ANGEL R

BL 248C

215-204-8578

BIOL 2001

Second Edition

Microbiology

Basic and Clinical Principles

Content Management: *Jeanne Zalesky, Lara Braun, Nida Ahamed*
Content Development: *Matthew Walker, Laura Bonazzoli, Hilair Chism*
Product Management: *Serina Beauparlant, Jennifer McGill Walker*
Content Production: *Mike Early, Titas Basu, Mireille Pfeffer, Lucinda Bingham, Nishant Bhaskar, Mary Tindle, Molly Montanaro*
Product Marketing: *Timothy Galligan, Kate Gittins, Rosemary Morton*
Rights and Permissions: *Matthew Perry, Zhalene Marian Vergara, Ben Ferrini*

Please contact https://support.pearson.com/getsupport/s/ with any queries on this content

Cover Image by Brandi Bleak/ Brandi Angel Photography

Library of Congress Cataloging-in-Publication Data

Names: Norman-McKay, Lourdes, author.
Title: Microbiology : basic and clinical principles / Lourdes Norman-McKay.
Description: Second edition. | New York, NY : Pearson, 2022. | Includes index.
Identifiers: LCCN 2021053587 | ISBN 9780136785750 (hardcover)
Subjects: MESH: Microbiological Phenomena
Classification: LCC QR46 | NLM QW 4 | DDC 616.9/041--dc23/eng/20211130
LC record available at https://lccn.loc.gov/2021053587

3 2022

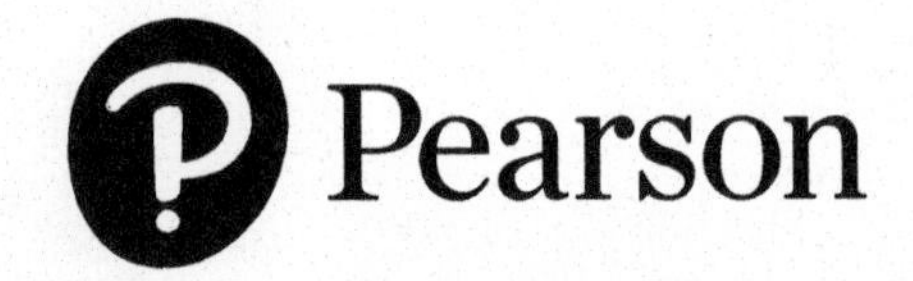

Rental
ISBN-10: 0136785751
ISBN-13: 9780136785750

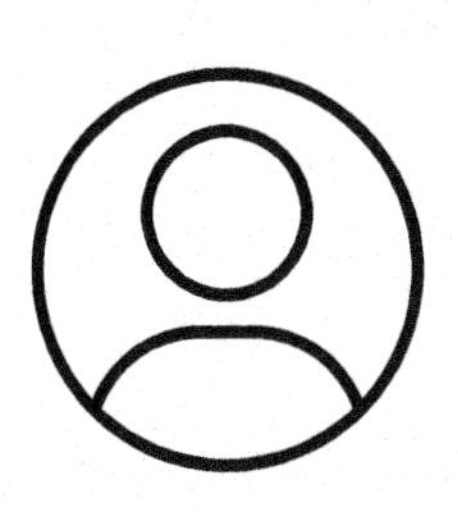

Pearson's Commitment to Diversity, Equity, and Inclusion

Pearson is dedicated to creating bias-free content that reflects the diversity, depth, and breadth of all learners' lived experiences.

We embrace the many dimensions of diversity, including but not limited to race, ethnicity, gender, sex, sexual orientation, socioeconomic status, ability, age, and religious or political beliefs.

Education is a powerful force for equity and change in our world. It has the potential to deliver opportunities that improve lives and enable economic mobility. As we work with authors to create content for every product and service, we acknowledge our responsibility to demonstrate inclusivity and incorporate diverse scholarship so that everyone can achieve their potential through learning. As the world's leading learning company, we have a duty to help drive change and live up to our purpose to help more people create a better life for themselves and to create a better world.

Our ambition is to purposefully contribute to a world where:

- Everyone has an equitable and lifelong opportunity to succeed through learning.
- Our educational content accurately reflects the histories and lived experiences of the learners we serve.
- Our educational products and services are inclusive and represent the rich diversity of learners.
- Our educational content prompts deeper discussions with students and motivates them to expand their own learning (and worldview).

Accessibility

We are also committed to providing products that are fully accessible to all learners. As per Pearson's guidelines for accessible educational Web media, we test and retest the capabilities of our products against the highest standards for every release, following the WCAG guidelines in developing new products for copyright year 2022 and beyond.

You can learn more about Pearson's commitment to accessibility at **https://www.pearson.com/us/accessibility.html**

Contact Us

While we work hard to present unbiased, fully accessible content, we want to hear from you about any concerns or needs with this Pearson product so that we can investigate and address them.

Please contact us with concerns about any potential bias at **https://www.pearson.com/report-bias.html**

For accessibility-related issues, such as using assistive technology with Pearson products, alternative text requests, or accessibility documentation, email the Pearson Disability Support team at **disability.support@pearson.com**

About the Author

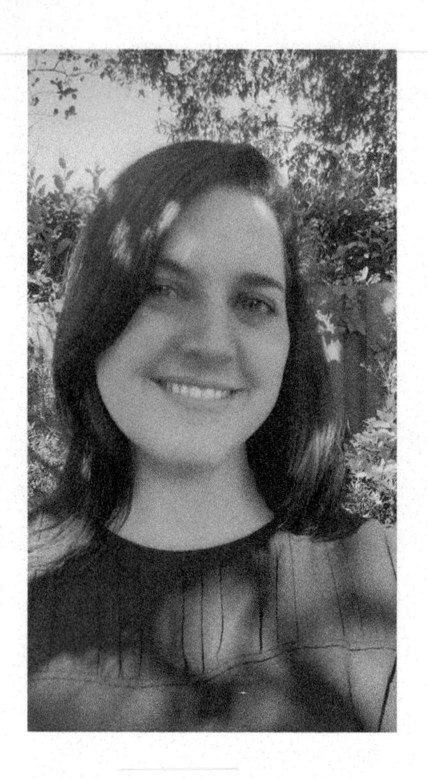

DR. LOURDES NORMAN-MCKAY is a professor at Florida State College Jacksonville, where her peers and students have recognized her with the Outstanding Faculty Award. She earned her B.S. in microbiology and cell science from the University of Florida and her Ph.D. in biochemistry and molecular biology from the Pennsylvania State University College of Medicine. Her postdoctoral specialization in microbiology and immunology—also at PSU College of Medicine—was funded through a competitive fellowship award from the National Institutes of Health and focused on studying the role of viruses in cancer. During her nearly two decades as a scientist-educator, she has trained thousands of undergraduate and graduate students pursuing healthcare careers and secured extensive federal funding to promote STEM education and empower underrepresented groups in STEM.

Her considerable STEM program development experience ranges from designing and launching a biomedical sciences baccalaureate program to serving as a curriculum designer and subject matter expert for the Florida Space Research Institute and Workforce Florida. To date, a highlight of her international experience includes work as a speaker for the U.S. Department of State's International Information Programs (IPP). As a part of the IIP program, Dr. Norman-McKay traveled to Tajikistan and Uzbekistan to meet with students, faculty, government officials, community leaders, U.S. diplomats, and other stakeholders to bolster STEM education, empower women and youth in STEM, and help build STEM capacity in central Asia.

In 2022, the National Academies of Sciences, Engineering, and Medicine, in coordination with the U.S. Department of State, selected Dr. Norman-McKay as a Jefferson Science Fellow—a prestigious award granted to only a handful of tenured faculty at U.S. institutions of higher education. In this role she will serve as a Senior Science Policy Advisor to the State Department's Office of the U.S. Global AIDS Coordinator and Health Diplomacy.

Dr. Norman-McKay also currently serves as a section editor for the American Society for Microbiology's (ASM) *Journal of Microbiology and Biology Education* and is an active participant in ASM's Microbiology in Nursing and Allied Health Committee. In addition to authoring the first and now second editions of her Pearson text *Microbiology Basic and Clinical Principles*, she is a lead author of the *Microbiology Laboratory Theory and Applications* series (Morton Publishing).

Dedication

Although I often get emails from former students, in March of 2020, I received more than usual—and they all had a different tone. These emails were from professionals in medicine who recalled their microbiology training and our discussions of emerging pathogens, especially viruses. They recollected discussions about transmission precautions, biosafety level ratings, vaccines and how they work, and antiviral drugs (to include the limited nature of our arsenal of these drugs). They also knew that the last major pandemic, the 1918 flu pandemic, had killed more Americans in 9 months than World Wars I and II, the Vietnam War, and the Korean War combined. They were scared. They had been mobilized in full force, briefed on new triage protocols, given tips for conserving their personal protective equipment, and advised to keep their distance from family and friends who were at risk for severe disease from "the virus." I replied to each email with a reassurance that while they would likely be pushed to the breaking point, they had the training and constitution to rise to the challenge, and in the balance of it all, they would save lives and ease suffering.

In the first year of the pandemic, COVID-19 killed over 3,600 U.S. healthcare workers. I'm not sure how many former students of mine have become ill or how many have died in this pandemic, but after nearly two decades of teaching *thousands* of prospective nurses and other allied health workers, I must assume a number greater than zero.

This book is dedicated to my students, to the healthcare teams of today and tomorrow, and to the over 115,000 healthcare workers around the world who lost their lives to COVID-19.

With gratitude,
Lourdes Norman-McKay
and all of Pearson Education

Preface

If you're reading this, there's a good chance that you either train tomorrow's healthcare team or you are training to become one of tomorrow's healthcare heroes. In both cases, this pandemic has offered a new lens through which you should view your work. This course is about employable education, learning to think critically and clinically, and seeing how topics integrate so that you are ready to meet and overcome the challenges you'll inevitably face as you advance in your training and eventually your career. This course is about more than a grade—this is a core component of your career readiness. The faculty teaching you are highly trained professionals who will push you to realize your potential. They are your mentors, your cheer squad, and your drill sergeants all in one.

It is through this lens that I and our editorial team have continued the work of setting *Microbiology: Basic and Clinical Principles* apart from other texts as the premier teaching and learning resource built from the ground up, specifically for training tomorrow's healthcare team.

Because this revision was written in the throes of the COVID-19 pandemic, every day has been a vivid reminder to stay true to the goal of supporting an employable educational experience that supports integrated learning as well as clinical and critical thinking. This lens focused *every* aspect of this new edition, literally from cover to cover, on that mission.

When you look at the chapter openers and the cover of the text, you'll see real-life healthcare professionals who, like you, overcame barriers to get through their training. Despite being overworked throughout the pandemic, they took the time to send us pictures of themselves in action so that students could see diverse races and ethnicities, all body types, genders, and ages and know that there is a place for *everyone* in STEM and medicine, regardless of where you come from or what you look like. As you undertake your microbiology journey, you are preparing to join thousands of professionals who have embraced the challenges of always learning, pushed themselves to be kind even when they were exhausted, and in many cases even made the ultimate sacrifice.

To support your important journey, we have designed new e-learning tools called *Interactive Content Reviews* that provide active learning experiences that help you explore and learn and test your knowledge attainment on the most challenging topics in this course. We also added over half a dozen more *Concept Coaches*—an expansion of an already well-loved feature that brings the art to life and helps you see how concepts interconnect. Lastly, the e-revision adds even more quiz and test questions to help you practice the content and master thinking critically and clinically.

We also did a "deep dive" into the text and made important updates across a wide variety of topics to, of course, include extensive coverage of SARS-CoV-2/COVID. Check out the "New to This Edition" section to learn more.

New to This Edition

Chapter 1

- *New Asset:* Concept Coach—The Gram Stain
- In section 1.1, expanded discussion of laws versus theories; revised Figures 1.3 and 1.4 for clarity.
- Revised section 1.2 on classifying microbes for clarity; moved content on biofilms to follow content on the normal microbiota, which was expanded.
- In section 1.3, clarified terminology for primary dye and counterstain; simplified discussion of phase contrast in Table 1.3; clarified discussion of lipopolysaccharide; added a Clinical Vocabulary Note on septic shock.

Chapter 2

- In section 2.2, expanded the discussion of valence electrons, polar covalent bonds, and dipoles; clarified the discussion of Van der Waals interactions and hydrophobic, hydrophilic, and amphipathic molecules; added a Chem Note with line art to clarify the chemical shorthand used in molecular line drawings.
- In section 2.4, added a discussion of polysaccharides to the content on carbohydrate structure.

Chapter 3

- *New Asset:* Interactive Content Review (combined with Chapter 4 to build connections between prokaryotic versus eukaryotic cells)
- New Concept Coach: The Gram Stain
- In section 3.1, increased detail in Figure 3.2 showing prokaryotic cell.
- In section 3.2, added Chem Notes on ether and ester bonds, isoprenoids, and glycans; added a brief explanation of L-form bacteria and the role of Emmy Klieneberger-Nobel; added Figure 3.10 on the structure of peptidoglycan and Figure 3.12 on medically important Gram-negative and -positive bacteria; added details to Figures 3.11 and 3.15.
- In section 3.3, rearranged Figure 3.27 to match presentation in text.

Chapter 4

- *New Asset:* Interactive Content Review (combined with Chapter 3 to build connections between prokaryotic versus eukaryotic cells)
- In section 4.1, provided the word origin for *eukaryotic* to emphasize that these cells are nucleated; revised Figure 4.2 showing eukaryotic cell, adding a prokaryotic cell for comparison.
- In section 4.4, revised Figure 4.20 of the endoplasmic reticulum for clarity; added a new Chem Note on toxic oxygen intermediates.

Chapter 5

- *New Asset:* Interactive Content Review
- In section 5.1, revised Figure 5.1 for clarity and match changes in Chapter 3.
- In section 5.2, explained the naming convention for DNA polymerases; revised DNA replication Figures 5.7 and 5.8 for clarity.
- In section 5.3, revised structure to improve flow of narrative; revised Figure 5.9 to improve readability; changed orientation of tRNA in Figure 5.10 to match revised Figure 5.12 on translation; revised Figures 5.12 and 5.13 for clarity.
- In section 5.5, added explanation of why induced mutations are often harmful to cells.

Chapter 6

- *New Asset:* Interactive Content Review
- Throughout the chapter, added content on SARS-CoV-2.
- In section 6.1, replaced the Training Tomorrow's Health Team with one on coronavirus vaccines; replaced the TEM of herpes virus in Figure 6.2 with one of a coronavirus.
- In section 6.2, added SARS-CoV-2 to Figure 6.8; altered Figure 6.9 to improve size comparison.
- In section 6.3, modestly expanded the discussion of T-even/T4 bacteriophages; revised Figures 6.13 through 6.15 for clarity; rearranged Figure 6.16 to match the order of presentation in the narrative.
- In section 6.4, added content on chloroquine and hydroxychloroquine and the limitation of *in vitro* findings in predicting drug efficacy *in vivo*; replaced the Training Tomorrow's Health Team on egg-free influenza vaccines with one on detecting SARS-CoV-2.

- Moved discussion of prions to new section 6.5; expanded to discuss neurodegenerative diseases associated with misfolded proteins.

Chapter 7

- New Concept Coach: The Bacterial Growth Curve and an Overview of Aerotolerance
- Changed chapter title to Fundamentals of Microbial Growth and Decontamination.
- In section 7.1, reorganized the learning outcomes to better reflect the order of topics.
- In section 7.2, expanded the discussion of antioxidants, and added the term to the glossary; altered Table 7.1 for clarity.
- In section 7.3, added a discussion of hemolysins, and added the term to the glossary; deleted the Training Tomorrow's Health Team on the use of genetics in disease outbreaks.
- Changed the title of section 7.4 to Basics of Microbial Growth Reduction and Decontamination.
- Replaced the Training Tomorrow's Health Team on triclosan with one on decontaminating N95 masks during the COVID-19 pandemic.

Chapter 8

- *New Asset*: Interactive Content Review
- In section 8.1, redrew Figure 8.1 of anabolism and catabolism for clarity; made the chemical drawing of ATP a numbered figure; revised Figure 8.3 (former 8.2) for clarity.
- In section 8.2, expanded the discussion of allosteric regulation; revised the layout of Figure 8.13 to improve comparison of concepts.
- In section 8.4, revised Figures 8.17 through 8.22 to improve visual and textual accessibility, readability, and clarity; revised the discussion of the electron transport chain for clarity.
- In section 8.5, in Figure 8.24 of fermentation, improved visual accessibility and representation of carbon; redrew Figure 8.25 to better align with cellular respiration tabbed headings and to serve as a more detailed summary.
- In section 8.6, revised the discussion of lipid biosynthesis for clarity; revised Figures 8.27 through 8.30 on building organic molecules to better align with cellular respiration tabbed headings and improve visual accessibility and clarity.
- Condensed section 8.8 and simplified Figure 8.31.
- Replaced end-of-chapter review question 15 with a matching-style question.

Chapter 9

- Extensive SARS-CoV-2 content infused throughout the chapter.
- New Concept Coach: Modes of Transmission and the Stages of an Active Infection
- In section 9.1, revised the discussion of Koch's postulates for clarity; deleted Table 9.1 because it repeats material in Chapter 1.
- In section 9.2, expanded the discussion of airborne transmission to compare droplet and aerosol transmission.
- Reorganized section 9.3 for clarity.
- In section 9.4, added discussion of double-blinded placebo-controlled studies; updated Figure 9.6; replaced Figure 9.7 with a graph of COVID-19 hospitalization; added a Training Tomorrow's Health Team on graphing inequality in access to healthcare.
- In section 9.5, updated Figure 9.10 of most common healthcare-acquired infections.
- In section 9.6, updated Figure 9.12, replacing avian flu with COVID-19; expanded the discussion of ethical considerations in vaccination.

Chapter 10

- Throughout the chapter, added content on SARS-CoV-2.
- Reorganized section 10.1 for clarity; added photo of 1918 flu pandemic.

- In section 10.2, revised headings for clarity; expanded discussions of virulence and basic reproduction number (R-naught or R_0) and effective reproduction number (R_e); added a micrograph.
- In section 10.3, revised the discussion of quorum sensing and Figures 10.5, 10.8, and 10.10 for clarity.
- In section 10.4, expanded the discussion of airborne precautions; added a figure on PPE for COVID-19; revised Figure 10.14 (formerly 10.13) on NIOSH-approved respirators.

Chapter 11

- *New Asset:* Interactive Content Review (combined with Chapter 12 to integrate the branches of immunity)
- Revised section 11.1 to introduce immune self-tolerance earlier; revised Figure 11.1 to emphasize innate and adaptive immunity; updated other orienting diagrams to match.
- In section 11.2, mentioned the evolutionary conservation of antimicrobial peptides.
- In section 11.5, added a discussion of cytokine storms; revised Figure 11.11 to improve visual accessibility and clarity.
- In section 11.6, corrected Figures 11.13 through 11.15 and improved visual accessibility and readability.

Chapter 12

- *New Asset:* Interactive Content Review (combined with Chapter 11 to integrate the branches of immunity)
- In section 12.1, added a definition of the term *thymocytes* and expanded the discussion of T cell receptors; revised Figure 12.1 to match the related figure in Chapter 11; updated orienting diagrams to match; revised Figure 12.5 of T cell receptor for clarity.
- In section 12.2, revised discussion of superantigens and T cell activation for clarity.
- In section 12.3, revised the discussion of antibody structure and isotopes to cover constant versus variable regions; revised Figure 12.19 for clarity; added visuals to T-dependent and T-independent descriptions; added visual of full antigen with different epitopes.
- In section 12.4, added a discussion of convalescent plasma and monoclonal antibody therapies; expanded the Training Tomorrow's Health Team on antibody titers to include their role in COVID-19 and plasma donations for emerging infectious diseases; expanded the Bench to Bedside on antibody therapies to explain naming.

Chapter 13

- *New Asset:* Interactive Content Review
- In section 13.1, emphasized the role of vaccine hesitancy in measles resurgence; revised the content introducing autoimmune disorders and the possible role of infection in etiology.
- Reorganized the introduction to section 13.2 for clarity.
- In section 13.3, updated information on allergy management and desensitization.

Chapter 14

- New Concept Coach: An Introduction to Vaccines
- New Concept Coach: Key Molecular Diagnostics 101
- To better represent chapter content, changed chapter title to Biomedical Applications: Vaccines, Diagnostics, Therapeutics, and Molecular Methods.
- In section 14.2, restructured the discussion of vaccine formulations and altered Figure 14.3 on herd immunity for clarity; replaced former Table 14.3 on vaccine formulations with Figure 14.4.
- In section 14.3, deleted discussion of immunoprecipitation reactions and Figures 14.7 and 14.8; combined former Figures 14.10 and 14.11 to improve comparison of indirect, direct, and sandwich ELISA; added Figure 14.12 showing Western blot test result.
- In section 14.4, revised discussion of CRISPR-Cas9 and Figure 14.15 for clarity.

Chapter 15

- New Concept Coach: Antimicrobial Drug Basics
- New Concept Coach: Understanding Antimicrobial Drug Resistance and Drug Stewardship
- In section 15.1, revised and renamed the discussion on modifying antimicrobials.
- In section 15.2, expanded the discussion of the order Enterobacterales.
- In section 15.3, added a discussion of the use of remdesivir for COVID-19 and replaced the Bench to Bedside on antivirals for Zika with one on remdesivir.
- In section 15.5, updated the discussion of resistant microbes.

Chapter 16

- In section 16.1, revised the section on the respiratory microbiome, including Figure 16.4.
- In section 16.2, thoroughly revised the discussions of respiratory syncytial virus and influenza; distinguished between New World and Old World hantaviruses.
- Added COVID-19 content throughout the chapter; the main coverage occurs in section 16.2 and includes new learning outcomes, key terms, table, Disease Snapshot, and Build Your Foundation questions. The SARS map (former Figure 16.6) and Disease Snapshot have been removed.
- In section 16.3, expanded the discussion of diphtheria; updated Figure 16.2 on pertussis; replaced the Bench to Bedside on flu vaccines with one on ventilator-associated pneumonia in COVID-19 patients.

Chapter 17

- Added 40+ photos that show how skin conditions present in people of color; an often-overlooked aspect of dermatology (e-book photo carousel).
- In section 17.2, updated shingles vaccine information to reflect the discontinuation of the Zostavax vaccine and the shift to Shingrix.

Chapter 21

- In section 21.1, expanded the discussion of sepsis and added information on the latest Sequential Organ Failure Assessment (SOFA) for scoring sepsis in patients.
- In section 21.2, replaced map of Ebola outbreak with new Figure 21.10 on Ebola virus transmission; added information about the new Ebola vaccine, Erbevo.
- In section 21.3, updated plague incidence map.
- Restructured section 21.5 on malaria; added that the GlaxcoSmithKline RTS,S/AS01vaccine is now the first ever WHO recommended malaria vaccine to protect children.

Global changes:

- Updated references as needed to communicate the most current developments in microbiology (e.g., new COVID vaccines and Zostavax discontinuation)
- All statistics updated to reflect the most recent data available at time of publication.
- Latest naming conventions applied throughout:
  - *Clostridium difficile* updated to *Clostridioides difficile*
  - Carbapenem-resistant Enterobacterales rather than carbapenem-resistant Enterobacteriaceae. The taxonomic order name, Enterobacterales, is now recommended in place of the family name, Enterobacteriaceae, to reflect that carbapenem resistance is no longer isolated to one family but now spans multiple families within the order.
- Revised and updated section learning outcomes to align with new and/or updated content
- Infused extensive SARS-CoV-2 information throughout the text and media assets.

Continuing to Support DE&I

- Graphing Inequality in Access to Healthcare: TTHT, Chapter 9, page 290
- Novel Vaccines for a Novel Coronavirus (Training Tomorrow's Health Team feature) and Dr. Kizzmekia Corbett: page176
- Encouraging students to build their STEM identity (chapter openers and cover)
- Chapter 17 dermatological images expansion
- Expanded language notes for non-native speakers of English

Acknowledgments

Although my name is on the cover, writing a textbook is not a one-person effort. It takes a team of editors, media specialists, talented artists, and expert reviewers to make a high-quality educational resource that is *thoroughly* peer-reviewed, student-centered, and employability driven. Our team, which was curated from across the globe, is full of talented people.

I especially want to thank Jennifer McGill-Walker for her leadership and willingness to take the "road less traveled" to do things like host a live photoshoot that brought the visages of *real-life* healthcare heroes to every chapter of this text, as well as to the cover. Lara Braun was an instrumental part of the development process—she brought her tireless work ethic, acumen for development, and ever-patient personality to the table. Laura Bonazolli served as our incredibly talented development editor; she is probably the only person on the planet who has read this book more times than me! Matt Walker provided excellent content strategy leadership that brought 2e to the next level. Titas Basu, our content producer, went beyond the call of duty to make the cover-to-cover healthcare heroes vision a reality—overseeing an endless stream of permissions documents and releases, all while coordinating the underlying machinations of seeing a major text revision to market. Hilair Chism, returned to this edition to continue building the unique art style that sheds clutter for clarity and treats art as an avenue to learning.

Mireille Pfeffer, served as our rich-media guru who navigated the perils of the ever-shifting technology world to bring our e-assets to fruition. Kassi Foley worked for over a year with me, Jen, and Lara to shape and pilot the first ever *Interactive Content Reviews* that are an exemplar of how technology can support best practices in teaching. Behind the scenes, Serina Beauparlant is always cheering us on as we shape teaching and learning tools for the world. And then there is Jeanne Zalesky who has been unflagging in supporting our team's work to innovatively conquer the biggest teaching and learning challenges in science higher education. A special thanks to Yez Alayan, Brett Freitas, Kelly Galli-Reynolds, Kate Gittins, Adam Goldstein, Tim Galligan, Jennifer Key, and Tim Wilson--your outreach and support are indispensable. Brandi Bleak, I am grateful for your talented photographer's eye on our cover and many of the chapter openers. Mary Tindle, thanks for being an amazing partner in the production process.

Lastly, I couldn't have finished this work without my husband and best friend, Andrew McKay—nor without the understanding and patience of our amazing daughters, Lourdes Catherine (18) and Delia (13). I love you all more than you can know! Thanks also to my parents for pitching in with a healthy meal when I was up against tight deadlines. And much love to my friend and fellow authoress, Dr. Erin Amerman.

Reviewers

First Edition

Alejandro Vazquez
El Paso Community College
Alicia Carley
Northwest Technical College
Alicia Musser
Lansing Community College
Amanda Mitchell
Hinds Community College
Amee Mehta
Seminole State College of Florida
Amy Goode
Illinois Central College
Andrea Castillo
Eastern Washington University
Andrew O'Brien
Hagerstown Community College
Andrew Thompson
College of Central Florida
Anne Heise
Washtenaw Community College
Ben Hanelt
University of New Mexico
Ben Whitlock
University of St. Francis
Benjamin Rowley
University of Central Arkansas
Bob Iwan
Inver Hills Community College
Bridget Joubert
Northwestern State University
Brinda Govindan
San Francisco State University
Carol Cleveland
Northwest Mississippi Community College
Carrie Arnold
Mount Wachusett Community College
Carron Bryant
East Mississippi Community College
Cassy Cozine
University of Saint Mary
Charlotte Barker
East Texas Baptist University
Christian Eggers
Quinnipiac University
Christine Cutucache
University of Nebraska Omaha
Christine Fleischacker
University of Mary
Christine Kirvan
California State University, Sacramento
Clifton Franklund
Ferris State University
Corrie Andries
Central New Mexico Community College
Craig Smith
Washington University in St. Louis
Cynthia Anderson
Mt. San Antonio College
Dale Amos
University of Arkansas-Fort Smith
Dale Horeth
Tidewater Community College
Daniel Aruscavage
Kutztown University of Pennsylvania
David Ansardi
Calhoun Community College
Dena Berg
Tarrant County College
Diane Dixon
Southeastern Oklahoma State University
Diane Lewis
Ivy Tech Community College
Don Dailey
Austin Peay State University
Edith Porter
California State University, Los Angeles
Elizabeth McPherson
University of Tennessee
Elizabeth Mitchell
Central Piedmont Community College
Elizabeth Yelverton
Pensacola State College
Emily Booms
Northeastern Illinois University
Eric Ford
East Mississippi Community College
Felicia Goulet-Miller
Florida Gulf Coast University
Fernando Monroy
Northern Arizona University
Gail Stewart
Camden County College
Georgia White-Epperson
Washtenaw Community College
Geraldine Rimstidt
Daytona State College
Gina Holland
Sacramento City College
H. Kathleen Dannelly
Indiana State University
Heather Seitz
Johnson County Community College
Helene Ver Eecke
Metropolitan State University of Denver
Holly Walters
Cape Fear Community College
Jack Shurley
Idaho State University
Jacqueline Brown
Angelo State University
Jacqueline Spencer
Thomas Nelson Community College
James Rago
Lewis University
Janice Dorman
University of Pittsburgh
Janice Speshock
Tarleton State University
Jason Furrer
University of Missouri
Jason Hitzeman
Georgia Highlands College
Jaya Dasgupta
Hudson Valley Community College
Jeanne Ferguson
Ivy Tech Community College
Jeffrey Morris
University of Alabama at Birmingham
Jennifer Bess
Hillsborough Community College
Jennifer Metzler
Ball State University
Jerald Hendrix
Kennesaw State University
Joan Boyd
Florida State College at Jacksonville
Joanne Scalzitti
Aurora University
John Dahl
University of Minnesota Duluth
John Jones
Calhoun Community College
John Myers
Owens Community College
John Treml
Fort Scott Community College
John Whitlock
Hillsborough Community College
Judith Coston
Bossier Parish Community College
Judy Kaufman
Monroe Community College
Julie Huggins
Arkansas State University

Justin Hoshaw
Waubonsee Community College
Kabeer Ahammad Sahib
Lansing Community College
Kalina White
Community College of Allegheny County
Karen Braley
Daytona State College
Karen Huffman
Genesee Community College
Karen Meysick
University of Oklahoma
Karen Sellins
Front Range Community College
Kari Cargill
Montana State University
Katy Jamshidi
Santa Rosa Junior College
Kelley Black
Jefferson State Community College
Kelly Johnson
Somerset Community College
Ken Malachowsky
Florence-Darlington Technical College
Kenneth Teveit
Baker College
Kevin Sorensen
Snow College
Kimberly Harding
Colorado Mountain College
Kristin Burkholder
University of New England
Laurie Freeman
Fulton-Montgomery Community College
Lawrence Aaronson
Utica College
Layla Khatib
Moraine Valley Community College
Lee Couch
University of New Mexico
Lisa Blankinship
University of North Alabama
Lisa Shimeld
Crafton Hills College
Liza Mohanty
Olive-Harvey College
Lori Smith
American River College
Lorraine Findlay
Nassau Community College
Lynne Hinkey
Trident Technical College
Manjushri Kishore
Heartland Community College
Marcia Watkins
Eastern Kentucky University
Mari Aaneason
Western Illinois University
Marian Kristie Bompadre
Reading Area Community College
Mark Randa
Cumberland County College
Mark Roos
Daytona State College
Mary Stewart
University of Arkansas at Monticello
Matthew Sattley
Indiana Wesleyan University
Mehrdad Tajkarimi
California State University, Los Angeles
Melissa Elliott
Butler Community College
Melissa Schreiber
Valencia College
Michael Buoni
Delaware Technical Community College
Michael Pressler
Delta College
Michael Shea
Hudson Valley Community College
Michael Womack
Gordon State College
Michelle Hughes
Clovis Community College
Mike Land
Northwestern State University of Louisiana
Misty Wehling
Southeast Community College
Mustapha Lahrach
Hillsborough Community College
Naina Phadnis
University of Utah
Nilesh Sharma
Western Kentucky University
Pamela Coker
Pima Community College
Patricia Alfing
Davidson County Community College
Paul Johnson
University of North Georgia
Peggy Mason
Brookhaven College
Peter Kish
Northampton Community College
Pramila Sen
Houston Community College
Ramaraj Boopathy
Nicholls State University
Randall Pegg
Florida State College at Jacksonville
Richard Karp
University of Cincinnati
Richard Knapp
University of Houston
Rita Moyes
Texas A&M University
Robert Leunk
Grand Rapids Community College
Robin Cotter
Phoenix College
Ruhul Kuddus
Utah Valley University
Samiksha Raut
University of Alabama at Birmingham
Sandra Horikami
Daytona State College
Scott Mittman
Essex County College
Sean Rollins
Fitchburg State University
Shannon Mackey
St. Ambrose University
Shannon McCallister
Old Dominion University
Sharon Gusky
Northwestern Connecticut Community College
Shaun Bowman
Clarke University
Sheela Vemu
Waubonsee Community College
Sheri Miraglia
City College of San Francisco
Sherry Stewart
Navarro College
Stacey Lettini
Gwynedd Mercy University
Stephanie Daugherty
University of Texas at Tyler
Steven Scott
Merritt College
Susan Bornstein-Forst
Marian University
Susan Skelly
Rutgers University
Suzanne Keller
Indian Hills Community College
Suzanne Long
Monroe Community College
Teresa Cowan
Baker College of Auburn Hills
Terrence Ravine
University of South Alabama
Terry Miller
Central Carolina Community College
Tracey Dosch
Waubonsee Community College

Vanessa Rowan
Palm Beach Atlantic University
Vijay Sivaraman
North Carolina Central University
Wei Wan
Texas A&M University
Wendy Hadley
Allan Hancock College

Second Edition

Tonya Bates, *University of North Carolina*
John Battista, *Louisiana State University*
Tilman Baumstark, *University at Buffalo SUNY*
Doug Bernstein, *Ball State University*
Jennifer Bess, *Hillsborough Community College*
Jennifer Bourcier, *Grand Rapids Community College*
Brad Bowser, *San Francisco State University*
Kari Cargill, *Montana State University*
Francisco Cruz, *Georgia State University*
Terry Conley, *Cameron University*
Clinton Copp, *University of Mississippi*
Kevin Costa, *Nassau Community College*
Emerson Crabill, *Angelo State University*
Deborah Dardis, *Southeast Louisiana University*
Cheryl Doughty, *University of Colorado*
Janet Dowding, *Miami Dade College*
Amy Goode, *Illinois Central College*
Natalie Hendrix, *Bossier Parish Community College*
Paul Himes, *University of Louisville*
Gina Holland, *Sacramento City College*
Heather Klenovich, *Community College of Allegheny County*
David Laborde, *Hillsborough Community College*
Mustapha Lahrach, *Hillsborough Community College*
Diane Lewis, *Ivy Tech Community College*
Lauren Logsdon, *University of Tampa*
Catarino Morales, *Southwest Texas Junior College*
Karen Neal, *Reynolds Community College*
Holly Paquette, *College of Western Idaho*
Marcia Pierce, *Eastern Kentucky University*
Robert Pollack, *Nassau Community College*
Laraine Powers, *East Tennessee State University*
Andrea Rediske, *Valencia College*
Benjamin Rowley, *University of Central Arkansas*
Vinaya Sampath, *Long Island University*
Sue Sawyer, *Kellogg Community College*
Julie Bogart, *University of South Florida*
Jack Shurley, *Idaho State University*
Lisa Smith, *Hillsborough Community College*
Jacqueline Spencer, *Thomas Nelson Community College*
Sherry Stewart, *Navarro College*
Denis Trubitsyn, *Longwood University*
Shannon Ulrich, *St. Petersburg College*
Debra Vanhouten, *Chemeketa Community College*
Roxanne Vargas, *Laredo College*
Sheela Vemu, *Waubonsee Community College*
Roland Vieira, *Green River Community College*
Richard Watkins, *Jacksonville State University*
Richard Wells, *Ozarks Technical Community College*
Kalina White, *Community College of Allegheny County*
Michael Womack, *Gordon State College*

Contributors

First Edition

Amy Reese, St. Louis College of Pharmacy; Amy Siegesmund, Pacific Lutheran University; Andrea Rediske, Valencia College; Christopher Thompson (Clinical Cases author), Loyola University Maryland; Damian Hill; Justin Hoshaw, Waubonsee Community College; Heather Klenovich, Community College of Allegheny County; Ines Rauschenbach, Union County College; Jennifer Walker, University of Georgia; Julie Oliver, Cosumnes River College; Kate Donnelly; Katherine Rawls, Florida State College at Jacksonville; Lanh Bloodworth, Florida State College at Jacksonville; Marcia Pierce, Eastern Kentucky University; Mary Miller, Baton Rouge Community College; Rebekah Ward, Georgia Gwinnett College; Sandra Gibbons, Moraine Valley Community College; Shannon Ulrich, St. Petersburg College; Sharron Crane, Rutgers University; Stephanie Havemann, Alvin Community College; Terri Lindsey, Tarrant County College; Tracey Dosch, Waubonsee Community College; Warner Bair III, Lone Star College

Second Edition

Franscisco Cruz, Georgia State University; Janet Dowding, Miami Dade College; Natalie Hendrix, Bossier Parish Community College; Marcia Pierce, Eastern Kentucky University; Andrea Rediske, Valencia College; Julie Bogart, University of South Florida; Jacquelyn Spencer, Thomas Nelson Community College; Heather Wilson, Florida State College; Deborah Dardis, Southeastern Louisiana University and Dusty Stutts, Roane State Community College

Contents

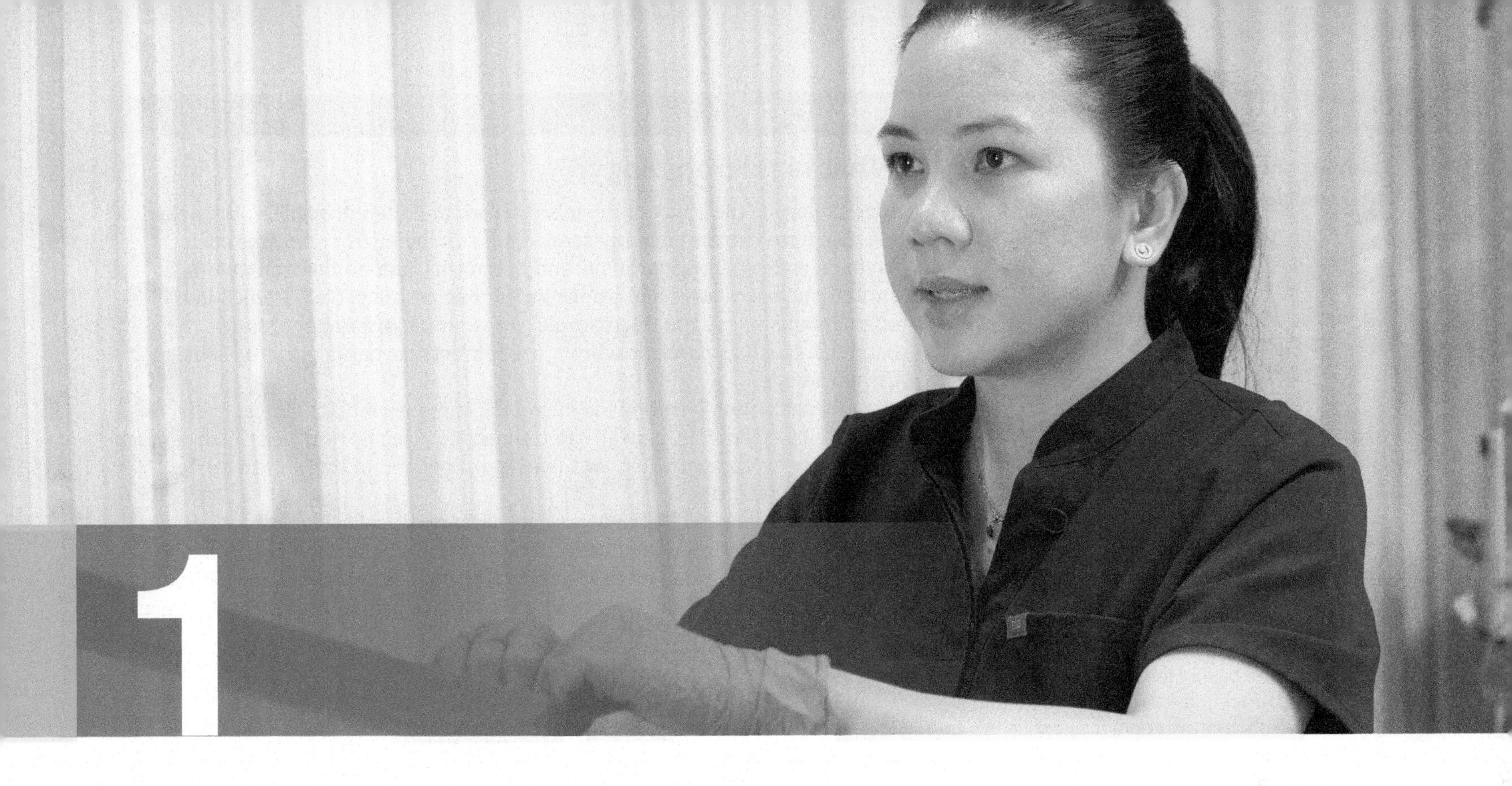

Introduction to Microbiology

What Will We Explore? You are venturing into a new world, one consisting of extremely small organisms so abundant that they are estimated to comprise at least half of the living biomass on Earth.[1] For thousands of years people were unaware of this invisible microbial landscape. Our veil of ignorance started to lift in the late 1600s, when microscopes were invented, and we saw for the first time the diversity of life-forms coexisting around us, in us, and on us. In 1665 Robert Hooke first formally described microbial life in his book *Micrographia*, stating, "By the means of telescopes, there is nothing so far distant but may be represented to our view; and by the help of microscopes, there is nothing so small as to escape our inquiry[.]" Since Hooke's time, we have continued expanding our knowledge. This chapter introduces a brief history of the field, then explains how we classify microbes and their interactions. Finally, we discuss the modern microscopes and staining techniques essential to the practice of microbiology today.

Why Is It Important? Understanding microbes is central to understanding human health and disease. These days, even toddlers are taught the connection between microbes and illness. But in recent decades, it has become clearer that our relationship with microbes is more complex than just "germs make us sick." In a way, each one of us is an ecosystem of many hundreds of species of microbes, some of which perform important functions in our bodies like making essential vitamins, training our immune system, and helping us digest food. Microbes also play important roles in sustaining ecosystems, cleaning our global environment, and producing foods, drugs, and even fuels.

Clinical CASE NCLEX HESI TEAS

The Case of the Mystery Pathogen

Visit the **Mastering Microbiology** Study Area to watch the case and find out how microbiology can explain this medical mystery.

Hong Phung
PTA; Jacksonville, FL

[1] Kluyver, A. J., & van Niel, C. B. (1956). *The microbe's contribution to biology*. Cambridge, MA: Harvard University Press, p. 3.

1.1 A BRIEF HISTORY OF MICROBIOLOGY

Learning Outcomes

After reading this section, you should be able to:

1.1 Define the term *microorganism*, and give examples of microbes studied in microbiology.

1.2 Explain the distinction between a pathogen and an opportunistic pathogen.

1.3 Compare the theories of biogenesis and spontaneous generation, and summarize Louis Pasteur's role in proving biogenesis.

1.4 Describe how Robert Koch helped shape the germ theory of disease, and list his postulates of disease.

1.5 Identify the goals of aseptic technique, and explain why it is important in healthcare facilities and laboratories.

1.6 Discuss how Semmelweis, Lister, and Nightingale contributed to health care.

1.7 Outline the basic steps of the scientific method, distinguish an observation from a conclusion, and compare a scientific law to a theory.

What is microbiology?

Microbiology is the study of **microbes**, which are often invisible to the naked eye. The term *microbe* encompasses cellular, living microorganisms such as bacteria, archaea, fungi, protists, and helminths, and nonliving/noncellular entities such as viruses and prions (infectious proteins) (TABLE 1.1). Some microorganisms are not microscopic. For example, a number of fungi, helminths such as parasitic worms, and protists such as algae are visible to the naked eye.

At least half of Earth's life is microbial. Microbes inhabit almost every region of our planet, from deep-sea trenches to glaciers. And we still have much to discover, as it is estimated that less than 1 percent of the world's microbes are currently identified.

Evolving about 3.5 billion years ago, **prokaryotic cells** (PRO-care-ee-ah-tic) are Earth's earliest lifeforms. They include unicellular bacteria and archaea (are-KEY-uh), which are structurally and functionally simpler than **eukaryotic cells** (YOU-care-ee-ah-tic). Eukaryotic cells make up all multicellular organisms and a number of unicellular microorganisms such as amoebae and yeast. The **endosymbiotic theory** states that eukaryotic cells evolved from prokaryotic cells. (For more on prokaryotic cells, eukaryotic cells, and endosymbiotic theory, see Chapters 3 and 4.)

Microbiology spans a wide variety of fields, including health care, agriculture, industry, and environmental sciences. Humans rely on microbes for many things such as food production, making medications, and breaking down certain environmental hazards.

A great deal of microbiology research focuses on **pathogens**—microbes that cause disease. However, most microbes are not pathogens. Just over 1,400 pathogenic microbes are known to infect humans;[2] but overall, less than 1 percent of all microbes are likely to be pathogenic. Among pathogens, we find so-called

TABLE 1.1 Living and Nonliving Agents Studied in Microbiology

Microbe	Cell Type	Notes
Bacteria	Prokaryotic	Unicellular;* pathogenic and nonpathogenic
Archaea	Prokaryotic	Unicellular; nonpathogenic; most live in extreme environments
Protists	Eukaryotic	Unicellular and multicellular; pathogenic and nonpathogenic (unicellular example: amoebae; multicellular example: algae)
Fungi	Eukaryotic	Unicellular and multicellular; pathogenic and nonpathogenic (unicellular example: yeast; multicellular example: mushrooms)
Helminths	Eukaryotic	Multicellular;* parasitic roundworms and flatworms
Viruses	Not cells; nonliving	Infect animal, plant, or bacterial cells; can have a DNA or RNA genome
Prions	Not cells; nonliving; infectious proteins	Not discovered until the 1980s; transmitted by transplant or ingestion; some prion diseases are inherited

*__Unicellular__ = one-celled organism; **multicellular** = organism made of many cells

S.M.A.R.T. Strategy

CONCEPT COACH

NCLEX HESI TEAS

Visit the **Mastering Microbiology** Study Area to watch the Concept Coach and learn a useful strategy for thinking clinically and critically.

[2] Woolhouse, M. E. J., & Gowtage-Sequeria, S. (2005). Host range and emerging and reemerging pathogens. *Emerging Infectious Diseases*, 11(12), 1842–1847.

true pathogens, which in theory can cause disease in any otherwise healthy host, and we find **opportunistic pathogens**, which tend to cause disease only in a weakened host. Typically, microbiology laboratories have the task of identifying the specific microbe causing a patient infection.

Great advances occurred in and around the golden age of microbiology.

The golden age of microbiology (approximately 1850–1920) was sparked by innovations in microscopes and new techniques to isolate and grow microbes (FIG. 1.1). Many of the techniques that facilitated this turning point in microbiology are still used today.

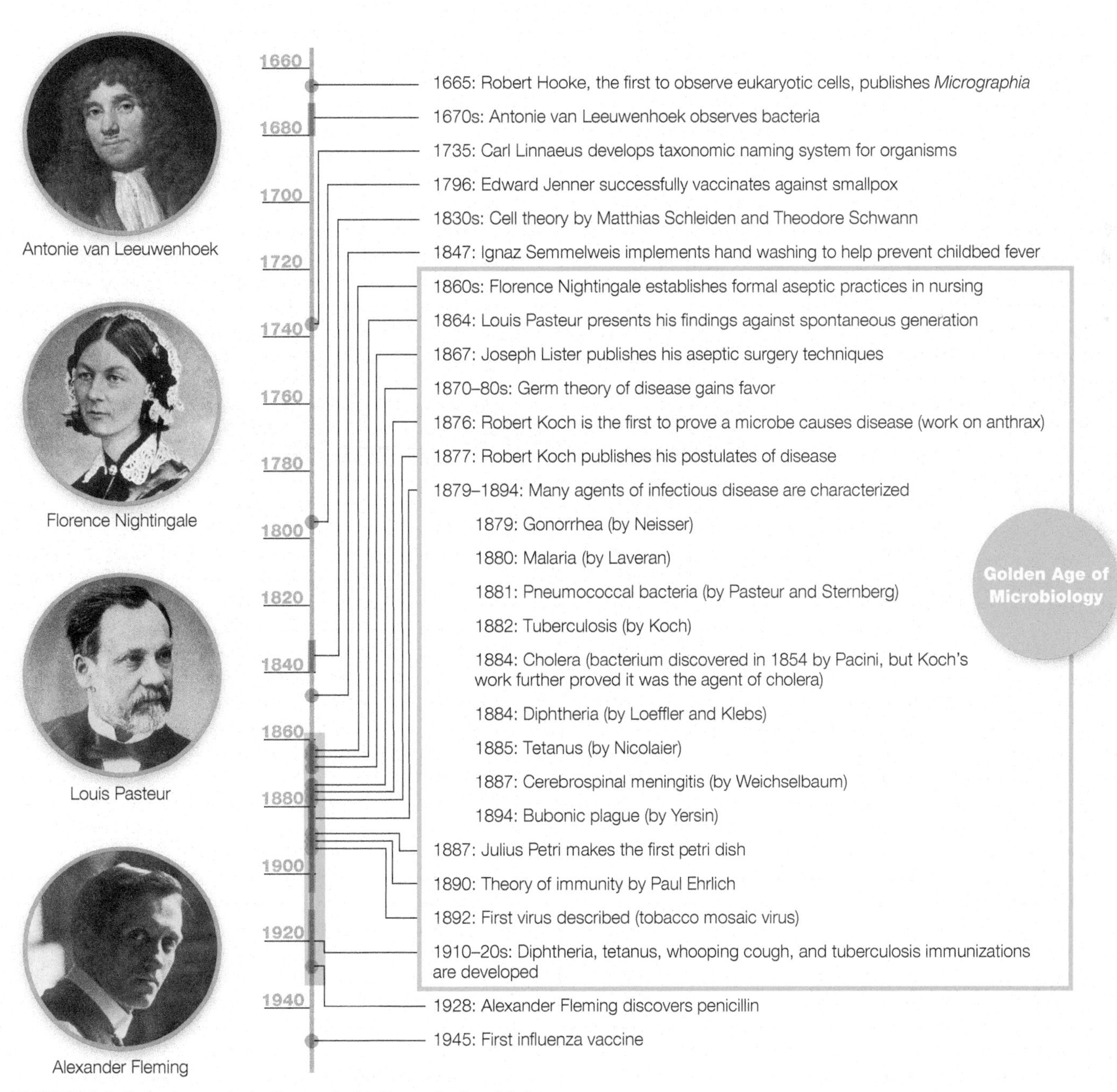

FIGURE 1.1 Select events in the early history of microbiology

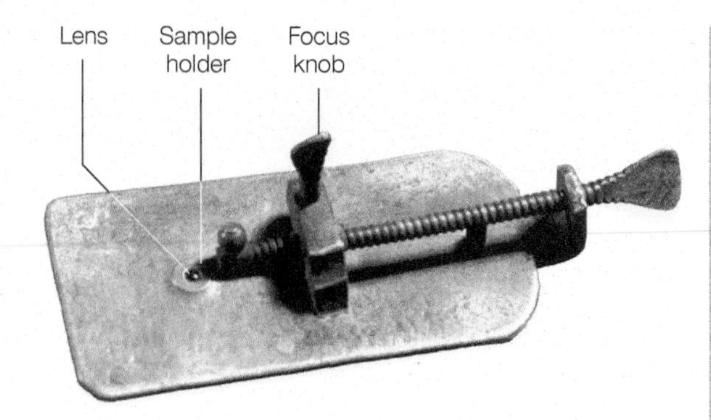

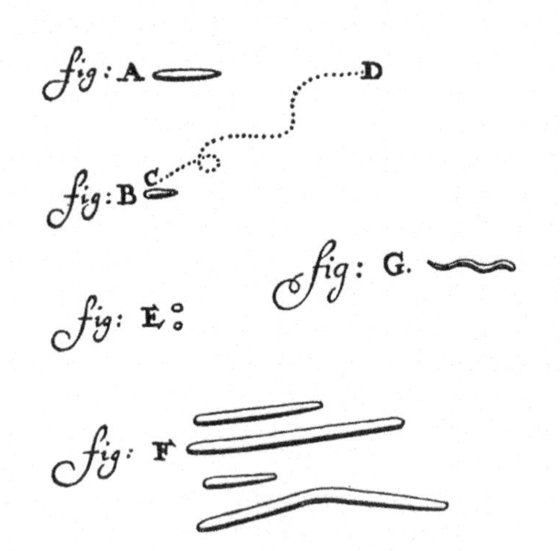

FIGURE 1.2 First views of bacteria Antonie van Leeuwenhoek was the first to report descriptions of bacterial cells. *Top*: Antonie van Leeuwenhoek used a small handheld microscope that had at best a 300 × magnification capability. *Bottom*: Leeuwenhoek's drawings of "very little animalcules."

CHEM • NOTE

Fermentation is a chemical process that often leads to the production of acids and/or alcohols. Fermentation has long been exploited by people who tap microbes to create wine, beer, and vinegar, among other food products.

Spontaneous Generation versus Biogenesis

In the mid-1600s, Robert Hooke used crude microscopes to view a variety of tiny structures, from fleas to snowflakes, and became the first scientist to publish descriptions of cells. Antonie van Leeuwenhoek, a contemporary of Hooke, refined earlier versions of the microscope and became the first to see bacteria (FIG. 1.2). Like many of their contemporaries, these scientists participated in the debate about **spontaneous generation**, an idea that life comes from nonliving items, and **biogenesis**, the idea that life emerges from existing life.

Several other 17th-century scientists, including Francesco Redi, performed experiments to test the hypothesis of spontaneous generation. One piece of evidence often cited as proof of spontaneous generation was that rotting meat gave rise to maggots. To further explore this evidence, Redi placed one piece of meat in an uncovered jar and a second piece of meat in a jar with a gauze-covered top. The uncovered meat gave rise to maggots because flies could lay their eggs on it. The meat in the covered jar did not give rise to maggots, because flies were unable to touch the meat.

One would think Redi's experiments would have debunked spontaneous generation, but the theory persisted for another 200 years until the late 1800s, when Louis Pasteur showed that biogenesis is responsible for the propagation of life. Before his experiments, stored wine often fermented and turned bitter. A common explanation for the bitterness was spontaneous generation of yeast during the fermentation process. Pasteur proved that yeast *performed* fermentation—they were not spontaneously generated by it. He showed that by heating new wine to 50–60°C, he could kill off the yeast within it and delay spoilage for years. We now know this heating process as **pasteurization**. It is most commonly used to treat milk and juices to render these foods safe for consumption and to slow spoilage.

Convinced that air contained contaminating microbes, Pasteur further investigated his hypothesis by performing an experiment with a specialized S-necked flask that was partially filled with broth (FIG. 1.3). He boiled the broth and showed

FIGURE 1.3 Pasteur's experiment Louis Pasteur's S-necked flask experiment disproved spontaneous generation.

Critical Thinking *Why was it important that the broth was heated in the same flask as it was cooled?*

that it remained unspoiled because microbes in the air were trapped in the bent portion of the flask, unable to reach the liquid below. When the flask was shaken, broth encountered the microbes previously trapped in the curved neck, and the broth would then spoil. Pasteur's work went beyond disproving spontaneous generation. He also developed the first vaccines to protect against anthrax and rabies (two deadly diseases that can be transmitted to humans from other animals) and he had a significant role in solidifying the germ theory of disease.

Germ Theory of Disease

The **germ theory of disease** states that microbes cause infectious diseases. From the late 1800s forward, determining the specific **etiological**, or causative, agent of an infectious disease became an important role of microbiology labs. Despite over a century of research, we are still nowhere near discovering the specific etiological agent of every infectious disease. We'll probably never have a complete catalog of every infectious agent, because even as we make progress identifying them, new diseases emerge. Further complicating matters is the fact that current laboratory techniques enable us to grow only about 2 percent of the bacterial species found in our environment.

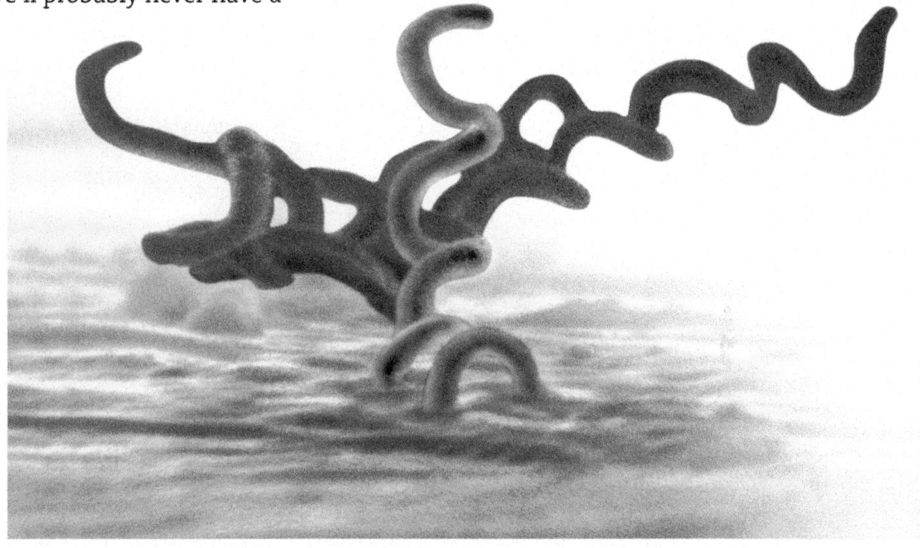

T. pallidum

Although laboratory limitations and the evolving nature of microbes make characterizing the etiological agents of disease more challenging, it does not make it impossible. For example, the bacterium *Treponema pallidum* was discovered to be the cause of syphilis over 100 years ago, yet it was only in 2018 that the bacterium was successfully cultured *in vitro* (meaning "in glass," or in an artificial setting).[3] Prior to 2018, it had only been sustained *in vivo* (meaning "in the living"—in animal models) using rabbits.

Louis Pasteur and his contemporary Robert Koch (pronounced "coke") both reinforced the germ theory of disease. Koch was a German physician who developed staining techniques and ways to grow and isolate bacteria. Some of Koch's most groundbreaking work started with the study of anthrax, which primarily infects grazing animals but, as mentioned earlier, can infect humans. (For more on anthrax, see Chapter 17.) Koch discovered that anthrax is caused by a bacterium, which he named *Bacillus anthracis*. He was able to isolate *Bacillus anthracis* from diseased animals, then introduced the purified bacteria into mice to establish an infection.

Koch's Postulates of Disease

Koch's work on anthrax led to the development of **Koch's postulates of disease** (FIG. 1.4). These four principles establish the criteria for determining the causative agent of an infectious disease. (Koch's postulates are briefly listed here, but are discussed in more detail in Chapter 9.)

1. The same organism must be present in every case of the disease.
2. The organism must be isolated from the diseased host and grown as a pure culture.
3. The isolated organism should cause the disease in question when it is introduced (inoculated) into a **susceptible host** (a host that can develop the disease).
4. The organism must then be re-isolated from the inoculated, diseased animal.

---

[3] Edmondson, D. G., Hu, B., & Norris, S. J. (2018). Long-term in vitro culture of the syphilis spirochete *Treponema pallidum* subsp. pallidum. *MBio*, 9(3), e01153-18.

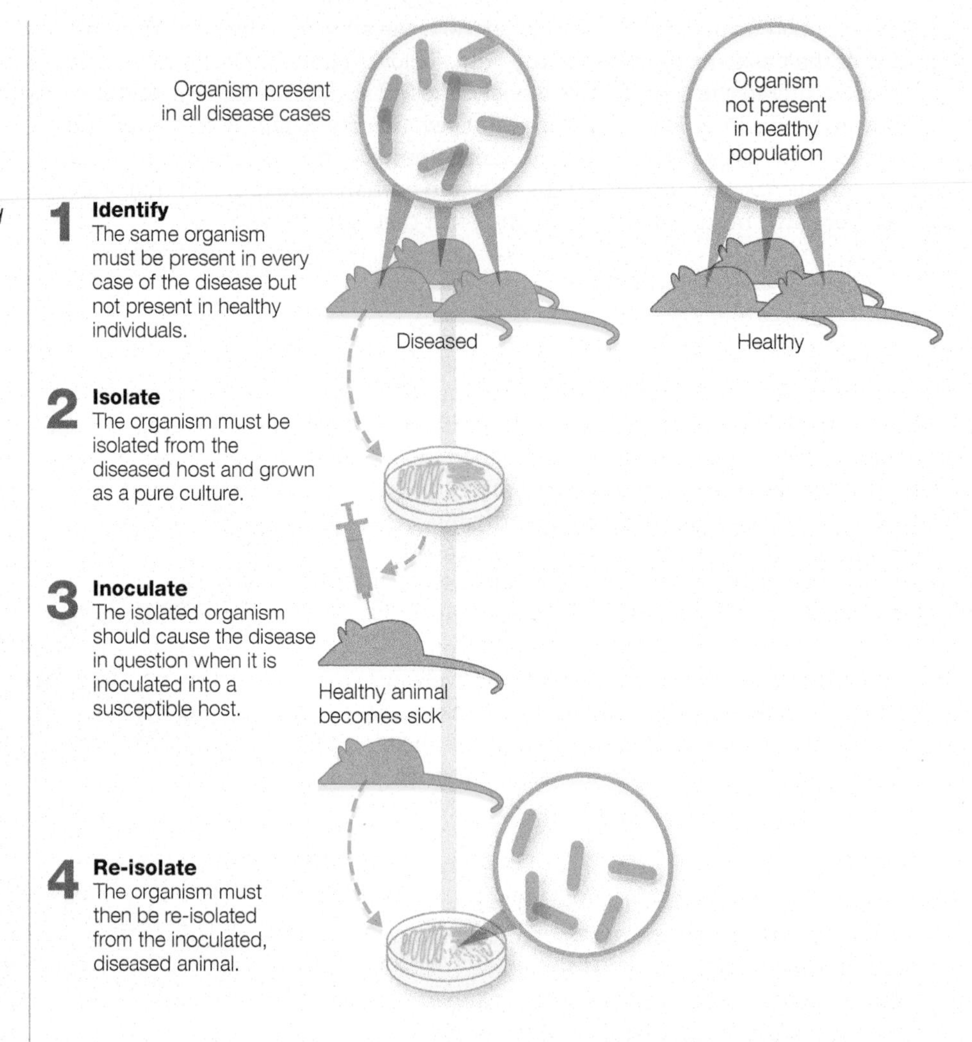

FIGURE 1.4 Koch's postulates of disease Koch's postulates can help identify the causative agent of certain infectious diseases.

Critical Thinking *It is not ethical to purposefully expose people to a suspected pathogen to fulfill Koch's third postulate. With this in mind, how could the third step be ethically observed if the disease being studied only occurred in humans and an animal model could not be used?*

In his studies, Koch confirmed that not all infections cause evident disease. This is why the third postulate states that disease *should* result, but avoids the term *must* as is found in his other statements.

Hand Hygiene and Aseptic Techniques

From the 1800s through early 1900s, as the biogenesis theory and the germ theory of disease were being developed and debated, several medical professionals were emphasizing the importance of aseptic techniques. In a medical setting, **aseptic technique** entails preventing the introduction of potentially dangerous microbes to a patient; it doesn't mean that everything in the environment needs to be **sterile** (absent of all microbes). While most surfaces in an operating room or other healthcare setting are disinfected to limit potentially dangerous microbes, they are not sterile. Aseptic procedures are central to health care because they prevent **healthcare-acquired infections** or **HAIs** (also known as **nosocomial infections** or **healthcare-associated infections**) and limit the spread of diseases.

Aseptic procedures also allow us to safely study microbes in the laboratory; moreover, they are necessary for maintaining pure samples so that we can study one microbe at a time. The type of aseptic technique used depends on the situation, but aseptic procedures usually include washing hands, wearing gloves, sterilizing instruments, and decontaminating surfaces. (See Chapter 10 for more on healthcare biosafety precautions.)

A Hungarian physician named Ignaz Semmelweis first developed aseptic techniques in the 1840s. He recommended that physicians and other care providers practice hand washing to decrease mortality rates from childbed fever (puerperal sepsis), an infection that killed many women in childbirth before the antibiotics era.

Semmelweis saved many lives, yet his work was not fully appreciated until about 20 years later, when Pasteur disproved spontaneous generation. Work by Semmelweis and Pasteur inspired the British surgeon Joseph Lister to investigate processes for aseptic surgery. Lister's work in the 1860s proved that sterilizing instruments and sanitizing wounds with carbolic acid encouraged healing and prevented pus formation. Around the same time, Florence Nightingale established the use of aseptic techniques in nursing practices, which, along with other patient-care innovations, led to her being recognized as the founder of modern nursing.

CHEM • NOTE

Carbolic acid (C_6H_6O), also known as phenol, is an organic molecule with antiseptic (degerming) and anesthetic (numbing) properties. It can be found in sore throat sprays, lip balms, and various household cleaners, and is also used for making plastics. Phenol can be commercially produced but also occurs naturally in many foods. Concentrated phenol is toxic to humans and is regulated.

The scientific method is the guiding investigative principle for microbiology.

Before the modern age, illness was often attributed to evil spirits or sinfulness. As such, early medical treatments almost always included some form of penance, pilgrimage, or protective charm. Early physicians thought illness derived from an imbalance of the body's humors that could be relieved by bloodletting or applying leeches to the body (FIG. 1.5).

Today, by contrast, we explore questions about the origins of diseases and potential treatments through the **scientific method**. In its most basic form, the scientific method starts with a question that can be investigated. Next, a **hypothesis**—a prediction based on prior experience or **observation**—is proposed as a potential answer to the question. The researcher collects and analyzes observations (data) and uses them to formulate a **conclusion** that states whether the data supported or contradicted the hypothesis.

Observations versus Conclusions

Failure to recognize the difference between observations and conclusions leads to errors and confusion. An observation is any data collected using our senses or instrumentation, whereas a conclusion interprets observations. For example, suppose you witnessed a robber driving a getaway car with a Florida license plate. If you tell the detective that the robber is from Florida instead of saying the robber's car had a Florida plate, you might mislead the investigation. This scenario is an example of *inference–observation confusion*—more commonly known as "jumping to conclusions." In contrast, science requires a collection of observations and many different methods of testing to draw accurate conclusions. It is essential that healthcare workers avoid inference–observation confusion because it can lead to an inaccurate assessment of the patient. We should also recognize the limits of experimental design and what we can truly conclude from our observations.

Law versus Theory

The difference between a scientific law and a scientific theory also confuses many students. A **law** is a precise statement, or mathematical formula, that predicts a specific occurrence. Laws only hold true under carefully defined and limited circumstances. By contrast, a **theory** is a hypothesis that has been proven through many studies with consistent, supporting conclusions. Laws predict *what* happens, while theories explain *how* and *why* something occurs. Furthermore, theories encompass laws. For example, the theory of evolution, which explains how and why organisms change over time, includes a number of guiding laws such as natural selection promoting the survival of the fittest. Unlike a hypothesis, which focuses on a specific problem, theories are comprehensive bodies of work that are useful for making generalized predictions about natural phenomena. Theories unite many different hypotheses and laws.

FIGURE 1.5 **Bloodletting** For about 3,000 years, bloodletting was practiced as a primary medical therapy for practically every ailment.

Sometimes people think that laws must be superior to theories—likely because of the way laws are defined and enforced within governments. But one should not equate the social definition of a law with the scientific one. In science, laws and theories have completely different goals and facilitate discovery in different ways. Thus, the idea that a theory would be "elevated" to a law is inaccurate; to do so would be like turning an apple into an orange. Moreover, neither laws nor theories are considered final. Scientists continue to retest laws and theories as our technology and knowledge increase.

Build Your Foundation

Build Your Foundation (BYF) Quick Quiz: Visit the **Mastering Microbiology** Study Area to quiz yourself.

1. Microbiology is the study of living and nonliving microscopic entities. Explain. (LO 1.1)
2. What is a pathogen, and how is it different from an opportunistic pathogen? (LO 1.2)
3. Distinguish between biogenesis and spontaneous generation, and discuss how Pasteur disproved spontaneous generation. (LO 1.3)
4. List Koch's postulates of disease, and describe how they contributed to the germ theory of disease. (LO 1.4)
5. The term *aseptic* does not mean 100 percent sterile. Explain why. (LO 1.5)
6. How did Lister, Semmelweis, and Nightingale contribute to health care? (LO 1.6)
7. In a lab report, a student wrote that a red color developed in the test tube being used. Is this an observation or a conclusion? Explain. (LO 1.7)
8. How is a scientific law different from a theory? (LO 1.7)

1.2 CLASSIFYING MICROBES AND THEIR INTERACTIONS

Learning Outcomes

After reading this section, you should be able to:

1.8 Summarize the taxonomic hierarchy from domain to species.

1.9 Define the term *strain*.

1.10 Describe the binomial nomenclature system and the information it provides about an organism.

1.11 Define the terms *parasitism*, *mutualism*, and *commensalism*.

1.12 Define the term *normal microbiota*, and discuss its establishment and roles.

1.13 Describe a host–microbe interaction that influenced human evolution.

1.14 State how a biofilm forms, and discuss the healthcare implications of biofilms.

1.15 Provide examples of how microbes affect industry and the environment.

Taxonomy groups organisms.

Microbial classification is important both for study and for clear communication among researchers and healthcare providers. **Taxonomy** is the study of how organisms can be grouped by shared features—to include physical features (**morphology**) and physiological characteristics. In the mid-1700s, Carl Linnaeus established criteria for classifying and naming organisms. He is now recognized worldwide as the father of taxonomy.

Taxonomic Hierarchy

There are eight rankings within the taxonomic hierarchy. From broadest to narrowest they are: *d*omain, *k*ingdom, *p*hylum, *c*lass, *o*rder, *f*amily, *g*enus, *s*pecies. The mnemonic device, "*D*elightful *K*ing *P*hilip *c*ame *o*ver *f*or *g*reat *s*paghetti" may help you remember the taxonomic hierarchy.

Domains The broadest category, **domain**, primarily groups organisms by cell type. There are three domains: Bacteria, Archaea, and Eukarya. Domains Bacteria and Archaea encompass prokaryotic organisms—all of which are unicellular and lack a nucleus (an intracellular structure that houses a cell's genetic information). Most of the prokaryotes you will learn about in this text belong to the Domain Bacteria—this grouping includes potential pathogens.

Members of Domain Archaea are best known for living in extreme environments: high-temperature deep-sea vents, areas of bitter cold, or environments with harsh chemical conditions usually devoid of other lifeforms. However, they can also live in normal environments and are also found in the human gut and on our skin. To date, Archaea members have not been shown to be pathogens.

TABLE 1.2 Six-Kingdom Classification System

Kingdom	Archaea	Bacteria	Fungi	Plantae	Animalia	Protists*
Example	*Sulfolobus*	*S. aureus*	*Candida albicans*	(Plants)	(Animals)	*Paramecium*
Domain	Archaea	Bacteria	Eukarya	Eukarya	Eukarya	Eukarya

*Not a true kingdom; a catchall category for lifeforms formerly grouped in Kingdom Protista

The Domain Eukarya encompasses unicellular and multicellular organisms that are made of eukaryotic cells—that is, cells that have a distinct nucleus. There are four main types of eukaryotic cells: animal, plant, fungal, and protist; you'll see versions of these names in the next section on kingdoms.

Kingdoms Beneath the umbrella of domains are a variety of **kingdoms**, the designated number of which has fluctuated from five to eight. The older five-kingdom classification scheme includes the eukaryotic kingdoms Animalia, Plantae, Fungi, Protista, and the prokaryotic kingdom Monera. The trouble with the five-kingdom scheme is that it fails to assign separate kingdoms for Domain Archaea and Domain Bacteria, instead lumping these prokaryotic domains together into Kingdom Monera.

The six-kingdom schematic, which is what this text employs, replaces Kingdom Monera with Kingdom Archaea and Kingdom Bacteria (TABLE 1.2). This scheme also isn't perfect: Traditionally, Protista was a sort of miscellaneous catchall kingdom for organisms that couldn't be described as plants, animals, or fungi. Genetics research now shows that protists can't logically be lumped into a single kingdom. However, the term "Kingdom Protista" persists, and for convenience's sake, this book also continues to informally refer to these diverse organisms as protists. To the ardent taxonomist, this is perhaps a heretical approach, but the goal here is not for you to become a taxonomy expert, just to explore these important lifeforms without feeling overwhelmed by the constantly changing world of taxonomy.

The disagreement on kingdoms emerges because taxonomy has become more than just the grouping of organisms with shared features. Because of new genetic analysis techniques, taxonomy is now often used as a tool to understand how organisms are evolutionarily related. There are a variety of styles for grouping organisms by relatedness in "the tree of life" and there likely will never be a unanimous agreement as to which tree is the best.

Phyla Through Species Regardless of how many kingdoms people acknowledge, each can be further subdivided into smaller and smaller groups: first phylum (phyla = plural); then class, order, family, and genus (genera = plural); and finally into the most precise grouping, species (FIG. 1.6).

Although each of these categories is useful, genus and species are arguably the most common groupings you'll encounter in an introductory microbiology course. You can think of a **genus** as a group of related species. For example, the genus *Canis* includes wolves (*Canis lupus*) and dogs (*Canis familiaris*). Traditionally, a species is defined as a group of similar organisms that sexually reproduce together, or breed. Such a definition doesn't work for prokaryotic cells like bacteria because they reproduce asexually. Instead, **prokaryotic species** are defined as cells that share physical characteristics, but also have at least 70 percent DNA similarity (based on the degree to which their DNA can stably pair

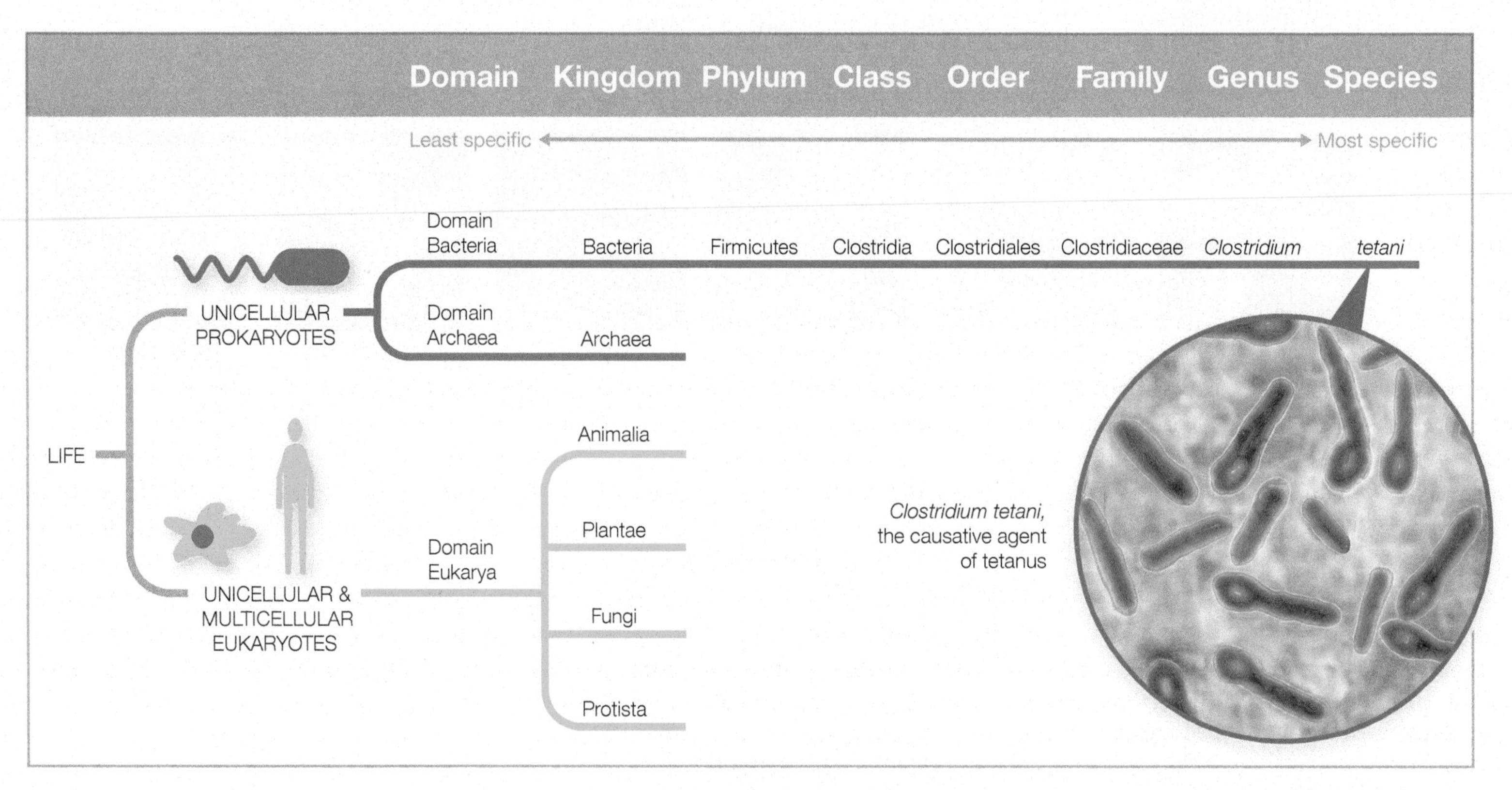

FIGURE 1.6 Classifying organisms This figure shows the taxonomic hierarchy based on a three-domain and six-kingdom system. Domains can be systematically broken down until the most specific ranking, species, is reached. The classification hierarchy for the causative agent of tetanus, *Clostridium tetani*, is shown as an example. *Note that Protista isn't a kingdom in the classical sense, but a catchall category that persists as a matter of convenience.

with each other in solution).[4] Also, prokaryotic species are at least 97 percent identical when genetic material in their ribosomes (organelles that build proteins) is compared.[5]

Strains Many microbiologists argue that trying to systematically identify prokaryotic species might be an exercise in futility because prokaryotes transfer genes and can take up genetic material from their environment. Therefore, recognizing genetic variants is helpful to microbiologists. The term **strain** is used to recognize genetic variants of the same species. Different strains of a species typically have a hallmark genetic trait that warrants a special designation. Mutations and gene transfer often lead to new strains.

As you can see, the parameters for bacterial classification are diverse. Fortunately, the Society of American Bacteriologists (now the American Society for Microbiology, or ASM), has worked to unify the classification criteria for bacteria. Their efforts laid the foundation for *Bergey's Manual of Determinative Bacteriology*. *Bergey's Manual*, as it is often called, has evolved through numerous updates, but remains a cornerstone reference for bacterial identification and classification.

---

[4] Wayne, L. G., Brenner, D. J., Colwell, R. R., et al. (1987). Report of the ad hoc committee on reconciliation of approaches to bacterial systematics. *International Journal of Systematic Bacteriology*, 37(4), 463–464; Tindall, B. J., Rosselló-Móra, R., Busse, H. J., Ludwig, W., & Kämpfer, P. (2010). Notes on the characterization of prokaryote strains for taxonomic purposes. *International Journal of Systematic and Evolutionary Microbiology*, 60(1), 249–266.

[5] Ribosomes are made of RNA (ribonucleic acid) and proteins. Ribosomal RNA (rRNA) sequences change very little over time, which is why they can be used to determine evolutionary relatedness. Prokaryotic organisms that are closely related will be almost identical in their 16S rRNA sequences—one of several types of rRNA in prokaryotic ribosomes. In Chapter 5, Genetics, you will learn more about rRNA and ribosomes. Stackebrandt, E., & Goebel, B. M. (1994). Taxonomic note: A place for DNA–DNA reassociation and 16S rRNA sequence analysis in the present species definition in bacteriology. *International Journal of Systematic Bacteriology*, 44(4), 846–849.

Scientific Names

As a part of his work in taxonomy, Linnaeus also established a **binomial nomenclature system**, or two-name system, that is still used today. In this system, an organism's first name is capitalized and reflects the genus, whereas the second name is lowercase and designates the species. Scientific names are also italicized (or underlined if handwritten). When first referring to an organism in writing, the genus and species names are written out in full, and thereafter the name is usually abbreviated by using the capitalized first letter of the genus name, followed by a period, and then the lowercase species name in full. For example, *Staphylococcus aureus* can be written as *S. aureus* once the full name has been noted.

Often scientific names are also Latinized, meaning the names are Latin-sounding; in most cases they are nothing near real Latin. For example, in the name of the bacterium *Escherichia coli*, the genus name derives from the discoverer's name, Theodor Escherich; the species name *coli* reflects that the organism is abundant in the colon. Often a microbe's name refers to its discoverer, cell shape, cell arrangement, or other distinct traits that the person who coined the name found noteworthy. Occasionally, even narrower categories such as strain or subspecies may be noted after the species name. Strain names typically include numbers and/or letters after the species name. For example, a strain of *Escherichia coli* commonly found in laboratories is *E. coli* K-12.

Microbes may be friends or foes.

Some people think of all microbes as a pestilence to be eliminated. However, microbes constitute such a huge part of the Earth's biomass that eliminating them would be an ecological disaster. It is suspected that there are several million species of microbes in our world, but so far just over 7,000 microbes have been characterized.[6] One study showed that there are about 150 different species of bacteria just on the palms of our hands.[7] Fortunately, most microbes are helpful or neutral to human health. Only a small minority are pathogens.

Host–Microbe Interactions

A **symbiotic relationship** exists when two or more organisms are closely connected. Humans and dogs, for example, have a symbiotic relationship. Microbes and their human hosts have evolved a variety of symbiotic relationships, including those that hurt the host (**parasitism**), help the host (**mutualism**), or have no perceived benefit or cost to the host (**commensalism**).

Pathogens are described as having a parasitic relationship with their host. However, the term **parasite** is most commonly used to refer to specific organisms such as helminths (worms) and protozoans (certain protists), which are eukaryotic infectious agents with complex life cycles.

The relationship between humans and our **normal microbiota** (the collection of all the microbes that reside in and on the human body) was traditionally described as commensal. But as we learn more, it is increasingly obvious that our ecological relationship with our normal microbiota is often mutualistic—we benefit from our microbiota and they benefit too.

Indeed, our close ecological relationships with microbes—from our normal microbiota to other microbes encountered in our environment—have led humans and microbes to coevolve. The same is true of the interaction between microbes and other organisms. A classic example of pathogen influence on human evolution is seen with malaria. This mosquito-borne, tropical disease has plagued humans for at least 100,000 years. According to the World Health Organization (WHO),

---

[6] Achtman, M., & Wagner, M. (2008). Microbial diversity and the genetic nature of microbial species. *Nature Reviews Microbiology*, 6(6), 431–440.

[7] Fierer, N., Hamady, M., Lauber, C. L., & Knight, R. (2008). The influence of sex, handedness, and washing on the diversity of hand surface bacteria. *Proceedings of the National Academy of Sciences of the USA*, 105(46), 17994–17999.

TRAINING TOMORROW'S HEALTH TEAM

E. coli Strains: Little Differences Have Big Consequences

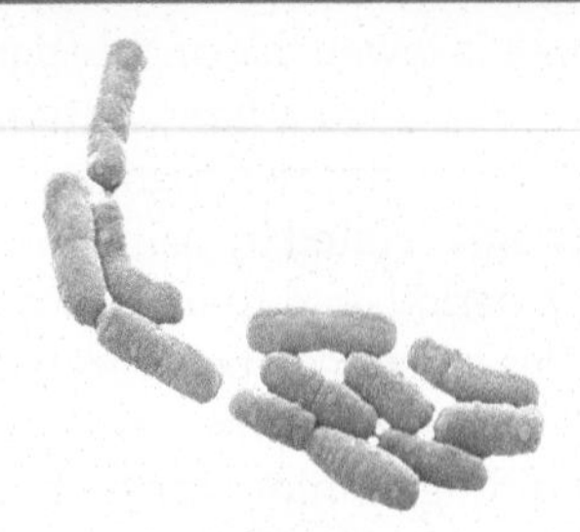

SEM of *E. coli* O157:H7, one of the causative agents of HUS.

Escherichia coli, the agent of much-publicized foodborne disease outbreaks, is also a normal resident of the human gut. Given that, why aren't more people suffering from the terrible symptoms generally associated with an *E. coli* infection?

The answer is that there are hundreds of *E. coli* strains, and most are harmless. But some, such as *E. coli* O157:H7, cause severe illness. The first recognized *E. coli* O157:H7 outbreak occurred in 1982 and was linked to undercooked hamburgers. However, more recent outbreaks have been associated with fruits and vegetables rather than meat.

The CDC estimates that up to 8 percent of *E. coli* O157:H7 infections result in hemolytic uremic syndrome (HUS), which leads to kidney failure. *E. coli* O157:H7 also makes a toxin known as Shiga-like toxin thanks to a gene that it acquired from *Shigella* species that cause dysentery. From a healthcare standpoint, rapidly identifying the culprit strain of an infection can be the difference between life and death for the patient. For example, studies have shown that treating *E. coli* O157:H7 with antibiotics may actually increase the risk of developing HUS by up to 17 times. The reason for this effect might be that the antibiotic damages the bacteria, causing a sudden and concentrated burst of toxin to be released. Instead of prescribing antibiotics, patients infected with *E. coli* O157:H7 should receive early administration of intravenous fluids, which may help protect the patient from HUS, or limit the damage of HUS.

QUESTION 1.1

Imagine you are a nurse caring for a child with an E. coli *O157:H7 infection. The patient's mother is concerned that an antibiotic is not being prescribed. She says that last year she received antibiotics for a urinary tract infection due to* E. coli. *How would you explain why the child needs different treatments against* E. coli?

malaria kills over 600,000 people every year. People who carry the gene for sickle cell anemia, a blood disorder characterized by a mutation in the gene for hemoglobin, are less likely to develop serious disease if infected by the protozoan pathogens that cause malaria. For this reason, carriers of the sickle cell gene have a survival advantage in areas where malaria is common. This helps explain why in some malaria-plagued regions of sub-Saharan Africa, up to 40 percent of tribal people carry the gene. People originating from parts of the world where malaria incidence is low rarely carry the sickle cell gene.

Normal Microbiota and the Human Microbiome

As previously mentioned, our normal microbiota—also referred to as our normal flora or the human microbiome—consists of all the microbes that reside in and on the human body. Our normal microbiota include bacteria, archaea, and eukaryotic microbes. They train our immune system, produce vitamins for us, and help us digest foods. Studies suggest that they may even impact our moods and brain function.[8] The *Human Microbiome Project* (HMP), an effort sponsored by the U.S. National Institutes of Health, was initiated to study the human microbiome and shed light on the links between these resident microbes and pathological conditions such as obesity, food allergies, heart disease, multiple sclerosis,

[8] Rea, K., Dinan, T. G., & Cryan, J. F. (2020). Gut microbiota: A perspective for psychiatrists. *Neuropsychobiology*, *79*(1–2), 50–62.

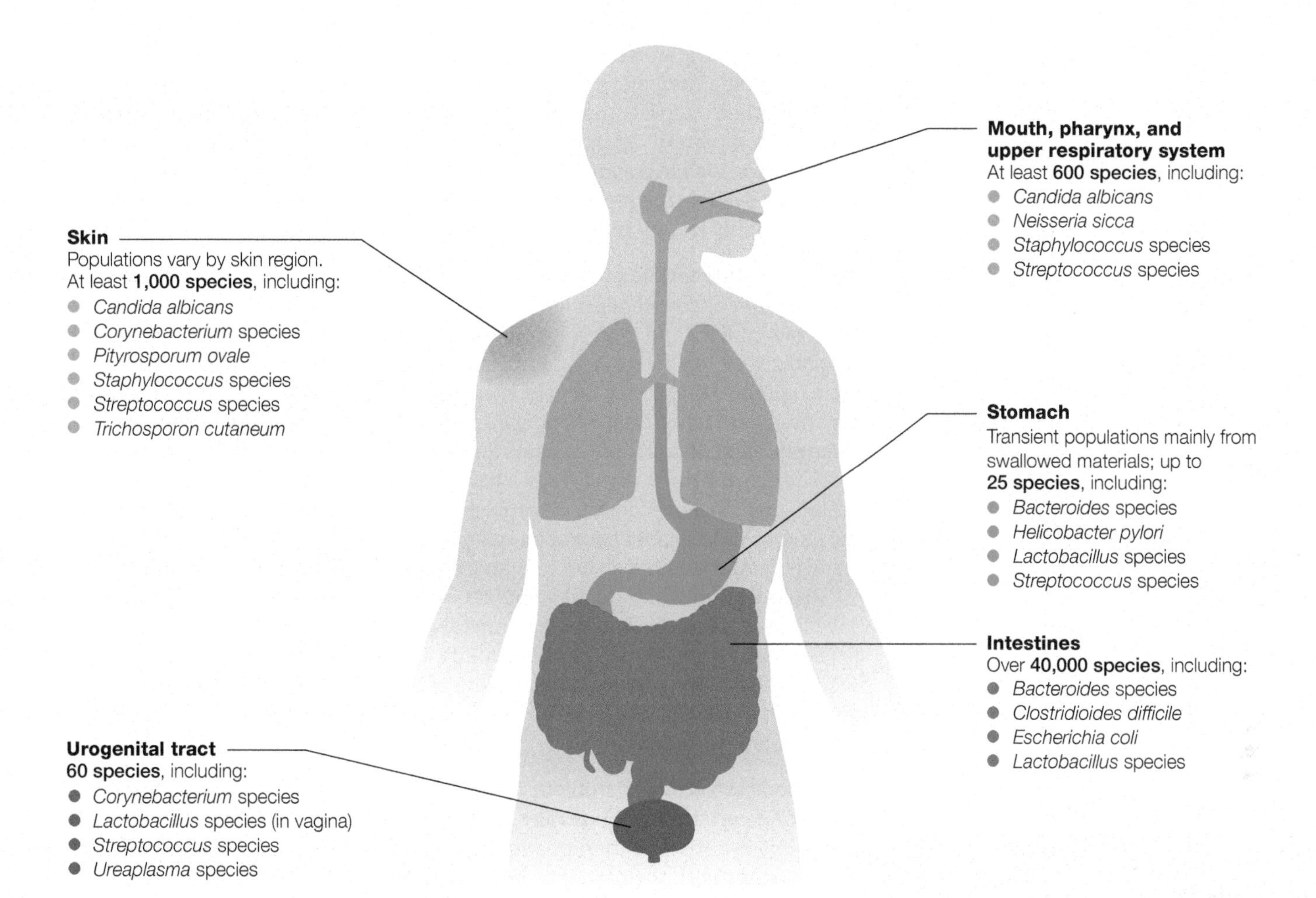

FIGURE 1.7 Our normal microbiota The human microbiome project is characterizing the microbes that call us home. Note the diversity and staggering variety of microbes in and on our bodies.

Critical Thinking *Based on the data presented in the figure, you can see that certain groups of microbes prevail in one body area over another. For example, the streptococci are much more abundant in the mouth than in other body regions. Why do you think microbes dominate a particular region of the body over another?*

and diabetes. Many parts of the human body teem with microbial life; there are at least as many microbial cells in and on us as there are human cells.[9]

Our skin, nose, mouth, intestinal tract (primarily the large intestine), and genital/urinary tract tend to harbor the most microbes; however, certain species are favored in each biological niche of our body (FIG. 1.7). For example, the skin's salinity limits the types of microbes that can thrive there to those that can tolerate salt. Likewise, the low-oxygen environment in the intestinal tract limits which microbes call the gut home. However, there is still tremendous diversity of life in and on us. Our gut microbiota alone is thought to include almost 5,000 bacterial species.[10] That said, not all parts of our body house normal microbiota. For example, the brain and spinal cord of healthy individuals are both microbe-free zones. (The normal microbiota of each body system is discussed in Chapters 16–21.)

What makes microbiota "normal" sometimes has to do with the location of the microbe rather than the species itself. For example, our normal microbiota often includes pathogens—27 percent of adults asymptomatically carry *Staphylococcus aureus* on their skin. However, this bacterium can cause skin infections by entering a cut, or food poisoning when toxins made by certain strains are ingested. It can even cause life-threatening sepsis if it enters the bloodstream. Fortunately, the majority of normal microbiota is harmless and tends to protect us by "crowding out" potential pathogens that might otherwise grow on or

[9] Sender, R., Fuchs, S., & Milo, R. (2016). Revised estimates for the number of human and bacteria cells in the body. Preprint on *bioRxiv*. Online only; doi: http://dx.doi.org/10.1101/036103.

[10] Almeida, A., Nayfach, S., Boland, M., et al. (2021). A unified catalog of 204,938 reference genomes from the human gut microbiome. *Nature Biotechnology*, 39, 105–114.

within us. Our understanding of how normal microbiota impacts health and disease is just emerging.

Initial data suggests that some people's microbiome profile may increase the chance of certain chronic diseases or disorders, whereas other profiles may be protective. We have a long way to go in exploring potential links between the microbiome and human physiology, but a better understanding of these links may eventually lead to retooling of normal microbiota to treat certain diseases.

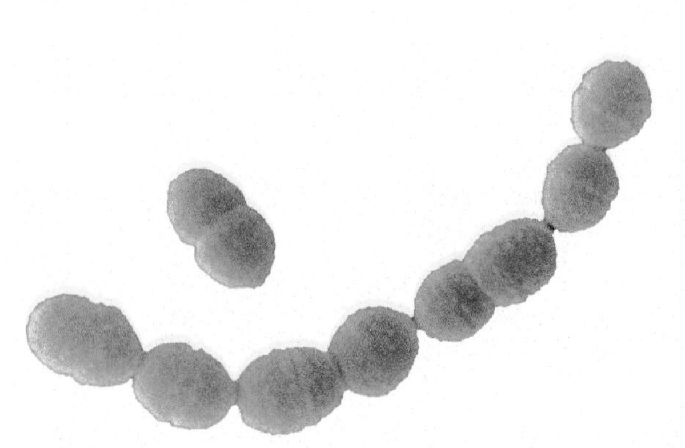

Streptococcus cremoris, SEM can be found as part of the normal microbiota.

Establishing Normal Microbiota Babies are colonized in their first days of life by the microbes they encounter during childbirth and through early interactions with their environment and caregivers. However, data suggests that microbes may start to colonize us even before birth. Some researchers have found that the placenta, an organ unique to pregnancy that transfers oxygen, nutrients, and certain antibodies from mother to fetus, may harbor low levels of microbes, although the existence of a true placental microbiome remains debatable.[11] While additional research still needs to be performed, some studies have documented that the bacterial profile of a baby's first stool (meconium) resembles bacterial profiles documented for amniotic fluid.[12] Also, bacteria have been isolated from fetal lung samples, further suggesting that the womb may not be sterile after all.[13]

Regardless of when initial colonization by normal microbiota takes place, it is well established that our normal microbiota expands and continues to develop throughout the early weeks of life and evolves as we approach adulthood. The developing normal microbiota of an infant is greatly influenced by whether the baby is born by cesarean section or vaginal birth, as well as by the choice of breast milk or formula for infant feeding.

Disruptions in Normal Microbiota When our normal microbiota is perturbed, our risk for certain infections may increase. A common disruption to normal microbiota is antibiotic therapy, which may kill many types of benign resident bacteria along with the pathogen being targeted. With normal microbiota reduced, opportunistic pathogens are more likely to thrive and establish infections. A common example of this is when a woman takes antibiotics to treat a urinary tract infection (UTI), only to develop a vaginal yeast infection (vulvovaginal candidiasis) soon after the antibiotic therapy concludes. The yeast *Candida albicans* is an opportunistic pathogen that is usually present in the vagina. Its growth is normally kept in check by vaginal microbiota—especially *Lactobacillus* bacteria, which make lactic acid that keeps the vaginal pH low. As antibiotics kill off these beneficial bacteria, the vaginal pH increases to a level that allows *Candida albicans* to thrive and cause vulvovaginal candidiasis. Similarly, diarrhea is a common side effect of antibiotic therapies because they disrupt the gut microbiome. Hormonal changes, diet, age, and our general environment also contribute to shifts in our normal microbiota.

Transient Microbiota Some microbes are just temporary passengers that do not persist as stable residents of our bodies. These **transient microbiota** (or *transient flora*) can be picked up through a handshake or contact with environmental surfaces. Most acquired pathogens are transient microbiota. Unlike normal microbiota, transient microbiota can be removed through hygiene—especially via proper hand-washing technique.

Probiotics and Health

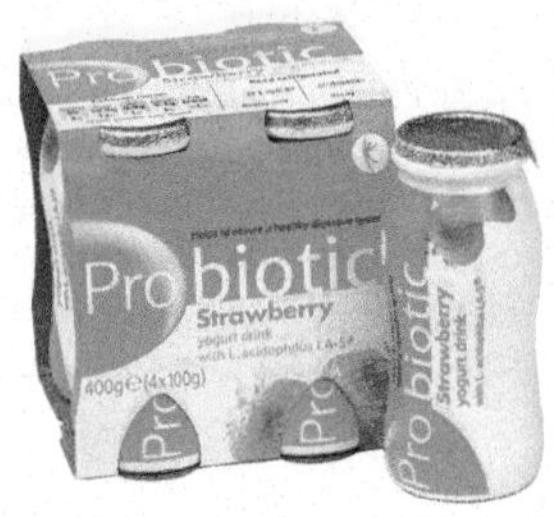

Probiotics are defined as microorganisms that provide a health benefit when ingested. In the past few decades, nutritional supplements containing living bacteria such as *Lactobacillus* and *Bifidobacterium* have become popular in the United States. These organisms can be swallowed in pill form or added to foods like yogurt or even baby formula in a powder form. Studies indicate that certain probiotics are helpful in combating or even preventing some forms of diarrhea (especially antibiotic-induced diarrhea), vaginal yeast infections, and urinary tract infections. There is also data to suggest that probiotics may help ease eczema and irritable bowel disease. There is still much research to be done, but the outlook is promising.

QUESTION 1.2

Why do you think probiotics may prevent or alleviate antibiotic-associated diarrhea?

[11] Kuperman, A. A., Zimmerman, A., Hamadia, S., et al. (2020). Deep microbial analysis of multiple placentas shows no evidence for a placental microbiome. *BJOG: An International Journal of Obstetrics & Gynaecology*, 127(2), 159–169.
and
Sterpu, I., Fransson, E., Hugerth, L. W., et al. (2021). No evidence for a placental microbiome in human pregnancies at term. *American Journal of Obstetrics and Gynecology*, 224(4), 296.e1–296.e23.

[12] He, Q., Kwok, L. Y., Xi, X., et al. (2020). The meconium microbiota shares more features with the amniotic fluid microbiota than the maternal fecal and vaginal microbiota. *Gut Microbes*, 12(1), 1794266.

[13] Al Alam, D., Danopoulos, S., Grubbs, B., et al. (2020). Human fetal lungs harbor a microbiome signature. *American Journal of Respiratory and Critical Care Medicine*, 201(8), 1002–1006.

Biofilms

Biofilms are sticky communities made up of single or diverse microbial species (FIG. 1.8). They allow microbes to coordinate responses within an environment, making the community much more durable than single, **planktonic** (free-floating) bacteria. The cells that seed a biofilm often have adhesion factors such as fimbriae (short, hair-like projections discussed in Chapter 3) to help them attach to a target surface. They then secrete a sticky substance that forms a protective matrix in which the bacteria grow. Multiple layers tend to develop in the biofilm as populations expand, with the residents of the innermost layers being highly protected. Periodically, microbes in the film are released as free-growing planktonic cells, allowing them to spread to other areas and serve as a source for chronic infection.

Dental plaque is an example of a biofilm. Each day we brush our teeth to remove it and prevent dental caries (cavities) from forming, but it persists on teeth despite brushing, so the plaque quickly grows back. Biofilms can develop on nearly any surface, including contact lenses, water filtering units, cutting boards, and catheters. The National Institutes of Health estimate that between 60 and 80 percent of infectious diseases in humans are due to biofilm-creating microbes. Whereas dental plaque can be scrubbed off regularly to limit cavities, internal biofilms are not so readily managed. They are more resistant to antibiotic treatments and more difficult for our immune system to destroy than planktonic bacteria. Numerous conditions, including chronic lung infections in cystic fibrosis patients, kidney stones, inner ear infections, atherosclerosis, endocarditis, and urinary tract infections involve biofilms. Scientists are working to understand how bacteria establish them and what allows them to persist, in the hope of creating better treatments against them. (See more on biofilms in Chapter 5.)

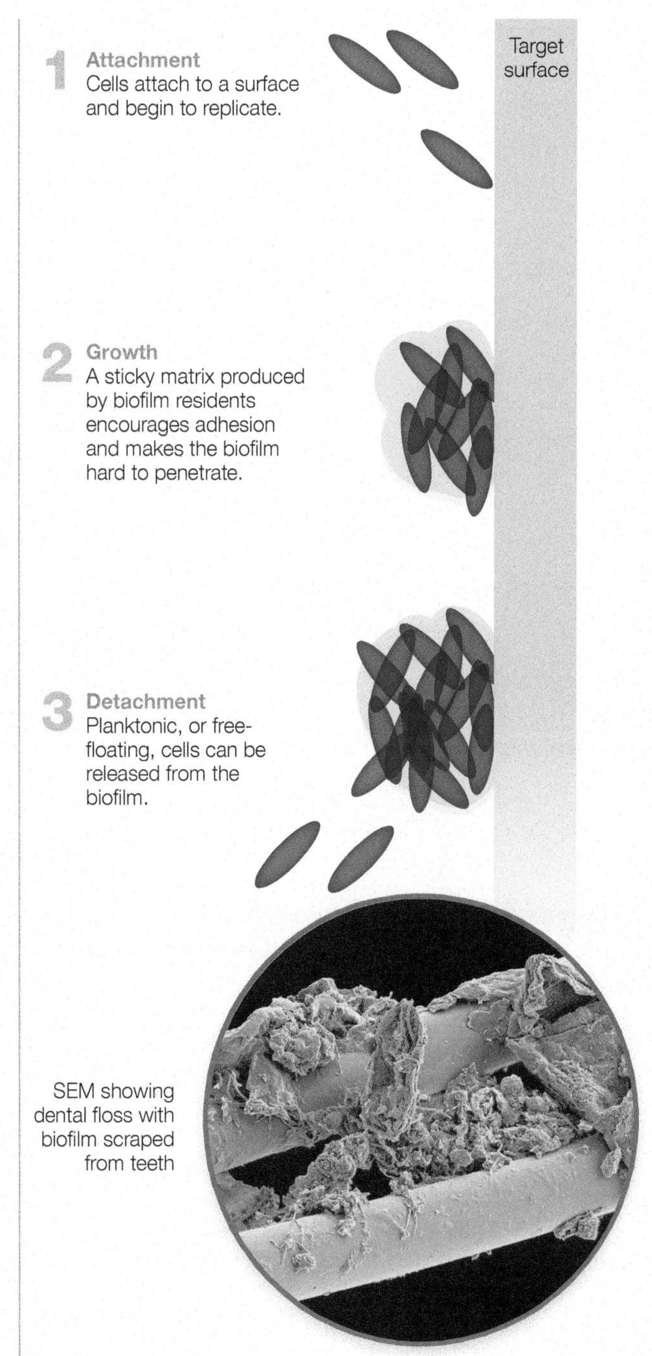

FIGURE 1.8 Biofilms Microbes form biofilm communities. Biofilms are medically concerning because they are difficult to treat, can affect medical implants and devices, and can serve as a chronic source of infection.

Critical Thinking *Drug doses sufficient to kill planktonic cells may be ineffective against biofilm microbes. Explain why this makes sense.*

Environmental and Industrial Uses for Microbes

It is well known that environmental toxins affect human health. The Environmental Protection Agency (EPA) documents thousands of chemical spills per year in the United States alone. **Bioremediation** harnesses the power of microbes to help clean up toxic waste. In bioremediation, certain nonpathogenic microbes are used to metabolize toxic substances into harmless intermediates. For example, hundreds of bacteria, archaea, and fungi species degrade petroleum oil spills into carbon dioxide. Bioremediation typically involves introducing nitrogen, sulfur, phosphate, and sometimes iron supplements to a spill environment along with dispersants, which break oil into smaller droplets to encourage microbes to more effectively break down the oil slick.

No cure is without a cost, and even bioremediation has its downsides, such as disturbing the microbe balance in the area, which can decrease the dissolved oxygen levels of the local aquatic environment and adversely impact the health of other organisms. The dispersants used to aid bioremediation might also harm the environment. There is no easy or obvious answer during environmental disasters, and ultimately, decisions come down to a cost/benefit analysis that tries to pick the course of action that will do the least harm. Oil slicks are not the only hazards to consider: Arsenic, mercury, and selenium are also dangerous contaminants that, in part, can be removed from contaminated soil and water through bioremediation.

In addition to cleaning up our environment, microbes have major roles in other industries. Bacteria and yeast serve as key ingredients in recipes for vinegar, beer, wine, chocolate, cheese, sauerkraut, kimchee, yogurt, sour cream, buttermilk, and bread. We tap microbes to help us produce drugs, including bacitracin, erythromycin, penicillin, and streptomycin. Microbes make important precursors for consumer products and are a source for biodegradable plastics. For example, *Xanthomonas campestris* (ZAN-tho-moe-nas kam-PES-tris) bacteria make a natural polysaccharide commonly called xanthan gum, which is used in a variety of cosmetics and foods as a thickening agent. Microbes are even a potential source for biofuels—researchers are working to engineer *E. coli* that can metabolize sewage to produce diesel fuel.

Build Your Foundation

9. List the taxonomic rankings in order from the broadest classification to the most specific. (LO 1.8)
10. What is a strain? (LO 1.9)
11. Explain the parts of a scientific name and the basic rules for writing one. (LO 1.10)
12. Compare and contrast parasitism, commensalism, and mutualism. (LO 1.11)
13. Describe the term *normal microbiota*. Give examples of how it could be disrupted, and discuss why disruption of normal microbiota is medically concerning. (LO 1.12)
14. Why has the sickle cell trait been perpetuated in populations exposed to malaria? (LO 1.13)
15. What are biofilms, and why are they a medical concern? (LO 1.14)
16. What are the pros and cons of bioremediation? (LO 1.15)
17. List three industrial uses for microbes. (LO 1.15)

Build Your Foundation (BYF) Quick Quiz: Visit the **Mastering Microbiology** Study Area to quiz yourself.

1.3 GROWING, STAINING, AND VIEWING MICROBES

Learning Outcomes

After reading this section, you should be able to:

1.16 Discuss the main formats of culture media used in the laboratory.

1.17 Describe the goal of aseptic culture technique, and note elements that are central to it.

1.18 Explain the goal of the streak plate technique, and state why it is important in microbiology.

1.19 Summarize simple versus structural staining techniques, and discuss what information they provide about a sample.

1.20 Describe the Gram stain procedure, why it works, and sources of error in the procedure.

1.21 Explain how the acid-fast stain is performed, why it works, and its clinical uses.

1.22 Correctly label the parts of the compound light microscope, and describe the general features of bright field microscopy.

1.23 Define the term *resolution*, and describe how immersion oil and wavelength impact it.

1.24 Compare and contrast transmission electron, scanning electron, and fluorescence microscopy.

We culture microbes so we can study them.

The first step in studying a microbe is to try to grow it in the laboratory setting. This is easier said than done for the majority of known species. Unfortunately, microbes often require complex growth environments that are poorly understood, or perhaps not replicable with today's culturing techniques.

Introduction to Growth Media

Growth media or **culture media** (singular: *medium*) are mixtures of nutrients that support organismal growth in an artificial setting. Robert Koch, the scientist known for his disease postulates, assembled the first research team that used agar, a polysaccharide from seaweed, as a growth medium. Agar proved to be an ideal solidifying agent for bacterial culture, allowing for isolation and identification of individual species in a way that earlier nutrient mixtures in broth or gelatin form had not. A research associate of Koch's, Julius Richard Petri, later developed the petri dish, which when filled with the agar-solidified media, made it even easier to isolate and observe bacteria.

Media come in a wide variety of consistencies and formulations (recipes), the most common of which are nutrient agar and nutrient broth. There are also a variety of formats: Most culture media are poured as plates (a petri dish filled with solidified medium), deeps, slants, or broths (FIG. 1.9). Each type of media format has its advantages and disadvantages. Choosing the appropriate recipe and media format depends on the goal at hand. (Types of media are further explored in Chapter 7.)

Aseptic Culture Techniques

In nature, microbes do not tend to grow in **pure culture**—that is, they rarely exist as single-species groups. As such, it is often desirable to isolate a specific type of microbe from a diverse multitude in a collected sample. Characterizing cultures depends on **aseptic culturing techniques**, where conditions are maintained to limit contaminants, so that only the desired microbes in a given sample are grown. As a part of aseptic culture technique, the media used to grow the specimen is sterile, as are all the instruments and lab ware (tubes, plates, inoculation instruments) that directly touch specimens. (Sterilization methods are covered in Chapter 7.) Commonly, surrounding surfaces are also decontaminated before and after handling cultures; gloves and other protective clothing may also be required depending on the specimen being studied.

In some labs, samples are handled in a biological safety cabinet to minimize the chances of contaminating the culture and also to protect the researcher. A **biological safety cabinet** is a large piece of equipment that maintains a specific flow of filtered air and also is readily decontaminated using UV light and surface cleaning with an antimicrobial solution (FIG. 1.10).

The **streak plate technique** helps to isolate a specific species of microbe for study. The general goal is to spread the

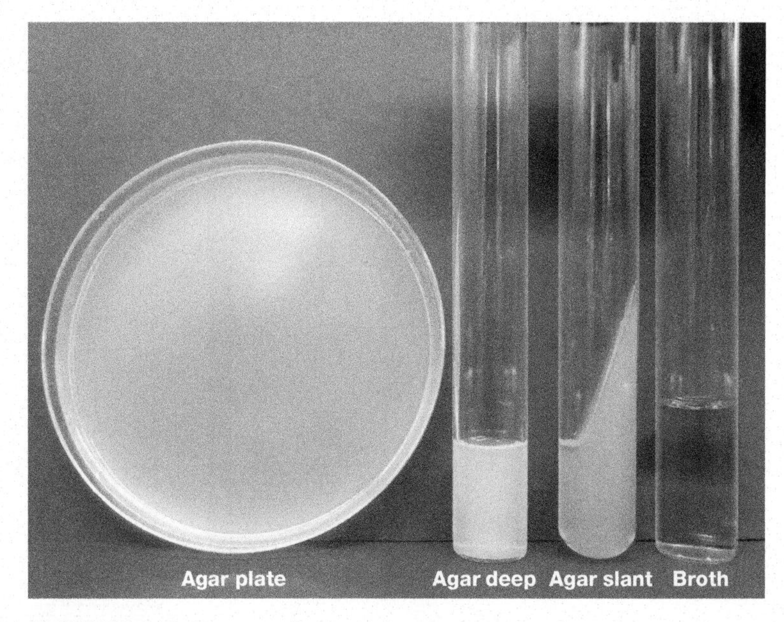

FIGURE 1.9 **Culture media** Culture media are typically poured as plates, deeps, slants, or broths. An agar plate, which is a petri dish filled with agar-solidified media, is used to isolate bacterial species. Culture media can be poured into test tubes, and agar can be added to the recipe to prepare solidified deeps and slants, or used without solidifying agar as a liquid broth.

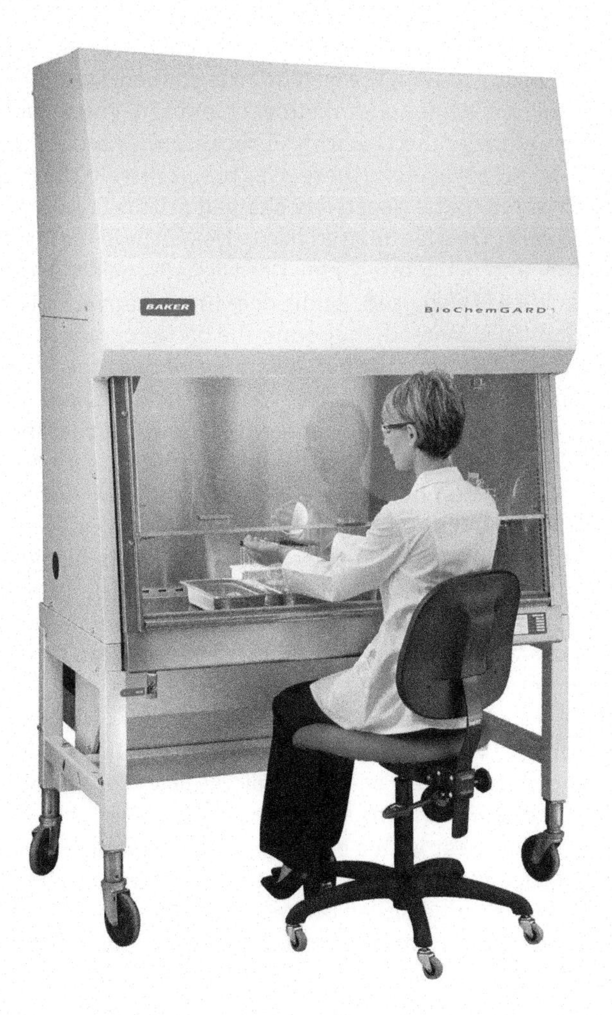

FIGURE 1.10 **Safety in the lab** Biological safety cabinets are used to practice aseptic culturing techniques of certain samples. UV lights to decontaminate the air and surfaces in the cabinet, along with controlled airflow, make biosafety cabinets an ideal way to limit sample contamination while also protecting the researcher.

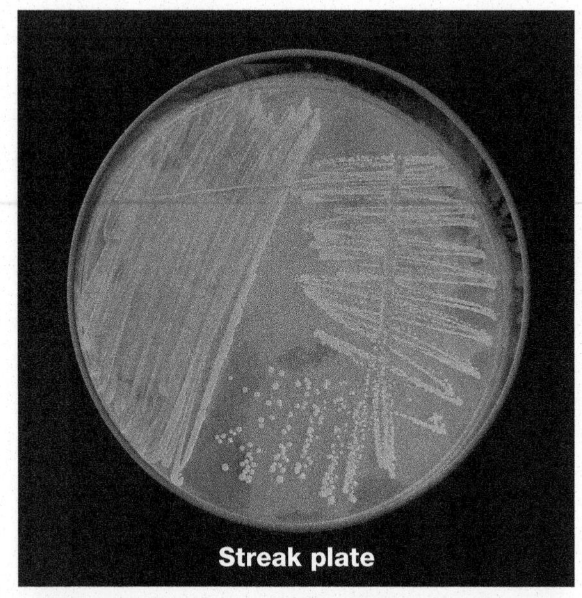

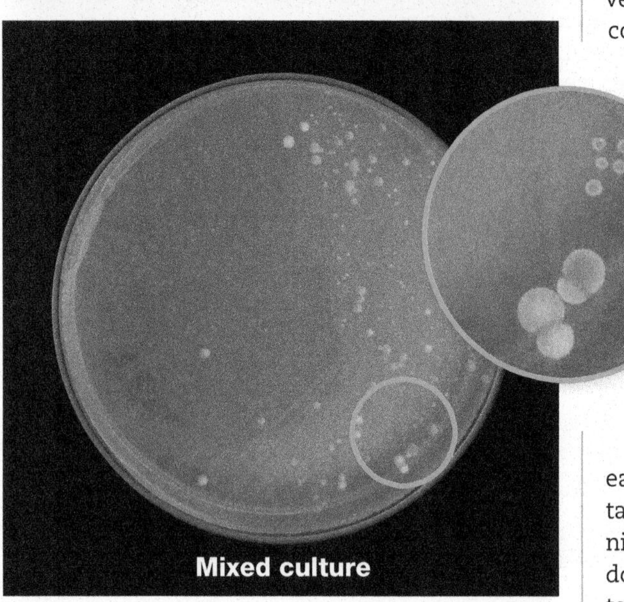

FIGURE 1.11 Isolating colonies The streak plate technique allows for isolation and characterization of individual colonies. *Top*: The goal of the streak plate technique is to spread the sample out thinly enough so that individual cells are isolated on the surface of the agar-solidified media. After an incubation period, colonies are visible on the plate. Each individual colony has developed from a single cell. *Bottom*: You can see from the variety of colonies on this plate that this sample is a mixed culture, not a pure culture.

CHEM • NOTE

The **pH scale** describes how acidic or basic a substance is based on hydrogen ion (H^+) concentration. A pH of 7 is neutral; above 7 is basic; below 7 is acidic.

sample thinly enough on an agar plate so that the various cells in the sample are sufficiently separated and can give rise to individual colonies (FIG. 1.11, *top*). A **colony** is a grouping of cells that developed from a single parent cell. The cells in a colony are genetically identical to the parent cell; they are clones. The characteristics of the colonies on a plate can be helpful for identifying a microbial species. For example, some bacteria grow in different colors, and some have characteristic margins to their colonies. When grown on a plate, a pure culture has colonies that appear somewhat uniform in shape and color, even when the colonies vary in size. By contrast, a plate with a **mixed culture** will have at least two characteristically different colonies (FIG. 1.11, *bottom*). (Additional culture techniques are reviewed in Chapter 7.)

Specimens are often stained before viewing with a microscope.

Imagine trying to find a clear marble at the bottom of a deep swimming pool—it would be a challenging task. However, if the marble were colored, it would be much easier. The same is true of specimens viewed under the microscope. The very first **stains**, or dyes, used in microbiology were added simply to increase contrast so the sample was easier to see. Eventually certain stains became an integral part of classifying microbes.

Most bacterial staining techniques involve first making a **smear** of the specimen, which involves placing a small amount of the sample on a glass slide. In most staining techniques the smear is then **fixed** by exposing it to heat or by adding a chemical reagent. Fixation adheres the sample to the slide, so that it is not as easily washed away during the staining process, and it kills most of the cells on the slide, making the stained specimen safer to handle.

After fixation, the specimen is stained. Dyes used are typically organic molecules with distinct coloration. The most common ones are sometimes called **basic dyes** because they are mildly basic (alkaline) on the pH scale. Also, these dyes are positively charged—therefore, the stain is attracted to the negatively charged cell surface of microbes and easily enters the cells. Frequently used basic dyes include methylene blue, crystal violet, safranin, and malachite green. Occasionally, **acidic dyes** such as nigrosin or India ink are also used. Acidic dyes are negatively charged, so they do not easily enter cells. Instead, they stain the background of a specimen in a technique called **negative staining**. An advantage of negative staining is that it doesn't require heating or chemical fixation, and the dye is not absorbed by the sample. This means the sample has a more true-to-life appearance, with fewer distortions of delicate cellular features. When two dyes are used in the same staining procedure, the first dye is often called the primary dye and the second dye is often called the **counterstain**.

Mordants are chemicals that may be required in certain staining procedures to interact with a dye and fix, or trap, it on or inside a treated specimen. There are numerous types of mordants such as iodine, alum, and tannic acid. Microbiologists use many staining techniques, but most can be classified as either simple, structural, or differential.

Simple Stains

Simple staining techniques use one dye. *Typically only size, shape, and cellular arrangement can be determined using simple stains.* A common protocol involves covering a heat-fixed smear with a basic dye such as methylene blue for about a minute, so that the dye has enough time to penetrate the cells. Then the smear is gently rinsed with distilled or deionized water to remove excess dye. A cover slip is placed over the smear, and it is ready for viewing with a light microscope (FIG. 1.12).

Structural Stains

Many microbes have specialized structures that can be seen using staining techniques that bring out these structural features so they may be easily viewed under a microscope.

Flagella Staining Some cells use whip-like extensions called **flagella** for motility. While eukaryotic microbes tend to have only one flagellum located at a single pole of the cell, prokaryotes can have single or multiple flagella with diverse arrangements. The number and position of flagella can help identify the microbe, making staining techniques to view flagella useful. Because flagella are very thin and difficult to see, staining methods coat samples with chemicals that act as mordants (substances that help dyes stick to a target) and also use dyes designed to thicken the appearance of flagella for easier viewing (FIG. 1.13, *left*).

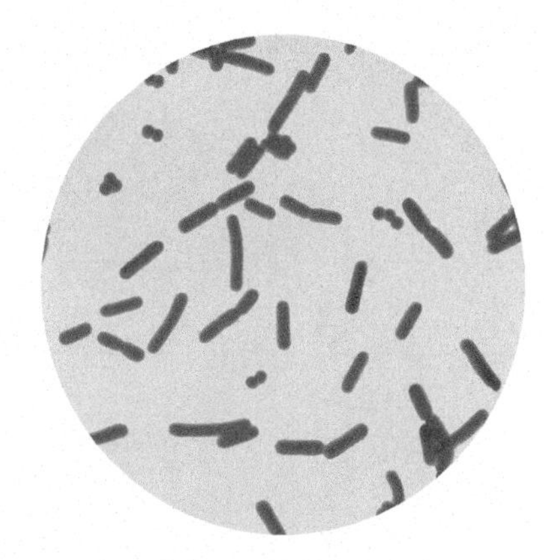

FIGURE 1.12 **Simple stain** Methylene blue stain of *E. coli* reveals small rods with a single arrangement.

Capsule Staining **Capsules** are sticky carbohydrate-based structures that some bacteria produce as a form of protection and also to help them adhere to surfaces. Certain pathogens require a capsule to establish an infection. Because capsules are easily dissolved in water and do not readily take up basic dyes, a negative staining technique utilizing an acidic dye is usually used in coordination with a basic dye. The basic dye stains the actual cell, while the acidic dye stains the background to enhance contrast. A capsule then appears as a clear halo around a stained cell (FIG. 1.13, *middle*).

Bacterial Endospore Staining Certain bacteria, including *Bacillus* species such as *B. anthracis*, make specialized dormant structures called **endospores** in response to stressful or harsh environmental conditions. Very few bacteria form them, so detecting endospore presence is useful for classification. Because endospores have a tough spore coat that resists staining, the technique involves heating the specimen to drive a dye, usually malachite green, into the spores. The excess dye is rinsed off, and a counterstain, usually safranin, is added to make it easier to see cellular material that surrounds the spore, or nonsporulating cells (those that do not form spores). As a result, endospores appear green, and other cellular features appear pink (FIG. 1.13, *right*). (Endospore formation and characteristics are reviewed in Chapter 3.)

Differential Stains: Gram and Acid-Fast

Differential staining highlights differences in bacterial cell walls, anatomical structures that contribute to cell rigidity. (Bacterial cell wall structure and function are discussed further in Chapter 3.) These differences in cell-wall structure

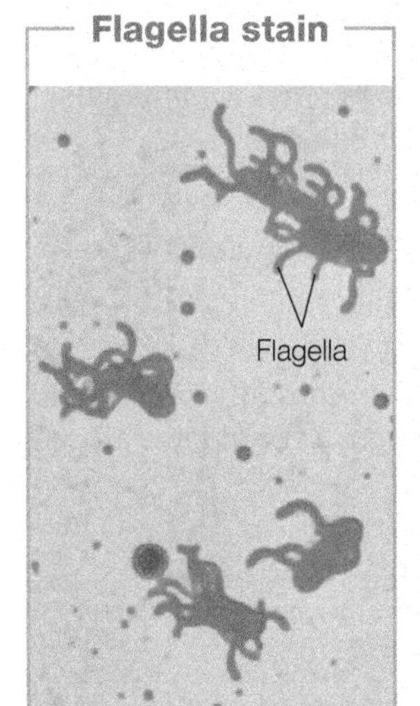

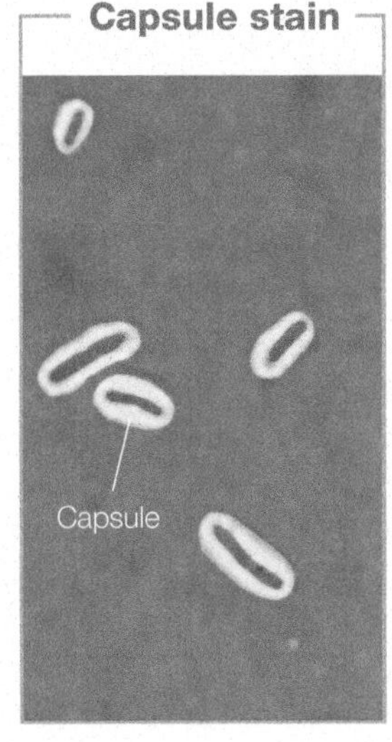

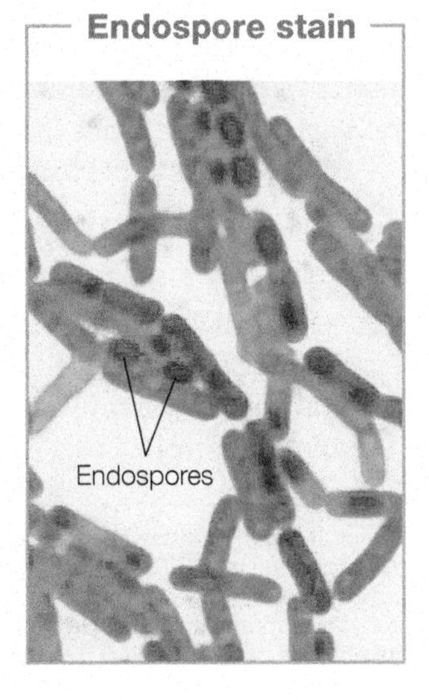

FIGURE 1.13 **Structural stains** Flagella staining reveals flagella at each pole of this bacterial cell. Capsule staining shows a clear halo surrounding pink-staining bacterial cells. Endospores appear green following the endospore stain.

Critical Thinking *Technically, any basic dye could be used as a secondary stain in the endospore stain, so why is pink (safranin) an ideal choice?*

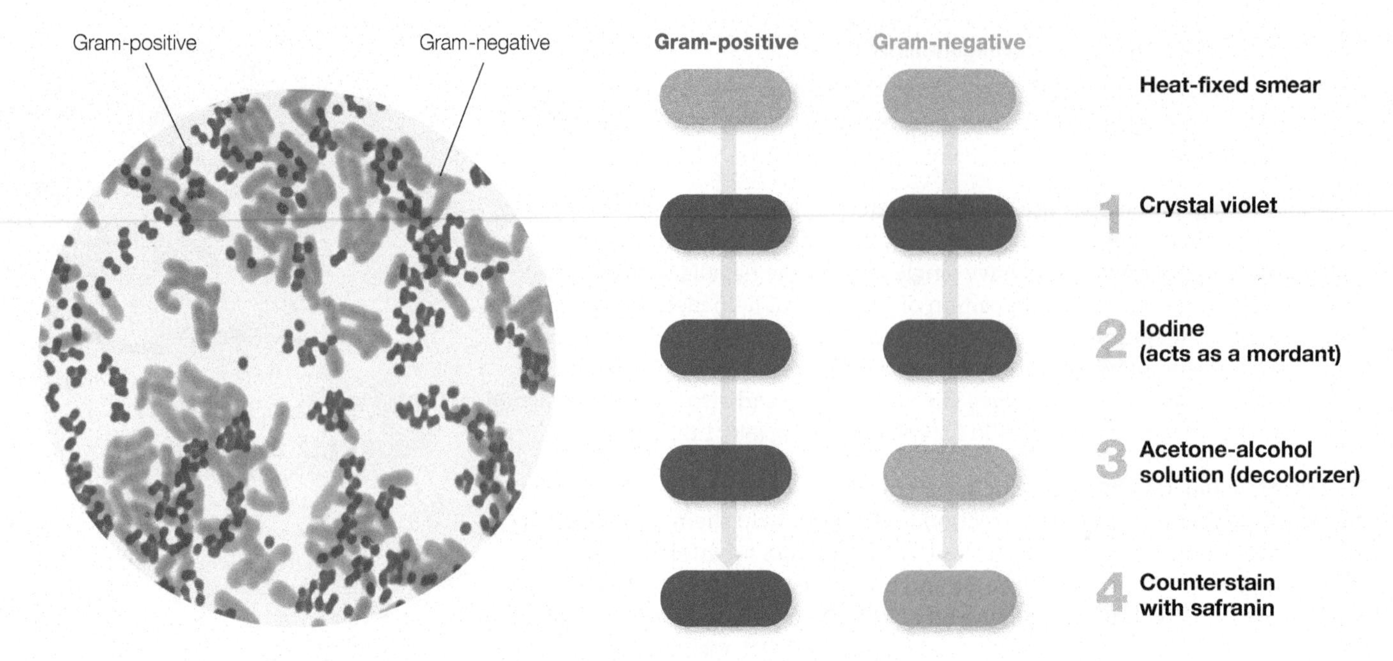

FIGURE 1.14 Gram staining *Left*: Gram-positive bacteria appear purple at the end of the Gram staining procedure while Gram-negative bacteria appear pink. *Right*: Summary of the Gram stain; notice that Gram-positive bacteria remain purple throughout the procedure, whereas Gram-negative bacteria are decolorized upon addition of the decolorizing solution, and then counterstained pink.

Critical Thinking *What color would Gram-positive bacteria be if the iodine step was omitted?*

Gram Stain

Bring the art to life! Visit the **Mastering Microbiology** Study Area to watch the Concept Coach and master how the Gram stain works and why it matters.

are clinically significant. The two most common differential stains used in microbiology are the Gram stain and the acid-fast stain.

Gram Staining The **Gram stain** is one of the most important stains in microbiology. This process classifies bacteria as either Gram-positive or Gram-negative. The Gram property of a specimen has important clinical implications, including the potential pathogenic features of the organism, and what antibiotics might be most effective in combating it. For example, penicillin is generally more effective at combating Gram-positive bacteria than Gram-negative bacteria.[14] Polymyxin antibiotics are typically effective against Gram-negative bacteria, but will not control infections caused by Gram-positive bacteria. Although bacterial cell wall structure isn't the only feature that influences whether a bacterium will be susceptible to a given drug, it is a key factor for defining general susceptibilities to certain classes of drugs.

The Gram stain works by revealing chemical and structural features of cell walls. It is important that you understand that the Gram stain procedure simply highlights differences in cell wall structures; it does *not* chemically turn a bacterium into a Gram-positive or Gram-negative specimen, any more than a pregnancy test makes a woman pregnant. *By the end of the Gram stain, Gram-positive cells appear purple, while Gram-negative cells appear pink* (FIG. 1.14).

The Gram stain technique is as follows:

1. A heat-fixed bacterial smear is stained with **crystal violet**, giving all the cells a deep purple color. This step is called the *primary stain*.
2. After briefly rinsing with water to remove excess crystal violet dye, an **iodine** solution is added to the sample. This acts as a mordant, forming an insoluble crystal violet-iodine complex (CV-I complex). At this point all cells remain purple.

[14] Not all Gram-positive bacteria are susceptible to penicillin family drugs. Some acquire drug-resistance factors such as the ability to break it down. In contrast, Gram-negative bacteria are naturally resistant to certain penicillin family drugs because they tend to struggle to get past the outer membrane structure. Because Gram-positive bacteria lack an outer membrane, barring the acquisition of resistance factors, they will generally respond well to penicillin-based drugs.

3. The sample is briefly rinsed with an **acetone-alcohol** solution. Although this is known as the *decolorizing step*, Gram-positive bacteria remain purple; only Gram-negative bacteria are decolorized and thereby left colorless after this step.
4. Because the prior step left Gram-negative bacteria colorless and hard to see on a glass slide, the last step of the Gram stain involves covering the smear with **safranin**, a counterstain, which stains the decolorized cells pink. Even if you know the Gram property, all samples are subjected to step 4. Before the sample is viewed under the microscope, the excess safranin is gently rinsed away with water, the slide is allowed to dry, and a cover slip is adhered to the slide.

Although Hans Gram developed the technique in the late 1800s, the mechanism for how the staining technique distinguishes between Gram-positive and Gram-negative bacteria (during decolorization, step 3) was not understood until the 1980s. What occurs in the different cell types has to do with cell wall composition.

Gram-positive bacterial cell walls contain a thick layer of a protein-carbohydrate substance called **peptidoglycan**, whereas Gram-negative bacterial cell walls have only a thin peptidoglycan layer. Additionally, Gram-positive bacteria do not have an outer membrane, whereas Gram-negative bacteria have one lying just on top of their thin peptidoglycan layer. This outer membrane is important to detect because it contains **lipopolysaccharide (LPS)**, of which the lipid portion is toxic to animals and can trigger septic shock. Although the outer membrane is rich in lipids, it is vulnerable. During the few seconds that the acetone-alcohol is in contact with Gram-negative bacterial cells, it dissolves the outer membrane of the cell wall to expose the thin peptidoglycan layer. The chemistry of this step renders the Gram-negative cell wall leaky, such that the purple CV-I complex formed during step 2 easily washes out of the Gram-negative bacteria. In contrast, Gram-positive bacteria remain purple because the acetone-alcohol solution dehydrates the thick peptidoglycan layer causing it to pack down and trap the CV-I complex inside Gram-positive cells.

The trickiest part of the Gram stain is the decolorizing step (step 3). If you leave the decolorizer on a sample too long, then even the thick peptidoglycan layer of Gram-positive cell walls will be damaged to the point of becoming leaky, allowing the CV-I complex to be readily rinsed out of the cells. Thus, over-decolorizing a sample leads to inaccurate results in which all cells appear pink irrespective of their true Gram property.

Other things can make interpreting results difficult, even when the procedure is done correctly. For example, *Acinetobacter* species have cell walls that are especially resistant to the action of the decolorizing solution, so these bacteria may appear Gram-positive using standard Gram staining methods despite the fact they are actually Gram-negative bacteria.[15] Another example of unexpected Gram staining result is found in *Mycobacterium* species, which include pathogens that cause tuberculosis and leprosy. These Gram-positive bacteria have a waxy cell wall that resists staining with crystal violet, making them appear Gram-negative.[16] Other bacteria appear to consist of both Gram-positive and Gram-negative cells, depending upon the life stage of the cells. For example, sporulating bacteria such as *Bacillus* species, which include *B. anthracis*, the cause of anthrax, and *Clostridium* species[17], which include the pathogen

[15] Bazzi, A. M., Al-Tawfiq, J. A., & Rabaan, A. A. (2017). Misinterpretation of Gram stain from the stationary growth phase of positive blood cultures for Brucella and Acinetobacter species. *The Open Microbiology Journal*, *11*, 126.

[16] Kuroda, H., & Hosokawa, N. (2019). Gram-ghost bacilli. *Journal of General and Family Medicine*, *20*(1), 31–32.

[17] *C. difficile*, a bacterium that cause pseudomembranous colitis, also forms endospores. This bacterium was fairly recently shifted from the genus *Clostriduium* and reclassified into the genus *Clostridioides*. As seen for *Bacillus* and *Clostridium*, sporulating bacteria in the genus *Clostridioides* also exhibit variability in Gram stain results.

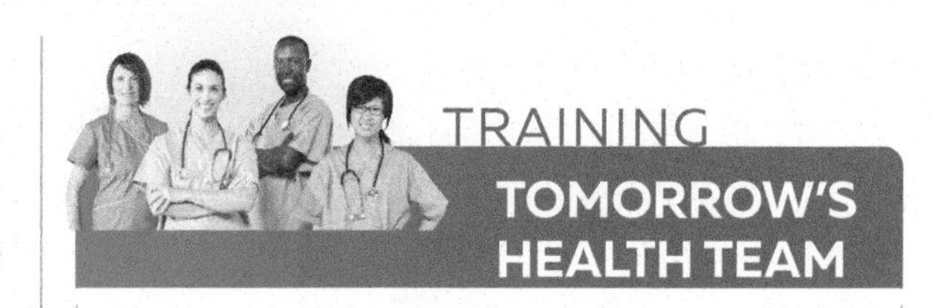

In the Gram Scheme of Things

One example of the vital role of microbiology labs in patient care is the diagnosis and treatment of bacteremia—a bacterial infection in the bloodstream. These infections have a mortality rate of about 20 percent in hospitalized patients. The first indication of bacteremia is usually based on Gram staining data for a blood sample. Gram-positive or Gram-negative bacteria can cause bacteremia, but knowing which is the culprit is important, as the Gram property influences the choice of antibiotic used.

When a physician receives the microbiology lab report, the empirical treatment (the treatment initiated prior to a definitive diagnosis) is often changed based on the Gram property data received.

QUESTION 1.3

Most Gram typing errors involve Clostridium *and* Bacillus *species. Based on your readings, why are these genera of bacteria particularly problematic in terms of Gram typing accuracy?*

CLINICAL VOCABULARY

Septic Shock A serious complication of a system-wide infection that can lead to organ failure and death. This complication is often, though not exclusively, associated with elevated levels of LPS that Gram-negative bacteria make. Signs and symptoms are usually nonspecific and can include fever, chills, nausea, low blood pressure, and difficulty breathing.

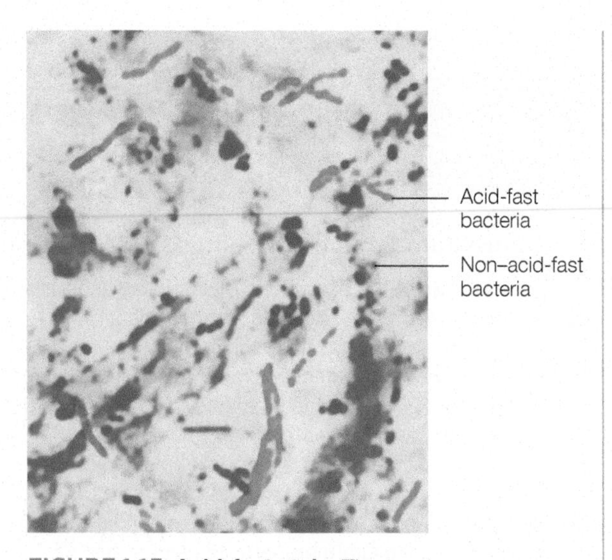

FIGURE 1.15 Acid-fast stain The most common acid-fast staining procedure is the Ziehl–Neelsen method. Acid-fast bacteria appear pink, and non–acid-fast bacteria are blue.

Critical Thinking *What color would non–acid-fast bacteria be if the decolorizing step with acid-alcohol was omitted?*

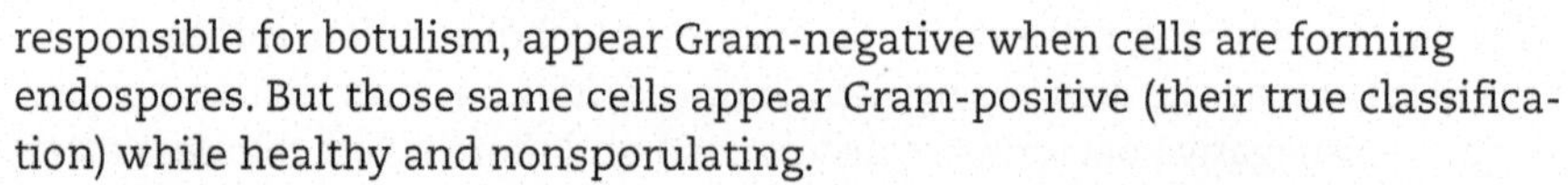

responsible for botulism, appear Gram-negative when cells are forming endospores. But those same cells appear Gram-positive (their true classification) while healthy and nonsporulating.

To minimize Gram property errors, fresh cultures between 24 and 48 hours old should be used, so that the cells tested are mostly healthy and undamaged. The older the sample cells, the more variable the Gram results may be. It is also a good practice to include a positive control (something you know is Gram-positive) on all test slides to ensure that the procedure is being properly performed.

Acid-Fast Staining The **acid-fast stain** is a differential stain that distinguishes between cells with and without waxy cell walls. Acid-fast bacteria have waxy cell walls that are rich in a substance called **mycolic acid**; they retain the primary dye, which is red, even after exposed to an acid wash. In non–acid-fast cells, the acid wash strips away the red stain (the primary stain). Clinically, the acid-fast stain is an important diagnostic tool for detecting *Mycobacterium* species such as the causative agents of tuberculosis and leprosy. The acid-fast stain also detects *Nocardia* species, bacteria that are usually in soil and water and can act as opportunistic pathogens of immune-compromised patients.

There are several acid-fast staining procedures, but the most common method is the Ziehl–Neelsen (Zeel NEEL-sen) method, where the primary dye, *carbol-fuchsin* (CAR-bowl-FEWK-sin), is layered onto a heat-fixed smear, and the sample is steamed for several minutes to drive the red dye into the bacteria. After a thorough water rinse, the sample is then briefly treated with an acid-alcohol solution, which acts as a decolorizing agent. This is the differentiating step; overdecolorizing with the acid-alcohol rinse can lead to false-negative results. Finally, a counterstain, *methylene blue*, is added. Because acid-fast bacteria have a waxy cell wall that resists decolorization by the acid-alcohol rinse, they appear a bright pink-red at the end of the procedure. Non–acid-fast bacteria appear blue at the end of the procedure (FIG. 1.15).

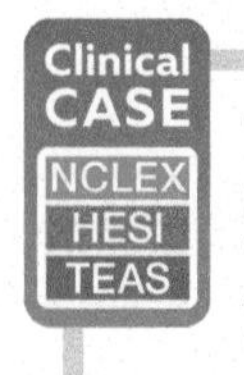

Clinical CASE

NCLEX HESI TEAS

The Case of the Mystery Pathogen

Practice applying what you know clinically: visit the **Mastering Microbiology** Study Area to watch Part 2 and practice for future exams.

Microscopy is central to microbiology.

Most early microscopes only magnified a specimen about 10 times $(10\times)$, which is not powerful enough to see most bacteria. Van Leeuwenhoek's microscopes of the 1600s reached about a $300\times$ magnification, allowing him to see and describe basic bacterial shapes and arrangements. Today's microscopes allow us to see substances that are 20 million times smaller than the visibility threshold of the naked eye. Our technology allows us to glimpse viruses, most of which are much smaller than most bacteria. We can also visualize atoms and even the bonds between atoms. **Micrographs**, or pictures taken through a microscope, allow us to document and share microscopy observations.

Light Microscopy

Light microscopy uses visible light to illuminate the specimen. This is one of the simpler ways to view a sample. As photons in a light wave interact with the specimen, they are channeled up to the viewer's eyes through a series of lenses. The type of lens used determines the final magnification observed. Also, the quality of the glass used to make the lens, as well as the shape of the lens itself, contribute to the quality of the final image. The **compound light microscope** is the most common type of optical microscope and is a basic tool found in microbiology labs (FIG. 1.16).

Parts of the Compound Light Microscope The **objective lens** is near the specimen, whereas the **ocular lens** sits at the top of the microscope near the viewer's eyes. The final magnification of the specimen is determined by multiplying the magnification of the ocular and objective lenses. For example, if the ocular lens has a $10\times$ magnification and the objective lens provides $40\times$ magnification, then the final magnification is $400\times$ $(10 \times 40 = 400\times)$. Most ocular lenses magnify a specimen by 10×, though some are 15 ×. Objective lenses come in varieties that usually include 4×, 10×, 40× and 100×.

Light microscopes also depend on a **condenser**, consisting of lenses that sharpen light into a precise cone to illuminate the specimen. The condenser also has an **iris diaphragm** that allows the viewer to modulate how much light is aimed at the specimen in order to improve contrast. The light that interacts with the specimen is channeled through the objective lens, which magnifies the image and also enhances the resolution. After the light passes through the objective lens, it continues up to the ocular lens, where the image is further magnified before reaching the viewer's eyes. The **coarse focus knob** on the microscope allows the viewer to roughly focus the image by affecting the distance of the objective lens from the specimen. The **fine focus knob** does the same, but in much smaller increments than the coarse knob, allowing for precision focusing.

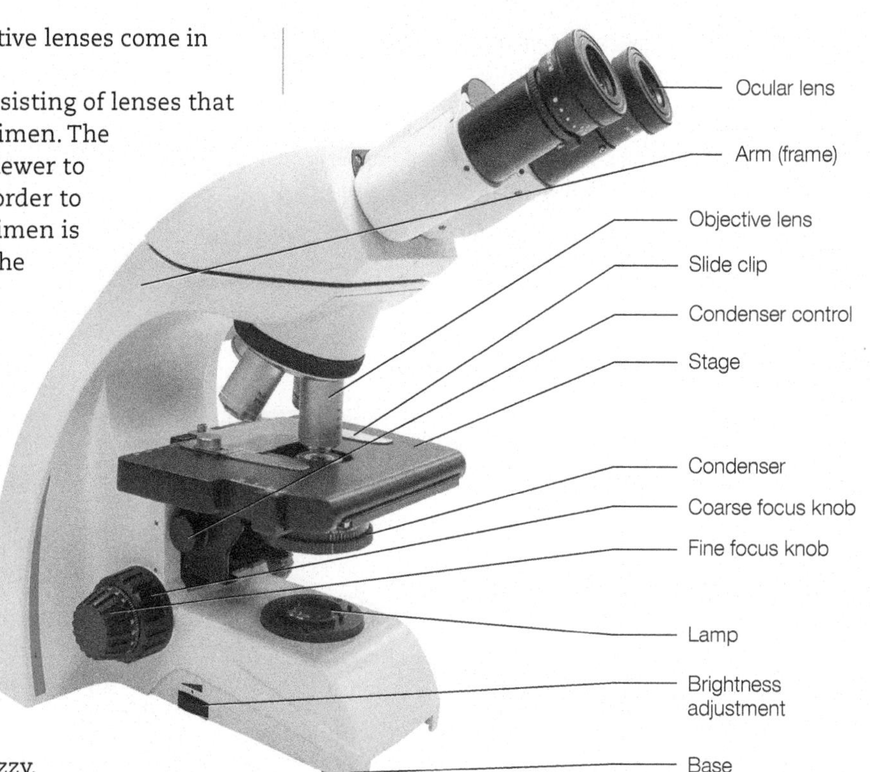

FIGURE 1.16 Parts of a compound light microscope

Resolution The ability to distinguish two distinct points as separate is **resolution**. You can think of resolution as being how sharp your vision is. A toy microscope may have decent magnification, but because its lenses are usually of low quality with poor resolving power, the images generated will be large but fuzzy.

The naked eye has a resolution of about 0.1 mm (100,000 nm).[18] Most compound light microscopes magnify a maximum of 1,000–1,500×. While greater magnification can be achieved, it wouldn't improve the image observed because the greatest resolution achievable with a light microscope is about 0.0002 mm (200 nm), or 500 times better than the resolving power of the naked eye. This limit for resolution is due to the fact that the smallest wavelength of visible light is 400 nm.

Oil immersion Another factor in the resolution of microscopes relates to **refractive index**, which is the degree to which a substance bends light. Air has a lower refractive index than glass. As such, when light passes through a slide and into the air above the slide, the light waves bend and are not channeled directly through the objective lens toward the eye. At low-power magnification, this light bending doesn't really impact resolution. However, to get a sharp image at high-power magnification (100× objective lens), **immersion oil** must be applied to the slide's surface. This specialized oil is formulated to have the same refractive index as glass. Therefore, the oil ensures that the light that interacts with the specimen on the glass slide is smoothly funneled up toward the high-power objective lens instead of scattering when it reaches the air between the slide and lens (FIG. 1.17). Oil immersion is used in a variety of light microscopy techniques—including bright field microscopy, which we'll cover next.

Bright Field Microscopy The simplest and most common form of all microscopy is **bright field microscopy**. In this light microscopy technique, a solid cone of light illuminates a sample from below, and the image is magnified through a series of lenses in a compound light microscope. The method's name, bright field, derives from the fact that the visualized sample appears as a darker contrasting image on a light background—that is, on a bright field of light. Bright field microscopy images are a product of how light is absorbed by a sample. Therefore, unless the sample has natural coloration to enhance light absorption, such as in photosynthetic cells, the specimen must be stained in

[18] One millimeter (mm) is equivalent to 1,000,000 nanometers (nm).

FIGURE 1.17 Oil immersion technique Immersion oil is used with the high-power objective lens (100×) of a light microscope to improve resolution. Because the oil has a similar refractive index to glass, light is directly channeled up through the objective lens instead of being refracted (bent) and lost to the surroundings.

Critical Thinking *Why would funneling as much light as possible through the objective lens improve resolution?*

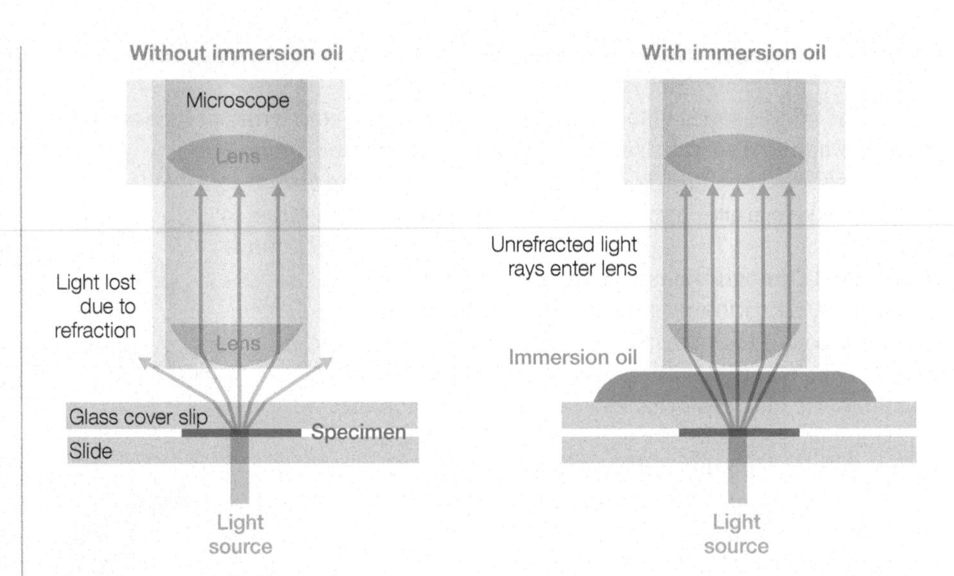

order to be seen with bright field microscopy. Because most staining techniques kill cells, bright field microscopy is less than ideal for viewing live specimens. Dark field, phase contrast, and differential interference contrast microscopy, all of which are summarized in TABLE 1.3, are better for observing live samples.

TABLE 1.3 Comparing Light Microscopy Techniques: Amoeba *Proteus* Viewed with Different Light Microscopy Techniques

Microscopy Technique	Image	Equipment	Notes
Bright Field	Darker contrasting image on a bright background	Compound light microscope	Illuminates sample with solid cone of light; image formed based on how light is absorbed; sample must be stained or have natural coloration
Dark Field	Negative image, where the sample appears light on a darker background	Modified condenser in a compound light microscope	Illuminates sample with hollow cone of light; image formed based on how light is scattered as opposed to how light is absorbed, so staining is not necessary; negative image made by dark field microscopy should not be confused with negative staining; visualizes unstained specimens (live or dead) and stained specimens
Phase Contrast	Negative image, where the sample appears light on a darker background	Modified condenser in a compound light microscope	Illuminates sample with hollow cone of light; a device in the microscope (i.e., a phase plate) interacts with light that passes through the viewed sample—thereby enhancing image brightness, shading, and contrast; visualizes unstained specimens (live or dead) and stained specimens
Differential Interference Contrast (or Nomarsky)	One side of specimen appears brighter than the other side, providing a false three-dimensional appearance	Modified compound light microscope	Illuminates specimen with polarized light (uniformly oriented light) as opposed to a hollow cone of unorganized nonpolarized light used in phase contrast microscopes

TABLE 1.4 Comparison of Electron Microscopy to Light Microscopy

Light Microscopes	Electron Microscopes
Use light waves to image the specimen	Use an electron beam to image the specimen
Small, portable, and affordable	Large, requires special designated space, expensive
Simple, cheap, and easy sample preparation that requires minimal training	Lengthy and complex sample preparation requires substantial training
Color images possible	Only black-and-white images (though color may be added later, as an aftereffect)
Most microscopes provide a maximum of 1,000×	Can magnify over 500,000×
Resolution of 200 nm	0.2 nm or about 1,000 times better than the best compound light microscopes
Specimens can be living or dead	Specimens are all dead
Stains often used, but certain forms can be done without staining and can visualize live cells	Specimens often must be stained with an electron-dense substance like osmium or gold

Electron Microscopy

As stated before, the smallest wavelength of visible light is 400 nm. *Resolution improves with smaller wavelengths.* Because an electron beam has a much smaller wavelength of about 1 nm, it is ideal for probing ultra-small structures like viruses, which are too small to be seen with light microscopes.

There are two main classes of electron microscopes: transmission electron microscopes and scanning electron microscopes. Both shoot electrons at a specimen and then generate an image by registering how the electrons interact with the specimen. While electron microscopes provide high-magnification and high-resolution images, they are very expensive, and they require considerable training to use. A comparison between electron microscopy and light microscopy is provided in TABLE 1.4. Electron micrograph examples are shown in FIG. 1.18.

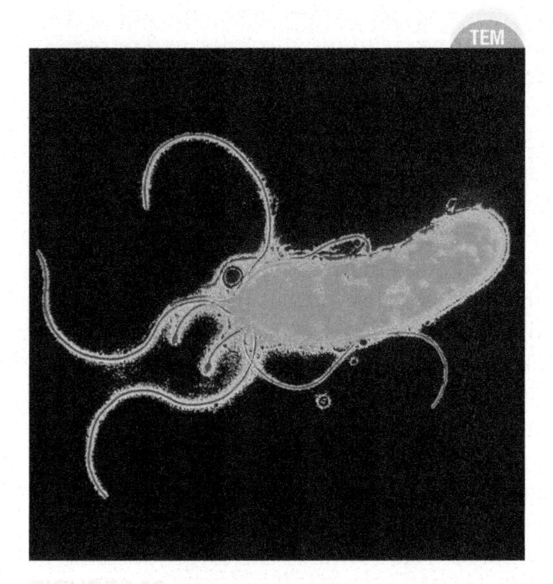

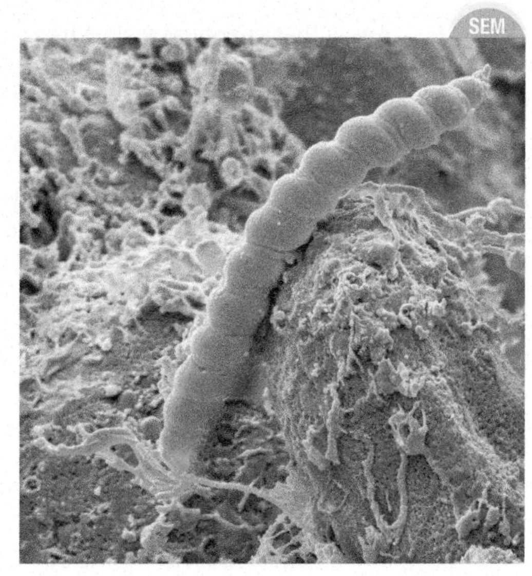

FIGURE 1.18 Examples of electron micrographs *Left*: Transmission electron micrograph (TEM) of *Helicobacter pylori*, a bacterium that causes ulcers. Note the numerous flagella. *Right*: Scanning electron micrograph (SEM) of *H. pylori* shows surface details based on how electrons are reflected off of an electron-dense sample surface. When you see colors in electron micrographs, they have been falsely generated for artistic effect.

Critical Thinking *Why are all electron micrographs naturally black-and-white images?*

Transmission Electron Microscopy (TEM) TEM is the most common form of electron microscopy. It provides up to 1 million times magnification and resolution, about 1,000 times better than that of light microscopy. Unfortunately, TEM samples must be extensively pretreated, and the maximum thickness of a specimen viewable with this type of microscope is 285 times thinner than a human hair. In TEM the electron beam passes through the specimen and hits a detector to generate two-dimensional black-and-white images of internal structures of the specimen. Dark areas represent regions where electrons were poorly transmitted, whereas light areas represent less dense regions where electrons were more readily transmitted. TEM is important to many fields, including biomedical research and nanotechnology. TEM has been central to understanding viruses and how they interact with host cells, as viruses are too small to be viewed by light microscopes.

Scanning Electron Microscopy (SEM) Hard, dry specimens like bone usually require no preparation for SEM. But soft tissues or organisms being viewed by SEM are usually preserved, dehydrated, and coated in an ultra-thin layer of conductive metal such as gold or a gold/palladium alloy. Then an electron beam scans over the specimen. Specialized detectors sense how the electrons interact with the surface of the specimen, and related signal information generates a black-and-white three-dimensional image. The overall resolution and magnification potential of SEM is lower than TEM, but SEM is useful because it provides excellent information about surface structures, whereas TEM provides information about internal structures.

SEM techniques have dramatically aided our understanding of the surface features of a variety of cells and viruses. They revealed that dramatic surface structure changes can occur in tissues with certain pathological conditions like kidney disease or cancer. SEM is also useful to study biofilms. Forensics uses SEM to examine crime scene evidence, and nanotechnology uses SEM for quality control. Certain SEM techniques can also help identify the chemical components of a substance.

Clinical CASE
NCLEX HESI TEAS

The Case of the Mystery Pathogen

Practice applying what you know clinically: visit the **Mastering Microbiology** Study Area to watch Part 3 and practice for future exams.

Using Fluorescence in Microscopy

Fluorescence is a naturally occurring phenomenon in which a substance absorbs energy, usually ultraviolet (UV) light that is invisible to the human eye, and then emits that energy as visible light. Many marine creatures, microbes, and minerals exhibit natural fluorescence. There are also a number of fluorescent dyes, called **fluorochromes**, which can be used to stain samples so they will fluoresce when illuminated by a UV light microscope. Fluorochromes can be natural or synthetic, and many chemically interact with certain cellular features. For example, Hoechst dyes bind to DNA and will glow blue when activated by UV light. The auramine-rhodamine fluorescent stain has a strong attraction for acid-fast bacteria and emits a distinct reddish-yellow glow when bound to its target, making it helpful for confirming the presence of mycobacterial species, such as those that cause tuberculosis. A fluorescing dye called calcofluor-white associates with cellulose and chitin found in the cell walls of fungi and other organisms and is helpful in rapid screening for yeasts and various pathogenic fungi.

Fluorescent dyes can also be linked to antibodies, immune-related proteins that help identify and target "nonself" cells such as bacteria. In **immunofluorescence**, a sample is exposed to fluorescent-tagged antibodies that recognize a specific target—perhaps a protein that is present only on the surface of certain pathogens. If the sample contains the target protein (or other molecule), then the antibodies will bind to the sample and glow when viewed under UV light. Immunofluorescence is used in many applications, such as rapid identification of bacteria in blood cultures, virus identification in patient samples, and fast screening for pathogenic bacteria in food-processing plants. TABLE 1.5 summarizes some forms of UV-based microscopy and also explores non–light-based microscopy techniques.

TABLE 1.5 UV Light Microscopy and Probe Microscopy Techniques

	Microscopy Technique	Image	Type of Sample	Notes
Fluorescence Imaging	**General Fluorescence** 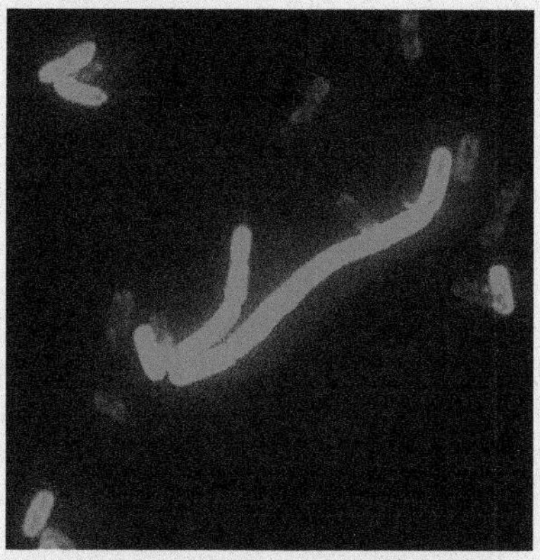	Flat image with coloration based on fluorochrome used	Live or fixed	The UV waves cause visible light to be released from fluorochromes; advantage over light microscopy is not improved resolution or magnification, but easy and sensitive detection; allows for the detection of even a single molecule in a sample
	Confocal 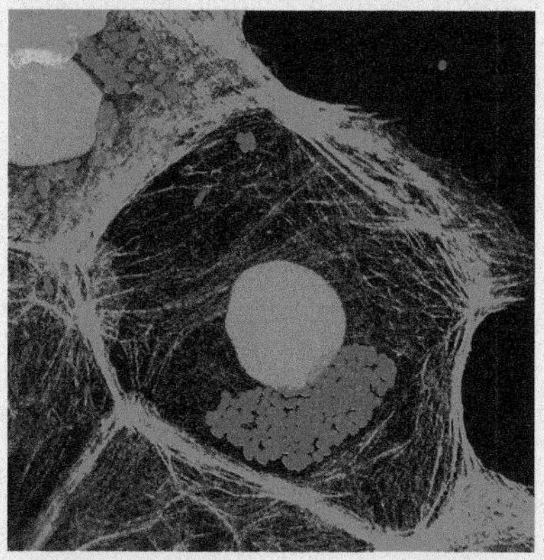	3D image	Live or fixed	Eliminates blurriness associated with standard fluorescence microscopes; images are taken at different planes of focus, and then these photo "slices" are compiled to generate a 3D image
Probe Techniques	**Scanning Tunneling** 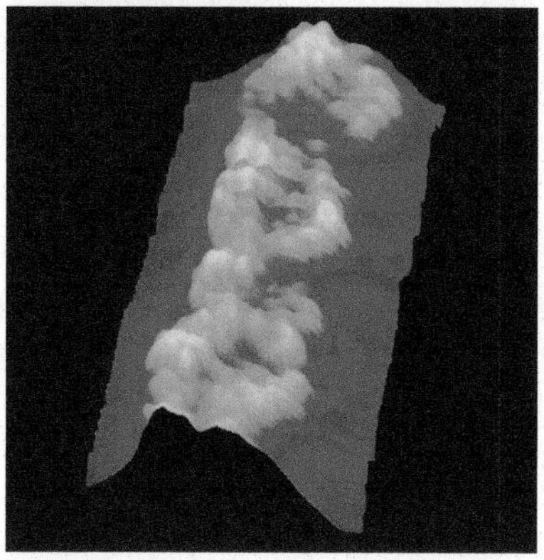	3D image	Atoms can be visualized; sample must conduct electricity, which limits what can be visualized	A probe sharpened to a single atom at the tip shoots electrons at the sample surface; elevations or dips in sample surface are registered to make the image; not as good as scanning electron microscopy for detecting steep rises or deep valleys in sample
	Atomic Force 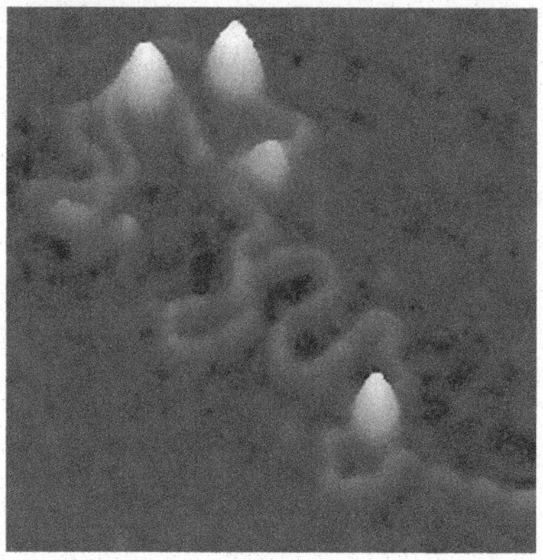	3D image	Atoms can be visualized; live samples under physiological conditions or fixed samples	Probe is dragged or tapped along specimen surface; not as good as scanning electron microscopy for detecting steep rises or deep valleys in sample

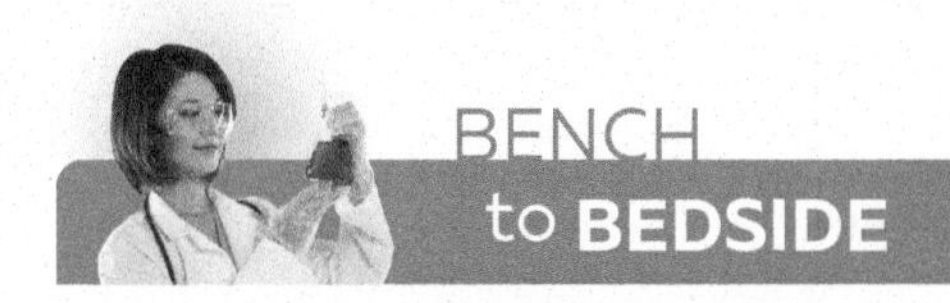

A Possible Link between Lost Normal Microbiota and Immune System Disorders

Autoimmune disorders arise when a patient's immune system attacks the person's own tissues. There are over 100 different recognized autoimmune disorders that impact over 23 million Americans. Examples include lupus, multiple sclerosis, type I diabetes, and inflammatory bowel disease. Statistically speaking, people in developed nations are at a greater risk for autoimmune disorders than are people living in developing nations. Nobody knows precisely why this is the case, or how autoimmune diseases emerge. But the hygiene hypothesis aims to explain the matter.

The hygiene hypothesis proposes that the modern quest for a microbe-free environment may actually encourage our immune systems to malfunction. The hypothesis states that because humans coevolved with parasitic worms and other microbes all around them, these creatures may have a symbiotic role in the evolution of our immune response and general health. In particular, intestinal helminths (multicellular, parasitic roundworms and flatworms) remain common ailments in many developing nations but are rare in developed countries thanks in large part to stringent water sanitation practices and food safety regulations.

Under most circumstances, humans tolerate parasitic worms (helminths) fairly well. Various studies show that geographical areas with the highest incidence of helminthic infections tend to have the lowest rates of allergy and autoimmune diseases. Studies also demonstrate that when people migrate from areas where helminth infections are common to places where they are rare, incidence of autoimmune disease rises in subsequent generations living in the new location. Interestingly, other studies show that some genes predisposing a person to asthma are possibly linked to resistance to certain helminthic infections.[1] Based on animal and human studies, it seems that helminthic infections may moderate the activity of certain immune system cells, called regulatory T cells.[2] Immune systems that mature without the presence of helminthic infections may be more likely to "overreact," leading to inappropriate immune responses against self cells or benign foreign substances, such as pollen.

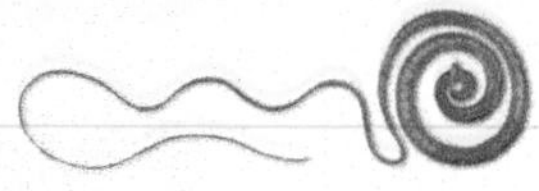

If immune system disorders can be stifled by helminths, it may be that reintroducing these worms into the body could lessen the severity of autoimmune disorders. Since 2003, over a dozen clinical trials have tested this approach. In one experiment, patients ingested the eggs of pig whipworm in an effort to alter or divert the host immune system response. Pig whipworms can't complete their life cycle in humans, so the spread of the worm to other people is unlikely, and sustained infection by the worms is only accomplished by periodically reinfecting the patient. While none of these studies has progressed to phase III, the final stage before approval for clinical use, the results have been promising in treating multiple sclerosis as well as Crohn's disease and other immune-mediated disorders.[3] That said, helminth therapy has a long way to go before it hits the clinic, but perhaps one day there will be probiotic worms to treat certain disorders linked to a hyperreactive immune response.

---

[1] Jesus, T. D. S., Costa, R. D. S., Alcântara-Neves, N. M., Barreto, M. L., & Figueiredo, C. A. (2019). Variants in the CYSLTR2 are associated with asthma, atopy markers and helminths infections in the Brazilian population. *Prostaglandins, Leukotrienes and Essential Fatty Acids*, 145, 15–22.

[2] Maizels, R. M. (2020). Regulation of immunity and allergy by helminth parasites. *Allergy*, 75(3), 524–534.

[3] Ryan, S. M., Eichenberger, R. M., Ruscher, R., Giacomin, P. R., & Loukas, A. (2020). Harnessing helminth-driven immunoregulation in the search for novel therapeutic modalities. *PLoS Pathogens*, 16(5), e1008508.

Build Your Foundation

18. What format of media is recommended for isolating bacteria? (LO 1.16)

19. Why are aseptic culturing techniques important in the microbiology lab? (LO 1.17)

20. Define the term *colony*, and describe a technique that allows for colony isolation. (LO 1.18)

21. What information do simple stains reveal about a bacterial specimen? (LO 1.19)

22. What stain reveals if a bacterium forms endospores? Why would knowing if a bacterium forms endospores be helpful? (LO 1.19)

23. What is the Gram stain, how does it work, and what are potential sources of error in the procedure? (LO 1.20)

24. How is the acid-fast stain performed, and what are its clinical applications? (LO 1.21)

25. Identify the parts of the compound light microscope, and describe the general features of bright field microscopy. (LO 1.22)

26. How does immersion oil improve the high-power resolution of a light microscope? (LO 1.23)

27. Compare and contrast SEM and TEM. (LO 1.24)

28. What is fluorescence microscopy, and what are its advantages? (LO 1.24)

Build Your Foundation (BYF) Quick Quiz: Visit the **Mastering Microbiology** Study Area to quiz yourself.

Microbiology's Golden Age

Biogenesis was accepted and the germ theory of disease embraced. Central figures were Pasteur with his S-necked flask study and Koch with his disease postulates.

Pasteur's study

1 Sterilized broth remains pure indefinitely unless...

2 ...flask is tilted, shaken, or S-neck broken...

3 ...allowing microbes to contaminate broth.

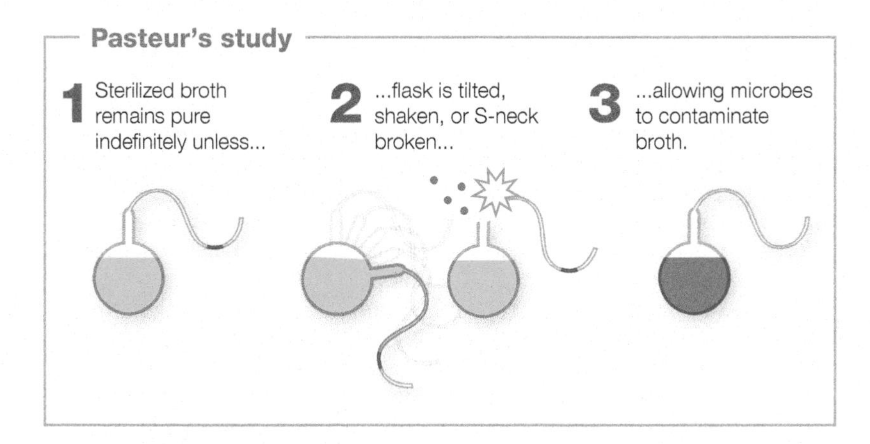

Koch's postulates

1 The same organism must be present in every case of the disease but not present in healthy individuals.

2 The organism must be isolated from the diseased host and grown as pure culture.

3 The isolated organism should cause the disease in question when it is inoculated into a susceptible host.

4 The organism must then be re-isolated from the inoculated, diseased animal.

Classifying Microbes

Organisms are classified by their shared features. A symbiotic relationship exists when two or more organisms are closely connected. Symbiotic relationships can be described as hurting the host (parasitism), helping the host (mutualism), or having no perceived benefit or cost to the host (commensalism).

CLASSIFICATION

Domain (least specific)
Kingdom
Phylum
Class
Order
Family
Genus
Species (most specific)

SYMBIOTIC RELATIONSHIPS

Host **Microbe**

Parasitic

Mutualistic

Commensal

Microscopy

Light microscopy relies on a compound light microscope. There are many forms of light microscopy, but bright field is the most common. Electron microscopy (SEM and TEM) provides high magnification and resolution.

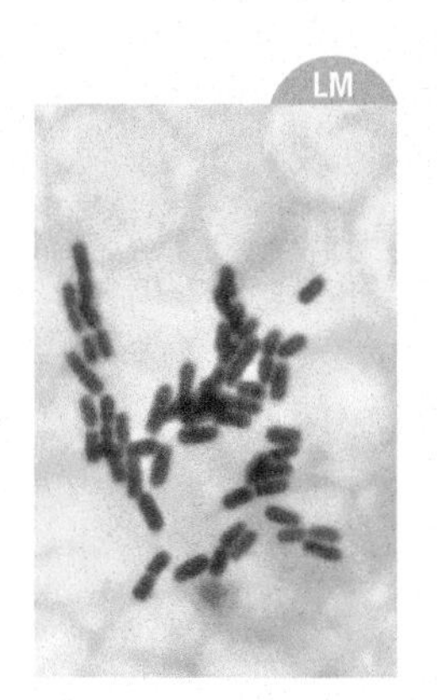

E. coli

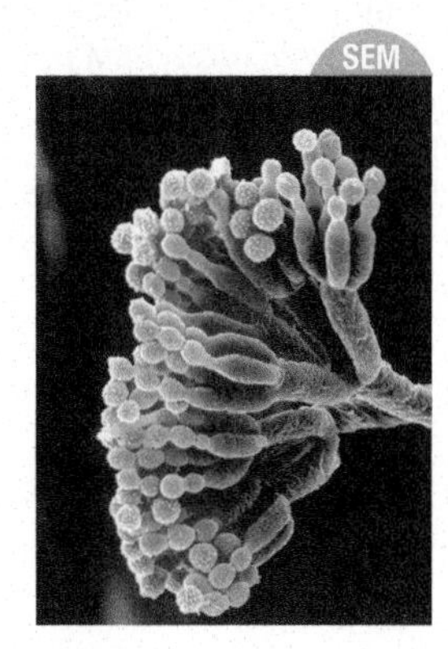

Penicillium mold

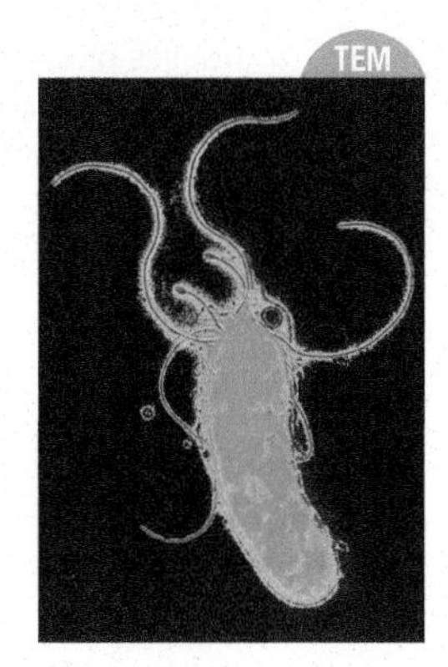

H. pylori

Staining Microbes

Simple stains give size, shape, and arrangement information. Structural stains are performed to reveal capsules, flagella, and endospores. Differential stains such as Gram and acid-fast stains distinguish bacteria based on cell wall components.

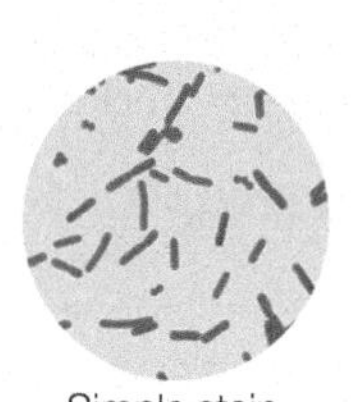

Simple stain

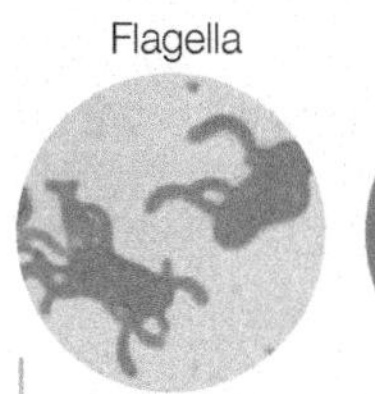

Flagella

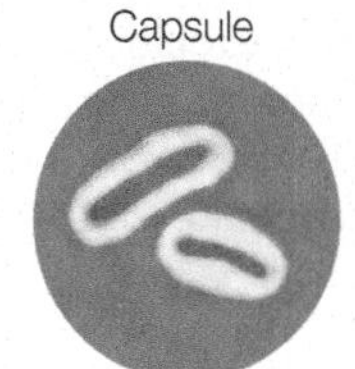

Capsule

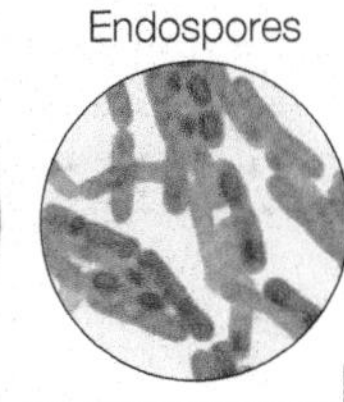

Endospores

Structural stains

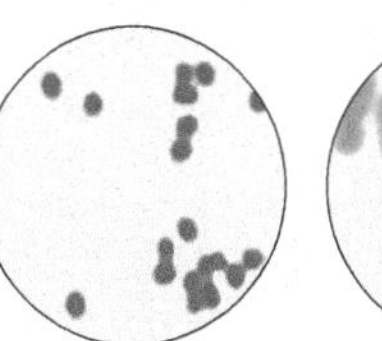

Gram-positive

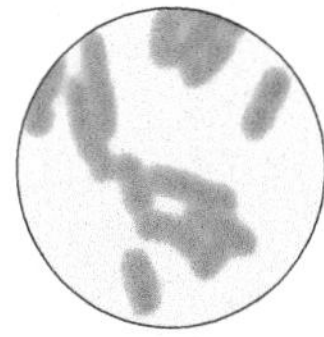

Gram-negative

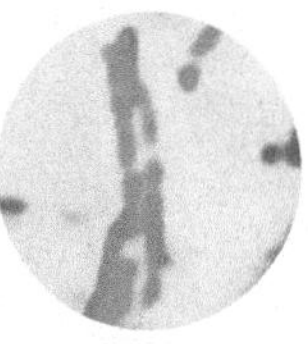

Acid-fast

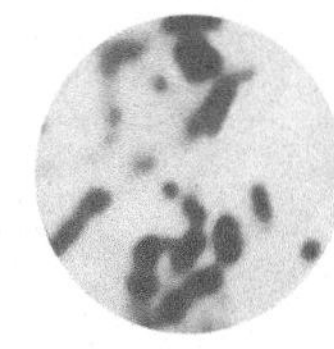

Non–acid-fast

CHAPTER 1 OVERVIEW

1.1 A Brief History of Microbiology

- Microbiology is the study of bacteria, archaea, fungi, protists, helminths, algae, viruses, and prions. Acellular nonliving microbes include viruses and prions (infectious proteins). Cellular microbes include prokaryotes and eukaryotes.
- Endosymbiotic theory states that eukaryotic cells evolved from prokaryotic cells. Prokaryotic cells include bacteria and archaea, which are always unicellular. Eukaryotes include single-celled organisms and multicellular organisms. Animals, plants, fungi, and protists are eukaryotic.
- The golden age of microbiology encompassed important advancements, including the theory of biogenesis and the development of the germ theory of disease. Pasteur's S-necked flask experiment was central to disproving spontaneous generation. The germ theory of disease states that microbes cause infectious diseases.
- Koch's postulates of disease outline a systematic process to directly link an infectious agent to a disease.
- Aseptic practices in surgery and nursing emerged as people embraced biogenesis and the germ theory of disease.
- The scientific method is the guiding investigative principle of microbiology; it involves formulating and testing a hypothesis by making observations and drawing conclusions. Observations include any data collected. Conclusions interpret observations.
- Hypotheses and laws are united in a theory; laws are not superior to theories. Theories seek to explain how and why something occurs. A law is a concise statement or mathematical formula to predict what will occur.

1.2 Classifying Microbes and Their Interactions

- Taxonomy identifies, names, and classifies organisms using a hierarchical system.
  - The three domains are Bacteria, Archaea, and Eukarya. Just under domains are kingdoms. Species is the most specific taxonomic grouping. Strains are genetic variants of the same species.
  - The binomial nomenclature system involves naming organisms by their genus and species. Scientific names are italicized and Latinized; the first name is capitalized and reflects the genus, and the second name is lowercase and designates the species.
- Symbiotic relationships such as parasitism, mutualism, and commensalism are ecological relationships between two or more closely connected organisms.
- Our normal microbiota consists of microbes that stably reside in and on us; our normal microbiota affects our physiology—including susceptibility to opportunistic pathogens.
- Biofilms are communities of microbes that coordinate their physiological responses.
- Microbes have major roles in bioremediation, food and drug production, making important precursors for consumer products, and serving as a source for biodegradable plastics, and may even make biofuels.

1.3 Growing, Staining, and Viewing Microbes

- Growth media are mixtures of nutrients that support the growth of cells or simple organisms in an artificial setting. Media can be poured as broths, plates, slants, or deeps. Petri plates of agar-solidified media are used for isolating bacteria and generating pure cultures. Pure cultures consist of only one species.
- The streak plate technique is often used to isolate individual colonies and determine culture purity; the characteristics of the colonies on a plate can help identify a microbial species. Characterizing cultures depends on aseptic culturing techniques to limit contaminants.
- Staining specimens is central to observing and classifying microbes. Simple stains use a single dye to reveal specimen size, shape, and arrangement. Structural stains reveal flagella, capsules, and endospores. Differential stains highlight differences in bacterial cell walls in order to discriminate between distinct classes of cells.
- The Gram stain is a differential stain that highlights differences in Gram-positive versus Gram-negative bacterial cell walls.
- Gram-positive bacteria appear purple at the end of the Gram stain because their cell walls are rich in peptidoglycan and lack an outer membrane, allowing for crystal violet retention after exposure to an alcohol-acetone solution. Gram-negative bacteria have a thin peptidoglycan layer and an outer membrane that allows crystal violet to be rinsed from the cell wall when exposed to an acetone-alcohol wash. Gram-negative cells are pink after the Gram stain.
- The acid-fast stain distinguishes between cells that have mycolic acid in their cell wall (acid-fast bacteria) and those that do not (non–acid-fast bacteria). Acid-fast bacteria retain the carbol-fuchsin dye after being exposed to an acid wash and appear pink at the end of the staining procedure. Non–acid-fast bacteria lack a waxy cell wall and are stripped of the carbol-fuchsin by an acid-containing solution; these cells appear blue at the end of the staining procedure.
- The compound light microscope is the most common type of optical microscope.
- Resolution is the ability to distinguish two distinct points as separate. Immersion oil enhances resolution in light microscopy by ensuring that the light that interacts with the specimen is smoothly funneled toward the high-power objective lens instead of being scattered.
- TEM is the most common form of electron microscopy; it generates images of internal cell structures. SEM produces three-dimensional images of surface structures; it has lower magnification and resolution than TEM.
- Fluorescence microscopy requires a fluorochrome and a specialized microscope that illuminates samples with UV light.

COMPREHENSIVE CASE

The following case integrates basic principles from the chapter. Try to answer the case questions on your own. Don't forget to be S.M.A.R.T.* about your case study to help you interlink the scientific concepts you have just learned and apply your new understanding of microbiology principles to a case study.

***The five-step method is shown in detail below. Refer back to this example to help you apply a SMART approach to other critical thinking questions and cases throughout the book.**

• • •

Cholera is an acute, infectious diarrheal illness that affects millions and kills up to 120,000 people every year. It likely originated in ancient India. Starting in the 1800s, it spread worldwide through ships' bilgewater. The first occurrence in Europe and the Americas spurred a flurry of theories about the cause of the disease. The miasma, or "harmful air," theory became a popular explanation. This theory was based on the observation that air smelled bad in poverty-stricken areas where cholera was common. The miasma theory, though inaccurate, pressed public officials to address sewage issues that plagued poor areas. As sewage management improved, the stench that miasma proponents felt caused cholera decreased; the number of cases also decreased. This was interpreted as further support for the miasma theory.

In 1849, English physician John Snow observed cholera cases clustered around a local well. He ordered the pump handle removed so water could not be drawn from it, and soon after, the local outbreak ended. Despite Snow's success, the miasma theory of cholera remained popular. In 1884 Robert Koch isolated a unique, comma-shaped bacterium from stool samples of cholera patients. Koch demonstrated that this bacterium, now called *Vibrio cholerae*, was always present in cholera patients. He described how to isolate and grow it and proposed it made a toxin that caused pathology. It was later determined that *V. cholerae* is a Gram-negative, flagellated bacterium that indeed makes a potent toxin.

Despite Koch's convincing evidence, many people continued to believe the miasma theory of cholera. One especially vocal advocate was Max von Pettenkofer, a renowned chemist and public hygienist who claimed it was an oversimplification to think that *V. cholerae* alone could cause disease. He noted that Koch was unable to infect animals with the isolated bacteria and therefore failed to fulfill his own third postulate of disease. Pettenkofer argued that the bacteria were harmless until they had a maturation period in the soil and released a toxic airborne miasma that had to be inhaled in order for infection to be established. To dramatically uphold his point, in the presence of several witnesses, Pettenkofer drank a preparation of *V. cholerae* that Koch himself had isolated from the stool sample of a dead cholera patient. The 76-year-old Pettenkofer won a victory for miasma theory when he did not develop full-blown cholera. While this minor victory sustained the miasma theory a little longer, it was eventually debunked, and the theory of biogenesis was embraced.

We now recognize that *V. cholerae* is a naturally occurring waterborne bacterium that invades humans through contaminated drinking water and foods. There are hundreds of *V. cholerae* strains, but the O1 and O139 strains are responsible for most outbreaks. The O139 strain has a capsule; the O1 strain does not. *V. cholerae* is not native to the human intestinal tract and tends to inhabit the human gut only temporarily, making a long-term carrier state extremely rare. While humans can spread cholera in feces, the source of initial infection that jump-starts outbreaks comes from the natural marine environment.

Copepods are small crustaceans found worldwide in estuaries, rivers, and ponds; they naturally harbor *V. cholerae* in their gut and as biofilms on their surface. These crustaceans appear to be important in the natural ecology of the bacteria. In the presence of live copepods, the population of *V. cholerae* rapidly increased by 100-fold, while in the presence of dead copepods the population showed a modest initial increase and then rapidly declined. An unrelated bacterium, a species of *Pseudomonas*, had a comparable increase in population in the presence of dead or live copepods. Electron microscopy has shown that *V. cholerae* adheres to the surface of these tiny zooplankton creatures, especially around the oral region and egg sacs. The bacteria secrete a substance that helps the egg sac rupture to release the eggs that are ready to be fertilized, and the bacteria benefit by having an environment rich in food and also safe from protozoans that normally graze on marine bacteria. Additional studies have shown that in response to starvation, *V. cholerae* colonies growing on agar plates change from their typical smooth morphology to wrinkled (or rugose) colonies. Bacteria in rugose colonies produce a slimy substance that makes them better adapted to forming biofilms on a variety of surfaces. Despite over 100 years of studies, we still have an incomplete picture of the pathophysiology of these bacteria and how biofilms play a role in infection.

• • •

CASE-BASED QUESTIONS

1. Population studies show type O blood is practically extinct in populations from the Ganges River delta in India. Based on this chapter, provide a possible evolutionary explanation for this.
2. Provide a possible explanation for why Koch experienced difficulties in establishing an animal model for cholera.
3. What type of electron microscopy was most likely used to obtain the data about *V. cholerae*'s colonization of copepod external surfaces? Explain your reasoning.
4. We now know the miasma theory is wrong, so how did proper sewage management help reduce cholera cases, if not by reducing the stench?
5. How would you define the symbiotic relationship between copepods and *V. cholerae*? Explain your definition.
6. How would you describe the biota classification or status of *V. cholerae* in humans? How about in copepods?

7. What is the most likely explanation for why Max von Pettenkofer did not develop a classic case of cholera after drinking the *V. cholerae* bacteria?
8. Why might *V. cholerae* cells from rugose colonies be better able to survive harsh conditions than cells from smooth colonies? Explain your reasoning.

Let's use our S.M.A.R.T. approach to address this case.

Now we'll walk through the cholera case questions using the SMART steps. Refer back to this section as a model for using this methodology elsewhere in the book.

1. **Summarize what is known and what is unknown.** *You may find it helpful to organize your thinking via a table similar to the one shown below.*
2. **Make connections between information in the case and your own knowledge to answer the questions.** *The case questions frequently require information from both the case itself and the chapter as a whole.*
3. **Avoid distractors; not all information in a case is relevant, and not all unknowns need to be known to solve the case!** *Stay focused by excluding information after you determine it doesn't impact the questions at hand.*
4. **Re-read the case and the questions to make sure you didn't miss anything.** *Do we have all the information we need to move forward with writing thorough answers? Is there anything that we need to look up or expand upon?*
5. **Thoroughly and thoughtfully address the questions being asked in the case.** *Apply logic, and organize your information into coherent answers.*

Working Through the Comprehensive Case on Cholera

Question	Relevant Information from Case (or question) and Unknown Details	Relevant Information from Chapter	Sample Answer
1. Population studies show type O blood is practically extinct in populations from the Ganges River delta in India. Based on this chapter, provide a possible evolutionary explanation for this.	• Cholera is an ancient disease that originated in India.	• Malaria is an ancient disease that has impacted human evolution. • People with the sickle cell trait are resistant to malaria. • The trait is common in populations hailing from malaria-prone regions.	Just as the sickle cell trait protects against malaria and therefore became common in populations of malaria-prone areas, so too could a trait that confers a *disadvantage* become rare in a population. If people with type O blood don't fight cholera infections as successfully as those with other blood types, then type O blood would present an evolutionary disadvantage and become rare in people hailing from cholera-prone regions.
2. Provide a possible explanation for why Koch experienced difficulties in establishing an animal model for cholera.	• Koch could not establish an animal model for cholera. Aside from this, no further information is provided in the case.	• Cholera causes either classic disease or a quiet infection. This is why the third postulate states that disease should (rather than must) result. • Koch couldn't establish animal models for some diseases. This led him to correctly assume that some diseases only affect humans.	The bacterium may have mutated in culture and lost its pathogenic capabilities. But, based on the chapter information, it's more likely that cholera is strictly a human disease and is not naturally found in other animals (at least not those Koch tested).
3. What type of electron microscopy was most likely used to obtain the data about *V. cholerae*'s colonization of copepod external surfaces? Explain your reasoning.	• Copepods have *V. cholerae* in their gut and on their outer surfaces. Electron microscopy shows the bacteria on external copepod surfaces.	• SEM provides information about surface structures. • TEM provides information about internal structures.	TEM is used to visualize internal structures, and samples have to be very thinly sliced before viewing. TEM may have been helpful in identifying *V. cholerae* in the gut of copepods. However, the question asks about colonization of *external surfaces* of copepods, not the gut. SEM is the recommended tool for visualizing relatively large surface areas and is most likely what was used.
4. We now know the miasma theory is wrong, so how did proper sewage management help reduce cholera cases, if not by reducing the stench?	• *V. cholerae* is in feces-contaminated water. • Snow observed cholera cases clustered around a local well, and closing it stopped the outbreak. • Marine environments are the initial outbreak source of cholera.	• Crowded living conditions, poor sanitation, and lack of waste management caused epidemics. • These outbreaks stimulated the formation of public health departments to improve sanitation and living conditions.	By properly managing the raw sewage in cholera-prone areas, the air smelled better and the drinking water was less likely to become contaminated with the cholera bacterium, which is present in the feces of infected patients.

Question	Relevant Information from Case (or question) and Unknown Details	Relevant Information from Chapter	Sample Answer
5. How would you define the symbiotic relationship between copepods and *V. cholerae*? Explain your definition.	• *V. cholerae* is in copepods and on their surface. • *V. cholerae* helps copepod reproduction, and the bacteria benefit by having an environment rich in food and safe from protozoans.	• Symbiosis is an ecological relationship between two or more closely connected organisms. • Parasites benefit at the host's expense. • In mutualism both participants benefit. • In commensalism, one participant benefits, and the partner is not harmed or benefited.	Because both the copepods and the bacteria benefit in this ecological relationship, this represents a form of mutualism, or mutual symbiosis.
6. How would you describe the microbiota classification or status of *V. cholerae* in humans? How about in copepods?	• *V. cholerae* temporarily inhabit the human gut; a carrier state is rare. • Copepods naturally harbor *V. cholerae*. Copepods have a role in the natural ecology of the bacteria.	• Normal microbiota consists of microbes that stably reside in and on us. • Transient microbiota are just temporary passengers and do not persist as residents. • Most acquired pathogens are transient biota. • *V. cholerae* is a naturally occurring waterborne bacterium; it invades humans through contaminated water and foods.	Most pathogens are transient biota. *V. cholerae* bacteria do not usually stably colonize the human gut, so are therefore most accurately described as transient flora, or transient biota, of humans. In contrast, *V. cholerae* are naturally found as stable colonizers of copepods and are therefore a part of the latter's normal microbiota.
7. What is the most likely explanation for why Max von Pettenkofer did not develop a classic case of cholera after drinking the *V. cholerae* bacteria?	• Hundreds of *V. cholerae* strains exist, but the O1 and O139 strains are responsible for most outbreaks.	• Koch confirmed that not all infection events lead to disease; certain diseases exist in a latent or quiet form that may not manifest symptoms.	Assuming the bacterium remained infectious even after being grown in the lab, von Pettenkofer may not have been a susceptible host or he may have had an asymptomatic infection.
8. Why might *V. cholerae* cells from rugose colonies be better able to survive harsh conditions than cells from smooth colonies? Be sure to explain your reasoning.	• When starving, *V. cholerae* colonies growing on agar plates change from smooth to rugose colonies. • *V. cholerae* from rugose colonies are better at forming biofilms.	• Biofilms are microbial communities that coordinate their responses to improve survival. • Microbes in a biofilm are protected against host immune response or the action of antibiotics. • The sticky matrix made by cells in a biofilm is protective and hard to eliminate; innermost biofilm residents are highly protected.	Bacteria in rugose colonies make a sticky substance that helps them form biofilms. This is a survival advantage because biofilms resist many environmental challenges. The sticky biofilm matrix is a protective barrier to chemical agents like antibiotics. Layers deep in a biofilm have added protection due to being isolated from the local environment.

END OF CHAPTER QUESTIONS

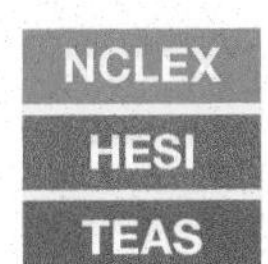

Think Critically and Clinically

Questions highlighted in **orange** are opportunities to practice NCLEX, HESI, and TEAS-style questions.

1. Indicate which form of electron microscopy is being described: SEM, TEM, or both.
   a. Images the external surface of a specimen
   b. Sample must be thinly sliced prior to viewing
   c. Provides details about internal structures of the sample
   d. Can be used to image viruses
   e. Provides black-and-white images
   f. Specimen is dead

2. Match the following people to their scientific/medical contribution:

   Note: Choices may be used more than once or not at all.

Antonie van Leeuwenhoek	Disproved spontaneous generation using an S-necked flask
Francesco Redi	Developed postulates of disease
Louis Pasteur	The first to observe bacteria
Robert Koch	The first to show that fermentation was caused by microbes
Ignaz Semmelweis	Implemented aseptic techniques in surgery
Joseph Lister	Practiced aseptic techniques in nursing
Florence Nightingale	Showed that hand washing decreased puerperal sepsis
Robert Hooke	First to publish descriptions of cells

3. Indicate the true statements **and** then correct the false statements so that they are true.
   a. The Gram stain is a simple stain.
   b. Simple stains reveal information about size, shape, and arrangement.
   c. Bright field microscopy requires a stained sample.
   d. Dark field microscopy requires a stained sample.
   e. The acid-fast stain detects peptidoglycan in the cell walls of certain bacteria.
   f. Gram-positive bacteria have a thin peptidoglycan layer in the cell wall.

4. Select the most accurate statement.
   a. The Gram stain is a simple stain.
   b. Bacterial endospores appear green from the acid-fast stain.
   c. The acid-fast stain is used to detect bacteria that have a capsule.
   d. Bacterial flagella can only be seen with electron microscopes.
   e. The mordant in the Gram stain is iodine.

5. Assume you are asked to view and draw all samples at a final magnification of 1000×. If the ocular lens is 10×, what objective lens should be in place as you draft your drawings?

6. Identify the following statements as observations or conclusions.
   a. The solution turned red.
   b. The bacterium is a rod shape.
   c. The cell died due to lack of nutrients.
   d. The bacterium is *E. coli*.
   e. The cell is Gram-positive.
   f. The bacterium is a pathogen.
   g. There are small green structures present upon performing the endospore stain.
   h. The solution remained clear after 10 hours of incubation.
   i. Fermentation occurred.
   j. Life comes from life.

7. Fill in the blanks: Bacteria are ________________ cells in the domain ________________. In contrast, the domain ________________ includes unicellular and multicellular organisms that are made of ________________ cells, or cells that have a nucleus and membrane-bound organelles.

8. Assume that you isolated a unicellular, non-nucleated cell from a deep-sea vent. Select the statement that is least likely regarding the cell you found.
   a. The cell is a prokaryote.
   b. The cell is a fungus or yeast.
   c. The cell is an archaea.
   d. The cell is not a pathogen.
   e. The cell is in Domain Archaea.

9. Fill in the blanks: In taxonomy, the broadest groupings are called ________________, which are further subdivided into six different ________________. The most specific or narrowest grouping is ________________, which is the ________________ name in the binomial nomenclature system.

10. How is an opportunistic pathogen different from a pathogen?

11. A(n) ________________ is a genetic variant of the same species.

12. Label and describe the function of each indicated part of the compound light microscope.

13. Fill in the blanks: ________________ are dormant structures that certain bacteria can make. These structures can be seen using the structural stain called the ________________.

14. Classify the following as bright field, dark field, phase contrast, or differential interference contrast (DIC).
   a. Sample must be stained or have its own coloration to be seen:
   b. Specimen is illuminated with polarized light:
   c. Generates a falsely three-dimensional image:
   d. The simplest and most common form of light microscopy:
   e. Generates a negative image using a hollow cone of light:
   f. Generates a negative image in which light shifts are converted into visible changes in brightness and contrast:

15. Which of the following is/are true regarding our normal microbiota? Select all that apply.
   a. Our normal microbiota is easily disrupted by hygiene practices like hand washing.
   b. Our normal microbiota can include pathogens.
   c. Normal microbiota compete with pathogens.
   d. Normal microbiota may colonize us before we are even born.
   e. Normal microbiota changes over time.
   f. The normal microbiota that we have as adults is the same as we had as children.
   g. The normal microbiota of the gut is similar to that of the skin.

16. Immersion oil improves resolution by:
   a. limiting light refraction.
   b. magnifying the specimen.
   c. improving specimen contrast.
   d. making light waves shorter.
   e. making light waves faster.

17. List the following taxonomic groupings in order from general to specific.

    Phylum, Kingdom, Genus, Family, Class, Domain, Species, Order

18. Why is it clinically helpful to know the Gram property of an organism?

19. Over time, Koch's disease postulates have been rephrased in a variety of ways, but they still reflect the same process outlined by Robert Koch. The following are reworded versions of Koch's postulates. Based on the original postulates, put the following items in the correct order.
    a. Use the purified agent to cause infection in a test animal.
    b. Isolate an infectious agent from a diseased animal.
    c. Grow the infectious agent as a pure culture in the lab.
    d. From the test animal, re-isolate the infectious agent that was originally grown in pure culture.

20. Acid-fast bacteria contain ________________ in their cell wall.

21. Which of the following is true?
    a. SEM is higher resolution than TEM.
    b. Most viruses can be seen using a light microscope.
    c. Simple staining involves an acidic and a basic dye in combination.
    d. Samples can be stained or unstained for light microscopy.
    e. These are all false statements.

22. What best describes the relationship between host and pathogen?
    a. Mutualism
    b. Parasitism
    c. Commensalism
    d. Endosymbiosis

CRITICAL THINKING QUESTIONS

1. Suppose you wanted to quickly detect a specific bacterium that was present at low levels in a sample. What microscopy technique would be the best, and why?

2. Assume that you forgot to add the iodine in the Gram staining procedure. Describe what color Gram-positive and Gram-negative bacteria would be at each step in the procedure.

3. How might an asymptomatic carrier state interfere with the application of Koch's postulates?

4. How many *different* species do you think are present in the culture pictured to the right? What led you to your conclusion?

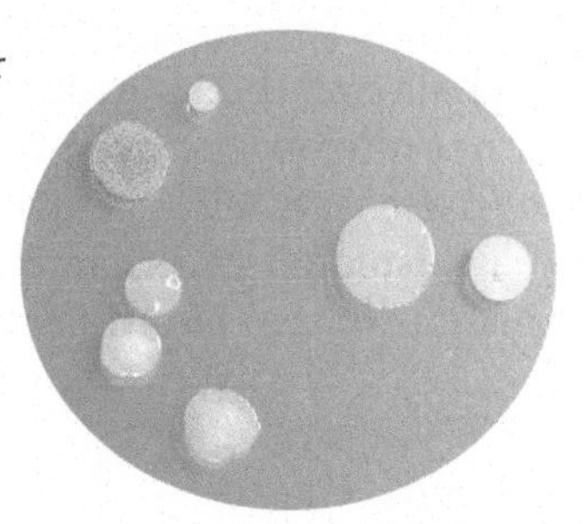

5. Biofilms are a concern in the medical field. If you were trying to design a drug to inhibit biofilm formation, what would be some potential targets for your drug?

6. An experienced microbiologist performed a Gram stain on an older culture of bacteria that was isolated from a blood sample. The cells were shown to be Gram-negative, endospore-forming, rod-shaped bacteria. The sample was then freshly cultured in the lab and then reanalyzed using Gram and endospore staining techniques. The second trial on the fresh culture showed rod-shaped, Gram-positive cells with no obvious endospores present. Assuming no errors were made, how can you logically explain the differences in the data from one trial to the next?

7. Based on the image (*right*) of Gram stained bacteria, how many different colonies (not total colonies, but *different* colonies) would you expect to grow if you were using a culture medium that only supported the growth of Gram-positive bacteria?

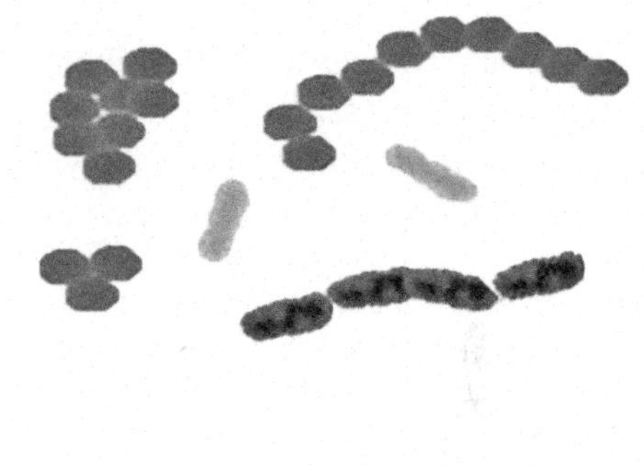

8. Imagine you have a sample that contains acid-fast and non–acid-fast bacteria. Describe two *separate* errors that could be made in the acid-fast staining procedure. Detail the exact consequences of each error on the final result of the acid-fast procedure on the acid-fast and non–acid-fast cells in your sample, and describe *why* each error would lead to the results you proposed.

Mount Sinai Hospital
National Nursing Day Celebration on May 6, 2021; New York

2

Biochemistry Basics

What Will We Explore? In this chapter we review the basics of atoms, molecules, and chemical reactions. We also review important macromolecules that will be discussed throughout the rest of the book.

Why Is It Important? Cells carry out millions of chemical reactions every second. We rely on chemistry for our very existence, as does everything around us, from the tiniest bacterium to the batteries in our electronic devices. Moreover, biochemical discoveries are among our most promising weapons for fighting infectious and noninfectious diseases. For example, many microbes emit chemical signals into their surroundings that inhibit the growth of other microbes or even kill them, yet only a small number of these naturally occurring antimicrobial chemicals have been isolated and characterized. Similarly, a number of plants contain pharmacologically active ingredients, but we have a poor understanding of the nature of these compounds. In essence, biochemistry is humanity's door to understanding how pathogens work, the intricacies of our own physiology, and the mechanism of action of many medical therapies.

Clinical CASE NCLEX HESI TEAS

The Case of the Confused Cattle Farmer

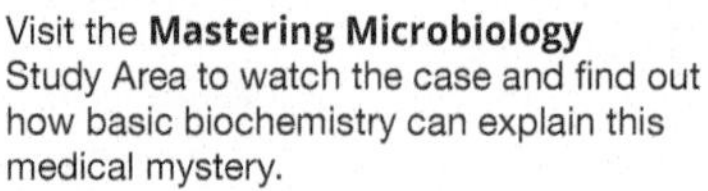

Visit the **Mastering Microbiology** Study Area to watch the case and find out how basic biochemistry can explain this medical mystery.

2.1 FROM ATOMS TO MACROMOLECULES

What are atoms?

Chemistry is the branch of science that studies atoms and molecules and how they interact. Biochemistry is the study of chemistry in living systems; it underpins how microbes live, how they impact us, and how we diagnose and treat the infections they cause.

Ordinary matter is everything around you. It exists in five different states: solids, liquids, gases, plasmas, and Bose-Einstein condensates (discovered in 1995). **Atoms** are the smallest units of **elements**, which are pure substances—such as carbon, oxygen, and iron—that make up ordinary matter. At the center of an atom is a nucleus that contains **protons** and **neutrons** (FIG. 2.1).

Protons are positively charged particles, whereas neutrons are non-charged or neutral particles. Around the nucleus is a cloud of negatively charged particles called **electrons**. Atoms can vary their number of neutrons and/or number of electrons. However, *the number of protons in an atom of a given element remains constant* and is a defining feature of that element. Therefore, elements are identified by their **atomic number**, or number of protons. Each element has a unique atomic number that is used to organize the elements into the periodic table.

Learning Outcomes

After reading this section, you should be able to:

2.1 Define the term *atom*, and identify an atom's parts.

2.2 Determine the atomic mass, atomic number, and chemical symbol of an element using the periodic table.

2.3 Explain the difference between an anion and a cation, and state how they are formed.

2.4 Describe isotopes and their importance in medicine.

2.5 Distinguish between a molecule, compound, and isomer.

2.6 Interpret and write a molecular formula.

2.7 Differentiate between organic and inorganic compounds, and identify selected functional groups.

2.8 Compare acids and bases, and discuss their effects on pH.

2.9 Explain what the pH scale reflects, and list its features.

2.10 Define the term *buffer*, and state why buffers are important in biological systems.

The Periodic Table

To date, 118 elements have been identified, most of which are naturally occurring. All of the known elements are organized by atomic number, electron arrangements, and chemical properties into a table format called the **periodic table of elements** (FIG. 2.2). Each element is noted by its **chemical symbol**, an abbreviated letter notation that derives from the name of the element, which is often Greek or Latin. For example, gold is "aurum" in Latin and is noted by the symbol Au on the periodic table. Above the chemical symbol is the atomic number, which is equal to the number of protons in an atom of that given element. Underneath the chemical symbol is a number noted in smaller font—usually it is a decimal; this number is the **atomic mass**. Because electrons have negligible mass, the atomic mass is mainly determined by the mass of the protons and neutrons in the atom. The atomic mass is the average mass of 6.022×10^{23} atoms, or one **mole**, of the given element.[1]

Ions and Isotopes: Variations of Atoms

Two forms of atoms are ions and isotopes. All elements exist as a variety of isotopes, but only certain elements form ions.

Ions In their elemental state, atoms have an equal number of positively charged protons and negatively charged electrons and are therefore neutral. However, under certain conditions some atoms form **ions**, which are charged atoms that have an unequal number of protons and electrons (FIG. 2.3). A positive ion, or **cation**, is an atom that has lost electrons and consequently has an overall positive charge. A negative ion, or **anion**, has a negative charge as a result of gaining that specifies the ion's charge. A calcium ion is noted as

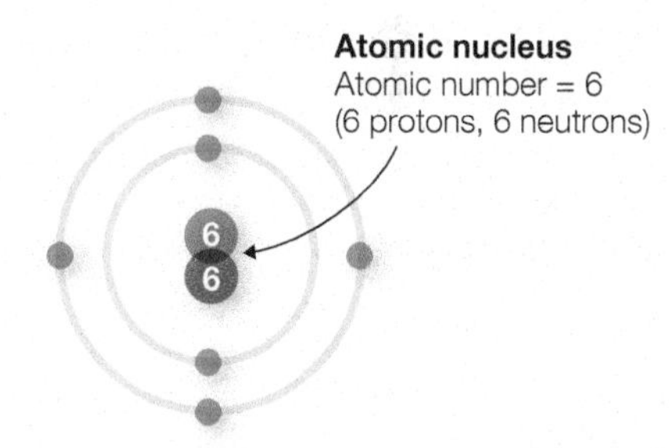

FIGURE 2.1 Basic structure of an atom
A generalized carbon atom is shown. The nucleus of an atom contains protons and neutrons; electrons are found in shells around the nucleus. These shells are not physical structures; rather, atoms have a cloud of electrons around their nucleus. The mass of an atom is mainly due to the number of protons and neutrons it contains.

[1] A mole is technically a collection of 6.022×10^{23} of a defined entity, or Avogadro's number of the substance being considered. If we had 6.022×10^{23} dollar bills, we could say we had a mole of dollars. However, this number is so large it would be unrealistic; you'd have to spend one billion dollars *per second* for 19 million years to spend it all. As such, chemists and physicists use this standard unit to refer to the quantity of very small things like atoms, subatomic particles, and molecules. If there were 6.022×10^{23} molecules of water in a sample, then we would say there was one mole of water in the sample.

1 — Atomic number
H — Chemical symbol
Hydrogen — Name
1.008 — Atomic mass

1	2	3	4	5	6	7	8	9	10	11	12	13	14	15	16	17	18
1 H Hydrogen 1.008																	2 He Helium 4.003
3 Li Lithium 6.997	4 Be Beryllium 9.012											5 B Boron 10.821	6 C Carbon 12.011	7 N Nitrogen 14.007	8 O Oxygen 15.999	9 F Fluorine 18.998	10 Ne Neon 20.180
11 Na Sodium 22.990	12 Mg Magnesium 24.305											13 Al Aluminium 26.982	14 Si Silicon 28.086	15 P Phosphorus 30.974	16 S Sulfur 32.076	17 Cl Chlorine 35.457	18 Ar Argon 39.948
19 K Potassium 39.098	20 Ca Calcium 40.078	21 Sc Scandium 44.956	22 Ti Titanium 47.867	23 V Vanadium 50.942	24 Cr Chromium 51.996	25 Mn Manganese 54.938	26 Fe Iron 55.845	27 Co Cobalt 58.933	28 Ni Nickel 58.693	29 Cu Copper 63.546	30 Zn Zinc 65.382	31 Ga Gallium 69.723	32 Ge Germanium 72.631	33 As Arsenic 74.922	34 Se Selenium 78.972	35 Br Bromine 79.907	36 Kr Krypton 83.798
37 Rb Rubidium 85.468	38 Sr Strontium 87.621	39 Y Yttrium 88.906	40 Zr Zirconium 91.224	41 Nb Niobium 92.906	42 Mo Molybdenum 95.941	43 Tc Technetium 98	44 Ru Ruthenium 101.072	45 Rh Rhodium 102.905	46 Pd Palladium 106.421	47 Ag Silver 107.868	48 Cd Cadmium 112.414	49 In Indium 114.818	50 Sn Tin 118.710	51 Sb Antimony 121.760	52 Te Tellurium 127.603	53 I Iodine 126.904	54 Xe Xenon 131.293
55 Cs Cesium 132.905	56 Ba Barium 137.327	71 Lu Lutetium 174.967	72 Hf Hafnium 178.492	73 Ta Tantalum 180.948	74 W Tungsten 183.841	75 Re Rhenium 186.207	76 Os Osmium 190.233	77 Ir Iridium 192.217	78 Pt Platinum 195.084	79 Au Gold 196.966	80 Hg Mercury 200.592	81 Tl Thallium 204.385	82 Pb Lead 207.210	83 Bi Bismuth 208.980	84 Po Polonium [209]	85 At Astatine [210]	86 Rn Radon [222]
87 Fr Francium [223]	88 Ra Radium [226]	103 Lr Lawrencium [262]	104 Rf Rutherfordium [261]	105 Db Dubnium [262]	106 Sg Seaborgium [266]	107 Bh Bohrium [264]	108 Hs Hassium [277]	109 Mt Meitnerium [268]	110 Ds Darmstadtium [281]	111 Rg Roentgenium [272]	112 Cn Copernicium [285]	113 Nh Nihonium [286]	114 Fl Flerovium [289]	115 Mc Moscovium [289]	116 Lv Livermorium [293]	117 Ts Tennessine [294]	118 Og Oganesson [294]
		57 La Lanthanum 138.905	58 Ce Cerium 140.116	59 Pr Praseodymium 140.907	60 Nd Neodymium 144.242	61 Pm Promethium 145	62 Sm Samarium 150.36	63 Eu Europium 151.964	64 Gd Gadolinium 157.253	65 Tb Terbium 158.925	66 Dy Dysprosium 162.500	67 Ho Holmium 164.930	68 Er Erbium 167.259	69 Tm Thulium 168.934	70 Yb Ytterbium 173.045		
		89 Ac Actinium [227]	90 Th Thorium 232.038	91 Pa Protactinium 231.036	92 U Uranium 238.029	93 Np Neptunium [237]	94 Pu Plutonium [244]	95 Am Americium [243]	96 Cm Curium [247]	97 Bk Berkelium [247]	98 Cf Californium [251]	99 Es Einsteinium [252]	100 Fm Fermium [275]	101 Md Mendelevium [258]	102 No Nobelium [259]		

FIGURE 2.2 The periodic table The periodic table of elements is organized by the chemical properties and atomic number of each element. Numbers in brackets indicate the **mass** of that element's most stable isotope.

Critical Thinking *Make a periodic table box for a fictitious element. Include a name, chemical symbol, atomic number, atomic mass, and describe how many protons, neutrons, and electrons are in your element. For simplicity, assume that your element does not have any isotopes and exists in only one form.*

CHEM • NOTE

Hydrogen is the only element with isotopes that have names that are different from their parent atom. Deuterium is ^{2}H and tritium is ^{3}H.

Ca^{2+} because it has an overall charge of +2 as a result of losing two electrons; chlorine often exists as an anion and is noted as Cl^{-} to reflect its overall charge of –1 that results when a chlorine atom gains a single electron.

Isotopes All elements exist as a mixture of **isotopes**, which are atoms with the same number of protons but different numbers of neutrons. Isotopes are denoted by their total number of protons and neutrons. For instance, about 99 percent of carbon atoms have six protons and six neutrons and are therefore known as carbon-12 (also written as ^{12}C). But two other isotopes exist for carbon: Carbon-13 (^{13}C) has seven neutrons and six protons, whereas carbon-14 (^{14}C) has eight neutrons and six protons. Carbon-12 and carbon-13 are stable nonradioactive isotopes, but carbon-14 is a radioactive isotope that decays over time. Radioactive atoms release energy (radiation energy) as their unstable nucleus breaks down.

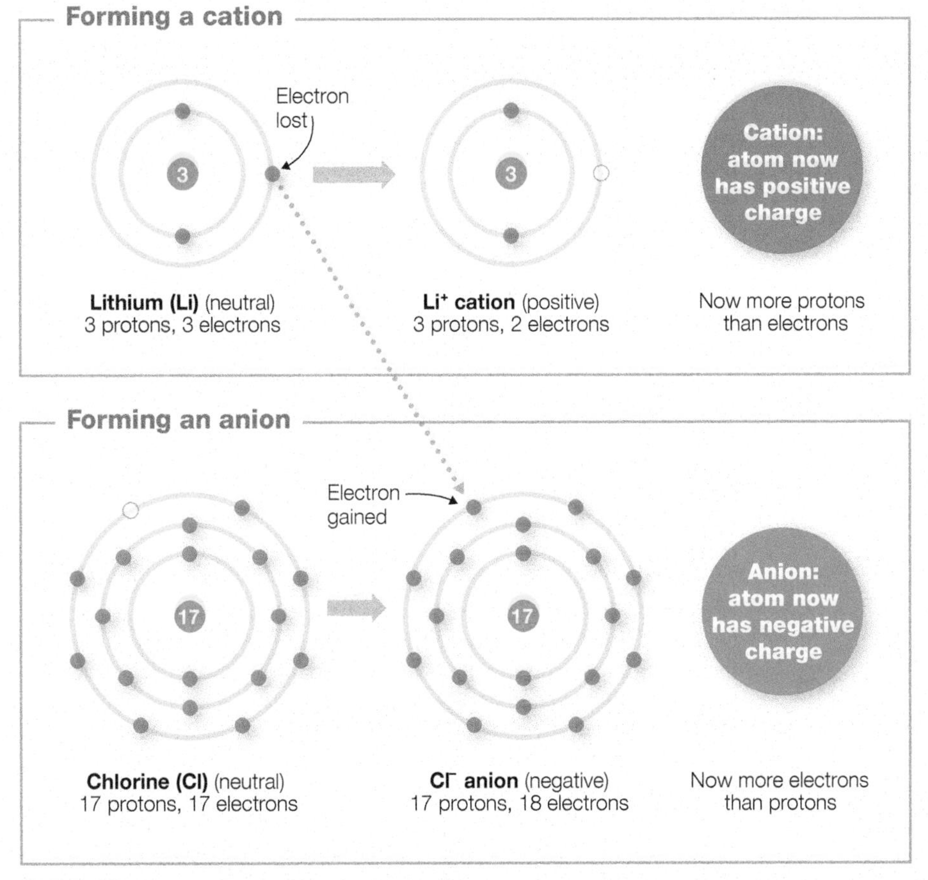

FIGURE 2.3 Ion formation Ions form when an atom gains or loses one or more electrons. Cations form when an atom loses one or more electrons; anions form when an atom gains one or more electrons.

Critical Thinking *Assume two elements, X and Y, interact to transfer electrons. In the interaction, three electrons are transferred from X to Y. Based on this information, write the correct ion representation for X and Y.*

Again, only certain elements form ions. There are cases where an atom of such elements simultaneously exists as both an ion and an isotope. For example, ^{3}He is an isotope of helium, He^{2+} is an ion of helium, and $^{3}He^{2+}$ is simultaneously an ion and an isotope of helium.

Isotopes have many applications. The rate of carbon-14 decay is used to determine the age of an organic sample in a process called *radiocarbon dating*. A field called *nuclear medicine* has also become increasingly important in modern health care. This branch of health care uses *radiopharmaceuticals*, or drugs that contain specific isotope formulations, to diagnose and treat certain diseases. Examples of nuclear medicine techniques include positron emission tomography (PET) scans, iodine-123 imaging to assess the thyroid gland, and radionuclide therapy to treat certain cancers.

What are molecules?

A **molecule** is formed when two or more atoms bond together. An elemental hydrogen molecule contains two atoms of hydrogen (H_2), while a molecule of water contains two hydrogen atoms and one oxygen atom (H_2O). Sometimes the word **compound** is used to describe molecules that are made of more than one type of element. For example, water is a molecule, but more specifically, water is a compound.

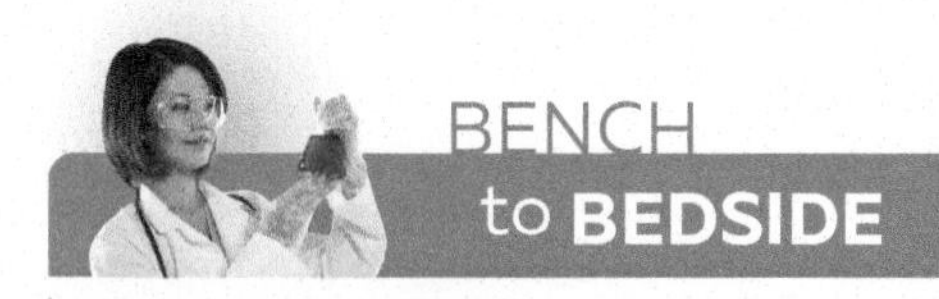

Neutrons to Nuke Cancer

When you hear the term "radiation," words like *toxic*, *deadly*, and *bomb* may spring to mind. Many see radiation as frightening and dangerous, and it certainly is not to be taken lightly. However, when applied clinically, radiation can be a key diagnostic tool and a lifesaving medical therapy. Nuclear medicine is a branch of health care that uses over 100 different isotopes to detect and treat disease, especially cancers. Most of the time the isotopes are used as oral, inhaled, or injected "tracers" to help visualize physiological processes, such as to trace blood flow in the brain, kidneys, and cardiovascular system. Technetium-99m (^{99m}Tc) is the most common isotope used in diagnostic nuclear medicine. This isotope has even been used to make radiolabeled antibiotics to help detect active bacterial infections in bone, which are otherwise challenging to definitively diagnose. In contrast, iodine-131 therapy is a widely used treatment for thyroid cancer and hyperthyroidism (overactive thyroid).

A new therapy, boron neutron capture therapy (BNCT), targets cancer cells with boron-10, a *nonradioactive* isotope. Most BNCT clinical trials have focused on treating cancers that are inoperable and have not responded well to conventional chemotherapy or radiotherapy. Although a wide variety of cancers may fall into this difficult-to-treat category, most BNCT studies have focused on treating head and neck cancers, malignant mesothelioma (a specific type of lung cancer), glioblastoma (a type of cancer that affects the brain and/or spinal cord), or melanoma (a type of cancer that affects the pigment-producing cells in the skin). In BNCT, after the boron isotope accumulates in cancer cells, the area is hit with a beam of neutrons, which are intensely absorbed by the boron-10. As the neutrons are absorbed, high-energy particles are emitted that kill the cancer cells.

The precise mechanisms by which BCNT triggers cell death remain poorly understood, but it's appreciated that the mechanism of death likely varies according to the type of boron-10-containing substance used to target the cancer cells.

Neutrons

Brain tumor

High-energy particles

Cells in brain tumor uptake boron-10. A beam of neutrons causes boron-10 to emit high-energy particles that kill the cells.

For BNCT to be effective, the boron-10 drug needs to specifically target and sufficiently accumulate inside cancer cells so that when the cells are bombarded with neutrons, enough energy is released to kill them. One boron-based drug being studied includes a hydrophobic molecule that has been delivered to cancer cells in liposomes, which are essentially lipid bilayer bubbles. Research shows that these liposome vehicles are able to penetrate deeply into tumors that are not directly served by the bloodstream and deliver their deadly cargo directly to the cancer cells. The use of boron-loaded nucleotides is also being investigated. In this technique, the boron-10 nucleotides bind specifically to nucleic acid targets in cancer cells.

Source: Dymova, M. A., Taskaev, S. Y., Richter, V. A., & Kuligina, E. V. (2020). Boron neutron capture therapy: Current status and future perspectives. *Cancer Communications*, *40*(9), 406–421.

CHEM • NOTE

Line drawings are a common way to represent an organic molecule's structure. Various simplified formats exist, and you'll see them in this text. This simplification helps to focus attention on parts of the molecule relevant to the learning goal. As an example, here are three ways to represent a molecule of glucose:

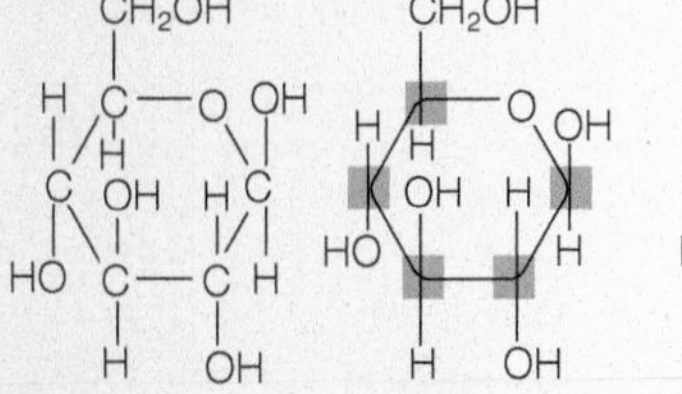

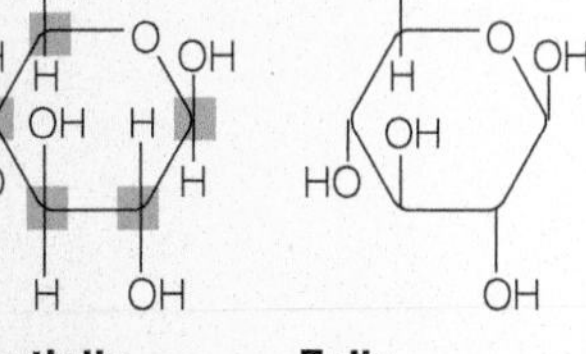

No abbreviation Write out all atoms.

Partially abbreviated Assume carbons present at any corner in drawing (yellow).

Fully abbreviated Take the partially abbreviated format and omit hydrogens that stabilize carbon (fill a carbon atom's valence shell).

Molecular Formulas

Molecules are often noted by their **molecular formula** (or chemical formula), which is like an atomic recipe, revealing what elements they contain as well as their ratios. For example, H_2 and H_2O are molecular formulas. To maintain consistency, there are rules for writing out molecular formulas. If carbon is present, it is typically listed first by its chemical symbol, "C," followed by hydrogen, and then any other elements are noted by their chemical symbol in alphabetical order. If carbon is not present, then alphabetical order by chemical symbol is usually followed. For ionic compounds, the positive ion is ordinarily listed first, followed by the negative ion. For example, hydrochloric acid is always formulaically presented as HCl, but when this molecule is in an aqueous (water) solution it ionizes to form H^+ cations and Cl^- anions. As with all rules, there are exceptions, but understanding the methodology of molecular formulas can help you note molecular characteristics.

It is possible to have different molecule structures with the same molecular formula. These are called **isomers**. The majority of biological molecules have at least one isomer. For example, the formula $C_6H_{12}O_6$ is the molecular formula for three structurally distinct sugars: glucose, fructose, and galactose. These isomers have the same number of atoms of the same three elements, but in different arrangements. Because fructose and galactose are isomers of glucose, they can easily be converted to glucose by our body and by many microbes as well.

TABLE 2.1 Selected Biologically Important Functional Groups

Functional Group	Formula	Notes (R = the remainder of a molecule; often a carbon-based addition to the molecule)
Alcohol	or R-OH	Found in all alcohols and added to steroids to make sterols; the suffix "ol" on a molecule name often means an alcohol group is present (examples: cholesterol, ethanol, glycerol); alcohol groups (also called hydroxyl groups) are OH groups tagged onto organic molecules. These are not to be confused with the hydroxide ion OH^-, which is an inorganic ion that is not bonded to carbon and instead tends to be free in a solution.
Amine	or $R\text{-}NH_2$	Important in many organic molecules including amino acids and the nitrogen bases of nucleotides
Carboxyl	or R-COOH	Found in a variety of organic acids such as amino acids and fatty acids; it's considered an acid because it ionizes to form $R\text{-}COO^-$ and release H^+
Ester	or R-COO-R'	In biology, esters tend to be formed by the condensation of an alcohol and an acid, by removing water (dehydration synthesis); lipids contain esters; phospholipids in bacteria and eukaryotic cell membranes have ester linkages.
Ether	or R-O-R'	Common linkage in carbohydrates; found in plasma membranes of archaea
Methyl	or $R\text{-}CH_3$	Common in many organic molecules, especially in hydrocarbon chains; added to DNA to regulate gene expression (DNA methylation)
Phosphate	or $R\text{-}PO_4^{2-}$	Found in DNA, RNA, ATP, and added to lipids, carbohydrates, and proteins; phosphate is denoted as PO_4^{3-} in an inorganic form or as PO_4^{2-} when bonded to an organic molecule as a functional group
Sulfhydryl	or R-SH	In cysteine and methionine (amino acids); important in building disulfide bonds in organic molecules

Organic versus Inorganic Molecules

Organic molecules contain carbon *and* hydrogen. **Inorganic molecules** may contain carbon, but will lack the associated hydrogen. A classic example of an inorganic molecule that contains carbon is carbon dioxide (CO_2). Organic molecules are typically more complex than inorganic molecules, but that doesn't mean they are more important to life than inorganic compounds. Both organic and inorganic molecules are necessary for life. After all, we wouldn't last long without the inorganic molecule oxygen (O_2), nor could we exist without complex organic molecules like our genetic material, DNA (deoxyribonucleic acid).

Functional Groups We classify and name organic molecules based in part on the functional groups they contain. **Functional groups** are molecules with shared chemical properties. Of the hundreds of functional groups, some of the most biologically important are presented in TABLE 2.1. Functional groups often participate in chemical reactions, and the presence of certain ones can be used to predict chemical properties of a molecule. For example, alcohols—like ethanol, isopropanol, and methanol—all contain an alcohol group (a hydroxyl group denoted as R-OH) and have shared chemical properties based on the shared chemical features of this group.[2]

CHEM • NOTE

Chemists often use R or **R-group** to denote the remainder of an organic molecule. This shorthand approach focuses us on the part of the molecule being discussed instead of including cumbersome chemical structures.

[2] Do not confuse hydroxyl groups, R-OH, with hydroxide ions (OH^-); they are chemically distinct.

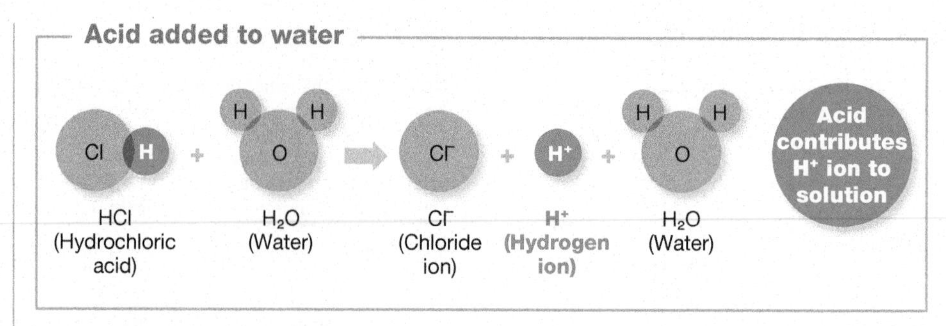

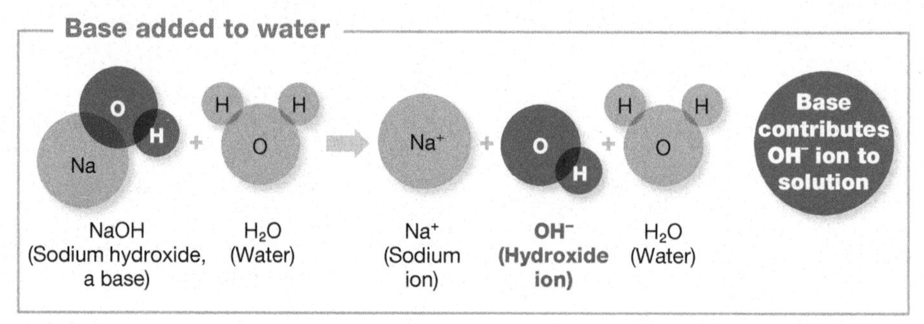

FIGURE 2.4 Acids and bases Acids contribute H^+ ions to a solution, whereas bases contribute OH^- ions.

Critical Thinking *Why would combining HCl (hydrochloric acid) and NaOH (sodium hydroxide; a base) produce water (H_2O) and a salt?*

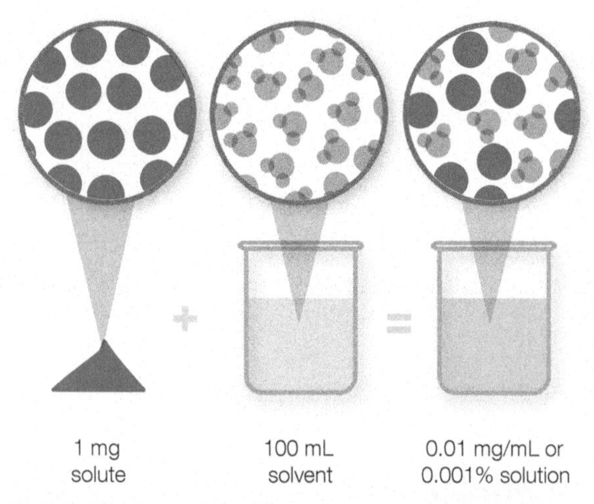

FIGURE 2.5 Solutes, solvents, solutions Solutes are dissolved in solvents to make solutions. A solution's concentration can be by molarity, a percentage, or a weight–volume proportion.

Critical Thinking *Assume you have been instructed to give a child an oral medication with a weight–volume concentration of 0.05 mg per mL (0.05 milligrams of the drug are dissolved per milliliter of solution). The child is to receive 0.1 mg of the drug every 6 hours. How many milliliters of the medication would you give the child every 6 hours? How many total milligrams would be administered over a 24-hour period?*

Acids, Bases, and Salts Most cellular chemistry occurs in an **aqueous solution**—that's a liquid mixture where water is the **solvent** (dissolving agent), and the dissolved substance is called the **solute**. **Acids** are molecules that contribute hydrogen ions (H^+) to an aqueous solution,[3] whereas **bases** are molecules that contribute hydroxide ions (OH^-) (FIG. 2.4). **Salts** form when acids and bases react with each other; the acid contributes the anion of a salt, while the base contributes a cation.

The concentration of a solution is determined by the amount of solute dissolved in a specific volume of solvent (FIG. 2.5). In clinical settings you will encounter several measures of concentration, so let's review them briefly:

Molarity is a measure of the concentration of a given solute in a liter of solvent (mol/L). Many blood chemistry values, like potassium levels, are reported in the smaller unit millimoles (mmol/L) to reflect the concentration of a given substance in a liter of blood. A patient's blood glucose level is also measured as a concentration and is usually presented as mg/dL (milligrams per deciliter); this is an example of a *weight–volume proportion*.[4] Intravenous solutions are typically labeled as having a particular *percentage* of a given solute. For example, saline solutions are labeled with a percentage: 0.9 percent normal saline is a solution that contains 9 grams of NaCl per liter of water.

[3] In reality, H^+ ions do not exist free in a solution. Instead, they interact with water to form H_3O^+, or hydronium ions. For this reason H_3O^+ ions, instead of H^+ ions, are often discussed when reviewing pH. Chemists have a number of different ways to define acids and bases. In this text we apply the more simplified (although limited) Arrhenius definition of acids and bases.

[4] A milligram (mg) is 1/1000 of a gram (g); a deciliter (dL) is 1/10 of a liter (L) or 100 milliliters (mL).

pH: A Measure of Acidity The balance of H^+ and OH^- ions is what determines the overall acidity or basicity of a solution, also called the **pH** (potential hydrogen) of a solution. We use the **pH scale** to describe the acidity and basicity of a solution (FIG. 2.6). Typically, pH values fall between zero and 14, but it is entirely possible to have pH values lower than zero and higher than 14, even in natural environments. For example, the hot springs around the Ebeko volcano in Russia contain hydrochloric acid and sulfuric acid and have a pH as low as –1.7.[5]

The logarithmic ($\log_{10}$) nature of the pH scale means there is a tenfold difference in H^+ ions for every whole-number increment on the scale. Referring to Figure 2.6, for example, sea water is 10 times more basic than pure water, and hydrochloric acid is 100 times more acidic than lemon juice.

When a solution contains a mixture of OH^- and H^+ ions, they combine to form water in what is called a **neutralization** reaction. Pure water has an equal concentration of H^+ and OH^- and therefore is chemically neutral ($pH = 7$). Basic, or alkaline, solutions have a higher concentration of OH^- compared to H^+ ions and exhibit a pH greater than 7. Acidic solutions have a higher concentration of H^+ than OH^- ions and have a pH less than 7. When an acid is added to a basic solution, every H^+ released by the acid neutralizes an OH^- from the base, and the pH will consequently decrease. Likewise, a base can be added to an acidic solution to raise pH. A classic example of the latter is taking an antacid medication to ease the discomfort of heartburn, which occurs as a result of acid reflux from the stomach into the esophagus.

Because pH can impact a number of biochemical reactions, it is often monitored in a variety of medical and industrial applications. In microbiology, pH indicators are chemicals frequently added to growth media to monitor if microbes make acidic, neutral, or basic products in certain biochemical reactions.

One common pH indicator is phenol red, which changes from red to yellow in acidic conditions. The pH of a solution can be roughly estimated with an indicator, whereas a pH meter can very accurately measure the concentration of H^+ ions in a solution if a precise pH value is required.

Most microbes grow best at a pH between 6.5 and 8.5. Our own physiology is also impacted by pH; our arterial blood has a tightly regulated pH of about 7.35–7.45. If blood strays even slightly outside of this pH range, a variety of pathologies result. **Acidosis** (lower-than-normal blood pH) and **alkalosis** (higher-than-normal blood pH) are both dangerous conditions. Because pH can greatly impact physiology, organisms tend to depend on **buffers**, which are compounds that stabilize pH by absorbing or releasing H^+ ions. Our blood pH is stabilized thanks to the presence of a number of buffers, including carbonic acid (H_2CO_3), which releases H^+ ions to lower pH, and bicarbonate (HCO_3^-), which absorbs H^+ ions to raise pH.

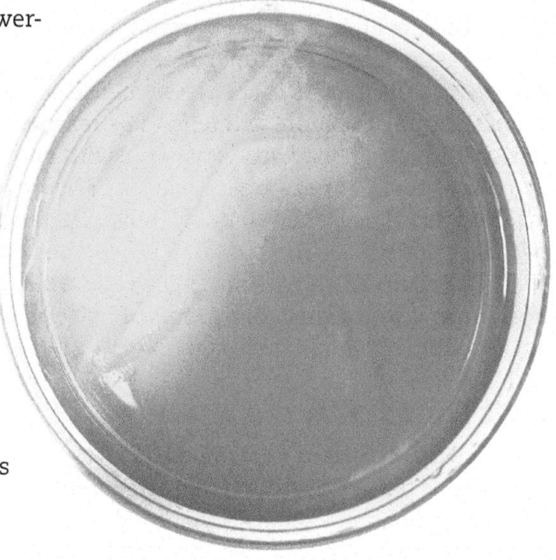

Phenol red, a pH indicator in certain media, turns yellow when bacteria make acids.

[5] Nordstrom, D. K., Alpers, C. N., Ptacek, C. J., & Blowes, D. W. (2000). Negative pH and extremely acidic mine waters from Iron Mountain, California. *Environmental Science and Technology*, 34 (2), 245–258.

CHEM • NOTE

pH is calculated based on the concentration of H^+ ions: $pH = -\log_{10}[H^+]$. The higher the concentration of H^+, the lower the pH.

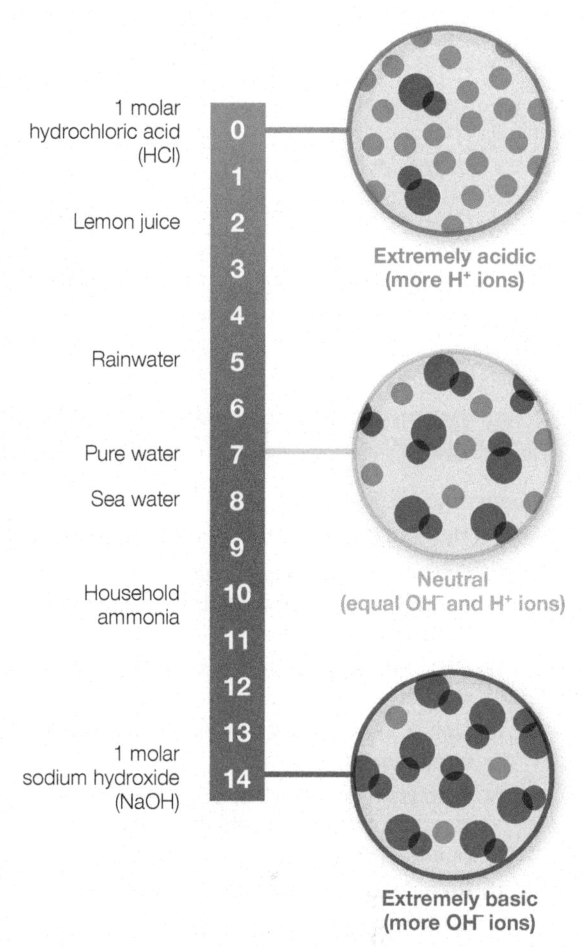

FIGURE 2.6 The pH scale The pH scale reflects the concentration of H^+ ions to OH^- ions.

Critical Thinking *The pH scale is based on a $\log_{10}$ scale. Knowing this, how many times more H^+ would a solution with a pH of 6 have as compared to a solution with a pH of 8?*

Build Your Foundation

1. What is an atom, and what are the parts of an atom? (LO 2.1)
2. Where on the periodic table are atomic mass and atomic number noted? (LO 2.2)
3. What is the difference between a cation and an anion? How are ions formed? (LO 2.3)
4. What are isotopes, and why are they medically useful? (LO 2.4)
5. Is O_2 considered a molecule, compound, or isomer? Explain your reasoning. (LO 2.5)
6. Write the molecular formula for methane, which contains four hydrogen atoms and one carbon atom. (LO 2.6)
7. State what makes a molecule organic, and name two organic and two inorganic functional groups. (LO 2.7)
8. How do acids and bases differ, and what effect does each have on pH? (LO 2.8)
9. What does the pH scale reflect, and what are its features? (LO 2.9)
10. How do buffers stabilize pH? (LO 2.10)

Build Your Foundation (BYF) Quick Quiz: Visit the **Mastering Microbiology** Study Area to quiz yourself.

2.2 CHEMICAL BONDS

Learning Outcomes

After reading this section, you should be able to:

2.11 Define the term *valence electron*, and explain how valence electrons relate to bonding.

2.12 Compare and contrast ionic and covalent bonds.

2.13 Describe electrolytes and their importance in biological systems.

2.14 Discuss the process of polar covalent bonding and how it sets the stage for hydrogen bonding.

2.15 Identify the characteristics of hydrogen bonds.

2.16 Describe van der Waals interactions.

2.17 Define the terms *hydrophobic*, *hydrophilic*, and *amphipathic*, and explain how these qualities relate to micelle formation.

Electrons determine what bonds can form.

Chemical bonds are the forces that "glue" atoms into molecules. The types of bonds present in a given molecule depend on how the electrons of the participating atoms interact.

The electrons of most atoms are organized in **electron shells** around the atomic nucleus.[6] Each shell has a maximum number of electrons it can hold, with those closest to the nucleus tending to hold fewer electrons than the shells farther from the nucleus. The **valence shell** is an atom's outermost shell. Valence electrons are the electrons found in the valence shell; for simplicity, they can be thought of as the electrons that typically participate in chemical reactions.[7] Ultimately, the type of valence electron interaction that occurs between atoms dictates what kind of chemical bond is formed.

When the valence shell of an atom is full, then the electron configuration is stable, and the atom tends to be nonreactive or inert. The noble gases (including helium, neon, argon, krypton, xenon, and radon) are all examples of elements that are usually inert because their valence shells are full. When the valence shell is not full, the atom will tend to be reactive—meaning that electrons in this shell can be gained, lost, or shared with another atom to stabilize the participating atoms. For example, an atom of hydrogen has only one electron, whereas its valence shell would be full—and the atom would be stable—with two electrons (FIG. 2.7). In contrast, oxygen has eight electrons, two in its first shell, and six in its valence shell. Oxygen would be stable if it had eight electrons in its valence shell. Two atoms of hydrogen therefore tend to "fill" one oxygen atom's valence shell, stabilizing all three atoms in a molecule of water.

CHEM • NOTE

Electrostatic forces are the attraction forces that exist between positive and negative atoms or molecules. Just as a magnet is attracted to metal, so too are positive and negative atoms or molecules attracted to one another.

Ionic Bonds

With ionic bonds, the saying "opposites attract" holds true. An **ionic bond** is the electrostatic force of attraction that exists between oppositely charged ions (between cations and anions). Ionic bonds form when electrons are transferred

[6] Electron shells are not physical structures; rather, this term is used here to describe regions where electrons are found. Each electron shell is organized into sub-shells and orbitals.

[7] With most elements it is fine to refer to the valence electrons as exclusively occupying the valence shell. However, the transition metals (see the d-block of a standard periodic table) do not always follow the classic definition of valence electrons because their most reactive electrons are not always exclusively in a single, outermost shell.

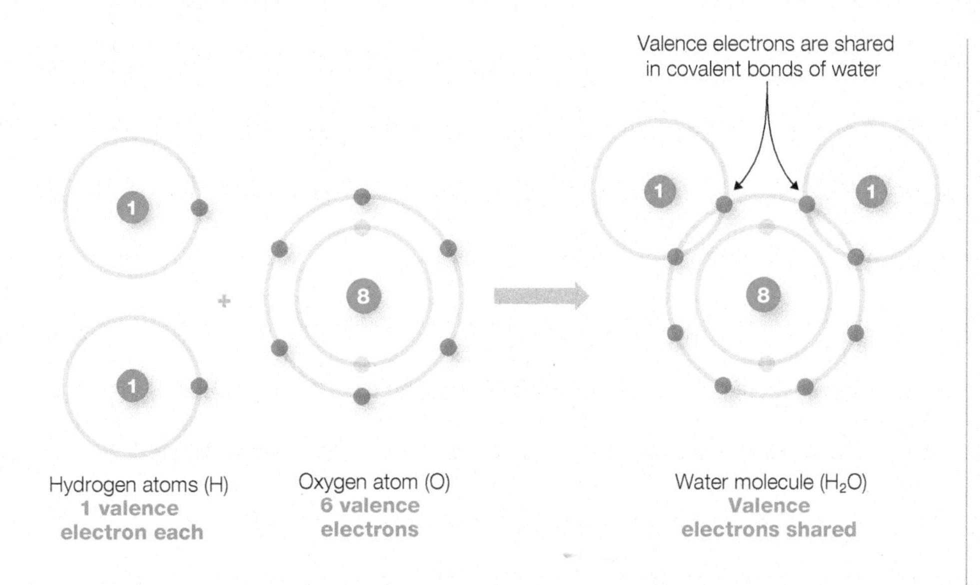

FIGURE 2.7 Valence electrons Valence electrons tend to be situated in an atom's outermost shell. They can be lost, gained, or shared to form chemical bonds between atoms.

from one atom to another to make ions. Sometimes students have a misconception that the atoms of ionic compounds only exist as ions in a solution, but this is not true. The atoms of ionic compounds *always* exist as ions irrespective of their physical state, be it solid, liquid, or gas. In a solid state the ions are just closely packed together into a matrix framework that restricts their movement (FIG 2.8, *top*).

When ionic compounds dissolve in a solution, the ions are freer to move around and are often called **electrolytes** (FIG. 2.8, *bottom*). The most common electrolytes are acids, bases, and salts. Physiological processes rely on many

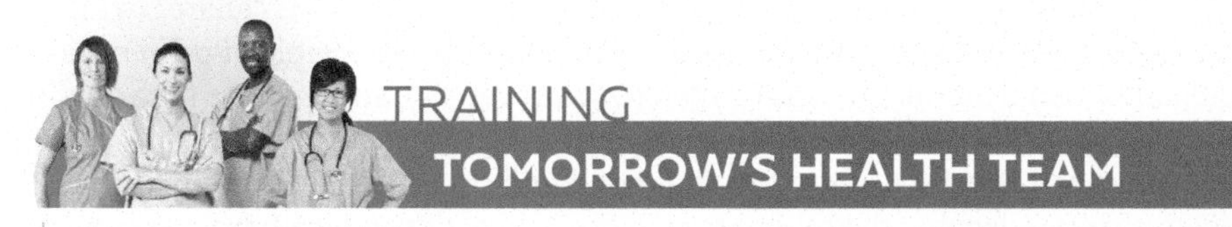

Electrolytes Save Lives

Rotaviruses are a significant cause of diarrhea in children. Before the introduction of the rotavirus vaccine in 2006, almost all children in the United States suffered from this virus before their first birthday. Worldwide, especially in poorer countries where the vaccine is still not widely available, it is estimated that rotavirus kills over 400,000 children per year.

Rotaviruses cause a watery diarrhea that leads to dehydration and loss of electrolytes. The virus triggers this pathology by generating a spike in calcium ions (Ca^{2+}) inside certain cells of the small intestine. The calcium ion spike leads to a cascade of tissue effects that ultimately impairs water and electrolyte absorption. The virus also causes intestinal cells to release chloride ions into the bowel, which leads to an even greater loss of water and electrolytes. Most deaths from rotavirus can be prevented by **oral rehydration therapies**, which consist of a simple water-based solution enriched with electrolytes.

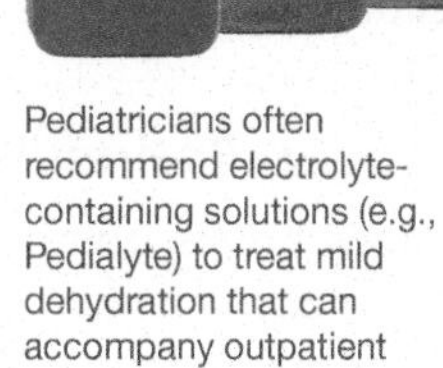

Pediatricians often recommend electrolyte-containing solutions (e.g., Pedialyte) to treat mild dehydration that can accompany outpatient gastrointestinal distress.

Most oral rehydration therapies include a specific formulation of electrolytes, such as sodium chloride (NaCl), glucose, potassium chloride (KCl), and citrate. If a solution is not effective at stabilizing the patient, or if severe dehydration is present, then intravenous rehydration therapies can be used.

QUESTION 2.1

Why is plain water insufficient for treating dehydration that can develop from gastrointestinal distress?

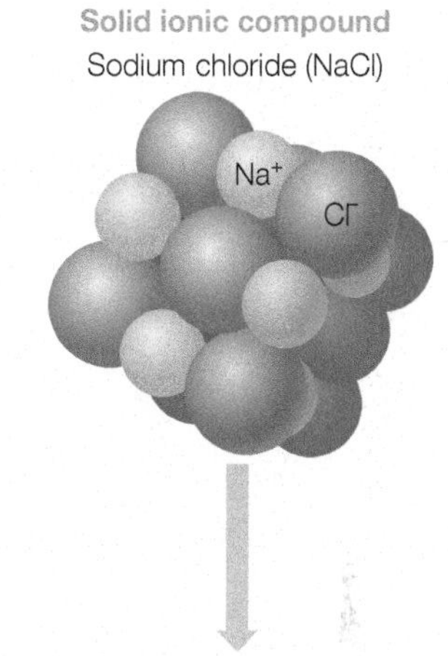

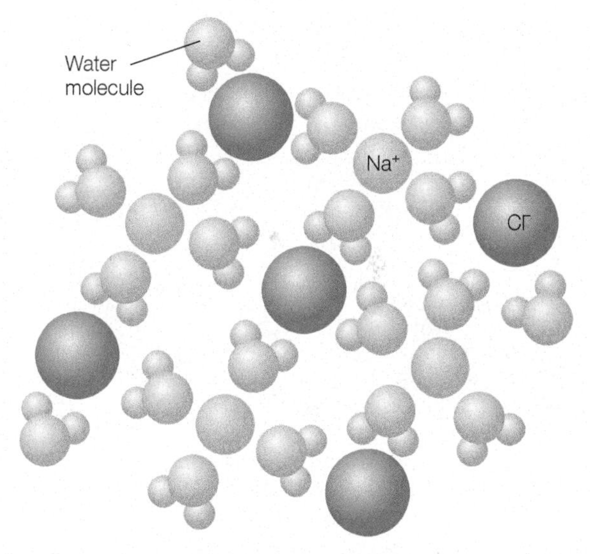

FIGURE 2.8 Ionic compounds Whether they are solids or dissolved in a solution, ionic compounds always consist of ions that are strongly attracted to each other by their opposite electrical charges. *Top:* In ionic solids, ions are locked in a matrix that restricts their movement. *Bottom*: When ionic solids dissolve in an aqueous solution, the ions are surrounded by water and are freer to move about in the solution.

Critical Thinking *Generation of an electrical current requires a flow of electrical charges. Most solid ionic compounds do not conduct electricity as effectively as aqueous solutions of the same ionic compounds. Explain why this is the case. (Hint: Compare ion mobility in one state versus the other.)*

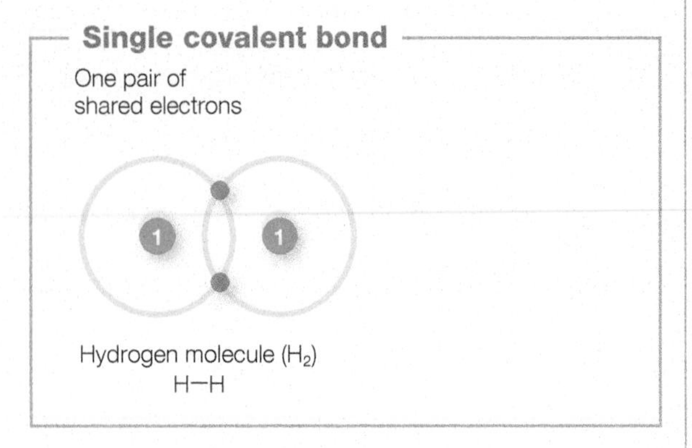

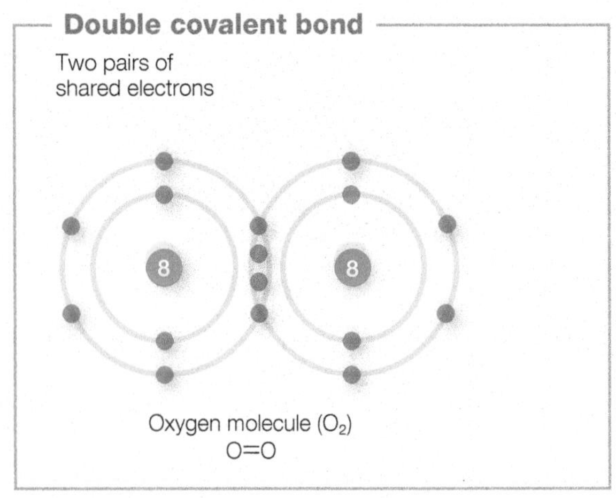

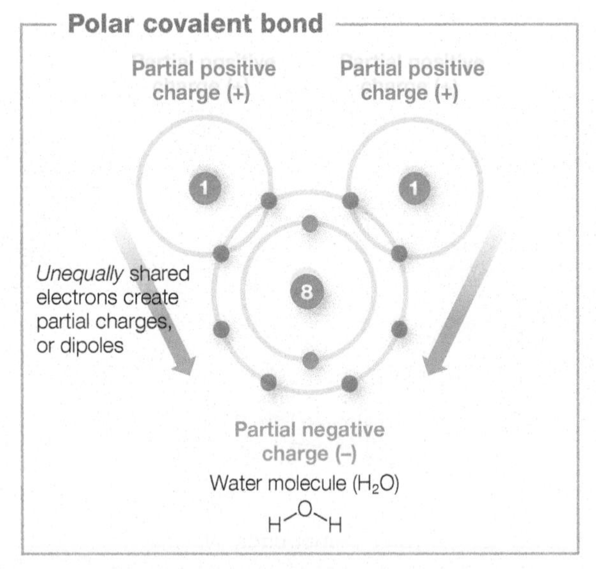

FIGURE 2.9 Covalent bonds Covalent bonds form when atoms share electrons. Atoms that share one pair of electrons have formed a single covalent bond. A double covalent bond entails sharing two pairs of electrons. Atoms like oxygen (O), nitrogen (N), and fluorine (F) form polar covalent bonds when bonded to certain atoms like hydrogen. In polar covalent bonds, unequal electron sharing leads to partial charges, or dipoles, within the molecule.

Critical Thinking *The molecule O_2, elemental oxygen, does not have a dipole despite oxygen being involved in the bond—explain why. (Hint: Consider what is fundamental to a dipole.)*

electrolytes, such as sodium (Na^+), calcium (Ca^{2+}), magnesium (Mg^{2+}), potassium (K^+), chloride (Cl^-), bicarbonate (HCO_3^-), and hydrogen phosphate (HPO_4^{2-}). Electrolyte imbalances can arise from excessive sweating, diarrhea, vomiting, kidney disease, or other pathologies. Electrolytes have important roles in regulating the nervous system, our heartbeat, skeletal muscle contraction, overall blood volume, and general water balance. Severe electrolyte imbalances can be deadly and require prompt medical attention.

Covalent Bonds

We just noted that, in ionic bonds, atoms *transfer* electrons. In contrast, a **covalent bond** exists between atoms that *share* electrons. An example has already been shown in Figure 2.7. Atoms that share one pair of electrons (one from each atom) are said to have a single covalent bond, whereas atoms that share two pairs of electrons (two from each atom) have a double covalent bond (FIG. 2.9).

Carbon Bonds Carbon's covalent bonding properties allow biological systems to make complex organic molecules. Carbon is commonly found as the core atom of organic molecules in part because it can form four covalent bonds and is one of the few elements capable of **catenation**, which is the ability of atoms of the same element to form long chains. We'll see examples of these long chains of carbon atoms when we explore the biologically important macromolecules later in this chapter.

Polar Covalent Bonds Sometimes atoms *within* a molecule have a charge; for example, the atoms of ionic compounds like the salt NaCl exist as charged ions, Na^+ and Cl^-. In such compounds the atoms are charged because electrons have been transferred. However, electrons do not have to be fully transferred in order for the atoms within a molecule to take on an electrical charge. Atoms participating in **polar covalent bonds** take on a charge because they share the molecule's electrons unequally. If you have ever seen how 3-year-olds "share" (which is to say, *not very well*), then you can understand the basic principle of polar covalent bonds. Some atoms, such as oxygen (O), nitrogen (N), and fluorine (F), are especially greedy when it comes to electron sharing. These highly electronegative atoms hog[8] the electrons of the covalent bond and consequently take on a partial negative charge, leaving the other atom in the interaction, which hardly ever gets to hold the bonding electron(s), with a partially positive charge. This asymmetric charge distribution between the participants in the covalent bond creates two poles (a **dipole**) within the molecule—one pole of the molecule being slightly negative and the other pole being slightly positive. Think of a pole as an opposing region, just as the Earth has a north and south pole. Molecules that contain dipoles are described as **polar** (FIG. 2.9, bottom). The presence of dipoles in polar molecules is significant because they lay the foundation for hydrogen bonds.

Hydrogen Bonds: Noncovalent Interactions

Hydrogen bonds do not bind atoms into molecules. Instead, hydrogen bonds are a noncovalent electrostatic attraction between two or more molecules or two or more regions within a single large molecule (e.g., hydrogen bonds within a protein). The name *hydrogen bond* is a bit misleading because the "bond" is only an electrostatic interaction between dipoles. What does that mean? Recall that a dipole develops when there is unequal sharing of electrons, which often occurs when hydrogen is covalently bonded to oxygen, fluorine, or nitrogen. In a polar covalent bond, hydrogen takes on a partial positive charge because it doesn't get to interact very much with the shared

[8] Conversational note for non-native speakers: "hog" or "hogging" means to act in a greedy, grasping, or possessive manner versus a generous or sharing manner.

electrons. In contrast, the O, F, or N atoms take on a partial negative charge because they hog the shared electrons. A hydrogen bond forms when the partially positive hydrogen of one polar molecule is electrostatically attracted to the partially negative charge of a nearby O, F, or N atom to which it is not already covalently bound (FIG. 2.10).

As an imperfect yet hopefully memorable analogy, you can think of hydrogen as being in a stable relationship with an O, F, or N atom. The hydrogen is happy enough with its partner that it won't break the bond—but it nevertheless has a weak attraction to another O, F, or N that is already taken (already covalently bonded to another hydrogen). Again, the attraction is not enough to make the atoms leave their respective partners, but they are attracted to each other and associate when near each other.

Hydrogen bonds that exist between two or more individual molecules are called **intermolecular** hydrogen bonds. The hydrogen bonding that occurs between separate H_2O molecules is an example. The abundant hydrogen bonding between water molecules is what gives water many of its unique and biologically important properties, such as its high surface tension, its high specific heat that allows it to remain at a fairly stable temperature unless a large amount of heat is added or removed, and its solvent properties.

Hydrogen bonds that occur within a single large molecule are known as **intramolecular** hydrogen bonds. As we'll see later in this chapter, large macromolecules like proteins and nucleic acids often rely on intramolecular hydrogen bonds to stabilize their structure.

CHEM • NOTE

Electronegativity is the tendency of an atom to attract electrons. Oxygen (O), nitrogen (N), and fluorine (F) are examples of highly electronegative elements that hog electrons when covalently bonded with less electronegative atoms. As such, O, F, and N atoms are commonly involved in polar covalent bonds.

Van der Waals Interactions

Some molecules have temporary dipoles that are not a result of hydrogen atoms bonding to O, F, or N atoms. These molecules do not form hydrogen bonds; however, they still exhibit a force of attraction called **van der Waals interactions** (van der Waals forces). Like hydrogen bonds, van der Waals interactions do not bind atoms into molecules; instead, they are considered electrostatic attractions between atoms that are near one another, but not covalently or ionically bonded to one another. Although van der Waals interactions are weaker than hydrogen bonds and much weaker than ionic and covalent bonds, they are significant stabilizers of molecular structures; for example, they contribute to the stability of a folded protein's higher-order structure.

Water molecules prefer to interact with other polar molecules.

Perhaps you have heard the tale of Narcissus, a mythological character who was so self-impressed that he couldn't stop staring at his own reflection in a pool of water. In many ways, water molecules love themselves so much that they, too, are reluctant to interact with other molecules unless those molecules have similar properties to water—especially in terms of polarity. Because water is so abundant in living systems, it influences their overall chemistry, including the structure of biologically important macromolecules. As such, a basic appreciation of polarity is central to your training.

Polarity and Solubility in Water

Just as their strong dipole causes water molecules to form hydrogen bonds with each other, they also readily form hydrogen bonds with other polar molecules. This makes water a great solvent for a broad collection of polar substances, such as sugars, but it also means water is not particularly good at dissolving nonpolar substances, such as fats. Like dissolves like; polar solutes dissolve in polar solvents, whereas nonpolar solutes dissolve in nonpolar solvents.

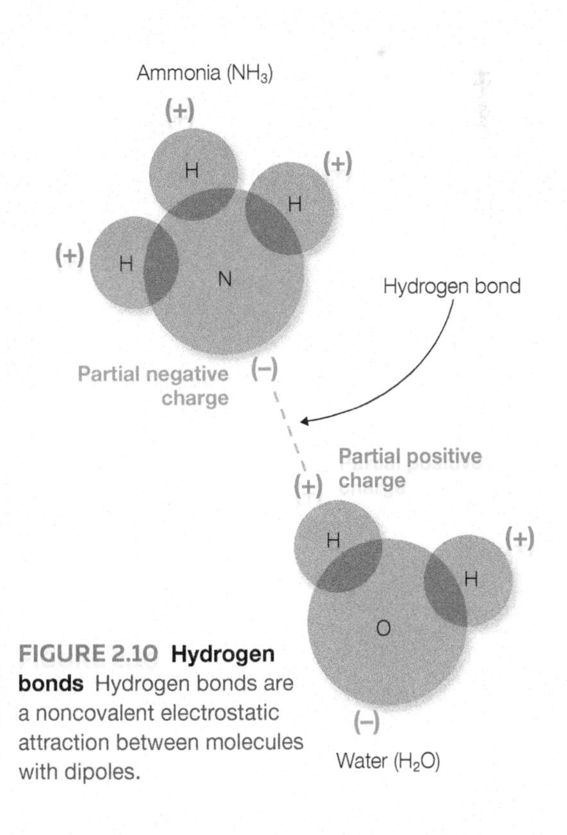

FIGURE 2.10 Hydrogen bonds Hydrogen bonds are a noncovalent electrostatic attraction between molecules with dipoles.

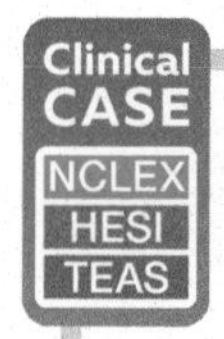

The Case of the Confused Cattle Farmer

Practice applying what you know clinically: visit the **Mastering Microbiology** Study Area to watch Part 2 and practice for future exams.

Consider, for example, fat-soluble versus water-soluble vitamins. Vitamins A, D, E, and K are nonpolar and thus not soluble in water. Instead, they are fat soluble, and any excess amount you consume can be stored in your body's fatty tissues. In contrast, the water-soluble B vitamins and vitamin C are polar. Consequently, these vitamins are easily dissolved in water and not readily stored; instead, any excesses are excreted in urine. The polarity of a drug molecule also impacts its absorption and distribution in the body, as well as how it is metabolized and excreted. Drugs that are administered topically are nonpolar, so they can be absorbed into the skin and cross fatty tissues to enter the body. For example, the nonpolar drug fentanyl, a potent painkiller, can be delivered by a transdermal patch because the drug readily crosses fatty tissues.

Hydrophobic, Hydrophilic, and Amphipathic Molecules

Substances that readily form hydrogen bonds with water are said to be **hydrophilic**, or "water loving." Substances that do not readily form hydrogen bonds with water are described as **hydrophobic**—"water fearing" (FIG. 2.11). Usually, polar substances are hydrophilic, and nonpolar substances are hydrophobic.[9] For example, if you add oil to water, you will notice that, because the oil is hydrophobic, it gathers into globules and eventually settles out as a layer on top of the water. This effect illustrates the principle that nonpolar molecules (like oil) do not readily form hydrogen bonds with water; thus, oil cannot form an aqueous solution—oil and water don't mix. In contrast, most (though not all) polar molecules will readily form hydrogen bonds with water and form an aqueous solution (a solution in which water is the solvent).

Some molecules are not simply polar or nonpolar, but instead have both hydrophobic and hydrophilic regions. These **amphipathic** molecules are capable of forming **micelles**—three-dimensional structures in which the hydrophobic regions of the component amphipathic molecules are positioned toward the center of the structure and hydrophilic regions face the aqueous (water-containing) environment. An example of an amphipathic molecule that can form micelles is detergent. The hydrophobic portion of the molecule surrounds the grease on your dishes, while the hydrophilic portion allows the bound grease to be washed away by water (FIG. 2.12, *left*). A cell's plasma membrane is a bilayer made up of amphipathic lipids called phospholipids. The phosphate-containing region of the lipid is hydrophilic and faces the aqueous environments outside and inside the cell, whereas the hydrophobic tail regions of the lipid are sequestered in the middle of the lipid bilayer (FIG. 2.12, *right*).

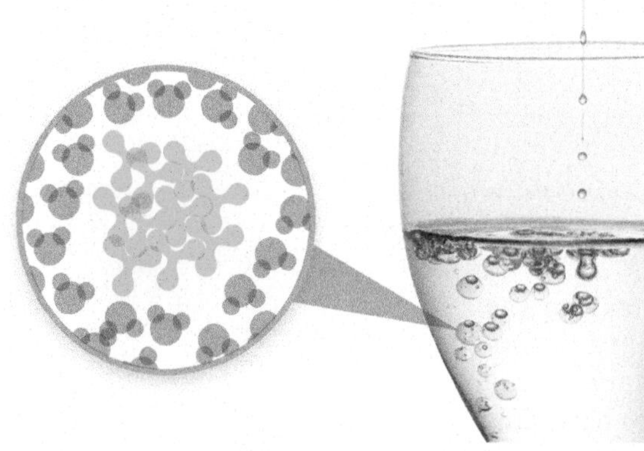

FIGURE 2.11 **Polar and nonpolar molecules** Polar substances dissolve in water, while nonpolar substances do not. Polar molecules are described as being hydrophilic because they readily interact with water. Nonpolar molecules are said to be hydrophobic because they do not interact well with water.

[9] In general, polar substances are hydrophilic and nonpolar substances are hydrophobic—but, as with most things in science, there are exceptions. There are examples of polar substances (e.g., certain nitriles and nitro compounds) that do not readily form strong hydrogen bonds with water and are hydroneutral. Therefore, the terms *polar* and *hydrophilic* are not technically synonyms.

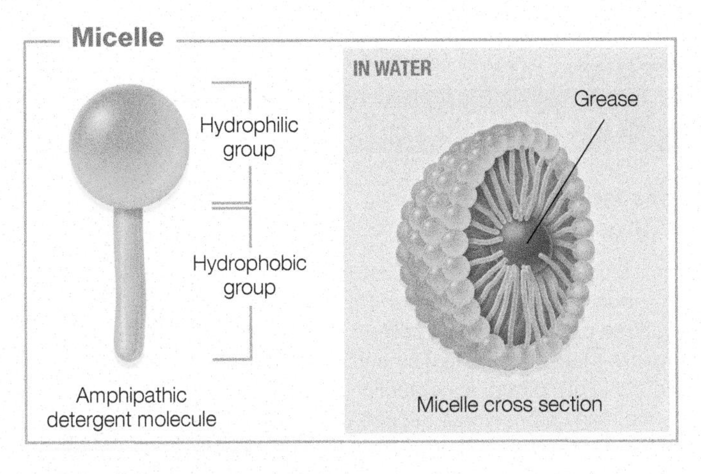

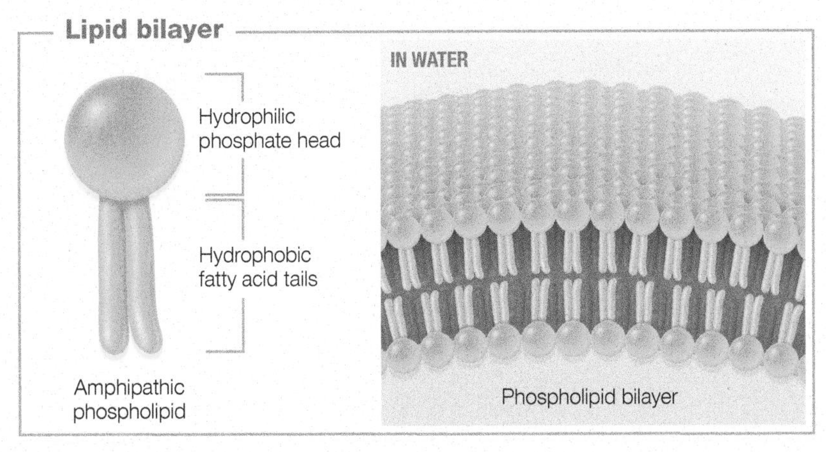

FIGURE 2.12 Amphipathic molecules This class of molecules has both polar and nonpolar properties. Micelles (*left*) and lipid bilayers (*right*) are examples of how amphipathic molecules orient themselves to prevent their hydrophobic tails from contacting water.

Critical Thinking *Emulsifiers are often added to salad dressings to help vinegar and oil blend better. Describe how an emulsifier might work.*

Build Your Foundation

11. Why are valence electrons important to consider in chemistry? (LO 2.11)
12. What sort of bond holds sodium (Na^+) and chloride (Cl^-) ions together in the salt sodium chloride? (LO 2.12)
13. What distinguishes a covalent bond from an ionic bond? (LO 2.12)
14. What are electrolytes, and why are they biologically important? (LO 2.13)
15. What are polar covalent bonds, and why are they central to hydrogen bond formation? (LO 2.14)
16. How are hydrogen bonds and van der Waals interactions similar? How are they different? (LO 2.15 and 2.16)
17. Define the terms *hydrophobic*, *hydrophilic*, and *amphipathic*. (LO 2.17)
18. How would micelles be organized if they formed in a nonpolar solution as opposed to in water? (LO 2.17)

Build Your Foundation (BYF) Quick Quiz: Visit the **Mastering Microbiology** Study Area to quiz yourself.

2.3 CHEMICAL REACTIONS

Chemical reactions make and break chemical bonds.

Chemical reactions involve making and/or breaking chemical bonds. By doing so, they often change the substances that participate in the reaction. The ingredients of a chemical reaction are called **reactants**, and the substances generated as a result of the reaction are called **products**. Every day you encounter chemical reactions—from the combustion of gasoline in a vehicle to cooking an egg on your stove. Many chemical reactions are accompanied by observable changes, such as when cut apples start to brown because of an oxidation reaction.

Learning Outcomes

After reading this section, you should be able to:

2.18 Identify the reactants and products in a chemical equation.

2.19 Describe catalysts and their importance in biological systems.

2.20 Explain synthesis, decomposition, and exchange reactions.

2.21 Distinguish between dehydration synthesis and hydrolysis reactions.

2.22 Describe activation energy and how it can be lowered in biochemical reactions.

2.23 Compare and contrast endergonic and exergonic reactions.

2.24 Explain what is meant by a reversible reaction and the concept of equilibrium.

Chemical Equations: Recipes for Reactions

A chemical equation is a written representation of a chemical reaction. It identifies the reactants and products of the reaction as well as any specific conditions that must be met to facilitate the reaction.

The reactants are listed to the left of a horizontal arrow; the products are listed to the right of the arrow; and any special considerations are listed above the reaction arrow. For example, if heat needs to be added to the reaction, it is often depicted as a small triangle, a symbol called delta (Δ), above the reaction arrow. If a catalyst is required, it also is written above the reaction arrow. A **catalyst** is an organic or inorganic substance that increases the rate of a reaction, but is not used up in the reaction. Proteins called **enzymes** act as catalysts in many biochemical reactions. The physical state of the reactants and products may also be indicated as (s) for solid, (l) for liquid, (g) for gas, or (aq) for ions in an aqueous solution. Examples of some key chemical reactions and their equations are discussed next and shown in (FIG. 2.13).

CHEM • NOTE

Oxidation reactions remove electrons from an atom. **Reduction reactions** add electrons to an atom. Oxidation and reduction reactions occur together as partners and are often termed redox reactions (red = reduction; ox = oxidation). Redox reactions make ions. We will refer to redox reactions again in the metabolism chapter.

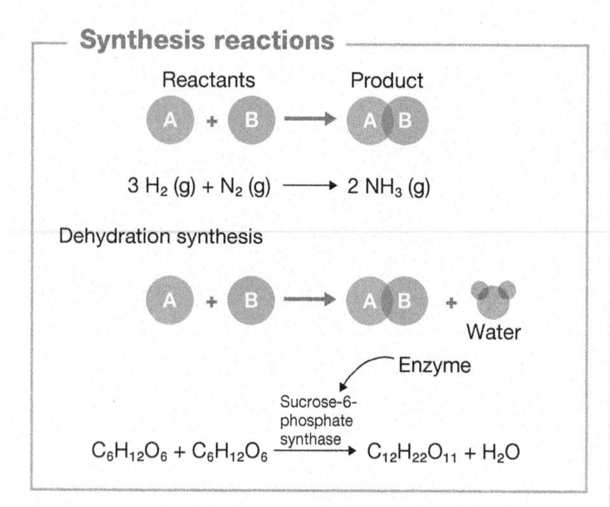

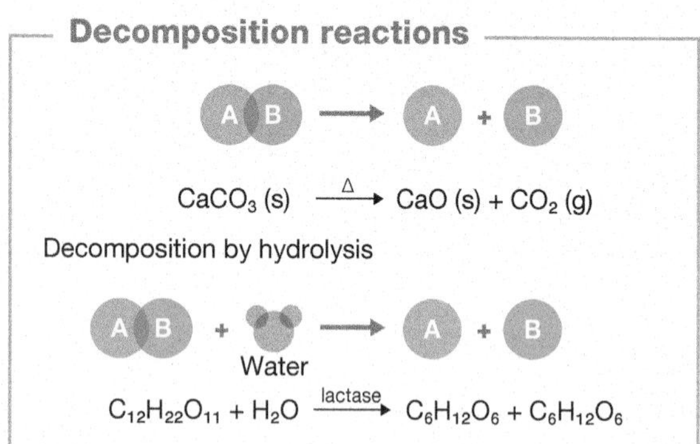

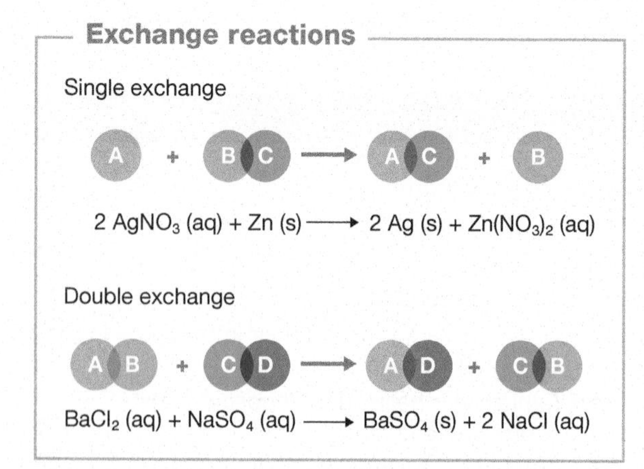

FIGURE 2.13 Examples of chemical reactions and their equations

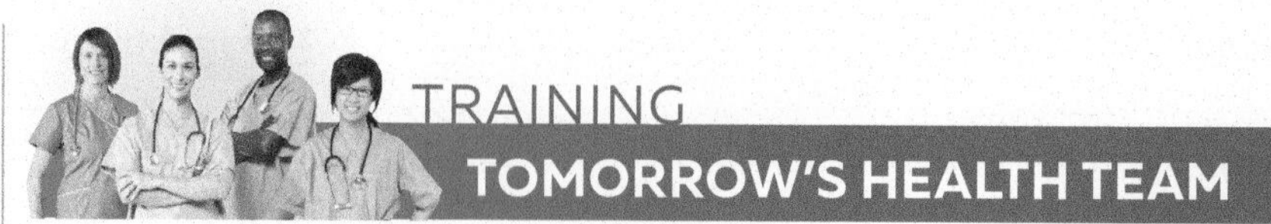

Enzymes That Make Isomers Might Also Make Good Drug Targets

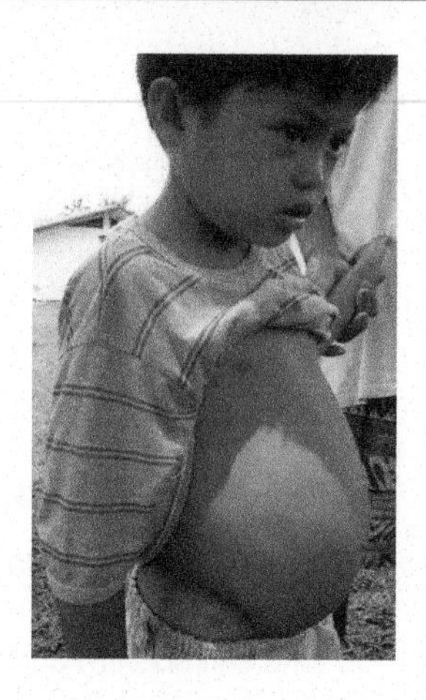

In many cellular processes one isomer of a molecule is converted to another by **isomerization reactions**. These reactions require specialized cellular enzymes called **isomerases**. An example is triosephosphate isomerase (TPI). This enzyme is found in glycolysis, a metabolic pathway that humans and many prokaryotes depend on to break down sugars in order to release energy. Researchers are currently investigating the use of compounds that inhibit TPI as potential drugs to combat a number of pathogens that cause parasitic infections such as giardiasis, leishmaniasis, schistosomiasis, or even the bacterial infection tuberculosis.

QUESTION 2.2

What potential negative effect would a drug that inhibits TPI have in humans if the drug is not engineered to be highly specific for TPI in the target pathogen?

Synthesis Reactions

Synthesis reactions build substances by combining two or more reactants. Typically, a synthesis reaction leads to the formation of more complex molecules from simpler building blocks. When cells build proteins, nucleic acids, carbohydrates, or fats, they use an elaborate series of synthesis reactions. Sometimes building a complex organic molecule requires bringing reactants together in such a way that water is released when a covalent bond is formed. These reactions are referred to as **dehydration synthesis** reactions (FIG. 2.14, *left*). Cells use dehydration synthesis reactions to build macromolecules like proteins, polysaccharides, lipids, and nucleic acids. A specific example of a dehydration synthesis reaction is the formation of a peptide bond to build proteins, which we will review later in this chapter in our discussion of proteins. Generically, a synthesis reaction can be represented by:

$$A + B \rightarrow AB$$

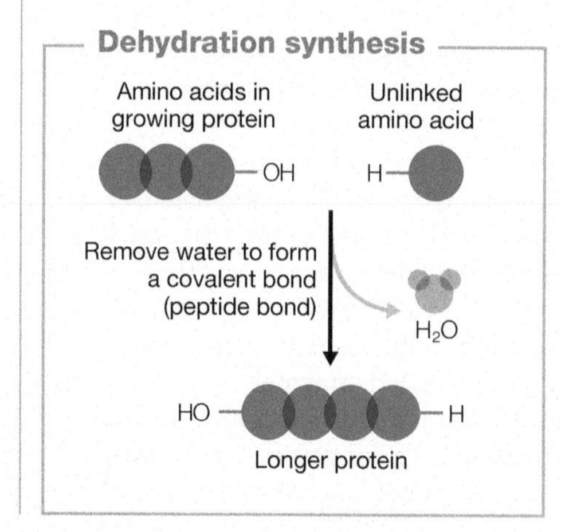

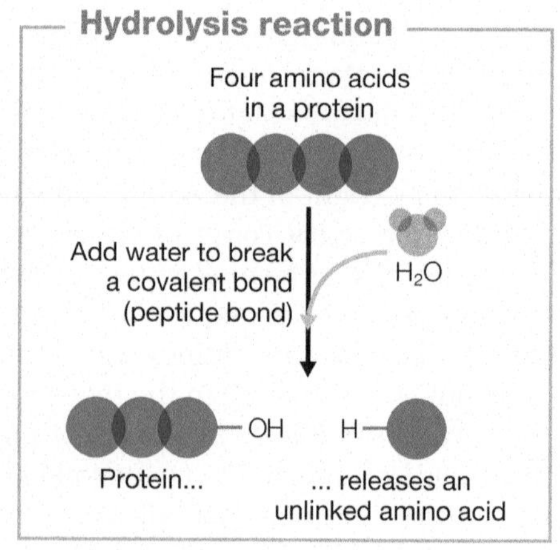

FIGURE 2.14 Dehydration and hydrolysis reactions

Critical Thinking *The reaction $2\ H_2O_2 \rightarrow 2\ H_2O + O_2$ is a decomposition reaction in which hydrogen peroxide breaks down to water and oxygen, but it is not a hydrolysis reaction; explain why.*

Decomposition Reactions

Decomposition reactions break a substance down into simpler components. In biochemical pathways it is common to add water to break the covalent bonds in complex molecules in a process called a **hydrolysis reaction** (FIG. 2.14, *right*). Hydrolysis reactions are used to break down macromolecules like proteins, polysaccharides, lipids, and nucleic acids. A standard depiction of a decomposition reaction would be:

$$AB \rightarrow A + B$$

Exchange Reactions

Exchange reactions involve swapping one or more components in a compound (see Figure 2.13, *bottom*). Sometimes these are called replacement or displacement reactions. The following are examples of exchange reactions:

$$A + BC \rightarrow AC + B \text{ (a single exchange reaction)}$$

or

$$AB + CD \rightarrow AD + CB \text{ (a double exchange reaction)}$$

Chemical reactions consume or release energy.

Reactions involve collisions between atoms or molecules. For a reaction to occur, the collisions between the reactants have to be energetic enough to lead to a change, and the reactants have to be properly oriented to interact with each other. As such, there are energetic barriers to overcome for a chemical reaction to occur. Even if the reactants are properly oriented, they won't react with each other if they don't collide with sufficient energy.

The minimum amount of energy needed to get a reaction started is called the **activation energy** (FIG. 2.15). Under physiological conditions, all reactions have an activation energy, even if so miniscule as to be almost zero. (In Chapter 8, we will discuss how enzymes act as catalysts to help reactions occur under physiological conditions by lowering activation-energy barriers.)

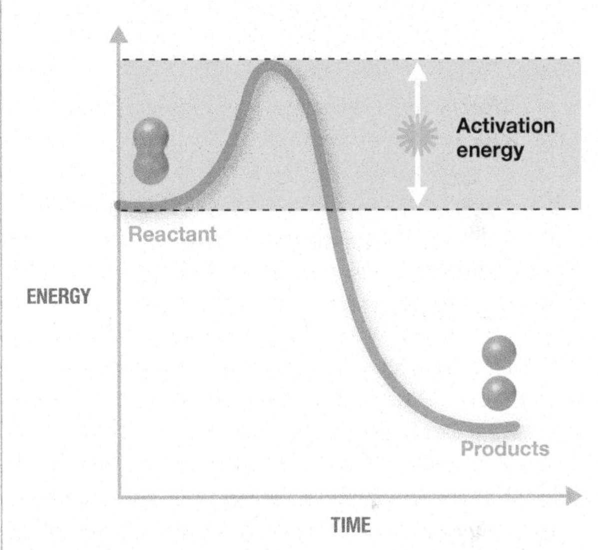

FIGURE 2.15 **Activation energy** This is the minimum amount of energy needed to get a reaction started.

Critical Thinking *Using the pictured energy diagram as a model, draw a diagram that shows an activation energy of near zero.*

Endergonic versus Exergonic Reactions

Some reactions will ultimately release more energy than is spent to start the reaction. These **exergonic reactions** make products with a lower final energy than the reactants. Other reactions use more energy than is released. These **endergonic reactions** require an energy investment to make products that have a higher final energy than the reactants. In biological systems the energy released by exergonic reactions is used to fuel endergonic reactions.

Reaction Reversibility

Some chemical reactions, like fuel combustion, are irreversible and are written with a unidirectional arrow that points strictly from the reactants to the products. However, in a **reversible reaction**, the forward and reverse reactions are both possible. Initially one reaction may occur at a higher rate than the other, but eventually the forward and reverse reactions occur at the same rate, and the reaction is said to be at **equilibrium**. Contrary to a common student misconception, equilibrium is *not* a static situation where the reaction just stops. Rather, there is a dynamic and continuous forming of products and the reforming of reactants at an equal rate. This does not mean that there is an equal amount of products and reactants; it just means that the total amount of products and reactants is no longer changing. A reaction at equilibrium is depicted by a set of arrows where one arrow points toward the products and the second arrow points toward the reactants, as shown here:

$$CH_3CO_2H + H_2O \rightleftharpoons CH_3CO_2^- + H_3O^+$$

Build Your Foundation

19. Label the reactant(s), product(s), and enzyme in the following reaction, and state what type of reaction it is: (LO 2.18)

$$2H_2O_2 \xrightarrow{\text{catalase}} 2H_2O + O_2$$

20. What is a catalyst, and why are catalysts useful in biological systems? (LO 2.19)
21. Name the class of reaction that this equation represents: (LO 2.20)

$$Na_2CO_3(aq) + CaCl_2(aq) \rightleftharpoons CaCO_3(s) + 2NaCl(aq)$$

22. Compare dehydration synthesis reactions and hydrolysis reactions. (LO 2.21)
23. In terms of energetics, why are enzymes necessary in biological systems? (LO 2.22)
24. What are endergonic and exergonic reactions, and why are they often coupled in biological systems? (LO 2.23)
25. How can a reaction at equilibrium still have formation of products, but not exhibit a net change in the amount of products? (LO 2.24)

Build Your Foundation (BYF) Quick Quiz: Visit the **Mastering Microbiology** Study Area to quiz yourself.

2.4 BIOLOGICALLY IMPORTANT MACROMOLECULES

Learning Outcomes

After reading this section, you should be able to:

2.25 Identify the four main groups of biomolecules and their building blocks.
2.26 Describe glycosidic bonds, peptide bonds, and phosphodiester bonds.
2.27 Explain the structural and functional characteristics of carbohydrates, lipids, nucleic acids, and proteins.
2.28 Summarize how saturation affects a lipid's characteristics.
2.29 Compare and contrast deoxyribonucleotides and ribonucleotides.
2.30 Describe the four levels of protein structure.
2.31 State what chaperone proteins do and why they are important.

There are four main classes of biomolecules.

Carbohydrates, lipids, nucleic acids, and proteins are the four main classes of biomolecules. All are essential to life (TABLE 2.2). Biomolecules often contain multiple functional groups that contribute to their chemical properties. Because these complex organic molecules tend to be large, they are called **macromolecules**. Biological macromolecules are built by a series of synthesis reactions and broken down by a series of decomposition reactions. **Polymerization**, the process of covalently bonding together smaller foundational units (**monomers**), is used to build many macromolecules.

Carbohydrates include simple sugars and polysaccharides.

The term **carbohydrate** refers to organic molecules consisting of one or more sugar monomers. These polar molecules come in many different formats and have diverse structures and functions. Single sugars are foundationally built

TABLE 2.2 The Four Main Classes of Biomolecules

Biomolecule	Examples	Building Blocks	Notes
Carbohydrates	Glucose	Simple sugars	Glucose is a monosaccharide (a simple sugar)
	Sucrose		Sucrose is a disaccharide (built by bonding two simple sugars)
	Glycogen		Glycogen is a polysaccharide (polymer of many simple sugars)
Nucleic Acids	Deoxyribonucleic acid (DNA)	Nucleotides	DNA is a polymer of deoxyribonucleotides
	Ribonucleic acid (RNA)		RNA is a polymer of ribonucleotides
Proteins	Enzymes Antibodies	Amino acids	Enzymes are proteins that serve as catalysts; antibodies are proteins made as part of an immune response
Lipids	Fats and oils	Glycerol and up to 3 fatty acids	Monoglycerides = Glycerol + one fatty acid Diglycerides = Glycerol + two fatty acids Triglycerides = Glycerol + three fatty acids
	Waxes	Long-chain alcohol and fatty acid	Mycolic acid is a type of wax found in acid-fast bacterial cell walls
	Steroids	Fused hydrocarbon rings	Cholesterol is incorporated into certain cell plasma membranes

from carbon, hydrogen, and oxygen and follow a general molecular formula $(CH_2O)_n$ where there is always a 2 to 1 ratio of hydrogen and oxygen, and *n* is often 3, 5, or 6. These single sugars can be polymerized to make larger molecules.

Carbohydrate Structures

Carbohydrate monomers are also known as **monosaccharides** ("one sugar"). The smallest unit of a carbohydrate, they include glucose, galactose, and fructose. **Disaccharides** ("two sugars") consist of two monosaccharides linked together. Common disaccharides include lactose and sucrose (see Table 2.2). **Polysaccharides** ("many sugars") consist of many monosaccharides (FIG. 2.16). Examples include starches, which are long chains of glucose found in grains and many other plant-based foods, and **glycogen**, a polymer of glucose that the body stores mainly in the liver and skeletal muscles.

Sugar monomers are linked by a **glycosidic bond**, a covalent bond formed between monosaccharides to build more complex sugars. The chemical reaction that leads to the formation of a glycosidic bond is a dehydration synthesis reaction because water is removed from the reactants when they come together. Glycosidic bonds can be broken by enzymatically adding water (hydrolysis). The synthesis and decomposition of carbohydrates requires the action of specific enzymes.

Carbohydrate Functions

Carbohydrates are the chief energy sources in biological systems, important structural biomolecules, and mediators of cellular adhesion, communication, and environmental sensing. Glucose, a monosaccharide, is a favorite energy source for many cell types, including bacteria and humans. (See more on glucose's role in metabolic pathways in Chapter 8.) Other examples of key monosaccharides include ribose and its closely related derivative, deoxyribose. Both of these *pentoses*, or five-carbon sugars, are important structural components of genetic material.

Cellulose is among the most abundant polysaccharides on the planet because it is a major structural macromolecule in plant and algal cell walls. Chitin (KYE-tin) is a structural polysaccharide in fungal cell walls. **Peptidoglycan**, a macromolecule that contains protein and carbohydrate components, is structurally important in bacterial cell walls. (See Chapter 3 for more on peptidoglycan.)

Many bacteria produce a carbohydrate-based structure called a **capsule**, which they use to adhere to surfaces in order to establish an infection. Furthermore, capsules afford bacteria some protection against our immune system. Bacteria also can secrete carbohydrates to form a sticky matrix that is important in forming biofilms. Last, in order for cells to survive, they must recognize and respond to changes in their environment and be able to communicate with other cells. The term **cell signaling** is used to refer to this collection of cellular communications that occur within a cell and between cells to manage cellular activities. Carbohydrates are central to many cell-signaling pathways.

Lipids include fats, oils, waxes, and steroids.

The term **lipid** refers to a collection of biomolecules that includes fats, oils, waxes, and steroids. All lipids are organic molecules made up of carbon, hydrogen, and oxygen, but they lack the 2-to-1 ratio of hydrogen to oxygen seen in carbohydrates. Lipids are predominantly hydrophobic; they do not readily dissolve in water. Some lipids contain modifications, such as added functional groups like phosphate or alcohol groups, to make them amphipathic, so that they have both polar and nonpolar properties.

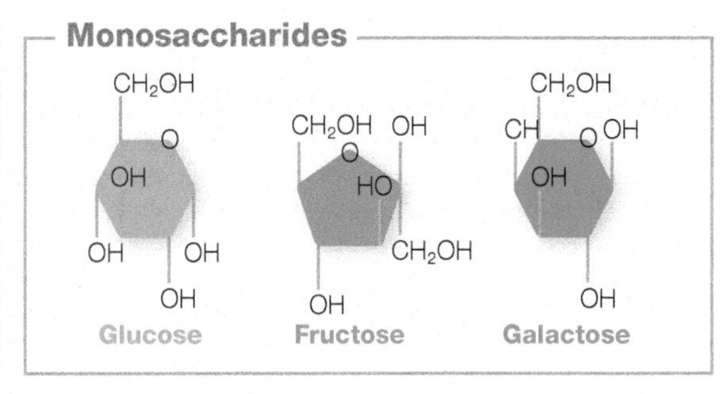

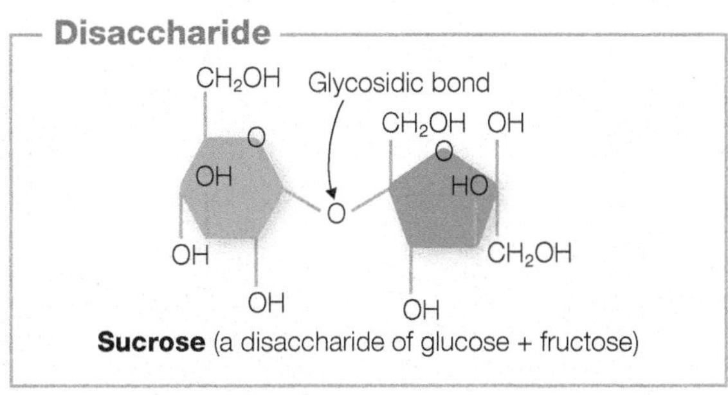

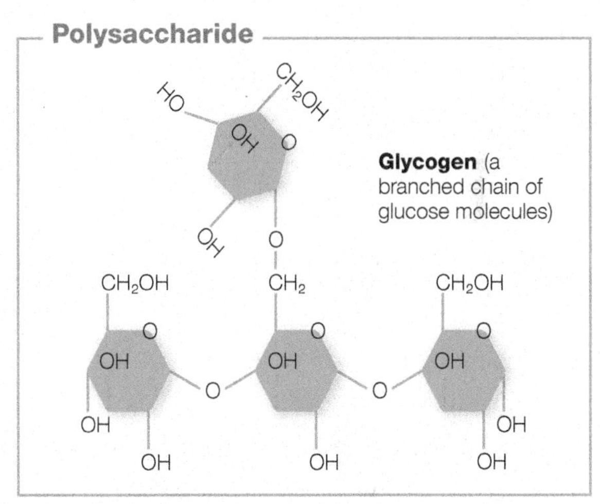

FIGURE 2.16 Types of carbohydrates
Monosaccharides are the simplest carbohydrates. Two monosaccharides bond to form a disaccharide. Polysaccharides are made by the polymerization of many monosaccharide subunits.

CHEM • NOTE

Carbon can form up to four covalent bonds, which means it can't ever bond to more than four atoms. Double covalent bonds in a long hydrocarbon chain can be **hydrogenated**, or reduced with hydrogen, to saturate the double bond.

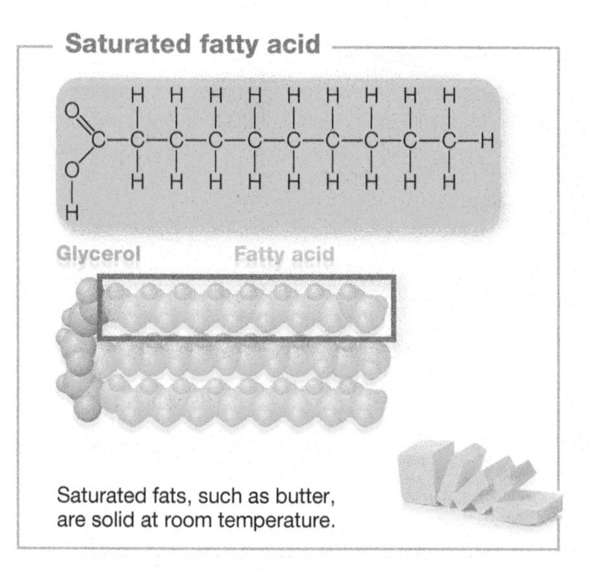

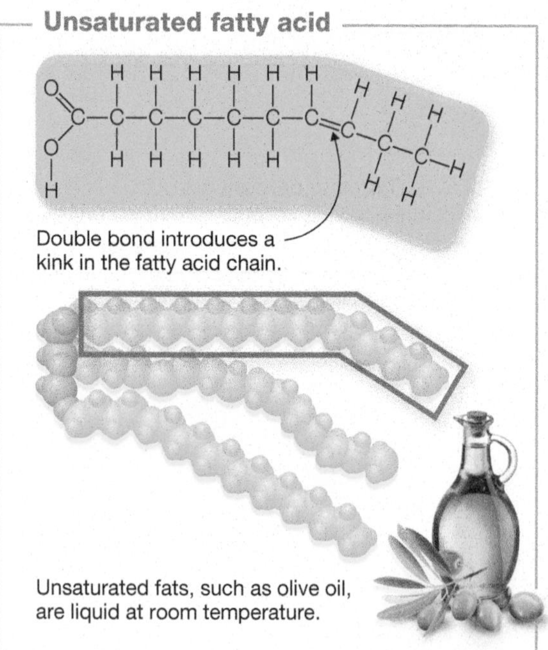

FIGURE 2.17 **Saturated versus unsaturated fats** The degree of saturation in the hydrocarbon tails of fatty acids impacts lipid fluidity. *Top*: Saturated lipids do not have double bonds in their fatty acid chains and tend to be less fluid at room temperature. *Bottom*: Unsaturated fats have at least one double bond in one or more of the fatty acid chains.

Lipid Structures

Lipids have diverse structures: fats, oils, and waxes have long fatty acid hydrocarbon chains, whereas steroids have complex cyclical structures. The fluidity of fats, oils, and waxes depends in part on their degree of saturation, or the number of double covalent bonds that exist in their fatty acid chains. **Saturated** lipids do not have double bonds in their fatty acid chains and therefore have the maximum number of hydrogen–carbon bonds (FIG. 2.17, *top*). Because saturated fatty acids pack tightly together, lipids high in saturated fatty acids, like butter and lard, exist as solids or semisolids at room temperature. Because a diet high in saturated fats has been associated with an increased risk for cardiovascular disease, all forms of saturated fat should be minimized in a healthy diet.

In contrast, **unsaturated** lipids have double bonds in one or more of the hydrocarbon chains of their fatty acids; *the more double bonds present, the less saturated the fat* (FIG. 2.17, *bottom*). Lipids rich in unsaturated fatty acids, like corn oil and olive oil, tend to have kinks and bends in their chains that prevent them from neatly stacking. This allows them to exist as liquids at room temperature. **Monounsaturated** fats have one double bond in their fatty acid chains. **Polyunsaturated** fats have more than one double bond.

Structure of Fats and Oils Again, fats tend to be solid or semisolid at room temperature, whereas oils are liquids at room temperature. Fats and oils are both built by combining one molecule of a short-chain alcohol called **glycerol** with up to three fatty acids through an **ester bond**—a type of covalent bond that forms between acids and alcohols. **Fatty acids** are long hydrocarbon chains with a carboxylic acid functional group, which is why they are called acids (FIG. 2.18). If only one fatty acid is linked to glycerol the lipid is called a **monoglyceride**. **Diglycerides** contain two fatty acid chains linked to glycerol, and **triglycerides** have three fatty acid chains bonded to glycerol. *Sphingolipids*, important fats that comprise parts of the nervous system, use a molecule called sphingosine instead of glycerol as the backbone to link fatty acids.

Structure of Waxes Waxes are lipids that contain a chemical link between a long-chain alcohol and fatty acids instead of a link between glycerol and fatty acids (Figure 2.18).[10] Waxes tend to be solids or flexible semisolids at room temperature. However, some, like the polyunsaturated wax commonly known as jojoba oil, are liquids at room temperature.

Structure of Steroids Belonging to a unique class of lipids, steroids are structurally distinct from other lipids in that they have four fused hydrocarbon ring structures instead of long hydrocarbon chains (Figure 2.18). One common steroid is cholesterol. Cholesterol is a modified steroid called a **sterol**. Sterols are steroid molecules with an attached alcohol functional group (R-OH). The addition of the alcohol group makes cholesterol and other sterols amphipathic.

Lipid Functions

Fats and oils serve as energy sources and structural components of cells, and they help mediate cell signaling. Important structural lipids include phospholipids, which are common cell membrane components. The phosphate group on phospholipids makes these diglycerides amphipathic, thereby allowing them to organize into lipid bilayers. Lipids and oils are also commonly linked to carbohydrates to form **glycolipids**, and to proteins to form **lipoproteins** (discussed

[10] Glycerol is a short-chain alcohol of only three carbons. Long-chain alcohols have from a dozen to over 30 carbons.

Fats and oils

Glycerol

Fatty acids

Triglyceride (glycerol + 3 fatty acids)

Waxes

Beeswax

Long-chain alchohol

Fatty acid

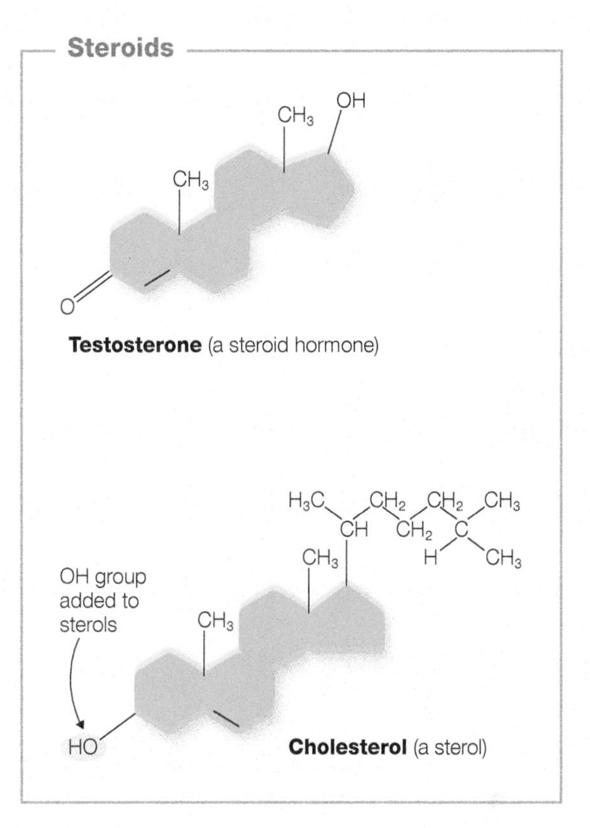

FIGURE 2.18 **Types of lipids** Lipids include fats, oils, waxes, and steroids. **Fats and oils** can be formed from glycerol and fatty acids; triglycerides have three fatty acids attached to glycerol. **Waxes** consist of a long-chain alcohol linked to fatty acids instead of a link between glycerol and fatty acids. **Steroids** have a fused ring structure; sterols, such as cholesterol, are steroids that have an alcohol functional group added to the ring structure.

Critical Thinking *Waxes are often shapeable semisolids, but liquid waxes also exist; Jojoba "oil" is actually a liquid wax. Based on what you have read regarding how fatty acid chain saturation impacts lipid fluidity, what can you assume about the level of saturation of jojoba oil's fatty acids?*

later). Glycolipids and lipoproteins are central to cell-to-cell recognition and cell signaling. Gram-negative bacteria make a glycolipid called **lipopolysaccharide** (LPS), which is toxic to animal cells but serves as a protective barrier for the bacterium, and helps it adhere to surfaces.

Waxes often serve as protection. They are found as a hydrophobic barrier on the surface of leaves, as a coating on certain insect structures, as a water barrier on bird feathers, and as a protective layer of the ear canal of many animals, including humans. Waxes also act as energy stores in microscopic marine life called plankton. Acid-fast pathogens like the bacteria *Mycobacterium tuberculosis* and *M. leprae* make a wax called mycolic acid, which is deposited in the bacterial cell wall, where it acts as a protective barrier for the cell and increases bacterial pathogenicity.

Steroids are important in cell-signaling pathways and also have structural roles. Some steroids act as hormones—lipid-based cell-signaling molecules that impact a wide variety of physiological functions. Eukaryotic organisms universally rely on a class of steroids called sterols as structural components of cellular membranes. For example, cholesterol is abundant in animal cell plasma membranes, and it's found in small

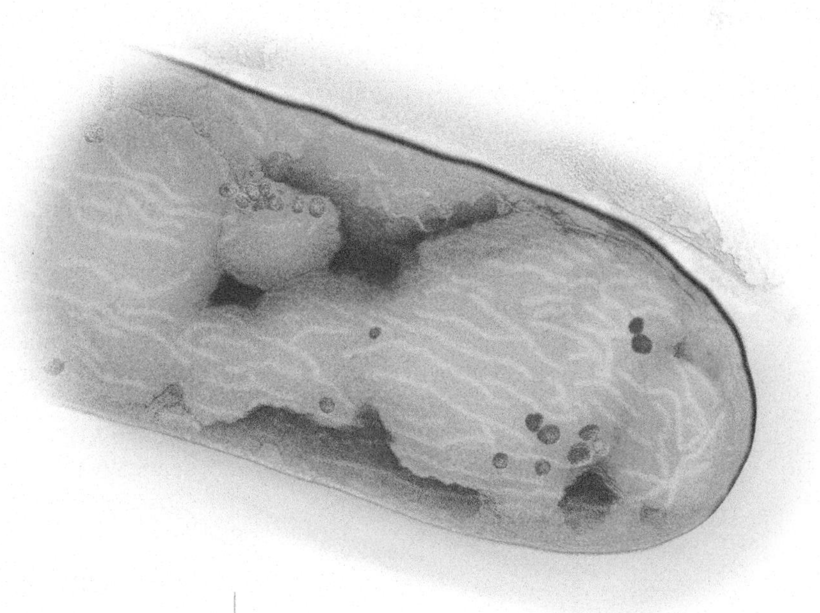

Rope-like mycolic acid (light orange) in the cell wall of *M. tuberculosis*.

quantities in certain plant membranes to keep membranes fluid.[11] A few species of prokaryotes can make steroids, but most do not.[12] Some prokaryotes, such as *Mycoplasma* bacteria, require cholesterol for their cell membranes but they cannot make it. Instead, they rely on host eukaryotic cells that they invade for providing them with cholesterol to reinforce their plasma membranes.[13]

Nucleic acids include DNA and RNA.

Nucleic acids are macromolecules that serve as the genetic material of cells and viruses. There are two main categories of nucleic acids: **deoxyribonucleic acid** (DNA) and **ribonucleic acid** (RNA). DNA exists as a double-stranded helical molecule. RNA is usually single-stranded, but it can fold in on itself to form higher-order loop and stem structures; also, some viruses have double-stranded RNA genomes (see Chapter 6 for more information). The following sections represent a brief introduction to these macromolecules. (A more detailed review of the structures, functions, and synthesis of DNA and RNA is provided in Chapter 5.)

CHEM • NOTE

Phosphate groups (PO_4^{3-}) are a type of functional group found in phospholipids, nucleic acids, ATP, and many other biologically important molecules.

Nucleic Acid Structures

Nucleic acids (FIG. 2.19) are polymers of nucleotides consisting of three ingredients:

1. A five-carbon sugar (deoxyribose in DNA; ribose in RNA)
2. One to three phosphate groups
3. A single nitrogen base (adenine, guanine, cytosine, thymine, or uracil)

The nucleotides that make up DNA are called **deoxyribonucleotides**. They differ from the **ribonucleotides** that make up RNA in their sugar type (deoxyribonucleotides contain deoxyribose, whereas ribonucleotides contain ribose). The nitrogen bases in deoxyribonucleotides also vary a bit from those found in ribonucleotides. Both contain adenine, guanine, and cytosine, but thymine is only found in DNA, and uracil is only in RNA.

DNA and RNA are both built by polymerization of their respective nucleotides. The sugar in one nucleotide bonds with a phosphate of another via **phosphodiester bonds** (FIG. 2.20). This creates the alternating pattern of sugar and phosphate seen in the backbone of DNA and RNA.

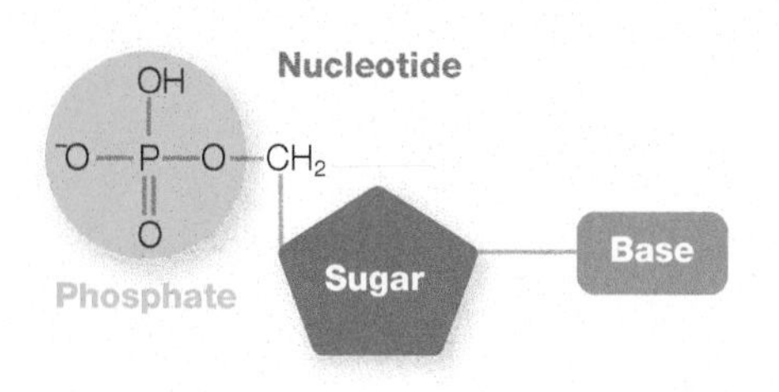

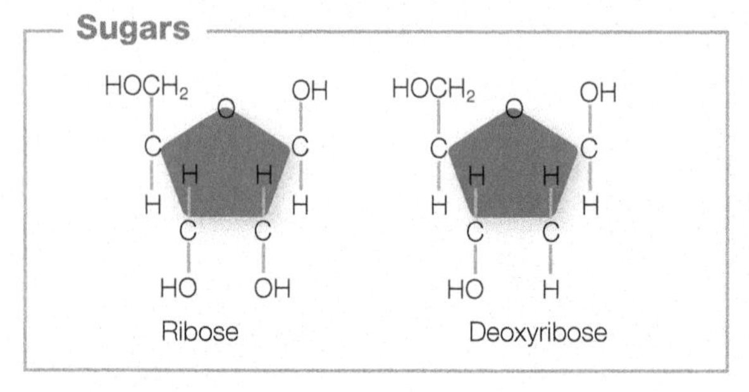

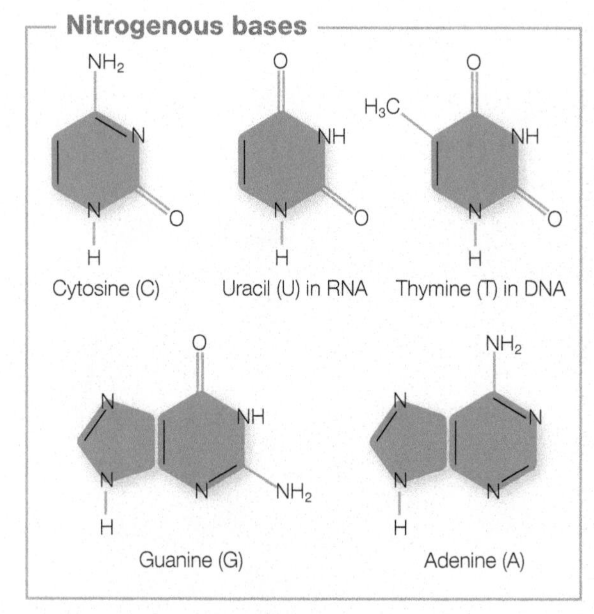

FIGURE 2.19 **Nucleotides** Nucleic acids are polymers of nucleotides. Each nucleotide consists of a sugar, at least one phosphate, and a nitrogen base.

ATP: A Special Ribonucleotide

Adenosine triphosphate (ATP) is a vitally important ribonucleotide (FIG. 2.21). Not only is it used to make RNA, but it is also the energy currency in cells. Energy can be extracted from foods to make ATP in a series of biochemical reactions. When phosphates are cleaved off ATP, energy is released. This energy can be used to fuel endergonic reactions that allow the cell to function, maintain functions, or divide. (The metabolic pathways that make ATP are discussed in our review of metabolism in Chapter 8.)

Nucleic Acid Functions

DNA is the genetic blueprint of all cells, while viruses can use DNA or RNA as their genetic blueprint. Not only does RNA direct the production of proteins in cells and viruses, but also specialized RNAs called ribozymes can catalyze particular reactions. Most ribonucleotides can serve as energy sources in cells—although ATP is the most widely used. Some ribonucleotides can even be chemically modified to function as messengers in cell-signaling pathways.

[11] Behrman, E. J., & Gopalan, V. (2005). Cholesterol and plants. *Journal of Chemical Education, 82*(12), 1791–1792.

[12] Bode, H. B., Zeggel, B., Silakowski, B., Wenzel, S. C., Reichenbach, H., & Müller, R. (2003). Steroid biosynthesis in prokaryotes: Identification of myxobacterial steroids and cloning of the first bacterial 2,3(S)-oxidosqualene cyclase from the myxobacterium *Stigmatella aurantiaca*. *Molecular Microbiology, 47*(2), 471–481.

[13] Jonathan, D., Kornspan, J. D., & Rottem, S. (2012). The phospholipid profile of mycoplasmas. *Journal of Lipids, 2012*, article ID 640762, doi:10.1155/2012/640762.

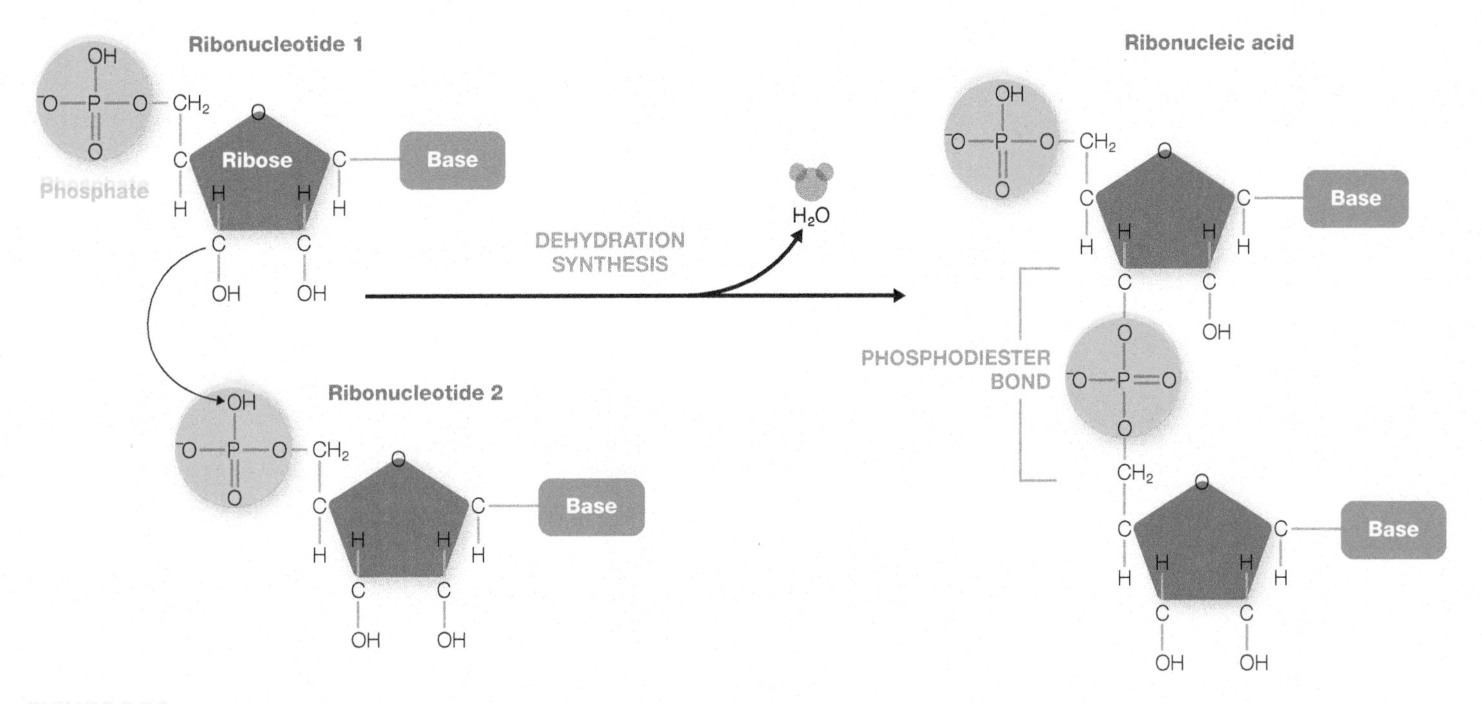

FIGURE 2.20 **Nucleic acids** Nucleic acids are polymers of nucleotides linked by phosphodiester bonds. A phosphodiester bond is formed between a hydroxyl (OH) group of the sugar of one nucleotide and the phosphate group of another nucleotide.

Critical Thinking *Dideoxynucleotides do not have any OH groups on their ribose sugar. These nucleotides are called "stop nucleotides" and can be added to stop the building of nucleic acids. Explain how these "stop nucleotides" block DNA production.*

Proteins are cellular workhorses.

Genes in DNA encode proteins, so the varieties of proteins in biology are as diverse as life itself. At some level every cellular process involves a protein; thus, proteins really are the workhorses of life. Here we will discuss the basics of protein structure and function. (See Chapter 5 for more on how cells build proteins.)

Protein Structures

Proteins are built from monomeric units called amino acids. Each amino acid has a general structure that consists of two functional groups: an amine group (NH_2) and a carboxyl group (COOH). Amino acids also have different side groups, or R groups (refer to the top left of FIG. 2.22). Some of the R groups are small, others are large; some are polar and others are nonpolar. Some R groups even have acidic properties, whereas others are bases. The properties of each amino acid contribute to the overall structure and function of the final protein.

Proteins are polymers of amino acids that range from just a few amino acids to thousands. The average protein has about 300 amino acids. There are hundreds of distinct amino acids, but only 22 are genetically encoded. Twenty of these 22 are called standard amino acids, whereas the other two (selenocysteine and pyrrolysine) are considered nonstandard, because of the way they are encoded. Figure 2.22 shows the genetically encoded amino acids. (Standard and nonstandard amino acids are discussed further in Chapter 5.)

Amino acids covalently bond to each other by peptide bonds between the amino group of one amino acid and the carboxyl group of the neighboring amino acid. When a peptide bond forms, water is removed, which means the reaction is a dehydration synthesis. Specialized enzymes called proteases can

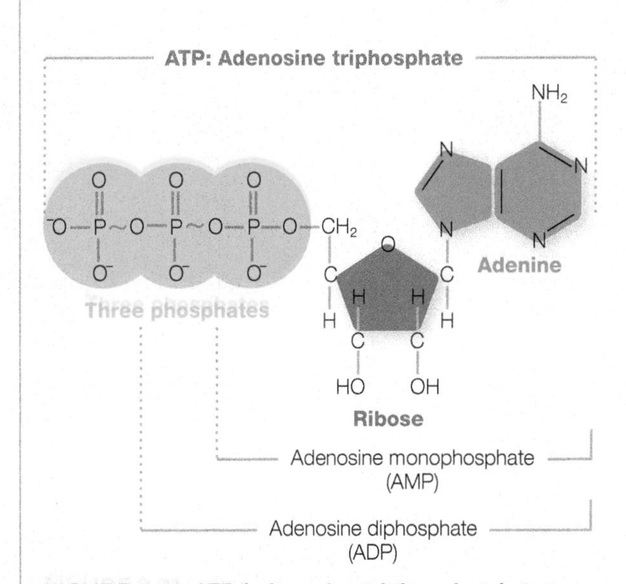

FIGURE 2.21 **ATP (adenosine triphosphate)** Cells use the ribonucleotide ATP to fuel cellular reactions. ATP is also used to build RNA.

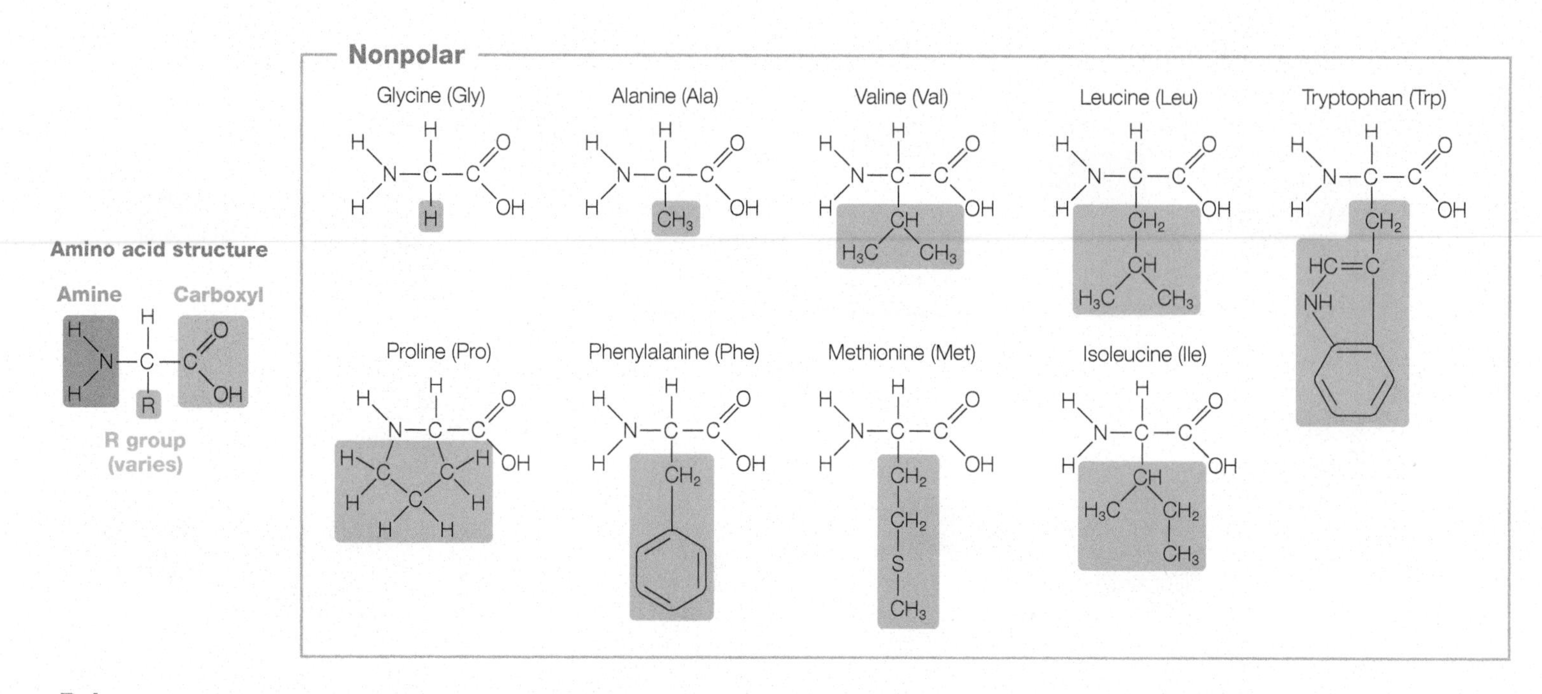

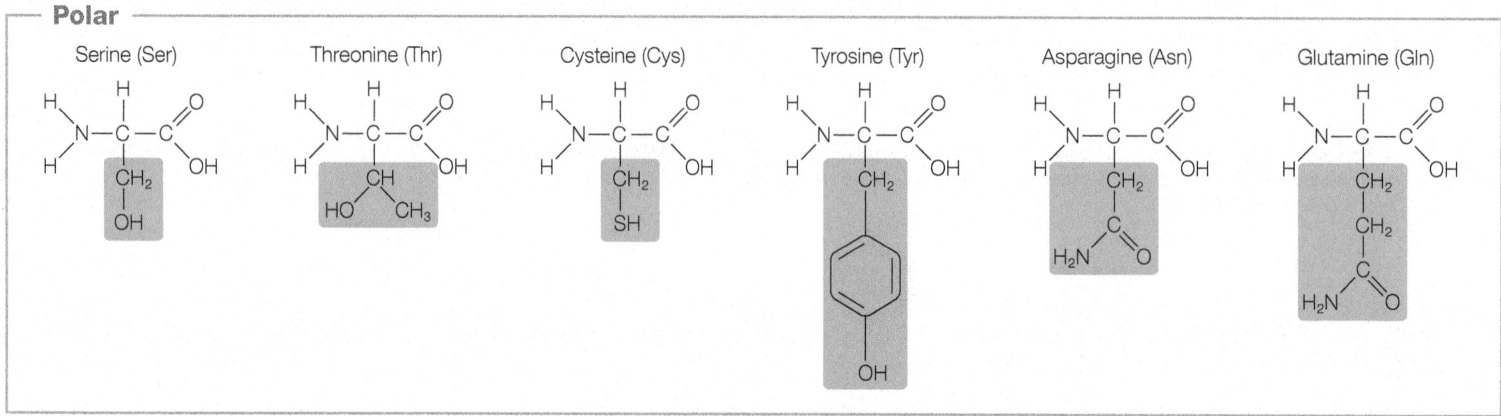

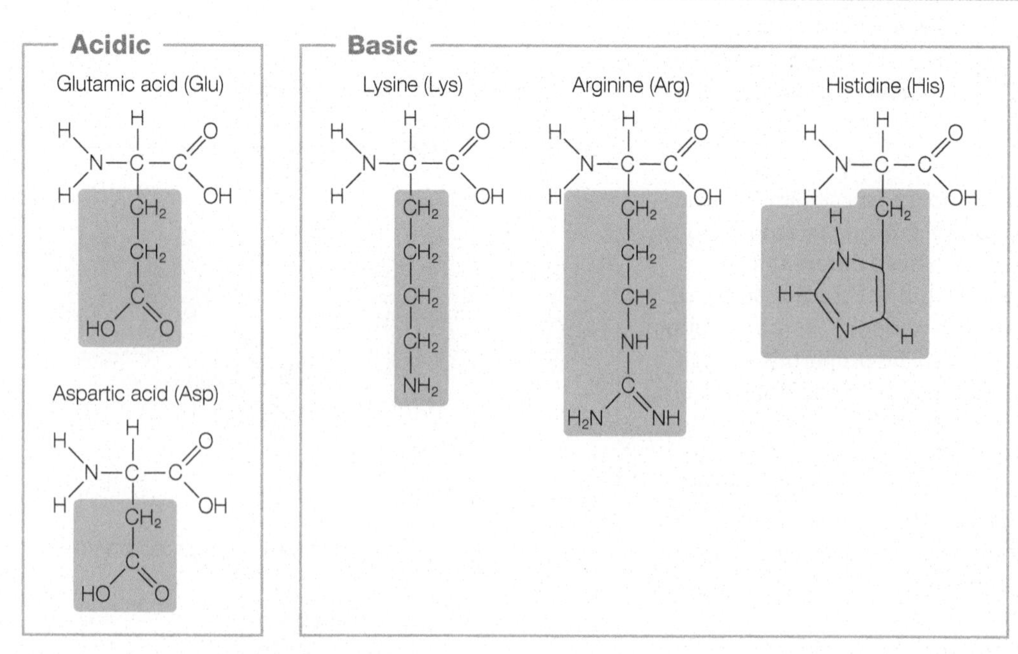

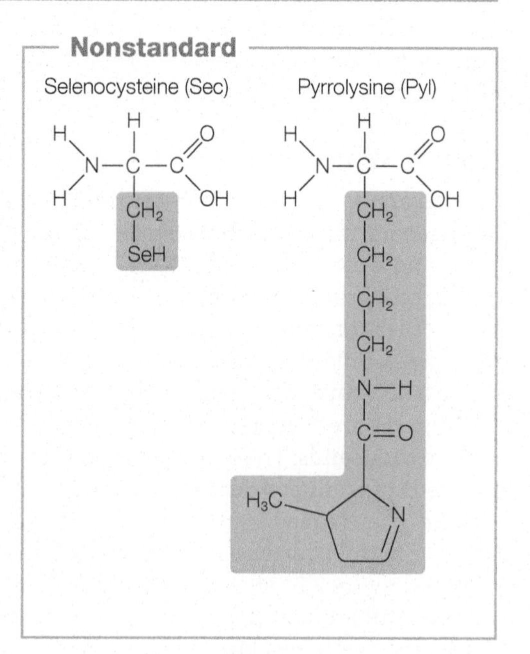

FIGURE 2.22 **Genetically encoded amino acids**

break peptide bonds by catalyzing hydrolysis reactions. Short chains of amino acids are called **peptides**, whereas longer amino acid chains are called **polypeptides**. The cutoff length between a peptide and a polypeptide is somewhat arbitrary, but usually peptides are fewer than 55 amino acids long. Human insulin, which is considered a peptide hormone, contains 51 amino acids.

There are four levels of protein structure: primary, secondary, tertiary, and quaternary (FIG. 2.23). Changes to any one of these levels of structure can affect protein function.

Primary Structure A protein's primary structure is simply the linear sequence of amino acids in that protein. You can think of the primary structure as a string of amino acid "beads" in a specific order, linked together by peptide bonds. A protein's primary structure is unique for that particular type of protein and is dictated by a genetic sequence. This means the primary structure can be predicted from the gene that encodes the protein. If the gene that encodes a

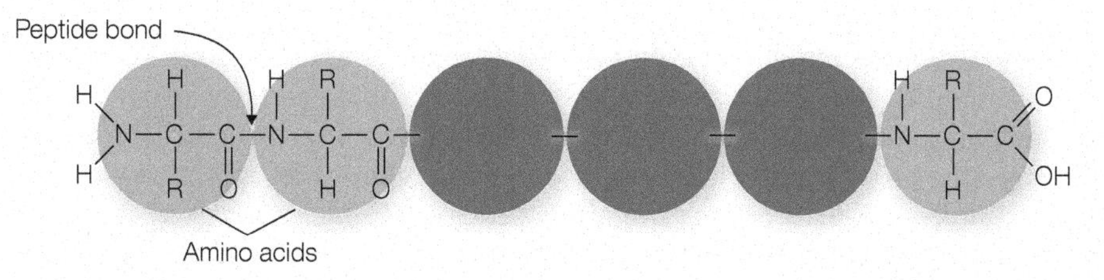

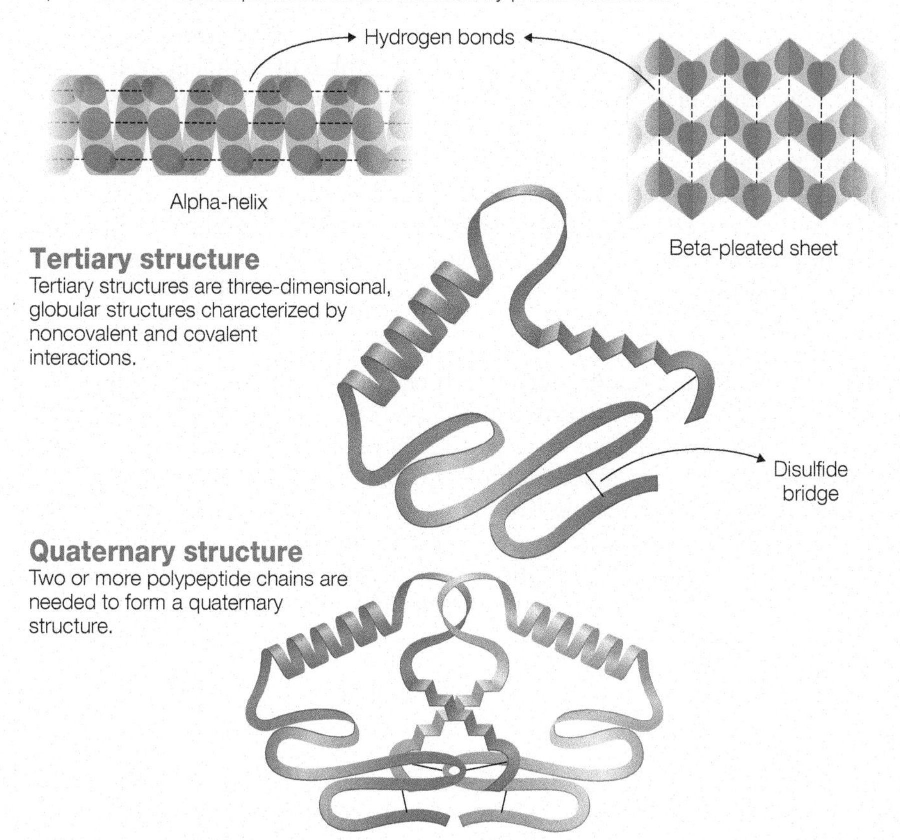

FIGURE 2.23 **Protein structure** There are four levels of protein structure.

Critical Thinking *A change in primary structure can lead to changes in all other levels of protein structure. Explain this statement.*

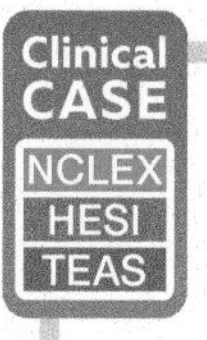

The Case of the Confused Cattle Farmer

Practice applying what you know clinically: visit the **Mastering Microbiology** Study Area to watch Part 3 and practice for future exams.

protein is not known, then a protein's primary structure can be directly determined in the laboratory. This is accomplished by sequentially removing amino acids and then characterizing them. **Mutations**, or errors that arise in genetic material, can lead to changes in a protein's primary structure. Because the primary structure lays the foundation for all higher-order structure of a protein, even small changes in a protein's primary structure can have devastating effects on overall protein structure and function.

Secondary Structure To picture secondary structure, imagine taking the string of beads that is the primary structure of a protein and organizing select segments of it into coils and folds. There are two forms of secondary structure in proteins, **alpha-helices** (or α-helices) and **beta-pleated sheets** (β-pleated sheets). Alpha-helices are spiral structures, whereas beta-pleated sheets are accordion-like folds; both have characteristic dimensions and are not random structures. Secondary structures are found in specific areas of the amino acid chain. They are mainly stabilized by hydrogen bonding between amino acids of the primary structure. Proteins often contain a mixture of these two secondary structures.

Tertiary Structure Imagine taking a string of beads containing coils and pleats as described earlier, and then folding it in upon itself. In tertiary structure, the protein chain does just this: fold into a three-dimensional structure. **Protein folding**—the process in which proteins take on higher-order structure—can be quite complex. A variety of factors affect protein folding, so much so that there is a whole subspecialty in biochemistry focused on studying how proteins fold and unfold. Some protein structures are so sophisticated that they require a variety of **chaperones**, or smaller proteins that the cell makes to help fold larger proteins. Because correct folding is essential to function, and proteins regulate most of a cell's activities, people who have errors in genes that encode chaperone proteins suffer from a number of pathologies. In fact, protein misfolding is what leads to Alzheimer's disease, as well as the neurodegeneration caused by infectious proteins called prions. (Prions are explored more in Chapter 6.)

Tertiary structures are stabilized by noncovalent interactions such as ionic bonds, hydrogen bonds, and van der Waals interactions. They are also stabilized by **disulfide bridges**, which are covalent bonds that form between sulfur-containing functional groups (thiol groups) of amino acids and add structural stability to proteins.

Quaternary Structure In quaternary structure, two or more separate polypeptide chains with tertiary structure combine. Proteins with quaternary structure are often called *multimers* because they have more than one subunit, or polypeptide chain, that interact. Multimers can be described by the number of subunits they have. For example, multimers consisting of two subunits are called *dimers*; *trimers* have three protein subunits, *tetramers* have four subunits, and so on. The subunits of a multimeric protein are held together by noncovalent and covalent interactions. Quaternary structures are essential to the function of many complex proteins such as cell surface receptors, antibodies, enzymes, and many structural proteins.

Protein Functions

There is tremendous diversity in proteins—our bodies make at least 30,000 different types. Proteins are essential to pretty much every process in a cell. They serve as structural scaffolds in cells, enzymes that facilitate thousands of chemical reactions needed for cell survival, cellular transporters shuttling a variety of substances into and out of cells, and hormones and other compounds central to cell recognition and communication.

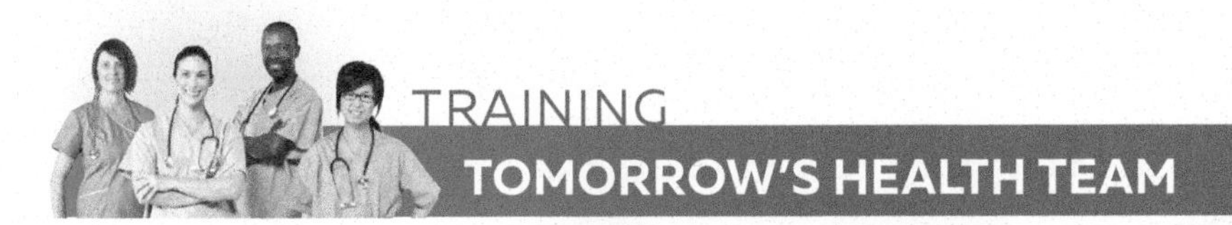

The Misfortune of Protein Misfolding

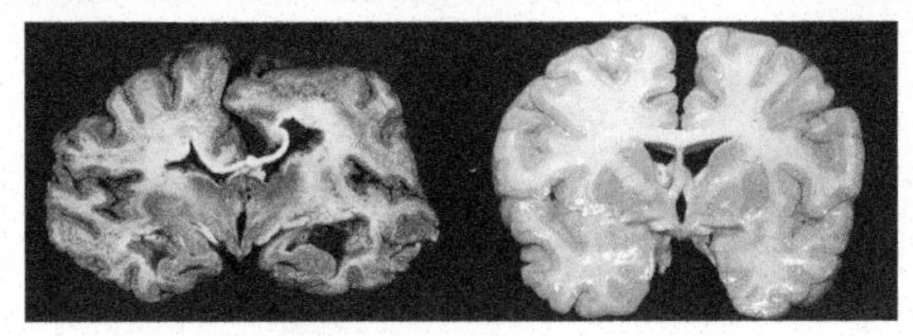

Brain damaged by Creutzfeldt-Jakob disease — Normal brain

Misfolded proteins cause hundreds of diseases in humans. In protein aggregation disorders, the misfolded proteins clump together. Eventually the clumps (aggregates) damage the tissue where they reside. Sometimes these disorders are inherited, as is the case with certain cataracts. Other diseases such as Alzheimer's are acquired over time. Some disorders of protein folding are caused by infectious proteins called prions. These include Creutzfeldt-Jakob disease—a fatal neurological disease related to mad cow disease.

To date there are no cures for these protein aggregation diseases. But research on how proteins fold, and how to fix misfolded proteins or eliminate them from afflicted cells, could lead to therapies. In particular, drugs that encourage the production of chaperones are being clinically tested, as are therapeutics that aim to enhance the body's ability to "clean up" misfolded protein aggregates. This research is especially accelerating in the area of Alzheimer's disease—there are currently over two dozen potential Alzheimer's disease therapeutics in phase 3 clinical trials, which is the final stage of testing before FDA approval can be solicited and marketing can begin in the United States.

QUESTION 2.3

Why would a mutation (an error) in a gene potentially lead to protein misfolding?

Build Your Foundation

26. Name the building blocks used to make each of the four classes of biomolecules, and state how these monomers are bonded. (LO 2.25 and 2.26)
27. What are the structural and functional features of each of the four classes of biomolecules? (LO 2.27)
28. What are triglycerides, and how does saturation affect their structure and physical properties? (LO 2.28)
29. How are deoxyribonucleotides and ribonucleotides similar? How are they different? (LO 2.29)
30. A protein's quaternary structure is ultimately dependent on its primary structure; explain why. (LO 2.30)
31. What are chaperone proteins, and why are they important? (LO 2.31)

Build Your Foundation (BYF) Quick Quiz: Visit the **Mastering Microbiology** Study Area to quiz yourself.

Atoms and Molecules

Atoms combine to make molecules. Molecular formulas like H_2O reveal what elements are in a molecule, as well as their ratios.

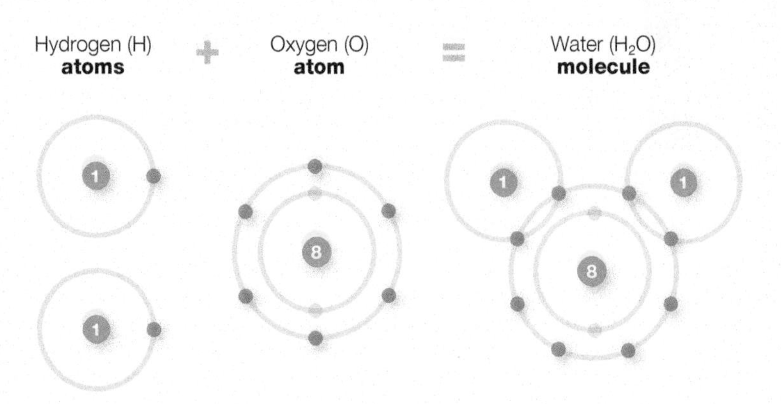

Ions and Isotopes

Ions are charged atoms. Ions are important electrolytes. Isotopes are atoms with the same number of protons but different numbers of neutrons. Some isotopes are used in medical treatments.

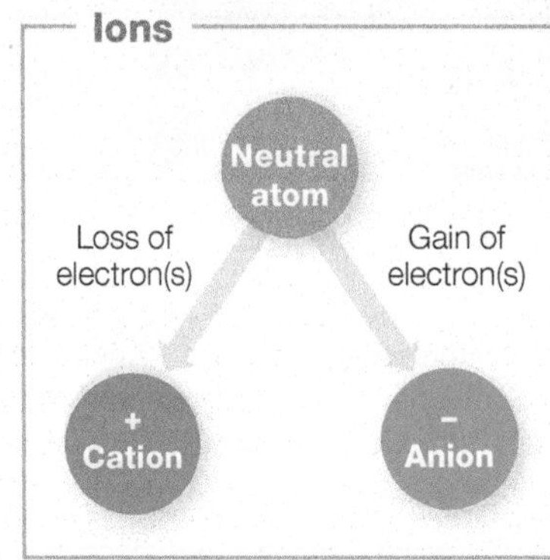

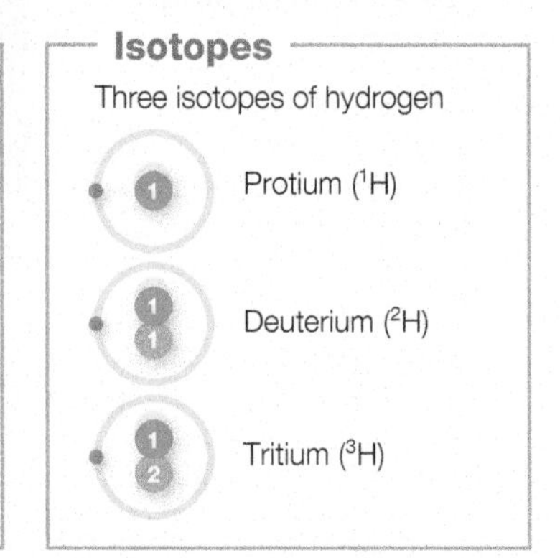

Acids and Bases

Acids contribute hydrogen ions (H^+) to an aqueous solution. Bases release hydroxide ions (OH^-). The pH scale measures acidity and basicity.

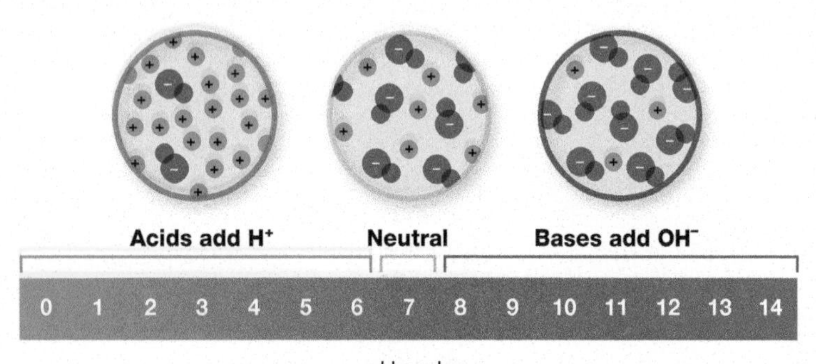

Four Classes of Biomolecules

Cells make their biomolecules from building blocks.

Building Blocks	Polymers
Simple sugars	Polysaccharides
Glycerol + Fatty acids	Lipids
Nucleotides	Nucleic acids
Amino acids	Proteins

Bonds and Interactions

Bonds are formed when valence electrons are shared (covalent bonds) or transferred (ionic bonds). Hydrogen bonds are electrostatic interactions between polar molecules. Many of water's properties are due to abundant hydrogen bonds.

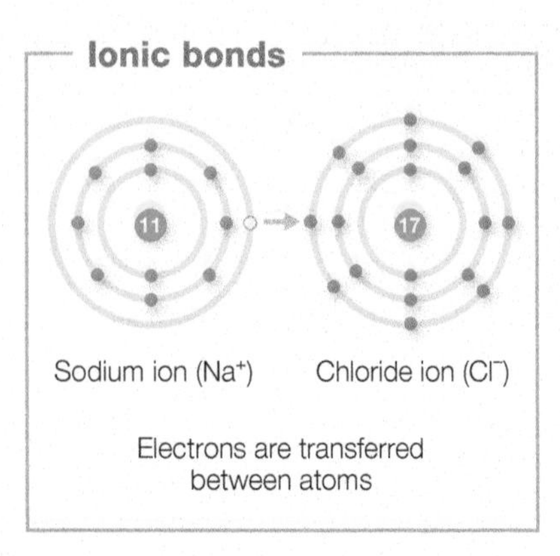

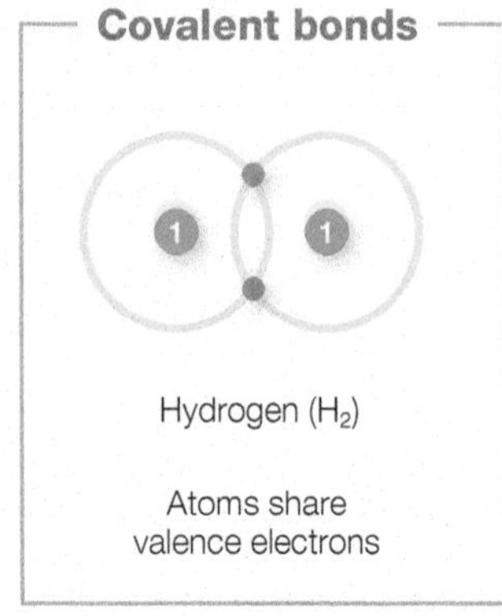

Chemical Reactions

Activation energy is the minimum amount of energy needed to start a chemical reaction. Exergonic reactions release more energy than they use to overcome the activation energy. Endergonic reactions require more energy than they release.

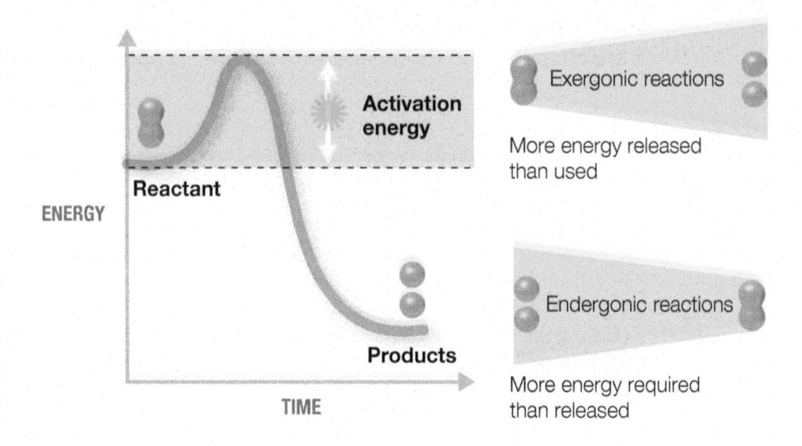

2 CHAPTER 2 OVERVIEW

2.1 From Atoms to Macromolecules

- Atoms are the smallest units of elements, which make up matter. Atoms have protons (positive charge) and neutrons (neutral) in their nucleus; electrons have a negative charge and exist in a cloud around the atomic nucleus.
- Ions are atoms that lost or gained one or more electrons as a result of forming ionic bonds. Cations are positive ions that form when at least one electron is lost. Anions are negative ions that form by gaining at least one electron.
- Isotopes are atoms of the same element with the same number of protons, but different numbers of neutrons. Radioactive isotopes release energy as their nucleus breaks down.
- A molecule forms when two or more atoms bond. Compounds are molecules made of more than one type of element.
- Isomers are molecules with the same molecular formula but different structures.
- Organic molecules contain carbon and hydrogen, while inorganic molecules may contain carbon, but lack the associated hydrogen.
- Acids donate H^+ (hydronium ion; also written H_3O^+) to aqueous solutions, which are water-based solutions. Bases donate OH^- (hydroxide ion) to aqueous solutions.
- Salts form when acids and bases react with each other; the acid contributes the anion of a salt, and the base contributes a cation.
- A solution's pH is determined by the concentration of H^+ and OH^-. Pure water has a pH of 7. At a pH below 7, there is more H^+ than OH^-, and the solution is acidic. At a pH above 7, there is less H^+ than OH^-, and the solution is basic.

2.2 Chemical Bonds

- Valence electrons, which are in the outermost electron shell, are gained, lost, or shared with another atom in a chemical reaction. Atoms with a full valence shell are stable.
- Ionic bonds are formed when electrons are transferred from one atom to another.
- Covalent bonds form when atoms share electrons. Carbon forms up to four covalent bonds. Polar covalent bonds arise when there is unequal sharing of electrons.
- Hydrogen bonds are electrostatic noncovalent interactions that form between molecules that have polar covalent bonds.
- Van der Waals interactions are weak electrostatic interactions between atoms that are not ionically or covalently bonded to one another.
- Molecules can be polar, nonpolar, or amphipathic. Polar molecules are generally considered hydrophilic because they interact well with water (form hydrogen bonds with water); nonpolar molecules are hydrophobic because they do not form hydrogen bonds with water; amphipathic molecules have polar and nonpolar properties and form structures such as micelles and lipid bilayers.

2.3 Chemical Reactions

- Ingredients of a chemical reaction are reactants, and the substances generated are products. Some reactions require a catalyst, which increases the rate of a reaction, but it isn't used up in the reaction.
- A chemical equation is a representation of a chemical reaction; it depicts the reagents and products and any specific conditions required for the reaction.
- Synthesis reactions build molecules; in biological systems, dehydration synthesis reactions remove water from reactants to build macromolecules like proteins, nucleic acids, and polysaccharides.
- Decomposition reactions break a substance down into simpler components. Hydrolysis reactions are a type that adds water across covalent bonds to break down macromolecules like proteins, polysaccharides, lipids, and nucleic acids.
- Exchange reactions involve swapping one or more components in a compound. Sometimes these are called replacement reactions.
- Activation energy is the minimum amount of energy needed to get a reaction started.
- In biological systems, exergonic reactions can be used to fuel endergonic reactions. Exergonic reactions release more energy than they consume; in an exergonic reaction the products are lower energy than the reactants. Endergonic reactions consume more energy than they release; the products of endergonic reactions have a higher energy than the reactants.
- Reversible reactions occur in the forward or reverse direction. The reaction is at equilibrium when the forward and reverse reactions occur at the same rate.

2.4 Biologically Important Macromolecules

- There are four main families of biomolecules: carbohydrates, lipids, nucleic acids, and proteins.
- Carbohydrates are organic molecules built from one or more sugar monomers. They include monosaccharides, disaccharides, and polysaccharides. Glycosidic bonds covalently link monosaccharides. Carbohydrates are the chief energy sources in biological systems and important structural biomolecules.
- Lipids include fats, oils, waxes, and steroids. Saturated lipids are less fluid at room temperature than unsaturated lipids due to how fatty acid chains stack. Fats and oils are built from glycerol and one to three fatty acids. Fatty acids can be saturated (no double bonds in the hydrocarbon chain) or unsaturated (one or more double bonds in the hydrocarbon chain). Fats and oils serve as energy sources for cells, they are integral to cell structures, and they help mediate cell signaling.
- Waxes are protective barriers or energy stores that contain a chemical link between a long-chain alcohol and fatty acids. Steroids have four fused hydrocarbon ring structures instead of long hydrocarbon chains. They work in cell-signaling pathways and have structural roles. Some steroids act as hormones.
- Nucleic acids, deoxyribonucleic acid (DNA) and ribonucleic acid (RNA), are the genetic material of cells and viruses. Nucleic acids are built by polymerizing nucleotides. The nucleotides that make up DNA are called deoxyribonucleotides; ribonucleotides make up RNA. Phosphodiester bonds covalently link nucleotides. ATP is a ribonucleotide that fuels cellular work.
- Proteins are made of amino acids that are bonded together by peptide bonds. There are 20 standard and two nonstandard amino acids.
- There are four levels of protein structure: primary, secondary, tertiary, and quaternary.

SMART

THINK CLINICALLY: | *Be S.M.A.R.T. About Cases*

COMPREHENSIVE CASE

The following case integrates basic principles from the chapter. Try to answer the case questions on your own. Don't forget to **be S.M.A.R.T.*** about your case study to help you interlink the scientific concepts you have just learned, and apply your new understanding of chemical principles to a case study.

*The five-step method is shown in detail in the Chapter 1 Comprehensive Case on cholera. See pages 31–33. Refer back to this example to help you apply a SMART approach to other critical thinking questions and cases throughout the book.

• • •

At 2:00 a.m., three-month-old Henry was rushed to the nearby children's hospital by ambulance. Upon arriving at the hospital Henry was unconscious, exhibited impaired breathing, and had a low body temperature. Once Henry was stabilized, the physician spoke with the parents. They said it was their first night of trying to limit nighttime feedings for Henry. Since birth, he had been eating on a two-hour, around-the-clock schedule that was exhausting for his mother. Because the baby's older sister had gone for six-hour stretches between feedings by the same age, his parents felt it was time to try a new feeding schedule.

The mother reported that Henry awoke after midnight, but that she did not feed him immediately, hoping that he would fall back asleep. The mother checked on Henry every 10 minutes during his crying spell to make sure he was all right. After about 45 minutes of nonstop crying, the mother decided to feed him. Walking toward Henry's room, she heard his cries suddenly stop, and upon entering the room found Henry convulsing in his crib.

The physician examined the baby and reported that Henry had an enlarged liver (hepatomegaly). His blood work was also concerning (select data table). The parents shared that when Henry was born, his blood sugar was extremely low (hypoglycemia), but he was readily stabilized with glucose. It was determined that Henry's low blood glucose level likely triggered that night's seizure. Henry was eventually released, and his parents were told to follow up on his condition with a specialist. After seemingly endless tests, Henry was finally diagnosed with von Gierke's disease, or type I glycogen storage disease (GSD).

The parents learned that GSDs are rare genetic disorders with an overall incidence of about 1 in every 43,000 live births. There are over 12 classes of GSDs, but about 90 percent of all cases are characterized as type I. In GSD I, the patient can't break down glycogen because of a deficiency of the enzyme glucose-6-phosphatase. This enzyme catalyzes the last chemical reaction before the release of glucose into the bloodstream: the hydrolysis of glucose-6 phosphate ($C_6H_{11}O_9P$) into glucose and inorganic phosphate (P_i). This enzyme is made of 357 amino acids and is a transmembrane protein (spans a lipid bilayer). GSD I patients have trouble stabilizing their blood glucose levels when fasting because they cannot make use of their glycogen energy stores.

GSD I patients suffer from an enlarged liver and kidneys, low blood sugar within a few hours of eating, chronic fatigue, muscle cramps, stunted growth, delayed puberty, and high levels of lactate, triglycerides, cholesterol, and uric acid in the blood. Currently GSD I can be controlled, but not cured. Longer-term complications for GSD I include liver tumors, kidney failure, and cardiovascular problems. The main goal of therapy is to regulate blood glucose levels to limit the likelihood that the patient will suffer from dangerous complications brought about by the accumulation of glycogen, lactate, pyruvate, fatty acids, cholesterol, and triglycerides.

Cornstarch is used as a dietary supplement for regulating blood glucose levels in GSD patients; it is a polysaccharide that, like glycogen, is a large polymer of glucose. Because cornstarch is broken down slowly, it helps stabilize blood glucose levels. A patient with GSD I can expect to get 30 to 45 percent of his or her daily calories from raw cornstarch. The cornstarch is mixed with water; it should not be heated or mixed with highly acidic juices, as the chemical composition of the cornstarch can be altered, making it a less effective treatment.

Henry's Blood Work Results

	Observed Levels	Normal Ranges for Children Under 1 Year
Glucose	38 mg/dL	60–105 mg/dL
Lactate (plasma)	3.9 mmol/L	1 to 3.3 mmol/L
Pyruvate	0.4 mmol/L	0.05 to 0.10 mmol/L
Free fatty acids	3.2 mmol/L	Below 2.3 mmol/ L
Triglycerides	200 mg/dL	20–150 mg/dL
Cholesterol	150 mg/dL	50–120 mg/dL
Blood pH	7.28	7.34 to 7.44

• • •

CASE-BASED QUESTIONS

1. Is glycogen an organic or inorganic compound? Explain your response.
2. Write the chemical equation for the reaction catalyzed by glucose-6-phosphatase (be sure to use molecular formulas, but don't worry about writing a balanced equation). After writing the reaction out, name the elements that make up the reactants, and specify how many atoms of each element are present in the reactants.
3. Explain what the notations mg/dL and mmol/L mean in the blood work results.
4. How would you describe Henry's blood pH in comparison with normal ranges? In relation to his observed blood pH, what would he be described as manifesting?
5. How would you describe the polarity of the enzyme glucose-6-phosphatase, knowing it is a transmembrane protein?

6. Why do patients with GSD exhibit hepatomegaly?
7. Hundreds of mutations in the enzyme glucose-6-phospahatase have been discovered; most result in small errors in amino acid sequence. Based on your introductory exposure to protein structure, explain why this would impact the enzyme's function.
8. What kind of bond do you think is broken to digest cornstarch (be specific)?
9. In the cornstarch therapy, what is the solvent, and what is the solute?
10. Henry's blood work reveals increases in a number of compounds. Using the diagram (right), explain the increases in cholesterol, triglycerides, free fatty acids, lactate, and pyruvate.
11. Based on the diagram to the right, explain why GSD type I patients are discouraged from consuming carbohydrates such as fructose, galactose, and lactose.

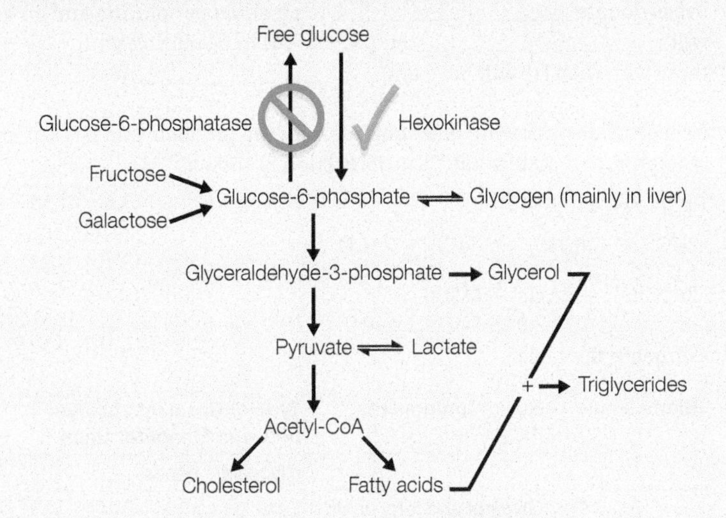

END OF CHAPTER QUESTIONS

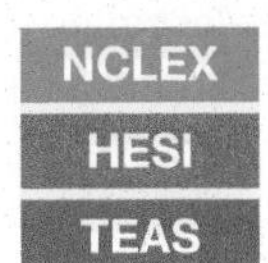

Think Critically and Clinically

Questions highlighted in orange are opportunities to practice NCLEX, HESI, and TEAS-style questions.

1. Using the periodic table, answer the following questions:
   a. What is the atomic number for lithium (Li)?
   b. What is the atomic mass of oxygen (O)?
   c. What is the chemical symbol for potassium?
   d. How many protons does nitrogen have?
   e. How many neutrons does lead have?
2. What ions result when hydrogen donates one electron to fluorine?
3. Select all of the *compounds* from the following list. If it is not a compound, then state what it is.
   a. H_2O
   b. HCO_3^-
   c. O_2
   d. H_2
   e. Li^{2+}
   f. $C_6H_{12}O_6$
   g. H^+
4. Determine the following:
   a. The molarity of a solution with 0.5 moles of glucose per liter of water.
   b. The concentration (in weight/volume percent) of a solution that contains 20 grams of sodium chloride per liter of water.
   c. The concentration (in mg/dL) of a solution with 1 gram of lactic acid per 100 mL of solution.
   d. The molarity of a solution with 1 mmol of solute in 1 L of water.
5. Select the *false* statement about salts.
   a. Salts are ionic compounds.
   b. Salts are formed when an acid and a base react with each other.
   c. Salts consist of an anion and a cation component.
   d. Salts may be inorganic.
   e. Salts are usually acids.
   f. Salts are usually hydrophilic.
6. Write the molecular formula for a substance that contains two oxygen molecules, two carbons, and four hydrogen atoms. Be sure to follow the standard conventions of writing molecular formulas.
7. Indicate the true statements **and** then correct the false statements so that they are true.
   a. Isotopes are atoms with differing numbers of protons and the same number of neutrons.
   b. A cation is a positive ion.
   c. Redox reactions create ions.
   d. Equal sharing of electrons leads to polar covalent bonds.
   e. Ions are charged atoms.
   f. CO_2 is an inorganic molecule.
   g. Isomers have the same molecular formula but different structures.
   h. Adding a base to a solution will decrease the pH.

8. Fill in the blanks:

Acids donate ______________ to an aqueous solution, which will lead to a(n) ______________ in pH. In contrast, bases donate ______________ to an aqueous solution and will ______________ the pH. The pH of a solution with more OH^- than H^+ will have a(n) ______________ pH.

9. In each of the following reactions, identify the products and reactants, and state what class of reaction is shown.

$2Ca + O_2 \rightarrow 2CaO$

$2HCl_2 + Ca(OH)_2 \rightarrow CaCl_2 + 2\,H_2O$

$2AgCl(s) \xrightarrow{\text{Light}} 2Ag(s) + Cl_2(g)$

10. Complete the table:

Biomolecule	Basic Components	Type of Covalent Linkage between Monomer Units
		Glycosidic bonds
	Glycerol and fatty acids	Ester bonds
RNA		
Proteins		
	Nucleotides (dNTPs)	

11. The notation ^{18}O denotes a(n)

a. isomer.
b. isotope.
c. dipole.
d. ion.
e. reaction.

12. Label the features of the periodic table box:

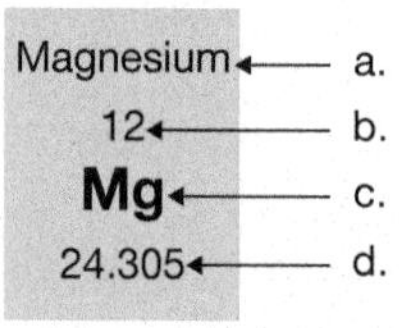

13. How many more times acidic is a solution with a pH of 9 versus a solution with a pH of 12?

14. Which of the following is *false?*

a. ATP is used to fuel cellular work.
b. ATP is a specialized ribonucleotide.
c. Energy is released from ATP by breaking off phosphate groups.
d. ATP is used to fuel exergonic reactions.

15. Cholesterol is best described as

a. a lipid.
b. a sterol.
c. an alcohol.
d. a fat.
e. a wax.

16. CIRCLE two covalent bonds and two hydrogen bonds in this image; label them. BOX a polar covalent bond.

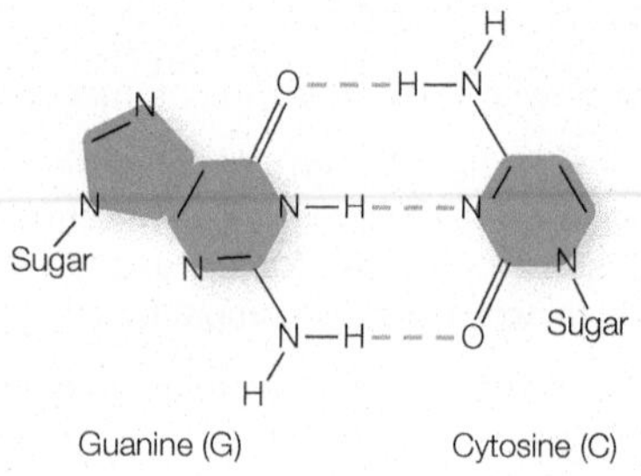

17. a. b. c. d. Label the following reactions as a neutralization reaction, a hydrolysis reaction, or a dehydration synthesis reaction.

a. 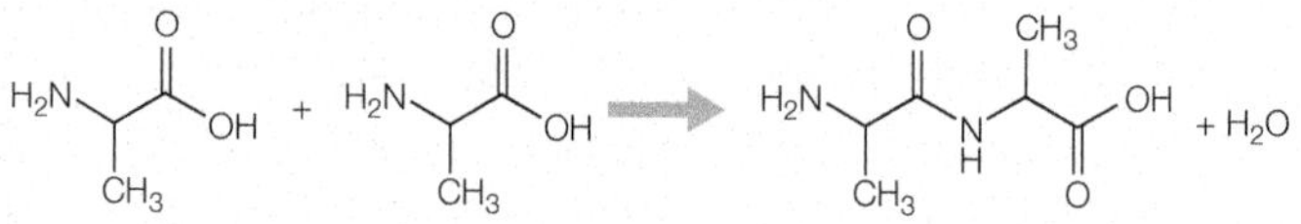

b. $HI + KOH \rightarrow H_2O + KI$

c. 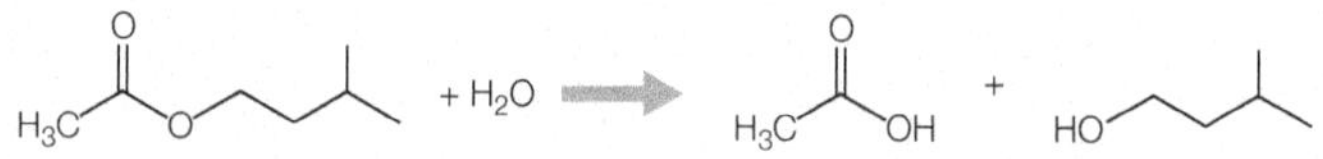

d. $H_2SO_4 + Mg(OH)_2 \rightarrow MgSO_4 + 2\,H_2O$

18. Which of the following is/are true regarding proteins? Select all that apply.

a. Proteins are made of amino acids.
b. Proteins can have higher-order structure.
c. Proteins are made by hydrolysis reactions.
d. Peptides are large proteins.
e. A protein's secondary structure is independent of the primary structure.
f. If the protein's primary structure is altered, it will not impact the protein's tertiary structure.

19. Label the following fatty acids as saturated, monounsaturated, or polyunsaturated.

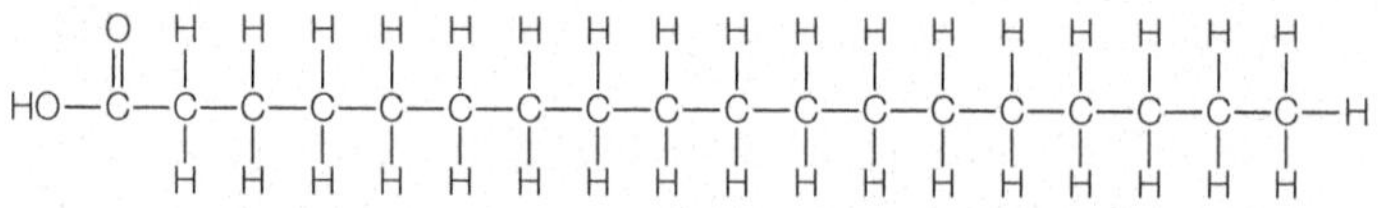

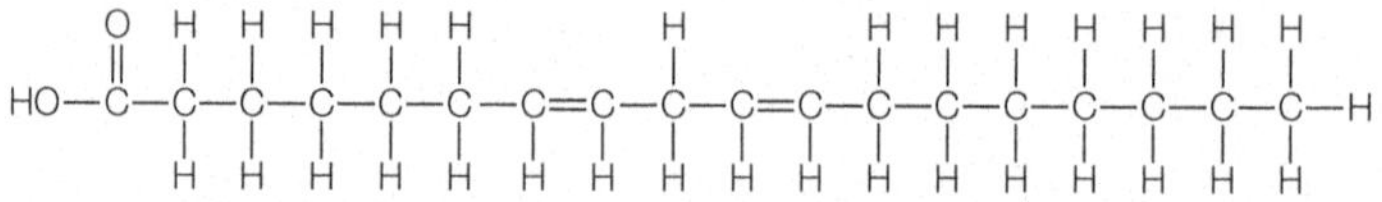

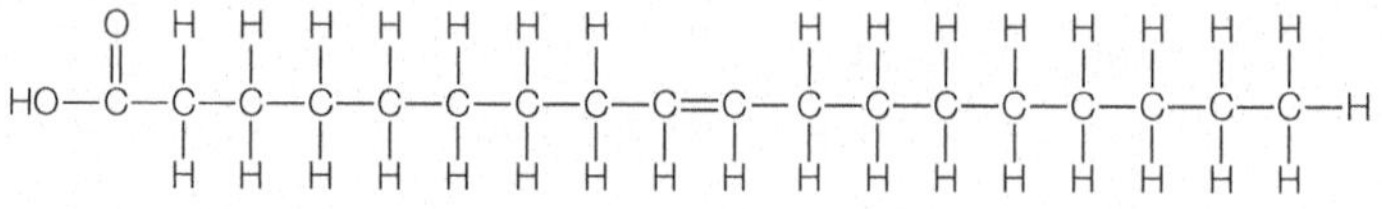

CRITICAL THINKING QUESTIONS

1. Cells make structures called vesicles to transport substances. They are basically double-layered micelles. How would vesicles be organized if they formed in a nonpolar solution as opposed to in water?

2. You may have noticed that before you get an injection, your skin is wiped with an alcohol solution. Alcohol solutions are frequently used as degerming agents because they interfere with hydrogen bonding in proteins. Describe what effect disruption of hydrogen bonding may have on a protein's structure and function.

3. Hydrophilic molecules do not have to contain polar covalent bonds. What other bonds could generate a hydrophilic substance?

4. The pH scale that is typically presented in textbooks ranges from 0 to 14, yet a pH higher than 14 or lower than 0 is possible. How is a pH lower than 0 mathematically possible if you consider that $pH = -\log_{10}[H^+]$? Hint: Consider that pure water has an agreed-upon pH of 7. At a pH of 7, the concentration of OH^- and H^+ are equal; each has a concentration of 10^{-7} mol/liter.

5. Microbes that are classified as psychrophiles live in extremely cold conditions. The bacterium *Psychromonas ingrahamii*, which lives in the ice of the Arctic Ocean at a frigid –20°C, holds the record in terms of bacteria growth at low temperatures. Microbes that live in such extreme conditions have a variety of adaptations; changes in membrane lipids and the cell's proteins are especially important adaptations. What adaptations would you suspect this bacterium has regarding its membrane fatty acids that help maintain membrane fluidity in such cold conditions? Provide a possible explanation as to why most enzymes from these bacteria do not function at room temperature.

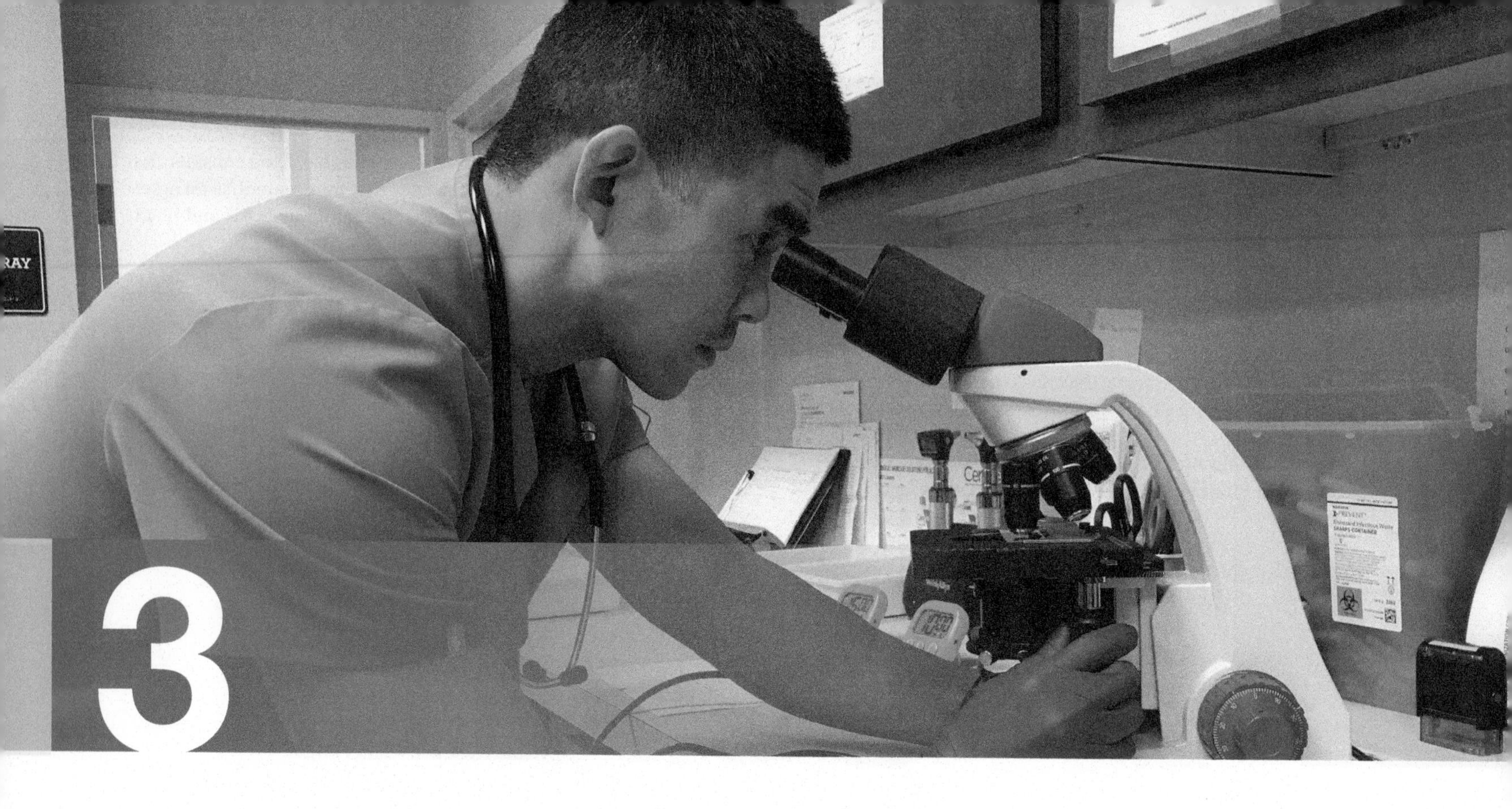

3 Introduction to Prokaryotic Cells

What Will We Explore? Life on Earth likely originated as prokaryotic cells about 3.8 billion years ago. Despite being the smallest lifeform, prokaryotes have been so successful that they constitute the most populated taxonomic domains—Domain Archaea and Domain Bacteria. Amazingly, some prokaryotes that make hearty cell structures called endospores can survive for millions of years in a dormant state. For example, researchers were able to grow bacteria found in the gut of a 30-million-year-old bee found fossilized in amber. Researchers also grew bacteria that had been trapped in salt crystals for 250 million years. In this chapter, we'll explore the extracellular and intracellular structures of prokaryotes, which vary tremendously from one genus to the next, allowing these organisms to thrive in pretty much every type of environment on Earth.

Why Is It Important? Researchers have only explored about 1 percent of the microbial life on Earth so far. As our understanding of prokaryotic cells grows, we will be better able to prevent and treat bacterial diseases. Furthermore, a firm understanding of how bacteria and our own cells differ allows for new drugs to be developed that target these differences.

Clinical CASE
NCLEX
HESI
TEAS

The Case of the Paralyzed Gardener

Visit the **Mastering Microbiology** Study Area to watch the case and find out how prokaryotic cells can explain this medical mystery.

Joseph Hwang
MLS, PA-C; Watkinsville, GA

3.1 PROKARYOTIC CELL BASICS

Bacteria and archaea are different types of prokaryotic cells.

As discussed in Chapter 1, cells are the smallest unit of life and can be described as being prokaryotic or eukaryotic. Prokaryotic cells are classified into two separate domains: Archaea (pronounced ahr-KEE-uh) and Bacteria. Eukaryotic cells, which evolved from prokaryotic cells, make up the third domain of life—Eukarya. FIG. 3.1 shows the relationship of these three domains of life. (We discuss eukaryotes in Chapter 4.) **Prokaryotes** are all unicellular and lack a membrane-bound nucleus. They also lack other membrane-bound organelles (intracellular structures with discrete functions) and have a much simpler genetic makeup than eukaryotic cells. FIG. 3.2 illustrates the key prokaryotic cell structures we will review in this chapter.

When discussing prokaryotes in this text, we primarily focus on the bacteria because they include many pathogens that healthcare providers must identify and treat in patients. So far, the archaea haven't been linked to human diseases, and so this book doesn't focus on them as much. However, archaea inhabit diverse environmental niches, from extreme environments like hot deep-sea vents to the surface of human skin. It could be that in the future, links between archaea and human health (either beneficial or harmful) may be discovered.

Learning Outcomes

After reading this section, you should be able to:

3.1 Name two prokaryotic domains, and state one way that these domains differ and one way they are alike.

3.2 Provide a basic description of a prokaryotic cell, and explain why prokaryotes tend to be small.

3.3 Describe the various shapes and arrangements that prokaryotes assume.

3.4 Define the term *pleomorphic*, and discuss how this property may impact an organism's ability to cause infection.

3.5 Define and summarize the process of binary fission.

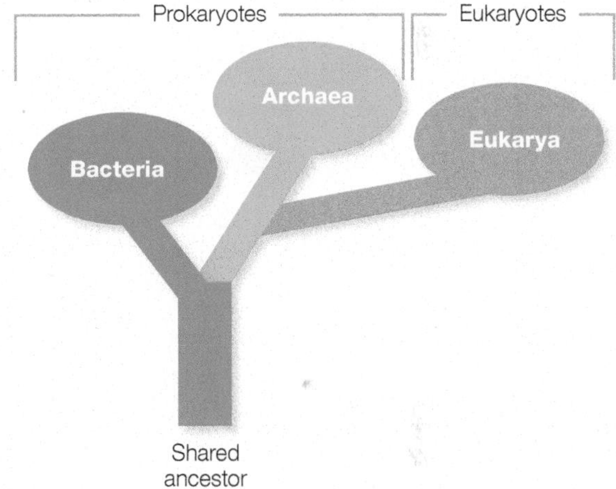

FIGURE 3.1 **Prokaryotes and the three domains of life** Members of the domains Bacteria and Archaea are all prokaryotes. Despite this shared grouping, genetic studies reveal that archaea are more closely related to eukaryotes than they are to bacteria. In other words, despite archaea and bacteria all being prokaryotes, they fall on different "boughs" of the evolutionary "tree."

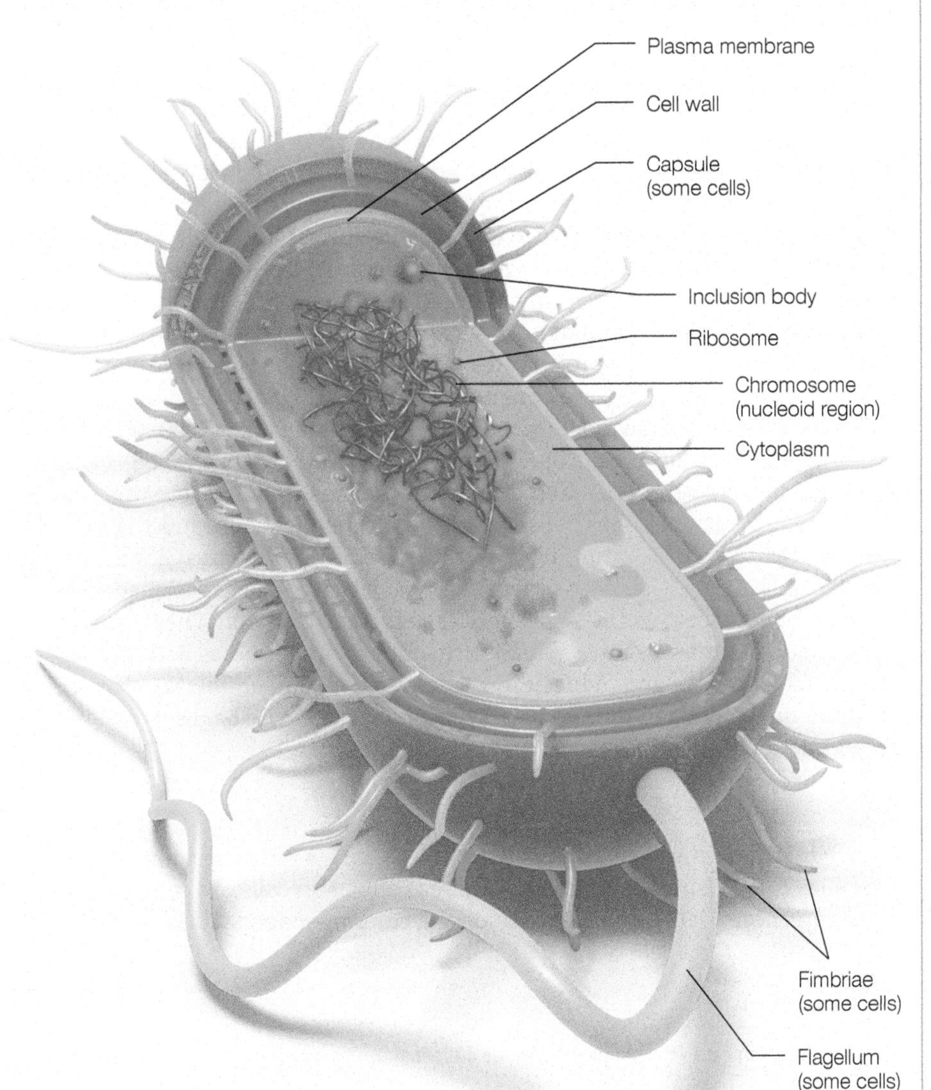

FIGURE 3.2 **Prokaryotic cell structure** For simplicity, this figure does not illustrate all possible cell structures.

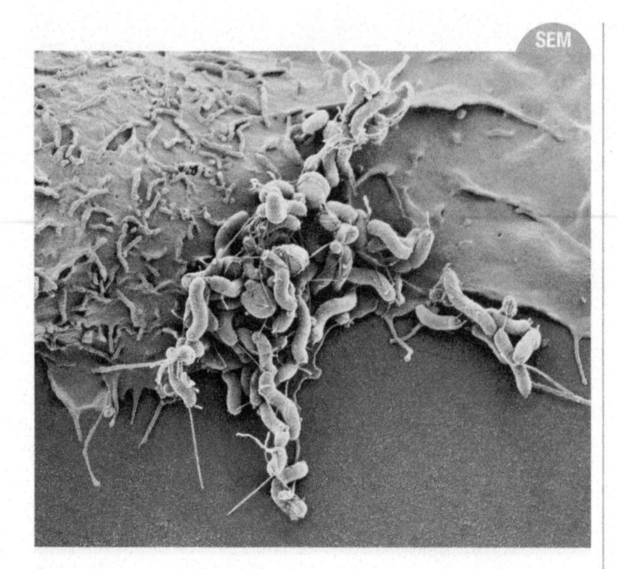

FIGURE 3.3 Pleomorphic *Helicobacter pylori* Scanning electron micrograph of *H. pylori*, the causative agent of stomach ulcers.

CHEM • NOTE

Diffusion is the movement of molecules from an area of higher concentration to an area of lower concentration without an energy investment.

Agar cubes that were soaked in dye and then cross sectioned.

Length of one edge of cube (inches)	2	4	6
Surface area (in^2)	24	96	216
Volume (in^3)	8	64	216
Surface area/volume	3.0	1.5	1.0

FIGURE 3.4 Diffusion and prokaryotes A high surface area-to-volume ratio favors efficient diffusion. As the agar cube (or a cell) gets larger, the surface area-to-volume ratio decreases. Soaking agar cubes of various sizes in dye demonstrates how surface area-to-volume ratio affects diffusion and helps explain why prokaryotes tend to be small. Recall that volume is calculated by multiplying length × width × height; as such, the units for this example will be cubic inches (in^3). In contrast, a cube's surface area (here presented as square inches, or in^2) is calculated with the formula $s^2 \times 6$, where "s" is the length of the cube's side.

Critical Thinking *What would the surface area-to-volume ratio be of a 1-inch cube. (Hint: If you are unsure of how to calculate this ratio directly, look at the pattern for surface area-to-volume of the 2-inch versus the 4-inch cube.)*

Prokaryotes have unique sizes, shapes, and arrangements.

Archaea and bacteria are genetically and biochemically distinct. However, they are practically indistinguishable from one another based on size, shape, and cellular arrangements. For a long time, scientists have wondered how prokaryotes evolved their diverse sizes and shapes and whether certain shapes hold an evolutionary advantage over others. Growing evidence indicates that life stage, environmental features, available nutrients, and even the presence of certain antimicrobial drugs affect prokaryotic morphological features such as shape and arrangement. This suggests that shape and arrangement may be directly linked to survival.

Clinically speaking, understanding prokaryotic shapes and arrangements is important because it is one of the criteria clinical microbiologists use to identify certain pathogens. From a patient care standpoint, it's also important to understand that a particular bacterium may change its shape and arrangement to enhance its ability to cause disease under different circumstances. The morphological changes bacteria undergo also impact their ability to form biofilms, which are an ongoing concern in medical settings (see Chapter 1 for more on biofilms). For example, the bacterium *Helicobacter pylori* (HEL-eh-ko-*bak*-ter pie-LOR-ee), which causes gastric and duodenal ulcers,[1] is **pleomorphic**—meaning it can take on different forms (FIG. 3.3). This enhances its survival and appears to be important for transmission to a new human host.

Sizes

Prokaryotes range in size from 0.2 micrometers (μm) in diameter, which is very small—close in size to some viruses—all the way up to 750 μm (or 0.75 mm), which is big enough to see with the naked eye.[2] However, most prokaryotes are between 0.5 and 2.0 μm, necessitating a light microscope to view them at the cellular level. The *Mycoplasma* (my-ko-PLAZ-muh) species are some of the smallest prokaryotes. The archaea *Thiomargarita namibiensis* (THIGH-o-*mar*-ga-*ree*-tah nah-mih-bee-EN-sis), which grows in ocean sediments, are among the largest prokaryotes characterized to date.

The discovery of large prokaryotes was surprising, because most of a prokaryotic cell's nutrients are obtained through diffusion. FIG. 3.4 represents diffusion in cells by showing how agar cubes of different sizes soak up dye (representing nutrient diffusion from the outer environment into the cell). Within 1 minute, the dye penetrates to the center of the smallest cube, but this is not the case for the larger cubes. This is because the smallest cube has the highest surface area-to-volume ratio. Likewise, nutrient diffusion is most efficient for smaller cells. As such, larger bacteria like *T. namibiensis* have special adaptations that allow them to thrive despite less efficient diffusion. These adaptations include numerous intracellular inclusions (storage bodies within a cell) that house nutrients and reduce the active intracellular volume to levels seen for more typically sized prokaryotes. Types of diffusion will be discussed in greater detail later in the chapter, under cell transport.

Shapes

Prokaryotes have diverse shapes (FIG. 3.5). Cells with a rod or cylindrical shape are called **bacilli** (singular: bacillus), whereas spherical cells are called **cocci** (singular: coccus). Other shapes include the comma-shaped **vibrio**, star-shaped

---

[1] Gastric ulcers appear in the stomach, whereas duodenal ulcers appear in the first segment of the small intestine.

[2] Schulz, H. N., & Jørgensen, B. B. (2001). Big bacteria. *Annual Review of Microbiology*, 55, 105–137.

stella, and ovoid **coccobacilli**, as well as spiral and club shapes. **Spirochetes** resemble spiral-shaped bacteria, but move in a corkscrew-rotary motion that is due to a specialized periplasmic flagellum, a tail-like extension we'll discuss later in this chapter. Spirochetes include the pathogens *Treponema pallidum*, the causative agent of syphilis, and *Borrelia burgdorferi*, the causative agent of Lyme disease (FIG. 3.6).

Cell shape has long been used to group microbes; it is so central to taxonomy that some genus names are based on the shape of member species. For example, the genus *Bacillus* includes a variety of rod-shaped bacteria, whereas the genus *Vibrio* includes *Vibrio cholerae*, a vibrio-shaped (comma-shaped) bacterium that causes cholera. Cell wall and cytoskeleton components, which are discussed later, are central to building and maintaining prokaryotic cell shapes. Some prokaryotes, such as the *Mycoplasma*, lack cell walls and take on diverse and flexible shapes.

Arrangements

Although all prokaryotes are unicellular, many species form arrangements as a result of their cell division patterns. For example, some cocci divide to produce a pairing of cells called a **diplococcal** arrangement (or diplococci). Longer chains of spherical cells called **streptococci** (*strep* = chain; *cocci* = sphere) may also form. You may be familiar with the infection "strep throat," which is caused by *Streptococcus pyogenes*—these bacteria have a beads-on-a-string appearance. Some bacilli can also divide in a way that makes **diplobacilli** or **streptobacilli** arrangements. When cocci divide to form grapelike clusters of cells, they are said to have a **staph** arrangement; members of the genus *Staphylococcus* exhibit such an arrangement. Bacilli can also form clusters of cells called a **palisade** arrangement. Some prokaryotes even take on a filamentous form that resembles long hair-like strands of cells.

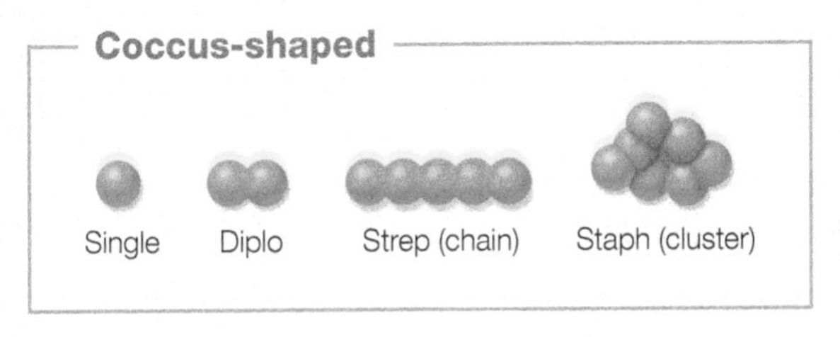

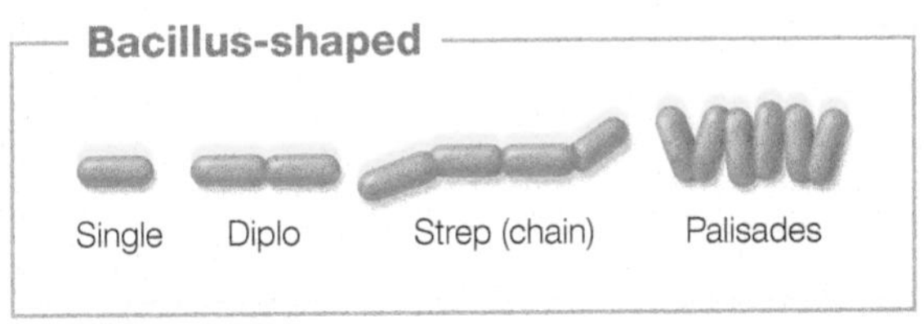

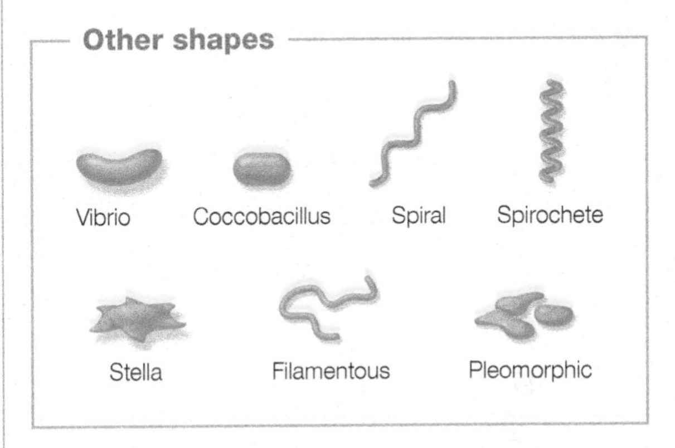

FIGURE 3.5 Prokaryotic shapes and arrangements Prokaryotes assume a variety of shapes and arrangements.

Critical Thinking *When viewing a sample of streptococci under a microscope, some of the cells appear to have a staph arrangement. Assuming you are not viewing a contaminant, how can this be explained?*

Prokaryotic cells primarily divide by binary fission.

Most prokaryotic cells reproduce asexually by **binary fission**. This process is simpler than mitosis, which is the form of asexual cell division that most eukaryotic cells use. Recall that sexual reproduction (meiosis) introduces genetic variation from parent to offspring; in asexual reproduction the offspring are genetically identical to the parent.

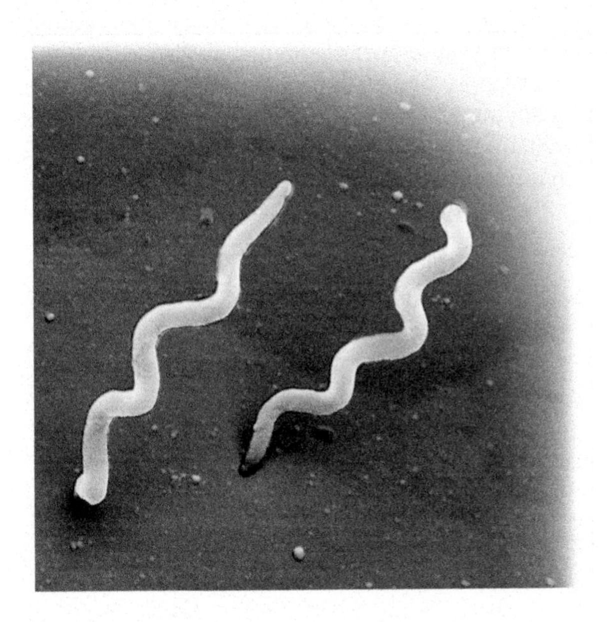

FIGURE 3.6 Spirochetes This colorized scanning electron micrograph of *Borrelia burgdorferi* shows the corkscrew shape that spirochetes assume as their periplasmic flagellum rotates.

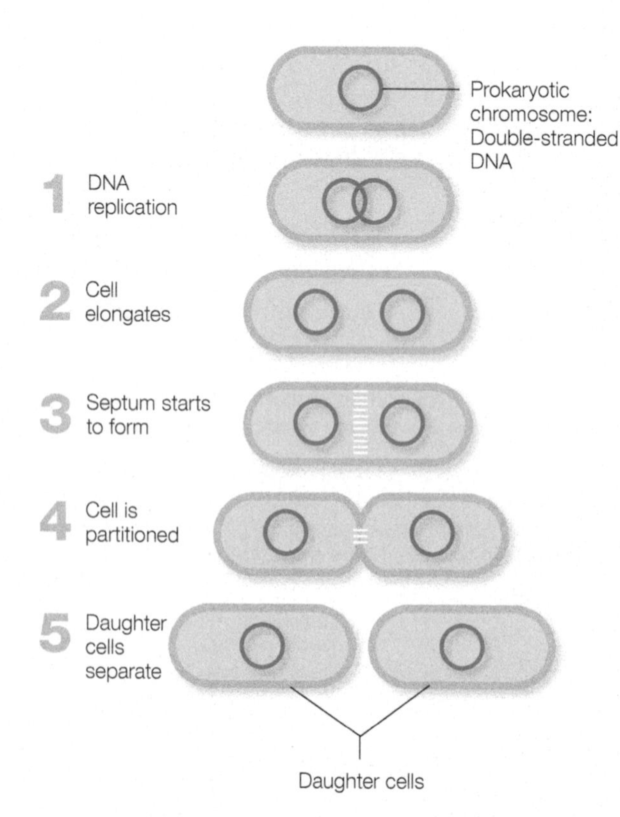

FIGURE 3.7 Binary fission Most prokaryotes reproduce by an asexual form of division called binary fission.

Critical Thinking *How would this process be slightly different for cells growing with a strep arrangement?*

Build Your Foundation (BYF) Quick Quiz: Visit the **Mastering Microbiology** Study Area to quiz yourself.

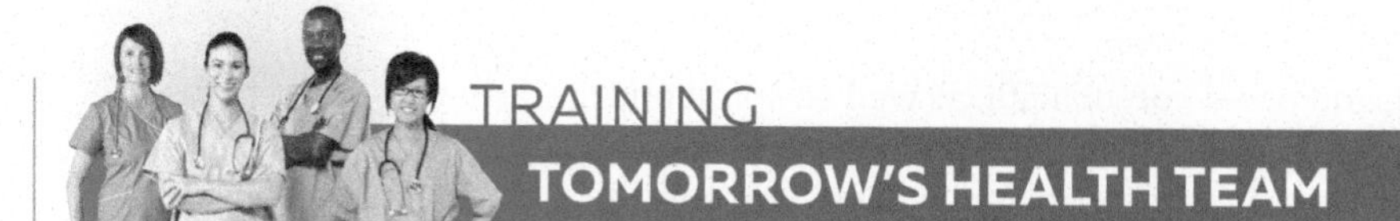

E. coli's Shape Shifting in UTIs

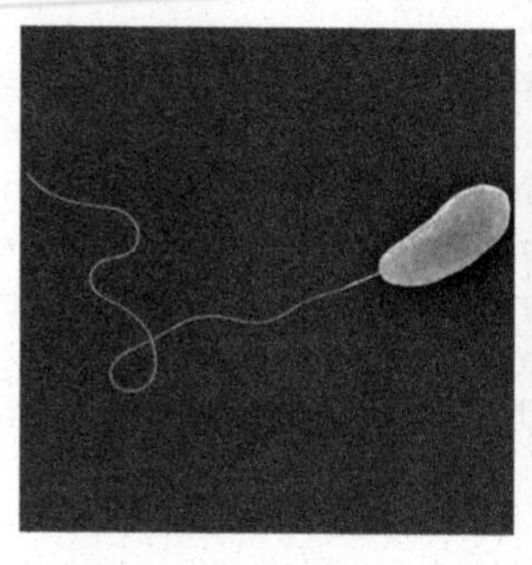

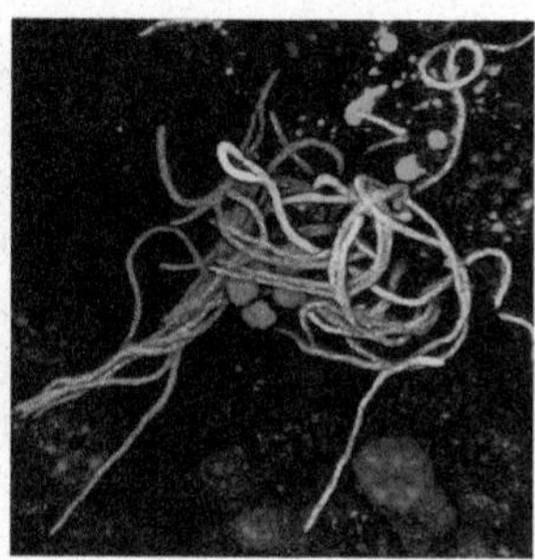

Usually *E. coli* is a single rod with a flagellum, as shown in the scanning electron micrograph on the left. The scanning confocal fluorescent micrograph (shown on the right) reveals that *E. coli* has a filamentous shape and loses its flagellum when grown on mouse bladder cells.

Free-living *Escherichia coli* grow as small flagellated bacilli (rods) with a single arrangement. However, studies suggest that the morphological features of this bacterium change in the course of urinary tract infections (UTIs), for which *E. coli* are frequently responsible. Using mouse models to study UTI progression, researchers learned that *E. coli* infecting the urinary system transition through four distinct morphological phases. Within the first few hours of invading host tissues, *E. coli* cells lose their flagella. This makes them unable to swim, so they exist as nonmotile rods within host cells. As infection progresses they pass through a stage where they appear as cocci (spherical), followed by a return to being motile rods that readily swim away from biofilms to spread the infection. In later stages of infection, *E. coli* assumes a filamentous form that is believed to help the bacteria avoid the host immune response, as well as provide a ready source of bacteria to perpetuate infection.

QUESTION 3.1

If you were to look at E. coli that was freshly grown in the lab, what sort of morphology would you expect?

Binary fission occurs in five main steps (FIG. 3.7). First, the cell that is preparing to divide copies its DNA-containing chromosome, so that each daughter cell will receive a complete chromosome. (Chapter 5 explores DNA replication.) As the cell grows, the copied chromosomes separate and are drawn to opposite ends of the cell. Eventually the cell reaches a critical size and a *septum* (partition) begins to form near the midpoint. The septum, which is essentially the newly forming cell wall, eventually walls off the resulting two daughter cells from each other. The daughter cells may completely separate or they can remain associated to form a variety of cellular arrangements, as discussed earlier.

The frequency of binary fission differs from species to species and is greatly affected by environmental conditions such as nutrient availability. (In Chapter 7, we will discuss the many factors that influence prokaryotic population growth and how to control their growth.)

Build Your Foundation

1. Name the two prokaryotic domains. List one way these domains differ and one way they are similar. (LO 3.1)
2. List three key features of prokaryotic cells. (LO 3.2)
3. Explain why the surface area-to-volume ratio impacts prokaryotic cell size. (LO 3.2)
4. Describe three prokaryotic cell shapes and three arrangements. (LO 3.3)
5. What is pleomorphism, and how can it improve a bacterium's ability to cause disease? (LO 3.4)
6. Describe what binary fission is, and explain the steps of this process. (LO 3.5)

3.2 EXTRACELLULAR STRUCTURES

Prokaryotic cells rely on their plasma membrane and cell wall as barriers.

Learning Outcomes

After reading this section, you should be able to:

3.6 Discuss structural and functional features of prokaryotic plasma membranes.

3.7 Compare bacteria and archaea plasma membranes and cell walls.

3.8 Compare and contrast Gram-positive and Gram-negative cell walls.

3.9 Name two acid-fast genera, state what makes them so, and explain why the acid-fast stain is clinically useful.

3.10 Compare L-forms to *Mycoplasma* bacteria.

3.11 Identify the types of passive and active transport mechanisms that exist in cells.

3.12 Define osmosis, and predict the outcome of placing a bacterial cell in hypertonic, hypotonic, or isotonic solutions.

3.13 Discuss the roles of flagella, fimbriae, pili, and glycocalyx structures.

3.14 Identify various flagella arrangements, and discuss how periplasmic flagella differ from regular flagella.

All cells, whether they are prokaryotic or eukaryotic, are surrounded by an outer boundary called a **plasma membrane** (also called the cytoplasmic membrane or cell membrane). All structures outside of the plasma membrane are considered *extracellular structures*, while structures that lie within the boundary defined by the plasma membrane are *intracellular structures*. Most prokaryotic cells also have a **cell wall** that lies just outside the plasma membrane. While most cells are able to repair minor damage to their plasma membrane and cell wall, drastic damage can be fatal. A number of antimicrobial drugs, detergents, and disinfectants target these cellular structures to limit bacterial growth.

Plasma Membrane Structure and Function

The plasma membrane is a thin, flexible, phospholipid bilayer that serves as a selective barrier separating the cell's cytoplasm from the external environment (FIG. 3.8). It is the cell's main platform for interacting with the environment. The plasma membrane is also a site for metabolic reactions that prokaryotes rely on to make ATP.

The plasma membrane is built like a sandwich. It includes a layer of "bread" on both the top and bottom that consists of water-loving hydrophilic phosphates that directly touch the water-based cytoplasm within the cell and the watery environment beyond it. The "filling" part of the membrane sandwich lies between the bread layers. It contains hydrophobic fatty acids that dislike interacting with water and are therefore chemically content to be sheltered within this water-free environment.

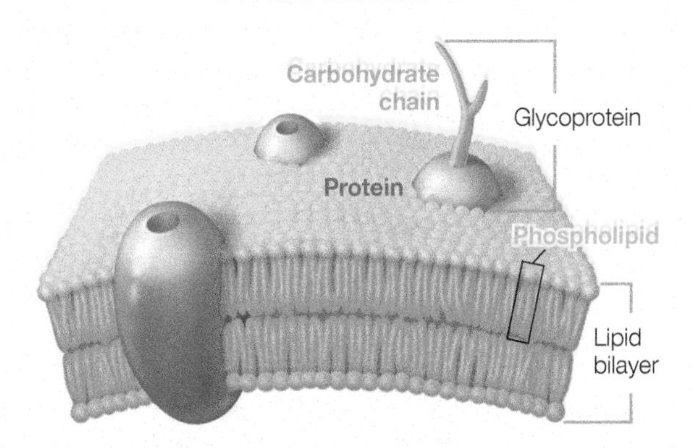

FIGURE 3.8 **Plasma membrane**

The types of lipids present in the plasma membrane differ considerably by species, and many prokaryotes can alter their membrane lipid profiles in response to certain stresses or environmental changes. Additionally, in many cases proteins constitute at least half of the plasma membrane mass. Membrane proteins serve as transporters, anchors, receptors, and enzymes. The types of proteins found in plasma membranes are tremendously diverse and confer specific capabilities to a cell. In many cases these membrane proteins are linked to sugar groups (*glycoproteins*) or lipid groups (*lipoproteins*). These added groups contribute important structural and functional properties to the membrane.

While the cell must be separated from its general surroundings, it must also obtain nutrients and gases and release waste products into the environment. Because the plasma membrane is selectively permeable, gases, water, and other small, noncharged substances can diffuse into and out of the cell without assistance. However, ions and larger polar substances like sugars often require specific protein transporters found in the membrane to enter or exit a cell. Many of these membrane proteins also detect changes in the environment and help coordinate cellular responses to these changes.

Membrane Fluidity The lipid bilayer is not a static structure, but is a *fluid-mosaic model*, where the membrane lipids and proteins move around within the bilayer. This fluidity enables proteins to relocate to areas of the membrane where they are most needed. Maintaining a certain level of membrane fluidity is essential to the cell and can impact its physiological functions. As such, it is not surprising that cells control membrane fluidity in response to environmental stimuli and cellular needs.

Warmer temperatures increase membrane fluidity, while colder temperatures reduce it. A higher proportion of unsaturated fats in the fatty acid portion of phospholipids improves membrane fluidity in cold temperatures by preventing

CHEM • NOTE

Polar molecules tend to interact with water, while nonpolar molecules avoid water. Phospholipids, the main components of plasma membranes, are built from glycerol, phosphate, and two fatty acids. The phosphate-containing region of a phospholipid is polar, while fatty acids are nonpolar.

CHEM • NOTE

Saturated fatty acids lack double bonds in their hydrocarbon backbone. Unsaturated fatty acids have double bonds in their hydrocarbon backbone, which limits how tightly they pack together, thereby keeping them fluid at colder temperatures. See Chapter 2 for figures and more review of these biomolecules.

CHEM • NOTE

Ester bonds (O-C=O) feature a carbon atom covalently bound to two oxygen atoms, with one of the oxygen atoms double bonded to the carbon. **Ether bonds** (O-C-O) also feature a carbon bound to two oxygen atoms, but a double bond is not present.

CHEM • NOTE

Isoprenoids are organic molecules made up of isoprene units—saturated hydrocarbons with the chemical formula C_5H_8.

tight packing of the fatty acids in the bilayer. In contrast, saturated fatty acids are more rigid at low temperatures because they tightly pack within the bilayer. Microbes that thrive in extremely cold climates tend to incorporate a higher proportion of short unsaturated fatty acids into their phospholipids to preserve membrane fluidity. Similarly, organisms in high-temperature environments lean toward longer and more saturated fatty acids in their phospholipids to keep membranes from becoming too fluid. Integrating steroid-based lipids like cholesterol in membranes can also improve fluidity; however, aside from certain *Mycoplasma* bacteria, cholesterol is rarely seen in prokaryotic cells. In contrast, it is standard in eukaryotic plasma membranes.

Archaea Plasma Membranes The chemical linkages in bacterial plasma membranes differ from those in archaea. In bacteria we find ester bonds joining linear fatty acid tails to glycerol, whereas archaea tend to use ester bonds to link glycerol to long-branched lipids called isoprenoids. Certain archaea that live in extreme heat build lipid monolayers as opposed to lipid bilayers. These monolayer membranes contain unique lipids (called tetraether lipids) that are basically long lipid molecules capped on each end with a polar head group (FIG. 3.9). These unique membrane adaptations help archaea thrive in harsh environments.

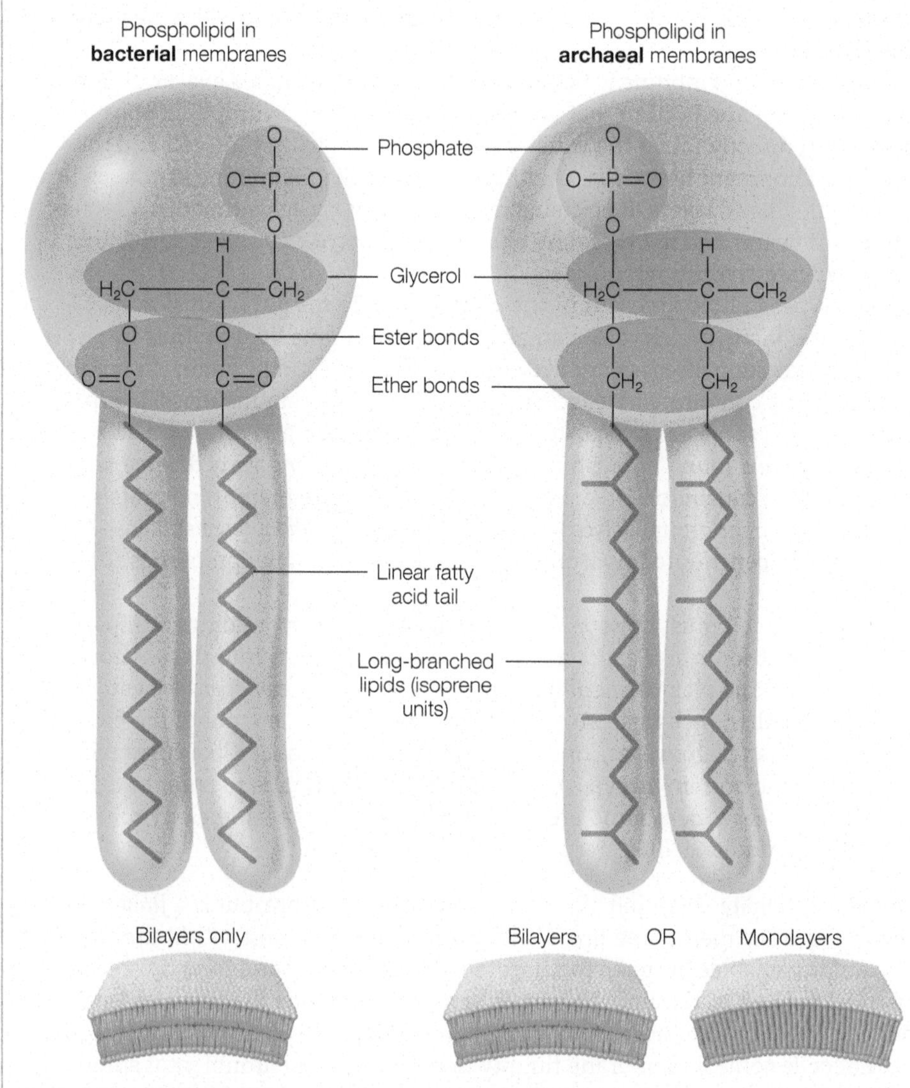

FIGURE 3.9 Bacterial and archaeal plasma membranes In bacterial membranes, ester bonds join fatty acids and glycerol; in contrast, archaea use ether bonds for this purpose. Archaeal membranes can be monolayer or bilayer formats— bacterial membranes are strictly bilayers.

Cell Wall Structure and Function

The cell wall confers a rigid structure to most prokaryotes and also serves as an extra layer of protection from the environment. Prokaryotic cells live in diverse environments, so it makes sense that they have developed diverse cell walls to survive specialized conditions. In general, archaea exhibit more diversity in their cell walls than do bacteria.

Bacteria use **peptidoglycan** (or murein) as a core component of their cell walls. Peptidoglycan contains protein ("peptid") and sugar ("glycan") components. The glycan chains are built of alternating N-acetyl-D-glucosamine (NAG or GlcNAc) and N-acetylmuramic acid (NAM or MurNAc) residues—molecules that somewhat resemble glucose. These long strands of alternating sugar residues are then cross-linked by short peptides to create a mesh-like structure much like a chain-link fence (FIG. 3.10). As we'll review in Chapter 15, drugs from the penicillin family work by interfering with peptidoglycan construction. Archaea lack peptidoglycan; instead, their cell walls tend to consist of one or more layers of *pseudopeptidoglycan* (or pseudomurein).

CHEM • NOTE

Glycans are chain-like structures made of covalently linked monosaccharide subunits (sugars).

Knowing a bacterium's Gram and acid-fast properties is clinically useful.

The Gram stain technique allows us to classify cells based on cell wall structure. (See Chapter 1 for how to perform the Gram stain and interpret results.) Gram-negative bacteria have a cell wall with a thin peptidoglycan layer surrounded by an outer membrane. The outer membrane structurally resembles the plasma membrane in that it is a lipid bilayer rich in phospholipids. Unlike the plasma membrane, however, the outer membrane is enriched with a glycolipid called **lipopolysaccharide (LPS)**. LPS contains a lipid portion (called lipid A or endotoxin) that is toxic to animals. (For more on endotoxins, see Chapter 10.) The space

Peptidoglycan (in bacterial cell walls)

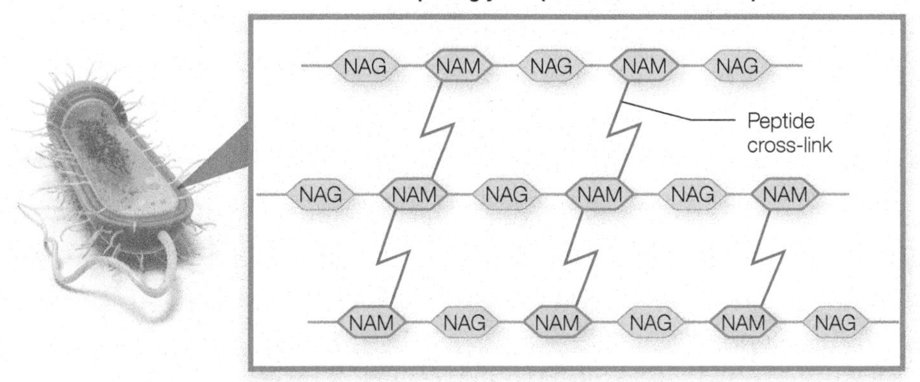

Glycan chains made of alternating N-acetyl-D-glucosamine (NAG) and N-acetylmuramic acid (NAM) residues

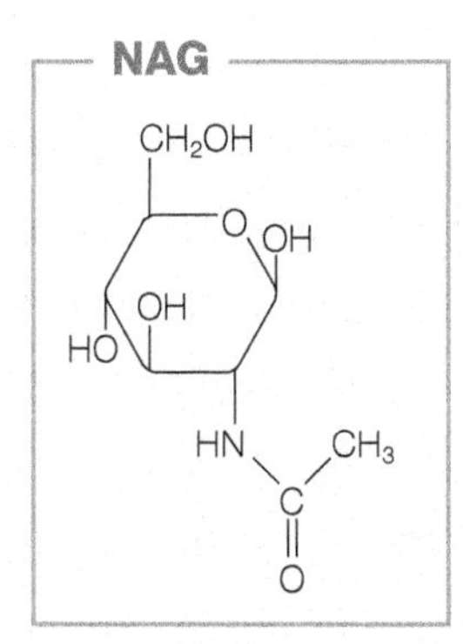

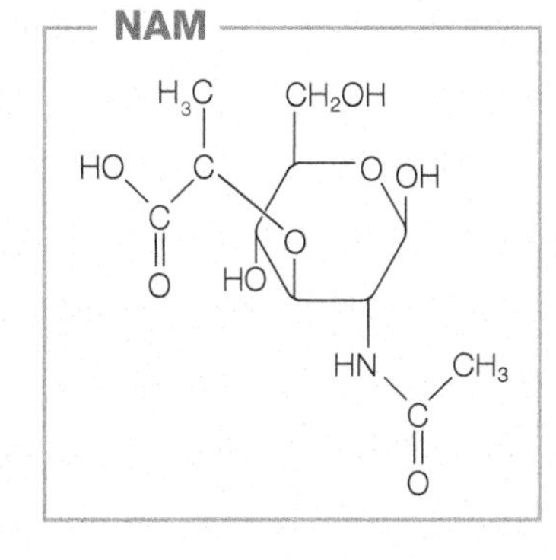

Glucose, included to show structural similarity

FIGURE 3.10 Peptidoglycan Most bacterial cell walls contain peptidoglycan, which is made of protein and sugar components.

CHEM • NOTE

Glycolipids are lipids that have added sugar groups. (*Glyco* means sugar.)

between the outer membrane and the plasma membrane is called the **periplasmic space** (FIG. 3.11). This space is not only the region where we find the peptidoglycan layer of the Gram-negative cell wall, but it is also filled with a gel-like fluid that is enriched with various factors that have important roles in helping the bacterium obtain nutrients and neutralize potentially toxic substances. Gram-negative cell walls stain red/pink in the Gram staining procedure.

FIG. 3.12 introduces you to examples of medically important Gram-negative and Gram-positive bacteria. Notice that the figure indicates not only Gram property, but also acid-fast property, endospore formation, and cell morphology (cell shapes and arrangements). Some of the pathogens shown may cause diverse infections; in such instances, only the main infections are identified. Note that other criteria, such as metabolic properties, are also key in classifying bacteria; you'll learn more about that aspect of classification in later chapters.

Gram Stain — CONCEPT COACH — NCLEX HESI TEAS

Bring the art to life! Visit the **Mastering Microbiology** Study Area to watch the Concept Coach and master how the Gram stain works and why it matters.

A number of the Gram-negative bacteria shown in Figure 3.12 belong to the **family *Enterobacteriaceae*** (EN-teh-row-back-teer-ee-AY-see-ee). This family includes over 100 species, many of which inhabit the intestinal tract of mammals and therefore are commonly referred to as **enteric bacteria**. These bacteria can include harmless bacteria as well as pathogens such as *Salmonella* species, *Escherichia coli*, *Proteus* species, and *Shigella* species. We'll explore the enteric bacteria more in later chapters.

In contrast to Gram-negative bacteria, the Gram-positive bacteria have a cell wall with a thick peptidoglycan layer, and they lack an outer membrane. Because there is not an outer membrane, the periplasmic space in Gram-positive bacteria is simply the small region between the plasma membrane and the thick peptidoglycan layer.[3] The Gram-positive cell wall stains purple in the

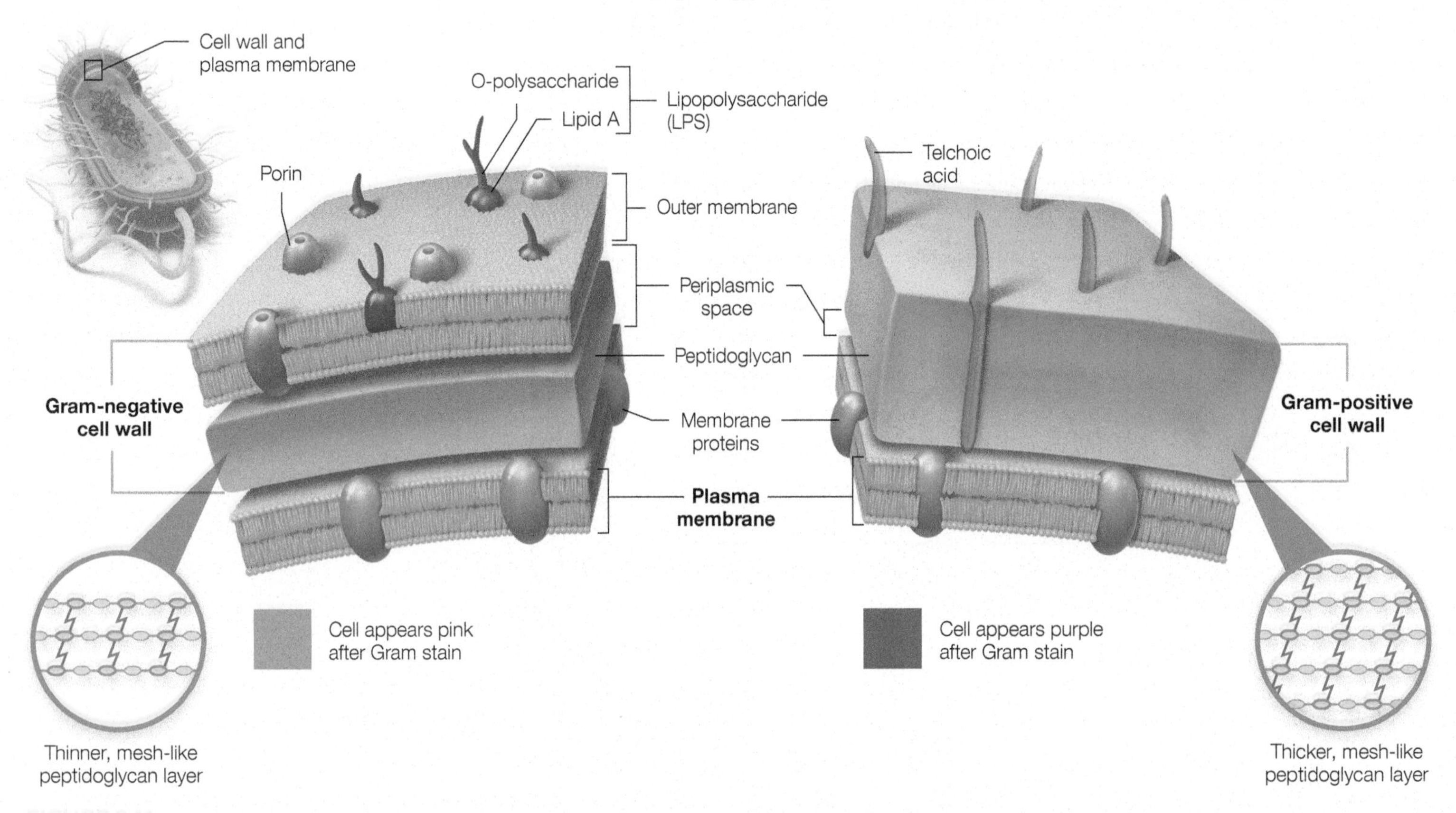

FIGURE 3.11 **Gram-negative and Gram-positive bacterial cell walls** Gram-negative bacteria have an outer membrane along with a thin peptidoglycan layer in their cell walls. Gram-positive bacteria lack outer membranes and have a thick peptidoglycan layer in their cell walls.

[3] Zuber, B., Haenni, M., Ribeiro, T., et al. (2006). Granular layer in the periplasmic space of Gram-positive bacteria and fine structures of *Enterococcus gallinarum* and *Streptococcus gordonii* septa revealed by cryo-electron microscopy of vitreous sections. *Journal of Bacteriology*, 188(18), 6652-6660.

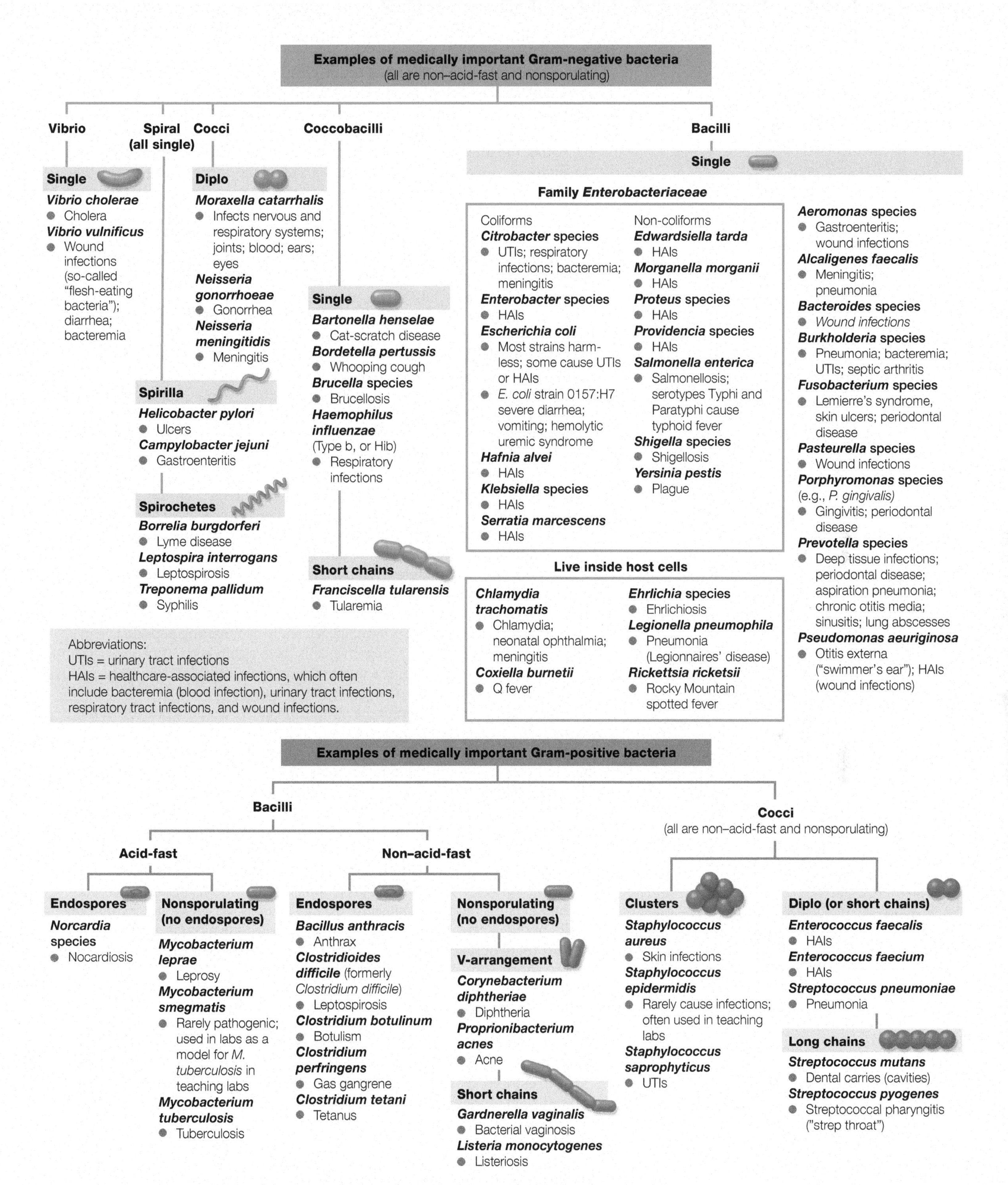

FIGURE 3.12 **Examples of medically important Gram-negative and Gram-positive bacteria**

Gram stain procedure. Note that Gram-positive cells with damaged cell walls may appear to be Gram-negative. In a given sample of cells, it's very likely that some may have cell wall damage. Thus, it's typical that a culture of Gram-positive cells will contain some cells that appear Gram-negative. As you can see in Figure 3.12, the Gram-positive bacteria include a wide variety of pathogens, including acid-fast bacteria and bacteria that make endospores—properties we'll discuss later in this chapter.

Clinical Implications of Gram Status

You might assume that the thick peptidoglycan layer in Gram-positive cell walls protects better than the thin peptidoglycan layer of Gram-negative cell walls. However, when going into battle, it's probably better to wear a thin bulletproof vest rather than a thick pillow. In other words, it's the quality, not the quantity, of the cell wall components that determines a cell's hardiness. When comparing two actively growing bacteria types, a Gram-negative species with a thin peptidoglycan layer (2–8 nm) is typically harder to kill with chemical agents than a Gram-positive species with a thick peptidoglycan layer (20–80 nm).

Hardiness of Gram-negative bacteria is in part due to the outer membrane, a somewhat selective barrier that guards against damage by certain agents, including lysozyme (an antibacterial enzyme abundant in tears, mucus, and saliva), a variety of drugs, and some detergents and disinfectants. That said, the Gram-negative cell's outer membrane is *not* as selective a barrier as the plasma membrane. Many substances cross the outer membrane that cannot cross the plasma membrane. This is thanks to **porins**—protein channels that form a pore—in the outer membrane. Most porins are nonspecific channels that allow substances such as amino acids, vitamins, and other nutrients to pass through the outer membrane, while excluding large molecules and a variety of substances that may be harmful to the cell (like certain drugs). As mentioned previously, penicillin-based drugs damage cell walls by targeting peptidoglycan construction. Because Gram-negative cells have a thin layer of peptidoglycan and the added protection of the outer membrane, they tend to be less sensitive to these compounds than are Gram-positive bacteria.

Although the outer membrane of Gram-negative bacteria provides some advantages, in certain situations, a Gram-positive cell wall is better suited to what the cell encounters. For example, the thick peptidoglycan layer of Gram-positive cell walls helps them retain moisture longer, making them better at surviving in dry environments. The thick peptidoglycan layer also protects Gram-positive cells from mechanical stresses like abrasion and crushing. Gram-positive bacteria also have teichoic acids, which stabilize the cell wall, help maintain shape, transport cations into the cell, and aid regulation of cell division. Teichoic acids also help certain Gram-positive bacteria cause disease by promoting adhesion to host tissues and protecting them from various antimicrobial compounds. *Staphylococcus aureus* (STAF-uh-low-*cock*-us OR-ee-us), a common resident of our skin, uses teichoic acids to colonize the nasal passages of about 30 percent of humans, increasing their risk of developing invasive infections.[4]

CHEM • NOTE

Cations are positively charged ions. Examples include magnesium (Mg^{2+}) and sodium (Na^{+}) ions.

In short, Gram-negative and Gram-positive cell walls each have unique features to help a cell survive in its preferred environment. TABLE 3.1 compares Gram-negative and Gram-positive bacteria.

Archaea and Gram Staining Archaea can be Gram stained, but it isn't a useful classification tool for these prokaryotes. Archaea assume a wide variety of cell wall structures, even within a given species, and may alter their cell wall based on transient environmental conditions. This is different from bacteria, which

[4] van Dalen, R., Peschel, A., & van Sorge, N. M. (2020). Wall teichoic acid in *Staphylococcus aureus* host interaction. *Trends in Microbiology*, *28*(12), 985–998.

TABLE 3.1 Comparing Gram-Negative and Gram-Positive Bacteria

Cell Feature	Gram-Negative	Gram-Positive
Outer membrane	Yes	No
Lipid A (endotoxin)	Yes	No
Porins	Yes	No
Teichoic acids	No	Yes
Peptidoglycan	Thin layer (10–20% of cell wall)	Thick layer (70–80% of cell wall)
Gram staining color	Pink	Purple
Resistant to physical disruption and drying?	No	Yes
Susceptibility to anionic (negatively charged) detergents	Low	High
Penicillin susceptibility	Low	High

typically assume one of two possible cell wall structures and reliably maintain the same structure across a given species.

Clinical Implications of Acid-Fast Staining

The acid-fast staining technique detects a waxy lipid called **mycolic acid** in cell walls. (See Chapter 1 for how to perform the acid-fast stain and interpret results.) This waxy layer makes these bacteria colorfast even when rinsed with acid. Acid-fast cells appear red/pink following the acid-fast staining procedure.

Genera *Nocardia* and *Mycobacterium* are the best-known examples of acid-fast bacteria. Most other bacteria are non–acid-fast. Clinically speaking, the acid-fast stain is especially important for identifying the causative agents of leprosy (*Mycobacterium leprae*) and tuberculosis (*Mycobacterium tuberculosis*). *Nocardia* species typically live in soil and may cause skin infections, or in the case of immune-compromised patients, lung infections.

Acid-fast bacteria stain weakly with the Gram stain, but are nonetheless considered Gram-positive. A weak Gram stain occurs because it is difficult for the crystal violet reagent to cross the waxy cell wall, but the crystal violet that does enter gets retained. Just as it is hard for crystal violet to enter these cells, nutrients and gases also face challenges in entering acid-fast cells—as a result, acid-fast bacteria grow slowly. Drugs also face challenges in crossing the waxy cell wall, thereby necessitating the use of drawn-out, multidrug therapies in patients fighting an infection caused by an acid-fast pathogen (FIG. 3.13).

Mycoplasma Bacteria and L-Forms

Unlike most bacteria, the very tiny members of the genus *Mycoplasma* entirely lack a cell wall. Instead, *Mycoplasma* species have a sterol-enriched plasma membrane. While these bacteria can be grown on specialized culture media outside a host cell, they normally live inside other cells and benefit from the protection conferred by their host cell. *Mycoplasma* species live in diverse hosts, from humans to plants. Because they are not shape-restricted by a rigid cell wall, they take on varied shapes (are *pleomorphic*).

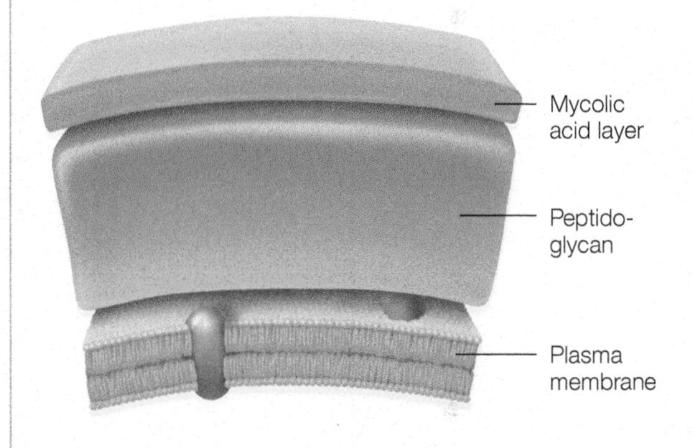

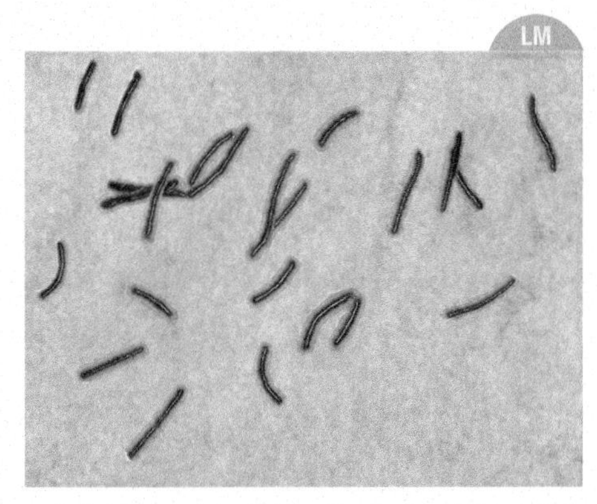

Acid-fast stained *Mycobacterium tuberculosis*

FIGURE 3.13 Acid-fast cell wall Acid-fast cell walls contain an external layer of waxy mycolic acids.

Critical Thinking *Explain why acid-fast cells do not stain well with the conventional Gram stain technique, yet are appropriately considered to be Gram-positive bacteria.*

Whereas *Mycoplasma* are wall-less throughout their life cycle, some bacteria that normally have a cell wall can lose it during their life cycle, or as a result of a mutation. Upon losing their rigid cell wall, these bacteria assume various morphologies. The term **L-form** is usually reserved for bacteria that had a cell wall and then lost it. These bacteria were first described by Emmy Klieneberger-Nobel, a Jewish scientist who was forced to flee to London to escape the Nazis in the 1930s. In London, she established the field of *Mycoplasma* research at the Lister Institute (the "L" of L-forms stands for "Lister"). Because *Mycoplasma* bacteria do not have a cell wall to lose, they are not technically classified as L-forms. It is suspected that L-forms may contribute to persistent infections because many antibiotics target cell wall structures. Research has also shown that L-forms are resistant to certain environmental stresses, including boiling and autoclaving, which are common processes for sterilizing medical devices.[5]

Prokaryotic cells transport substances across their cell wall and plasma membrane.

Although separated from their environment by their cell wall and plasma membrane structures, prokaryotic cells must still interact with their environment to obtain nutrients, maintain water balance, exchange gases, and dispose of waste products. Cells employ diverse mechanisms to accomplish these goals. Some substances are transported without an energy investment by the cell (**passive transport**), while other types of transport require energy to operate (**active transport**).

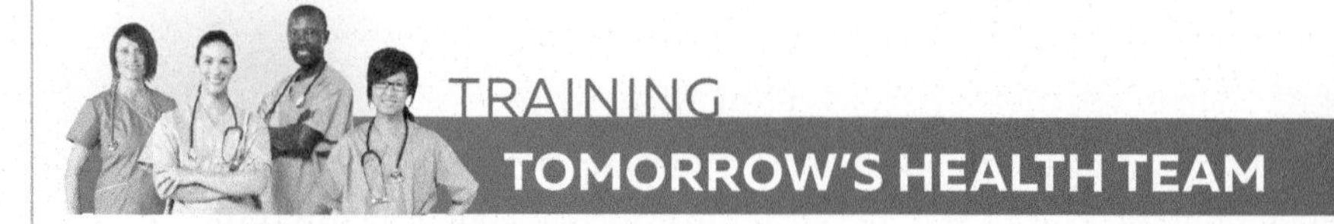

Walking Pneumonia

Mycoplasma pneumoniae's lack of a cell wall and its exceptionally small size had scientists initially classifying it as a virus. This bacterium causes a lung infection commonly called "walking pneumonia" because although patients may feel unwell, they are unlikely to be so ill as to be confined to bed. *M. pneumoniae* can infect anyone, but it is more common in older children and adults under age 40. The typically self-limiting disease often resolves without antibiotic therapy, although recovery can take about a month if antibiotics are not taken.

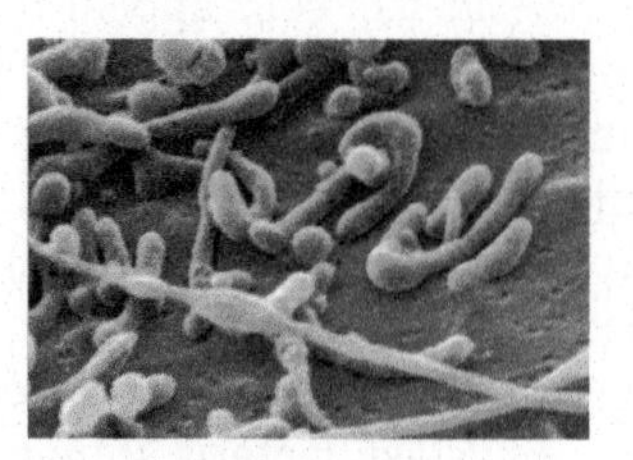

Scanning electron micrograph of *Mycoplasma pneumoniae.*

Despite the mildness of walking pneumonia for most, it is not to be lightly dismissed by healthcare workers assessing patients. There are about 2 million cases of this form of pneumonia per year in the United States, and in people with other health problems or compromised immune systems, the infection can be much more serious. It accounts for an estimated 20 percent of pneumonia-related hospitalizations annually.

QUESTION 3.2

Why are penicillin-based drugs ineffective at treating M. pneumoniae *infections? (Hint: Consider what cell structure penicillin-based drugs target.)*

---

[5] Ma, A., Glassman, H., & Chui, L. (2020). Characterization of *Escherichia coli* possessing the locus of heat resistance isolated from human cases of acute gastroenteritis. *Food Microbiology*, *88*, 103400.

Diffusion

The simplest form of exchange between cells and their environment is accomplished by **diffusion**, which is the passive movement of substances from areas of high concentration to areas of low concentration (that is, down or along a concentration gradient) until they are uniformly distributed. Diffusion occurs in solids, liquids, and gases (like air). If you've ever baked, then you've experienced diffusion—through random molecular motion, aromas diffuse through the air to fill the house with the scent of apple pie. Diffusion is also important in living systems. There are two general forms: simple and facilitated (FIG. 3.14). Because both forms are driven by concentration differences and both mechanisms can only move substances from higher to lower concentration, once the concentration is equalized (reaches equilibrium), there is no more *net* diffusion. That is, while molecules continue to move, they do so in a manner that doesn't ultimately shift their overall concentration. This is one of the limitations of diffusion: It doesn't allow a cell to concentrate a particular substance within it. In order to concentrate a substance on one side of a membrane, energy must be invested to drive active transport (discussed later).

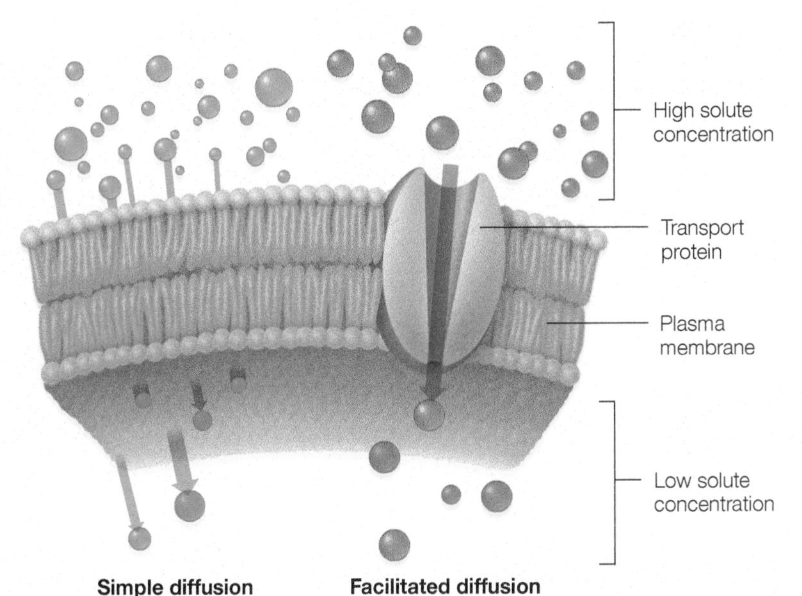

FIGURE 3.14 Simple and facilitated diffusion Simple diffusion occurs without the help of transport proteins. Facilitated diffusion requires them. Both are passive transport mechanisms that don't require an energy investment, and both move substances from high to low concentration across the plasma membrane.

Critical Thinking *Why can small lipid-soluble substances often enter cells by simple diffusion, while small charged substances require a transporter? (Hint: Consider the chemical composition of the plasma membrane.)*

Simple Diffusion This process does not require an energy investment, nor does it require assistance from transporters or carriers. Small noncharged molecules, gases, and lipid-soluble substances tend to enter and exit cells by simple diffusion.

Facilitated Diffusion Like simple diffusion, facilitated diffusion does not require energy. It moves substances along their concentration gradient with the help of one or more transport proteins in the plasma membrane (Figure 3.14). The transport proteins that mediate facilitated diffusion are specific for the substances they transport. Examples include channels that span the membrane and protein transporters that can flip-flop within the plasma membrane to translocate a given substance into or out of the cell.

CHEM • NOTE

Concentration gradients form when there is an unequal distribution of a dissolved substance in an environment. Electrochemical gradients are generated when ions (charged atoms) are unequally distributed. These specialized concentration gradients drive many biological processes, including the transmission of signals in our nervous system.

Osmosis

Water is one of the few substances that readily crosses the cell wall and plasma membrane. It does this when it is attracted to a solute—a dissolved substance such as sugar or salt. You may have heard the term "water follows salt"; this phrase emphasizes the point that water is drawn to, or moves toward, areas of higher solute concentration. The practice of using salt to kill garden slugs is a prime, although somewhat gruesome, example. The salt draws water out of the slug, eventually killing it by dehydration.

Water continues to move from areas of low solute concentration to areas of high solute concentration until the solute concentrations on either side of a selectively permeable membrane (a membrane permeable to water but not to solutes) are equal. This passive process is known as **osmosis**, the *net movement of water* across a selective membrane (FIG. 3.15). Sometimes osmosis is called the diffusion of water, but this is not an especially accurate description of the process. Diffusion doesn't necessarily involve a selective barrier, whereas osmosis, by definition, always does. Osmosis occurs until there is no difference in solute concentration on either side of the membrane. Even at equilibrium, when solute concentrations are equal on each side of the membrane, water molecules still move back and forth across the membrane. But, because the rate of water movement in each direction is equal, there is no *net* movement of water from one side of the membrane to the other. Therefore, osmosis stops at equilibrium, but water movement does not.

CHEM • NOTE

Solutes are dissolved substances; a 10 percent glucose solution has a higher solute concentration than a 5 percent glucose solution. As such, a 10 percent solution is hypertonic as compared to a 5 percent solution.

Isotonic environment

- Solute concentrations equal outside and inside cell

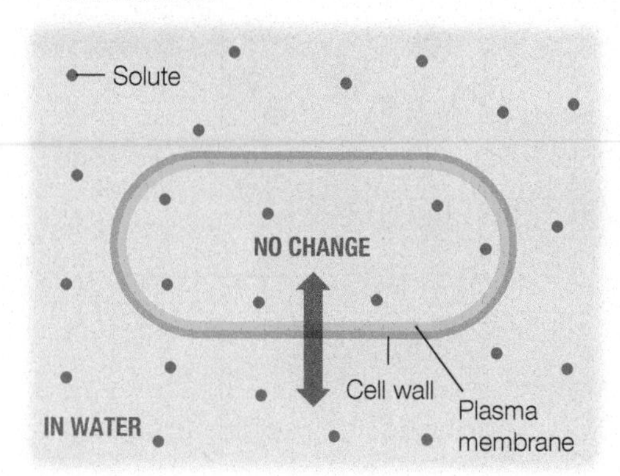

- No net water loss or gain (cytoplasm volume unchanged)
- Cell does not experience an effect

Hypertonic environment

- Higher solute concentration outside cell than inside cell

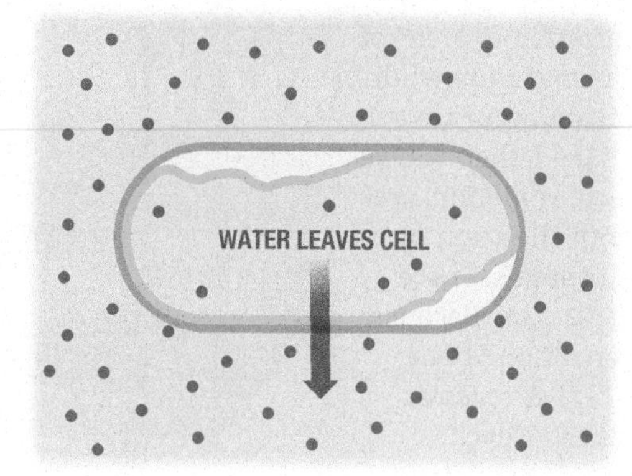

- Water drawn out of cell
- Results in plasmolysis (cytoplasm volume decreases and plasma membrane shrinks away from cell wall)

Hypotonic environment

- Lower solute concentration outside cell than inside cell

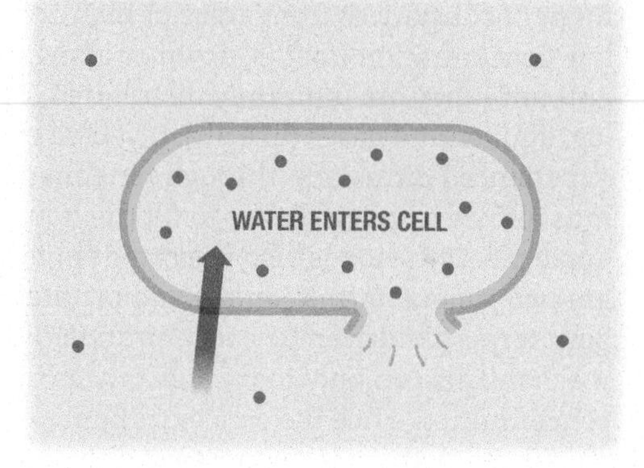

- Water drawn into cell (cytoplasm volume increases)
- Results in cell swelling
- Can result in lysis (cell bursting) if cell wall is damaged

FIGURE 3.15 **Osmosis** In osmosis water moves across a selective membrane from an area of low solute concentration to an area of higher solute concentration.

Critical Thinking *Does osmosis occur in an isotonic environment? Explain your answer.*

To envision osmosis at a microscopic level, imagine taking a bacterial cell and dropping it into a *hypertonic* solution (a solution with a higher solute concentration than inside the cell). This causes the cell to lose water to the environment. Because most bacteria have a cell wall, the general shape of the cell would be maintained, but within the cell, the plasma membrane will shrink away from the cell wall as water is drawn out; in this case the cell is said have undergone *plasmolysis*. Most bacteria are stressed by hypertonic environments, which is why salty and sugary foods have an extended shelf life; at sufficiently high concentrations, salt and sugars naturally act as preservatives by limiting bacterial growth. Some prokaryotes called *halophiles* (salt loving) thrive in salty environments. (See Chapter 7 for more on halophiles.)

Now imagine putting a bacterial cell in a *hypotonic* solution (a solution with a lower solute concentration than inside the cell). In such a case, the cell would rapidly take on water. If the bacterium's cell wall is damaged, or if the bacterium lacks a cell wall, then the cell would eventually burst (lyse). Most bacteria live in a hypotonic environment and rely on their cell wall to keep them from lysing. A number of antibiotics damage the bacterial cell wall, dooming the cell to lysis. If a bacterial cell is in an *isotonic* solution (a solution in which the solute concentration is equal to that of the cell interior), there is no net loss or gain of water from the cell; because the solute concentrations are equal inside and outside the cell, water is not drawn in either direction.

CHEM • NOTE

Hypertonic solutions contain more concentrated solute than a cell; **hypotonic** solutions contain less concentrated solute than the cell; **isotonic** solutions contain equal solute concentrations to those of the cell.

Active Transport

Active transport requires energy to transport specific substances through specialized channels or carrier proteins. *Active transport can move a substance along or against a concentration gradient*. Most cellular exchanges involve active transport mechanisms. They are vitally important to cells because they allow them to concentrate useful substances like nutrients in their cytoplasm. Prokaryotic cells use three main classes of active transport. Each class is defined by how it is fueled (FIG. 3.16).

Primary Active Transport The simplest model of active transport, this mechanism uses the molecule adenosine triphosphate (ATP) to drive the transport of specific substances. Primary active transport is often used to build ion gradients and to shuttle a variety of nutrients across plasma membranes (FIG. 3.16, *left*).

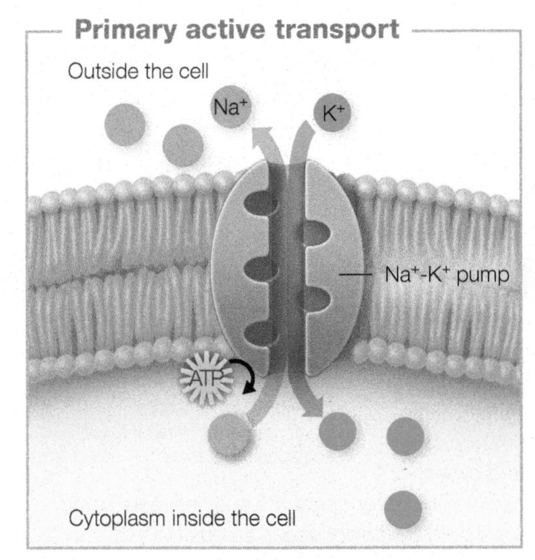

ATP fuels the transport of substances against their concentration gradients. The sodium-potassium pump is a common example.

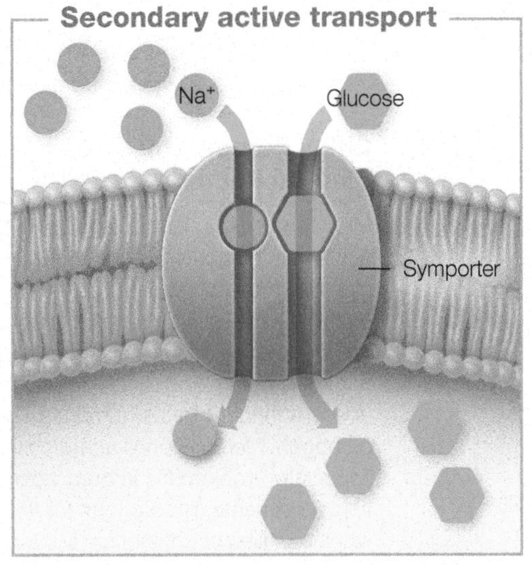

As an ion (Na^+) flows *with* its concentration gradient, it fuels the transport of another substance (glucose) *against* its concentration gradient. (A symporter is shown.)

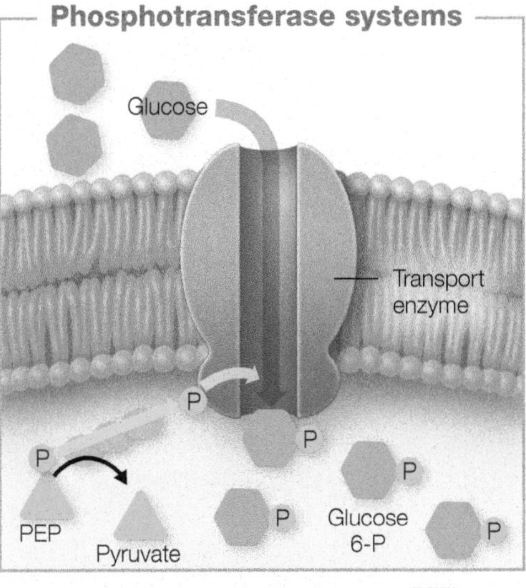

Intermediates like phosphoenol pyruvate (PEP) phosphorylate the transported substance (glucose in this case) and concentrate it within the cell.

FIGURE 3.16 Classes of active transport Active transport mechanisms are classified by how they are fueled.

Critical Thinking *Why is secondary active transport described as depending on primary active transport? (Hint: Consider how an ion gradient is generated.)*

Secondary Active Transport This mechanism relies on an ion gradient to drive transport. Secondary active transport is central to moving a variety of amino acids and sugars across plasma membranes. The energy released by the flow of an ion such as sodium from high to low concentration (with its concentration gradient) fuels the transport of an unrelated substance from low to high concentration (against its concentration gradient). Because transport of a target substance is linked to the transport of ions, this type of active transport is also known as *coupled transport*, or *co-transport*. Ions can flow in the same direction as the target substance (**symport**) or they can flow in the opposite direction of the target substance's movement (**antiport**) (FIG. 3.16, *center*).

Phosphotransferase Systems Also known as group translocation, these systems "transfer" a high-energy phosphate from a molecule other than ATP onto the substance being transported. From a chemical standpoint, this phosphorylation event favors the substance being retained inside the cell, thereby enabling the cell to concentrate the imported substance intracellularly. Phosphoenolpyruvate (PEP) is often the source of phosphates. Phosphotransferase systems are commonly used to import glucose into the cell and directly shuttle it into metabolic pathways (FIG. 3.16, *right*).

Many prokaryotic cells have external structures for adhesion, movement, and protection.

Prokaryotes have a variety of structures that allow them to move, adhere to surfaces, and even confer protection from environmental hazards. Many pathogens rely on such structures to effectively establish an infection.

Flagella

Certain bacteria make use of one or more **flagella** (singular: flagellum) for motility, or movement. About half of the bacteria that have been characterized have at least one flagellum. Flagella are filament-like extracellular structures built primarily from a protein called *flagellin*. In general, bacterial flagella can be described as working like a rotary propeller that spins from a rod-and-ring structure (basal body) embedded in the cell wall. In Gram-positive bacteria, two

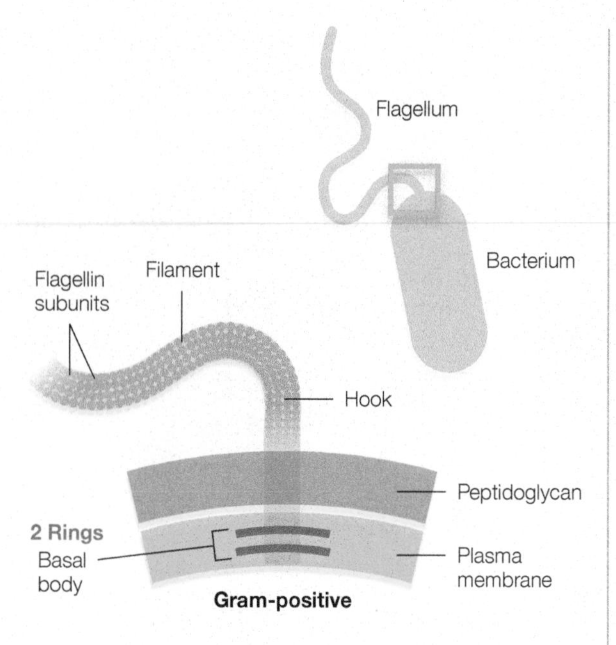

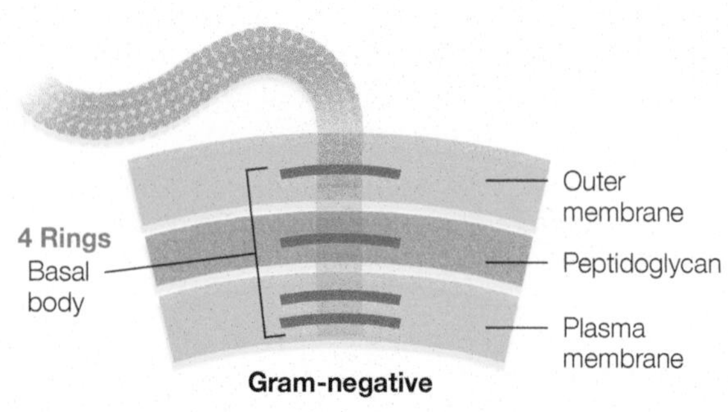

FIGURE 3.17 Bacterial flagella In Gram-positive bacteria, two rings secure the flagellum, while in Gram-negative bacteria, the flagella are anchored by four rings.

Critical Thinking *Would you expect archaea to have the same flagella structure as bacteria? Explain your answer.*

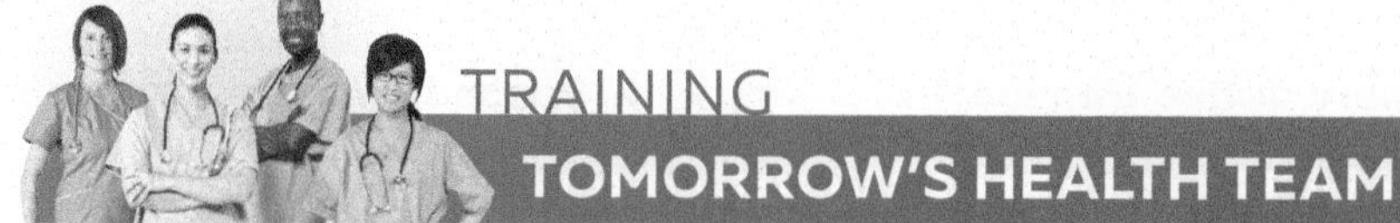

Efflux Pumps: One Active Transport Method Bacteria Use to Thwart Antibiotics

Derivatives of piperine from the black pepper plant decrease antibiotic resistance.

Many bacteria resist the damaging effects of antibiotics by actively pumping them out of their cytoplasm using efflux pumps (EPs). Because all bacteria have active transport machinery, this implies that, theoretically, any cell could expand its active transport repertoire to include pumping out antibiotics. (A scary thought for us humans.) The opportunistic pathogen *Pseudomonas aeruginosa*, which is especially problematic in burn victims and cystic fibrosis patients, is notorious for its EPs, which have rendered the species resistant to whole families of antibiotics. Methicillin-resistant *Staphylococcus aureus* (MRSA), which is infamous for hard-to-treat skin infections, also uses EPs to avoid drugs.

Research is ongoing to develop EP inhibitors that would allow drugs to stay inside bacteria long enough to damage and kill them. So far, a number of inhibitors have been discovered in plant extracts. For example, piperine, a compound found in a type of black pepper plant, has been shown to inhibit EPs. Initial studies with derivatives of piperine suggest they may render *Mycobacterium tuberculosis* more susceptible to rifampin, the first-choice drug for tuberculosis treatment. Similarly, garlic extracts seem to inhibit certain EPs. Because the majority of research on EP inhibitors has been in laboratory experiments that don't involve living systems, there is still *a lot* of work to be done to determine how effective EP inhibitors would be as disease therapies in the real world. Provided that these compounds prove to be effective in humans, the plan is to administer them in conjunction with antibiotics to decrease resistance.

QUESTION 3.3

Why does increasing the drug dose sometimes help overcome the effect of EPs?

rings secure the flagellum, while in Gram-negative species, four rings are used for anchoring (FIG. 3.17). Flagella propel prokaryotes at amazing speeds; it is not uncommon for a cell to "swim" 50 times its length in one second. Scaled up, that's equivalent to a six-foot-tall person being able to attain bursts of speed of up to 204 miles per hour.

While we use our eyes to see what direction we move in, prokaryotes rely on environmental sensing techniques to ensure they are swimming in a desirable direction. A *run-and-tumble* system allows the cell to sense its environment and change direction whenever necessary. In a run state, the cell moves in a concerted direction toward or away from a stimulus. After a short run phase, the cell momentarily stops and slightly tumbles around in its environment to sense its surroundings and redirect its motion accordingly (FIG. 3.18).

Prokaryotic cells move in response to a wide variety of stimuli, including chemical gradients, light, and oxygen concentrations. Movement in response to a chemical stimulus is **chemotaxis**. Movement in response to light is **phototaxis**. **Aerotaxis** is movement in response to oxygen levels.

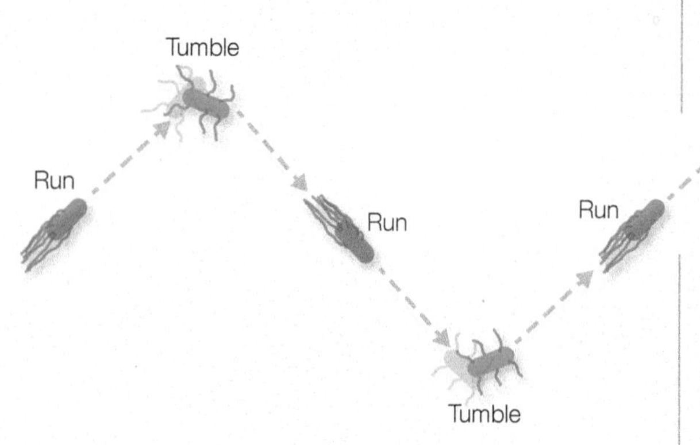

FIGURE 3.18 Run-and-tumble A run-and-tumble system allows the cell to sense its environment and ensure it is moving in the desired direction.

Flagella Arrangements The length and arrangements of flagella vary by species. Cells with a single flagellum are **monotrichous** (*mono* = one). Some cells have a tuft or cluster of flagella at one pole of the cell and are said to be **lophotrichous** (*lopho* = tuft). When one or more flagella are present at both poles of the cell, then the flagella arrangement is described as **amphitrichous** (*amphi* = both).

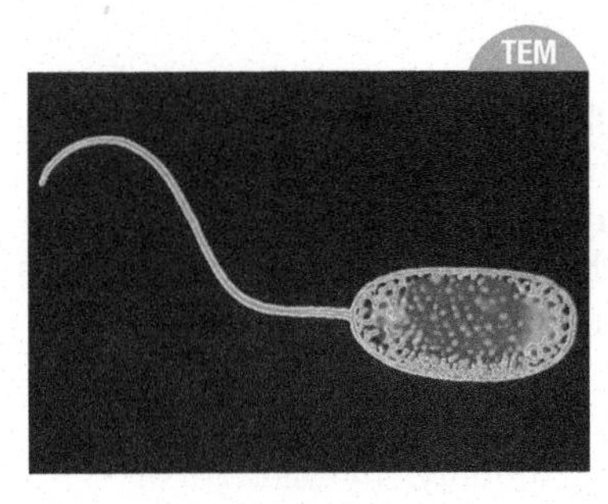

Monotrichous flagellum

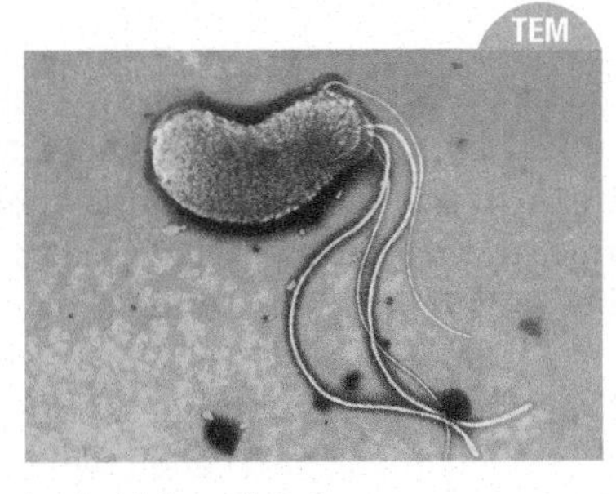

Lophotrichous flagella

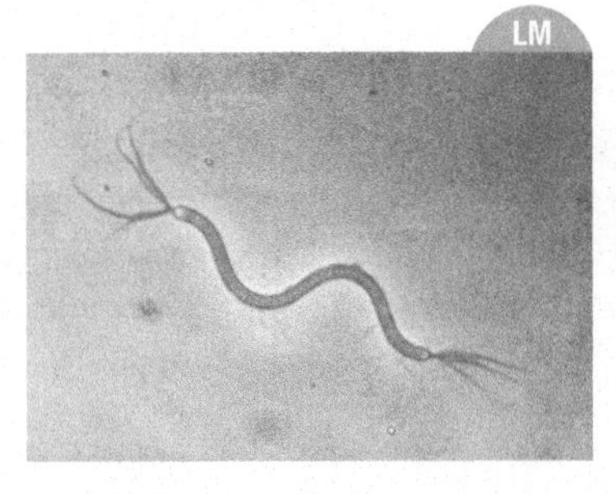

Amphitrichous flagella

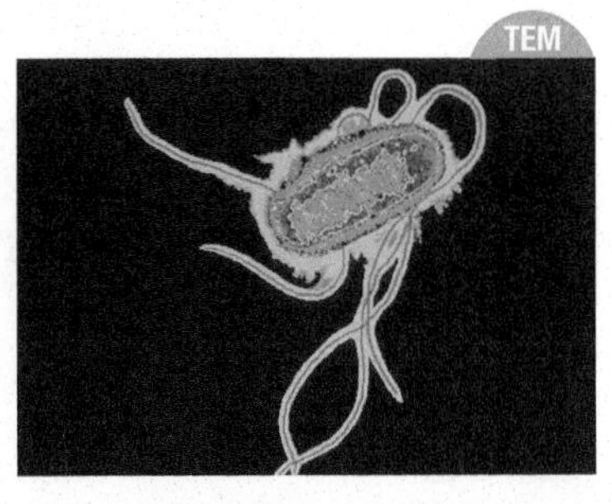

Peritrichous flagella

FIGURE 3.19 Flagella arrangements Flagella may exhibit different arrangements.

Critical Thinking *Would you expect all cells of a given species to have precisely the same number of flagella? Explain your response.*

In a **peritrichous** arrangement (*peri* = around/surround), the flagella are distributed all over the cell surface (FIG. 3.19).

Periplasmic Flagella (Axial Filaments) Unlike most flagella, which extrude from the cell wall, **periplasmic flagella** are found in the periplasmic space of certain Gram-negative bacteria (FIG. 3.20). These unique flagella allow spirochetes to move with their distinct corkscrew motion.

Fimbriae

Fimbriae (pronounced fim-BREE-ah or fim-BREE) are short, bristle-like structures that extrude from the cell surface. Made of protein, they tend to be numerous and cover the cell (FIG. 3.21). Their adhesive properties help prokaryotes stick to surfaces or to each other for establishing biofilms or for invading a host. Fimbriae are common in Gram-negative bacteria and are occasionally found on certain archaea and Gram-positive cells. A number of pathogens, such as certain *Salmonella* species and uropathogenic *E. coli* (a leading cause of urinary tract, bladder, and kidney infections), are unable to establish infections without use of their fimbriae. Consequently, drugs that specifically target fimbriae could act as potential preventive therapies.

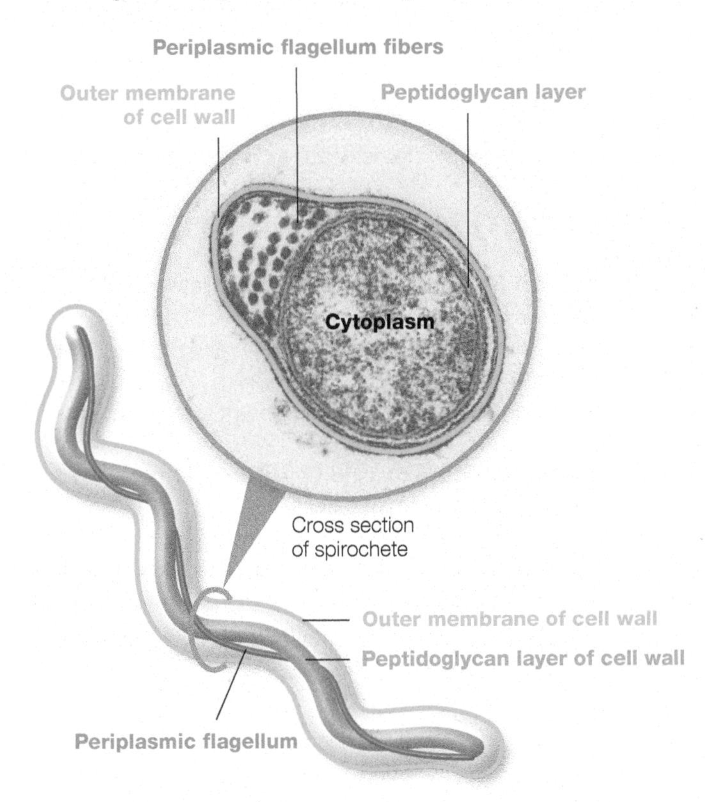

FIGURE 3.20 Periplasmic flagella Certain Gram-negative bacteria have these specialized flagella in the periplasmic space of their cell wall.

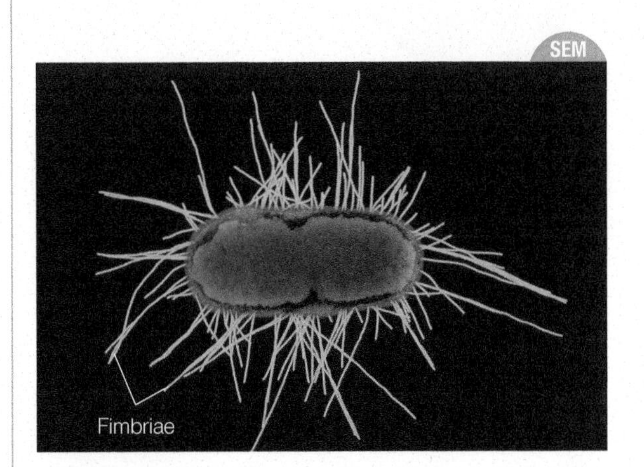

FIGURE 3.21 Fimbriae Scanning electron micrograph of *E. coli*'s fimbriae.

Critical Thinking *Why would a strain of* E. coli *that has lost its ability to make fimbriae be less pathogenic than a strain that can make these structures?*

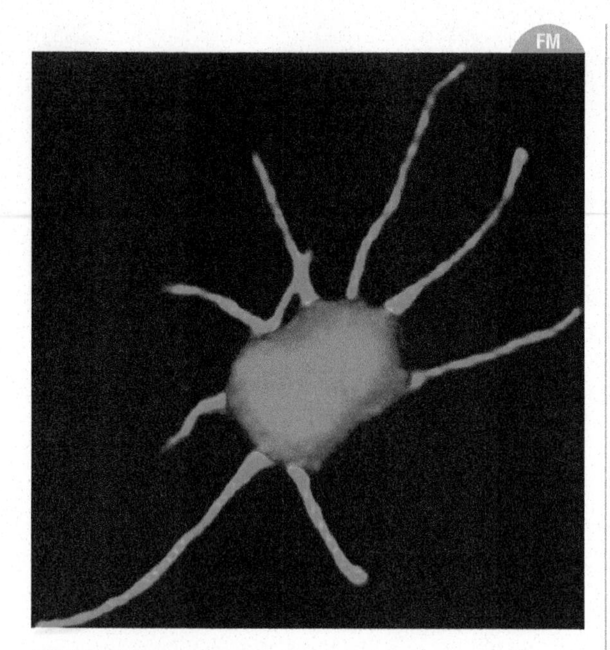

FIGURE 3.22 Pili Fluorescence micrograph of *Neisseria meningitidis* showing type IV pili in green.

Pili

Pili (singular: pilus) are similar to fimbriae, except that they tend to be longer, more rigid, and less numerous. They are used to adhere to surfaces, move, and they can aid in gene transfer through conjugation (discussed further in Chapter 5). *Neisseria gonorrhoeae*, the pathogen that causes the sexually transmitted disease gonorrhea, relies on pili to adhere to the urogenital tract to establish an infection (FIG. 3.22).

Type IV pili are unique in that they facilitate movement in prokaryotes through a gliding mechanism, also called twitching motility.[6] In this form of motility, pili attach the cell to a surface and then retract in a manner that "slingshots" the bacteria forward. Numerous pathogens use pili to move along surfaces, including *Pseudomonas aeruginosa*, which is notorious for causing infections in burn wounds and for pulmonary infections in cystic fibrosis patients.

Glycocalyx

Certain prokaryotes have an additional layer of protection external to the cell wall. A sticky carbohydrate-enriched layer, the **glycocalyx** (meaning "sugar coat"), can be loosely or tightly associated with the cell wall (FIG. 3.23). The exact composition of the glycocalyx varies greatly by species and environmental conditions. The glycocalyx helps bacteria stick to host tissues to establish infections, and it encourages adhesion for biofilm formation. It also protects cells from drying out and offers some protection against antibiotics and typical disinfection measures.

A **slime layer** is one type of glycocalyx that is fairly unorganized and loosely associated with the cell wall. In contrast, a **capsule** is a well-organized glycocalyx that is tightly associated with the cell wall. The presence of a capsule often increases pathogenicity (ability to cause disease) because it promotes adhesion to host tissues and provides some protection against host immune cells by interfering with phagocytosis (see Chapter 4 for more on phagocytosis). For example, encapsulated strains of *Streptococcus pneumoniae*, a leading cause of bacterial pneumonia, are about 100,000 times more virulent than nonencapsulated strains.

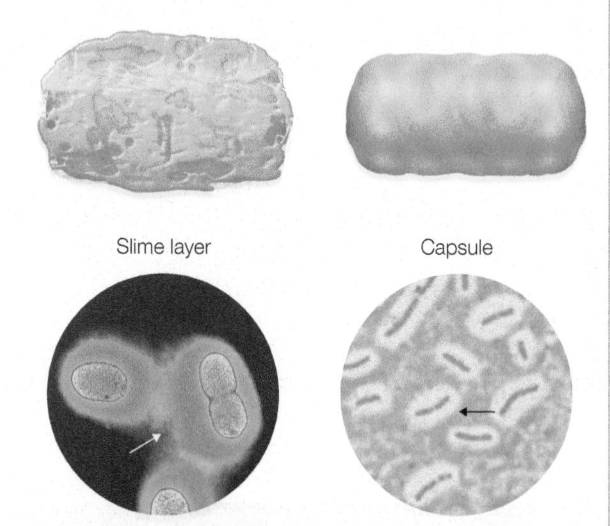

FIGURE 3.23 Glycocalyx The glycocalyx can be loosely linked to the prokaryotic cell wall as a slime layer or tightly associated as a capsule.

Critical Thinking *Explain why slime layers tend to be associated with biofilms.*

Build Your Foundation

7. Describe the structural and functional features of prokaryotic plasma membranes. (LO 3.6)
8. Compare and contrast bacterial and archaeal plasma membranes and cell walls. (LO 3.7)
9. Compare and contrast Gram-positive and Gram-negative bacterial cell walls. (LO 3.8)
10. Name two acid-fast bacteria genera, state what makes them so, and explain their clinical importance. (LO 3.9)
11. Explain how L-forms and *Mycoplasma* are different from other prokaryotic cells. (LO 3.10)
12. Describe the passive and active transport mechanisms that allow for the exchange of substances in prokaryotic cells. (LO 3.11)
13. Define osmosis, then state the outcomes of putting a bacterium with a damaged cell wall in a very salty solution versus pure water. (LO 3.12)
14. Name the structures prokaryotes use for motility and adhesion. (LO 3.13)
15. Describe prokaryotic flagella arrangements, and discuss how periplasmic flagella differ from regular flagella. (LO 3.14)

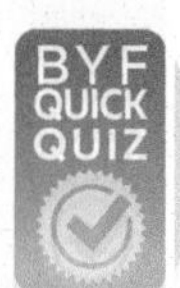

Build Your Foundation (BYF) Quick Quiz: Visit the **Mastering Microbiology** Study Area to quiz yourself.

[6] Nieto, V., Kroken, A. R., Grosser, M. R., et al. (2019). Type IV pili can mediate bacterial motility within epithelial cells. *MBio*, *10*(4), e02880-18.

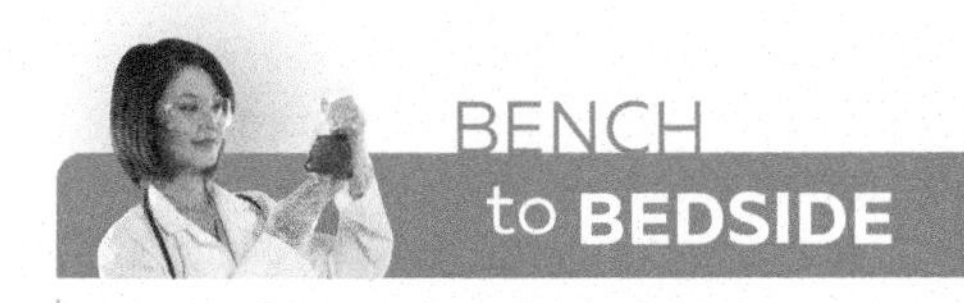

Harnessing Ancient Peptides for Modern Medicine

Antimicrobial peptides (AMPs) are small proteins made by all cellular organisms. Bacteria make AMPs in an effort to destroy competing microbes in their environment, while our cells produce AMPs as an immune defense. AMPs use a complex range of approaches for destroying their targets, including damaging plasma membranes, targeting cytoplasmic structures like ribosomes, preventing cell wall synthesis, blocking biofilm formation, and even breaking up existing biofilms.

Because AMPs represent an ancient defense used across lifeforms, it is reasonable to assume that bacteria have been exposed to AMPs since the dawn of time. Despite this, the development of bacterial resistance to AMPs is rare. In a world where antimicrobial drug resistance is a chronic and growing problem, this characteristic makes AMPs particularly promising as drugs.

Human-derived AMP drugs selectively attack bacterial cells without damaging our own cells. This selective toxicity is an ideal trait in drugs, and few synthetic compounds have managed to be so specific. By modeling natural AMPs, a number of synthetic AMPs have been developed.

At this point, well over a dozen of these are in clinical trials, with many currently in the last testing stage before marketing to the public begins. Researchers hope continued work in this area may lessen side effects, expand the spectrum of activity against pathogens, and make the drugs last longer in biological systems before they break down. Provided these goals can be met, it is likely that in the near future these ancient peptides will make their way into the pharmacopeia of modern medicine.

Some of the most potent AMPs have come from horseshoe crab blood. (Blood samples are taken, and then the animals are released back into their natural environment.)

Source: Magana, M., Pushpanathan, M., Santos, A. L., et al. (2020). The value of antimicrobial peptides in the age of resistance. *The Lancet Infectious Diseases*, *20*(9), e216–e230.

3.3 INTRACELLULAR STRUCTURES

Prokaryotes lack membrane-bound organelles, but still have intracellular structures.

Learning Outcomes

After reading this section, you should be able to:

3.15 Define the term *nucleoid*.

3.16 Describe the basic structure and function of the prokaryotic cytoskeleton.

3.17 Describe the function of prokaryotic ribosomes, and discuss how their structural features support the endosymbiotic theory.

3.18 Discuss the role of inclusion bodies, and provide examples of them.

3.19 Explain the basic structure and function of endospores, and discuss why they present challenges in healthcare settings.

Within the confines of the cell's plasma membrane, we find the internal structures of the cell bathed in a watery **cytoplasm** (or cytosol). Because prokaryotic cells lack the sophisticated, membrane-bound organelles that compartmentalize the intracellular environment of eukaryotes, most of their biochemical reactions occur in the cytoplasm, making this a busy place, crowded with all of the tools needed to carry out cellular processes. Though they lack membrane-bound organelles, prokaryotes do have membranous inclusions that they use for storage and non–membrane-bound organelles, such as ribosomes. Some microbiologists also consider the simple inclusions used for photosynthesis in certain prokaryotes to be primitive organelles.

Nucleoid

A cell's instructions for life processes come from its genetic material, DNA. Prokaryotic DNA is typically organized into a single, circular chromosome located in the **nucleoid** region of the cell—a somewhat centralized region, though its boundaries are not distinct because it isn't enclosed by a membrane. The nucleoid region of prokaryotic cells can be seen with electron microscopy (FIG. 3.24). In addition to the DNA, the region also includes some RNA and proteins. (Prokaryotic chromosomes are discussed further in Chapter 5.)

Ribosomes

All cells must make proteins to survive—proteins are integral parts of many cell structures, and certain proteins (enzymes) perform essential chemical reactions in cells. **Ribosomes**, organelles made of RNA and protein, build proteins by linking together amino acids. (See more on protein synthesis in Chapter 5.) Prokaryotic ribosomes have a lower overall mass and diameter than eukaryotic ribosomes. Antibiotics often target structural differences between prokaryotic and eukaryotic ribosomes—thereby exerting effects on bacteria, while leaving our own ribosomes alone.

CHEM • NOTE

Organic molecules contain carbon and hydrogen. Organisms that can fix carbon convert inorganic carbon like CO_2 into organic carbon by combining it with hydrogen.

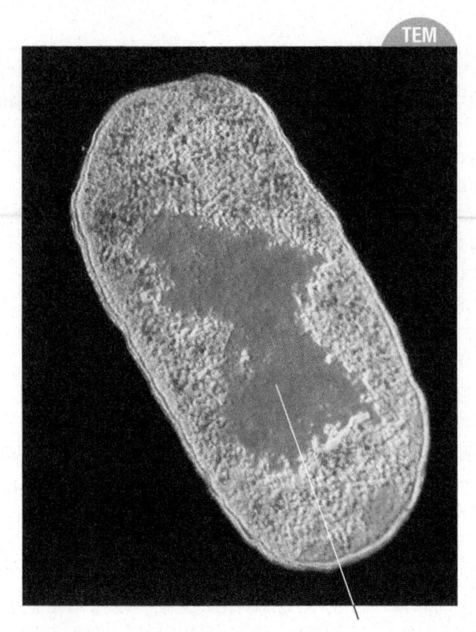

FIGURE 3.24 Nucleoid region Pseudo-colored transmission electron micrograph of *E. coli*'s nucleoid region.

Critical Thinking *Why is the nucleoid region not a true nucleus?*

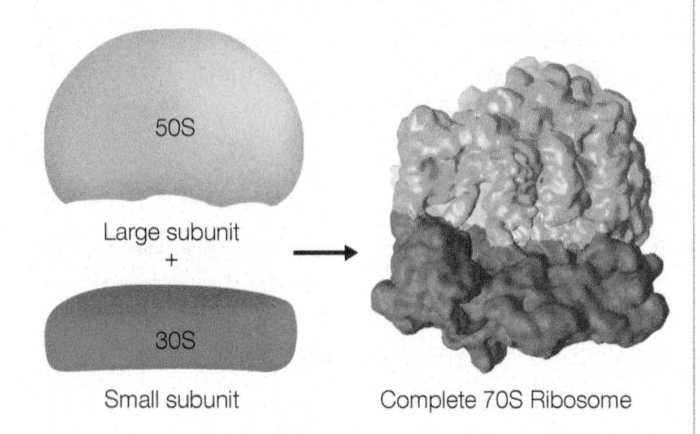

FIGURE 3.25 Prokaryotic ribosome An electron micrograph representation of a prokaryotic ribosome.

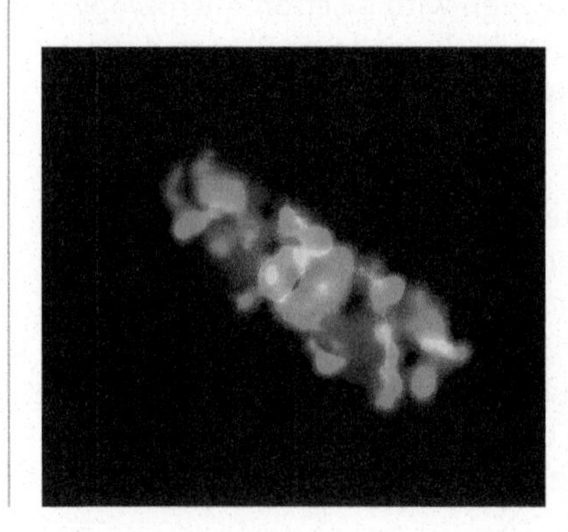

FIGURE 3.26 Prokaryotic cytoskeleton This micrograph shows the various proteins of the bacterial cytoskeleton in green and red. (The image appears slightly blurry due to the nature of fluorescence imaging.)

Prokaryotic ribosomes are also easily differentiated from eukaryotic ribosomes based on a property called sedimentation rate, which is how fast various particles in a solution settle out when spun in a centrifuge (a machine that rapidly spins samples). Sedimentation rate is measured in Svedberg (S) units. Ribosomes are made of two subunits; in prokaryotic cells, the larger ribosome subunit is called 50S, and the smaller is 30S. When subunits combine, the complete ribosome exhibits a sedimentation rate of 70S (FIG. 3.25). You likely noticed that 50 plus 30 is *not* 70; factors impacting sedimentation rate are complex, and so the combined sedimentation rate is not determined via simple addition of the two subunit values. This scenario holds true for eukaryotic ribosomes too, where 60S and 40S subunits combine to create an 80S ribosome.

Interestingly, the ribosomes of two eukaryotic organelles—chloroplasts and mitochondria—are 70S, just like prokaryotic ribosomes. This observation supports endosymbiotic theory, which hypothesizes that eukaryotic cells developed from prokaryotic cells that lived symbiotically with an ancient shared ancestor. (See Chapter 4 for more on eukaryotic ribosomes and endosymbiotic theory.)

Cytoskeleton

Composed of long protein filaments, the **cytoskeleton** gives prokaryotes structure and support (FIG. 3.26). The cytoskeleton acts as scaffolding to organize cell division (binary fission), directs the construction of the rigid cell wall, and provides an overall general organization to the cytoplasm for a variety of biochemical processes. Prokaryotes do *not* contain the same proteins (actin and tubulin) that eukaryotic cytoskeletons have, but their cytoskeleton proteins behave similarly to actin and tubulin and have some shared structural features of these proteins.

Inclusion Bodies

Inclusion bodies are distinct collections of substances inside prokaryotic cells. They consist of insoluble granules, or are sometimes bound in a membrane. Bacteria typically build inclusions in times of excess, such as when grown on nutrient-rich media. The stored materials can then be used in leaner times for a variety of tasks, including metabolic processes. Glycogen is a commonly stored molecule; in times of need, cells break it down one glucose molecule at a time to make ATP. Another commonly stored substance is the polymer poly-β-hydroxybutyrate (PHB), which acts as a carbon and energy reserve for times when more preferable nutrients like glucose run low.

Certain prokaryotes, such as cyanobacteria, fix carbon and help sustain life on Earth by serving as major participants in the carbon cycle. Just as plants reduce carbon dioxide in our air, cyanobacteria recycle carbon in our oceans. These cells carry out this important task with the help of protein-coated inclusions called *carboxysomes* that are packed with carbon-fixing enzymes.

Some prokaryotes even accumulate magnetic iron into membranous inclusions called *magnetosomes*. Prokaryotes like the aquatic organism *Aquaspirillum magnetotacticum* use these specialized inclusions as a magnetic compass to align with geomagnetic field lines to direct their path to deeper waters where they tend to thrive in low-oxygen environments. The iron in these membranous inclusions also seems to confer some protection against oxygen radicals (highly reactive oxygen-containing molecules that tend to damage DNA and proteins), suggesting that magnetosomes may protect against certain forms of oxidative stress (stress from toxic forms of oxygen). Examples of inclusion bodies are shown in FIG. 3.27.

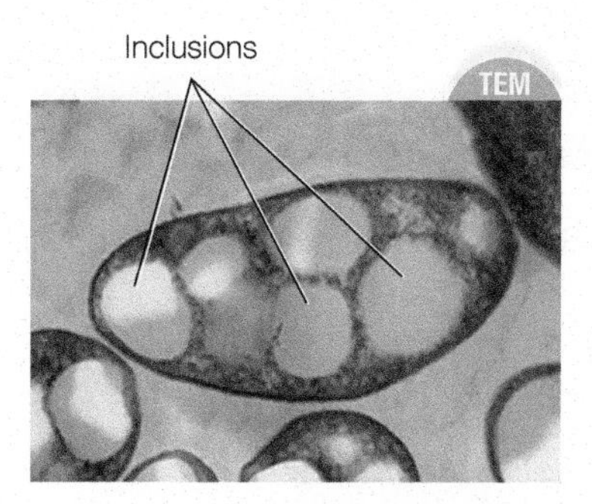

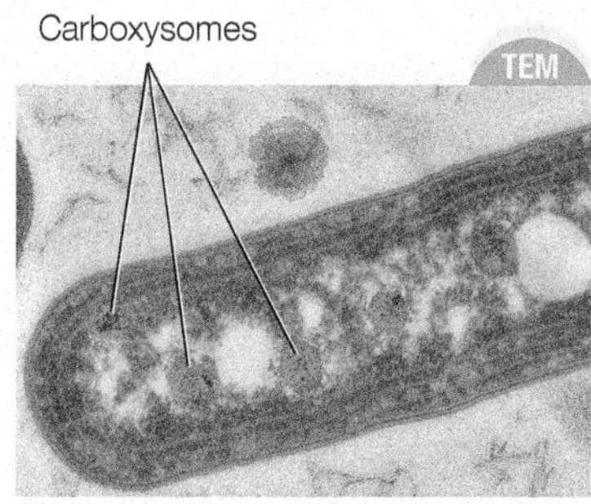

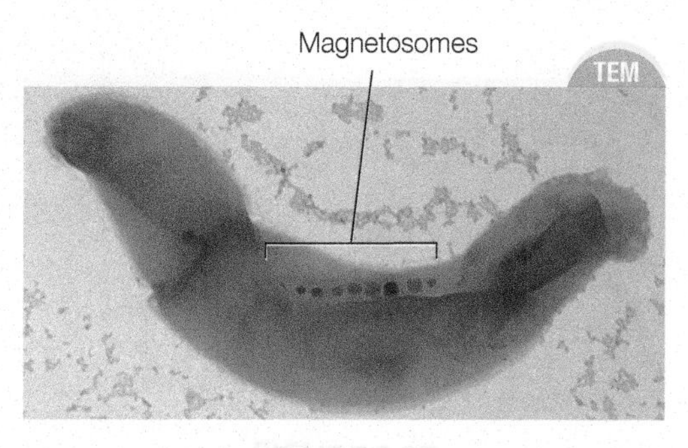

FIGURE 3.27 Prokaryotic inclusion bodies Prokaryotic cells use inclusion bodies to store useful substances.

Critical Thinking *Would you expect a cell to have inclusion bodies after being grown in nutrient-poor conditions for a prolonged period?*

Some bacterial species make endospores to survive harsh conditions.

Endospores (or bacterial spores) are metabolically inactive structures that allow certain cells to enter a dormant state. Once the endospore is released, it is simply called a spore. Bacterial spores are highly resistant to environmental stress such as starvation, heat, drying, freezing, radiation, or various chemicals—so much so that scientists have been able to grow bacteria from 250-million-year-old spores that were encased in salt crystals.[7] When environmental conditions become favorable, the spores germinate back into **vegetative cells** (actively growing cells). Bacterial endospores are not considered reproductive structures because in most species one stressed cell gives rise to one endospore, and therefore they do not increase the total population (FIG. 3.28, *top*).

Clinically Important Spore-Forming Bacteria

While most spore-forming prokaryotes are Gram-positive, there are Gram-negative cells that can form endospores, such as *Sporomusa ovate*. The most medically relevant species that make endospores are in the *Bacillus*, *Clostridium*, and *Clostridioides* genera. All members of these genera are considered Gram-positive, although when undergoing spore formation their cell wall may be compromised, leading these cells to appear Gram-variable (a mixture of Gram-positive and Gram-negative staining cells).

The capacity to make endospores allows dangerous pathogens like *Clostridium tetani* (the causative agent of tetanus), *Clostridium botulinum* (the causative agent of botulism), *Clostridium perfringens* (the causative agent of gas gangrene), and *Clostridioides difficile* (the causative agent of a severe diarrhea) to survive for extended periods on surfaces, even in healthcare facilities that work hard to eradicate them. *Bacillus anthracis* is another medically important bacterium. Historically, *B. anthracis* was more of a danger to grazing animals than humans, but it gained infamy as a potential bioterrorism weapon when spores were malevolently mailed in 2001. This act resulted in 22 victims developing anthrax, five of whom died. Upon entering a host, bacterial spores experience a nutrient-rich environment and therefore readily germinate into vegetative cells within a couple of hours and establish infection.

When a sporulating sample is stained using the endospore staining technique, bacterial spores appear as small, spherical, green structures on the slide. In an introductory microbiology lab, students frequently prepare and view slides showing endospore structures made by harmless bacteria such as *Bacillus subtilis* (FIG. 3.28, *bottom left*).

Sporulation

During **sporulation**, the process of forming an endospore, a cell copies its genetic material and then packages it—along with the ribosomes and enzymes

[7]Vreeland, R. H., Rosenzweig, W. D., & Powers, D. W. (2000). Isolation of a 250-million-year-old halotolerant bacterium from a primary salt crystal. *Nature*, 407(6806), 897–900.

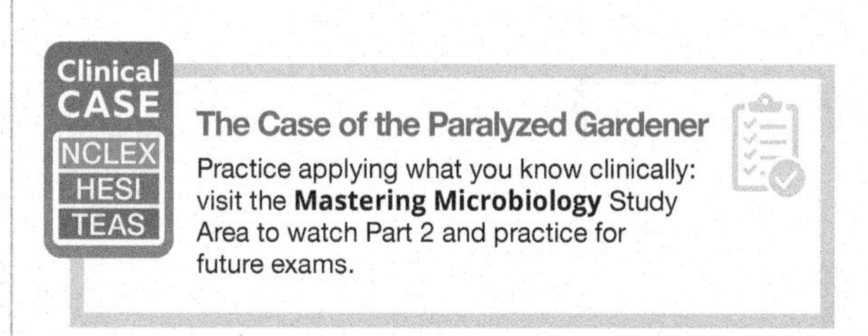

The Case of the Paralyzed Gardener
Practice applying what you know clinically: visit the **Mastering Microbiology** Study Area to watch Part 2 and practice for future exams.

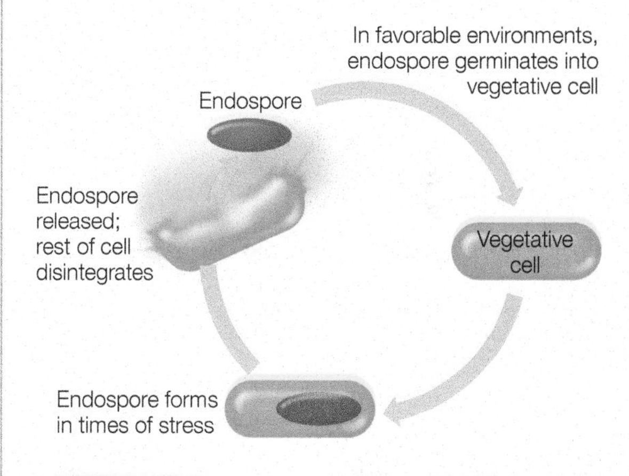

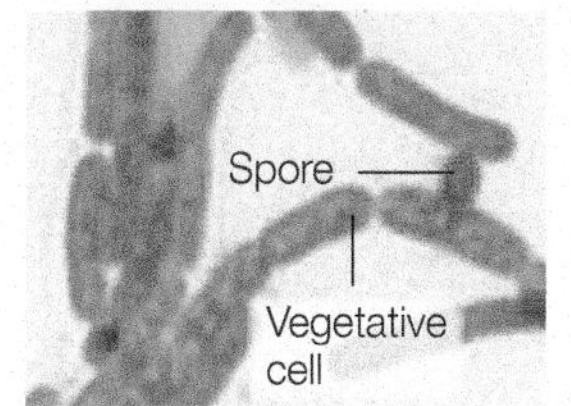

Endospore staining

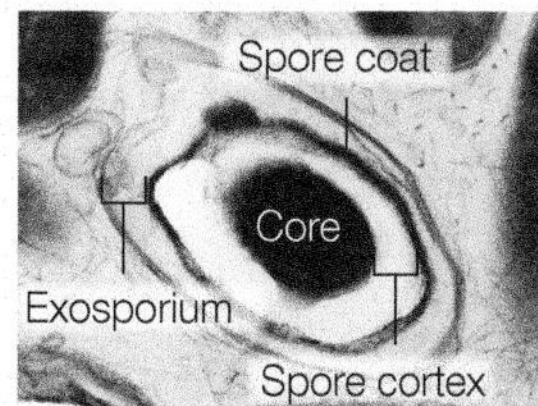

Endospore

FIGURE 3.28 Bacterial endospores *Top:* Cycle for a cell that can make spores. *Bottom left:* Endospore staining colors the spores green. *Bottom right:* A bacterial spore's coat consists of several layers that protect it from harsh environmental conditions.

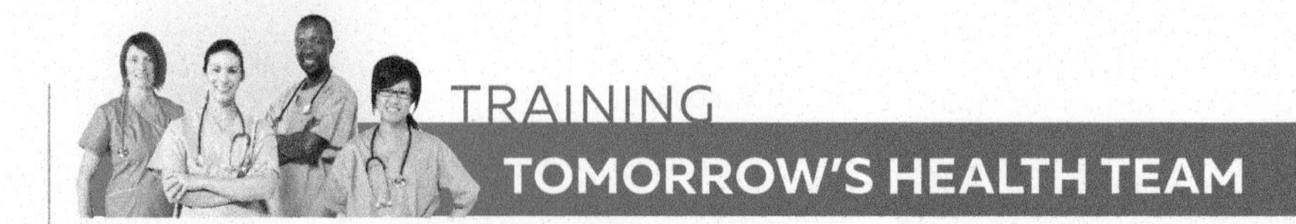

Cleaning up *C. difficile* in the Healthcare Setting

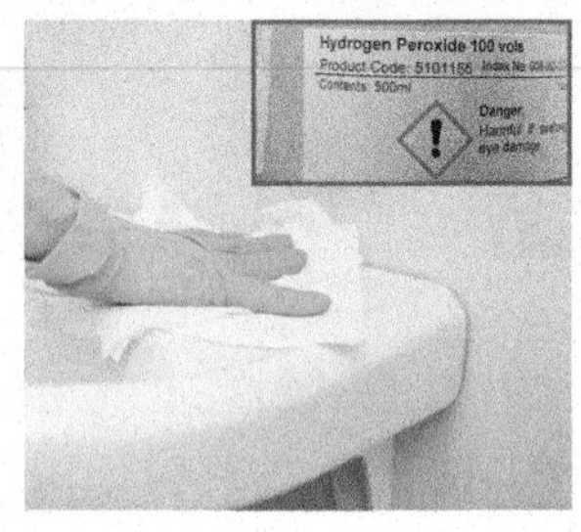

Cleaning up *C. difficile*.

Clostridioides difficile (klos–TRID–ee–OY-dees dif–uh–SEEL) (formerly *Clostridium difficile* and more commonly known as *C. diff*) is increasingly responsible for hospital-acquired infections. Not only are the vegetative forms of this bacterium increasingly antibiotic resistant, but they can also form endospores that are especially hard to eliminate from surfaces.

Patients with active cases of *C. difficile* experience profuse diarrhea and shed vegetative cells and spores in their feces. Thus, the pathogen can easily contaminate frequently touched hospital surfaces such as bed rails, call buttons, and doorknobs. It is estimated that one in four healthcare workers who directly assist a patient suffering from a *C. difficile* infection get the bacterium on their hands, and up to 75 percent of rooms occupied by patients suffering from *C. difficile* become contaminated with the bacterium. Up to 29 percent of rooms that are not even occupied by *C. difficile* patients become cross-contaminated with the pathogen.

A number of disinfectants have broad-spectrum antibacterial activity and can inactivate bacterial spores. These agents are commonly used for sanitizing rooms, beds, and surfaces potentially contaminated by spore-forming bacteria. The most effective disinfectants contain good old-fashioned bleach. However, bleach is corrosive and has an overpowering smell. So some facilities use agents enriched with hydrogen peroxide instead. When used properly, peroxide-based agents are as effective as bleach-based disinfectants.

QUESTION 3.4

Why don't antibiotics eliminate endospores? (Hint: Consider that antibiotics tend to target active cellular processes like protein synthesis.)

it needs to return to a vegetative cell—into a compact structure called the *spore core*. The cell then surrounds the spore core with several heat- and chemical-resistant layers. The lower water content of spores combined with the presence of this dense coat makes spores heat resistant (**FIG. 3.28**, *bottom right*). In addition, spores contain a high amount of *dipicolinic acid* (averaging about 10 percent of a spore's dry weight), which is thought to contribute to their hardiness by stabilizing the DNA packaged into the spore and adding heat resistance to the overall spore structure. Once the endospore is fully formed, it is released, becoming a free spore, and the rest of the cell deteriorates as a waste product. Because bacterial spores are so hardy, healthcare settings must apply special sanitation measures to limit spore-associated infections.

Clinical CASE — NCLEX HESI TEAS

The Case of the Paralyzed Gardener

Practice applying what you know clinically: visit the **Mastering Microbiology** Study Area to watch Part 3 and practice for future exams.

Build Your Foundation

16. What is in the nucleoid region of prokaryotes, and how is it different from the eukaryotic nucleus? (LO 3.15)
17. What is the prokaryotic cytoskeleton made of, and what is its general function? (LO 3.16)
18. What is a 70S ribosome, and what does it do? What structural feature of prokaryotic ribosomes supports the endosymbiotic theory? (LO 3.17)
19. State the role of inclusion bodies inside prokaryotic cells, and provide some examples of substances found in inclusion bodies. (LO 3.18)
20. What are endospores and why do certain bacteria make them? (LO 3.19)
21. Why are *Clostridium, Clostridioides*, and *Bacillus* species a challenge in healthcare sanitation practices? (LO 3.19)

Build Your Foundation (BYF) Quick Quiz: Visit the **Mastering Microbiology** Study Area to quiz yourself.

Interactive CONTENT REVIEWS Interactive Content Reviews: Visit the **Mastering Microbiology** Study Area to quiz yourself.

Prokaryotic Cell Basics

Domains Archaea and Bacteria are prokaryotes. They are unicellular and lack a nucleus and membrane-bound organelles.

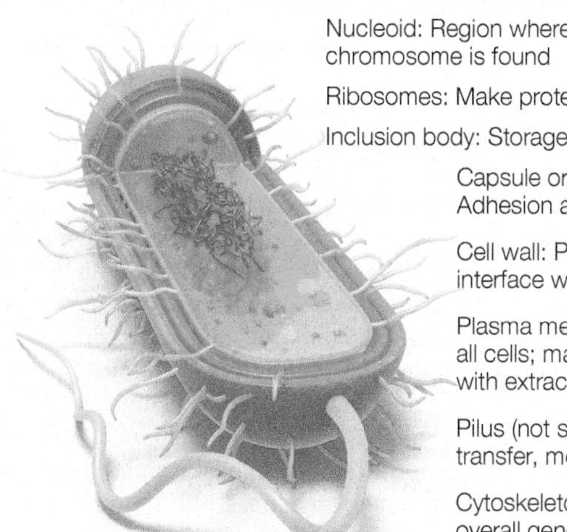

Gram Property

Gram-positive and Gram-negative bacteria have different cell walls; they are differentiated by the Gram stain.

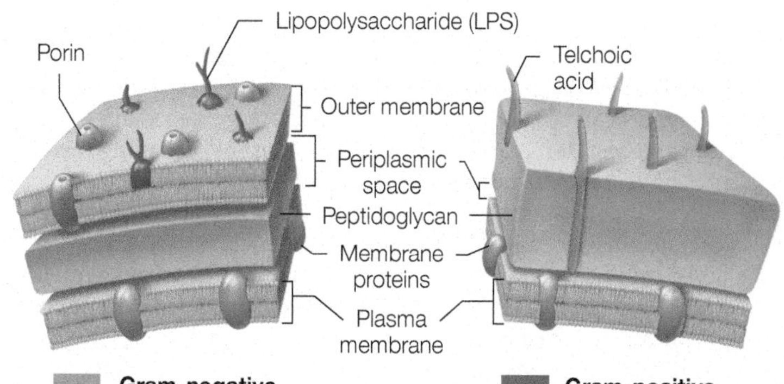

Prokaryotic Arrangements

Prokaryotes are usually smaller than eukaryotes and have diverse shapes and arrangements.

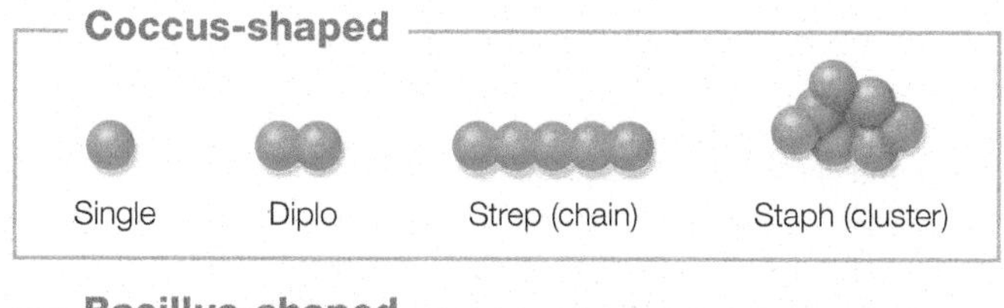

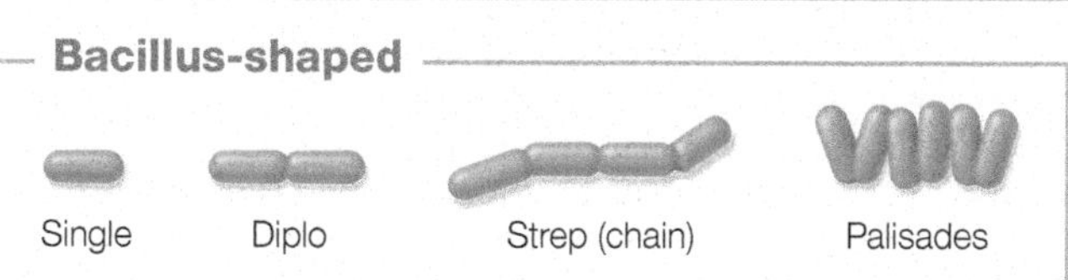

Endospores

Some bacteria, including the genera *Bacillus, Clostridium* and *Clostridioides*, make endospores to survive harsh conditions.

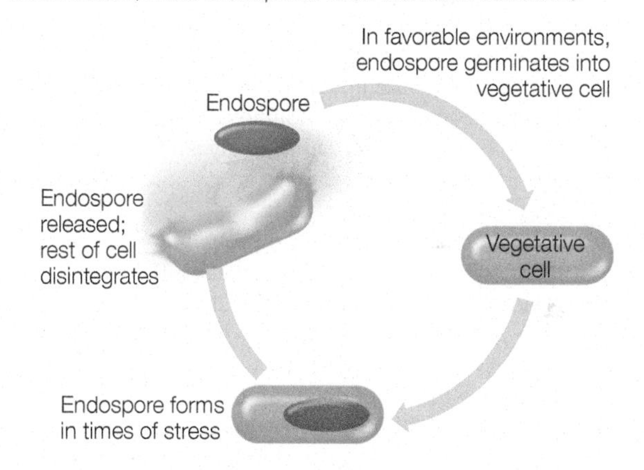

Binary Fission

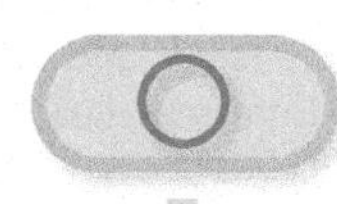

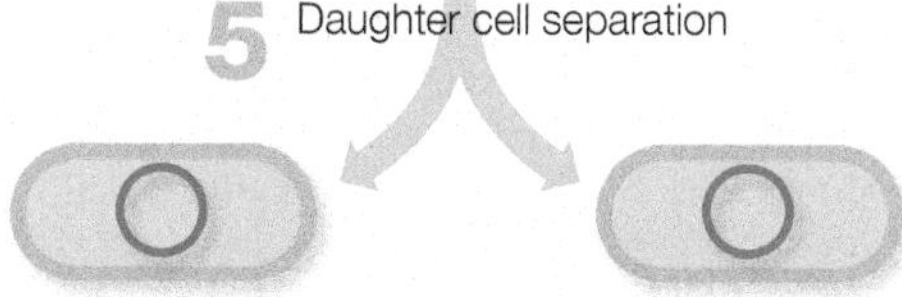

Passive and Active Transport

Passive transport: no energy required

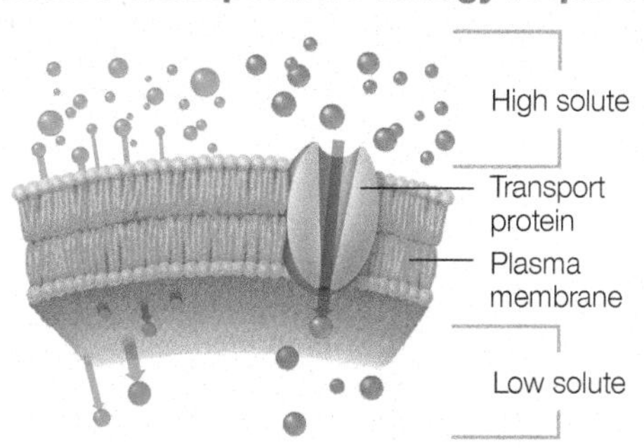

Osmosis is the net movement of water from an area of low solute concentration to an area of higher solute concentration across a selectively permeable membrane.

Active transport: requires energy

Primary active transport ATP fuels transport against a concentration gradient.

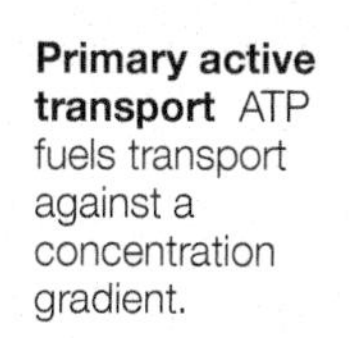

Secondary active transport An ion flowing along its concentration gradient fuels the transport of a substance against its gradient (Na^+-K^+ pump shown here).

Phosphotransferase systems Energy in intermediates phosphorylates the transported substance and concentrates it within the cell.

3 CHAPTER 3 OVERVIEW

3.1 Prokaryotic Cell Basics

- Archaea, Bacteria, and Eukarya are the three taxonomic domains. Prokaryotes are classified under the domains Archaea and Bacteria.
- Prokaryotic cells are unicellular, lack a nucleus (and other membrane-bound organelles), and are genetically simpler than eukaryotes.
- Most prokaryotes are small because a high surface area-to-volume ratio promotes efficient diffusion.
- Prokaryotes can be described by their morphology (cell shapes and arrangements). Some prokaryotes are pleomorphic, meaning they can change forms to aid survival and establish infections.
- Prokaryotic cells assume diverse shapes—for example, they can be cocci (spheres), bacilli (rods), spirals, or vibrio-shaped (comma-shaped cells). Cell arrangements include chains (strep arrangement) and clusters of cells (staph arrangement); rods in clusters are said to be arranged as palisades.
- Most prokaryotes use binary fission to reproduce asexually; this generates two genetically identical cells from a parent cell.

3.2 Extracellular Structures

- The plasma membrane is a selective barrier, a key structure through which cells interact with the environment, and a site for certain metabolic reactions that prokaryotes rely on to make ATP. This flexible lipid bilayer is composed of phospholipids. The polar phosphate-containing part of the lipid faces the water-containing environments of the cytoplasm and the extracellular environment. The nonpolar (or hydrophobic) tails of the phospholipids are sandwiched in the center of the bilayer. As described by the *fluid-mosaic model*, plasma membrane lipids and proteins migrate laterally through the bilayer. In bacteria, phospholipids are built from generally linear fatty acids that are linked to glycerol by an ester bond. In contrast, the archaea tend to have long-branched fatty acids that are linked to glycerol by an ether bond.
- External to the plasma membrane of most prokaryotes is a rigid cell wall made of peptidoglycan—a mesh-like molecule made of glycan chains and peptide cross linkages. Bacterial cell walls are usually described as Gram-positive or Gram-negative. Gram-positive cell walls have a thick peptidoglycan layer and teichoic acids; Gram-negative bacteria have a thin peptidoglycan layer, an outer membrane, and porins. The Gram-negative outer membrane is enriched with lipopolysaccharide (LPS). The lipid A portion of LPS is toxic to animal cells. Acid-fast bacteria, such as *Mycobacterium* species, contain mycolic acid in their cell walls and appear pink after the acid-fast staining procedure. Some bacteria, like *Mycoplasma* species, do not make cell walls. Other bacteria that normally have a cell wall can lose their cell wall as a part of their life cycle or in response to certain stresses and become L-forms.
- Passive transport does not require energy; it moves substances from high to low concentration. Key types of passive transport are simple diffusion (which does not need a protein transporter), facilitated diffusion (which requires a protein transporter), and osmosis.
- Osmosis is the net movement of water across a selective membrane from low to high solute; it is a passive process that is driven by unequal solute concentrations. Placing a cell in a hypertonic environment leads to plasmolysis. Cells in hypotonic environments take on water and may lyse if their cell wall is damaged.
- Active transport requires energy and moves specific substances through carrier proteins either along or against a concentration gradient. Transport against a concentration gradient allows cells to concentrate useful substances within the cell. Primary active transport uses ATP to drive transport. Secondary active transport uses ion gradients to drive transport. Symport and antiport systems are common examples. Phosphotransferase systems use high-energy molecules like phosphoenolpyruvate (PEP) to drive transport.
- Certain prokaryotes have flagella for motility. Flagella arrangements can be described as monotrichous, lophotrichous, amphitrichous, or peritrichous based on the number and position of the flagella. Spirochetes have periplasmic flagella, which propel them in a corkscrew-like motion.
- Fimbriae, pili, and glycocalyx structures such as capsules and slime layers aid in adhesion. Pili can also be used for a specialized gliding motion and exchange of genetic material.

3.3 Intracellular Structures

- Intracellular structures exist within the confines of the plasma membrane.
- The cytoplasm is a watery medium that surrounds intracellular structures and serves as the location for most prokaryotic biochemical processes.
- Prokaryotic DNA is typically organized into a single circular chromosome located in the nucleoid region of the cell.
- Prokaryotes have a cytoskeleton made up of proteins that resemble eukaryotic actin and tubulin. The cytoskeleton aids in cellular organization, coordinates cell division, and supports cell shape.
- Prokaryotic ribosomes are 70S structures that build proteins.
- Prokaryotes can store a variety of substances such as enzymes, iron, glycogen, and poly-β-hydroxybutyrate in inclusion bodies.
- Some bacteria make endospores to survive harsh conditions. Sporulation forms a spore, while germination causes the spore to regenerate a vegetative cell. The chemical and structural characteristics of bacterial spores make them resistant to numerous environmental stresses. They are hard to destroy and present a unique challenge in healthcare settings in terms of sanitation practices.

THINK CLINICALLY | Be S.M.A.R.T. About Cases

COMPREHENSIVE CASE

The following case integrates basic principles from the chapter. Try to answer the case questions on your own. Don't forget to be S.M.A.R.T.* about your case study to help you interlink the scientific concepts you have just learned and apply your new understanding of prokaryotic cells to a case study.

***The five-step method is shown in detail in the Chapter 1 Comprehensive Case on cholera. See pages 31–33. Refer back to this example to help you apply a SMART approach to other critical thinking questions and cases throughout the book.**

• • •

Delia, a 5-year-old, woke up one morning complaining that she had a sore throat, headache, and stomachache. Her dad noticed that her tonsils appeared swollen, and a tonsillar exudate, appearing as a cream-colored pus, was evident when he examined Delia's oropharynx. Delia's dad also noted she was running a low-grade fever. Suspecting strep throat (streptococcal pharyngitis), he made an appointment with you, the pediatrician.

You first performed a rapid strep test, which works by detecting certain cell surface proteins on group A streptococci (GAS). This test is inexpensive and can detect GAS in a matter of minutes. However, rapid strep tests that come back negative are not especially reliable; about 5 out of every 100 patients with streptococcal pharyngitis will have a negative rapid strep test result. Therefore, despite Delia's rapid strep test being negative, you ordered a bacterial culture. This was a wise decision, because Delia had signs and symptoms that pointed to a case of streptococcal pharyngitis caused by the Gram-positive, nonmotile, encapsulated prokaryote *Streptococcus pyogenes*.

When cultures to detect *S. pyogenes* are performed, the patient's sample (in this case a swab of the throat) is inoculated onto a specialized nutrient-rich agar called blood agar. Actively growing *S. pyogenes* can break down red blood cells (a process called β-hemolysis) and therefore is readily detectable on blood agar plates because a clear zone develops around its colonies. Based on the microbiology data that came back, you treated Delia for streptococcal pharyngitis. After 48 hours on an antibiotic, she felt much better and returned to school.

• • •

CASE-BASED QUESTIONS

1. What cell shape and cellular arrangement would the clinical microbiologist observing Delia's throat culture expect to find if the agent is indeed *S. pyogenes*?
2. Assuming that Delia is not allergic to penicillin-based drugs, would a penicillin-family drug (example: amoxicillin) typically be effective against *S. pyogenes*? Why or why not?
3. Would you expect to find endospores if you conducted an endospore stain of the patient's cultured infectious agent? Explain your reasoning.
4. Would an acid-fast stain be informative for diagnosis? Explain your reasoning.
5. Based on the data in the case, what intracellular and extracellular structures could you expect to find for *S. pyogenes* that was isolated from the patient? What cell structures can you safely assume are absent? Be thorough in your listings.
6. Would *S. pyogenes* be classified in the domain Archaea? Explain why or why not.

END OF CHAPTER QUESTIONS

NCLEX | HESI | TEAS

Think Critically and Clinically
Questions highlighted in **orange** are opportunities to practice NCLEX, HESI, and TEAS-style questions.

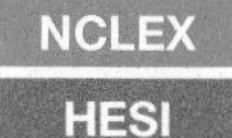

1. Which of the following options include prokaryotic cells? Select all that apply.
   a. Eukarya
   b. Archaea
   c. Protista
   d. Bacteria
   e. Fungi

2. Bacteria cell walls tend to contain:
   a. peptidoglycan.
   b. lipid bilayers.
   c. cholesterol.
   d. pseudomurein.
   e. flagellin.

3. Select any of the following characteristics that would NOT apply to prokaryotes.
   a. Generally simpler than eukaryotes
   b. Multicellular
   c. Lack a true nucleus
   d. Tend to have a single circular chromosome
   e. Often lack a cell wall
   f. All make endospores
   g. Divide by mitosis
   h. Includes the Domain Archaea
   i. Includes the Domain Bacteria
   j. Includes the Domain Eukarya

4. Indicate the true statements about prokaryotic cells, **and then** reword the false statements so that they are true.
   a. They have 80S ribosomes.
   b. They sexually reproduce by meiosis.
   c. Cell walls underlie the plasma membrane.
   d. They synthesize proteins with the help of ribosomes.
   e. They can store nutrients in inclusion bodies.
   f. Fimbriae are used for motility.
   g. Archaea and Bacteria can be classified using the Gram stain.

5. Archaea cell walls tend to contain:
   a. lipid bilayers.
   b. pseudopeptidoglycan.
   c. cholesterol.
   d. flagellin.
   e. peptidoglycan.

6. What is the flagella arrangement for each of the pictured bacteria?
   a. 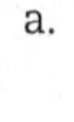
   b. 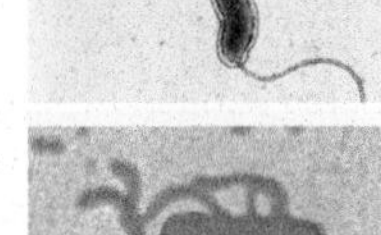
   c. 

7. State the shape **and** arrangement of the pictured bacterial samples.
   a. 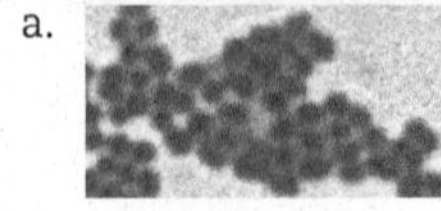
   b.
   c. 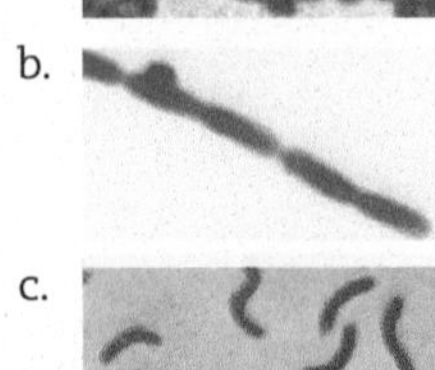

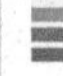

8. Which of the following would you expect to find in acid-fast cell walls? Select all that apply.
   a. Peptidoglycan
   b. Phospholipids
   c. Lipopolysaccharide
   d. Lipid A
   e. Mycolic acid

9. Match the cellular structures to their functions. Many statements will be used more than once; some structures may be matched to more than one functional description.

Structure	Function
Ribosome	Motility
Cytoskeleton	Make(s) proteins
Cell wall	Adhesion
Pili	Protection
Flagella	Exchange(s) genetic information
Fimbriae	Nutrient storage
Capsule	Aid(s) in cellular organization
Inclusion bodies	Support(s) cell shape

10. Complete the Venn diagram that compares and contrasts Gram-positive and Gram-negative bacteria.

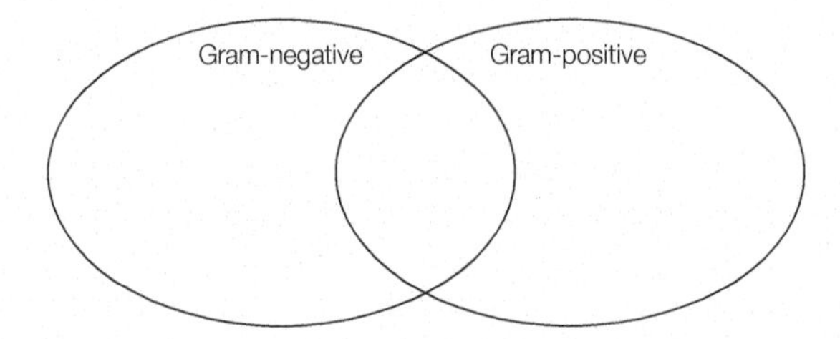

11. Complete the table by answering *yes* or *no* for each question.

	Simple Diffusion	Facilitated Diffusion	Active Transport
Requires energy?			
Requires a transporter of some kind?			
Moves substances from high to low concentration?			
Move substances from low to high concentration?			

12. Certain bacteria can form endospores in order to ______________.
   a. reproduce
   b. survive harsh conditions
   c. cause recurrent infections
   d. adhere to host tissues
   e. avoid the immune system of a host

13. Describe what would occur if a cell with a damaged cell wall was placed in each of the following environments:
   - Isotonic environment
   - Hypertonic environment
   - Hypotonic environment

14. Which of the following characteristics regarding prokaryotic ribosomes supports the endosymbiotic theory?
    a. Prokaryotic ribosomes are 80S just like eukaryotic ribosomes.
    b. Prokaryotic ribosomes are 70S just like mitochondrial ribosomes.
    c. Prokaryotic ribosomes make proteins in a similar manner to eukaryotic ribosomes.
    d. Prokaryotic ribosomes have two subunits like eukaryotic ribosomes.
    e. Prokaryotic ribosomes are intracellular structures like eukaryotic ribosomes.

15. Prokaryotic flagella are made of a protein called _______________.

16. Which of the following are *true* regarding osmosis? Select all that apply.
    a. It is driven by differences in solute concentration.
    b. It is the same as simple diffusion except it involves water.
    c. It involves a selectively permeable membrane.
    d. It requires energy to occur.
    e. It requires specific channels.
    f. Water moves from areas of high solute to areas of low solute.
    g. Salt moves from high-water to low-water areas.
    h. Water moves from areas of low solute to areas of high solute.

17. Which of the following are *true* regarding *Mycoplasma* bacteria? Select all that apply.
    a. They have cholesterol in their plasma membranes.
    b. They are among the largest bacteria characterized to date.
    c. They are pleomorphic.
    d. They are also called L-forms.
    e. They were initially thought to be viruses.
    f. Some can act as pathogens.
    g. They are classified in domain Archaea.
    h. They are prokaryotes.
    i. They are sensitive to penicillin-based drugs.

CRITICAL THINKING QUESTIONS

1. You observe a rod-shaped bacterium that forms endospores. Predict how this specimen most likely would stain with the Gram stain and the acid-fast stain. Support your reasoning.

2. Secondary active transport cannot occur without facilitated diffusion. Explain why.

3. Imagine you are a microbiologist who isolates a never-before-characterized prokaryotic cell. What characteristics could you look for to help you decide if you should classify the new microbe in the domain Archaea or Bacteria?

4. Assume you want to design a prokaryotic cell that can survive in extremely cold environments, forms sticky biofilms, can resist drying, is able to carry out conjugation, and is magnetotactic (moves in response to magnetic fields). What structures/features would you give the organism, and why?

5. What Gram property (positive or negative) would you expect *Mycoplasma* bacteria to be, and why?

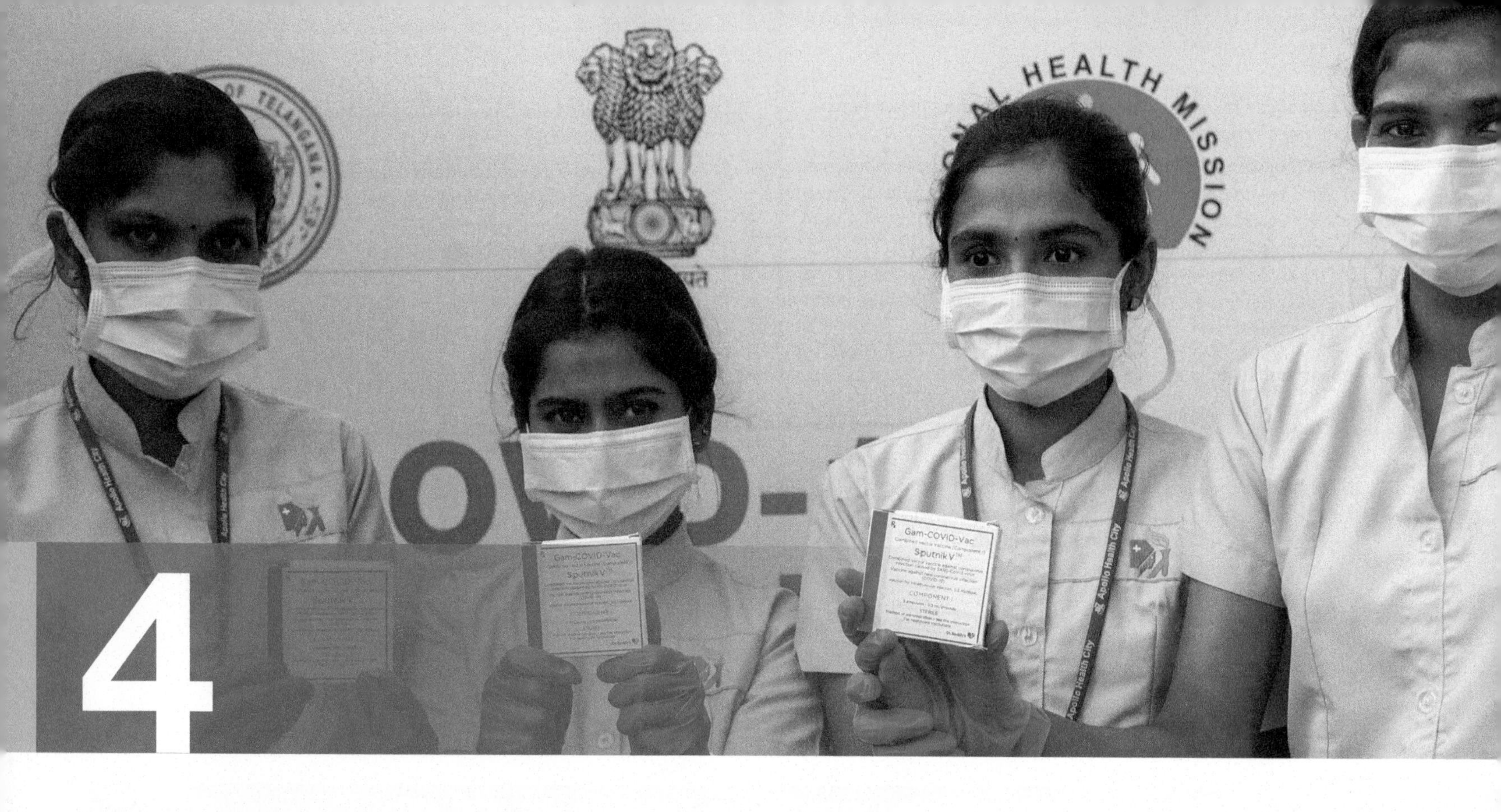

4

Introduction to Eukaryotic Cells

What Will We Explore? Eukaryotic cells make up plants, animals, fungi, and protists. It is projected that over eight million different eukaryotic species call Earth home. Amazingly, we have characterized only about 16 percent of land-dwelling eukaryotes and about 9 percent of water-dwelling eukaryotes.[1] This chapter reviews eukaryotic cell features and physiology.

Why Is It Important? In addition to the fact that humans are made up of eukaryotic cells, a variety of eukaryotic organisms are important in medicine. Various pathogenic fungi and protists as well as parasitic worms (helminths) afflict humans. As you read, you will notice variations among eukaryotes as well as between prokaryotes and eukaryotes. These differences are central to medicine because they tend to be targets of drug therapy. Fungal and parasitic infections can be especially challenging to treat, often requiring long courses of treatment with a fairly limited drug arsenal. This is, in part, because there are fewer differences to exploit between eukaryotic pathogens and our own eukaryotic cells. As such, the continuous exploration of basic cellular differences between "them" and "us" is key to developing new therapies.

Clinical CASE NCLEX HESI TEAS

The Case of the Infectious Hike

Visit the **Mastering Microbiology** Study Area to watch the case and find out how eukaryotes can explain this medical mystery.

Indian nurses displaying Sputnik V coronavirus vaccine
Hyderabad, India

[1] Mora, C., Tittensor, D. P., Adl, S., Simpson, A. G. B., & Worm, B. (2011). How many species are there on Earth and in the ocean? *PLoS Biology*, 9(8), e1001127.

4.1 OVERVIEW OF EUKARYOTES

The endosymbiotic theory proposes how eukaryotes evolved.

Life on Earth started about 3.5 billion years ago with the evolution of prokaryotes. Eukaryotic cells developed about 2.5 billion years ago. The **endosymbiotic theory** describes the evolution of eukaryotes as a series of sequential, cell-merging events between an ancient eukaryotic ancestor and certain prokaryotes.[2] The prefix *endo* means inside, while *symbiotic* refers to collaboration between organisms. Therefore, the word *endosymbiotic* reflects the idea that these cell-merging events came about through a mutually beneficial relationship among the participants.

There are several variations of the endosymbiotic theory. The most widely accepted version states that nonphotosynthetic prokaryotes, possibly an ancient *Rickettsia*-like species, merged with an ancestral cell to produce a *protoeukaryote*.[3] It is still debated whether the protoeukaryote was a full-fledged, nucleus-containing eukaryote or not. However, it's generally agreed that the engulfed bacteria eventually lost the ability to live outside of host cells, becoming what we now know as mitochondria (FIG. 4.1). Later, some of these protoeukaryotes engaged in a second merging event with a photosynthetic prokaryote—possibly a cyanobacterium. These engulfed cyanobacteria also lost their ability to live freely outside of their hosts, becoming what we now know as chloroplasts. The sequence of merging events also explains why the cells of plants and other photosynthetic eukaryotes have both mitochondria and chloroplasts.

Abundant evidence supports the endosymbiotic theory. Mitochondria and chloroplasts both have their own circular DNA and 70S ribosomes that are similar to those found in bacteria. These organelles also have a double-membrane structure, are similar in size to bacteria, and replicate by a process similar to binary fission. Furthermore, many mitochondrial genes resemble select proteobacteria genes, while the genes in chloroplasts resemble those found in certain photosynthetic bacteria. These clues all point to mitochondria and chloroplasts having once been independent, prokaryotic lifeforms. Even if the exact logistics are still debated, scientists widely accept the basic premise that eukaryotes evolved from prokaryotes.

Learning Outcomes

After reading this section, you should be able to:

4.1 Describe the endosymbiotic theory as it relates to the evolution of eukaryotes.

4.2 Provide a basic description of a eukaryotic cell, and state how eukaryotes differ from prokaryotes.

4.3 Compare and contrast mitosis and meiosis.

4.4 Describe and compare the main transport mechanisms used by eukaryotic cells.

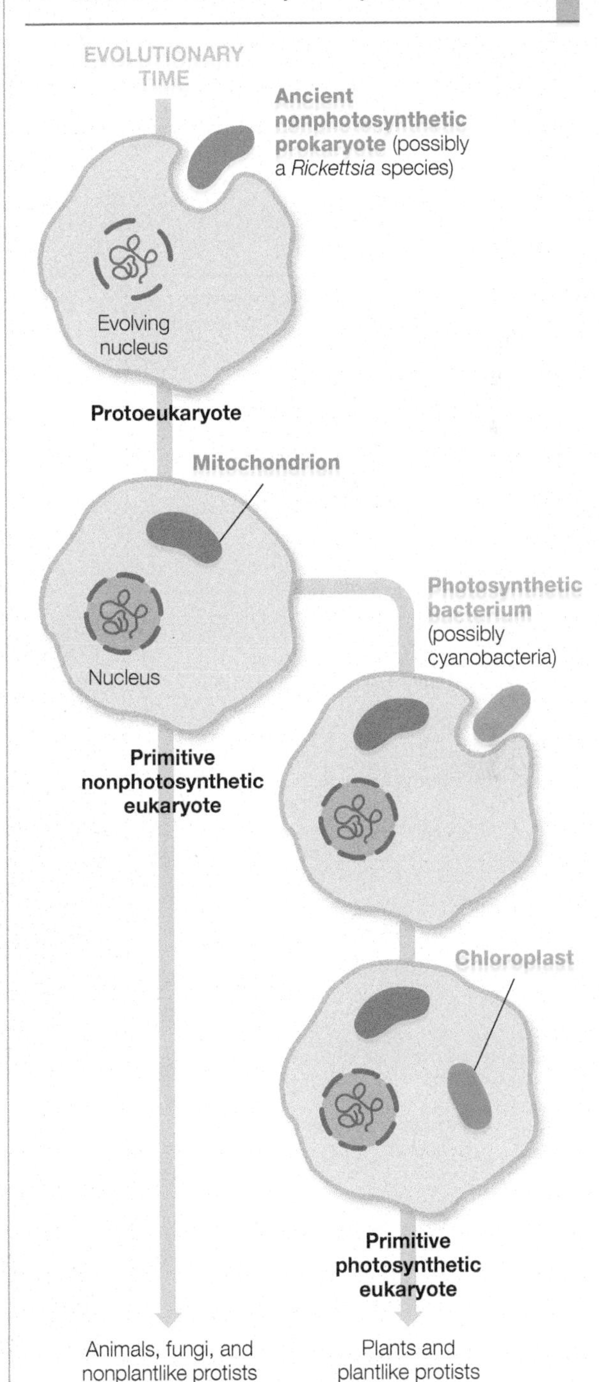

FIGURE 4.1 Endosymbiotic theory Eukaryotes evolved by cell-merging events between a protoeukaryote and prokaryotes.

Eukaryotic cell structures, as well as processes for cell division and transport, differ from prokaryotic cells.

Plants, animals, protists, and fungi are all eukaryotes. Eukaryotic cells are usually larger in size and structurally more complex than prokaryotic cells. They also tend to have larger genomes that are spread across multiple linear chromosomes, while most prokaryotes have a single, circular chromosome. TABLE 4.1 compares and contrasts features of eukaryotic and prokaryotic cells.

Membrane-Bound Organelles

The word *eukaryotic* derives from the ancient Greek for "true nut" (Eu = true/genuine; karyon = kernel/nut). This word dissection may help you remember that eukaryotic cells have a defined, or true, nucleus. They also have a variety of other membrane-bound organelles, such as mitochondria

[2] Margulis, L. (2004). Serial endosymbiotic theory (SET) and composite individuality. *Microbiology Today*, 31(4), 172–175.

[3] *Rickettsia* bacteria invade eukaryotes. A number of these bacteria are human pathogens that cause diseases like Rocky Mountain spotted fever (*R. rickettsii*) and typhus (*R. prowazekii*). Emelyanov, V. V. (2001). Evolutionary relationship of Rickettsiae and mitochondria. *FEBS Letters*, 501(1), 11–18.

TABLE 4.1 Eukaryotic versus Prokaryotic Cells

Characteristic	Eukaryotes	Prokaryotes
Organisms	Unicellular: Protists and yeast (a type of fungi) Multicellular: Animals, plants, and most fungi	Unicellular archaea and bacteria
Size	Usually much larger than prokaryotes	Usually much smaller than eukaryotes
Cell division	Asexual (mitosis) and sexual (meiosis)	Asexual (binary fission)
Plasma membrane	Often contains sterols	Rarely contains sterols
Cell wall	Only in plants, fungi, and certain protists	Most (except *Mycoplasma* and L-forms)
Nucleus	Yes	No
Ribosomes	80S: Cytoplasm, rough endoplasmic reticulum 70S: Mitochondria and chloroplasts	70S only
Genetic material	DNA	DNA
Chromosomes	Multiple linear chromosomes	Usually a single circular chromosome
Membrane-bound organelles	Yes	No (but may have membranous inclusions)

and chloroplasts. Eukaryotic cells are amazingly diverse in both structure and physiological capabilities. A sample eukaryotic cell is shown in FIG. 4.2.

Later in this chapter we'll review the organelles of eukaryotic cells and, where applicable, point out similarities and differences between prokaryotes and eukaryotes. This information is important because it helps explain why some drugs affect bacteria, but do not kill or damage our own eukaryotic cells. For example, the drug penicillin targets peptidoglycan production in the bacterial cell wall, so it won't impact eukaryotic cells, which don't make peptidoglycan.

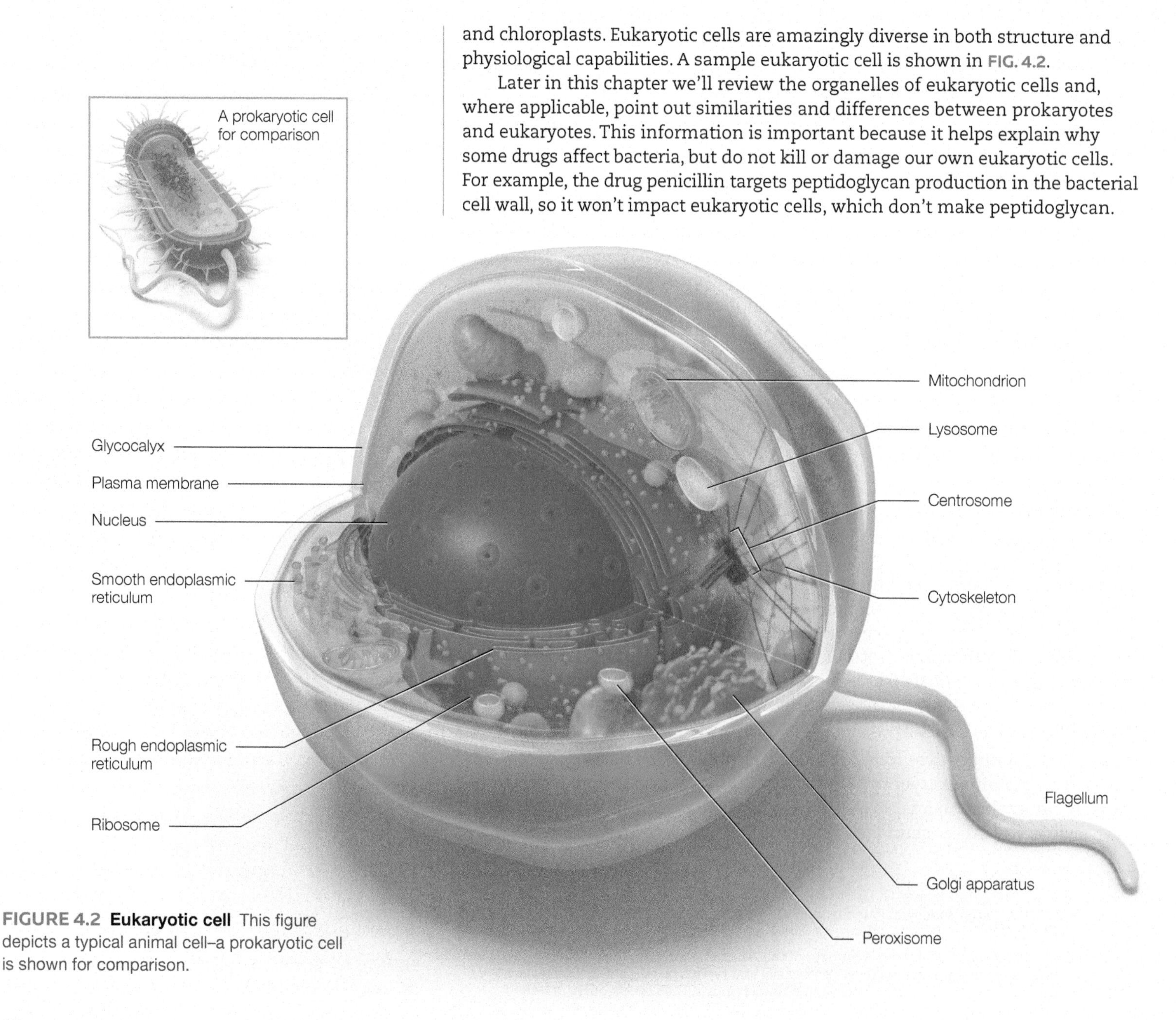

FIGURE 4.2 Eukaryotic cell This figure depicts a typical animal cell–a prokaryotic cell is shown for comparison.

This basic information on differences between cell types lays the foundation for the discussion of antibiotic therapies found later in the book (see Chapter 15).

Cell Division in Eukaryotes

Unlike prokaryotes, which do not conduct sexual reproduction, eukaryotic cells can exhibit sexual and/or asexual reproduction. Before a cell performs any form of cell division, it must copy its genetic material (see Chapter 5). Because eukaryotes have a larger genome and membranous organelles to replicate and separate into daughter cells, these processes are much more involved and take longer than binary fission in prokaryotes.

Mitosis This form of asexual reproduction is the most common way eukaryotic cells divide. **Mitosis** generates two genetically identical offspring from one parent cell (FIG. 4.3, *left*). The offspring cells maintain the same number of chromosomes as the parent cell. All human cells, except for egg and sperm cells, divide using mitosis.

Meiosis As you are aware, sexual reproduction (**meiosis**) in humans requires two contributors: sperm from a male and an egg from a female. These specialized cells, called *gametes*, combine to make a genetically unique zygote. Fungal sexual spores are another example of gametes. In some cases, two separate partners must contribute gametes in meiosis, but other times the gametes may come from a single parent, so that interaction with another organism of the same species is not necessary. Regardless of the number of parents involved, the gametes are all made by meiosis.

Meiosis consists of two cell division stages (FIG. 4.3, *center*). Consequently, one parent cell produces four gametes (daughter cells). Due to a genetic recombination event called *crossing over*, each gamete is genetically unique. The gametes are also haploid, which means they have half the number of chromosomes of their diploid parent, which originally had its chromosomes arranged in pairs before meiosis started. When two complementary gamete cells combine, the resulting zygote has the full number of chromosomes necessary to mature into the next generation of that particular organism.

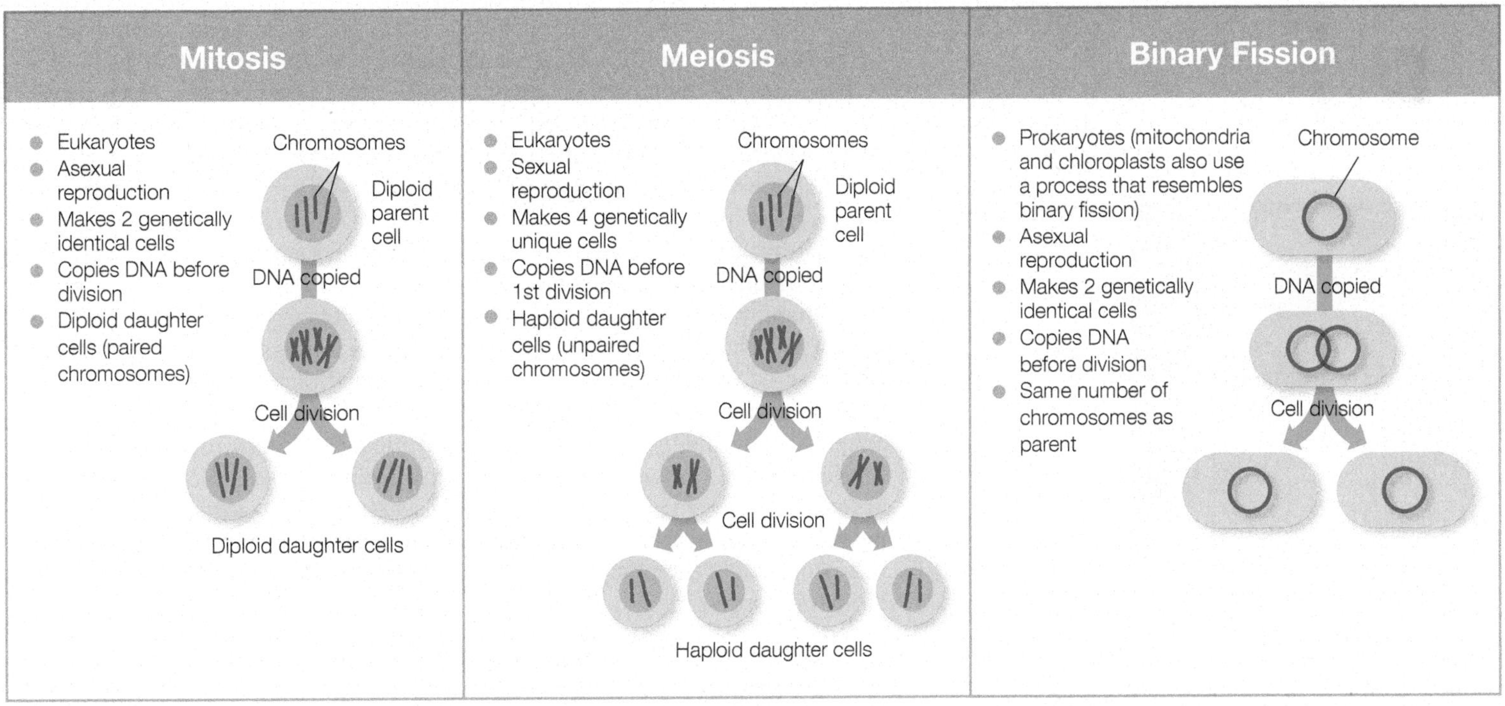

Mitosis	Meiosis	Binary Fission
• Eukaryotes • Asexual reproduction • Makes 2 genetically identical cells • Copies DNA before division • Diploid daughter cells (paired chromosomes)	• Eukaryotes • Sexual reproduction • Makes 4 genetically unique cells • Copies DNA before 1st division • Haploid daughter cells (unpaired chromosomes)	• Prokaryotes (mitochondria and chloroplasts also use a process that resembles binary fission) • Asexual reproduction • Makes 2 genetically identical cells • Copies DNA before division • Same number of chromosomes as parent
Chromosomes Diploid parent cell DNA copied Cell division Diploid daughter cells	Chromosomes Diploid parent cell DNA copied Cell division Cell division Haploid daughter cells	Chromosome DNA copied Cell division

FIGURE 4.3 **Comparing types of cell division**

Binary Fission As a reminder, prokaryotic cells divide by binary fission. However, the eukaryotic organelles mitochondria and chloroplasts replicate using a process very similar to binary fission (FIG. 4.3, *right*). The fact that these eukaryotic organelles use a binary fission-like method of division distinct from how the eukaryotic cell itself divides is additional evidence supporting the endosymbiotic theory.

Eukaryotic Cell Transport: Endocytosis and Exocytosis

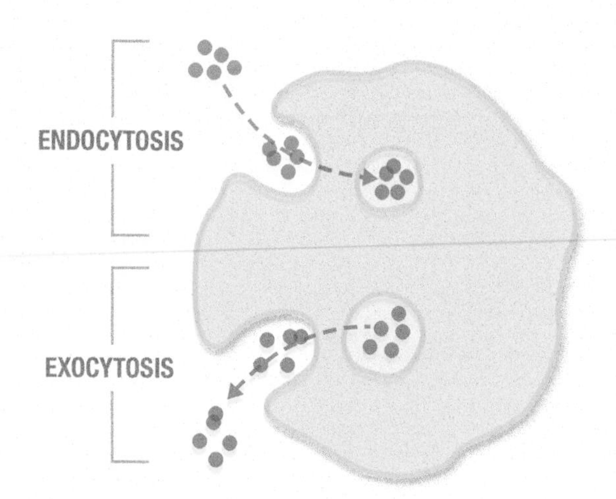

FIGURE 4.4 **Endocytosis and exocytosis**

Eukaryotes use many of the same transport processes as prokaryotes. (For more on osmosis, diffusion, and active transport, see Chapter 3.) But they also employ **endocytosis** to import things into the cell and **exocytosis** to remove things from the cell (FIG. 4.4). *Endocytosis and exocytosis both require ATP.* Both are used for generalized (bulk or mass) transport, as well as for specialized transport regulated by receptors on the cell's surface.

Endocytosis was once thought to be exclusive to eukaryotes, but relatively recently it was discovered in aquatic prokaryotes belonging to the bacterial phylum Planctomycetes.[4] These bacteria blur the lines between prokaryotic and eukaryotic cells in other ways aside from their ability to perform endocytosis. For example, they have some basic membrane compartmentalization and even separate their DNA in a membranous inclusion.[5]

During endocytosis, substances from the extracellular environment enter the cell in membranous *endocytic vesicles* that form as the cell's plasma membrane folds inward (invaginates) and pinches off. Nutrients, macromolecules, dissolved substances, and even viruses or other cells enter eukaryotic cells by endocytosis. The three main mechanisms of endocytosis are pinocytosis, phagocytosis, and receptor-mediated endocytosis.

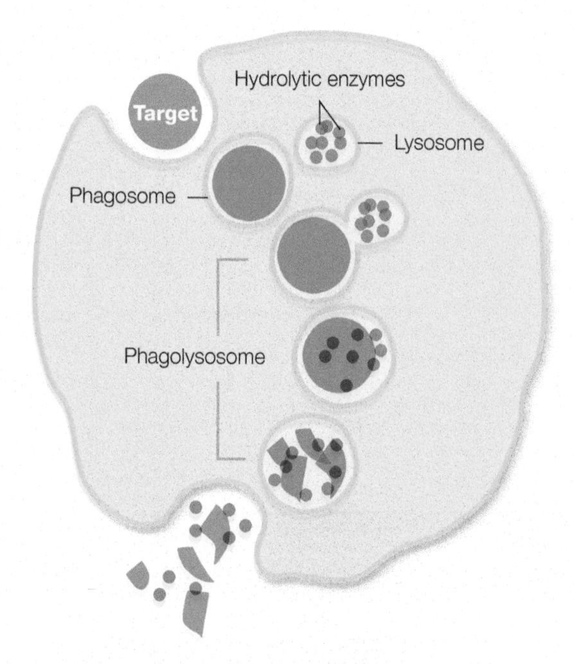

1 Nondissolved target (like a cell or virus) is engulfed.

2 Phagosome containing the target enters the cytoplasm.

3 Phagosome fuses with lysosome, forming a phagolysosome.

4 Hydrolytic enzymes usually destroy contents of the phagolysosome.

5 Waste products are released.

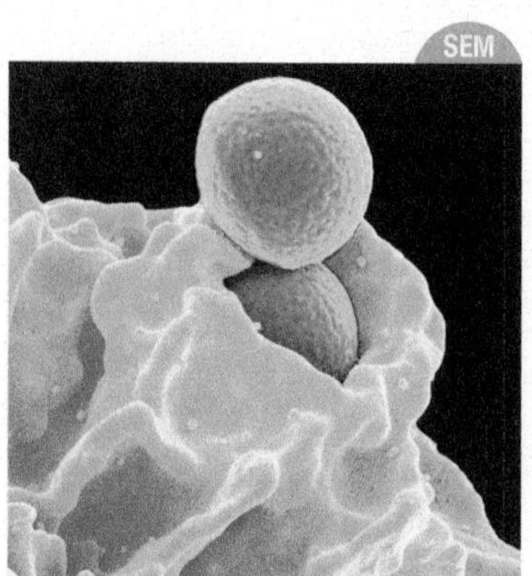

A human neutrophil engulfiing two *Staphylococcus aureus* bacteria.

FIGURE 4.5 **Phagocytosis** This multistep process starts with engulfing the target and ends with expelling waste products.

Critical Thinking *How would this diagram look for a pathogen that manages to escape into the cytoplasm before the phagolysosome stage?*

Pinocytosis The term **pinocytosis** means "cell drinking" and describes endocytosis of dissolved substances in small vesicles. Most eukaryotic cells are constantly carrying out pinocytosis as a form of nonspecific mass transport that is important for cell survival.

Phagocytosis The term **phagocytosis** means "cell eating" and describes endocytosis of undissolved substances. Eukaryotic cells often engulf whole cells or viruses. Because the undissolved cargo imported via phagocytosis is larger than what is imported by pinocytosis, it typically involves larger vesicles. Phagocytosis is often mediated by specific receptors on the cell surface. Features on the agent being engulfed often bind to these receptors. Immune system proteins such as antibodies may also coat pathogens to help target them for clearance by **phagocytes**—specialized immune system cells, such as macrophages, that engulf and destroy their targets. Amoebas and a variety of other protists also use phagocytosis to obtain food from their environments.

Generally speaking, a phagocytic cell engulfs its target in an endocytic vesicle known as a *phagosome*. Soon after the phagosome enters the cell, it fuses with a **lysosome**—a vesicle-like organelle packed with **hydrolytic enzymes**. The fusion of the phagosome and a lysosome creates a **phagolysosome**. The hydrolytic enzymes delivered by the lysosome destroy most cells and viruses engulfed by phagocytosis, and then waste products are expelled (FIG. 4.5). However, certain pathogens have evolved mechanisms for escaping the phagolysosome and/or neutralizing the enzymes. Once these agents escape destruction, they replicate in the cytoplasm of the phagocytic cell and cause disease. (Escaping phagocytosis is discussed further in Chapter 10.)

---

[4] Lonhienne, T. G., Sagulenko, E., Webb, R. I., et al. (2010). Endocytosis-like protein uptake in the bacterium *Gemmata obscuriglobus*. *Proceedings of the National Academy of Sciences of the United States of America*, 107(29), 12883–12888.

[5] Lindsay, M. R., Webb, R. I., Strous, M., et al. (2001). Cell compartmentalisation in planctomycetes: Novel types of structural organisation for the bacterial cell. *Archives of Microbiology*, 175(6), 413–429.

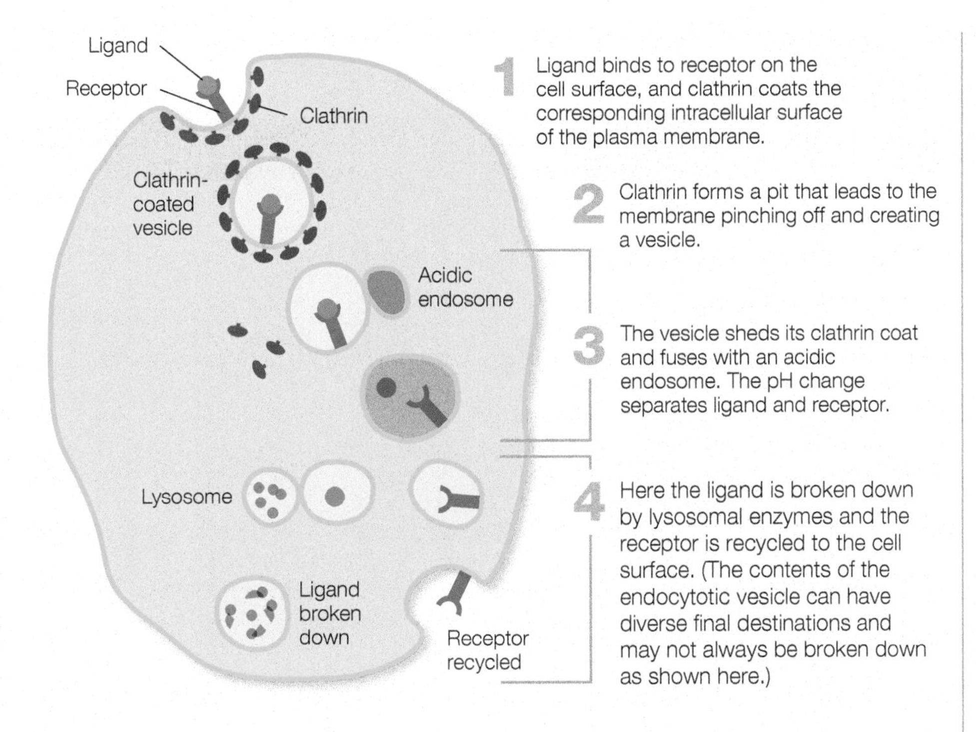

FIGURE 4.6 Receptor-mediated endocytosis Clathrin-mediated endocytosis, shown here, is the most common form of receptor-mediated endocytosis. The imported ligand and receptor can have diverse final destinations. This schematic shows the receptor being recycled and the ligand being broken down.

Critical Thinking *How would this image look different if the ligand and the receptor were both broken down?*

Receptor-mediated endocytosis A highly specific importation tool is **receptor-mediated endocytosis**. Here *ligands*, which are target substances such as hormones, nutrients, or pathogens, bind to specific cell-surface receptors. The hormone insulin enters cells by this process, as do a variety of viruses, including the one that causes polio.

There are several forms of receptor-mediated endocytosis, but the most common form is **clathrin-mediated endocytosis** (FIG. 4.6). When a ligand binds to a receptor on the cell surface, the *inner* surface of the plasma membrane where the receptor–ligand complex is located becomes coated with a protein called **clathrin**. The clathrin *polymerizes*, causing a pit to form. Eventually this *clathrin-coated pit*, which contains the receptor–ligand complex, pinches off from the inner plasma membrane surface, forming a *clathrin-coated vesicle*.

Soon after the vesicle enters the cell, it sheds its clathrin coat and fuses with an **endosome**, a small vesicle with an acidic interior. Fusing with the endosome alters pH so that the ligand and receptor separate from each other. Finally, the ligand and receptor are sorted and delivered to their respective final destinations (either somewhere within the cell or back to the cell surface). Alternatively, the sorting process could lead the endosome to fuse with a lysosome—forming an *endolysosome*—that breaks down the contents. This mirrors the process in phagocytes when a phagosome fuses with a lysosome to form a phagolysosome.

Exocytosis The cellular exportation process called **exocytosis** involves vesicles delivering their contents to the plasma membrane. These *exocytic vesicles* are formed inside the cell—often by budding from cellular organelles like the Golgi apparatus (discussed more later). The cell transports the vesicles to the plasma membrane, with which they then fuse in order to expel their contents from the cell (FIG. 4.7). Exocytosis can rid the cell of unwanted waste products, as occurs in the last step of phagocytosis. The process can also be used to secrete specific substances, like signaling factors, into the cell's environment. For example, our own neurons secrete neurotransmitters via exocytosis. Also, as exocytic vesicles fuse with the plasma membrane, they replace membrane that was removed by endocytosis. Highly phagocytic cells like macrophages are estimated to endocytose the equivalent of their entire plasma membrane in just half an hour; thus, they must constantly repair and restore their plasma membrane by exocytosis.

CHEM • NOTE

Hydrolytic enzymes use water to break chemical bonds. They are important for breaking down organic molecules in cells.

CHEM • NOTE

Polymerization is a process in which building block subunits are linked to make a larger molecule, or polymer.

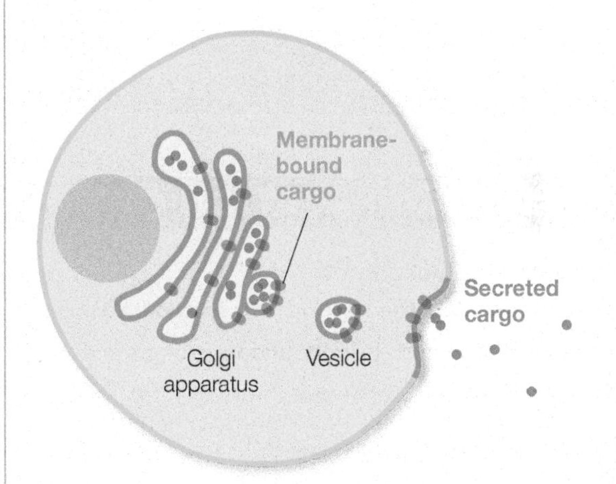

FIGURE 4.7 Exocytosis Vesicles transport substances to the cell surface. This removes waste, replenishes cell membrane lost to endocytosis, and allows secretion of substances like cell-signaling factors.

Critical Thinking *If there was more endocytosis than exocytosis, what do you think would generally happen to the cell's plasma membrane?*

TRAINING TOMORROW'S HEALTH TEAM

Endocytosis and HIV

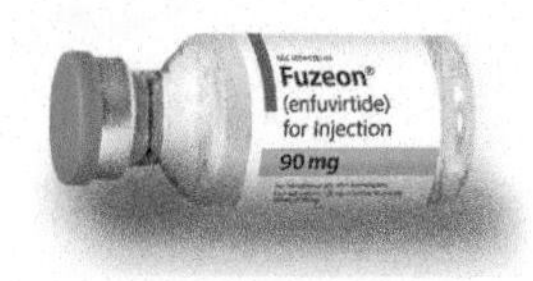

Fuzeon blocks endocytosis to limit HIV entry into host cells.

Many viruses rely on endocytosis to invade host cells, which means blocking endocytosis is one possible approach to treating a variety of viral infections, including human immunodeficiency virus (HIV). Drugs that use this mode of action are called fusion inhibitors.

Normally, HIV invades cells by binding to specific proteins on the plasma membrane of T cells, a type of human immune system cell. When HIV binds to the T cell's surface proteins, it triggers endocytosis, allowing the virus to enter the cell. An HIV fusion inhibitor drug called enfuvirtide limits HIV endocytosis by blocking HIV's binding to T cells. Because it doesn't actually eradicate infection, it is not a cure for HIV. However, it lowers the amount of HIV present in the blood, which is key for preserving better health for the patient.

QUESTION 4.1

While specifically blocking endocytosis is useful as an anti-HIV therapy, an overall generalized block on endocytosis would be dangerous for a cell. Based on what you read in previous sections of this chapter, why would a general block on all cellular endocytosis be hazardous to eukaryotic cells?

Build Your Foundation

Build Your Foundation (BYF) Quick Quiz: Visit the **Mastering Microbiology** Study Area to quiz yourself.

1. How did eukaryotic mitochondria and chloroplasts likely develop? (LO 4.1)
2. Name at least three ways that eukaryotes differ from prokaryotes. (LO 4.2)
3. List the ways that mitosis and meiosis differ and some ways they are similar. (LO 4.3)
4. What are the two main classes of endocytosis, and how do they differ? What is exocytosis, and why is it important? (LO 4.4)

4.2 CLASSIFICATION OF EUKARYOTES

Learning Outcomes

After reading this section, you should be able to:

4.5 Name and describe the four kingdoms of eukaryotes.

4.6 Name and describe the two main groups of parasitic helminths.

4.7 Explain what fungal hyphae are, and name the two forms of hyphae.

4.8 Name and describe the five main classes of fungal spores.

4.9 Define the term *mycosis*, and give examples of human mycoses.

4.10 Explain why Protista is sometimes described as a catchall kingdom.

4.11 Define the term *protozoan*, list the four main groupings of protozoans, and state how they are classified.

Eukaryotic organisms fall into four different kingdoms.

Eukaryotic cells make up single-celled and multicellular organisms that fall into four different kingdoms (TABLE 4.2). We will review the general features of these different kingdoms. All of these kingdoms (except for plants) include potential pathogens or parasites.

Animals

Animals are multicellular organisms that do not carry out photosynthesis, so they must obtain their organic carbon from nutrients. It is estimated that there are over 7.5 million animal species on our planet, making this the largest of the eukaryotic kingdoms. In their mature form, animals are not typically microscopic, yet we cover them in microbiology because certain parasitic worms, or **helminths**, are usually spread in a microscopic form. The term *helminth* refers to a broad collection of organisms that spans roundworms and flatworms. In general, such organisms act as parasites, meaning they live in or on a host. They tend to have complex life cycles that can involve different host species. Most helminthic parasites in humans spend at least some part of their life cycle in the gastrointestinal tract. (See more discussion of these pathogens in Chapter 19.)

TABLE 4.2 Summary of Eukaryotic Organisms

Kingdom	Animalia	Plantae	Fungi	Protista
Examples	Birds, helminths, reptiles, mammals, fish, amphibians, sponges, arthropods	Plants	Yeasts, molds, mushrooms	Euglena, diatoms, amoebas, paramecia, algae, slime molds
Organization	Multicellular	Multicellular	Some unicellular (yeasts), but most are multicellular (nonyeast fungi)	• Animal-like protists (protozoans): Unicellular • Plant-like protists: Unicellular or multicellular • Fungus-like protists: Unicellular or multicellular
Reproduction	Sexual and asexual	Sexual and asexual	Sexual and asexual	Sexual and asexual
Cell wall	No	Yes	Yes	Some (Examples: Slime molds at certain life stages, and algae)
Chloroplasts (or plastids)	No	Yes	No	Some (Example: Algae)
Mitochondria	Yes	Yes	Yes	Most (some have only remnants of mitochondria called mitosomes)
Medical examples	Parasitic worms; many arthropods like ticks and mosquitoes transmit infectious diseases	No pathogens, but some may produce toxins dangerous to animals	*Candida albicans* causes yeast infections; *Pneumocystis jirovecii* causes infections in the immune compromised	Various *Plasmodium* species cause malaria; *Entamoeba histolytica* causes amoebic dysentery; certain algae make toxins (such as the ones that cause red tide blooms)

Helminths represent a significant medical burden around the world. Improved water and food sanitation practices in developed countries reduce helminthic infections, but they are by no means eliminated. The World Health Organization (WHO) estimates that at least *half* of the world's population is infected with some sort of helminth at any given time. The general characteristics of the main groups of medically important helminths are presented in TABLE 4.3.

Parasitic worms are not the only members of the animal kingdom explored in microbiology. A number of arthropod species, especially ticks and mosquitos,

TABLE 4.3 Overview of Parasitic Helminths

Phylum	Roundworms	Flatworms	
Subtypes	Nematodes Hookworm	Tapeworms (cestodes) Tapeworm	Flukes (trematodes) Liver fluke
Structure	Nonsegmented; elongated, cylindrical	Segmented, flat, ribbon-like	Nonsegmented; flattened leaf shaped
Size range	Microscopic–1 meter	1 millimeter–10 meters	1 millimeter–7 centimeters
Reproduction	Sexual reproduction; two sexes	Sexual reproduction; hermaphroditic (male and female reproductive organs in same individual)	Sexual reproduction; except for blood flukes, all are hermaphroditic
Examples in humans	Hookworm, pinworm, *Ascaris*, filarial worms, *Trichinella*, *Strongyloides*, whip worm	Six tapeworms infect humans (Examples: *Diphyllobothrium latum* from fish, *Taenia saginata* from beef)	Blood flukes: *Schistosoma* species; Lung fluke: *Paragonimus westermani*, Liver fluke: *Fasciola hepatica*; Intestinal flukes: *Fasciolopsis buski* and *Heterophyes heterophyes*
Transmission mechanisms	Fecal/oral (eggs consumed in contaminated food such as undercooked meat, or in water or soil); some species' larvae burrow into skin, migrate to lungs, get coughed up and swallowed to arrive at the targeted intestines	Fecal/oral through contaminated water or food such as undercooked meat or fish from an infected animal	Embryonated eggs from host feces enter water and hatch; released larvae mature in snails and then are either ingested in contaminated food/water or burrow into human host

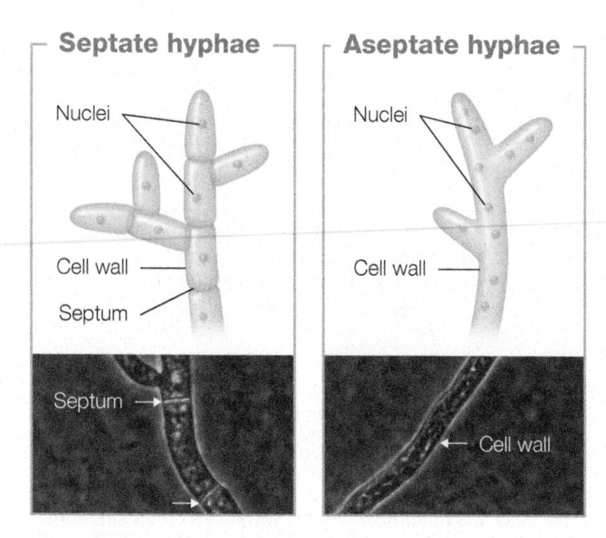

FIGURE 4.8 Fungal hyphae Fungal hyphae can be described as septate or aseptate.

are medically important due to their capacity to serve as disease vectors. (See Chapter 9 for more on vectors.)

Plants

Plants are multicellular organisms that carry out photosynthesis to make their own organic carbon using light energy. There are over 290,000 different plant species, none of which cause infectious disease—although vegetation can serve as a vehicle for infectious pathogens, such as improperly cleaned fruits and vegetables that retain microbes on their surfaces. Plant cells contain chloroplasts, organelles that we noted earlier are thought to be derived from photosynthetic bacteria. A variety of photosynthetic eukaryotic organisms rely on chloroplasts to perform photosynthesis. Chloroplasts will be explored later in this chapter.

Fungi

The Kingdom Fungi is believed to include over 600,000 different species, although most have not yet been characterized. Most fungi (singular: fungus) are multicellular or colonial; only yeasts are unicellular. Fungi do not carry out photosynthesis and instead rely on extracting carbon from the nutrients they absorb from their environment. Aside from pathogenic fungi, most fungi are **saprobes**, meaning they absorb nutrients from dead plants and animals in the environment.

Hyphae Most fungi (aside from yeasts) grow as a collection of tubular structures called **hyphae** (singular: hypha). Fungal hyphae can be described as *septate* or *aseptate* (or coenocytic). Septate hyphae include divisions between each cell in the filament and appear as a string of individual cells. Aseptate hyphae do not have divisions and appear as a long continuous chain with many nuclei (FIG. 4.8). Some fungi cycle between having hyphae and living as a yeast-like form and are called *dimorphic fungi* (having two forms). Many pathogenic fungi are dimorphic, exhibiting a yeast-like growth in humans and a hyphae growth form in the environment. (See Chapter 16 for more on fungal respiratory infections and dimorphism.)

Fungal Spores Fungi have a variety of reproductive strategies, but the most prevalent approach to reproduction involves the production of spores. Fungal spores are *not* like spores made by bacteria. They are used for reproduction, whereas bacterial spores are used to survive harsh conditions. A variety of features such as morphology, coloration, genetics, physiological features, and the type of reproductive spores made are all used to classify fungi. Fungal spores are classified as either asexual or sexual. Fungal spores are summarized in TABLE 4.4, and examples are shown in FIG. 4.9. **Asexual spores** arise from mitosis and do *not* result in genetic variation. Two classes of asexual fungal spores are *conidiospores* and *sporangiospores*. By contrast, **sexual spores** arise from the union of complementary mating strains of fungi generated by meiosis and do result in genetic variation. Three types of sexual fungal spores exist: *zygospores*, *ascospores*, and *basidiospores*. The type of sexual spore made is used to taxonomically group fungi into a given phylum.

Fungal Diseases Diseases caused by fungi are called **mycoses**. Of the fungi described to date, the vast majority are *not* pathogens. Most mycoses are seen in individuals with a weakened immune system, such as *Pneumocystis* pneumonia that occurs in AIDS patients. In other cases, people who develop fungal infections first experienced a disruption of their normal

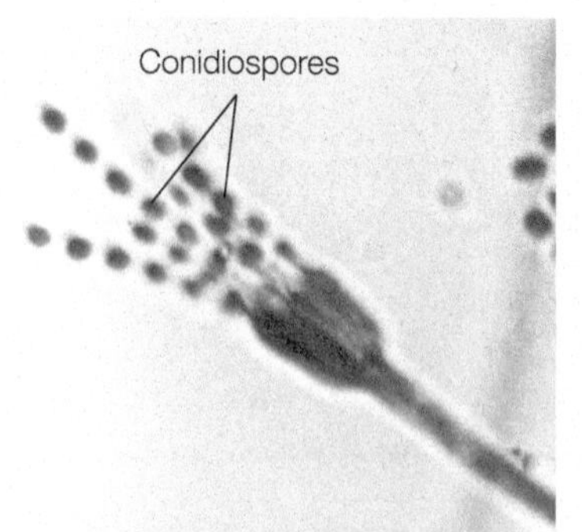

Conidiospores from *Penicillium*

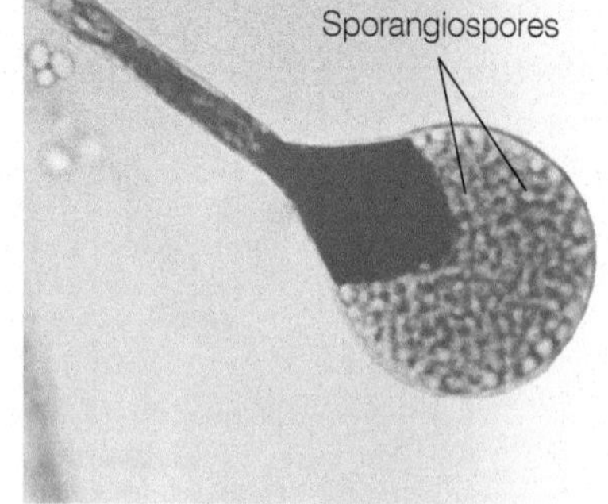

Sporangiospores from *Absidia*

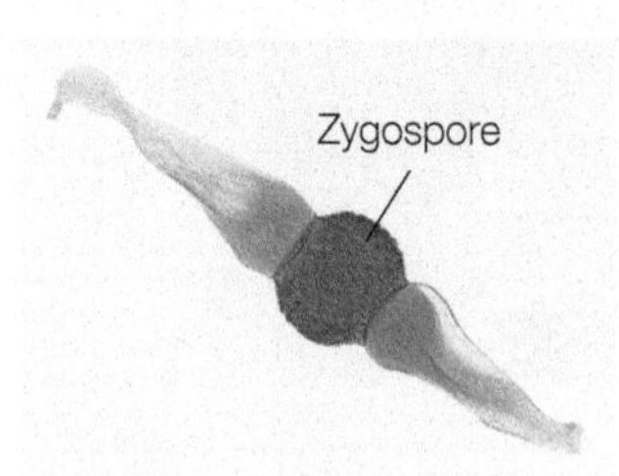

Zygospores from *Rhizopus*

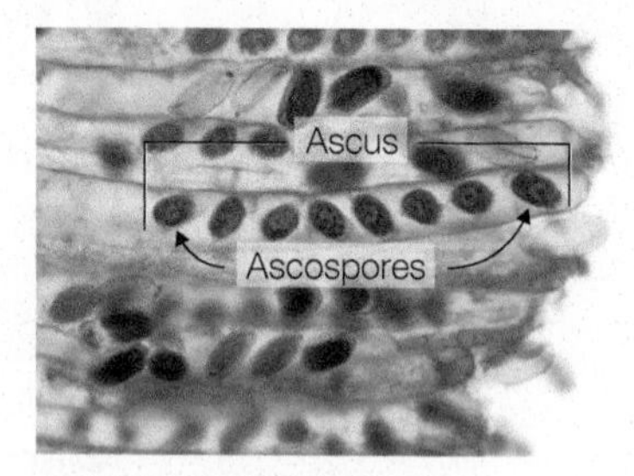

Ascospores from cup fungus

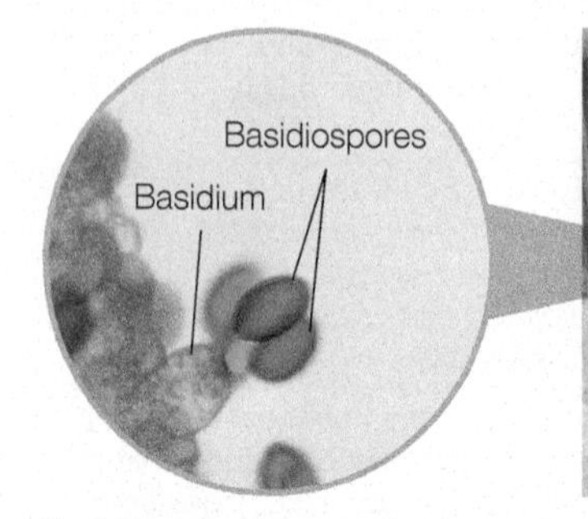

Basidiospores from mushrooms

FIGURE 4.9 Examples of fungal spores

TABLE 4.4 Fungal Spores

Type	Asexual Fungal Spores		Sexual Fungal Spores		
Name	Conidiospores	Sporangiospores	Zygospores	Ascospores	Basidiospores
Form	Chains of spores; not enclosed in a sac	Spores formed within a sac called a sporangium	Haploid gametes found at the tips of hyphae	Haploid gametes form within a sac called an ascus	Bud off of a pedestal structure called the basidium
Examples	*Penicillium* (source of penicillin) and *Aspergillus* species	*Absidia* species (the cause of mucormycosis in humans)	Phylum Zygomycota; includes black bread molds *(Rhizopus* species)	Phylum Ascomycota; includes truffles, morels, many yeasts, and cup fungi	Phylum Basidiomycota; includes mushrooms

microbiota through interventions like antibiotic therapies; this is the case for many vaginal yeast infections caused by *Candida* species.

However, some fungi are true pathogens that infect even a typically healthy host without a disruption in the normal microbiota. Such infections include histoplasmosis and coccidioidomycosis (cock-SID-ee-oh-doh-*my*-co-sis), also called valley fever. These infections are discussed later, in Chapter 16. Fungi called **dermatophytes** are also true pathogens; they infect the skin, hair, and nails and break down the protein keratin in these structures (FIG. 4.10). Common dermatophytic infections are **tinea**, or ringworm infections. Although the term *ringworm* suggests that a worm is responsible, these infections are actually caused by fungi. Tinea infections tend to be named for the body region they affect. For example, tinea pedis, or athlete's foot, develops when fungi such as *Trichophyton* species infect the feet.

Even if a fungus doesn't cause an infection, it could cause other clinical issues by stimulating allergies or by producing potentially deadly toxins called **mycotoxins**. The fungus *Claviceps purpurea* produces ergot toxin, a potent neurotoxin that can lead to seizures, psychosis, nausea, vomiting, and even death.

While we will not review agriculturally important fungi in this text, you should know that they represent a major challenge in farming, as they can readily infect and destroy crops. Lastly, a number of fungi are medically important not for causing human suffering, but for curing it. Thanks to their production of diverse antibiotics, a number of fungi save lives. Perhaps as more fungal species are discovered, we will naturally expand our armory of antimicrobial compounds. The roles of fungi in making antibiotics are discussed more in Chapter 15.

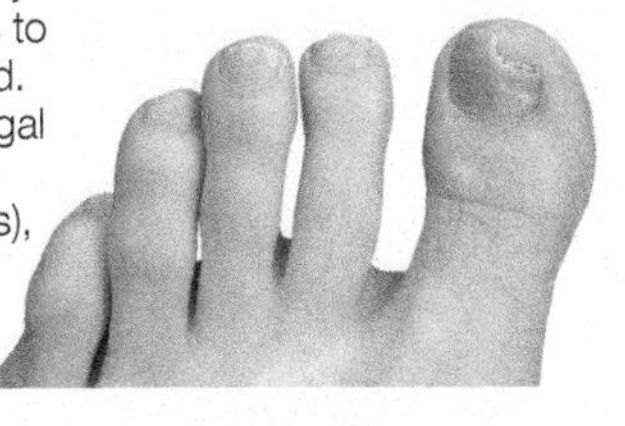

Tinea unguium, or dermatophytic onychomycosis, causes nails to become brittle and discolored. Some risk factors for this fungal infection include an infection with athlete's foot (tinea pedis), nail damage (as can occur when applying acrylic nails), and immunosuppression.

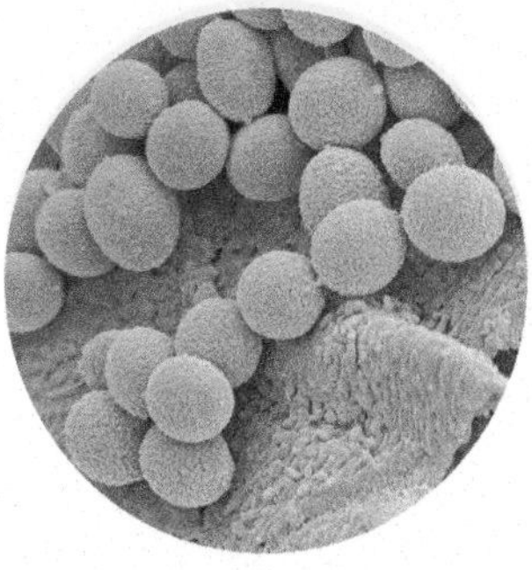

Tinea pedis, or athlete's foot, is a common infection that often starts between the toes and causes itching and burning as well as scaling skin. This colored SEM shows fungal spores (yellow) growing on skin of a human foot (pink).

FIGURE 4.10 Tinea unguium and tinea pedis

Protists

The miscellaneous nature of protists makes it difficult for scientists to agree on standard features that should be used to group them.[6] Although most protists are unicellular, some, like algae, are multicellular, and others, such as slime molds, form multinucleated cell masses when stressed (FIG. 4.11). The earliest eukaryotes probably resembled some of our modern-day protists; thus, these organisms are thought to represent an evolutionary link to plants, fungi, and animals. As evidence of this evolutionary link, some protists such as algae have plant-like features, including the ability to conduct photosynthesis and the presence of a cell wall. Some, like slime molds, are fungi-like saprobes that have cell walls and don't carry out photosynthesis. Others are called **protozoans**, which means "first animal." The term *protozoan* is not an official taxonomic rank; rather it is a term of convenience to describe animal-like protists that are unicellular, lack a true cell wall, exhibit asexual and sexual reproduction, and typically live by heterotrophic means (which means most don't conduct

Kelp are considered protists. These photosynthetic organisms can grow enormously tall and form kelp forests that are central to marine ecosystems.

This harmless "dog vomit slime mold," named for its texture and coloration, can often be spotted growing on mulch in gardens. It typically starts off yellow and eventually becomes a brownish-red before it hardens to a dry mass filled with spores.

FIGURE 4.11 Algae and slime molds

[6] In the past, protists were all placed in Kingdom Protista, but genetic analysis has since shown that this kingdom is actually a collection of organisms that are not all related. However, rather than review the complicated modern protist taxonomy here, in this text we opt to continue grouping these organisms together in the traditional manner. It's hoped that this approach will help you to appreciate the amazing diversity of these organisms without feeling overwhelmed by the taxonomic details that evolutionary biologists debate.

FIGURE 4.12 **Protozoan groups** These animal-like protists are grouped by their means of motility.

Amoeboid	Flagellated	Ciliated	Spore-forming
Pseudopods	Flagella	Cilia	
Entamoeba histolytica	*Giardia lamblia*	*Tetrahymena thermophila*	*Plasmodium sp.* (sporozoite stage)

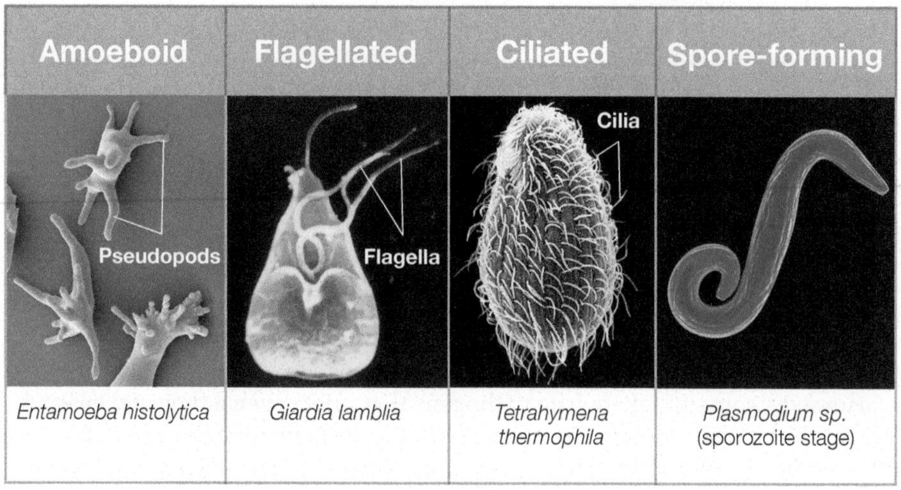

photosynthesis, although some, like *Euglena*, do). Protozoans that cause disease are always nonphotosynthetic.

People in developed countries often think of protozoan pathogens as only infecting those who live in or visit developing countries. This is a dangerous misconception that leads to misdiagnosis and underdiagnosis of protozoan infections in the developed world. In reality, protozoan pathogens are problematic in every nation. For example, the U.S. Centers for Disease Control and Prevention (CDC) estimates that at least 11 percent of people in the United States age 6 or older have had toxoplasmosis—an infection caused by the protozoan *Toxoplasma gondii*. Furthermore, the CDC estimates that in any given year about 3.7 million people in the United States are infected with the sexually transmitted protozoan *Trichomonas vaginalis*, and there are roughly 2 million cases of giardiasis, caused by various species of the *Giardia* protozoan, in the United States each year. (Each of these protozoan parasites is described in more detail in later sections.)

Protozoans include a number of pathogens, making them noteworthy in human and veterinary medicine. In general, they fall into four key groups based on their means of motility in their mature form. These groups are amoeboid, flagellated, ciliated, and spore-forming (FIG. 4.12).

Amoeboid Protozoans (Sarcodina) Amoeboid (uh-MEE-boid) protozoans use "false feet"—extensions of their cytoplasm called **pseudopods** (SUE-doh-pods) (also known as pseudopodia)—for movement. Many amoeboid protozoans are free living and don't need a host—for example, *Amoeba proteus*. However, others are parasitic. A number of amoebas such as *Naegleria fowleri*, *Acanthamoeba*, *Hartmannella*, and *Balamuthia* are human pathogens that cause rare forms of encephalitis, a potentially fatal inflammation of the brain. *Acanthamoeba* also causes corneal keratitis, a painful inflammation of the transparent front portion of the eye called the cornea, which can lead to blindness.

Entamoeba histolytica (EN-tuh-mee-buh HIS-toh-lit-ik-ah) is the most common amoeboid infection in humans. It infects people when they ingest food or water contaminated with the cyst form of the parasite, a form in which the parasite is encased in a tough protective layer. The cysts are excreted in feces of infected humans and other primates. About 90 percent of infections are asymptomatic; the other 10 percent of infections lead to amoebic dysentery (amoebiasis)—a form of severe diarrhea that includes the passing of mucus and blood. The WHO estimates that worldwide, there are about 50 million cases of amoebiasis every year, with about 70,000 deaths due to this parasite. *E. histolytica* infections can be treated, but it is challenging to destroy the cyst forms of the amoeba. Complications of amoebiasis can include liver abscess, intestinal perforation (tearing), and colitis. Amoebiasis is most common in Africa, Latin America, Southeast Asia, and India.

Amoeboid Infection and Neti Pots

Normally free-living protists, *Naegleria fowleri* exist in moist soil and warm waters like hot springs, lakes, or rivers. In humans, *N. fowleri* causes primary amoebic meningoencephalitis, an infection with a 97 percent mortality rate. This protozoan has been associated with the use of a neti pot, which many allergy and cold sufferers use to alleviate nasal congestion. A neti pot is a small basin with a narrow funnel through which warm water is introduced into one nostril. After rinsing the nasal passages, the water drains through the other nostril.

You cannot get primary amoebic meningoencephalitis by drinking contaminated water, because stomach acid kills the microbe. Disease only develops if the amoeboid enters through the nose and then migrates to the brain—hence its nickname, the "brain-eating amoeba." Headache, fever, nausea, or vomiting begin within a week of infection and closely resemble bacterial meningitis. The patient may then develop neck stiffness, compromised balance, hallucinations, cognitive impairment, and seizures. Death usually occurs within 12 days of initial symptoms.

Overall, the risk of infection is extremely low—according to the U.S. Centers for Disease Control and Prevention, in a span of over 50 years (1962–2019), only 148 infections were reported in the United States, and most were traced to swimming. Only a few cases resulted from nasal irrigation with contaminated tap water in neti pots. But given the high mortality rate, health officials urge neti pot users to take precautions. Standard tap water should never be placed in a neti pot; sterile or distilled solutions purchased at pharmacies are recommended. Homemade saline solution is also safe if the tap water is boiled for 3 minutes and cooled before use.

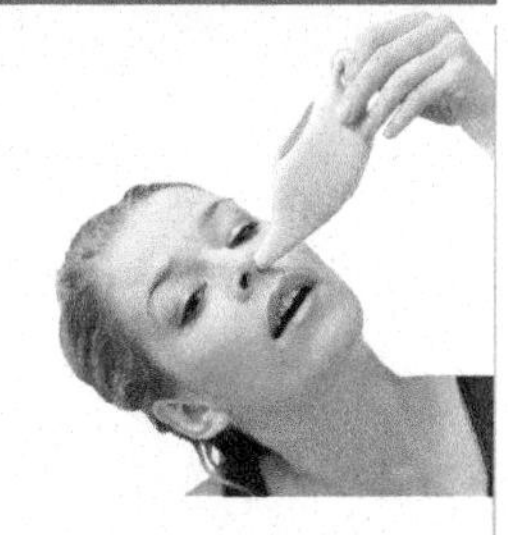

Improper use of neti pots has been associated with fatal *N. fowleri* infections.

QUESTION 4.2

Based on your readings, if you were to look at N. fowleri *under the microscope, what key features would you expect to see?*

Flagellated Protozoans (Mastigophora) Flagellated protozoans have one or more flagella for motility (more on eukaryotic flagella later). These protozoans include free-living organisms like *Euglena* and parasitic organisms such as *Trichomonas vaginalis*, *Trypanosoma* species, and *Giardia lamblia*—which are all human pathogens. *Trichomonas vaginalis* is a sexually transmitted protozoan commonly responsible for infections in industrialized and developing nations alike. African sleeping sickness, which is caused by *Trypanosoma* species, is a dangerous protozoan disease transmitted by tsetse flies. *Giardia lamblia* (also referred to as *Giardia intestinalis* or *Giardia duodenalis*), has five flagella that propel it through water and help it adhere to the intestinal epithelium of animals. The cyst stage of *Giardia* has a protective outer coat that can resist chemicals like chlorine, which are commonly used to treat recreational and drinking water. Outbreaks of giardiasis often occur when water sanitation processes are compromised, as is common following hurricanes, earthquakes, and other natural disasters. (See Chapter 18 for more on African sleeping sickness, Chapter 19 for giardiasis, and Chapter 20 for trichomoniasis.)

Ciliated Protozoans (Ciliophora) Ciliated (SIL-ee-ay-tid) protozoans use hair-like appendages called cilia for motility (more on cilia later). Ciliates are common in aquatic environments and include diverse species from free-living *Paramecium* species to parasites like *Balantidium coli*—the only ciliated protozoan known to cause human disease. Balantidiasis is a rare form of dysentery that can develop when the host ingests the cyst form of *Balantidium coli* in contaminated food or water. The protozoan can then grow in the colon and cause persistent diarrhea and abdominal pain, and in severe cases it can lead to perforation (tearing) of the colon.

Spore-Forming Protozoans (Apicomplexa) The phylum Apicomplexa (formerly called Sporozoa) is one of the largest phyla of protozoans. In their mature form, these microbes have no flagella, cilia, or pseudopodia and instead move by gliding. Most apicomplexans are obligate intracellular parasites, meaning they cannot survive outside of a host cell. These parasites have complex life

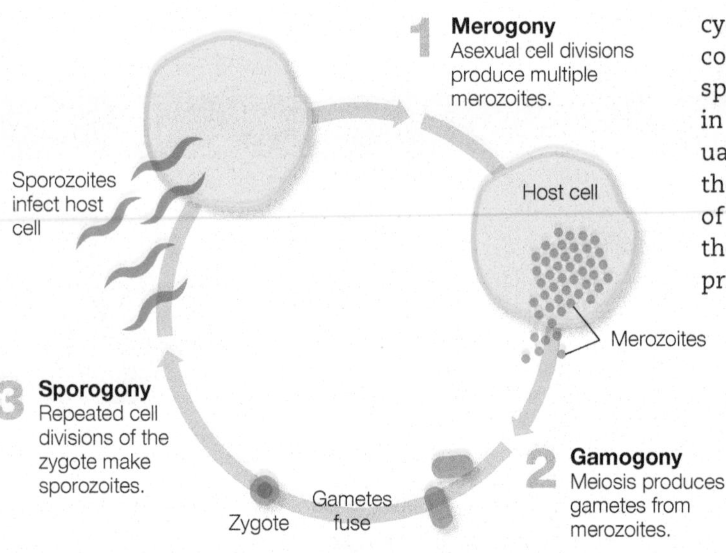

FIGURE 4.13 Generalized life cycle of spore-forming protozoans In general, apicomplexans go through three life stages: merogony, gamogony, and sporogony.

Critical Thinking *At what stage would you expect the organism to invade a new host?*

cycles that include sexual and asexual stages. In general, the Apicomplexans go through three phases: merogony, gamogony, and sporogony (FIG. 4.13). **Merogony** is an asexual stage of reproduction, in which daughter cells called *merozoites* are made by repeated asexual cell division. This occurs by a modified form of mitosis in which the nucleus divides multiple times and is later followed by division of the remaining cell components into discrete cells. **Gamogony** is the sexual phase of reproduction in which the merozoites from the prior stage produce male and female haploid gametes by meiosis. Male and female gametes then randomly fuse. In **sporogony**, the zygote made by gamete fusion divides repeatedly to make *sporozoites*. Typically, the sporozoites are the stage that invades a new host to repeat the reproductive cycle.

Apicomplexans are notorious human pathogens, causing diseases such as toxoplasmosis (often spread by cats), malaria, and gastrointestinal illnesses such as cryptosporidiosis and microsporidiosis. Some of the members of this phylum require arthropod vectors like mosquitoes or ticks for one part of their life cycle. For example, *Plasmodium* species, which cause malaria, have their sexual phase of reproduction in mosquitoes. In contrast, their asexual phase of reproduction occurs in humans. (Protozoan pathogens are discussed more in the disease chapters of this text—Chapters 16 through 21.)

Build Your Foundation

5. What eukaryotic kingdom(s) include unicellular organisms? Which include multicellular organisms? Which have photosynthetic organisms? (LO 4.5)
6. What are the two main groupings of helminths? (LO 4.6)
7. What are the tubular extensions that make up most fungi called, and what are their two forms? (LO 4.7)
8. Name the types of sexual and asexual fungal spores, and state how they are made. (LO 4.8)
9. Define the term *mycosis*, and give an example from this chapter. (LO 4.9)
10. Why is it challenging to develop specific criteria for describing protists? (LO 4.10)
11. What are animal-like protists called, and how are they primarily classified? (LO 4.11)

BYF QUICK QUIZ

Build Your Foundation (BYF) Quick Quiz: Visit the **Mastering Microbiology** Study Area to quiz yourself.

4.3 EXTRACELLULAR STRUCTURES

Learning Outcomes

After reading this section, you should be able to:

4.12 Discuss the basic structural and functional features of eukaryotic plasma membranes and how they differ among kingdoms.
4.13 Give an example of eukaryotes with cell walls, and discuss how cell walls differ among eukaryotic kingdoms.
4.14 Describe the eukaryotic glycocalyx, and list some of its roles.
4.15 Discuss the basic structure of eukaryotic flagella, and compare eukaryotic and prokaryotic flagella.
4.16 Describe the structure and function of cilia.

All eukaryotes have a plasma membrane.

All cells have a plasma membrane, which serves as a selective barrier and interface through which the cell interacts with its outside environment. Typically, cells have a plasma membrane with a **phospholipid** bilayer structure that's crowded with a variety of receptors and channels for normal cell functions, including cellular communication and transport processes. (See Chapter 3 to review phospholipid bilayers.)

Although it's rare for prokaryotes to contain **sterols** in their membranes, eukaryotic membranes contain many sterols, which have central roles in maintaining membrane stability and fluidity. Cholesterol is especially abundant in animal cell membranes, whereas plant cell membranes contain a wide variety of phytosterols as well as low levels of cholesterol. Fungi and protists have a wide range of sterols in their membranes, with *ergosterol* being a key example. Certain azole drugs and other antifungals such as amphotericin B and nystatin target the ergosterol in plasma membranes of a wide range of fungal and protozoan pathogens. Because animal cells lack ergosterol, these drugs do not target human cells.

Certain eukaryotes have a cell wall.

Fungi, plants, and certain protists have cell walls, but animal cells don't. When a cell wall is present, it's external to the plasma membrane. This structure helps maintain cell shape and protects the cell against mechanical and **osmotic stress**. Despite the chemical diversity seen among eukaryotic cell walls, none of them contain peptidoglycan or pseudopeptidoglycan. This feature makes them chemically distinct from prokaryotic cell walls. In general, *chitin* (KITE-in) is a core compound in many fungal cell walls, while cellulose is abundant in plant cell walls. Protists have diverse cell walls that can include cellulose, calcium carbonate, xylan, silica, and a variety of other protein- and carbohydrate-based compounds.

CHEM • NOTE

Phospholipids are amphipathic molecules with a hydrophilic (water-loving) phosphate head group and hydrophobic (water-hating) fatty acid chains. See Chapter 2 for more on phospholipids.

CHEM • NOTE

Sterols, or steroid alcohols, are organic ring-structured compounds; the most common example is cholesterol. These compounds are hydrophobic, which means they do not like interacting with water. See Chapter 2 for more on sterols.

CHEM • NOTE

Osmotic stress may occur when a cell tries to maintain its water balance by counteracting the movement of water down its concentration gradient. Such a situation would occur when the solute concentration in the cell and its environment are unequal. Cells in hypotonic environments will burst as they take on water unless they have an intact cell wall. Hypertonicity and hypotonicity are also reviewed in Chapters 2 and 3.

Many eukaryotes have structures for protection, adhesion, and movement.

Structures external to the plasma membrane and cell wall have roles in cellular protection, adhesion, and motility. Although the general functions of these eukaryotic cellular tools don't differ greatly from those found in prokaryotes, there are some structural differences to consider.

Eukaryotic Glycocalyx

Like many prokaryotes, most eukaryotic cells have a sticky extracellular layer called a **glycocalyx** (which means "sugar husk") as their outermost coat. It contains a diverse collection of carbohydrates, glycoproteins, and glycolipids. Each of these sugary components can have a different role in protecting the cell from a variety of stresses, promoting or preventing cell adhesion where appropriate, and assisting with cellular communication (FIG. 4.14). In multicellular organisms, the eukaryotic glycocalyx is also central to proper tissue development. In contrast, prokaryotic cells do not form higher-order tissues, so although their glycocalyx is important to their physiology and their ability to make biofilms, it has a slightly less sophisticated role.

The composition of the glycocalyx can vary, and some of these variations may help cancer to progress or infectious agents to establish disease. For

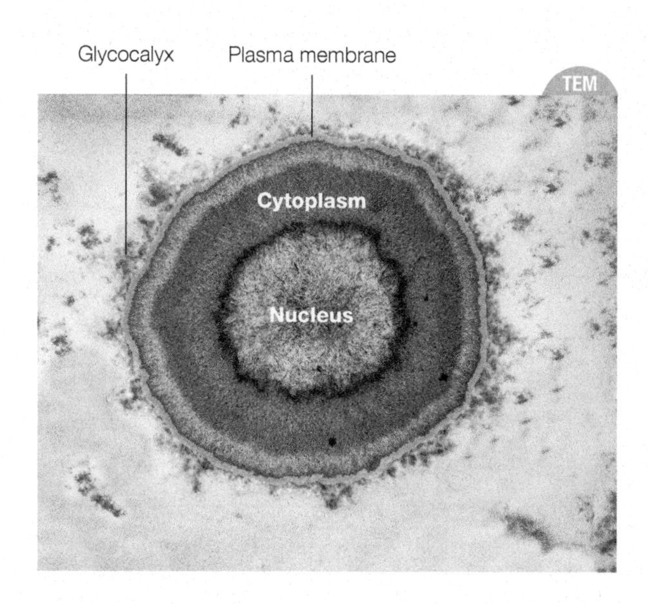

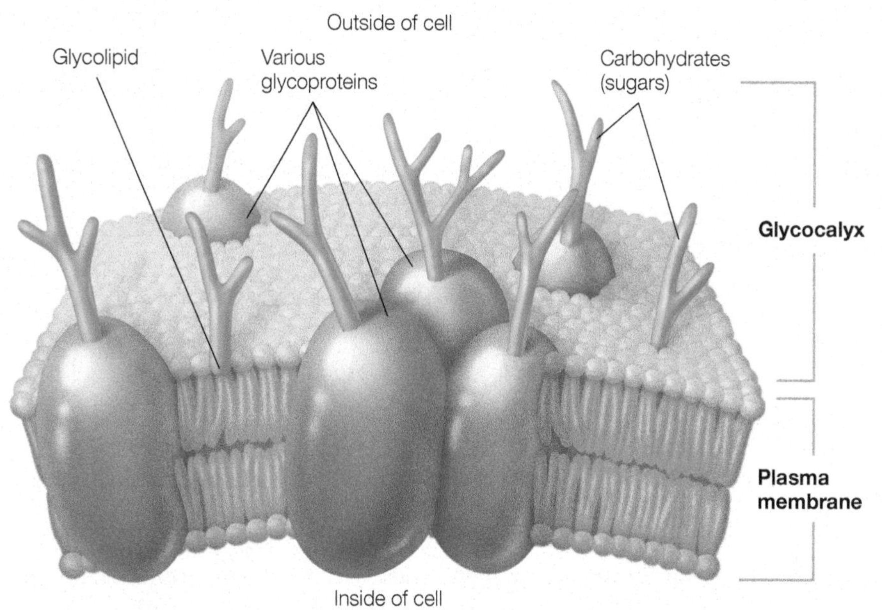

FIGURE 4.14 Eukaryotic glycocalyx The eukaryotic glycocalyx serves roles in cell protection, adhesion, and assisting with cellular communication.

Critical Thinking *You read that pathogens often rely on components of their glycocalyx to adhere to target cells, but how might a host cell's glycocalyx impact what pathogens can invade it?*

example, cancer cells with an abnormally thickened glycocalyx tend to more readily metastasize, or spread.[7] In terms of infectious agents, the glycocalyx of pathogenic fungi and protozoans can differ in ways that influence infectivity and immune responses, making this structure important to consider in drug and vaccine development. Some glycocalyx variations reduce pathogenicity: Studies done with the protozoan pathogen *Leishmania major*, which causes skin lesions, showed that mutations to one of the organism's glycocalyx components decreased the pathogen's ability to invade host cells and also made it more susceptible to immune system attack.[8]

One of nine sets of fused microtubule pairs
Central microtubules
Flagellum
Plasma membrane
Microtubules sprouting from a centriole

FIGURE 4.15 Eukaryotic flagella Eukaryotic flagella have a nine-plus-two arrangement of microtubules.

Flagella

Flagella are long, tail-like structures used for motility. Eukaryotes, like prokaryotes, may have one (called a flagellum) or multiple flagella. Many gametes essential to eukaryotic reproduction and a variety of protists are flagellated. Structural differences between prokaryotic and eukaryotic flagella are identified in TABLE 4.5. Not only is the eukaryotic flagellum thicker and typically longer than that of prokaryotes, but also it is wrapped in a membrane derived from the cell's plasma membrane. Therefore, the inner part of the flagellum isn't a sealed-off compartment—instead, it is continuous with the rest of the cell's cytoplasm. Eukaryotic flagella are made of the protein *tubulin* and have what is called a *nine-plus-two arrangement* $(9 + 2)$, in which nine fused pairs of tube-like protein structures called microtubules (described later in this chapter) form the border of the flagellum, and two nonfused microtubules are at the center. The eukaryotic flagellum is anchored to the cell by a basal body—a structure consisting of a cylindrical collection of microtubules sprouting from a centriole (the centriole is an intracellular structure described shortly) (FIG. 4.15). As you may recall from Chapter 3, prokaryotic flagella have a rotary, or propeller-like, motion; in contrast, eukaryotic flagella have a wave-like, back-and-forth motion (FIG. 4.16, *top*).

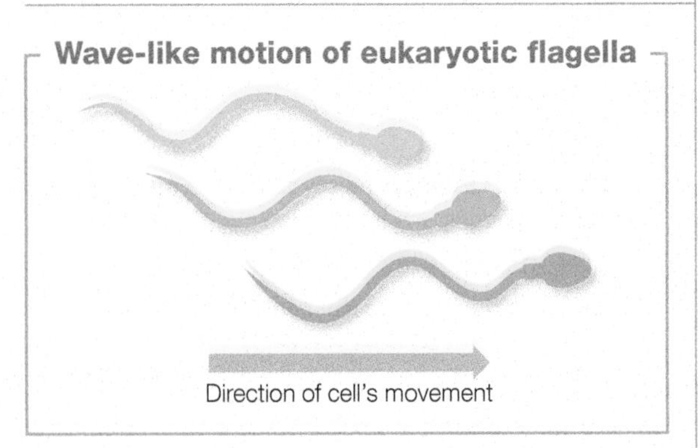

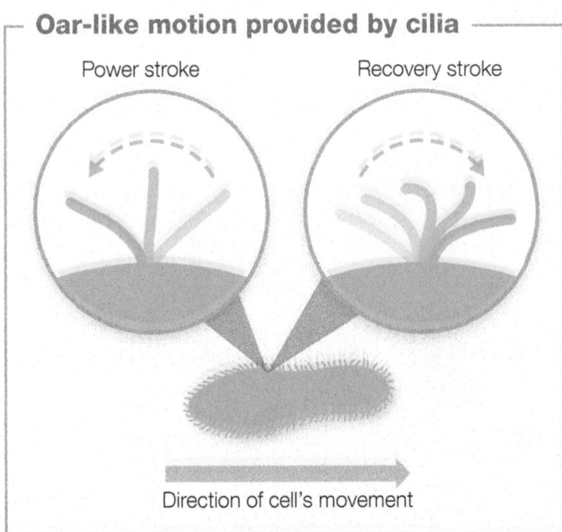

FIGURE 4.16 Flagella and cilia motion Flagella and cilia differ in how they propel an organism.

Cilia

Cilia, the Latin word for "eyelashes," structurally resemble flagella except that they are much shorter and far more numerous on a cell. Only eukaryotes have cilia. The synchronized rowing motion they provide helps *Paramecium* species and other ciliophora protists move in their aquatic environments (FIG. 4.16, *bottom*). Cilia are also found on certain animal cells. One example is the epithelial cells of our upper airway, which sweep mucus and pathogens up and away from the lungs. (For more on the mucociliary escalator, see Chapter 11.) Like flagella, cilia are surrounded by a membrane and sprout from centrioles associated with the plasma membrane.

TABLE 4.5 Prokaryotic versus Eukaryotic Flagella

Flagella Type	Prokaryotic	Eukaryotic
Built from	Flagellin protein	Tubulin protein
Microtubules	No	Yes; $9 + 2$ arrangement
Membrane enclosed	No, except for periplasmic flagella	Yes
Anchor	Hook-and-filament structure anchored by rings	Microtubules sprout from a centriole
Motion	Rotary (propeller)	Wave-like (whips back and forth)

[7] Buffone, A., & Weaver, V. M. (2020). Don't sugarcoat it: How glycocalyx composition influences cancer progression. *The Journal of Cell Biology*, *219*(1).
[8] Abaza, S. (2020). Virulence factors. *Parasitologists United Journal*, *13*(2), 76–92.

Build Your Foundation

12. How are fungal and protist plasma membranes distinguishable from other eukaryotic plasma membranes? (LO 4.12)
13. List some eukaryotes that have cell walls. (LO 4.13)
14. What is the eukaryotic glycocalyx, and what does it do? (LO 4.14)
15. Name three differences between prokaryotic and eukaryotic flagella. (LO 4.15)
16. What are cilia, and where can they be found? (LO 4.16)

Build Your Foundation (BYF) Quick Quiz: Visit the **Mastering Microbiology** Study Area to quiz yourself.

4.4 INTRACELLULAR STRUCTURES

Ribosomes can be free or bound to a membrane.

Learning Outcomes

After reading this section, you should be able to:

4.17 Describe the structure of eukaryotic ribosomes, and state where they are located.

4.18 Explain the structure and function of the cytoskeleton, and name the organelle that builds microtubules.

4.19 Describe the general structure and function of the nucleus.

4.20 Discuss the basic structural and functional features of the endoplasmic reticulum.

4.21 Describe the general structural and functional features of the Golgi apparatus.

4.22 Outline the types of vesicles and vacuoles that exist in eukaryotic cells, and state their general functions.

4.23 Review the structure and function of mitochondria, and state what features make them similar to bacteria.

As discussed in Chapter 3, both prokaryotic and eukaryotic cells have **ribosomes**, organelles essential for making proteins. All ribosomes are made up of protein and ribosomal RNA (rRNA). Eukaryotic cells have 80S ribosomes that are made up of a small subunit (the 40S subunit) and a large subunit (the 60S subunit) (FIG. 4.17). (Refer back to Chapter 3 for a review of Svedberg units, which are used to denote the sedimentation rates of the subunits.) Eukaryotic ribosomes can be found free in the cytoplasm or bound to the membrane of an organelle called the endoplasmic reticulum (ER).

The main difference between bound and free ribosomes is not in the ribosomes themselves, but in what type of proteins they make. Ribosomes that are bound to the ER tend to produce proteins that are destined for secretion from the cell, whereas ribosomes free in the cytoplasm tend to make proteins that will be used inside the cell. That said, free ribosomes can attach to the ER if they are making proteins bound for secretion. Similarly, bound ribosomes can become free ribosomes as needed. This means that ribosomal populations are not static and can change from free to bound, based on the protein production demands of the cell.

Lastly, inside mitochondria and chloroplasts, eukaryotes have ribosomes that resemble 70S ribosomes found in prokaryotes. These mitochondrial and chloroplastic ribosomes are separate from the membrane-bound and free ribosomes previously discussed.

Many toxins target eukaryotic ribosomes. For example, ricin, a deadly toxin made by castor beans, binds to the 60S subunit of eukaryotic ribosomes and blocks protein production. Similarly, toxins made by certain *Shigella* species and some *Escherichia coli* strains (bacteria that can cause severe dysentery) also inactivate the 60S subunit of the eukaryotic ribosome and disrupt protein synthesis in affected cells.

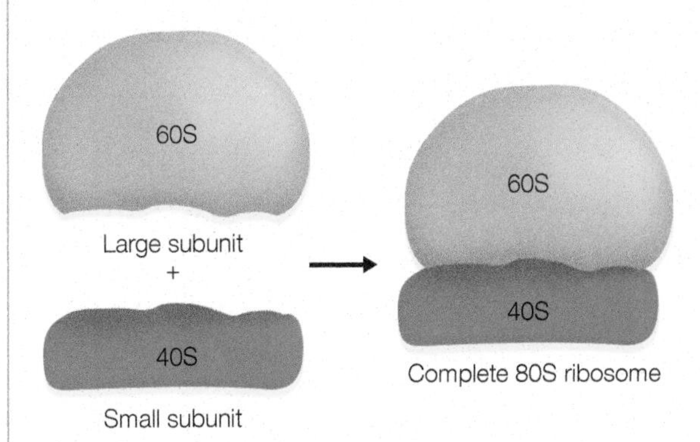

FIGURE 4.17 Eukaryotic ribosomes Eukaryotes have 80S ribosomes in their cytoplasm and attached to the rough endoplasmic reticulum.

Critical Thinking *Why don't antibacterial drugs necessarily target eukaryotic ribosomes?*

The cytoskeleton shapes cells and coordinates cell cargo movement.

The **cytoskeleton** is a dynamic and responsive intracellular network of protein fibers that helps maintain shape, facilitates movement, protects against external forces that may otherwise deform the cell, and directs transport of vesicles, organelles, and other cellular cargo. It also coordinates cell division by moving chromosomes and organelles to developing daughter cells. A number of pathogens interact with the eukaryotic cytoskeleton as a part of their toolbox for causing disease. For example, the bacterium *Salmonella enterica*, which causes a severe form of gastroenteritis, releases substances that affect the host cell's cytoskeleton in a way that helps the pathogen invade gut epithelial cells. *Listeria monocytogenes* bacteria (the causative agent of listeriosis, a foodborne infection) hijack certain fibers in the host cytoskeleton to jump from one host

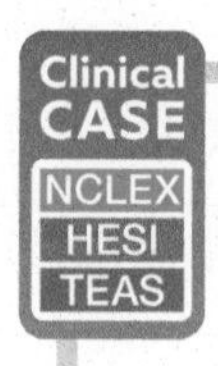

The Case of the Infectious Hike

Practice applying what you know clinically: visit the **Mastering Microbiology** Study Area to watch Part 2 and practice for future exams.

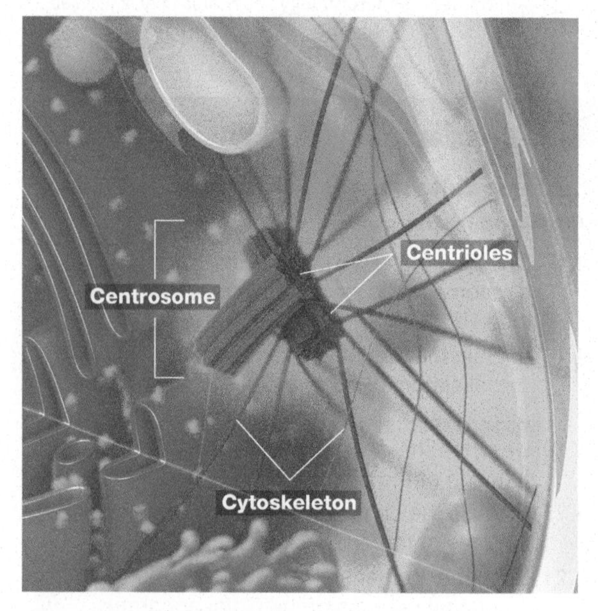

FIGURE 4.18 The centrosome and the cytoskeleton The cytoskeleton is made of fibrous proteins that stem from an organelle called the centrosome. The centrosome is made up of two centrioles and surrounding unstructured material.

cell to another.[9] Additionally, the cytoskeleton is an important target for anti-cancer drugs such as Taxol and vincristine.

The cytoskeleton is made up of three main fibers: microtubules, intermediate filaments, and microfilaments. We'll discuss each of these fibers next.

Microtubules

Microtubules are hollow tubes made of a protein called **tubulin**. They tend to be about 25 nm in diameter, making them the thickest of the three cytoskeleton fibers. (To give you an idea of scale, a helium atom is about 1 nm in diameter.) Microtubules serve as roadways in the cell. Just as a car takes a particular street to reach a destination, ferrying proteins such as *kinesin* and *dynein* move cargo along microtubules. During cell division, microtubules polymerize to form a structure called the *spindle*. The spindle helps separate chromosomes and deliver organelles to the newly forming daughter cells. Microtubules usually arise from and are attached to a specialized organelle called the **centrosome**, which is normally located near the nucleus. The centrosome is made up of two *centrioles* (barrel-shaped structures made of microtubules) and surrounding unstructured material (FIG. 4.18).

Intermediate Filaments

Intermediate filaments are rope-like fibers about 10 nm in diameter. They mainly contribute tensile strength to the cytoskeleton in order to oppose external mechanical forces. Many different proteins have been found in intermediate filaments; their composition varies from one type of cell to another. For example, the intermediate filaments in a skin cell are likely to be made of keratin proteins, while those in a muscle are more likely to be made of desmin proteins.

Microfilaments

Microfilaments are fine fibers made of the protein **actin**. They tend to be 3–6 nm in diameter and associate with the motor protein **myosin** to facilitate movement. The collaboration between actin and myosin allows our muscles to contract and helps cells divide by pinching apart the newly formed daughter cells made in mitosis and meiosis. Actin microfilaments also act as roadways in cells, used by myosin motor proteins to ferry cargo from one point in the cell to another. Actin fibers also polymerize and depolymerize, thereby forming and then deconstructing pseudopodia that facilitate the cell movement that amoeboid protozoans use.

Eukaryotes have a variety of membrane-bound organelles.

Membranous organelles are unique to eukaryotes. They compartmentalize cells and allow them to concentrate specific substances in areas where needed. The term means "little organ," and indeed you can think of these intracellular structures as having specific roles in the health and survival of the cell—much like organs have specialized roles in maintaining life in a complex, multicellular organism. The intracellular and extracellular structures we review in this chapter are summarized for you in TABLE 4.6.

Nucleus

As alluded to at the beginning of the chapter, the exact origins and development of the **nucleus** are still debated. Nevertheless, this distinct,

[9] Souza Santos, M., & Orth, K. (2015). Subversion of the cytoskeleton by intracellular bacteria: Lessons from *Listeria, Salmonella* and *Vibrio*. *Cellular Microbiology*, *17*(2), 164–173.

TABLE 4.6 Summary of Extracellular and Intracellular Structures

Location	Structure	Function/Notes
Extracellular	Plasma membrane	Present in all cells; phospholipid bilayer is a selective barrier and interfaces with extracellular environment
	Cell wall	In most prokaryotes and some eukaryotes; adds rigidity and protects cell from mechanical and osmotic stress
	Glycocalyx	Sticky layer present in prokaryotes and eukaryotes; important in cell adhesion, protection, and cell communication
	Flagella	Motility of certain prokaryotes and eukaryotes; eukaryotic and prokaryotic flagella have different general structures
	Cilia	Only in eukaryotes; short and numerous hair-like extensions used for motility and also found on the surface of various human cells—including cells of the upper airway as part of the mucociliary escalator
Intracellular	Ribosomes	Found in prokaryotes and eukaryotes; build proteins; in eukaryotes, 80S ribosomes are membrane bound or free in the cytoplasm, while 70S ribosomes are in mitochondria and chloroplasts
	Cytoskeleton	Made of microtubules, intermediate filaments and microfilaments; provides cell shape, plays important roles in cell movement, facilitates cargo transport, and protects the cell against external mechanical stress forces; prokaryotes also have a cytoskeleton, although it is different in structure and in various functions from that of eukaryotes
	Nucleus	Only in eukaryotes; houses the cell's DNA and serves as the cell's command center; dense region of the nucleus, called the nucleolus, is enriched with RNA
	Endoplasmic reticulum	Only in eukaryotes; rough ER has ribosomes on surface and is mainly involved in protein production, modification, and folding; smooth ER lacks ribosomes on surface and is central to lipid production and detoxification
	Golgi apparatus	Only in eukaryotes; modifies cellular proteins, builds lipids, and further sorts and distributes the finished products; central to secretion
	Mitochondria	Only in eukaryotes; generate most cell ATP; contain own circular genome and 70S ribosomes
	Chloroplasts	Only in photosynthetic eukaryotes; harvest energy from light through photosynthesis; contain own circular genome and 70S ribosomes
	Vesicles and vacuoles	Bud from membranous organelles and/or plasma membrane; roles in storing and transporting substances; certain vesicle structures like lysosomes and peroxisomes are only in eukaryotes

membrane-enclosed structure is a hallmark feature that evolved billions of years ago and forever separated eukaryotic cells from their prokaryotic cousins (FIG. 4.19). Because of its fairly large size, the nucleus is typically visible when viewing eukaryotic cells with a light microscope. This command center of the cell houses DNA, which ultimately orchestrates all cellular activities. Unless the cell is dividing, the DNA of the nucleus is loosely organized as *chromatin* (a collection of DNA and protein) and floats in the liquid *nucleoplasm*. A dense region of the nucleus, called the *nucleolus*, is enriched with RNA. In particular, the nucleolus is the site where ribosomal subunits begin their development; the ribosomal subunits are eventually exported to the cytoplasm for maturation and assembly into complete ribosomes.

The nucleus is enclosed by a double-membrane structure called the nuclear envelope. This envelope has pores that control the traffic of materials into and out of the nucleus. The nuclear envelope gives rise to the endoplasmic reticulum, which is discussed next.

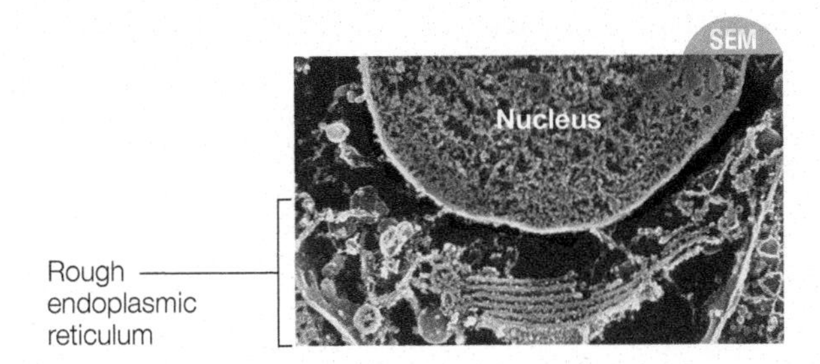

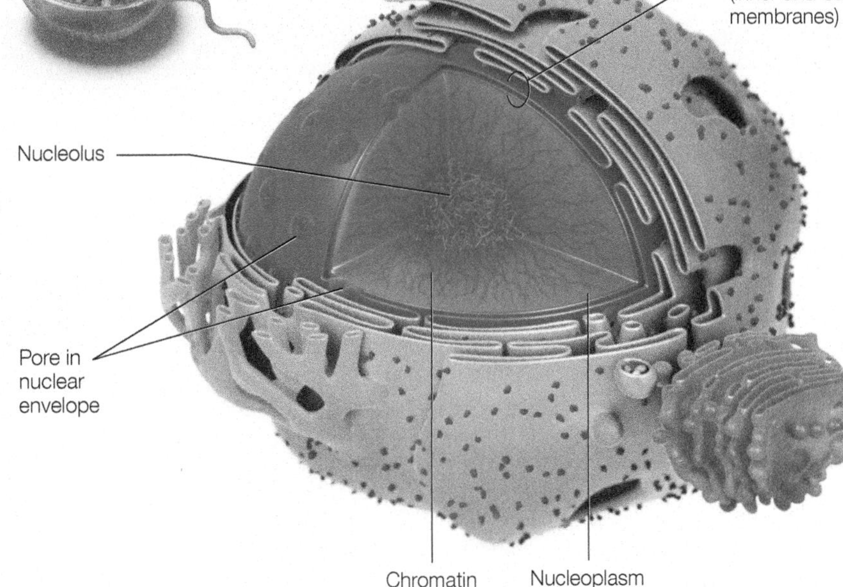

FIGURE 4.19 Nucleus The nucleus is a double-membrane-enclosed organelle that houses the cell's genome. The nuclear envelope, a double-membrane structure surrounding the nucleus, gives rise to the endoplasmic reticulum.

Critical Thinking *Explain why nuclear pores are important in a cell's nucleus.*

Endoplasmic Reticulum

The **endoplasmic reticulum (ER)** is an undulating series of interconnected membranous enclosures that originate from the outer membrane of the nuclear envelope (FIG. 4.20). It has essential roles in protein and lipid production. Many of the products made in the ER are destined for transport to other organelles or for secretion outside the cell.

The ER can be either rough or smooth. The *rough ER* has millions of ribosomes on the outer surface. Proteins made by these membrane-bound ribosomes can be directly inserted into the ER membrane or transported into the internal area of the ER, called the ER lumen. As a protein matures in the rough ER, it may be clipped, folded, and modified with organic and/or inorganic factors. The rough ER also serves as a quality control manager by making sure that proteins that are abnormally folded or formed are broken down.

In contrast to the rough ER, the *smooth ER* is not associated with ribosomes and has a more interlaced tubular appearance. It also tends to be more involved in lipid production than in protein modification and folding. Cells that specialize in lipid production, such as liver cells that make cholesterol and cells that are central for making lipid-based hormones, tend to have a more pronounced smooth ER than cells that do not specialize in this sort of biosynthesis. The smooth ER also has an important role in detoxifying substances like drugs, alcohol, and various metabolic byproducts.

Both the rough and the smooth ER also coordinate with the Golgi apparatus (discussed next) to sort, package, and ship proteins and lipids around the cell, as well as to ferry cargo to the cell surface for secretion or for incorporation into the plasma membrane.

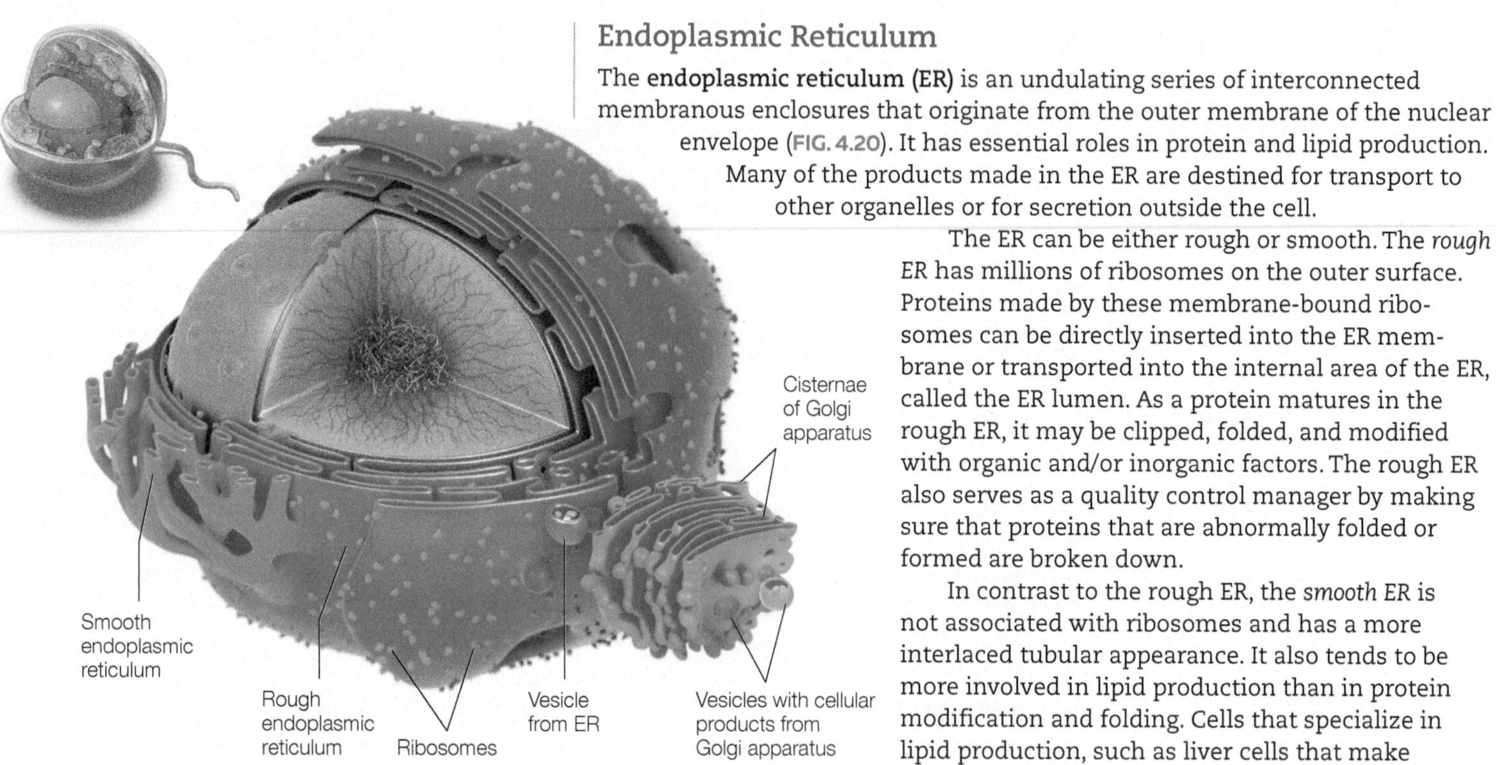

FIGURE 4.20 Endoplasmic reticulum (ER), Golgi apparatus, and vesicles *Left:* The smooth ER lacks ribosomes, while the rough ER is studded with them. *Right:* The Golgi apparatus modifies cellular proteins, builds lipids, and sorts and distributes a variety of products for the cell with vesicles that bud off the organelle.

Critical Thinking *Why does it make sense to have ribosomes associated with the ER?*

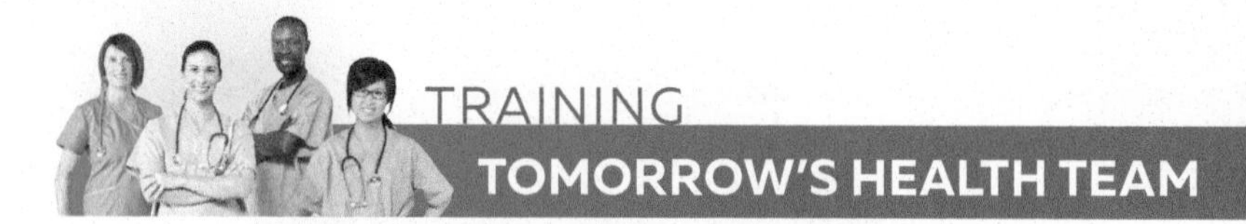

The Link between the Endoplasmic Reticulum Stress Response and Disease

When viruses and other pathogens invade eukaryotic cells, they often trigger a stress response in the endoplasmic reticulum. Normally the ER stress response limits pathogen spread because it helps stimulate an immune response. But the ER stress response also has a downside. If not resolved within a certain time frame, it triggers cell suicide (apoptosis). If extensive, apoptosis can lead to tissue damage.

The protozoan pathogen *Toxoplasma gondii* (which causes toxoplasmosis) and a number of bacterial pathogens induce a sustained ER stress response that promotes tissue damage. Other pathogens may induce an ER stress response that correlates with the development of certain cancers. Sustained ER stress responses also contribute to the pathology of a wide variety of nervous system disorders, including Alzheimer's, Parkinson's, and Huntington's diseases. Given the associations of the ER stress response to the progression of *many* infectious and noninfectious diseases, the ER has become a hot focus of research. Suppression of the ER stress response may one day become a viable option in the treatment plan for these diseases.

QUESTION 4.3

What potential side effects would ER stress inhibitors have? (Hint: Consider the normal role of a controlled ER stress response.)

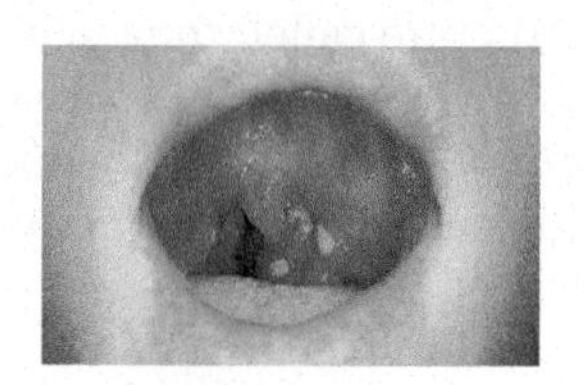

S. pyogenes, the causative agent of streptococcal pharyngitis (strep throat), is thought to benefit from the amino acids and other nutrients released into the surrounding environment by stressed host cells.

Golgi Apparatus

Vesicles that bud off the ER are often shuttled to the **Golgi apparatus** (also called the Golgi body). The Golgi apparatus resembles a series of disc-like, flattened sacs called *cisternae* that stack upon one another. It coordinates with the ER to modify cellular proteins, build lipids, and further sort and distribute the finished products. A variety of vesicles, small membranous pods that bud from the Golgi surface, are used to ship materials from the Golgi to their final destination, which is often the cell's plasma membrane, for secretion from the cell (FIGURE 4.20).

TEM of a Golgi apparatus.

Vesicles and Vacuoles

Vesicles are lipid bilayer sacs packed with substances ranging from simple organic and inorganic molecules to complex enzymes. There are several types. **Transport vesicles** move substances around the cell to diverse cellular destinations—including the plasma membrane. Vesicles destined for the plasma membrane fuse with the membrane, thereby replenishing the lipid bilayer and delivering membrane proteins and other necessary surface molecules. **Secretory vesicles** are a specific type of transport vesicle. They shuttle materials to the cell surface for discharge from the cell. In some instances the discharged materials are waste products; in other cases the secreted substances are bioactive molecules (for example, hormones and neurotransmitters secreted by certain cells in animal tissues).

Other important vesicles to consider are lysosomes and peroxisomes. **Lysosomes** contain a wide variety of hydrolytic enzymes that break down substances engulfed by the cell during phagocytosis and receptor-mediated endocytosis. They also act as garbage disposal tools for the cell, removing damaged organelles and other defunct cell components. **Peroxisomes** are found in most eukaryotic cells. They contain enzymes that break down certain fats and amino acids and protect the cell from hydrogen peroxide and other toxic oxygen intermediates.

Vacuoles are a conglomerate of many vesicles that have merged to make a larger membranous sac. Vacuoles contain mainly water and various organic and inorganic substances such as nutrients, toxins, and even waste products. Vacuoles vary in size and shape. They are common in plants and fungi, but may also be found in certain animal cells, prokaryotes, and protists.[10] For example, contractile vacuoles are commonly used by protists such as *Paramecium* and *Amoeba* species to regulate osmotic pressure to avoid lysing when in a **hypotonic environment**—an environment with a lower solute concentration than the cell.

CHEM • NOTE

Toxic oxygen intermediates—also called reactive oxygen species (ROS)—are chemically reactive molecules that contain oxygen. Examples include superoxide (O_2^-), peroxides (O_2^{-2}), and hydroxyl ions (OH^-). ROS can damage cellular proteins and nucleic acids.

CHEM • NOTE

When a cell is placed in a **hypotonic (low-solute) environment**, water is drawn into the cell. This occurs because the concentration of solute is greater in the cell than in the surrounding environment. If the cell doesn't have a cell wall or some other tool for managing osmotic stress, it will lyse (burst).

Mitochondria and Chloroplasts

Two groups of organelles, mitochondria and chloroplasts, are double-membrane structures with their own 70S ribosomes and circular chromosome. As mentioned earlier, the remarkable similarity of mitochondrial and chloroplastic ribosomes and genomes to prokaryotic cells supports the theory that these organelles were once free-living prokaryotes that were engulfed by an ancient ancestral cell, eventually losing their independence.

Mitochondria The **mitochondria** (singular: mitochondrion) make most of a cell's adenosine triphosphate (ATP), which is the preferred molecule cells use to meet their energy needs. As such, mitochondria's nickname is the "powerhouse" of the cell. (Chapter 8 discusses adenosine triphosphate, ATP, and the metabolic pathways that make it.) However, mitochondria are now known to be central to

The Case of the Infectious Hike

Practice applying what you know clinically: visit the **Mastering Microbiology** Study Area to watch Part 3 and practice for future exams.

[10] Sulfur bacteria such as *Thiomargarita* and *Thioploca* are known to have vacuoles. Aquatic prokaryotes may also use gas vacuoles to regulate their buoyancy.

Mitochondria and Medicine

It has long been known that mitochondria are a key component in a cell's ability to fuel the chemical reactions that keep it alive. More recent research shows these fascinating organelles are not just ATP generators, but are also important cell-signaling hubs that influence at least 50 different disease states. Every year, about one in every 4,000 children in the United States is born with some sort of mitochondria-based disease. Mitochondria also play a part in modulating normal aging. Some so-called diseases of aging, including Parkinson's, cancer, atherosclerotic heart disease, and type 2 diabetes, are also linked to mitochondria.

An average human cell has anywhere from hundreds to thousands of mitochondria, with cells that have higher energy needs (such as muscle or neurons) containing more mitochondria than those that are less active, like skin cells. Mitochondria are unique organelles in that they have their own genome, called mitochondrial DNA (mtDNA). Mitochondria copy their DNA as they divide, and sometimes mutations (errors in DNA) arise. All of us have some sort of mutated mtDNA, but the physiological effects aren't noticeable until these mutations start to accumulate.

One hypothesis about why mutations accumulate is that the reactive oxygen species* made as a natural byproduct of ATP production damage mtDNA. Our mitochondria have ways to manage these reactive oxygen species, but sometimes they cannot fully neutralize them. As mtDNA mutations accumulate, mitochondrial function diminishes, energy output decreases, and the tendency toward programmed cell death (apoptosis) increases.

As mitochondria decline in function and cells die, physiological symptoms develop. Thus, the accumulation of mtDNA mutations has been dubbed the "aging clock." There is strong evidence that supports this mitochondrial aging clock mechanism. For example, mouse models that have been engineered to have a higher-than-normal rate of mtDNA mutation exhibit accelerated aging.

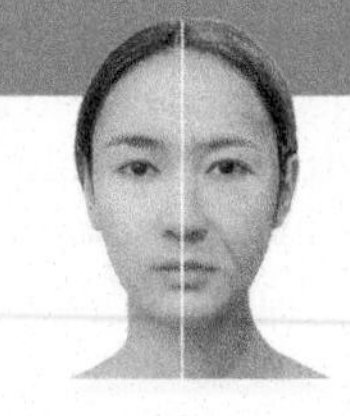

Mitochondrial stress is associated with aging and a variety of disease pathologies.

Also concerning is the fact that even useful substances like antibiotics can negatively affect our mitochondria. At therapeutically useful levels, aminoglycosides and penicillin family–based drugs have all been shown to increase reactive oxygen species in mitochondria and cause tissue stress. Other drugs, such as the fluoroquinolones (ciprofloxacin, levofloxacin, and ofloxacin), which are used to treat pneumonia, gonorrhea, and other common bacterial infections, are also associated with mitochondrial oxidative stress. Since 2004 the FDA has required a printed warning disclosing the association of fluoroquinolones with peripheral nerve damage, which has been linked to mitochondrial stress. The damage can occur suddenly and even as a result of short-term therapy (5 days or less) and may last for years after stopping drug use; in some cases it is permanent.

Since the first mitochondrial disorder was diagnosed and characterized in 1962, mitochondria have become a focus of medical research, and there are even specializations in mitochondrial medicine. As our understanding of the role of mitochondria in health and disease expands, it is likely that we will see new therapies designed to reduce mitochondrial stress or, in some cases, specifically induce it to kill off cancer cells.

*Reactive oxygen species (ROS) is a term used to collectively refer to molecules that contain oxygen and may damage cells.

Sources: Xiao, Y., Xiong, T., Meng, X., Yu, D., Xiao, Z., & Song, L. (2019). Different influences on mitochondrial function, oxidative stress and cytotoxicity of antibiotics on primary human neuron and cell lines. *Journal of Biochemical and Molecular Toxicology*, *33*(4), e22277.

For a detailed review, read Zhang, J., Duan, D., Song, Z. L., Liu, T., Hou, Y., & Fang, J. (2021). Small molecules regulating reactive oxygen species homeostasis for cancer therapy. *Medicinal Research Reviews*, *41*, 342–394.

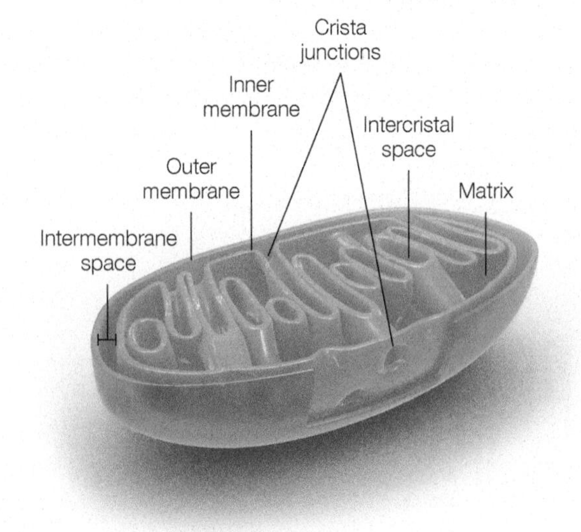

FIGURE 4.21 Mitochondria A smooth outer membrane makes up the surface of the mitochondrion. The inner membrane has a large surface area thanks to specialized cristae.

other cellular processes, including making amino acids and vitamins, regulating cell division, and carrying out programmed cell death (*apoptosis*). Mitochondria impact a number of disease states and have roles in cancer and aging.

Again, mitochondria have a double membrane (FIG. 4.21). In general, the smooth outer membrane makes up the surface of the mitochondrion. Underlying this is the inner membrane, which has a large surface area thanks to specialized *cristae*[11]. Traditionally, cristae were thought to be deep invaginations of the inner membrane. But by the 1990s advanced microscopy techniques had confirmed the presence of small pores in the inner membrane that lead into chambers of sorts, making the cristae more like little compartments than simple undulating folds. Surrounding the cristae is the matrix, which houses the mitochondrion's DNA and ribosomes.

The number of mitochondria in a cell can differ. For example, a typical yeast cell contains 5–10 mitochondria, a tobacco plant cell has 500–600, and a

[11] Many textbooks continue to show cristae as deep invaginations of the inner membrane, but it is now known that this "baffle model" is outdated. The newer crista junction model reflects that cristae are more like compartments within the mitochondrion than simple undulating folds. Surrounding the cristae is the matrix, which houses the mitochondrion's DNA and ribosomes.

human liver cell may contain up to 2,000 mitochondria. In general, highly active cells tend to have more mitochondria than less active cells. Whereas only photosynthetic cells have chloroplasts, most eukaryotic cells, *including plant cells*, have mitochondria.

Chloroplasts Structurally similar to mitochondria, **chloroplasts** allow the cells that have them to harvest energy from sunlight using light-collecting pigments like chlorophylls and carotenoids. Moving from the inside of the chloroplast outward, you would find a series of membranous structures called thylakoids (FIG. 4.22). These are interconnected by lamella (thin membrane extensions) and stacked into structures called grana (singular: granum). The grana are bathed in a colorless fluid called stroma. The entire contents of the chloroplast are encased in an inner membrane, which is in turn encased in a second, outer membrane.

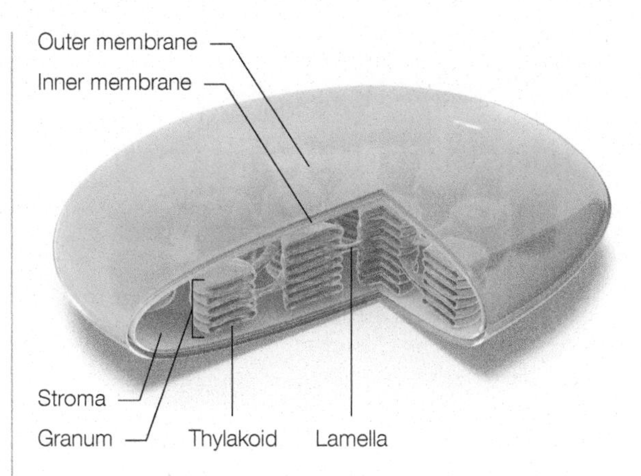

FIGURE 4.22 **Chloroplasts**

Build Your Foundation

17. How are eukaryotic ribosomes different from prokaryotic ribosomes? Where are they located in eukaryotic cells? (LO 4.17)
18. What is the cytoskeleton made of, and what does it do? Describe the centrosome and its function. (LO 4.18)
19. Provide three features of the nucleus. (LO 4.19)
20. Name two functions of the endoplasmic reticulum. (LO 4.20)
21. Name two functions of the Golgi apparatus. (LO 4.21)
22. What are lysosomes, and what do they do? What are secretory vesicles? (LO 4.22)
23. What features of mitochondria are similar to bacteria? Why are they called the powerhouse of the cell? (LO 4.23)

Build Your Foundation (BYF) Quick Quiz: Visit the **Mastering Microbiology** Study Area to quiz yourself.

VISUAL SUMMARY | *Eukaryotic Cell Basics*

Interactive CONTENT REVIEWS Interactive Content Reviews: Visit the **Mastering Microbiology** Study Area to quiz yourself.

Endosymbiotic Theory

Eukaryotes evolved through cell-merging events.

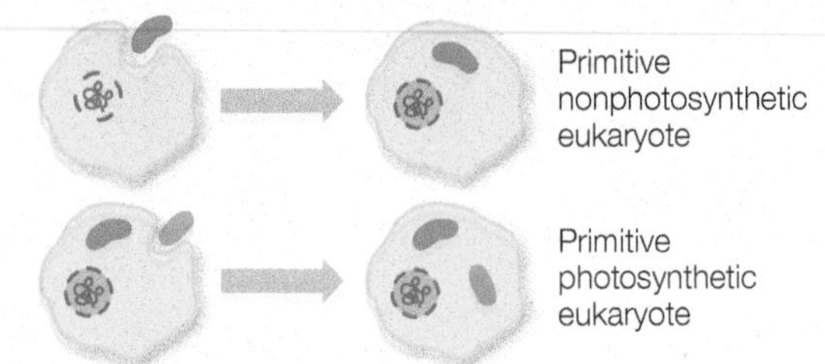

Transport

Both endocytosis and exocytosis require energy.

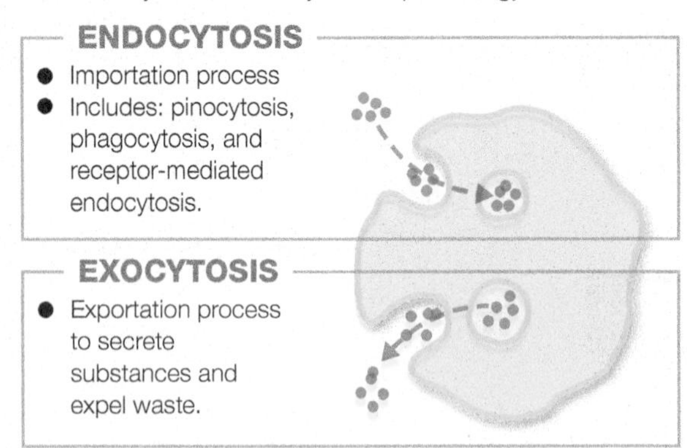

Intracellular Structures

Eukaryotes may be unicellular or multicellular; they have a nucleus and membrane-bound organelles. They divide asexually by mitosis; some divide sexually by meiosis.

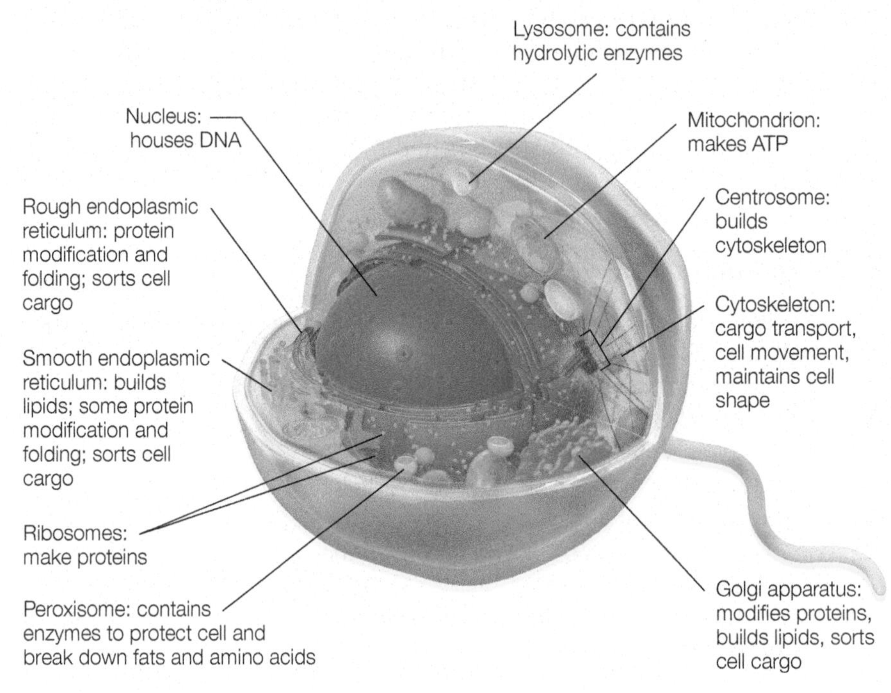

Extracellular structures

All eukaryotes have a plasma membrane. Plants and fungi have a cell wall just outside of their plasma membrane. Some eukaryotes use flagella or cilia to move.

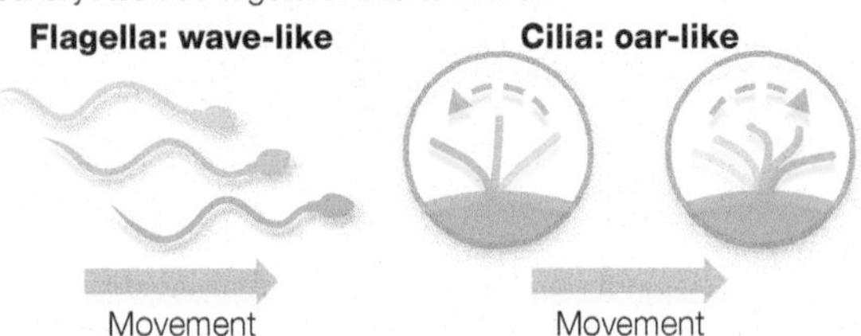

The eukaryotic glycocalyx is used for protection, adhesion, and communication.

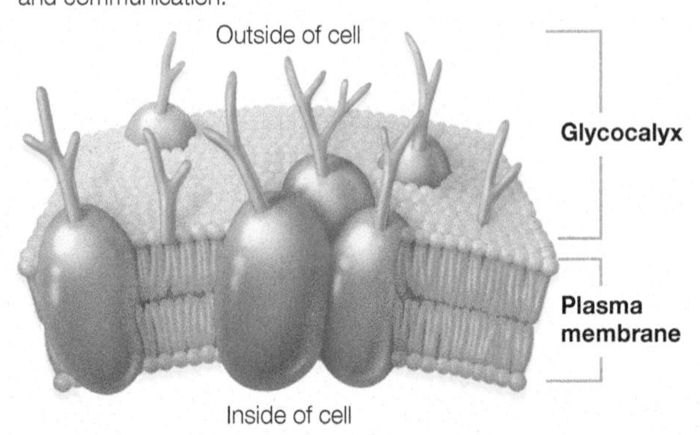

Domain Eukarya

Eukaryotes include plants, animals, fungi, and protists.

Protozoans (animal-like protists) are grouped by how they move in their mature form.

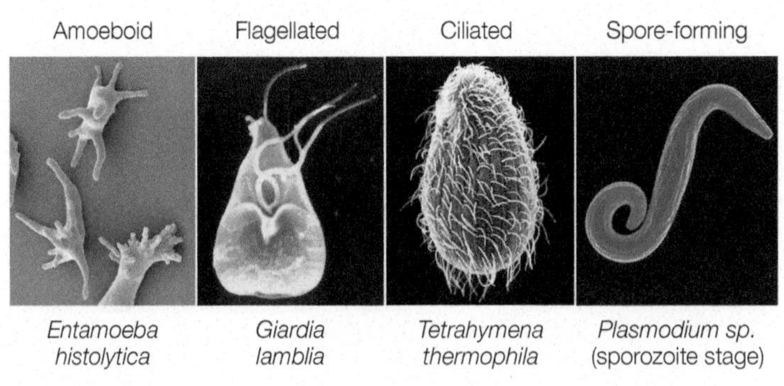

Entamoeba histolytica | *Giardia lamblia* | *Tetrahymena thermophila* | *Plasmodium sp.* (sporozoite stage)

Fungi can be grouped into phyla by the type of sexual spore they form.

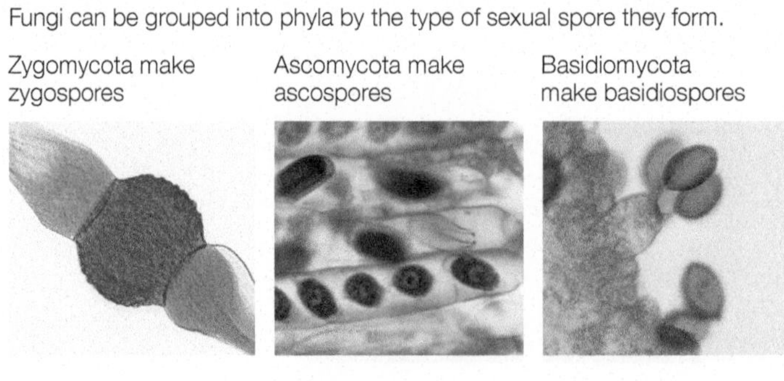

4 CHAPTER 4 OVERVIEW

4.1 Overview of Eukaryotes

- The endosymbiotic theory describes eukaryotic evolution as starting when an ancestral cell engulfed a nonphotosynthetic prokaryote, which lost its self-sufficiency and became mitochondria. Chloroplasts came about when photosynthetic bacteria were engulfed by an ancestral cell. Similarities among prokaryotes, mitochondria, and chloroplasts support the theory.
- Mitochondria and chloroplasts have a double-membrane structure, are similar in size to bacteria, and replicate by a process similar to binary fission. Mitochondria and chloroplasts have their own prokaryote-like genome (single circular chromosome), and they have prokaryote-like ribosomes (70S ribosomes).
- Eukaryotes conduct sexual and asexual cell division. Mitosis is asexual cell division that produces two genetically identical daughter cells from one parent cell. Meiosis is sexual cell division where four daughter cells are made from one parent cell. The daughter cells from meiosis have half the number of chromosomes as the parent cell and are genetically distinct from the parent cell and from each other. When haploid cells made by meiosis combine with complementary haploid cells (example: egg and sperm uniting in a fertilization event), they produce a zygote.
- Eukaryotes use many of the same cellular transport processes as prokaryotes (osmosis, diffusion, active transport, etc.), but they also employ endocytosis and exocytosis. Endocytosis is an *importation* process that can engulf solids (phagocytosis) or dissolved substances (pinocytosis). Exocytosis is an *exportation* process in which vesicles deliver their contents to the plasma membrane.

4.2 Classification of Eukaryotes

- There are four eukaryotic kingdoms. Animalia are multicellular, heterotrophic, and nonphotosynthetic, and they lack a cell wall. Plantae are multicellular, photosynthetic, and autotrophic, and they have a cell wall. Fungi are unicellular or multicellular, heterotrophic, and nonphotosynthetic, and they have a cell wall. Protista are unicellular or multicellular, some are heterotrophic and others are autotrophic, some have a cell wall, and some are photosynthetic.
  - Most fungi rely on spores for reproduction. Sexual fungal spore types are used to group fungi into different phyla; they include zygospores, ascospores, and basidiospores. There are two classes of asexual spores made by fungi: conidiospores and sporangiospores.
  - Fungal diseases are called mycoses; most mycoses occur in people with compromised immune function or disrupted normal microbiota.
  - Animal-like protists are called protozoans. Protozoans are grouped by their means of motility into the following: amoeboid (Sarcodina), flagellated (Mastigophora), ciliated (Ciliophora), and spore-forming (Apicomplexa).

4.3 Extracellular Structures

- All cells have a plasma membrane. The plasma membrane serves as a barrier to the external environment. Animal cell membranes include cholesterol, while a wide variety of phytosterols and low levels of cholesterol are found in plants. Fungi and protists have a wide range of sterols in their membranes, with ergosterol being a key example.
- Eukaryotes like plants, fungi, and some protists have cell walls. Animal cells do not have cell walls.
- The eukaryotic glycocalyx is a carbohydrate-based extracellular layer. It protects the cell from a variety of stresses, has roles in cell adhesion, and assists with cellular communication.
- Eukaryotic flagella are used for motility. Eukaryotic flagella are made of the protein tubulin and have what is called a nine-plus-two arrangement. Eukaryotic flagella move in a back-and-forth, wave-like motion.
- Cilia provide an oar-like motion. They structurally resemble flagella, but they are shorter and more numerous than flagella. Only eukaryotes have cilia.

4.4 Intracellular Structures

- Eukaryotic cells have 80S ribosomes made up of a small subunit (the 40S subunit) and a large subunit (the 60S subunit); these ribosomes can be free in the cytoplasm or bound to the endoplasmic reticulum. The 70S ribosomes are present in mitochondria and chloroplasts.
- The cytoskeleton is a network of protein fibers that maintains cell shape, facilitates cell movement, and coordinates the movement of vesicles, organelles, and other cellular cargo. It is made of microtubules, intermediate filaments, and actin microfilaments.
- The centrosome helps form microtubules. It is made up of two centrioles and surrounding unstructured material.
- The nucleus is a large double-membrane structure that houses the cell genome. Nuclear pores regulate nuclear transport. A dense region of the nucleus, called the nucleolus, is enriched with RNA; it is the site where ribosomal subunits begin their development.
- The endoplasmic reticulum (ER) is a series of interconnected membranous sacs that originate from the outer membrane of the nuclear envelope. Rough ER has ribosomes on the outer surface; smooth ER is not associated with ribosomes. The ER assists with protein modifications. The rough and smooth ER coordinate with the Golgi apparatus to sort, package, and ship proteins and lipids around the cell and to the cell surface.
- The Golgi apparatus is a series of interconnected flattened sacs that modifies cellular proteins, builds lipids, and sorts and distributes the finished products. Diverse vesicles ship materials from the Golgi to their final destination.
- Transport vesicles move substances around the cell. Secretory vesicles shuttle materials to the cell surface for discharge from the cell.
- Lysosomes contain a wide variety of hydrolytic enzymes that break down substances engulfed by the cell during phagocytosis.
- Peroxisomes contain enzymes that break down certain fats and amino acids and protect the cell from hydrogen peroxide and other toxic oxygen intermediates.
- Vacuoles are membranous enclosures formed by the merging of many vesicles; they contain mainly water and dissolved organic and inorganic substances.
- Mitochondria make ATP, produce amino acids and vitamins, regulate cell division, and carry out programmed cell death (apoptosis).
- Chloroplasts conduct photosynthesis; while only photosynthetic cells have chloroplasts, most eukaryotic cells (including plant cells) have mitochondria.

THINK CLINICALLY | Be S.M.A.R.T. About Cases

COMPREHENSIVE CASE

The following case integrates basic principles from the chapter. Try to answer the case questions on your own. Don't forget to be **S.M.A.R.T.*** about your case study to help you interlink the scientific concepts you have just learned and apply your new understanding of eukaryotic cells to a case study.

*The five-step method is shown in detail in the Chapter 1 Comprehensive Case on cholera. See pages 31–33. Refer back to this example to help you apply a SMART approach to other critical thinking questions and cases throughout the book.

• • •

Emma recently suffered from bacterial pneumonia and just completed the prescribed antibiotic. She had been feeling much better, but recently started experiencing mild vaginal itching. She thought it might be due to irritation from a new body wash she got for her 25th birthday. Emma stopped using the product, but the itching got worse over the next couple of days and was accompanied by a white vaginal discharge. Concerned, Emma made an appointment with her nurse practitioner (NP). The NP noted that Emma takes oral contraceptives and reported having a regular menstrual cycle. Her vital signs were normal, and she appeared well aside from her noted symptoms. The NP then asked Emma about her sexual history. Emma said that over the past couple of months she had become intimate with her boyfriend. Before him, she had been with one other sexual partner, but that was at least two years ago. In light of the symptoms Emma had, in combination with her recent antibacterial therapy, the NP ordered a vaginal swab for wet mount for immediate inspection under the light microscope.

After viewing the wet mount, the NP came in to discuss her findings. Emma had a yeast infection (vulvovaginal candidiasis, usually caused by the nonmotile yeast *Candida albicans*) and also was infected with the protozoan *Trichomonas vaginalis*. The NP explained that *T. vaginalis* is a common curable sexually transmitted infection (STI) and that about 70 percent of cases in women are asymptomatic. She stated that Emma's symptoms of pruritus (itching) and the white vaginal discharge were most likely due to the yeast infection. Emma was also informed that the candidiasis likely developed as a sequel to the antibiotic therapy. Emma was prescribed a single 150-mg oral dose of fluconazole (brand name Diflucan) for the yeast infection.

Emma was shocked about the STI and worried about what it meant for her relationship. The NP said that most men are asymptomatic—and there was no way to tell if she'd contracted it from her current or previous partner. Emma was prescribed a single 2-g oral dose of metronidazole (there are many brand names, including Flagyl) to treat the *T. vaginalis* infection. She was advised that her boyfriend also needed to see his doctor and get treated. Emma asked if she and her boyfriend could still be intimate during treatment if they used condoms. The NP explained that, although proper condom use greatly reduces the chance of *T. vaginalis* infection, they both should wait to have sex for a week after taking the metronidazole to ensure that the infection has cleared in both of them. Emma was told to come back to see the NP if all of her symptoms didn't resolve after the therapy or if new symptoms developed.

• • •

CASE-BASED QUESTIONS

1. Imagine you are Emma's healthcare provider, and she asks you why the antibacterial drug she took for her recent case of pneumonia did not prevent the yeast infection. How would you reply?
2. The following is a light microscopy image of *T. vaginalis*. Based on your observation, to what group of protozoans does this organism belong? Explain your answer.

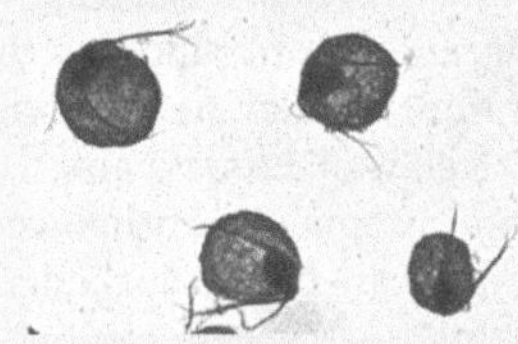

3. What cellular features would you expect to be shared between *C. albicans* and *T. vaginalis*?
4. How does the azole drug fluconazole damage yeast? Why does this drug not target human cells?
5. What cellular features would you expect to be different between *C. albicans* and *T. vaginalis*?
6. A variety of bacteria normally inhabit the vagina. How could a practitioner differentiate between these normally present bacteria and *C. albicans* just by looking at a wet mount sample under the microscope without differential staining procedures?

END OF CHAPTER QUESTIONS

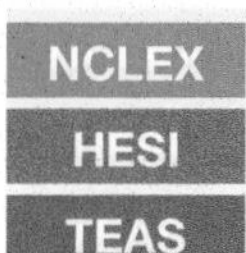

Think Critically and Clinically

Questions highlighted in orange are opportunities to practice NCLEX, HESI, and TEAS-style questions.

1. Which of the following include eukaryotic cells? Select all that apply.
   a. Fungi
   b. Yeasts
   c. Protista
   d. Bacteria
   e. Helminths
   f. Animalia
   g. Archaea
   h. Plantae

2. Which of the following are *not* characteristics of eukaryotes? Select all that apply.
   a. They are generally simpler than prokaryotes.
   b. They can be multicellular.
   c. They all have a nucleus.
   d. They tend to have multiple chromosomes.
   e. They can have a cell wall.
   f. They include pathogens.
   g. They divide by mitosis.
   h. They make up the Domain Archaea.
   i. They make up the Domain Bacteria.
   j. They make up the Domain Eukarya.

3. Indicate the true statements about eukaryotic cells **and then** reword the false statements so that they are true.
   a. Eukaryotic cells have 70S ribosomes on the rough endoplasmic reticulum.
   b. Eukaryotic cells sexually reproduce by mitosis.
   c. Eukaryotic cells can make up unicellular or multicellular organisms.
   d. Eukaryotic cells always have a cell wall.
   e. Eukaryotic cells can be photosynthetic.
   f. Eukaryotic cells use fimbriae for motility.
   g. Yeast is a multicellular eukaryote.

4. Which of the following are characteristics of the Kingdom Fungi? Select all that apply.
   a. They have a cell wall.
   b. They are mainly unicellular.
   c. They are mostly pathogens.
   d. They contain peptidoglycan in their cell wall.
   e. Some are photosynthetic.
   f. They lack mitochondria.
   g. They are prokaryotic and eukaryotic.

5. In what group of protozoans would you place the following organism?

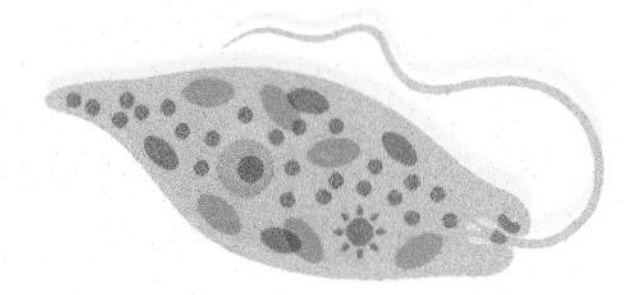

6. ________________ is an exportation process used in eukaryotes and involves vesicles that may bud off the Golgi apparatus. In contrast, ________________ is an importation process by which a solid *or* a dissolved substance enters the cell by being engulfed by invaginations of the plasma membrane.

7. ________________ is also called cell drinking and is an example of an endocytic process.

8. The presence of which of the following would be helpful in distinguishing a prokaryote from a eukaryote? Select all that apply.
   a. Peptidoglycan
   b. Phospholipids
   c. A cell wall
   d. A nucleus
   e. Chloroplasts
   f. Ribosomes
   g. Ability to carry out active transport
   h. DNA

9. Match the organelle to the function

Structure	Description
Nucleus	Energy production
Endoplasmic reticulum	Builds microtubules
Golgi apparatus	Initial site of ribosome subunit formation
Cilia	Adhesion and cell signaling
Mitochondria	Modifies and packages proteins; can be rough or smooth
Glycocalyx	Builds and packages lipids
Ribosome	Motility
Centrosome	Makes proteins
Nucleolus	Houses the cell's DNA

10. Complete the Venn diagram that compares and contrasts prokaryotes and eukaryotes.

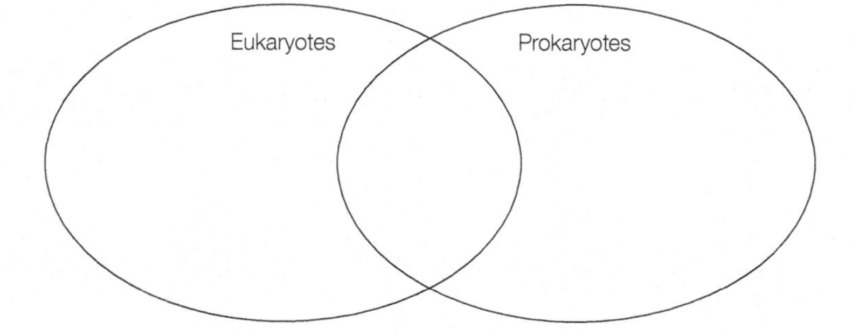

11. Complete the table by answering *yes* or *no* for each question.

	Endocytosis	Exocytosis
Requires energy?		
Starts at the plasma membrane?		
Transports substances out of the cell?		
Transports substances into the cell?		
Includes phagocytosis?		
Includes pinocytosis?		

12. Fungal spores can be made by __________. Select all that apply.
    a. binary fission
    b. mitosis
    c. meiosis
    d. asexual reproduction
    e. sexual reproduction
    f. sporogony

13. Eukaryotic flagella differ from prokaryotic flagella in that:
    a. eukaryotic flagella use a rotary motion to propel the cell.
    b. eukaryotic flagella are made of proteins.
    c. eukaryotic flagella are not enclosed in a membrane.
    d. eukaryotic flagella have a 9 + 2 arrangement of microtubules.

14. Which of the following characteristics regarding eukaryotic organelles supports the endosymbiotic theory? Select all that apply.
    a. Eukaryotic ribosomes are 70S just like prokaryotic ribosomes.
    b. Mitochondria and chloroplasts have 70S ribosomes like prokaryotes.
    c. Mitochondria and chloroplasts are double-membrane organelles similar in size to bacteria.
    d. Mitochondria and chloroplasts have their own circular chromosome.
    e. Eukaryotes can have cell walls like prokaryotes.

15. Most fungi grow as a collection of tubular structures called ________________.

16. Which of the following statements are *true* regarding protozoans? Select all that apply.
    a. They are unicellular.
    b. They have cell walls.
    c. They are usually nonphotosynthetic.
    d. They are animal-like protists.
    e. Some are pathogens.
    f. They usually have simple life cycles.
    g. They often use sexual and asexual reproduction.
    h. They usually lack nuclei.
    i. They all have mitochondria.

CRITICAL THINKING QUESTIONS

1. Imagine you have just discovered an organism that is multicellular, is nonphotosynthetic, and lacks a cell wall. What kingdom does it most likely belong to? Explain your reasoning.

2. You are viewing a single-celled organism under the microscope. It has a cell wall, is unicellular, has a nucleus, is nonmotile, and does not have chloroplasts. In what kingdom would this organism most likely be classified? Explain how you came to your answer.

3. Assume you were engineering a eukaryotic cell that could survive in a dark environment, would be resistant to osmotic stress, and also would be motile in an aquatic environment. It also would need to be able to adhere (stick) to certain animal hosts in its environment. What features would you give your cell, and what organelles might you leave out of the cell? Explain your choices.

4. Why is it unlikely for a photosynthetic microbe to be a pathogen in humans?

5. What would be a possible adverse side effect of a drug that blocked fusion of the lysosome and phagosome in humans?

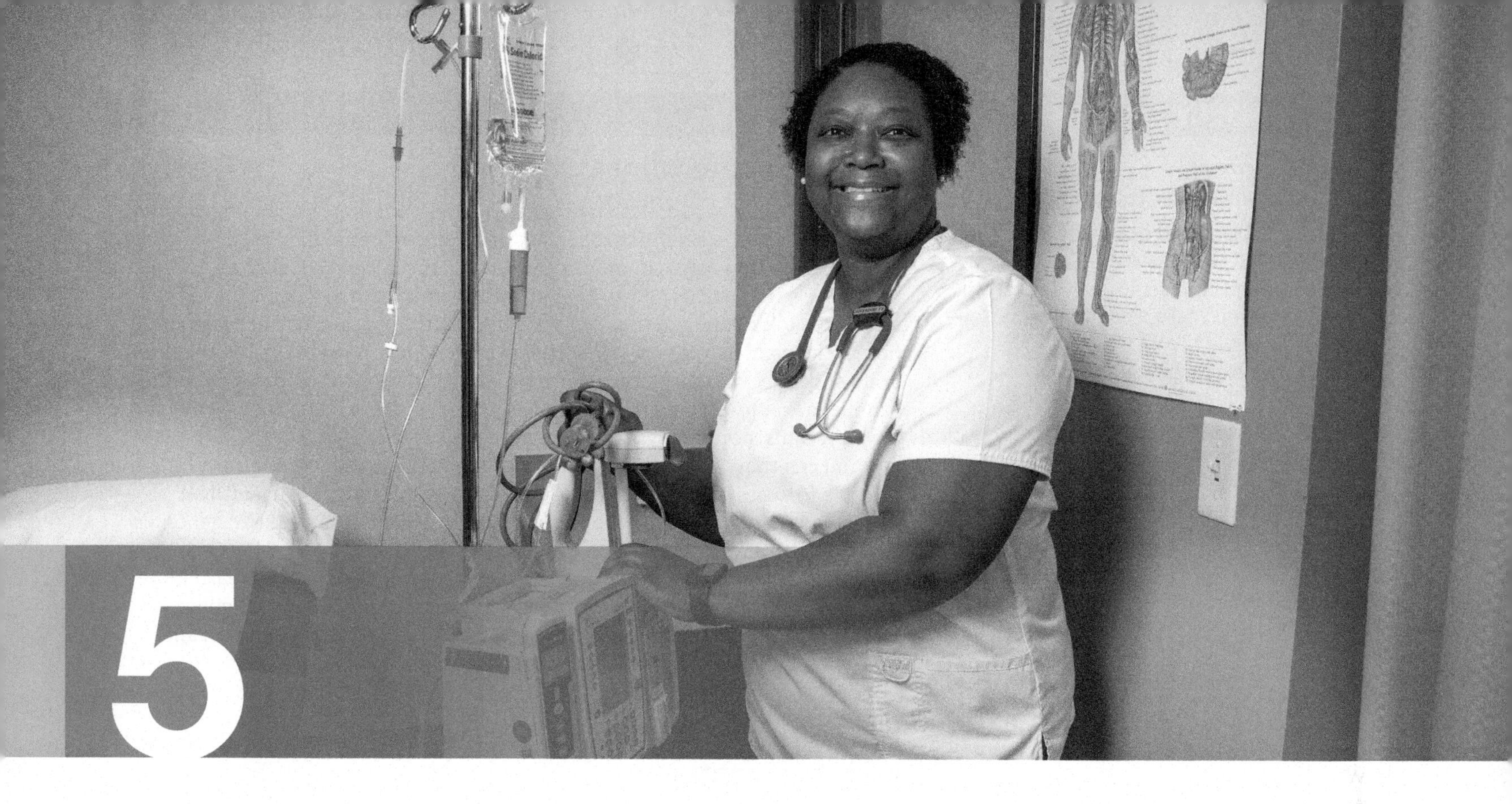

5 Genetics

What Will We Explore? This chapter introduces you to basic genetics topics, including an overview of heredity, how cells copy their genetic material and build proteins, how mutations (changes in genetic material) occur, and how bacteria share genetic information without dividing.

Why Is It Important? These topics are essential for understanding a number of medical concepts, including antibiotic resistance, bacterial growth, modes of drug action, personalized medicine, gene therapy, and the development of therapies to manage viral epidemics, from Ebola to SARS-CoV-2—the causative agent of COVID-19.

Genes impact our risk for diseases, including our susceptibility to certain infections. For example, an extremely rare genetic alteration protects a very small segment of the population against HIV infection. Of course, not all genetic changes confer an advantage; at least 6,000 medical disorders, including cystic fibrosis and sickle cell anemia, result from genetic changes.

The study of genetics is propelling us toward an age of personalized medicine that promises customized preventive therapies, earlier diagnosis, and treatments targeted to our individual genetic makeup. Hundreds of diseases are now directly linked to certain genes and can be diagnosed through genetic testing, and the effectiveness of numerous drugs has been studied in the context of patient genetic profiles.

In recognition of the increasing role of genetics in medical therapies, the FDA has developed standards for using genetic information for personalized medicine, and a number of FDA-approved genetic-based therapies are in use. In the future, healthcare providers will increasingly rely on genetic profiles to diagnose and treat patients. Thus, tomorrow's healthcare team will need a working knowledge of basic genetics principles and how those principles relate to patient care.

Renata JM Henderson
RN, Professor of Nursing; Jacksonville, FL

> **Clinical CASE** NCLEX HESI TEAS
> **The Case of the Spreading Superbug**
> Visit the **Mastering Microbiology** Study Area to watch the case and find out how genetics can explain this medical mystery.

5.1 HEREDITY BASICS

Learning Outcomes

After reading this section, you should be able to:

5.1 Define genotype and phenotype and discuss their relationship.

5.2 Compare and contrast the ways eukaryotic and prokaryotic cells organize their genomes.

5.3 Describe the structural features of DNA and RNA nucleotides.

5.4 Relate the directionality of nucleotides to the formation of phosphodiester bonds.

5.5 Describe the structural features of RNA, and state how RNA differs from DNA.

5.6 Explain the flow of genetic material as it is presented by the "central dogma."

Genotype determines phenotype.

The modern field of genetics studies genes, their function, and how variations arise in **genomes**—the entire collection of genetic material in a cell or virus. Gregor Mendel, an Austrian monk who lived in the mid-1800s, is often called the father of modern genetics. His pioneering work with pea plants paved the way for us to understand that traits are **heritable**, or passed from one generation to the next. By the early 1900s the term **gene** was used to describe heritable units of genetic material that determine a particular trait. The theory that the genetic makeup, or **genotype**, of an organism influences its physiological and physical traits, or **phenotype**, then became widely accepted.

You can think of genomes as instruction manuals that determine all the possible features of a cell or virus. Cells have deoxyribonucleic acid (DNA) genomes, whereas viruses have either DNA or ribonucleic acid (RNA) genomes (FIG. 5.1). This chapter focuses on cellular genetics. (See Chapter 6 for viruses.) DNA and RNA were first introduced in Chapter 2, but here you will learn more about the structure and function of these molecules that are the language of life. Let's begin by reviewing DNA as the substance of cellular genomes, and then we'll explore how DNA is built and read.

Prokaryotic and eukaryotic genomes differ in size and organization.

In general, the more complex an organism, the more genes it has. Prokaryotic cells tend to have smaller genomes compared with eukaryotic cells, although there is amazing variety in the sizes of genomes among organisms. For example, the nonpathogenic bacterium *Escherichia coli* strain K12 has around 4,400 genes, while a pathogenic strain of *E. coli* called O157:H7 has about 5,500 genes. For a eukaryotic comparison, consider that a yeast cell has about 6,000 genes, while a human cell has about 24,000 genes.

CHEM • NOTE

DNA and **RNA** are nucleic acids made of repeating subunits called nucleotides. Nucleotides are made up of a phosphate, a sugar (deoxyribose in DNA/ribose in RNA), and one of five possible nitrogen bases (A, T, G, C, or U).

FIGURE 5.1 Genomes The genome is all the genetic material in a cell or virus. Illustrations are not to scale.

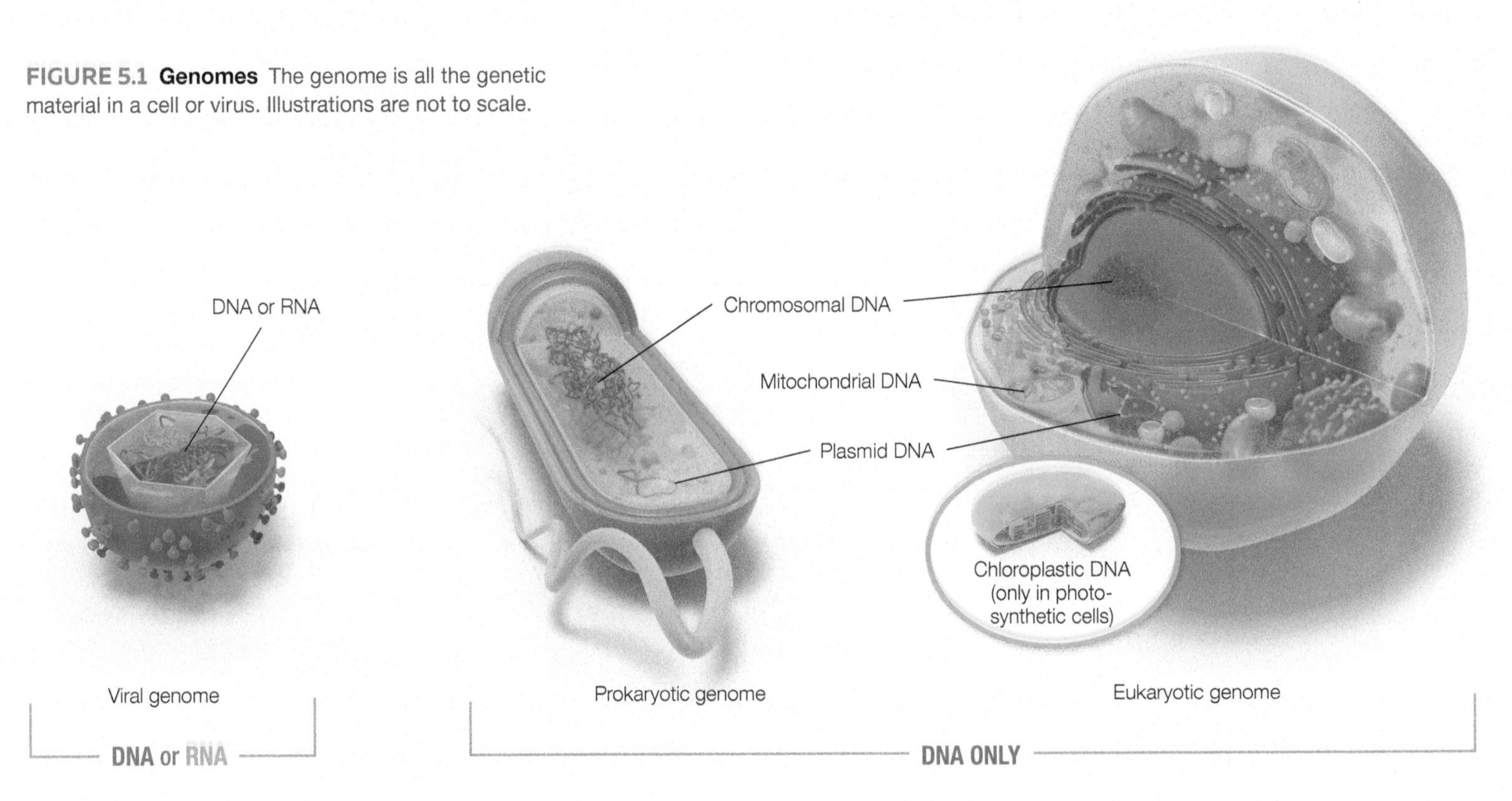

All cells organize their genomes into **chromosomes**, which are carefully packaged strands of DNA associated with organizational proteins. The number of chromosomes does not influence organism sophistication, or even its total number of genes. For example, human cells have 46 chromosomes; chimpanzee cells have 48; and single-celled amoebas have hundreds of chromosomes.

Most prokaryotic cells organize their genome in one to three chromosomes located in the nucleoid region of the cell (recall that prokaryotic cells lack a nucleus). Most bacteria have a single circular chromosome, but some have linear chromosomes—*Streptomyces* species, many of which make useful antibiotics, and the bacterium that causes Lyme disease, *Borrelia burgdorferi*, are examples of bacteria that have linear chromosomes. In contrast, eukaryotic cells tend to have numerous linear chromosomes housed in the cell's nucleus. Eukaryotic chromosomes also have organizational proteins called **histones** that help keep their DNA from getting tangled. Just as thread is wound around a spool to keep it organized, DNA wraps around these histones. Prokaryotes also make use of several histone-like proteins to organize their DNA into chromosomes.

Although the bulk of a cell's genome is found in chromosomes, prokaryotic and eukaryotic genomes also may include **plasmids**, which are pieces of DNA that tend to be circular and exist outside of the chromosomal DNA. Plasmids usually confer a survival advantage to a cell; for example, some include genes related to antibiotic resistance. Mitochondria and chloroplasts also contribute to eukaryotic genomes because these organelles have their own DNA. The prokaryote-like features of mitochondrial and chloroplastic genomes suggest that these organelles evolved from a bacterial ancestor. (See Chapter 4 for more on the endosymbiotic theory). TABLE 5.1 summarizes key differences between prokaryotic and eukaryotic genomes.

The sequencing of the human genome took 13 years and was completed in 2003 at a cost of about a billion dollars. Now individuals can have their entire genome sequenced in a few days for about $1,000. Researchers have also decoded the genomes of thousands of other organisms and viruses. Our progress in *genomics*, a specialized area in genetics that focuses on characterizing entire genomes, occurred rapidly, considering that the structure of DNA was not confirmed until the mid-1950s.

The nucleic acids DNA and RNA govern cell life.

As reviewed in Chapter 2, **nucleic acids** include DNA (deoxyribonucleic acid) and RNA (ribonucleic acid). DNA and RNA are built from nucleotides, which have three basic parts: a phosphate, a sugar (*deoxyribose* in DNA or *ribose* in RNA), and one **nitrogen base** (FIG. 5.2). In DNA, the possible nitrogen bases are adenine (A), guanine (G), cytosine (C), or thymine (T). In RNA, T is replaced by uracil (U), so we see the nitrogen bases A, G, C, and U. The five nitrogen bases are classified as either **purines** or **pyrimidines**, depending on their chemical structure (see TABLE 5.2). The nitrogen bases in nucleic acids pair up according to specific rules—G always pairs with C, while A pairs with T in DNA versus U in RNA (AT pairs in DNA/AU pairs in RNA). These nitrogen bases can occur in any

TABLE 5.2 DNA and RNA Nitrogen Bases

Nitrogen Base	Family	Pairs with	Found in
Adenine (A)	Purine	Thymine (T)	DNA and RNA
Guanine (G)	Purine	Cytosine (C)	DNA and RNA
Cytosine (C)	Pyrimidine	Guanine (G)	DNA and RNA
Thymine (T)	Pyrimidine	Adenine (A)	*Only DNA*
Uracil (U)	Pyrimidine	Adenine (A)	*Only RNA*

TABLE 5.1 Genome Characteristics

Factor	Prokaryotic Genomes	Eukaryotic Genomes
Complexity:	Simple	More complex
Genome can include:	Chromosomal DNA and plasmids	Chromosomal DNA, plasmids, and DNA in mitochondria and chloroplasts
Chromosomes:	Few (generally only one); usually circular	Many; nuclear chromosomes are linear
Location of chromosome(s):	Nucleoid region	Nucleus
DNA organized by:	Histone-like proteins	Histones

CHEM • NOTE

Numbering carbons in DNA and RNA All nucleotides have a five-carbon sugar. The first carbon in the sugar is called the 1′ ("*one prime*") carbon; the second is the 2′ carbon, and so on. The 1′ carbon links to a nitrogen base, the 3′ carbon always bonds to a hydroxyl (OH) group, and the 5′ carbon links to phosphate.

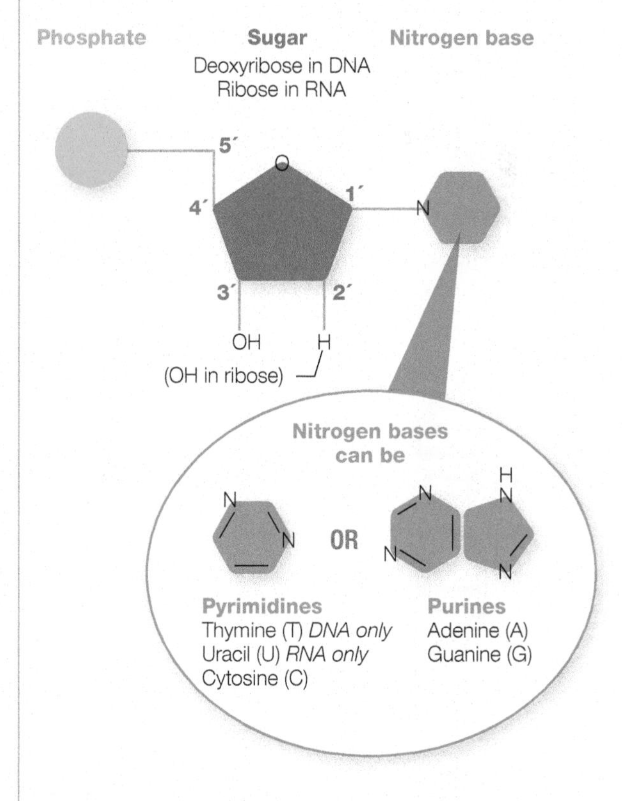

FIGURE 5.2 **Nucleotides** DNA and RNA are made of nucleotides.

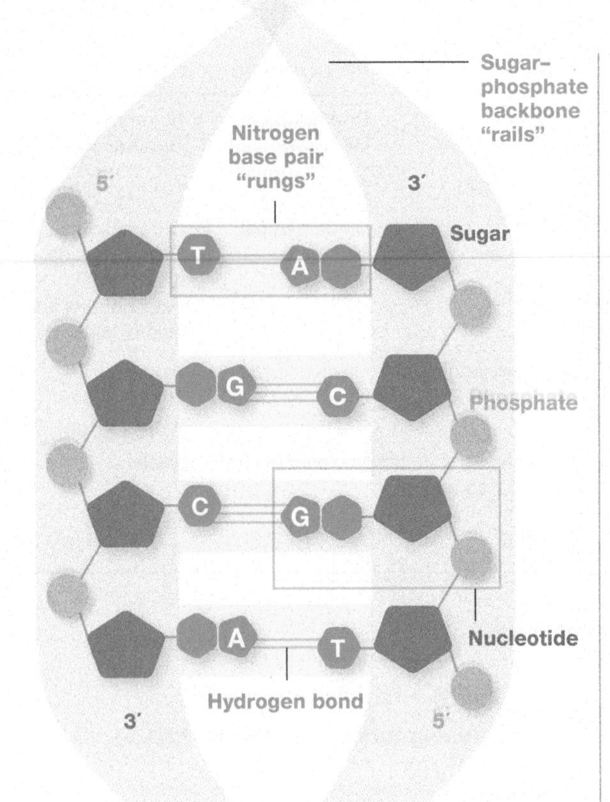

FIGURE 5.3 DNA's structure DNA has an antiparallel arrangement. The sugar–phosphate backbone makes up the "rails" of the DNA ladder, while the nitrogen bases form the "rungs" of the ladder. Note that T pairs with A and G pairs with C.

Critical Thinking *What would the 5′–3′ sugar–phosphate backbone look like if DNA had a parallel arrangement instead of antiparallel?*

CHEM • NOTE

Phosphodiester bonds are covalent bonds that link nucleotides together in DNA and RNA.

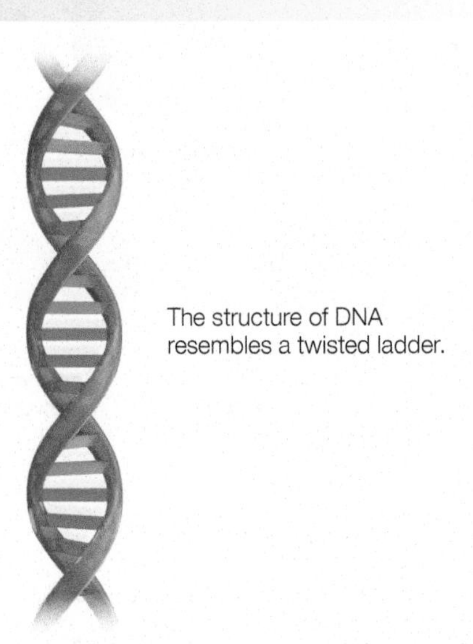

The structure of DNA resembles a twisted ladder.

sequence, thereby providing a limitless "code" that determines a cell's structural and physiological features.

DNA's Structure and Function

As previously mentioned, all cells have DNA as their genomic material—meaning DNA functions as the source code for all the RNAs and proteins a cell can make. DNA is a long, double-stranded molecule that curves into a helix; in many ways it resembles a twisted ladder (FIG. 5.3). DNA's "rungs" consist of the nitrogen bases (A, T, G, and C). The set pairings of the bases (A with T and G with C) ensure there is always a purine across from a pyrimidine, so that DNA has a consistent and predictable dimensionality, even though it has an unpredictable sequence of bases. Base-pairing rules also establish that one strand of DNA directs what the partner strand looks like; therefore, the strands are said to be *complementary* of each other, or balanced.

The "side rails" of the DNA molecule "ladder" are an alternating pattern of sugar and phosphate molecules held together by phosphodiester bonds (FIG. 5.4). These covalent bonds give DNA strands directionality—just as we read from left to right, DNA is built and read with a specific orientation. DNA has 5′ to 3′ directionality (pronounced "five-prime to three-prime") because it is built by linking a 5′ phosphate to a 3′ hydroxyl group. FIG. 5.5 presents an analogy to help you understand DNA's chemical directionality.

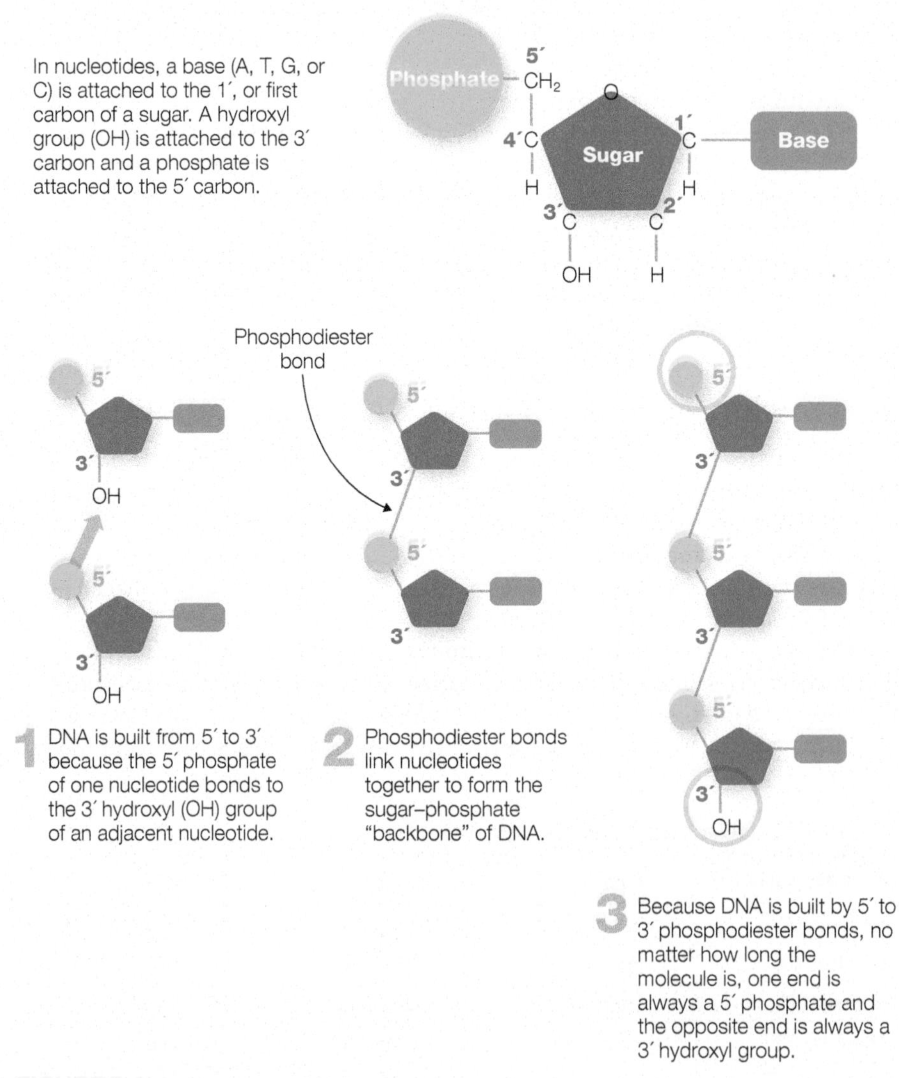

FIGURE 5.4 Phosphodiester bonds

Like two lanes of traffic headed in the opposite direction, double-stranded DNA has an **antiparallel** arrangement. In DNA, one strand runs 5′ to 3′ and the partner strand runs 3′ to 5′. Antiparallel directionality allows the complementary base pairs of DNA to properly associate, much like the cogs of a gear must be properly positioned to interlock.

RNA's Structure and Function

Unlike DNA, RNA doesn't serve as the genome in cells (although, it can in certain viruses). Instead, in cells RNA mainly serves as an essential intermediate for building proteins. The nucleotides in RNA, called **ribonucleotides**, are built using the sugar ribose, as opposed to the deoxyribose sugar found in DNA's nucleotides. Also, recall that in RNA the nitrogen base uracil (U) replaces thymine (T), leading to AU base pairs in RNA instead of AT base pairs (see Table 5.2). Like DNA, RNA also has 5′ to 3′ directionality. Unlike DNA, which tends to exist primarily as a double-stranded helix, RNA is often single stranded. However, it does fold in on itself to form helical and loop structures. Types of RNA include messenger RNA, transfer RNA, and ribosomal RNA. All of them participate in making proteins. Later, in the protein synthesis section, we will review the structure and function of each RNA type.

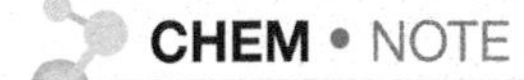

CHEM • NOTE

Phosphate groups are noted as PO_4^{3-}. **Hydroxyl groups** are designated as R–OH, where R is often a carbon-based molecule. Hydroxyl groups are different from hydroxide ions (OH^-).

Genetic information typically flows from DNA, to RNA, to protein.

The general flow of genetic information is usually from DNA to RNA to protein. That is, DNA directs production of RNA, and RNA then directs assembly of proteins, which are central to most cellular functions (FIG. 5.6). This flow of information is sometimes called the **central dogma** of molecular biology.[1] We now know,

1 Imagine a paper clip where half the clip is yellow and half the clip is purple. The yellow half represents the 5′ phosphate and the purple half represents the 3′ hydroxyl group.

2 Imagine the purple end of one paper clip can only link to the yellow end of another paper clip.

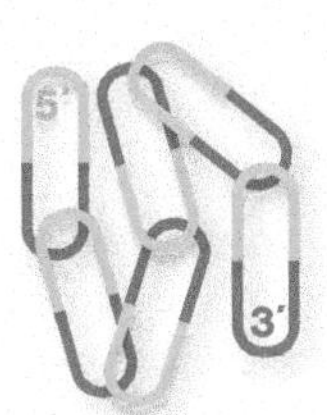

3 No matter how long the chain, one end will be yellow and the other will be purple.

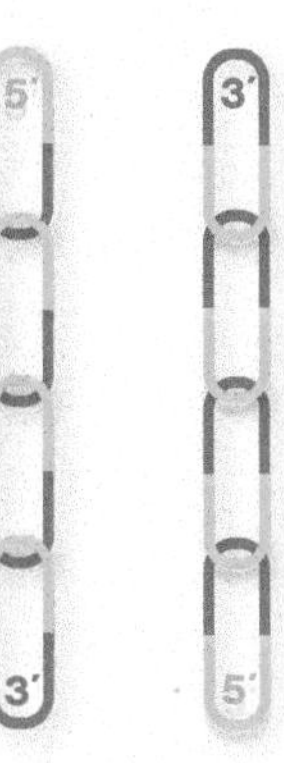

4 If you made two separate paper clip chains you could lay them antiparallel to each other.

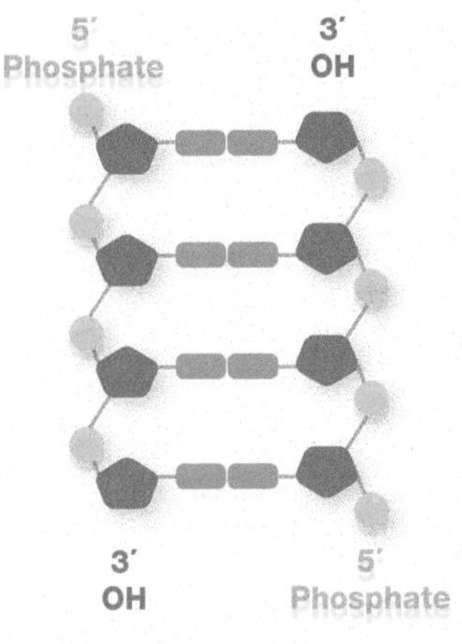

5 DNA works like these paper clip chains. Nucleotides are added in a 5′ to 3′ manner so that no matter how long the DNA strand is, one end will be a 5′ phosphate and the other end will be a 3′ hydroxyl group.

FIGURE 5.5 Chemical directionality of DNA

[1] The term *dogma* means an unquestionably true principle—something that really has no place in science, which embraces questioning to advance understanding. Scientist Francis Crick reportedly coined the phrase "central dogma" without truly understanding the definition of the word—he just thought it was catchy. The phrase persists in scientific literature and, as such, is included here for historical perspective.

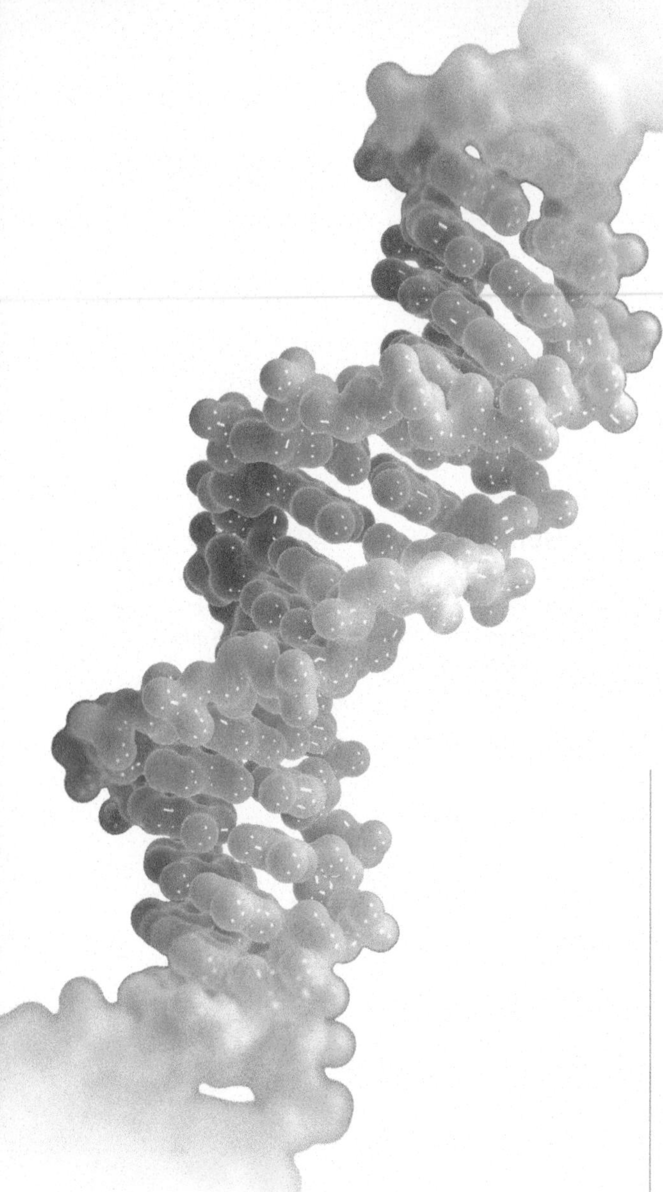

however, that genetic information does *not* flow strictly from DNA to RNA and then to protein; that is, RNA is sometimes used as a template to make DNA. As a result, the "central dogma" has been revised to accommodate the fact that the flow of genetic information is not always unidirectional. We will discuss this modified flow of genetic information further when we cover transcription.

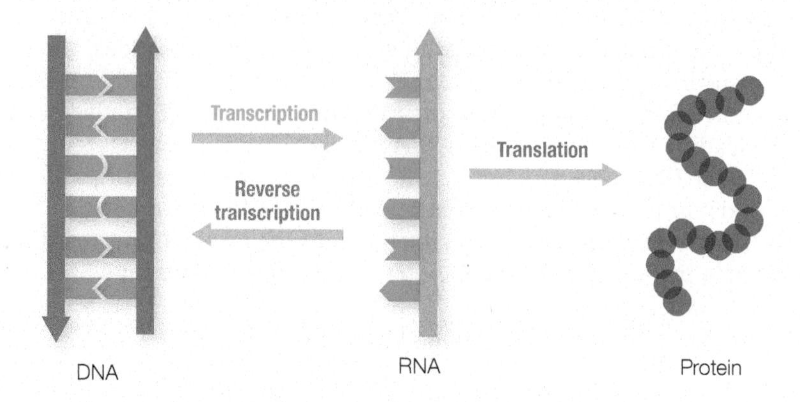

FIGURE 5.6 Flow of genetic information The primary flow of genetic information in cells is from DNA to RNA to protein; this is referred to as the "central dogma" of molecular biology.

Critical Thinking *RNA could be considered a more versatile nucleic acid than DNA. What information in this figure supports that idea?*

Purines and Gout

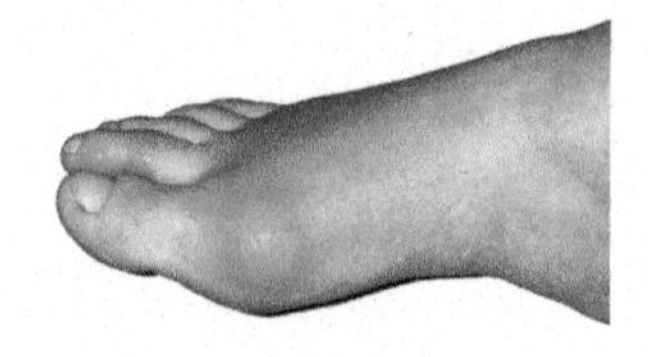

In humans, uric acid is the product of metabolizing dietary purines. When uric acid accumulates in the blood (hyperuricemia), it may cause gout, a painful form of arthritis in which uric acid crystallizes and settles in joints—especially joints of the extremities—causing inflammation and pain. The small joint at the base of the big toe is the most commonly affected, followed by knees and ankles. Gout tends to be more common in men than in women, in part because men have higher levels of uric acid in their blood than women. There is also a genetic component to gout.

People with gout should follow a diet that is low in meat, poultry, shellfish, and fish—all of which have high purine content. However, dairy products are a great way for gout patients to meet their protein needs while minimizing dietary purines. In addition to dietary management to reduce hyperuricemia, medications like allopurinol (Aloprim, Zyloprim, Lopurin, etc.) or febuxostat (Uloric) are often prescribed.

QUESTION 5.1

Why do you think meat contains higher levels of purines than dairy products?

Build Your Foundation

1. What is the primary difference between genotype and phenotype? (LO 5.1)
2. Compare and contrast how DNA is packaged in prokaryotic and eukaryotic cells. (LO 5.2)
3. Describe the building blocks of nucleic acids, and describe how they differ between DNA and RNA. (LO 5.3)
4. Explain how DNA's chemical 5′ to 3′ directionality is established. (LO 5.4)
5. What are the basic functions of DNA and RNA, and how do these nucleic acids structurally compare to each other? (LO 5.5)
6. What does the "central dogma" state? (LO 5.6)

Build Your Foundation (BYF) Quick Quiz: Visit the **Mastering Microbiology** Study Area to quiz yourself.

5.2 DNA REPLICATION

DNA replication allows cells to copy their DNA.

Learning Outcomes

After reading this section, you should be able to:

- **5.7** Discuss the role of DNA replication and factors that contribute to its accuracy.
- **5.8** Name the enzymes required for DNA replication, and discuss their specific functions.
- **5.9** Discuss how DNA replication begins.
- **5.10** Describe how the directionality of DNA impacts replication.
- **5.11** Compare and contrast DNA replication on the leading and lagging strands.
- **5.12** Identify differences in how prokaryotic and eukaryotic cells conduct DNA replication.

DNA replication is the process by which a cell copies its genome before it divides. In general, DNA replication involves unwinding the original DNA, copying it, and then rewinding the parent DNA and the newly made DNA. To make a new DNA strand, a typical bacterial cell like *E. coli* must copy about four million nucleotides, whereas a human cell must copy about three *billion* nucleotides. To give you an idea of the enormity of these numbers, if you were to read the human genome sequence at the rate of one nucleotide per second, it would take you over 101 years to read the entire sequence out loud without stopping.[2]

Fortunately, DNA replication is incredibly fast and accurate; relatively few mutations (errors) are made. For example, a bacterium like *E. coli* can copy about 500 nucleotides per second and makes an error only about once every 10–100 billion base pairs.[3] This high accuracy rate is due to complementary base-pairing rules, as well as proofreading mechanisms of certain DNA replication enzymes. While there are definite differences in how prokaryotic and eukaryotic cells replicate DNA, the processes also have a lot in common. The next section focuses on how the bacterium *E. coli* copies its DNA.

Enzymes are proteins that facilitate chemical reactions in a cell; usually enzyme names end in the suffix "ase." (For more details on enzymes, see Chapter 8.) As shown in TABLE 5.3, many enzymes participate in DNA replication, to include **DNA polymerases**—enzymes that help build DNA. In *E. coli*, at least five different DNA polymerases have been characterized, but only two (DNA polymerases I and III) directly participate in DNA replication—the other polymerases tend to have roles in DNA repair. The DNA polymerases were named based on their order of discovery, not the order in which they operate or even the extent to which they contribute to DNA replication. For example, **DNA polymerase III** (not DNA polymerase I) is the main enzyme that copies DNA; this fact is a bit counterintuitive because of the "III" after its name, so be aware.

CHEM • NOTE

Hydrogen bonds hold complementary DNA strands together. These bonds are weaker than covalent and ionic bonds, making it easier to separate the DNA strands for replication. In DNA, three hydrogen bonds form between GC pairs, while two hydrogen bonds form between AT pairs. Therefore, the more GC pairs in a DNA molecule, the harder it is to separate the DNA strands.

DNA replication begins with helix unwinding.

Just as you have to open a book before photocopying it, the DNA helix must be opened before duplication occurs. In prokaryotes, DNA replication starts when the **primosome**, which is a collection of enzymes and other factors, is recruited to a specific point in the prokaryotic chromosome called the *origin of replication*. The primosome includes **helicase** to unwind the DNA strands and **primase** to

[2] One year has about 31,563,000 seconds. If you divide 3.2 billion by the number of seconds in a year, you'll see it is about 101 years (101.47, to be precise).

[3] Makiela-Dzbenska, K., Maslowska, K. H., Kuban, W., et al. (2019). Replication fidelity in *E. coli*: Differential leading and lagging strand effects for *dnaE* antimutator alleles. *DNA Repair*, *83*, 102643.

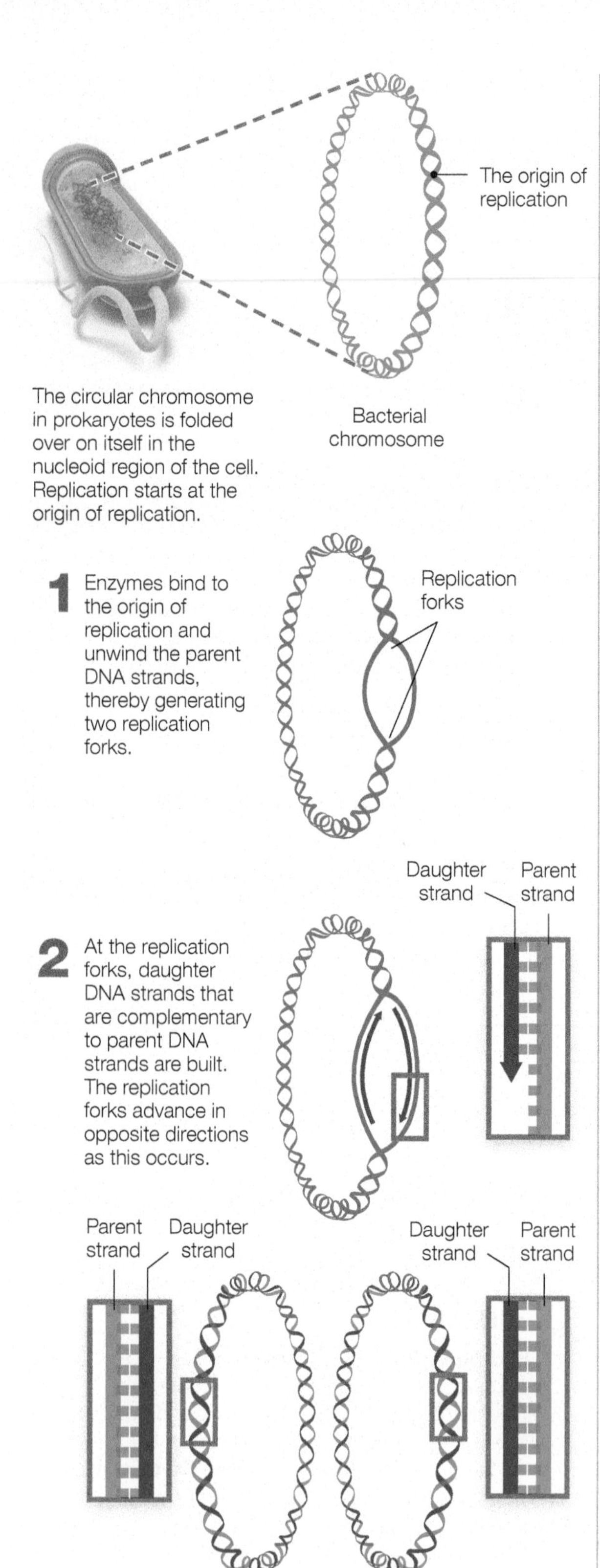

FIGURE 5.7 DNA replication forks in a circular chromosome

TABLE 5.3 Key Enzymes in DNA Replication

Enzyme	Function
Helicase	Unwinds DNA helix
Primase	Builds RNA primers; multiple primers are required to build the lagging strand
DNA polymerase III	Main enzyme that copies DNA on the leading and lagging strand
DNA polymerase I	Replaces RNA primers with DNA; also has a role in DNA repair
Ligase	Forms phosphodiester bonds to seal nicks in the DNA sugar–phosphate backbone; important in DNA replication and DNA repair
Gyrase and Topoisomerases	Relieve torsion stress that develops ahead of helicase as DNA unwinds

lay down **RNA primers**—short segments of RNA that DNA polymerase III requires to start replication.

Helicase breaks the hydrogen bonds between the nitrogen bases of the double-stranded helix to unwind the DNA and form a single-stranded bubble in the DNA (FIG. 5.7, STEP 1). DNA polymerase III then binds DNA within this bubble region, and replication starts. The original DNA strands are often called "parent" strands, while the newly made DNA strands are called "daughter" strands.

The immediate point where unwinding occurs is called a **replication fork.** As the replication bubble grows, the DNA helix unwinds to the right and left of the origin of replication, and two replication forks develop. Thus, DNA is replicated bidirectionally (FIG. 5.7, STEP 2). For a circular chromosome, which is what *E.coli* have, the replication forks advance in opposite directions around the chromosome until they eventually run into one another, and the completed DNA copies separate (FIG. 5.7, STEP 3).

During replication, **single-strand DNA-binding proteins** bind the DNA strands to keep them separated until they are copied. Specialized detangling enzymes called **gyrase** and **topoisomerases** relieve the coiling tension that develops as the helix unwinds. Another enzyme, primase, lays down RNA primers to jump-start replication. DNA polymerase III relies on these RNA primers as a springboard to start DNA replication. This is because DNA polymerase III can only extend a nucleic acid strand. It cannot start a new strand from scratch because it requires a 3′ OH group to build upon. FIG. 5.8 shows an overview of DNA replication.

Leading and lagging strand replication differs.

DNA polymerase III, the primary enzyme that copies DNA, only builds DNA in a 5′ to 3′ direction. This, along with DNA's antiparallel arrangement, means one side of the parent DNA molecule is copied in the *same* direction as helix unwinding and is therefore called the **leading strand**. In contrast, the other side of the DNA helix (the complementary DNA strand) is copied in the *opposite* direction of unwinding and is therefore called the **lagging strand**. Every replication fork has a leading and a lagging strand.

DNA Replication on the Leading Strand

On the leading strand, DNA polymerase III launches off the single RNA primer laid down by primase. As DNA polymerase III advances, it lays down new nucleotides that correctly pair with the parent DNA template to form AT and GC pairs. For example, if the parent DNA strand has an adenine (A) nucleotide, then the polymerase lays down a complementary thymine (T) nucleotide. Overall, the goal is to make the new DNA strand look exactly like the parent DNA. Because the direction of replication on the leading strand is the same as unwinding, replication is **continuous** along the template, with one nucleotide after another paired in sequence without interruption.

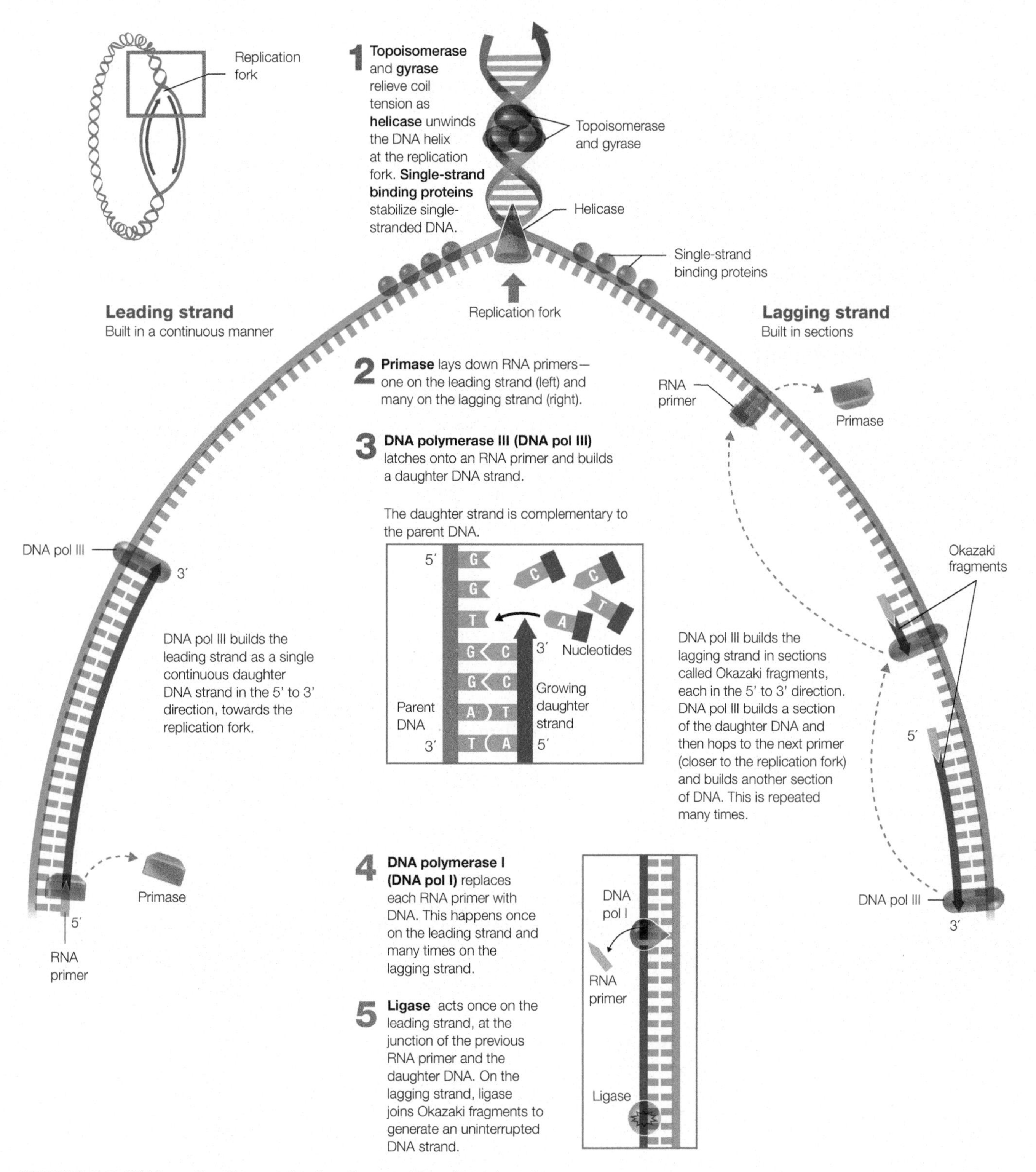

FIGURE 5.8 DNA replication on the leading and lagging strands

The final DNA copy cannot have any RNA in it, so the RNA primer that was laid down to jump-start replication must be removed before the leading strand is complete. To accomplish this, **DNA polymerase I** replaces the RNA primer with DNA. The final step is for the enzyme **ligase** to glue together the junction between this replacement DNA and the rest of the leading strand.

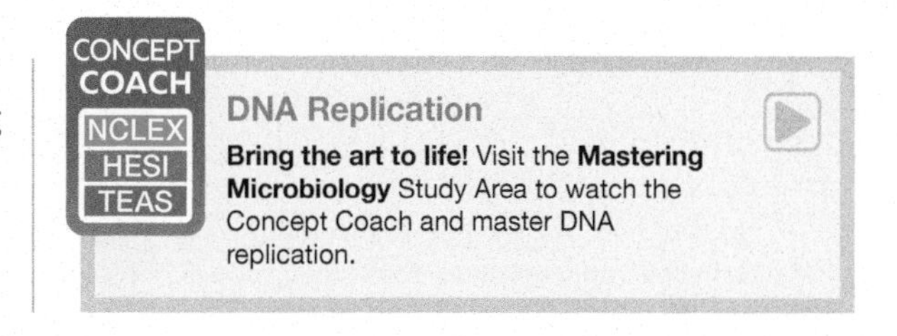

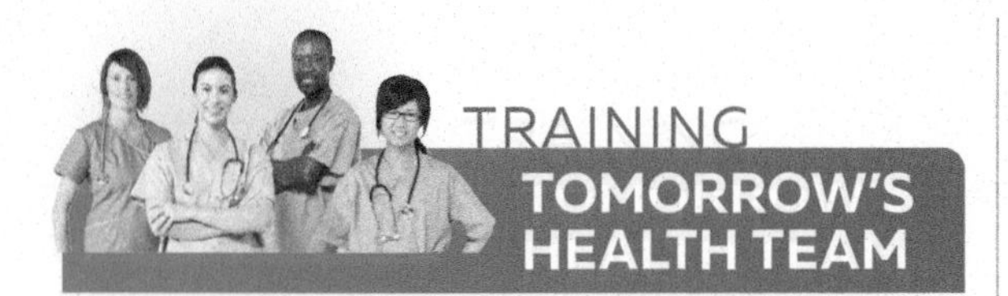

Hampering Herpes by Inhibiting DNA Replication

Many drugs, from antibiotics to anticancer drugs, work by inhibiting DNA replication. Acyclovir is an example of an antiviral drug used to treat (not cure) outbreaks of genital herpes, shingles, cold sores, and chickenpox.

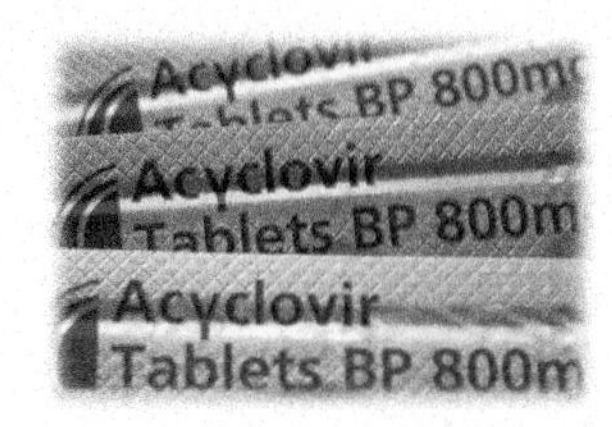

Acyclovir drugs limit the breakout of sores associated with chickenpox, shingles, and genital herpes.

Acyclovir is a chemical compound that resembles the guanine nucleotide found in DNA, except it does not have a sugar group and will therefore cause DNA replication to stop once it is added to the growing DNA strand. Once DNA replication stops, the viral genome cannot be copied, leading to a dead end for the virus, which must copy its genome to make new viral particles.

This drug also kills the infected host cell because it not only blocks viral DNA replication, but it also interferes with host DNA and RNA. Acyclovir is only toxic to cells infected with certain herpes viruses that encode an enzyme that activates acyclovir into a toxic form. Mammalian cells that are not infected with certain herpes viruses will not transform the compound into a toxic form, meaning they can persist in the presence of the drug without harm.

QUESTION 5.2

Based on what you have learned about DNA replication, why would adding acyclovir to the growing DNA strand ultimately result in a dead end for DNA replication?

DNA Replication on the Lagging Strand

DNA polymerase III builds new DNA on the lagging strand, too. However, because the lagging strand builds in the opposite direction of the helix unwinding, replication is a **discontinuous** process, in which enzymes construct chunks of the lagging strand in a sort of "back-filling" action. These DNA chunks are called **Okazaki fragments**, after the scientist who discovered them. Unlike on the leading strand, where primase synthesizes a single RNA primer, on the lagging strand primase lays down *many* RNA primers at intervals. DNA polymerase III latches onto the end of an RNA primer and builds DNA in a 5′ to 3′ direction, until it encounters the next primer. This results in DNA fragments that are capped at their start by a short RNA sequence. Then DNA polymerase I replaces the RNA primers with DNA. In the final step, ligase glues the multiple Okazaki fragments together, producing an uninterrupted strand.

By the end of replication, there will be two separate DNA helices. Each helix contains one original parent DNA strand paired with a newly formed complementary strand. Because DNA replication produces a hybrid molecule that is half new and half parent DNA, it is described as a **semiconservative** process.

Prokaryotes and eukaryotes replicate DNA somewhat differently.

Although eukaryotic DNA replication tools differ in structure, number, and sometimes name from what bacteria use, collectively the eukaryotic replication machinery works similarly to that of prokaryotes. That being said, DNA replication in eukaryotes takes longer and involves more protein factors, as eukaryotic genomes tend to be significantly larger and spread across more chromosomes than prokaryotic genomes. An average animal (eukaryotic) cell has about 50 times more DNA to copy than the average bacterial cell.

To manage their large genomes, eukaryotic cells depend on higher-order packaging of their DNA, so that the genome can fit into the nucleus. This higher-order organization takes time to undo and then reconstruct, so it slows replication. Also, because eukaryotes have a lot of DNA to copy, they have multiple replication initiation sites on their chromosomes (thousands) versus the single origin of replication seen in most bacteria. However, even the huge number of replication sites doesn't enable eukaryotic cells to copy their DNA as fast as prokaryotes. Certain bacteria can double their population about every 20 minutes, while eukaryotic cells often take between 18 and 24 hours to complete cell division.

Build Your Foundation

7. When do cells carry out DNA replication, why is it essential, and what contributes to its accuracy? (LO 5.7)
8. What is the main enzyme that builds DNA? Describe at least one of its rules or limitations. (LO 5.8)
9. What enzyme builds RNA primers, and what role do RNA primers have in DNA replication? (LO 5.8)
10. What is the role of DNA polymerase I in DNA replication? (LO 5.8)
11. How is DNA replication started? (LO 5.9)
12. How does the directionality of DNA impact replication? (LO 5.10)
13. How are the leading and lagging strands defined, and how is replication performed differently on these strands? (LO 5.11)
14. List three differences in DNA replication in prokaryotes versus eukaryotes. (LO 5.12)

BYF QUICK QUIZ

Build Your Foundation (BYF) Quick Quiz: Visit the **Mastering Microbiology** Study Area to quiz yourself.

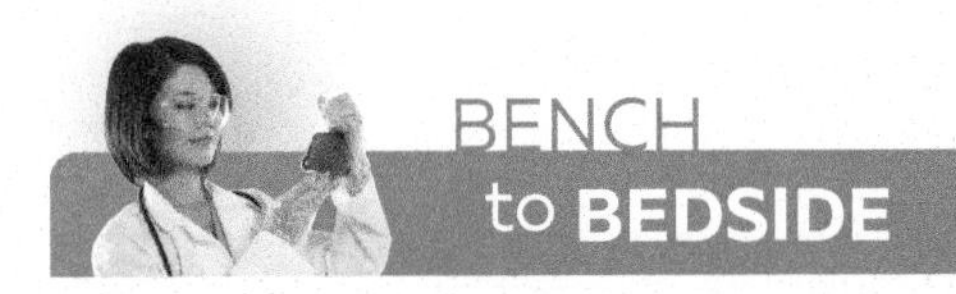

BENCH to BEDSIDE

Genetically Engineered Bacteria and Medical Therapies

The first modern **genetic engineering** (the manipulation of genes to produce specific proteins) was accomplished in the 1970s when scientists figured out how to selectively cut and paste sections of DNA to make recombinant gene constructs. *E. coli* was engineered to produce human insulin (Humulin). In 1983, Humulin became the first recombinant DNA human healthcare product on the market. Since then, bacteria have been actively employed as drug factories for recombinant gene products. Other recombinant DNA products include many antibiotics and vaccines, blood-clotting factors, anti-anemia drugs, drugs to stimulate bone marrow growth following a bone marrow transplant, and various diagnostic tests for HIV, tuberculosis, and several cancers.

Whereas the products made by recombinant microbes have been used in medical applications since the early 1980s, the direct use of engineered bacteria in medical therapies is still developing. There has been promising progress in this area, especially in treating inflammatory bowel diseases (IBDs), which include ulcerative colitis and Crohn's disease. IBD illnesses are characterized by chronic inflammation of the intestines. The administration of an anti-inflammatory protein, interleukin-10 (IL-10), has been shown to alleviate the inflammation in IBD. Barriers to using IL-10 in therapy are that it cannot be taken orally because stomach acid destroys it before it reaches the intestines, and injections are poorly tolerated.

Engineered bacteria are a possible alternative to deliver IL-10 treatments. Lactococci, which naturally live in acidic environments, can survive the stomach and temporarily grow in the bowel, making them ideal for delivering compounds such as IL-10 directly to the intestines. Plus, these bacteria are safe to consume; they are abundant in dairy products like yogurt and cheeses. Seizing on this collection of knowledge, scientists have engineered *Lactococcus lactis* to secrete human IL-10. Results suggest that these IL-10 engineered bacteria are effective at treating irritable bowel disease (IBD) in tested mouse models, and they have been effective in phase I clinical trials with humans.

Sources: Steidler, L., Hans, W., Schotte, L., et al. (2000). Treatment of murine colitis by *Lactococcus lactis* secreting interleukin-10. *Science*, *289*, 1352–1355. Braat, H., Rottiers, P., Hommes, D. W., et al. (2006). A phase I trial with transgenic bacteria expressing interleukin-10 in Crohn's disease. *Clinical Gastroenterology and Hepatology*, *4*(6), 754–759. Del Carmen, S., De Moreno de LeBlanc, A., Perdigon, G., et al. (2012). Evaluation of the anti-inflammatory effect of milk fermented by a strain of IL-10-producing *Lactococcus lactis* using a murine model of Crohn's disease. *Journal of Molecular Microbiology and Biotechnology*, *21*(3–4), 138–146.

5.3 PROTEIN SYNTHESIS (GENE EXPRESSION)

Protein synthesis entails reading the genomic instruction manual using transcription and translation.

In the last section you learned how cells copy their entire instruction manual (genome) and transfer it to a new generation of cells using DNA replication. Here we discuss **protein synthesis**, also known as **gene expression**, a process in which a cell's genetic information is read and used to create gene products—proteins. Whereas DNA replication copies the whole genome, in protein synthesis, only parts of the genome are read at any given time. What is read (or what genes are expressed) depends on the cell's circumstances. As an analogy, you may have access to a full library of information, but you won't read a book on Shakespeare if your task is to write a microbiology paper. Similarly, cells read the parts of their genetic instruction manual that will help them with a necessary task.

Protein synthesis is central to life, as it is responsible for most physical and functional cell features. Making proteins is so important that if it is blocked the cell will eventually die. Many antibiotics work by blocking one or more steps in the process. Protein synthesis has two main stages: transcription and translation. During **transcription**, genes in DNA are copied into a new format, RNA. The cell's DNA is *not* changed in this process. The second stage, **translation**, entails ribosomes decoding RNA to build proteins.

Learning Outcomes

After reading this section, you should be able to:

5.13 Explain what protein synthesis (gene expression) is, describe how it impacts phenotype, and state why it is essential to cell survival.

5.14 Explain transcription and translation, and point out differences in these processes in prokaryotes versus eukaryotes.

5.15 Define reverse transcription.

5.16 Describe structures and functions of the three RNA types.

5.17 Summarize mRNA splicing in eukaryotes.

5.18 Discuss the features of the genetic code and how these features benefit a cell.

5.19 Compare and contrast standard genetically encoded amino acids to nonstandard genetically encoded amino acids.

5.20 Define and give examples of post-translational modifications.

Transcription makes RNA.

In transcription, the enzyme **RNA polymerase** uses DNA as a template to build RNA. Transcription occurs in the nucleus in eukaryotes and in the cytoplasm of

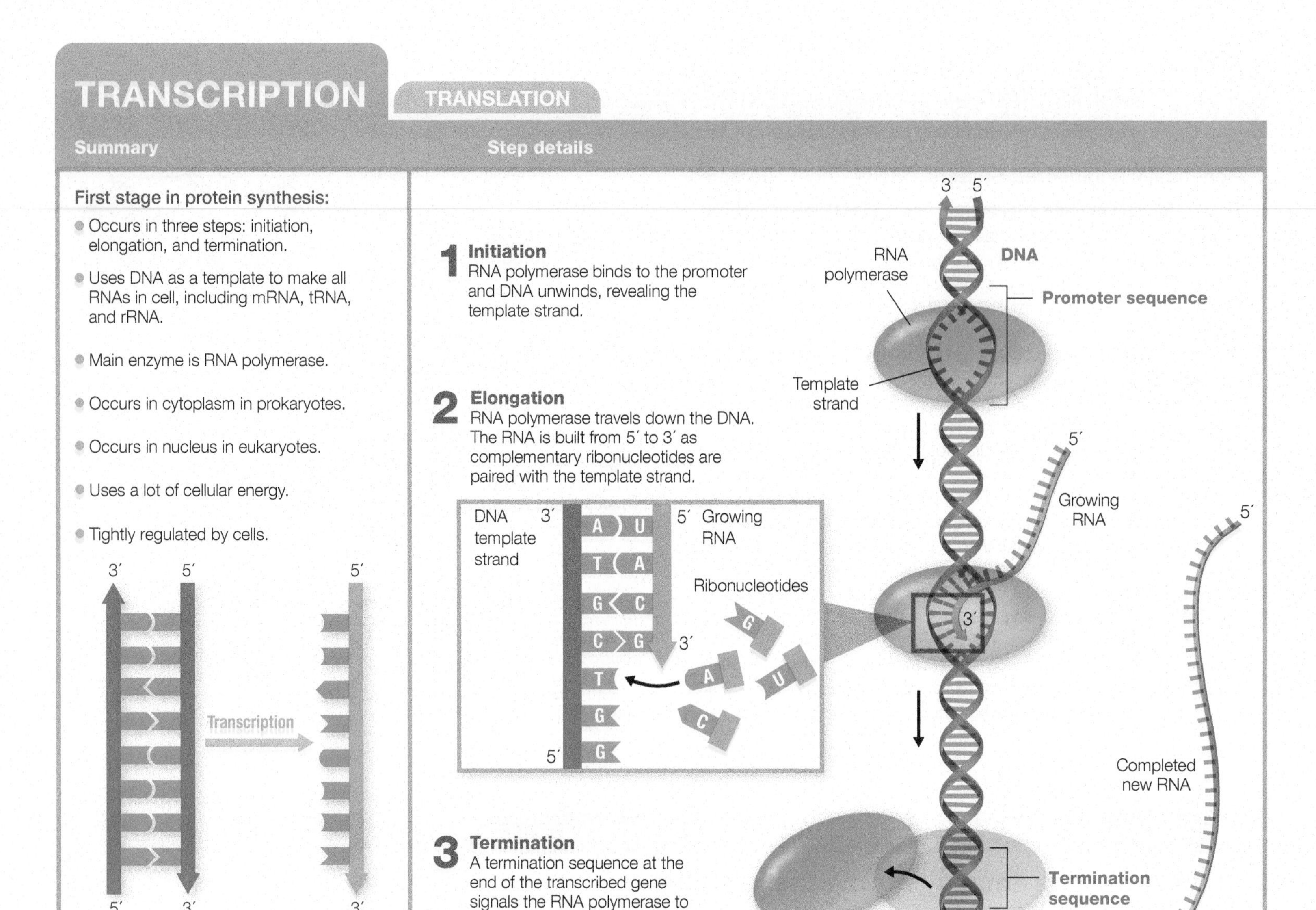

FIGURE 5.9 Transcription Transcription is the first stage of protein synthesis.

Critical Thinking *Transcription generates a different version of the same information. Explain this statement.*

prokaryotes and involves three main steps: (1) initiation, (2) elongation, and (3) termination (FIG. 5.9).

(1) In initiation, RNA polymerase binds to the *promoter*—the beginning region of a gene in DNA. Once bound to the promoter, RNA polymerase unwinds the DNA to reveal a single-stranded template of the gene that will be transcribed into RNA.

(2) During elongation, the enzyme advances along the single-stranded DNA, laying down complementary ribonucleotides as it goes. (Recall that in RNA A pairs with U rather than T, and G pairs with C.)

(3) The termination step occurs when the RNA polymerase reaches the end of the gene. At this point, a termination sequence signals the polymerase to fall off the DNA, and the newly made RNA transcript is released.

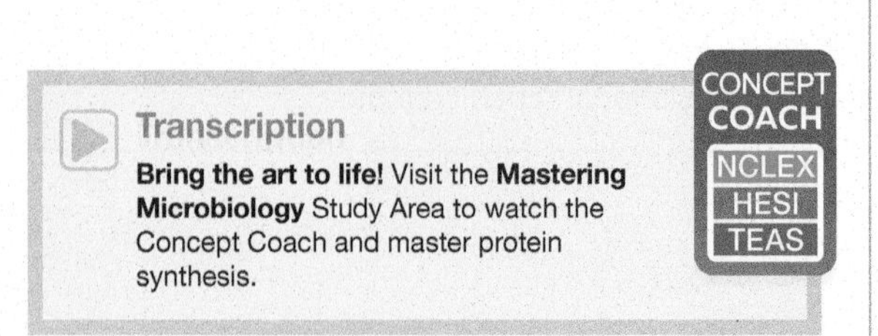

Reverse Transcription

As previously reviewed, genetic information usually flows one way—from DNA to RNA to proteins. However, some cells and viruses carry out **reverse transcription**, a process where an enzyme called **reverse transcriptase** reverses the flow of genetic information by using RNA as a template to build DNA. The DNA that reverse transcriptase builds is called *copy DNA*—or *cDNA*—to differentiate it from genomic DNA. Retroviruses such as HIV (human immunodeficiency virus) and certain human cells carry out reverse transcription.

Three main types of RNA are involved in protein synthesis.

Using DNA as the guiding instructions, transcription makes all RNA molecules in a cell. Although there are numerous types of RNA in a cell, three main types are involved in protein synthesis (FIG. 5.10). Here you will simply learn the names and basic roles of these RNA molecules. Then, in the translation section, we'll cover these molecules in greater detail.

1. Messenger RNA (mRNA): These generally linear molecules carry the genetically coded messages stored in DNA to ribosomes, the organelles that build proteins. In the upcoming section on translation, you'll explore features of the triplet code in mRNA and learn how ribosomes decode (or "read") mRNA to build proteins.
2. Transfer RNA (tRNA): These cloverleaf-shaped RNA molecules bring the correct amino acid to a ribosome to build proteins. There are two key regions in tRNA that you should note: a 3′ end, which can be bound to a specific amino acid, and a region called an **anticodon loop**, which, as you'll learn shortly, has a role in ensuring the correct amino acid is added to the growing protein during translation.
3. Ribosomal RNA (rRNA): These RNA molecules fold up into elaborate three-dimensional structures and combine with proteins to form ribosomes, the organelles that carry out translation. A ribosome's rRNA helps it to "read" or scan an mRNA molecule and form peptide bonds between amino acids to build a protein. There are several types of rRNA in ribosomes, and there is impressive diversity regarding rRNA sequences among species. This makes rRNA sequencing a useful tool for taxonomically grouping organisms.

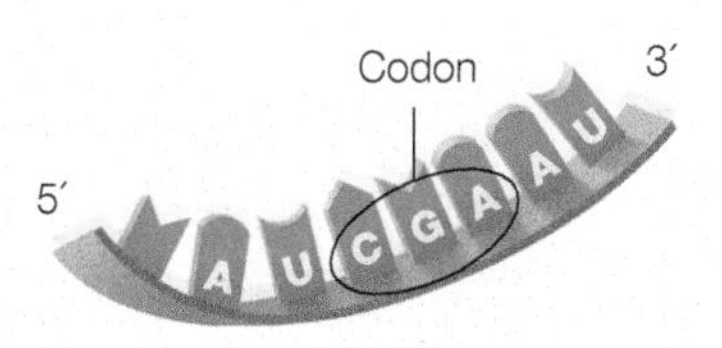

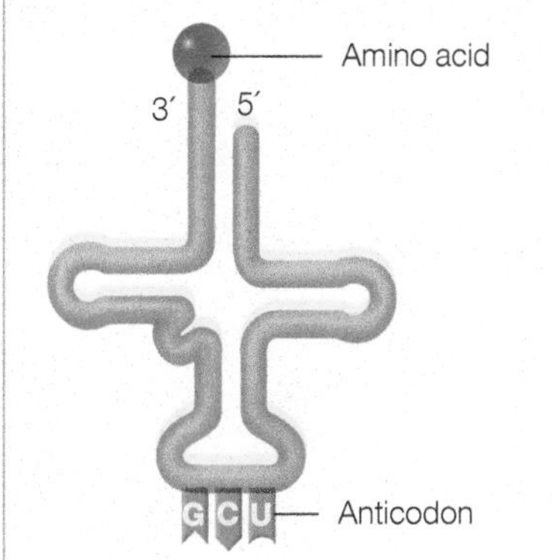

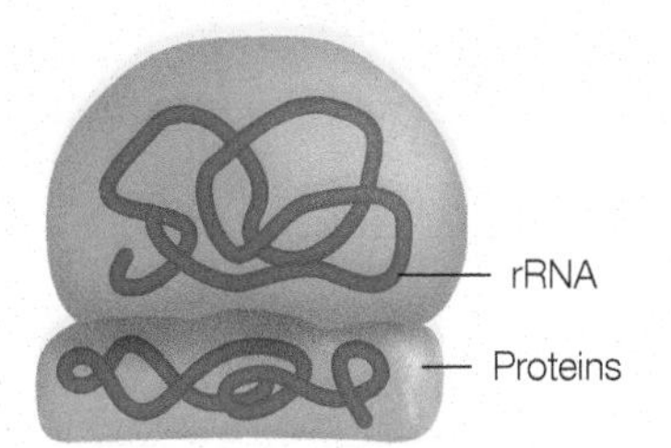

FIGURE 5.10 **RNAs** Three classes of RNA participate in translation.

Some RNAs must undergo splicing before translation.

It is important to note that certain RNAs must undergo splicing before they can be translated. **Splicing** involves clipping out specific sequences in RNA and joining the remaining parts of the molecule. This process occurs in many types of RNA in prokaryotes and eukaryotes, but here we focus on mRNA splicing in eukaryotes.

In eukaryotic cells, not all segments of mRNA are decoded to build a protein—these coding sequences are called **exons**. The intervening sequences in mRNA that are not decoded to build the protein are called **introns**. Even though intron sequences aren't decoded to contribute to the protein, they are still important tools in protein synthesis. Examples of intron function include regulating how much of a protein is made and determining what versions of the protein are present in a cell.

While introns were originally thought to be strictly in eukaryotes, they have been found in prokaryotic RNAs[4]—although not in prokaryotic mRNA. Consequently, prokaryotic mRNA does not need to be spliced before it is translated into protein. However, in eukaryotic cells, mRNA must be spliced and transported out of the nucleus before it can be translated. Eukaryotic cells remove the introns in their mRNAs and join the exons so that a coherent message is decoded by the ribosome. This mRNA editing is performed in the nucleus by a complex called the **spliceosome** (FIG. 5.11). The advantage of mRNA splicing is that there is the potential for *alternative splicing*, or new combinations of exons. Therefore, introns shouldn't be perceived as junk or a nuisance; rather, they provide an opportunity for genetic diversity.

CHEM • NOTE

Amino acids are the building blocks of proteins. They have a core structure consisting of an amine group (NH_2) and a carboxyl group (COOH).

CHEM • NOTE

Peptide bonds are a type of covalent bond in which the amine group of one amino acid covalently bonds with the carboxyl group of an adjacent amino acid. Peptide bonds link amino acids together to form polypeptides, or proteins. This chemistry happens between amino acids in the P and A sites of the ribosome.

[4] Four main classes of introns have been identified in nature. Prokaryotic introns fall into "self-splicing" classes and do not require the spliceosome machinery seen in eukaryotes.

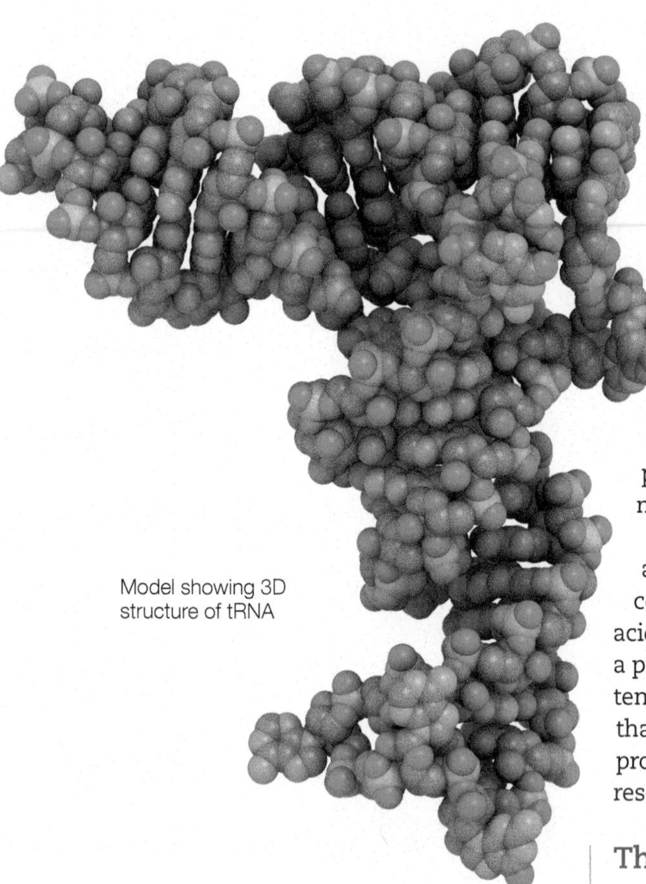

Model showing 3D structure of tRNA

Ribosomes translate mRNA to build proteins.

Again, translation is the process in which ribosomes decode mRNA to build proteins—that is, ribosomes translate the language of nucleic acids (the nucleotides in RNA) into the language of proteins (amino acids). In keeping with the language analogy, consider that letters form words; words build sentences; punctuation indicates the end of each sentence, and each new sentence is started with a capitalized word. A cell's RNA nucleotides (A, U, G, and C) can be thought of as its alphabet. These letters are organized into three-letter words—the **codons** in mRNA. Because codons are three letters long, they are often referred to as a *triplet code*. Codons are only found in mRNA. And while most codons simply code for an amino acid, some serve as punctuation that tell the ribosome where to end a sentence (stop building the protein). Just as we read words and punctuation and extract the meaning from a sentence, ribosomes read mRNA and use the information to assemble the specified protein.

Because there are only four letters in a cell's alphabet (A, U, G, and C), and all of a cell's codons are three letters long, there are only 64 possible codons in a cell's vocabulary ($4 \times 4 \times 4 = 64$). These 64 codons encode 20 standard amino acids (the words in the sentence), a start signal (the "capital letter" for the start of a protein sentence), and three stop signals (punctuation to end the protein sentence). Codons that encode amino acids are called *sense codons*, whereas those that encode stop signals are called *nonsense codons*. The codons in mRNA guide the production of all proteins—and the set of rules that determines which codon represents a particular amino acid or start/stop signal is called the **genetic code**.

The Genetic Code and Amino Acids

As described in Chapter 2, amino acids are the building blocks of proteins. In nature there are hundreds of amino acids, but 20 are called *standard genetically encoded amino acids*. These 20 amino acids are found in the proteins of all cells and are typically depicted in genetic code tables like that in TABLE 5.4.[5]

TABLE 5.4 The Genetic Code

	Second Base								
First Base	U		C		A		G		Third Base
U	UUU	Phenylalanine	UCU	Serine	UAU	Tyrosine	UGU	Cysteine	U
	UUC		UCC		UAC		UGC		C
	UUA	Leucine	UCA		UAA	STOP	UGA	STOP	A
	UUG		UCG		UAG	STOP	UGG	Tryptophan	G
C	CUU	Leucine	CCU	Proline	CAU	Histidine	CGU	Arginine	U
	CUC		CCC		CAC		CGC		C
	CUA		CCA		CAA	Glutamine	CGA		A
	CUG		CCG		CAG		CGG		G
A	AUU	Isoleucine	ACU	Threonine	AAU	Asparagine	AGU	Serine	U
	AUC		ACC		AAC		AGC		C
	AUA		ACA		AAA	Lysine	AGA	Arginine	A
	AUG	Methionine START	ACG		AAG		AGG		G
G	GUU	Valine	GCU	Alanine	GAU	Aspartate	GGU	Glycine	U
	GUC		GCC		GAC		GGC		C
	GUA		GCA		GAA	Glutamate	GGA		A
	GUG		GCG		GAG		GGG		G

[5] There are technically 22 genetically encoded amino acids—however, only 20 are considered *standard*. The other two (selenocysteine and pyrolysine) are described as *nonstandard* genetically encoded amino acids. We'll review these special amino acids later in this chapter.

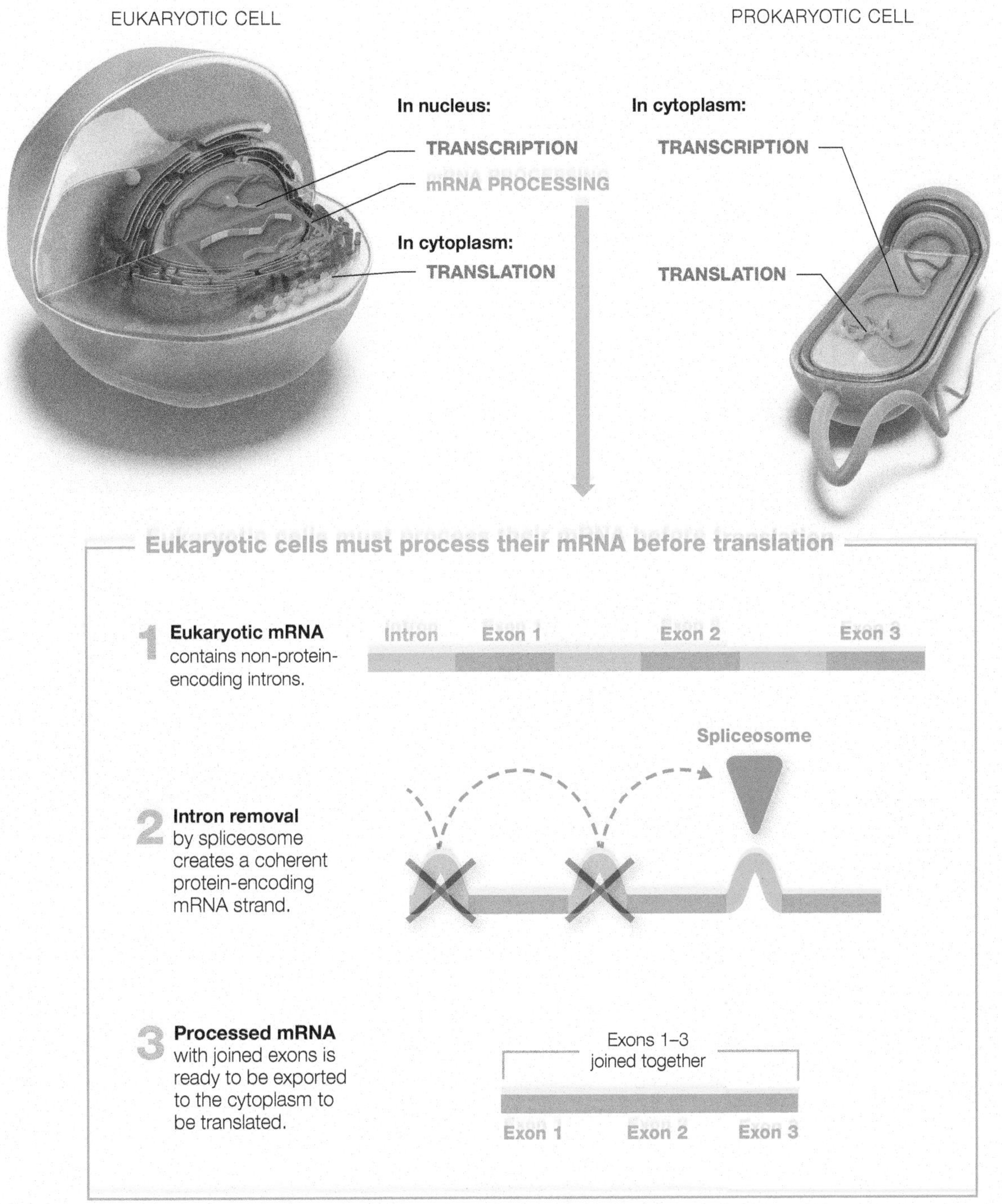

EUKARYOTIC CELL

PROKARYOTIC CELL

In nucleus:

TRANSCRIPTION

mRNA PROCESSING

In cytoplasm:

TRANSLATION

In cytoplasm:

TRANSCRIPTION

TRANSLATION

Eukaryotic cells must process their mRNA before translation

1 **Eukaryotic mRNA** contains non-protein-encoding introns.

Intron Exon 1 Exon 2 Exon 3

Spliceosome

2 **Intron removal** by spliceosome creates a coherent protein-encoding mRNA strand.

3 **Processed mRNA** with joined exons is ready to be exported to the cytoplasm to be translated.

Exons 1–3 joined together

Exon 1 Exon 2 Exon 3

FIGURE 5.11 Eukaryotic mRNA processing The spliceosome removes introns and joins exons.

Critical Thinking *How could an alternative protein be made from the portrayed mRNA?*

Looking at the table, you will notice that a single amino acid can be represented by more than one codon (for example, there are six codons for leucine). This *redundancy* in the genetic code protects cells from genetic changes (mutations). In redundant codons the first two nucleotides tend to remain constant, while the third nucleotide, sometimes referred to as the *wobble position,* differs.

TRANSCRIPTION

TRANSLATION

Summary

Second stage in protein synthesis:

- Occurs in three steps: initiation, elongation, and termination.
- With the help of tRNA, ribosomes decode/read mRNA to make a protein.
- Process occurs in the cytoplasm in both prokaryotes and eukaryotes.
- Uses a lot of cellular energy.
- Tightly regulated by cells.

5′ Translation 3′ mRNA Protein

Step details

1 Initiation

- Ribosome attaches to mRNA and scans until it reaches a start codon (usually AUG).
- Initiator tRNA carrying amino acid methionine (met) then enters the ribosome's P site.

met
Initiator tRNA
P site (peptidyl site)
E site (exit site)
A site (acceptor site)
mRNA
5′
E P A
Ribosome
3′
A U G
Start codon

- The start codon on the mRNA base pairs with the the anticodon on the initiator tRNA.

3′ 5′
U A C
A U G
5′ 3′
Anticodon
Start codon

FIGURE 5.12 Translation overview

Critical Thinking *Describe two possible ways that tRNA could introduce a mutation in translation.*

Step 1 of Translation: Initiation

Earlier you learned that transcription builds RNA in three steps, *initiation*, *elongation*, and *termination*. Here you'll see those same three stepwise terms again. However, this time they describe the steps in translation. The ribosome serves as an efficient translation machine. Recall from Chapters 3 and 4 that ribosomes have two subunits, a large and a small subunit. These subunits combine to make an active ribosome, which has an exit site (E), a peptidyl site (P), and an acceptor site (A). These sites coordinate translation (FIG. 5.12).

In prokaryotes, translation is initiated even before transcription concludes. This is possible because unlike eukaryotes, prokaryotes do not have to splice their mRNA before translating it, and their mRNA does not have to be shipped out of a nucleus. In the initiation step of translation, a ribosome attaches to the mRNA and scans it until it encounters a start codon, where it adds the first amino acid. Several protein factors help initiation in prokaryotes, while over a dozen factors help initiation in eukaryotes, but we will discuss the process in simplified terms. Usually the start codon is AUG, and the first amino acid added by the initiator tRNA in prokaryotes is *formyl methionine* (fMet). In eukaryotes, the initiating amino acid is methionine (Met). In either case, the initiator tRNA enters the ribosome at the P site to deliver the first amino acid of the protein being built.

Once one ribosome has initiated translation, the next ribosome can latch on the mRNA and launch initiation so that eventually multiple ribosomes are reading (translating) the same mRNA. These so-called *polysomes*, or collections

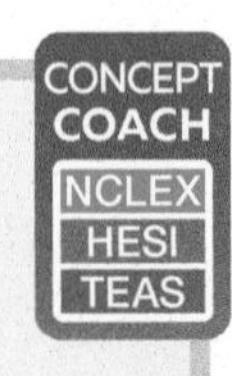

Translation

Bring the art to life! Visit the **Mastering Microbiology** Study Area to watch the Concept Coach and master protein synthesis.

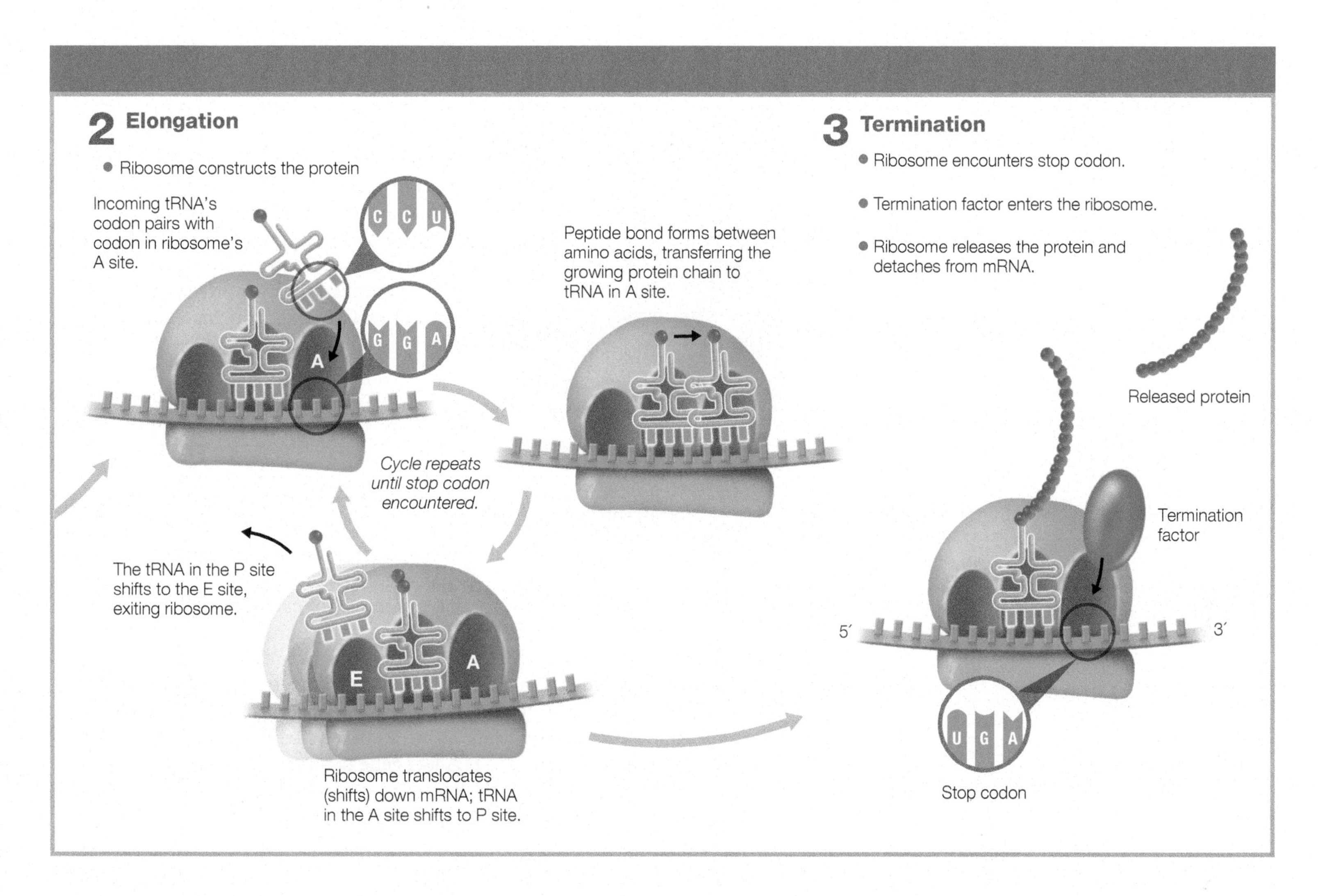

of many ribosomes on a single mRNA, allow prokaryotes and eukaryotes to rapidly produce proteins from mRNA transcripts (FIG. 5.13).

Step 2 of Translation: Elongation

During elongation, the ribosome constructs or elongates the protein. Once ribosomes attach to the mRNA, transfer RNAs (tRNAs) shuttle amino acids to them. Each genetically encoded amino acid has at least one specific tRNA molecule.

A tRNA that is carrying its specific amino acid is said to be "charged." One at a time, charged tRNAs enter the ribosome at the A site and try to pair with a codon on the mRNA. If the tRNA's anticodon loop is complementary to the mRNA codon (by base-pairing rules), then the amino acid is added to the growing protein chain. The codon/anticodon pairing process ensures that the manufactured protein follows the genetic instructions in mRNA.

Assuming the mRNA's codon and the tRNA's anticodon are complementary, the ribosome forms a peptide bond between adjacent amino acids. The ribosome then shifts along the mRNA, kicking the uncharged tRNA out of the ribosome (via the E site) to make room for the next charged tRNA to enter the ribosome.

Step 3 of Translation: Termination

A ribosome continues the elongation process, adding amino acids to the growing protein chain at each codon on the mRNA, until it reaches a stop codon (or nonsense codon) that signals termination—the end of translation. Three codons serve as termination or stop signals (see Table 5.4) that encourage termination factors to enter the ribosome, end translation, and release the completed protein. Eventually the ribosome detaches from the mRNA, enabling it to then reattach to the start codon of the same (or another) mRNA and begin the translation process anew.

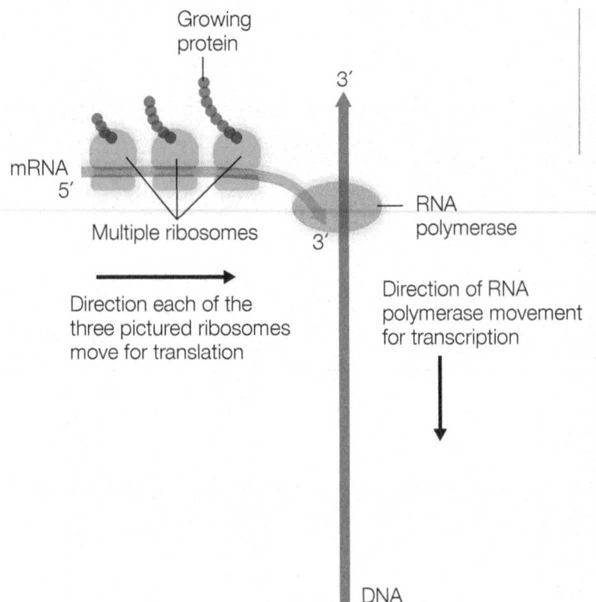

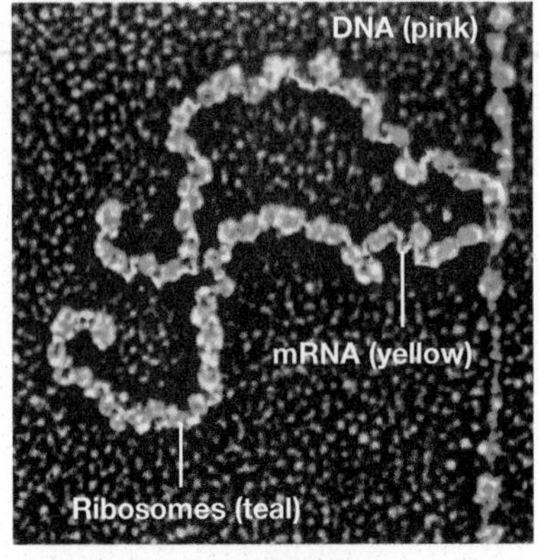

TEM showing simultaneous transcription and translation.

FIGURE 5.13 Polysome formation In bacteria, polyribosomes (or polysomes) can develop before transcription finishes. In eukaryotes (not shown), the mRNA must be processed and exported out of the nucleus before translation can start. However, once the mRNA is in the cytoplasm, multiple ribosomes can translate it in parallel.

In eukaryotes, most mRNA molecules are **monocistronic**, meaning they only encode a single protein. But this is not the case in prokaryotes, whose mRNAs are commonly **polycistronic**, meaning they can encode multiple types of proteins.[6] Thus, in prokaryotes the ribosome may not immediately detach from the mRNA when it encounters a stop codon. Instead, part of the ribosome often remains on the mRNA and continues reading it in search of the next start codon. Having ribosomes remain on the mRNA to check for additional start codons is very efficient for prokaryotic cells.

Translation termination does not necessarily mark the end of the protein's development. Once released from the ribosome, proteins often undergo modifications such as trimming. For example, the first amino acid added to a protein (methionine in eukaryotes or formyl methionine in prokaryotes) can be trimmed off so the protein no longer starts with the initiator amino acid. Proteins may also have various organic factors (e.g., sugars and lipids) and inorganic factors (e.g., phosphate and metal ions) added to them. These **post-translational modifications** are often required for a protein to function and provide a way for cells to regulate gene product functionality.

Nonstandard Genetically Encoded Amino Acids and Stop Codons

For many years it was thought that stop codons always terminated translation and that no tRNAs interacted with stop codons to bring in amino acids. The discovery of two nonstandard genetically encoded amino acids changed this thinking and is reframing how scientists look at genetic messages and the process of translation. Under specific circumstances, the stop codon UGA encodes a nonstandard amino acid called selenocysteine. Initially it was thought that UGA's ability to code this was unique to bacteria and archaea, but we now know that this also occurs in many eukaryotes, including humans. Errors and/or deficiencies in proteins that contain selenocysteine have been associated with cancer, male infertility, Keshan disease (a heart condition), thyroid hormone abnormalities, and diminished immune response. Although the role of selenocysteine in human health and disease is still being investigated, and there is no scientific evidence that selenocysteine supplements promote health, supplement manufacturers sell a variety of selenocysteine concoctions.

Like selenocysteine, the nonstandard amino acid pyrrolysine can be incorporated into proteins under specific circumstances. This step is facilitated by the stop codon UAG. So far, only certain prokaryotes (mainly methanogenic archaea) seem to use pyrrolysine.

---

[6] The existence of polycistronic messages in eukaryotes has been known for over 20 years, yet many sources continue to refer to polycistronic messages as unique to prokaryotes. This topic is nicely reviewed by Brunet, M. A., Leblanc, S., & Roucou, X. (2020). Reconsidering proteomic diversity with functional investigation of small ORFs and alternative ORFs. *Experimental Cell Research*, 393(1), 112057.

Build Your Foundation

15. What is protein synthesis, and why is it important in cells? (LO 5.13)
16. Describe ways that transcription and translation differ in prokaryotes versus eukaryotes. (LO 5.14)
17. What is transcription, and what enzyme is primarily responsible for it? (LO 5.14)
18. What is the general function of reverse transcriptase? (LO 5.15)
19. Name and describe three different types of RNA that are made by transcription. (LO 5.16)
20. What is an anticodon, where is it found, and how does it factor into protein synthesis? (LO 5.16)
21. What are introns, and how do they impact protein synthesis? (LO 5.17)
22. Explain the term *redundancy* as it relates to the genetic code. (LO 5.18)
23. What is meant by "triplet code"? (LO 5.18)
24. Name two nonstandard genetically encoded amino acids, and describe how they are incorporated into a protein. (LO 5.19)
25. Why are post-translational modifications important? (LO 5.20)

Build Your Foundation (BYF) Quick Quiz: Visit the **Mastering Microbiology** Study Area to quiz yourself.

5.4 REGULATING PROTEIN SYNTHESIS

Controlling protein synthesis is essential for all cells.

Protein synthesis through transcription and translation is energy and resource intensive, so it is essential for cells to control when these processes occur. As mentioned previously, cells do not constantly express every gene in their genome. Typically, less than 20 percent of a cell's genes are expressed at any given time. Some are used for everyday cell work and are referred to as *house-keeping genes*—these genes encode proteins that are reliably found in a cell and are thus considered *constitutive genes*. In contrast, proteins from *facultative genes* are made selectively, when a cell encounters a specific environmental change, or has a special job to do. Of *E. coli*'s approximately 4,200 protein-encoding genes, only about 250 are considered constitutive.[7] Even when a gene is switched on, the amount of a gene product made is still carefully regulated. Protein synthesis is usually regulated at the pre-transcription and post-transcriptional levels (FIG. 5.14).

Learning Outcomes

After reading this section, you should be able to:

5.21 Describe the difference between constitutive and facultative genes.

5.22 Summarize the general effect of pre-transcriptional regulation and the basic role of transcription factors in regulation.

5.23 Define the parts of an operon, and compare and contrast inducible and repressible operons.

5.24 Explain what epigenetic regulation is, and how it can impact protein synthesis.

5.25 Describe the role of quorum sensing in protein synthesis, and how this enhances bacterial survival.

5.26 Define post-transcriptional regulation, and provide specific examples of it.

Pre-transcriptional regulation impacts when and how often transcription occurs.

Pre-transcriptional regulation helps cells control when and how often transcription occurs, thereby regulating protein synthesis by affecting how much RNA is made from a given gene. Most forms of pre-transcriptional regulation are managed by *transcription factors*—specialized proteins that bind to DNA and help recruit RNA polymerase to start transcription. By controlling the presence and action of transcription factors, cells exert amazing control over protein synthesis.

Bacterial cells experience frequent and dramatic changes in their environment. To survive, they must rapidly adapt to such changes. As such, bacteria are excellent models for understanding how protein synthesis is regulated. Next,

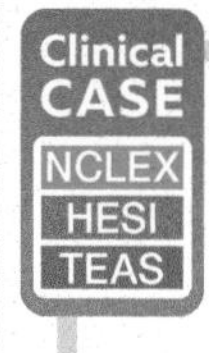

The Case of the Spreading Superbug

Practice applying what you know clinically: visit the **Mastering Microbiology** Study Area to watch Part 2 and practice for future exams.

[7] Puigb, P., Romeu, A., and Garcia-Vallv, S. (2008). HEG-DB: A database of predicted highly expressed genes in prokaryotic complete genomes under translational selection. *Nucleic Acids Research*, vol. 36, pp. D524–D527.

FIGURE 5.14 Regulating protein synthesis by controlling transcription and/or translation Cells regulate gene expression at a variety of levels. Two key points of regulation are at the pre-transcriptional and post-transcriptional levels.

Critical Thinking *Why could pre-transcriptional regulation be considered more energy efficient than post-transcriptional regulation?*

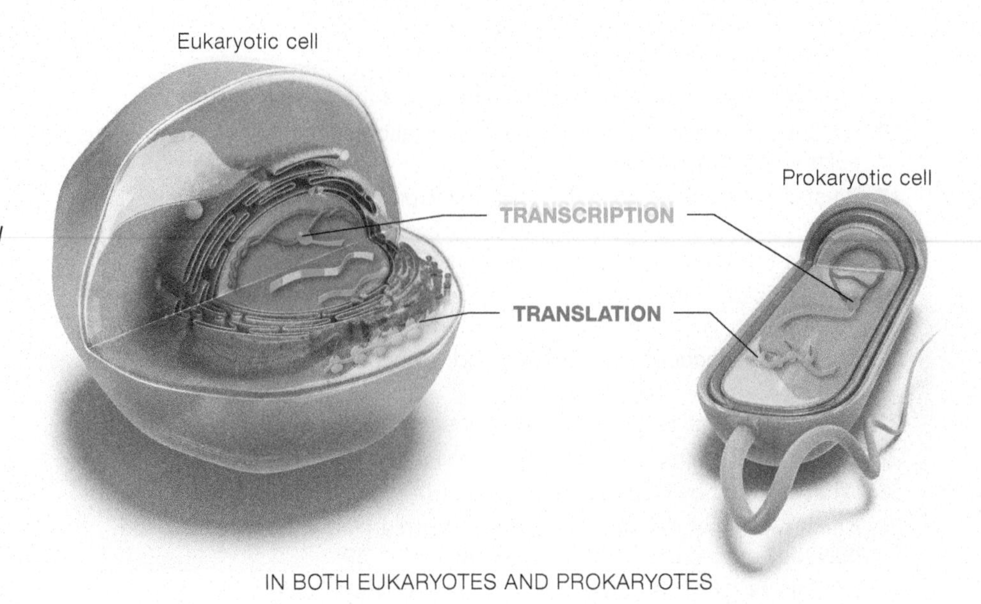

IN BOTH EUKARYOTES AND PROKARYOTES

Pre-transcriptional regulation allows cell to control gene expression by regulating how much mRNA is made from a given gene.

Examples:

- Operons
- Quorum sensing
- Epigenetic control (such as DNA methylation)
- Recruitment of transcription factors

Post-transcriptional regulation allows cell to control gene expression by regulating how often mRNA is translated into protein.

Examples:

- Recruiting ribosomes to mRNA
- Controlling mRNA stability
- Small noncoding RNAs
- Riboswitches
- Eukaryotes also can control RNA processing and nuclear export. (This is not used by prokaryotes.)

we will examine operons—one important way to regulate protein synthesis at a pre-transcriptional level.

Operons are one form of pre-transcriptional regulation.

An operon is a collection of genes controlled by a shared regulatory element. The genes in an operon participate in a joint goal, such as building a specific substance, or breaking something down. Think of operons as an efficient way to coordinate an assortment of sports team members who must work together to make a winning play. It was originally thought that only prokaryotic cells used operons, but operons have also been found in certain eukaryotes.[8]

Operons include four key parts that combine to make an on–off switch to regulate transcription. These four parts are (1) a promoter, which RNA polymerase associates with to start transcription; (2) two or more genes that encode proteins that work together toward a shared task; (3) a repressor that blocks transcription; and (4) an operator, the part of the operon that the repressor binds to in order to block transcription. When the repressor binds the operator, transcription is blocked, and the operon is "off." When it falls off the operator, the genes in the operon are transcribed and the operon is "on."

There are two main classes of operons. Inducible operons are *off* by default unless certain conditions arise under which they are activated (induced) to allow transcription. In contrast, repressible operons are *on* by default, meaning

[8] Blumenthal, T., Davis, P., & Garrido-Lecca, A. (2018). Operon and non-operon gene clusters in the C. *elegans* genome. In *WormBook: The Online Review of* C. *elegans Biology [Internet]*. WormBook.

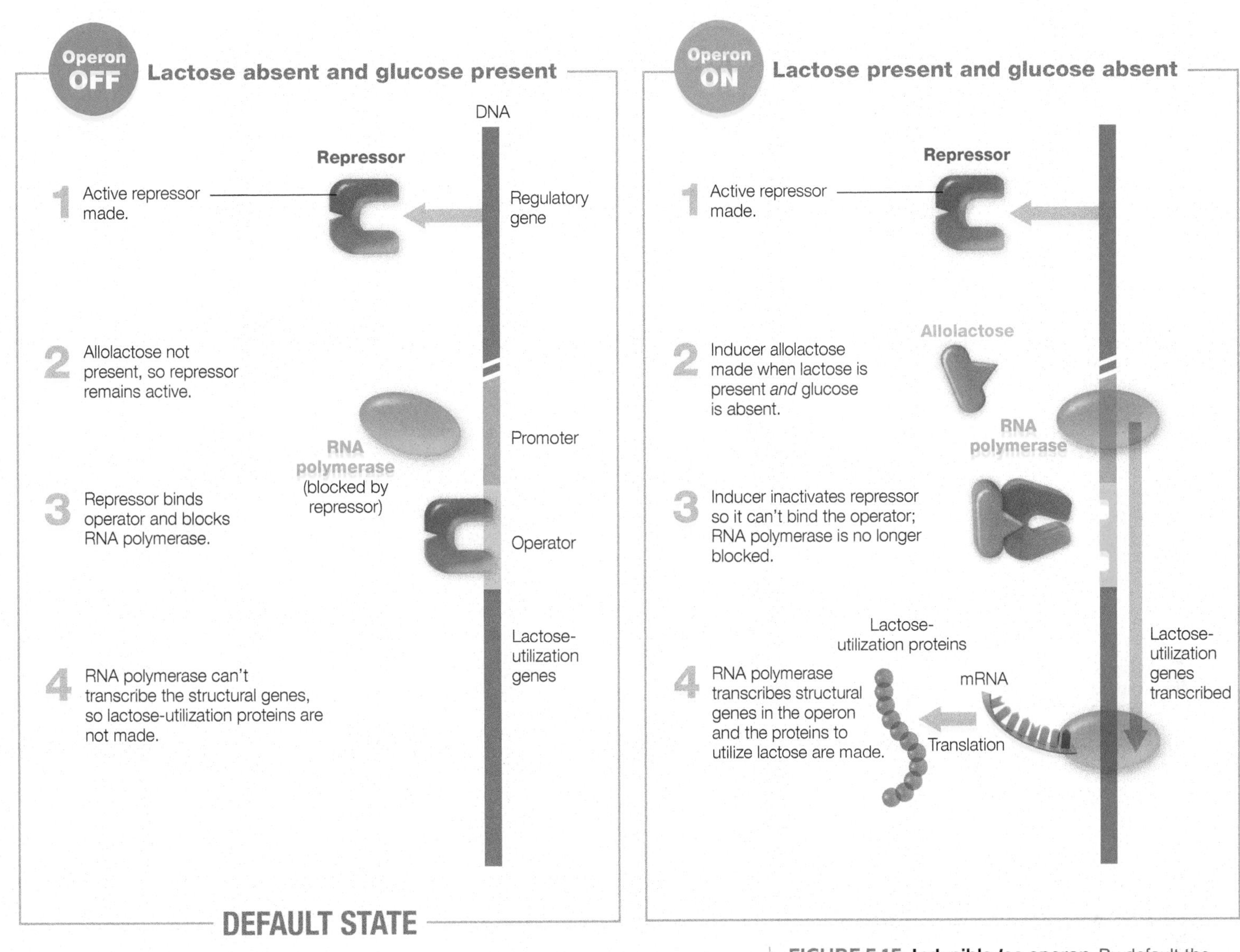

FIGURE 5.15 Inducible *lac* operon By default the repressor is active. Unless the repressor is inactivated, the operon is OFF.

they are actively transcribed until they are switched off (repressed). As an analogy for inducible operons, think of a car's headlights, which are by default off unless you are *induced* to turn them on by dark conditions. Repressible operons can be thought of as akin to the electricity supply to your home, which is always on and ready to power devices or appliances, unless a power outage results in the power being shut off, or repressed.

Inducible Operons

An example of an inducible operon is the lactose (*lac*) operon of *E. coli*. There are three structural genes in the *lac* operon that help the cell digest lactose. This operon is induced, or actively transcribed, *only* when lactose is present *and* the cell's preferred food, glucose, is absent (FIG. 5.15). Unless both conditions are met, the repressor remains attached to the operator, blocking transcription.

When lactose is present *and* glucose is absent, the cell makes an inducer called *allolactose*, which inactivates the repressor and prevents it from binding to the operator. With the repressor removed, RNA polymerase's path is no longer blocked, and it can transcribe the structural genes in the operon. If all the lactose is digested, or if glucose becomes available again, the cell will stop making allolactose, and the repressor will reassociate with the operator, once again blocking transcription.

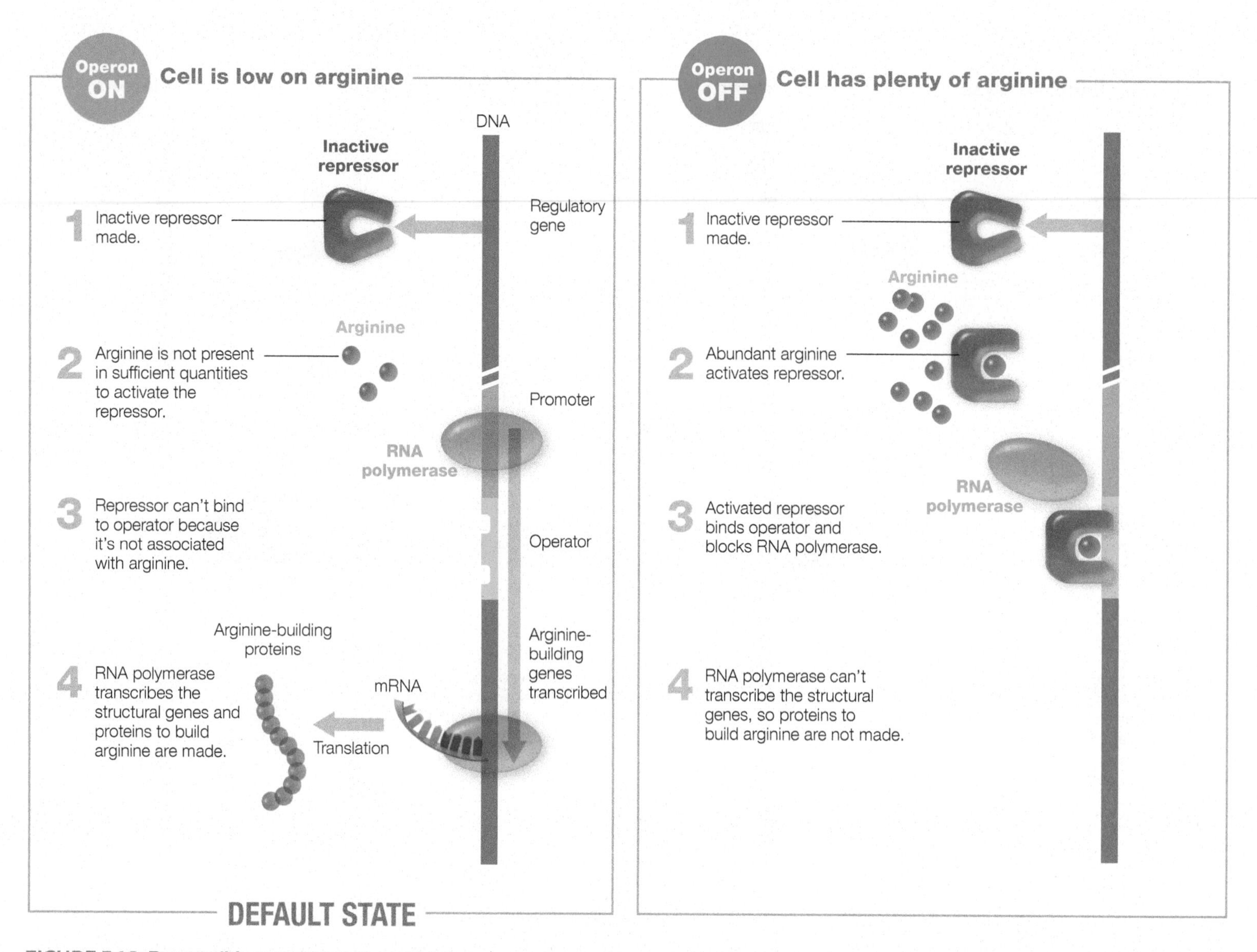

FIGURE 5.16 **Repressible *arg* operon** By default the repressor is inactive. Unless the repressor is activated by arginine, the operon is ON.

Repressible Operons

Repressible operons allow cells to halt production of substances that are already abundant.

One common example is the arginine (*arg*) operon, which regulates the production of the amino acid arginine. Of course, the cell doesn't need to make arginine if there is already plenty of it available. The *arg* operon is by default transcribed (is always "on") unless the repressor gains the ability to bind to the operator (FIG. 5.16).

The repressor can only associate with the operator when it is bound to arginine. Therefore, when arginine is present in sufficient quantities, it will enable the repressor to bind to the operator to block transcription. When the supply of arginine diminishes to a level at which the cell needs more, the repressor dissociates from arginine and falls off the operator, allowing transcription to resume.

Cells alter their epigenome as a means of pre-transcriptional regulation.

We have already learned that a *genome* is all the genetic material in a cell. An **epigenome**, by contrast, is the collection of all the chemical changes to the genome. Epigenetic regulation is a way to control protein synthesis by directly altering the appearance of DNA without changing its sequence. One way eukaryotic and prokaryotic cells do this is by **DNA methylation**, which involves adding methyl groups (CH_3 groups) to specific regions in DNA enriched with cytosine (C). Methylating a gene silences it by preventing transcription (FIG. 5.17). DNA methylation patterns have important roles in the development of different tissues in multicellular organisms. This is because a cell's phenotype is based not only on the genes that were inherited, but also on which genes are expressed.

All the cells in a multicellular organism have the same genome, yet they can be structurally and functionally diverse. How can skin or liver cells be so different in appearance and function, yet both have the same genome? As it turns out, DNA methylation patterns "teach" cells how to develop and function by affecting protein synthesis. DNA methylation patterns can also be inherited, or passed from one cell generation to the next; this heritability of methylation is thought to be just as important to the function of cells as inheritance of the genome itself.

Although the particulars of DNA methylation are not completely understood, research has shown that when methylation patterns are changed, there are dramatic affects. Certain cancer cells have altered DNA methylation patterns compared with noncancerous cells in the same tissue.[9] Also, many agents that cause birth defects have been shown to alter DNA methylation patterns.

In bacteria, DNA methylation patterns have diverse roles, from coordinating DNA repair and responses to environmental stress to affecting a bacterium's ability to cause disease.[10] Methylation patterns of bacterial genomic DNA also protect bacteria from infection by bacteriophages, which are viruses that infect bacteria (bacteriophages are discussed further in Chapter 6).

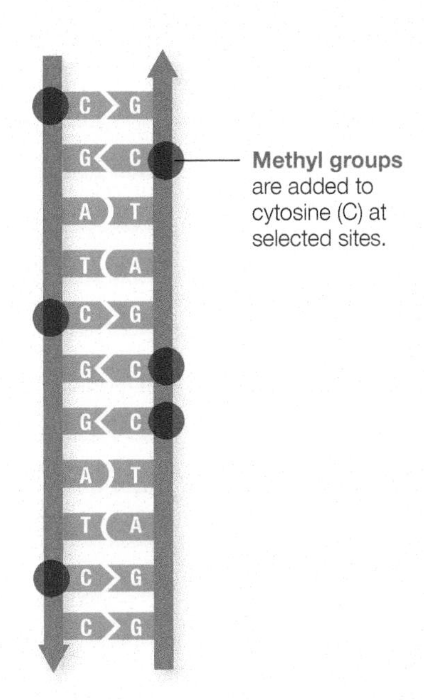

FIGURE 5.17 **DNA methylation** Adding methyl groups (CH_3) to selected cytosine in DNA silences genes by blocking transcription.

Many bacteria use quorum sensing to pre-transcriptionally control protein synthesis.

Although bacteria are unicellular organisms, they often exist in communities where they communicate using chemical messengers called *autoinducers*. These molecules allow bacteria to sense what is occurring in their community and respond accordingly to enhance their chance of survival. The collective sensing and responding to changes within a bacterial community is called **quorum sensing**. Quorum sensing allows bacteria to alter their protein synthesis in response to changes in the density of the population. In a way, quorum sensing allows unicellular organisms to coordinate their processes and behave more collectively, almost like a primitive multicellular organism. Coordinated protein synthesis through quorum sensing is the foundation for forming biofilms. (See Chapter 1 for more on biofilms.)

Post-transcriptional regulation impacts how often mRNA is translated into protein.

In contrast to all the previously mentioned pre-transcriptional regulation methods, which act before mRNA is even made, post-transcriptional regulation impacts how often mRNA is translated into protein.

Cells can regulate translation by affecting how readily ribosomes associate with mRNA to read it. Cells may also alter mRNA stability; mRNA with a fleeting existence has a limited amount of time to be translated, while stable mRNA exists longer in the cell, providing more opportunities for translation. As previously mentioned, in eukaryotes mRNA is spliced and exported out of the

[9] Yau, H. L., Ettayebi, I., & De Carvalho, D. D. (2019). The cancer epigenome: Exploiting its vulnerabilities for immunotherapy. *Trends in Cell Biology*, 29(1), 31–43.

[10] Sánchez-Romero, M. A., & Casadesús, J. (2019). The bacterial epigenome. *Nature Reviews Microbiology*, 1–14.

nucleus before translation occurs. Regulating the rate of these processes is another way for eukaryotes to regulate translation.

Small Noncoding RNAs

In addition to the various RNAs you have already learned about, cells produce a variety of small noncoding RNAs that impact protein synthesis. In eukaryotic cells, small noncoding RNAs that work in post-transcriptional regulation include *microRNAs* and *short interfering RNAs*. In prokaryotes there are similar-acting RNAs called *small RNAs*. These regulatory RNAs limit protein synthesis by binding to mRNAs using complementary base pairing, thereby decreasing the rate of mRNA translation. When these small RNAs associate with mRNA, the mRNA is usually either tagged for destruction, or struggles to have effective interactions with ribosomes. Either way, this reduces how much protein is made from a particular mRNA. In eukaryotic cells, small noncoding RNAs can also reduce the rate of mRNA processing and nuclear export, ultimately decreasing translation of the affected mRNA. The world of small noncoding RNAs is an active area of research and could give us further insight into certain mechanisms of bacterial and viral pathogenesis, as well as the progression of various cancers and inherited disorders.

Riboswitches

There are over 20 classes of **riboswitches**, but in general they are segments within particular mRNAs that can fold into higher-order structures to increase or decrease gene expression by influencing transcription or translation. Riboswitches contain "sensor" regions that bind to a specified small molecule—often a cellular metabolite like a vitamin derivative. The riboswitch's folded structure changes when the sensor region associates with the target metabolite. These changes in structure serve as on-off signals that determine how the riboswitch will affect transcription or translation.[11] For example, in *E. coli*, riboswitches seem to mainly work by preventing the ribosome from accessing the mRNA start codon.[12] Riboswitches are common gene regulation tools in many bacteria, and similar RNA-based switches have been discovered in certain archaea and eukaryotes, too. Diverse riboswitches detect and respond to a wide variety of signals. As a whole, they act as effective tools for regulating transcription at the pre-transcriptional and/or post-transcriptional level.

Build Your Foundation

26. How are facultative genes different from constitutive genes? (LO 5.21)
27. What is meant by *pre-transcriptional regulation*, and why are transcription factors key to such regulation? (LO 5.22)
28. Compare and contrast inducible and repressible operons. (LO 5.23)
29. What is DNA methylation, and how can it affect protein synthesis? (LO 5.24)
30. How is quorum sensing accomplished, and what effect does it have on bacteria? (LO 5.25)
31. What is post-transcriptional regulation? (LO 5.26)
32. Explain what small noncoding RNAs are and how they affect protein synthesis. (LO 5.26)
33. What are riboswitches, and how can they impact protein synthesis? (LO 5.26)

BYF QUICK QUIZ

Build Your Foundation (BYF) Quick Quiz: Visit the **Mastering Microbiology** Study Area to quiz yourself.

---

[11] Bédard, A. S. V., Hien, E. D., & Lafontaine, D. A. (2020). Riboswitch regulation mechanisms: RNA, metabolites and regulatory proteins. *Biochimica et Biophysica Acta (BBA)-Gene Regulatory Mechanisms, 1863*(3), 194501.

[12] Bastet, L., Chauvier, A., Singh, N., et al. (2017). Translational control and Rho-dependent transcription termination are intimately linked in riboswitch regulation. *Nucleic Acids Research, 45*(12), 7474–7486. https://doi.org/10.1093/nar/gkx434

5.5 MUTATIONS

There are three main categories of mutations: substitutions, insertions, and deletions.

A mutation is a change in the genetic material of a cell or virus. Mutations are essential to evolution and variation within species. In single-celled organisms, a mutation is more likely to be passed to daughter cells, but in a multicellular organism most mutations are not inherited by offspring. This is because in multicellular organisms the mutation would have to be in a germ line cell (a sex cell like egg and sperm) as opposed to a somatic cell (body cell, such as liver or skin). This is just one reason why evolution rates tend to be faster in unicellular organisms than in multicellular organisms. The three main classes of mutations are substitutions, insertions, and deletions.

1. Substitution mutations occur when an incorrect nucleotide is added—for example, adding a nucleotide that contains guanine instead of thymine. Cells may make this sort of error when replicating or repairing DNA. Sometimes these are called *point mutations* because they occur at a specific point, affecting a single nitrogen base at a time. As you'll read later, these mutations can cause the wrong amino acid to be incorporated into a protein or lead to a premature stop in translation.
2. Insertion mutations occur when a cell adds one or more nucleotides to its genome sequence. These mutations tend to lead to shifts in the mRNA reading frame, which we'll review later.
3. Deletion mutations occur when one or more nucleotides are removed from a genome sequence. As with insertion mutations, deletions tend to lead to frameshifts in the mRNA nucleotide sequence that can drastically alter the protein sequence.

Learning Outcomes

After reading this section, you should be able to:

5.27 Describe how mutations contribute to evolution.

5.28 Describe substitution, insertion, and deletion mutations, and discuss how the impact of each can be minimized.

5.29 Compare silent, missense, and nonsense mutation effects.

5.30 Explain what a spontaneous mutation is, and discuss the primary contributors to spontaneous mutations.

5.31 Discuss the various classes of mutagens and how they can induce mutations.

5.32 Summarize what the Ames test is, how it works, and why it is useful.

5.33 Explain what excision repair is and how it protects cells from mutations.

Mutation effects differ.

Just as making changes to a recipe can lead to good, bad, or neutral effects in the resulting food, changes in the genome lead to diverse outcomes. Some mutations are helpful to a cell and increase its likelihood of survival—for instance, the mutation might help the cell better manage a certain environmental stress. Other mutations can be detrimental to a cell and compromise survival. Still others are neutral, having no impact on the cell's phenotype. The effect of a base substitution on a cell's phenotype depends on what substitution is made and where it occurs. The three potential effects of substitution mutations are shown in FIG. 5.18. The impact of insertions and deletions on the cell's phenotype depends on how large the insertion or deletion is and where the change is made.

Silent Mutation (No Effect on the Protein Amino Acid Sequence)

Substitution mutations that do not change the amino acid sequence of a protein are called **silent mutations**. Recall that the genetic code relies on codons to direct the production of a protein during translation. The genetic code is redundant—meaning several different codons encode the same amino acid. Redundant codons tend to differ at the third, or so-called wobble, position of their triplet sequence. Therefore, a nucleotide change at the wobble position tends to be silent because it is unlikely to change which amino acid is added to the protein being made. For example, the amino acid arginine would be added to the growing protein if the ribosome encountered the codons CG<u>U</u>, CG<u>C</u>, CG<u>A</u>, or CG<u>G</u>

FIGURE 5.18 Three potential effects of base substitution mutations A base substitution mutation results in a single nucleotide change in the DNA. The DNA is eventually transcribed into mRNA, and the mRNA is translated into a protein. In a **silent mutation**, the resulting amino acid is not changed. In a **missense mutation** the new codon encodes the wrong amino acid. A **nonsense mutation** occurs if the change leads to a stop signal instead of an amino acid.

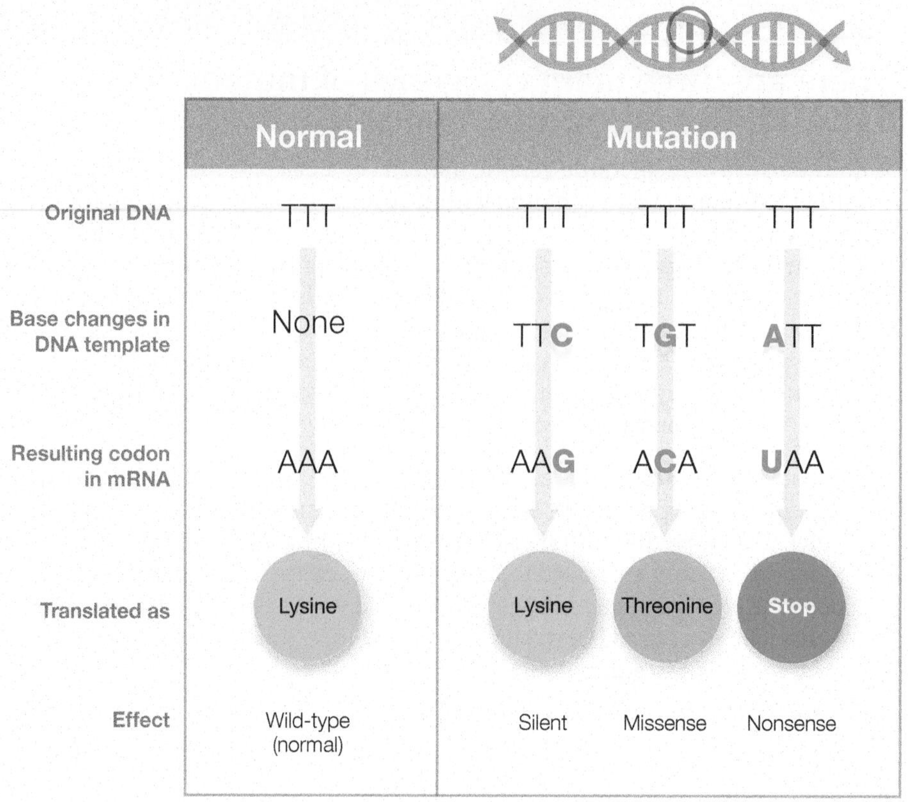

	Normal	Mutation		
Original DNA	TTT	TTT	TTT	TTT
Base changes in DNA template	None	TTC	TGT	ATT
Resulting codon in mRNA	AAA	AAG	ACA	UAA
Translated as	Lysine	Lysine	Threonine	Stop
Effect	Wild-type (normal)	Silent	Missense	Nonsense

(see Table 5.4). In this case, a mutation in the third position of the codon would have no effect on the amino acid encoded and therefore would not alter the protein being assembled. This built-in redundancy limits the impact of base-substitution mutations and is an important protective feature of the genetic code.

Occasionally a base substitution mutation can be corrected by another base substitution; in effect, the error is corrected by another error. This is called a **reversion mutation**, or back mutation, because the second mutation caused the DNA to revert to the original sequence. This may be one of the few situations in which two wrongs really do make a right!

Nonsense and Missense Mutations (Premature Translation Stop or Erroneous Amino Acidic Incorporation)

A mutation that modifies the meaning of a codon can be a dangerous scenario for a cell. For example, if a base substitution mutation changed the codon CGA (which encodes arginine) to UGA (a stop signal), the ribosome would prematurely stop building the protein when it came across the mutated codon. As a result, the cell would end up with an incomplete protein that likely can't perform its job. A mutation that causes a codon to go from encoding an amino acid to encoding a stop signal is called a **nonsense mutation**. In studies using *E. coli*, it was estimated that about 3 percent of all substitution mutations are nonsense mutations.[13] In comparison, a **missense mutation**, in which the meaning of the codon is changed in a way that the wrong amino acid is added to the growing protein, is far more common. For example, if the codon CGU (arginine) was changed to UGU (serine), the protein's amino acid sequence would be altered.

[13] Lee, H., Popodi, E., Tang, H., & Foster, P. (2012). Rate and molecular spectrum of spontaneous mutations in the bacterium *Escherichia coli* as determined by whole-genome sequencing. *PNAS, 109*(41), E2774–E2783.

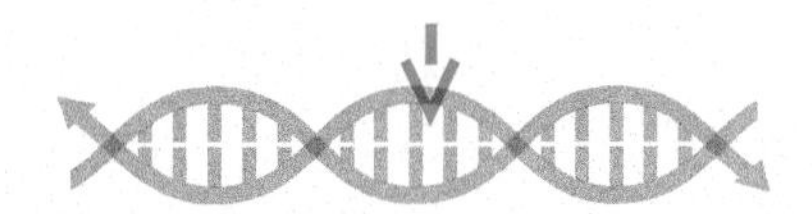

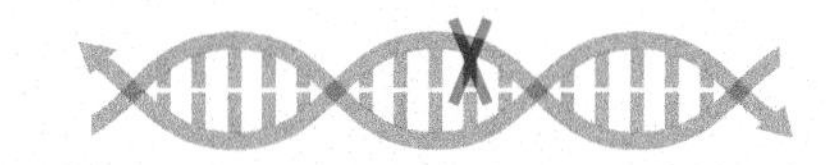

	Normal	Single insertion	Single deletion
Original DNA	TACTTTGAATATACT	TACTTTGAATATACT	TACTTTGAATATACT
Mutation	None	Insert Adenine (A)	Remove Thymine (T)
Resulting DNA	TACTTTGAATATACT	TACATTTGAATATACT	TACTTTGAATATACT
Resulting mRNA	AUG AAA CUU AUA UGA	AUG UAA ACU UAU AUG A	AUG AAC UUA UAU GA
Read as	Methionine Lysine Leucine Isoleucine Stop	Methionine Stop	Methionine Asparagine Leucine Tyrosine
Effect	Wild-type (normal)	Premature stop	Completely altered protein

FIGURE 5.19 Insertion and deletion mutations Insertions and deletions can shift the mRNA reading frame and lead to drastic changes in the resulting protein.

Critical Thinking *Why are base substitution mutations usually less troublesome to a cell than an insertion or deletion?*

Frameshift Mutations (Insertions and Deletions)

FIG. 5.19 shows the potential impact of inserting or deleting bases from the coding region of a genetic sequence. The resulting **frameshift mutation** can be devastating for a cell because it tends to lead to a useless protein.

For example, if you were told to decode the message THEBOYRANFORTHETOY by breaking it into sequential three letter chunks, the message would be decoded as: THE BOY RAN FOR THE TOY.

Now, let's insert one letter into the sequence. Doing this will give us: THEBOYARANFORTHETOY, which would decode as: THE BOY ARA NFO RTH ETO—gibberish.

Similarly, cells follow decoding rules; they start decoding mRNA at a start codon and read it in triplet code to build a protein. If the reading frame is shifted by insertions or deletions, then the mRNA will direct the building of a jumbled protein that is usually useless to the cell. The impact of insertions and deletions is minimized if these mutations occur in multiples of three. For example, if we were to add three letters to the sequence THEBOYARABCNFORTHETOY it would be decoded as: THE BOY ARA BCN FOR THE TOY. While this sentence is not as coherent as the original, it still conveys part of the original meaning. The triple insertion changed some words, but the reading frame itself was not entirely shifted. Likewise, in a gene, the insertion or deletion of bases in multiples of three (i.e., three, six, nine, etc.) can alter the final protein by adding or eliminating amino acids. Again, the consequences of such mutations always depend on the location and extent of the mutation.

Mutations can be spontaneous or induced.

Spontaneous mutations typically result from errors in normal biological processes of DNA replication and/or transcription, whereas induced mutations are caused by environmental factors.

Spontaneous Mutations

Naturally occurring mutations are often referred to as **spontaneous mutations**. Most spontaneous mutations are either neutral or harmful to a cell. In the rare instance that a mutation is beneficial, the *mutant strain* (or the cells carrying the mutation) may have a survival advantage over nonmutated, or *wild-type,* strains in a particular environment. This survival advantage increases the chances for the mutant strain to pass the mutation to other generations. Over time, the mutation may become a standard trait in cell lines that are able to thrive in a particular environment.

Spontaneous mutations are an important mechanism for evolution because they introduce genetic variation even in organisms that replicate asexually, like viruses and bacteria. It is estimated that in bacteria there is a spontaneous mutation in 1 out of every 10 billion base pairs. This sounds like a rare occurrence, but when you consider that under ideal conditions *E. coli* can divide about every 20 to 30 minutes, it means that one cell can generate a population of over 1 million cells in about 10 hours. Because *E. coli* has about 4 million base pairs in its genome, this means that in a population of about a million cells, at least 400 mutations will have occurred. This represents a tremendous opportunity for evolution in just 10 hours. Over time, enough mutations occur that eventually a cell will make an unexpectedly lucky mistake that confers a fabulous advantage.

Induced Mutations

Induced mutations are prompted by factors in the organism's environment. These factors, called **mutagens**, are chemical, physical, or biological agents that increase the rate of mutation. Induced mutations tend to rapidly accumulate, so their effects are usually harmful to a cell. This is because any rare "stroke of luck" by which a mutation may confer an advantage is easily outweighed by the more numerous detrimental errors.[14] *Chemical mutagens* include organic or inorganic agents. These agents cause mutations by a variety of methods, such as inducing breaks in DNA, modifying bases, or promoting frameshift mutations by directly inserting themselves into the DNA. Examples of chemical mutagens include arsenic, asbestos, alcohol, and a variety of compounds in tobacco smoke. *Physical mutagens* induce DNA damage in a similar fashion as chemical mutagens, but they include radiation such as ultraviolet (UV) light, X-rays, and radioactive gamma rays. *Biological mutagens* are agents that can introduce genetic change through **recombination**, which is an exchange of genetic material that leads to new genetic combinations. Transposons and certain viruses can act as biological mutagens (these agents are discussed further in the last section of the chapter). Many mutagens cause a rate of mutation that promotes the development of cancers; such mutagens are called **carcinogens**.

Being able to test if a substance could be mutagenic is increasingly important given the CDC estimates that about 49 percent of Americans take at least one prescription drug in any given month, and there are over 85,000 industrial chemicals currently in use that people may encounter, including flame retardants in upholstery or clothing, plastic drink and food containers, pollution in

[14] A note for nonnative English speakers: A "stroke of luck" is a common phrase that describes an unplanned yet fortunate event or series of events.

the air we breathe, etc. The **Ames test** can quickly identify mutagens that alter DNA by either base substitutions or frameshift mutations. It relies on strains of *Salmonella* bacteria that cannot make the amino acid histidine (called *his*$^-$ strains). This amino acid is essential for growth; cells that cannot make their own histidine must get it from their environment. Like all cells, *his*$^-$ bacteria experience spontaneous mutations, and as a result they may regain their ability to make histidine (revert to *his*$^+$). Normal cells have repair mechanisms (discussed next) that would address these mutations. However, the mutations in the *his*$^-$ Ames strains cannot be repaired because these strains lack DNA repair tools. This deficiency allows for the detection of reversion mutations because only *his*$^+$ reversion mutants will grow on the histidine-free agar used in the test (FIG. 5.20).

If *his*$^-$ strains are grown in the presence of a mutagen, they experience more mutations, and consequently the rate of *his*$^+$ reversion is greater than that produced by spontaneous mutation. Therefore, when *his*$^-$ bacteria are exposed to a chemical and experience a reversion rate above that seen for spontaneous mutation, the tested chemical is regarded as a mutagen.

DNA proofreading and repair mechanisms protect the stability of the genome.

Although DNA polymerases proofread the DNA they build and fix detected errors, they may still make mistakes that don't get corrected. Errors made by DNA polymerases during replication are an important contributor to the total spontaneous mutation rate, which is estimated to be 1 error in every 10 billion base pairs. However, errors are also introduced outside of replication, whenever DNA is damaged by a variety of agents such as mutagens or reactive intermediates from cellular metabolism. If cells did not have the ability to find and repair errors, then the rate of spontaneous mutations could be as much as one thousand times higher.[15]

There are diverse repair mechanisms in cells; one important example is **excision repair**. During excision repair, specialized enzymes clip out damaged or mismatched nucleotides, and then DNA polymerase I lays down new nucleotides to repair the DNA. The enzyme ligase, which was used to glue Okazaki fragments together in DNA replication, also glues the nicks in the DNA sugar–phosphate backbone, so that the repair patch is seamless (FIG. 5.21). Excision repair is especially active in fixing UV radiation damage in DNA. UV light can cause neighboring thymine bases in DNA to cross-link and form a bulge in the DNA. These *thymine dimers* interfere with DNA replication and transcription unless they are removed and repaired. The human genetic disorder *xeroderma pigmentosum* is a testimony to the importance of DNA repair mechanisms. When patients with this disorder are exposed to sunlight, a common and powerful source of UV radiation, their skin cells experience DNA damage that cannot be repaired, ultimately leading to skin cancer and death at an early age.

Whenever the rate of DNA damage outpaces the rate of DNA repair, mutations accumulate in a cell and have detrimental effects. In single-celled organisms, this usually results in death. In animals, accumulated mutations often manifest as cancers. Although DNA repair in higher eukaryotes is more involved than in prokaryotes, these repair mechanisms are closely related in terms of goals and how they are performed across prokaryotes and eukaryotes.

FIGURE 5.20 **Ames test** In the Ames test a sample of *his*$^-$ *Salmonella* is exposed to a test chemical; the control sample is not exposed to the test chemical. The addition of liver enzymes provides a more accurate simulation of how the chemical being tested may "appear" once it is chemically processed by the body. Both samples are plated onto a medium that is lacking the amino acid histidine. If the tested chemical causes the bacteria to mutate at a rate higher than spontaneous mutation, the test plate will have significantly more colonies on it than the control plate. In such a case, the test chemical is a potential mutagen and possibly a carcinogen.

Critical Thinking *Would you expect this experiment to work if the bacteria were grown on media that contained the amino acid histidine? Explain your reasoning.*

[15] Kunkel, T. A., & Erie, D. A. (2005). DNA mismatch repair. *Annual Review of Biochemistry*, *74*, 681–710.

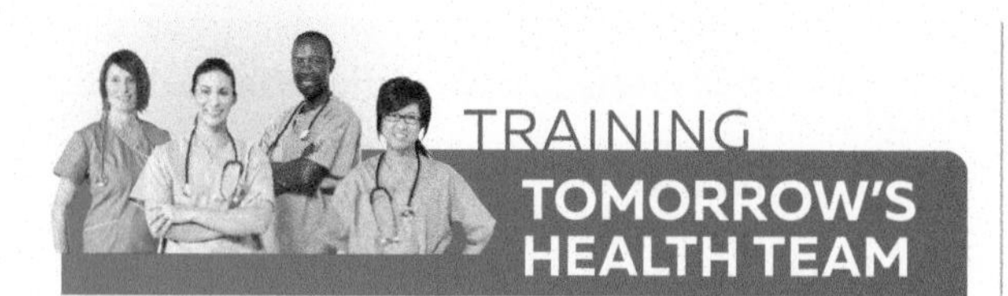

The New Field of Genetics Nursing

Advances in genetics and the beginning of personalized medicine led to the development of a new specialization in health care called genetics nursing. A genetics nurse is a licensed professional with additional training in genetics. These nurses work closely with physicians and genetics counselors in practices that diagnose or manage genetically based disorders. Genetics nurses help assess the contributions of genetic factors to disease, collect and analyze samples for genetic testing, and educate patients and their families about how genetic factors contribute to disease.

Some genetics nurses specialize in prenatal care, assisting with genetic screenings and sharing results of such tests with patients. They also help patients understand the risk of passing a specific genetic disorder to a child and how to care for a child with a genetic disorder.

Most genetics nurses have a bachelor's of nursing degree, and after being licensed to practice nursing, they gain their genetics certification through the Genetic Nursing Credentialing Commission (GNCC).

Build Your Foundation (BYF) Quick Quiz: Visit the **Mastering Microbiology** Study Area to quiz yourself.

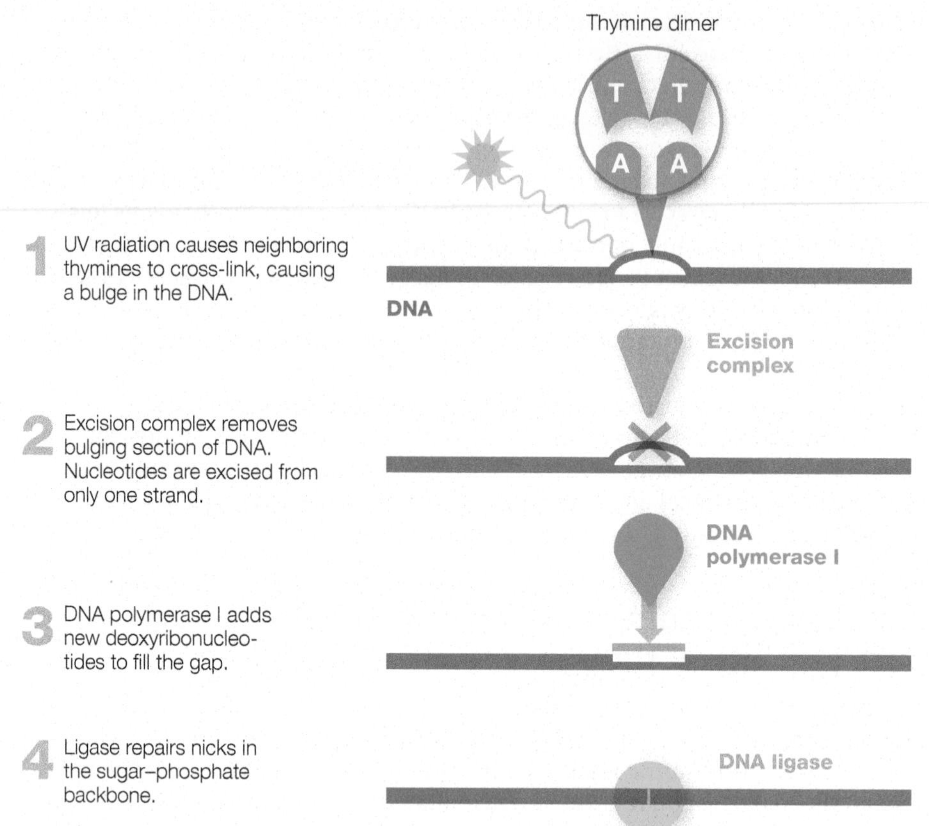

FIGURE 5.21 Excision repair UV radiation can lead to the formation of thymine dimers, which form a bulge in the DNA. Excision repair factors are recruited to the bulge to clip out the damaged section of the DNA strand, leaving the opposite strand untouched. DNA polymerase I and ligase then repair the gap.

Critical Thinking *People with xeroderma pigmentosum have faulty DNA repair mechanisms and can't go in the sun because their skin cells can't repair DNA damage inflicted by UV radiation. However, these patients have a short life span even if they stay out of the sun. How might this be explained?*

Build Your Foundation

34. How do mutations come about, and how do they contribute to evolution? (LO 5.27)

35. Why would a substitution mutation likely have less impact on a cell than a frameshift mutation? (LO 5.28)

36. How do silent, missense, and nonsense mutation effects differ from one another? (LO 5.29)

37. What are spontaneous mutations, and what primarily leads to their occurrence? (LO 5.30)

38. Name three classes of mutagens, and discuss how they induce mutations. (LO 5.31)

39. How is a *his*$^-$ strain of *Salmonella* used in the Ames test to help detect mutagens? (LO 5.32)

40. What are thymine dimers, and how do cells deal with them? What happens if thymine dimers are not removed? (LO 5.33)

5.6 GENETIC VARIATION WITHOUT SEXUAL REPRODUCTION

Horizontal gene transfer allows bacteria to share genes without a cell division event.

Learning Outcomes

After reading this section, you should be able to:

5.34 Compare horizontal and vertical gene transfer.

5.35 Explain what plasmids are, and describe why they are important in nature and in the laboratory.

5.36 Describe conjugation, and explain how it contributes to genetic diversity.

5.37 Discuss the features of cells that can conduct conjugation.

5.38 Define recombination, and explain how this process leads to the production of Hfr strains.

5.39 Describe transformation, and explain how Griffith's experiments demonstrate this process.

5.40 Compare and contrast specialized and generalized transduction, and explain how they lead to genetic variation in bacteria.

5.41 Describe the two main categories of transposons, and state how they can impact a cell's genetic landscape.

Vertical gene transfer occurs when cells pass their genetic information to the next generation (from parent cell to offspring) as a result of sexual or asexual cell division. In contrast, **horizontal gene transfer** passes genetic information between cells by a process independent of cell division, and therefore *separate* from reproduction. It enables cells that are related or not related to share genetic information. Horizontal gene transfer may even occur across different phylogenetic domains. For example, there is evidence that the bacterium *Neisseria gonorrhoeae*, which causes gonorrhea in humans, has acquired some human gene sequences.[16] The integration of bacterial DNA into human cells has also been documented and may promote the development of certain cancers.[17] This section focuses on horizontal gene transfer mechanisms that plasmids, bacteriophages, and transposons facilitate.

Plasmids and Horizontal Gene Transfer

Cells commonly use plasmids to share genetic information by horizontal gene transfer. Recall from earlier in this chapter that plasmids are nonchromosomal DNA segments found in bacteria and a number of eukaryotic cells. These small, usually circular segments of double-stranded DNA tend to carry only a few genes and are rarely essential to the cell; instead, they typically confer a survival advantage such as the ability to make toxins or defend against antibiotics. Plasmids that confer antibiotic resistance are often called **R plasmids**, which stands for *resistance plasmids*. Molecular biologists can use plasmids to generate cell lines that make useful products for medical applications. (See Chapter 14 and the Bench to Bedside in this chapter for more on plasmids in biomedical applications.)

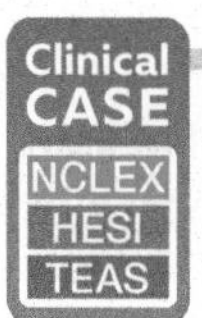

The Case of the Spreading Superbug

Practice applying what you know clinically: visit the **Mastering Microbiology** Study Area to watch Part 3 and practice for future exams.

Conjugation

Some plasmids allow bacteria to conduct a process called **conjugation**, a specialized form of horizontal gene transfer. Conjugation is the closest thing to sex that bacteria have, but it is *not* sexual reproduction. In conjugation, a bacterium that carries a **fertility plasmid** (F plasmid or F factor) forms a small hollow tube called a **pilus**, which attaches to a neighboring bacterial cell that lacks a fertility plasmid. The pilus serves as a bridge for transferring a copy of the fertility plasmid to the cell lacking it (FIG. 5.22, *left*). Then the pilus is dismantled, and the cells separate. Assuming a complete copy of the fertility plasmid transfers, the recipient cell can then initiate conjugation with cells that lack fertility plasmids.

From a medical standpoint, the troubling part of conjugation is that fertility plasmids often carry genes that endow cells with drug resistance or the ability to produce toxins. Of further concern is that the partners in conjugation do not have to be the same bacterial species. This characteristic enables the widespread sharing of antibiotic-resistance genes and other potentially dangerous genetic tools across bacterial species.

Occasionally the fertility plasmid will merge with the cell's chromosome and cease to be an independent plasmid in the cytoplasm. When this occurs, the resulting bacterial cells are called **high-frequency recombination (Hfr)**

---

[16] Anderson, M., & Steven, H. S. (2011). Opportunity and means: Horizontal gene transfer from the human host to a bacterial pathogen. *mBio*, *2*(1), e00005–e00011.

[17] Emamalipour, M., Seidi, K., Vahed, S. Z., et al. (2020). Horizontal gene transfer: From evolutionary flexibility to disease progression. *Frontiers in Cell and Developmental Biology*, *8*.

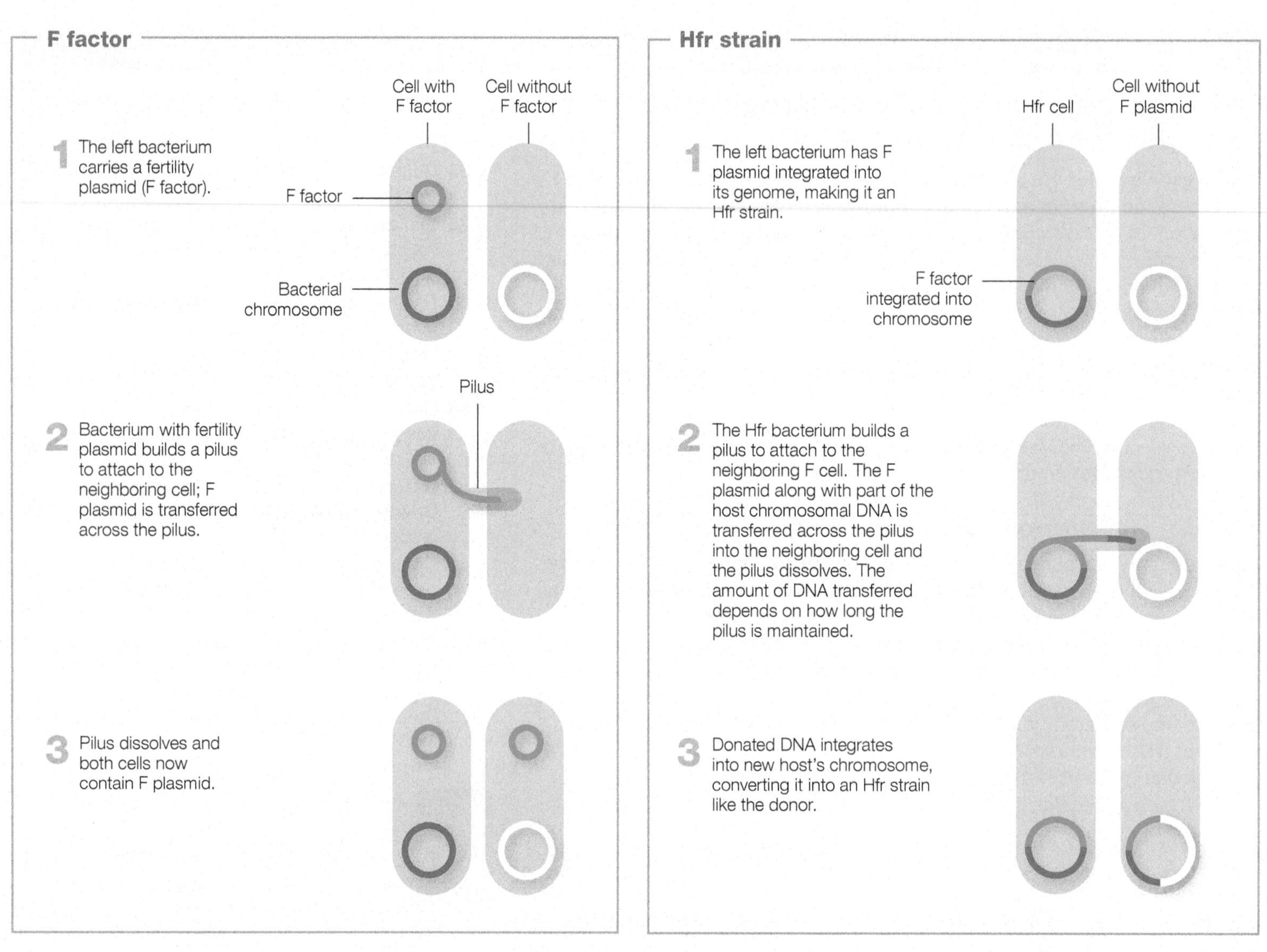

FIGURE 5.22 Conjugation Regarding Hfr strain conjugation (right panel of figure), a new Hfr strain only results if the entire F plasmid is transferred into the new host cell.

strains (FIG. 5.22, *right*). This process, called **recombination**, is important in nature and is also used as a molecular biology tool that allows researchers to alter the genetic landscape of a cell.

Hfr strains share genetic material with cells that lack a fertility plasmid. The amazing thing about Hfr strains is that they can pass on copied segments of DNA beyond the usual set of genes found on the fertility plasmid. In fact, if the pilus is maintained for enough time (about 100 minutes in *E. coli*), the Hfr strain can pass a complete copy of its genome to the recipient cell. This has tremendous implications for how bacteria evolve.

Conjugating bacteria

Transformation

Another horizontal gene transfer process is called **transformation**. Here bacteria are genetically altered when they take up DNA from their environment. Unlike conjugation, a pilus is not involved; instead, the DNA crosses the cell wall through an absorption process. This so-called exogenous (or non-self) DNA can be from practically any source and ranges from laboratory purified plasmids to fragments of DNA from a dead bacterial cell. Some researchers have even demonstrated that ancient DNA, such as 43,000-year-old DNA from a woolly mammoth bone, could be used to transform bacteria. This suggests that fresh DNA is not the only potential influence on bacterial evolution because ancient DNA can persist in certain environments for many thousands of years, and even badly fragmented ancient DNA can be integrated into modern-day bacterial cells.[18]

The process of transformation is natural to over 80 species of bacteria, but also can be induced by certain treatments in other species. The transforming DNA that is taken into the cell can be maintained in the cytoplasm as a plasmid, or it can integrate itself into the cell's chromosomal DNA to generate a stably transformed cell. Cells that can be transformed are called *competent cells*. Competent cells are characterized by a specialized physiological state in which they express a variety of genes that make DNA uptake and processing possible. Under certain circumstances eukaryotic cells can also take up DNA from their environment, although the process is referred to as *transfection* instead of transformation. (Note that when animal cells are called "transformed," it means the cell is abnormal, with precancerous or cancerous features.) Both transformation and transfection are important tools to molecular biology and are routinely used in a variety of biomedical applications. (See Chapter 14 for more.)

Griffith's Experiments Classic experiments performed in the 1920s by Fredrick Griffith showed that bacteria could be transformed to gain certain pathological features. Griffith exposed mice to four different combinations of *Streptococcus pneumoniae,* which causes bacterial pneumonia in humans (FIG. 5.23). The experiment involved four scenarios:

1. First, Griffith infected mice with a living strain of *S. pneumoniae* that tends to be dangerously virulent because of its capsule. (See Chapter 3 for more on capsules.) As expected, the mice infected with the live encapsulated *S. pneumoniae* died.

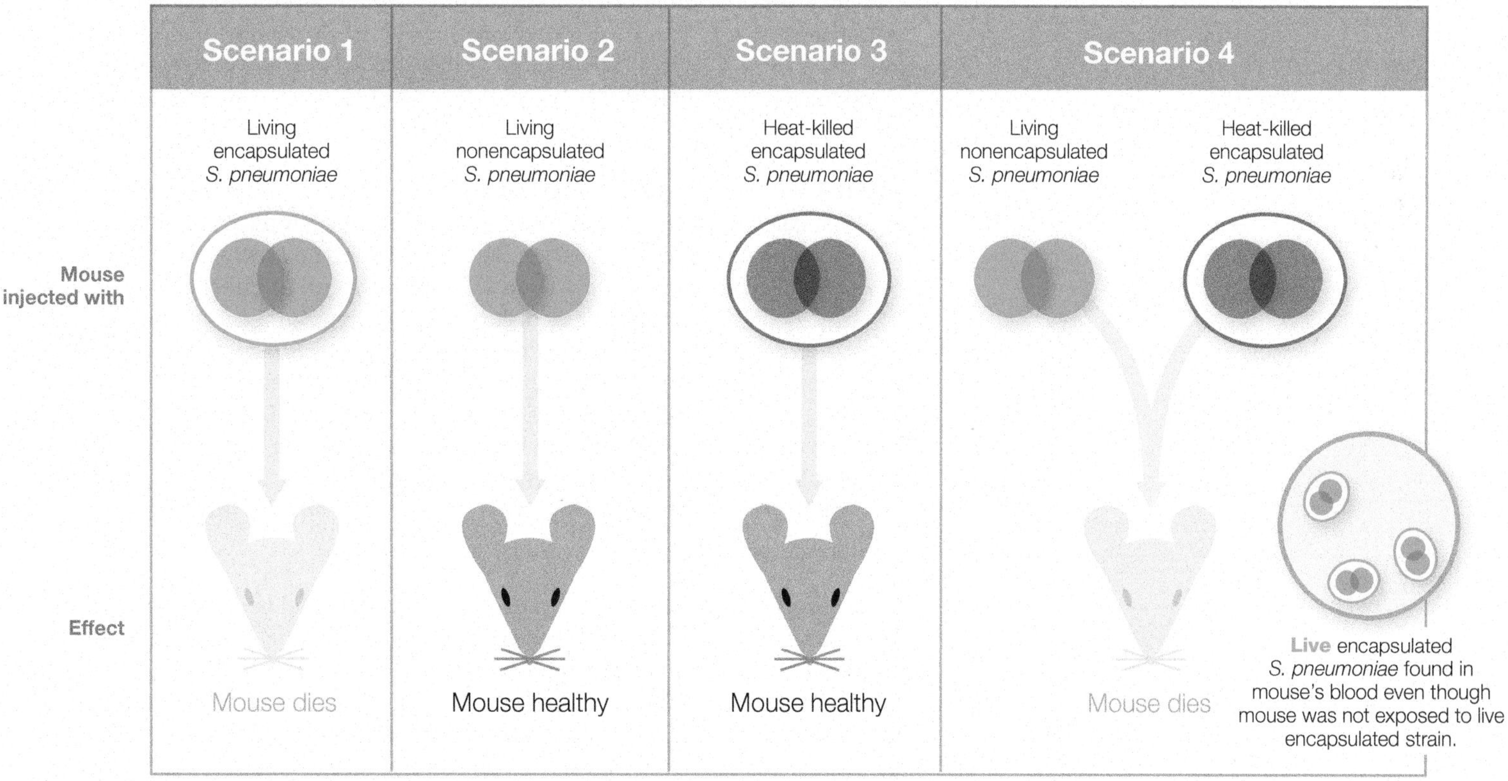

FIGURE 5.23 Griffith's experiments and transformation

Critical Thinking *What outcome would you expect if a fifth experiment that combined scenarios 3 and 4 was performed? Explain your reasoning.*

[18] Overballe-Petersen, S., Harms, K., Orlando, L., et al. (2013). Bacterial natural transformation by highly fragmented and damaged DNA. PNAS, 110(49), 19860–19865.

2. Next, Griffith infected mice with a nonvirulent, living strain of S. *pneumoniae* that does not make a capsule. Again, he got expected results: The mice remained healthy.
3. In the third scenario, he infected mice with heat-killed, encapsulated S. *pneumoniae*. Because the pathogenic bacteria were dead, Griffith did not expect that the mice would be killed by the treatment. And as expected, the mice remained healthy.
4. Finally, Griffith gave mice the combined treatments from scenarios two and three. Both of these scenarios previously resulted in healthy mice and presumably, when combined, would also lead to healthy mice. However, when Griffith infected the mice with heat-killed encapsulated S. *pneumoniae* along with the live nonencapsulated S. *pneumoniae,* the mice died. Griffith was then able to isolate living encapsulated S. *pneumoniae* from the dead mice.

We now know that the nonencapsulated S. *pneumoniae* were able to take up DNA from the killed encapsulated S. *pneumoniae* via transformation. This turned the formerly nonpathogenic bacteria into the virulent variety. To this day, Griffith's experiments are highly regarded because they opened the door to our understanding that environmental DNA could transform bacteria and help pathogens evolve.

Transduction

The introduction of new genetic material into a bacterial cell by a virus is called **transduction**. The viruses that infect bacteria are called **bacteriophages**, or simply phages. (These viruses are discussed in detail in Chapter 6.) Bacteriophages must infect bacterial cells to multiply. When a bacteriophage injects its DNA into a host bacterial cell, there are two possible outcomes: either the phage DNA immediately directs production of new phage particles, or the phage DNA integrates (combines itself) into the host cell's genome and postpones making new phages. The first outcome sets the stage for generalized transduction, while the second scenario can lead to specialized transduction.

Generalized Transduction As new phage particles are built in the host, a part of the host cell's fragmenting genome can accidentally get packaged into a phage particle in place of the necessary phage DNA. Once the newly made bacteriophages burst out of the infected bacterium, they go on to infect other bacteria. When they do so, it occasionally occurs that the next targeted bacterium gets lucky, and rather than being infected with a standard phage that will kill it, the bacterium is instead infected with a phage carrying genes from the prior host (FIG. 5.24, *top*). Once this bacteriophage injects the donor DNA into the new host, the DNA can incorporate into the host cell's genome and confer new properties to the cell. This process, called **generalized transduction**, is random and does not specify any particular gene being carried or transduced from the prior host to the new host.

Specialized Transduction Certain bacteriophages, called **temperate phages,** do not necessarily direct the production of new phage particles immediately upon infecting the host bacterium. Instead, they may integrate their DNA into the host cell's genome, often in a specified region, and bide their time before they make new phages. At some point the phage can become reactivated, excise itself from the host genome, and begin to build new phage particles. When it does so, it may take surrounding bacterial genes with it. Because the bacteriophage takes very specific genes along with it into the new phage particles, the process is called **specialized transduction** (FIG. 5.24, *bottom*).

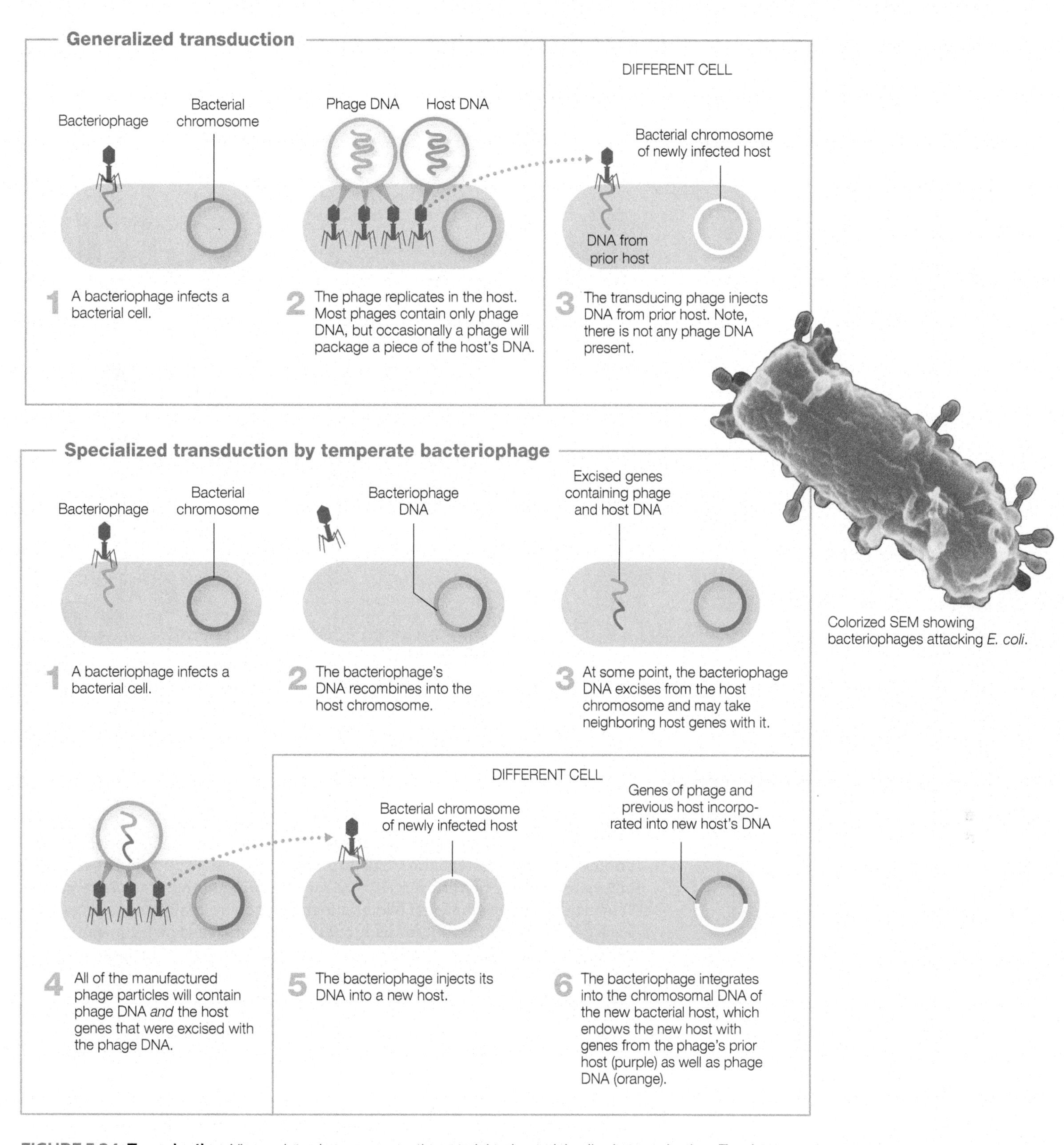

Colorized SEM showing bacteriophages attacking *E. coli*.

FIGURE 5.24 Transduction Viruses introduce new genetic material to bacterial cells via transduction. *Top*: In generalized transduction a bacteriophage accidentally packages DNA from its prior host into a new phage particle and delivers the randomly packaged DNA to the next host. *Bottom*: In specialized transduction a bacteriophage takes very specific genes along with it to the next host. This is possible because temperate phages integrate their DNA into specific regions of the host cell's genome, and when they excise from the host genome to build new phage particles, they often take adjacent bacterial genes with them.

"It's trans-something... if only I could remember which one!"

Silly but effective memory devices for keeping similar genetics terms straight

Protein Synthesis Vocabulary

Transcription Trans = change; script = a written document. A medical transcriptionist takes information dictated by the doctor and puts it in written form without changing the actual information. Similarly, cells use the information provided by DNA to build RNA without altering the DNA.

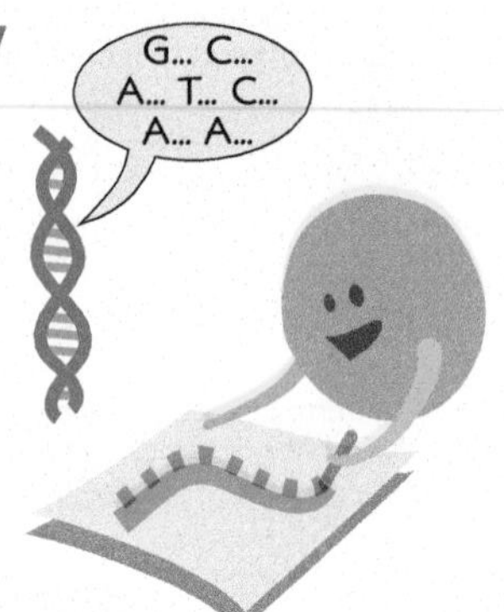

Cell transcribing DNA to RNA

Translation Ribosomes translate mRNA to build protein. Silly tip: Translation rhymes with nation and people in one nation may need a translator to help them understand the other's language. Ribosomes translate the message in mRNA to a new language, the language of proteins.

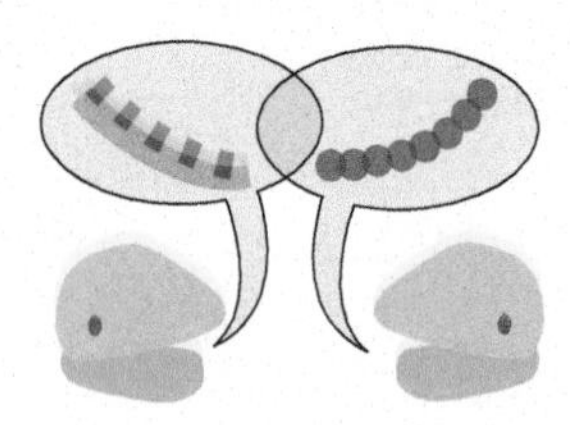

Ribosomes translating

Horizontal Gene Transfer Vocabulary

Transformation Bacterial cells can be transformed when they pick up DNA from their environment. Consider this silly story: Eric Coli was a shy guy, but when he moved to the city he was transformed by his environment and even found a great new pair of "genes" that made him look brand-new.

"Genes" transforming Eric

Transduction Bacteriophages can steal pieces of bacterial cell DNA in the cells they invade and then they can shuttle this DNA to another bacterial cell. You can remember this by thinking of the word abduction, which is the kidnapping of a person. In a way, the bacteriophage is a gene kidnapper. It takes genes from one bacterial cell and carries them away to another bacterial cell.

Genes getting kidnapped

Transposon *Trans* = change, *pos* = position. Transposons are simply genes that can change their position in the genome.

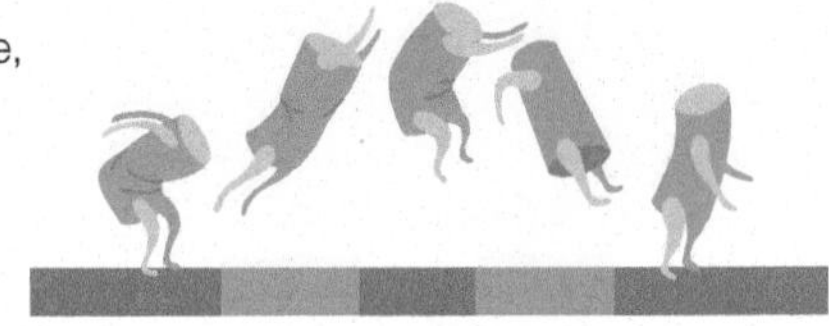

Jumping gene changing position

FIGURE 5.25 Keeping genetics terminology straight

The ability of bacteriophages to shuttle genes from one bacterial cell to another through transduction has important medical implications because it allows related species of bacteria that may be living far apart to share genes. There are many examples of bacteria learning how to make toxins after they have gained genes through transduction. Eukaryotic cells also can gain new genetic material through transduction by a bacteriophage. This point is currently being explored as a potential means of gene therapy, in which a normal copy of a gene is delivered to mutated cells in an effort to restore a normal phenotype. See FIG. 5.25 for helpful tips on keeping certain genetics terms straight.

Transposons

In the 1940s the geneticist Barbara McClintock described "jumping genes," or transposable elements, that led to the varied colors seen in the kernels of maize (FIG. 5.26). At first, other scientists were skeptical about her findings—her data suggested that genomes could be naturally remodeled at any time and change a cell's phenotype. Now it is well accepted that transposable elements, or **transposons**, exist not only in plants, where they were first discovered, but that they also make up a significant portion of eukaryotic, prokaryotic, and viral genomes. Eventually, Dr. McClintock was awarded a Nobel Prize for her

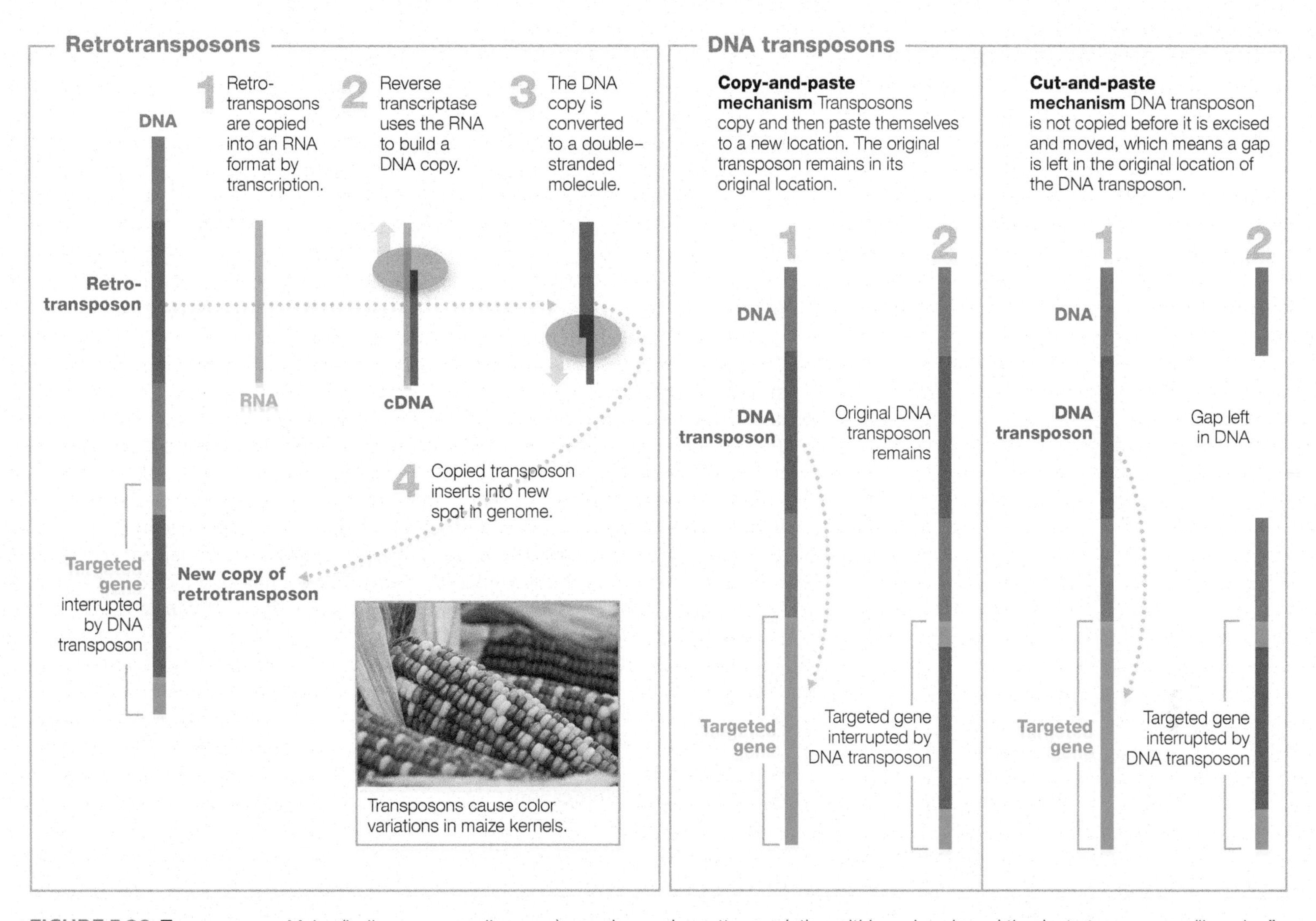

FIGURE 5.26 Transposons Maize (Indian corn or calico corn) can show color pattern variation within a given kernel thanks to transposons "jumping" around the genome to activate and inactivate certain pigment genes. *Left*: Retrotransposons are copied into an RNA format by transcription. They are then converted into cDNA by reverse transcriptase. The new double-stranded DNA transposon can then insert itself into a new spot in the genome. *Right*: DNA transposons can "jump" around in the genome using either a cut-and-paste or copy-and-paste strategy.

discovery. It is estimated that about 45 percent of the human genome is composed of transposable elements; however, fewer than 0.05 percent are actively mobile in the human genome.[19] As these mobile gene elements jump from one spot in a genome to another, they can alter protein synthesis patterns, change genetic sequences, introduce new genes, and even cause mutations such as deletions and insertions that cause disease. The fact that our genome is capable of widespread rearrangement is both fascinating and frightening. Dozens of human diseases have been directly linked to transposon insertions, and it is likely that many more will eventually be linked to these genomic elements.

Many biologists argue that transposons must confer at least modest benefits, or they would not have been retained so vigorously across lifeforms. If they do more harm than good, then natural pressures would have weeded them out. Alternatively, maybe the cells that don't fall prey to these restless genetic

[19] Mills, R. E., Bennett, E. A., Iskow, R. C., & Devine, S. E. (2007). Which transposable elements are active in the human genome? *Trends in Genetics*, *23*(4), 183–191. https://doi.org/10.1016/j.tig.2007.02.006

parasites have just figured out how to keep transposon jumping in check, so that their effects are limited. Now we have evidence suggesting that transposons are not actively jumping around in most normal cells unless some sort of stress or spontaneous mutation occurs. In contrast, cancer cells often have highly active transposons, which likely contribute to their abnormal and diverse phenotypes. Whatever the cause, transposable elements remain abundant in all genomes studied to date—and we still have a lot to learn about them.

There are two main classes of transposons (FIG. 5.26, *left*). The first are *retrotransposons*, segments of DNA that get transcribed into RNA; their RNA is then *reverse transcribed* into cDNA (copy DNA or complementary DNA) by reverse transcriptase. The resulting cDNA then inserts into a new place in the genome, while the original element remains in its home location.

The second class of transposons, *DNA transposons*, are mobilized as DNA and do not require an RNA intermediate. They can use a "copy-and-paste" or a "cut-and-paste" mechanism to move in the genome (FIG. 5.26, *right*). Transposons that operate using a "copy-and-paste" mechanism (replicative transposition approach) copy themselves and then paste the copy into a new location. In such a scenario the original transposon remains in its original location, but it gave rise to a new transposon that can insert elsewhere in the genome. In a "cut-and-paste" mechanism (or nonreplicative transposition), the DNA transposon is *not* copied before it is excised and moved, which means a gap is left in the original location of the DNA transposon. This gap can usually be repaired by DNA repair mechanisms.

Both classes of transposons are found in prokaryotes and in eukaryotes. However, in eukaryotes retrotransposons are more common, while DNA transposons are more common in prokaryotes. Some of the best-known transposons in bacteria encode antibiotic-resistance genes. Because these genes are mobile, they can insert themselves into plasmids, which in turn can be passed to other bacteria through conjugation.

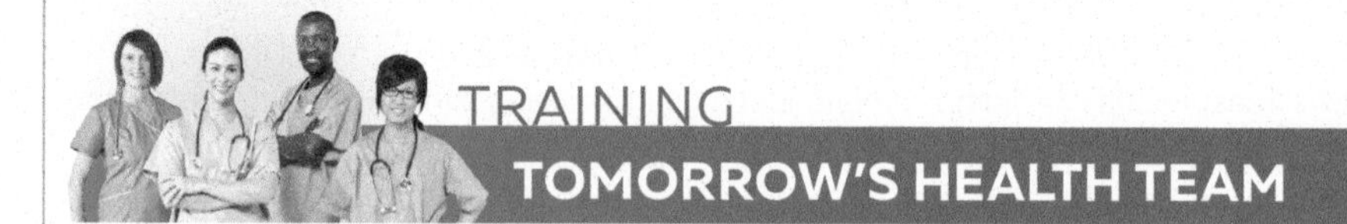

The Power of Sharing Genes: From Antibiotic Resistance to Bacterial Toxins

Conjugation, transformation, and transduction represent powerful evolutionary tools that create pathogens with increased virulence and antibiotic resistance. Unfortunately, it is fairly standard for antibiotic resistance to emerge in bacteria within 3 years of an antibiotic being released to market. Rapid and widespread resistance is largely due to conjugation.

Most bacterial toxins are also encoded by mobile gene elements that are shared by horizontal gene transfer, especially conjugation and transduction. Similarly, toxins such as Shiga toxin, diphtheria toxin, and cholera toxin are housed in genes encoded by temperate bacteriophages.

QUESTION 5.3

Compounds that can inhibit conjugation may represent a way to inhibit or at least slow the progression of antibiotic resistance and the emergence of new pathogens. What do you think may be a good target to prevent conjugation? Explain your decision. (Hint: Think about what bacteria must do to successfully carry out conjugation.)

Build Your Foundation

41. Define horizontal gene transfer, and explain how it is different from vertical gene transfer. (LO 5.34)
42. What are plasmids, and why are they important in the laboratory and in nature? (LO 5.35)
43. What is conjugation, and how does it contribute to genetic diversity? (LO 5.36)
44. How does a fertility plasmid facilitate conjugation? (LO 5.37)
45. How does a bacterium with a fertility factor differ from an Hfr strain? (LO 5.38)
46. What are competent cells, and how can they alter their genetic makeup? (LO 5.39)
47. Why was the result of Griffith's fourth experimental scenario surprising, and what did the scenario demonstrate? (LO 5.39)
48. How are specialized and generalized transduction different? (LO 5.40)
49. Why were transposons such a revolutionary discovery in genetics? (LO 5.41)
50. How do retrotransposons and DNA transposons differ? (LO 5.41)

Build Your Foundation (BYF) Quick Quiz: Visit the **Mastering Microbiology** Study Area to quiz yourself.

VISUAL SUMMARY | *Genetics*

Interactive CONTENT REVIEWS Interactive Content Reviews: Visit the **Mastering Microbiology** Study Area to quiz yourself.

Heredity Basics

The genome is all the genetic material in a cell or virus. Genotype dictates phenotype. DNA makes up the genome of the cells. DNA is a complementary double helix with an antiparallel arrangement made of nucleotides.

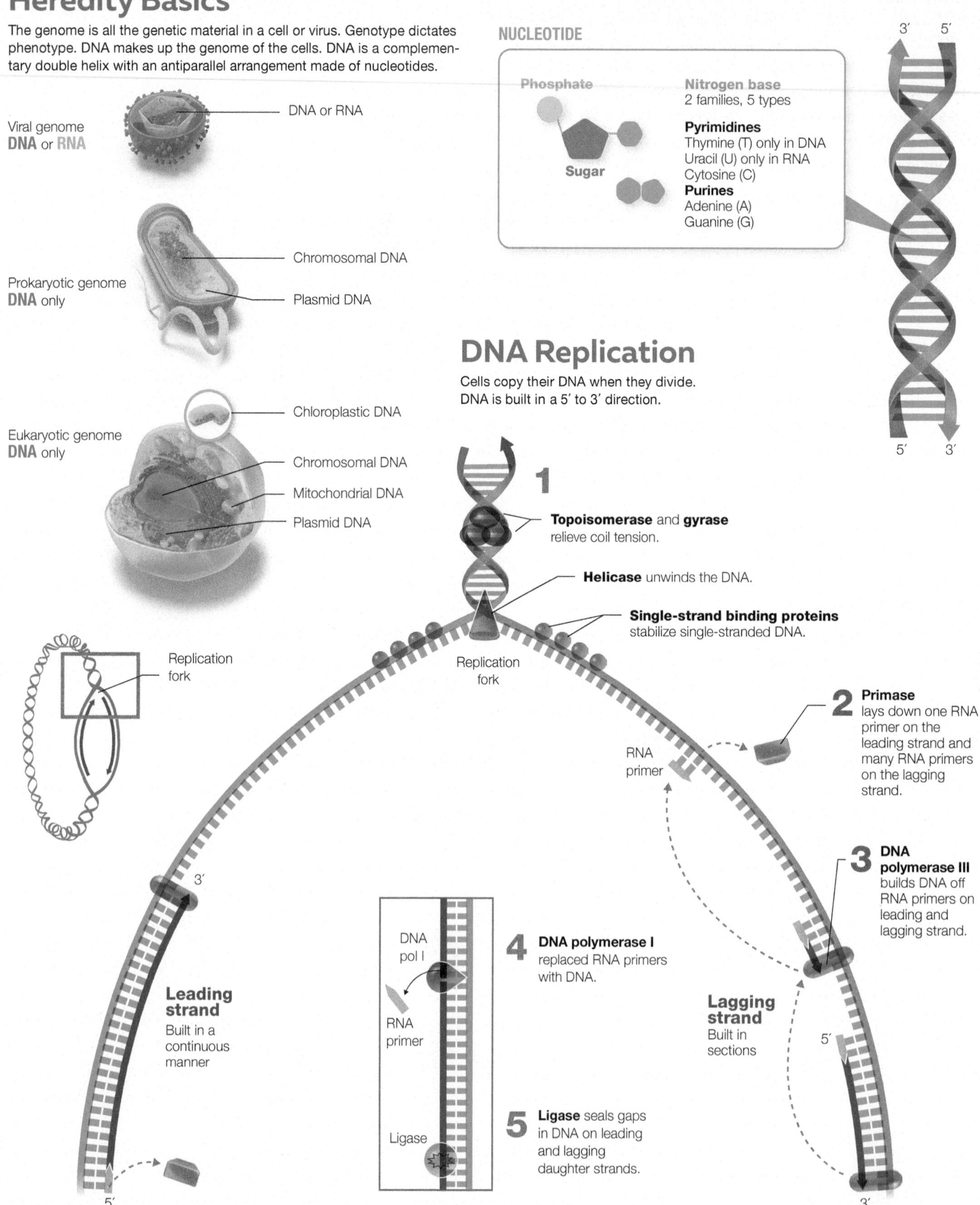

Protein Synthesis and Regulation

Genes in DNA direct the production of proteins, which leads to manifestation of a trait (phenotype).

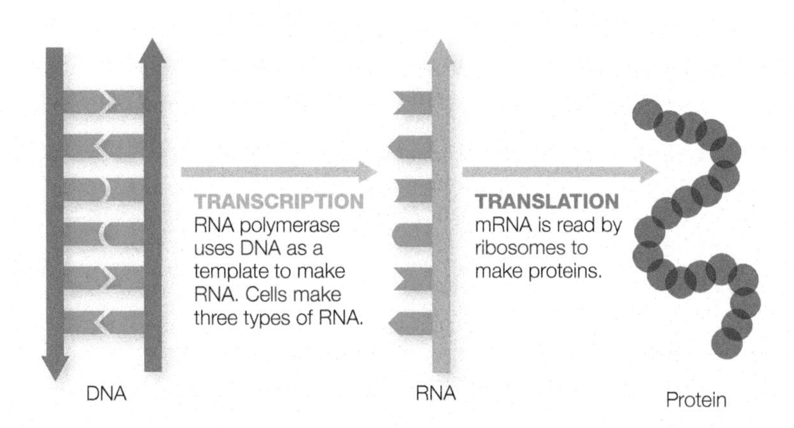

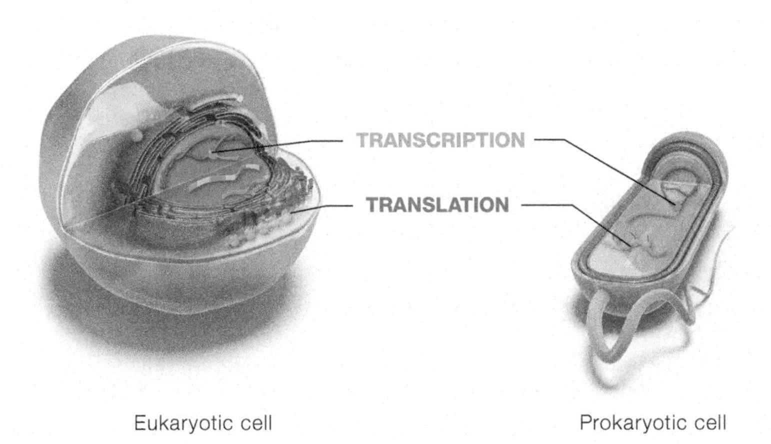

IN BOTH EUKARYOTES AND PROKARYOTES

Pre-transcriptional regulation allows cell to control gene expression by regulating how much mRNA is made from a given gene.

Examples:

- Operons
- Quorum sensing
- Epigenetic control (such as DNA methylation)
- Recruitment of transcription factors

Post-transcriptional regulation allows cell to control gene expression by regulating how often mRNA is translated into protein.

Examples:

- Recruiting ribosomes to mRNA
- Controlling mRNA stability
- Small noncoding RNAs
- Riboswitches
- Eukaryotes also can control RNA processing and nuclear export. (This is not used by prokaryotes.)

Genetic Change: Mutation and Horizontal Gene Transfer

A mutation is a change in the genetic material of a cell or virus. Horizontal gene transfer allows bacteria within the same generation to share genes without a cell division event.

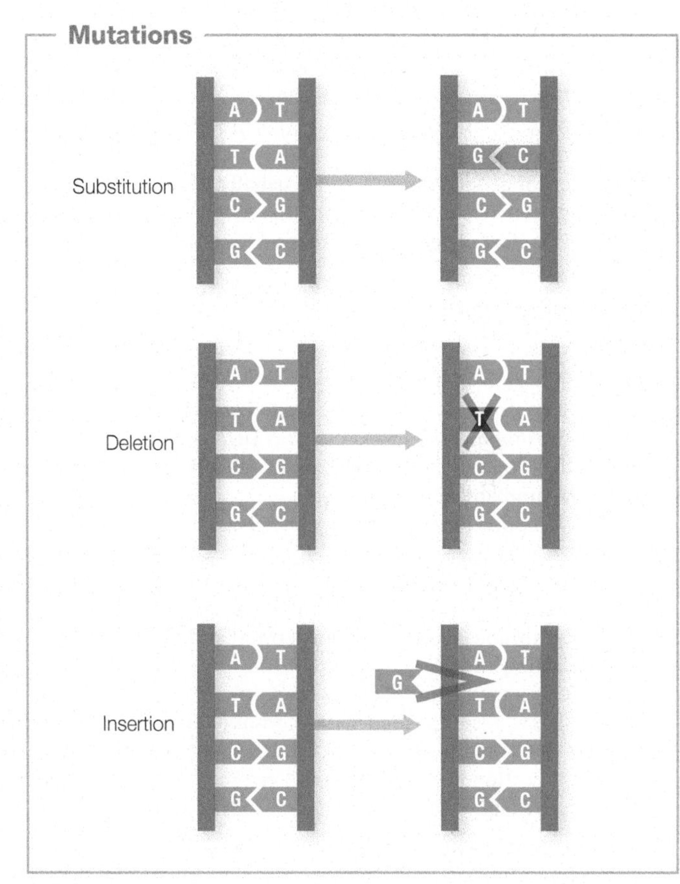

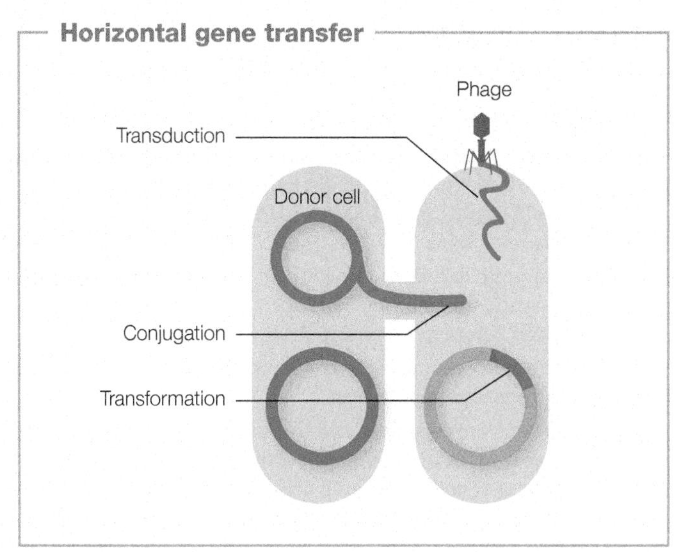

5 CHAPTER 5 OVERVIEW

5.1 Heredity Basics

- A genome is all the genetic material in a cell or virus; cells have DNA genomes, while viruses can have DNA or RNA genomes.
- The genotype (inherited set of genetic instructions) of an organism influences its phenotype, or collection of traits.
- Eukaryotes have many linear chromosomes housed in a nucleus; prokaryotes usually have a single circular chromosome in a nucleoid region.
- DNA exists as a complementary double-stranded helix with antiparallel arrangement and 5′ to 3′ directionality.
  - DNA and RNA are made of nucleotides. Nucleotides contain phosphate, a sugar (deoxyribose in DNA or ribose in RNA), and a purine or pyrimidine nitrogen base.
  - In DNA, AT and GC pairs are made. In RNA, AU and GC pairs are formed.
- The flow of genetic information is usually from DNA to RNA to protein. In reverse transcription, RNA is used as a template to make DNA. This flow of genetic information is called the "central dogma" of molecular biology.

5.2 DNA Replication

- DNA replication copies DNA before cells divide. In bacteria, DNA replication begins at the origin of replication.
  - The primosome, which includes helicase and primase, is recruited to the origin of replication to start DNA replication. The replication fork is the immediate area where the DNA is being unwound and actively copied.
  - DNA replication is primarily carried out by DNA polymerase III (DNA pol III).
    - DNA pol III requires an RNA primer to start DNA replication and it can only build DNA in a 5′ to 3′ direction. DNA is made continuously on the leading strand and discontinuously on the lagging strand.
    - The lagging strand is built in chunks called Okazaki fragments. This is accomplished by the coordinated actions of primase, ligase, DNA polymerase I, and DNA polymerase III.
- DNA replication is a semiconservative process because it produces a molecule that is half new and half parent DNA.

5.3 Protein Synthesis (Gene Expression)

- Protein synthesis is responsible for all the physical and functional features of a cell. It entails using information in genes to direct the formation of proteins. Protein synthesis occurs in two main stages: transcription and translation.
- In transcription, DNA is used as a template to make RNA.
  - RNA polymerase builds new RNA one ribonucleotide at a time in a 5′ to 3′ direction. The three types of RNA used in protein synthesis are mRNA (messenger RNA), tRNA (transfer RNA), and rRNA (ribosomal RNA).
  - Prokaryotic mRNA is immediately ready to be translated, but eukaryotic mRNA must be spliced and then transported out of the nucleus before it is translated.
    - The spliceosome is a complex in the nucleus that removes introns and joins exons.
    - Introns are segments in a transcribed gene that do not encode amino acids in the protein; exons are stretches of a transcribed gene that code for amino acids in the protein.
- In translation, tRNAs and ribosomes work together to decode mRNA and build a protein.
  - Each amino acid in a protein is encoded by a triplet code—a collection of codons that are three nucleotides long in mRNA.
  - The amino acid specified by a codon on mRNA is shuttled by an adaptor-like molecule, called tRNA, to the ribosome.
  - The anticodons of tRNA molecules interact with codons on mRNA to ensure that the genetic message in mRNA directly dictates the sequence of amino acids in the protein being made.
  - An mRNA encodes start signals that are required to initiate translation as well as stop signals to end translation.
- The genetic code is the decoding platform for translating mRNA into a protein. It consists of 64 different codons that encode 20 different standard amino acids, two stop signals, and one start signal. The code is described as redundant because a single amino acid can be represented by many different codons.
- Most proteins will undergo post-translation modifications before they are fully functional.

5.4 Regulation of Protein Synthesis

- Protein synthesis is an energy-intensive process that is tightly regulated.
- Pre-transcriptional regulation impacts when and how often transcription occurs; transcription factors are central regulators of pre-transcriptional regulation.
  - Operons are a collection of genes that are controlled by a shared regulatory element; they regulate gene expression at a pre-transcriptional level. Operons have four key parts: a promoter, a repressor, an operator, and two or more structural genes.
    - Inducible operons are by default off, unless certain conditions arise in which they are activated or induced to allow transcription. The *lac* operon is induced, or actively transcribed, only when lactose is present *and* the cell's preferred food, glucose, is absent.
    - Repressible operons such as the *arg* operon are by default on or actively transcribed until they are switched off, or repressed.
  - Epigenetic regulation, such as methylation, is a form of pre-transcriptional regulation that alters the appearance of DNA without changing its sequence.
  - Bacteria can regulate protein synthesis by quorum sensing, which lays the foundation for forming biofilms.
- Post-transcriptional regulation such as riboswitches and small noncoding RNAs impact the degree of mRNA translation.

5.5 Mutations

- A mutation is a change in the genetic material of a cell or virus.
- The three main classes of mutations are substitutions (also called point mutations), insertions, and deletions.
- Naturally occurring mutations are called spontaneous mutations.
- Mutagens are chemical, physical, or biological agents that increase the rate of mutation.
- The Ames test uses *his*$^-$ strains of *Salmonella* as a quick and effective way to identify mutagens.
- Cells rely on DNA proofreading and repair mechanisms to protect the stability of their genome.
- UV radiation can cause thymine dimers to form in DNA. In excision repair, specialized enzymes clip out damaged or mismatched nucleotides in damaged DNA.

5.6 Genetic Variation Without Sexual Reproduction

- Horizontal gene transfer enables cells that are not related, or even in the same phylogenetic domain, to share genetic information.
- The primary forms of horizontal gene transfer are conjugation, transformation, transduction, and transposons.

COMPREHENSIVE CASE

The following case integrates basic principles from the chapter. Try to answer the case questions on your own. Don't forget to be S.M.A.R.T.* about your case study to help you interlink the scientific concepts you have just learned and apply your new understanding of genetics principles to a case study.

***The five-step method is shown in detail in the Chapter 1 Comprehensive Case on cholera. See pages 31–33. Refer back to this example to help you apply a SMART approach to other critical thinking questions and cases throughout the book.**

• • •

Returning from the hospital cafeteria with coffee in hand, David was anxious to get back to the maternity unit to visit with his wife, Patricia, and their newborn daughter. The spring in his step evaporated when he noticed the concerned expression on Patricia's face and heard the sympathetic tone of the neonatologist who was speaking to her. The neonatologist was explaining that, in a routine blood screening performed on all newborns in the United States, their daughter had tested positive for the genetic disease cystic fibrosis (CF). The doctor explained that more than 10 million Americans (about 1 in every 31 people) are symptomless carriers of the mutation that causes the life-threatening disease.

Cystic fibrosis only develops when a person inherits two mutated copies (one from each parent) of the cystic fibrosis transmembrane regulator (CFTR) gene. The family was referred to a local cystic fibrosis care center to follow up. Unfortunately, the follow-up tests, which consisted of a sweat test and DNA screening, were positive for CF.

In 1989 it was discovered that cystic fibrosis resulted from mutant forms of the CFTR gene. To date about 2,000 different mutations have been found in the gene; most of the mutations do not lead to cystic fibrosis. The CFTR gene has 27 exons and 26 introns and encodes a protein channel that spans the cell membrane and allows chloride ions (Cl^-) to exit cells. In CF patients, this channel is defective, so chloride ions accumulate inside cells. As a result, people with CF have high levels of salt (sodium chloride) in their sweat, and they make abnormally thick and sticky mucus that accumulates in mucous membranes throughout the body, especially in the lungs and GI tract. This thick mucus is hard to cough up. It consequently settles in the lungs and serves as a breeding ground for bacteria. CF patients often battle hard-to-treat lung infections because the bacteria in their lungs establish a biofilm. *Pseudomonas aeruginosa*, a relatively rare infection in otherwise healthy people, is especially problematic in CF patients. Chronic infection, inflammation, and obstruction eventually lead to lung failure. Progress is being made in treating CF, but there is still no cure. An exciting advance in CF treatment is gene therapy. In gene therapy the goal is to infect lung cells with a normal copy of the CFTR gene. Due to early diagnosis and improved therapies, including lung transplantation, the average life span for a CF patient is now about 35 years.

The good news was that the baby's genotype was associated with less severe CF. The genetics counselor speculated that her phenotype should allow her to lead a fairly normal life, although certainly with challenges. Thanks to research and outreach programs, David and Patricia have access to a lot of great resources to help their daughter remain as healthy as possible.

• • •

CASE-BASED QUESTIONS

1. What class of mutations do you think would be *less* likely to lead to CF? Explain your reasoning.
2. Describe at least one way that a cell could make a shorter protein from the CFTR gene.
3. What is the most likely effect of a mutation in an intron of the CFTR gene?
4. Why does the establishment of biofilms in the lungs complicate treating bacterial infections in CF patients?
5. How can you envision viruses being potential vehicles for CF gene therapy?
6. Even though the CFTR gene is in every cell of the body, only certain cells contribute to the CF phenotype due to CFTR mutations. Discuss a possible explanation for this.
7. David and Patricia learned that their baby's CFTR genotype was one that typically generated a milder disease phenotype. How is it possible for one mutant genotype to have a milder phenotypic effect than another mutant genotype?

END OF CHAPTER QUESTIONS

NCLEX
HESI
TEAS

Think Critically and Clinically
Questions highlighted in **orange** are opportunities to practice NCLEX, HESI, and TEAS-style questions.

1. List three features of the genetic code.

2. Indicate the true statements and then correct the false statements so that they are true.
   a. DNA is replicated in a 3′ to 5′ direction.
   b. DNA has a parallel arrangement.
   c. RNA primase is required on the leading and the lagging strand.
   d. DNA ligase forms phosphodiester bonds between Okazaki fragments on the lagging strand.
   e. Prokaryotic mRNA requires processing before it is translated.
   f. In RNA, A bonds to U.
   g. In DNA, C bonds with G.
   h. RNA contains deoxyribonucleotides.

3. Select the false statement:
   a. DNA is made of deoxyribonucleotides.
   b. RNA is made of ribonucleotides.
   c. RNA is built in a 5′ to 3′ direction.
   d. DNA is built in a 5′ to 3′ direction.
   e. RNA primase builds RNA in transcription.

4. During __________ a pilus forms between an F^+ and an F^- cell and allows for the exchange of genetic material. By the end of the process, the previously F^- cell is converted to a(n) __________ cell.

5. Match the following:

Primase	Keeps DNA strands separated during replication
Gyrase	Unwinds DNA helix
Helicase	Important for removing RNA primers on the lagging strand
DNA polymerase I	The primary enzyme that copies DNA
DNA polymerase III	Lays RNA primers to start DNA replication
Single-strand DNA binding protein	Serves as a jump-start platform for DNA polymerase I
RNA primer	Relieves coiling tension in DNA as it unwinds

6. Codons are __________ nucleotides long and are in __________, which is transcribed from DNA. During __________, tRNAs serve as adapter molecules to bring __________ to the ribosome to build a protein. Once the ribosome reaches a(n) __________ on the mRNA, translation ends.

7. Assume you have the DNA sequence 3′-ACGTATCCAGCAGCTCCACCAA-5′.

   Use the genetic code table found in the chapter to answer the following questions:
   a. What would the complementary DNA sequence be?
   b. What would the corresponding mRNA sequence be?
   c. Could the mRNA sequence you generated be translated? Why or why not?

8. Label the following as a biological, chemical, or physical mutagen:

UV radiation:	X-rays:
Transposons:	Plasmids:
Cigarette smoke:	Alcohol:
Viruses:	

9. How is a ribonucleotide different from a deoxyribonucleotide?

10. For the lactose operon to be "on" and actively transcribed, __________ must be present and __________ must be absent.

11. Why are the terms *gene expression* and *protein synthesis* often used interchangeably?

12. What would the likely consequence be if a gene's promoter was deleted or severely mutated?

13. Which of the following are involved in pre-transcriptional regulation? Select all that apply.
    a. Methylation
    b. Riboswitches
    c. Operons
    d. Short interfering RNAs (siRNAs)
    e. Transcription factors

14. In eukaryotic cells, mRNA must be __________ before it is __________ into protein. In this process __________ sequences are removed from the mRNA and __________ are joined. A complex called the __________ performs this process.

15. Does the statement apply to DNA, RNA, or both?
    a. Contains uracil
    b. Usually double stranded
    c. Found in the cytoplasm of eukaryotic cells
    d. Contains thymine
    e. Is made by transcription
    f. Contains adenine
    g. Made of nucleotides
    h. Contains ribose
    i. Contains phosphodiester bonds
    j. Built in a 5′ to 3′ direction

16. Classify the effects of the following mutations as missense, nonsense, or silent (use the genetic code table in the chapter to help you):
    a. mRNA codon AUG is mutated to AUC __________
    b. mRNA codon UAC is mutated to UAA __________
    c. mRNA codon GGC is mutated to GGG __________
    d. mRNA codon UAA is mutated to UAG __________
    e. mRNA codon UGG is mutated to CGA __________

17. Which of the following helps to prevent mutations? Select all that apply.
    a. Conjugation
    b. Transposons
    c. Transduction
    d. DNA proofreading
    e. Specialized transduction
    f. Excision repair

18. Select all the true statements about repressible operons:
    a. By default they are on until turned off.
    b. An example is the lactose operon.
    c. An example is the arginine operon.
    d. A repressor must bind to the operator in order for the operon to be turned off.

19. Quorum sensing helps cells __________.
    a. mutate
    b. form biofilms
    c. carry out transduction
    d. copy their DNA
    e. perform conjugation

20. Cells that can be transformed are said to be __________.

21. In Griffith's classical experiments on transformation, which of the following scenarios led to a dead mouse? Select all that apply.
    a. Infecting the mouse with a living strain of *S. pneumoniae* that makes a capsule
    b. Infecting the mouse with a heat-killed strain of *S. pneumoniae* that makes a capsule
    c. Infecting the mouse with a heat-killed strain of *S. pneumoniae* that makes a capsule and a living strain of *S. pneumoniae* that cannot make a capsule
    d. Infecting the mouse with a living strain of *S. pneumoniae* that cannot make a capsule

22. Protein synthesis occurs in two main stages: __________ and __________.

23. Which of the following is produced by transcription?
    a. mRNA
    b. Protein
    c. DNA
    d. None of the above

24. Use the genetic code table in the chapter and the DNA sequence below to answer the following questions:
    3′-TACATAAAATAATGGCGTTCTATT-5′
    a. What would the mRNA sequence be, based on the provided DNA sequence?
    b. What would the corresponding polypeptide sequence be for this DNA sequence?
    c. What tRNA anticodon loop would correspond to the third codon of the mRNA?
    d. What would the mRNA and polypeptide sequences be if the second adenine in the DNA was deleted?

25. The following is a schematic of the parent DNA that is about to be replicated. Which side of the pictured DNA molecule (A or B) would be the leading side? Explain your answer.

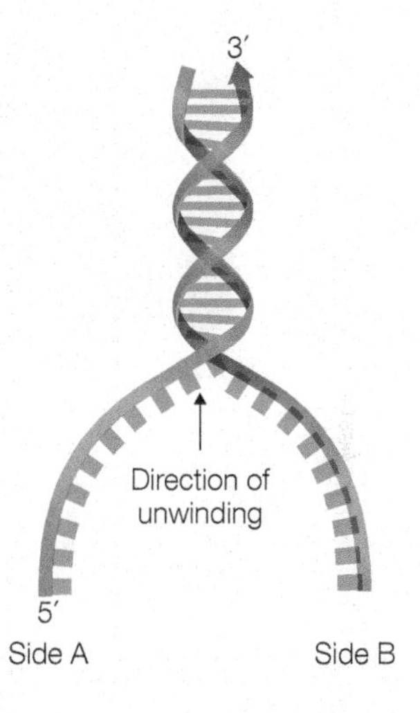

CRITICAL THINKING QUESTIONS

1. Working backward with the help of the genetic code table in the chapter, provide a DNA sequence that would encode the following polypeptide sequence. Why might the DNA sequence you provided be different from that of another student, yet equally correct?

   Methionine-Proline-Arginine-Lysine-Tyrosine-Valine-Histidine-Stop

2. Assume a particular drug inhibited transcription. What are the likely effects of this drug on the cell, and why?

3. Assume a mutation occurs in one strand of the parent DNA molecule. If the mutation was not fixed, what proportion of the next generation would inherit the mutation?

4. Imagine a cell that makes a fictitious compound called X. Making this compound is energetically costly for the cell, and the compound is only needed under specific circumstances. Design an operon that would control the production of compound X and explain how your operon would work and under what circumstances it would be on versus off.

5. Identical twins are genetically identical, but they still are different in certain regards. For example, one twin may develop a certain form of cancer that the other twin may never develop. Based on what you have read and learned in this chapter, explain why identical twins are not always identical in their phenotype despite being identical in their genotype.

6. The central dogma of molecular biology suggests that the flow of genetic information is from DNA to RNA and then to protein, and occasionally from RNA to DNA by reverse transcription. Assuming reverse translation was possible, what might it entail? Be creative in your design, but be sure to consider the tools that cells already have and include them in your proposed reverse translation process.

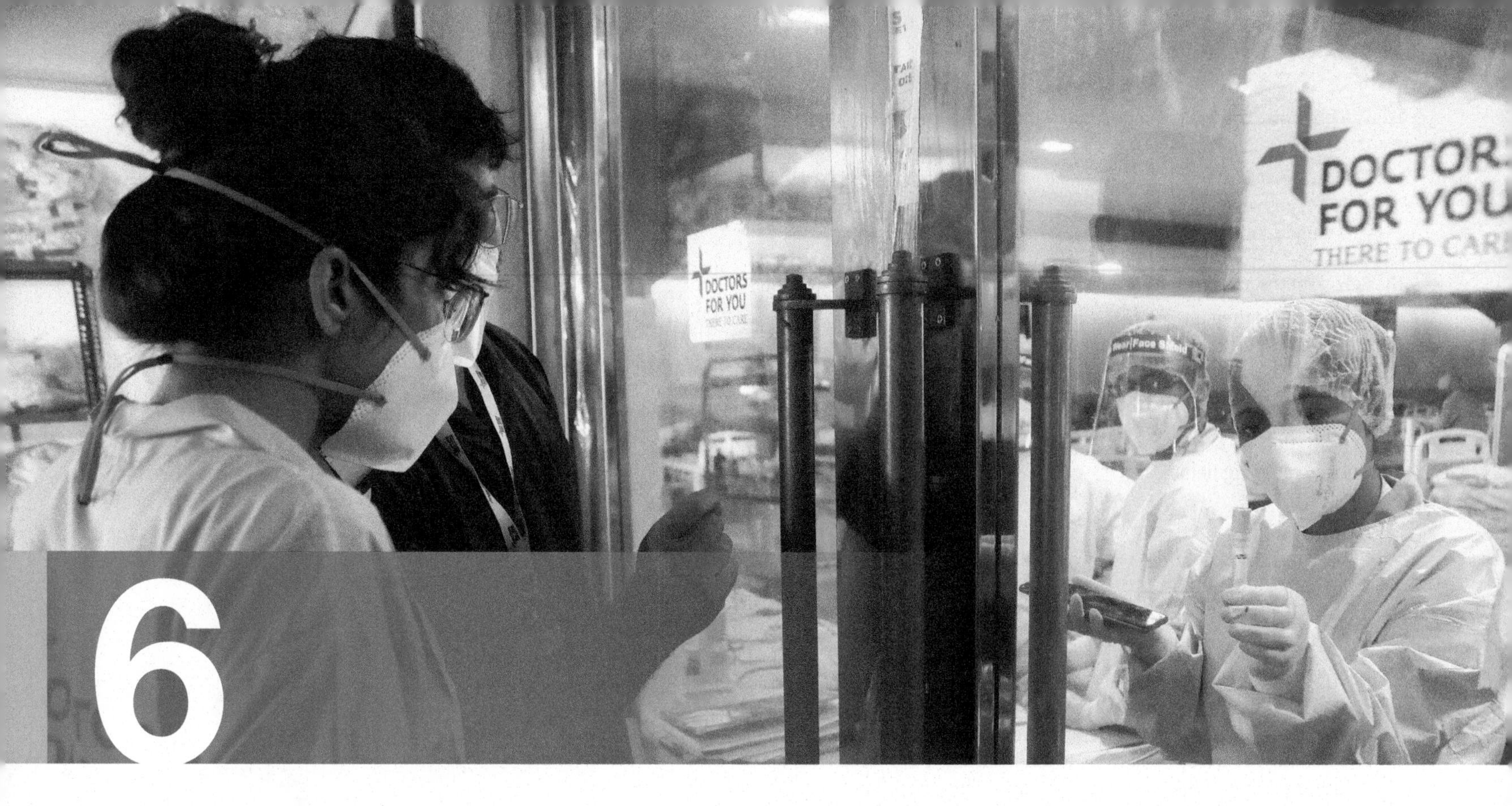

6

Viruses and Prions

What Will We Explore? In this chapter we investigate the foundations of virology, the study of viruses. We discuss general characteristics of viruses, their classification, and the pathways they use to replicate. We also examine how viruses and prions (infectious proteins) cause disease, discuss antiviral drugs, explain how viruses evolve to cause pandemics, and review ways to detect viral infections.

Why Is It Important? Virology is a tremendously important area of research and medicine because most emerging diseases—including COVID-19 (**Co**rona**vi**rus **D**isease 20**19**), Ebola, and Zika—are viral. Plus, the ability of healthcare workers to quickly identify and isolate infected patients remains crucial for limiting the scope and impact of outbreaks. As happened with SARS-CoV-2 (the virus that causes COVID-19), widespread and rapid travel means that isolated viral outbreaks can become worldwide pandemics at an alarming speed.

Unlike other microbes we study, viruses and prions are nonliving pathogens. Most viruses are smaller than bacteria, outnumber cellular life by more than 10 to 1, appear in all ecosystems, and can infect every type of life. In humans, viruses cause everything from minor ailments such as the common cold to the deadly disease rabies. Other viruses, such as the human papilloma viruses (HPV), can cause cancer. For millennia, viruses have been a major source of human disease and suffering. Eradication of the smallpox virus through vaccination was one of the greatest public health victories ever. Following decades of research and development, we finally have drugs that eliminate certain hepatitis C viruses that cause severe liver damage. Likewise, vaccines for influenza protect millions each year, and vaccines that protect against HPV, chickenpox (varicella-zoster), and rotaviruses (a common cause of viral diarrhea in infants) are widely used. And new immunization strategies paved the way for the development of vaccines that prevent COVID-19.

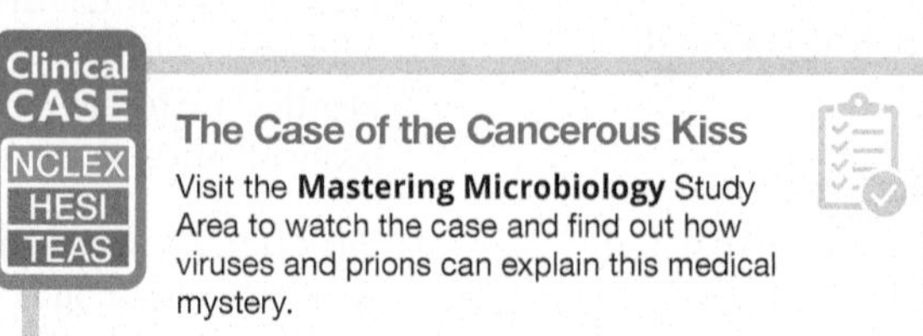

The Case of the Cancerous Kiss

Visit the **Mastering Microbiology** Study Area to watch the case and find out how viruses and prions can explain this medical mystery.

Healthcare worker collecting a blood sample from a Covid-19 patient
India

6.1 GENERAL VIRUS CHARACTERISTICS

Viruses are nonliving pathogens.

Two important characteristics of viruses are that they are especially small, and they are nonliving pathogens. The extremely small scale of viruses was discovered in the late 1800s when it was noted that sap from an infected tobacco plant remained infectious to healthy tobacco plants even after it was passed through a specialized filter designed to remove prokaryotic and eukaryotic cells. The infectious agent, which was too small to be seen via light microscopy, was given the name **virus** from the Latin term for a venomous secretion. Viruses were described as *filterable* infectious agents (agents that pass through a filter), reflecting their usually tiny size compared with cells. Today over 5,000 mammal-infecting viral species have been described. Of these, at least 220 infect humans.[1] It is estimated that at least 320,000 mammalian viruses remain uncharacterized. About 70 percent of viruses that infect humans are harbored in other animals, representing a significant pool for the emergence of new viral diseases in humans.[2] Clearly, an introduction to the study of viruses, or **virology**, is essential for any healthcare provider.

A distinguishing feature of viruses is that they are nonliving infectious agents. Most biologists consider cells to be the smallest units of life, so anything not made of cells (anything *acellular* like viruses) is classified as nonliving. Viruses are also considered nonliving because they cannot synthesize their own components, such as nucleic acids or proteins, without the help of the host cells they infect. As such, viruses are called *obligate intracellular pathogens*—disease-causing microbes that must invade living cells and hijack their biochemical and cellular tools to replicate. Although a number of bacteria are also intracellular pathogens, all bacteria have their own biochemical processes to extract energy from nutrients; in contrast, viruses lack these metabolic processes. TABLE 6.1 compares viruses with prokaryotic and eukaryotic cells.

Learning Outcomes

After reading this section, you should be able to:

6.1 Explain why viruses are classified as nonliving microbes.

6.2 Compare viruses to prokaryotic and eukaryotic cells.

6.3 Describe features and functions of viral structures, including capsids, envelopes, and spikes.

6.4 Describe the genomic variations seen among viruses.

6.5 Summarize the various ways that DNA and RNA viruses use their genome to make mRNA for protein production.

6.6 Describe the key contributors to and consequences of viral genome evolution, and state why RNA viruses evolve faster than DNA viruses.

6.7 Compare and contrast antigenic shift and antigenic drift, and state how they impact influenza virus evolution and outbreaks.

TABLE 6.1 Comparing Viruses, Prokaryotes, and Eukaryotes

Characteristic	Viruses	Prokaryotes	Eukaryotes
Cells?	No	Yes	Yes
Considered alive?	No	Yes	Yes
Relative size	Generally smaller than prokaryotes; most require electron microscopy to be seen	Most are bigger than viruses and smaller than eukaryotes; usually visible via light microscopy	Usually bigger than prokaryotes and viruses; often visible via light microscopy
Pass through a filter and are therefore described as filterable	Yes	Not usually (some exceptionally small bacteria such as *Mycoplasma* species are filterable)	No
Structure	Protein capsid coating and nucleic acid	Cells without nuclei or other membrane-bound organelles	Cells with nuclei and membrane-bound organelles
Replication	Host cell energy and machinery are hijacked to replicate the virus	Binary fission (asexual)	• Mitosis (asexual) • Meiosis (sexual)
Exhibit metabolism?	No	Yes	Yes
Genome composition	DNA or RNA	DNA	DNA

[1] Woolhouse, M., Scott, F., Hudson, Z., Howey, R., & Chase-Topping, M. (2012). Human viruses: Discovery and emergence. *Philosophical Transactions of the Royal Society of London. Series B: Biological Sciences*, 367(1604), 2864–2871.

[2] Anthony, S. J., Epstein, J. H., Murray, K. A., et al. (2013). A strategy to estimate unknown viral diversity in mammals. *mBio*, 4(5), e00598-13 (online only).

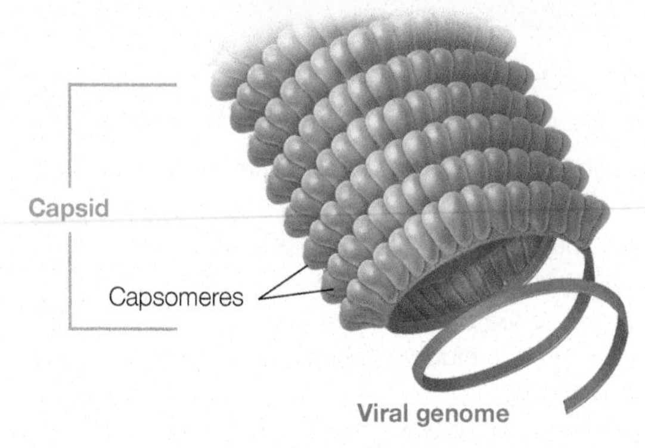

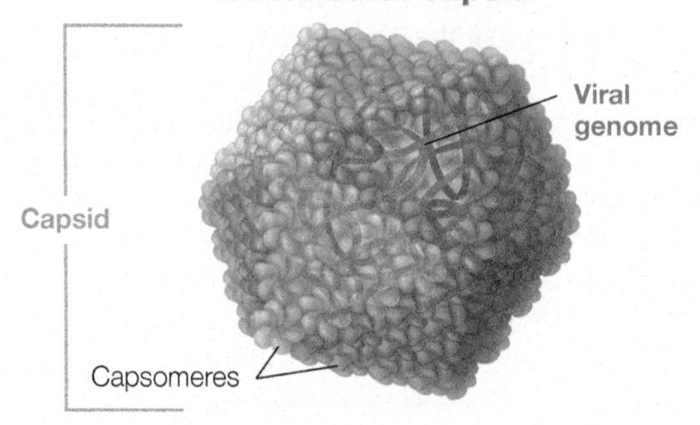

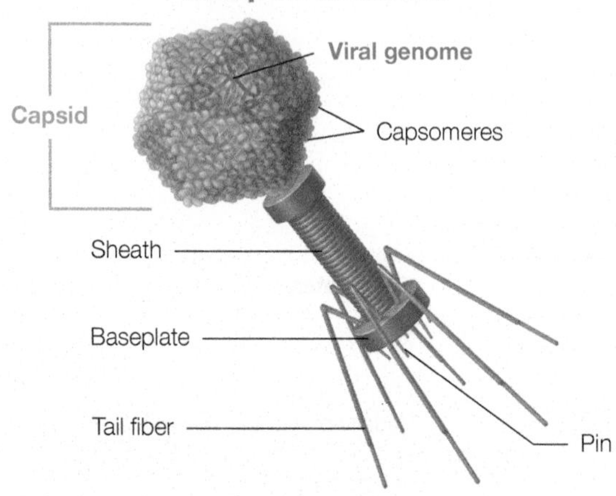

FIGURE 6.1 Capsid structures Most animal viruses have capsids with helical or icosahedral symmetry. Bacteriophages and certain animal viruses have complex structures.

Critical Thinking *How can the basic structure of a virus be compared to the nucleus of a eukaryotic cell?*

CHEM • NOTE

Protein self-assembly As described in Chapter 2, most proteins aren't physiologically active until they are properly folded into a higher-order structure. To achieve proper folding, most proteins enlist the help of specialized cellular tools called chaperones. Proteins that self-assemble, like some capsid proteins, achieve their final form without the help of chaperones.

Viruses exhibit diverse structural and genomic features.

A virus's structural and genomic features dictate which cells the virus can infect, as well as the progression of the infection. Viruses can infect every branch in the tree of life. However, in this chapter we'll focus on **bacteriophages** (or phages), which are viruses that infect bacteria, and **animal viruses**, which include viruses that infect humans. A single, infectious virus particle is called a **virion**. All virions have an exterior protein shell packed with genetic material (DNA or RNA), and some have an outer lipid envelope.

Viral Capsids

The protein shell that packages and protects a virion's genome and also accounts for the bulk of a virion's mass is called a **capsid** (FIG. 6.1). A capsid's shape is based on how the individual three-dimensional subunits, called **capsomeres**, are arranged. The capsids of most animal viruses have either helical or icosahedral symmetry. **Helical** capsids look like a hollow tube. **Icosahedral** capsids look like three-dimensional polygons, but may appear fairly spherical—just as a soccer ball is spherical, yet made of multiple hexagon and pentagon shapes. The novel Coronavirus SARS-CoV-2 has a helical capsid. Viruses with less conventional capsids are lumped into the catchall **complex** structure category. Some animal viruses, such as the *Poxviridae* family that includes the smallpox virus, have ovoid or brick-shaped capsids and are therefore described as having a complex structure. Bacteriophages also exhibit a complex structure; although their capsids often have icosahedral symmetry (see the bottom of Figure 6.1), they are associated with additional complex structures that enable them to act like hypodermic syringes that inject the viral genome into target bacterial cells.

The capsomeres that build capsids may be made from a single type of protein or a collection of different proteins. Unlike most cellular proteins, viral capsids often exhibit self-assembly, meaning the amino acid sequence determines how the final proteins fold and come together to make the larger capsid structure. Other more complex viruses use the host cell machinery to put the capsids together properly. Because capsid assembly is such a central part of virion formation, it represents a potential target for antiviral drugs.

Viral Envelopes

Some animal viruses have a lipid-based **envelope** that surrounds the capsid. Virions that lack an envelope are said to be **naked** (or nonenveloped) (FIG 6.2). Enveloped viruses develop by budding from the host, taking a portion of the cell membrane with them as a coating when they go. In contrast, naked viruses lyse (burst) out of host cells and do not coat themselves in an envelope. Because most bacteriophages lyse out of host cells that have a cell wall, these complex-shaped viruses are almost always naked.[3] Animal viruses may be either enveloped or naked. Enveloped animal viruses are numerous and include coronaviruses (e.g., SARS-CoV-2), influenza viruses, and herpes viruses. Examples of nonenveloped animal viruses are also numerous and include the human papilloma viruses, poliovirus, and rotavirus.

Viral Spikes (Peplomers)

Many viruses have glycoprotein extensions called **spikes** (or peplomers). These spikes may protrude from the viral capsid or, if present, from the viral envelope (Figure 6.2).

The coronaviruses—a large family of RNA viruses that includes SARS-CoV-2—have numerous spikes that protrude through the viral envelope, conferring a "corona" or crown-like appearance. Spikes help viruses attach and gain entry to

[3] Lyytinen, O. L., Starkova, D., & Poranen, M. M. (2019). Microbial production of lipid-protein vesicles using enveloped bacteriophage phi6. *Microbial Cell Factories*, 18(1), 1–9.

host cells. Because they only bind to specific factors on a given host cell, spikes have an important role in determining what species and tissues the virus can infect, similar to how a lock and key must match. The host immune system may also recognize spikes and mount an immune response to them. Unfortunately, if the host develops an immune response to the spikes and the spikes then change—something certain viruses do rapidly and routinely in their genetic evolution—then the newly evolved virus may escape immune detection. Viral spike evolution is a major concern regarding the longer-term efficacy of vaccines against SARS-CoV-2—immunity conferred by a vaccination designed to protect against today's circulating strains might not protect against strains that are likely to evolve. This could necessitate a seasonal vaccination program much like the program we have for influenza management.

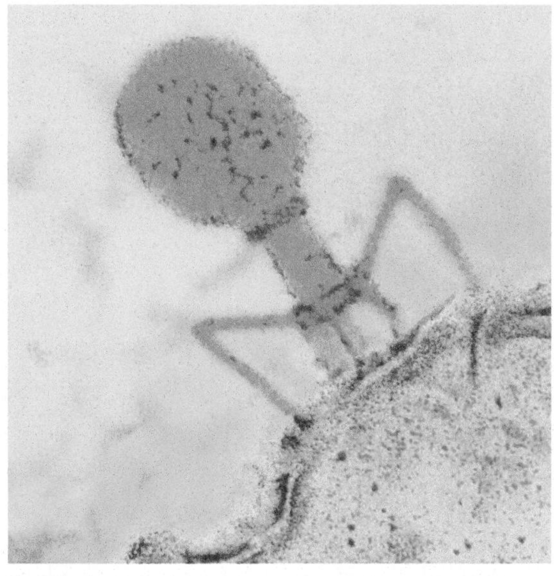

Bacteriophage

As with coronaviruses, influenza viruses frequently mutate and experience small changes in their spike proteins. This is such an important issue for influenza A (a collection of influenza viruses that cause seasonal flu and have flared up to cause flu pandemics) that the viral **hemagglutinin (HA)** and **neuraminidase (NA) spikes** (referred to as HA and NA spikes) are a key part of the virus subtype naming convention (FIG. 6.3). For example, swine flu's scientific name is H1N1, whereas bird flu is called H5N1. There are at least 18 known variations of HA spikes and 11 characterized variations of NA spikes in influenza A. (Later in this chapter we revisit how changes in influenza spikes impact influenza outbreaks, and Chapter 16 also explores influenza pathology and outbreaks.)

Viral Genomes

Viruses do not require many genes to exist because they do not build organelles, nor do they have metabolic processes to mediate. Viral genes generally encode capsomere proteins, various enzymes needed for viral replication (e.g., enzymes needed to copy the viral genome), and structural factors. Most viruses therefore have fewer than 300 genes. The smallest have fewer than 2,000 base pairs in their whole genome—this is much smaller than the genomes typically found in cells. By comparison, the human genome has about 25,000 genes and about 3.2 billion base pairs.

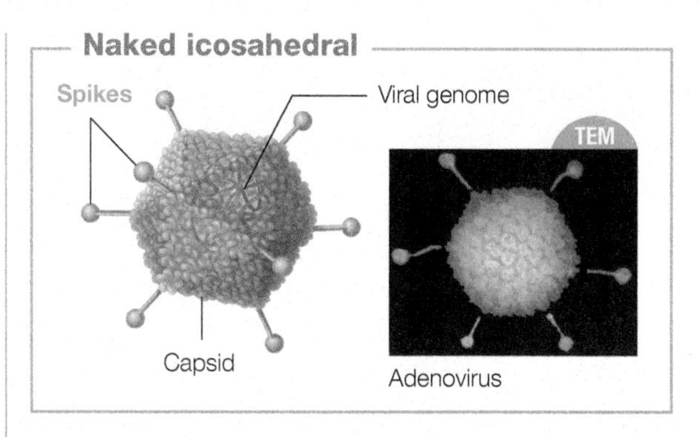

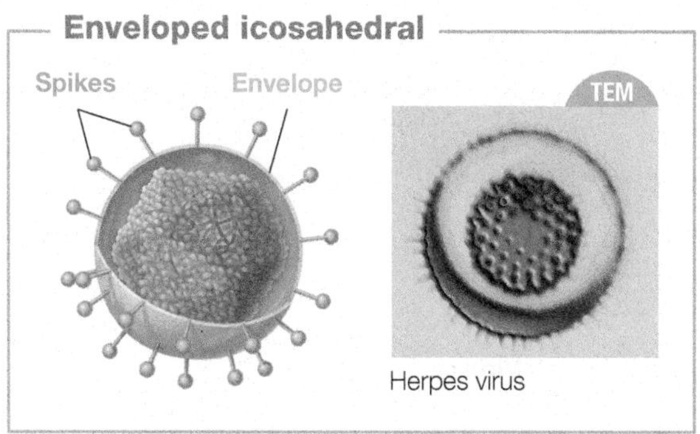

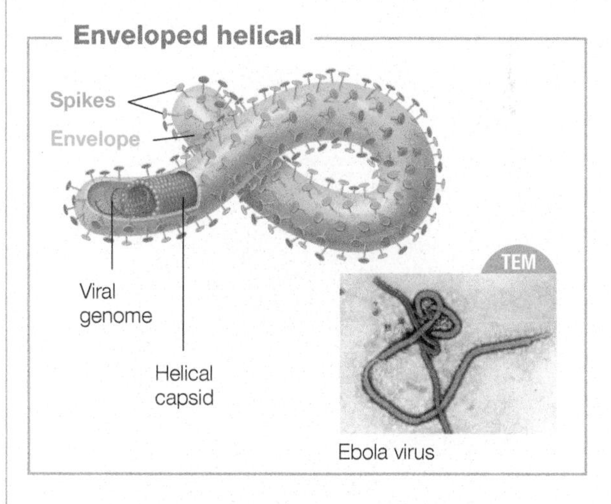

FIGURE 6.2 **Viral envelope and spikes** Animal viruses can be enveloped or naked. Many viruses have spikes (peplomers) that extend from the capsid or envelope.

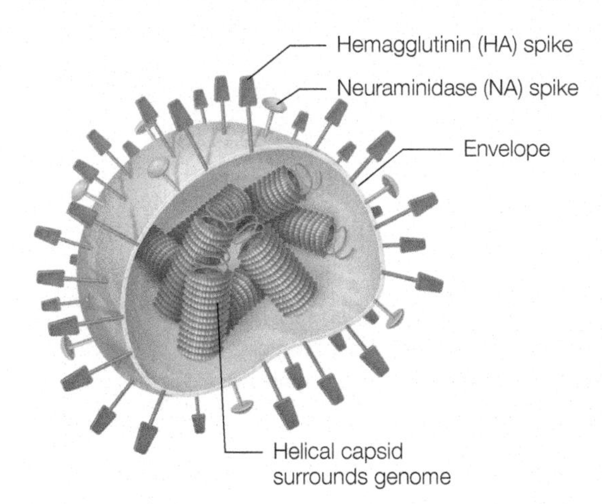

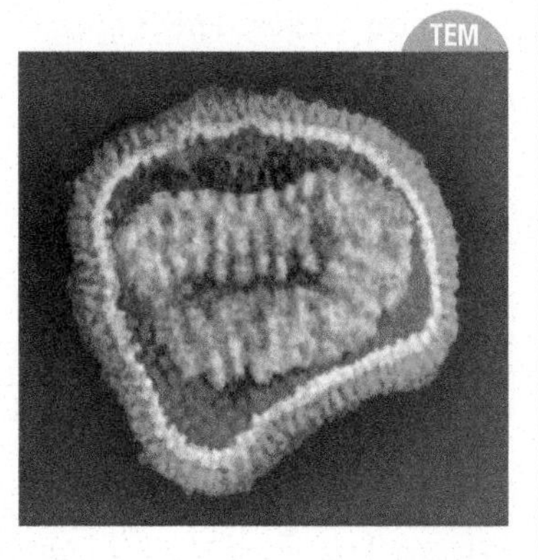

FIGURE 6.3 **Influenza's glycoprotein (HA and NA) spikes** Influenza's hemagglutinin (HA) and neuraminidase (NA) spikes affect how these viruses interact with host cells and how the host's immune system recognizes these viruses. The colored transmission electron micrograph shows influenza's HA and NA spikes (red) covering the surface envelope (yellow).

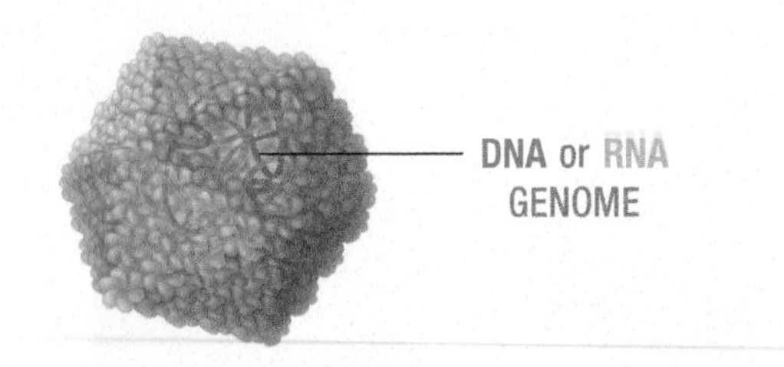

Genome arrangements:

Circular Linear Segmented

DNA viral genomes
- Circular or linear
- Often double stranded
- May also be single stranded

RNA viral genomes
- Linear or segmented
- Often single stranded
- May also be double stranded

FIGURE 6.4 **Viral genomes**

A 3D model of an enveloped icosahedral virus.

Viruses (to include animal viruses and bacteriophages) can have either an RNA- or DNA-based genome (but usually not both),[4] and the nucleic acid present can be either single- or double-stranded. In contrast, cells always have a double-stranded DNA genome. Furthermore, viral genomes can exist in diverse arrangements; some viruses spread their genome over multiple segmented sections, while others have a single circular or linear molecule. FIG. 6.4 shows possible nucleic acid variations for viruses.

Ultimately, all viruses require a host cell to make viral proteins and thereby build more virions. As you learned in Chapter 5, DNA is transcribed to make messenger RNA (mRNA), which in turn is translated by ribosomes to build proteins. So, no matter what the viral genome may be made of, it will direct the production of mRNAs that the host cell will then translate into proteins. The bulleted points that follow and FIG. 6.5 review how viruses do this and how the process varies depending upon the virus's genome architecture. Although these general processes apply to all viruses, here we focus on animal viruses and bacteriophages as our case-in-point examples.[5]

- **Double-stranded DNA viruses (dsDNA):** For viruses that have double-stranded DNA (dsDNA), the process of making mRNA follows what we reviewed in Chapter 5. As shown near the top of Figure 6.5, the viral DNA is transcribed using host cell RNA polymerases, and the resulting mRNA is then translated into protein. Double-stranded DNA viruses include most bacteriophages as well as numerous medically important animal viruses such as the papillomaviruses, hepatitis B virus, and herpes viruses.
- **Single-stranded DNA viruses (ssDNA):** If the virus has a single-stranded DNA (ssDNA) genome, then it converts its genome to a double-stranded form before conducting transcription. Examples of ssDNA viruses include parvovirus B19 and bacteriophages from the *Microviridae* family.
- **Single-stranded sense RNA viruses (ssRNA+):** In these viruses the RNA genome functions as mRNA that host cell ribosomes directly translate to make viral proteins. Examples of viruses with an ssRNA+ genome include the causative agents of COVID-19 (SARS-CoV-2), polio, rubella, West Nile encephalitis, and dengue fever. Some bacteriophages such as the MS2 phage also fall into this category.
- **Single-stranded antisense RNA viruses (ssRNA−):** These *single-stranded, antisense,* or *negative-stranded RNA* (ssRNA−) viruses have an RNA genome that is *complementary* to mRNA. Consequently, their RNA genome must be transcribed into a readable mRNA format before the host cell can perform translation to make viral proteins. This is accomplished by virally encoded enzymes called **RNA-dependent RNA polymerases (RdRPs)**. Unlike host cell RNA polymerases that build new RNA from a DNA template, RdRPs build new RNA from an existing RNA template. This group of RNA viruses encompasses a wide variety of pathogens, including the causative agents of influenza, measles, Ebola, and rabies.
- **Retroviruses:** These specialized RNA viruses conduct reverse transcription. Recall from Chapter 5 that in reverse transcription, a virally encoded enzyme called **reverse transcriptase** builds a double-stranded DNA using the single-stranded viral RNA genome as a template. Once completed, the viral double-stranded DNA is usually inserted into the host genome. The integrated viral DNA is then transcribed in the nucleus, and the resulting

---

[4] In 2012 a virus with a hybrid RNA/DNA genome was isolated from a boiling springs lake in Lassen Volcanic National Park. To date, no medically important viruses have been found to have a hybrid genome, and the discovered DNA/RNA hybrid virus seems to be a rare exception.

[5] Genome architecture is the premise of the Baltimore classification system—a viral classification model initially developed by virologist David Baltimore in the 1970s. The Baltimore model dominated virology for a while, but because it does not fully address the challenge of developing a hierarchical taxonomy for viruses that drilled down to the family level, in the 1990s the International Committee on Taxonomy of Viruses (ICTV) added a few other criteria for classification. The newer ICTV classification model is reviewed later in this chapter.

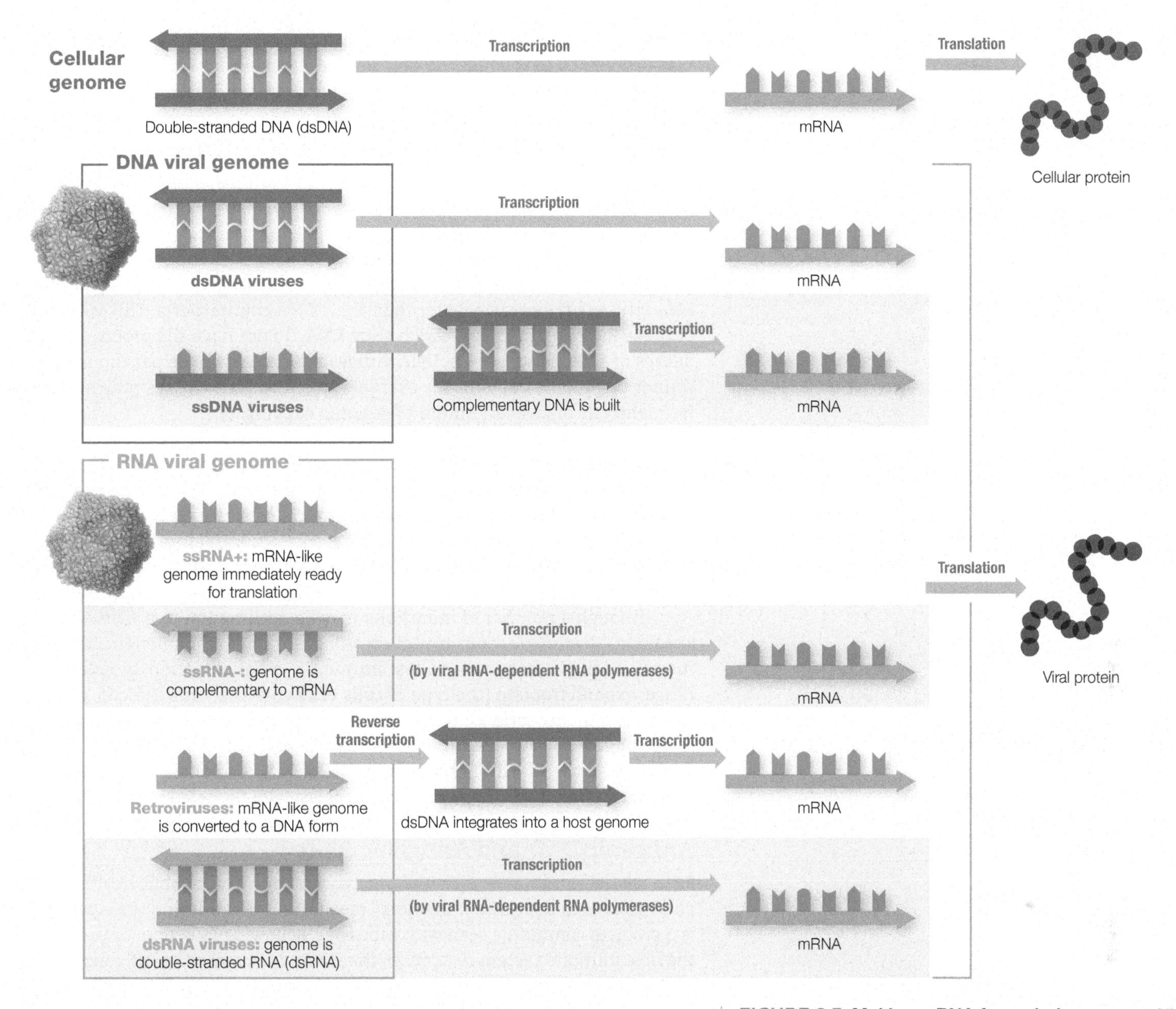

mRNA is translated. HIV (the virus that causes AIDS) as well as human T-lymphotropic viruses (also called human T cell leukemia viruses or HTLVs) carry out reverse transcription and operate through a DNA intermediate. Retroviruses are covered in more detail later in this chapter.

- **Double-stranded RNA viruses (dsRNA):** As shown at the bottom of Figure 6.5, if the virus has a double-stranded RNA (dsRNA) genome, then the RNA must be unwound, so that RNA polymerases can transcribe it into an mRNA format. The process resembles that of a double-stranded DNA virus, but instead of using host cell RNA polymerases, virally encoded RNA-dependent RNA polymerases are required. Rotaviruses, which cause severe diarrhea, are an example of dsRNA viruses. Certain bacteriophages, such as *Pseudomonas aeruginosa* phage phiYY, also have a double-stranded RNA genome and employ RNA-dependent RNA polymerases to make mRNA.

FIGURE 6.5 Making mRNA from viral genomes All viruses must be able to make mRNA that can be translated by host cell ribosomes to make proteins. DNA viruses tend to closely resemble cells in mRNA production. RNA viruses have four general pathways they may use to get to mRNA.

Viral genomes change over time.

In general, viruses evolve faster than cells. Two factors explain this phenomenon. First, as compared to cells, viruses tend to replicate faster and more prolifically—the large quantity of virions released within a host during an infection provide many opportunities for nature to fine-tune viral transmission and infection

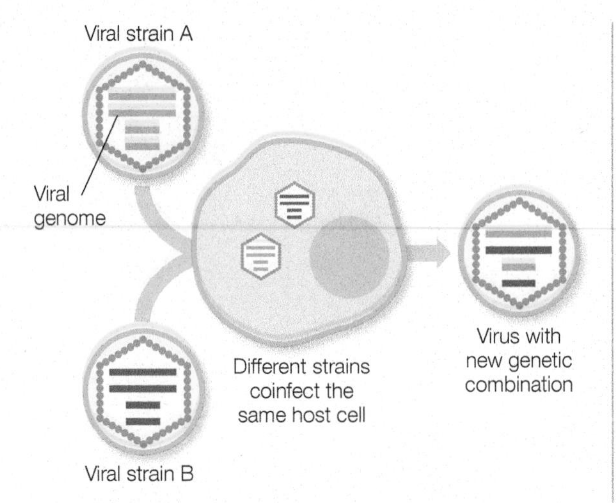

FIGURE 6.6 Viral genome reassortment When different viral strains coinfect a host cell, their genomes mix and can generate new viral combinations.

Critical Thinking *Genetic reassortment is typically between related viral strains. Why do you think this is the case?*

strategies. Second, reassortment events and random mutations foster viral evolution—especially in the RNA viruses. Here we'll review how viruses change over time and the potential effects of that evolution on outbreaks and pandemics.

Reassortment Events and Random Mutations

Genome reassortment events can contribute to viral evolution when two different viral strains coinfect a single host cell (FIG. 6.6), providing an opportunity for the viral genomes to mix and generate new viral strains.

Random mutations also cause viral genomes to change over time. RNA genomes tend to mutate more frequently than DNA genomes, and therefore RNA viruses often evolve faster than their DNA counterparts. This is mainly because RNA polymerases, which copy RNA, do not have the proofreading capabilities of DNA polymerases. DNA viruses may mutate once per thousand rounds of genome copying, for example, whereas RNA viruses may mutate as frequently as once every round of genome copying.

As reviewed in Chapter 5, mutations can have neutral, beneficial, or negative effects, depending on where they occur in the genome. Genetic changes that limit infectivity, for example by impairing a virus's ability to invade host cells and replicate, are unfavorable. Such mutants are called **attenuated strains**. Although these strains don't cause disease in a host with a normal immune system, they do stimulate an immune response. This makes them ideal candidates for vaccine development. The oral polio vaccine is formulated with attenuated viral strains.

Although a number of mutations may be detrimental to a virus, statistically speaking, an advantageous mutation will eventually occur. Beneficial mutations may allow the virus to escape host immune system detection, broaden host range, expand **tropism** (the type of cells or tissues the virus infects), or make the virus more infectious so that it is more easily spread from one host to the next. For example, SARS-CoV-2 evolved enhanced infectivity traits that further complicated the COVID-19 pandemic. The effects of spontaneous mutations and reassortment in viral evolution are classically modeled in influenza viruses.

Antigenic Drift and Antigenic Shift

Even relatively minor virus mutations may greatly impact public health, as they allow viruses to thwart the "memory" mechanism of our immune response. This is a problem commonly seen with influenza. Following infection (or vaccination), the host immune system recognizes the HA and NA spikes on influenza's surface. These spikes, which the virions use to attach to host cells before entry, also serve as *antigens* against which the immune system creates specific *antibodies*. To clarify, an antigen is any substance that can trigger the immune system to make antibodies, proteins that defend against foreign agents. In Chapter 12 we discuss how the immune system makes antibodies, but suffice it to say here that during an infection, the immune system recognizes a particular antigen profile and makes antibodies that help clear the infection and combat the pathogen if it re-enters the body later. However, because influenza's RNA genome mutates frequently, minor changes to the HA and NA spikes also routinely occur. These minor changes, referred to as **antigenic drift**, allow the virus to evade a quick antibody response by making the new strains different enough to go unrecognized by the immune system—even if the host had a prior interaction with a related strain (FIG. 6.7). Antigenic drift is the reason we all need a new, different flu shot every year. Furthermore, antigenic drift means there is always a portion of the population susceptible to the latest virus, which in turn leads to seasonal outbreaks.

Occasionally, influenza viruses undergo a major genetic reassortment that dramatically changes HA and NA spikes. These broader mutations, termed **antigenic shift**, often lead to viral strains with new features, such as increased infectivity or expanded host range. Antigenic shift is what occurred when avian and swine influenza strains "jumped species," becoming able to infect humans.

FIGURE 6.7 Antigenic shift and antigenic drift

Critical Thinking *Why is antigenic shift more likely to lead to expanded host range than antigenic drift?*

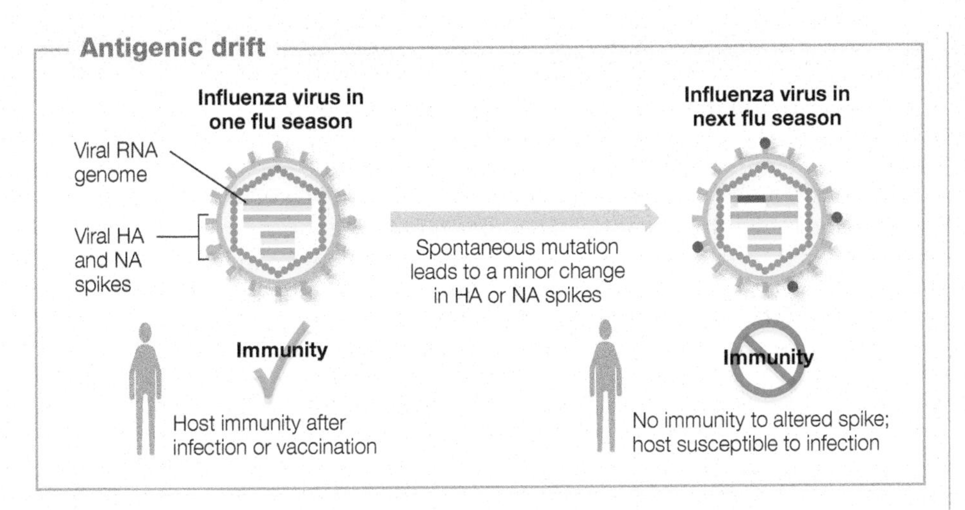

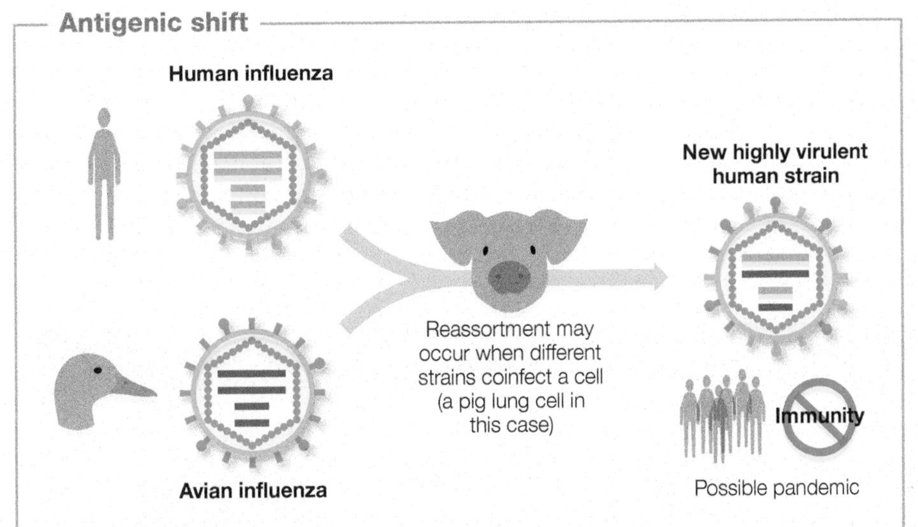

Not only is a vaccine unlikely to exist at the outset of this sort of strain emergence, but the drastic change in the virus usually means that people have no residual immune protection from prior influenza infections or vaccinations. These factors set the stage for a worldwide outbreak or **pandemic**. Because of the limited preexisting immunity in the population, pandemics may have a dramatically increased mortality rate compared with epidemics caused by the earlier, progenitor virus strains. (For more on antigenic drift, antigenic shift, and influenza, see Chapter 16.)

Build Your Foundation

1. What is the primary reason that viruses are not considered alive? (LO 6.1)
2. How do viruses differ from prokaryotes and eukaryotes? (LO 6.2)
3. Describe the features and functions of viral capsids, spikes, and envelopes. (LO 6.3)
4. Why are bacteriophages almost always naked? (LO 6.3)
5. What are the possible genomic variations for viruses? (LO 6.4)
6. How do DNA viruses use their genomes to make mRNA? (LO 6.5)
7. What are four ways that RNA viruses may make mRNA? (LO 6.5)
8. What are the leading contributors to viral genome evolution, and what are their potential effects on viral features? (LO 6.6)
9. Why do RNA viruses evolve faster than DNA viruses? (LO 6.6)
10. Define antigenic shift and antigenic drift, and describe how they influence influenza evolution and outbreaks. (LO 6.7)

Build Your Foundation (BYF) Quick Quiz: Visit the **Mastering Microbiology** Study Area to quiz yourself.

6.2 CLASSIFYING AND NAMING VIRUSES

Learning Outcomes

After reading this section, you should be able to:

6.8 Identify the properties considered in classifying viruses.

6.9 Give examples of viral families that are medically important in humans.

6.10 Explain the significance of a virus's host range and tropism.

6.11 Summarize the conventions for naming viruses.

Diverse features are used to classify and name viruses.

The International Committee on Taxonomy of Viruses (ICTV) works to develop criteria and refine the naming conventions for viruses. Before this committee took over viral taxonomy, naming was haphazard and often misleading. For example, hepatitis A, B, and C viruses were all given a common name indicating that they all infect and damage the liver ("hepato" means liver; and "itis" means inflammation); however, they are not otherwise related to each other.

Viruses are now grouped by the following properties:

1. Type of nucleic acid present (DNA or RNA)
2. Capsid symmetry (helical, icosahedral, or complex)
3. Presence or absence of an envelope
4. Genome architecture (ssDNA, ssRNA, etc.)[6]

FIG. 6.8 shows the medically important families of viruses that infect humans and how they are grouped into families using the aforementioned criteria. Additional features such as virus size, host range, tropism (the type of cells or tissues the virus infects), and disease features are often reflected in virus names because they help to refine the grouping of viruses into species.

TRAINING TOMORROW'S HEALTH TEAM

Novel Vaccines for a Novel Coronavirus

The COVID-19 pandemic is an example of how antigenic shift can trigger widespread disease outbreaks. In the case of SARS-CoV-2, the RNA virus expanded its host range from bats to another animal host (possibly palm civets), and then evolved the capacity to infect humans. Ultimately, the virus rapidly spread because humans lacked residual immune protection to the novel virus.

SARS-CoV-2 vaccines represent two new immunization strategies—mRNA vaccinations and vector vaccines. As you learned in Chapter 5, cells translate mRNA to make proteins. This is a natural process that viruses leverage for their own replication. When a virus infects a host cell, it hijacks the cell's protein-building machinery to make viral proteins. SARS-CoV-2 mRNA vaccines exploit this strategy, but with the following important differences: Unlike most vaccines on the market, the mRNA vaccines do not expose patients to an inactivated form of the virus (e.g., a common influenza vaccine strategy) nor an attenuated form (e.g., an approach used for oral polio vaccines). Instead, the patient's immune system only encounters a harmless bit of mRNA that encodes part of a SARS-CoV-2 spike protein. Containing only mRNA, salts, a small amount of sugar (sucrose), lipids, and some pH stabilizers (buffers already widely used in medical therapies), the vaccine is arguably the closest to "all natural" as it can get. The mRNA in the vaccine is coated in lipids that help the patient's cells take up the mRNA so it is translated to make the segment of the viral spike protein. The immune system then mounts a response against the produced viral spike protein segment. As the immune system reacts, the original ingredients in the vaccine are all eliminated. The mRNA vaccines designed by pharmaceutical companies Moderna and Pfizer were the first two vaccines approved for use in the United States to prevent COVID-19.

The vector vaccine approach, which the vaccine designed by Johnson and Johnson uses, also relies on host cells making a segment of the SARS-CoV-2 spike protein. However, instead of delivering mRNA to host cells, the vaccine contains a harmless virus that delivers a snippet of DNA. The delivery virus (or vector virus) is an adenovirus (a common cold virus) that has been genetically modified so that it can't replicate (make more of itself) in the host cells it infects. Instead, the vector virus simply delivers a piece of DNA that is then transcribed and translated to make a segment of the viral spike protein. As described for the mRNA immunization strategy, the resulting protein is detected by the patient's immune system, and it, along with the original vector virus, is eliminated from the body. Neither of these vaccination strategies alter a person's DNA.

Dr. Kizzmekia Corbett's work led to Moderna's mRNA vaccine.

As with influenza viruses, SARS-CoV-2 and related viruses will be closely monitored for strain variations, and seasonal vaccines are likely to become a preventative strategy.

QUESTION 6.1

Some patients have expressed a concern that the new vaccines cause COVID—pointing to the achiness associated with the second booster dose of the mRNA vaccine as evidence. What information would you share with them to alleviate their concern?

---

[6] Recall that genome architecture is the premise of the Baltimore classification system—a viral classification model developed by virologist David Baltimore in the 1970s and replaced in the 1990s by the ICTV recommendations described in this chapter.

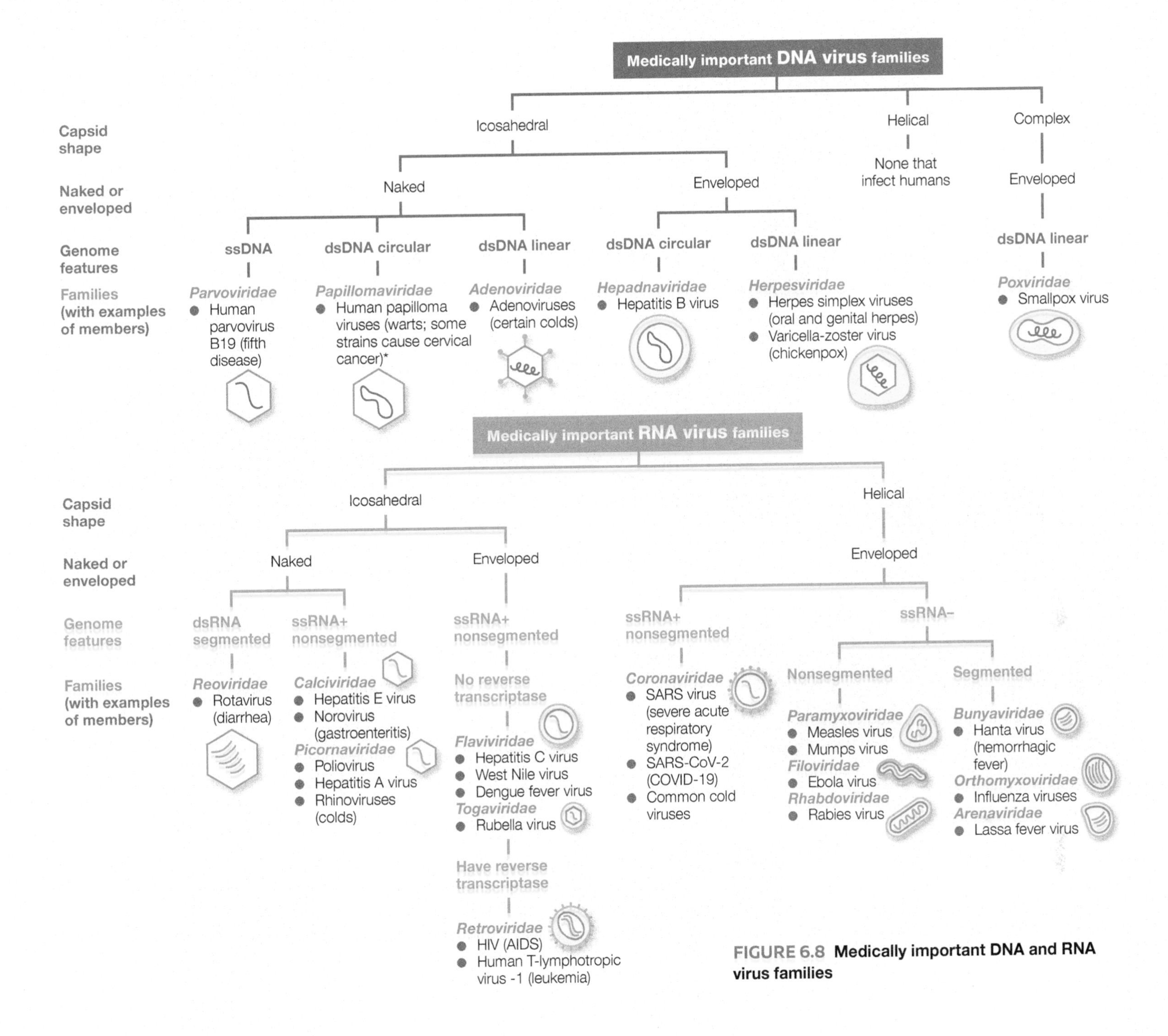

FIGURE 6.8 Medically important DNA and RNA virus families

Host Range and Tropism

Viruses infect every type of lifeform on Earth; however, a given virus will exhibit a specific **host range** or collection of species that it can infect. Some viruses infect more than one species, whereas others infect only one species. For example, the measles virus is only able to infect humans. Most of the viruses that infect humans can also infect some other animal. Indeed, most human viruses evolved in other species and then through genetic changes they gained the capacity to infect us. For example, avian influenza (bird flu or H5N1), which primarily infected certain waterfowl species, especially geese and ducks, underwent genetic reassortment that expanded its host range to include humans. The same holds true for swine flu, which initially infected pigs and can now infect humans.

- Poliovirus, 30 nm
- Rhinovirus, 30 nm
- HIV, 120 nm
- Bacteriophage T4, 225 nm

Ebola virus, 970 nm long

Pandoravirus, 1,000 nm long

Pithovirus, 1,500 nm long

E. coli bacterium, 2,000 nm long

Human red blood cell, 8,000 nm diameter

FIGURE 6.9 Virus sizes vary greatly

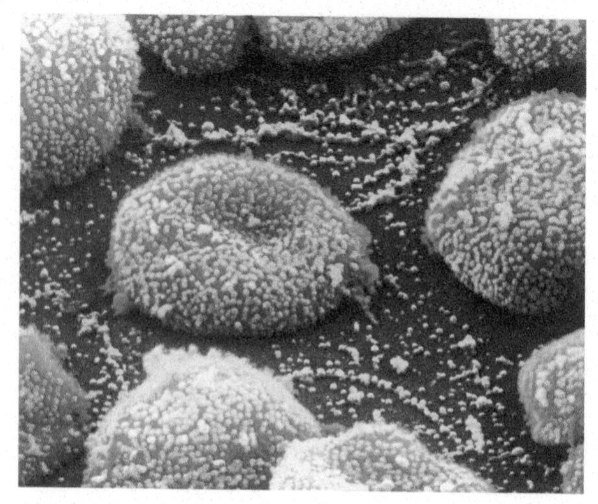

Influenza viruses infect red blood cells.

Viruses also exhibit specificity regarding what tissues or cells they will infect in their given host. This tropism is due to virus surface factors only being able to bind to specific surface molecules on certain host cells. Some viruses, like Ebola, can infect a wide range of host cells or tissues and are said to have a *broad tropism*. Other viruses have a *narrow tropism* and may specifically target only one type of cell or tissue in the given host. For example, the hepatitis viruses target the liver, so they are described as hepatotropic. Human T-lymphotropic viruses get their name as a result of their partiality for infecting human T cells, which are specialized white blood cells of the immune system—this virus name reveals the virus's narrow tropism and host range.

Virus Sizes

Viruses are usually smaller than bacteria, but there is tremendous size diversity. Some viruses, such as rhinoviruses (which typically cause colds) and polioviruses (the cause of polio), may have a diameter as small as 30 nanometers (nm). By comparison, a molecule of hemoglobin, the oxygen carrier in human blood, is around 5 nm. In contrast, Ebola viruses and the pandoraviruses, are relatively large and can have lengths nearing 1,000 nm. Pithovirus, which was discovered in 2014 in ice cores from Siberia, is one of the largest viruses discovered to date; its dimensions of 1,500 nm in length by 500 nm in diameter make it close to the size of *E. coli* (FIG. 6.9). Furthermore, some bacteria, such as *Mycoplasma, Rickettsia,* and *Chlamydia* species, are just as small as certain viruses. Considering these points, it's easy to appreciate why size is not a stand-alone criterion for classifying viruses.

Viruses are named using standardized rules.

Unlike organisms, viruses are not assigned to domains or kingdoms—the phylum level is the highest taxon for viruses. Under the phylum ranking, we find the order level followed by family (and occasionally subfamily), genus, and species.[7] Viral taxa and naming conventions are summarized in TABLE 6.2. Classification rankings below the species level are not overseen by the ICTV.

Build Your Foundation

11. What criteria are used to classify viruses? (LO 6.8)
12. Name the virus families that are medically important in humans. (LO 6.9)
13. What is meant by host range and tropism? (LO 6.10)
14. What are the naming conventions for viruses? (LO 6.11)

BYF QUICK QUIZ

Build Your Foundation (BYF) Quick Quiz: Visit the **Mastering Microbiology** Study Area to quiz yourself.

TABLE 6.2 Naming Conventions for Viral Taxonomy

Taxon	Examples	Notes
Order	*Herpesvirales*	Italicized with first letter capitalized; always ends in "virales"
Family	*Herpesviridae*	Italicized with first letter capitalized; always ends in "viridae"
Subfamily	*Alphaherpesvirinae*	Italicized with first letter capitalized; always ends in "*virinae*"
Genus	*Simplexvirus*	Italicized with first letter capitalized; always ends in "virus"
Species	*Human herpesvirus-1,* also known as *Herpes simplex virus-1*	Italicized and first name as well as proper nouns are capitalized; should not be abbreviated
Common name	human herpes virus-1 (HHV-1), also known as herpes simplex virus-1 (HSV-1)	Often the same as the species name except not italicized and only proper nouns are capitalized; may be abbreviated after an initial use of the full name

[7] Kuhn J. H. (2020). Virus taxonomy. *Reference Module in Life Sciences*, B978-0-12-809633-8.21231-4. https://doi.org/10.1016/B978-0-12-809633-8.21231-4

6.3 VIRAL REPLICATION PATHWAYS

Viruses hijack host cell machinery to multiply.

Once inside a host cell, a virus commandeers the cell's energy, enzymes, organelles, and molecular building blocks such as amino acids to build new virions. In this section, we explore the general pathways that bacteriophages use to replicate in bacterial cells and then the replication pathways that animal viruses use.

Learning Outcomes

After reading this section, you should be able to:

6.12 Explain the features of bacteriophage lytic and lysogenic replication.

6.13 Define the term *phage conversion*, and discuss why it is medically important.

6.14 Describe the generalized steps for animal virus replication.

6.15 Compare and contrast how enveloped and naked animal viruses differ in their replication pathways.

6.16 Identify key mechanisms by which viruses can cause chronic or latent persistent infections.

6.17 Explain what makes a virus oncogenic, and name examples of oncogenic viruses and the cancers they may cause.

Bacteriophages and Their Replication Processes

Bacteriophages are viruses that infect bacteria. These nonliving infectious agents are described as having a protein capsid that encloses a genome consisting of double- or single-stranded DNA or RNA—with double-stranded DNA being the most common. And although bacteriophages don't infect human cells, they're still medically important because they serve as a means for bacteria to evolve. Bacteriophages facilitate specialized and generalized transduction, which enable bacteria to develop new genetic combinations despite their inability to sexually reproduce. (See Chapter 5.) Also, bacteriophages can influence bacterial population levels—an important consideration, in light of the role that microbiome bacteria play in human physiology.

The ICTV recognizes over 2,000 bacteriophage species. Despite the incredible diversity of bacteriophages, they can all be described as performing one of two possible replication pathways—lytic or lysogenic replication.

Lytic replication Although many bacteriophages perform lytic replication, here we review the generalized process as performed by one species—the T4 phage.[8] In lytic replication, a bacteriophage infects the host bacterial cell and immediately builds new virions using a **lytic replication pathway**. This pathway kills the host cell as newly made bacteriophages are released. The bacteriophage T4 is just one of many bacteriophages that perform lytic replication. However, it is often presented as a model for this pathway because it was one of the first bacteriophages characterized. The lytic replication pathway involves five key steps, which are described next and depicted in FIG. 6.10.

1. **Attachment** (adsorption[9]): Because viruses can't move on their own, the bacteriophage and host cell usually come together from random contact. Bacteriophage tail fibers help the virus adhere to specific proteins on the bacterial cell wall surface. The specificity of these interactions means that certain bacteriophages only infect certain bacteria.
2. **Penetration** (entry): Like a hypodermic needle, the bacteriophage injects genetic material into the cell, through the host's cell wall and plasma membrane. The empty capsid remains outside the cell.
3. **Replication** (synthesis): Once the viral genome is inside, the bacteriophage commandeers host cell factors to transcribe and translate viral genes. Among the earliest viral proteins made are **DNAases** (DNA-degrading enzymes) that break up the host cell's DNA. The viral genome also encodes proteins to build new phage particles and enzymes that will copy the viral genome.
4. **Assembly** (maturation): Once all the parts of the bacteriophage are replicated, viral factors pack the viral genome into the capsid, and the remaining phage parts are assembled to build completed virions. Hundreds to thousands of new bacteriophages may be generated in this step.

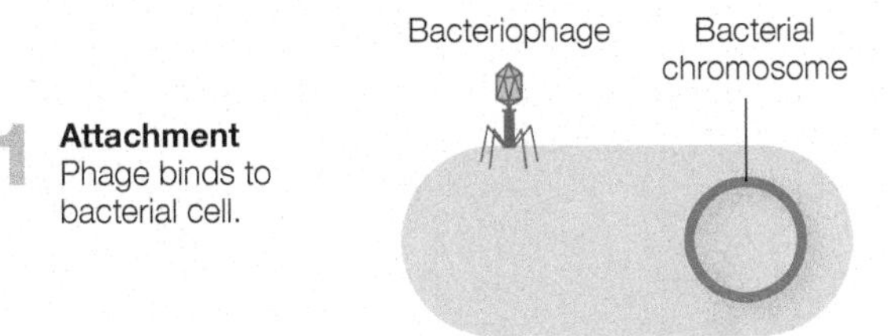

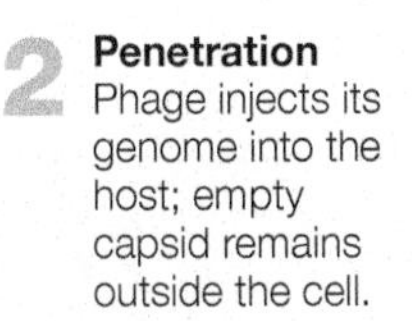

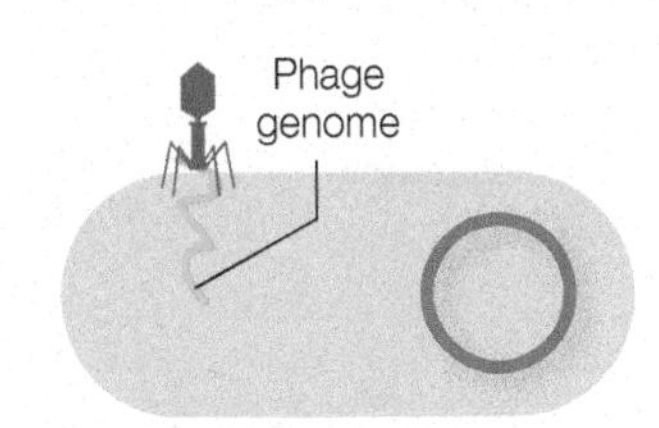

3 Replication
Protein synthesis makes phage parts and genome is replicated; host cell DNA is broken down by bacteriophage DNAases.

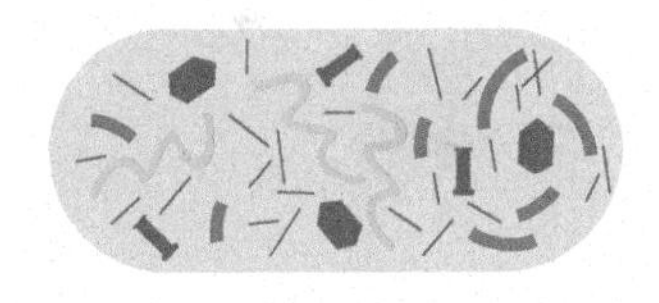

4 Assembly
Genome is packed into capsid, and phage structures are assembled.

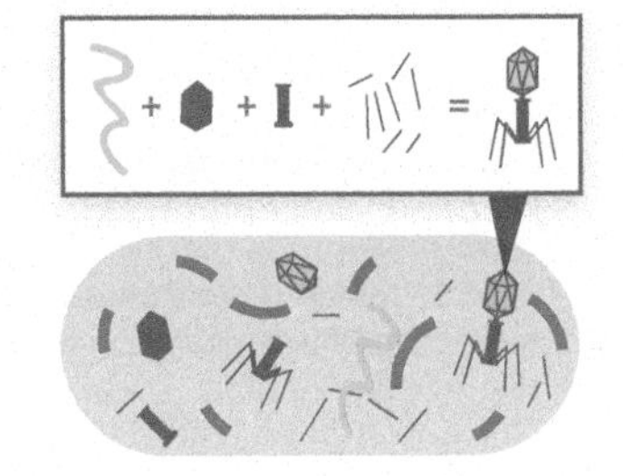

5 Release
Bacterial cell lyses, and new phages are released.

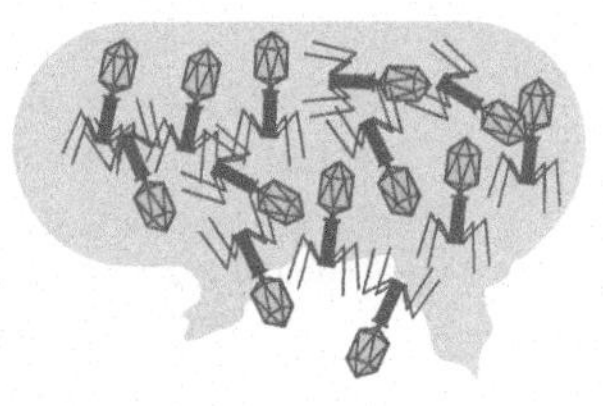

FIGURE 6.10 **Bacteriophage lytic replication**

[8] Bacteriophages types 1–7 (abbreviated as T1, T2, etc.) were some of the earliest-described phages. The bacteriophages with even numbers were called the T-even phages, and those with odd numbers were the T-odd phages. All seven of these phage types can infect *Escherichia coli*; however, numerous other bacteriophages can as well. The T4 phage (one of the original three T-even phages) has been embraced as a sort of prototypical phage—much like *E. coli* is so commonly referenced as a prototypical bacterium.

[9] Language note: It can be easy to confuse a**b**sorption with a**d**sorption; one letter changes the word's meaning. A**b**sorption is the process of soaking up/absorbing something while a**d**sorption is the process of sticking to something/adhering to something.

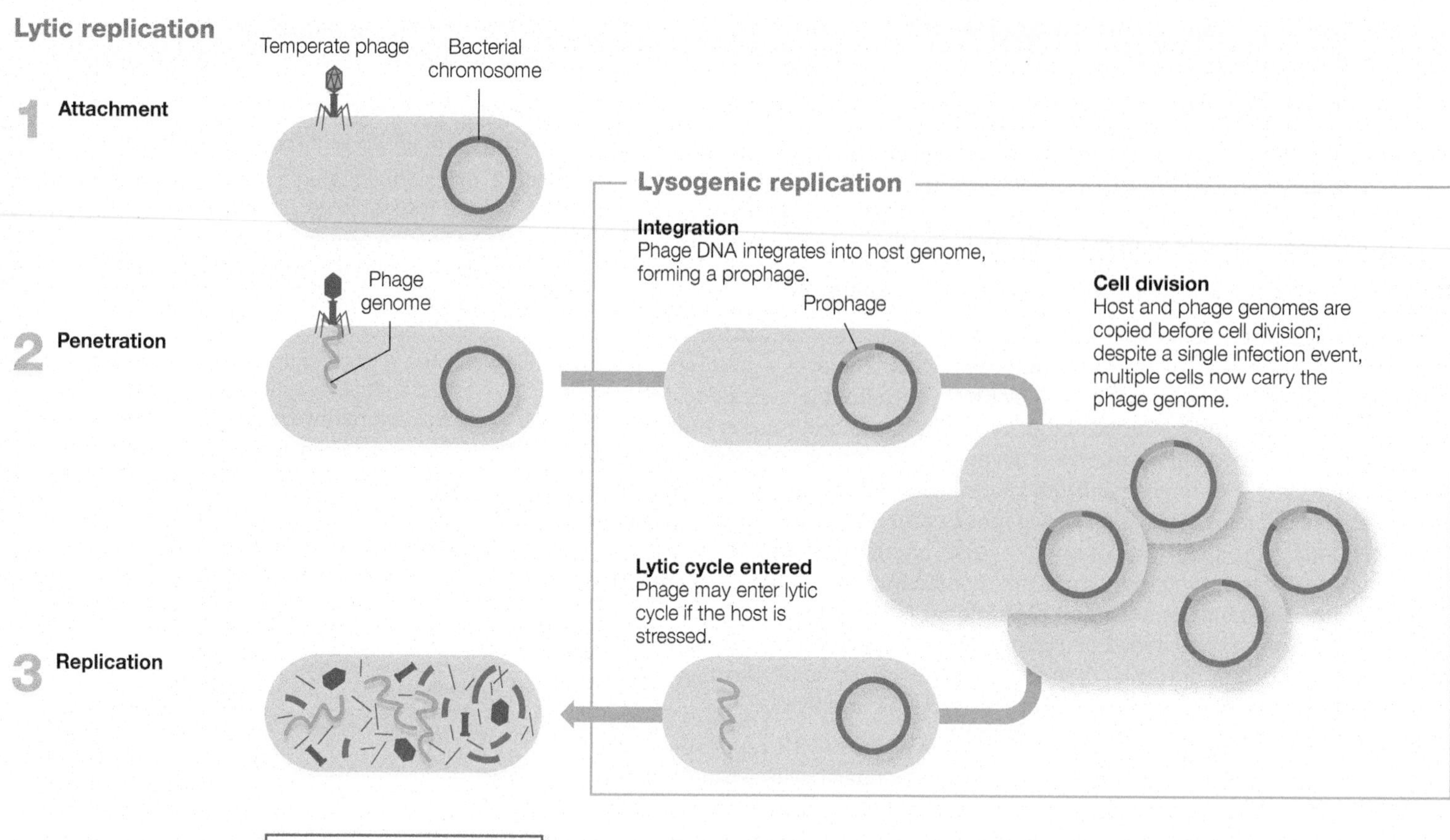

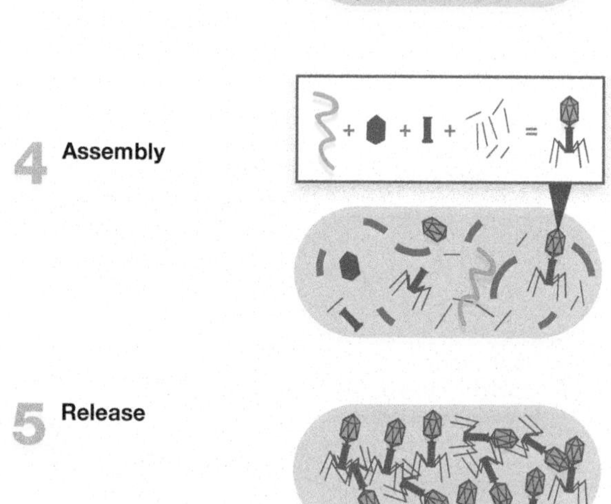

FIGURE 6.11 Temperate bacteriophage lytic and lysogenic replication

Critical Thinking *What evolutionary advantages does the lysogenic pathway confer on temperate phages?*

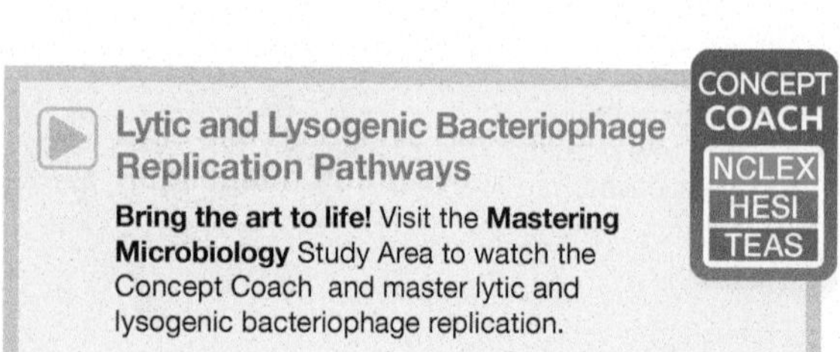

5. **Release:** Bacteriophages encode an enzyme called **lysozyme**, which breaks down host cell walls and causes bacterial cell lysis (bursting) once the newly assembled phages are mature. The released bacteriophages can then infect other cells.

Lysogenic replication Bacteriophages such as lambda phages (λ phages) are called **temperate phages**. These bacteriophages can replicate by either a lytic or a **lysogenic pathway**. The first two steps of the lysogenic cycle, attachment and penetration, are the same as in the lytic pathway. However, following penetration, the phage genome is incorporated into the host cell genome, forming a **prophage**. As the infected bacterial cell divides, it copies its own genome as well as the prophage's genome. Therefore, a single infection event ultimately results in many cells carrying the bacteriophage's genome. If a host cell carrying a prophage is stressed, the prophage may excise itself from the host genome and enter the lytic replication pathway. This "abandon ship" approach allows the bacteriophage to replicate and find a new host before its current host cell dies. The lytic and lysogenic paths that temperate phages can exploit are illustrated in FIG. 6.11.

Prophages are medically important because of their ability to confer new pathogenic properties to bacterial cells, a situation called **phage conversion**. The capacity to make certain toxins, for example, can result from phage conversion. The bacterium that causes diphtheria, *Corynebacterium diphtheria*, gained its ability to make the potent diphtheria toxin from a prophage. Similarly, the botulinum toxin made by the bacterium that causes botulism, *Clostridium botulinum*, is encoded by a prophage.

Generalized Animal Virus Replication

The generalized animal virus replication pathway has six main steps: attachment, penetration, uncoating, replication, assembly, and release. Except for the uncoating step, the names of these steps are exactly the same as for bacteriophages—however, the precise details differ. FIG. 6.12 summarizes the process.

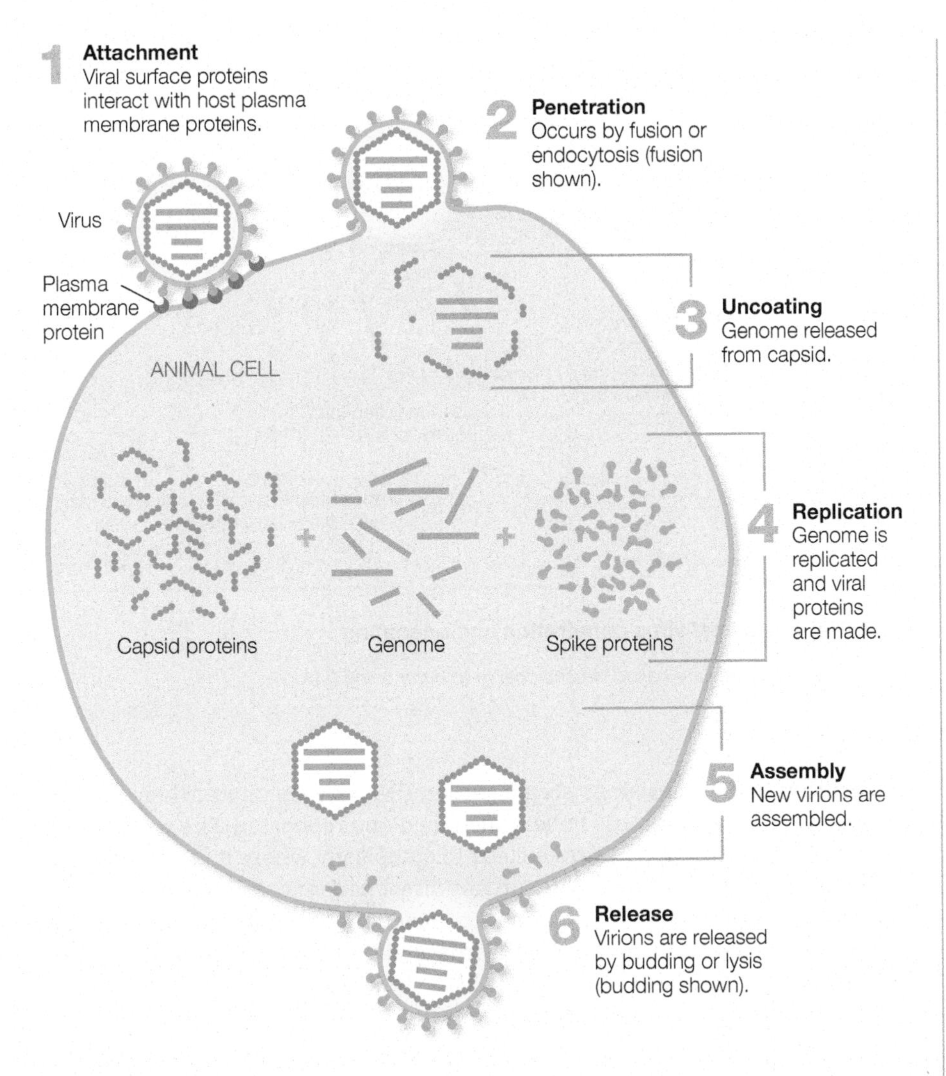

FIGURE 6.12 Overview of animal virus replication Animal viruses use six general steps in their replication. The finer specifics of these steps vary from one viral family to the next.

1. **Attachment:** Many naked viruses attach to host cell membranes through capsid proteins. Other viruses (both naked and enveloped) use spikes (FIG. 6.13). As described earlier, specificity of this binding is why viruses exhibit host range and tissue tropism. This makes attachment proteins a target for drug therapy to limit or prevent infection. One drug called maraviroc (Selzentry) works by blocking HIV's attachment to host cell proteins.
2. **Penetration (Entry):** Enveloped animal viruses enter the host cell through membrane fusion or endocytosis, whereas naked viruses mainly penetrate via endocytosis (see Chapter 4 for more on endocytosis). In membrane fusion, the cell's plasma membrane and the viral envelope blend together, and the viral capsid is released into the cytoplasm (FIG. 6.14). In endocytosis, the virus binds to host cell surface receptors, triggering uptake of the virus into vesicles.
3. **Uncoating:** Unlike bacteriophage capsids, animal virus capsids enter the host cells. The capsid is then entirely or partially broken down, releasing the viral genome. This **uncoating** varies by process and location. A virus entering by fusion may have its capsid degraded by enzymes in the host cell cytoplasm (Figure 6.14). A virus entering by endocytosis usually has its capsid digested away by enzymes in the endocytic vesicle. Poliovirus is an exception: It enters via endocytosis, but spits out its genome through a pore that forms in the capsid and the surrounding endocytic vesicle.[10] Many DNA viruses don't undergo uncoating until their capsid is safely delivered to the host nucleus.

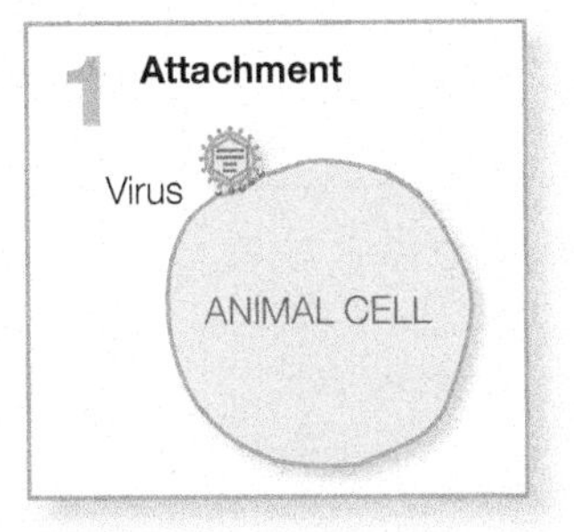

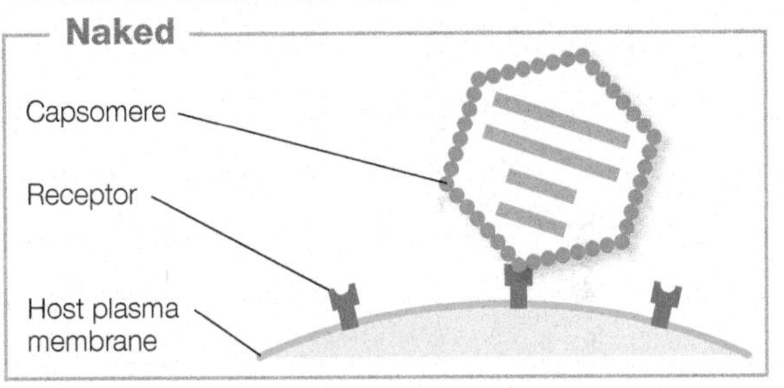

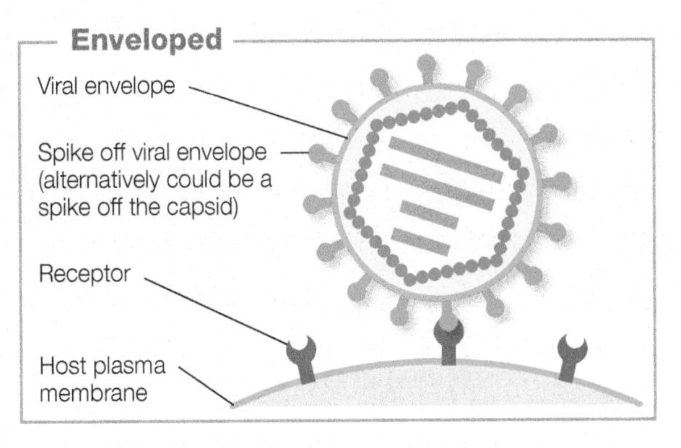

FIGURE 6.13 Viral attachment to host cells Viral envelope or capsid proteins attach to host cell proteins to facilitate attachment.

Critical Thinking *Some antiviral drugs block viral attachment. Blocking viral proteins is typically preferred to blocking host proteins. Why do you think this is the case?*

[10] Brandenburg, B., Lee, L. Y., Lakadamyali, M., et al. (2007). Imaging poliovirus entry in live cells. *PLoS Biology*, 5(7), e183. doi: 10.1371/journal.pbio.0050183 (online only).

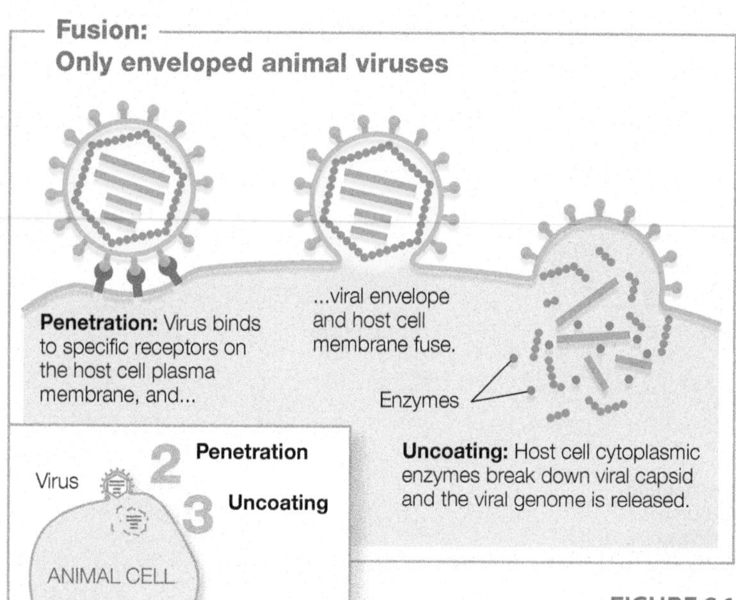

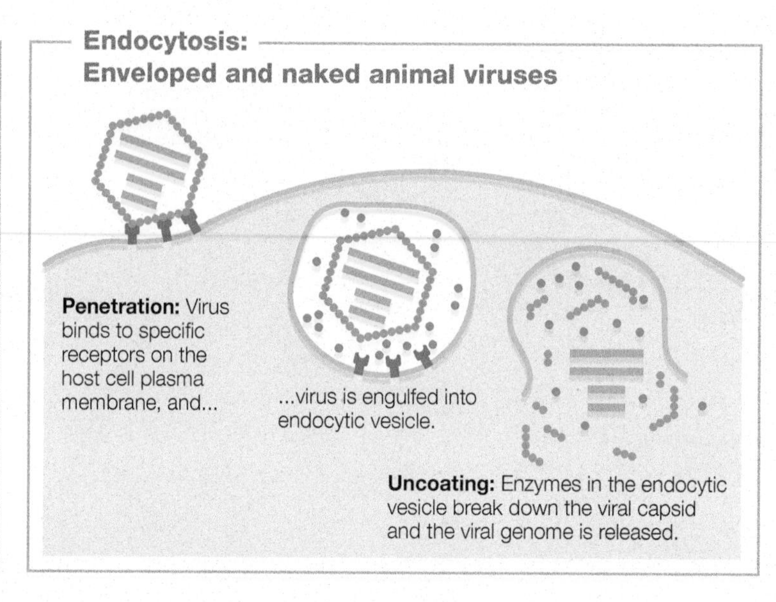

FIGURE 6.14 Animal virus penetration and uncoating

Critical Thinking *Why are naked viruses unable to enter using fusion?*

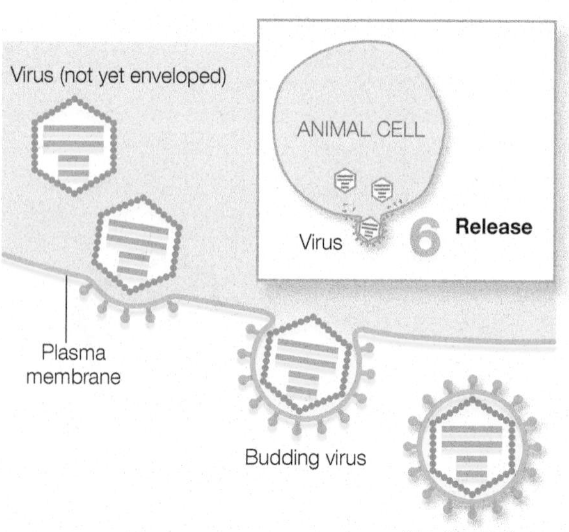

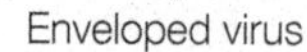

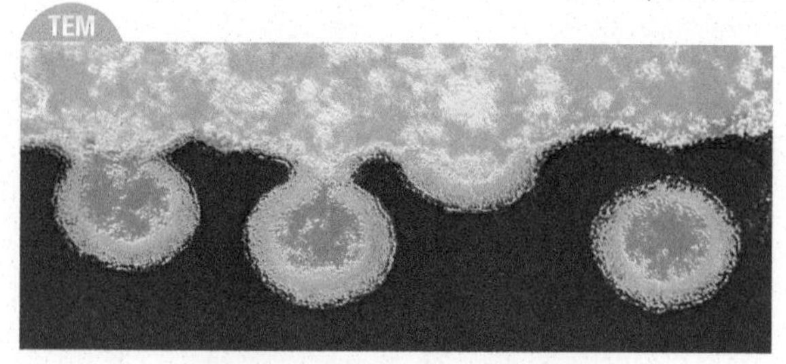

HIV budding off a white blood cell

FIGURE 6.15 Release of enveloped animal viruses Enveloped viruses are released by budding. Viral envelope proteins must be embedded in the host cell plasma membrane before budding occurs.

Critical Thinking *Although viruses that bud from the host cells do not lead to cell lysis, they are no less dangerous to the host. Explain this statement.*

4. **Replication (Synthesis):** As stated earlier, DNA viruses import their genome into the host cell nucleus to be transcribed and replicated. The resulting mRNA is then shipped from nucleus to cytoplasm, where it is translated to make viral proteins. Most RNA viruses direct genome replication and protein synthesis from the cytoplasm, never entering the nucleus. Notable exceptions include the orthomyxoviruses (which cause influenza) and retroviruses such as HIV, which are RNA viruses but have a replication process that takes place in the nucleus.
5. **Assembly:** Sometimes new capsids are partially built and packed with the viral genome before they are finished and sealed. However, usually the capsid assembles around the genome.
6. **Release:** Enveloped viruses often require viral proteins to be embedded in the host cell plasma membrane before virion release. Then, the newly assembled enveloped viruses bud off the host cell. As they do so, they usually take a portion of the cell's plasma membrane, enriched with the viral surface proteins, with them (FIG. 6.15). Naked viruses rupture the host cell during release, usually killing the cell.

Some animal viruses have unique replication mechanisms that cause persistent infections.

The steps of animal virus replication just described apply to a *productive infection*—one in which a virus infects a host cell and immediately starts making new virions. Viruses that employ this replication strategy cause **acute infections**, which run their course and are cleared by the host immune system. The common cold and influenza are examples of acute viral infections.

In contrast, some viruses have replication strategies that allow them to avoid immune system clearance. They cause **persistent infections**. Persistent viruses tend to remain in the host for long periods—from many weeks to a lifetime. Most persistent infections can be described as chronic or latent (FIG. 6.16).

Chronic Persistent Infections

Chronic infections are characterized by continuous release of virions over time and a slow progression of disease. In some cases, a period of quiet infection for months or years precedes a period of active viral replication. In the quiet period, host cells produce a small number of virions. Disease is not evident, but the patient can still pass the virus to others. Human immunodeficiency virus (HIV), which harms the immune system by killing T cells, follows this slow progressing chronic model. Those infected in the quiet stage are said to be "HIV positive." Without treatment, eventually the quiet stage ends. Virus count increases, and healthy T cell levels drop, causing loss of immune function. The end stage of HIV infection is called acquired immune deficiency syndrome, or AIDS. (For more on HIV/AIDS, see Chapter 21.)

Some viruses that cause persistent infections will integrate their genome into the host cell, forming a **provirus** that resembles the prophage made by temperate bacteriophages that we discussed earlier. The key difference is that proviruses don't excise themselves from the genome to employ the "abandon ship" approach seen in lysogenic bacteriophages. Retroviruses like HIV are notorious for forming a provirus in infected cells and then remaining silent for years before releasing sufficient virions to destroy the host's immune system (FIG. 6.17). Similarly, the papillomaviruses, which are perhaps best known for causing cervical cancer, form proviruses and generate persistent infections.

Still other viruses, like those that cause hepatitis B and C, cause persistent infections, but they do not form proviruses.

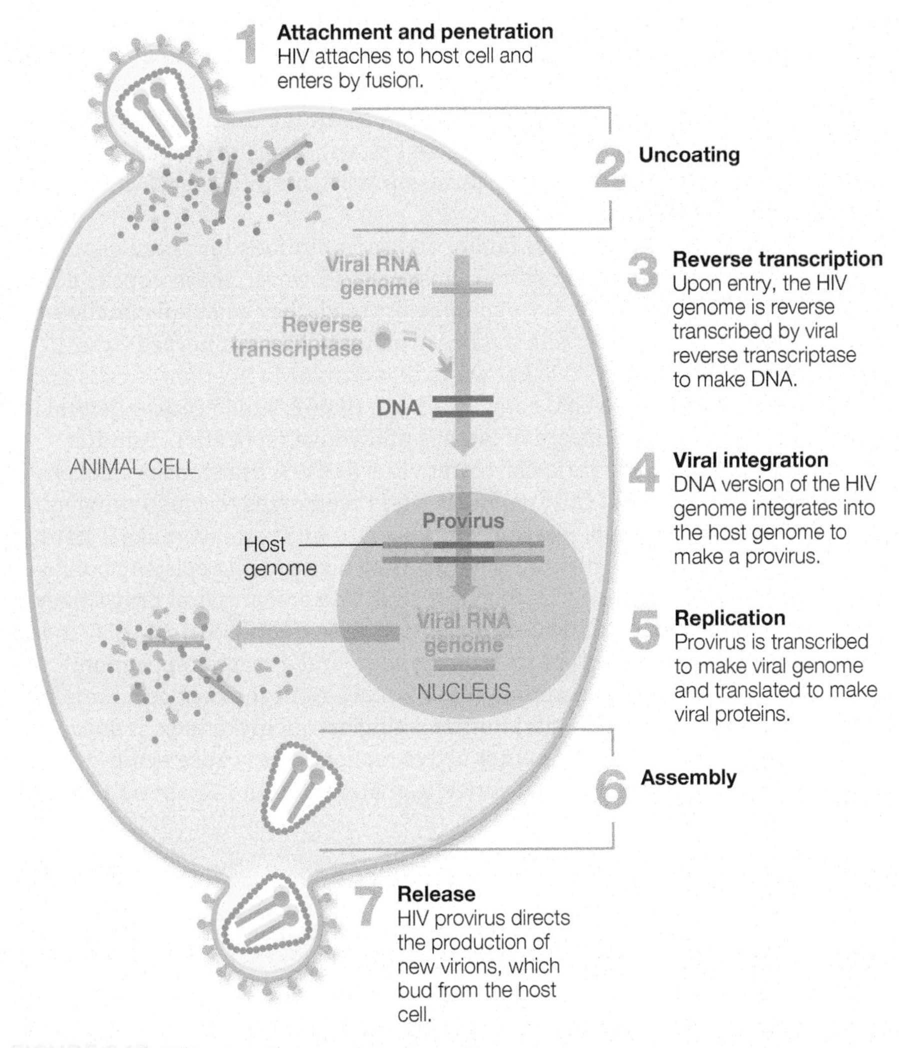

FIGURE 6.17 HIV retroviral provirus formation

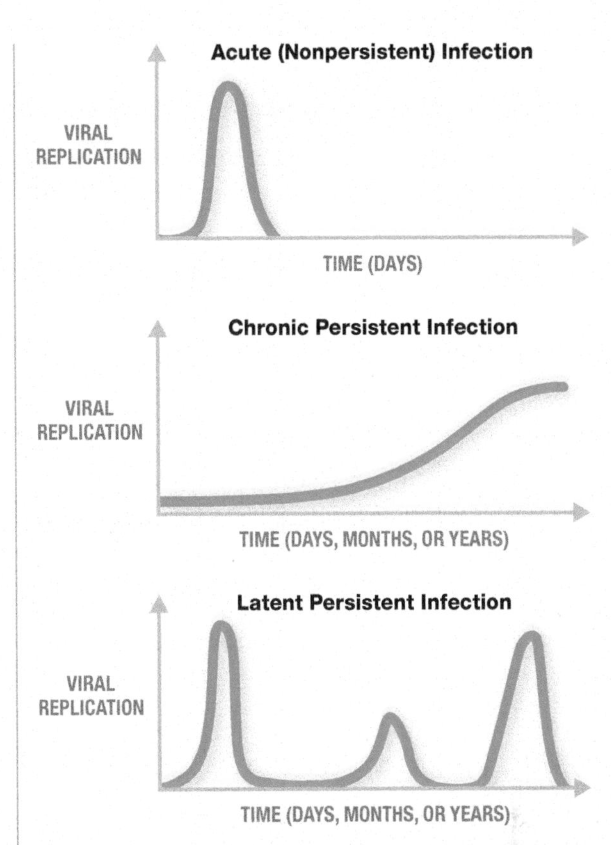

FIGURE 6.16 Acute, chronic, and latent infections (generalized graphs) In an acute (nonpersistent) infection, viral replication peaks, and then is followed by immune clearance of the virus. Chronic and latent infections evade immune clearance and persist in the host. Chronic infections exhibit steady viral production that may increase over time. Latent infections exhibit bursts of viral replication with intermittent (silent) periods.

Critical Thinking *It is impossible to make a single graph that would accurately model the precise pattern of viral replication for all latent persistent infections. Explain why.*

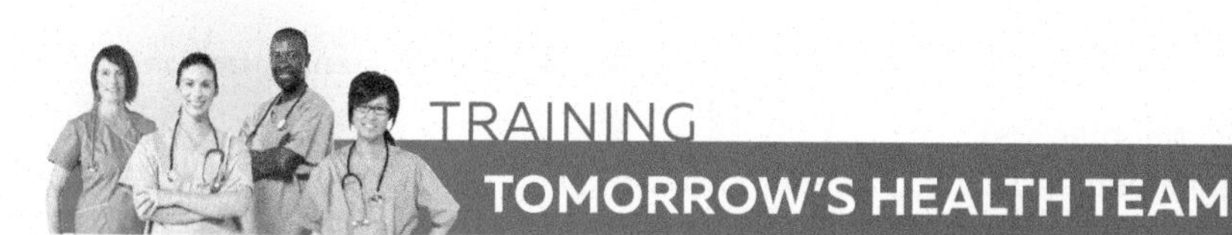

Keeping HIV Quiet

In the 1980s, when HIV first emerged worldwide, patients usually only lived for about 12 years following infection. Today, HIV-positive patients may have an almost normal life expectancy thanks to daily "cocktails" of antiviral medications.

Some HIV drugs slow viral replication by targeting the viral genome. Others directly attack viral replication proteins. One class of drug stops HIV from inserting itself into the genome of T cells. Ultimately, all these drugs aim to extend the "quiet," symptom-free stage of the disease, so that an infection doesn't progress to AIDS.

Other anti-HIV drugs interfere with the virus's ability to infect new cells. One prevents the maturation of essential viral proteins to limit the infectivity of newly made virions. Another prevents viral attachment by blocking the receptors on human immune system cells. The final drug class prevents HIV entry by blocking viral fusion with host plasma membranes.

For healthcare providers, one of the most important advances in HIV treatment is postexposure prophylaxis (PEP), or preventing the infection from starting. This is administered to people after an incident like an accidental needle stick. PEP must start within 72 hours of the initial exposure. It consists of three different drugs that are taken for at least a month.

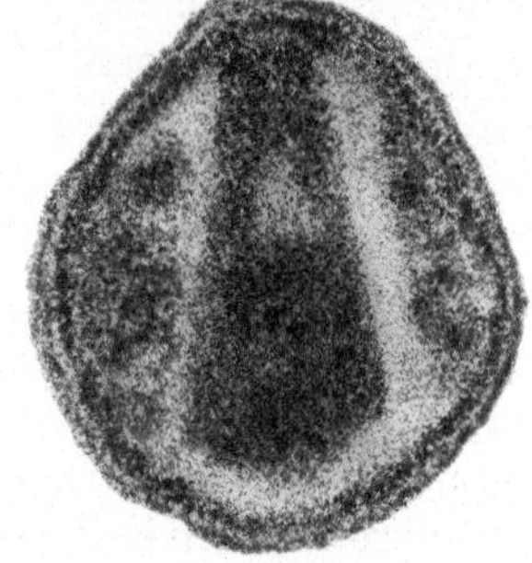

TEM of HIV

QUESTION 6.2

Based on the previous sections, what step in viral replication might a new class of anti-HIV drugs target? How might they work?

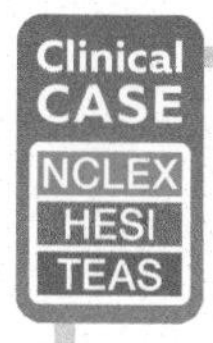

The Case of the Cancerous Kiss

Practice applying what you know clinically: visit the **Mastering Microbiology** Study Area to watch Part 2 and practice for future exams.

Latent Persistent Infections

Latent infections are distinguished by flare-ups with intermittent periods of dormancy (latency). During the flare-up, virions are shed, and the infected person experiences symptoms. When the flare-up concludes, the virus retreats into a period of inactivity, during which virion levels fall drastically and may be difficult to detect. Flare-ups can be triggered by any stress to the host, including an infection with another pathogen, fever, sunburn, hormone level changes, and immune suppression.

Members of the *Herpesviridae* family are also notorious for causing latent infections by going dormant in certain host cells. However, these viruses do not integrate themselves into the host genome; instead, they exist **episomally**—that is, their genome remains outside the host's genome. Human herpes virus-1 (HHV-1, also known as HSV-1) is known to lie dormant in host nerve cells and then reemerge to generate cold sores (FIG. 6.18). HHV-2, which causes genital herpes, follows a similar pattern of latency and active replication. Another *Herpesviridae* member, the varicella-zoster virus (HHV-3) that causes chickenpox, persists episomally in host nerve cells and reemerges to cause shingles, a painful, blistering skin rash. Intermittent flare-ups and retreats make it hard for the immune system to eliminate the virus. They also provide episodic bursts of virions that allow for transmittal to a new host. The oral antiviral drug valacyclovir (trade names Valtrex and Zelitrex) is a drug that limits HHV-1, HHV-2, and HHV-3 lytic cycle progression by interfering with viral genome replication. Although the drug doesn't eliminate these viruses from the host and therefore is not a cure for cold sores, genital herpes, shingles, or chickenpox, it does reduce the extent to which they enter active replication to cause symptoms. (See Chapter 17 for more on varicella-zoster virus and Chapters 17 and 20 for more on herpes viruses.)

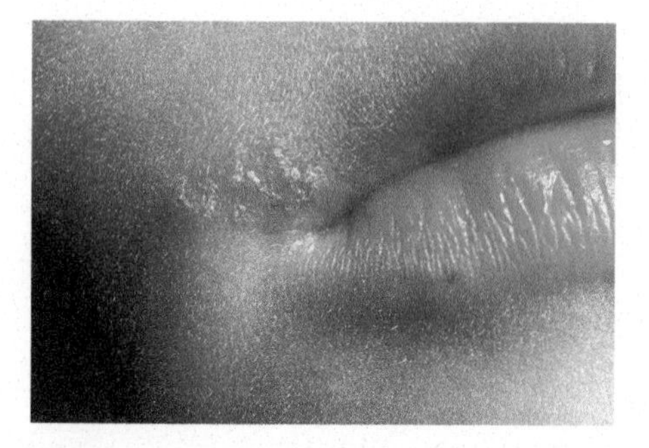

FIGURE 6.18 HSV-1 cold sores HSV-1 causes cold sores when it transitions from a latent to an active replication state. Most people have been infected with this virus by young adulthood, but most people will not have repeated outbreaks of cold sores.

Persistent Infections That Can Lead to Cancer

Viruses that cause persistent infections are often associated with cancer. Those that can cause cancer are called **oncogenic viruses** (or oncoviruses). They include a number of DNA viruses as well as the RNA viruses. Some of them integrate into the host genome, while others are maintained episomally—either in the nucleus or cytoplasm. The six human viruses that are associated with cancers are

TABLE 6.3 The Six Main Human Viruses Associated with Cancers

Virus	Viral Genome	Integrates with Host Genome	Cancer Link	Cancer-Causing Mechanism
Human papilloma viruses (HPVs)	DNA; *Papillomaviridae* family	Yes	Cervical, oropharyngeal, anal, and rare vaginal and penile cancers	Viral genes cause uncontrolled cell division
Human herpes virus-8	DNA; *Herpesviridae* family	No	Kaposi sarcoma	
Epstein–Barr virus (also causes mononucleosis)	DNA; *Herpesviridae* family	No	Associated with a number of malignancies to include B cell and T cell lymphomas and Hodgkin's disease	
Human T-lymphotropic viruses (HTLVs)	RNA; *Retroviridae* family	Yes	Adult T cell leukemia	
Hepatitis B virus	DNA; *Hepadnaviridae* family	No	Liver cancer (hepatocellular carcinoma)	Hypothesized that chronic inflammation from the virus triggers host cell DNA damage and mutations, leading to cancer
Hepatitis C viruses	RNA; *Flaviviridae* family	No		

reviewed in TABLE 6.3. Oncogenic viruses are thought to cause between 10 and 15 percent of cancers. In general terms, oncogenic viruses cause cancer by stimulating uncontrolled host cell division and/or decreasing host cell responsiveness to death signals. Cells that ignore death signals are said to be "immortalized." Unlike normal cells that have a limited life span and can be stimulated to undergo apoptosis (cell suicide), immortalized cells may undergo repeated division and survive indefinitely. It should be noted that infection with an oncogenic virus increases risk, but does not guarantee cancer will develop.

Some of the most well-known oncogenic viruses are the **human papilloma viruses** (HPVs). There are over 200 different types of HPVs, at least 40 of which spread through sexual contact. The vast majority cause benign warts. About a dozen HPV types have been linked to cancer, with HPV-16 and HPV-18 causing the majority of cases. Even the HPV viruses most associated with cancer are usually cleared by the immune system within 1 or 2 years without further problems. Nonetheless, the U.S. Centers for Disease Control and Prevention (CDC) reports that about 90 percent of all cervical cancer cases are linked to HPV. The CDC therefore recommends that all children 11–12 years old receive the Gardasil vaccine that protects against the HPV types most frequently associated with cancer. (For more on HPV and cervical cancer, see Chapter 20.)

Other oncogenic viruses include the **human T-lymphotropic viruses** (HTLV). HTLV-1 and HTLV-2 are clinically important. These retroviruses form a provirus and can quietly persist in host cells for more than a decade before emerging to cause leukemia or lymphoma. Worldwide, up to 20 million people are infected with these viruses.[11] The primary modes of transmission are through breast milk, sexual contact, and blood contact.

Build Your Foundation

15. Compare and contrast lytic and lysogenic bacteriophage replication. (LO 6.12)
16. How does phage conversion impact bacterial pathogens? (LO 6.13)
17. What are the steps in animal virus replication, and what occurs in each step? (LO 6.14)
18. How does the presence or absence of an envelope impact viral entry and release? (LO 6.15)
19. Compare and contrast latent and chronic infections. (LO 6.16)
20. Give three examples of oncogenic viruses and the cancers they cause. (LO 6.17)
21. In general, how do oncogenic viruses generate cancer? (LO 6.17)

Build Your Foundation (BYF) Quick Quiz: Visit the **Mastering Microbiology** Study Area to quiz yourself.

[11] Amar, L., Le, M., Ghazawi, F. M., et al. (2019). Prevalence of human T cell lymphotropic virus 1 infection in Canada. *Current Oncology (Toronto, Ont.)*, 26(1), e3–e5.

6.4 CLINICAL ASPECTS OF VIRUSES

Learning Outcomes

After reading this section, you should be able to:

6.18 Discuss the various laboratory methods for growing bacteriophages and animal viruses.

6.19 Explain what the plaque assay is and why it is useful.

6.20 Describe several methods for detecting viral proteins and genetic material, and state their advantages and limitations.

6.21 Explain the different drug approaches to managing viral infections, and name several antiviral drugs.

Viruses can be grown in the laboratory.

In order to develop vaccines and drugs to combat viruses, researchers must be able to grow viruses in a laboratory setting. Here we review some of the basic techniques for cultivating viruses.

Growing Bacteriophages

Bacteriophages can be grown in the laboratory with relative ease. While specific techniques vary, generally all that's needed is a combination of the desired bacteriophage, an appropriate bacterial host cell, and the right medium to support the growth of the bacteria. The bacteria can be grown and infected with bacteriophages in a liquid broth culture or using solid agar in a petri plate. The advantage of the solid agar approach is that it allows for a technique called the **plaque assay**. When lytic bacteriophages lyse out of the host cell at the end of their replication pathways, they kill the host cell. This killing leaves a clear zone called a *plaque* on the growing plate of bacteria (FIG. 6.19). A higher initial virus level will lead to more plaques following incubation. In theory, each plaque represents a single bacteriophage in the initial sample. Thus, the quantity of bacteriophages in an initial volume of sample can be noted in **plaque-forming units** (PFUs). The plaque assay is also adaptable to animal viruses and is a useful way of determining how much of a given animal virus is present. In addition to determining **viral titer**, or quantity of virus present in a given volume of sample, the plaque assay can be used to purify specific viruses. Some limitations of the plaque assay are that the virus must cause cell lysis. Also, the cells used as the host must grow in a plate format—this is a potential barrier because some cells require a broth media format and will not grow on plates.

Growing Animal Viruses

Animal viruses are more difficult to cultivate than bacteriophages. Most animal viruses are grown using tissue culture techniques. A variety of commercially available human and animal cell lines can be used to support viral replication. For example, HeLa cells, which are cancer cells that were derived from a patient named Henrietta Lacks, are among the most common commercially available human cells used in research today. Unfortunately, commercially available cell lines don't always support the growth of a given virus. This is because cells that are well adapted to being cultured outside of the animal they originally came from may lose the ability to make specific surface factors the virus requires for attachment. In this event, primary cell lines derived directly from the preferred host can often be used. For example, white blood cells from donated blood can serve as a primary cell line, as can samples from tissue biopsies. A lot of specialized equipment, technical experience, and reagents go into supporting tissue culture, making this an expensive and time-consuming endeavor. Live animal hosts such as mice, rats, guinea pigs, or other animals may also be required to support the study of certain viruses. Animal models are essential to clinical trials for antiviral drug and vaccine development and are unlikely to ever be fully replaced by tissue culture methods.

Embryonated eggs (usually fertilized chicken eggs) are also useful for growing certain viruses. Active virions are injected into the egg, and then the

1 The initial virus stock is sequentially diluted.

0.1 ml

0.1 ml

Concentrated virus stock

Dilute virus stock

2 A portion of the diluted samples is mixed with bacteria and melted agar and poured into a petri plate.

3 Clear zones, or plaques, form where host cells are killed by viruses.

4 Following incubation, plaques are counted and the initial viral titer is calculated and presented as PFU/ml.

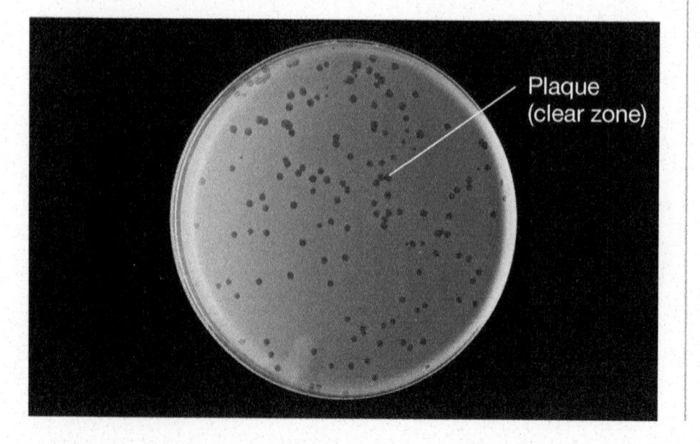

FIGURE 6.19 Plaque assay Only diluted samples are plated because high concentrations would kill all cells.

Critical Thinking *Would the plaque assay be useful for calculating lambda phage titers? Explain your reasoning.*

egg is incubated to allow for viral replication. The site of injection in the egg determines what cells in the embryonated egg become infected; this in turn affects viral replication. Thus, the region of the egg that should be injected depends on the viruses being grown (FIG. 6.20). The resulting virions are purified from the eggs and used for a number of applications, for example, in certain vaccines. Because it is possible that miniscule amounts of egg protein will remain in the purified viral preparation, people allergic to eggs can't be vaccinated with this type of vaccine. This is why patients are asked about allergies to eggs when getting certain vaccines. (See Chapter 14 for more on vaccines and their development.)

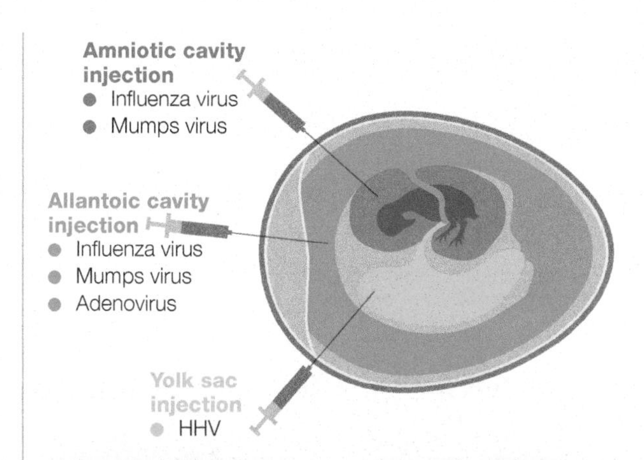

Holding a light up to the egg allows for specific location of the areas for injection.

FIGURE 6.20 Using embryonated eggs to grow viruses

Critical Thinking *Not all viruses can be grown in chicken eggs. In general terms, give a reason why this is the case.*

Diagnostic tests determine the presence of certain viruses.

The ability to detect the presence of a virus is central to studying them, but it's also important for determining appropriate patient care. Accurate diagnostics are necessary to ensure safe, virus-free transplant tissues and pharmaceutical products; and analyzing clinical samples (tissue, blood, saliva, etc.). With such diversity in what needs testing, it stands to reason there are many methods available for identifying viruses.

Because viruses do not have their own biochemical processes and aren't viewable via standard light microscopy, most detection techniques use molecular methods to identify viral genetic material, viral proteins, or antibodies that a patient may have against viral proteins. When a clinical sample (sputum, cerebrospinal fluid, blood, etc.) is sent to the lab, it is usually filtered or centrifuged (rapidly spun in a specialized machine) so that only particles as small as viruses are left in the test fluid. Then the fluid is subjected to diagnostic tests or used to inoculate cell lines.

The most clinically useful detection tests are specific, are sensitive, and have short turnaround times. **Specificity** means that the diagnostic test reliably detects only the virus(es) of interest without producing false-positive results. **Sensitivity** means the test detects very low levels of the target to limit false-negative results. Finally, it goes without saying that a highly specific and sensitive test that returns results after the patient is permanently impaired or dead is not the best clinical tool. The shortest turnaround time possible for receiving results is important for any diagnostic test.

Detecting Viral Proteins

Some of the most common virus detection methods involve searching for viral proteins in a sample. These techniques tend to rely on purified antibodies, which can recognize and bind to even miniscule amounts of a target viral antigen. When properly developed and quality control evaluated, tests based on antigen–antibody interactions can be highly sensitive as well as specific. Here we'll review some of the most common of these tests, also called immune assays, for detecting viruses. There are commercially available kits using these methods for most common viruses. These types of tests are also adapted for detecting other pathogens and are explored further in Chapter 14.

Agglutination tests Purified antibodies can be linked to tiny latex beads that are no larger than a fine grain of sand. These antibody-coated beads can then be mixed with a sample. If the sample contains the viral antigen being sought, then the antibodies will bind to the antigen and the beads will clump, or **agglutinate**. The top of FIG. 6.21 shows how this agglutination occurs. This version of the latex agglutination test is useful for determining if a sample contains a specific virus. It can be used to screen bodily fluids such as urine, cerebrospinal fluid, or blood for specific viruses. However, this form of the test is useless if virions are not actively being produced in a patient.

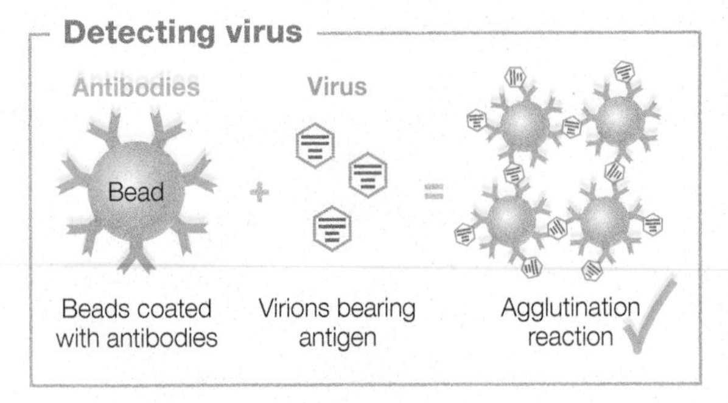

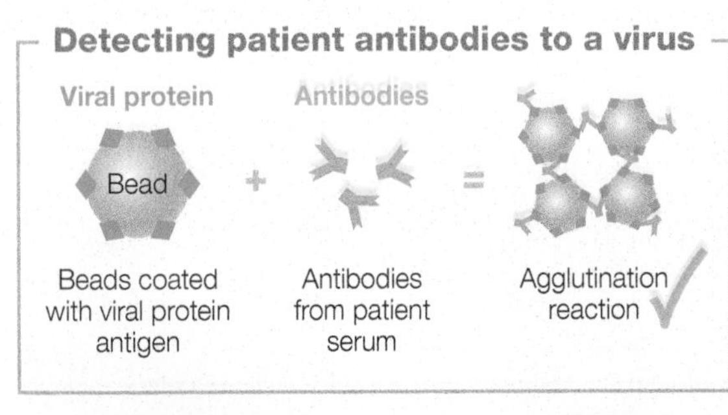

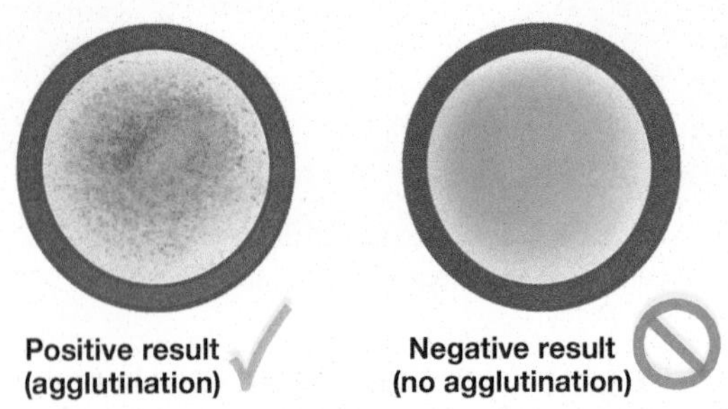

FIGURE 6.21 **Latex agglutination test to detect antigens or antibodies** Latex beads linked to antibodies can be used to detect the presence of viruses in samples. Modifying the beads so that they are linked to viral proteins (viral antigens) is useful for determining if a patient has had an infection with the virus in question.

Critical Thinking *Which version of the latex agglutination test would be used to determine if a patient who is no longer exhibiting symptoms had been exposed to the varicella-zoster virus?*

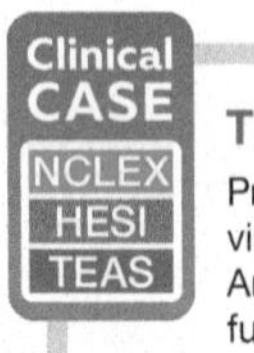

The Case of the Cancerous Kiss

Practice applying what you know clinically: visit the **Mastering Microbiology** Study Area to watch Part 3 and practice for future exams.

Fortunately, a minor modification of the agglutination test makes it possible to determine if a prior exposure to a virus occurred—even if the virions in question aren't actively present in the patient. Our immune system "remembers" antigens and can make the same antibodies against them later, even if the virus has long since been cleared from the host or is latent. So, testing for an antibody response to a specific antigen serves as evidence of a prior infection. The **latex agglutination test** is one commonly performed test that can do this. To perform this test, latex beads are coated with viral proteins. If the patient has antibodies to the viral proteins coating the beads, then the beads will agglutinate when exposed to the patient's serum (the part of blood that contains antibodies). The bottom of Figure 6.21 shows how this agglutination occurs.

Enzyme-linked immunosorbent assays (ELISAs) Another clinical mainstay in pathogen detection is the ELISA assay. Like agglutination assays, ELISA methods exploit the specificity and sensitivity of antibody–antigen interactions, and they can be adapted to detect either antigens or antibodies in a sample. Instead of clumping beads, ELISAs adhere the antigen or antibody to a surface and usually change color if there is binding. (ELISA methods are further discussed in Chapter 14.)

Limitations of agglutination tests and ELISAs Most viruses that cause human disease can be detected using an agglutination assay or ELISA. As useful as these tests are, they have some important limitations. In both cases the sample being tested must be a liquid, and the antigens being detected must be fairly well characterized to ensure detection specificity. Another major consideration is that if a virus undergoes an antigenic shift, or even antigenic drift in just the right antigen, it may no longer be detectable using existing agglutination or ELISA methods.

Furthermore, tests that detect patient antibodies are ineffective during early infection because it takes at least a couple of weeks following infection for detectable antibodies to develop, a time period called the **seroconversion window**. For HIV, the seroconversion window can be several weeks to months. This makes methods that detect patient antibodies against HIV an unreliable tool to assess infection status. For this reason, it is helpful to use a combination of detection methods, which brings us to our next topic: detecting viral genetic material.

Detecting Viral Genetic Material

Genetic sequencing has broadened our ability to detect viral infections before seroconversion. Tests for detecting viral nucleic acids are a growing trend in diagnostics because they are more sensitive and sometimes more rapid than antigen–antibody-based tests. Also, they can detect new viruses and early-stage infections that antigen–antibody tests are likely to miss. This is why the earliest tests for SARS-CoV-2 were based on detecting viral genetic material.

To perform the test, nucleic acids are extracted from a clinical sample such as sputum, blood, cerebrospinal fluid, or tissue. Then very specific segments of viral nucleic acid, usually those coding for a unique viral gene, can be detected by using fluorescent-labeled probes, by sequencing the nucleic acids, or by carrying out a process called PCR (polymerase chain reaction) that can amplify specific parts of a genetic sequence. A modification of the PCR procedure called RT-PCR (reverse transcription PCR) can detect viral RNA and remains the gold standard for detecting active SARS-CoV-2 infections. (Nucleic acid detection methods are discussed in more detail in Chapter 14.)

Antiviral drugs treat infections, but don't typically cure them.

Most viral diseases are self-limiting and do not require treatment. But others, like HIV or certain influenza strains, cause serious disease and warrant therapy. Because viruses are not cells, viral infections are not cured by antibacterial

TRAINING
TOMORROW'S HEALTH TEAM

Detecting Active SARS-CoV-2 Infections

Two categories of testing dominate the public health options for detecting active SARS-CoV-2 infections: reverse transcription-PCR (RT-PCR) tests and antigen tests. The RT-PCR test is designed to detect the genetic material of SARS-CoV-2 (viral RNA). In contrast, antigen tests are designed to detect SARS-CoV-2 proteins. Note that neither test can detect a prior infection. Instead, these tests are only useful for detecting an active infection.

Early in the pandemic, most health departments require a negative RT-PCR test result as evidence that an individual was not infected and could be released from quarantine; a negative antigen test result was insufficient to clear quarantine. On the other hand, either a positive RT-PCR test result *or* a positive antigen test result constituted sufficient evidence to require a quarantine period. If the antigen test was satisfactory to prove infection status and require quarantine, then why wasn't it considered good enough to release a person from quarantine? The answer lies in the sensitivity differences for each test.

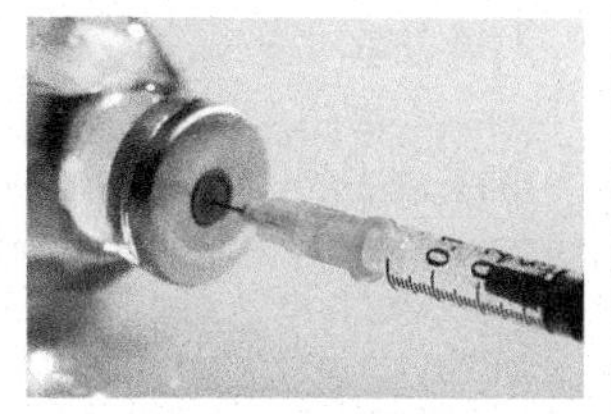

In general, antigen tests are less sensitive than RT-PCR tests. As compared to a RT-PCR test an antigen test is more likely to yield a false-negative result (suggesting you don't have the virus when you actually do). Conversely, a positive result from either test can justify quarantine because the tests are comparably specific. A positive RT-PCR result or a positive antigen test result are equally likely to reflect true infection status. The bottom line is that any patient with COVID signs and symptoms should assume they could have COVID—even if they had a negative antigen test. The RT-PCR test remains the gold standard of clearing a person as COVID-free.

QUESTION 6.3

As noted briefly in the discussion, neither RT-PCR tests nor antigen tests can detect a prior infection. Why does this make sense?

drugs (e.g., antibiotics), which are designed to target bacterial cell structures and biochemical processes. (See Chapter 15 for more on the dangers of using antibiotics to treat viral infections.) However, any step in the viral replication pathway is a potential drug target. As such, basic research into how a given virus gains entry into a host cell and then replicates is essential to developing antiviral drugs. Studying how our own immune systems successfully end certain viral infections also gives us a model for developing antiviral treatments. Research has led to the introduction of more and more antivirals to treat diverse viral infections; however, most antiviral drugs only limit rather than cure infections. To date, the one exception is hepatitis C, which is now curable with antiviral drugs. Special difficulties come into play when designing antiviral drugs. Because viral pathogens exist inside host cells much of the time, drugs cannot always reach them. And to be an effective drug rather than just a poison, antivirals must be selectively toxic, stopping the pathogen but leaving host cells unharmed. Ideally, a drug targets processes that are unique to the pathogen—but that goal is difficult to achieve when viruses use the cell's own machinery and metabolism. Lastly, being so simple in their structure, viruses have fewer chemically distinct targets than living pathogens like bacteria, protists, or fungi. The difficulty in developing antiviral drugs is the main reason we have so few effective antiviral agents and why prevention of serious viral diseases through vaccination is so incredibly important. Vaccines train the immune system to recognize viruses and are effective means to limit infection. (Vaccines are covered in more detail in Chapter 14.)

Drugs that Block Viral Attachment, Penetration, and Uncoating

A number of antiviral drugs prevent viral entry into cells by blocking attachment or penetration (FIG. 6.22). Antibodies can block these early steps in viral replication and can therefore be administered to combat active infections. For example, during the COVID-19 pandemic, antibodies from survivors were transfused into severely ill patients to improve patient outcomes. Antibody therapies are also commonly administered to prevent infection. For example, a laboratory-prepared mixture of injectable antibodies to the rabies virus (known as

Attachment blocked by:
- HRIG (Rabies)
- Maraviroc (HIV)

Penetration blocked by:
- Interferon-alfa (HBV, HCV)
- Enfuvirtide (HIV)
- Docosanol (HHV-1)
- Palivizumab (RSV)

Uncoating blocked by:
- Amantadine (influenza)
- Rimantadine (influenza)
- Vapendavir (rhinoviruses)

Replication and assembly blocked by:
- NRTIs/reverse transcriptase inhibitors like AZT (HIV)
- Ribavirin (RSV, HCV, and certain hemorrhagic fevers)
- Protease inhibitors (HIV)

Release blocked by:
- Oseltamivir (influenza)
- Zanamivir (influenza)

FIGURE 6.22 **Mechanisms of action of select antiviral drugs** Every aspect of viral replication is a potential target. Ideally, antiviral drugs should not induce significant damage to host cells. Note the acronyms for hepatitis B virus (HBV), hepatitis C virus (HCV), human herpes virus (HHV), and respiratory syncytial virus (RSV).

Critical Thinking *Most influenza A strains are now resistant to amantadine. What do you think contributed to this evolutionary change in the virus?*

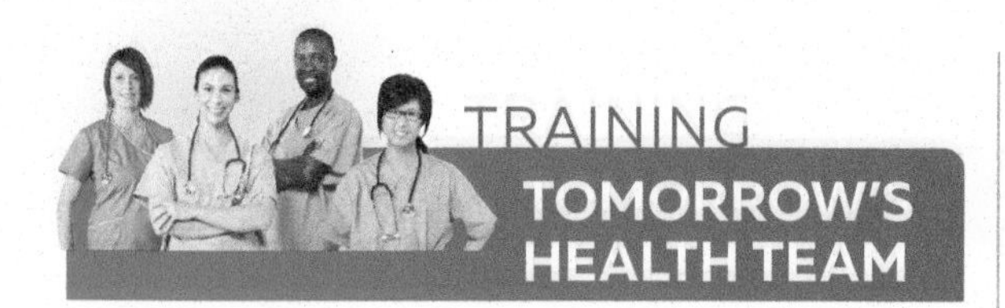

The Virome

When we hear the word "microbiome," we tend to think about bacteria. But viruses are part of the microbiome, too. Our "virome" includes bacteriophages that infect the bacteria that live within us and viruses that infect our own cells.

Bacteriophages are best known for conferring virulence genes to bacteria. For example, a phage gives *Corynebacterium diphtheriae* the ability to produce the toxin that makes human diphtheria infections so deadly. But phages may indirectly affect a human host. Some studies suggest they kill off bacteria that rapidly overtake a niche in the human body, serving as a buffer against changes in our bacterial microbiomes. Also, phages bind to the epithelial layers of human gastrointestinal tracts, potentially preventing pathogenic bacterial adhesion. Some preliminary studies also show that when combined with antibiotics, phages may help control the emergence of antibiotic-resistant bacteria.

Furthermore, viral genes make up as much as 1.5 percent of the human genome. Some of these viral elements stimulate the immune system, while others increase our risk for certain cancers.

QUESTION 6.4

Given what you know about the evolution of antibiotic resistance, why would it be helpful to administer bacteriophages alongside an antibiotic?

Active form of acyclovir that when added to DNA will halt replication

Guanine nucleotide that is normally added to DNA

FIGURE 6.23 Active form of acyclovir Inside infected cells, viral enzymes activate acyclovir to the represented form. The activated drug is very similar to guanine nucleotides that are the "G" of the G-C pairs in DNA.

HRIG, human rabies immunoglobulin) prevents the virus from binding to and entering host cells. This therapy is a key **postexposure prophylaxis**, a prevention treatment applied after an exposure to limit infection. It is administered if a rabid animal bites a person who has not been vaccinated against rabies. While rabies is vaccine preventable, the vaccine is not routinely used except in people at an increased risk for exposure, such as veterinarians or people who research the rabies virus. Even if the rabies vaccine were administered immediately upon a suspected exposure to a rabid animal, the patient's immune system would need weeks to make sufficient antibodies to protect the patient from infection—and in the meantime, the patient would likely die. Having a viral attachment blocker is therefore an important part of the medical toolkit against rabies. (See Chapter 18 for more on rabies.)

Blocking viral penetration is a key strategy in postexposure prophylaxis against HIV. The injected peptide drug enfuvirtide is administered to people who suspect they have been exposed to HIV, such as a healthcare worker who gets an accidental needle stick while caring for an AIDS patient. The drug binds to proteins essential for HIV to enter new cells. It works by blocking fusion of the viral envelope with host cell membranes.

Docosanol is not a postexposure prophylactic, but, like enfuvirtide, it blocks viral entry into host cells. As a topically applied agent that blocks viral fusion, its main clinical use is to treat cold sores caused by HHV-1. Lastly, palivizumab is an injectable antibody preparation that blocks fusion of the respiratory syncytial virus (RSV)—a virus that causes potentially dangerous respiratory distress in premature babies.

If viral entry can't be targeted, then targeting other steps that precede viral replication may be useful. One drug that is in late-stage clinical trials to prevent the common cold is vapendavir. This orally administered drug includes a small compound that binds to capsid proteins of rhinoviruses to inhibit viral uncoating. Thus, even if the virus enters host cells, the viral genome is not accessible for replication. Other agents, such as chloroquine and hydroxychloroquine, which are currently used to treat malaria and certain autoimmune disorders, affect endosomal environments (see Chapter 4 to review endosomes) and may limit viral penetration. Because cell-based studies (*in vitro* studies) conducted by several laboratories supported the potential usefulness of these drugs against coronaviruses, in 2020 a number of *in vivo* studies (organism-level studies) were established to test them as potential therapies for treating SARS-CoV-2 infection. Unfortunately, these agents do not statistically improve patient outcomes, nor do they prevent infection. It is not unusual for a drug to appear promising *in vitro* only to fail *in vivo*. This is why experts caution that *in vivo* studies needed to be done before conclusions about any drug's efficacy can be made.

Drugs that Target Viral Replication, Assembly, and Release

The most common drugs that block replication are **nucleoside analogs**. There are at least a dozen drugs in this class, but all of them tend to work by blocking successful nucleic acid production to limit viral replication. In certain virus-infected cells, nucleoside analogs are activated into compounds that mimic normal *nucleotides* (adenine, guanine, cytosine, thymine, and uracil). However, these analogs are chemically different from natural nucleotides and represent a chemical dead end for nucleic acid replication. If a DNA or RNA polymerase uses an analog in place of a natural nucleotide, nucleic acid replication will be interrupted.

One of the best-known nucleoside analogs is the compound acyclovir, the active ingredient in brand-name antiviral drugs such as Zovirax and Valtrex (FIG. 6.23). This compound blocks DNA replication in cells infected with HHV-1, HHV-2, or varicella-zoster virus (HHV-3). Blocking DNA replication sounds like a dangerous prospect, and it definitely would be if this drug did not specifically target cells infected with these viruses. Fortunately, only infected cells chemically modify the drug to activate it. The activated drug mimics guanosine nucleotides that DNA polymerases use to make DNA. If the activated analog is incorporated, it prematurely ends DNA replication. The activated form of the drug preferentially interacts with viral DNA polymerases over host DNA polymerases, but that preference is not absolute, so the drug can lead to collateral host cell damage—killing some infected host cells, but not enough to generate a level of tissue damage that renders the cost-benefit of the drug null. All drugs have the potential to damage host cells and/or cause undesirable side effects, which is why extensive efficacy and safety testing is necessary. Of course, some drug therapies are ultimately found to inflict too much damage on host cells and are deemed overly toxic for practical use (although some may be reserved for last resort therapies that are administered under tightly regulated, in-hospital dosing). We discuss drug testing and safe dosing levels in Chapter 15.

Whereas acyclovir targets DNA-based viruses, another nucleoside analog, ribavirin, targets RNA polymerases and is used to combat certain RNA viruses. It is mostly effective against RSV and hepatitis C virus; it may also have some efficacy in early infections with Lassa virus, hantavirus, and some others that cause certain hemorrhagic fevers. Unfortunately, it is not effective against hemorrhagic fevers like Ebola. Another nucleoside analog drug that targets viral RNA polymerases is remdesivir (trade name Veklury). This drug was initially developed to treat hepatitis C, but was not effective. It was then repurposed to treat Ebola, but was less effective than other options. Starting in 2019, healthcare providers began using remdesivir to treat severely ill COVID-19 patients. In October of 2020, the FDA formally approved it for treating COVID-19.

Nucleoside reverse transcriptase inhibitors (NRTIs) target reverse transcriptase enzymes. Recall that reverse transcriptase is an enzyme that retroviruses require to convert their RNA genome into a DNA format so that the virus can integrate into the host genome and perpetuate viral replication. HIV treatment usually includes administration of the NRTI drug called azidothymidine (also known as AZT, or by brand names like Retrovir and Zidovudine). Originally, AZT was developed as a potential anticancer drug, but it failed in that capacity, only to be resurrected later as a successful drug for fighting HIV. Because AZT has about a 100-fold preference for binding to viral reverse transcriptase as opposed to human polymerases, at therapeutic doses it has minimal effects on noninfected host cells.

CHEM • NOTE

Nucleotides are the basic building blocks of the nucleic acids DNA and RNA. They include a five-carbon sugar, a nitrogenous base, and phosphate groups. In contrast, a **nucleoside** is basically a nucleotide that is missing the phosphate groups. Most antiviral drugs are administered as a nucleoside form and then virally encoded enzymes that are only found in infected cells convert the drug to the active nucleotide form. Nucleoside analogs only mimic natural nucleosides, so they may not contain all the same parts as a naturally occurring nucleoside.

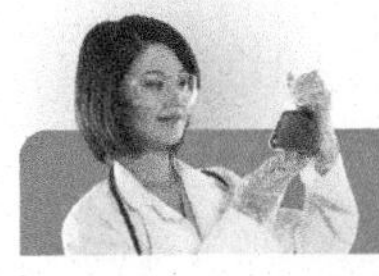

BENCH to BEDSIDE

Safeguarding Healthcare Workers Who Treat Ebola Patients

An article in the journal *The Lancet* evaluated the effect of postexposure prophylaxis for healthcare workers after exposure to Ebola. Study participants were given a combination of favipiravir, a viral RNA-polymerase inhibitor; TKM-Ebola, which binds and inactivates the viral mRNA; and ZMapp, immunotherapy with specific combinations of three monoclonal anti-Ebola virus antibodies. None of the treated patients developed Ebola.

Despite the availability of potentially useful postexposure prophylaxis treatments, the goal is to prevent an exposure in the first place. Despite the fact that Ebola has landed on U.S. shores, most healthcare workers still have only limited training on how to manage patients infected with this deadly biolevel 4 agent. The need for enhanced training on this front became all too clear when two American nurses, Nina Pham and Amber Vinson, contracted Ebola while caring for patients. The CDC requires that people treating Ebola patients wear coveralls with a hood and integrated face shield as well as a respirator. Extensive training and diligent observation to ensure proper donning and doffing of gear as well as equipment maintenance is central to prevention.

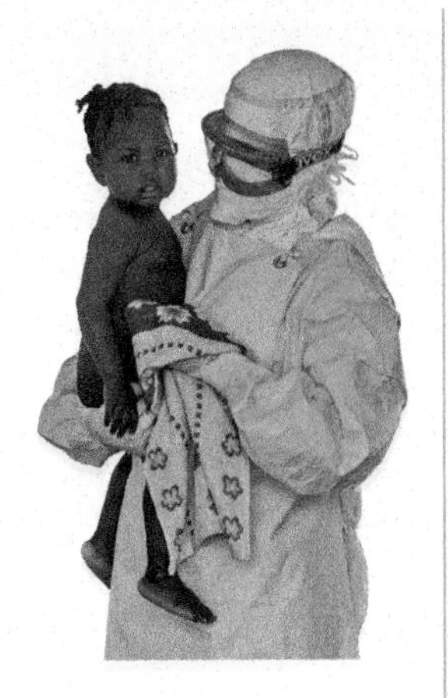

Sources: Jacobs, M., Aarons, E., Bhagani, S., et al. (2015). Post-exposure prophylaxis against Ebola virus disease with experimental antiviral agents: A case-series of health-care workers. *The Lancet Infectious Diseases*, *15*(11), 1300–1304. doi:10.1016/s1473-3099(15)00228-5.

Other enzymes besides reverse transcriptase can also be targeted to limit HIV replication. For example, HIV requires a number of proteases (enzymes that cut proteins) to make mature infectious virions. Over 25 protease inhibitors have been developed to target HIV proteases to limit viral assembly.

Lastly, viral replication can be blocked by **antisense antivirals**. These agents are short sequences of nucleotides that are complementary to the RNA transcribed by specific viruses. After viral mRNA is made, antisense antivirals can bind to it, preventing the host ribosome from translating the viral mRNA. Additionally, the targeted RNA is destroyed by cellular enzymes. So far, the only antisense drug that has ever been approved to treat viral-associated pathologies is Vitravene. In 1998 the FDA approved Vitravene for the local treatment of cytomegalovirus (CMV)-induced retinitis. However, this drug has now been replaced by safer and more effective therapies and is no longer manufactured. Nevertheless, the fact that an antisense antiviral drug was ever approved by the FDA represents a huge breakthrough, and research into other such drugs is ongoing.

Naturally occurring substances called **interferons** are released by cells in response to viral infections. Interferons signal the presence of a virus, causing neighboring, uninfected cells to make defensive changes that limit viral entry and/or replication. These molecules can be produced in the lab and administered to help limit the progression of certain viral infections. Interferon therapies have been used to treat hepatitis B and C infections in combination with other drugs. Unfortunately, they tend to generate flu-like symptoms as side effects, making them a fairly unpleasant treatment option. (Chapter 11 reviews the role of interferons in immune responses.)

Lastly, if viral entry or replication can't be blocked, there is always the last step—virus release. Drugs like oseltamivir (Tamiflu) and zanamivir (Relenza) prevent influenza A and influenza B virions from budding off the host cell surface.

Build Your Foundation

Build Your Foundation (BYF) Quick Quiz: Visit the **Mastering Microbiology** Study Area to quiz yourself.

22. How are bacteriophages and animal viruses grown in the laboratory? (LO 6.18)

23. What is the plaque assay, and what information does it reveal? (LO 6.19)

24. What are the different main methods for detecting if a patient has been exposed to a virus or if a virus is present in a sample? What are the limitations of the methods you named? (LO 6.20)

25. What are three different antiviral drugs and their modes of action? (LO 6.21)

6.5 PRIONS

Learning Outcomes

After reading this section, you should be able to:

6.22 Describe prions and the diseases they cause in humans.

6.23 Identify four ways that Creutzfeldt–Jakob disease can be transmitted.

6.24 Explain why neurological disorders such as Alzheimer's disease, Parkinson's disease, and amyotrophic lateral sclerosis are described as prion-like neurological disorders.

Prions are infectious proteins.

Even though viruses are acellular and therefore nonliving, they arguably act like a lifeform in some ways (albeit indirectly, via a host). But another type of acellular infectious agent, **prions**, are merely proteins. In fact, the name *prion* reflects their characteristics as "proteinaceous infectious particles." Unlike viruses, prions do not have any genetic material, and they do not replicate. Thus, prions prove that causing a transmissible disease doesn't require much biochemical sophistication.

Prions cause spongiform encephalopathies.

Prions cause a class of diseases called **spongiform encephalopathies**, which destroy brain tissue. These extremely rare diseases have a worldwide incidence of about one case per million people.

The normal prion protein is found throughout the nervous system, but it is especially prevalent in the brain. In their normal form, prions have a neuroprotective role, but when misfolded, these proteins cause serious damage to the nervous system. The current theory is that the misfolded, infectious version of the prion protein makes contact with the normal version, causing changes to the normal protein's shape. This leads to a chain reaction of protein misfolding that spreads throughout the brain's prion proteins. As misshapen prion proteins clump together, they cause brain tissue degeneration through an unknown mechanism. As the brain tissue deteriorates, spongy holes are left in the tissue—hence the name "spongiform" encephalopathy (FIG. 6.24). Detecting these sponge-like features in brain tissue collected upon autopsy remains the standard for definitively diagnosing a spongiform encephalopathy.

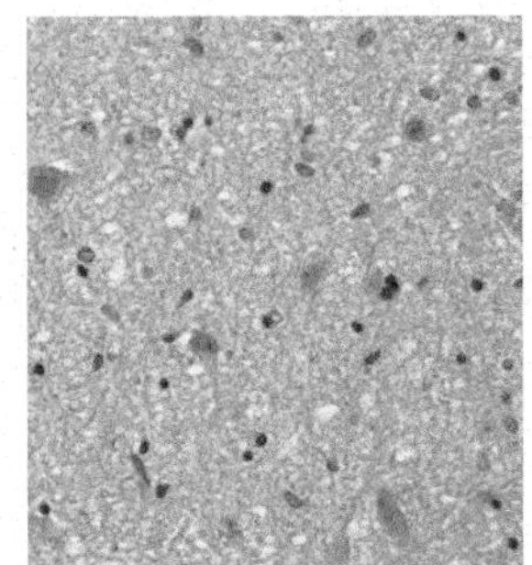
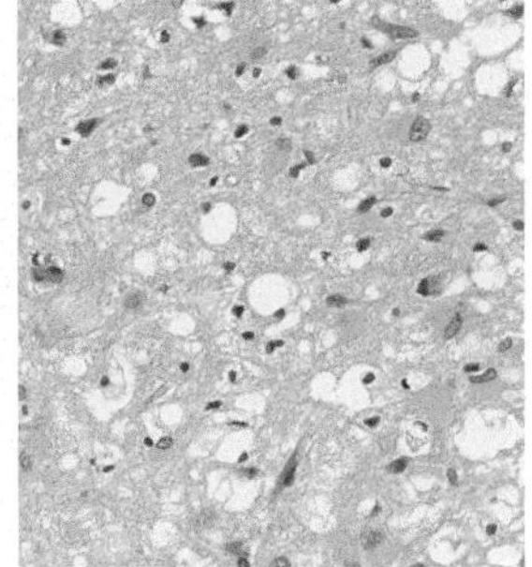

FIGURE 6.24 Prions Prions infect nervous tissue and cause sponge-like holes. (*Left:* Normal brain tissue; *Right:* Brain tissue showing prion induced damage)

Spongiform encephalopathies can be inherited or acquired. Gerstmann–Sträussler–Scheinker syndrome and a condition called fatal familial insomnia are both inherited when a parent passes on a gene that encodes a mutated form of the prion protein to their children. In contrast, Creutzfeldt–Jakob disease (CJD), which is probably the best-known spongiform encephalopathy in humans, has acquired (i.e., variant, sporadic, and iatrogenic) as well as inherited etiologies. Variant CJD is caused by a prion that has a different protein sequence than other CJDs. It was given the infamous moniker "mad cow disease" after a study linked human infections to consumption of meat contaminated with the form of the prion that primarily affects cattle. Sporadic CJD is linked to a spontaneous mutation in the normal cellular prion protein and affects 200–400 people in the United States every year. Inherited CJD is uncommon, accounting for only 15 percent of human prion disease cases. Iatrogenic CJD, which is accidental transmission of CJD to a patient as a result of a medical intervention, is much rarer and has been detected in only 400 people worldwide. Most iatrogenic CJD cases can be traced to either contaminated surgical instruments or tissue transplants derived from infected cadavers. Dura mater grafts used for head injury and brain surgery patients and corneal transplants have been associated with iatrogenic CJD. In the past, human growth hormone treatments were also a risk factor for iatrogenic CJD. This is because the hormone was formerly derived from cadaver pituitary tissue. As of 1985, human growth hormone is no longer derived from cadavers and is instead produced in genetically engineered cells in a lab, eliminating the risk of iatrogenic CJD in this treatment.

CLINICAL VOCABULARY

Etiology Refers to the origin of an infectious or noninfectious disease. Note, this term is not necessarily interchangeable with the term *causative agent*. For example, here prions are the causative agent of disease, but etiology refers to how the prion disease originated (i.e., inherited or acquired).

Most spongiform encephalopathies are characterized by dementia at an older age, but patients may demonstrate psychological and behavioral symptoms earlier. Unfortunately, there are currently no cures, treatments, or vaccines for spongiform encephalopathies. (A more detailed discussion of spongiform encephalopathies can be found in Chapter 18: Nervous System Infections.)

Some neurodegenerative diseases exhibit prion-like features.

Like prion-induced spongiform encephalopathies, a number of other neurodegenerative diseases have been associated with misfolded proteins in the brain. These include Alzheimer's disease (AD), which is a severe form of dementia;

Parkinson's disease (PD), a movement disorder characterized by muscle tremors and rigidity; and amyotrophic lateral sclerosis (ALS), which causes muscle weakness, dysfunction, and eventually paralysis. Interestingly, a number of studies show that exposing healthy mice to the misfolded proteins found in AD, PD, or ALS can trigger the development of the same neurological problems seen in mice with these diseases. These findings suggest that AD, PD, and ALS may be prion-like in their progression, meaning that misfolded proteins may cause otherwise normal versions of the same protein to change their conformation, deposit in the brain, and thereby cause disease. It is important to note that AD, PD, and ALS are not classified as spongiform encephalopathies and are not considered prion diseases; they are merely described as exhibiting a prion-like pathophysiology mechanism.[12]

Build Your Foundation

Build Your Foundation (BYF) Quick Quiz: Visit the **Mastering Microbiology** Study Area to quiz yourself.

26. What are prions, and what diseases do they cause in humans? (LO 6.22)

27. How is iatrogenic Creutzfeldt–Jakob disease (CJD) contracted versus variant CJD? (LO 6.23)

28. Why are Alzheimer's disease, Parkinson's disease, and amyotrophic lateral sclerosis described as having a prion-like pathophysiology? (6.24)

[12] Ma, J., Gao, J., Wang, J., & Xie, A. (2019). Prion-like mechanisms in Parkinson's disease. *Frontiers in Neuroscience*, 13, 552.

VISUAL SUMMARY | *Viruses and Prions*

Interactive CONTENT REVIEWS — Interactive Content Reviews: Visit the **Mastering Microbiology** Study Area to quiz yourself.

Viral Structure

Naked

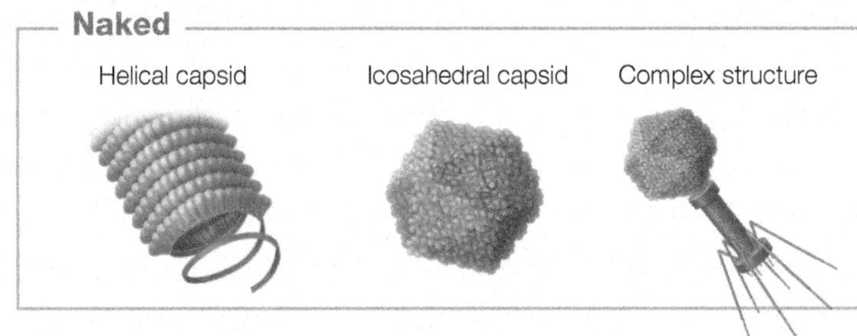

Enveloped

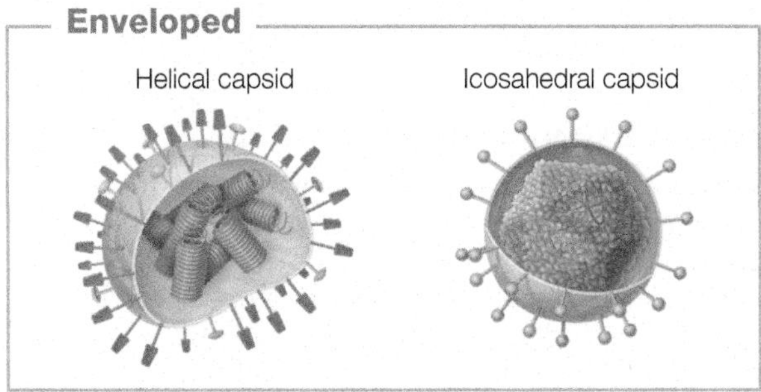

Viral Genomes

Viruses can have double- or single-stranded **DNA** or **RNA**.

DNA viral genomes
- Circular or linear
- Often double stranded
- May also be single stranded

RNA viral genomes
- Linear or segmented
- Often single stranded
- May also be double stranded

Viruses can have special enzymes.

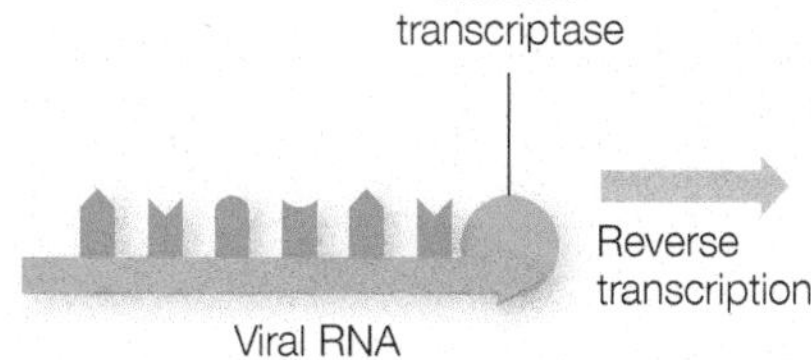

Reverse transcription

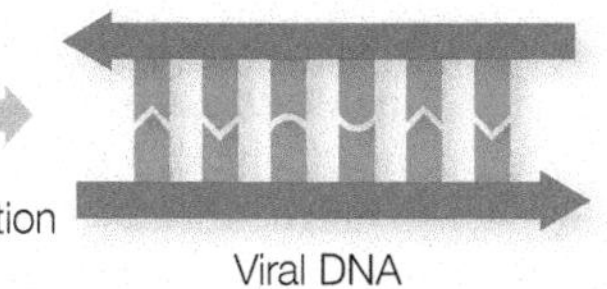

Viral Replication

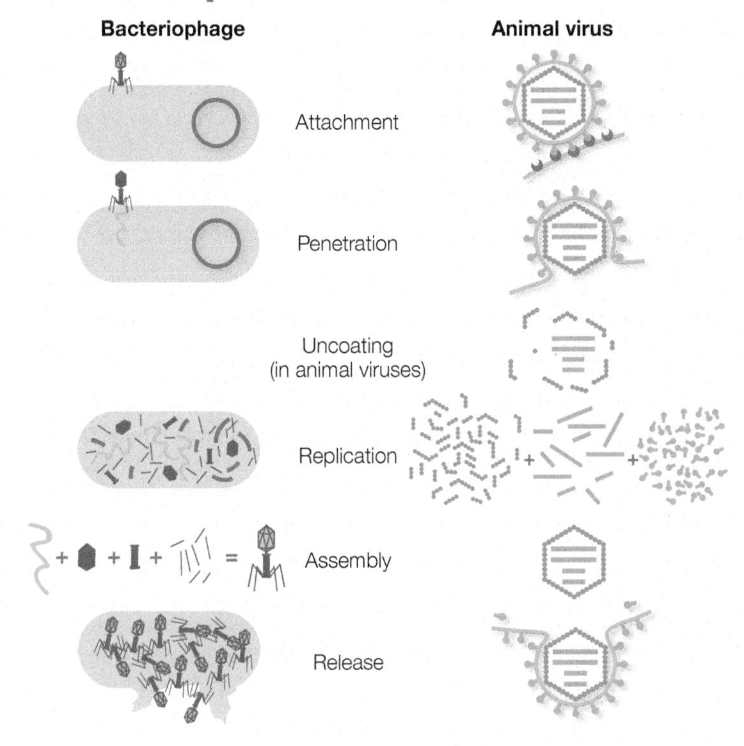

Some viruses integrate their genome into the host cell genome.

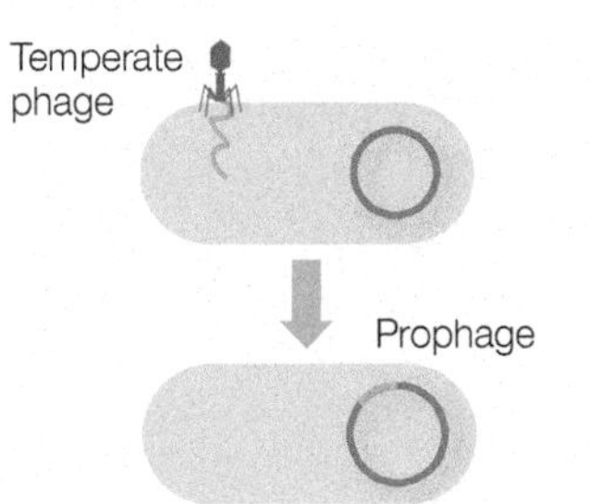

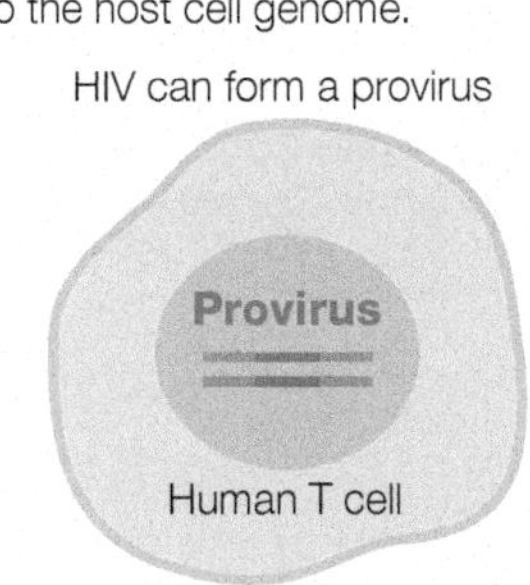

Antiviral Drugs

Antiviral drugs can block any step in viral replication.

- Attachment
- Penetration
- Uncoating
- Replication
- Assembly
- Release

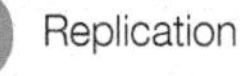

Detecting Viral Infections

Common diagnostic tools for detecting viral infections in patients include:

Detect virus

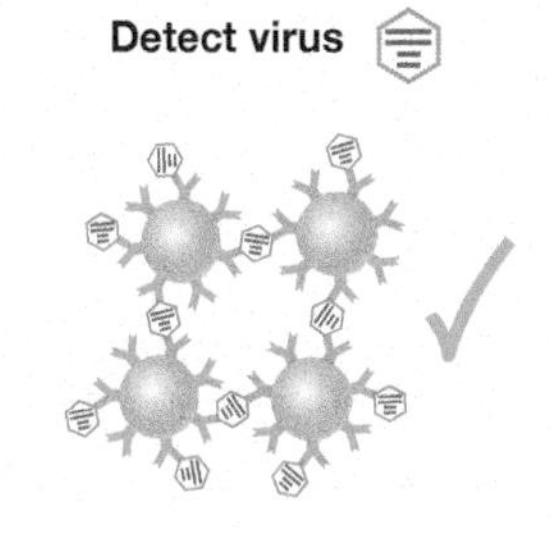

Detecting patient antibodies for virus

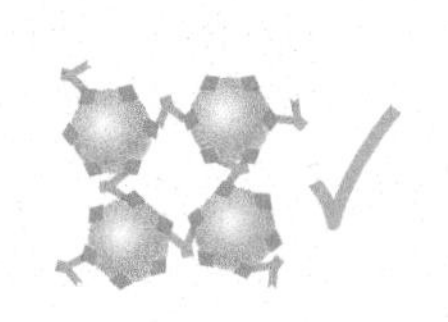

Prions

Prions are infectious proteins that cause transmissible spongiform encephalopathies.

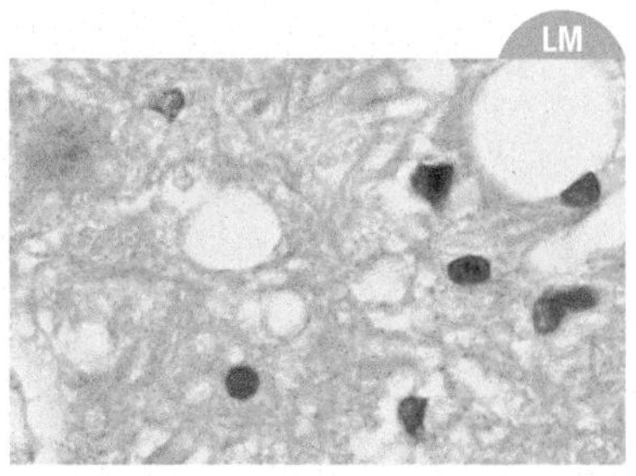

Light micrograph of prion-induced sponge-like holes in brain tissue.

6 CHAPTER 6 OVERVIEW

6.1 General Virus Characteristics

- Viruses are infectious, acellular particles that must invade living cells to replicate.
- The viral genome is encased in a capsid that's made of repeating protein units called capsomeres. Capsids can be helical, complex, or icosahedral. Outside the capsid, some viruses have a lipid-based envelope. Those without an envelope are called naked. Many viruses have spikes (peplomers) that help them infect a host cell.
- Viruses can have DNA or RNA genomes. RNA viruses have different ways of making viral mRNA:
  - Single-stranded RNA genomes that are *positive/sense-stranded RNA* ($ssRNA+$) can be used directly by the host's ribosomes to make protein.
  - Single-stranded RNA genomes that are *antisense/negative-stranded RNA* ($ssRNA-$) use RNA-dependent RNA polymerases to make mRNA.
  - Retroviruses convert their RNA genome into DNA and integrate into the host genome.
  - Double-stranded RNA genomes use virally encoded RNA-dependent RNA polymerases after unwinding the strands in order to make mRNA.
- Rapid replication promotes viral evolution. Because they use error-prone polymerases to copy their genome, RNA viruses tend to mutate more frequently than DNA viruses. Antigenic drift occurs due to small mutations in antigen-coding genes; it is responsible for seasonal influenza variations. Antigenic shift occurs when two different strains infect the same cell and undergo reassortment to make a new virus strain; this is often responsible for pandemic strains.

6.2 Classifying and Naming Viruses

- Viruses are taxonomically organized by order, family (and occasionally subfamily), genus, and species. They are mainly classified by the type of nucleic acid present (DNA or RNA); capsid symmetry; the presence or absence of an envelope; and genome architecture (ssDNA, ssRNA, etc.). Size, host range, tissue/cell tropism, and disease features help to refine viral groupings into species.
- Viruses exhibit characteristic host ranges and tissue tropisms.

6.3 Viral Replication Pathways

- Viruses use a host cell's energy, enzymes, organelles, and molecular building blocks to build new virions.
- Bacteriophages are viruses that infect bacteria. They perform either lytic or lysogenic replication pathways.
  - Lytic pathways kill the host cell. They involve attachment, penetration, replication, assembly, and release from the host.
  - Lysogenic replication involves insertion of the phage genome into the host's chromosome to form a prophage. Through phage conversion, prophages confer new properties onto a host bacterium. At any point, the prophage can excise itself from the host genome to continue into the lytic pathway.
- Animal virus replication has six general steps: (1) Attachment (often aided by spikes or other capsid proteins); (2) penetration (naked viruses mainly enter by endocytosis, while enveloped viruses use membrane fusion or endocytosis); (3) uncoating; (4) replication; (5) assembly; and (6) release (enveloped viruses bud out of host cells, while naked viruses lyse the cell).
- Acute viral infections have a rapid increase in virions and then resolve. Persistent infections last longer. Some are caused by viruses that integrate into the host genome, forming a provirus. Some persistent infections are caused by oncoviruses, viruses that can cause cancer. Latent persistent infections involve viruses that lie dormant for long periods of time between outbreaks with large numbers of free virions. Chronic persistent infections have a slow buildup of virions over time, eventually leading to a more rapid increase in viral replication.

6.4 Clinical Aspects of Viruses

- The number of infectious virions, or viral titer, can be determined using a plaque assay. Animal viruses may be grown in animals, cell culture, and fertilized chicken eggs.
- Detection tests are evaluated based on specificity, sensitivity, and turnaround time. Specificity means that the diagnostic test reliably detects only the virus(es) of interest without producing false-positive results. Sensitivity means the test detects very low levels of the target to limit false-negative results.
- The two major classes of tests for detecting viruses are antigen–antibody based or nucleic acid based. Agglutination assays involve the clumping of antibodies and antigens after they bind. ELISA is similar to agglutination tests except the antigen or antibody is bound to a surface. Using sequencing, fluorescent probes, or polymerase chain reaction, special genes unique to a certain virus are used to determine if the virus is present in a host.
- Antiviral drugs target all phases of the viral infection cycle. Because viruses live inside the host cells, the best antivirals target processes specific to the pathogen. Preventing infection can sometimes be done using postexposure prophylaxis.
- Commercially available antibodies against a virus, or drugs that bind viral proteins, can prevent successful attachment. Uncoating can be blocked by drugs that bind the capsid. Replication can be disrupted by nucleoside analogs that are modified and then incorporated into the viral genome, interrupting replication. Many antivirals target special enzymes that viruses use to replicate their genomes. Drugs like Tamiflu that block release of new virions, as well as immune system molecules called interferons, can prevent the spread of the virus to new cells.

6.5 Prions

- Misfolded prion proteins cause spongiform encephalopathies, which destroy brain tissue. These diseases can be inherited or acquired.
- There are many types of prion diseases, including Creutzfeldt–Jakob disease.
- Alzheimer's disease, Parkinson's disease, and amyotrophic lateral sclerosis (ALS) are not classified as spongiform encephalopathies and are not considered prion diseases; however, they do involve misfolded proteins and are therefore sometimes described as having a prion-like pathophysiology.

COMPREHENSIVE CASE

The following case integrates basic principles from the chapter. Try to answer the case questions on your own. Don't forget to be S.M.A.R.T.* about your case study to help you interlink the scientific concepts you have just learned and apply your new understanding of virology to a case study.

***The five-step method is shown in detail in the Chapter 1 Comprehensive Case on cholera. See pages 31–33. Refer back to this example to help you apply a SMART approach to other critical thinking questions and cases throughout the book.**

• • •

Jimena was quietly enjoying her lunch when an unexpected knock on the door startled her. She found two people in suits standing on her front porch. They explained they were medical scientists from Columbia University in New York, and they were visiting her to talk about some old blood samples from a transfusion she had back in the 1970s. They showed Jimena a consent form that she had signed back then; it had allowed scientists to store and study her blood to help advance our medical knowledge. Now she was worried. Did she have some terrible disease? She invited them in and sat down in case the news was bad.

The researchers explained that they were part of a team that searches for viruses in old blood transfusion samples. Their work was aimed at finding new transfusion-transmissible viruses that ought to be screened for in the donated blood supply to further protect patients. They were looking at samples from the 1970s because, unlike today, the only virus screened for back then was hepatitis B. These lightly screened samples might lead to discovery of previously unrecognized viruses. Sadly, lack of broader screening for bloodborne viruses meant that thousands of people were infected with hepatitis C and HIV via contaminated blood transfusions during the 1970s and 1980s.

People with hemophilia, a blood-clotting disorder, had an unusually high prevalence of hepatitis C virus and human immunodeficiency virus (HIV). This is because managing hemophilia in the 1970s and 1980s often involved administering clotting factors derived from human blood. Before the stricter screening went into effect, at least 6,000 hemophiliacs in the United States are believed to have contracted hepatitis C from blood transfusions. Of course, hemophiliacs were not the only ones who suffered; anyone who got a transfusion for any reason at that time was at high risk for exposure to these viruses.

In 1976, during her last trimester of pregnancy, Jimena had developed severe anemia. She remembered receiving a blood transfusion a few weeks before she delivered her son, Jose. The transfusion was done to ensure that her blood levels could support both her and her child during delivery. She knew this must have been the reason the doctors were visiting her now. She asked them if they were here to tell her she had HIV or hepatitis—she had been feeling very tired lately, and wondered if this was the cause.

The doctors hastened to explain that the blood she received was not infected with hepatitis C or HIV. But they had found something else: a new virus never seen before in humans. They examined the medical records of all the people who had received blood with this virus, and it did not seem to cause disease. However, they wanted to investigate more. They were here to get a more detailed medical history from Jimena, along with a new blood sample. They also wanted to reach out to Jose, Jimena's son, and see if he'd be willing to take part in the study, too.

Jimena's eyes became moist as she explained that Jose had died of cancer eight years ago. The doctors expressed their sympathy, but pressed Jimena for the type of cancer he'd developed. She said it was liver cancer. "Are you sure we didn't catch hepatitis C from the transfusion?" she asked the researchers. "I know that can cause liver cancer, right?" "Yes, it can," they replied.

• • •

CASE-BASED QUESTIONS

1. How could old samples of donated blood be screened to search for new viruses?
2. What information would be needed about this new virus in order to classify it, and how could researchers get that information?
3. How could the researchers isolate and grow a particular virus from Jimena's blood?
4. Assume the doctors determined the virus had an RNA genome. What kinds of antivirals would likely be prescribed to treat her infection? How would the drugs possibly work?
5. A small number of genes in the unnamed virus appear similar to hepatitis C virus. When doctors expose animals that had been injected with a hepatitis C vaccine to the unnamed virus, their blood shows a strong immune reaction to the unnamed virus (using a latex agglutination test). The blood of unvaccinated animals shows no reaction to the same test. What does this suggest about the shared genes?
6. The researchers discovered the new virus in six other people who had received transfusions around the same time as Jimena. The doctors monitored all the patients over a period of years. They found that viral titers remained very low in all, with only a small increase in only a few patients. What does this suggest about the infection?
7. Besides Jose, no one suspected to be infected with the unnamed virus has died of cancer. Does this information rule out the possibility that this is an oncogenic virus? Why or why not?

END OF CHAPTER QUESTIONS

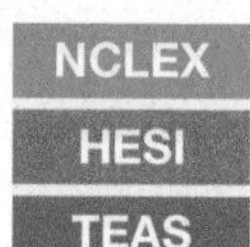

Think Critically and Clinically
Questions highlighted in orange are opportunities to practice NCLEX, HESI, and TEAS-style questions.

1. An RNA virus with an antisense or negative strand must have which of the following enzymes to replicate?
   a. Reverse transcriptase
   b. Host DNA polymerase
   c. RNA-dependent RNA polymerase
   d. Host RNA polymerase

2. A young mother who has hepatitis B is under your care. Your main concern is to prevent the spread of the virus from infected cells to healthy ones. You would recommend
   a. acyclovir.
   b. Retrovir.
   c. interferon.
   d. Valtrex.

3. Assume a new enveloped RNA virus is causing an epidemic. Which of the following may be useful to manage the disease? Select all that apply.
   a. Injectable antibodies
   b. Interferons
   c. Antibiotics
   d. Nucleoside analogs

4. Compare and contrast production of viral proteins for dsDNA viruses and ssRNA+ viruses.

5. An accidental needle stick occurs during routine treatment of an HIV-positive patient who arrives at your clinic. Immediate administration of ____________ can block viral entry.
   a. enfuvirtide
   b. vapendavir
   c. AZT
   d. amantadine

6. How might phage conversion provide a bacterium with an evolutionary advantage?

7. The patient has a viral titer of 200 on day four after infection, 30 at 4 weeks after infection, 600 at 8 months after infection, and 23 after 1 year of infection. This is likely a(n) ____________ infection.

8. Match the following terms.

a. Retrovirus	T4 bacteriophage
b. Lytic replication	Rhinoviruses
c. Lysogenic replication	Lambda phage
d. Persistent infection	Human papilloma virus (HPV)
e. Acute infection	Human immunodeficiency virus (HIV)

9. Which of the following could inform you if your patient had a previous infection with a nonpersistent virus and recovered? Select all that apply.
   a. A PCR test
   b. A plaque assay
   c. An ELISA test
   d. An agglutination assay

10. Which of the following is a potential feature of an animal virus? Select all that apply.
    a. Presence of an envelope
    b. Presence of a naked icosahedral capsid
    c. Ability to inject naked RNA into the host cell
    d. Ability to integrate into the host cell's genome
    e. Ability to cause host cell lysis
    f. Ability to build DNA from an RNA template

11. Why don't bacteriophages undergo an "uncoating" step during replication?

12. A young man comes into your clinic and asks for an HIV test. He had unprotected sex two weeks ago and is nervous. His best option would be a(n)
    a. ELISA of anti-HIV antibodies.
    b. test that detects HIV genes.
    c. latex agglutination test.
    d. cell culture of blood sample.

CRITICAL THINKING QUESTIONS

1. Viruses are not considered alive. However, some virologists believe that viruses should be included in the category of living things. How would they argue for this position?

2. Many public health experts believe there might be a relationship between farming practices and increased influenza A antigenic shift. Why might they think this?

3. The realization that many cancers are associated with viral infections is relatively recent. Why do you think that is the case?

7 Fundamentals of Microbial Growth and Decontamination

What Will We Explore? Understanding where pathogenic microbes dwell, how they grow in their selected environments, and how to control their numbers has saved many lives. This chapter covers the basics of microbial growth, with an emphasis on bacterial growth. We'll discuss nutritional requirements and environmental factors and explore how microbes may survive in some of the most inhospitable places on Earth. We'll also cover methods used in industry and medical settings to limit microbial growth and remove microbes from surfaces.

Why Is It Important? A particular concern with infectious diseases is the spread of pathogens among both healthcare workers and patients in healthcare settings. Gloves, masks, and other personal protective equipment combined with decontamination efforts can help prevent such transmission. Even the simple act of hand washing reduces pathogen transmission—for example, during the COVID-19 pandemic, frequent hand washing was emphasized as something everyone could do to help reduce the spread of SARS-CoV-2. Despite the attention placed on decontamination and transmission precautions, the U.S. Centers for Disease Control and Prevention (CDC) states that on any given day, 1 in every 31 hospitalized patients is battling at least one healthcare-associated infection. Healthcare workers stand on the front lines of the battle to control these infections. Identifying the places where microbes can linger and grow in healthcare settings is crucial for limiting transmission and saving lives.

Of course, microbial control and decontamination strategies are also important in nonhealthcare settings, such as in the food industry. The CDC estimates that every year, one in six people suffers from a foodborne illness. Although most cases resolve within a day or two, about 128,000 Americans are hospitalized and 3,000 die of foodborne illness each year. Pasteurization and other food safety measures prevent food spoilage and limit foodborne illnesses.

Finally, studying where and how microbes grow has revealed that they can survive in conditions that we previously assumed could not support life. If microbes can survive in deep-sea vents, within glaciers, or even in jet fuel tanks, then there is a possibility that they could survive the inhospitable environments of planets like Mars.

Kimberly Horn
RN, Director of Nursing; Decatur, GA

Clinical CASE
NCLEX HESI TEAS

The Case of the Sickly Soaker

Visit the **Mastering Microbiology** Study Area to watch the case and find out how microbial growth can explain this medical mystery.

7.1 MICROBIAL GROWTH BASICS

Learning Outcomes

After reading this section, you should be able to:

7.1 Discuss basic differences in features of microbial growth in a laboratory versus in nature.

7.2 Define the term *binary fission*, and compare it to budding and spore formation.

7.3 Calculate generation time for a bacterium.

7.4 Outline the features of the four stages of bacterial growth in a closed pure batch system.

Microbes show dynamic and complex growth in nature, and often form biofilms.

When nutritional requirements are met, a microbe will enlarge in size and eventually divide. **Microbial growth** is cell division that produces new (daughter) cells and increases the total cell population. We observe distinct microbial growth and reproduction stages in the laboratory. But in nature, with its more diverse environments and mix of microbial species, growth is not always so straightforward. In this section we highlight bacterial examples because we understand a great deal about their growth. However, it should be noted that most of this knowledge comes from studying species that can be cultured in the laboratory—and those species amount to just 1 percent of the total bacterial species on our planet.

In the laboratory, bacteria are usually grown as pure, single-species cultures. But in nature, bacteria intermingle and live side by side with archaea and eukaryotes. Even a basic morphological property like cell shape can be altered by environmental factors. For example, *Escherichia coli* converts from its motile bacillus shape to a filamentous nonmotile form during urinary tract infections.[1] Cells living communally, as in biofilms where they communicate and collaborate to survive, are a more realistic representation of how microbes normally live and grow than are the pure cultures used in a lab. (See Chapters 1 and 5 for more information on biofilms.)

In healthcare settings, biofilms are a major concern because they make treating infections difficult and contribute to persistent infections. Although the top layer of cells may be readily killed (or in the case of the tooth biofilm *plaque*, scrubbed off), cells deeper in a biofilm are highly protected. Methicillin-resistant *Staphylococcus aureus*, vancomycin-resistant *Enterococcus* species, *Clostridioides difficile* (formerly *Clostridium difficile*), and *Pseudomonas aeruginosa* are some examples of bacteria for which biofilm formation complicates treatments.

Biofilms form when free-floating (planktonic) bacteria adhere to a surface. Indwelling devices such as catheters and heart valves are therefore potential havens for biofilms. For example, once a urinary catheter is placed in the urethra, bacteria can adhere to it and produce a sticky, polysaccharide matrix that allows other microbes to adhere to the growing biofilm. Eventually a mixed microbial community develops. As the biofilm assembles, channels form within it to allow nutrients to flow in and waste to move out. Some of the biofilm may even break off and attach to another location—for example, a bit of biofilm from a urinary catheter might break off and adhere to the bladder (FIG. 7.1). Because urine flow is insufficient for flushing out a biofilm in a catheter, the bacteria there readily grow undisturbed.

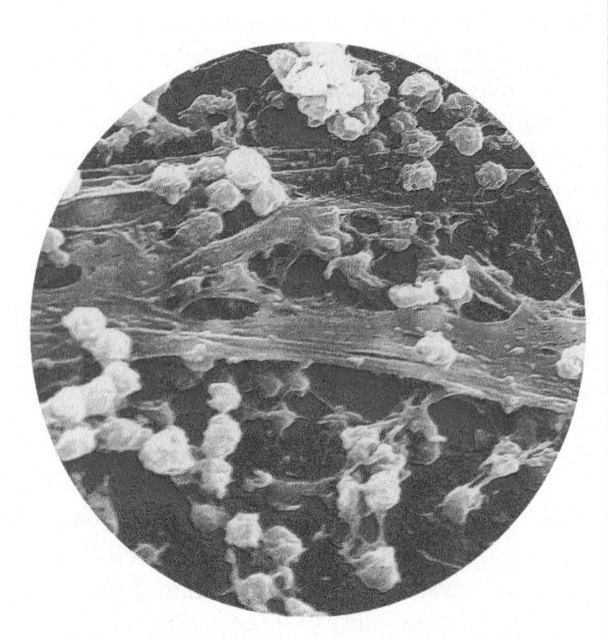

FIGURE 7.1 **Biofilm on a catheter** The inside of a catheter was cut open to reveal a biofilm. The rod-shaped cells are *Alcaligenes* species, Gram-negative bacteria that can cause blood infections.

Bacteria usually divide by binary fission, but some may use budding or spore formation.

As cells grow, they eventually reach a point where they divide. For most prokaryotes, this involves dividing a single cell into two cells via the asexual process called **binary fission**. (See Chapter 3 for a step-by-step overview of this process.) However, there are some exceptions to this tendency. For instance, the giant bacterium *Epulopiscium* produces several sporelike cells that are released when the parent cell lyses.

To review, before dividing by binary fission, a prokaryotic cell replicates its chromosome so that each new daughter cell receives a complete copy of the cell's genetic information. Then the parent cell begins to pinch off at the middle, with

[1] This research paper was groundbreaking because it established that bacterial shape affects pathogenesis. Other studies have shown similar effects with diverse bacteria. Justice, S. S., Hunstad, D. A., Cegelski, L., & Hultgren, S. J. (2008). Morphological plasticity as a bacterial survival strategy. *Nature Reviews Microbiology*, 6(2), 162–168.

ribosomes and other cell components migrating to each end. Finally, the partition (septum) in the center becomes complete, creating two genetically identical daughter cells. Cellular arrangements such as chain or cluster morphologies are determined by the direction (division plane) in which binary fission occurs. For example, chains (strep arrangements) form if binary fission occurs in one linear plane, whereas clusters (staph arrangements) develop from fission occurring in multiple planes. FIG. 7.2 shows bacteria performing binary fission.

Another method of asexual reproduction is **budding**. Certain fungi and some bacteria such as *Hyphomicrobium* reproduce this way. Budding in bacteria involves the original cell elongating and developing a small outgrowth on one side. Once the chromosome is duplicated and placed in the bud, separation from the mother cell occurs. Budding is fairly similar to binary fission, except that the bud is not an equal-sized division of the original cell. Both binary fission and budding are asexual, leading to genetically identical cells.

Some fungi and even bacteria use **spore formation** for reproduction. In fungi, spore formation can be sexual or asexual, whereas in bacteria it is asexual. For example, spore-forming *Streptomyces* bacteria have an unusual replication mechanism that mimics asexual fungal spore formation. These common soil bacteria form spores called conidia that hang from long hyphae extensions. Eventually the conidia, which are the bacterial daughter cells, break off (FIG. 7.3). These spores are *not* to be confused with bacterial endospores, which are the thick-walled, nongrowing structures that some bacteria like *Bacillus*, *Clostridium*, and *Clostridioides* species form to survive unfavorable growth conditions.[2]

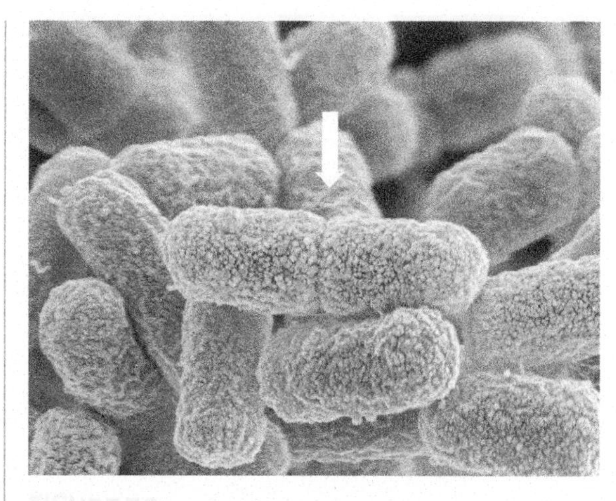

FIGURE 7.2 Binary fission A scanning electron micrograph of *Escherichia coli.* The white arrow shows a cell undergoing binary fission.

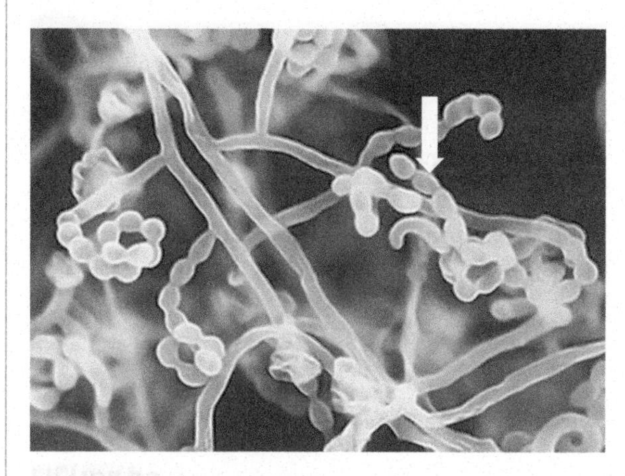

FIGURE 7.3 *Streptomyces* spores This scanning electron micrograph of *Streptomyces* shows spore formation. The arrow indicates a string of spores (conidia) extending out of a mass of hyphae.

Critical Thinking *What cellular components do the spores need to have before breaking off?*

Bacterial species differ widely in their generation time.

The time it takes for a particular species of cell to divide is its **generation time**. Bacterial generation times are diverse and can range from about 15 minutes to 24 hours or more, depending on the species and growth conditions. As bacteria divide by binary fission, one cell begets two, the two yield four, those four become eight, and so on. This type of growth is described as **exponential**. FIG. 7.4 explains how to calculate generation time.

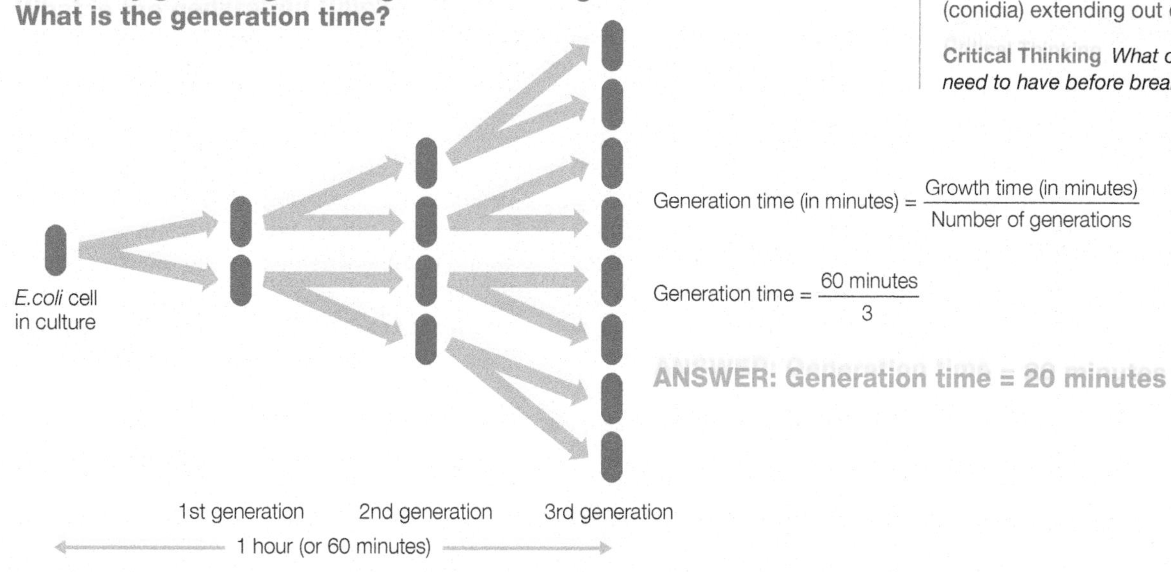

FIGURE 7.4 Calculating generation time: A worked example

Critical Thinking *What would the generation time be for a bacterium that grew for 3 hours, but required 1 hour to become acclimated to its environment before it entered exponential growth to produce four generations?*

[2] *The genus Clostridioides was established in 2016. This genus name means "Clostridium like" and, to date, includes only two species—Clostridioides difficile (formerly known as Clostridium difficile) and Clostridioides mangenotii (formerly known as Clostridium mangenotii).* Both the U.S. CDC and the Clinical and Laboratory Standards Institute (CLSI) have formally acknowledged this new genus and embraced the resulting classification changes.

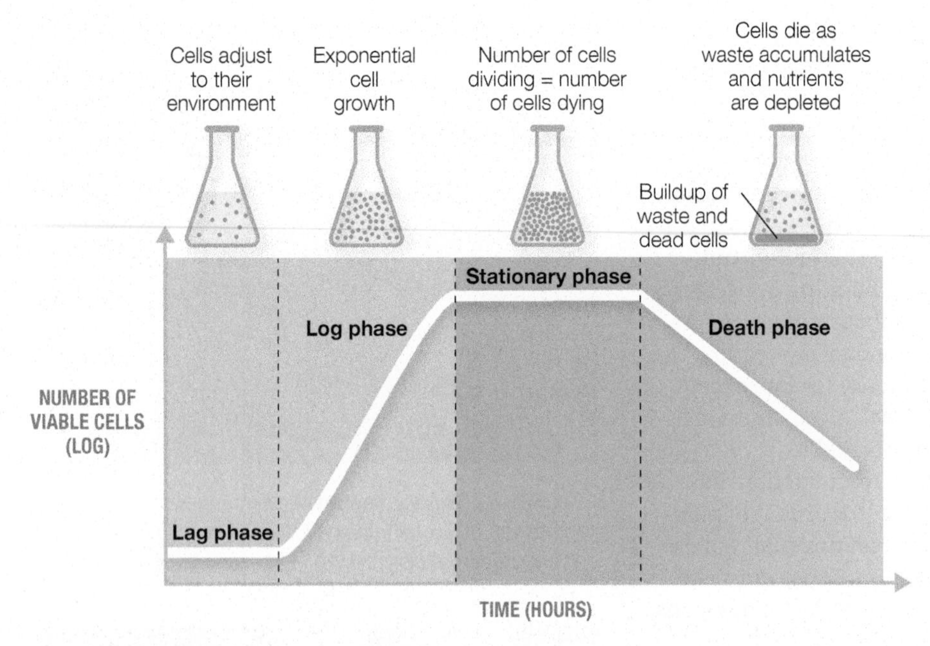

FIGURE 7.5 Bacterial growth curve The logarithmic graph displays the stages of bacterial growth over time in a batch culture. The exact time frame and cell numbers present at each phase depend on many factors, including the species and growth conditions.

Critical Thinking *In what phase would genome replication incidence be the highest?*

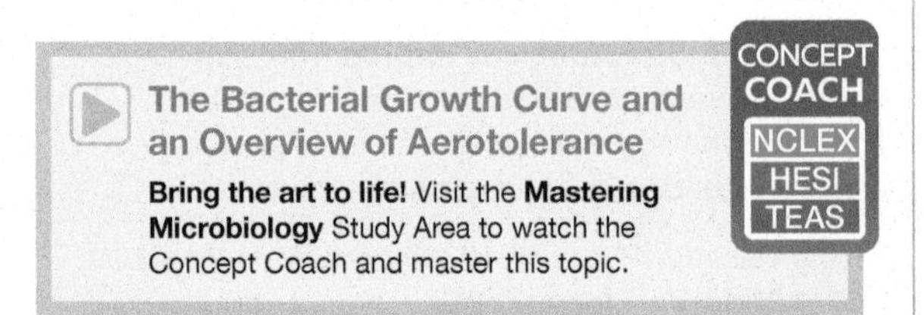

The nutrients available impact how fast a microbial population increases. In optimal growth conditions, the generation time for many common bacteria is less than an hour. For example, under ideal conditions, *E. coli* has a generation time of about 20 minutes. This means that every 20 minutes the population would double. Provided unlimited nutrients and unchecked growth, if we started with just one *E. coli* cell, in just under 20 hours there would be enough *E. coli* to cover the entire surface of the Earth.[3] If you consider this point, it's easy to appreciate how a patient's condition can rapidly deteriorate if a bacterial infection is not effectively managed. Fortunately for us, *E. coli* grows much more slowly in the human intestines because of competition for nutrients with other microbes. Some bacteria have fairly slow generation times, even under optimal growth conditions. For example, *Mycobacterium tuberculosis* cultured under optimum lab conditions has a 15- to 20-hour generation time, mainly because it is hard for nutrients to cross the waxy mycolic acid layer in these bacteria. (Chapter 3 reviews acid-fast bacteria and their cell wall features.)

Bacteria have four distinct growth phases when cultured using a closed pure batch system.

In the lab, bacteria are usually isolated and grown in closed pure batch cultures. First, bacteria are deposited into a flask of sterile growth medium with a set amount of organic and inorganic nutrients. The flask is then covered to prevent outside contamination. (See Chapter 1 for more on culturing techniques.) In such carefully controlled conditions, the bacterium undergoes distinct growth phases that can be detected by counting the number of viable cells in the culture. However, walk outside the lab into the environment where the bacteria being studied normally live, and you'll observe a different type of growth altogether. FIG. 7.5 shows a generalization of the four growth phases associated with bacteria in a closed pure batch culture. The length and magnitude of each phase is highly dependent on the type of bacteria being cultured, the general growth conditions, and the number of cells that were initially inoculated into the medium.

Phase One: Lag Phase

Imagine you're hiking through the desert with an empty canteen and belly, only to be suddenly dropped into an air-conditioned supermarket. With such a sudden change, you might have to sit for a few minutes to adjust to your new surroundings. Similarly, bacterial growth may pause when suddenly moving from nutrient-poor to nutrient-rich conditions. This first phase of delay that occurs while cells adjust to their new environment is called the **lag phase**. During this phase, cells alter their gene expression in response to their new setting. For example, they may make new enzymes and transporter proteins so that they can take up and metabolize nutrients provided in their new environment. They may also stop making certain products that are no longer necessary to survive. These modifications take time—especially if the cells' prior environment was particularly limiting for growth.

The first part of the graph in Figure 7.5 shows the lag phase—a flat line where the number of viable cells remains the same for a period of time. The length of the

[3] An average *E. coli* cell covers about 1 square μm (0.5 μm width by 2 μm length). The Earth's surface area is 510 million square kilometers (or 5.1×10^{17} μm^2). Assuming a generation time of 20 minutes, after 59 generations (or about 19.7 hours) there would be 5.8×10^{17} bacteria. Having a "footprint" of about 1 μm^2 per cell, that would be enough *E. coli* to cover all the land and water surfaces of the planet.

lag phase depends on a variety of factors, including the species being cultured, the type of media used, and the general healthiness of cells in the initial inoculum.

Phase Two: Log Phase

If the growth conditions are optimized for nutrients, pH level, and temperature, then once the cells have adjusted to their new environment, they will enter a phase of rapid exponential growth. This second growth phase is called the **logarithmic (log) phase.** It is characterized by an upward-sloped line that results when the number of viable cells is plotted on a logarithmic scale as a function of time. This phase lasts as long as sufficient nutrients are available and metabolic wastes are not appreciably accumulating. The generation time can be calculated during this phase; it impacts how quickly the population increases. Cells with a short generation time (rapid rate of cell division) will exhibit a steeper slope on a growth graph for this phase than cells with a longer generation time (slower rate of cell division).

Phase Three: Stationary Phase

Whether grown in a closed flask or in a natural, more limiting environment, at some point available nutrients are depleted, and waste products accumulate. Over time, the population growth rate slows and eventually levels off as the number of cells dying matches the number of cells dividing—this is the **stationary phase.** As with the other phases, the length of this third phase also depends on factors such as the amount and type of nutrients in the culture and the species being grown. From an industrial standpoint, the stationary phase is a useful period because many important metabolic products, like certain antibiotics, are made in this phase. The stationary phase is also the stage in which bacteria that can form endospores, such as *Bacillus, Clostridium,* and *Clostridioides* species, will gear up to do so. (Bacterial endospore formation is covered in Chapter 3.)

Phase Four: Death Phase

At a critical point of waste buildup and decreasing nutrients, the cells in a closed batch culture system begin to die. During the **death phase** the rate of cell death is exponential, but the rate of population decline varies based on numerous starting factors and the species being grown. Note that the entire population will not be killed. Over a period of hours, days, or even months, a small number of the cells survive by adapting to the new waste-filled environment and by feeding off dead cells.

In industry, maintaining cells at a specific growth phase is often necessary for optimizing the production of useful products like ethanol or pharmaceuticals. Therefore, instead of the closed batch culture method, a **chemostat** is typically used. In a chemostat system, fresh growth medium is added at one end of the culturing device, while waste, nutrient-depleted medium, and excess cells are removed at another end of the system to maintain a constant culture volume. This maintains cellular growth at a constant rate and can be used to hold the culture in a desired growth phase—typically either the exponential or stationary growth phase. A chemostat is often maintained for days or weeks (FIG. 7.6).

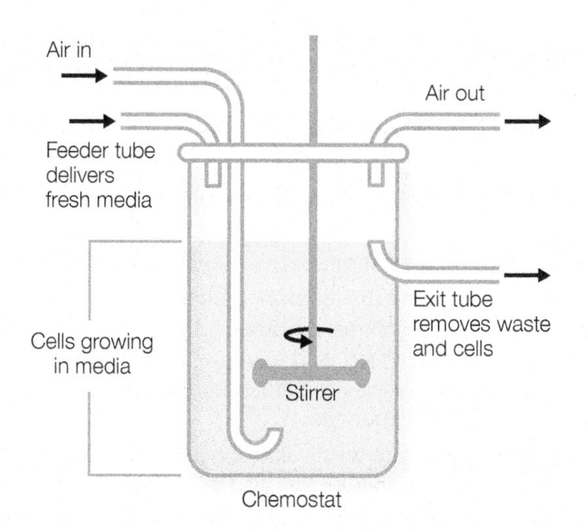

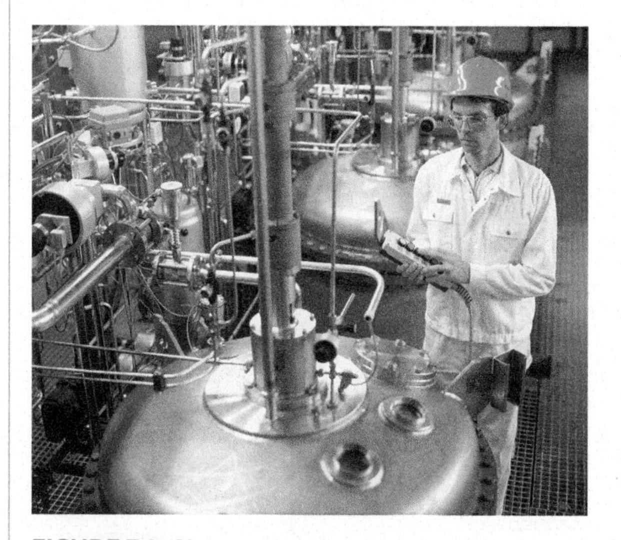

FIGURE 7.6 Chemostat cultures Bacteria and yeast cultures can be grown in a chemostat so that their growth rate is maintained at a constant level. A feeder tube delivers fresh media, and an exit tube siphons off extra media to remove waste and excess cells. The large stainless steel container shown here is an industrial-sized chemostat.

Build Your Foundation

1. State some differences between how cells grow in a lab versus in nature. (LO 7.1)
2. Define the term *binary fission*, and compare it to budding and spore formation. (LO 7.2)
3. What would the generation time be for an organism that goes through five generations in 150 minutes? (LO 7.3)
4. Name and describe the phases of bacterial growth in a closed pure batch system. (LO 7.4)

Build Your Foundation (BYF) Quick Quiz: Visit the **Mastering Microbiology** Study Area to quiz yourself.

7.2 PROKARYOTIC GROWTH REQUIREMENTS

Learning Outcomes

After reading this section, you should be able to:

7.5 Define the terms *optimal*, *minimum*, and *maximum* as they apply to temperature and pH conditions.

7.6 Describe the temperatures at which psychrophiles, psychrotrophs, mesophiles, thermophiles, and extreme thermophiles would thrive, and state the grouping for most pathogens.

7.7 Define the terms *acidophiles*, *alkaliphiles*, and *neutralophiles*, and describe ways that microbes survive in pH extremes.

7.8 Define the term *halophile*, and state how these microbes combat osmotic stress.

7.9 Name the various classes of microbes according to their oxygen use and tolerance.

7.10 Define the terms *essential nutrient* and *growth factor*, then describe the energy sources required by phototrophs versus chemotrophs.

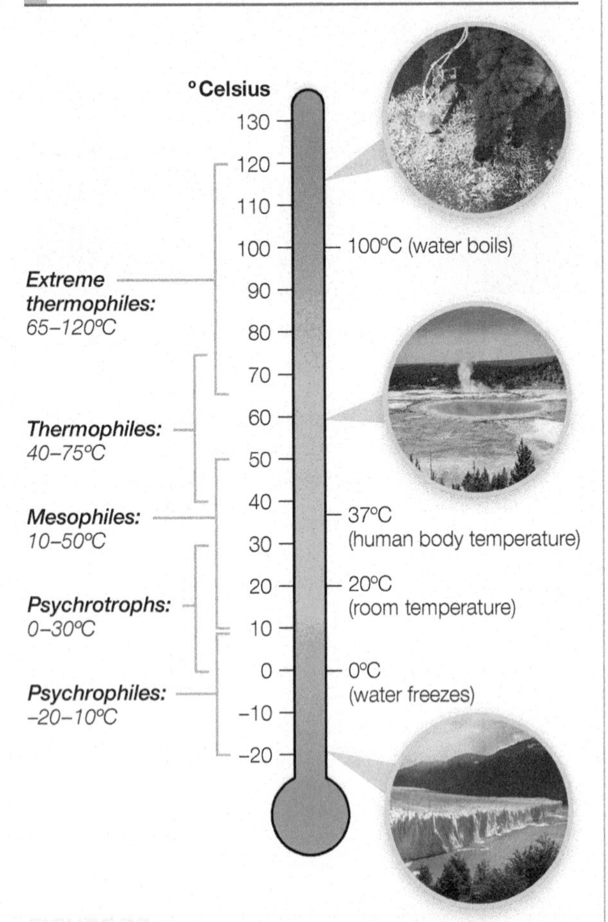

FIGURE 7.7 Prokaryote temperature growth ranges Prokaryotes thrive at diverse temperatures and can be grouped by their preferred growth temperature range.

Critical Thinking *At what temperature would an intestinal pathogen grow best? What term would be used to describe such a pathogen?*

Prokaryotes adapt to various growth conditions.

From the bottom of the ocean to an implanted arterial stent, microbes live in practically every natural and synthetic environment on this planet. They thrive by adapting to the conditions in those environments, no matter how extreme. Archaea in particular are known for their ability to thrive in harsh conditions, although they can live in hospitable places too—like our own gastrointestinal tract.[4] The fact that archaea are often able to adapt to extreme conditions suggests that microbial life may very well exist on other planets. In the meantime, researchers scour our own globe to find new bacterial species. It is estimated that less than 1 percent of the Earth's microbial population has been characterized.

All microbes find a niche by adapting to specific conditions. These include temperature, pH, salinity, and levels of oxygen.

Temperature

Temperature is an important component of a microbe's environment. Increased temperature can speed up enzymatic reactions and thereby increase growth rate, but once temperature rises too high, cell proteins denature, killing the cell. Conversely, temperatures that are too low tend to reduce enzyme activity and slow growth. Because microbes cannot regulate their internal temperature in response to their surroundings, they must instead create proteins and membranes that function within a particular temperature range. In terms of microbial growth, there are three principal temperatures to consider: the minimum, maximum, and optimum temperatures. The **maximum** and **minimum temperatures** represent the upper and lower temperatures that support a given microbe's growth. Within the minimum and maximum range there is an **optimal temperature**; this is the temperature at which cellular growth rate is highest. Amazingly, prokaryotes thrive at almost every imaginable temperature—whether it is under glacier ice, or inside a steaming, deep-sea volcanic vent. Temperature is such a fundamentally important factor that it's also one of the ways we can classify microbes. FIG. 7.7 shows the main classifications of microbes by temperature growth range.

Psychrophiles can thrive between about −20°C and 10°C. These organisms tend to live in environments that are consistently cold, like the Arctic. The highly saturated lipids in their cell membranes help these prokaryotes maintain a fluid bilayer, even at cold temperatures. Other cold-tolerant organisms called **psychrotrophs** grow at about 0–30°C, and are associated with foodborne illness because they grow at room temperature as well as in refrigerated and frozen foods. *Listeria monocytogenes* is an example; this pathogen readily grows in refrigerated meats, fruits, vegetables, and unpasteurized juices and dairy products. **Mesophiles** prefer moderate temperatures and tend to grow best around 10–50°C, a range that includes body temperature. Thus, most pathogens are part of the mesophilic temperature group. Mesophiles cover a broad range of the planet, dwelling in soils, streams, and eukaryotic organisms.

Thermophiles prefer warm temperatures of roughly 40–75°C. They dwell in compost piles and hot springs. A subset of thermophiles that prefer extremely hot temperatures are the **extreme thermophiles**. They prefer growth temperatures from 65–120°C. These organisms can live in boiling water and volcanic vents. Some strains of *Methanopyrus kandleri* can even survive temperatures of up to 122°C (about 251°F). In light of the high temperatures

[4] Nkamga, V. D., Henrissat, B., & Drancourt, M. (2016). Archaea: Essential inhabitants of the human digestive microbiota. *Human Microbiome Journal*, 3, 1–8.

where thermophiles thrive, they have uniquely adapted proteins that resist thermal denaturing. Also, many thermophiles make specialized chaperone proteins to help refold proteins and keep them in their functional state (see Chapter 5 for more on chaperones). Lastly, their cell membranes are adapted to high temperatures. One way that plasma membranes can be made thermostable is by building them from lipids that are highly branched and saturated.

High-temperature environments frequently also have extremes in pressure that microbes must manage. For example, bacteria that live near thermal vents are not only extreme thermophiles, but also **barophiles** that can withstand the high-pressure environment of the deep sea.

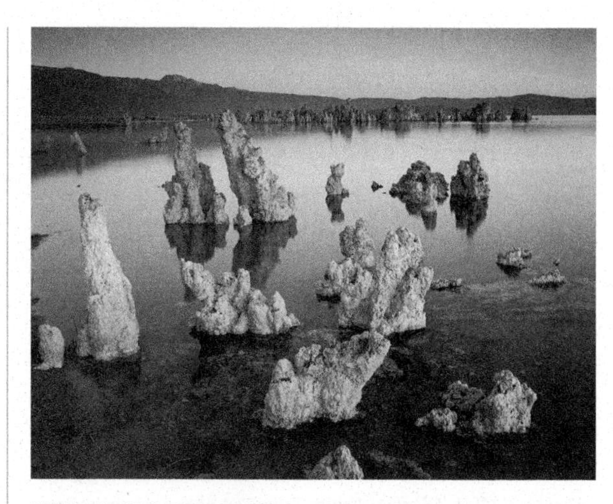

FIGURE 7.8 **Alkaliphiles** Mono Lake in California has a pH of 10. The extremely basic pH of this lake favors the growth of alkaliphiles. The white columns seen in the image are limestone formations that rise above the extremely salty and alkaline waters.

Critical Thinking *Mono Lake has a high pH and a high salt concentration. What terms describe a microbe that grows in both of these conditions?*

pH

Although microbes can't regulate their internal temperature, most can adjust their intracellular pH in response to slight environmental pH shifts. This is an important capability for microbes to have because even small changes in extracellular pH, which will occasionally occur in any natural setting, can adversely affect cellular processes. In contrast to microbes that live in neutral (or nearly neutral) pH environments, those that live in pH extremes must constantly work to stabilize their intracellular pH and therefore have evolved specialized adaptations. As with temperature, each microbe's survivable pH range includes a minimum, optimum, and maximum pH for growth, with **optimum pH** being the pH at which a microbe grows the best.

Acidophiles can grow at pH 1 (or less) to pH 5 and live in areas such as sulfur hot springs and volcanic vents. Often these organisms are archaea that use the inorganic elements around them for energy and carbon sources. Despite existing in an acidic environment, acidophiles often maintain a fairly neutral cytoplasmic pH. This can be accomplished in a number of ways, such as using proton pumps to export excess protons from the cytoplasm to raise pH. Some archaea resist low pH by building special monolayer plasma membranes that H^+ ions cannot easily cross to enter the cell. Ultimately, these adaptations are important because most enzymes do not function well when intracellular pH drops below 5. Because most bacteria do not have these unique adaptations to withstand consistently acidic environments, the vinegar used in pickling is a great way to preserve foods.

Neutralophiles grow best in a pH range of 5–8. They make up the majority of microorganisms, especially pathogens, known today. These include *E. coli*, *Salmonella* bacteria, and *S. aureus*.

Alkaliphiles are microbes that grow in the basic pH range of 9–11. They live in areas of the world that have extremely basic pH conditions, such as soda lakes that contain high concentrations of sodium carbonate. Mono Lake in California hosts bacteria and archaea, such as *Spirochaeta americana*, that are alkaliphiles (FIG. 7.8). Just as acidophiles have ways to maintain a relatively neutral cytoplasmic pH, so too do alkaliphiles. By enriching their plasma membrane with acidic compounds like phosphoric acid and gluconic acid, these microbes attract H^+ ions to their cell surface where they are then absorbed, thereby lowering cytoplasmic pH.

CHEM • NOTE

Saturated lipids lack double bonds in their hydrocarbon tail structures. Saturated lipids stack and pack together, which means membranes enriched with them are solid at cooler temperatures and fluid at warmer temperatures.

CHEM • NOTE

pH is a measure of acidity or alkalinity. It is based on the concentration of H^+ (hydrogen proton concentration). The lower the pH value, the higher the H^+ concentration. A pH below 7 is acidic; a pH above 7 is basic; and a pH of exactly 7 is neutral. Acids lower pH, and bases raise pH. When cells are described as "pumping protons" to manage pH, they are actually moving H^+ ions.

High-Salt Conditions

Organisms that thrive in high-salt environments are called **halophiles**. These microbes tolerate environments that are up to 35 percent salt and can thrive in places like the Dead Sea and the Great Salt Lake of Utah. To do so, they must overcome the stress that a high-salt environment puts on their water balance. Because bacterial cytoplasm is 80 percent water, cells in a high-salt environment will undergo *plasmolysis*—a situation in which water is drawn out of their

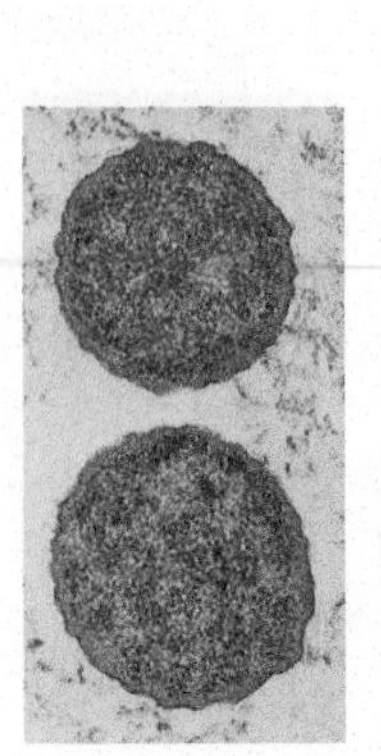
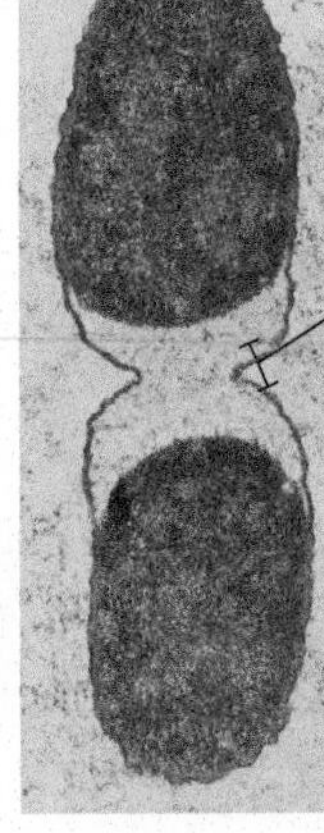

FIGURE 7.9 **Plasmolysis** This TEM reveals the effect of high-solute concentration. *Left*: These cells show a normal cellular structure. *Right*: These cells, which are just wrapping up binary fission, have undergone plasmolysis as evidenced by the space between the cell wall and the plasma membrane. This would have occurred as water rushed out of the cells into the higher salt environment.

cytoplasm into the solute-rich environment (FIG. 7.9). This is due to the basic principles of osmosis (see Chapter 3 for a review of osmosis and plasmolysis). This osmotic stress is one reason why so few organisms can survive growing in sugary jellies and jams or on dried and salted meats like jerky.

One way that halophiles avoid plasmolysis and even flourish in high-salt environments is by keeping a high concentration of organic materials and ions in their cytoplasm. This allows them to retain their water in the high-solute concentration of their surroundings. Organisms such as S. *aureus* that tolerate higher solute concentrations but may not grow especially well in them are called *facultative halophiles*. These bacteria that thrive on our skin despite salty sweat can withstand solute concentrations of up to 15 percent.

Most **extremophiles**, or organisms that live in extremes of pH, temperature, and/or salt, are exposed to a combination of stresses. For example, some psychrophiles live in saltwater veins found in arctic ice because the high salt prevents the water from freezing. Consequently, these microbes would be halophiles as well as psychrophiles, so they are called *halo-psychrophiles*. An archaeal cell living in a soda lake would simultaneously have to manage stresses caused by high salt and high pH, so it would be called a *halo-alkaliphile*. Pathogens are not usually found in extreme environments and instead prefer body temperature and neutral tissue and blood pH. Thus, pathogens are typically considered *mesophilic neutralophiles*. In every environmental niche, there are specific conditions to which microbes adapt for survival.

CHEM • NOTE

Osmosis is the net movement of water from an area of low-solute (salt, sugar, or other dissolved substance) concentration to an area of higher solute concentration across a selectively permeable membrane. It does not require energy to occur. Cells experience osmotic stress when they are placed in environments where their water balance is disrupted.

Oxygen Requirements

Because humans depend on oxygen for survival, the thought of life without abundant oxygen rarely comes to mind. However, oxygen can be toxic to certain cells, and many microbes have evolved ways to live either entirely without it or with minimal amounts. For example, oxygen levels are much lower beneath the soil or within silt deposits deep in lakes and oceans than they are at Earth's surface, and many microbes thrive in these environments. Most pathogenic microbes also thrive in low-oxygen environments within their host.

In contrast, many microbes thrive in environments where they are continuously exposed to oxygen. To understand how they do it, let's look at how oxygen harms cells. Atmospheric oxygen (O_2) easily diffuses across cell plasma membranes. Once inside the cell, some of the oxygen is converted into **reactive oxygen species** (ROS).[5] These reactive intermediates include superoxide ions (O_2^-) and hydrogen peroxide (H_2O_2), both of which can rapidly damage cellular proteins and DNA. Many microbes have evolved ways to detoxify ROS so that they can safely use oxygen in their metabolism; such organisms are said to be **aerobes**.

Within the aerobes there are varying degrees of oxygen dependence and tolerance, which reflect the degree to which an organism can detoxify ROS. Most aerobes rely on **antioxidants**—compounds and enzymes that reduce the effects of ROS—to survive in aerobic environments. For example, many aerobic microbes use antioxidant enzymes called superoxide dismutases to convert reactive superoxide ions to hydrogen peroxide; they then convert the hydrogen peroxide to water and oxygen using the enzyme catalase. Some organisms (including humans) are called **obligate aerobes**—they have an absolute dependence on oxygen for cellular processes and will die unless it is abundant. In contrast, **microaerophiles** use only small amounts of atmospheric oxygen and live in low-oxygen settings where they can limit their exposure to ROS while still meeting their oxygen needs.

Organisms that do not use oxygen in their metabolic processes are **anaerobes**. One subtype is the **aerotolerant anaerobes**; they tolerate atmospheric oxygen, even though they do not use it in their metabolic processes. Like aerobes, these microbes have ways to deactivate ROS. On the other hand, **obligate anaerobes** do not use oxygen in their metabolism, and they tend to die in aerobic environments

[5] Reactive oxygen species (ROS) is a term used to collectively refer to molecules that contain oxygen and may damage cells.

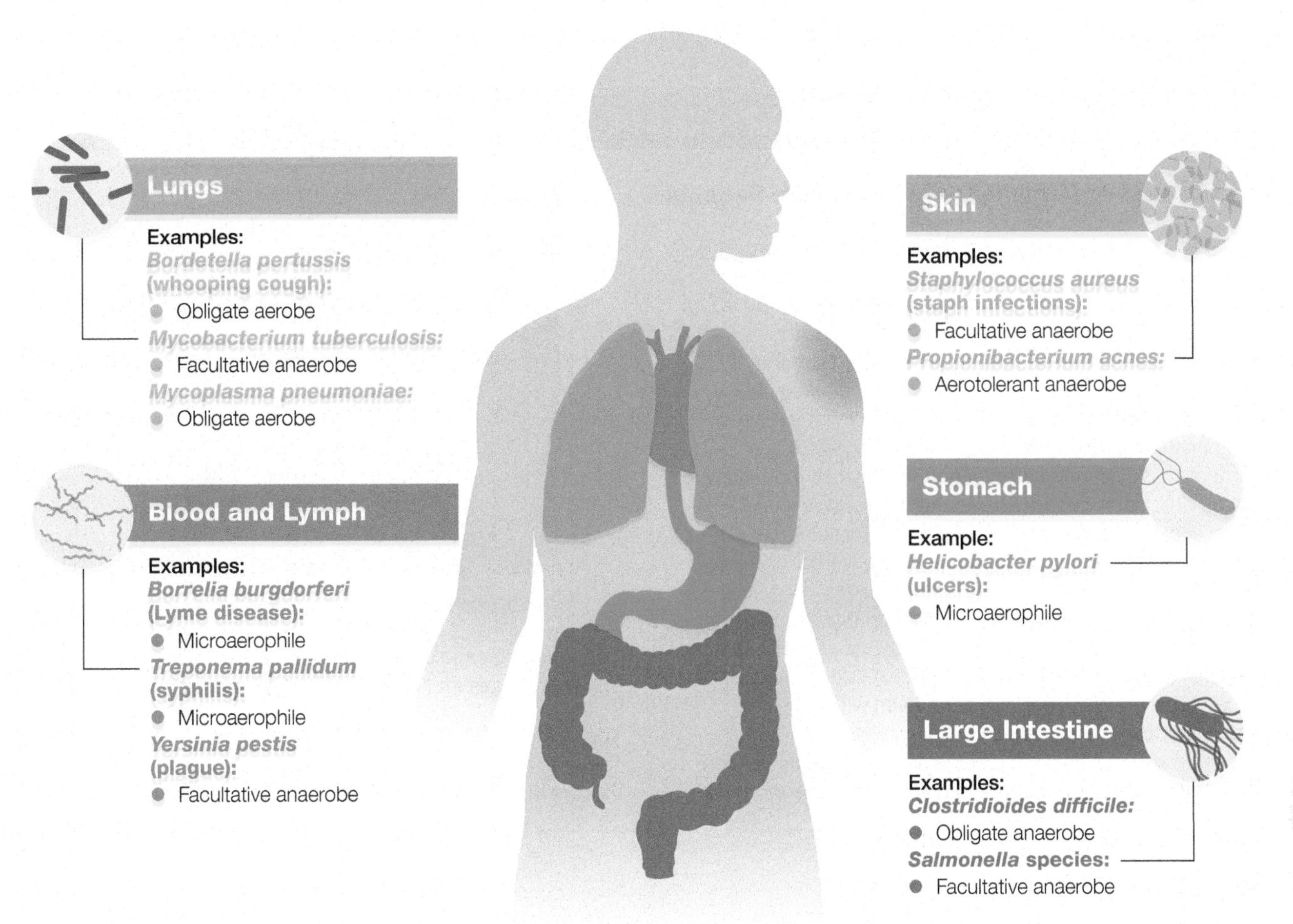

FIGURE 7.10 Select pathogens and their oxygen profiles Bacterial pathogens are shown. Note that most pathogens are not obligate aerobes.

Critical Thinking *Why would bacterial pathogens that infect the lungs evolve to be facultative anaerobes when they infect an area rich in oxygen?*

because they struggle to eliminate ROS. A final group of microbes that spans both aerobic and anaerobic environments are **facultative anaerobes**. They can be compared to the "switch hitter" in baseball who can bat right-handed or left-handed. Facultative anaerobes have enzymes to detoxify ROS, and they can switch between using oxygen for metabolism or employing fermentation as an anaerobic form of metabolism. Ultimately, facultative anaerobes tend to grow best when oxygen is present, but can switch to anaerobic metabolism if necessary. FIG. 7.10 presents examples of pathogens and their oxygen-tolerance profiles. (For more on aerobic and anaerobic cellular metabolism, see Chapter 8.) TABLE 7.1 summarizes terminology related to how microbes use and tolerate atmospheric oxygen.

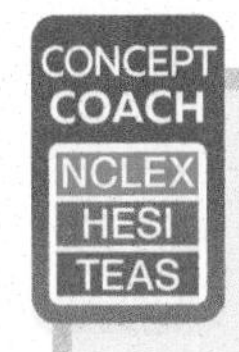

The Bacterial Growth Curve and an Overview of Aerotolerance

Bring the art to life! Visit the **Mastering Microbiology** Study Area to watch the Concept Coach and master this topic.

TABLE 7.1 Oxygen Use and Tolerance Classifications

	Obligate Aerobe	Obligate Anaerobe	Microaerophile	Aerotolerant Anaerobe	Facultative Anaerobe
Appearance of growth reflects oxygen tolerance/use. Medium is semisolid thioglycolate, which generates an oxygen gradient ranging from an oxygen-rich environment at the very top to an anaerobic environment at the bottom.					
Oxygen use in metabolism	Absolute dependence	Not used	Small amounts	Not used	Prefer using oxygen but can survive without it
Can effectively manage reactive oxygen species?	Yes	No	Yes (but only low amounts)	Yes	Yes

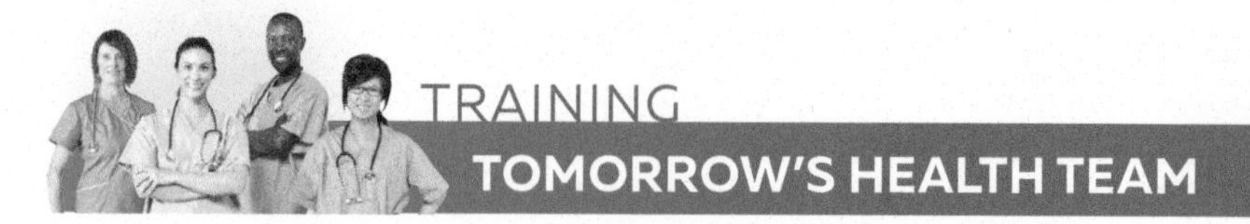

Pseudomonas and Its Persistence in Healthcare Settings

The hardy bacterium *Pseudomonas aeruginosa* can survive on very limited nutrients in many environments. Deemed an environmental pathogen, it has become a problem for healthcare facilities and is one of the top microbes that cause healthcare-acquired infections (HAIs).

P. aeruginosa can grow as a biofilm on the interior surface of hospital water pipes, as well as within humidifiers, catheters, and endoscopes. It can live on bandages, on operating room surfaces, and in contaminated disinfectants. Like most pathogens, it is a chemoheterotroph and secretes a collection of enzymes that scavenge for available organic compounds. Able to infect almost any tissue of the body, it is responsible for a wide array of infections that usually begin with accidental transmission by a healthcare worker or through inadequate equipment sanitation.

P. aeruginosa infections are common in burn patients and cystic fibrosis patients, as well as in the elderly and anyone with reduced immune function. An additional clinical concern is that *Pseudomonas* isolates are increasingly multidrug-resistant. Even with antibiotic treatment, only the top layer of cells in a biofilm would be killed, allowing deeper layers to survive and serve as a continued infection source. New methods are being developed to combat *Pseudomonas* contamination and infections, such as the use of *Pseudomonas*-specific phages and drugs that disrupt the biofilm formation process.

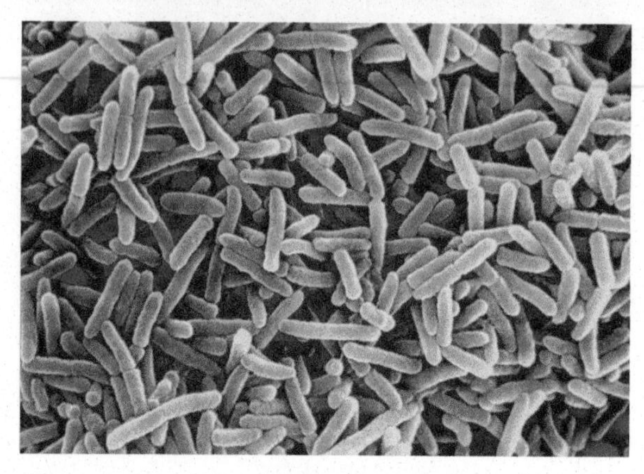

Pseudomonas aeruginosa easily thrives in multiple environments.

QUESTION 7.1

If you wanted to develop a novel drug to combat Pseudomonas biofilm formation, what is one feature of the microbes that could be targeted?

Microbes require nutrients, growth factors, and a source of energy.

Microbes use nutrients from the environment to divide and to build structural components, enzymes, and other factors. In order to grow microbes in a laboratory setting, we must understand what nutrients and growth factors they need.

Essential Nutrients

About 90 percent of a cell's dry weight is carbon, hydrogen, nongaseous oxygen, and nitrogen. Other important elements include sulfur, phosphorus, potassium, sodium, calcium, magnesium, chlorine, and various metal ions (ions of copper, zinc, cobalt, or iron, for example). These essential nutrients are required to build new cells and can be found in the organic and inorganic compounds of a microbe's environment. Essential nutrients that an organism needs in large quantities, such as the elements that make up the bulk of a cell's dry weight, are often called *macronutrients*. Nutrients used in very small amounts are referred to as trace nutrients or *micronutrients*.

Carbon alone constitutes at least half of a cell's dry weight and is found in virtually every component of the cell. Microbes can be categorized according to how they obtain organic carbon. **Heterotrophs** require an external source of organic carbon such as sugars, lipids, and proteins. These cells then extract the carbon from these nutrients for use in various cell parts. **Autotrophs** are "self-feeding" organisms that use a process called **carbon fixation** to convert inorganic carbon into organic carbon. Consequently, these organisms, which include photosynthetic microbes and plants, do not require an external source

CHEM • NOTE

Organic carbon is carbon that is bonded to hydrogen. Methane (CH_4), sugars, lipids, and proteins are all examples of organic carbon sources. Inorganic carbon such as that found in carbon dioxide (CO_2) and carbon monoxide (CO) is not bonded to hydrogen.

of organic carbon. (For more on photosynthesis and carbon fixation, see the Chapter 8 online appendix.)

Nitrogen and phosphorus are key ingredients of nucleotides, which are used as building blocks for nucleic acids (DNA and RNA) and as sources of energy in a cell (ATP, for example) (see Chapter 2 for more on nucleotides). Nitrogen is also a major component of amino acids, which are the building blocks of proteins. Most of a cell's nitrogen and phosphorus are extracted from organic nutrients. However, some cells get their nitrogen directly from the atmosphere in a process called **nitrogen fixation**. This process is an important step in converting atmospheric nitrogen from a gas to a nongaseous form, like ammonia, that certain species may then metabolize.

CHEM • NOTE

Nitrogen bases (adenine, guanine, cytosine, uracil, and thymine) are used as a component of nucleotides. Nucleotides in turn are used to build nucleic acids (DNA and RNA), and they also serve as energy-rich molecules (such as ATP) in a cell.

Growth Factors

Some microbes can't build all their organic precursors like amino acids, vitamins, or nitrogenous bases from scratch. In such cases, cells import these required substances from their environment. The necessary substances that a cell can't make on its own are called **growth factors**—the organism will not grow if these factors are missing from the environment.

Many intracellular bacteria such as *Bordetella pertussis*, which causes whooping cough, have become so dependent on their host that they have lost the ability to make most of their own amino acids and nucleotides and instead must get these substances from their environment. This is much like people becoming so dependent on matches that the skill of building a fire without them is lost. Organisms that need multiple growth factors are said to be **fastidious**. When growing fastidious microbes in the laboratory, amino acids, vitamins, and/or nitrogenous bases must be supplied in the growth medium—this is in addition to supplying all the other essential nutrients and an energy source.

Energy Sources

In order to carry out certain functions and construct various cellular parts, cells require energy. Organisms that use light energy are **phototrophs**. Organisms that break down chemical compounds for energy are **chemotrophs**. FIG. 7.11 shows carbon sources for phototrophs and chemotrophs.

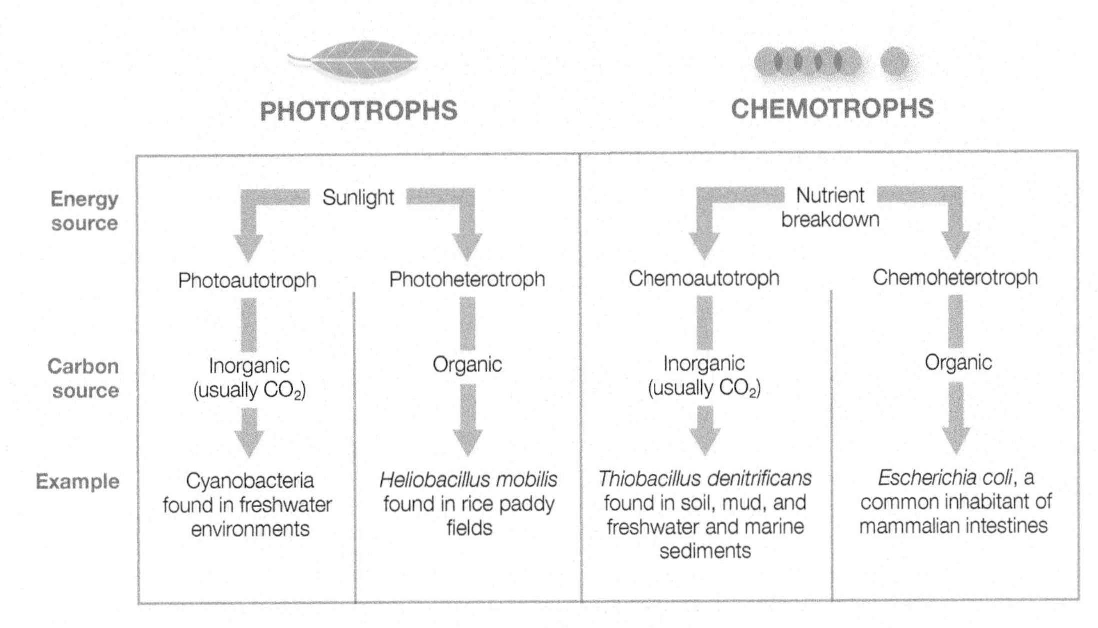

FIGURE 7.11 **Carbon sources for phototrophs and chemotrophs**

Build Your Foundation

5. How do the terms *optimal*, *minimum*, and *maximum* apply to temperature and pH conditions? (LO 7.5)
6. Using temperature classification terms, describe the types of microbes that would thrive in the following ecosystems: a deep-sea hydrothermal vent, the human intestinal tract, or beneath a glacier. (LO 7.6)
7. Define the terms *acidophile*, *alkaliphile*, and *neutralophile*. Explain ways microbes may adapt to pH extremes. (LO 7.7)
8. What are halophiles, and what adaptations do they possess to manage osmotic stress? (LO 7.8)
9. Name the various classes of microbes according to their oxygen use and tolerance. Explain why oxygen is toxic to anaerobes. (LO 7.9)
10. Define the terms *essential nutrient* and *growth factor*, and describe the energy sources required by phototrophs versus chemotrophs. (LO 7.10)

Build Your Foundation (BYF) Quick Quiz: Visit the **Mastering Microbiology** Study Area to quiz yourself.

7.3 GROWING, ISOLATING, AND COUNTING MICROBES

Learning Outcomes

After reading this section, you should be able to:

7.11 Describe the three formats of media, and state when each would be used.

7.12 Explain the features and uses of complex, defined, selective, and differential media.

7.13 Provide examples of how to culture anaerobic microbes.

7.14 Describe the considerations made in collecting clinical samples.

7.15 Summarize the streak plate technique, and state its purpose.

7.16 Describe direct and indirect methods for cell enumeration.

Microbes are grown using various media.

An assortment of culture media helps us grow microbes (FIG. 7.12). We classify media by their physical state (liquid, solid, or semisolid formats), their chemical composition, and their function.

Physical State: Liquid, Solid, and Semisolid Media

The physical state, or format, of media used depends on the application. Liquid formats, also called broth media, are ideal for growing large batches of microbes. However, they are also used to study certain metabolic properties of an isolated bacterium. Solid media are useful for isolating colonies and observing specific culture characteristics. They are also used to isolate bacteria into pure cultures using the streak plate isolation technique. (See more on the streak plate technique in Chapter 1.) The most common application of semisolid media is for the motility test to determine if an isolated specimen is able to move.

Broth media are made by adding various nutrients to purified water. The prepared media are then poured into flasks or tubes and sterilized. Solid and semisolid media are made by adding a powdered polysaccharide called agar to liquid media; most liquid media can be easily converted to a solid or semisolid format if needed for the given laboratory application. After being mixed with agar, the medium is poured into flasks or bottles to be heat sterilized; the heating process also melts the agar. While still hot, the liquefied agar-containing medium is poured into petri plates or test tubes and allowed to cool so that it sets (much like Jell-O gelatin must cool to set). The medium hardens as it cools to create a firm yet gelatin-like consistency. The more agar that is used, the firmer the medium—so semisolid media contain less agar than solid media. Gelatin itself is not a standard solidifying agent in most microbiological media because many microbes can break it down, which would cause the media to liquefy.

Occasionally solid media are poured into test tubes and allowed to cool at an angle to create a slanted surface for inoculation. These preparations are often described as **slants**. The Simmons citrate test and triple sugar iron (TSI) media are common examples of slants. If the test tube is placed in an upright position while the medium solidifies, it is called a **deep**. The motility test uses a semisolid medium that is poured into test tubes and allowed to set as a deep (FIG. 7.13).

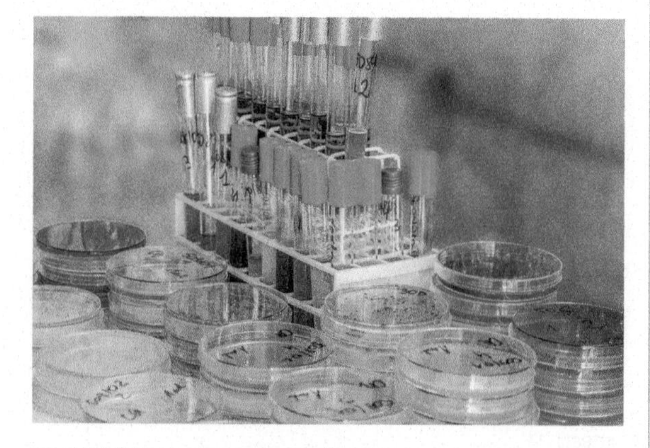

FIGURE 7.12 **Growth media** There are hundreds of different types of media to cultivate microbes. Liquid media are poured into flasks or test tubes; solid media are poured into plates or test tubes.

Chemical Composition (Complex and Defined Media)

After determining what media formats would best serve the laboratory goal, the next feature to consider is what macro- and micronutrients to add to the media. Based on their chemical composition, media can be described as either defined or complex (TABLE 7.2). **Defined media** (also called synthetic media) have a chemically defined or precisely known composition; each organic and inorganic component is completely known and quantified. A carbon source, such as a sugar like glucose, is added along with other macro- and micronutrients (calcium chloride and potassium phosphate, for example). Additional growth factors, like specific amino acids and vitamins, may also be added. We use defined media to grow microbes in a specific environment under exactly measured conditions, as is required in certain microbiological assays. An example of a microbiological assay that uses defined media supplemented with the amino acid histidine is the Ames test. (See more on the Ames test in Chapter 5.) Fastidious organisms that have complex nutrient requirements and most pathogens isolated directly from patients don't usually grow well in defined media. But defined media are useful for growing certain autotrophs and some heterotrophs.

Complex media (also called enriched media)[6] contain a mixture of organic and inorganic nutrients present in complex ingredients like blood, milk proteins, or yeast extracts. These are chemically sophisticated mixtures for which we don't know the precise quantity of every vitamin, carbon nutrient, or other ingredient. Consequently, there are variations among media batches. For example, the level of glucose in one blood sample to the next can vary, as can the amount of a given amino acid or a particular vitamin. Complex media are often used to grow fastidious organisms with complex growth requirements. They are also useful for growing a broad collection of heterotrophs and are ideal for isolating pathogens from clinical samples.

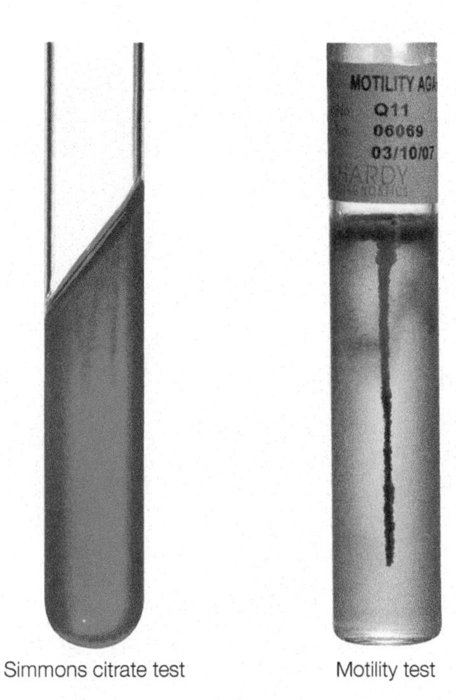

Simmons citrate test   Motility test

FIGURE 7.13 **Slant and semisolid growth media** *Left*: When agar solidifies at an angle, it forms a slant; the Simmons citrate test is an example of such media. *Right*: The motility test is an example of a semisolid media that sets as a deep.

Function (Differential and Selective Media)

A urine sample, throat swab, and fecal (stool) sample are all examples of clinical specimens commonly analyzed in a microbiology lab to determine if a particular pathogen is present. These clinical samples are not pure cultures; they contain a wide variety of microbes, most of which are harmless, normal microbiota. Therefore, a clinical microbiologist must be able to separate out pathogens from among the normal microbiota. Some of the most helpful tools to do this are differential and selective media. These media are often a solid format poured into petri plates. Although we'll introduce differential and selective media properties

TABLE 7.2 Complex and Defined Media Examples

Complex Media **Example: Luria–Bertani Media** **(Ingredients in 1 Liter of Liquid Medium)**	**Defined Media** **Example: Glucose Minimal Salts Media** **(Ingredients in 1 Liter of Liquid Medium)**
• 10 grams of tryptone (broken-down milk protein) • 5 grams of yeast extract (ground-up yeast cells) • 10 grams of sodium chloride (NaCl) • Water added to bring level to 1L	• 20 mL of a 20% glucose solution • 2 mL of 1 molar magnesium sulfate $MgSO_4$ • 0.1 mL of 1 molar calcium chloride $CaCl_2$ • 12.8 g of disodium phosphate Na_2HPO_4 • 3.0 g of potassium phosphate KH_2PO_4 • 0.5 g of sodium chloride (NaCl) • 1.0 g of ammonium chloride NH_4Cl • Water added to bring level to 1L
Although plenty of amino acids and sugars are provided in the complex yeast and protein preparations, we don't know the exact levels of each.	Growth factors like vitamins and amino acids are not added to minimal salts media, so any organism grown on it must be able to take the nitrogen supplied in the ammonium chloride and use it to build all necessary amino acids.

[6] Sometimes people add specialized ingredients (such as specific amino acids) to defined media to help certain bacteria grow. In such cases, the person may call the media *enriched*; however, it would be more correct to say that the media is *supplemented*. The term *enriched* implies that a complex ingredient was added.

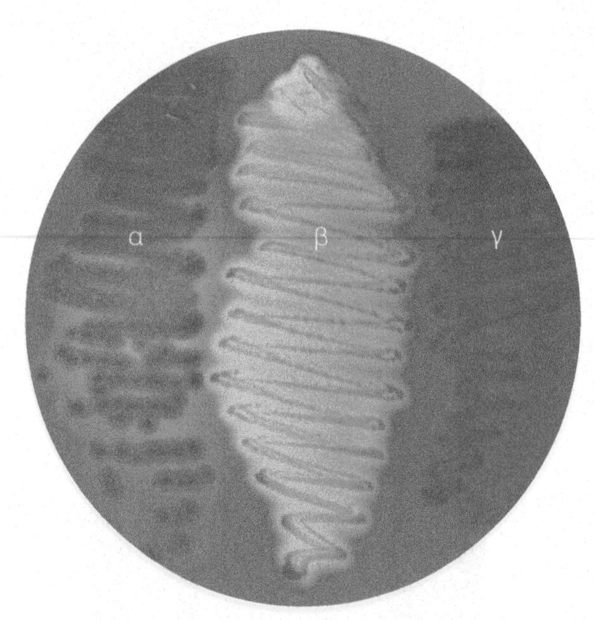

Blood agar inoculated with bacteria that exhibit alpha, beta, and gamma hemolysis.

FIGURE 7.14 Hemolysis Alpha (α) hemolysis is characterized by a green zone around colonies; it results from hemoglobin oxidation. Beta (β) hemolysis completely breaks down red blood cells and hemoglobin thereby generating a yellow zone in the media. In gamma (γ) hemolysis, there is no red blood cell breakdown, nor is there hemoglobin oxidation.

Critical Thinking *Is blood agar complex or defined media? Explain your reasoning.*

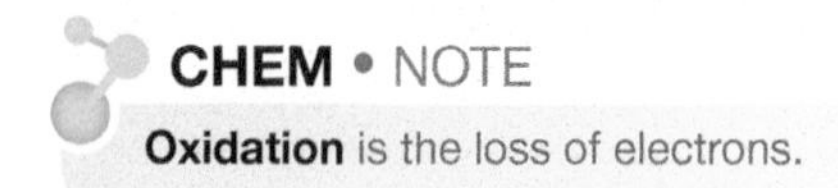

CHEM • NOTE

Oxidation is the loss of electrons.

separately, it's important to understand that media are often formulated to be simultaneously differential and selective.

Differential media Say that a patient, 4-year-old Michael, has had a terrible sore throat and fever. As his healthcare provider, you suspect streptococcal pharyngitis, or "strep throat," so you swabs his throat. At the microbiology lab, a technician checks the collected sample for the presence of the pathogen *Streptococcus pyogenes*. This is done using **differential media**—specialized media that are formulated so that we can *visually* distinguish one microbe from another based on how they metabolize media components. The formulation of differential media makes bacterial colonies with different physiological properties from one another appear as different colors on the medium—or the medium itself may change color in response to the presence of certain bacteria.

A common example of a differential medium is **blood agar**. It contains sheep red blood cells that serve as both a nutrient and a differentiating indicator. Some pathogenic bacteria lyse (break down) red blood cells for nutrients by releasing enzymes called **hemolysins**—specialized proteins that can affect hemoglobin, red blood cells, white blood cells, and other cells. When grown on blood agar, bacteria that make hemolysins (such as the pathogen *S. pyogenes*) lyse red blood cells in the media and break down the hemoglobin in the red blood cells generating a clear and slightly yellow-appearing zone around colonies. Bacteria that can do this are said to be **beta hemolytic** (usually written as β-hemolytic). Therefore, if Michael's throat swab reveals β-hemolytic bacteria, the lab can confirm he has streptococcal pharyngitis.

Some bacteria do not fully lyse red blood cells and instead just oxidize hemoglobin, the oxygen-carrying component of blood. Bacteria with this property are often described as conducting partial hemolysis or as being **alpha hemolytic** (usually written as α-hemolytic); they turn blood agar a green color. Examples of α-hemolytic bacteria are *Streptococcus pneumoniae* and several types of streptococci found among the normal oral microbiota. Bacteria that do not lyse red blood cells nor oxidize hemoglobin are described as **gamma hemolytic** (usually written as γ-hemolytic) (FIG. 7.14).

Selective media To become a healthcare provider, students must take certain courses and pass licensing exams. Not everyone ultimately makes the cut, because it's a selective process. Similarly, **selective media** single out bacteria that have specific properties. This is accomplished by including ingredients in the media that foster the growth of certain bacteria while suppressing the growth of others.

Mannitol salt agar (MSA) is selective as a result of its high salt content. Most bacteria can't grow on this medium. Notable exceptions are *Staphylococcus* species, which are Gram-positive bacteria that tolerate high salt and are commonly found on our skin. Certain other bacteria, like *Bacillus subtilis,* can also grow on this medium, so MSA is not considered strictly selective for *Staphylococcus* species—rather, MSA is best described as selective for bacteria that can tolerate high salt concentrations. As previously mentioned, many media are simultaneously selective and differential (FIG. 7.15). While MSA selects for the growth of bacteria that can tolerate high salt levels, it also differentiates organisms based on their ability to ferment a sugar called mannitol. Pathogenic *Staphylococcus* species like *S. aureus* are mannitol fermenters, whereas most nonpathogenic *S. epidermidis* cannot ferment mannitol. If an organism growing on MSA ferments mannitol, then it will turn the normally red MSA medium a bright yellow color.

Eosin methylene blue agar (EMB) is another common medium with selective and differential capabilities. It contains the dyes eosin and methylene blue, which limit Gram-positive bacterial growth, while allowing Gram-negative bacteria to grow. It also helps us to differentiate among Gram-negative bacterial species based on their ability to ferment the sugar lactose. Most lactose-fermenting bacteria appear as dark pink colonies on EMB, though a metallic

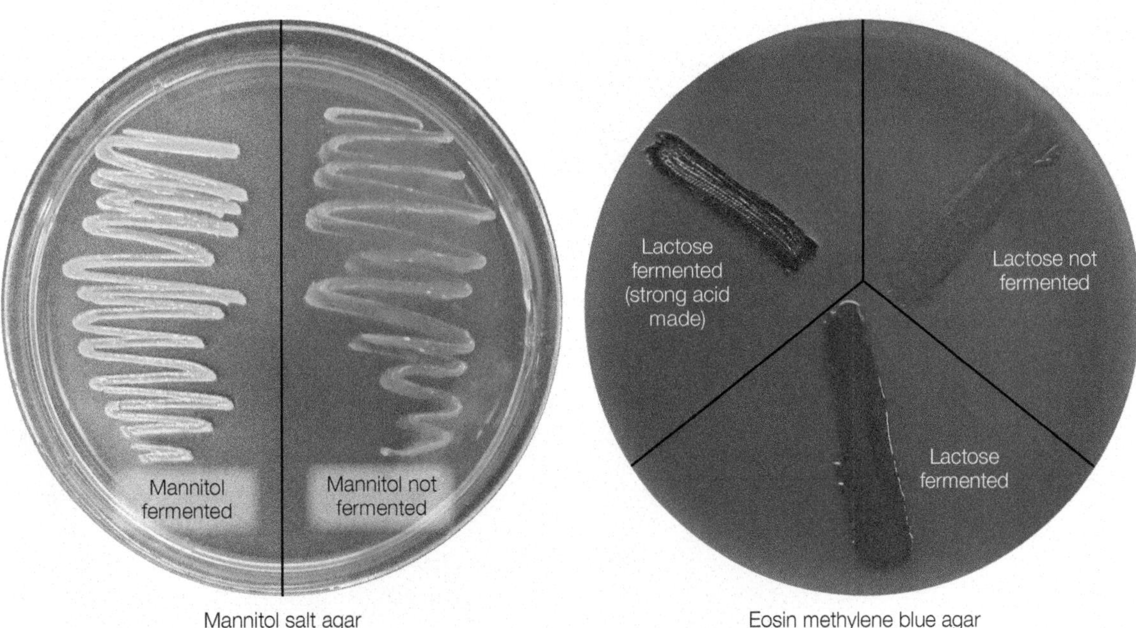

FIGURE 7.15 Selective and differential media Most selective media also have differential properties. *Left*: Mannitol salt agar (MSA) selects for bacteria that tolerate high salt, like *Staphylococcus* species. It differentiates based on mannitol fermentation. *Right*: Eosin methylene blue agar (EMB) is selective for Gram-negative bacteria. It differentiates Gram-negative bacteria based on lactose fermentation.

Critical Thinking *Assume you observe pink growth on an EMB plate. What two conclusions can you make about the growing bacterium?*

green coloration is often seen for bacteria such as *E. coli* that make strong acids while fermenting lactose. Gram-negative bacteria that don't ferment lactose, such as *Pseudomonas aeruginosa* and *Salmonella enterica*, appear colorless on EMB agar.

Anaerobic Media

Because our environment is naturally aerobic, there are challenges in isolating, transporting, and culturing anaerobic organisms. On average, anaerobes make up about 38 percent of the bacterial population in a wound.[7] Clinicians must be careful when collecting anaerobic samples from deep wounds, blood, the vagina, cerebrospinal fluid, or other sites because exposure to oxygen can kill the harvested microbes before they can be cultured for identification. To prevent this, samples must be quickly deposited and capped in vials or tubes in media that are depleted of molecular oxygen. Molecular oxygen is removed from media in a number of ways. One common method adds a reducing agent like **thioglycolate**, which converts molecular oxygen to water. Culturing anaerobic organisms such as *Bacteroides*, often associated with abdominal infections, requires blood agar plates made anaerobically and maintained inside an anaerobic chamber.

Sometimes a small anaerobic jar or chamber is used. The sample is dropped into it, and then a packet of oxygen-reacting chemicals is opened inside it, creating the necessary anaerobic (oxygen-free) conditions. An anaerobic chamber is usually a large box that is continually maintained anaerobically. Gloves are inserted into the chamber to allow handling of anaerobic organisms within the chamber. Samples and equipment are placed in a side compartment, and then nitrogen and/or CO_2 are piped into the chamber to displace all oxygen before materials are moved into the main part of the anaerobic chamber (FIG. 7.16).

Collecting, isolating, counting, and identifying microbes are important in microbiology.

Aseptic techniques are methods designed to prevent the introduction of contaminating microbes to a patient, a clinical sample, or others in the healthcare setting. They protect people and promote sample integrity. Healthcare workers

[7] Although newer resources also cover this content, this is one of the best review articles available on the topic: Bowler, P. G., Duerden, B. I., & Armstrong, D. G. (2001). Wound microbiology and associated approaches to wound management. *Clinical Microbiology Reviews*, 14(2), 244–269.

FIGURE 7.16 **Methods to grow anaerobic cultures** *Left*: Anaerobic jar and media. *Right*: Anaerobic chamber with air lock port to remove oxygen before placing in the chamber.

Critical Thinking *How could biofilms promote infections by anaerobic bacteria even in an aerobic setting?*

apply these techniques as a routine part of patient care when administering an injection or collecting a clinical sample for microbiological analysis. Clinical analyses often involve pathogen isolation, identification, and enumeration. While this chapter focuses on bacteriological methods, proper aseptic technique is just as important to studying viruses. (Methods for viral identification and enumeration were covered in Chapter 6.)

Collecting Samples for Clinical Microbial Analysis

A patient's preliminary diagnosis, as well as the stage and site of infection, affect how and when a representative sample is obtained for microbiological analysis. All clinical samples must be collected and transported aseptically to prevent contamination with microbes from the environment or from the healthcare worker collecting the sample. This requires that samples are collected using sterile materials (that is, swabs and containers must be devoid of all microbes before they contact the sample). The specific containers used depend on the tests that need to be done. Specimen containers often include a transport medium to protect sample viability. Proper hand washing before and after specimen collection, wearing gloves, quickly sealing samples in containers, and carefully sampling only the site in question without touching other body structures or surface sites (for example, avoiding contact with the tongue during a throat swabbing) all help to ensure that a "clean" and representative sample is obtained.

The properties of the suspected causative agent must also be considered during sample collection. If an anaerobic organism is suspected, the sample must be collected and transported using anaerobic methods. Another consideration is the type of equipment used. Even something as simple as the wood of a wooden-handled swab may inactivate certain viruses or interfere with certain bacterial cultures. For example, if *Chlamydia trachomatis* is the suspected pathogen, the clinical sample must be collected using a swab with a plastic or wire shaft and a noncotton tip because wood and cotton interfere with the bacterium's isolation. Once a sample is collected, it may have to be immediately refrigerated or placed on dry ice to ship to a laboratory. If samples are improperly collected, or left at the wrong temperature, and/or transport to the lab is delayed, then lab data are very likely to be compromised, leading to the wrong diagnosis and improper treatment.

Isolating Microbes Using the Streak Plate Technique

The first step in identifying a pathogen in a sample is to separate it from other organisms in the sample. The most common way to accomplish this is the **streak plate technique**, in which a culture is diluted on an agar plate in such a way that individual cells are separated from one another over the medium's surface. As cells divide, their population increases to form a mound of cells called a **colony**. Following an incubation period, numerous isolated colonies are easily seen growing on the agar plate (FIG. 7.17). If we assume that a single cell generates a single colony, then we can say that a colony represents a pure line of a given microbe. In practice, this assumption is commonly made, but it should be noted that it is impossible to guarantee that a couple of cells or more did not stick together and end up in the same area of the plate to generate a colony. If multiple species are growing on the plate, it is possible to see a mixture of colonies that exhibit a range of sizes, shapes, colors, elevation, sheen, textures, and other features.

When performing the streak plate technique, the proper solid medium must be chosen first. The choice depends upon the agent to be cultured. If the nutritional requirements of the bacterium to be isolated are well characterized, then one type of medium may be used; if not, then diverse media formulations would be used to optimize the chances of isolating the bacterium of interest. Next, a flame-sterilized inoculating loop is used to transfer bacteria from a sample to the chosen plated medium. By making a series of streaks across the plated medium's surface and sterilizing the loop between each streak, the culture will be sufficiently thinned to allow for colony isolation.

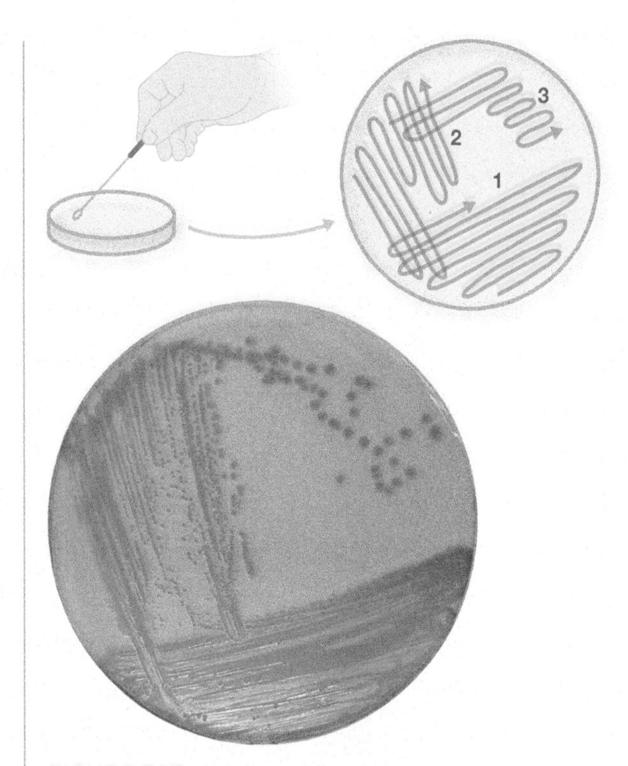

FIGURE 7.17 Streak plate method Some labs streak the plate over three sections, while others may use four quadrants. Whatever the method used, the basic goal is the same—thin the culture on the plate so individual colonies are easily observed.

Critical Thinking *If you see a variation in colony sizes on a streak plate of a pure culture, does that mean the culture was contaminated? Explain your reasoning.*

Methods for Counting Microbes

Sometimes it's important to determine how many microbes are in a sample. Food and beverage manufacturers, water treatment plants, and other industries must conduct microbial counts to ensure product quality. Clinical laboratories may use bacterial growth rates from patient samples to determine antibiotic susceptibility. Microbes in either mixed or pure cultures can be counted using direct or indirect methods—all of which estimate population size, rather than counting every single cell in a sample.

Direct methods These methods involve counting individual cells or colonies (plate counts). As the name implies, a **cell count** enumerates the number of cells in a small portion of the sample. Cell counts can be done using automated or manual procedures. Manual cell counting requires a microscope and a specialized counting chamber that has a volumetric grid etched on it (FIG. 7.18). A known volume of the sample to be counted is mixed with a dye and placed in the counting chamber. Individual cells within the etched grid are then counted and averaged to calculate how many cells are in a total sample. This method is easy and requires minimal reagents and equipment, which makes it relatively inexpensive. The disadvantages are that dead cells are not differentiated from living cells, and a fairly dense culture suspension is needed to make the count statistically accurate (at least 10^7 cells/mL). It's also labor intensive if numerous samples need to be analyzed.

For faster, highly reproducible results, automatic counting methods can be used to count the cells in a culture. A **Coulter counter** is a machine that counts the number of cells as they pass through a thin tube. An electronic counter detects the passing cells and keeps a tally. However, all cells, both living and dead, get counted. Another machine called a **flow cytometer** works similarly to the Coulter counter, but uses a laser light to detect cells passing through a

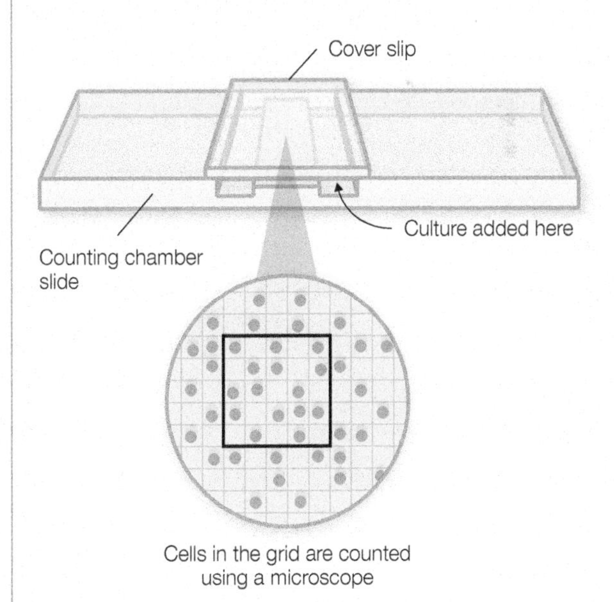

FIGURE 7.18 Direct cell count using a microscope A counting chamber is loaded with a sample of dyed cells, and the cells are directly counted using a microscope.

Critical Thinking *A manual cell count is still an estimate of the total population of cells in a sample. Explain why.*

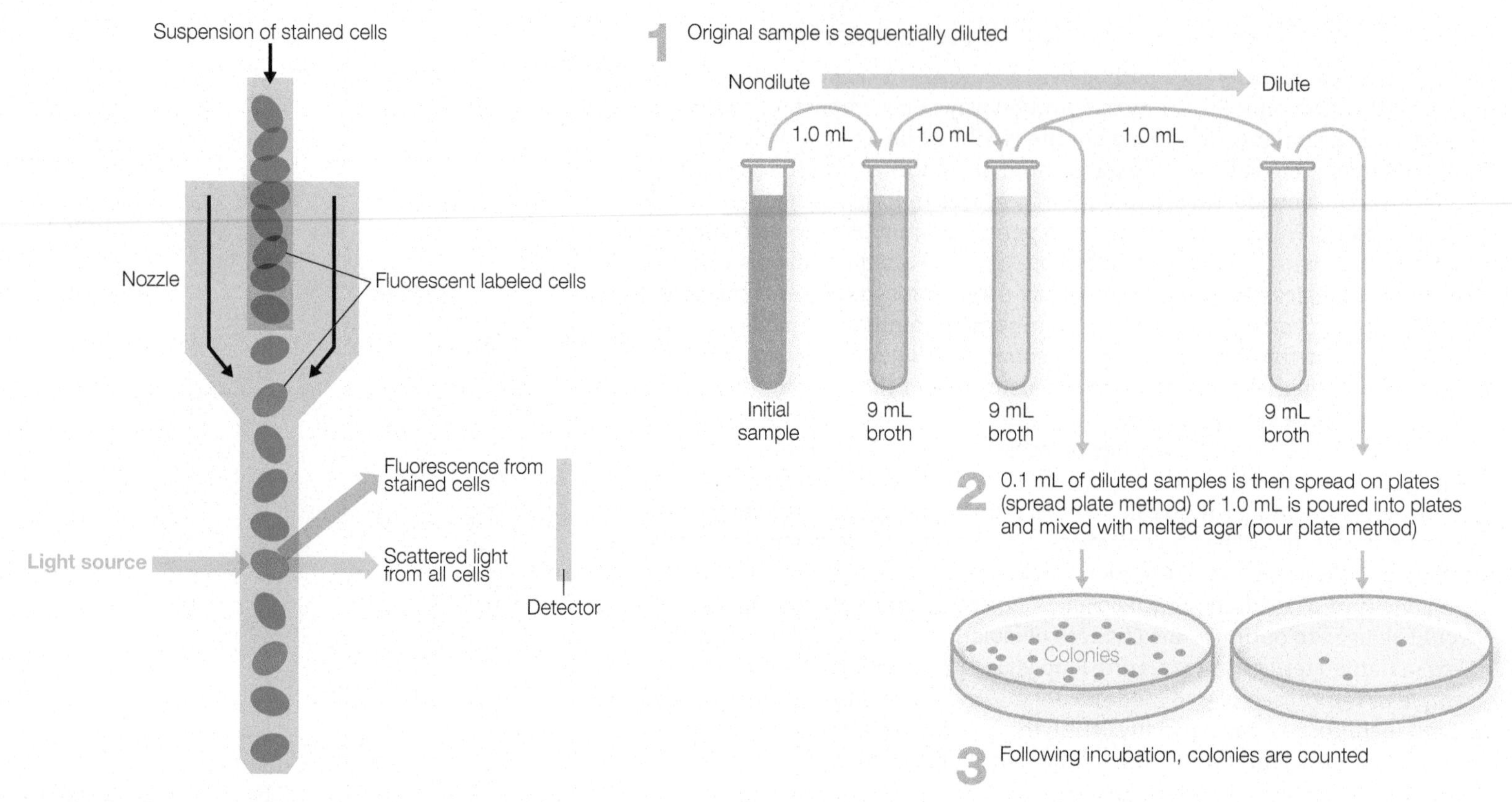

FIGURE 7.19 Flow cytometer A flow cytometer detects cells tagged with a fluorescent label. The type of tag selected for cell labeling determines which cells are counted. The device's detector is designed to differentiate between the scattered light from nontagged cells versus the fluorescence emitted from tagged cells.

Critical Thinking *How could this method be adapted to count two different species of bacteria in the same sample?*

FIGURE 7.20 Viable plate count for direct population enumeration

Critical Thinking *What would be an advantage of the pour plate method over the spread plate method in terms of the types of microbes that could be grown?*

narrow channel (FIG. 7.19). Cells must be fluorescently labeled before counting, but the advantage is the ability to differentiate one cell type from another by using different colored labels. Additionally, the flow cytometer can distinguish between living and dead cells, depending on the type of dye or tag used for the cells.

Another important method for directly enumerating a population is the **viable plate count** (FIG. 7.20). Samples from a liquid broth culture are serially diluted and either spread on solid agar (*spread plate method*) or poured into a petri plate with melted agar media (*pour plate method*). After an incubation period, colonies are visible and can be counted. If the spread plate method was used, then all the colonies will be on the surface of the media. If the pour plate method was used, then colonies are embedded in the media and on the surface. Taking the dilution factor into consideration, the number of counted colonies is then used to calculate the total number of living cells in the original undiluted sample. Numerical data for plate counts are usually represented as **colony-forming units** (CFU) per milliliter (or per gram). This verbiage reflects the fact that, although a single cell *may* give rise to a single colony, this is not guaranteed. This is because some cells will stick together and may even naturally grow as clumps or chains that won't be separated and could theoretically codivide to generate the colony being observed. The advantage of the viable plate count method is that it is very sensitive, detecting even small numbers of cells. A disadvantage is that only cells that can thrive on the media chosen for the application will grow, meaning noncultivable cells will not be enumerated.[8]

Indirect methods Whereas direct methods count colonies or cells, **indirect methods** rely on secondary reflections of overall population size.

[8] As discussed in other chapters in this text, noncultivable bacteria are those bacteria that fail to grow in cell culture, but they are known to be present because they can be detected in other ways (e.g., often via molecular methods that detect DNA).

TABLE 7.3 Direct and Indirect Cell-Counting Methods

Method	Name	Description
Direct	Microscopic count	Manual count of cells using a microscope and counting chamber; doesn't differentiate between living or dead cells
	Coulter counter or flow cytometer	Both are automated methods; Coulter counter doesn't differentiate between living or dead cells. Flow cytometer distinguishes cells by detecting fluorescent tags, and can differentiate between living and dead cells
	Viable plate count	Colonies that grow from a diluted cell sample account for original cell count
Indirect	Turbidity measurement	Spectrophotometer is used to measure cloudiness of a liquid culture
	Dry weight	Dry mass measurement
	Biochemical activity	Byproducts of cell growth are measured

One of the fastest and easiest ways to indirectly measure cell numbers is to measure the cloudiness, or **turbidity**, of a liquid culture. The more cells in the sample, the more turbid it is. A spectrophotometer, a machine that aims light at the sample, measures either how much light can pass through the liquid (percent light transmission) or how much light is absorbed (absorbance or optical density/OD) (FIG. 7.21). The downside to this type of measurement is that both living and dead cells account for turbidity in the growth medium.

Another indirect method for population enumeration is assessment of total dry weight. First a culture is centrifuged and all liquid withdrawn from the cell pellet, then the dried mass is weighed. Lastly, biochemical activity can indirectly reflect population growth. For example, an indirect test might detect an increase in the amount of DNA present, or an increase in wastes made as cellular metabolism byproducts. TABLE 7.3 summarizes direct and indirect enumeration methods.

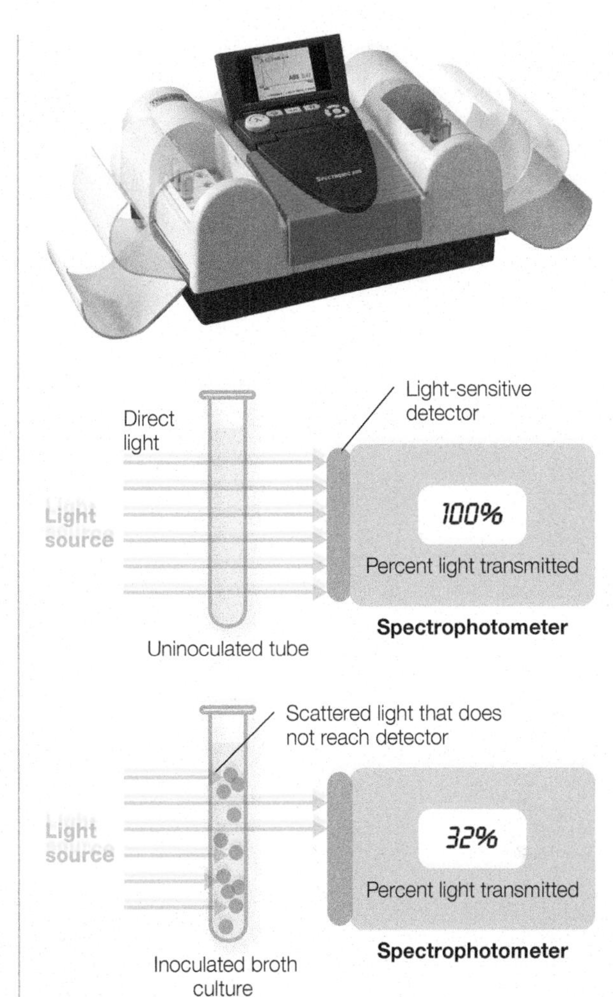

FIGURE 7.21 Spectrophotometer *Top*: This simple spectrophotometer measures absorbance and transmittance of light through a sample. *Bottom*: Inside the spectrophotometer, light is directed through the culture sample and measured.

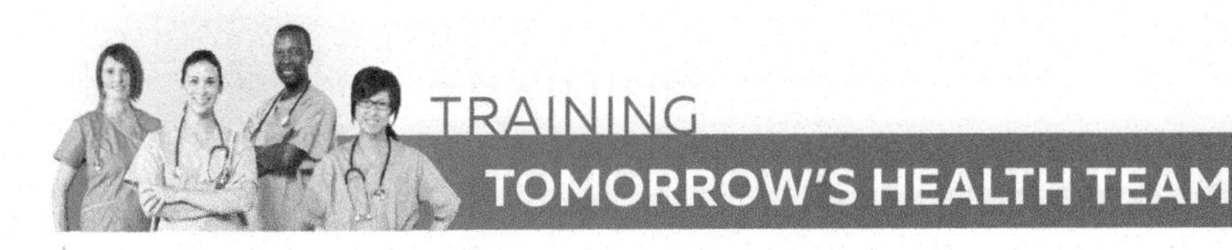

Decontaminating N95 Masks During COVID-19

During the COVID-19 pandemic, personal protective equipment (PPE) was often in short supply—especially N95 respirators, which are high-level filtration masks designed to protect the wearer from inhaled pathogens (including viruses). At the pandemic's outset, there were no data on how mask decontamination and reuse could alter filtration efficacy—no one knew if healthcare workers would truly be protected if their masks, which were designed for single use, were decontaminated and reused. As the pandemic surged on and N95 masks remained in high demand and short supply, mask reuse became necessary. The idea that something might be better than nothing was embraced while studies were launched as to which decontamination measure would be the most effective at reducing N95 contamination while still preserving the masks' filtering properties.

Four main decontamination strategies were tested: the application of 70% ethanol solutions, exposure to dry heat (70°C), UV light (260–285 nm) irradiation, or sterilization using vaporized hydrogen peroxide (VHP). Studies revealed that ethanol treatments eroded filtration efficacy the fastest and should be avoided. VHP treatments proved to be the best approach for preserving mask filtration efficacy while still achieving excellent decontamination results. The researchers concluded that, when faced with a critical PPE shortage, VHP is useful for decontaminating N95 masks up to 10 times, depending on the mask brand, without compromising mask filtration efficacy. Of course, PPE reuse is not a best practice, so once supply kept pace with demand, reuse was discouraged.

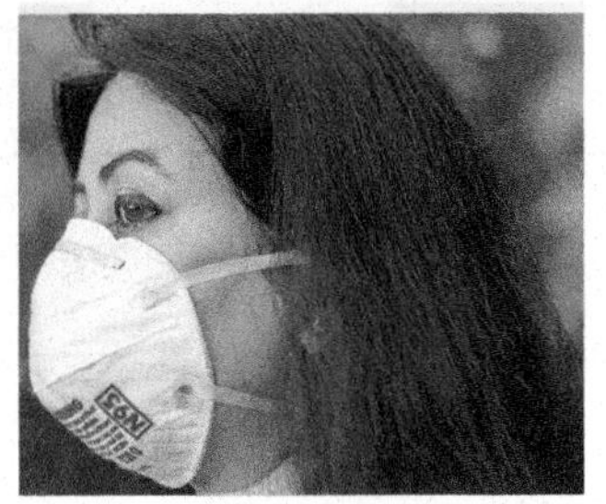

A person wearing an N95 respirator—a mask designed to protect the wearer from inhaled pathogens.

QUESTION 7.2

How might each of the tested decontamination approaches work to inactivate SARS-CoV-2, the virus that causes COVID-19?

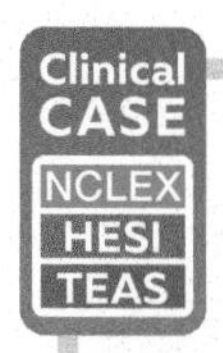

Clinical CASE NCLEX HESI TEAS

The Case of the Sickly Soaker

Practice applying what you know clinically: visit the **Mastering Microbiology** Study Area to watch Part 2 and practice for future exams.

Methods for Identifying Microbes

Physical, biochemical, and genetic analysis are all commonly used tools to identify microbes. Physical analysis usually involves microscopy and staining techniques, such as the Gram stain, to observe morphological features. Specific fluorescent labels can also be used to identify features that are known to exist on a particular species. (Staining and microscopy techniques are covered in Chapter 1.) Biochemical analysis involves a wide collection of media that assess metabolic properties of an isolated microbe. Many biochemical test media have selective and differential properties. (Because understanding biochemical tests and how to read them requires a basic understanding of metabolic pathways, we'll review these methods in more detail in Chapter 8.)

Genetic methods also help to quickly identify microbes. Probes that bind to specific genetic sequences can quickly classify pathogen species. Polymerase chain reaction (PCR) can reveal a genetic sequence indicative of a specific strain, even if it is present in small amounts. DNA "fingerprinting" is another genetic method that uses special enzymes to cut up genomic DNA of bacterial samples. Using electrophoresis separation methods, the genomic DNA fragments reveal a pattern unique to a species. Molecular methods such as these are more closely reviewed in Chapter 14.

Build Your Foundation

11. Describe a scenario for use of each of the three media formats. (LO 7.11)
12. To isolate Gram-negative organisms in a mixed clinical sample, what media format and formulation would you recommend? Explain why. (LO 7.12)
13. What techniques may be used to grow anaerobic microbes? (LO 7.13)
14. If you suspect a patient has septicemia (bacterial blood infection), what considerations would you need to take into account before collecting a sample for microbial analysis? (LO 7.14)
15. What is the underlying goal of the streak plate method, and what information does it provide? (LO 7.15)
16. Which enumeration method is best if time is not an issue and viable cell numbers are needed? Which is best if time is limited and viable cell numbers are needed? (LO 7.16)

BYF QUICK QUIZ

Build Your Foundation (BYF) Quick Quiz: Visit the **Mastering Microbiology** Study Area to quiz yourself.

7.4 BASICS OF MICROBIAL GROWTH REDUCTION AND DECONTAMINATION

Learning Outcomes

After reading this section, you should be able to:

7.17 Define the terms *decontamination*, *sterilization*, *disinfection*, *microbiostatic*, *microbiocidal*, *disinfectant*, and *antiseptic*.

7.18 Describe different heat treatments used to control microbe levels. State when each is used, and define the terms *decimal reduction time*, *thermal death point*, and *thermal death time*.

7.19 Describe radiation and filtration controls, and state when each is used.

7.20 Name and describe the various classes of germicides, and discuss what goes into selecting the right one for a given situation.

7.21 Name the levels of germicide activity, and identify the types used for critical, semicritical, or noncritical equipment.

7.22 Discuss the factors that should be considered when selecting an appropriate germicide.

7.23 Provide examples of how *Mycobacteria*, endospore, virus, protozoan, and prion levels are each controlled.

Control strategies aim to reduce or eliminate microbial contamination.

Microbes versus humans: We constantly wage a war to reduce or eliminate pathogens and food-spoiling microbes. In the United States, we often take for granted all the measures enacted to reduce the overall microbial load in our environments. Control measures occur in every area of our lives, from water sanitation to hospital and restaurant cleaning standards.

Decontamination measures remove or reduce microbial populations to render an object safe for handling. These measures fall into two key categories based on their final outcome: **Sterilization** eliminates *all* bacteria, viruses, and endospores, whereas **disinfection** just reduces microbial numbers. Sterilization is often required for drugs, objects used for invasive medical procedures, and media and glassware used to culture microbes in a lab. Disinfection is usually sufficient for most cosmetics, foods, surfaces, and medical equipment that doesn't contact internal body tissues. Sterilization and disinfection are achieved using a variety of chemical and physical control mechanisms. The method used depends on whether sterilization or disinfection is the goal, the nature of the substance being treated, and what microbes are present.

Temperature, radiation, and filtration are all physical methods to control microbial growth.

Physical controls most commonly include temperature changes (heat/cold), radiation, and filtration; they do not involve chemicals. These controls lead to disinfection or sterilization, depending on how they are applied.

Temperature Changes

Both cold and heat have important roles in controlling microbial growth. Refrigeration (4°C) and freezing (0°C) do not eliminate microbes, but they slow the growth of many associated with food spoilage and foodborne illnesses. In the laboratory, cold temperatures preserve specimen isolates and increase the shelf life of media and other reagents. Refrigeration also preserves clinical samples (blood, tissue biopsies, or swabs containing suspected pathogens) until they are delivered to the clinical lab for analysis. Sometimes freeze-drying (also called lyophilization or cryodesiccation) is used to increase food and medication shelf life. In this process the substance being preserved is subjected to freezing temperatures under low pressure to remove water.

Most microbes are sensitive to heat, so decontamination methods often involve heat processing. Depending on how it is applied, heat can be used to achieve either sterilization or decontamination. Different microbes exhibit varying heat tolerances, so protocols that decontaminate foods, drugs, and medical equipment take into account the temperature and time needed to either reduce or fully eliminate microbes. In some cases, sterilization isn't necessary to render a product safe, but microbe numbers need to be low. In such instances a **decimal reduction time** (DRT or *D value*) is an important measure. DRT is the time in minutes that it takes to kill 90 percent of a given microbial population at a set temperature.

Sterilization has other temperature-related parameters. The **thermal death time** is the shortest period of time that a given temperature must be held to kill all microbes in a sample. This information is useful if you know that a product can't be heated above a certain temperature without damaging or altering it. Another measure, the **thermal death point**, is the minimum temperature needed to kill all microbes in a sample within 10 minutes. Both the thermal death time and thermal death point are impacted by many criteria, including type and concentration of microbe(s) being targeted for reduction or elimination, the ingredients in the treated product, packaging, and if the treated substance is a solid or liquid.

Autoclaving, boiling, pasteurization, and dry heat are all heat-related control methods we will discuss next.

Autoclaving An **autoclave** is a machine that applies steam heat along with pressure to sterilize microbiological media and assorted medical or lab equipment. Most substances are rendered sterile within 20 minutes using standard autoclave settings, which exert a pressure of 15 pounds per square inch and steam heat at 121°C (294°F) (FIG. 7.22). Modified autoclave settings that increase pressure and temperature are needed to effectively destroy endospores. A downside of autoclaving is that only materials that withstand moisture, high heat, and pressure can be treated. Some plastics easily melt, and bottles may collapse. Certain media and drugs also can't be autoclaved without affecting their chemical integrity.

Canned goods, which are excellent environments for germination of *Clostridium botulinum* endospores, are sterilized using a large autoclave called a retort. The prepared food is placed in cans and sealed airtight using a vacuum. The cans are then subjected to sterilization temperatures in the retort that ensure endospore destruction.

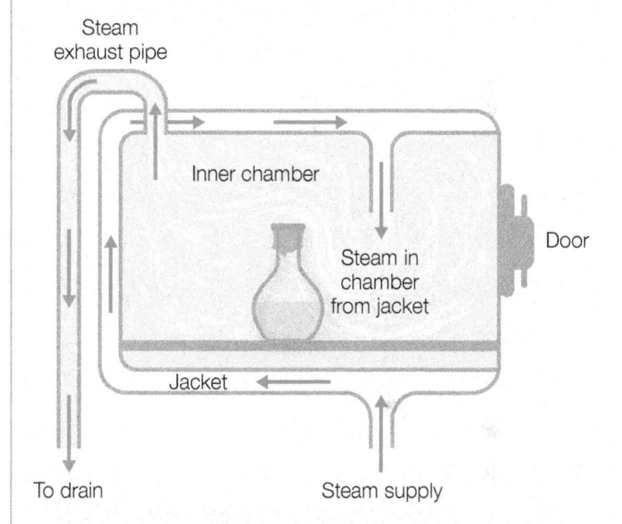

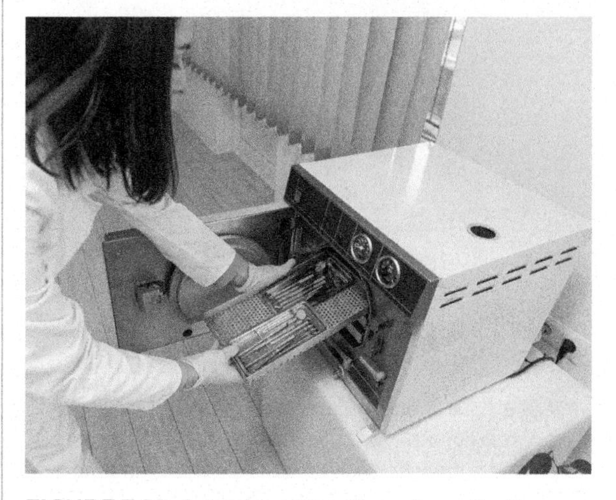

FIGURE 7.22 Autoclave *Top:* The schematic diagram shows how high heat, steam, and pressure are created in an autoclave. *Bottom:* A benchtop autoclave effectively sterilizes medical equipment.

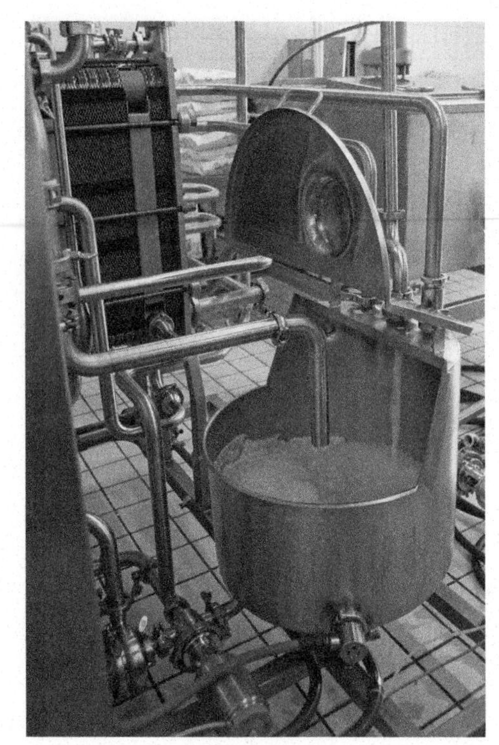

Milk being prepared for pasteurization

FIGURE 7.23 **Pasteurization**

Boiling Another way to reduce microbial numbers is through boiling. Municipalities often issue a "boil water advisory" when drinking water is contaminated. Boiling water for 5 minutes eliminates most pathogenic bacteria, protozoans (such as *Cryptosporidium* and *Giardia* that are found in most bodies of water), and viruses, but some endospores can withstand hours of boiling. Thus, boiling is a highly effective decontamination strategy, but not necessarily an efficient sterilization method. Heat-resistant objects like glassware or metal tools can be decontaminated (but not sterilized) by submerging them in boiling water for 10 to 20 minutes.

Pasteurization A number of pathogens such as *Listeria, Salmonella, E. coli* O157:H7, and *Coxiella burnetii* can be found in milk, so the dairy industry eliminates these bacteria using **pasteurization**. Application of moderate heat, well below the liquid's boiling point, eliminates pathogens and reduces (though does not fully eliminate) harmless microbes that cause milk spoilage (FIG. 7.23). Once pasteurization is complete, milk is refrigerated to prevent growth of the milk-spoiling microbes that remain.

Some dairy products like individual coffee creamers in restaurants and Parmalat milk have undergone ultra-high-temperature treatments and are described as *ultra-pasteurized*. Because they are sterile, these milk products are safe to store at room temperature for extended periods (months), provided their airtight containers haven't been opened. Other liquids, including juices, liquid eggs (e.g., Egg Beaters), soy milk, wine, and beer, can also be pasteurized or ultra-pasteurized to make them safe and extend shelf life.

Dry heat Incineration or hot-air ovens can also be used for sterilization or disinfection. Common examples of dry heat sterilization include heating an inoculating loop to red hot in a Bunsen burner flame and incinerating waste. Placing an object at 170°C (338°F) for about 2 hours in a dry heat oven is also a means of sterilization.

Radiation

Some physical decontamination methods involve **radiation**, or high-energy waves. Radiation can serve as a disinfection or sterilization tool depending on the protocol applied. Based on its features, radiation is described as ionizing or non-ionizing.

Ionizing radiation Gamma rays released by radioactive substances and X-rays, which are electron beams, are classified as **ionizing radiation**. These high-energy waves generate reactive ions that kill microorganisms and inactivate viruses by damaging their nucleic acids. Ionizing radiation passes though packaging and is useful in food and pharmaceutical industries. It can also sterilize medical supplies that can't be autoclaved and has been approved for eliminating microbes on dried spices, meats, and vegetables. Growing concern over foodborne illness outbreaks has brought food irradiation greater attention.

Non-ionizing radiation Ultraviolet (UV) rays are a form of **non-ionizing radiation** that can change the bond structure of DNA, leading to severe mutations that ultimately result in cell death. (See Chapter 5 for more on UV light, formation of thymine dimers, and DNA mutations.) UV light boxes in air-handling systems can reduce the number of airborne microbes; they also can be used to sanitize drinking water and swimming pools. In hospitals, UV light is often used to disinfect surfaces in operating rooms, while labs commonly use UV light to disinfect biosafety cabinet surfaces (FIG. 7.24).

FIGURE 7.24 **Non-ionizing radiation** UV light is a form of non-ionizing radiation used to decontaminate air, water, and surfaces—such as the inside of a laboratory biological safety cabinet, as shown here.

Critical Thinking *A student places a sealed petri plate of bacteria into a biological safety cabinet before turning off the UV light. What will the effect be on the microbe?*

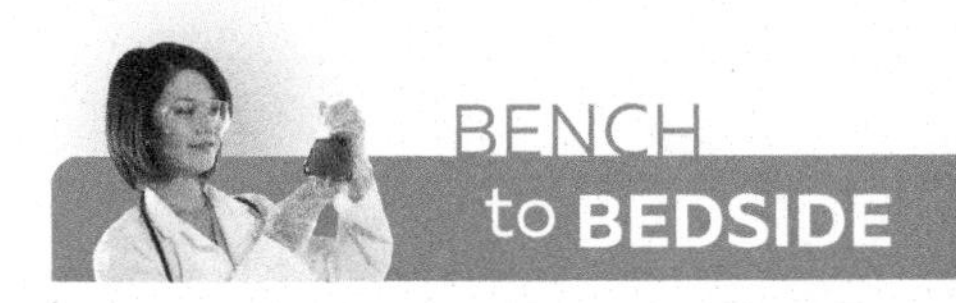

Irradiating Food: Public Fears and Practice

Traditionally, U.S. consumers have viewed food irradiation with suspicion, and the food industry has not embraced it widely, either. Some outspoken citizen groups protest the practice as "unnatural" and claim it may be dangerous. For industry, there are additional costs for processing foods at an irradiation facility. This makes the process less desirable if companies are trying to keep costs as low as possible—especially if they feel consumers may shy away from the treated product. But scientifically speaking, is food irradiation actually dangerous? The data say no.

In the United States, food irradiation has been reviewed by the Food and Drug Administration (FDA) for over 30 years. The practice appears to be safe and effective for removing foodborne pathogens without compromising food taste or nutritional properties. The CDC, U.S. Department of Agriculture, and World Health Organization all approve irradiation as a food safety practice. In the United States, meats, shellfish, wheat, fruits, and vegetables are all approved for irradiation. Military personnel and NASA astronauts regularly eat irradiated food. The FDA also recommends irradiated foods for severely immune-compromised patients, to reduce the risk of foodborne infection. And whether they realize it or not, consumers already eat foods every day that contain irradiated spices or herbs.

The greatest benefits of irradiation are that raw foods can be treated with the heat-free process, and non-spore-forming bacteria and pests are eliminated from even packaged food without chemical additives. Despite these advantages, it should be mentioned that not all foods can be irradiated. The process turns some foods, like melons and cucumbers, into a mushy mess.

At this point, food irradiation has been tested on thousands of different food products. Data show that nutritional value is preserved, and few or no reactive by-products are made. The trace amounts of radiolytic products found after irradiation are similar in level to what is found in nonirradiated food. But until the public and food industries embrace food irradiation more, the benefits of irradiated food will primarily remain with soldiers and space explorers.

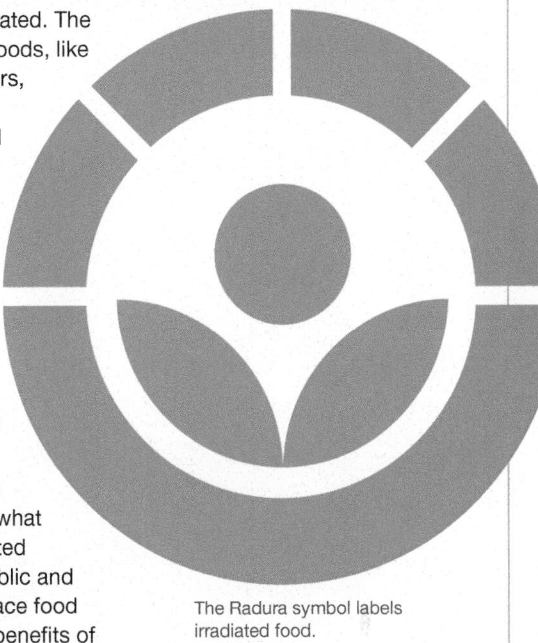

The Radura symbol labels irradiated food.

Source: Zehi, Z. B., Afshari, A., Noori, S. M. A., Jannat, B., & Hashemi, M. (2020). The effects of X-ray irradiation on safety and nutritional value of food: A systematic review article. *Current Pharmaceutical Biotechnology*, *21*(10), 919–926.

UV light treatments cannot pass through packaging, and they are ineffective at decontaminating objects covered in organic material. Temperature and distance from the UV light sources can also affect decontamination efficacy.

Filtration

Large volumes of liquid or air that must be quickly decontaminated can easily be passed through microbe-capturing filters. Filter pore sizes can even be made small enough to remove viruses. In clinics, hospitals, labs, and homes, high-efficiency particulate air (HEPA) filters remove bacteria, environmental allergens, fungi, and even larger viruses from the air. HEPA filters, as shown in FIG. 7.25, are made of randomly arranged fibers that screen out microbes to remove 99.97 percent of airborne substances that are 0.3 micrometer (μm) or larger. Because HEPA filters do not remove 100 percent of microbial agents, they aren't a means of sterilizing air.

Liquids can be sterilized using superfine membrane filters made from a variety of polymers such as cellulose nitrate and polyester. Pore sizes can range from 0.1 micrometer (μm), to filter out bacteria and protists, to 0.01 μm to remove viruses. Large volumes of liquid can be pulled through filters using a vacuum mechanism, whereas smaller volumes can simply be pushed through syringes with filters attached to the end. Special filter straws like "LifeStraws" have 0.2 μm filters that remove pathogens from drinking water and are used by people venturing into remote areas, or in communities that lack municipal water treatment. These straws do not sterilize water because viruses are not fully removed (FIG. 7.26).

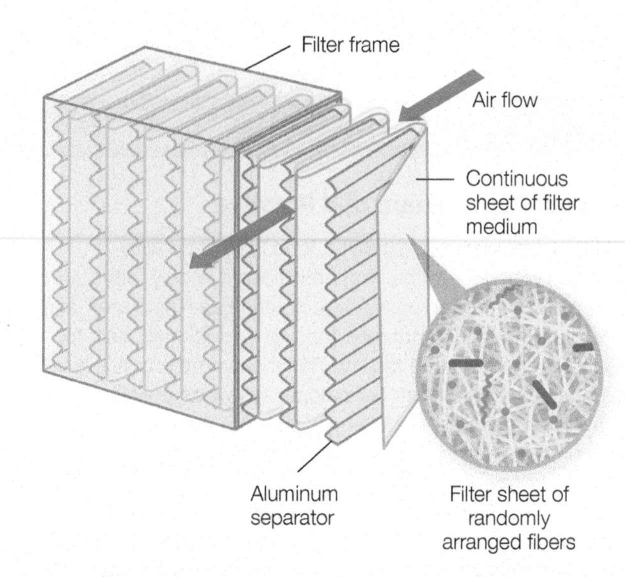

FIGURE 7.25 Air filtration HEPA filters are made of matted fibers that trap particles, including microbes, as air passes through.

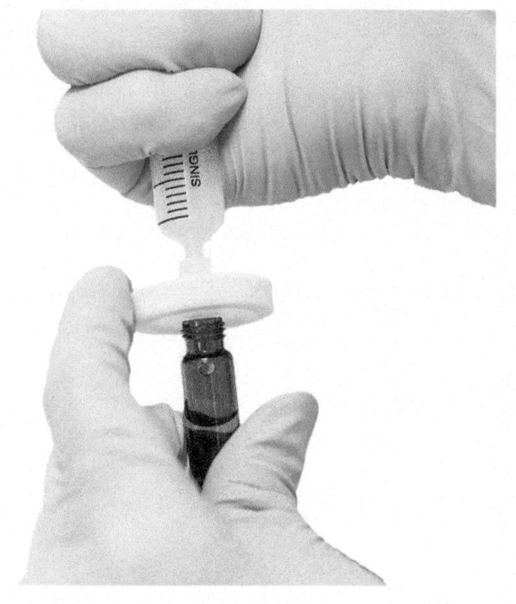

FIGURE 7.26 Liquid filtration *Top*: Filters can be attached to the end of a syringe to sterilize small volumes of liquids. *Bottom*: Special filter straws can render untreated water safe to drink.

Critical Thinking *What potential problem could arise in using filter straws? (Hint: Think about large particles that are being trapped.)*

Germicides are chemical controls that limit microbes.

Chemical control of microbial growth involves chemicals called **germicides**. Two key classes are **disinfectants**, which are used to treat inanimate objects, and **antiseptics**, which are applied to living tissue such as skin. Germicides are ranked across three tiers. High-level agents destroy all microbes (bacteria, fungi, and viruses) and endospores—provided they are at a low concentration. Intermediate-level agents destroy all bacteria (including *Mycobacterium tuberculosis*), fungi, and viruses, but not endospores. Low-level agents destroy bacteria (but not *M. tuberculosis*), fungi, and some viruses, but not endospores. Notably, these rankings do not consider efficacy against prions or protozoans. Additionally, germicides that kill microbes are classified as **microbiocidal**, whereas those that only inhibit microbial growth are **microbiostatic**. At one concentration a germicide may be "cidal" for one microbial species and "static" for another.

Medical equipment that requires regular decontamination is classified in three tiers as well. **Critical equipment** comes into contact with sterile body sites or the vascular system and must therefore be sterilized. Examples include surgical tools, implants, and urinary and cardiac catheters. High-level disinfectants are usually recommended for critical equipment. **Semicritical equipment** comes in contact with mucous membranes or nonintact skin and should be free of bacteria, fungi, and viruses, but low numbers of endospores are not a threat. Tubing for endoscopes or anesthesia or respiratory therapy is an example of semicritical equipment. High-level germicides are usually preferred for this type of gear, but occasionally intermediate-level disinfectants are used. Special concern remains over endoscopes, which are considered semicritical equipment. They have been linked to the most healthcare-associated outbreaks compared with other medical equipment.[9] The problem is that endoscopes are expensive and must be reused. However, their long and flexible structure can prove cumbersome to thoroughly clean, and after a single use, endoscopes can carry large amounts of microbial contamination. Finally, **noncritical equipment** such as stethoscopes, blood pressure cuffs, and most surfaces in patient rooms

[9] Endoscopes are long, flexible, fiber optic–based imaging tools that are most commonly used to view gastrointestinal tract structures.

TABLE 7.4 Germicides for Reducing or Eliminating Microbes

Level*	Germicide	Mode of Action	Pros/Cons
Low	Detergents	Target lipid membranes	*Pros:* Cheap, low toxicity, pleasant scent, useable as disinfectant and antiseptic *Cons:* Activity is decreased in hard water, easily contaminated by *Pseudomonas* bacteria
Intermediate	Alcohols: • Isopropanol • Ethanol	Target proteins and lipid membranes	*Pros:* Cheap, easily applied, useable as disinfectant and antiseptic *Cons:* Flammable, can react with plastics
	Phenols	Target proteins and lipid membranes	*Pros:* Easy to apply, effective in hard water, useable as disinfectant and antiseptic *Cons:* Leave residue, irritants, harsh on surfaces, medicinal scent, sensitive to water hardness
High	Aldehydes: • Formaldehyde • Glutaraldehyde	Target proteins, nucleic acids	*Pros:* Achieve sterility at certain concentrations *Cons:* Toxic, irritants, and leave a residue
	Halogens: • Chlorine • Iodine	Oxidizing agents, mainly target proteins, nucleic acids	*Pros:* Sterilants at higher concentrations, cheap, useable as disinfectant and antiseptic *Cons:* Rapidly inactivated by organic material, corrosive, discolor fabrics
	Peroxygens: • Hydrogen peroxide • Peracetic acid	Oxidizing agents that mainly target nucleic acids and proteins	*Pros:* Effective sterilization at high concentrations, useable as disinfectant and antiseptic; peracetic acid is effective despite organic material present and has no residue *Cons:* Most readily inactivated by organic matter, corrosive, irritants
	Ethylene oxide	Target proteins, nucleic acids	*Pros:* Can treat items that can't withstand heat or moisture, gentle on equipment *Cons:* Toxic and flammable

*Denotes the highest possible germicide level of the agent. Any germicide that is greatly diluted or improperly applied will have a low effect. Higher concentrations usually provide a higher disinfection potential (or even sterilize).

only contact patients' intact skin and therefore require less stringent disinfection than semicritical and critical equipment. Low-level disinfectants are sufficient to remove microbes from noncritical surfaces.

TABLE 7.4 summarizes the level, modes of action, and pros and cons of the various germicides we'll review next.

Alcohols

Ethanol and isopropanol are two intermediate-level disinfectants commonly used in healthcare settings. They denature proteins and attack lipid membranes. Their optimal concentration for use is a 60 to 90 percent solution. **Alcohols** are most commonly used to disinfect small equipment like thermometers, scissors, and stethoscopes. Alcohol wipes also clean small surfaces, such as stoppers, and the outside of equipment, such as ventilators. Wipes are also used as skin antiseptics.

Unfortunately, alcohols can react with plastic tubing and swell or harden rubber components of medical equipment, so they should be used only on approved surfaces. Additionally, alcohols are flammable, so cool, dry storage must be available. Because alcohols evaporate readily, exposure to this germicide is brief unless the item is immersed in a solution.

CHEM • NOTE

Alcohols are organic molecules that have a polar hydroxyl (R-OH) functional group that confers most of their chemical properties. Functional groups are reviewed in Chapter 2.

Aldehydes

Formaldehyde and glutaraldehyde act as high- or intermediate-level disinfectants based on their concentration. Formaldehyde is sold in both liquid and gas states, which makes it a versatile option for equipment cleaning. Both types of **aldehydes** work by reacting with proteins and nucleic acid. Aldehydes are used to sterilize surgical instruments, endoscopes, dialyzers, and anesthesia and respiratory equipment.

The biggest drawback of aldehydes is their toxicity. Formaldehyde generates irritating fumes and is a suspected carcinogen. Glutaraldehyde is preferred over formaldehyde, but it, too, irritates skin and mucous membranes. Both aldehydes leave residues on equipment, so rinsing is required after these germicides are used. A newer chemical, ortho-phthalaldehyde, is now available. It is a high-level disinfectant with similar bactericidal (bacteria killing) action as formaldehyde and glutaraldehyde, but it is safer, less irritating, and stable at a variety of pH levels.

Phenols

A number of **phenols** and phenol derivatives (chemically modified phenols) are used in personal hygiene items such as mouthwashes and soaps and in common disinfectants such as Lysol. They have long been used as intermediate-level germicides. Phenols destroy bacterial cell walls and interact with proteins. They are often used in medical settings for surface disinfection (bed rails, tables, and floors) and to disinfect semicritical and noncritical equipment.

The drawback to phenols is that they are easily absorbed by porous surfaces and can be irritating to skin. If reusable equipment is disinfected by a hospital-grade phenol, all residue must be rinsed off before the equipment can be used for another patient.

Halogens (Chlorine and Iodine)

Halogens include chlorine and iodine compounds. These germicides work by oxidizing cell components, especially cellular proteins and nucleic acids. Chlorine bleach (sodium hypochlorite) is one of the most widely used halogen disinfectants. Other chlorine-based germicides include chlorine dioxide and sodium dichloroisocyanurate. The concentration of chlorine determines the level of disinfection. Chlorines can be applied to a variety of surfaces, including medical equipment and floors. Most notably, chlorine can be added to drinking water to remove potential pathogens such as *Legionella* bacteria that are normal inhabitants of water and can cause severe respiratory infections, especially in the immune compromised. Chlorine-based disinfectants are usually cheap, don't leave residue, and act fast. Unfortunately, chlorine germicides are easily inactivated by organic material and can be corrosive or discoloring to equipment and fabrics. They also can be irritants, especially as concentrated solutions.

Traditionally, iodine tinctures (iodine compounds dissolved in alcohol and water) are used as antiseptics to decontaminate skin before an incision. Iodophors are another iodine-containing germicide that can be used as an antiseptic and disinfectant against bacteria, mycobacteria, and viruses. They are ineffective against endospores. Povidone-iodine is a widely used iodophor that can be applied as an antiseptic or used as a disinfectant for medical equipment such as thermometers and endoscopes. Iodophors must be diluted, with lower concentrations used as antiseptics and higher concentrations suitable as disinfectants. A benefit of using iodophors over iodine tinctures is that they are nonstaining and nonirritating.

Peroxygens

Peroxygens are a group of chemicals with strong oxidizing properties that easily convert into reactive oxygen species that attack proteins, nucleic acids, and other biomolecules. Hydrogen peroxide and peracetic acid are two commonly used peroxygens. They can be used as antiseptics and disinfectants. At high concentrations, these germicides serve as high-level disinfectants.

When hydrogen peroxide concentrations are high enough, they overcome bacteria that produce the hydrogen peroxide-neutralizing enzyme, catalase. (Refer back to the earlier chapter section on oxygen requirements to review reactive oxygen species and catalase.) In comparison with glutaraldehyde, hydrogen peroxide is less toxic, is odorless, and shows the same effectiveness in disinfectant activity. It has been used for disinfecting soft contact lenses, ventilators, fabrics, and endoscopes. Some chemical irritation can occur at high concentrations, and organic matter tends to rapidly inactivate hydrogen peroxide.

Peracetic acid is effective even in the presence of organic material such as blood, pus, or fecal matter. It leaves no residue, which allows objects to be used immediately after decontamination. In the United States, many medical and dental clinics use an automated washing and sterilizing machine with peracetic acid (FIG. 7.27). The waste wash can be poured down the sink. Newer disinfectants that combine both hydrogen peroxide and peracetic acid have been used in dialysis centers for decontaminating hemodialyzers, machines that remove waste products from blood.

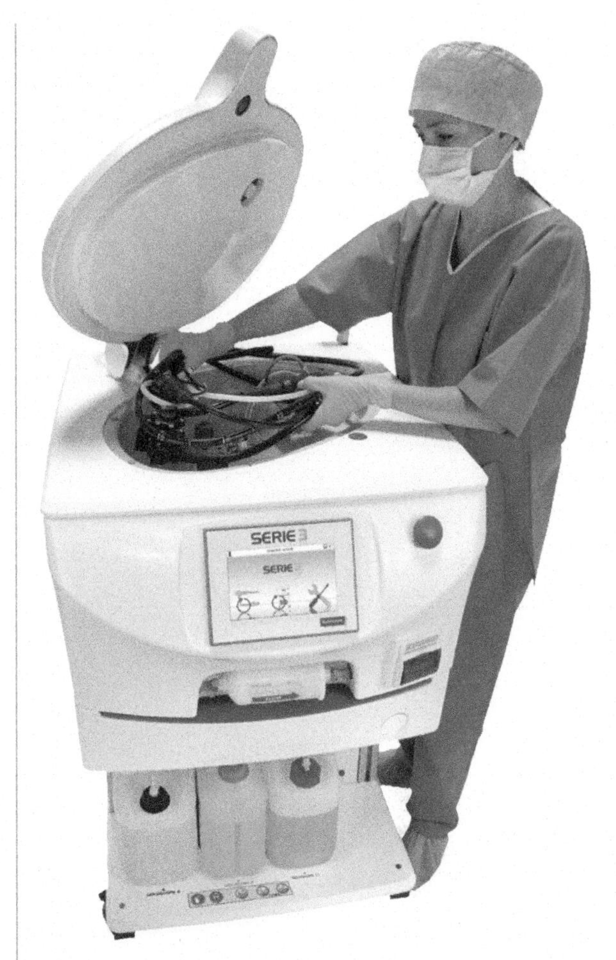

FIGURE 7.27 Automated cleaning and disinfecting machine Using a peracetic acid–based germicide, this machine washes and disinfects medical equipment, such as the endoscope seen here.

Ethylene Oxide

For temperature-sensitive materials and equipment susceptible to moisture, the colorless **ethylene oxide** gas is a good sterilization method. It primarily works by damaging proteins and nucleic acids. Sterilization efficacy is influenced by temperature, exposure time, humidity, and gas concentration. Ethylene oxide is often applied to implant devices such as pacemakers that contain electronic parts or plastic components, but it can also be used on tissue such as heart valves being prepared for transplantation.

Although beneficial in removing all microbes including viruses and endospores, ethylene oxide does have drawbacks. It is highly toxic, flammable, costly, and time consuming to use. The gas requires a large pressurized chamber in which devices are placed for treatment. Gas is pumped into the chamber, and the objects must be exposed for a number of hours. After treatment, the chamber must be flushed with air for several hours because the gas penetrates the exposed object and must be fully removed for safety. Finally, organic material and inorganic salts contaminating the device or tool can reduce the effectiveness of ethylene oxide. Thus, a prewash is often recommended before treatment.

Detergents

Detergents are *amphipathic* molecules; this chemical feature allows them to remove water-soluble and water-insoluble substances, which makes them good cleaning agents. Some detergents damage the lipid envelope of certain viruses and also the lipid membrane of certain bacterial cells, but most detergents have limited microbiocidal activity and instead work mostly as cleaning agents that reduce microbial counts by simply washing them away.

Based on chemical structure, there are four classes of detergents. Two common categories are *anionic* and *cationic detergents*. Anionic detergents have a negative charge and include soaps. An anionic detergent (or soap) that you may have heard of is sodium laureth (or laurel) sulfate. It is a common ingredient in hand soaps, cosmetics, shampoos, laundry detergents, and household cleaning agents. These agents were often recommended for routine hand washing to prevent SARS-CoV-2 transmission during the COVID-19 pandemic.

Cationic detergents have a positive charge; one common class of cationic detergents in healthcare settings is **quaternary ammonium compounds** (QACs). QACs have bactericidal activity and are sporostatic (inhibit spore germination). They are used as antiseptics on unbroken skin and certain mucous membranes

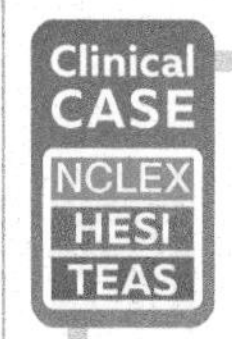

The Case of the Sickly Soaker

Practice applying what you know clinically: visit the **Mastering Microbiology** Study Area to watch Part 3 and practice for future exams.

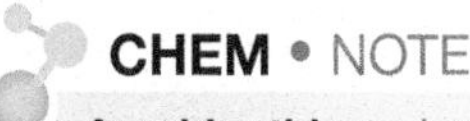

CHEM • NOTE

Amphipathic molecules have a region that interacts well with water (hydrophilic) and a region that interacts well with lipids or other water-repellent (hydrophobic) molecules. Soaps and detergents are common examples of amphipathic molecules.

CHEM • NOTE

Cations are positively charged ions that result from the loss of one or more electrons. **Anions** are negatively charged ions that are created when atoms gain electrons.

and as disinfectants on noncritical equipment and surfaces. Given their low microbiocidal potential, detergents and soaps can easily become contaminated and spread microbes to new surfaces. Therefore, the FDA recommends discarding detergent solutions after a given time frame and frequently replacing them with freshly prepared detergent.

Many factors must be considered to select an appropriate germicide.

A germicide's concentration and application method impact the achievement of sterility or decontamination. Choosing the right germicide depends on a number of factors, some of which we discuss next.

Item Uses

How the object is used, and therefore how likely it is to transmit an infectious agent, must be considered. If the object contacts sterile tissue, then it must be sterilized. For example, surgical instruments must be sterilized, whereas bedpans and bed rails can simply be cleaned with disinfectants.

Germicide Reactivity

The chemical makeup of the object being treated should also be considered. The object may chemically react with certain germicidal agents; for example, some germicides corrode metals, and harsh chemicals like sodium hypochlorite (bleach) burn living tissues. Other items, such as bandages, are treated with sterilizing gases because liquids distort their physical structure.

Germicide Concentration and Treatment Times

Many germicides come as concentrates that require dilution. Manufacturers' recommended solution concentrations depend on whether the goal is sterilization or disinfection. For example, a 2.4 percent glutaraldehyde solution should be applied for 20 minutes at 20°C to achieve sterilization. The number of microbes to be killed also impacts treatment times and concentrations needed; the more microbes present, the longer the required treatment time and/or the greater the required concentration.

Types of Infectious Agents Being Controlled

Awareness of what infectious agents are most likely present is key in selecting the proper chemical agent for control of growth. Some infectious agents, like prions and endospores, require much more aggressive treatments to eliminate.

Presence of Organic and Inorganic Matter on Treated Item

Organic material such as blood, soil, or fecal matter can inactivate chlorine and iodine disinfectants. The presence of excessive inorganic substances like minerals and salts in hard water can chemically react with germicides and lessen their effectiveness.

Impact of Germicide Residues on Equipment Use

Objects such as endoscopes must be cleaned with germicides to remove microbes, but they then need to be flushed with distilled water to remove any chemical residue before they are used in another patient. This prevents potential irritation of mucous membranes by the applied germicide.

Germicide Toxicity

Germicides can be hazardous to humans and the environment. Gloves should be worn to protect the skin, and a face mask may be necessary to prevent inhalation of fumes. Germicide waste must also be properly discarded. Some chemicals such as formaldehyde and some phenols cannot be poured down the sink; they are disposed of in labeled containers that must be shipped off for treatment before disposal to reduce environmental impact.

Different control methods work for different microbes.

One strategy does not fit all in reducing or eliminating microbes. Thanks to inherent structural features, many microorganisms are tough to eliminate even with harsh treatments. Here we'll highlight examples of microbes' best efforts to survive.

Mycobacterium Control

Mycobacterium species cause several serious diseases such as tuberculosis (caused by *Mycobacterium tuberculosis*) and leprosy (caused by *Mycobacterium leprae*). These bacteria have cell walls rich in waxy mycolic acids, which provide protection against many disinfectants. *M. tuberculosis* is transmitted by airborne droplets, so control measures target reducing airborne particles from infected individuals. In healthcare settings, quickly identifying those infected with *Mycobacterium* is key. Protocols for isolating tuberculosis patients helps reduce spread. Isolation rooms for tuberculosis patients have special ventilation equipped with HEPA filters that remove bacteria from the air. Disinfectants used on reusable medical equipment such as endoscopes or dental devices must be strong enough to kill mycobacterial cells. Such chemical concoctions usually contain glutaraldehyde or hydrogen peroxide.

Endospore Control

In response to harsh environmental conditions such as low nutrients or radiation, certain bacteria form endospores, dormant structures that can revert to growing (vegetative) cells once favorable growth conditions are restored. Thanks to their low water content and durable shell, endospores survive drying, radiation, boiling, chemicals, and heat treatments that would kill vegetative cells. *Bacillus anthracis*, the causative agent of anthrax, and *Clostridiodes difficile,* an intestinal pathogen, are some medically important bacteria that form endospores. (Chapter 3 reviews medically important spore-forming bacteria and the process of spore formation.) The most effective way to ensure elimination of endospores is by autoclaving. Other methods include use of hydrogen peroxide vapor at high heat or application of specific disinfectants termed sporicides that usually contain bleach.

Viral Control

Because they are inactive outside of host cells, viruses can be resistant to some measures of microbial control. Lipids in the viral envelope are sensitive to heat, drying, and detergents; therefore, most enveloped viruses are easier to destroy than naked (nonenveloped) viruses. In clinical settings, enveloped viruses such as HIV, hepatitis B, SARS-CoV-2, and hepatitis C must be eliminated from reusable equipment to ensure patient safety. Several types of glutaraldehyde-based detergents are effective at inactivating these viruses. Naked viruses like the intestinal pathogen norovirus are usually inactivated by chlorine-based agents.

Protozoan Control

Different stages of a protozoan's life cycle can resist certain control methods. For example, the common parasite *Cryptosporidium parvum* contaminates drinking water and has a spore phase that resists chlorine treatment. Therefore, water management agencies use a variety of treatments such as filtration, carbon dioxide, and ozone treatments, as well as exposure to ultraviolet light, to eliminate these protozoa from drinking water. *Entamoeba histolytica* and *Giardia lamblia* are other intestinal protozoan pathogens that can be found in natural waters like lakes and rivers. For hikers who need a drink, boiling or filtering water from a nearby stream will remove these pathogens.

Prion Control

A growing concern is the presence of infectious proteins called prions on surgical equipment. (See Chapter 6 for details on prions.) Surgical devices are reused after autoclaving or chemical sterilization. Prions, however, are known to withstand standard autoclaving times and temperatures. Prions can be eliminated through a combination of chemical treatments and increased temperature and pressure during autoclaving.

Build Your Foundation

17. What do the terms *decontamination*, *sterilization*, *disinfection*, *microbiostatic*, *microbiocidal*, *disinfectant*, and *antiseptic* mean? (LO 7.17)
18. How does heat control microbial levels? Which heat treatments sterilize versus disinfect? (LO 7.18)
19. Define the terms *decimal reduction time*, *thermal death point*, and *thermal death time*. (LO 7.18)
20. How are filtration and radiation used to reduce microbial levels? Describe which treatments are sterilizing and which are disinfecting. (LO 7.19)
21. What are the various classes of germicides, and what goes into selecting the right one? (LO 7.20)
22. What are the levels of germicide activity, and which ones are used for critical, semicritical, or noncritical equipment? (LO 7.21)
23. What factors should be considered when selecting an appropriate germicide? (LO 7.22)
24. What features help *Mycobacteria*, endospores, viruses, protozoa, and prions resist germicide actions? (LO 7.23)

Build Your Foundation (BYF) Quick Quiz: Visit the **Mastering Microbiology** Study Area to quiz yourself.

VISUAL SUMMARY | Fundamentals of Microbial Growth and Decontamination

Microbial Growth

Bacteria mainly use binary fission to divide. Generation time is the time it takes for a particular species of cell to divide.

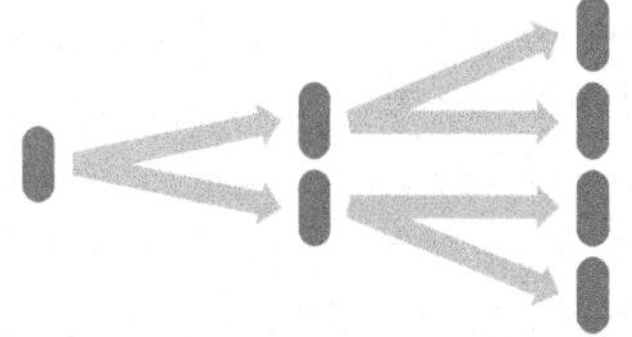

In closed culture, bacterial populations have four distinct growth phases, with exponential growth in the log phase.

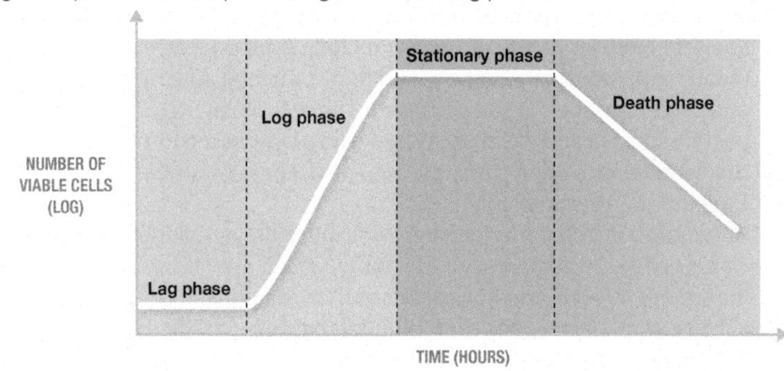

Growing and Isolating Microbes

Media can be solid, liqud, or semisolid and have defined or complex formulations. Some are selective and/or differential.

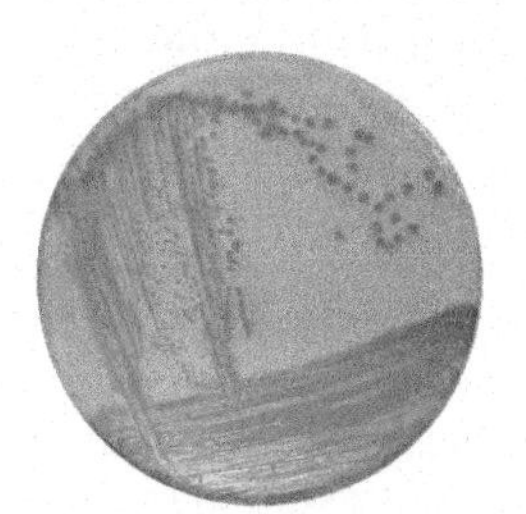

The streak plate isolation technique helps to isolate a pure culture.

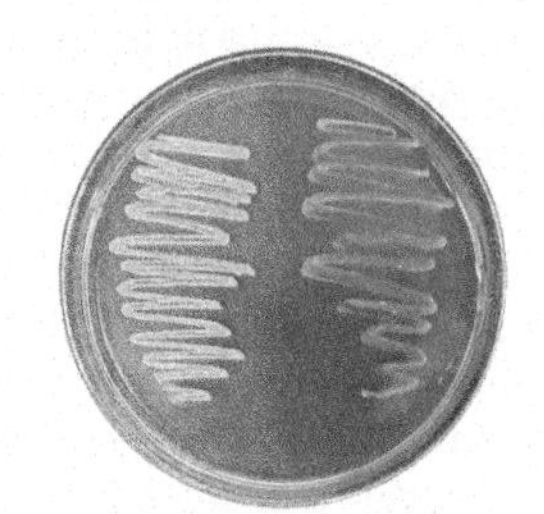

Mannitol salt agar is one type of selective and differential media.

Population Enumeration

While direct methods count colonies or cells, indirect methods rely on secondary reflections of overall population size.

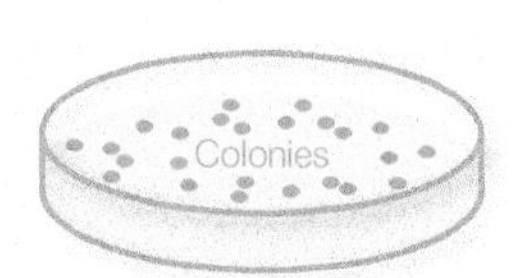

Viable plate counts

Direct methods
- Viable plate count
- Manual cell counts
- Automated cell counts (coulter counter/flow cytometer)

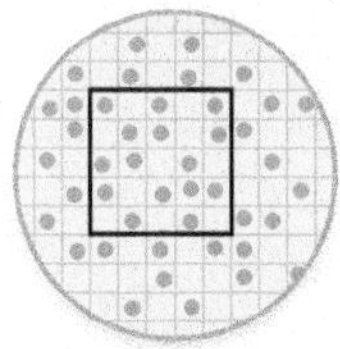

Manual cell counts

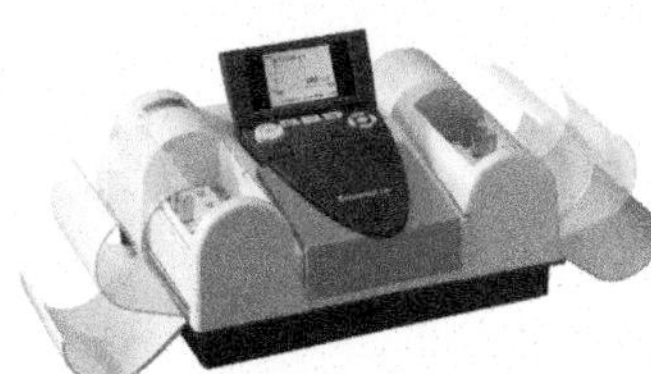

Spectrophotometry

Indirect methods
- Turbidity (spectrophotometry)
- Dry weight
- Biochemical activity

Environmental Growth Conditions

Temperature, pH, salinity, and oxygen levels impact microbial growth. Bacteria also require various organic and inorganic nutrients and growth factors.

Temperature ranges for growth

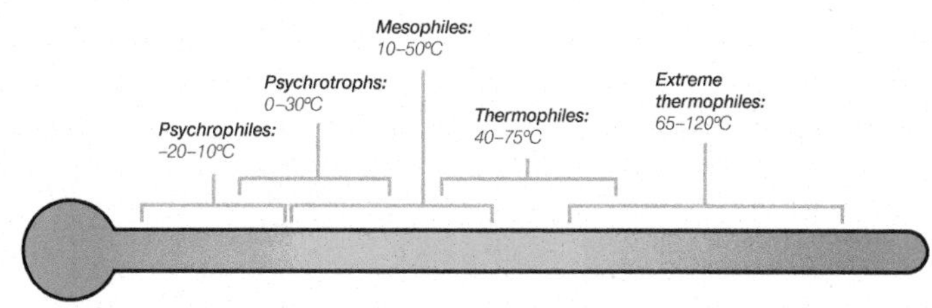

Oxygen use/tolerance

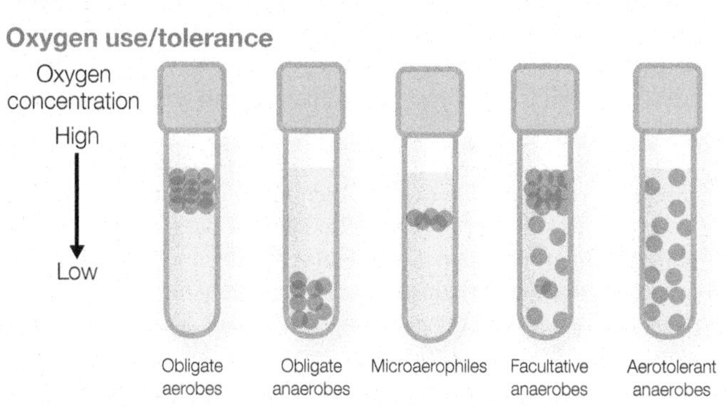

Controlling Microbial Growth

Sterilization and disinfection are achieved using chemical or physical methods.

Physical methods

Heat
- Autoclaving
- Boiling
- Pasteurization
- Dry heat

Radiation
- Ionizing (gamma rays/X-rays)
- Non-ionizing (UV light)

Filtration
- Air filters (HEPA filter)
- Liquid filters (membrane filters)

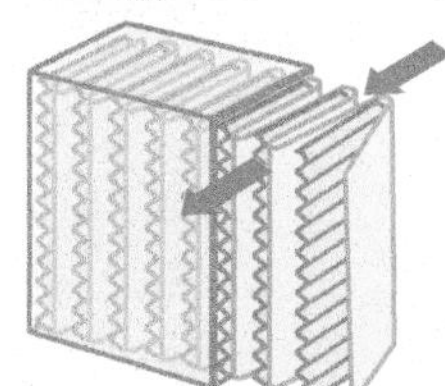

Chemical methods (germicides)

Disinfectants and antiseptics
- Alcohols
- Aldehydes
- Phenols
- Halogens
- Peroxygens
- Ethylene oxide
- Detergents

7 CHAPTER 7 OVERVIEW

7.1 Microbial Growth Basics

- In nature, bacteria have complex associations with other microbes. Biofilms often pose a hazard in healthcare facilities.
- Most bacteria reproduce asexually by binary fission. Some rare examples use asexual budding or spore formation. Fungi reproduce by sexual and asexual means, often through spore formation.
- Bacterial populations grow at an exponential rate. Generation time is the time required for a given species to divide.
- When placed in a closed culture system, bacteria exhibit four distinct phases of growth: lag, log, stationary, and death phases.

7.2 Prokaryotic Growth Requirements

- Temperature, pH, and salinity levels impact prokaryotic growth. Each species has an optimal, minimum, and maximum level for each of these factors. Psychrophiles thrive from −20–10°C; mesophiles include pathogens and thrive in temperatures from 10–50°C; thermophiles grow at roughly 40–75°C; extreme thermophiles prefer growth temperatures from 65–120°C. Acidophiles grow in pH levels below 5; neutralophiles grow between pH 5 and pH 8; alkaliphiles prefer a pH above 9. Halophiles withstand salt concentrations of 9 percent or higher.
- The presence of gaseous oxygen can lead to the production of toxic reactive oxygen species (ROS), which kill cells that can't eliminate them. Organisms can be categorized as aerobes, microaerophiles, obligate anaerobes, aerotolerant anaerobes, and facultative anaerobes based on their use and tolerance of oxygen.
- Cells need nutrients, growth factors, and a source of energy to grow. Heterotrophs require a source of organic carbon, while autotrophs make their own organic carbon. Chemotrophs use chemical energy from breaking down nutrients, while phototrophs get their energy from sunlight. Fastidious microbes need growth factors provided in their environment.

7.3 Growing, Isolating, and Counting Microbes

- Growth media are classified by their physical state (liquid, solid, or semisolid formats), their chemical composition (complex/defined), and their function (differential/selective).
- Anaerobic organisms must be grown in oxygen-free conditions; there are special media and equipment to facilitate this.
- Aseptic techniques prevent the introduction of contaminating microbes to a patient, a clinical sample, or others in the healthcare setting. The streak plate technique is a simple way to isolate colonies so that a pure culture can be obtained.
- Microbes can be enumerated using direct or indirect methods. Direct methods include manual cell counts, viable plate counts, and automated counts by Coulter counters and flow cytometers. Indirect enumeration can be done by spectrophotometry, dry weight, and assessing biochemical activity.
- Microbes can be identified using biochemical tests, microscopy techniques, and genetic analysis.

7.4 Basics of Microbial Growth Reduction and Decontamination

- Decontamination measures remove or reduce microbial populations to render an object safe for handling; these measures fall into two key categories, sterilization or disinfection, based on their final outcome. Sterilization eliminates all bacteria, viruses, and endospores, while disinfection just reduces microbial numbers.
- Physical controls most commonly include temperature changes (heat/cold), radiation (ionizing and non-ionizing), and filtration (air and fluid filtration). Heat methods include autoclaving, boiling, pasteurization, and dry heat. When using heat as a control measure, the decimal reduction time, thermal death point, and thermal death time are important to consider.
- Chemical control of microbial growth relies on germicides. Two key classes of germicides are disinfectants, which are used to treat inanimate objects, and antiseptics, which are applied to living tissue such as skin. Germicides that kill microbes are classified as microbiocidal, while those that only inhibit microbial growth are microbiostatic. Germicides are ranked across three tiers (high-, intermediate-, or low-level agents) based on the microbes they eliminate. Medical equipment also is classified across three tiers (critical, semicritical, and noncritical) based on the level of decontamination required.
- The main classes of germicide are alcohols, aldehydes, phenols, halogens, peroxygens, ethylene oxide, and detergents. These germicides are summarized in Table 7.4.
- Choosing the right germicide depends on a number of factors, such as how the item being treated is used; how the item will react to the germicide (for example, corrosives may harm metal equipment); the concentration of germicide and time needed to achieve sterility versus disinfection; the type of microbial agent(s) being controlled; the presence of organic matter; the impact of residue on equipment use; and germicide toxicity.
- One strategy does not fit all in reducing or eliminating microbes. Special decontamination conditions are often required to effectively eliminate *Mycobacteria*, bacterial endospores, naked viruses, certain protozoans, and prions.

COMPREHENSIVE CASE

The following case integrates basic principles from the chapter. Try to answer the case questions on your own. Don't forget to be **S.M.A.R.T.*** about your case study to help you interlink the scientific concepts you have just learned and apply your new understanding of microbial growth to a case study.

*The five-step method is shown in detail in the Chapter 1 Comprehensive Case on cholera. See pages 31–33. Refer back to this example to help you apply a SMART approach to other critical thinking questions and cases throughout the book.

• • •

Within a week, you, a nurse practitioner, saw three different patients ranging from 16 to 18 years old with infections from recent piercings of the upper ear cartilage (auricular cartilage of the ear helix). The infections showed redness, swelling, and pus. Patients did not report swimming since their piercings and had covered their ears while showering. Curious about this unusual uptick in piercing-related infections, you asked the third patient when and where they had their ear piercings done. They said it was done at a piercing kiosk at the local mall just a few days before. The patient stated they had their lower ear lobes pierced at the same kiosk 6 months earlier, and those had healed without infection.

You decided to contact the other two patients you saw earlier that week. In doing so, you learned that their piercings were also done within the same week at the same kiosk. The three patients had selected different earrings, and different workers did the procedures. Surprisingly, all three patients knew each other from school.

Taking pus samples from the ear of the third patient revealed a *Pseudomonas aeruginosa* infection (Pseudomonal chondritis). This bacterium is common in the environment and can be found on skin. Oral antibiotics were prescribed to all three patients, but because of the lack of blood flow in cartilage, healing was not successful with antibiotics alone. Each patient ultimately needed surgical removal of the infected tissue to help clear the infection in their cartilage.

Upon learning that all three of your patients went to the same piercing kiosk, you called the local health department to report your concerns. Health officials investigated. Safety procedures at the kiosk dictated that a worker would first decontaminate their hands with an alcohol-based hand sanitizer and then put on gloves. Next, they cleaned the area of the ear to be pierced with an antiseptic-soaked cotton ball. The worker would slide the earring into the back of a manual tool and then squeeze the two ends of the tool as it was placed over the ear, so that the earring stud pierced the cartilage. The back of the earring would be automatically pushed onto the earring stud in the same process. To account for anatomical differences between customers' ears, the tool was adjustable—but any adjustments were supposed to occur while workers wore gloves.

In an effort to track down the causative agent, health authorities sampled various surfaces and items at the kiosk. Each of the three different workers who performed the piercings on the three recent customers who developed infections claimed to follow all safety procedures. The antiseptic agent used to clean customer ears was also discovered in a half-empty bottle that had been opened at least a month earlier.

• • •

CASE-BASED QUESTIONS

1. What items and/or surfaces in the kiosk were probably sampled, and why were they selected for sampling?
2. The kiosk workers claimed they couldn't have transferred the pathogen from their hands to the customers because they wore gloves. Is this a valid conclusion? What are other ways *P. aeruginosa* could have been transferred to the customers during their piercing experience?
3. Initially, you were concerned that the causative agent could have been *S. aureus*, a Gram-positive bacterium, or *P. aeruginosa*, a Gram-negative bacterium; both are common culprits in piercing infections. What culture methods would allow for isolation and differentiation of these two bacteria? Would an anaerobic culture condition be needed? Be sure to explain your reasoning for all answers.
4. What role (if any) could the antiseptic have played in pathogen transmission?
5. Which tier would the piercing tool be classified for decontamination purposes? What precautions/protocols would the health authorities likely have recommended to limit future infections from piercings at the kiosk?
6. What is the most likely explanation for why the patient who had their lower earlobe pierced 6 months ago did not develop a *P. aeruginosa* infection?

END OF CHAPTER QUESTIONS

NCLEX
HESI
TEAS

Think Critically and Clinically
Questions highlighted in **orange** are opportunities to practice NCLEX, HESI, and TEAS-style questions.

1. A *Salmonella* species is grown for 8 hours. In this time, each cell divides about four times. What is the generation time?

2. If a bacterium that normally lives in the gastrointestinal tract is plunged into a salty solution, ____________ would occur.

   a. halophilic adjustment
   b. osmotic concentration
   c. lysis
   d. plasmolysis
   e. nothing

3. After performing the streak plate method, what feature(s) would you look for on the agar plate to determine if you have a pure culture?

4. Which of the following pathogens would hyperbaric oxygen therapy, a treatment that increases the level of dissolved oxygen in tissues, most likely ward off? Select all that apply.

   a. Anaerobic thermophile
   b. Anaerobic mesophile
   c. Facultative anaerobic thermophile
   d. Facultative anaerobic mesophile
   e. Microaerophilic mesophile

5. Choose the *false* statement about binary fission.

   a. It generates genetically diverse daughter cells.
   b. It is an asexual form of reproduction.
   c. It is the most common way that prokaryotes divide.
   d. It leads to exponential population growth.
   e. If it occurs in a single plane, it could generate chains of bacteria.

6. Match the term to the proper description(s). A term can have more than one description.

Terms	Descriptions
Death phase	A period of exponential population growth
Log phase	Cells die and are generated at an equal rate
Stationary phase	No population growth observed
Lag phase	Rate of cell death exceeds cell division
	Cells are acclimating to their environment
	Cells that can form spores prepare to do so

7. In a closed batch system, not all cells are expected to die even as the death phase advances. Why?

8. Which direct enumeration method differentiates living from nonliving cells?

   a. Manual cell counts
   b. Measuring dry weight of cells
   c. Viable plate count
   d. Measuring biochemical activity
   e. Coulter counter

9. Match the term to the proper description(s). A term can have more than one description.

Terms	Descriptions
Sterilization	A form of decontamination
Antisepsis	Wiping bed rails with a low-level germicide
Disinfection	Using a germicide on living tissue
	Autoclaving surgical equipment
	Treatment needed for critical equipment

10. You are collecting a clinical sample for microbiological analysis. Which of the following is the *most* important thing you must do?

    a. Follow aseptic protocols.
    b. Refrigerate the samples immediately after collection.
    c. Determine if the potential pathogen is an aerobe or strict anaerobe.
    d. Determine if normal flora have been removed before the sample is collected.

11. Label each statement that follows as true or false, and *correct the false statements* so they are true:

    a. Most pathogens would be considered mesophilic alkaliphiles.
    b. Sterilization is a form of decontamination.
    c. Disinfection is a form of decontamination.
    d. High-level germicides achieve sterilization.
    e. Ionizing radiation is a form of chemical microbial control.
    f. Standard pasteurization is a way to sterilize milk.

12. You have a patient who is suffering from a *Clostridioides difficile* infection. Which of the following would most likely be recommended to decontaminate small heat-stable equipment used for the patient?

    a. Autoclave all equipment at 121°C for 15 minutes.
    b. Boil equipment for 3 minutes.
    c. Place the equipment in a hot-air oven at 121°C for 15 minutes.
    d. Treat all equipment with a detergent solution.
    e. None of the above.

13. Choose the *false* statement about turbidity as an enumeration method.

    a. It is an indirect enumeration method.
    b. It is performed using a spectrophotometer.
    c. It must be done using a liquid culture.
    d. It differentiates between live and dead cells.
    e. It is a rapid enumeration method.

14. ____________ microbes use oxygen in metabolism, while ____________ do not.

15. If you were a manufacturer of electronic pacemakers for heart implantation, which agent would you most likely use to treat your product?
    a. Ethanol
    b. Iodophor
    c. Glutaraldehyde
    d. Autoclave
    e. Ethylene oxide

16. You are collecting a sample from a deep wound for analysis by the clinical microbiology lab. Which of the following is *not* a consideration as you undertake this process?
    a. Avoiding the skin as the wound is swabbed
    b. Using an anaerobic culture tube
    c. Disinfecting the tube before collecting the sample
    d. Washing your hands before and after sample collection
    e. Using complex media

17. The ____________ is the time needed to kill 90 percent of a given microbial population at a set temperature. The ____________ is the lowest temperature needed to kill all microbes in a sample within 10 minutes.

18. Which of the following is (are) true? Select all that apply.
    a. Scalpels are critical equipment.
    b. Endoscopes are noncritical equipment.
    c. Stethoscopes are noncritical equipment.
    d. Anesthesia tubing is semicritical equipment.
    e. Critical equipment contacts intact skin.

19. A bacterial specimen exhibits the following growth on blood agar. What can you most reasonably conclude about the bacterium? Select all that apply.
    a. It is not *S. pyogenes*.
    b. It is alpha hemolytic.
    c. It is beta hemolytic.
    d. It is gamma hemolytic.
    e. It is Gram-positive.

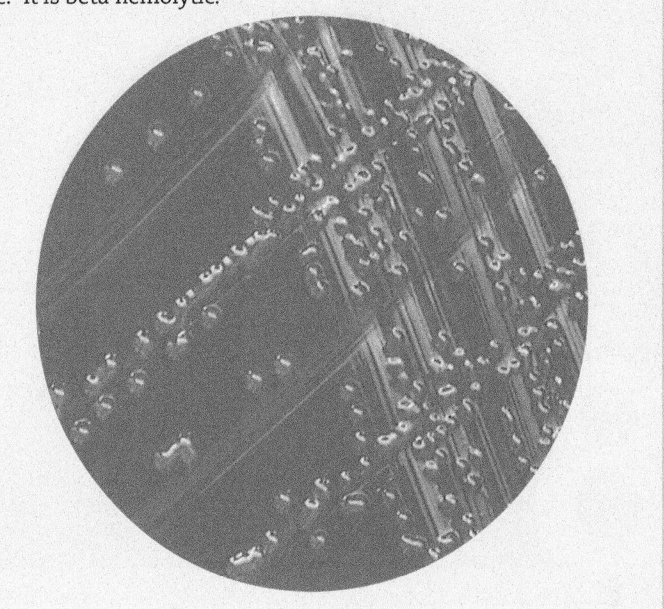

20. While hiking, Huda wants to ensure that the river water is clean enough to drink and will not cause any intestinal infections. How could she treat water from a stream so that it is safe to drink?

CRITICAL THINKING QUESTIONS

1. Using DNA analysis, a new bacterial species was found growing on the surface of heart stents. Unfortunately, the bacterium that was fairly numerous on stents could never successfully be grown as a pure culture in the lab. This was despite using complex media and simulating the oxygen, pH, pressure, temperature, and salinity features of the bacterium's normal habitat. Provide two possible explanations for why the bacterium could not be grown in a pure culture in the lab.

2. If you were growing a closed (or batch) culture of *Vibrio cholerae* and wanted to kill it as fast as possible with a cell wall–targeting antibiotic such as vancomycin, during which phase of growth should the antibiotic be added? Explain your answer.

3. A patient is suspected of having bacterial meningitis. When taking a spinal fluid sample, what environmental conditions must be maintained in the sample to ensure that all candidate microbes remain viable for microbiological analysis? If the conditions for sample collection are *not* met, does that mean laboratory confirmation of the pathogen is impossible?

4. In a clinical microbiology lab, you observe a bacterium growing on mannitol salt agar (MSA) with a yellow zone. Name the genus and species of the microbe you are most likely observing; state its likely oxygen tolerance and use; and predict if you will observe it on the eosin methylene blue (EMB) agar you are about to take out of the incubator. Explain the reasoning behind each of your conclusions and your prediction.

5. You are working in the emergency department (ED). One day a construction worker comes in to be seen for a puncture wound sustained from a nail gun. After cleaning up the wound, you administer a tetanus shot to prevent tetanus by an anaerobic bacterium called *Clostridium tetani*. A student on rotation through the ED pulls you to the side and states that they thought tetanus was only a risk if the wound was from a rusty object because the rust would be necessary to protect the anaerobic bacterium from the air. Explain why rusty objects may pose a greater threat for tetanus, but are definitely not the only threat for transmission.

Microbial Metabolism

What Will We Explore? Microbes, like all living cells, need energy to survive. They extract energy from food sources and use it to move, multiply, maintain cell structures, and perform other functions. In this chapter, we examine the key metabolic pathways microbes use to extract energy from nutrients. We also examine how metabolic features are useful for classifying microbes.

Human cells have a lot in common with microbes when it comes to harvesting energy from organic molecules. The fact that key metabolic pathways work in the same general way in both prokaryotic and eukaryotic cells supports the theory that all cells evolved from a common ancestor.

Despite shared metabolic features, there is still great metabolic diversity across lifeforms, especially among prokaryotes. Some bacteria utilize distinctive pathways that allow them to thrive in unique and diverse environments, ranging from your body to deep-sea vents.

Why Is It Important? The pathways you are going to learn about are the very reason we breathe. We need oxygen to carry out cellular respiration, an important process used to fuel cell work in many eukaryotes and prokaryotes alike. The basic metabolic pathways we explore in this chapter are essential for cell survival.

Hundreds of human medical disorders stem directly from failed metabolic pathways. In bacteria and eukaryotic microbes, damaged metabolic pathways also have disastrous results, including cell death. Metabolic differences between microbes can also help us identify bacterial species in the laboratory. Finally, an understanding of microbial metabolism and enzymes can support the design of antimicrobial drugs.

Clinical CASE
NCLEX
HESI
TEAS

The Case of the Drunk Nursing Student

Visit the **Mastering Microbiology** Study Area to watch the case and find out how microbial metabolism can explain this medical mystery.

Healthcare heroes being celebrated on National Nursing Day 2021
Leesburg, VA

8.1 DEFINING METABOLISM

Catabolic and anabolic reactions are the yin and yang of metabolism.

You have probably heard and even used the term *metabolism*. But what does it really mean? **Metabolism** collectively refers to the chemical reactions that organisms use to break down substances to release energy, as well as to reactions that use the released energy to build new substances. Often combinations of reactions, rather than single ones, help cells achieve a desired result. We call these organized sets of chemical reactions *metabolic pathways*. The final product of a metabolic pathway is called an *end product*, whereas the molecules made between the start and the end are called *intermediates*.

There are three categories of metabolic pathways:

- **Catabolic pathways** break down substances and release energy.
- **Anabolic pathways** combine energy and molecules to build new substances.
- **Amphibolic pathways** have a dual role and can be used for both breaking down and building substances.

The prefix "cata" comes from the Greek word meaning "down." You see this prefix in many common words, like "catastrophe"—meaning the breakdown of the natural order of things. Catabolic reactions are generally *hydrolytic* and *exergonic*. A cell breaking down sugars into carbon dioxide and water is an example of catabolism.

Anabolic reactions, also known as **biosynthetic reactions**, use energy to build molecules. "Ana" is the Greek word for "up." To remember this term, think of anabolic steroids, which some athletes illegally use to help build muscles. Anabolic processes often involve *dehydration synthesis* reactions and tend to be *endergonic*. Anabolic reactions include the building of large macromolecules from building-block precursors. For example, amino acids build proteins, nucleotides build nucleic acids, and simple sugars build polysaccharides.

You have probably already noticed the complementary symmetry to metabolism. Catabolism splits bigger molecules into smaller components, releasing energy in the process. Anabolism reuses that energy—and in many cases, the smaller components—to build new, more complex molecules. Because catabolic and anabolic reactions are opposite in nature, they tend to balance each other and are often *coupled*. As such, they are a sort of biological yin and yang that cells rely on to change their chemical components and maintain energy balance (FIG. 8.1).

A key molecule supplies energy for these reactions: **adenosine triphosphate (ATP)** (FIG. 8.2). (See Chapter 2.) All cells require ATP to do cellular work. If a cell loses its ability to make ATP, it will die. ATP is made by catabolic reactions and provides the energy for anabolic reactions.

Other nucleoside triphosphates, including guanosine triphosphate (GTP) and uridine triphosphate (UTP), have a role in certain energy transfers, but ATP is the most widely used by cells and, therefore, the key nucleoside triphosphate that we will discuss throughout this chapter.

ATP is like metabolic money.

To help understand what ATP is and how cells use it, think of money. Many things such as cars, homes, and personal belongings are worth money. However, these items are not *spendable* until they are converted into cash. ATP can be thought of as the cash that cells get when they break down nutrients. Organic

Learning Outcomes

After reading this section, you should be able to:

8.1 Define metabolism.

8.2 Describe the differences between anabolism and catabolism.

8.3 Describe how and why catabolic and anabolic reactions depend on each other.

8.4 Define adenosine triphosphate, state why it is an ideal energy molecule, and describe how it is recharged by ADP/ATP cycling.

CHEM • NOTE

Hydrolytic reactions use water to break apart molecules.

CHEM • NOTE

Exergonic reactions release more energy than they use. **Endergonic reactions** use more energy than they release.

CHEM • NOTE

Dehydration synthesis reactions remove water to form chemical bonds.

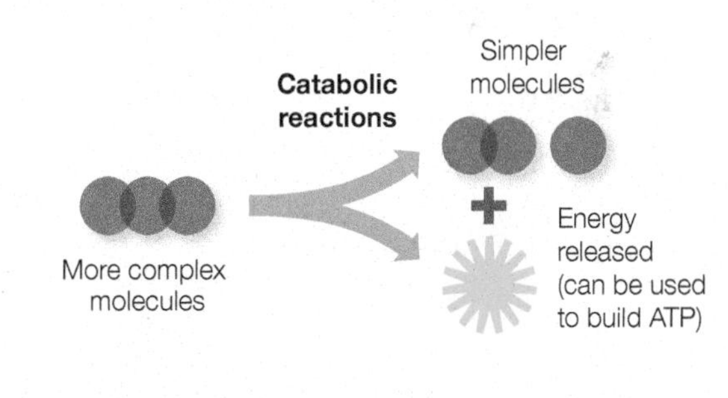

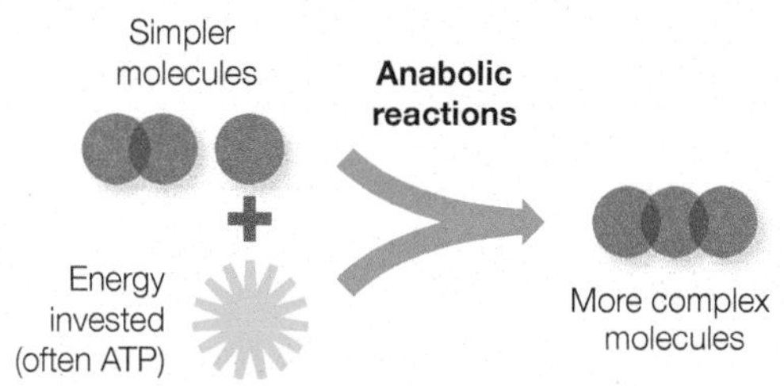

FIGURE 8.1 **Anabolism and catabolism** Cells use catabolic reactions to harvest energy from complex molecules by breaking them down into simpler, lower-energy molecules. Anabolic reactions use the ATP generated in catabolic reactions to make building-block molecules and macromolecules.

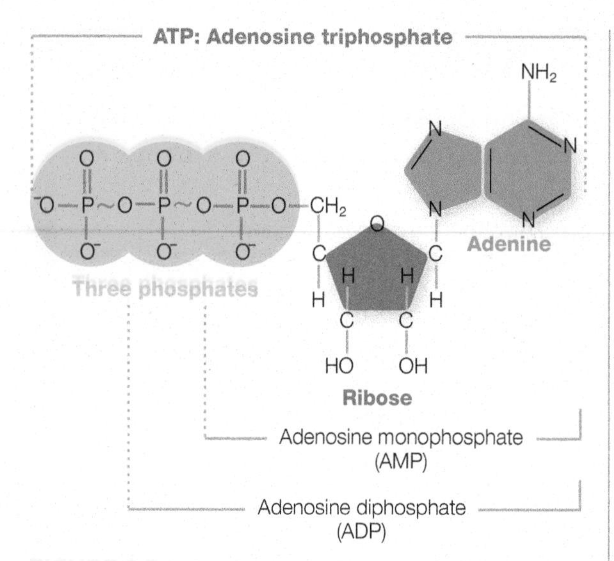

FIGURE 8.2 ATP molecule

CHEM • NOTE

ATP Catabolic reactions make adenosine triphosphate (ATP); anabolic reactions use ATP to fuel synthesis of other substances.

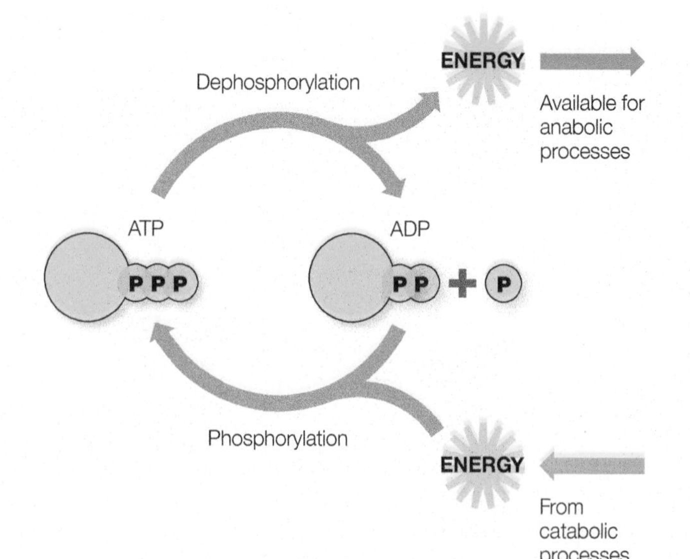

FIGURE 8.3 ATP/ADP cycling Energy is released when adenosine triphosphate (ATP) is dephosphorylated. Cells use phosphorylation reactions to "recharge" adenosine diphosphate (ADP) to ATP.

Critical Thinking *Why do you think cells mostly cycle between ADP and ATP as opposed to stripping ATP down to AMP and then recycling?*

molecules such as proteins, lipids, and carbohydrates are valuable because they store energy in their bonds. The higher the energy of a starting molecule, the more ATP will be made when that molecule is cashed in through metabolic pathways. For example, fats contain more chemical energy in their bonds than do carbohydrates, so catabolizing one lipid molecule would produce more ATP for a cell than catabolizing one glucose molecule.

It is important to realize that ATP is *not* something cells can save up, like dollars in a bank account. ATP is made on demand because cells use so much that they could never effectively store enough. A single *Escherichia coli* cell is estimated to use about 12 billion molecules of ATP in 40 minutes. That's about 50 times its own weight in ATP.[1] In humans, an average adult burns through about 60 kilograms (roughly 132 pounds) of ATP every day, and elite marathon runners use that much ATP in just 2 hours of racing.[2] Therefore, cells store energy in fats, proteins, and polysaccharides and cash in these more complex molecules to create ATP as needed.

As the name implies, adenosine triphosphate is made of adenine, ribose, and *three* phosphate groups. Removing the last (terminal) phosphate group from ATP by dephosphorylation releases energy, along with **adenosine diphosphate (ADP)**, which contains *two* phosphate groups. When a cell needs more ATP, it can add back a phosphate group to ADP via phosphorylation. This adding and removing of the terminal phosphate group is called an **ATP–ADP cycle** (FIG. 8.3). A Nobel Prize–winning biochemist estimated that every minute, the terminal phosphate of each ATP molecule in a cell is added and removed three times. This is especially remarkable considering that at any given time, one cell contains about a billion molecules of ATP.[3] The energy to recharge ADP back to ATP comes from catabolic reactions, while the energy released when ATP becomes ADP can fuel anabolic reactions.

Whereas ATP is often described as a "high-energy" molecule, the energy in ATP is not especially large. Just as 20-dollar bills are more practical to have on hand for daily spending versus 100-dollar bills, so too do we find that it is easier for cells to "spend" energy in the form of ATP as compared to higher-energy molecules. This is mainly because enzymes make it possible for cells to readily dephosphorylate ATP, making it an excellent molecule to fuel a cell's work. It should be noted that cells can also dephosphorylate ADP to make **adenosine monophosphate (AMP)**. (See Chapter 2.) However, an ATP–ADP cycle is more typical.

Later in this chapter we will discuss the three general processes that allow cells to recharge ADP to ATP. But now we will turn to discussing enzymes, the specialized proteins that allow cells to conduct and monitor chemical reactions. Enzymes not only facilitate the ATP–ADP cycle, but also guarantee that nutrients are not needlessly funneled into catabolic pathways to make ATP when the cell would be better served by storing them for later use as carbohydrates, fats, or proteins.

[1] Farmer, I. S., & Jones, C. W. (1976). The energetics of *E. coli* during aerobic growth in continuous culture. *European Journal of Biochemistry*, *67*(1), 115–122. Note that this calculation assumes an *E. coli* cell is about 1×10^{-13} g, and 12 billion molecules of ATP would weigh about 1×10^{-11} g (because one mole of ATP weighs approximately 507 g).

[2] Buono, M. J., & Kolkhorst, F. W. (2001). Estimating ATP resynthesis during a marathon run: A method to introduce metabolism. *Advances in Physiology Education*, 25, 142–143.

[3] Kornburg, A. (1989). *For the love of enzymes: The odyssey of a biochemist*. Cambridge, MA: Harvard University Press, p. 65.

Build Your Foundation

1. Give a general description of metabolism. (LO 8.1)
2. Describe how anabolic and catabolic reactions are interconnected, yet different. (LO 8.2)
3. Describe how ATP acts as a go-between, or intermediate, in catabolic and anabolic reactions. (LO 8.3)
4. Even though ATP is not very high energy, why is it an ideal molecule for fueling cellular processes? (LO 8.4)

Build Your Foundation (BYF) Quick Quiz: Visit the **Mastering Microbiology** Study Area to quiz yourself.

8.2 ENZYMES

Cells rely on enzymes for metabolism.

Learning Outcomes

After reading this section, you should be able to:

8.5 Describe the structural and functional features of enzymes, and state why cells need them.

8.6 Describe how enzymes lower activation energy.

8.7 Give two examples of coenzymes and their functions.

8.8 Describe what ribozymes are and how they are different from typical enzymes.

8.9 List several factors that affect enzyme activity, and differentiate the various modes of enzyme inhibition.

Enzymes are protein catalysts that help chemical reactions occur under cellular conditions. A **catalyst** is something that is only needed in small amounts to make a reaction happen faster; it increases the **reaction rate**. Catalysts are not consumed or permanently changed by a reaction. (Think of a hammer, which is not consumed or changed when it drives down nails.) TABLE 8.1 shows general characteristics of enzymes. Enzymes do *not* facilitate reactions that would be chemically impossible without them. Instead, they just help chemical reactions occur under physiological conditions and within a time frame conducive to life.

Put simply, enzymes facilitate chemical reactions by assisting the molecules' positioning. Recall that all chemical reactions involve the making or breaking of chemical bonds. (See Chapter 2.) Atoms and molecules constantly move and often hit one another. The energy transferred during these collisions can disturb the electron structures of atoms and molecules enough to make or break chemical bonds. This phenomenon is called **collision theory**. However, just because reactants make contact with each other does not mean there will be a reaction. Molecules that collide must also be correctly positioned. Think of two pieces of Velcro: Not only must each piece be combined with enough force to stick together, but the different sides of the fastener must also be properly positioned. Enzymes are tools that cells use to hold reactants in their proper orientation and lower the energy required for starting the reaction.

TABLE 8.1 General Enzyme Characteristics

- Biological catalysts
- Effective in small amounts
- Act on specific substrates to generate specific products
- Increase the rate of a reaction
- Provide a route from reactants to products that has a lower activation energy
- Not consumed or permanently changed in a reaction
- Can be regulated to control metabolic pathways
- Some require cofactors to be active
- Genetically determined
- Often named after its substrate and/or the type of reaction it performs
- Name often ends in "ase" (Example: sucr*ase*)

Enzyme Names

Enzyme names usually end in the suffix *-ase*. In general, enzymes fall into six different classes based on the reactions they catalyze (TABLE 8.2). For example, *oxidoreductases* assist in oxidation–reduction reactions (discussed later in this chapter). We can further catalog enzymes by the reactions they conduct. Some enzymes have very specific names to better reflect their function. For example, alcohol dehydrogenase removes hydrogen from alcohol.

The molecule(s) that an enzyme acts upon is called the **substrate**. Enzymes are so specialized as to the substances they act upon that they are often named after their substrate. For example, the enzyme that breaks down the sugar sucrose is called sucrase. The fact that enzymes are so specialized is beneficial, as specialization provides cells with a way to control when, where, and how often specific chemical reactions occur.

CHEM • NOTE

Enzymes Cells use these protein catalysts to decrease energy needed to start a reaction.

Enzyme–Substrate Interactions

Because of their specificity, enzymes were originally described by the **lock-and-key model**, which stated that an enzyme is like a lock that can only be opened by a specific key: its substrate. The "keyhole" in this analogy, or the site where

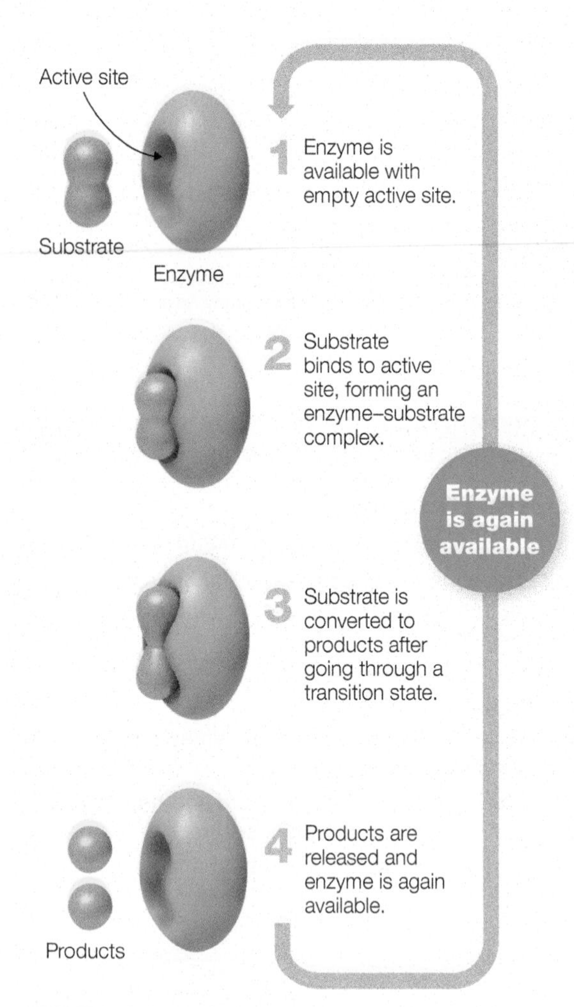

FIGURE 8.4 **General mechanism of enzyme–substrate interaction**

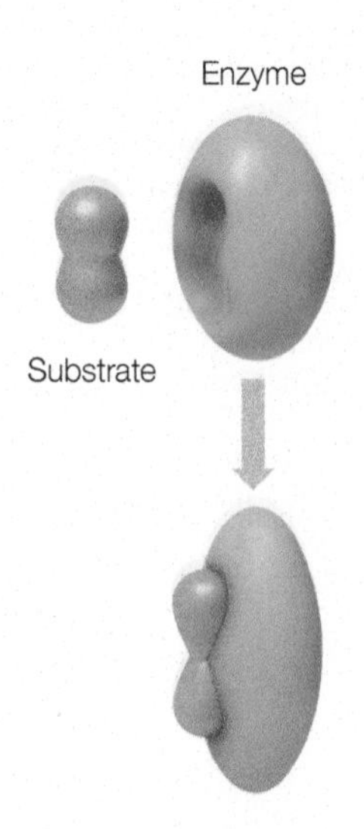

FIGURE 8.5 **Enzyme–substrate induced fit model** Enzymes and their substrates both change shape slightly as they interact.

Critical Thinking *How would a decrease in enzyme flexibility likely affect enzyme activity? Explain your reasoning.*

TABLE 8.2 Enzyme Classification Based on Type of Chemical Reaction Catalyzed

Class	Reactions Catalyzed	Examples
Oxidoreductase	Oxidation–reduction	• Cytochrome oxidase • HMG-CoA reductase • Alcohol dehydrogenase
Transferase	Transfer of functional groups from one molecule to another	• DNA methyltransferase • Pyruvate kinase • Alanine transaminase
Hydrolase	Hydrolysis (addition of water to break bonds)	• Lipase • Sucrase • Carboxypeptidase
Lyase	Removal of groups of atoms without hydrolysis	• Pyruvate decarboxylase • Isocitrate lyase
Isomerase	Rearrangement of atoms within a molecule	• Triose phosphate isomerase • Alanine racemase
Ligase	Joining of two molecules (usually using energy)	• Acetyl-CoA synthase • DNA ligase

the substrate and enzyme interact to generate a chemical reaction, is called the enzyme's **active site** (FIG. 8.4). The active site is a result of the enzyme's three-dimensional protein structure and is chemically and structurally defined. If an enzyme's active site is chemically or physically changed, then the enzyme will struggle to interact with its substrate, decreasing enzyme activity.

Although the lock-and-key model helps us appreciate the basics of enzyme–substrate interactions, it is outdated. Enzymes and substrates are actually somewhat flexible, and both can change shape slightly upon interacting. This allows enzymes to slightly mold and position the substrate in a way that will encourage a reaction. This model of describing how substrates and enzymes interact in the active site is called the **induced fit model** (FIG. 8.5).

When an enzyme and its substrate come together, we call it an **enzyme–substrate complex**. This complex is important because it allows two criteria to be met for the reaction to occur. First, the minimal amount of energy required to start a chemical reaction—the **activation energy**—is met. You can think of activation energy as the energy needed to "get the ball rolling" (get the reaction started). Second, the reactants are properly oriented for the reaction to proceed. Just as Velcro strips have to be properly positioned with respect to each other to stick together, so too do reactants have to be properly positioned to react.

Not only does the enzyme–substrate complex bring reactants together in the right orientation, but it also stabilizes the transition state of the reaction. A **transition state** is a brief yet critical point when reactants are chemically becoming products, but the reaction is not yet completed. It is somewhat like the time between glue being applied and setting, except infinitely faster. Just as it is helpful to stabilize pieces being glued together so the glue sets properly, stabilizing a transition state facilitates chemical reactions. The combined effect of efficiently bringing reactants together and stabilizing transition states is how enzymes lower activation energy (FIG. 8.6). In many cases, it would be impossible to meet the noncatalyzed activation energy needed to carry out vital chemical reactions while also maintaining normal cellular conditions. Consider the breakdown of a protein into its component amino acids: It has been estimated that it would take about 500 years for a peptide bond to break down in a neutral water solution under noncatalyzed conditions. Enzymes accomplish the same task in the same environment in about 10 milliseconds—that's about *one trillion* times faster than the noncatalyzed reaction.[4]

[4] Wolfenden, R. (2006). Degrees of difficulty of water-consuming reactions in the absence of enzymes. *Chemical Reviews*, 106, 3379–3396.

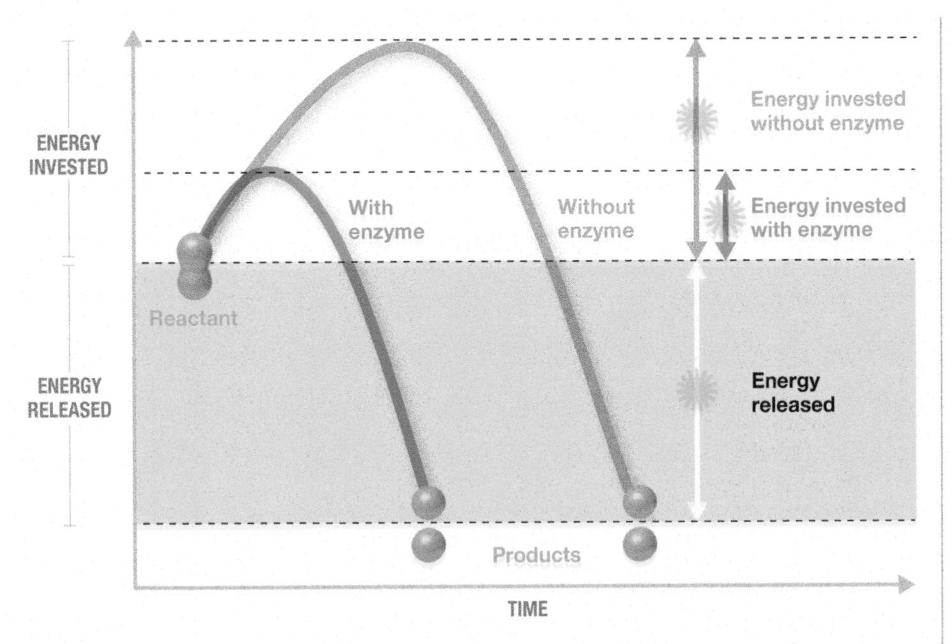

FIGURE 8.6 Enzymes and reactions Enzymes lower activation energy and speed up reactions. This graph shows the progress of a catabolic, exergonic reaction with and without an enzyme.

Critical Thinking *What would the graphs look like if enzymes did not increase the reaction rate (didn't make products form faster)?*

Enzyme Cofactors

Some enzymes need additional components to function; these helpers are called **cofactors**. An enzyme without its necessary cofactor is inactive and is called an **apoenzyme**. The functional form of the enzyme that includes the enzyme and any necessary cofactors is called a **holoenzyme** (FIG. 8.7). Unlike enzymes, which are made of proteins, cofactors are nonprotein substances that can be inorganic or organic. Inorganic cofactors are often metal ions (examples include ions of zinc, iron, magnesium, and calcium). Organic cofactors, which are typically called **coenzymes**, range from free molecules that can move about to factors anchored to the enzyme they assist. Coenzymes are often vitamins or are made from vitamins (TABLE 8.3).

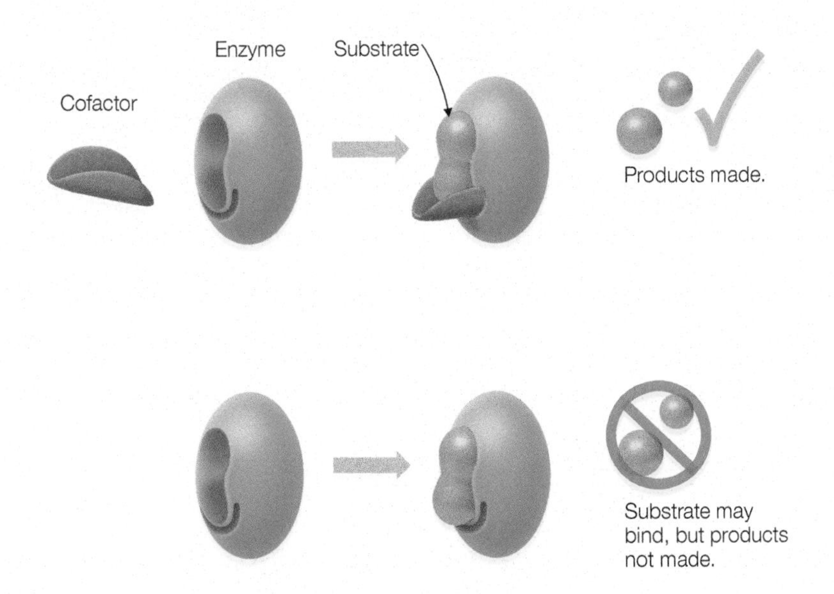

FIGURE 8.7 Enzyme cofactors Cofactors are nonprotein agents that some enzymes require to carry out a reaction. Cofactors do not necessarily affect whether a substrate can bind to the enzyme, but they are often required for the actual chemical reaction to occur.

TABLE 8.3 Selected B Vitamin Coenzymes

Vitamin	Common Name	Coenzyme	Function	Sources	Deficiency Symptoms
B_1	Thiamine	Thiamine pyrophosphate (TPP)	Helps break carbon–carbon bonds	Whole or enriched grains, lean meats	Nervous system disorders (e.g., beriberi)
B_2	Riboflavin	Flavoproteins (FAD and FMN)	Electron transfers	Enriched grains, dairy, liver, spinach	Anemia, mouth/skin sores, neuropathy
B_3	Niacin	NAD^+ and $NADP^+$ molecules*	Electron transfers	Enriched grains, meat, fish, mushrooms	Pellagra
B_5	Pantothenic acid	Coenzyme A (CoA)	Metabolism of pyruvic acid and lipids	Many foods	Extremely rare because in so many foods
B_7	Biotin	Carboxylase and decarboxylase enzymes	Transfer CO_2 groups	Made by intestinal bacteria, many foods	Dermatitis, neurological symptoms
B_9	Folic acid	Tetrahydrofolate (THF), dihydrofolate (DHF)	Used to make purines and pyrimidines	Fortified cereals, liver, leafy vegetables	Spina bifida in baby if low during pregnancy
B_{12}	Cobalamin	Methylcobalamin	Transfer of methyl groups; amino acid metabolism	Animal products (vegans should consider vitamin supplement)	Pernicious anemia, nerve degeneration

*NAD^+ nicotinamide adenine dinucleotide (common in catabolic reactions that release energy); $NADP^+$ nicotinamide adenine dinucleotide phosphate (common in anabolic reactions).

Coenzymes NAD^+ and FAD Certain coenzymes act as **electron carriers.** When electrons are removed from a substrate during a reaction, these coenzymes act as taxicabs—that is, they pick up electrons from one reaction and drop them off for use in another reaction. Electron carrier coenzymes can do this over and over again, just as a taxi repeatedly picks up and drops off passengers. Like a taxi, if the electron carrier coenzyme is "occupied" (has electron passengers), it must drop off those electrons before it can pick up more. Additionally, just as there are many taxicab companies, there are different types of cofactors whose job it is to transport electrons. Even though the names and structures of these coenzymes vary, they tend to operate in the same general way.

Common coenzymes that serve as electron carriers include **nicotinamide adenine dinucleotide (NAD^+)** and **nicotinamide adenine dinucleotide phosphate ($NADP^+$).** These coenzymes are made from the B vitamin niacin (nicotinic acid). Other electron carrier coenzymes include **flavin mononucleotide (FMN)** and **flavin adenine dinucleotide (FAD).** These are made from the B vitamin riboflavin. Unlike NAD^+ and $NADP^+$ the flavin coenzymes do not exist as free molecules in a cell. Instead, they are anchored to enzymes and are sometimes called *prosthetic groups*. Usually NAD^+ and FAD are found in catabolic pathways, while anabolic pathways often involve $NADP^+$.

CHEM • NOTE

Electron-carrier coenzymes In cells, coenzymes often collect electrons from one reaction and shuttle them to other reactions in the cell. Common coenzymes in metabolism include NAD^+, $NADP^+$, FMN, and FAD.

Coenzyme A (CoA) The last coenzyme we will discuss is **coenzyme A (CoA),** which contains a derivative of pantothenic acid (vitamin B_5). Coenzyme A is important in carbohydrate and fat metabolism, as we will see later in the chapter.

Ribozymes

Unlike protein enzymes, **ribozymes** are made of the nucleic acid RNA, and they have a more limited range of substrates. So far, ribozymes have only been shown to act on other RNA molecules. There are many naturally occurring ribozymes, but man-made ribozymes have also been developed to carry out a variety of research applications. Preliminary clinical research suggests that man-made ribozymes hold promise as drug therapies, especially in genetic disorders and as antiviral therapies.

Many factors affect enzyme activity.

Cells can control biological processes by regulating how much of an enzyme they make and by managing how active enzymes are. Because cells are so dependent on enzymes for their survival, we can control microbial growth through drugs and environmental factors that impact enzyme activity. Cofactors, temperature, pH, how much substrate is present, the phosphorylation state, and the presence of inhibitors are some factors that can affect enzyme activity.

Temperature

In most cases, lowering temperature will lower enzyme activity, causing cells to grow more slowly. This is why keeping foods at cold temperatures usually slows food spoilage (see Chapter 7). Warmer temperatures can increase enzyme activity. However, in living systems, there is a limit to how much we can safely raise temperature and still support metabolism. Temperatures too much above an enzyme's **optimal temperature**, or the temperature at which a given enzyme's activity is highest, significantly reduce enzyme activity (FIG. 8.8). Exposing cells to high temperatures can cause enzymes and other proteins to become **denatured**, meaning they lose their three-dimensional structure and become nonfunctional (FIG. 8.9). Denaturation can be reversible or irreversible. If you have cooked an egg, you have witnessed irreversible protein denaturation. As an egg is cooked, its protein albumin starts out clear and gelatinous and ends up permanently transformed into a more solid, white substance.

All proteins have a temperature at which they will denature. Some proteins have uniquely high denaturation temperatures, such as those found in the hot-springs bacteria *Thermus aquaticus*. For most pathogenic bacteria and human cells, temperatures above 41°C (105.8°F) can lead to protein denaturation. This is partially why a high fever can be dangerous. Enzymes can also denature through contact with strong acids, bases, salts, detergents, alcohol, radiation, and heavy metals such as ions of lead, arsenic, or mercury.

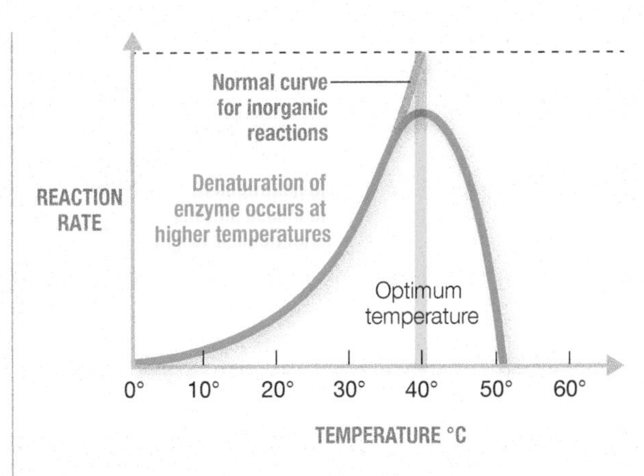

FIGURE 8.8 Optimal conditions for enzyme-catalyzed reactions Reaction rates are highest under optimal conditions. The reaction rate will decrease above or below the optimal temperature, as shown here, or outside the optimal pH range (not shown).

Critical Thinking *What would you predict this graph would look like for a thermophile (organism that thrives at high temperatures)?*

pH

A pH above or below **optimal pH** will also alter enzyme structure and lead to a reduced reaction rate (Figure 8.9). Changes in pH alter an enzyme's structure because H^+ (acid) and OH^- (base) affect the hydrogen and ionic bonds that support the three-dimensional structure of a protein. Extreme changes in pH can also lead to protein denaturation. Amylase, an enzyme in human saliva, is active in the mouth, but it is quickly deactivated when it reaches the low pH of the stomach. Stomach acid interferes with the activity of many enzymes, which limits bacterial growth in that organ. Most pathogens prefer a neutral pH, but there are examples of bacteria with special adaptations that allow them to live in pH extremes. *Helicobacter pylori* can thrive in the low-pH environment of the stomach and cause stomach ulcers.

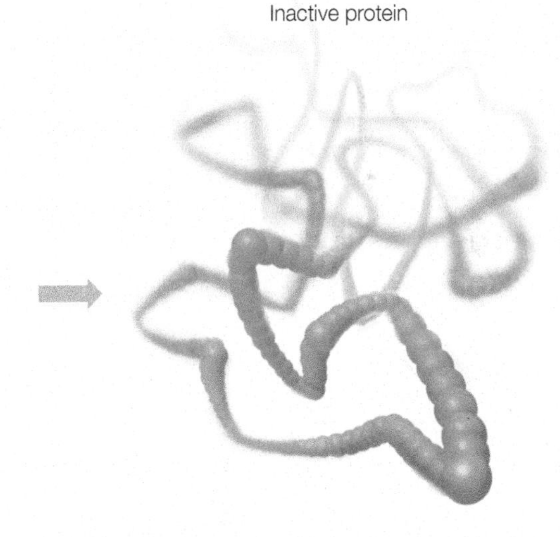

FIGURE 8.9 Denatured proteins Denaturing changes a protein's structure and therefore impacts its function.

Regulating Enzyme Activity

Cells can regulate enzyme activity by controlling how much of an enzyme is physically present and by adjusting how active a given enzyme is via phosphorylation state, inhibitors, and allosteric regulatory factors.

Enzyme and Substrate Concentrations Recall that for an enzyme to make a product, the substrate must enter the enzyme's active site. Therefore, the rate of product formation depends on how many active sites are available (the enzyme concentration) *and* how much substrate is present. If the amount of substrate increases while the amount of enzyme remains unchanged, then we will

CHEM • NOTE

pH scale Low pH (below 7) is acidic; high pH (above 7) is basic; and a pH of exactly 7.0 is neutral. Acids lower pH, and bases raise pH.

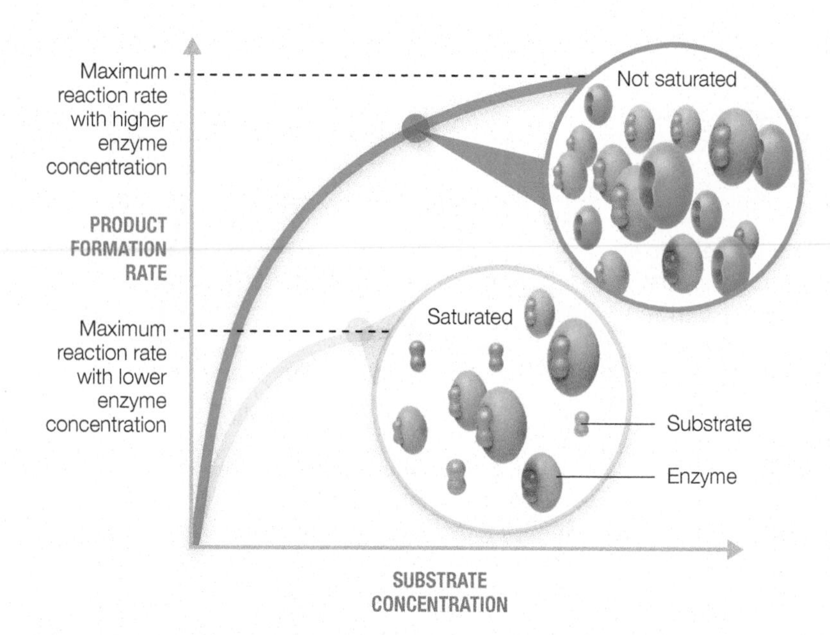

FIGURE 8.10 Enzyme saturation Enzyme saturation occurs when all enzyme active sites are occupied by substrate. This only happens when substrate concentration is much higher than the enzyme concentration. Notice that the saturation point and the maximum reaction rate are both higher when more enzyme is present.

eventually observe a *saturation point* at which every enzyme active site is occupied by a substrate (FIG. 8.10). To understand how enzyme and substrate concentrations affect the rate of product formation, think about checking out at a grocery store. In busy times, a store may open additional registers to speed the rate of checkout. Likewise, if a cell needs more products formed in a shorter amount of time, it can increase the amount of a particular enzyme available for action.

Enzyme Regulation by Phosphorylation State Even small structural changes can affect an enzyme's function. Cells can change the structure of certain enzymes by adding or removing phosphate groups. Cells use specialized enzymes called **kinases** and **phosphatases** to accomplish this type of regulation. Kinases use phosphorylation reactions to add phosphates to their targets, while phosphatases use dephosphorylation reactions to remove phosphate groups from their targets.

Some enzymes are more active when they are phosphorylated, whereas others are less active; it depends on the enzyme in question. The bacterium *E. coli* uses phosphorylation to regulate the production of lipopolysaccharide, a toxin in Gram-negative bacteria that can accumulate in an infected person and cause septic shock. In humans, a key enzyme used in making cholesterol, called HMG-CoA reductase, is inactivated by phosphorylation and activated by dephosphorylation.

CHEM • NOTE

Phosphorylation and dephosphorylation reactions The element phosphorus (P) is important in cells. When phosphorus is bonded to three oxygen atoms, it is called a phosphoryl group (PO_3^{2-}). When it is bonded to four oxygen atoms, it is a phosphate group (PO_4^{3-}). Phosphoryl groups are added in phosphorylation reactions. They are removed by dephosphorylation reactions. When a phosphoryl group is added to a molecule by bonding to another oxygen, it becomes a phosphate group (as occurs when ADP is recharged to ATP).

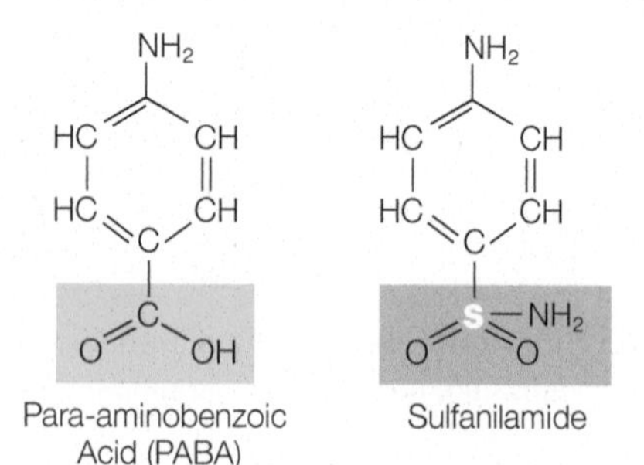

Sulfanilamide, which is similar in structure to PABA, blocks bacterial folic acid synthesis.

Enzyme Inhibitors Penicillin-based antibiotics inhibit an enzyme that helps bacteria produce their cell walls. The statin drug Lipitor inhibits a human liver enzyme that makes cholesterol. These common drugs are examples of how modern medicine has employed enzyme inhibitors to improve human health. However, cells also use naturally occurring inhibition to regulate enzyme activity.

Competitive Inhibitors Competitive inhibitors slow reactions by competing with a substrate for the target enzyme's active site. Just as the wrong key may fit into the keyhole but not turn the lock, in competitive inhibition the affected enzyme cannot carry out the reaction until the competitive inhibitor leaves the active site and the substrate can enter. Competitive inhibition can be overcome if the concentration of substrate is greater than the concentration of the inhibitor (FIG. 8.11). This is because it is harder for the inhibitor to "compete" for the enzyme's active site if it is outnumbered by the substrate. Statistically, it is more likely that the enzyme will interact with an abundant substrate instead of a less prevalent competitive inhibitor.

An example of a competitive inhibitor is the antibiotic sulfanilamide. Because *para*-aminobenzoic acid (PABA) and sulfanilamide have similar structures (see line drawing), they compete with each other for the active site of an enzyme important in folic acid production.[5] If sulfanilamide enters the enzyme instead of the enzyme's normal substrate, PABA, then folic acid production decreases. The affected bacterial cell soon suffers from a deficiency in folic acid, stops growing, and eventually dies. Because we get folic acid from our diet and our cells cannot make it, sulfanilamide selectively affects bacterial cells without causing harm to human cells. (For more on how sulfa drugs work, see Chapter 15.)

[5] Note that in some nutrition texts you may find that folate and folic acid are treated as entirely different, with folate being described as a "natural" form of the molecule, and folic acid being described as the form manufactured for nutritional supplements. Biochemists make no such distinction. In chemistry the suffixes "-ic acid" and "-ate" simply reflect the organic acid and its conjugate base—both of which exist in nature. Thus, biochemists consider folic acid and folate to be two sides of the same coin. In this text, organic acid names (e.g., folic acid) are used instead of conjugate base names (e.g., folate).

Noncompetitive Inhibitors Unlike competitive inhibitors, **noncompetitive inhibitors** do not compete with a substrate for the enzyme's active site. Instead, they decrease enzyme activity by binding to the enzyme at a site other than the active site. Sometimes noncompetitive inhibitors are described as deforming an enzyme's structure so that substrates cannot interact with the active site. However, in most cases, the noncompetitive inhibitor and the substrate can simultaneously bind to the enzyme—but the altered enzyme shape reduces the enzyme's activity, often preventing product formation (FIG. 8.12). Some inhibitors bind to enzymes reversibly, while others bind irreversibly. Cyanide is an example of a noncompetitive inhibitor. Cyanide inhibits enzymes in the electron transport chain, a pathway that makes ATP.

Allosteric Regulation In the prior sections, we discussed how competitive and noncompetitive inhibitors reduce enzyme activity. Here we touch on a specific mechanism of enzyme activity regulation called allosteric regulation—a key way that cells regulate metabolic pathways. The word *allosteric* means "other form." This is an appropriate name because this regulatory mechanism involves a change in enzyme shape in response to a regulatory molecule. This change in shape either increases or decreases enzyme activity. The change in the enzyme's shape occurs when the regulatory molecule binds to a specific site in the enzyme called the **allosteric site** (or regulatory site). **Allosteric activation** occurs when a regulatory molecule (an allosteric activator) *increases* enzyme activity. In contrast, in **allosteric inhibition**, a regulatory molecule (an allosteric inhibitor) *decreases* the enzyme's activity.

It can seem that allosteric inhibitors are the same as noncompetitive inhibitors—neither is competitive, and both decrease enzyme activity by binding to a site other than the enzyme's active site. However, although they produce a similar end result, they achieve that

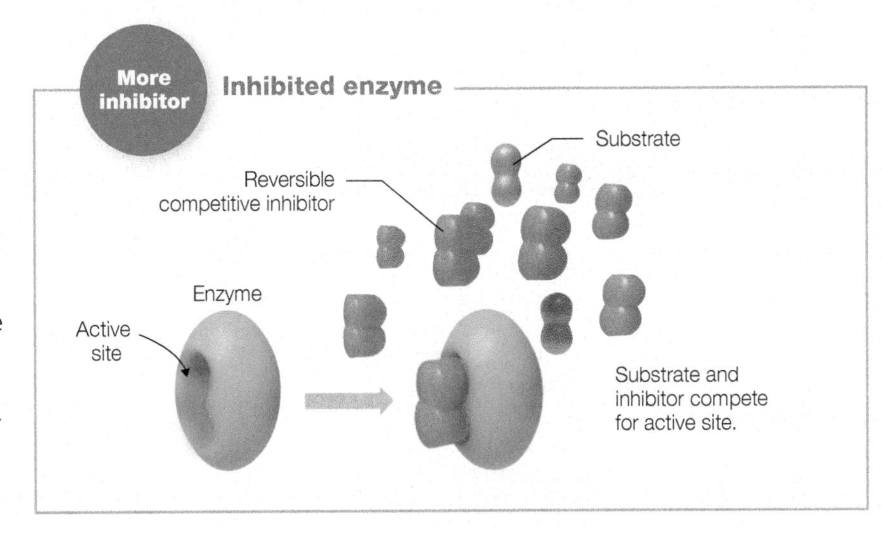

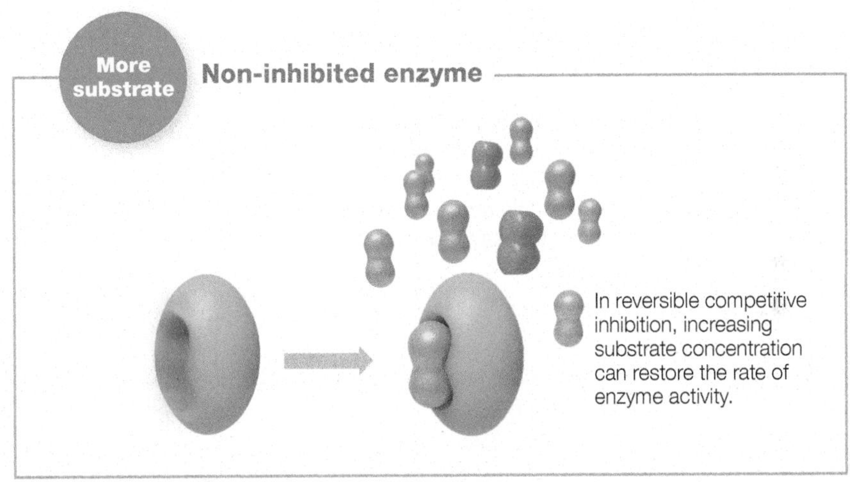

FIGURE 8.11 **Competitive inhibition** Competitive inhibitors compete with substrate for access to an enzyme's active site.

Critical Thinking *In the figure, the enzyme has an equal affinity (liking) for its substrate and for the competitive inhibitor. How would the concentration of substrate needed to overcome competitive inhibition be affected if the enzyme preferred (bound better or more readily to) the competitive inhibitor as compared to the substrate?*

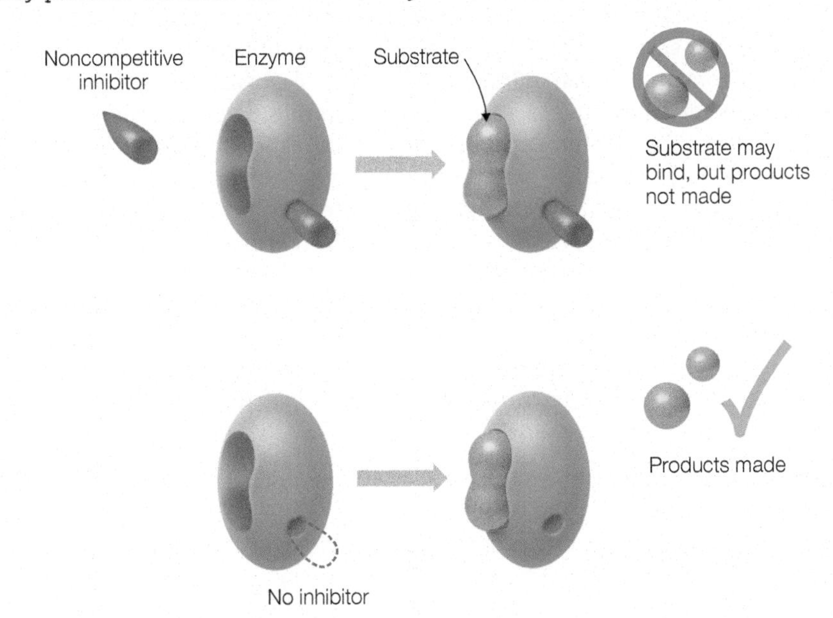

FIGURE 8.12 **Noncompetitive inhibition** Noncompetitive inhibitors bind to a site other than an enzyme's active site. Normally these inhibitors don't fully prevent substrates from interacting with enzymes, but they do impact the progress of the reaction. The noncompetitive inhibitor can bind irrespective of how much substrate is present. This results in an overall *decrease* in reaction rate even with very little inhibitor present.

Critical Thinking *What is a potential advantage of a noncompetitive inhibitor over a competitive inhibitor for regulating enzyme reaction rates? (Hint: Consider how these two different inhibitors work.)*

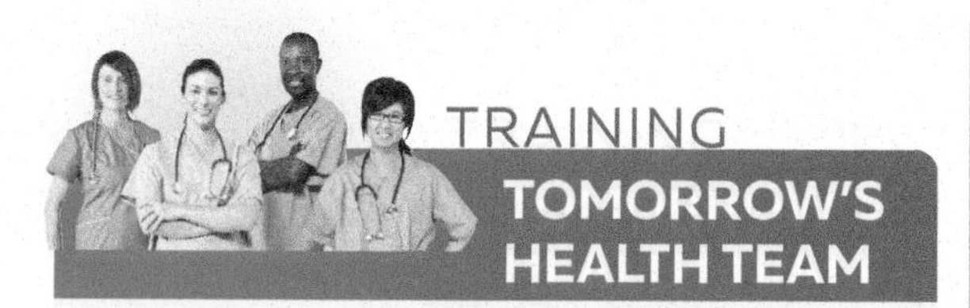

Statins

Many drugs prescribed by healthcare providers work by inhibiting enzymes. This is true for the class of drugs known as statins, which act as competitive inhibitors of HMG-CoA reductase (3-hydroxy-3-methyl-glutaryl-CoA reductase), a liver enzyme important for cholesterol production.

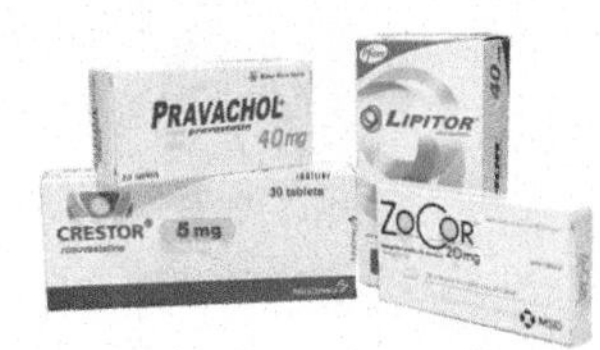

Statin drugs are marketed under a variety of names, including Lipitor, Crestor, Pravachol, and Zocor.

These drugs are commonly prescribed to lower blood cholesterol levels. According to the National Institutes of Health, about one-half of U.S. men ages 65–74 take statin drugs, while just over one-third of women in that age group take them.

QUESTION 8.1

Describe the general way that these drugs act upon the enzyme HMG-CoA reductase to lower a patient's cholesterol. Hint: Look at Figure 8.11.

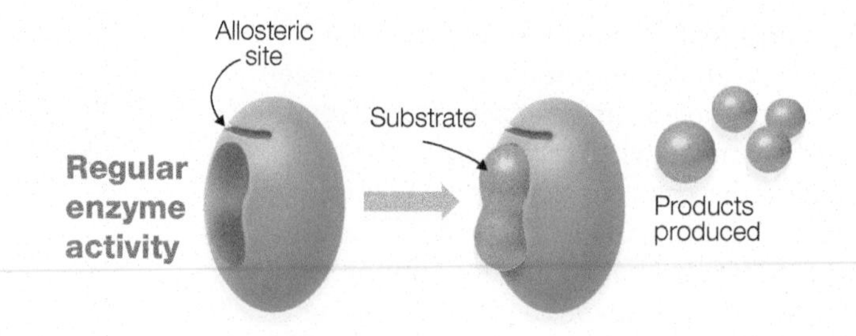

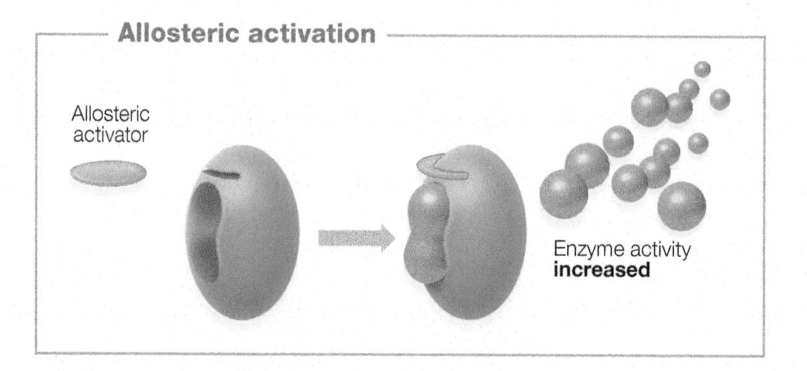

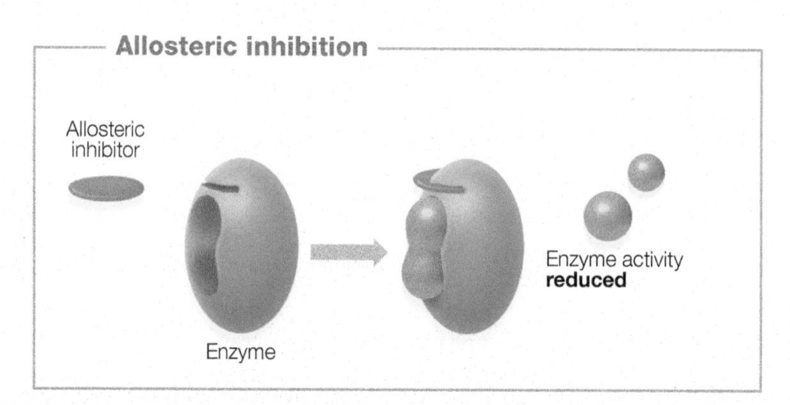

FIGURE 8.13 **Allosteric activators and inhibitors** Enzyme activity can be enhanced by an allosteric activator or decreased by an allosteric inhibitor.

Critical Thinking *An allosteric activator for an enzyme in one pathway could act as an allosteric inhibitor of another enzyme in a separate pathway. Why would coordinating inhibition and activation in this way benefit a cell?*

result through a different approach. An allosteric site is a specific location in the enzyme that serves a regulatory function by specifically interacting with allosteric factors. In contrast, noncompetitive inhibitors are not necessarily picky about where they bind to enzymes. For example, heavy metals such as lead and mercury do not have specific allosteric sites for binding to particular enzymes and therefore cause generalized enzyme inactivation that leads to the symptoms of heavy metal poisoning. Also, allosteric inhibitors do not permanently inhibit an enzyme, whereas certain noncompetitive inhibitors can. Another difference is that feedback inhibition, which is discussed next, can be accomplished through a few mechanisms, one of which is allosteric inhibition. In summary, allosteric regulation is a physiological process that regulates enzyme activity much as a volume knob controls a stereo's sound level; binding of the allosteric activator can "turn up" the enzyme's activity, while an allosteric inhibitor "turns it down" (FIG 8.13).

Feedback Inhibition Cells often use **feedback inhibition** to regulate biochemical pathways. This form of regulation slows down (or can even turn off) metabolic pathways to help cells operate more efficiently. Biological pathways often involve making many intermediates on the way to a particular end product. Therefore, cells normally only run these complex pathways when needed. If the

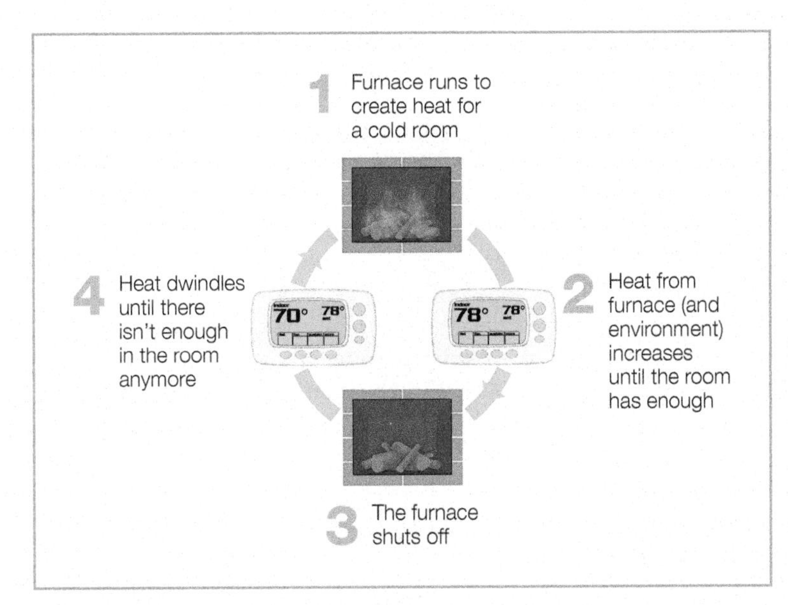

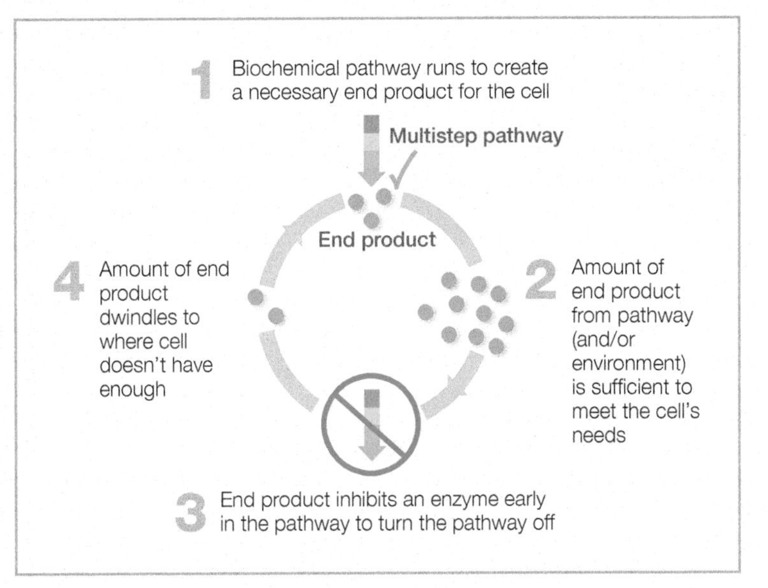

FIGURE 8.14 Feedback inhibition Feedback inhibition in a cell works similarly to how a furnace and thermostat function in a room. *Left*: When heat (end product) is needed, the furnace (reaction pathway) makes it for the room (cell). Once there is enough heat, the furnace is blocked from making more until the room gets cold again. *Right*: Likewise, a cell makes a product to the point where sufficient quantities exist; then the pathway is blocked from making any more. When product decreases to less than what the cell needs, the block goes away, allowing the cell to make more again.

cell has enough of a specific end product, it would be a waste of energy and resources to continue to produce it at high levels. A cell won't invest in making more of a product if it already has enough, or in certain cases, if it has a readily available supply in its environment.

Feedback inhibition is reversible and is usually accomplished by an end product acting as either a competitive inhibitor or as an allosteric inhibitor of an enzyme earlier in the pathway. When the level of end product drops to a point that the cell needs more, the production pathway resumes (FIG. 8.14). It is important to understand that feedback inhibition usually takes into account *total* end product (what the cell has made *and* the amount of the product obtainable from the cell's environment). Cells use feedback inhibition to manage their production of many biologically important compounds, such as amino acids and nucleotides.

One example of feedback inhibition is the production of the amino acid isoleucine in *E. coli*. Five biochemical steps convert the amino acid threonine to isoleucine. If isoleucine is plentiful, either because the cell made enough or there is a large amount in the environment, then the first step in the production pathway will be inhibited. As a result, *E. coli* will stop making isoleucine until more is needed.

Build Your Foundation

5. What is an active site, and what role does it have in enzyme specificity? (LO 8.5)
6. Why is enzyme specificity an advantage to a cell? (LO 8.5)
7. How do enzymes help reactions occur more quickly and efficiently? (LO 8.6)
8. Describe the functions of NAD^+ and FAD coenzymes. (LO 8.7)
9. What are ribozymes, and how are they different from typical enzymes? (LO 8.8)
10. List at least four different factors that affect enzyme activity, and describe how they do so. (LO 8.9)
11. How does feedback inhibition help cells live more efficiently? (LO 8.9)

Build Your Foundation (BYF) Quick Quiz: Visit the **Mastering Microbiology** Study Area to quiz yourself.

8.3 OBTAINING AND USING ENERGY

Redox reactions fuel the recharging of ADP to ATP.

Learning Outcomes

After reading this section, you should be able to:

8.10 Explain what oxidation–reduction reactions are and why they are considered coupled reactions.

8.11 List three general mechanisms for phosphorylating ADP to make ATP.

Now that we understand the basics of enzymes and their regulation, we can turn to the mechanisms that recharge ADP to ATP. As discussed earlier, the energy to recharge ADP back to ATP comes from catabolic reactions, while the energy released when ATP becomes ADP can be used to fuel anabolic reactions and other cellular work. Cells use oxidation and reduction reactions to extract energy from nutrients in order to recharge ADP to ATP:

CHEM • NOTE

Redox reactions Remember "Oil Rig" to keep track of what happens to electrons in these reactions: **O**xidation **I**s **L**oss of electrons; **R**eduction **I**s **G**ain of electrons.

- In an **oxidation reaction**, an atom or molecule loses electrons (e^-).
- In a **reduction reaction**, an atom or molecule gains electrons.

Oxidizing agents carry out oxidation reactions. *Reducing agents* carry out reduction reactions. Because electrons are negative, when a molecule gains electrons, its total oxidation state, or hypothetical atomic charge, is *reduced*; which is why it is appropriate to call such reactions reduction reactions.

Because oxidation and reduction reactions *always* occur as partners, we refer to them as **redox reactions**. Much as somebody making money means that someone else is spending money (you can't have one without the other), whenever a molecule or atom loses electrons and is oxidized, some other molecule or atom gains those electrons and is reduced (FIG. 8.15). In biochemical reactions, oxygen is one common oxidizing agent, whereas hydrogen often acts as a reducing agent.

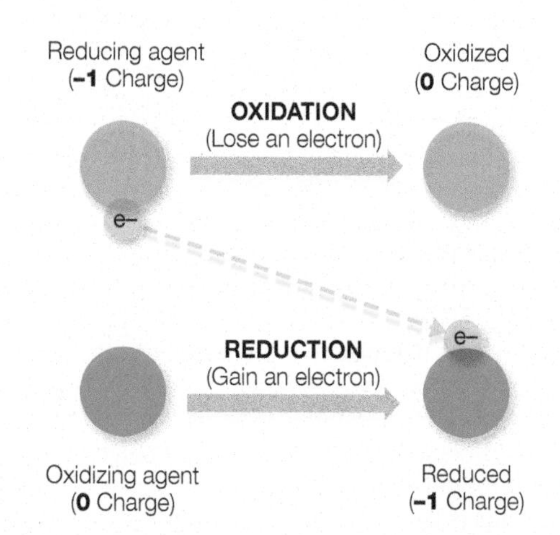

FIGURE 8.15 Oxidation–reduction reactions Oxidation–reduction reactions are coupled. When an atom or molecule loses an electron (or is oxidized), another entity absorbs that electron and is reduced.

Critical Thinking *From the following reaction, describe what is being reduced and what is being oxidized (Fe is iron, and Cu is copper).*

$$\text{Fe (metal)} + \text{Cu}^{2+} \rightarrow \text{Fe}^{2+} + \text{Cu (metal)}$$

Oxidation of Nutrients

Cells harvest energy from nutrients by oxidizing them—that is, by stripping electrons from them. Molecules that contain abundant hydrocarbon bonds (C—H bonds) are highly reduced; this means they can be readily oxidized to release energy. For example, carbohydrates and fats are highly reduced molecules that cells can extract energy from using redox reactions. Proteins are less reduced and consequently will make less ATP than carbohydrates and fats. Therefore, cells tend to use carbohydrates and fats first, and then will turn to proteins as energy sources. The more completely cells can oxidize these molecules, the more energy they can harvest from these nutrition sources.

Coenzymes and Redox Reactions

Cells have evolved metabolic pathways that extract energy from carbohydrates, fats, and proteins using oxidation–reduction (redox) reactions. The enzymes that carry out these redox reactions usually rely on coenzymes like NAD^+ and FAD to act as electron taxis. These coenzymes pick up electrons and are *reduced* as sugars, fats, or other nutrient sources are *oxidized*. NADH is the reduced form of NAD^+, while $FADH_2$ is the reduced form of FAD (FIG. 8.16). In this chapter, we will focus on the oxidation of glucose, which is a primary energy source for many cells. Note that cells also oxidize fats and proteins to release energy, and that these macronutrients are also funneled into the pathways we will discuss. The redox reactions of catabolic pathways are often tied to phosphorylation reactions that recharge ADP to ATP (discussed next).

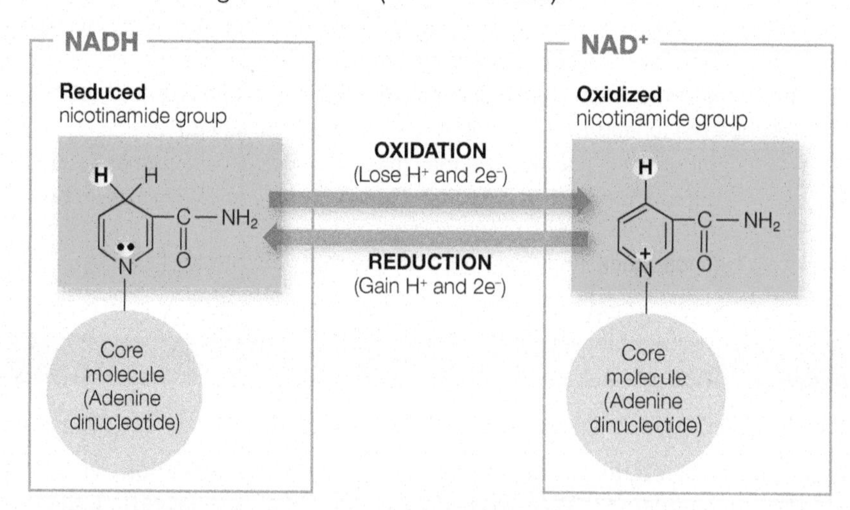

FIGURE 8.16 Reduction and oxidation of coenzymes When NAD^+ (nicotinamide adenine dinucleotide) is reduced it is called NADH. NADH is returned to an oxidized state (NAD^+) when it donates its electrons—indicated by the dots over the nitrogen (N) atom—in a reduction reaction.

TABLE 8.4 Three Phosphorylation Mechanisms to Recharge ADP to ATP

	Substrate-Level Phosphorylation	Oxidative Phosphorylation	Photophosphorylation
How ATP is made	A high-energy phosphate from a pathway intermediate is transferred directly to ADP to make ATP.	ADP recharged to ATP using electron transport chains powered by nutrients	ADP recharged to ATP using electron transport chains powered by solar energy
Electron transport chain used?	No	Yes	Yes
Used in	Glycolysis, Krebs cycle, fermentation*	Aerobic and anaerobic electron transport chains of cellular respiration	Light-dependent reactions of photosynthesis
Cell types that use it	Prokaryotic and eukaryotic	Prokaryotic and eukaryotic	Only photosynthetic cells; prokaryotic and eukaryotic

*Some fermentation pathways directly make ATP via substrate level phosphorylation (e.g., acetate fermentation).

Three general mechanisms make recharging ADP to ATP possible.

We previously mentioned that cells carry out ATP–ADP cycling, a situation where a phosphorylation reaction recharges ADP to ATP. Cells use three general mechanisms to drive the phosphorylation of ADP to ATP (TABLE 8.4). All three mechanisms participate in the metabolic pathways we will discuss later in this chapter.

Substrate-Level Phosphorylation

Substrate-level phosphorylation occurs when an enzyme transfers a phosphoryl group (PO_3^{2-}) from a donor substrate directly to ADP to make ATP. Sometimes these phosphoryl group donors are called *high-energy intermediates*.

Oxidative Phosphorylation

Oxidative phosphorylation involves a collection of redox reactions that strip electrons from a food source (such as carbohydrates, fats, or proteins) and eventually hand off those electrons to an **electron transport chain** to fuel the phosphorylation of ADP to ATP. Sometimes electron transport chains are called respiratory chains. (We will discuss respiratory chains in more detail under cellular respiration.)

Photophosphorylation

Photophosphorylation, like oxidative phosphorylation, relies on the redox reactions of an electron transport chain. However, instead of using energy and electrons derived from the oxidation of a macronutrient to run the electron transport chain, light energy is used to activate electrons. Only photosynthetic cells can perform photophosphorylation.

Build Your Foundation

12. What are redox reactions, and how are NAD^+ and FAD central to them in cells? (LO 8.10)
13. Why is glucose important to cells? (LO 8.10)
14. Describe the three general ways that ADP is recharged to ATP. (LO 8.11)
15. Describe the link between redox reactions and the phosphorylation reactions. (LO 8.11)

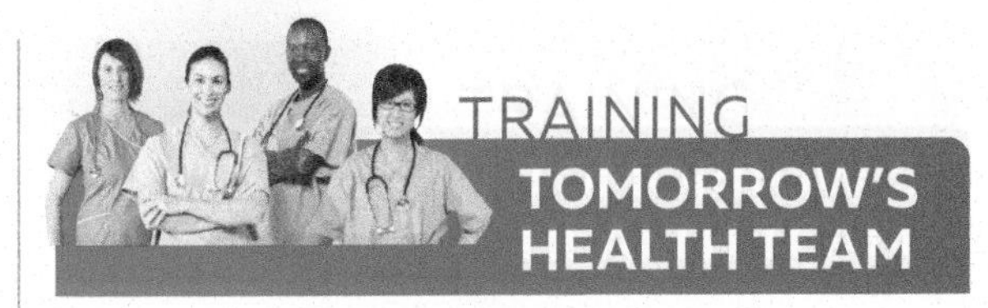

Niacin Deficiency

Niacin (vitamin B_3) is important for several cellular activities, including extracting energy from foods. Bacteria can make niacin, but we primarily get it from our diet. In humans, the disease pellagra can develop from sustained niacin deficiency. Pellagra may result from malnutrition or alcoholism. Symptoms include abdominal cramps, scaly skin lesions, dementia, headaches, diarrhea, fatigue, anxiety, depression, loss of strength, and nausea (with or without vomiting). Pellagra is fatal if not treated. The typical treatment is oral or intravenous niacin.

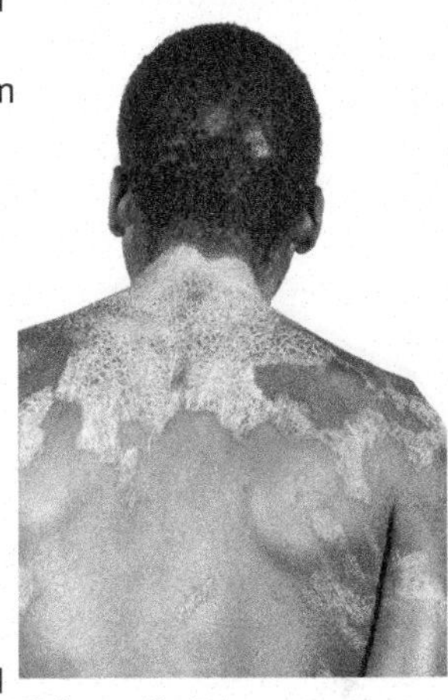

Pellagra skin lesions.

QUESTION 8.2

What is it about niacin that makes it so important for extracting energy from foods? Hint: Refer to Table 8.3.

Build Your Foundation (BYF) Quick Quiz: Visit the **Mastering Microbiology** Study Area to quiz yourself.

8.4 THE CATABOLIC PROCESS OF CELLULAR RESPIRATION

Learning Outcomes

After reading this section, you should be able to:

8.12 Define cellular respiration.

8.13 Describe what occurs in glycolysis, including general reactants and products.

8.14 Explain what occurs in the Krebs cycle, including general reactants and products.

8.15 Describe how chemiosmosis drives ATP production.

8.16 List the similarities and differences between aerobic and anaerobic respiration.

8.17 Compare and contrast aerobic and anaerobic electron transport chains.

Cellular respiration is one way cells harvest energy from nutrients.

Glucose is the most common carbohydrate energy source used by cells. Therefore, **carbohydrate catabolism**, or the breakdown of carbohydrates to release energy, is central to a cell's survival. Cells primarily extract energy from carbohydrates through cellular respiration and fermentation. FIG 8.17 shows a simplified schematic of these two processes. Looking at the diagram, you may notice that cellular respiration has more stages than fermentation and requires an electron transport chain. There are also shared features between these processes—for example, both rely on glycolysis and electron carriers (e.g., NAD^+). We'll begin by describing cellular respiration, and later we'll review fermentation.

Cellular respiration is a collection of reactions that use redox reactions to extract energy from nutrients and then *transfer* that energy into the bonds of ATP. Cellular respiration occurs through the combined efforts of multiple pathways. In glycolysis, the intermediate step, and the Krebs cycle, coenzymes NAD^+ and FAD collect electrons from the cell's nutrients and then deposit

CHEM • NOTE

First Law of Thermodynamics This law states that energy is not created or destroyed; it just changes forms. So, although cellular respiration is sometimes described as "making" energy, it's more accurate to say these pathways "transfer" energy. Specifically, the energy in the bonds of the nutrients is transferred to the bonds of ATP.

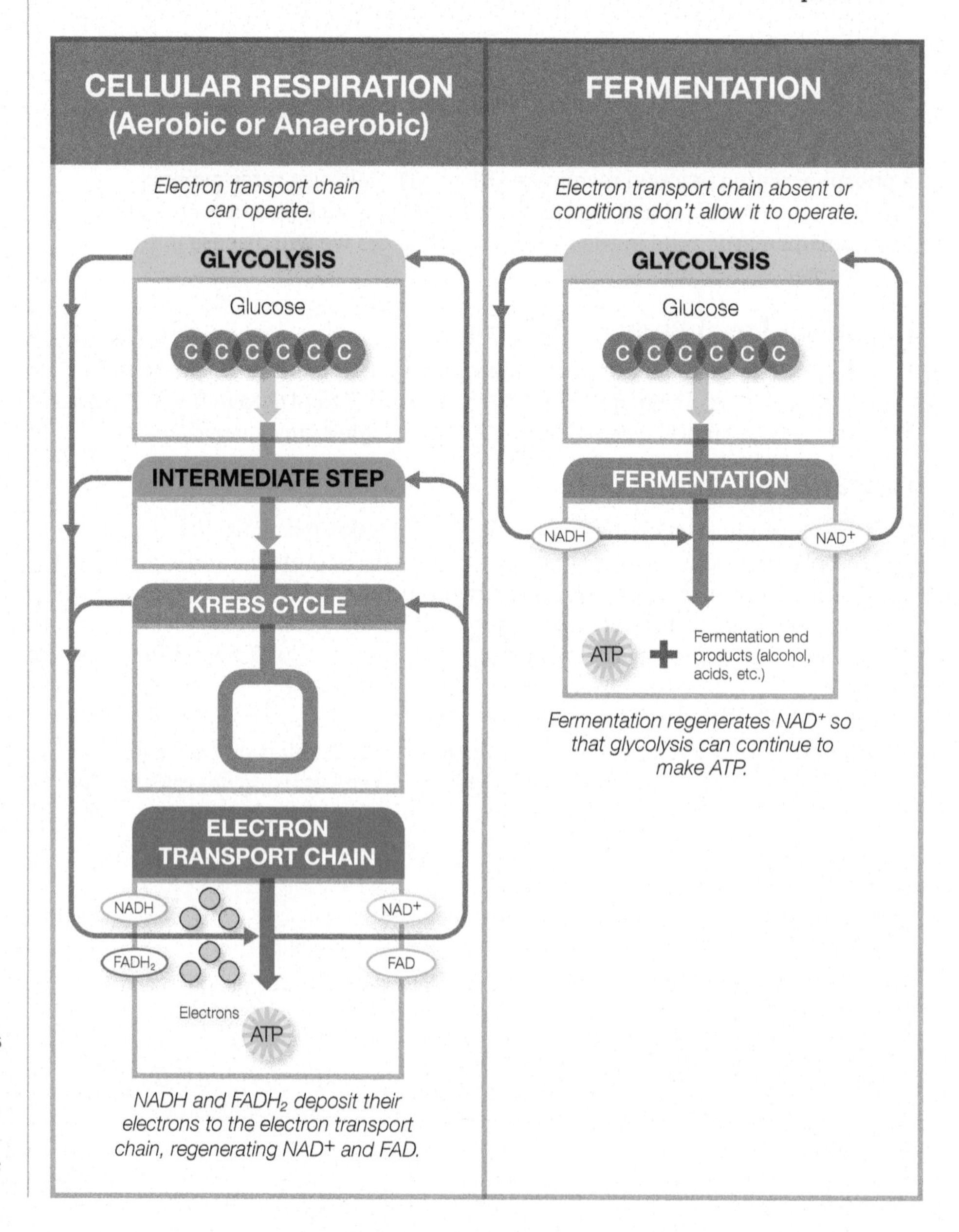

FIGURE 8.17 Comparing carbohydrate catabolism pathways Cellular respiration and fermentation both catabolize carbohydrates. Although most fermentation pathways do not directly make ATP, there are exceptions (e.g., acetate fermentation).

Critical Thinking *Many students confuse fermentation with anaerobic respiration. Based on this simple diagram, describe three ways that fermentation differs from anaerobic respiration.*

those electrons to the electron transport chain to drive ATP production (see Figure 8.17). As we proceed through the next few pages, we'll take a closer look at each part of cellular respiration. Although cellular respiration has a role in the catabolism of a variety of nutrients, let's first learn how cells use cellular respiration to harvest energy from carbohydrates (FIG. 8.18).

You may be wondering why cells have evolved such complex pathways for energy extraction; couldn't they harvest energy faster, with fewer reactions? To answer this question, you need to consider the fact that some energy released in these reactions is in the form of heat. All living things release lost energy as heat. In fact, up to 60 percent of the total energy available in the original nutrients is lost as heat, with the rest making its way into ATP. In multicellular animals, this contributes to "body heat." If energy were extracted from nutrients in one step, even more of the energy would be lost as heat.

The *many* steps of the metabolic pathways we'll examine can be loosely compared to those that transform energy from 400,000 volts as it leaves a power plant to the 120–240 volts used by your home appliances. To transform the electricity into a usable form, an extensive series of transformers progressively reduces the voltage to a level that your appliances can easily use. Likewise, in cells a series of biochemical reactions transforms energy-rich nutrients into readily usable packets of energy: ATP.

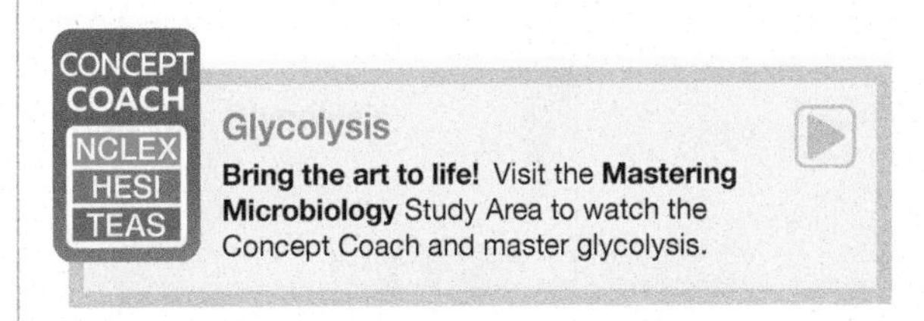

Glycolysis

For most organisms, **glycolysis** (also called the *Embden–Meyerhof–Parnas pathway*) is the first stage in aerobic *and* anaerobic carbohydrate catabolism (FIG. 8.19). It extracts energy from complex carbohydrates such as starches, from disaccharides like sucrose and lactose, and from simple sugars such as mannose, fructose, and of course, glucose. The word *glycolysis* means splitting of sugar.

Glycolysis consists of 10 reactions that occur in two basic stages: an energy investment stage (steps 1– 5 in Figure 8.19) and a payoff stage (steps 6–10 in Figure 8.19). This pathway, which occurs in the cytoplasm of prokaryotic and eukaryotic cells, does not require oxygen.

We start the pathway with glucose, a six-carbon sugar, and invest energy (2 ATP) to get the reaction going. Eventually the original six-carbon glucose is split to make two molecules of pyruvic acid. No carbons are removed during glycolysis: We start with six in one molecule of glucose and end with two molecules of

CHEM • NOTE

Carbohydrates Also known as saccharides, carbohydrates consist of oxygen, hydrogen, and carbon. The simplest carbohydrates consist of just one sugar molecule and are called monosaccharides. Glucose is a monosaccharide. Carbohydrates are broken into monosaccharide units as they enter catabolic pathways.

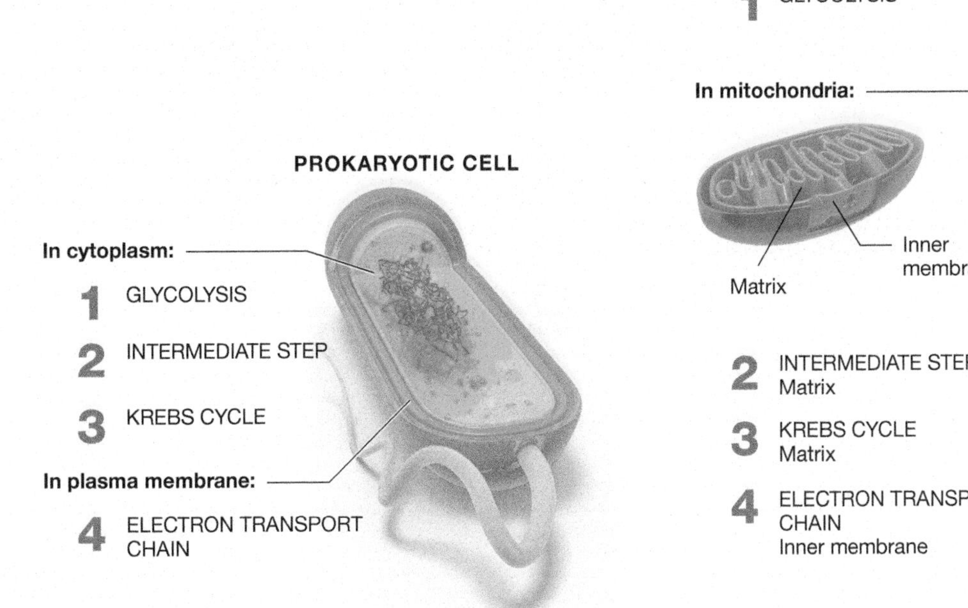

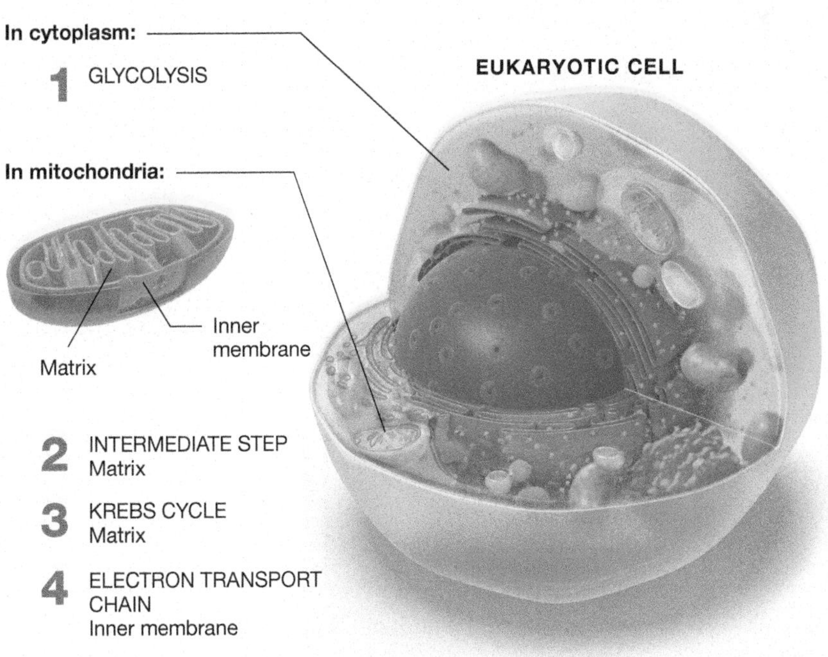

FIGURE 8.18 **Where cellular respiration occurs**

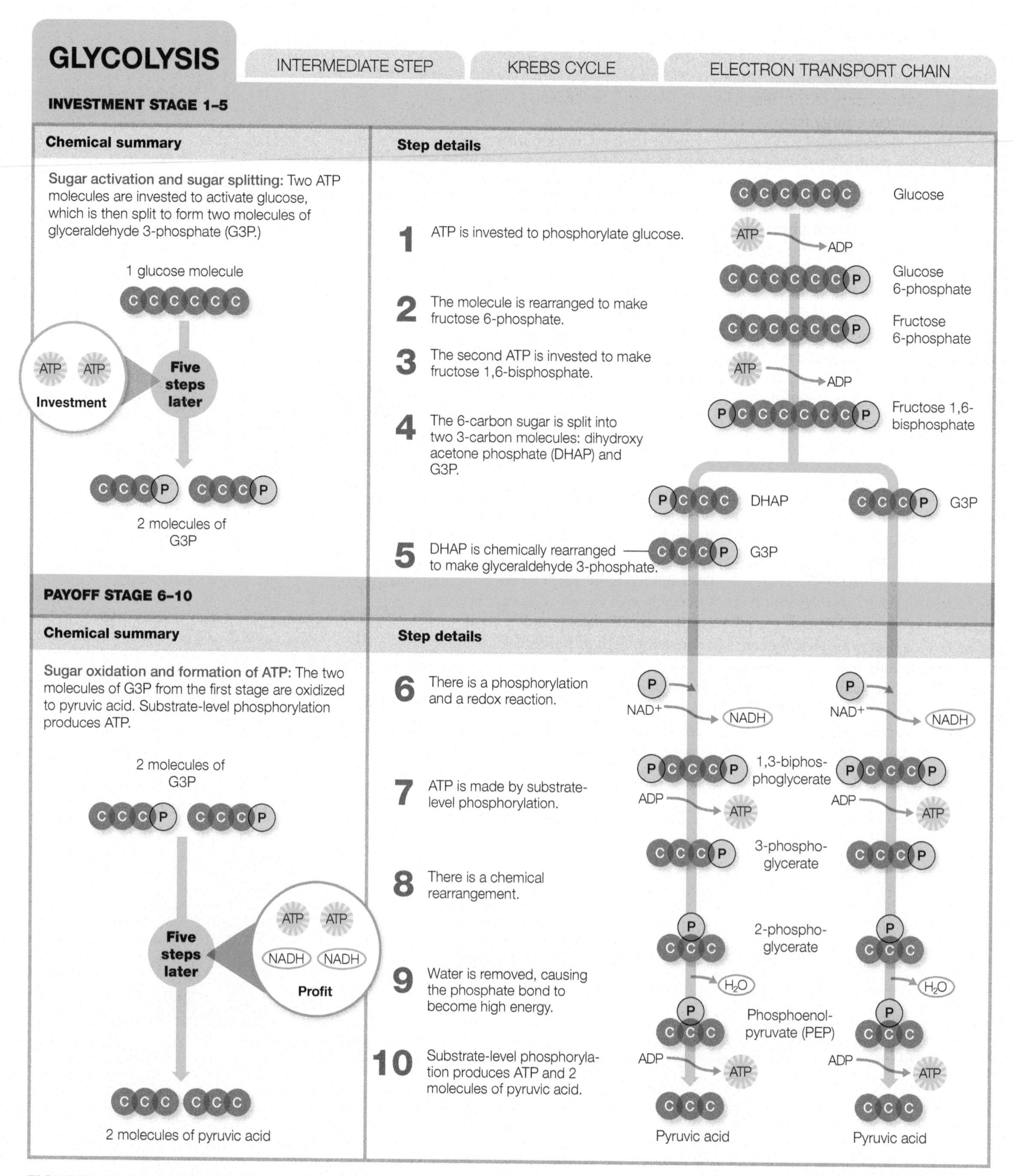

FIGURE 8.19 Dissecting cellular respiration: Glycolysis

Critical Thinking *It is often said, "It takes money to make money." How can this phrase similarly apply to harvesting energy from nutrients? (Hint: Consider how glycolysis is started.)*

three-carbon pyruvic acid. In addition, NAD^+ is reduced to NADH, which contains energy that will be harvested later using the electron transport chain. Because two ATP molecules served to jump-start glycolysis, and the entire process produces four molecules of ATP, *there is a net gain of only two molecules of ATP for each molecule of glucose that is oxidized during glycolysis.*

Intermediate Step

In cellular respiration, each pyruvic acid made in glycolysis proceeds to the intermediate step (or preparatory step). In this step, we see the first **decarboxylation reaction** in cellular respiration—a reaction in which carbon dioxide (CO_2) is removed from a molecule. Once CO_2 is removed from pyruvic acid, a two-carbon acetyl group remains. This acetyl group is bonded to coenzyme A (CoA) to make acetyl-CoA, which is then funneled into the Krebs cycle (FIG. 8.20). We will discuss the other possible pathways that metabolize pyruvic acid, including fermentation, later in this chapter.

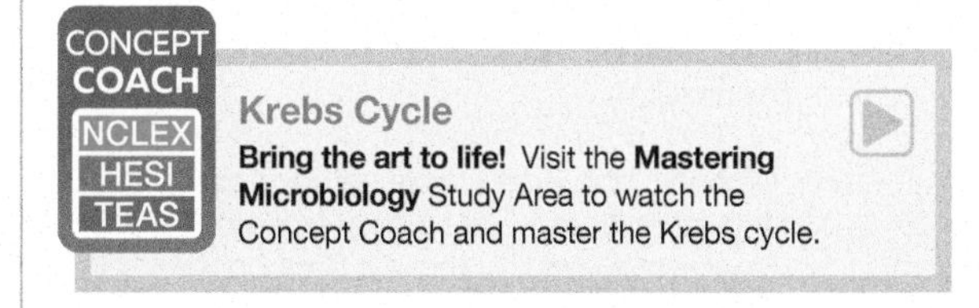

Krebs Cycle

The **Krebs cycle** (also called the *citric acid cycle* or the *tricarboxylic acid cycle*) is a series of redox reactions and decarboxylation reactions. It begins with the formation of citric acid from oxaloacetic acid and the acetyl-CoA made in the intermediate step. As in glycolysis, specific enzymes catalyze the various steps. The Krebs cycle produces some ATP[6] and a lot of the reduced cofactors NADH and $FADH_2$ (FIG. 8.21). NADH and $FADH_2$ contain valuable potential energy that

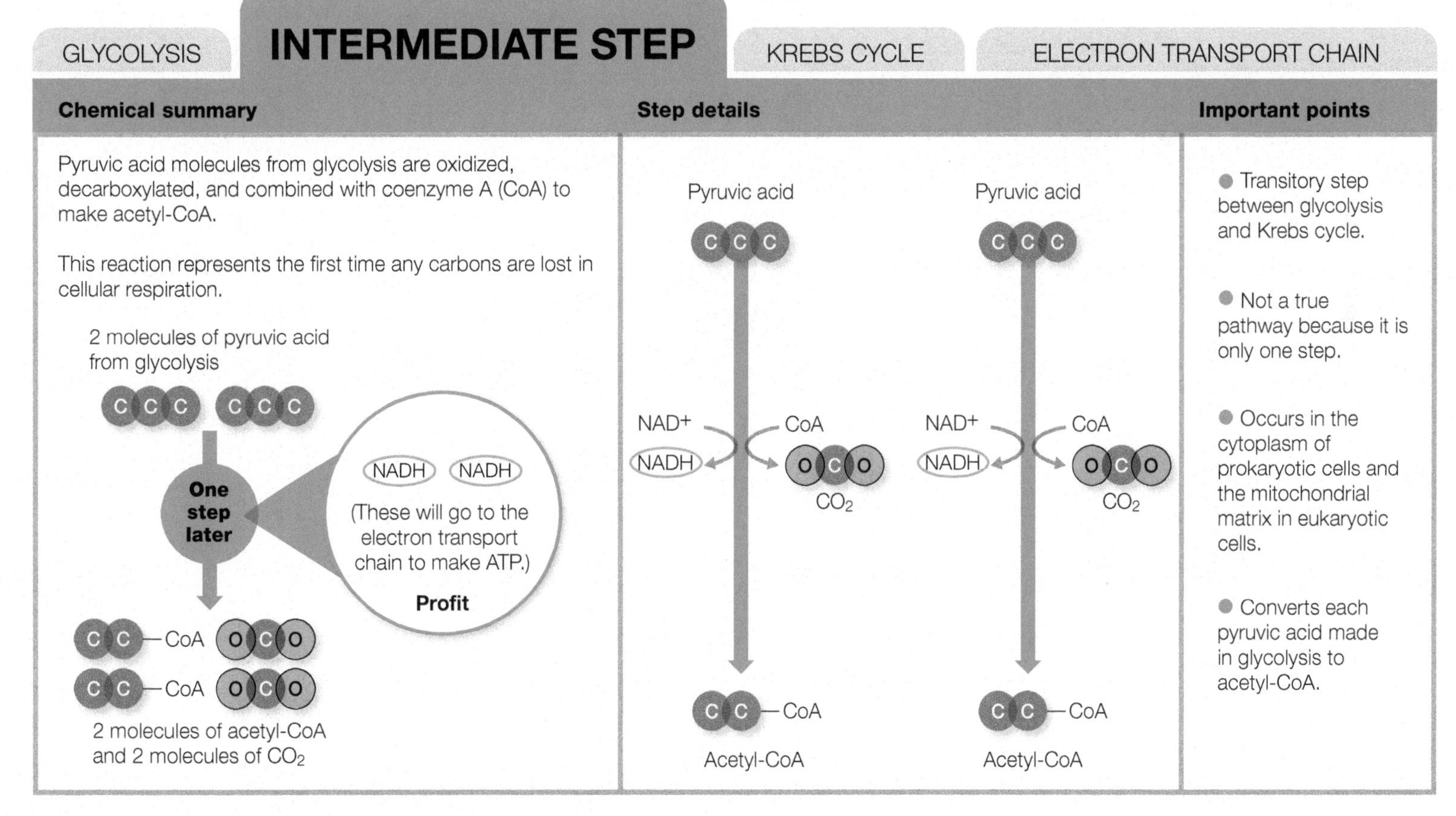

FIGURE 8.20 **Dissecting cellular respiration: Intermediate step**

[6] In the Krebs cycle, you may notice that ATP is produced via a GTP (guanosine triphosphate) intermediate (see Figure 8.21). Recall from earlier in this chapter that GTP, like ATP, is used for energy transfers, but it tends to have a more restricted usage than ATP.

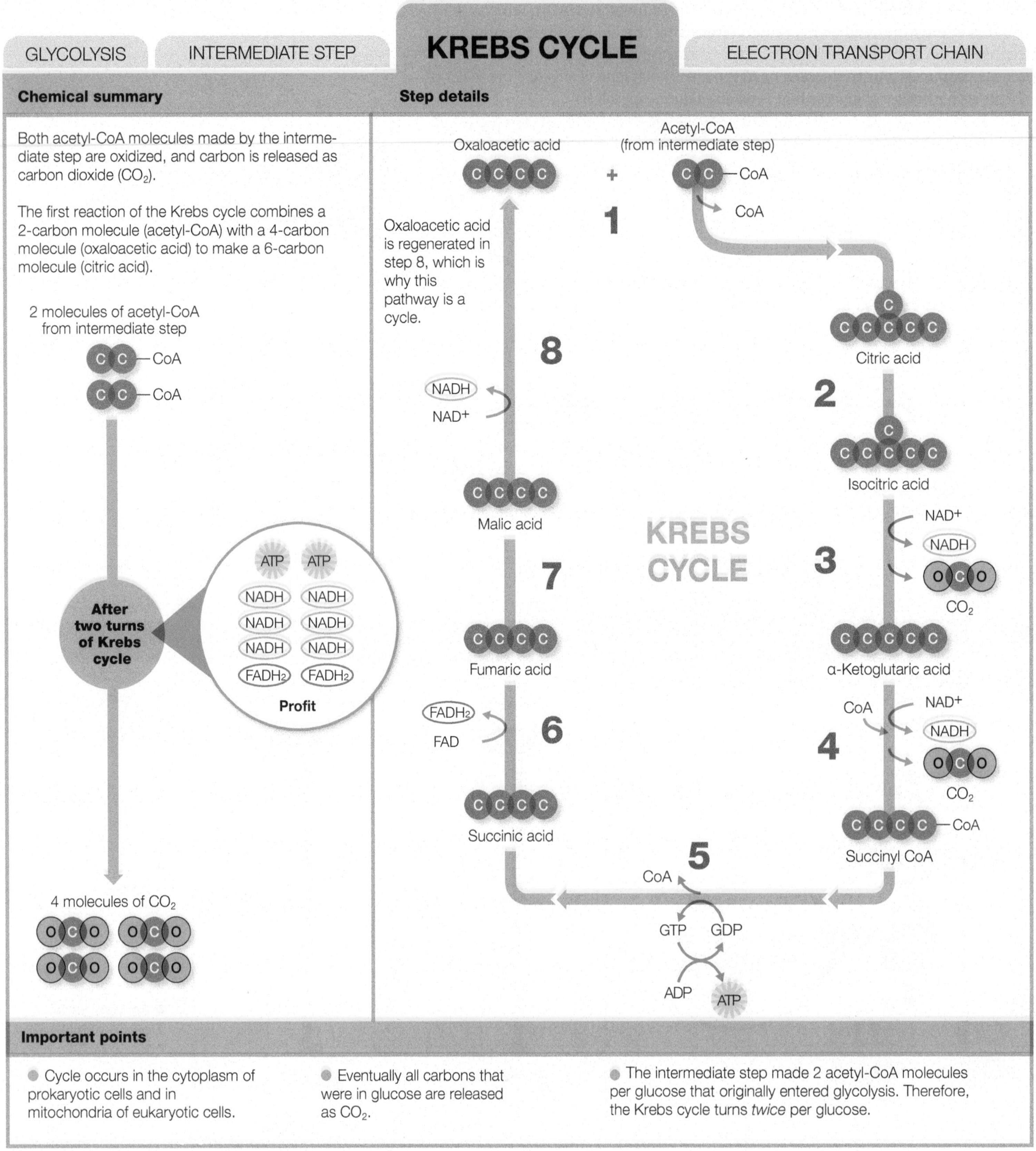

GLYCOLYSIS | INTERMEDIATE STEP | **KREBS CYCLE** | ELECTRON TRANSPORT CHAIN

Chemical summary

Both acetyl-CoA molecules made by the intermediate step are oxidized, and carbon is released as carbon dioxide (CO_2).

The first reaction of the Krebs cycle combines a 2-carbon molecule (acetyl-CoA) with a 4-carbon molecule (oxaloacetic acid) to make a 6-carbon molecule (citric acid).

2 molecules of acetyl-CoA from intermediate step

C C—CoA

C C—CoA

After two turns of Krebs cycle

ATP ATP
NADH NADH
NADH NADH
NADH NADH
$FADH_2$ $FADH_2$

Profit

4 molecules of CO_2

O C O O C O
O C O O C O

Step details

Oxaloacetic acid
C C C C
+
Acetyl-CoA (from intermediate step)
C C—CoA
1
CoA

Oxaloacetic acid is regenerated in step 8, which is why this pathway is a cycle.

C C C C C C
Citric acid

2

C C C C C C
Isocitric acid

3
NAD^+
NADH
O C O
CO_2

C C C C C
α-Ketoglutaric acid

4
CoA
NAD^+
NADH
O C O
CO_2

C C C C—CoA
Succinyl CoA

5
CoA
GTP GDP
ADP ATP

C C C C
Succinic acid

6
$FADH_2$
FAD

C C C C
Fumaric acid

7

C C C C
Malic acid

8
NADH
NAD^+

KREBS CYCLE

Important points

- Cycle occurs in the cytoplasm of prokaryotic cells and in mitochondria of eukaryotic cells.
- Eventually all carbons that were in glucose are released as CO_2.
- The intermediate step made 2 acetyl-CoA molecules per glucose that originally entered glycolysis. Therefore, the Krebs cycle turns *twice* per glucose.

FIGURE 8.21 **Dissecting cellular respiration: Krebs cycle**

Critical Thinking *Why must the Krebs cycle run twice for each glucose molecule that enters glycolysis?*

will be harvested in the electron transport chain. The Krebs cycle runs once for each acetyl-CoA made in the intermediate step; therefore, the cycle eventually runs twice for each glucose molecule.

By the end of the Krebs cycle, every carbon originally in glucose is converted into carbon dioxide. In microbes, this CO_2 simply diffuses into the environment. For humans and many other organisms, CO_2 is a waste product to be released—in our case, the bloodstream collects CO_2 and sends it to the lungs, where it is exhaled.

Cells can use the Krebs cycle to catabolize nutrients, but many of the intermediates of the cycle have important roles in other pathways, especially in amino acid production. Some photosynthetic bacteria can even run the Krebs cycle *backward* to build carbohydrates. In prokaryotic cells, the Krebs cycle occurs in the cytoplasm. In eukaryotic cells, it occurs in the matrix of the mitochondria.

While some ATP is made in the Krebs cycle, the main energy benefit of the Krebs cycle is the mass production of the reduced cofactors NADH and $FADH_2$. These reduced cofactors are like electron taxis—they ferry the electrons that were harvested in glycolysis, the intermediate step, and the Krebs cycle and deliver those electrons to the next stage of cellular respiration, the electron transport chain.

Electron Transport Chains

Electron transport chains (or respiratory chains) drive two of the three mechanisms for making ATP: oxidative phosphorylation and photophosphorylation. Here we focus on oxidative phosphorylation, which is the last stage of cellular respiration *and the step where the most ATP is made.*

Electron transport chains are called "chains" because they are essentially a chain reaction of reduction and oxidation events (redox reactions) that transfer electrons from one membrane-associated electron carrier to another, until a final electron acceptor is encountered. As an analogy, envision a ball bouncing down a staircase. The ball would start its journey at the highest step (the first membrane-associated electron carrier) and then bounce down to the next step (the second electron carrier), and the next. This step-by-step descent would continue until the ball reached the ground (the final electron acceptor). **Aerobic respiratory chains**, which are used in aerobic cellular respiration, require oxygen as a final electron acceptor (the ground in our analogy). In contrast, **anaerobic respiratory chains** are used in anaerobic cellular respiration and require an inorganic substance *other than oxygen* as the final electron acceptor (e.g., nitrite or nitrate). Regarding the electron carriers in respiratory chains (the stairs in our analogy), they can vary in number as well as type from one organism to another—much as the number and style of stairs in a stairway can vary. TABLE 8.5 presents examples of electron carriers.

Building on our analogy, just as a ball would bounce from a higher stair to a lower one, respiratory chains are organized such that electrons are passed from

TABLE 8.5 Examples of Electron Carriers

Carrier Type	General Description	Electron Transfers	Found In
Flavin groups	Made from the vitamin riboflavin (vitamin B_2); FAD (flavin adenosine dinucleotide) and FMN (flavin mononucleotide) are examples	Can participate in single or double electron transfers	Flavoproteins
Ubiquinone	A nonprotein hydrophobic molecule made of repeating isoprene units (soluble in lipid membranes)	Can participate in single or double electron transfers	Coenzyme Q
Iron–sulfur complexes	Switches between oxidized (Fe^{3+}) and reduced (Fe^{2+}) states	Involved in single electron transfers	Non-heme iron–sulfur proteins
Heme groups	The iron contained in a heme group can switch between oxidized (Fe^{3+}) and reduced (Fe^{2+}) states	Involved in single electron transfers	Cytochromes
Copper	Switches between oxidized (Cu^{2+}) and reduced (Cu^{+}) states	Involved in single electron transfers	Copper proteins

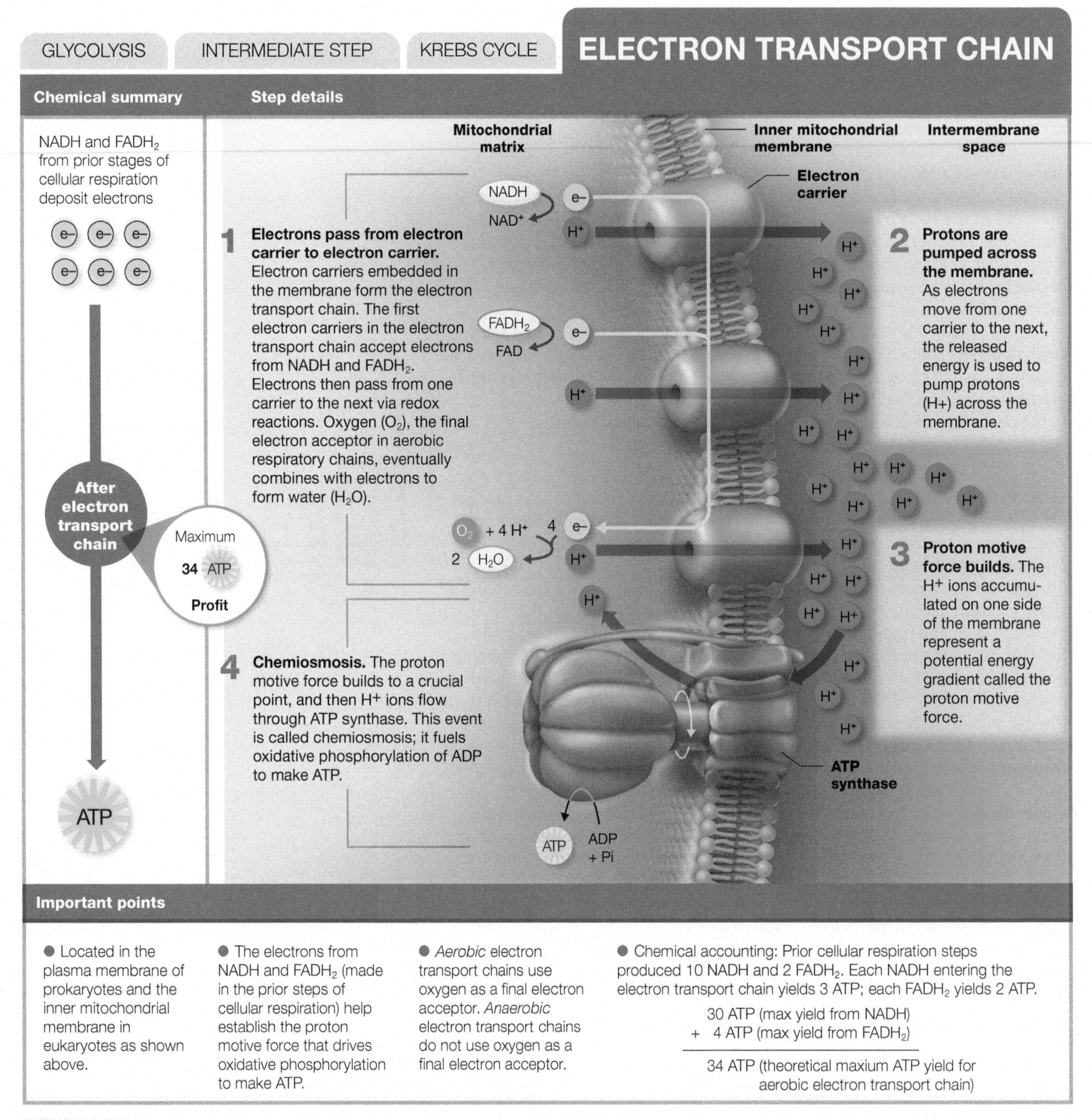

FIGURE 8.22 Dissecting cellular respiration: Electron transport chain

Critical Thinking *Is the pictured respiratory chain aerobic or anaerobic? How can you tell?*

a higher to a lower energy state. As electrons pass from carrier to carrier, energy is released. This released energy is used to pump protons across the membrane that houses the carrier molecules (FIG. 8.22). The significance of these protons will be explained shortly. In eukaryotic cells, electron transport chains are associated with the inner mitochondrial membrane. In prokaryotic cells, they are located in the plasma membrane.

The organization of electron transport chains can differ between prokaryotic and eukaryotic cells, and even one bacterial cell could have several types of electron transport chains. This diversity in electron transport chain carrier molecules is sometimes used to identify bacteria. For example, the **oxidase test** can detect if the electron carrier called cytochrome c oxidase is present. *Neisseria gonorrhoeae*, the bacterium that causes the sexually transmitted disease gonorrhea, has cytochrome c oxidase and is said to be "oxidase positive." *Salmonella* species, which are infamous for causing foodborne illnesses, do not have cytochrome c oxidase and are described as "oxidase negative." Interestingly, cyanide, which is a deadly poison, kills cells by targeting cytochrome c oxidase and essentially shutting off respiratory chains.

Oxidative phosphorylation uses chemiosmosis to recharge ADP to ATP.

Respiratory chains depend on **chemiosmosis** to drive the phosphorylation reactions that ultimately recharge ADP to ATP. Respiratory chains set the stage for chemiosmosis as follows:

- The reduced coenzymes NADH and $FADH_2$, which were made in earlier pathways, drop their electron passengers off to electron carriers in respiratory chains. Through a series of redox reactions, the electrons pass from one electron carrier in the chain to the next.
- Electrons moving through the chain go from a high-energy state to a lower-energy state (or, per our earlier analogy, bouncing from the top stair of a staircase to the ground). The energy released in this cascade is used to pump protons (H^+) across the membrane that contains the carrier proteins. In eukaryotes, H^+ ions are pumped across the inner mitochondrial membrane into the intermembrane space (Figure 8.22). In prokaryotes, the protons are pumped across the plasma membrane, and they accumulate between the cell wall and the plasma membrane.
- This pumping action causes protons (H^+ ions) to accumulate on one side of the membrane, where they form a **proton motive force**. It is the proton motive force that sets the stage for chemiosmosis.
- Eventually, by *chemiosmosis*, the accumulated protons flow through an enzyme in the membrane called ATP synthase. In eukaryotes, **ATP synthase** is found in the inner mitochondrial membrane; prokaryotes have this enzyme in their plasma membrane. It is ATP synthase that captures the energy of the flowing protons and uses it to recharge ADP to ATP.

Chemiosmosis can be compared to the production of hydroelectric power. In such a system, water is collected on one side of a dam and funneled through a specific point. The kinetic energy of the flowing water turns a turbine, converting kinetic energy into electrical energy. Similarly, in cells the proton motive force builds up on one side of a membrane and then funnels through ATP synthase, an enzyme in the membrane. The ATP synthase acts like the turbine in the dam—it even uses a rotational motion like a turbine. The structure of ATP synthase allows cells to capture energy released by the flow of the protons to recharge ADP to ATP.

In chemiosmosis, protons funnel through ATP synthase from an area where they are concentrated to an area where they are less concentrated (FIG. 8.23). This flow of protons allows a cell to make an average of three ATP molecules for each NADH that delivers electrons to the electron transport chain and about two ATP molecules for each $FADH_2$ that drops electrons off to the chain. The reason less ATP is made for each $FADH_2$ is that it drops off its electrons later in the electron transport chain than NADH does, thereby resulting in fewer protons building up across the membrane.

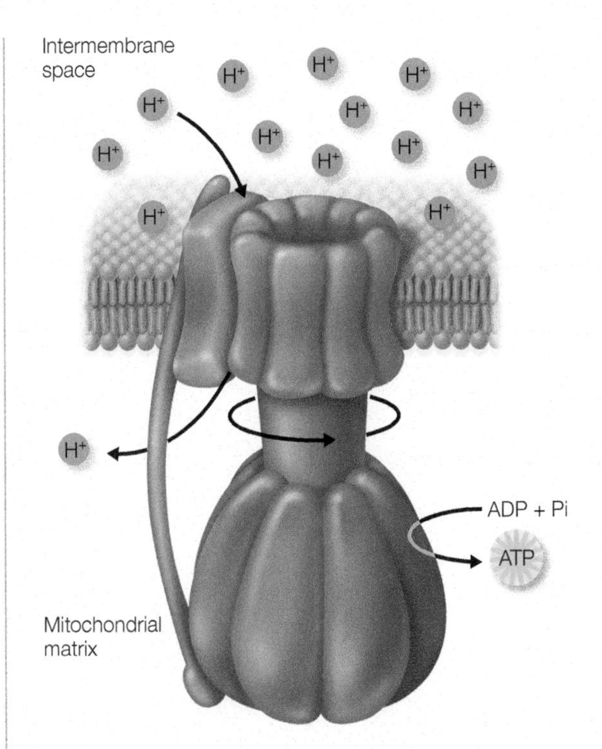

FIGURE 8.23 Proton motive force and chemiosmosis The accumulation of protons on one side of the membrane is known as the proton motive force. In chemiosmosis, these accumulated protons are funneled through the membrane channel ATP synthase to make ATP.

TABLE 8.6 Examples of Final Electron Acceptors in Anaerobic Respiratory Chains

Bacteria Type	Importance	Examples of Final Electron Acceptors
Nitrogen fixing	Nitrogen cycle	Nitrate (NO_3^-) acts as the final electron acceptor and is reduced to nitrogen gas (N_2).
Methanogens	Carbon cycle	Carbonate (CO_3^{2-}) acts as a final electron acceptor and is reduced to methane (CH_4).
Sulfate reducing	Sulfur cycle	Reduced sulfate (SO_4^{2-}) acts as a final electron acceptor and is reduced to hydrogen sulfide (H_2S).

Aerobic cellular respiration uses oxygen as the final electron receptor, while anaerobic cellular respiration does not.

Aerobic electron transport chains reduce oxygen to water. Some examples of final electron acceptors in anaerobic respiration are shown in TABLE 8.6. Not only do anaerobic organisms *not* require oxygen as a final electron acceptor, but they may even be unable to grow in the presence of oxygen. This is because anaerobic organisms are not as well equipped as aerobic bacteria are to deal with toxic oxygen intermediates such as oxygen radicals.

Diverse species in a number of biological niches where oxygen is limited use anaerobic respiration. These environments range from your own intestinal tract to the organic sludge at the bottom of marshes and ponds. The various microbes that can carry out anaerobic respiration are centrally important to the carbon, nitrogen, and sulfur cycles that sustain our planet. The fact that these microbes are so diverse means there are hundreds of different anaerobic respiration chains. Some closely resemble the pathways of aerobic cellular respiration that we have just discussed, while others look like long-lost evolutionary cousins. (See Chapter 7 to review how environmental oxygen levels impact aerobic, anaerobic, and facultatively anaerobic bacterial growth.)

ATP Production Yields for Aerobes and Anaerobes

In summary, aerobic respiration, through the collective chemical reactions of glycolysis, an intermediate step, the Krebs cycle, and the electron transport chain, can theoretically make a *maximum* of 38 ATP from one molecule of glucose (see the Visual Summary at the end of the chapter). In general, anaerobic respiration makes less ATP than aerobic respiration. (A theoretical maximum ATP yield for anaerobic respiration is not practical to quote, as there are hundreds of anaerobic respiratory mechanisms, and overall ATP yields can differ dramatically.) Most scientists are reluctant to specify a *fixed* ATP profit for aerobic or anaerobic cellular respiration because it varies among cell types, and the total ATP made is influenced by a number of factors. Note that four of the ATP molecules came from substrate-level phosphorylation that occurred in glycolysis and the Krebs cycle; the other 34 ATP molecules came from oxidative phosphorylation (TABLE 8.7). We can summarize the overall reaction for *aerobic respiration* as follows:

$$\underset{\text{Glucose}}{C_6H_{12}O_6} + \underset{\text{Oxygen}}{6\ O_2} + 38\ ADP + 38\ PO_3^{2-} \rightarrow \underset{\text{Carbon dioxide}}{6\ CO_2} + \underset{\text{Water}}{6\ H_2O} + 38\ ATP$$

Notice that in this summarized equation for aerobic respiration, oxygen is required. This is because, as already mentioned, *oxygen is the final electron acceptor in aerobic respiration*. Indeed, this is why we require oxygen and do not live long without it. The oxygen that is so central to aerobic respiration is eventually reduced to water (H_2O).

TABLE 8.7 Maximum ATP Yields for Aerobic Cellular Respiration in Prokaryotes

Pathway	ATP Invested	ATP Made	Pathway's Net ATP Yield
Glycolysis	2	4	2
Intermediate Step	0	0	0
Krebs Cycle	0	2	2
Electron Transport Chain	0	34	34
Combined Net Totals	**2**	**40**	**38**

Build Your Foundation

16. What is cellular respiration, and what pathways comprise cellular respiration? (LO 8.12)
17. Summarize the chemical steps in the investment and payoff stages of glycolysis. (LO 8.13)
18. Describe what occurs in the Krebs cycle (pay particular attention to the roles of NAD^+ and FAD). (LO 8.14)
19. How do electrons get to the electron transport chain, and what happens to them once there? (LO 8.15)
20. List three ways anaerobic and aerobic respiration are alike and three ways they are different. (LO 8.16 and LO 8.17)

Build Your Foundation (BYF) Quick Quiz: Visit the **Mastering Microbiology** Study Area to quiz yourself.

8.5 OTHER CATABOLIC PATHWAYS FOR OXIDIZING NUTRIENTS

Glycolysis is not the only pathway to oxidize sugars.

Glycolysis has been evolutionarily conserved and is present in most cells. However, cells also have alternates to the standard glycolytic pathway for carbohydrate metabolism.

Pentose Phosphate Pathway

The **pentose phosphate pathway** uses a carbon-shuffling process to convert pentoses (five-carbon sugars) into trioses (three-carbon sugars) and hexoses (six-carbon sugars) that can be directly funneled into glycolysis. Alternatively, the pathway can shift glucose 6-phosphate away from the glycolytic pathway, repurposing it to make pentoses, which are building blocks for nucleotides and certain amino acids. Lastly, this pathway produces the coenzyme NADPH (nicotinamide adenine dinucleotide phosphate), which is used in anabolic pathways.

As a testimony to how well this pathway is evolutionarily conserved, a number of the steps in the pentose phosphate pathway are even used by photosynthetic cells to build glucose from carbon dioxide. Many bacteria, including *Bacillus subtilis* and *E. coli*, use the pentose phosphate pathway, as do some eukaryotes—including human red blood cells.

Entner–Doudoroff Pathway

The Entner–Doudoroff pathway is primarily found in Gram-negative bacteria that require oxygen to grow (obligate aerobes); however, it is also found in certain facultative anaerobes, select Gram-positive bacteria, and some eukaryotic microbes. Most likely it is an ancient pathway that evolved early on to catabolize glucose. It is less efficient than glycolysis, producing a profit of only one molecule of ATP per glucose (versus the two-ATP profit seen in glycolysis). However, unlike glycolysis, which makes NADH, this pathway makes NADPH, which is used in anabolic reactions. This pathway is not found in human cells.

Learning Outcomes

After reading this section, you should be able to:

8.18 Describe the roles of the pentose phosphate and Entner–Doudoroff pathways.

8.19 Summarize the process of fermentation.

8.20 Describe three different fermentation processes.

8.21 Describe how lipids and proteins are broken down by extracellular and intracellular catabolism.

TABLE 8.8 Comparing Fermentation to Aerobic and Anaerobic Respiration

	Fermentation	Aerobic Cellular Respiration	Anaerobic Cellular Respiration
Oxygen Needed?	No	Yes	No
Final Electron Acceptor	Organic molecule (often pyruvic acid)	Oxygen	Inorganic molecule other than oxygen (commonly nitrite, nitrate, carbonate, and sulfate)
ATP Amount Produced	Typically, 2–3 ATP gain, depending on the type of fermentation pathway*	The most efficient of these three processes (maximum of 38 ATP made per glucose)	Typically less than 38 ATP and more than 2 ATP per glucose
Method of Phosphorylation Used to Make ATP	Substrate level	Substrate level *and* oxidative	Substrate level *and* oxidative
Electron Transport Chain?	No	Yes	Yes

*In most cases the ATP noted comes from glycolysis. However, a range of ATP is provided here because some fermentation pathways directly make ATP such that the ATP yield is not just that provided by glycolysis (the acetate fermentation pathway is an example).

Interestingly, this pathway enhances the survival and biofilm-forming capabilities of some *Campylobacter jejuni* isolates associated with campylobacteriosis, one of the most common foodborne infections.[7]

Fermentation catabolizes nutrients without using a respiratory chain.

The most efficient way to harvest energy from glucose is for the pyruvic acid molecules created through glycolysis to be further oxidized through the remaining steps of aerobic cellular respiration (the intermediate step, the Krebs cycle, and the electron transport chain). However, some organisms do not have respiratory chains. Others may have respiratory chains, but may be unable to use them because the necessary final electron acceptor is missing. In such instances a cell might rely on **fermentation** pathways. Again, these are not nearly as efficient as aerobic or even anaerobic respiration.

Bacteria can affect human health through fermentation pathways. For example, *Lactobacillus* bacterial species ferment the carbohydrate glycogen and produce acids that lower the vaginal pH to about 4. This acidic pH inhibits the growth of microbial pathogens such as the pathogenic yeast *Candida albicans*, which causes vulvovaginal candidiasis (vaginal yeast infections).

Although carbohydrate fermentation works in combination with glycolysis, fermentation is not considered part of cellular respiration. The main goal of fermentation is *not* to make ATP, but rather to *sustain* ATP production by glycolysis when respiratory chains are not available.[8] Why would cells need fermentation to do this? Recall that cells rely on coenzymes like NAD^+ to oxidize nutrients by acting as electron taxis. These coenzyme taxis typically drop their electron passengers off at respiratory chains and then go back to catabolic pathways, like glycolysis, for more electron passengers. However, if respiratory chains are not functioning, then the taxis have to find somewhere else to drop off their electron passengers. In fermentation, NADH passes its electron passengers to an organic molecule like pyruvic acid. In so doing, the NADH is restored to its oxidized form, NAD^+, and can go pick up more electron passengers, thereby allowing glycolysis to continue. You can refer back to Figure 8.17 as well as to TABLE 8.8 to compare fermentation to aerobic and anaerobic respiration.

Antibiotics

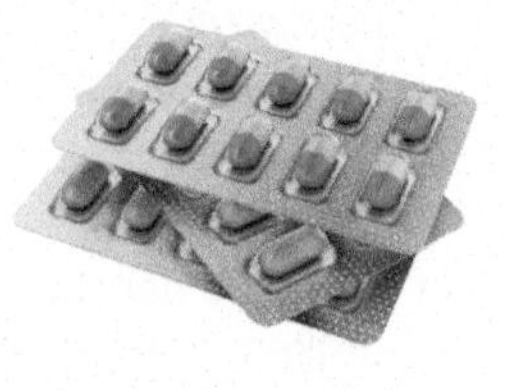

Ideally, antibiotics will have minimal side effects on human cells and will instead specifically target microbes. This is one of many reasons why it is so important for us to study the basic structures and functions of a wide variety of human and nonhuman cells. Basic science, not just clinical research, is essential if we are to continue to open doors for drug therapies that can ultimately improve our quality of life.

QUESTION 8.3

Which would be the best pathway to target in an antibiotic made for people: glycolysis, the pentose phosphate pathway, or the Entner–Doudoroff pathway? Why? Hint: Reread the preceding section, Other Catabolic Pathways for Oxidizing Nutrients.

[7] Vegge, C. S., van Rensburg, M. J. J., Rasmussen, J. J., et al. (2016). Glucose metabolism via the Entner–Doudoroff pathway in *Campylobacter*: A rare trait that enhances survival and promotes biofilm formation in some isolates. *Frontiers in Microbiology*, *7*, 1877.

[8] Although most of the fermentation pathways studied to date do not make ATP outside of what is generated in glycolysis, some do—acetate fermentation is one example.

Many types of fermentation pathways occur in eukaryotic cells, such as yeast and muscle cells, as well as in a variety of prokaryotes (FIG. 8.24). Fermentation pathways are often named for their end products, which vary by organism. Consequently, fermentation end products can also be used in a laboratory setting to identify microorganisms (more on this later). Just like aerobic and anaerobic respiration, fermentation can catabolize a range of organic nutrients. For simplicity, introductory texts tend to focus exclusively on how glucose is fermented, but fermentation of nonglucose nutrients, such as amino acids and fatty acids, is also important in nature, in food microbiology, and in a variety of industrial applications.

Examples of common carbohydrate fermentation reactions are summarized in the next few sections.

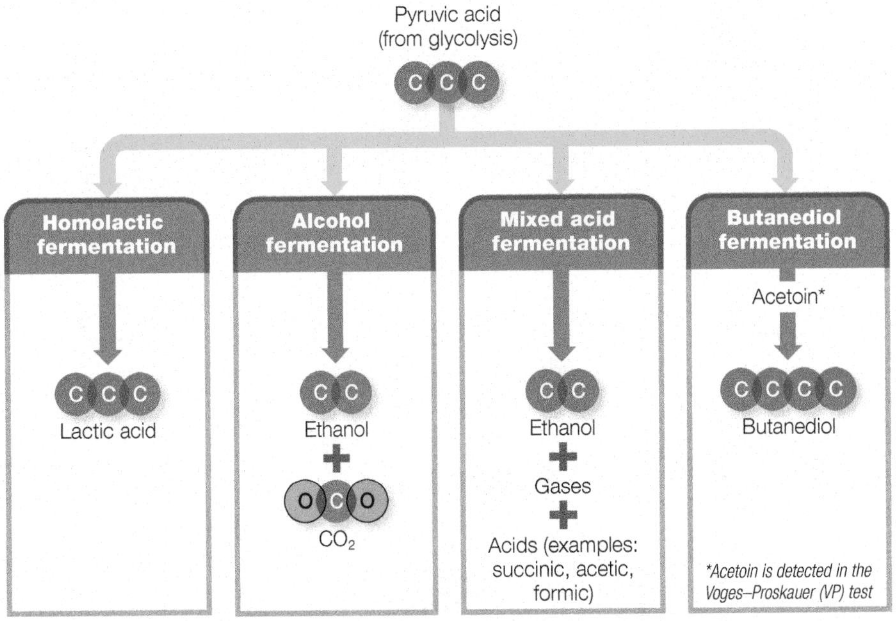

FIGURE 8.24 Fermentation reactions As shown in this simplified diagram, pyruvic acid made in glycolysis can enter diverse fermentation pathways. Although carbohydrate fermentation works in combination with glycolysis, fermentation is not considered a part of cellular respiration.

Critical Thinking *A cell may carry out several different fermentation reactions. Why is this advantageous to a cell?*

Lactic Acid Fermentation

Lactic acid fermentation can be described as being either homolactic or heterolactic. In **homolactic fermentation**, the pyruvic acid made in glycolysis is only reduced to lactic acid. The two ATP molecules that are gained are made in glycolysis. Bacteria used to make yogurt (*Lactobacillus bulgaricus* and *Streptococcus thermophiles*) and sauerkraut (*Pediococcus* and certain *Lactobacillus* species) often carry out homolactic fermentation. *Streptococcus mutans* causes dental caries (cavities) when the acid it makes via homolactic fermentation erodes tooth enamel. Lastly, our muscle cells also use homolactic fermentation when oxygen is limited, as during anaerobic exercise (short bursts of intense activity, such as lifting a heavy weight or sprinting). Human cells that can carry out fermentation can only use this process for brief periods.

Heterolactic fermentation results in the production of equal quantities of lactic acid, ethanol, and carbon dioxide, with minor amounts of acidic end products such as acetic acid and formic acid. The pentose phosphate pathway is involved in heterolactic fermentation. A wide variety of bacteria and fungi can carry out heterolactic fermentation; some common examples can be found among the genera *Lactobacillus*, *Leuconostoc*, and *Weissella*. During heterolactic fermentation there is usually a gain of only one ATP molecule from glycolysis, making heterolactic fermentation slightly less efficient than homolactic fermentation.

Alcohol Fermentation

In alcohol fermentation, the pyruvic acid from glycolysis is converted to ethanol and carbon dioxide. The carbon dioxide produced by yeast such as *Saccharomyces cerevisiae* causes bread to rise, while the ethanol that these cells make is central to beer and wine production. Although some bacteria, fungi, and even certain plant cells can carry out alcohol fermentation, human cells cannot. (If they could, then alcohol could potentially be made during anaerobic exercise . . . which would certainly be an interesting way to motivate people to exercise regularly!)

CHEM • NOTE

Anaerobic Respiration versus Fermentation
They are both anaerobic (do not use oxygen), but fermentation does not use a respiratory chain. Also, fermentation occurs in the cytoplasm of a cell, whereas respiratory chains are associated with a membrane—the plasma membrane in prokaryotic cells and the inner mitochondrial membrane in eukaryotic cells.

Mixed Acid and Butanediol Fermentation

Certain bacteria, especially **enteric bacteria** (inhabiting the gut), can use mixed acid fermentation pathways to convert the pyruvic acid made in glycolysis

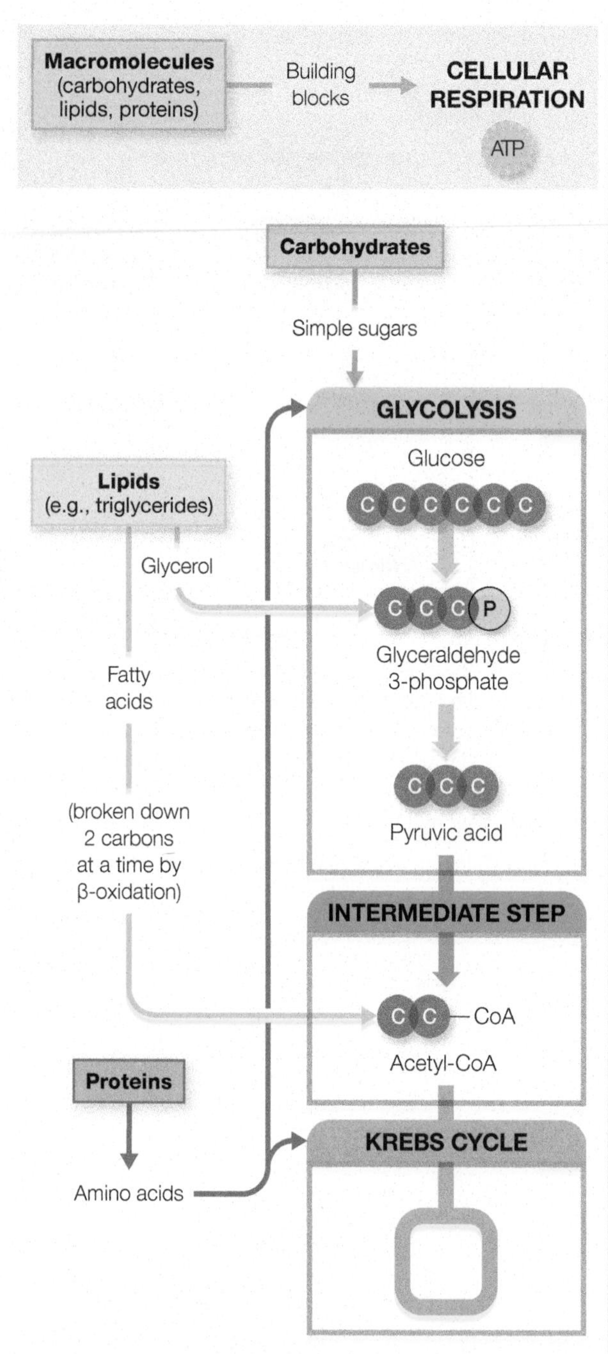

FIGURE 8.25 Funneling diverse nutrients into cellular respiration Larger macromolecules are broken down into their building-block monomers, which are then catabolized via cellular respiration.

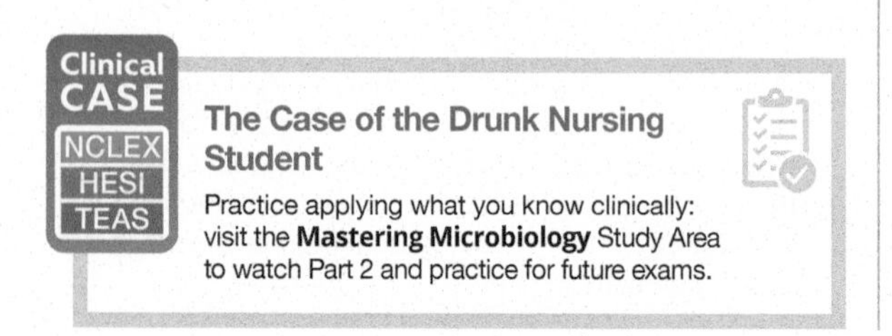

The Case of the Drunk Nursing Student

Practice applying what you know clinically: visit the **Mastering Microbiology** Study Area to watch Part 2 and practice for future exams.

into a variety of acidic end products and gases that include lactic acid, formic acid, acetic acid, succinic acid, ethanol, CO_2, and H_2. In **butanediol fermentation**, pyruvic acid is converted to many of the acidic products seen in mixed acid fermentation, but to a lesser extent. Mostly neutral end products such as butanediol and ethanol are made. A bacterium's ability to carry out mixed acid fermentation or butanediol fermentation can help identify the organism. (See the upcoming section on identifying bacteria by metabolic properties.)

In summary, all cells depend on redox reactions to make ATP.

Even though cells rely on diverse pathways to harvest energy from food or sunlight, an important common feature unites them. Every pathway has an absolute dependence on redox reactions to make ATP, and there is always some sort of final electron acceptor. In cellular respiration, the final electron acceptor is an inorganic molecule that runs electron transport chains (oxygen in aerobic respiration, or some molecule other than oxygen for anaerobic respiration). In fermentation, the final electron acceptor is an organic molecule that is usually a metabolic intermediate made by the cell. In photosynthesis, the final electron acceptors are coenzymes and photosynthetic pigments.

As electrons are shuffled through a variety of redox reactions toward a final electron acceptor, ATP is made. The byproducts created and the amount of ATP made in these diverse metabolic processes depends on the type of cell as well as its needs.

Cells also catabolize lipids and proteins for energy.

So far we have focused on how cells harvest energy from carbohydrates, especially from the simple sugar glucose. However, most cells also harvest energy from lipids and proteins. The initial catabolism of large macromolecules such as starches, lipids, and proteins is often **extracellular**, occurring outside the cell. Before large macromolecules can enter catabolic pathways, they must first be broken down into smaller molecules that can be transported across the cell's plasma membrane into the cell. For example, the carbohydrate starch is broken down into glucose subunits before it can enter the catabolic pathways we have discussed.

Likewise, lipids must be broken into smaller building-block molecules; for example, triglycerides are broken down into glycerol and fatty acids. Extracellular catabolism can be compared to taking apart a large piece of furniture in order to get it through a narrow doorway. Bacteria secrete enzymes into their local environment (**exoenzymes**) to do this. One class of exoenzymes called *lipases* dismantles lipids into glycerol and fatty acids. *Proteases* and *peptidase* dismantle proteins into smaller peptides and eventually amino acids. Once large macromolecules have been broken down by exoenzymes into smaller molecules, the building-block materials can be recycled, or funneled into intracellular catabolic pathways to harvest their energy. FIG. 8.25 provides a generalized overview of how nutrients like larger carbohydrates, lipids, and proteins can be funneled into cellular respiration so the cell can use them for energy.

Different bacterial species can make different exoenzymes. Thus, detecting the presence of certain exoenzymes can be helpful in identifying an unknown organism. In fact, a variety of bacterial growth media are routinely used to detect exoenzymes (FIG. 8.26). These growth media, in combination with other biochemical tests, are commonly used to identify microbes.

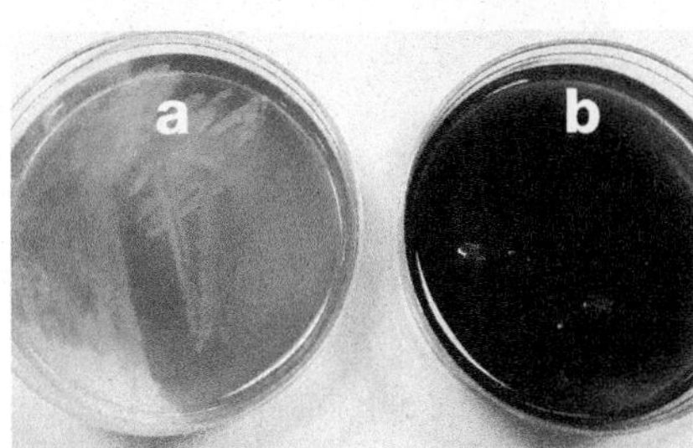

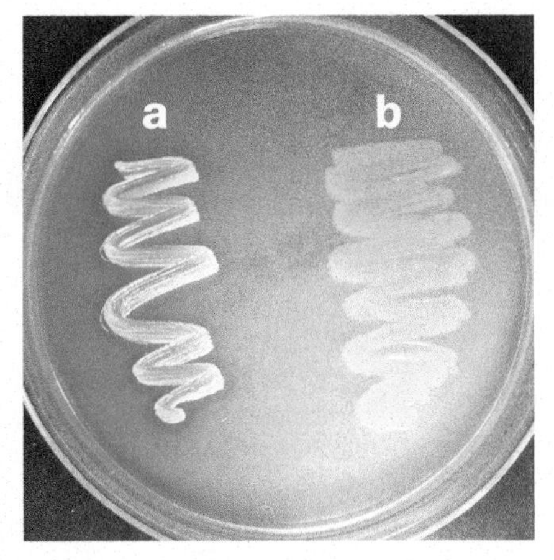

FIGURE 8.26 Exoenzyme profiles *Top left*: Media solidified with the protein gelatin can be used to detect the exoenzyme gelatinase, which breaks down gelatin. Organism **a** made gelatinase and liquefied the media during the incubation period. Organism **b** did not make gelatinase, so the media remained solid.

Bottom left: Bacteria are grown on starch agar to detect the exoenzyme amylase. The level of starch hydrolysis is determined by adding iodine to the medium after incubation. Iodine turns a dark blue/black if starch is present and remains yellow if starch was broken down. Organism **a** made amylase; organism **b** did not.

Right: Spirit blue agar detects exoenzymes that break down lipids. When lipids in the agar are broken down, the pH of the medium around the bacterial growth drops. This drop in pH causes the dye in the agar to precipitate and a dark blue halo develops around lipase-positive organisms. Organism **a** is lipase positive; organism **b** is lipase negative.

Lipid Catabolism

Lipids are a rich and diverse source of energy for cells. Here we focus on triglycerides, which are a common example of a fat (a simple lipid) that cells metabolized for energy. Once lipases break triglycerides into glycerol and fatty acids, these building blocks can be further catabolized to extract energy. The glycerol component is chemically converted to dihydroxyacetone phosphate (DHAP) and enters glycolysis. The fatty acids are broken down, two carbons at a time, into acetyl-CoA molecules that can enter the Krebs cycle in a process called **beta-oxidation** (or **β-oxidation**).

CHEM • NOTE

Polymers These macromolecules consist of repeated smaller units (monomers).

Protein Catabolism

Proteases and peptidases break proteins down into smaller peptides and eventually into amino acids. Amino acids can then be recycled to make new proteins, chemically modified to make new amino acids, or further broken down to extract energy. Before an amino acid can be broken down for energy, it must be stripped of its amine group (NH_2) by a **deamination** reaction. Most amino acids are catabolized via the Krebs cycle, though some enter in earlier stages of cellular respiration.

Nucleic Acid Catabolism

Nucleic acid polymers, DNA and RNA, are also broken into their building blocks—nucleotides. Recall that nucleotides consist of a phosphorylated sugar (deoxyribose in DNA; ribose in RNA), and a purine (adenine/guanine) or pyrimidine (cytosine/uracil/thymine) base. Organisms can differ in their overall ability to catabolize various nucleotides, but in general nucleotides are not considered a key dietary energy source because energy can only be extracted from the sugar component of nucleotides. In most cases, organisms will salvage what they can from nucleotides rather than break them down for energy.

Build Your Foundation

21. Even though the pentose phosphate and Entner–Doudoroff pathways do not produce much ATP, they are useful. Explain why. (LO 8.18)
22. Describe three different examples of fermentation, and state how fermentation differs from anaerobic respiration. (LO 8.19 and LO 8.20)
23. Why do organisms that survive using fermentation require an abundance of sugars? (LO 8.20)
24. Describe the general processes of protein and lipid catabolism. (LO 8.21)

Build Your Foundation (BYF) Quick Quiz: Visit the **Mastering Microbiology** Study Area to quiz yourself.

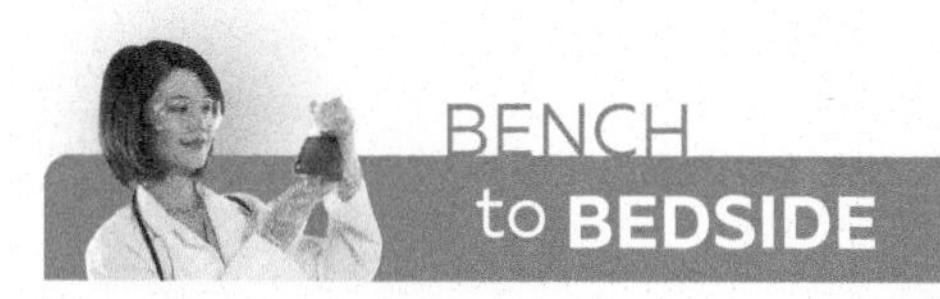

Do Gut Microbes Contribute to Obesity?

Our gastrointestinal tract is home to over 100 trillion microbial cells representing between 1,000 and 36,000 species. Research has shown that gut microbes are important for many functions, including assisting us with harvesting energy from food, making vitamins, and stimulating our immune system. In fact, they are so central to our health that some researchers have referred to our gut microbiota as our "forgotten organ."

In 2004, a landmark study on germ-free mice (mice that are born and raised in an aseptic environment and therefore lack normal flora) suggested that the gut microbiome could directly influence obesity. In the experiment, germ-free mice were colonized with microbiota from normal counterparts. Within 14 days they experienced a 57% increase in total body fat and showed signs of insulin resistance (a risk factor for diabetes). This increase in body fat was not linked to increased food consumption or decreased activity levels.

More recent research reveals that obese adults have reduced gut microbial diversity as compared to nonobese adults. It's also increasingly clear that the gut microbiome generates metabolites that can enter the bloodstream from the intestinal tract and influence appetite, insulin resistance, and fat development. Although mounting evidence suggests that normal microbiota differ between lean and obese populations, it is still unclear how diet, genetics, and age impact the gut microbiota.

It has been proposed that in combination with other obesity risk factors, gut microbes could contribute to obesity by increasing the amount of energy harvested through the digestion of food, causing low-level systemic inflammation, increasing fat deposition, and influencing appetite. Research on how the microbiome affects human health and disease could result in a variety of probiotic and prebiotic therapies. Such therapies may still be a while off, though, because many questions remain about how we can shape the microbiome to improve human health.

Sources: Bäckhed, F., Ding, H., Wang, T., et al. (2004). The gut microbiota as an environmental factor that regulates fat storage. *PNAS, 101*(44), 15718–15723.

Lee, C. J., Sears, C.L., & Maruthur, N. (2020). Gut microbiome and its role in obesity and insulin resistance. *Annals of the New York Academy of Sciences, 1461*(1), 37–52.

Ejtahed, H. S., Angoorani, P., Soroush, A.-R., et al. (2020). Gut microbiota-derived metabolites in obesity: A systematic review. *Bioscience of Microbiota, Food and Health, 39*(3), 65–76.

8.6 ANABOLIC REACTIONS: BIOSYNTHESIS

Learning Outcomes

After reading this section, you should be able to:

8.22 List three categories of molecules produced by anabolism, and briefly describe the biosynthetic pathways that make them.

8.23 Describe how cells balance anabolism and catabolism.

Polysaccharide biosynthesis starts with simple sugars.

Although cells rely on ATP to fuel many processes, such as active transport and cellular movement (e.g., flagellar motion), a key use of ATP is to fuel **biosynthesis**, the construction of biological molecules. Biosynthetic pathways are made up of anabolic reactions that require ATP and the reducing power found in coenzymes like NADPH. By building carbohydrates, lipids, amino acids, purines, and pyrimidines, these pathways help cells grow, multiply, and repair damage. Therefore, they are just as central to a cell's survival as the catabolic pathways we just discussed.

Polysaccharides are polymers of simple sugars. They are common structural components of cells and act as important energy reserves because cells do not store ATP.

Polysaccharide assembly begins with glucose. Most eukaryotic and prokaryotic cells can funnel off intermediates from glycolysis, the Krebs cycle, and lipid or protein catabolism to make glucose or other simple sugars. The chemical process of building glucose from nonsugar starting materials is called **gluconeogenesis** (*gluco* = sugar, *neo* = new, *genesis* = creation). Once glucose molecules (or other simple sugars) are built, cells can assemble them into polysaccharides such as glycogen and starch. The production of glycogen, a branched polymer of glucose, is called **glycogenesis**.

Bacteria make glycogen by chemically modifying glucose 6-phosphate (an intermediate of glycolysis) (FIG. 8.27). Animals carry out a similar pathway to make glycogen, but they fuel the process with UTP (uridine triphosphate) instead of ATP. UTP is similar to ATP, but UTP has a more specialized role in energy transfers in cells. Bacteria build peptidoglycan, which is found in bacterial cell walls, using reactions that involve fructose 6-phosphate (also an intermediate of glycolysis).

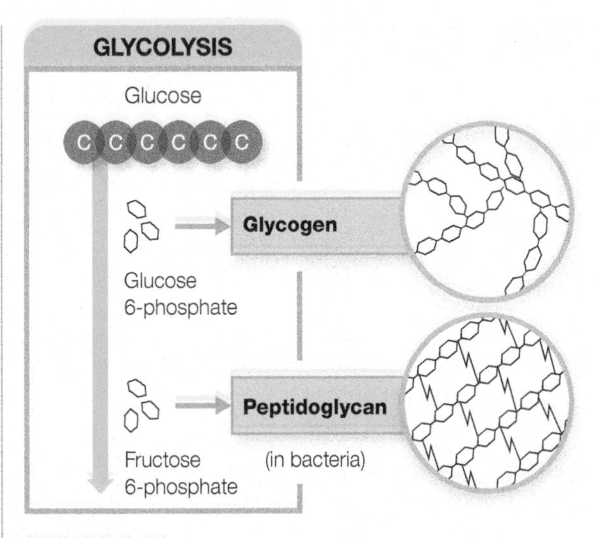

FIGURE 8.27 **Building carbohydrates** Prokaryotic and eukaryotic cells use intermediates from glycolysis to make carbohydrate polymers.

Critical Thinking *A bacterial cell surviving on a diet of lipids can still make carbohydrate polymers. How is this possible?*

Lipid biosynthesis starts with carbohydrate catabolism intermediates.

As reviewed in Chapter 2, lipids include fats, oils, waxes, and steroids. Cells build lipids using diverse metabolic pathways, but most cells can make lipids from carbohydrate catabolism intermediates using an anabolic process that requires an energy investment.

Cells often build triglycerides (an important and common example of a fat), to store energy. Triglycerides are built by combining glycerol and three fatty acids. Glycerol can be made from dihydroxyacetone phosphate (DHAP), which is an intermediate in glycolysis. Fatty acids are long chains of hydrocarbons that are made by linking acetyl-CoA molecules end to end. Recall that acetyl-CoA is made in the intermediate step before the Krebs cycle (FIG. 8.28). When cells have sufficient ATP, they will shift acetyl-CoA away from the Krebs cycle and channel it toward making fats, to store energy.

Phospholipids are the most abundant lipid in cell membranes. They are made up of glycerol, two fatty acids, and a phosphate-containing compound. In animal cells, cholesterol (a steroid lipid), maintains the fluidity of phospholipid cell membranes. Cholesterol can be built from the acetyl-CoA produced in the catabolism of fatty acids or, as mentioned earlier, via the intermediate step in carbohydrate catabolism. Acetyl-CoA is also a key building block for making certain waxy lipids in bacteria. For example, it is a precursor for mycolic acid, a waxy lipid found in the cell wall of the acid-fast bacterium *Mycobacterium tuberculosis*.

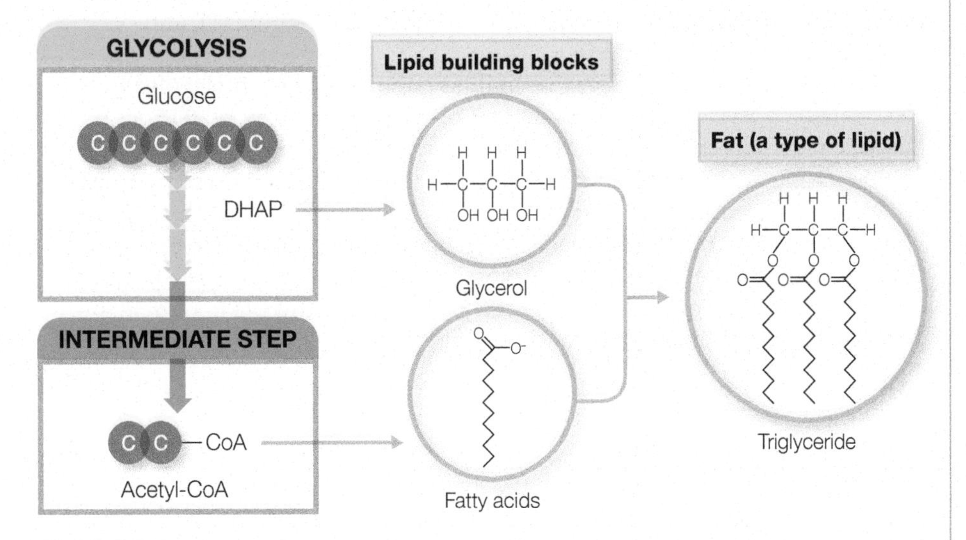

FIGURE 8.28 **Building lipids** The building blocks of triglycerides (fats), the type of lipid cells use to store energy, are glycerol and fatty acids. Glycerol can be made from dihydroxyacetone phosphate (DHAP), which is an intermediate of glycolysis. Fatty acids can be made by linking acetyl-CoA molecules created during the intermediate step of cellular respiration.

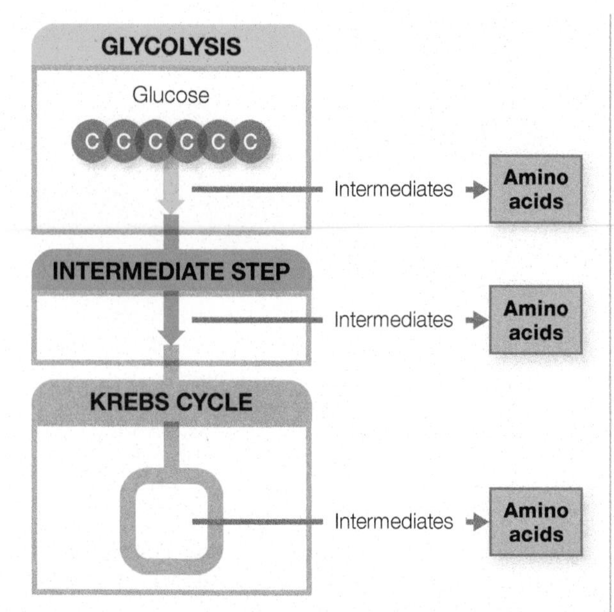

FIGURE 8.29 Building nonessential amino acids Cells make nonessential amino acids using anabolic reactions and intermediates from glycolysis, the intermediate step, and the Krebs cycle.

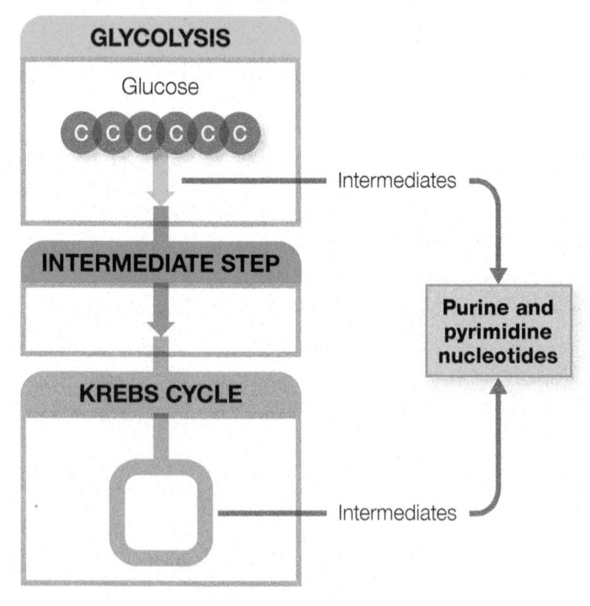

FIGURE 8.30 Building nucleotides Purine and pyrimidine nucleotides can be salvaged from dietary sources or built using intermediates of glycolysis, the Krebs cycle, and the pentose phosphate pathway. In certain microbes the Entner–Doudoroff pathway is central to nucleotide production.

Critical Thinking *Even if a cell is not replicating its DNA, it still needs nucleotides to survive. Thinking back to your genetics chapter, explain why this is the case.*

Cells use anabolic reactions to make amino acids.

You learned earlier that enzymes are proteins that are vital to cell physiology. In addition to serving as enzymes, proteins are found in many cell structures. In certain cells, they even serve as toxins. In the genetics chapter (Chapter 5), we learned that cells build proteins by forming peptide bonds between amino acids, which are the building blocks of proteins.

Most cells can make some (if not all) of the amino acids they need using intermediates from other pathways as start materials (FIG. 8.29). The amino acids that cells cannot make are called **essential amino acids**. Cells must obtain essential amino acids from their environment. In contrast, the amino acids they can make are called **nonessential amino acids**. Both prokaryotic and eukaryotic cells need amino acids, but the degree of dependence on essential amino acids is as diverse as life itself. For example, most strains of *E. coli* can make all amino acids. However not all prokaryotes are as independent as *E. coli*; *Lactobacillus* species and a number of intracellular bacteria like *Mycoplasma pneumoniae* require most amino acids from their environment to survive. Cells make nonessential amino acids by adding an amine group (NH_2) to a biological intermediate. This is a process called **amination**. Often the aminated intermediate is from a catabolic pathway such as glycolysis or the Krebs cycle.

Purines and pyrimidines are not usually made from scratch.

Purines and pyrimidines are essential ingredients in nucleic acids (DNA and RNA) and in energy-carrying molecules like ATP. Collectively, purines and pyrimidines are sometimes called *nitrogen bases* because they are weakly basic, nitrogen-containing compounds. The purines **adenine (A)** and **guanine (G)** are used to make DNA and RNA. In addition to these purines, there are other naturally occurring purines such as caffeine and uric acid, which is found in kidney stones. Pyrimidines are structurally and functionally related to purines and include molecules such as **uracil (U)**, **thymine (T)**, and **cytosine (C)**.

Purines and pyrimidines can be built as completely new molecules from scratch (which is called *de novo synthesis*), or they can be recycled (FIG. 8.30). The anabolic reactions that build purines and pyrimidines from scratch depend on certain amino acids as key starting materials. Building purines and pyrimidines *de novo* uses a great deal of energy. Therefore, these pathways are highly regulated and mainly run when recycling pathways are not meeting the needs of the organism. Unlike bacterial cells, most human cells cannot make purines and pyrimidines, so our cells mainly rely on salvage pathways to recycle the purines and pyrimidines obtained from food.

Build Your Foundation

25. Amino acids are used to build proteins, but where do cells get their amino acids? (LO 8.22 and 8.23)
26. How are lipids linked to the catabolism of glucose? (LO 8.22 and 8.23)
27. Why are purines and pyrimidines usually recycled instead of being built from scratch? (LO 8.22 and 8.23)

BYF QUICK QUIZ

Build Your Foundation (BYF) Quick Quiz: Visit the **Mastering Microbiology** Study Area to quiz yourself.

8.7 THE INTERCONNECTED WEB OF METABOLISM

Amphibolic pathways simultaneously function in catabolism and anabolism.

Learning Outcomes

After reading this section, you should be able to:

8.24 Explain the features of an amphibolic pathway.

We have seen how cells carefully balance catabolism and anabolism so that they run as efficiently as possible. In many cases, anabolic reactions start with intermediates of catabolic pathways. For example, if a cell has an adequate supply of ATP, it can repurpose certain intermediates in catabolic pathways to build new molecules like amino acids and fats (refer back to Figure 8.25). Some pathways, such as the Krebs cycle and the pentose phosphate pathway, simultaneously have a role in anabolism and catabolism. Pathways that simultaneously function in anabolism and catabolism are called **amphibolic pathways**.

Cells maintain a delicate balance between anabolism and catabolism by regulating enzymes, which serve as the key facilitators and moderators of *all* metabolic pathways in a cell. Recall that a number of things can affect enzyme activity, including the availability of cofactors like NAD^+ and $NADP^+$. Cells rely on these two separate cofactors to run anabolic and catabolic reactions. NAD^+ is mostly found in catabolic pathways, while NADPH is usually used in anabolic pathways. If a certain cofactor is limited, then the pathway that depends on that cofactor will be less active. This means the amount of NAD^+ available can affect the degree of catabolic pathway activity, while the amount of $NADP^+$ can affect the activity of anabolic pathways. Enzyme activity is also regulated by mechanisms such as feedback inhibition, allosteric activation/inhibition, and phosphorylation/dephosphorylation, which allow cells to slow down or speed up certain pathways as needed. TABLE 8.9 summarizes the features of anabolism versus catabolism.

TABLE 8.9 Summary of Anabolism and Catabolism

Catabolic Pathways	Anabolic Pathways	All Metabolic Pathways
Breakdown of molecules	Building molecules	Tightly regulated (occur only as needed)
Overall release energy	Overall consume energy	Necessary for cell survival
Rely on oxidized coenzymes (especially NAD^+ to oxidize intermediates)	Rely on reduced coenzymes (especially NADPH) to reduce intermediates	Require enzymes
Example: cellular respiration	Example: lipid biosynthesis	–

Build Your Foundation

28. Define what an amphibolic pathway is, and provide an example. (LO 8.24)

Build Your Foundation (BYF) Quick Quiz: Visit the **Mastering Microbiology** Study Area to quiz yourself.

8.8 CLASSIFYING ORGANISMS BY CARBON SOURCE AND MODE OF ATP PRODUCTION

Autotrophs fix carbon; heterotrophs cannot.

Learning Outcomes

After reading this section, you should be able to:

8.25 Categorize organisms according to their carbon source and how they obtain energy to make ATP.

There is great nutritional diversity in nature, so it is useful to classify organisms according to their nutritional patterns—especially their source of organic carbon. **Autotrophs** are considered *self-feeders* because they make their own organic carbon-containing molecules from inorganic start materials using a process called **carbon fixation**. To clarify, organic molecules contain carbon associated with hydrogen. Examples include methane (CH_4) and carbohydrates like glucose ($C_6H_{12}O_6$). In contrast, inorganic carbon molecules lack associated hydrogen (for example, CO_2). All plants are autotrophs that can "fix" inorganic carbon by incorporating it into sugars. By contrast, **heterotrophs** cannot fix carbon—to live and grow they require an external source of organic carbon, which they obtain from the nutrients they break down. Multicellular animals, such as humans, are heterotrophs.

How a cell obtains energy to make ATP defines phototrophs and chemotrophs.

In addition to considering an organism's source of organic carbon, it is also useful to classify organisms based on how they make ATP. **Phototrophs** use photosynthesis to harvest energy from light to make ATP, whereas **chemotrophs** rely on cellular respiration and/or fermentation to harvest energy from their nutrients to make ATP.

The Case of the Drunk Nursing Student

Practice applying what you know clinically: visit the **Mastering Microbiology** Study Area to watch Part 3 and practice for future exams.

To more precisely describe an organism's metabolic features, the terms *phototroph* and *chemotroph* can be combined with the terms we discussed earlier (*autotroph* and *heterotroph*) to produce the following descriptors: photoautotroph, photoheterotroph, chemoautotroph, and chemoheterotroph (FIG. 8.31). Plants are examples of photoautotrophs, whereas pathogenic bacteria, animals, and fungi are examples of chemoheterotrophs.

We have yet to identify a carbon source that cannot be broken down by some sort of prokaryote, which explains why bacteria and archaea can live in such diverse environments. Their diverse metabolic capabilities are useful to us. For example, bioremediation processes like breaking down of old tires, cleaning up oil spills, and detoxifying mining waste all rely on prokaryotes.

Some organisms are harder to pin down in terms of their metabolic pathways because they can switch between modes of metabolism; these microbes are called **mixotrophs**. Mixotrophs can be eukaryotic or prokaryotic; they use a variety of carbon sources and are also diverse in how they obtain energy. These microbes manage to survive dramatic changes in their surroundings by switching between metabolic modes. For example, *Euglena* (a type of protist) can live as a phototroph or as a chemotroph.

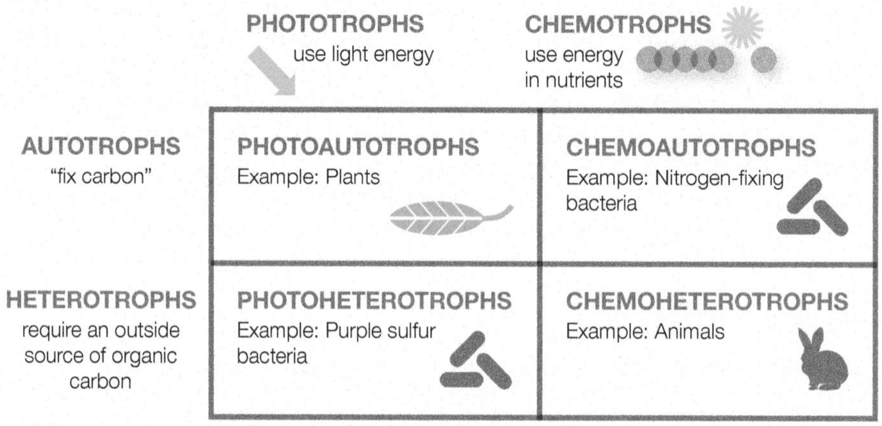

	PHOTOTROPHS use light energy	CHEMOTROPHS use energy in nutrients
AUTOTROPHS "fix carbon"	PHOTOAUTOTROPHS Example: Plants	CHEMOAUTOTROPHS Example: Nitrogen-fixing bacteria
HETEROTROPHS require an outside source of organic carbon	PHOTOHETEROTROPHS Example: Purple sulfur bacteria	CHEMOHETEROTROPHS Example: Animals

FIGURE 8.31 **Classification by metabolic properties** As shown here, organisms can be classified by their source of organic carbon and how they make ATP. Although prokaryotes are key examples of chemoautotrophs and photoheterotrophs, they are technically found among all four of the presented categories. Like us and other animals, pathogenic bacteria and pathogenic protists are chemoheterotrophs.

Build Your Foundation (BYF) Quick Quiz: Visit the **Mastering Microbiology** Study Area to quiz yourself.

Build Your Foundation

29. Based on metabolic properties, how are most nonviral pathogens classified? (LO 8.25)

8.9 USING METABOLIC PROPERTIES TO IDENTIFY BACTERIA

Learning Outcomes

After reading this section, you should be able to:

8.26 Describe why biochemical tests are useful.

8.27 List at least three examples of different biochemical tests and what they detect.

8.28 State what tools or techniques, aside from biochemical tests, are useful for identifying a bacterial specimen.

8.29 Give an example of a rapid analysis technique and how it works.

Biochemical tests help us identify bacterial specimens.

Given the diversity of cellular life, it is not surprising that species can vary in their metabolic profiles. In fact, metabolic profiles can act as biochemical fingerprints, making them useful for identifying microbes. In order to make a definitive diagnosis, clinical samples are often sent to a microbiology lab so that potential pathogens can be identified. **Biochemical tests** that allow us to detect metabolic end products, intermediates, or particular enzymes are extremely useful in identifying microbes.

There are over 1,000 different tests available to identify microbes. Tests range from specialized media to grow and isolate microbes to procedures that identify molecular, genetic, and metabolic characteristics of a given microbe. (In addition to some of the tests discussed here, there are also media that combine differential and selective properties, discussed in Chapter 7.) Typically, when working to identify an unknown sample, a pure culture is grown and then subjected to a panel of tests rather than a single biochemical test. Aside from biochemical tests, microscopy and observing general culture characteristics are also key tools for specimen identification. (Chapter 1 provides information on how microscopy and staining are useful in bacterial identification.)

Amino Acid Catabolism Tests

Amino acid catabolism tests reveal which amino acids a bacterium can breakdown. If a bacterium has the necessary enzymes—such as a particular deaminase or decarboxylase—to catabolize the test amino acid present in the growth media, then the bacterium will grow, and the media will develop a detectable color change. These tests are useful for identifying certain enteric bacteria based on their amino acid catabolism profiles. Common amino acid catabolism tests include the phenylalanine deaminase test and the ornithine decarboxylase test. In another test, *Salmonella* species are differentiated from *E. coli* based on their ability to make hydrogen sulfide (H_2S) by catabolizing amino acids that contain sulfur (FIG. 8.32). Several types of media are used to detect the production of H_2S; in these media H_2S combines with iron to form a black precipitate.

FIGURE 8.32 **Amino acid catabolism and hydrogen sulfide (H_2S) production** *Salmonella* and *Proteus* species commonly make H_2S. As shown in this image, the H_2S made by the bacteria combines with iron in the growth medium to make the black precipitate.

CHEM • NOTE

Amino acids All amino acids contain an amine functional group ($R\text{-}NH_2$) and a carboxyl functional group (R-COOH) group. Deaminases are enzymes that remove the amine group from an amino acid, releasing ammonia (NH_3) in the process. Decarboxylases are enzymes that remove the carboxyl group from an amino acid, releasing carbon dioxide (CO_2) in the process.

Fermentation Tests

Fermentation tests are biochemical tests commonly used to identify an unknown specimen. Fermentation test media usually contain protein, a single carbohydrate (usually a monosaccharide or disaccharide), a pH indicator, and sometimes an inverted Durham tube to capture gas. Most fermentation tests involve inoculating a pure culture of bacteria into the fermentation test media and then observing the medium after the specimen has had time to grow.

Often fermentation pathways produce acids, although the type and amount of acid made depends on the fermentation pathway the bacterium uses. The accumulation of acidic fermentation end products will lower the pH of the medium. A pH indicator, which changes color in an acidic sample, is used to detect the drop in pH. Some organisms make gas *and* acid when they ferment carbohydrates; in such a case the gas formed would be captured in the inverted Durham tube, and a bubble would be seen (FIG. 8.33). Sometimes fermentation tests can be used to distinguish pathogens from harmless bacteria. For example, commensal *E. coli* (bacteria that normally live in the large intestine) ferment the carbohydrate sorbitol. However, the pathogenic *E. coli* O157 strain does not ferment sorbitol. Another example of fermentation profiles being used to identify a bacterium is found in comparing *Shigella* species, some of which cause dysentery, to *E. coli*; *Shigella* species usually do not ferment lactose, while most *E. coli* strains do.

As mentioned in the fermentation section of this chapter, the type of fermentation reaction performed is also helpful in identifying certain enteric bacteria. The **methyl red/Voges–Proskauer (MRVP)** test is designed to distinguish between mixed acid fermentation and butanediol fermentation. When bacteria that carry out mixed acid fermentation are grown in broth containing glucose, they make an abundance of acidic products that lower the media's pH. The drop in pH can be detected by adding a pH indicator called methyl red. This is the **methyl red (MR)** test; again, it is often used in combination with the **Voges–Proskauer (VP)** test, which is designed to detect acetoin, an intermediate

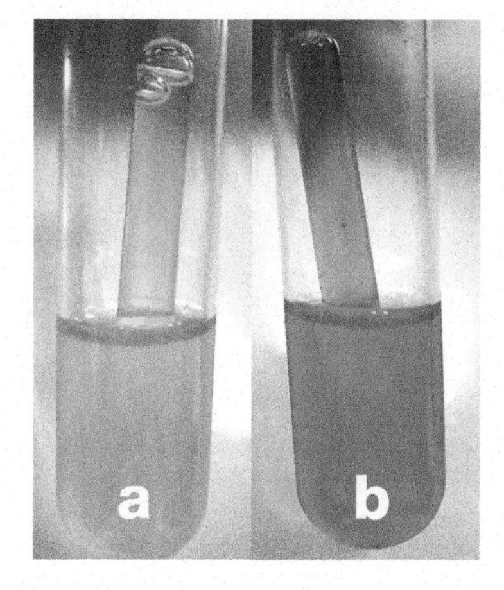

FIGURE 8.33 **Fermentation tests** In these biochemical tests for bacterial identification, fermentation end products cause a drop in the pH of the test medium. A pH indicator (e.g., phenol red) in the medium changes color from red to yellow if a fermentation reaction occurred. (**a**) Specimen A fermented the sugar lactose; a bubble in the Durham tube also tells us that the bacteria made gas during the fermentation process. (**b**) Specimen B did not ferment the sugar lactose.

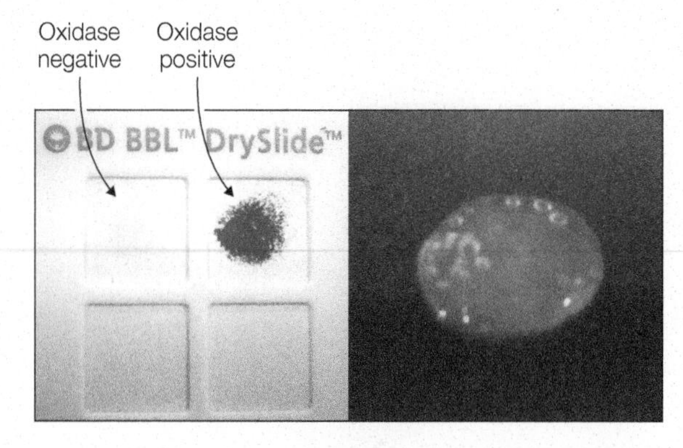

FIGURE 8.34 Oxidase and catalase tests These tests give clues about how a bacterial specimen uses oxygen. *Left*: A bacterium that is oxidase positive will generate a purple/blue pigment when smeared onto a DrySlide card. An oxidase-negative specimen will not generate a color change. *Right*: This is a catalase-positive organism, as evidenced by the bubbling in the puddle of hydrogen peroxide.

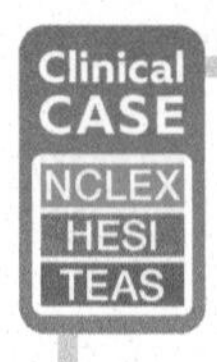

The Case of the Drunk Nursing Student

Practice applying what you know clinically: visit the **Mastering Microbiology** Study Area to watch Part 4 and practice for future exams.

of butanediol fermentation. In a clinical microbiology lab, the MRVP test can help identify the causative agent of a foodborne illness. For example, the MRVP test can differentiate *E. coli* and *Salmonella*, which carry out mixed acid fermentation, from species of *Enterobacter*, which carry out butanediol fermentation. (Refer to Figure 8.24.)

Oxidase and Catalase Tests

Because the electron transport chains of bacteria are so diverse, biochemical tests that look for differences in electron carriers can be used to identify certain specimens. The **oxidase test** can be used to determine if a bacterium has cytochrome c oxidase in its respiratory chains (FIG. 8.34). Organisms that have cytochrome c oxidase can use oxygen as a final electron acceptor. Note that the oxidase test detects only *one* type of electron carrier involved in aerobic respiration; it does not detect all oxidases that participate in aerobic respiration. This means that bacteria that have a negative oxidase test result could still be aerobic—they may just use an oxidase other than cytochrome c oxidase. A negative oxidase test result reveals only that the tested bacterium lacks cytochrome c oxidase; it does not mean a bacterium is anaerobic. The oxidase test can help identify *Neisseria gonorrhoeae* (the causative agent of the sexually transmitted disease gonorrhea) because this bacterium contains cytochrome c oxidase. The oxidase test is also useful for differentiating certain Gram-negative rods; *Pseudomonas* species like *P. aeruginosa* (which commonly causes infections in burn patients) are oxidase positive, while *E. coli* is oxidase negative.

Another test commonly used in introductory microbiology labs is the **catalase test**. This test detects if an organism has the enzyme catalase, which protects organisms from toxic oxygen radicals. The test is performed by adding hydrogen peroxide to the test specimen. If the enzyme catalase is present, it will break the hydrogen peroxide down into water and gaseous oxygen. The oxygen is released as bubbles and is easily seen in the puddle of hydrogen peroxide on the sample smear. If bubbles are made, the organism is catalase positive. An organism that is catalase negative will be unable to break the hydrogen peroxide down into oxygen and water and will not generate bubbles (Figure 8.34). Most aerobes and facultative anaerobes are catalase positive. Most anaerobes are catalase negative. This test is useful for differentiating *Staphylococcus* species (catalase positive) from *Streptococcus* species (catalase negative).

Rapid Identification Techniques

In clinical and food microbiology labs with high volumes of samples, it is crucial that specimens are identified as quickly, reliably, and inexpensively as possible. One way to accomplish this is to use specialized test strips like the Analytical Profile System (API® system). The test strips are essentially miniaturized versions of the biochemical tests that students perform in a typical introductory microbiology lab session. Depending on the strip type, there are about 20 mini-tests of media. The media are inoculated with a pure culture, and the entire strip is then incubated for 18–24 hours. An index number can then be generated using an automated reader or by manually observing and recording the test results (FIG. 8.35). These test strips are sold in a variety of formats and are routinely used to identify Enterobacteriaceae (a large family of Gram-negative bacteria that includes many pathogens), clinical staphylococci (which can cause "staph" infections) and micrococci (usually only pathogenic in immune-compromised patients, such as AIDS patients), and Gram-negative non-Enterobacteriaceae, to name a few examples.

1 After the incubation period, visually interpret the test strip (or place in an automated reader).

2 Add the numbers that align with positive test results in each segment. (For example, in the first segment, only tests in columns associated with 1 and 4 were positive, so we add 1 + 4 to get 5).

3 Following step 2 for each segment will generate an index number.

4 Use the API database to look up which bacteria correspond to the index number. Based on the index number, the most likely bacterial candidate is *Klebsiella pneumoniae*.

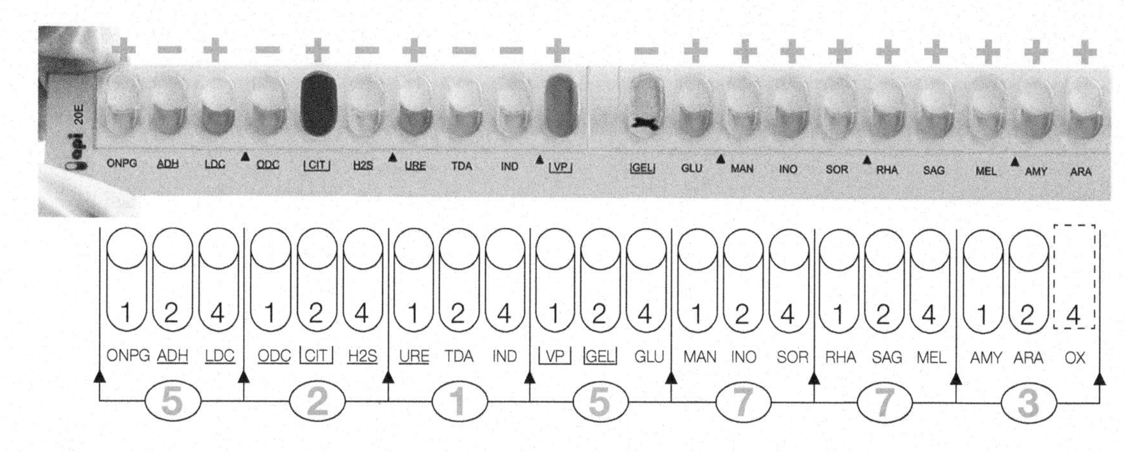

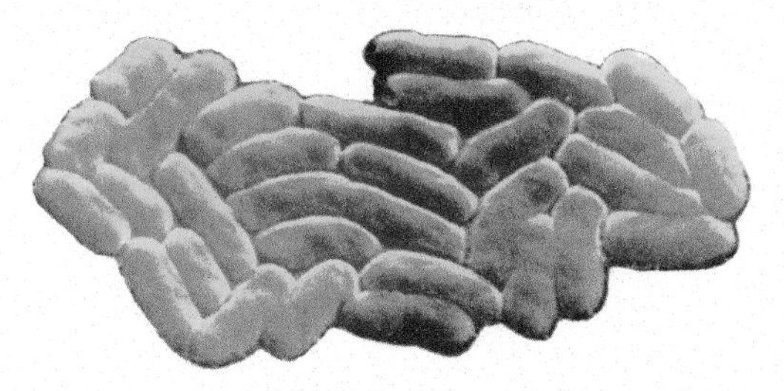

Klebsiella pneumoniae

FIGURE 8.35 **Automated identification of clinical and food microbiology specimens** The API system (Analytical Profile Index) is a rapid, semi-automated tool commonly used for the identification of clinical and food microbiology specimens. This system uses a combination of numerical coding and test result data to generate unique identifiers for over 500 bacterial species.

Critical Thinking *To identify a specimen using biochemical tests, a pure culture of the suspected pathogen must be obtained. Why must a pure culture be used for these tests?*

Build Your Foundation

30. Describe the basic underlying principle that makes biochemical testing useful for identifying an unknown specimen. (LO 8.26)

31. Describe three different categories of biochemical tests and how they work. (LO 8.27)

32. In addition to standard biochemical tests, what else is often used to identify a specimen? (LO 8.28)

33. Give an example of a rapid ID tool, and state when such a tool would be helpful. (LO 8.29)

Build Your Foundation (BYF) Quick Quiz: Visit the **Mastering Microbiology** Study Area to quiz yourself.

VISUAL SUMMARY | *Microbial Metabolism*

Interactive CONTENT REVIEWS

Interactive Content Reviews: Visit the **Mastering Microbiology** Study Area to quiz yourself.

Metabolism

Metabolism is the collection of anabolic and catabolic reactions in a cell.

Anabolic reactions build molecules, *using* energy in the process.

Catabolic reactions break down molecules *releasing* energy in the process.

ATP

ATP is essential for all cells to do cellular work. Cells do not store ATP. Instead, they use energy from catabolic pathways to continuously recharge or recycle ADP to ATP. The ATP made by catabolic pathways is used to fuel anabolic pathways to build organic molecules such as lipids, proteins, nucleic acids, and complex carbohydrates.

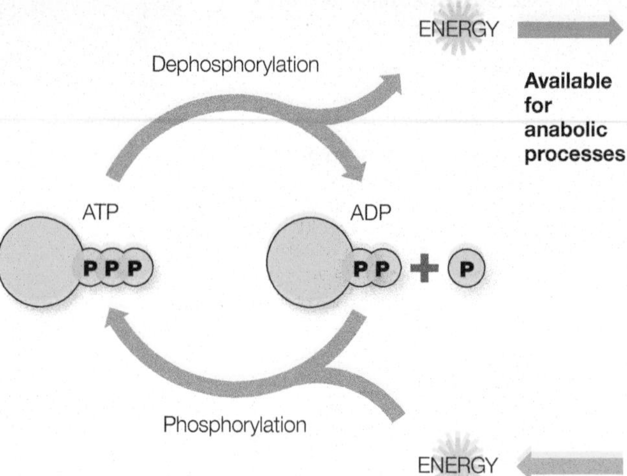

Enzymes

Enzymes are necessary for metabolism and are not consumed in reactions.

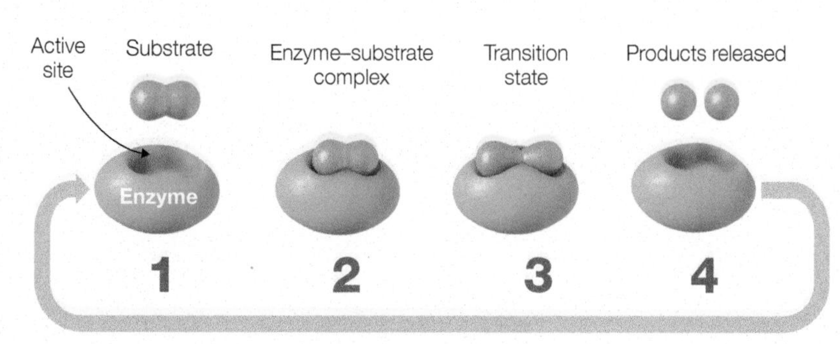

Enzymes are also

- specific for their substrate.
- protein catalysts that are not consumed by the reactions they perform.
- used by cells to help reactions occur under physiological conditions.
- sensitive to changes in pH and temperature and can become denatured if conditions stray outside optimum ranges.
- often detected by biochemical tests that are used to characterize bacterial samples.

Enzyme Inhibitors

There are several types of enzyme inhibitors.

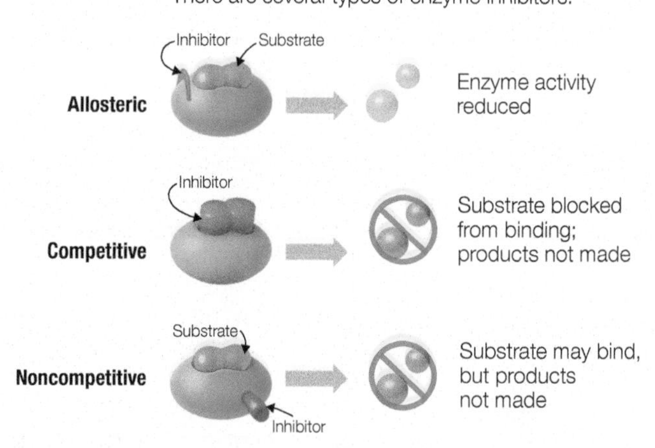

Cofactors

Organic cofactors (or coenzymes) such as NAD^+ and FAD are central to catabolic pathways and act as electron carriers.

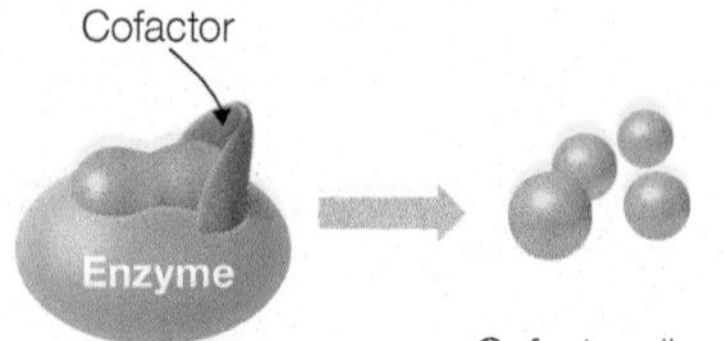

Cofactor allows chemical reaction that converts substrate to products

Regulation

Cells can reduce enzyme activity if the amount of product is sufficient. Feedback inhibition is one example of regulation.

ATP Production in Cellular Respiration

Most of the ATP made in cellular respiration comes from oxidative phosphorylation, which occurs in the electron transport chain.

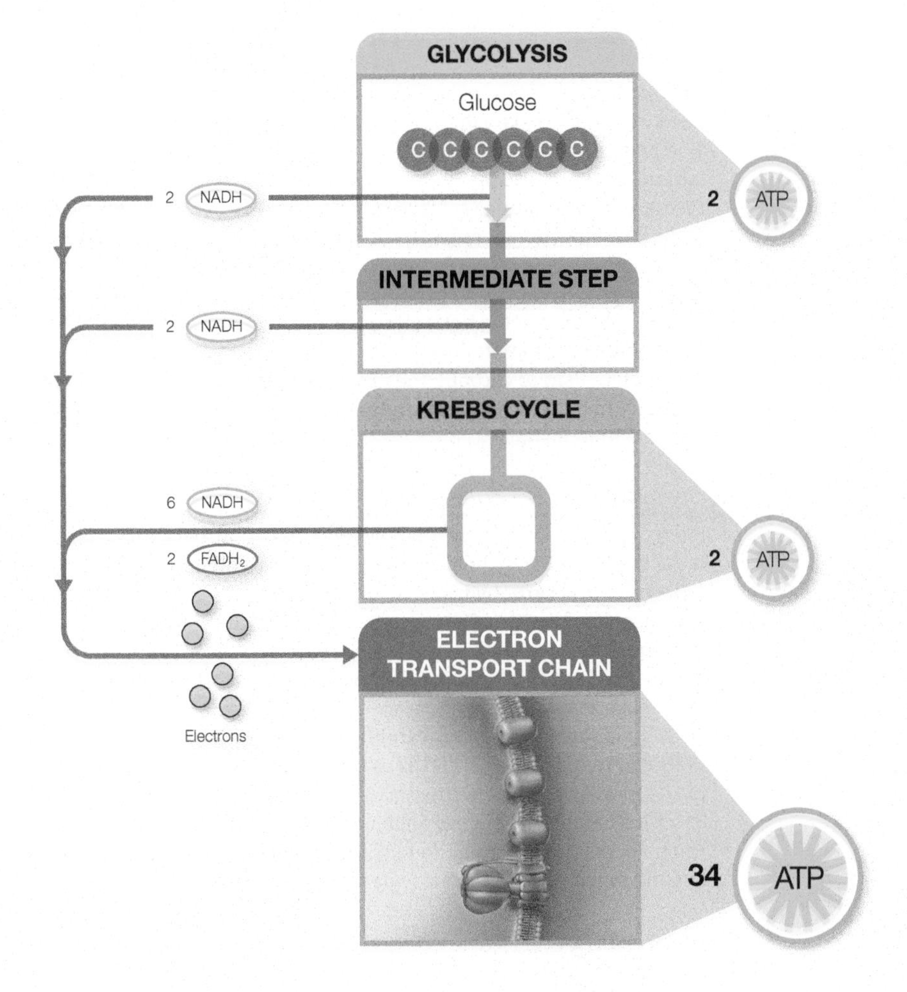

Substrate level phosphorylation
4 ATP per glucose
(2 in glycolysis and 2 in Krebs cycle)

+

Oxidative phosphorylation
34 ATP per glucose
(3 ATP per NADH and 2 ATP per $FADH_2$ that enters aerobic electron transport chains)

= **38** ATP

Theoretical maximum per glucose catabolized in aerobic cellular respiration

Biochemical Tests

Biochemical tests provide a metabolic profile or biochemical fingerprint that is useful to identify microbes. There are many types of biochemical tests. Some directly detect a particular enzyme. Others detect products of a particular metabolic pathway. Aside from biochemical tests, microscopy and observing general culture characteristics are used for specimen identification.

Gelatin hydrolysis test

Detecting amylase with starch agar

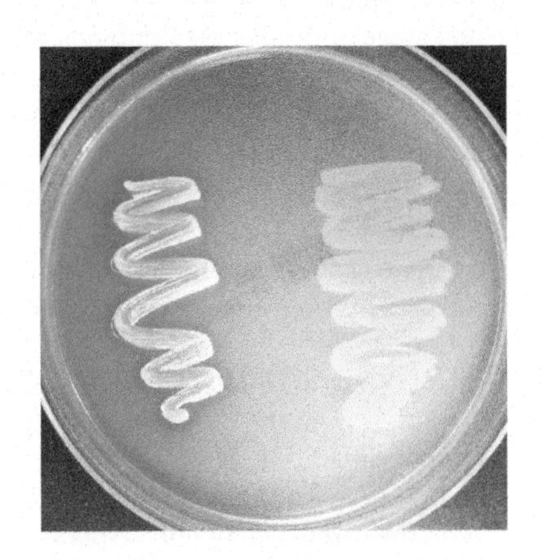

Spirit blue agar to detect lipase

CHAPTER 8 OVERVIEW

8.1 Defining Metabolism

- Metabolism is the collection of reactions that break down substances to release energy (catabolic reactions) and the reactions that use energy to build substances (anabolic reactions).
- Energy-rich nutrients are "cashed in" to make ATP (adenosine triphosphate).
- Cells carefully balance ATP production and use.

8.2 Enzymes

- All metabolic pathways depend on enzymes.
- Enzymes are highly specific protein catalysts that lower the activation energy of a reaction and increase the rate of a reaction.
- Enzymes are named based on their function.
- Enzymes are not consumed by the reactions they facilitate.
- The induced fit model describes how enzymes interact with substrates through the enzyme's active site.
- Inorganic cofactors and organic coenzymes (especially $NAD^+/NADH$ and $FAD/FADH_2$) are often needed for an enzyme to function.
- Ribozymes are specialized catalytic RNA molecules.
- Cofactors, temperature, pH, amount of substrate present, phosphorylation state, and the presence of inhibitors can all affect enzyme activity.
- Cells carefully regulate enzyme activity.

8.3 Obtaining and Using Energy

- Cells do not store ATP.
- The production of ATP relies on coupled oxidation–reduction (redox) reactions.
- Cells re-phosphorylate ADP to form ATP.
- Cells use three general mechanisms to drive the phosphorylation of ADP to ATP: (1) substrate-level phosphorylation, (2) oxidative phosphorylation, and (3) photophosphorylation.

8.4 The Catabolic Process of Cellular Respiration

- Cellular respiration is a collection of reactions that *extract* energy from nutrients to make ATP.
- Cellular respiration involves glycolysis, an intermediate step, the Krebs cycle, and the electron transport chain.
- The first stage of cellular respiration is glycolysis.
  - Glycolysis does not require oxygen.
  - It occurs in the cytoplasm of prokaryotes and eukaryotes.
  - Glycolysis oxidizes and splits glucose.
  - The process makes two ATP profit (four ATP gross) and some NADH.
  - Two pyruvic acid molecules are the final products of glycolysis.
- An intermediate step converts pyruvic acid from glycolysis to acetyl-CoA.
- The second stage of cellular respiration is the Krebs cycle.
  - Acetyl-CoA from the intermediate step is further oxidized and decarboxylated.
  - The Krebs cycle occurs in the mitochondrial matrix of eukaryotes and in the cytoplasm of prokaryotes.
  - By the end of the Krebs cycle, all of the carbons originally found in glucose are removed.
  - The Krebs cycle produces ATP and reduced cofactors NADH and $FADH_2$.
- The final stage of cellular respiration relies on electron transport chains.
  - Electron transport chains are found in the inner mitochondrial membrane of eukaryotic cells and in the plasma membrane of prokaryotic cells.
  - Aerobic respiratory chains use oxygen as a final electron acceptor.
  - Anaerobic respiratory chains use an inorganic substance other than oxygen as a final electron acceptor.

8.5 Other Catabolic Pathways for Oxidizing Nutrients

- The pentose phosphate pathway operates parallel to glycolysis and allows for the breakdown of five-carbon sugars (e.g., ribose).
- The Entner–Doudoroff pathway is an ancient relative of glycolysis; certain microorganisms use this pathway to catabolize sugars.
- Fermentation is a way to catabolize nutrients without a respiratory chain.
  - In fermentation an organic molecule serves as the final electron acceptor.
  - Examples of common carbohydrate fermentation reactions include homolactic, heterolactic, alcohol, mixed acid, and butanediol fermentation.
- The metabolism of carbohydrates, lipids, nucleic acids, and proteins is interconnected. The Krebs cycle is especially important in protein and fat catabolism.

8.6 Anabolic Reactions: Biosynthesis

- Biosynthetic reactions are anabolic reactions that require ATP and reducing power. They help cells grow, multiply, and repair damage.
- Cells can build carbohydrates.
  - Gluconeogenesis is the process of building glucose from noncarbohydrate start materials.
  - The production of glycogen, a polymer of glucose, is called glycogenesis.
- Cells build fats from glycerol and fatty acids.
- Cells can make nonessential amino acids by adding an amine group (NH_2) to a biological intermediate. Often the aminated intermediate is from a catabolic pathway such as glycolysis or the Krebs cycle.
- Anabolic reactions that build purines and pyrimidines from scratch depend on certain amino acids as key starting materials.

8.7 The Interconnected Web of Metabolism

- Amphibolic pathways have roles in anabolism and catabolism.
- The Krebs cycle and the pentose phosphate pathway are examples of amphibolic pathways.

8.8 Classifying Organisms by Carbon Source and Mode of ATP Production

- Organisms can be classified by their ability to fix carbon; autotrophs can fix inorganic carbon (CO_2) into organic carbon molecules, while heterotrophs cannot fix carbon.
- Organisms can extract energy from nutrients (chemotrophs) or from light (phototrophs) to make ATP.
- Most pathogens are chemoheterotrophs.

8.9 Using Metabolic Properties to Identify Bacteria

- Biochemical tests that detect metabolic end products, intermediates, or particular enzymes in a bacterium can help identify a bacterial sample.
- When a high volume of samples must be identified, rapid identification tools like the Analytical Profile System (API® system) are useful.

SMART

THINK CLINICALLY | *Be S.M.A.R.T. About Cases*

COMPREHENSIVE CASE

The following case integrates basic principles from the chapter. Try to answer the case questions on your own. Don't forget to be **S.M.A.R.T.*** about your case study to help you interlink the scientific concepts you have just learned and apply your new understanding of metabolism principles to a case study.

*The five-step method is shown in detail in the Chapter 1 Comprehensive Case on cholera. See pages 31–33. Refer back to this example to help you apply a SMART approach to other critical thinking questions and cases throughout the book.

• • •

A 40-year-old male called 911 stating that he was having difficulty breathing and felt too nauseated to drive himself to the hospital. He said he had type I diabetes and had not been able to afford insulin on a consistent basis, especially over the past week, since he lost his job. Fortunately, the emergency medical team promptly got to the patient and took him to the hospital. His blood work revealed an exceedingly high blood glucose level, a finding consistent with the fact that the body relies on insulin to get glucose into cells. The patient was diagnosed as suffering from ketoacidosis. This is a metabolic state in which the body perceives that it is starving for glucose and so switches to mainly catabolizing fats. In ketoacidosis, ketones, which are byproducts of fat catabolism, rapidly build up in the body and dangerously lower the blood's pH level. After stabilizing the patient, the physician performed a thorough physical examination.

Upon physical examination, the physician noticed the man had an infected ulcer on his foot. A bacterial specimen was collected, and analysis showed Gram-positive rods that form endospores. The doctor said it was fortunate that the man had sought medical assistance because he could have gone into a coma and died from ketoacidosis. Also, they told the patient that if ketoacidosis had not killed him, the foot wound could have. The ulcer was infected with an anaerobic bacterium, *Clostridium perfringens,* which causes gas gangrene.

The patient stated that he often had slow-healing sores on his feet and asked if it was something to be concerned about. The doctor explained that people who have diabetes and do not consistently regulate their blood glucose often suffer from a number of complications, including circulatory system vessel damage. When blood vessels are damaged, blood delivery to the tissues is decreased, which can affect healing. This pathology is particularly common in the extremities, and especially the feet. Follow-up care for all wounds is important to prevent future infections with dangerous bacteria like *C. perfringens*. Fortunately, the patient's infection was localized and readily responded to basic wound care and a regimen of clindamycin and penicillin.

• • •

CASE-BASED QUESTIONS

1. Blood work revealed that the patient had high blood glucose levels. However, ketoacidosis results when the body is "starving" for glucose. Why can these two conditions simultaneously exist in a diabetic patient?
2. Why would the body preferentially metabolize fats before metabolizing proteins?
3. Explain why low blood pH, like that seen in ketoacidosis, is potentially deadly.
4. Why are diabetics at risk for wound infections by anaerobic bacteria like *Clostridium perfringens*?
5. Name two biochemical tests that were discussed in this chapter that would most likely be *negative* for *Clostridium perfringens*. Explain why.

END OF CHAPTER QUESTIONS

NCLEX
HESI
TEAS

Think Critically and Clinically

Questions highlighted in orange are opportunities to practice NCLEX, HESI, and TEAS-style questions.

1. Label each reactant *and* product in the reactions as reduced or oxidized (X is a theoretical molecule or atom).

X^{+2} + NADPH → $NADP^+$ + XH^+
X + NAD^+ → NADH + X^+
Cu + 2 Ag^+ → Cu^{2+} + 2 Ag
$C_6H_{12}O_6$ + 6 O_2 → 6 CO_2 + 6 H_2O

2. Indicate the true statements about ATP, and then reword the false statements so that they are true.
   a. ATP is made using anabolic reactions.
   b. Substrate-level phosphorylation converts ATP to ADP.
   c. ATP is commonly used by cells to store energy.
   d. ATP is used to jump-start cellular respiration.
   e. Catabolic reactions are used to make ATP.
   f. In cellular respiration, the most ATP is made by glycolysis.
   g. ATP can be made by phosphorylating ADP.

3. The graph shows an enzyme-catalyzed reaction.
   a. Draw a line that would correspond to a noncatalyzed reaction.
   b. Is this an endergonic or exergonic reaction? How can you tell?
   c. Label the point on your graph that corresponds with the activation energy.

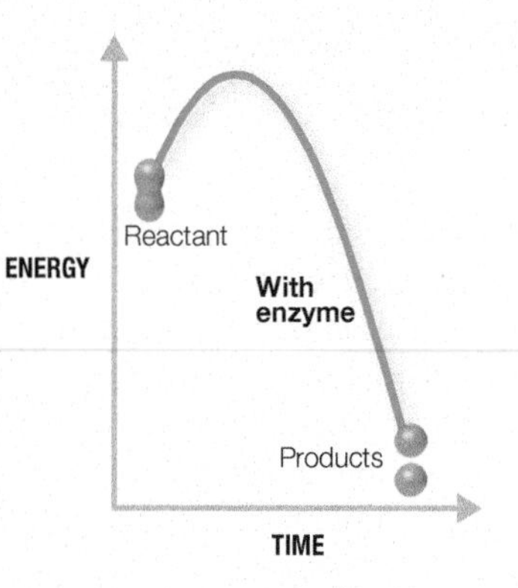

4. Match the term to the statement (some statements may be used more than once; others may not be used at all).

Electron transport chain	Produces two pyruvic acid molecules from glucose
Glycolysis	Results in the production of acetyl-CoA from fatty acids
Krebs cycle	Makes acetyl-CoA from pyruvic acid
Fermentation	Uses light energy to make ATP and NADPH and carbohydrates
Beta-oxidation	Produces $FADH_2$
Intermediate step	Decarboxylation and reduction reactions combine to catabolize acetyl-CoA
	Requires an investment of two ATP
	An organic molecule acts as the final electron acceptor
	Requires oxygen as a final electron acceptor
	Uses a proton gradient to make ATP

5. Complete the table.

Process	Cellular location in prokaryotic cells	Cellular location in eukaryotic cells	Uses oxygen?
Glycolysis			
Intermediate step			
Krebs cycle			
Aerobic electron transport chain			

6. Indicate which statements about fermentation are true, and then correct the false statements.
   a. Fermentation is an anaerobic process that can be used by prokaryotic or eukaryotic cells.
   b. Sugars are the only nutrients that can be fermented.
   c. Fermentation is a low ATP yielding process.
   d. There are only five types of fermentation: homolactic, heterolactic, alcohol, mixed acid, and butanediol fermentation.
   e. Fermentation is the same as anaerobic respiration.

7. Draw and then fill out a Venn diagram (see sample version here) that compares and contrasts anaerobic respiration to aerobic respiration. The middle section of the diagram is where you should list the features these processes have in common. Include as many details as possible in this diagram to make this a useful exercise.

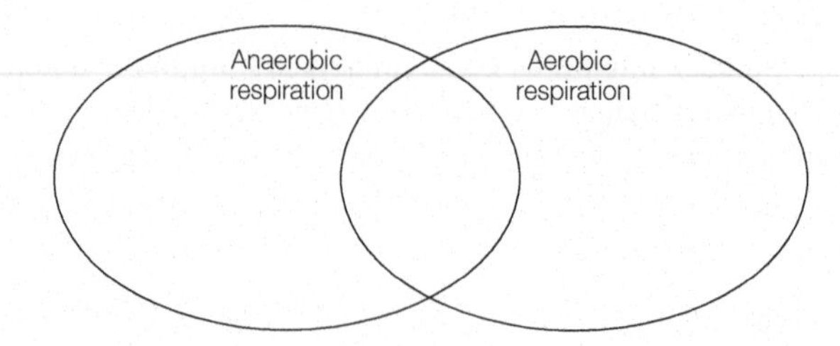

8. Why do human cells require oxygen? Select all relevant statements.
   a. To carry out glycolysis
   b. To carry out fermentation
   c. To carry out cellular respiration
   d. To oxidize fats
   e. To carry out substrate-level phosphorylation
   f. To carry out oxidative phosphorylation
   g. To carry out photophosphorylation

9. ________________ is a process that relies on a phosphorylated intermediate to directly convert ADP to ATP.

10. ________________ is a process that uses light energy to drive an electron transport chain to make ATP.

11. ________________ is a process that uses energy from nutrients to fuel an electron transport chain to make ATP.

12. The ________________ test detects if an organism can convert hydrogen peroxide to water and oxygen. The test result is usually ________________ in anaerobic microbes and usually ________________ in aerobic microbes. You know the test is positive by ________________.

13. The photo shows an organism growing in glucose fermentation broth. What can you conclude about this specimen with regard to its ability (or lack thereof) to ferment the sugar sucrose? Explain your answer.

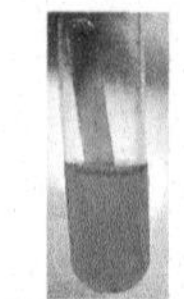

14. Use this pathway schematic to answer questions a through d.
   a. Which enzyme carries out a redox reaction? In this reaction, what is being reduced and what is being oxidized?
   b. Which enzyme would be the most likely to be regulated by feedback inhibition?
   c. What is the end product of this pathway?
   d. Which enzyme carries out substrate-level phosphorylation?

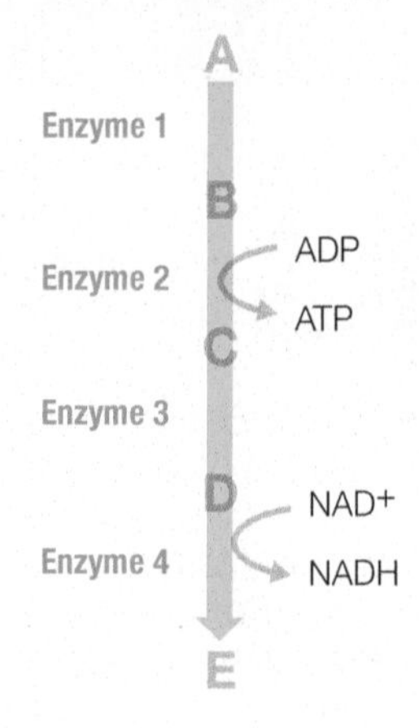

15. Match the term to the statement. (Some terms will be used more than once.)

Photoautotrophs	Require an external source of organic carbon
Chemoheterotrophs	Harvest energy from light to make ATP
Mixotrophs	Can switch between modes of metabolism
	Extract energy from nutrients to make ATP
	Make their own organic carbon-containing molecules from inorganic start materials

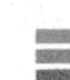

16. Rank the following from the *most* ATP that could be made to the *least* ATP that could be made:
    - 1 glucose molecule processed via a fermentation pathway (consider that glycolysis is the first stage of the process)
    - A lipid made of glycerol and three 10-carbon fatty acid chains entering cellular respiration
    - 1 glucose molecule entering the Entner–Doudoroff pathway
    - 1 glucose molecule entering cellular respiration

CRITICAL THINKING QUESTIONS

1. Two samples were both analyzed for the activity of enzyme A. Both samples were incubated under the same temperature and with the same amount of substrate and the same type of inhibitor. What is the most likely explanation for the difference seen between sample 1 and sample 2?

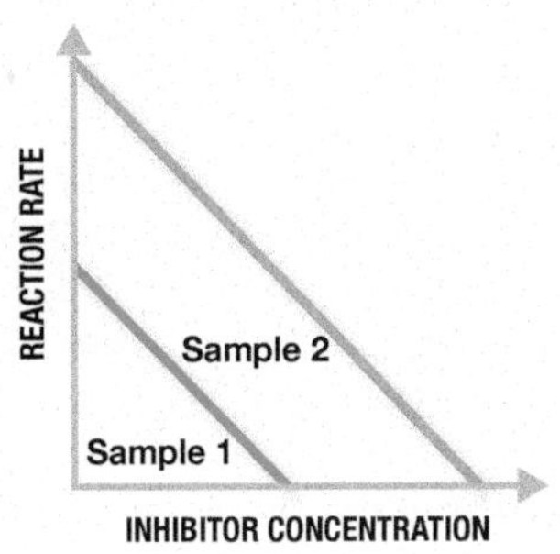

2. An organism was found to be insensitive to levels of oxygen in its environment and made about 20 ATP molecules per glucose molecule. What type of metabolic pathway does this organism most likely depend upon? Explain your answer.

3. Draw a diagram to represent how an inhibitor of a kinase could affect a pathway that converts compound A to compound C in two steps.

4. The following is a set of observations about an organism. What type of environment do you think it lives in? What type of metabolic pathway does it most likely use to meet its energy and ATP needs? Support your conclusions using the graphical data.

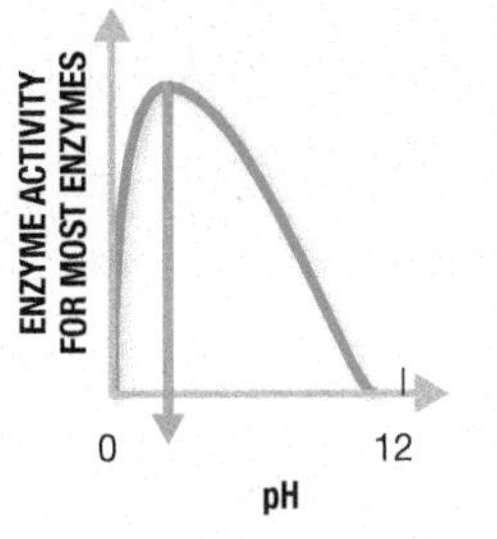

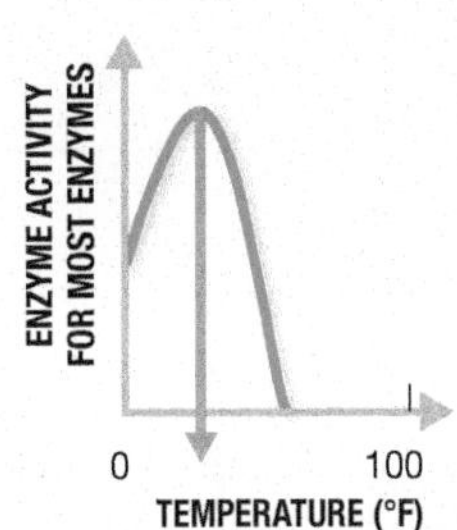

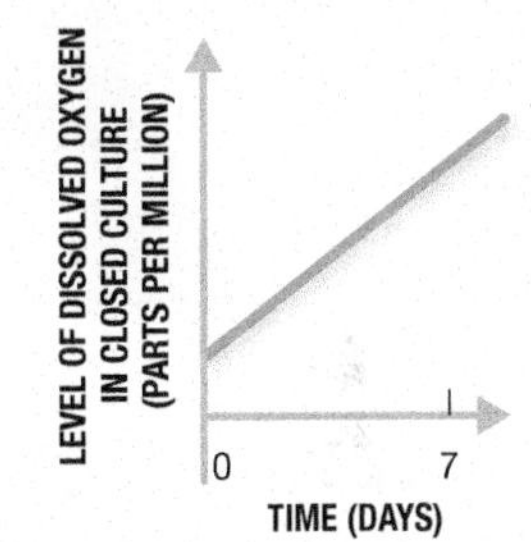

5. Cellular respiration is not the same as breathing, although they do interconnect with each other. Explain this statement.

6. An enzyme can convert an exergonic reaction into an endergonic reaction. Is this statement true or false? Support your viewpoint.

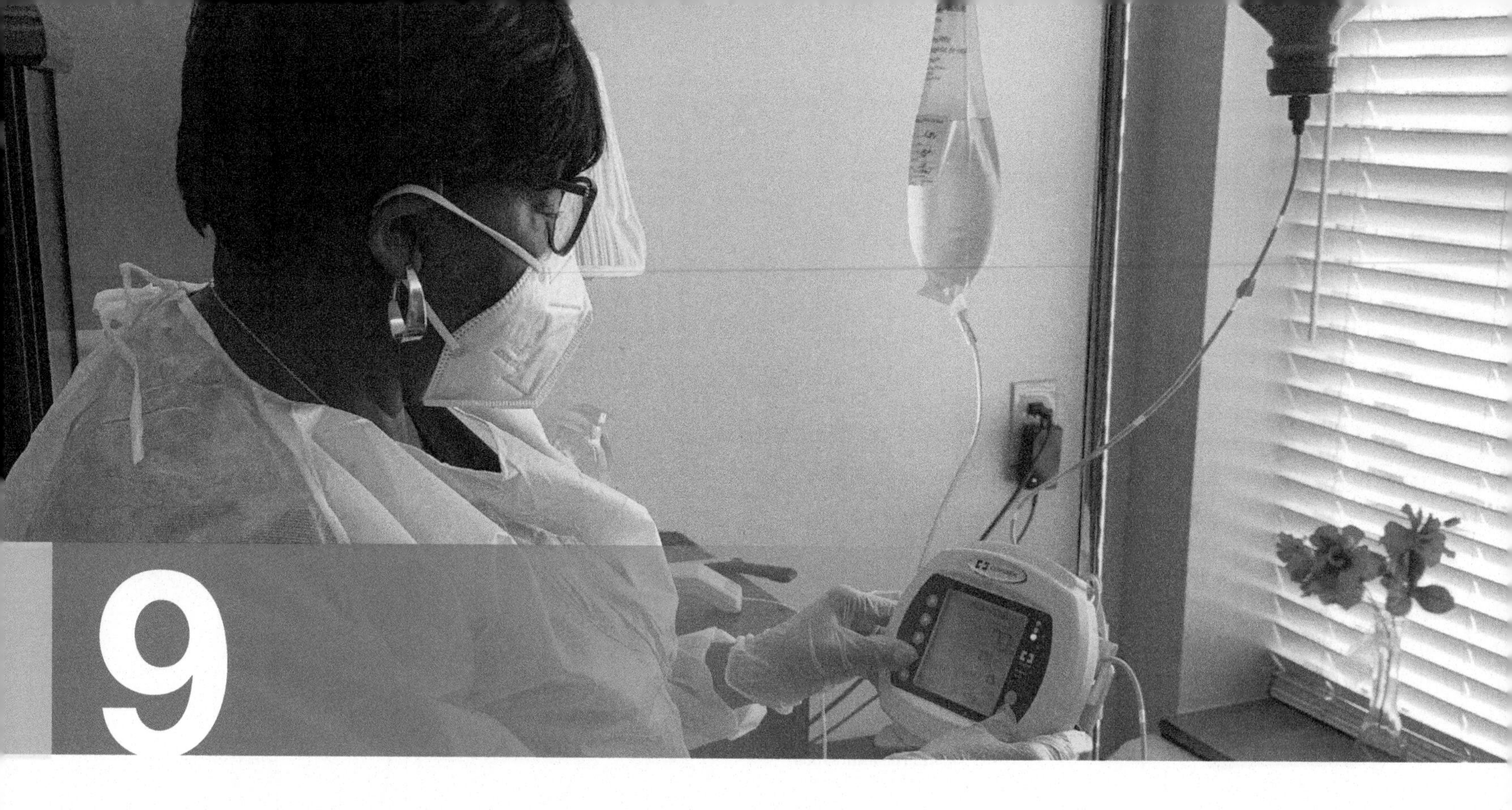

9

Principles of Infectious Disease and Epidemiology

What Will We Explore? Epidemics of infectious disease have evolved right along with humans. Ancient texts and artwork depict various epidemics contributing to wars, the collapse of empires, and other historical upheavals. Plague, a scourge that wiped out huge segments of the world population in the 14th century, still grips our imagination. Even before the outbreak of COVID-19, dangerous disease outbreaks served as central themes for modern movies, TV programs, books, music, and art.

Like us, our ancestors were anxious to discover and understand the causes of disease and to protect themselves from such dangers. Before the germ theory of disease was accepted, illnesses were often regarded as supernatural phenomena—the work of "bad air" or "evil spirits." English physician John Snow helped dispel such notions and is now considered one of the first modern epidemiologists. Cholera's cause—the waterborne bacterium *Vibrio cholerae*—was unknown to Snow when he worked in the 1850s. Yet by carefully tracking infection cases on a map, he pinpointed a public well in London as the likely origin of a cholera outbreak. Snow had the well handle removed, and cholera cases immediately declined in the area.

This chapter explains basic principles of infection and disease, along with the evolution and practices of epidemiology. We will learn how the search to understand "what plagues the people" has shaped human history and modern health policies.

Why Is It Important? Today the field of epidemiology, which studies patterns in disease incidence as well as control and prevention, is inseparable from modern medicine. It is the basis of what we call public health. Modern equipment and techniques allow us to identify the causes and progression of infection in individuals. Comprehensive data collection also allows us to track incidence of certain diseases within a population. Such epidemiological studies influence public health policies and help limit disease transmission. Healthcare providers are essential partners in epidemiology, as they are responsible for reporting disease incidence and breaking cycles of transmission.

Rita Griffiths
RN, Unit Manager; Decatur, GA

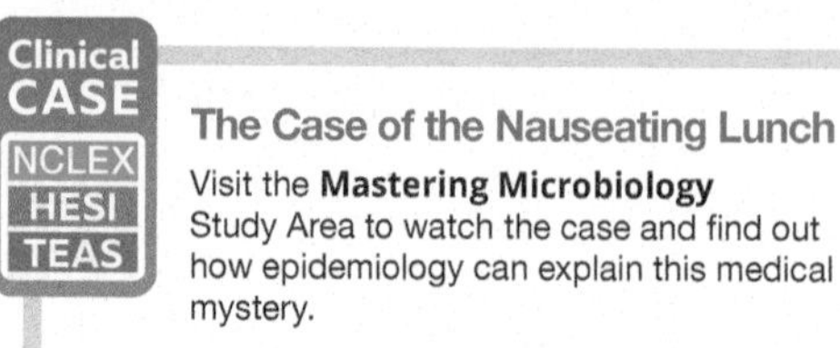

The Case of the Nauseating Lunch

Visit the **Mastering Microbiology** Study Area to watch the case and find out how epidemiology can explain this medical mystery.

9.1 CAUSES OF INFECTIOUS DISEASES

Disease terminology is the foundation of understanding modern health care and epidemiology.

In this chapter we discuss basic terms and principles of infectious disease, as well as epidemiology, which is focused on monitoring and controlling the occurrence of diseases to promote public health. (Chapter 10 discusses *pathogenesis*, the disease-causing mechanisms of microbes.) An **infectious disease** is an illness caused by a pathogen. The six categories of pathogens are viruses, bacteria, protozoans, helminths, fungi, and prions. **Opportunistic pathogens** tend to only cause disease when their **host**, the targeted organism, is weakened in some way—often through shifts in the normal microbiota or weakening of the immune system. A **true pathogen** does not require a weakened host to cause disease. **Sporadic** cases of a disease are isolated infections in a particular population. By contrast, **endemic** infections are routinely detected in a population or region. For example, cold viruses are endemic, whereas Ebola cases are sporadic.

An **epidemic** is a widespread disease outbreak in a particular region during a specific time frame. If an epidemic spreads to numerous countries, it is called a **pandemic**. Emerging and reemerging pathogens are especially likely to cause epidemics or pandemics. **Emerging pathogens** include newly identified agents such as SARS-CoV-2 as well as pathogens that previously caused only sporadic cases, but are increasingly common and/or exhibit an expanded geographical distribution. The Zika virus is an example of the latter. Although it has caused sporadic cases since its discovery about 70 years ago, the outbreak in 2015/2016 combined with an expanded geographical footprint make Zika virus an important emerging pathogen. In contrast, a **reemerging pathogen** is an infectious agent that was under control as a result of prevention or treatment strategies and is now resurfacing. Bacterial diseases that were previously treatable with antibiotics but have developed resistance to available drugs, such as multidrug-resistant tuberculosis, are examples of reemerging pathogens.

About 60 percent of the emerging infections characterized since the 1970s are **zoonotic diseases**, which spread from animals to humans. Many zoonotic pathogens are **noncommunicable**, meaning they do not spread from person to person. Unfortunately, noncommunicable pathogens can mutate into **communicable** forms that do transmit from human to human. Communicable diseases that are easily transmitted from one host to the next (such as cold viruses and influenza viruses) are often described as **contagious** diseases.

We can also describe how an infection is experienced by its host. During an **active infection**, the patient is symptomatic, meaning they have signs and symptoms. **Signs** are objective indicators of disease that can be measured or verified. Common signs include fever, rash, or blood in stool. **Symptoms** are sensed by the patient and are subjective rather than precisely measurable; pain, fatigue, and nausea are examples. In a **latent infection** the host does not have signs or symptoms (they are *asymptomatic*). Finally, infections are sometimes characterized by onset and duration. **Acute** infectious diseases have a rapid onset and progression, whereas **chronic** infectious diseases have slower onset and progression.

Koch's postulates reveal the cause of some infectious diseases, but have limitations.

Koch's postulates of disease (Koch is pronounced "coke") are the four criteria used to determine what pathogen is the causative agent of a particular disease.

Learning Outcomes

After reading this section, you should be able to:

9.1 List the six different categories of pathogens.

9.2 Compare the following sets of terms: opportunistic versus true pathogen; endemic versus sporadic disease; communicable versus noncommunicable disease; and acute versus chronic disease.

9.3 Explain the differences between an epidemic and a pandemic.

9.4 State how an emerging disease differs from a reemerging disease.

9.5 Explain Koch's postulates of disease and how they are used.

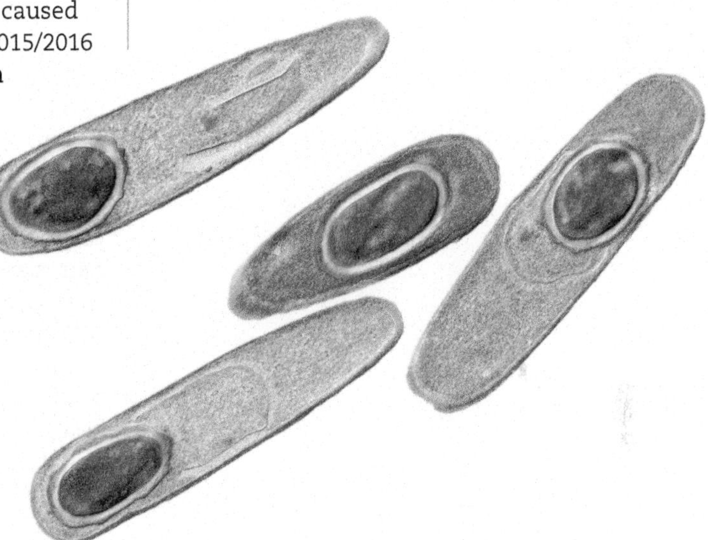

Bacillus anthracis causes anthrax and is a true pathogen.

Chapter 1 contains a brief introduction to them. To review, Koch's postulates are:

1. The same organism must be present in every case of the disease.
2. The organism must be isolated from the diseased host and grown as a pure culture.
3. The isolated organism should cause the disease in question when it is introduced (inoculated) into a **susceptible host** (a host that can develop the disease).
4. The organism must then be re-isolated from the inoculated, diseased animal.

Koch's postulates are important because they allowed us to identify the causative pathogens of many infectious diseases. However, it's important to understand their limitations. They do not apply to **noninfectious diseases** because those illnesses are not directly caused by pathogens. Even certain infectious diseases cannot be characterized using the postulates in their original form. For example, many infectious agents won't grow in the laboratory as a pure culture with existing techniques, meaning that the second postulate is unattainable. Other infectious agents called **obligate intracellular pathogens**, which include viruses and certain bacteria and protozoans, only replicate inside a host cell and are therefore impossible to grow as independent pure cultures. In still other cases, growing a microbe as a pure culture may cause it to attenuate, or lose its ability to cause disease. Attenuated pathogens would not fulfill the third postulate.

Furthermore, there are many human-specific infectious agents. In such cases, laboratory animal test subjects cannot be used because they are not susceptible hosts. For example, human immunodeficiency virus (HIV) does not cause disease in nonhuman hosts. In such instances, confirmation that an isolated agent causes disease would require inoculating a human (postulate three). Clearly, such an experiment should never be performed. In such cases, scientists may use records of accidental exposures (such as an accidental prick with a contaminated needle) to demonstrate confirmation of the third and fourth postulates.

Lastly, Koch's postulates are not easily applied to infectious agents that primarily cause latent disease. Not every infection by a microbe leads to a readily detectible disease state—for latent infections, signs and/or symptoms of illness might not emerge for years. Long periods of latency and/or a high instance of asymptomatic infections can make the third postulate difficult to fulfill—after all, science depends on replicable results, and if replicating a disease state is difficult, then linking a pathogen to a specific disease becomes challenging. As you can see, the four steps for linking an infectious agent to a specific disease may seem simple enough in theory, but in practice there are many considerations and exceptions.

Build Your Foundation

Build Your Foundation (BYF) Quick Quiz: Visit the **Mastering Microbiology** Study Area to quiz yourself.

1. Name the six different categories of pathogens. (LO 9.1)
2. What epidemiological terms describe a disease that is commonly found in a population, tends to only infect people with weak immune systems, shows sudden onset, and transmits from person to person? (LO 9.2)
3. What is an epidemic, and how is it different from a pandemic? (LO 9.3)
4. What differentiates an emerging disease from a reemerging disease? (LO 9.4)
5. List Koch's postulates of disease, and state why they are useful. (LO 9.5)
6. What step(s) in Koch's postulates of disease could prove impossible to fulfill in the case of a disease that only affects humans? (LO 9.5)

9.2 INFECTIOUS DISEASE TRANSMISSION AND STAGES

Learning Outcomes

After reading this section, you should be able to:

9.6 Define the terms *reservoir*, *source*, and *transmission*, and contrast endogenous and exogenous sources.

9.7 Describe the various direct and indirect modes of transmission, and provide examples of each.

9.8 Compare and contrast biological and mechanical disease vectors.

9.9 Explain the five stages of infectious disease, and summarize how these stages vary from one pathogen to another and the challenges that such variations present in epidemiology.

9.10 Compare chronic carriers and asymptomatic carriers.

Pathogens come from different sources, some of which are reservoirs.

The **reservoir** of an infectious agent is the animate or inanimate habitat where the pathogen is naturally found. Often a reservoir can harbor the infectious agent without damage to itself. In contrast, a **source** disseminates the agent from the reservoir to new hosts. Sometimes the reservoir and source are the same; for example, humans are the only reservoir and source for varicella-zoster virus—the chickenpox virus. Other times, the reservoir and source of a pathogen differ. For example, soil is the natural reservoir for the bacterium *Clostridium botulinum*, the causative agent of botulism. Although this bacterium has its natural habitat in soil, most people do not get botulism from eating soil; rather, the source of infection is usually a contaminated food. In epidemiology it is important to know the typical source of an infection as well as the reservoir. An **endogenous source** means the pathogen came from the host's own body. A source that is external to the host is an **exogenous source**. Examples of exogenous and endogenous sources are provided in TABLE 9.1.

Transmission is the spread of a pathogen from a source to a new host.

The *mode of transmission* is the means by which the pathogen spreads to a host. In the case of a common cold, the source may have been your sick friend, and the mode of transmission was the airborne droplets you inhaled after she sneezed. There are many modes of transmission for infectious agents, but they are generally categorized as direct contact or indirect contact transmissions (FIG. 9.1). Be aware that some pathogens can spread by direct *and* indirect means. As you read the following discussion, you may develop an impression that classifying transmission as direct or indirect is straightforward and not open to interpretation. In reality, there is no single agreed-upon classification system for direct versus indirect transmission. Here you are encouraged to use the nature of the *contact with the source* of the infectious agent as your guide.

Direct Contact Transmission

With **direct contact transmission**, the host comes into physical contact with the source of the pathogen. A source could be an animal, another human, or the environment. Examples include being bitten by a rabid animal; touching, kissing, or sexual contact among people; or taking a dip in a pond and getting swimmer's ear. In the case of swimmer's ear, the microbe naturally lives in and replicates in the water—the water is both the reservoir and the source.

Vertical transmission, a specialized form of direct contact transmission, occurs when the pathogen passes from mother to offspring during pregnancy

TABLE 9.1 Examples of Exogenous and Endogenous Sources of Infection

Exogenous Sources	Endogenous Sources
Environmental: contaminated food, medical equipment, soil, or water **Animals:** transmit zoonotic diseases to people **Humans:** transmit communicable infections from one person to another	**Misplaced normal microbiota:** Bacteria living harmlessly on skin can enter surgical incisions to cause postoperative infections. **Disrupted microbiota and opportunistic pathogens:** For example, yeast in the vagina may proliferate and cause infection after antibiotics kill off bacterial neighbors.

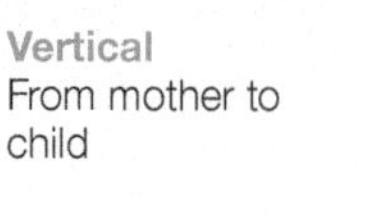
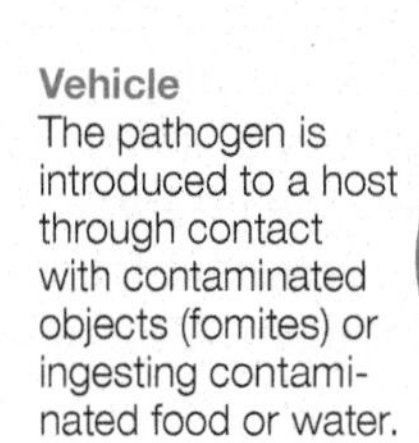
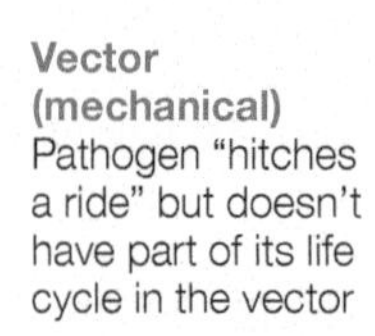

FIGURE 9.1 Infectious disease transmission modes Pathogens can be transmitted by direct or indirect contact. In some cases, a pathogen can be transmitted both ways.

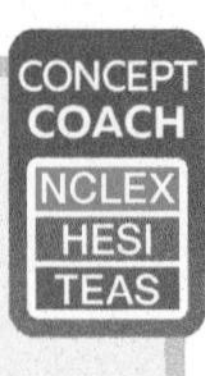

Modes of Transmission and the Stages of an Active Infection

Bring the art to life! Visit the **Mastering Microbiology** Study Area to watch the Concept Coach and master the basic science behind getting sick.

(*in utero*), during childbirth (transcervical), or through breast milk (postpartum). Any pathogen that can cross the placenta or be secreted into breast milk can potentially be spread by vertical transmission. HIV is one example.

Indirect Contact Transmission

Indirect contact transmission means a pathogen spreads without direct physical contact with the source (again, the source could be an animal, a human, or the environment). There are three main categories:

1. **Airborne transmission**: The pathogen enters through the respiratory route. This usually occurs by inhaling pathogen-laden **respiratory droplets** and/or finer **aerosols**. Most respiratory infections that are transmitted from person to person (e.g., influenza, colds, and COVID-19) can spread via respiratory droplets—larger liquid-laden particles (generally 5 μm or more in diameter) that are expelled from the mouth as people exhale, talk, cough, sing, or sneeze. Because of their size and weight, droplets tend to fall to the ground soon after being expelled and do not travel far from their source. Consequently, social distancing (maintaining a space of at least 6 feet from others) combined with wearing a mask can reduce respiratory droplet transmission.

   Compared to droplets, aerosols are extremely small (less than 5 μm in diameter) and light (due to a lower moisture content than droplets). This means they can remain suspended in the air for hours and easily travel more than 6 feet from their source. The same actions that produce droplets can expel aerosols. In addition, wind, water-based cooling systems,[1] and even

[1] Aerosols from air-conditioning cooling towers, water systems, whirlpool spas, and humidifiers can transmit Legionnaires' disease—an atypical pneumonia caused by the bacterium *Legionella pneumophila* (covered in Chapter 16).

actions like cleaning, digging, and demolishing can generate aerosols. Notably, because aerosols are so small, they tend to contain far fewer virions and/or bacteria than droplets. Diseases like measles and tuberculosis can be established at a very low exposure dose (1 virion for measles and as few as 10 *Mycobacterium tuberculosis* cells for tuberculosis) and are readily transmitted via aerosols. It is also now known that SARS-CoV-2 can be transmitted through aerosols—though, as of this text's publication, the percentage of infections attributable to aerosol transmission as compared to droplet transmission is unknown. Because aerosols carry relatively fewer agents, anything that can spread via aerosols is also likely to be transmissible via droplets, whereas the opposite is not necessarily true.

As you'll read in Chapter 10, in healthcare settings there are *droplet precautions* and then there are the more stringent *aerosol precautions*. It's important to note that aerosol precautions are often called *airborne precautions*, a name that better emphasizes that these precautions protect against *both* forms of airborne particles—droplets and aerosols.

2. **Vehicle transmission** (also called **fomite transmission**): Contaminated objects or substances (fomites) such as doorknobs, needles, sheets, food, or drinking water convey a pathogen from a source to a new host. Note, in such cases the pathogen source could be an infected human who touched the fomite and has since left. Or consider a cook who forgot to wash their hands before preparing a meal. In this case, the food the cook prepared would be the vehicle for transmitting a pathogen from the source (the cook's hands) to the new host (the diner).
3. **Vector transmission**: **Vectors** are organisms such as arthropods and rodents that spread infectious agents to other susceptible hosts. Many zoonotic diseases require a vector for transmission to human hosts.

Many arthropods such as mosquitoes and ticks act as **biological vectors**, meaning the vector organism has a role in the pathogen's life cycle. For example, the protozoan that causes malaria conducts the first part of its life cycle in mosquitoes and completes it in humans. (For more on malarial parasites, see Chapter 20.) In contrast, a **mechanical vector** spreads disease without being integral to a pathogen's life cycle. For example, a fly could pick up bacteria on its legs when it lands on garbage, then transfer those bacteria to your picnic lunch. Flies, rodents, and cockroaches are notorious mechanical vectors.

Five general stages of disease occur during infections.

Once a pathogen contacts a potential host, it aims to establish infection. In some cases, it will fail to do so. The term **infectivity** describes how good an infectious agent is at establishing an infection. Even when an infection occurs, the host may not develop disease. The general ability of an infectious agent to cause disease is described as its **pathogenicity**, whereas **virulence** describes the severity of the disease it causes. (Pathogenicity and virulence factors are discussed in Chapter 10.)

The progression of many infectious diseases follows five basic stages (FIG. 9.2): the incubation period, prodromal phase, acute phase, period of decline, and convalescent phase. These five stages are generalizations; the characteristics of each stage vary by pathogen and from one patient to another. It's possible for diseases to be transmitted to other susceptible hosts at any point in these five stages. Some infectious diseases like genital herpes can transmit to others throughout all five phases. In other diseases, such as measles and hepatitis A, the pathogen spreads most easily during the incubation and prodromal periods.

Incubation Period

The time between infection and the development of disease symptoms is called the **incubation period**. It could be hours for one disease and years for another.

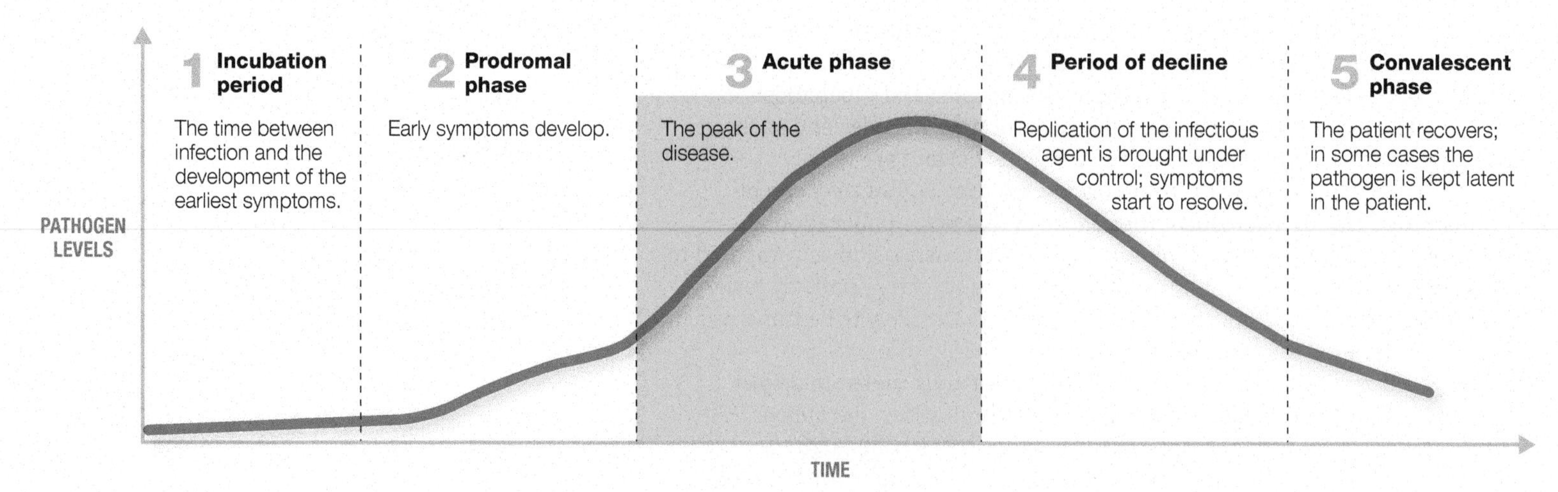

FIGURE 9.2 Five stages of infectious disease This graph is generalized; not all five phases occur in all patients. Also, the duration of each phase can vary from what is presented here.

Critical Thinking *In a subclinical (or asymptomatic) case, would you expect that there was a successful infection event? Explain your reasoning.*

CONCEPT COACH NCLEX HESI TEAS

Modes of Transmission and the Stages of an Active Infection

Bring the art to life! Visit the **Mastering Microbiology** Study Area to watch the Concept Coach and master the basic science behind getting sick.

Even the same pathogen will have a range because host factors and the nature of the exposure can influence how quickly symptoms develop following infection. Accurately determining incubation time for a newly observed infectious disease can be challenging because it may be difficult to pinpoint the exact timing of pathogen exposure. A disease like influenza, with a short incubation period of 1–4 days, is usually easier to track and monitor in a population than a disease like HIV, which has a highly variable incubation period of several months to many years. COVID-19 can be epidemiologically challenging to track because the incubation period could extend as far out as 2 weeks after the initial infection event.

TABLE 9.2 Estimated Subclinical Cases of Disease

Disease	Asymptomatic Cases by Age (in Years)
Polio	All children 99%
Mononucleosis (Epstein–Barr virus)	1–5—99% 6–15—90% 16–25—50%
Hepatitis A	Under 5—95% 5–9—90% 10–15—85% Adults—50%
COVID-19*	Pooled age-group data—16%
Rubella	5–20—50%
Influenza	20–40 (among nonvaccinated)—35%
Mumps	10–14—10%
Measles	5–20—1%
Rabies	All ages—0%

*He, J., Guo, Y., Mao, R., & Zhang, J. (2021). Proportion of asymptomatic coronavirus disease 2019: A systematic review and meta-analysis. *Journal of Medical Virology*, *93*(2), 820–830.

Based on 50,155 confirmed COVID-19 patients from 41 studies. The percentage of asymptomatic cases varies greatly by age and even by study. Based on pooled data, about 27% of confirmed COVID-19 cases in children were described as asymptomatic.

Prodromal and Acute Phases

During the **prodromal phase** the patient starts to feel run down and may have mild symptoms. After the prodromal phase, the patient could progress to an **acute phase** and experience the full-blown classic symptoms of the disease. This is known as a **symptomatic case** (or a *clinical infection*). Often someone with symptoms can transmit the agent causing their symptoms, though in some cases transmission can occur before the onset of telltale symptoms. For example, chickenpox is most easily spread about 1–2 days before the hallmark rash appears. It's not hard to imagine how tough it is to prevent the spread of a disease that is most contagious before the classic symptoms develop.

To further complicate things, some pathogens infect the host, but cause an **asymptomatic case** (also called a *subclinical case*) in which symptoms are mild or nonexistent. In these instances, the prodromal and acute phases often go unnoticed or are dismissed as a simple cold. Most mild diseases are not public health concerns unless evidence suggests that they are linked to something more dangerous. For example, sometimes a pathogen will cause a deadly disease in one host, but remain asymptomatic or mild in others. Poliovirus is an example. Most infections result in subclinical cases, but sometimes the infection progresses to cause permanent paralysis or death. Many diseases "hide" in populations by causing subclinical infections (TABLE 9.2). For example, asymptomatic cases of COVID-19 can spread the SARS-CoV-2 virus.

Period of Decline

During the **period of decline**, pathogen replication decreases, and the patient begins to feel better. Of course, not all patients experience a period of decline—some patients are killed by the pathogen. If patients do experience a decline phase, they may still be infectious even though they feel better. Sometimes people who were taking an antibiotic to treat a bacterial infection may prematurely stop taking their medication during the period of decline. In such cases, the pathogen might return in full force, causing a patient to relapse. Even worse, the pathogen may have developed antibiotic resistance as a result of the patient failing to follow the required course of treatment.

Convalescent Period

The **convalescent period** usually involves elimination of the pathogen from the body, but sometimes the host harbors a pathogen indefinitely; this is especially true for certain viruses. (See Chapter 6 for more on latent and chronic viral infections.) In such cases, the pathogen can exist in a dormant (or latent) state in the host and reactivate later. The resulting **chronic carrier** may remain asymptomatic for long periods, only to have symptoms reemerge from time to time. In some cases, even when chronic carriers are asymptomatic, they may be capable of infecting others. Human herpes simplex virus-2, which causes genital herpes, is an example of a virus that can induce a chronic carrier state. Note, viruses are not the only pathogens that can lead to a chronic carrier state; many bacteria, fungi, and parasites also take advantage of this mechanism.

Asymptomatic carriers harbor certain pathogens for extended periods without experiencing symptoms—this differentiates asymptomatic carriers from chronic carriers, who occasionally experience signs and/or symptoms of infection (e.g., episodic outbreaks of herpes lesions). The asymptomatic carrier condition can develop after an initial infection that was either symptomatic or asymptomatic. *Streptococcus pyogenes*, the cause of streptococcal pharyngitis (strep throat), *Neisseria meningitides* (the pathogen that causes meningococcal meningitis), and many other pathogens and parasites can spread through asymptomatic carriers. A classic example of an asymptomatic carrier is Mary Mallon, known as "Typhoid Mary." In the early 1900s, she was the first person identified as an asymptomatic carrier of typhoid fever in the United States (FIG. 9.3). The causative agent, *Salmonella enterica* serovar Typhi, remained in Mallon's gallbladder and served as a source of infection to those around her. Working as a cook, she unwittingly spread typhoid fever to over 50 people, at least three of whom died. Public health authorities traced outbreaks to Mallon and prohibited her from working as a cook any longer. Unfortunately, Mallon was poor and her options for earning a living were limited—so she assumed a fake name and returned to cooking. This resulted in more cases of typhoid fever, and the discovery that she was once again the source. Because she had disobeyed the order not to cook for others, she was deemed a danger to society and quarantined for life on North Brother Island in New York's East River. (We will read more about quarantines in the next section on epidemiology essentials.)

FIGURE 9.3 Typhoid Mary Mary Mallon (1869–1938) was quarantined on an island near Manhattan. She was an asymptomatic carrier of *Salmonella enterica* Typhi.

Build Your Foundation

7. What do the terms *reservoir*, *source*, and *transmission* mean, and how do exogenous and endogenous pathogen sources differ? (LO 9.6)
8. What are the various modes of direct and indirect contact transmission? Be sure to give examples of each. (LO 9.7)
9. How do mechanical and biological vectors differ? (LO 9.8)
10. Name the five stages of infectious disease, and state their features. (LO 9.9)
11. Why would a long incubation period be a particular challenge to epidemiologists? (LO 9.9)
12. How do chronic carriers differ from asymptomatic carriers? (LO 9.10)

Build Your Foundation (BYF) Quick Quiz: Visit the **Mastering Microbiology** Study Area to quiz yourself.

9.3 EPIDEMIOLOGY ESSENTIALS

Learning Outcomes

After reading this section, you should be able to:

9.11 Define the term *epidemiology*, and state its two primary goals.

9.12 Give examples of host and environmental factors that form an epidemiological triangle, and provide several examples of how the epidemiological triangle can be broken.

9.13 Define the term *quarantine*, and explain when it may be an effective tool to limit disease.

9.14 In your own words, describe at least three roles of public health.

Now that we've discussed some basic principles of infectious disease in individuals, we can discuss how these factors relate to disease in populations. The word **epidemiology** literally means the "study of what is upon the people." Hippocrates, a physician in ancient Greece, is considered the father of Western medicine and the first epidemiologist. He observed that malaria and yellow fever mainly occurred in warm, swampy areas. He also noted that diet and personal habits influenced health, that some diseases occurred in seasonal cycles, and that certain sources of water were linked to disease. Hippocrates appreciated that individual patient treatment improves when one understands how disease spreads in populations, and what is generally "upon the people" when a particular case of illness arises.

Epidemiology focuses specifically on diseases (infectious and noninfectious) in populations. It is sometimes referred to as *population medicine* because, unlike general medicine, which treats individuals, epidemiology aims to understand and prevent illness in communities. The two general goals of epidemiology are to: (1) describe the nature, cause, and extent of new or existing diseases in populations and (2) intervene to protect and improve health in populations.

The epidemiological triangle links host, etiological agent, and environment.

The epidemiological triangle represents the who, what, and where of a particular disease: the host, etiological agent, and environment (FIG. 9.4). We've already discussed how Koch's postulates of disease can help determine the **etiological agent** (causative agent) of an infectious disease. However, in some cases knowing the host and environmental factors that lead to the disease may be even more important to saving lives than knowing the etiological agent, especially if the disease cannot be cured. In this book we discuss only infectious etiological agents of disease, but the epidemiological triangle is equally useful for describing noninfectious diseases. Noninfectious etiological agents may include chemicals, physical factors, a dietary insufficiency, and even social factors.

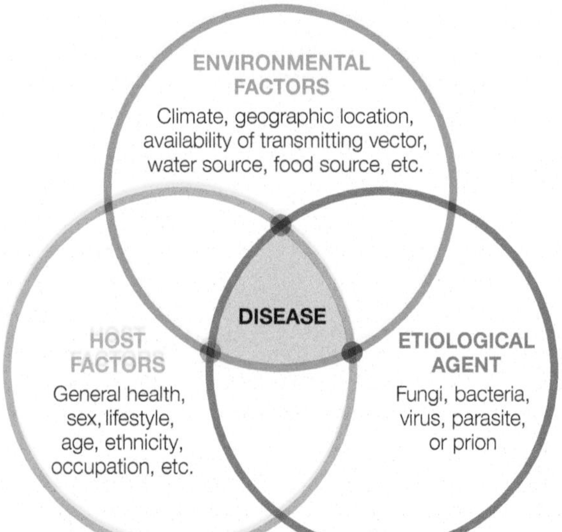

Many pathogens have a limited *host range*—the type of host (or cell within a host) they can infect. Because pathogens continually evolve, the host range can change over time. For example, an influenza virus may start out infecting only birds or pigs, but later mutate into a strain that can also infect people. Certain host factors can affect how a disease progresses or if the disease can even be established. A person's overall health, age, nutrition, and underlying medical conditions all influence risk for infection, as well as the *prognosis*, or likely outcome if disease develops. If the host is immune compromised, they are at risk for a greater number of infections overall, including those caused by opportunistic pathogens. Unfortunately, an infectious disease that is barely noticeable in a normally healthy host could kill an immune-compromised host. Preventing disease in certain populations such as babies, the elderly, or even organ transplant recipients (who take immunosuppressive drugs to decrease the risk of rejecting the transplant) is especially important.

Some factors that influence health are not within an individual's control, but certain personal habits and activities can lower the risk of infectious disease. Washing hands frequently cuts down on many infections. So does avoidance of intravenous drug use, not walking barefoot in mud, and not drinking contaminated water. Understanding risk factors for a particular infectious disease allows health officials to recommend preventive behaviors. Preventing a

ENVIRONMENTAL FACTORS
Climate, geographic location, availability of transmitting vector, water source, food source, etc.

DISEASE

HOST FACTORS
General health, sex, lifestyle, age, ethnicity, occupation, etc.

ETIOLOGICAL AGENT
Fungi, bacteria, virus, parasite, or prion

FIGURE 9.4 **Epidemiological triangle** When host, agent, and environment overlap, there is the possibility of disease; often removing one member of the epidemiological triangle is enough to prevent disease.

disease is always better than having to treat a disease. Despite our significant medical advances, not all infectious diseases are curable.

Finally, environmental factors are central to epidemiology. In general, disease can be established whenever an environment that allows an etiological agent to thrive overlaps with an environment in which a susceptible host lives. Epidemiologists are keenly interested in preventing such overlaps.

Public health strategies can break the epidemiological triangle.

In many cases, taking away one of the links in the epidemiological triangle is enough to prevent disease spread. Some basic public health strategies to target disease include education, quarantine, and vector control.

Public Education

Accurate public education is central to breaking the epidemiological triangle. For example, among the Maasai tribes of East Africa, an unusually high number of newborns used to die of tetanus. Researchers realized that this was partly due to a Maasai cultural tradition of coating the baby's umbilical stump with cow manure as a symbol of the tribe's connection to cattle herding. Once the Maasai people were educated about how tetanus was being transmitted, they worked with researchers to modify certain cultural celebrations and birthing practices, and the cases decreased. Other common examples of public health education include vaccination campaigns, promotion of prenatal care and breastfeeding, and programs to prevent sexually transmitted diseases.

Public health education can also limit disease by helping people understand their role in preventing development of drug-resistant pathogens. According to the Centers for Disease Control and Prevention (CDC), about two million people per year in the United States become infected with antibiotic-resistant bacteria. Drug resistance occurs naturally over time, as pathogens encounter our drugs. This selective pressure favors the survival of pathogens that have evolved strategies to escape a given antimicrobial drug. However, through public health education, we can decrease the rate of drug-resistance development by decreasing overuse of antibiotics. At least 30 percent of antibiotics prescriptions issued in the United States are unnecessary—for instance, those prescribed when a patient complained about cold symptoms, which are generally caused by viruses, not the bacteria that antibiotics treat.[2] In fact, the issue of antibiotic resistance is so concerning that specific areas of epidemiology are dedicated to monitoring and preventing antibiotic resistance of certain pathogens. (Mechanisms of drug resistance are discussed more in Chapter 15, on antimicrobial drugs.)

Quarantine

Some infectious diseases with short incubation times can be controlled with **quarantine**, a period of confinement away from the general population for a

[2] Shively, N. R., Buehrle, D. J., Clancy, C. J., & Decker, B. K. (2018). Prevalence of inappropriate antibiotic prescribing in primary care clinics within a Veterans Affairs health care system. *Antimicrobial Agents and Chemotherapy*, 62(8), e00337-18. doi:10.1128/AAC.00337-18; Fleming-Dutra, K. E., Hersh, A. L., Shapiro, D. J., et al. (2016). Prevalence of inappropriate antibiotic prescriptions among US ambulatory care visits, 2010–2011. *Journal of the American Medical Association*, 315(17), 1864–1873.

Animal Quarantine

Many existing and emerging diseases are zoonotic. To prevent their spread, animals that enter the United States are subject to inspection and possible quarantine and can also be denied entry. Because Hawaii is the only U.S. state that does not have endemic rabies, bringing a pet to Hawaii from the U.S. mainland or from another country is strictly monitored (and certain pets are not allowed to enter).

For animals that can develop rabies, proof of a rabies vaccine being administered within the last 30 days is usually a *minimum* requirement before entering any state in the United States. (Hawaii has very strict rules, and not following them can lead to a costly 120-day animal quarantine.) Importation regulations also apply to service dogs and even to pets that were taken out of the United States and are returning.

The animals that are regulated by the CDC are dogs, cats, bats, civets, African rodents, birds from certain countries, turtles, and monkeys. If you plan to travel with your pet, it is advisable to check the requirements at least 6 months before you plan your trip so your pet is not subject to extended quarantine.

QUESTION 9.1

Why do you think that specific animals are subject to importation regulations by the CDC, whereas others, like horses, are not?

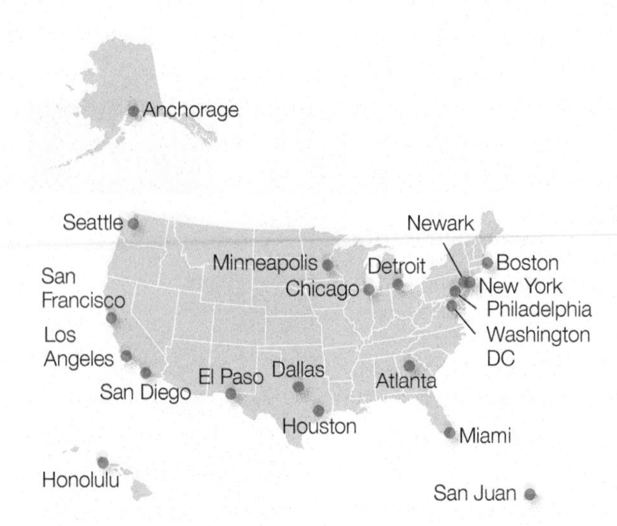

FIGURE 9.5 U.S. quarantine locations In the United States there are 20 quarantine stations situated at airports and land-border crossings where 85 percent of international travelers arrive.

Critical Thinking *Why is diphtheria on the quarantinable disease list when it is preventable by routine childhood vaccination?*

person or animal that may be infected. If the typical incubation period passes without symptoms arising, the person or animal is released. If disease develops, the quarantine can be maintained throughout the duration of illness, to prevent further transmissions. Quarantines are a useful tool. However, if the pathogen in question has a long incubation period, quarantine may be an impractical or inhumane way to prevent disease spread.

The CDC can quarantine any person entering or already in the United States who may have an infectious disease that is deemed quarantinable. There are currently nine diseases/disease categories on the U.S. quarantinable disease list: cholera, diphtheria, infectious tuberculosis, plague, smallpox, yellow fever, viral hemorrhagic fevers, influenza strains that could cause pandemics, and severe acute respiratory syndromes (viral infections that currently include severe acute respiratory syndrome caused by SARS-CoV-1, Middle East respiratory syndrome caused by MERS-CoV, and COVID-19 caused by SARS-CoV-2). Infectious agents can be added to or removed from the list by presidential executive order on recommendation by the Health and Human Services Secretary (FIG. 9.5). Many countries have quarantine laws related to diseases of particular concern. For example, Australia maintains very strict animal quarantine measures to prevent rabies, which is not endemic to the island continent, from gaining a foothold there.

Vector Control

Fleas, mosquitoes, and ticks are infamous for spreading parasitic, viral, and bacterial diseases. Arboviruses (arthropod-borne viruses) such as West Nile virus, yellow fever, and dengue fever, as well as bacterial diseases such as plague and Lyme disease, and the protozoan illness malaria, are all spread by arthropods. Limiting the numbers of biological vectors in an environment helps prevent the spread of many dangerous infectious diseases.

Mosquito Control Efforts

Mosquito control, which is usually organized and paid for by local governments, typically entails aerial spraying and ground fogging of pesticides that limit mosquito vectors in a community. If you live in a mosquito-prone area of the country, you may have seen a fogging truck. Larvicide pellets containing chemicals such as methoprene and biopesticides (natural pest control agents) containing the bacterium *Bacillus thuringiensis* can also be used to control mosquito populations by poisoning mosquito larvae without harming other wildlife.

Media campaigns may also urge residents to minimize standing water sources such as buckets, children's swimming pools, or birdbaths because mosquitoes use these locations to breed and grow. In some communities, governments may also provide mosquito-larvae-eating fish to residents who have ponds on their property.

QUESTION 9.2

Describe three specific things that you can do to protect yourself from vectorborne illnesses.

Public health aims to improve overall health in a population.

The U.S. public health system investigates, diagnoses, prevents, and works to reduce health problems in the community. This entails monitoring within the population to identify problems, developing policies and plans that support good community health, ensuring a competent public and personal healthcare workforce, and mobilizing health resources (including public education) to meet needs. Many public health goals could not be fulfilled if the nature, extent, and cause of a disease were ignored. Therefore, epidemiology is a necessary partner with public health.

You see the role of public health in almost every aspect of life in the United States, from the "Employees must wash hands" signs in restaurant restrooms, to vaccination mandates for children enrolling in most publicly funded schools, to quarantine, travel restrictions, and mask-wearing mandates. These efforts are usually overseen by local, regional, and state agencies following national guidelines. In the United States, the **Centers for Disease Control and Prevention (CDC)** is a federal health agency that falls under the U.S. Department of Health and Human Services (HHS). One major role of the CDC is to serve as a central source of epidemiological information and public health recommendations. Another critical agency of the HHS is the National Institutes of Health (NIH), which is among the world's foremost medical research centers. The CDC and NIH are just two examples of U.S. public health agencies. Worldwide, approximately 40 countries also have established healthcare systems and public health agencies.

Build Your Foundation

13. What is epidemiology, and what are its two key goals? (LO 9.11)
14. What does the epidemiological triangle describe, and how can it be broken to limit disease? (LO 9.12)
15. What is quarantine, and when might it be a useful control measure? (LO 9.13)
16. Name a few roles of public health, and explain why public health and epidemiology are necessary partners. (LO 9.14)

Build Your Foundation (BYF) Quick Quiz: Visit the **Mastering Microbiology** Study Area to quiz yourself.

9.4 EPIDEMIOLOGICAL MEASURES AND STUDIES

Learning Outcomes

After reading this section, you should be able to:

9.15 Explain why epidemiological measures are useful.

9.16 Describe how rates, proportions, and ratios may be used to describe disease.

9.17 Describe the difference between measures of frequency and measures of association.

9.18 Calculate the prevalence rate and incidence rate of a given disease, and explain what factors impact prevalence rates, and why.

9.19 Define the terms *population*, *morbidity*, and *mortality*.

9.20 Define the term *correlation*, and state why correlation is not necessarily evidence of causation.

9.21 Give examples of descriptive and analytical epidemiology study designs, and explain when each type of study is useful.

Epidemiological measures are often presented as ratios, proportions, or rates.

To determine the impact and risk factors of an infectious disease, epidemiologists rely on numerical measures. We hear versions of these all the time. When the media reports "1 in 4 people currently has a sexually transmitted disease," or "6 out of 10 deaths due to stroke are in women," these statistics are the result of epidemiological data collection and study. The measures often present data as ratios, proportions, or rates. A **rate** is used to measure the occurrence of an event over time. A **ratio** presents the occurrence of an event or condition in one group as compared to another group (in the United States in people age 65 and older, the ratio of males to females is 77:100—read as 77 to 100). A **proportion** is a percentage of a whole (80 out of 100 women with gonorrhea do not have symptoms, or 80 percent of women with gonorrhea do not have symptoms). Measuring specific aspects of health provides insight as to where public health campaigns or prevention efforts may be needed most.

Measures of frequency include disease prevalence and incidence data.

Measures of frequency give information about the occurrence of a disease in a population during a certain period of time—they reveal the degree of **morbidity** (existence of disease) in a population. A **population** is any defined group of people; for example, a population could be nonsmoking women over age 50 in the United States, or all students taking microbiology classes this semester. In order to appreciate how a certain disease affects a population, we try to learn how many cases exist in that population at a given time. This measure is known as the **prevalence rate** (sometimes just called prevalence). To calculate the prevalence rate, you simply take the total number of disease cases during a given time and divide it by the total number of people in the defined population during that same time.

$$\text{Prevalence Rate} = \frac{\text{Total cases in a defined population during a given time}}{\text{Total people in the defined population during the same time}}$$

When interpreting prevalence statistics, it is important to understand the characteristics of the population being described. For example, is the prevalence being reported for all people living in the United States, or for infants, the elderly, low-income families, smokers, or adults who are obese? Such factors can influence susceptibility to certain diseases; for example, the prevalence of a disease may be high among people over age 65, but rare in other age groups. Furthermore, prevalence is not a measure of how dangerous a disease is. A very dangerous disease could have a low prevalence rate (like Ebola).

Two main factors impact prevalence rate: the **incidence rate** of a disease (the number of *new* cases in a defined population during a defined time frame)

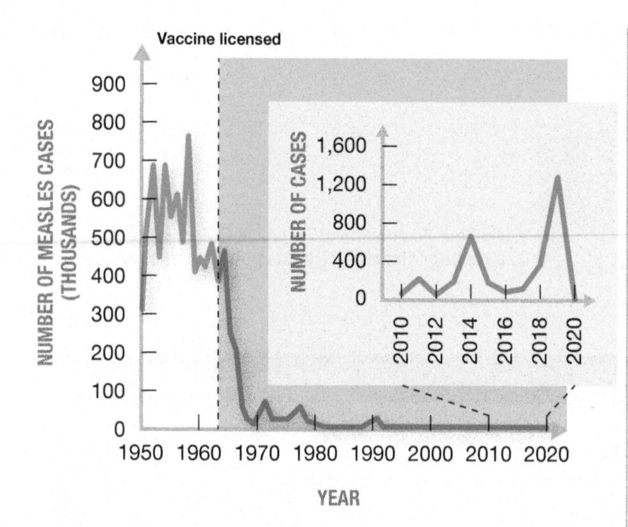

FIGURE 9.6 Number of U.S. measles cases from 1950–2019 If a preventive measure works, the disease incidence will decrease after the preventive measure begins. The incidence of measles greatly decreased once children were vaccinated, suggesting it was an effective preventive measure. The periodic outbreaks occurring after 1970 have been linked to decreased vaccination rates in children and failure to have booster shots.

(Data from: Hamborsky, J., & Kroger, A. (Eds.). (2015). *Epidemiology and prevention of vaccine-preventable diseases, The Pink Book*. Public Health Foundation, CDC.)

Critical Thinking *Based on this incidence graph, when do you think measles was the most prevalent? Explain your reasoning.*

TABLE 9.3 Mortality Rates

Term	Definition
Crude mortality rate	General death rate in a population, not determined based on specific causes of death
Cause-specific mortality rate	Deaths due to a specific cause, in a given population, during a specified period
Infant mortality rate	Death rate of children under age 1, as compared to number of live births
Maternal mortality rate	Maternal deaths per 100,000 live births, from any cause related to pregnancy or management of pregnancy
Case fatality rate	Percentage of people with a particular diagnosis who die in a specified time period after diagnosis

and the **duration** of the disease (how long the infection lasts). A disease that lasts a long time and has many new cases will also have a high prevalence.

To calculate incidence rate, the total number of *new* cases in a population during a specified time is divided by the total number of people in the population who are susceptible hosts.

$$\text{Incidence Rate} = \frac{\text{Total new cases in a defined population during a specified time}}{\text{Total susceptible hosts in the population during the same time}}$$

The incidence rate alone is a useful measure; if a preventive measure such as a vaccine works, then you'd expect the disease incidence to decrease during the period after the preventive measure begins (FIG. 9.6).

It can be challenging to accurately determine incidence rate for diseases with a high percentage of subclinical (asymptomatic) cases. For example, many people who have HIV do not know it. A pathogen that mutates to become more infective and/or virulent may lead to a higher incidence of disease. (Factors that affect pathogen infectivity and virulence are discussed in Chapter 10.) This remains a key concern with SARS-CoV-2; the evolution of more easily transmissible strains is bound to trigger surges in cases. This is also, in part, why epidemiologists are interested in monitoring the number of new influenza cases: A jump in incidence could point to a new strain of influenza. In order to decrease the burden of a disease, it helps to understand what risk factors are associated with the disease. This is where measures of association are helpful.

Measures of association may reveal risk factors for a disease.

As incidence and prevalence data are collected, **measures of association** can be determined. Measures of association tell us what factors may be linked with cases of the disease, and this in turn shows who might be at risk for developing it. For example, *Vibrio* gastroenteritis is associated with eating raw oysters. Knowing disease associations allows public health officials to warn people about how to avoid certain infections. The data also help healthcare providers arrive at the correct diagnosis more quickly for individual patients, even when signs or symptoms are rather general. For instance, when people living in certain geographic areas complain of joint pain and fatigue, a treating doctor will ask if the patient spends much time outdoors, or has noticed any tick bites recently. This line of questioning is not random; it comes from data that strongly associates tick bites with Lyme disease.

Epidemiologists analyze data in a variety of ways to uncover potential trends. For example, graphing the number of disease cases and comparing it to the number of sexual partners people with the disease had could show if the disease may be sexually transmitted. Generally speaking, analyzing data by ethnicity, age, or other host factors may help public health researchers identify interventions that may directly benefit populations suffering the greatest burden of a given disease (FIG. 9.7). Similarly, organizing data based on the geographical location of cases could point to a shared environmental factor linked to disease.

The most common association measure is **mortality rate**, the number of deaths during a specific time period. Several types of mortality rates are described in TABLE 9.3; all help us understand the association between a disease or a particular occurrence and death.

Correlation does not imply causation.

Looking at associations can uncover disease risk factors. But the fact that two things are related to each other—for example, because they occur around the

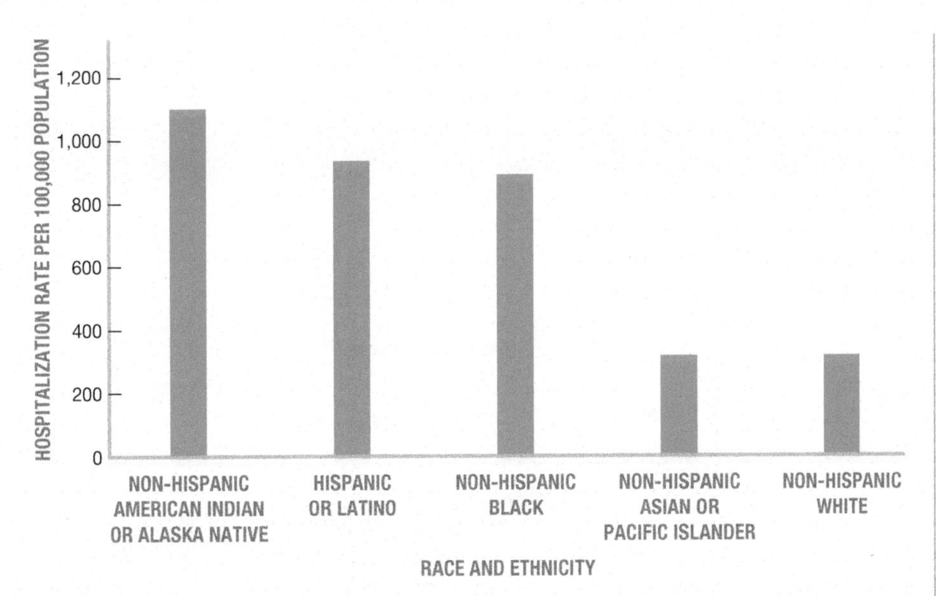

FIGURE 9.7 **Age-adjusted COVID-19-associated hospitalization rates by race/ethnicity** The data in this graph are from the CDC's COVID-NET (March 1, 2020-April 10, 2021). Hospitalization rates were statistically adjusted to account for differences in age distributions within race/ethnicity groups. For the time period these data span, race and ethnicity data were reported to COVID-NET for about 98.5% of COVID-19-related hospitalizations. The CDC data reveal that several minority populations disproportionately experience higher rates of severe COVID-19 disease that requires hospitalization.

same time or place—does not mean that one of the things *causes* the other. This is the meaning of the cautionary phrase "correlation does not imply causation." Unfortunately, in the middle of a disease outbreak, it can be easy to confuse correlation and causation. For example, during the late 1940s, public health data showed increased ice cream consumption in the United States during the same period as a polio outbreak. It seems silly today, but at the time, some people were convinced that ice cream consumption was somehow linked to the polio cases. In actuality, the polio outbreak occurred during the summer, a period during which ice cream consumption usually increases. The peaks in polio cases and ice cream consumption were correlated by their time of occurrence, but there was no causal relationship between them. A similar situation occurred in the 1990s, when the overlap in timing of the emergence of autism symptoms in early childhood and the administration of childhood immunizations led to unprofessional and unethical research suggesting a causal relationship between the two things. Unfortunately, the incorrect association between autism and vaccines spread like wildfire and has persisted. Even today, some parents avoid or delay vaccination of their children based on false understanding of the data. This behavior, in turn, has directly promoted the reemergence of vaccine-preventable diseases in several populations.

Epidemiological studies can be categorized as either descriptive or analytical.

Epidemiologists use several approaches for research. **Descriptive epidemiology** uncovers *who* is infected, *where* cases occur, and *when* cases occur. The goal is to describe the occurrence and distribution of disease so that hypotheses related to causes, prevention, or treatment can be developed and tested. Data collected from correlation studies, case reports, and cross-sectional studies—described next—are reviewed from numerous angles to pull out possible trends.

TRAINING TOMORROW'S HEALTH TEAM

Graphing Inequality in Access to Healthcare

This graph makes some people uncomfortable for two reasons—neither of which are the *right* reasons. First, some people mistakenly assume the data reflect differences in sexual behavior. They do not. When the data are disaggregated to consider sexual behavior patterns across groups (e.g., same number of sexual partners), the seroprevalence for HSV-2 among non-Hispanic Black people remains disproportionately high. Second, some people are uncomfortable with discussing STIs (sexually transmitted infections) because they are still wrongfully stigmatized. If the data here were about COVID-19 or breast cancer mortality (other scenarios that disproportionality burden minorities), the reactions would likely be different. So let's analyze this graph as a small step toward recognizing systemic racism and destigmatizing STIs.

The graph reveals that, as compared to the two other groups represented, non-Hispanic Black people suffer the highest burden of HSV-2 infection. The graph tells us nothing, however, about the sexual behavior of non-Hispanic Black people. Indeed, other CDC data reveal that a non-Hispanic Black person who has had 2–4 sexual partners is more than three times as likely to suffer from HSV-2 as compared to members of other groups with the same number of partners. In fact, a non-Hispanic White person who has had *10 or more sexual partners* is less likely to suffer from HSV-2 than a non-Hispanic Black person with only 2–4 partners.

What *should* make people uncomfortable is that these data are evidence of systemic racism in our healthcare system—especially in access of non-Hispanic Blacks to diagnosis and treatment. Groups with reduced access to healthcare are less likely to receive preventive health education and, if they become infected, to receive a diagnosis and treatment. Several drugs can reduce the spread of HSV-2 (e.g., Valtrex [valacyclovir] has been available since 1995), but reduced access to healthcare means reduced access to these drugs. Also, if patients don't know they are infected, they may not take necessary measures to reduce transmission to others and are more likely to see HSV-2 become increasingly burdensome in their communities. In short, the HSV-2 seroprevalence data in this graph underscore the healthcare access gaps that Black populations face. And the data suggest those gaps are larger for non-Hispanic Black people as compared to certain Hispanic groups (e.g., Mexican Americans).

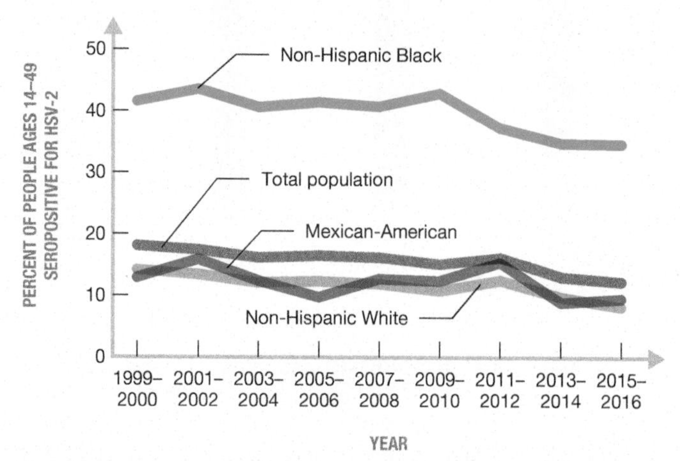

Percentage of people ages 14–49 who are seropositive for herpes simplex 2 virus (HSV-2). The data are disaggregated by race/Hispanic ethnicity. Total population includes all races and ethnicities in the United States, including those not shown separately. Data for the Asian subpopulation are only available since 2011, so this subpopulation is not shown separately, but included in the total population. (Data presented in: Centers for Disease Control and Prevention. (2018). *NCHS Data Brief No. 304,* February 2018.)

Inequality similarly affects cancer mortality. According to the American Cancer Society, Black people have the shortest survival time and highest death rate of any racial or ethnic group for most cancers.

According to the CDC, "Even when health care is readily available to racial and ethnic minority populations, fear and distrust of health care institutions can negatively affect the health care-seeking experience. Social and cultural discrimination, language barriers, provider bias, or the perception that these may exist, likely discourage some people from seeking care. Moreover, the quality of care can differ substantially for minority patients."

This graph is an important conversation starter. Future healthcare providers must be able to discuss scenarios that may feel uncomfortable and learn from those conversations.

QUESTION 9.3

Provide examples of how healthcare inequality could lead to a high burden of HSV-2 in minority populations in the United States.

Once descriptive epidemiology yields a testable hypothesis, analytical techniques are used to test its validity. **Analytical epidemiology** investigates *what* caused the disease (etiological agent), *why* people get the disease, and *how* the disease can be prevented or treated.

Descriptive Epidemiology: Correlation Studies, Case Reports, and Cross-Sectional Studies

Descriptive epidemiology helps characterize health problems, identify at-risk populations, and provide a map for allocating resources. However, it does not reveal the cause of the disease. Descriptive epidemiology includes correlation studies, case reports, and cross-sectional studies. This type of research is usually cheaper and less time consuming than analytical techniques.

Correlation studies (also called ecological studies) search for possible associations between an exposure and the development of a disease in a population. A correlation study might examine the association between fluoridated water and the number of dental visits for tooth decay. **Case reports** are individual or group records of a disease. They are an important connection between clinical medicine and epidemiology. The value of these reports is demonstrated in the epidemiological studies that eventually led to the discovery of the causative agent of acquired immune deficiency syndrome (AIDS). Case reports spanning October 1980 through May 1981 revealed five previously healthy homosexual men, all close in age and all living in the Los Angeles area, who developed pneumocystis pneumonia (PCP). The shared features of the cases pointed to the possibility of a shared etiological agent. Analytical studies were then launched to reveal the agent as a new virus, which we now know as human immunodeficiency virus (HIV).

Cross-sectional studies, also called prevalence studies, evaluate exposure to a possible hazard (e.g., infection with a particular pathogen, exposure to a specific environmental factor, etc.) and the development of disease (e.g., development of cancer in people exposed to asbestos) across a defined population at a single point in time. These studies often use surveys to determine prevalence of a disease. Because these studies determine the prevalence of a risk factor and the prevalence of disease at the same time, it is sometimes hard to determine if the exposure to the risk preceded the disease. Links between lung cancer and smoking were first suggested by cross-sectional studies and later confirmed by analytical approaches. Cross-sectional studies also revealed the association between certain human papilloma viruses (HPVs) and cervical cancer (FIG. 9.8).

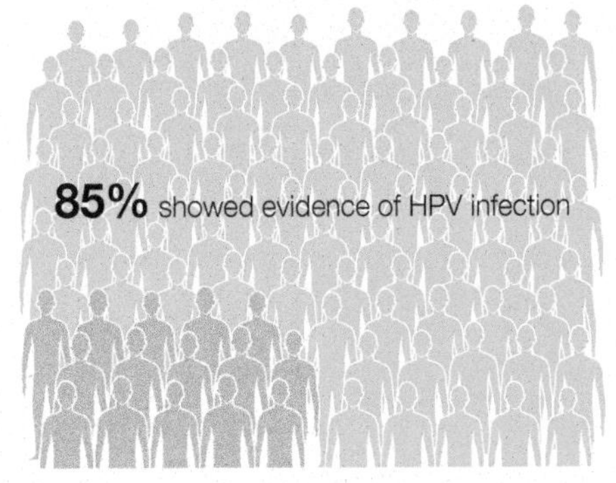

FIGURE 9.8 **Associations between human papilloma virus (HPV) and cervical cancer** In the largest cross-sectional study of HPV and cervical cancer, 10,575 cases of invasive cervical cancer were analyzed. Eighty-five percent of the samples were positive for HPV DNA. The two most commonly identified HPV types in samples were HPV 16 and 18. These data, along with other studies, were helpful in deciding which HPV strains to vaccinate against for preventing cervical cancer.

Analytical Epidemiology: Observational and Experimental Studies

Analytical studies tend to be more challenging and expensive than descriptive studies. This is because they are designed to test the validity of hypotheses that come out of descriptive studies and therefore often require expensive lab equipment and/or long-term patient data tracking. There are two types of analytical studies.

Observational studies do not involve administering an intervention, a treatment, or an exposure for a disease; they simply watch and collect data on cases that exist, used to exist, or develop over time in a tracked/monitored group—often a large population. The researchers do not expose the participants to the potential risk factor; they just develop the study groups based on the existence of the risk factor (example: groups of smokers versus nonsmokers). These studies are designed to directly measure risk for developing a certain disease, such as lung cancer, when certain risk factors, such as smoking, are present.

Experimental studies allow the researcher to change variables (factors, traits, or conditions) and determine the effect of the change on the outcome. In contrast to observational studies, experimental studies are used to determine the effectiveness of a treatment or preventive measure (drug, vaccine, etc.). Experimental study designs include clinical trials. An important feature of experimental studies is their use of strategies that can reduce biased interpretations. For example, if a drug or vaccine is being tested for efficacy, the study may be "double-blinded," meaning neither the patients nor the care provider knows if the patient enrolled in the study is receiving the intervention or a placebo (something that seems like a real medical intervention, but in reality has no pharmacological properties—e.g., a saline injection versus a trial vaccine injection). This strategy allows the researchers who analyze the data to distinguish between effects that are truly attributable to the experimental intervention and effects that could develop as a result of the patient and/or healthcare professional having certain expectations for the intervention.

Build Your Foundation

17. Why are epidemiological measures useful? (LO 9.15)
18. What would be the best way to express the data (rate, proportion, or ratio) of heterosexual males who have HIV compared with all heterosexual males? (LO 9.16)
19. How do measures of frequency differ from measures of association? Give examples of each type of measure. (LO 9.17)
20. If there were 600 cases of a disease in a given year and 10,000 people in a susceptible host population in that same year, what is the prevalence rate? What factors might impact the prevalence rate of a disease, and why? (LO 9.18)
21. What is the incidence rate for a disease that has three new cases per year out of 100 in the at-risk population? (LO 9.18)
22. Define the terms *population*, *morbidity*, and *mortality*. (LO 9.19)
23. What does *correlation* mean, and why is correlation not equivalent to causation? (LO 9.20)
24. What study design would you recommend to test the validity of the statement, "An apple a day keeps the doctor away"? Explain your reasoning. (LO 9.21)

Build Your Foundation (BYF) Quick Quiz: Visit the **Mastering Microbiology** Study Area to quiz yourself.

9.5 EPIDEMIOLOGY IN CLINICAL SETTINGS

Learning Outcomes

After reading this section, you should be able to:

9.22 Describe the historical context of hospital epidemiology, and describe why healthcare epidemiology is essential to quality care.

9.23 Characterize healthcare-acquired infections (HAIs), and discuss how they may be prevented.

9.24 List the most common sources and classes of HAIs, and give examples of pathogens that cause them.

9.25 State why healthcare settings are hot zones for antibiotic-resistant pathogens and the role of epidemiology in managing such resistance.

In the 1840s, Ignaz Semmelweis first showed that hand washing prevents infectious disease.

Florence Nightingale, the founder of modern nursing, said, ". . . the very first requirement in a hospital [is] that it should do the sick no harm." For thousands of years, in addition to being centers for healing, hospitals have been inadvertent sources of infection. Limiting infections in hospitals is an important role of healthcare epidemiology.

One of the earliest hospital epidemiology studies occurred in Vienna in the 1840s. Dr. Ignaz Semmelweis worked at two maternity wards. He noticed that the one staffed by midwives had significantly lower maternal death rates from puerperal sepsis (childbirth fever) than the ward served by physicians and medical students (FIG. 9.9). In those days, physicians and students performed autopsies and then went directly to the maternity ward to deliver babies—without washing up. By contrast, the midwives never performed autopsies. As a precaution, Semmelweis started washing his hands with a liquid chlorine bleach solution before examining patients. This significantly reduced childbed fever cases, and for the 2 years that this practice was in place at his hospital, there was no difference between the maternal death rates of the two wards. Sadly, Semmelweis's work was not universally appreciated in his lifetime. It was not until the 1890s, when the germ theory of disease finally took root, that people started to realize the importance of **aseptic** (or germ-free) techniques to prevent **healthcare-acquired infections** (HAIs)[3], diseases that develop from a healthcare intervention. Also known as *nosocomial infections*, they may occur at hospital or long-term care facilities, at outpatient care facilities, or even at home, from professionally administered home health care.

[3] The terms healthcare-associated infection and healthcare-acquired infection are often used interchangeably.

Healthcare-acquired infections are dangerous, expensive, and an increasing problem.

Hospital epidemiology involves the surveillance, prevention, and control of HAIs. These activities are critical because, despite shorter average hospital stays and better infection control and prevention measures, HAI incidence is still problematic—between 5 and 10 percent of all acute care patients contract at least one HAI during their stay.[4] According to the CDC, on a typical day, 1 in every 31 patients hospitalized in the United States develops an HAI, and about 75,000 people die from HAIs every year. HAIs are estimated to add at least an extra $10 billion dollars per year to U.S. healthcare costs, with the majority of those costs falling directly on patients and the healthcare facility.[5]

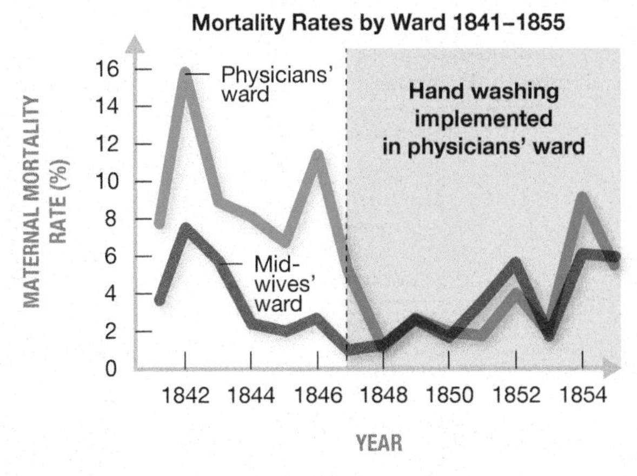

FIGURE 9.9 Semmelweis's hand hygiene intervention Until Semmelweis implemented hand washing in the physician ward, the midwife ward consistently had lower mortality rates due to puerperal sepsis than did the physician ward.

Critical Thinking *Even though the mortality rate due to puerperal sepsis in the midwife ward was consistently lower than that seen in the ward staffed by the physicians and medical students, there are still occasional peaks in cases in both wards. Provide a theory for these occasional peaks in the ward attended by the midwives and why these peaks roughly mirror what occurs in the physician-tended ward.*

Common HAI Sources and Pathogens

HAIs transmit by direct and indirect contact. Contaminated medical devices and healthcare workers' hands are the most common sources. Medical devices can either be directly contaminated with an infectious agent, or they can serve as a way for a pathogen to enter the body without encountering many of the natural barriers that otherwise would have to be overcome (such as getting through the skin). Furthermore, medical devices inside a patient (such as a catheter or central line) can serve as a breeding ground for infections. (Biofilms are discussed more in Chapter 10.) Hospital personnel can also inadvertently spread infectious agents, which is why hand washing between patients is so important.

HAIs can be **localized** (restricted to a specific part of the body) or **systemic** (spread throughout the body), and they can affect any body system or tissue. According to HAI tracking data from the CDC, the most common HAIs are *Clostridioides difficile*[6] gastrointestinal infections; urinary tract infections associated with catheterization of the urethra; surgical wound infections; methicillin-resistant *Staphylococcus aureus* bloodstream infections (MRSA bacteremia); intravascular device–related bloodstream infections (usually associated with a central line/catheter in a vein); and ventilator-associated events (respiratory infections such as pneumonia that can develop during mechanical ventilation) (FIG. 9.10).[7]

Bacteria are the most common cause of HAIs. Some of the most common bacterial pathogens associated with HAIs are *Staphylococcus aureus*, *Escherichia coli*, *Clostridioides difficile*, and *Pseudomonas aeruginosa*. All of these bacteria are demonstrating increased antibiotic resistance, which means they pose an even greater risk. Examples of key pathogens and HAIs are presented in TABLE 9.4.

Role of Healthcare Epidemiology Programs

Healthcare epidemiology programs reduce transmission of infectious diseases in the healthcare setting through surveillance, prevention, and control programs. It is estimated that up to 55 percent of HAIs are preventable; therefore, healthcare epidemiology programs have a real opportunity to save lives.[8]

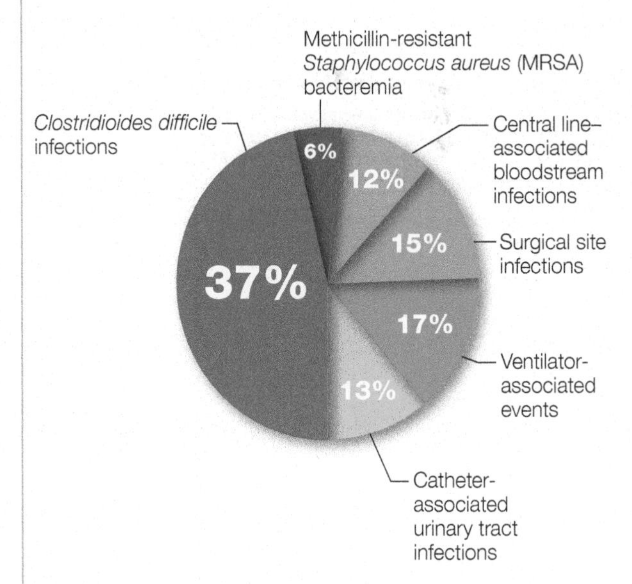

FIGURE 9.10 Most common healthcare-acquired infections (HAIs) Graph based on data from the 2019 Annual National and State HAI Progress Report. The data apply to acute care hospitals. Due to rounding, percentages are approximate. These CDC-generated summary reports are updated every 5 years and used to track progress in decreasing HAIs.

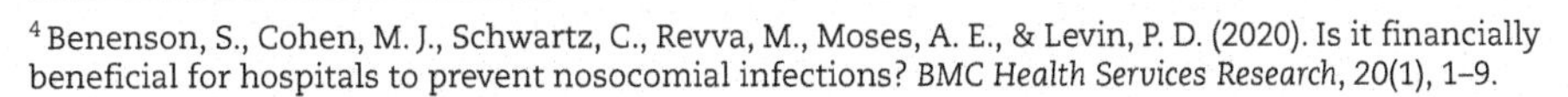

[4] Benenson, S., Cohen, M. J., Schwartz, C., Revva, M., Moses, A. E., & Levin, P. D. (2020). Is it financially beneficial for hospitals to prevent nosocomial infections? *BMC Health Services Research*, 20(1), 1–9.

[5] Kollef, M. H., Torres, A., Shorr, A. F., Martin-Loeches, I., & Micek, S. T. (2021). Nosocomial infection. *Critical Care Medicine*, 49(2), 169–187.

[6] Previously known as *Clostridium difficile*

[7] Centers for Disease Control and Prevention. (2019). *2019 National and State Healthcare-Associated Infections Progress Report*. This report presents data collected by the National Healthcare Safety Network (NHSN) (https://www.cdc.gov/hai/data/portal/progress-report.html) for the calendar year 2019. To account for delays in reporting, 2019 data reported to NHSN through July 1, 2020, is included.

[8] Markwart, R., Saito, H., Harder, T., et al. (2020). Epidemiology and burden of sepsis acquired in hospitals and intensive care units: A systematic review and meta-analysis. *Intensive Care Medicine*, 1–16.

TABLE 9.4 Key HAI Pathogens

Microbe	Examples	Notes
Bacteria	*Clostridioides difficile*	Causes infectious colitis (inflammation of the colon); any surface contaminated by feces can serve as a source of infection; endospores allow it to survive for extended periods on surfaces, making it difficult to kill
	E. coli	Urinary tract infections and bacteremia (blood infection); common bacteria of the gastrointestinal tract
	Methicillin-resistant *Staphylococcus aureus* (MRSA)	Causes dangerous infections of surgical wounds, bedsores, and central lines; primarily transmitted by touch; mainly limited by hand washing
	Pseudomonas aeruginosa	Infections in burn patients, catheterized patients, and people on ventilators; found in water and soil; resistant to many antibiotics; grows in low-nutrient environments, including soaps, disinfectants, and distilled water
	Vancomycin-resistant enterococci (VRE)	Causes wound, bloodstream, and urinary tract infections; enterococci are normally present in the intestines, but may cause disease when introduced to other areas of the body; strains resistant to the antibiotic vancomycin are of increasing concern; hand washing is an essential prevention
Viruses	Hepatitis B	Causes acute illness that is followed by a high risk of chronic infection that causes severe liver damage and increases the risk of liver cancer; in healthcare settings the virus is mainly transmitted through sharp instruments—especially accidental needle sticks; vaccination prevents infection
	Human immunodeficiency virus (HIV)	Causes acquired immune deficiency syndrome (AIDS); in healthcare settings it is mostly transmitted by sharps, so proper sanitization of surgical instruments, single use of needles, and consistent use of protective personal wear and proper disposal of biologically contaminated sharps is essential; screening blood and tissue donations limits transmission to patients
	Influenza	Especially dangerous for hospitalized patients; often transmitted by sick healthcare workers or visitors; limit spread through vaccination; wear a face mask when entering rooms with possible influenza cases; place patients with suspected infection in private room or with patients who also have the virus
	Viral gastroenteritis (mainly norovirus and rotavirus)	Characterized by gastrointestinal upset; transmission reduced through proper hygiene and decontamination measures; place patients with suspected infection in private room or with patients who also have the virus
Fungi	*Aspergillus* mold species	A threat to severely immune-compromised patients, especially bone marrow transplant or leukemia/lymphoma patients; specialized airflow and air filtration systems limit environmental exposure; prohibit flowers and plants in areas with high-risk patients
	Candida yeast species	Causes urinary tract, bloodstream (candidemia), and wound infections; antibiotic resistance is increasing; proper management of indwelling devices and hand washing limits infections

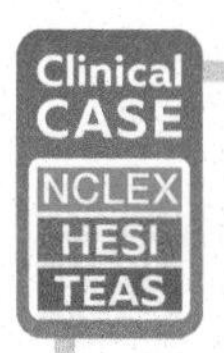

The Case of the Nauseating Lunch

Practice applying what you know clinically: visit the **Mastering Microbiology** Study Area to watch Part 2 and practice for future exams.

Preventing HAIs The first step to reducing HAIs in a healthcare facility is to establish surveillance and monitoring systems that collect information on the prevalence, incidence, and nature of infections occurring in the facility. Infection control personnel and committees monitor transmission patterns and investigate specific risk factors for HAIs in the facility. The data are then used to prevent HAIs moving forward. Basic preventive measures include hand washing, consistent use of personal protective wear (gloves, masks, and gowns), environmental sanitization, and equipment sterilization. Additionally, limiting patient transport unless medically necessary, dedicating equipment for a single patient's use whenever possible (for example, a blood pressure cuff), placement of the patient in an airflow-controlled room when there is a risk for airborne transmission, and patient isolation (as needed) all cut down on HAI incidence. Infections through catheters and central lines can be limited by strictly adhering to aseptic technique during insertion, dressing, changing, and removal of such devices. Proper handling and disposal of sharps helps to prevent the spread of bloodborne pathogens like HIV and hepatitis B.

Simple and inexpensive, proper hand washing is one of the most important yet overlooked factors for HAI prevention. Hand washing may be improperly performed when healthcare staff are caring for a high number of patients, or when caregivers are overly fatigued. Certain healthcare workers may wrongly assume that glove use can replace hand hygiene. Gloves should *never* be considered a replacement for hand washing. Unnoticed tears in gloves can be big enough for a pathogen to pass from a caregiver to a patient. The chance of contaminating the outside of gloves is also higher when hands are not washed.

Preventing "superbug" HAIs Drug-resistant pathogens often originate in healthcare settings as a result of extensive antibiotic use and the abundance of susceptible hosts. Therefore, an important aspect of hospital epidemiology is monitoring antibiotic-resistant strains so that proper drug therapies are used. Hospital microbiologists routinely test isolated pathogens for their susceptibility to commonly prescribed drugs. These susceptibility rates are then shared with the healthcare team to help them make informed decisions about antibiotic therapies.

Build Your Foundation

25. Describe Semmelweis's efforts, and state why they were not initially accepted. (LO 9.22)

26. What is the general goal of healthcare/hospital epidemiology, and why is it important? (LO 9.22)

27. What are healthcare-acquired infections, and how can they be prevented? (LO 9.23)

28. What are the most common categories of HAIs and common agents identified in HAIs? (LO 9.24)

29. Why are hospitals hot zones for antibiotic resistance, and how does epidemiology help address antibiotic resistance? (LO 9.25)

Build Your Foundation (BYF) Quick Quiz: Visit the **Mastering Microbiology** Study Area to quiz yourself.

9.6 SURVEILLANCE, ERADICATION, AND ETHICS IN EPIDEMIOLOGY

Learning Outcomes

After reading this section, you should be able to:

9.26 Explain how surveillance for nationally notifiable diseases works.

9.27 Compare and contrast emerging and reemerging diseases, and describe factors that contribute to each.

9.28 Define the term *eradication*, and name factors to consider in an eradication program.

9.29 Provide an explanation of how ethical issues can emerge in epidemiology.

Surveillance programs monitor, control, and prevent disease.

In many countries public health agencies have surveillance programs designed to monitor the prevalence and incidence of certain diseases. In the United States, the CDC develops surveillance recommendations and then collects and shares surveillance data with healthcare providers and the general public. Surveillance networks help prevent disease by enabling earlier detection of outbreaks caused by existing and emerging pathogens, and by providing feedback on how well preventive measures work.

Notifiable and Reportable Diseases

The CDC's **National Notifiable Diseases Surveillance System (NNDSS)** is a network that depends upon local hospitals, laboratories, and private healthcare providers to monitor and report certain infectious and noninfectious diseases. Most states have laws requiring healthcare providers to report cases of certain diseases to a local health authority. The diseases on a state or local tracking list are called **reportable diseases**. Usually the list of reportable diseases includes the diseases that the CDC has an interest in monitoring, called **notifiable diseases**, as well as diseases that the local authorities want to monitor. There are usually about 60 different infectious diseases (and multiple noninfectious diseases) on the CDC's notifiable diseases list.

When a healthcare provider diagnoses a reportable disease, they document the case per state or local rules. The case reports trickle up the local and state reporting chain to the CDC, which can then gather the data for national

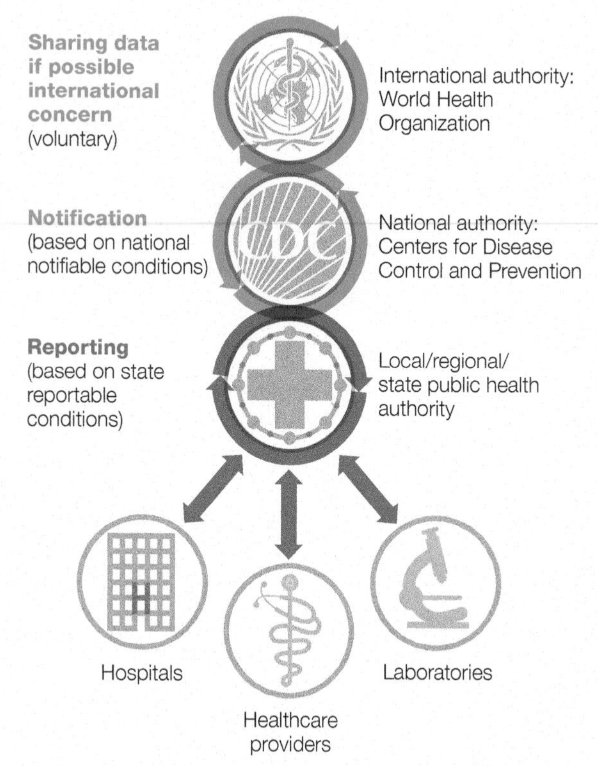

FIGURE 9.11 Effective public health surveillance systems In the United States, all healthcare providers and laboratories are required to report certain disease cases to a local, state, or regional health authority. Data sharing at the national and international level is voluntary.

statistics (FIG. 9.11). Although no federal rule requires every state to report such data, all states have willingly participated in the NNDSS since it was started in 1951. In addition to the NNDSS, there are also nationally coordinated surveillance systems for monitoring antibiotic resistance, HAIs, foodborne illnesses, and bioterrorism.

The CDC publishes the *Morbidity and Mortality Weekly Report* (MMWR) to update health officials and care providers on U.S. health issues. This report provides summaries of health watches and a running report for selected notifiable diseases.

Surveillance of Emerging and Reemerging Diseases

In addition to surveilling a multitude of ancient human diseases, epidemiologists are constantly monitoring populations for the emergence or reemergence of diseases. Surveillance by location, causative agent, virulence, and size of outbreak is crucial for formulating effective action plans when outbreaks hit.

Emerging diseases are new or newly identified infections in a population. Since the 1970s at least 40 never-before-described diseases that afflict humans have been identified. These include Ebola, HIV, H1N1 influenza ("swine flu"), Zika, and COVID-19. The rate of disease emergence has quadrupled in the past 50 years.

Emerging disease hot spots are often places with population crowding, poverty, tropical or subtropical climates, diverse wildlife, and new shifts in the environment—like deforestation (FIG. 9.12). Climate change allows vector populations to move into new territories, while deforestation and urbanization may lead to once rarely encountered arthropods or animals more frequently encountering people. Changes in sociocultural practices (such as injectable illicit drugs, widespread global travel, and increased food processing and handling) also accelerate disease emergence. As an example, in 2007 the first outbreak of the chikungunya virus occurred in Europe and has since spread to the Caribbean and subtropical parts of the United States. This virus is a mosquito-borne tropical disease that was previously only found in tropical regions of Africa and Asia.

Reemerging diseases were previously under control, but are now showing increased incidence. These include infections caused by well-known pathogens that have evolved increased virulence. For example, plague was once easily curable with modern drug therapies. Unfortunately, the bacteria that cause plague (*Yersinia pestis*) are developing resistance to antibiotics. Similar trends of drug resistance are seen in numerous other pathogens, including tuberculosis. We also have a much larger population of immune-suppressed people in the world today: AIDS patients, cancer patients, organ transplant recipients, and the elderly. More people with poor immune function means an increased number of hosts in whom opportunistic pathogens can develop increased virulence. It is also hypothesized, though not confirmed, that diseases that normally wouldn't be transmitted from human to human might be able to use an immune-compromised human host as an intermediary to evolve into a true human pathogen. Add to this the human element of bioterrorism, and it is hard to imagine a future in which surveillance systems are not needed.

Eradication is the ultimate triumph over an infectious disease.

Eradication of an infectious disease means that there are no longer any cases of it anywhere in the world. Eradication is a triumph over infectious disease that has been sought for many different diseases, but so far has only succeeded once, with the eradication of smallpox in 1977. Fortunately, polio and guinea worm (dracunculiasis) are close to being eradicated. (Polio is discussed in Chapter 17.)

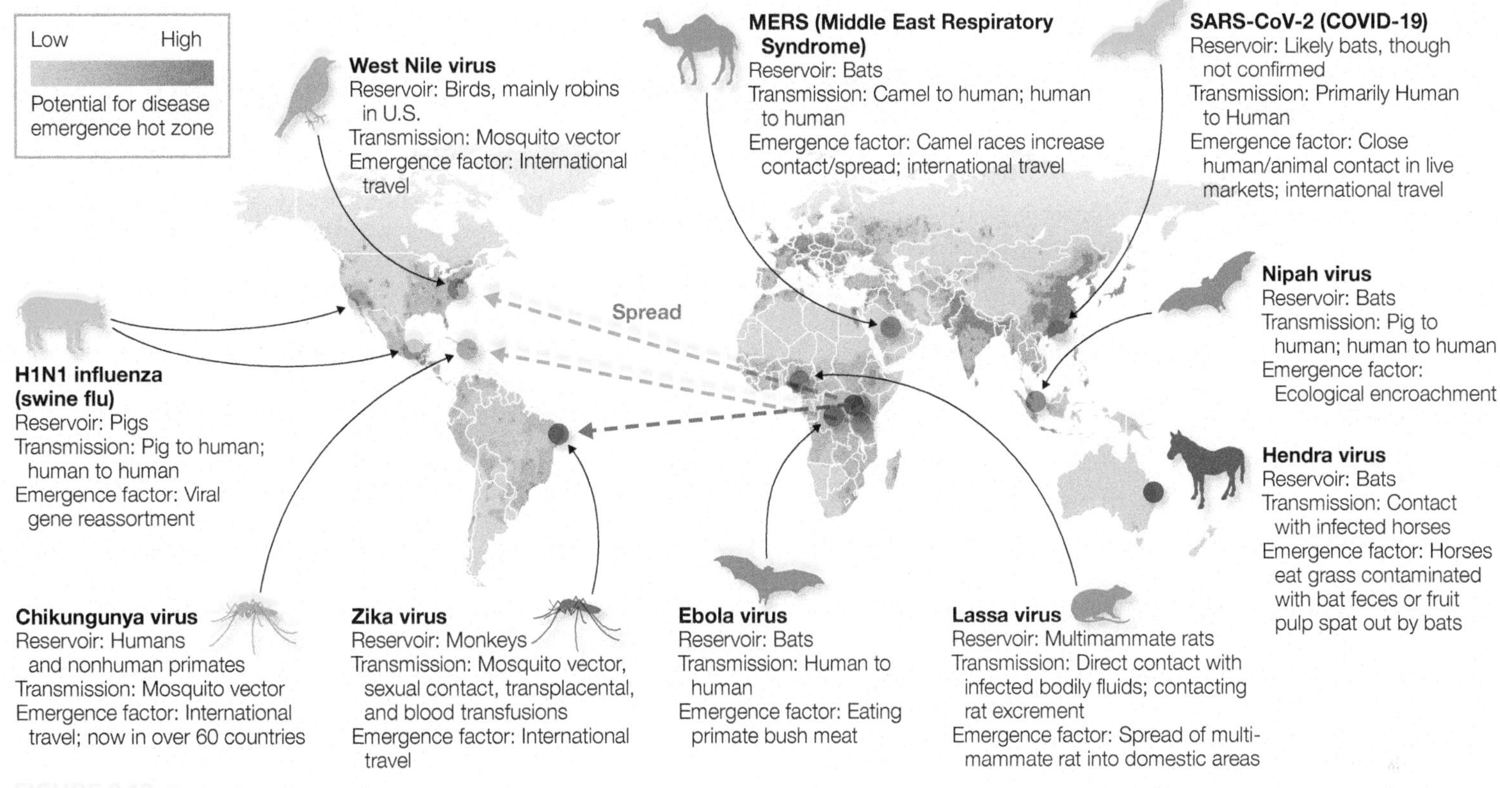

FIGURE 9.12 Emerging disease hot spots

Critical Thinking *About 70 years ago the hot spots for disease were mainly in poor countries. Increasingly, hot spots are found in dense metropolitan areas in developed nations. What are some reasons for this observed shift?*

The best candidates for eradication are easily identifiable, treatable, or preventable infectious diseases that only humans transmit and catch. Infectious diseases with longer incubation periods, or pathogens that can cause latent infections, are harder (but not impossible) to eradicate because surveillance is more difficult. In addition to biological considerations, there also must be strong social and political support for the eradication program. Disease eradication efforts are costly in terms of money, time, personnel, and government resources. The good news is that eradication of serious diseases pays off financially and in terms of human lives saved. For example, the smallpox eradication program cost a total of about $321 million, of which the United States contributed $23 million. Yet the Center for Global Development estimates that every 26 days, the United States directly saves *at least* $23 million, thanks to not having to vaccinate against or treat smallpox. Furthermore, UNICEF estimates that smallpox eradication saves about 5 million lives per year.

Conducting epidemiology involves weighing ethical issues.

Effects of epidemiological research are not always equally positive for all involved. Even something as overwhelmingly beneficial as disease eradication requires patients to surrender certain individual rights in favor of the common good. As such, weighing competing interests is an important part of practicing epidemiology. Ultimately, we should try to base public health priorities on evidence, rather than self-interests or political agendas. And the means by which research is conducted should always be as benign as possible.

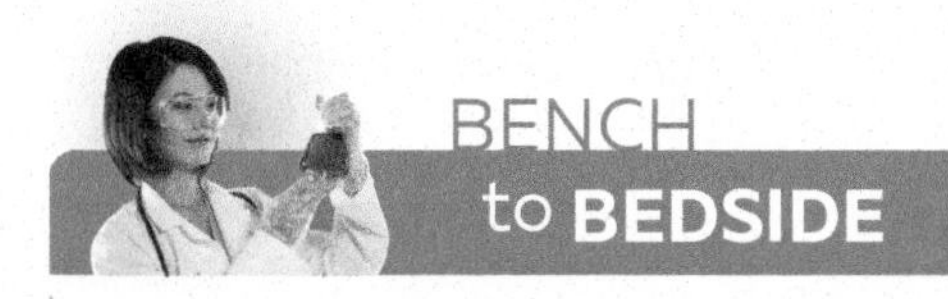

Will Polio Be the Second Eradicated Human Disease?

The polio virus, which can lead to permanent paralysis or death, fits the criteria of a disease worthy of eradication. However, eliminating polio requires more effort than smallpox did; the following table explains some of the reasons for this.

Factor	Smallpox	Polio
Population of affected areas	500 million	4 billion
Vaccine stability	Stable and easy to ship	Unstable, easily inactivated in tropical climates
Vaccine dose	One dose	Multiple doses
Case monitoring	Easy because of severe symptoms	Difficult because many cases are asymptomatic

Thanks to the World Health Organization's (WHO's) Global Polio Eradication Initiative, most countries have been polio free since 2003. In 1988, when the polio eradication program began, about 1,000 children were newly infected with polio *every day*. According to the CDC, in 2020, there were approximately 500 polio cases worldwide.

One barrier in the race to eradicate polio is that there are three serotypes of polio virus. Unfortunately, the original oral polio vaccine (OPV) was not as effective against two of the three virus serotypes as would be needed for eradication to succeed. In 2009, a new vaccine was released. The new bivalent oral polio vaccine (bOPV) has proven to be about 30 percent more effective than the original OPV. The new vaccine contributed to India—formerly an epicenter for polio—being declared "polio free" in 2014. The Americas, Europe, and the western Pacific are also polio-free areas. However, it is essential that vaccination programs remain in place in all areas because polio is still endemic in Nigeria, Afghanistan, and Pakistan. If vaccination is stopped before eradication, then polio is likely to reemerge. This was demonstrated by the 2013 outbreak of polio in Syria that occurred when vaccination workers were unable to reach children in war-torn areas of the county. Prior to this outbreak, Syria had been polio free for 14 years. The COVID-19 pandemic has also set back polio eradication efforts by disrupting vaccination administration in many countries, leaving children there vulnerable to infection.

In order to declare polio eradicated, there cannot be a new case of polio anywhere in the world for three consecutive years.

Source: Data for this feature is from the WHO's Global Polio Eradication Initiative.

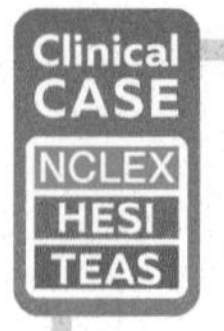

The Case of the Nauseating Lunch

Practice applying what you know clinically: visit the **Mastering Microbiology** Study Area to watch Part 3 and practice for future exams.

Ethical Considerations and Discrimination

Unfortunately, high ethical standards have not always been followed with every U.S. health study. The most notorious example of unethical health research in modern America is the Tuskegee syphilis experiment, which took place in Alabama between 1932 and 1972. Researchers devised the study to look at long-term effects of syphilis in a group of poor African-American men. At the time it began, no cure existed, and incidence was high; tracking disease progression was arguably a valid area of study. However, study participants with syphilis were never informed they had the disease and were continually misled about the reasons for ongoing tests and physical examinations. When penicillin proved to be a cure for syphilis in the mid-1940s, researchers actively prevented participants from receiving it so the study could continue. As a result, many men taking part in the research needlessly suffered and died. A number of the men's spouses and some of their children also contracted the disease during the study.

When news of the Tuskegee experiment broke in the 1970s, the resulting scandal brought many ethical questions to the forefront of research. Today, informed consent is the cornerstone of analytical research. Before enrollment in any study, potential participants should know their disease status, along with any other important details that are part of the proposed research. Likewise, no study, epidemiological or otherwise, should ever be constructed in a manner that puts people at serious risk. In fact, drug studies are stopped early if data show evidence of harm.

Discrimination is also an ethical concern in genetic research. In particular, there are concerns over how genetic information could be used to discriminate against patients and how this data should be kept private. While the Genetic Information Nondiscrimination Act (GINA) of 2008 provides some protection against discrimination, many argue that it does not go far enough. For example, it states that someone cannot be denied health insurance on the basis of their genetics, but it does not prevent denial of long-term care insurance, life insurance, or disability insurance.

Ethical Considerations and Vaccination

In the past, the majority of citizens in the United States and other countries agreed that vaccination was worthwhile for both individuals and the general population. Thus, they willingly participated in immunization programs in great numbers. Getting as many people with normal immune systems to take part in vaccination is key because herd immunity occurs only when a high percentage of the population is immune. **Herd immunity** means a pathogen won't find enough susceptible people in the community to persist, even if a small number of individuals there remain unvaccinated. Herd immunity is the only protection available to those who are unable to receive immunizations themselves, including newborns, immune-compromised people, and those with certain other medical conditions. However, most infectious agents require at least 85 percent of the population to be immune before herd immunity ensues—hence the conflict between the group and the individual that can take place under these circumstances.

Earlier in this chapter, we mentioned unprofessional and unethical research, published in the 1990s, that suggested a link between vaccination and the development of autism in children. This situation prompted unfounded fears among some parents, leading them to skip or delay vaccination of their children. Even after public health officials and pediatricians began debunking the false link between autism and vaccines, fears persisted. The resulting drop in vaccinations was a factor in certain long-controlled diseases such as measles reemerging.

Now some public health officials have gone beyond pro-vaccination information campaigns in their quest to bring vaccination levels back up. Personal exemptions from vaccination requirements to attend public schools have been revoked in some areas. That means only children with medical issues that make vaccination dangerous can attend public school without having up-to-date immunizations. Communities opting to do this consciously place the safety of people who cannot receive a vaccine above the rights of healthy people who can but don't wish to.

Is it acceptable to curtail some freedom of choice to maintain the safety of others? These sorts of questions come up any time a public health issue is addressed, and they should be thoughtfully and thoroughly considered from all angles before a policy is implemented. Unfortunately, public policies rarely make everyone happy. (For more on vaccines, see Chapter 14.)

Build Your Foundation

30. What would be the general route of communications followed if a physician diagnosed a case of standard influenza versus a novel influenza strain? (LO 9.26)

31. What is the NNDSS, and why is it important to public health? (LO 9.26)

32. How is an emerging disease different from a reemerging disease, and what factors contribute to pathogen reemergence? (LO 9.27)

33. Why are there certain hot spots for emerging diseases, and why do people who live outside of these developing disease zones still need to be concerned? (LO 9.27)

34. What is eradication, and what factors should be considered in the development of an eradication program? (LO 9.28)

35. Discuss some of the ethical issues that can emerge in epidemiology. (LO 9.29)

Build Your Foundation (BYF) Quick Quiz: Visit the **Mastering Microbiology** Study Area to quiz yourself.

Infectious Disease Transmission

A reservoir is the animate or inanimate habitat where the pathogen is naturally found. A source disseminates the agent from the reservoir to new hosts. Sometimes the source and reservoir are the same. The mode of transmission is how the pathogen is spread to a host.

Direct contact transmission

The host is exposed through direct contact with the source.

Person to person | Animal | Environment | Vertical

Indirect contact transmission

The host is exposed without direct physical contact with the source.

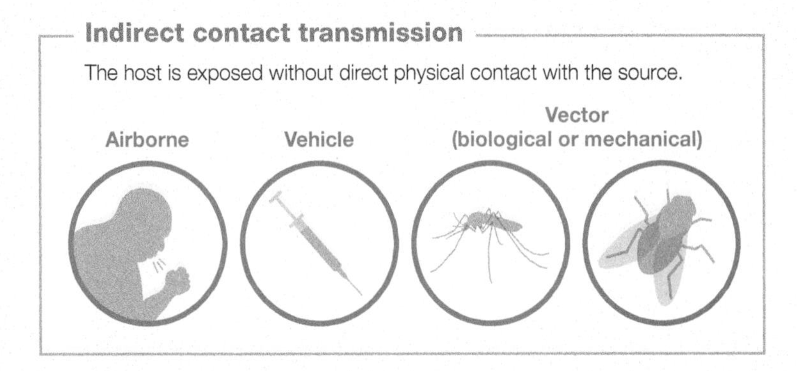

Infectious Disease Stages

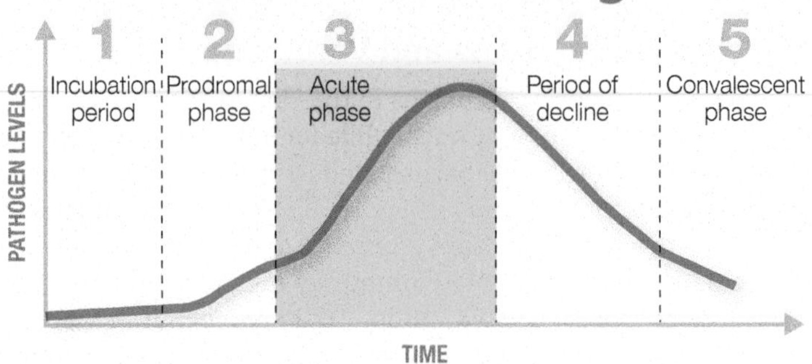

Epidemiology

Epidemiology aims to understand and prevent illness in populations. Education, quarantine, and vector control are some ways to break the epidemiological triangle to prevent disease.

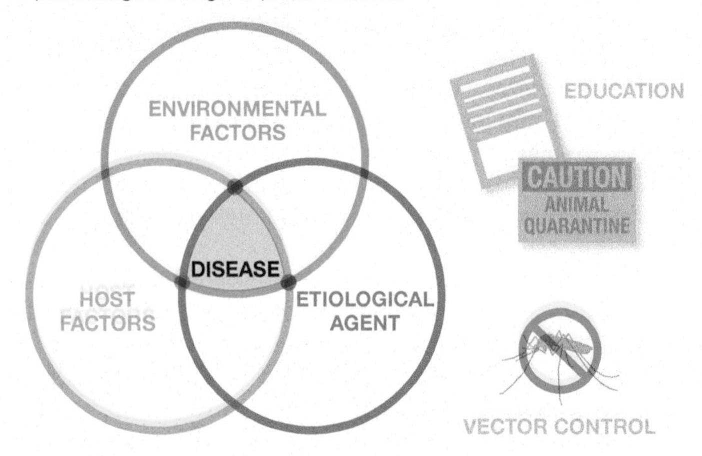

Healthcare Epidemiology

Healthcare-associated infections (HAIs) are infections that develop as a result of healthcare intervention. Contaminated hands and medical devices often spread HAIs. Bacteria are the leading cause of HAIs.

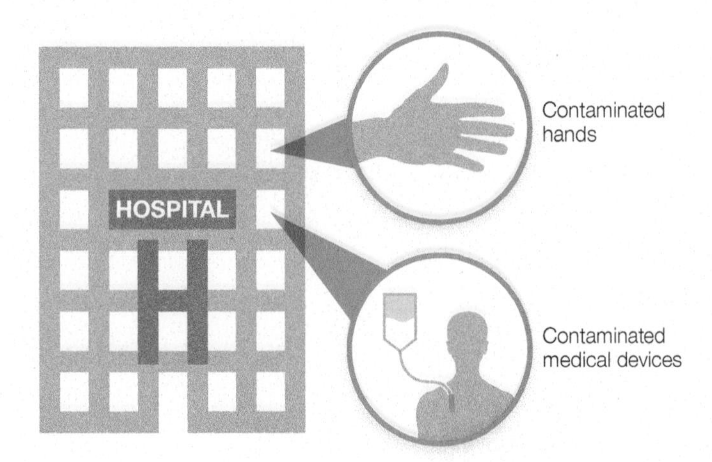

Disease Reporting Scheme

Surveillance programs help monitor, control, and prevent disease. The CDC monitors notifiable diseases; diseases on a state or local tracking list are called reportable diseases.

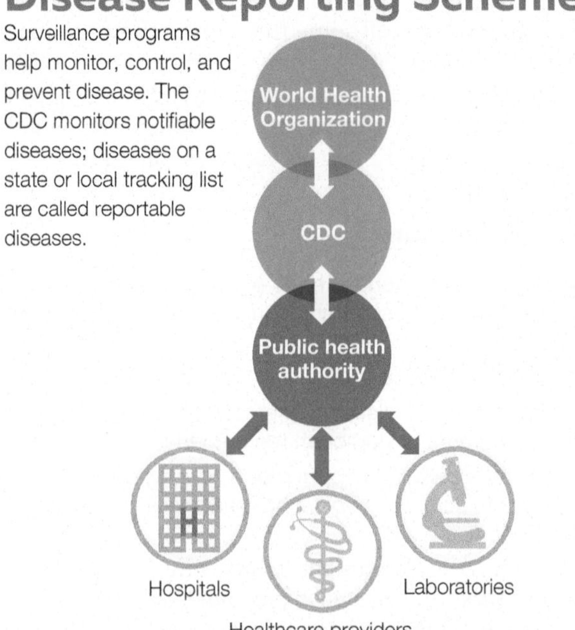

CHAPTER 9 OVERVIEW

9.1 Causes of Infectious Diseases

- Pathogens such as viruses, prions, bacteria, protozoans, helminths, and fungi can cause disease. True pathogens cause disease in a regularly healthy host; opportunistic pathogens require a weakened host or a disruption in the normal flora.
- Vocabulary related to infectious diseases:
  - Sporadic cases of disease are scattered or isolated infections in a particular population, while endemic diseases are routinely found in a specified population or region.
  - An epidemic is a widespread outbreak of disease in a particular region during a specified time frame; an epidemic that spreads to numerous countries is called a pandemic.
  - Zoonotic diseases are transmitted from animals to humans.
  - Infectious diseases can be communicable (spread from person to person) or noncommunicable (not spread from person to person). A contagious disease is a communicable disease that is easily spread in a population.
  - Pathogens can cause latent infections (asymptomatic) or active infections in which the patient has signs and symptoms of disease.
  - Signs are objective indicators of disease that can be witnessed or verified. Symptoms are subjective and mainly sensed by the patient.
  - Acute diseases have a rapid onset and rapid progression; chronic diseases are characterized by slower onset and progression.
- Koch's postulates of disease help identify the causative agent of infectious diseases. Koch's postulates are
  - The same organism must be present in every case of the disease.
  - The organism must be isolated from the diseased host and grown as a pure culture.
  - The isolated organism should cause the disease in question when it is introduced (inoculated) into a susceptible host (a host that can develop the disease).
  - The organism must then be re-isolated from the inoculated, diseased animal.

9.2 Infectious Disease Transmission and Stages

- To cause disease, an infectious agent must be transmitted to a susceptible host. A reservoir is the animate or inanimate habitat where the pathogen is naturally found. In contrast, a source disseminates the agent from the reservoir to new hosts.
- Transmission of an infectious disease can be through direct or indirect contact.
  - Direct contact transmission: host is infected by direct physical contact with the source of the pathogen. Indirect contact transmission: the pathogen spreads from a source to a host without direct physical contact. There are three general types of indirect transmission: airborne, vehicle, and vector.

    The five general stages of infectious disease are the incubation period, the prodromal phase, an acute phase, a period of decline, and a convalescent phase.

9.3 Epidemiology Essentials

- Epidemiology seeks to understand and prevent illness (infectious and noninfectious) in populations.
- The two general goals of epidemiology are to: (1) describe the nature, cause, and extent of new or existing diseases in populations and (2) intervene to protect and improve health in populations.
- The epidemiological triangle represents the *what* (etiological agent), *who* (host), and *where* (environment) of disease.
  - In general, disease can be established whenever an environment in which the etiological agent can thrive overlaps with the susceptible host's environment.
  - Public education, quarantine, and vector control are ways to break the epidemiological triangle.
- Epidemiology is the foundation of public health. The CDC is a U.S. health agency that oversees national epidemiological endeavors such as disease surveillance. Public health agencies affect human health at a variety of levels—from health awareness campaigns to research. Epidemiology is a necessary partner with public health.

9.4 Epidemiological Measures and Studies

- Measures of frequency and association are useful in epidemiology.
  - Measures of frequency describe the occurrence of a disease in a population during a certain period of time; they reveal the degree of morbidity (existence of disease) in a population. Examples include prevalence and incidence rates.
  - A prevalence rate describes the overall occurrence of a disease in a population in a given period of time. An incidence rate describes new occurrences of a disease in a population in a specified period of time.
  - Measures of association reveal risks associated with a particular disease. The most common association measure is mortality rate.
- Epidemiological studies are descriptive or analytical.
  - Descriptive studies describe the occurrence and distribution of disease so that hypotheses can be developed and then tested using analytical approaches; data is often collected through correlation studies, case reports, and cross-sectional studies.
  - Analytical studies investigate the etiological agent of a disease, what specific factors lead to disease, and how the disease can be treated or prevented. Analytical epidemiology studies are either observational or experimental.

9.5 Epidemiology in Clinical Settings

- Healthcare-acquired infections (HAIs) are infections that develop as a result of a healthcare intervention; they are a major concern in healthcare settings because they are dangerous and expensive.
  - HAIs are transmitted by direct and indirect contact. Contaminated medical devices and healthcare workers' hands are the most common sources. Bacteria are the most common cause of HAIs (although viruses, fungi, and parasites can also be culprits).
  - Hand washing is the easiest and most significant way to reduce HAIs.
- Hospital epidemiology involves the surveillance, prevention, and control of HAIs and monitoring antibiotic-resistant strains.

9.6 Surveillance, Eradication, and Ethics in Epidemiology

- Surveillance programs are used to monitor, control, and prevent disease. The CDC develops surveillance recommendations and collects, then shares, surveillance data with healthcare providers and the general public; diseases that the CDC recommends monitoring are called notifiable diseases.
- Emerging diseases are linked to changes in the environment and changes in sociocultural practices.
- When a disease is declared eradicated, there are no longer any cases of the infectious disease anywhere in the world.
- Ethics in epidemiology and public health includes research ethics, protecting the well-being of a population while respecting individual freedoms, guaranteeing fair distribution of public resources, and ensuring that public health priorities are based on evidence.

THINK CLINICALLY | *Be S.M.A.R.T. About Cases*

COMPREHENSIVE CASE

The following case integrates basic principles from the chapter. Try to answer the case questions on your own. Don't forget to be **S.M.A.R.T.*** about your case study to help you interlink the scientific concepts you have just learned and apply your new understanding of epidemiology to a case study.

*The five-step method is shown in detail in the Chapter 1 Comprehensive Case on cholera. See pages 31–33. Refer back to this example to help you apply a SMART approach to other critical thinking questions and cases throughout the book.

• • •

Michelle, a Colorado resident in her 20th week of pregnancy, had been running a fever and experiencing muscle aches and diarrhea for the past 2 days. When she noticed bright red blood on her underwear, she called her obstetrician, who instructed her to go to the hospital immediately. Unfortunately, during her exam, no fetal heartbeat was detected. Michelle was most likely experiencing a miscarriage. Two sets of aerobic and anaerobic blood cultures were taken, and Michelle was then given a combination antimicrobial therapy consisting of ampicillin and gentamycin—a choice driven by her attending physician's clinical suspicion. The microbiological report was received the next day and revealed that Michelle had a *Listeria monocytogenes* infection. Laboratory analysis of fetal and placental tissue also revealed the presence of *L. monocytogenes*.*

Once stabilized, Michelle was interviewed about the foods she had eaten for the 4 weeks prior to symptoms developing. It was difficult for her to recall all food choices over such a lengthy time period, but her husband mentioned she had been craving fruit, especially cantaloupe. The interviewer asked a few more questions about the cantaloupe and said that someone from the department of health might follow up with Michelle. Local health authorities were starting to investigate the possibility of an outbreak of listeriosis.

Listeriosis is a nationally notifiable foodborne illness caused by *Listeria monocytogenes*, a Gram-positive bacterium. The bacterium can be found in soil, water, and in a number of animals. About 10 percent of people carry *L. monocytogenes* asymptomatically in their intestinal tract. Forty other mammals and about 17 bird species also carry the bacterium. Any food can become contaminated with *Listeria*, but cooking and pasteurization kill the bacterium, so raw foods and/or foods that become contaminated after cooking are more commonly sources. Deli meats and hot dogs processed by contaminated packing equipment are prime sources for *Listeria*, as the bacteria can grow at refrigerator temperatures, and these foods are often insufficiently heated before serving. Unpasteurized milk products and cheeses are also common sources. Less frequently, raw vegetables, fruits, prepared deli salads, smoked fish spreads, and meat pâté are sources of infection.

According to the CDC, about 800 laboratory-confirmed cases of listeriosis occur in the United States every year. Most are asymptomatic. Clinical cases are mainly seen in pregnant women and the elderly. Pregnant women are 20 times more likely to become infected than nonpregnant healthy adults, likely because of the reduced immunity (selective immune suppression) that occurs during pregnancy. Listeriosis during pregnancy is especially dangerous because about 22 out of 100 perinatal listeriosis cases result in miscarriage, stillbirth, or neonatal death. The incubation period for listeriosis ranges from 3 to 70 days, but symptoms usually appear within a month and can last several days to several weeks. Michelle was one of 147 people affected by the *Listeria* outbreak in 2011 that spread to at least 24 states and killed 33 people. Among the 140 outbreak victims who were able to provide information on what they ate, 131 (94%) reported consuming cantaloupe within the month before becoming ill. Eventually all of the cases were linked to cantaloupes from Jensen Farms in Colorado.

This case is based on a real outbreak of listeriosis that occurred in 2011. The patient's name and the specifics of her case have been altered, though the numerical data have not been changed. This foodborne outbreak ranks among the deadliest in U.S. history.

---

*There is always a cost/benefit analysis in healthcare—the risk of withholding an antimicrobial drug until microbiological data are returned could have irreversible negative consequences. In contrast, it's easy to swap out a drug or change the dose if the data suggests this is the best course of action. As such, in acute cases an empiric treatment is usually prescribed and then revisited as needed.

• • •

CASE-BASED QUESTIONS

1. In general, what is the source of the pathogen *L. monocytogenes*?
2. What is the mode of transmission for *L. monocytogenes*?
3. Is listeriosis considered an infectious disease? Explain your reasoning.
4. Based on the information in the case, do you think listeriosis is a communicable disease? Explain your reasoning.
5. What was the case fatality rate in the 2011 *Listeria* outbreak (express your answer as a percentage)?
6. What is the overall incidence rate of listeriosis in the United States (assume a population of 315,505,000, and express your answer per 1,000,000 in the population)?
7. Even if you calculated the incidence rate correctly in the previous question, it is probably not a true reflection of the number of *Listeria* infections that occur in the United States every year. Why?
8. From the case, identify at least one rate, one proportion, and one ratio.
9. What features of listeriosis present epidemiological challenges?
10. What type of epidemiological study design was most likely used in order to recommend a food recall?
11. Would Michelle have been interviewed even if there were not a suspected outbreak? Explain your reasoning.

END OF CHAPTER QUESTIONS

NCLEX
HESI
TEAS

Think Critically and Clinically
Questions highlighted in orange are opportunities to practice NCLEX, HESI, and TEAS-style questions.

1. In developed nations, which of the following are considered endemic diseases, and which are considered sporadic diseases?

   Influenza: (endemic or sporadic)
   Tetanus: (endemic or sporadic)
   Plague: (endemic or sporadic)
   Common cold: (endemic or sporadic)
   Streptococcal pharyngitis: (endemic or sporadic)
   Botulism: (endemic or sporadic)
   Pneumonia: (endemic or sporadic)

2. Indicate the true statements and then correct the false statements, so they are true.
   a. Zoonotic diseases pass from humans to animals.
   b. Communicable diseases spread from person to person.
   c. Noncommunicable diseases are contagious.
   d. Koch's postulates of disease are mainly used to study noninfectious diseases.

3. Match the following:

Emerging	A disease with rapid onset and rapid progression
Epidemic	Worldwide epidemic
Chronic	A disease with a common occurrence in a population
True pathogen	A disease that has a slow progression and a slow onset
Pandemic	A pathogen that causes disease in a normally healthy host
Acute	An outbreak of a disease
Endemic	A pathogen that usually only causes disease in a compromised host
Opportunistic pathogen	A new disease in a population

4. List three functions of public health.

5. State what type of mortality rate is applicable.

   Scenario 1: Out of 6,000 live births last week, 10 of the women died. Type of mortality rate?

   Scenario 2: 300 patients had disease X last year, 10 of whom died. Type of mortality rate?

   Scenario 3: Of the 120,000 live births in a particular community last year, 15 of the babies died before their first birthday. Type of mortality rate?

   Scenario 4: Out of 3,000 people in a given population, 100 died of pneumonia. Type of mortality rate?

   Calculated mortality rate (expressed per 100 in the population):

6. Draft a Venn diagram to compare and contrast descriptive and analytical epidemiology.

7. Diseases that the CDC collects information on through collaboration with state and local health authorities are called
   a. communicable diseases.
   b. reportable diseases.
   c. nationally notifiable diseases.
   d. investigative diseases.
   e. case report illnesses.

8. An epidemiological study design that is commonly used for determining the efficacy of a drug therapy is a(n)
   a. experimental study.
   b. case report.
   c. cross-sectional study.
   d. correlation study.
   e. observational study.

9. A(n) ____________ groups the study populations by exposure versus nonexposure to a certain risk factor to see if either group develops the outcome in question.
   a. experimental study
   b. case report
   c. cross-sectional study
   d. correlation study
   e. observational study

10. Label the following modes of transmission as either direct or indirect. For all *indirect* transmissions, also specify which of the three categories of indirect transmission is involved.

    Transmission of HIV across the placenta:
    Transmission of a pathogen through drinking contaminated water:
    Transmission of malaria by a mosquito to a human host:
    Transmission of a pathogen through breast milk:
    Transmission of rabies by a dog bite:
    Transmission of a pathogen by touching a doorknob:
    Transmission of a pathogen by a contaminated needle:
    Transmission of a respiratory pathogen through respiratory droplets:

11. From the following choices, select *all* of the factors that impact prevalence rate.
    a. Duration of a disease
    b. The type of pathogen responsible (such as if the pathogen is viral or bacterial)
    c. Cure rates for a disease
    d. The pathogenicity of the microbe that causes the disease
    e. The effectiveness of preventive measures
    f. The incidence rate of a disease
    g. The quality of diagnostic tools
    h. The severity of the disease

CRITICAL THINKING QUESTIONS

Data related to a fictitious nationally notifiable disease, microbiotitis, are presented here. *Use the provided data to answer questions 1–5.*

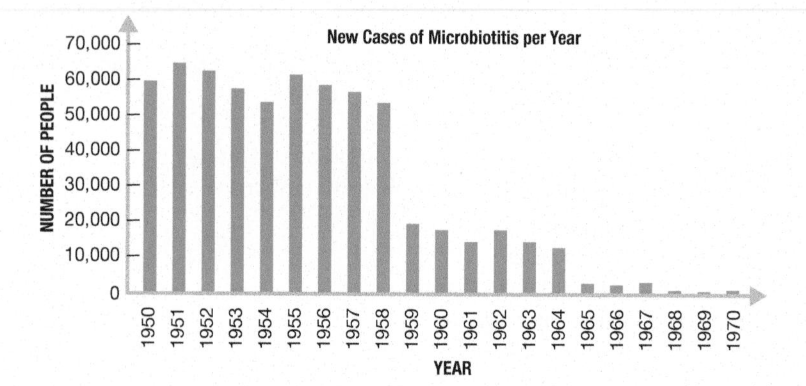

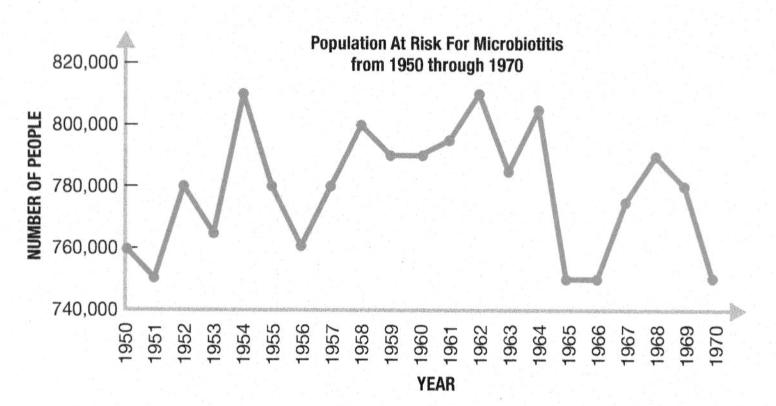

1. When microbiotitis was first characterized, a number of measures were used to lower the incidence of the disease. Based on your analysis of the graphs, in what year was the first successful incidence-decreasing measure applied to the population?

2. Some time after the first measure to reduce the incidence of microbiotitis was introduced, another successful measure was introduced to the population. In what year was the second successful measure applied?

3. Based on the data, were the measures that were introduced to lower the incidence of microbiotitis preventive measures or treatment measures? Explain your reasoning.

4. Assuming that all factors except incidence remained constant for microbiotitis, how would the prevalence rate of 1950 compare in general with the prevalence rate of 1969 (a calculation is not needed to answer this)?

5. Would the presented epidemiological data collected for microbiotitis be described as descriptive or analytical? Why?

6. Draw two lines on the following plot area of a graph: (1) draw a line in red that shows a direct association between developing disease X and drinking water from well A, and (2) draw a second line in green that represents a direct association between drinking water from the same well and *preventing* the development of disease Y.

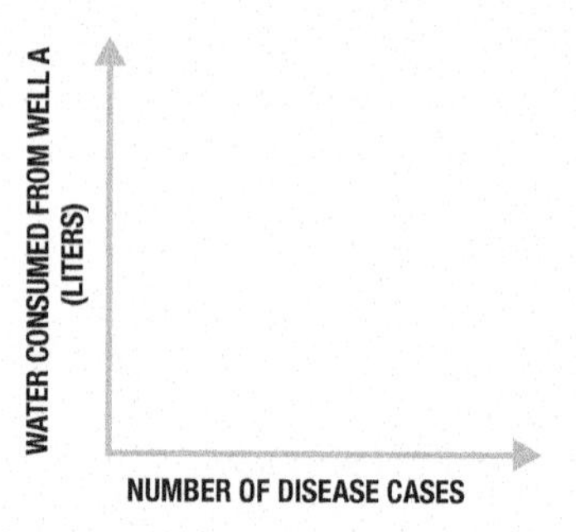

7. Health authorities are investigating a fictitious disease, Fever B. The first case was observed in a 30-year-old male, Marcus, who is an ecological scientist with routine occupational exposure to various areas in and around the Amazon River basin. Two months after returning from a 10-day trip to Bolivia, Marcus developed a high fever, a painful vesicular rash over about 30 percent of his body, and debilitating muscle aches and fatigue. He wasn't really himself until about 6 weeks after his early symptoms developed. He reported starting to feel weak and exhausted with mild aches about 2 days before the high fever and rash developed. Upon investigating the case, it was discovered that two of the three other people who had been traveling with Marcus also became ill about 4 days after Marcus did. One of the patients died of the illness. No one else developed the illness other than these three people. The same virus was isolated from blood samples taken from all three patients; it appears to be a new virus that has never been reported in humans. Upon questioning the patients, it was found that none of them had ever traveled together before. Also, while none of them recalled a specific encounter with an animal, they all recalled being bitten by ticks while in the jungle (including the one individual who did not develop Fever B).

Use the information in the case report about Fever B to answer the following questions:

a. Is Fever B a communicable disease? Explain your reasoning.
b. What evidence suggests that Fever B is caused by a zoonotic pathogen?
c. Would Koch's original postulates of disease have been useful in characterizing the pathogen that causes Fever B? Explain why or why not.
d. What is the *approximate* incubation period for Fever B?
e. What is the duration of Fever B?
f. What is the case fatality rate for Fever B?
g. How long does the prodromal phase of Fever B seem to be?
h. What are the acute-phase symptoms of Fever B?

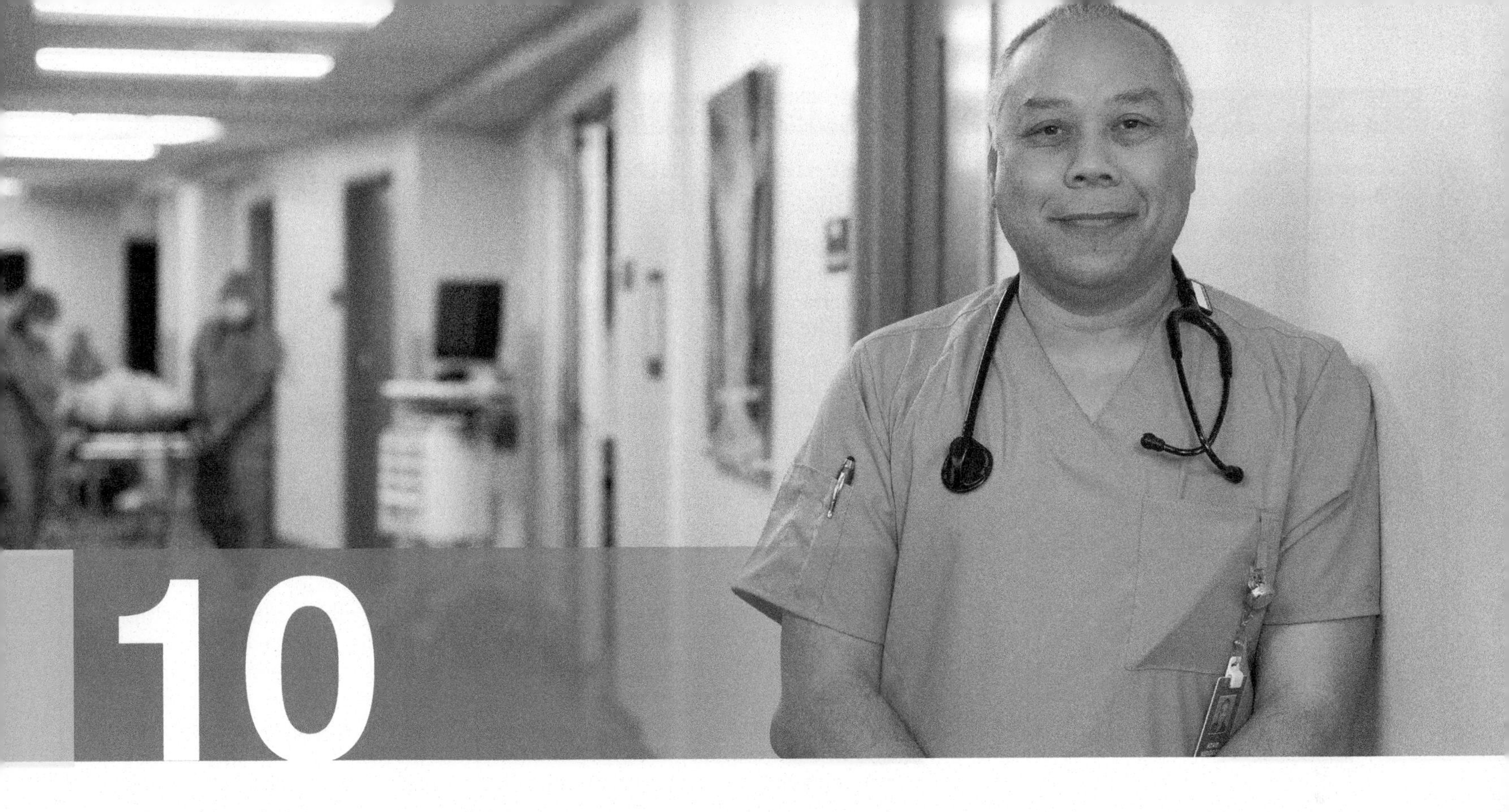

Host-Microbe Interactions and Pathogenesis

What Will We Explore? The **human microbiome project**, which was completed in 2016, was a major initiative to catalog the microbes that comprise the normal human microbiota. This effort revealed that at least 10,000 distinct microbial species call us home. Although we've discussed the microbiome throughout this text, in this chapter we'll further explore its features. We'll also review qualities that distinguish the approximately 1,400 microbes documented as human pathogens from the other 99 percent of their fellow microbes that do not cause disease.[1] Finally, we'll review the assignment of pathogens to biosafety levels and describe precautions that limit infectious disease transmission.

Why Is It Important? Antibiotics, vaccines, and a variety of public health measures have all revolutionized medicine. Yet infectious diseases remain a major global health issue. Antibiotic resistance and healthcare-acquired infections are growing problems, and since 1980 at least 87 new pathogens have been described.[2] Although we're better equipped to fight pandemics than our ancestors were, modern travel has also enabled emerging diseases to spread around the globe more rapidly than ever—COVID-19 was a reminder to us all of that fact. The steady emergence of new infectious diseases means that a firm grasp of biosafety levels and their associated precautions is now the responsibility of *all* healthcare providers, not just researchers. Communities largely rely on first-line healthcare responders to diligently follow proper transmission precautions in order to limit transmission of emerging and reemerging diseases. Ultimately, our ability to fight pathogens relies on understanding how they fight us.

Clinical CASE NCLEX HESI TEAS

The Case of the Deadly Mistake

Visit the **Mastering Microbiology** Study Area to watch the case and find out how host-microbe interactions can explain this medical mystery.

Adam Mendoza
MSN, RN, CCRN, NE-BC Professor of Nursing; Jacksonville, FL

[1] This is an older reference, but still an excellent summary: Woolhouse, M., & Gaunt, E. (2007). Ecological origins of novel human pathogens. *Critical Reviews in Microbiology*, 33(4), 231–242.
[2] Ibid.

10.1 BASICS OF HOST–MICROBE INTERACTIONS

Learning Outcomes

After reading this section, you should be able to:

10.1 Describe host–microbe interactions, and explain how they can foster health or lead to disease.

10.2 Discuss how shifts in normal microbiota levels or location can promote disease.

10.3 Explain why a commensal organism in one host could be a pathogen in another host.

10.4 Explain what tropism is, and discuss how it can influence pathogen emergence.

Host–microbe interactions are not always harmful.

Although this chapter mostly focuses on disease-related exchanges between microbes and people, it's important to remember that **host–microbe interactions** are a dynamic give-and-take, and what results is not necessarily damaging to the host.

Normal microbiota (flora) colonize virtually every part of the human body—there is even evidence of microbiota in tissues previously thought to only harbor microbes during an infection (e.g., blood and lungs). The densest microbiota populations tend to exist on our skin as well as in certain areas of the digestive, genital, urinary, and respiratory systems. In return for a place to live, most members of our microbiota are helpful to us. For example, they can compete with potential pathogens, make vitamins for us, or even promote the maturation of our immune system. Given the reciprocal benefit of this relationship, these microbes are said to have mutualistic relations with us.[3]

Disrupting normal microbiota balance can compromise a patient's health. One example of this microbiota disruption, or **dysbiosis**, occurs when a course of antibiotics kills off some of the normal microbiota in the intestines. This decrease in the normal "gut" microbiota levels reduces local competition and can allow a pathogen like *Clostridioides difficile* (previously known as *Clostridium difficile*) to successfully establish an infection in the large intestine and flourish. Just as it's difficult to find a parking space in a packed parking lot, if the intestinal tract is full of thriving normal microbiota (full of parked cars), *C. difficile* is less likely to find a "space" to park itself (establish disease), unless some of the existing "cars" (normal microbiota) are removed. Patients in hospitals or long-term care facilities are generally at a greater risk for antibiotic-associated diarrhea caused by *C. difficile* (FIG. 10.1).

As already mentioned, normal microbiota are generally helpful to us, but differences in host factors can render a harmless species in one host pathogenic in another—that is, the host–microbe interaction can shift. An example of this is seen in Group B streptococci (GBS) infections. Up to 30 percent of women harbor GBS as normal commensals in the vagina, but the same bacterium in newborns is associated with sepsis, meningitis, and pneumonia. The reason for this is that the GBS doesn't encounter the same host factors in a newborn as it does in the mother—namely, the baby has reduced immunity and lacks a firmly established normal microbiota to compete with the GBS. Most GBS infections in newborns are due to exposure to the bacterium during a vaginal birth. Before pregnant women were routinely screened for GBS, the U.S. Centers for Disease Control and Prevention (CDC) noted that about 2 to 3 babies per 1,000 live births developed GBS disease. Today, pregnant women who are GBS carriers receive intravenous antibiotics during labor—a practice that the CDC states has reduced infant GBS incidence in the United States to about 0.22 cases per 1,000 live births per year.

The immune system recognizes our normal microbiota's presence and even mounts a moderate response to it, but a balanced communication between our immune system and these resident microbes normally prevents

[3] Recall from Chapter 1 that mutualism is a type of symbiotic relationship in which both participants benefit. In comparison, commensalism is a relationship in which one organism benefits and the other is not affected. It was formerly thought that our normal microbiota had a commensal relationship with us, but based on the aforementioned benefits that normal microbiota confer on us, it is increasingly evident that we have a mutualistic relationship.

FIGURE 10.1 ***C. difficile* and the balance of host–microbe interactions in the large intestine** The normal microbiota of the large intestine contains diverse species that help to keep pathogenic members in check. When the normal microbiota are killed off with antibiotics, a pathogen such as *C. difficile* encounters decreased competition and may flourish. The pathogen benefits, while the host is harmed.

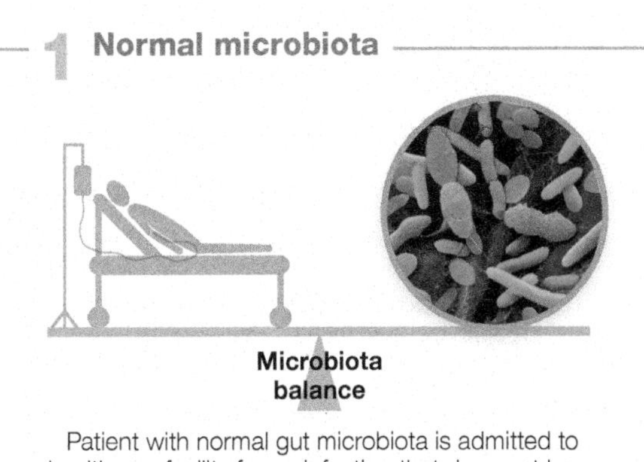

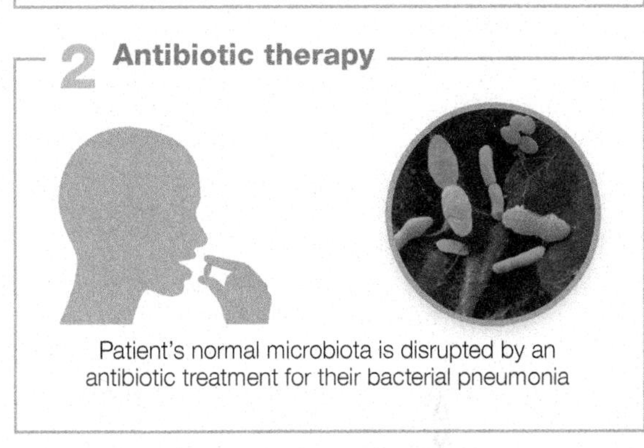

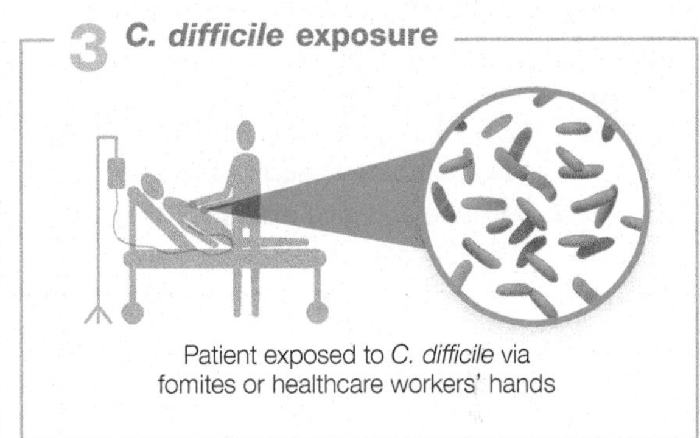

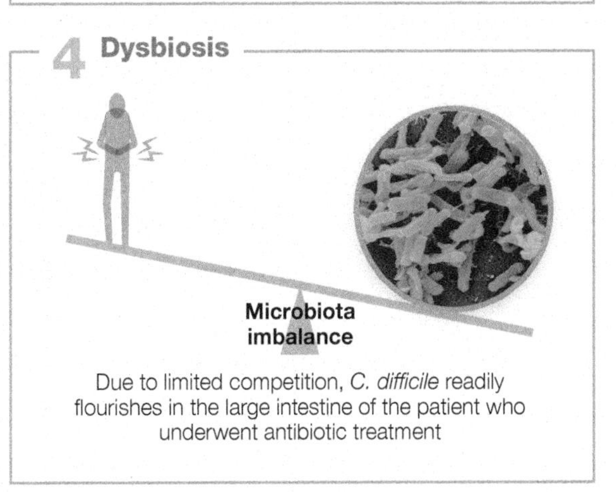

a full-blown immune attack. A sort of cease-fire treaty exists between them and us. However, occasionally microbes living under this peace treaty break their end of the deal and invade other tissues to cause disease. For instance, *E. coli* lives harmlessly in the appendix, but if it invades the abdominal cavity, as occurs when the appendix ruptures, then it's attacked by the immune system.

Under certain circumstances, a species harbored among the normal microbiota can become an *opportunistic pathogen*. For instance, the yeast *Candida albicans* is found among normal microbiota, most notably in the mouth, in the vagina, and on the skin. But these fungi may cause disease if they encounter a weakened immune system or face reduced competition as a result of a disruption in the normal microbiota from something like antibiotic usage. For example, when a woman takes an antibiotic that disrupts the balance of lactobacilli bacteria in the vagina (common microbiota in that tissue), then the resident *C. albicans* can thrive and cause vulvovaginal candidiasis (a vaginal yeast infection).

Why does the body perceive some microbes as friends and allow them to persist, while tagging others as foes? There are several interesting hypotheses on this topic, but the truth is that we don't yet fully understand why some microbes are tolerated as normal microbiota and others are not. However, it is clear that the host–microbe interaction is delicately balanced and depends on host factors just as much as microbial features.

Tropism is the preference of a pathogen for a specific host.

All microbes typically require specific host features to establish infection. The preference of a pathogen for a specific host (and even a specific tissue within the host) is called **tropism**. For example, *Shigella flexneri*, a bacterium that causes dysentery, preferably infects the intestines of humans and other primates. Most microbes exhibit some level of tropism, but it's important to understand that this factor can change over time. In fact, an estimated 60–80 percent of new infectious diseases in humans emerged as a result of expanded host tropism.[4] Bats, rodents, birds, and nonhuman primates are all common sources of emerging human pathogens. HIV, which jumped from nonhuman primates into humans about a century ago, and SARS-CoV-2, which is thought to have originated in bats, are examples.[5]

The fact that a pathogen has successfully invaded its preferred host and made its way to its favored tissue doesn't necessarily mean it will cause severe disease—or any disease at all. Factors specific to each host (e.g., age, gender, and overall health and fitness) interact with pathogen-specific factors, which we'll discuss in the next section, to influence the likelihood and severity of disease.

[4] Morens, D. M., & Fauci, A. S. (2013). Emerging infectious diseases: Threats to human health and global stability. *PLoS Pathogens*, 9(7), e1003467.

[5] dos Santos Bezerra, R., Valença, I. N., de Cassia Ruy, P., et al. (2020). The novel coronavirus SARS-CoV-2: From a zoonotic infection to coronavirus disease 2019. *Journal of Medical Virology*, 92(11), 2607–2615.

Build Your Foundation

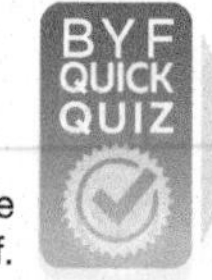

Build Your Foundation (BYF) Quick Quiz: Visit the **Mastering Microbiology** Study Area to quiz yourself.

1. What are host–microbe interactions, and how do they impact health and disease? (LO 10.1)
2. What effects could changing the level or location of normal microbiota have? (LO 10.2)
3. Why would a commensal microbiota in one host be potentially dangerous in another host? (LO 10.3)
4. Define tropism as it relates to pathogens. (LO 10.4)

10.2 INTRODUCTION TO VIRULENCE

Learning Outcomes

After reading this section, you should be able to:

10.5 Define *pathogenicity* and *virulence*.

10.6 Describe various virulence factors, and provide examples of how host–microbe interactions impact pathogen virulence and transmission.

10.7 Discuss what R_0 and R_e values reveal, and state when these values may be useful in epidemic management.

10.8 Explain why virulence is best viewed as an evolving property.

10.9 Define the term *attenuated pathogen*.

10.10 Distinguish between ID_{50} and LD_{50}.

10.11 Compare and contrast endotoxins and exotoxins, and explain how endotoxins can contribute to septic shock.

10.12 Describe and give examples of the three main types of exotoxins.

Host–microbe interactions influence virulence.

Pathogenicity is the ability of a microbe to cause disease. It is a term that is used to describes an all-or-nothing feature. In contrast, **virulence** describes the *degree or extent* of disease that a pathogen causes. The various strains of influenza, for example, share pathogenicity, but some tend to cause mild disease, whereas others are associated with widespread morbidity and mortality. **Virulence factors** are the mechanisms pathogens use to overcome host defenses. Making virulence factors requires an energy investment, so most pathogens only develop and keep virulence factors that benefit them. Beneficial virulence factors often include an improved ability to adhere to host cells, an enhanced ability to invade host tissues, more efficient ways to acquire nutrients, and mechanisms to evade host immune defenses (FIG. 10.2). Virulence factors can also include toxins—substances with diverse ways of thwarting the immune system and/or damaging cells.

Host Factors and Virulence

It was previously thought that a microbe's features were the sole determinants of virulence. But it's now understood that how adept a microbe is at causing disease (its overall virulence) also relates to host properties, including immune system fitness. Whereas some virulence factors directly damage host cells, others provoke an immune response that may ultimately be more dangerous. For

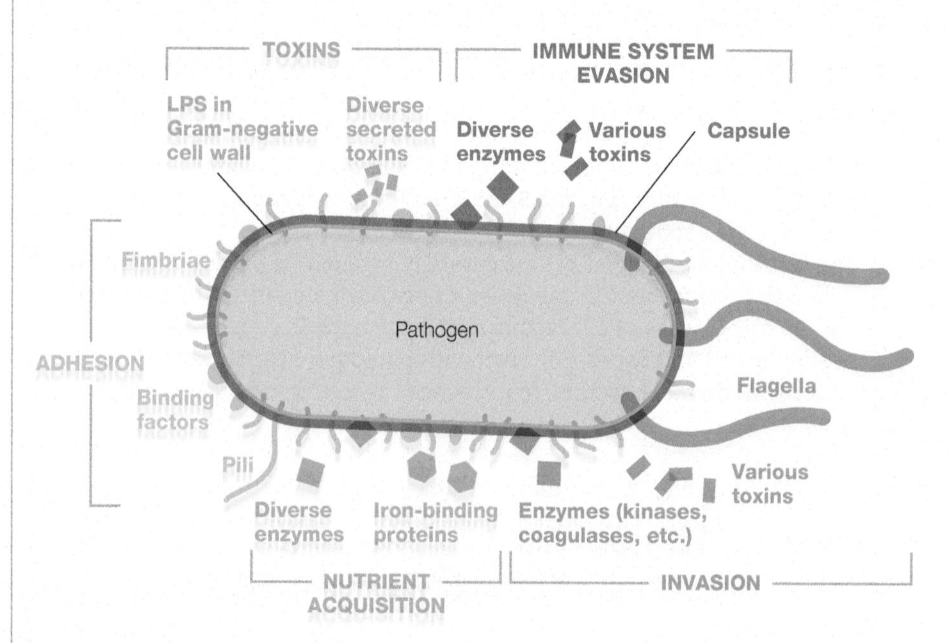

FIGURE 10.2 **Overview of virulence factors**

example, the influenza pandemic of 1918 killed millions and was notable for being more virulent in young adults than in older patients. Notably, at least 99 percent of the deaths during this pandemic occurred in patients younger than age 65.[6] This is the opposite of typical flu outbreaks, in which the elderly usually suffer higher mortality rates. Many decades later, researchers hypothesize that the particular influenza strain involved probably stimulated an overreaction of the immune system in healthy young adults, leading to fatal tissue and organ damage. By contrast, older adults with less agile immune systems likely survived in greater numbers precisely because they didn't mount the same overzealous response.

Of course, the ability to establish an infection doesn't always mean a pathogen is particularly virulent in every host. For example, poliovirus infections are asymptomatic in most hosts, but some infections cause paralysis. And SARS-CoV-2 infections are more likely to be asymptomatic in children as compared to adults.[7] SARS-CoV-2 also exhibits a differential virulence in men versus women—despite having the same infection prevalence as women, men are statistically more likely to suffer severe COVID-19 and die (case fatality ratio is about 2.4 times higher for men as compared to women).[8] The differences in disease manifestations for polio and COVID-19 from one host to another are likely due to diverse host factors, not just immune system fitness.

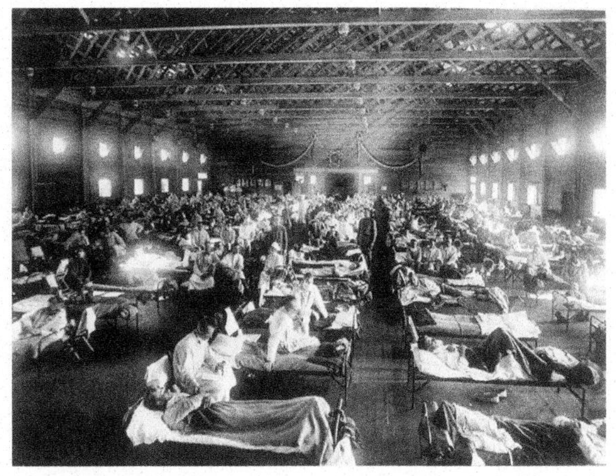
Photo of 1918 flu pandemic

Transmission and Virulence

To endure over time, a pathogen's ability to overcome host defenses and live within host cells and tissues has to be balanced with its ability to promote transmission to others. Agents that trigger a rapidly progressing symptomatic infection that quickly debilitates the host are more likely to cause isolated outbreaks. In contrast, viruses that have a less aggressive onset and can transmit readily from one person to another, even if the host is asymptomatic (as seen for SARS-CoV-2), are more likely to cause pandemics. Because evolving viruses can lead to pandemics, emerging disease surveillance is central to protecting public health. Indeed, virus tracking programs are actively monitoring SARS-CoV-2 evolution (e.g., monitoring for the emergence of new strains with enhanced transmissibility) as well as a variety of other pathogens like MERS-CoV, a coronavirus that causes Middle East Respiratory Syndrome.

From an evolutionary standpoint, it makes sense that virulence factors are often linked to transmission. For example, someone who feels terrible isn't likely to spread a sexually transmitted disease, so it doesn't make sense for such a pathogen to quickly incapacitate or rapidly kill the host. Accordingly, most sexually transmitted pathogens evolve virulence factors that allow them to be infectious, yet minimally symptomatic, and/or quietly persistent for long periods in the host before becoming symptomatic. For example, someone could be infected with HIV and not know for months or even years, during which time they may unknowingly transmit the virus to others. Two of the most common sexually transmitted infections, chlamydia and gonorrhea, are asymptomatic in most women and even in many men, making it easy for this bacterium to spread through the human population.

If you learn how a pathogen is transmitted, you can often also deduce what virulence factors it has (or could develop, in the case of emerging diseases). The

[6] Taubenberger, J. K. (2006). The origin and virulence of the 1918 "Spanish" influenza virus. *Proceedings of the American Philosophical Society*, 150(1), 86.

[7] The true rate of asymptomatic SARS-CoV-2 infection is emerging. So far, the data supports the conclusion that children are statistically more likely than adults to have an asymptomatic infection. He, J., Guo, Y., Mao, R., & Zhang, J. (2021). Proportion of asymptomatic coronavirus disease 2019: A systematic review and meta-analysis. *Journal of Medical Virology*, 93(2), 820–830.

[8] Sharma, G., Volgman, A. S., & Michos, E. D. (2020). Sex differences in mortality from COVID-19 pandemic: Are men vulnerable and women protected? *Case Reports*, 2(9), 1407–1410.

bacterium that causes cholera transmits through diarrhea, so strains of *Vibrio cholerae* that generate profuse diarrhea have a better chance of spreading. Similarly, most colds spread via coughing and sneezing. The cold viruses that stimulate the most coughs and sneezes while still allowing the host to be mobile and interact with others have an evolutionary advantage. Waterborne and vectorborne diseases, infections from endospore-forming pathogens, and antibiotic-resistant healthcare-acquired infections all tend to feature pathogens that don't rely on host mobility for transmission. As such, they can increase their virulence without compromising the chance of transmission. This explains why waterborne cholera or vectorborne malaria can be so debilitating to their hosts, yet evolutionarily persist.

Ultimately, pathogens that are more easily transmitted from one host to a new host become more prevalent in populations. This is in keeping with the evolutionary tenet of natural selection—that those organisms that "survive" over time are the ones best equipped to propagate, or reproduce. A pathogen's **basic reproduction number**, or $\mathbf{R_0}$ **(R-naught)** value, is a measure of a pathogen's transmissibility, or contagiousness—it represents the number of people that a single infected person is, on average, expected to infect in a population where all people are susceptible to infection and no prevention strategies have been applied. For example, if R_0 is 2.0, then one infected person is expected to infect an average of two other people in a fully susceptible population (no one is vaccinated against the pathogen, and no one has had the disease before) that is not subjected to any preventions against the pathogen (e.g., no public health measures/precautions are in place). A pathogen's R_0 value is calculated using complex mathematical models; it is an *estimate* of pathogen contagiousness based on human behaviors as well as pathogen features. R_0 values are useful for assessing the possible trajectory of an outbreak caused by a newly identified pathogen. For example, pathogens with R_0 values above 1.0 are likely to spread in a population and could progress to an epidemic. Those with R_0 values less than 1.0 are not especially contagious and are unlikely to progress to an epidemic. An R_0 closer to 1 suggests the disease may remain stable in a population for some time without interventions, but is unlikely to progress to an epidemic. R_0 estimates for SARS-CoV-2 ranged from 2.24 up to 3.58—a range that, coupled with the mortality data available at the time, suggested huge death tolls if countries did nothing to reduce spread.[9]

Because R_0 is mainly used to describe disease spread in a fully susceptible population, the **effective reproduction number**, or $\mathbf{R_e}$ (also called R_t or simply the R value), is more appropriate to consider in the midst of epidemics and pandemics. R_e values can change as host–pathogen interactions change. For example, as more people take precautions against transmission, or as an effective vaccination is made available, R_e is likely to decrease. On the other hand, if the pathogen evolves to become more easily transmitted, as has occurred for some SARS-CoV-2 strains, then (provided all other factors that could affect transmission remain the same) R_e may increase.

A pathogen's environment influences virulence.

Through interacting with their environment, including responding to the selective pressures we impose on them, pathogens evolve new virulence factors—a point that is illustrated by antibiotic resistance. Ampicillin used to be a first-line treatment against *Salmonella typhi* until widespread drug resistance discouraged practitioners from prescribing it. Many years after we abandoned ampicillin

[9] Rabi, F. A., Al Zoubi, M. S., Kasasbeh, G. A., Salameh, D. M., & Al-Nasser, A. D. (2020). SARS-CoV-2 and coronavirus disease 2019: What we know so far. *Pathogens*, 9(3), 231.

therapy against it, the bacterium is again developing susceptibility to the drug.[10] Upon removing the selective pressure (ampicillin), the virulence factor (ampicillin resistance) became inconsequential for survival. Consequently, those *Salmonella typhi* strains with virulence factors that enabled them to survive a more immediate environmental challenge—perhaps some other drug—became concentrated in the population as compared to strains that were less equipped to survive the most pressing environmental selective pressure.

Another example of environmental factors influencing virulence can be observed in pathogens grown in cell culture. In the harmless conditions of a petri dish, pathogens aren't fighting the host's immune response. Consequently, they often lose their former virulence factors, becoming **attenuated**—still infectious, but weakened to the point that they do not cause disease in an immune-competent host. In some cases, attenuated pathogens are used in vaccines. (See Chapter 14 for more on vaccines.) In short, a pathogen's virulence is best viewed as an evolving property because it changes in response to host factors as well as environmental factors.

The dosage of pathogen and toxin exposure affects host health outcomes.

A threshold exposure dose is required for a pathogen to successfully establish an infection in a new host. The **infectious dose-50 (ID_{50})** describes how many cells (bacterial, fungal, or parasitic) or virions are needed to establish an infection in 50 percent of exposed susceptible hosts. The more infectious (easy to "catch") the pathogen is, the lower the ID_{50}. Fortunately, the fact that a pathogen is highly infectious doesn't necessarily mean it is especially dangerous; cold viruses are highly infectious but rarely deadly.

As with the infection dosage, there is a threshold dosage for toxins to trigger health effects in a host. That dosage, called the **lethal dose-50 (LD_{50})** describes the amount of toxin needed to kill 50 percent of affected hosts who are not treated. For toxin-producing pathogens, you may see an ID_{50} for the infectious agent and an LD_{50} for its toxin. Note that the lethality of all pathogens (not just toxin-producing pathogens) tends to be expressed as mortality rate, not as an LD_{50}.

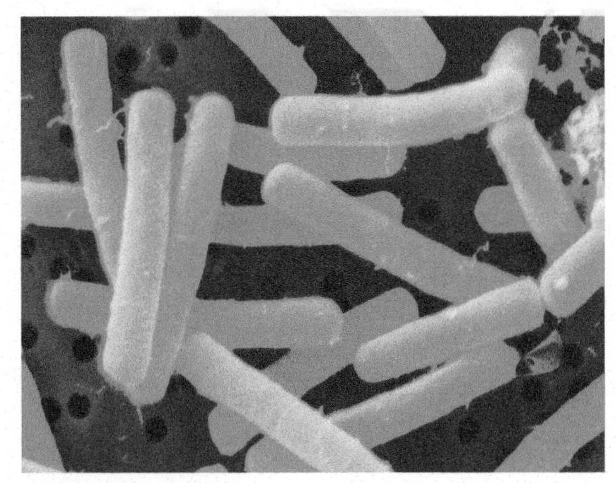

SEM of *Bacillus cereus*

ID_{50} and LD_{50} can change based on the species affected, the host's immune fitness, and the route of exposure (TABLE 10.1). For example, the lethal dose of botulinum toxin is lower when the toxin is injected than when it is inhaled. Also, LD_{50} and ID_{50} measures usually come from animal studies, so they are not a perfect predictor of what would occur in people. In general, you should view these numbers as estimates, not law; they just help us compare the potential hazard associated with a pathogen or toxin.

TABLE 10.1 Examples of Infectious Dose and Lethal Dose

Agent	Disease	ID_{50}	LD_{50}
Bacillus anthracis	Anthrax	Cutaneous anthrax: 10–50 bacterial spores Respiratory anthrax: 10,000–20,000 spores Gastrointestinal anthrax: 250,000–1 million spores	8.4 $\mu g/kg$ body weight (for *B. anthracis* lethal toxin in canines)
Botulinum toxin A	Botulism (neurotoxicity; toxin from *Clostridium botulinum*)	Not applicable (Toxins do not have an ID_{50} because although they are often made by infectious agents, they are not themselves infectious.)	0.001 $\mu g/kg$ body weight (injected or oral) 0.07 $\mu g/kg$ body weight (inhaled) (Based on studies in nonhuman primates)

[10] Kaushik, D., Mohan, M., Borade, D. M., & Swami, O. C. (2014). Ampicillin: Rise, fall, and resurgence. *Journal of Clinical and Diagnostic Research*, 8(5), ME01–ME03.

TABLE 10.2 Comparing Endotoxins and Exotoxins

Properties	Endotoxins	Exotoxins
Made of	Lipid	Protein
Made by	Gram-negative bacteria	Gram-negative and Gram-positive bacteria
Released from	Gram-negative cell wall when bacteria divide or die	Actively growing bacteria
Vaccines	No	Yes (some)
Fever	Yes	Sometimes (certain superantigens)
Can be neutralized in patient	No	Yes (some)
Toxicity level	Lower (relatively high LD_{50})	Higher (many have a low LD_{50})

Toxins are major virulence factors.

Toxins are molecules that, in small amounts, generate a range of adverse effects in the host. Examples of toxin effects include tissue damage and suppressed immune response. A microbe that makes toxins is said to be **toxigenic**. Although toxins may start out acting locally in the host, they cause serious problems when they enter the bloodstream and spread throughout the body—a situation called **toxemia**. You should note that a patient does not necessarily have to have an infection with a toxigenic bacterium to be exposed to its toxin(s). For example, *Bacillus cereus* makes a heat-stable toxin that causes food poisoning. Even if the food is heated to a point at which the bacteria are killed, the heat-stable toxin is not inactivated and can affect anyone who ingests it. Some fungi and algae make toxins, but bacteria tend to make most of the toxins seen in clinical scenarios. Therefore, we'll focus on bacterial toxins here. Endotoxins and exotoxins are the two main classes; they are compared in TABLE 10.2.

Endotoxins

All Gram-negative bacteria have an outer membrane rich in lipopolysaccharide (LPS). LPS consists of a lipid portion (**lipid A**) and associated sugars. The lipid A region of LPS is called **endotoxin**; it is toxic to us and other animals. Gram-negative bacteria mainly release endotoxin when they die, although a small amount can be released as the bacteria divide (see FIG. 10.3). Gram-positive bacteria do *not* make endotoxin.

As the immune system and/or antibiotics kill Gram-negative pathogens, the circulating endotoxin levels increase in the body and cause symptoms, including fever, chills, body aches, hypotension (pathologically low blood pressure), tachycardia (increased heart rate; pronounced TACK ee *car* dee uh), increased respiratory rate, inflammation, a feeling of disorientation, nausea, and vomiting. If present in sufficient quantities, endotoxin causes septic shock (or *endotoxic shock*) and may kill the host as organs fail. It is estimated that septic shock kills at least 11 million people every year globally, placing it among the leading causes of death worldwide.[11]

Endotoxin enters the bloodstream (*endotoxemia*) from localized or systemic infections, or when Gram-negative bacteria that are part of the normal microbiota are introduced to areas where they don't belong. A number of conditions, including surgery complications, can lead to bacterial misplacement, but as mentioned previously, this scenario is classically demonstrated with appendicitis,

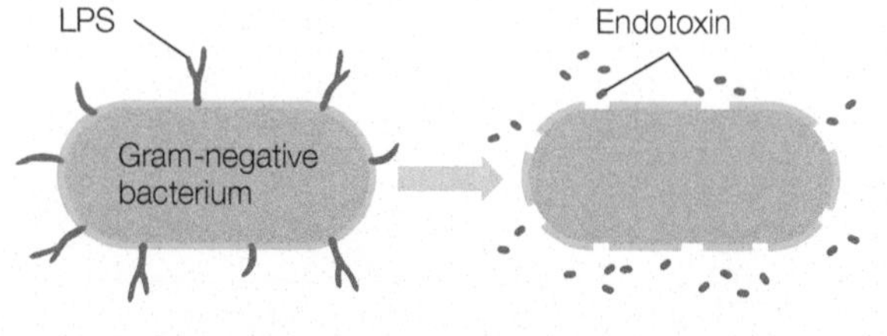

Lipid A (endotoxin) is a component of LPS in the outer membrane of cell wall.

Endotoxin is released when cell wall breaks apart

FIGURE 10.3 **Endotoxins**

[11] Kempker, J. A., & Martin, G. S. (2020). A global accounting of sepsis. *The Lancet*, 395(10219), 168–170.

(an inflammation of the appendix). Appendicitis requires prompt medical attention because if the appendix ruptures, Gram-negative bacteria leak into the abdomen, leading to a high endotoxin load that may lead to septic shock.

Lipid-based endotoxins are not readily neutralized, and no vaccines or other effective therapies exist to protect against them. Because of this, it is important to ensure that endotoxins don't contaminate anything used in patient care (drugs, implanted devices, etc.). Microbial control measures, decontamination practices, and aseptic methods, all of which were reviewed in Chapter 7, limit such contamination.

Exotoxins

Both Gram-positive and Gram-negative bacteria make **exotoxins**—toxic, soluble proteins that affect a wide range of cells. These protein toxins are often named based on the organism that makes the toxin or the type of cells the toxin targets. Neurotoxins affect the nervous system, enterotoxins target the gastrointestinal (GI) tract, hepatotoxins affect the liver, nephrotoxins damage the kidneys, and so on. To date, over 200 exotoxins have been described. Exotoxins are often classified into three main families based on their mode of action (FIG. 10.4).

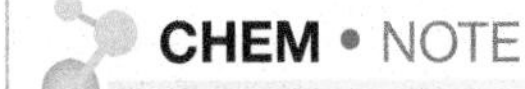

Solubility refers to a substance's ability to dissolve in a solvent, especially water.

Type I These membrane-acting extracellular toxins bind to specific receptors on the host cell's plasma membrane. They are often referred to as surface-acting toxins because, rather than entering the cell, they propagate a signaling cascade via the receptor they bind. This signaling cascade can evoke changes in gene expression in the cell that lead to diverse outcomes ranging from temporarily altered cell physiology to cell death. Examples of toxins found in this grouping include **superantigens**, which act by overstimulating the immune system to cause massive inflammation that harms the host. Superantigens also tend to cause fever (they are pyrogens). (The mechanism that superantigens use to activate our immune system is further discussed in Chapter 12.)

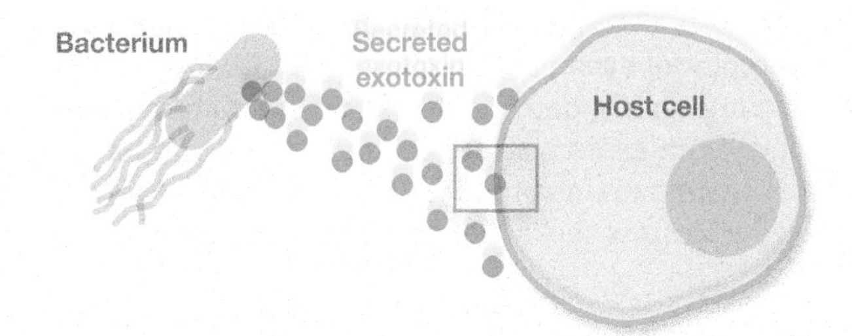

Type I

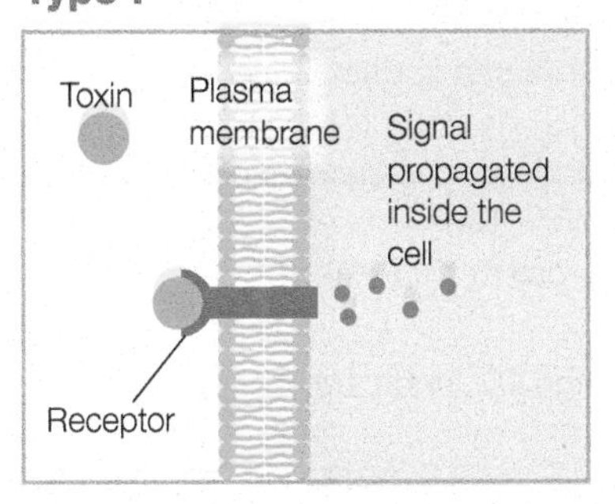

Toxin binds at host plasma membrane to generate a signal that generates effects; toxin doesn't enter cell.

Type II

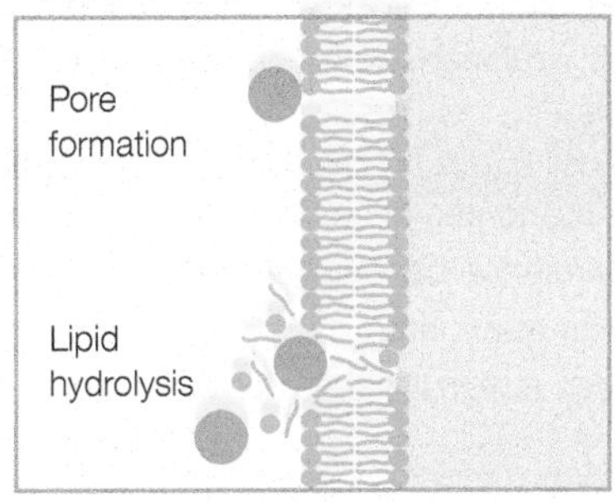

Toxins disrupt host cell membranes by forming pores or breaking down membrane lipids.

Type III

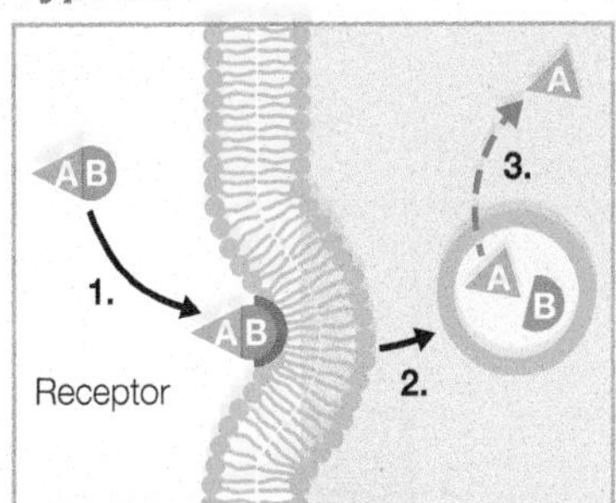

1. Binding portion (B) of toxin binds plasma membrane. **2.** Toxin enters cell, often by endocytosis. **3.** Active portion (A) enters the host cell and exerts an effect.

FIGURE 10.4 **Exotoxin families by modes of action**

TABLE 10.3 Exotoxin Examples

Exotoxin Family	Examples
Type I: Membrane-acting extracellular toxins	• Superantigens made by certain *Streptococcus* and *Staphylococcus* species cause toxic shock syndrome (TSS). • *Staphylococcus aureus* enterotoxins cause food poisoning. • Certain *E. coli* strains make heat-stable enterotoxins that cause diarrhea. • Erythrogenic toxins made by *Streptococcus pyogenes*; damage skin capillaries to produce a red rash characteristic of scarlet fever.
Type II: Membrane-damaging toxins	• Hemolysins: hemolytic toxins that lyse red and white blood cells to interfere with host immune response. Streptolysins are hemolysins made by certain streptococcal species. • Cytolysins interfere with host immune response by lysing white blood cells; they also may target general host cells to damage tissues; pneumolysins made by *Streptococcus pneumonia* (pneumonia, septicemia, meningitis) are an example. • Phospholipases are made by many bacteria, including *Clostridium perfringens* (gas gangrene), *Pseudomonas aeruginosa* (wound infections and pneumonia), and *Staphylococcus aureus* (skin/wound infections; pneumonia; sepsis).
Type III: Intracellular toxins	• Diphtheria toxin; cytotoxin made by *Corynebacterium diphtheria* (diphtheria); blocks protein synthesis. • Pertussis toxin: cytotoxin made by *Bordetella pertussis*, which causes whooping cough; suppresses host immune response. • Cholera toxin: enterotoxin made by *Vibrio cholerae*; induces a watery diarrhea. • Botulinum toxin: neurotoxin made by *Clostridium botulinum*; causes flaccid paralysis. • Tetanospasmin: neurotoxin made by *Clostridium tetani*; causes contractile paralysis that gives tetanus its common name, "lock-jaw."

Type II These membrane-damaging toxins disrupt the host cell plasma membrane by forming pores or by removing phosphate head groups from phospholipids of the lipid membrane (phospholipases). Removing the phosphate head groups destabilizes the membrane and causes cell lysis.

Type III These intracellular toxins bind to a receptor and then enter the cell to exert an effect. Most exotoxins in this group are **AB toxins**. The B (binding) part of the toxin binds a cell receptor, while the A (active) portion exerts effects inside the cell.

Examples of each of these three toxin families are presented in TABLE 10.3. Neutralizing agents, detection tests, and certain routine childhood vaccines protect us from exotoxins. For example, the DTaP is a childhood vaccine that protects against toxins produced by diphtheria, tetanus, and pertussis (whooping cough). Despite these measures, exotoxins are still associated with deaths.

Build Your Foundation

5. Define the terms *pathogenicity* and *virulence*. (LO 10.5)
6. What are virulence factors, and what host factors can impact virulence? (LO 10.6)
7. What do R_0 and R_e reveal about a pathogen, and when is each value most appropriate to use in epidemic management? (LO 10.7)
8. Why is virulence best viewed as an evolving property? (LO 10.8)
9. What is an attenuated pathogen? (LO 10.9)
10. What does a high ID_{50} mean? What would a high LD_{50} indicate versus a low LD_{50}? (LO 10.10)
11. Compare and contrast endotoxins and exotoxins, and explain how untreated endotoxemia can lead to septic shock. (LO 10.11)
12. Describe and give examples of the three main types of exotoxins. (LO 10.12)

Build Your Foundation (BYF) Quick Quiz: Visit the **Mastering Microbiology** Study Area to quiz yourself.

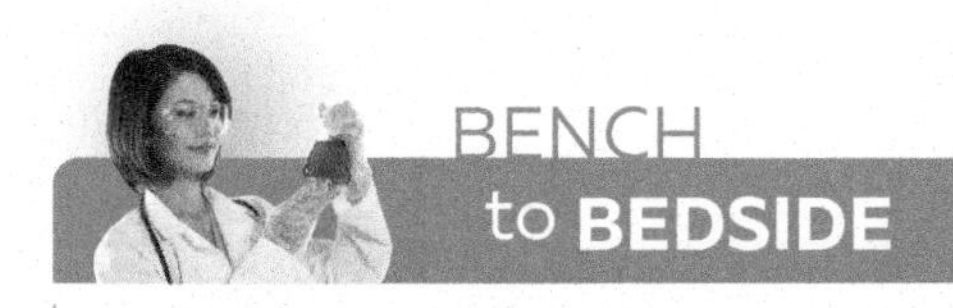

Time to Take Your Toxin

At first, it may be hard to imagine how poisonous bacterial toxins could do humans any good. However, bacterial toxins influence protein synthesis, cellular communication, and our immune response, abilities that can sometimes be tapped by humans for beneficial effects.

The best-known toxin therapy is use of the botulinum exotoxin (Botox®) to reduce facial wrinkles, but this toxin is useful in diverse clinical situations. For example, the paralysis that the toxin causes can be targeted at specific muscles to reduce dystonia—the involuntary, unregulated muscle contractions that accompany neurodegenerative disorders such as Parkinson's and Huntington's diseases. The botulinum exotoxin can also reduce painful muscle spasms that occur in multiple sclerosis and cerebral palsy. Moreover, the toxin is a potential therapy for certain forms of epilepsy; in experimental models, it has an antiseizure effect on the brain.

Another approach is to engineer bacterial toxins to specifically target cancer cells. Several clinical trials on engineered forms of diphtheria toxin and *Pseudomonas* exotoxin (PE) have shown these agents to be effective at killing a variety of leukemia cells. Also, an engineered form of lethal factor (LF) toxin from *Bacillus anthracis* is being tested as an antitumor agent.

Additional emerging uses for bacterial toxins include fighting viruses, enhancing memory, and treating pain. Pertussis toxin, which is made by *Bordetella pertussis*, the whooping cough bacterium, has been shown in cell culture to block certain HIV strains from entering cells. Additionally, this toxin keeps the HIV latent in already infected cells. Research with mice suggests that cytotoxic necrotizing factor 1 (CNF1), an exotoxin made by certain strains of *E. coli*, changes neuronal synapses in a way that improves learning and memory. In one study, a single dose of CNF1 improved memory in Alzheimer's dementia mouse models. Lastly, mouse models on inflammation-induced pain suggest that CNF1 may serve as a pain reliever. A great deal of research is still needed to bring these potential therapies into clinical use, but one day the phrase "time to take your toxin" may not be so far-fetched.

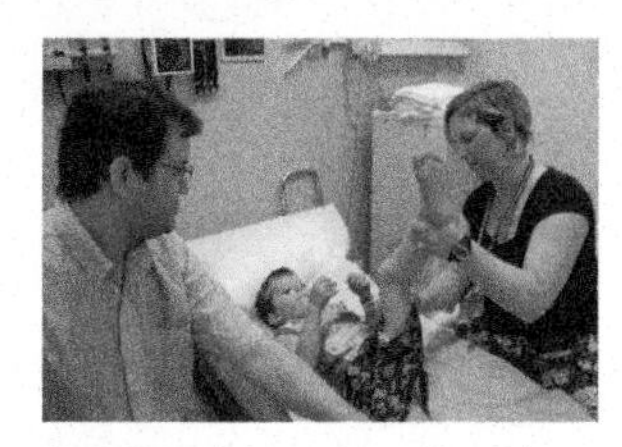

Botox can be used to relieve the painful spasms associated with cerebral palsy.

Sources: Rasetti-Escargueil, C., & Popoff, M. R. (2021). Engineering botulinum neurotoxins for enhanced therapeutic applications and vaccine development. *Toxins*, *13*(1), 1.

Sawant, S. S., Patil, S. M., Gupta, V., & Kunda, N. K. (2020). Microbes as medicines: Harnessing the power of bacteria in advancing cancer treatment. *International Journal of Molecular Sciences*, *21*(20), 7575.

Travaglione, S., Loizzo, S., Ballan, G., Fiorentini, C., & Fabbri, A. (2014). The *E. coli* CNF1 as a pioneering therapy for the central nervous system diseases. *Toxins*, *6*(1), 270–282.

10.3 FIVE STEPS TO INFECTION

To establish an infection, a successful pathogen must complete five general tasks: enter the host, adhere to host tissues, invade tissues and obtain nutrients, replicate while warding off immune defenses, and transmit to a new host. Next, we'll discuss various methods pathogens can employ to succeed at each of these tasks.

Learning Outcomes

After reading this section, you should be able to:

10.13 Identify the five tasks a pathogen must complete to successfully infect a host.

10.14 Identify various portals of entry and exit.

10.15 Name some adhesins and invasins, and describe their roles in establishing infections.

10.16 Discuss the most common tools that pathogens use to obtain nutrients.

10.17 Describe the various mechanisms that pathogens may use to avoid immune detection and elimination.

10.18 Explain the relationship between symptoms and mode of transmission.

10.19 Define the term *reservoir*, and provide examples of environmental and organismal reservoirs.

First, a pathogen must enter a host.

The first task involves a *portal of entry,* which is any site that a pathogen uses to enter the host. Because mucous membranes line every entrance of the body, they are key **portals of entry**. As you'll read in the upcoming sections, our mucous membranes have specialized defenses to protect against pathogen entry, and successful pathogens have virulence factors that overcome these barriers.

The portal of entry is often the site where disease develops, but it is not necessarily the only or even the main site affected. For example, systemic infections result when a pathogen proceeds from its initial portal of entry to enter the cardiovascular or lymphatic system. (Chapter 21 reviews these infections.) FIG. 10.5 shows potential portals of entry and lists examples of pathogens that use them. The portal of entry is often determined by the mode of transmission—such as fecal–oral, arthropod-borne, and sexual.

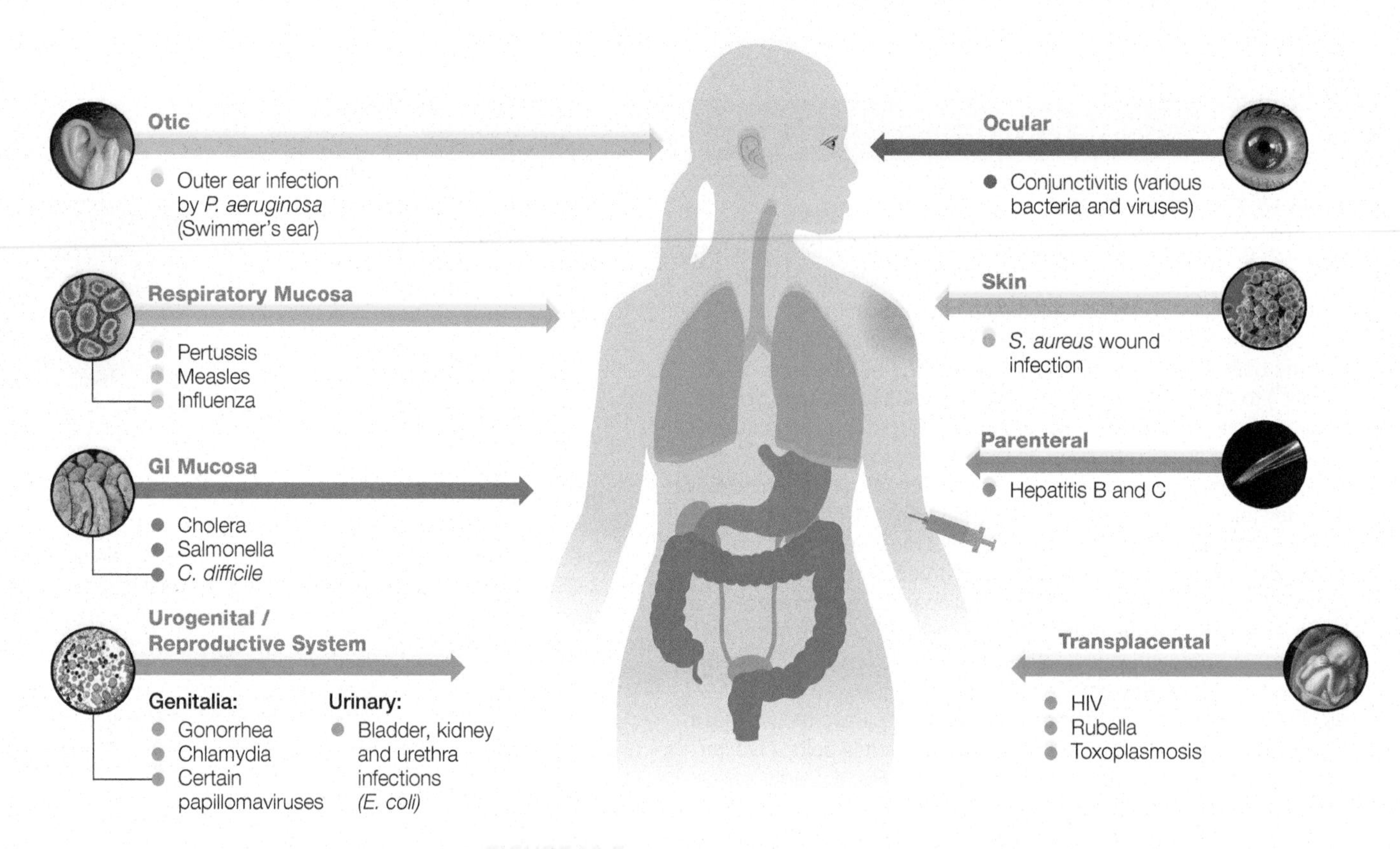

FIGURE 10.5 **Portals of entry** The balance between host–microbe interactions determines if an agent is received as normal flora or targeted for immune system attack and clearance.

We can think of pathogens as having keys that open various doors into the body. Some agents hold a key for only a single doorway, whereas others possess multiple keys and thus can enter by diverse portals. Influenza invades only through the respiratory tract, for example, but *Bacillus anthracis* can enter via the skin, GI tract, or respiratory system. The portal of entry used often impacts virulence. For example, *B. anthracis* entering through the skin causes cutaneous anthrax, which has a lower mortality rate than gastrointestinal or respiratory anthrax.

Skin, Ocular, Otic, and Parenteral Entry

Based on overall weight and surface area, the **integumentary system**, consisting of skin, hair, nails, and associated glands, is the largest body system. It blocks most microbes, but some have virulence factors that penetrate this essential barrier. Some pathogens undertake *otic entry*, meaning they enter via the ear. Pathogens may also exploit a skin wound, such as a burn or abrasion, to gain entrance—and some, like certain parasites, can physically bore through skin.

Continuous with the skin is an ocular mucous membrane called the **conjunctiva**. This thin membrane lines our eyelids and covers exposed eye surfaces. Pathogens that invade the conjunctiva cause inflammation called conjunctivitis. (See more on infectious skin and eye diseases in Chapter 17.)

A pathogen that successfully breaches the skin barrier may enter the bloodstream and spread systemically. Alternatively, a pathogen can bypass the skin and directly invade the underlying subcutaneous tissues, muscles, or bloodstream using **parenteral entry**. Bites, injections, and surgical incisions facilitate parenteral entry.

Respiratory Tract Entry

The respiratory tract is the most common portal of entry. As people cough and sneeze, pathogens travel through the air as respiratory droplets and aerosols, which another person may inhale. Multiple pathogens, including SARS-CoV-2, influenza viruses, and common cold viruses enter this way. Infectious agents may also settle in dust or live in soil and be inhaled when they are stirred up. We've noted that the portal of entry is not necessarily the site where a pathogen establishes the infection. For example, chickenpox virus (varicella-zoster virus) is usually transmitted through the mucous membranes of the respiratory tract, but the virus ultimately infects neurons. (See Chapter 9 for more on how diseases are transmitted and Chapter 16 for more on respiratory system infections.)

Gastrointestinal (GI) Tract Entry

Digestive system pathogens frequently spread through fecal–oral transmission; that is, they exit the body in infected stool and enter a new host via the mouth—the entrance of the GI tract. They then invade the mucosal surfaces throughout the GI tract and sometimes spread from there. Poliovirus enters through the GI mucosa and in some cases spreads to the nervous system, where it may cause paralysis. In contrast, many pathogens that don't enter via the GI tract eventually cause GI symptoms. (See Chapter 19 for more on digestive system infections.)

Urogenital Tract and Transplacental Entry

Most sexually transmitted pathogens enter through the mucosal lining of the vagina or cervix in women or the urethra in men. Some sexually transmitted pathogens invade through the skin of the genitalia. Pathogens that cause urinary tract infections invade the urethra in men and women. As reviewed in Chapter 9, some pathogens exhibit vertical transmission (passed from mother to child) by **transplacental entry**, meaning they cross the placenta to infect the fetus. (See Chapter 20 for urinary and reproductive system infections.)

Second, a pathogen must adhere to host tissues.

After entering a host, the pathogen must next adhere to host tissues. The initial adhesion is often nonspecific, such as through hydrophobic interactions. Following nonspecific anchoring, the agent may target an exact surface molecule on the host cell. Sometimes pathogens trick host cells into interacting with them by mimicking something to which the cell would normally bind. Species and tissue tropism is largely due to the specificity a pathogen has for a particular host cell surface marker. Because adhesion is an early step necessary for infection, it also represents a potential drug target to prevent disease.

Adhesion Factors

Adhesins are a major class of virulence factors that bacteria, fungi, protists, and viruses all use to stick to host cells in a specific or nonspecific manner. Some pathogens make a variety of adhesins that facilitate binding to different tissues. The thousands of bacterial adhesins we currently know of include cell wall components, capsules, extracellular appendages (such as fimbriae and pili), and a variety of plasma membrane–associated molecules. Because adhesion factors tend to be on the surface of pathogens, they make ideal vaccine development targets. (See Chapter 14 for more on vaccine formulations.) TABLE 10.4 presents examples of adhesins.

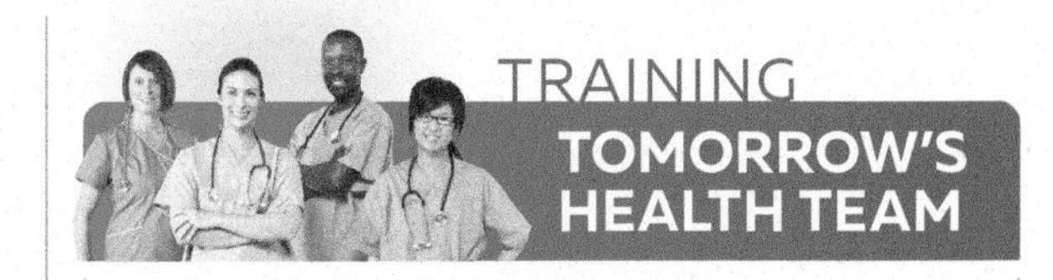

Don't Pass the TORCH

TORCH stands for toxoplasmosis, other agents, rubella, cytomegalovirus, and herpes simplex virus-2. These infectious agents can all pass to a developing fetus by transplacental entry to cause congenital (present from birth) infections. They all may also cause birth defects in the developing fetus, kill the fetus *in utero*, or cause death of the infant soon after birth.

TORCH pathogens cause a collection of clinical signs that include fever, poor feeding, jaundice, hepatosplenomegaly (enlarged liver and spleen), hearing impairment, eye abnormalities, and a characteristic "blueberry muffin rash" of dark purple blotches. Other manifestations vary according to the specific agent and the time during gestation that the fetus was exposed. Thankfully, vaccines against rubella and other agents like varicella zoster (chickenpox virus) have limited the number of TORCH syndrome babies. (For more on TORCH, see Chapter 20.)

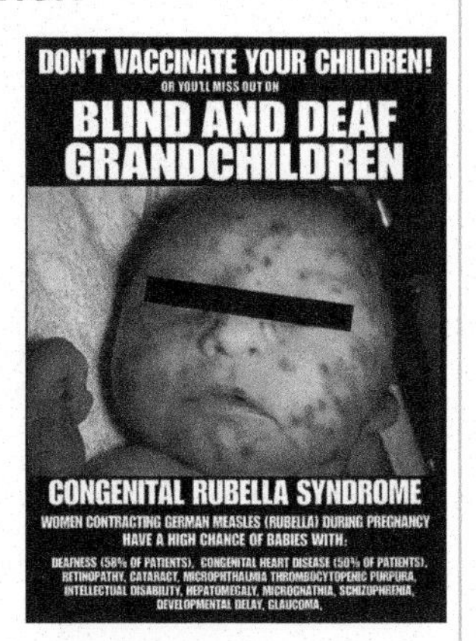

This image was developed by pro-vaccination advocates and circulated via social media. It is included here as a reminder of the emotional aspects that tend to accompany the vaccine debate. Women infected with rubella in their first trimester of pregnancy have up to a 90 percent chance of having a baby with congenital rubella syndrome, provided the baby survives.

QUESTION 10.1

TORCH infections cause systemic disorders in the developing fetus. What would you predict is the portal of entry into the fetus once these pathogens cross the placenta?

CHEM • NOTE

Hydrophobic interactions occur between nonpolar molecules in a watery (aqueous) environment. Oil droplets joining together in water is an example.

TABLE 10.4 Some Examples of Pathogen Adhesins

Adhesin	Mechanism	Examples
Fimbriae or pili	Extracellular hair-like appendages that bind carbohydrates on host cells	• *E. coli* GI and urinary system infections • *Pseudomonas aeruginosa* respiratory and wound infections • *Neisseria gonorrhoeae* (gonorrhea) • *Vibrio vulnificus* deep wound infections like cellulitis and necrotizing fasciitis
Sialic acid binding factors	Surface molecules that bind to sialic acid (diverse acidic sugar molecules on host cells)	• Rotaviruses (GI infections) • Influenza viruses • *Aspergillus fumigatus* (fungal lung infection) • Plasmodium falciparum (malarial parasite)
Heparan and heparin sulfate binding factors	Factors that target host cell heparan and heparin sulfate (acidic sugary molecules that are present in many cells and tissues)	• *Borrelia burgdorferi* (Lyme disease) • *Helicobacter pylori* (gastric ulcers) • Human immunodeficiency virus (HIV/AIDS) • Herpes simplex viruses
Fibronectin binding factors	Assorted surface molecules that bind to the protein fibronectin in host epithelial tissues	• *Streptococcus pyogenes* ("strep" throat) • *Staphylococcus aureus* ("staph" infections) • *Mycobacterium tuberculosis* (tuberculosis) • *Clostridioides difficile* (infectious diarrhea) • *Candida albicans* (yeast infection) • *Leishmania* species (protozoan infection leishmaniasis)

Biofilms and Quorum Sensing

Bacteria form biofilms on almost any natural or artificial surface, and, according to the National Institutes of Health, 60 to 80 percent of human infections originate from them. (For more on biofilm formation, see Chapter 1.) Common places for healthcare-acquired biofilm formation include implanted devices such as venous catheters that serve as portals to deliver medications, and implanted prosthetic joints. Rocky deposits in the kidneys (kidney stones) or gallbladder (gallstones) can also be biofilm formation sites (FIG. 10.6). Established biofilms—for instance, dental plaque, which may contain over 300 different bacterial species—often serve as an easy adhesion platform for agents that may otherwise struggle to stick to noncolonized surfaces.

Pathogens that are growing in biofilms often have enhanced virulence over their *planktonic* (free-floating) counterparts. This shift in virulence is primarily due to *quorum sensing*—a form of bacterial communication that occurs in biofilms. Quorum sensing affects gene expression patterns to increase community survival, and in many cases, it also enhances the expression of pathogen virulence factors. For example, most of the virulence factors that *Staphylococcus aureus* makes, including toxins that lyse white blood cells and other factors that help the bacterium evade host immune responses, are induced upon the bacterium growing in a biofilm.[12] Quorum sensing can also dictate when bacteria living in a biofilm will break free from it (returning to a planktonic state) and spread to other regions in a host.

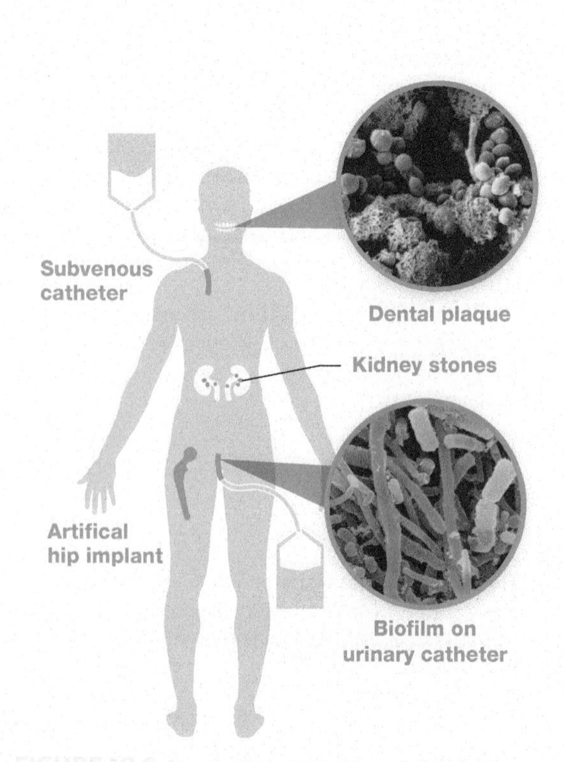

FIGURE 10.6 Examples of biofilm sites Biofilms can form on most natural and synthetic surfaces in the body, serving as a platform for pathogen adhesion.

Third, a pathogen must invade tissues and obtain nutrients.

Once a pathogen enters the host and adheres to tissues, there is a third task to fulfill: invading tissues further and, in the case of cellular pathogens, obtaining nutrients. Following adhesion, the pathogen has several options: It may stay on

[12] Cheung, G. Y., Bae, J. S., Liu, R., Hunt, R. L., Zheng, Y., & Otto, M. (2021). Bacterial virulence plays a crucial role in MRSA sepsis. *PLoS Pathogens*, 17(2), e1009369.

the surface of the host cell, pass through cells (or the junctions between cells) to invade deeper tissues, or enter cells to reside as an intracellular pathogen (FIG. 10.7). As pathogens accomplish invasion, they tend to damage host tissues and generate cytopathic effects (host cell damage).

Invasins and Motility

Invasins allow pathogens to invade host tissues. These local-acting factors are usually extracellular enzymes that pathogens secrete (release into their local environment) or have on their surface. Some break down host tissues, whereas others form blood clots or other structures that wall off the pathogen from host defenses that would otherwise prevent the pathogen from replicating. Sometimes, the pathogen has nonenzymatic surface molecules that have a dual role in adhesion and invasion. Such molecules cause changes in the host cell's cytoskeleton, inducing the cell to take up the pathogen. *Helicobacter pylori* (which causes gastric ulcers) and *E. coli* are examples of pathogens that use these dual-functionality adhesive/invasive factors to stick to and then enter host cells. Examples of invasins are presented in TABLE 10.5. Finally, motility (the ability to move) is another important invasion strategy that helps a pathogen spread. In addition to making movement possible, molecular features of flagella can also aid adhesion.

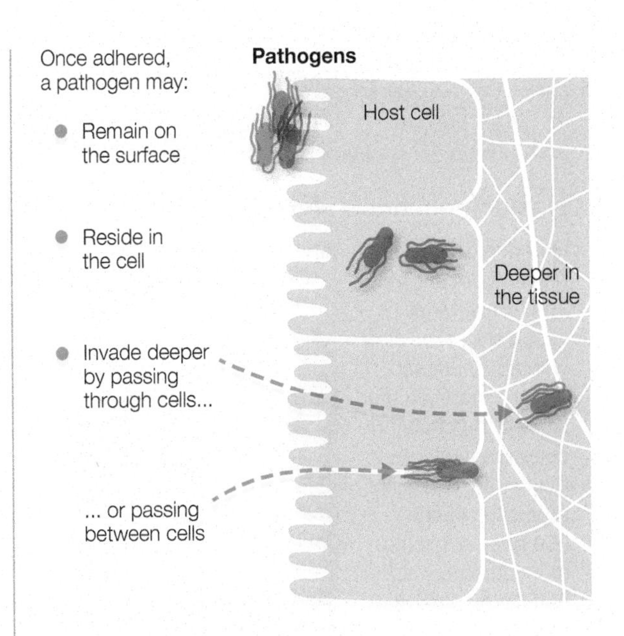

FIGURE 10.7 **Three main options for pathogen invasion**

Tools to Obtain Nutrients: Siderophores and Extracellular Enzymes

Cellular microbes—prokaryotes and eukaryotes—must acquire nutrients to survive. (This doesn't apply to viruses and prions; as you will recall from Chapter 6, those agents are classified as nonliving because they are acellular.) Most cellular pathogens require iron to survive, yet typically there is very little iron freely circulating in our tissues and blood. That's because a protein called *transferrin* binds to iron and shuttles it to tissues for incorporation into different molecules, including hemoglobin, the oxygen carrier found in red blood cells. This mechanism not only limits potentially toxic effects associated with large quantities of free iron in the body, but it also serves as a host defense by making our inner environment less nutrient-rich for microbial growth. To get around this defense, many bacteria produce iron-binding complexes called **siderophores**, which snatch iron from transferrin for their own use.

Many bacteria and fungi also make extracellular enzymes (cell-surface or secreted enzymes) that break down nutrients in the local environment. Doing so allows pathogens to scavenge nutrients as they damage host tissues.

TABLE 10.5 Invasin Examples

Invasin	Mechanism	Examples
Flagella	Motility enhances spread; molecular features of flagella may assist adhesion	• *E. coli* (GI and urinary system infections) • *Vibrio vulnificus* (deep wound infections like cellulitis and necrotizing fasciitis)
Collagenases	Enzymes that break down collagen, an important structural protein in tissues, allowing for invasion of new host areas	• *Clostridium perfringens* (gas gangrene) • *Vibrio vulnificus* (deep wound infections like cellulitis and necrotizing fasciitis) • *Streptococcus mutans* (dental caries; periodontal disease)
Neuraminidases	Enzymes that break down neuraminic acid, which has important roles in regulating host cell communications and membrane transport. Damage disrupts normal and immune cell functions.	• *Vibrio cholerae* (cholera) • Influenza virus • *Hemophilus influenza* (pneumonia/epiglottitis/Hib disease meningitis)
Coagulases	Enzyme that promotes blood clotting to form a protective layer around the pathogen	• *Staphylococcus aureus* (skin/wound infections; pneumonia; sepsis) • *Streptococcus faecalis* (urinary tract infections; meningitis; endocarditis)
Kinases	Enzymes that break down blood clots to allow pathogens to spread out of clots that may trap them	• *Streptococcus pyogenes* ("strep" throat) • *Staphylococcus aureus* (skin/wound infections; pneumonia; sepsis)

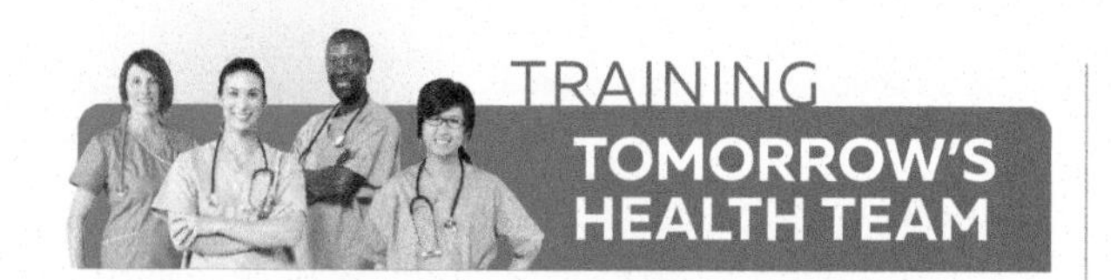

Hemochromatosis and Infections

Hemochromatosis results from elevated iron levels. Most cases are associated with an inherited genetic mutation that causes the patient to absorb and store too much iron. Affecting about 1 in every 250 people, it's the most common genetic disorder that healthcare providers in the United States are likely to encounter.

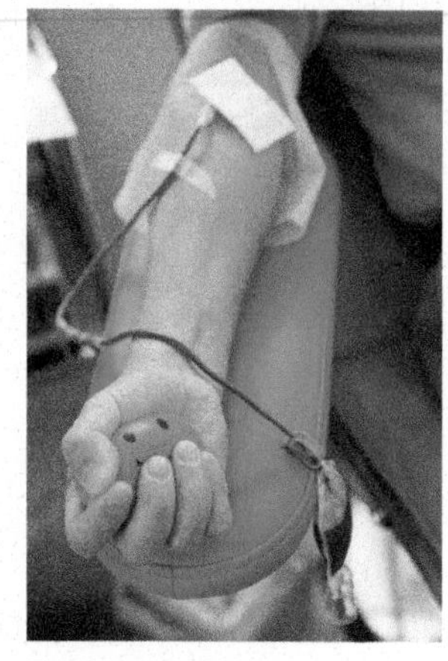

Therapeutic phlebotomy is the only treatment for hemochromatosis.

The total amount of iron in an adult's blood is normally between 2.5 and 3.5 grams, but patients with hemochromatosis may have total blood iron levels as high as 50 grams. This high iron load can cause skin bronzing and organ damage. It can also increase the patient's risk for bacterial infections because free iron is an essential—but normally scarce—bacterial nutrient. All international treatment guidelines agree that excess iron should be treated with venesection (removing blood by therapeutic phlebotomy) and that the usefulness of other therapies like strict dietary management and chelation therapies (treatments that involve intravenous infusion of metal chelating substances such as EDTA—ethylenediaminetetraacetic acid—into the bloodstream to bind free iron) is limited.

QUESTION 10.2

Why is it not surprising that people with hemochromatosis have an increased risk of bacterial infections?

Sources: Oliver, D. (2021). Hereditary hemochromatosis and iron overload disorders: A clinical review. *Lynchburg Journal of Medical Science, 3*(1), 51.

Powell, L. W., Seckington, R. C., & Deugnier, Y. (2016). Haemochromatosis. *The Lancet, 388*(10045), 706–716.

Extracellular enzymes like lipases (which break down lipids) and proteases (which break down proteins) are primarily used to obtain nutrients, but they also serve a secondary role in promoting invasion.

Cytopathic Effects in the Host

As pathogens establish themselves in the host, they can damage host cells and generate **cytopathic effects**. These effects can kill the cell (*cytocidal*), or simply damage it (*noncytocidal*). Bacteria induce cytopathic effects as they invade host cells, release toxins, and/or exploit host nutrients. Viral pathogens generate cytopathic effects when they hijack cellular machinery and disrupt normal host cell function. Viral release also tends to kill or damage host cells. *Oncogenic pathogens* are viruses or bacteria that transform normal cells into cancer cells. (See Chapter 6 for more on viruses.)

As noted earlier, the host's immune system also inflicts damage on the body as a byproduct of the tactics it uses to fight infections. For example, most of the tissue damage that develops in tuberculosis is due to the immune system's attack on cells infected with the bacterium *Mycobacterium tuberculosis*, rather than to effects of the bacteria itself. Similarly, most viral infections result in the immune system killing virus-infected cells. While this immune-mediated killing ends the ability of the virus to replicate in the targeted cell, it also can lead to tissue damage. For example, as the immune system targets SARS-CoV-2-infected lung cells, it can cause serious lung tissue damage. (Chapters 11 and 12 review how the immune system combats infections and the potential collateral host tissue damage that may accompany such efforts.)

Fourth, a pathogen must evade host immune defenses so that it can replicate.

Assuming a pathogen has made it this far (which most do not), it now must evade the immune system so it can safely replicate. The human body has evolved many defenses to prevent and overcome pathogen infections. (Immune defenses are discussed in Chapters 11 and 12.) But pathogens continually evolve countermeasures to escape host immune defenses. It is not uncommon for pathogens to employ multiple strategies at once. In general, these strategies either help the microbe evade detection (hide) or undermine pathogen-fighting tools of the immune system. TABLE 10.6 summarizes ways that pathogens avoid host defenses.

TABLE 10.6 Key Mechanisms for Escaping Host Immune Defenses

Approach	Examples
Hide	• Antigen masking, mimicry, and variation • Latency • Intracellular lifestyle
Undermine	• Suppress immune function ▪ Break down antibodies ▪ Infect immune system cells ▪ Block immune system signals ▪ Inhibit production of immune system factors • Avoid phagocytosis ▪ Make a capsule ▪ Block phagosome–lysosome fusion ▪ Neutralize hydrolytic enzymes in phagocytes ▪ Secrete toxin that damages phagocytic cells ▪ Evolve to thrive inside the phagolysosome

Hiding from Host Immune Defenses

To destroy a pathogen, first the immune system must find it. As already mentioned, pathogens may hide within biofilms that immune defenses struggle to penetrate and eliminate. Other common hiding techniques include living as intracellular pathogens, entering a latent state, and changing antigenic factors either to trick the immune system into thinking the pathogen is a harmless body cell that should be left alone or to evade circulating antibodies.

Intracellular Pathogens Rather than residing on or between cells, these pathogens spend the majority of their time inside host cells. All viruses, many protozoans, and some bacterial pathogens lead intracellular lifestyles. Examples of intracellular bacteria include *Listeria monocytogenes* (causative agent of listeriosis), *Mycobacterium tuberculosis*, and *Salmonella* species. Intracellular pathogens have special adaptations that allow them to reside inside host cells. Microbes that live inside phagocytic cells, for example, must avoid destruction during phagocytosis. These adaptations are discussed shortly.

Latency The ability of a pathogen to quietly exist inside a host is called **latency**. Pathogens with this capability cause persistent or recurrent disease. By remaining quiet inside host cells, latent pathogens stay under the immune system's "radar" and then reemerge to cause signs and symptoms at opportune moments.[13] Such pathogens frequently use a strike-and-retreat approach where they enter the host, perpetrate damage, and then retreat into a period of latency or dormancy during which the host immune system doesn't detect them. Numerous viruses, especially the herpes viruses that cause intermittent flare-ups in the host, are infamous for this approach. Other viruses like HIV invade host cells and act as genetic hitchhikers by integrating themselves into the host genome and waiting to emerge later. Bacteria can also exhibit latency. *M. tuberculosis* achieves this by walling itself off in the lungs to generate a latent infection that may or may not progress to active tuberculosis. The latent state protects the pathogen not only from the immune system, but also from drug therapies because most antibiotics target actively replicating bacteria.

Antigenic masking, mimicry, and variation An *antigen* is a foreign substance that triggers an immune response—for example a specific feature of a pathogen. Upon entering the host, the pathogen may conceal its antigenic features so that the immune system doesn't detect it. How this **antigen masking** is accomplished varies, but one common theme is that the pathogen coats itself with host molecules, allowing it to masquerade as part of the body. Another approach is **antigenic mimicry**, which involves imitating host molecules. Numerous bacteria make capsules that resemble host carbohydrates and thus do not stimulate an immune response. Viruses can also exhibit immune system evasion strategies—for example, SARS-CoV-2 viral RNA is associated with specialized proteins that allow the viral RNA to mimic host cell RNA and thereby evade immune detection and destruction.[14]

Other pathogens undergo **antigenic variation**, periodically altering the surface molecules that host immune cells could recognize and target as foreign. This variation delays the immune system's response. For example, *Trypanosoma brucei*, a protozoan that causes African sleeping sickness, switches its surface appearance. It covers itself in a thick coat of proteins, and just when the immune system is ready to mount an attack against that particular surface profile, it shifts production to a new protein, dodging the immune attack (FIG. 10.8). There are also other ways pathogens achieve antigenic variation.

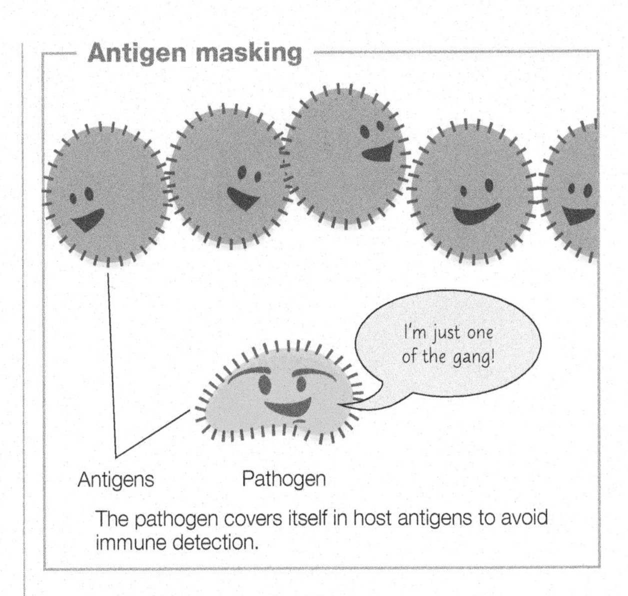

The pathogen covers itself in host antigens to avoid immune detection.

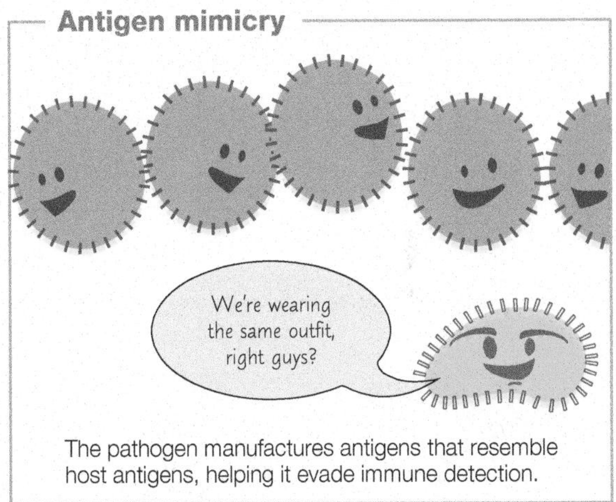

The pathogen manufactures antigens that resemble host antigens, helping it evade immune detection.

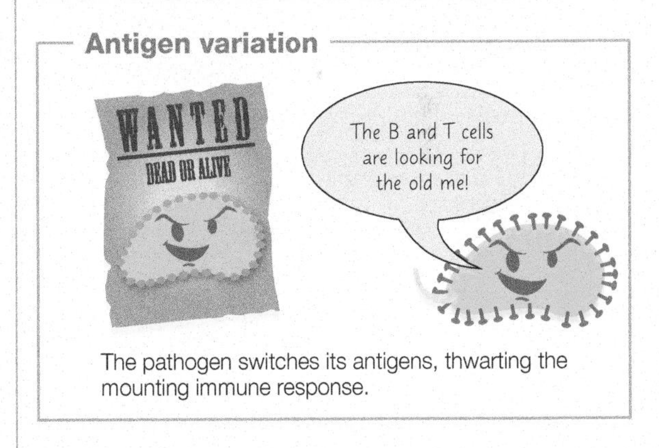

The pathogen switches its antigens, thwarting the mounting immune response.

FIGURE 10.8 Keeping terminology straight: Antigen masking, mimicry, and variation Silly but memorable ways to remember that pathogens use antigen masking, mimicry, and variation to evade immune detection.

---

[13] Conversational note for non-native speakers: The phrase "staying under the radar" is a direct reference to a plane remaining undetected by radar devices. Most commonly, this phrase is used to refer to remaining unnoticed.

[14] Astuti, I. (2020). Severe Acute Respiratory Syndrome Coronavirus 2 (SARS-CoV-2): An overview of viral structure and host response. *Diabetes & Metabolic Syndrome: Clinical Research & Reviews*, 14(4), 407–412.

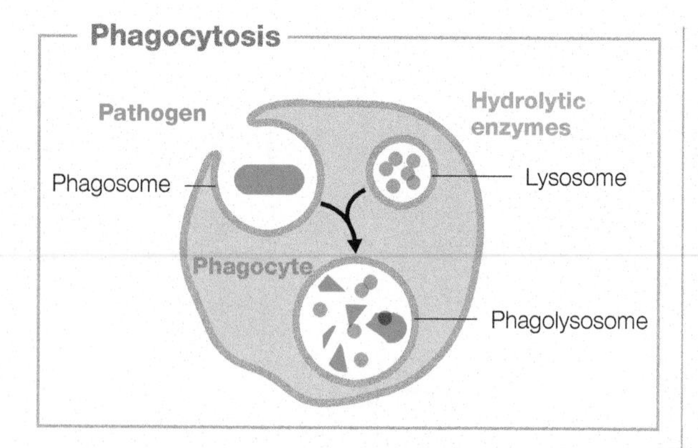

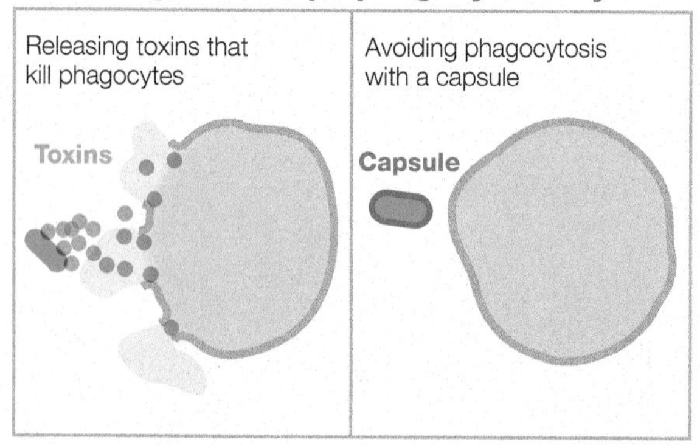

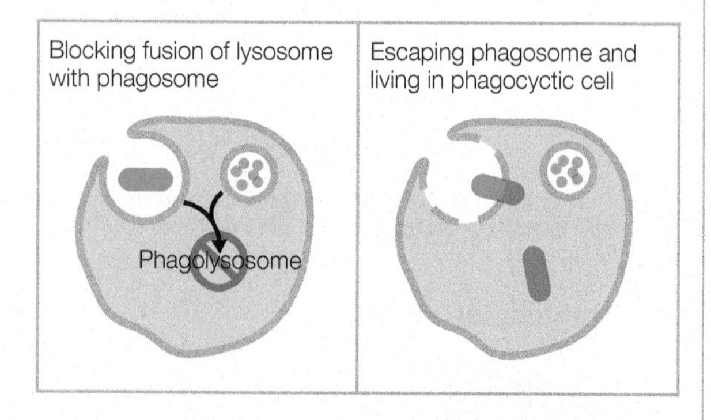

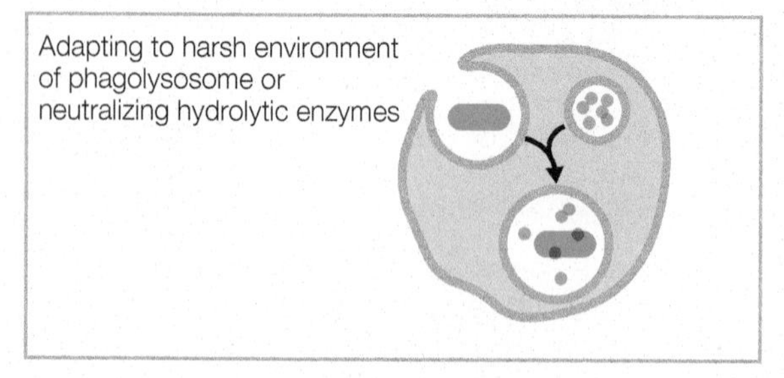

FIGURE 10.9 Escaping phagocytosis Pathogens have evolved many ways to escape phagocytosis.

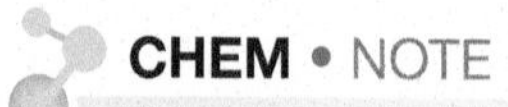

CHEM • NOTE

Hydrolytic enzymes use water to break chemical bonds. They are important for breaking down organic molecules in cells.

Sometimes, as with influenza viruses, a pathogen's genome is prone to mutation, leading to modified proteins. As previously mentioned, this tendency to undergo antigenic drift is why we create different flu vaccine formulations every year (see Chapters 6 and 16). SARS-CoV-2 is also adept at mutating—we'll eventually see the emergence of new SARS-CoV-2 strains that can evade immune protections conferred by a prior infection or vaccination. To limit cases, updated COVID-19 vaccines that are based on newer circulating SARS-CoV-2 strains will be necessary.

Undermining the Host Immune Response

Even if a pathogen can't avoid immune detection, it could limit the immune system's actions. Two key examples are interfering with phagocytosis and suppressing the immune response.

Interference with phagocytosis Immune system cells include phagocytes that engulf pathogens and then use hydrolytic enzymes to destroy them. But many pathogens have evolved mechanisms to escape phagocytosis (FIG. 10.9). Some make a capsule that protects them from phagocytic cells. Others, especially intracellular pathogens living inside macrophages, burst free of the phagosome (the vesicle that contains the phagocytized target) or block fusion of the phagosome with the lysosome to avoid destruction by hydrolytic enzymes. Certain pathogens can even neutralize the enzymes phagocytes use to kill bacteria or they can make toxins that directly damage phagocytic cells. Lastly, the pathogen may simply have evolved to thrive inside the harsh environment of the phagolysosome; *Coxiella burnetii*, the bacterium that causes Q fever (a flu-like disease that is on the notifiable disease list), is one such pathogen.

Immune suppression Pathogens like HIV suppress immune function by directly targeting immune system cells and thereby decimating the immune system. Other pathogens make proteases that break down host proteins. For example, certain bacteria make IgA proteases, enzymes that break down host antibodies (immune system proteins). Some pathogens, such as *Leishmania* parasites, suppress immune function by interfering with transcription of immune system factors called interleukins (factors we'll discuss in Chapter 11). Some agents, like the smallpox virus, interfere with the molecular signaling that activates parts of the human immune response.

Fifth, a pathogen must be transmitted to a new host to repeat the cycle.

The last crucial step in a pathogen's success is transmission to a new host. (Chapter 9 reviews modes of transmission.) Sometimes the symptoms a pathogen generates facilitate transmission to others. For example, the parasite pinworm causes anal itchiness, which may lead the patient to scratch the anal area. As the patient scratches, sticky pinworm eggs attach to the fingers. Young children are not inhibited by social norms that limit this activity—plus they tend to touch everything in their environment (including doorknobs). This makes it likely that a new host will eventually pick up and ingest the pinworm eggs. Perhaps a more palatable example is observed in agents transmitted by the respiratory route. Pathogens like cold and influenza viruses often cause sneezing or coughing, making it easy for them to be inhaled by a new host. Other times, it is the very lack of symptoms that favors transmission; this occurs in HIV, where only the end stage of the disease has obvious signs and symptoms.

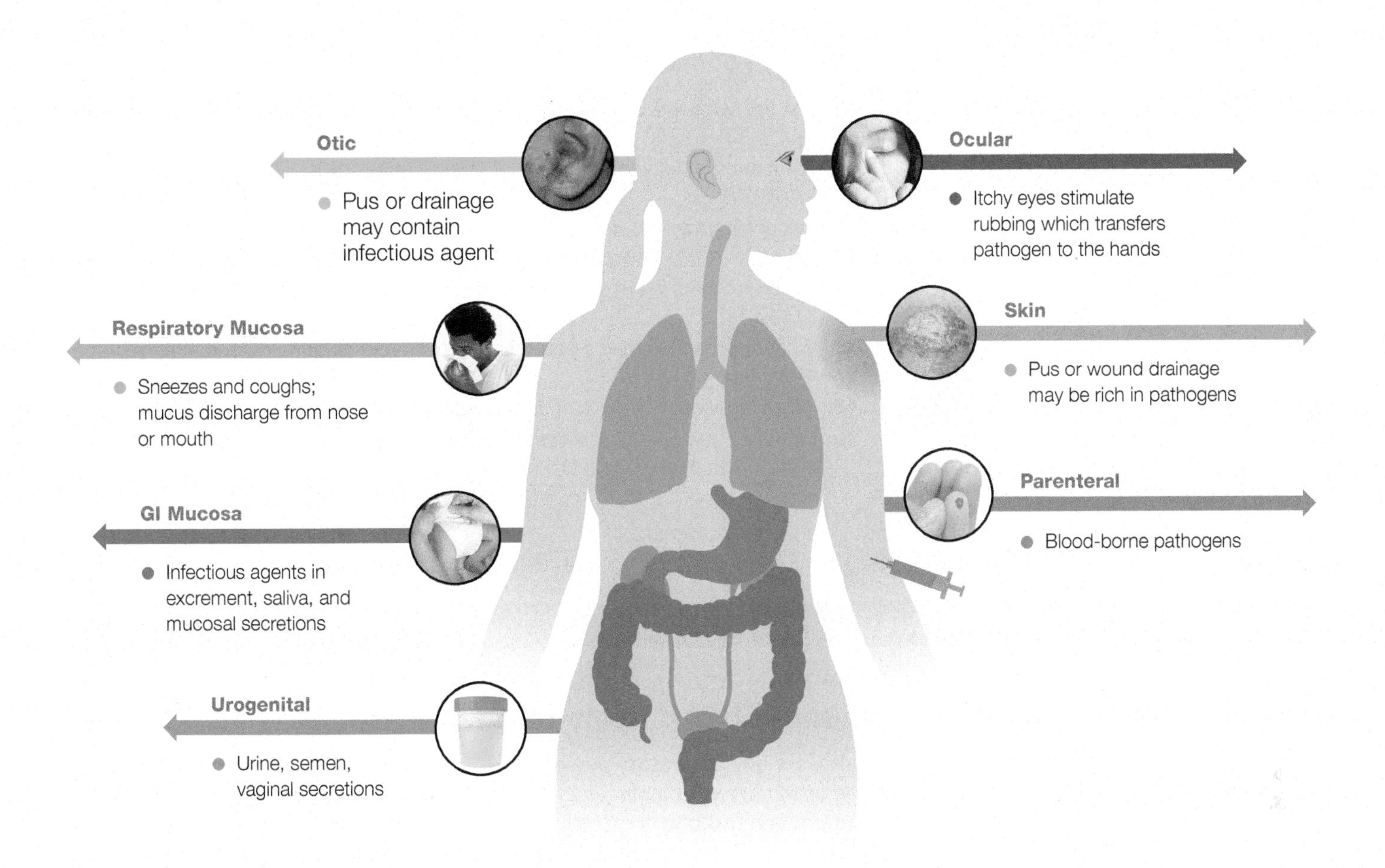

FIGURE 10.10 Portals of exit Many pathogens use the same portal of exit as used for entry.

Exiting the Host

Any route a pathogen uses to exit its host is a **portal of exit**. Feces, urine, fluids drained from wounds, vomit, saliva, mucus, or semen are all possible ways for a pathogen to exit the host. The portal of entry used by a pathogen is often the same as the portal of exit, but there are exceptions, especially in the case of agents that employ multiple entry portals like *B. anthracis*. FIG. 10.10 shows potential portals of exit and symptoms that encourage transmission through the given portal.

Maintaining a Reservoir

As discussed in earlier chapters, a *disease reservoir* is any environment or organism in which a pathogen tends to naturally thrive. Some pathogens have only human reservoirs. In zoonotic diseases, animals other than humans are the reservoir. Usually, nonhuman reservoirs are minimally affected by human pathogens. For example, bats are reservoirs for many viruses that are dangerous in us yet harmless in them.

Some agents survive on inanimate objects (or fomites), for prolonged periods; others must quickly enter a new host or perish. *C. difficile*, the causative agent of acute infectious diarrhea and pseudomembranous colitis, forms endospores that can persist on hospital room surfaces for prolonged periods, making transmission more likely. As stated in Chapter 9, a reservoir could be a source of infection, but *not all sources of infection are reservoirs*. For example, marine environments are a reservoir for *Vibrio* species, including the bacterium that causes cholera, but the source for most cholera cases is infected people.

Build Your Foundation

13. List the five tasks a pathogen must complete to successfully infect a host. (LO 10.13)
14. What are the different portals by which microbes enter and exit the body? (LO 10.14)
15. List at least two examples of adhesins and two of invasins, and identify their role in facilitating infection. (LO 10.15)
16. What are some ways that pathogens may obtain nutrients in the host? (LO 10.16)
17. What approaches may a pathogen use to avoid immune detection and elimination? (LO 10.17)
18. How are disease symptoms and transmission associated? (LO 10.18)
19. What is a reservoir, and what are some examples of environmental and organismal reservoirs? (LO 10.19)

Build Your Foundation (BYF) Quick Quiz: Visit the **Mastering Microbiology** Study Area to quiz yourself.

10.4 SAFETY AND HEALTH CARE

Learning Outcomes

After reading this section, you should be able to:

10.20 Identify the basic criteria for assigning pathogens to a biosafety level.

10.21 Describe features of each biosafety level (BSL), and give examples of pathogens for each BSL.

10.22 Discuss what standard precautions entail.

10.23 Name the three main categories of contact precautions, and explain what each entails.

10.24 Explain how healthcare providers can help prepare for and manage outbreaks of infectious disease.

Biosafety levels dictate appropriate on-the-job behaviors in healthcare.

Healthcare and laboratory personnel have jobs that regularly bring them into close proximity with biological hazards. As such, people working in these fields must understand how to protect themselves and others. Not all agents present the same level of danger—some may not even cause disease. Knowing the risks that a particular agent presents helps workers apply appropriate safety protocols. **Biosafety level (BSL)** assignment is based on numerous criteria, with the following being some key considerations:

- Level of infectivity (not necessarily level of contagiousness)
- Extent of disease caused and mortality rates
- Mode of transmission
- Availability of preventive measures and treatments for the disease

There are four biosafety levels. Laboratories and hospitals refer to BSL rankings to help them properly apply safety practices, such as appropriate safety gear and facility features, that limit exposure risks (see TABLE 10.7). Most laboratory and hospital facilities are set up to manage BSL-1 and -2 agents. Some laboratory facilities are approved to manage BSL-3 or -4 agents.

Biosafety Levels 1 and 2

BSL-1 agents are well characterized and rarely cause disease in healthy people; they pose limited, if any, risk. *Bacillus subtilis*, *E. coli* K-12 strains, and *Staphylococcus epidermidis* are common BSL-1 agents.

BSL-2 pathogens are known to cause human disease but mostly cause infections that are preventable by vaccines, or at least treatable. Most infectious agents fall into BSL-2, so the majority of hospitals and research laboratories operate at the BSL-2 level. *Staphylococcus aureus*, herpes simplex viruses, most influenza strains, *Clostridium tetani*, and *Salmonella* species are examples of BSL-2 pathogens.

Some agents, like HIV, are considered BSL-2-plus pathogens (BSL-2+) because they are dangerous, incurable, and not vaccine preventable. These pathogens are managed in BSL-2 facilities because they are not airborne, but they require additional safety practices or worker monitoring.

Clinical CASE
NCLEX
HESI
TEAS

The Case of the Deadly Mistake

Practice applying what you know clinically: visit the **Mastering Microbiology** Study Area to watch Part 2 and practice for future exams.

TABLE 10.7 U.S. General Biosafety Level Precautions

Level	Minimum Personal Protection Equipment (PPE) Required	Examples of Lab Facility Considerations (not comprehensive)
BSL-1	None	• Hand-washing sinks must be available • Work may be done on open lab bench • No food, beverages, or chewing gum in the lab
BSL- 2	• Lab coats and gloves (safety glasses/face shield if splash risk) • BSL-2+ agents like dengue virus, hepatitis C virus, *Salmonella enterica* serotype Typhi, rabies virus, Zika virus, and HIV require additional precautions.	**All BSL-1 measures, plus:** • Most agents worked on at open lab bench spaces • Biological safety cabinet needed when working with *certain* BSL-2 agents or samples, like tissues or bodily fluids, that may contain such agents • Limit lab access • Biohazard signage (signs indicating biosafety level, agents used, emergency contact personnel, etc.) • Eye-wash stations must be present • Lab design should allow easy cleaning and decontamination (no carpets, upholstery, etc.) • Autoclave (a specialized machine for sterilizing)
BSL-3	• Protective lab covering • Gloves • Respirators if indicated • Must wear PPE at all times • PPE worn for BSL-3 work should not be worn in other areas • Monitoring to ensure the worker has not been infected	**All BSL-1 and -2 measures, plus:** • Agents manipulated in biological safety cabinet • Controlled access/authorized personnel • People entering area warned of risks, vaccinated (if possible), and monitored for infection • Decontaminate all waste • Decontaminate lab wear before laundering • Special airflow management • Self closing, double door access
BSL-4	• Airtight, pressurized, full-body hazardous material suits; air is piped into the suit through a specialized air delivery system • Clothing changes and showers before entering and leaving facility	**All BSL-1, -2, and -3 measures, plus:** • Specialized facility design and engineering • Highly restricted/lockdown access

Unless there is reasonable suspicion that the patient is infected with a BSL-3 or -4 agent, human bodily fluids and tissues are treated as BSL-2 hazards. The main risk of exposure to BSL-2 hazards is through accidental parenteral exposure (such as a needle stick) or mucous membrane contact.

Biosafety Level 3

BSL-3 agents cause serious or lethal human diseases. Some of these agents have airborne transmission. Certain pathogens in this class are treatable, but because of the severity of disease, they are classified at a higher level. *Coxiella burnetii*, *Mycobacterium tuberculosis*, certain influenza strains, and St. Louis encephalitis virus are examples of BSL-3 pathogens. At the time of this text's publication, SARS-CoV-2 is also classified as a BSL-3 agent, but availability of effective vaccines may eventually lead to a BSL reclassification.[15] There are close to 2,000 BSL-3 facilities across the United States; these include laboratories and hospitals with BSL-3 areas used to care for patients with certain contagious BSL-3 diseases. Entry to BSL-3 areas is restricted to authorized personnel and specialized **personal protective equipment (PPE)** (FIG. 10.11).

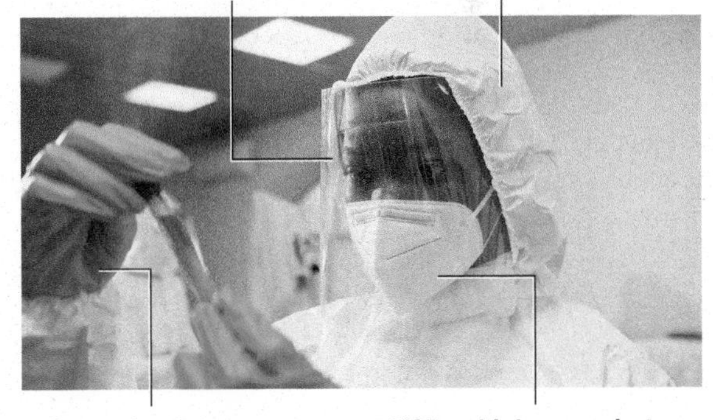

FIGURE 10.11 **Standard PPE for healthcare providers caring for patients infected by SARS-CoV-2—a BSL-3 pathogen**

[15] Kaufer, A. M., Theis, T., Lau, K. A., Gray, J. L., & Rawlinson, W. D. (2020). Laboratory biosafety measures involving SARS-CoV-2 and the classification as a Risk Group 3 biological agent. *Pathology*, 52(7), 790–795.

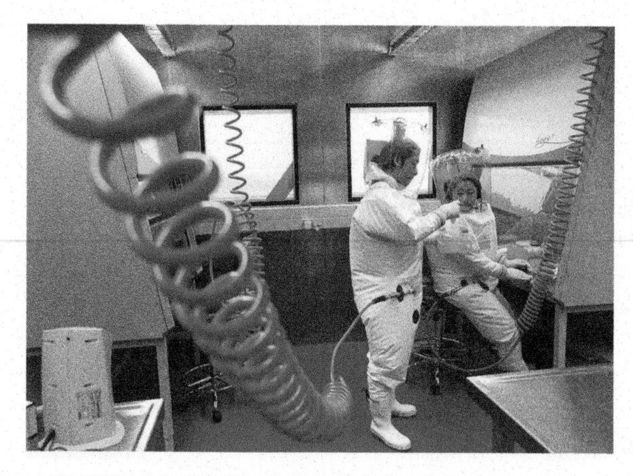

FIGURE 10.12 **Full containment suit for working with BSL-4 agents in the laboratory**

Biosafety Level 4

BSL-4 agents include dangerous and so-called exotic pathogens that tend to be lethal in humans and do not have cures or treatments. Ebola and Marburg viruses, as well as agents with unknown modes of transmission (for example, the Nipah virus), fall into this group.[16] Currently there are about 15 BSL-4 facilities in the United States (FIG. 10.12).

All biosafety levels and their related protocols are summarized in Table 10.7.

Infection control practices protect both workers and patients in healthcare facilities.

The rise of healthcare-associated infections and the frequency of emerging and reemerging disease outbreaks make it essential that healthcare workers learn how to protect themselves, their patients, and the community. Most healthcare facilities have an infection control team that strives to limit infection risks for healthcare workers and patients. Standard precautions and transmission precautions are the two main classes of infection control practices used by healthcare workers (FIG. 10.13).

Standard precautions
(used with all patients)

- Hand hygiene
- Gloves if risk of exposure to wounds, bodily fluids, or mucous membranes
- Barrier gowns and face shields when splash risk
- Disinfection

Transmission precautions

Contact Precautions	Droplet Precautions	Airborne Precautions
• Wound/skin infection • Resistant infection • Infectious diarrhea	• Most respiratory infections • Influenza • Pertussis	• Tuberculosis • Rubeola (measles) • Varicella (chickenpox)
Limit patient transport	Limit patient transport	Limit patient transport
Special attention to hand washing	Procedural mask at all times	N95 or comparable respirator
Gloves at all times		Place patient in AIIR facility
Gown at all times		
Single patient use equipment		

FIGURE 10.13 **Transmission precautions and standard precautions** One or more of these contact precautions may be needed for a patient.

[16] The new vaccine, Erbevo, may eventually result in a BSL ranking change for Ebola viruses against which the vaccine is protective.

Standard Precautions

In the wake of the first HIV cases in the 1980s, healthcare facilities adopted **universal precautions** to limit transmission of bloodborne pathogens such as HIV and hepatitis B and C. These guides applied when dealing with patients suspected or known to have a bloodborne infection. They required workers to prevent contact with blood or bodily fluids contaminated with blood. The CDC later expanded the universal precautions and called them **standard precautions**. Under the newer rules, infection status became irrelevant—all patients are treated as potential sources of bloodborne or other infectious agents. Now, handling precautions exist for all bodily fluids, including blood, urine, feces, sputum, and vomitus; for membranes, nonintact skin, and fresh tissues; and for all excretions or secretions except for sweat. Thus, the phrase "standard precautions" is technically the more comprehensive term, but people often use it synonymously with "universal precautions."

Under standard precautions, you should follow proper hand washing before and after each patient contact, even dry skin contact. Wear new gloves whenever you may encounter blood, mucous membranes, nonintact skin, or other bodily fluids. Change gloves between tasks or procedures. Wear barrier clothing and face shields or masks if there is a splash risk. Properly manage biosharps waste. Lastly, surfaces must be disinfected, and any contaminated laundry or garments must be removed and laundered as soon as possible.

Transmission Precautions

Transmission precautions are taken in addition to standard precautions to prevent direct contact, droplet, and airborne disease transmission. These precautions apply when a specific infectious agent is suspected or known to be present. When transmission precautions are in place, signage is posted to inform healthcare providers and visitors (if allowed) of necessary precautions.

Contact precautions Used in conjunction with standard precautions, contact precautions help minimize transmission of infectious agents spread by fomites and healthcare workers' hands. Barrier gowns and gloves are worn in the patient's room, disinfection practices increase, patient transport is limited, and noncritical equipment (stethoscope and blood pressure cuff) should be dedicated for single-patient use. Patients infected with MRSA (methicillin-resistant *Staphylococcus aureus*) or other wound infections, and those with *C. difficile*, or antibiotic-resistant Gram-negative bacteria, should all be cared for using contact precautions. When possible, these patients should have a private room. When necessary, **cohorting**, which is rooming patients with the same active infection and no other infections, may be acceptable if approved by the infection control team.

Droplet precautions When the agent spreads through large moist respiratory droplets that do not remain suspended in the air, and therefore require closer contact for transmission, droplet precautions are appropriate. Droplet precautions involve standard precautions plus wearing a procedural mask when in the patient's room and limiting patient transport. When transport does occur, the patient is asked to wear a mask, but the individual(s) performing the transport needn't wear a mask. Rubella, influenza, and pertussis (whooping cough) are examples of diseases that require droplet precautions. Whenever possible, patients should have a private room. If this is impossible, then acceptable approaches include cohorting, or providing at least three feet of separation between beds and using drawn-curtain divisions.

Airborne precautions When an agent spreads via fine respiratory aerosols that can remain airborne for longer periods and travel greater distances than respiratory droplets, airborne precautions are implemented. Patients should be placed

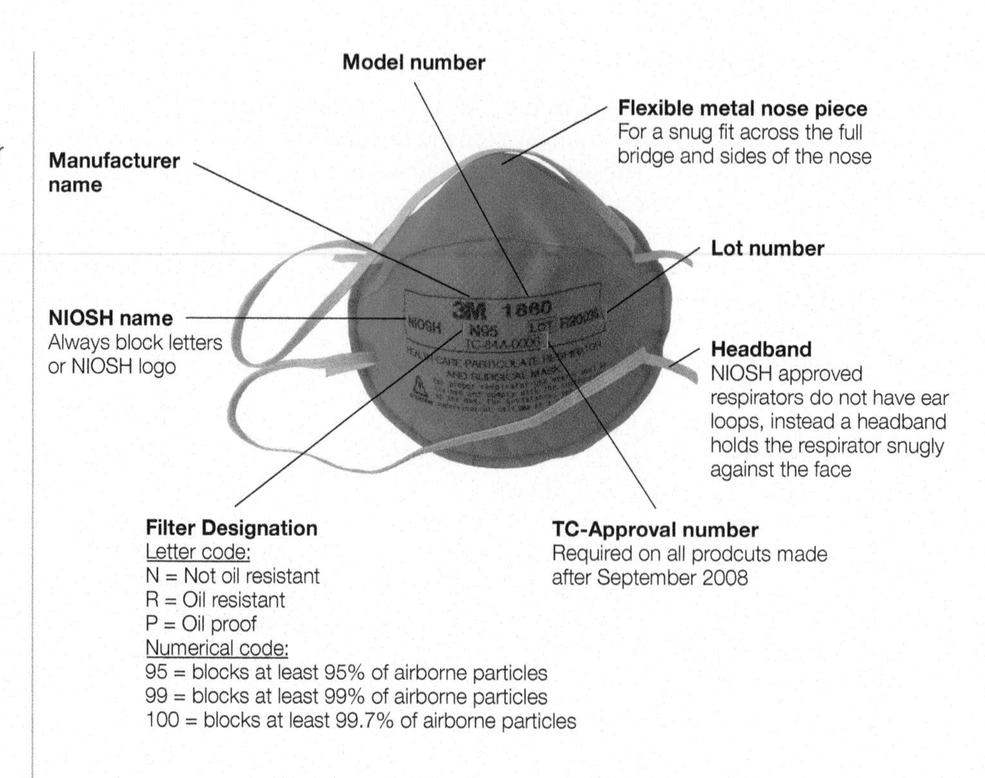

FIGURE 10.14 **Markings of NIOSH approved respirators** All NIOSH respirators should be worn snugly against the skin of the face. They are not guaranteed to provide NIOSH-rated efficacy if worn over facial hair. The following are the only NIOSH approved respirators: N95, N99, N100, R95, P95, P99, P100, and surgical 95 (approved by NIOSH as a respirator and by the FDA as a surgical mask).

in an airborne infection isolation room (AIIR). AIIRs come equipped with specialized pressure systems that keep air in the room from entering surrounding areas. If an AIIR is not available, the patient should be transported to a facility with these rooms. Unless immune to the infectious agent in question, people entering AIIRs must wear air-purifying respirators that effectively block 95 percent or more of airborne particles from being inhaled. In the United States, the National Institute for Occupational Safety and Health (NIOSH) tests and certifies a variety of high-efficiency respirators for airborne precautions. In general, numbers and letters represent different respirator features—for example, an N95 designation means the mask is a non-oil-resistant respirator that blocks at least 95 percent of airborne particles (FIG. 10.14). N95 respirators are the most commonly recommended for airborne precautions in medical settings—they are less obstructive to breathing than higher filtering respirators (e.g., N99), yet sufficiently protective. Note, masks with "breathing valves" are not NIOSH approved. Tuberculosis, chickenpox, and measles are examples of cases managed with airborne precautions. In some settings, airborne precautions for managing tuberculosis are called *AFB precautions* (for acid-fast bacilli precautions).

COVID-19 precautions Transmission precautions for managing COVID-19 in healthcare settings are a blend of the aforementioned contact, droplet, and airborne transmission precautions. It is likely that the recommended precautions will change as people are more widely vaccinated and perhaps as new SARS-CoV-2 strains emerge. At the time of this text's publication, the CDC has defined three PPE use and distribution strategies: conventional, contingency, and crisis. Because contingency and crisis strategies are complex and depend on diverse case-by-case factors, here we focus on conventional strategies that apply when PPE supply chains are keeping pace with demand.

When managing confirmed or suspected COVID-19 patients, healthcare providers should follow standard precautions with splash risk (wear a face shield) and a NIOSH-approved fitted respirator (N95, N99, P100, etc.). Whenever possible, patients with confirmed COVID-19 should be placed in private rooms with individual-access bathrooms, or if space is limited, confirmed COVID-19 patients can be cohort roomed. COVID-19 patients should only be transported out of designated COVID spaces when medically necessary, and patients should be masked when transported (unless the patient cannot medically tolerate masking). Surrounding areas should also be cleared for patient transport, and departments receiving COVID-19 transfers should be notified ahead of time whenever possible. When AIIR space is limited, patients suspected or confirmed to have COVID-19 should only be placed in AIIR rooms when undergoing aerosol-generating procedures (e.g., endotracheal intubation/extubation, laryngoscopy, positive pressure ventilation, airway suctioning).

Learning from Emerging Disease Outbreaks

Ebola, swine flu, SARS-CoV-1 (Severe Acute Respiratory Syndrome), MERS-CoV (Middle East Respiratory Syndrome), and SARS-CoV-2 (COVID-19) are examples of emerging diseases that require extra concern and precautions. The 2013–2016 Ebola outbreak in West Africa killed over 11,000 people. Swine flu caused a pandemic in 2009 that is estimated to have killed over 500,000 people by 2012. Thanks to air travel, the 2003 SARS-CoV-1 outbreak took only 1 day to spread from China to Canada. MERS-CoV, which has a 35 percent mortality rate and was first reported in Saudi Arabia in 2012, has now spread throughout the Arabian Peninsula and has resulted in imported cases in the United States and other countries. And the story of COVID-19 continues to unfold in nations around the world.

Outbreaks of emerging diseases are here to stay. First-line healthcare responders like emergency technicians and nurses are central to the management of such outbreaks. These groups are at the greatest risk for exposure, as they tend to spend the most time with patients before diagnosis is confirmed. The very real risks that healthcare providers face are demonstrated in the case of nurse Nina Pham, who was the first person to ever contract Ebola while on U.S. soil. It's suspected she became infected while removing the extensive PPE that she and others caring for Ebola patients must wear. The PPE required for BSL-4 agents is cumbersome to properly remove. Each layer must be taken off in a specific order and with particular attention to technique, or else microbes on the PPE could be introduced to the wearer. Proper technique to remove BSL-4 PPE requires careful and extensive training—something that most healthcare providers had never received before Ebola came to the United States. And, as history will note, even the wealthiest nations across the world were poorly equipped to handle widespread BSL-3 containment and safety precautions needed to manage COVID-19 cases.

Recent epidemics and pandemics have made it clear that all healthcare providers must have a firm grasp of standard and transmission precautions needed to limit their own risk and that of others in their care. A good start is to *never* be casual about the standard and transmission precautions discussed earlier, as you never know what you may be dealing with—it could be some new pathogen that has yet has to be characterized. Learn how to properly put on and remove PPE, speak up if potential dangers are spotted, and advocate for refresher training and updates in the workplace.

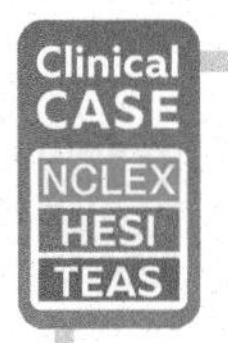

The Case of the Deadly Mistake

Practice applying what you know clinically: visit the **Mastering Microbiology** Study Area to watch Part 3 and practice for future exams.

Build Your Foundation

20. What features of a pathogen are used to assign it to a biosafety level? (LO 10.20)

21. What are the features of each biosafety level? (LO 10.21)

22. If you are told to follow standard practices with a patient, what does that mean? (LO 10.22)

23. Compare and contrast contact, droplet, and airborne transmission precautions. (LO 10.23)

24. What can individual healthcare providers do to prepare for and respond to an outbreak of infectious disease? (LO 10.24)

Build Your Foundation (BYF) Quick Quiz: Visit the **Mastering Microbiology** Study Area to quiz yourself.

VISUAL SUMMARY | *Host–Microbe Interactions and Pathogenesis*

1 Entry

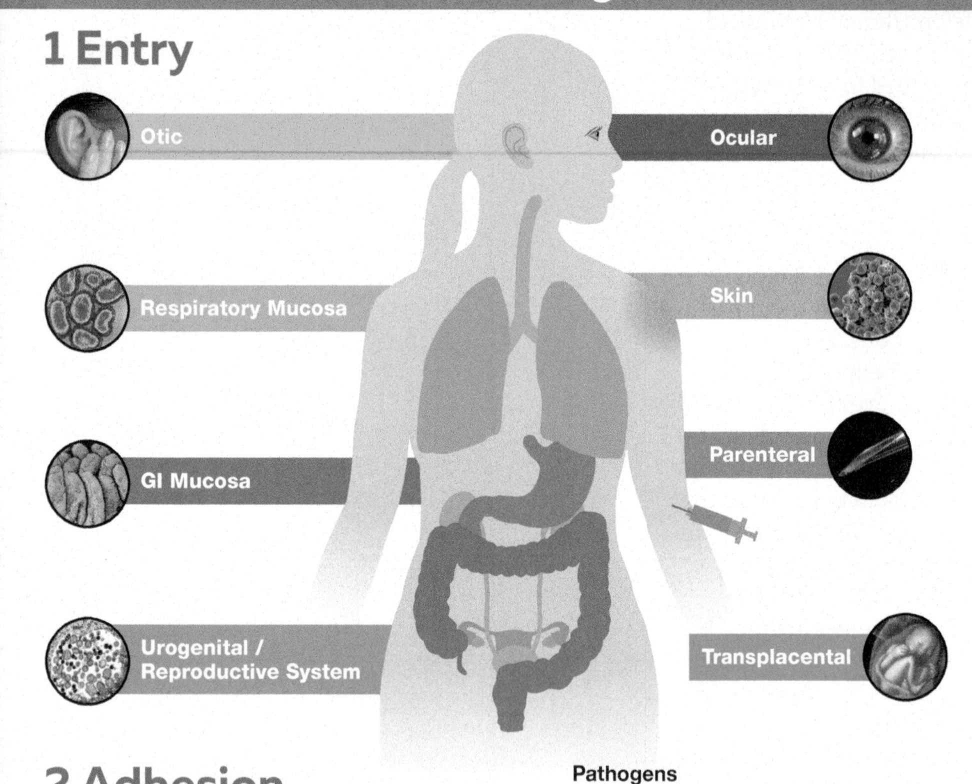

2 Adhesion

Pathogens

Host cell

3 Invasion

Invasins help the pathogen spread deeper into host tissues. Siderophores and extracellular enzymes help cellular pathogens obtain nutrients so they can propagate.

Deeper in the tissue

4 Evasion

Hide from immune defenses
- Intracellular lifestyle
- Latency
- Antigen masking, mimicry, and variation

Host Cell

Antigen masking

Antigen mimicry

Antigen variation

Undermine immune defenses
- Suppress immune function
- Avoid phagocytosis

Phagocyte

5 Exit and Transmission

Portals of exit are determined by transmission mode and are often (though not always) the same as the portal of entry.

Virulence Factors

Virulence factors help a pathogen accomplish the five steps at left.

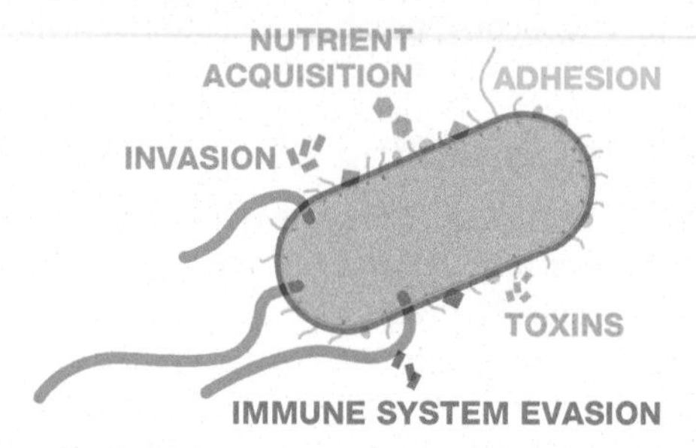

Lethal dose-50
The lethal dose-50 is the amount of toxin needed to kill 50% of affected hosts that are not treated.

Infectious dose-50
The infectious dose-50 describes how many cells or virions are needed to establish an infection in 50% of exposed susceptible hosts.

Exotoxins

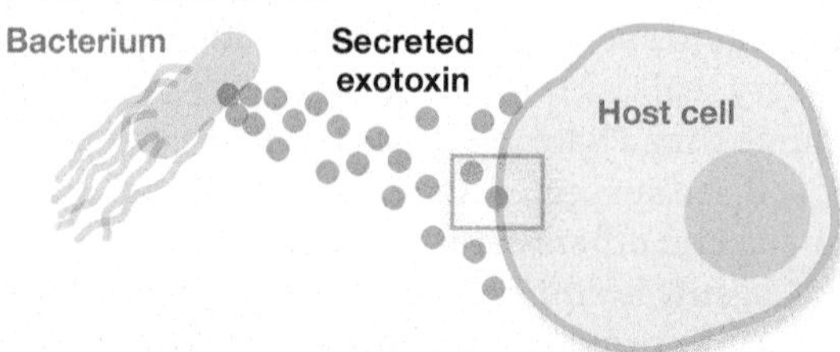

- **Type I:** Toxin binds at host plasma membrane.
- **Type II:** Toxin disrupts host cell membranes.
- **Type III:** Toxin enters cell, often by endocytosis.

Safety

Limit infection! Always follow standard precautions and transmission precautions.

Hand hygiene

Gloves if risk of exposure to wounds, bodily fluids, or mucous membranes

Barrier gowns and face shields when splash risk

Disinfection

Biosafety levels are assigned based on pathogenic features of the agent in question.

Biosafety levels

BSL-1 **Minimal risk**
BSL-2
BSL-3
BSL-4 **Highest risk**

10 CHAPTER 10 OVERVIEW

10.1 Basics of Host–Microbe Interactions

- The host–microbe interaction does not always lead to disease and often has roles in health.
  - Mutualism is a type of symbiotic relationship in which both participants benefit. In comparison, commensalism is a relationship in which one organism benefits and the other is not affected. It was formerly thought that our normal microbiota had a commensal relationship with us, but based on the benefits discussed in this chapter, it is increasingly evident that we have a mutualistic relationship.
  - *Opportunistic pathogens* often live among the normal microbiota, but do not cause disease unless an opportunity (weakened immunity or disruption in the normal microbiota) develops.
- Tropism is the specificity a pathogen has for infecting a defined host or host tissue; it can change as pathogens evolve.

10.2 Introduction to Virulence

- Virulence describes the *degree or extent* of disease that a pathogen causes. Pathogenicity is the ability of a microbe to cause disease. Key virulence factors include toxins and tools that help pathogens adhere to host cells, invade tissues, acquire nutrients, and evade immune defenses. Host properties, host–microbe interactions, and environmental factors all influence virulence and pathogenicity.
- Not all pathogens are equally virulent. Pathogens balance maximally exploiting their host with promoting transmission. A pathogen's R_0 value describes disease spread in a fully susceptible population when no interventions have been applied; the R_e value describes transmission in the midst of an epidemic where not all hosts are susceptible and/or interventions to combat transmission may have been applied.
- Infectious dose-50 (ID_{50}) describes how many cells or virions are needed to establish an infection in 50 percent of exposed susceptible hosts. The more infectious the pathogen is, the lower the ID_{50}. Just because a pathogen is highly infectious doesn't mean it is especially dangerous.
- Lethal dose-50 (LD_{50}) is the amount of toxin needed to kill 50 percent of affected hosts that are not treated. The lethality of an infectious agent is usually expressed as mortality rate rather than as LD_{50}.
- Two main classes of bacterial toxins are endotoxins and exotoxins. Gram-negative bacteria make endotoxin. In high quantities, endotoxin causes endotoxic shock (sepsis). Exotoxins are made by Gram-positive and Gram-negative bacteria and are often classified into three main families based on their mode of action.

10.3 Five Steps to Infection

- First, pathogens must enter a host. A *portal of entry* is any site that a pathogen uses to enter the host. It is usually determined by the mode of transmission.
- Second, pathogens must adhere to host tissues. Adhesins are virulence factors that help infectious agents stick to host cells. Biofilms facilitate pathogen adhesion and serve as a source of infection.
- Third, pathogens must invade host tissues and obtain nutrients. Invasins are virulence factors that break down host tissue to help with this. Host cell invasion by pathogens often causes cytopathic effects.
- Fourth, pathogens must avoid host defenses. They may hide from immune defenses using latency, living intracellularly, or applying antigenic variation, mimicry, or masking. Other pathogens undermine host immune defenses by interfering with phagocytosis and suppressing the immune response.
- Fifth, pathogens must transmit to new hosts. Portals of exit are often (though not always) the same as the portals of entry. A disease reservoir aids in transmission by making it possible for a pathogen to exist outside a host.

10.4 Safety and Health Care

- Four biosafety levels (BSLs) are assigned based on pathogenic features of the agent in question. BSL-1 agents present minimal risk to people, while BSL-4 agents have the highest risk.
- Managing biological risks is central to patient and healthcare worker safety.
  - Under standard precautions, blood, all bodily fluids, excretions and secretions (except sweat), mucosal membranes, nonintact skin, and fresh tissues are handled as though they are a source of bloodborne or other infectious agents irrespective of presumed or confirmed infection status.
  - Transmission precautions are taken in addition to standard precautions to prevent direct contact, droplet, and airborne disease transmission and are also applied when a specific infectious agent is suspected or known to be present. COVID-19 transmission precautions are a blend of contact, droplet, and airborne transmission precautions.

COMPREHENSIVE CASE

The following case integrates basic principles from the chapter. Try to answer the case questions on your own. Don't forget to be S.M.A.R.T.* about your case study to help you interlink the scientific concepts you have just learned and apply your new understanding of microbial pathogenesis to a case study.

***The five-step method is shown in detail in the Chapter 1 Comprehensive Case on cholera. See pages 31–33. Refer back to this example to help you apply a SMART approach to other critical thinking questions and cases throughout the book.**

• • •

It was a hot summer Sunday in Fort Myers, Florida, and Max was anxious to get some last-minute surfing in before the end of the weekend. A slip of the knife while prepping dinner the previous Monday had him nursing a minor hand wound that had kept him out of the water for the past week. The wound was not very deep and was healing well—it was well scabbed over and itchy, but painless.

After enjoying a full morning back in the surf, Max was famished, and stopped at a local burger stand for lunch. While devouring greasy fries and a burger, he noticed that his lengthy time in the water had softened the scab on his hand to no more than a small pinkish layer. The wound looked much better than it had in days.

However, by 5:00 p.m. that evening, Max felt achy, and his upper arm and hand were sore. He noticed the area that was previously covered by the small pink remnants of a scab had reddened and become tender, swollen, and warm to the touch. He figured the soreness was normal after a day in the surf and that the hand abrasion was just irritated from sea and sand. He felt exhausted and feverish, and also a bit nauseated, though he assumed that might be due to the greasy food he had eaten for lunch. He skipped dinner and went to bed early.

His mom called him around 8:30 p.m. She noticed that he sounded terrible, and upon learning of his symptoms, told him to go to the emergency room right away. Because his mom was a nurse, he decided he'd better listen to her, even though he thought she was probably overreacting.

Max was 33 and, aside from this acute situation, was in perfect health. Despite this, the emergency department physician was concerned about Max's condition, especially because he was running a fever, and admitted him to the hospital for wound management, intravenous antibiotic therapy, and monitoring. By the morning, Max had a heartbeat of 105 beats per minute (tachycardia), a temperature of 101.4°F (38.4°C), remained nauseated, and was disoriented. Despite ongoing intravenous volume resuscitation (IV fluid administration), Max was hypotensive (had a pathologically low blood pressure). In addition, his arm was looking much worse: it was severely swollen, and pulses in the arm were difficult to detect. The skin, which had previously been mostly spared, now appeared eccyhmotic (took on a deep bluish color due to the escape of blood from ruptured blood vessels into the surrounding tissue), and hemorrhagic bullae (large blood-filled blisters) were evident. Max was also in excruciating pain that morphine barely dulled.

As the attending physician, you suspected Max had necrotizing fasciitis (informally known as "flesh-eating bacteria"), a soft tissue infection that is usually caused by Gram-positive group A streptococci. You noticed that Max had some mild sunburn and asked him if he'd been out swimming lately. Max confirmed he'd been surfing the morning before falling ill. The microbiology report confirmed your suspicions: Max was fighting off a Gram-negative bacterium called *Vibrio vulnificus*. Knowing that *V. vulnificus* has many virulence factors (such as a capsule; extracellular collagenases, proteases, and lipases; motility; siderophores; and toxins that act as cytolysins and hemolysins), you were worried about Max's rapid decline.

Max was immediately taken into surgery for wound debridement—the removal of infected, damaged tissue. Following surgery Max was moved to the intensive care unit (ICU). He endured several additional wound debridement procedures and a skin graft. Max was told that his age and general overall health would likely lead him to a full recovery. The nurse explained to Max that, had his fever and hypotension not resolved after the debridement procedures, he likely would have needed an amputation. Fortunately, that was not the case. Although he would require a good deal of rehabilitation, he would eventually regain full use of the affected arm.

• • •

CASE-BASED QUESTIONS

1. Based on the microbiology report and Max's signs and symptoms, what toxin-based complication was Max's healthcare team most likely concerned could develop? Explain your reasoning.
2. Explain how *V. vulnificus*'s virulence factors contributed to the pathology described in the case.
3. In Max's illness, what was the likely reservoir and source of the pathogen *V. vulnificus*?
4. What portal of entry did *V. vulnificus* most likely use? Explain your reasoning.
5. What infection control precautions were most likely used when managing Max's health in the ICU? Discuss how you came to your conclusions.
6. To which biosafety level is *V. vulnificus* most appropriately assigned? Explain your reasoning.

END OF CHAPTER QUESTIONS

NCLEX
HESI
TEAS

Think Critically and Clinically

Questions highlighted in **orange** are opportunities to practice NCLEX, HESI, and TEAS-style questions.

1. Assume your patient has a superantigen circulating in their blood. Select the single statement that is most likely to apply to your patient.
   a. They are at risk for endotoxic shock.
   b. They are not up to date on their vaccinations.
   c. They are infected with a Gram-positive microbe.
   d. They do not have a fever.
   e. They have a viral infection.

2. Which of the following is a true statement?
   a. If a pathogen establishes an infection, it is described as virulent.
   b. Pathogenicity is the extent of disease caused by a microbe.
   c. Normal microbiota are not usually affected by host factors.
   d. A pathogen's virulence factors change over time in response to selective pressures.
   e. Attenuated pathogens cause disease in a normal host.

3. Define the class of each listed exotoxin as type I, II, or III:
   a. Superantigen
   b. Hemolysins
   c. *Staphylococcus aureus* enterotoxins that cause food poisoning
   d. AB toxin
   e. Membrane-damaging toxins
   f. Phospholipases

4. Which of the following is true regarding tropism?
   a. It is the preference of a pathogen for a given tissue.
   b. It is constant for a given microbe.
   c. It limits a pathogen to infecting only one host.
   d. It is determined by portal of entry.
   e. It is independent of host factors.

5. Indicate the true statements **and** then correct the false statements so that they are true.
   a. HIV is transmitted by a parenteral route.
   b. *Candida albicans* is an opportunistic pathogen that can cause disease in an immune-competent host if the normal microbiota are disrupted.
   c. Gram-positive bacteria may produce endotoxin.
   d. Siderophores help pathogens obtain calcium.
   e. Emerging pathogens tend to exhibit expanded tropism.
   f. The more toxic a substance is, the higher its LD_{50}.
   g. Virulence is the ability of a microbe to cause disease.
   h. Gram-negative bacteria may produce exotoxins.

6. Select the *false* statement about normal microbiota.
   a. They compete with pathogens.
   b. They do not include potential pathogens.
   c. They make vitamins for the host.
   d. They train the immune system.
   e. A disruption in their balance can lead to disease.

7. Fill in the blanks:

   Pili, fimbriae, and sialic acid binding factors are examples of ________, which are virulence factors that allow pathogens to ________ host tissues—an essential early step in pathogenesis. In contrast, flagella, collagenases, and coagulases tend to act as ________, which help pathogens spread deeper into host tissues.

8. What is a reservoir, and why can *C. difficile* use a fomite as an effective environmental reservoir?

9. Fill in the blanks:

   Toxigenic microbes produce ________. A high ID_{50} would suggest ________, and a low LD_{50} would suggest ________.

10. A pathogen that makes endotoxin, enters through the fecal–oral route, and lacks a nucleus is most likely a
    a. virus.
    b. Gram-positive bacterium.
    c. Gram-negative bacterium.
    d. protozoan pathogen.
    e. There is not enough information to answer this question.

11. Complete the table:

Description of portal(s) of entry	Microbe/pathology example
	Conjunctivitis
	Influenza
	Congenital infections
	HIV

12. Which of the following is *false* regarding biofilms?
    a. They tend to consist of one species of microbe.
    b. They are platforms on which pathogens may adhere.
    c. They form on indwelling devices.
    d. They may harbor pathogens.
    e. They can form on natural and manufactured surfaces.

13. Make a Venn diagram to compare and contrast endotoxins and exotoxins.

14. What three main options can a pathogen pursue following adhesion?

15. Which of the following is *false* regarding toxemia?
    a. It can be caused by bacteria or fungi.
    b. It is localized in the patient's body.
    c. It can be caused by endotoxins.
    d. It can be caused by exotoxins.
    e. Some forms are vaccine preventable.

16. What precautions or actions would apply to an HIV/AIDS patient? Select all that apply.
    a. Droplet precautions
    b. Standard precautions
    c. BSL-4 precautions
    d. Universal precautions
    e. AFB precautions
    f. Isolation practices

17. Antigen ________ is a scenario in which pathogen antigens resemble host antigens. Antigen ________ is a scenario in which the pathogen changes its antigens. These are just a couple of ways that pathogens may avoid host immune system detection.

18. Which of the following is/are features of endotoxic shock? Select all that apply.
    a. Fever
    b. Confusion
    c. Hypertension
    d. Bradycardia
    e. Decreased respiratory rate
    f. Achiness

19. What is IgA protease, and what effect would it possibly have on host immune function?

20. Place the following steps for infection in order from first to last:
    - Invade tissues and obtain nutrients
    - Adhere to host tissues
    - Enter the host
    - Exit the host
    - Evade immune defenses

21. What BSL would an airborne pathogen that causes potentially deadly, but treatable, disease be placed into? Explain your answer.

CRITICAL THINKING QUESTIONS

1. Pathogens often evolve virulence factors that are associated with their mode of transmission. Explain this statement, and provide a theoretical example.

2. Proteases and lipases can be said to have a dual role in obtaining nutrients and invasion. Explain why.

3. Imagine you are a nurse in an emergency room, and a patient comes in late at night with signs of dehydration, a high fever, and sneezing—but no cough. Assuming you will be in close direct contact with this patient, what standard and transmission precautions would you follow? Which precautions, if any, are unnecessary? Explain your reasoning.

4. You are volunteering as a healthcare aid worker in Bambari, in the Central African Republic. Lately, forest fires have caused animals and people to relocate. Water is fouled with feces due to poor waste management and crowded conditions exist in refugee areas. The mosquitoes are also terrible. You were called in to help manage an outbreak of a newly emerged illness that causes fever, bloody diarrhea, achiness, and fatigue. Cases seem limited to people with suppressed immunity. Currently available drugs do not cure affected patients, and 20 percent die from dehydration. Based on these details, what BSL level would this pathogen likely be placed in, and why? What do you think is the most likely portal of entry of the newly emerged pathogen? Explain your reasoning.

5. Diseases that quickly debilitate and kill patients typically have less capacity to cause pandemics than do pathogens that cause slowly progressing diseases. Explain why this is the case, and also explain why, despite this fact, they are still considered important to globally monitor.

11 Innate Immunity

What Will We Explore? Although microbes have many tools to establish infection, we are not helpless in the face of microbial invasions. When a threat is detected, our immune system tirelessly works to protect us. Its defense mechanisms are divided into two broad categories, innate and adaptive immunity. Here, we explore innate immunity, the first group of defenses activated against microbial threats.

Why Is It Important? Pathogens reside in us and on us and occupy every part of our environment, yet thanks to our immune defenses, we rarely become ill. Innate immunity is so central to our health that if we overly suppress this branch of our immune system with anti-inflammatory drugs, we become immune compromised and are less able to fight infections and ward off cancer. Indeed, many of the foundational standards of patient care such as hand washing, avoiding unnecessary antibiotic use, considering white blood cell levels, letting a low-grade fever run its course untreated, and prescribing targeted inflammation interventions all relate to innate immunity. Lastly, understanding innate immunity is central to developing vaccines and immunotherapies—treatments that can bolster natural immune defenses so they are more efficient at fighting certain cancers and infections.

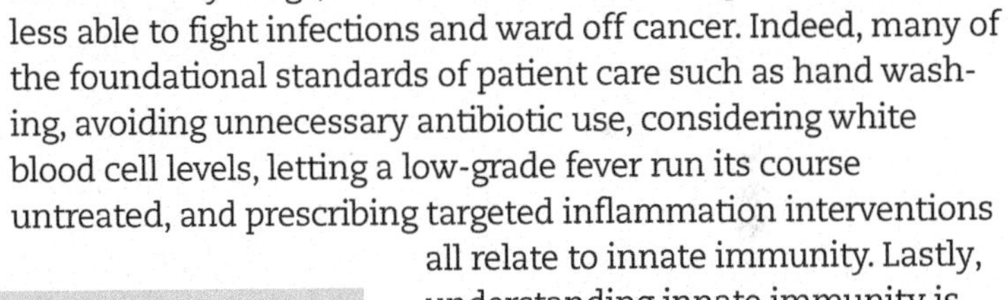

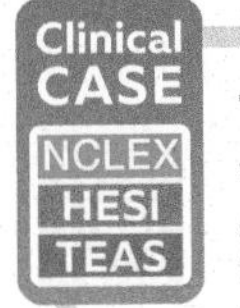

The Case of the Missing Bleach

Visit the **Mastering Microbiology** Study Area to watch the case and find out how innate immunity can explain this medical mystery.

Jordyn Henderson and Laurie Iliff
RNs, school clinic nurses; Saint Johns, FL

11.1 OVERVIEW OF THE IMMUNE SYSTEM AND RESPONSES

Learning Outcomes

After reading this section, you should be able to:

11.1 Describe the general features of innate and adaptive immunity.

11.2 Describe how normal microbiota may impact immune responses and limit pathogens.

Immune responses are classified as either innate or adaptive.

An **immune response** is a physiological process coordinated by the immune system to eliminate foreign substances (**antigens**). Our immune system includes two key branches: innate and adaptive immunity. The three common features of both branches are that they (1) recognize diverse pathogens, (2) eliminate identified invaders, and (3) demonstrate **self-tolerance** (discriminate self from foreign and do not attack self).

Innate immunity is an inborn, ancient protection existing in one form or another in all eukaryotic organisms. It includes barrier defenses, such as the skin or stomach acid, as well as specialized molecules, cells, and tissues. Its responses are generalized; that is, they don't vary based on the pathogen being fought. Because of this, innate immunity is sometimes called *nonspecific immunity*.

Evolutionarily speaking, **adaptive immunity** is newer and exists only in vertebrate animals. It includes specialized immune cells and their products—most importantly, **antibodies**. This set of defenses matures over time, tailoring its responses to the pathogens it encounters. This feature gives adaptive immunity its other names, *specific immunity* and *acquired immunity*. Adaptive immunity typically requires a 4- to 7-day window to fully activate, making it a much slower process for fighting infection than innate immunity, which is more immediate. However, adaptive immunity can recognize pathogens it previously fought and respond more quickly and aggressively upon later exposures to the same agent. Vaccines work because of adaptive immunity. They introduce harmless antigens to train the adaptive immune system to recognize and neutralize a particular pathogen. (Adaptive immune responses are covered in Chapter 12, and vaccines are covered in Chapter 14.)

When we say we are **immune** to a particular pathogen, we are referring to the specific protection conferred by adaptive immune responses. In contrast, when we say someone is **susceptible**, it means they aren't immune to a given pathogen, and it may infect them. TABLE 11.1 compares innate and adaptive immunity.

You can think of immunity as three collaborating lines of defense (FIG. 11.1). The first-line barrier defenses and the second-line molecular, cellular, and tissue defenses fall under innate immunity. They are the subject of this chapter. The third and final line of defense is adaptive immunity. It is covered in Chapter 12. Although the full explanation of immune responses is spread across two chapters, it's essential to understand that we have a *single* immune system, with all components working together to confer immune protection. This

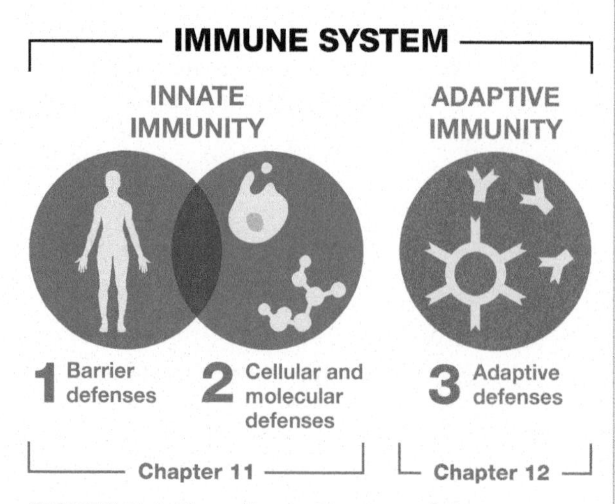

FIGURE 11.1 **Three lines of immune defense**

TABLE 11.1 Comparing Innate and Adaptive Immunity

Feature	Innate Immunity Tools	Adaptive Immunity Tools
Response time	Immediate	4–7 days
Organisms that have it	All eukaryotes (multicellular and unicellular eukaryotic organisms)	Only vertebrates
Distinguishes self from foreign	Yes	Yes
Kills invaders	Yes	Yes
Effective against diverse threats	Yes	Yes
Tailors response to specific antigen	No	Yes
Remembers antigen and amplifies response upon later exposure	No	Yes

interdependency between immune responses isn't surprising when we consider that the adaptive immune system evolved in the context of a successful and well-established innate system.

Deficiencies in innate or adaptive defenses will render the host immune compromised—meaning the host is more susceptible to opportunistic and true pathogens. A variety of situations are associated with immune deficiencies—infancy, old age, pregnancy, taking certain medications (e.g., chemotherapy and medications to prevent transplant rejection), having an infection with pathogens such as HIV that target the immune system, suffering severe burns, and even genetic disorders can all lead to a compromised immune system.

Normal microbiota has a role in shaping immune responses and conferring protection.

The immune system allows certain microbes to live as symbiotic partners in and on our bodies while excluding others. This ongoing interaction with our normal microbiota allows our immune system to fine-tune its preparedness to fight pathogens while also training it to tolerate nonpathogens, food, and self-tissues. (Chapter 10 discusses the microbiota–host balancing act at length.)

When our normal microbiota community changes in terms of the types of microbes present and/or their overall numbers, our immune system may become confused and overreact against harmless agents, and even target self-tissues. Over the years, antibiotic use and changes in our diets and lifestyles have shifted our normal microbiota. Some scientists suggest that these shifts could be linked to our rising incidence of allergies and autoimmune diseases (a pathological condition in which the immune system attacks self-tissues). The *hygiene hypothesis* proposes that decreasing the diversity and levels of microbes in our normal microbiota may negatively affect immune responses. (See Chapter 13 for specifics on immune system disorders and the hygiene hypothesis. Also, refer back to the Chapter 1 Bench to Bedside feature, "A Possible Link Between Lost Normal Microbiota and Immune System Disorders," for more information on this topic.)

Our understanding of how normal microbiota affects our immune responses is in its infancy, but we do know that eliminating normal microbiota or shifting its constituents or density can lead to disease. Studies performed on germ-free animals that were born and raised in microbe-free environments reveal that these animals have an underdeveloped immune system and struggle to combat pathogens. Interestingly, these animals lack certain gastrointestinal (GI) tract immune system factors. This is probably because the gut represents a major interface between our normal microbiota and immune system. Colonizing germ-free mice with normal GI tract microbiota stimulates the development of their gut immune structures and demonstrates the direct role that normal microbiota may have in immune system development. Furthermore, people with chronic inflammatory disorders like irritable bowel disease (IBD) have clear shifts in the normal gut microbiota. In mouse models for IBD, Clostridia bacteria populations are especially decreased.[1] It has been proposed that these bacteria may stimulate immune system regulators to limit gut inflammation.[2]

As discussed in Chapter 10, members of our normal microbiota compete with would-be invaders and/or establish environments that limit pathogen growth. For example, pathogens face robust competition from the normal microbiota of the GI tract and may fail to find a niche to survive in unless the microbiota population is decreased. As a reminder, pathogens that require a disrupted microbiome or an otherwise compromised host to establish infection

---

[1] Clostridia is a taxonomic class that falls under the order Firmicutes.

[2] Belkaid, Y., & Hand, T. (2014). Role of the microbiota in immunity and inflammation. *Cell*, 157(1), 121–141. Palm, N. W., de Zoete, M. R., & Flavell, R. A. (2015). Immune–microbiota interactions in health and disease. *Clinical Immunology*, 159(2), 122–127.

are called *opportunistic pathogens*. *Clostridioides difficile* is an example of an opportunistic pathogen. Patients who have decreased gut microbiota due to an antibiotic therapy are at increased risk for infection because *C. difficile* can thrive unchecked by competitors. Our normal microbiota also makes substances that may directly damage pathogens or generate an environment that limits their survival. The acidic pH of the vagina is a prime example of this eco-limitation. Vaginal pH is kept low by lactic acid–producing bacteria called lactobacilli. A reduction in lactobacilli levels, as can occur following antibiotic therapies, leads to an increase in vaginal pH, which in turn allows the opportunistic pathogen *Candida albicans* (a yeast) to grow and cause a vaginal yeast infection.

Build Your Foundation (BYF) Quick Quiz: Visit the **Mastering Microbiology** Study Area to quiz yourself.

Build Your Foundation

1. Compare and contrast innate and adaptive immunity. (LO 11.1)
2. Explain the basic role of normal microbiota in shaping immune responses and limiting pathogens. (LO 11.2)

11.2 INTRODUCTION TO FIRST-LINE DEFENSES

Learning Outcomes

After reading this section, you should be able to:

11.3 State the general function of first-line defenses.

11.4 Describe mechanical, chemical, and physical barriers, and provide examples of each.

11.5 Discuss the features that make skin a useful barrier.

11.6 Explain what lysozyme is, what it does, and where it is found.

11.7 Describe antimicrobial peptides, and review their basic role.

First-line defenses limit pathogen entry.

All **first-line defenses** help to prevent pathogen entry. For convenience, we traditionally subcategorize these defenses as *mechanical*, *chemical*, and *physical barriers*. However, they work together and often have roles that blur the lines of these groupings (FIG. 11.2).

Mechanical Barriers

Mechanical barriers rinse, flush, or trap pathogens to limit their spread into the body. Examples include tears washing debris or pathogens from the eyes and

INNATE IMMUNITY

1 Barrier defenses

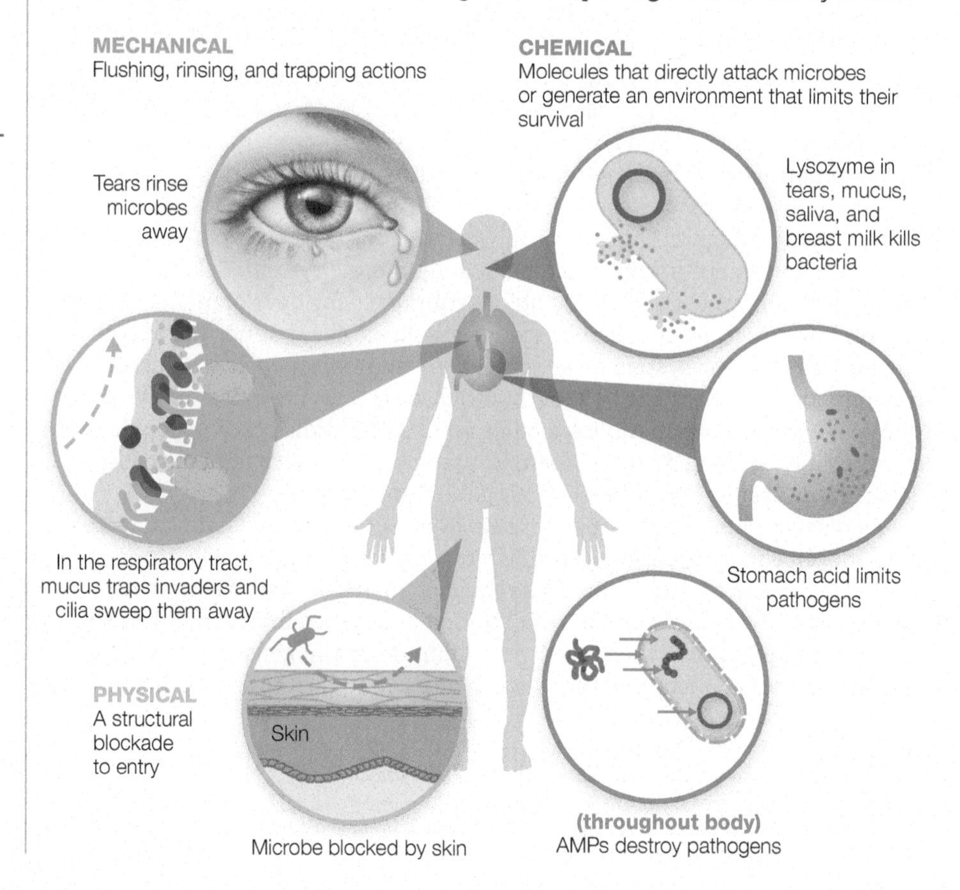

FIGURE 11.2 **Summary of first-line defenses**

urine flushing microbes out of the body. Even the constant rinsing of saliva in the mouth limits what microbes adhere to oral tissues.

Mucous membranes are an excellent example of trapping as a mechanical barrier. Mucous membranes line all body entrances as well as the stomach, intestines, lungs, and bladder. These specialized tissues make mucus, a sticky and viscous substance that traps invaders, microbial and nonmicrobial alike. Within airways, mucus's trapping action is complemented by the **mucociliary escalator**, where ciliated cells sweep the mucus away from the lungs and toward the mouth. (For more on the mucociliary escalator, see Chapter 16.) Mucus also contains chemical factors that are discussed in the next section. Lastly, earwax is another sticky substance that traps microbes and makes tissue invasion more difficult.

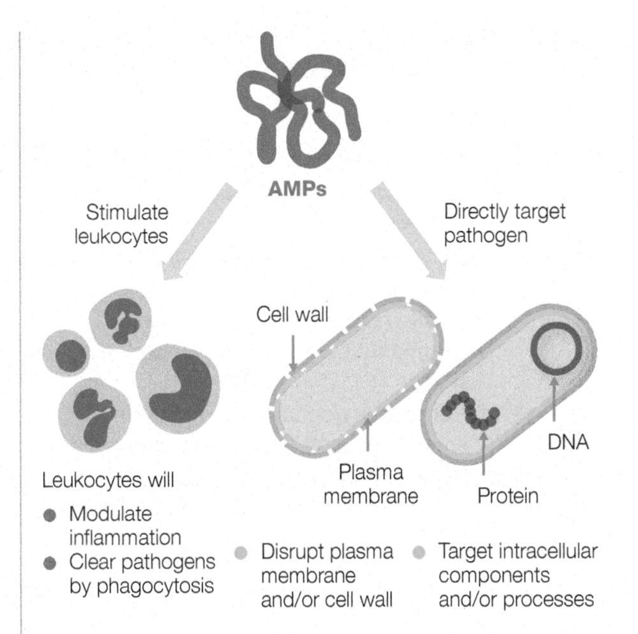

FIGURE 11.3 **Examples of how antimicrobial peptides (AMPs) work**

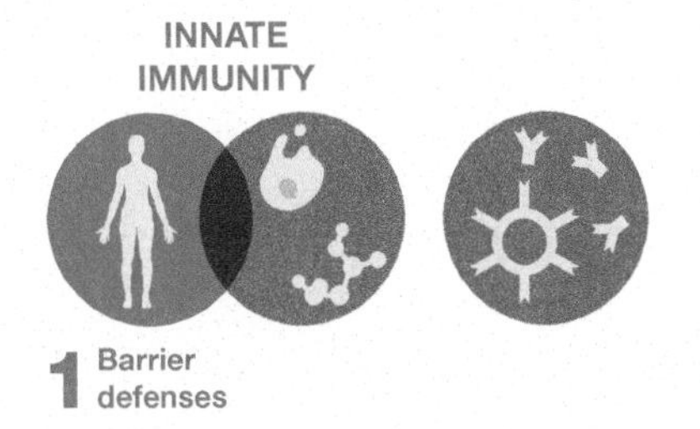

Chemical Barriers

In addition to the mechanical protections that mucus, saliva, and tears confer, these secretions are also rich in substances that serve as **chemical barriers**. These chemical factors may directly attack invaders, or they can establish environments that limit pathogen survival in or on a particular tissue. The enzyme **lysozyme**, which is found in all the aforementioned secretions, acts as a chemical barrier by breaking down bacterial cell walls. Chemical barriers of the stomach and skin also exist. Most pathogens can't survive the corrosive actions of hydrochloric acid in the stomach, while the skin's relatively dry, low-nutrient, salty, and slightly acidic environment inhibits the growth of many microbes. Certain fatty acids in sweat and earwax serve as additional chemical barriers. (See Chapter 17 for more on skin's protective properties.)

Antimicrobial peptides (AMPs) are proteins that act as chemical barriers by destroying a wide spectrum of viruses, parasites, bacteria, and fungi. There are thousands of different AMPs, and they have diverse modes of action (FIG. 11.3). AMPs are evolutionarily conserved—studies suggest every organism from prokaryotes to humans makes some form of these proteins. In humans, leukocytes (immune system cells we'll discuss under second-line defenses) and certain skin and mucous membrane cells make them. AMPs also appear in earwax, mucus, milk, tears, and saliva. Unlike antibiotics, to which microbes frequently become resistant, it's rare for microbes to develop resistance to AMPs. **Defensins** are one important class of mammalian AMPs that rapidly kill invaders by inserting themselves into the invader's cell membrane—we'll learn more about defensins later.

Physical Barriers

Innate **physical barriers** include structures that physically block pathogen entry. **Epithelial tissue** is a main physical barrier in animals. It is made of epithelial cells bound together in one or more layers, and it lines every body cavity and body entrance, including the GI tract, the urinary tract, and the airways. The epidermis, the outermost layer of the skin, is also an epithelial tissue. Its tightly compacted dead epithelial cells are enriched with specialized proteins (such as keratin) and lipids to serve as a water-resistant layer. Aided by the chemical factors already mentioned, the skin is one of our most important physical barriers. (For more on skin's structural features, see Chapter 17.)

Build Your Foundation

3. What is the main function of first-line defenses? (LO 11.3)
4. List examples of mechanical, chemical, and physical barriers, and describe how they work. (LO 11.4)
5. What skin features make it one of the most important first-line barriers? (LO 11.5)
6. What is lysozyme, and where is it found? (LO 11.6)
7. What are antimicrobial peptides, and what do they do? (LO 11.7)

Build Your Foundation (BYF) Quick Quiz: Visit the **Mastering Microbiology** Study Area to quiz yourself.

11.3 INTRODUCTION TO SECOND-LINE DEFENSES AND THE LYMPHATIC SYSTEM

Learning Outcomes

After reading this section, you should be able to:

11.8 Name the two categories of second-line defenses.

11.9 Describe lymph, and explain how it is collected and filtered.

11.10 Name the primary and secondary lymphoid tissues, and describe their basic roles.

11.11 Name the two main categories of leukocytes, and describe how they differ.

11.12 Explain several roles of molecular second-line defenses in innate immunity.

Second-line defenses kick in when first-line defenses are breached.

Despite the general effectiveness of our first-line barriers, microbes still gain entry into the body. They may come through a cut in the skin, or have special adaptations that help them circumvent first-line defenses. For example, the bacterium that causes stomach ulcers, *Helicobacter pylori*, neutralizes stomach acid. No matter how microbes breach first-line barriers, they still face **second-line defenses**. These primarily consist of assorted molecular factors and **leukocytes** (white blood cells)—specialized cells of the immune system. Second-line defenses call on the body's lymphatic system, which we discuss next.

The lymphatic system collects, circulates, and filters body fluids.

Our immune system is supported by a variety of cells, tissues, and organs from many different body systems, but it is especially interconnected with and dependent upon the lymphatic system. The **lymphatic system** is a collection of tissues and organs that have roles in collecting, circulating, and filtering fluid in body tissues before such fluid is returned to the bloodstream.

Lymph and Lymphatic Vessels

As blood is delivered to our tissues, some **plasma**—the liquid portion of blood—exits the capillaries and seeps into the small spaces between tissue cells. The cells of our tissues are constantly being bathed in this **interstitial fluid** (or extracellular fluid), which is water-based and rich in proteins and leukocytes. Lymphatic capillaries, which tend to run parallel to blood capillaries, take up interstitial fluid and return it to our bloodstream. This action prevents **edema**—tissue swelling. Once the interstitial fluid enters lymphatic vessels, we call it **lymph**. The lymphatic vessels transport lymph to lymph nodes, where it is screened for pathogens and filtered. From there, the fluid is eventually channeled into veins, where it rejoins the blood plasma (FIG. 11.4).

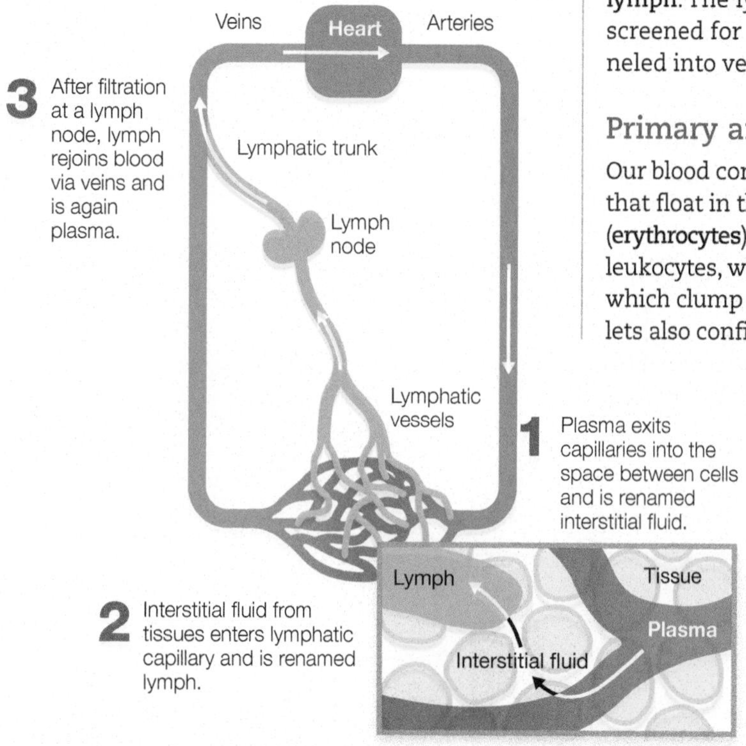

Primary and Secondary Lymphoid Tissues

Our blood consists of numerous cells and cell fragments known as *formed elements* that float in the blood plasma. Most cells in our blood are red blood cells (**erythrocytes**) that deliver oxygen to tissues. Other formed elements in blood include leukocytes, which are essential immune cells, and cell fragments called **platelets**, which clump together and help stem blood loss when vessels are damaged. Platelets also confine invading agents in mesh-like clots. **Primary lymphoid tissues**, including the thymus and red bone marrow, have roles in the production and maturation of erythrocytes, leukocytes, and platelets.

Secondary lymphoid tissues filter lymph and sample surrounding body sites for antigens. As they do this, the leukocytes that reside in secondary lymphoid tissues are brought into contact with antigens to stimulate an immune response. Secondary lymphoid tissues include the lymph nodes, spleen, and mucosa-associated lymphoid tissue (MALT) (FIG. 11.5).

Thymus and bone marrow The **thymus** is a butterfly-shaped lymphoid organ located just behind the sternum (breastbone) near the heart. **Bone marrow** is spongy tissue that is located inside bones and is a key site for red and white blood cell production. The long bones of the arms and legs, the sternum, and the pelvis

Veins Heart Arteries

3 After filtration at a lymph node, lymph rejoins blood via veins and is again plasma.

Lymphatic trunk

Lymph node

Lymphatic vessels

1 Plasma exits capillaries into the space between cells and is renamed interstitial fluid.

Lymph Tissue Plasma Interstitial fluid

2 Interstitial fluid from tissues enters lymphatic capillary and is renamed lymph.

FIGURE 11.4 **Collection and flow of lymph**

(hips) are especially rich in bone marrow. As primary lymphoid tissues, both the thymus and bone marrow make leukocytes and help them mature before they enter circulation. Specifically, these tissues are central to maturation of B cells and T cells, which are both classes of leukocytes involved in adaptive immunity. T cell precursors are made in the bone marrow and then make their way to the thymus to mature into T cells. In contrast, B cells are made in the bone marrow and also mature there. (Chapter 12 covers B cell and T cell maturation.)

Lymph nodes Throughout the body we have 500–700 **lymph nodes** clustered in six main areas, including the neck, underarm, and groin. These small, bean-shaped, secondary lymphoid organs serve as filtering and screening centers for lymph before it is returned to the bloodstream. Upon detecting an invading microbe, leukocytes residing in a node rapidly multiply in order to expand the population available to combat it. This abundant cell division often causes the normally small nodes to become swollen and tender to the touch. A healthcare provider may gently probe the neck region to check for swollen superficial lymph nodes as an indicator of infection. Filtering lymph before it is returned to the bloodstream is useful, as it prevents most microbial invaders and cancer cells from gaining access to the whole body through the bloodstream.

Spleen The **spleen** is a fist-sized secondary lymphoid organ located in the upper left part of the abdomen, just under the diaphragm. Like lymph nodes, the spleen is populated by leukocytes looking for invaders. However, lymph nodes filter lymph, whereas the spleen filters blood. The spleen also has non-immune system functions, including disposal of damaged erythrocytes.

Mucosa-associated lymphoid tissue (MALT) The majority of secondary lymphoid tissue is **MALT**. This diffuse system of lymphoid tissue is found in all mucosal linings of our body and at all body entryways. Tonsils, the appendix, and Peyer's patches (lymphoid tissues found in the small intestine) are common examples. MALT is often more specifically named based on its location in the body. For example, **GALT** refers to gut-associated lymphoid tissue. As in other secondary lymphoid tissues, an abundance of diverse leukocytes patrol MALT tissue searching for invaders. Because the places MALT exists are all common portals of entry for pathogens, these tissues play a key role in finding and fighting harmful microbes.

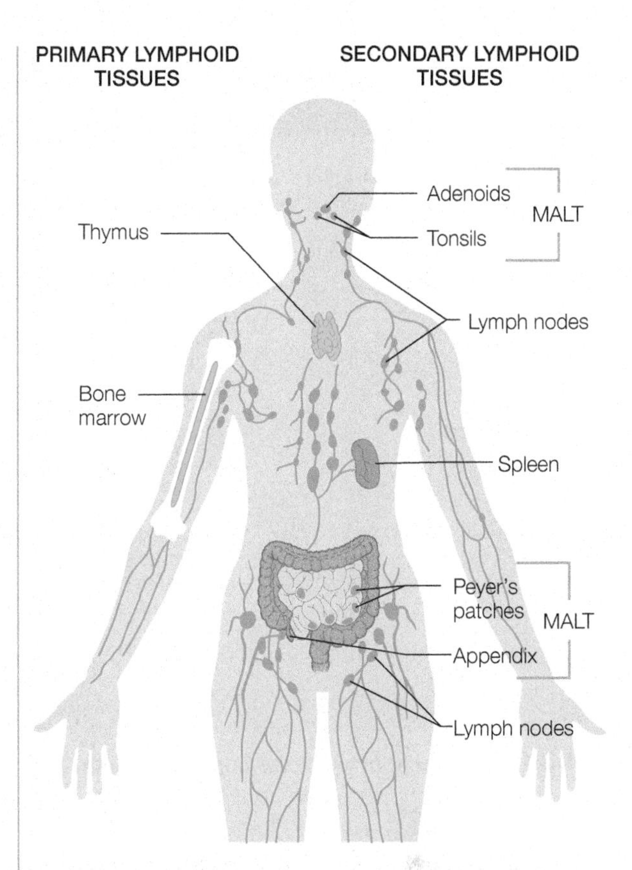

FIGURE 11.5 **Primary and secondary lymphoid tissues** The thymus and bone marrow are primary lymphoid tissues. Most secondary lymphoid tissue is MALT. Sometimes MALT is more specifically named based on its location in the body. For example, the acronym GALT is sometimes used to refer to MALT tissues such as Peyer's patches and the appendix that are associated with the gut.

Leukocytes are essential in all immune responses.

As mentioned earlier, leukocytes are made in primary lymphoid tissues and are then mobilized to seek and destroy microbes that have breached first-line defenses. Leukocytes have diverse functions, but generally they can be structurally classified as either granulocytes or agranulocytes. FIG. 11.6 summarizes the most important leukocyte groups of our immune defenses. **Granulocytes** are cells with granules in their cytoplasm that are visible when stained and then viewed by light microscopy. **Agranulocytes** do not have granules in cytoplasm that are readily detectable with light microscopy.

All granulocytes have roles in the innate immune response, and they tend to have important roles in activating other leukocytes involved in adaptive immunity—the third-line defenses (see Chapter 12). In contrast, agranulocytes include both innate immune responders as well as B cells and T cells that are central to adaptive immunity. As we discuss leukocytes throughout this chapter, it's important to remember that given the complex nature of immune interactions, our categorizations of whether a cell or other factor "belongs" in the innate or adaptive immune system can become a bit blurry. While it is useful for us to differentiate between the innate and adaptive systems for study and research, it's also important to realize that ultimately, humans have one single, highly coordinated and integrated immune system.

	Granulocytes				Agranulocytes		
	Neutrophils	**Eosinophils**	**Basophils**	**Mast cells**	**Monocytes**	**Dendritic cells**	**Lymphocytes**
							• NK cells • B cells • T cells
Appearance	Multilobed nucleus	Bilobed nucleus; red-orange staining granules	Bilobed nucleus obscured by dark purple granules	Circular nucleus; dark staining granules around nucleus	Large horseshoe-shaped nucleus	Ruffled membrane with long cytoplasmic extensions	Small cells with a large rounded nucleus and limited cytoplasm
Notes	Highly phagocytic; fight many invaders, especially bacteria and viruses	Moderately phagocytic; attack allergens and parasites	Attack allergens and parasites	Moderately phagocytic; attack bacteria, allergens, and parasites; reside in tissues	Highly phagocytic once they mature into macrophages (which can be fixed or wandering); activate adaptive immune responses	Highly phagocytic; activate adaptive immune responses	*NK cells*: innate immunity to viruses, bacteria, parasites, and tumor cells *B and T cells*: adaptive immunity

FIGURE 11.6 **Leukocytes of the immune system**

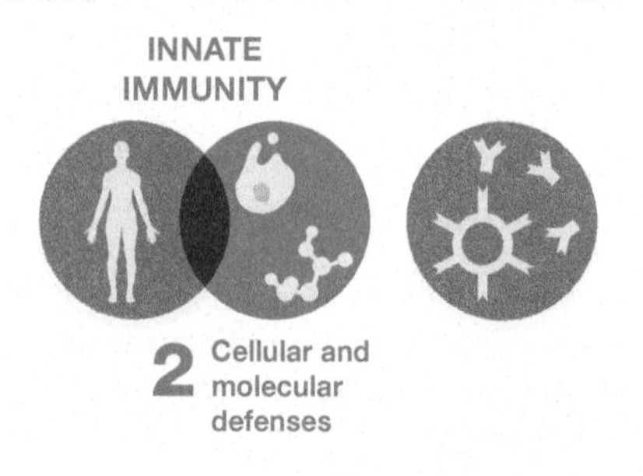

TABLE 11.2 Examples of Leukocytoses

Name	Leukocyte Increased	Typical *Noncancerous* Causes
Neutrophilic leukocytosis	Neutrophils	Acute (sudden onset) bacterial infections
Eosinophilia	Eosinophils	Allergy, asthma, parasitic infections
Basophilia	Basophils	None; usually only occurs with certain rare blood cancers
Monocytosis	Monocytes	Chronic infections/ inflammation
Lymphocytosis	Lymphocytes (usually T or B cells)	Chronic infections/ inflammation; viral infections

White Blood Counts as a Clinical Measure

Given their key role in battling infections, it's not surprising that monitoring the number and type of leukocytes present in a patient is useful for diagnosing infections. A **differential white blood cell count** (WBC differential) is a rapid, inexpensive test that healthcare providers commonly use to help with diagnosis. It determines if any leukocytes are over- or underrepresented in a patient's blood. An increase in leukocytes is called **leukocytosis** and is often caused by infection or, less commonly, by cancer. Examples of different leukocytoses and the clinical scenarios they may indicate are described in TABLE 11.2.

The "normal" or reference ranges for various leukocytes are broad and influenced by numerous patient factors that include age, sex, pregnancy status, smoking status, and ethnicity. As such, what is considered within normal range in a child may be abnormal in an adult. Even if patient factors are normalized, laboratory instrumentation and methodologies also impact reference ranges, which is why these ranges are always noted on the individual lab results.

Cooperation Between Leukocytes and Molecular Factors

When leukocytes are called to action, they often exert their effects by releasing physiologically active molecules into their local environment. These molecules have diverse functions such as recruiting other leukocytes to the site of infection, restricting pathogen growth, triggering fever, or stimulating inflammation. They also buy valuable time by sometimes slowing, if not outright killing, the pathogen, while the body begins to mount its third-line, adaptive immunity response.

Build Your Foundation

8. What are the two main categories of second-line defenses? (LO 11.8)
9. What is lymph, and how is it collected and filtered? (LO 11.9)
10. What are the primary and secondary lymphoid tissues, and what are their roles? (11.10)
11. What are the two main categories of leukocytes, and how do they differ? (11.11)
12. What are some roles of molecular second-line defenses? (11.12)

Build Your Foundation (BYF) Quick Quiz: Visit the **Mastering Microbiology** Study Area to quiz yourself.

11.4 CELLULAR SECOND-LINE DEFENSES

Granulocytes include neutrophils, eosinophils, basophils, and mast cells.

Learning Outcomes

After reading this section, you should be able to:

11.13 Describe structural and functional features of neutrophils, eosinophils, basophils, and mast cells.

11.14 Discuss the features of monocytes, macrophages, and dendritic cells.

11.15 Name the three main categories of lymphocytes, and discuss their general features.

Most granulocytes and agranulocytes exhibit some degree of **phagocytosis**, a specialized form of endocytosis in which nondissolved, extracellular targets are engulfed by a cell and digested. (Refer back to Chapter 4 for details on phagocytosis.) Phagocytic cells (phagocytes) of the immune system target bacterial cells, viral particles, or general debris. Enzymes and chemical agents in the lysosome typically destroy the targeted material. However, some microbes have virulence factors to help them avoid or thwart steps in phagocytosis. (See Chapter 10 for more on how some microbes thwart phagocytosis.)

Neutrophils

Comprising 40 to 70 percent of the leukocyte population, **neutrophils** are the most numerous white blood cells in circulation. Their multilobed segmented nucleus makes them fairly easy to identify by light microscopy. Neutrophils are the first leukocytes recruited from the bloodstream to injured tissues. They release potent antimicrobial peptides (AMPs) and enzymes from their granules into the surrounding environment. These released factors destroy many microbes, especially bacteria, and stimulate inflammation—a process we will review later. Neutrophils also participate in the formation of web-like structures called neutrophil extracellular traps (NETs). As the name implies, these net-like structures trap and immobilize pathogens, allowing neutrophils to more efficiently target the invading agents.

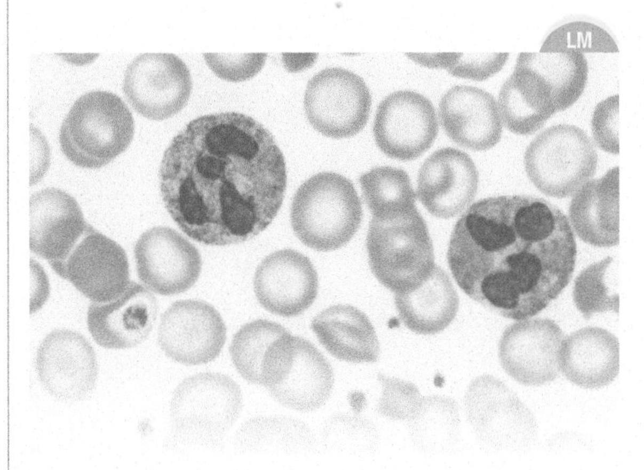

Neutrophils are our most numerous white blood cells.

Neutrophils also phagocytize foreign cells and viruses. An elevated neutrophil count in a blood sample may mean the patient is suffering from an acute bacterial infection. A lower-than-normal neutrophil count, **neutropenia**, may be caused by certain viral infections. Lastly, because neutrophils are the initial cellular responders, they are key "alarm sounders" that recruit other white blood cells to the scene and stimulate their activity upon arrival.

Eosinophils

Eosinophils account for less than 5 percent of the total white blood cell population. They have relatively large cytoplasmic granules that stain red-orange, and their nucleus appears to have two lobes connected by a thin band—a bit like an

TRAINING TOMORROW'S HEALTH TEAM

Counting Immature Neutrophils to Diagnose Bacterial Infections

An increase in leukocytes (leukocytosis) can signal physiological conditions such as infection, autoimmune disorders, allergies, steroid use, certain cancers, or an inflammatory response from injury. In addition to counting normal mature leukocytes, a differential white blood cell count detects immature leukocytes. In particular, it identifies immature neutrophils, or *bands*, to help assess whether or not the patient has a bacterial infection.

Mature neutrophils have a segmented multilobed nucleus and are therefore sometimes called "segs" (for segmented nucleus) on differential white blood cell reports. "Bands," by contrast, are immature neutrophils that have not yet developed a segmented nucleus.

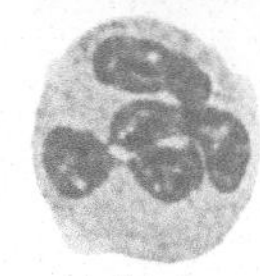
Seg

Band

Banded neutrophils usually constitute less than 6 percent of the total neutrophil population. However, during a bacterial infection, there is an increase in bands as the body works to rapidly increase the neutrophil population. Sometimes this skewing toward banded neutrophils is called a "left shift" because of how these cells used to be noted in older pathology reports.

QUESTION 11.1

Why aren't mast cells counted as a part of a differential white blood cell count and monitored for increases or decreases like other granulocytes? (Hint: Consider where mast cells are found.)

old-fashioned telephone receiver. Eosinophil granules contain diverse enzymes and antimicrobial toxins. These mediators are expelled into surrounding tissues in response to certain allergens and microbes—especially parasites. Eosinophils also exhibit moderate phagocytic activity. An elevated eosinophil count (eosinophilia) suggests a parasitic infection, or possibly asthma or seasonal allergies.

Basophils

At less than 1 percent of the leukocyte population, **basophils** are the least abundant of our white blood cells. They have a double-lobed nucleus (bilobed nucleus) and numerous cytoplasmic granules. Upon staining, the granules take on a dark purple color that tends to obscure the view of the nucleus. Basophil granules are packed with many defense molecules, but of particular note is the presence of **histamine**, a molecule that stimulates inflammation. Like eosinophils, basophils also combat parasitic infections and have roles in allergic responses, but their overall circulating levels are rarely elevated except in certain blood cancers.

Mast Cells

Mast cells were once thought to be basophils that resided in tissues instead of circulating in blood. However, it is now known that mast cells are a distinct population of leukocytes. That said, they are similar to basophils in that they release histamine and have roles in allergies and fighting parasites. Their ability to conduct phagocytosis, along with diverse enzymes and AMPs in their granules, also make them important in fighting bacteria. Furthermore, mast cells have key roles in activating the adaptive immune response.

Mature mast cells are common in tissues near body openings, such as the skin and mucous membranes of the airway and GI tract. Because they stay put in tissues rather than circulate, mast cells act as local lookouts where pathogens are most likely to enter the body. Many factors in mast cell granules act as the initial chemical alarm that recruits neutrophils and other leukocytes to the scene and simultaneously promotes early phases of inflammation.

Agranulocytes associated with innate immunity include monocytes (macrophage precursors), dendritic cells, and certain lymphocytes.

Agranulocytes are leukocytes that don't have visible granules in their cytoplasm when stained and then analyzed by light microscopy (Figure 11.6), though when viewed with electron microscopes, granules can be seen in some of these cells. This group includes cells with roles in both innate and adaptive immunity.

Key Phagocytes: Monocytes (Macrophages) and Dendritic Cells

Like neutrophils, macrophages and dendritic cells use phagocytosis to clear invaders (FIG. 11.7). They also activate the adaptive branch of immunity. (See Chapter 12 for their roles as antigen-presenting cells.)

FIGURE 11.7 **Phagocytosis** Colorized scanning electron micrograph showing the intracellular detail as a macrophage (gray) phagocytizes *Mycoplasma tuberculosis* bacteria (pink).

Monocytes, the largest agranular white blood cells, have horseshoe-shaped nuclei and account for up to 10 percent of our circulating leukocytes. Monocyte levels can become increased as a result of chronic infections and inflammation, autoimmune disorders, and certain cancers. As they migrate out of the circulatory system into tissues, they mature into **macrophages**, which are highly phagocytic cells that destroy a wide range of pathogens. Macrophages are described as either fixed or wandering. **Fixed macrophages** reside in specific tissues—for example, Kupffer cells in the liver, or alveolar dust cells in the lungs. **Wandering macrophages** do as their name implies: roam through tissues.

Dendritic cells are named for their long cytoplasmic extensions that resemble the spindly dendrite extensions of neurons (nerve cells). These highly phagocytic cells patrol most body tissues, but are especially abundant in areas of the body that have contact with the environment, such as skin and body openings. In addition to phagocytizing a broad range of antigens, dendritic cells play roles in preventing our immune system from attacking self and from overreacting to nonthreatening substances.

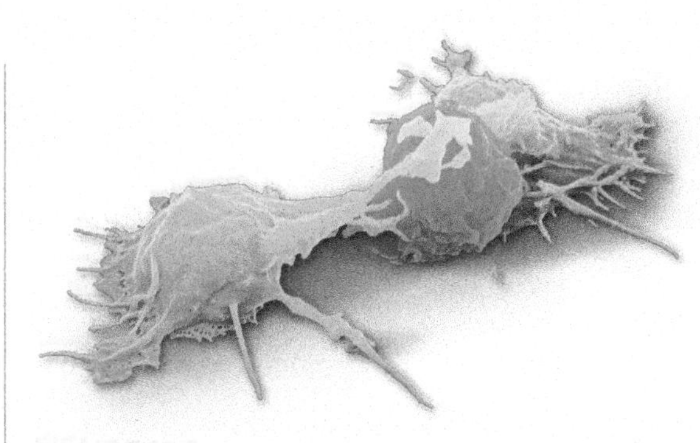

FIGURE 11.8 **Natural killer cells** Colorized scanning electron micrograph of NK cells attacking a cancer cell.

Lymphocytes: NK cells, B cells, and T cells

Lymphocytes are a subgroup of leukocytes that include natural killer (NK) cells, B cells, and T cells. About 25 percent of circulating leukocytes are **lymphocytes**. These cells tend to be relatively small compared with other leukocytes. When stained and viewed with a light microscope, they have a large, rounded nucleus and limited cytoplasm. **Natural killer (NK) cells** are abundant in the liver and have important roles in innate immunity. Their name derives from the fact that these cells directly kill virus-infected cells and tumor cells; however, they also protect against bacteria and parasites (FIG. 11.8). Interestingly, research is increasingly revealing that NK cells play at least some role in adaptive immunity. Nevertheless, they are still classically considered lymphocytes of innate immunity because they lack antigen-specific cell surface receptors. T cells and B cells are still considered the main coordinators of the adaptive immune response. (B and T cells are covered in more detail in Chapter 12.)

Build Your Foundation

13. Discuss the structural and functional features of neutrophils, eosinophils, basophils, and mast cells. (LO 11.13)

14. What are features of monocytes, macrophages, and dendritic cells? (LO 11.14)

15. What are the three main categories of lymphocytes, and which are involved in innate immunity versus adaptive immunity? (LO 11.15)

Build Your Foundation (BYF) Quick Quiz: Visit the **Mastering Microbiology** Study Area to quiz yourself.

11.5 MOLECULAR SECOND-LINE DEFENSES

A number of defense molecules mediate innate immune responses.

Learning Outcomes

After reading this section, you should be able to:

11.16 List at least three general roles of second-line molecular defenses.

11.17 Describe why leukocytes are so central to second-line molecular defenses.

11.18 Define the term *cytokine*, and provide examples of classes of cytokines and their respective functions.

11.19 Discuss the role of iron-binding proteins, and provide examples of how pathogens can overcome this defense mechanism.

11.20 Explain the three main pathways for complement activation, and describe their final outcomes.

When leukocytes are called to action, they often release molecules into the local environment. These molecules recruit more immune system cells to sites where antigens are detected, restrict pathogen growth, generate fever, and induce inflammation. Their collective functions can be distilled into one key effort: eliminating the invader, or when that is not possible, limiting its propagation and further invasion until the adaptive immune system responds. The main molecular second-line defenses are summarized in TABLE 11.3.

Cytokines

Cytokines are signaling proteins that help cells communicate with each other, initiating and coordinating immune actions. Most cells in our body can make and release at least one type of cytokine. Hundreds have been identified, and many more are still to be characterized. Cytokines are often grouped by functions. For example, cytokines that induce inflammation are collectively called proinflammatory cytokines.

Because cytokines are such important immune system mediators, they are increasingly used as diagnostic, therapeutic, and even prognostic markers. For example, an exaggerated cytokine response, called a **cytokine storm**, is associated with a poorer prognosis. This is because excessive cytokine levels trigger a self-destructive immunological response that may lead to irreversible tissue damage. Some of the sickest COVID-19 patients develop cytokine storms, and treatment with drugs that suppress cytokine actions, such as the rheumatoid

TABLE 11.3 Summary of Key Molecular Second-Line Defenses

Molecular Defense	Function(s)
Cytokines	• Stimulate inflammation • Generate fever • Recruit leukocytes to fight infection • Stimulate tissue and blood vessel repair • Promote leukocyte and lymphatic tissue development • Antiviral effects (interferons) • Immune system regulation/activation
Iron-Binding Proteins	• Limit availability of free iron to reduce bacterial growth
Complement Proteins	• Stimulate inflammation • Tag targets for elimination (opsonization) • Directly kill targeted cells (cytolysis)

CLINICAL VOCABULARY

Prognosis A prediction regarding a patient's outcome as it relates to a medical condition. Prognostic tools help predict the course of disease and possible outcomes. Levels and types of cytokines present can provide clues for diagnosing disease and determining the possible outcome.

arthritis drug Actemra (tocilizumab), may be useful to improve patient outcomes. (See Chapter 16 for more on cytokine storms and their pathophysiological effects in respiratory infectious diseases such as COVID-19 and influenza.) Given their increasing role in clinical applications, it is useful for allied health students to have a basic understanding of these molecules (TABLE 11.4).

Chemokines A subcategory of cytokine, **chemokines** act as signaling proteins that attract white blood cells to areas where they are needed. Their ability to induce **chemotaxis**, cell movement in response to a chemical stimulus, is reflected in their name. The over 40 known chemokines have important roles in wound healing, blood vessel formation and repair, lymphoid tissue development, and activation of innate and adaptive immune responses.

Interleukins (ILs) Diverse cytokines called **interleukins** have roles in activating adaptive and innate immune responses and stimulating the production of new blood cells and platelets—a process called **hematopoiesis** (hem-a-toe-POY-ee-sis). Interleukin families are usually denoted as "IL" followed by a number. The *IL-1 family* is especially noteworthy, as all cells in the innate immune system make at least one of the 11 different IL-1 family members and/or are affected by them. These signaling proteins have diverse effects on the body, but their primary immune system roles are to regulate inflammation, stimulate innate and adaptive immune responses, and generate fever (each of these processes is reviewed later). Genetic defects in IL-1 family interleukins can lead to an assortment of diseases, including autoimmune disorders, inflammatory disorders, insulin resistance/diabetes, and atherosclerosis.

Another key group of interleukins, which you will encounter in Chapter 12, is the *IL-2 family*. For now, suffice it to say that these interleukins are central to T cell development and training the immune system in self-tolerance—that is, the ability to differentiate self from foreign and only attack foreign substances. The IL-2 family also has roles in triggering programmed cell death—a process known as **apoptosis** (ay-pah-TOE-sis).[3]

TABLE 11.4 Examples of Cytokines

Cytokine Type	Examples	Notes
Chemokines	40+ types (Monocyte chemoattractant protein-1 is an example)	Chemotactic molecules that recruit white blood cells to areas of injury or infection; roles in wound healing, blood vessel formation/repair, lymphoid tissue development, and activation of innate and adaptive immune responses
Interleukins	IL-1 family	Regulate inflammation, stimulate innate and adaptive immune responses, and generate fever
	IL-2 family	Influence T cell development; have roles in generating immune system self-tolerance (see Chapter 12)
Interferons	Interferon-alpha (IFN-α) Interferon-beta (IFN-β)	IFN-α and IFN-β both made by virus-infected cells to signal neighboring cells to mount antiviral defenses; stimulate innate and adaptive immune responses to viruses, bacteria, and parasites; IFN-α can also trigger fever
	Interferon-gamma (IFN-γ)	Mainly made by NK cells and certain T cells to activate macrophages and stimulate innate and adaptive immune responses to viruses, bacteria, and parasites
Tumor necrosis factors	Tumor necrosis factor-alpha (TNF-α)	Mainly made by macrophages; induce inflammation and kill tumor cells; can induce fever

[3] Kuska, B. (1997). You say tomato and I say tomahto: Getting a handle on pronouncing apoptosis. *Journal of the National Cancer Institute*, *89*(5), 351.

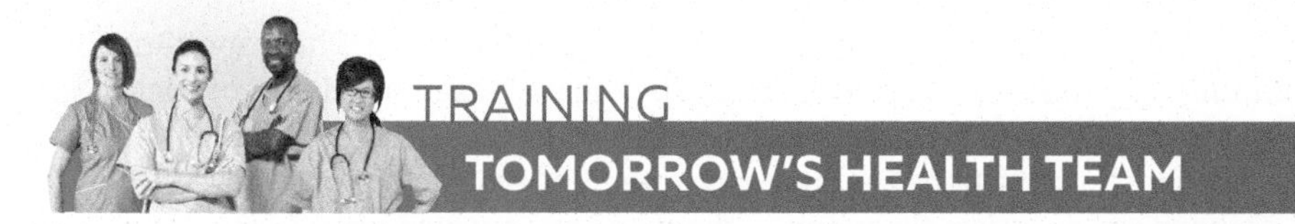

Interferons as Therapies

Because of their ability to stimulate the immune system to target cancer- and virus-infected cells, interferons have been used to treat a number of diseases. Drugs of this class tend to end in the suffix "ron" (e.g., Betaseron, Intron-A, Sylatron). Interferon alpha (IFN-α) has been used in combination with other drugs to treat hepatitis B and C viruses, as well as a variety of blood and lymphatic system cancers such as chronic myeloid leukemia, hairy cell leukemia, and nodular lymphoma.

Multiple sclerosis is often treated with IFN beta (IFN-β), which is thought to lessen autoimmune reactions in these patients by increasing interleukin 10, a molecule that reduces inflammation. Unfortunately, interferon therapies have a lot of side effects, including flu-like symptoms of fever, achiness, and fatigue.

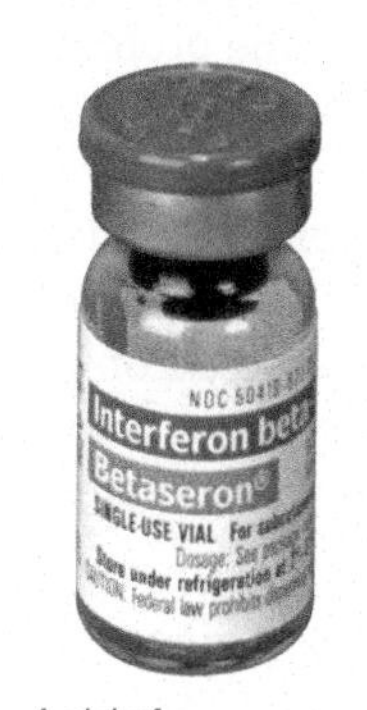

A vial of recombinant IFN-α for injection.

QUESTION 11.2

Why isn't it surprising that IFN therapies induce fever as a side effect?

Interferons (IFNs) A collection of signaling molecules called **interferons** give the alarm when pathogens or tumor cells are detected. They are especially well known for antiviral effects and derive their name from their ability to "interfere" with viral replication. Initially thought only to battle viruses, they are now also known to be important activators of innate and adaptive immune responses against bacteria and parasites. Many classes of interferons exist, but interferon alpha (IFN-α), interferon beta (IFN-β), and interferon gamma (IFN-γ) are among the better-understood types. Virus-infected cells make IFN-α and IFN-β as chemical alarms that stimulate nearby uninfected cells to mount antiviral defenses (FIG. 11.9). IFN-γ is made by certain lymphocytes, especially NK cells and certain T cells, and stimulates a range of innate and adaptive immune system effects that help combat viruses, bacteria, and parasites.

Unfortunately, many viruses block interferon signaling in infected cells. This is a factor with Ebola viruses, some adenoviruses, influenza A viruses, polioviruses, and human papillomaviruses, to name just a few.

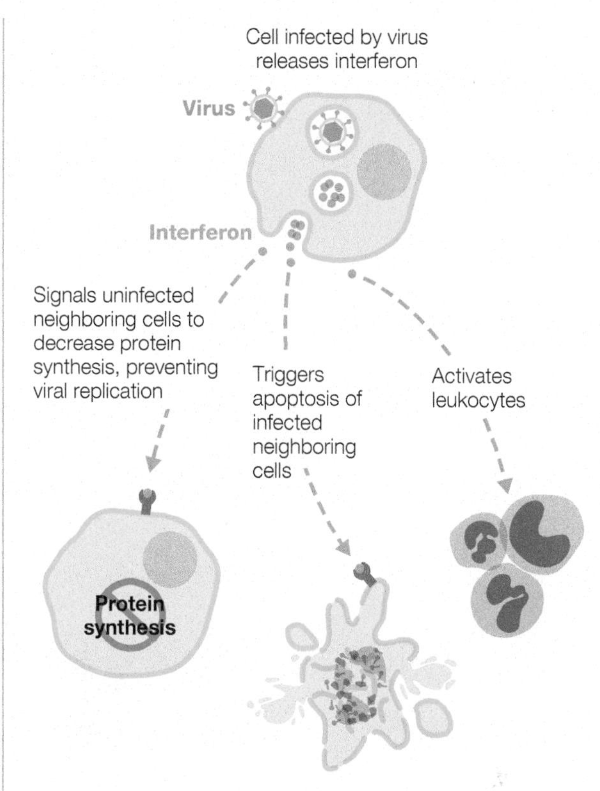

FIGURE 11.9 Summary of key interferon actions

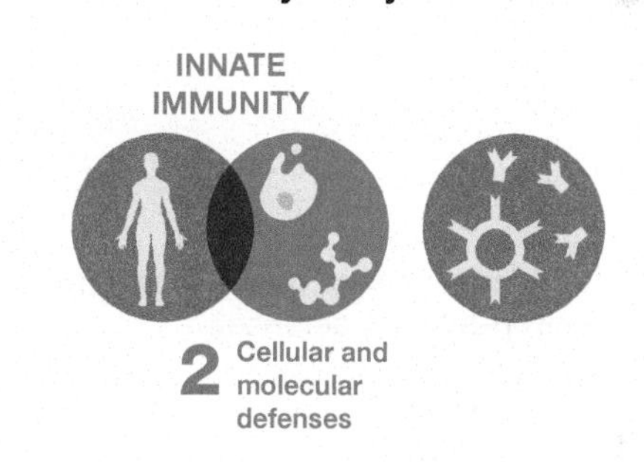

Tumor necrosis factors (TNFs) Like the other families of cytokines already discussed, the tumor necrosis factors (TNFs) are signaling proteins. There are a lot of TNFs, but **tumor necrosis factor alpha** (TNF-α), a factor made primarily by macrophages, is one that has received much clinical attention for its ability to stimulate inflammation and kill tumor cells. TNF-α also stimulates fever. TNF-α inhibitors treat certain immune disorders, such as rheumatoid arthritis and inflammatory bowel disease.

Iron-Binding Proteins

Because of its role in a variety of metabolic and other physiologically essential pathways, iron is a vital nutrient for most cells. Consequently, if access to iron is limited, then so is cell growth and survival. Normally, the amount of freely available iron in circulation and within our tissues is well below the necessary amount microbes require for survival. This is thanks to our **iron-binding proteins**. Oxygen-transporting *hemoglobin* in our erythrocytes is one example. Others include *ferritin*, found in most cells; *lactoferrin*, seen in milk, tears, saliva, mucus, and neutrophil granules; and *transferrin*, found in blood plasma and extracellular fluids. They all sequester iron, ensuring that our own cells have an adequate supply when

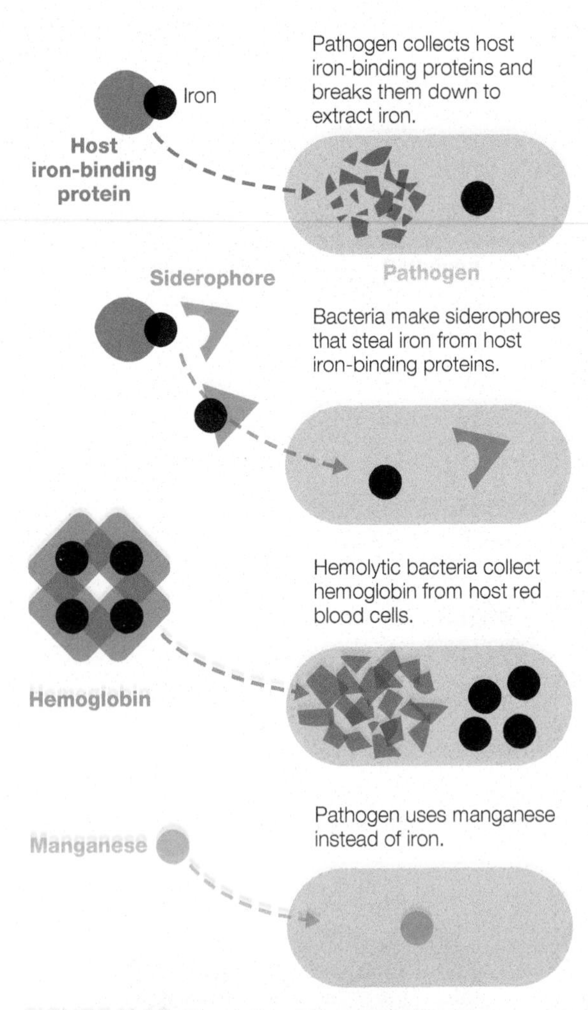

FIGURE 11.10 Examples of mechanisms for overcoming iron-binding defenses

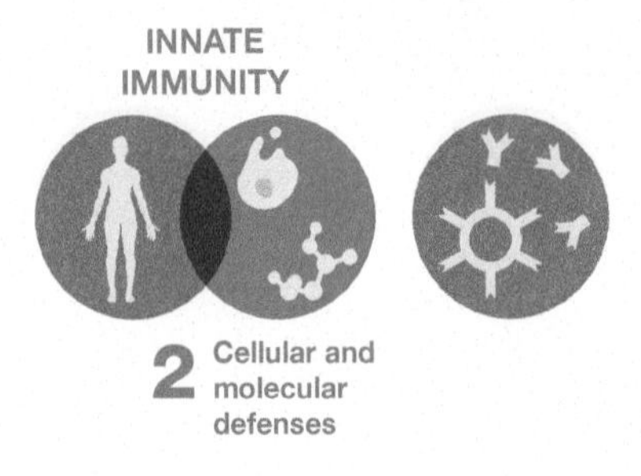

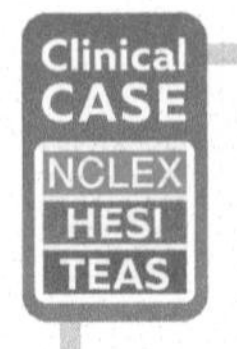

The Case of the Missing Bleach
Practice applying what you know clinically: visit the **Mastering Microbiology** Study Area to watch Part 2 and practice for future exams.

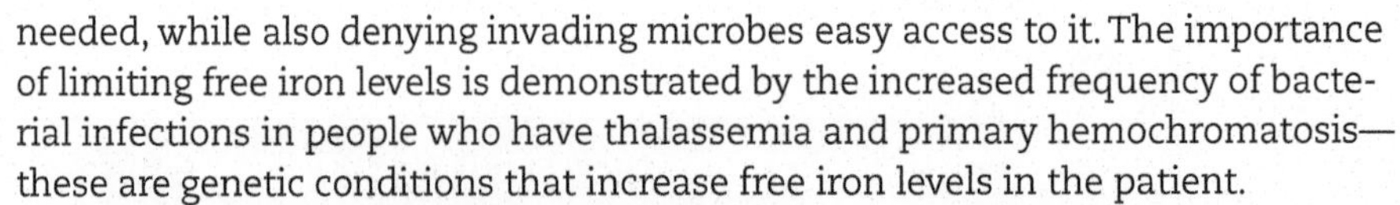

needed, while also denying invading microbes easy access to it. The importance of limiting free iron levels is demonstrated by the increased frequency of bacterial infections in people who have thalassemia and primary hemochromatosis—these are genetic conditions that increase free iron levels in the patient.

Perhaps not surprisingly, many pathogens have evolved ways to steal iron from us (FIG. 11.10). Some bacteria such as *Neisseria gonorrhoeae* capture our iron-binding proteins and break them down to obtain iron for their own use. Another iron-acquisition approach involves **siderophores**, which are organic molecules that extract iron from our iron-binding proteins without going through the step of breaking the iron-binding protein down (see Chapter 10). Some pathogens, such as group A streptococci, conduct hemolysis (the breakdown of erythrocytes) to access iron-rich hemoglobin. Other pathogens such as *Borrelia burgdorferi*, the bacterium that causes Lyme disease, use manganese in their metal-requiring enzymes instead of iron, thereby circumventing our iron-sequestering defenses.

Complement cascades boost the effectiveness of other innate immune responses.

The **complement system** consists of over 30 different proteins that work together in a cascade fashion to support other immune defenses. These proteins tend to be labeled with a "C" followed by a number (such as C1, C2, etc.). This number is based on the order of protein discovery, not the order in which the proteins act in the cascade. For example, C3 is the key trigger point for cascade activation and regulation. Complement proteins are mostly made by the liver and then circulated in our blood plasma in an inactive form. When the proteins are activated, they generate a cascade of events that boost our response against infectious agents.

Complement proteins are activated in diverse ways, such as being triggered by macrophages and neutrophils, or by certain blood-clotting proteins. Here we will focus on three of the main pathways for activating complement proteins: the classical pathway, the alternative pathway, and the lectin pathway (FIG 11.11).

When discussing complement cascades, it's important to realize that they are distinguished by how they are activated, not by their end results. *All three of these complement pathways have the same three outcomes:*

1. **Opsonization**, which tags the invader with complement proteins so it stands out and is more readily cleared by phagocytic cells
2. Formation of a **membrane attack complex** (MAC), which drills into cells, causing them to burst (undergo cytolysis) as water and ions rush into the targeted cell
3. Inflammation (discussed in more detail later)

Classical Pathway

The **classical complement cascade** was the first pathway scientists discovered. It is most commonly triggered by **antibodies** bound to an invading agent.[4] Antibodies are soluble proteins that play a key role in adaptive immunity when they bind to specific antigens, tagging them for elimination (see Chapter 12 for a more detailed review of antibodies). Their role in triggering the classical pathway illustrates how one component of the immune system can straddle the line between adaptive and innate processes. Although this pathway involves antibodies, it is still considered part of the innate system. This is because, whereas the antibodies themselves target specific antigens, the complement cascade proteins always remain the same and do not exhibit a memory component.

---

[4] Until the late 1990s it was assumed that the classical pathway was only initiated by antibodies. We now know that other body proteins that act as "danger signals" can activate the classical pathway, too. For example, C-reactive protein, which is released when tissues are injured or inflamed, directly activates the classical pathway. Thus, the classical pathway is not as "adaptive immunity dependent" as was previously thought.

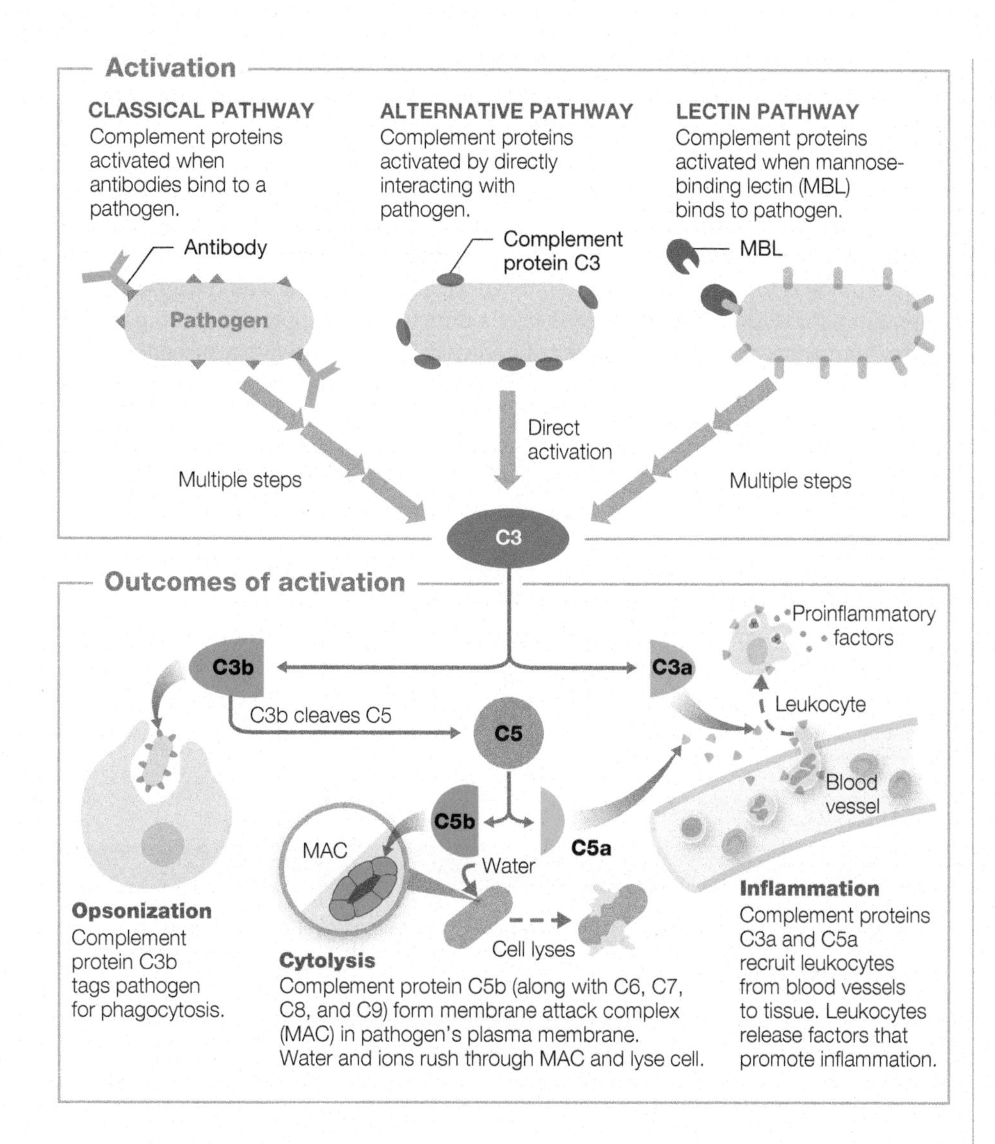

FIGURE 11.11 Complement cascades and outcomes The three main pathways for complement activation all have the same general outcomes: opsonization, cytolysis, and inflammation.

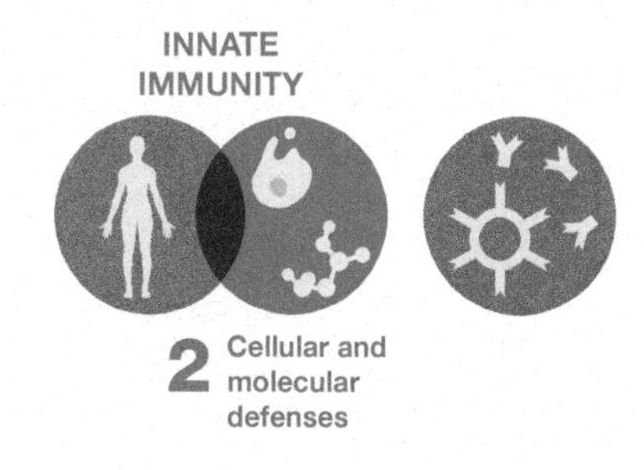

Alternative Pathway

The **alternative pathway** of complement activation was so named because it was the second complement activation scheme to be discovered and thereby represented an "alternative" to the classical pathway. Here, complement proteins activate by directly interacting with the invading agent—there is no need for an intermediary antibody or other protein to modulate the activation.

Lectin Pathway

Like the alternative pathway, the **lectin pathway** of complement activation is independent of antibodies. Instead, it becomes activated when a protein in our blood called **mannose-binding lectin** associates with certain sugars (mannose or other sugars) on a microbe's surface. Other than how it is initiated, the lectin pathway for complement activation is identical to the classical pathway.

Complement Evasion and Regulation

Complement proteins are potent defenses that, like all immune responses, will damage our own tissues if they are not properly regulated. One form of regulation is a "self-destruct feature": Because the activated forms of complement proteins are highly unstable, they will deteriorate if not stabilized by other activated proteins in the cascade. This stabilization requirement ensures a coordinated cascade and limits the risk of a rogue protein triggering cascades when they're not needed and inadvertently causing host tissue damage. The body also makes **regulators of complement activation** (or RCAs), a collection of proteins that turn off complement cascades after a threat passes.

As they have done with most of our defenses, some pathogens have evolved ways to evade our complement systems. One of the most widespread mechanisms of evasion is for a pathogen to make proteins that mimic our naturally occurring RCAs; this puts the brakes on complement activation. Another strategy some bacteria employ is to build a capsule to hide surface features that would trigger complement activation. Other pathogens take a direct assault approach and break down complement proteins and/or antibodies that could activate a cascade.

Build Your Foundation

Build Your Foundation (BYF) Quick Quiz: Visit the **Mastering Microbiology** Study Area to quiz yourself.

16. What are three main roles of molecular second-line defenses? (LO 11.16)

17. Why are molecular second-line defenses considered closely related to cellular second-line defenses? (LO 11.17)

18. Define the term *cytokine*, and then name four different classes of cytokines and state their functions. (LO 11.18)

19. What role do iron-binding proteins have in immunity, and how do pathogens defeat this defense tool? (LO 11.19)

20. Compare and contrast the three key complement activation pathways. (LO 11.20)

11.6 INFLAMMATION AND FEVER

Learning Outcomes

After reading this section, you should be able to:

11.21 List the three primary functions of inflammation, and describe its three phases.

11.22 State the four cardinal signs of inflammation, and describe how they come about.

11.23 Discuss roles of several key chemical mediators in inflammation.

11.24 Describe the features of chronic inflammation.

11.25 State what fever is, list its potentially useful effects, and describe how it is generated and treated.

11.26 Correctly use clinical terminology to classify different fever patterns.

Inflammation and fever are key protective innate immune responses.

When a threat is detected, tissue cells and local leukocytes (like mast cells and fixed macrophages) release chemical mediators that signal the need for further immune actions. As cellular defenses converge on the scene, cytokines, complement proteins, and other pro-inflammatory factors coordinate defense efforts. The combined actions of cellular and molecular second-line defenses trigger inflammation and sometimes fever.

Inflammation is essential to healing and immunity, but if unregulated, it damages our own tissues.

Inflammation is an innate immune response that tends to develop when our tissues are damaged, either from physical factors like trauma or burns, or from infectious agents. Although a physical injury often introduces an infectious agent, such as when a skin cut allows a pathogen to gain entry, the injury could also be *aseptic*, meaning it doesn't introduce an infectious agent. A sprained ankle, bruise, or torn ligaments are examples of aseptic injuries. Either way, a tissue injury initiates blood-clotting cascades. Blood clots curb blood loss or seepage into surrounding tissues and limit pathogen spread—at least when dealing with microbes that don't circumvent our clotting protections. Although inflammation is generally perceived as a negative process that can damage the associated tissues, so long as it doesn't get out of control it is actually an important part of our innate immune defense and is essential to healing. The three main functions of inflammation are to:

1. Recruit immune defenses to the injured tissue
2. Limit the spread of infectious agents
3. Deliver oxygen, nutrients, and chemical factors essential for tissue recovery

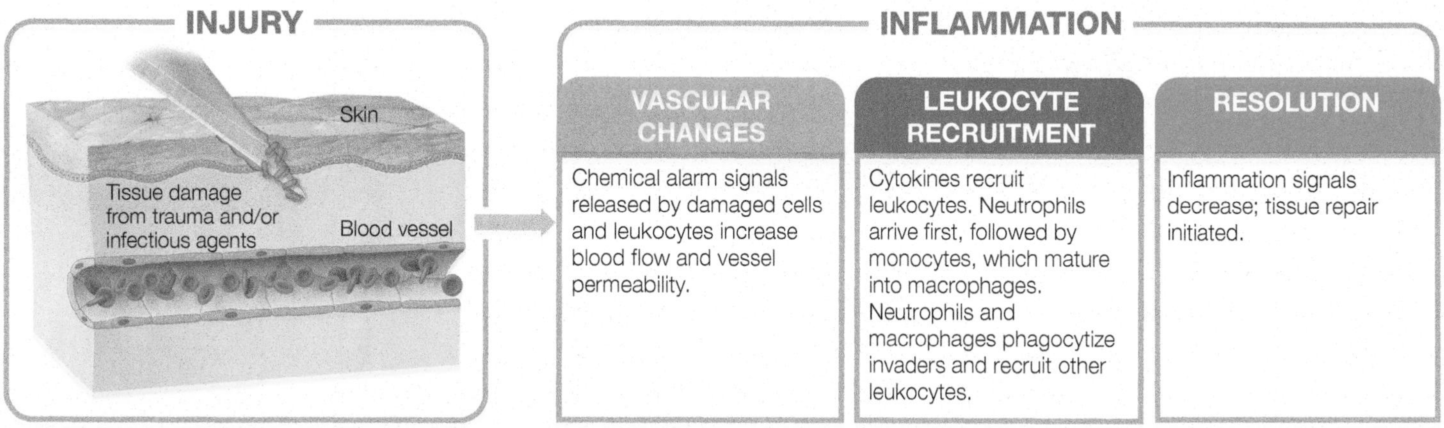

INJURY

Skin

Tissue damage from trauma and/or infectious agents

Blood vessel

INFLAMMATION

VASCULAR CHANGES

Chemical alarm signals released by damaged cells and leukocytes increase blood flow and vessel permeability.

LEUKOCYTE RECRUITMENT

Cytokines recruit leukocytes. Neutrophils arrive first, followed by monocytes, which mature into macrophages. Neutrophils and macrophages phagocytize invaders and recruit other leukocytes.

RESOLUTION

Inflammation signals decrease; tissue repair initiated.

FIGURE 11.12 **Injury and the three general phases of inflammation**

INNATE IMMUNITY

2 Cellular and molecular defenses

Cardinal signs of inflammation are redness, pain, localized heat (not fever), and swelling. Occasionally a fifth feature, loss of function, also develops; this occurs with severe inflammation that generates intense pain and/or swelling that limits use of the affected body part.

Inflammation Phases

Inflammation occurs in three general phases: vascular changes, leukocyte recruitment, and resolution (FIG. 11.12). Each is governed by chemical signals including complement proteins, cytokines, and molecules stored in granulocytes—especially histamine in mast cell granules.

Vascular changes phase Early inflammation is characterized by vascular changes, in which undamaged blood vessels increase in diameter, so more blood flows to injured tissues. This process is called **vasodilation**. At the same time, **vessel permeability** increases, which causes vessels to be slightly "leaky," to allow blood plasma and the proteins it carries (including complement proteins) to enter tissues (FIG. 11.13). This fluid, which is called **exudate**, accumulates in the tissues. Many of the proteins in exudate help finalize blood clot formation. As such, inflammation and blood-clotting cascades are tightly connected.

Vascular changes are caused by diverse chemical signals, the most important of which include histamine, kinins, and eicosanoids. **Histamine** is a small amine molecule and potent inducer of vascular changes; as such it is said to be a **vasoactive** molecule. Many cells make it, including mast cells, basophils, and eosinophils, as do platelets—the cell fragments in blood that are central to clot formation. **Antihistamines** block histamine's actions and therefore serve as anti-inflammatory drugs. (Histamines in allergy are discussed more in Chapter 13.) Mast cells, which stay localized in tissues, are one of the most important histamine releasers early in the inflammation process. Mast cells and other leukocytes also release cytokines, such as TNF-α to enhance inflammation.

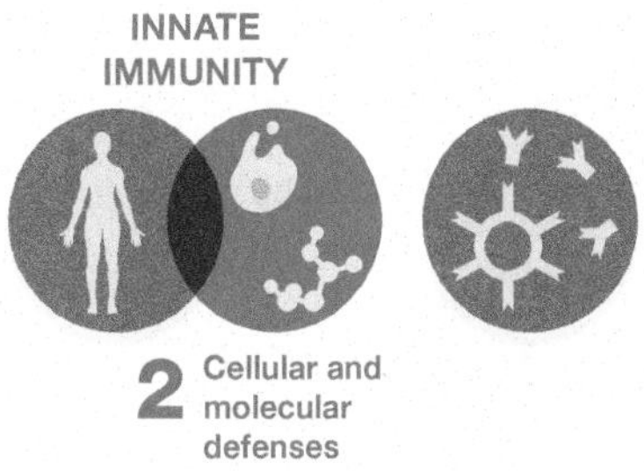

CHEM • NOTE

Amines are organic molecules that have NH_2 groups.

Kinins are another class of pro-inflammatory, vasoactive factors. These blood plasma proteins induce vascular changes, stimulate pain receptors, and assist in blood-clotting cascades. They also amplify inflammation by triggering the production of numerous downstream signaling molecules, including eicosanoids.

Eicosanoids (or icosanoids) are vasoactive signaling molecules that participate in many physiological events, from inducing uterine contractions in childbirth to regulating stomach acid secretions. During inflammation, they induce vascular changes and stimulate pain receptors. Certain eicosanoids also generate fever when released in the hypothalamus of the brain. Eicosanoids are made from *arachidonic acid*, a fatty acid derived from cell membrane phospholipids. Examples of eicosanoids include prostaglandins, leukotrienes, and thromboxanes. Nonsteroidal anti-inflammatory drugs (NSAIDs) such as aspirin, ibuprofen (Advil/Motrin), and naproxen (Aleve) and steroidal anti-inflammatory

INFLAMMATION

VASCULAR CHANGES

LEUKOCYTE RECRUITMENT | RESOLUTION

Summary

- An Injury or infection triggers the release of vasoactive molecules from damaged tissue cells and resident leukocytes such as mast cells.
- Vasoactive molecules induce nearby, undamaged vessels to dilate and become more permeable.
- Increased blood flow and vessel permeability promote swelling as plasma seeps into the tissues. Complement proteins also move into the tissues from the blood.

Key players

Mast cells 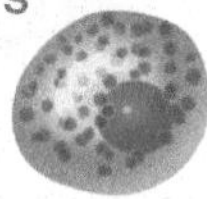

Vasoactive molecules (mainly kinins, eicosanoids, and histamines)

Exudate (fluid and associated plasma proteins, such as complement proteins)

Details

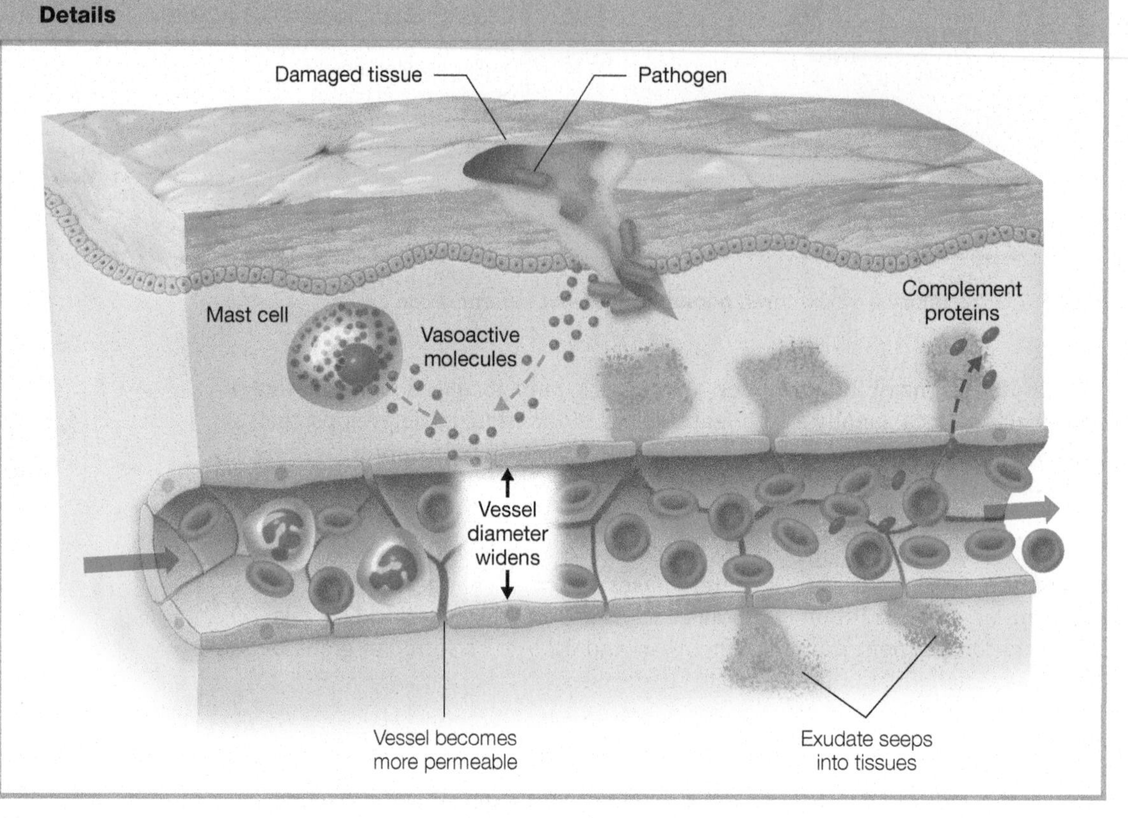

FIGURE 11.13 **Vascular changes in early inflammation** Chemical signals released by injured tissues, and resident leukocytes cause blood vessels to vasodilate and become more permeable.

How Inflammation Works in the Body

Bring the art to life! Visit the **Mastering Microbiology** Study Area to watch the Concept Coach and master inflammation.

CHEM • NOTE

Phospholipids are molecules that make up the plasma membrane. They have a phosphate region linked to fatty acid tails.

drugs (SAIDs) such as glucocorticosteroids (cortisone, prednisolone, etc.) reduce eicosanoid production and are useful for managing inflammation and related pain.[5]

Both vasodilation and increased vessel permeability contribute to the cardinal signs of inflammation. Vasodilation stimulates blood to enter the affected area—and because blood is red and warm, it's logical that this event is linked to the symptoms of local heat and redness. If you have "blushed" in response to a situation, then you know firsthand how a rush of blood to the face instantly makes your cheeks appear reddened and feel warm. Swelling results as fluids exit leaky blood vessels to enter tissues; the pressure generated by swelling then activates pain receptors. To reduce swelling, the exudate that accumulated in the tissues must be removed; we'll review that process later.

Leukocyte recruitment phase The recruitment phase of inflammation relies on **chemoattractants**, which are a diverse collection of cytokines and other signaling molecules that draw leukocytes to the injured tissue. This phase is also aided by the vascular changes that occurred earlier in inflammation. Increased blood flow conveys more leukocytes to the vicinity, while increased vessel permeability makes it easier for the leukocytes to escape blood vessels and migrate into the tissues where they are needed.

Cells exit blood vessels in two steps (FIG. 11.14). In **margination**, leukocytes slow down as they roll along vessel walls; they eventually adhere to the vessel wall and stop rolling. The leukocyte then dramatically changes shape, a

[5] Acetaminophen (Tylenol) is a fever-reducing drug (has antipyretic capabilities), and it is an analgesic (pain reliever), but it is not technically an anti-inflammatory drug.

INFLAMMATION

VASCULAR CHANGES | **LEUKOCYTE RECRUITMENT** | RESOLUTION

Summary

- Chemoattractants recruit leukocytes to the inflamed tissue. Neutrophils and monocytes are the first recruits.
- Leukocytes undergo margination and diapedesis to exit capillaries.
- Monocytes mature into macrophages as they migrate through the tissue.
- As neutrophils and macrophages carry out phagocytosis they release cytokines to recruit other leukocytes.

Key players

Chemoattractants (cytokines and other signaling molecules released by local tissues and leukocytes such as mast cells)

Leukocytes:

Details

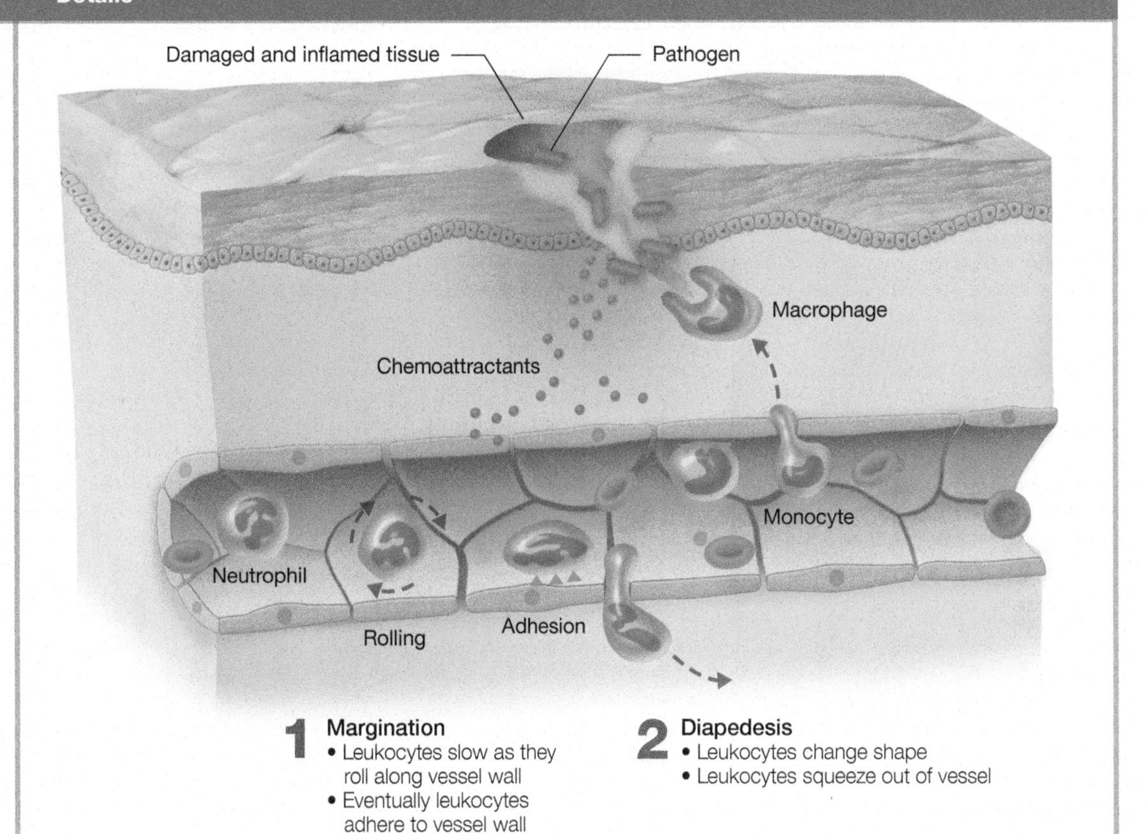

FIGURE 11.14 **Leukocyte recruitment phase of inflammation** Chemoattractants signal leukocytes to leave the blood vessels and migrate to injured tissue.

TRAINING TOMORROW'S HEALTH TEAM

NSAIDs and SAIDs

Eicosanoids are significant pro-inflammatory factors, so blocking their production is one important approach to limiting inflammation. Nonsteroidal anti-inflammatory drugs (NSAIDs) target an enzyme that converts arachidonic acid to the eicosanoids known as prostaglandins and thromboxanes. NSAIDs aren't effective for managing inflammation in asthma, in part because they do not reduce the production of leukotrienes, eicosanoids that stimulate airway inflammation seen in asthma. Indeed, NSAIDs may even *enhance* leukotriene production; this is why aspirin usage can trigger asthma episodes in some people.

Steroid anti-inflammatory drugs (SAIDs) are more potent anti-inflammatory drugs than NSAIDs. This is because steroid drugs decrease the production of many pro-inflammatory factors, including cytokines and a broader range of eicosanoids; with regular use they also decrease tissue mast cell numbers. SAIDs block the production of arachidonic acid, which is an earlier step in the inflammation cascade than NSAIDs target.

QUESTION 11.3

If administered over a prolonged period, SAIDs lead to suppressed immunity and put people at risk for infections. Based on what you have read in this feature and in the chapter, why does this side effect make sense?

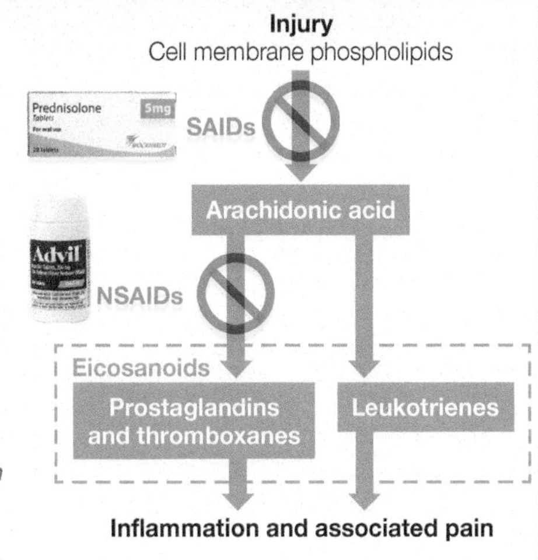

INFLAMMATION

VASCULAR CHANGES | LEUKOCYTE RECRUITMENT | **RESOLUTION**

Summary

- The resolution phase begins as the threat passes. Local capillaries return to normal while leukocytes and tissue cells in the area release chemical signals that reduce inflammation and promote healing.
- Leukocytes that are no longer needed die off through apoptosis, contributing to the pus that may form in this stage.
- Swelling is reduced as exudate is collected by nearby lymphatic capillaries.

Key players

Cytokines and growth factors (reduce inflammation and encourage healing)

Leukocytes (especially neutrophils and macrophages)

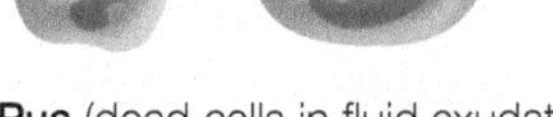

Pus (dead cells in fluid exudate)

Details

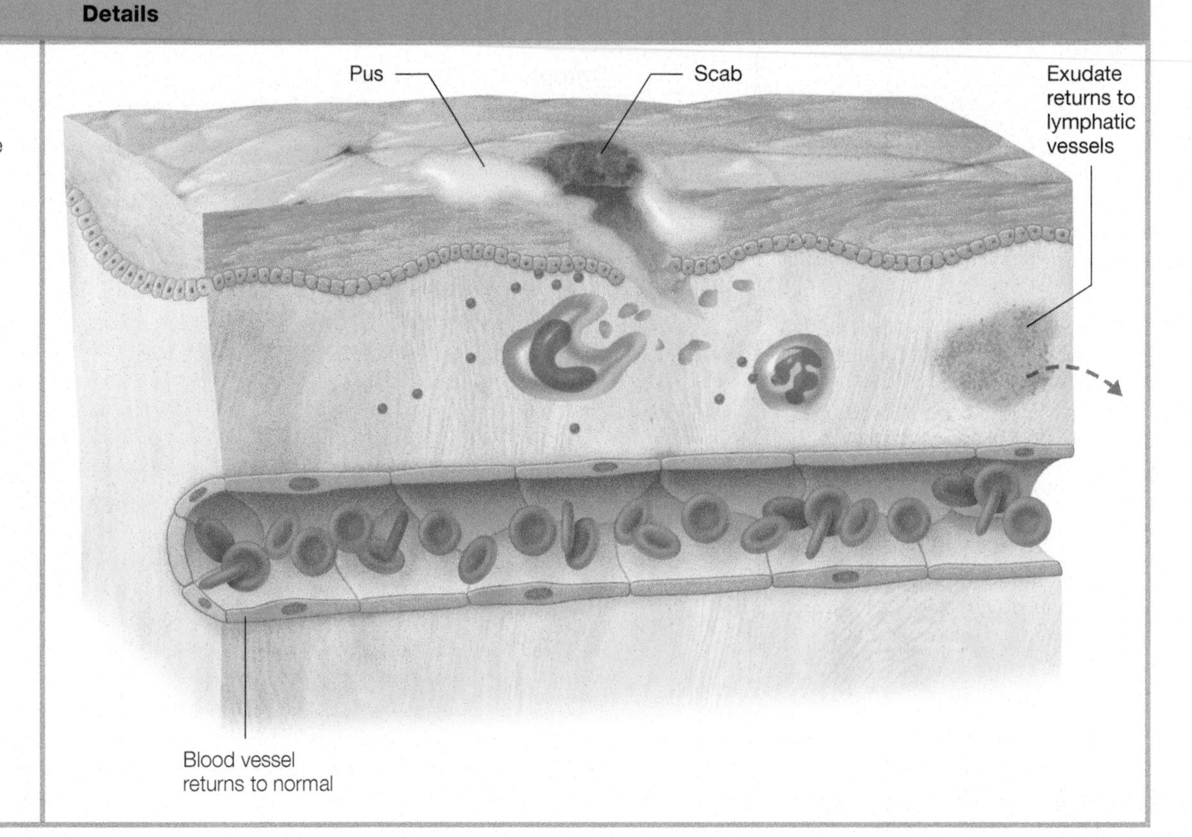

FIGURE 11.15 Resolution phase of inflammation During the resolution phase, inflammation signals decrease, and tissue begins to heal.

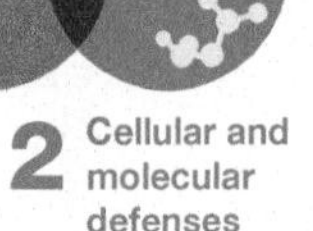

process called **diapedesis** (also called transmigration or extravasation), to squeeze out of the blood vessel. The first leukocytes to exit vessels are neutrophils, followed by monocytes. Monocytes mature into macrophages as they migrate through tissues.

Once in tissue, neutrophils release substances from their granules to eliminate invaders (especially bacteria); they also phagocytize pathogens. Macrophages clear invaders mainly by phagocytosis and also work closely with other leukocytes to stimulate adaptive immune responses.

Resolution phase As the threat passes, the resolution phase works to tone down inflammation so that host tissues don't experience unnecessary collateral damage (FIG. 11.15). The resolution phase is organized by assorted chemical signals, including cytokines released by leukocytes and tissue cells. During resolution, activated leukocytes in the tissue decrease, and the surrounding blood vessels begin to revert to a normal state. Swelling decreases as exudate in the tissues is collected by nearby lymphatic capillaries. Late in the resolution phase, wound healing begins: Damaged tissues are repaired, and a process called **angiogenesis** builds new blood vessels. Blood clots dissolve, leukocytes that are no longer needed in an area are destroyed by apoptosis (programmed cell death), and damaged host cells are cleared. **Pus** may form as dead tissue cells and leukocytes are collected into the fluid exudate made by earlier phases of inflammation.

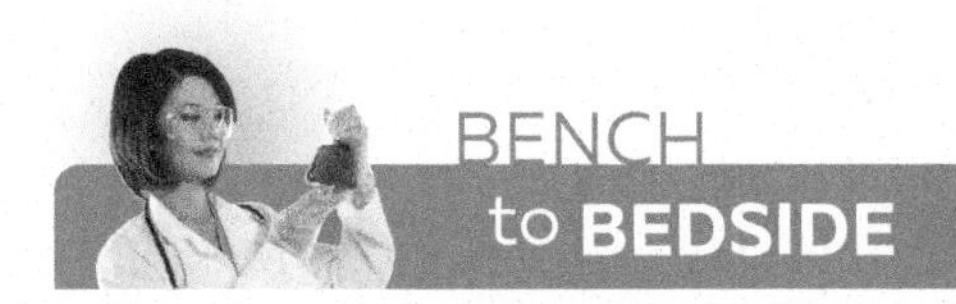

Getting Closer to a Cure for Harmful Inflammation

Resolution of inflammation is an active process that the body implements once the threat that triggered acute inflammation passes. Unfortunately, inflammation doesn't always resolve, resulting in chronic inflammation and more tissue damage. Medically speaking, we manage inflammation by limiting its progression, but available drugs are currently limited in usefulness.

Most anti-inflammatory drugs on the market today block inflammation resolution signals in addition to blocking the inflammation cascades that they target. Aspirin and glucocorticosteroids are exceptions, but they aren't robust stimulators of the resolution phase. Furthermore, routine glucocorticosteroid use causes immunosuppression, while aspirin is associated with gastrointestinal damage and is often contraindicated for certain patients (children and those with asthma are not usually put on aspirin therapies). An ideal therapy would block inflammation progression *and also* directly activate inflammation resolution factors in the body.

Lipid molecules called *specialized pro-resolving lipid mediators (SPM)*, which are made from dietary omega-3 fatty acids, stimulate resolution of inflammation, making them potential cures for chronic inflammation. Lipoxins, maresins, resolvins, and protectins are all SPMs. They work by recruiting noninflammatory macrophages and neutrophils to clear dead or damaged cells and also neutralize pro-inflammatory factors in the area.

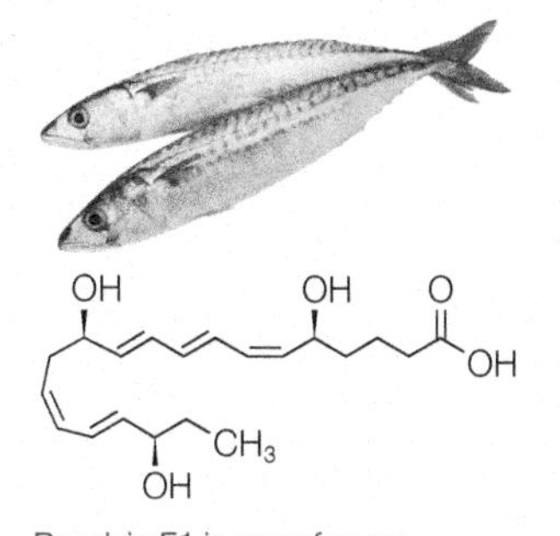

Resolvin E1 is one of many inflammation resolution lipids derived from omega-3 fatty acids that are abundant in fish oils.

Studies in animals demonstrate that administering synthetic versions of SPMs reduces and resolves inflammation and promotes tissue healing without generating immune suppression. These resolution-phase molecules also promote pathogen clearance by enhancing phagocytic activity against bacteria, and certain protectins have been shown to inhibit influenza virus replication. Clinical trials using synthetic SPMs to treat inflammation-based disorders such as dry eye syndrome and periodontal disease have had promising results. However, as of the time of this publication, no SPMs have FDA approval for routine use. Perhaps inflammation linked to disorders such as arthritis, cystic fibrosis, asthma, neurodegenerative disorders, and cardiovascular disease will one day be cured by these novel approaches.

Fattori, V., Zaninelli, T. H., Rasquel-Oliveira, F. S., Casagrande, R., & Verri, Jr., W. A. (2020). Specialized pro-resolving lipid mediators: A new class of non-immunosuppressive and non-opioid analgesic drugs. *Pharmacological Research*, *151*, 104549.

Although wound healing starts during the resolution phase of inflammation, complete healing can take anywhere from days to years, depending on the location and severity of the injury. The degree to which tissue repair is possible depends on the tissue affected and the extent of the damage. Often, instead of functional tissue developing as wounds heal, scar tissue results.

Chronic Inflammation

Chronic inflammation may develop when an inflammatory response goes on too long. Unlike acute inflammation, chronic inflammation is not useful or protective. It exacerbates tissue injury by inflicting further damage. It is now widely accepted that chronic inflammation promotes atherosclerosis (narrowing of blood arteries), a variety of cancers, progressive neurodegenerative disorders such as multiple sclerosis and Alzheimer's disease, and numerous other diseases.

Fever is a systemic innate immune response.

Fever (pyrexia) is an abnormally high systemic body temperature, different from the "local heat" that characterizes inflammation. Infectious agents trigger fevers, but certain drugs and a variety of noninfectious pathologies also induce it. Collectively, fever-inducing agents are called **pyrogens** (*pyro* = fire/heat; *gen* = genesis/creation). Many bacterial toxins act as pyrogens—particularly endotoxin (lipopolysaccharide) found in the outer membrane of Gram-negative bacteria. Pyrogens trigger the release of cytokines, especially interleukin 1, tumor necrosis factor, and interferon alpha, which signal the hypothalamus of the brain to raise the body's baseline temperature from 37°C (98.6°F) to a higher temperature (FIG. 11.16). Upon thermostat reset, the body's temperature increases as a result of hormone level changes,

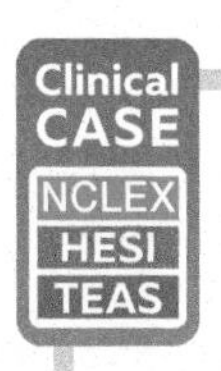

The Case of the Missing Bleach

Practice applying what you know clinically: visit the **Mastering Microbiology** Study Area to watch Part 3 and practice for future exams.

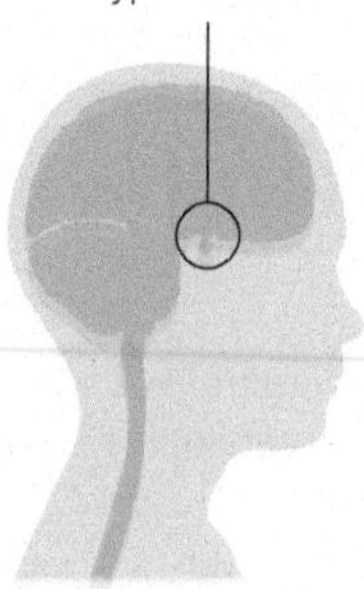

FIGURE 11.16 Fever Fever develops as pyrogens trigger the release of cytokines that signal the hypothalamus to reset the body's thermostat.

INNATE IMMUNITY

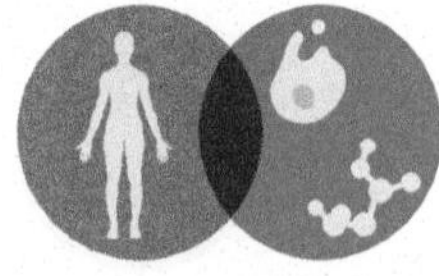

2 Cellular and molecular defenses

Build Your Foundation (BYF) Quick Quiz: Visit the **Mastering Microbiology** Study Area to quiz yourself.

TABLE 11.5 Fever Classifications

Term	Description
Fever of undetermined origin (FUO)	Fever of at least 38.3°C (101°F) that persists more than 3 weeks and isn't linked to a specific cause within a week of inpatient study; usually due to infections or neoplasms (malignant cancers or benign growths)
Intermittent fever	Body temperature elevates, but falls to normal at some point in the day
Pel–Ebstein fever	Fever lasting 3–10 days followed by nonfever state of similar length; characteristic of Hodgkin's disease and certain other lymphomas
Relapsing fever	Recurrent episodes spaced by days or weeks of normal body temperature; characteristic of certain tick-transmitted bacteria such as *Borrelia* species
Remittent fever	Elevated body temperature that fluctuates, but doesn't reach normal in the course of the fluctuations
Sustained fever	Consistently elevated body temperature with limited fluctuation (0.3°C or less) during a 24-hour period
Tertian fever	Fever that occurs on 1st and 3rd days; common in malaria caused by *Plasmodium vivax*
Quartan fever	Fever that occurs on 1st and 4th days; common in malaria caused by *Plasmodium malariae*

shivering, increased metabolism, and limiting heat loss by constricting blood vessels that lead to the skin. Terms that describe different fever classifications are summarized in TABLE 11.5.

Low-grade fever, which in adults is typically defined as an oral temperature of 37.5°C to 38.3°C (99.5–101°F), is considered protective. All effects of fever are not entirely understood, but studies suggested it enhances the antiviral effects of interferons, increases phagocyte efficiency, enhances leukocyte production, limits growth of certain pathogens, and promotes tissue repair. In light of these protective effects, some healthcare providers argue that a low-grade fever should be allowed to run its course because fever is usually self-limiting. That said, these same healthcare providers also tend to agree that if the patient is uncomfortable and not resting much, it makes sense to treat fever with an antipyretic (fever-reducing drug). Antipyretic drugs like aspirin, ibuprofen (Advil/Motrin), and acetaminophen (Tylenol) are all commonly used to reduce fever. They work by limiting the production of prostaglandins in the hypothalamus, which is ultimately what "resets" the body's thermostat. A fever that reaches 40.5°C (105°F) and does not decrease with treatment is a life-threatening medical emergency because essential cellular enzymes and other proteins will begin to denature and stop working at that temperature range. A body temperature above 43°C (109.4°F) is fatal.

Build Your Foundation

21. What are the three primary functions of inflammation? (LO 11.21)
22. What are the three phases of inflammation, and what occurs in each? (LO 11.21)
23. What are the four cardinal signs of inflammation, and what process are they directly linked to in inflammation? (LO 11.22)
24. Name at least three chemical mediators of inflammation, and state their roles. (LO 11.23)
25. What is chronic inflammation, and what may cause it? (LO 11.24)
26. What is fever, and how is it generated, treated, and potentially useful? (LO 11.25)
27. List and describe at least three fever classifications. (LO 11.26)

VISUAL SUMMARY

The Visual Summary and Interactive Content Review (ICR) for this chapter is combined with Chapter 12 on pages 384–385 to provide a holistic review of immune response.

CHAPTER 11 OVERVIEW

11.1 Overview of the Immune System and Its Responses

- Immune responses recognize diverse pathogens, kill identified invaders, and discriminate between self and foreign.
- The immune system has two collaborative branches. Innate immunity is a nonspecific, immediate defense that doesn't require specialized training. Adaptive immunity acts more slowly, but is trainable, has a memory component, and is highly specific.
- Normal microbiota have central roles in training and tempering immune responses.

11.2 Introduction to First-Line Defenses

- First-line defenses prevent pathogen entry into the body; they include mechanical, chemical, and/or physical barriers. They are rarely fully separable from each other.
- Mechanical barriers (mucus, tears, flushing action of urine, etc.) limit pathogen entrance by rinsing, flushing, or trapping them. Chemical barriers (stomach acid, lysozyme, antimicrobial peptides, etc.) either directly attack invaders or establish an environment that limits pathogen ability to survive in a particular tissue. Physical barriers provide a physical blockade to entry; skin is the most important physical barrier.

11.3 Introduction to Second-Line Defenses and the Lymphatic System

- Second-line defenses are categorized as cellular or molecular.
- The lymphatic system collects, circulates, and filters body fluid to prevent edema and screens for foreign agents. Lymphatic capillaries collect lymph from tissue, convey it to nodes for filtering, and send lymph back into the blood supply (where it is called plasma) via veins.
- Primary lymphoid tissues (thymus and bone marrow) produce and mature leukocytes and other formed elements in blood. Secondary lymphoid tissues (lymph nodes, spleen, and mucosa-associated lymphoid tissue) bring leukocytes into contact with antigens to stimulate immune responses.
- Blood consists of plasma, leukocytes, and other formed elements. Granulocytes (basophils, eosinophils, neutrophils, and mast cells) are innate immunity leukocytes. Innate-system agranulocytes include macrophages (immature version: monocytes), natural killer cells, and dendritic cells. Other agranulocytes, T cells and B cells, are adaptive actors.
- A differential white blood cell count determines if any leukocytes are over- or underrepresented in a patient's blood. An increase in leukocytes is called leukocytosis.

11.4 Cellular Second-Line Defenses

- Phagocytosis is a special form of endocytosis. Phagocytes engulf nondissolved targets such as bacterial cells, viral particles, or general debris. Neutrophils, macrophages, and dendritic cells are key phagocytes.
- Neutrophils, which have multilobed nuclei and lightly stained cytoplasmic granules, are the most numerous leukocytes in circulation. These phagocytic first responders mostly combat bacteria.
- Eosinophils have double-lobed nuclei and large, red-orange staining cytoplasmic granules; they are moderately phagocytic cells with roles in asthma, allergy, and combating parasites.
- Basophils, rarest of the leukocytes, have double-lobed nuclei obscured by dark-purple staining cytoplasmic granules packed with histamine. They fight parasites and mediate allergies.
- Mast cells reside in tissues and act as local lookouts. These moderately phagocytic cells attack bacteria, allergens, and parasites and activate adaptive immune responses.
- Monocytes have a large horseshoe-shaped nucleus; they mature into highly phagocytic macrophages as they migrate to tissues. Macrophages eliminate a wide range of invaders and activate adaptive immune responses.
- Dendritic cells are highly phagocytic, important activators of adaptive immunity and are found in tissues.
- Natural killer cells are innate immunity lymphocytes that protect against viruses, bacteria, parasites, and tumor cells.

11.5 Molecular Second-Line Defenses

- Leukocytes exert effects in part by releasing physiologically active cytokines into their local environment. Cytokines (which include chemokines, interleukins, interferons, and tumor necrosis factors) recruit other immune system cells to sites where they are needed, restrict pathogen growth, generate fever, and induce inflammation.
- Iron-binding proteins (ferritin, lactoferrin, hemoglobin, and transferrin) sequester iron to limit pathogen access to this essential nutrient.
- Complement proteins are activated in a cascade fashion and promote opsonization, cytolysis using a membrane attack complex (MAC), and inflammation.
- Examples of pathways for activating complement proteins are the classical pathway (initiated by antibodies), the lectin pathway (initiated by binding sugars on a pathogen), and the alternative pathway (triggered by a direct association of complement proteins with a pathogen).

11.6 Inflammation and Fever

- Combined actions of cellular and molecular second-line defenses trigger inflammation and sometimes fever. Inflammation is a protective response initiated by tissue injury. Cardinal signs are redness, pain, local heat, and swelling. In extreme inflammation a fifth feature, loss of function, may occur.
- Inflammation recruits immune defenses to the injured tissue, limits spread of infectious agents, and delivers oxygen, nutrients, and chemical factors for tissue recovery.
- Inflammation occurs in three general phases: vascular changes, leukocyte recruitment, and resolution.
- Vascular changes of inflammation (vasodilation and increased vessel permeability) bring cellular and chemical defenses to the injury site. Histamines, kinins, and eicosanoids are central. Drugs that block pro-inflammatory mediators include antihistamines, SAIDs, and NSAIDs. Vasodilation and increased vessel permeability both contribute to the cardinal signs of inflammation.
- The leukocyte recruitment phase of inflammation relies on chemoattractant molecules that stimulate leukocytes to exit the vasculature and migrate to injured tissue. Leukocytes undergo margination and diapedesis to exit vessels.
- As the initial threat passes, the resolution phase lessens inflammation so that host tissues do not experience unnecessary collateral damage. Wound healing starts during this phase. If resolution is not effective, chronic inflammation may develop.
- Fever (pyrexia) is abnormally high body temperature. Pyrogens such as endotoxin or other bacterial toxins trigger the brain's hypothalamus to raise the body's temperature.
- Low-grade fever is considered protective; fever approaching 40.5°C (105°F) that doesn't decrease with treatment is a life-threatening medical emergency; a body temperature above 43°C (109.4°F) is fatal. Body temperature levels and patterns of temperature fluctuation help classify fevers.

THINK CLINICALLY | *Be S.M.A.R.T. About Cases*

COMPREHENSIVE CASE

The following case integrates basic principles from the chapter. Try to answer the case questions on your own. Don't forget to be S.M.A.R.T.* about your case study to help you interlink the scientific concepts you have just learned and apply your new understanding of innate immunity to a case study.

* The five-step method is shown in detail in the Chapter 1 Comprehensive Case on cholera. See pages 31–33. Refer back to this example to help you apply a SMART approach to other critical thinking questions and cases throughout the book.

• • •

After watching their son run around all Saturday afternoon at his fifth birthday party, Jackson's parents, Erin and Nia, weren't surprised that, by bedtime, he was exhausted. He also complained of ear discomfort, so Nia gave him children's Advil. A few hours later Erin checked on Jackson and found he was feverish, with some foul-smelling fluid draining from his ear. Jackson had experienced an inner ear infection (otitis media) 10 days earlier and had recently finished a course of amoxicillin. His parents assumed the infection was back. He'd always been prone to bacterial ear and respiratory infections, and since beginning prekindergarten, he'd been constantly ill.

Jackson's oral temperature was 102°F (38.9°C), despite taking Advil recently. Erin called the pediatrician, who sent a new antibiotic prescription to a 24-hour pharmacy. She told Erin to start the medication, keep Jackson on Advil, and bring him in for an appointment Monday. She also mentioned that if Jackson developed other symptoms or if his current symptoms or fever got worse, they should take him to the emergency room. By Sunday afternoon Jackson complained of a stiff neck (nuchal rigidity) and seemed disoriented. His parents drove him to the hospital.

Laboratory tests on Jackson's cerebrospinal fluid (CSF) and fluid drained from his ear revealed Gram-positive, encapsulated cocci in chains. Jackson had bacterial meningitis, an infection of membranes surrounding the brain and spinal cord. *Streptococcus pneumoniae* was cultured from the ear fluid and CSF samples, confirming diagnosis. Nia asked a physician how Jackson developed a *S. pneumoniae* infection when he was up to date on all routine vaccines and had been vaccinated against this bacterium. The physician clarified that the vaccine works against the most commonly seen invasive strains of the bacterium, but not all of them.

Jackson spent 3 weeks in the hospital for treatment but fully recovered. However, over the next few years he continued to suffer from pyrogenic bacterial infections, which prompted his pediatrician to check for an immune deficiency and hemochromatosis. The tests revealed that Jackson had abnormally low levels of mannose-binding lectin (MBL). Unfortunately, there is currently no therapy for MBL deficiencies. Jackson's family can only be vigilant about early infection symptoms and seek immediate treatment. His pediatrician recommended that Jackson receive the meningococcal vaccine earlier than most children to prevent infection by *Neisseria meningitidis*—a bacterium that causes a form of meningitis that Jackson had not yet contracted.

• • •

CASE-BASED QUESTIONS

1. What do you predict a differential white blood cell count on Jackson's blood would reveal? Be sure to support your conclusion with information from the chapter.
2. Given Jackson's diagnosis of MBL deficiency, what do you think his prognosis (forecast) is for a normal life? Consider what immune system defenses would be most directly affected in Jackson versus defenses that are less directly impacted.
3. Despite Jackson's MBL deficiency, his pediatrician is convinced that Jackson's immune system will respond to the meningococcal vaccine and provide protection against *N. meningitidis*. Why?
4. Describe the general molecular mechanism by which Jackson developed a fever, and explain why Advil was expected to help manage Jackson's discomfort and fever. Also, explain why his pediatrician recommended emergency care if the Advil didn't limit Jackson's fever.
5. Describe some first-line defenses that were likely overcome by the pathogen to successfully invade Jackson's body, and explain what second-line defenses were likely called to action to help fight the infection.
6. Why did Jackson's doctors check for hemochromatosis?
7. What leukocyte(s) would you expect would be increased in Jackson's cerebrospinal fluid sample? Explain your reasoning.

END OF CHAPTER QUESTIONS

NCLEX
HESI
TEAS

Think Critically and Clinically
Questions highlighted in orange are opportunities to practice NCLEX, HESI, and TEAS-style questions.

1. Classify each defense as either first-line, second-line cellular, or second-line molecular:

Inflammation	Lysozyme	Mucus
Neutrophils	Stomach acid	Iron-binding proteins
Skin	Eosinophils	Phagocytosis
Antimicrobial peptides	Fever	
	Complement proteins	

2. Pick which statements are true, then correct all false statements, so they are also true.
   a. Redness, pain, fever, and swelling characterize inflammation.
   b. Granulocytes include monocytes and lymphocytes.
   c. Pyrogens induce fever.
   d. Adaptive and innate immune responses are completely independent from one another.
   e. The innate immune responses occur faster than adaptive responses.
   f. Monocytes are highly phagocytic cells.
   g. Complement cascades share the same outcomes: opsonization, cytolysis, and fever.

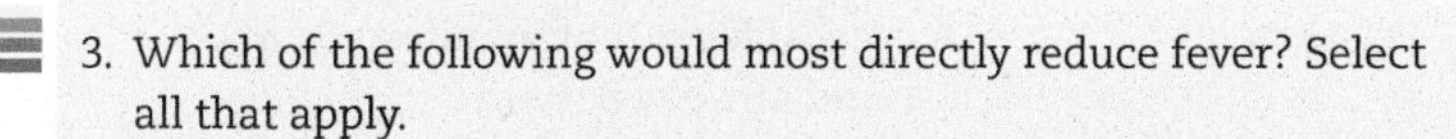

3. Which of the following would most directly reduce fever? Select all that apply.
   a. Limiting the number of circulating white blood cells
   b. Reducing eicosanoid production
   c. Inhibiting pyrogenic cytokines
   d. Stimulating the action of prostaglandins
   e. Administering antihistamines

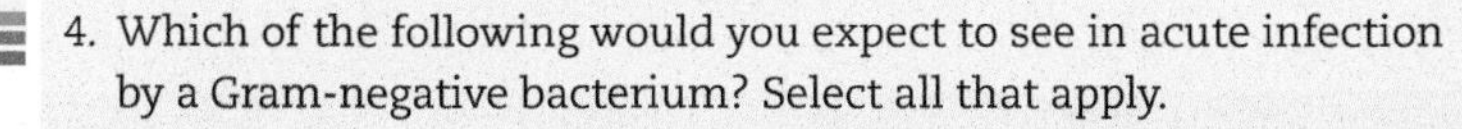

4. Which of the following would you expect to see in acute infection by a Gram-negative bacterium? Select all that apply.
   a. Pyrexia
   b. Decreased lymphocytes
   c. Neutrophilic lymphocytosis
   d. Decreased monocytes
   e. Increased release of pro-inflammatory cytokines

5. Which of the following would you expect to see increased in circulation in a patient suffering from allergies? Select all that apply.

a. 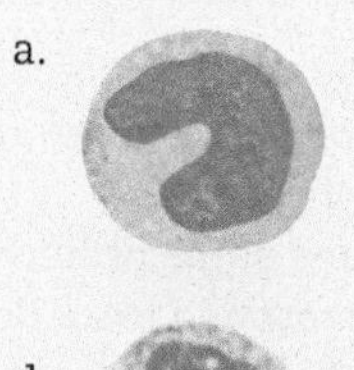
b. 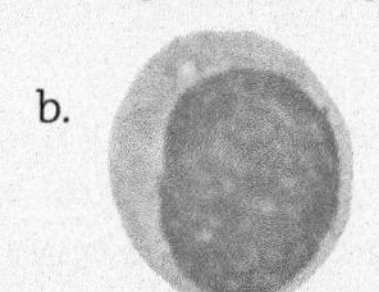
c. 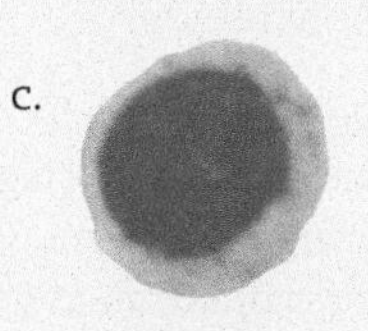
d. 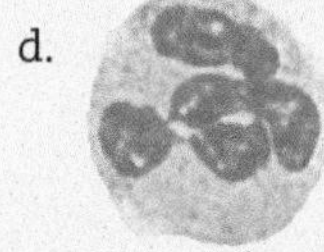
e. 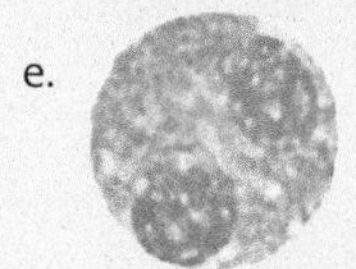

6. The __________ cascade of complement activation is initiated by antibodies. In contrast, the __________ cascade is activated by a direct interaction with complement proteins, and the __________ cascade is activated by MBL associating with a pathogen.

7. Which of the following would you not expect to see in the first stage of inflammation?
   a. Histamine
   b. Kinins
   c. Macrophages
   d. Increased blood vessel permeability
   e. Eicosanoids

8. Label the following as granulocytes or agranulocytes *and* classify them as innate or adaptive cellular responders.

Basophil	Lymphocyte	Mast cell
Monocyte	Neutrophil	NK cell
Macrophage	Eosinophil	T cell

9. Make a Venn diagram to compare and contrast innate and adaptive immunity.

10. Which of the following would be the most likely immediate consequence of an aseptic tissue injury?
    a. Monocytosis
    b. Complement activation
    c. Eosinophilia
    d. Fever
    e. Inflammation

11. Which of the following are considered cytokines? Select all that apply.
    a. Eicosanoids
    b. TNF-α
    c. Interferon β
    d. Histamine
    e. Chemokines

12. __________ are innate molecular defenses that collectively limit free iron in the blood. Examples of these factors in humans include __________, __________, __________, and __________.

13. Which would be expected to contribute to chronic inflammation? Select all that apply.
    a. A reduced innate defense
    b. Fever
    c. Persistent tissue injury
    d. Glucocorticosteroids
    e. Antihistamines

14. Which of the following is not a feature of innate immunity?
   a. Better protection upon later exposure to a given pathogen
   b. Recognition of diverse pathogens
   c. Discrimination between self and foreign
   d. Killing of identified invaders
   e. Stimulation of adaptive immunity

15. Label the following as either primary or secondary lymphoid tissues:

   Spleen, lymph node, adenoids, thymus, tonsils, bone marrow

16. Why are vascular changes in early inflammation considered central to generating inflammation's cardinal signs?

17. Which of the following is false regarding histamine?
   a. Histamine is a vasodilator.
   b. Histamine increases vascular permeability.
   c. Histamine is a pro-inflammatory factor.
   d. Histamine is a pyrogen.
   e. Histamine is released by leukocytes.

18. Why is innate immunity considered a generalized defense?

19. Which of the following shows a correct chronological order of events in inflammation?
   a. Neutrophil recruitment, macrophage recruitment, vascular changes, resolution
   b. Vascular changes, resolution, neutrophil recruitment, macrophage recruitment
   c. Vascular changes, macrophage recruitment, neutrophil recruitment, resolution
   d. Vascular changes, macrophage recruitment, resolution, neutrophil recruitment
   e. Vascular changes, neutrophil recruitment, macrophage recruitment, resolution

20. Select all the false statements about fever.
   a. It is generated by pyrogens.
   b. It is an innate immune defense.
   c. It is accompanied by a decrease in metabolism.
   d. It can be reduced by anti-inflammatory drugs.
   e. It can accompany inflammation.

21. Which of the following is a chemical defense found in tears? Select all that apply.
   a. Water
   b. Lysozyme
   c. Antimicrobial peptides
   d. Neutrophils

22. Which of the following are formed elements of the blood? Select all that apply.
   a. Platelet
   b. Leukocyte
   c. Plasma
   d. Erythrocyte
   e. Complement proteins

23. Match the following fever terms to their proper definition; not all definitions will be used.

Fever terms:	Definitions:
Relapsing	Fever lasting several days followed by nonfever state of similar length
Quartan	Body temperature elevated, but falls to normal at some point each day
Sustained	Characteristic of certain tick-transmitted bacterial infections
Pel–Ebstein	A cyclical fever that manifests in a 1st-and 4th-day pattern
Remittent	A stably elevated body temperature with limited fluctuation

CRITICAL THINKING QUESTIONS

1. Imagine you are a healthcare provider in an immunology practice. You recently learned that one of your patients has a C9 complement protein deficiency. What health issues would you anticipate for the patient, and what immune defense would most immediately be impacted? Be sure to support your answer with information from the chapter.

2. Imagine that a new drug under development blocks inflammation by stopping production of leukotrienes, but not thromboxanes or prostaglandins. Draw a schematic that best summarizes which branch (or branches) in the inflammation "tree" are blocked by the drug.

3. A friend in class is confused about why innate and adaptive immunity are considered separate when they are not fully independent of one another. Based on this chapter, explain this fundamental concept to your friend. Be sure to include specific examples that show the interdependence of these two branches of immunity.

4. Why is applying ice a recommended therapy for acute inflammation?

5. Design a white blood cell that is efficient at destroying bacteria and inducing early stages of inflammation, and also induces fever. Your designed cell should also limit the opportunity for chronic inflammation to develop. Be sure to explain some of the cellular and molecular features the imaginary cell may respond to and/or have.

Adaptive Immunity

What Will We Explore? In Chapter 11 we learned about the first two lines of immune defense, which are innate responses. Here, in Chapter 12, we review the third and final line of immune defense—adaptive responses. When the innate system fails to prevent infection, the adaptive system kicks in, formulating a custom-designed attack against the pathogen at hand. Another crucial, unique component of the adaptive immune system is memory. After defeating a pathogen, the adaptive immune system will remember its particular antigen profiles and then mount a more rapid reaction should it invade the body again later.

Why Is It Important? The adaptive response is what trains our immune system to act against certain microbes we naturally encounter. It's also the set of responses that vaccinations activate to confer protection against a specific pathogen. Although the first modern version of a vaccine was not developed until the late 18th century, the protective effect of surviving certain infections was recognized as far back as ancient Greece and Egypt. In the 10th century, Chinese physicians had healthy people inhale small doses of smallpox scabs to induce a mild form of the disease, which protected against the more severe form brought about by natural infection.

In addition to protecting us from pathogenic bacteria, viruses, fungi, and parasites, our adaptive immune responses eliminate cancer cells. In fact, cancer is largely considered a failure of the adaptive immune response, which is tasked with eliminating abnormal self-cells, as well as foreign material detected in the body.

> **Clinical CASE** NCLEX HESI TEAS
>
> **The Case of the Coughing Newborn**
>
> Visit the **Mastering Microbiology** Study Area to watch the case and find out how adaptive immunity can explain this medical mystery.

Jane Galligan
RN, Director of Community Relations; New Jersey

12.1 INTRODUCTION TO THIRD-LINE DEFENSES

Learning Outcomes

After reading this section, you should be able to:

12.1 Name the branches of adaptive immunity, and compare adaptive to innate immunity.

12.2 Compare and contrast T and B cells.

12.3 Compare and contrast T helper versus T cytotoxic cells, and discuss the various T helper cell subclasses.

12.4 Define antigens, epitopes, and haptens, and discuss the concept of immunogenicity.

12.5 Explain the role of T and B cell receptors in antigen recognition, and state how receptor diversity comes about.

12.6 Define self-tolerance, and describe how and why the body screens immature T and B cells for this feature.

The adaptive response is the body's third and final line of defense.

In Chapter 11 we reviewed the innate immune response, which consists of first- and second-line defenses. Here we introduce the **adaptive immune response**, which comprises the third and final line of our immune defenses (FIG. 12.1). Adaptive immunity is a set of defenses that the body acquires as it responds to a specific antigen and learns to remember it. And although we focus on adaptive responses in this chapter, it is important that you appreciate that adaptive defenses are not completely independent of innate defenses. Let's take a closer look.

When innate first- and second-line defenses fail to overcome a foreign antigen, adaptive responses are called into action. As discussed earlier, innate responses are immediate and always the same, regardless of the antigen encountered. By contrast, adaptive responses typically take a few days to more than a week to detect and respond to an antigen during the *primary exposure* (first encounter with the antigen). This extra time is necessary because adaptive responses are *specific* to a particular antigen. Once a specific immune response is triggered, the second feature of adaptive immunity, **immunological memory**, ensures that a *secondary exposure*[1] to the same antigen is greeted by reactions that are so rapid and effective that we frequently will not even experience symptoms.

The adaptive immune system is subdivided into two branches: the **cellular response** (also called T cell-mediated immunity) and the **humoral response** (also called B cell-or antibody-mediated immunity). However, both of these branches are highly dependent on each other for full functionality. Moreover, the function of both branches is the same: *Eliminate an identified antigen, and remember it so that next time adaptive responses are faster.*

Cellular and humoral responses both progress through four main stages (FIG. 12.2).

Stage 1 – Antigen Presentation: Dendritic cells and certain other white blood cells act as **antigen-presenting cells** (APCs) that take up antigens, process them, and show them to T cells. B cells do not require APCs to show them antigens; instead, they can directly interact with an antigen.

Stage 2 – Lymphocyte Activation: Upon successful antigen presentation, lymphocytes are activated by a collection of released cytokines. Activated T cells influence B cell activation.

Stage 3 – Lymphocyte Proliferation and Differentiation: Activated B and T cells undergo multiple rounds of cell division to proliferate (increase their population); this process is sometimes called **clonal expansion**. Some of the cells, designated **effector cells**, will engage in the response against the antigen, while others will become **memory cells** that remain in lymphatic tissues to rapidly recognize the antigen if it's encountered again later.

Stage 4 – Antigen Elimination and Memory: Cellular and humoral responses collaborate to eliminate the antigen against which they were activated. Once the threat passes, effector cells die off, while memory cells endure for years in lymphatic tissues.

We'll review each of these stages of the adaptive immune response in greater detail as we progress through this chapter. However, before diving into the details, you first need to know a bit more about T and B cells.

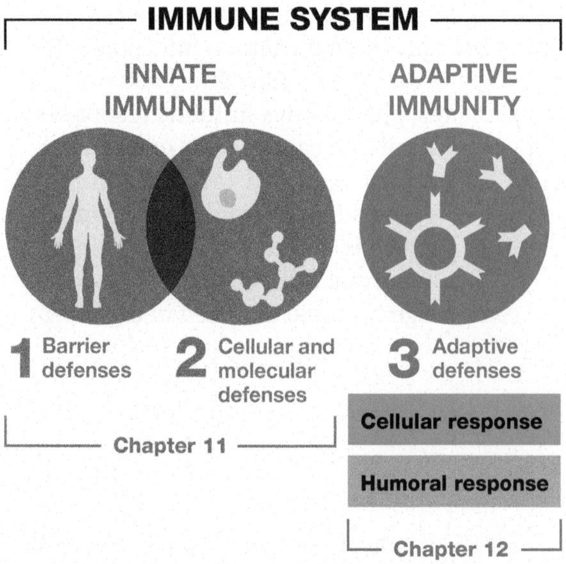

FIGURE 12.1 **Three lines of defense in immunity** Adaptive immunity, which constitutes our third line of defenses, includes cellular and humoral responses.

[1] The term *secondary exposure* does not exclusively refer to just a second exposure. Rather, it is *any* exposure that occurs subsequent to a first/primary exposure.

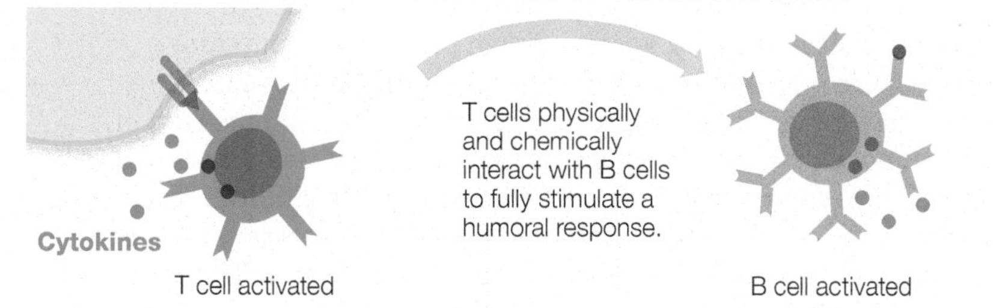

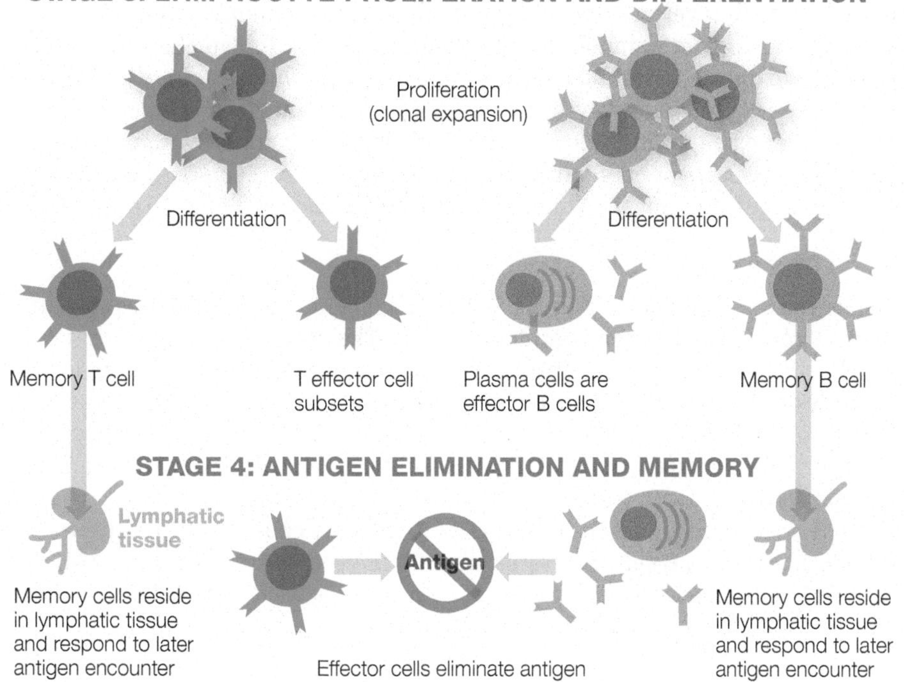

FIGURE 12.2 **Overview of adaptive immunity** The adaptive immune response is divided into two branches: cellular and humoral. This figure serves as a road map for what we'll be exploring in this chapter. Many of the details presented in the figure have not yet been discussed, so it's normal to look at a big-picture overview like this and have questions. Don't worry, we'll explore each of these stages in more detail, and it will all come together for you as you progress through the chapter.

CONCEPT COACH

NCLEX HESI TEAS

An Overview of Adaptive Immunity

Bring the art to life! Visit the **Mastering Microbiology** Study Area to watch the Concept Coach and master adaptive immunity.

T cells and B cells, the main lymphocytes of adaptive immunity, can recognize practically any natural or man-made antigen.

The most important adaptive immunity leukocytes (white blood cells) are *lymphocytes* called **T cells** and **B cells**. Soon after their production in the bone marrow, **thymocytes** (immature T cells) migrate to the thymus gland to mature, which is how they came to be named T cells. In contrast, B cells are produced in the bone marrow and also mature there. Mature T and B cells tend to be present

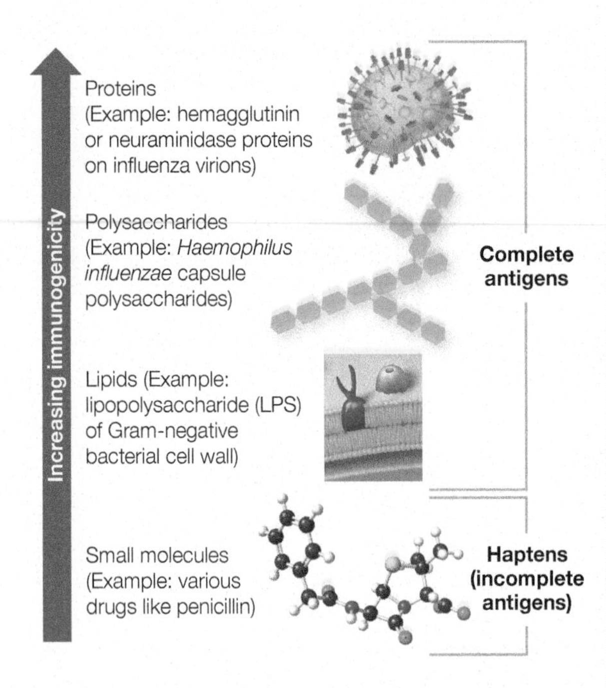

FIGURE 12.3 Antigen immunogenicity The degree to which the antigen provokes an immune response varies and depends on the antigen's biochemical features.

CHEM • NOTE

Proteins are made of amino acids. Proteins can have primary, secondary, tertiary, and quaternary structure. Refer back to Chapter 2 for an in-depth review of this concept.

CHEM • NOTE

Polysaccharides are carbohydrate molecules made up of long chains of single sugars (monosaccharides). Polysaccharides can be linear or highly branched.

at relatively low levels in circulation; instead, they mainly reside in lymphoid tissues throughout the body. (Refer back to Chapter 11 to review primary and secondary lymphoid tissues.)

T cells have roles in both the humoral and cellular branches of adaptive immunity, whereas B cells coordinate the humoral response by making antibodies. Our immune system generates a vast array of T cells and B cells, which have the capacity to recognize virtually any antigen.

Immunogenicity

An antigen is any substance that, if presented in the right context, may trigger an immune response. In terms of composition, most antigens are proteins or polysaccharides that come from a bacterium, virus, fungus, or protist. Cancer cells also frequently make proteins and/or polysaccharides that the body recognizes as abnormal, allowing them to be flagged as something to attack. In abnormal situations, the body may develop an immune response against self-antigens, which leads to an autoimmune disorder. Finally, certain nonpathogenic environmental substances, such as pollen or dust, are sometimes antigenic. (See more on allergies and autoimmunity in Chapter 13.)

In general, any antigen that can successfully trigger an immune response is said to be **immunogenic**. Immunogenicity is influenced by a combination of antigen size, overall molecular complexity, and chemical composition. Proteins tend to be more immunogenic than polysaccharides, which on average tend to be more immunogenic than lipids. However, this ranking is a generalization and not a hard-and-fast rule. It depends on the molecules being compared and how the patient is exposed to the antigen (injected, ingested, inhaled, etc.). Sometimes immunogenic antigens are called *complete antigens* to differentiate them from antigens that are not immunogenic on their own (FIG. 12.3). **Haptens**, or *incomplete antigens*, are unable to stimulate an immune response unless they are linked to a more complex protein or polysaccharide. A number of medications may act as haptens. For example, in some patients, penicillin may combine with proteins and trigger allergies or other adverse immune responses. (Haptens and their role in triggering various adverse immune reactions are reviewed more in Chapter 13.)

Antigen Features

Just as you recognize your friends by their facial features, lymphocytes rely on distinct portions of antigens for identification. The parts of an antigen that B and T cells recognize and mount an immune response to are called **epitopes**, or *antigenic determinants*. Often in conversation the broader term *antigen* is used instead of *epitope*, but you should understand that when we say a lymphocyte "recognizes" an antigen, what we mean is that the cell can bind to a particular epitope on that antigen (FIG. 12.4).

The antigen recognition receptors on T cells are called **T cell receptors (TCRs)**, whereas on B cells they are called **B cell receptors (BCRs)** (FIG. 12.5). A given T cell has thousands of TCRs on its surface. However, the TCRs on a given

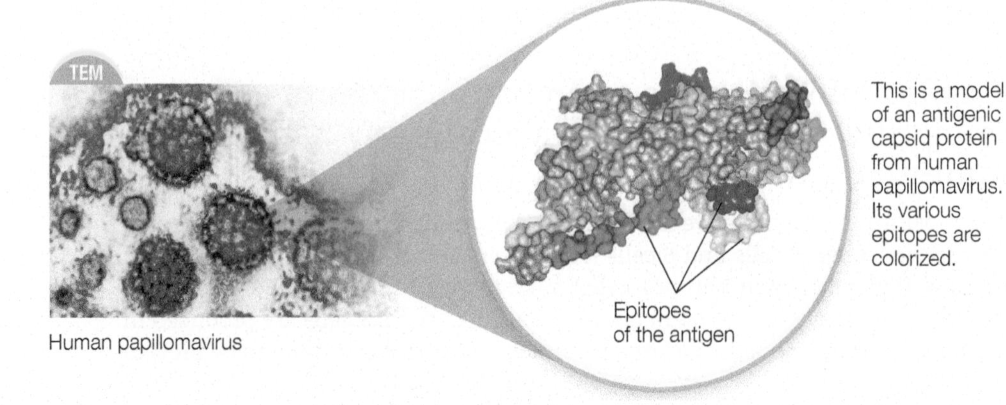

FIGURE 12.4 Antigens and their epitopes As demonstrated by this space-filling model of a papillomavirus capsid protein, a given antigen will usually have multiple epitopes. TCRs and BCRs bind to a specific epitope.

cell are typically specific for the same epitope, which means most T cells are *monospecific* (geared at recognizing one type of epitope).[2] The situation is similar for B cells—a given B cell is covered in thousands of BCRs, and each BCR on the cell will typically target the same type of epitope. Even so, antigen recognition capacity is essentially unlimited because the body makes such a diverse variety of B and T cells.

When a T or B cell binds to a specific epitope of an antigen (through its TCR or BCR), the cell becomes activated. Lymphocyte activation will be reviewed later, but basically the chemical signals that cause the T or B cell to undergo proliferation (also called clonal expansion) entail many rounds of cell division to make genetically identical cells (**clones**). As the activated lymphocyte proliferates, some of the generated clones develop into effector cells, while others become memory cells. Activated B cells **differentiate**, or become specialized, into effector cells called **plasma cells** and a small group of memory cells. Plasma cells make antibodies, which are a secreted form of the BCR that binds to the antigen that stimulated the activation event. Activated B cells and their released antibodies mediate the humoral response of adaptive immunity, discussed later in this chapter. T cell activation leads to cell proliferation and differentiation into memory cells and diverse effector cell lineages that govern the cellular branch of adaptive immunity. The various T cell lineages are reviewed next.

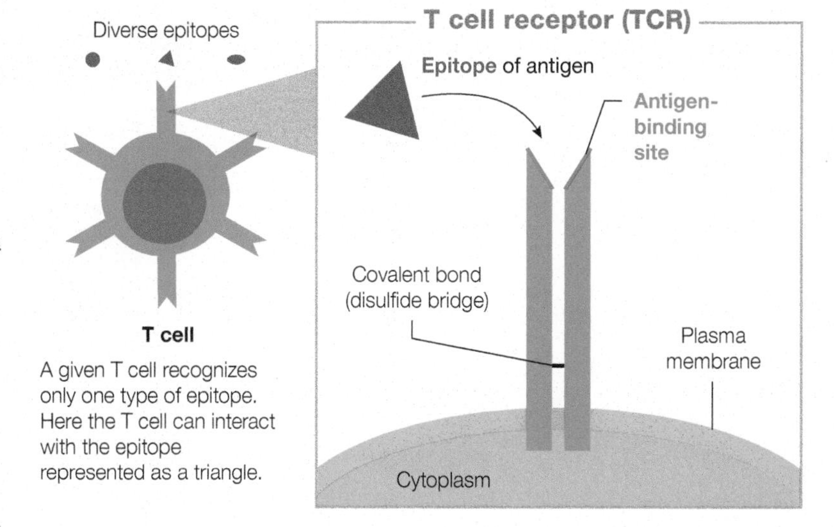

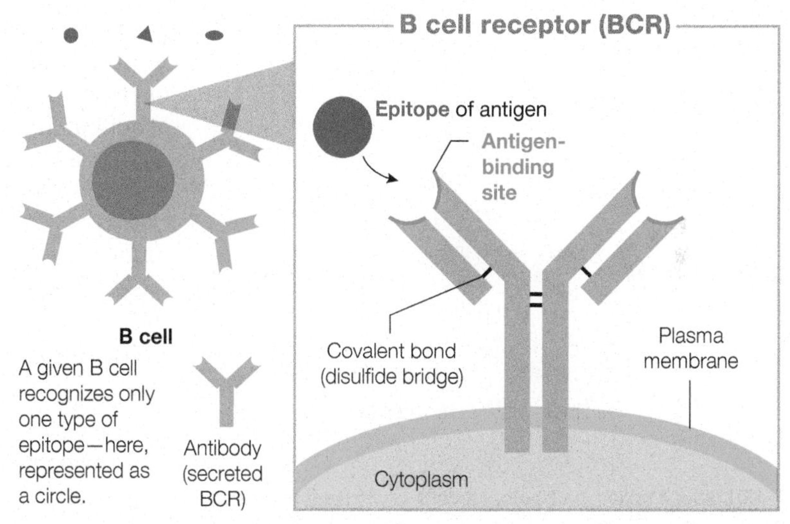

FIGURE 12.5 T cell receptor (TCR) and B cell receptor (BCR) Each TCR on a given T cell recognizes the same type of epitope on an antigen. The same applies to the BCRs of a given B cell.

Critical Thinking *Although a given T cell or B cell typically recognizes one type of epitope, the immune system can still recognize a virtually limitless repertoire of antigens. Explain why.*

T cells fall into two main classes: helper or cytotoxic cells.

There are two main lineages, or classes, of T cells. **T cytotoxic cells** (T_C cells) can directly destroy infected or cancerous body cells.[3] In contrast, **T helper cells** (T_H cells) do not directly seek and destroy their target(s)—instead, they coordinate an adaptive immune response by stimulating other white blood cells. We can tell these two classes of T cells apart by the presence of specialized glycoproteins on their surfaces called **cluster of differentiation (CD) proteins**. More than 300 CD proteins have been characterized, but we'll concern ourselves with only two of them: CD4 and CD8.

T Helper Cells (T_H)

T helper cells are the most abundant T cells. You can use the phrase "call 4 help" to remember that $CD4^+$ cells and T helper cells are the same. Upon activation, they "help" coordinate the adaptive immune response by releasing cytokines that boost activity of other white blood cells, especially macrophages, B cells, and T cytotoxic cells. T helper cells can be thought of as the generals of your immune system. They are the main organizers of both the cellular and humoral branches of adaptive immunity (FIG. 12.6).

T helper cells are so essential that if killed off, the host is unable to mount an effective adaptive immune response. This is what occurs during human immunodeficiency virus (HIV) infection. Healthcare providers monitor T helper

CHEM • NOTE

Glycoproteins are proteins that are covalently bonded to sugars. The sugars can be diverse, and they often participate in cell signaling, binding, or other cellular actions.

[2] Dual-TCR T cells have been described and are probably physiologically important. However, this type of T cell remains poorly understood, and research suggests dual-TCR T cells are the minority of T cells. As such, for simplicity, here we focus on monospecific T cells.

[3] T cytotoxic cells have also been found to directly target bacteria, fungi, and parasites. However, that mechanism of action remains poorly understood and will not be reviewed in this text.

FIGURE 12.6 **T helper cells bridge the cellular and humoral branches of adaptive immunity**

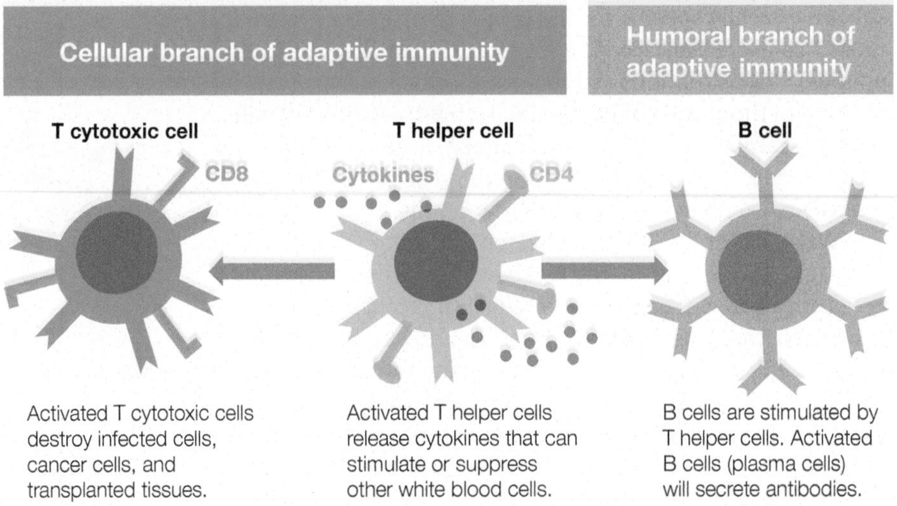

Cellular branch of adaptive immunity

Humoral branch of adaptive immunity

T cytotoxic cell

T helper cell

B cell

CD8

Cytokines

CD4

Activated T cytotoxic cells destroy infected cells, cancer cells, and transplanted tissues.

Activated T helper cells release cytokines that can stimulate or suppress other white blood cells.

B cells are stimulated by T helper cells. Activated B cells (plasma cells) will secrete antibodies.

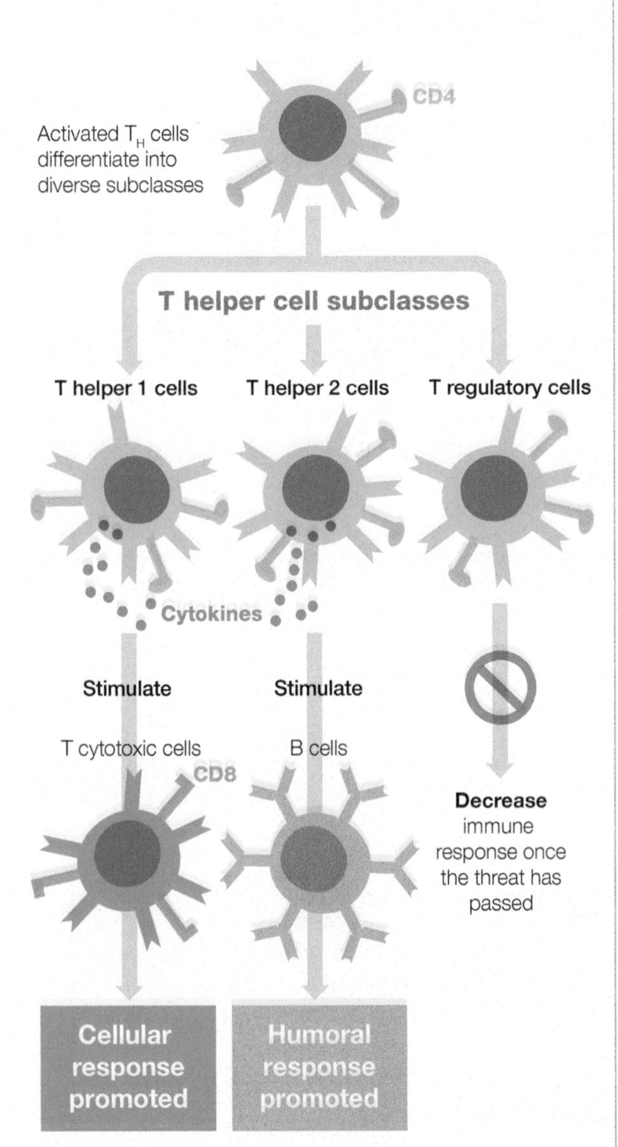

FIGURE 12.7 **Three examples of T helper cell subclasses**

cell population levels in HIV-positive patients as a part of disease management. When the number of $CD4^+$ cells drops to a critical point (below 200 T_H cells/mm^3 of blood), the HIV-positive patient is then classified as having developed acquired immunodeficiency syndrome (AIDS).

Once activated, T helper cells may differentiate into a variety of subclasses that have specific functions (FIG. 12.7). We currently know of at least seven subclasses of T helper cells, all of which are important in adaptive immunity and communicate with one another.[4] To keep this discussion at an introductory level, we'll focus on the three T helper cell subclasses that laid the foundation for our current understanding of the complex interactions among T helper cell populations.[5]

- **T helper 1 (T_H1) cells** mainly activate T cytotoxic cells, macrophages, and natural killer cells (see Chapter 11) to destroy pathogens inside of host cells. Excessive T_H1 activity is linked to autoimmune and inflammatory disorders. Any population of T_H1 cells will include effector and memory cells.
- **T helper 2 (T_H2) cells** primarily stimulate B cells to make antibodies and are therefore key stimulators of humoral immune responses. Abnormal activity of these cells is linked to allergy and certain autoimmune and inflammatory disorders. A given population of T_H2 cells will consist of effector and memory cells.
- **T regulatory (T_{reg}) cells** control functions of other white blood cells, including dendritic cells, mast cells, B cells, and other T cells to ensure that immune responses taper off once a threat subsides. This reduces risk for auto-inflammatory disorders and collateral host tissue damage from sustained inflammation. A given population of T_{reg} cells will consist of effector and memory cells.

T Cytotoxic Cells (T_C)

T cytotoxic cells can be thought of as the foot soldiers of the cellular branch of adaptive immunity. They directly destroy cells that are virus infected, damaged, foreign/transplanted, or cancerous. You can remember that T_C cells are $CD8^+$ by noting that the number 8 is composed of a series of "c" shapes. Like T helper cells, T_C cells also differentiate into specialized subsets. However, so far T_C subset features and functions remain poorly understood compared with the T_H

[4] The known T helper cell subclasses are: T_H1, T_H2, T regulatory (also occasionally called T_H3 or T suppressor cells), T_H9, T_H17, T_H22, and T follicular helper cells (T_{FH}). The interactions of these subclasses in inflammatory disorders and autoimmunity are being intensely studied and may present avenues for therapies. You could expect to review all of these T_H subclasses in an immunology course.

[5] Raphael, I., Nalawade, S., Eagar, T. N., & Forsthuber, T. G. (2015). T cell subsets and their signature cytokines in autoimmune and inflammatory diseases. *Cytokine*, *74*(1), 5–17.

subsets. As seen for the T helper cell subclasses, T cytotoxic subclasses also will consist of memory and effector cells.

The body screens T cells and B cells for self-tolerance.

T and B cells recognize a wide variety of antigens. This is largely due to **gene shuffling** mechanisms that generate an incredibly diverse repertoire of antigen receptors (TCRs and BCRs) during lymphocyte development. The random process inevitably gives rise to some immune cells that have receptors that could bind to normal body cells. If allowed to mature, these lymphocytes would attack self-tissues. To prevent that, the body has screening mechanisms that select for immune cells with **self-tolerance**, meaning they will not attack normal self-cells (FIG. 12.8).

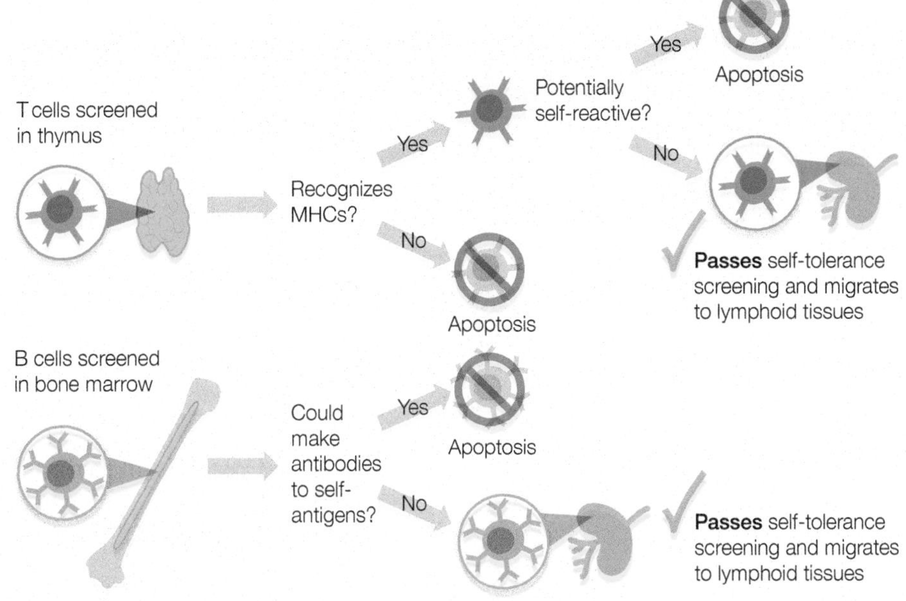

FIGURE 12.8 **Self-tolerance screening** T and B cells are screened for self-tolerance. Lymphocytes that would damage healthy self-tissues undergo apoptosis.

For T cells, ensuring self-tolerance involves screening them for their ability to recognize **major histocompatibility complex (MHC) proteins**. MHCs are specialized "self-proteins" also known as human leukocyte antigens (HLAs). You can think of MHCs as the body's uniform. As a T cell matures in the thymus, it's tested to confirm it recognizes the "self" uniform but won't attack "self" cells. If the maturing T cell can't pass these tests, then it's banned from the team because it may go after its own teammates. MHCs also have crucial roles in activating T cells; we cover this later in the chapter.

The mechanism for B cell self-tolerance screening differs from that of T cells and occurs in the bone marrow. Because B cells mainly exert effects by releasing antibodies, their selection process is designed to ensure that any future antibodies they make won't cross-react with self-antigens and damage host tissues.

T and B cells that don't exhibit self-tolerance receive signals to undergo **apoptosis**, which is cell suicide (programmed cell death). It's thought that certain autoimmune diseases, which develop when our immune system attacks self, may be linked to errors in self-tolerance screening and/or the apoptosis process. TABLE 12.1 compares T and B cells.

TABLE 12.1 Comparing T Cells and B Cells

	T cells	B cells
Adaptive immunity mediated:	Cellular branch	Humoral branch
Include:	• T helper (CD4+ or T_H) cells: stimulate B cells and other leukocytes • T cytotoxic (CD8+ or T_C) cells: seek and destroy cancer cells and cells infected with intracellular pathogens	Plasma cells (activated B cells that make antibodies)
Site of maturation:	Thymus	Bone marrow
Found in:	Mainly in lymphatic tissues; low levels in circulation	Mainly in lymphatic tissues; low levels in circulation
Antigen recognition receptors:	T cell receptors (TCRs)	B cell receptors (BCRs), which are secreted as antibodies following B cell activation
Require antigen-presenting cell to become activated?*	Yes	No
Memory cells made after activation?	Yes	Yes
Major histocompatibility complex (MHC) proteins present on cell surface:*	MHC I	MHC I and II
Considered antigen-presenting cells?*	No	Yes

*Concept reviewed later.

Matching Uniforms: Tissue Typing for Transplants

For a successful organ transplant, the MHCs of the donor and recipient must be closely matched. This task is made especially challenging by the huge genetic diversity seen among MHCs. If the cellular uniforms are not similar enough for the transplanted tissue to blend in with the self-tissue, the immune system of the transplant recipient will attack the transplant, and the patient will experience graft rejection. Administering immunosuppressive drugs for the rest of the patient's life also limits the chances of transplant rejection.

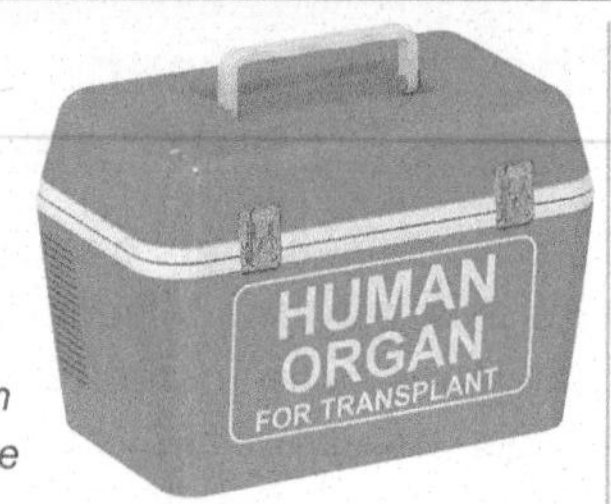

QUESTION 12.2

Would a patient who has transplanted tissue from an identical twin need to take immunosuppressive drugs? Explain your response.

Build Your Foundation

1. What are the branches of adaptive immunity, and how does adaptive immunity differ from innate immunity? (LO 12.1)
2. Compare and contrast the main lymphocytes of adaptive immunity. (LO 12.2)
3. Name the different classes and subclasses of T cells, state how they're differentiated from one another, and describe their general roles in adaptive immunity. (LO 12.3)
4. What do the terms *immunogenicity*, *hapten*, *antigen*, and *epitope* mean? (LO 12.4)
5. What antigens tend to be most immunogenic? (LO 12.4)
6. What are TCRs and BCRs, what role do they play in antigen recognition, and how does their diversity come about? (LO 12.5)
7. Define self-tolerance, and explain how the body screens T and B cells for this capability; include a discussion of MHCs in self-tolerance screening. (LO 12.6)

Build Your Foundation (BYF) Quick Quiz: Visit the **Mastering Microbiology** Study Area to quiz yourself.

12.2 CELLULAR BRANCH OF ADAPTIVE IMMUNITY

Learning Outcomes

After reading this section, you should be able to:

12.7 Describe the cellular branch of adaptive immunity, and name its key effector cells.

12.8 Describe how the two types of MHCs present antigens, and summarize how MHCs impact transplant rejection.

12.9 Explain the two-signal mechanism of T cell activation, and discuss the factors that affect subclass differentiation.

12.10 Summarize how superantigens activate T helper cells.

12.11 Discuss how T helper and T cytotoxic cells eliminate antigens, and summarize memory T cell roles.

In the cellular response, T cells mobilize against diverse antigens.

The cellular branch of adaptive immunity is mainly organized by T helper cells and carried out by T cytotoxic cells. As described back in Figure 12.2, both the cellular and humoral branches can be described as going through four general stages: (1) antigen presentation, (2) lymphocyte activation, (3) lymphocyte proliferation and differentiation, and (4) antigen elimination and memory. Here we'll review each of the four stages for the cellular branch of adaptive immunity.

Stage 1: Antigen-presenting cells use major histocompatibility complexes I or II to present antigens to T cells.

Cellular response
STAGE 1: ANTIGEN PRESENTATION
STAGE 2: LYMPHOCYTE ACTIVATION
STAGE 3: LYMPHOCYTE PROLIFERATION AND DIFFERENTIATION
STAGE 4: ANTIGEN ELIMINATION AND MEMORY

We previously discussed the role of major histocompatibility complexes (MHCs) in lymphocyte self-tolerance screening, but they're important for another reason: They present antigens to T cells. There are two main classes of MHCs important for antigen presentation: MHC I and MHC II.

MHC I is found on the surface of all nucleated body cells (including antigen-presenting cells)[6]; it acts like the body's uniform. Our immune system recognizes transplanted cells because they wear a different uniform—they have different MHC I molecules on their cell surface. In contrast, MHC II is only on antigen-presenting cells (APCs). As a reminder, APCs are white blood cells that collect antigens (usually by phagocytosis), process them, and then present them to T cells. Although macrophages and B cells are important APCs, *dendritic cells are considered the most active APCs in adaptive immunity*. (For more on dendritic cells, which are innate immunity leukocytes, see Chapter 11.)

Having two classes of MHCs makes sense if we consider that antigens can exist in two locations: either inside a host cell (intracellular antigen) or outside of a host cell (extracellular antigens). Examples of intracellular antigens include viral proteins made inside an infected host cell, an antigen from a bacterium or protist with an intracellular lifestyle, or even abnormal proteins made by cancer cells. Extracellular antigens include any agent that exists outside of host cells, such as extracellular bacteria, parasitic worms, and various fungi.

Discerning the source of an antigen is essential because the immune system requires a different set of immunological tools to eliminate a pathogen that is outside versus inside host cells. Intracellular antigens are eliminated by killing the host cell. In contrast, extracellular pathogens are directly attacked without the need to kill host cells. The type of MHC used to present an antigen to T cells helps the immune system understand exactly what sort of antigen is being fought. MHC I presents intracellular antigens to T cytotoxic cells, and MHC II presents extracellular antigens to T helper cells.

MHC I: Presenting Intracellular Antigens to T Cytotoxic Cells

All nucleated body cells have the capacity to present intracellular antigens in the context of MHC I. This is an important point because any of our cells could become infected with an intracellular pathogen or mutate into a cancer cell; so, they all need a mechanism for being detected by the immune system and destroyed for the body's greater good. As an example of how intracellular antigens are presented to T cytotoxic cells via MHC I, let's consider antigen presentation during a viral infection (FIG. 12.9).

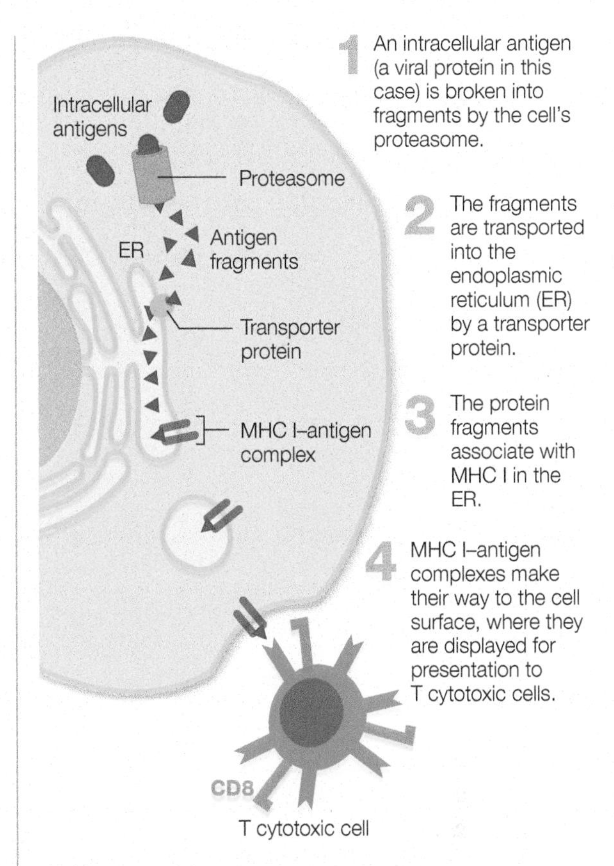

FIGURE 12.9 Intracellular antigen presentation by MHC I

When a virus infects a cell, it hijacks the cell's machinery and builds viral proteins inside the host cell. Viral proteins are chopped up into small segments by a specialized cell component called the **proteasome**. This barrel-shaped structure consists of enzymes called proteases that break down proteins. Sometimes proteasomes are referred to as "protein garbage disposals" because they break down proteins that have been tagged for destruction. A protein could be tagged for destruction because it is damaged/abnormal, is foreign, or has fulfilled its useful life span. The viral protein snippets are then shipped into the cell's endoplasmic reticulum—the organelle that has an important role in modifying and folding cellular proteins. There, thousands of MHC I molecules act as quality-control tools that are constantly sampling the contents of the endoplasmic reticulum and binding proteins to make **MHC I–antigen complexes**. After binding a protein snippet, the MHC I–antigen complex migrates to the cell surface and displays the bound antigens extracellularly. It is the extracellular portion of the MHC I–antigen complex that ultimately beckons T cytotoxic cells and signals them to kill the cell harboring the displayed antigen.

Notably, MHC I will bind to a diverse collection of proteins in the ER—including self-proteins—and display them on the cell surface. Therefore, it is up to patrolling T cytotoxic cells to determine if the protein being displayed is a normal self-protein or something worth reacting against, like a mutated protein that a cancer cell may be producing, or a viral protein. *However, only T cytotoxic cells that have been activated by APCs to recognize the given antigen can effectively*

[6] Note that mature red blood cells (erythrocytes) and platelets (cell fragments important for blood clotting) are not nucleated and therefore lack MHC I.

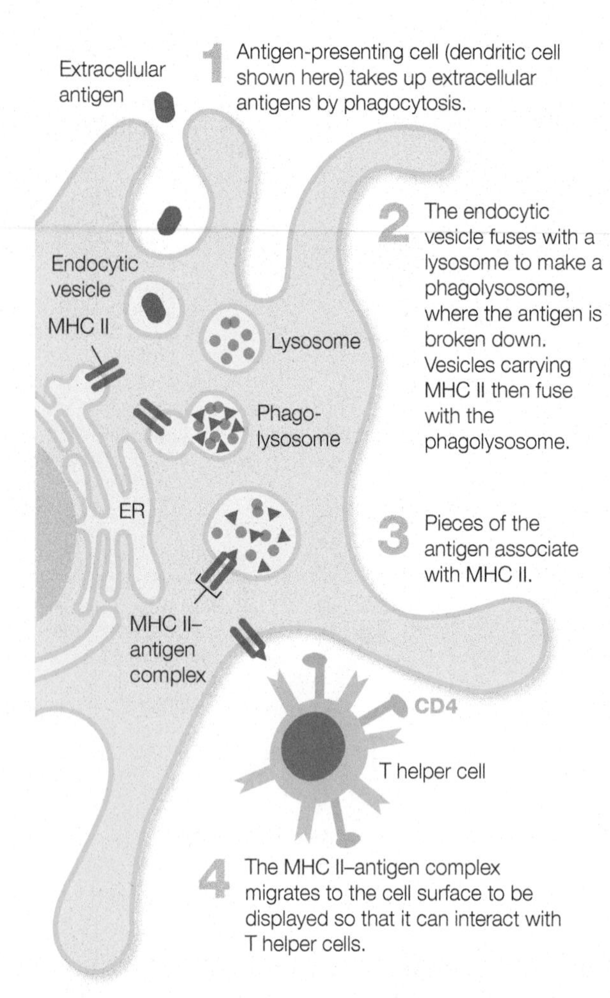

FIGURE 12.10 Extracellular antigen presentation by MHC II

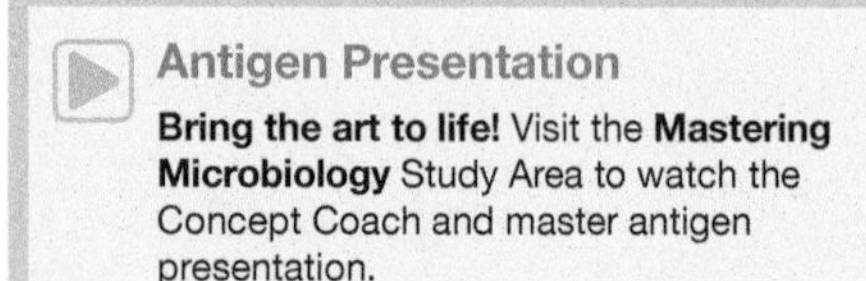

Antigen Presentation

Bring the art to life! Visit the **Mastering Microbiology** Study Area to watch the Concept Coach and master antigen presentation.

TABLE 12.2 Major Histocompatibility Complex Features

Type	MHC I	MHC II
Location	On all body cells except red blood cells; serves as the body's uniform	Only on antigen-presenting cells (APCs); main APCs are dendritic cells
Interacts with:	CD8 on T cytotoxic cells	CD4 on T helper cells
Antigens presented:	Intracellular antigens	Extracellular antigens

Escaping Self-Destruction

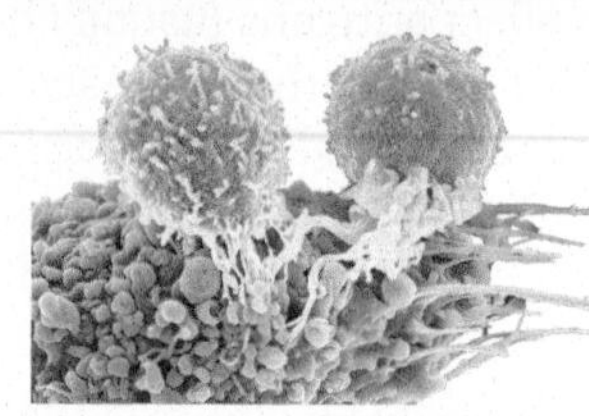

Colorized scanning electron micrograph of T_C cells attacking a virus-infected cell.

Many viruses have developed ways to evade T_C cell detection by interfering with MHC I antigen presentation. For example, certain adenoviruses block MHC I from leaving the endoplasmic reticulum (ER) to migrate to the cell surface; herpes viruses block transporters that shuttle proteins from the cytoplasm into the ER to allow MHC I to bind to them for display; and cytomegalovirus triggers MHC I–antigen complex destruction to prevent antigen display on the cell surface. This blocking of antigen exhibition by MHC I allows infected cells to escape the cytotoxic actions of patrolling T_C cells.

QUESTION 12.1

In the case of HIV, the virus infects T_H cells and lacks the ability to evade MHC I presentation. As such, the immune system signs and executes its own death warrant. Explain the meaning of this statement.

patrol the body and eliminate cells displaying suspicious antigens. As such, elimination of intracellular antigens is not as independent of APCs as it may first appear.

To activate T cytotoxic cells, APCs obtain viral antigen samples by being infected with the virus in question or by phagocytizing an infected host cell. If the APC is directly infected, it simply grabs viral peptides with MHC I and displays them on the cell surface using the process we just reviewed. In the phagocytosis scenario, as virus-infected cells die, they are engulfed by APCs—usually dendritic cells. Consequently, viral antigens inside the cell are also ingested. The APC then binds viral antigens with MHC I for presentation to T cytotoxic cells. The interaction between the APC (usually a dendritic cell) and a naïve (not yet activated) T cytotoxic cell mainly occurs in lymphatic tissues—especially lymph nodes. Activated T cytotoxic cells then roam the body and can target cells bearing the antigen in question.

Although viruses are important intracellular pathogens, many eukaryotic parasites, bacteria, and fungi also exist intracellularly. Tuberculosis, chlamydia, listeriosis, malaria, and toxoplasmosis are all caused by nonviral intracellular pathogens that will require MHC I antigen presentation to be eliminated. We'll cover T cytotoxic cell activation and antigen elimination soon, but let's first review how MHC II molecules present extracellular antigens to T helper cells.

MHC II: Presenting Extracellular Antigens to T Helper Cells

Only APCs make MHC II; they use it to present extracellular antigens to T helper cells. To envision this process, imagine a patient who steps on a piece of glass while walking on the beach. As the dirty piece of glass penetrates the skin, it damages local cells. The damaged cells immediately sound a "chemical alarm" as they release pro-inflammatory factors (Chapter 11). These pro-inflammatory factors recruit innate immunity cells such as neutrophils, macrophages, and dendritic cells to the area. Dendritic cells, which are key APCs, phagocytize dead self-cells as well as potential invaders—perhaps bacteria on the glass. As the dendritic cells break down the phagocytized antigen, protein fragments derived from the antigen associate with MHC II proteins. These **MHC II–antigen complexes** then make their way to the APC's cell surface (FIG. 12.10). The APC, which is now displaying a collection of MHC II–antigen complexes on its surface, migrates from the inflammation site to lymphoid tissues and seeks out T helper cells to activate them. TABLE 12.2 summarizes MHC I and MHC II features.

A Note on MHCs and Tissue Transplants

Because MHCs have a central role in helping the immune system differentiate self from foreign, they are the main proteins that must be matched between a tissue donor and recipient. If they are not closely matched, then the recipient's immune system will recognize the transplanted tissue as foreign and mount an immune response against it. The process that lymphocytes use to differentiate self from foreign MHCs is called **allorecognition**.

In the case of bone marrow transplants, the transplanted tissue will eventually reconstitute the recipient's immune system cells as well as their red blood cells, even converting the recipient's blood type to that of the bone marrow donor. If donor and recipient's MHCs are not closely matched, then the developing B and T cells mount an immune response against their new host, which leads to a scenario called **graft-versus-host disease** (GVHD).

When we hear about waiting lists for transplant patients, it is because finding an adequate match is often difficult. Unless the patient has an identical twin there is no one else in the world that will be a perfect MHC match; the best that can be hoped for is a very close match between the donor and recipient so that T cells and B cells are not activated against the transplanted tissue. (Chapter 13 reviews the various categories of tissue grafts and pathology associated with graft rejection.)

Stage 2: T cells are activated by antigen-presenting cells in lymphatic tissues.

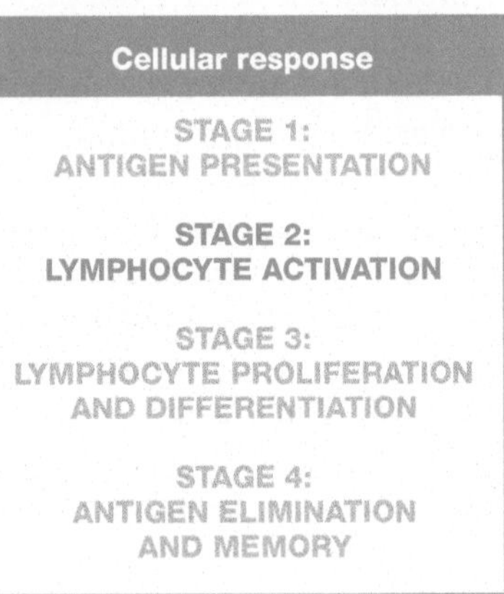

An APC bearing MHC-antigen complexes on its cell surface migrates to lymphoid tissues to interact with T helper and T cytotoxic cells. The APC arriving in lymphoid tissues is a little like someone walking into a crowded room with their hand extended, hoping to receive at least one handshake in return. Chemically speaking, a handshake happens when a T cell with a compatible receptor binds to one of the antigens presented by the APC. T cytotoxic cells will interact with antigens presented in the cleft of MHC I molecules, while T helper cells will bind to antigens present in the cleft of MHC II molecules (FIG. 12.11).

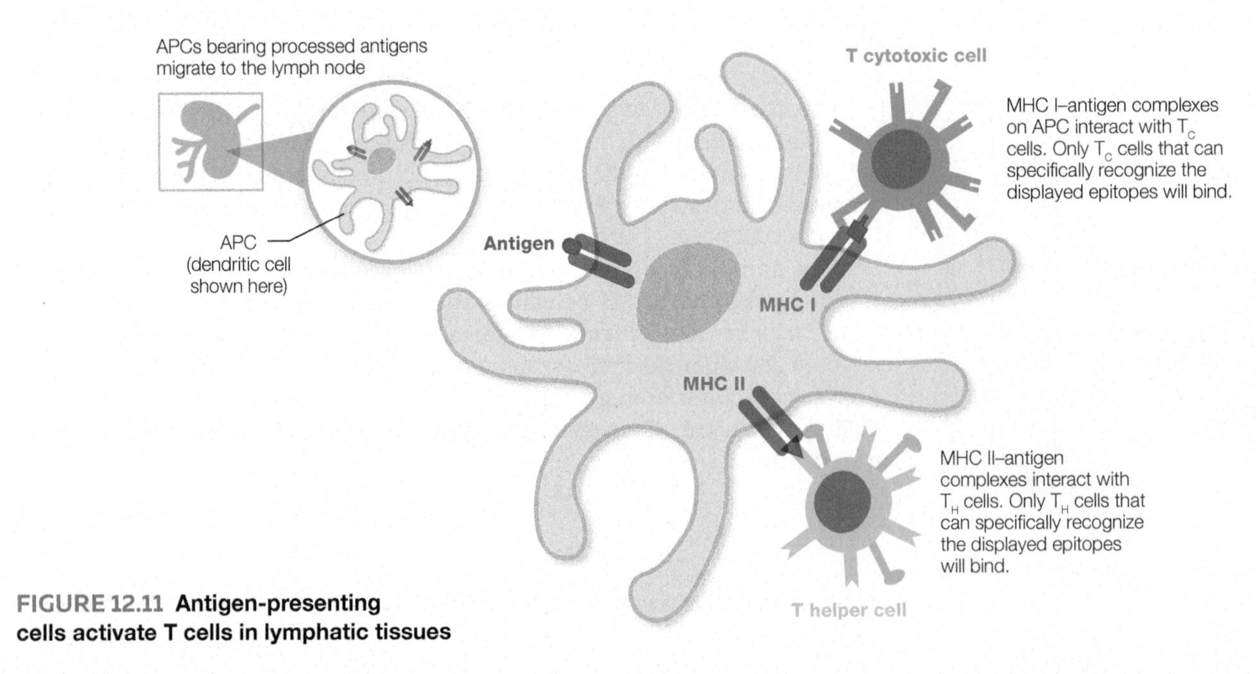

FIGURE 12.11 Antigen-presenting cells activate T cells in lymphatic tissues

Two-Signal Activation

It is important to recall that due to TCR specificity, each T cell can only recognize one type of antigen (or more accurately, one epitope of the antigen). Because an APC presents diverse epitopes from diverse antigens, it is possible for more than one T cell to interact with the APC and become activated. If a given T cell successfully binds to one of the MHC–antigen complexes on the APC's surface, then the first activation signal is initiated. However, one signal isn't enough; T cells require two signals for full activation (FIG. 12.12).

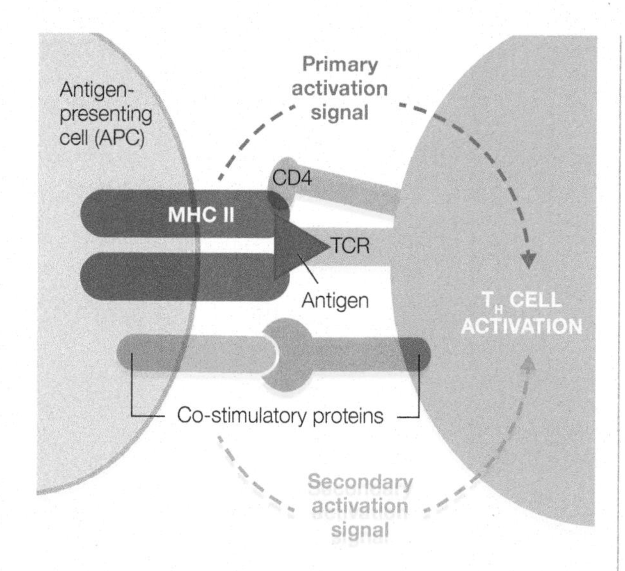

The primary activiation signal for T cytotoxic cells is a little different from what is shown above for T helper cells. Here the primary activation signal involves CD8 interacting with an MHC I–antigen complex.

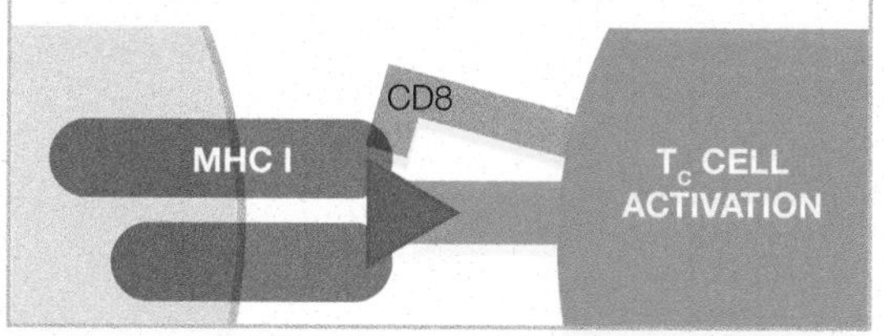

FIGURE 12.12 **Two-signal activation of T cells**

1. The **primary activation signal** involves the T cell's TCR interacting with the MHC–antigen complex. Recall that T cytotoxic cells will only interact with MHC I–antigen complexes, and T helper cells will only interact with MHC II–antigen complexes. This TCR/MHC/antigen interaction is further assisted by CD8 on the surface of T cytotoxic cells and by CD4 on the surface of T helper cells.
2. The **secondary activation signal** involves **co-stimulatory proteins** on the surface of the APC binding to co-stimulatory proteins on the T cell's surface. A number of cell membrane proteins on the APC and the T cell can interact to provide the secondary signal.

Requiring two activation signals prevents immune system misfires. Also, it provides a way for T cell populations to become specialized. As mentioned before, there are seven known subclasses of T helper cells as well as less understood T cytotoxic cell subclasses. Just as certain team members have specific roles in making a winning play, so too do various T cell subclasses have specific roles that must be coordinated for best results. Of course, to make the winning play, the participating members have to know what their roles are—which is where the second activation signal is helpful.

When a co-stimulatory protein on an APC binds to a co-stimulatory protein on a T cell, a signaling cascade is sparked. The type of cascade initiated, which varies according to the co-stimulatory proteins involved, determines what class of interleukins and other factors the activated T cell will make. This in turn defines what specialized subclass the T cell joins.

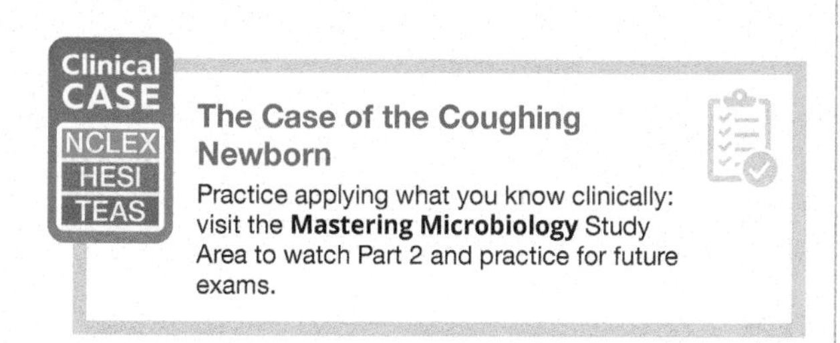

Superantigens and T Helper Cell Activation

Some antigens are especially potent T helper cell activators; these so-called **superantigens** include a variety of bacterial toxins such as staphylococcal enterotoxins, staphylococcal toxic shock toxin, and streptococcal exotoxins that generate the features of scarlet fever. T cells are normally activated when an APC presents them with an antigen in the context of MHC. However, superantigens circumvent the entire antigen-presentation process and instead trigger a nonspecific activation by essentially forcing an interaction between MHC II and TCRs (FIG. 12.13).

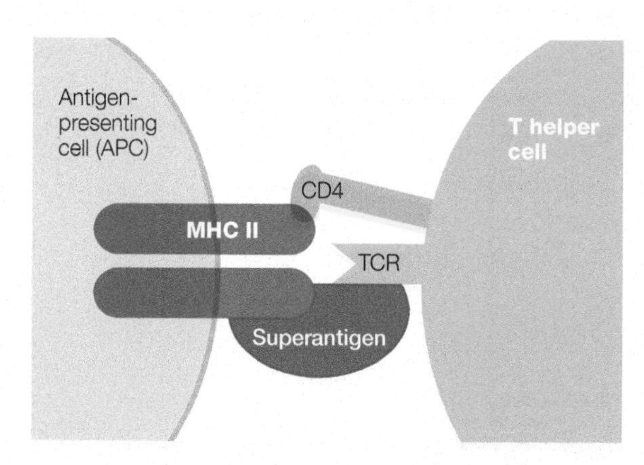

FIGURE 12.13 **Superantigens** Note how superantigens directly cross-link TCRs to MHC II without a processed antigen.

This nonspecific interaction leads to a broad activation of T cells—some data suggest that up to 20 percent of the T helper cells in the lymphatic tissues may become activated if superantigens are in the bloodstream. In response, T helper cells release dangerous levels of cytokines—especially interleukin 2, interferon gamma (INF-γ), and tumor necrosis factor alpha (TNF-α), which are all reviewed in Chapter 11—that can lead to shock and even death.[7]

[7] Fraser, J. D. (2011). Clarifying the mechanism of superantigen toxicity. *PLoS Biol*, 9(9), e1001145.

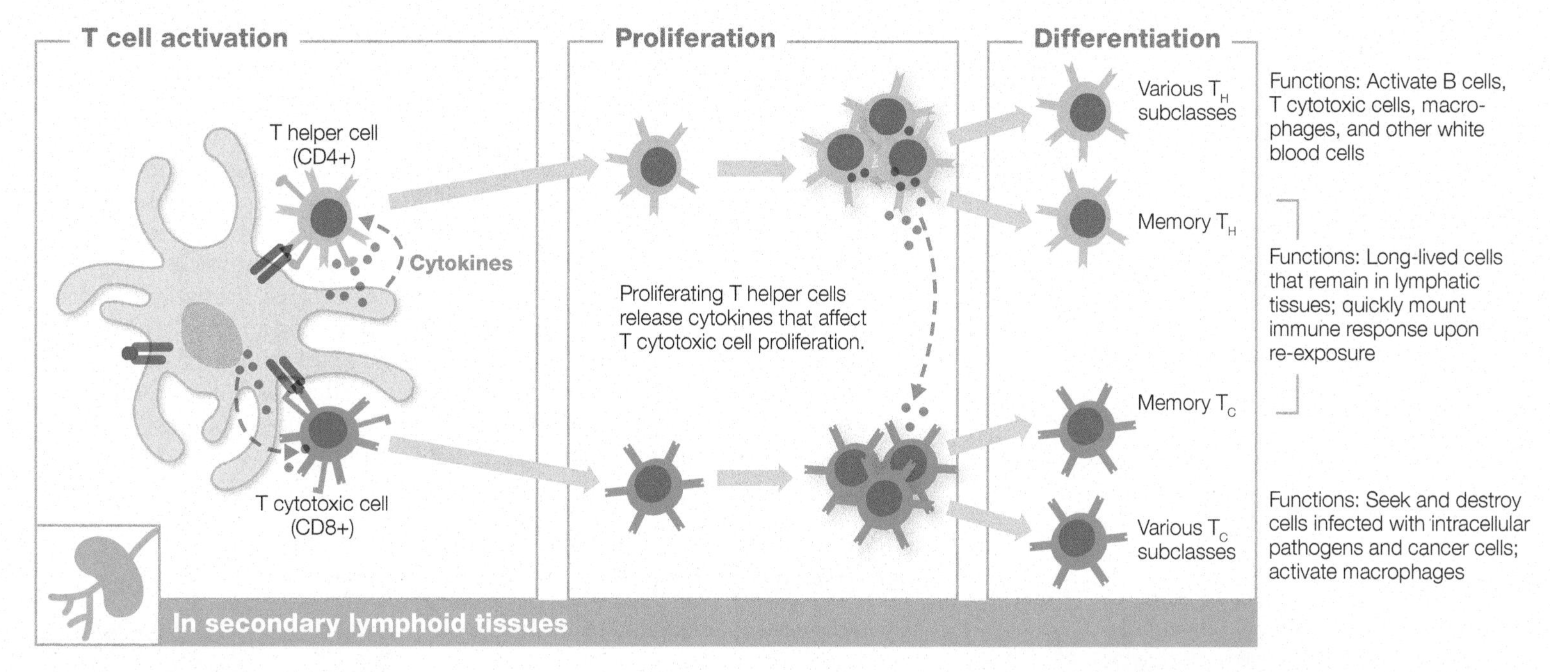

FIGURE 12.14 T cell proliferation and differentiation Antigen-presenting cells interact with T cells to stimulate T cell proliferation and differentiation.

Stage 3: Activated T cells undergo proliferation and differentiation.

Cellular response

STAGE 1:
ANTIGEN PRESENTATION

STAGE 2:
LYMPHOCYTE ACTIVATION

**STAGE 3:
LYMPHOCYTE PROLIFERATION AND DIFFERENTIATION**

STAGE 4:
ANTIGEN ELIMINATION AND MEMORY

In secondary lymphoid tissues, T cytotoxic and T helper cells bind to an epitope presented by an APC. This triggers the T cells to undergo proliferation (or clonal expansion) by repeated rounds of mitosis (FIG. 12.14). During these cell division events, chemical signals influence cell differentiation—that is, what T helper and T cytotoxic cell subclasses develop. Each subclass made includes effector and memory cells. A clone made in the proliferation stage, no matter what class or subclass it belongs to, will recognize the same epitope that activated the original parent cell. Proliferation and differentiation signals that lead to various cell subclasses are best understood for T helper cells.

As APCs and T helper cells interact in lymphatic tissues, APCs release various cytokines. As just noted, the types of cytokines released influence what T helper cell subclasses develop (FIG. 12.15). For example, if the APC releases interleukin 12 and interferon gamma (INF-γ), then T_H1 cells tend to develop.[8] In contrast, interleukin 2 and interleukin 4 favor T_H2 cell development.[9] The types of cytokines released depend on numerous factors that include the antigen's nature and the amount of antigen present. The T helper cell subclasses generated as a result of the initial APC interaction are important in dictating the nature of the developing immune response. If T_H1 cells develop, the cellular

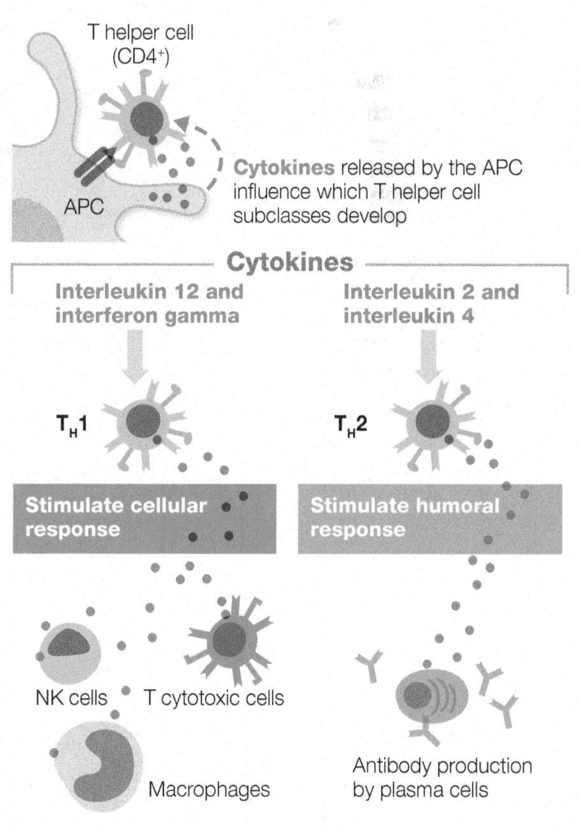

FIGURE 12.15 T helper cell differentiation into T_H1 and T_H2 subsets

[8] Trinchieri, G., Pflanz, S., & Kastelein, R. A. (2003). The IL-12 family of heterodimeric cytokines: New players in the regulation of T cell responses. *Immunity*, *19*(5), 641–644.

[9] Zhu, J. (2015). T helper 2 (Th2) cell differentiation, type 2 innate lymphoid cell (ILC-2) development and regulation of interleukin 4 (IL-4) and IL-13 production. *Cytokine*, *75*(1), 14–24.

[10] T_H1 cells mainly make interferon gamma (IFN-g) and interleukin 2; T_H2 cells make a wide array of interleukins such as IL-4, IL-5, IL-6, IL-10, and IL-13. The different cytokine profiles influence what type of antibodies B cells make when they are stimulated by T helper subclasses.

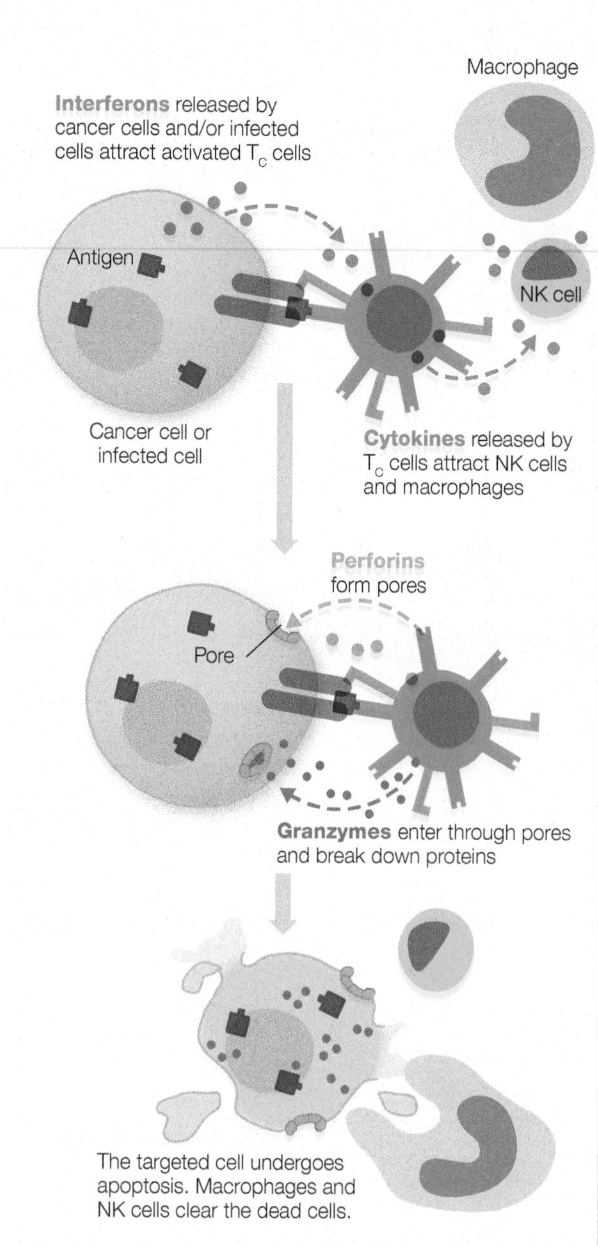

FIGURE 12.16 T cytotoxic (T_C) cell efforts to eliminate antigens

response tends to be favored because T_H1 cells release diverse cytokines that are potent T cytotoxic cell stimulators.[10] However, if T_H2 cells develop, then the humoral response is usually favored because T_H2 cells release cytokines that promote B cell maturation and discourage T cytotoxic cell actions.

The type of T helper cell subclass that develops impacts the immune system's overall defense strategy. An example of this is seen in leprosy. If the patient develops a T_H2 response against the bacteria that cause leprosy (*Mycobacterium leprae*), then a lepromatous form of the disease develops. Lepromatous leprosy is disfiguring and deadly. In contrast, if a T_H1 response develops, the patient manifests a tuberculoid form of the disease that is not disfiguring and not lethal. In this case (and in other diseases, too), the decision toward one T helper cell subclass or another impacts the progression and outcome of infectious as well as noninfectious diseases. We revisit this concept of T_H1 and T_H2 balance in Chapter 13 when we discuss the development of allergy and autoimmunity.

Stage 4: Effector T cells eliminate antigens, and memory T cells remain in lymphatic tissues.

As you have read, T helper cells influence the direction of immune responses based on what cytokines they release. The diverse T helper and T cytotoxic cell subclasses that exist and the complex interactions among them are geared toward eliminating the target antigen. As the threat subsides, memory cells of these various subclasses remain in lymphatic tissues ready to rapidly proliferate and differentiate upon a subsequent exposure.

Cellular response
STAGE 1: ANTIGEN PRESENTATION
STAGE 2: LYMPHOCYTE ACTIVATION
STAGE 3: LYMPHOCYTE PROLIFERATION AND DIFFERENTIATION
STAGE 4: ANTIGEN ELIMINATION AND MEMORY

T Cytotoxic Cell Roles in Antigen Elimination

T cytotoxic subclasses of effector cells will "seek and destroy" cells that display the activating antigen in the grip of MHC I (FIG. 12.16). When a cell is infected with a virus or is cancerous, various interferons are released, recruiting activated T cytotoxic cells to the area.[11] Released interferons also enhance MHC I production inside host cells, making the likelihood of grabbing and displaying a viral protein to flag down activated T cytotoxic cells even greater. Interferons and a variety of other inflammatory signals put the immune system on high alert, causing the screening process to become even more intensive when there is an invader in the vicinity. When the TCR of a patrolling T cytotoxic cell binds to an MHC I–antigen complex, the T cytotoxic cell releases **perforins** that form pores in the target cell and **granzymes**, which enter through the pore to break down host cell proteins and induce apoptosis.

T Helper Cell Roles in Antigen Elimination

T helper cells do not directly attack invaders, cancer cells, or infected host cells. Rather, they support the action of the cells that will actually do the work in the immune response—B cells, T cytotoxic cells, and innate immunity leukocytes such as macrophages and natural killer cells. The T helper cell subclass made during the proliferation and differentiation stage impacts the

[11] Zaidi, M. R., & Merlino, G. (2011). The two faces of interferon-g in cancer. *Clinical Cancer Research*, 17(19), 6118–6124.

immune response. For example, T_H1 cells favor the action of T cytotoxic cells, while T_H2 cells promote the humoral response, which is mediated by B cells. (Refer back to Figure 12.14.)

Build Your Foundation

8. What is the cellular response, and what are its key effector cells? (LO 12.7)
9. What are the main types of MHCs, and how do they present antigens to T cells? (LO 12.8)
10. How do MHCs impact transplant rejection? (LO 12.8)
11. How are T cells activated, and what factors affect T cell subclass differentiation? (LO 12.9)
12. How do superantigens activate T helper cells? (LO 12.10)
13. How do T cytotoxic and T helper cells eliminate and remember antigens? (LO 12.11)

Build Your Foundation (BYF) Quick Quiz: Visit the **Mastering Microbiology** Study Area to quiz yourself.

12.3 HUMORAL RESPONSE OF ADAPTIVE IMMUNITY

Learning Outcomes

After reading this section, you should be able to:

12.12 Explain how T-dependent and T-independent antigens activate B cells.

12.13 Describe the basic structural and functional features of antibodies.

12.14 Explain what isotype switching is, and state why it's advantageous.

12.15 Discuss the structural and functional features of each of the five antibody classes.

Stage 1: B cells are antigen-presenting cells.

Humoral response
STAGE 1: ANTIGEN PRESENTATION
STAGE 2: LYMPHOCYTE ACTIVATION
STAGE 3: LYMPHOCYTE PROLIFERATION AND DIFFERENTIATION
STAGE 4: ANTIGEN ELIMINATION AND MEMORY

B cells mediate humoral responses; when activated, they become plasma cells that secrete antibodies. Unlike T cells, B cells act as antigen-presenting cells, so they don't require another antigen-presenting cell to process and present antigens. Instead, they have two different paths to activation; the route taken depends on the type of antigen. **T-independent antigens** can fully activate B cells without T cell stimulatory signals (activation is *independent* of T cells). In contrast, for B-cells to be fully activated by **T-dependent antigens**, T helper cells (especially T_H2 cells) are necessary.

Stage 2: B cells are activated by T-dependent and T-independent antigens.

Humoral response
STAGE 1: ANTIGEN PRESENTATION
STAGE 2: LYMPHOCYTE ACTIVATION
STAGE 3: LYMPHOCYTE PROLIFERATION AND DIFFERENTIATION
STAGE 4: ANTIGEN ELIMINATION AND MEMORY

Most antigens that interact with B cells are described as T-dependent (FIG. 12.17). This activation pathway is a two-signal mechanism—much like we saw with T cells. The first activation signal is the binding of the antigen to the B cell receptor (BCR). Recall that B cells act as APCs, which means they possess MHC II. So, a B cell with a bound receptor then internalizes the antigen, processes it like any other antigen-presenting cell, and displays the processed antigen on its cell surface in the context of MHC II. T helper cells then interact with the MHC II–antigen complex on the B cell surface. When co-stimulatory proteins on the B cell and T helper cell surface interact, the T helper cell releases cytokines that provide the second required activation signal for the B cell. Examples of cytokines that stimulate B cells are interleukins 2, 4, and 5.

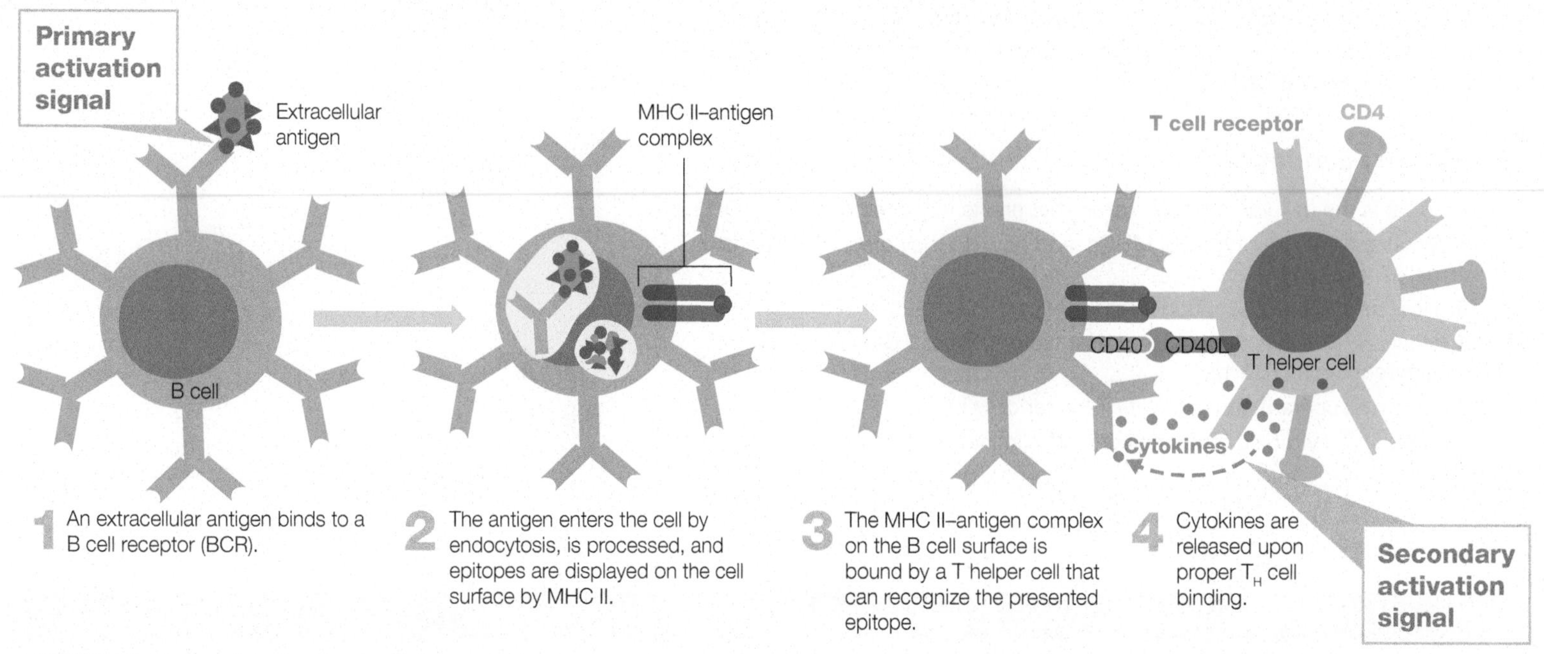

FIGURE 12.17 **B cell activation by T-dependent antigens**

Antigens with repeating features (also known as repetitive antigens) may act as T-independent antigens. These antigens are usually, although not exclusively, polysaccharides (examples: bacterial capsule polysaccharides). In T-independent activation, multiple BCRs on the given B cell directly bind to the antigen. It is hypothesized that the binding of the BCRs to the antigen provides the first signal for B cell activation and that perhaps some other sort of immune system cell other than a T_H cell (for instance, a dendritic cell) provides a second signal for activation. That said, the specifics regarding a second activation signal, or if a second signal is always needed, remains a topic of investigation. In response to activation by T-independent antigens, the B cell will undergo proliferation and differentiation to make plasma cells, but they only have a limited capacity for memory, and they do not tend to confer the same long-term protection as T-dependent antigens (FIG. 12.18).

FIGURE 12.18 **B cell activation by T-independent antigens**

Stage 3: Activated B cells proliferate and differentiate into plasma cells and memory cells.

Just as occurred for T cells, fully activated B cells undergo proliferation by repeated rounds of mitosis and eventually differentiate into effector cells and memory cells. All of the resulting B cell clones recognize the exact same epitope of the antigen. Most of the generated clones become antibody-producing plasma cells. A small number become memory B cells that will reside in lymphatic tissues along with memory T helper and memory T cytotoxic cells made in the cellular response (FIG. 12.19).

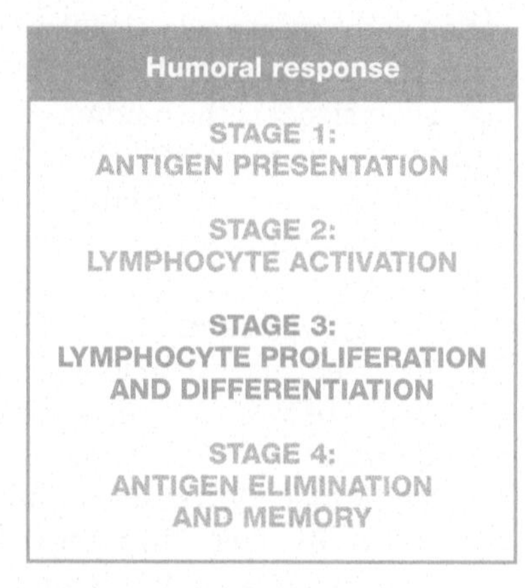

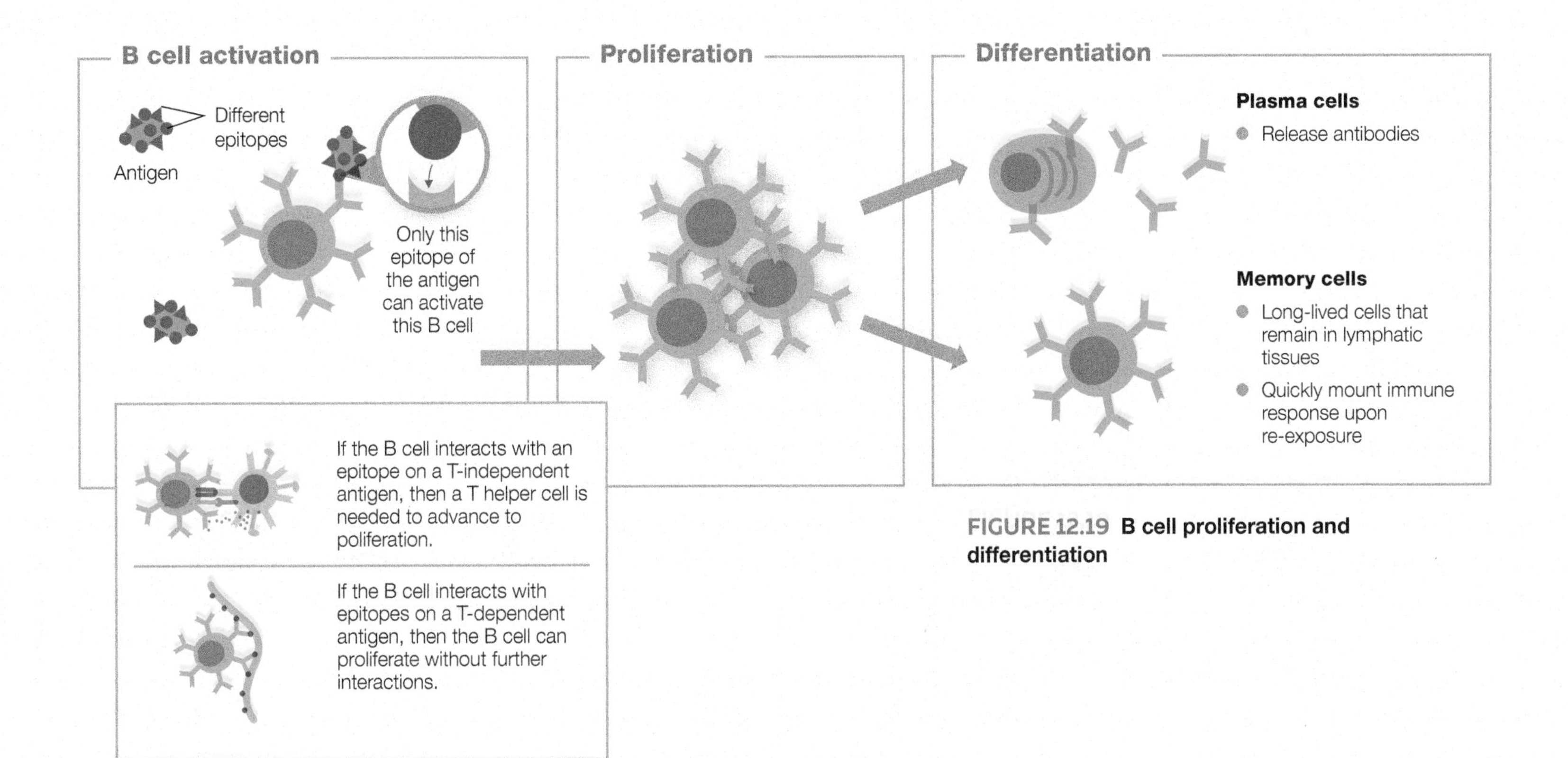

FIGURE 12.19 B cell proliferation and differentiation

Stage 4: Antibodies help eliminate antigens.

Humoral response
STAGE 1: ANTIGEN PRESENTATION
STAGE 2: LYMPHOCYTE ACTIVATION
STAGE 3: LYMPHOCYTE PROLIFERATION AND DIFFERENTIATION
STAGE 4: ANTIGEN ELIMINATION AND MEMORY

As noted earlier, plasma cells secrete antibodies, also known as **immunoglobulins (Ig)**, that bind to the antigen that triggered the B cell's activation. Three key antibody functions help eliminate antigens (FIG. 12.20).

First, antibodies directly neutralize antigens to prevent them from interacting with target host cells. For example, toxins, viruses, and bacteria often bind specific host cell receptors. When bound by antibodies, these entities can't interact with host cell receptors.

Second, the classical complement cascade is activated when antibody–antigen complexes bind to certain complement proteins in the pathway. (Review complement cascades in Chapter 11.) Activated complement proteins lead to cytolysis, inflammation, and opsonization (tagging antigens to enhance their chances of being phagocytized).

Lastly, antibodies increase phagocytosis. They accomplish this by:

- Precipitating small, soluble antigens to make them readily detectible by phagocytes
- Causing agglutination of large antigens, such as bacterial cells
- Serving as *opsonins*, factors that bind a target antigen to tag it for phagocytosis.

Antibody Structure and Isotypes

An antibody's single-unit, *monomeric* structure consists of two heavy chains and two light chains held together by covalent bonds—specifically, disulfide bonds. If we were to view an antibody from a two-dimensional perspective, it would have a "Y" shape. At the tips of the "Y"-shaped molecule we find **antigen-binding sites**

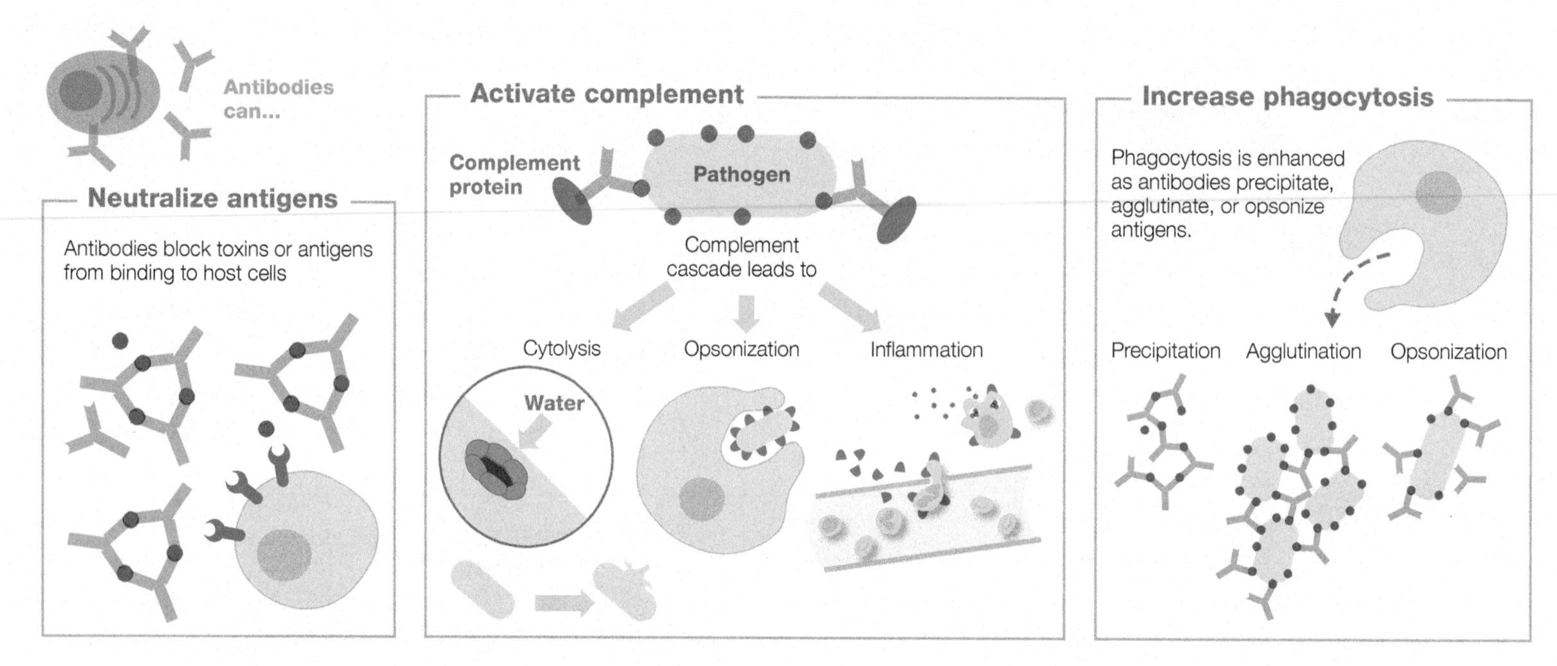

FIGURE 12.20 **Key antibody functions**

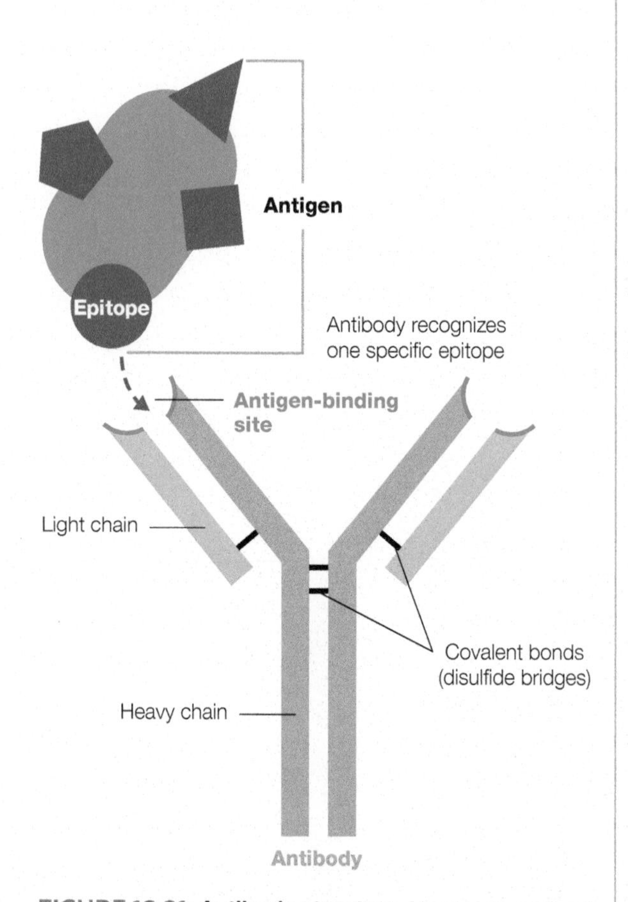

FIGURE 12.21 **Antibody structure** Here a generalized antibody is shown. Antibodies consist of heavy and light protein chains that are connected by disulfide bridges. All the antibodies made by a given plasma cell will recognize the exact same epitope.

(FIG. 12.21). An antibody's antigen-binding site varies based on the epitope that activated the B cell. For this reason, the antigen-binding site is sometimes referred to as the antibody's **variable region**. In contrast, an antibody's **constant region** is the stem portion of the "Y"-shaped molecule. The constant region does not directly participate in antigen binding and is the same (held constant) for a given class (or **isotype**) of antibody.

Based on constant region structures, five antibody isotypes have been described—IgG, IgA, IgM, IgE, and IgD. You can remember "GAME Day" to recall the five antibody isotypes. The first three types listed in this memory device also reflect general abundance in our antibody pool, with IgG being the most common, IgA the second most prevalent, and IgM the third most abundant. The isotype made depends on numerous factors, including the cytokines that stimulate the B cell and the time frame of antigen exposure (early versus late infection or primary versus secondary exposure).

Although a given B cell tends to be monospecific (primed to recognize a single specific epitope), B cells can perform **isotype switching** to alter the isotype of antibody made. As such, a given plasma cell may start off making IgM against a specific epitope and then later switch to making IgG against that same epitope. Although isotype switching won't expand what sort of epitope is recognized, it does expand the operational capacity of the antibodies made. Different isotypes have different specializations and also predominate in different areas of the body. The features of the five antibody isotypes, which are summarized in TABLE 12.3, include the following.

Immunoglobulin G (IgG) The most abundant antibody in human blood, IgG constitutes up to 85 percent of total antibodies. Found in all bodily fluids, it is especially abundant in blood, lymph, cerebrospinal fluid, and peritoneal fluid (which lubricates organs and tissues of the abdominal cavity). IgG has a half-life of about 21 days in circulation before breaking down. Detecting IgG to a particular antigen indicates the patient is either in the late stages of an initial exposure, or the patient had a prior exposure to that antigen by either an infection or through vaccination. Structurally speaking, this antibody has only one of the "Y"-shaped molecules described in the prior section, so is described as a monomer. IgG is one of the most versatile antibodies we make. It is the only antibody that can cross the placenta to protect the developing fetus. IgG is a key activator of complement proteins, neutralizes antigens very effectively, and is a powerful opsonin.

TABLE 12.3 Five Antibody Isotypes

Isotype	IgG	IgA	IgM	IgE	IgD
Structure	Monomer	Monomer or dimer	Monomer or pentameter	Monomer	Monomer
Proportion of antibody pool	Most abundant	Second most abundant	Third most abundant	Rare	Rare
Neutralization	Strong	Strong	Some	Negligible	Negligible
Complement activation	Strong	Some	Strong	Negligible	Negligible
Opsonization	Strong	Some	Negligible	Negligible	Negligible
Agglutination/precipitation	Some	Negligible	Strong	Negligible	Negligible
Half-life	21 days	6 days	10 days	2 days	2 days
Notes	Crosses placenta; made later in infection	Main antibody in breast milk; resistant to destruction by stomach acid	Made early in infection; large structure limits where it migrates	Fights parasites; mediates allergic responses	Bound to B cells; poorly understood

Immunoglobulin A (IgA) IgA represents up to 15 percent of our total antibodies and has a half-life of about 6 days before it breaks down. Prevalent in mucus, IgA is therefore also abundant on mucous membranes of the gut, respiratory tract, and urogenital tract. It is also found in most bodily secretions such as tears, saliva, sweat, and breast milk. IgA can exist as a monomer or as a dimer—meaning it has two "Y"-shaped subunits. The dimer form is especially resistant to harsh environmental conditions. This makes it an ideal antibody to protect nursing infants because it survives stomach acid and reaches the intestines to confer protection. IgA isn't very good at activating complement cascades, but it has excellent neutralizing capabilities and some opsonization capabilities.

Immunoglobulin M (IgM) IgM is mainly in blood and accounts for up to 10 percent of total antibody population. It has a half-life of about 10 days before it breaks down. Usually made early in infection upon a primary antigen exposure, its presence tends to indicate a recent exposure to a given antigen. IgM can exist as either a monomer or a snowflake-shaped pentamer with five "Y"-shaped subunits. The large pentamer structure is central to agglutination and precipitation reactions that promote phagocytosis. It also has important roles in complement activation. It has some neutralization ability, but is not as effective at neutralizing antigens as IgA or IgG. It is not a strong opsonin.

Immunoglobulin E (IgE) IgE is a monomer with a half-life of only 2 days before it breaks down. It is present in very low concentrations and mostly in lungs, skin, and mucous membranes. IgE is not central to complement activation or agglutination, and it isn't a good opsonin; however, it is central to fighting parasites. It also plays a key role in allergies by encouraging mast cells and basophils to release histamine, leukotrienes, and other allergy mediators. It is common to find elevated IgE in patients with allergies, asthma, eczema, dermatitis, and certain parasitic infections, such as those caused by roundworms or tapeworms. (The role of IgE in hypersensitivities is covered more in Chapter 13.)

Immunoglobulin D (IgD) IgD is a sparsely represented antibody with a 2-day half-life. It is a monomer mainly found on the surface of B cells; only small quantities can be found circulating in blood. Its precise function remains unknown.

We've now summarized the basic activities of both branches of the adaptive immune system. FIGURE 12.22 provides some silly memory devices to help keep track of the various cells involved.

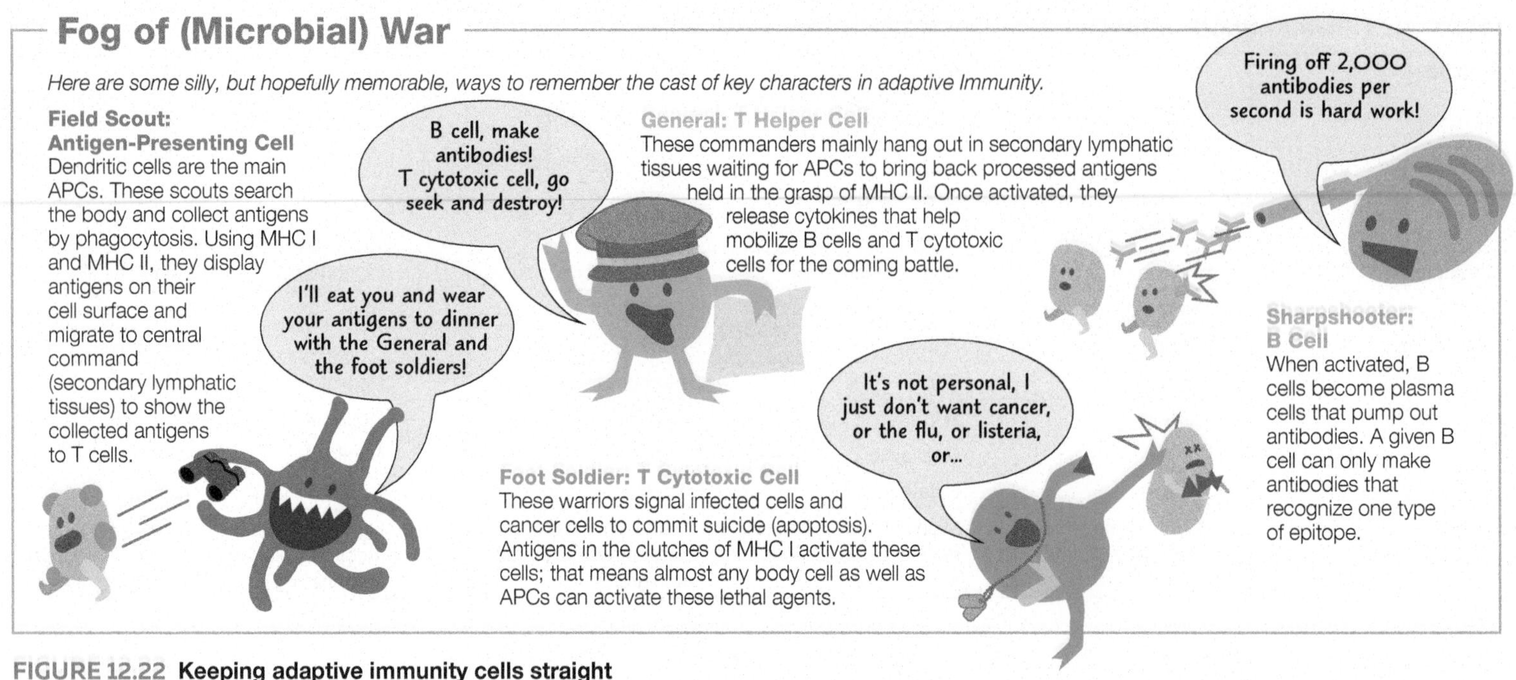

FIGURE 12.22 **Keeping adaptive immunity cells straight**

Build Your Foundation (BYF) Quick Quiz: Visit the **Mastering Microbiology** Study Area to quiz yourself.

Build Your Foundation

14. Compare and contrast T-dependent and T-independent antigens and their modes of activating B cells. (LO 12.12)

15. What are the basic structural and functional features of an antibody? (LO 12.13)

16. What is isotype switching, and how is it advantageous? (LO 12.14)

17. Name the five antibody isotypes, and provide details on their structures and functions. (LO 12.15)

12.4 A DEEPER EXPLORATION OF HUMORAL MEMORY

Learning Outcomes

After reading this section, you should be able to:

12.16 Describe immunological memory, and compare it to a primary response.

12.17 Define antibody titer, and state how it differs in a primary versus secondary antigen exposure.

12.18 Name and describe the four categories of adaptive immunity, and state which confer long-term protection and why.

Memory cells allow for fast, amplified response upon re-exposure to an antigen.

As you learned in earlier sections, the proliferation events that follow T cell and B cell activation generate memory cells and effector cells. Effector cells die off once the threat subsides, but memory cells are long-lived. They are primarily found in lymphoid tissues, where they provide immunological memory. That is, by "remembering" antigens encountered before, our memory cells allow for a rapid reactivation of the cellular and humoral adaptive response if the same antigen is encountered again later. This **secondary immune response** requires the coordinated activity of memory B and T cells. Here we specifically focus on humoral memory because it is a measurable feature that you need to understand to appreciate how vaccines work and how we determine what degree of immunity a patient may have to a pathogen.

Primary exposure to a given antigen generates IgM antibodies first, then IgG. In a secondary response to the same antigen, activated memory cells generate a surge of IgG and only a small amount of IgM. The antibody surge is not only faster to develop in the secondary response, but the **antibody titer**, or amount of antibody present in the blood, is also greater (FIG. 12.23). Furthermore, the antibodies made in a secondary response can bind the antigen even better than their predecessor antibodies—a feature that's referred to as enhanced *affinity*. In fact, this reactivation is so rapid and effective that we usually don't develop any symptoms and may not even realize we were re-exposed to the antigen.

CLINICAL VOCABULARY

Serology The study of blood serum (the liquid portion of blood) for the presence of certain antibodies.

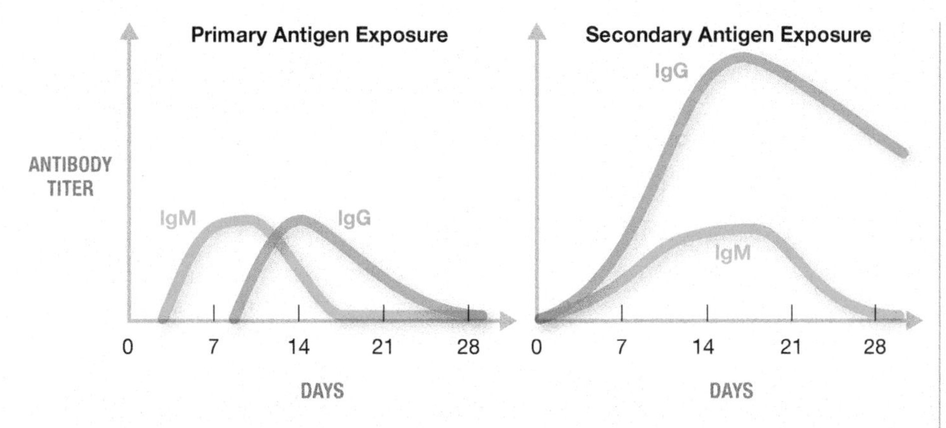

FIGURE 12.23 Generalized antibody production in primary versus secondary responses In a secondary exposure, the surge in antibodies is faster and greater than in a primary exposure.

A healthcare provider may order **serological testing** to assess a patient's antibody titers to help determine if a person has been exposed to a disease or was effectively vaccinated against a particular pathogen. For example, checking if the patient has antibodies to HIV, dengue, Zika, Chikungunya, SARS-CoV-2, or other infectious agents assists with diagnosis. Because antibodies are not made immediately upon infection, such tests can't reliably detect the earliest stage of infection and are generally better suited for detecting later stages of infection and/or determining if a patient has developed immunity to an infectious agent.

Vaccinations optimize immunological memory and antibody titers. Some vaccines require multiple doses to confer optimal protection, while others confer lifelong protection with one dose. Some pathogens, like influenza viruses, frequently mutate. This necessitates new vaccines to protect patients.

In the case of emerging infectious diseases for which there are no approved therapies or vaccinations, **convalescent plasma**—or plasma from a person with a protective antibody titer against the infectious agent—have been experimentally used to treat severely ill patients. For example, convalescent plasma was used to treat SARS (severe acute respiratory syndrome) in the 2003 outbreak and MERS (Middle East respiratory syndrome), which was first identified in 2012. In 2020, the U.S. Food and Drug Administration (FDA) approved using convalescent plasma as an experimental therapy for COVID-19. Convalescent plasma has also been used to treat Ebola patients, although with mixed success. In terms of antibodies as therapeutics, **monoclonal antibodies** (antibodies that are highly specific for a single epitope) are approved for use to combat certain cancers as well as a variety of infectious agents, such as respiratory syncytial virus (RSV) and rabies. (See the Bench to Bedside feature at the end of the chapter for more about these useful therapeutics.)

Humoral immunity is acquired naturally or artificially and is either passive or active.

There are four classifications for adaptive immunity: naturally acquired active immunity, artificially acquired active immunity, naturally acquired passive immunity, and artificially acquired passive immunity. The first part of these classifications depends upon whether the immunity process occurred via a natural process such as an infection or via a human (artificial) intervention like vaccines. The next part of the classifications refers to whether the person benefiting from the antibodies made them (active) or got them from another source (passive). Active immunity—whether naturally or artificially acquired—provides long-term immunological protection because the host actively makes memory cells and antibodies. In contrast, naturally and artificially acquired passive immunity confer only temporary protection because the host doesn't invest in making memory cells or antibodies to the antigens in question and instead gets the antibodies from a secondary source. You can use a dichotomous key to decide which of the four categories of humoral immunity applies to a given situation (FIG. 12.24).

Antibody Titers

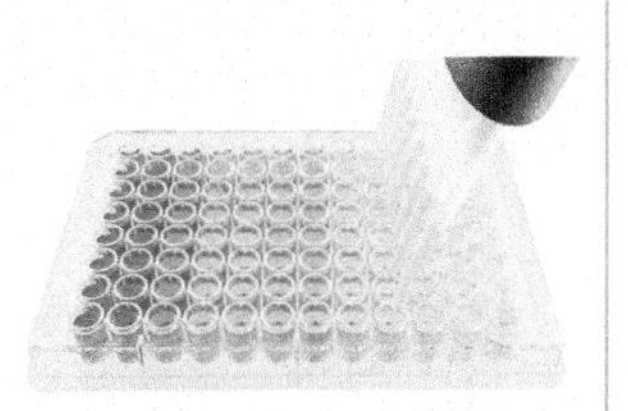

Antibody titer assays are run in microwell plates.

Serological tests provide information about antibody titers (TIE ters)—a measure of how much of a specific antibody is in a blood sample. In general, antibody levels correlate to the strength of the immune response against an antigen and can reveal if a person has developed immunity to a pathogen. For example, assessing antibody levels to SARS-CoV-2 is useful for determining if a person has developed immunity to the virus that causes COVID-19. Titer evaluation is also at the heart of assessing vaccination efficacy because it reveals whether or not a vaccine can trigger a sufficient humoral response to confer protection.

Titers are expressed as a ratio, based on the degree to which the sample is diluted to eliminate an antibody–antigen interaction. The greater the dilution needed to eliminate a reaction, the higher the antibody concentration. For example, a 1:10 ratio indicates a fairly low antibody concentration. This is because the blood sample was only diluted tenfold before the antibody was undetectable. In contrast, a titer of 1:400 would be high because the sample was diluted 400-fold before the antibody was undetectable. You can think of having a sugary drink that must be diluted to be enjoyable. The more concentrated the sugar, the greater the necessary dilution.

The titer cutoffs and their meanings depend on the situation. For example, antirubella antibody titers of at least 1:8 are protective against rubella, a disease vaccinated for during childhood. For the hepatitis B vaccine, titer levels should be at least 1:10 for optimal protection. In the case of emerging infectious diseases, which are not vaccine preventable, a person with a protective antibody titer against the infectious agent may be encouraged to donate plasma to help treat infected patients in critical condition. This strategy was used to combat the Ebola outbreak in 2014, although the success was minimal.

QUESTION 12.3

If IgG to a particular infectious agent is present in substantial quantities, what could you conclude about the chronological stage of the infection?

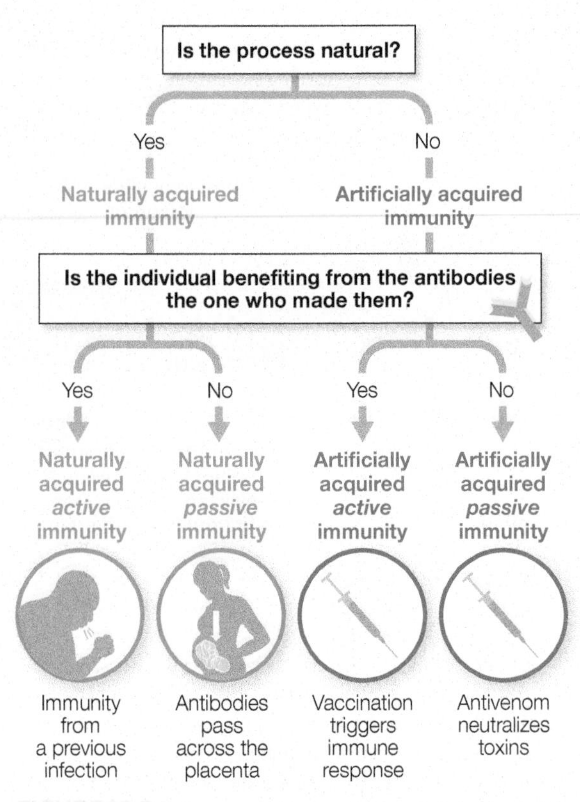

FIGURE 12.24 Four categories of humoral immunity

Natural Active Immunity

Naturally acquired active immunity involves contracting an infection that triggers the patient's immune system to make memory cells and antibodies that confer long-term protection thereafter. This form of immunity can be developed from either symptomatic or asymptomatic infections.

Artificial Active Immunity

Artificially acquired active immunity involves using vaccines to trigger an immune response in the patient. Vaccinations train the immune system to recognize specific antigens and result in the creation of the patient's own memory cells and antibodies, and thereby confers long-term protection. This is why the term "immunized" is often used interchangeably with the term "vaccinated."

Natural Passive Immunity

Naturally acquired passive immunity involves receiving antibodies through nonmedical means. The best example is maternal antibodies passing across the placenta to a baby in utero or passing into an infant through colostrum and breast milk (colostrum is an antibody- and mineral-rich fluid that precedes breast milk). In this form of immunity, the baby isn't making the antibodies to the antigens it is being protected against—the baby may not even have encountered these antigens. Because the baby isn't developing their own memory cells and antibodies to these antigens, they won't receive long-term protection from this form of immunity. Rather, naturally acquired passive immunity is a temporary way to protect babies until their own immune system is fully functional.

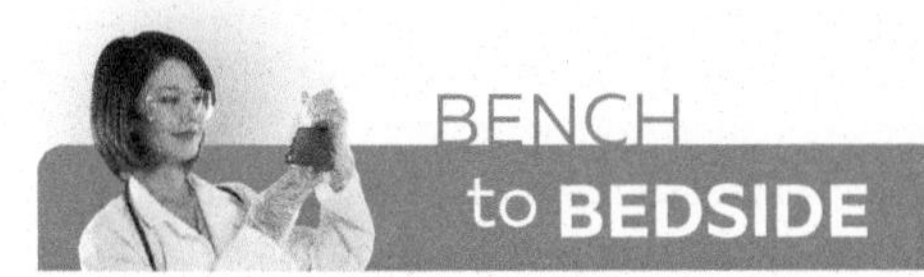

BENCH to BEDSIDE

Antibodies to Revolutionize Medicine

Imagine a drug that tags cancer cells for destruction while leaving healthy cells alone, or one with the capacity to hunt down and neutralize infectious agents in the body before they cause disease. Monoclonal antibody (or mAb) therapies, which use antibodies that target a single antigen epitope, are making some of these medical goals a reality.

In 1986, the first commercially available monoclonal antibody therapy was designed to prevent kidney transplant rejections. Since then, monoclonal antibodies have been developed to treat a huge range of medical disorders, including cancers, asthma, and rheumatoid arthritis. They're also being researched as a means to block mother-to-child HIV transmission, as well as dengue, Ebola, Zika, and SARS-CoV-2 infections.

Effective antibody treatments for cancer are highly desirable because traditional chemotherapy drugs typically cause harsh side effects. The advantage of monoclonal antibodies is their specificity: They can tag just the cancer cells for destruction by the immune system, or deliver toxic payloads to cancer cells with relatively minimal side effects. Despite their great potential, these therapies only confer a passive immunity, so effects are not long term. Earlier versions of these antibody therapies were made with mouse antibodies, which triggered adverse side effects, because people's immune systems mounted attacks on the foreign antibodies. To reduce side effects, most of the monoclonal antibodies now used in therapeutics come from genetically engineered mice or cell lines that make humanized antibodies—that is, the antibodies share a high degree of protein-sequence similarity (homology) with human antibodies.

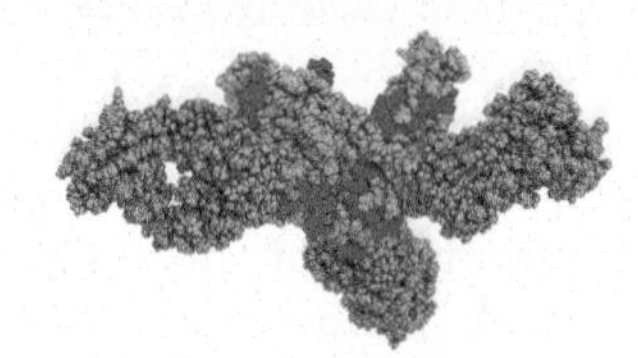

A molecular model reveals how antibodies (brown) bind to select Ebola proteins (blue). The orange and red areas denote new mutations that developed in the course of the 2014–2016 Ebola outbreak.

A monoclonal antibody therapy's generic name reveals how humanized it is. Typically, therapies with names ending in "**u**mab" (e.g., adalimumab—brand name Humira) are more humanized than therapies with names ending in "**zu**mab" (e.g., palivizumab—brand name Synagis). Therapies with names ending in "**xi**mab" (e.g., basiliximab—brand name Simulect) are usually the least humanized. Note that this naming convention categorizes monoclonal antibody therapies by their protein sequence similarity to human antibodies, *not* by treatment efficacy. As such, you should never assume that "umab" therapies are safer or more effective than "ximab" therapies—each is humanized and represents an improvement over fully mouse antibodies ("**o**mab" agents).

With an average of four or five new monoclonal antibodies being approved each year for clinical use, these protein molecules represent one of the fastest-growing areas of pharmacotherapeutics. An important limitation is that they are delivered intravenously, which makes these therapies harder to administer outside of healthcare settings.

Source: AminJafari, A., & Ghasemi, S. (2020). The possible of immunotherapy for COVID-19: A systematic review. *International Immunopharmacology*, 106455.

Artificial Passive Immunity

In **artificially acquired passive immunity** the patient receives protective antibodies as a medical treatment. The external source of the antibodies is often a horse, rabbit, or goat. **Antiserum**, a preparation of antibodies developed to neutralize specific toxins or venoms, is an example. Antisera include antivenom administered after a venomous snakebite and antitoxin that neutralizes a bacterial toxin, such as that made by *Clostridium botulinum*. Protection is nearly immediate. However, because the patient didn't make their own antibodies to the venom or toxin, they won't be protected from future exposures to the agent. There are also antibody preparations that can be administered to protect against actual pathogens, not just their toxins. An example is the monoclonal antibody palivizumab. This antibody therapy protects premature infants from respiratory syncytial virus, a common virus that is potentially deadly if it infects the baby's underdeveloped lungs.

Theoretically, a person could develop naturally acquired active immunity to toxins and venoms, but that assumes they survive long enough to do so. Certainly some people can claim this hard-earned form of immunity to a variety of rare and deadly agents. These survivors sometimes donate antibodies to treat patients or submit serum samples to researchers working to develop neutralizing antibody preparations.

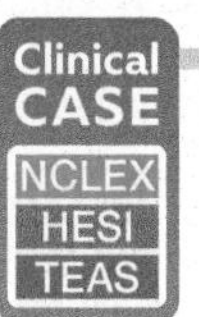

The Case of the Coughing Newborn

Practice applying what you know clinically: visit the **Mastering Microbiology** Study Area to watch Part 3 and practice for future exams.

Build Your Foundation

18. Define immunological memory, and compare it to a primary response. (LO 12.16)

19. What is an antibody titer, and how would a secondary exposure alter it? (LO 12.17)

20. Name the four categories of adaptive immunity, provide an example of each, identify those that generate long-term immunity, and explain why. (LO 12.18)

Build Your Foundation (BYF) Quick Quiz: Visit the **Mastering Microbiology** Study Area to quiz yourself.

VISUAL SUMMARY | *Innate and Adaptive Immunity*

Interactive CONTENT REVIEWS

Interactive Content Reviews: Visit the **Mastering Microbiology** Study Area to quiz yourself.

Three Lines of Immune Defense

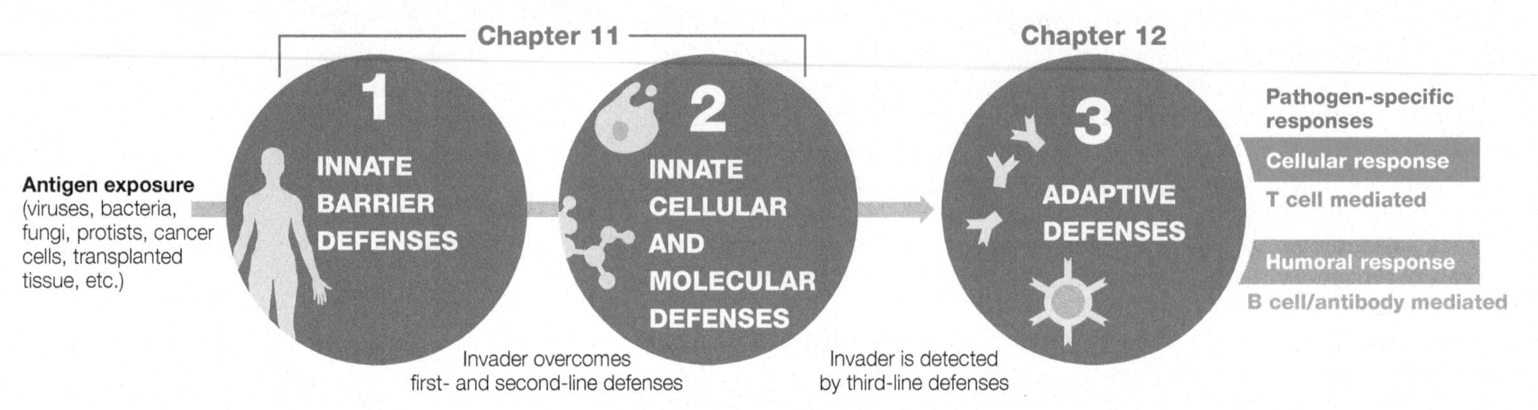

First-Line Defenses

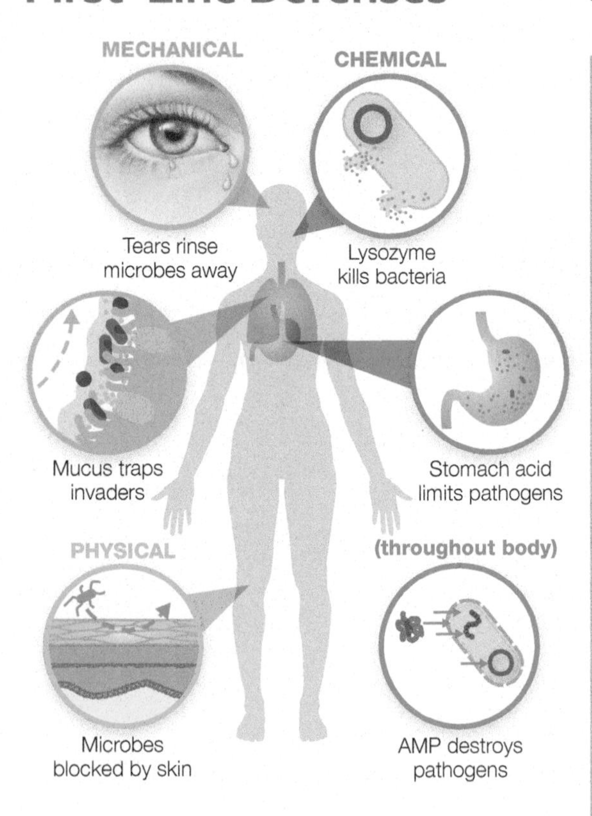

Second-Line Defenses

Molecular defenses

Cytokines: Stimulate inflammation and tissue repair; recruit leukocytes; generate fever; promote leukocyte and lymphatic tissue development; antiviral; immune system regulation

Iron-binding proteins: Limit availability of free iron to reduce bacterial growth

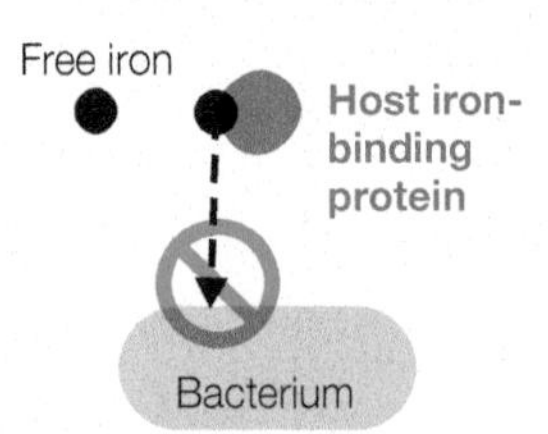

Complement proteins:

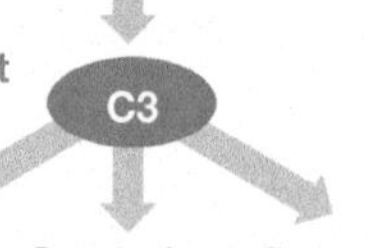

Inflammation

VASCULAR CHANGES	LEUKOCYTE RECRUITMENT	RESOLUTION
Chemical alarm signals increase blood flow and vessel permeablility.	Cytokines recruit leukocytes. Neutrophilis and macrophages phagocytize invaders.	Inflammation signals decrease; tissue repair initiated.

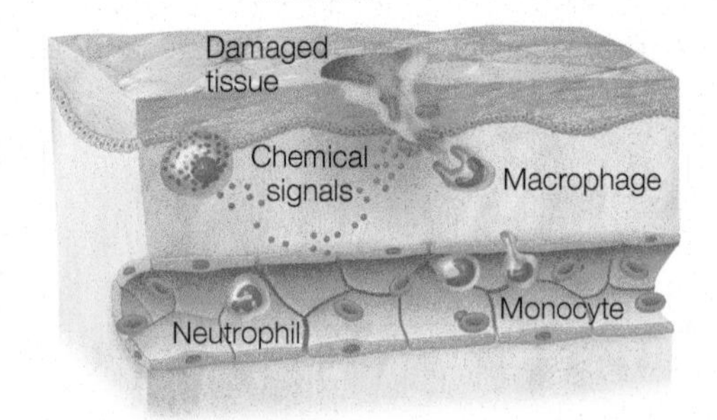

Fever

Hormone level changes, shivering, and increased metabolism raise body temperature—also, blood vessels constrict such that heat loss through the skin is reduced.

Lymphoid Tissues

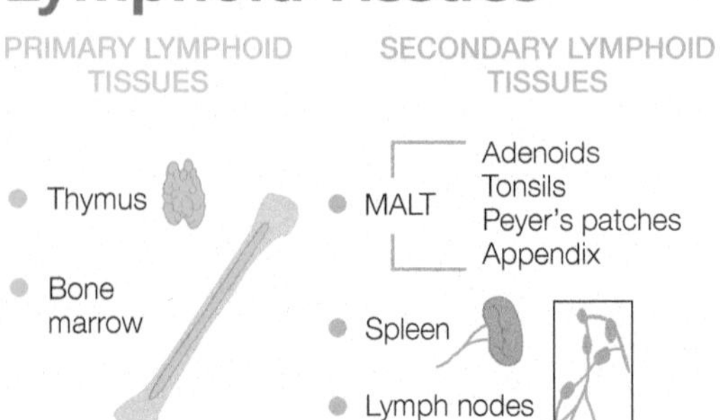

Leukocytes of Innate Immunity

Granulocytes

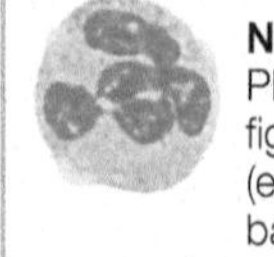

Neutrophilis: Phagocytic; fight pathogens (especially bacteria)

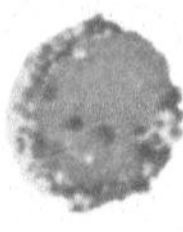

Basophils: Fight parasites; roles in allergy

Eosinophils: Fight parasites; roles in allergy

Mast cells: Fight parasites and bacteria; roles in allergy; reside in tissues

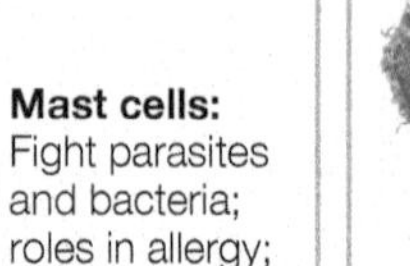

Agranulocytes

Monocytes (mature into macrophages): Phagocytes; serve as antigen-presenting cells

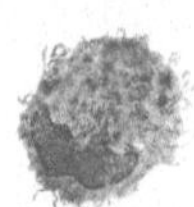

Dendritic cells: Phagocytes; serve as antigen-presenting cells

Natural killer cells: Innate immunity lymphocytes; fight pathogens and cancer cells

Third-Line Defenses (Adaptive Immunity)

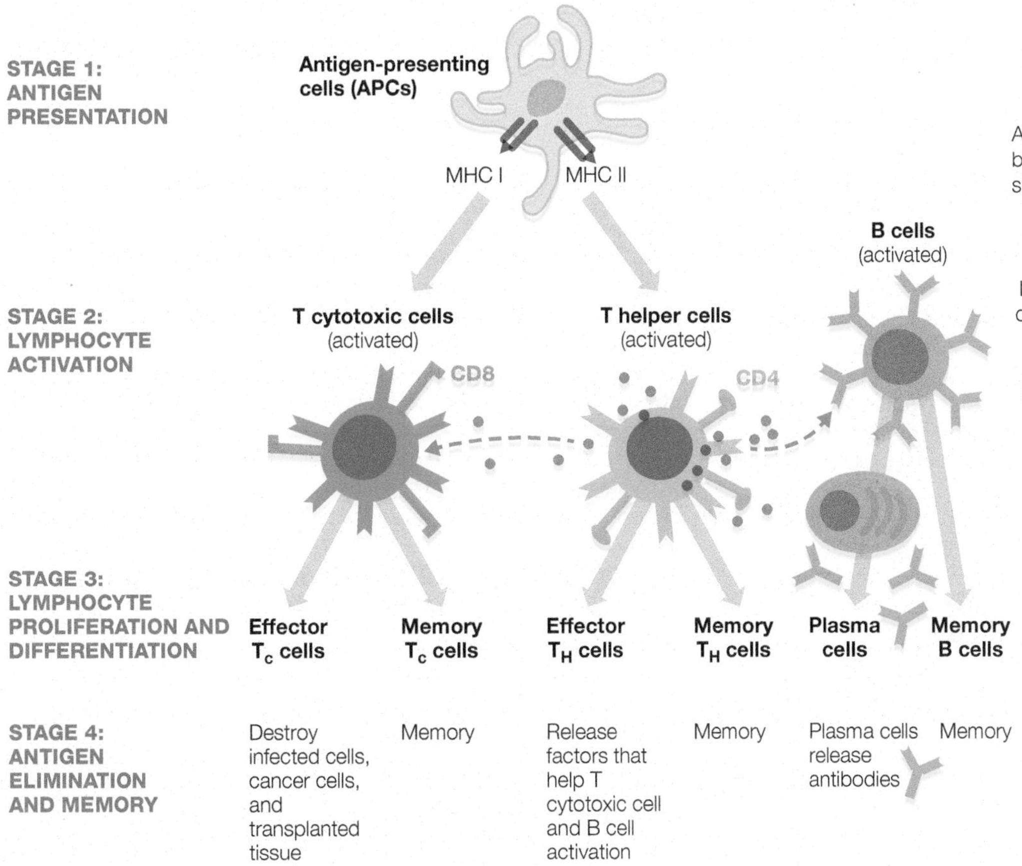

STAGE 1: ANTIGEN PRESENTATION

Antigen-presenting cells (APCs)

MHC I

MHC II

STAGE 2: LYMPHOCYTE ACTIVATION

T cytotoxic cells (activated)

CD8

T helper cells (activated)

CD4

B cells (activated)

STAGE 3: LYMPHOCYTE PROLIFERATION AND DIFFERENTIATION	**Effector T_c cells**	**Memory T_c cells**	**Effector T_H cells**	**Memory T_H cells**	**Plasma cells**	**Memory B cells**
STAGE 4: ANTIGEN ELIMINATION AND MEMORY	Destroy infected cells, cancer cells, and transplanted tissue	Memory	Release factors that help T cytotoxic cell and B cell activation	Memory	Plasma cells release antibodies	Memory

Antibodies

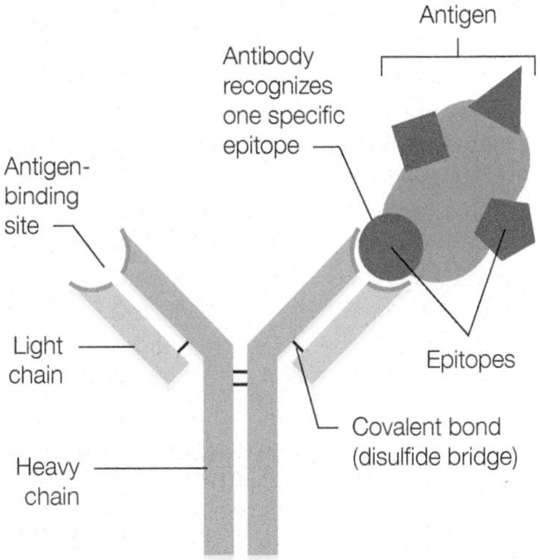

Antigen

Antibody recognizes one specific epitope

Antigen-binding site

Light chain

Heavy chain

Epitopes

Covalent bond (disulfide bridge)

Antibodies can:

- **Neutralize antigens**
  Antibodies block toxins or antigens from binding to host cells.
- **Activate complement**
  Complement activation leads to cytolysis, opsonization, and inflammation.
- **Increase phagocytosis**
  Phagocytosis is enhanced as antibodies agglutinate, precipitate, or opsonize antigens.

Five antibody isotypes:

IgG	Crosses placenta; made later in infection
IgA	In secretions
IgM	Made early in infection
IgE	Fights parasites; mediates allergic responses
IgD	Bound to B cells; poorly understood

Lymphocytes of Adaptive Immunity

	Activated cell types	Site of maturation	Antigen recognition	Require APC to become activated?	Major histo-compatibility complex (MHC)
T cells (cellular response)	**T cytotoxic cell** (CD8⁺ or T_C)	Thymus	T cell receptors (TCRs)	Yes	T cytotoxic cells: Interact with antigens presented in the context of MHC I
	T helper cell (CD4⁺ or T_H)				T helper cells: Interact with antigens presented in the context of MHC II
B cells (humoral response)	**Plasma cells** (activated B cells that make antibodies)	Bone marrow	B cell receptors (BCRs), which are secreted as antibodies following B cell activation	No	Directly interact with antigens

12 CHAPTER 12 OVERVIEW

12.1 Introduction to Third-line Defenses

- The adaptive immune response comprises the third and final line of our immune defenses. Adaptive immunity is activated slower than innate defenses, but it's specific and exhibits immunological memory.
- Adaptive immunity is divided into cellular and humoral branches—both progress through four key stages (antigen presentation, lymphocyte activation, lymphocyte proliferation and differentiation, and antigen elimination and memory).
- T cells and B cells are lymphocytes central to adaptive immunity. T cells are generated in the bone marrow and mature in the thymus. B cells are made in the bone marrow and mature there. T cells have roles in the humoral and cellular branches of adaptive immunity; B cells mainly coordinate the humoral response by making antibodies.
- An antigen is any substance that, if presented in the right context, may trigger an immune response; most antigens are proteins or polysaccharides. The parts of an antigen that B and T cells recognize and mount an immune response to are called epitopes. The antigen recognition receptors on T cells are called T cell receptors (TCRs); on B cells they are called B cell receptors (BCRs).
- T cytotoxic cells ($CD8^+$ cells) and T helper cells ($CD4^+$ cells) are the two main lineages (or classes) of T cells. Activated T helper cells may differentiate into a variety of subclasses that have specific functions. Activated T cytotoxic cells directly destroy cells that are virus infected, damaged, foreign/transplanted, or cancerous.
- T and B cells are screened for self-tolerance. Those that would attack normal self-tissue are eliminated. Major histocompatibility proteins (MHCs) have roles in screening T cells for self-tolerance and in antigen presentation.

12.2 Cellular Branch of Adaptive Immunity

- The cellular branch of adaptive immunity is mainly organized by T helper cells and T cytotoxic cells that have been activated by antigen-presenting cells (APCs). Dendritic cells, macrophages, and B cells all can act as APCs.
- Major histocompatibility complex (MHC) proteins are important for showing antigens to T cells. MHC I (found on all nucleated cells) presents intracellular antigens to T cytotoxic cells; MHC II (only on APCs) presents extracellular antigens to T helper cells. MHCs have a central role in helping the immune system differentiate self from foreign—these are therefore the main proteins that must be closely matched for a successful tissue transplant.
- T helper and T cytotoxic cells require two signals for full activation. Upon activation they undergo proliferation and differentiation to make memory cells and effector cells. Superantigens can non-specifically activate T cells.
- Activated T helper cells coordinate the cellular immune response as well as promote the humoral response (next section). Activated T cytotoxic cells "seek and destroy" cells that display the activating antigen in the grip of MHC I.

12.3 Humoral Response of Adaptive Immunity

- B cells mediate humoral responses; when activated by an encounter with a foreign antigen (either T-dependent or T-independent antigens), they become plasma cells that secrete antibodies. As with T cell activation, B cells also require multiple signals to become fully activated.
- Activated B cells (plasma cells) make diverse antibodies. Based on the constant region of an antibody, there are five different antibody isotypes (IgG, IgA, IgM, IgE, and IgD)—each isotype has distinct structural and functional properties (see Table 12.3). Antibodies activate complement cascades, neutralize antigens, and promote phagocytosis of targeted antigens.

12.4 A Deeper Exploration of Humoral Immunity

- Memory B and T cells are long-lived and reside in lymphoid tissues to serve as historians for immunological memory. They rapidly reactivate the cellular and humoral responses if the same antigen is encountered again.
- A primary antigen exposure is hallmarked first by the production of IgM antibodies followed by IgG. A secondary response generates a rapid surge in IgG antibodies and only a minor production of IgM antibodies. The total antibody amount made in a secondary response is greater (has a higher titer) than in a primary response, and the generated antibodies bind targeted antigens more effectively (exhibit a higher affinity).
- Humoral immunity can be naturally acquired active, artificially acquired active, naturally acquired passive, and artificially acquired passive. Naturally acquired active immunity and artificially acquired active immunity are characterized by long-term immunological protection. In contrast, naturally acquired passive immunity and artificially acquired passive immunity only confer temporary protection.

THINK CLINICALLY | *Be S.M.A.R.T. About Cases*

COMPREHENSIVE CASE

The following case integrates basic principles from the chapter. Try to answer the case questions on your own. Don't forget to be **S.M.A.R.T.*** about your case study to help you interlink the scientific concepts you have just learned and apply your new understanding of adaptive immunity principles to a case study.

*The five-step method is shown in detail in the Chapter 1 Comprehensive Case on cholera. See pages 31–33. Refer back to this example to help you apply a SMART approach to other critical thinking questions and cases throughout the book.

• • •

Hassan, a middle schooler, complained of feeling tired and uncomfortable because of pain in his legs. His dad gave him Tylenol and encouraged him to go to bed earlier—he assumed Hassan was probably just suffering from growing pains, overexertion at soccer practice, or adjustment to a heavy course load at school. A few weeks later Hassan started to run a fever (oral temperature of 101°F, or 38.3°C) and complained that he felt short of breath (was experiencing dyspnea). Because it was late on a Friday night, his dad took him to the local urgent care center. Hassan was treated for a suspected bacterial infection. Further tests were not performed, and the doctor advised Hassan's dad that should symptoms worsen, or if new symptoms developed, then he should immediately follow up with Hassan's pediatrician.

Hassan's fever resolved within a few days of starting the prescribed antibiotics, and he returned to school. Over the next few weeks, however, his leg pain worsened, and he could barely muster the energy to go to school, let alone soccer practice. One morning Hassan called his dad to come and get him from school because he had spiked another fever and the pain in his legs was causing him to limp. Hassan's dad got him into the pediatrician's office for an appointment that same day.

Because Hassan reported bone pain and had an enlarged spleen (splenomegaly), the pediatrician ordered several new tests, including a complete blood count (CBC), a peripheral blood smear, tests to profile the coagulation features of Hassan's blood, and various serum chemistry tests. The test results suggested that Hassan could be suffering from acute lymphoblastic leukemia (ALL), the most common cancer in children.

The pediatric oncologist that Hassan was referred to explained that patients with ALL usually have an increased number of lymphocytes in their circulating blood, but that they are abnormal and can't perform their normal immune functions. When these lymphocytes divide uncontrollably in the bone marrow, they can cause pain as they crowd out healthy cells and normal tissue in the bones. Patients tend to develop a low red blood cell count, which causes anemia that leads to the persistent fatigue that Hassan was experiencing. More tests led Hassan's medical team to diagnose a type of ALL that causes immature B cells (pre-B cells) to divide uncontrollably.

Hassan's cancer did not respond well to chemotherapy. His oncologist recommended eliminating Hassan's immune system cells and reestablishing them through a bone marrow stem cell transplant. Fortunately, Hassan's brother was a compatible donor. This was a stroke of luck, because there's only about a 25 percent chance that a sibling will be a compatible tissue donor. Fortunately, a year out from the transplant, Hassan had not experienced significant or progressive GVHD—a scenario in which the white blood cells that develop from the donated bone marrow attack the tissues of the new body they inhabit because they are seen as "foreign."

After years of therapy and monitoring, Hassan was eventually declared cured. Based on his antibody titers, he had to get vaccinated with a number of routine childhood vaccines that he had already completed well before his cancer onset. After his revaccination, his titer levels were sufficiently high to be considered protective. This was another good indicator that Hassan had re-established a normal immune system that would improve his quality of life.

• • •

CASE-BASED QUESTIONS

1. Predict some of the basic abnormalities that would have been detected upon microscopic evaluation of Hassan's blood before treatment. Explain how these observed features relate to his symptoms.
2. Despite an increase in B lymphocytes in his blood *before* his treatment, why would Hassan's antibody titers to illnesses that he had been previously vaccinated against be so low as to necessitate revaccination?
3. What category of adaptive immunity would Hassan develop upon revaccination? Explain your reasoning.
4. Assuming Hassan did have a bacterial infection when he went to the urgent care center, what would his likely antibody isotypes and general titer profiles possibly have looked like compared with those of a healthy patient?
5. Based on your chapter readings, explain what the phrase "compatible tissue donor" means, and discuss why this reduces the risk for GVHD.
6. Explain what type of immunological response would develop if an incompatible bone marrow tissue were used for a transplant; you should also describe how such a response would progress.
7. Hassan's dad asks you why siblings aren't guaranteed to be good tissue matches. As a member of Hassan's healthcare team, what would you say?

END OF CHAPTER QUESTIONS

NCLEX
HESI
TEAS

Think Critically and Clinically
Questions highlighted in **orange** are opportunities to practice NCLEX, HESI, and TEAS-style questions.

1. Indicate the true statements *and* correct the false statements so they are true.
   a. B cells are activated by antigen-presenting cells.
   b. T cytotoxic cells are activated by antigens bound to MHC I.
   c. Upon activation, T helper cells stimulate T cytotoxic cells and B cells.
   d. IgG is the first antibody made during a primary response.
   e. T-dependent antigens rely on T_H cells to activate B cells.

2. In which of the following scenarios would administering immunoglobulins be useful? Select all that apply.
   a. To neutralize a toxin
   b. To aid a patient who is immune compromised
   c. After venom exposure
   d. To block IgM from crossing the placenta
   e. To protect a premature infant from respiratory syncytial virus

3. Which of the following most directly cause tissue graft rejection?
   a. Antigen-presenting cells
   b. B cells
   c. T helper cells
   d. T cytotoxic cells
   e. Plasma cells

4. Which of the following would you anticipate seeing in a patient who suffers from allergies? Select all that apply.
   a. Decreased IgG titers
   b. Increased histamine release upon allergen exposure
   c. Decreased capacity for naturally acquired active immunity
   d. Increased IgM titers
   e. Increased IgE titers

5. Complete the table to indicate which MHCs are present for each cell.

	MHC I	MHC II
B cell		
Macrophage		
Dendritic cell		
Red blood cell		
T_H cell		
T_C cell		

6. Why are packed red blood cells not tissue typed? Select all that apply.
   a. Red blood cells lack MHC II.
   b. Red blood cells lack MHC I.
   c. Red blood cells can't stimulate an immune response.
   d. Red blood cells aren't transferred to others.
   e. Red blood cells don't make antibodies and therefore do not need to be typed.

7. Which of the following does not generate long-term immunological memory? Select all that apply.
   a. Antivenom
   b. Antitoxins
   c. Vaccinations
   d. Breast-feeding
   e. Antigens

8. T helper cells activate B cells to become ______________, which make ______________.

9. Compare and contrast T-dependent and T-independent antigens.

10. Choose the false statement about T cytotoxic cells.
    a. They stimulate B cells.
    b. They destroy virus-infected cells.
    c. They destroy cancer cells.
    d. They are activated by MHC I bound to antigens on APCs.
    e. They mediate the cellular branch of adaptive immunity.

11. Where do T cells undergo self-tolerance selection?

12. Which of the following is not a function of antibodies?
    a. Opsonization
    b. Activating complement proteins
    c. Activating T helper cells
    d. Enhancing phagocytosis
    e. Antigen neutralization

13. The distinct feature of an antigen that stimulates an adaptive immune response is called a(n) ______________.

14. Match the cell to its stated feature. Some features will be assigned more than once.

Plasma cells	Have CD4
T helper cells	Have CD8
T cytotoxic cells	Activate T cells
APCs	Make antibodies
B cells	Make memory cells upon activation
	Activated by MHC II bound to antigen
	Activated by MHC I bound to antigen
	Present antigens to T cells
	Lymphocytes

15. What is opsonization, and which antibodies have opsonizing activity?

16. Why is a second signal useful in T cell activation?

17. Select all the false statements about artificially acquired immunity.
    a. It can be passive.
    b. It can be active.
    c. It may be generated by vaccines.
    d. It is a form of autoimmunity.
    e. It may generate memory cells.
    f. An example includes the transfer of antibodies across the placenta.

18. List the antibody isotype(s) that exhibit the stated feature. Some features will be assigned to more than one antibody.

    Feature:
    - Most abundant antibody in serum
    - Made as a dimer
    - Stimulates allergic responses
    - Does not cross the placenta
    - Considered a complement activator
    - Rare antibody that's poorly characterized
    - Main antibody in breast milk and mucus
    - Dominates the secondary immune response
    - Made early in the course of infection
    - Made in a primary immune response

19. Label the indicated parts of the antibody.

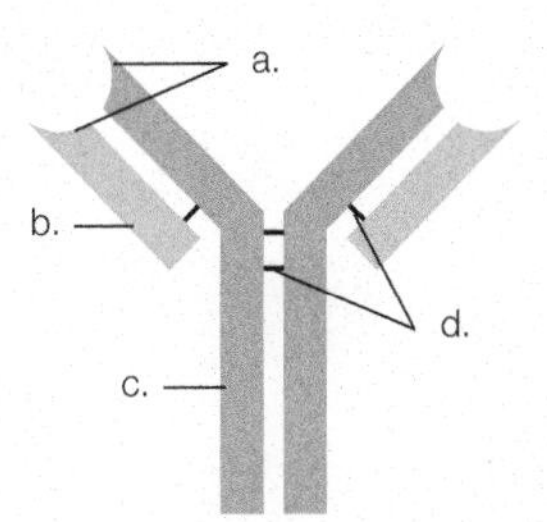

20. Match the T helper cell subset to its function. Some choices may not be used, and some may be used more than once.

T_{reg}	Activates T cytotoxic cells
T_H1	Activates B cells
T_H2	Decreases the immune response upon threat removal
	Mainly supports the humoral response
	Activates natural killer cells
	Have MHC I
	Serve as APCs

CRITICAL THINKING QUESTIONS

1. Cancer is more common in people with a weak immune system, including the very young, the elderly, or people taking drugs that suppress immunity. Explain why this is the case.

2. A patient has a herpes virus infection. Herpes viruses are infamous for their ability to go dormant inside host cells. Describe why antibodies alone would be insufficient to clear such an infection.

3. A child's parents want to delay vaccination until their child is no longer breast-feeding. They are convinced that delaying the vaccination schedule until at least 3 months after the child stops nursing is safe. Develop a professional and compassionate response to the parent to address their misconception.

4. At time point A, the patient's IgG titer against antigen X is 1:6; the patient's IgM titer to the same antigen is zero. Later, at time point B, the same patient encounters antigen X. The IgG titer to X is zero, and the IgM titer to X is 1:8. Describe a scenario that could explain these data.

13 Immune System Disorders

What Will We Explore? In Chapters 11 and 12 we explored the immune system's potential to respond to essentially any antigen, natural or manmade. The immune system's might is easily appreciated when we consider that an immune reaction gone astray, such as certain severe allergies, can kill a patient in a matter of minutes. Here we explore the scenarios where the immune system goes awry—from allergy and various autoimmune disorders to immunosuppression.

Why Is It Important? The immune system is essential to our survival, but people who suffer from autoimmune disorders and allergies are victims of their own most powerful defense system. The American Autoimmune Related Diseases Association estimates that at least 50 million Americans suffer from more than 100 known autoimmune disorders—an epidemic that in the United States alone costs about $100 billion dollars per year to treat. The impact of allergies is every bit as devastating. The Asthma and Allergy Foundation of America states that allergies (which include the most common forms of asthma) are the fifth-leading cause of chronic disease in the United States; allergies are now estimated to affect at least one in every five Americans. Add to this that all cancers are technically a failure of the immune system to clear abnormal cells, and it's easy to appreciate why a survey of immune disorders is an important part of your training.

Clinical CASE
NCLEX
HESI
TEAS

The Case of the Terrible Turkey Surprise

Visit the **Mastering Microbiology** Study Area to watch the case and find out how immune system disorders can explain this medical mystery.

Jennifer de la Cruz
MMSc, PA-C, Director of Clinical Education; Decatur, GA

13.1 IMMUNE DEFICIENCIES AND AUTOIMMUNITY

Learning Outcomes

After reading this section, you should be able to:

13.1 Compare and contrast primary and secondary immunodeficiencies.

13.2 Explain why cancer can be considered a failure of the immune system.

13.3 Describe and provide examples of autoimmune disorders.

13.4 Discuss the general approaches to diagnosing and managing autoimmune disorders.

Genetic defects may lead to primary immunodeficiencies.

The risk of developing certain infections, as well as the safety of receiving certain vaccines and treatments, hinges on the patient's immune system status. **Immunodeficiency** is the lack of a properly functioning immune system. The immune system is so complex (see Chapters 11 and 12) that it's not surprising that there are many ways for it to become deficient or react inappropriately.

A **primary immunodeficiency** (congenital immunodeficiency) is an inborn error that affects one or more immune system factors and leads to deficient immunity. These genetic immunodeficiencies span a collection of more than 300 disorders that have a broad spectrum of effects.[1] Some of these disorders are manageable and survivable, whereas others have limited treatment options and result in a decreased life span. Fortunately, they are relatively rare.

Primary immunodeficiencies can develop from defects in humoral immunity, cellular immunity, or innate immunity (FIG. 13.1). According to the *Merck Manual*, about 50 percent of primary immunodeficiencies are due to B cell issues, about 30 percent are linked to T cell defects, roughly 18 percent are associated with errors in phagocytic cells, and 2 percent relate to complement cascade deficiencies. An example of a primary immunodeficiency is DiGeorge syndrome (also called 22q11.2 deletion syndrome). This deletion in part of chromosome 22 causes diverse developmental defects, including impaired thymus development. Because T cells mature in the thymus, a thymic deficiency greatly reduces the patient's ability to mount cellular immune responses.

In some primary immunodeficiencies, there may be a family history of susceptibility to infections. Other times, the patient's immunodeficiency results from a spontaneous mutation that was not inherited.

Patients with primary immunodeficiencies tend to experience recurring, persistent, and severe infections, often caused by uncommon agents—especially opportunistic pathogens. You can think of "SPUR" (severe, persistent, uncommon, recurring) to help recall the hallmark characteristics of infections in patients with primary immunodeficiencies. Therapies for primary immunodeficiencies include bone marrow transplants, intravenous or subcutaneous antibody administration, cytokine therapies, and experimental treatments like stem cell transplants, thymus transplantation, and gene therapy.

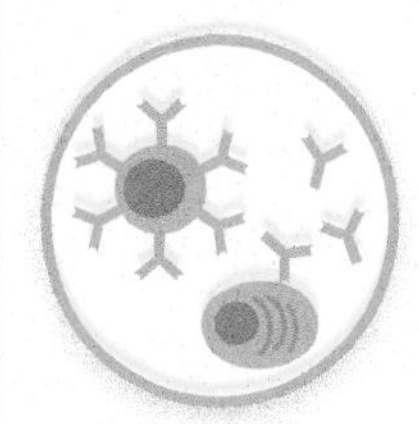

Humoral immunodeficiencies
B cell and/or antibody deficiencies or defects

- Selective IgA or IgG deficiencies
- X-linked agammaglobulinemia
- Immunodeficiency with hyper-IgM

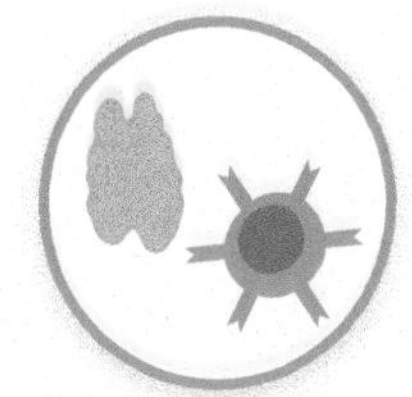

Cellular immunodeficiencies
Underdeveloped or absent thymus; low T cell numbers; defective T cell responses

- DiGeorge syndrome
- Congenital thymic dysplasia
- Ataxia-telangiectasia

Combined humoral and cellular immunodeficiencies

- Severe combined immunodeficiency (SCID)
- Wiskott-Aldrich syndrome
- Common variable immunodeficiency

Innate immunodeficiencies
Phagocyte defects and/or deficiencies

- Chronic granulomatous disease
- Cyclic neutropenia
- Chédiak-Higashi syndrome

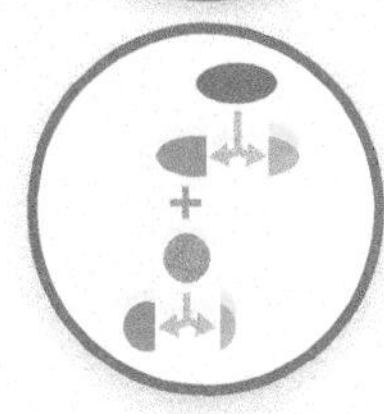

Complement protein deficiencies or defects

- C2 deficiency
- C4 deficiency
- C9 deficiency

FIGURE 13.1 Examples of primary immunodeficiencies Primary immunodeficiencies are genetic disorders that limit immune function. Innate and/or adaptive immunity factors can be affected.

Critical Thinking *A defect in an innate immune system factor is unlikely to only impact innate immunity. Explain why.*

Aging, chronic disease, and various external factors can cause secondary immune deficiencies.

Secondary immunodeficiencies (acquired immunodeficiencies) are much more common than primary immunodeficiencies. A person with this type of immunodeficiency doesn't have an inborn error in the immune system, but starts out with a normal immune system and then experiences a decline in immune system rigor. For example, as we age our immune function declines because the number of lymphocytes we make gradually decreases over the years, as does the effectiveness of other immune system factors. Secondary immunodeficiencies also can result from certain infections, medical interventions, or systemic disorders like diabetes, malnutrition, alcoholism, or hepatitis. When caring for an elderly patient or someone who is chronically ill, it is common to assume that the person has reduced immune function.

The exact immune system factor or factors that a secondary immunodeficiency impacts depends upon the underlying issue. For example, someone with

[1] Kitcharoensakkul, M., & Cooper, M. A. (2020). Autoimmunity in primary immunodeficiency disorders. In N. R. Rose & I. R. Mackay (Eds.), *The autoimmune diseases* (6th ed., pp. 513–532). London: Academic Press.

chronic kidney disease cannot filter blood properly, and this leads to the loss of certain blood proteins, including antibodies. People suffering from liver disease may experience a reduced production of complement proteins. And sometimes disease or physical trauma requires removal of the spleen (splenectomy), an important lymphatic organ. Any of these scenarios can increase a patient's susceptibility to infections.

Secondary Immune Deficiencies Caused by Prescribed Drugs

Aside from natural aging, medical interventions are the leading cause of secondary immunodeficiencies. Cancer treatments such as radiation and chemotherapy, steroid anti-inflammatory drugs (corticosteroids), and antiseizure medications (anticonvulsants) are all medical interventions that suppress immunity, thereby increasing the risk of infections or cancers. Usually drug-induced immunosuppression is reversed when the patient stops taking the drug. However, it isn't always possible for a patient to stop taking medication. For instance, an organ transplant recipient will likely always stay on immunosuppressive drugs; the trade-off—dealing with decreased immunity—is better than rejecting a lifesaving transplanted organ. Examples of immunosuppressive drugs are presented in TABLE 13.1.

Secondary Immune Deficiencies Caused by Infectious Agents

Many pathogens have virulence factors that directly inhibit host immune defenses. (Virulence factors were discussed in Chapter 10.) Some microbes break down antibodies; others interfere with cellular signaling that is central to coordinating immune defenses. Certain infectious agents may directly infect immune system cells. This is the case with the notorious **human immunodeficiency virus (HIV)**, which infects T helper cells, the generals of the immune system that coordinate adaptive immunity actions. As HIV causes T helper cells to decline, immune responses also become deficient. Eventually, HIV patients lose enough T helper cells to have their condition be classified as acquired immunodeficiency syndrome (AIDS). A patient may be HIV positive, but may not have advanced to having AIDS (Chapter 21 covers HIV/AIDS in more detail).

HIV is not the only virus that infects immune system cells. Human T cell lymphotropic viruses and Epstein–Barr virus can infect lymphocytes and lead to blood cancers like leukemia and lymphoma, which in turn decrease immune function. The measles virus also causes reduced B cell and T cell function that can endure for months after infection. During this period of suppressed immunity, measles patients are at increased risk for other infectious diseases, including tuberculosis and various parasitic infections. Measles is much less common than it was prior to the 1960s, when vaccines dramatically reduced its incidence in industrialized nations. However, as more people in developed nations decide to forgo vaccinating their children, resurgences have been increasingly common—so much so that in 2019 the U.S. Centers for Disease Control and Prevention warned that if the patterns continue, the United States may lose its measles elimination status in the world. Unfortunately, in developing countries routine vaccines are not always accessible, and measles remains a significant cause of childhood mortality. According to the World Health Organization, measles killed more than 140,000 children in 2018—that's about 16 deaths per hour. (For more on measles, see Chapter 17.)

Immune system deficiencies may lead to cancer.

In addition to protecting us against infections, our immune system defends against cancer. In fact, cancer can be considered a failure of the immune system to clear abnormal cells. Ultimately any defect or deficiency in our immune system can lead to an increased cancer risk, but in particular, T cytotoxic cells are important for destroying cancer cells (FIG. 13.2). This point is made abundantly

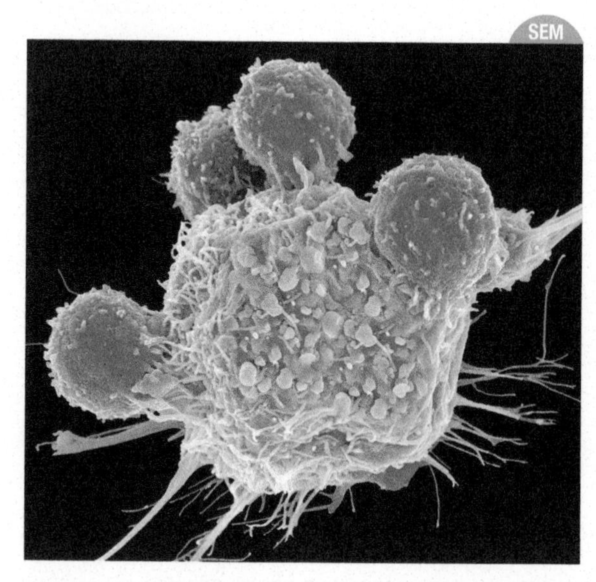

FIGURE 13.2 T cytotoxic cells and cancer This colorized SEM shows T cytotoxic cells (red) attacking a cancer cell.

TABLE 13.1 Examples of Immunosuppressive Drugs

Drug Class	Treats	Examples
Anticonvulsants	Seizures	Carbamazepine, valproate, phenytoin
Corticosteroids	Inflammation disorders, including autoimmune disorders	Prednisone, methylprednisolone, hydrocortisone
Immunosuppressants	Limits transplant rejection risk and/or autoimmune disorders, or chronic inflammation disorders	Azathioprine, cyclosporine, sirolimus, adalimumab, daclizumab

clear when we consider AIDS patients. As HIV kills off their T cell populations, they are at increased risk for rare cancers like Kaposi sarcoma and various blood cancers. Similarly, transplant patients, because they take immunosuppressant antirejection drugs, are almost three times more likely to develop cancer than nontransplant patients.[2] They are also about 35 percent more likely to die from their cancer compared with immune-competent, nontransplant patients.[3]

Because the immune system is central to combating cancers, a number of anticancer treatments, called **immunotherapies**, are geared at boosting immune defenses. These therapies can be *prophylactic*, meaning they try to prevent cancer, or *therapeutic*, meaning they treat existing cancer. One well-known prophylactic immunotherapy is Gardasil, the vaccine that prevents infection by strains of human papilloma virus (HPV) that most commonly cause cervical cancer. Monoclonal antibody treatments can also fight certain cancers; examples include ipilimumab and nivolumab, used in cases of inoperable, metastatic melanoma. Other immunotherapies for cancer include interferon and interleukin treatments. Investigational T cell therapies also aim to train T cells to identify and destroy cancer cells that would otherwise remain undetected by the immune system.

Lack of self-tolerance leads to autoimmune disorders.

Let's now shift gears to **autoimmunity**, a specific immune system attack against healthy self-tissues that should normally be left alone. **Autoimmune disorders** are chronic conditions that develop from these damaging self-tissue attacks. Although it's still not fully understood why or how autoimmune disorders develop, it is known that autoimmune disorders all involve B and/or T cells that lack self-tolerance. (Refer to Chapter 12 to review self-tolerance.) Autoimmune disorders are generally described as inflicting damage via one or more hypersensitivity mechanisms—responses that we'll review later.

More than 100 different autoimmune disorders have been described (FIG. 13.3). Among them are celiac disease, lupus, type I diabetes mellitus, rheumatoid arthritis, multiple sclerosis, and Graves' disease. Autoimmune disorders can affect any part of the body. **Systemic** autoimmune disorders like lupus affect diverse tissues throughout the body. **Localized** autoimmune disorders like rheumatoid arthritis tend to target a specific type of tissue. The American Autoimmune Related Diseases Association estimates that at any given moment, about 50 million Americans are living with an autoimmune disorder. For unknown reasons, women make up 80 percent of those afflicted.

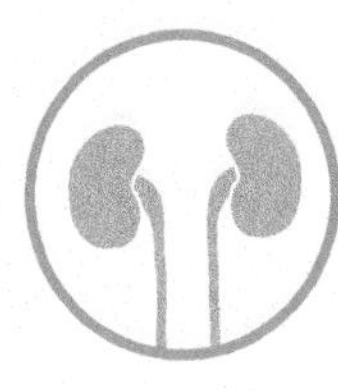

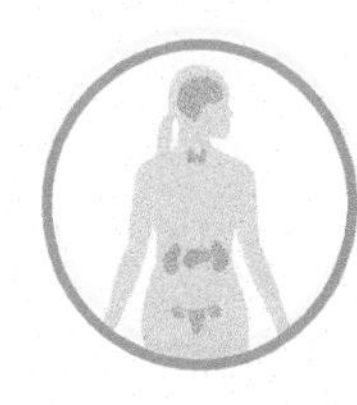

FIGURE 13.3 **Examples of autoimmune disorders** These autoimmune disorders are grouped by the primary tissue affected, but in many cases, there are effects in different tissues.

[2] Huo, Z., Ge, F., Wang, R., et al. (2020). Cancer risk in solid organ transplant recipients: A systematic review and meta-analysis. *Transplantation*, 104(S3), S51–S53.

[3] Benoni, H., Eloranta, S., Ekbom, A., Wilczek, H., & Smedby, K. E. (2020). Survival among solid organ transplant recipients diagnosed with cancer compared to nontransplanted cancer patients—A nationwide study. *International Journal of Cancer*, 146(3), 682–691.

TABLE 13.2 Some Autoimmune Disorders with Possible Infectious Agent Association

Disease	Features	Implicated Infectious Agents
Type 1 diabetes	Immune system attacks insulin-producing cells of the pancreas	Coxsackievirus B (cause of infectious myocarditis or inflammation of the heart muscle) is one possibly linked agent.
Guillain–Barré syndrome	Peripheral nerves are attacked, and muscle weakness develops	*Campylobacter jejuni* (leading cause of bacterial diarrhea) may be involved.
Rheumatic heart disease	Heart inflammation and scarring; antibiotics make this disorder very rare in developed countries	*Streptococcus pyogenes* (cause of streptococcal pharyngitis, "strep throat") may have a role.
Multiple sclerosis (MS)	Loss of insulating sheath (myelin sheath) on nerves leading to delayed nerve impulse transmission and pain	Over 20 different viral agents have been proposed as linked to MS; examples include human herpesvirus 6 and Epstein–Barr virus.

Etiology of Autoimmune Disorders

As previously stated, researchers still aren't sure why autoimmune disorders occur, but exposure to certain infectious agents is known to influence their development. For example, it is hypothesized that some of the complications that develop in severely ill COVID-19 patients may result from SARS-CoV-2 triggering autoimmune events.[4,5] Several theories attempt to explain the link between autoimmunity and various pathogens. One theory proposes that if a pathogen has antigens that resemble host factors, it could trigger the production of antibodies that cross-react with host tissues. Another theory proposes that certain pathogens may release superantigens that inappropriately activate certain T cells against self-factors. Still another theory suggests that the cytopathic (cell-damaging) effects generated by a pathogen could encourage host antigen-presenting cells to process and present self-antigens to T cells.

So far, the exact role of infectious agents in autoimmunity is still being investigated, but we do know that a previous infection with such an agent is not sufficient in and of itself to cause a disorder. This suggests that there is some other underlying factor involved. Genetic factors are the most likely predisposing elements, but the exact genes that put one at risk for an autoimmune response remain elusive. TABLE 13.2 summarizes examples of autoimmune disorders and some of the infectious agents suspected—but not confirmed—to contribute to their development.

Diagnosing Autoimmune Disorders

Joint and muscle pain, fatigue, rash, organ dysfunction, and low-grade fever are some common, general signs and symptoms of autoimmune disorders. Of course, presentation differs based on what tissues are affected—for instance, celiac disease damages the lining of the small intestine, so it tends to generate abdominal pain, intestinal cramping, vomiting, and/or diarrhea. In contrast, someone with lupus-related kidney damage might experience edema (swelling).

It can be challenging for clinicians to identify the autoimmune disorder causing the patient's condition because of the large number of options, variability of symptoms, and the fact that signs and symptoms can take a long time to fully develop and may change over time. Diagnosis often involves detecting self-reactive immune system cells and/or **autoantibodies**, which are antibodies that bind to self-tissues. Depending on the suspected disorder, the clinician may also order other hematological tests (blood tests) that look for signs of inflammation and certain metabolic factors. An autoimmune disorder is rarely definitively diagnosed using a single test; instead, a collection of tests and careful assessment of signs and symptoms are required. Some autoimmune disorders have scoring criteria to help clinicians make a diagnosis. As a generic example, if the patient met five out of a dozen criteria for a particular disorder, then they might have an 85 percent likelihood of being affected by it, based on collective historical data from numerous patients with the same diagnosis. Typically, the more criteria a patient meets in such a scoring system, the greater the chance that the diagnosis is accurate.

Basics of Managing Autoimmunity

Managing autoimmunity usually involves suppressing the immune response and reducing inflammation that damages tissues. Thus, treatments for autoimmunity usually lead to secondary immunodeficiency, including the host of complications just discussed. So far, there are no cures for autoimmune

[4] Ehrenfeld, M., Tincani, A., Andreoli, L., et al. (2020). Covid-19 and autoimmunity. *Autoimmunity Reviews, 19*(8), 102597.

[5] Franke, C., Ferse, C., Kreye, J., et al. (2020). High frequency of cerebrospinal fluid autoantibodies in COVID-19 patients with neurological symptoms. *Brain, Behavior, and Immunity, 93*, 415–419.

disorders, and no way to prevent them. Sometimes a patient will benefit from replacement therapies—a common example is the administration of insulin in people with type 1 diabetes to replace what their body doesn't make.

In some cases, a hematological test called the **erythrocyte sedimentation rate (ESR)** may be used in combination with other measures to determine whether or not the prescribed therapy is effectively reducing inflammation. ESR reflects how long it takes for red blood cells to separate out from plasma, the straw-colored fluid of our blood. It is reported as the volume of plasma in millimeters present at the top of the tube after 1 hour (mm/hr). The ESR tends to increase when the patient is experiencing inflammation; however, this is a nonspecific test, and ESR may also increase during pregnancy or as a result of infection, anemia, or aging.

Build Your Foundation

1. Compare and contrast primary and secondary immunodeficiencies. (LO 13.1)
2. Why can cancer be considered a failure of the immune system? (LO 13.2)
3. What is an autoimmune disorder, and what is an example of one? (LO 13.3)
4. What are some general means for diagnosing autoimmune disorders, and what is the basic goal of therapy for most of these disorders? (LO 13.4)

Build Your Foundation (BYF) Quick Quiz: Visit the **Mastering Microbiology** Study Area to quiz yourself.

13.2 INTRODUCTION TO HYPERSENSITIVITIES

Learning Outcomes

After reading this section, you should be able to:

13.5 Define the term *hypersensitivity*.

13.6 Describe the hygiene hypothesis and its relevance to hypersensitivities.

13.7 Name the four types of hypersensitivity reactions.

13.8 Name some features and examples of each type of hypersensitivity reaction.

Hypersensitivities are defined by an inappropriate secondary immune response.

Recall (from Chapter 12) that secondary immune responses involve the coordinated action of B and T cells. In most cases, these adaptive, secondary immune responses are appropriate and helpful. In some cases, however, the response can be exaggerated and even autoimmune in nature. Inappropriate secondary immune responses, which include allergy and autoimmune disorders, are classified as **hypersensitivities**. Hypersensitivity reactions can be localized and therefore restricted to a given tissue. Or they may be systemic, affecting multiple tissues and organ systems.

To be clear, all the hypersensitivities we discuss in this section develop because of an inappropriate *secondary* immune response mediated by antibodies or T cells. Recall, however, that secondary immune responses commonly recruit primary immune-response factors such as granulocytes to attack flagged targets (e.g., antigens coated with antibodies) and to amplify pathways like inflammation (e.g., antibody-initiated complement cascades can fuel inflammation). Ultimately all of these hypersensitivities are specific in their etiology and therefore should not be confused with generalized inflammation disorders or other *nonspecific* primary immune responses—responses that in their own right can go awry but are not technically classified as hypersensitivities.

The hygiene hypothesis suggests a link between our microbiome and hypersensitivities.

There is evidence that genetics and environmental factors have roles in hypersensitivities, but it still isn't fully understood what puts a person at risk for developing them. Studying the global incidence of these immune-based pathologies has revealed that autoimmunity and allergy are both more common in residents of developed nations than in residents of developing countries. The **hygiene hypothesis** suggests this differential risk may relate to our microbiota profiles. It proposes that in developed countries, ultra-clean water and food combined with common antibiotic usage and decreased infection incidence has decreased the biodiversity of people's normal microbiota. This in turn may affect how our immune system develops and therefore increase our risk for developing autoimmune disorders and allergies.

Supporting the hygiene hypothesis is increasing evidence that our normal gut microbiota profiles impact our T cell populations, which are key mediators of our secondary immune responses (the responses that go awry in hypersensitivities).[6] For example, upsetting the balance of T regulatory cell, T helper 1 cell, and T helper 2 cell populations has been linked to the development of hypersensitivities in mice.[7] Although disruptions to normal gut microbiota have been associated with various hypersensitivities, it's still not clear if microbial population shifts are the cause or an effect of these pathologies.[8] The details of what constitutes a healthy microbiome versus one that puts us at risk for various diseases, including immunopathologies like allergy and autoimmunity, are still poorly understood, but as our knowledge expands, the microbiome may become a new target for managing hypersensitivities.

There are four classes of hypersensitivities.

The **Gell and Coombs classification system** is helpful for organizing the main types of hypersensitivity reactions that a patient may experience. This system, which is not the only one that exists but tends to be the easiest for introductory students to follow, describes four classes of hypersensitivity reactions—types I, II, III, and IV. Antibodies are key participants in the first three classes of hypersensitivities, whereas T cells mediate type IV reactions. On the other hand, type I hypersensitivities, which are allergies, are the only one of the four classes that is not associated with autoimmunity.

As a clarification, autoimmune disorders with well-characterized pathophysiology mechanisms can be described as type II, III, or IV hypersensitivities—and in some cases, more than one of these tissue-damaging mechanisms is involved. However, because we do not have a complete knowledge of all autoimmune disorders and how they cause tissue damage, we cannot definitively state that every autoimmune disorder has pathology mechanisms that fall within the Gell and Coombs classification system—the possibility of exceptions exists, even if we do not yet know of them. TABLE 13.3 introduces you to the four classes of hypersensitivities that we will review in the remaining parts of this chapter. Notice that, in most cases, hypersensitivities cannot be cured. Instead, the goal of treatment is symptom management.

TABLE 13.3 Four Classes of Hypersensitivity Reactions

Class	Type I	Type II	Type III	Type IV
Description*	Allergies	Cytotoxic (often, but not exclusively so)	Immune complex	Delayed hypersensitivity
Driven by	IgE interacting with soluble antigens	IgG or IgM interacting with nonsoluble antigens on cell surfaces or extracellular antigens (i.e., connective tissue proteins)	IgG or IgM interacting with soluble antigens	T cells interacting with soluble or cell- or matrix-bound antigens
Drugs can trigger the response	Yes	Yes	Yes	Yes
Associated with autoimmunity?	No	Yes	Yes	Yes
Selected examples	Localized or systemic allergies	Hemolytic disease of the newborn; blood transfusion reactions; autoimmune disorders such as Goodpasture syndrome, Graves' disease, and rheumatic heart disease	Serum sickness; autoimmune disorders like lupus, rheumatoid arthritis, and poststreptococcal glomerulonephritis	Nickel, latex, or poison ivy reactions; tuberculin skin test reaction, chronic graft rejection, and certain autoimmune disorders such as multiple sclerosis and Hashimoto thyroiditis

**Use the mnemonic "ACID" to remember these.*

Note: Although autoimmune disorders may be assigned to one category of hypersensitivity as their main mode of development, many autoimmune disorders are perpetuated through more than one type of hypersensitivity reaction.

[6] Di Gangi, A., Di Cicco, M. E., Comberiati, P., & Peroni, D. G. (2020). Go with your gut: The shaping of T-cell response by gut microbiota in allergic asthma. *Frontiers in Immunology*, *11*, 1485.

[7] Round, J. L., & Mazmanian, S. K. (2009). The gut microbiota shapes intestinal immune responses during health and disease. *Nature Reviews Immunology*, 9(5), 313–323.

[8] Aguilera, A. C., Dagher, I. A., & Kloepfer, K. M. (2020). Role of the microbiome in allergic disease development. *Current Allergy and Asthma Reports*, 20(9), 1–10.

Build Your Foundation

5. What does the term *hypersensitivity* mean? (LO 13.5)
6. How might the hygiene hypothesis connect to the development of immune hypersensitivities? (LO 13.6)
7. Describe the following as type I, II, III, or IV reactions: (1) Delayed hypersensitivity reactions, (2) allergy, (3) immune complex, and (4) cytotoxic reactions. (LO 13.7)
8. What are some features and examples of each class of hypersensitivity reaction? (LO 13.8)

Build Your Foundation (BYF) Quick Quiz: Visit the **Mastering Microbiology** Study Area to quiz yourself.

13.3 TYPE I HYPERSENSITIVITIES

Allergy and certain forms of asthma are also called type I hypersensitivities.

Learning Outcomes

After reading this section, you should be able to:

13.9 Describe type I hypersensitivities, and provide examples.

13.10 Define the term *allergen*, discuss the mechanism by which allergens trigger an allergic response, and identify the factors that can influence the signs, symptoms, and severity of an allergic response.

13.11 Define the term *anaphylaxis*, name some key signs and symptoms of systemic anaphylaxis, and state the types of allergens most commonly associated with it.

13.12 Explain how allergies can be diagnosed.

13.13 Differentiate seasonal allergies from atopic asthma, and summarize the various strategies for managing each.

13.14 Discuss desensitization immunotherapy, and state what types of allergies it is mainly used to treat.

An **allergen** is any antigen that triggers IgE production and leads to **allergy**, a scenario in which the immune system fights off a perceived threat that would otherwise be harmless. Type I hypersensitivity reactions include all allergies, **atopic asthma** (allergy-based asthma), and **atopic dermatitis**—an inflamed and itchy skin condition also known as *atopic eczema* or *allergy-based eczema* (FIG. 13.4). These are the most common hypersensitivities that healthcare providers assess and treat.

As you learned previously (see Chapter 12), the body must encounter the antigen (or something that mimics the antigen) before it can make antibodies to it. The allergen exposure that triggers the immune system to produce IgE is called the **sensitizing exposure** (FIG. 13.5, *top*). Sensitization can occur on the very first exposure to the antigen, or after years of many uneventful exposures. There is no way to predict if or when someone will develop an allergy to a given substance. However, one important risk factor is a family history of allergies, atopic asthma, or atopic eczema.

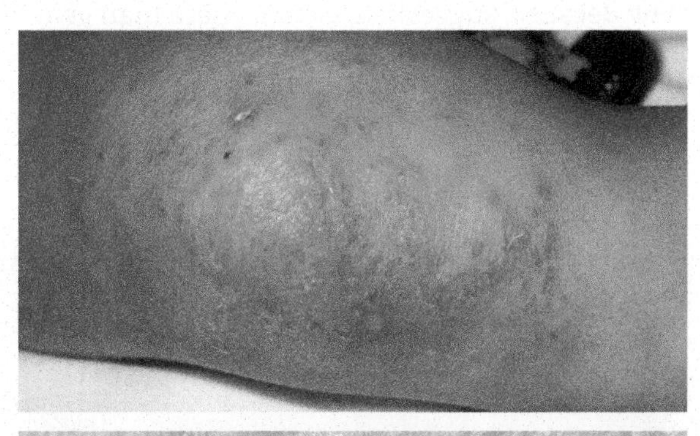

FIGURE 13.4 **Atopic dermatitis (eczema)** A cycle of inflammation, itching, and scratching generates reddened scaly skin lesions. In most people it is a chronic condition that has intense flare-ups interspersed with periods during which symptoms may be less noticeable.

How Allergies Develop

Bring the art to life! Visit the **Mastering Microbiology** Study Area to watch the Concept Coach and master how allergies develop.

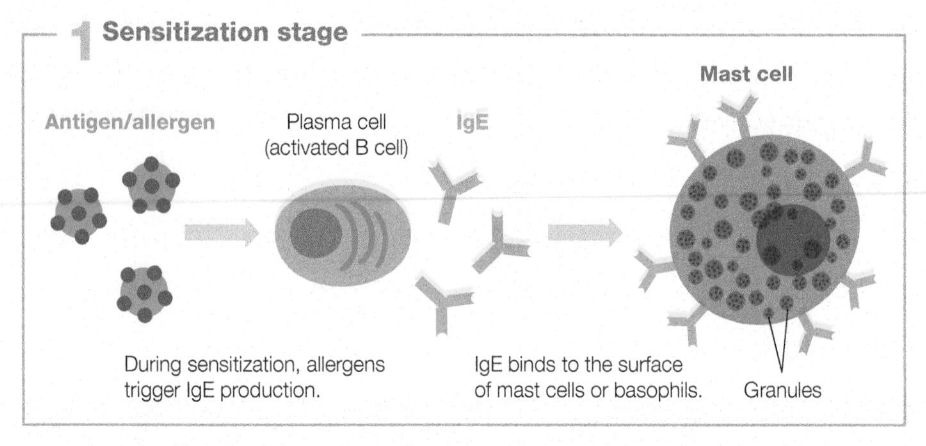

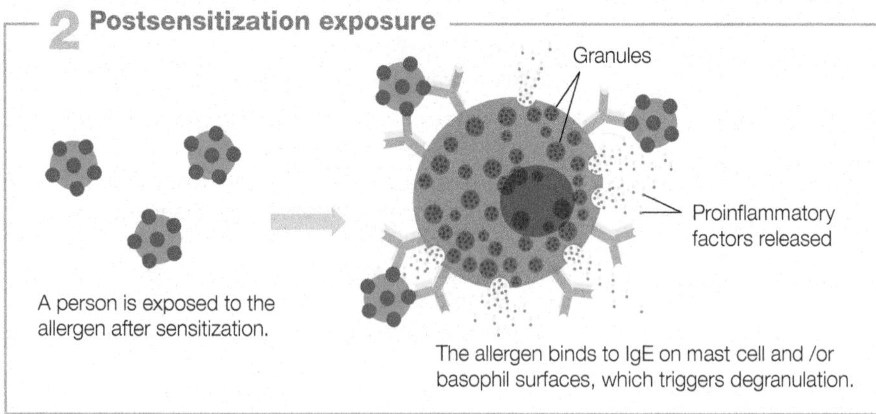

FIGURE 13.5 **Type I hypersensitivity**

Critical Thinking *Numerous websites recommend that parents feed a single-ingredient food to their baby "for a few days before introducing the next new food, to check for possible allergic reactions." Is this sound advice? Explain your reasoning.*

Following sensitization, any exposure to that allergen can trigger symptoms. In a postsensitization exposure, the allergen binds to the numerous previously made IgE antibodies that are anchored to the plasma membrane of mast cells and basophils. This IgE–allergen interaction causes the antibody-coated mast cells and basophils (also called IgE-primed mast cells or basophils) to release proinflammatory factors, such as histamine and leukotrienes, from their cytoplasmic granules. This process is called **degranulation**. As proinflammatory factors are released, they generate symptoms (Figure 13.5, *bottom*).

Route of exposure can impact allergy symptoms. Inhaled allergens tend to cause respiratory issues such as swollen airways and coughing. Ingested allergens such as foods and drugs may cause digestive system distress, skin manifestations like hives, and respiratory distress. TABLE 13.4 summarizes various type I hypersensitivities and the corresponding signs and symptoms. Not only does the route of exposure influence symptom manifestation, but also typically the more IgE a person makes against an allergen, the more severe the symptoms. For this reason, a healthcare provider will often order blood tests to assess a patient's IgE level (or titer) to a particular allergen as a way to predict allergy severity.

In most developed countries, the number of patients affected by type I hypersensitivities has greatly increased in the relatively short time frame of just a few decades, suggesting there is more than genetics at play. The food allergy epidemic seems to have its roots in the 1990s—data show that between 1997 and 2011, incidence of food allergies in the United States

TABLE 13.4 Type I Hypersensitivities: Signs and Symptoms

Type I Categories	Signs/Symptoms
Food and drug allergies	• Hives/rash/skin swelling • Diarrhea/stomach pain • Odd taste or feeling in mouth • Tingling or itchy throat • Itchy, watery, red eyes • Sneezing • Cough/congestion
Atopic asthma	• Cough • Shortness of breath (dyspnea) • Chest tightness • Wheezing
Seasonal allergies	• Cold-like symptoms (sneezing, cough, congestion, runny nose, postnasal drainage/drip) • Itchy, watery, red eyes • Itchy sinuses • Itchy or sore throat • Ear congestion
Atopic dermatitis	• Persistent, dry, itchy, scaly skin rash

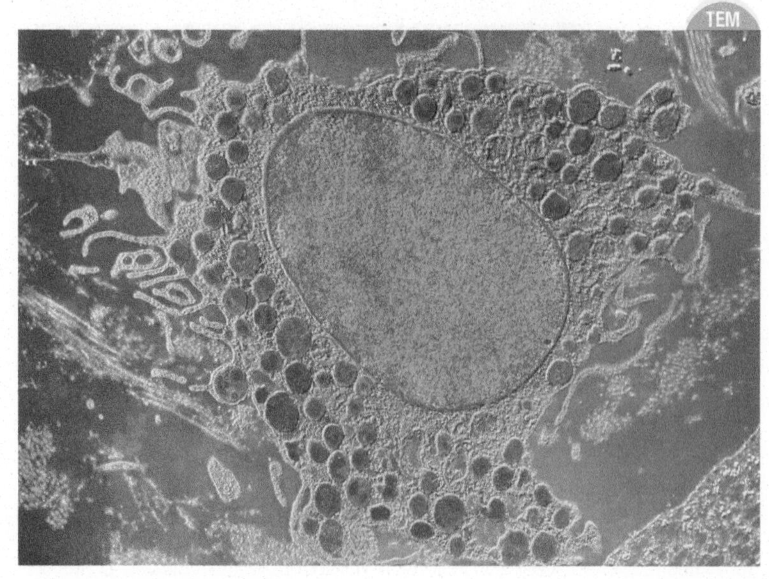

A mast cell packed with granules.

Central Nervous System

10–15% of Reactions
- Uneasiness/anxiety
- Dizziness
- Fainting (syncope)
- Confusion
- Intense headache

Skin

80–90% of Reactions
- Hives (urticaria)
- Itching (pruritus)
- Flushing
- Swelling (angioedema)

Cardiovascular System

10–45% of Reactions
- Chest pain
- Decreased blood pressure
- Rapid heart rate (tachycardia)
- Weak pulse

Respiratory System

70% of Reactions
- Swollen throat
- Itchy throat
- Hoarseness (dysphonia)
- Shortness of breath (dyspnea)
- Wheezing (bronchospasm)

Gastrointestinal System

30–45% of Reactions
- Nausea
- Vomiting
- Cramping/abdominal pain
- Diarrhea

FIGURE 13.6 **Recognizing systemic anaphylaxis**

Critical Thinking *Can an anaphylactic reaction occur upon a primary exposure? Explain your answer.*

increased by 50 percent.[9] Worldwide, roughly 1 in every 13 children and 1 in every 10 adults has a food allergy.[10] Interestingly, just eight foods—peanuts, tree nuts, milk, eggs, fish, shellfish, soy, and wheat—account for the majority of food allergies. Moreover, according to the American Academy of Allergy, Asthma & Immunology, worldwide up to 30 percent of people suffer from *allergic rhinitis,* an allergy to inhaled agents leading to nasal inflammation and sinus pressure.

Localized and Systemic Anaphylaxis

Anaphylaxis is a term used to describe an allergic response to an antigen. Anaphylactic reactions fall into two categories: localized and systemic. **Localized anaphylaxis** tends to feature isolated symptoms such as the watery eyes, sneezing, and runny nose seen with seasonal allergies, or a confined rash due to a contact allergy. Localized reactions are not nearly as dangerous as a system-wide response called **systemic anaphylaxis,** which if not treated, can progress to anaphylactic shock. This is a life-threatening medical emergency. Systemic anaphylaxis may manifest with a variety of symptoms, which are summarized for you in FIG. 13.6. Any allergen can trigger this dangerous response, but usually it is caused by ingested or injected allergens such as foods, drugs, and insect venoms. Treatment usually involves administration of epinephrine using an autoinjector (such as an EpiPen) followed by emergency medical care.

Diagnosing Allergies

Allergy diagnosis can be challenging because the symptoms vary and can mimic those of other disorders. In most cases, the patient should be diagnosed by a clinical specialist called an allergist–immunologist. Adding to the challenge of diagnosis are common misconceptions that only children develop allergies, and that something previously tolerated cannot later trigger an allergy. Allergy diagnosis is typically based on signs and symptoms along with blood and/or skin tests.

The Case of the Terrible Turkey Surprise

Practice applying what you know clinically: visit the **Mastering Microbiology** Study Area to watch Part 2 and practice for future exams.

[9] Jackson, K. D., Howie, L. D., & Akinbami, L. J. (2013). Trends in allergic conditions among children: United States, 1997–2011. *NCHS Data Brief, 121,* 1–8.

[10] Messina, M., & Venter, C. (2020). Recent surveys on food allergy prevalence. *Nutrition Today,* 55(1), 22–29.

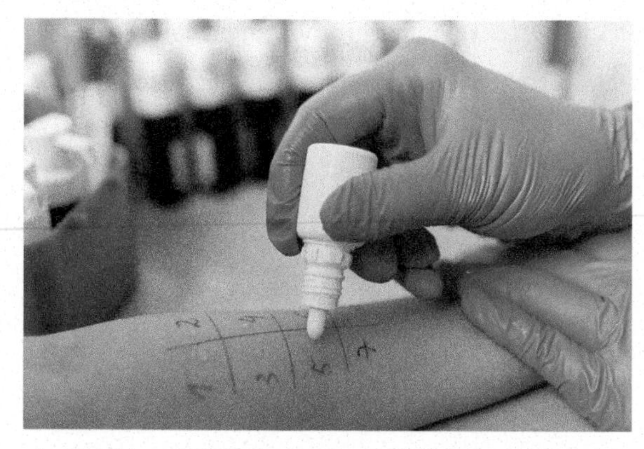

1 Solutions of different suspected allergens are dropped on the skin and then pricked with a sterile needle so they may enter the skin.

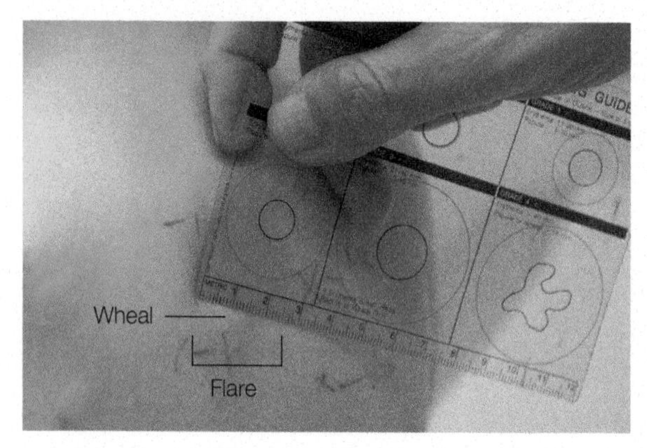

2 The extent of the wheal (raised lesion) and flare (reddened area) is measured for each tested allergen.

FIGURE 13.7 Skin prick test for allergies

Critical Thinking *Could the skin prick test detect an allergy to a food that the patient never previously ate?*

Signs and Symptoms Again, common signs and symptoms of various allergies are identified in Table 13.4, and the symptoms of systemic anaphylaxis are shown in Figure 13.6.

Sometimes patients are encouraged to eliminate single foods from the diet and monitor how they feel. Although this strategy can help determine the source of food allergy, it can be challenging when so many processed foods contain multiple ingredients and traces of potential allergens. Allergens may also share antigenic features. This means that a sensitizing exposure to one allergen may generate a cross-reactivity to other related allergens. For example, a person may only ever have been exposed to pecans, but the generated IgE antibodies may cross-react with antigens in walnuts, almonds, and cashews, rendering the patient allergic to all of those foods *even if they have never been directly exposed to them*. For this reason, a person who reacted against one tree nut would be advised to avoid all tree nuts as a precaution.

It should also be clarified that an "intolerance," such as lactose intolerance, is *not* the same as a food allergy. Intolerances do not involve the immune system. Lactose intolerance, for example, is due to inadequate secretion of the enzyme lactase by the small intestine. Also, the onset of symptoms in intolerances tends to be delayed (sometimes taking up to 48 hours to manifest), whereas allergic reactions occur rapidly, with symptoms usually manifesting within a few minutes of an allergen exposure.

In most cases it's impossible to fully remove inhaled allergens from one's environment, so pinpointing the cause by elimination can prove difficult. That said, using air purifiers, implementing more frequent household cleaning (with an emphasis on dusting), and removing pets from environments are strategies often suggested to see if symptoms diminish as exposures decrease. Regarding potential contact allergens such as soaps and laundry detergents, patients are typically encouraged to implement stepwise product eliminations to see if signs and symptoms decrease as a result.

Blood and Skin Tests An important part of allergy care is determining the specific allergen that is generating the patient's symptoms. Blood tests look for IgE levels, or titers, against a variety of allergens. They can provide numerical (quantitative) data that can help the clinician monitor the escalation or remission of an allergy. Reports generally rank the patient's IgE titers to help make interpretation easier.

Skin tests are used to screen for allergies against numerous airborne, food, and contact allergens. They are inexpensive and fast, but can cause itchy, unsightly lesions. Skin tests include scratch/prick tests, intradermal tests, and patch tests, each of which involves exposing the patient's skin to the allergen, waiting a few minutes, then assessing for a skin lesion at the exposure site. If the patient has an allergy to the tested allergen, then the skin develops a **wheal and flare lesion**. The wheal is the raised, inflamed area of the lesion, and the flare is the flattened, reddened area (FIG. 13.7). Typically, the extent of the lesion is measured for each allergen tested.

Allergy Management

The mainstay approach to allergy management is to avoid the allergen. This sounds simple enough in theory, but in practice it can be quite challenging—environmental allergens such as pollen and dust mites are practically unavoidable.

Seasonal allergies (allergic rhinitis) are the most common example of environmental allergies. Patients usually suffer from a collection of one or more of the following upper respiratory signs and symptoms that resemble a common cold: watery or itchy eyes, sneezing, sinus congestion, runny nose, and sore throat (often due to postnasal drip). Seasonal allergies are usually managed with over-the-counter medications, such as antihistamines and decongestants. Some antihistamines, like diphenhydramine in Benadryl, work by reversing the effects of histamine on target cells—making them useful even as postallergen exposure treatments.[11] Others, such as loratadine in Claritin and fexofenadine in Allegra, are H1 receptor antagonists—drugs that block histamines from interacting with histamine receptors (H1 receptors) on target cells. Oral decongestants such as pseudoephedrine (e.g., Sudafed) can help alleviate congestion as can glucocorticoid nasal sprays (steroid anti-inflammatory sprays). In addition, prescription antileukotriene drugs that block the production or actions of proinflammatory factors called leukotrienes can reduce airway inflammation. For example, zafirlukast (Accolate) and montelukast (Singulair) interfere with leukotriene action, whereas zileuton (Zyflo) inhibits leukotriene production. In severe seasonal allergies, injected or oral steroid drugs and inhaled bronchodilators (e.g., albuterol and levalbuterol)—drugs that quickly open the airways in deeper regions of the lungs (bronchioles)—may be useful. Because injected and oral steroid drugs can lead to immunosuppression, they are a last resort for seasonal allergy management.

Unlike seasonal allergies in which symptom severity usually mirrors the seasonal release of different environmental allergens (e.g., different plants release pollen in different seasons), atopic asthma (allergic asthma) is a chronic and more severe condition. Most common in children, it is characterized by bouts of coughing, wheezing, chest tightness, and shortness of breath (dyspnea). One of the biggest problems in atopic asthma is the hypersecretion of extremely viscous mucus that can form mucous plugs that block the airways. These plugs combined with bronchoconstriction can be deadly. Therefore, to keep the airways open, inhaled bronchodilators and inhaled steroids are cornerstone therapies for atopic asthma. Antileukotriene drugs can occasionally help manage atopic asthma, but antihistamines are largely ineffective. Atopic asthma that isn't effectively managed with the aforementioned therapies may also be treated with an injected antibody preparation (e.g., omalizumab—trade name Xolair) that ties up the patient's IgE antibodies. If IgE antibodies are blocked from interacting with allergens, then they can't trigger mast cells and basophils to release proinflammatory factors like histamines and leukotrienes.

Atopic dermatitis (eczema) is usually treated topically with ointments that may contain anti-inflammatory steroids. If the skin becomes infected, antibiotics are often required.

Other types of allergies, like food allergies and insect bite/sting allergies, are managed post-exposure using antihistamines like Benadryl, rather than with ongoing drug therapies. Again, if an individual starts to show signs of systemic anaphylaxis, then epinephrine should be administered, and emergency medical care must be sought.

Desensitization and Allergies

Perhaps you or someone you know had an allergy as a child and "outgrew" it? What *really* occurred was a process called **allergen desensitization**. This occurs when the body shifts its immune responses to the allergen. Patients with severe allergies may undergo desensitization immunotherapy in hope of fostering desensitization (FIG. 13.8).

Desensitization is believed to occur when the body moves from a T helper 2–dominated immune response to a T helper 1–dominated response. This shift in T cells then alters the balance of antibodies made to the allergen from

[11] Because diphenhydramine can reverse the effects of histamines, it is useful for treating patients who have sporadic exposures—as occurs in food allergies and insect bite/sting allergies.

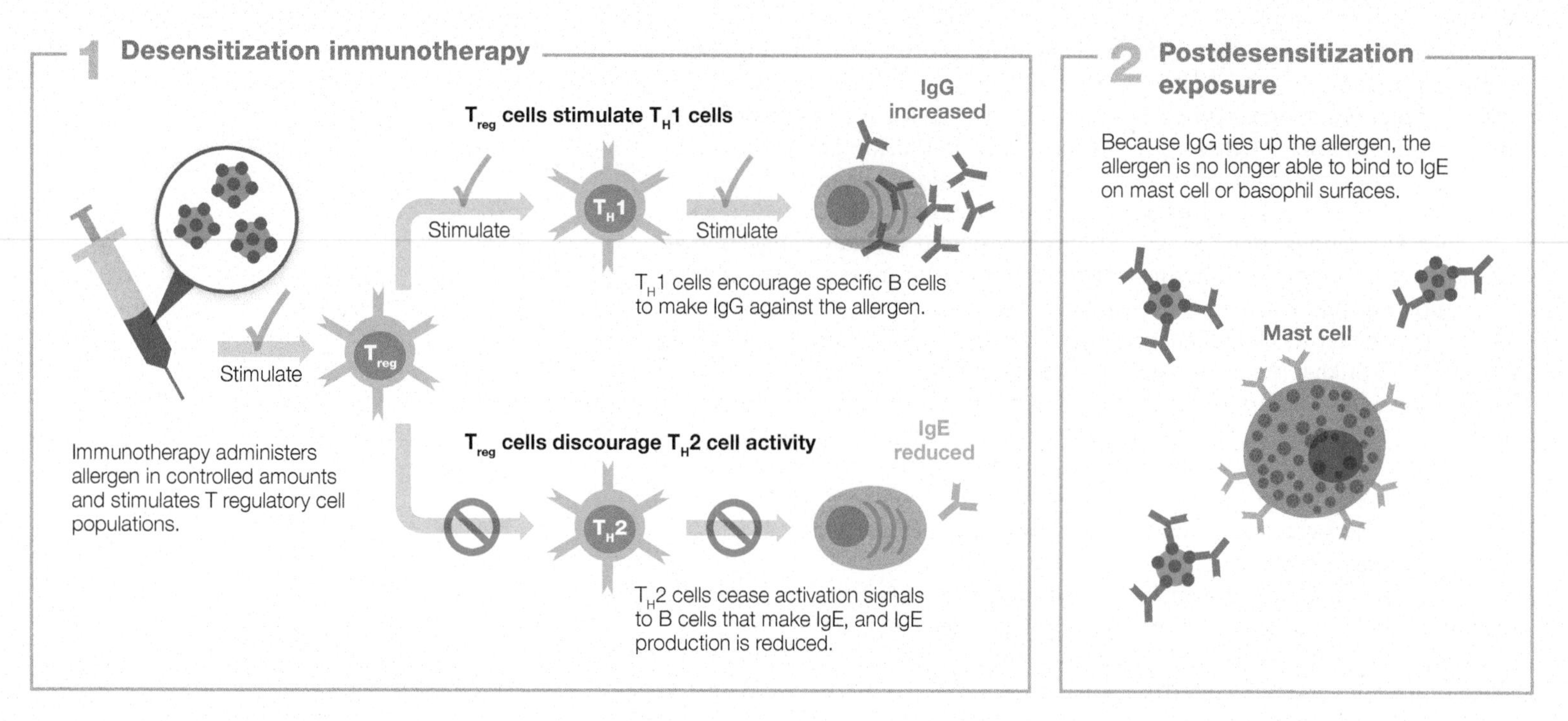

FIGURE 13.8 Desensitization immunotherapy Immunotherapy alters the patient's immune response to an allergen by stimulating regulatory T cells (T_{reg}). These T_{reg} cells promote a T helper 1 (T) cell response and discourage T helper 2 cells (T). T_H1 cells encourage IgG production so that upon a later exposure to the allergen, IgE–allergen complexes are limited, and mast cells and basophils are less likely to be activated.

Critical Thinking *What potential side effects could immunotherapy cause, and how could these effects be managed?*

IgE to IgG.[12],[13] IgE is the antibody class that activates mast cells and basophils, causing inflammation and the other hallmark symptoms of allergy. IgG is much more abundant in serum than IgE—so when the dominant antibody class made against a particular antigen shifts to IgG, it will readily outcompete IgE for allergen binding. This can limit allergy symptoms to the point that they may not even be noticeable.

As an immunotherapy, desensitization involves exposing the patient to greater and greater amounts of the allergen over time until the patient's immune response is altered. This is usually done using injections or oral administration. Desensitization immunotherapies are most effective to treat allergies to insect venoms (bees, fire ants, wasps, etc.) and environmental allergens such as dust mites, animal dander, and pollen. Unfortunately, most food allergies are only weakly responsive to traditional desensitization therapies, though some people will naturally become desensitized to these agents as they age. Data suggests that about 65–75 percent of patients with milk allergies and about 50 percent of people with egg allergies will become tolerant or "outgrow" their allergies. In contrast, people rarely "outgrow" peanut allergies. In the past, desensitization strategies for peanut allergies were not very effective—only about 20 percent of patients with peanut allergies became desensitized to the point that they did not have an anaphylactic reaction upon encountering peanut allergens. However, in 2020, desensitization strategies using peanut allergen powder gained FDA approval. The results of clinical trials are promising: about 67 percent of enrolled participants experienced a reduction in their allergic response to peanuts.[14]

[12] Akdis, M., & Akdis, C. A. (2014). Mechanisms of allergen-specific immunotherapy: Multiple suppressor factors at work in immune tolerance to allergens. *Journal of Allergy and Clinical Immunology*, 133(3), 621–631.

[13] van de Veen, W., & Akdis, M. (2020). Tolerance mechanisms of allergen immunotherapy. *Allergy*, 75(5), 1017–1018.

[14] Dougherty, J. A., Wagner, J. D., & Stanton, M. C. (2020). Peanut allergen powder-dnfp: A novel oral immunotherapy to mitigate peanut allergy. *Annals of Pharmacotherapy*, 1060028020944370.

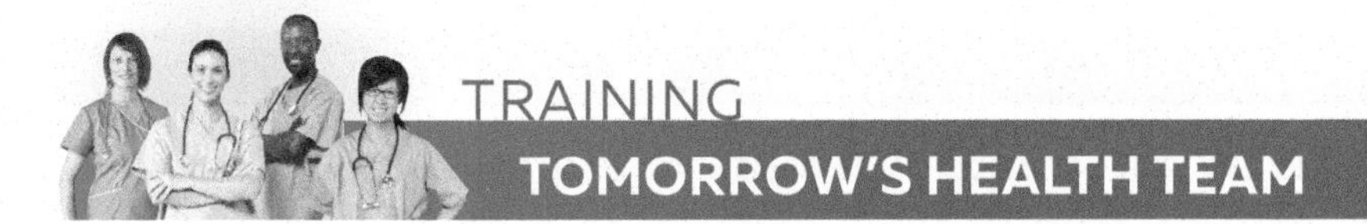

A New Approach to Reduce Peanut Allergy Severity

For millions of patients with severe peanut allergies, being on a plane where someone opens a snack pack or accidentally eating something with even just a trace of peanut butter could lead to a deadly reaction in a matter of minutes. Traditional desensitization immunotherapies have not been widely effective for people with severe peanut allergies, nor do children with this allergy tend to outgrow it. Fortunately, a new desensitization therapy called peanut allergen powder-dnfp (PAP; Palforzia) seems promising as a new approach for helping patients who suffer from severe peanut allergies.

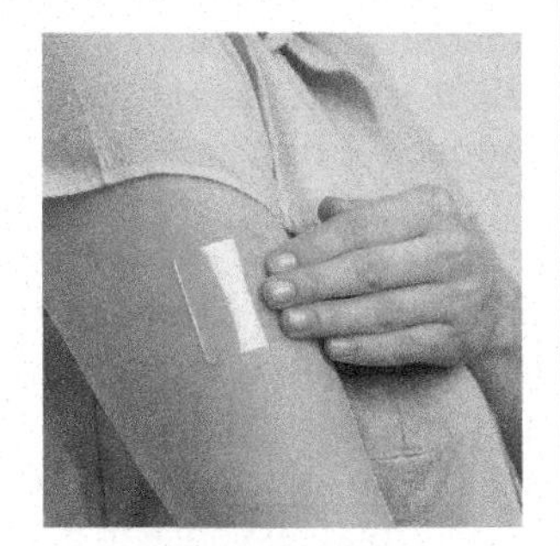

The peanut patch is a small patch that delivers immunotherapy via the skin to reduce the severity of peanut allergy reactions.

In 2020, the FDA approved the orally administered immunotherapy to treat peanut allergies in children 4 to 17 years old. The dosing regimen occurs in three stages: an initial dose escalation (the first day of exposure), up-dosing (the periodic increasing of dosages), and eventually maintenance dosing (a daily dose of Palforzia to maintain immune tolerance). The maintenance stage is reached when the patient can tolerate 300 mg of the product (equivalent of about one peanut). Because the therapy has been reported to trigger anaphylactic shock in some patients, Palforzia is only available through the FDA's Risk Evaluation and Mitigation Strategy—an FDA program for medications that require close monitoring to ensure the medical benefits outweigh the known safety risks.

Palforzia is not considered a cure for peanut allergies; instead, it reduces risks associated with severe allergy upon accidental allergen exposure. As such, patients who undergo therapy must still maintain a peanut-avoidant diet. Following the immunotherapy, about two-thirds of patients in clinical trials experienced reduced peanut allergy severity—with some patients even tolerating doses equivalent to eating two peanuts. This level of desensitization could improve the quality of life for many patients, especially children who are required to sit at "peanut-free" lunch tables at school or have had to forgo foods that are not prepared in a 100 percent peanut-free facility.

Source: Hise, K., & Rabin, R. L. (2020). Oral immunotherapy for food allergy—a US regulatory perspective. *Current Allergy and Asthma Reports*, *20*(12), 1–7.

QUESTION 13.1

Why is a low-dose exposure administered over time essential for desensitization therapies?

Build Your Foundation

9. What are type I hypersensitivities? (LO 13.9)
10. What are allergens, how do they trigger an allergic response, and what factors may impact allergic reaction severity, signs, and symptoms? (LO 13.10)
11. What is anaphylaxis, and what allergens are most likely to cause systemic anaphylaxis? (LO 13.11)
12. What are the signs and symptoms of systemic anaphylaxis? (LO 13.11)
13. How are allergies diagnosed? (LO 13.12)
14. How do atopic asthma and allergic rhinitis differ in manifestation and management? (LO 13.13)
15. How do desensitization immunotherapies work, and what allergens tend to be ideal desensitization candidates? (LO 13.14)

Build Your Foundation (BYF) Quick Quiz: Visit the **Mastering Microbiology** Study Area to quiz yourself.

13.4 TYPE II HYPERSENSITIVITIES

Learning Outcomes

After reading this section, you should be able to:

13.15 Describe how type II cytotoxic and noncytotoxic hypersensitivity reactions work.

13.16 Provide examples of type II hypersensitivities, and state if they are mediated by cytotoxic or noncytotoxic mechanisms.

13.17 Explain how the ABO and Rh antigens impact blood type, and determine if a transfusion is compatible or not.

13.18 Outline the development of hemolytic disease of the newborn, and describe how it is prevented.

Type II hypersensitivities are often characterized by cytotoxic reactions.

As with type I reactions, antibodies mediate **type II hypersensitivities.** Type II reactions involve IgG or IgM binding to *nonsoluble* antigens on the surface of a cell or within the extracellular environment, such as extracellular collagen in connective tissues.

There are three mechanisms by which type II reactions cause harm. One route to damage relies on complement cascade activation to lyse cells and recruit phagocytes to destroy target cells. Another involves antibodies directly recruiting leukocytes (especially natural killer cells) to lyse tagged extracellular substances and/or cells (FIG. 13.9). Both of these mechanisms tend to trigger cell lysis, so they are known as **cytotoxic reactions** (or cytolytic reactions). Cytotoxic reactions can be a normal response to a foreign antigen, as occurs in blood transfusion reactions and in hemolytic disease of the newborn, both of which are discussed shortly. Or they can be autoimmune based, as occurs in *Goodpasture syndrome* (connective tissues of the kidney and lungs are attacked), *autoimmune hemolytic anemia* (red blood cells are attacked when bound to drugs like cephalosporins and penicillins), and *rheumatic heart disease* (antibodies made to fight the strep throat bacterium *Streptococcus pyogenes* cross-react with the patient's heart valves).

A third way for type II reactions to inflict damage occurs *without* causing cytolysis. This route usually has an autoimmune basis, with antibodies interacting with a cell-surface receptor on self-cells. Rather than causing cell lysis, the antibody–receptor interaction causes either a receptor inactivation or an overactivation (FIG. 13.10). A neuromuscular condition called *myasthenia gravis* is one example of a disease caused by inactivation of a receptor. It causes muscle weakness and fatigue—for example, drooping of the eyelids and difficulty swallowing. *Graves' disease,* which leads to an overly active thyroid (hyperthyroidism), is an example of receptor overactivation.

FIGURE 13.9 **Type II hypersensitivity cytotoxic reactions**

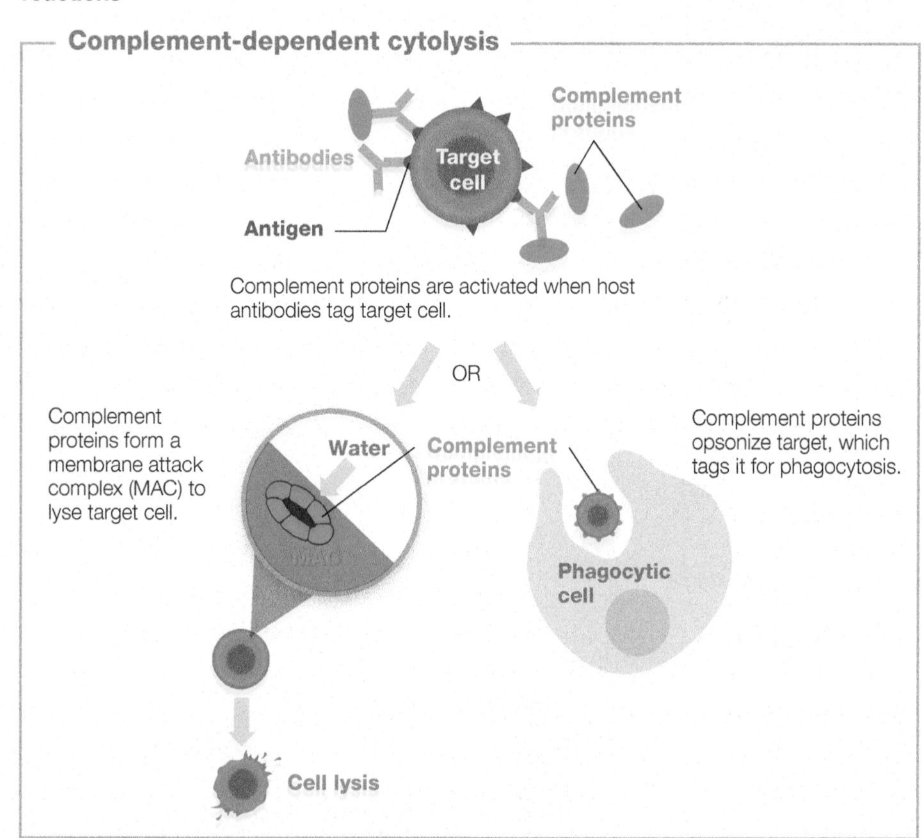

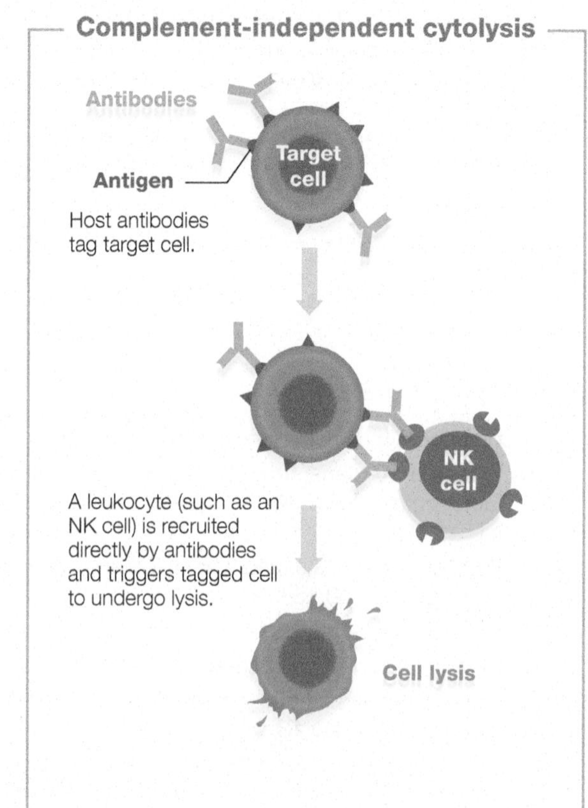

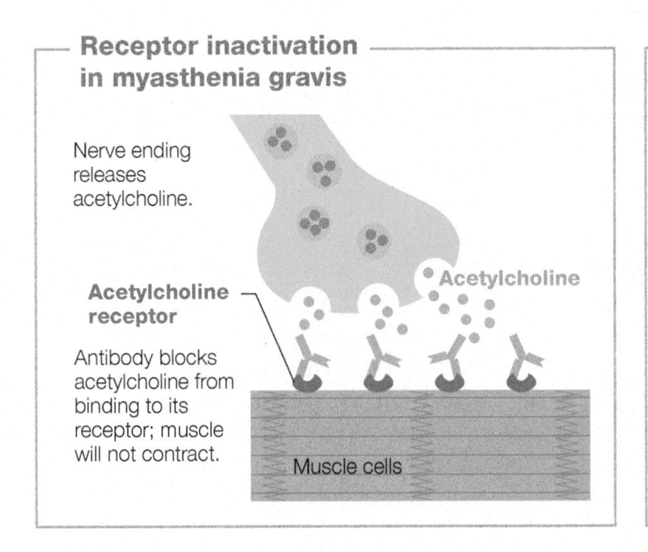

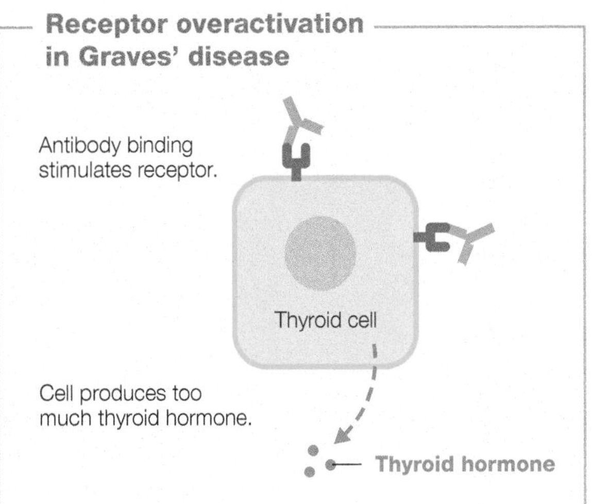

FIGURE 13.10 Type II hypersensitivity noncytotoxic reactions Rather than causing cell lysis, the antibody–receptor interaction causes either a receptor inactivation *or* an overactivation.

Blood group incompatibility may lead to a transfusion reaction.

If you have donated blood, then there is a good chance that you know your blood type. Blood banks keep careful records of this information because blood transfusion compatibility depends on the antigens found on the surface of the transfused red blood cells. If the recipient's immune system considers the red blood cell surface antigens to be foreign, then a transfusion reaction may occur.

Blood Groups

When we discuss blood types, we are referring to the presence of particular antigens on the red blood cell's surface. **A**, **B**, **O**, and **Rh** (rhesus factor) are the most widely analyzed blood antigens. The A, B, and O antigens are carbohydrates, whereas the Rh antigen is a protein.

The convention for denoting blood type is to first list the carbohydrate antigens (A, B, O) found on the blood cells, followed by a superscripted plus or minus sign to reflect the Rh status; a plus sign means the Rh factor is present and a minus sign means it's absent. Our genes determine the types of antigens on our red blood cells. If a person has genes for the A, B, and Rh antigens, then their red blood cells will be denoted as AB^+ ("*AB positive*"). In contrast, if someone has genes for the A antigen, but not the B or Rh antigens, their blood will be denoted as A^- ("*A negative*").

Someone who doesn't have genes to make A, B, or Rh antigens and instead only has genes for the O antigen is designated as having type O^- ("*O negative*") blood (FIG. 13.11). All told, there are eight possible blood types that result from combinations of the A, B, O, and Rh antigens: They are A^+, A^-, B^+, B^-, AB^+, AB^-, O^+, and O^-. The most common blood types are O^+ (38 percent of people) and A^+ (34 percent of people).

Transfusion Compatibility and Reactions

The A and B antigens are not unique to our red blood cells; various foods and microbes also have these same molecules, and so everyone is exposed to them through eating and other environmental contact. Thus, anyone whose immune system categorizes A or B antigens as foreign develops antibodies against these

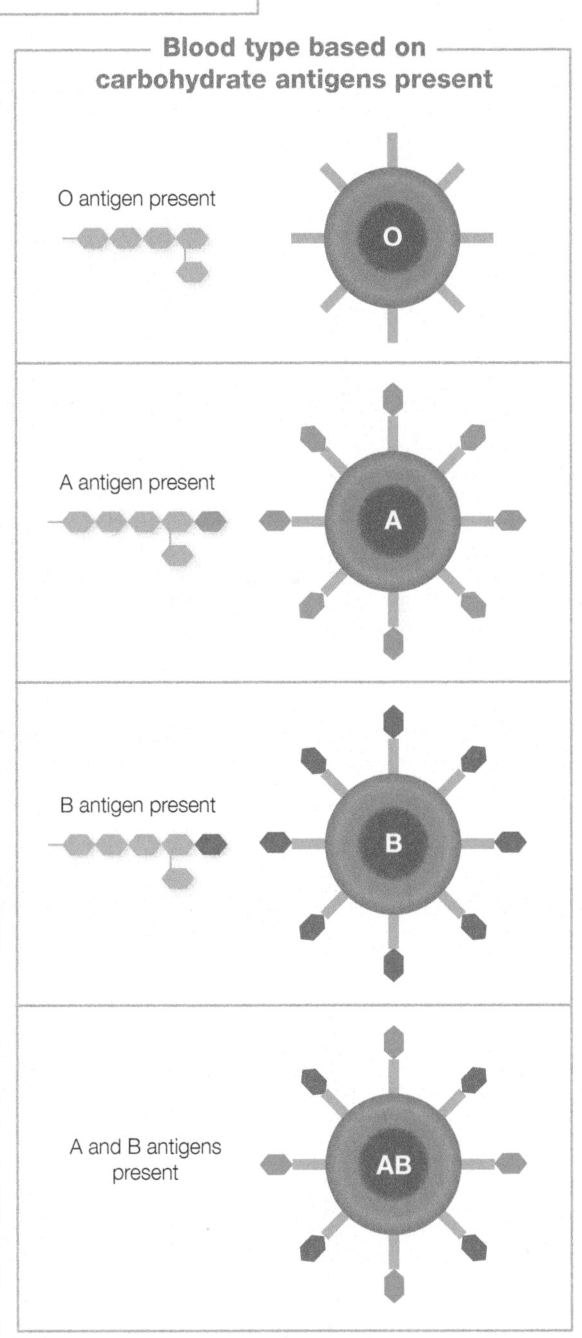

FIGURE 13.11 O, A, and B carbohydrate antigens in blood The O carbohydrate is a foundational part of A and B carbohydrates. This shared structure is why people with A or B antigens can tolerate type O blood transfusions, while those with type O blood treat A and B antigens as foreign.

Critical Thinking *There is another blood type called Bombay O that will treat type A, B, and O blood as foreign. Propose a possible structure for this antigen based on the schematic presented here.*

CHEM • NOTE

Carbohydrate molecules are made up of chains of simple sugars (monosaccharaides). The **A**, **B**, and **O blood antigens** are carbohydrates. The A and B antigens share the same core structure as the O antigen, which is why people with A, B, or AB blood tolerate O blood.

O carbohydrate antigen

A carbohydrate antigen

B carbohydrate antigen

factors—even if the patient never encounters foreign blood. In contrast, exposure to the Rh antigen is not routine, so an Rh-negative person would only develop antibodies to the Rh antigen after exposure to Rh-positive blood (usually from a transfusion or through exposure during a pregnancy with an Rh-positive fetus—a situation we'll review shortly).

The immune response against incompatible transfused red blood cells causes a **hemolytic transfusion reaction**, which lyses red blood cells and could kill the patient. Signs and symptoms tend to occur within hours of incompatible transfusion and may include fever, chills, lower back pain, constricting chest pain, a rapid heart rate (tachycardia), reduced blood pressure, and muscle aches. Unfortunately, there are no therapies to reverse a transfusion reaction or block it once it starts. Aside from stopping the transfusion, the treatment plans for acute reactions aim to reduce the risk of kidney failure, systemic blood clots (disseminated intravascular coagulation, or DIC), and hypotension (low blood pressure).

Fortunately, transfusion reactions are largely avoidable thanks to techniques that identify red blood cell antigens and ensure that individuals only receive compatible blood. Many factors must be considered to ensure compatibility, but for most patients A, B, O, and Rh are the critical antigens to check. A patient cannot safely receive blood that contains antigens their own blood type doesn't naturally have. TABLE 13.5 summarizes how a person's blood type relates to their compatibility for transfusions with other blood types.

As shown in the table, those with AB^+ blood are considered "universal recipients," meaning they can be given any of the eight major ABO blood types. In contrast, people with type O^- blood (about 7 percent of the population) do not have A, B, or Rh factors on their red blood cells and are therefore considered the universal donor that can give blood to all eight of the major ABO blood types. Unfortunately, that benefit is one sided; people with O^- blood cannot be exposed to A, B, or Rh antigens and therefore can only receive O^- blood.

Rh factor incompatibility during pregnancy may lead to hemolytic disease of the newborn (HDN).

So far we have focused on blood group incompatibilities between blood donors and recipients, but these incompatibilities can also exist between a mother and her fetus. ABO incompatibilities in pregnancy do not usually cause serious problems, although they can cause the newborn to become jaundiced—a condition in which bilirubin (a waste product) increases as red blood cells are lysed. Normally the liver chemically modifies bilirubin so it can be easily cleared from the body. Although excessively high bilirubin levels in the newborn can lead to brain

TABLE 13.5 Key Blood Types and Transfusion Compatibility

Blood Type	Antigen(s) present	Antigen(s) missing	Patient can receive transfusions from
AB^+	A, B, and Rh	None of the A, B, or Rh antigens are missing	All ("universal recipient")
AB^-	A and B	Rh	A^-, B^-, AB^-, O^-
A^+	A and Rh	B	A^+, A^-, O^+, O^-
A^-	A	B and Rh	A^-, O^-
B^+	B and Rh	A	B^+, B^-, O^+, O^-
B^-	B	A and Rh	B^-, O^-
O^+	Rh	A and B	O^+, O^-
O^-	None of the A, B, or Rh antigens are present	A, B, and Rh	Only O^- (but a "universal donor" to other types)

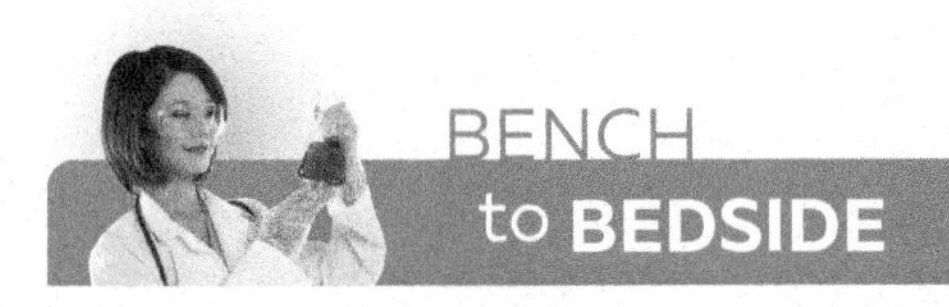

BENCH to BEDSIDE

Turning All Blood Into the "Universal Donor"

According to the Red Cross, every 2 seconds someone in the United States needs a blood transfusion. In an emergency situation there may not be time to get a patient blood type, so to ensure patient safety the "universal donor," type O$^-$, will be given. Although blood donations of all types are needed, the special status of O$^-$ blood makes it an extremely valuable commodity in the healthcare setting. Unfortunately, only 7 percent of the U.S. population is O$^-$, so stockpiling adequate amounts of the blood is always difficult. Adding to the challenge is the point that a pint of blood has a shelf life of only about 42 days.

In light of these statistics, a method to convert all blood donations to a "universal donor" would be useful. For decades researchers have been investigating the possibility of removing the end portion of type A and B carbohydrate antigens to make them look like the O antigen. Over 2,500 enzymes have been screened, and so far, only a few are viable candidates.

The ideal blood-converting enzyme must specifically modify the A and B carbohydrate antigens and be concentrated to a level that leads to adequate conversion. Another issue is that the enzyme needs to work at roughly room temperature, and at a neutral pH. These restrictive parameters disqualify most enzymes, and even promising candidates, like one derived from unroasted green coffee beans, ended up being less efficient than hoped. Another limitation is that to date the enzymes investigated do not remove the Rh factor, which means the process to make O$^-$ blood would have to start by using Rh$^-$ blood.

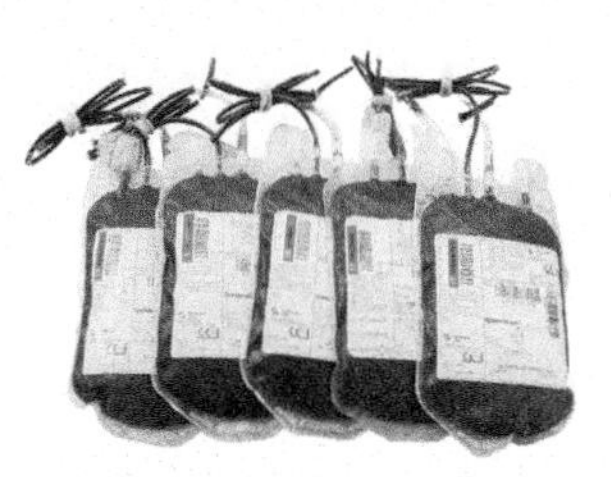

In 2019, new carbohydrate-cleaving enzymes were isolated from gut microbiota. The researchers who discovered these enzymes determined that a mixture of the isolated enzymes could convert type A blood to type O. Interestingly, the analysis of the blood conversion was confirmed using MTS™ Gel Cards—an industry standard for determining blood type. Furthermore, the researchers were able to remove the conversion enzymes using procedures that are already approved and routinely used for preparing blood cells for transfusion. In 2020, another collection of enzymes was identified as efficiently converting types B and AB blood to type O. Much work remains to determine if these blood type conversion systems will be practical, safe, and cost effective, but researchers in the field remain optimistic.

Sources: Rahfeld, P., Sim, L., Moon, H., et al. (2019). An enzymatic pathway in the human gut microbiome that converts A to universal O type blood. *Nature Microbiology*, *4*(9), 1475–1485. Rahfeld, P., & Withers, S. G. (2020). Toward universal donor blood: Enzymatic conversion of A and B to O type. *Journal of Biological Chemistry*, *295*(2), 325–334.

damage, treatments for newborn jaundice are easy and effective. However, Rh factor incompatibilities between mother and fetus are much more serious.

In **hemolytic disease of the newborn (HDN)** an Rh$^+$ baby's red blood cells are lysed in response to maternal anti-Rh antibodies that cross the placenta. This can occur when a mother who is Rh$^-$ has been sensitized to the Rh factor—usually when a mother had an earlier pregnancy with an Rh$^+$ baby. Note that fetal and maternal blood do not normally mix during pregnancy. Thus, exposure to fetal blood and the development of maternal antibodies against the Rh factor occur either following a miscarriage, or abortion, or after the baby is

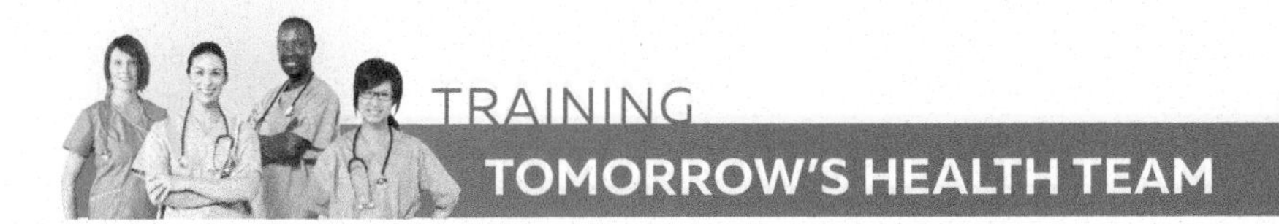

TRAINING TOMORROW'S HEALTH TEAM

Shining a Light on Jaundice

If only all that ails us could be cured with a ray of sunshine! Fortunately, jaundice, a condition that develops in up to 50 percent of normal-term newborns and 80 percent of premature babies, is treatable with inexpensive and safe phototherapy. Jaundice presents as a yellowing of the whites of the eyes and skin. It develops when bilirubin, a byproduct of red blood cell breakdown, accumulates in the blood—a situation called hyperbilirubinemia. Blue light (420 to 470 nm wavelength), which is delivered through so-called bili lights, converts bilirubin to a product that is easily excreted in urine and feces. This light is not ultraviolet light, so it will not cause sunburn, but it can damage the newborn's eyes. To avoid eye exposure, biliblankets are useful—these portable phototherapy devices comfortably wrap around the newborn's torso like a blanket to safely deliver the light therapy.

In newborns, hyperbilirubinemia is associated with numerous conditions such as incompatibility between the mother and baby's blood types, hemorrhage, anemia, certain infections, and liver malfunction.

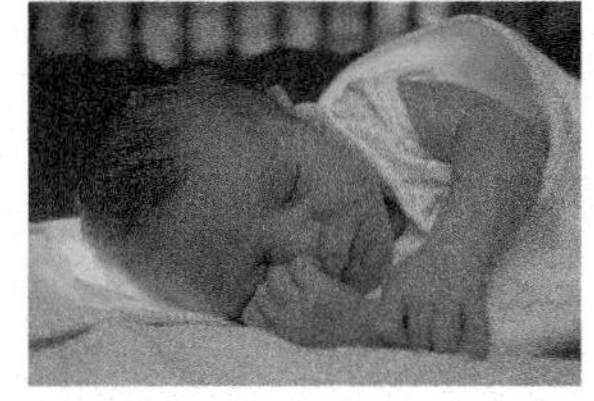

Biliblankets can be used for at-home phototherapy.

QUESTION 13.2

Would persistent jaundice be expected if there was a blood type incompatibility between mother and fetus? Explain your response.

FIGURE 13.12 Hemolytic disease of the newborn (HDN)

Critical Thinking *Why isn't HDN a concern in Rh+ mothers?*

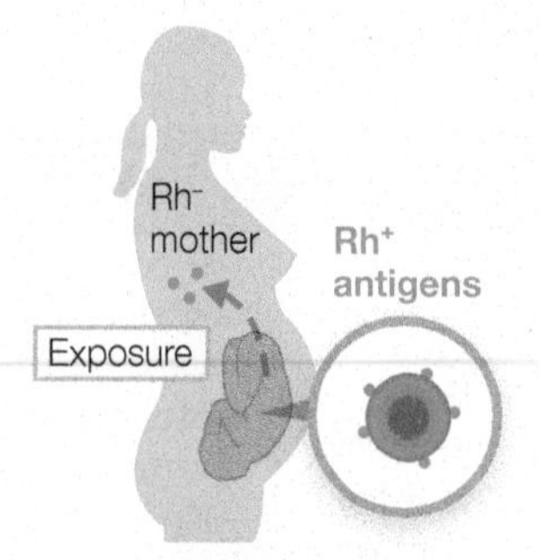

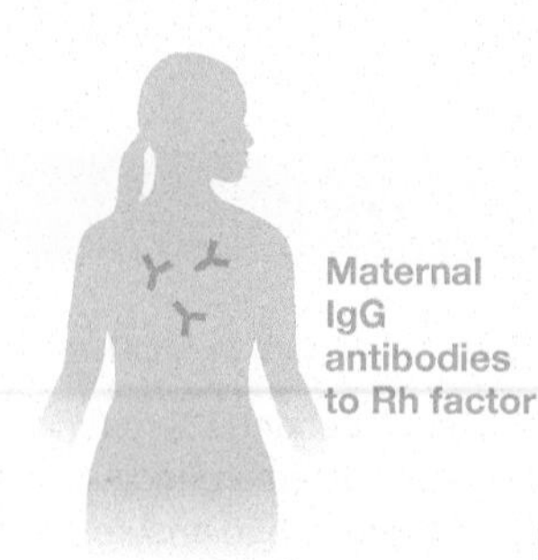

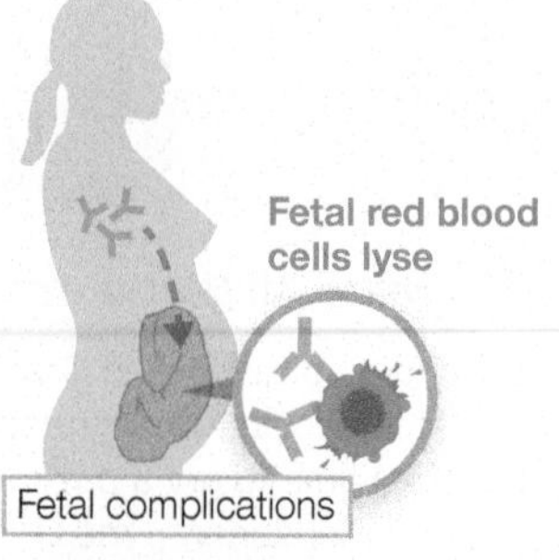

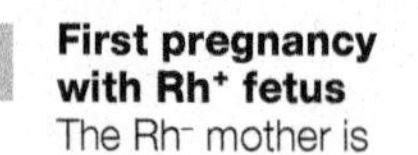

1 First pregnancy with Rh+ fetus
The Rh− mother is exposed to the Rh+ blood of the fetus during pregnancy or birth.

2 Between pregnancies
After exposure, the mother develops IgG antibodies against the Rh factor.

3 Later pregnancy with Rh+ fetus
Maternal IgG antibodies cross the placenta and target fetal Rh+ red blood cells, leading to red blood cell lysis and severe anemia in the fetus.

born and the placenta separates from the uterine wall. If the mother is Rh sensitized, then during a subsequent pregnancy with an Rh+ fetus, maternal IgG antibodies against the Rh factor cross the placenta and target the fetus's Rh+ red blood cells. This leads to fetal red blood cell lysis and induces a severe and possibly fatal anemia. FIG. 13.12 illustrates the progression of HDN. If the fetus is Rh−, there is no risk for HDN developing because maternal antibodies will not find an Rh factor to bind to on the fetal red blood cells. *In utero* signs of HDN may include yellowing of amniotic fluid, fetal tissue edema (swelling), and enlargement of the fetal liver, heart, or spleen. Assuming the baby survives, postnatal symptoms may include severe jaundice, enlarged spleen and liver, anemia, breathing difficulties, and widespread tissue edema.

Unfortunately, if a mother is already sensitized against the Rh factor, she already makes antibodies to the Rh antigen, so there is no effective way to prevent HDN. In such cases, the pregnancy is closely monitored, and interventions such as fetal blood transfusions may be required to prevent fetal death. The only way to prevent HDN from occurring involves preventing Rh− women from ever being sensitized to the Rh factor. This can be accomplished by administering **Rh(D) immunoglobulin** (for example, RhoGAM) to a pregnant Rh− mother. This drug prevents the mother's body from making antibodies to the Rh factor that may be present on fetal red blood cells. The injected immunoglobulin preparation is administered to an Rh− mother around the 28th week of pregnancy. The drug should also be given after amniocentesis or chorionic villus sampling; following an episode of vaginal bleeding during pregnancy; subsequent to miscarriage or abortion; and within 72 hours of childbirth if the baby is Rh+. These are all periods when the mother might be exposed to Rh+ fetal blood. Rh− women who do not receive adequate prenatal care are much more likely to become sensitized to the Rh factor during a first pregnancy with an Rh+ fetus and then experience HDN complications in a subsequent pregnancy with an Rh+ fetus.

Build Your Foundation

16. What are the features of type II cytotoxic and noncytotoxic hypersensitivity reactions? (LO 13.15)

17. Name several examples of type II hypersensitivities that are mediated by cytotoxic reactions and two that are mediated by noncytotoxic mechanisms. (LO 13.16)

18. List the compatible blood types that an A− patient could receive. How about someone who is O+? (LO 13.17)

19. If a person is A+, who can safely receive their blood? (LO 13.17)

20. How does hemolytic disease of the newborn develop, and how can it be prevented? (LO 13.18)

Build Your Foundation (BYF) Quick Quiz: Visit the **Mastering Microbiology** Study Area to quiz yourself.

13.5 TYPE III HYPERSENSITIVITIES

Type III hypersensitivities are characterized by immune complexes depositing in tissues.

Learning Outcomes

After reading this section, you should be able to:

13.19 Explain the underlying process that leads to inflammation in type III hypersensitivities.

13.20 Provide examples of autoimmune type III hypersensitivities.

13.21 Describe how serum sickness develops and how it is treated.

Type III hypersensitivities develop when IgG or IgM antibodies bind to soluble antigens to make an excessive number of antibody–antigen complexes. This distinguishes these reactions from type II reactions, in which antibodies bind to *nonsoluble* antigens. As type III reactions progress, relatively large antigen–antibody complexes form. These nonsoluble complexes are deposited in tissues where they attract complement factors, triggering massive inflammation (FIG. 13.13). The antibodies involved in type III reactions can be made as part of an autoimmune response or formed as a normal response to foreign antigens.

Autoimmune Type III Hypersensitivities

Some autoimmune disorders are characterized by type III reactions. Examples include systemic lupus erythematosus, rheumatoid arthritis, scleroderma, Sjögren's syndrome, and poststreptococcal glomerulonephritis. The features of these autoimmune disorders are identified in TABLE 13.6.

Nonautoimmune Type III Hypersensitivities

Antivenoms (or antivenins) are antibody preparations that neutralize the effects of venom from snakes, spiders, scorpions, or other venomous creatures. They are administered to patients who have been bitten or stung. Similarly, **antitoxins** are made using antibodies against toxins. Botulism, anthrax, and tetanus are all bacterial diseases for which antitoxin treatments exist.

TABLE 13.6 Examples of Autoimmune Type III Hypersensitivities

Disease	Auto-Antibodies	General Features
Systemic Lupus Erythematosus	DNA, histones, ribosomes, and ribonucleoproteins	Systemic (gastrointestinal, lung, kidney, and thyroid issues); often manifests with rash across cheeks and nose, fatigue, joint pain, fever, or hair loss
Rheumatoid Arthritis	Rheumatoid factor (mainly in the lining of joints)	Severe arthritis; mainly in wrists and hands; can cause bone erosion that deforms joints
Scleroderma	Mainly against centromeres and topoisomerases (enzymes for DNA replication)	Attack on connective tissues; all organs may be affected; external manifestation includes hardened, thickened, and tightened skin
Sjögren's Syndrome	Rheumatoid factor (with or without rheumatoid arthritis); antibodies to nuclear proteins	Systemic; affects up to 4% of population; early signs include enlarged parotid gland, dry eyes and mouth; often develops with other autoimmune disorder—especially lupus
Poststreptococcal Glomerulonephritis	Antibodies against *Streptococci* cross-react with proteins in the kidney.	May develop after untreated *Streptococcus pyogenes* infection; antibiotics make it rare in developed countries; usually resolves in weeks to months but may progress to renal failure

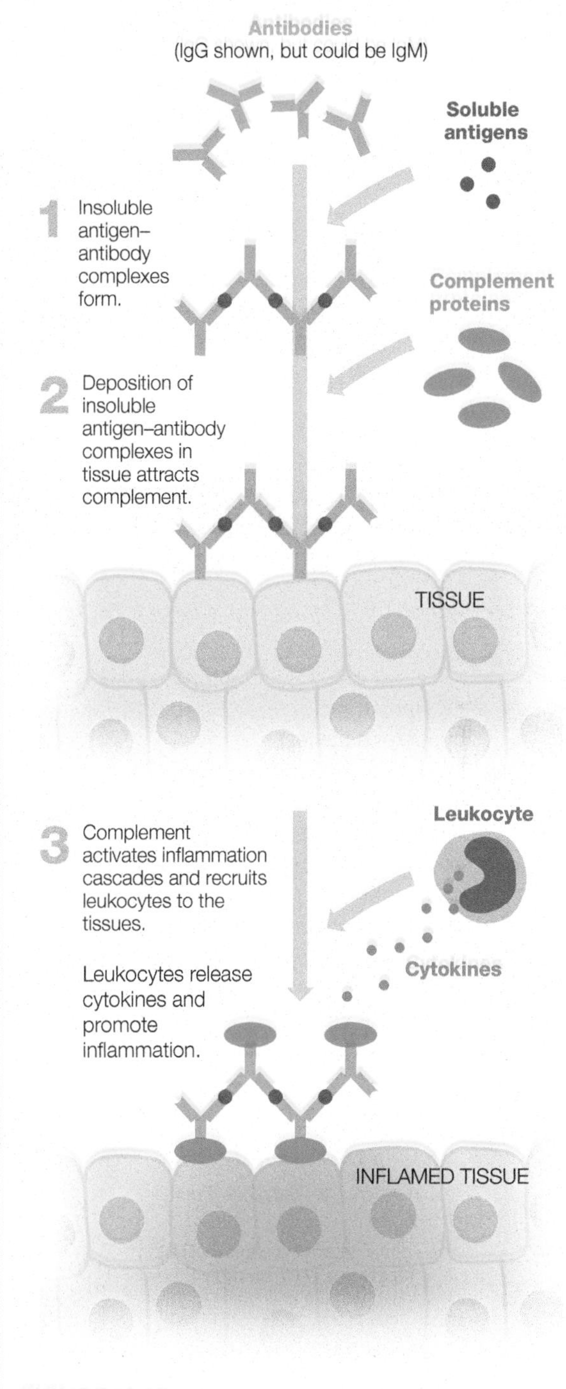

FIGURE 13.13 **Type III hypersensitivity reactions**

CHEM • NOTE

Substances that dissolve in water (the main liquid component of blood and tissues) are described as **soluble**. Substances that do not dissolve in water are said to be **nonsoluble**.

Antivenoms and antitoxins save lives, but are not without potential risks. One is **serum sickness**—a situation in which the patient's immune system recognizes the administered substance as foreign. About 3 weeks after initially receiving the drug, antivenom, or antitoxin, the patient forms antibodies against it. As the antibodies bind to their targets, the resulting immune complexes become lodged in blood vessels or joints, producing symptoms such as rash, fever, fatigue, achiness in joints and muscles, headache, labored breathing (dyspnea), and abdominal pain. Serum sickness usually resolves without treatment, although anti-inflammatory drugs (steroidal or nonsteroidal) and antihistamines can be administered. Fortunately, most patients fully recover within a week and have no further symptoms once exposure to the causative substance is stopped. Monoclonal antibody preparations (e.g., rituximab, infliximab) have also been known to occasionally cause serum sickness.

It is also important to mention that certain medications may cause reactions that are like serum sickness, but do not involve immune complex formation. Some of the most common causes of serum sickness-like reactions are penicillins, cephalosporins, and sulfonamide drugs.[15]

BYF QUICK QUIZ

Build Your Foundation (BYF) Quick Quiz: Visit the **Mastering Microbiology** Study Area to quiz yourself.

Build Your Foundation

21. What process triggers inflammation in type III hypersensitivities? (LO 13.19)
22. Name several examples of type III autoimmune disorders. (LO 13.20)
23. How does serum sickness develop, and how it is treated? (LO 13.21)

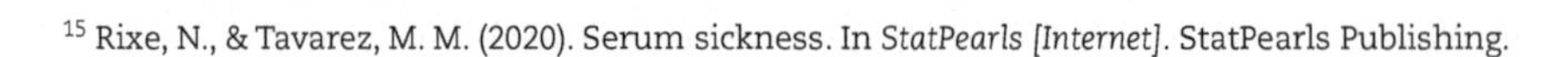

[15] Rixe, N., & Tavarez, M. M. (2020). Serum sickness. In *StatPearls [Internet]*. StatPearls Publishing.

13.6 TYPE IV HYPERSENSITIVITIES

Learning Outcomes

After reading this section, you should be able to:

13.22 Explain how type IV reactions are mediated, how they progress, and why they are called delayed hypersensitivities.
13.23 Provide examples of autoimmune and nonautoimmune type IV hypersensitivities.
13.24 Discuss the different classes of transplants, and explain the meaning of the term *immune-privileged site*.
13.25 Describe graft-versus-host disease and the circumstances in which it is most likely to develop.

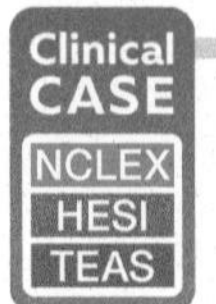

The Case of the Terrible Turkey Surprise

Practice applying what you know clinically: visit the **Mastering Microbiology** Study Area to watch Part 3 and practice for future exams.

Type IV hypersensitivities are mediated by T cells.

Any immune response, whether it's mediated by antibodies (humoral response) or by T cells (cellular response), is considered a hypersensitivity if it is directed at self-tissues or is aimed at an otherwise harmless substance. Unlike the other three hypersensitivities we have reviewed to this point, antibodies do *not* mediate **type IV hypersensitivities**. Instead, these are T cell–mediated responses against self-antigens or otherwise harmless antigens. Because type IV reactions manifest slowly over 12–72 hours after the stimulating antigen is encountered, they are often called **delayed hypersensitivity reactions**.

As reviewed in Chapter 12, antigen-presenting cells normally activate T cells by showing them MHC-associated foreign antigens. The activated T cells then mount a response against the foreign antigen. However, if the antigen presented is a self-antigen, resembles a self-antigen, or is an otherwise harmless substance (such as latex, a transplanted tissue antigen, or a drug), then a type IV hypersensitivity reaction may develop.

Autoimmune Type IV Hypersensitivities

Type IV reactions are responsible for pathology seen in a variety of autoimmune disorders, including Guillain–Barré syndrome, Hashimoto thyroiditis, type I diabetes, multiple sclerosis, and celiac disease. These disorders are marked by the destruction of specific tissues rather than by the systemic inflammation seen in some type III autoimmune disorders like lupus and Sjögren's syndrome.

Guillain–Barré (GEE yawn *bur* ray) syndrome is a nervous system disorder that develops as T cells attack nerves that regulate muscle contractions, resulting in loss of motor function or paralysis; sometimes the nerve damage is reversible, but other times not. This syndrome has been linked to a prior infection with *Campylobacter jejuni*, *Mycoplasma pneumoniae*, cytomegalovirus, or

TRAINING TOMORROW'S HEALTH TEAM

"Do You have Any Drug Allergies?"

It's a routine question that every patient has been asked: "Do you have any drug allergies?" Regrettably, there is no way to tell if or even when a person may develop an adverse immune reaction to a drug, so a patient's answer to this question is never set in stone. And although people may refer to adverse drug reactions as "drug allergies," in reality type I reactions mediated by IgE—true allergies—are just one segment of the reactions patients mean when they say "allergy."

Most drugs are too small to trigger an immune response on their own. However, they can act as haptens. Type I and IV reactions are the most common immune drug reactions. Most type I reactions are against amoxicillin, ampicillin, penicillin, and tetracycline. Type IV reactions tend to generate a characteristic drug rash and are most commonly linked to beta-lactam and sulfa drugs. Type II and III drug hypersensitivities are very rare. Type II reactions include drug-induced immune hemolytic anemia (DIIHA). The disorder is linked to at least 125 drugs, but the antimicrobial drugs cefotetan, ceftriaxone, and piperacillin are the most common culprits. Serum sickness, a type III reaction, is most commonly associated with beta-lactam-based drugs, sulfonamides, and carbamazepine.

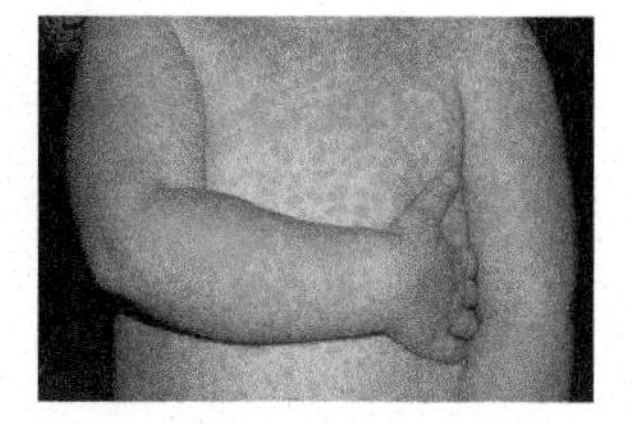

Rash in penicillin allergy.

QUESTION 13.3

Unfortunately, an adverse reaction to one drug may render multiple classes of drugs unsafe for a patient. Why is this logical from an immune system standpoint?

Epstein–Barr virus. There is also evidence to suggest SARS-CoV-2 can elicit an autoimmune response that leads to Guillain–Barré syndrome.[16] It's thought that these pathogens trigger a cross-reactive immune response in which T cells recognize pathogen antigens that resemble self-antigens found on nerve cells.

Hashimoto thyroiditis results from T cell-mediated attacks on the thyroid, resulting in the gland becoming underactive (hypothyroidism). In type I diabetes, insulin-producing cells in the pancreas are destroyed, so that the body can no longer regulate blood glucose levels. With multiple sclerosis, it's the myelin-producing cells of the nervous system that are damaged by the immune system, leading to compromised nerve-signal transduction and impaired movement (FIG. 13.14).

Celiac disease is sometimes referred to as a "gluten allergy;" however, it isn't an allergy because the pathophysiology is not IgE mediated. Instead, the patient's T cells attack the lining of the small intestine within 2–3 days of the patient consuming gluten—a protein found in wheat, rye, and barley. Extreme inflammation and tissue damage result, interfering with the ability of the intestines to absorb essential nutrients from food. Consequently, patients with celiac disease may experience significant malnutrition and weight loss.

Therapies for type IV reactions are aimed at quieting the T-cell response and reducing inflammation. Corticosteroid anti-inflammatory drugs and substances that block the actions of pro-inflammatory cytokines like TNF are standard interventions for many type IV hypersensitivities. However, not all people with type IV autoimmune disorders require such drugs. Most people with type 1 diabetes, for example, can manage their symptoms with insulin injections, a careful diet, and exercise—and the best therapy for people with celiac disease is a strict gluten-free diet. For all autoimmune type IV hypersensitivities, there is also hope that one day T cells can be induced toward tolerance.

FIGURE 13.14 Autoimmune type IV hypersensitivity reaction in multiple sclerosis

[16] Rahimi, K. (2020). Guillain-Barré syndrome during COVID-19 pandemic: An overview of the reports. *Neurological Sciences*, 1–8.

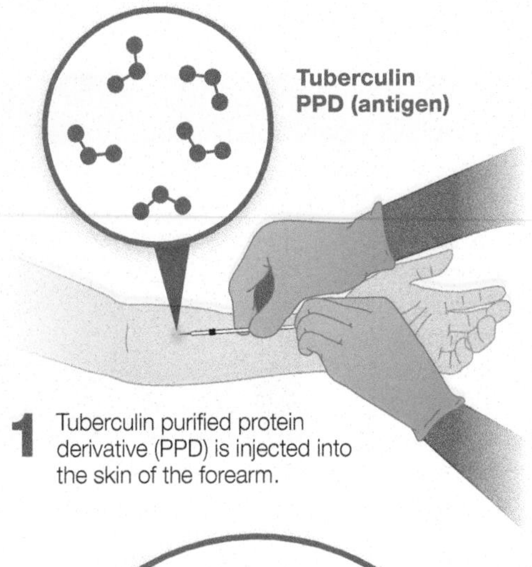

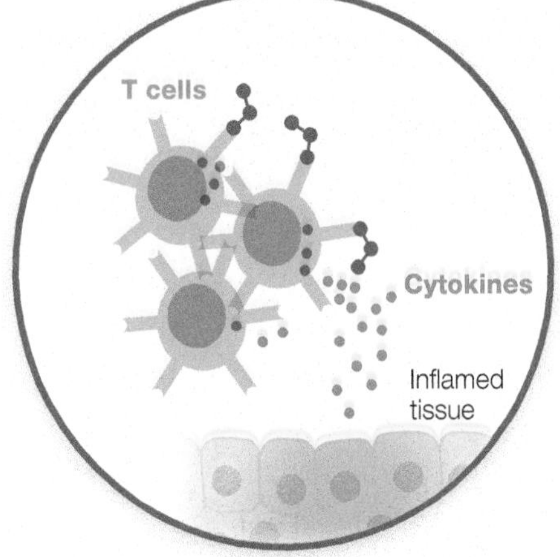

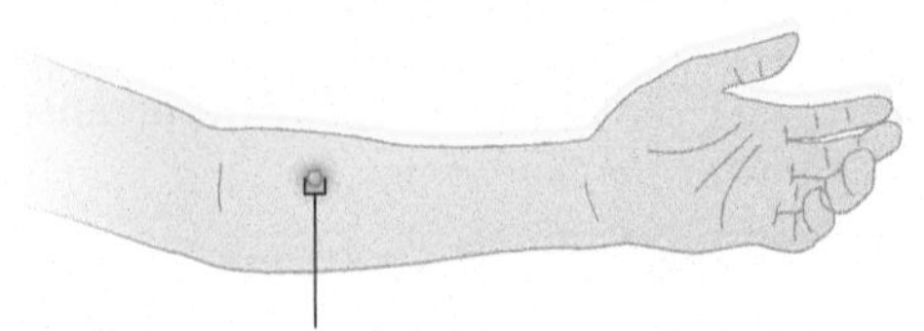

FIGURE 13.15 Type IV reactions and the tuberculin skin test

Critical Thinking *Does a positive test result mean the patient has a current tuberculosis infection? Explain your answer.*

Nonautoimmune Type IV Hypersensitivities

Most nonautoimmune type IV reactions are triggered by **haptens**—molecules that on their own are too small to trigger immune responses, but bind to host proteins, thereby becoming immunogenic and provoking an immune response. (See Chapter 12 for more on complete antigens versus haptens.) Most pharmaceutical drugs that cause hypersensitivities do so by acting as haptens. Other haptens include metals like nickel and chromate. Key examples of type IV nonautoimmune reactions are the tuberculin skin test, contact dermatitis, transplant rejection, and graft-versus-host disease.

Tuberculin skin test (PPD test) A classic example of a type IV delayed hypersensitivity reaction is the tuberculin skin test (also called the PPD test, or the Mantoux test). This test detects exposure to *Mycobacterium tuberculosis*. To perform the test, tuberculin purified protein derivative (PPD) is injected into the skin of the forearm. The injection site is then observed within 48 to 72 hours. If an area of induration (a localized hardened and reddened skin inflammation) develops at the injection site and the lesion meets a certain size criterion, then the test is declared positive—meaning the patient has had some prior exposure to *M. tuberculosis* antigens (FIG. 13.15). Unfortunately, the test cannot differentiate between natural exposures or vaccination against tuberculosis, so the test is not useful in areas of the world where people are routinely vaccinated against *M. tuberculosis*. (This test is covered in more detail in Chapter 16.)

Contact dermatitis If you've ever had a poison ivy or poison oak rash or suffer from a latex sensitivity, then you've had a type IV **contact dermatitis** reaction. Sometimes contact dermatitis develops against a complete antigen, and other times it is against haptens like pharmaceutical drugs, nickel, chromate, or the poison ivy toxin (pentadecacatechol) that have combined with proteins in skin cells. Once T cells are sensitized, upon a secondary exposure to the same antigen, skin inflammation generates an extremely itchy (pruritic) red rash (FIG. 13.16).

Latex sensitivities are an increasing concern in healthcare settings because not only are some patients likely to be sensitive to this natural rubber agent, but also an increasing number of healthcare providers have developed sensitivities. The increased incidence of latex sensitivity in healthcare providers is largely due to increased exposure. A contributing factor to the issue among healthcare workers is that wearing gloves on a regular basis often causes dry, cracked skin that may promote entry of the latex into the skin to trigger sensitization.

Transplant rejection and graft-versus-host disease Transplanted tissue is rejected if the recipient patient's T cytotoxic cells detect that the tissue is foreign. If the transplanted tissue's major histocompatibility molecules (MHCs) are not closely matched to the MHCs of the patient receiving the graft, the graft is likely to be rejected. (See more on MHCs in Chapter 12.) Grafts can be classified based on their degree of similarity to the host:

- **Autografts** are transplants from self, like a self-skin graft from one part of the body to another location. No rejection reactions will occur.
- **Isografts** are transplanted tissue from an identical twin. Genetically identical to the host, these grafts are accepted as "self" and typically safe from immune rejection.
- **Allografts** are similar to the host, but not genetically identical. The closer the MHC match to the host, the higher the chance they will be tolerated.
- **Xenografts** are interspecies transplants. For example, pig heart valves have successfully been used in humans.

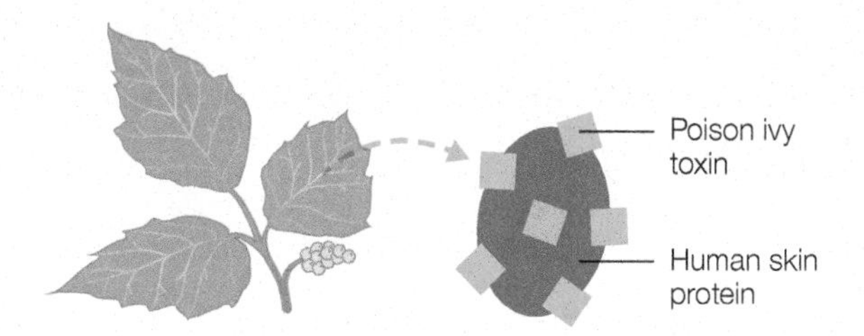

Poison ivy toxin is a hapten that combines with human skin proteins to make a complete antigen that stimulates an immune response.

Primary contact (no dermatitis)
On first exposure, T memory cells specific for the poison ivy toxin are created.

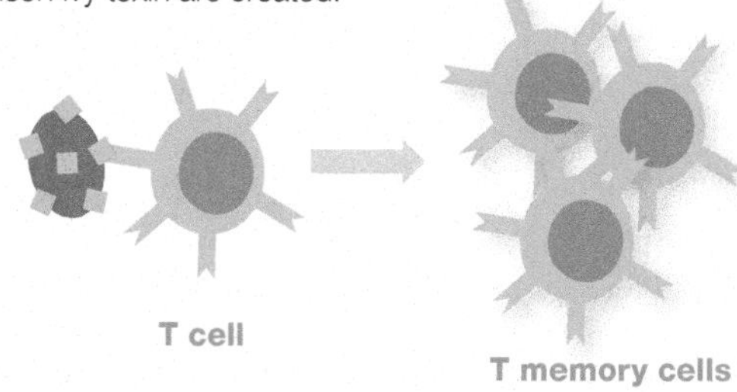

Secondary contact (dermatitis)
Active T cells sensitized against the toxin release cytokines and attract phagocytes, promoting inflammation and causing visible lesions.

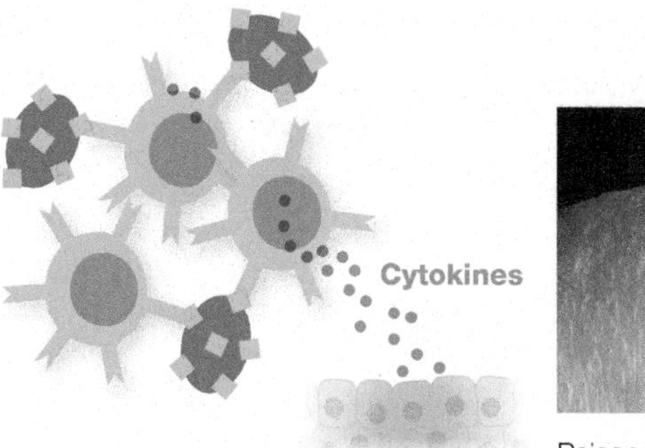

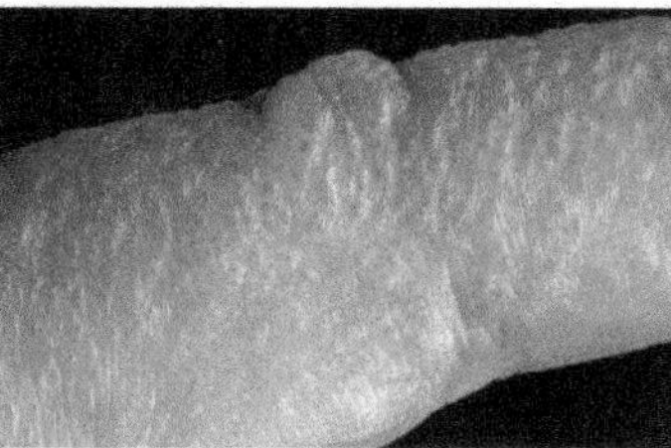

Poison ivy lesions

FIGURE 13.16 Contact dermatitis A schematic of how type IV reactions may develop to haptens like the poison ivy toxin pentadecacatechol.

Transplants that are made in an **immune-privileged site** in the body are the least likely to be rejected. Immune-privileged sites include the eye, brain, uterus, and testicles.

Just as the body can reject a transplant, the reverse situation, called **graft-versus-host disease (GVHD)**, is also possible. In GVHD, which may occur in bone marrow transplants, the graft attacks host tissues. Because a bone marrow transplant is basically an immune system transplant, the white blood cells made in the transplanted bone marrow could attack the new body they find themselves inhabiting.

Build Your Foundation

24. How are type IV hypersensitivities mediated, how do they progress, and why are they called delayed hypersensitivities? (LO 13.22)
25. What are some examples of autoimmune and nonautoimmune type IV hypersensitivities? (LO 13.23)
26. Define the terms *autograft*, *isograft*, *allograft*, and *xenograft*. (LO 13.24)
27. What does the term *immune-privileged site* mean? (LO 13.24)
28. What is graft-versus-host disease, and when is it most likely to develop? (LO 13.25)

Build Your Foundation (BYF) Quick Quiz: Visit the **Mastering Microbiology** Study Area to quiz yourself.

Interactive CONTENT REVIEWS Interactive Content Reviews: Visit the **Mastering Microbiology** Study Area to quiz yourself.

Immune Deficiencies

Primary immunodeficiencies are genetically based disorders that limit immune function and can affect the main lines of immune defense.

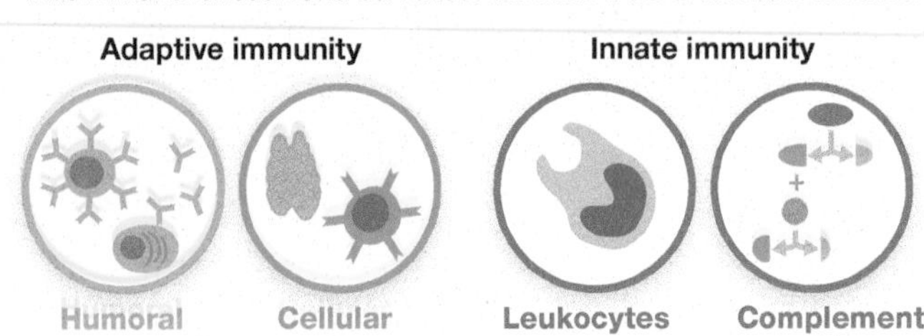

Secondary immunodeficiencies are a decline in normal immune system rigor.

Autoimmunity

Autoimmunity is an immune system attack against healthy self-tissues. Type II, III, and IV hypersensitivities include autoimmune responses.

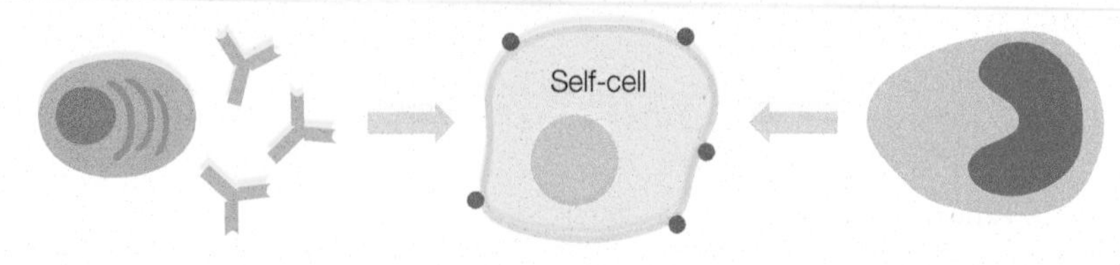

A person makes antibodies against self or their own T cells attack self.

Hypersensitivities

Hypersensitivities are inappropriate immune responses that can be localized or systemic. The acronym **ACID** reminds us of the four hypersensitivity classes (allergy, cytotoxic, immune complex, delayed).

Allergy (Type I hypersensitivity)

- IgE mediated
- Can lead to systemic anaphylaxis
- Desensitization therapy may decrease patient's response to allergen
- Not autoimmune reactions

Allergen binds to IgE anchored to a mast cell or basophil surface and triggers degranulation.

Wheal

Flare

The skin prick test for allergies

Cytotoxic (Type II hypersensitivity)

- Usually involves cytolysis triggered by IgG or IgM bound to nonsoluble antigens on the surface of a cell or within the extracellular environment
- May be autoimmune (examples: autoimmune hemolytic anemia, Goodpasture syndrome)
- Includes blood transfusion reactions and hemolytic disease of the newborn (HDN)

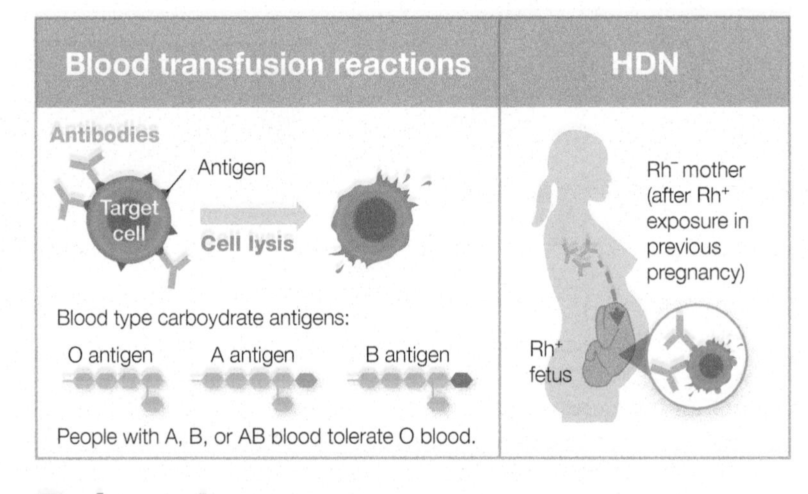

Immune Complex (Type III hypersensitivity)

- IgG or IgM antibodies bind to soluble targets to make antigen–antibody complexes that precipitate and lodge in tissues
- May be autoimmune (examples: lupus, rheumatoid arthritis)
- Also applies to serum sickness

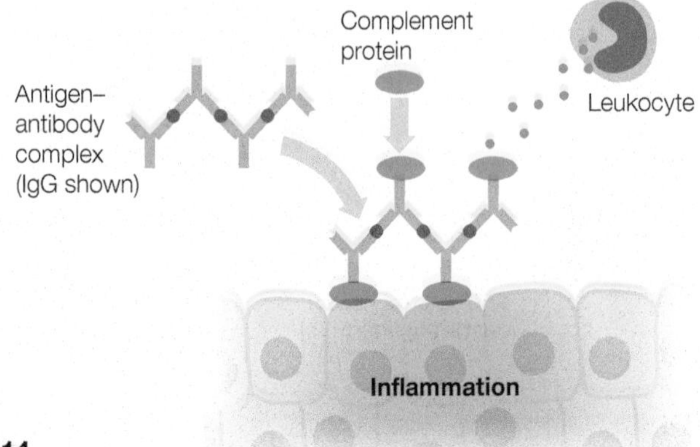

Delayed (Type IV hypersensitivity)

- Reactions are mainly mediated by T cells and are delayed in their time course
- May be autoimmune (examples: type I diabetes, multiple sclerosis)
- Also applies to tuberculin skin test and poison ivy/poison oak reactions and latex sensitivities

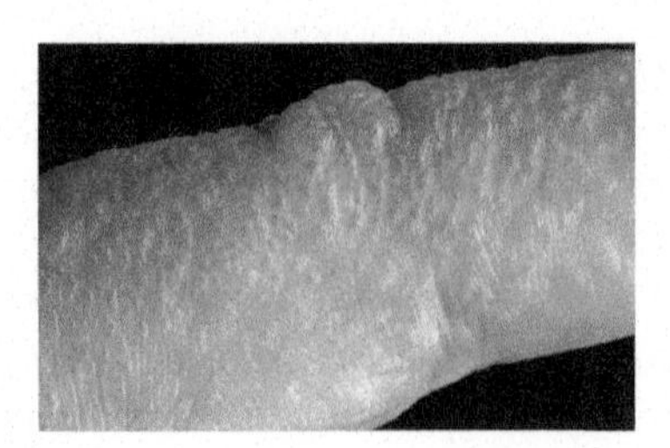

13 CHAPTER 13 OVERVIEW

13.1 Immune Deficiencies and Autoimmunity

- Primary immunodeficiencies are congenital/inherited, while secondary immunodeficiencies are acquired. Cancer is more common in immune-compromised patients. Immunotherapies aim to boost immune defenses.
- Autoimmune disorders develop when the immune system attacks self-tissues. Diagnosis of these disorders often involves autoantibody detection. Treatments involve suppressing the immune response and reducing inflammation.

13.2 Introduction to Hypersensitivities

- Hypersensitivities are exaggerated immune responses that can cause systemic or localized pathology. Genetic and environmental factors contribute to developing hypersensitivities.
- The Gell and Coombs classification system describes four classes of hypersensitivities. Pharmaceutical drugs can trigger all four classes of hypersensitivities.

13.3 Type I Hypersensitivities

- Type I hypersensitivities include all allergies, atopic asthma, and atopic dermatitis. IgE antibodies mediate type I reactions. Of all four hypersensitivities, type I reactions are the only ones that are not autoimmune.
- To have an allergy, a patient must first experience a sensitizing exposure, which may or may not be the first antigen exposure. In an exposure after sensitization, IgE antibodies that are anchored to mast cells and basophils bind the allergen and trigger symptoms.
- The hygiene hypothesis proposes that decreased normal microbiota diversity may lead to an increased risk of allergy and/or autoimmunity.
- A potentially deadly system-wide allergic response is called systemic anaphylaxis, or anaphylactic shock. Any allergen can cause a systemic anaphylactic response, but usually injected or ingested allergens are the culprits.
- Assessing IgE levels against allergens and skin prick tests help diagnose allergies. Most nonasthma allergies are managed on a contact basis using antihistamines. An epinephrine autoinjector may be needed for patients at risk for anaphylactic shock. Inhaled drugs that open up the bronchi and nonsteroid allergy medications are often prescribed to treat atopic asthma. Desensitization therapies offer a means of limiting some allergic responses.

13.4 Type II Hypersensitivities

- Type II hypersensitivities develop when IgG or IgM bind to nonsoluble antigens on the surface of a cell or within the extracellular environment.
- Damage can develop through cytotoxic or noncytotoxic mechanisms. Transfusion reactions from blood group incompatibilities and hemolytic disease of the newborn are examples of cytotoxic nonautoimmune reactions. Goodpasture syndrome, autoimmune hemolytic anemia, rheumatic heart disease, Graves' disease, and myasthenia gravis are examples of autoimmune type II reactions.
- Blood type is based on what A, B, or Rh antigens are present on the surface of red blood cells. Type O^- blood is considered the universal donor, while type AB^+ blood is called the universal recipient.
- Hemolytic disease of the newborn (HDN) may develop if an Rh^- mother has been sensitized to the Rh factor and carries an Rh^+ fetus. Provided an Rh^- mother is not already sensitized to the Rh factor, Rh(D) immunoglobulin can be administered to prevent HDN.

13.5 Type III Hypersensitivities

- Type III hypersensitivities develop when IgG or IgM antibodies bind to soluble targets and make an excessive number of antibody–antigen complexes. When the complexes deposit in tissues, they cause massive inflammation and tissue destruction.
- Systemic lupus erythematosus, rheumatoid arthritis, scleroderma, Sjögren's syndrome, and poststreptococcal glomerulonephritis are examples of type III autoimmune diseases. Serum sickness is a nonautoimmune type III reaction that can develop to antivenoms, antitoxins, or certain therapeutic monoclonal antibody preparations. Some drugs cause serum sickness-like reactions, but they do not involve immune complex formation seen in classic serum sickness.

13.6 Type IV Hypersensitivities

- Type IV hypersensitivity reactions are primarily due to the actions of T cells against antigens that are self-antigens, resemble self-antigens, or are otherwise harmless (such as transplanted tissue or latex).
- Called delayed hypersensitivity reactions, they usually develop over 12–72 hours after the stimulating antigen is encountered.
- They require a sensitizing exposure in which antigen-presenting cells activate T cells. The activated T cells expand their population and form memory cells. Symptoms do not develop unless memory T cells are activated by a subsequent exposure.
- Type IV reactions mediate contact dermatitis such as poison ivy or latex sensitivities. They are also characteristic of certain autoimmune disorders like Guillain–Barré syndrome, Hashimoto thyroiditis, type I diabetes, multiple sclerosis, and celiac disease.
- Graft rejection is mediated by this type of reaction. Transplanted tissues may be autografts, isografts, allografts, or xenografts. Graft-versus-host disease may occur in bone marrow transplants when the graft attacks host tissues.

THINK CLINICALLY | *Be S.M.A.R.T. About Cases*

COMPREHENSIVE CASE

The following case integrates basic principles from the chapter. Try to answer the case questions on your own. Don't forget to be **S.M.A.R.T.*** about your case study to help you interlink the scientific concepts you have just learned and apply your new understanding of immune system disorders to a case study.

*The five-step method is shown in detail in the Chapter 1 Comprehensive Case on cholera. See pages 31–33. Refer back to this example to help you apply a SMART approach to other critical thinking questions and cases throughout the book.

• • •

After working as a surgical nurse for 5 years, Xavier noticed that increasingly often, he had an itchy red and scaly rash on his hands. Suspecting he was developing a latex sensitivity, he started to wear nitrile gloves and to avoid items he knew contained latex. Although it is estimated that less than 6 percent of the general population has a latex hypersensitivity, it is estimated that up to 18 percent of healthcare providers develop this hypersensitivity. Despite his efforts, Xavier still regularly encountered latex because it is in many household products, as well as in the tourniquets, blood pressure cuffs, and stethoscope tubes he routinely used at work. By the time Xavier had been practicing for 7 years, he was quite sensitive to latex and had to be very careful to avoid it, or else he would develop a terrible blistering rash within about 12 hours of exposure.

One day a trauma patient was rushed into surgery. The surgeon Xavier was working with hastily put on latex gloves—which was their preference for fine vascular work because the latex provided excellent dexterity. In the rush to glove up, the surgeon accidentally broke the glove, and a fine powder misted the area near Xavier. Thinking nothing of it, the surgeon grabbed a new glove. A few minutes later, Xavier started to feel his throat tightening up; it became a stranglehold. He became dizzy and collapsed right there in the operating room. Eventually he was stabilized, and it was determined he had experienced a systemic anaphylactic reaction.

Months later Xavier noticed that his mouth would tingle and his throat become scratchy whenever he ate bananas, which he regularly enjoyed with his cup of morning coffee. A few months after he first experienced issues with bananas, Xavier was eating guacamole and had a similar scratchy and tingling feeling. Concerned that he was developing food allergies, he went to see an immunologist, who confirmed Xavier had an avocado and banana allergy and also was allergic to kiwi—a real surprise to Xavier because he had never even eaten kiwi. Over time, Xavier was able to manage his hypersensitivities and was able to remain working in healthcare, although he did relocate to another hospital that was working to limit latex in all departments.

• • •

CASE-BASED QUESTIONS

1. When Xavier first noticed his latex sensitivity, what class of hypersensitivity had he developed? Support your response with evidence from the case and from your readings.
2. When Xavier collapsed in the operating room, what class of hypersensitivity was he manifesting, and how is it different from the hypersensitivity reactions he had previously suffered in response to latex?
3. When Xavier experienced a reaction to bananas and guacamole, what class of reaction was he most likely experiencing? Support your response with information from the case and your readings.
4. Predict what Xavier's immunologist did to arrive at a diagnosis for Xavier, and describe key aspects of the patient care plan the doctor most likely recommended.
5. Xavier had allergies to avocado, banana, and kiwi, yet he had never eaten kiwi. Why was Xavier so surprised when the doctor told him about the kiwi allergy, and how is this scenario even possible?
6. When his allergist took Xavier's health history, he remembered that as a child, he had a reaction to penicillin. About 3 weeks after he started taking the medication, he developed a rash, achiness, and a fever. His symptoms fully resolved within about 7 days. What sort of reaction was this sensitivity, and why might Xavier have a propensity toward hypersensitivities?

END OF CHAPTER QUESTIONS

NCLEX
HESI
TEAS

Think Critically and Clinically
Questions highlighted in orange are opportunities to practice NCLEX, HESI, and TEAS-style questions.

1. Which type of hypersensitivity is not antibody mediated?
   a. Type I hypersensitivities
   b. Type II hypersensitivities
   c. Type III hypersensitivities
   d. Type IV hypersensitivities
   e. Antibodies mediate all of these.

2. A patient has developed a type III reaction to a drug. Which of the following is the most immediate action required?
   a. Lower the patient's fever.
   b. Stop administration of the drug.
   c. Treat the patient's skin rash to avoid possible infections.
   d. Hook the patient up to an IV for rehydration therapy.
   e. Administer antihistamines to limit the response.

3. Indicate the true statements, **and then** reword the false statements so they are true.
   a. Immunodeficiencies are associated with a decreased cancer risk.
   b. Type I reactions are mediated by IgG antibodies.
   c. Autoimmune disorders are not caused by type I hypersensitivities.
   d. Systemic lupus is mainly mediated by a type III hypersensitivity.
   e. A person with type O$^-$ blood is called a universal donor.
   f. Immunotherapies are useful to reduce type II reactions.

4. Which of the following would be recommended as a means to diagnose a type I sensitivity? Select all that apply.
   a. Skin prick test
   b. IgG titers
   c. IgM titers
   d. IgE titers
   e. Histamine levels

5. Which of the following is the most suggestive of an anaphylactic reaction?
   a. Sudden fever
   b. A localized rash
   c. Nasal congestion
   d. Hemorrhage
   e. Respiratory distress

6. Fill in the table.

Blood type	Can safely donate to:	Can safely receive blood from:	Antigens on red blood cell surface	Antibodies present in serum
				B
AB$^-$				
		O$^-$ and O$^+$		
				A, B, and Rh

7. Which scenario presents the greatest risk for HDN developing?
   a. O$^-$ mother with O$^+$ fetus
   b. A$^+$ mother with O$^+$ fetus
   c. B$^-$ mother with AB$^-$ fetus
   d. A$^-$ mother with O$^-$ fetus
   e. AB$^+$ mother with AB$^-$ fetus

8. Which patient would be the most likely to benefit from desensitization immunotherapy?
   a. A person suffering from serum sickness
   b. A person at risk for HDN during pregnancy
   c. A person with an allergy to pollen
   d. A transplant patient
   e. An asthmatic patient

9. List the applicable hypersensitivities (type I, II, III, IV) as they relate to the description. Note, more than one type of hypersensitivity may be listed for a given description.
   - IgG antibodies can mediate type ______________ hypersensitivities.
   - T cells mediate type ______________ hypersensitivities.
   - Type ______________ hypersensitivities may be generated in response to pharmaceutical drugs.
   - Type ______________ hypersensitivities may be associated with autoimmunity.
   - Type ______________ hypersensitivities require a sensitizing exposure.
   - IgE antibodies can mediate type ______________ hypersensitivities.

10. Which of the following is (are) *true* regarding type III hypersensitivity reactions? Select all that apply.
    a. They involve IgG.
    b. They involve IgM.
    c. They are considered delayed reactions.
    d. They include autoimmune disorders like multiple sclerosis.
    e. They are rare compared with type I reactions.

11. Which of the following is the most likely to lead to graft-versus-host disease?
    a. An allogeneic bone marrow transplant
    b. An allogeneic liver transplant
    c. A xenogeneic heart valve transplant
    d. An isogenic bone marrow transplant
    e. A xenogeneic skin graft

12. Imagine that one of your patients is an emergency medical technician who has a family history of autoimmune disorders and allergy, but she does not currently suffer from either. Which of the following is (are) most likely true of your patient? Select all that apply.
    a. She is at an increased risk for cancer.
    b. She is at an increased risk for type I hypersensitivities.
    c. She is at an increased risk for type III hypersensitivities.
    d. She is at an increased risk for latex hypersensitivity.
    e. She is at an increased risk for immunosuppression.

13. Which types of white blood cells are most likely to be involved in type I hypersensitivities? Select all that apply.
    a. T helper cells
    b. T cytotoxic cells
    c. Basophils
    d. Macrophages
    e. Mast cells

14. What is the general goal of treatments used for autoimmune disorders?

15. Which of the following is (are) *true* regarding primary immunodeficiencies? Select all that apply.
    a. These disorders tend to be present from birth.
    b. These disorders are also called congenital immunodeficiencies.
    c. These disorders are more common in elderly patients.
    d. Patients with these disorders may not have a familial history of immunodeficiencies.
    e. Patients with these disorders tend to experience recurring, persistent, and severe infections, often caused by uncommon agents.

CRITICAL THINKING QUESTIONS

1. Packed blood cells are one form of transfusion, but often plasma and not erythrocytes must be transfused. What blood type would be considered the universal plasma donor, and why?

2. Your 30-year-old female patient, Hilair, was given a blood transfusion 3 days ago and was started on a new antibiotic a day ago. The patient has a history of latex sensitivity and tree nut allergy. She had started to feel better, so she was feeling up to solid foods and visitors. A few hours ago, Hilair's friend brought her one of her favorite pastries from the local bakery. About 4 hours after their visit Hilair developed a fever, itchy rash, and abdominal pain. What hypersensitivity is Hilair most likely experiencing? Support your response with information from the chapter.

3. What hypersensitivities would a patient with a complement deficiency have the lowest risk of developing? Explain your reasoning.

4. A routine tuberculin skin test was performed on a 25-year-old patient. Within 30 minutes the patient had an extensive red itchy rash. How can this development be best explained?

5. You have a patient who is exhibiting persistent and severe infections caused by opportunistic pathogens. What clinical suspicion should you have, and why?

14

Biomedical Applications: Vaccines, Diagnostics, Therapeutics, and Molecular Methods

What Will We Explore? In the early 1960s, a huge rubella epidemic moved across the United States. Thousands of pregnant women who contracted it gave birth to children with congenital rubella syndrome, which causes smaller birth size and a variety of congenital birth defects—with blindness and deafness being among the most common consequences. This heartbreaking event became known as the "rubella bulge."

Never heard of this fearsome episode? You aren't alone. The end of rubella as a public health threat in the United States came soon after a vaccine became available in 1966. Life in the vaccine era means that many dangerous infections like congenital rubella are now rare, or in the case of smallpox, eradicated.

Vaccines are just one type of biomedical application. During the last century, advances in our understanding of microbiology, immunology, and genetics have also led to the development of techniques for diagnosing disease more quickly and accurately, and for treating it with medications and other therapies based on manipulation of genes.

In this chapter we'll explore the basics of vaccine history and how vaccines are made and work. We'll then discuss medical applications of various immune- and genetics-based technologies.

Why Is It Important? Every day we're exposed to diverse microbes, yet we can rest easy knowing that some of the most dangerous ones can't harm us, thanks to vaccines that have trained our immune system. When we can't prevent infectious disease, other immune- and genetics-based technologies can help us diagnose and treat it. Because they are increasingly central to modern medicine, it's important for tomorrow's clinician to understand the biotechnology tools reviewed in this chapter.

Clinical CASE
NCLEX
HESI
TEAS

The Case of the Deadly Delay

Visit the **Mastering Microbiology** Study Area to watch the case and find out how vaccines and biotechnology-based diagnostics and therapeutics can explain this medical mystery.

Patricia Galvin
G7 Intensive Care Unit (ICU) nurse; Houston, TX

14.1 A BRIEF HISTORY OF VACCINES

Learning Outcomes

After reading this section, you should be able to:

14.1 Review the early history of vaccines from variolation to Jenner's advancements.

14.2 Describe factors that contribute to the reemergence of vaccine-preventable diseases.

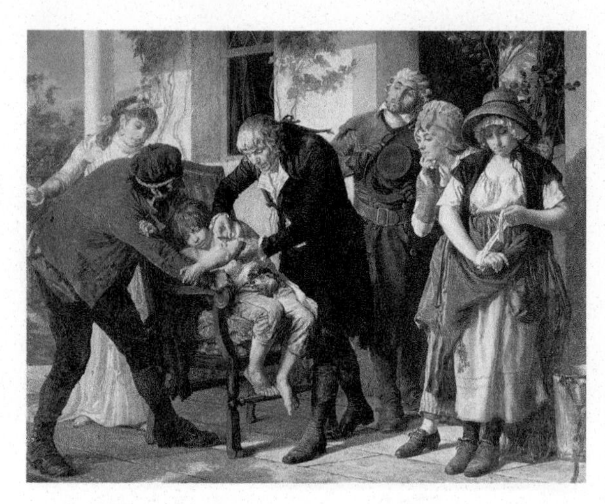

FIGURE 14.1 **Jenner's first vaccination** This painting depicts Edward Jenner inoculating a boy with cowpox. Jenner hoped that by exposing the boy to cowpox, the child would be protected against smallpox.

Critical Thinking *What did Jenner unknowingly depend on for this experiment to be a success?*

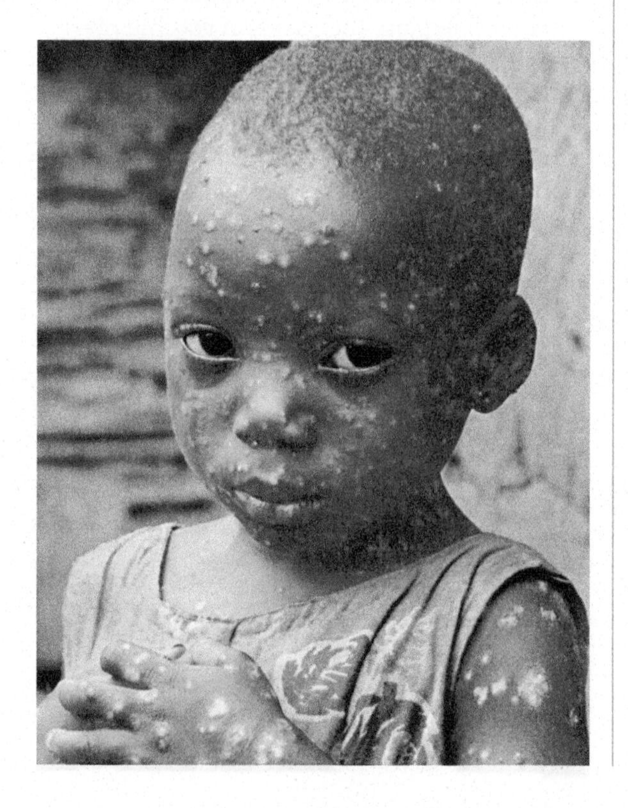

FIGURE 14.2 **Smallpox eradication** Smallpox is an extremely contagious disease with a high mortality rate (the more common form of the disease, variola major, has a mortality rate of 30%; variola minor mortality rate is about 1%). Thanks to worldwide efforts to vaccinate, smallpox was last diagnosed in 1977 in Somalia and is now eradicated.

Critical Thinking *Polio is currently targeted for eradication. What challenges may exist for polio eradication?*

Vaccine history includes triumphs as well as controversies.

Efforts to *vaccinate*, or trigger the body's immune response against infections, go back hundreds of years to when the Chinese used a procedure called **variolation** to combat smallpox. In variolation, the practitioner blew a powder made from the dried scabs of smallpox lesions into a healthy individual's nose. The resulting smallpox infections tended to be milder, with only a 1–2 percent mortality rate compared to the 30 percent mortality associated with naturally acquired infections. Additionally, after recovery, inoculated people were immune to smallpox for the rest of their lives. The practice spread across Asia but was slower to be accepted in the Western world. In the 1700s it gained a foothold in England after Lady Mary Wortley Montagu, the wife of a British ambassador, learned of variolation while living in Turkey. Severely disfigured by scars from her own experience with smallpox, Lady Montagu had her children inoculated against the disease and convinced members of the British royal family to do the same. Despite the apparent success of variolation, fear and distrust of the practice endured in the West, preventing it from becoming common for decades.

In 1796, the British physician Edward Jenner created a new method of purposely infecting people with pathogens to spur immunity. He noticed that milkmaids weren't affected by smallpox epidemics like the rest of the population. Upon investigation, he learned that most of them contracted cowpox, a less severe and nonfatal bovine version of the disease. Jenner suspected that a prior cowpox infection somehow was protective against smallpox. To test his hypothesis, he purposely inoculated the arm of an 8-year-old boy with pus from a cowpox lesion (FIG. 14.1). The boy contracted cowpox, but quickly recovered. Jenner then infected the boy with smallpox, but the child showed no symptoms. These early experiments gave us the word *vaccination*, which derives from *vacca*, the Latin word for cow.

Smallpox vaccination was soon mandated for British soldiers and sought out by others. Thereafter, other vaccines were created. A leader in this effort was Louis Pasteur, who in the late 1800s developed an early version of the rabies vaccine to protect humans, as well as a vaccine to protect cattle against anthrax. As knowledge of pathogens increased, scientists developed more vaccines. Currently, at least 25 different infections are vaccine preventable (TABLE 14.1). In addition to eradicating smallpox, global vaccination programs have saved millions of lives—it is estimated that between 2000 and 2030, routine vaccines against 10 pathogens (including measles, rotavirus, and hepatitis B) will prevent 69 million deaths in low- and middle-income countries (FIG. 14.2).[1]

Although centuries of historical evidence and formal scientific research back the lifesaving benefits of vaccines, the practice has been a frequent source of controversy. Antivaccination societies formed in England during the 1800s to protest the growing use of vaccinations. In early colonial India, many people opposed taking the smallpox vaccine because it was derived from cows, an

[1] Li, X., Mukandavire, C., Cucunubá, Z. M., et al. (2021). Estimating the health impact of vaccination against ten pathogens in 98 low-income and middle-income countries from 2000 to 2030: A modelling study. *The Lancet*, 397(10272), 398–408.

TABLE 14.1 Vaccines Licensed for Use in the United States

Vaccine	Administration	Formulation	Notes
Adenovirus vaccine	Oral administration; not routine for public	Live attenuated	Protects against adenoviruses 4 and 7 that cause colds
Anthrax vaccine	Intramuscular injection for pre-exposure prophylaxis; subcutaneous for post-exposure prophylaxis; not routine for public; adults only	Purified subunit	*Bacillus anthracis* bacterium; potential biological weapon due to the dangerous exotoxins it makes
BCG (Bacille Calmette–Guérin) vaccine	Intramuscular injection; not routine for public	Live attenuated	*Mycobacterium tuberculosis*; vaccine prevents severe forms of tuberculosis in children, but is variably effective in adults
Cholera vaccine	Oral; for people 18–64 years old traveling to cholera-afflicted areas	Live attenuated	Protects against *Vibrio cholerae* serotype O1
COVID-19 vaccines*	Injected; age for administration depends on vaccine brand	mRNA and vector formats available	Protect against COVID-19 disease caused by SARS-CoV-2
Dengue vaccine	Subcutaneous injection; for use in children 9–16 years old who live in endemic areas and who have had a laboratory-confirmed previous dengue infection	Live attenuated	Prevention of severe secondary dengue disease
Diphtheria, tetanus, and acellular pertussis (DTaP and Tdap) vaccines	Intramuscular injection; DTaP is routine pediatric vaccine; Tdap is formulated with fewer diphtheria and pertussis agents and is a booster vaccine	Subunit/toxoid combination	Protects against diphtheria caused by *Corynebacterium diphtheria*, tetanus caused by *Clostridium tetani*, and whooping cough caused by *Bordetella pertussis*
Ebola Zaire vaccine	Intramuscular injection; for use in people 18 years of age and older	Recombinant vector vaccine	Prevents Ebola disease caused by Zaire ebolavirus (first approved vector vaccine in U.S.)
Haemophilus B (Hib) vaccine	Intramuscular injection; routine pediatric vaccine	Conjugate	Protects against *Haemophilus influenzae* type b bacteria
Hepatitis A vaccine	Intramuscular injection; routine pediatric vaccine	Whole-agent inactivated	Protects against hepatitis A virus
Hepatitis B vaccine	Intramuscular injection; routine pediatric vaccine	Recombinant subunit	Protects against hepatitis B virus
Human papillomavirus (HPV) vaccines	Intramuscular injection; routine vaccine used in pediatric and adult patients	Recombinant subunit	Protects against the main strains of HPV associated with genital warts and cancer
Influenza vaccines (many types available)	Intramuscular injection; routine vaccine used in pediatric and adult patients; annual vaccine	Whole-agent inactivated and purified subunit available	Nasal mist made with live attenuated strains is no longer recommended in the U.S. due to low efficacy
Japanese encephalitis virus vaccine	Subcutaneous injection; recommended for travelers to endemic areas	Whole-agent inactivated	Protects against Japanese encephalitis virus
Measles, mumps, and rubella (MMR) vaccine	Subcutaneous injection; routine pediatric vaccine	Live attenuated	Protects against viruses that cause measles, mumps, and rubella
Meningococcal vaccine	Subcutaneous injection or intramuscular injection depending on the brand and manufacturer	Recombinant subunit and conjugate formulations available	Prevents meningococcal disease caused by select *Neisseria meningitidis* serogroups
Pneumococcal vaccine (PCV)	Subcutaneous injection or intramuscular injection; PCV13 is routine pediatric vaccine; PCV23 is a routine adult vaccine and is also for pediatric patients over 2 years old who have an increased risk for pneumococcal disease	Conjugate	Protects against the main serogroups of *Streptococcus pneumoniae* that cause pneumonia; PCV13 protects against 13 serotypes; PCV23 protects against 23 serotypes
Inactivated poliovirus vaccine (IPV)	Intramuscular injection; routine pediatric vaccine	Whole-agent inactivated	Protects against poliovirus
Rabies vaccine	Intramuscular injection; only administered to at-risk groups	Whole-agent inactivated	Pre-exposure and post-exposure prophylaxis for rabies virus
Rotavirus	Oral; routine pediatric vaccine	Live attenuated	Protects against rotavirus gastroenteritis
Smallpox vaccine	Skin puncture with bifurcated needle	Live attenuated	Formulated using the vaccinia virus—a virus closely related to the cowpox virus; confers protection against variola virus (smallpox virus)
Typhoid vaccine	Oral for live attenuated or subcutaneous for conjugate formula; reserved for at-risk groups age 2 years and older and travelers to endemic areas	Live attenuated and conjugate formulations available	Protects against typhoid fever caused by the bacterium *Salmonella typhi*

(*Continued*)

TABLE 14.1 (*Continued*)

Vaccine	Administration	Formulation	Notes
Varicella-zoster virus vaccine	Subcutaneous injection; routine pediatric vaccine	Live attenuated	Protects against chickenpox virus
Yellow fever vaccine	Subcutaneous injection; reserved for at-risk groups age 9 months and older and travelers to endemic areas	Live attenuated	Protects against mosquito-transmitted flavivirus
Zoster vaccine	Subcutaneous injection; recommended for people over age 50	Live attenuated	Boosts immunity to varicella zoster to reduce the risk of herpes zoster (shingles)

Table based on 2021 Food and Drug Administration (FDA) recommendations

*As of this text's publication, some COVID-19 vaccines are being administered under FDA emergency use authorization.

animal sacred to Hindus.[2],[3] In the 1900s, when many U.S. states enforced vaccination among citizens to prevent outbreaks of diseases such as polio, some Americans protested that such policies violated their individual right to choose medical interventions. However, over time, most state governments in the United States opted to value protection of public health over the individual right to opt out of vaccinations.

In the 20th century, the World Health Organization helped bring vaccines to developing nations and served as an important partner in eradicating smallpox. The next disease targeted for eradication is polio. One feature that makes polio eradication feasible is that the virus is found only in humans. However, the challenge of maintaining consistent vaccination access, especially in war-torn areas, makes eradicating polio difficult.

In addition to existing vaccines, new vaccines are always on the horizon. For example, in addition to the newest vaccines that protect against COVID-19, there are vaccines in development to protect against hookworm infections, HIV, and other infectious agents.

Vaccination fears continue to contribute to the persistence or reemergence of preventable diseases.

Fears about vaccination re-erupted in 1998, after a paper published in *The Lancet* (a medical journal) described a study of just 12 patients and claimed a correlation between the measles, mumps, and rubella (MMR) vaccine and the development of autism. Shortly after publication, many parents in the United States and the United Kingdom started to decline MMR and many other vaccinations for their children.

In 2010, *The Lancet* fully retracted the MMR/autism study as bad science. The study's authors had been partially funded by lawyers of parents with autistic children who were filing lawsuits against vaccine companies—a clear conflict of interest—and the main study author has since lost his medical license amid charges of fraud and malpractice. Furthermore, subsequent studies with thousands of participants have found no link between vaccines and autism. For example, a 2015 study of records for over 95,000 children showed no link between the MMR vaccine and autism.[4] Despite this, a 2019 Gallup poll found that 10 percent of U.S. adults still erroneously believe vaccines can cause autism—up from 6 percent in 2015.[5] It seems vaccine misinformation can be as difficult to eradicate as real pathogens!

---

[2] Davey, S. (2018). Smallpox vaccination in Early Colonial India: Diversity in resistance. *South Asia Research*, *38*(2), 130–139.

[3] This is an older reference but provides primary source information. Arnold, D. (1993). Smallpox and colonial medicine in nineteenth century India. *Institutions and Ideologies: A SOAS South Asia Reader*, *24*(10), 136–142.

[4] Jain, A., Marshall, J., Buikema, A., Bancroft, T., Kelly, J. P., & Newschaffer, C. J. (2015). Autism occurrence by MMR vaccine status among US children with older siblings with and without autism. *JAMA*, *313*(15), 15341540. (Note, the large sample size for this study may lead one to think this is a metanalysis, a statistical study that combines data from a collection of smaller studies. However, this was not a metanalysis—it was a single retrospective cohort study.)

[5] Gallup Poll Social Series, December 2-15, 2019. Retrieved from: https://news.gallup.com/poll/276929/fewer-continue-vaccines-important.aspx

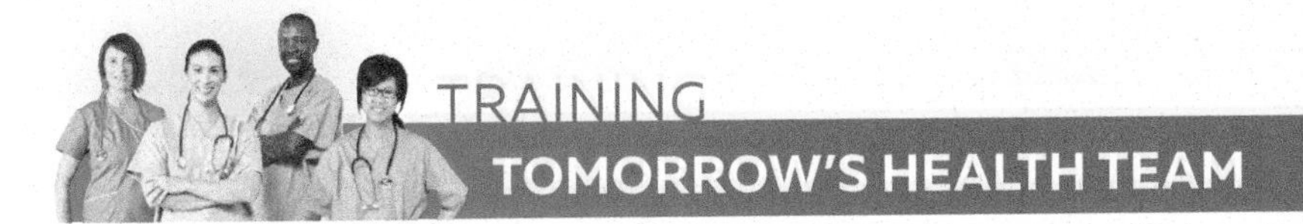

National Vaccination Injury Compensation Program

Like any medicine, vaccines are carefully tested for safety and potential side effects. Usually, side effects are limited to injection-site soreness and/or a low-grade fever. The CDC estimates that one in four children who get the DTaP (diphtheria, tetanus, and pertussis) vaccine as well as hepatitis A and B vaccines experience inflammation at the injection site and fully recover in a few days.

A very small percentage of vaccinated patients—representing one in one million doses administered—experience potentially severe symptoms such as an allergic response. The National Vaccine Injury Compensation Program (NVICP) provides federal funds to defray medical costs for individuals who claim severe complications from the following vaccines: DTaP, MMR (measles, mumps, rubella), polio, hepatitis B, *Haemophilus influenzae* type b (Hib), varicella (chickenpox), rotavirus, and pneumococcal conjugate vaccines.

To qualify for compensation, the petitioner must either show that the patient experienced an injury that is listed in the "Vaccination Injury Table" or, if their injury is not in the table, then they must either prove the vaccine caused it or prove that the vaccine significantly aggravated a pre-existing condition.*

So, if you see headlines publicizing awards through this program, don't assume that the vaccine was *proven* to cause the complication. Most petitions claim the patient suffered a "Table Injury" because in such a claim the petitioner does not have to definitively prove the vaccine was the cause of the injury. Also, even petitions that are deemed "noncompensable" still receive payment to cover attorney's fees; this is sometimes incorrectly presented by the media as a court award.

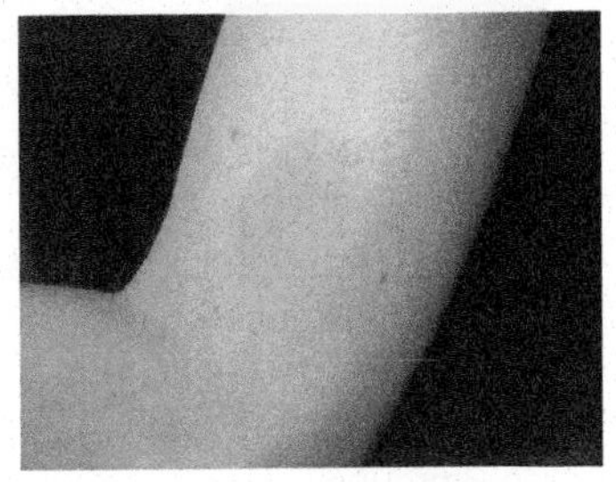

As of the time of this text's publication, any complications from COVID-19 vaccines (as well as new treatments, devices, protocols, etc. used to "diagnose, mitigate, prevent, treat, or cure COVID-19, or the transmission of SARS-CoV-2 or a virus mutating therefrom") are covered under the Countermeasures Injury Compensation Program (CICP), not the NVICP.

QUESTION 14.1

How would you explain the benefits of immunization in light of potential side effects to someone who is skeptical of vaccines in general?

*The vaccine injury table is maintained and updated by the Health Resources and Services Administration, an agency of the U.S. Department of Health and Human Services.

In both the United States and the United Kingdom, a drop in childhood vaccination rates has led to outbreaks of vaccine-preventable diseases that had been rare for over half a century. Measles shows the strongest reemergence. CDC tracking data showed that the 2019 measles outbreak resulted in nearly 1,300 cases that spanned 31 states—the greatest number of measles cases in the United States since 1992. In response to increasing outbreaks, some areas changed school enrollment rules that previously allowed parents to opt out of having their children vaccinated for personal reasons. For example, California, which formerly recognized nonmedical vaccination exemptions, now requires that children receive at least 10 immunizations to attend school (public or private) or daycare. Furthermore, the California public health department can revoke vaccination exemptions it deems medically unsound.

Build Your Foundation

1. What is variolation, and how does it compare to the way Jenner protected his patients from smallpox? (LO 14.1)
2. What prompted people to stop immunizing their children starting in the late 1990s, and how have dips in immunization rates affected public health? (LO 14.2)

Build Your Foundation (BYF) Quick Quiz: Visit the **Mastering Microbiology** Study Area to quiz yourself.

14.2 OVERVIEW OF VACCINES

Learning Outcomes

After reading this section, you should be able to:

14.3 Discuss the general immunology principles underlying vaccinations.

14.4 Explain herd immunity, and describe how it protects nonimmunized people.

14.5 Describe the various types of vaccine formulations.

14.6 State what adjuvants are, and describe their purpose.

14.7 Discuss how RNA vaccines and recombinant vector vaccines stimulate immunity.

Broadly speaking, immunity can be acquired by either natural or artificial means. (See Chapter 12 for more on acquired immunity.) In this section we focus on vaccines—important tools for stimulating artificially acquired active immunity. To be effective, vaccines must stimulate immunological memory without causing the disease they aim to prevent.

Vaccines do not provide immediate protection; on average it takes 2 weeks for antibody levels to reach a peak. However, because vaccines stimulate immunological memory, if the vaccinated patient encounters the real microbe weeks, months, or even years later, memory cells act quickly to prevent the pathogen from establishing an infection.

Immunization programs aim to create herd immunity.

Although the risks posed by routine vaccinations are extremely low for most people, certain vaccines may not be recommended for certain patients, such as newborns, pregnant women, or immune-compromised patients. Even so, provided that a sufficient percentage of the rest of the population is vaccinated, nonvaccinated individuals still reap the protective benefits of immunization. This phenomenon is called **herd immunity** (FIG. 14.3). The fewer disease-susceptible people in a community, the harder it is for a pathogen to be transmitted to a susceptible host who could then spread the disease.

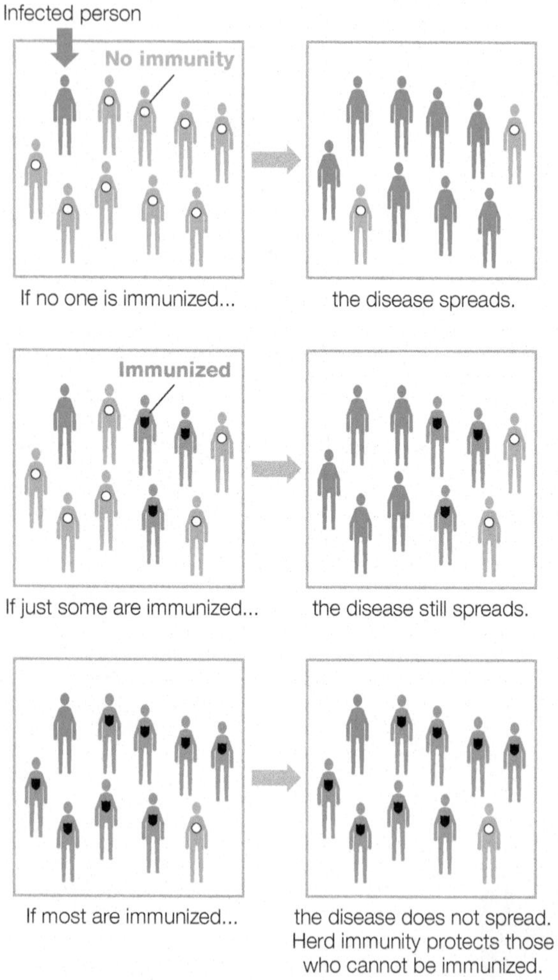

The percentage of a population that must be vaccinated for herd immunity to be effective varies based on the pathogen. For example, highly infectious pathogens that have an airborne transmission, such as SARS-CoV-2, usually require a higher vaccination rate. For most pathogens that we routinely vaccinate against, at least 85 percent of the total population must be vaccinated in order to achieve effective herd immunity. Some pathogens, like measles and whooping cough (pertussis), require a higher percentage of vaccination for herd immunity—closer to 95 percent. Anyone who can be vaccinated should be vaccinated because herd immunity protects the most vulnerable people in the population—babies who have not yet been fully vaccinated and immune-compromised people. When a community's herd immunity dips below certain thresholds, then outbreaks are likely. Unfortunately, any time an agent causes an infection, it has the potential to mutate and render our current vaccines and/or drugs ineffective, putting vaccinated people at risk, too. Indeed, a concern in the fight against SARS-CoV-2 is that newly emerging strains may develop mutations that allow the virus to evade the immune protection current vaccines evoke.

Public health immunization initiatives aim to create herd immunity. In the United States, the Centers for Disease Control and Prevention (CDC) provides vaccination recommendations and publishes an annual pediatric immunization schedule (TABLE 14.2). Routine childhood vaccines protect children against more than 15 different pathogens. To stimulate optimal immunological memory, two or more subsequent doses (boosters) of a vaccine are often needed. The doses are spaced apart to give the adaptive immune system time to respond and create more memory B and T cells against the vaccine agent. The recommended time between boosters may be anything from a month to years, depending on the particular vaccine. If a child is behind on their vaccination schedule (i.e., has missed doses), then the attending healthcare provider can implement a CDC recommended "catch-up" schedule. These schedules take the child's age, the types of vaccines that need to be administered, and other medical factors into consideration. In most cases, catch-up schedules shorten the interval between boosters.

In recent years, some parents have preferred to space out vaccinations, allowing just one per doctor visit, rather than following a schedule that calls for babies and young children to receive as many as four to five vaccines at a time. The problem with this approach is that it can lead to a breakdown of herd

FIGURE 14.3 Herd immunity Vaccines limit the number of susceptible hosts in a population and generate herd immunity by breaking the chain of pathogen transmission.

Critical Thinking *Name at least two ways herd immunity can break down in a community.*

TABLE 14.2 CDC's Recommended Routine Pediatric Immunization Schedule for 2021

Vaccines	Birth	1 mo	2 mos	4 mos	6 mos	9 mos	12 mos	15 mos	18 mos	19–23 mos	4–6 yrs	11–12 yrs	16 yrs
Hepatitis B	1st dose	2nd dose					3rd dose						
Rotavirus*			1st dose	2nd dose									
Diphtheria, tetanus, and pertussis (DTaP)**			1st dose	2nd dose	3rd dose			4th dose			5th dose		
***Haemophilus influenzae* type b (Hib)^**			1st dose	2nd dose			3rd dose						
Pneumococcal (PCV13)			1st dose	2nd dose	3rd dose		4th dose						
Polio (IPV)			1st dose	2nd dose			3rd dose				4th dose		
Influenza							Vaccine recommended annually starting at 6 months						
Measles, mumps, and rubella (MMR)							1st dose				2nd dose		
Varicella (VAR)							1st dose				2nd dose		
Hepatitis A							2 doses 6 to 18 months apart						
Meningococcal§												1st dose	2nd dose
Diphtheria, tetanus, and pertussis booster (Tdap)**												Single dose	
Human papillomavirus (HPV)^^												2 doses	

Immunization doses are often given in a range of time. The ranges here represent the recommended time frames assuming the patient is on time for their vaccine dosing and does not have underlying conditions that require special adjustments.

* Rotarix version shown; RotaTeq version requires a third dose at 6 months.

** DTaP is a combined vaccine against diphtheria, tetanus, and pertussis. Tdap protects against the same pathogens as the DTaP, but is formulated differently and is recommended as a single booster at age 11 (or older if not received at age 11) and in the third trimester of each pregnancy.

^ PedvaxHIB version; ActHIB, MenHibrix, Hiberix, or Pentacel consists of four doses given at ages 2, 4, 6, and 12–15 months.

§ Menactra/Menveo vaccine schedule shown; other versions may be recommended for certain high-risk groups and have a different dosing schedule.

^^ 11- to 12-year-olds get two doses that are at least 6 months apart; the three-dose series is recommended for people with a weakened immune system and patients age 15 or older.

immunity. Children who aren't up to date with vaccines are not fully protected. This puts them and others who are too young or too immunologically fragile to be vaccinated at risk. Therefore, keeping to the CDC immunization schedule is best.

Immunization is often considered a childhood rite of passage, but the need for vaccines continues into late adolescence and adulthood. For example, the bacterial meningitis vaccine is typically given to people 16 through 23 years of age. And even once a person is fully vaccinated, immunological memory can wane over time, necessitating routine booster shots for adults. For example, the recent resurgence of whooping cough has led the CDC to add a one-time adolescent/adult tetanus/diphtheria/acellular pertussis (Tdap) booster at age 11 or older.[6] Pregnant women also get the Tdap in the third trimester of each pregnancy. This boosts antibodies that are passed to the baby to protect it until it is vaccinated. To best match circulating virus strains, the influenza vaccine formulation is altered every year; therefore, it is recommended that everyone over 6 months old receive

Clinical CASE NCLEX HESI TEAS

The Case of the Deadly Delay

Practice applying what you know clinically: visit the **Mastering Microbiology** Study Area to watch Part 2 and practice for future exams.

[6] Note that the adolescent/adult Tdap vaccine is formulated differently from the DTaP vaccine, which is for children under age 11. These two vaccines are not interchangeable, even though they both protect against the same infectious agents: tetanus, diphtheria, and pertussis (whooping cough).

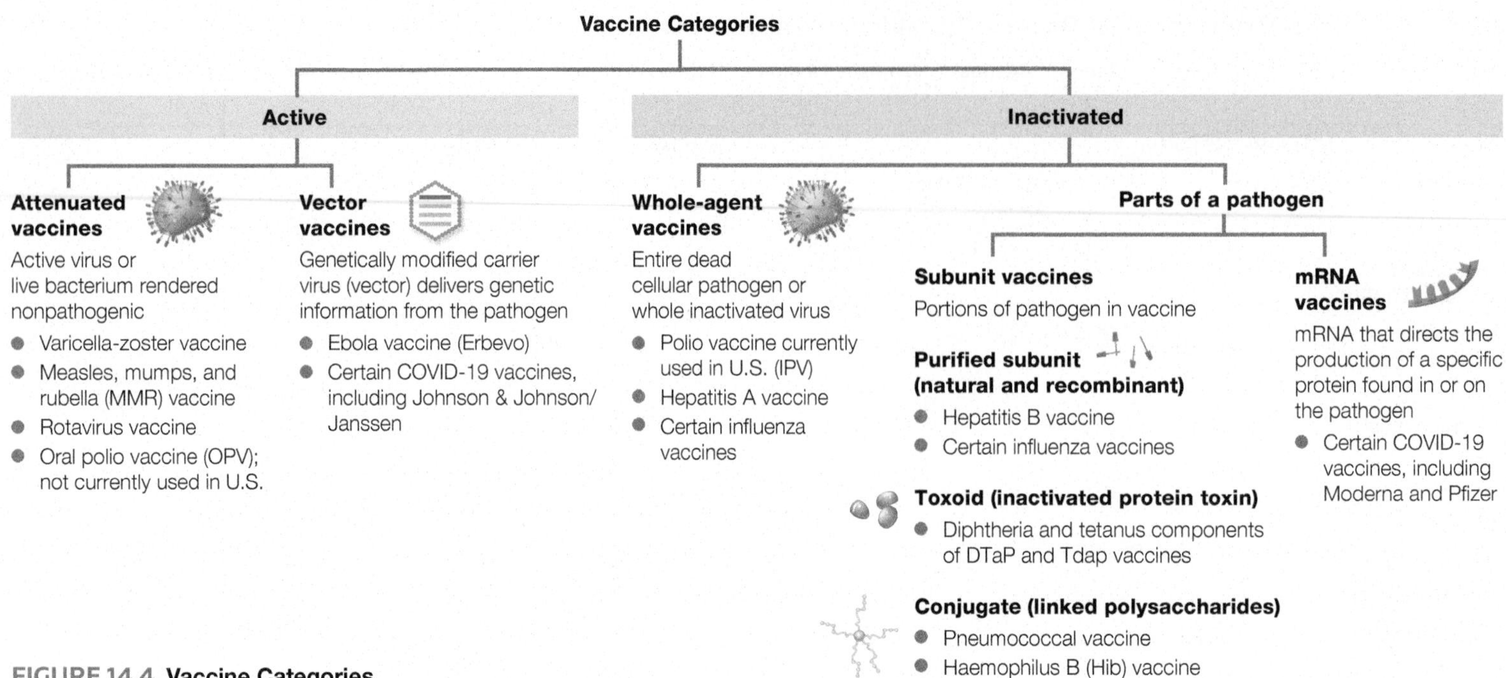

FIGURE 14.4 **Vaccine Categories**

an annual "flu shot." Senior citizens also need routine vaccines. Classic examples are the administration of vaccines to prevent bacterial pneumonia and shingles. (Shingles as a sequela to chickenpox is discussed at length in Chapter 17.) There are many vaccine types, all with different pros and cons.

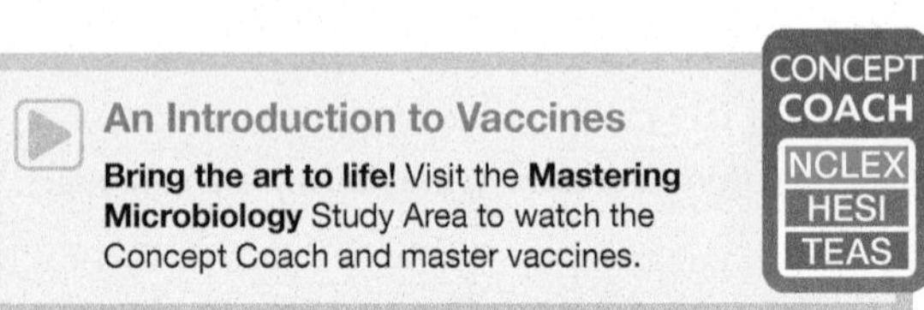

Vaccine formulations are varied.

Vaccines can be injected, inhaled, or ingested, and they come in a range of formulations. They are mainly categorized by how they are made. Some vaccines contain active agents (either live bacteria or active viruses) that have been weakened (attenuated) or genetically modified so they do not cause disease in a normal healthy host. Others consist of inactive agents—killed bacteria, inactivated viruses, fragments of the microbe, a genetically manufactured portion of the microbe, or mRNA that encodes a protein (FIG. 14.4). Whether a virus, bacterium, or parasite is responsible for the disease, scientists develop vaccines based on the features of the responsible agent so that B and T cell-mediated immunity are both activated. Typically, the more closely the vaccination agent resembles the agent encountered in nature, the more likely it is to elicit a strong adaptive immune response that specifically recognizes the pathogen and induces long-term immunological memory.

Live Attenuated Vaccines

Live attenuated vaccines contain pathogens that have been altered so that they do not cause disease (are not pathogenic), but are still infectious. Attenuated agents can be developed in a number of ways that include cultivating the pathogen in cell culture in such a way that it loses its pathogenicity or by genetic manipulation. The chickenpox vaccine is an example of a live attenuated vaccine; it contains varicella-zoster virus that has been weakened by culturing first in guinea pig cell lines and then in human cell lines.

The main benefit of live attenuated vaccines is that they are the closest to the pathogen encountered in nature and therefore tend to simulate potent immunological responses that are accompanied by long-lived memory. Furthermore, different B and T cells are activated against multiple antigens on the pathogen, conferring broader protection. Live attenuated agents remain in the body longer than other vaccine agents, giving them ample opportunity to stimulate an effective immune response.

A drawback of live attenuated vaccines is that the agents in them could cause disease in an immune-compromised host. Another drawback is that, sporadically, the live attenuated pathogen can mutate to an infectious form and cause disease. For example, in rare circumstances, the virus in the oral polio vaccine, which is no longer administered in the United States, mutated back into a pathogenic form, causing paralysis in a very small number of recipients. Live attenuated vaccines can also potentially transmit other infectious agents, such as other viruses, if the tissue culture is contaminated. There has also been some concern from the public about secondary transmission—the possibility that a person might spread live attenuated organisms to others after receiving the vaccine. There are some documented cases of secondary transfer, particularly to immunocompromised individuals. Finally, live attenuated vaccines often require refrigeration right up until they are administered, making them difficult to transport and use in regions where electricity may be unreliable or absent.

Inactivated Vaccines: Whole-Agent and Subunit

Inactivated vaccines can consist of whole inactivated pathogens, such as killed bacteria or inactivated viruses, or parts of pathogens. Many inactivated vaccines are stable at room temperature, which makes them easy to ship and store, plus their noninfectious nature means they are safe for immune-compromised patients. However, their lack of infectivity also means they are quickly cleared from the body, which limits exposure to the antigens and necessitates booster doses to achieve full immunity. There are two general categories of inactivated vaccines: whole-agent and subunit vaccines.

As the name implies, **whole-agent vaccines** contain the entire pathogen, which has been rendered inactive by heat, chemicals, or radiation. The benefit of whole-agent vaccines is that the agent is essentially the same as what would be encountered in nature, but can't cause disease in a weak host or mutate to cause disease. Examples of whole-agent inactivated vaccines include the inactivated polio vaccine (IPV), the rabies vaccine, and the hepatitis A vaccine. Some COVID-19 vaccines (e.g., several being used in China and others that are in development in other parts of the world) are whole-agent inactivated vaccines.

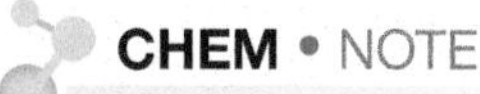

CHEM • NOTE

Killing pathogens through heat is possible because high temperature denatures proteins and enzymes required for pathogen infectivity.

Subunit vaccines do not include whole pathogens; instead, they consist of purified parts of the pathogen (antigens from the actual pathogen) or antigens produced by genetic engineering (*not* directly purified from the pathogen). Because subunit vaccines have fewer antigens than whole-agent or live attenuated vaccines, they usually require **adjuvants** to stimulate a strong immune response. Adjuvants are pharmacological additives that enhance the body's natural immune response to an antigen. The precise mechanisms that adjuvants use to accomplish this are not yet fully understood, but in general they encourage antigen uptake and processing by antigen-presenting cells (especially dendritic cells), and they stimulate the release of certain cytokines that stimulate immune system activation.[7]

The three main groups of subunit vaccines are purified subunit vaccines, toxoid vaccines, and conjugate (or polysaccharide) vaccines.

- **Purified subunit vaccines (natural and recombinant)** Because not all proteins on a pathogen stimulate an immune response—that is, not all pathogen components are *immunogenic*—the best antigens for stimulating the immune response have to be identified and purified. As noted earlier, the pathogen components (antigens) that go into a **purified subunit vaccine** can either be harvested from a natural pathogen or the antigens can be produced using a genetically engineered expression system and then purified from that system. Those that are purified from a genetically

[7] Awate, S., Babiuk, L. A., & Mutwiri, G. (2013). Mechanisms of action of adjuvants. *Frontiers in Immunology*, 4(114), http://doi.org/10.3389/fimmu.2013.00114.

engineered expression system are called **recombinant subunit vaccines**. To make a recombinant vaccine, a gene that encodes a protein antigen (such as a viral surface protein) is inserted into a cell that will make large quantities of the protein—usually a bacterial or mammalian cell culture line is used for this purpose. The desired protein is then harvested and purified for inclusion in the vaccine. The Novavax/Sanofi vaccine against COVID-19 relies on a recombinant purified subunit strategy—as of the date of publication, this vaccine had not yet been FDA approved. We'll explore recombinant DNA methods later in this chapter.

- **Toxoid vaccines** Many bacteria make protein toxins. These naturally occurring toxins can be purified and inactivated to make a class of subunit vaccine called **toxoid vaccines**. The tetanus and diphtheria components of the DTaP and Tdap vaccines are routine toxoid vaccinations. A number of recombinant toxoid vaccines are also in development. Recombinant toxoid vaccines are made by genetically engineering an expression system to make a protein toxin that is then purified, inactivated, and used as a vaccine ingredient.
- **Conjugate (or polysaccharide) vaccines** This class of vaccines uses polysaccharides from bacterial cell parts, such as capsule-associated polysaccharides. These are not as immunogenic as protein antigens. To help polysaccharide antigens stimulate a sufficient immune response, they may be conjugated (or linked) to a more immunogenic protein antigen; therefore, these vaccines are often called **conjugate vaccines**. Routine conjugate vaccines include those that protect against bacterial meningitis (meningococcal vaccines), pneumococcal vaccines against *Streptococcus pneumoniae*, and Hib vaccines that prevent infections by *Haemophilus influenzae* type B bacteria.

Inactivated Vaccines: mRNA Vaccines

This vaccination strategy relies on delivering mRNA—the genetic instructions for assembling a protein—to host cells, whereupon the host cells translate the mRNA to build an antigenic protein that triggers an immune response. Although mRNA immunizations have been in development for decades, they only recently became well known because the first FDA-approved vaccines to prevent COVID-19 (Moderna and Pfizer vaccines) are mRNA vaccines. In general, these vaccines contain mRNA molecules that encode a specific protein from the pathogen of interest—in the case of the COVID-19 vaccines, the mRNA encodes a SARS-CoV-2 spike protein. The purified mRNA is then encased in lipids, which protect the fragile mRNA. The resulting lipid nanoparticles are close in size to a typical virus, and because the lipid encasement is chemically compatible with the cell plasma membrane, host cells readily take up the mRNA-laden particles. (Note, mRNA vaccines using lipid nanoparticles are technically different from nanoparticle vaccines, which are reviewed in the Bench to Bedside Feature.)

Once the mRNA enters host cells, their ribosomes translate the nucleic acid to build the target protein(s) that will stimulate an immune response. The vaccine does *not* alter host cell DNA; it simply allows host cells to *temporarily* make the protein that the mRNA encodes. Because the mRNA is only temporarily present in host cells, booster shots are usually necessary to allow for a second exposure that will bolster the immune response and enhance the development of immunological memory. (Refer to Chapter 12 to review secondary immune response features.) Because these vaccines rely on host cells for making the antigen, adjuvants are not required to enhance vaccine efficacy.

Recombinant Vector Vaccines

For this immunization method, select genes from a pathogen are packed inside a harmless virus or bacterium, which is referred to as the vector. The engineered vector is then purified and introduced into the body (FIG. 14.5). If an

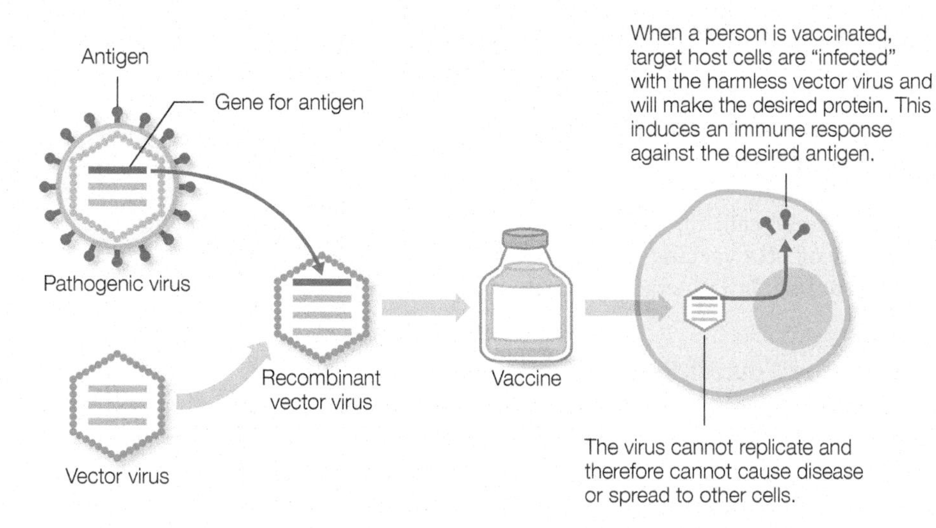

FIGURE 14.5 Recombinant vector vaccine To review this concept, let's consider a recombinant vector vaccine against influenza. In this case, a copy of the hemagglutinin gene from a pathogenic influenza virus is inserted into a harmless virus (also called a vector). The resulting recombinant vector is introduced to the human host. Once it infects host cells, it triggers the production of the immunogenic antigen (the hemagglutinin surface protein in this case).

Critical Thinking *What type of information must be known about the pathogen before a recombinant vector vaccine can be made?*

engineered viral vector is used, it will cause the host cells to make pathogen antigens to stimulate an immune response. In scenarios that employ engineered bacteria as the vector, the harmless bacteria are genetically modified to make selected pathogen antigens that will stimulate an immune response.

In 2019 the first recombinant vector vaccine, Erbevo, was FDA approved to prevent Ebola virus disease in adults (age 18 and older). Then, in 2021, the FDA issued an emergency use authorization for Johnson and Johnson's Janssen COVID-19 vaccine. The Janssen vaccine relies on adenovirus type 26 (Ad26) as the viral vector to deliver a piece of DNA that encodes a SARS-CoV-2 spike protein. As host cells produce the viral spike protein, the immune system is activated to target the foreign protein.

New vaccines are in development for persistent and emerging diseases.

As new technologies develop, so do new approaches for vaccines. A relatively new vaccine approach is to use a pathogen's DNA to stimulate an immune

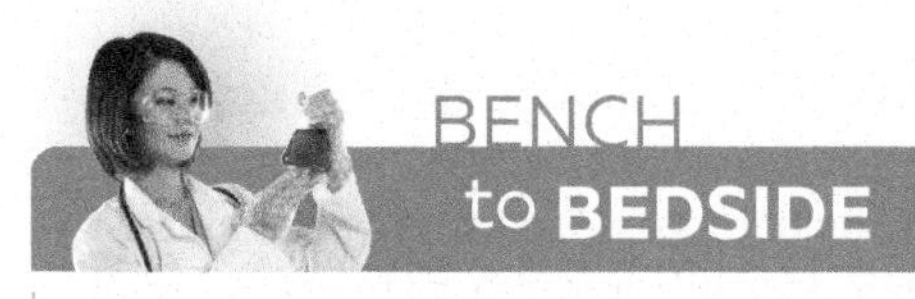

BENCH to BEDSIDE

Nanoparticle Vaccines

Scientists are searching for new vaccine formats that are easy to deliver and trigger a strong immune response and immunological memory. Nanoparticle vaccines are one candidate formulation. At just 1–100 nanometers in diameter, nanoparticles are exceptionally tiny. To give you some perspective, a nanometer is one *billionth* of a meter—that's about the span of a couple of silicon atoms or approximately half the diameter of a DNA molecule. Vaccines could be made by attaching antigens to nanoparticles made of gold, carbon, phospholipids, or even engineered organic polymers. Unlike current mRNA vaccines that use lipid-based nanoparticles to deliver mRNA into target cells for antigen production, nanoparticle vaccines deliver the actual antigen to the body—that is, they do not rely on host cells to make the antigen via mRNA translation. Nanoparticles can be taken up by antigen-presenting cells, thus allowing for effective antigen processing and presentation. (Refer back to Chapter 12 to review antigen processing and presentation as an essential part of activating adaptive immunity.) Another key benefit is that nanoparticles elicit a stronger immune response than the antigen would stimulate on its own.

This emerging vaccine technology has also made a debut in the fight against COVID-19. For example, Novavax developed a COVID-19 vaccine by linking SARS-CoV-2 spike protein to a lipid carrier. In clinical trials, the Novavax vaccine was highly effective against newly emerging UK and South African strains of SARS-CoV-2. Researchers are also investigating nanoparticle vaccines as a strategy to protect against a variety of other infectious agents such as influenza, respiratory syncytial virus, and rotavirus—and even as cancer immunotherapies.

Sources: Diaz-Arévalo, D., & Zeng, M. (2020). Nanoparticle-based vaccines: Opportunities and limitations. In R. Shegokar (Ed.), *Nanopharmaceuticals* (pp. 135–150). Amsterdam: Elsevier. Mahase, E. (2021). Covid-19: Novavax vaccine efficacy is 86% against UK variant and 60% against South African variant. *BMJ*, *372*, n296. Muluh, T. A., Chen, Z., Li, Y., Xiong, K., Jin, J., Fu, S., & Wu, J. (2021). Enhancing cancer immunotherapy treatment goals by using nanoparticle delivery system. *International Journal of Nanomedicine*, *16*, 2389.

response. These **DNA vaccines** require identification of genes that encode highly immunogenic antigens. The genes are then placed into a plasmid.[8] The engineered plasmid is then injected into a human host. Some cells of the human body take up the plasmid and transcribe and translate the pathogen genes to create the corresponding protein antigens. Essentially, the recipient's cells become the antigen producers. When the antigen is secreted or displayed by the human cell, a humoral and a cellular immune response can ensue. The majority of DNA vaccines being developed focus on HIV or cancer. Other targets being studied for DNA vaccines include West Nile virus and influenza.

Build Your Foundation

3. What is the general immunological premise behind how vaccines protect patients? (LO 14.3)
4. What is herd immunity, how is it generated, and how does it protect a nonimmunized person? (LO 14.4)
5. Compare and contrast the various categories of vaccine formulations. Be sure to review the pros and cons of each. (LO 14.5)
6. What are adjuvants, and when would they be used? (LO 14.6)
7. How do DNA vaccines and recombinant vector vaccines stimulate immunity? (LO 14.7)

Build Your Foundation (BYF) Quick Quiz: Visit the **Mastering Microbiology** Study Area to quiz yourself.

[8] Plasmids, introduced in the genetics chapter (Chapter 5), are further discussed later in this chapter as tools for genetic engineering. For now, suffice it to say that a plasmid is a small circular piece of DNA that cells can take up and use to direct protein production.

14.3 IMMUNOLOGICAL DIAGNOSTIC TESTING

Learning Outcomes

After reading this section, you should be able to:

14.8 Describe the advantages of immunological over biochemical testing.

14.9 Describe agglutination reactions, and explain how they can be used to determine blood type.

14.10 Describe how plaque reduction neutralization tests work and how they are clinically useful.

14.11 Compare and contrast direct, indirect, and sandwich ELISA techniques.

14.12 Explain how fluorescent-tagged antibodies may be used in diagnostic testing.

14.13 Discuss how interferon gamma release assays work and why they are an important development in the fight against tuberculosis.

As reviewed in Chapter 8, biochemical tests are useful for identifying bacteria that are responsible for an infection. However, these methods can take up to 24 hours or more to perform, and they are useless if the pathogen can't be cultured in the lab. They also can't identify noncellular pathogens like viruses that lack their own biochemical processes. Immunological diagnostics are therefore essential tools for identifying a variety of cellular as well as viral pathogens and determining if a patient has been exposed to a particular pathogen or has developed immunity to it. Immunological tests are typically a part of **serology**, the study of what is in a patient's serum. Recall that serum is the liquid portion of blood that remains once all the formed elements (platelets and all blood cells) are removed by centrifuging (spinning) the sample at high speed using a machine called a **centrifuge**. Often the goal of serology is to determine if a patient has certain antigens and/or antibodies in their blood. Because antibodies bind to specific antigens, antibody–antigen reactions can be used to quickly identify specific pathogens in a patient sample or to determine if a patient has made antibodies to a particular antigen. This section describes select antibody–antigen reactions that are useful for diagnostic purposes.

Agglutination tests reveal antigen–antibody interactions.

Antibodies can attach to more than one antigen because each antibody molecule has two antigen-binding sites. (See Chapter 12 to review antibody structure.) Therefore, antibodies can bind antigens into a clump. This effect is called

agglutination. Agglutination reactions can be seen in the lab when antibodies interact with cells that display multiple antigens on their surface or with tiny synthetic beads coated with antigens. These tests are usually run on a slide or in the wells of a microtiter plate.

Agglutination reactions are usually used for blood typing, to identify infections, and to diagnose noninfectious immune disorders such as autoimmune hemolytic anemia (a disease where autoantibodies are generated against red blood cells). Testing can identify either the antibody or the antigen in patient samples such as serum, urine, or cerebrospinal fluid. For example, the *Treponema pallidum* particle agglutination assay (TPPA) test confirms a syphilis diagnosis by detecting patient antibodies against the bacterium *T. pallidum*. The patient's serum is collected from a blood sample, diluted, and added to a small well that contains tiny sand-sized beads coated in *T. pallidum* antigens. The beads agglutinate (or clump) if there are antibodies in the patient's serum that bind the antigens on the beads (FIG. 14.6).

Blood typing can also be determined by agglutination tests (FIG. 14.7). As described in Chapter 13, a person's blood type is determined by what antigens are on their red blood cells. By adding antibodies that detect A or B antigens, a person's red blood cell antigens can be determined. Type A blood will only agglutinate in the presence of antibodies that detect the A antigen. Type B blood will only agglutinate when antibodies to the B antigen are added. Type AB blood will exhibit agglutination when antibodies to A or B are added. Type O blood lacks both A and B antigens and will not agglutinate when either one is added. The Rh status of a patient's blood (Rh^+ or Rh^-), is also determined by agglutination tests. Determining Rh status relies on anti-D antibodies, which agglutinate red blood cells bearing the Rh factor.

Neutralization reactions are useful for detecting immunity to certain viruses.

A young mother has just gotten upsetting news. An ultrasound revealed that the baby she is carrying has microcephaly, an abnormally small skull and brain. Because the patient frequently travels to Brazil to see family, the doctor

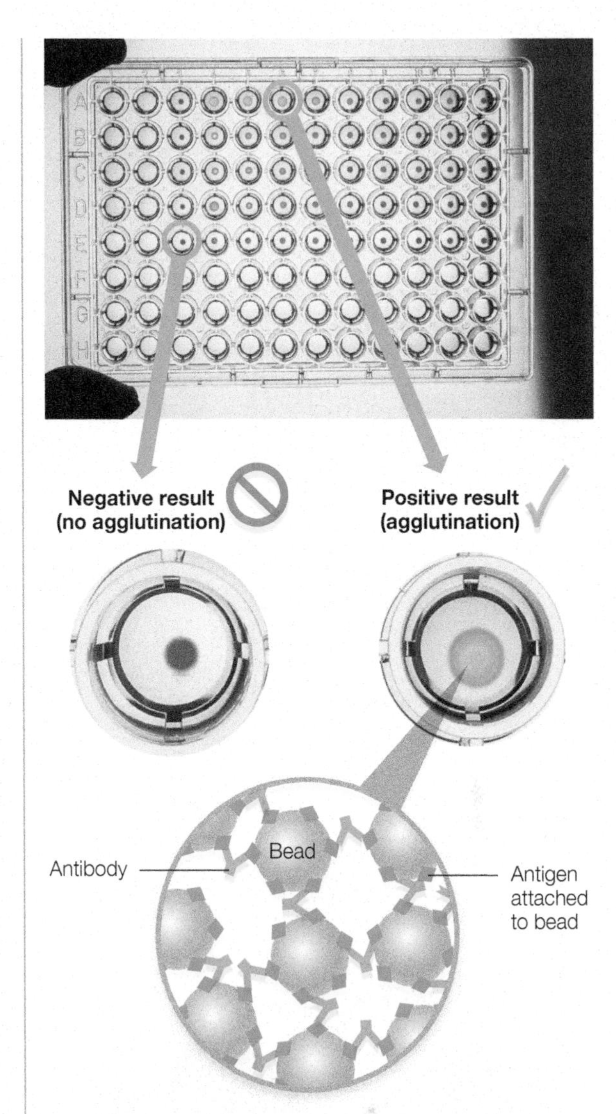

FIGURE 14.6 Agglutination assay for syphilis A patient's diluted serum is added to wells in a plate containing antigen-coated beads. If antibodies to *T. pallidum* antigens are present in the patient's serum, they will cross-link the beads, producing a broad patch of agglutination (a positive result). If antibodies are absent, the beads won't agglutinate, and instead they settle at the bottom of the well as a small spot (a negative result).

Critical Thinking *Why does the antigen have to be attached to beads for this test?*

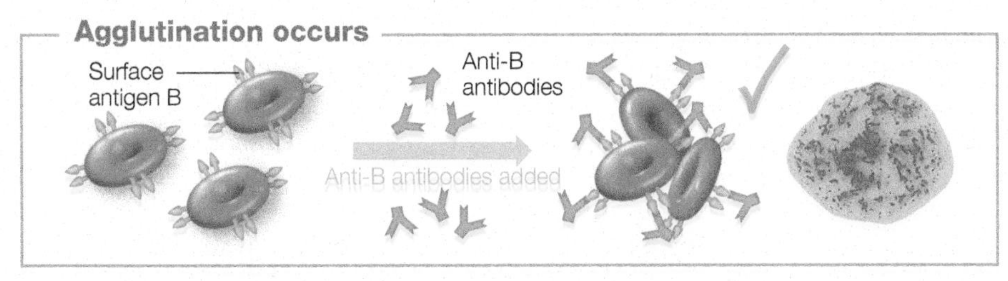

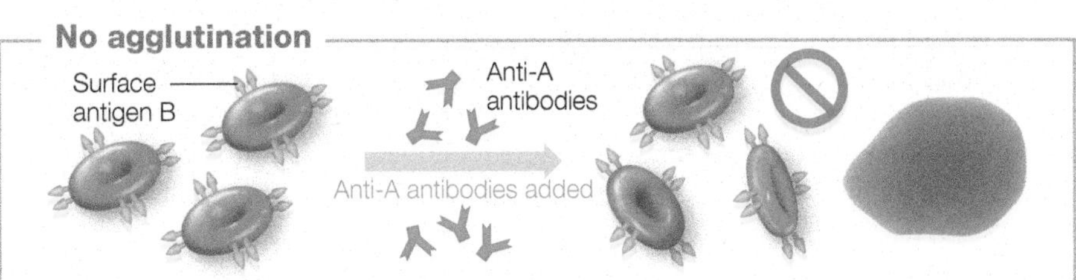

FIGURE 14.7 Agglutination to determine blood type The reactions for type B blood in response to anti-A and anti-B antibodies are shown. Agglutination will occur only if antibodies interact with antigens on the red blood cell surface.

Critical Thinking *Technically, type O blood contains O antigens. However, detecting this antigen would not differentiate between A, B, or O blood type. Based on the discussion of the O antigen in Chapter 13, why is this the case?*

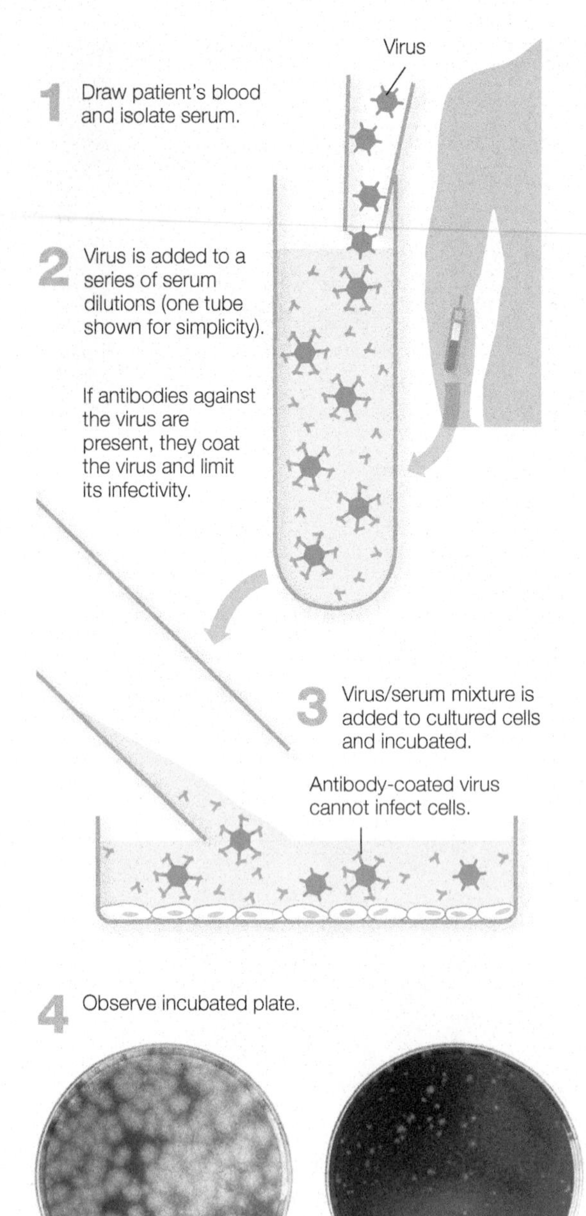

FIGURE 14.8 Plaque reduction neutralization test (PRNT) This test detects patient antibodies against a specific virus.

Critical Thinking *Could the PRNT be used to diagnose a patient in the prodromal phase of infection? (Hint: Refer to Chapter 9 to review the prodromal phase.)*

suspects the case is due to an emerging virus—Zika. How can it be determined if the mother was exposed to Zika virus when the virus is no longer in her system? This is a scenario where a test called a **plaque reduction neutralization test (PRNT)** is helpful (FIG. 14.8).

In this test, the patient's serum is extracted from a blood sample and serially diluted. Next, a purified preparation of the suspected virus is added to the various tubes of diluted serum. Each serum/virus mixture is then added to separate petri plates of cultured cells and incubated for a few days. If the serum lacks antibodies to the virus (or if antibody levels are very low), then the virus infects the cultured cells at a level similar to that seen for a control sample lacking antibodies to the virus. As cells in the culture are infected, **plaques** form—the plaques develop as cells are lysed by the virus.[9] If there are sufficiently high levels of virus-specific antibodies in the tested serum, then they bind to the virus and *neutralize* it. Neutralized virions can't infect the cells in the culture because they are coated with antibodies and are blocked from binding to host cells; this leads to a reduced number of plaques.

Because the PRNT is highly specific and allows for differentiation between infections with viruses that have antigen overlap (like the closely related flaviviruses Zika and dengue), this method is often preferred over other faster and easier methods like the enzyme-linked assays that we'll review next.

Enzyme-linked immunosorbent assays (ELISAs) leverage antigen–antibody interaction for rapid diagnosis.

Some of the most versatile and rapid diagnostic tests used today are **enzyme-linked immunosorbent assays** (ELISAs). With applications that include pregnancy tests, procedures for certifying a food as "allergen free," tests for illicit drugs in urine or blood, and diagnostics for hundreds of infectious and noninfectious diseases, ELISA methods have revolutionized medicine and industry. Like the tests we've already reviewed, ELISA techniques rely on antibody–antigen interactions. However, unlike the techniques described so far, these tests rely on a reporter enzyme that is attached to a detecting antibody—this is where the tests get the "enzyme-linked" part of their name. For best results, the detection antibody should be a **monoclonal antibody**—an antibody that recognizes just one epitope of a particular antigen. (See Chapter 12 for more on epitopes.)

The reporter enzyme (often horseradish peroxidase or alkaline phosphatase) chemically modifies an added substrate in a way that can produce a color change or release light.[10] An instrument called a plate reader is used to measure the color changes or detect emitted light.

Although there are three main formats for ELISA (direct, indirect, and sandwich, all reviewed next), there are hundreds of commercially available ELISA kits that quickly and accurately detect either a specific antigen, such as a protein or hormone, or a specific antibody in a patient sample. ELISA tests require only a small amount of a liquid sample and can analyze many samples at once because they are usually performed on 96-well plastic microtiter plates. Also, depending upon the way the test is designed, ELISAs provide quantitative information such as how much of a given antigen or antibody is present in the tested sample.

[9] Plaques were also covered in Chapter 6 in reference to bacteriophages lysing bacterial cells in a culture; refer to that content to review tests that allow us to enumerate viruses in a sample. You can see Chapter 10 for more on cytopathic effects, but cell lysis (bursting) is one of many potential cytopathic effects associated with viral infections.

[10] Fluorescently tagged antibodies are used in immunofluorescence assays (IFAs). Such assays do *not* require a reporter enzyme reaction to detect bound antibodies and are therefore mechanistically distinct from ELISAs. IFAs are reviewed later in this chapter.

Direct ELISA

Direct ELISA methods allow for the identification of antigens or antibodies in a sample. The technique is mostly used for research applications, but it is presented here because it is the simplest version of ELISA and represents the technique upon which other more commonly used ELISA methods are built. In this test a solution containing antigens is added to microtiter plate wells in a manner such that the antigens will stick to the bottom of the wells. Next, detection antibodies that are linked to a reporter enzyme and only bind to the antigen of interest are added. If the antigen of interest is in the wells, then the detection antibodies will directly bind it—which is why this test has the term "direct" in its name. Any unbound antibodies are then washed out of the wells. Bound antibodies are detected by adding a substrate to the wells. The substrate interacts with the antibody-linked reporter enzyme and either changes color or emits light. The plate is then inserted into a plate reader to measure the signal in each well (FIG. 14.9, *top*). This test is not recommended for crude samples because less abundant antigens may be outcompeted by more abundant antigens for binding to the bottom of the plastic wells and therefore may not be detected.

Direct ELISAs can also be modified to determine if a particular antibody is present in a test sample—such as a patient's or an animal's blood. To do this, a known purified antigen, such as an inactivated virus, is used to coat the microtiter plate well bottoms. Then, serum from the test subject is treated such that the antibodies in the sample are conjugated to reporter enzymes. The newly enzyme-linked antibodies are then added to the antigen-coated wells. Excess antibody is washed away, and indicator substrate is added. The downside of this form of the direct ELISA is that it is laborious and challenging to label diverse antibody isotypes, making this a fairly impractical approach. The indirect ELISA that we review next makes detecting antibodies in a sample much easier and is therefore far more commonly used as compared to direct ELISA methods.

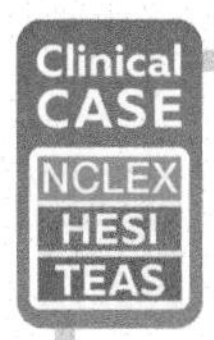

The Case of the Deadly Delay

Practice applying what you know clinically: visit the **Mastering Microbiology** Study Area to watch Part 3 and practice for future exams.

Indirect ELISA

A more common tool for determining if a patient has antibodies to a particular pathogen is the **indirect ELISA**. This form of ELISA requires two antibodies. The first antibody recognizes the antigen bound at the bottom of the plate wells, and the second antibody is linked to an enzyme and serves as the detection antibody. To perform the test, the patient's serum is added to plates that have wells that were precoated with known antigens, such as an inactivated form of the suspected pathogen. Any patient antibodies that recognize the antigens coating the wells will bind. The antibodies that don't bind get rinsed away. Next, the detection antibody is added; it is designed to bind to specific isotypes of human antibodies present so that specific isotype titers can be determined.[11] After unbound detection antibodies are rinsed away, a substrate that interacts with the reporter enzymes is added; the resulting signal levels are then measured (FIG. 14.9, *middle*).

Sandwich ELISA

Like the direct ELISA, the sandwich ELISA format allows us to detect a specific antigen in a sample, but it is more sensitive and overcomes the downsides of the direct ELISA. In its simplest form, the **sandwich ELISA** requires two antibodies: a capture antibody and a detection antibody (FIG. 14.9, *bottom*). To perform the test, a sample is added to plates that have well bottoms coated with capture antibody. Antigens in the sample that bind the capture antibody will do so, and

[11] We reviewed antibody isotypes and the importance of knowing their levels (titers) in Chapter 12. Recall that IgM isotype titers are highest in an initial infection, while IgG isotype is higher in subsequent infections as well as in patients considered immune to a given pathogen due to a resolved infection or successful vaccination. It is also clinically useful to determine how much IgE isotype is present against a particular antigen for allergy assessments.

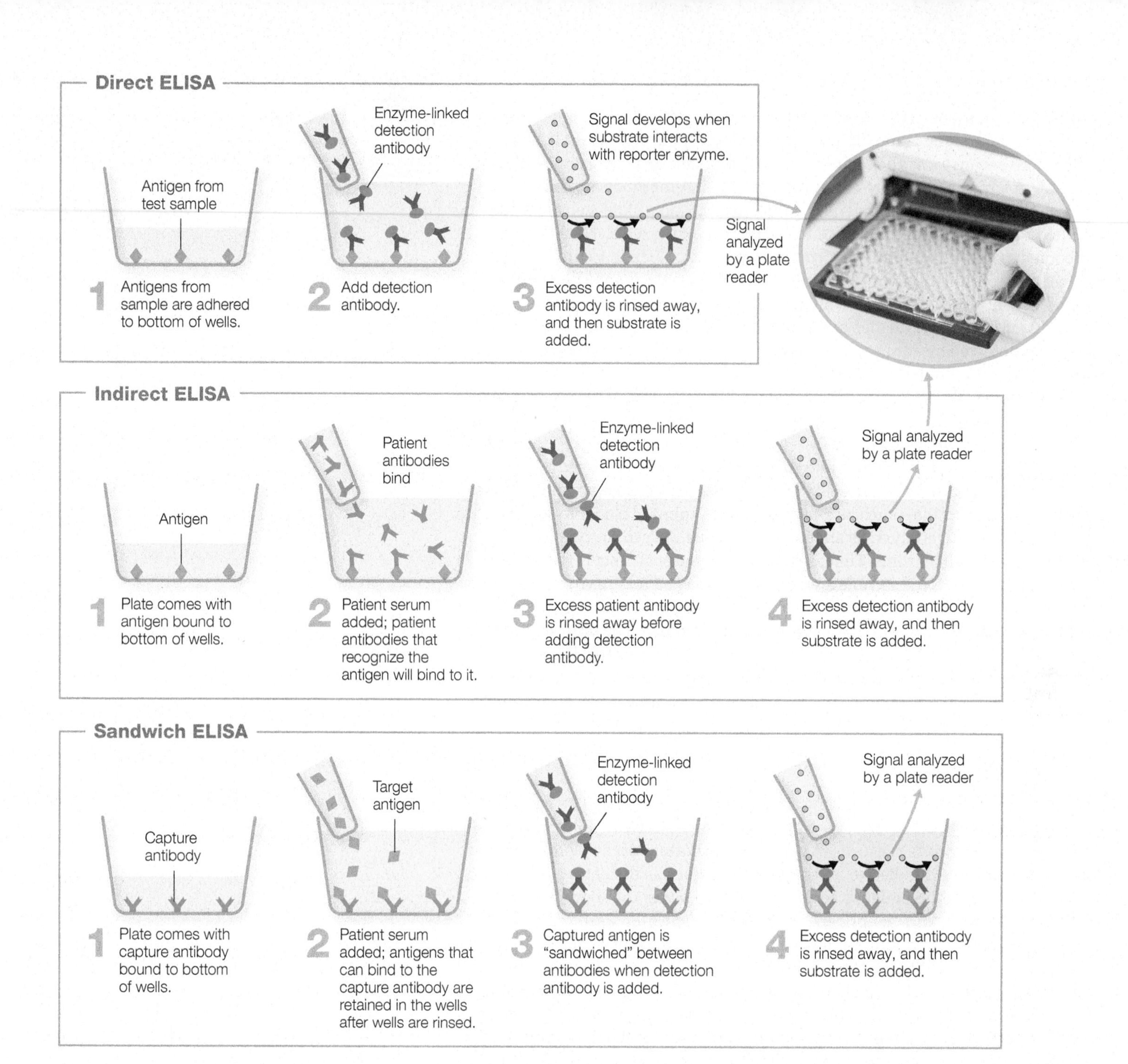

FIGURE 14.9 Direct, Indirect, and Sandwich ELISA

Critical Thinking *Assume the technician performing a direct ELISA on a patient sample forgot to rinse the plate wells after the detection antibody step. What would you expect in terms of test results?*

Key Molecular Diagnostics 101

Bring the art to life! Visit the **Mastering Microbiology** Study Area to watch the Concept Coach and master the basics of molecular diagnostics.

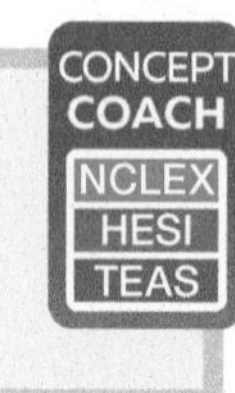

then unbound components are washed away. A second antigen-specific antibody is added and allowed to bind, thus "sandwiching" the antigen between two antibodies. This second antibody is linked to a reporter enzyme. Sometimes the sandwiching antibody is *not* linked to a reporter enzyme, in which case a third enzyme is needed to serve as the detection antibody; the third antibody is designed to bind the sandwiching antibody. Tests that require the third antibody for detection are sometimes called *indirect sandwich ELISAs*, while those that use two antibodies are occasionally called *direct sandwich ELISAs*. After rinsing away any unbound detector antibodies, an indicator substrate is added to each well, and the degree of signal is measured.

Some sandwich ELISA tests, like home pregnancy tests, are designed so they don't require specialized plate readers (FIG. 14.10). Instead, a positive test result is indicated by the development of a colored line if the urine contains the pregnancy hormone hCG (human chorionic gonadotropin).

Fluorescent-tagged antibodies can detect antigens or antibodies in a sample.

Imagine that public health authorities suspect that a dog is rabid. After euthanizing it, they test brain tissue from the animal to determine if the rabies virus is present. This is accomplished by **immunofluorescence microscopy**, a protocol in which fluorescent-tagged antibodies that recognize a specific antigen are incubated with tissue or cell samples, after which the sample is observed with a specialized microscope. In this case, fluorescent-tagged antibodies will bind to rabies antigens in infected cells (FIG. 14.11). As reviewed in the microscopy section of Chapter 1, when viewed with an ultraviolet-light microscope, the fluorescent-tagged antibodies glow, making it easy to decide if the animal was rabid. Immunofluorescence microscopy can also detect certain cancer cells in biopsied tissue and a variety of infections, including genital herpes and syphilis.

Fluorescent antibodies are also used in **immunofluorescence assays (IFAs)**. Like ELISAs, IFAs detect antigens or antibodies in a patient sample, but the detection antibody is linked to a fluorescent tag instead of an enzyme. This eliminates the need to add substrate for the detection step and minimizes sample processing.

Lastly, cells can be sorted using a process called **flow cytometry**. As discussed in Chapter 7, this method relies on the use of a special piece of equipment called a **fluorescence-activated cell sorter (FACS)**. The overall process is useful for enumerating specific cells, such as detecting T helper cell levels in HIV patients. To perform flow cytometry, fluorescent-tagged antibodies are incubated with a patient blood sample. After unbound antibodies are removed, the sample is loaded into the FACS machine, which counts tagged cells and sorts them from nontagged cells (see Figure 7.19).

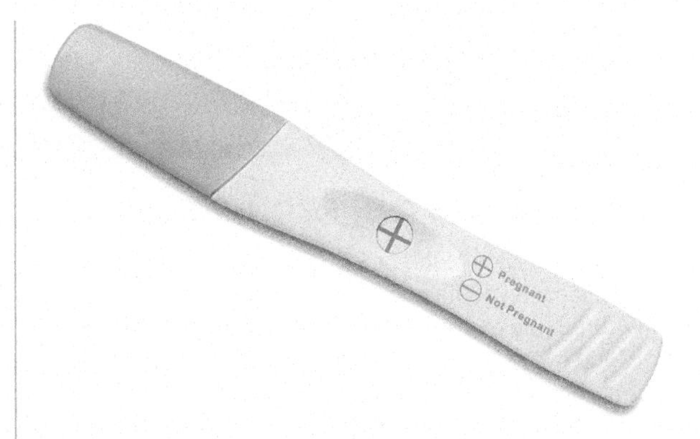

FIGURE 14.10 A home pregnancy test A built-in control generates one line on the pregnancy test—this line informs the user that the test is working as designed. A second line only develops if the urine contains the pregnancy hormone hCG (human chorionic gonadotropin).

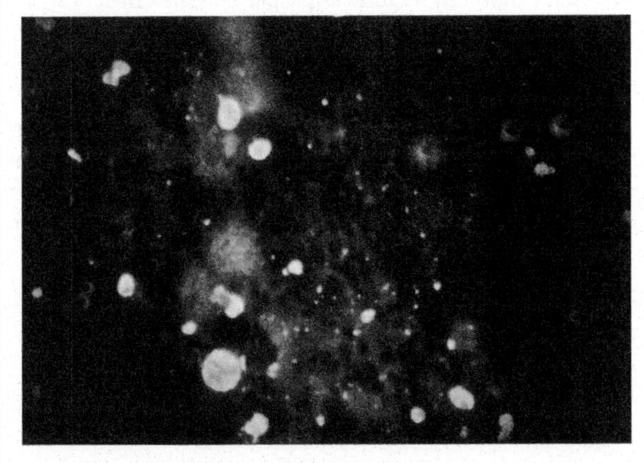

FIGURE 14.11 Immunofluorescence microscopy Here a brain tissue section glows green wherever rabies antigens are bound by fluorescently tagged antibodies.

Interferon gamma release assays (IGRAs) detect tuberculosis infections.

According to the CDC, about a third of the world's population is infected with tuberculosis (TB), and in 2019 there were about 1.4 million TB-related deaths worldwide. Why is a bacterium that is usually treatable with antibiotics such a problem? One answer lies in the challenge of diagnosing an infection before lung damage arises. In the United States, a cheap and easy tuberculin skin test makes monitoring for TB simple. However, not all countries can use the skin test to screen their residents because it is only reliable if a person has not been vaccinated against TB. Because the vaccine is only recommended for populations at high risk for TB exposure, the United States and a few other developed countries do not routinely administer it; however, most countries do. After over 100 years of having only the tuberculin skin test and a variety of unreliable serology tests as TB screening tools, in 2001 **interferon gamma release assays (IGRAs)** finally provided a fast and relatively reliable way to detect TB in the early stages in vaccinated populations.

IGRAs work by measuring how a patient's T cells respond to *Mycobacterium tuberculosis* antigens. T cells from a person who has TB (either a latent or an active infection) release more interferon gamma (IFN-γ) than T cells from someone who does not have TB. (See Chapter 11 to learn more about the cytokine IFN-γ.) To perform the test, a patient's blood is mixed with *M. tuberculosis* antigens, and the level of IFN-γ released is measured. IGRAs are sensitive, require only a single patient visit, and can provide results within a day. However, they can't differentiate between an active versus latent TB infection. To optimize treatment, follow-up tests can determine if a patient with a positive IGRA has an active versus latent infection.

TRAINING TOMORROW'S HEALTH TEAM

Point-of-Care Tests for Quick Diagnosis

Point-of-care tests, or POCTs, are now a mainstay in most doctors' offices. If you have ever gone to your doctor complaining of a sore throat, he or she most likely swabbed the back of your throat and left you in the room for 10 minutes. After returning, your doctor might have said, "Yes, you are positive for strep!" During those 10 minutes, most likely your physician used an immunoassay-based POCT that detects antigens associated with the strep throat bacterium *Streptococcus pyogenes*. POCTs provide rapid diagnosis and easier monitoring for a variety of conditions, facilitating precise treatments without the higher cost or wait time associated with outsourced laboratory tests. Urine tests for pregnancy and drug levels, and finger-prick tests for glucose and cholesterol monitoring are other common POCTs.

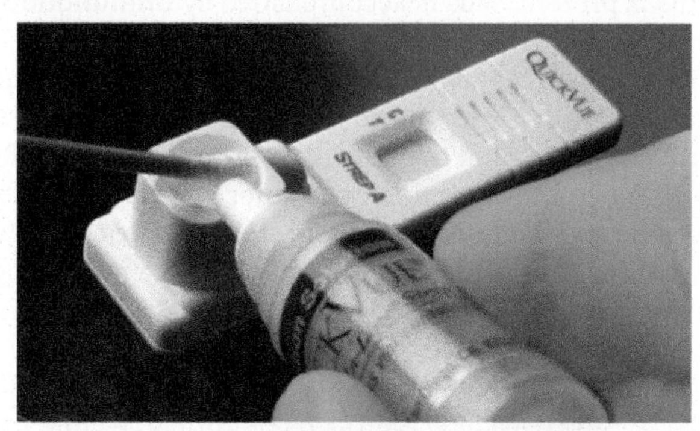

Image courtesy of Quidel Corporation.

Like any medical test, POCTs are subject to potential sources of error that may include improper sample collection or processing, sample or reagent contamination, faulty equipment (especially incorrect calibration), and expired or improperly stored reagents. Furthermore, it is important that healthcare providers and patients understand the limits of POCTs, such as the relative sensitivity and reliability of these tests compared with tests performed in a lab. In order for any diagnostic to be useful, it must be sensitive, specific, and accurate—but no test is 100 percent infallible. One measure of test reliability is the rate of false negatives (tests that read negative when they should not) and false positives (tests that read positive when they should not). (See Chapter 6 for a review of this concept.) Generally speaking, POCTs have a higher margin of error than outsourced lab tests, but the benefits of POCTs far outweigh any downsides, and patient care is ultimately improved by these faster and cheaper test options.

QUESTION 14.2

Regarding the POCT for strep throat, what factors could lead to a false-negative result?

Western blotting and complement fixation assays are less common immunodiagnostic techniques.

Western blotting is a protocol that detects specific proteins in a sample, whereas **complement fixation assays** detect patient antibodies in a serum. (See Chapter 12 for a review of complement cascades.) Although these methods are still occasionally used in clinical labs, faster and more sensitive techniques like ELISAs and some of the genetics methods we'll review next have made these tests much less prevalent in modern diagnostic testing.[12] Western blotting is still an extremely useful protocol in scientific research, and there is a good chance that if you were to pursue a career in basic research, you would perform plenty of Western blots.

Like ELISA methods, Western blots can reveal what proteins are present as well as their levels (although Western blots are not currently as sensitive as ELISA for quantifying proteins). Unlike ELISA, Western blots provide added information about the size of the protein being detected. Western blotting starts with separating proteins by size using a technique called electrophoresis. The samples are loaded onto a polymer gel (usually a polyacrylamide gel), and then an electrical current is applied to the gel. In response to the applied current, the proteins basically wiggle their way through the gel matrix. Small

[12] Balmer, C. (2015). The past, present, and future of Western blotting in the clinical laboratory. *Clinical Laboratory News*. Professional communication article published by the American Association for Clinical Chemistry (AACC). https://www.aacc.org/publications/cln/articles/2015/october/the-past-present-and-future-of-western-blotting-in-the-clinical-laboratory

proteins make their way through the gel faster than larger proteins. After the gel has run, the proteins are then transferred or "blotted" from the gel onto a specialized membrane. This membrane is then exposed to detection antibodies, which lead to a dark blot on the film wherever the target protein is located (FIG. 14.12).

FIGURE 14.12 **A Western Blot** Dark blots on the film are areas where the target protein was detected.

Build Your Foundation

8. What advantages do immunological tests hold over biochemical tests? (LO 14.8)
9. How might agglutination reactions be used to determine blood type? (LO 14.9)
10. How does the plaque reduction neutralization test work, and why is it clinically useful? (LO 14.10)
11. What are the main similarities and differences between the various ELISA formats, and how is each clinically useful? (LO 14.11)
12. How are fluorescent-tagged antibodies used in diagnostic procedures? (LO 14.12)
13. What are IGRAs, how do they work, and why are they an advancement over prior tuberculosis diagnostics? (LO 14.13)

Build Your Foundation (BYF) Quick Quiz: Visit the **Mastering Microbiology** Study Area to quiz yourself.

14.4 SELECTED GENETICS APPLICATIONS IN MEDICINE

The polymerase chain reaction (PCR) can help diagnose infections and genetic disorders.

The American Red Cross collects about 13.6 million units of blood per year. All donated blood is tested for bloodborne pathogens like HIV to ensure it is safe for transfusion. Immunological techniques, like those reviewed in the last section, are powerful tools for ensuring a safe blood supply, but what if a blood donor is infected with a virus like HIV or hepatitis B and gives blood during the **seroconversion period**—the weeks between infection and the production of antibodies? Without techniques like the **polymerase chain reaction (PCR)** to detect viral genetic material, our blood supply wouldn't be nearly as safe as it is.

PCR methods create billions of copies of a target gene in just a few hours. This makes PCR ideal for facilitating gene sequencing for genetic disorders or diagnosing infections, even at their earliest stage—PCR is sensitive enough to detect even a single viral particle in a sample. PCR's speed and sensitivity have made it an essential tool in clinical labs.

Learning Outcomes

After reading this section, you should be able to:

- **14.14** Describe the polymerase chain reaction (PCR), real-time PCR, and reverse transcription PCR (RT-PCR), and state the clinical applications of each.
- **14.15** Describe the three general steps for producing a recombinant DNA (rDNA) vector, state how rDNA can be introduced into cells, and discuss the clinical applications of rDNA.
- **14.16** Discuss the features and functions of restriction enzymes.
- **14.17** Explain how the CRISPR-Cas9 protein system edits genetic material.
- **14.18** Describe the general process of gene therapy, and state how viruses are used in this process.
- **14.19** Explain what genome maps are, and state why they are useful.
- **14.20** Describe DNA microarrays, and explain how they are applied to clinical diagnostics.

PCR Technique

The process is performed using a machine called a **thermocycler** and requires the following reagents: (1) At least one copy of the template DNA to be copied; (2) two single-stranded DNA primers that are usually about 20 nucleotides long and are complementary to the DNA sequences that flank a target gene; (3) DNA polymerase that is heat resistant, such as ***Taq* polymerase**—an enzyme from the hot springs bacterium *Thermus aquaticus*; and (4) deoxynucleotide triphosphates (dNTPs)—the building blocks of DNA (FIG. 14.13).

A small tube with reagents mixed in the proper concentrations is placed in the thermocycler machine, which is programmed to cycle through a series of temperature changes. First, a short period of high temperature (about 95°C) is applied to separate double-stranded DNA into single strands that can be copied. Next, the machine quickly brings the tube contents to a lower temperature (usually between

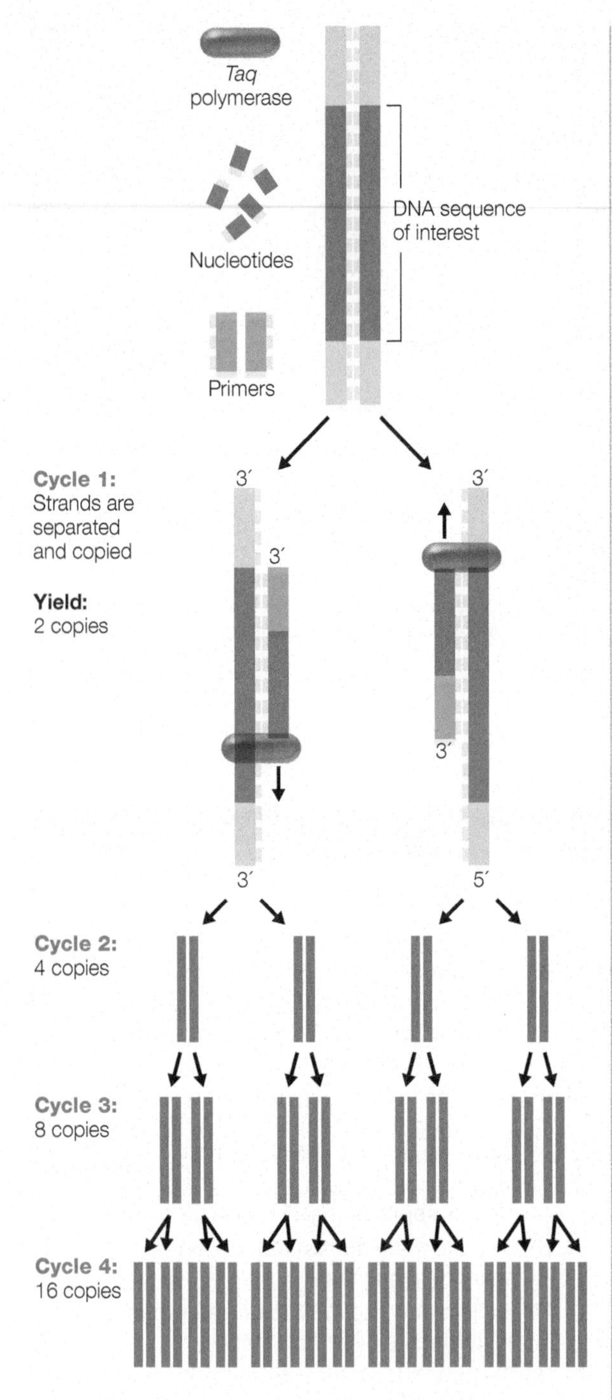

FIGURE 14.13 The polymerase chain reaction

Critical Thinking *Do you need to know the whole sequence of the gene being copied in order to perform PCR?*

50°C and 65°C),[13] which allows the primers to *anneal* (or pair) with the template DNA to serve as a start point for the DNA polymerase to copy the template. During the *extension period* of a given cycle, the tube contents are held at a temperature that allows DNA polymerase to copy the target DNA. The extension temperature used depends on the polymerase, but is usually between 65°C and 75°C.

During each series of heating and cooling cycles, new copies of the gene are made, each of which becomes the subsequent template for the next round of copies. Therefore, the number of copies of a desired gene doubles with each cycle and exhibits an exponential increase of 2^n, where n is the number of copying cycles. Based on this formula, if the reaction started with *just one copy* of a target gene, after 30 cycles of copying we'd have over 10 billion copies.

Real-time PCR

The amplified DNA made in traditional PCR can be analyzed using a method called **gel electrophoresis**, in which molecules are separated according to their size. This process can take several hours. Fortunately, a faster PCR method called **real-time PCR** is available. Real-time PCR uses fluorescence imaging to visualize DNA copies as they are made—making the data "real time," or immediate. This method is sometimes called *quantitative PCR* (or qPCR) because it can quantify, or measure, how many copies of a particular gene were present in a sample from the start of the reaction.

Reverse Transcription PCR (RT-PCR)

Another form of PCR, called **reverse transcription PCR (RT-PCR)**, is useful for detecting RNA in a sample—such as the genome of an RNA virus like SARS-CoV-2. In RT-PCR, the enzyme **reverse transcriptase** is used in conjunction with primers to build DNA that is complementary to RNA molecules in a sample. As can be done for PCR of a DNA target, there are real-time RT-PCR procedures that facilitate immediate data visualization.

PCR for Genetic Testing

Screening for genetic disorders is nothing new; since the early 1960s, babies born in the United States have been tested for genetic disorders, and all 50 states require newborn screening.[14] The sequencing of the human genome that was completed in 2003 has advanced our ability to detect genetic diseases. It is no longer rare for clinicians to order DNA-based tests; over 2,000 genetics-based tests are available, and many of them rely on PCR to work. Tests can range from looking for genetic mutations that cause cystic fibrosis or Huntington's disease to screening a person for genes that influence their cancer risk. The specificity of the designed PCR primers allows for certain genes in a patient's DNA to be analyzed rather than their entire genome.

Ethical considerations in genetics testing With genetic information being increasingly relied upon in health care and the surge of people submitting their genetic information to companies like 23andMe or Ancestry.com for genetics-

[13] Temperature ranges used for PCR protocols depend on various factors such as the types of primers used. The temperatures stated here are generalizations.

[14] State newborn screening programs screen nearly all of the four million babies born each year in the United States. The federal government provides a list of tests that should constitute the core testing panel. In general, states test for 29–40 disorders in newborn screening. All 50 states have laws that require testing and outline specific rules about opting out of testing. Most parents don't even realize their child has been tested for genetic disorders; they assume they have to ask for testing or that it is optional. To explore what tests your state performs and the laws governing newborn testing, you can visit http://babysfirsttest.org/newborn-screening/states. To review legal and ethical aspects of newborn screening, you may find the following dissertation interesting: Dayno, A. (2020). Unwinding the ethical concerns of newborn screening in the age of genomic medicine (Doctoral dissertation, Temple University Libraries).

based analysis, there are some valid concerns about protecting patients. The **Genetic Information Nondiscrimination Act of 2008 (GINA)** was passed to protect patients from discrimination based on genetic information; however, the act is fairly weak and leaves various loopholes for insurers and employers. And there is the psychosocial aspect to consider, too. It could be emotionally devastating to test positive for a particular gene—especially if the gene causes a deadly disorder like Huntington's disease for which there is no cure.

Drug development often relies on recombinant DNA techniques.

Our ongoing search for effective antibiotics and other drugs has taught us that pharmacologically useful proteins are found in a wide variety of sources, including plants, fungi, bacteria, snake venom, frog skin, and even cockroaches. However, in many cases it is impractical if not impossible to rely on a natural source for the large quantities of the protein needed for pharmaceutical applications. Recombinant DNA techniques solve this issue by providing a way to insert a desired gene into an expression system to produce large amounts of a particular protein quickly and easily. The term **recombinant DNA (rDNA)** refers to DNA that is generated or engineered by combining DNA from different organisms. Here we'll review the overall process of building and using recombinant DNA in the context of making a viral protein in *E. coli* for use in a subunit vaccine.

Recombinant DNA Step 1: Gene Isolation and Copying

Building a recombinant DNA construct starts with isolating the gene that encodes the desired protein. There are numerous ways to do this, but here we'll review a method that relies on PCR. By designing primers that flank the desired gene in a template, researchers can amplify a specific gene. For our case-in-point illustration (FIG. 14.14), one of the primers would be designed to bind just ahead of the targeted viral gene (perhaps a gene that encodes a viral surface protein), and the second primer would be designed to bind the complementary DNA strand at the tail end of the desired gene. Following the process described earlier, the desired gene is copied through a series of thermocycles.

Recombinant DNA Step 2: Inserting the Desired Gene into a Plasmid

Once the desired gene is copied, the goal is to insert it into a cloning vector, which is usually a plasmid. Recall from Chapter 5 that plasmids are small circular DNA structures that are distinct from chromosomes and are found in diverse eukaryotes and prokaryotes. There are many commercially created plasmids that are used for building recombinant DNA constructs.

The gene of interest and the plasmid that the gene is going to be joined with must be specially clipped so that their ends are complementary to

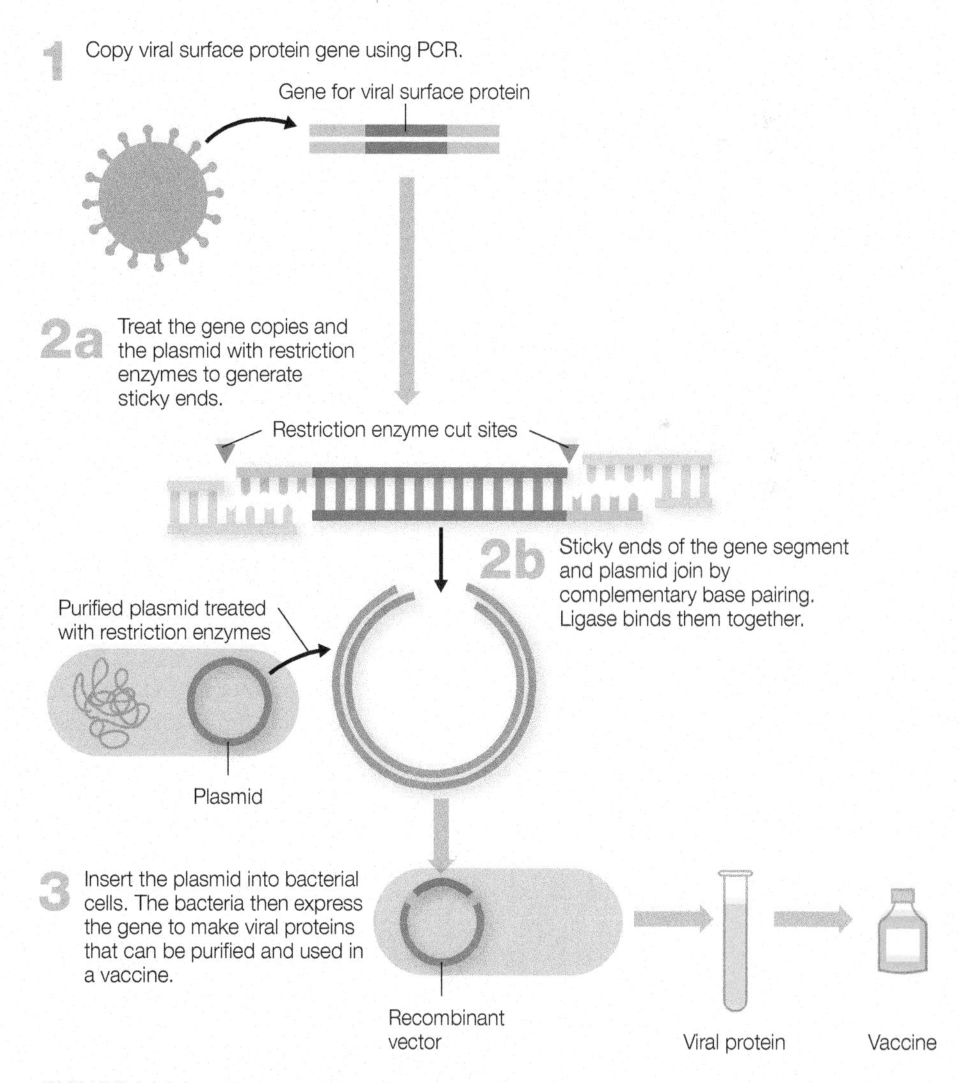

FIGURE 14.14 **Recombinant DNA (rDNA)** Engineered expression vectors are used to make a variety of medically useful products. Here we follow the use of rDNA for making a viral surface protein for inclusion in a vaccine.

TABLE 14.3 Common Restriction Enzymes Used in Molecular Cloning

Restriction enzyme and source	DNA recognition sequence and cut point
*Eco*RI *Escherichia coli* (generates sticky ends)	Cut point; 5′ G AATTC 3′; 3′ C TTAA G 5′; Sticky end; Recognition sequence
*Bam*HI *Bacillus amyloliquifasciens*	5′ G GATCC 3′; 3′ C CTAG G 5′
*Hind*III *Haemophilus influenza*	5′ A AGCTT 3′; 3′ T TCGA A 5′
*Sma*I *Serratia marcescens*	5′ CCC GGG 3′; 3′ GGG CCC 5′

one another.[15] To generate these compatible ends with nucleotide overhangs, or **sticky ends**, both the plasmid and the DNA construct are incubated with **restriction enzymes**. These enzymes occur naturally in a variety of bacterial species and constitute a defense mechanism that allows a bacterium to cut up bacteriophage genomes when they enter a cell. Restriction enzymes cut DNA at specific target sequences; see TABLE 14.3 for a few examples of restriction enzymes and the sequences they use for identifying where to cut the DNA.

Once the plasmid and the target gene have been clipped by the same restriction enzyme, they will have compatible sticky ends and are ready to be joined—like connecting compatible puzzle pieces. An enzyme called ligase, which was reviewed in Chapter 5, forms phosphodiester bonds between the plasmid DNA and the DNA being inserted into the plasmid to leave us with a completed recombinant vector.

Recombinant DNA Step 3: Transforming the Plasmid into Cells for Expression

The last step is to purify the newly made rDNA construct and insert it into host cells. Host cells are diverse and can include just about any cell line, provided it is easy to culture, readily takes up the rDNA, and produces large quantities of the desired protein. Often prokaryotic cell lines are used because they are among the cheapest cultures to maintain, and their growth is easily scalable for mass production. However, eukaryotic cells may be needed in certain cases because they modify and process proteins in a manner that may be required for the final protein to be active in humans.

As reviewed in Chapter 5, *transformation* is a process by which cells take up DNA from their environment. Some cells do this naturally (especially certain bacteria), but most cells must be coaxed into taking up rDNA through chemical treatments or exposure to a brief electrical current. Cells that take up the rDNA will transcribe and translate it to make the desired protein, which is then harvested. The viral protein we have been using as a case-in-point model would need to be harvested from lysed cells. However, some proteins are secreted (or can be engineered to be secreted); in such a case the secreted protein could be purified from the liquid media used to grow the cells.

Additional Challenges in Recombinant Drug Production

Recombinant DNA technology is fairly straightforward in theory but often complex in practice. For instance, in some cases more than one gene is required for effective protein production. Or, if a eukaryotic gene is to be expressed in a prokaryotic cell, introns (intervening genetic sequences that do not code for proteins) must first be removed. Even when the vector integrates with the host cell successfully, protein production may not go as planned. The host cell might contain particular metabolic pathways that limit production of the desired protein. Alternatively, the protein may fail to correctly fold in the host cell, rendering it useless.

CRISPR can edit any genetic material.

CRISPR-Cas9 is a gene-editing tool that has taken the world by storm.[16] The system basically allows researchers to perform gene surgery—CRISPR-Cas9 can locate a specific DNA sequence and cut it out with surgical precision so that new DNA can be plugged into the cut site. This is highly beneficial because dysfunctional genes cause thousands of medical disorders, such as cancer, muscular dystrophy, and cystic fibrosis. Replacing mutated or missing genes with normal genes could

---

[15] Recall that in DNA, A (adenine) bonds to T (thymine) to make AT pairs, while G (guanine) bonds to C (cytosine) to make GC pairs.

[16] CRISPR-Cas9 stands for "clustered regularly interspaced short palindromic repeats (CRISPR)-associated protein 9 nuclease."

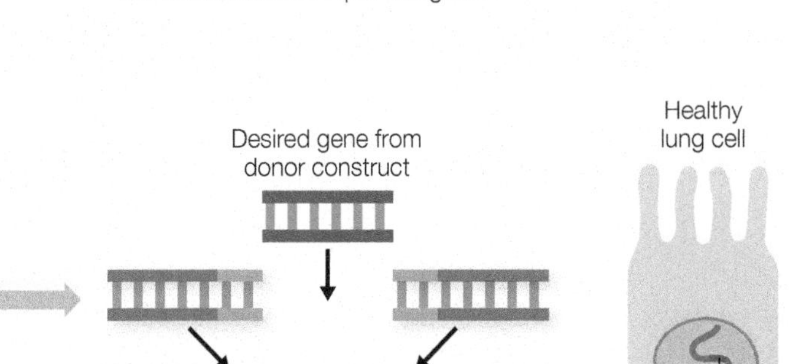

FIGURE 14.15 **CRISPR-Cas9 editing system** In the future, the CRISPR-Cas9 system may make it possible to cure genetic diseases such as cystic fibrosis by replacing mutated genes in the patient's cells with correct gene sequences.

help to control or even cure such diseases; however, simply injecting or ingesting a gene won't result in the desired gene being taken up by human cells.

The gene-surgery system consists of two key components. The CRISPR part of the system includes a guide RNA that acts like a molecular GPS for finding the desired genetic sequence that is to be cut out—the guide RNA nucleotide sequence is complementary to the target DNA nucleotide sequence. Once the guide RNA locates and pairs with the complementary target DNA, the Cas9 enzyme serves as the scalpel that cuts the targeted DNA sequence. The Cas9 enzyme cuts both strands of the targeted DNA, generating a double strand break. The cell will then specifically insert the desired sequence into the area the CRISPR-Cas9 system clipped (FIG. 14.15).

The CRISPR-Cas9 system has been used in all types of cells, both prokaryotic and eukaryotic. Even human cells can be manipulated with it. The possibilities for gene editing are endless. CRISPR-Cas9 applications being investigated include modifying plants for better crop production, genetically altering livestock to produce more product using less feed, and human applications to fix genetic diseases such as hemophilia and sickle cell anemia. However, ethical concerns have spurred caution in the use of this system, particularly with human cell lines.

Viruses can deliver genes to human cells.

Another technique used to introduce genetic material into host cells involves viruses. This process, when used as a treatment for human diseases, is known as **gene therapy**. Over 1,800 gene therapy clinical trials are currently in progress across 30 countries; most are aimed at treating cancer.

Engineered adenoviruses and retroviruses are the most common viruses used in gene therapy clinical trials. The viruses are genetically engineered to be nonpathogenic, but they must remain infectious in order to deliver their genetic cargo to host cells. The desired human gene is then packaged into the delivery virus. The gene-carrying virus is then introduced into the patient, or cells from the patient can be removed, cultured with the virus, and then returned to the body.

When the virus enters a human cell, the normal gene is delivered to the host cell nucleus. The human host cell then transcribes and translates the normal gene (FIG. 14.16). Retroviruses can insert their genetic material into host cell chromosomes. In contrast, adenoviruses rarely integrate into host cell genomes, so they present limited potential in effecting permanent genetic editing in standard gene therapy. However, by loading an adenoviral vector with newer CRISPR-Cas9 systems, it is entirely possible to edit out mutated genes in human cells and replace them with normal genes.

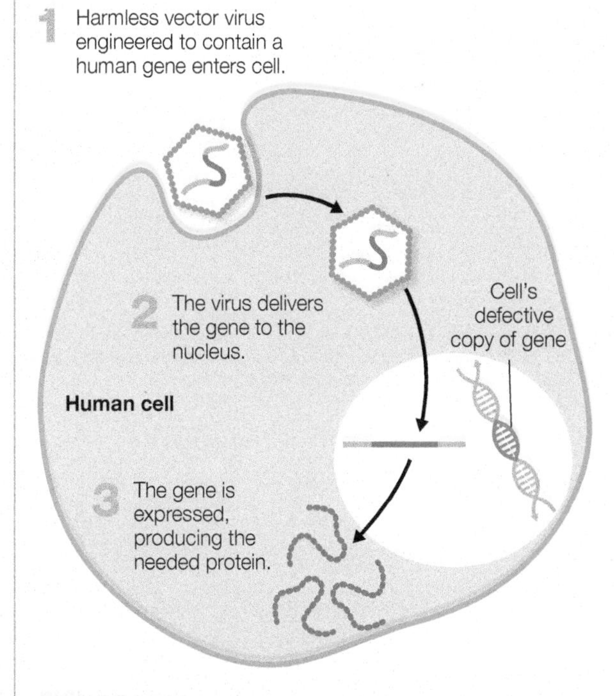

FIGURE 14.16 **Human gene therapy using viral gene vectors**

Genome maps reveal valuable information.

Mapping an organism's genome requires documenting the position of every nucleotide and identifying possible genes and their locations. The human genome was mapped in 2007, but three decades before, researchers mapped their very first microbial genome, which was for a bacteriophage. The 1990s showed rapid progress, with maps created for *Haemophilus influenzae* and then *Saccharomyces cerevisiae*. Now the genomes of many pathogens have been sequenced and mapped.

Similar to a blueprint of a building that identifies the location of pipes and wires, genome maps serve as a blueprint for identifying the proteins and regulatory RNAs required for every cellular function. Genome maps can thus reveal how microbes cause disease and how they may have acquired virulence factors. Certain regions in pathogen genomes encode toxins, virulence factors, and resistance mechanisms. These special gene groupings, called **pathogenicity islands**, also reveal how some virulence factors are acquired through horizontal gene transfer such as via bacteriophages, transposons, and conjugation events. (Horizontal gene transfer mechanisms are reviewed in Chapter 5.)

Gene maps also reveal information about genes that pathogens may need for day-to-day living—that is, their normal cellular functions. Such information is useful for developing drugs that can inhibit or kill pathogens. A genome map can also help to quickly identify the gene that encodes a desired protein for recombinant DNA and molecular cloning methods. FIG. 14.17 provides an example of a genome map of an *E. coli* plasmid.

Gene microarray technology provides a global view of cellular functions.

Gene microarrays are useful tools for investigating differences between healthy cells and diseased ones, such as cancer cells. Microarrays are made by adhering single-stranded segments of DNA to a slide. Some companies make

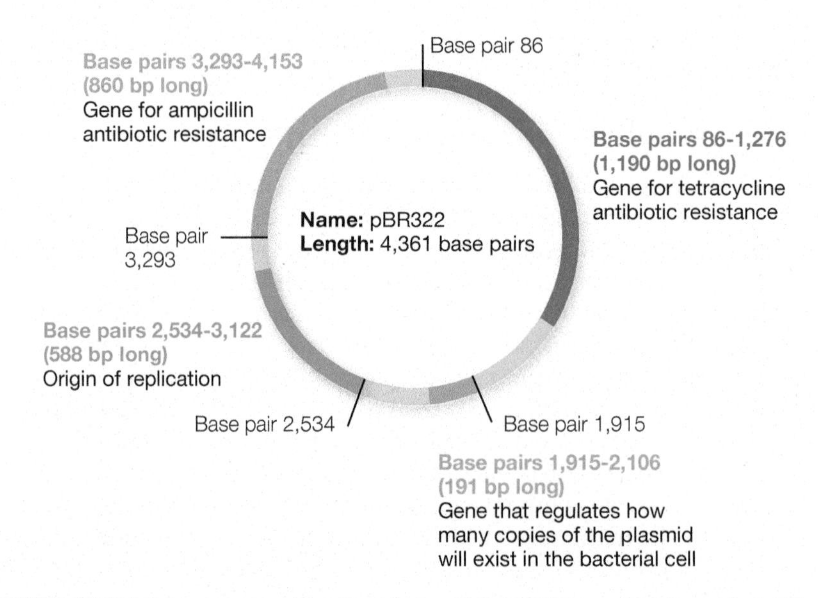

FIGURE 14.17 Genome map The *E. coli* plasmid pBR322 was developed in 1977. The name includes the initials of the researchers who engineered it (p = plasmid; B = Bolivar; R = Rodriguez). Researchers use maps like this to determine where genes are located. Gene maps for numerous animals (including humans) and diverse microbes are also available, but are too large and complex to be shown here.

FIGURE 14.18 Gene microarrays

Critical Thinking *How could gene arrays be used to study the effects of a drug on human cells?*

arrays that represent the entire human genome, whereas others make arrays for selected profiling, such as "cancer arrays" that include the most common genes associated with certain cancers. For example, the oncotypeDX test profiles breast cancers based on 21 different genes and can give information about the likelihood that the cancer will recur, whether chemotherapy would be beneficial, and the possible impact of radiation therapy on the cancer. The fact that most insurance companies cover the oncotypeDX test reflects just how far microarray technology has come in terms of moving into mainstream clinical diagnostics.

The concept of complementary base paring between nucleotides underlies microarray technology (FIG. 14.18). To envision the process, let's imagine that we want to profile a breast cancer cell to determine its genetic properties so that we can apply the best possible therapy. Cells from a biopsy of the tumor are first lysed to release the DNA and RNA in the cells. Because proteins are encoded by mRNAs, the mRNA present in the cell lysates represents a snapshot of what genes the cells were expressing before they were killed. However, because the microarray slide is loaded with DNA, the mRNA from the patient cells must be converted to single-stranded complementary DNA (cDNA) using reverse transcriptase. The patient's cDNA is then fluorescently tagged, then added to the microarray slide. If cDNAs bind to a particular DNA sequence on the array, then a fluorescent signal will be detected with a specialized array reader. In general, the brighter the "glow" associated with a given spot on the chip, the more that particular gene was expressed in the tested cells. Microarrays can therefore reveal not only what genes are being expressed in biopsied cancer cells, but also the degree to which specific genes are expressed. This information can help clinicians make important treatment decisions.

Microarray technology can also be used in the detection of pathogens. The beauty of the technique is that multiple pathogens can be detected on a single chip, even when relatively few cells are present. Food-associated pathogens, biological warfare agents such as plague and anthrax, and viral pathogens are among the types of microbes that can be identified.

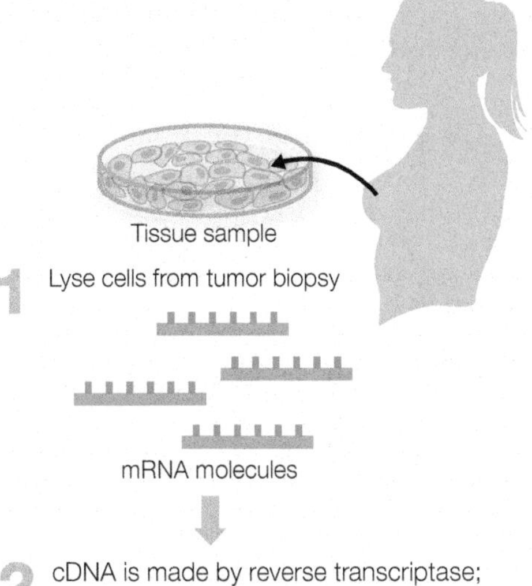

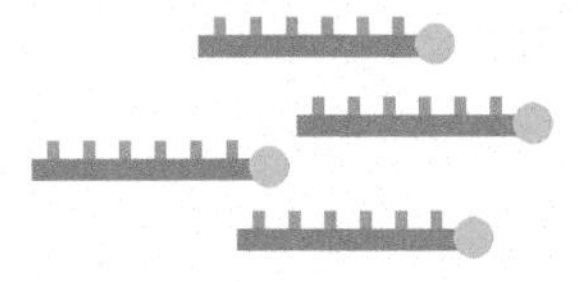

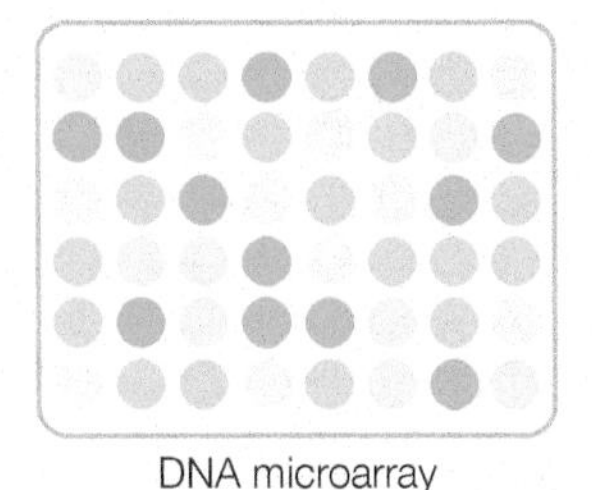

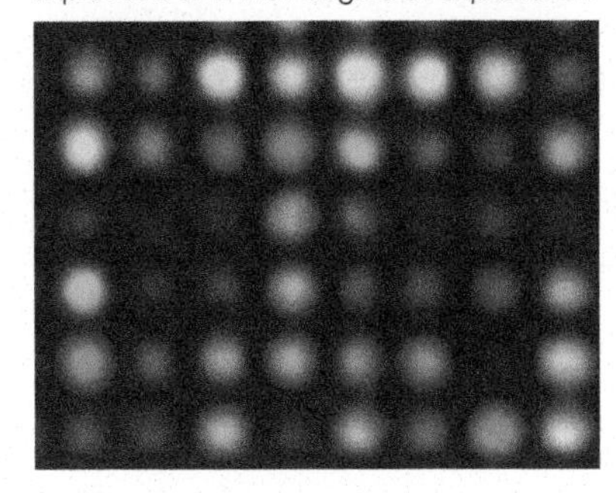

Build Your Foundation

14. What are the various PCR methodologies, how do they each work, and what are their clinical applications? (LO 14.14)
15. What is the general process for making recombinant DNA constructs, how can they be introduced into cells for gene expression, and what are the clinical applications of the process? (LO 14.15)
16. What are restriction enzymes, and how are they used in recombinant DNA methods? (LO 14.16)
17. How does the CRISPR-Cas9 system perform genome editing? (LO 14.17)
18. What is gene therapy, and how can viruses be used for it? (LO 14.18)
19. What type of information is provided by a genome map, and why are such maps useful? (LO 14.19)
20. What are microarrays, and how are they clinically useful? (LO 14.20)

Build Your Foundation (BYF) Quick Quiz: Visit the **Mastering Microbiology** Study Area to quiz yourself.

Herd Immunity

Herd immunity protects those who cannot be vaccinated. If the vast majority are vaccinated, infectious disease will not spread through the population.

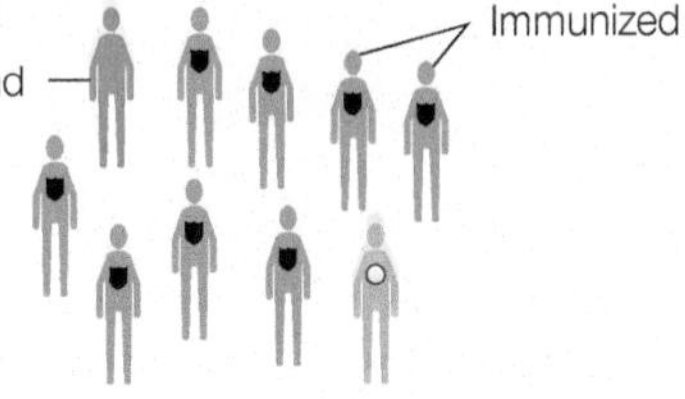

Vaccine Types

Active Agent	**Attenuated Vaccines:** Live weakened pathogen
Active Agent	**Vector Vaccines:** Genetically modified carrier virus (vector) delivers genetic information from the pathogen
Inactivated Agent	**Whole-Agent Vaccines:** Inactivated pathogen 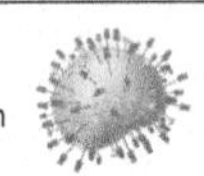
Inactivated Agent	**Subunit Vaccines:** Portion of pathogen is used to stimulate an immune response **Purified Subunit Vaccines (Natural and Recombinant)** Includes natural or engineered parts of the pathogen **Toxic Vaccines** Inactivated protein toxin **Conjugate (or Polysaccharide) Vaccines** Polysaccharides are conjugated or linked to a component that enhances immunogenicity
Inactivated Agent	**mRNA Vaccines** mRNA that directs the production of a specific protein found in or on the pathogen

Polymerase chain reaction (PCR):
Copies genes for genetic engineering and to detect DNA in a sample; reverse transcription PCR (RT-PCR) detects RNA in a sample; both PCR and RT-PCR can be performed using real-time visualization methods.

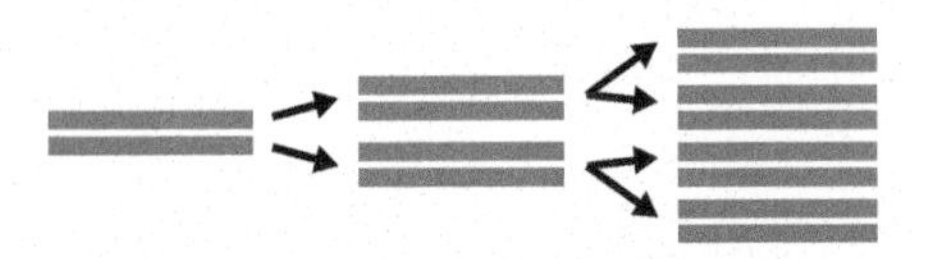

Agglutination- and Neutralization-Based Diagnostics

Agglutination reactions:
Used in blood typing, to identify infections, and to diagnose noninfectious immune disorders.

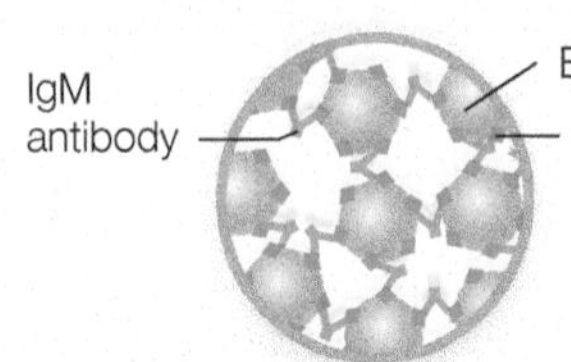

Neutralization reactions:
Tests like the plaque reduction neutralization test detect patient antibodies against a specific virus.

Antibody-coated virus cannot infect cells.

Enzyme-Linked Immunosorbent Assays (ELISAs)

Sensitive rapid diagnostic tests that detect antigens or antibodies in a sample; rely on a reporter enzyme linked to an antibody.

Direct ELISA **Indirect ELISA** **Sandwich ELISA**

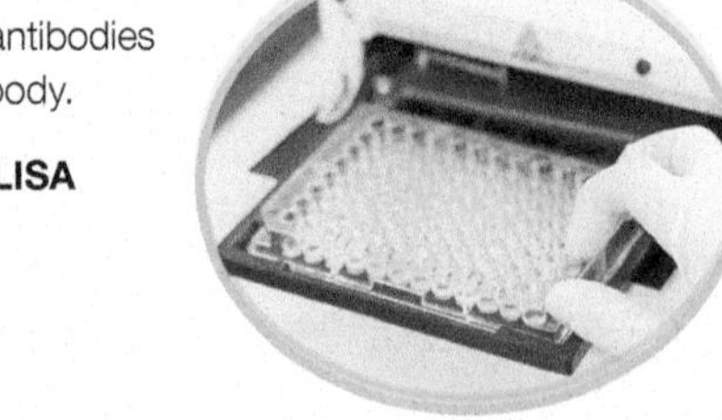

Genetics Techniques for Diagnostics and Therapeutics

Recombinant DNA techniques:
An expression system is engineered to make large quantities of a desired protein.

Desired gene is isolated, amplified by PCR, and cut with restriction enzymes to generate sticky ends.

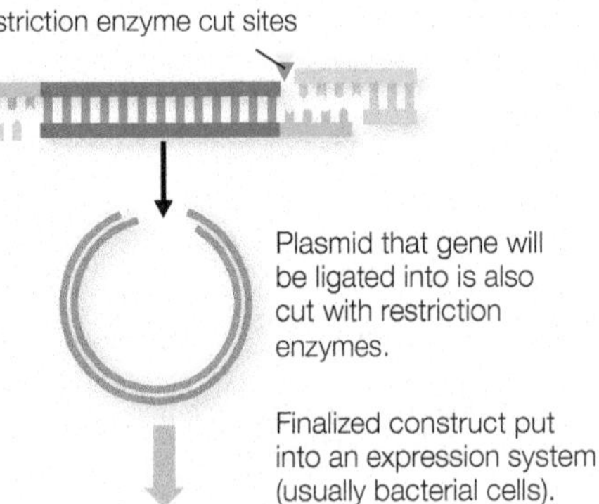

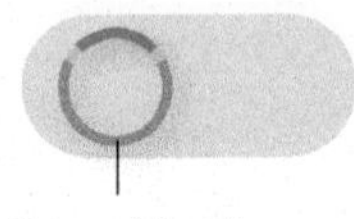

CRISPR-Cas9 editing system:
A gene-editing tool that locates a specific DNA sequence and cuts it out so that a new sequence can be inserted.

Gene therapy:
The process of using viruses to introduce genetic material into human cells to treat disease.

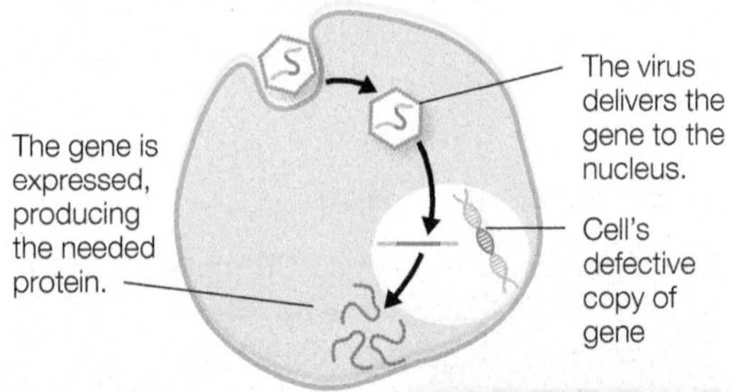

Gene microarray technology:
Provides a global view of cellular functions; useful for detecting pathogens and for informing clinical decisions for cancer therapies.

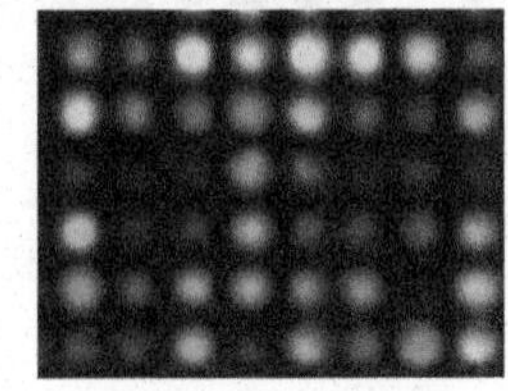

14 CHAPTER 14 OVERVIEW

14.1 A Brief History of Vaccines

- Efforts to immunize people against pathogens originated as variolation against smallpox. Jenner made vaccination more widespread when he developed an attenuated cowpox vaccine to protect against smallpox. Later, Pasteur developed vaccines to protect humans against rabies and cattle against anthrax.
- Smallpox was eradicated through vaccination campaigns, and the same approach is being applied to polio.
- Many parents in the United States and the United Kingdom stopped vaccinating their children in response to a 1998 study claiming a link between the measles/mumps/rubella vaccine and autism. The paper was later retracted, and several studies have since debunked the claim.
- Herd immunity occurs when a large enough percentage of a population is vaccinated so that a disease cannot spread. Herd immunity protects people who can't get vaccinated for medical reasons. The percentage of the population that must be vaccinated to achieve herd immunity depends on the agent, but it is at least 85 percent for most infectious diseases.

14.2 Overview of Vaccines

- Live attenuated vaccines are made up of weakened whole pathogens that do not cause disease in immune-competent patients. Because they closely mimic the real pathogen, they tend to elicit robust immunity.
- Inactivated vaccines include whole-agent and subunit vaccines. Subunit vaccines are divided into three groupings: purified subunit (natural and recombinant), toxoid, and conjugate (or polysaccharide) vaccines. Adjuvants are commonly added to subunit vaccines to improve the strength of the immune response they generate.
- Vector vaccines and mRNA vaccines are the newest vaccine strategies and have been used for COVID-19 vaccines.

14.3 Immunological Diagnostic Testing

- Immunological diagnostic tests rely on antigen–antibody interactions such as agglutination and neutralization. Agglutination occurs when antibodies cause cells or antigen-coated beads to clump. Neutralization reactions are typified by the plaque reduction neutralization test (PRNT), in which antibodies coat virions and thereby limit host cell infection.
- Enzyme-linked immunosorbent assays (ELISAs) detect the presence of a particular antigen or antibody in a sample. The main ELISA formats are direct, indirect, and sandwich.
- Interferon gamma release assays (IGRAs) examine whether a patient's T cells release interferon when exposed to *Mycobacterium tuberculosis* antigens. IGRAs provide a sensitive and accurate means of detecting TB at early stages in patients—even those that have been vaccinated against TB, which have traditionally been a hard group to quickly screen because they are seropositive for TB and have false-positive tuberculin skin tests.
- Fluorescent-tagged antibodies are useful for immunofluorescence microscopy to detect antigens in tissue samples. They are also central to immunofluorescence assays, which work similarly to ELISA methods, but without the need to add a substrate for the detection step.

14.4 Selected Genetics Applications in Medicine

- Recombinant DNA combines DNA from two different sources. Microbes containing recombinant DNA are used to produce some drugs and other products.
- The polymerase chain reaction (PCR) is a sensitive and fast way to determine if a particular gene is present in a sample. It's also used to copy DNA for the production of recombinant DNA constructs. Standard PCR has been modified to real-time PCR and reverse transcription PCR (RT-PCR). Real-time PCR uses a fluorescent signal to visualize data immediately rather than relying on time-consuming electrophoretic analysis to determine if a specified PCR product was made. RT-PCR is diagnostically useful for detecting RNA.
- Making a recombinant DNA construct entails isolating the gene of interest, inserting it into an expression vector, and then coaxing cells to undergo transformation—take up the rDNA and express it. Restriction enzymes help scientists build rDNA constructs because they clip DNA to yield sticky ends of compatible base pairs that help insert DNA segments into plasmids cut with the same restriction enzyme.
- The CRISPR-Cas9 system consists of a guide RNA (the CRISPR part of the system) and the Cas-9 enzyme; the system cuts DNA in a location defined by the guide RNA and can be used for DNA editing.
- Gene therapy delivers new genetic information to cells to treat disease. Retroviruses and adenoviruses can be engineered to deliver gene therapy to human cells.
- Genome maps record the location of genes in an organism's genome and provide information that helps us understand what genes an organism has.
- Gene microarrays use complementary binding of single-stranded DNA to identify which genes are being expressed in a cell or tissue. This technique can be used to get expression profiles of dysfunctional cells, such as cancer cells, and identify pathogens in a sample.

SMART THINK CLINICALLY | *Be S.M.A.R.T. About Cases*

COMPREHENSIVE CASE

The following case integrates basic principles from the chapter. Try to answer the case questions on your own. Don't forget to be **S.M.A.R.T.*** about your case study to help you interlink the scientific concepts you have just learned and apply your new understanding of vaccines and biotechnology-based diagnostics and therapeutics to a case study.

*The five-step method is shown in detail in the Chapter 1 Comprehensive Case on cholera. See pages 31–33. Refer back to this example to help you apply a SMART approach to other critical thinking questions and cases throughout the book.

• • •

Mariam, a school nurse, was examining 5-year-old Sam, who was sent to her office an hour after school began. He complained of a bad cough and felt extremely tired. After talking with Mariam for a few minutes, Sam erupted into a coughing fit with loud wheezing sounds between coughs. Although Mariam had never seen a case of whooping cough in a school setting, she suspected this disease immediately. To obtain more information, she called Sam's parents.

When the boy's parents arrived, Mariam asked about recent changes in Sam's health. Sam's mother said she noticed he had a runny nose and felt warmer than usual a week or so earlier, but thought it was just a lingering cold. Two nights before, Sam had broken into a coughing fit and then vomited, but he appeared to be fine the next morning, except for the cough.

Mariam told the parents that based on Sam's signs and symptoms, she suspected he had whooping cough (pertussis), a bacterial infection caused by *Bordetella pertussis* in the lungs. She also added that the disease is highly contagious, prolonged, severe, and preventable by a vaccine.

The parents were unconvinced. They believed that Sam was suffering from a bad cold or perhaps the flu that had been circulating among children at the school. They were not aware of anyone with whooping cough symptoms who had been in contact with their son. Mariam urged them to seek a doctor to make a final diagnosis of the boy's condition. They assured Mariam that they would, and left with Sam.

After the parents left, Mariam looked through Sam's medical records. Not surprisingly, he did not have any vaccinations listed, because his parents had been granted an exemption. Mariam walked straight to the principal's office to discuss Sam's condition and what steps should be taken by the school to warn other parents, if he was indeed diagnosed with whooping cough.

• • •

CASE-BASED QUESTIONS

1. Should Mariam be concerned for the other children who are in class with Sam? Why or why not?
2. Was Mariam being overly worried by telling the principal she suspected Sam had whooping cough?
3. If you were Sam's pediatrician, what type of diagnostic tool would you most likely use in the clinic to quickly identify whether or not he has whooping cough?
4. If Sam does indeed have whooping cough, should the entire school body get vaccinated, or would another action be preferable?
5. The parents seem convinced that their child contracted his respiratory infection, whatever it may be, at school. If the nurse is correct that Sam has whooping cough, would the school be a likely place for him to have contracted it? Be sure to explain your reasoning.
6. If another child was exposed to Sam and did not have the pertussis vaccination for whooping cough, should the child be vaccinated immediately?

END OF CHAPTER QUESTIONS

Think Critically and Clinically
Questions highlighted in orange are opportunities to practice NCLEX, HESI, and TEAS-style questions.

1. The human papillomavirus vaccine consists of surface proteins engineered from various HPV strains. How would you classify this vaccine?
   a. Whole inactivated vaccine
   b. Whole live attenuated vaccine
   c. Recombinant subunit vaccine
   d. mRNA vaccine
   e. Vector vaccine

2. Which of the following vaccines is most likely to be contraindicated for an immune compromised patient?
   a. An mRNA vaccine
   b. A whole inactivated vaccine
   c. A toxoid vaccine
   d. A conjugate vaccine
   e. A live attenuated vaccine

3. If parents in the United States decline or delay vaccinations for their children because they fear the refuted claim of a link between vaccinations and autism, which of the following is a possible result?
   a. A decrease in diagnosis of autism with a decrease of vaccinations
   b. An increase in diagnosis of autism with an increase of vaccinations
   c. An increase in the administration of subunit vaccines
   d. A decrease in herd immunity
   e. An increase in public demand for inactivated vaccines

4. How does an antibody neutralize a virus?
   a. By preventing the virus from binding to a receptor protein on a host cell
   b. By preventing the virus from injecting its nucleic acid into its host cell
   c. By degrading the capsid
   d. By inserting holes into the viral envelope
   e. By preventing the virus from replicating while inside the host

5. A direct ELISA test is utilized as a pregnancy test because it tests for:
   a. antibodies to the fetus in urine.
   b. antibodies to a pregnancy hormone in urine.
   c. an antigen that recognizes a pregnancy hormone in urine.
   d. T cells that recognize a pregnancy hormone in urine.
   e. the presence of a pregnancy hormone in urine.

6. What was the purpose of the practice of variolation?

7. What caused thousands of parents in the United States and the United Kingdom to stop vaccinating their children after 1998?

8. Is there a definitive scientific link between autism and vaccines? Support your response with information from the text.

9. ___________ active immunity creates ___________ lymphocytes that will remember a pathogen and quickly respond to the same pathogen later when exposed.

10. Match the pathogen part to the inactivated vaccine:

a. Whole agent	An inactivated toxin from a pathogen
b. Subunit	The entire intact pathogen killed by heat or chemical methods
c. Recombinant	An antigen that elicits a strong immune response is joined to a pathogen part
d. Toxoid	Parts of a pathogen are made by another host cell via genetic engineering
e. Conjugate	A portion or piece of the pathogen such as a surface protein

11. Select the true statements about mRNA vaccines. Select all that apply.
    a. These vaccines contain a type of nucleic acid.
    b. These vaccines can alter cellular DNA.
    c. These vaccines often require a booster dose.
    d. These vaccines do not require adjuvants.
    e. These vaccines require an engineered virus to deliver the mRNA to a target cell.

12. Place the following steps in order for a neutralization test:
    a. Incubate cell culture for a few days.
    b. Add suspected infecting virus to the patient's serum.
    c. Inspect cell culture for viral infection.
    d. Extract serum from patient.
    e. Mix patient serum with virus and add the mixture to a cell culture.

13. Describe one difference between a direct ELISA and an indirect ELISA.

14. How many copies could PCR make if we started with one copy of a template and performed 15 cycles of copying?
    a. A few
    b. A few hundred
    c. A few thousand
    d. Millions

15. How could a CRISPR-Cas9 enter a human cell for genome editing purposes?

CRITICAL THINKING QUESTIONS

1. Why is it better to use a conjugate vaccine than to provide several boosters of a vaccine that elicits a weak immune response?

2. The MMR (measles, mumps, and rubella) vaccine must be given in two doses over a four-year period for children, while the DTaP (diphtheria, tetanus, and pertussis) vaccine is given five times over the same period. Explain why the DTaP vaccine needs to be given so many times.

3. How may inactivated vaccines be safer for the public than live attenuated vaccines?

4. If you wanted to create a subunit inactivated vaccine against *Salmonella enterica* serovar typhimurium, what general bacterial components would you choose to include?

5. What potential issues could arise with using the CRISPR-Cas9 system to edit human genomes?

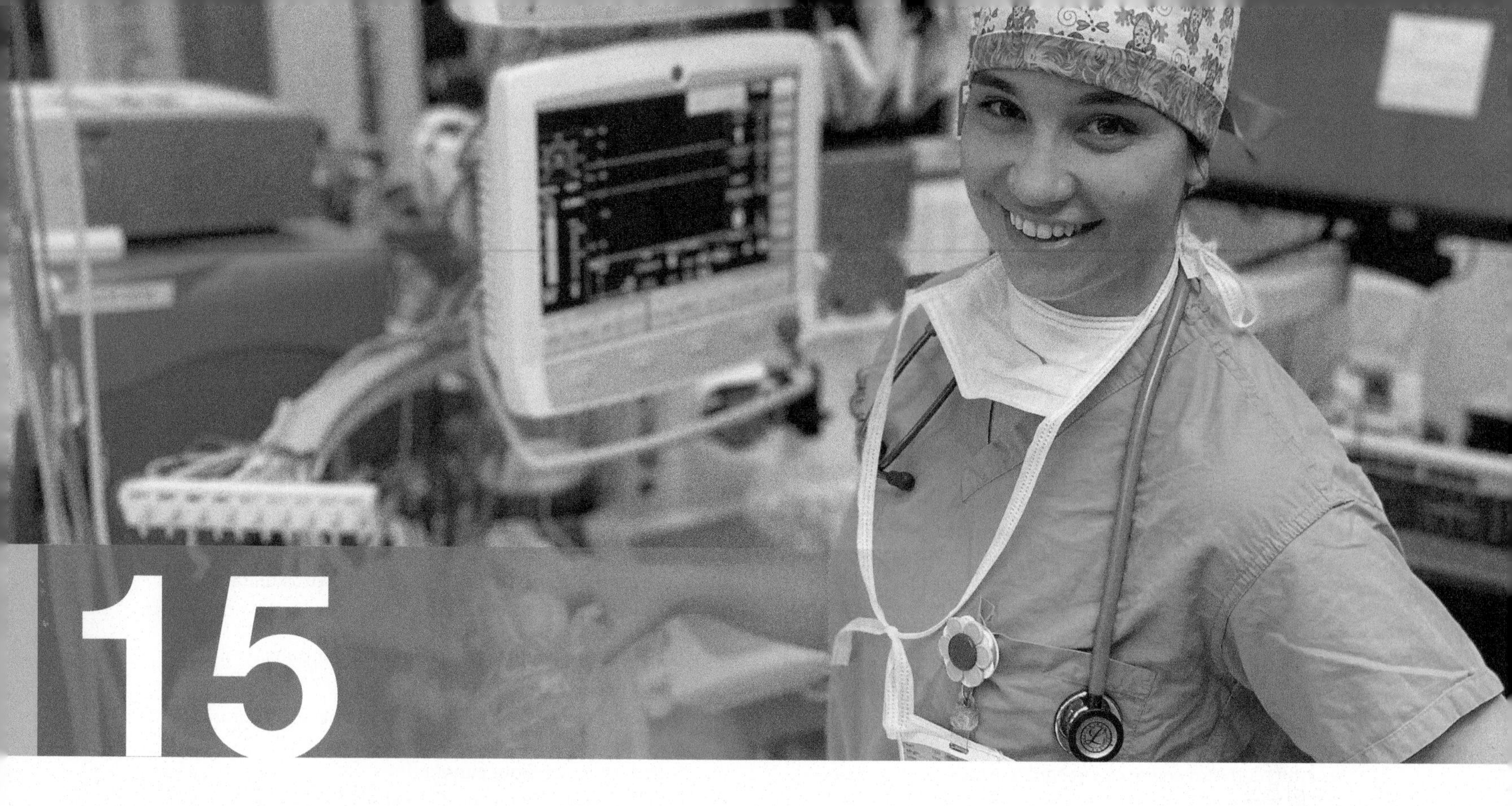

15 Antimicrobial Drugs

What Will We Explore? Antimicrobial drugs have been available for the last 80 years. They have saved countless lives and dramatically changed the way we approach infection control. In days past, a simple scratch or cut could turn into a life-threatening infection. Now, most nonviral infections are usually curable. In recent decades, however, many pathogens have evolved resistance to our most used antimicrobials, necessitating an ongoing search for new drugs.

We begin this chapter with a basic overview of the history of antimicrobial drugs and common terms for antimicrobials. Then we'll survey key antimicrobial families, including antibacterial drugs and drugs used to fight eukaryotic infections. Antiviral drugs are covered only briefly in this chapter, as they are discussed in detail in Chapter 6. We'll also explore the types of tests used to determine which antimicrobial would work best against a given pathogen. Finally, bacterial antimicrobial resistance will be discussed, along with ways to counter resistance.

Why Is It Important? Antimicrobials are among the most commonly prescribed drugs in the United States. According to the U.S. Centers for Disease Control and Prevention (CDC), U.S. physicians write more than 200 million prescriptions for antibiotics each year. Furthermore, annually in the United States there are about 2.8 million infections caused by antibiotic-resistant bacteria, and about 35,000 people per year die as a result.[1] As such, all healthcare providers require an understanding of the most common antimicrobial drugs, their targets and modes of action, and their common side effects. They also need to understand the pathogen mechanisms and human behaviors that contribute to the evolution of antimicrobial resistance.

Clinical CASE
NCLEX
HESI
TEAS

The Case of the Slapdash Self-Medication

Visit the **Mastering Microbiology** Study Area to watch the case and find out how antimicrobial drugs can explain this medical mystery.

Lauren Dyer
CAA, Anesthesiology specialist; Westerville, OH

[1] U.S. Centers for Disease Control and Prevention. (2019). *Antibiotic resistance threats in the United States, 2019.* U.S. Department of Health and Human Services, Centers for Disease Control and Prevention.

15.1 INTRODUCTION TO ANTIMICROBIAL DRUGS

Antimicrobial drugs are therapeutic compounds that kill microbes or inhibit their growth. They can be categorized according to the type of pathogen they target: **Antibacterial drugs** treat bacterial infections, **antiviral drugs** target viral infections, **antifungal drugs** work against fungal infections, and **antiparasitic drugs** treat protozoan and helminthic (worm) infections. Of these compounds, antibacterial drugs are the most commonly prescribed and are therefore a primary focus of this chapter.

Learning Outcomes

After reading this section, you should be able to:

15.1 Explain how Alexander Fleming discovered penicillin, and name the genus of the fungus that makes this antibiotic.

15.2 Distinguish between broad-spectrum and narrow-spectrum antimicrobials, and state why an empiric therapy may be intentionally broad-spectrum.

15.3 Compare bacteriostatic to bactericidal drugs, discuss scenarios where each may be useful, and explain why these terms are less concrete in clinical scenarios.

15.4 Contrast natural, semisynthetic, and synthetic antimicrobials, and state the potential value of drug modifications.

15.5 Distinguish between an antibiotic and an antimicrobial drug.

15.6 Define the following, and discuss how they relate to drug development: therapeutic index, selective toxicity, and drug half-life.

15.7 Discuss hepatotoxicity and nephrotoxicity as they relate to antimicrobials, and provide examples of drug classes associated with each effect.

15.8 Discuss how route of administration, drug interactions, and contraindications play into drug development.

Antimicrobial drugs radically changed modern medicine.

Before antimicrobial drugs, people routinely died from infections that today pose little threat. Strep throat could be fatal, and ear infections could spread to the brain. Infections were the main cause of war-related deaths throughout most of history. During the Spanish American and Mexican Wars, for instance, seven times as many soldiers died from wound infections and from diseases such as dysentery, typhoid fever, pneumonia, and tuberculosis as died on the battlefield.[2]

Tonics and potions claiming to cure infections existed for hundreds of years, but the first safe and effective antimicrobial drugs weren't developed until the 20th century.[3] During World War I, physician and bacteriologist Alexander Fleming frequently amputated soldiers' limbs as the only way to prevent death from infected wounds. Later, Fleming studied the bacterium *Staphylococcus aureus,* a pathogen that killed many of the soldiers Fleming treated. In 1928, he noticed that one of his *S. aureus* culture plates was contaminated with a mold. The bacteria were unable to grow near the mold on the culture plate. Fleming determined that the mold excreted a compound that could inhibit the bacteria. He named this chemical **penicillin**, after the *Penicillium* species of mold that had contaminated the original plate (FIG. 15.1). However, it wasn't until 1940, with the onset of World War II, that penicillin was mass-produced—mainly to combat streptococcal and staphylococcal infections that were responsible for high death tolls among soldiers.

Penicillin wasn't the only antimicrobial drug identified during this period. A red dye called Prontosil, the first sulfa drug, proved effective against streptococcal bacteria. Sulfa drugs and their derivatives are still prescribed today. Streptomycin, another bacteria-inhibiting compound, was isolated from a culture of the soil bacterium *Streptomyces griseus* and proved to be effective against tuberculosis (FIG. 15.2), although resistance developed quickly. In the following years, many more antimicrobial drugs were discovered. To this day, many of these early drugs remain important compounds to treat diverse pathogens.

FIGURE 15.1 **Penicillin** This photo shows a re-creation of the mold-contaminated culture plate that led to Alexander Fleming's discovery. Notice that the area around the mold is free of bacterial colonies.

Antimicrobial drugs are described by the pathogens they target and their mechanisms of action.

Antimicrobial drugs work via specific mechanisms and are selective about the microbes they affect. Antibacterial drugs tend to exhibit a *spectrum* of activity. **Broad-spectrum drugs** are effective against both Gram-negative and Gram-positive bacterial cells. Quinolones, which include levofloxacin and ciprofloxacin, are examples. **Narrow-spectrum drugs** target a limited range of bacteria.

[2]Murray, C. K., Hinkle, M. K., & Yun, H. C. (2008). History of infections associated with combat-related injuries. *Journal of Trauma*, 64, S221–S231.

[3]In 1910, German physician Paul Ehrlich created a drug called Salvarsan to treat syphilis. This technically predates the development of penicillin. However, Salvarsan was toxic to patients and fell out of use. Because penicillin had far fewer side effects, it is often considered the first major antibiotic.

FIGURE 15.2 ***Streptomyces griseus*** This soil bacterium naturally makes streptomycin. Several antibiotics are naturally made by *Steptomyces* species; these drugs tend to end in the suffix "mycin" to reflect their origin.

Critical Thinking *Why might a bacterium that makes an antimicrobial compound have a survival advantage in nature?*

For example, bacitracin and vancomycin are mainly effective against Gram-positive bacterial species. In most cases, narrow-spectrum drugs are preferred because they present less disruption to the normal microbiota than do broad-spectrum drugs. However, to prescribe a narrow-spectrum drug, the clinician needs to have a fairly firm idea of the causative pathogen's identity. Although microbiological samples are collected with the goal of specifically identifying the causative agent, definitive pathogen identification can take several days. Consequently, an **empiric therapy**, or treatment based on clinical presentation in the absence of definitive or complete clinical data, is commonly started to protect the patient. Such treatments are often intentionally broad spectrum. If a definitive bacterial identification is eventually made, then the clinician may narrow the therapy spectrum. As discussed in Chapter 10, microbiota disruption can put a patient at risk for *Clostridioides difficile* infection. Although any antibiotic may increase a patient's risk for *C. difficile*, most cases are preceded by a prolonged treatment with a broad-spectrum antimicrobial drug.[4]

Antimicrobial drugs also differ in whether they are **bacteriostatic**, preventing bacteria from growing, or **bactericidal**, actively killing bacteria. Bactericidal drugs tend to target bacterial cell walls or cell membranes and nucleic acids. In contrast, bacteriostatic drugs tend to target bacterial protein synthesis and metabolic pathways like folic acid production. It is important to note that in the clinical world there is not such a sharp line of distinction between these categories. An antimicrobial drug that is bactericidal for one pathogen may be bacteriostatic for another. Furthermore, a drug's bactericidal or bacteriostatic properties can change according to the drug dose, length of the drug regimen, pathogen load, and route of administration.

Bacteriostatic drugs, such as erythromycin, are typically effective for patients who have a healthy immune system that can destroy the bacteria during the drug course. Bactericidal drugs might seem more effective, but a potential drawback of these drugs, especially if their spectrum is broad, is that while targeting the pathogen they also tend to kill off normal microbiota. They can also lead to a spike in bacterial toxin release that can be deadly. For example, lipopolysaccharide (LPS, or endotoxin) made by Gram-negative bacteria is mainly released as cells die. Therefore, administering a bactericidal drug in a patient with a Gram-negative infection could trigger a dangerous surge in LPS levels in the patient. That said, bactericidal drugs have their place; bacterial endocarditis and bacterial meningitis are often treated with bactericidal drugs. Notably, when a patient has bacterial meningitis caused by Gram-negative bacteria, the bactericidal antibiotic is usually administered along with a steroid anti-inflammatory drug because as the Gram-negative bacteria die, the release of LPS can cause damaging inflammation.

Antimicrobial drugs may be natural, synthetic, or semisynthetic.

The development of penicillin from mold and streptomycin from bacteria led to a huge interest in characterizing more **antibiotics**, which are naturally occurring antimicrobial compounds.[5] Although there are many antibiotics, microbes have evolved an almost equal number of resistance mechanisms to combat them. Thus, **synthetic antimicrobials**, which are wholly manufactured by chemical

[4]In particular, carbapenems and third- and fourth-generation cephalosporin significantly increase a patient's risk for *C. difficile*. Slimings, C., & Riley, T. V. (2021). Antibiotics and healthcare facility-associated *Clostridioides difficile* infection: Systematic review and meta-analysis 2020 update. *Journal of Antimicrobial Chemotherapy*, *76*(7), 1676–1688.

[5]The terms *antimicrobial* and *antibiotic* are often used interchangeably. However, *antibiotic* is a more specific term because it refers to naturally occurring antimicrobial compounds and would therefore technically exclude synthetic and semisynthetic compounds.

processes, represent one avenue for making drugs that can overcome antibiotic-resistance mechanisms that naturally evolve as pathogens encounter our pharmacopeia of drugs. Developing synthetic drugs is a tedious process that involves screening thousands of possible candidates. Prontosil, the sulfa drug mentioned earlier, is a synthetic antimicrobial. Sometimes new antimicrobial drugs don't need to be built from scratch; instead, naturally occurring antibiotics can be chemically modified to improve their pharmacological actions and/or stability. These drugs are collectively called **semisynthetic antimicrobials**.

Antimicrobial drugs are rarely purified from their natural source, and many are chemically modified to enhance desirable properties.

Although antibiotics may be abundant in nature, they are often difficult to isolate from their original source; therefore, scientists often rely on genetic expression models for their mass production. For instance, although something like frog skin might yield a possible polypeptide antimicrobial, it would be difficult to isolate enough of the compound from frogs for clinical testing, let alone mass production. Instead, the gene encoding the compound could be identified and then genetically engineered into a new host, such as a bacterial cell, that could be used for mass production. In addition, the gene might be edited to produce a more potent drug or one better suited to production. (Refer to Chapter 14 for details on how genetic engineering may be used to make drugs.)

Researchers can also modify an antimicrobial compound's molecular structure by chemical means, such as by adding or changing R groups on the molecule. Ampicillin and amoxicillin are two semisynthetic antimicrobials derived in this way from the antibiotic penicillin (FIG. 15.3).

It is also often the case that naturally occurring antibiotics must be chemically altered to enhance their stability and usefulness. Just as new iPhones get generation numbers, so do drugs that are modified from a core molecule. For example, drugs that result from a first round of chemical modification would be called **first-generation drugs**, while those resulting from a second round of chemical modification would be called **second-generation drugs**. Drugs in later generations have expanded capabilities over their predecessors. Examples of bonus features can include an extended spectrum of activity, increased stability in the body that requires less frequent dosing, or even the ability to circumvent certain drug-resistance mechanisms. Aside from cephalosporins, which we'll discuss later, most antimicrobials are not formally grouped into generations. However, in conversation people may refer to "later-generation drugs" as a way to casually communicate the drug's general timeline of entry into medical use.

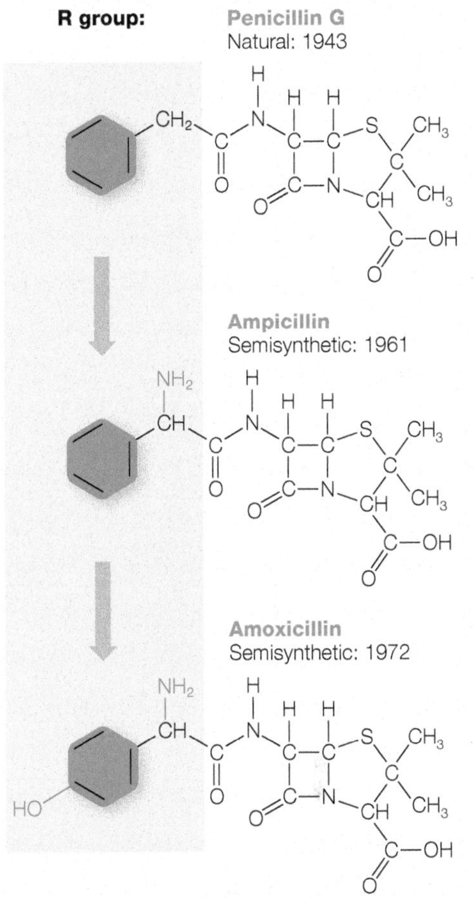

A number of factors impact antimicrobial drug development.

Researchers continue to look for new drugs to combat pathogens, but the search is difficult because a potential antimicrobial must have a variety of features. This section is intended to serve as a basic introduction to key clinical considerations in drug development and is by no means exhaustive.

Drug Safety

Every medical intervention has a risk/benefit consideration, and even "safe" drugs have the potential to generate side effects. A core principle of antimicrobial development is that the drug must exhibit **selective toxicity**, meaning that it inhibits or kills the targeted microbe without damaging host cells. This is one

R group:

Penicillin G
Natural: 1943

Ampicillin
Semisynthetic: 1961

Amoxicillin
Semisynthetic: 1972

FIGURE 15.3 **Examples of penicillin derivatives** Ampicillin and amoxicillin are derived by modifying penicillin G—one of many naturally occurring penicillin antibiotics. While penicillin G primarily targets Gram-positive bacteria, ampicillin and amoxicillin have a broader activity spectrum due to added chemical groups (noted here in red) that allow them to better penetrate Gram-negative cell walls.

Critical Thinking *What structural feature of Gram-negative cell walls makes them harder for certain drugs to penetrate as compared to Gram-positive bacterial cell walls?*

CHEM • NOTE

Chemists often use R or **R group** to describe chemical additions to a core structure (that is, to denote the remainder of an organic molecule). This shorthand approach focuses us on the part of the molecule being discussed instead of including cumbersome chemical structures.

feature that sets antimicrobials apart from disinfectants. A disinfectant, like bleach, is not selectively toxic; it effectively eliminates pathogens, but it also would damage host cells. Also, side effects should be limited and nonpermanent; rash, nausea, vomiting, and diarrhea are common antimicrobial side effects. If an antimicrobial is the best available treatment for a dangerous infection, some potentially dangerous side effects may be deemed acceptable as part of a total risk/benefit analysis. For example, in 1910, when arsphenamine (Salvarsan), the first antimicrobial drug that could treat syphilis, became available, it was embraced despite potentially severe side effects that included nerve damage and even death. At the time, the risk of potential side effects from Salvarsan was worth the benefit of avoiding the well-known outcomes of untreated syphilis—irreversible nervous system and cardiovascular system damage, dementia, and eventual death. With the arrival of penicillin, Salvarsan was discontinued.

Therapeutic index One measure of a drug's general safety is the **therapeutic index** (or therapeutic ratio). This is the ratio of the maximum tolerated or safe dose to the minimum effective or therapeutic dose.

$$\text{Therapeutic index} = \frac{\text{Maximum safe dose}}{\text{Minimum effective dose}}$$

In general, a drug with a high therapeutic index is effective well below the dose at which it is potentially toxic. It is therefore considered safer than a drug with a narrow therapeutic index, which has a therapeutic dose that is close to the toxic dose. Drugs with a narrow therapeutic index must be carefully dosed, and patients must be closely monitored for drug toxicity effects. In some cases, **therapeutic drug monitoring** (TDM) is used to ensure patient well-being and/or assess the therapeutic benefit of a drug. TDM may involve measuring drug concentrations in the bloodstream as well as monitoring other patient parameters such as renal and hepatic function. TDM is also used in late-phase clinical trials to ensure that the drug being developed is not adversely impacting the patient. The amount of active drug in circulation is another important metric to establish for many drugs. It is influenced by diverse factors, including how well the drug is absorbed (for noninjected drugs[6]), as well as how quickly the drug is metabolized and eliminated from the body.

Toxicity considerations Because the kidneys and liver are key organs that metabolize and eliminate drugs, they may be particularly susceptible to damage by certain drugs. **Nephrotoxic** (or kidney-toxic) antimicrobial drugs include a long list of agents, with aminoglycosides being prevalent on the list (these are drugs that tend to end in "mycin" or "micin," such as gentamicin, vancomycin, or streptomycin). Antimicrobials are a leading cause of drug-associated nephropathy, but fortunately, adverse effects tend to resolve upon stopping use of the drug. Notably, it's not just antimicrobials that can be nephrotoxic—the ongoing use of many nonsteroidal anti-inflammatory drugs like aspirin, contrast dyes used in MRIs, several blood pressure medications, and even Chinese herbals containing aristolochic acid may be nephrotoxic.

Hepatotoxic (or liver-toxic) antimicrobial drugs can induce liver damage and are leading agents of **drug-induced liver injury (DILI)**. Fortunately, the liver is a relatively resilient organ that tends to recuperate after the drug regimen is stopped. Several antimicrobial drugs are associated with DILI, but amoxicillin–clavulanate, which is known as Augmentin, Clavulin, and other

[6]Orally administered drugs, suppositories (vaginal and rectal), and medications applied to the skin (such as skin patches and ointments) all must have their absorption properties characterized as part of developing drug-dosing parameters.

trade names, is the most prevalent cause. But antimicrobial drugs are certainly not the only agents that are hepatotoxic; drugs like Tylenol, statins that lower cholesterol, and some dietary and herbal supplements are also known to affect the liver.[7] In fact, DILI is the most commonly cited reason for discontinuing drug development or removing a drug from the market.[8]

Administration Route

Another property that matters in a candidate antimicrobial is how it is administered. Oral administration is preferred because it's the easiest. Orally administered drugs must be stable in the acidic environment of the stomach, withstand digesting enzymes, and be sufficiently absorbed in the intestines to reach a therapeutic dose. Unfortunately, the chemical composition of some antimicrobials prevents oral administration. **Parenteral administration** (injection or infusion) is another option. The drug may be injected into a vein (intravenously), into the muscle tissue (intramuscularly), or under the skin (subcutaneously). The downside of the parenteral method is that needles and/or intravenous lines must be used, meaning that either healthcare workers must administer the drug, or the patient or caregiver must receive some basic medical training. "Needle phobia" and injection discomfort are also considerations for some patients.

Drug Stability and Elimination

The length of time the drug remains active in the body is another consideration. A drug's **half-life** is the time it takes for half of a dose to be eliminated or deactivated by the body—often by the kidneys or liver, as already noted. For example, penicillin V (one of many naturally occurring penicillins) has a half-life of 30 minutes, while azithromycin has a half-life of up to 68 hours. A drug with a short half-life is usually administered frequently; a drug with a long half-life doesn't need to be taken as frequently to maintain a therapeutic level in the body. That is why some antimicrobials must be taken three times a day, whereas others are taken only once a day. Azithromycin's long half-life is why a "Z-Pak" prescription dosing only requires a once-per-day administration and the total therapy course is short—available as 3- or 5-day regimens.[9] In contrast, penicillin G is dosed every 4 hours and often requires 7–10 days of treatment.

Drug Interactions and Contraindications

Ideally, an antimicrobial drug will have few or no adverse interactions with other drugs a patient may be taking. In reality, nearly every drug has some sort of contraindication or warning listed.[10] For example, women on the antibiotic rifampin are warned that the drug can inactivate oral contraceptives. And digoxin, a drug used for treating heart failure, shows higher toxicity when a macrolide such as clarithromycin is also prescribed. Drugs with nephrotoxic effects or that are eliminated via the kidneys are usually avoided in patients with already reduced kidney function. Similarly, patients who have reduced liver function would not be given a potentially hepatotoxic drug as a first-line therapy. Lastly, some antimicrobials, such as tetracycline, are not recommended during pregnancy or for nursing mothers.

---

[7]Dietary supplements and herbal remedies are not regulated by the U.S. Food and Drug Administration (FDA) as medications. Supplements can be marketed without their manufacturers having to demonstrate efficacy or even safety to the FDA.

[8]Yokoi, T., & Oda, S. (2021). Models of idiosyncratic drug-induced liver injury. *Annual Review of Pharmacology and Toxicology*, 61, 247–268.

[9]Azithromycin, also referred to as Zithromax or Z-Pak, is a commonly prescribed antibacterial drug.

[10]A contraindication is a specific situation in which a given drug or medical intervention would not be recommended.

Build Your Foundation (BYF) Quick Quiz: Visit the **Mastering Microbiology** Study Area to quiz yourself.

Build Your Foundation

1. How was penicillin discovered, who discovered it, and what organism makes it? (LO 15.1)
2. What are the pros and cons of narrow- versus broad-spectrum drugs, why are these groupings less concrete in a clinical setting, and when should an empiric therapy be intentionally broad spectrum? (LO 15.2)
3. What are bacteriostatic versus bactericidal drugs, and what clinical scenarios would benefit from each class of drug? (LO 15.3)
4. How do natural, synthetic, and semisynthetic antimicrobials differ, and what is the value of drug modifications? (LO 15.4)
5. The term *antibiotic* is more specific than *antimicrobial*. Explain this statement. (LO 15.5)
6. What do the terms *therapeutic index*, *selective toxicity*, and *drug half-life* mean, and how do these features relate to antimicrobial drug development? (LO 15.6)
7. What is meant by hepatotoxicity and nephrotoxicity, and what antimicrobial drug classes are associated with each effect? (LO 15.7)
8. How do the route of drug administration, drug interactions, and contraindications play into drug development? (LO 15.8)

15.2 SURVEY OF ANTIBACTERIAL DRUGS

Learning Outcomes

After reading this section, you should be able to:

15.9 Describe the four main groups of beta-lactam drugs, and state how they work.

15.10 Give examples of glycopeptide drugs, state how they work, and describe how they differ from beta-lactam drugs.

15.11 Describe how quinolones and rifamycins work, and state when each may be recommended.

15.12 Explain how sulfa drugs work and why they don't target human cells.

15.13 State how macrolides, lincosamides, phenicols, tetracyclines, and aminoglycosides each work, and describe the potential adverse effects of each.

15.14 Give examples of polypeptide drugs, describe how they work, and state when they may be used.

Antibacterial drugs may be grouped by their cellular targets.

Ideally, an antibacterial drug should attack the infecting bacteria while sparing the patient's cells. Thus, the best drug targets tend to be structures and processes that bacteria rely on, but human cells do not—such as the bacterial cell wall. FIG. 15.4 provides an overview of the main antibacterial drug targets. Most antibiotics that inhibit protein synthesis or block metabolic pathways are considered bacteriostatic, whereas most drugs that target the bacterial cell wall, cell

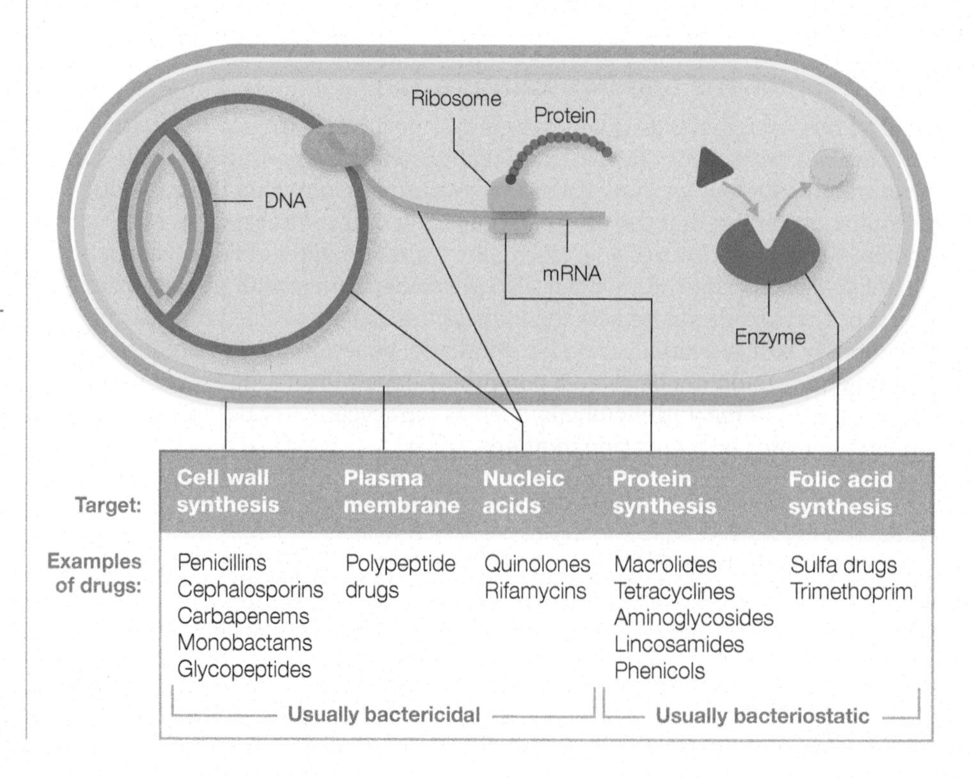

FIGURE 15.4 Key antibacterial drug targets

Critical Thinking *If an antimicrobial inhibits transcription, what are the repercussions for other cellular processes?*

TABLE 15.1 Summary of Key Antibacterial Drug Families

Target	Action	Drug Family	Examples	Activity Spectrum	
Cell wall synthesis	Usually bactericidal	Penicillins (end in "*cillin*")	Penicillin V Penicillin G Ampicillin	Natural penicillins = narrow spectrum Semisynthetic derivatives = broad spectrum	Beta-lactam superfamily
		Cephalosporins (start with "*cef*" or "*ceph*")	Cephalexin Cephapirin Fifth-generation drugs such as ceftaroline combat MRSA/ORSA strains	Broad (range of activity tends to increase for each successive drug generation)	Beta-lactam superfamily
		Carbapenems (often end in "*penem*")	Doripenem Imipenem Meropenem	Broad	Beta-lactam superfamily
		Monobactams	Aztreonam	Narrow	Beta-lactam superfamily
		Glycopeptides	Teicoplanin Vancomycin	Narrow	
		Miscellaneous non-beta-lactam antimicrobials	Bacitracin Isoniazid (targets mycobacteria)	Narrow	
Nucleic acids	Usually bactericidal	Quinolones (fluoroquinolones contain "*fl*" in their names)	Ciprofloxacin Levofloxacin Ofloxacin	Broad	
		Rifamycins	Rifampin (rifampicin) Rifabutin Rifapentine	Broad	
Folic acid synthesis	Usually bacteriostatic	Sulfa drugs (sulfonamides)	Trimethoprim-sulfamethoxazole* Sulfisoxazole Sulfasalazine	Broad	
Ribosomes (protein synthesis)	Usually bacteriostatic	Macrolides	Erythromycin, Azithromycin ("Z-Pak") Clarithromycin	Broad	
		Lincosamides	Clindamycin Lincomycin Pirlimycin	Broad	
		Phenicols	Chloramphenicol Thiamphenicol	Broad	
		Tetracyclines	Tetracycline, Demeclocycline Doxycycline	Broad	
		Aminoglycosides (end in "*mycin*" or "*micin*")	Neomycin Amikacin Streptomycin Gentamicin	Narrow	
Cell membrane	Usually bactericidal	Polypeptide drugs	Polymyxin B Colistin	Narrow	

*This is a combination therapy of a sulfa drug (sulfamethoxazole) and a nonsulfa antifolate drug (trimethoprim). This combination therapy is often called by its trade names Bactrim or Septra.

membrane, or nucleic acids (DNA or RNA) act as bactericidal drugs. TABLE 15.1 presents an overview of the antibacterial drugs reviewed in this chapter.

Many antimicrobial drugs disrupt bacterial cell wall production.

Most bacteria have a cell wall consisting of peptidoglycan. As discussed in Chapter 3, peptidoglycan is made up of sugar units (N-acetylmuramic acid and N-acetylglucosamine) and amino acids. The sugar molecules are joined into

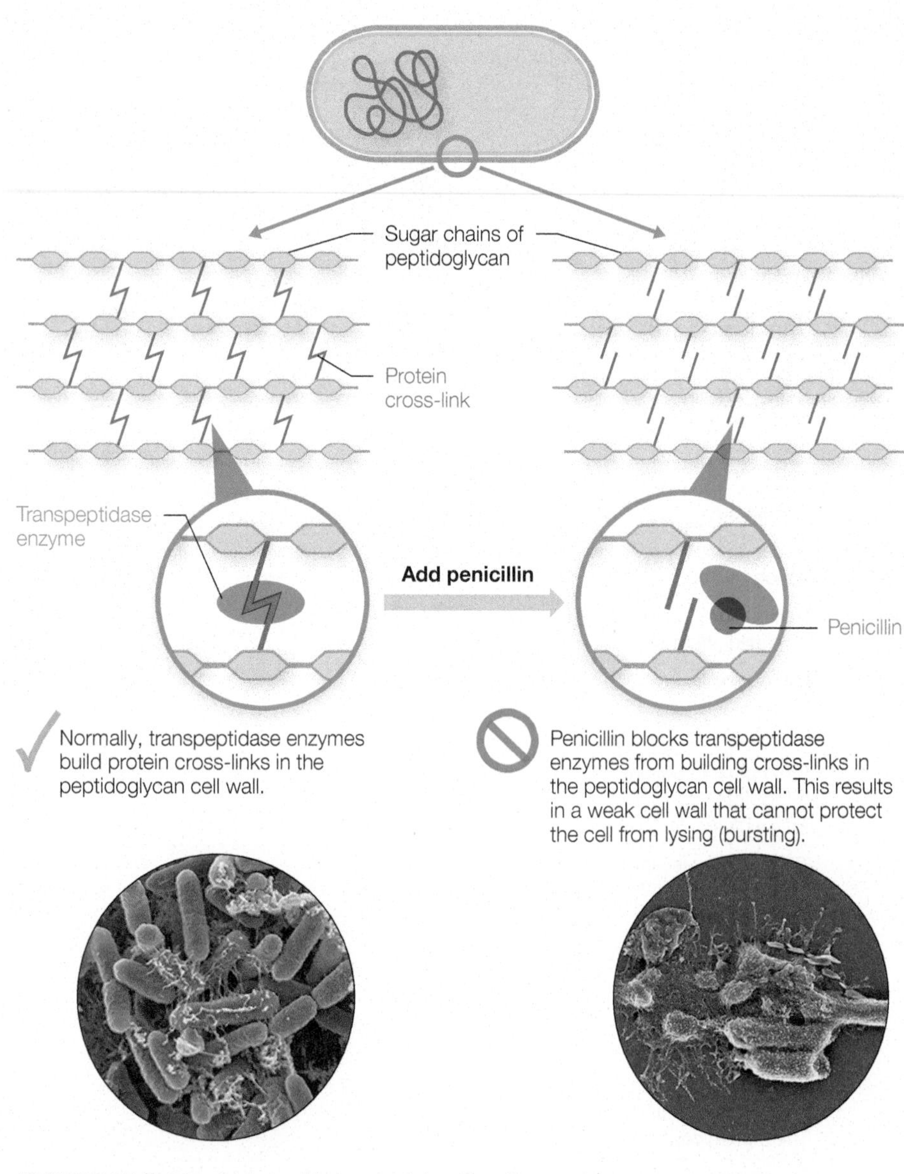

FIGURE 15.5 **Interfering with bacterial cell wall construction** Beta-lactam and glycoprotein antimicrobials interfere with cell wall synthesis by blocking transpeptidation—the formation of peptide linkages between the sugar chains in peptidoglycan. This weakens the cell wall and leads to bacterial cell lysis.

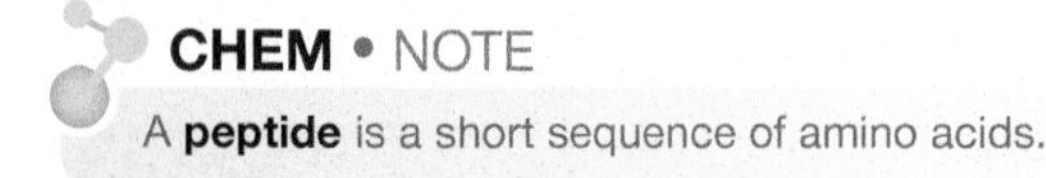

A **peptide** is a short sequence of amino acids.

long chains that are then interlinked by peptides. Drugs that block peptidoglycan production tend to do so by interfering with **transpeptidation**—the step in which reinforcing peptide linkages are made between sugar chains. This disruption weakens the cell wall and leads to cell lysis, making these drugs bactericidal (FIG. 15.5). Cell wall synthesis inhibitors tend to be most effective during the exponential growth phase, when the bacterial cells are actively building their cell walls as a part of cell division.

Penicillins, Cephalosporins, Carbapenems, and Monobactams: Beta-Lactam Antimicrobials That Target Bacterial Cell Walls

A superfamily of drugs called **beta-lactam antimicrobials** work by blocking cell wall construction. These drugs include penicillins, cephalosporins, carbapenems, and monobactams. All four of these drug families have a four-sided **beta-lactam ring** (or β-lactam ring) as part of their active chemical structure (FIG. 15.6). Beta-lactam antibiotics work by binding to *transpeptidase enzymes* that are central to building peptidoglycan. These enzymes form the protein cross-links that bind peptidoglycan's carbohydrate chains together in a chain link fence-like structure.

Many bacteria make **beta-lactamases** (or β-lactamases), which are enzymes that inactivate beta-lactam drugs. To combat these resistance enzymes, **beta-lactamase inhibitors** such as clavulanate, sulbactam, or tazobactam can be co-administered with certain beta-lactam antibiotics. For example, Augmentin is a combination of amoxicillin and clavulanate. Beta-lactamase inhibitors have a beta-lactam ring structure and bind strongly to beta-lactamase enzymes. This binding blocks beta-lactamase from deactivating the administered antimicrobial. Unfortunately, these decoys do not thwart all beta-lactamases that have evolved.

Penicillins The **penicillin drugs** are a family of natural and semisynthetic compounds that tend to end in the suffix *cillin*. They remain among the most widely prescribed and safest antimicrobials. Common side effects are minimal and tend to be rashes and gastrointestinal upset (FIG. 15.7). The main challenges in using penicillins are the prevalence of patient allergies to these drugs as well as the widespread occurrence of resistant bacterial strains.

Naturally occurring penicillin G was the first type of penicillin widely used. It is primarily effective against a wide range of Gram-positive bacteria, as well as select Gram-negative pathogens such as the causative agents of syphilis, diphtheria, meningitis, and gonorrhea. Penicillin G is still widely used, but it must be administered by injection because it is destroyed by stomach acid. In contrast, the natural antibiotic penicillin V can be taken orally.

As you learned earlier in this chapter, the penicillin G core molecular structure has been modified to create an array of penicillin drugs that have improved pharmacological properties. Some, like amoxicillin and ampicillin, are designed

FIGURE 15.6 Beta-lactam superfamily of antibacterial drugs Penicillins, cephalosporins, carbapenems, and monobactams all possess a beta-lactam ring. This four-sided structure is a target of antimicrobial-resistance enzymes called beta-lactamases.

Critical Thinking *Aside from making beta-lactamase, what is another mechanism that bacteria could employ to become broadly resistant to beta-lactam drugs? (Hint: Consider the enzyme involved in the inhibited reaction.)*

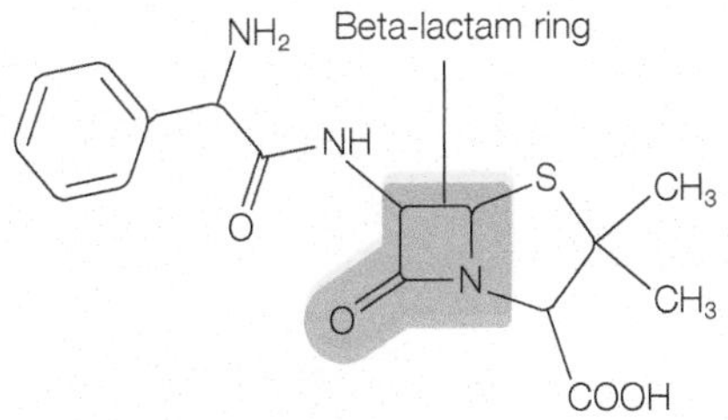

Beta-lactam ring

Penicillins (ampicillin)

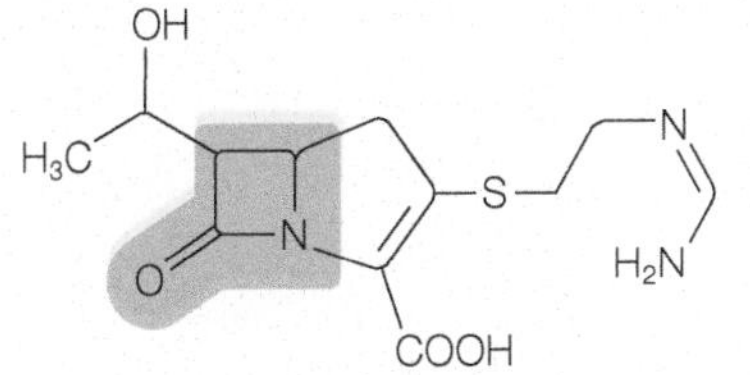

Carbapenems (imipenem)

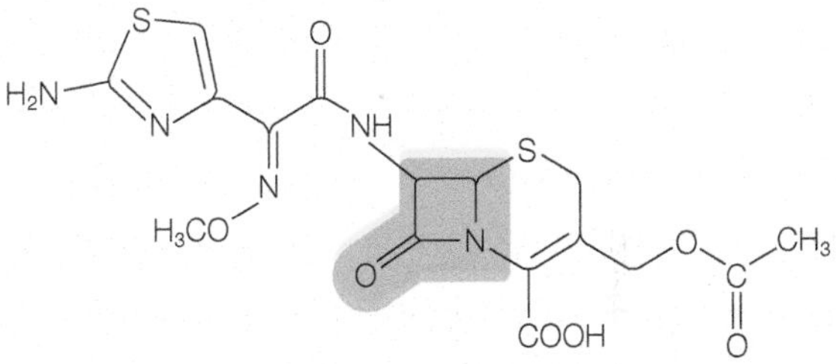

Cephalosporins (cefotaxime)

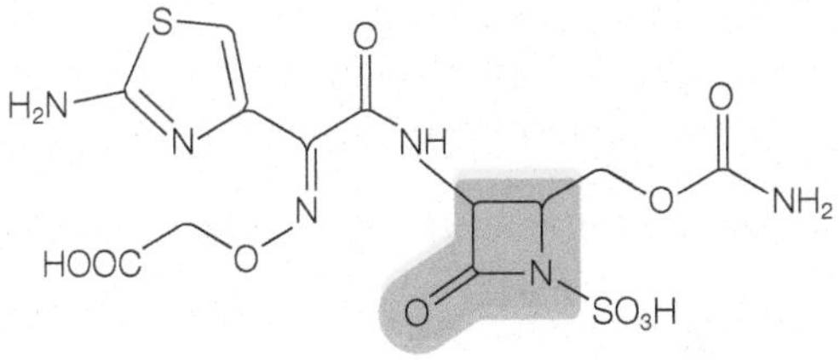

Monobactams (carumonam)

to have an extended spectrum of activity against Gram-negative bacteria. Although amoxicillin and ampicillin have extended spectra, they remain susceptible to beta-lactamases. Cloxacillin, oxacillin, nafcillin, dicloxacillin, and methicillin[11] were developed to combat bacteria that make beta-lactamases, but they lack the expanded spectrum of other penicillin derivatives. Just as offensive linemen may physically block access to a football team's quarterback, the bulky R groups on these drugs use a strategy called *steric hindrance* to physically block beta-lactamases from interacting with the drug's beta-lactam ring.

Inevitably, methicillin- and oxacillin-resistant S. *aureus* strains (**MRSA** and **ORSA**) eventually emerged and are now relatively common. Rather than trying to go after the beta-lactam drugs, these resistant strains adopted a disconcertingly effective strategy—they altered their transpeptidase enzymes so that the majority of beta-lactam drugs can no longer target it. This means that MRSA and ORSA strains are resistant to almost all beta-lactam drugs, not just methicillin and oxacillin, as their acronyms imply.

Cephalosporins These broad-spectrum, bactericidal drugs constitute the largest collection of beta-lactam drugs. **Cephalosporins** can be recognized by their *cef* or *ceph* prefixes. Like penicillins, they tend to be relatively safe and are commonly used, with minimal side effects such as rash and gastrointestinal upset. Patients with penicillin allergies can have a cross-reactive allergy to cephalosporins; when they do, it is almost always to first-generation cephalosporins that are the most structurally similar to penicillins—later-generation cephalosporins are rarely problematic in patients allergic to penicillin drugs.[12]

Cephalosporins exist naturally and have also been semisynthetically derived. As of publication of this book, semisynthetic versions span five different generations. In general, each successive generation has an expanded activity against Gram-negative bacteria, usually with a corresponding decrease in activity against Gram-positive bacteria. Only fifth-generation cephalosporins such as ceftaroline (Teflaro) combat MRSA/ORSA strains.

Carbapenems Like penicillins and cephalosporins, **carbapenems** are beta-lactam drugs that target bacterial cell wall construction. These drugs are usually well tolerated, with rash and gastrointestinal upset being the most reported side effects. However, people with renal dysfunction are at a greater risk for seizures when taking these drugs. The incidence of allergy to these drugs is low, and people who have penicillin allergies can usually tolerate carbapenems.[13] Members of this drug group tend to end in the suffix *penem* and are considered broad-spectrum drugs—acting against most Gram-negative bacilli as well as Gram-positive streptococci and non-MRSA staphylococci. Carbapenems are effective against a variety of **multidrug-resistant (MDR)** bacterial strains and are therefore reserved to fight against healthcare-acquired infections (HAIs) that

FIGURE 15.7 Penicillin allergy rash

[11]The mnemonic to remember these is CONDM; however, methicillin is no longer used because of its nephrotoxicity.

[12]Patients with a documented penicillin allergy have a four- to fivefold increased risk of having an allergic reaction to a first-generation cephalosporin drugs. Ward, S., Geurin, M. D., & Mikulic, A. (2021). Is it safe to use cephalosporins in patients who are allergic to penicillin? *Evidence-Based Practice*, 24(4), 16–17.

[13]Caruso, C., Valluzzi, R. L., Colantuono, S., Gaeta, F., & Romano, A. (2021). β-lactam allergy and cross-reactivity: A clinician's guide to selecting an alternative antibiotic. *Journal of Asthma and Allergy*, 14, 31.

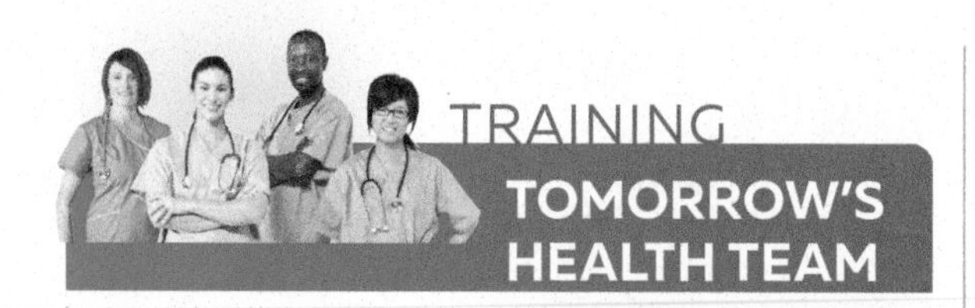

Carbapenem Resistance on the Rise

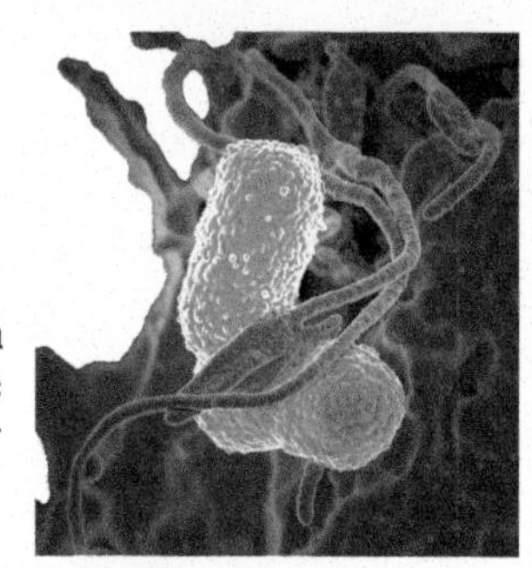
Carbapenem-resistant Enterobacterales such as *E. coli* are increasingly prevalent in healthcare settings.

You have been concerned about your patient in room 203—a 14-year-old with cystic fibrosis. You notice her mom is at her bedside anxiously awaiting the infectious disease doctor's visit. Your heart sinks as you read the antibiogram report that reveals the cultured bacteria are resistant to basically every drug available. Sadly, this is not an imagined scenario, but an increasingly common occurrence when dealing with carbapenem resistance. Blood infections with carbapenem-resistant Enterobacterales (CRE) and carbapenem-resistant *Acinetobacter* (CRA) strains have a roughly 50 percent mortality rate.

Bacteria have evolved several carbapenem-resistance mechanisms, but the most disconcerting approach is the production of carbapenemase enzymes that break down these drugs. CRE and CRA strains are especially threatening because their resistance genes are on plasmids that are easily shared with other bacteria in the environment by horizontal gene transfer. To limit the spread of carbapenem resistance, healthcare providers must practice proper hand hygiene *before* they gown up and put on gloves. They should then remove their gown and gloves and thoroughly wash their hands before exiting a patient's room. Also, patients infected with CRE or CRA should not be roomed with non-CRE/CRA patients—ideally, a facility will have CRE/CRA-designated spaces. During an outbreak, facilities are encouraged to designate staff to attend only CRE/CRA patients, in order to limit cross-contamination to non-CRE/CRA patients.

QUESTION 15.1

In addition to standard precautions, what transmission precautions would you apply if caring for a patient with CRE pneumonia? (Hint: Chapter 10 reviewed transmission precautions.)

CHEM • NOTE

The term **glycopeptide** describes a peptide molecule (a short sequence of amino acids) that is associated with sugar components.

are often resistant to two or more groups of antibiotics. Currently there are only injectable forms of these drugs, and they are mainly reserved for use in critically ill patients and patients who have infections that have not responded to first-line therapies. Examples include doripenem, ertapenem, imipenem, and meropenem. Unfortunately, over the past decade the rise of **carbapenem-resistant Enterobacterales** (CRE)[14] has rendered even these last-resort drugs useless against some infections. Enterobacterales is a large order of Gram-negative bacteria that taxonomically encompasses a variety of bacterial families, including the Enterobacteriaceae. *Klebsiella pneumonia* and *Escherichia coli* are among the more prevalent CRE. A number of bacteria that are not in the order Enterobacterales, such as *Pseudomonas aeruginosa* and carbapenem-resistant *Acinetobacter* (CRA), can also exhibit carbapenem resistance. *P. aeruginosa* is a notorious cause of healthcare-associated infections (HAIs)—especially pneumonia, wound infections, and blood infections. *Acinetobacter* species, particularly *Acinetobacter baumannii*, are also a cause of HAIs such as pneumonia, blood infections, wound infections, and urinary tract infections. Unfortunately, carbapenem-resistant bacteria tend to be resistant to most other antimicrobials, making them difficult if not impossible to treat with current drugs. As such, CRE and CRA bacteria are a pressing medical concern that the CDC is carefully monitoring.

Monobactams Unlike other beta-lactam drugs that usually have two rings in their structure, the **monobactams** have only one ring. Aztreonam is the lone example of a commercially available monobactam drug for human use. Aztreonam is most effective against Gram-negative bacteria and is therefore useful against certain Enterobacteriaceae family members such as *Pseudomonas*, *Klebsiella*, *Proteus*, and *Enterobacter* species. It is not especially useful against Gram-positive bacteria and therefore is considered a narrow-spectrum drug. Like other beta-lactam drugs, aztreonam blocks bacterial cell wall production and promotes cell lysis. Although aztreonam shares some structural features with penicillin, it can usually be used in penicillin-allergic patients. Side effects from aztreonam are similar to those for other beta-lactam drugs.

Glycopeptides: Non-Beta-Lactam Antimicrobials That Target Bacterial Cell Walls

Like beta-lactam drugs, glycopeptides interfere with bacterial cell wall construction. However, unlike the beta-lactams, **glycopeptide drugs** do not have a beta-lactam ring, so they are not susceptible to beta-lactamases. Glycopeptides, such as teicoplanin and vancomycin, are key glycopeptides. These narrow-spectrum drugs are effective against Gram-positive organisms and can be useful for fighting certain antibiotic-resistant bacteria. For example, vancomycin can be used against methicillin-resistant *S. aureus* (MRSA); it is also a preferred treatment for *C. difficile* infections.

The original glycopeptides are naturally occurring products made by *Streptomyces*. However, considering growing resistance to commonly used glycopeptides, including vancomycin-resistant *S. aureus* (VRSA) strains and vancomycin-resistant enterococci (VRE), newer-generation synthetic and semisynthetic glycopeptides are being developed. In 2017, a new drug was introduced that has been dubbed "vancomycin version 3.0" and uses a three-pronged approach to combat bacteria. Initial testing suggests that this newer vancomycin analog may be much more potent and far less susceptible to antibiotic resistance than classic vancomycin, which has been in use since the 1950s.[15]

[14]Formerly these bacteria were called carbapenem-resistant Enterobacteriaceae (CRE). The taxonomic order name, Enterobacterales, is now recommended in place of the taxonomic family name. This change was made to reflect that carbapenem resistance is no longer isolated to one family (Enterobacteriaceae), but now spans multiple families within the order.

[15]Akinori, O., Isley, N. A., & Boger, D. L. (2017). Peripheral modifications of vancomycin with added synergistic mechanisms of action provide durable and potent antibiotics. *PNAS*, *114*(26), E5052–E5061.

Because glycopeptides are not easily absorbed across the intestines, intravenous administration is used for systemic infections. In contrast, oral preparations are useful against intestinal tract infections, such as *C. difficile*. Most people tolerate glycopeptides well, and there is only a low incidence of allergy to these drugs. Some common glycopeptide side effects include taste disturbances, nausea, vomiting, headache, dizziness, and **red man syndrome**—a situation that can be mistaken for a classic drug allergy because it is characterized by a red flush spreading over the skin and itchiness (FIG. 15.8).

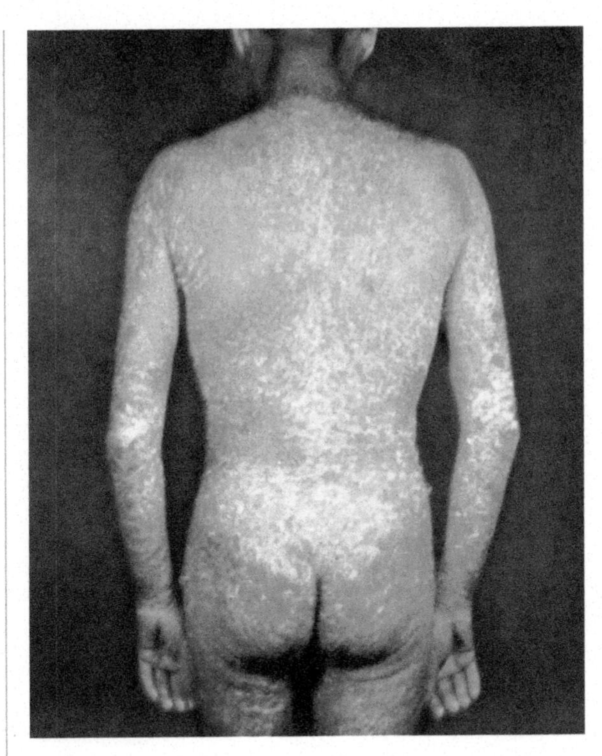

FIGURE 15.8 Red man syndrome This syndrome is the most common adverse reaction to vancomycin.

Miscellaneous Non-Beta-Lactam Antimicrobials That Target Bacterial Cell Walls

In addition to the beta-lactam drugs and glycopeptides, some other antimicrobials inhibit cell-wall synthesis. For example, **bacitracin** is a polypeptide drug that targets bacterial cell walls and acts as a narrow-spectrum agent against Gram-positive bacteria—especially staphylococci. It is a common antibacterial drug in topically applied ointments like Neosporin. Another drug called **isoniazid** is used in combination with other antibacterial drugs to treat tuberculosis. Isoniazid interferes with mycolic acid construction in cell walls of acid-fast bacteria.

Quinolones and rifamycins target nucleic acids.

Some drugs target nucleic acids. Because DNA and RNA production pathways are highly conserved among different bacteria and are necessary for survival, these drugs tend to be broad spectrum and bactericidal in nature.

Quinolones

Quinolones are synthetic antimicrobials that target DNA replication enzymes, namely DNA gyrase and topoisomerases. The more modern and most prescribed quinolones are called fluoroquinolones, so called because they contain a fluorine atom (a feature echoed by the inclusion of "fl" in their drug names). These drugs are not usually used as first-line therapies and instead are reserved for treating infections that show antimicrobial resistance. Ciprofloxacin is an example of a fluoroquinolone that is effective against *Mycobacterium,* the genus that includes the causative agents of tuberculosis and leprosy, and *Pseudomonas* species, which cause wound, burn, and other infections. Levofloxacin is a fluoroquinolone used to treat community-acquired pneumonia, including "walking pneumonia" caused by *Mycoplasma pneumoniae*, which is insensitive to all beta-lactam drugs because it lacks a cell wall. In addition to the broad spectrum of action, fluoroquinolones can be taken orally and have a relatively long half-life (patients only need to take them once or twice a day), which makes them especially helpful bactericidal antimicrobials. They are excellent in situations when there isn't time to identify the responsible pathogen. These drugs tend to have the standard side effects of gastrointestinal upset, such as nausea, vomiting, and diarrhea. They can also trigger headache, dizziness, and insomnia. These minor side effects tend to resolve upon finishing the drug course. In rare instances these drugs have been associated with more serious side effects such as nerve damage, tendonitis, and possible tendon rupture (especially of the Achilles tendon).

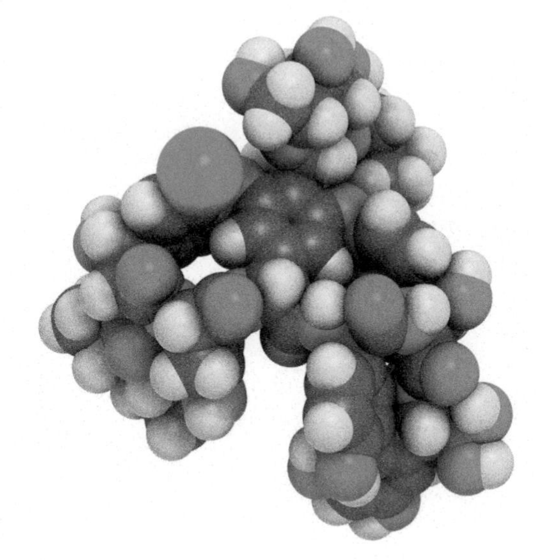

Molecular structure of vancomycin.

Rifamycins

The **rifamycins** are a group of bactericidal drugs that were originally isolated from bacteria. They are now mainly produced as synthetic and semisynthetic compounds. **Rifampin** (also called rifampicin) is a key representative of this group. It inhibits transcription by binding to RNA polymerase. This broad-spectrum

Ciprofloxacin is a fluoroquinolone drug.

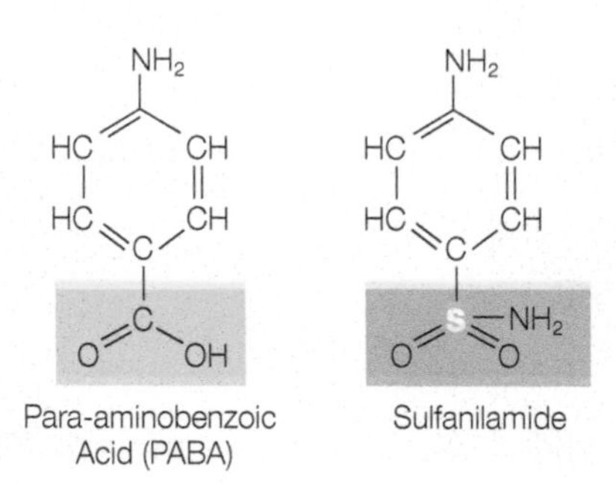

FIGURE 15.9 Sulfa drugs resemble PABA Sulfa drugs are structurally similar to para-aminobenzoic acid (PABA), the natural substrate of bacterial enzymes that build folic acid.

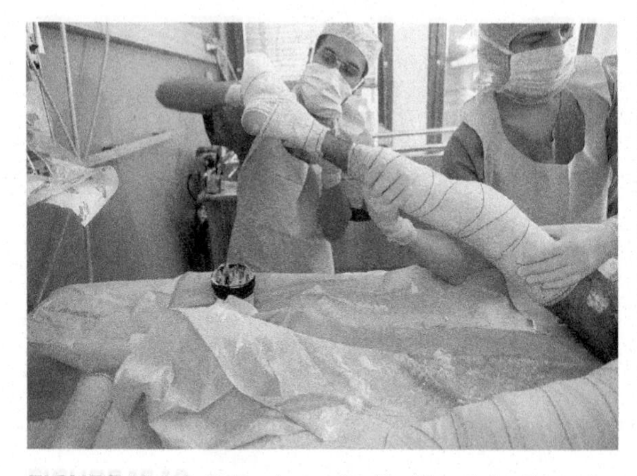

FIGURE 15.10 Sulfa drugs for limiting infections in burn patients Sulfa drugs, such as silver sulfadiazine, can be impregnated into dressings or applied as ointment to burns. In particular, *P. aeruginosa* is a common infection in burn patients that can be prevented with sulfa drug dressings.

Critical Thinking *Why are burn victims more susceptible to tissue infections by the common bacteria* P. aeruginosa*?*

antimicrobial is effective against most Gram-positive bacteria and some Gram-negative bacteria. It is especially useful to combat mycobacterial species, which have a waxy, mycolic acid-enriched cell wall that many drugs cannot cross. In particular, rifampin is often administered in conjunction with other antibiotics as a part of **combination therapy** to control tuberculosis and leprosy infections: Using two or more drugs in combination decreases the likelihood that a pathogen will develop resistance and survive the therapy. Most people tolerate rifampin without serious side effects. When side effects do occur, they tend to include heartburn, gastrointestinal upset, headache, drowsiness, fatigue, and dizziness. Care must be taken when prescribing rifampin with other antimicrobials, as its presence inhibits several drugs, including certain HIV medications, digoxin (used to treat certain heart conditions), oral contraceptives, certain blood pressure medications (such as beta-blockers), and anticoagulants ("blood thinners").

Antifolate drugs target a microbe's folic acid production.

Originally discovered in the 1930s, **sulfa drugs** (or sulfonamides) were among the first synthetic antimicrobials used to combat bacterial infections. Sulfa drugs act as competitive inhibitors of folic acid production. They chemically resemble **para-aminobenzoic acid (PABA)**, the natural substrate of a bacterial enzyme in the pathway that builds folic acid (FIG. 15.9). (Chapter 8 reviews competitive inhibitors.) These drugs do not affect mammalian cells because, unlike bacteria, mammals do not make their own folic acid and therefore do not have the enzyme that these drugs target. Sulfa drugs are bacteriostatic and have a broad spectrum of action. They can be administered orally, topically, intravenously, and as eyedrops or eardrops. Sulfonamides can also be combined with silver and applied to wound dressings for burn victims (FIG. 15.10).

Sulfa drugs like **sulfamethoxazole** are often administered in combination with another antifolate drug called **trimethoprim**, resulting in **trimethoprim-sulfamethoxazole (TMP/SMX)**. Leading trade names for this combination are Bactrim and Septra.[16] Trimethoprim is *not* a sulfa drug, nor does it resemble PABA. Instead, this antifolate drug targets a part of folic acid synthesis that is different from the step that sulfa drugs inhibit. Because these drugs target different steps of the same biochemical pathway, when used together, they tend to exhibit **synergism**—that is, their combined effectiveness is greater than the sum of their effectiveness when used alone. Common sulfa drug therapy side effects include increased skin sensitivity to UV light, rash, and gastrointestinal upset such as nausea, vomiting, and diarrhea. Unfortunately, sulfa drug allergies are among the more common antibiotic allergies that patients report.

Some drugs target prokaryotic ribosomes.

Ribosomes translate mRNA to build proteins and are essential organelles for cell growth. Because prokaryotic and eukaryotic ribosomes differ, it is possible for an antimicrobial to target prokaryotic ribosomes without affecting the ribosomes in human cells. Many drugs can bind to prokaryotic ribosomes to inhibit protein synthesis (FIG. 15.11). These drugs are often considered bacteriostatic, although as mentioned earlier, the line between bactericidal and bacteriostatic drugs can easily blur with shifts in the length of the drug course and the strength of the dose. Some drugs that inhibit protein synthesis are bactericidal at higher doses, and certainly a bacterium that is blocked from making proteins for a long enough period will eventually die. Here, we discuss five types of drugs that target prokaryotic ribosomes.

[16] The nonsulfa antifolate drug trimethoprim is often used in combination with sulfamethoxazole (TMP/SMX combination therapy). This combination therapy is also effective against certain protozoan infections and will be revisited later in this chapter.

Macrolides

Common examples of **macrolide drugs** include erythromycin, azithromycin ("Z-Pak"), and clarithromycin. Macrolides target the 50S subunit of prokaryotic ribosomes to block protein synthesis. These compounds are broad-spectrum drugs effective against various Gram-negative cocci as well as aerobic and anaerobic Gram-positive rods and cocci. Macrolides may be used to treat a variety of infections, including *Streptococcus pneumoniae*, *S. pyogenes* (strep throat), chlamydia, Legionnaires' disease, diphtheria carriers, and Lyme disease. Administration can be oral, parenteral, or topical.

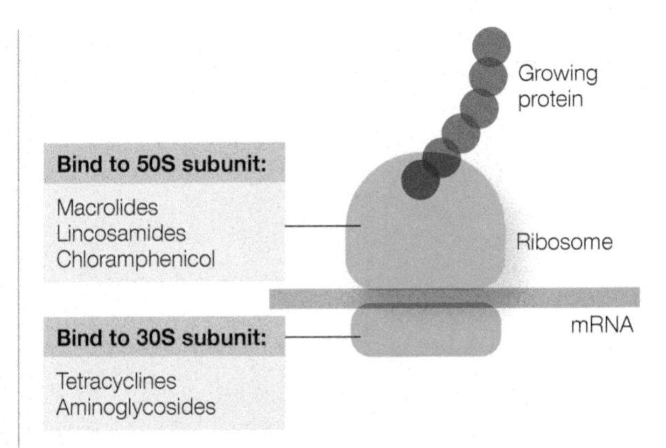

FIGURE 15.11 Drugs that block protein synthesis

Critical Thinking *Theoretically speaking, could a macrolide-resistant pathogen be successfully treated with a tetracycline antimicrobial? Explain your reasoning.*

Lincosamides

Lincosamides bind to the 50S subunit of prokaryotic ribosomes to block protein synthesis. These broad-spectrum drugs work against a wide collection of aerobic and anaerobic Gram-positive as well as anaerobic Gram-negative bacteria and even some protozoans. Clindamycin is one commonly used lincosamide and is also effective against MRSA. In some cases, clindamycin can also be used to treat macrolide-resistant infections, particularly those resistant to erythromycin. A downside of clindamycin is that it is one of the antimicrobial agents that is most commonly associated with pseudomembranous colitis caused by *C. difficile*.

Phenicols

The **phenicols** are structurally simple, broad-spectrum drugs that bind to the 50S ribosomal subunit. The main phenicol drug used in humans is **chloramphenicol**. It is effective against Gram-positive and Gram-negative cocci and bacilli that include *Rickettsia*, *Mycoplasma*, and *Chlamydia* species. However, this antimicrobial has a narrow therapeutic index and is associated with bone marrow toxicity that results in aplastic anemia (lack of red blood cell production). Because of these unfavorable side effects, chloramphenicol is not typically a first-line drug choice. Instead, it is usually reserved for treating severe infections caused by certain multidrug-resistant bacteria.

Tetracyclines

Tetracyclines are a family of chemically related compounds that include drugs such as tetracycline, demeclocycline, and doxycycline, with doxycycline being the most commonly used in a clinical setting. These drugs bind to the prokaryotic 30S ribosomal subunit. As with macrolides, tetracyclines are bacteriostatic and are relatively easily absorbed through the intestines. These broad-spectrum drugs have many uses, such as treating acne, chlamydia, cholera, *Mycoplasma* infections, Lyme disease, anthrax, plague, and syphilis, and even can be used to prevent malaria.[17] As seen with other broad-spectrum antimicrobial drugs, tetracyclines are associated with an increased risk of *C. difficile* infection. These drugs also induce photosensitivity, so patients taking them should stay out of the sun. Lastly, these compounds should not be given to children under 8 years old due to detrimental effects on bones and teeth (FIG. 15.12).

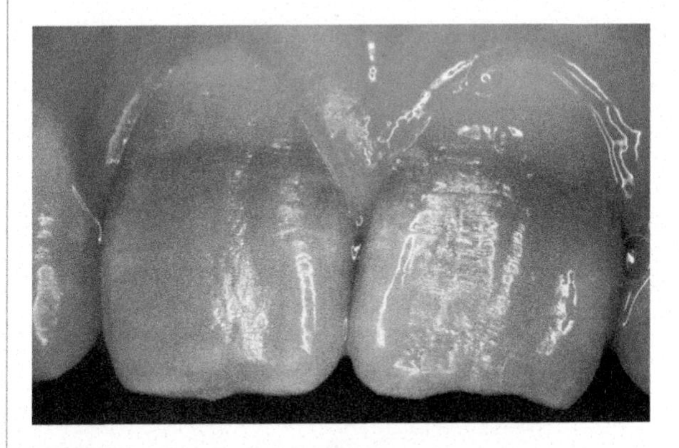

FIGURE 15.12 Tetracycline induces tooth discoloration Tetracycline can cause permanent discoloration of developing teeth.

Aminoglycosides

Aminoglycosides include a collection of drugs that bind to the 30S ribosomal subunit to block protein synthesis. Their drug names tend to end in "mycin" or "micin." Because they are poorly absorbed in the intestinal tract and have a

[17]The mechanism that tetracyclines use for preventing malaria, a protozoan infection, is different than the mechanism that blocks protein synthesis in bacteria. The details of their action against malarial parasites (*Plasmodium* species) are still under investigation, but are thought to involve targeting the parasite's apicoplast—an organelle found in the parasite.

short half-life of 2 to 3 hours, these drugs are usually given intravenously rather than orally. Aminoglycosides are mainly narrow spectrum, working against Gram-negative aerobic and facultative anaerobic bacilli—most notably *P. aeruginosa*, a major cause of wound, burn, and postoperative infections. Aminoglycosides are rarely used alone and are most often used as a partner in a combination therapy—often with beta-lactam drugs, which work synergistically with aminoglycosides. A common aminoglycoside is neomycin, an ingredient in over-the-counter triple antibiotic topical ointments that many people keep in their medicine cabinets. Nontopically administered aminoglycosides include amikacin, tobramycin, streptomycin, and gentamicin.

Nontopical aminoglycosides are known to cause irreversible hearing loss. For this reason, hearing may be tested before therapy and during therapy to make sure the drugs are stopped, or the doses are lowered if hearing becomes affected. These drugs also exhibit nephrotoxicity. These potentially serious side effects make nontopical aminoglycosides less useful as first-line therapies; instead, they tend to be used when other drugs have failed.

Certain polypeptide drugs target membrane structures.

The outer membrane of Gram-negative cells functions as a selective barrier that prevents many antimicrobials from entering these cells. Polypeptide drugs overcome that barrier. Two notable bactericidal polypeptide drugs are **polymyxin B** and **colistin** (also called polymyxin E). These drugs tend to be narrow-spectrum agents that target the outer membrane of facultative anaerobes and aerobic Gram-negative cells. Their structural features allow them to interact with lipopolysaccharide and destabilize the outer membrane of the Gram-negative cell wall. They can then further destabilize the plasma membrane to cause cytoplasmic leakage and cell lysis (FIG. 15.13).

Because they have a narrow therapeutic index, polymyxin B and colistin are mainly used topically. Most notably, polymyxin B is a common component of over-the-counter triple antibiotic ointments applied to scrapes and cuts. In life-threatening multidrug-resistant infections, especially those caused by *P. aeruginosa*, *Acinetobacter baumannii*, or *Klebsiella pneumoniae*, there are forms of these drugs that can be administered intravenously.

FIGURE 15.13 Polypeptide drugs Polymyxin B and colistin are polypeptide drugs that disrupt the outer membrane of Gram-negative bacterial cells and destabilize the cell's plasma membrane to induce lysis.

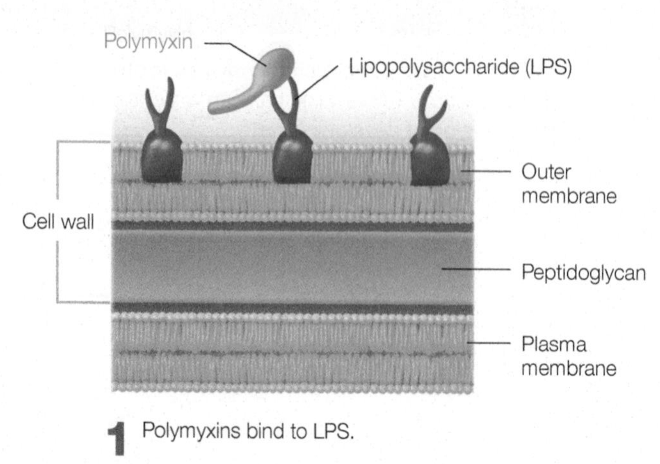

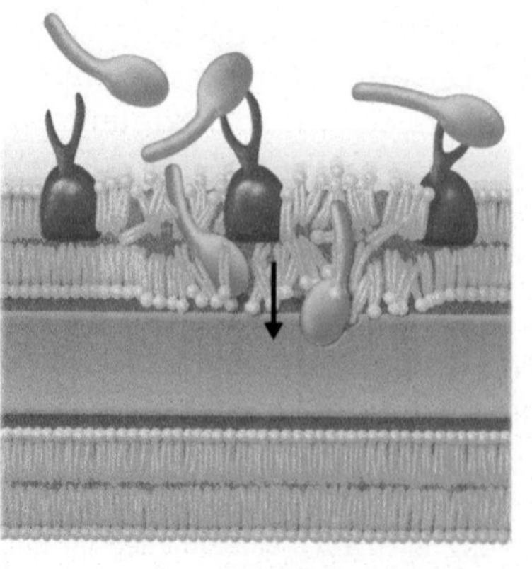

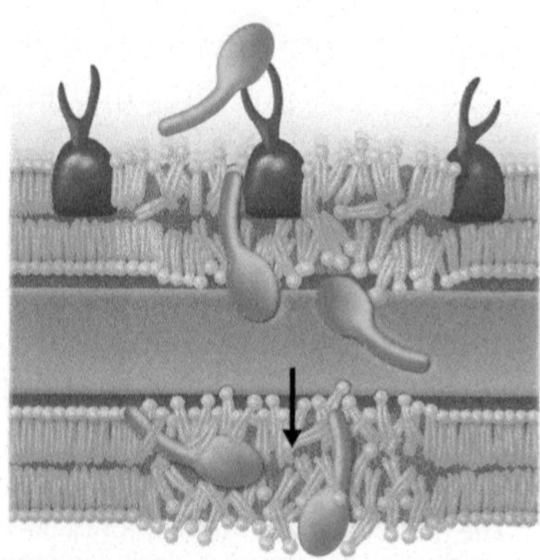

Build Your Foundation

9. What are the four groups of beta-lactam drugs, and how do they work? (LO 15.9)
10. Name a glycopeptide drug, summarize how it works, and state how it differs from the beta-lactam superfamily of drugs. (LO 15.10)
11. How do quinolones and rifamycins work, and when would each be recommended? (LO 15.11)
12. How do sulfa drugs work, and why don't they target human cells? (LO 15.12)
13. How do macrolides, lincosamides, phenicols, tetracyclines, and aminoglycosides each work, and what potential adverse effects may each present? (LO 15.13)
14. What are some examples of polypeptide drugs, how do they work, and when might they be used? (LO 15.14)

Build Your Foundation (BYF) Quick Quiz: Visit the **Mastering Microbiology** Study Area to quiz yourself.

15.3 DRUGS FOR VIRAL AND EUKARYOTIC INFECTIONS

Learning Outcomes

After reading this section, you should be able to:

15.15 Describe the difficulties of developing drugs against viruses and eukaryotic pathogens.

15.16 List potential points in viral replication that antiviral drugs may target.

15.17 Provide examples of antifungal agents, describe the fungal infections they may treat, and discuss how they target fungi.

15.18 Describe antiprotozoal drugs—including examples, mechanisms of action, and therapeutic applications.

15.19 Discuss the main antihelminthic drugs, their mechanisms of action, and medical applications.

Developing selectively toxic drugs against eukaryotic pathogens and viruses is challenging.

The collection of antibacterial drugs that we have is more extensive than the collection of drugs we have against viruses, fungi, protozoa, or helminths (worms). This is because it is difficult to develop drugs that specifically target viruses and eukaryotic pathogens such as fungi, protozoa, and helminths without inflicting collateral damage on our own cells.

Antiviral drugs target specific points in viral replication.

Viruses rely on host cell machinery to replicate, which means they are often difficult to target without affecting host cell processes. Antiviral agents and interferon therapies to stimulate immune responses against viral infections were reviewed in Chapter 6. As a reminder, antiviral drugs are classified into five main groups according to the viral activity they target: attachment, penetration, uncoating, viral replication and assembly, or viral release. A sixth category includes drugs such as **interferons** that stimulate immune responses against viruses. Antiviral drugs are mainly effective when viruses are actively replicating; thus, latent viruses are difficult to treat. To date, most antivirals have been against human immunodeficiency virus (HIV), certain herpes family viruses, viruses that cause hepatitis, and influenza viruses. In 2020, remdesivir became the first antiviral drug approved by the U.S. Food and Drug Administration (FDA) to treat COVID-19. This drug, which targets the RNA polymerase that copies the SARS-CoV-2 viral genome, was approved as a treatment for hospitalized COVID-19 patients age 12 and older who weigh at least 40 Kg (about 88 pounds).

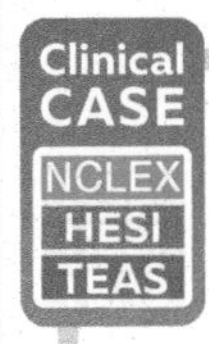

The Case of the Slapdash Self-Medication

Practice applying what you know clinically: visit the **Mastering Microbiology** Study Area to watch Part 2 and practice for future exams.

Antifungal drugs often target cell wall and membrane structures.

Like our own cells, fungi are eukaryotic; this makes it hard to find drugs that target features of pathogenic fungi that human cells do not share. Most antifungal drugs target fungal cell walls and plasma membranes; some affect nucleic acid synthesis.

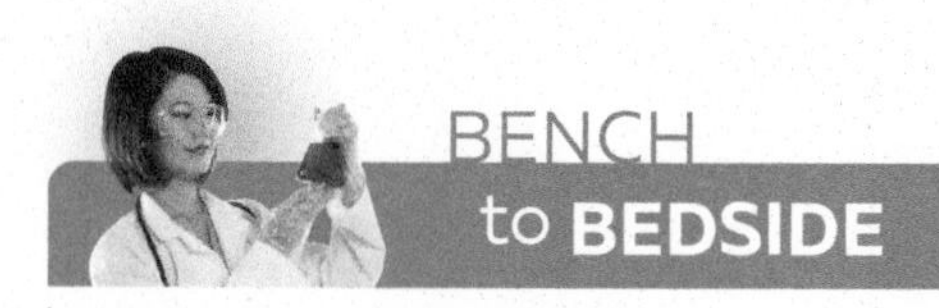

Remdesivir's Journey from the Bench to the Bedside to Combat SARS-CoV-2

Remdesivir's development started in 2009 when it was originally studied as a possible treatment for hepatitis C and respiratory syncytial virus (RSV). When activated, the drug works as a nucleotide analog that targets certain RNA polymerases and thereby interferes with RNA viral genome replication. Although remdesivir wasn't effective against the initially intended viral targets, it was later shown to modestly interfere with the replication of several single-stranded RNA viruses such as Ebola viruses, MERS-CoV-1 (the virus that causes Middle East respiratory syndrome), and SARS-CoV-1 (the virus that causes severe acute respiratory syndrome). Because early studies showed that remdesivir exhibited modestly positive action against SARS-CoV-1, the nearest genetic relative of SARS-CoV-2, the drug joined a short list of antiviral drugs to test in the fight against COVID-19. As with many other drugs that were tested against SARS-CoV-2, early trials for remdesivir provided a suggestion of potential, but clinical testing had to be expanded to determine whether or not the drug might provide any sort of statistically significant benefit.

In early 2020, researchers launched a large, randomized, controlled trial of remdesivir. In the study, 1,062 hospitalized COVID-19 patients were randomly assigned to a group that received an intravenous 10-day course of remdesivir or a control group that received the standard therapy and a placebo. The results indicated that remdesivir accelerated recovery for hospitalized patients with severe COVID-19. The median time to recovery for the remdesivir test group was 10 days as compared to 15 days for the placebo control group. Further analysis revealed that patients receiving supplemental oxygen were more likely to benefit from remdesivir as compared to patients receiving more rigorous respiratory support via high-flow oxygen, non-invasive, or mechanical ventilation—in other words, remdesivir does not seem to help the sickest COVID-19 patients. Mortality rates were also lower in the remdesivir treated group, though not to a statistically significant degree, which means the role of chance in the observed difference can't be ruled out.

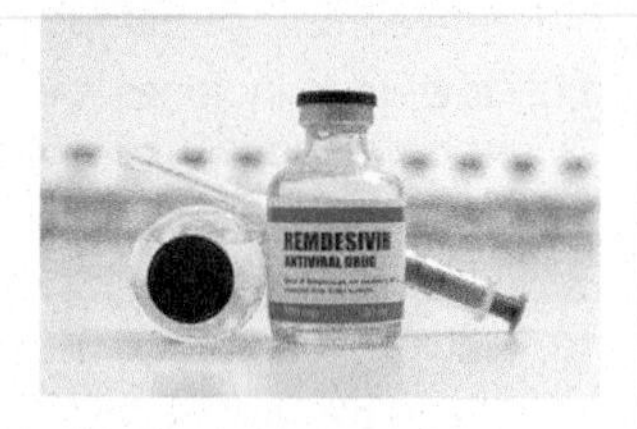

In the fall of 2020, remdesivir became the first FDA-approved drug to treat certain hospitalized COVID-19 patients, but it is far from being a pharmaceutical home run. Also, just as influenza viruses are endemic and constantly evolving, SARS-CoV-2 will continue to evolve and remain a lingering threat. Consequently, it's a global imperative that new broad-spectrum antiviral drugs be developed to combat SARS-CoV-2 and other emerging viruses. In short, the war on viruses is far from over.

Sources: Lin, H. X. J., Cho, S., Aravamudan, V. M., et al. (2021). Remdesivir in Coronavirus Disease 2019 (COVID-19) treatment: A review of evidence. *Infection*, 1–10. Madsen, L. W. (2020). Remdesivir for the treatment of Covid-19-final report. *The New England Journal of Medicine*, *338*(19), 1813–1826.

Targeting Fungal Plasma Membranes

Unlike animal cell membranes, which contain cholesterol, fungal cell membranes contain a sterol called ergosterol, which can serve as a selective target for antifungal agents. The three main classes of drugs that target fungal plasma membranes—azoles, allylamines, and polyenes—all act on ergosterol.

Azoles (examples: fluconazole and ketoconazole) and **allylamines** (examples: terbinafine and naftifine) are broad-spectrum agents commonly prescribed for athlete's foot, ringworm, and yeast infections. They work by inhibiting enzymes that build ergosterol. This leads to improperly built plasma membranes and fungal cell lysis.

Polyenes (examples: amphotericin B and nystatin) directly interact with ergosterols. This causes targeted plasma membranes to become leaky and leads to cell lysis. Unfortunately, polyenes also interact with cholesterol in animal cell membranes. This gives them a narrow therapeutic index and can result in adverse side effects, including nephrotoxicity, when administered systemically by oral or intravenous routes. Nystatin is too toxic to administer systemically and is instead used in topical preparations that treat skin infections—especially cutaneous candidiasis caused by *Candida albicans*. Amphotericin B is effective against most medically relevant fungi, but because of its toxicity, it is reserved for treating life-threatening systemic fungal infections.

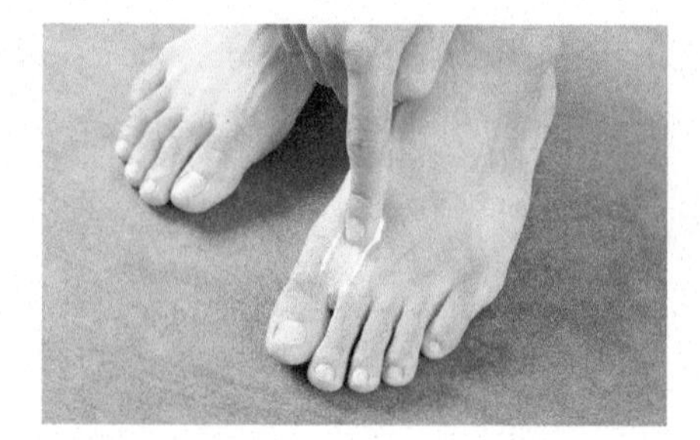

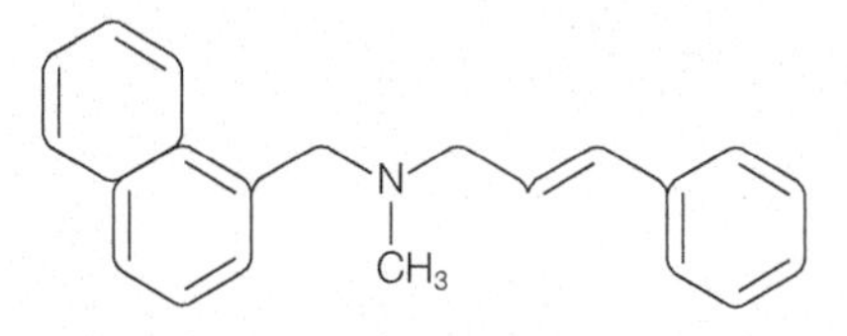

Naftifine is often applied as a topical cream.

Inhibiting Fungal Cell Wall Synthesis

Echinocandin drugs, such as caspofungin acetate, inhibit fungal cell wall synthesis by targeting an enzyme that makes beta-glucan (β-glucan), a component of the fungal cell wall. These injected drugs are mainly useful against systemic fungal infections in immune-compromised patients.

Inhibiting Nucleic Acid Synthesis

Flucytosine is a drug that targets fungal DNA replication and also blocks transcription. Upon entry into fungal cells, it is converted to a nucleotide analog that blocks DNA and RNA synthesis. Flucytosine is usually administered in combination with amphotericin B for severe fungal infections such as *Cryptococcus* meningitis and systemic *Candidiasis* infections.

Antiprotozoan and antihelminthic drugs often target intracellular components.

Relatively few drugs exist to treat protozoan and helminthic infections. Effective drugs are difficult to develop not only because these parasites are eukaryotes, but also because of their complex life cycles. A drug that targets one stage in a parasite's life cycle may be ineffective against other stages.

Antiprotozoan Drugs

Antiprotozoan drugs are mainly split into antimalarial drugs and nonmalarial antiprotozoan drugs. As their name implies, antimalarial drugs target the protozoans that cause malaria—a potentially deadly mosquito-transmitted infection that leads to high fevers and red blood cell lysis. (See Chapter 21 for more on malaria.) According to the World Health Organization, in 2019 there were about 229 million malaria cases worldwide, with an estimated 409,000 deaths. Although five *Plasmodium* species cause malaria, 95 percent of cases are caused by *P. falciparum* or *P. vivax*, with *P. falciparum* causing the most deaths.[18]

It is common for people who live in developed countries to assume that protozoan parasites are not a concern in their lives. However, protozoan infections still represent an important burden of disease in developed countries and should not be ignored or underestimated; millions of people in developed countries are infected with protozoan pathogens. According to the CDC, at any given time about 3.7 million people have trichomoniasis, caused by the sexually transmitted protozoan *Trichomonas vaginalis*. The CDC also states that roughly 30 million Americans are chronically infected with *Toxoplasma gondii*, a protozoan that is most commonly transmitted to people by domestic cats.

Antimalarial drugs Currently, **antimalarial drugs** fall into six main classes: aminoquinolines, arylaminoalcohols, artemisinins, respiratory chain inhibitors, antifolates, and certain antibacterial drugs such as doxycycline and clindamycin that have crossover efficacy against select protozoans. The drug treatment applied depends on the species that is causing the disease and the infection's severity. The aminoquinoline drugs **chloroquine** and **hydroxychloroquine** have been the most widely used antimalarial drugs, but malarial pathogen resistance to these drugs is increasingly common.[19] **Quinine**, an arylaminoalcohol compound, was one of the earliest treatments ever used for malaria. It remains an important drug to fight uncomplicated and severe forms of malaria—although you may be more familiar with it as the compound that gives tonic water its bitter taste. As seen with chloroquine, malarial protozoans are increasingly able to resist quinine and quinine derivatives such as mefloquine.

To combat resistant strains of *Plasmodium*, **artemisinin-based combination therapies (ACT)** are increasingly used. Artemisinin drugs originated from an herb called wormwood that is used in Chinese medicine. ACTs combine an

---

[18] *P. falciparum, P. vivax, P. ovale, P. malariae*, and *P. knowlesi* all cause malaria. *P. falciparum* is mainly in Africa; *P. vivax* dominates in Asia and South America; *P. ovale* is primarily in West Africa; *P. malariae* is rare and is mainly isolated to Africa; *P. knowlesi* is very rare and is mainly in parts of southeast Asia.

[19] Chloroquine and hydroxychloroquine are chemically related compounds that were both investigated for efficacy against SARS-CoV-2. Although these drugs both showed initial promise in laboratory studies, in clinical trials neither statistically reduced the severity or duration of COVID-19. Furthermore, these drugs were not found to prevent SARS-CoV-2 infection.

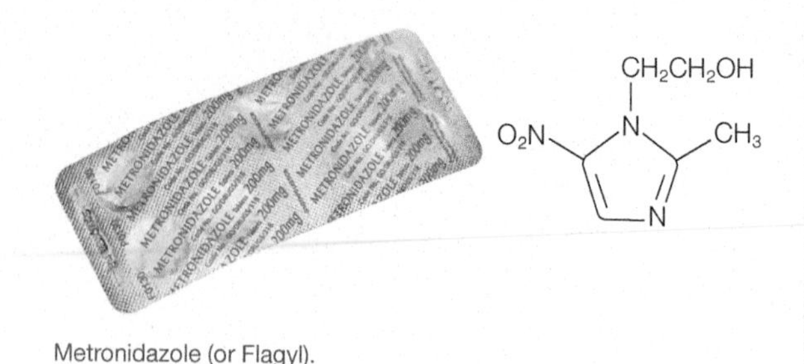

Metronidazole (or Flagyl).

artemisinin-class drug with one or more nonartemisinin drugs. The precise mechanism that artemisinin-class drugs use against malarial parasites remains unclear, but so far resistance to these drugs remains rare.

Nonmalarial antiprotozoan drugs One of the most prescribed drugs that treats nonmalarial protozoan infections is **metronidazole** (or Flagyl). A member of the nitroimidazoles, it targets nucleic acids and is effective against *Toxoplasma gondii* (toxoplasmosis), *Trichomonas vaginalis* (trichomoniasis), and the intestinal protozoans *Giardia lamblia* (giardiasis) and *Entamoeba histolytica* (amoebiasis). Metronidazole is also useful for treating certain anaerobic bacterial infections—it is often a therapy of choice in mild to moderate *C. difficile* infections that don't have complications.[20]

Trimethoprim-sulfamethoxazole (TMP/SMX), which is known by the trade names Septra or Bactrim, is an antifolate drug combination that we reviewed in the antibacterial drugs section. This drug combination works by blocking folate production in certain bacteria as well as in certain protozoans. It is effective against *Toxoplasma gondii* (the causative agent of toxoplasmosis), and it can treat *Giardia*, *Cryptosporidium*, and *Entamoeba*, which all cause severe diarrhea. Another antiprotozoan drug called **nitazoxanide** is mainly used to treat *Giardia* and *Cryptosporidium* as well as certain parasitic worms. Nitazoxanide blocks anaerobic energy metabolism in protozoa.

Antihelminthic Drugs

Two of the most common antihelminthic drugs are **albendazole** and **mebendazole**. These are broad-spectrum agents that interfere with glucose uptake in worms by targeting microtubules. These drugs treat a wide collection of roundworms such as *Ascaris*, hookworms, pinworms, and *Trichinella* (an infection mainly associated with eating undercooked pork). They are also used against certain tapeworms. **Praziquantel** (Biltricide) is another antihelminthic drug for fluke and tapeworm infections. It essentially paralyzes these parasites. Paralyzed intestinal worms and flukes release their attachment to the intestinal wall and are then expelled in the feces. (Roundworms, flatworms, and flukes are further reviewed in Chapter 19: Digestive System Infections.)

Build Your Foundation

15. Why is it especially challenging to develop drugs that target viruses, protozoans, or helminths? (LO 15.15)
16. What are the potential points in viral replication that antiviral drugs may target? (LO 15.16)
17. What three main classes of drugs target fungal plasma membranes, how do they each work, and what fungal infections are they each used against? (LO 15.17)
18. How do echinocandins and flucytosine work, and what fungal infections might they be used to treat? (LO 15.17)
19. Name two malarial antiprotozoan drugs and two nonmalarial antiprotozoan drugs, and state what specific protozoan infections each drug treats. (LO 15.18)
20. What is the purpose of artemisinin-based combination therapies? (LO 15.18)
21. What are two main antihelminthic drugs, how do they work, and what helminthic infections might they be prescribed to treat? (LO 15.19)

Build Your Foundation (BYF) Quick Quiz: Visit the **Mastering Microbiology** Study Area to quiz yourself.

[20]The antibacterial drug vancomycin is also a common empiric therapy against *C. difficile* infections.

15.4 ASSESSING SENSITIVITY TO ANTIMICROBIAL DRUGS

The CDC reports that each year more than 2 million people in the United States suffer from an antimicrobial-resistant infection, and that about 23,000 of these patients die as a result. Given these statistics, it's not surprising that an important part of using antimicrobials to treat infections is susceptibility testing. Clinical microbiology laboratories perform **susceptibility testing** on bacteria (and to a lesser extent on fungi) to assess if the pathogen is likely to be treatable with a particular antimicrobial drug. Although protozoan parasites also develop drug resistance, it's difficult and time consuming to test them for their drug response, which means the prescribed therapy tends to use a trial-and-error approach.

The most common forms of susceptibility testing are reviewed in this section. As you read about these tests you should note that they measure susceptibility in strictly defined contexts. For each test, a known and set amount of bacteria is used, and the concentration of the drug is well controlled. These controlled conditions are quite different from a real-life situation in which bacterial levels and active drug concentrations vary from one tissue to another. Therefore, it is entirely possible for a pathogen to appear susceptible in these tests and yet be resistant in the body.

Learning Outcomes

After reading this section, you should be able to:

15.20 Describe the premise of antibiotic susceptibility testing, and explain its general limitations in a clinical context.

15.21 Explain how the Kirby–Bauer test works and what information it provides.

15.22 Describe how the E-test works and what information it can provide.

15.23 Distinguish between minimum inhibitory concentration (MIC) and minimum bactericidal concentration (MBC), and discuss the various ways these measures may be determined.

15.24 Discuss how broth dilution tests work, and state what information they provide.

Agar diffusion tests can determine a bacterium's susceptibility to antimicrobial drugs.

Agar diffusion tests include the Kirby–Bauer test and the E-test. Both tests are used to determine a basic antimicrobial susceptibility profile for a specific bacterium (with modified protocols for testing fungi). These methods are relatively inexpensive and can be used to determine pathogen susceptibility to a wide variety of antimicrobial drugs. Here we'll review these tests in the context of testing bacterial susceptibility to antibacterial drugs.

Kirby–Bauer Test

The **Kirby–Bauer** test (or disk diffusion test) involves spreading a set amount of a pure test bacterium on the surface of solidified agar media. Usually **Muller Hinton agar** is used because it supports the growth of most common bacterial pathogens while also allowing for fairly uniform diffusion of drugs through the agar from their point of application. Once the bacterial inoculum is spread on the agar, filter paper disks that are infused with a set amount of a specific drug are placed on the agar surface.

Following an incubation period, a **zone of inhibition**, which is a clear zone around the disk, may be evident if the bacteria are prevented from growing. It is important to appreciate that a zone of inhibition could develop for either of two reasons: because the tested drug inhibits growth (as would occur if the drug is bacteriostatic) or because it kills the organism (as would occur if the drug is bactericidal). Unfortunately, agar diffusion tests don't differentiate between bacteriostatic and bactericidal actions.

If a zone of inhibition develops, its diameter is measured and then compared to a susceptibility table (FIG. 15.14). The bacterium is then described as *susceptible* to the drug if the measured zone of inhibition falls in the susceptible range or *resistant* if in the resistant range. Sometimes there is a borderline response, and the bacteria are said to have an *intermediate response* to the drug. The collected data is often presented in an **antibiogram**, which is basically a summary report of the organism's susceptibility to the tested antimicrobial drugs.

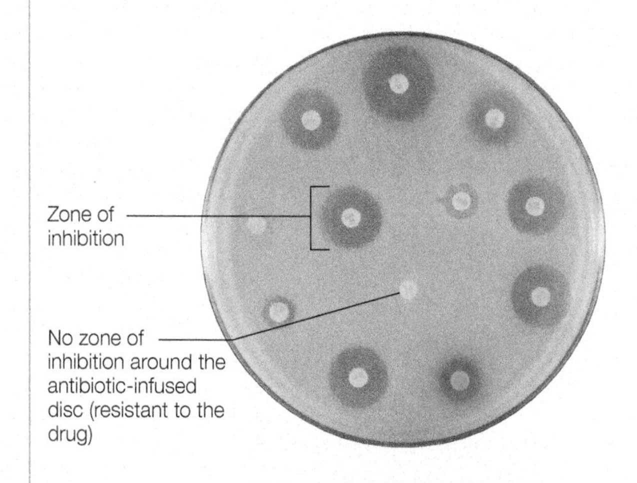

FIGURE 15.14 Kirby–Bauer test (disk diffusion test) The diameter of the zones of inhibition must be measured and compared to a standardized table to determine the bacterium's response to the tested drugs.

Critical Thinking *Can you assume the tested bacterium is susceptible to the drug pictured in the bottom photo? Explain your reasoning.*

E-Test

The **E-test** conceptually resembles the Kirby–Bauer test, but instead of applying round disks infused with a set amount of an antimicrobial drug to the agar plate, strips infused with a variable gradient of drug are placed on the agar surface. Because the test strip provides a gradient of drug concentrations, the E-test can reveal the **minimal inhibitory concentration (MIC)** of the drug—the MIC is the lowest concentration of the antimicrobial drug that inhibits the microbe's growth. After incubation, the MIC is the point on the strip that intersects the start of the zone of inhibition (FIG. 15.15). As with disk diffusion tests, an E-test can't reveal if the tested drug is bacteriostatic or bactericidal. Therefore, although we refer to "inhibition" in this test, the truth is that the bacterium may actually be inhibited from growing because it was killed.

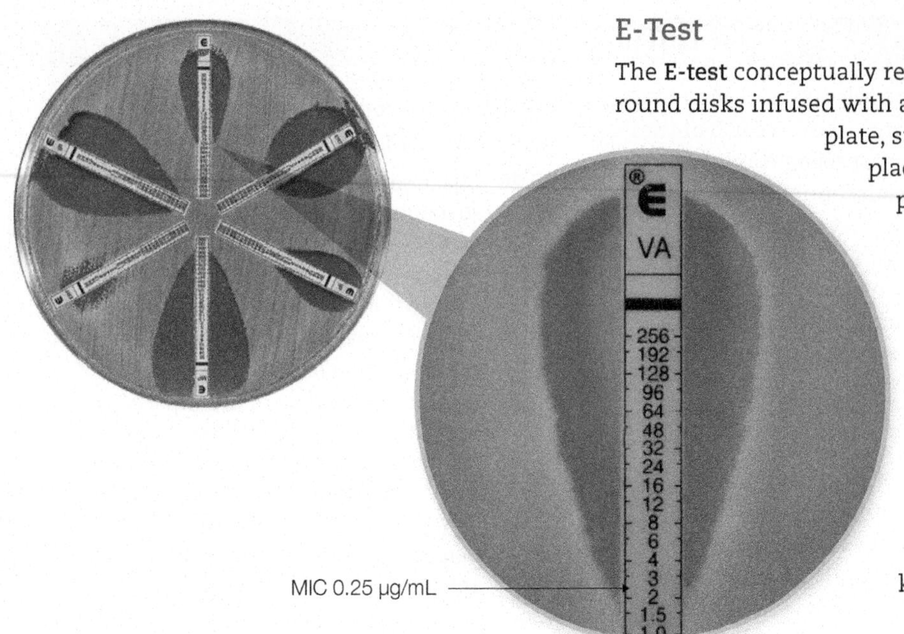

FIGURE 15.15 E-test The minimal inhibitory concentration (MIC) is the concentration where the zone of inhibition starts to develop.

Critical Thinking *Could the MIC determined by this test actually turn out to be a minimum bactericidal concentration? Explain your reasoning.*

Broth dilution tests can distinguish between bactericidal and bacteriostatic actions.

Unlike agar diffusion tests, **broth dilution tests** can help us to differentiate between bactericidal and bacteriostatic actions. In these tests each antibiotic is serially diluted and then added to a set amount of liquid growth medium containing a standardized amount of bacteria. Although the experiment can be set up in test tubes, in a clinical lab microtiter plates are usually preferable to save space and accommodate the testing of many antibacterial drugs (FIG. 15.16).

After an incubation period, the level of growth for each diluted drug series is measured by assessing turbidity—the cloudiness of the sample.[21] Alternatively, there are colorimetric indicators of population growth.[22] The MIC is the lowest drug concentration at which turbidity stops increasing; if a colorimetric version of this assay is used, it is the point where the color intensity levels off. The **minimum bactericidal concentration (MBC)** is the minimum concentration of the drug that kills at least 99.9 percent of the bacteria present. MBC is determined using a series of steps *after* the MIC is established. To determine the MBC, the dilution where the MIC was observed along with samples from at least two of the adjacent more concentrated preparations are plated on media that does not contain antibiotics. Because only live cells in the sample will be able to grow once the cells are no longer exposed to the antibiotic, the colonies that grow on each plate directly correlate to the number of live bacterial cells in the sample. Therefore, the plate that has a 99.9 percent colony reduction as compared to the MIC plate is deemed to be the MBC.

---

[21]As reviewed in Chapter 7, turbidity is the cloudiness of a liquid sample. The more bacteria present in a suspension, the more turbid the sample. A spectrophotometer determines turbidity by measuring how much light passes through a sample. Spectrophotometers can be designed to assess solution turbidity for samples in a microtiter plate or in a test tube.

[22]Gattringer, R., Nikš, M., Ostertág, R., et al. (2002). Evaluation of MIDITECH automated colorimetric MIC reading for antimicrobial susceptibility testing. *Journal of Antimicrobial Chemotherapy*, 49(4), 651–659.

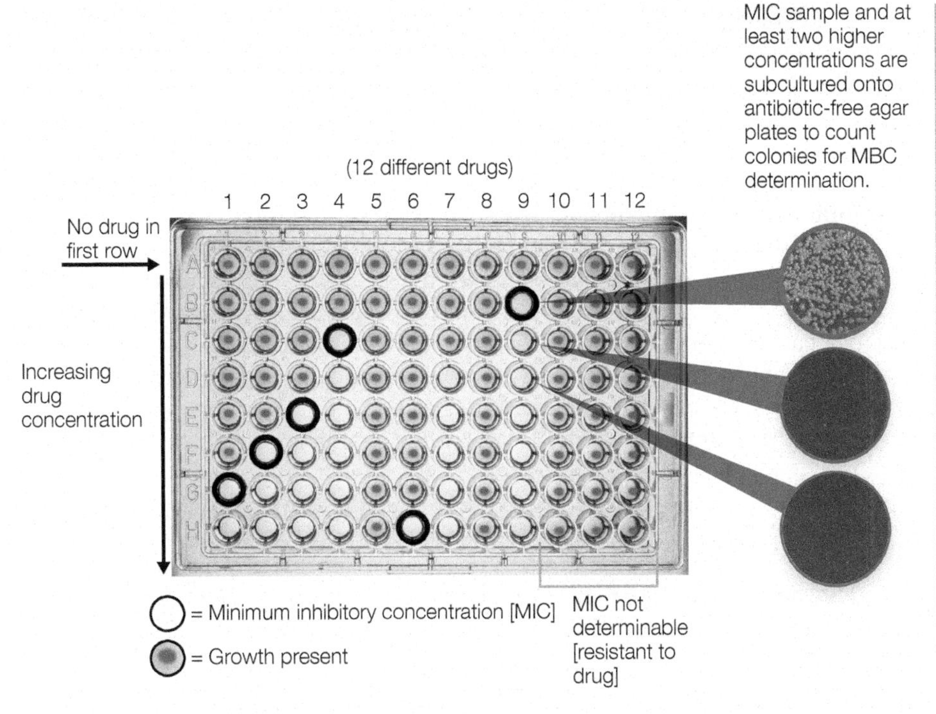

FIGURE 15.16 Broth dilution test for antimicrobial susceptibility The MIC is the lowest drug concentration that inhibits bacterial growth. In this test a colorimetric indicator was added; the indicator turns red if cells are growing. As you can see, drug #9 was effective at the lowest concentration, whereas drugs 5, 8, 10, 11, and 12 were not effective even at high concentrations. By subculturing the MIC sample and at least two samples exposed to higher drug concentrations onto antibiotic-free agar media, the MBC can be determined. The MBC is the lowest concentration that yields at least a 99.9 percent reduction in cell number. Because a colony grows from a single cell, the colonies on the subcultured plates can be counted to determine if the appropriate reduction was realized.

Critical Thinking *What could you reasonably conclude if all plates that were subcultured for MBC determination produced the same number of colonies as the MIC plate? (Assume that the MIC sample and all concentrations above the MIC were subcultured in this case.)*

The broth dilution procedure can be automated to expedite testing. Most clinical labs purchase microtiter plates that contain prediluted panels of antimicrobial drugs so that they only need to pipet a standardized amount of the test bacteria into each well and incubate the plate.

Build Your Foundation

22. Susceptibility testing reveals that a bacterium isolated from a patient's blood is susceptible to drug X, but the patient's infection fails to respond to the drug. Assuming the microbiology lab did not make a mistake, what explanations can you offer for this observation? (LO 15.20)
23. A zone of inhibition is seen around a disk on an agar plate. What test does this describe, and what (if any) conclusion can be made about the bacterium's response to the drug based on this information? (LO 15.21)
24. How does the E-test differ from the Kirby–Bauer test? (LO 15.21and 15.22)
25. Distinguish between a minimum inhibitory concentration (MIC) and a minimum bactericidal concentration (MBC). (LO 15.23)
26. If you needed to accurately assess the MBC and MIC of a new antimicrobial against a bacterium, what susceptibility test would you use, and why? (LO 15.23)
27. How does the broth dilution test work, and what can it reveal? (LO 15.24)

Build Your Foundation (BYF) Quick Quiz: Visit the **Mastering Microbiology** Study Area to quiz yourself.

15.5 DRUG RESISTANCE AND PROPER ANTIMICROBIAL DRUG STEWARDSHIP

Learning Outcomes

After reading this section, you should be able to:

15.25 Explain what is meant by antimicrobial resistance, and compare intrinsic to acquired resistance.

15.26 Name and describe the three categories of acquired antimicrobial resistance tools that microbes may use to thwart drug action.

15.27 Discuss how the use of antimicrobials selects for resistant microbes, and summarize how resistance may be spread by agricultural and clinical misuse.

15.28 Summarize what healthcare workers and patients can do to reduce the emergence of antimicrobial resistance.

15.29 List the top three urgent drug-resistant bacterial threats to health in the United States.

15.30 Describe the challenges and opportunities that exist in developing new antimicrobial drugs.

Antimicrobial resistance is defined as a situation in which a microbe, which could be a bacterium, virus, fungus, protozoan, or helminth, is not affected by a drug therapy that is intended to inhibit or eliminate the pathogen. Resistant microbes are sometimes called **superbugs** because they remain unaffected by the administered antimicrobial therapy and may readily increase their numbers in the patient, causing a **superinfection** that is difficult to treat (FIG. 15.17).

The problem of resistant pathogens started soon after antimicrobials became widely used. Within just a few years of penicillin being used in medical practice, resistant bacterial strains emerged. Whenever an antimicrobial drug is used, no matter how appropriate the use may be, microbes have an opportunity to develop drug resistance. That said, the epidemic increase in drug resistance, which has been dubbed the "antibiotic crisis," is largely due to antibiotic overuse and misuse.

Pathogens may have intrinsic and/or acquired antimicrobial resistance.

Before we dive deeper into resistance, we should first clarify that many microbes can have a natural inborn resistance, and some acquire resistance over time. Next, we'll review these two categories of resistance.

Intrinsic Resistance

Some bacteria have built-in qualities that help them naturally resist antimicrobial drugs. This natural resistance is often called **intrinsic resistance.** While intrinsic resistance rarely provides full protection from all antimicrobials, it does make certain pathogens harder to eliminate, and it tends to limit what drugs can be used to combat those pathogens. For example, *M. pneumoniae*, which causes walking pneumonia, lacks a cell wall and is therefore intrinsically resistant to drugs that target cell wall construction. Another example of intrinsic resistance is seen in *C. difficile*, a bacterium that can form endospores—dormant structures with a tough spore coat that blocks the entry of most antibiotics. Because *C. difficile* can undergo sporulation and go dormant in a patient's intestinal tract, a patient who has recovered from *C. difficile* pseudomembranous colitis (a severe infection often characterized by excessive diarrhea, flu-like symptoms, vomiting, and abdominal cramping) may still carry the bacterium and may experience a relapse. Recurrent *C. difficile* infection develops in about 18 to 35 percent of patients who recovered

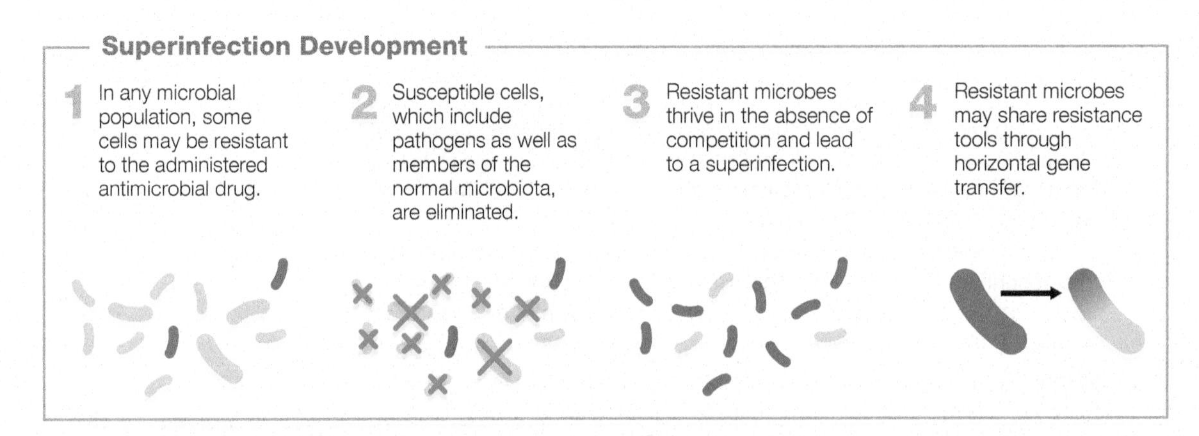

FIGURE 15.17 Superinfection development Whenever an antimicrobial drug is taken, it presents a selective pressure that weeds out bacteria that are not drug resistant. This increases the prevalence of resistant strains.

Critical Thinking *How could combination therapies help reduce the likelihood of this selection process?*

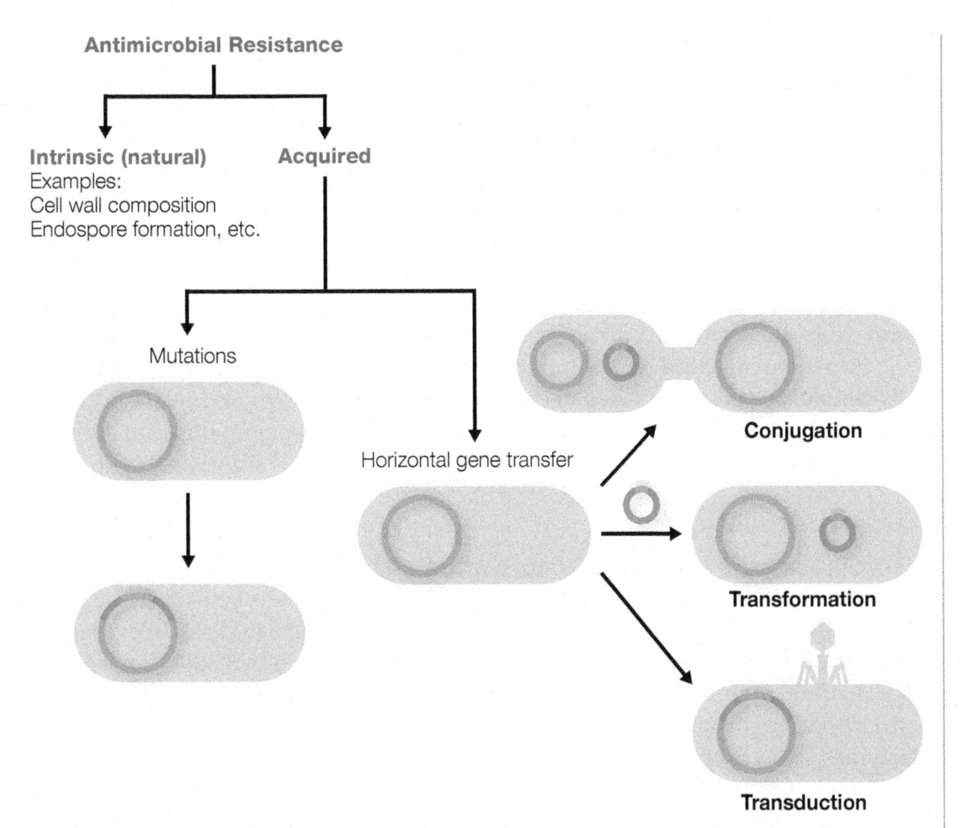

FIGURE 15.18 Intrinsic versus acquired resistance Recall from Chapter 5 that horizontal gene transfer mechanisms include conjugation, transformation, and transduction. Conjugation involves the donor bacterium building a conjugation pilus through which genetic information can be passed to a recipient bacterium; transformation involves a bacterium taking up DNA from the environment; in transduction a bacteriophage delivers new genetic material to a bacterial cell.

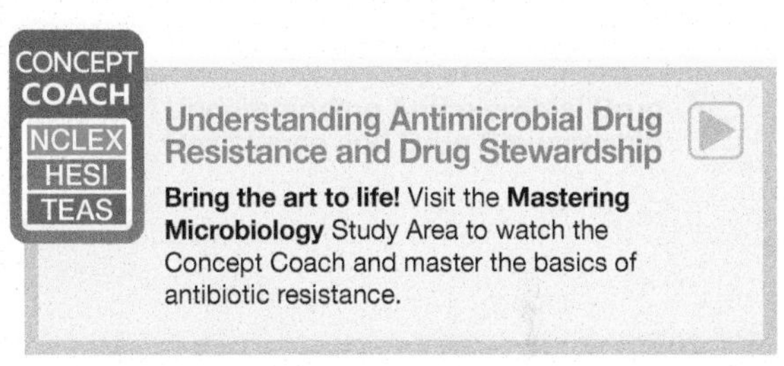

from a primary episode.[23] Similarly, mycobacteria such as *Mycobacterium tuberculosis* have a waxy cell wall enriched with mycolic acid. This layer is particularly difficult for drugs to cross in order to enter cells. Gram-negative bacteria are another example—they are naturally resistant to drugs that are unable to cross their lipid outer membrane. Lastly, microbes in biofilms exhibit intrinsic resistance because drugs often do not permeate deep into biofilms. Also, individual members of a biofilm community tend to have different gene expression profiles; such variation may alter their susceptibility to a given drug.

Acquired Resistance

In most cases, when we refer to the emergence of "resistant strains" we are really referring to **acquired resistance**. This form of resistance is due to genetic mutation or the acquisition of resistance genes through horizontal gene transfer from other strains. (Chapter 5 reviews mutations and horizontal gene transfer.) FIG. 15.18 compares intrinsic versus acquired resistance.

There are three main ways microbes evade antimicrobial drugs.

The three main ways microbes evade antimicrobial drugs are by: (1) altering the drug's target, (2) inactivating the drug, or (3) reducing drug concentrations inside the cell by blocking drug entry or by pumping the drug out of the cell (FIG. 15.19). Although we mainly focus on bacteria and their antimicrobial resistance, you should note that viruses, fungi, protozoans, and helminths also have evolved ways to resist antimicrobial drugs.

Target Alterations

One way microbes resist antimicrobials is by altering the drug's target; in many cases a mutation in the targeted protein prevents drug binding. There are

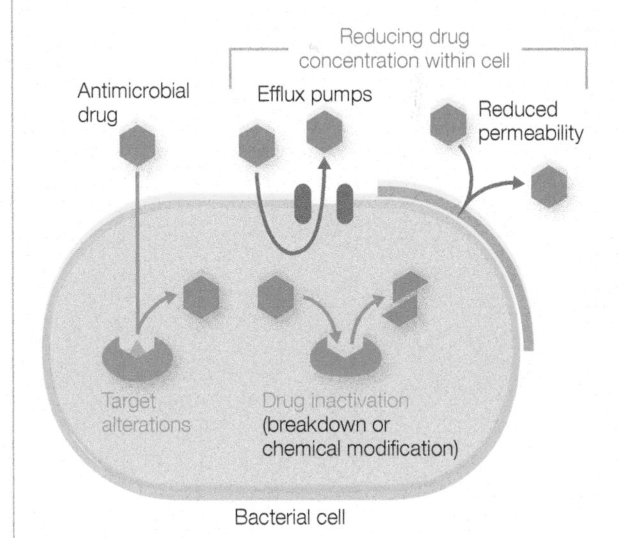

FIGURE 15.19 Three key mechanisms of antimicrobial resistance

Critical Thinking *What possible risks may a bacterium that employs target alteration take on in terms of its own survival and physiological processes?*

[23]De Roo, A. C., & Regenbogen, S. E. (2020). *Clostridioides difficile* infection: *Clostridium difficile* infection: An epidemiology update. *Clinics in Colon and Rectal Surgery*, 33(2), 49.

diverse examples of this mechanism. One is seen with rifampin, a drug used to treat tuberculosis. Rifampin targets RNA polymerases in *M. tuberculosis,* which has evolved resistance by altering its RNA polymerases in a manner that allows them to work perfectly well, yet avoid drug binding. Another example is seen in bacteria that have become resistant to antifolate drugs such as sulfa drugs (sulfonamides) and trimethoprim. Many resistant strains have altered folic acid production enzymes that limit drug binding while still allowing folic acid production.[24] As a third example, we can consider vancomycin-resistant *S. aureus* (VRSA) strains, which have altered their cell wall-building enzymes in ways that limit drug binding while still allowing cell wall synthesis.

Viruses are also infamous for employing this resistance strategy. Some herpes simplex virus strains have become resistant to acyclovir and related drugs such as famciclovir or valacyclovir by altering viral DNA polymerases and viral thymidine kinases.

Drug Inactivation

As we discussed earlier in this chapter, microbial beta-lactamases break the beta-lactam ring of drugs such as penicillins and other members of the beta-lactam superfamily. These enzymes represent some of the earliest recognized antimicrobial-resistance tools and are especially common in *Staphylococcus* species. To date, more than 1,000 different beta-lactamases have been characterized. Other important drug-inactivating enzymes are carbapenemases; these enzymes break down carbapenem drugs.

In other cases, an enzyme may inactivate the antimicrobial drug by adding a chemical group, such as a phosphate, to the molecule. Aminoglycosides are particularly vulnerable to such drug-modifying enzymes. Chloramphenicol can also be enzymatically inactivated by adding an acetyl group.

Plasmids commonly carry genes that encode drug-inactivation tools. This makes it easy for these resistance tools to spread by horizontal gene transfer. For example, genes that encode carbapenemases are typically carried on plasmids that are easily transferred to other bacteria. As such, not only are carbapenem-resistant Enterobacterales (CRE) resistant to last-resort drugs, but they also have the capacity to share their resistance tools with other pathogens.

CHEM • NOTE

Acetyl groups have the chemical formula C_2H_3O. These functional groups are common in biology and can be found in acetylcholine (a neurotransmitter) and acetyl-CoA (the molecule that is fed into the Krebs cycle).

Reducing Drug Concentrations Inside the Cell

Lowering the amount of a drug in a cell's cytoplasm to levels below the minimum inhibitory concentration (MIC) is an important resistance mechanism. Microbial cells can accomplish this by preventing a drug's entry or by pumping drugs out of their cytoplasm.

Limiting drug entry An effective form of drug resistance is to keep the drug out of the cell in the first place. Many drugs require transporters or porin channels to enter targeted cells. For example, fluoroquinolones, aminoglycosides, and carbapenems require specific porins to enter bacterial cells. Some bacteria have become resistant to these compounds by altering the number and/or character of their porins in a way that limits drug entry.

Pumping drugs out of cells Another way to lower drug levels in a cell is through **efflux pumps**. These plasma membrane-spanning proteins actively pump drugs out of cells. Because efflux pumps tend to remove diverse classes of drugs, they are important contributors to multidrug resistance. Interestingly, efflux pumps have been found in all cells where they have been sought—including in human cells, where they may thwart cancer chemotherapies.[25]

---

[24]Nunes, O. C., Manaia, C. M., Kolvenbach, B. A., & Corvini, P. F.-X. (2020). Living with sulfonamides: A diverse range of mechanisms observed in bacteria. *Applied Microbiology and Biotechnology*, 104(24), 10389–10408.

[25]Wang, J. Q., Yang, Y., Cai, C. Y., et al. (2021). Multidrug resistance proteins (MRPs): Structure, function and the overcoming of cancer multidrug resistance. *Drug Resistance Updates*, 54, 100743.

So, if all cells have these pumps, why aren't *all* cells drug resistant? It turns out that resistance is not simply due to the natural levels of these pumps in cells, but to the number of pumps present as well as their overall ability to bind antimicrobials (or chemotherapy compounds, as seen in human cancers).[26] Mutations that lead these pumps to be overexpressed (such as a mutation in the gene's promoter) and/or mutations that expand their drug-binding potential facilitate antimicrobial resistance. To be effective, these pumps do not need to completely remove a given antimicrobial; they only need to lower the effective concentration to a point at which the drug is harmless to the cell. Drug levels that fall below the MIC also promote the development of selective mutations that confer additional resistance mechanisms, such as a mutation that alters the drug's target.

Some efflux pump genes that exhibit increased expression and/or activity levels have been found on plasmids that are easily shared with other cells by horizontal gene transfer. Many bacteria use efflux pumps to resist antimicrobials, with *P. aeruginosa* being one of the most notoriously resistant pathogens, thanks to its prolific use of efflux pumps. These pumps also protect *P. aeruginosa* from disinfectants, making it challenging to eliminate from healthcare settings. Also, pathogenic fungi such as *Candida* species have developed azole drug resistance due to increased efflux pump activity.

Human behaviors can accelerate the emergence of drug resistance.

Antibiotic resistance is fueled by natural selection. Any time an antibiotic is used, a selective pressure is applied, and there is the potential for those bacteria that have resistance to the drug to dominate the exposed population. However, behaviors such as noncompliance with prescribed dosing parameters along with antimicrobial misuse are key factors that have accelerated the evolution of drug-resistant pathogens (FIG. 15.20).

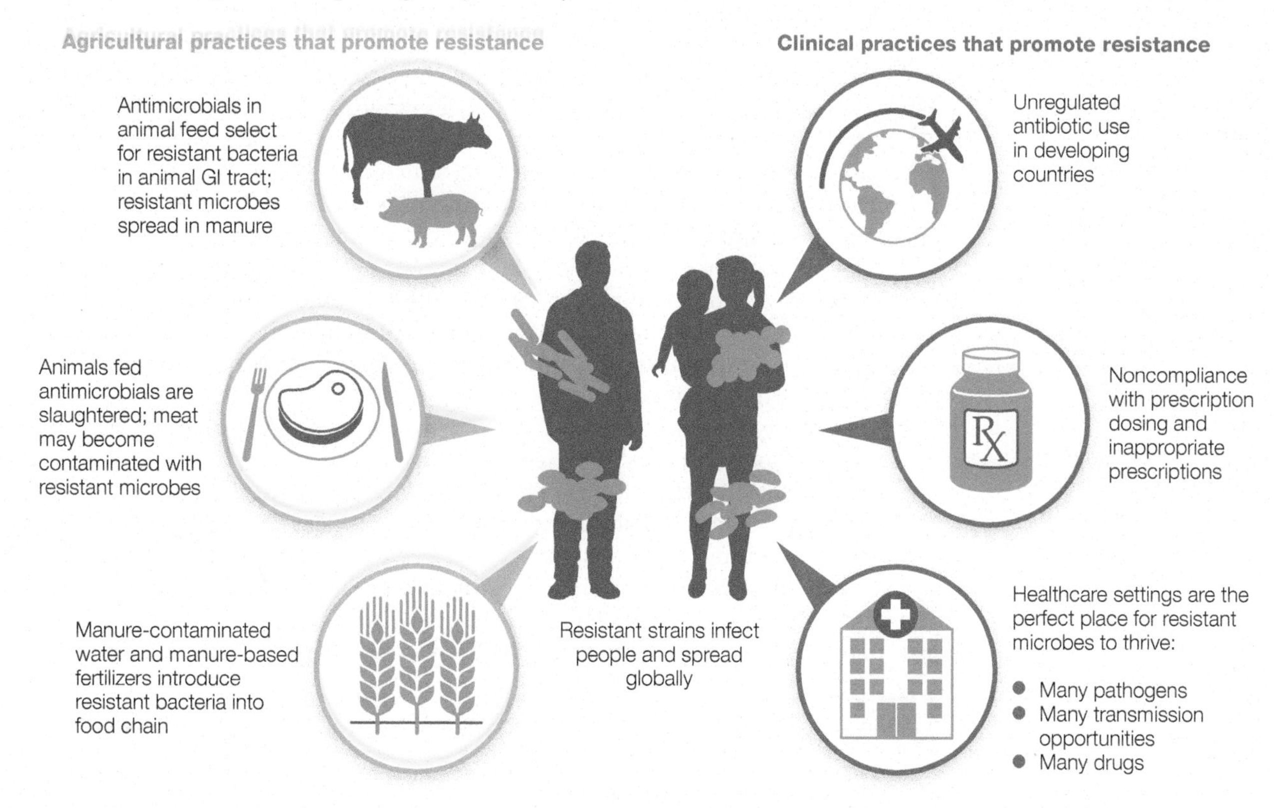

FIGURE 15.20 Factors that contribute to antimicrobial resistance

[26]Kumar, S., Lekshmi, M., Parvathi, A., Ojha, M., Wenzel, N., & Varela, M. F. (2020). Functional and structural roles of the major facilitator superfamily bacterial multidrug efflux pumps. *Microorganisms*, *8*(2), 266.

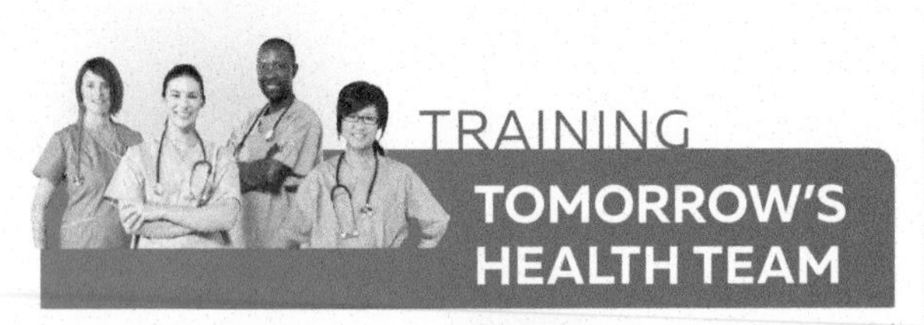

Tips for Saying "No" to Patients Who Want Antibiotics

All the professional societies agree that clinicians should "just say no" to patients who don't need antimicrobials for self-limiting (and viral) infections such as colds. But there is little practical advice on how to do so without seeming insensitive. Saying no to a patient can be made "ECE" (easy) by employing these three tips:

Empathy: Make it clear to your patient that you care about them, that you want to do the best thing for their short- and longer-term medical well-being, and that you know they are suffering and longing for relief. Take a moment to recall the last time you felt miserable and desperate to feel better. Share that feeling with your patient.

Competence: Help your patient feel confident in your competence. Give them a peek into the systematic thought process that led to the diagnosis. Try to infuse some information about how viral and bacterial infections differ, and share what features make it clear that they have a viral infection.

Education: Close the "no" conversation with some education about antibiotics. Explain that antibiotics don't cure viral infections, and then remind your patient that these drugs may cause adverse side effects like diarrhea and nausea that may be just as unpleasant as a cold—and if used unnecessarily antibiotics could be ineffective when they are really needed. Furthermore, you can mention that we have helpful bacteria in and on our bodies and that whenever we take an antibiotic, we disrupt those helpful bacteria, possibly putting us at risk for other complications—a risk only worth taking when antibiotic therapy is necessary. Let your patient know that if their condition worsens or if they develop new symptoms, you are there to help and will adjust the care plan as needed to best protect their health.

Patient Noncompliance and Antibiotic Resistance

Brandon was just diagnosed with streptococcal pharyngitis, or "strep throat." At first it was easy for him to remember to take the prescribed amoxicillin on the required dosing schedule—after all, his throat was dreadfully sore. A couple of days into his prescription he started to feel much better, so much so that he forgot to take his medication before bed. At lunchtime the next day he realized he had left his medication on the kitchen table. After a week of spotty adherence to his drug regimen, Brandon decided he felt so much better that he stopped taking the antibiotic, leaving about 3 days' worth of pills in the bottle. He decided to save the drug in case he got sick again later. Brandon's story of missed doses, an unfinished drug regimen, and self-medicating with a leftover prescription is not uncommon. Whenever pathogens are exposed to drug levels that do not meet the MIC, they have an opportunity to evolve resistance.

Suboptimal drug levels can result whenever a patient skips drug doses, doesn't adhere to the proper drug-dosing schedule, fails to take the antimicrobial drug for the full length of the prescription, uses an expired antibiotic, or incorrectly stores the prescription in a way that reduces drug potency. Low-dose exposures serve as a selective pressure that acts as a training opportunity for microbes with an intermediate resistance to optimize their resistance strategies and become fully resistant. Low drug levels also allow susceptible bacteria to become intermediate in their resistance as they undergo mutations that allow them to explore ways of thwarting a drug's actions.

Antimicrobial Misuse in Agricultural and Clinical Settings

Antimicrobial misuse is prevalent in agricultural and medical practices. In livestock and poultry production, antibiotics are commonly added to animal feed to promote growth. Globally, between 120 million and 480 million pounds of antibiotics are added to animal feed each year.[27] Although the FDA has banned adding antimicrobials used in human medicine to animal feed for growth promotion, it still allows the addition of antimicrobials that are not approved for humans. While this may seem like a way to prevent resistance to clinically useful antimicrobials, it is not necessarily a safe bet. For example, the veterinary glycopeptide drug avoparcin has been documented to induce cross-resistance to the glycopeptide vancomycin, which is used in humans.[28] Resistant strains can then be introduced into humans through meat, the environment, animal handling, or vegetable crops that are fertilized with manure or watered with reclaimed water contaminated with manure.

In clinical practice, inappropriately prescribing antimicrobial drugs for self-limiting infections, noninfectious conditions such as allergies, or viral infections that are not treatable with antibacterial drugs are common examples of drug misuse. The CDC reports that about 30 percent of antibiotic prescriptions are unnecessary.[29] Such inappropriate drug use can be due to misdiagnosis, but it is at least in part also due to patients putting pressure on clinicians to provide a prescription even if the clinician doesn't think one is needed.[30]

The unregulated use of antimicrobial drugs in many developing countries also contributes to antimicrobial resistance. In some countries, patients can obtain antimicrobial drugs without a prescription. Furthermore, they may not be able to afford the full drug regimen all at once and may buy only half of the ideal dosing regimen. And once they start to feel better, they are likely to save the drug for future use.

[27]Oliveira, N. A., Gonçalves, B. L., Lee, S. H. I., Oliveira, C. A. F., & Corassin, C. H. (2020). Use of antibiotics in animal production and its impact on human health. *Journal of Food Chemistry and Nanotechnology, 6*(01), 40–47.
[28]Marshall, B. M., & Levy, S. B. (2011). Food animals and antimicrobials: Impacts on human health. *Clinical Microbiology Reviews*, 24(4), 718–733.
[29]U.S. Centers for Disease Control and Prevention. (2019). *Antibiotic resistance threats in the United States, 2019*. U.S. Department of Health and Human Services, Centers for Disease Control and Prevention.
[30]Kohut, M. R., Keller, S. C., Linder, J. A., et al. (2020). The inconvincible patient: How clinicians perceive demand for antibiotics in the outpatient setting. *Family Practice*, 37(2), 276–282.

Healthcare Settings as Perfect Incubators for Antimicrobial Resistance

Nowhere are antimicrobial-resistant strains more troublesome than in healthcare facilities. Pathogen prevalence combined with the prolific use of diverse antimicrobial drugs provides the perfect environment for spawning antimicrobial resistance. Healthcare settings are the stomping grounds of multidrug-resistant pathogens that represent a serious threat to global health. Adding to the threat is the fact that in healthcare settings, healthcare workers and other staff often pick up resistant strains on their hands, scrubs, and personal electronic devices, making it fairly easy to transfer those microbes to themselves, to others in their community, or to patients—any of whom can also spread the pathogens further in the community. One study found that antibiotic-resistant bacteria such as MRSA could be isolated in up to 32 percent of sampled clinical personnel's clothing and/or handheld electronic devices.[31]

Resistant Microbes to Watch

The CDC compiles a list of the top drug-resistant pathogens in the United States and ranks them as urgent, serious, or concerning based on the level of risk they present and their potential for spread. At the time of this text's publication, the five microbes graded as urgent threats are *C. difficile*, drug-resistant *Neisseria gonorrhoeae*, drug-resistant *Candida auris* (a type of fungus), carbapenem-resistant Enterobacterales (or CRE), and carbapenem-resistant *Acinetobacter* (CRA). We reviewed *C. difficile* and carbapenem-resistant bacteria earlier in this chapter, but we have not discussed the sexually transmissible *N. gonorrhoeae* (Chapter 20) or *C. auris*. Each year in the United States there are about half a million cases of drug-resistant *N. gonorrhoeae*.[32] In women, these infections are often persistent and can result in pelvic inflammatory disease followed by infertility and long-term pelvic pain. The yeast *C. auris* is a concern because it is difficult to detect with standard laboratory protocols and is therefore prone to misdiagnosis and improper management that leads to a poor patient prognosis. Furthermore, *C. auris* is usually resistant to multiple drugs and is easily transmitted in healthcare settings and in nursing homes. Ranked just below these urgent threats are pathogens that the CDC considers serious concerns. These include, to name a few, multidrug-resistant strains of *Campylobacter, Candida, Pseudomonas, Mycobacterium, Staphylococcus*, and *Streptococcus*.

Combating drug resistance requires proper drug stewardship.

Health agencies around the world have been stressing the importance of proper antimicrobial drug stewardship to keep resistance in check. Analysts have projected that if we stay on our current course and do not improve our stewardship of antimicrobials, we can anticipate that by 2050 about 10 million people per year worldwide will be killed by antimicrobial-resistant pathogens; that's about 2 million more people than cancer kills.[33]

Calling Healthcare Workers and Patients to Action

It is especially important that healthcare workers faithfully follow proper hand hygiene practices and rigidly enforce contact precautions. Healthcare workers can also limit unnecessary antimicrobial prescriptions and use narrow-spectrum

Clinical CASE NCLEX HESI TEAS

The Case of the Slapdash Self-Medication

Practice applying what you know clinically: visit the **Mastering Microbiology** Study Area to watch Part 3 and practice for future exams.

---

[31] Haun, N., Hooper-Lane, C., & Safdar, N. (2016). Healthcare personnel attire and devices as fomites: A systematic review. *Infection Control & Hospital Epidemiology*, *37*(11), 1367–1373.

[32] U.S. Centers for Disease Control and Prevention. (2021). *Sexually transmitted disease surveillance 2019*. Atlanta, GA: U.S. Department of Health and Human Services.

[33] United Nations Interagency Coordination Group on Antimicrobial Resistance. (2019). *No time to wait: Securing the future from drug-resistant infections*. https://www.who.int/antimicrobial-resistance/interagency-coordination-group/final-report/en

drugs whenever possible. If a broad-spectrum drug is given as an empiric therapy, and then culture data reveals that a narrow-spectrum drug could be used, the therapy can be switched to limit the impact on normal gut microbiota that are important competitors of antimicrobial-resistant bacteria. Healthcare workers can also educate patients about the importance of following their drug-dosing regimen and make it clear when antibiotics are unnecessary.

Patients should follow all drug-dosing instructions and execute follow-up as recommended. Medications should also be stored under proper conditions to protect drug potency. Patients may also consider the implications of demanding an antibiotic and recognize that an unnecessary antibiotic could be harmful.

Developing New Drugs: Challenges and Opportunities

Identifying new antimicrobials is difficult, time consuming, and costly. For every antimicrobial that makes it to the clinic, 5,000–10,000 candidate compounds may be screened. The development of a successful drug may take 10–15 years and have a net cost (counting failed candidates) of about a billion dollars. Thus, the road to discovering new antimicrobials is rocky. In addition, the profit from a new antimicrobial is less than that from, say, a new heart medication. This is because in most cases the more a drug is used the more money a company will make, but in the case of antimicrobials, using the drug as little as possible is what reduces resistance development. For these reasons, there is little economic incentive to discover, test, and market antimicrobials. Fewer large pharmaceutical companies are developing antimicrobials; instead, discovery and development are left to smaller, newer companies. To incentivize the costly yet essential search for new antimicrobials, many nations have committed money to subsidizing antimicrobial development. In 2012, the United States also passed legislation that extends patent rights on antimicrobial drugs to 20 years before generic drugs can be sold.

Some companies are investing in multidrug approaches. Taking currently available drugs and combining them into optimized formulations against pathogens may reduce the chance of a pathogen escaping the drug therapy. Alternative types of antimicrobials, such as peptide drugs, are also being pursued. Peptide drugs act on cell membranes, as seen with polymyxin B. Some studies indicate that resistance to these drugs is slow to develop. Another new approach is to pair a redesigned antimicrobial with inhibitors that block bacterial-resistance mechanisms. For example, tetracyclines can be paired with efflux pump inhibitors. Lastly, other antimicrobial alternatives include phage therapy—using lytic bacteriophages to specifically target pathogens. The war is far from over in the fight to combat drug-resistant bacteria.

Build Your Foundation

28. What is the difference between intrinsic and acquired antimicrobial resistance? (LO 15.25)

29. What are the three main mechanisms that microbes use to resist antimicrobial drugs, and how does each mechanism work? (LO 15.26)

30. How does using antimicrobials select for the growth of resistant microbes? (LO 15.27)

31. What agricultural and clinical practices contribute to antimicrobial resistance emergence? (LO 15.27)

32. What can healthcare workers and patients do to reduce the emergence of antimicrobial resistance? (LO 15.28)

33. According to the CDC, what are the top five most urgent microbial threats? (LO 15.29)

34. What challenges and opportunities exist in developing new antimicrobial drugs? (LO 15.30)

Build Your Foundation (BYF) Quick Quiz: Visit the **Mastering Microbiology** Study Area to quiz yourself.

Antimicrobial Drugs Overview

Antimicrobial drugs are selectively toxic compounds that kill microbes or inhibit their growth. They can be broad- or narrow-spectrum agents, and they can be natural, synthetic, or semisynthetic.

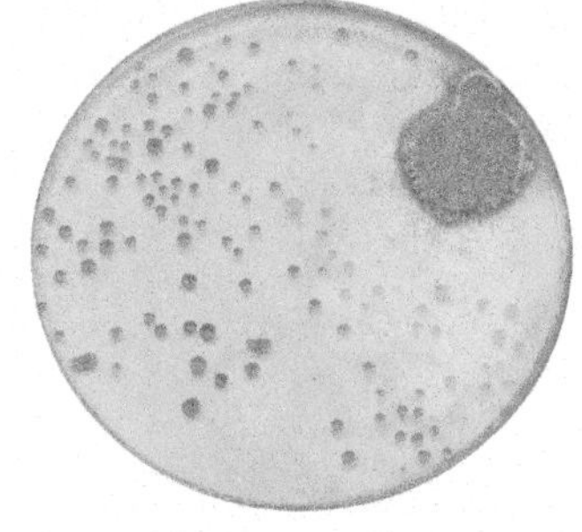

A high therapeutic index means a drug is effective well below its potentially toxic dose.

$$\text{Therapeutic index} = \frac{\text{Maximum safe dose}}{\text{Minimum effective dose}}$$

Antimicrobial Drugs Against Viruses, Fungi, Protozoans, and Helminths

- Antiviral drugs target steps in viral replication.
- Antifungal drugs usually target fungal cell walls, plasma membranes, or nucleic acid synthesis.

- Antiprotozoan drugs are mainly split into antimalarial drugs and nonmalarial antiprotozoan drugs.
- Common antihelminthic drugs are albendazole, mebendazole, and praziquantel.

Antimicrobial Stewardship

Drug resistance is promoted by:

- Including antimicrobials in animal feed
- Unregulated antibiotic use
- Prescription noncompliance
- Inappropriate prescriptions

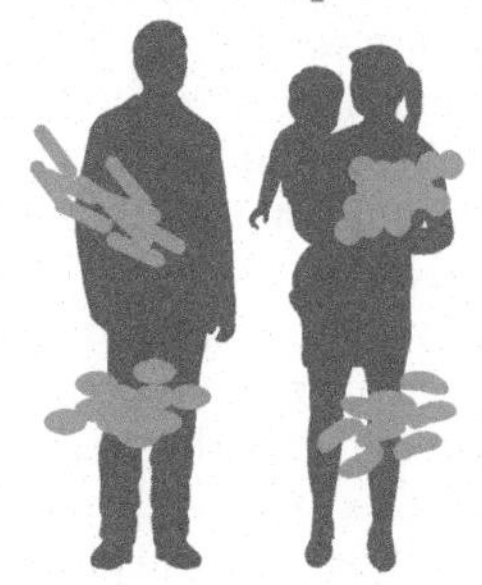

Resistant strains infect people and spread globally

Survey of Antibacterial Drugs

Antimicrobial drugs target structures and processes that bacteria rely on, but human cells do not.

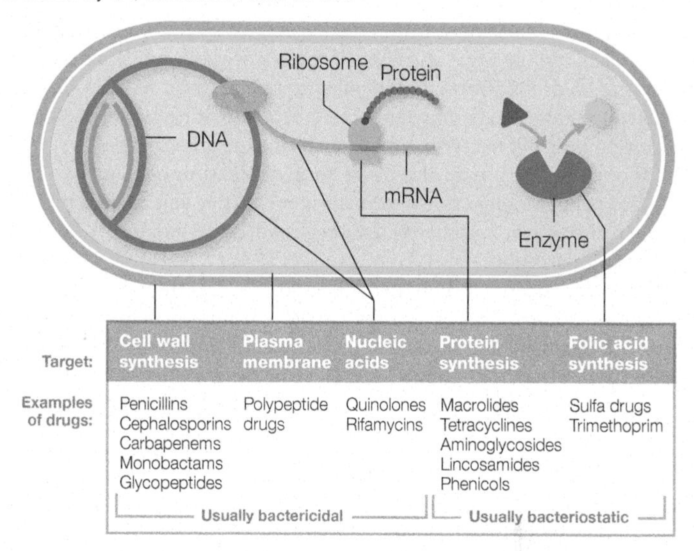

Target:	Cell wall synthesis	Plasma membrane	Nucleic acids	Protein synthesis	Folic acid synthesis
Examples of drugs:	Penicillins Cephalosporins Carbapenems Monobactams Glycopeptides	Polypeptide drugs	Quinolones Rifamycins	Macrolides Tetracyclines Aminoglycosides Lincosamides Phenicols	Sulfa drugs Trimethoprim
	Usually bactericidal			Usually bacteriostatic	

Sensitivity Testing and Drug Resistance

Disk diffusion testing, the E-test, and broth dilution tests reveal information about sensitivity or resistance to an antimicrobial drug.

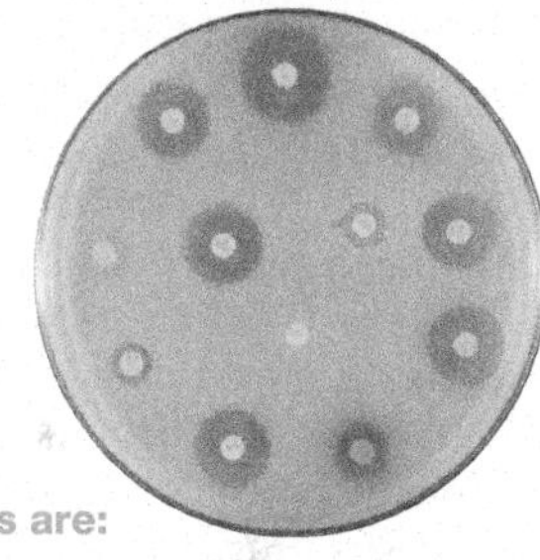

Three key drug-resistance mechanisms are:

- Altering the drug's target
- Inactivating the drug
- Reducing drug concentrations by blocking drug entry or using efflux pumps

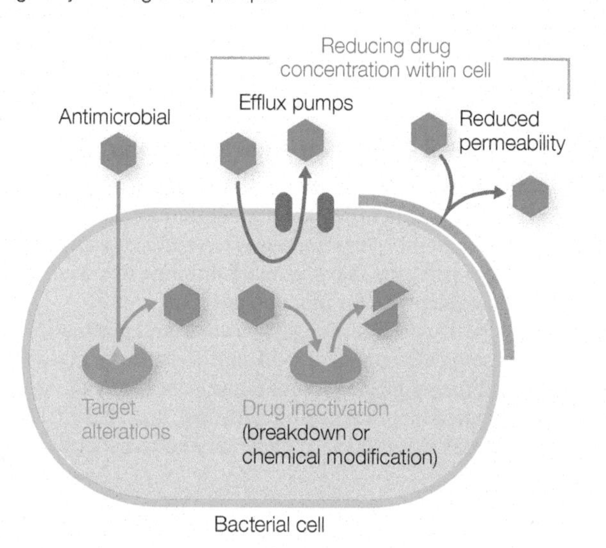

15 CHAPTER 15 OVERVIEW

15.1 Introduction to Antimicrobial Drugs

- Antimicrobials are classified as natural, semisynthetic, or synthetic and can further be described by their spectrum of activity and mechanism of action.
- In terms of antibacterial drugs, broad-spectrum agents tend to be effective against Gram-positive and Gram-negative bacteria, while narrow-spectrum drugs affect either Gram-positive or Gram-negative bacteria.
- Bacteriostatic drugs prevent bacteria from growing, while bactericidal compounds kill bacteria. An antimicrobial drug that is bactericidal for one pathogen may also be bactericidal for another. Furthermore, a drug's bactericidal or bacteriostatic properties can change based on the drug dose, length of the drug regimen, pathogen load, and route of administration.
- Ideally, an antimicrobial drug will exhibit a high therapeutic index and selective toxicity. Other features to consider in drug selection are how it is administered, its stability and elimination mechanisms, its toxicity to the kidneys and liver, and any drug interactions or contraindications.

15.2 Survey of Antibacterial Drugs

- Penicillins (*-cillin* suffix), cephalosporins (*cef-* or *ceph-* prefix), carbapenems (*-penem* suffix), and monobactams all fall under the beta-lactam superfamily of drugs. They inhibit cell wall construction. Beta-lactamases may inactivate these drugs. To combat these resistance enzymes, beta-lactamase inhibitors can be co-administered with beta-lactam drugs.
- Glycopeptides are non-beta-lactam drugs that interfere with bacterial cell wall construction and are especially effective against Gram-positive organisms.
- Bacitracin and isoniazid are other drugs that target cell walls. Bacitracin is mainly used against Gram-positive bacteria, and isoniazid is used against mycobacteria.
- Quinolones (fluoroquinolones) and rifamycins are broad-spectrum families of drugs that target nucleic acids.
- Antifolate drugs such as sulfa drugs and trimethoprim inhibit folic acid synthesis in prokaryotes and some protozoans.
- Macrolides, lincosamides, phenicols, tetracyclines, and aminoglycosides target ribosomes to inhibit protein synthesis. These drugs tend to be bacteriostatic and aside from aminoglycosides are broad spectrum.
- Polypeptide drugs disrupt the outer membrane of Gram-negative bacteria and are bactericidal.

15.3 Drugs for Viral and Eukaryotic Infections

- Antiviral drugs are classified into five main groups according to the viral activity they target: attachment, penetration, uncoating, viral replication and assembly, or viral release. A sixth category includes drugs such as interferons that stimulate immune responses against viruses. Review Chapter 6 for more on antiviral agents.
- Most antifungal drugs target fungal cell walls and plasma membranes; some affect nucleic acid synthesis. Three main classes of drugs target fungal plasma membranes: azoles, allylamines, and polyenes. Echinocandins are antifungals that target fungal cell walls. Flucytosine targets fungal DNA replication and transcription.
- Chloroquine, quinine, and artemisinin-based combination therapies are key drugs that treat malaria. Metronidazole is one of the most commonly prescribed drugs that treats nonmalarial protozoan infections. Trimethoprim-sulfamethoxazole (TMP/SMX) and nitazoxanide are other important nonmalarial antiprotozoan drugs.
- Two of the most common antihelminthic drugs are albendazole and mebendazole. These drugs treat a wide collection of roundworms by interfering with glucose uptake in the worms. Praziquantel is often used to treat fluke and tapeworm infections; it paralyzes these flatworms.

15.4 Assessing Sensitivity to Antimicrobial Drugs

- The Kirby–Bauer test and the E-test are agar diffusion tests. These tests don't differentiate between bacteriostatic or bactericidal actions. The Kirby–Bauer test reveals if a bacterium is susceptible, resistant, or intermediate to a drug based on the diameter of measured zones of inhibition. E-tests reveal the minimal inhibitory concentration of a drug (MIC). The MIC is the lowest concentration of the antimicrobial drug that inhibits the microbe's growth.
- Broth dilution tests can differentiate between bactericidal and bacteriostatic actions and can be used to determine a drug's MIC and minimum bactericidal concentration (MBC). The MBC is the minimum concentration of the drug that kills at least 99.9 percent of the bacteria present. These tests can be automated for rapid testing.

15.5 Drug Resistance and Proper Antimicrobial Drug Stewardship

- Bacteria can have intrinsic resistance to an antimicrobial drug or can acquire resistance through mutation or horizontal gene transfer.
- Common resistance mechanisms are (1) altering the drug's target, (2) inactivating the drug, or (3) reducing drug concentrations inside the cell by blocking drug entry or by pumping the drug out of the cell.
- Noncompliance with prescribed dosing parameters along with antimicrobial misuse accelerate the emergence of drug-resistant pathogens. Adding antibiotics to livestock feed can select for resistant bacteria. In clinical practice, resistance evolution is promoted by inappropriately prescribing antimicrobial drugs or failing to regulate antimicrobial use (as occurs in many developing nations). Healthcare workers and patients have important roles in preventing resistance emergence.
- The CDC has ranked *C. difficile*, *N. gonorrhoeae*, *C. auris*, carbapenem-resistant Enterobacterales (CRE), and carbapenem-resistant *Acinetobacter* (CRA) as the most urgent antibiotic-resistant bacterial threats.
- Patients and healthcare workers can reduce the emergence of drug-resistant pathogens by following good drug stewardship practices.
- Developing new antimicrobials is difficult, time consuming, and costly. Potential opportunities for new therapies include optimizing combination therapies, using peptides, pairing a redesigned antimicrobial with inhibitors that block bacterial-resistance mechanisms, and phage therapy.

COMPREHENSIVE CASE

The following case integrates basic principles from the chapter. Try to answer the case questions on your own. Don't forget to be **S.M.A.R.T.*** about your case study to help you interlink the scientific concepts you have just learned and apply your new understanding of antimicrobial drugs to a case study.

*The five-step method is shown in detail in the Chapter 1 Comprehensive Case on cholera. See pages 31–33. Refer back to this example to help you apply a SMART approach to other critical thinking questions and cases throughout the book.

• • •

Michael, a 25-year-old nursing student with asthma and a documented penicillin allergy, developed a dry cough that tended to get worse at night. He felt fatigued and lost his appetite. He decided that he probably just had a cold. About 3 days after his initial symptoms had developed, Michael experienced an increasing tightness in his chest that made breathing difficult. His partner reminded him that as an asthmatic he really shouldn't put off being seen for his respiratory symptoms. Michael reluctantly agreed and went to the nearby urgent care center, where the doctor prescribed him azithromycin—a "Z-Pak"—to treat what was presumed to be walking pneumonia caused by *M. pneumoniae*.

Within a few days Michael felt much better. Although he did have some asthma flare-ups, he never missed attending his nursing school rotations at the hospital. About a week after finishing the Z-Pak, Michael woke up at about midnight with flu-like symptoms and terrible abdominal cramping. He ended up vomiting and he suffered diarrhea every hour throughout the night. By 9:00 a.m. he was exhausted and feeling worse; he was running a fever and was dehydrated. He knew from his training that it was time to be seen for his symptoms.

His partner drove him to the emergency room. Based on Michael's patient history, the ER doctor suspected a *C. difficile* infection. A stool sample was taken, and then Michael was started on vancomycin and admitted to the hospital. The next day he was informed that rapid screening methods on the stool sample had confirmed a *C. difficile* infection. A couple of days later, Michael was released from the hospital. He was careful to stick to his vancomycin dosing and finished his whole 10-day prescription. He took things a bit easier than usual, but was concerned about falling behind at school.

Within a month Michael was back at the ER with the same symptoms. Rapid screening tests confirmed a *C. difficile* relapse, and he was once again prescribed vancomycin. He was discharged after a couple of days and again diligently took his vancomycin prescription at home. Fortunately, he did not experience a subsequent relapse. Moreover, Michael's nursing program faculty and the program director were able to help him get back on track and graduate just one semester behind his original cohort.

• • •

CASE-BASED QUESTIONS

1. Based on the presumed causative agent of Michael's pneumonia, what drugs would have been immediately ruled out of the empiric therapy, and why?
2. What are the pros and cons of a Z-Pak prescription?
3. What aspects of Michael's patient history made the ER doctor suspect *C. difficile*, and why were they clues for his diagnosis?
4. What is the most likely explanation for Michael's *C. difficile* relapse?
5. What transmission precautions would have been implemented for Michael's stay in the hospital?
6. Assume that a bacterial pathogen other than *C. difficile* was isolated from the patient stool sample and subjected to susceptibility testing that produced the following data:

Table of Zone Diameter Limits (mm)

Drug Name	Observed Diameter (mm)	Resistant	Intermediate	Susceptible
Ampicillin	20	≤ 13	14–16	≥ 17
Erythromycin	15	≤ 13	14–22	≥ 23
Ceftriaxone	25	≤ 19	20–22	≥ 23
Ciprofloxacin	23	≤ 15	16–20	≥ 21

(a) What susceptibility test was most likely done to obtain this data? Support your conclusion.

(b) Based on the data, what antimicrobial drug(s) would you exclude from Michael's treatment plan, and why?

END OF CHAPTER QUESTIONS

Think Critically and Clinically

Questions highlighted in **blue** are opportunities to practice NCLEX, HESI, and TEAS-style questions.

1. A broad-spectrum drug is best described as
   a. bactericidal against a wide range of species.
   b. bacteriostatic against a wide range of species.
   c. effective against a wide range of species.
   d. an empiric therapy.
   e. selectively toxic.

2. A patient who is not a healthcare worker is diagnosed with *C. difficile* pseudomembranous colitis. What most likely led to this infection?
   a. The patient was recently treated with a bacteriostatic drug.
   b. The patient was recently treated with a broad-spectrum drug.
   c. The patient was treated with a bactericidal drug.
   d. The patient recently became immune compromised and therefore had an increased risk for infection.
   e. The patient was recently exposed to someone with an active *C. difficile* infection.

3. Match the antimicrobial drug class to its action. You may assign more than one drug class to a given action and some actions may not be applicable.

Drug Class	Action
a. Macrolides	____ Inhibit DNA replication
b. Quinolones	____ Inhibit transcription
c. Rifamycins	____ Inhibit translation
d. Tetracyclines	____ Inhibit cell wall synthesis
e. Cephalosporins	____ Disrupt plasma membrane integrity
f. Glycopeptides	____ Block folic acid production
g. Polypeptide drugs	

4. A patient has an uncomplicated infection with a Gram-negative bacterium. He also has a history of penicillin allergy. Which drug is the best treatment option for this patient?
   a. Ampicillin
   b. A first-generation cephalosporin
   c. A carbapenem
   d. Isoniazid
   e. Azithromycin

5. Which drug family would be the most effective to treat a patient diagnosed with a MRSA infection?
   a. Penicillins
   b. Third-generation cephalosporins
   c. Carbapenems
   d. Lincosamides
   e. Aminoglycosides

6. What advantages might a semisynthetic antimicrobial drug have over an antibiotic?

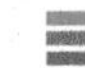

7. Choose the true statement(s) about therapeutic index (TI). Select all that apply.
   a. A drug with a high therapeutic index would be effective above the dose at which it is potentially toxic.
   b. A narrow TI is preferable.
   c. A drug for which the maximum safe dose is close to the minimum effective dose would have a high TI.
   d. It is one measure of a drug's general safety.
   e. A drug that is not selectively toxic would most likely have a high TI.

8. The difference between a synthetic drug and a semisynthetic drug is that
   a. the semisynthetic drug is a modified synthetic drug.
   b. the synthetic drug is a modified natural drug.
   c. the synthetic drug is a modified semisynthetic drug.
   d. the semisynthetic drug is a modified natural drug.

9. Which of the following antimicrobial properties would be the most crucial to consider in developing a new antimicrobial?
   a. Selective toxicity
   b. Ease of administration
   c. Lack of drug interactions
   d. Long half-life
   e. The drug's capacity to be bactericidal

10. Assume a bacterium makes beta-lactamase. Could you still use a glycopeptide drug to treat an infection caused by this bacterium? Explain your reasoning.

11. Assume a clinical sample yields a strain of *S. aureus* containing a plasmid that encodes two antimicrobial-resistance genes. How did the bacterium most likely acquire these new resistance genes?
    a. The strain was intrinsically resistant.
    b. The strain obtained the genes through horizontal gene transfer.
    c. The strain acquired the genes by a random mutation.
    d. The strain picked up the genes by an efflux pump.
    e. The strain acquired the genes through cell division events.

12. If a gene encoding a bacterial transpeptidase enzyme undergoes mutation, which of the following antimicrobials may no longer be effective against the mutated bacterium?
    a. Macrolides
    b. Polypeptide drugs
    c. Tetracyclines
    d. Penicillins
    e. Quinolones

13. Acquired antibiotic resistance can include all of the following *except*:
    a. altering an enzyme that a given drug may target.
    b. making endospores.
    c. altering a point of entry for a drug.
    d. making enzymes that inactivate a drug.
    e. increasing the number of efflux pumps that are active in a cell.

14. Mark the following as true or false, and then correct the false statements so they are true.
    a. Human cells make drug efflux pumps.
    b. The minimum bactericidal concentration is the minimum concentration of the drug that kills at least 50 percent of the bacteria present.
    c. The E-test can reveal if a drug is bactericidal or bacteriostatic.
    d. A drug that is bactericidal at one dose may be bacteriostatic at another dose.
    e. The antifolate combination therapy trimethoprim-sulfamethoxazole may be used to treat protozoan infections.

15. Why is it challenging to obtain selectively toxic drugs against fungi, protozoans, and viruses?

16. Choose the false statement(s). Select all that apply.
    a. Antifungal drugs may target cholesterol in fungal cell membranes.
    b. Azole and polyene drugs promote cell lysis by impacting fungal cell plasma membranes.
    c. Echinocandin drugs inhibit fungal cell wall synthesis.
    d. Antifungal drugs may target DNA replication.
    e. Antifungal drugs may target protein synthesis.

17. Match the antimicrobial drug to its feature. Some features may be used more than once, and some may not be used at all.

Drugs	Features
a. Chloroquine	___ Treats nonmalarial protozoan infections
b. Metronidazole	___ Antihelminthic drug
c. Quinine	___ Treats MRSA infections
d. Praziquantel	___ Antimalarial drug that treats nonresistant cases
e. Albendazole	___ Treats malaria that resists more common drugs
f. Artemisinin combination therapy (ACT)	

18. Which sensitivity test is best for determining the minimum bactericidal concentration *and* the minimum inhibitory concentration of a drug?

CRITICAL THINKING QUESTIONS

1. Multidrug-resistant bacteria tend to first emerge in healthcare settings and then spread to the community. Describe why hospital settings are epicenters for emerging antimicrobial drug resistance, and state what specific actions healthcare workers and patients can take to limit the spread of these agents into the community.

Hospitals and health departments often track drug susceptibility patterns for bacteria. This helps inform decisions about what drugs may be best for treating an infection in the given institution or region. Use this antibiogram to answer the following critical thinking questions.

General Hospital of Microbe County (GHMC)
% Susceptibility of Tested Isolates

	Pepracillin-tazo	Cefepime	Amikacin	Gentamicin	Ciprofloxacin	Levofloxacin	Imipenem	Meropenem	TMP/SMX
E. coli	88	88	100	95	79	81	100	100	62
K. pneumoniae	76	100	100	86	77	79	100	100	90
E. cloacae	55	73	100	95	85	81	100	100	55
P. aeruginosa	45	90	90	67	49	37	54	46	0

2. Within 2 days of being discharged from GHMC, a patient developed a urinary tract infection caused by *K. pneumoniae*. The patient was treated as an outpatient with an oral prescription of TMP/SMX to be taken for 1 week. However, the therapy was *not* effective against the infection. Assuming all susceptibility-testing protocols were properly followed, and the patient was compliant with the drug regimen, provide *three* possible explanations for this observed situation.

3. An adult, nonpregnant patient has a mixed infection caused by *K. pneumoniae* and *P. aeruginosa*. The patient reports living and working in Microbe County, no recent travel, and a severe penicillin allergy. Based on the antibiogram data, what is the best treatment option for this patient, and why?

4. The first drug listed in the table is a combination therapy. What does "tazo" stand for, and what drug-resistance mechanism is it designed to overcome? Name *three* drug-resistance mechanisms that a bacterium could develop to thwart tazo's action.

5. Which bacterium represented in the antibiogram is the most problematic in terms of overall drug resistance? If we were to assume a *single* resistance mechanism is being used by the bacterium you named, what is the most likely resistance mechanism at work? Explain your reasoning.

6. Describe the drug class for each drug listed in the antibiogram. Then, based strictly on the cellular features of *M. pneumoniae* and drug action mechanisms, explain which drugs in the antibiogram would not treat an infection caused by *M. pneumoniae*.

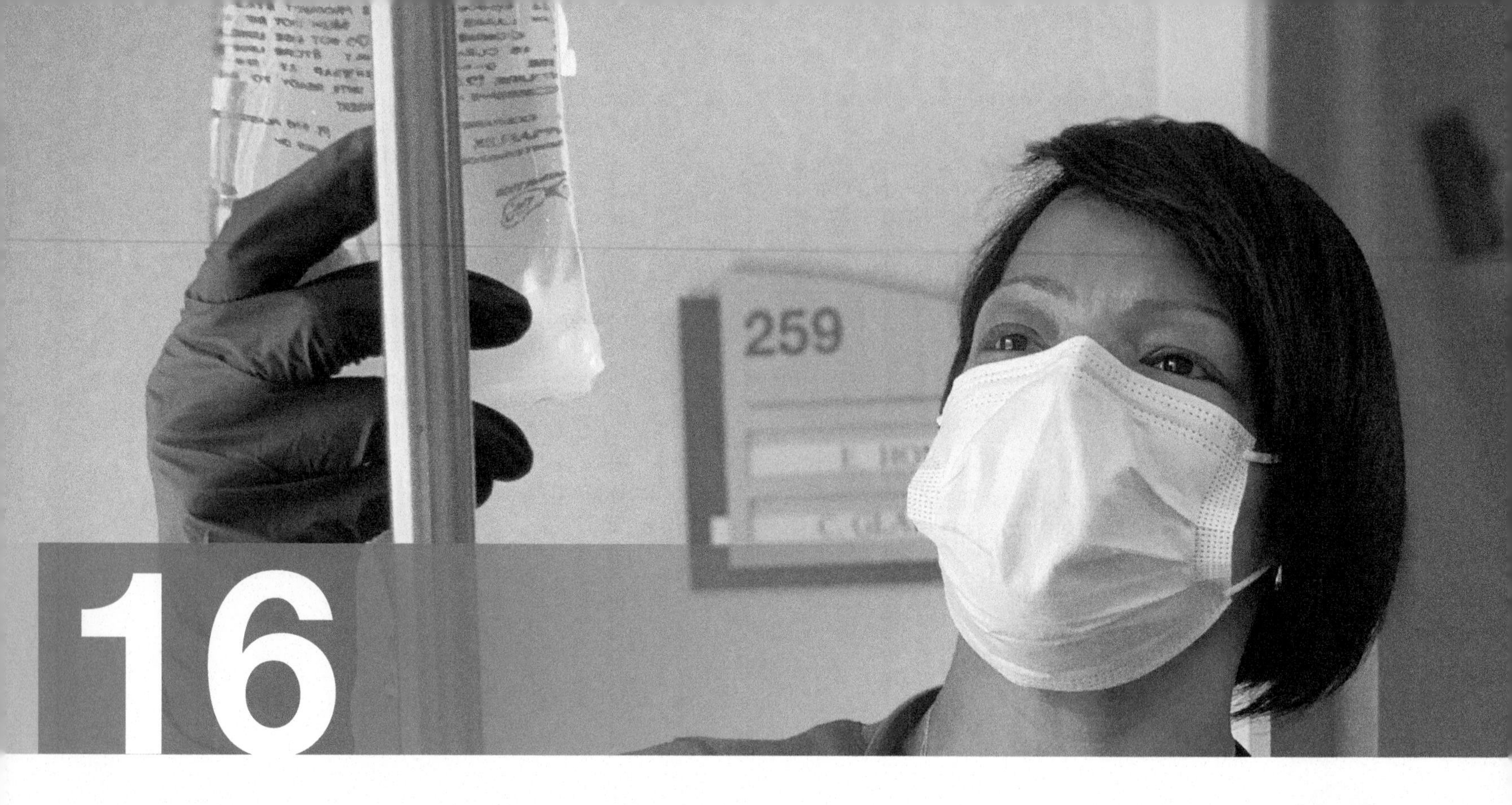

Respiratory System Infections

What Will We Explore? An average adult takes about 20,000 breaths per day, inhaling about 11,000 liters of air. This makes the respiratory tract the most common and readily accessible entryway for microbes to establish infection in the human body. The roll call of culprits that can infect the respiratory tract includes bacteria, viruses, fungi, and parasites. In this chapter, we explore the most important of these pathogens. Some have afflicted humans since ancient times, and others are more recent threats.

Why Is It Important? Bacterial and viral infections of the respiratory tract have had an especially profound impact around the world. Just a century ago, tuberculosis, typical pneumonia, and influenza were routinely among the top three causes of death. In the 1800s, tuberculosis was responsible for about 40 percent of deaths among working-class people in cities, and the influenza pandemic of 1918 infected at least a third of the global population, killing more people in a 9-month period than died throughout all of World War I.

Despite the modern medical advances of antibiotics, vaccines, and supportive therapy, respiratory infections are still consistently a worldwide threat today. Because diseases that spread via the respiratory route can be extremely easy to transmit in populations, they pose a significant risk for serious epidemics and pandemics. For example, the World Health Organization (WHO) estimates that each year influenza and its complications kill about 250,000 to 500,000 people worldwide. And in December 2019, COVID-19, an infectious disease caused by a novel coronavirus, emerged to cause a global pandemic that, to date, has claimed more than 4.55 million lives worldwide. Based on these data, it's not surprising that tackling the threat of respiratory pathogens, especially certain coronaviruses and influenza viruses, remains a top priority of public health agencies around the world.

Clinical CASE
NCLEX
HESI
TEAS

The Case of the Suffering Spelunker

Visit the **Mastering Microbiology** Study Area to watch the case and find out how respiratory infections can explain this medical mystery.

Jackie Lewis
RN, Infection Preventionist; Decatur, GA

16.1 OVERVIEW OF THE RESPIRATORY SYSTEM

The respiratory system is the most common portal of entry for microbes.

The human respiratory system includes organs and tissues that facilitate gas exchange—namely, bringing oxygen into the body so cells can carry out cellular respiration and the removal of carbon dioxide, a metabolic waste product. It is divided into two tracts. The upper respiratory tract includes the mouth, nasal passages, paranasal sinuses, pharynx (throat), and epiglottis, a cartilage structure associated with the pharynx. The lower respiratory tract includes the larynx (voice box), trachea (windpipe), bronchi, bronchioles, lungs, and alveoli, which are the air sacs in the lungs where gas exchange occurs (FIG. 16.1).

Our constant need to breathe presents an opportunity for pathogens, which are coughed and sneezed into the air, to infect anyone who happens to inhale the expelled respiratory droplets and aerosols. In fact, the respiratory route is the most common microbial portal of entry.

Once inside the respiratory tract, the microbe's impact and ultimate destination varies by pathogen. Well-known diseases that enter the body through the respiratory tract but affect other systems include chickenpox, mumps, and measles. There are also some pathogens that do not enter through the respiratory route but can migrate there and cause disease. For example, protozoans and helminths may affect the lungs, though usually this is a secondary manifestation that occurs when they migrate from other areas of the body, such as the gastrointestinal tract. (See Chapter 19 for more on protozoan and helminth infections.) This chapter focuses on the main infectious agents that enter through the respiratory tract and directly cause infections there. Because viral and bacterial respiratory infections are the most prevalent, this chapter concentrates mostly on them. You will also learn about selected fungal infections, the increasing incidence of which is due to a larger percentage of our population being immune deficient from medical interventions, aging, or HIV/AIDS.

Learning Outcomes

After reading this section, you should be able to:

16.1 Describe the parts of the upper and lower divisions of the human respiratory system.

16.2 Describe the factors that limit infection of the respiratory system.

16.3 Define common terminology associated with respiratory system inflammation and infections.

16.4 State where normal microbiota exists in the respiratory system, and give examples of respiratory microbiota phyla.

Upper Respiratory Tract Anatomy and Related Infection Terms

The upper respiratory tract warms, humidifies, and filters the air we breathe. These functions are partially accomplished by the paranasal sinuses, four pairs of interconnected hollow cavities lined with mucous membranes. Specialized cells and glands in these membranes secrete mucus, which warms and humidifies inhaled air and traps microbes and debris. The mucus forms a layer along **cilia**—tiny hairs lining the airway. The cilia sweep the debris-laden mucus toward the nose and mouth (FIG. 16.2). Many infectious agents and inhaled allergens like pollen can aggravate the sinuses and nasal passages, leading to inflammation of the sinus membranes (**sinusitis**). This can block drainage of mucus from the sinuses into the nose—a condition called sinus congestion. As mucus accumulates in the normally hollow chambers of the sinuses, it puts pressure on surrounding structures, causing a painful headache. If sinuses do not drain, the trapped mucus can also serve as a breeding ground for bacteria, which can lead to a sinus infection. The American Academy of Family Physicians estimates that 15 to 20 percent of all antibiotic prescriptions in outpatient care are

FIGURE 16.1 **The respiratory system**

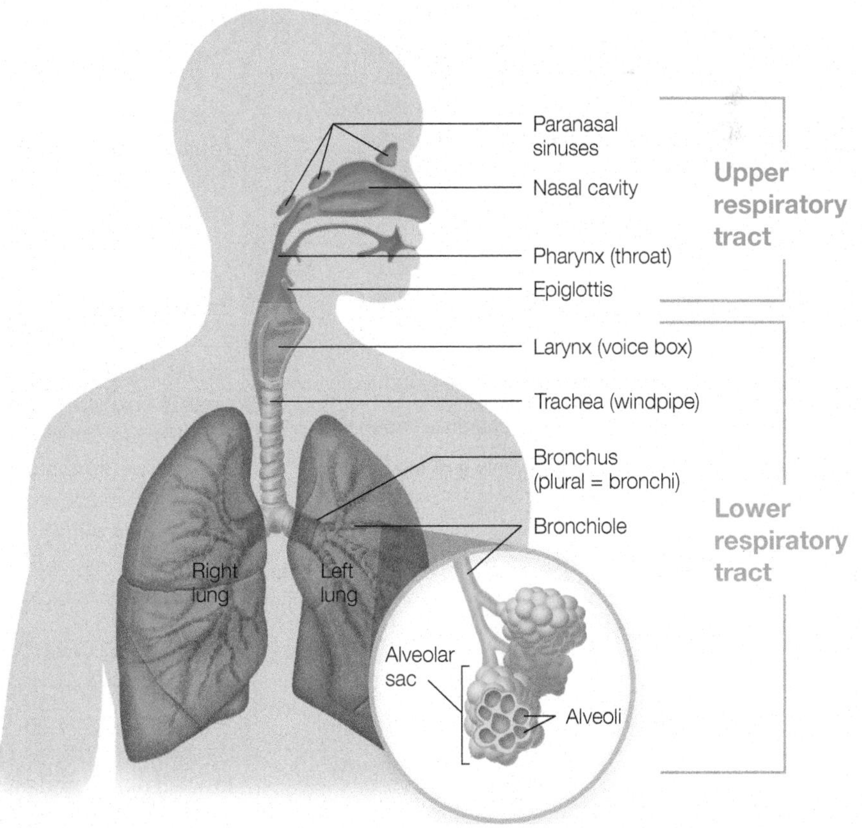

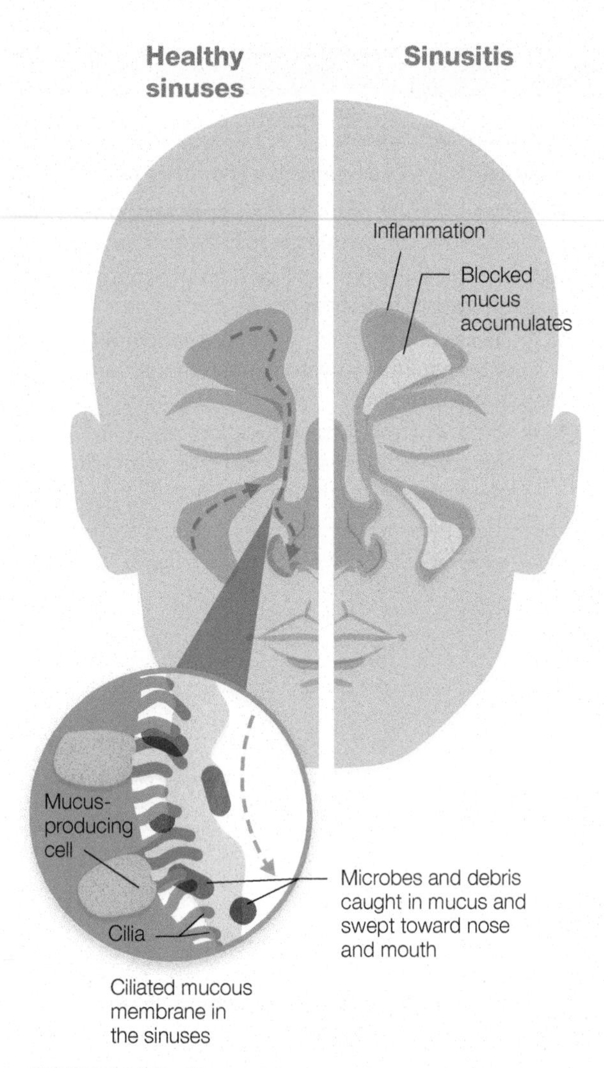

FIGURE 16.2 Sinuses In the upper respiratory tract, ciliated mucous membranes in the sinuses and airways trap debris and sweep it toward the mouth and nose. When the process is blocked, sinusitis and sinus infections may occur.

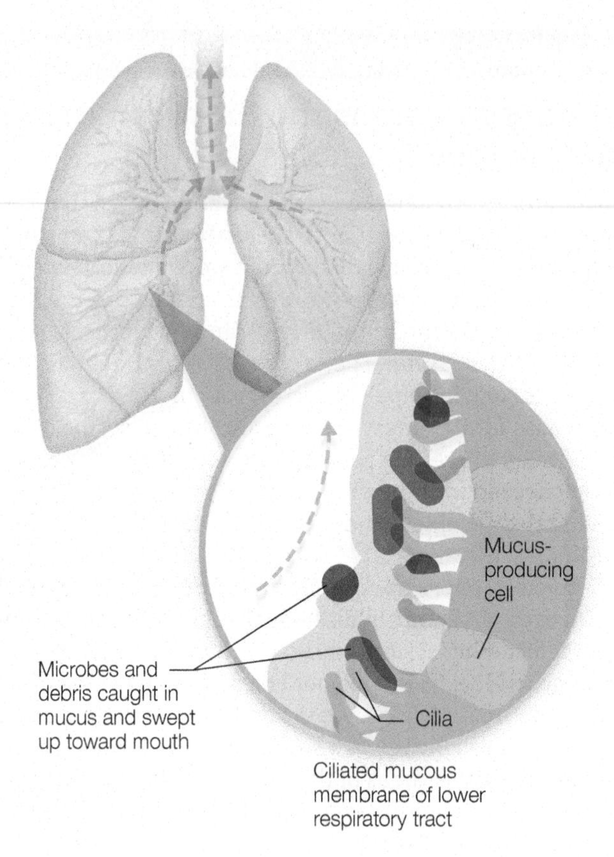

FIGURE 16.3 Mucociliary escalator In the lower respiratory tract, microbes and other debris are trapped in mucus and then swept away from the lungs by the mucociliary escalator.

Critical Thinking *Smoking can damage the cilia that sweep mucus away from the lungs. How do you think this relates to the fact that smokers are at greater risk for lung infections than nonsmokers?*

CLINICAL VOCABULARY

Sinusitis inflammation and swelling of the sinus membranes

Pharyngitis inflammation of the pharynx (throat)

Epiglottitis inflammation and swelling of the epiglottis; can quickly block the airway

prescribed to treat sinusitis. This is unfortunate because the most common cause of the condition is not bacteria, but viruses (especially cold viruses) and inhaled allergens like pollen and dust—none of which are treatable with antibiotics.

The pharynx (commonly called the throat) marks the end of the upper respiratory tract. It is the gatekeeper between the respiratory and digestive tracts. Bacteria, viruses, and allergens can all cause the pharynx to become inflamed—a condition called **pharyngitis**, which most patients describe as having a "sore throat." The epiglottis is a cartilage structure associated with the pharynx; it guards entry to the lower respiratory system and seals off the airway during swallowing to prevent food and beverages from entering the lungs. Of course, when a person is not swallowing, the airway needs to be open so that air can make its way through the larynx toward the lungs. **Epiglottitis** is an inflammation and swelling of the epiglottis, which can block the airway. It is a medical emergency caused by certain infections—especially *Haemophilus influenzae* type b (Hib), which is reviewed later in this chapter.

Lower Respiratory Tract Anatomy and Related Infection Terms

The primary function of the lower respiratory tract is to direct air to the lungs, where gas exchange occurs. The lower respiratory tract airways, which include the trachea, bronchi, and bronchioles, have ciliated mucous membranes that trap inhaled debris and sweep it toward the mouth to prevent it from entering the lungs. This mucus-sweeping mechanism is called the **mucociliary escalator** (FIG. 16.3).

Debris and pathogens that are not trapped by the mucociliary escalator can often be cleared out by macrophages residing in the alveoli of the lungs. (For more information on alveolar macrophages, see Chapter 11, which covers the innate immune system.)

Inhaled allergens and microbes can cause inflammation of lower respiratory tract tissues. When the larynx is inflamed, a person is said to have **laryngitis**, which can cause a temporary voice loss due to the swelling of the vocal cords. Inflammation of the trachea is called **tracheitis**, and inflammation of the bronchi and/or bronchioles is called **bronchitis**. A combined inflammation of the larynx, trachea, bronchi, and bronchioles is called laryngotracheobronchitis—also known as **croup**. Viruses cause most croup cases. Croup is characterized by a barking cough and respiratory **stridor** (wheezing or loud breathing associated with airway obstruction). Inflammation of the lung tissue, especially of the alveoli, which are specialized regions of the lung where gas exchange occurs, results in **pneumonia**—a generalized term that refers to inflammation of the lungs, especially of the alveoli. Inflammation of the lower respiratory tract can upset the delicate oxygen balance required by the body, making the condition dangerous if it is not mitigated. Because the lungs are highly vascularized—meaning there are extensive networks of blood vessels associated with them—infections in the lungs can readily spread to the bloodstream, increasing the risk that they can become systemic (systemic infections, or infections that affect the whole body, are discussed in Chapter 21).

Infectious agents that enter the body by the respiratory route often cause signs and symptoms that include one or more of the following: coughing, stridor, **dyspnea** (shortness of breath), fatigue, sneezing, sore throat, and fever. Runny nose, coughing and sneezing are not only uncomfortable, but also help spread the agent to new hosts through respiratory droplets and/or finer aerosols. (See Chapter 9 for a review of modes of transmission.) Note that respiratory allergies often imitate common respiratory infections; however, fever is absent in allergies (despite the common term "hay fever" that refers to seasonal allergies).

CLINICAL VOCABULARY

Laryngitis inflammation and swelling of the larynx (voice box); can cause temporary voice loss

Tracheitis inflammation and swelling of the trachea

Bronchitis inflammation and swelling of the bronchi and/or bronchioles

Croup laryngotracheobronchitis (or a combined inflammation of the larynx, trachea, bronchi, and bronchioles) caused mainly by viruses; characterized by a barking cough and stridor

Stridor wheezing or loud breathing associated with a blocked or narrowed airway

Pneumonia inflammation of the alveoli, which are the small air sacs in the lungs where gas exchange occurs

Dyspnea shortness of breath

Respiratory Tract Microbiome

As we inhale air, our airways are essentially brought into contact with elements of the outside environment, so it is not surprising that many microorganisms, including bacteria, viruses, fungi, and archaea, call our respiratory tract home. Although healthy lungs were previously thought to be a sterile environment, at this point numerous studies have demonstrated that the lungs are colonized by normal microbiota and, at least in terms of bacterial profiles, these microbes resemble what is found in the mouth (FIG. 16.4). Thus, normal microbiota are now known to reside throughout the respiratory system. These resident microbiota are typically more helpful than harmful to us because they compete with potential pathogens, and some secrete antimicrobial peptides to limit the growth of would-be pathogens. The normal microbiota profiles in the respiratory tract vary between individuals and also between people with healthy lungs and those who have underlying health issues like asthma, chronic obstructive pulmonary disease (COPD), cystic fibrosis, and even lung cancer. For example, data from several studies suggest that the pulmonary microbiota profile can affect the risk of developing certain lung cancers and even influence how well certain lung tumors respond to

FIGURE 16.4 Bacterial microbiome of the respiratory system The upper respiratory tract has a much greater diversity of bacterial species and density of microbial populations than the lungs.

Sources: Nasal cavity data are from Kumpitsch, C., et al. (2019). The microbiome of the upper respiratory tract in health and disease. *BMC Biol. 17*(87). Pharynx data are from Gao, Z., et al. (2014). Human pharyngeal microbiome may play a protective role in respiratory infections. *Genomics, Proteomics & Bioinformatics, 12*(3). Lung data are from Ramesh, M. Y., et al. (2021). Lung microbiome composition and bronchial epithelial gene expression in patients with COPD versus healthy individuals: A bacterial 16S rRNA gene sequencing and host transcriptomic analysis. *The Lancet Microbe* (online publication https://doi.org/10.1016/S2666-5247(21)00035-5).

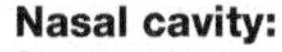

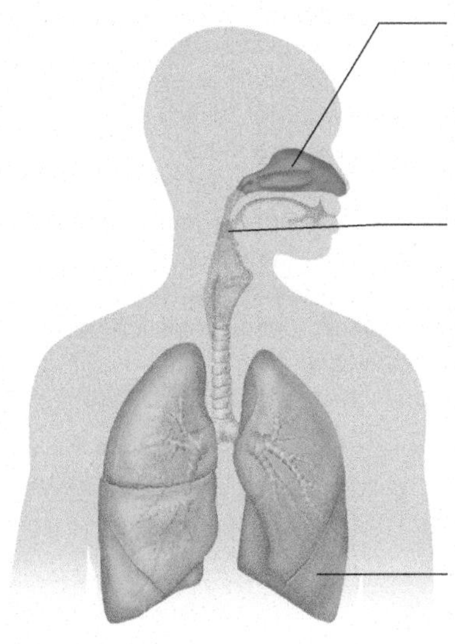

Nasal cavity:
Dominated by members of the phyla Firmicutes (especially members of *Staphylococcus* and *Streptococcus* genera), Actinobacteria, and Proteobacteria

Pharynx (throat):
Dominated by the phyla Firmicutes, Bacteroidetes, Proteobacteria, Actinobacteria, and Fusobacteria; more Bacteroidetes members live in the throat than any other body site. The most common genera, in descending order are *Prevotella, Capnocytophaga, Campylobacter, Veillonella, Streptococcus, Neisseria,* and *Haemophilus.*

Lungs:
Bacteroidetes, Firmicutes, Actinobacteria, Proteobacteria, and Fusobacteria are the most prevalent phyla; main genera in healthy people include *Prevotella, Veillonella, Streptococcus, Neisseria, Haemophilus,* and *Fusobacterium.*

chemotherapy.[1] Evidence has also emerged that changes in lung microbiota exist in asthmatic patients as compared to nonasthmatics.[2] A healthy respiratory microbiome has a high diversity of species and is mainly dominated by five phyla of bacteria: Bacteroidetes, Firmicutes, Actinobacteria, Proteobacteria, and Fusobacteria.[3]

Build Your Foundation

1. List the anatomical structures that make up the human upper and lower respiratory tract. (LO 16.1)
2. Describe two anatomical and/or physiological factors that help to limit infections of the respiratory tract. (LO 16.2)
3. Define the following terms: *sinusitis*, *pharyngitis*, *epiglottitis*, *laryngitis*, *tracheitis*, *bronchitis*, *croup*, *stridor*, *pneumonia*, and *dyspnea*. (LO 16.3)
4. State where normal microbiota exist in the respiratory tract, and name at least two bacterial phyla considered part of the healthy lung microbiota. (LO 16.4)

Build Your Foundation (BYF) Quick Quiz: Visit the **Mastering Microbiology** Study Area to quiz yourself.

TRAINING TOMORROW'S HEALTH TEAM

Listening for Respiratory Disease

When a patient complains of respiratory symptoms, healthcare providers reach for their stethoscopes and listen. A number of breath sounds can hint at respiratory infections.

Stridor is a noisy breathing or loud wheezing that's most audible during inhalation and more common in pediatric patients. It's caused by a narrowing of the airway and sounds like a high-pitched whistle that is often audible even without a stethoscope. Stridor can result from an inflammation of the pharynx, larynx, trachea, or bronchi and is therefore a clinical sign of many different upper respiratory conditions, both infectious and noninfectious. The most prevalent infectious cause of stridor is croup, which is a common childhood infection caused by human parainfluenza viruses.

Other sounds, such as wheezing, are due to narrowing of the airways and may be heard during inhalation and exhalation. Wheezing sounds are often categorized by pitch. Higher-pitched wheezes have a squeaky sound quality, whereas lower-pitched wheezes, which are sometimes referred to as rhonchi or sonorous wheezes, have a snore-like or gurgling quality. Higher-pitched wheezes are common in asthma, bronchitis, and chronic obstructive pulmonary disease (COPD) (i.e., emphysema and chronic bronchitis). Rhonchi are often heard when there is restricted airflow in the larger upper airways such as the bronchi. Ronchi are often more pronounced upon exhalation and may temporarily clear after coughing; they are common in pathologies like pneumonia, bronchitis, pulmonary edema (fluid in the lungs), and COPD.

Crackles (also called rales) are heard most clearly during inhalation and can be described as coarse or fine. Coarse crackles sound like Velcro fastener strips being pulled apart, whereas fine crackles are more like the sound of numerous hairs being rolled between the fingers. Crackles are often associated with inflammation of the bronchi, bronchioles, or alveoli. Crackles that don't resolve after a cough may be due to pulmonary edema.

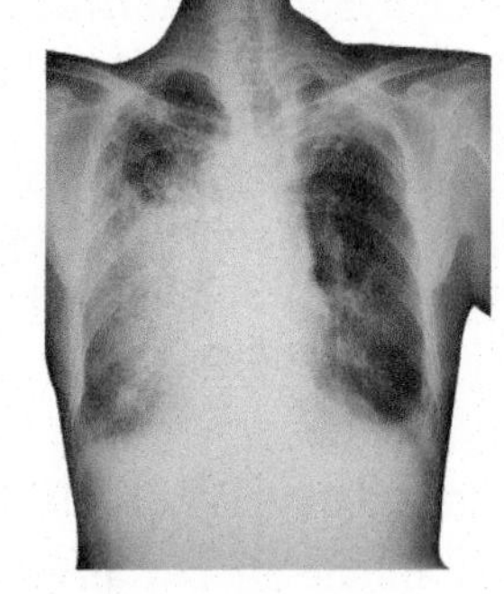

Pneumonia is sometimes seen as a cloudy region in a chest X-ray.

In addition to listening for these aforementioned sounds, clinicians may also perform a "percussion test" that can help to determine if there is fluid in the lungs. In this test, the clinician taps on the chest wall. This action generates distinct tones; normal lungs produce a resonant sound, whereas fluid-filled lungs produce a dull thud.

Pulmonary auscultation and percussion help healthcare providers narrow down which additional tests, if any, should be performed to definitively diagnose the cause of the patient's respiratory symptoms.

QUESTION 16.1

How might a percussion test help to suggest typical versus atypical pneumonia?

[1] Sommariva, M., Le Noci, V., Bianchi, F., et al. (2020). The lung microbiota: Role in maintaining pulmonary immune homeostasis and its implications in cancer development and therapy. *Cellular and Molecular Life Sciences, 77*(14), 2739–2749.

[2] Barcik, W., Boutin, R. C., Sokolowska, M., et al. (2020). The role of lung and gut microbiota in the pathology of asthma. *Immunity, 52*(2), 241–255.

[3] Yang, D., Xing, Y., Song, X., & Qian, Y. (2020). The impact of lung microbiota dysbiosis on inflammation. *Immunology*, 159(2), 156–166.

16.2 VIRAL INFECTIONS OF THE RESPIRATORY SYSTEM

Hundreds of different viruses can infect the respiratory tract. Even though the general signs and symptoms of viral respiratory illnesses overlap, the severity of the disease and the **prognosis** (likely outcome) vary according to the infectious agent as well as patient factors (age, general health, etc.). Although viruses cannot be treated with antibiotics, there are vaccines against some viral respiratory illnesses.

Learning Outcomes

After reading this section, you should be able to:

16.5 Compare and contrast the signs and symptoms of colds and influenza.

16.6 List special clinical features of illnesses caused by respiratory syncytial virus, human parainfluenza viruses, and adenoviruses.

16.7 Discuss how antigenic shift and antigenic drift affect viral evolution, and state the likely role of antigenic shift in the development of the influenza pandemic of 1918 (i.e., the "Spanish flu" pandemic).

16.8 Discuss characteristics of SARS-CoV-2, including how it is transmitted, how it invades host cells, and features of COVID-19—the disease it causes.

16.9 Define the terms *viral variant* and *viral strain*, and discuss the three categories used to group SARS-CoV-2 variants.

16.10 Describe what hanta pulmonary syndrome is and how it is transmitted.

Colds are the most common cause of respiratory infections.

Over 200 genetically distinct viruses from at least eight different genera cause the common cold, which is clinically known as an acute respiratory infection (often abbreviated ARI). Colds are the most common infections of the respiratory system. Although they usually resolve without any therapy (they are self-limiting), they still have a huge economic impact—lost productivity, healthcare expenditures, and other related expenses amount to billions of dollars every year.

Hundreds of **serovars** (genetically distinct variants of the same species that are distinguished according to their different surface antigens) of rhinoviruses and coronaviruses are estimated to cause between 60 and 80 percent of all colds. Parainfluenza viruses, adenoviruses, and nonpolio-type enteroviruses are a few examples of other viruses that can cause colds. The type of virus that typically causes a cold in any particular patient depends on host factors like the patient's age, environment, and existing medical conditions, as well as the time of year. Peak cold season is from fall through early spring, but nonpolio enteroviruses are a common cause of colds in warmer months, and cold viruses can be found in human populations year-round in all climates. Also, many studies show no statistically significant increase of colds in people exposed to cold and/or wet conditions, so the idea that you'll "catch a chill" if you get wet feet or go outside underdressed for cold weather seems to be untrue.

Adults average about three colds per year, whereas children have about twice as many colds each year. This means an average person will spend about two years of their life suffering from colds.

Cold viruses are highly infectious and are spread through personal contact, respiratory droplets expelled during coughs and sneezes, and fomites such as doorknobs, cell phones, pacifiers, and toys. Cold symptoms can include sudden onset of sore throat, runny nose, sneezing, fatigue, general body achiness, loss of appetite, and a cough. A low-grade fever can accompany a cold, although fever is more common in children than in adults. Thickened, opaque, and discolored (e.g., yellow or green) mucus can accompany the later stages of a cold and does not necessarily indicate an underlying bacterial infection.

Only about 1 out of 200 colds will result in a secondary bacterial infection such as a sinus infection, ear infection, or a lower respiratory tract infection like pneumonia. In such cases, antibiotic therapy may be necessary. Antibiotics are not usually prescribed for a patient suffering from a cold unless symptoms persist for more than 10 days and there are no improvements or if new symptoms develop (such as a fever), which could mean a bacterial infection has developed. Bacterial complications of colds are discussed further in the bacterial infections of the respiratory system section.

Because so many different viruses cause colds, it is unlikely that a vaccine will be developed to prevent our most common medical foe. Treatment is supportive and may include rest, increased fluids, and over-the-counter remedies to alleviate symptoms.

Disease**Snapshot** • Respiratory Syncytial Virus (RSV)

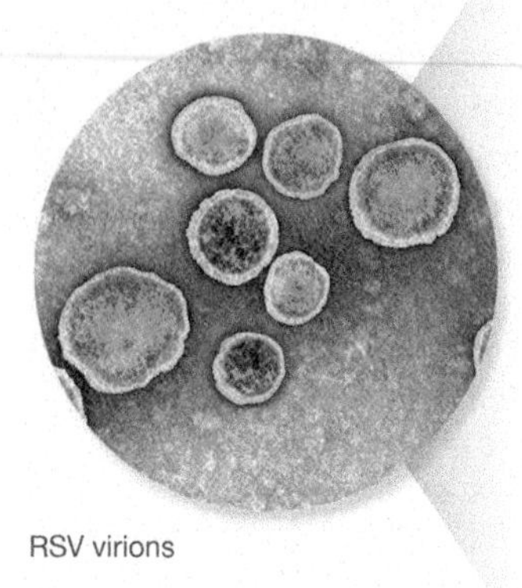
RSV virions

Causative agent	Respiratory syncytial virus (RSV): enveloped, single-stranded RNA genome, *Pneumoviridae* family
Epidemiology	Endemic among children under age 5; can be dangerous in premature infants and in patients who are immune compromised or suffer from chronic lung and/or heart conditions; most cases occur from fall through early spring; incubation period is usually 4–6 days
Transmission & prevention	Transmitted by respiratory droplets and fomites; typical preventions are sanitary practices and hand washing; antibody injections for high-risk groups (especially premature infants)
Signs & symptoms	Coughing, sneezing, and fever are typical symptoms; wheezing may also occur
Pathogenesis & virulence factors	Enters the upper respiratory tract and spreads to the lower respiratory tract within 1–3 days, causing inflammation in the airways; masses of multinucleated cell bodies, called syncytia, form; immune system cells can damage epithelial cells in the airway and contribute to pathology
Diagnosis & treatment	Antigen-detection tests are commonly used for diagnosis; supportive therapies alleviate symptoms; aerosolized antiviral drug ribavirin may be administered in severe cases

RSV, HPIV, and adenoviruses produce cold-like symptoms but have other clinical features worth noting.

The viruses discussed in this section cause cold-like illnesses in most people. Sometimes, however, they are accompanied by additional symptoms, or they can present a specific danger to certain at-risk groups. In particular, the viruses discussed here can lead to pneumonia—which, as previously mentioned, is not a single infection, but rather a generalized term that refers to inflammation of the alveoli. (Additional details on pneumonia are presented later.)

Respiratory Syncytial Virus (RSV) Infections

Respiratory syncytial virus (RSV) infections are caused by an enveloped RNA virus that belongs to the *Paramyxoviridae* family—the same viral family as measles, rubella, and parainfluenza viruses. The viral genome, which encodes 11 proteins, is about 15,000 nucleotides. There are two main RSV subtypes: A and B.

In older children and adults, RSV resembles a cold. However, in the elderly and in infants (especially those born before 36 weeks gestation), RSV can be quite serious. Worldwide, RSV is the leading cause of acute lower respiratory tract infection in children under age 5—it is the top cause of bronchitis and pneumonia in children under 1 year old and is the primary reason for hospitalization in this age group. Almost all children have battled this virus before their second birthday. Re-infection throughout life is common, although as just noted, these later infections tend to be milder. Globally, there are an estimated 33 million cases, 3 million hospitalizations, and 66,000 deaths from RSV each year, making it a priority for vaccine development.[4]

To date there are about 60 RSV vaccine candidates in development with over a dozen in various stages of clinical trials. Until a vaccine is approved, antibody preparations can be given to high-risk patients to prevent the disease. Unfortunately, the antibody preparations are expensive (roughly $2,000 U.S. dollars per injection), require monthly injections, do not provide long-term protection against RSV, and cannot be used to treat or cure a person who has already contracted RSV.

[4] Tam, C. C., Yeo, K. T., Tee, N., et al. (2020). Burden and cost of hospitalization for respiratory syncytial virus in young children, Singapore. *Emerging Infectious Diseases*, 26(7), 1489.

DiseaseSnapshot • Adenovirus Infections

Causative agent	Over 50 types of adenoviruses; most cause respiratory illnesses, others cause conjunctivitis, diarrhea, and cystitis: nonenveloped, double-stranded DNA genome, *Adenoviridae* family
Epidemiology	Endemic in populations worldwide, human reservoir; infections occur year-round, but outbreaks are more common in late winter through early summer; incubation period is usually 2–10 days
Transmission & prevention	Respiratory droplets and fomites for types that target the respiratory tract; vaccine against types 4 and 7 approved for military personnel only; sanitary practices and hand washing are most common prevention methods
Signs & symptoms	Most adenoviruses attack the respiratory tract and cause sore throat and cold-like symptoms, but occasionally they cause more severe and potentially deadly infections of the respiratory tract, including bronchitis, viral pneumonia, and croup
Pathogenesis & virulence factors	Attach to host cells and are engulfed by endocytosis; as the viral load increases, the production of host cell RNA and proteins kills the cell; make a special viral-associated RNA (VA RNA) that prevents the host cell from mounting a defense against the virus
Diagnosis & treatment	Usually identified using antigen detection or polymerase chain reaction; supportive therapies to alleviate symptoms

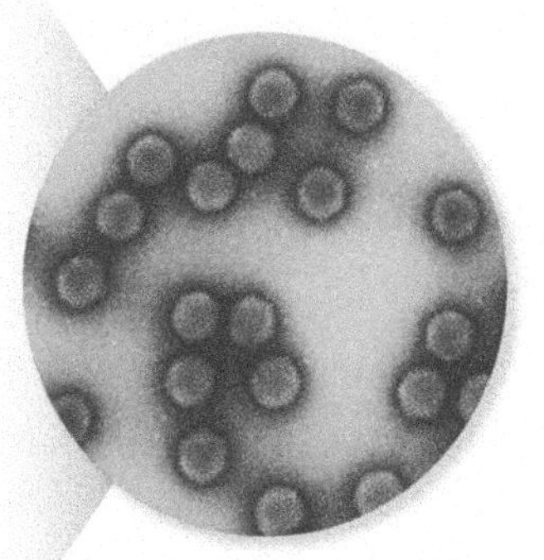

Adenovirus virions

Human Parainfluenza Virus (HPIV) Infections

The name parainfluenza virus means "influenza-like virus"; however, these single-stranded RNA viruses are in the family *Paramyxoviridae* and are genetically and clinically different from influenza viruses and should not be confused with them. **Human parainfluenza virus (HPIV) infections** are characterized by cold-like symptoms, but in infants or the elderly they can cause more severe illnesses like croup, bronchitis, and pneumonia. It is estimated that these viruses are responsible for up to 30 percent of respiratory infections in children under age 5. HPIV infections are second only to respiratory syncytial virus infections in terms of hospitalization due to viral respiratory tract infections.

There are four types of HPIV; all are transmitted by respiratory droplets and fomites, and they tend to cause infections in fall, spring, and summer. HPIV-1 and HPIV-2 are the most common causes of croup in children, while HPIV-3 tends to cause bronchiolitis, bronchitis, and pneumonia. HPIV-4 is less well characterized, but it seems to be clinically unremarkable and causes only mild cold-like symptoms. In developed nations, deaths due to HPIV are rare, but in developing nations where health care is often limited, HPIV poses a considerable risk for death from a viral respiratory tract infection in normally healthy preschool-age children. Vaccines against parainfluenza viruses are currently in clinical trials. The only treatments are supportive therapies to alleviate symptoms.

Adenovirus Infections

Adenovirus infections are caused by nonenveloped DNA viruses. Although these viruses infect patients of all ages, they are responsible for up to 10 percent of respiratory illnesses in children. Over 50 types of adenoviruses can infect humans. Most attack the respiratory tract and cause sore throat and cold-like symptoms, but occasionally they cause more severe and potentially deadly infections of the respiratory tract that include bronchitis, pneumonia, and croup. Certain adenoviruses cause viral conjunctivitis (pink eye), gastroenteritis (diarrhea), and bladder infections (cystitis). There is a vaccination against adenovirus types 4 and 7, which affect the respiratory tract; however, in the United States the vaccine is usually only given to military recruits.

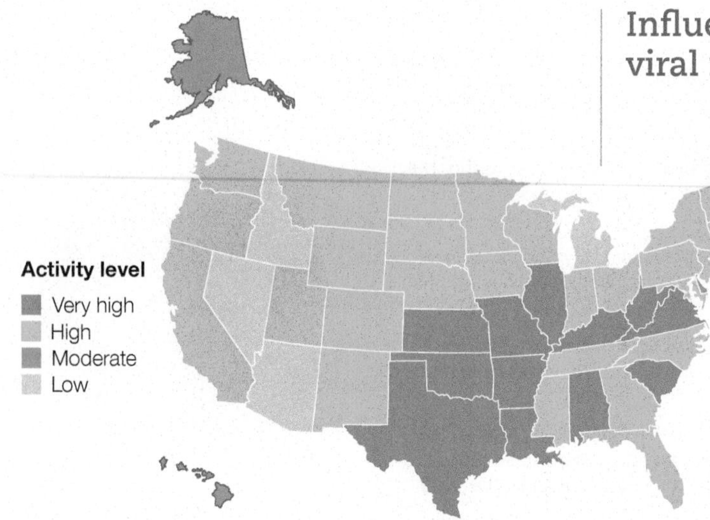

FIGURE 16.5 U.S. February 29, 2020, seasonal flu-like illness map This map from the week ending February 29, 2020, shows high levels of influenza-like illness. It is likely that this map, which is based on data from roughly 1 month after the first COVID-19 case was confirmed in the United States, includes COVID-19 cases that were not diagnosed because of limited testing capabilities early in the COVID-19 pandemic.

Influenza is among the most common viral respiratory illnesses in humans.

Influenza is an infectious respiratory disease caused by certain strains of enveloped, single-stranded RNA viruses in the *Orthomyxoviridae* family. Most strains that cause influenza ("the flu") infect the host by binding to ciliated cells in the upper respiratory tract. However, some strains (especially more virulent avian and swine strains) can directly affect the lungs. Symptoms tend to mimic those of a severe cold, and infection is frequently associated with dangerous complications like pneumonia. Influenza is always more debilitating than a cold, but for those with pre-existing conditions such as asthma or heart disease, it can be deadly. Disease and death rates vary greatly from year to year during seasonal outbreaks (epidemics), depending on the strains most commonly circulating. For example, the Centers for Disease Control and Prevention (CDC) reported that the mild flu season of 2011–2012 caused about 12,000 deaths in the United States, whereas the widespread 2017–2018 season was responsible for about 61,000 deaths (FIG. 16.5). And the infamous influenza pandemic of 1918 killed millions worldwide (see more on that pandemic later).

Types of Influenza Viruses

Three related types of influenza viruses can infect humans: types A, B, and C. Type A is the most common cause of human flu epidemics. Moreover, the CDC has found that when type A influenza strains are prominent, nearly three times as many deaths tend to occur, compared to seasons when type A strains are not the main ones circulating. Historically, type B influenza has been less impactful, but still important, in epidemics. Type C usually causes mild illness and is not associated with epidemics.

Influenza strain virulence is usually determined based on disease severity and the number of fatalities attributed to a given strain. Among several important virulence factors are influenza's hemagglutinin (HA) and neuraminidase (NA) proteins. HA and NA spikes are glycoproteins on the surface of the influenza virus. The virus uses HA spikes to attach to and invade target cells in the respiratory tract. NA spikes help newly formed viral particles escape from the host cell so they can go on to infect more cells. The HA and NA subtypes are numbered and used to name and group influenza type A viruses. To date, 18 different subtypes of the HA protein and at least 11 distinct variants of the NA protein have been identified, making at least 198 influenza A subtypes possible.[5] (Note that Chapter 6 gives more detail on many topics touched upon in this influenza discussion, including HA and NA spikes, flu naming conventions, RNA viral genomes, antigenic shift, and antigenic drift.)

Historically there have been three main influenza A subtypes that infect humans: H1N1, H2N2, and H3N2. Emerging influenza strains such as swine flu (H1N1) and avian or "bird" flu (H5N1 and H7N9) changed their HA and NA proteins in such a way that they gained the ability to invade human cells. Fortunately, "bird" flu strains have not yet demonstrated efficient spread from person to person. However, influenza's RNA genome mutates quickly and has a genetic mix-and-match ability, which means constant monitoring of influenza strains is required to identify new strains that might have new ways to infect the host.

Seasonal Outbreaks versus Worldwide Pandemics

Influenza's HA and NA spikes frequently undergo minor changes through random mutations. This process, referred to as *antigenic drift,* allows viruses that

[5] Jang, J., & Bae, S. E. (2018). Comparative co-evolution analysis between the HA and NA genes of influenza A virus. *Virology: Research and Treatment*, 9, doi.org/10.1177/1178122X18788328.

DiseaseSnapshot • Influenza

Causative agent	Influenza types A, B, and C (type A strains are the major concern in epidemics and pandemics): enveloped, single-stranded RNA genome, *Orthomyxoviridae* family
Epidemiology	Endemic, with ability to cause epidemics and pandemics; exhibits seasonality with most infections in late fall through early spring; 1- to 4-day incubation period
Transmission & prevention	Respiratory droplets and fomites; prevented by seasonal vaccines, sanitary practices, and hand washing
Signs & symptoms	Symptoms tend to be similar to those of a severe cold; especially fever, chills, and body aches. Severe complications like pneumonia can occur.
Pathogenesis & virulence factors	NS1 protein helps type A viruses dodge antiviral response of host cells through interferon; HA proteins impact how effectively the virus can invade cells; NA proteins help the virus escape host cells to infect more host cells
Diagnosis & treatment	Rapid immunoassays detecting either influenza A or B are routinely used for diagnosis; in high-risk patients, antivirals may be used to treat infections.

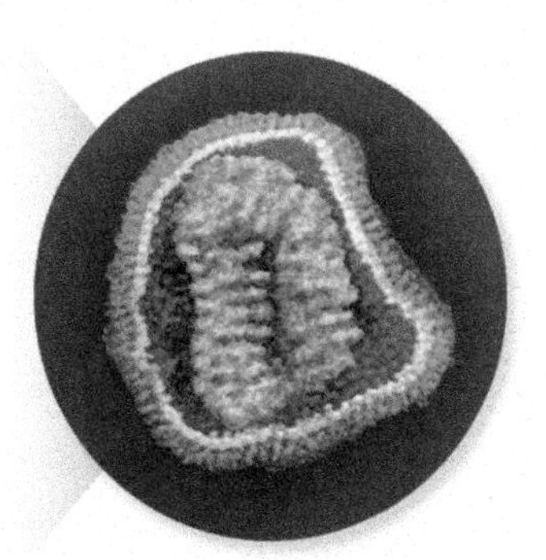

Influenza virion

are closely related to ones that previously infected us to evade recognition by the immune system, thus allowing seasonal outbreaks to occur. When an influenza strain undergoes a major genetic change, it may cause major alterations in viral antigens (*antigenic shift*). This may allow the virus to expand host range, or simply be different enough that few people possess immunity against it. This scenario leads to pandemics, or worldwide outbreaks. There have been numerous influenza pandemics in human history, but one of the worst was the so-called Spanish influenza outbreak of 1918. Although the virus that caused that pandemic probably did not originate in Spain, the outbreak was named for the country after 8 million people there died from the disease in just 1 month early in the flu season. Unlike most influenza strains, which mainly claim victims among the elderly or the very young, the 1918 strain showed the highest mortality rate in young adults (15- to 34-year-olds). This population had strong immune responses and experienced what is called a cytokine storm when exposed to the novel virus (Chapter 11 reviews cytokines). Consequently, these younger patients were more likely to experience an aggressive and out-of-control immune response that tended to do more harm than good.

Influenza Vaccines

Given the limited usefulness of antiviral medications against influenza, flu vaccines are still the best option to reduce influenza's burden in populations. We create new flu vaccines every year using strains that the World Health Organization (WHO) believes will most likely be circulating widely in the next season. Because the viral candidates usually change from year to year, for maximal protection people must get vaccinated every year with the new formulation. Billions of dollars are spent every year in a global effort to head off influenza outbreaks that could mirror the disaster of the 1918 influenza pandemic. Because we do have flu vaccinations and drugs to help reduce the severity of the infection, a devastating flu pandemic on the scale of that in 1918 is less likely, but not impossible or even improbable. Therefore, the annual scramble to predict which viral strains will cause the next seasonal flu outbreak, and the rush to develop appropriate amounts of the updated vaccine formulation, continues.

Types of influenza vaccines There are currently three main types of influenza vaccine available: inactivated formulations, recombinant preparations, and live attenuated formats. Inactivated vaccines contain at least three **inactivated viral strains**—viral particles that have been inactivated by heat and/or chemical treatments to render the virion noninfectious. Recombinant influenza vaccines contain purified parts of the virus. Attenuated vaccines usually contain four different **attenuated viral strains**— weakened viruses that do not cause disease in

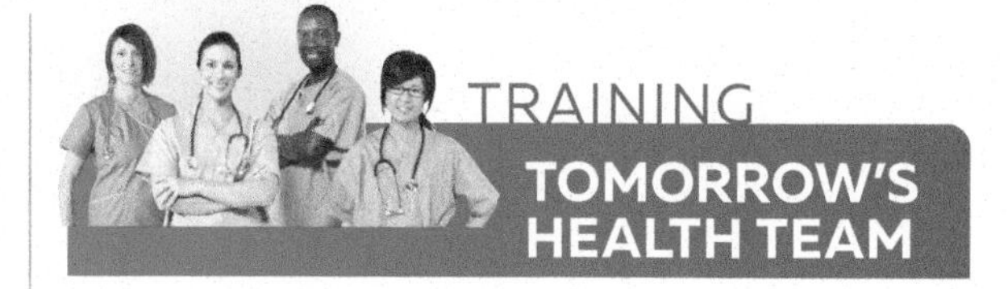

How to Fight the Flu

In the United States, four antiviral drugs are currently approved by the Food and Drug Administration (FDA) for treating influenza A and B:

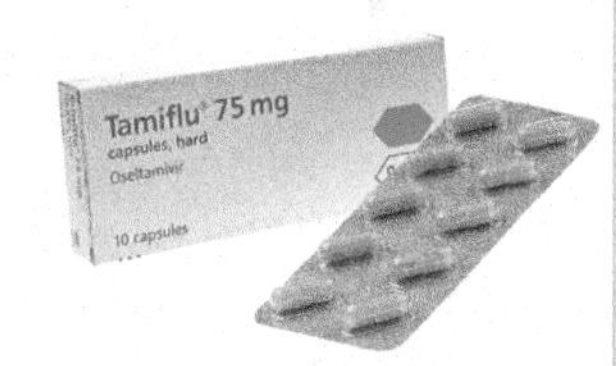

Tamiflu is one of several FDA-approved drugs that acts as a neuraminidase inhibitor.

- oseltamivir (Tamiflu®)
- zanamivir (Relenza®)
- peramivir (Rapivab®)
- baloxivir marboxil (Xofluza®)

Oseltamivir, zanamivir and peramivir all act as neuraminidase inhibitors that inhibit the release of new virions from infected host cells. Unfortunately, some influenza strains have developed resistance to drugs that target neuraminidase. In such cases, baloxivir marboxil, the newest of the anti-influenza drugs, may be useful. Baloxivir marboxil targets influenza's RNA polymerase and thereby blocks the virus from replicating its genome. These drugs are all most effective if administered within 48 hours of initial flu symptoms. However, even when they are administered early in infection, they have a nominal benefit and generally only reduce illness duration by about a day. As such, the CDC primarily recommends these antiviral drugs for patients who have a high risk for influenza-related complications (e.g., patients with heart disease, diabetes, or asthma).

QUESTION 16.2

Given the information discussed here, why is the flu vaccine still the best protection against influenza?

people with a normal immune system. (Vaccine formulations are reviewed in more detail in Chapter 14.)

From a public health standpoint, it is important to vaccinate a significant portion of the population so that **herd immunity** results. Herd immunity against influenza not only benefits at-risk groups, but it also limits circulation of the virus in humans, thereby directly reducing the opportunity for the virus to mutate and form new variants. So, getting a flu vaccine not only protects an individual, but is also a sort of community service.

Vaccine myths and reality Some people argue against getting a flu vaccine, saying they "got the flu from the vaccine." Decades of data collected from around the world show that available flu vaccine formulations are safe and effective preventive health measures that do not cause influenza. Sometimes patients complain, "I got the flu shot and still caught the flu." This scenario has more truth to it. Vaccine formulation is based on an educated guess regarding which strains will be circulating during the season; sometimes a strain begins circulating widely that wasn't included in that year's vaccination, reducing the vaccine's efficacy. Still, the vaccine is currently the best protection we have against this disease. Another consideration is that sometimes people are already infected with an influenza virus when they get vaccinated, or they may be exposed to a circulating strain before their immune system has had the chance to develop immunity from the vaccine; either scenario can lead to symptoms that the patient may incorrectly attribute to the vaccine. The CDC estimates that in seasons when the vaccine is well matched to circulating strains, the vaccine provides a 60 percent reduced risk of influenza illness across the overall population, and vaccinated patients who may develop flu tend to have only a mild case.

Novel coronaviruses include SARS-CoV-2, the cause of COVID-19.

Coronaviruses, which are a large and diverse family of enveloped RNA viruses, were initially characterized in the 1960s as causing colds, pneumonia, and bronchitis. However, during the last two decades, three novel coronaviruses that are far more dangerous than earlier described isolates have been characterized. In 2003, a new coronavirus was discovered to cause a highly infectious and dangerous respiratory illness, **severe acute respiratory syndrome (SARS)**. The SARS outbreak of 2003 was caused by **severe acute respiratory syndrome coronavirus-1 (SARS-CoV-1)**, which originated in China and rapidly spread to at least 24 other countries, infecting at least 8,000 people and killing almost 800 before it was contained by travel advisories and quarantine efforts.

Then, in 2012, **Middle East Respiratory Syndrome (MERS)** caused by another newly emerging coronavirus, **MERS-CoV**, triggered an outbreak in Saudi Arabia that spread through the Middle East. As of 2021, there have been about 2,600 confirmed MERS cases worldwide with infections detected in about 27 countries (including the United States). So far, most MERS cases diagnosed outside the Middle East have been travel related. MERS-CoV, which kills about 30 to 40 percent of those it infects, is primarily transmitted from dromedary camels to humans.[6] That said, human-to-human transmission has been documented—mainly from close contact scenarios (e.g., cases in healthcare workers who were not wearing appropriate personal protective equipment while providing care to MERS patients). Because human-to-human transmission is possible, the WHO

---

[6] This mortality rate figure is based on confirmed cases. Mortality rates may be lower; the true incidence and mortality of MERS is not known—current surveillance processes are unlikely to detect mild cases.

and other public health agencies around the world are carefully tracking MERS-CoV infections and evolution.

In December of 2019, a severe respiratory disease emerged in workers in a live-animal market of Wuhan, China. The causative agent, identified as **severe acute respiratory syndrome coronavirus-2 (SARS-CoV-2)**, is more closely related to SARS-CoV-1 than MERS-CoV. Highly contagious, SARS-CoV-2 spread globally to cause what has become known as the **coronavirus disease 2019 (COVID-19)** pandemic. As of September 2021, SARS-CoV-2 had infected over 220 million people and killed about 4.5 million.

SARS-CoV-2 Pathophysiology

SARS-CoV-2 has a single-stranded RNA genome of about 30,000 nucleotides that encodes 29 proteins. Among these proteins are RNA-dependent RNA polymerase, which copies the viral genome and is a target for the antiviral drug remdesivir, and spike proteins (S proteins) that extrude from the viral lipid envelope and help the virus infect host cells.

The virus spreads via respiratory droplets and aerosols. Once in the host respiratory tract, it uses its spike proteins to bind to angiotensin-converting enzyme 2 (ACE2) on host cells and then gain entry via clathrin-mediated endocytosis.[7] (See Chapter 4 for information on this endocytosis mechanism.) Some of the pathophysiology seen in COVID-19 is thought to directly relate to the virus binding ACE2, thereby blocking ACE2 from performing its normal function, which is to inactivate angiotensin II (ANG II).[8] Clinical studies have shown that severely ill COVID-19 patients have higher plasma concentrations of ANG II.[9] Increased ANG II levels are associated with inflammation, elevated blood pressure, and other pathologies. Broad ACE2 expression in human tissues (e.g., kidneys, heart, blood vessels, neurons, and lungs) may explain some of the wide-ranging COVID-19 complications like kidney damage and neurological manifestations—there is evidence that SARS-CoV-2 can directly infect the kidneys and brain.[10,11]

Not everyone who is infected with SARS-CoV-2 develops symptoms, but when symptoms do develop, they tend to do so within 2 to 14 days of infection. The degree of disease severity varies greatly according to patient factors—morbidity and mortality rates are significantly higher in people over age 65, especially those with underlying conditions. However, COVID-19 has killed young and otherwise healthy patients. Most SARS-CoV-2 deaths are linked to a severe viral pneumonia that develops as a result of the patient's inflammatory immune response to the virus. More specifically, COVID-19 kills patients by inducing a rapidly progressing form of inflammatory lung injury that leads to a sudden and marked reduction in blood oxygenation—a situation known as **acute respiratory distress syndrome (ARDS)**. Chest X-rays or CT imaging of patients experiencing ARDS show shadowy areas in both lungs (bilateral pulmonary infiltrates) that are not fully attributable to cardiac failure or intravenous fluid overload. The Disease Snapshot on COVID-19 further reviews disease features.

[7] Bayati, A., Kumar, R., Francis, V., et al. (2021). SARS-CoV-2 infects cells after viral entry via clathrin-mediated endocytosis. *Journal of Biological Chemistry*, 296(100306).

[8] Trougakos, I. P., Stamatelopoulos, K., Terpos, E., et al. (2021). Insights to SARS-CoV-2 life cycle, pathophysiology, and rationalized treatments that target COVID-19 clinical complications. *Journal of Biomedical Science*, 28(1), 1–18.

[9] Wu, Z., Hu, R., Zhang, C., et al. (2020). Elevation of plasma angiotensin II level is a potential pathogenesis for the critically ill COVID-19 patients. *Critical Care*, 24(1–3), 290.

[10] Wang, M., Xiong, H., Chen, H., et. al. (2020). Renal injury by SARS-CoV-2 infection: A systematic review. *Kidney Diseases*, 1–11.

[11] Song, E., Zhang, C., Israelow, B., et al. (2021). Neuroinvasion of SARS-CoV-2 in human and mouse brain. *Journal of Experimental Medicine*, 218(3), e20202135.

Disease**Snapshot** • **Coronavirus Disease (COVID-19)**

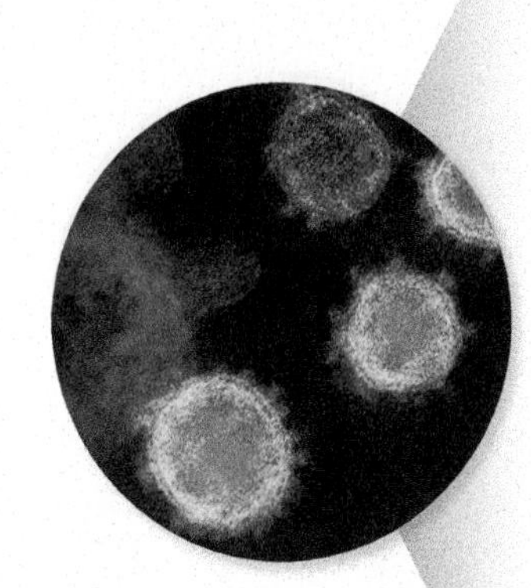

SARS-CoV-2 virions

Causative agent	Severe acute respiratory syndrome coronavirus-2 (SARS-CoV-2): enveloped, single-stranded RNA genome, *Coronaviridae* family
Epidemiology	Originated in Wuhan China late 2019; affects people of all ages, though children are less likely to become severely ill; case fatality rate varies, with older patients and patients with underlying conditions having a higher mortality rate—worldwide the average case fatality rate is about 2%; incubation period of 2–14 days; percentage of asymptomatic cases varies greatly by age and other patient factors, but pooled data suggest that about 16% of infections are asymptomatic*
Transmission & prevention	Transmitted by respiratory droplets and aerosols; COVID-19 vaccines are available as an effective prevention against severe COVID-19 (see Chapter 14 for details)
Signs & symptoms	Characterized by flu-like signs and symptoms, including one or more of the following: fever, chills, cough, shortness of breath, body aches, fatigue, headache, loss of smell (anosmia), sore throat, nausea or vomiting, diarrhea, congestion or runny nose. Emergency care should be sought if any of the following develop: trouble breathing, persistent pain or pressure in the chest, difficulty with waking or staying awake, confusion (new or abnormal for the patient), pale/gray/or blue-colored skin, lips, or nail beds
Pathogenesis & virulence factors	Can trigger an exaggerated inflammatory immune response and cytokine storm; broad tissue tropism can lead to systemic failure likely because of dysregulated ACE2/ANGII function
Diagnosis & treatment	RT-PCR to detect viral RNA remains the gold standard for diagnosis; rapid tests for viral antigen detection are also available; remdesivir (inhibitor of viral RNA polymerase) is FDA-approved for treating hospitalized COVID-19 patients ages 12 and older who weigh at least 40 kg; dexamethasone may also decrease mortality, with the largest benefit seen among patients receiving invasive mechanical ventilation

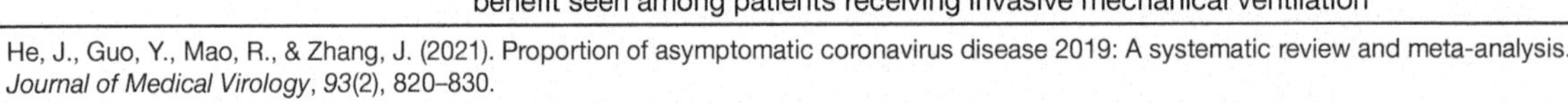

He, J., Guo, Y., Mao, R., & Zhang, J. (2021). Proportion of asymptomatic coronavirus disease 2019: A systematic review and meta-analysis. *Journal of Medical Virology*, *93*(2), 820–830.

*Based on 50,155 confirmed COVID-19 patients from 41 studies. The percentage of asymptomatic cases varies greatly by age and even by study. Based on pooled data, about 27% of confirmed COVID-19 cases in children were described as asymptomatic.

SARS-CoV-2 Variants and Strains

As mentioned in Chapter 6, because RNA polymerases do not have the same proofreading capabilities as DNA polymerase, RNA viruses tend to mutate more frequently than DNA viruses. Therefore, it is not at all surprising that new variants and strains of SARS-CoV-2 are emerging. A **viral variant** is an isolate of a virus that has at least one mutation that makes it genetically distinct from the original form of the virus. The mutation (or mutations) may not change how the virus behaves; however, if it does alter viral functions, the variant is more specifically considered a new **viral strain**. For example, the SARS-CoV-2 Delta variant (B.1.617.2 variant) has 14 mutations, three of which (a deletion and two amino acid substitutions) affect how the viral spike protein binds to ACE2. Compared to earlier forms of the virus, the Delta variant is more easily transmitted and may cause more severe disease.[12] Because it has altered features that trigger physiological consequences, the Delta variant is more specifically considered a new strain of SARS-CoV-2; however, to avoid public confusion, it is usually referred to using the less specific term, variant.[13]

Any active infection represents an opportunity for the virus to mutate into a new variant—this is one reason why even people who are not at high risk for severe COVID-19 should be vaccinated. Vaccines decrease the pool of susceptible hosts, lowering the incidence of active infections and thereby slowing the emergence of viral variants. The CDC is carefully monitoring the emergence of

[12] Faria, N. R., Claro, I. M., Candido, D., et al. (2021). Genomic characterisation of an emergent SARS-CoV-2 lineage in Manaus: Preliminary findings. *Virological*.

[13] Just as ice cream is a dessert, but not all desserts are ice cream, all strains are variants, but not all variants are strains. Using the more general term *variant* when one is discussing a strain is not technically incorrect, though it is less specific/descriptive.

TABLE 16.1 Categorical Classification of SARS-CoV-2 Variants

Category	Examples of Variant's Attributes	Examples of Public Health Actions
Variants of Interest	• Predicted to affect transmission, reduce efficacy of diagnostics, therapeutics, or promote evasion of vaccine-induced immunity • Linked to increased number of cases/new outbreak clusters • Limited circulation	• Enhanced sequence surveillance • Enhanced laboratory characterization • Continued evaluation of transmission features, efficacy of existing vaccines and therapeutics against the variant, and disease severity linked to the variant
Variants of Concern	• Some evidence of reduced efficacy of existing diagnostics, treatments, and/or vaccines • Some evidence of increased transmissibility • Some evidence of increased disease severity	• Notify WHO • Enhanced testing for the variant to facilitate tracking and reporting of cases caused by the variant to control spread • Additional research to assess efficacy of existing vaccines and therapeutics against the variant • Research and develop new vaccines and therapeutics that are effective against the variant
Variants of High Consequence	• Significant reduction in efficacy of existing diagnostics, treatments, and/or vaccines • Clear evidence of increased transmissibility • Clear evidence of increased disease severity	• Notify WHO • Announce strategies to prevent or contain transmission • Recommend treatment and vaccine updates

new variants. Emerging SARS-CoV-2 variants are categorized by the risk they present to public health as variants of interest, concern, or high consequence. TABLE 16.1 summarizes the features of these three categories.

As the pandemic ebbs, cases of COVID-19 are unlikely to entirely disappear—instead, SARS-CoV-2 is likely to become an endemic virus that will continue to circulate among humans in a seasonal pattern that mirrors what is seen for influenza viruses. We can anticipate that new SARS-CoV-2 strains will continue to evolve, and virus tracking efforts already used to monitor new influenza virus strains will be similarly applied to monitor SARS-CoV-2 evolution and inform future vaccine formulations.

Hantavirus pulmonary syndrome is a rare but dangerous illness.

There are currently over 40 known viruses assigned to the genus *Hantavirus*. Some hantaviruses, called New World hantaviruses, cause **hanta pulmonary syndrome (HPS)**. HPS is an acute and potentially fatal respiratory illness that humans can acquire upon inhaling airborne dust particles that contain hantavirus particles (sometimes called Sin Nombre virus) shed in rodent urine or feces. Other hantaviruses, called Old World hantaviruses, cause hemorrhagic fever with renal syndrome (HFRS). (Because it impacts the kidneys, HFRS and Old World hantaviruses are covered in more detail in Chapter 20.)

About 25 different New World hantaviruses located throughout North and South America can cause HPS. In 2012, there was an outbreak of the virus among visitors to Yosemite National Park in California; ultimately about 10 people, who all stayed in the same tent-cabin area, became infected. The infectious dose and infectivity rate for New World hantaviruses are not known, although the fact that up to 10,000 people stayed in the tent-cabins at Yosemite that season, yet only 10 developed HPS, suggests that exposure may require certain host factors and/or extensive exposure to the virus for illness to develop (FIG. 16.6).

Most people who develop HPS were exposed to rodent excrement when they entered or cleaned areas that had not been disturbed for some time (sheds, storage facilities, barns, or other enclosed areas where rodents may nest) or they encountered rodents in the wild (camping or in other rustic settings like cabins). Different rodent species can carry hantaviruses; in the United States the most common rodent carrier is the deer mouse.

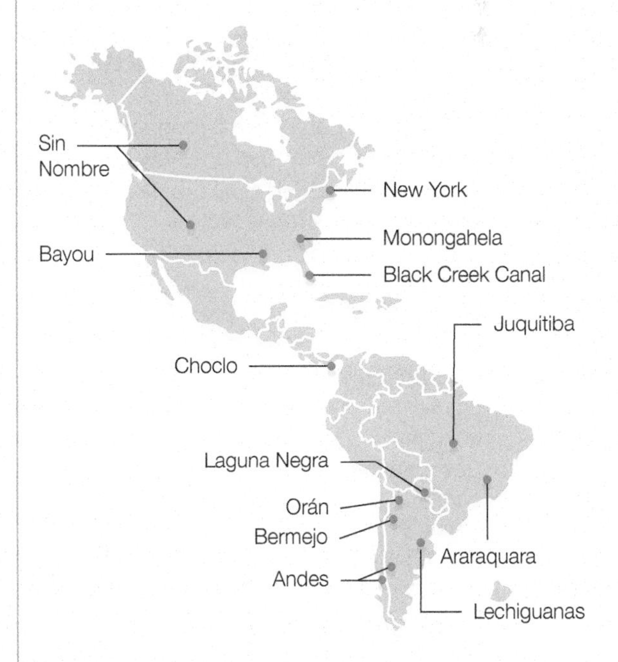

FIGURE 16.6 **Pathogenic New World hantaviruses found in the Americas**

According to the CDC, between 30 and 40 percent of people who develop HPS in the United States do not survive. Diagnosis is confirmed by detecting patient antibodies to the virus, or by confirming the presence of viral RNA or antigens in patient blood or tissues. Initial symptoms often lead to **pulmonary edema** (fluid accumulation in the lungs), which has a high mortality rate. There are no antiviral drugs to treat HPS, nor is there a vaccine to prevent it. These factors, combined with recent evidence that certain Andes virus strains found in South America are not just zoonotic, but can also be transmitted from person to person, make the hantaviruses important emerging pathogens to monitor.

Build Your Foundation

5. Why is it challenging to differentiate colds from influenza? (LO 16.5)
6. What are the primary features that differentiate respiratory syncytial virus, human parainfluenza viruses, and adenoviruses? (LO 16.6)
7. Explain the consequences of antigenic drift and antigenic shift on viral evolution and pandemics (especially regarding the 1918 influenza pandemic). (LO 16.7)
8. Name three characteristics of SARS-CoV-2 and three features of the disease it causes, COVID-19. (16.8)
9. When would the term *viral variant* be used over the term *viral strain*? (16.9)
10. What category of variant would an isolate of SARS-CoV-2 fall into if it exhibited widespread evasion of existing vaccinations? (16.9)
11. What is the carrier for hantavirus, and how can people decrease the risks of exposure? (16.10)

Build Your Foundation (BYF) Quick Quiz: Visit the **Mastering Microbiology** Study Area to quiz yourself.

16.3 BACTERIAL INFECTIONS OF THE RESPIRATORY SYSTEM

Learning Outcomes

After reading this section, you should be able to:

16.11 Describe why children are prone to developing otitis media as a complication of colds.

16.12 Give examples of diseases caused by *Streptococcus pyogenes*.

16.13 Describe the pathological progression of diphtheria and how it is prevented.

16.14 Describe the pathological progression of pertussis, and discuss factors contributing to its reemergence.

16.15 Name and describe the causative agent of tuberculosis and the forms of the disease.

16.16 Give examples of organisms that cause typical pneumonia, and describe their effects.

16.17 Describe the features of the most common causes of atypical bacterial pneumonia: *Chlamydia pneumonia*, *Mycoplasma pneumoniae*, *Legionella pneumophila, Chlamydophila psittaci, Coxiella burnetii*, and *Francisella tularensis.*

Viruses are the most common infectors of the respiratory tract, but bacteria are also important causes of respiratory tract infections. Bacterial infections are also a common secondary complication of the viral infections we discussed in the previous section. Unlike viral infections, where drug therapies are limited and minimally effective, most bacterial infections respond well to antibiotic therapies.

Otitis media is a common bacterial complication of colds.

During a cold, inflamed membranes can cause mucus to accumulate in the respiratory tract, providing an ideal environment for bacteria to grow. In children, **otitis media**, or a middle ear infection, is a frequent example of such an infection. The middle ear is connected to the pharynx by a small hollow structure called the eustachian tube. Although the ear is not considered a part of the respiratory system, this physical connection to the respiratory system allows for pressure to be equalized between the tympanic membrane (eardrum) and the environment.

By age 2 at least 80 percent of children have experienced otitis media.[14] Children are especially susceptible to ear infections because their eustachian tubes are shorter, narrower, and oriented such that draining into the pharynx is less efficient, even when they are not suffering from a cold. During a cold, the eustachian tubes can become inflamed, further blocking mucus drainage and setting the stage for a bacterial infection of the middle ear. The most common bacteria that cause pediatric otitis media are *Streptococcus pneumoniae* (mainly

[14] Wilson, M., & Wilson, P. J. (2021). Middle ear infections. In *Close encounters of the microbial kind* (pp. 233–242). Cham: Springer.

Disease**Snapshot** • **Otitis Media**

Causative agent	Most commonly *Streptococcus pneumoniae* (primarily types not included in the pneumococcal conjugate vaccine), *Moraxella catarrhalis*, and nontypable *Haemophilus influenzae*
Epidemiology	Widespread; 70% of children in the U.S. have a middle ear infection by age 3; common complication of colds and allergies; often develops 2–7 days after a cold
Transmission & prevention	Noncommunicable; vaccination with pneumococcal conjugate vaccine reduces occurrence due to specific bacteria; treating colds with decongestants may reduce mucus buildup in the eustachian tubes to reduce infection incidence
Signs & symptoms	May include earache, redness of the eardrum, leakage of pus or blood from the ear, and fever
Pathogenesis & virulence factors	Accumulation of mucus in the eustachian tube serves as a breeding ground for bacteria; some causative bacteria have capsules and/or fimbriae to help establish infection; disruption of normal barriers such as earwax removal through ear cleaning and tissue trauma (such as a ruptured eardrum) may promote otitis media
Diagnosis & treatment	Symptomatic diagnosis includes a swelling or effusion from the tympanic membrane combined with visible membrane redness and earache; amoxicillin is a drug of choice for treatment; in minor cases, decongestants to drain the ear canal and pain management are often recommended over antibiotics

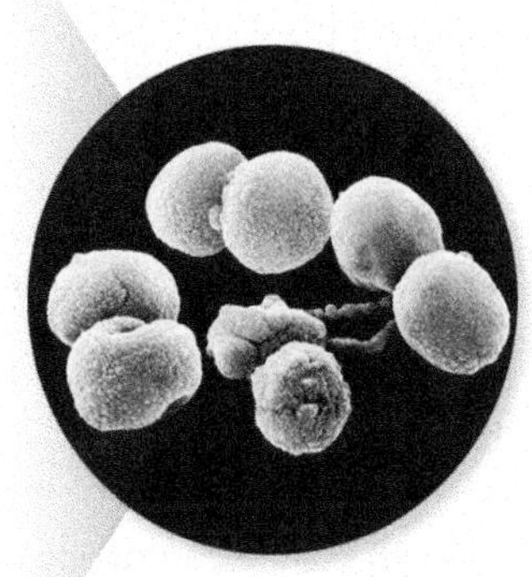

S. pneumoniae

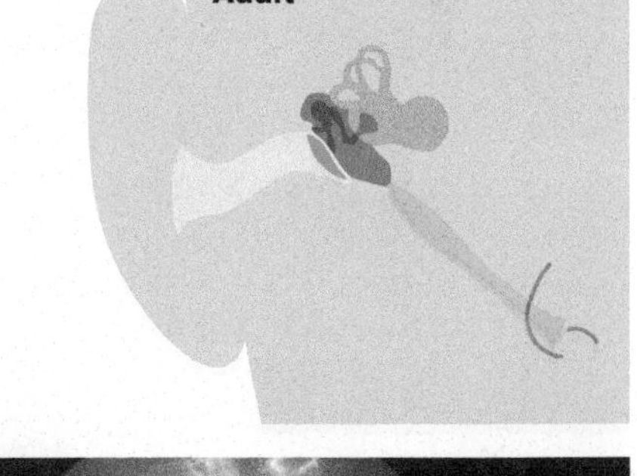

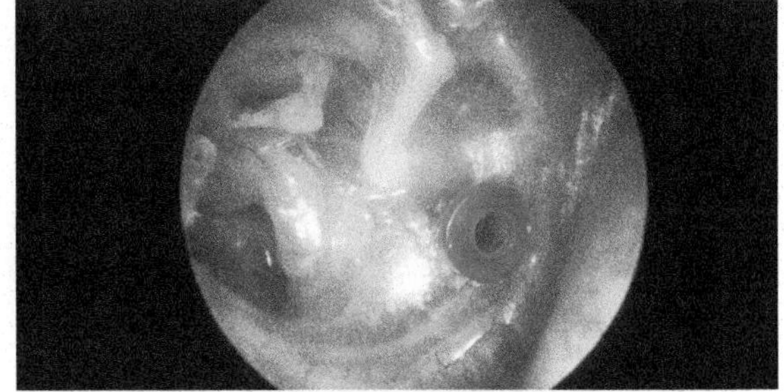

Tube inserted in child's tympanic membrane to lower incidence of infection.

FIGURE 16.7 **Draining fluid from the middle ear** Compared to adults, the eustachian tube in children is narrower and more horizontal, making fluid drainage from the middle ear to the throat less efficient. This increases the risk of middle ear infection in children. Temporary tubes can be inserted into the eardrum to drain fluid from the middle ear, which eases painful ear pressure and decreases infections.

Critical Thinking *Why do you think sounds may seem muted or muffled when you have a cold?*

types not included in the pneumococcal conjugate vaccine), *Moraxella catarrhalis*, and nontypable *Haemophilus influenzae*.[15] Some children experience repeated ear infections because of poor eustachian tube draining. In such cases, tubes may be surgically inserted into the tympanic membrane (eardrum) to help fluid drain and lower the incidence of infection (FIG. 16.7).

Streptococcus pyogenes primarily causes strep throat.

The genus *Streptococcus* includes over 40 species of Gram-positive cocci that grow in chains. In humans, streptococci are often found as normal microbiota of the upper respiratory tract and skin, but species in this genus can also cause endocarditis, pharyngitis (strep throat), meningitis, infections of the genitourinary system, sepsis, dental caries (cavities), skin infections, and pneumonia.

Most pharyngitis cases are viral, but when bacterial pharyngitis occurs, it is often **streptococcal pharyngitis (strep throat)** caused by the **group A streptococcus (GAS)**, *S. pyogenes*. (Refer to Chapter 7 for the Lancefield grouping of streptococcal bacteria.) According to the CDC, this bacterium causes up to 15 percent of acute pharyngitis cases in adults and up to 30 percent of such cases in children. However, *S. pyogenes* also can cause otitis, sinusitis, invasive pneumonia, joint or bone infections, meningitis, endocarditis, impetigo, necrotizing fasciitis, and streptococcal toxic shock syndrome. Most streptococcal pharyngitis cases occur in the winter and early spring. Humans are a natural reservoir for *S. pyogenes*, and many people are asymptomatic carriers. The bacterium is primarily transmitted by respiratory droplets, and prevention is mainly through hand washing

[15] Hu, Y. L., Lee, P. I., Hsueh, P. R., et al. (2021). Predominant role of *Haemophilus influenzae* in the association of conjunctivitis, acute otitis media and acute bacterial paranasal sinusitis in children. *Scientific Reports*, *11*(1), 1–6.

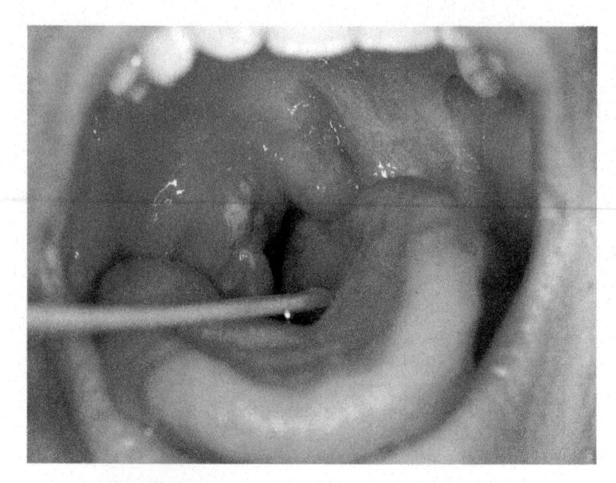

FIGURE 16.8 Streptococcal pharyngitis This infection is sometimes characterized by white patches of exudate on the tonsils.

and avoiding contact with infected patients. Typical symptoms usually include inflammation of the throat, swollen lymph nodes in the neck (cervical lymph nodes), a low-grade fever, and pus (**exudate**) in the throat or tonsils; a cough is absent in most cases (FIG. 16.8). Diagnosis is confirmed by a rapid strep test for bacterial antigens and/or a culture that identifies S. *pyogenes*. The main treatment is penicillin-based drugs. Patients who are allergic to penicillin can be given macrolide drugs such as erythromycin and azithromycin or cephalexin (a cephalosporin drug).

Virulence and Pathogenesis Factors

S. *pyogenes* has a multitude of factors that make it a dangerous pathogen, including at least 9 superantigens and 11 different adhesion factors. S. *pyogenes'* hyaluronic acid capsule resembles human connective tissue components, which helps the bacteria avoid immune detection. Also, various bacterial proteins and enzymes allow it to dodge complement cascades and phagocytosis. (See Chapter 11 for more on complement cascades and phagocytosis.) The M protein of S. *pyogenes* is a particularly important virulence factor that helps with adhesion to host cells and avoiding phagocytosis. Once the bacterium has avoided the immune system and successfully attached to host cells, it is then ready to invade tissues.

Scarlet Fever

As a result of infection by specific bacteriophages, some S. *pyogenes* strains have gained the ability to produce an erythrogenic toxin. (Bacteriophages and their ability to impart new features to bacteria are discussed in Chapter 6.) **Scarlet fever** is one well-known disease caused by such erythrogenic toxin-producing strains. Its easily recognizable clinical signs include a red sandpaper-like rash and a reddened tongue that resembles the surface of a strawberry (FIG. 16.9). The scarlet-colored rash is attributed to the toxin's ability to stimulate an inflammatory response in the host and dilation of capillaries. The rash begins on the face and neck and works its way down the body. After the rash resolves, the skin is sloughed off. Scarlet fever is fairly rare and occurs in less than 10 percent of streptococcal pharyngitis cases. It is most common in children under 10 years old.

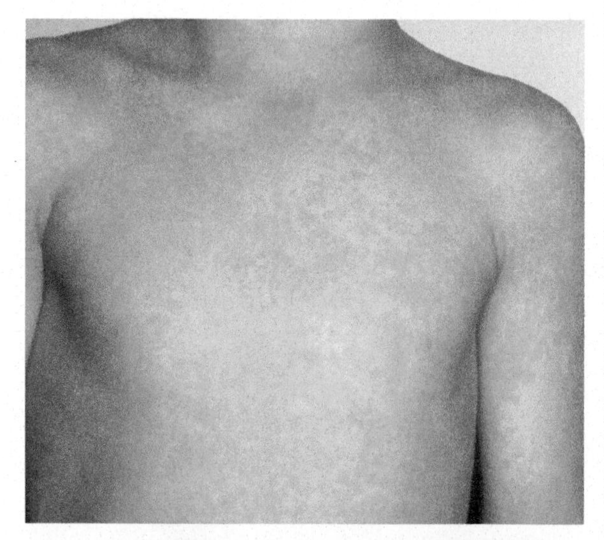

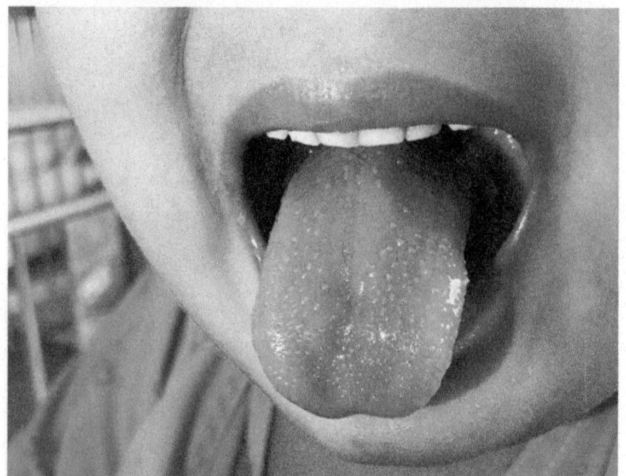

FIGURE 16.9 Scarlet fever rash and "strawberry tongue" A sandpaper-like rash begins on the face and neck and works its way down the body. After the rash resolves, the skin is sloughed off. "Strawberry tongue" is another manifestation of scarlet fever.

Autoimmune Complications of S. *pyogenes* Infection

There are more than 100 different strains of S. *pyogenes*. One way these strains are characterized is by a cell surface virulence factor M protein. Certain M proteins look a lot like proteins found in our heart valves, parts of the kidney, and in our joints. M proteins that resemble our own proteins can trigger the body to make antibodies that attack S. *pyogenes*, but unfortunately may also cross-react with our own proteins (FIG. 16.10). M proteins that can stimulate autoimmune complications are said to be rheumatogenic. The current understanding is that cross-reactive antibodies (antibodies that react to both the bacterial and human proteins) may be more likely to develop when a patient's case of strep throat is untreated. Cross-reactive antibodies are characteristic of rheumatic fever and can cause inflammation of various tissues to include the joints (arthritis), nervous system, heart (carditis), and skin. Young people in developing nations are more likely to be affected. It is estimated that as of 2021, at least 40 million people worldwide live with rheumatic fever-induced heart complications—also known as rheumatic heart disease.[16]

[16] Beaton, A., Kamalembo, F. B., Dale, J., et al. (2020). The American Heart Association's Call to Action for Reducing the Global Burden of Rheumatic Heart Disease: A Policy Statement from the American Heart Association. *Circulation*, 142(20), e358–e368.

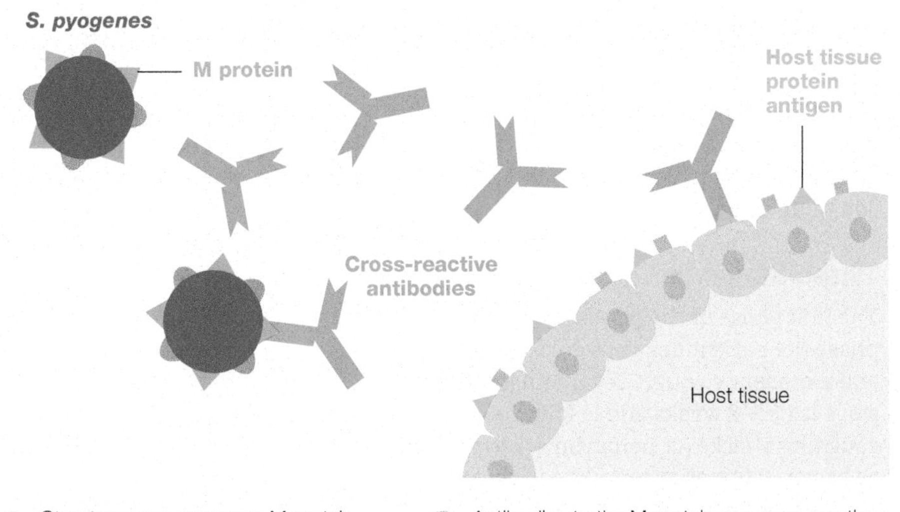

1 *Streptococcus pyogenes* M protein stimulates antibody production in the host.

2 Antibodies to the M protein are cross-reactive, meaning they bind to host tissues bearing proteins that resemble the M protein. These antibodies recruit immune factors that then attack the tagged host tissues.

FIGURE 16.10 M protein of *Streptococcus pyogenes* stimulating an autoimmune reaction M proteins of *S. pyogenes* stimulate the production of antibodies that cross-react with host tissues.

Critical Thinking *Why does prompt antibiotic treatment for an S. pyogenes infection limit autoimmune complications?*

Corynebacterium diphtheriae causes diphtheria.

The Gram-positive rod *Corynebacterium diphtheriae* causes **diphtheria**, a bacterial illness that can affect the upper respiratory tract or the skin (cutaneous diphtheria). The respiratory form of the illness is transmitted by respiratory droplets. It most commonly affects children under age 5 and causes cold-like symptoms that are accompanied by a sore throat and a low-grade fever. Often, cough and hoarseness develop. In acute disease, the neck region swells, which is often described as a "bull neck" (FIG. 16.11 *top*). Untreated, this disease can kill its victims in a few days. Historically, this much-feared illness had a variety of names such as "the putrid sore throat" and "the strangling angel of children."

The bacteria usually remain localized to the upper respiratory tract, so the broader pathophysiology that develops during infection is mainly attributed to the diphtheria toxin—a dangerous type A-B exotoxin that can enter the bloodstream. The toxin inhibits protein synthesis thereby limiting the ability of host cells to make new proteins. (See Chapter 10 for more on A-B toxins.) A patient who is suffering from diphtheria exotoxin exposure and is not treated usually develops a rapid pulse, paleness (pallor), progressive weakness, loss of consciousness, and coma, and dies within 6–10 days in about 10 percent of treated cases and in up to 50 percent of untreated cases. The toxin also kills cells in the upper respiratory tract, thereby contributing to the formation of a thickened leathery structure called a pseudomembrane in the upper airway (usually on the tonsils and throat) (FIG. 16.11 *bottom*). Treatment using antibiotics and an antiserum that neutralizes the exotoxin is usually started before a definitive culture diagnosis.

In the 1920s, a vaccine was made with inactivated diphtheria toxin (the diphtheria toxoid vaccine). In the United States, where most children are routinely vaccinated with the DTaP vaccine (diphtheria and tetanus toxoids and acellular pertussis vaccine), diphtheria is rare. However, there are still cases of this disease in developing nations, almost exclusively in unimmunized children. Even though diphtheria is rare in developed countries, it is still important to have children vaccinated because this bacterium continues to circulate worldwide in asymptomatic carriers in the population.

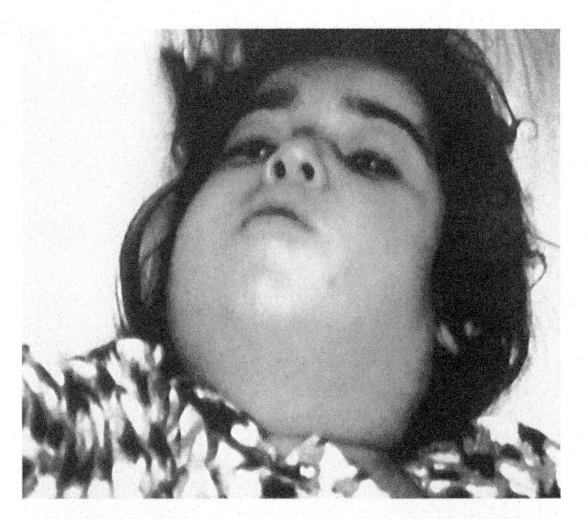

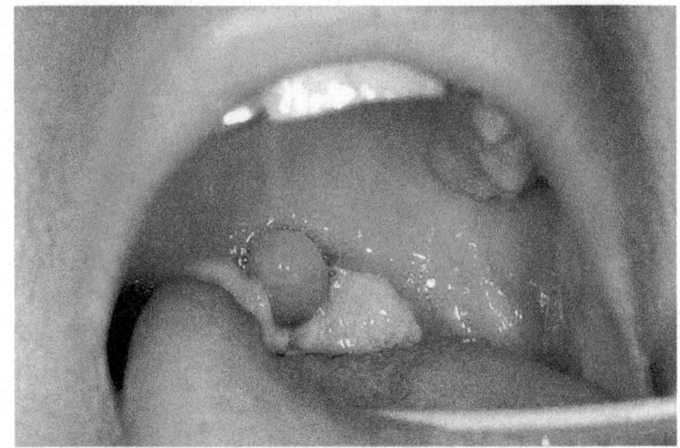

FIGURE 16.11 Key signs of diphtheria *Top*: The distinct swelling of the neck that can accompany diphtheria is called a "bull neck." *Bottom*: Diphtheria is also characterized by the formation of a leathery pseudomembrane in the upper airway.

Pertussis (whooping cough) is an acute infection of the respiratory tract.

Pertussis (whooping cough) is caused by the Gram-negative bacterium *Bordetella pertussis*. This disease is commonly referred to as whooping cough, due to the sound patients make as they try to catch a breath between long and intense coughing attacks.

Three Stages of Pertussis

The first phase of the disease is called the catarrhal phase. During the catarrhal phase the patient develops cold-like symptoms such as a runny nose, watery eyes, and a moderate cough. Fever and a sore throat are usually absent. The catarrhal stage lasts 1–2 weeks and is followed by the paroxysmal stage, where severe coughing attacks (or **paroxysms**) develop, and the disease continues to intensify as bacterial toxins accumulate in the patient's respiratory system. One of these, the pertussis toxin, is a type A-B toxin that triggers a generalized inflammation in the respiratory tract. It also helps the bacteria avoid the patient's immune response. (See Chapter 10 for more on A-B toxins.) The coughing fits can be so severe that they may cause vomiting, fractured ribs, and loss of bladder control. In children and babies, bleeding behind the eyes and in the brain can also occur.

After 2 to 6 weeks of the exhausting paroxysmal stage, the patient enters the **convalescent** stage. The convalescent stage lasts for about a month and is characterized by less frequent coughing spells. The entire course of pertussis is prolonged and typically lasts several months, which is why the Chinese name for pertussis translates to "100-day cough."

Reemergence

In 2012, the United States had the highest number of pertussis cases since 1955. Pertussis is one of the few vaccine-preventable diseases that is currently on the rise, especially in adolescents and adults (FIG. 16.12). Two key factors seem to be contributing the rise in pertussis incidence: waning immunity and bacterial evolution.

- **Waning Immunity.** Immunity from the routine childhood DTaP vaccine (which protects against diphtheria, tetanus, and pertussis) is not long lived; by age 12 the immunity originally conferred by the DTaP vaccination series declines. To renew immunity to pertussis, the CDC recommends that adolescents and adults get a booster shot (the Tdap). Adolescents should receive a single dose of Tdap at age 11 or 12. Adults should get the booster vaccine at least every 10 years (although a booster may be medically necessary more frequently). Unfortunately, Tdap booster vaccination rates are

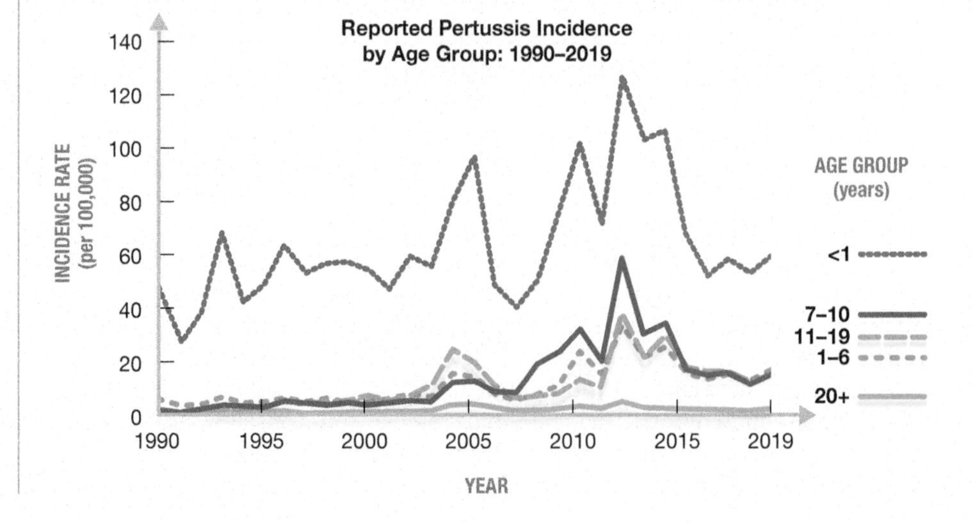

FIGURE 16.12 Curbing pertussis reemergence with booster vaccines As illustrated by this graph, the incidence of pertussis cases has increased in the past decade. To prevent this reemergence, adolescents and adults should get a booster pertussis vaccine (Tdap booster).

Critical Thinking *How do you think a reemergence of pertussis in adolescents and adults could affect the prevalence of the disease in infants under 1 year old?*

DiseaseSnapshot • Pertussis (Whooping Cough)

Causative agent	*Bordetella pertussis*; Gram-negative bacterium
Epidemiology	Main reservoir is adults and adolescents with atypical or undiagnosed infection; highly contagious (infects 80–90% of susceptible hosts); found worldwide; reemergence due to need for a booster vaccine and newly emerging *B. pertussis* strains; about 1.7 million cases per year worldwide (most are not diagnosed) with about 89,000 deaths (WHO statistic from 2019; most dangerous in infants; 7–10 days average incubation period
Transmission & prevention	Transmitted by respiratory droplets or direct contact; prevented by the DTaP vaccine (booster vaccine recommended for adolescents, adults, and pregnant women); to limit spread, antibiotics may be given to those with close/regular contact with a diagnosed case
Signs & symptoms	Catarrhal phase: cold-like symptoms Paroxysmal stage: intense coughing fits that leave patient gasping for air and can also induce convulsions Convalescent stage: lasts up to 3 months and is characterized by less frequent coughing spells
Pathogenesis & virulence factors	Tracheal cytotoxin damages the ciliated cells of the mucociliary escalator, which causes mucus to accumulate and settle in the lungs; pertussis toxin damages host airway cells and stimulates an inflammatory response
Diagnosis & treatment	Bacterial culture from a nasopharyngeal sample is the best diagnostic technique; serological tests to detect patient antibodies to *B. pertussis* and screening patient samples for bacterial DNA by PCR techniques are also used; treated with macrolide antibiotics (i.e., erythromycin, clarithromycin, or azithromycin)

B. pertussis

low among adults—according to the CDC only 8 to 28 percent of adults meet the recommended Tdap booster criteria. Because the most common and serious pertussis cases are in infants under 1 year old, it is especially important for people who are around infants (e.g., childcare workers, healthcare providers, and parents) to have their Tdap booster so that they do not develop pertussis and inadvertently spread it to a baby. To further protect infants, the CDC recommends that all pregnant women get vaccinated during each pregnancy (even if pregnancies are only a year or two apart). Vaccinating against pertussis during pregnancy is safe, and it enhances the level of antibodies that the mother can pass to the baby to better provide a natural passive immunity for the infant until the child is vaccinated against pertussis.

- **Bacterial Evolution.** The pertussis vaccine currently used in most developed nations is a subunit vaccine that contains purified inactivated pertussis toxin alone or in combination with other purified *B. pertussis* components. Unfortunately, there are newly emerging *B. pertussis* strains that have altered certain features and can thereby evade the immunity that our current subunit vaccines trigger. Consequently, next-generation pertussis vaccines may be needed to protect against newer *B. pertussis* strains.[17]

Tuberculosis (TB) is one of the top infectious disease killers in the world.

Tuberculosis (TB) is an ancient infectious disease that historically was known as "consumption" because of the increasingly thin and wasted appearance of victims in the terminal stage of the disease. TB is found throughout the world, but is especially prevalent in Latin America, Africa, eastern Europe, and throughout Asia. Until COVID-19, worldwide TB was the leading cause of death from an infectious agent. (Of course, TB may reclaim its standing as the leading cause of mortality from an infectious agent as COVID-19 vaccines are more widely applied.) According to the WHO, TB killed about 1.4 million people in

[17] Dewan, K. K., Linz, B., DeRocco, S. E., et al. (2020). Acellular pertussis vaccine components: Today and tomorrow. *Vaccines*, *8*(2), 217.

Disease**Snapshot** • Tuberculosis

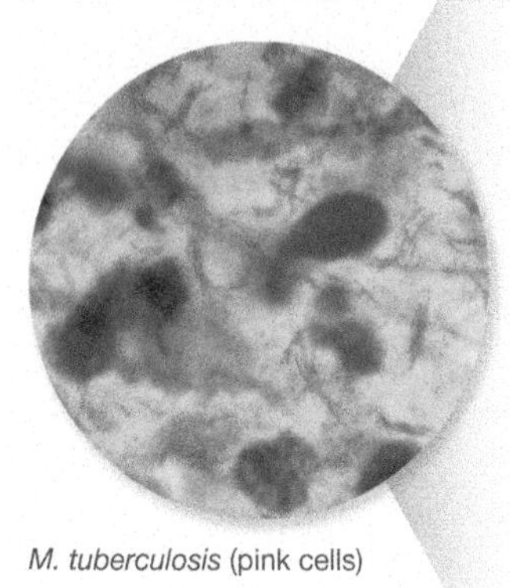

M. tuberculosis (pink cells)

Causative agent	Primarily *Mycobacterium tuberculosis*; acid-fast, Gram-positive bacterium
Epidemiology	Worldwide incidence, with Africa and Asia being the most affected; reservoir is humans and diseased animals; 4- to 8-week incubation period with 90–95% of infection events leading to latent TB; about 5% of these latent infections progress to active TB within 5 years; another 5% will progress to an active infection within the patient's lifetime
Transmission & prevention	Transmitted by respiratory droplets; highly communicable; prevented by the bacillus Calmette–Guérin (BCG) vaccine
Signs & symptoms	Active TB is characterized by cough (sometimes with blood-tinged sputum), fever, night sweats, fatigue, and weight loss
Pathogenesis & virulence factors	Specialized waxy cell wall makes these bacteria resistant to immune system defenses and protects the bacteria against antibiotics; bacteria avoid destruction within macrophages and can escape the phagolysosome and grow within macrophages
Diagnosis & treatment	Diagnostic tools include: the Mantoux tuberculin skin test; fluorescent microscopy or acid-fast staining to identify mycobacteria in sputum smears; culture techniques; Xpert MTB/RIF test; and immune response evaluation using interferon gamma (IGRA). Latent TB is usually treated with isoniazid and/or rifampin for 3–9 months; active TB is often treated with a combination of four drugs: isoniazid, rifampin, pyrazinamide, and ethambutol (or streptomycin) for the first 2 months, followed by isoniazid and rifampin for the next 4–7 months; treatment for resistant strains is much longer and varied based on drug-resistance profiles of the bacterium. Sirturo (bedaquiline) may be used as a part of combination therapy to treat MDR and XDR TB

2019. The good news is that global efforts to reduce TB cases are working—TB incidence is on average decreasing by about 2 percent per year.

TB is usually caused by *Mycobacterium tuberculosis* (also called tubercle bacillus). However, before pasteurization was used to make milk products safer, *Mycobacterium bovis* was also a common cause of tuberculosis and is still responsible for a small percentage of cases in developing nations. *Mycobacterium avium* causes tuberculosis in birds and can occasionally be transmitted from birds to people. *M. avium* is not normally a human pathogen, but when it does cause human illness, it is usually in HIV patients.

Not everyone infected with *M. tuberculosis* gets sick; most people exposed to the tubercle bacteria only develop a **latent infection**, which means they don't have symptoms of TB and are not contagious. Most latent TB cases never progress to an active case; however, some people—especially those with compromised immunity—do progress to active infection. In active pulmonary tuberculosis, the patient develops symptoms that include a cough (sometimes with blood-tinged sputum), fever, night sweats, fatigue, and weight loss. Patients with active TB can transmit the disease to others. FIG. 16.13 compares latent and active tuberculosis.

FIGURE 16.13 Latent tuberculosis (TB) progression to active infection

Critical Thinking *Why do you think the treatment for latent TB is so prolonged? (Hint: Consider the structure of the granuloma and the nature of how mycobacteria grow.)*

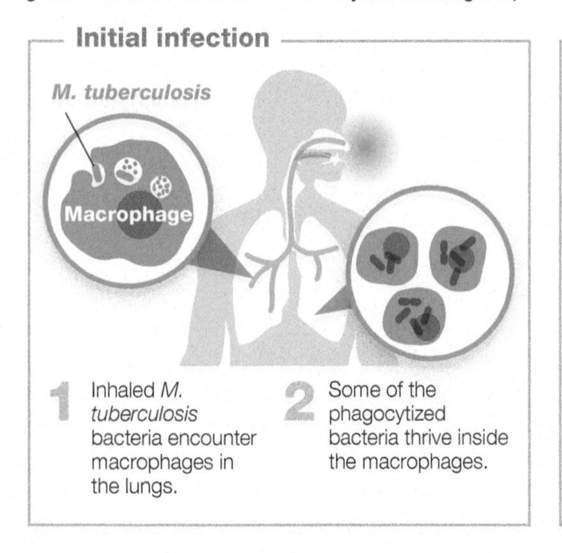

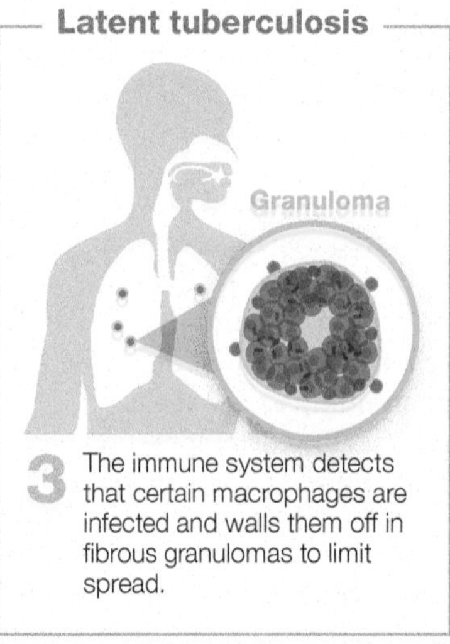

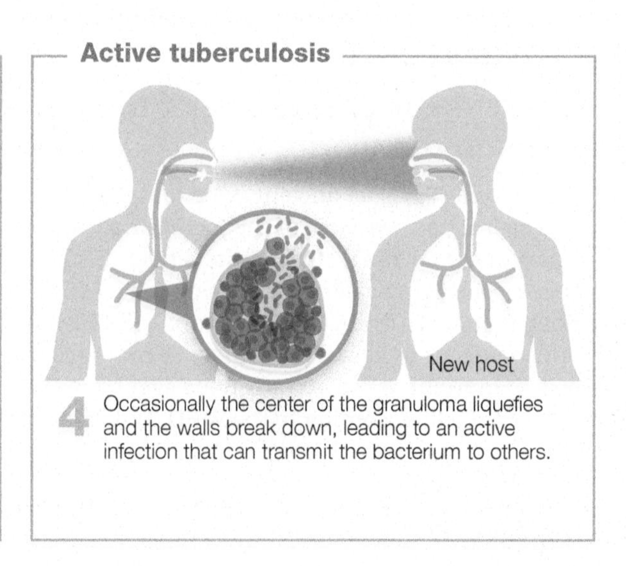

Vaccination and Screening

In over 157 countries, the live-attenuated bacillus Calmette–Guérin (BCG) vaccine is routinely given to children who are HIV negative and have a negative tuberculin skin test. (See Chapter 14 on live-attenuated vaccines.) However, there is wide variation in how well the vaccine protects adults, so it is only recommended for people at high risk for TB exposure (such as people who live in the many places where TB is endemic). In areas such as the United States, where people are not routinely vaccinated against TB, the tuberculin skin test is regularly used to screen at-risk groups such as healthcare workers and people who travel to or have guests from areas where TB is endemic.

The Mantoux tuberculin skin test is performed by an intradermal injection of tuberculin purified protein derivative (PPD) into the skin of the forearm (FIG. 16.14). Within 48–72 hours, the injection site is inspected for **induration**, which is the development of a hardened area at the injection site. Redness is also common around the induration, although only the size of the induration weighs into the test result. The size of the induration is measured to determine if the patient has a positive or negative result. Guidelines regarding the cutoff size for a positive result depend on various patient factors. Patients who have been vaccinated against TB can have a false-positive skin test result, so they should instead be tested with molecular diagnostics like the Xpert MTB/RIF test that detects the bacterium's DNA, or by an interferon gamma release assay (IGRA), which measures the immune response against *M. tuberculosis*. The tuberculin skin test does not distinguish between an active versus a latent TB infection; it just reveals a potential infection.

If screening tests and patient history suggest a TB infection, then the next step is to determine if the person has a latent or an active infection. In such cases, an X-ray or occasionally a CT scan (computed tomography scan) is recommended to look for lung damage. In latent TB, the person does not have symptoms, and a chest X-ray will usually appear normal. In an active TB case, the radiograph may reveal lung damage.

Antibiotic Resistance

Unfortunately, some strains of *M. tuberculosis* have developed resistance to the preferred antibiotic therapies isoniazid and rifampin. Such strains are called multidrug-resistant strains (MDR TB strains). MDR TB strains can be found in all countries, but these strains are especially common in the Russian Federation, India, and China. Some MDR TB strains are further described as extensively drug-resistant (or XDR TB strains). These are rare strains that, according to the WHO's 2021 updated definition, are resistant to first-line drugs (isoniazid and rifampin) as well as any fluoroquinolone drug and at least one of three injectable second-line drugs used to treat TB (amikacin, capreomycin, or kanamycin). The outcome of an XDR TB treatment varies dramatically and depends on the degree of bacterial resistance to available drugs, the overall health and age of the patient, patient compliance with treatment, the severity of the disease, and how early it was diagnosed.

Not only are cure rates for cases of MDR or XDR TB lower than typical TB cases, treatment can last for up to 2 years and involves more expensive drugs with more side effects. The WHO estimates that in 2019 there were about 206,000 cases of MDR TB worldwide (a roughly 10 percent increase from 2018). With current therapies, there is only about a 57 percent chance that a patient with MDR TB will be cured and about a 39 percent chance that someone with XDR TB will be cured.[18]

[18] *Global Tuberculosis Report 2020*. (2020). Geneva: World Health Organization.

Administer test

Purified tuberculin proteins are injected into the dermis of the forearm.

At 48 to 72 hours after injection, the diameter of the induration (raised area) is measured. The diameter of the red area surrounding the induration is not measured.

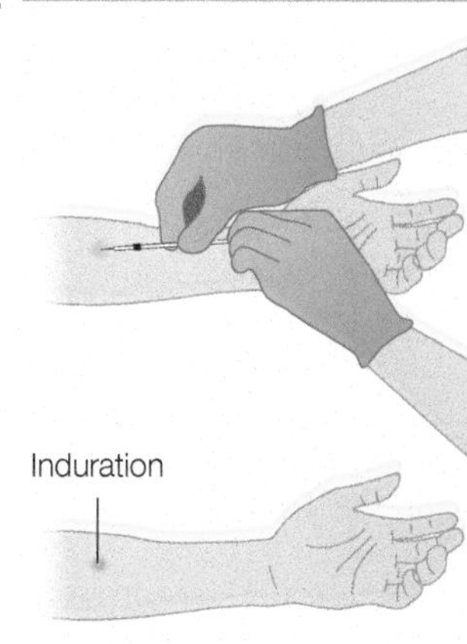

Interpret results

Induration **5 or more mm** is positive test for:
- Immune-compromised patients (example: HIV patient)
- Patients with a history of prior TB infection
- Patients with close contact with someone who has TB
- Patient with fibrotic chest X-ray

Induration **10 or more mm** is positive test for:
- Patients who visit, live in or work in TB-endemic areas
- Patients under age 4
- Patients with an underlying medical condition (example: diabetes)
- Residents of long-term facilities and shelters
- Injection drug users

Induration **15 or more mm** is positive test for:
- Anyone with or without risk for TB exposure

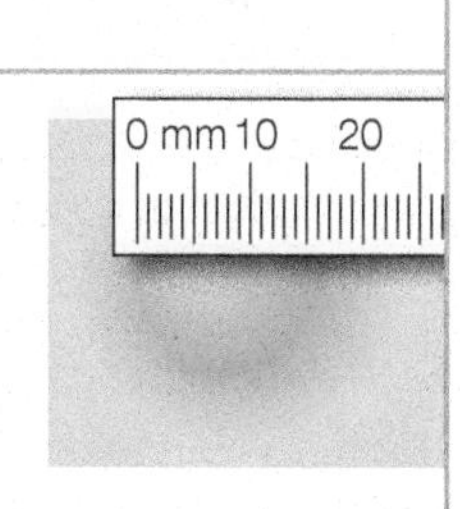

FIGURE 16.14 **Tuberculin skin test (TST)** The test is administered by injecting purified tuberculin proteins into the dermal skin layer of the forearm; the test is interpreted 48–72 hours later.

Critical Thinking *Why do you think there is a lower threshold set for a positive TST result in an immune-compromised patient?*

CLINICAL VOCABULARY

Exudate a fluid discharge (can be pus-like or clear fluid) that is present in a given tissue; usually produced in response to inflammation

Paroxysm a sudden and violent attack (as in a paroxysm of coughing)

Convalescent not fully better, but recovering

Latent infection the infectious agent is present, but not causing symptoms; if conditions are right, the agent can cause an active illness

Induration the hardening of a tissue often caused by inflammation; a hardened or fibrous mass in a normally elastic and soft tissue

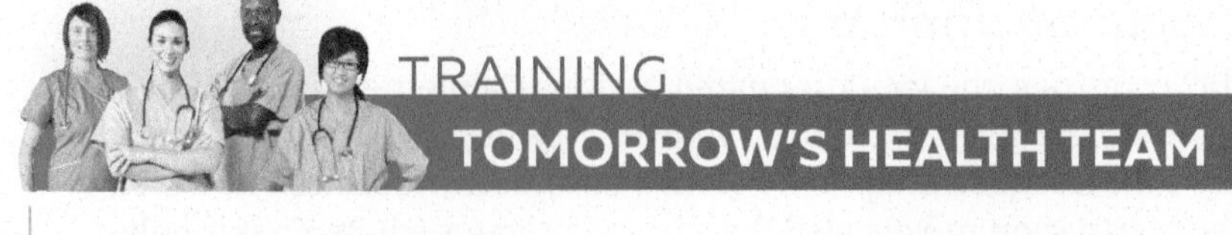

Testing for TB Exposure

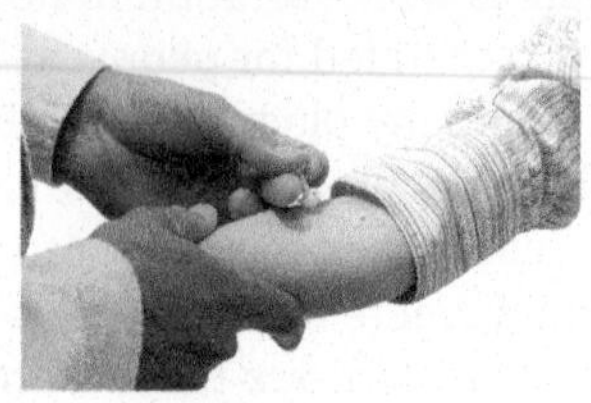
Mantoux tuberculin skin test can cause an induration, or hardened area, at the injection site.

Mycobacterium tuberculosis causes a type IV hypersensitivity reaction in a patient after administration of the tuberculin skin test. Although the resulting immune reaction is strong, it can take 48–72 hours to develop. For this reason, a type IV reaction is also called a delayed hypersensitivity. While all other hypersensitivity reactions (types I through III) involve antibodies, sensitized T cells and macrophages are the key players in delayed hypersensitivities. The skin test is the most common method of testing for TB because it is easy to interpret and inexpensive. However, there are some disadvantages to this test: It lacks sensitivity, does not differentiate between latent and active TB, and gives false positives for those who have received the TB vaccine.

There are three alternative TB tests that are used frequently: interferon gamma assays, microscopy and culture, and the Xpert MTB/RIF. The interferon gamma assay can be faster and require fewer visits; however, latent and active TB give the same positive result, and it can be expensive. Microscopy and culture are both accurate and inexpensive, but require expertise for analysis and cannot detect latent TB at all. Finally, the Xpert MTB/RIF is sensitive, can determine drug resistance, is simple to administer, and has fast results. Although this may seem an ideal test, it is expensive, requires specialized training and equipment, and also cannot differentiate latent from active TB.

All of these tests have strengths and weaknesses, and each is preferentially used under different circumstances. The skin test remains the most common worldwide.

QUESTION 16.3

Based on the preceding information, would reading the tuberculin skin test before the recommended 48-hour minimum more likely result in a false negative or a false positive? Explain your answer.

Pneumonia is among the leading causes of death in the United States.

CLINICAL VOCABULARY

Healthcare-acquired infection (HAI) an infection that a patient develops while receiving care for another condition in a healthcare setting

Community-acquired obtained from a human, animal, or environmental source in a nonhealthcare setting

Consolidation refers to the apparent merging of the lung air sacs (alveoli) when fluid accumulates in the lungs; evident as white opaque areas by X-ray imaging

Bacteremia presence of bacteria in the blood; usually confirmed by culturing bacteria from a blood sample

According to the CDC, in the United States pneumonia is one of the leading causes of death from an infectious agent; worldwide, it kills about a million children under age 5 every year. As already stated, the term *pneumonia* is general and describes an inflammation of the alveoli. Over 100 different microbes can cause pneumonia when they invade the lower respiratory tract, including certain bacteria, viruses, fungi, protists, and parasitic worms. In general, though, bacteria and viruses cause most pneumonia cases. Pneumonia is suspected in patients who have fever, cough, dyspnea (shortness of breath), and **tachypnea** (an abnormally high number of breaths per minute).

In the next section, we will review some of the most common causes of bacterial pneumonias. Here, we discuss classification of pneumonia as a healthcare-acquired infection or as community acquired, and as typical or atypical.

Pneumonia is among the most common healthcare-acquired infections (HAIs). There are three categories of HAI pneumonia. **Healthcare-associated pneumonia (HCAP)** occurs in ambulatory patients who are not hospitalized but who have had extensive healthcare contact within the last 3 months. The term **hospital-acquired pneumonia (HAP)** describes new pneumonia cases that develop in hospitalized, nonintubated patients at least 48 hours after admission. **Ventilator-associated pneumonia (VAP)** is a form of pneumonia that occurs more

TABLE 16.2 Clinical Presentation of Pneumonia

Clinical Feature	Typical Pneumonia	Atypical Pneumonia
General Presentation	• High fever • Sudden onset of severe chills • Shortness of breath • Cough with sputum (productive cough) • Chest pain when inhaling • Runny nose, muscle aches, joint aches usually absent	• Fever usually lower • Little to no chills • Shortness of breath • Cough with little or no sputum (nonproductive cough) • Usually no chest pain upon inhaling • Runny nose, muscle aches, joint aches usually present • Usually community acquired • Onset is usually more subtle than for typical pneumonia
Physical Examination	• Noticeable respiratory distress • Consolidation usually present (fluid accumulation in the lungs)	• Limited respiratory distress • Limited consolidation (minimal fluid in lungs) • Auscultation (or listening to the patient's breathing sounds with a stethoscope) reveals a crackling sound (or rales) as the patient breathes
Chest X-ray	• Reveals fluid accumulation in the lung air sacs in one or both lungs (consolidation) • Usually an opaque white area is visible in at least one lung lobe	• X-ray often looks worse than the patient appears or feels • May see white opaque areas in both lungs in a patchy or a diffuse pattern • Atelectasis (collapse of the lung air sacs) may be evident • Usually only a small amount of fluid in the lung air sacs • Consolidation (fluid in lungs) is not necessarily evident
Sputum Culture	Reveals Gram-positive or Gram-negative bacteria	Often an agent is not successfully cultured from sputum
Prevalence	4 out of 5 pneumonia cases	1 out of 5 pneumonia cases

than 48 hours after a hospitalized patient is put on a mechanical ventilator.[19] With a mortality rate ranging from about 27 to 76 percent (depending on patient factors), VAP represents a serious problem in intensive care units.[20] *Acinetobacter baumannii*, *Pseudomonas aeruginosa*, and methicillin-resistant *Staphylococcus aureus* (MRSA) are among the more common bacteria isolated in VAP.[21] These bacteria are notoriously resistant to antimicrobial drugs, which makes them challenging infections to treat. (For information on VAP in COVID-19 patients, see the Bench to Bedside feature.) **Community-acquired pneumonia (CAP)** is a case of pneumonia that does not meet any of the three aforementioned HAI pneumonia criteria (note that pneumonia which develops within 48 hours of admission is also considered CAP). According to the CDC, in the United States about a million people per year are hospitalized with pneumonia, and about 50,000 die.

Pneumonias are further classified as being either typical or atypical. Even before it was understood that a wide variety of agents cause pneumonia, it was recognized that cases of pneumonia could differ. Throughout the early to mid-1900s a patient was described as having typical pneumonia if they had an illness that resembled pneumococcal pneumonia (pneumonia caused by *Streptococcus pneumoniae*). If the patient had a pneumonia-like illness that presented and progressed differently from pneumococcal pneumonia, that patient was said to have an atypical pneumonia. A separate category exists for viral infections, *viral community-acquired pneumonias*. As previously mentioned, most of the viral agents we reviewed in the earlier section on viral respiratory infections can cause pneumonia.

Grouping pneumonia cases helps narrow down the likely etiological agent. Atypical pneumonia describes bacterial pneumonia that, unlike typical pneumonia, is not characterized by signs and symptoms of **consolidation**, a term meaning that the alveoli contain fluid instead of gas. Consolidation appears as a white or hazy opaque region on a chest X-ray. TABLE 16.2 shows a side-by-side comparison of some general features of typical versus atypical pneumonia.

Clinical CASE
NCLEX
HESI
TEAS

The Case of the Suffering Spelunker

Practice applying what you know clinically: visit the **Mastering Microbiology** Study Area to watch Part 2 and practice for future exams.

[19] If pneumonia develops before the 48-hour mark, then the causative microbe is likely to be *Streptococcus pneumoniae* or *Haemophilus influenzae*.

[20] Sadigov, A., Mamedova, I., & Mammmadov, K. (2019). Ventilator-associated pneumonia and in-hospital mortality: Which risk factors may predict in-hospital mortality in such patients? *Journal of Lung Health and Diseases*, 3(4), 8–12.

[21] Karakuzu, Z., Iscimen, R., Akalin, H., et al. (2018). Prognostic risk factors in ventilator-associated pneumonia. *Medical Science Monitor: International Medical Journal of Experimental and Clinical Research*, 24, 1321–1328.

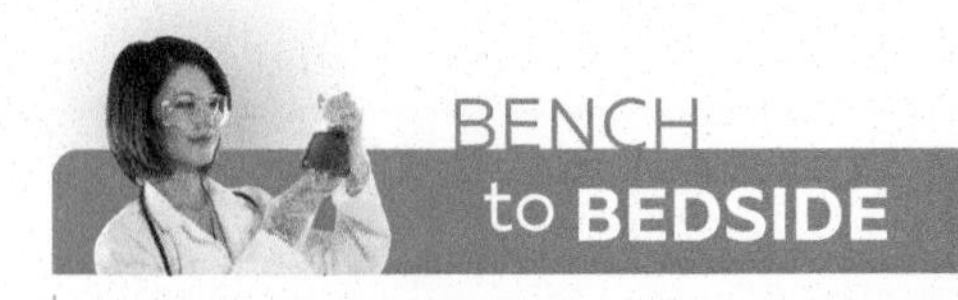

COVID-19 Patients and New Biomarkers for Ventilator-Associated Pneumonia

Approximately 10–20 percent of patients admitted to a hospital with severe COVID-19 require invasive mechanical ventilation (IMV). Furthermore, data reveal that COVID-19 patients have a much higher risk of developing ventilator-associated pneumonia (VAP) as compared to non-COVID-19 patients—there are about 28 VAP cases per 1,000 ventilator days for COVID-19 patients versus 13 VAP cases for non-COVID-19 patients. This greater risk for VAP in COVID-19 patients is detected even when the data are analyzed in a way that accounts for standard known risks for VAP such as longer IMV periods and a variety of patient factors (e.g., advanced age and preexisting cardiac or pulmonary diseases).

VAP is usually suspected when new signs and/or symptoms develop in the patient and new infiltrate becomes evident on a chest X-ray. However, no symptom, sign, or X-ray finding is specific for VAP—and none facilitates early diagnosis. As with most infectious diseases, an early and accurate diagnosis usually leads to a better prognosis. As such, clinical trials are currently underway to characterize biomarkers that could help clinicians detect VAP earlier and more reliably. Just as monitoring the levels of certain liver enzymes in blood can provide early clues about liver health, early indicators of VAP may also be detectable in blood serum. Diverse blood serum biomarkers are being investigated as possible early indictors of VAP. Here we'll discuss two biomarkers that have been investigated with respect to COVID-19 patients and VAP—procalcitonin (PCT) and pentraxin 3 (PTX3).

PCT is a small protein that is converted to the hormone calcitonin—a substance that regulates blood calcium levels. Normally, blood PCT levels are low; however, levels detectibly increase 6 to 12 hours following an initial bacterial infection. And when a proper host immune response and antibiotic therapy are applied, PCT levels usually decrease by about 50 percent over 24 hours. Thus, monitoring PCT levels can provide early clues about bacterial infections like VAP and possibly help clinicians assess treatment efficacy. However, a variety of noninfectious conditions (severe burns, certain carcinomas, and chronic kidney disease—to name a few) can raise blood PCT levels. Therefore, clinicians must rule out these noninfectious pathologies when interpreting PCT data. Initial studies suggest that PCT may be a useful early indicator of VAP development in COVID-19 patients. Unfortunately, even if monitoring PCT levels proves useful as an early bacterial infection indicator, it cannot serve as a sole metric for VAP because PCT levels can increase during any bacterial infection.

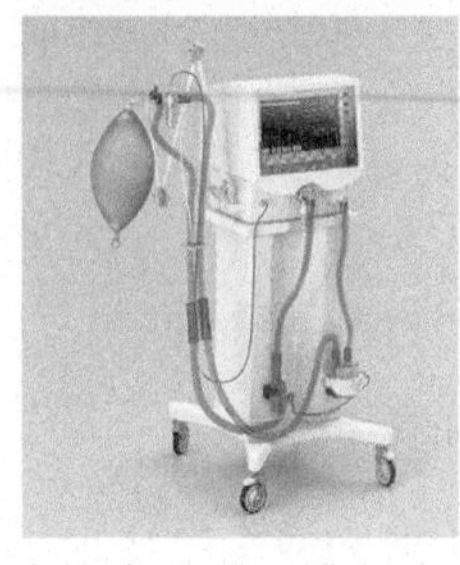
A mechanical ventilator

PTX3 is an innate immunity factor that some cells, like dendritic cells, release in response to the early stages of an infection (including SARS-CoV-2 infection and VAP). This molecule activates innate immunity protections such as complement cascades. Elevated PTX3 levels in COVID-19 patients have been found to independently predict disease severity and short-term (28-day) patient mortality. Although further research is needed, some medical scientists speculate that high PTX3 levels reflect a dysregulated inflammatory response that increases lung tissue damage in COVID-19. Detecting elevated PTX3 levels in patients' blood could provide clues about COVID-19 case severity and help clinicians identify patients who are more likely to need IMV and have a greater risk for VAP.

In short, biomarkers like PTX3 could help clinicians determine which COVID-19 patients are likely to become the sickest and need ventilation. In patients on IMV, data like PCT levels may empower clinicians to detect VAP in its earliest stages to improve patient outcomes.

Sources: Maes, M., Higginson, E., Pereira-Dias, J., et al. (2021). Ventilator-associated pneumonia in critically ill patients with COVID-19. *Critical Care*, *25*(1), 1–11. Côrtes, M. F., de Almeida, B. L., Espinoza, E. P. S., et al. (2021). Procalcitonin as a biomarker for ventilator associated pneumonia in COVID-19 patients. *Diagnostic Microbiology and Infectious Disease*, 115344. Brunetta, E., Folci, M., Bottazzi, B., et al. (2021) Macrophage expression and prognostic significance of the long pentraxin PTX3 in COVID-19. *Nature Immunology*, *22*(1), 19–24.

Many bacteria cause typical pneumonia.

Many bacterial agents cause typical pneumonia; in this section, we will explore two of the most common causative agents.

Pneumococcal Pneumonia

The most common form of bacterial pneumonia—and the standard for classifying typical pneumonia syndrome—is pneumococcal pneumonia caused by *Streptococcus pneumoniae*—a Gram-positive, encapsulated diplococcus of the Group B streptococci (GBS). (See Chapter 10 for more information on GBS.) Humans are the only known natural source of pneumococcal bacteria, which means we get these bacteria from each other. People under age 5 and over age 65 are at the greatest risk for pneumococcal pneumonia.

The bacterium spreads through respiratory droplets. When *S. pneumoniae* spreads to a new host, it usually colonizes the nasopharynx without causing disease. Occasionally it may spread from the nasopharynx to the middle ear and cause otitis media. Alternatively, it can migrate to the lungs and cause pneumonia. Sometimes the bacteria spread from the lungs to the bloodstream,

DiseaseSnapshot • Typical Pneumonia

Disease	Pneumococcal pneumonia	*Haemophilus* pneumonia
Causative agent	*Streptococcus pneumoniae*; Gram-positive bacterium	*Haemophilus influenzae*; Gram-negative bacterium
Epidemiology	Worldwide; most common cause of community-acquired pneumonia; people under age 5 and over age 65 are at the greatest risk; 1- to 3-day incubation	Nontypable *H. influenzae remains* a leading cause of community-acquired pneumonia in adults; 2- to 4-day incubation
Transmission & prevention	Respiratory droplets; transmitted from person to person; prevented by PCV-13 (Prevnar 13®) and PPSV-23 (Pneumovax®) vaccines	Respiratory droplets; contact with discharge from the nose and throat; commonly carried asymptomatically in the nasopharynx; infections caused by typable *H. influenzae* prevented by childhood Hib conjugate vaccine
Signs & symptoms	Symptoms are characterized by a sudden high fever and episodes of shaking chills; cough, chest pain, shortness of breath, disorientation, and rust-colored, blood-tinged sputum are also sometimes seen	Infections causing pneumonia present fever (but older people may have lower than normal body temperature), chills and sweating, cough, chest pain, and shortness of breath; sometimes headache, muscle pain, and fatigue are present as well
Pathogenesis & virulence factors	Capsule protects against phagocytosis; pneumolysin inhibits ciliary action of the mucociliary escalator; produces hydrogen peroxide, which can cause oxidative damage in host tissues	Capsule protects against phagocytosis and is used for adhesion to host tissues; endotoxin is present in the bacterium's outer membrane
Diagnosis & treatment	Diagnosed upon culturing bacterium from patient samples; urine samples can also be used to detect the C-polysaccharide antigen of *Streptococcus pneumoniae;* commonly treated with cephalosporins or penicillin	Diagnosis is confirmed by culture techniques or through detection of bacterial DNA or antigens; treatable with many different antibiotics (cephalosporins and quinolones are commonly prescribed)

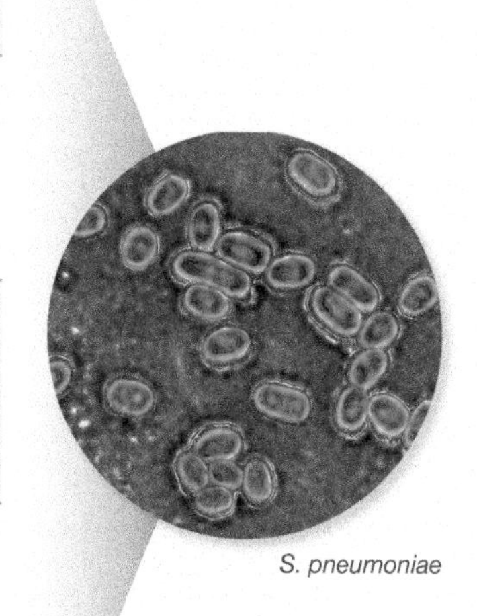

S. pneumoniae

leading to a dangerous situation called **bacteremia**. In pneumococcal bacteremia, the bacteria enter the bloodstream and spread throughout the body and cause complications like meningitis, an inflammation caused by an infection of parts of the nervous system. (See Chapter 18 for more on meningitis.) When *S. pneumoniae* does cause disease, the symptoms usually come on quite suddenly.

Vaccines against *S. pneumoniae* have greatly decreased pneumococcal disease incidence. The conjugate vaccine PCV-13 (Prevnar 13®) is used mainly in children, whereas the pneumococcal polysaccharide vaccine PPSV-23 (Pneumovax®) is recommended for people age 65 and older and for anyone age 2 or older who has an underlying condition that puts them at an increased risk of pneumococcal disease.

Haemophilus influenzae Type b (Hib) Pneumonia

Haemophilus influenzae **type b (Hib) pneumonia** is caused by *Haemophilus influenzae,* a Gram-negative bacterium that was first described in 1892, but until the 1930s was mistakenly thought to cause influenza (hence the potentially misleading species name). Based on analyses of polysaccharide capsules, there are six major serotypes (a through f) of *H. influenzae*—of these, type b (or Hib) is the most dangerous. Before vaccines against it, Hib was a notorious cause of severe epiglottitis, cellulitis, pneumonia, and meningitis in children. (Hib meningitis is reviewed in Chapter 18.)

Aside from the six *typable strains*, there are also strains that do not have polysaccharide capsules and are called *nontypable strains*. Nontypable

H. influenzae bacteria are among the leading causes of community-acquired pneumonia in adults.[22] Unfortunately, we do not yet have vaccines that protect against nontypable *H. influenzae*.

In the United States, five licensed vaccines are currently used to protect children against Hib pneumonia and the other Hib diseases named earlier. Not only have these vaccines reduced the instances of life-threatening Hib disease, but they also have reduced cases of Hib-induced otitis media. Thanks to the high efficacy of these vaccines, Hib disease in children has become rare in the United States—according to the CDC, since Hib vaccines have been widely used in the United States, there have been, on average, fewer than 50 cases reported per year.

There are six leading causes of atypical bacterial pneumonia.

In general, there are six leading causes of **atypical bacterial pneumonia**. Three—Q fever, psittacosis, and tularemia—are zoonotic illnesses, which means they are passed from animals to humans. The other three—illnesses caused by *Mycoplasma pneumoniae*, *Chlamydia pneumoniae*, and *Legionella* species—are non-zoonotic. Infections caused by *M. pneumoniae* and *C. pneumoniae* are by far the most common of these six illnesses. These two agents are thought to cause about 80 percent of all atypical pneumonia cases and up to 17 percent of all pneumonia (typical and atypical combined) in adults and young children.

Mycoplasma Pneumonia (Walking Pneumonia)

Even though ***Mycoplasma* pneumonia** is the most common cause of atypical community-acquired pneumonia, studies suggest that up to 20 percent of infections are asymptomatic. It is estimated that about 2 million cases occur in the United States every year, with about 100,000 cases requiring hospitalization.

The causative agent, *Mycoplasma pneumoniae,* is one of the smallest self-replicating organisms known. The virus-like characteristics of these bacteria (small size, lack of cell wall, parasitic nature) initially caused scientists to wrongly assume that they were viruses. These bacteria are so small that they are difficult to see with a standard light microscope. Additionally, unlike most bacteria, *Mycoplasma* species do not have a cell wall, so describing them by Gram property is not appropriate. They also have far fewer genes than most bacteria. Their smaller genome is thought to have evolved because *M. pneumonia* have a parasitic lifestyle inside host cells and rely on their host for many functions.

When illness from *M. pneumoniae* does develop, it has a slow onset and variable progression. In most cases the patient develops a mild form of pneumonia, although some people (especially the elderly, or those with lung disease) can be severely affected. Sometimes an infection with *M. pneumoniae* is called **walking pneumonia** because the clinical findings point to a more severe illness than the symptoms may suggest. Immunity to *M. pneumoniae* is not long lived, so a person can have walking pneumonia more than once.

Chlamydophila Pneumonia (formerly *Chlamydia* Pneumonia)

***Chlamydophila* pneumonia** is caused by *Chlamydophila pneumoniae,* a Gram-negative bacterium that lives inside host cells. It is now recognized as a common cause of sinusitis, pneumonia, bronchitis, and pharyngitis. The severity of the illness varies greatly, but typically starts with a sore throat and then progresses to a cough that lasts up to 6 weeks. Like *M. pneumoniae*, it prefers to

---

[22] Shoar, S., Centeno, F. H., & Musher, D. M. (2021). Clinical features and outcomes of community-acquired pneumonia caused by *Haemophilus influenza*. *Open Forum Infectious Diseases*, 8(4), 622.

Disease**Snapshot** • Atypical Pneumonia, Part 1

Disease	Walking Pneumonia	*Chlamydophila* Pneumonia
Causative agent	*Mycoplasma pneumoniae* (lack a cell wall)	*Chlamydophila pneumoniae*; Gram-negative bacterium
Epidemiology	Common worldwide, but more cases in temperate climates; outbreaks are mostly in late summer; most common in people ages 5 to 40 years; responsible for at least 40% of community-acquired pneumonia; incubation period is 4–23 days	Common worldwide; most cases occur in school-age children; about 10% of community-acquired pneumonia and 5% of bronchitis; incubation period is about 21 days
Transmission & prevention	Respiratory droplets person to person; mainly prevented by hand washing	
Signs & symptoms	Usually include cold-like symptoms, a persistent dry cough, and mild fatigue	Symptoms vary widely but typically start with a sore throat and then progress to a cough that lasts up to 6 weeks
Pathogenesis & virulence factors	Specialized attachment structure that is enriched with adhesins; induces damage to respiratory epithelium by hydrogen peroxide and superoxide radicals	Bacteria has mechanisms to thrive inside phagocytic cells and avoid destruction in phagolysosomes
Diagnosis & treatment	Diagnosed using serology or nucleic acid-based tests; culturing the bacteria is challenging, but is sometimes used for diagnosis; usually treated with macrolides; not susceptible to penicillin-based drugs because these bacteria lack a cell wall	Diagnosed by culture methods, antigen detection, serology, and PCR; macrolides, fluoroquinolones, or tetracycline drugs are common therapies

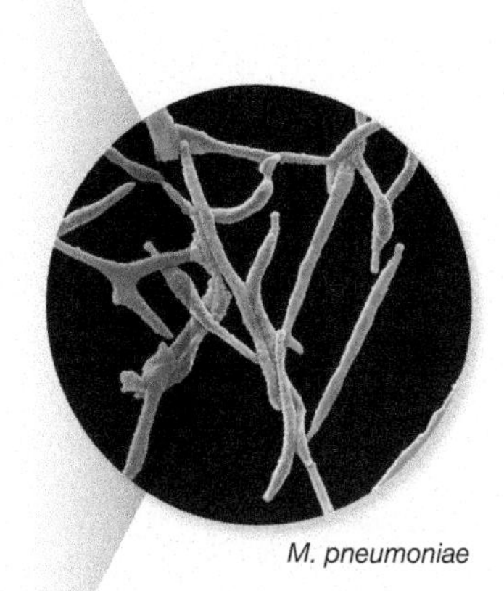

M. pneumoniae

thrive inside host cells, so it can be difficult to culture. It is estimated to cause 2–5 million cases of atypical pneumonia in the United States every year, with about 500,000 of these cases requiring hospitalization.

Although anyone can develop a *C. pneumoniae* infection, it is most common in school-age children. Serological data show that about half of the population worldwide has been infected with this bacterium by age 20. Immunity after infection is not long lived, so people are often infected more than once, and chronic (or lingering) infections are also possible.

Legionnaires' Disease (Mainly Caused by *Legionella pneumophila*)

In 1976, more than 2,000 military veterans gathered at a hotel in Philadelphia for a convention. Three days later, one of them developed a disease that later came to be known as Legionnaires' disease. Although it took almost 5 years to discover, the causative agent, *Legionella pneumophila*, was present in the conference hotel's air-conditioning system and was transmitted to attendees through the building's ventilation system. Mortality rates from **Legionnaires' disease** can be up to 30 percent if untreated.

Legionella species are Gram-negative coccobacilli that are usually motile. Over 50 species of *Legionella* have been characterized. About 90 percent of human cases are due to *L. pneumophila*, although at least 19 other species are known to cause illness in humans. *L. pneumophila* can cause two clinically distinct types of legionellosis: Legionnaires' disease and **Pontiac fever**. The Disease Snapshot table compares Legionnaires' disease to Pontiac fever.

Every year between 8,000 and 18,000 people in the United States are hospitalized with Legionnaires' disease. Because Legionnaires' disease outbreaks frequently trace back to contaminated buildings that house large numbers of people, it's important to track cases of legionellosis so the source of infection can be eliminated.

Disease**Snapshot** • Atypical Pneumonia, Part 2

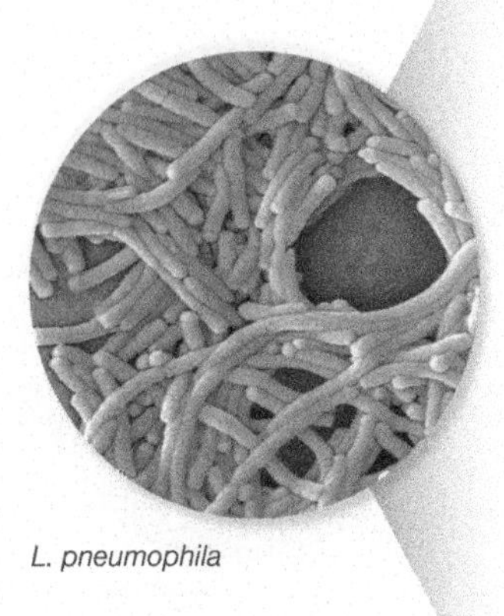
L. pneumophila

Disease	Legionnaires' Disease	Pontiac Fever
Causative agent	Usually *Legionella pneumophila*; Gram-negative bacterium	
Epidemiology	Occurs worldwide year-round, but has higher incidence from summer through early fall when air conditioners are in high use; men and smokers are disproportionately affected; around 5,000 cases per year in the U.S.; incubation period of 2–14 days	Found worldwide and occurs year-round; slightly more prevalent in people under age 40; incubation period of 1–3 days
Transmission & prevention	Aerosols or aspiration of contaminated water; common in industrial air-conditioner systems and other artificial water systems; prevent by proper air-conditioning system cleaning and maintenance	
Signs & symptoms	Symptoms usually include high fever, cough, and potentially severe atypical pneumonia	Symptoms include mild flu-like illness; only about 50% have a cough; vomiting and diarrhea have been described and, rarely, neurological symptoms are seen
Pathogenesis & virulence factors	Bacteria have mechanisms to thrive inside phagocytic cells and avoid destruction in phagolysosomes	Pathogenesis is different from Legionnaires' disease and is poorly understood
Diagnosis & treatment	Usually diagnosed using a urine antigen test for *L. pneumophila* serogroup 1; confirmed by culture methods; quinolones (example: ciprofloxacin) or macrolides (example: erythromycin) are common treatments	

Psittacosis or Ornithosis (Caused by *Chlamydophila psittaci*)

Chlamydophila psittaci are Gram-negative bacteria that live as a parasite of eukaryotic cells and cause a zoonotic atypical pneumonia that transmits from infected birds to humans. Almost any bird can harbor *C. psittaci*, but pigeons and birds in the parrot family (psittacines) such as parakeets, macaws, and cockatiels are especially likely to be carriers. Not surprisingly, human **psittacosis** (sit uh KOH sis) is most common among pet shop workers, zoo employees, taxidermists, veterinarians, poultry workers, and bird owners.

The usual route of transmission is when people inhale dust that is stirred up from dried bird droppings found in birdcages. Patients often report generalized muscle aches as well as stiffness and spasms of back and neck muscles, which can cause a clinician to erroneously suspect meningitis. In untreated cases the fever can come and go over several weeks, and the infection can be fatal in 10–40 percent of cases. Infection does not provide lifelong immunity, and there is no vaccine, so re-infection is possible.

Q Fever (Caused by Coxielle burnetti)

In 1935 an outbreak of an unknown respiratory illness occurred in Australia. It was called query (meaning question) fever—**Q fever**—for short. The Gram-negative bacterium that causes the illness is highly infectious (the infectious dose is just *one* bacterial cell) and is named *Coxiella burnetii*. *C. burnetii* causes a flu-like illness that can be accompanied by pneumonia. Livestock such as cattle, sheep, and goats become infected from ticks. People can become infected if they inhale dust or aerosols from feces, urine, or birth products (as would be encountered by farmers or veterinarians helping an animal deliver), or consuming unpasteurized milk products.

Tularemia (Caused by Francisella tularensis)

In 1911, the Gram-negative bacterium, *Francisella tularensis*, was named after Tulare County in California, where it was first isolated from ground squirrels. The causative agent of **tularemia** can exist as a free-living bacterium, but it usually prefers to live as a parasite of eukaryotic cells—especially macrophages.

Disease**Snapshot** • Atypical Pneumonia, Part 3

Disease	Psittacosis or Ornithosis	Q fever	Tularemia
Causative agent	*Chlamydophila psittaci*; Gram-negative bacterium	*Coxiella burnetii*; Gram-negative bacterium	*Francisella tularensis*; Gram-negative bacterium
Epidemiology	Zoonotic disease; found worldwide, mainly in pigeons and psittacines; most common among people who contact birds; incubation period of 5–14 days	Zoonotic disease; found worldwide except in Antarctica and New Zealand; most outbreaks linked to infected pets, livestock, or wild animals; high-risk groups include veterinarians, meat packers, dairy workers, and livestock farmers; incubation period of 13–28 days	Zoonotic disease; most prevalent in rural areas of North America, Europe, and northern Asia; host range includes humans, birds, over 100 species of wild animals (especially rabbits), and occasionally domesticated animals; incubation period of 3–5 days
Transmission & prevention	Most commonly transmitted via aerosols containing dust stirred up from infected bird droppings; not communicable; prevented by limiting exposure to bird droppings	Transmitted by airborne bacteria from animal excrement and contaminated animal products; amniotic fluid and placental tissue of infected animals is a common exposure source for veterinarians; not communicable; prevented by the Q-vax vaccine for at-risk groups (not approved for use in U.S.)	Transmitted by aerosolized bacteria; not communicable; prevented by an investigational vaccine for people at high risk for exposure
Signs & symptoms	Fever that increases over several days, chills, a severe headache, and hacking cough; often these are accompanied by generalized stiffness and back pain	Symptoms vary widely; include high fever, nonproductive cough, muscle aches, chills, sweats, weakness, nausea, vomiting, diarrhea, chest pain, severe headache, or abdominal pain	Symptoms include a sore throat, swelling of the lymph nodes, sudden fever, chills, muscle aches, headache, joint pain, a dry cough, and progressive weakness with sudden onset
Pathogenesis & virulence factors	Has mechanisms to thrive inside phagocytic cells and avoid destruction in phagolysosomes	Can form a spore-like resistant structure in host cells, allowing chronic infections; has mechanisms to thrive inside phagocytic cells and avoid destruction in phagolysosomes	Can escape phagosomes and live in macrophages; capsule is important for aerosolized forms; siderophores bind and collect iron; pili can attach to host tissues
Diagnosis & treatment	Diagnosed by serology and PCR of respiratory samples; antibiotic of choice is doxycycline	Mainly diagnosed by patient history and signs/symptoms; treated with doxycycline	Patient history of tick or deer fly bites, or contact with sick or dead animals; serology, PCR, and culture methods can confirm diagnosis; treated with aminoglycosides

F. tularensis

However, *F. tularensis* remains infectious for long periods of time outside of the host. Tularemia is an extremely infectious zoonotic disease; many scientists consider *F. tularensis* to be the most infectious pathogenic bacterium known to cause disease in humans, making it a potential biological weapon. Bacteria can become naturally airborne when dried grasses or hay contaminated with infected animal carcasses are stirred up; skinning or butchering contaminated animals also can put one at risk for inhaling aerosolized bacteria. When aerosolized, inhaled *F. tularensis* can cause pneumonic tularemia. Other forms of tularemia are discussed in Chapter 21, Systemic Infections.

A pneumonic tularemia outbreak that occurred in 1942 before the battle of Stalingrad was possibly caused by weaponized *F. tularensis* thought to have been released by the Soviet military (although this has never been confirmed, and a natural outbreak still has not been ruled out). It was reported that the outbreak among the German and Soviet troops affected tens of thousands of soldiers.

Build Your Foundation

12. Why is otitis media such a common problem in young children? (LO 16.11)
13. Why is it important to promptly treat a *Streptococcus pyogenes* infection? (LO 16.12)
14. How can diphtheria be prevented? (LO 16.13)
15. What factors are contributing to the reemergence of pertussis? (LO 16.14)
16. Describe the bacterium that causes tuberculosis, and state how it can be detected in a patient. (LO 16.15)
17. What features distinguish a latent tuberculosis infection from an active one? (LO 16.15)
18. What are some of the challenges in treatment of tuberculosis? (LO 16.15)
19. Compare and contrast typical and atypical bacterial pneumonia. (LO 16.16)
20. What are some of the features that differentiate infections by *Mycoplasma pneumoniae* from *Chlamydia pneumoniae*? (LO 16.17)
21. How would a person be exposed to *Legionella pneumophila, Chlamydophila psittaci, Coxiella burnetii,* and *Francisella tularensis*? (LO 16.17)

Build Your Foundation (BYF) Quick Quiz: Visit the **Mastering Microbiology** Study Area to quiz yourself.

16.4 FUNGAL RESPIRATORY SYSTEM INFECTIONS

Learning Outcomes

After reading this section, you should be able to:

16.18 Define the term *mycosis*, and explain why mycoses are becoming more common.

16.19 State why certain mycoses are called endemic, and others are not.

16.20 Provide examples of three endemic mycoses and their clinical features.

16.21 Describe the causative agents and clinical features of aspergillosis, mucormycosis, and *Pneumocystis* pneumonia.

At least 1.5 million species of fungi exist, but fewer than 200 are known to cause disease in animals and humans. As compared to bacterial and viral infections of the respiratory tract, fungal infections (**mycoses**) are fairly rare. However, they are becoming increasingly common as a result of increased urban development, deforestation, and climate change—all of which bring people into increased contact with fungal pathogens. Another factor in the rise of fungal diseases is the increasingly large number of people with weakened immunity, which can be due to advanced age, immune-related illnesses such as HIV/AIDS, or the use of immunosuppressant drugs after organ transplants.

Knowing the geographical distribution of certain pathogenic fungi can be useful for narrowing down a patient's diagnosis. **Endemic fungi** grow only in defined geographical locations because they require certain climate and soil conditions to thrive. A person is unlikely to develop an **endemic mycosis** (a fungal infection associated with a distinct geographical location) if they haven't been to the area where the fungus grows. On the other hand, **ubiquitous fungi** grow in varied climates and under diverse conditions; some are even common members of the normal microbiota of the human body. Knowing a patient's travel history is unlikely to help a healthcare provider rule out a **ubiquitous mycosis** (a fungal infection that is not confined to a specific geographical region). All of the respiratory mycoses you will read about here affect the lower respiratory tract and are contracted when fungal spores are inhaled. They are all forms of fungal pneumonia.

Endemic fungi are found in certain geographic regions and can cause endemic mycoses.

Endemic mycoses are caused by fungi that are considered true pathogens—meaning these fungi do not require a compromised host to establish an infection. Here we'll review three of the more common endemic respiratory mycoses.

Blastomycosis (Chicago Disease)

The endemic fungus *Blastomyces dermatitidis* causes the noncommunicable lung infection **blastomycosis** (FIG. 16.15). This illness is also known as Chicago disease and Namekagon River fever. It grows in the central, southern, and northern

United States and is especially common in the Mississippi and Ohio River valleys, the Great Lakes region, and parts of Canada near the Great Lakes and the St. Lawrence Seaway. Although most blastomycosis cases are isolated and not associated with an outbreak, one of the larger modern-day outbreaks in the United States occurred in 2015; 49 blastomycosis cases were associated with the Wisconsin Little Wolf River. Officials believe the fungal spores became airborne in areas near the river and were inhaled by people who went tubing during the summer. According to the CDC, Wisconsin tends to have the largest incidence of blastomycosis in the United States. In South America there is another form of blastomycosis caused by *Paracoccidioides brasiliensis*.

B. dermatitidis is a **dimorphic** fungus, meaning it's able to exist in two distinct forms. Like many pathogenic fungi, in natural environments it lives as a *saprophyte*, a mold form that thrives on decomposing plant matter in moist soil. It produces conidiospores that become airborne and are easily inhaled when soil is disturbed by digging, wood clearing, or construction. Once inside lungs, the spores can germinate into a unicellular, yeast-like form that causes disease.

Even when treated with antifungal drugs such as fluconazole, itraconazole, ketoconazole, or amphotericin B, blastomycosis mortality rates are up to 40 percent in immune-compromised patients and around 10 percent in otherwise healthy patients. (Antifungal drugs are discussed in Chapter 15.) The higher mortality rates seen in immune-compromised patients are due to the increased likelihood of the fungus spreading from the lungs to other parts of the body, a situation called disseminated blastomycosis.

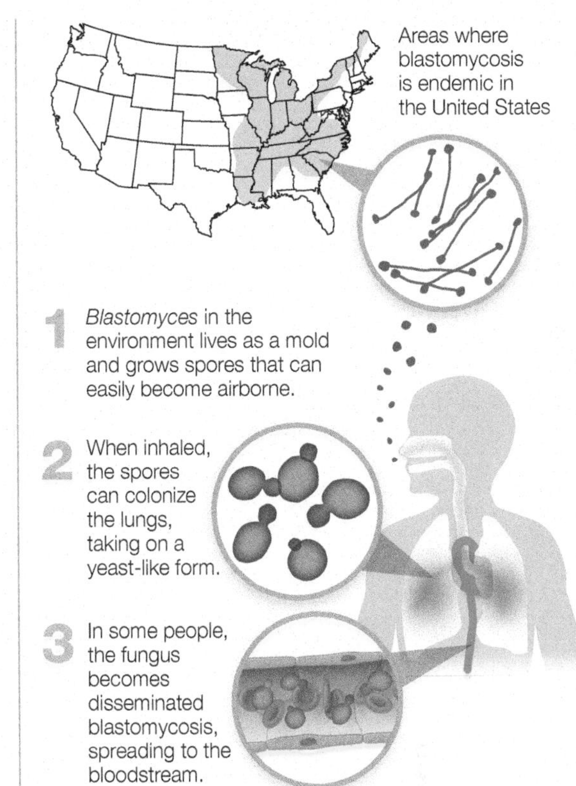

FIGURE 16.15 Blastomycosis

Critical Thinking *Dogs are also affected by blastomycosis. What natural behavior do dogs have that could make them prone to this mycosis? (Hint: Where in the environment does this fungus thrive?)*

Coccidioidomycosis (Valley Fever)

Coccidioidomycosis (cock SID ee *oy* doh my *koh* sis) is a noncommunicable lung disease first clinically described in the late 1800s. It is caused by the fungi *Coccidioides immitis* and *Coccidioides posadasii*. *C. immitis* is native to the semi-arid soil of California's San Joaquin Valley, which is why coccidioidomycosis is also known as **Valley fever**. The second fungus, *C. posadasii*, mainly grows in the dry soil of the southwestern United States, northern Mexico, and parts of Central and South America.

C. immitis and *C. posadasii* are dimorphic, growing as molds that make asexual spores called arthroconidia that spread with the dry winds common to arid climates (FIG. 16.16). Once in the lungs, the arthroconidia change form and develop into spherules, which are multinucleated structures that will produce

Disease**Snapshot** • Blastomycosis (Chicago Disease)

Causative agent	*Blastomyces dermatitidis* or *Paracoccidioides brasiliensis*
Epidemiology	*B. dermatitidis* is mostly found in North America, whereas *P. brasiliensis* causes most cases in Latin America. Isolated incidence worldwide. There are approximately 1–2 symptomatic cases per 100,000 in endemic areas. Disease is most common in immune-compromised people. Incubation period is 3 weeks to 3 months.
Transmission & prevention	*B. dermatitidis* transmits via conidia spores that are inhaled. Avoiding exposure is the best preventive method.
Signs & symptoms	Flu-like with a fever, headache, cough, muscle aches, chills, and joint pain; pneumonia-like symptoms are also common. Rarely, the fungus spreads throughout the body, causing disseminated blastomycosis, which has a poor prognosis.
Pathogenesis & virulence factors	*Paracoccidioides brasiliensis* has adhesins to bind to human cells and can form a biofilm. It is also capable of inducing apoptosis and host coagulation. *Blastomyces dermatitidis* also has adhesins and can deactivate some of the functions of macrophages in the lungs. Both of these fungi change into the parasitic yeast form upon exposure to human body temperature.
Diagnosis & treatment	Common methods include chest X-rays, microscopic examination of sputum samples, and culture methods. Immunoassays are also available. Treatments include antifungal drugs such as fluconazole, itraconazole, ketoconazole, or amphotericin B.

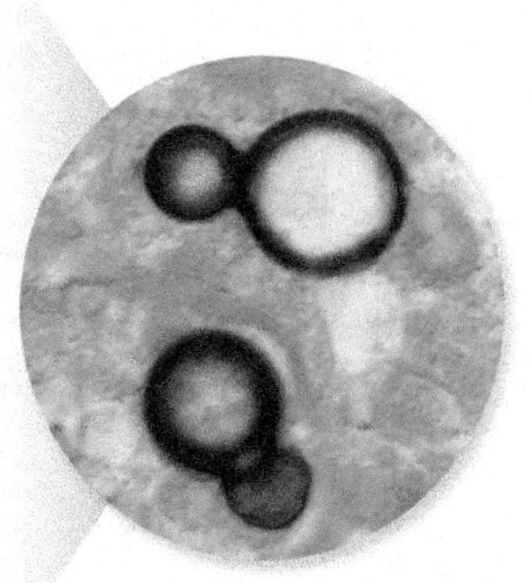

B. dermatitidis

DiseaseSnapshot • Coccidioidomycosis (Valley Fever)

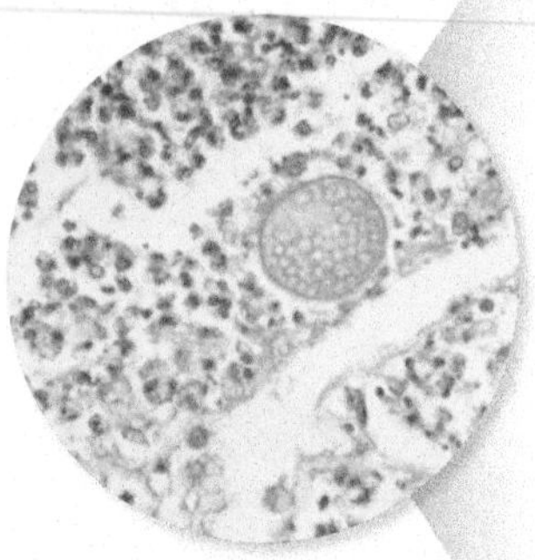

Coccidioides species

Causative agent	*Coccidioides immitis* and *Coccidioides posadasii*
Epidemiology	The fungi are usually found in semi-arid or arid soil in the southwest and California, as well as northern Mexico. Although around 150,000 people are infected each year, 65% of cases are asymptomatic. In the U.S., 50–100 people per year die from it. Incubation period is 1–4 weeks.
Transmission & prevention	Transmits via inhalation of fungal spores. No preventive measures, beyond avoiding inhalation.
Signs & symptoms	Body aches, fever, cough, headache, knee and ankle joint pain, and a rash on the torso and extremities. If the fungus causes a disseminated infection, skin lesions and meningitis also commonly occur.
Pathogenesis & virulence factors	A protein found on the outer wall of spherules helps the fungus hide from the immune system. *C. immitis* can also prevent macrophages from producing nitric oxide that kills invading cells.
Diagnosis & treatment	Most cases are diagnosed using an immunoassay to find related antibodies. Chest X-rays and tissue biopsies are common. Treatment involves 6 months of antifungal drugs such as fluconazole, itraconazole, ketoconazole, or amphotericin B in severe cases.
Prevention	None, beyond avoiding exposure.

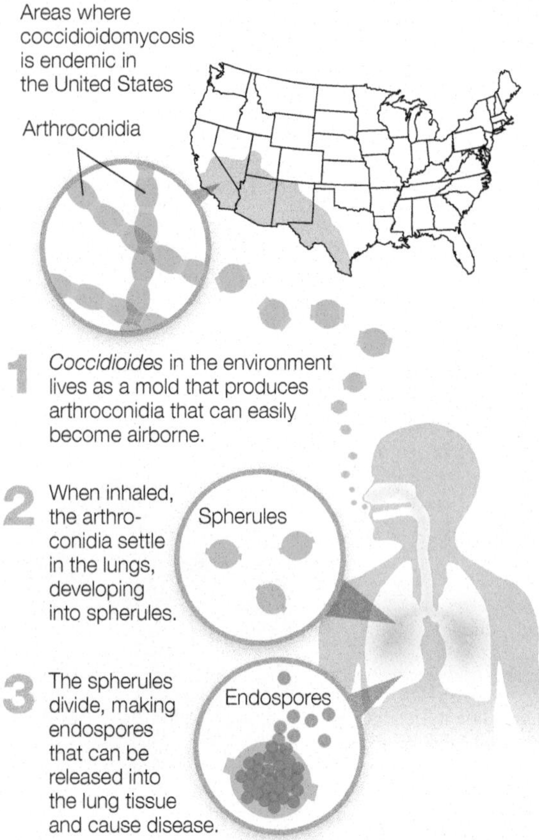

FIGURE 16.16 Coccidioidomycosis

endospores. The fungal endospores, morphologically and physiologically distinct from bacterial endospores, are then released into the lung tissue when the spherules rupture.

Histoplasmosis

Histoplasmosis (also called Darling's disease, cave disease, spelunker's lung, and Ohio Valley disease) is the most common endemic mycosis seen in humans. It is caused by the dimorphic fungus *Histoplasma capsulatum* (FIG. 16.17). The fungus, which grows as a mold mainly in soil enriched with bat and bird droppings, produces macroconidia and microconidia—types of asexual spores. The spores become airborne when dust or soil containing the fungus is disturbed. If inhaled into the lungs, spores germinate into the pathogenic, yeast-like form.

Exposure to the fungus is largely occupational, with poultry farmers and people who harvest guano (bat droppings) from caves for fertilizer at the greatest risk for infection. *H. capsulatum* is found in many parts of the world, but tends to be concentrated near river valleys. In the United States, *H. capsulatum* is abundant near the Mississippi and Ohio River valleys. Most people who live in areas where *H. capsulatum* is endemic have been exposed to the fungus; however, only about 5 percent of infections are symptomatic. Anyone can get histoplasmosis, but immune-compromised patients are more likely to develop the dangerous disseminated form, which occurs when the fungus spreads from the lungs to other parts of the body.

Ubiquitous fungi can cause serious infections in immune-compromised patients.

In addition to endemic fungi that tend to be isolated to specific geographical areas, there are several clinically important ubiquitous fungi, which are found throughout the world. The ubiquitous fungi discussed here are by no means the only ones that can affect the respiratory system, but they are some of the more common culprits. The vast majority of ubiquitous mycoses are caused by opportunistic fungi that require an immune-compromised host to establish an infection. Here we'll review three of the more common examples of ubiquitous respiratory mycoses—and even these are fairly rare.

DiseaseSnapshot • Histoplasmosis

Causative agent	*Histoplasma capsulatum*
Epidemiology	Much more common among populations with occupations that come into contact with bat or bird droppings. Where endemic, many people are exposed, but only around 5% are symptomatic. Immune-compromised patients are at greater risk. In Latin America, almost 30% of HIV patients coinfected with histoplasmosis die of the disease. Incubation period is 3–17 days.
Transmission & prevention	Transmits via inhaled spores. To prevent, avoid caves, or wear masks in areas known to harbor the fungus, including caves and chicken coops.
Signs & symptoms	Resembles atypical pneumonia (fever, chest pain, and dry, nonproductive cough). Fatigue and headache sometimes present. *H. capsulatum* can become disseminated into other parts of the body, with central nervous system infections being most severe.
Pathogenesis & virulence factors	Some of its cell wall polysaccharides can resist macrophage binding. Even when ingested by the macrophage, it can survive and scavenge iron from the cell.
Diagnosis & treatment	Detecting via immunoassays from a blood sample or urine is routine. Chest X-rays, microscopic examination of sputum or tissue, and culture methods are also used. The most common treatment is the antifungal itraconazole.

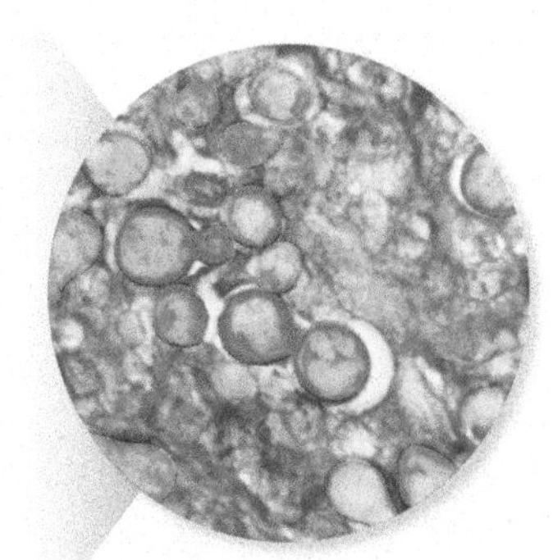

H. capsulatum (orange)

Aspergillosis

Aspergillus species are a common soil fungus throughout the world. The fungi make spores that are inhaled by most of us every day. There are five forms of **invasive aspergillosis**, with the pulmonary (or lung) and rhinocerebral (or brain-sinus) forms being the most common. Several species are associated with invasive disease, with *A. fumigatus* being the one mostly frequently isolated from patients. Invasive aspergillosis is almost exclusively seen in immune-compromised patients—especially among organ transplant patients, those undergoing chemotherapy for cancer, and people on high doses of corticosteroids. Unless invasive aspergillosis is treated early, the infection has a poor prognosis with a high mortality rate. As such, treatment is usually recommended before laboratory confirmation of the diagnosis. In the laboratory, the organism is usually cultured from a patient specimen or microscopically observed in the patient sample.

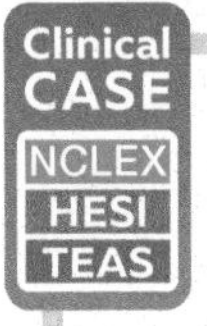

The Case of the Suffering Spelunker

Practice applying what you know clinically: visit the **Mastering Microbiology** Study Area to watch Part 3 and practice for future exams.

Mucormycosis

The fungi that cause **mucormycosis** are found in soil that is enriched with rotting wood material. Most of us encounter the fungi that can cause mucormycosis in our everyday lives and never develop illness. Several fungi can cause the disease; however, most infections are associated with inhaling spores made by *Rhizopus arrhizus*. There are five forms of invasive mucormycosis, but like aspergillosis, the most common are the pulmonary and rhinocerebral forms. The overall outcome for the patient can vary and is mostly influenced by how early the illness is detected, as well as the level of patient immunosuppression. Fortunately, mucormycosis is rare; in the United States fewer than 2 people per million in the population develop this illness.

Pneumocystis Pneumonia

Pneumocystis pneumonia (PCP) is a serious fungal infection seen in immune-compromised patients; it is very rare for a person with a normal immune system to develop PCP. The fungus *Pneumocystis jirovecii* causes it. Unlike all of the other fungi you have read about in this section, *P. jirovecii* is not native to the soil, and it does not make spores. In fact, the natural reservoir for this fungus is not entirely known, but it is thought that asymptomatic human carriers may transmit it. Despite the vagueness of a natural source, we do know that *P. jirovecii* is common enough that serological data show that most healthy children have been exposed to it by age 4.

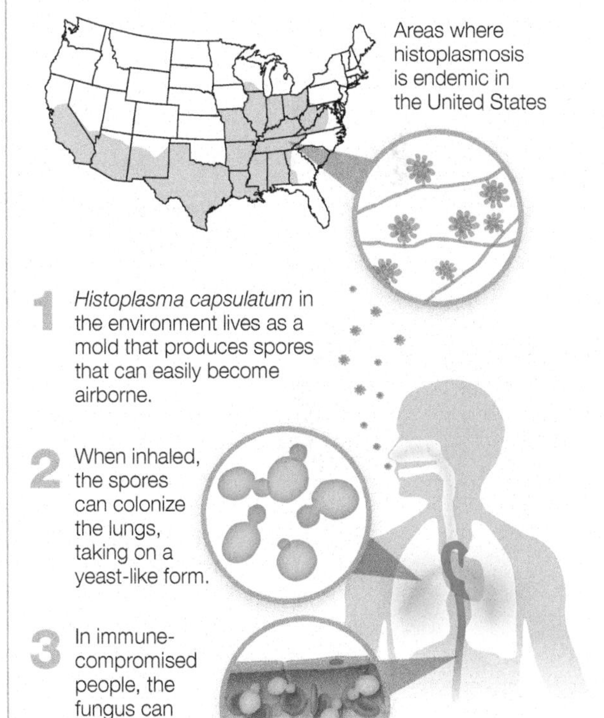

FIGURE 16.17 **Histoplasmosis**

DiseaseSnapshot • Aspergillosis

Aspergillosis species

Causative agent	Several species associated with invasive disease, but *A. fumigatus* is most common.
Epidemiology	Invasive aspergillosis is almost exclusively seen in immune-compromised patients, especially people undergoing cancer chemotherapy, those on high doses of corticosteroids, and those taking antirejection drugs after organ transplants. Incubation period is 2 days to 3 months.
Transmission & prevention	Transmits by inhalation of spores. There is no vaccine, and the only prevention is to avoid inhaling the spores.
Signs & symptoms	Fever, coughing up mucous plugs and possibly blood, and difficulty breathing.
Pathogenesis & virulence factors	RodA proteins and DNH-melanin on the surface of *A. fumigatus* help it avoid immune detection prior to changing to the yeast form. After the yeast form is engulfed by macrophages, it can prevent acidic digestion inside the phagolysosome. Infected cells are also prevented from undergoing apoptosis.
Diagnosis & treatment	Diagnosis is made by a positive skin test that detects immune reaction to *A. fumigatus*, serological tests to detect antibodies to *A. fumigatus*, or positive culture from sputum or tissue biopsy. Treatment starts immediately upon positive results, but the diagnosis is usually confirmed via culture from sputum. Intravenous voriconazole is the first-line treatment, sometimes in combination with other agents such as caspofungin, amphotericin B, and echinocandins. Itraconazole can also be used if first-line treatment fails. Mortality rate is 30–90% in symptomatic cases.

The natural source of this fungus is not the only thing that is baffling about this organism. For about 80 years it was widely thought to be a protozoan, as it has several characteristics of protozoan organisms. Also, *P. jirovecii* has a trophozoite stage of development—a common feature in protozoans. Furthermore, it is still not exactly understood how people become infected with *P. jirovecii*, but there is increasing evidence that it is transmitted from person to person in aerosols. High-risk patients such as those with HIV may be given low doses of trimethoprim-sulfamethoxazole daily or three times a week to help prevent PCP. About 9 percent of hospitalized HIV patients and about 1 percent of solid organ transplant patients develop PCP. Untreated cases are usually fatal, and even with treatment up to 40 percent of patients die.

DiseaseSnapshot • Mucormycosis

Rhizopus species

Causative agent	Most commonly *Rhizopus arrhizus*
Epidemiology	In the U.S., fewer than 2 people per million in the population develop this illness. However, in some countries, cases rose 70% in the past decade. The fungus is usually found in rotting wood, and human exposure is common. Infection is usually confined to immune-compromised patients. If the infection is detected late in immune-compromised patients, mortality can be as high as 50%. Incubation is 3–10 days.
Transmission & prevention	Transmits through respiratory route via inhaled fungal spores. No preventive vaccine or methods, beyond avoiding the spores.
Signs & symptoms	Fever, chest pain, cough, and shortness of breath are common pulmonary symptoms.
Pathogenesis & virulence factors	Proteins can bind and take up iron from the host, bind endothelial cells, and enter the bloodstream. Some strains can decrease the oxidative burst from phagocytes as well.
Diagnosis & treatment	Immunoassays from sputum is routine. Chest X-rays, microscopic examination of sputum or tissue, and culture methods are also used. Amphotericin B, often in lipid form to allow for high doses, is administered.

DiseaseSnapshot • *Pneumocystis* Pneumonia

Causative agent	*Pneumocystis jirovecii*
Epidemiology	Serological data show most healthy children have been exposed to the fungus by age 4. The majority of infections occur in immune-compromised patients. Untreated cases are typically fatal. Up to 40% of treated cases are fatal. Incubation period is 4–8 weeks.
Transmission & prevention	Transmission is unclear; it may be via direct contact (person to person) from asymptomatic carriers. Those at high risk (immune-compromised people), especially in healthcare facilities, are sometimes given low doses of trimethoprim-sulfamethoxazole to prevent infection.
Signs & symptoms	Symptoms include fever, fatigue, dyspnea, and a dry, nonproductive cough.
Pathogenesis & virulence factors	The fungal cells can vary major surface glycoproteins to evade the host immune system. The same surface proteins help with host cell attachment and binding of iron.
Diagnosis & treatment	Diagnosed by microscopic observation of the fungus in a patient's sputum sample; bronchoscopy or lung biopsy may also be useful for diagnosis. Treated with trimethoprim-sulfamethoxazole because traditional antifungals are ineffective.

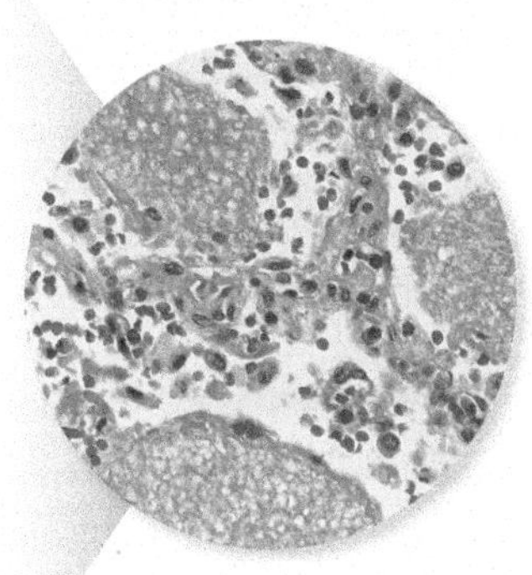

P. jirovecii (black)

Build Your Foundation

22. How could deforestation increase the number of mycoses in a population? (LO 16.18)

23. Compare and contrast endemic mycoses with ubiquitous mycoses. (LO 16.19)

24. List four shared features of blastomycosis, coccidioidomycosis, and histoplasmosis. (LO 16.20)

25. List four shared features of aspergillosis, mucormycosis, and *Pneumocystis* pneumonia. (LO 16.21)

Build Your Foundation (BYF) Quick Quiz: Visit the **Mastering Microbiology** Study Area to quiz yourself.

VISUAL SUMMARY | *Respiratory System Infections*

Otitis Media (Bacterial Infections)

- Usually *Streptococcus pneumonia*, *Moraxella catarrhalis*, and nontypable *Haemophilus influenzae*
- Colds can cause eustachian tubes to become inflamed and then mucus doesn't drain well (especially in children), which sets the stage for bacterial infection.

Colds (Viral Infections)

- Rhinoviruses and coronaviruses, but also parainfluenza viruses, adenoviruses, and nonpolio-type enteroviruses
- Self-limiting; can predispose patient to secondary bacterial infections

Fungal Lung Infections

- Infections that can impact otherwise healthy people include **blastomycosis** (*Blastomyces dermatitidis*), **coccidioidomycosis** (*Coccidioides immitis* and *Coccidioides posadasii*), and **histoplasmosis** (*Histoplasma capsulatum*).
- Infections that impact immune-compromised patients include **aspergillosis** (*Aspergillus* species), **mucormycosis** (*Rhizopusarrhizus*), and **pneumocystis pneumonia** (*Pneumocystis jirovecii*), which is most common in hospitalized HIV/AIDS patients.
- Fungal infections that reach the bloodstream have high mortality rates.

Viral Lung Infections

Respiratory Syncytial Virus (RSV)

- Leading cause of acute lower respiratory tract infection in children under age 5
- Antibody preparations available for high-risk patients

Influenza

- Influenza viruses types A, B, and C
- Antigenic drift associated with seasonal outbreaks; antigenic shift linked to pandemics

COVID-19

- Caused by SARS-CoV-2
- Remdesivir FDA approved for hospitalized patients weighing ≥40Kg and ≥12 years of age

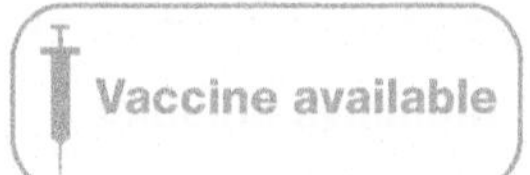

Bacterial Throat Infections

Strep Throat

- Streptococcal pharyngitis (strep throat) caused by *Streptococcus pyogenes*
- Features a fever, sore throat (pus may be evident), and swollen cervical lymph nodes
- Bacterial strains that make erythrogenic toxin cause scarlet fever
- Rheumatic fever is autoimmune complication

Diphtheria

- *Corynebacterium diphtheria*
- Usually in children under age 5; bacteria form a thickened leathery structure called a pseudomembrane on the tonsils and throat

Bacterial Lung Infections

Pertussis

- Pertussis (whooping cough) is caused by the bacterium *Bordetella pertussis*
- Considered reemerging
- Most serious cases in children under 1 year old

Tuberculosis

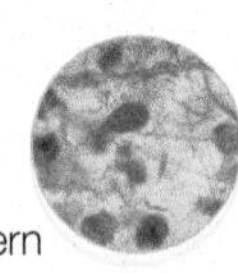

- Tuberculosis is caused by the bacterium *Mycobacterium tuberculosis*, an acid-fast bacillus; drug resistance is a growing concern
- Vaccine not routinely given in the U.S.
- Tuberculin skin test screens for infection in nonendemic areas
- Infection can be active or latent

Typical Bacterial Pneumonia

- Productive cough, high fever, severe chills, chest pain upon inhaling, fluid is usually present in lungs
- Most commonly *Streptococcus pneumoniae*

Atypical Bacterial Pneumonia

- Nonproductive cough, low-grade fever, no chest pain upon inhaling, and limited fluid in lungs
- Examples include walking pneumonia (*Mycoplasma pneumoniae*), *Chlamydophilia pneumonia* (*Chlamydophila pneumoniae*), Legionaires' disease (*Legionella pneumophila*), Psittacosis (*Chlamydophila psittaci*), Q fever (*Coxiella burnetii*), and Tularemia (*Francisella tularensis*)
- Often agent is not successfully cultivated from sputum sample

Pneumonia is not a single disease but a general term that describes inflammation of the alveoli in the lungs.

Alveoli

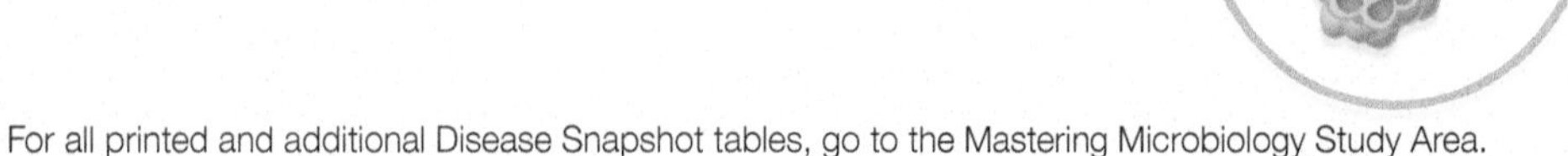

For all printed and additional Disease Snapshot tables, go to the Mastering Microbiology Study Area.

16 CHAPTER 16 OVERVIEW

16.1 Overview of the Respiratory System

- The respiratory system is the most common portal of entry for pathogens; most respiratory infections are transmitted by respiratory droplets.
- The respiratory tract is divided into upper and lower segments.
  - The upper respiratory tract warms, humidifies, and filters air; it consists of the nasal passages, paranasal sinuses, and the pharynx (throat).
  - The lower respiratory tract directs air to the lungs, where gas exchange occurs; it includes the larynx (voice box), trachea (windpipe), bronchi, bronchioles, and lungs.
- Ciliated mucous membranes in the airway use a mucociliary sweeping action to keep debris-laden mucus out of the lungs.
- Infectious agents or inhaled allergens can cause inflammation of respiratory tract structures.
  - Conditions caused by inflammation are named for the affected structure: sinusitis (sinus membranes, which can lead to a sinus infection), pharyngitis (pharynx or throat), epiglottitis (epiglottis), laryngitis (larynx), tracheitis (trachea), and laryngotracheobronchitis (larynx, trachea, bronchi, and bronchioles, also known as croup).
  - Pneumonia is an inflammation of the alveoli, which is where gas exchange occurs in the lungs.
- Normal microbiota of the upper airway include species of Proteobacteria, Firmicutes, Bacteroidetes, and Actinobacteria. These usually help to prevent infection. There are small numbers of bacteria (usually Proteobacteria) that colonize the lower respiratory tract.
- Respiratory allergies can be confused with respiratory infections because they share several symptoms and are both characterized by inflammation of respiratory tract structures.

16.2 Viral Infections of the Respiratory System

- Viruses are the most common cause of respiratory infections; they cannot be treated with antibiotics.
- Colds are caused by hundreds of viruses, but coronaviruses and rhinoviruses are the most common.
  - Colds are self-limiting but can predispose someone to a secondary infection by bacteria; otitis media, bronchitis, and sinus infections are common examples.
  - Respiratory syncytial virus (RSV), adenoviruses, and human parainfluenza viruses all cause cold-like illness that have additional clinical features.
- Influenza is an acute respiratory tract infection that shares characteristics with colds, but is more severe. Frequent changes in HA and NA spikes necessitate strain monitoring and booster vaccines.
- The pandemic influenza of 1918 exhibited extremely high morbidity and mortality; this pandemic shaped policies and practices regarding influenza management.
- The novel coronaviruses include SARS-CoV-2, which causes COVID-19. This RNA virus frequently mutates, so monitoring variants for booster vaccine development will be necessary.
- New World hantaviruses cause hanta pulmonary syndrome, a rare but dangerous infection.

16.3 Bacterial Infections of the Respiratory System

- Otitis media is a common complication of colds, especially in children.
- Group A streptococci (*Streptococcus pyogenes*) causes streptococcal pharyngitis and other infections.
  - Scarlet fever is caused by a strain of *S. pyogenes* that makes an erythrogenic toxin.
  - Autoimmune complications of streptococcal pharyngitis include rheumatic fever.
- *Corynebacterium diphtheriae* causes diphtheria; the DTaP vaccine and Tdap booster vaccine can prevent it.
- Pertussis (whooping cough) is a prolonged respiratory illness caused by *Bordetella pertussis*; the DTaP vaccine and Tdap booster vaccines are used for prevention—though newly emerging strains of the bacterium may be able to avoid vaccine-induced immunity.
- The acid-fast bacterium, *Mycobacterium tuberculosis*, is the most common cause of tuberculosis (or TB).
  - *M. tuberculosis* can cause active or latent infections, and a vaccine (bacillus Calmette–Guérin) is often used to protect children where TB is endemic.
  - Diagnostic tools for TB include: bacterial culture from patient sputum samples, microscopic examination of sputum for acid-fast bacilli, chest X-rays, the tuberculin skin test (TST), interferon gamma release assay (IGRA), and molecular screening techniques such as the Xpert MTB/RIF test.
  - TB treatment with antibiotics is prolonged (many months) and complicated by the existence of multidrug-resistant and extensively drug-resistant strains.
- Bacterial pneumonia can be described as typical or atypical.
  - Typical pneumonia includes *Haemophilus influenza* and pneumococcal pneumonia (which is caused by *Streptococcus pneumoniae*). Pneumococcal pneumonia is the most common type of pneumonia; it is the role model for describing typical pneumonia syndrome. Pneumococcal conjugate vaccines can prevent pneumococcal disease.
  - The six leading atypical bacterial pneumonias are: *Mycoplasma pneumoniae*, *Chlamydophila pneumoniae*, Legionnaires' disease, psittacosis, Q fever, and tularemia.

16.4 Fungal Respiratory System Infections

- Endemic mycoses such as blastomycosis, coccidioidomycosis, and histoplasmosis mainly cause illness in specific geographic regions.
- Mycoses such as *Pneumocystis* pneumonia, mucormycosis, and aspergillosis are found all over the world, but mainly cause disease in immune-compromised patients.

SMART

THINK CLINICALLY | *Be S.M.A.R.T. About Cases*

COMPREHENSIVE CASE

The following case integrates basic principles from the chapter. Try to answer the case questions on your own. Don't forget to be **S.M.A.R.T.*** about your case study to help you interlink the scientific concepts you have just learned and apply your new understanding of respiratory system infections to a case study.

*The five-step method is shown in detail in the Chapter 1 Comprehensive Case on cholera. See pages 31–33. Refer back to this example to help you apply a SMART approach to other critical thinking questions and cases throughout the book.

• • •

As an epidemiologist working for the World Health Organization for the past five years, Kelsey has lived in many parts of the world. For the past 2 years she served in the Emerging Disease Surveillance and Response Team stationed in Manila, Philippines. About 2 months ago she traveled to the United States for a meeting. Over the past month, she felt fatigued and had lost her appetite. Although she did not have a fever or chills, she did have a cough that occasionally produced yellow-tinged sputum. Kelsey went to a clinic in Manila, where she was prescribed a broad-spectrum antibiotic (clarithromycin) to treat what the doctor felt was probably a case of bronchitis.

After finishing the course of antibiotics, Kelsey still felt fatigued, and her cough persisted. Within 2 days of finishing her course of clarithromycin, she developed a fever, chills, night sweats, and occasionally coughed up sputum that was tinged with blood (hemoptysis). After 2 days of these worsening symptoms, she returned to the clinic where she meets you—the physician assistant on duty at the time.

After your routine exam, you discover that Kelsey has lost 10 pounds in the past 4–6 weeks. Questioning her further, you learn she is married and has been in good health until recently, she isn't diabetic, and she is HIV negative. Kelsey is a nonsmoker who is up to date on her standard vaccines and booster shots. Her vaccine records show she had the bacillus Calmette–Guérin vaccine as an infant. She did have a positive tuberculin skin test last year, but no signs of latent or active TB were found afterward.

You decide to order a chest X-ray, take a sample of her sputum for culture and microscopy analysis, and also order an interferon gamma release assay. You do not order a tuberculin skin test. While you wait for the test results, you admit Kelsey to the clinic and assign her to an isolation room.

• • •

CASE-BASED QUESTIONS

1. As the PA on duty, you have developed a working diagnosis to help you care for Kelsey. What disease do you suspect? What clues led you to your working diagnosis?
2. Provided your working diagnosis is *correct*, what results do you expect to see from her chest X-ray, microscopy and culture of the sputum sample, and interferon gamma release assay?
3. If your working diagnosis is *wrong*, what would those tests show? (That is, would her X-ray be normal or abnormal, and would her interferon gamma release assay be positive or negative?) What microscopy and culture of sputum findings might cause doubt about your initial diagnosis?
4. Explain why you did not order a tuberculin skin test. Was this a mistake on your part?
5. Why is the patient's HIV status important in this case?
6. Explain why you put Kelsey in isolation.
7. Why did the physician who first saw Kelsey assume she had bronchitis?
8. If the chest X-ray comes back abnormal, but bacteria are not observed in or cultured from the sputum, and the interferon gamma release assay is negative, what are some possible differential diagnoses (other illnesses the patient could have, given her signs and symptoms)?

END OF CHAPTER QUESTIONS

NCLEX
HESI
TEAS

Think Critically and Clinically
Questions highlighted in orange are opportunities to practice NCLEX, HESI, and TEAS-style questions.

1. Define the following key terms.
   a. Fomite
   b. Mucociliary escalator
   c. Alveolar macrophage
   d. Prognosis
   e. Serovar
   f. Otitis media
   g. Tympanic membrane
   h. Epidemic
   i. Pandemic
   j. Antigenic drift
   k. Herd immunity
   l. Antigenic shift
   m. Rheumatogenic
   n. Attenuated viral strain
   o. Inactivated viral strain
   p. Pulmonary edema
   q. Community-acquired pneumonia (CAP)
   r. Healthcare-associated pneumonia (HCAP)
   s. Typical pneumonia
   t. Atypical pneumonia
   u. Endemic mycoses
   v. Dimorphic fungus

2. List the anatomical structures of the human upper respiratory tract.

3. List the anatomical structures of the human lower respiratory tract.

4. Which of the following can lead to pneumonia?
   a. Viruses
   b. Bacteria
   c. Fungi
   d. All of the above

5. Most cases of croup are caused by
   a. bacteria.
   b. viruses.
   c. trauma injuries.
   d. fungi.
   e. allergens.

6. Select the FALSE statement:
   a. Typical pneumonia can be community acquired.
   b. *Streptococcus pneumoniae* is one cause of typical pneumonia.
   c. Viruses are the only cause of atypical pneumonia.
   d. Atypical pneumonia is not usually characterized by consolidation.

7. Rapid diagnosis and treatment of streptococcal pharyngitis (strep throat) are important mainly because
   a. streptococcal pharyngitis is painful.
   b. streptococcal pharyngitis is highly contagious.
   c. failure to treat streptococcal pharyngitis can compromise the efficiency of the DTaP vaccine.
   d. streptococcal pharyngitis can cause certain autoimmune complications.
   e. not treating streptococcal pharyngitis leads to antibiotic resistance.

8. The ________ vaccine is a childhood vaccine to protect against tetanus, diphtheria, and pertussis; in contrast, the ________ vaccine is a booster shot recommended for adolescents and adults to protect against tetanus, diphtheria, and pertussis.

9. List the three stages of pertussis, and provide some general characteristics of each stage.

10. Give two reasons why the bacillus Calmette–Guérin (BCG) vaccine is not routinely recommended in the United States to prevent tuberculosis.

11. Indicate the true statements, and then correct the false statements so that they are true.
    a. The most common viruses that cause human respiratory system infections are cold viruses.
    b. Viruses can be treated with antibiotics.
    c. People who have been exposed to cold and/or wet conditions are statistically more likely to develop a cold.
    d. Many respiratory illnesses that are not caused by influenza viruses have flu-like symptoms.
    e. The flu vaccine is not recommended for children under 2 years of age.
    f. Pneumonia is a rare healthcare-acquired infection.

12. Match the following:

Mycosis	An above-normal number of breaths per minute
Bacteremia	Often responsible for seasonal epidemics
Dyspnea	Disease caused by fungi
Stridor	Wheezing or loud breathing associated with a blocked or narrowed airway
Antigenic drift	Can lead to pandemics
Antigenic shift	A clinical condition characterized by bacteria in the bloodstream
Tachypnea	A clinical condition characterized by fluid in the alveoli
Consolidation	Shortness of breath

13. Choose the general agent of the disease (select ONE choice of agent from the options after each illness):
    a. Pneumococcal pneumonia is caused by a (virus, bacterium, fungus).
    b. Pneumocystis pneumonia is caused by a (virus, bacterium, fungus).
    c. Tuberculosis is caused by a (virus, bacterium, fungus).
    d. Histoplasmosis is caused by a (virus, bacterium, fungus).
    e. Influenza is caused by a (virus, bacterium, fungus).
    f. Colds are caused by (viruses, bacteria, fungi).
    g. Coccidioidomycosis is caused by a (virus, bacterium, fungus).
    h. Diphtheria is caused by a (virus, bacterium, fungus).
    i. Pertussis is caused by a (virus, bacterium, fungus).
    j. Scarlet fever is caused by a (virus, bacterium, fungus).
    k. Legionellosis is caused by a (virus, bacterium, fungus).
    l. Q fever is caused by a (virus, bacterium, fungus).
    m. Hanta pulmonary syndrome is caused by a (virus, bacterium, fungus).
    n. Streptococcal pharyngitis is caused by a (virus, bacterium, fungus).

14. Create a Venn diagram like the following one to compare and contrast latent and active tuberculosis.

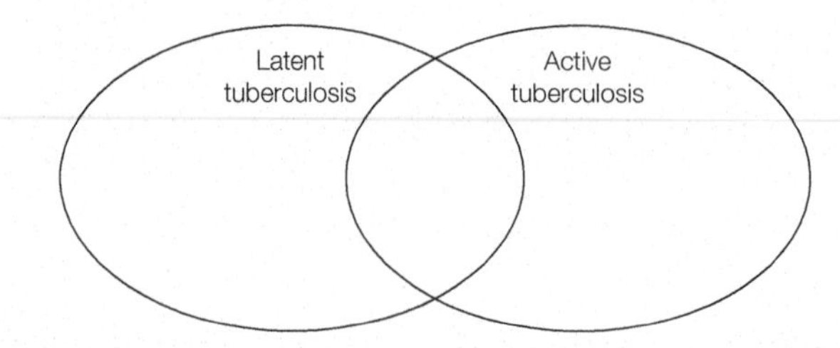

15. From the following list, select *all* of the diseases that antibiotics do not directly cure.
    a. Hanta pulmonary syndrome
    b. *Streptococcus pneumoniae*
    c. Colds
    d. Influenza
    e. *Haemophilus influenzae*
    f. COVID-19
    g. Legionellosis
    h. Tularemia

16. Select the true statements about SARS-CoV-2. (Select all that apply.)
    a. It causes COVID-19.
    b. It is vaccine preventable.
    c. The FDA has approved hydroxychloroquine to treat infections with this virus.
    d. It is considered a zoonotic infection.
    e. It causes a form of viral pneumonia.

17. Which of the following would you expect for a SARS-CoV-2 variant that is classified as a variant of interest? (Select all that apply.)
    a. The isolated virus is genetically unique as compared to earlier circulating versions of SARS-CoV-2.
    b. There is evidence that the variant completely evades vaccine-induced immunity.
    c. The variant is broadly detectable with current diagnostic tools.
    d. The variant can cause viral pneumonia.
    e. Infections caused by the variant are not treatable with existing approved drug therapies.

CRITICAL THINKING QUESTIONS

1. Explain why accurately diagnosing typical pneumonia as compared to influenza is an important clinical step to preventing antibiotic resistance.

2. Why is a universal vaccine possible for the flu but highly unlikely for colds?

3. The following graph shows data for two groups (A and B). Each group contained 1,000 study participants. Study participants were vaccinated with the given flu vaccine for that season and then followed over the flu season for that year to monitor if they developed true influenza or a flu-like illness (not truly the flu, but mimicking the flu). Use the graph to answer the following questions:

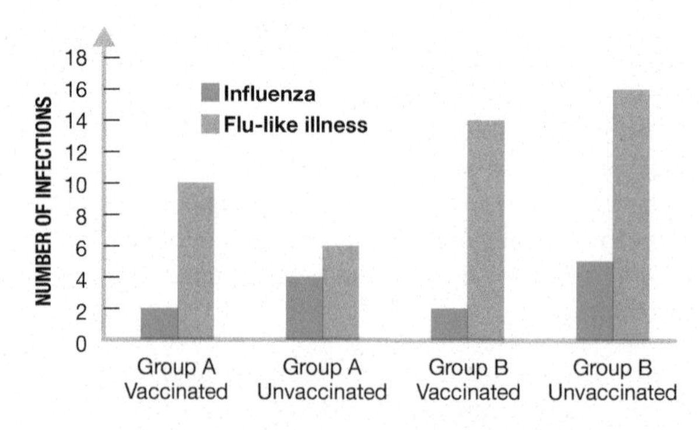

    a. Based on the data presented in the graph, was the influenza vaccine generally effective at preventing influenza in group A? What about group B?
    b. Based on the data presented in the graph, was the influenza vaccine generally effective at preventing influenza-like illness in group A? What about group B?
    c. How might people develop a misconception about the value of the flu vaccine if they only considered development of a flu-like illness as their criteria for determining vaccine efficacy? Use the data in the graph to develop your position.

Skin and Eye Infections

What Will We Explore? There are a myriad of diseases that impact the skin and eyes. This chapter addresses key structures and defenses of these important organs, and describes how these tissues may be compromised by infectious diseases, as well as the types of microbes that may afflict them.

Why Is It Important? Since the 20th century we've made great progress in our understanding and treatment of many microbial skin and eye diseases. Vaccines to prevent chickenpox, measles, and rubella have saved millions of lives. The eradication of smallpox, a deadly and feared ancient disease, is a hallmark of our modern medical era. But today's healthcare providers also face emerging problems with antibiotic-resistant bacterial skin diseases such as methicillin-resistant *Staphylococcus aureus* (MRSA), which can cause devastating tissue destruction and even death. Also, due to fewer children being vaccinated than recommended by federal guidelines, the reemergence of measles has become an increasing public health concern—showing that even if a disease has long been absent from an area, it can rapidly return and cause problems anew, if public safety protocols are not understood and embraced by the public.

Clinical CASE
NCLEX
HESI
TEAS

The Case of the Hidden Blinder

Visit the **Mastering Microbiology** Study Area to watch the case and find out how an eye infection can explain this medical mystery.

For organizational purposes, this chapter discusses diseases where a skin rash or lesion is the main manifestation or primary route to diagnosis, irrespective of the primary body system infected or the route of transmission.

Laly Silva
APRN, Family Medicine, Nurse Practitioner; Jacksonville, FL

17.1 OVERVIEW OF SKIN STRUCTURE, DEFENSES, AND AFFLICTIONS

Learning Outcomes

After reading this section, you should be able to:

17.1 Describe the parts of the skin and how they are protective.

17.2 Specify examples of normal skin microbiota and explain where they are found.

17.3 Define the following terminology of skin lesions and rashes: vesicle, macule, papule, pustule, cyst, vesicular rash, papular rash, and maculopapular rash.

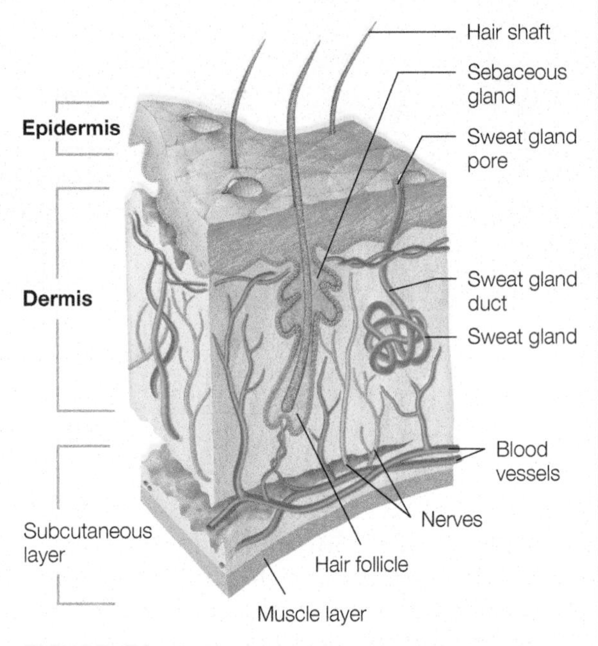

FIGURE 17.1 Skin's two distinct layers: the epidermis and the dermis The epidermis consists of several layers of dead skin. The thicker dermis layer further protects the subcutaneous fat layer below.

Critical Thinking *Explain why a lotion containing antimicrobial peptides could be a useful product to protect burned skin.*

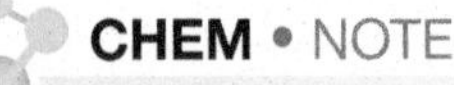

CHEM • NOTE

pH scale Low pH (below 7) is acidic, high pH (above 7) is basic, and a pH of exactly 7.0 is neutral. Acids lower pH and bases raise pH. Perspiration causes the pH of skin to be 4–5.5.

CHEM • NOTE

Fatty acid metabolism Fatty acids are broken down into a wide variety of organic molecules. Some bacteria use these byproducts in their metabolic processes, while others cannot—and may even be killed by them (as occurs when sebum is broken down).

The skin, our largest organ, has specialized defenses.

As the largest body organ, the skin's structure is designed to keep out infectious agents (FIG. 17.1). It's helpful to imagine the skin as a medieval, walled city's defenses, protecting the inner areas from the hostile outside world—in this case, from potentially damaging invaders such as ultraviolet light, chemicals, microbes, or anything else that might breach the surface.

First, would-be invaders come in contact with a city's outer wall. In the skin the outermost defensive structure is the **epidermis**, made up of several layers of cells. The outer layer consists of tightly packed dead cells. Below these layers, the inner basal level of cells is alive and protected. Most of the epidermis is made of keratinocytes, specialized cells that contain **keratin**, a waterproofing protein. While many microbes may come in contact with the body (city), so long as the epidermis (outer wall) remains intact, microbes are unable to gain access to the body.

If the epidermis is penetrated, invaders will come upon the thicker **dermis**. This is connective tissue comprised of dermal fibroblasts (cells that make connective tissue to help skin heal and replenish itself), collagen fibers, and immune system cells. This layer also contains blood vessels, nerves, hair follicles, sweat gland ducts, and sebaceous glands (oil glands). In our imaginary battle, the dermal layer can be seen as soldiers guarding the inner wall. Below the dermis are fat cells, nerves, and blood vessels that comprise the **subcutaneous layer**. This layer is closely associated with the skin, but is not technically considered a skin layer.

The Skin's Defense Mechanisms

Our dermis warriors have defensive weapons. While some sun exposure allows our bodies to manufacture vitamin D, too much exposure to ultraviolet light isn't healthy. **Melanin**, the brown pigment in our skin, has antimicrobial properties and protects against ultraviolet radiation that depletes the skin of folic acid and causes skin damage. **Perspiration** (sweat) is a particularly effective weapon against microorganisms—the pH is too acidic for some microorganisms, while the salt content is too high for others. **Lysozyme**, an enzyme present in sweat, breaks down the peptidoglycan found in bacterial cell walls. In addition to this antimicrobial action, sweat may also wash away microbes, or be wiped away, carrying microbes with it.

Sebum is an oily or waxy substance produced by the skin's **sebaceous glands** (oil glands) and is a combination of low-pH lipids and proteins that moisturize and further protect the skin. While some common skin bacteria metabolize sebum, the resulting breakdown products of fatty acids are toxic to other microbes, thereby limiting what organisms colonize the skin. The skin also manufactures a number of antimicrobial peptides for defense.

Skin Microbiome

Normal microbiota reside on the epidermis as well as in sweat glands, sebaceous glands, and hair follicles. Even when we wash our hands, many microbes still remain. Scrubbing just exposes new layers of the diverse species that reside under the more superficial skin microbiota.

Environmental exposure, occupation, antibiotic usage, age, cosmetics, moisturizers, choice of soaps or clothing, and many other factors impact our normal skin biota profiles. Hormone levels also affect the skin environment by altering the constitution and quantity of skin secretions, which is why acne is common when hormone levels change in pregnancy, puberty, or even with a woman's menstrual cycle. The increased sebum production that occurs in puberty is also associated with shifts in normal skin biota that may lead to chronic acne.

Gram-positive *Staphylococcus* species are abundant skin residents (FIG. 17.2). The Human Microbiome Project surprised many when it found that the Gram-negative rods *Pseudomonas* and *Janthinobacterium* species were the most prevalent on certain skin sites such as the forearm/inner elbow.[1] Skin moisture also affects which bacteria are supported in certain skin areas. Recent work has shown fungi to be abundant among the normal skin biota. Genetic analysis reveals that the fungi *Malassezia* are widespread skin residents. Interestingly, our heels (which harbor up to 80 different genera of fungi) seem to be the most complex in their fungal inhabitants.[2] The most important issue to keep in mind is that the skin is a complex ecosystem where species type varies greatly from site to site, so that the species found in warmer, moister areas such as the inguinal (groin) region are very different from the species found in dryer, cooler areas like the forearm or scalp.[3]

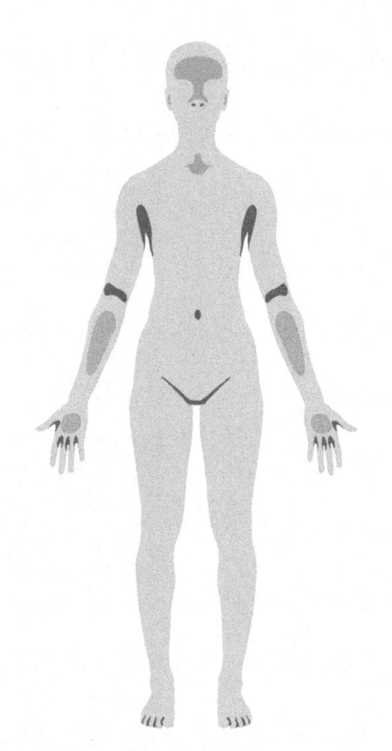

Skin type

Oily
Actinobacteria
Bacteroidetes
Staphylococcus
Malassezia

Moist
Actinobacteria
Bacteroidetes
Cyanobacteria
Proteobacteria
Staphylococcus
Malassezia
up to 80 different genera of fungi on heels

Dry
Actinobacteria
Bacteroidetes
Proteobacteria
Staphylococcus
Pseudomonas
Janthinobacterium

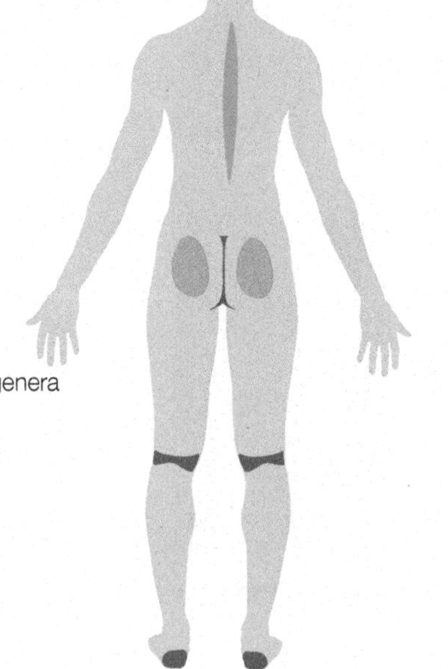

*Non-italicized names are phyla; italicized names are genera

FIGURE 17.2 **Skin microbiome** What microbes prevail on our skin depends on which regions are sampled and whether or not they are oily, moist, or dry. Sampled regions were those likely to exhibit bacterial infections. *Staphylococcus* are well-known skin inhabitants and have been easily cultured from diverse skin sites.

Rashes and lesions are typical skin afflictions.

Dermatologists are medical specialists concerned with defining and describing a variety of *dermatoses* (infections and noninfectious skin diseases) as well as the normal anatomical and physiological features of skin. However, knowledge of skin lesions and rashes is essential for any competent healthcare provider. The descriptions here are not exhaustive—rather, they serve as general guidelines.

Classification of Rashes and Lesions

Lesion is the clinical term for any observable abnormality of the skin. **Primary lesions** are directly associated with a disease and are considered key features for diagnosing a variety of infections. **Secondary lesions** are less strictly defined and have diverse origins. They may evolve from primary lesions, or from external forces like trauma or scratching; they are also seen during the progression of skin damage from certain infectious agents or allergic reactions. Examples of primary and secondary lesions are presented in TABLE 17.1 and TABLE 17.2. You should become familiar with these terms, as they are central to diagnosis and care.

CLINICAL VOCABULARY

Lesion change or abnormality in the skin that is usually in a defined area; may be harmless or serious

Primary lesion closely associated with a specific disease process; useful for diagnosis

Secondary lesions have diverse origins and are less obviously associated with a specific disease; may develop from a primary lesion

Rash a more widespread eruption of lesions; may be symptomatic or asymptomatic

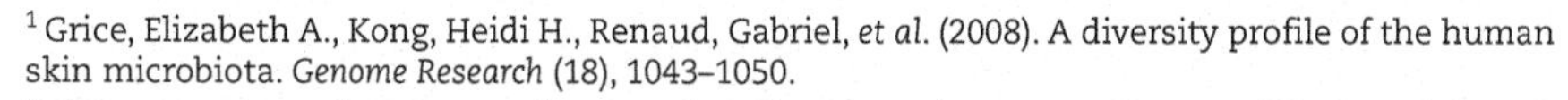

[1] Grice, Elizabeth A., Kong, Heidi H., Renaud, Gabriel, *et al.* (2008). A diversity profile of the human skin microbiota. *Genome Research* (18), 1043–1050.

[2] NIH press release, "NIH researchers conduct first genomic survey of human skin fungal diversity." Embargoed for Release: Wednesday, May 22, 2013, 1 p.m. EDT from http://www.nih.gov/news/health/may2013/nhgri-22.htm

[3] Byrd, A. L., et al. (2018). The human skin microbiome. *Nature Reviews Microbiology*, 16(3), 143–155.

Build Your Foundation

1. List the parts of the skin and explain how each is protective.
2. Describe the types of skin microbiota and how they are protective.
3. Compare and contrast different types of skin lesions and rashes.

Build Your Foundation (BYF) Quick Quiz: Visit the **Mastering Microbiology** Study Area to quiz yourself.

TABLE 17.1 Primary Lesions

Name and Description	
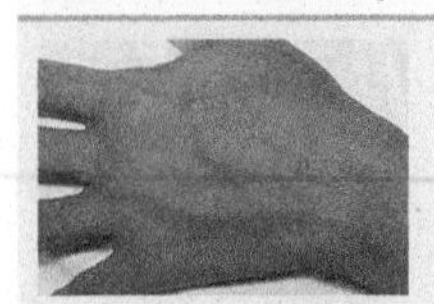	**Vesicle/Vesicular rash:** Small, elevated lesion filled with clear fluid *(Example: chickenpox)*
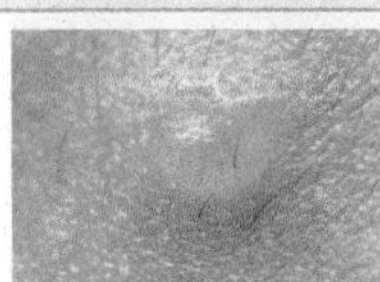	**Bulla (plural bullae):** Vesicle more than 0.5 cm in diameter *(Example: blister)*
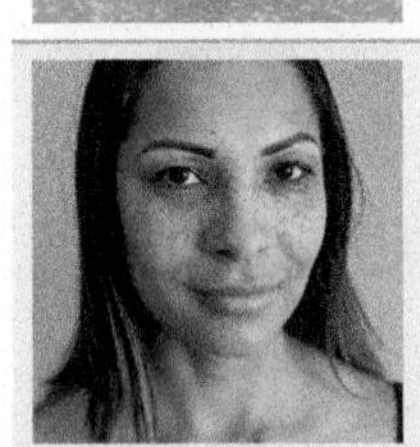	**Macule:** Flat discolored area of the skin; doesn't alter thickness or texture of skin *(Examples: freckles, café-au-lait spots)*
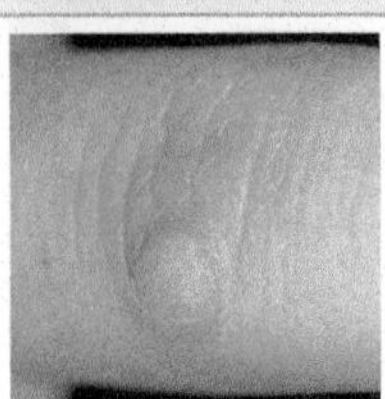	**Papule (papular rash):** Raised solid lesion (not fluid filled); may be discolored; has distinct borders; less than 0.5 cm in diameter *(Example: warts)*
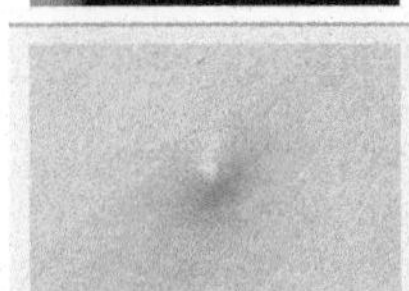	**Pustule:** Raised lesion with pus below the surface *(Example: inflammatory acne)*
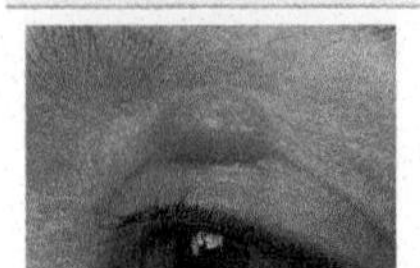	**Cyst:** Closed fluid filled sac; deeper in skin; usually painless unless infected or ruptured *(Example: nodular cystic acne)*
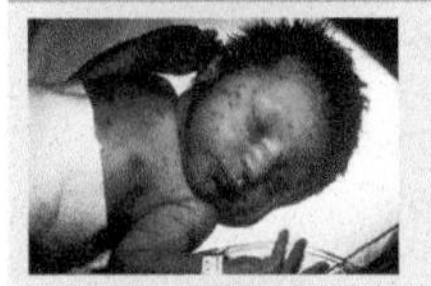	**Maculopapular rash:** Small slightly raised papule lesions that overlay or are interspersed with macules *(Examples: rubella, rubeola)*

TABLE 17.2 Secondary Lesions

Name and Description	
	**Crust:** Dried exudate or plasma on the skin *Honey-colored crusted lesions seen in staphylococcal impetigo*
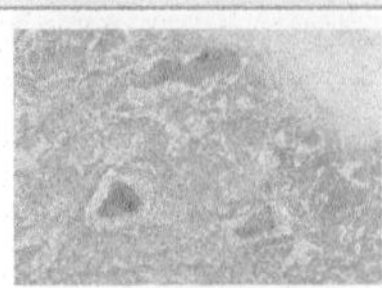	**Scale:** Outer epidermal layers flaking off of lower skin layers as a result of tissue damage and/or infection *Dermatophytic infections such as athlete's foot and skin conditions like psoriasis*
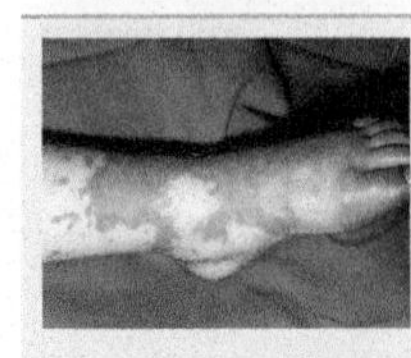	**Purpura:** Rash of large purple spots that evolve from bleeding in skin layers (subcutaneous bleeding); does not turn white when skin is depressed (no blanching); measure 3–10 mm *Bloodstream infections (septicemia/bacteremia); meningococcal meningitis; also seen in many noninfectious scenarios*
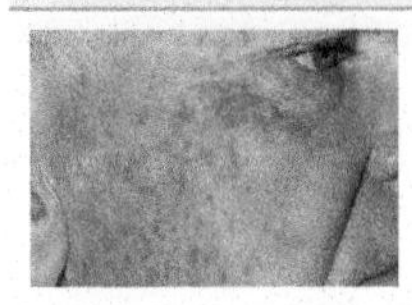	**Petechiae:** Same as purpura but smaller (less than 3 mm); often described as pinpoint spots; arise when tiny capillaries burst *Congenital cytomegalovirus infection, scarlet fever, bacterial endocarditis, excessive strain from vomiting or coughing or childbirth; drug reactions*
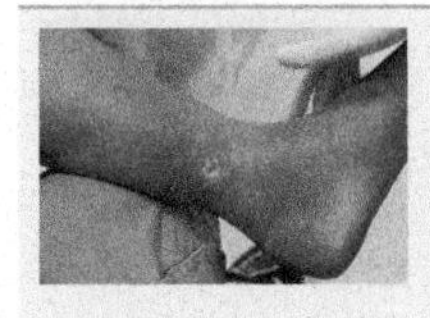	**Ulcer:** A sore of irregular size and shape that results as epidermal and dermal skin layers are destroyed *Often develop when primary skin lesions become infected with secondary infectious agents (example: bacteria infect a chickenpox vesicle due to child scratching the rash) and begin to involve deeper skin layers; cutaneous anthrax*

17.2 VIRAL SKIN INFECTIONS

Learning Outcomes

After reading this section, you should be able to:

17.4 Explain how the chickenpox vaccine has impacted the number and severity of cases and how chickenpox is linked to shingles.

17.5 Compare and contrast chickenpox with smallpox.

17.6 Discuss how HSV-1 infections are latent and can reactivate.

17.7 Describe how the MMR vaccine has impacted the number and severity of measles and rubella cases.

17.8 Differentiate the following diseases from each other: fifth disease, roseola, measles, rubella, and hand, foot, and mouth disease.

17.9 Describe how viruses cause warts.

Vesicular or pustular rashes characterize a variety of viral infections.

It is important to note that not every skin rash or lesion indicates infection. Drug side effects, allergic reactions, and even excessive exertion (such as development of petechiae during childbirth) can lead to some of the lesions detailed in Tables 17.1 and 17.2. Similarly, just because a patient exhibits a skin manifestation doesn't necessarily mean they have a skin-based infection. As stated before, a number of systemic infections (e.g., bacteremia), nervous system (e.g., meningococcal meningitis), and respiratory infections have skin manifestations. The diverse causes of skin lesions makes diagnosis challenging.

Some of the most classic skin manifestations are due to viruses that infect via the respiratory tract and migrate to the nervous system. In this section we explore a number of viruses that are primarily diagnosed by their skin manifestation, yet they are not actually skin-based diseases.

Chickenpox

1 Virus enters via the respiratory tract, migrates to the blood, and gains access to the liver, spleen, and lymph nodes.

2 In the next wave of viral replication (secondary viremia), viruses move from blood to skin, producing lesions on the head, trunk, or limbs.

3 The immune system destroys most of the viruses, but some may migrate to peripheral nerves and become latent.

Chickenpox lesions

Shingles

4 If the virus reactivates, it travels down nerve cells toward the skin, resulting in a banded rash in the area served by those particular nerves.

Shingles rash

FIGURE 17.3 The connection between chickenpox and shingles Once an individual is infected with chickenpox (varicella virus), some of the viruses may become latent in nerves and may reactivate later.

Critical Thinking *When someone experiences a shingles outbreak, it boosts the immune system and decreases the chances of another outbreak. Why would this lead to a boost in immunity?*

Chickenpox and Shingles

Before the development of a vaccine in 1995, chickenpox (caused by the *Herpesviridae* family virus varicella-zoster) was a standard childhood disease that was mild for most, but potentially life threatening for immunocompromised individuals and newborns. Since usage of the vaccine, cases are much rarer. Adults who contract chickenpox for the first time tend to have more severe cases than children, probably because the adult immune system mounts a stronger response to it.

The virus is highly contagious, typically spreading through respiratory droplets and occasionally from direct contact with pox lesions. The progression of chickenpox is detailed in FIG. 17.3. Chickenpox is on the list of nationally notifiable diseases (see Chapter 9 for more on how this impacts healthcare provider reporting).

Some varicella-zoster viruses may travel to peripheral nerves and become latent (dormant). **Shingles** is the reactivation of the virus within the nerves of

DiseaseSnapshot • Chickenpox and Shingles

Disease	Chickenpox	Shingles
Causative agent	Varicella-zoster virus: Enveloped, double-stranded DNA genome, *Herpesviridae* family	
Epidemiology	Humans are only reservoir; highly contagious; most dangerous in infants and immune compromised; incubation period 14 to 16 days	Reactivation of virus, variable incubation period
Transmission & prevention	Transmitted by respiratory droplets or direct contact with lesions; prevented by routine childhood vaccine	Reactivation of dormant virus; prevented by Shingrix* vaccine for older adults
Signs & symptoms	Fever with itchy vesicular rash	Headache, flulike symptoms without fever, painful, vesicular rash on one side of body
Pathogenesis & virulence factors	Virus enters via the respiratory system or skin abrasions; virions go latent in nerves during chickenpox infection and reactivate later in life as shingles	
Diagnosis & treatment	Diagnosed by lesion appearance and medical history; no cure; calamine lotion and oatmeal baths reduce itch; acyclovir for immune compromised individuals	

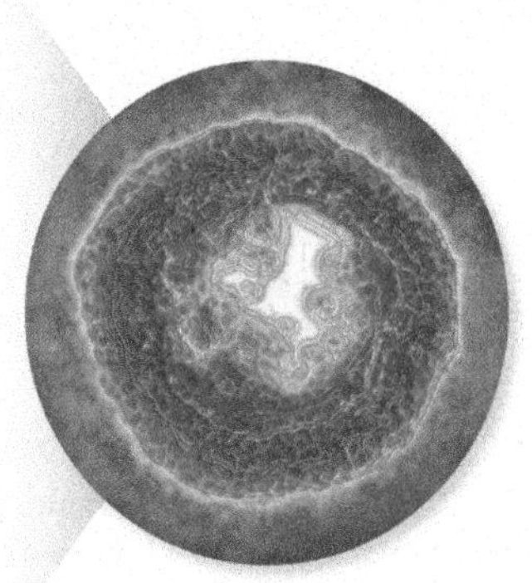

Varicella-zoster virion

*Zostavax was discontinued in the United States in 2020. Patients who previously received Zostavax are encouraged to get the Shingrix vaccine series.

CLINICAL VOCABULARY

Malaise overarching sense of feeling unwell or uncomfortable for reasons that may be unclear

Pruritic itchy

Postherpetic neuralgia (PHN) pain that persists after a shingles rash heals; due to skin nerve damage caused by a virus in the herpes family

Supportive therapy treatment to manage discomfort or symptoms, such as rest, fluids, or general pain medication

someone who previously had chickenpox. Unlike chickenpox, the shingles rash typically occurs over a smaller area of the body and tends to appear on the back, or in other nerve-rich areas, such as the face. While the chickenpox rash is usually itchy, the shingles rash is usually described as painful and burning. The rash appears as a band of blisters on one side of the body, rarely crossing the midline. This gives shingles its other name: **herpes zoster**, from the Greek word for belt, indicating that the rash develops along a belt of nerves.

According to the Centers for Disease Control and Prevention (CDC), one in three individuals who had chickenpox could experience shingles. While most shingles sufferers see the rash and pain subside after a few weeks, the disease can lead to a chronic pain condition called **postherpetic neuralgia** (PHN). The chance of developing shingles increases with age and as immune function declines; it is rare in individuals under 40 years old.

The Shingrix vaccine is recommended for people age 50 and older. This recombinant subunit vaccine, which is 90 percent effective at preventing shingles, is administered in two doses that are 2 to 6 months apart. Antiviral medications may limit a shingles outbreak if begun soon after lesions appear and topical ointments may alleviate discomfort.

Someone with shingles can't transmit the disease to anyone who previously had chickenpox. It is possible (though rare) for someone with shingles to transmit chickenpox to an individual who has never had it. As such, shingles patients should stay clear of newborns, immune-compromised patients, pregnant women, and nonvaccinated individuals.

Smallpox

Before worldwide eradication through vaccination, smallpox—with a 30 percent mortality rate—was for thousands of years one of the most feared diseases. The variola major virus is a highly contagious pathogen acquired from inhaling respiratory droplets containing the virus or touching contaminated fomites.

Around two weeks after exposure, an infected individual runs a high fever and feels very fatigued. After a few days, the fever drops and a rash appears, first on the face and then on the arms and legs. The lesions evolve from fluid-filled vesicles to pustules, which then dry up, crust over, and eventually fall off (FIG. 17.4). Those who survived had lifelong immunity, and often had skin scarring.

An extensive campaign by the World Health Organization (WHO) eradicated smallpox by 1980. By international agreement, now the only virus stocks are kept in special containment in authorized laboratories in the United States and Russia. There has been discussion as to whether it would be best to destroy all stocks, but they've been maintained in case they are needed for research purposes to develop a counterdefense in the event of bioterrorism or the use of smallpox as a biological weapon.

Because of smallpox's potential to be weaponized, it remains important for healthcare providers to be able to distinguish the disease from other skin rashes, particularly chickenpox. Smallpox lesions tend to develop on the palms of the hands and soles of the feet, and lesions appear more on the extremities versus the torso as seen in chickenpox. Smallpox lesions tend to appear domed, are solid to the touch, extend deep into the dermis, and can have a center dimple.

While no longer routinely administered, a vaccine is still available and is required for certain U.S. military personnel. Due to potentially serious complications, the immunization is not recommended for general usage. Besides supportive therapy, there is not a specific treatment for smallpox.

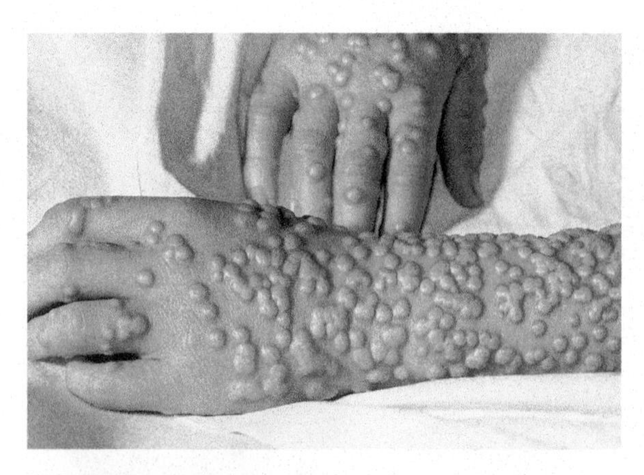

FIGURE 17.4 **Smallpox lesions** Although smallpox is eradicated, it is valuable for healthcare workers to be able to identify differences between chickenpox and smallpox.

Critical Thinking *How are smallpox lesions different from those in chickenpox?*

Disease**Snapshot** • **HSV-1**

Causative agent	Herpes simplex virus 1: Enveloped, double-stranded DNA genome, *Herpesviridae* family
Epidemiology	600,000 new cases/year in U.S.; virus reemerges as cold sores in about 30% of patients; incubation period averages 3 to 6 days
Transmission & prevention	Transmitted by respiratory droplets/saliva or direct contact with infected person; virus shed even in absence of active lesions; do not share items that contact the mouth (i.e. cups, lip balm, etc.); wrestlers should insist on mat-cleaning protocols and not practice with open lesions
Signs & symptoms	Painful, itchy, vesicular lesions on lips
Pathogenesis & virulence factors	Surface glycoproteins aid entry into host cells; viral replication produces lesions; virus suppresses host immune responses; some virions go latent in the trigeminal nerve ganglia and reemerge later
Diagnosis & treatment	Diagnosis is by lesion features and medical history; may detect viral DNA or host antibodies to virus; no cure; topical or oral antiviral drugs reduce outbreaks

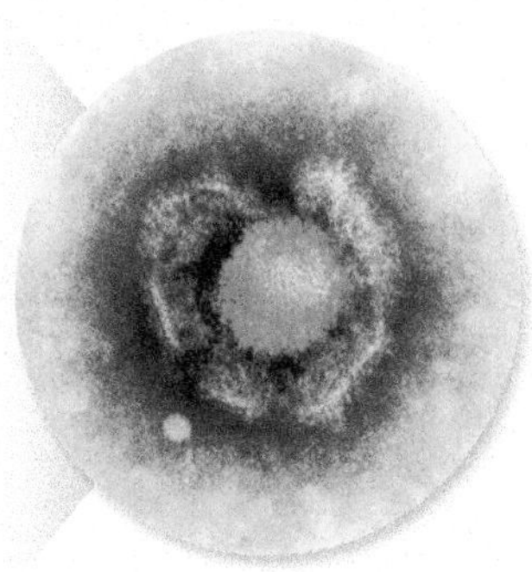

HSV-1

Human Herpes Viruses 1 and 2 (Herpes Simplex Viruses)

There are over 80 types of herpes viruses—only eight of which infect people—including human herpes viruses 6 and 7 (discussed later as causes of roseola) and varicella-zoster virus (previously discussed). In this section we focus on human herpes viruses 1 and 2, which are sometimes also called herpes simplex viruses (HSV). As seen for chicken pox and smallpox, HSV-1 and 2 generate a vesicular skin lesions. About 90 percent of people have antibodies to at least one of the simplex viruses (HSV-1 and/or HSV-2). HSV-2 is usually sexually transmitted and is therefore discussed further with sexually transmitted diseases in Chapter 20.

HSV-1 primarily causes oral herpes, characterized by cold sores (or fever blisters). It is transmitted through saliva—kissing or sharing eating utensils with an infected person commonly leads to infection. Contact with contaminated towels, sheets, or clothing or casual direct skin contact with an infected individual can also lead to infection.

Painful, itchy, vesicular lesions may develop on the lips about a week after infection (FIG. 17.5). These cold sores or fever blisters are not the same as canker sores (mouth sores sometimes associated with gum trauma, stress, or certain dietary deficiencies). Initial infection may also be accompanied by a sore throat or flu-like symptoms that make the individual feel generally unwell. Most infected individuals are asymptomatic, or the symptoms are so minor they go unrecognized. After an initial stage, HSV-1 migrates to the trigeminal nerve, which connects facial tissues to the central nervous system. There it becomes latent and may later cause flare-ups in about two-thirds of patients.

Various events, such as ultraviolet radiation, menstruation, hormonal changes, or other stressful conditions, trigger reactivation of the virus at the nerve endings near the lips. Fortunately, flare-ups tend to be shorter and less severe than initial infection because of the immune system's recognition of the virus.

HSV-1 may also cause lesions in other body locations. For example, **herpes gladitorum** is common in wrestlers who contact the virus through contaminated wrestling mats or direct contact with an infected wrestler. **Herpes whitlow** is a rare condition (2 out of 100,000 individuals) that can occur when the virus enters fingers via an abrasion and has been known to occur in healthcare providers who contact the virus.

1 HSV-1 enters the body and may cause painful oral and lip lesions.

2 The first outbreak heals, but the virus goes dormant in nerves.

Trigeminal nerve

3 If reactivated, the virus travels from the site of latency back down the same nerve path toward the skin, resulting in a new outbreak of lesions.

FIGURE 17.5 HSV-1 lesions Just as the chickenpox virus can be reactivated to cause shingles, an individual with HSV-1 can experience a flare-up as the virus reactivates from nerve cells. Viral reactivation can be triggered by stress, illness, sunlight, trauma, or other factors.

Critical Thinking *What is meant by the phrase "latent viral infection"?*

HSV-1 diagnosis is often made based on signs and symptoms, such as the recurring lesions. Microscopy of infected tissues reveals host cells that have fused and become larger (**syncytia**); these tissue changes are due to viral cytopathic effects on host cells. (See Chapter 10 for more on cytopathic effects.) Detection of the virus by PCR analysis is the most definitive way to diagnose infection. While there is not a cure for herpes infections, treatments may aid in managing symptoms. Topical or oral antiviral medications like acyclovir can reduce the frequency and severity of flare-ups as well as decrease viral shedding, but these drugs do not cure the infection because they do not clear the virus from the nerve cells. Researchers have been working on vaccines for years, but currently there is not a vaccine against HSV-1.

Maculopapular rashes are typical of several viral infections.

Recall that from Table 17.1 we know that a maculopapular rash consists of small, slightly raised, papule lesions (not fluid filled, possibly discolored, and without distinct borders) that overlay or are interspersed with macules (flat discolored areas of the skin). These rashes are typical in measles, rubella, rubeola, fifth disease, and hand, foot, and mouth disease.

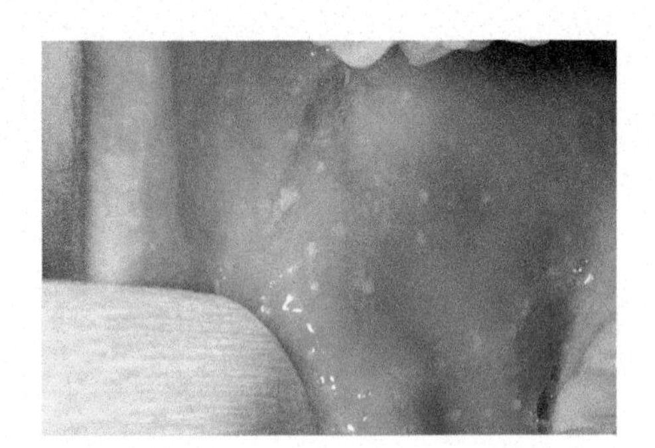

Koplik's spots

Maculopapular rash

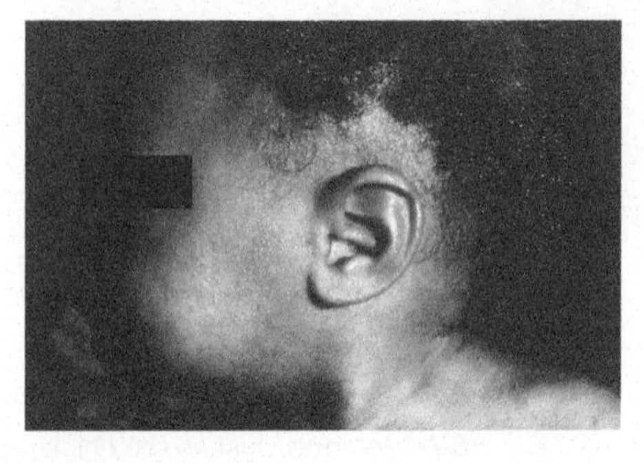

Maculopapular rash

FIGURE 17.6 Measles rash After fever, sore throat, and a dry cough develop, Koplik's spots (red spots with white dots in the center) may appear in the mouth. The measles maculopapular rash is the last sign of infection to develop. It often begins at the face, spreads to the trunk, and then the extremities.

Measles (Rubeola)

According to the CDC, before the public health measles vaccination campaign began in 1963, between 3 million and 4 million people in the United States got **measles** each year. Measles used to annually cause about 48,000 hospitalizations, 1,000 cases of chronic disabilities due to measles encephalitis, and up to 500 deaths from complications. Thanks to vaccinations, measles was rare in the United States for decades. But after the Disneyland outbreak in 2014 and recent outbreaks across many states, health officials became alarmed and legislation was proposed in several states to remove or limit exemptions for vaccination. These laws are based on the science that demonstrates vaccination efficacy and the dangers created by leaving pockets of susceptible individuals unvaccinated.

The measles virus is highly contagious and spreads by the respiratory route. About ten days after exposure, infected individuals may develop a fever, sore throat, dry cough, spots in the mouth, and finally the characteristic maculopapular rash that begins on the face and spreads to the trunk and then to the extremities (FIG. 17.6). Individuals are contagious before they show signs of measles, making quarantining ineffective and disease control difficult in outbreaks.

Once inside host cells, the measles virus causes cells to fuse together, allowing the viruses to hide from antibodies that the immune system makes to battle the infection. The virus multiplies in the respiratory tract and then spreads through the body through lymph and blood. The immune system kills the infected host cells, but this results in the lesions and other symptoms. The infection also dampens the immune system's ability to respond, putting patients at risk for secondary infections—particularly conjunctivitis, ear infections, and bacterial pneumonia. This is partially what makes measles so dangerous and even deadly. Immunocompromised individuals are at even greater risk for these secondary infections.

The best prevention is the **measles, mumps, and rubella (MMR) vaccine**. Unfortunately, in recent years, some parents have chosen not to vaccinate their children for various reasons, including erroneous concerns that the vaccination could cause autism or diabetes, an incorrect assumption that vaccination is unnecessary because the disease is no longer around, or an opinion that the disease is not severe enough to outweigh potential side effects of the vaccine. The most common vaccine side effects are fever, rash, or swelling at the injection site. In very rare cases, there have been incidents of joint pain, febrile

Disease **Snapshot** • Measles (Rubeola)

Causative agent	Measles (rubeola) virus: Enveloped, single-stranded RNA genome, *Paramyxoviridae* family
Epidemiology	Highly contagious; incubation period of 8 to 12 days
Transmission & prevention	Respiratory droplet transmission; prevented by MMR vaccine
Signs & symptoms	Fever, sore throat, dry cough, Koplik's spots (red spots with white dots in the center) in the mouth, red raised rash begins at face and spreads downward
Pathogenesis & virulence factors	Virus causes cells to fuse (form syncytia) allowing it to hide from host immune defenses; cytotoxic T-cells kill infected host cells, which promotes symptoms and lesions; immune system suppression increases risk for secondary infections
Diagnosis & treatment	Diagnosed by symptoms and medical history (with travel or exposure information); may detect viral DNA, patient antibodies to the virus, or virions in respiratory secretions or urine; no cure, supportive therapy only

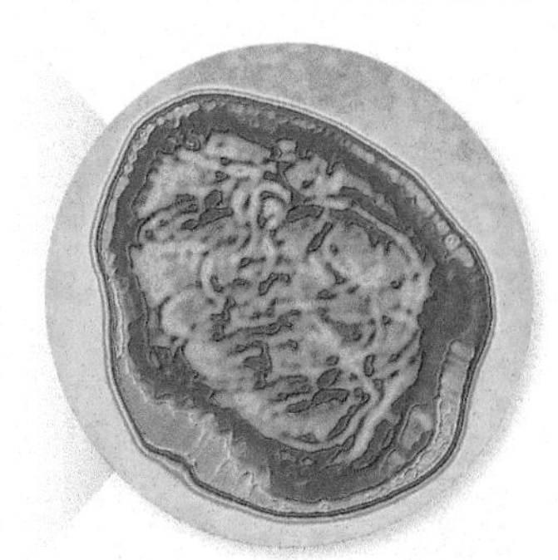

Measles (Rubeola) virion

seizures (convulsions caused by fever), severe allergic reactions (less often than 1 in a million doses), or a decrease in platelet count. As made apparent by recent outbreaks, the disease is in fact still around and is much more of a threat to the average person than getting the vaccine. Children under age 5 are at particular risk. According to the CDC, about 1 in 4 people who get measles will require hospitalization, 1 out of 1,000 will experience swelling of the brain, which could lead to permanent brain damage, and 1 or 2 out of every 1,000 who have measles will die—despite care. In short, the chances of serious complications from measles are considerably greater than the chances of severe reactions to the vaccine.

The vaccine is a live attenuated virus, so it's inappropriate for infants under the age of 1. As such, herd immunity is critical for protecting infants and others for whom the vaccine may be contraindicated. There is no treatment for measles; vaccination is the only protection.

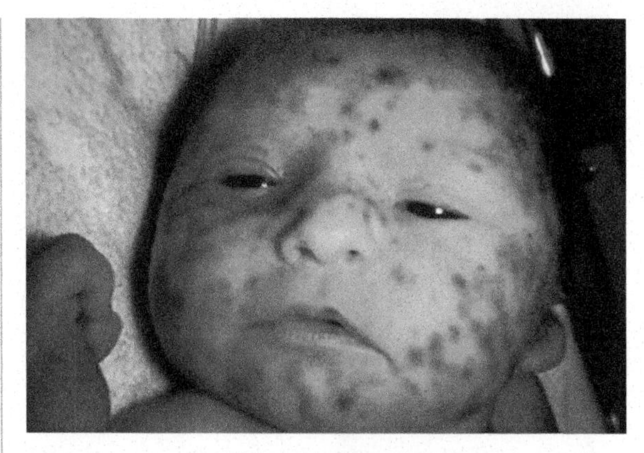

FIGURE 17.7 Congenital rubella syndrome Rubella, or German measles, is most dangerous when contracted during pregnancy. Transmission to the fetus leads to congenital rubella syndrom (CRS), shown here.

Critical Thinking *What is the most effective way to prevent CRS?*

Rubella (German Measles)

The rubella virus is an enveloped, single-stranded RNA virus of the *Togaviridae* family. **Rubella** is sometimes called German measles, but it should not be confused with measles, which is much more serious. The virus enters by the respiratory tract, spreads to the lymph nodes, and eventually spreads to the rest of the body by the bloodstream. Infection is characterized by a red rash that lasts for three to seven days. The rash starts at the face and then spreads downward. Patients may also experience fever, swollen and tender lymph nodes, muscle and joint aches, and a runny or stuffy nose. Rubella infections in pregnant women, especially during the first trimester, can result in **congenital rubella syndrome (CRS)**, which can lead to stillbirth or miscarriage or a variety of birth defects, including, blindness, deafness, heart defects, and growth or mental disabilities (FIG. 17.7).

Rubella is prevented with the measles, mumps, and rubella (MMR) vaccine. This vaccine shouldn't be given to pregnant women, children under the age of 1, or immune-suppressed individuals. Thanks to an effective vaccine, CRS is extremely rare in the United States, but according to the CDC it remains a significant global issue, with more than 100,000 children suffering CRS every year (particularly in Africa, southeast Asia, and the western Pacific).

CLINICAL VOCABULARY

Febrile seizures convulsions, twitching, or shaking induced by fevers that could result in loss of consciousness; last from a few minutes to 15 minutes or longer; most are harmless, though scary to those witnessing them

Congenital rubella syndrome condition in infants born of mothers who contract rubella while pregnant (usually during the first trimester) may result in serious birth defects, including deafness, blindness, and heart conditions

Fifth Disease; Roseola; and Hand, Foot, and Mouth Disease

The unusual name of fifth disease comes from 1899 when this infection was listed as the fifth most common childhood rash/illness. Fifth disease causes a mild illness characterized by a red facial rash that makes the patient look as if they had been slapped on the cheeks—hence its common name, "slapped cheek

MMR Vaccine Timing

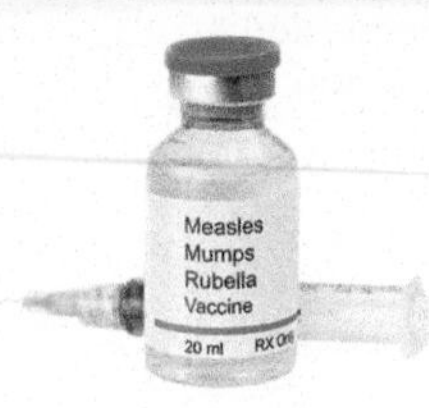

The measles, mumps, and rubella (MMR) vaccine has been in use since 1971, while the chickenpox or varicella (Var) vaccine has been available since 1995. Both have been tremendously effective in reducing the incidence of disease associated with the agents against which they protect, which has been a major victory for public health.

The MMR and Var vaccines are both administered beginning at 12 months of age, whereas other vaccines are administered earlier—right after birth or at 2, 4, and 6 months. The MMR and Var vaccines are administered later because they include live attenuated (weakened) viruses and should not be given before the immune system has time to develop. Many other vaccines are made of dead pathogens or only parts of pathogens and therefore are safe to administer even if the immune system is still maturing.

QUESTION 17.1

Based on this information about the formulation of these two vaccines, why are they not administered until the child is a year old?

syndrome" (FIG. 17.8). It has also been referred to as *erythema infectiosum*. In 1980, parvovirus B19 was discovered to be the causative agent of fifth disease. This is not the same parvovirus that affects dogs. Typical symptoms include sore throat, low fever, joint aches, stomach discomfort, and fatigue. Following the facial rash, a lacy rash may develop over the trunk and limbs. While fifth disease most commonly affects children between ages 4 and 10, adults can also contract it. By adulthood at least half of the population has been exposed. About 20 percent of cases are asymptomatic. Most cases of fifth disease are self-limiting, resolving without medical treatment. However, this virus is known to cause complications in pregnancy (such as miscarriage) as well as in patients who have sickle-cell anemia or immunodeficiency disorders. A vaccine is in clinical trials.

At some point most children have a mild or asymptomatic infection of **roseola (roseola infantum, exanthem subitum**, or sixth disease) that goes undiagnosed. Roseola, which is caused by human herpes viruses 6 and 7, is a very common viral disease that is often asymptomatic. If it does present with symptoms, these are marked by a sudden high fever (up to 104°F) for a few days followed by a rash that covers most of the body (Figure 17.8). After a day or two the rash subsides. In a few cases, the fever triggers febrile seizures.

Hand, foot, and mouth disease (HFMD) is common in infants and children but can also occur in adults. Uncomfortable blisterlike sores in the mouth (**herpangina**) and lesions on the hands and feet distinguish this disease from others. The disease is mild, but it's common in childcare settings, and it is important to distinguish it from other skin rashes (Figure 17.8). HFMD in humans is mainly caused by coxsackievirus A16 and enterovirus 71 viruses.

Certain viruses cause warts.

Viruses that cause warts are very common. They replicate in the cells they infect and deregulate host cell division to cause abnormal cell growth. Most warts are harmless, but because of their location, some may hurt, itch, or seem unsightly. They are spread by direct contact, virus-contaminated fomites, and by autoinoculation from one infected site to other parts of the body; the elbows, knees, trunk, and face are often affected (FIG. 17.9).

Fifth disease

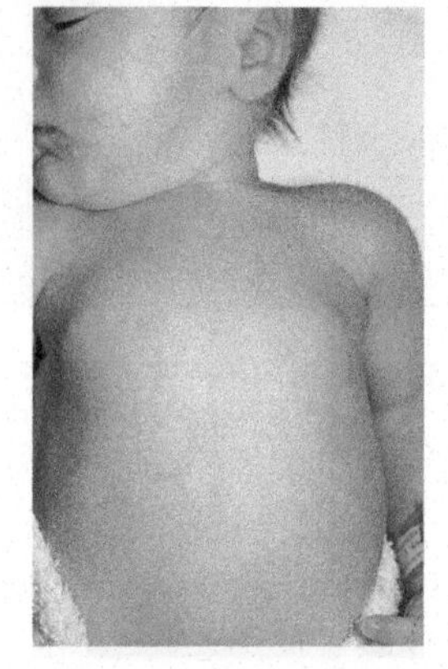

Roseola

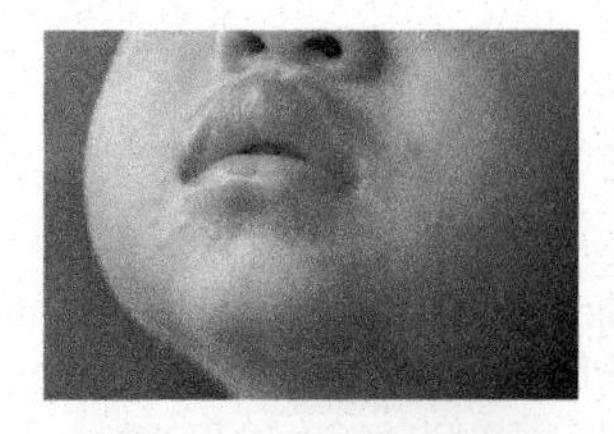

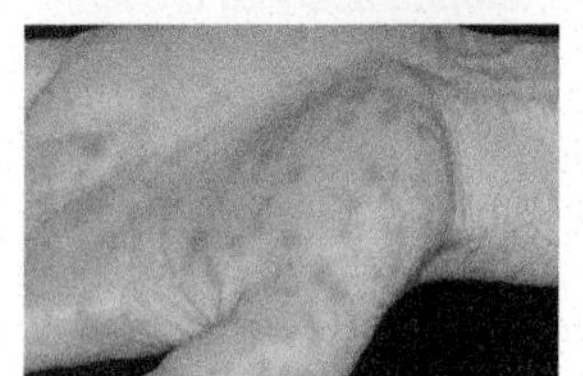

Hand, foot, and mouth disease

FIGURE 17.8 **Human parvovirus B19, roseola, and hand, foot, and mouth disease viruses** These three viruses cause different rash and lesion patterns, though none is considered particularly dangerous.

Critical Thinking *Compare and contrast the rashes and lesions for each of these common childhood infections.*

Papillomaviruses

Over 50 different types of **papillomaviruses** cause warts. These viruses enter abrasions by direct or indirect contact and through contaminated fomites. Generally, warts are painless benign growths on the skin or mucous membranes. They are commonly found on the feet (plantar warts), hands and toes (seed warts), genital region (genital warts), or rest of the body (flat warts). Most warts self-resolve, but affected tissue may be frozen off (cryogenic methods), burned off (cauterization), laser removed, or removed chemically. These treatments may not permanently clear the causative viruses and the wart may redevelop later.

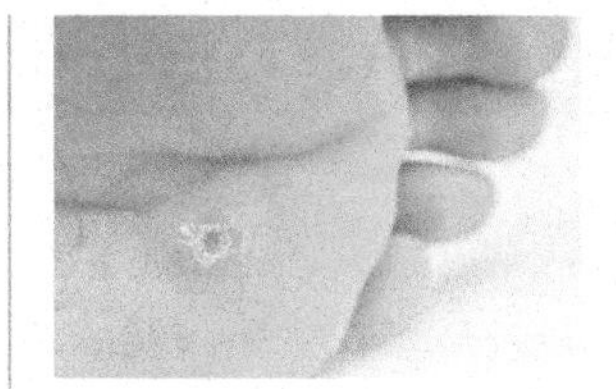

Plantar wart

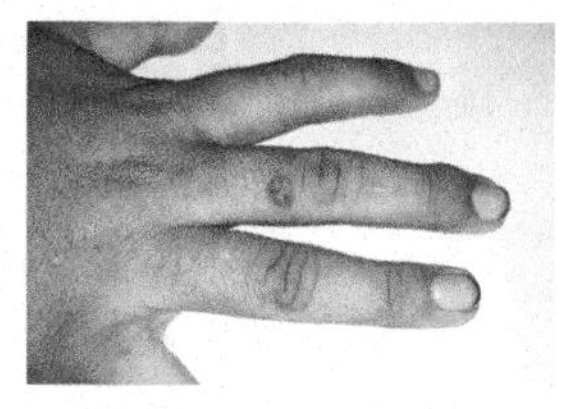

Seed wart

FIGURE 17.9 Warts Plantar warts of the feet and seed warts of the hand can be uncomfortable or unsightly, but are not usually serious.

Critical Thinking *What are some common procedures for treating warts like these?*

Build Your Foundation

4. Describe the connection between chickenpox and shingles, and how cases can be prevented.
5. Describe smallpox and explain why we don't routinely vaccinate against it.
6. Describe the process that HSV-1 uses to cause fever blisters around the mouth.
7. Describe why preventing measles and rubella with the MMR vaccine is important.
8. Compare and contrast measles, rubella, fifth disease, roseola, and hand, foot, and mouth disease.
9. Discuss the role of viruses in skin warts.

Build Your Foundation (BYF) Quick Quiz: Visit the **Mastering Microbiology** Study Area to quiz yourself.

DiseaseSnapshot • Fifth Disease, Roseola, and Hand, Foot, and Mouth Disease

Disease	Fifth Disease	Roseola	Hand, Foot, and Mouth Disease
Causative agent	Human parvovirus B19: Non-enveloped, single-stranded DNA genome, *Parvoviridae* family	Human herpes viruses 6 and 7 (HHV-6, HHV-7): Enveloped, double-stranded DNA genome, *Herpesviridae* family	Mainly coxsackievirus A16 and enterovirus 71 viruses: Non-enveloped, single-stranded RNA genome, *Picornaviridae* family
Epidemiology	Very contagious; common in childcare settings; most cases in children 5 to 15 years old; peaks in winter and early spring; incubation period of 4 to 14 days	Humans are the only hosts; no seasonal preference; universal incidence, mainly seen in children 9 months to 2 years old; incubation period of 9 to10 days	Spreads easily among young children in school/daycares; more common in summer and fall; incubation period of 3 to 6 days
Transmission & prevention	Respiratory droplet transmission; not contagious once the rash appears; only prevention is to avoid infected patient	Respiratory droplet transmission; keep infected children home to limit spread	Transmitted by respiratory route, fecal/oral route, direct contact, and fomites; prevented by general hygiene and fomite disinfection; keep infected children home to limit spread
Signs & symptoms	Cold-like symptoms, occasionally followed by a maculopapular "slapped cheek" rash that can spread to other parts of the body, 20% of patients are asymptomatic	Sudden high fever (3 to 5 days), runny nose, mild diarrhea, swollen lymph nodes, following fever a non-itchy rash of flat, small, pink spots develops on trunk and arms	Fever, reduced appetite, malaise, sore throat, mouth sores, non-itchy but sometimes blistering rash on hands and feet within a couple of days of the fever
Pathogenesis & virulence factors	Mild in most; arthritis or anemia in some; pregnant women may miscarry	Virus replicates in salivary glands and leukocytes; invasion of nervous system may cause seizures and other CNS complications	Virus mutates rapidly to make new strains, evades interferon and other innate immune responses, complications include viral meningitis and/or encephalitis
Diagnosis & treatment	Diagnosed mainly by telltale rash; self-resolves; supportive therapy is only treatment	Diagnosed mainly by telltale rash that follows fever, host antibodies to virus confirms exposure; treat fever with tepid baths or non-aspirin NSAIDs	Mainly diagnosed by symptoms, hydration therapy if children with mouth sores are not taking in enough fluids, non-aspirin NSAIDs for fever and pain reduction

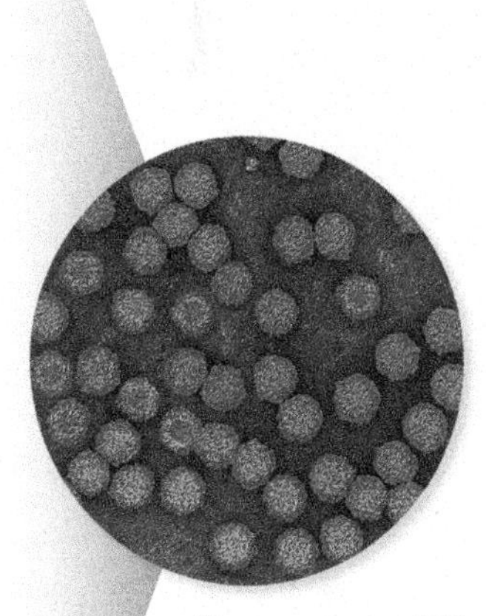

Human parvovirus B19

17.3 BACTERIAL SKIN INFECTIONS

Learning Outcomes

After reading this section, you should be able to:

17.10 Describe *Propionibacterium acnes's* role in acne.

17.11 Provide examples of *Staphylococcus aureus's* virulence factors and describe the skin diseases it causes.

17.12 List and describe *Streptococcus pyogenes's* virulence factors and describe the skin diseases it causes.

17.13 Describe the laboratory characteristics of *Staphylococcus aureus* and *Streptococcus pyogenes*.

17.14 Detail the virulence factors of *Pseudomonas* species and describe the diseases they cause.

17.15 Name and describe the bacterial agents of gas gangrene and cutaneous anthrax and describe the progression of these diseases.

CHEM • NOTE

Fermentation is a broad term that describes catabolic processes that certain cells use to break down nutrients (often carbohydrates) into simpler substances. Acids and alcohols are common products of fermentation reactions.

Acne is a common skin infection mainly caused by *Propionibacterium acnes* bacteria.

Acne is one of the most common skin afflictions, affecting 70–80 percent of teenagers and young adults at some point. There are three main types: mild **comedonal acne**, moderate **inflammatory acne**, and the most severe and scarring form **nodular cystic acne**. While mild and moderate forms of acne are common, their causes are not uniform or fully understood. The overarching cause of acne relates to the clogging of skin pores and hair follicles. A range of factors contribute to the severity of acne in different individuals, including, but not limited to, an individual's sebum production, genetics, hormone levels, cosmetics, and usage of certain drugs or medications. Despite what people might have once thought, diet, dry skin, and hygiene have not been shown to affect acne. *Propionibacterium acnes* bacteria, however, have been shown to play a major role in at least some types of acne.

P. acnes bacteria are Gram-positive rods that are normal residents of the skin. When these bacteria break down carbohydrates, they are capable of producing propionic acid as a fermentation product, thus their name.

Under normal circumstances, the sebaceous glands produce sebum to oil, soften, and protect the skin. Sebum normally is released to the skin's surface through pores. In acne, sebum and dead skin cells block the skin's pores. Hormones (particularly in young adult males) increase sebum production, making matters worse. How the skin pores are blocked impacts what type of acne develops (FIG. 17.10).

As previously mentioned, sebum has a generally antibacterial effect. However, *P. acnes* have enzymes that allow them to use sebum as a nutrition source. When pores become clogged with sebum, oxygen levels in the pores decrease, leading to an ideal environment for *P. acnes*. As the bacteria proliferate in the pores they can generate an inflammatory response, leading to the characteristic pustules and cysts of inflammatory acne. Over time, biofilms can develop in pores and sebaceous glands, adding to the challenge of treating acne.

Because of its prevalence on skin, *P. acnes* is difficult to manage. A variety of retinoid drugs such as isotretinoin (Accutane) decrease sebum production.

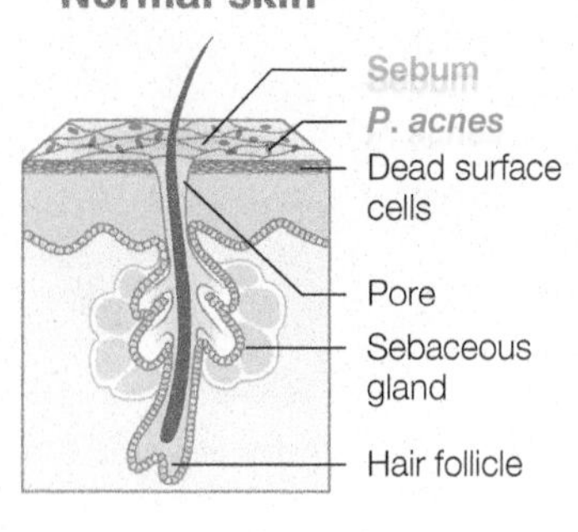

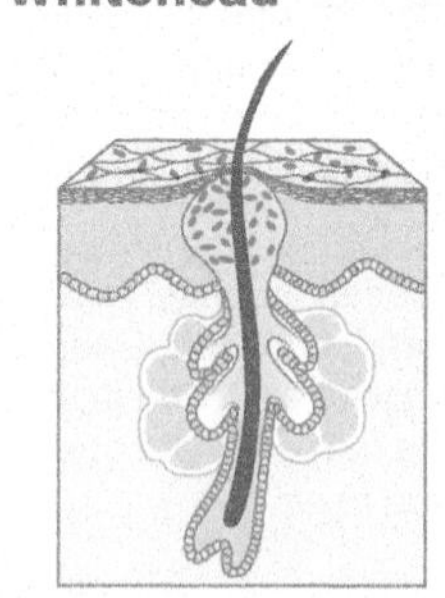

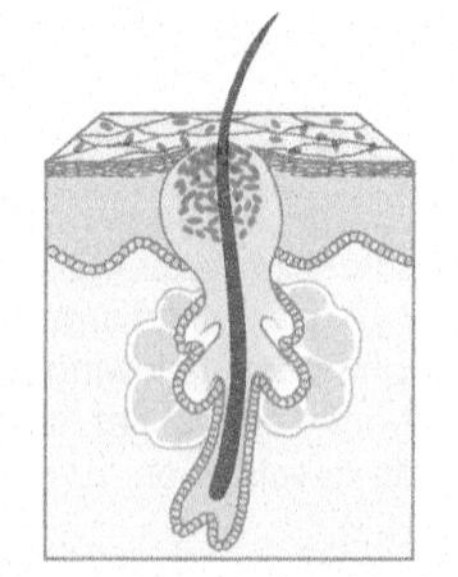

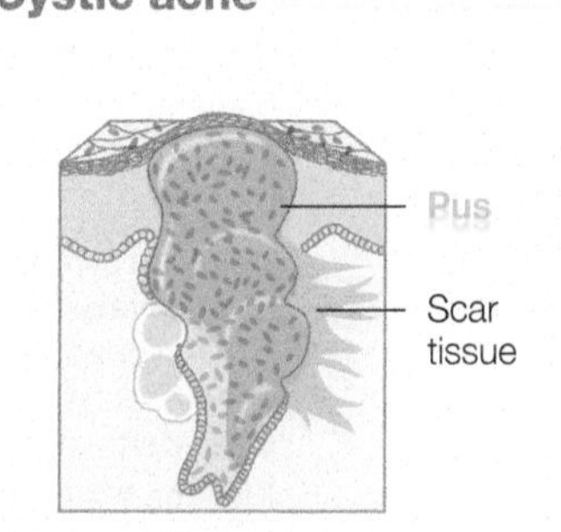

FIGURE 17.10 **Different types of acne** In a clogged pore, or comedo, diverse substances such as sebum, keratin, or dead skin cells may cause the obstruction, which may either close the pore or allow it to remain open. The type of acne lesions that develop depend on the nature of the comedo.

Critical Thinking *Antibiotics are prescribed in some severe cases of acne. What negative side effects could taking these antibiotics have?*

Unfortunately, a number of retinoid drugs are also associated with birth defects. Topical or oral antibiotics also limit *P. acnes*, but these change the biota balance, making the patient susceptible to opportunistic fungi or bacteria. Topical agents containing benzoyl peroxide exfoliate skin cells and reduce the load of *P. acnes* on the skin. FDA-approved treatments such as blue light wavelengths or photodynamic therapy (ClearLight), laser light (Smoothbeam), or intense heat pulses (ThermaClear) kill *P. acnes* and decrease the size of sebaceous glands.

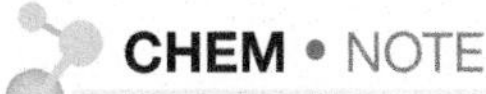

Blue light has a short wavelength compared with other visible light spectrum colors and is therefore a higher energy wave; it is closest to ultraviolet light in its energy. Energy in the blue light range activates porphyrin, a light-absorbing molecule in *P. acnes*, to make toxic oxygen species that kill the bacterium.

Staphylococcus aureus causes a spectrum of skin diseases.

So-called "staph infections" are a collection of skin diseases caused by various *Staphylococcus* species, but *Staphylococcus aureus* is the leading cause of such infections. *S. aureus* and other staphylococcal strains, such as *S. epidermidis*, are normal skin residents. However, these organisms can shift from being normal microbiota to invading pathogens if the skin is damaged. (Chapter 10 explores how normal microbiota shift to invading agents.)

The word "staphylococcus" describes the arrangement and shape of the bacterial cells, as they are clusters ("staph") of Gram-positive spheres ("cocci"). The word "aureus" is Latin for golden, and describes the yellow color of most *S. aureus* colonies. This coloring seems to protect the bacterium from ultraviolet light while it resides on the sun-exposed skin, and it chemically protects it from neutrophils (innate immunity cells).

Most *S. aureus* strains are catalase positive, mannitol fermenters, beta hemolytic (meaning they break down red blood cells), and coagulase positive. *S. aureus* produces a number of virulence factors that make it an effective pathogen; these are described in TABLE 17.3. *S. aureus* also has surface factors that play key roles, including a protective polysaccharide capsule and protein A on its cell surface. Both of these factors protect the bacterium from phagocytosis. In addition to skin infections, *S. aureus* is associated with certain gastrointestinal conditions, systemic diseases of the cardiovascular and lymphatic systems, central nervous system infections, and pneumonia (see Chapters 16, 18, 19, 21 for other conditions caused by *S. aureus*).

Staphylococcal Impetigo, Erysipelas, Cellulitis, and Folliculitis

Although *S. pyogenes* (discussed later) may also cause the skin and soft tissue infections we are about to discuss, here we'll focus on *S. aureus*.

TABLE 17.3 Select *Staphylococcus aureus* and *Streptococcus pyogenes* Virulence Factors

Virulence Factor	*S. aureus*	*S. pyogenes*
Coagulase: clots blood to help hide bacteria from immune system	Yes	No
Catalase: breaks down reactive oxygen species	Yes	No
Enzymes to break down blood clots	Yes	Yes
Pyrogenic toxin: stimulates fever, rash, and shock	Some strains	Some strains
Exfoliative toxin: separate upper epidermis from rest of skin	Some strains	No
Hyaluronidase: breaks down connective tissues	Yes	Yes
Hemolysins: lyse red blood cells	Yes	Yes
Lipases: digest lipids in tissues	Yes	Yes
M-protein: anti-phagocytosis factor	No	Yes

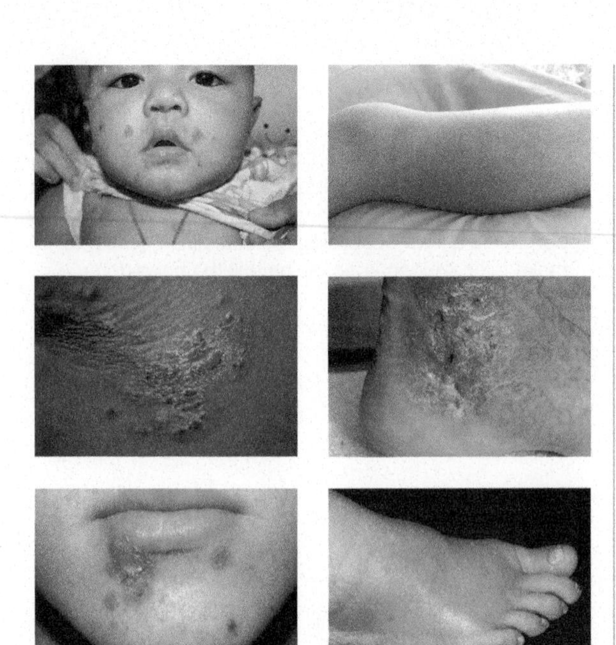

FIGURE 17.11 Impetigo and cellulitis *Staphylococcus aureus* may cause relatively mild and superficial impetigo or the more serious dermal infection cellulitis.

Critical Thinking *What virulence factors may allow S. aureus to invade deeper tissues to cause cellulitis?*

CLINICAL VOCABULARY

Leukocytosis increase in white blood cells (leukocytes), often associated with infection

Lymphangitis inflammation of the lymphatic vessels; may present as red streaks below the skin

Impetigo is characterized by superficial pus-filled vesicles on reddened skin usually on the face, lips, or extremities (FIG. 17.11). The occasionally itchy vesicles rupture and ooze, and eventually crust over into honey-colored lesions. Impetigo is highly contagious, though not generally serious unless secondary bacterial infections occur. The bacteria are easily transmitted by direct contact or through contaminated fomites. The condition most often occurs in 2- to 5-year-olds and is often associated with outbreaks in childcare centers. Impetigo may be treated with topical antibiotics.

If impetigo spreads to surrounding skin and lymph nodes and involves extensive inflammation and pain, it is referred to as **erysipelas**. This infection is more common in children and the elderly. It is a much more serious condition than superficial impetigo and is accompanied by fever, chills, and an increase in white blood cells (**leukocytosis**) to fight off the infection; if untreated, erysipelas may be fatal.

Unlike impetigo and erysipelas, which affect the upper portion of the dermis, **cellulitis** is a deeper infection of the lower dermal and subcutaneous fat. This infection is mainly seen in adults and is characterized by red, swollen, and painful skin and may be accompanied by fever, leukocytosis, and/or **lymphangitis** (inflamed lymphatic vessels). While any part of the body can be affected, the legs are the most commonly infected. The infection may quickly spread to lymph nodes and the bloodstream and is potentially fatal if not treated (Figure 17.11). Most cases of cellulitis are effectively treated with antibiotics—although treatment has been complicated by the emergence of methicillin-resistant *Staphylococcus aureus* (MRSA; discussed later).

S. aureus may also cause **folliculitis**, which is hallmarked by swollen, red, pus-filled hair follicles. Sometimes the infected region spreads beyond the hair follicle to create a raised painful nodule called a *furuncle*. These are more likely to occur in thicker skin, including the back of the neck.

Proper hand hygiene and wound care are the best ways to prevent *S. aureus* infections. Limiting contact with infected patients and, in the case of healthcare workers, observing all posted precautions reduces transmission. Diagnosis is usually made by skin lesion appearance as well as microscopic and biochemical analysis of bacteria cultured from infection sites. The coagulase test is especially useful in diagnosis, since *S. aureus* typically makes this virulence factor. Infections caused by other *Staphylococcus* species that do not produce coagulase may be referred to as coagulase-negative *Staphylococcus* (CoNS) species.

Methicillin-Resistant *Staphylococcus aureus* (MRSA)

Certainly, these days **MRSA** (methicillin-resistant *Staphylococcus aureus*) has become a large concern. Most bacteria produce enzymes that assemble peptidoglycan in bacterial cell walls. Some of these enzymes are also called penicillin-binding proteins because they can bind to and be inactivated by penicillin family drugs. When this happens the affected bacterium will not develop a strong cell wall and it will lyse (burst). Many bacterial strains produce penicillinases (also called beta lactamases) that break down beta lactam–based drugs (i.e., penicillin family drugs).

To combat penicillinase-producing bacteria, a modified synthetic form of penicillin, called methicillin, was developed in the 1950s. Like traditional penicillin drugs, methicillin still targets cell wall building enzymes—it is just chemically resistant to be being broken down by penicillinases. Unfortunately, by the early 1960s, MRSA strains were becoming common in hospitals, and now they are commonplace in healthcare settings and in the general community. MRSA strains have modified their cell wall construction enzymes in such a way that they do not associate with methicillin, penicillin, or a variety of other penicillin based drugs. For this reason, non-penicillin-based drugs like vancomycin, rifampin, tetracyclines, and other drugs are commonly used to treat MRSA infections.

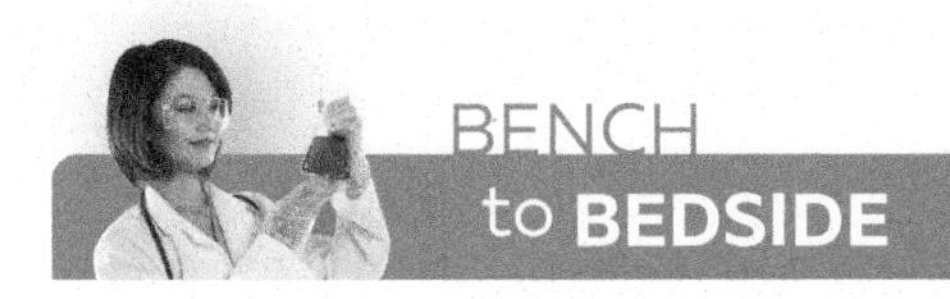

Diagnosing Illness by Studying Immune Responses

Since the invention of the polymerase chain reaction technique in the early 1980s, medical diagnostics have taken off. Now clinicians have a variety of rapid tools with which to detect specific pathogens, or even panels of "usual suspect" pathogens that may be present in patient samples. While useful, these techniques can be expensive to perform. And though these methods speed up detection of known pathogens, they usually require a fairly firm idea of the suspected agent so that the appropriate test is used. That means many techniques remain ineffective at identifying new pathogens or unsuspected agents.

A new approach that researchers are investigating is using host response to tell us what ails a patient. Our bodies respond differently when fighting infections caused by different pathogens. This approach of studying immune responses has also allowed researchers to develop host expression profiles for people infected with acute respiratory infection agents, in comparison to the signatures of healthy, uninfected individuals. Furthermore, researchers are finding ways to successfully predict if symptoms are associated with viral or bacterial infections. These approaches have even successfully distinguished between host responses to Gram-positive bacteria versus Gram-negative bacteria—something that has traditionally been challenging to determine. Human immune responses to *Staphylococcus aureus* were distinguishable from responses to *Escherichia coli*, suggesting this approach could allow labs to quickly detect the presence of certain bacteria, and more quickly recommend appropriate treatment.

Our immune system responds differently to various pathogens. Future medicine may take our individual immune responses into consideration for disease treatment.

These methods are still in the early research stages and will require further study and development to better account for patient variations. However, technologies are advancing quickly, and this approach holds great promise for future personalized rapid diagnostics and treatment.

Sources: Ravichandran, S., et al. (2021). VB10, a new blood biomarker for differential diagnosis and recovery monitoring of acute viral and bacterial infections. *EBioMedicine, 67*, 103352.

Thair, S., et al. (2021). Gene Expression–Based Diagnosis of Infections in Critically Ill Patients—Prospective Validation of the SepsisMetaScore in a Longitudinal Severe Trauma Cohort. *Critical Care Medicine, 49*(8), e751–e760.

Unfortunately, **vancomycin-intermediate** S. *aureus* **(VISA)** and **vancomycin-resistant** S. *aureus* **(VRSA)** strains are emerging, making S. *aureus* even more challenging to treat. Healthcare workers must carefully follow hand-washing and antisepsis precautions to limit the spread of these healthcare-associated infections (HAIs).

Scalded Skin Syndrome

Staphylococcal scalded skin syndrome is a potentially serious condition; it looks as if the skin has been burned in boiling water (FIG. 17.12). Infants may present as fussy and tired, with malaise and/or fever. Red blistering skin is the sign for significant concern. The fluid in the blisters does not contain bacteria, but rather the **exfoliative** toxins that are made by some strains of S. *aureus*. The toxin causes the outer layer of epidermal cells to peel in sheets and leave the skin red and exposed. These toxins originate at the initial site of infection and are carried by the bloodstream to other areas as the disease develops. Depending on the severity of a case, treatment may require IV fluids for rehydration, IV antibiotics to kill the bacteria, treatments for skin wounds, and protection from secondary skin infections and to allow the skin to heal. Scarring should not occur, since this condition only affects the outer dermal layer.

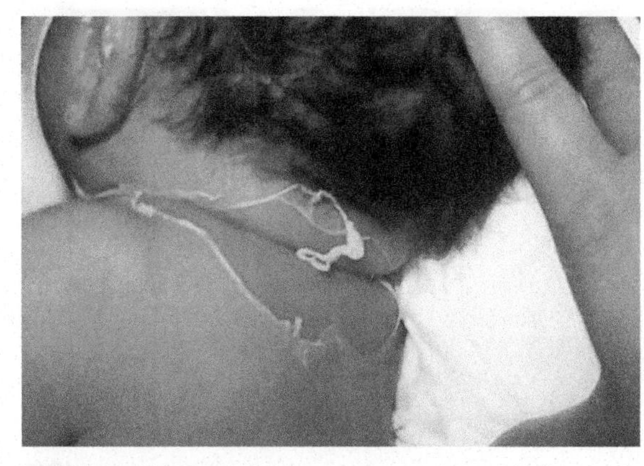

FIGURE 17.12 **Staphylococcal scalded skin syndrome** Some strains of *S. aureus* produce exfoliative toxins that cause outer skin layers to peel off.

Streptococcus pyogenes primarily causes strep throat, but can cause skin infections.

Some of the same skin infections caused by S. *aureus* can also be caused by *Streptococcus pyogenes*. The word "streptococcus" describes the beads-on-a-string appearance of these Gram-positive bacteria. The word "pyogenes" comes from the Greek for "pus-forming"—an appropriate name, given that this organism often causes pus-filled lesions. These bacteria are catalase negative (separating

TRAINING TOMORROW'S HEALTH TEAM

Precautions Against Skin Infections in Newborns

Infant skin is thin and still developing, making it more at risk for injury and dryness. Premature infants in particular have extremely thin and easily damaged skin. Additionally, their skin is not yet colonized with normal, harmless bacteria; this means that both true and opportunistic pathogens can easily take root and cause serious skin infections. Also, infants are often handled by many, particularly in a hospital setting. Because potentially pathogenic staphylococci and streptococci are part of the resident microbiota of many of these caregivers, it sets up a scenario where passing an infection to a child can be surprisingly easy.

While some hospitals are looking into the effectiveness of screening healthcare workers and patients to find those colonized with bacteria that may be hazardous to newborns, good old-fashioned hand-washing remains the fundamental way to prevent these infections. Other recommendations for caregivers in neonatal units include the removal of personal items such as rings or watches before entering regions with infants, since these items can harbor bacteria. Keeping nails trimmed short and using alcohol-based degerming products right before any physical contact also cuts down on infant skin infections.

QUESTION 17.2

Explain the specific importance and value of each of these recommendations for this at-risk neonatal population.

them from staphylococci) and beta hemolytic on blood agar plates (separating them from other streptococci). *S. pyogenes* produces many virulence factors that make it a formidable pathogen; these factors are outlined in Table 17.3. In addition to skin infections, *S. pyogenes* is also associated with certain upper respiratory infections (Chapter 16) and systemic infections of the cardiovascular and lymphatic systems (Chapter 21).

Streptococcal Impetigo and Cellulitis

Streptococcus strains, particularly *S. pyogenes*, are commonly present as normal skin residents, making it easy for them to enter the body through broken skin. As with *S. aureus*, there are several skin and soft tissue diseases caused by *S. pyogenes*. While impetigo may be caused by either *S. aureus* or *S. pyogenes*, the latter is more likely to cause infections in newborns who have not yet fully developed their normal skin microbiota or immune system. *S. pyogenes* is also a frequent cause of cellulitis.

Necrotizing Fasciitis and Streptococcal Toxic Shock Syndrome

S. pyogenes strains are often clinically referred to as **group A streptococci** (GAS). GAS can be invasive and deadly, thanks to their diverse virulence factors (see Table 17.3) that break down tissues and facilitate spread. *S. pyogenes* also has key surface factors including a hyaluronic acid capsule to camouflage the bacteria from host and M protein on the surface that serves as a binding site plasminogen to trigger a degradation cascade. According to the CDC, few people who come in contact with GAS will develop invasive GAS disease; however, individuals with chronic conditions or a weak immune system are more susceptible. The two most severe, yet rare, diseases associated with *S. pyogenes* are **necrotizing fasciitis** (FIG. 17.13) and **streptococcal toxic shock syndrome**. Necrotizing fasciitis has been referred to by the media as "flesh-eating disease" because of the way the bacteria break down flesh with their specialized tissue-degrading enzymes. CDC tracking data suggests that there are only 650–850 cases of necrotizing fasciitis per year.

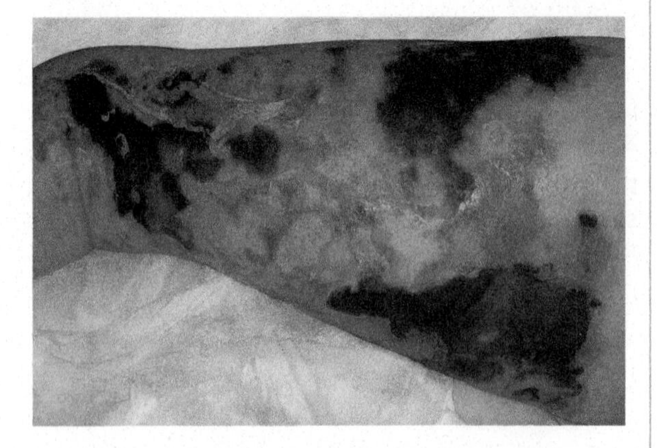

FIGURE 17.13 **Necrotizing fasciitis** The most common causes of necrotizing fasciitis are invasive group A streptococci, such as *Streptococcus pyogenes*, that produce enzymes that degrade connective tissues.

Critical Thinking *Using TABLE 17.3, describe several specific virulence factors that could help GAS invade tissues.*

In streptococcal toxic shock syndrome, the bacterial toxins released into the bloodstream cause a drop in blood pressure and organ failure. Cases may manifest with a red skin rash, which is why this disease is covered in this skin

Disease**Snapshot** • **Necrotizing Fasciitis**

Causative agent	Usually *Streptococcus pyogenes:* Gram-positive bacterium
Epidemiology	Enters via damaged skin; incubation period is minimal (tissue destruction starts within hours of infection)
Transmission & prevention	Usually transmitted by direct contact with an environmental source; prevent spread to others by using universal precautions with special attention to wound hygiene
Signs & symptoms	Skin swelling and heat at infection site, pain that is disproportionate to the initial injury; flulike symptoms, fever, nausea, malaise, dizziness, rash, or drop in blood pressure
Pathogenesis & virulence factors	Numerous bacterial enzymes and toxins (including deoxyribonucleases, streptokinases, hyaluronidase, exotoxin A, and streptolysin S) facilitate tissue invasion
Diagnosis & treatment	Diagnosed by wound appearance, medical history, skin biopsy, and culture confirmation; treated by surgical debridement (removal of affected tissue) and various intravenous antibiotics (often penicillin and clindamycin); mortality rate is about 15% for invasive GAS infection and around 40% for necrotizing fasciitis

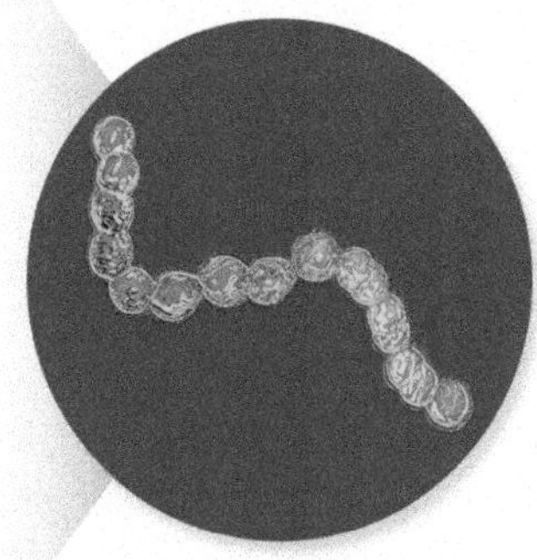

S. pyogenes

chapter. It is worth noting that this infection is different from staphylococcal toxic shock syndrome, which is associated with improper tampon usage and systemic infections (see Chapter 21).

Pseudomonads can cause opportunistic infections as well as serious wound infections.

Pseudomonas bacteria (collectively called pseudomonads) are Gram-negative rods that are normal skin residents and abundant in soil and water. They can metabolize a wide range of substances for food and energy, allowing them to grow in some strange places. Their ability to thrive on plants and in floral vase water is one reason why certain clinical spaces do not allow fresh flowers or plants. A particularly interesting and alarming feature is their ability to break down certain detergents, making them difficult to eliminate from surfaces. In addition, they are increasingly resistant to many antibiotics.

Pseudomonas aeruginosa is the most medically relevant member of the pseudomonads; it is a common cause of healthcare-acquired infections. As an opportunistic pathogen it readily establishes infections in people with weak immune systems, damaged skin, or other underlying health conditions (i.e., cystic fibrosis). Its ability to form biofilms on indwelling devices such as catheters, as well as in a variety of tissues, allows it to cause a number of infections. Examples include respiratory conditions—particularly in cystic fibrosis patients (Chapter 16)—systemic diseases of the cardiovascular and lymphatic systems (Chapter 21), swimmer's ear, and keratitis eye infections (discussed later).

Wound Infections

Burn and other skin wound patients are particularly at risk for *P. aeruginosa* infections; up to two-thirds of burn patients develop *P. aeruginosa* infections. As skin wounds heal, the bacteria can grow underneath scabs and directly access the bloodstream, leading to systemic infections. Such cases are more serious and challenging to treat. Meticulous wound care, antibiotics, and protective creams that contain silver (see Chapter 15 on antibiotics) help manage infections. Treatment may also require debridement of affected scabs or tissue to remove bacteria.

Many strains also make *pyocyanin*, a greenish-blue pigment that generates reactive forms of oxygen to further damage tissue. This pigment may cause pus to take on a blue/green coloration, which is often a feature that aids diagnosis along with clinical laboratory findings (FIG. 17.14).

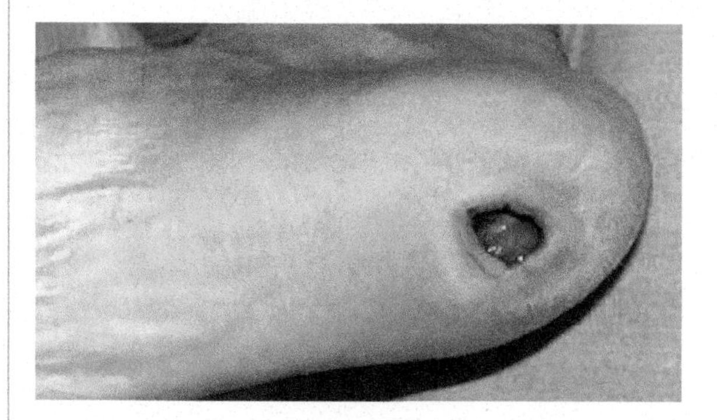

FIGURE 17.14 *P. aeruginosa* **wound infections** Wounds, especially burns, put patients at risk for *P. aeruginosa* infections. The pyocyanin pigment made by some strains increases tissue damage.

Critical Thinking *Describe P. aeruginosa's virulence factors and state how they contribute to disease.*

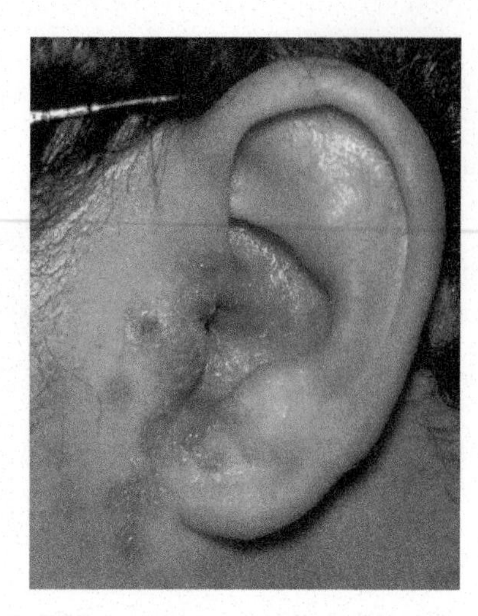

FIGURE 17.15 *P. aeruginosa* **otitis externa** Exposure to *P. aeruginosa* in recreational water is a common cause of otitis externa (swimmer's ear).

Critical Thinking *In addition to surviving in inadequately treated recreational water, what metabolic features do P. aeruginosa have that allow them to survive in a broad range of conditions?*

CLINICAL VOCABULARY

Gangrene a specific type of necrosis at an extremity

Necrosis tissue death due to loss of blood flow; may be caused by infections, injury, or diseases that affect the vasculature (i.e., diabetes)

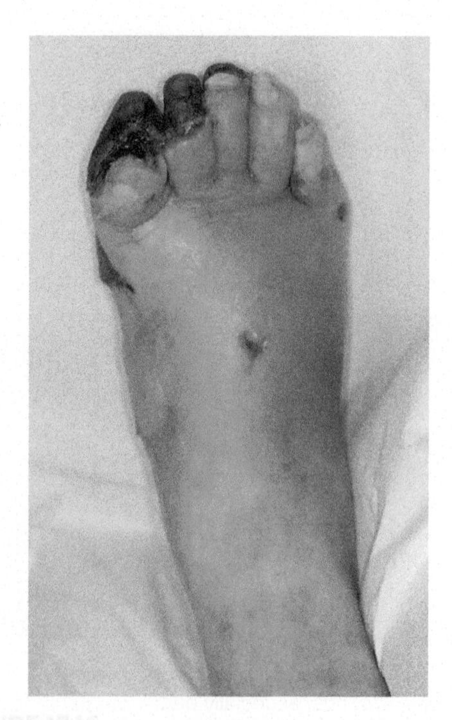

FIGURE 17.16 *Clostridium perfringens* **caused by gas gangrene**

Critical Thinking *Why is hyperbaric oxygen therapy a potentially useful treatment for gas gangrene?*

Otitis Externa (Swimmer's Ear)

Saunas, pools, or hot tubs with improperly maintained water may have higher pH levels and lower-than-ideal chlorine levels, allowing *P. aeruginosa* to thrive. Swimmers in such waters can suffer **otitis externa** (swimmer's ear), in which the outer ear canal is infected (FIG. 17.15); in some cases the pinna (the flaplike tissue that most think of as the ear) can also become inflamed and exhibit pus-filled lesions.

Gas gangrene and cutaneous anthrax are both bacterial infections characterized by tissue necrosis.

Necrosis (tissue death) is associated with decreased blood flow to tissues. Wounds, certain underlying conditions (such as diabetes), and a variety of infectious agents may cause necrosis as blood vessels are damaged and tissues starve for oxygen and nutrients. When the extremities are affected, the condition is usually called gangrene (FIG. 17.16). The bacterial infections discussed in this section are infamous for their ability to cause tissue necrosis—although they do so to varying degrees.

Gas Gangrene (*Clostridium perfringens*)

Clostridium perfringens are anaerobic Gram-positive, endospore-forming rods that naturally live in in the soil. These bacteria may infect deep wounds and cause gas gangrene—a condition named for the foul-smelling gases emitted as the bacteria destroy infected tissues. In addition to causing rapid and extensive tissue necrosis, if untreated, gas gangrene quickly progresses to shock, kidney failure, and death.

Cutaneous Anthrax

Anthrax is a collection of infections caused by the Gram-positive endospore-forming bacterium *Bacillus anthracis*. These bacteria cause rare infections of the respiratory tract, digestive system, or the skin. The skin form, **cutaneous anthrax**, is the mildest form of anthrax and accounts for about 95 percent of all anthrax infections (the more severe respiratory form is discussed in Chapter 16).

Cutaneous anthrax develops when *B. anthracis* enters a wound through soil or by direct contact with infected animals or contaminated animal products. Upon entering a skin abrasion, the bacteria form a solid nodule on the skin and start to kill superficial and then deeper skin cells. Eventually the nodule progresses to a blackened, swollen ulcer of necrotic tissue that is painless, but may itch (FIG. 17.17). Due to various exotoxins that these bacteria make, the infection is potentially deadly if not promptly treated with antibiotics.

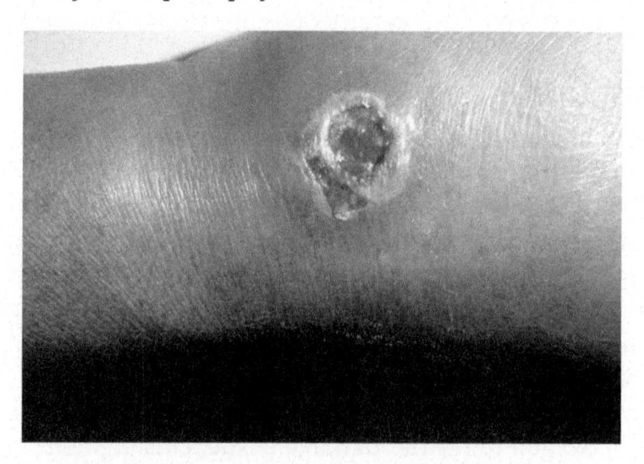

FIGURE 17.17 **Cutaneous anthrax caused by** *Bacillus anthracis*

Disease**Snapshot** • *Pseudomonas* Wound and Ear Infections

Infection	Wound Infection	Otitis Externa (Swimmer's Ear)
Causative agent	*Pseudomonas aeruginosa:* Gram-negative bacterium	
Epidemiology	Burned or wounded skin; 1–3 day incubation period	Typically affects frequent swimmers; pain usually develops within 24–48 hours of infection
Transmission & prevention	Environment or healthcare inoculation, autoinoculation from normal skin biota; keep wounds clean	Recreational water (minimize risk by properly maintaining pools and hot tubs); not spread person to person; keep ears dry and administer a 50/50 solution of vinegar and rubbing alcohol to the ears after each swim; don't use cotton swabs to remove earwax—a natural protectant
Signs & symptoms	Wound inflammation and green pigment may be apparent; if enters bloodstream, individuals may experience fever, chills, and septic shock	Painful outer ear; drainage
Pathogenesis & virulence factors	Enzymes that break down tissues and combat immune system factors; make exotoxins and endotoxin; excellent at biofilm formation; use diverse organic compounds for nutrients (even detergents) allowing them to grow in most environments; broad antibacterial and disinfectant resistance	
Diagnosis & treatment	Diagnosed mainly by wound appearance and culturing techniques; treated by removing affected tissue and by antibiotics (examples: aminoglycosides and beta-lactam drugs)	Diagnosed by symptoms, patient history of swimming, and observation of inflamed outer ear; treated with antibiotic eardrops (examples: aminoglycosides, quinolones, and polymyxin B)

P. aeruginosa

Disease**Snapshot** • Gangrene

Causative agent	Usually *Clostridium perfringens:* Gram-positive bacterium
Epidemiology	Present in soil, water, and intestinal tract of animals (including humans); diabetics and others with reduced circulation are at greatest risk; 1–3 day incubation period
Transmission & prevention	Usually associated with deep wounds or tissue trauma; prevented by meticulous wound care
Signs & symptoms	Severe pain and swelling at injured site, fever, darkened skin, foul-smelling wound drainage, gas bubbles under the skin, and a drop in blood pressure (hypotension); may progress to shock, coma, and death
Pathogenesis & virulence factors	Endospores allow for survival in harsh conditions
Diagnosis & treatment	Diagnosed by symptoms/wound appearance and staining or culture confirmation of agent; treated with intravenous antibiotics (example: penicillin), removal of dead tissue (debridement), or amputation if extensive damage; hyperbaric oxygen therapy (breathing oxygen while in a pressurized chamber) limits anaerobic *Clostridium* bacteria growth by increasing oxygen levels in tissues

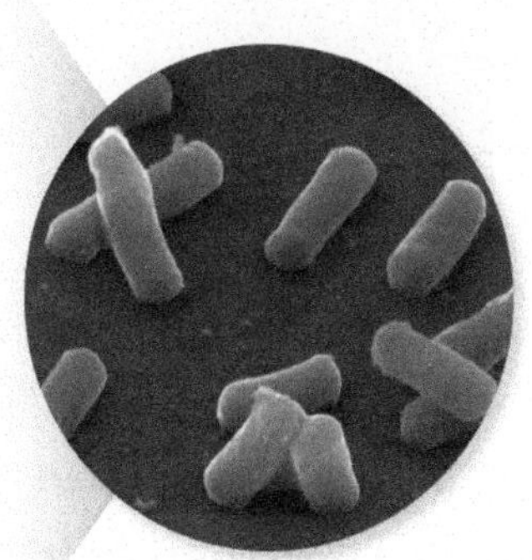

C. perfringens

Build Your Foundation

10. What are the different types of acne and what role does *P. acnes* have in acne?

11. Describe *Staphylococcus aureus* virulence factors and some of the skin diseases it causes.

12. Describe *Streptococcus pyogenes* virulence factors and some of the skin diseases it causes.

13. Compare and contrast the laboratory characteristics of *Staphylococcus aureus* and *Streptococcus pyogenes*.

14. Describe *Pseudomonas aeruginosa* virulence factors and some of the skin diseases it causes.

15. Name and describe the causative agents of cutaneous anthrax and gas gangrene and discuss the general progression of these diseases.

Build Your Foundation (BYF) Quick Quiz: Visit the **Mastering Microbiology** Study Area to quiz yourself.

17.4 FUNGAL SKIN INFECTIONS

Learning Outcomes

After reading this section, you should be able to:

17.16 Describe the types and severity of skin mycoses.

17.17 Describe cutaneous candidiasis.

17.18 List and describe dermatophytic infections.

CLINICAL VOCABULARY

Dermatophytes a collection of fungal organisms that cause conditions of the skin, hair, or nails

Fungal skin infections are usually superficial.

Like bacteria, many fungi live on our skin as normal microbiota that may cause disease when skin barriers are breached or our microbiota balance shifts. Fungal skin infections are called **cutaneous mycoses**. Common agents include *Candida* species and organisms that cause athlete's foot or ringworm (discussed later). Most cutaneous mycoses are superficial, meaning the fungi do not invade deep into the skin. Some, such as *Piedraia*, *Trichosporon*, and *Malassezia* species, cause mild skin, hair, or nail discolorations. Others, called dermatophytes, break down keratin and cause skin, hair, and nails to become brittle and flaky.

In most cases removing affected hair or using topical or oral antifungals will clear cutaneous mycoses, although prolonged treatment is common and normal pigmentation may not return for months. Immunocompromised individuals are at risk for more widespread discoloration and deeper tissue or systemic infections.

Subcutaneous mycoses involve deeper dermal or muscle infections and can be more serious. Subcutaneous mycoses are often associated with wounds or abrasions; an example is *Sporothrix schenckii*, which can be inoculated into the skin by the prick of a rose thorn.

Yeast Infections of the Skin (Cutaneous Candidiasis)

The opportunist pathogen *Candida albicans* is a common culprit of cutaneous mycoses. While this unicellular yeast is a normal resident of the gastrointestinal tract and skin, a shift in normal microbiota, decreased immune function, or inoculation of the yeast into areas where it is not a normal resident can lead to disease.

C. albicans is most well known for causing diaper rash and afflicting damp, friction-prone skin folds of the underarm, groin, and under the breasts (FIG. 17.18). It is important to thoroughly dry these body areas to limit infection. Frequent diaper changes help keep this region dry to reduce the risk of diaper rash, and use of topical diaper creams helps form a protective barrier to reduce skin contact with *C. albicans*, which is naturally found in feces.

Cutaneous candidiasis manifests as a bright red macular rash that may be accompanied by small white pustules, especially at the edges of the rash. The rash tends to be intensely itchy or burning and may exhibit scaling. Diabetics have a higher risk for cutaneous candidasis, due to high blood sugar levels that promote yeast growth and vascular damage that reduces circulation.

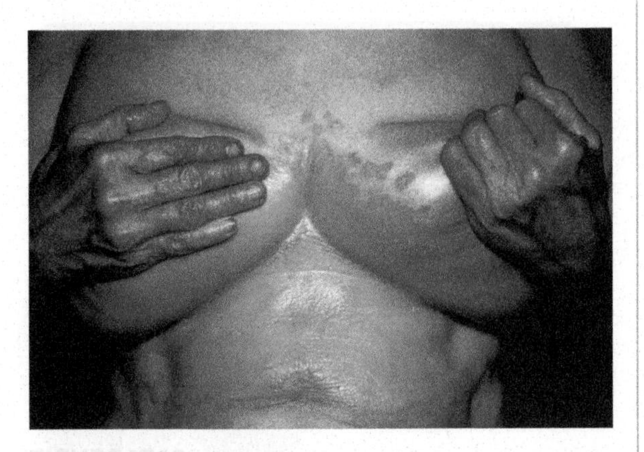

FIGURE 17.18 **Cutaneous candidiasis**

Critical Thinking *Why are friction-prone areas at greater risk for cutaneous candidiasis?*

DiseaseSnapshot • Cutaneous Candidiasis

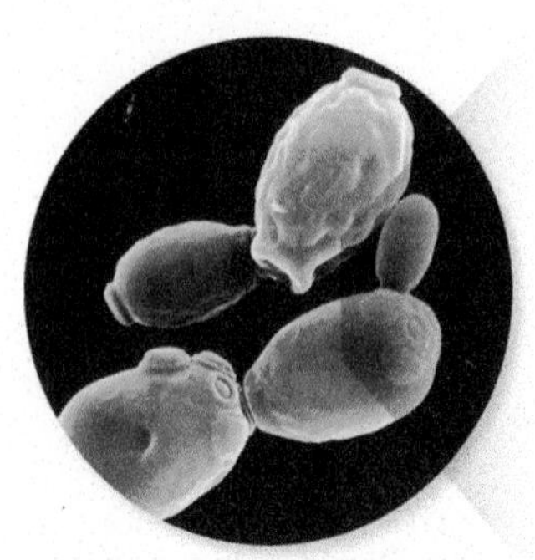

C. albicans

Causative agent	Yeast species, usually *Candida albicans*
Epidemiology	Prior antibiotic usage, moist skin conditions, diabetes, and reduced immune system function; incubation period is usually 1–5 days, but varies based on area affected
Transmission & prevention	Opportunistic infection by normal flora; prevent by keeping skin dry (especially skin folds and diaper region); diabetics should carefully regulate blood sugar levels
Signs & symptoms	Red macular rash that is intensely itchy or may burn
Pathogenesis & virulence factors	*C. albicans* makes diverse adhesion factors to assist with attachment to host tissues and breaks down keratin to aid skin invasion
Diagnosis & treatment	Diagnosed by patient history and rash appearance/ location; culture and microscopic analysis can help to confirm diagnosis; infection often self-resolves if skin is kept dry; antifungal creams/ointments or oral medications may be used to treat advanced cases

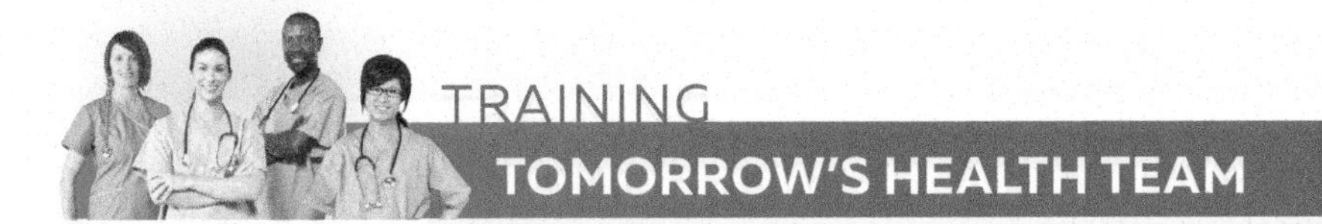

An Array of Topical Anti-Fungals

Antifungal medications come in many different formats and formulations. For instance, the brand name drugs Desenex, Tinactin, and Lamisil all have different active ingredients, yet all are marketed to treat athlete's foot. This is not just a marketing ploy; the diversity of fungi makes the availability of varied active ingredients important.

There are multiple species that cause cutaneous mycoses; one species may respond to a particular antifungal drug better than another. Many antifungals are low dose and used over several days. This is partly because fungi are eukaryotes like us, and therefore targeting their cells in ways that do not also damage our own is challenging. Also, because they grow slower than bacteria and drugs tend to target actively growing cells, a longer treatment course is required to kill them.

QUESTION 17.3

Miconazole is often used to treat vaginal yeast infections. It is sold in variants that include 1-, 3-, and 7-day doses. Why might a longer treatment course be more effective?

Tinea Infections (Ringworm)

Ringworm is not actually a worm, but a fungus that in some cases produces a round skin lesion that appears wormlike, with a scaly red margin and a central lighter red coloration. The Latin name tinea (meaning "worm") is combined with other names that refer to the afflicted region, such as tinea pedis (athlete's foot), tinea cruris (groin or "jock itch"), tinea capitis (head or scalp), tinea unguium (nails), or tinea corporis (body). These names are not italicized because they refer to clinical conditions rather than specific organism names.

Tinea infections are mainly caused by *Trichophyton*, *Microsporum*, and *Epidermophyton* species, although infections are usually treated without determining the exact causative agent. All three of these genera can cause skin and nail infections; *Trichophyton* can also cause hair infections. Infected skin may become scaly, blistered, discolored, or inflamed; infected nails may thicken, darken, or be disfigured; and infected hair may undergo discoloration, and short- or long-term bald patches can develop. Skin abrasions, fomite contamination, and intimate contact with fungal cells, spores, or lesions can enhance transmission. Some dermatophytes can be picked up from infected people or animals.

Because they are among our normal skin biota, it is impossible to completely avoid dermatophytes. Diagnosis is typically made based solely on lesion appearance (FIG. 17.19). Topical antifungal creams and ointments that contain tolnaftate, miconazole, itraconazole, thiabendazole, or terbinafine are available over the counter. Medical attention is needed for persistent, expanding, or reoccurring cases, or anything that looks further infected. Oral drugs like terbinafine or griseofulvin may be necessary in severe cases.

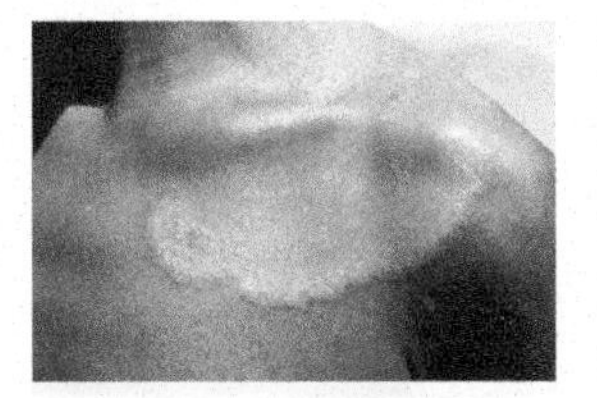
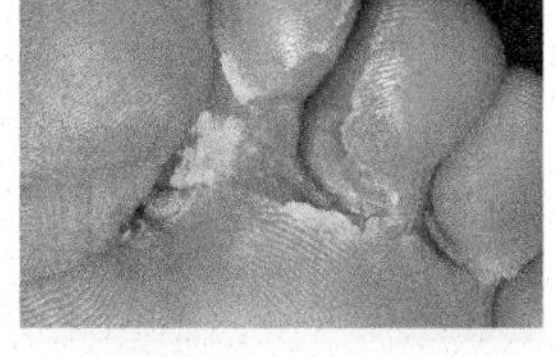
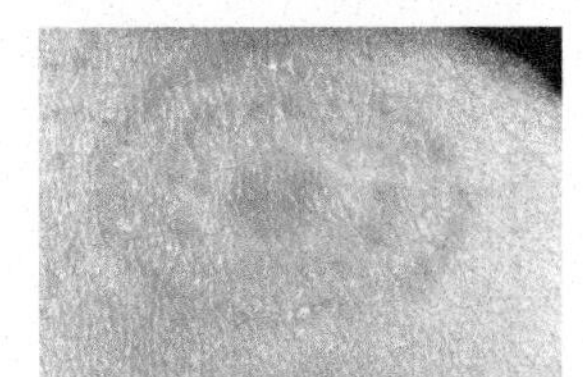
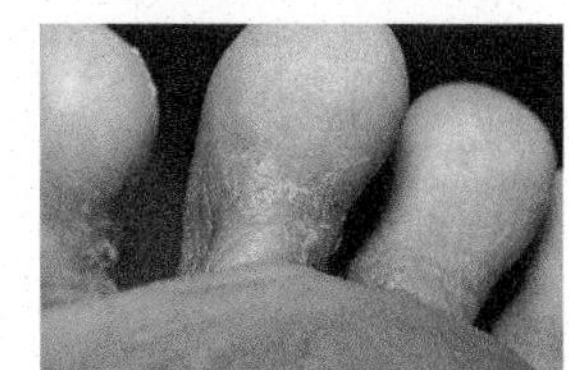

Tinea corporis (body ringworm) Tinea pedis (athlete's foot)

FIGURE 17.19 Body ringworm and athlete's foot These are common dermatophytic conditions caused by various fungi.

Build Your Foundation

16. Describe several different types of skin mycoses and differentiate between cutaneous versus subcutaneous infections.
17. Describe skin infections caused by *Candida albicans*.
18. Describe different types of dermatophytic skin infections.

Build Your Foundation (BYF) Quick Quiz: Visit the **Mastering Microbiology** Study Area to quiz yourself.

17.5 PARASITIC SKIN INFECTIONS

Learning Outcomes

After reading this section, you should be able to:

17.19 Describe the causative agent of leishmaniasis and discuss general disease features.

Cutaneous leishmaniasis is a protozoan infection.

The protozoan infection **leishmaniasis** is classified as a neglected tropical disease caused by *Leishmania* species of protozoan that are transmitted by the bite of infected sand flies. *Leishmania* parasites are found on every continent except Australia and Antarctica, but they are most prevalent in the tropics, subtropics, Middle East, and southern Europe. As climate change alters habitats it is likely that leishmaniasis will spread to new areas and/or increase in prevalence as sand fly vectors expand into new areas. The CDC estimates that there are 0.7–1.2 million cases of cutaneous leishmaniasis worldwide each year. U.S. healthcare workers may encounter this infection in people who travel to endemic areas or in military personnel stationed in such areas.

There are three types of leishmaniasis caused by over 20 different *Leishmania* species. The responsible protozoan species, transmission vector, and the host immune response all impact disease severity. In **cutaneous leishmaniasis**, skin ulcers, which are usually painless, form at bite sites and can persist for months or even years; scars remain after healing due to dermal damage (FIG. 17.20). Secondary bacterial infections of these lesions is also common, and can be dangerous if not properly treated. In **mucocutaneous leishmaniasis**, the lesions develop in mucous membranes of the nose or mouth. It can lead to severe and permanent disfigurement as the protozoan destroys these structures. In **visceral leishmaniasis**, the protozoan spreads throughout the body. According to the World Health Organization, in developing countries where proper treatment is difficult, visceral leishmaniasis is usually fatal within two years.

The only prevention is to avoid sand fly bites by covering exposed skin and using DEET-containing insect repellants. Treatment approaches depend on the *Leishmania* species responsible for the infection, but pentavalent antimonial compounds and sodiumstibogluconate are common treatments. Intravenous liposomal amphotericin B (usually considered an antifungal) and other medications, such as orally administered miltefosine, may also be used.

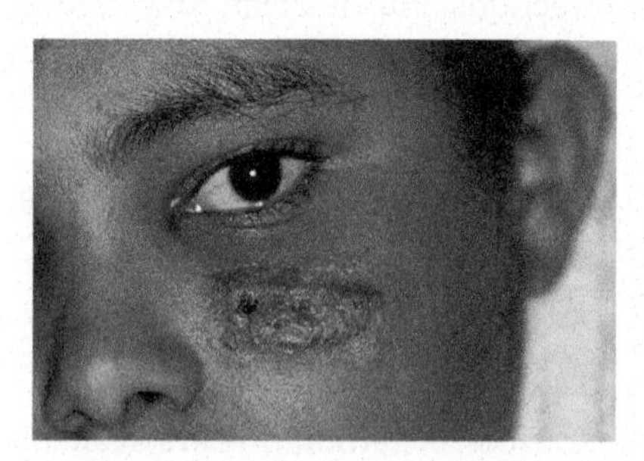

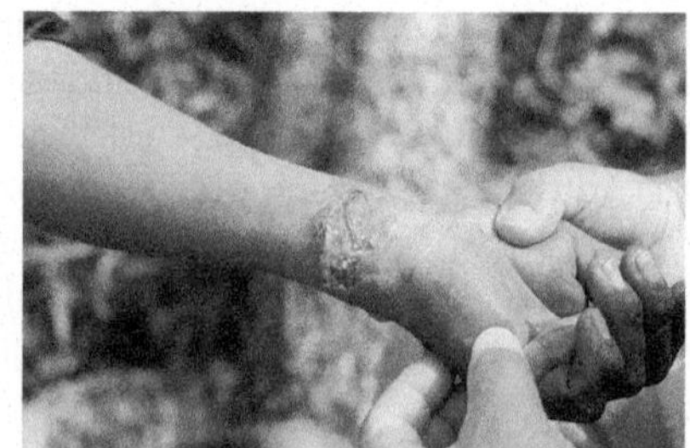

FIGURE 17.20 **Cutaneous leishmaniasis**

BYF QUICK QUIZ

Build Your Foundation (BYF) Quick Quiz: Visit the **Mastering Microbiology** Study Area to quiz yourself.

Build Your Foundation

19. Describe leishmaniasis and its global impact.

17.6 STRUCTURE, DEFENSES, AND INFECTIONS OF THE EYES

The eye has specialized structures and defense mechanisms.

Learning Outcomes

After reading this section, you should be able to:

17.20 Describe structures, protective factors, and normal microbiota of the eyes.

17.21 Compare and contrast bacterial versus viral conjunctivitis, including causative agents, prevention, and treatments.

17.22 Describe trachoma and review its causative agent, prevention, and treatments.

17.23 Explain keratitis and list common causative agents.

While it isn't hard to imagine the skin contacting many possible threats, the threats our eyes are up against may be less obvious. Bacteria and viruses are the main causes of microbial eye infections, but fungi and even parasites can also infect the eyes. Specialized eye structures and even normal microbiota protect our eyes against pathogens.

Basic Structure of the Eyes

The eye is an amazing and complex structure. FIG. 17.21 is a simplified view of the eye structures and the **lacrimal** system that produces and collects tears. Pathogens generally target the conjunctiva or cornea, so these are our main areas of focus.

Defense Mechanisms of the Eyes

The **cornea** is a transparent layer at the front of the eye that covers the iris. It has five to six layers of epithelial cells that can quickly be replaced if superficially damaged. The **conjunctiva** is the epithelial membrane that covers the eyeball and lines the eyelids. It does not cover the cornea but surrounds it. The eyeball itself has a thick outer wall that helps prevent microbial penetration. The lacrimal gland produces tears that constantly rinse away foreign objects, dirt, or dust from the eye. Tears are made up of water and protective factors including oils, mucus, sugars, lysozyme, and lactoferrin. Lysozyme breaks down bacterial cell walls, and lactoferrin binds up free iron (which certain microbes need). The collective goal of these defenses is to protect our eye structures without obstructing our vision.

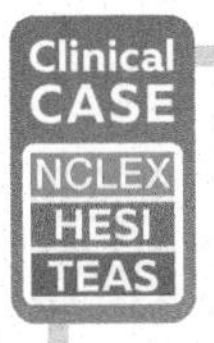

The Case of the Hidden Blinder

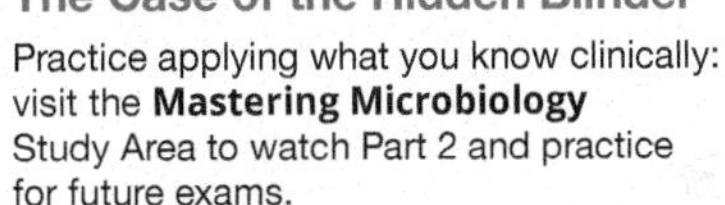

Practice applying what you know clinically: visit the **Mastering Microbiology** Study Area to watch Part 2 and practice for future exams.

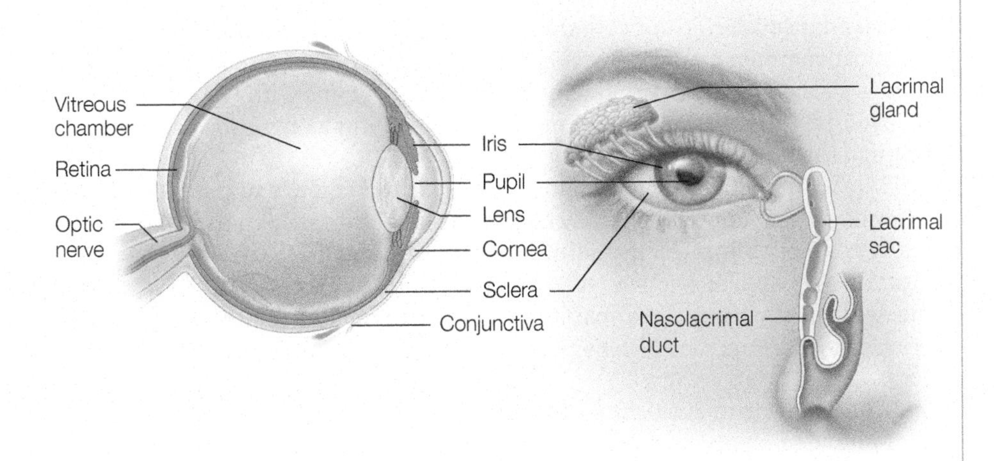

FIGURE 17.21 Structures of the eye and lacrimal system The eye is a complex structure that is linked both to skin structures (through the conjunctiva) and to the brain (through the optic nerve). Structures and defenses provide protection against infections of the conjunctiva and cornea.

Critical Thinking *Why would blocking tear ducts potentially lead to eye infections?*

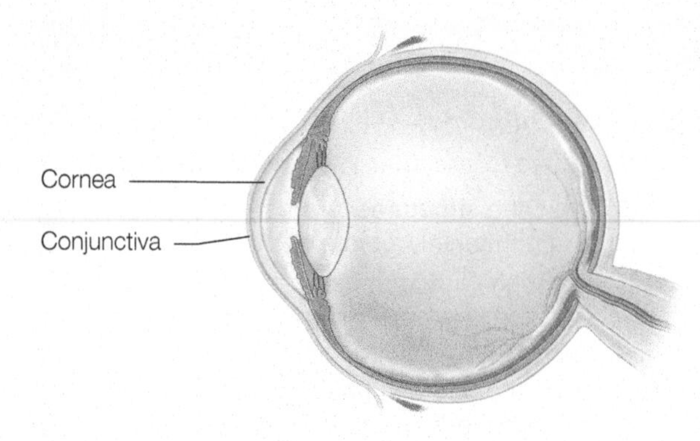

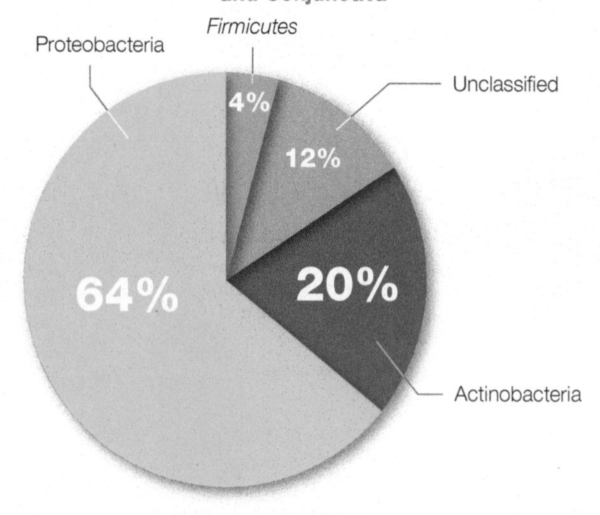

Eye microbiome Researchers are still learning about the full range of microbes that reside on the eye. Recent studies show the conjunctiva and cornea are home to over a dozen genera, spanning over 200 bacterial species, in healthy adults. Most of these bacteria fall into just a few phyla. (Note: Data shown in the chart are rounded.)

Eye Microbiome

Students are often surprised to learn that the eyes have any natural biota at all, and until recently, even scientists didn't consider the eyes to be normally colonized by microbes. We knew there were some limited *Staphylococcus* and *Corynebacterium* species that live on the eyes. However, the Ocular Microbiome Project has determined that we have numerous bacteria on exposed epithelial layers of the eyes. The work suggests that a dozen or so species dominate the conjunctiva and a different set of a dozen or so species dominate the cornea. Some of the species were not yet familiar to scientists, and not all were bacteria. The findings are still early, but it is intriguing to consider what has been right under (or in) our eyes![4]

Bacteria, viruses, fungi, and parasites can all cause eye infections.

Almost every known class of microbe can cause eye infections. In the developed world, there has been a general increase in eye infections that tracks with the increasing number of people who wear contact lenses. Contact lens wearers should remove and clean contacts as recommended, and never use homemade saline solutions that could introduce microbes into their eyes. Eye infections are also associated with daycares and schools, as young children tend to share objects and wipe their hands on their faces, and are not practiced in hygiene.

Conjunctivitis

Conjunctivitis is inflammation of the conjunctiva or epithelial membrane over and around the eye. The white sclera portion of the eye may become red or pink from inflamed blood vessels, so conjunctivitis is often referred to as "pink eye" (FIG. 17.22). Conjunctivitis from allergens, chemical irritation, or the introduction of foreign objects is not contagious. However, most microbial forms of conjunctivitis are contagious.

Viral conjunctivitis Some forms of conjunctivitis are viral; upper respiratory viruses that cause colds or sore throats are common culprits. Adenoviruses are the most common causes of viral eye infections, but other viruses, including herpes viruses, cause conjunctivitis. Neonatal infections with HSV may develop following transmission from the mother to the infant as the baby passes through the birth canal.

Viral conjunctivitis often begins in one eye and spreads to the other within a few days. Most forms are highly contagious and are easily transmitted to others by contact or fomites. In addition to red or pink color to the eyes, the conjunctiva may appear swollen (even more common in bacterial conjunctivitis), the eyes usually feel itchy or painful, and the conjunctiva may tear. Upper respiratory symptoms may also be present. A watery (but not thick) discharge is common.

The best way to avoid viral conjunctivitis is to avoid infected individuals or contaminated objects and wash hands thoroughly and frequently. Using hand sanitizers may be valuable in institutional, healthcare, or travel situations. Infected individuals should not share eye drops, towels, bedding, or eye makeup, and they should avoid touching or rubbing their eyes. When conjunctivitis develops, disposable contacts should be discarded and nondisposable long-wear contacts should be thoroughly disinfected and not worn until the conjunctivitis is cleared up in order to avoid re-infection and promote proper

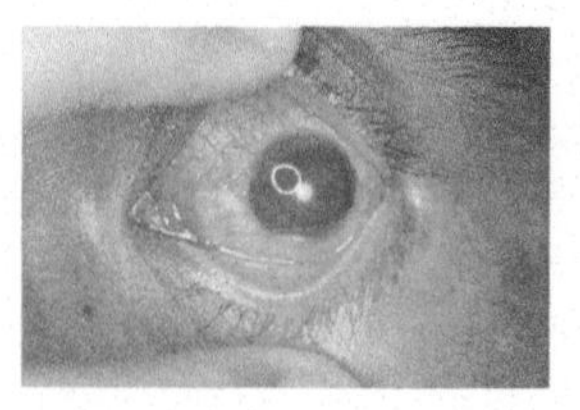

Viral conjunctivitis

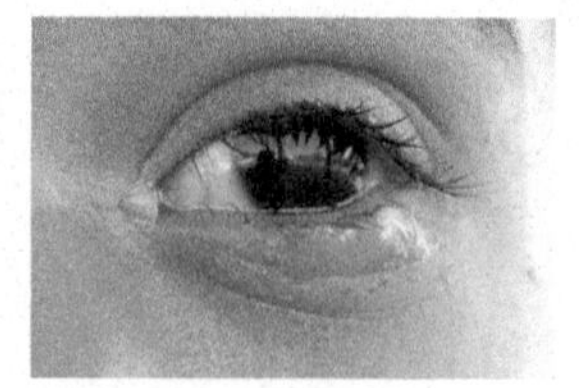

Bacterial conjunctivitis

FIGURE 17.22 **Conjunctivitis (non-neonatal)**

Critical Thinking *What clinical features allow for the differentiation of bacterial from viral conjunctivitis?*

[4] Shaikh-Lesko, Rina. (2014). *The Scientist*, retrieved June 2014 from http://www.the-scientist.com/?articles.view/articleNo/39945/title/Visualizing-the-Ocular-Microbiome/ and http://www.microbiota.org/cgi-bin/ocular/index.cgi.

DiseaseSnapshot • Conjunctivitis (Non-neonatal)

Causative agent	Viral (usually adenoviruses) and bacterial (often *Streptococcus, Staphylococcus, Haemophilus*, or *Moraxella* species)
Epidemiology	Highly contagious, common in contact lens wearers and school-aged children; incubation period of 24 hours to a few days
Transmission & prevention	Transmitted by contact with infected individuals or fomites; prevented by frequent hand washing, proper contact lens use and care; do not share sunglasses, makeup, or other items that encounter the eyes
Signs & symptoms	Red or pink eyes, swollen conjunctiva, itchy or painful eyes; watery discharge accompanies viral forms, while bacterial forms may exhibit pus with crusty eyelids or lashes
Pathogenesis & virulence factors	Infectious agents have adherence factors for the conjunctiva that provide resistance to the rinsing action of tears
Diagnosis & treatment	Mainly diagnosed by symptoms; bacterial forms can be confirmed by culture methods although such confirmation is not usually cost effective or necessary for routine cases; most viral cases self-resolve; antibiotic eye drops or ointments are used to treat bacterial cases (often aminoglycosides or fluoroquinolone drugs)

Adenovirus

healing. Swimming pools should be avoided to prevent further irritation or spread. Ordinarily, viral conjunctivitis is mild, resolving on its own without treatment. However, care should be taken to prevent spread or re-infection. Since the source and severity of conjunctivitis is often not known, physician consultation is recommended. This may be particularly necessary for daycare or school-aged children.

Bacterial conjunctivitis Viral and bacterial conjunctivitis have the same symptoms, but can usually be differentiated by the nature of the fluid discharge from the eye. A puslike discharge (often green or yellow) is usually associated with bacterial eye infections. The eyes may feel "stuck shut" or grainy in the morning, and eyelids and lashes may be crusted over may be crusted over; in contrast, most viral conjunctivitis is accompanied by a watery discharge (Figure 17.22). The conjunctiva is more likely to be swollen in bacterial conjunctivitis than in viral forms.

Bacteria that cause upper respiratory infections can also cause conjunctivitis. *Haemophilus influenzae, Staphylococcus* species, *Streptococcus* species, and *Moraxella* species are some of the most common causative agents of bacterial conjunctivitis in children. These agents are also often associated with ear infections.

Neonatal bacterial conjunctivitis is usually caused by *Neisseria gonorrhoeae* or *Chlamydia trachomatis*, and is acquired by vertical transmission at birth. Neonatal bacterial conjunctivitis can lead to serious eye damage if not treated. This is why newborns in the United States routinely receive antibiotic drops to prevent infections. Adults infected with these organisms may inadvertently transmit the bacteria from the genital tract to eyes by rubbing the eyes with contaminated hands.

Bacterial and viral conjunctivitis are prevented in the same ways. Similar to viral conjunctivitis, bacterial cases are often mild and resolve in a few days without treatment. However, some bacterial cases may last a few weeks and may be treated with antibiotic eye drops or ointment. Again, it is recommended that children be evaluated by a medical professional when signs of conjunctivitis appear.

Trachoma

According to the CDC, trachoma is the leading cause of preventable microbial blindness; worldwide about 6 million people are blind due to trachoma. The World Health Organization has targeted this treatable infection for elimination by 2030. The word "trachoma" comes from the Greek word for rough,

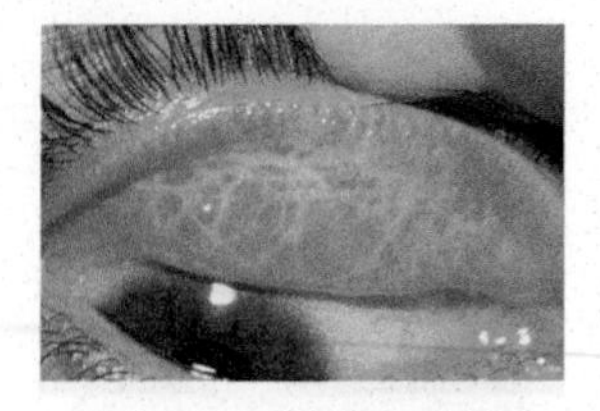

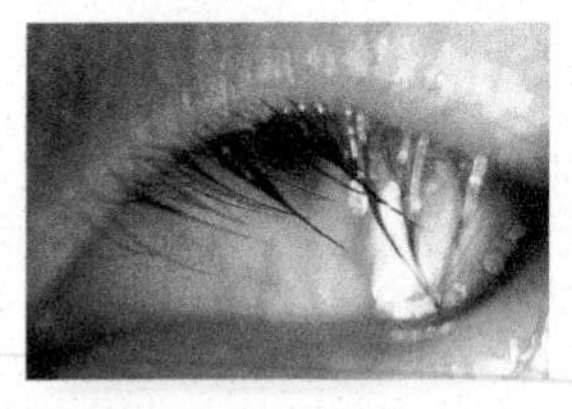

FIGURE 17.23 Trachoma The left image shows conjunctiva scarring while the image to the right shows changes in eyelid structure that lead the lashes to scratch the cornea.

Critical Thinking *The World Health Organization is using a public health strategy called S.A.F.E. to address trachoma. S.A.F.E. stands for surgery to correct advanced blindness, antibiotics for active infections, facial cleanliness, and environmental improvements to reduce transmission. Explain why each approach is needed for an effective strategy to combat this global concern.*

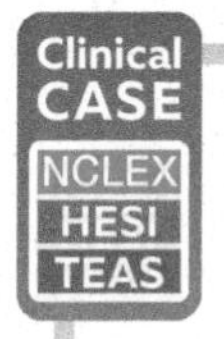

The Case of the Hidden Blinder

Practice applying what you know clinically: visit the **Mastering Microbiology** Study Area to watch Part 3 and practice for future exams.

which describes the appearance of the affected conjunctiva. Trachoma is caused by certain serotypes of *Chlamydia trachomatis*. Untreated or repeat infections scar the conjunctiva lining the eyelid, causing the eyelid to turn inward so that the eyelashes scratch and permanently damage the cornea (FIG. 17.23). The disease mainly afflicts the impoverished in developing nations and remains an issue in areas with little access to clean water and general sanitation systems.

Trachoma can be acquired by sharing contaminated fomites like towels, by touching the eyes with contaminated hands, or by contact with contaminated flies. In summary, inoculation occurs by fomites, fingers, and flies. Prevention measures include facial cleaning, environmental improvements, and public education campaigns. Diagnosis is usually made by patient signs and symptoms and through clinical inspection using a magnifying lens to detect inflammation and damage of eye structures. The preferred treatment is a single oral dose of azithromycin; ophthalmic ointments or eye drops can also be used, but they require multiple doses and patient compliance tends to be lower. Surgery may be required to fix eyelids so that lashes do not scar the cornea.

Keratitis

Severe inflammation of the cornea is called **keratitis**. This disease is more serious than conjunctivitis and warrants immediate medical attention to prevent serious and irreversible damage to the eyes. In keratitis, the protective layers that cover the cornea are destroyed, leaving the cornea accessible to bacteria, viruses, parasites, or fungi. Contact lens wearers should follow all recommendations for the use, care, and cleaning of their contact lenses.

Viral keratitis The number-one cause of viral keratitis is the herpes simplex 1 virus (herpetic keratitis). Individuals who have had prior HSV-1 outbreaks can have reactivation of the virus in the eye. Instead of reactivation through the trigeminal nerve (as discussed earlier in this chapter regarding viral skin infections caused by herpes viruses), the reactivation occurs via the ophthalmic nerve. HSV-2 keratitis is also possible. Symptoms can include conjunctivitis, eye pain, blurred vision, sensitivity to light, and watery discharge. Prevention includes hand-washing hygiene, especially with herpes blisters, and proper eye care. Diagnosis is based on eye exams and treatment is with trifluridine, acyclovir, or both. Because herpes can reoccur, so can herpetic keratitis, making it the leading cause of infectious blindness in the United States.

Bacterial keratitis According to the CDC, *Pseudomonas aeruginosa* and *Staphylococcus aureus* from our surroundings or our own skin are the most common causes of bacterial keratitis. These infections occur most often in those who wear contacts overnight, improperly disinfect their lenses, or share contacts. Bacterial keratitis can also develop following eye injury, or other eye conditions, or as a result of a weakened immune system. Symptoms parallel those for viral keratitis, but also tend to include eye discharge and excessive tearing. Prevention includes hand hygiene and the proper use, cleaning, and storage of contacts. Diagnosis is made based on an eye exam and may involve a sample for analysis. Treatment includes antibiotic eye drops.

Fungal keratitis Individuals that have suffered eye trauma or have had eye surgery (including radial keratotomy to correct nearsightedness) have the highest risk for fungal keratitis. Other risk factors include a weakened immune system, contact lens use, and underlying eye disease. The most common fungal culprits are *Fusarium, Aspergillus,* and *Candida* species; the first two are common in the environment, and the third is common in and on our bodies. Symptoms are similar to those for bacterial keratitis. While fungal keratitis is rare, it can be very serious. Prevention includes protective eyewear for workers at risk for eye injury. Diagnosis is parallel to bacterial keratitis, but treatment is more intense, as fungal infections can be difficult to treat. Antifungal drugs may be needed for several months, and surgery may be required.

Parasitic keratitis This form of keratitis can be caused by helminths or protozoans. *Acanthamoeba* is a protozoan form of keratitis while river blindness is a helminthic form. Both of these forms of keratitis are discussed next.

Acanthamoeba (Protozoan Keratitis)

Parasitic keratitis caused by *Acanthamoeba* is rare but very serious when it occurs. This disease is referred to as *Acanthamoeba* keratitis and is again connected to improper handling, storage, or disinfection of contact lenses or swimming with contacts in place. Tap water should not be used to clean contact lenses or contact lens cases. *Acanthamoeba* species are single-celled protozoans found in natural bodies of water as well as in tap water, heating or cooling systems, and whirlpools. Transmission is by direct introduction of the protozoan from contaminated water into the eye; the infection is not contagious nor can it be acquired by ingesting the protozoan. Symptoms are similar to viral keratitis. Diagnosis is by physical examination and medical testing, and treatment varies by diagnosis, individual, and severity. Early diagnosis and treatment is essential to avoid advanced damage to the cornea, which may necessitate a corneal transplant.

River Blindness or Ocular Onchocerciasis (Helminthic Keratitis)

Onchocerciasis develops when an infected blackfly bites someone and introduces *Onchocerca volvulus* larvae to the skin. The larvae of this parasitic helminth (worm) then migrate throughout the body to cause a variety of symptoms that depend on larval dose and location. As the larvae die in the skin, an itchy "leopard skin" rash may develop. If the larvae migrate to the eye and cause inflammation that damages the optic nerve and/or cornea, the infection is called ocular onchocerciasis or "river blindness" (FIG. 17.24). This name comes from the habitat of the flies— they live and breed near rivers. Adult worms, which mature from the larvae in 3 months to a year after initial infection, often cause skin nodules to form.

This neglected tropical disease mostly affects people in sub-Saharan Africa. The WHO estimates that at least 25 million are infected; of these, about 300,000 are blinded by the larvae while another 800,000 are visually impaired. After trachoma, it is the second leading cause of infectious blindness. River blindness is diagnosed by the slit-lamp eye exam, which magnifies the eye structures so that the worm's larvae may be seen; skin infections are diagnosed by detecting larval DNA or live larvae in skin biopsies. Orally administered ivermectin kills the larvae. The antibiotic doxycycline kills adult worms by killing the *Wolbachia* bacteria that the worms rely on to survive.

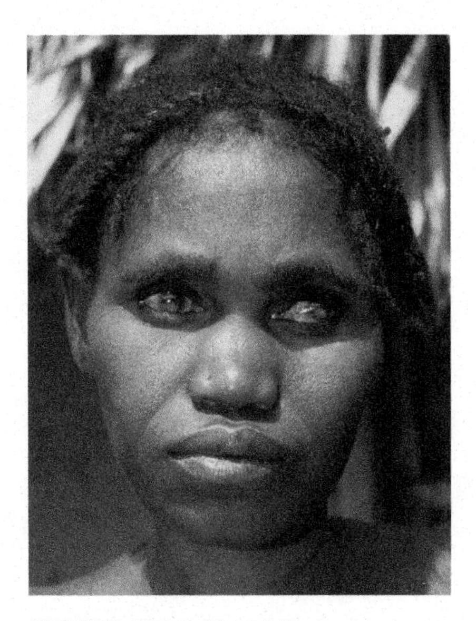

FIGURE 17.24 **River blindness**

Build Your Foundation

20. What are the parts of the eye and some of its defenses?
21. What is conjunctivitis and how is it caused, treated, and prevented?
22. Describe trachoma and state how it is caused, treated, and prevented.
23. Describe keratitis and give examples of causative agents.

Build Your Foundation (BYF) Quick Quiz: Visit the **Mastering Microbiology** Study Area to quiz yourself.

VISUAL SUMMARY | *Skin and Eye Infections*

SKIN INFECTIONS

EYE INFECTIONS

Bacterial

S. aureus

Staphylococcus aureus
- Normal microbiota, fomite, or HAI transmission
- Range from mild impetigo to cellulitis, issues with resistant strains

Streptococcus pyogenes
- Normal microbiota, fomite, or HAI transmission
- Range from mild impetigo to necrotizing fasciitis

Pseudomonas aeroginosa
- Normal microbiota, fomite, or HAI transmission
- Wounded or burned skin at particular risk

Clostridium perfringens
- Naturally lives in soil
- Gas gangrene
- Tissue necrosis can lead to amputation or death

Bacillus anthracis
- Introduction to wound from soil or infected animal
- Skin nodules progress to blackened ulcers

Viral

HSV

Chickenpox (varicella-zoster virus)
- Respiratory droplets or direct contact transmission
- Itchy vesicular rash that can lead to complications

Shingles (varicella-zoster virus)
- Reactivation of previous chickenpox virus
- Painful vesicular rash, often occurs in a band on one side of the body

Smallpox (variola major virus)
- Respiratory droplet or direct contact transmission
- Pustular lesions with dimple, 30% fatal, now eradicated

Herpes simplex virus 1
- Respiratory droplet or direct contact transmission
- Painful or itchy vesicular lesions on the lips, recurrent

Measles (rubeola virus)
- Respiratory droplet transmission, highly contagious
- Manifests as rash, risk of measles encephalitis

German measles (rubella virus)
- Respiratory droplet transmission, highly contagious
- Teratogenic to fetus

Fifth disease (human parvovirus B19)
- Respiratory droplet transmission
- Childhood mild illness with "slapped cheek" rash

Roseola (human herpes virus 6)
- Respiratory droplet transmission
- High fever followed by rash in young children, most dangerous for unborn fetus

Hand, foot, and mouth disease (coxsackievirus A16 and enterovirus 71)
- Respiratory droplet transmission
- Rash on hands and feet often with mouth sores, common in children

Human papillomaviruses
- Direct or indirect contact transmission
- Warts
- No vaccines for cutaneous forms
- Vaccination only against certain sexually transmitted HPVs

Fungal

C .albicans

Candida albicans
- Normal microbiota, opportunistic pathogen
- Degrades skin keratin

Tinea infections
- Normal microbiota, but may also be transmitted by contact with contaminated surfaces or infected individuals
- Ringworm, athlete's foot, and other dermatophyte conditions

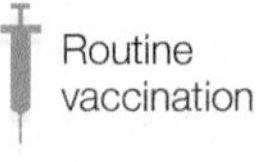

Routine vaccination

Non-routine vaccination

SKIN INFECTIONS

EYE INFECTIONS

Conjunctivitis

Viral conjunctivitis

Viral
- Pink or red eye appearance, watery discharge, hygiene important
- Often mild and resolves in a few days, caused by common respiratory viruses

Bacterial
- Pink or red eye appearance, pus-like discharge likely, hygiene important, neonatal infection is more serious
- Caused by common respiratory agents
- Neonate infections caused by *Chlamydia* or gonorrhea bacteria

Keratitis

River blindness

Viral
- Herpes simplex 1 virus reactivation in eye
- Herpetic keratitis is the leading cause of infectious blindness in U.S.

Bacterial
- Associated with improper use or care of contact lenses or eye injury
- Often *P. aeruginosa* or *S. aureus* autoinoculation

Fungal
- Associated with eye injury
- Similar to bacterial keratitis but more difficult to treat

Protozoan
- Associated with improper handling of contacts or swimming
- *Acanthamoeba* keratitis is rare but serious

Helminthic
- *Onchocerca volvulus* causes river blindness

Trachoma

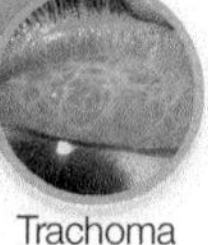

Trachoma lid

Bacterial
- Caused by *Chlamydia trachomatis*
- Transmitted by fomites, fingers, or flies
- Inversion of eyelids and corneal scarring, leading cause of infectious blindness worldwide

For all printed and additional Disease Snapshot tables, go to the Mastering Microbiology Study Area.

CHAPTER 17 OVERVIEW

17.1 Skin Structures, Defenses, and Afflictions

- Many systemic diseases have primary skin manifestations central to diagnosis. Rashes and lesions are characterized by their color, elevation, and whether or not they contain pus.
- Protective skin barriers include multiple layers of cells, normal microbiota, and chemical factors like salinity and pH that limit microbial growth.

17.2 Viral Skin Infections

- The varicella-zoster virus causes chickenpox and shingles; vaccines exist for both. In chickenpox, a vesicular rash develops. The virus may become latent in nerve cells and may reactivate later in life to cause shingles. Immunocompromised individuals are at high risk for secondary infections from chickenpox.
- Smallpox is eradicated, making routine vaccinations unnecessary. It is important for healthcare providers to distinguish smallpox from chickenpox in the event of a weaponized attack.
- Herpes simplex virus type 1 can cause fever blisters (oral herpes) and is usually contracted by direct contact with infected individuals or indirect contact with contaminated objects. Initial infection may be asymptomatic or cause painful itchy lesions a week later. Stress can cause the virus to reactivate at nerve endings near the mouth.
- Measles, caused by the rubeola virus, is a vaccine-preventable infection that causes a sore throat, dry cough, Koplik's spots in the mouth, and a characteristic maculopapular rash. Immunocompromised patients are at high risk for secondary complications.
- Rubella (German measles) is caused by the rubella virus. In pregnant women it can cause congenital rubella syndrome, resulting in severe infant defects or death.
- Fifth disease is usually a mild rash in children that can give a characteristic "slapped cheek" rash. Roseola can be asymptomatic or result in a high fever for a few days, followed by a rash. Hand, foot, and mouth disease may result in uncomfortable mouth blisters or lesions on the hands and feet.
- Warts are caused by human papillomaviruses acquired by direct contact or contact with contaminated objects. Skin warts may resolve on their own or require other removal efforts.

17.3 Bacterial Skin Infections

- Acne results when skin pores or hair follicles are plugged by bacteria, sebum, pus, or white blood cells and cause inflammation. *Propionibacterium acnes* sometimes increases inflammation.
- *Staphylococcus aureus* causes a spectrum of skin diseases. Its enzymes impact clotting, degrade tissue, and lyse cells. Its toxins overstimulate the immune system. *S. aureus* bacteria can cause mild (impetigo) to severe skin problems (erysipelis, cellulitis, folliculitis, or scalded skin syndrome). Strains that produce beta lactamase (penicillinase) break down penicillin so it is no longer effective. These methicillin-resistant *Staphylococcus aureus* (MRSA) strains are major problems in today's healthcare facilities.
- *Streptococcus pyogenes* can cause skin infections because of virulence factors similar to those of *S. aureus*, including enzymes and toxins. It can cause mild (impetigo) to severe skin problems (cellulitis, necrotizing fasciitis, or streptococcal toxic shock syndrome).
- Pseudomonads (most often *Pseudomonas aeruginosa*) cause opportunistic infections, otitis externa, and serious wound or burned skin infections. *Pseudomonas* can grow using unusual metabolites and produce virulence factors including adhesion factors, toxins, and tissue-degrading enzymes and pigments.
- Bacterial gangrene is caused by *Clostridium perfringens*, anaerobic endospore-forming Gram-positive rod bacteria that live in soil and thrive in deep wounds where oxygen levels are low. The bacterium's metabolism generates gases under the skin and tissue necrosis. Without treatment, shock and death can result.
- Cutaneous anthrax can result from contact exposure to animals or skins that are contaminated with *Bacillus anthracis* Gram-positive endospore-forming rod bacteria or their toxins. Resulting black painless ulcers should be treated promptly.

17.4 Fungal Skin Infections

- Mycoses (fungal skin diseases) include superficial skin and hair discoloration, cutaneous infections of the hair, nails, or skin, and subcutaneous dermal wounds associated with punctures.
- Cutaneous candidiasis, usually caused by *Candida albicans*, may occur in moist skin folds, regions of high friction, diaper regions, or around the mouth (thrush). Risk factors include long-term moist skin, diabetes, antibiotic usage, and immunosupression. *Candida* species are also associated with vaginal yeast infections.
- Dermatophytic infections are superficial hair, skin, or nail infections such as ringworm or athlete's foot. *Trichophyton*, *Microsporum*, and *Epidermophyton* are common dermatophytic agents that may be acquired by contact with animals, contaminated fomites, or fungal spores or cells.

17.5 Parasitic Skin Infections

- Leishmaniasis is caused by *Leishmania* protozoan species; it is spread by the bites of infected sandflies in tropical regions and is classified as a neglected tropical disease. There are three leishmaniasis forms: Cutaneous leishmaniasis results in ulcers at bite sites; mucocutaneous leishmaniasis results in disfiguring scars; and visceral leishmaniasis affects the full body and is fatal if not treated.

17.6 Structures, Defenses, and Infections of the Eyes

- The regions at most risk for eye infections are the conjunctiva and cornea. Many eye infections are associated with improper contact lens use or poor hand hygiene and eye contamination.
- Eye defenses include the conjunctiva lining the lens and surrounding the eyeball except for over the cornea, lacrimal system tears that keep eyes moist and clean, and resident microbes.
- Conjunctivitis (pink eye) is an inflammation of the conjunctiva that may be itchy or painful. It is most often caused by viruses or bacteria. This condition is contagious and often associated with daycares or schools. Viruses increase tear formation and bacterial agents are more likely to cause a pus discharge.
- Neonatal conjunctivitis is a severe condition and requires prompt treatment. It is associated with *Neisseria gonorrhoeae* or *Chlamydia trachomatis* bacteria or HSV acquired during birth. Prevention is by standard application of antibiotics in newborns' eyes.
- Trachoma is the leading cause of microbial blindness in the world. The disease results from repeated infections by *Chlamydia trachomatis*, causing scarring of the conjunctiva that makes the eyelids turn inward and induces cornea damage by eyelashes. Transmission of the bacteria is by fomites, fingers, and flies.
- Keratitis is severe inflammation of the cornea. Viral keratitis is most often caused by herpes simplex virus 1. Bacterial and fungal keratitis is most often from our own skin microbiota, the environment, and hygiene issues. Parasitic keratitis includes *Acanthamoeba* keratitis, caused by an amoeboid protozoan from contaminated water in the eye, and river blindness (onchocerciasis) caused by the worm *Onchocerca volvulus*, which is transmitted by infected blackflies.

THINK CLINICALLY | *Be S.M.A.R.T. About Cases*

COMPREHENSIVE CASE

The following case integrates basic principles from the chapter. Try to answer the case questions on your own. Don't forget to be **S.M.A.R.T.*** about your case study to help you interlink the scientific concepts you have just learned and apply your new understanding of skin and eye infections to a case study.

*The five-step method is shown in detail in the Chapter 1 Comprehensive Case on cholera. See pages 31–33. Refer back to this example to help you apply a SMART approach to other critical thinking questions and cases throughout the book.

• • •

A couple in their early forties have twin two-year-old girls, Elisa and Elizabeth, as well as a five-year-old daughter, Eowin. The twins attended daycare and all children were current in their vaccinations. Both parents had standard vaccinations as children. One day Elizabeth seemed a little bit under the weather. By the next day, she was complaining that the inside of her mouth hurt. Elisa had a bit of a fever and was crabby. It was the weekend, so the girls weren't in daycare for a few days. No one was ill enough to take to the pediatrician, and the girls seemed fine a few days later.

Over the weekend, their mother noticed a maculopapular rash on her hands. She was prone to sensitive-skin conditions; she was extremely sensitive to poison ivy, and when the twins were born, she had experienced severely itchy feet and had to seek help from a dermatologist to resolve the problem (dyshidrosis). The oldest daughter did not appear to exhibit any symptoms and neither did the father.

By early the following week, a health notice was posted at the daycare to report that cases that sounded similar to what the family was experiencing were documented in some of the other children attending the daycare.

• • •

CASE-BASED QUESTIONS

1. What skin infections can you rule out quickly and why?
2. What types of infections could fit the described signs and symptoms?
3. What agents are on your most likely suspect list? Do you think the etiologic agent is most likely bacterial, viral, fungal, protozoan, or helminthic (worm)? Support your answer. What additional clinical information do you wish you had?
4. The family members were probably all infected with the same agent. Why might they have experienced different symptoms or no symptoms at all? The symptoms appeared in the children just before the weekend. If they had occurred on Monday, should they have gone to daycare? How can the spread of this infection be halted?

END OF CHAPTER QUESTIONS

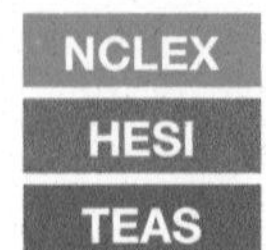

Think Critically and Clinically

Questions highlighted in **orange** are opportunities to practice NCLEX, HESI, and TEAS-style questions.

1. Which of the following is not considered a skin defense?
   a. Perspiration
   b. Melanin
   c. Sebum
   d. Lysosomes
   e. Antimicrobial peptides

2. Which lesion/rash is mismatched with the disease?
   a. Ulcer – inflammatory acne
   b. Honey-colored crusted lesion – impetigo
   c. Papule rash – wart
   d. Maculopapular rash – measles
   e. Vesicular rash – chickenpox

3. Which of the following is associated with chronic neuralgia?
   a. Herpes simplex 1 virus
   b. *Streptococcus pyogenes*
   c. Measles virus
   d. Varicella-zoster virus
   e. Papilloma virus

4. Choose the false statement about HSV-1:
   a. Viruses may be transmitted via wrestling mats.
   b. Viruses may be transmitted via contact with lesions.
   c. Viruses may be transmitted via the saliva of others with the virus.
   d. Viruses may reactivate under stress-inducing conditions.
   e. Viruses can be easily cured with common antivirals.

5. Which genus is primarily associated with acne?
   a. *Staphylococcus*
   b. *Propionibacterium*
   c. *Clostridium*
   d. *Streptococcus*
   e. *Pseudomonas*

6. All of the following are commonly associated with *Streptococcus pyogenes* EXCEPT:
   a. impetigo.
   b. cellulitis.
   c. scalded skin syndrome.
   d. necrotizing fasciitis.
   e. strep throat.

7. Which of the following is not a virulence factor of *P. aeruginosa?*
   a. Protein A
   b. Exotoxins
   c. Enzymes that damage host tissues
   d. Endotoxins
   e. Factors that enhance biofilm formation

8. Select ALL the true statements about dermatophytes:
   a. They are a group of bacteria that cause cutaneous infections.
   b. They may be treated with antifungals without knowing the exact causative agent.
   c. They are easily treated with antibiotics.
   d. They are commonly acquired from the soil, environment, or animals.
   e. They produce enzymes that digest keratin found in hair, nails, and skin.

9. Select the false statement about cutaneous candidiasis:
   a. The most common causative species is *Candida albicans.*
   b. *Candida albicans* can be part of normal microbiota.
   c. Usage of antibiotics can increase the chance of cutaneous candidiasis.
   d. Changes in pH can permit overgrowth.
   e. The causative agents are naturally found as mold filaments.

10. All of the following defend the eyes EXCEPT:
    a. tears.
    b. lysozyme.
    c. several corneal epithelial layers.
    d. a hard external layer encasing the entire eyeball, including the cornea.
    e. lactoferrin.

11. Conjunctivitis is caused by the following agent types (select ALL that apply):
    a. bacteria.
    b. viruses.
    c. protozoa.
    d. fungi.
    e. helminths.

12. Which best describes conjunctivitis (select ALL that apply)?
    a. Itchy eyes
    b. Scarred cornea
    c. Red eyes
    d. Inverted eyelashes
    e. Scarred conjunctiva

13. *Acanthamoeba* protozoa species are associated with:
    a. keratitis.
    b. conjunctivitis.
    c. river blindness.
    d. trachoma.
    e. all of the above.

14. Select the false statement about trachoma:
    a. The causative agent is bacterial.
    b. It is the leading cause of infectious blindness in the United States.
    c. It is transmitted by unhygienic items, such as flies, fingers, and fomites.
    d. Uncomplicated cases can be resolved with antibiotics.
    e. Severe cases require surgery.

15. Your diabetic patient has a foot wound that has developed into serious necrosis and the recommended treatment has been sessions in the hyperbaric oxygen chamber. You are explaining to him what has contributed to this condition, including the causative agent, which is:
    a. *Bacillus anthracis.*
    b. *Clostridium perfringens.*
    c. *Corynebacteria diphtheriae.*
    d. *Streptococcus pyogenes.*
    e. *Pseudomonas aeroginosa.*

16. A frantic mother comes into your clinic because her two-year-old child has had a very high fever for two days now, with mild diarrhea and coldlike symptoms. As you consider the differential list, which is the most likely causative agent if the fever breaks and is replaced with a rash by tomorrow?
    a. Human parvovirus B19
    b. Hand, foot, and mouth disease
    c. Roseola
    d. Rubella
    e. Measles

17. Your patient is reminiscing about her childhood diseases while reviewing her medical history. She claims she had a really bad case of rubella (German measles) when she was a child and that she was very sick. You think she is confusing it with rubeola (measles) because:
    a. measles can cause congenital rubella syndrome.
    b. German measles causes a mild rash and is not likely to make a patient very sick.
    c. the largest concern is for secondary infections like pneumonia.
    d. she didn't mention Koplik's spots in the mouth or raised lesions.
    e. the raised red rash usually begins on the trunk and spreads from there.

18. A child comes into your clinic with impetigo. The lab cultures a sample for further analysis. If the sample is *S. aureus*, which lab results would you expect?
    a. Gram-positive cocci in clusters, catalase and coagulase positive
    b. Gram-negative diplococci, catalase positive, and coagulase negative
    c. Gram-positive cocci in clusters, catalase negative, and coagulase positive
    d. Gram-positive cocci in chains, catalase positive, and coagulase negative
    e. Gram-positive cocci in chains, catalase negative, and coagulase positive

19. A 65-year-old patient calls the nursing hotline to ask about some painful blisters arranged in a band on one side of his waist. After hearing about the lesions, you ask him if he had chickenpox as a child. This is because you feel you are hearing about a case of:
    a. candidiasis.
    b. measles.
    c. herpes.
    d. shingles.
    e. rubella.

20. **Concept Mapping:**

    Using the following terms, create a concept map to organize and review microbial diseases of the eyes.

Acanthamoeba	*Haemophilus influenzae*	*Streptococcus* species
Adenoviruses	Herpes simplex 1 virus	Turning of lashes and further scarring
Aspergillus	Keratitis	
Candida	*Moraxella*	
Chlamydia trachomatis	River blindness	
Flies, fomites, fingers	Scarring of eyelid	
Fusarium	*Staphylococcus* species	

CRITICAL THINKING QUESTIONS

1. Some pathogens are normal residents of the body, but can also cause disease when the opportunity arises. Explain some circumstances that could favor this shift from normal resident to invading pathogen.

2. Leishmaniasis and river blindness are classified as neglected tropical diseases (NTDs). In fact, most developing countries have five or more NTDs circulating in their population. Discuss some of the challenges in addressing these global health issues.

3. You are working in a free health clinic and have noticed that the same middle-aged woman keeps returning for what seems like fairly mild but persistent cutaneous candidiasis. On her most recent visit you notice that she has a gangrenous foot sore with a significant amount of necrosis around the big toe. You are starting to suspect that a systemic disorder is at the root of her recurrent and persistent skin infections. What systemic condition is the woman likely experiencing that is contributing to these skin manifestations? Support your conclusion with evidence from the chapter.

4. Your best friend has a six-month-old baby and you and your mother would like to visit her. You are concerned because your mother has been experiencing an outbreak of shingles, so you call to reschedule the visit. Your friend says it will be fine and you both should come by. She argues that her baby can't get anything from your mom because shingles is an older person disease and there is no risk for the baby. Who has the safer plan and why?

5. The measles, mumps, and rubella (MMR) vaccine has been available since 1963 and has been tremendously effective at preventing these diseases. Assume you're speaking with a parent at the pediatrics practice where you work. The parent is arguing that because "almost everyone else is vaccinated" her child is unlikely to get any of the diseases that the MMR vaccine protects against. Draft a response to this parent making sure that you address how they are in some ways correct, but in other ways they are misguided. Be sure to include a discussion of the risks associated with failing to vaccinate her child with the MMR vaccine.

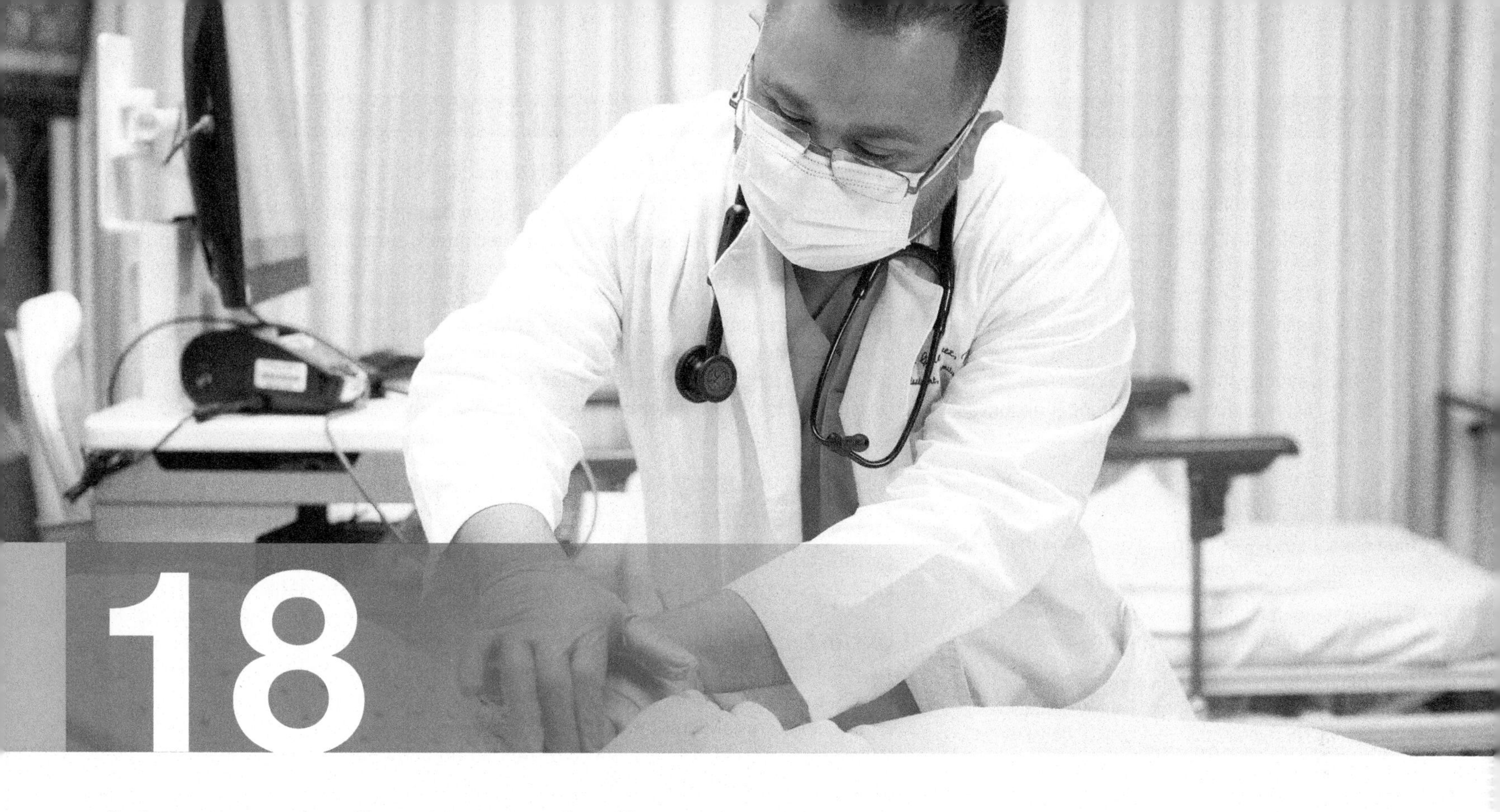

Nervous System Infections

What Will We Explore? Due to its essential functions, the human nervous system has specialized structures that protect it against infection. Despite that, infections still sometimes occur, and their results can be devastating and deadly. In this chapter we examine major viral, bacterial, and fungal infections that impact the nervous system, along with disease signs, symptoms, diagnostic methods, and treatments.

Why Is It Important? Some nervous system infections are on the rise in the United States, making it especially important for healthcare providers to be able to identify and respond to signs and symptoms. For example, in 2015 chikungunya fever, a mosquito-borne illness that can cause horrible joint pain and headaches, became a nationally notifiable disease in the United States. Until a few years ago there were only a dozen or so reported cases annually in the United States, and those traced to people who had contracted it outside the country. But almost 3,000 cases of this disease, which causes life-threatening encephalitis, were reported in the United States in 2014. The expansion of this disease throughout the United States has serious implications; some strains of the virus kill one in every ten people who are infected.

Clinical CASE
NCLEX
HESI
TEAS

The Case of the Pain in the Neck

Visit the **Mastering Microbiology** Study Area to watch the case and find out how nervous system infections can explain this medical mystery.

Pedro Gutierrez
DMS PA-C; Jacksonville, FL

18.1 OVERVIEW OF NERVOUS SYSTEM STRUCTURE AND DEFENSES

Learning Outcomes

After reading this section, you should be able to:

18.1 Explain the differences between the central and peripheral nervous systems.
18.2 Describe the function of a neuron.
18.3 Define the functions of cerebrospinal fluid and the meninges.
18.4 Describe the blood–brain barrier.
18.5 Distinguish meningitis from encephalitis.

CLINICAL VOCABULARY

Meninges three layers of specialized tissue that encase the central nervous system

Cerebrospinal fluid the colorless, watery fluid that cushions and nourishes the central nervous system

Concussion injury to the brain and/or spinal cord caused by jarring physical force; usually causes temporary CNS dysfunction that may or may not be accompanied by a brief period of unconsciousness

Coma a prolonged state of unconsciousness that resists waking

Meningitis inflammation of the meninges

Encephalitis inflammation of the brain

Meningoencephalitis inflammation of both the brain and meninges

Lumbar puncture a diagnostic procedure that removes cerebrospinal fluid for analysis using a needle inserted between two vertebrae in the lower back

Immunocompromised an individual who doesn't have a fully functioning immune system; can be associated with organ transplants, blood transfusion, and certain diseases including, but not limited to, cancer and HIV/AIDS

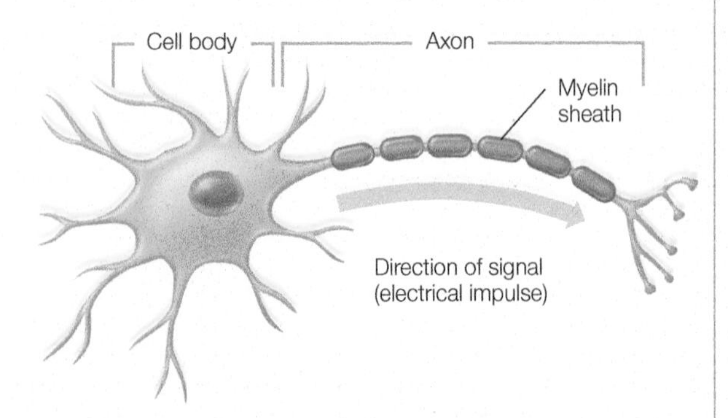

FIGURE 18.2 A neuron The nervous system has specialized cells called neurons that send signals to communicate between the CNS and PNS.

The nervous system includes two main segments.

The human nervous system is divided into two major segments that accomplish two different tasks within the body. The **peripheral nervous system (PNS)** inputs and transmits information, while the **central nervous system (CNS)** integrates information received, and sends back an "action plan." Nerves comprise the peripheral nervous system, while the spinal cord and brain make up the central nervous system (FIG. 18.1). Everything you do, from drinking your morning coffee to studying for your next microbiology exam, requires many complex signals passing back and forth between your peripheral and central nervous systems.

The nervous system contains specialized cells for transmitting signals.

Signals for the CNS and PNS are transferred by a collection of specialized cells called **neurons**. Neurons can be categorized as sensory (specializing in input, or afferent), motor (specializing in output, or efferent), or interneurons (allowing CNS and PNS communication). Nerves are bundles of neurons that extend from the CNS and make up the PNS. The neurons in a given nerve sense internal and external stimuli, and send out responses in the form of chemical messengers called neurotransmitters. Neurotransmitters cause changes in neighboring cells—for example, muscle cells contract when activated by neurotransmitter molecules released from a nearby motor neuron.

As a stimulus activates a neuron, an electrical wave is generated. This wave travels down the **axon**, or length of a neuron (FIG. 18.2). Axons are covered in myelin, a fatty layer that helps electrical impulses travel quickly. Once the electrical wave reaches the end of the axon, neurotransmitters are released to convey the signal to a neighboring neuron. This signaling cascade continues from neuron to neuron until it reaches the CNS. Similarly, the CNS relates information to the PNS to support effective two-way communication.

FIGURE 18.1 The two divisions of the nervous system The peripheral and central nervous systems work together to receive, interpret, and respond to conditions both inside and outside the body. Specialized nervous system cells called neurons facilitate communication between and within the CNS and PNS.

Critical Thinking *When a baby's hand touches something hot and quickly pulls it back, is it the central nervous system, peripheral nervous system, or both at work? Why do you think so?*

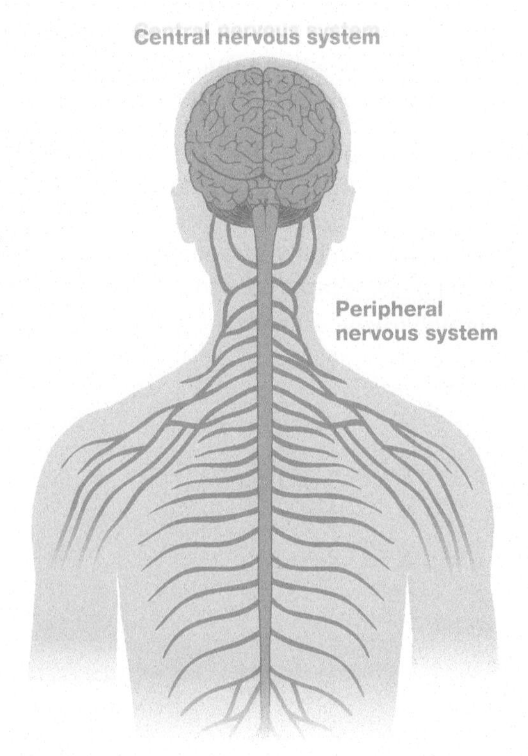

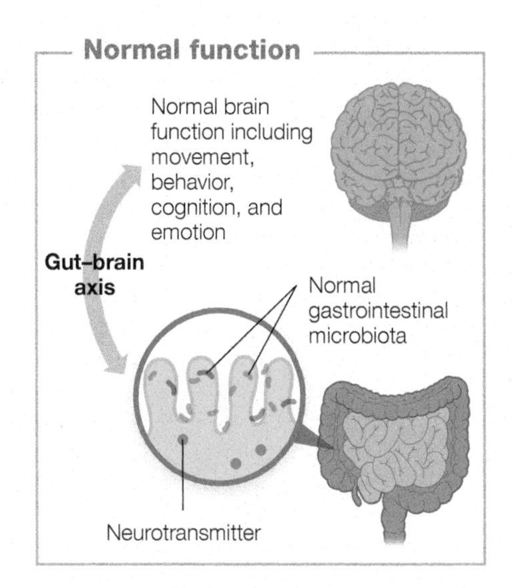

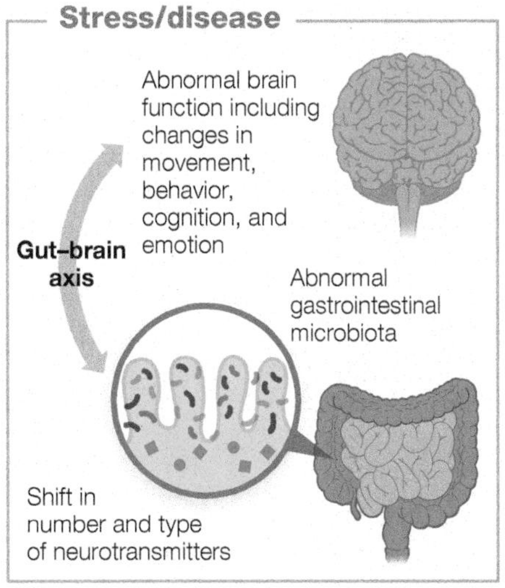

FIGURE 18.3 Relationship between resident microbes of the GI tract and nervous system There are no normal nervous system microbiota. However, research indicates resident microbes in the gastrointestinal system may impact nervous system functioning.

Source: Figure adapted from Cryan, John and Dinan, Timothy. "Mind-altering microorganisms: The impact of gut microbiota on brain and behavior," *Nature Reviews Neuroscience*, Volume 13, 701–712, 2012.

The nervous system contains no normal microbiota, but organisms living in the GI tract may impact this system.

The brain, spinal cord, and nerves are easy to damage and hard to heal. Disruption of crucial functions can have deadly consequences. Perhaps not surprisingly, the nervous system contains no resident microbiota (normal flora) to compete with infectious agents that manage to invade this body system. Therefore, discovery of any microbes within the nervous system is considered abnormal.

That said, resident microbes in the digestive system may actually exert some influence on nervous system functions (FIG. 18.3). The term "gut–brain axis" describes the complex, back-and-forth communication that occurs between the endocrine, immune, and nervous systems and the gastrointestinal tract. The resident bacteria of the gut have already been linked to certain autoimmune disorders (see Chapter 19). Research now indicates that similar ties may exist between gastrointestinal flora and some nervous system disorders that impact mood, thinking, and movement.

A recent study found that people experiencing depression had lower levels of Bacteroidetes, Proteobacteria, and Actinobacteria in their GI tracts compared with people who were not depressed. Likewise, another study showed that the GI tracts of those suffering from the neurodegenerative disorder Parkinson's disease have decreased levels of *Prevotella* bacteria, and greatly increased levels of Enterobacteriaceae species compared with healthy individuals.[1] Modes of action remain poorly understood, and further research must take place to discover the nature of these relationships. And, as with all complex diseases and disorders, there are likely many contributing factors. However, knowledge of links between gut microbes and the nervous system holds promise for new diagnostic screenings—even new treatments—for a variety of neurological diseases.

Unique defenses protect the nervous system from infection.

Given how important nervous system processes are, this system has unique defenses that protect against both physical damage and infectious agents.

[1] Dinan, Timothy G., & Cryan, John F. (2015). The impact of gut microbiota on brain and behavior: Implications for psychiatry. *Current Opinion in Clinical Nutrition and Metabolic Care* 18, (6), 552–558. Scheperjans, Filip, et al. (2015). Gut microbiota are related to Parkinson's disease and clinical phenotype. *Movement Disorders* 30, (3), 350–358.

FIGURE 18.4 **Meninges** The meninges form a protective layer around the central nervous system.

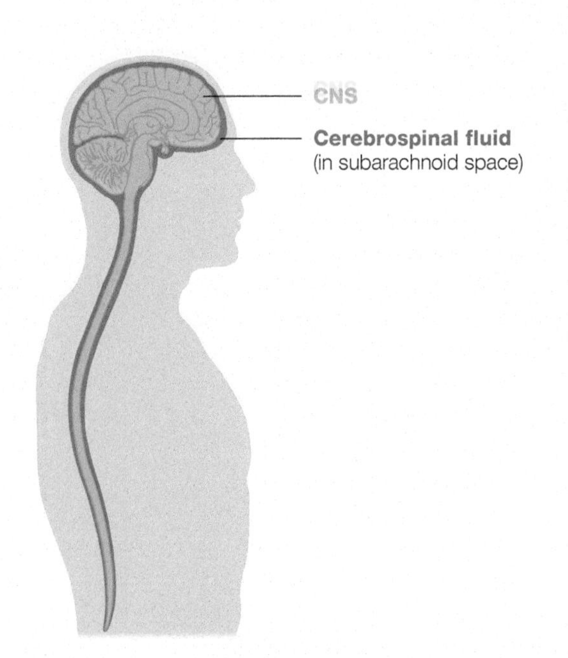

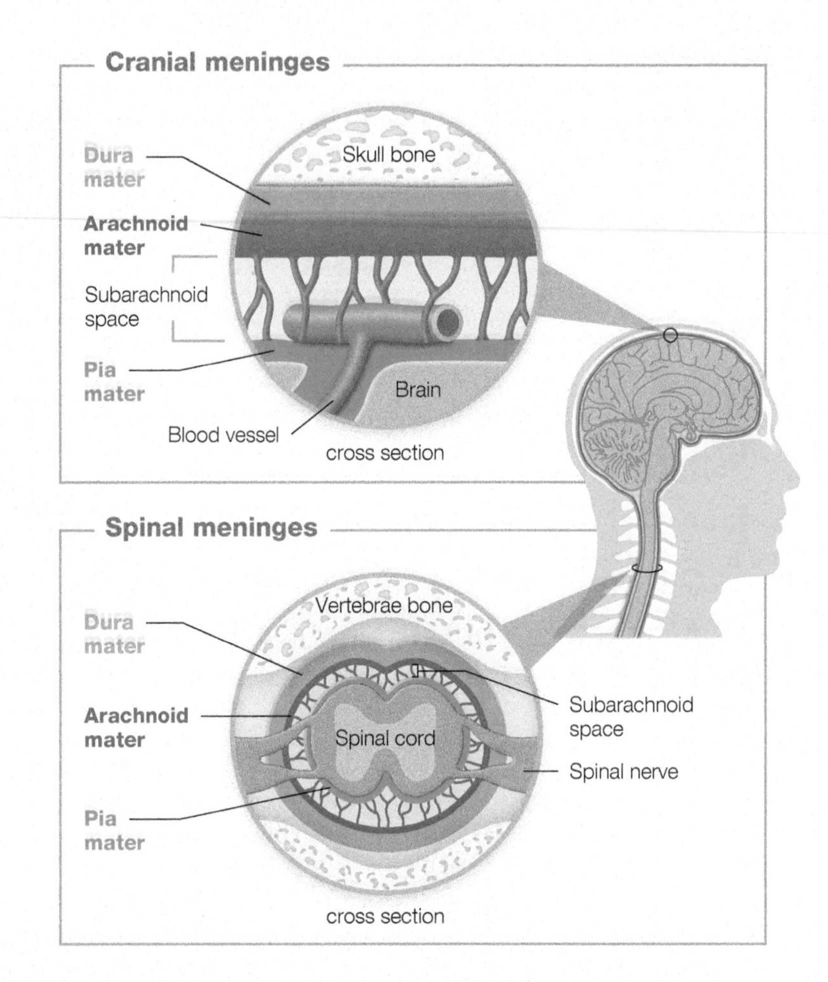

FIGURE 18.5 **Cerebrospinal fluid** CSF insulates and nourishes the brain and spinal cord.

Critical Thinking *What would be the consequences of lacking cerebrospinal fluid?*

The Meninges and Cerebrospinal Fluid

There are three layers of tissue that surround our brain and the dense collection of nerves in our spinal cord. Together, these tissues are known as the **meninges** (meh NIN geez; FIG. 18.4). Layers include the dura mater, arachnoid mater, and pia mater. These Latin-derived names roughly translate as "tough mother," "spider mother" (named for the shape of the cells), and "tender mother," respectively. All of them act as mothers in that they support and care for the CNS by supplying nutrients, removing waste, and protecting the CNS from physical shocks.

Structures within the brain produce **cerebrospinal fluid (CSF)** as the cushion that flows in a space between the meninges called the subarachnoid space (FIG. 18.5). Some infectious agents invade this fluid and attack the CNS. Physical damage to any of the CNS tissues can cause **concussion** or even **coma**.

Blood–Brain Barrier

The blood vessels that deliver nutrients and oxygen to the CNS are structurally specialized to protect the CNS from pathogens. The **blood–brain barrier** allows only a few types of molecules to pass from capillaries into the CNS (FIG. 18.6). This occurs because blood vessel cells are sealed much more tightly in the areas around our CNS than they are in other areas of the body. Very small molecules and lipophilic (fat-soluble) molecules can pass through these capillary cells. Larger essential molecules such as glucose have special transporters to help them cross. But most other molecules, including those that are hydrophilic, are blocked from entering the cerebrospinal fluid.

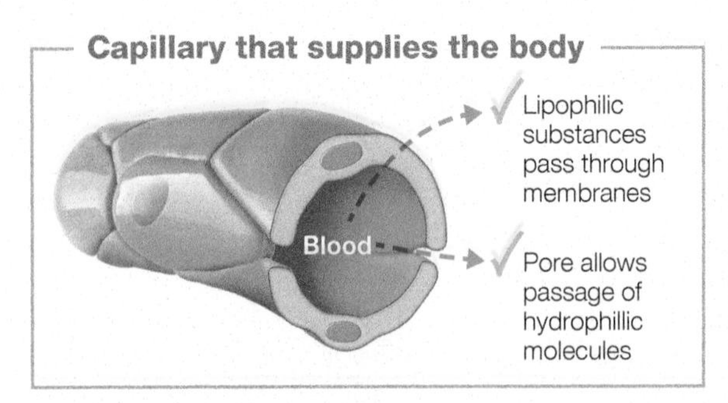

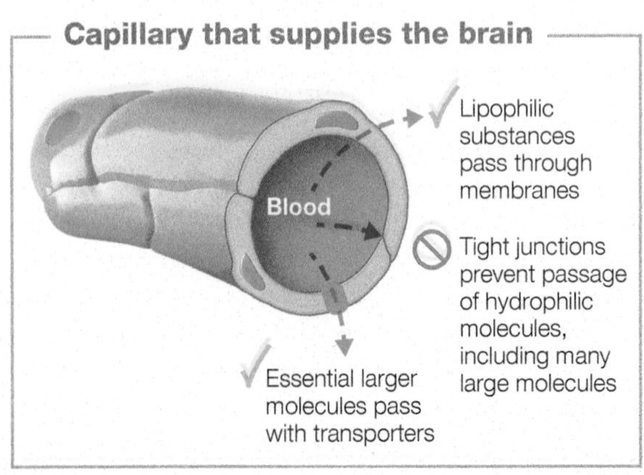

FIGURE 18.6 **Blood-brain barrier** The blood-brain barrier describes differences between the structure and function of capillaries that feed and nourish the brain and other capillaries in the body. These special capillaries protect the brain from infection by limiting pathogen access to this sensitive tissue.

TRAINING
TOMORROW'S HEALTH TEAM

Diagnosing Infections Using Lumbar Puncture

The "triad" symptoms of fever, neck stiffness (nuchal rigidity), and headache are warning signs of a serious nervous system infection: meningitis. Most patients exhibit at least one of these triad symptoms, although less than half will exhibit all three, so symptoms alone are insufficient for diagnosis. One of the most useful diagnostic tools for meningitis is the **lumbar puncture** (or spinal tap). In this procedure, a needle is inserted between the lumbar vertebrae to collect CSF for examination. The results can reveal either purulent (pus-containing, usually associated with bacterial infection) or aseptic (without cultivable organisms, usually viral infections) meningitis.

QUESTION 18.1

Based on the information in the section about nervous system anatomy, do you think a lumbar puncture could pinpoint the location of a CNS infection? Explain your reasoning.

Bacteria require a very special set of virulence factors in order to slip through this barrier. Viruses, being significantly smaller than most bacteria, have an easier time invading the CNS, but still encounter challenges. Because of these defenses, nervous system infections are more likely to follow some injury to the system or develop if there is a problem with the immune system.

Both the CNS and PNS may become infected when defenses break down.

When any of the meninges are inflamed, the condition is called **meningitis** (*men* in JY tis). **Encephalitis** (en *sef* ah LY tis) refers to inflammation of the brain (TABLE 18.1). When both are affected, the condition is called **meningoencephalitis** (*men* in JOE en *sef* ah LY tis); it displays a variable combination of meningitis and encephalitis symptoms. Viruses are the most common cause of both meningitis and encephalitis; however, bacteria (and less commonly parasites and fungi) are also culprits. Viral forms of meningitis are serious, but usually have a better prognosis than bacterial forms.

Peripheral nervous system infections can occur as well. Several viruses, such as herpes viruses and varicella-zoster virus (the causative agent of chickenpox), hide inside neurons and cause occasional flare-ups. But since the symptoms of these pathogens are rarely neurological, they're not considered nervous system diseases. Furthermore certain pathogens make powerful toxins that can disable the PNS.

CHEM • NOTE

Lipophilic molecules are "fat-loving" molecules that dissolve easily in fats or lipids. The hydrocarbon tails on fatty acids are a common example of lipophilic molecules. These are often synonymous with hydrophobic, or "water hating," molecules.

CHEM • NOTE

Hydrophilic molecules are "water-loving" molecules that dissolve easily in water. Sugars, like glucose, are common hydrophilic molecules.

TABLE 18.1 Comparing Meningitis and Encephalitis

Condition	Meningitis	Encephalitis
Location	The meninges—membranes surrounding the brain and spinal cord	The brain (when the spinal cord is also involved it's called encephalomyelitis; here we focus only on encephalitis)
Causative agents	Usually viral; nonpolio enteroviruses and herpes simplex viruses common. Bacterial agents (less common) include *Streptococcus pneumoniae*, *Neisseria meningitidis*, and *Haemophilus influenzae*.	Usually viral. Viral: herpes simplex virus 1, and arboviruses such as West Nile viruses are common. Bacterial agents (less common) include *Listeria monocytogenes*, *Borrelia burgdorferi* (covered in Chapter 21). Rare: Certain fungi and parasites.
Signs and symptoms	Viral meningitis: fever, headache, stiff neck (nuchal rigidity), nausea, photophobia, and vomiting; tend to be alert. Bacterial meningitis: same as viral, but also exhibit lethargy, body aches, chills, and sometimes a rash.	Fever, headache, and often one or more of the following: disorientation, abnormal behavior, or seizures.
Routine diagnostics	Lumbar puncture for CSF analysis (culturing and staining); in certain cases, computed tomography (CT scan) or magnetic resonance imaging (MRI) may be used to decide if it is safe to perform a lumbar puncture; because normal imaging data is common in early stages of these pathologies, CT and MRI are usually more useful for ruling out other possible diagnoses than for confirming a meningitis or encephalitis diagnosis.	
Prevention and treatment	Viral: There are no vaccines against the most common culprits; most recover in 7–10 days without specific antiviral treatment. Bacterial: Vaccines are available to prevent forms caused by *Neisseria meningitidis*, *Streptococcus pneumoniae*, and *Haemophilus influenzae* type b; treated with antibiotics.	

Build Your Foundation

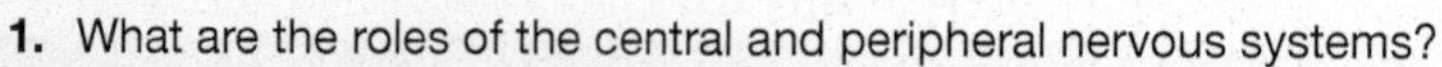

1. What are the roles of the central and peripheral nervous systems?
2. What is the structure of a neuron and how does it transmit information?
3. What roles do the meninges play in keeping the CNS healthy?
4. What is the blood–brain barrier and how does it confer protection?
5. How is meningitis distinguished from encephalitis?

Build Your Foundation (BYF) Quick Quiz: Visit the **Mastering Microbiology** Study Area to quiz yourself.

18.2 VIRAL NERVOUS SYSTEM INFECTIONS

Learning Outcomes

After reading this section, you should be able to:

18.6 Explain how polio affects the nervous system and describe prevention strategies.

18.7 Describe the pathogenesis of rabies and note methods for preventing infection.

18.8 Give examples of arboviruses and then describe their associated epidemiology and how they cause nervous system infections.

CLINICAL VOCABULARY

Cytolytic an agent that causes the lysis, or rupture, of a cell

Zoonosis a disease that is transmitted from animals to humans

Systemic infection an infection that affects the whole body, usually carried in the blood

Paraesthesia a neurological symptom associated with a prickling or "pins-and-needles" sensation

Flaccid paralysis paralysis caused by an inability to contract muscles

Symmetric paralysis paralysis that affects both sides of the body equally

Asymmetric paralysis paralysis that affects one side of the body more than another

Viruses cause the most common nervous system infections.

As mentioned, because of their small size, viruses can more easily pass through the blood–brain barrier and infect the nervous system. This section explores the three viral infections that most commonly cause neurological symptoms.

Enteroviruses such as coxsackieviruses and poliovirus are the most common causes of viral meningitis. Despite what their name implies, enteroviruses do not routinely cause enteric (gastrointestinal) disease; instead, they are named for their fecal–oral transmission route. Coxsackieviruses commonly infect children less than one year old. Unless the infection progresses to meningoencephalitis, it often resolves itself. Coxsackieviruses also cause the childhood skin infection hand, foot, and mouth disease discussed in Chapter 17. Also, in the early years of the HIV epidemic, a large number of HIV-positive patients exhibited neuropathologies. Although HIV itself can cause encephalitis, this is a marginal source of mortality for HIV-positive patients (for more on HIV/AIDS, see Chapter 21). Antiretroviral treatments for HIV have dramatically decreased the number of these cases that were mostly caused by secondary infections (see the section on fungal infections in this chapter).

Poliomyelitis

While **poliomyelitis**, more commonly known as **polio**, no longer plagues the developed world, for centuries outbreaks brought suffering, permanent disability, and death throughout the globe. To date, the disease persists in certain war-torn developing nations, including Afghanistan and Pakistan. According to the U.S. Centers for Disease Control (CDC), the United States had 35,000 cases of polio per year in the early 1950s. Thanks to vaccine development, there were zero cases of polio in the United States by 1979. The massive decrease in the incidence of polio is considered to be one of the triumphs of modern medicine.

Polio is caused by the *poliovirus*, which belongs to the *Picornaviridae* family of very small RNA viruses that have only a protein capsid without an envelope. The name *Picornaviridae* actually comes from *pico* ("small" in Latin) and *RNA*. The tough protein coat protects it from stomach acid, allowing it to reach and bind to intestinal cells. From there, the poliovirus travels to skeletal muscle cells, replicates, and progresses up motor neurons to the CNS.

Poliovirus is **cytolytic**, causing rapid cell rupture after infection and viral replication. This lysis of infected neurons causes severe inflammation, giving the disease its name: inflammation of the myelin, or myelitis. Very few polio patients progress to this stage of disease. Most cases only present with flulike symptoms. However, if poliovirus reaches the CNS, it can attack different areas of the spine and brain, causing muscle weakness or paralysis.

Recovery from polio depends on the number and type of neurons destroyed by the virus. If the nerves coming out of the brain are destroyed, bulbar poliomyelitis occurs. This causes muscle weakness that, depending on the severity, can cause respiratory failure. Some patients experience complete **flaccid paralysis** if the motor cortex of the brain is damaged (FIG. 18.7). This means the muscles are unable to contract, and are permanently relaxed. Polio can result in **symmetric paralysis**, in which the affected regions may be on both sides of the body, but **asymmetric paralysis** is more common. There was a time in U.S. history when all hospitals had machines called iron lungs that simulated breathing in polio patients who suffered from flaccid paralysis.

Approximately 25 percent of those who recover from polio will experience postpolio syndrome within 10–40 years of initial infection. Deterioration of muscle function and serious pain can occur in the body region where infection originally took place. This syndrome is progressive but is not usually life threatening. Postpolio syndrome is not infectious, and incidence has decreased as polio vaccines have become more widespread.

Two polio vaccines are available. One, the Salk vaccine, was initially released to the public in 1957 and is made of inactivated poliovirus particles; the Salk vaccine is also known as the inactivated polio vaccine (IPV). It's injected into muscle and requires several boosters. This vaccine gives lifelong immunity in the blood, but only temporarily allows secretion of antipolio antibodies into the saliva, stopping the virus before it enters the blood. The other vaccine was released in 1963 and is made of live, attenuated poliovirus. It is known as either the Sabin or oral polio vaccine (OPV). This version is made up of a live but weakened (attenuated) viral strain. Patients who receive the OPV continue to make protective antibodies against poliovirus in both their blood and secretions for their entire life. The OPV's oral administration and its lower production cost make it the polio vaccine of choice in poorer, underdeveloped nations. Another advantage of the OPV over the IPV is that the attenuated virus is shed from people vaccinated with the OPV, effectively immunizing those around them. However, there are disadvantages to the OPV as well. **Immunocompromised** individuals—those without fully functioning immune systems—can develop illness from the attenuated virus in the OPV. In rare circumstances, patients who receive the OPV can also develop paralytic polio. This is caused by viral reversion, in which the virus evolves back into a virulent form after entering the body as a harmless strain.

The IPV is used in the United States, whereas the OPV has been historically preferred by global health agencies in areas with wild-type polio. This was motivated by the added benefit of the OPV; those who receive the attenuated virus effectively increase the number of those vaccinated. The fecal–oral route of transmission, combined with poor water sanitation facilities, expose the local community to the nonvirulent virus, priming the immune system. For those who contract polio, reverse transcriptase polymerase chain reaction (RT-PCR) is used to detect the viral mRNA. Pleconaril, a new antiviral drug, has been effective if administered early in the disease. Although a complete recovery from some types of polio is possible, it can require up to two years of medical attention.

Rabies

In the 1980s, horror writer Stephen King wrote a bestselling novel called *Cujo* that was adapted into a popular film. It was the story of a family pet that contracted rabies and the terrifying result of its neurological symptoms. Unfortunately, **rabies** truly is a horror story, infecting and killing about 55,000 people worldwide every year—mainly in developing nations of Asia and Africa. Very few of those deaths occur in developed countries, largely because of successful vaccination campaigns for dogs, which are the most common domesticated rabies reservoir. Rabies is considered a classical **zoonosis**, a disease passed from animals to humans. The virus is usually introduced into humans through an animal bite, although aerosolized bat droppings have also led to infection.

The rabies virus is a member of the *Rhabdoviridae* (Greek for "rod-shaped") family of enveloped, single-stranded RNA viruses. Its RNA genome codes for only five different proteins, allowing for faster mutation of individual genes. Therefore, each rabies vaccine must be targeted for a single host species.

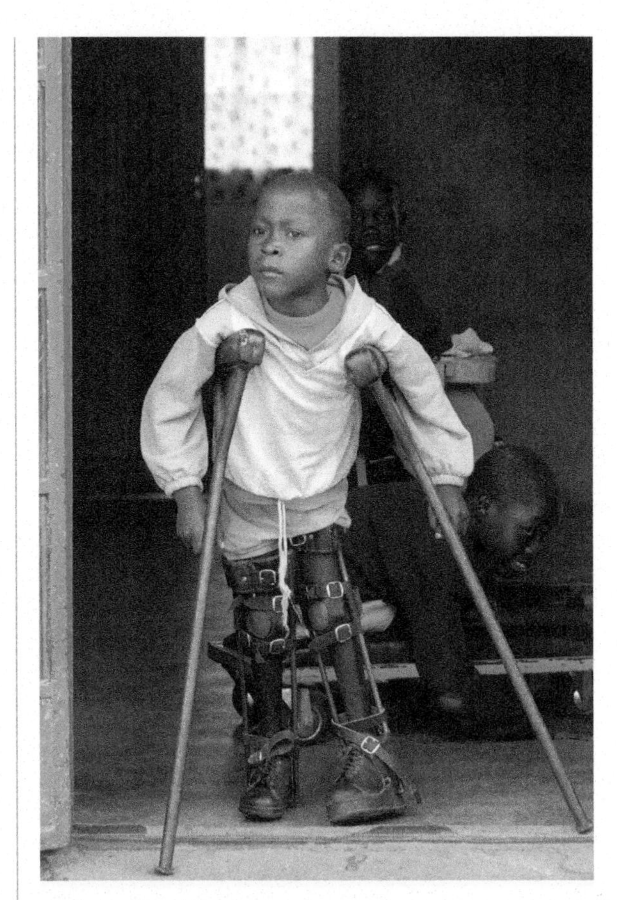

FIGURE 18.7 **Flaccid paralysis in an individual with poliomyelitis**

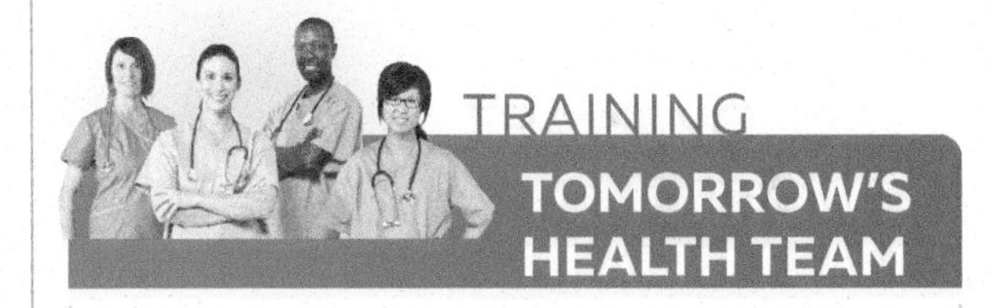

Why is Polio so Difficult to Eradicate?

In 1988, the World Health Organization made polio eradication a public health priority. Progress has been made: Formerly endemic in over 100 countries, by 2015 only two countries continue to have endemic cases. However, unlike smallpox, which was famously eradicated, polio has proven very difficult to completely remove from the global population.

Many of the regions that still have sporadic polio outbreaks are war-torn areas that are hard for public health workers to safely access and administer the vaccine. Religious or cultural practices may also influence acceptance of vaccination, as they sometimes do in the United States.

QUESTION 18.2

Several public health organizations suggested introduction of the IPV in countries with infrequent polio outbreaks, to prepare for the eventual discontinuation of the OPV. Why do you think this was the recommendation?

DiseaseSnapshot • Rabies

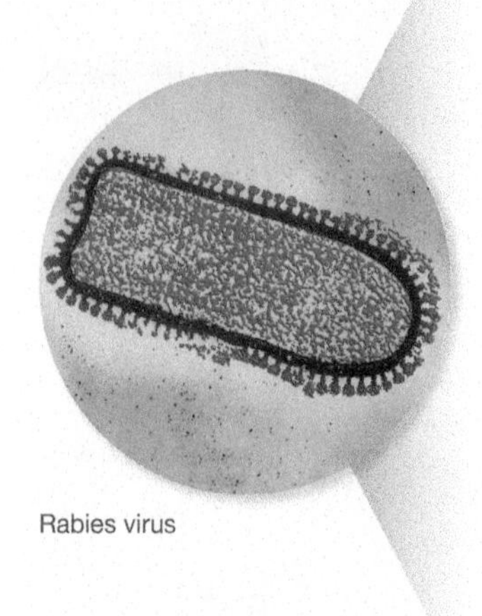

Rabies virus

Causative agent	Rabies virus: Enveloped, single-stranded RNA genome, *Rhabdoviridae* family
Epidemiology	2–5 human cases per year in the United States. 55,000 deaths per year globally, mostly in Africa and Asia; Incubation is 2–12 months
Transmission & prevention	Zoonosis, contracted through an infected animal's bite or by breathing aerosolized droppings. In the United States, the most common animal reservoirs are bats, skunks, and raccoons. The inactivated (killed-virus) rabies vaccine is recommended for animal handlers and those traveling to places with high rabies incidence.
Signs & symptoms	Fever, headache, disorientation, hydrophobia (fear of water), excessive salivation, hallucinations, and partial paralysis
Pathogenesis & virulence factors	Virus enters muscle cells and replicates slowly, eventually traveling up the peripheral nervous system to invade the central nervous system. Pathogenesis is not well understood but does not involve cytolysis. One of the virus's five proteins, the G protein, is implicated in adhesion and cell penetration.
Diagnosis & treatment	Infected animals are often tested using direct antibody fluorescence after a human is bitten. Diagnosis in humans usually involves reverse transcriptase PCR. Seeing Negri bodies in brain tissue is a postmortem diagnosis method. Administering human or equine antirabies antibodies can give short-term protection to those exposed to the virus. Individuals are also vaccinated to develop a long-term immune response. No treatments consistently prevent death once symptoms appear.

At first the virus slowly replicates in the muscle cells leading to a long incubation period that lasts from two months to a year. However, once it reaches the peripheral nervous system, it rapidly travels through neurons, replicating and spreading to the CNS. The virus hijacks a cellular transport system down the axon that usually carries neurotransmitters, which helps speed its spread from neuron to neuron. Unlike poliovirus, rabies virus does not lyse host cells. In fact, little or no damage is seen upon postmortem examination of the CNS. However, a diagnostic sign of rabies is the formation of Negri bodies, small clusters of virus inside the neurons. Recent work suggests that rabies symptoms stem from the virus's ability to shut down normal neuron functioning. Apoptosis, or forcing the cell to commit "suicide," is seen very late in the disease, which means that the immune system can't react in time to stop the virus. After invading brain cells, rabies spreads back out into the PNS. It replicates and is shed from tissues that are well supplied with nerves, including the salivary glands. This explains why rabies is so easily transmitted through bites.

If treatment is received before symptoms appear, the likelihood of survival is high. Patients who suspect exposure to rabies are injected with both antirabies antibodies and the inactivated vaccine for rabies, a regime called rabies postexposure prophylaxis. This allows the body to produce an effective immune response to fight off the initially slow-progressing virus. However, once fever, nausea, and a pins-and-needles sensation (paraesthesia) around the wound develops, the disease rapidly progresses. These initial symptoms last 2–10 days, followed by a collection of neurological symptoms that usually last no longer than a week. These final neurological symptoms include pharyngeal spasms, confusion, loss of coordination, delirium, and the characteristic hydrophobia. (Patients are terrified of drinking water—not because water harms them, but probably because of the extreme pain involved in swallowing.) Coma, cardiac arrest, and respiratory failure follow neurological symptoms. After the onset of neurological symptoms, rabies is almost always fatal.

Disease**Snapshot** • Arboviral Encephalitis: West Nile and La Crosse

Causative agent	West Nile virus: Enveloped, single-stranded RNA genome, *Flaviviridae* family La Crosse virus: Enveloped, single-stranded RNA genome, *Bunyaviridae* family
Epidemiology	Around 1,300 cases of neuroinvasive disease each year in the United States. West Nile incubation is 2–6 days, while La Crosse incubation is 5–15 days.
Transmission & prevention	Zoonosis, usually passed from birds to humans via mosquito or tick bites. The most common mosquito vectors for West Nile virus are part of the *Culex* genus, while the La Crosse virus is usually carried by members of the *Aedes* genus. Preventions include wearing insect repellent and spraying for mosquitoes.
Signs & symptoms	Fever, headache, flulike symptoms, meningitis, and encephalitis
Pathogenesis & virulence factors	Special glycoproteins embedded in the viral envelope allow the viruses to bind different human tissue, triggering them to take up the virus by endocytosis. Once inside the targeted cells, the virus replicates and then lyses the infected host cells to go on to infect other cells. Cytolysis is the primary form of damage caused by arboviruses. Both West Nile and La Crosse virus use proteins (NS4B and NSs, respectively) to block the interferon-mediated immune system response.
Diagnosis & treatment	Many antibody-based methods exist to detect arboviruses, but these can fail to distinguish between closely related species. Reverse transcriptase PCR and genome fingerprinting can also be used for viral identification. Once infected, the only treatment is supportive.

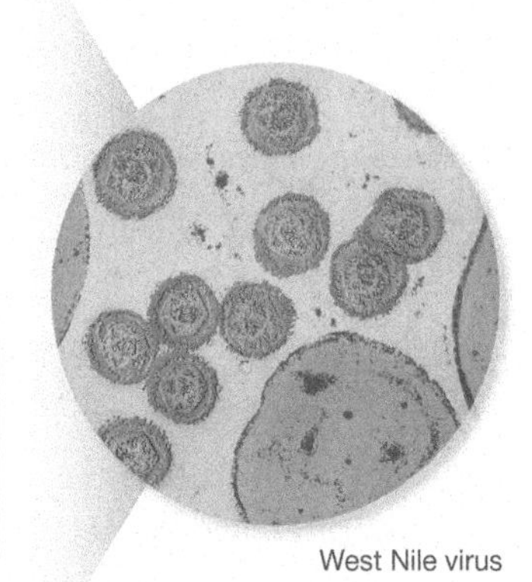
West Nile virus

Selected Arboviral Encephalitis and Meningitis

West Nile virus is one of the most common and well-known members of a group of viruses called arboviruses (named for **ar**thropod **bo**rne). Arboviruses come from different families, but they share key similarities: They are all enveloped, single-stranded RNA viruses and are transmitted by arthropods such as mosquitoes or ticks. Different arboviruses have different reservoir species and geographic distribution. For West Nile virus, mosquitoes become infected with the virus upon taking a blood meal from an infected reservoir, usually an infected bird. When infected mosquitoes feed on animals, some of the viruses already in their stomachs gets regurgitated into the bite of the latest victim. This passes the virus from one animal to another. (See Chapter 9 for details on vectors and reservoirs.) Humans are considered a "dead-end host," because the low number of virus particles in the blood makes it very unlikely that a mosquito would become infected from us in a blood meal.

All identified arboviruses can cause flulike symptoms and invade white blood cells for replication. Upon becoming **systemic**, or widespread throughout the body, some of these viruses invade and destroy the blood vessel cells that form the blood–brain barrier. They can then infect the brain and cause arboviral encephalitis, inflicting severe neurological damage and even causing death. However, the most common arbovirus on a global level, dengue virus, only rarely causes neuroinvasive disease (meningitis, encephalitis, or a combination of the two). The neuroinvasive arboviruses West Nile virus and La Crosse virus are the most common in the United States.

According to the CDC, in 2019 there were about 1,000 cases of West Nile in the United States. West Nile virus damages neurons by lysing them as they exit after replication or by triggering apoptosis. One way to diagnose the disease postmortem is to look for patches of microglial nodules—dead tissue in the brain that is surrounded by immune cells.

The **La Crosse virus** causes around 50–80 cases with neurological symptoms each year. Unlike West Nile, the reservoir for La Crosse is usually chipmunks or squirrels. Another difference is that this virus can survive for long periods of time inside the mosquito vector and in mosquito eggs. La Crosse also shows microglial nodules in postmortem patients; however, they develop in different regions of the brain than those caused by West Nile virus.

Build Your Foundation (BYF) Quick Quiz: Visit the **Mastering Microbiology** Study Area to quiz yourself.

Build Your Foundation

6. How does the poliovirus invade and damage nervous tissue?
7. How do the two polio vaccines compare to each other?
8. How does the rabies virus affect the CNS?
9. What are some of the treatment options after infection with rabies?
10. How do arboviruses invade the CNS?
11. What are good prevention strategies for arboviral encephalitis?

18.3 BACTERIAL NERVOUS SYSTEM INFECTIONS

Learning Outcomes

After reading this section, you should be able to:

18.9 Identify key features of *Haemophilus* meningitis.

18.10 Explain how meningococcal meningitis is prevented, detected, and treated.

18.11 Describe the cause of pneumococcal meningitis and note the differences between the vaccines for it.

18.12 Name the risk factors for *Listeria* meningitis and describe how the disease spreads in the body.

18.13 Explain why leprosy is confined to the PNS and describe its clinical features.

18.14 Describe how botulism is contracted and state the effect of the botulinum toxin on the nervous system.

18.15 Detail how tetanus is contracted and how it causes paralysis and sometimes death.

Bacteria can infect the central nervous system, causing meningitis.

While bacteria are a rarer cause of meningitis than viruses, they cause a more dangerous form of the disease. Because bacterial meningitis progresses quickly, treatment must be sought immediately upon experiencing symptoms (Table 18.1 reviewed symptoms of bacterial meningitis). Twenty percent of bacterial meningitis patients suffer long-term disabilities, including brain damage and hearing loss. According to the National Meningitis Association, bacterial meningitis has an 11 percent mortality rate, even with treatment.

Bacterial meningitis differs from the viral diseases already covered, even though the signs of disease are often similar (TABLE 18.2). These diseases occur when bacteria invade the CSF, where they can grow and divide in the nutrient-rich environment. One way to determine the cause of meningitis is to measure CSF glucose levels and compare them to blood glucose levels. Viral meningitis does not affect glucose levels in the CSF. But bacteria in the CSF lower the usual glucose amount because they use it for energy and as a carbon source. Bacterial meningitis can also be determined by growing the bacterium from a CSF sample or Gram staining such samples.

One form of bacterial meningitis presents as acute meningitis in infants within the first 28 days of being born. **Neonatal meningitis** occurs when the bacteria are passed to a neonate from their mother during delivery. Several different bacteria cause neonatal meningitis, including Group B *Streptococcus*, *Escherichia coli*, and *Listeria monocytogenes*. Both streptococcal meningitis and *Listeria* meningitis are covered in this section. *Escherichia coli*, which primarily causes gastrointestinal disease, is covered in Chapter 19 (Digestive System Infections).

Haemophilus Meningitis

Scientists working to identify the cause of the 1918 Spanish flu pandemic found small, Gram-negative bacteria in many of the respiratory samples of

CLINICAL VOCABULARY

Spastic paralysis an inability to relax muscles

Toxemia presence of bioactive toxins in the blood; often these toxins are made by bacterial pathogens

Serotype a group of related microbes that share common antigens (i.e., specific proteins on their surface) that are often used for detection and diagnosis

TABLE 18.2 Patient Observations: Bacterial Versus Viral Meningitis

Test Results and Observations	Bacterial Agents	Viral Agents
Lumbar puncture	High pressure released on puncture	Low or no pressure released
White blood cell counts	High (usually much greater than 300/mm^3), mostly neutrophils except in *Listeria* meningitis, where lymphocytes predominate	Elevated, but lower than seen in bacterial infections (usually less than 300/mm^3), mostly lymphocytes
Cerebrospinal fluid composition	Higher protein, lower glucose (except in *Listeria* meningitis, where glucose is normal in over 60% of cases)	Higher protein, normal glucose

those who died. Because of its discovery in flu victims, the bacterium was named *Haemophilus influenzae* (HEE mo *fill* us *in* flu EN zee). It was later learned that *H. influenzae* does not cause influenza and is a common resident of the normal microbiota of healthy people. There are many serotypes of *H. influenzae* (strains that share cell surface features that react the same way with the immune system), but the most serious invasive strain as it relates to the nervous system is type b.

According to the CDC, before the Hib vaccine against *H. influenzae* type b was common in the United States, each year this bacterium caused over 12,000 cases of meningitis in children under five years old. Each year up to 600 children died and as many as 4,000 became permanently disabled due to *H. influenzae* meningitis. (Serotypes, respiratory *H. influenzae* infections, and their complications are discussed in Chapter 16.)

Thanks to the Hib vaccine, what was once the leading cause of bacterial meningitis in children in the United States is now, on average, responsible for fewer than 20 cases of meningitis per year. However, in developing countries where vaccination campaigns can be difficult because of a lack of infrastructure, *H. influenzae* type b is still a serious danger to children. Some estimates say over 3 million cases of severe disease and over 700,000 deaths occur every year due to this pathogen.

The symptoms of *Haemophilus* meningitis have a sudden onset and include fever, headaches, stiff neck, and confusion or disorientation. In patients with untreated *H. influenzae* meningitis, mortality approaches 100 percent. Even with appropriate antibiotics, 3 to 6 percent of children who exhibit symptoms will die.

Meningococcal Meningitis

Neisseria meningitidis (ny SEH *ree* ah men *ing* jih TY dis), which causes **meningococcal meningitis**, is a Gram-negative aerobe that has a capsule that contributes to pathogenesis. There are several different serotypes based on the polysaccharides in the capsule. Serotypes A, B, C, W, X, and Y are the groups implicated in

DiseaseSnapshot • Meningococcal Meningitis

Causative agent	*Neisseria meningitidis*: Gram-negative bacterium
Epidemiology	Most common in dry months; potential to cause epidemics; most cases are in 11–34 year olds; bacterium can be found as a normal resident of the mouth, nose, and throat. Average incubation period is 3–4 days.
Transmission & prevention	Transmits during contact between carriers and susceptible hosts (sharing food, kissing, or living in crowded quarters). Vaccines for serotypes A, B, C, Y, and W. Adolescents should receive the quadrivalent conjugate vaccine. Only the polysaccharide vaccine (Menomune®) is effective in older adults.
Signs & symptoms	Fever, chills, disorientation and confusion, agitation, headache, sensitivity to light, petechial pinpoint rash that may progress to a more marked purple bruise-like rash, stiff neck and back
Pathogenesis & virulence factors	It has a capsule, pili that aid in adherence, and makes endotoxin. It can evade the immune system through antigenic variation, molecular mimicry, and an enzyme that destroys IgA. The endotoxin, along with the related immune response, causes tissue damage associated with the disease. The most effective immune response is antibody-mediated complement attack; this is why those with complement deficiency disorders are routinely vaccinated against all forms of it.
Diagnosis & treatment	Gram stains of blood and CSF are common detection methods, as is culturing bacteria from the CSF. If a patient has received some antibiotic treatment, commercial kits that detect *N. meningitidis* capsular antigens may be used. Antibiotics must be administered as soon as possible to prevent tissue damage. Ceftriaxone is given first, until the susceptibility of the organism to penicillin can be tested. If appropriate, the patient can be switched to penicillin. Antibiotics like rifampicin are given as prophylaxis to family members, roommates, and romantic partners of those diagnosed with *N. meningitidis*.

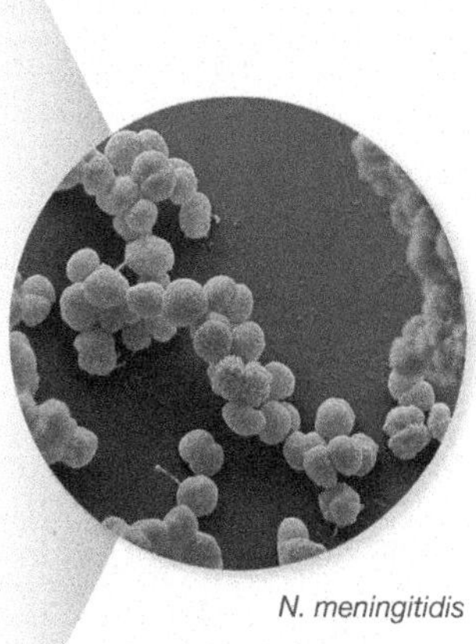

N. meningitidis

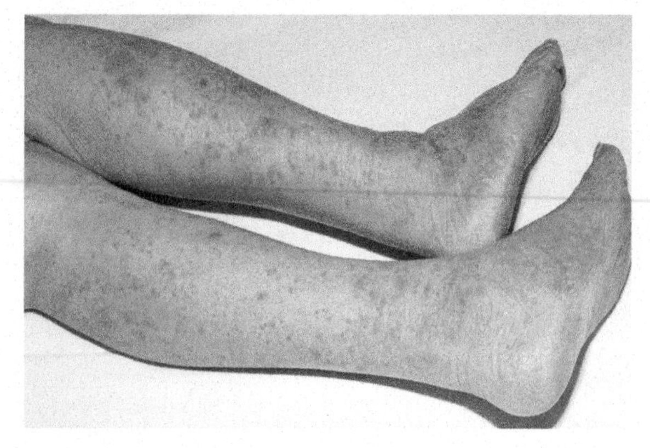

FIGURE 18.8 **Rash associated with meningococcal meningitis**

TABLE 18.3 Common Causes of Bacterial Meningitis by Age Group

Age Group	Causes
Newborns	Group B *Streptococcus, Escherichia coli, Listeria monocytogenes*
Infants and Children	*Streptococcus pneumoniae, Neisseria meningitidis, Haemophilus influenzae* type b
Adolescents and Young Adults	*Neisseria meningitidis, Streptococcus pneumoniae*
Older Adults	*Streptococcus pneumoniae, Neisseria meningitidis, Listeria monocytogenes*

meningococcal meningitis (this organism can also cause generalized bacteremia). The epidemiology of each serotype is distinct; types B, C, and Y are most common in the United States. Serogroup B causes a majority of the cases in children younger than one year old, while other serogroups are more common in older patients.

N. meningitidis is capable of "capsular switching," allowing organisms to exchange genes coding for the polysaccharide capsule. Almost 10 percent of the population carries *N. meningitidis* as a part of their resident nose and throat microbiota, and sporadic disease outbreaks (especially among people living in close quarters such as dormitories or barracks) occasionally occur following "capsular switching."

The symptoms of meningococcal meningitis are similar to those of other types of bacterial meningitis. However, a distinctive petechial rash that progresses to larger bruiselike lesions is found in cases where *N. meningitidis* invades the bloodstream (FIG. 18.8). Also, meningococcal meningitis has a particularly rapid onset and course—if not treated, death can occur within hours of fever onset.

Before effective antibiotic treatments, meningococcal meningitis killed about 80 percent of those affected. Now mortality rates are down to around 15 percent. But the most dramatic advances have been made in the area of prevention. There are only about 0.18 cases per 100,000 people versus the prevaccine incidence of about 300 cases per 100,000 people. As of 2016, there were three types of vaccines available in the United States: meningococcal conjugate vaccines (Mentactra®, MenHibrix®, and Menveo®), meningococcal polysaccharide vaccine (Menomune®), and serogroup B meningococcal vaccines (Bexsero® and Trumenba®). The CDC recommends vaccination of all adolescents (and a booster shot in late adolescence) with the conjugate vaccine. The group B vaccine is only strongly encouraged for high-risk groups. Certain groups of older adults (over 56 years old) are vaccinated using the polysaccharide vaccine, the only one that is immunogenic in this age group.

Pneumococcal Meningitis

According to the National Institutes of Health, most bacterial meningitis cases in the United States are **pneumococcal meningitis**, caused by *Streptococcus pneumoniae* (*strep* toe KOK us new MOAN ee *ay*) (TABLE 18.3). This Gram-positive diplococcus is a facultative anaerobe that was first isolated over 100 years ago.

DiseaseSnapshot • Pneumococcal Meningitis

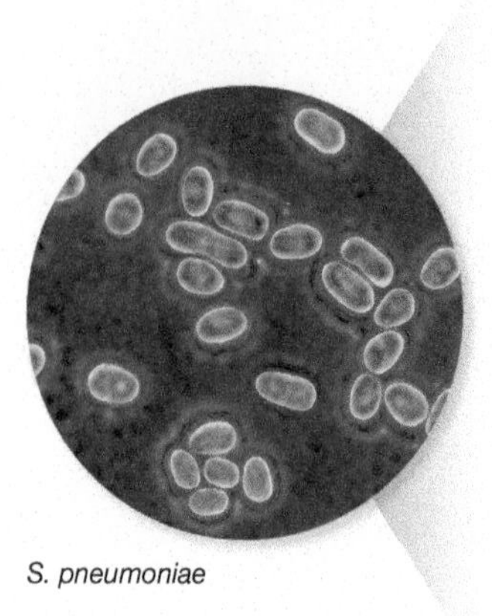

S. pneumoniae

Causative agent	*Streptococcus pneumoniae*: Gram-positive bacterium
Epidemiology	3,000–6,000 cases every year in the U.S.; mostly affects nonimmunized people under two or over 50 years of age. Incubation period is 1–3 days.
Transmission & prevention	Usually an opportunistic infection; person-to-person transmission is rare. PPSV23 (or Pneumovax®) vaccine recommended for all people over age 65 and children older than 2 years. PCV13 (Prevnar®) vaccine protects against 13 strains and is recommended for children under age 2 and adults with compromised immunity.
Signs & symptoms	Fever, chills, headache, confusion, disorientation, agitation, vomiting, stiff neck and back, sensitivity to light
Pathogenesis & virulence factors	A capsule helps it adhere to airway mucous membranes and evade phagocytosis; a toxin called pneumolysin allows the pathogen to lyse phagocytic cells. Toxins and antigens stimulate inflammation and tissue damage, which allows the bacteria access to the blood (bacteremia) and changes the permeability of the blood–brain barrier, causing meningitis.
Diagnosis & treatment	Microscopic examination and culture of the bacteria from CSF; antigen-based tests such as ELISA also detect pneumococcal C polysaccharide. Penicillin is commonly effective; however, bacterial resistance is increasing. Alternative antibiotics include cephalosporin; vancomycin must sometimes be used.

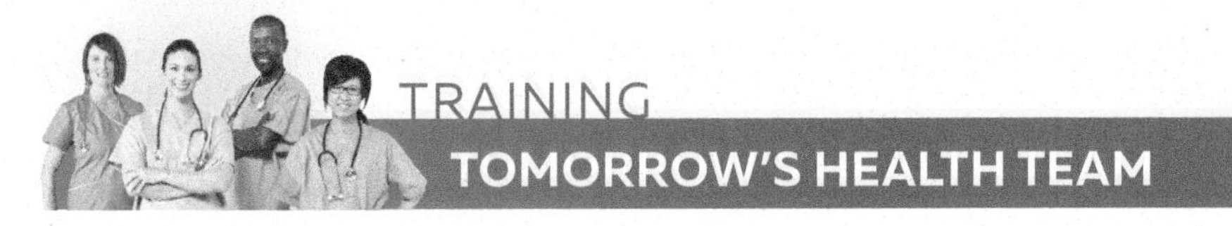

Meningitis Outbreaks on College Campuses

In March 2013, Princeton University was the center of a debate about bacterial meningitis. According to the CDC, over the course of the previous two years, nine cases of serogroup B meningococcal meningitis were associated with the university. This form of the disease had no vaccine on the market in the United States at the time, although one, Bexsero®, was approved for use in Europe and Australia. In this situation the CDC recommended that the students be given Bexsero to prevent future outbreaks, and the Food and Drug Administration (FDA) approved the import of the vaccine specifically for use at Princeton. By February of 2014 more than 5,000 Princeton students had received their second dose of this vaccine; fortunately, serious side affects were not reported. There are now two serogroup B vaccines licensed for administration in the United States, Bexsero and Trumenba®.

Thirty-six states require students to receive a meningitis vaccine before attending college. This often applies whether or not the college has on-campus housing and is enforceable by law irrespective of if the school is public or private. This means that if you are reading this, you may have been mandated to get a meningococcal vaccine (Menactra®, Menveo®, or MenHibrix®).

QUESTION 18.3

Based on the meningitis sections you just read, why do you think the FDA approved the use of the Bexsero vaccine?

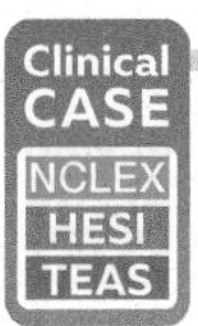

The Case of the Pain in the Neck

Practice applying what you know clinically: visit the **Mastering Microbiology** Study Area to watch Part 2 and practice for future exams.

S. pneumoniae commonly causes pneumonia, but it can also result in sinusitis, ear infection (otitis media), blood infections (bacteremia), and meningitis.

Pneumococcal meningitis usually begins with the spread of resident bacteria from the nose and throat into the bloodstream. This allows rapid bacterial growth and host tissue damage that can break down the blood–brain barrier. The bacteria then grow inside the cerebrospinal fluid and may invade the brain (FIG. 18.9). The immune response to the bacterial infection is often the most damaging aspect of the disease. *S. pneumoniae* makes an exotoxin that stimulates severe inflammation and provokes host tissue damage as immune cells release oxygen radicals and enzyme-destroying proteins in response to the toxin. Like many other causes of bacterial meningitis, pneumococcal meningitis is preventable by vaccination. The routine immunization of young children with the PCV7 vaccine around the year 2000 decreased average annual rates of pneumococcal meningitis hospitalizations in children under 2 by over 60 percent.

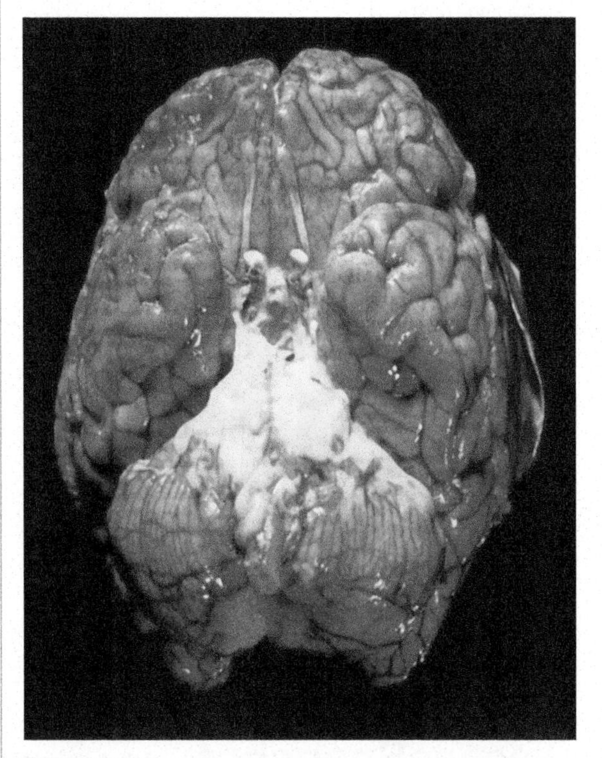

FIGURE 18.9 Pneumococcal meningitis in the brain This image shows the tissue damage caused by infection with *S. pneumoniae.*

Critical Thinking *What might cause normal microbiota to become invasive and cause disease?*

Listeria meningitis (Listeriosis)

Following the introduction of the Hib vaccine, which greatly reduced *H. influenzae* meningitis cases, *Listeria monocytogenes* (Lis TEER ee ah mon *oh* sy TOE jen *eez*) became the fourth most common cause of bacterial meningitis. It accounts for 20 percent of all meningitis in neonates and in people over 60 years old. *Listeria* infections, or listerioses, frequently make headlines as a food-borne illness when tainted cheese, lunchmeats (deli meats), or even mangoes have been discovered carrying the pathogen.

Listeria meningitis can develop from the preliminary gastrointestinal symptoms of listeriosis. The Gram-positive rods that cause this disease have a few very special adaptations that make them difficult to fight unless a patient's cellular immunity is strong. *L. monocytogenes* evades the immune system by adhering to cells, being phagocytized, and then breaking out of the phagosome into the cytoplasm. With this intracellular lifestyle, the bacteria can grow and divide hidden from antibody-mediated immune responses—*L. monocytogenes*

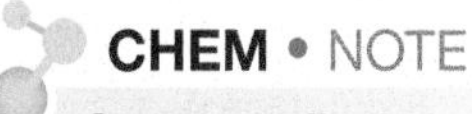

CHEM • NOTE

Oxygen radicals have an unpaired electron in their outer shell that makes them reactive with nearby molecules. Hydrogen peroxide and nitric oxide are examples of molecules that contain reactive oxygen. The immune system uses these molecules to combat invaders since they lethally damage targeted DNA and proteins.

Listeria

Host cell

Phagosome

Tail

1 *Listeria* enters a cell through phagocytosis and then breaks out of the phagosome, entering the cytoplasm.

2 *Listeria* reproduces within a host cell, safe from antibody-mediated immune response.

3 The bacterium creates a tail for propulsion using actin filaments from the host cell's cytoskeleton. The tail propels the bacterium into a neighboring cell, where it again enters the cytoplasm.

4 New cells are infected in the same way.

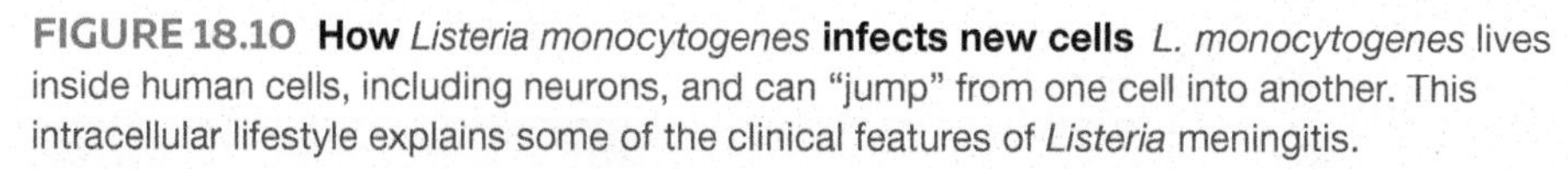

FIGURE 18.10 How *Listeria monocytogenes* infects new cells *L. monocytogenes* lives inside human cells, including neurons, and can "jump" from one cell into another. This intracellular lifestyle explains some of the clinical features of *Listeria* meningitis.

Critical Thinking *Why does Listeria's lifestyle protect it from certain immune defenses?*

doesn't even have to enter the intercellular space to spread the infection. It forces a protein called actin in the cytoskeleton of the eukaryotic cell it's infecting to form a "tail" and propel it like a rocket into an adjacent cell (FIG. 18.10). In addition to this intracellular lifestyle, *L. monocytogenes* is capable of growing at a wide range of pH and temperature, including refrigeration temperatures. If it's introduced to undercooked meat or dairy products, it continues to grow while the food is refrigerated. Once contaminated food is consumed, it causes a range of symptoms, depending on the host.

In healthy adults, *L. monocytogenes* may cause few symptoms or just appear like a mild cold. In some patients, acute but short-lived gastrointestinal problems arise, causing watery diarrhea, fever, nausea, and headache. But in elderly patients, immunocompromised individuals, neonates, and pregnant women, the infection can be much more serious and develop into nervous system infections. AIDS patients are twice as likely as the general population to develop neurological symptoms. All of these patients share a weak cell-mediated immune response. Because *L. monocytogenes* is resistant to the antibody-mediated immune system, the cell-mediated component is much more important in defeating the infection. When cellular immunity is compromised, the body can't control the spread of *L. monocytogenes*. The bacteria can invade macrophages, spread throughout the body, and cross the blood–brain barrier to infect the meninges. In pregnant women, the disease can cause profound damage to the fetus, sometimes inducing spontaneous abortion or causing stillbirth. It can also cause neonatal meningitis. Incidence and mortality are higher in the elderly: 25 percent of patients over age 50 with *L. monocytogenes* infection die (this includes both bacteremia and CNS forms). Roughly 15 percent of *L. monocytogenes* cases involve pregnant women; in about a quarter of these cases, the fetus dies.

Bacteria can infect the peripheral nervous system.

While several bacteria can enter the peripheral nervous system, it is very rare for this tissue to act as a long-term host to any bacteria. When it does occur, such

DiseaseSnapshot • *Listeria* Meningitis (Listeriosis)

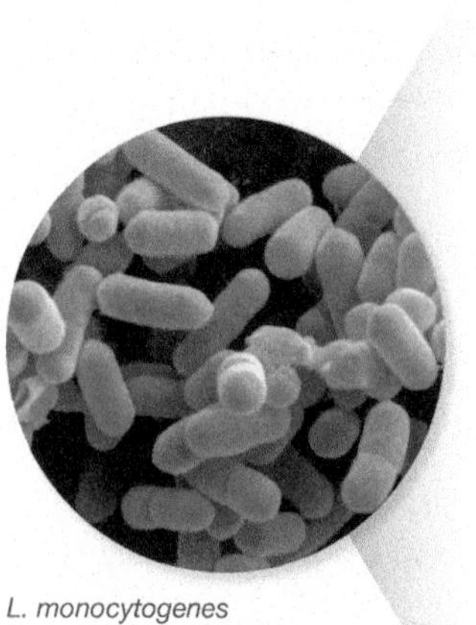

L. monocytogenes

Causative agent	*Listeria monocytogenes*: Gram-positive bacterium
Epidemiology	1,600 cases with 260 deaths annually in the U.S., mostly involving pregnant women or the elderly. Up to one quarter of cases develop nervous system infections. Average incubation period for invasive disease is 30 days.
Transmission & prevention	Transmits from contaminated food or *in utero*; best prevented by avoiding raw or undercooked meat, unwashed vegetables, and soft cheeses.
Signs & symptoms	Fever, headache, stiff neck and back, confusion, agitation, loss of balance, and convulsions
Pathogenesis & virulence factors	Internalin A helps bacteria adhere to cells and become phagocytized. Lisetriolysin allows escape from the phagosome so the bacteria can grow inside the cytoplasm of a host cell. The key protein involved in cell-to-cell transmission is called ActA; this protein uses the host cell's cytoskeleton to spread throughout the surrounding cells. Cell-mediated immunity is crucial to clearing the disease because of its intracellular lifestyle.
Diagnosis & treatment	Due to the small numbers of extracellular bacteria, microscopic examination of CSF is not sufficient for diagnosis. Culture-based methods, however, are very sensitive, but may require cold incubation, or growth on differential media. Usually a combination of gentamycin with ampicillin treats serious illness. Cephalosporins, which are often first-choice therapies for pyrogenic (or fever inducing) meningitis, are not effective.

disease leaves the central nervous system intact but cripples sensory and motor neurons in the human body. For example, *Mycobacterium leprae* (my koh bak TEER ee um LEP ray) uses this unusual lifestyle to sustain itself inside the human body over a period of years. The slow growth of this organism contributes to the drawn-out progress of the disease known as **Hansen's disease,** or **leprosy**.

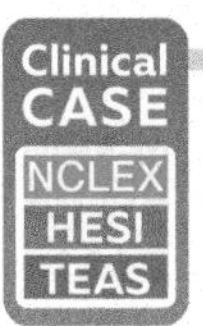

The Case of the Pain in the Neck

Practice applying what you know clinically: visit the **Mastering Microbiology** Study Area to watch Part 3 and practice for future exams.

Hansen's Disease (Leprosy)

M. leprae is an acid-fast bacterium closely related to the causative agent of tuberculosis (*M. tuberculosis*). When it infects humans, it mainly causes damage to the skin, peripheral nerves, and mucosal tissues in the eyes and respiratory tract. Symptoms may emerge decades after the infecting exposure.

Armadillos are natural carriers of the bacterium, probably due to their cooler body temperature (30–35°C), which is necessary for *M. leprae*'s growth.

According to the World Health Organization, globally there were about 212,000 cases diagnosed in 2015, with the majority found in India. Historically, disease sufferers who experienced disfiguring effects were persecuted and isolated in so-called leper colonies. Today multidrug therapy, available since the 1980s, can cure the disease. Transmission is not fully understood, but it's thought that the bacteria may be contracted via the respiratory route. The disease does not spread easily among people.

The bacteria's preference for a lower growth temperature explains why *M. leprae* thrives in the extremities rather than migrating to deeper, warmer tissues within the human body. Hansen's disease comes in two general forms: *tuberculoid* (paucibacillary), and *lepromatous* (multibacillary). Tuberculoid forms have fewer and more isolated skin plaques associated with infected nerves. Lesions tend to be flattened and less pigmented than nonaffected skin (FIG. 18.11). Although this form often causes significant localized sensory loss, it's not very contagious and it's fairly easily treated. It therefore tends to cause less significant long-term health impacts.

The lepromatous form is more serious. It can spread to others and causes irreversible tissue damage in the form of macules, papules, and nodules in many places on the body. (See Chapter 17 for more on skin lesions.) Bone and cartilage also decay, especially in the extremities. Without treatment, lepromatous leprosy is often fatal. For unknown reasons, people have different immune responses to *M. leprae*—which form of the disease develops is dictated by which parts of the immune system respond to the infection. Fortunately, the more dangerous and disfiguring lepromatous form is less common than the milder tuberculoid form.

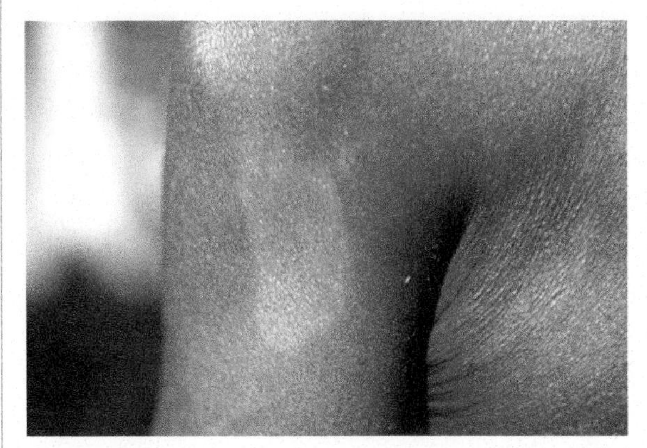

Tuberculoid form

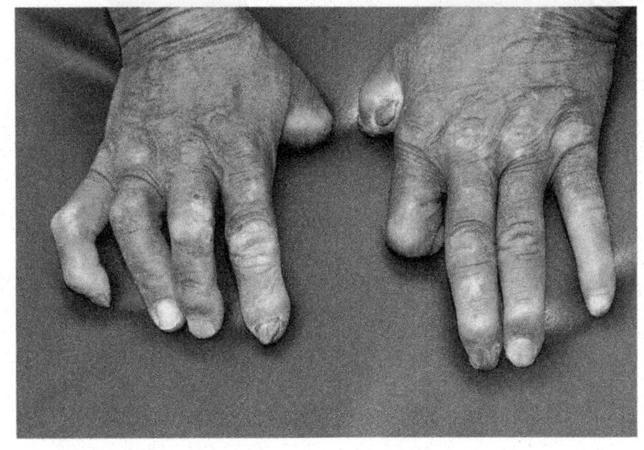

Lepromatous form

FIGURE 18.11 The effects of Hansen's disease (leprosy) The more common tuberculoid form of Hansen's disease, or leprosy, is characterized by flattened lesions. The lepromatous form that affects extremities is more contagious, however.

Critical Thinking *Why does the acid-fast property of this bacterium cause the bacterium to grow slowly? (Hint: Reflect on acid-fast property/revisit Chapters 1 and 3.)*

Bacterial toxins can damage the nervous system.

Some bacteria have a pathogenic effect primarily because they secrete toxins. Although many bacteria secrete deadly toxins, those of *Clostridium botulinum* and *Clostridium tetani* are among the most famous causes of **toxemia**, or blood poisoning from toxic compounds. (For more on toxins, see Chapter 10.) Some of the most dangerous toxins known to mankind are neurotoxins made by bacteria. Surprisingly, some of these same neurotoxins have commercial and medical uses when administered in exceedingly small doses. For example, the injectable treatment Botox has cosmetic and medical uses. It causes mild facial paralysis, temporarily reducing the appearance of wrinkles, and has also been used to treat a number of neurological conditions including, but not limited to, chronic migraines, muscle spasticity in multiple sclerosis, and overactive bladder dysfunction.

Botulism

Clostridium botulinum (klos TRID ee um bot yu LIN um) is a Gram-positive, rod-shaped anaerobe that makes a collection of powerful neurotoxins. The effects of botulinum toxins produce the disease **botulism**. The organism gets its name from the Latin word for "sausage" because, during the Middle Ages, sausages were commonly contaminated with the bacteria. Botulinum toxins are so powerful that as little as one nanogram can kill an adult. The bacterium that produces it is found globally in soils. This begs the question, why don't more people get botulism?

The answer is that *C. botulinum* has very specific growth needs. It doesn't tolerate high pH well, and can't grow unless the oxygen level is quite low. In nonconducive growth environments, it exists as a dormant endospore; but when favorable growth conditions are met, it germinates, actively grows, and produces the botulinum toxin. Endospores are highly resistant bacterial structures that can survive harsh conditions, including cooking. Canned vegetables such as beets, carrots, and spinach provide the perfect growth conditions for *C. botulinum* endospores to germinate, since salt, pH, and oxygen are all low. Cans contaminated with *C. botulinum* appear bloated, with bulging sides or tops. They may even produce a foamy spray when opened and release a foul odor (gases produced from fermentation). Improperly canned vegetables accounted for 38 percent of the 116 cases of foodborne botulism from 1996–2008 in the United States. All forms of botulism are nationally notifiable and are immediately reported to the CDC to limit other infections. Wound botulism can also occur, although it is relatively rare. This form is most common among IV drug users who use needles contaminated with *C. botulinum* endospores.

Due to greater precautions in home canning and the use of sterile needles in the United States, the most common form of botulism is infant botulism. This form of botulism is generally found in children younger than 1 year old. These cases stem from ingesting *C. botulinum* endospores. Infants encounter the endospores through wind and dust, or more commonly by eating foods, like honey, that contain endospores. The gastric acids in the digestive tract trigger endospore germination, allowing a temporary production of the toxins as the *C. botulinum* travels through the GI tract. A survey of U.S. honey products showed that from 2 to 24 percent of them contained *C. botulinum* endospores. For this reason, children under 1 year old should not be fed honey. In older children and adults, a mature immune system and the healthy microbes that live in the gastrointestinal tract usually prevent *C. botulinum* from finding a place to grow. But infants are not fully colonized by helpful bacteria, and their immune system is still maturing so *C. botulinum* encounters little competition, and will readily grow and produce its toxins.

Disease**Snapshot** • Botulism

C. botulinum

Causative agent	*Clostridium botulinum*: Gram-positive bacterium
Epidemiology	Annually about 200 cases in the United States (with most cases in California and Pennsylvania). Incubation period for infants is about 3 days; for foodborne botulism incubation is 12–72 hours.
Transmission & prevention	Not considered contagious; infant botulism is commonly due to ingesting endospores in foods such as honey; foodborne botulism is most commonly associated with canned vegetables. Wound botulism is most commonly linked to dirty needles. Proper cooking destroys toxin in foods. Proper canning techniques prevent endospore germination and bacteria growth. Vaccine is available, but usually only administered to researchers, emergency first responders, and military personnel.
Signs & symptoms	Nausea, vomiting, muscle cramps, difficulty breathing, drooping eyelids, and difficulty speaking
Pathogenesis & virulence factors	Infant botulism involves temporary colonization of the GI tract by *C. botulinum*, which produces paralytic botulinum toxins resulting in flaccid paralysis. In foodborne botulism, the toxin is ingested without requiring growth of the organism in the GI tract.
Diagnosis & treatment	Ninety percent of cases are diagnosed in the U.S. (under-reported elsewhere). Characteristic symptoms and clinical suspicion are essential to early diagnosis. Botulinum toxins can be detected in the patient's blood or feces; and the bacteria may also be cultured from feces. Treatment with antitoxin and breathing support may be needed; penicillin is administered to kill any live *C. botulinum* and the GI tract may be cleared. Because the toxin binds to neurons, full recovery can be slow or limited, depending on how well neurons can regenerate.

Botulinum toxin causes **flaccid** (relaxed) **paralysis**. This condition occurs because the toxin blocks the release of the neurotransmitter acetylcholine from neurons and thereby prevents muscle contraction. In foodborne botulism cases, this causes severe constipation, possibly because of local paralysis of the muscles that move food through the GI tract. If timely treatment is provided, patients usually recover from this form of botulism. Affected nerves of the intestinal tract are regenerated in about 3–4 weeks, but some other affected neurons may not regenerate. The long-term effects of botulism largely depend on what nerves were impacted and how soon treatment was administered.

FIGURE 18.12 **An individual with tetanus** Tetanospasmin produced by *Clostridium tetani* prevents muscle tissue from relaxing.

Tetanus

Tetanus is caused by *Clostridium tetani* (teh TAN eye). A patient with tetanus experiences the effects of a toxin almost as powerful as botulinum: **tetanospasmin** (*tet* an oh SPAZ min). The common name "lock jaw" comes from one of the effects of the toxin: Patients' muscles contract powerfully and then do not relax. This prevents patients from being able to open their mouths to speak or even cry out from the pain (FIG. 18.12).

C. tetani is found in soils all over the world. Tetanus develops when the bacterium grows in deep puncture wounds. People often associate wounds from rusty objects like a nail as a source of tetanus, but it is the spore-containing soil or dust clinging to the object rather than rust that transfers the pathogen. Because the bacterium only grows and makes its toxin in low-oxygen conditions, deep wounds are more conducive to tetanus than shallow abrasions. Thanks to routine vaccination against tetanus, in the United States there are fewer than 50 cases per year. Cases that do develop tend to be in the undervaccinated and in the elderly, whose immunity is less effective. Globally, tetanus is still a terrible threat, with over a million cases occurring each year.

C. tetani causes irreversible damage to neurons. The toxin, tetanospasmin, is taken up by peripheral motor neurons and transported to the spinal cord. Once there, it becomes active and blocks the release of inhibitory neurotransmitters, preventing muscle relaxation. This leads to **spastic paralysis**, which is an inability to relax muscles. This in turn causes intense muscle spasms, drooling, sweating, and irritability (see FIG. 18.13).

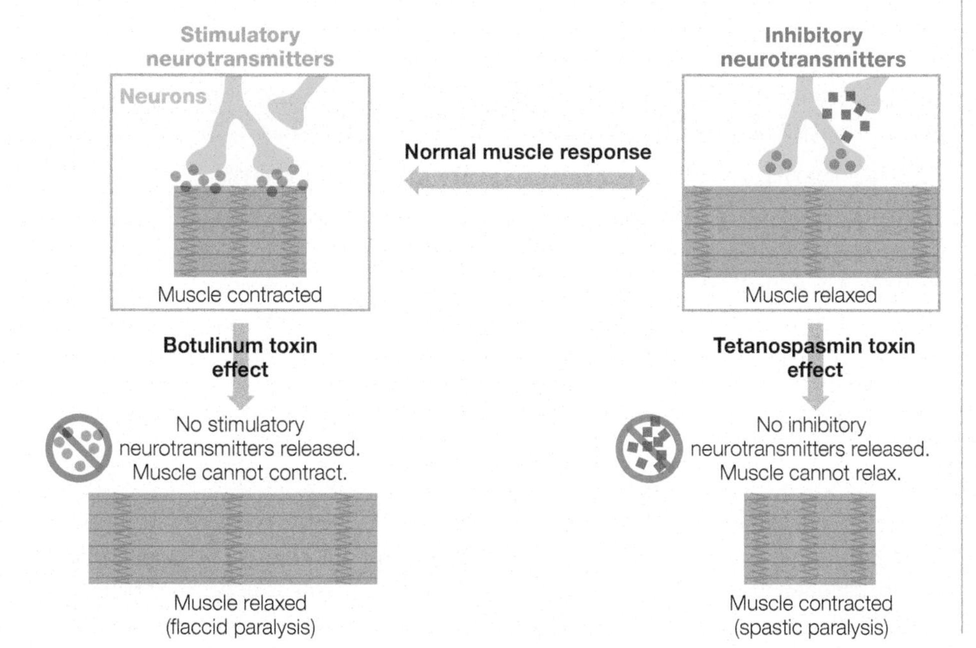

FIGURE 18.13 **Neurotoxin effects on motor neurons** Normal muscle contractions occur when stimulatory neurotransmitters are released from a motor neuron and they end when inhibitory neurotransmitters turn them off. The tetanospasmin toxin blocks normal inhibition while botulinum toxin blocks normal stimulation.

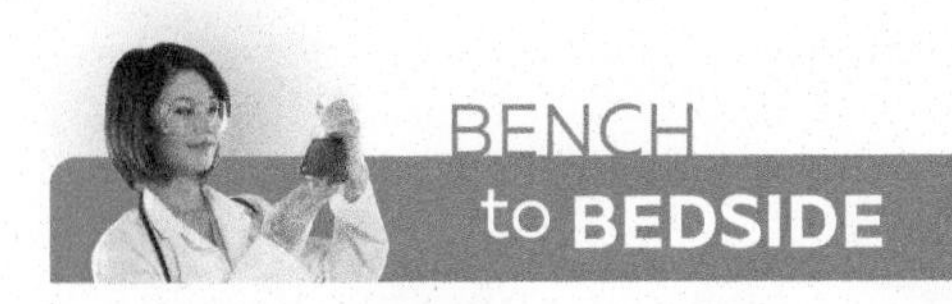

Searching for a Treatment for Herpes Simplex Encephalitis

Herpes simplex virus 1 (HSV-1) is most commonly known as the cause of cold sores. But in some cases the virus can reactivate and travel up the neurons in the face to the brain, causing herpes simplex encephalitis (HSE). In fact, it is the most common identified cause of sporadic fatal encephalitis worldwide.

HSE can attack the frontal and temporal lobes of the brain, leading to profound personality changes, hallucination, seizures, memory loss, and speech problems. HSV-1 is found in approximately 50 percent of the population in the United States. The virus accounts for around 10 percent of the encephalitis cases in the United States.

The normal antivirals that prevent effective viral DNA replication have selected for drug-resistant strains of HSV-1, especially in immunocompromised patients. Recent work by some researchers suggests that injection of an antibody made to target a glycoprotein on the surface of HSV-1 can prevent some of the worst symptoms from occurring. Mice given a lethal dose of HSV-1 lived after receiving the antibody as well. Because the glycoprotein is involved in the spread of the virus from cell to cell, these antibodies may be the key to ensuring that HSV-1 cannot leave the PNS and invade the CNS to cause encephalitis.

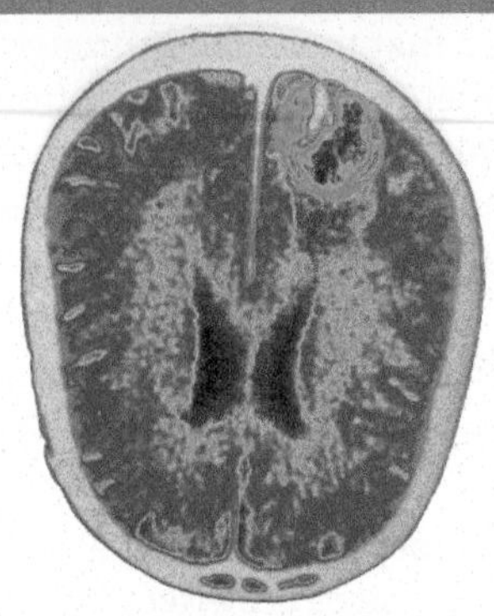

Sources: Krawczyk, A., Dirks, M., Kasper, M., Buch, A., Dittmer, U., Giebel, B., Wildschütz, L., et al. (2015). Prevention of herpes simplex virus induced stromal keratitis by a glycoprotein B-specific monoclonal antibody. *PLoS One* (10), 1.

Death, primarily due to respiratory failure, occurs in about 50 percent of affected adults who don't get adequate and timely treatment; the rate is lowered to 10 percent where proper supportive care is provided. Neonatal tetanus kills over 90 percent of those infected. This form of the disease is almost exclusively found in developing nations and is usually caused by soil contamination of the umbilical cord after birth.

Build Your Foundation

12. Who is most likely to contract *Haemophilus influenzae* and why?
13. What are the different types of vaccines against meningococcal meningitis?
14. Who should get vaccinated for meningococcal meningitis?
15. How do people contract pneumococcal meningitis?
16. Who is most at risk for serious disease by *Listeria monocytogenes*? Why?
17. Why is Hansen's disease usually confined to the PNS?
18. What is the most common type of botulism and how is it treated?
19. How do botulinum and tetanus toxin compare to each other?
20. How is tetanus prevented?

Build Your Foundation (BYF) Quick Quiz: Visit the **Mastering Microbiology** Study Area to quiz yourself.

18.4 OTHER NERVOUS SYSTEM INFECTIONS

Learning Outcomes

After reading this section, you should be able to:

18.16 Discuss how the fungus *Cryptococcus neoformans* invades and damages the nervous system.
18.17 Describe trypanosomes and detail how they cause African sleeping sickness.
18.18 Name the cause, effects, and health impact of primary amoebic meningoencephalitis.
18.19 Describe the disease toxoplasmosis and the risk factors for getting the disease.
18.20 Explain the role of prions in the neurological progression of spongiform encephalopathies.

Fungi can infect the central nervous system.

When we hear the word fungi, we think about the mushrooms in our pasta sauce or perhaps the mold on our bread—not meningoencephalitis. That's because fungal infections are relatively rare, and only a few fungi are primary pathogens that cause disease in a healthy host. But fungal infections in immunocompromised patients are much more common, and have serious health impacts. Because fungal spores (fungal reproductive structures) are found in the dirt, air, and water all around us, prevention of all fungal diseases is almost impossible.

Cryptococcosis

Bird droppings, especially from pigeons, can contain dangerous pathogens that get aerosolized when the dry droppings are disturbed. This is especially true at construction or industrial work sites, where dust is constantly being generated

Disease**Snapshot** • Cryptococcosis

Causative agent	*Cryptococcus neoformans* and *C. gattii* fungi
Epidemiology	Over 1 million cases of cryptococcal meningitis occur annually, mostly in immune compromised people, including HIV/AIDS patients. The disease results in 624,700 deaths annually, mostly in sub-Saharan Africa. Incubation period is unknown.
Transmission & prevention	Fungal spores get inhaled from stirred-up dried bird droppings. No effective preventative measures exist.
Signs & symptoms	Cough, flulike symptoms, fever, headache, agitation, and disorientation.
Pathogenesis & virulence factors	*Cryptococcus* species have a thick polysaccharide capsule that shields the fungi from many of the normal immune responses. The capsule also contains melanin, which may protect the fungi from oxidative bursts from neutrophils and NK cells.
Diagnosis & treatment	Culture and microscopic methods are used to test for *Cryptococcus* in the blood, sputum, or CSF. Antigen-based tests are also used for identification of the species, although *C. neoformans* and *C. gattii* require additional testing to differentiate. Severe infections with neurological symptoms are treated with two antifungal drugs, amphotericin B and flucytosine.

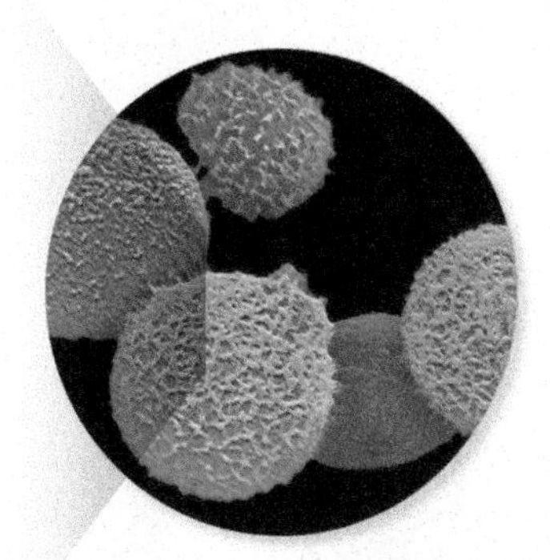

C. neoformans

and people have ample opportunity to inhale stirred-up pathogens from soil or bird droppings. One of the diseases that is spread this way is **cryptococcosis** (KRIP toh kok *oh* sis), which is caused by infection with *Cryptococcus neoformans*.

This yeastlike fungus forms tough, resistant spores. Once they enter the lungs, they invade macrophages and grow inside them. These macrophages migrate out of the lungs to the lymphatic system, carrying the fungi with them. From there, the organism can infect the blood and invade the CNS.

C. neoformans primarily infects immunocompromised hosts; it is the leading cause of death for people who have HIV/AIDS. However, the emerging disease caused by the related fungus, *C. gattii*, usually occurs in healthy hosts. In the United States there were over 100 cases reported in the last 10 years, almost all in the Pacific Northwest. While *C. gattii* is usually contracted from natural settings, *C. neoformans* is associated with urban environments.

CLINICAL VOCABULARY

Parasites organisms that benefit themselves at the expense of their host (in medical contexts this usually refers to protozoans and helminths)

Protozoans cause rare but serious nervous system infections.

Protozoans are a type of animal-like protist; many of them are single-celled eukaryotic parasites. Certain members of the flagellated protozoans and amoeboid protozoans are known for causing neurological symptoms in humans. Like most other protozoans, these pathogens have complex life cycles and go through different stages of development that correspond to different hosts and/or environments. They are facultative anaerobes, and reproduce through sexual as well as asexual means, depending on their life stage (Chapter 4 reviews protists in more detail).

African Sleeping Sickness

African sleeping sickness is a tropical disease found in Africa. The name refers to the second stage in the disease progression, one characterized by hallucinations and sleep disturbances. Those infected cannot sleep at night and cannot stay awake during the day. If the disease goes untreated, it is fatal.

The flagellated protozoan *Trypanosoma brucei* (*try* pan oh SOH mah BROO see eye) causes this debilitating neurological disease. *T. brucei* is carried by the blood-sucking tsetse fly from host to host. Two morphological forms alternate, based on infective stage. One form has a flagellum that runs the entire length of the parasite's body, whereas the other stage has a shorter flagellum and is only found in the salivary glands of the fly. In addition to direct cytotoxic effects from *T. brucei*, the organism also appears to induce damage to the brain through the immune response itself. It induces an immune system reaction so strong that host neutrophils can harm host neurons.

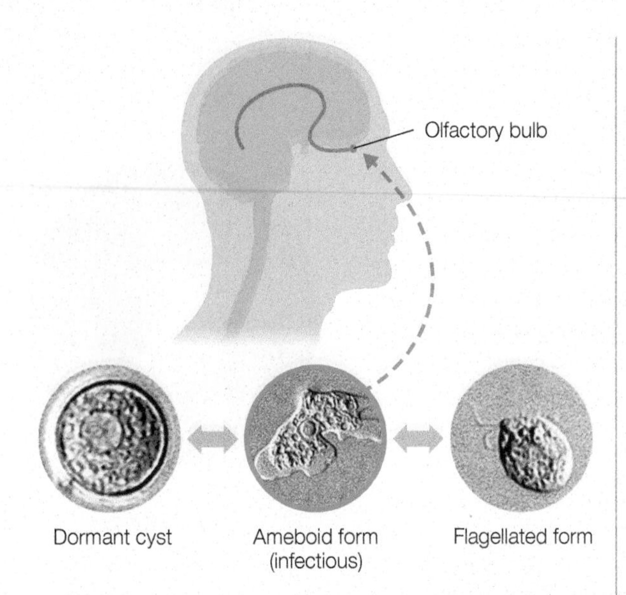

FIGURE 18.14 Morphological forms of *Naegleria fowleri* *N. fowleri* has flagellated and amoeboid trophozoite forms, as well as a more dormant cyst form. The amoeboid trophozoite form can invade the olfactory nerves and cause primary amoebic meningoencephalitis (PAM).

Critical Thinking *Would you expect more cases of PAM in cold climates or warm climates? Explain your reasoning.*

There are two medically important subspecies of *T. brucei*: *rhodesiense* and *gambiense*. These protists have different features regarding geographic distribution (although there is growing overlap), reservoirs, and disease progression in humans. Subspecies *T. b. rhodesiense* is rare and causes a more rapid and acute disease, with patients often dying from cardiac arrest before neurological symptoms emerge. Subspecies *T. b. gambiense* causes a more chronic disease. Because *T. b. gambiense* accounts for 98 percent of all African sleeping sickness cases, it is the focus of the remainder of this section.

T. b. gambiense symptoms begin with a sore, or chancre (SHAN ker), surrounding the site of the tsetse bite. Fever, headache, and swollen lymph nodes follow this, although infection with *T. b. gambiense* can also be asymptomatic during the initial phase. In this phase of infection, there are very few white blood cells and almost no trypanosomes are found in the CSF. After entering the blood, the parasite can cross the blood–brain barrier and infect the CNS.

Primary Amoebic Meningoencephalitis (PAM)

Although very rare, **primary amoebic meningoencephalitis (PAM)**, caused by the amoeba *Naegleria fowleri*, is almost always fatal. Often called "the brain-eating amoeba," the dramatic effects on the CNS after infection are truly terrifying. *N. fowleri* has different life stages, only one of which is infective. It's a thermophilic protist that lives in streams and lakes all around the world as either a dormant, protected cyst or an active infectious trophozoite (troh foh ZOE ite). Under normal circumstances, it feeds solely on bacteria in its aquatic environment. If the trophozoites encounter a lack of resources, they change morphology into a flagellated form that makes them motile and therefore better able to relocate to a new habitat (FIG. 18.14).

Humans very rarely come into contact with *N. fowleri* in a way that allows for infection. When someone does contract the disease, it's usually by swimming in warm, stagnant water. In a few cases, infections were traced to using nonboiled tap water in a neti pot (see the Training Tomorrow's Health Team on *N. fowleri* in Chapter 4). Over the past fifty years, there have only been 128 confirmed cases in the United States.

Once *N. fowleri* enters the nasal passages of a human host, the trophozoite can burrow into the nasal mucosa. From there, it travels up the olfactory nerves into the brain. Initial symptoms include intense headaches, sore throat, vomiting, and fever. As the disease progresses, stiff neck, seizures, hallucinations, and coma can occur. Patients usually die within a week or two after infection. Postmortem examination shows hundreds of trophozoites growing in the brain. The CSF is also filled with the protozoans and dead white blood cells.

A similar set of symptoms can also be caused by another disease usually called granulomatous amebic encephalitis. This disease is caused by other protozoans, *Acanthamoeba* species and *Balamuthia mandrillaris*, and is distinguished by a postmortem autopsy of the brain tissue, although DNA-based assays are most definitive. One of the most important differences between these protozoans and *N. fowleri* is the speed with which they affect the patient. These protozoans can reside in the host for anywhere from three months to two years before symptoms are recognized. However, *Acanthamoeba* species and *B. mandrillaris* are, like *N. fowleri*, almost always fatal.

Toxoplasmosis

It is common to hear that pregnant woman shouldn't clean the litter box for household cats. A protozoan called *Toxoplasma gondii*, which causes **toxoplasmosis**, is the reason behind this avoidance. Cats, both feral and household pets, often carry *T. gondii* and act as the definitive host, shedding the protozoans in their feces, making the disease toxoplasmosis a zoonosis. Although rare, there are reports of dogs shedding the parasite. Cats are usually infected by exposure

DiseaseSnapshot • Toxoplasmosis

Causative agent	*Toxoplasma gondii*, a protist
Epidemiology	Up to 23 percent of the U.S. population has been exposed, with mostly asymptomatic cases. Global exposures are likely up to 50 percent of all people. Serious illness is almost always associated with immunocompromised patients or pregnant woman. Incubation is 5–23 days.
Transmission & prevention	Most transmission is likely foodborne from undercooked meat containing *T. gondii* cysts; cat feces exposure is another source of infection; may transmit from mother to fetus. Prevention includes careful handling and complete cooking of meat, avoiding untreated drinking water, changing cat litter often, and avoiding cat litter if pregnant or immunocompromised.
Signs & symptoms	Healthy people tend to be asymptomatic or have mild flulike symptoms. Congenital toxoplasmosis can cause miscarriage, stillbirth, or neurological problems. Immunocompromised patients may have fever, headache, confusion, seizures, and psychological symptoms.
Pathogenesis & virulence factors	*T. gondii* contains genes for several proteins that help it to create a special compartment called a parasitophorous vacuole. This vacuole inside a host cell offers protection for the protozoan from the immune system. The protozoan also has special secretory organelles called rhoptries that release proteins that disrupt cellular functions.
Diagnosis & treatment	ELISA or other serological tests detect *T. gondii* antibodies. PCR to detect *T. gondii* DNA in amniotic fluid is used to diagnose fetal infections. Pyrimethamine and sulfadiazine are routine treatments, although these drugs do not remove all cysts from the tissues.

T. gondii

to other cats' feces, or through hunting. *T. gondii* can transmit through contaminated drinking water to a number of other animals, including birds, rats, and mice. These animals do not shed *T. gondii* in feces but do contain infected cysts within their muscle tissue. When a cat consumes these infected small animals, it develops an infection and begins to shed oocysts (daughter cells inside a thick, resistant protein coat). In addition to exposure to cat feces, other important risk factors include eating raw or undercooked meat such as wild game, pork, and shellfish, because these organisms may have cysts in their tissues.

Once *T. gondii* oocysts are consumed, they may develop into a pathogenic form of the protozoan called tachyzoites. This is a motile form that can invade tissues. Usually, tachyzoites return to the cyst form in the heart, muscles, eyes, and brain.

Surveys of humans with antibodies to *T. gondii* suggest that about 9 percent of the population has been exposed to the protozoan. Other estimates suggest that exposure may be even higher, with almost 25 percent of people over 12 years old having come into contact with *T. gondii*. But most people do not develop illness; symptoms are almost always seen in immunocompromised individuals. HIV/AIDS patients account for almost 85 percent of the adult cases seen in the United States. In these cases, the disease is usually caused when cysts already in the host begin to rupture and form tachyzoites that damage brain tissue, inducing encephalitis. Symptoms of this disease stage include confusion, headache, and fever. If untreated, seizures, psychiatric symptoms, coma, and death may follow. But toxoplasmosis can be treated to prevent encephalitis, although the drugs do have some side effects, such as rash.

Another medically important form of toxoplasmosis in the United States is congenital, passed from mother to fetus. Congenital toxoplasmosis impacts approximately 500 to 5,000 of 4.2 million live births per year in the United States. If a pregnant mother has an acute infection or if the cysts in her tissue reactivate to become tachyzoites, the protozoan can cross the placenta and infect the fetus. There it invades nervous tissue and can cause convulsions, deafness, or neurological disabilities. Sometimes, a fetal toxoplasmosis infection can cause miscarriage or stillbirth.

DiseaseSnapshot • Human Transmissible Spongiform Encephalopathy

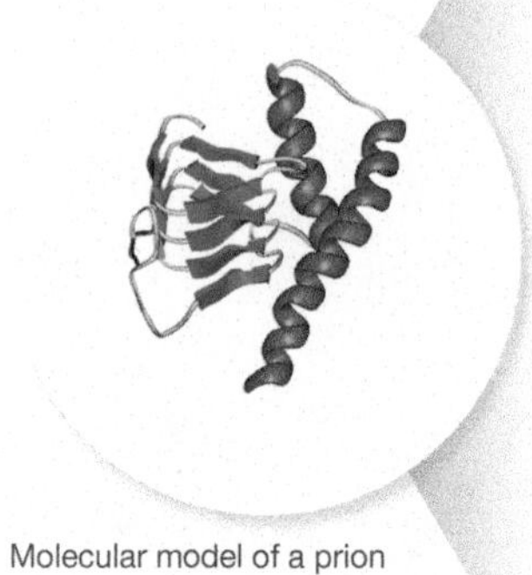

Molecular model of a prion

Causative agent	Prions
Epidemiology	For Creutzfeldt-Jakob disease (CJD), the median age of death is 68 years old, while for the variant form (vCJD), the median age of death is 28 years; about 300 cases per year in the United States. Incubation can range from 5 to 50 years.
Transmission & prevention	vCJD transmits through exposure of the nervous system to abnormal prions through medical procedures or ingesting contaminated meat. CJD is considered spontaneous, although there is a genetic component. No known preventative measures.
Signs & symptoms	Personality changes, memory loss, blurred vision, problems with speaking and swallowing, and sudden, jerky movements
Pathogenesis & virulence factors	Prions cause infection by changing the conformation of normally folded proteins on the surface of cells called PrP^C. These proteins become misfolded versions of themselves. Once patients begin to show symptoms, deterioration is inevitable and death usually occurs in less than a year for CJD and 12–14 months for vCJD.
Diagnosis & treatment	No treatments available. CJD and vCJD have no diagnostic test, although there is ongoing research in this area. An MRI (magnetic resonance image) and an EEG (electroencephalogram) of the brain can be used for preliminary diagnosis. CJD and vCJD can only be confirmed by autopsy.

Infectious proteins called prions can damage the central nervous system.

In 1974, a 55-year-old woman began showing serious neurological symptoms. Memory loss, difficulty speaking, and tremors all indicated a neurological disease that eventually led to death. An autopsy was performed at Columbia University in New York and revealed brain tissue literally full of holes, sponge-like in its appearance. Eighteen months before death, the patient had received a corneal transplant. An autopsy was performed on the donor and the same spongelike brain tissue was seen. This was then categorized as a case of **transmissible spongiform encephalopathy, or TSE**.

Although TSEs were identified as early as 1730, it wasn't until the 1960s that the cause of this strange neurological disease began to emerge. The idea was very controversial—infectious proteins called **prions** caused TSEs. This textbook is full of examples of how cellular pathogens such as bacteria, fungi, and protozoans cause disease. It's also well understood how viruses can damage the host. But the idea that exposure to a single protein can cause a debilitating neurological disease was confusing to scientists at the time. This bold hypothesis and the work to prove it earned two scientists, Daniel Carleton Gajdusek and Stanley Prusiner, Nobel prizes.

There are many different TSEs. For instance, European sheep have a long history of a disease called scrapie. This disease is named for the way that affected sheep rub their skin raw on fences. There is also bovine spongiform encephalopathy (BSE), also known as mad cow disease. In addition, in the year 1996, a human prion disease called vCJD (variant Creuztfeldt-Jakob disease [KROOTZ felt YAH kob]) was linked to eating beef contaminated with BSE prions. Hundreds of people in the United Kingdom died in this outbreak.

A concern for public health officials in the United States is chronic wasting disease (CWD). Deer and elk in certain northern areas of the United States have the species-specific prion disease CWD. Given the cases of cross-species infection through consumption of contaminated meat, there are fears that venison from these areas could cause a new outbreak. But the majority of human prion cases are due to non-variant CJD, or Creuztfeldt-Jakob disease. CJD normally occurs among the elderly and is considered spontaneous. These spontaneous cases may be caused by mutations during cell division in neuronal stem cells.

In all mammals with TSEs examined to date, the disease is tied to a normal cellular prion protein (PrP^C) that is found mostly on the surface of neurons. As with all proteins, PrP^C has a shape that is essential to its cellular function. Scientists are still working to understand the normal function of this protein. When the normal prion protein encounters an abnormally shaped version of itself, PrP^{SC} (scrapie-like prion protein), the normal prion's shape is changed and it becomes an infectious PrP^{SC}. (As discussed in Chapter 2, protein conformation is determined by the bonds that form between different amino acids.) The newly made PrP^{SC} goes on to transform more prions on adjacent neurons, causing a fatal cascade. The abnormal prions clump together and kill the affected neurons. These pockets of dead cells account for the spongelike appearance of brain tissue found in TSE autopsies. Scientists are still exploring how prions induce cytotoxicity.

Build Your Foundation

21. Who usually gets cryptococcosis and how do they usually get it?
22. Describe *Trypanosoma brucei* and how it causes disease.
23. How does *Naegleria fowleri* cause such severe damage to patients so quickly?
24. What are three of the life stages of *Toxoplasma gondii* and how do they interact with human hosts?
25. What are prions and how do they damage the nervous system?

Build Your Foundation (BYF) Quick Quiz: Visit the **Mastering Microbiology** Study Area to quiz yourself.

VISUAL SUMMARY | *Nervous System Infections*

CNS INFECTIONS

PNS INFECTIONS

Bacterial

N. meningitidis

Haemophilus influenzae
- Direct contact or respiratory droplet transmission
- Disorientation and irritability are common

Neisseria meningitidis
- Associated with close-quarters living
- Petechial pinpoint rash and endotoxin-related pathology

Streptococcus pneumoniae
- Mainly an opportunistic infection
- Toxin-induced inflammation and tissue damage helps it to cross the blood–brain barrier

Listeria monocytogenes
- Often from contaminated food
- Intracellular lifestyle

Viral and Prions

Rabies virus

Rabies Virus
- Zoonosis; bite from an infected animal
- Prognosis poor after neurological symptoms manifest

West Nile Virus
- Insect vector transmission
- Limiting mosquito population is key prevention

Prions
- Infectious versions of neuron proteins
- Cause spongiform encephalopathies such as Creuztfeldt-Jakob disease

Fungal and Protozoan

T. brucei

***Cryptococcus neoformans* and *C. gattii* (fungal)**
- Inhaled spores infect lungs and from there spread to CNS
- Much more common among immunocompromised patients

***Trypanosoma brucei* (protozoan)**
- Tsetse fly insect vector
- Changes surface antigens to evade immune system

***Naegleria fowleri* (protozoan)**
- Nasal exposure can lead to brain infection; most cases from swimming in warm lakes or ponds
- Causes massive cytolysis

***Toxoplasma gondii* (protozoan)**
- Often from undercooked meat or a zoonosis
- Usually asymptomatic but can cause cytolysis in certain tissues including those of the CNS

Comparing two specific CNS infections

Infection	Meningitis	Encephalitis
Description	Inflammation of the meninges—membranes surrounding the brain and spinal cord	Inflammation of the brain; when spinal cord is also involved it's called encephalomyelitis
Causative agents	Usually viral; less commonly from bacteria	Usually viral, with bacterial agents less common, and certain fungi and parasites causing the rarest varieties
Signs & symptoms	Viral meningitis: fever, headache, stiff neck, nausea, photophobia, and vomiting; patients tend to be alert Bacterial meningitis: same as viral, but also exhibit lethargy, body aches, chills, and sometimes a rash	Fever, headache, and often one or more of the following: disorientation, abnormal behavior, or seizures
Routine diagnostics	Lumbar puncture for CSF analysis (culturing and staining); in certain cases, computed tomography (CT scan) or magnetic resonance imaging (MRI) may be used to decide if it is safe to perform a lumbar puncture	

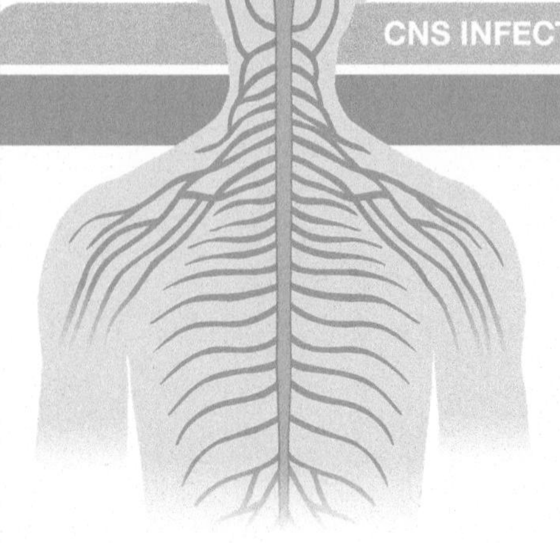

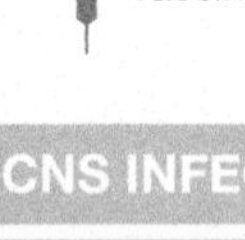

Vaccine available

CNS INFECTIONS

PNS INFECTIONS

Viral

Polio virus

Polio Virus
- Transmitted by fecal–oral route
- Flaccid paralysis in some cases

Bacterial

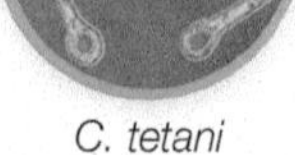

C. tetani

Mycobacterium leprae
- Direct contact or respiratory droplet transmission
- Lepromatous or tuberculoid forms; cooler temperatures of extremities served by the PNS often confine growth

Clostridium botulinum
- Often in improperly canned foods; honey is a common means of transmission in infant botulism
- Botulinum toxins prevent muscle contraction

Clostridium tetani
- Often site of entry is a puncture wound
- Tetanospasmin toxin prevents muscle relaxation

For all printed and additional Disease Snapshot tables, go to the Mastering Microbiology Study Area.

18 CHAPTER 18 OVERVIEW

18.1 Overview of Nervous System Structure and Defenses

- The brain and spinal cord make up the central nervous system (CNS); the peripheral nervous system (PNS) includes nerves that branch off the spinal cord.
- Neurons pass chemical signals to and from one another to convey information between the PNS and CNS.
- The blood–brain barrier is an important CNS defense. Three layers of meninges and the cerebrospinal fluid (CSF) also protect and nourish the CNS.
- The meninges have special structures that make it difficult for pathogens to enter the CSF and brain. Meningitis is an inflammation of the meninges, while encephalitis is an inflammation of the brain.
- A lumbar puncture, or spinal tap, allows for CSF analysis to differentiate between viral and bacterial meningitis.

18.2 Viral Nervous System Infections

- Viruses are the most common nervous system pathogens; their small size helps them pass through the blood–brain barrier.
- Nonpolio enteroviruses are the most common cause of viral meningitis.
- Polio has been eradicated in the United States, although it is still not globally eradictated. It has two effective vaccines: One is oral (OPV) and the other injected (IPV). Polio invades cells, replicates inside them, and lyses them, physically destroying certain parts of the CNS.
- Rabies is a deadly viral disease, but widespread vaccination of domesticated animals makes it rare in the United States. Rabies virus is introduced through an animal bite and travels up the PNS into the CNS to cause damage.
- Certain arboviruses (arthropod-borne viruses) such as West Nile virus cause encephalitis. West Nile virus relies on a mosquito vector and is the most common of the arboviral diseases in the United States.

18.3 Bacterial Nervous System Infections

- Bacterial meningitis or encephalitis can induce long-term disabilities such as brain damage and/or hearing loss.
- CSF samples taken from patients with bacterial meningitis typically have a high white blood cell count, a low glucose level, and a milky appearance.
- *Haemophilus influenzae* type b meningitis (Hib) used to be one of the most common forms of meningitis; it is now rare thanks to the Hib vaccine. Many people harbor *H. influenzae* among the normal throat microbiota.
- Meningococcal meningitis is more common among young adults and people who live in close quarters. There are vaccines for most of the serotypes and vaccination can be mandatory under certain conditions. *Neisseria meningitidis* grows inside phagocytes and spreads to the CSF.
- Certain *Streptococcus pneumoniae* serotypes cause pneumococcal meningitis as well as other respiratory, ear, and throat infections. *S. pneumoniae* causes meningitis when it enters and grows in the bloodstream and then invades the CNS. The immune response to the bacteria causes many of the pathogenic effects.
- *Listeria monocytogenes* is mainly a foodborne infection that causes *Listeria* meningitis. The bacterium grows inside cells, allowing it to avoid host antibodies. Pregnant women are at risk for spontaneous abortion or passing the bacteria to their infant during birth or in utero. The elderly have decreased cellular immunity and are at greater risk.
- Hansen's disease, or leprosy, is caused by *Mycobacterium leprae*. This bacterium infects the PNS. When it infects peripheral nerves, the immune system reaction can cause flattened skin lesions (tuberculoid leprosy) or disfiguring and often deadly manifestations (lepromatous leprosy). The bacteria damage peripheral nerves, causing muscle weakness and loss of sensation.
- *Clostridium botulinum* and *Clostridium tetani* are both endospore-forming bacteria that normally reside in soil and make neurotoxins when they geminate in growth-conducive environments. The neurotoxins they make enter the blood and generate widespread effects.
- *C. botulinum* secretes botulinum toxins, which cause flaccid paralysis by preventing the release of the neurotransmitter that causes muscle contractions. The infection is usually foodborne (especially from canned goods). Most botulism cases are in infants.
- Tetanus is caused by the introduction of *C. tetani* into an anaerobic environment, such as a deep wound; an effective vaccine makes this disease rare in the United States. The tetanospasmin toxin blocks muscle relaxation.

18.4 Other Infections of the Nervous System

- Fungal nervous system infections, such as cryptococcosis, are rare in people with a normal immune system. *Cryptococcus neoformans* and *C. gattii* are the leading cause of death in HIV/AIDS patients. The fungal spores are found in bird droppings and soil. Upon inhalation, the fungi may spread to the CNS.
- The protozoan parasite *Trypanosoma brucei* causes African sleeping sickness, a CNS disease. It is transmitted by the tsetse fly.
- Primary amoebic meningoencephalitis is a fatal neurological disease caused by the amoeba *Naegleria fowleri*. Infections result when water containing the amoeba enters nasal passages and travels up olfactory nerves to the brain.
- Prions are infectious proteins that cause spongifiorm encephalopathies.

THINK CLINICALLY | *Be S.M.A.R.T. About Cases*

COMPREHENSIVE CASE

The following case integrates basic principles from the chapter. Try to answer the case questions on your own. Don't forget to be S.M.A.R.T.* about your case study to help you interlink the scientific concepts you have just learned and apply your new understanding of nervous system infections to a case study.

***The five-step method is shown in detail in the Chapter 1 Comprehensive Case on cholera. See pages 31–33. Refer back to this example to help you apply a SMART approach to other critical thinking questions and cases throughout the book.**

• • •

After working at Piedmont Hospital in the ER for 40 years, LaTonia retired and finally had some time for herself. She had enjoyed her going-away party a lot, maybe a little too much. Two days later, she woke up feeling awful. She was burning up and had a nasty headache. As the day progressed, she became nauseous and started to vomit. When her granddaughter Denisha called her that night, LaTonia sounded confused and her speech was halting and slow. LaTonia also complained about muscle cramps. Denisha knew something was wrong and drove her grandmother to the emergency clinic. As the physician assistant on duty at the time you admit her and show her to a private room. After talking with LaTonia and Denisha, you discover she:

- Is up to date on the polio, meningococcal, PCV13, and tetanus vaccines
- Has not lost or gained weight recently, nor has she had a decrease in appetite over the last month
- Is taking medication for high cholesterol
- Has no family history of neurological disorders
- Has gone through menopause
- Has a slightly elevated temperature, but not a high fever
- Is not diabetic, HIV positive, or a recent organ transplant recipient
- Had attended a retirement party two days before
    - The party was in a gazebo at a park.
    - The park was near a lake where she used a paddle boat.
    - The party lasted from 11:00 am to 1:00 pm.
    - She ate a roast beef sandwich, potato salad, green bean salad, and banana pudding.

Because she has neurological symptoms, confusion and agitation, and muscle cramping, you suspect meningitis. You order a lumbar puncture to examine her CSF. The lab order includes the following tests on the CSF: an ELISA for West Nile virus, a Gram stain, culture methods, white blood cell count, and glucose level. When you return to her room after ordering these tests, you notice her eyelids are beginning to droop. You realize the tests you just ordered are not the most immediate concern.

• • •

CASE-BASED QUESTIONS

1. As the physician assistant on duty, you have narrowed down the possible causes of LaTonia's illness. What did you think was the most likely cause of her illness when you ordered the initial tests? Why?
2. Why did LaTonia's drooping eyelids change the differential diagnosis you had originally developed? What is your new working diagnosis?
3. Did you make a mistake in ordering the tests on her CSF? Support your answer.
4. Provided your newest working diagnosis is correct, what do you expect the tests you ordered will reveal?
    a. Differential white blood cell and glucose level of the CSF sample
    b. ELISA for West Nile virus
    c. Gram stain and culturing of the CSF
5. To confirm your diagnosis, what item(s) will you ask Denisha to bring to you and how will you explain the need for these items?
6. If your final diagnosis is confirmed, what public health steps, if any, must you take?

END OF CHAPTER QUESTIONS

NCLEX HESI TEAS

Think Critically and Clinically

Questions highlighted in orange are opportunities to practice NCLEX, HESI, and TEAS-style questions.

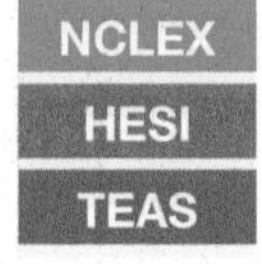

1. A patient is admitted who is complaining of headache, disorientation, and numbness in his left arm. He was bitten by a raccoon two days before. The FIRST step taken to help this patient would be
    a. retrieving and testing the raccoon's brain tissues for the rabies virus.
    b. collecting a CSF sample from the patient.
    c. performing a tissue biopsy on the patient to look for Negri bodies.
    d. administering an anti-rabies vaccine as well as anti-rabies antibodies.

2. Viral meningitis is more common than other forms of meningitis because
   a. these infectious agents are all intracellular, evading the immune system.
   b. there are more types of viruses than bacteria because they evolve more quickly.
   c. the small size of viruses makes them able to pass more easily through the blood–brain barrier.
   d. viruses are spread person to person more easily than bacteria, fungi, or protozoans.

3. A patient's lab data report has come back with a positive ELISA for pneumococcal C polysaccharide from CSF samples. The MOST appropriate next step for a health professional to recommend is to
   a. culture from the CSF sample to test for antibiotic resistance.
   b. administer tetracycline.
   c. administer cephalosporin.
   d. wait for RT-PCR confirmation before antibiotic treatment.

4. A patient under one year old is vomiting, drowsy, floppy (flaccid), and unresponsive. Both infant botulism and infant meningitis caused by other bacteria are suspected. Which piece of information would best help narrow down the possible diagnosis?
   a. The child has spent almost no time outside the home.
   b. The child is being breast-fed.
   c. The child was born one month ago.
   d. The child has a very high fever.

5. CSF glucose levels are often assessed to differentiate between bacterial and viral meningitis. This is because
   a. bacteria cells use glucose and lower the overall concentration in the CSF.
   b. viral meningitis causes inflammation that blocks glucose transport into the CSF.
   c. viral infection of the meninges causes cell lysis, releasing glucose into the CSF.
   d. None of the above.

6. During a lumbar puncture procedure, the nurse observed that the patient's CSF pressure was high. However, when the analysis of the CSF sample comes back, the glucose levels are normal. The nurse would most likely suspect:
   a. arboviral meningitis.
   b. *Neisseria meningitidis*.
   c. *Listeria monocytogenes*.
   d. poliovirus.

7. A vaccine against *Listeria monocytogenes* is a challenge to develop because
   a. very few people get *Listeria* meningitis.
   b. the antibody-mediated immune response is not effective against *L. monocytogenes*.
   c. better prevention approaches exist, such as food preparation precautions.
   d. vaccinations are not used for foodborne illness.

8. Tetanospasmin and botulinum toxins affect muscles by
   a. blocking the action potential traveling from the CNS to the PNS.
   b. allowing the bacteria to enter neurons.
   c. blocking the relaxation or contraction of the muscles.
   d. damaging the axon itself.

9. A patient who has just immigrated to the United States after serving in the Iraqi army is displaying a bruiselike rash, headache, and sensitivity to light. The doctor will immediately recommend administration of
   a. ceftriaxone.
   b. penicillin.
   c. gentamycin.
   d. rifampicin.

10. The protists *Trypanosoma brucei*, *Naegleria fowleri*, and *Toxoplasma gondii*
    a. invade host cells to cause cellular damage.
    b. are only infective in the flagellated form.
    c. have a nucleus and are eukaryotic.
    d. have all of the above characteristics.

11. An HIV/AIDS patient is presenting with flulike symptoms, headache, fever, and disorientation. The MOST important piece of information a nurse must obtain for the patient is his or her
    a. travel history, in and outside the United States.
    b. diet for the last 3 days.
    c. exposure to lakes or streams.
    d. exposure to bats.

12. Which of the following symptoms would be most informative in terms of ruling out a prion disease as a differential diagnosis?
    a. nuchal rigidity
    b. hallucinations
    c. fatigue
    d. headache

CRITICAL THINKING QUESTIONS

1. Why is CSF useful for diagnosing encephalitis despite the point that this fluid is extracted from the subarachnoid space and not the brain? What clinical signs and symptoms would have to be present in order to determine that the patient had encephalitis over meningitis?

2. What pathogenic features and physiological scenarios could help a bacterium cross the blood–brain barrier?

3. Toxoplasmosis sometimes causes neurological symptoms in rats, making them less afraid of cats. Why would this trait be advantageous to *T. gondii*?

4. In theory, could the botulinum toxin be used to treat tetanus? Explain your reasoning.

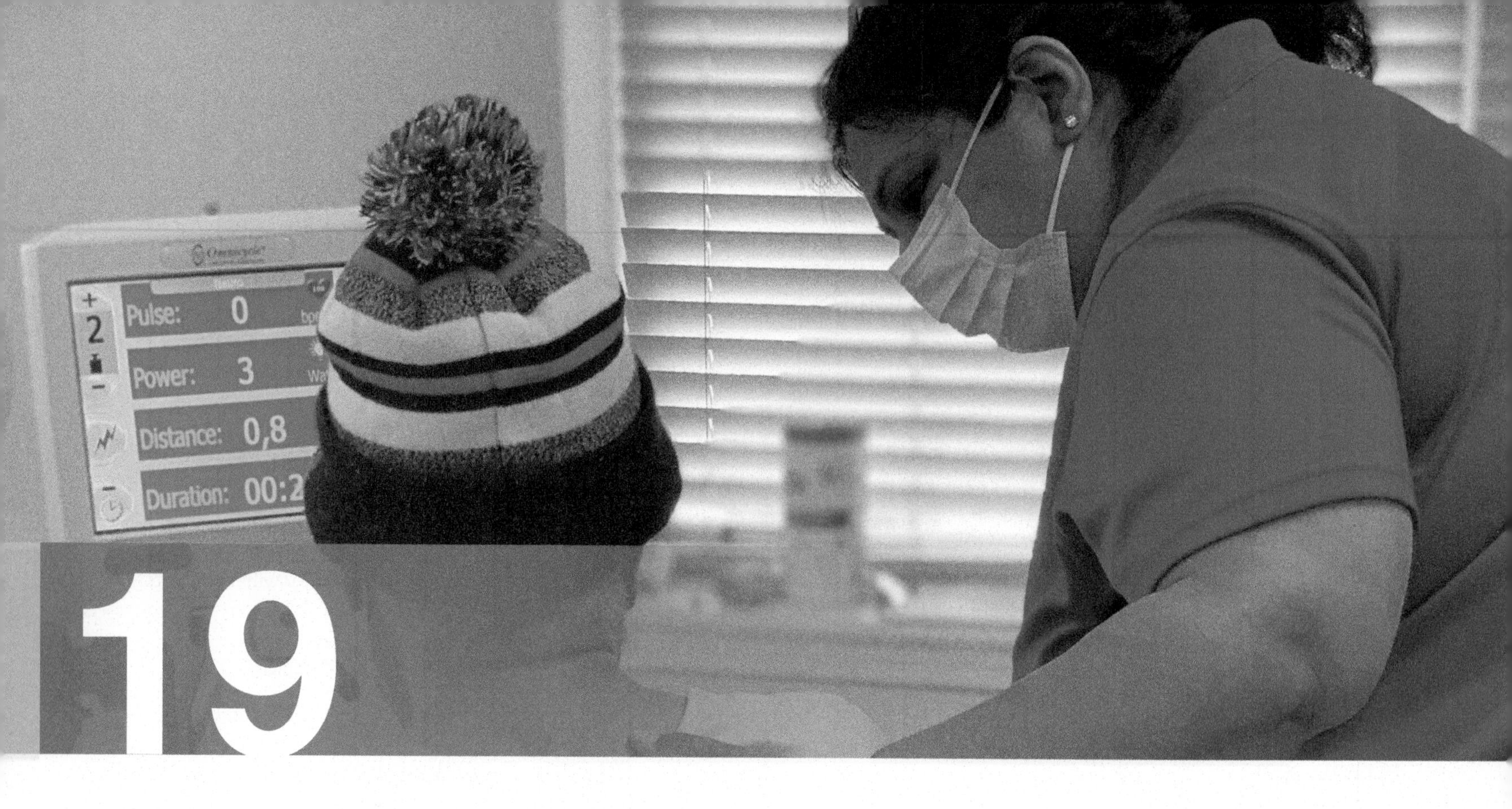

19 Digestive System Infections

What Will We Explore? The human digestive system includes accessory organs and 30 feet (9 meters) of tubing that stretches from the mouth to the anus. Every chewed fingernail, sip of poorly sanitized water, or improperly prepared bite of food may introduce viruses, bacteria, or parasites to the digestive system. While many of these infectious agents pose little risk to human health, a number do cause illness. In this chapter, we'll explore the various microbes that wreak havoc on the digestive system and review how these agents produce disease.

Why Is It Important? The digestive system is home to the most abundant and diverse normal microbiota of the entire body. Our gut microbiota not only make certain vitamins for us, but they also impact a number of physiological processes, from our mental health to our immune system functions.

Pathogenic microbes also remain a troublesome malady throughout the world. Despite public health measures that limit food and water contamination, the U.S. Centers for Disease Control and Prevention (CDC) estimates at least 200 million people in the United States suffer from some form of gastrointestinal (GI) infection annually. Because patients in developed countries have relatively easy access to health care, GI infections are rarely fatal. However, in developing countries, the burden of GI infection is even higher, due to less rigorous public health measures, and fatalities are more common due to limited healthcare access.

Clinical CASE NCLEX HESI TEAS

The Case of the Hurricane Health Hazard

Visit the **Mastering Microbiology** Study Area to watch the case and find out how digestive system infections can explain this medical mystery.

Vaani Bhatia
PT, Director of Rehab; Decatur, GA

19.1 DIGESTIVE SYSTEM ANATOMY AND DEFENSES

The digestive system includes the GI tract and accessory organs.

Learning Outcomes

After reading this section, you should be able to:

19.1 Name the organs and describe the segments of the gastrointestinal (GI) tract.

19.2 Describe the role of mucosa-associated lymphoid tissues in relation to the digestive system.

19.3 Identify where normal microbiota exist in the digestive system and describe the factors that limit microbial growth in certain parts of the GI tract.

19.4 State some of the roles of our normal gut microbiota in health and disease.

If you're eating a snack while reading this, you probably give little thought to your intricate digestive system working to acquire water and nutrients from the food you just swallowed. The digestive system consists of the **gastrointestinal (GI) tract** (or alimentary canal) and accessory organs. The GI tract is what comes into direct contact with our food. It extends from the mouth to the anus and is divided into two sections. The **upper GI tract** includes the mouth, pharynx, esophagus, and stomach. The **lower GI tract** consists of the small and large intestines, rectum, and anus. The digestive system's **accessory organs**, which include the salivary glands, liver, gallbladder, and pancreas, do not directly contact our food, but have essential roles in digestion. Additionally, the GI tract is associated with lymphatic tissues such as tonsils, appendix, and Peyer's patches. The placement of immune system tissues along the GI tract is important because it's a major entry portal for pathogens (FIG. 19.1).

Upper GI Tract with Salivary Glands

Food enters the upper GI tract at the mouth. The salivary glands, of which the parotid glands are the largest, secrete saliva that is rich in enzymes that start lipid and carbohydrate digestion. The chewed food is swallowed and makes its way to the stomach, where it mixes with gastric juices, and protein digestion begins. The harshly acidic stomach environment limits microbial growth. The stomach itself is an expandable structure that is lined with alkaline mucus, which protects the stomach tissue from the acidic gastric juices. The food and gastric juices mix to form **chyme**, which then enters the small intestine.

Lower GI Tract with Liver, Pancreas, and Gallbladder

As chyme enters the small intestine, the gallbladder secretes bile to help aid in fat digestion. Bile is a greenish-yellow fluid made by the liver and then concentrated and stored in the gallbladder. The salts in bile inhibit the growth of many bacteria, especially Gram-positive bacteria.

The pancreas also secretes substances to aid in digestion. The liver, which stores certain nutrients and metabolizes most drugs and toxins, directly receives nutrient-rich blood from vessels of the small intestine. Additionally, the liver removes substances like bilirubin, a byproduct of red blood cell breakdown. Finally, whatever is not absorbed in the small intestine makes its way to the large intestine, where most of our water and water-soluble vitamins are absorbed. Any undigested matter remains in the rectum until excreted through the anus as feces.

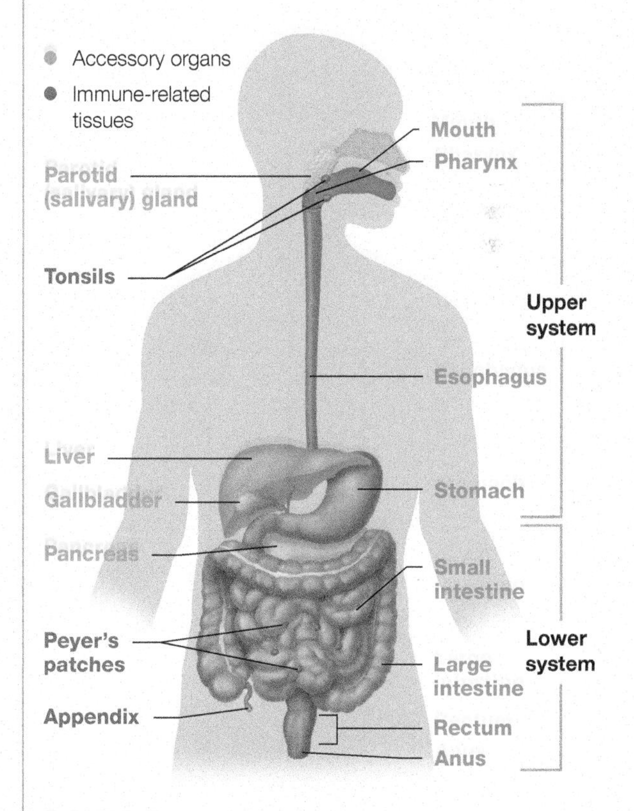

FIGURE 19.1 The digestive system

Critical Thinking *What is the purpose of lymphatic tissues such as tonsils and Peyer's patches in relation to the digestive tract?*

A variety of GI tract features limit digestive system infections.

A number of mechanical, chemical, and physical barriers limit what microbes may colonize the GI tract. Factors like lysozyme in saliva, the acidity of gastric juices, and the salts in bile kill many bacteria passing through the system (Chapter 11 reviewed mechanical and chemical barriers as first-line defenses). Additionally, mucus lining the GI tract serves as a physical barrier that prevents many microbes from attaching to epithelial cells.

Specific cells of the digestive system also aid in immune responses in the event that a pathogen takes advantage of this major portal of entry. As discussed in Chapter 11, lymphatic tissue is found all along the GI tract. Tonsils at the back of the throat, the appendix, and Peyer's patches of the small intestine are all examples of **mucosa-associated lymphoid tissue (MALT)** found along the GI tract. Cells in these lymphatic tissues "sample" the environment by phagocytosis, and present their findings to lymphocytes that may or may not need to initiate an immune response. If a response is warranted, then inflammation is an early defense, followed by activation of the adaptive immune response. This is why the doctor feels your neck and under your jaw if you complain of a sore throat. Enlarged and inflamed lymph nodes in the neck region signal a potential infection detected by tonsils of the MALT system.

Digestive System Microbiome

The digestive tract is home to the most diverse and densely populated collection of microbes in the body. According to studies performed as a part of the Human Microbiome Project, there are thousands of different species that call our digestive tract home. Examples of the most common residents in the mouth and on the tongue and gums are the *Streptococcus* species. The teeth are mainly colonized by a variety of *Streptococcus, Neisseria, Fusobacterium,* and *Actinomyces* species. (Think about that before you kiss someone!) Meanwhile, stool is rich in *Bacteroides* species. Fungi, such as *Candida,* and a few protozoa also live in the digestive tract, although bacterial populations vastly outnumber them (FIG. 19.2).

These normal gut microbiota have essential roles in both our health and in disease. Similar to a crowded parking lot, resident bacteria can prevent pathogenic bacteria from settling into areas of the intestine by simply taking up surface space that would otherwise be available for potential colonization. They also compete for surrounding nutrients and excrete antimicrobial products to kill off contending newcomers. In addition to warding off pathogens, normal gut microbiota assist in digesting food and provide nutrients for the body. An important source of vitamin K in the intestine comes from none other than *Escherichia coli.*

In addition to making vitamins and competing with pathogens, research reveals there is a complex association between our gut microbiota and our health. Our GI microbiota may impact metabolism and obesity; affect the development of depression and diseases such as diabetes; and train and modulate immune responses. (See the Bench to Bedside feature in Chapter 8 for more details.)

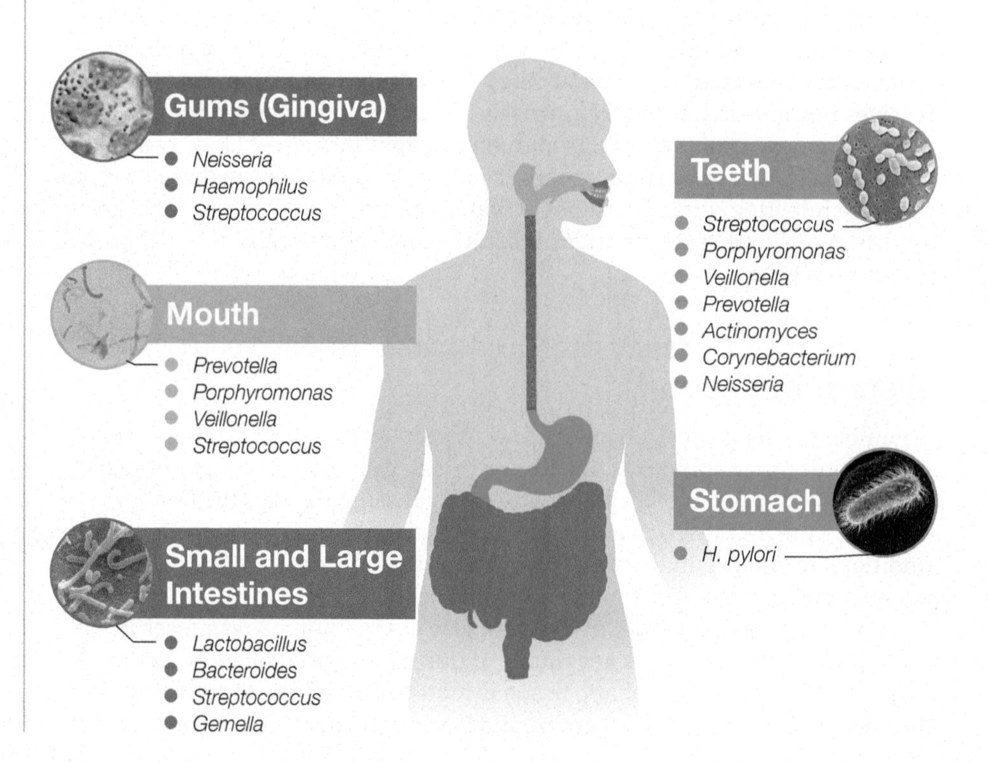

FIGURE 19.2 Microbiome of the digestive system Microbes reside on mucous membranes throughout the digestive tract. Different locations along the GI tract harbor different microbial genera.

Build Your Foundation

1. What anatomical structures make up the upper and lower GI tract and what are the accessory organs of the digestive system?
2. Using Figure 19.1, name and locate three different mucosa-associated lymphoid tissues of the digestive tract.
3. Where do resident microbes exist along the digestive tract?
4. Name several roles of our normal gut microbiota.

Build Your Foundation (BYF) Quick Quiz: Visit the **Mastering Microbiology** Study Area to quiz yourself.

19.2 GASTROINTESTINAL INFECTION SYMPTOMS AND DIAGNOSTIC TOOLS

Digestive system infections may result in dysentery, gastroenteritis, or other GI symptoms.

A huge variety of pathogens invade the digestive tract, but the symptoms of the infections they cause are very similar. They include abdominal pain and loose stools that may contain blood or mucus. In addition, there may be fever and/or vomiting. **Diarrhea** is described as a loose or watery stool. When a pathogen infects the intestinal lining, an inflammatory response called **enteritis** develops. In **gastritis** the stomach is inflamed. Inflammation of both the stomach and intestines is called **gastroenteritis**. Abdominal pain also commonly develops—which isn't surprising, since pain is a cardinal sign of inflammation.

Destruction of infected intestinal cells may also lead to excess fluid in the stool and bleeding. **Dysentery** may develop in severe cases of gastroenteritis; it is similar to diarrhea, but painful, and typically characterized by blood and mucus in the stool. Both diarrhea and dysentery can lead to excessive water and electrolyte loss (**dehydration**). Severe dehydration can cause potentially deadly **hypovolemic shock**, as organs begin to shut down due to lowered blood pressure.

Most intestinal pathogens that cause diarrhea or dysentery transmit via the fecal–oral route—meaning pathogens found in feces are ingested. The most common means of fecal–oral transmission is contaminated food or water. All of us are exposed to some degree of fecal contamination in our daily lives. Hand washing, sewer management, water sanitation practices, and food processing/handling regulations all help limit fecal–oral transmission.

The CDC estimates about one in six Americans will suffer from a foodborne illness each year. In developing nations, poor water sanitation and limited food-handling regulations make these fecal–oral-transmitted illnesses even more prevalent (FIG. 19.3). Worldwide, diarrhea and dysentery kill about 2 million people each year—with the majority of deaths occurring in children under five years old who live in developing nations.

Learning Outcomes

After reading this section, you should be able to:

19.5 Describe the symptoms and related clinical terms associated with GI tract infections.

19.6 Explain some of the basic tools used to diagnose GI tract infections.

CLINICAL VOCABULARY

Diarrhea frequent passing of loose or watery stool

Enteritis inflammation of the intestines

Gastritis inflammation of the stomach

Gastroenteritis inflammation of the stomach and intestines

Dysentery diarrhea accompanied by pain, blood, and/or mucus

Dehydration excessive loss of body fluid; may develop due to severe diarrhea and/or vomiting

Hypovolemic shock low blood volume due to loss of blood or severe dehydration; can lead to organ failure

FIGURE 19.3 Sanitation to reduce intestinal disease In parts of Sierra Leone, sewage and animal waste contaminate drinking water and food. Up to 69 percent of the residents do not have access to toilets. When public toilets are available, the cost of using one is equivalent to paying for a meal.

Diagnostic Tools

When a patient presents with intestinal symptoms, fecal samples are usually collected in an attempt to identify the causative agent. From the clinical microbiology lab perspective, a diverse collection of selective and differential culture media, such as eosin methylene blue and MacConkey agar, are used to detect GI bacterial pathogens[1]. *E. coli* O157 strains can be cultured using sorbitol-MacConkey agar (SMAC). Often SMAC agar is supplemented with an antibiotic called cefixime and an oxide mineral called tellurite. These additives allow for the

[1] The following review article is recommended for more insight into culture techniques for diagnosing bacterial GI infections: Humphries, R. M., & Linscott, A. J. (2015). Laboratory diagnosis of bacterial gastroenteritis. *Clinical microbiology reviews*, 28(1), 3–31.

CLINICAL VOCABULARY

Upper endoscopy insertion of a fiber-optic camera through the mouth to view the esophagus, stomach, and small intestine

Lower endoscopy an endoscope is inserted via the anus to view the large intestine (colon)

specific pathogenic *E. coli* strain to grow and be identified while preventing the growth of commensal *E. coli* and other harmless GI bacteria that are commonly present in fecal samples. Bacterial culture is the most common tool for pathogen identification, but specialized molecular tests that detect antigens on a pathogen or characterize pathogen genes are also occasionally used. These molecular techniques tend to be used to detect *C. difficile*, Shiga toxin-producing *E. coli*, and certain *Campylobacter* species. Furthermore, microscopic examination of stool samples may identify parasites or their eggs.

Tissue damage, such as ulcers or inflammation, can be assessed using an **upper GI endoscopy**, but this diagnostic tool won't reveal the precise microbe responsible. During this procedure an endoscope, a long flexible tube with a light and lens attached to the end, is inserted in the mouth and used to view the esophagus, stomach, and small intestine. A **lower endoscopy** (also referred to as a colonoscopy) involves a colonoscope, which resembles an endoscope. The colonoscope is inserted into the anus and is used to view the rectum and the large intestine (colon). Large patches of mucous membranes sloughing off can provide immediate diagnosis for diseases such as *Clostridioides difficile*-associated dysentery, but again, this technique can't confirm the specific microbial agent responsible for the pathology.

BYF QUICK QUIZ

Build Your Foundation (BYF) Quick Quiz: Visit the **Mastering Microbiology** Study Area to quiz yourself.

Build Your Foundation

5. Compare and contrast diarrhea and dysentery, and define gastritis, enteritis, and gastroenteritis.
6. What approaches exist to diagnose GI tract infections and assess tissue damage?

19.3 VIRAL DIGESTIVE SYSTEM INFECTIONS

Learning Outcomes

After reading this section, you should be able to:

19.7 Describe the features of mumps virus infections.

19.8 Compare and contrast rotavirus and norovirus.

19.9 Distinguish the features of hepatitis A, B, C, D, and E.

19.10 Compare and contrast acute and chronic hepatitis infections.

Mumps is a viral infection of the salivary glands.

Viral digestive tract infections can range in symptoms from a mildly upset stomach to severe diarrhea and dehydration. Even salivary glands can be infected, which is what occurs in mumps. Before 1967, mumps was a common childhood illness in the United States. Today, it's extremely rare in developed countries thanks to the measles, mumps, and rubella (MMR) vaccine. Usually infections only arise in people who are not immunized and travel outside the United States to areas where mumps is endemic.

Mumps is an RNA virus that spreads among humans through infected saliva. When the mumps virus enters the nose or mouth, it quickly replicates in the upper respiratory tract. Thereafter, the virus enters the bloodstream to infect salivary glands, particularly the parotid glands. The incubation period of two to three weeks is often followed by dry mouth, fever, headache, and fatigue. Subsequent swelling under the jaw is due to inflammation of the parotid gland, termed *parotitis* (FIG. 19.4). Because the virus is released in saliva almost a week before symptoms develop, preventing spread to others can be difficult.

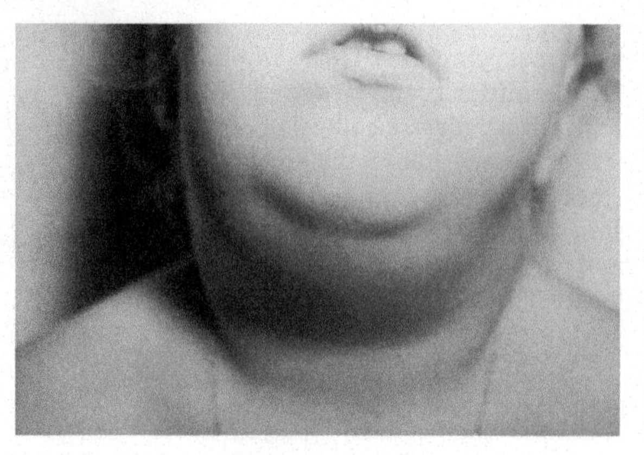

FIGURE 19.4 Mumps Swelling under the chin and around the jaw is typical of a mumps infection.

Critical Thinking *A mumps infection may be confirmed by the presence of IgM in the patient's blood. Occasionally blood is assessed for the IgG antibodies to the virus, but this diagnostic is less useful for detecting early stages of disease. Why? (Hint: Refer back to Chapter 12.)*

DiseaseSnapshot • Mumps

Causative agent	Mumps virus: enveloped, single-stranded RNA genome, *Paramyxoviridae* family
Epidemiology	Endemic in areas with low MMR vaccination rates such as Africa, southern Asia, and the Middle East. Incubation period is 2–3 weeks.
Transmission & prevention	Exposure to respiratory droplets or direct contact with infected saliva. Two doses of MMR vaccine; avoid close contact with patient and do not share utensils and cups.
Signs & symptoms	Swelling of the parotid salivary glands, pain under the jaw, dry mouth, fever, headache, fatigue
Pathogenesis & virulence factors	Targets the parotid salivary glands; virus utilizes a V protein to block interferon and interleukin signaling within the host cell
Complications	Meningitis and orchitis (inflammation of the testes) that can lead to sterility
Diagnosis & treatment	Usually reverse transcription polymerase chain reaction is performed on saliva samples; blood may also be assessed to detect IgM antibodies to the mumps virus. Nonsteroidal anti-inflammatory drugs such as ibuprofen manage fever and parotid gland swelling.

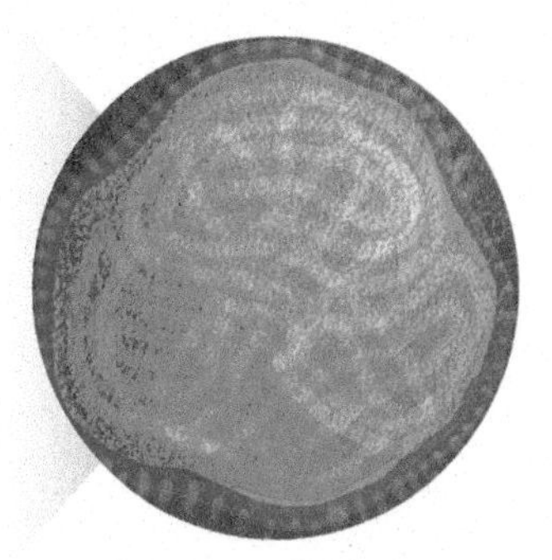
Mumps virus virion

Many viruses cause gastroenteritis.

Most people have experienced what they call "the stomach flu." In reality, the symptoms of diarrhea and nausea have nothing to do with the influenza virus, but rather are caused by a number of other viruses that target the intestines and stomach. The immune system tends to mount a rapid and effective response to GI-tract viral infections. Therefore, these illnesses are usually *acute* rather than chronic—meaning symptoms have sudden onset but quickly resolve (usually within a few days) rather than being sustained for long periods in the host. Rotavirus and norovirus are the most common viruses that target the intestines to cause acute illness.

CLINICAL VOCABULARY

Acute rapid onset and progression of disease

Rotavirus

Worldwide, **rotavirus** is the leading cause of enteritis in children under five years old. This RNA virus also infects adults, but symptoms are usually mild and may go unnoticed. Most cases are acquired through fecal–oral transmission, in which contaminated food or water is ingested. However, contaminated inanimate objects, or *fomites*, can also transmit the virus—for example, a baby's pacifier. Infected patients may have trace amounts of feces on their hands after going to the bathroom. If they don't wash their hands, they can transfer the fecal material to the food they prepare or to common surfaces. After touching a contaminated surface, such as a doorknob or countertop, people who do not wash their hands before eating may ingest the virus. The main concern with rotavirus infection is the possibility of dehydration. TABLE 19.1 compares mild and severe dehydration signs and symptoms.

Vaccinations combined with maintaining good hygiene practices, such as washing hands before food preparation and eating, and access to sanitized water limit transmission. The rapid identification of rotavirus antigens in stool samples is a conclusive diagnostic test.

TABLE 19.1 Dehydration Signs and Symptoms

Type	Signs and Symptoms
Mild Dehydration	Dry mouth and skin Decreased urine; no wet diaper for 3 hours Little or no tears when crying Headache
Severe Dehydration	Extreme thirst Little urination, darker urine Decreased skin turgor Sunken eyes, sunken fontanels on babies Low blood pressure that may lead to hypovolemic shock and organ failure

Norovirus

Norovirus (or Norwalk virus) is the leading cause of acute viral gastroenteritis in the United States. Over half of all outbreaks occur in long-term-care facilities. The RNA virus is extremely contagious, as only 20 particles are required for infection. Feces and vomit serve to spread the virus via a fecal–oral route. Symptoms develop within 12 to 48 hours of ingesting the virus. If you have ever experienced the symptoms of norovirus, you know that this infection is usually not a quick 24-hour "bug"; it can last for three days or more. Acute diarrhea

DiseaseSnapshot • Rotaviral Gastroenteritis

Rotavirus

Causative agent	Rotavirus: nonenveloped, double-stranded RNA genome, *Reoviridae* family
Epidemiology	Endemic worldwide. Due to inadequate sanitation and lack of health services, poor countries have more cases and higher mortality rates. Incubation period is 48 hours.
Transmission & prevention	Fecal–oral (primarily through contaminated food or water). Vaccinations include RotaTeq and Rotarix. Proper hand washing and water sanitation practices also aid prevention.
Signs & symptoms	Children tend to experience fever and vomiting followed by watery diarrhea that lasts 3–8 days. Adults tend to have less obvious signs and symptoms.
Pathogenesis & virulence factors	The viral capsid consists of three layers, allowing the virus to survive on surfaces and pass through the stomach unharmed. Once ingested, the virus invades and damages the tight junctions between intestinal cells in a manner that causes water and electrolyte loss.
Diagnosis & treatment	Detect viral RNA in stool or vomit samples using reverse transcription polymerase chain reaction (RT-PCR), a specialized form of PCR. Supportive therapy (particularly hydration); intravenous fluid administration is used to treat severe dehydration.

lasting 36 to 72 hours and projectile vomiting can be alarming symptoms. Most healthy individuals fully recover within a few days without medical intervention. However, patients that become dehydrated, have blood in their stool, and/or experience excessive vomiting that makes oral rehydration impossible need medical attention.

There are many types of norovirus, and new strains arise every few years. Thus, suffering through a bout of norovirus will usually not provide immunity to the next norovirus. The illness induces profuse vomiting and diarrhea that easily contaminate surfaces and make it hard to limit disease transmission (FIG. 19.5). Definitive diagnosis requires detecting norovirus RNA in patient stool samples.

FIGURE 19.5 **Simulation of norovirus-induced vomiting** The Health and Safety Lab in Buxton, U.K. developed a vomiting simulation to help safety officials understand how far norovirus spreads through vomiting. Particles were detected up to 3 meters (almost 10 feet) from "Larry" the simulator. Information gathered from this simulation can help develop protocols for effective decontamination during norovirus outbreaks.

Source: The Health and Safety Lab in Buxton, U.K.

Hepatitis is a liver infection most commonly caused by three unrelated viruses.

Hepatitis is a general term meaning inflammation of the liver. There are infectious and noninfectious sources of hepatitis. Here we focus on the primary cause of infectious hepatitis: hepatitis A (*Picornaviridae* family), hepatitis B (*Hepadnaviridae* family), and hepatitis C viruses (*Flaviviridae* family). Although they all are all hepatotropic (target the liver), they are *not* genetically or structurally related viruses.

DiseaseSnapshot • Norovirus Gastroenteritis

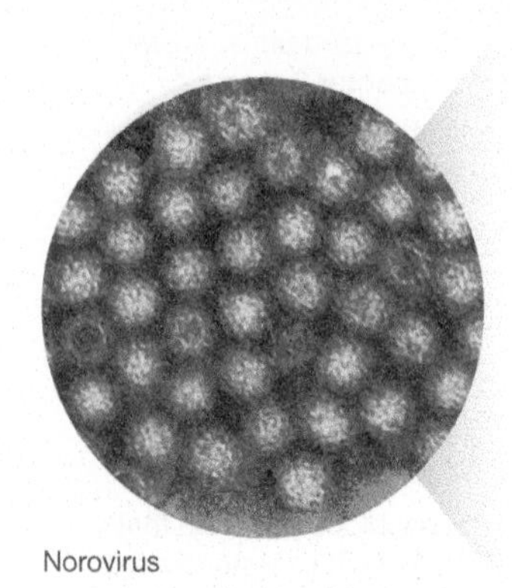

Norovirus

Causative agent	Norovirus (formerly called Norwalk-like virus): Nonenveloped, single-stranded RNA genome, *Caliciviridae* family
Epidemiology	Worldwide with periodic outbreaks as new strains emerge. Incubation period is 12–48 hours.
Transmission & prevention	Fecal–oral transmission (usually on contaminated surfaces). No vaccines currently available, so prevention involves decontamination of surfaces with chlorine bleach solution and hand washing.
Signs & symptoms	Quick onset of profuse vomiting and diarrhea lasting 36–72 hours. Low-grade fever, and malaise. Young, elderly, and immunocompromised people are at risk for moderate to severe dehydration.
Pathogenesis & virulence factors	As few as 20 viral particles are needed for infection. Little known about how the virus infects, but studies show virus-infected cells of microvilli undergo cell death.
Diagnosis & treatment	Reverse transcription polymerase chain reaction detects RNA in stool or vomit samples. Treated with oral rehydration therapies or, in cases of moderate to severe dehydration, intravenous fluids.

Hepatitis A: Foodborne and Waterborne Hepatitis

Hepatitis A virus (HAV) is a nonenveloped, single-stranded RNA virus. HAV infection rates decreased dramatically in the United States since a vaccine was introduced in 1995. The virus is typically spread by the fecal–oral route, so infected restaurant workers who fail to wash up after using the bathroom may contribute to an outbreak. Unvaccinated individuals who travel to countries where HAV is common are at the greatest risk for infection. Most children under age 6 will have an asymptomatic infection. Patients older than age 6 have about a 70 percent chance of developing signs and symptoms such as a low-grade fever, nausea, vomiting, joint pain, abdominal pain, and fatigue. Diagnostic symptoms include clay-colored feces and jaundice (FIG. 19.6), which presents as a yellowing of the skin and of the **sclera** (whites of the eyes).

Once ingested, HAV multiplies within intestinal cells. From there the virions enter the bloodstream, a situation called **viremia**, and eventually they infect liver cells (hepatocytes). As hepatocytes become infected, inflammation develops and impacts normal liver functions, including limiting the clearing of bilirubin from the blood. Bilirubin, a yellow-tinted substance, is a natural byproduct of red blood cell breakdown—a normal physiological process that occurs regularly, not just during a hepatitis infection. As bilirubin accumulates in the bloodstream it causes a yellowing of the skin and sclera. When liver function is restored and bilirubin is cleared, this yellowing resolves.

The virus is fairly stable and can remain infectious for months outside a host's body; however, it's readily inactivated by high temperature. Another factor contributing to HAV spread is that viral particles are most concentrated in the feces two weeks before symptoms develop, allowing for someone to unknowingly spread the infection. Severe manifestations are rare, and usually HAV infections are self-limiting and resolve within a few weeks without drug therapies. For those with a preexisting liver condition, a hepatitis A infection can be severe and even life threatening.

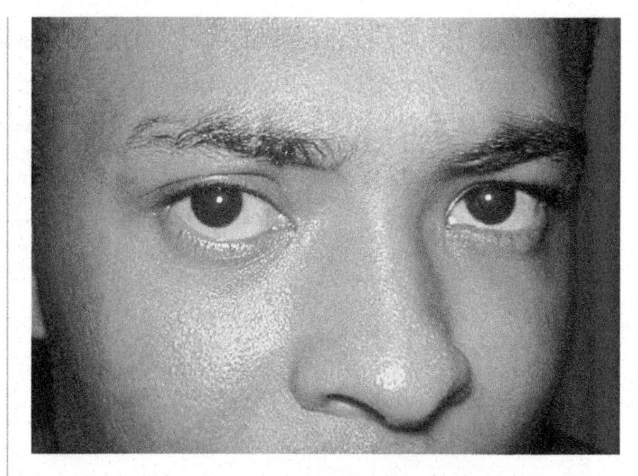

FIGURE 19.6 Hepatitis-associated jaundice This hepatitis A patient shows the typical symptom of jaundice. Note the yellowing of the eyes due to bilirubin buildup in the blood.

Critical Thinking *Explain what causes bilirubin to build up in the bloodstream during a hepatitis infection.*

Hepatitis B: Acute and Chronic Infections

Hepatitis B virus (HBV) is a double-stranded enveloped DNA virus. The CDC records that, since 1991, HBV vaccinations have dramatically reduced infections by about 82 percent. However, as of 2016, chronic HBV still affects an estimated 800,000 to 2.2 million people living in the United States. Transmission usually occurs through direct contact with bodily fluids or open sores from an infected individual—sexual contact with an infected partner, sharing HBV-contaminated needles, razors, or even sharing toothbrushes can lead to infection. Accidental **percutaneous** (through the skin) exposures due to needle sticks or cuts from HBV-contaminated sharps are responsible for the higher incidence among healthcare workers. While HBV does not cross the placenta, vertical transmission from an infected mother to her baby can still occur during delivery. If the mother is infected, then within 12 hours of birth, the baby is given the HBV vaccine and an antibody preparation called HBIG (hepatitis B immune globulin) that reduces the chance of the baby becoming infected. Although HBV is found in breast milk of infected mothers, breast-feeding has *not* been shown to transmit the virus (this was true even before immune prophylaxis options such as HBIG and vaccines were available).

Viral incubation takes an average of 120 days before symptoms similar to those seen in HAV develop; however, up to about 50 percent of cases are asymptomatic. For most patients, the virus causes an acute infection and clears within weeks. However, in up to 10 percent of patients age 6 and up the virus is not cleared and it establishes a chronic infection. Children under age 5 are far more likely to become chronically infected than adults. Over time, with intermittent inflammation, **cirrhosis** (scarring of the liver) occurs and eventually leads to liver failure. HBV also can induce hepatocellular carcinoma, a form of liver cancer (FIG. 19.7).

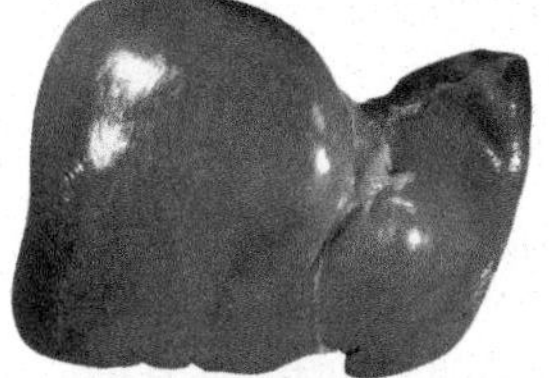

FIGURE 19.7 HBV and liver cirrhosis Inflammation in chronic HBV infection promotes liver damage and cirrhosis.

Critical Thinking *What pathogenic factors expressed in acute HBV infections limit progression to cirrhosis?*

DiseaseSnapshot • Hepatitis A, B, and C

Hepatitis B virus

Disease	Hepatitis A	Hepatitis B	Hepatitis C
Causative agent	Hepatitis A virus: enveloped, single-stranded RNA genome, *Picornaviridae* family	Hepatitis B virus: enveloped, double-stranded DNA genome, *Hepadnaviridae* family	Hepatitis C virus: enveloped, single-stranded RNA genome, *Flaviviridae* family
Epidemiology	Thanks to vaccine, less common in U.S. now; incubation is 15–50 days (average 28 days)	Decreasing U.S. incidence due to vaccine; 45–160-day incubation (average 120 days)	Most common chronic bloodborne infection in the U.S.; 14–180-day incubation (average 45 days)
Transmission & prevention	Mainly fecal–oral, with contact transmission rarely. Prevented with vaccine and proper hygiene/hand washing.	Bloodborne, sexual, and vertical (mother to child) transmission. Prevented with vaccine as well as condom usage and not sharing needles. If pregnant woman is infected, administer vaccine to baby within 12 hours of birth.	Bloodborne, sexual, and vertical (mother to child) transmission. No vaccine, so prevention involves condom use and not sharing needles.
Signs & symptoms	All types of hepatitis may generate fever, loss of appetite, joint pain, fatigue, abdominal pain, nausea, clay-colored bowel movements, and jaundice		
Pathogenesis & virulence factors	Inhibits host interferon response; virus cloaks itself with host cell membrane to avoid antibody detection as it circulates in the blood; virus can remain infectious for months outside the host's body	Protein secreted from infected liver cells helps the virus evade immune detection in acute infections. In chronic infections, T cytotoxic cells attack infected hepatocytes, causing cirrhosis. Virus survives outside the body for up to 7 days.	Antigen switching through constant viral mutation helps the virus evade immune detection; initial infection is usually asymptomatic
Chronic infections	None	90% of those under age 1; up to 50% in ages 1–5, and up to 10% in ages 6 and older	75–85% of patients
Diagnosis & treatment	Blood tests detect HAV antibodies. Treatment is rest, proper nutrition, and fluids. Antibodies given within 2 weeks of primary exposure can sometimes limit infection.	Blood tests detect HBV viral DNA, surface antigens, or antibodies. Treatment is rest, proper nutrition, and fluids. For chronic cases, antivirals can reduce viral load and help to clear infections.	Blood tests detect HCV antibodies and surface antigens. Most cases are now curable with antivirals such as Harvoni (2014) and Epclusa (2016).

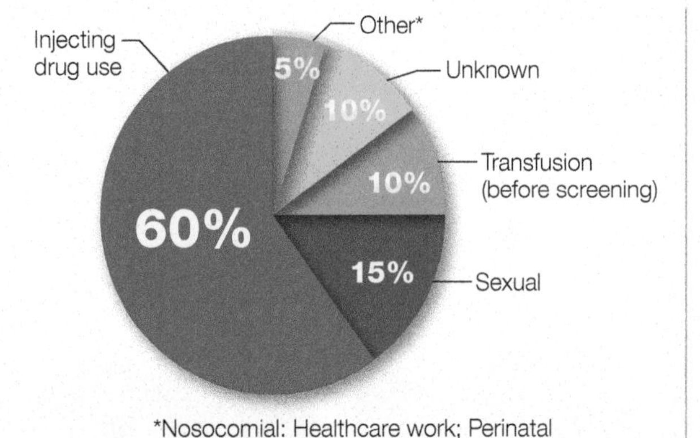

FIGURE 19.8 Sources of hepatitis C infection According to the CDC's 2014 studies, injected drug use, which often involves shared needles, is the most common source of HCV infection. The cases from transfusions were acquired before blood was routinely screened for HCV.

Critical Thinking *Propose some possible explanations for why breast-feeding has not been shown to be a means of HCV transmission despite the presence of this virus in breast milk from infected mothers.*

Hepatitis C: The Silent Killer

Hepatitis C virus (HCV) is enveloped, with a single-stranded RNA genome. As of 2020, the CDC estimates that about 2.4 million Americans are infected with HCV. However, the number of infections is difficult to accurately estimate because most people don't even know they have the virus until the later stages of a chronic infection, when liver damage develops.

The virus wasn't characterized until 1989, and it was not until sensitive HCV detection tests became available in 1992 that transmission via blood transfusions became rare. After that point, donated blood was screened for the pathogen before being used. Today, most HCV cases are associated with shared needles, although some cases develop following direct contact with blood and bodily fluids from an HCV-infected patient (FIG. 19.8). Vertical transmission from mother to child is also possible. The virus has been shown to cross the placenta. The absence of an approved drug for use during pregnancy, combined with the lack of a vaccine, make vertical transmission a concern. Like HBV, HCV can be found in breast milk of infected mothers, but breast-feeding has not been shown to cause infection.

HCV continually mutates, making it difficult for the immune system to recognize and even harder for an effective vaccine to be developed. Approximately 75–85 percent of HCV patients develop a chronic form of the infection. Early infection with HCV is usually asymptomatic and chronic infections are not usually diagnosed until liver problems begin—often years after the initial acute

DiseaseSnapshot • Hepatitis D and Hepatitis E

Disease	Hepatitis D	Hepatitis E
Causative agent	Hepatitis D virus: enveloped, single-stranded RNA genome, no family designation to date	Hepatitis E virus: nonenveloped, single-stranded RNA genome, *Hepeviridae* family
Epidemiology	Worldwide, though HDV is only found with those who are coinfected with HBV. Incubation is 45–160 days (average 90 days).	Found worldwide, but more prevalent in areas with poor sewer and water management/ sanitation, such as south and east Asia. Incubation is 2–10 weeks (average 5–6 weeks).
Transmission & prevention	Primarily bloodborne transmission; sexual and vertical transmission also possible. Since HDV requires HBV to cause infection, HBV vaccine protects against HDV, too. Other prevention methods include condom use and not sharing needles.	Mainly fecal–oral transmission (usually through contaminated food and water); some cases due to blood transfusions, ingesting infected meat or undercooked shellfish, and vertical transmission. No vaccine, so prevention methods include proper water sanitation and hygienic practices.
Chronic infections	Both acute and chronic infections occur	Both acute and chronic infections occur
Signs & symptoms	All forms of viral hepatitis may generate fever, loss of appetite, joint pain, abdominal pain, fatigue, nausea, clay-colored bowel movements, and jaundice	
Pathogenesis & virulence factors	Enters the blood and targets liver cells for replication; requires HBV replication machinery to replicate	Enters the blood and targets liver cells for replication; virulence factors for HEV remain poorly understood
Diagnosis & treatment	Detecting HDV antibodies or antigens in a patient's blood; detecting HDV viral RNA using reverse transcriptase polymerase chain reaction. Treated with antiviral therapies (ribavirin or interferon).	HEV antibodies detected in the patient's blood; RT-PCR to detect HEV RNA in blood or stool samples. Treated with antiviral therapies (ribavirin or interferon).

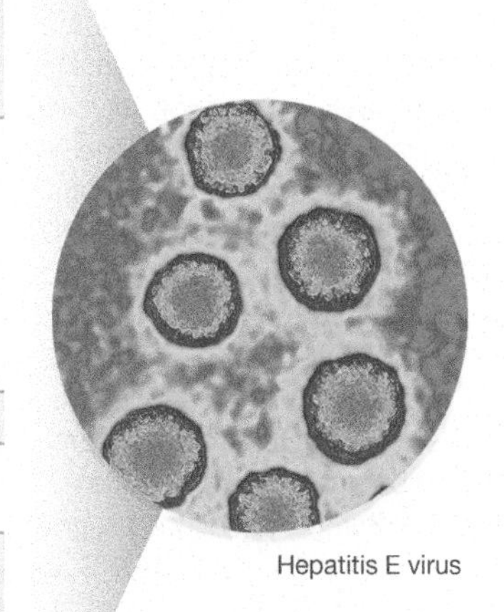

Hepatitis E virus

infection. Liver cancer and cirrhosis are complications of chronic infection. Fortunately, newer drug treatment can cure 90 percent of those infected. Treatments consist of protease, polymerase, and other viral protein inhibitor medications. While side effects from these medications can be severe, they are clearly preferable to cirrhosis or liver cancer. The CDC estimates that up to 20 percent of infected individuals spontaneously clear the virus without treatment. If the body cannot clear the virus, then repeated immune responses will eventually damage the liver. The combined HAV and HBV vaccine is recommended for those with HCV in order to avoid coinfection and greater viral load in the liver.

Hepatitis D and E: Less Prevalent Forms of Hepatitis

In the United States, hepatitis D (HDV) and hepatitis E (HEV) infections are rare, but they are a serious concern in many developing nations. HDV infection cannot occur without HBV being present. As such, an HDV infection can either occur simultaneously with HBV infection as a "coinfection," or it can occur as a "superinfection" in which the patient has HBV and then later contracts HDV in addition. Patients with an HBV-HDV coinfection or superinfection tend to exhibit more severe symptoms than HBV patients.

HEV primarily has a fecal–oral transmission. In some cases, HEV outbreaks have been linked to eating undercooked meat and shellfish from endemic areas. Countries with poor sanitation and lack of clean drinking water have higher incidences of HEV infections. According to the World Health Organization, HEV is estimated to infect about 20 million people per year, with about 3.3 million of those cases being accompanied by symptoms. About 56,000 people each year die of HEV-associated complications—so it's not inconsequential. There are no vaccines currently available in the United States for either HEV or HDV viruses.

TRAINING

TOMORROW'S HEALTH TEAM

Reducing Hepatitis Transmission Through Syringe Exchange Programs

In the early 1990s, health authorities in major U.S. cities began offering syringe exchange programs (SEPs) to people who injected illicit drugs. These programs allow anyone to swap used syringes for new, sterile needles free of charge. The hope was that these publicly funded programs would reduce HIV and hepatitis B and C infections—all diseases where sharing needles during drug use is a common factor in transmission.

The CDC estimates that about 60 percent of hepatitis C infections are acquired through illicit drug use. Today there are about 351 SEPs offered in 42 states, plus D.C. and Puerto Rico. Clinics that conduct needle exchange may additionally offer HIV tests, hepatitis B and C screening, tuberculosis screening, health education, and referrals for substance abuse treatment.

Syringe exchange programs have been a topic of political debate in the United States. The U.S. Congress has a long history of switching its stance on federal funding for SEPs. As of this text's publication federal funds are permitted for SEP support. The World Health Organization encourages use of syringe exchange programs in all countries because they reduce the collateral harm of injected drug use. Ultimately, harm reduction strategies like SEPs are much cheaper than the costs associated with public health care for HIV/AIDS or HCV patients, and there is ample evidence that these programs reduce sharing and reusing needles and thereby reduce bloodborne pathogen transmission, especially HIV. The barriers to widespread implementation of SEPs are largely political because these harm-reduction programs tend to be seen as being at odds with a zero-tolerance stance on drug abuse.

Sources: Motie, I., Carretta, H. J., & Beitsch, L. M. (2020). Needling Policy Makers and Sharpening the Debate: Do Syringe Exchange Programs Improve Health at the Population Level?. *Journal of Public Health Management and Practice, 26*(3), 222–226.

Adams, J. M. (2020). Making the case for syringe services programs. *Public Health Reports, 135*(1_suppl), 10S–12S.

QUESTION 19.1

Why might dollars spent on SEPs actually reduce public health expenditures in the longer term?

Build Your Foundation (BYF) Quick Quiz: Visit the **Mastering Microbiology** Study Area to quiz yourself.

Build Your Foundation

7. What are the features of mumps and how is it transmitted?
8. Compare and contrast rotavirus and norovirus.
9. Compare and contrast hepatitis A, B, C, D, and E.
10. Which hepatitis viruses may establish chronic infections?

19.4 BACTERIAL DIGESTIVE SYSTEM INFECTIONS

Learning Outcomes

After reading this section, you should be able to:

19.11 State the signs and symptoms of dental caries and periodontal disease and explain the role of biofilms in these conditions.

19.12 Explain how *Helicobacter pylori* causes stomach ulcers.

19.13 Compare and contrast foodborne infection to food poisoning.

19.14 Describe the features of infections caused by *Campylobacter* species, *Shigella* species, *E. coli*, *Salmonella* species, and *Listeria monocytogenes*.

19.15 Discuss the virulence factors of *Shigella* and *E. coli* O157:H7.

19.16 State the features of *Vibrio cholerae* and discuss the disease it causes.

19.17 Describe the disease caused by *Clostridioides difficile* and explain why it's an increasing concern in healthcare settings.

Dental caries are prevalent in children while periodontal disease is prevalent in adults.

According to the CDC's Division of Oral Health, 67 percent of adolescents age 16–19 years old report they have at some point had **dental caries** (cavities) and 70 percent of patients over 65 years old report suffering from periodontal disease. Periodontal disease starts as an infection of the gums (**gingivitis**) and, if not treated, may progress to further inflammation and eventual erosion of the bone that surrounds the tooth (**periodontitis**). Periodontal disease is a leading cause of tooth loss in adults.

Dental Caries

In babies, the normal oral microbiota becomes established very soon after birth through exposure to bacteria from caregivers and feedings. Just as the normal microbiota of the intestinal tract can vary and impact health and disease, so too can members of the oral microbiota vary. In general, *Streptococcus* and *Actinomyces* species are among the first to begin attaching to the tooth enamel. Other bacterial genera such as *Viellonella* and *Haemophilus* may attach next, creating the complex biofilm called **dental plaque**. The presence of plaque promotes tooth decay because it holds resident bacteria close to the tooth enamel. As bacteria,

1 Bacteria in plaque biofilms ferment sugars. The acid byproducts they release erode tooth enamel, forming caries.

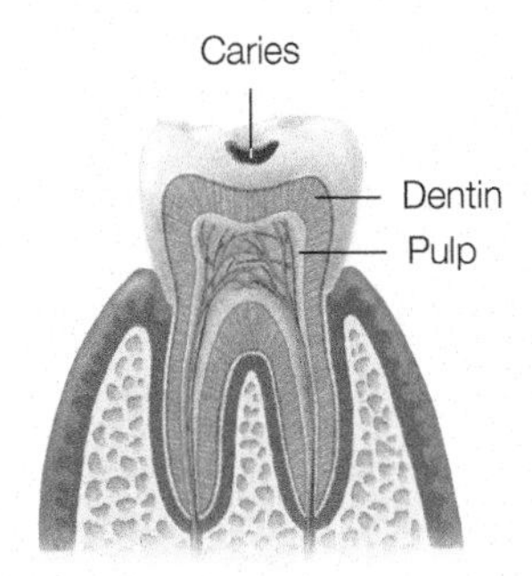

2 As enamel erodes, the underlying dentin is damaged.

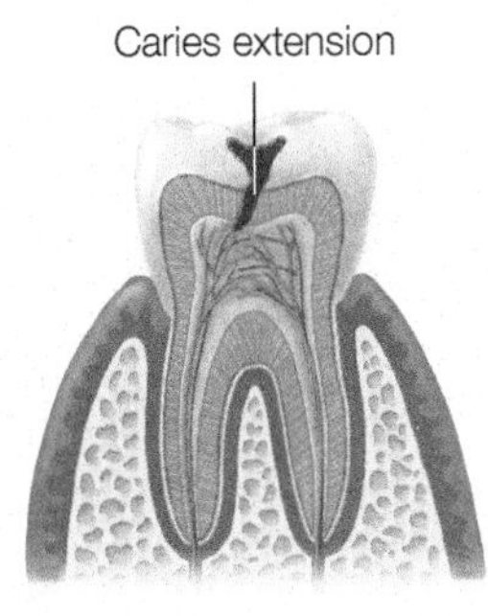

3 If left untreated, the pulp becomes infected and an abscess may form.

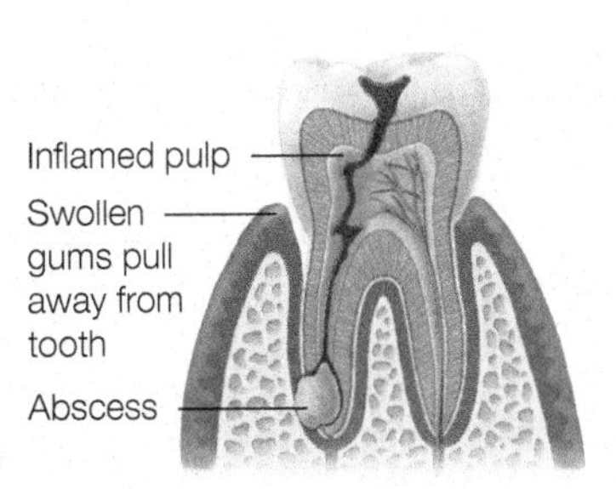

FIGURE 19.9 **Progression of dental caries**

especially *Streptococcus mutans*, metabolize sugars, they create acids that can eat away at the enamel to cause dental caries. As enamel erodes, the underlying dentin is exposed. Eventually, if the cavity is not filled, the pulp, where blood vessels and nerves are located, can become infected. Infection can then generate an abscess, or pus-filled lesion, in the surrounding gum or at the root of the tooth. Dentists repair cavities by cleaning out the affected area (frequently requiring drilling) and then filling the hole with resin composite or amalgam. Severe cases may require removal of the entire tooth. See FIG. 19.9 for progression of dental caries and the disease snapshot table for more details on dental caries.

Plaque's dense, mat-like structure along the gums and in tooth crevices prevents salivary enzymes from penetrating to the layers of bacteria that lie against the tooth enamel. As most people who regularly visit their dentist know, brushing teeth twice daily and flossing will physically remove superficial plaque layers. However, bacterial growth within the deeper biofilm layers and new bacteria entering the mouth quickly reestablish what was removed. Also, when plaque calcifies into **tartar** (or calculus), it must be scraped off the teeth. Dentists recommend a twice-yearly dental cleaning and checkup to limit caries and periodontal disease. Sealants are another preventive measure that involves plastic coatings that cover crevices in back teeth where plaque could grow. Additionally, limiting dietary sugars (especially sucrose) reduces lactic acid production by the bacteria in dental plaque and decreases the risk for dental caries. In the United States, federal guidelines recommend addition of fluoride to community water systems; this ion is also added to many toothpastes and mouthwashes. Fluoride hardens teeth and promotes mineralization, making them less susceptible to decay. The American Dental Association, which conducts surveys on the effectiveness of fluoridated water, states that tooth decay has been reduced in 25 percent of the American population due to fluoridated water.

Periodontal Disease

Periodontal disease is characterized by tender, swollen gums with a bright red or purplish coloration. The gums may also pull away from teeth. Bad breath and tooth loss are also frequent manifestations. As periodontal diseases progress, the swollen gums pull away from the tooth, forming gaps that can be colonized by anaerobic microbes such as *Porphyromonas gingivalis, Tannerella forsythia,* and *Fusobacterium nucleatum*. If not treated, the pronounced inflammation that these bacteria trigger leads to bone damage and eventual tooth loss (FIG. 19.10).

Regular brushing and flossing, along with regular dental visits and professional cleanings, reduce plaque and therefore limit periodontal disease. If periodontal disease is suspected, the groove between gums and teeth is measured to assess the extent of the problem. Dental X-rays are also taken to check for bone

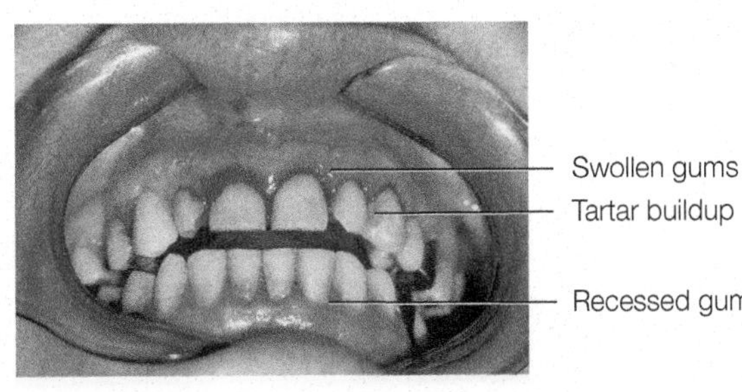

FIGURE 19.10 **Periodontal disease**

Critical Thinking *What is the common shared cause of periodontal disease and dental caries?*

DiseaseSnapshot • Periodontal Disease and Dental Caries

Dental plaque

Disease	Periodontal Disease	Dental Caries
Causative agent	Anaerobic bacteria biofilm along gum line, notably *Porphyromonas gingivalis* and *Tannerella forsythia*	Diverse bacteria, but especially *Streptococcus mutans*
Epidemiology	Almost half of adults over age 30 in the U.S. have some stage of periodontal disease	More than half of the U.S. population has their first cavity by age 15
Transmission & prevention	Oral bacteria transmit to infants from family members and caretakers. Preventing dental disease involves brushing and flossing teeth at least twice a day to reduce plaque. Twice-yearly cleanings at the dentist, adding fluoride to community water, and reducing sugar intake also prevent disease.	
Signs & symptoms	Swollen, bright red and bleeding gums; loose adult teeth; receding gum from base of tooth	Small discoloration may be visible; tooth sensitivity; mild to severe pain develops as the cavity progresses deeper into the tooth
Pathogenesis & virulence factors	Plaque biofilm structure holds bacteria against the tooth enamel and protects from brushing and saliva. Biofilm residents coordinate their gene expression to enhance survival.	
Diagnosis & treatment	Signs and symptoms are main diagnostic criteria, along with X-rays to assess bone loss. Treated with antibiotics and by scraping away plaque at and just below gum line. Severe gum loss requires tissue grafts, and bone loss may require bone grafts.	X-rays reveal white areas where caries exist. Treatment involves drilling away decay and filling the remaining tooth. Extensive decay may require replacing the top of a tooth with a porcelain crown. If infection reaches the inner tooth, a root canal drills out the pulp and replaces it with filling. In severe cases, teeth may be removed.

loss. If the disease isn't advanced, then plaque removal from the teeth and below the gums may be all that is required. Topical or oral antibiotics may also be prescribed. Advanced periodontitis requires surgery to help restore gums by either grafting new tissue to the affected region or deep cleaning below the gum line to remove plaque near the tooth root. For bone degeneration, bone grafts or application of a special matrix to encourage bone regrowth can be used.

Helicobacter pylori can cause gastritis and stomach ulcers.

Before 1980, most ulcers were attributed to stress or diet. Scientists believed the stomach was sterile due to its low-pH environment. However, two Australian scientists, Drs. Barry Marshall and Robin Warren, identified small curved rods in a number of ulcer biopsies. They cultured and identified a new species of bacteria they named *Helicobacter pylori*. In their publication, they strongly suggested that it was an infection that causes gastritis and ulcers. However, their colleagues weren't convinced that the bacteria *caused* ulcers—perhaps they simply colonized the ulcers after formation. To prove their discovery and follow Koch's postulates, Dr. Barry Marshall, who did not suffer from ulcers, drank a beaker of *H. pylori* himself! Sure enough, within 10 days, endoscopy revealed he was suffering from gastritis.

According to the CDC, at any given time the microaerophilic, Gram-negative bacterium *Helicobacter pylori* infects two-thirds of the world's population, although many never suffer symptoms. When symptoms do develop, they may include belching, vomiting, abdominal pain due to gastritis, and stools tinged with blood from bleeding stomach ulcers. Most likely, transmission is via the fecal–oral route.

Ulcers form in areas where the stomach lining is damaged by the bacterium and the host inflammatory response (FIG. 19.11). On the other hand, some benefits to *H. pylori* colonization of the stomach have

FIGURE 19.11 Stomach ulcer formation

Critical Thinking *If a drug were created that cleaved the flagella off of Helicobacter pylori, how would the bacterium's ability to cause disease be affected?*

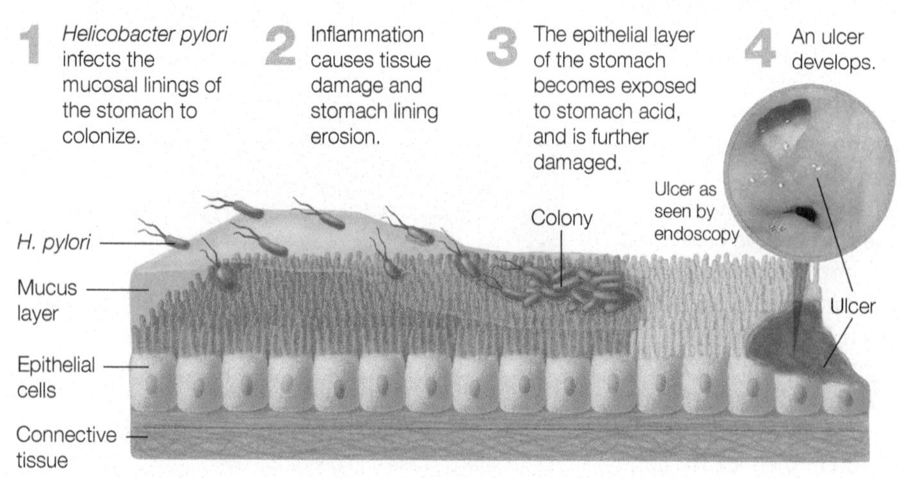

DiseaseSnapshot • Stomach Ulcers

Causative agent	*Helicobacter pylori*; Gram-negative bacterium
Epidemiology	About two-thirds of all people are infected, with incidence higher in developing nations. Incubation is months to years, depending on overall patient health.
Transmission & prevention	Fecal–oral transmission suspected. The bacteria can also be found in the mouth, so direct contact may also transmit. Proper water sanitation and food handling may prevent disease.
Signs & symptoms	Most infections are asymptomatic; those who have gastric ulcers suffer with nausea, vomiting, and abdominal pain; bloody stool may indicate a bleeding ulcer
Pathogenesis & virulence factors	Polar flagella help bacteria burrow into mucosal lining to escape stomach acid. Enzymes decrease the acidity around them. They release toxin VacA, which triggers host cell death.
Diagnosis & treatment	*H. pylori* antigens can be detected in stool, or antibodies to the bacterium can be found in blood samples. A urea breath test can indirectly identify the presence of the bacterium by detecting urea made by the organism. Biopsy of the stomach lining during endoscopic evaluation can also confirm diagnosis. Treatment is dual antibiotic therapy (often clarithromycin and amoxicillin) along with a proton pump inhibitor to reduce stomach acidity so the ulcer may heal.

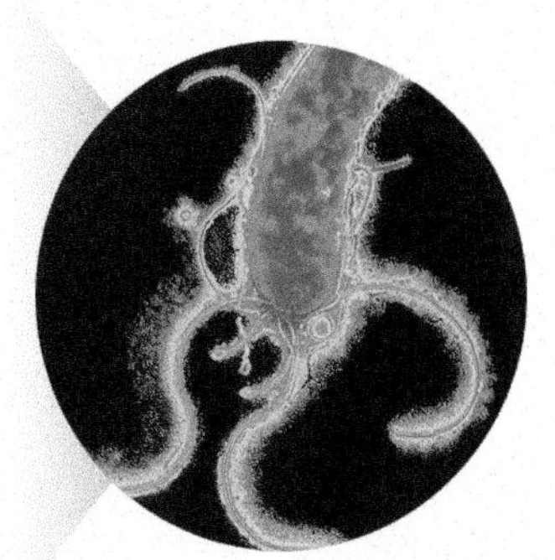

H. pylori

emerged. A few studies suggest that conditions such as asthma and acid reflux are reduced by the presence of the bacterium. These findings continue to probe the complex interaction between gut microbiota and human health.[2]

H. pylori infections are curable with antibiotics. In addition, a proton pump inhibitor is often prescribed to reduce stomach acid levels so that the stomach lining can more readily heal.

Bacteria are common causes of foodborne illnesses.

The CDC estimates that each year one in six Americans suffers from a foodborne illness. Of the 31 characterized pathogens that are known to cause foodborne illness, norovirus is the most common causative agent. However, bacteria cause a significant proportion of foodborne illnesses. The majority of foodborne illnesses are due to fecal contamination of food or water. Foodborne illnesses can be sorted into two categories, foodborne infection and food poisoning (also

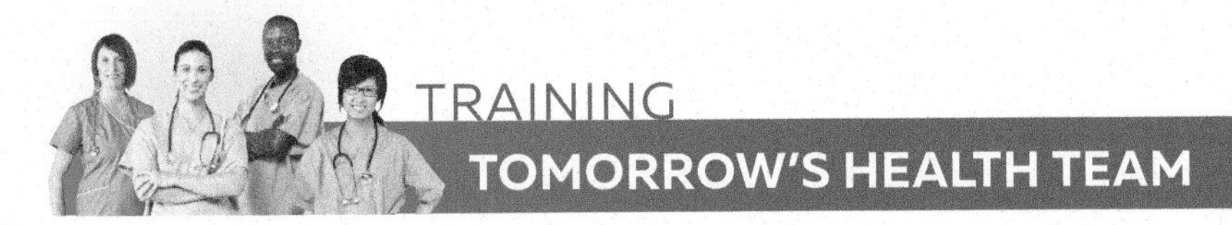

TRAINING TOMORROW'S HEALTH TEAM

The Link Between *Helicobacter pylori* Infections and Gastric Cancer

Gastric cancer is the fourth most common cancer worldwide. In recent years, a link between stomach cancers and *H. pylori* infections has become evident. In one study, the rate of stomach cancer was six times higher in infected people compared to noninfected people. The International Agency for Research on Cancer, a suborganization of the World Health Organization, has categorized the bacteria as a carcinogen.

We still do not fully understand how *H. pylori* causes cancer. Researchers suspect that the prolonged inflammation of the stomach turns cells cancerous. While there is evidence that a long-standing *H. pylori* infection is one factor that promotes stomach cancer, diet and environment are also likely to play a significant role in disease development. Another factor may be the specific *H. pylori* strain that is causing the infection.

QUESTION 19.2

Helicobacter pylori was the first bacterial agent termed a carcinogen. Can you name another cancer-causing agent found in this chapter? What is a common factor between them? (Hint: Think about the infection time course.)

Endoscopic view of gastric cancer

Sources: Piscione, M., et al. Eradication of *Helicobacter pylori* and gastric cancer: a controversial relationship. *Frontiers in Microbiology, 12.*

[2] Testerman, T. L., & Morris, J. (2014). Beyond the stomach: An updated view of *Helicobacter pylori* pathogenesis, diagnosis, and treatment. *World Journal of Gastroenterology*, 20(36), 12781–12808.

TABLE 19.2 Foodborne Infection Versus Food Poisoning

Illness	Foodborne Infection	Food Poisoning
Description	After ingestion, pathogen establishes infection in the host	No infection; an ingested toxin causes the illness symptoms
Common bacterial agents	*Campylobacter jejuni*, *Shigella* species, *E. coli*, *Salmonella* species, and *Listeria monocytogenes*	Toxins from *Staphylococcus aureus*, *Bacillus cereus*, and *Clostridium perfringens*
Symptom onset	1–5 days (or longer) after exposure	30 minutes to 6 hours after exposure
Signs & symptoms	May include fever, headache, muscle aches, nausea, diarrhea, abdominal pain	Most commonly nausea and vomiting and occasionally diarrhea and abdominal pain
Recovery	Days to weeks	Usually within 24 hours; up to several days in severe cases

called food intoxication), both of which are entirely preventable. The incidence of these illnesses can be reduced by understanding the sources of foodborne illness and educating food processors and the public on ways to reduce food contamination. TABLE 19.2 compares foodborne infections with food poisoning.

Leading Causes of Foodborne Bacterial Infections

Foodborne bacterial infections are the result of ingesting food that harbors live bacterial pathogens that infect the GI tract. The most common symptom of foodborne infections is diarrhea. The severity of the disease depends on the type of pathogen that establishes the infection. The most common bacterial causes of foodborne infections are *Campylobacter* species, *Shigella* species, *E. coli*, *Salmonella*, and *Listeria monocytogenes*. (*L. monocytogenes* is reviewed in the Training Tomorrow's Health Team feature, "Is *Listeria monocytogenes* lurking in your refrigerator?" and in Chapter 18; the other four bacteria are reviewed in later sections of this chapter.)

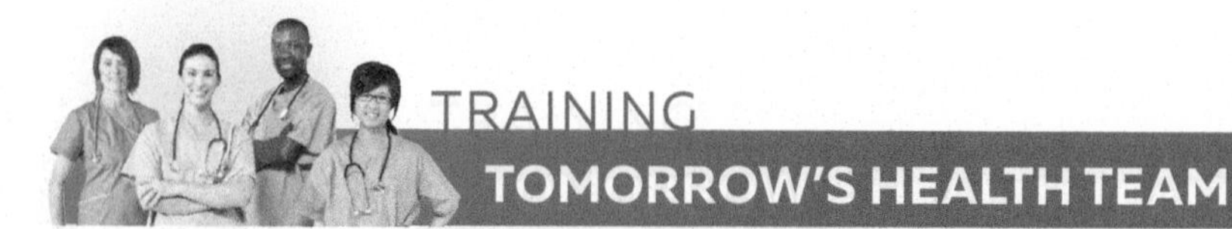

TRAINING TOMORROW'S HEALTH TEAM

Is *Listeria monocytogenes* Lurking in Your Refrigerator?

In March of 2016, the largest ever food recall was issued due to *Listeria monocytogenes* contamination of foods distributed by CFR Frozen Foods. It impacted millions of frozen food items across all 50 states, Canada, and Mexico. While the chilly environment of a refrigerator or freezer is one of the last places we think microbes will thrive, the Gram-positive bacterium *Listeria monocytogenes* can thrive even when food is properly refrigerated, or even frozen. *L. monocytogenes* is naturally found in soil and water, and animals can harbor it too, which is why it can easily contaminate produce and is often associated with meats and unpasteurized dairy products. Most people with a healthy immune system experience a 24- to 48-hour self-limiting infection with symptoms such as diarrhea, vomiting, and fever. However, people with reduced immunity and pregnant women may die or suffer serious complications such as meningitis.

Infection in the first trimester of pregnancy can cause a miscarriage, while infections later in pregnancy can cause pre-term labor, stillbirth, neonatal meningitis, or a collection of impairments including paralysis, seizures, intellectual disabilities, blindness, and organ damage. Given these severe risks, it is extremely important that pregnant women avoid eating high-risk food such as hot dogs and lunch meats (unless cooked to steaming), unpasteurized milk products, refrigerated meat pâtés, or refrigerated smoked seafood.

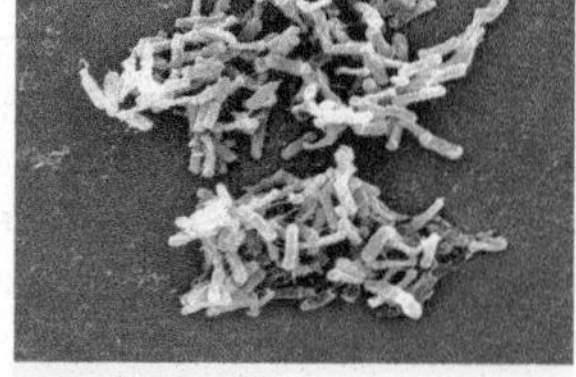

L. monocytogenes (colorized SEM)

QUESTION 19.3

While there are certain "high-risk" foods for Listeria, fresh and even frozen fruits and vegetables have also been linked to outbreaks. Keeping this in mind, what specific food precautions would you discuss with a patient who is trying to become pregnant?

Leading Causes of Food Poisoning

Unlike foodborne infections, which require bacteria to infect the host, in food poisoning the bacteria don't actively grow in the host. Instead, symptoms are triggered by ingested exotoxins that inflame intestinal cells and prevent water from absorbing into the intestine, causing diarrhea and abdominal pain. Toxins that target the intestines are called **enterotoxins**. They may also be **emetic**, meaning they trigger vomiting. Food poisoning can be linked to any food, but it is often associated with prepared foods that require a lot of handling and processing. Leaving foods out at room temperature for too long is a key contributing factor to most food poisoning cases because it allows bacteria in food to multiply and release toxins. Because many enterotoxins are not affected by heat, cooking or reheating the food may kill the bacteria, but it doesn't necessarily destroy the enterotoxin.

A few bacterial pathogens that are the leading causes of food poisoning are *Staphylococcus aureus, Clostridium perfringens,* and *Bacillus cereus*. *S. aureus* food poisoning is often linked to creamy salads (tuna, chicken, potato, macaroni, etc.) and dairy-based foods. Poultry and meat products also have been associated with outbreaks due to improper food handling. Since *S. aureus* can become part of normal nose and skin microbiota, transmission is easy if hands are not washed before preparing food.

C. perfringens and *B. cereus* are both spore-forming organisms. *C. perfringens* is commonly found in animal intestines, and is readily found on raw meat. Most *C. perfringens* food poisoning cases are linked to gravy and meat products that are left too long at room temperature. *B. cereus* is present in soil and can contaminate produce. *B. cereus* can contaminate most foods, including meat, fish, and milk products, as well as starchy foods such as rice, potatoes, or pasta. Cooking does not always kill its spores, so when food is left out, the spores become actively growing bacteria and produce toxin (FIG. 19.12).

Botulism is another type of food poisoning from the spore-forming *Clostridium botulinum*. This exotoxin enters the blood and targets nerves. (See botulism coverage in Chapter 18 for more details.)

Another type of food poisoning can be associated with toxin-producing fungi that grow on crops. Corn and peanuts must be treated to remove the *Aspergillus flavus* toxin, alfatoxin. Meat, eggs, and milk may become contaminated with alfatoxin due to animals ingesting contaminated feed. In the past, large outbreaks occurred in countries such as India, Taiwan, and Uganda due to poorly stored food supplies.

Campylobacter jejuni is a leading cause of bacterial foodborne illness.

The CDC estimates that up to 1.6 million people in the United States suffer from a *Campylobacter jejuni* (cam PILLOW bac ter juh JUNE ee) infection every year, yet you probably haven't heard of this bacterium. This is likely because it tends to cause isolated and sporadic infections, not the sizable outbreaks and food recalls that are associated with *E. coli* and *Salmonella*. Still, *C. jejuni* is certainly one of the top bacterial causes of foodborne infection in the United States and may even be the leading culprit in terms of the total number of cases. It's just that cases are underreported.

C. jejuni is a Gram-negative, flagellated, spiral-shaped bacterium that grows best in microaerophilic (limited oxygen) conditions. The bacterium is commonly found in the intestinal tract of healthy birds—most notably in chickens. As such, *C. jejuni* infections are primarily associated with eating undercooked poultry or cross-contaminating foods with raw poultry juices. Once ingested, the bacteria burrow through the mucosal layer of the intestine and migrate through intestinal epithelial cells to multiply just beneath the epithelial layer.

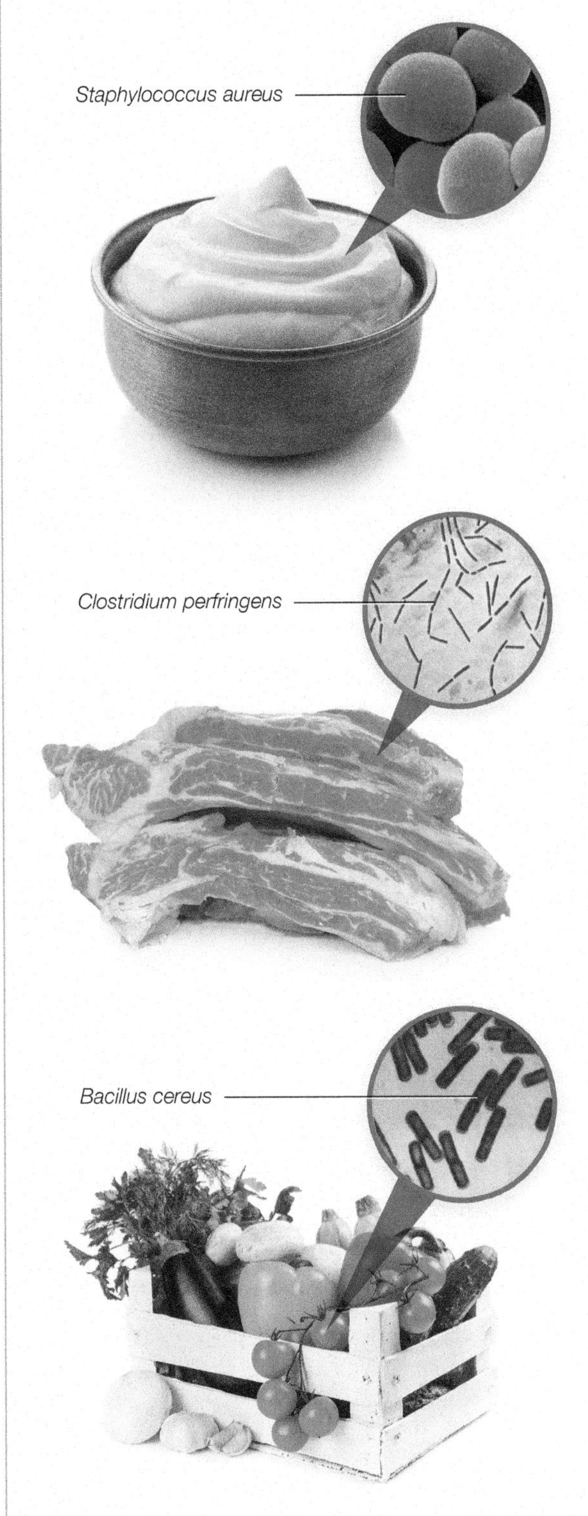

FIGURE 19.12 **Food poisoning** Most food poisoning cases are associated with *S. aureus*, *C. perfringens*, and *B. cereus*.

Critical Thinking *Is thoroughly cooking meat guaranteed to prevent C. perfringens food poisoning? Explain your reasoning.*

Disease**Snapshot** • Campylobacteriosis

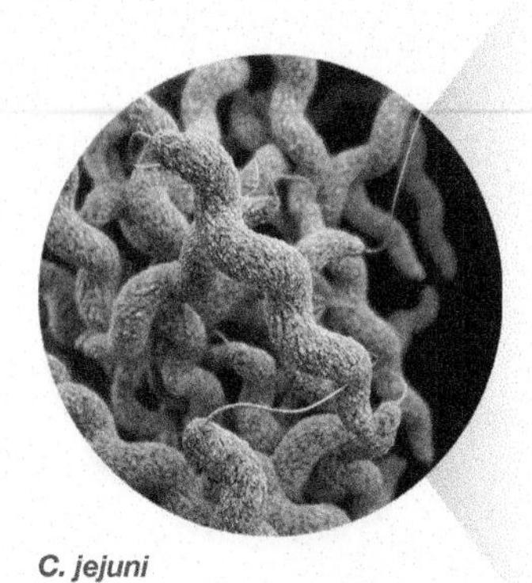

C. jejuni

Causative agent	*Campylobacter jejuni*; Gram-negative bacterium
Epidemiology	The most common diarrheal illness in the U.S, with isolated and sporadic cases. More prevalent in summer months than other times of the year. Incubation is 2–5 days.
Transmission & prevention	Chickens naturally harbor the bacteria in their intestines, so cross-contamination by raw chicken is the main transmission mode, along with fecal–oral. Hygienic handling of raw chicken, washing hands and surfaces to prevent cross-contamination, and fully cooking poultry prevents infections.
Signs & symptoms	Diarrhea, abdominal cramps, vomiting, and fever
Pathogenesis & virulence factors	Flagella and helical shape are thought to enhance burrowing ability; a capsule and numerous adhesion factors aid host cell invasion. The bacteria make cytolethal distending toxin, which blocks intestinal cell division and eventually kills intestinal epithelial cells.
Complications	Guillain-Barré syndrome, an autoimmune disorder that causes muscle weakness/paralysis
Diagnosis & treatment	Stool sample cultures or PCR to identify *C. jejuni* DNA in samples are diagnostic techniques. Most recover with just supportive therapy and fluid restoration. Immune-compromised patients may be treated with azithromycin and fluoroquinolones like ciprofloxacin.

CLINICAL VOCABULARY

Bacteremia bacteria found in the blood

Sepsis a body-wide immune response to persistent or large numbers of microbes in the blood; may progress to septic shock

Septic shock a dangerously advanced stage of sepsis in which the body's organs start to shut down; has a 20–30 percent mortality rate

Within 2 to 5 days, an inflammatory response leads to symptoms of diarrhea, abdominal cramps, and fever. Nausea and vomiting may also occur. Symptoms usually resolve within a week, but in immune-compromised patients the bacteria may spread into the blood (a condition called bacteremia), which may in turn lead to sepsis and possibly septic shock.

Aside from thoroughly cooking poultry, using safe food-handling practices while handling raw poultry is one of the best ways to prevent *C. jejuni* infection. According to the CDC, even one drop of juice from raw chicken may have enough bacteria to cause infection. Using a different cutting board for meat and vegetables is recommended, and hands should be thoroughly washed after handling raw poultry.

A rare but troubling complication associated with *Campylobacter* infection is the development of **Guillain-Barré syndrome (GBS)** (GHEE lan *bar* ay). It's thought that a *C. jejuni* infection may trigger this autoimmune neurological disease; 25 to 50 percent of reported GBS cases are preceded by a *C. jejuni* infection.[3] Weeks after diarrhea clears, the patient develops a tingling sensation in the legs and feet. Eventually this sensation advances to paralysis and may spread to other parts of the body. The symptoms are temporary, but individuals must be hospitalized. The prevailing thought is that antibodies that recognize *C. jejuni*'s lipopolysaccharides cross-react with sugars on neurons. This triggers an autoimmune response against the nervous system, creating the paralysis.

Dysentery and fever may occur during *Shigella* infections.

Shigella (shih GEHL uh) are Gram-negative bacteria that include four pathogenic species: *S. flexneri*, *S. sonnei*, *S. boydii*, and *S. dysenteriae*. In the United States, *S. sonnei* accounts for about 80 percent of *Shigella* cases, while *S. flexneri* predominates in developing countries. According to the CDC, *Shigella* affects about half a million people in the United States every year. It spreads from person to person and also through human fecal contamination of food, water, or environmental surfaces that are touched (such as countertops, banisters, or light switches). Children between 2 and 4 years old are most likely to become infected. Within a day or two of ingesting *Shigella*, symptoms of frequent watery or bloody

[3] Willison, H. J., Jacobs, B. C., & van Doorn, P. A. (2016). Guillain-Barré syndrome. *The Lancet, 388*(10045), 717–727.

diarrhea, abdominal pain, and fever are typical. *S. dysenteriae* is usually seen in Africa and Central America; of the four *Shigella* species, it causes the most severe form of dysentery and according to the CDC carries a case fatality rate of 5–15 percent.

Shigella species have numerous virulence factors. Some key virulence factors are: (1) a system that induces host cells to endocytose the bacteria; (2) the manufacture of several toxins that damage intestinal cells and induce fluid efflux; (3) the capacity to pass between infected cells using an actin propulsion system; and (4) the ability to escape from phagocytes (FIG. 19.13). The endocytosis induction system involves the pathogen injecting specialized proteins into the would-be host to induce it to internalize the bacterium by endocytosis. Regarding toxins, most *Shigella* species can make at least one of two enterotoxins that damage intestinal cells and promote fluid efflux into the intestines, generating diarrhea.

A third toxin, the **Shiga toxin**, is only associated with *S. dysenteriae*; it targets ribosomes to block protein synthesis and kill host cells. Shiga toxin's cytotoxic effects are largely why *S. dysenteriae* has such a high mortality rate compared with other *Shigella* species. In some patients, the Shiga toxin can damage the kidneys and induce **hemolytic uremic syndrome (HUS)**. HUS symptoms include blood in the urine (hematuria), bruising (purpura), and generalized swelling (edema) in the extremities. Also, certain strains of *E. coli* that have

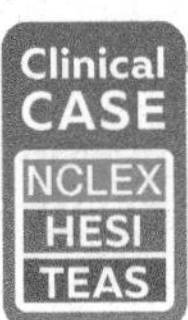

The Case of the Hurricane Health Hazard

Practice applying what you know clinically: visit the **Mastering Microbiology** Study Area to watch Part 2 and practice for future exams.

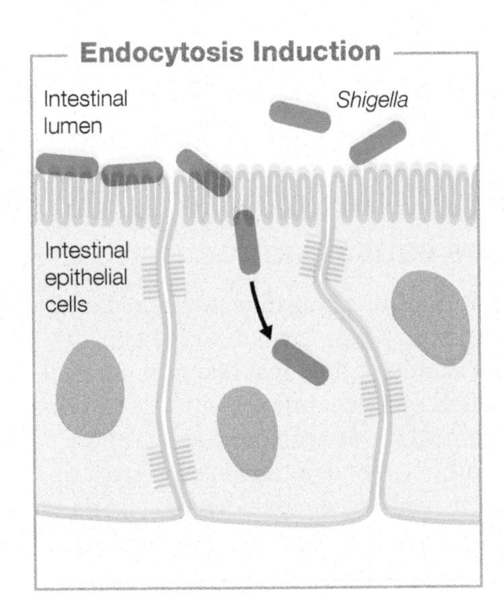

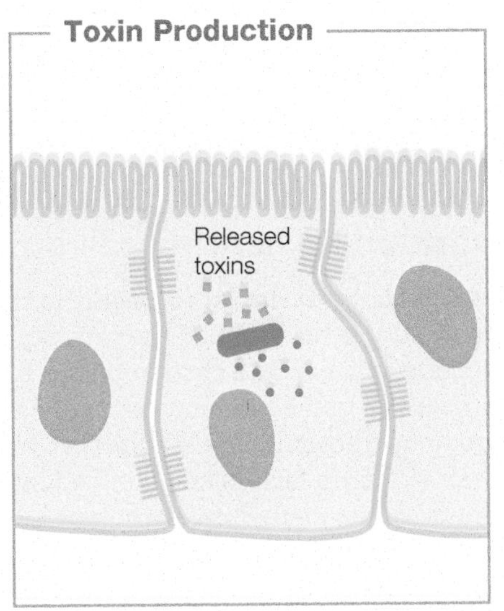

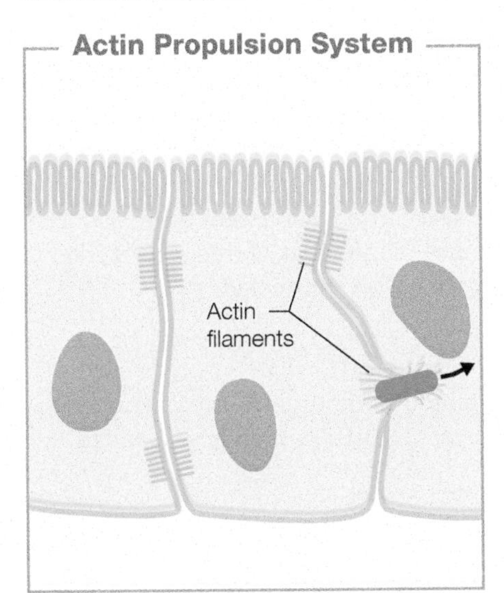

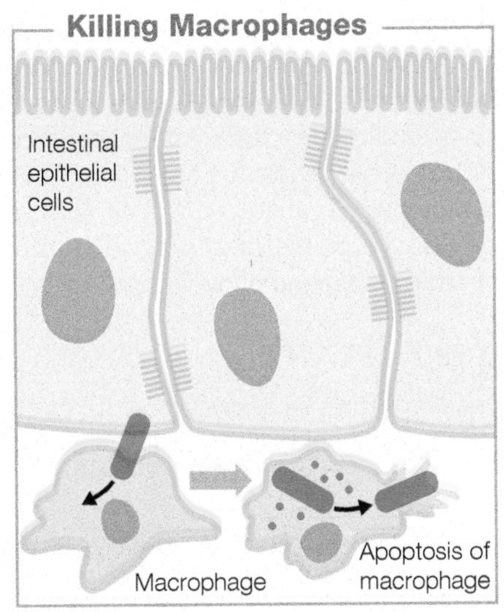

FIGURE 19.13 **Key *Shigella* virulence factors**

Critical Thinking *What symptoms are directly linked to intestinal epithelial cell death that Shigella may cause?*

Disease**Snapshot** • **Shigellosis**

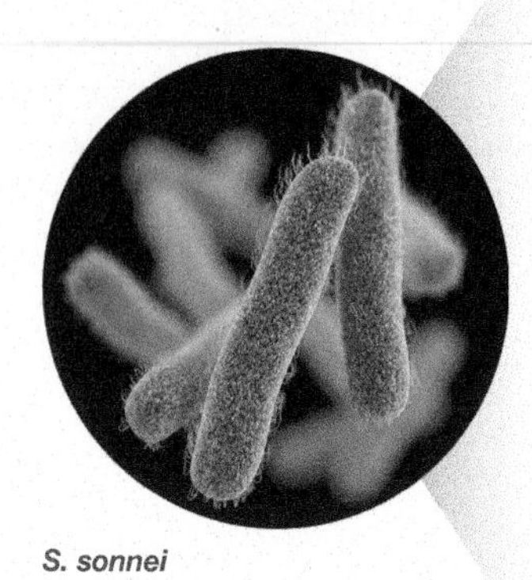

S. sonnei

Causative agent	*Shigella sonnei* most common in the U.S.; *S. flexneri* is more prevalent in developing countries. Both are Gram-negative bacteria.
Epidemiology	Toddlers are most likely to become infected. International travelers who consume feces-contaminated food or water are also at risk. Incubation period is 1–2 days.
Transmission & prevention	Highly contagious, with both fecal–oral and direct contact (human to human) transmission. Careful hand washing can prevent it. Also, when traveling to areas without fully sanitized water, drink only boiled water, and eat only well-cooked foods, or fruit you peel yourself.
Signs & symptoms	Bloody diarrhea, abdominal cramping, and fever
Pathogenesis & virulence factors	Induces endocytosis and produces various toxins (Shiga toxin is only made by *S. dysenteriae*). It can pass between infected cells using an actin propulsion system, and can escape phagocytosis.
Complications	Hemolytic uremic syndrome (HUS) induced by Shiga toxin
Diagnosis & treatment	Specialized culture and biochemical tests identify the pathogen in patient stool samples. Most individuals recover without treatment. For severe disease, antibiotics such as ampicillin and trimethoprim-sulfamethoxazole are first-line therapies, but many strains are now drug resistant. Antidiarrheal medications that decrease intestinal motility such as loperamide (Imodium) or diphenoxylate-atropine (Lomotil) are not recommended for patients with bloody diarrhea or diarrhea and fever, because these drugs may increase risk for invasive disease.

gained the capacity to make the Shiga toxin are associated with HUS. In fact, most HUS cases in the United States are due to *E. coli* rather than *S. dysenteriae*.

The ability of the bacterium to propel itself from one host cell to the next by commandeering host cytoskeletal actin allows the pathogen to remain hidden from the immune system. Even when host macrophages manage to phagocytize *Shigella*, the bacterium releases substances that cause the macrophage to undergo apoptosis (cell suicide)—allowing the pathogen to escape.

CLINICAL VOCABULARY

Hematuria blood in the urine

Purpura a rash of purple spots/bruising caused by red blood cell lysis and/or capillary destruction rather than by external physical trauma

Edema generalized swelling, often in the extremities, in response to fluid accumulation in tissues

Various *Escherichia coli* strains cause gastroenteritis.

Escherichia coli is similar to *Shigella* in that it is a Gram-negative motile rod and some strains have gained the capacity to make a Shiga toxin due to the introduction of genes by a bacteriophage. *E. coli* is among the most common normal microbiota in the human intestines as well as other mammals, but most strains are nonpathogenic. The strains that are pathogenic are categorized based on their virulence factors and grouped into pathotypes (also called pathovars). There are six main *E. coli* pathotypes that cause diarrhea (diarrheagenic) and other pathotypes that cause urinary tract infections, bacteremia, and meningitis. *E. coli* pathotypes can be grouped by their O and H antigens—molecules on the bacterial cell. Probably the most infamous pathotype is *E. coli* O157:H7, which is reviewed in the next section.

All six diarrheagenic *E. coli* pathotypes are transmitted via the fecal–oral route and are therefore commonly associated with poor water sanitation and unsafe food-handling practices. Animal contact, environmental contact, and direct contact with infected people are other ways that these bacteria spread. Although these six pathotypes all generate diarrhea, some may also cause fever and/or dysentery. The specific symptoms, disease severity, time course of infection, and recommended treatments depend on what pathotype the patient has contracted.[4]

E. coli O157:H7

Deadly outbreaks associated with contaminated hamburgers, unpasteurized juices, and even petting zoos have made *E. coli* O157:H7, which is also called

[4] Croxen, M. A., Law, R. J., Scholz, R., Keeney, K. M., Wlodarska, M., & Finlay, B. B. (2013). Recent advances in understanding enteric pathogenic *Escherichia coli*. *Clinical Microbiology Reviews*, 26(4), 822–880.

DiseaseSnapshot • Enterohemorrhagic *Escherichia coli*

Causative agent	*Escherichia coli* O157:H7 (Shiga toxin–producing *E. coli*); Gram-negative bacterium
Epidemiology	Outbreaks are associated with contaminated meat, unpasteurized dairy and juice products, and produce washed or watered with feces-contaminated water. Incubation period is 1–10 days
Transmission & prevention	Transmits via fecal–oral route through food or water, or by direct person-to-person contact, as well as environmental or animal carrier exposure. Proper food and water sanitation, hand washing, thorough cooking of meat, and avoidance of unpasteurized dairy and juice products are the main preventive measures
Signs & symptoms	Watery diarrhea that may turn bloody, abdominal pain, and vomiting; usually there is not a fever
Pathogenesis & virulence factors	Shiga toxin kills host cells by inhibiting protein synthesis. Resists stomach acid, making for a low infectious dose
Complications	Hemolytic uremic syndrome in up to 10% of infections, especially among children, immune-compromised patients, and the elderly
Diagnosis & treatment	Diagnosed via stool samples using bacterial culture techniques. Molecular methods analyze DNA and the O and H antigens to identify specific pathotype. Treatment is usually rest and rehydration. Antibiotics do not reduce disease and may precipitate HUS, due to sudden burst in toxin levels. Antidiarrheal medications that decrease intestinal motility such as loperamide (Imodium) or diphenoxylate-atropine (Lomotil) are not recommended, as they slow clearing of bacteria

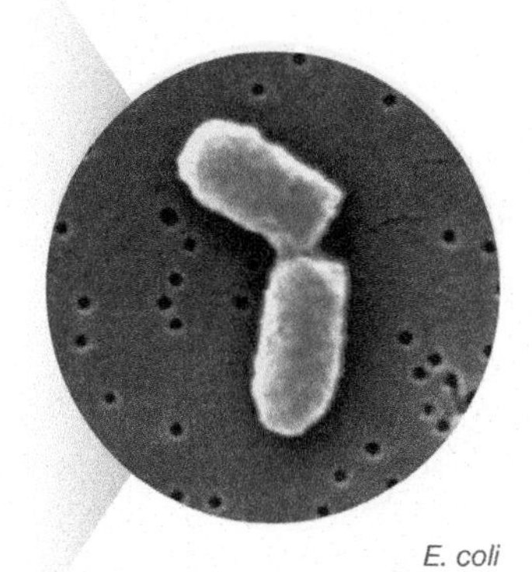

E. coli

Shiga toxin-producing *E. coli* (or STEC), an infamous pathogen—it is the most common dysentery-associated serovar in the United States. Although it is among the normal gut microbiota of many animals, its presence in the normal gut microbiota of healthy cattle is the main way that this microbe makes its way into humans. When meat is processed and the intestines are not sufficiently removed before the meat is ground up, bacteria can contaminate the meat. Vegetables become contaminated when watered or washed with cow manure-contaminated water. Dairy and juice products can also harbor the agent, which is why pasteurization is such an important safety measure. Person-to-person transmission also occurs, especially among young children who may touch feces-contaminated surfaces and then put their hands in their mouth. Because the organism is resistant to gastric acids, ingesting just 10 to 100 bacterial cells may be sufficient to establish disease.

Symptoms include abdominal pain and watery diarrhea followed by the development of bloody diarrhea; occasionally the patient develops a low-grade fever, but usually fever is absent. The illness typically resolves within a week, but in up to 10 percent of cases it progresses to hemolytic uremic syndrome (HUS). Children, the elderly, and immune-compromised patients are at the greatest risk for developing HUS as a complication of infection.

Some *Salmonella* cause common foodborne gastroenteritis, while others cause typhoid fever.

Salmonella bacteria are among the most common causative agents of diarrhea and dysentery. The CDC estimates that every year these Gram-negative rods are responsible for at least 1.2 million foodborne infections, called **salmonellosis**, per year in the United States. As high as this number is, even this estimate is conservative because, like most other foodborne illnesses, there tends to be a huge number of unreported cases; the CDC estimates that there are about 29 *Salmonella* cases that go unreported for every case that is reported.

The genus *Salmonella* is divided into two species, of which only *S. enterica* includes pathogens that can infect humans. Using O and H antigens for classification schemes (as is done for *E. coli* strains), there are over 2,600 immunologically distinct varieties, or **serovars**, of *S. enterica*. The two most common diarrheagenic serovars in the United States are Enteritidis and Typhimurium. Another serovar, Typhi, which causes typhoid fever, is no longer common in the United States, but it remains endemic in many developing nations.

Disease**Snapshot** • Nontyphoid Salmonellosis

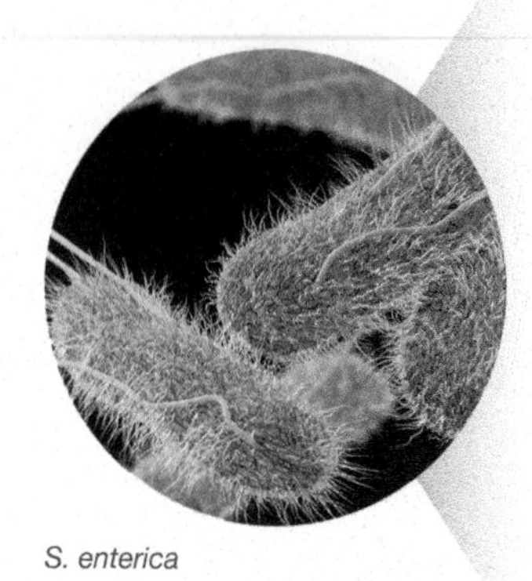

S. enterica

Causative agent	*Salmonella enterica* nontyphoid serovars such as Enteritidis and Typhimurium; Gram-negative bacteria
Epidemiology	Outbreaks are usually associated with consumption of undercooked meat and raw eggs, handling reptile pets (especially turtles), or live chickens. Most cases are isolated, rather than part of mass outbreaks. Cases occur throughout the year, but seasonality is observed, with peaks in summer and fall. Incubation period is 6 hours to 3 days.
Transmission & prevention	Fecal–oral transmission and direct contact with infected people. Prevent by thoroughly cooking meat and eggs, and using proper hand-washing and surface sanitation practices when handling meat and poultry, or after handling pets.
Signs & symptoms	Fever that tends to last a couple of days, vomiting, chills, headache, abdominal cramps, and diarrhea that may contain blood or mucus. Symptoms lasts 2–7 days and usually self-resolve.
Pathogenesis & virulence factors	Bacteria that survive acid barrier of the stomach induce intestinal epithelial cells to take them up by endocytosis. Pathogen can escape phagocytosis and multiply inside macrophages, killing them.
Diagnosis & treatment	Culture bacterium from stool samples; various biochemical media and molecular methods that identify bacterial O and H antigens. Supportive therapy for mild cases; antibiotics given if progresses to bacteremia and in immune-compromised patients; antimotility agents like loperamide (Imodium) or diphenoxylate-atropine (Lomotil) may be used as long as dysentery, *Shigella*, and *E. coli* O157:H7 have been ruled out

Salmonella enterica Serovars Enteritidis and Typhimurium

There are numerous *S. enterica* serovars that cause diarrhea and dysentery, but based on CDC tracking data, Enteritidis and Typhimurium consistently cause the most cases. Widespread recalls on foods including peanut butter, cantaloupe, chicken products, and spinach make it clear that practically any food can serve as an infection source. This is not surprising, given that a wide array of animals harbor the bacterium in their gut; it's found in the intestinal tract of birds, mammals, and some reptiles. That said, poultry products are most likely to harbor *Salmonella* serovars that infect people. The bacterium is commonly detected in raw chicken and eggs—both inside and on the outside of shells. Raw eggs, undercooked meat or seafood, and produce that has been washed or watered with feces-contaminated water are all potential sources of infection. Additionally, pet reptiles (especially turtles) have been associated with *Salmonella* infections.

Symptoms appear within a day or two of consuming contaminated food or water. Fever, abdominal cramps, and diarrhea can last four to seven days. Blood and mucus may also appear in the stool, indicating a progression to dysentery. Vomiting, chills, and headache are also commonly reported.

Salmonella bacteria induce endocytosis in intestinal epithelial cells. This mechanism allows the pathogen to readily enter the intestinal epithelium to avoid immune factors such as complement cascades and antibodies. *Salmonella* also escape phagocytosis and multiply inside macrophages.

Salmonella enterica Serovar Typhi

Typhoid fever is caused by *S. enteritidis* serotype Typhi, which is much more dangerous than the nontyphoid salmonellosis caused by the more common Enteritidis and Typhimurium serovars. The increased virulence of this serovar is primarily due to a toxin that its members make: the **typhoid toxin**. Humans are the only known source and host for the bacterium, so typhoid fever is only spread from human to human. This happens when human sewage contaminates water or food or when someone who is a carrier transmits the bacteria through unsanitary food-handling practices. Although rare today in the United States, typhoid fever is common in developing countries that suffer from poor sewage management and water sanitation.

Symptoms begin between one and two weeks following exposure. Patients tend to exhibit fever, headache, a rose-colored spotted rash, constipation, and generalized abdominal pain that can last weeks; vomiting is usually absent.

Intestinal rupture, internal bleeding, and shock are potentially deadly complications. While typhoid fever can be treated with antibiotics such as azithromycin, resistance to multiple antibiotics is an increasing concern. The *Salmonella* Typhi vaccine is recommended for people traveling to endemic areas.

People who get typhoid fever can and often do fully recover, but up to 5 percent may become asymptomatic carriers and continue to shed the bacteria in their feces and serve as a source of infection. The famous incident of Mary Mallon, dubbed "Typhoid Mary," of New York in the early 1900s brought to light this occurrence of *Salmonella* carriers. She claimed innocence in the 53 cases of typhoid fever suffered by people who ate where she served as a cook. (See Chapter 9 for more information on Mary Mallon.)

Poor sanitation contributes to cholera outbreaks.

Cholera cases in the United States tend to be sporadic and mainly linked to eating raw shellfish. However, in Africa, southeast Asia, and now Haiti, the disease is endemic and mainly due to poor water sanitation and person-to-person transmission. The World Health Organization states that worldwide each year there are as many as 4.3 million cases, with 142,000 deaths.

Cholera is caused by *Vibrio cholerae*, a comma-shaped Gram-negative bacterium with a single flagellum. Acute symptoms appear within hours to a few days after ingesting contaminated food or water. The infection can cause mild diarrhea, or can be severe with profuse watery diarrhea, vomiting, and leg cramps brought on by the loss of vital electrolytes. In severe cases, up to 20 liters (just under five gallons) of fluid a day can be excreted. Without prompt fluid replacement, severe dehydration develops, which in turn can lead to the cascade of decreased blood volume, hypovolemic shock, organ failure, and death.

V. cholerae are not as resistant to stomach acid as are *E. coli* and *Shigella*, so infection requires exposure to at least a million of the bacteria. The bacteria that survive the acidity of the stomach burrow past the mucous layer of the small intestines and adhere to the surface of intestinal epithelial cells. As they establish themselves, the bacteria release **cholera toxin**. This potent exotoxin triggers intestinal epithelial cells to release electrolytes and water into the intestines, generating a watery, starchy-colored diarrhea or "rice water" stool. This stool may also contain flecks of intestinal mucosa—of course, it's also loaded with the cholera bacteria that will get passed on to caregivers and the environment to perpetuate the spread of this disease.

Proper sewage and safe water management are key cholera preventions. Higher incidence occurs during rainy seasons and following hurricanes, typhoons, and other natural disasters that may compromise municipal water sanitation. As mentioned with *Shigella*, travelers should be careful about eating or drinking anything that has not been boiled, cooked, or peeled.

Various vaccines have been used in over 60 different countries, but to date they have only offered an incomplete protection. A newer vaccine, Vaxchora™, was approved by the U.S. Food and Drug Administration in 2016. Clinical trials data suggest this oral vaccine confers about a 90 percent rate of protection, but data regarding its levels of protection for people who live in endemic areas is yet to be determined.

The gold standard for diagnosis is culturing *V. cholera* from the patient's feces. Oral rehydration is usually sufficient for most patients to fully recover, but unfortunately access to oral rehydration therapy is limited in many developing nations. To combat this, the WHO recommends the establishment of cholera treatment centers in countries experiencing outbreaks or seasonal occurrences of cholera to ensure timely treatment. In cases that persist even after rehydration therapies or in immune-compromised patients, antibiotics can be prescribed. Doxycycline is the first choice for use in adults, while azithromycin is preferred for pregnant women and children.

Disease**Snapshot** • **Cholera**

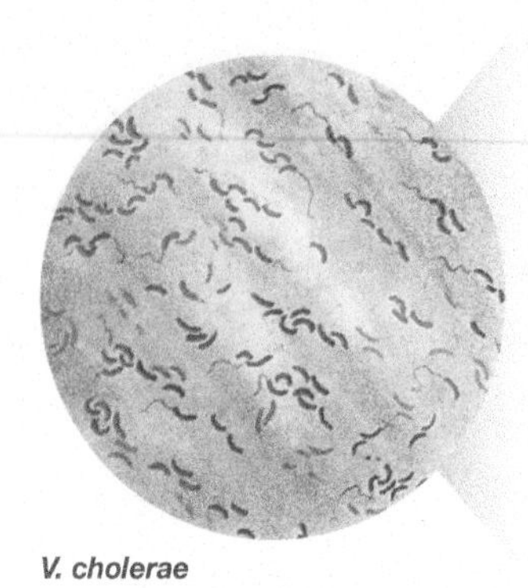

V. cholerae

Causative agent	*Vibrio cholerae*; Gram-negative bacterium
Epidemiology	Rare in developed countries, endemic in African and southeast Asian countries and in areas lacking proper sanitation. Incubation period is hours to a few days.
Transmission & prevention	Transmits via fecal–oral route through contaminated food and water as well as via human-to-human contact. Vaxchora, a single-dose live vaccine, is available to 18–64-year-olds traveling to areas with active cholera transmission. The vaccine offers incomplete protection with unknown longevity, so the best prevention remains proper water sanitation and hygienic food-handling practices.
Signs & symptoms	Vomiting, profuse and watery diarrhea
Pathogenesis & virulence factors	Cholera toxin triggers intestinal epithelial cells to release water and ions into the intestines, generating watery diarrhea
Diagnosis & treatment	Diagnosis is through bacterial culture from stool samples. Treated with oral rehydration therapy for mild symptoms, and intravenous fluids and antibiotics (doxycycline for adults and azithromycin for pregnant women and children) for severe cases.

Pseudomembranous colitis

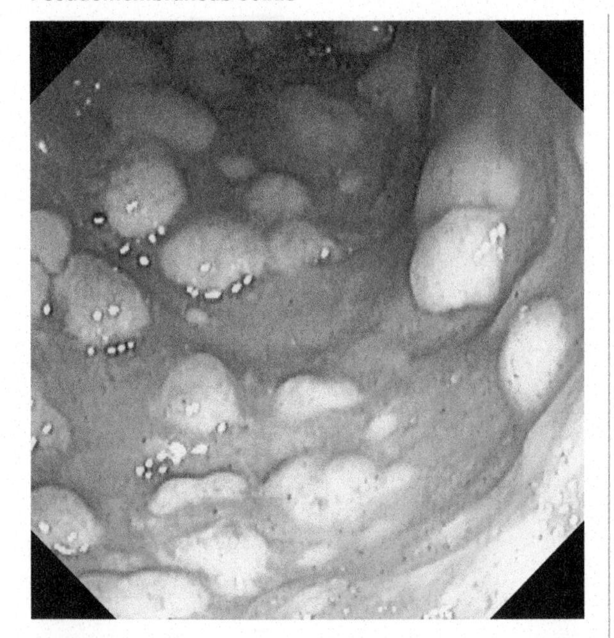

Healthy colon

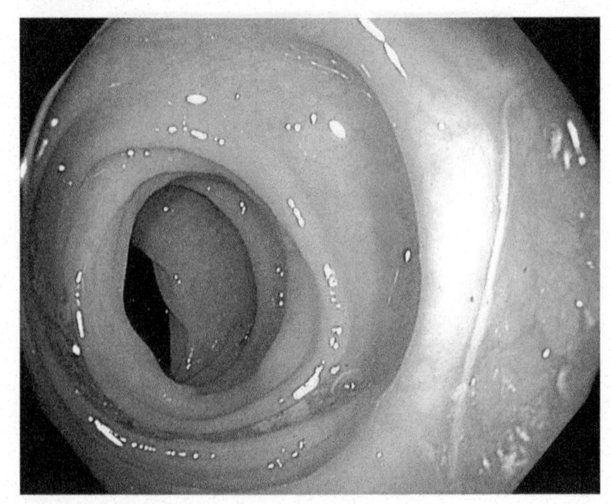

FIGURE 19.14 **Pseudomembranous colitis** Pseudomembranous colitis is usually associated with *C. difficile*. In this form of colitis, pus-filled lesions called pseudomembranous plaques form along the colon lining as *C. difficile* toxins cause tissue damage.

Clostridioides difficile increasingly causes serious, healthcare-acquired infections.

The CDC has put *Clostridioides difficile* (klos TRID ee OY-dees dif uh SEEL) at the top of the list of emerging public health threats. According to the CDC, every year up to half a million people develop *C. difficile* infections, and as many as 29,000 of them die. About 75 percent of cases occur in older people with a history of a hospital stay, or residents of long-term-care facilities. It is also increasingly common for children, people who do not report prior antibiotic therapy, and people who do not have a hospitalization history, to be affected.[5] *C. difficile* is among the most common healthcare-associated infections—surpassing even MRSA (methicillin-resistant *Staphylococcus aureus*).[6]

This spore-forming, Gram-positive anaerobe spreads via the fecal–oral route, through contact with contaminated fomites, or from contaminated hands. It is impossible to fully prevent exposure, since it exists almost everywhere in our environment, and colonizes many people. However, being colonized by *C. difficile* does not necessarily cause disease. People with nonpathogenic strains that don't make toxins, and even those colonized with pathogenic, toxin-manufacturing forms, are usually protected by their normal intestinal microbiota. However, *C. difficile* forms dormant endospores and can therefore naturally resist antibiotics—so if the normal microbiota are removed via a course of antibiotics, *C. difficile* may suddenly have the opportunity to thrive. As such, the CDC's strategy for *C. difficile* prevention is largely focused on limiting unnecessary antibiotic prescriptions. (The CDC estimates that up to half of antibiotic prescriptions are for illness not caused by bacteria, or for disease that will resolve without any medication.) Every unwarranted antibiotic prescription, be it in a hospital or community setting, provides an opportunity for *C. difficile* to cause disease.

When *C. difficile* flourishes, it causes anything from mild diarrhea to **pseudomembranous colitis**, which is an inflammation of the colon that is accompanied by pus-filled nodules (pseudomembranous plaques) (FIG. 19.14). Symptoms usually develop toward the end of an antibiotic treatment—or soon after the antibiotic therapy is completed. Mild cases feature diarrhea and moderate abdominal pain. Serious cases are characterized by severe abdominal pain, fever, nausea, and abundant watery diarrhea that may progress to dysentery. In extreme but rare cases, **toxic megacolon** may develop. In this condition the

[5] Gupta, A., & Khanna, S. (2014). Community-acquired *Clostridium difficile* infection: An increasing public health threat. *Infection and Drug Resistance, 7*, 63–72.

[6] Fu, Y., Luo, Y., & Grinspan, A. M. (2021). Epidemiology of community-acquired and recurrent *Clostridioides difficile* infection. *Therapeutic Advances in Gastroenterology, 14*, 1–11.

Disease**Snapshot** • ***Clostridioides difficile* pseudomembranous colitis**

Causative agent	*Clostridioides difficile*; Gram-positive bacterium
Epidemiology	Most cases in elderly who have had an extended stay in a healthcare facility and recent antibiotic therapy. May be a resident microbe that emerges as an opportunistic pathogen after antibiotics.
Transmission & prevention	Fecal–oral transmission; contact with contaminated fomites or hands. Isolation of infected individual and protective gloves and gown for healthcare workers prevents transmission, as does reducing unnecessary antibiotic prescriptions that destroy protective gut microbiota.
Signs & symptoms	Diarrhea, fever, abdominal pain, and nausea
Pathogenesis & virulence factors	Spore-forming bacteria that naturally resist detergents and drugs. Releases A and B toxins that damage the colon and induce inflammation. Emerging strains make a third toxin (binary toxin) that reduces eosinophil function and may suppress the innate immune response.
Diagnosis & treatment	Diagnosed by culturing toxigenic *C. difficile* from the patient's stool. Molecular methods detect *C. difficile* toxin genes or actual toxins in stool, or look for glutamate dehydrogenase (an enzyme made by the bacterium) in stool. Colonoscopy reveals pseudomembranous colitis. First-line therapy for moderate cases is the antibiotic metronidazole. Rehydration therapies and the antibiotic vancomycin for severe cases.

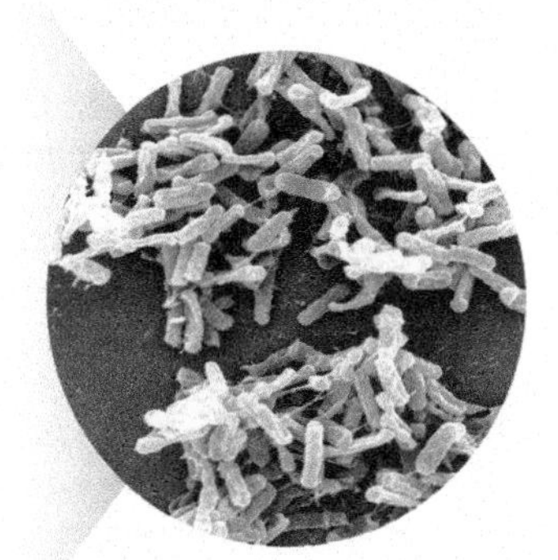

C. difficile

large intestine can't expel gas and feces so it becomes distended and stressed; and eventually a *perforation*, or a tear in the colon, may develop. As bacteria seep into the normally sterile abdominal cavity and cause inflammation, the patient may progress to septic shock and die.

Most pathogenic *C. difficile* strains release two exotoxins (A and B toxins). An emerging and more dangerous strain makes a third toxin (binary toxin). The A and B toxins promote a vigorous inflammatory response and also damage the cytoskeleton of the colon's epithelial cells. As the A and B toxins damage the integrity of the colon's lining, pus-filled pseudomembranous lesions develop. These lesions eventually slough off and may occasionally be seen in the stool. The effect of binary toxin is still being investigated, but it has been shown to interfere with the innate immune response by suppressing eosinophil functions.[7] Strains that make binary toxin cause more severe disease and are associated with a higher mortality rate.

The threat of newly emerging hypertoxic strains is compounded by the fact that the bacterium's spores can survive outside the body on diverse fomites for months. Even after bed rails, call buttons, bathroom areas, and door handles have been cleaned, they often still harbor *C. difficile* spores. Not only does the bacterium have natural resistance to several antibiotics and cleaning agents due to its ability to retreat into a dormant spore form, but it readily acquires additional resistance mechanisms through genetic variation. Research indicates a large number of transposons containing antibiotic-resistance genes exist in the genome.

In short, the stage is set for a perfect storm. First, there is the challenge of managing exposure to a bacterium that can naturally escape drugs and disinfectants by retreating into a resistant spore state. Second, antibiotic misuse and overuse needlessly put patients at risk for infection. And third, there are emerging superstrains that make a third type of toxin and increasingly resist our drug arsenal through newly acquired genes. To combat this storm we'll need new drugs and interventions. Fortunately, probiotics and fecal transplants from healthy individuals provide hope in the battle against *C. difficile*. So far, these therapies have proven effective in restoring normal intestinal microbiota and reducing disease reoccurrence. (For more on fecal transplants, see the Bench to Bedside feature.)

CLINICAL VOCABULARY

Colitis inflammation of the colon (the large intestine)

[7] Cowardin, C. A., Buonomo, E. L., Saleh, M. M., Wilson, M. G., Burgess, S. L., Kuehne, S. A., et al. (2016). The binary toxin CDT enhances *Clostridium difficile* virulence by suppressing protective colonic eosinophilia. *Nature Microbiology*, 1, 16108.

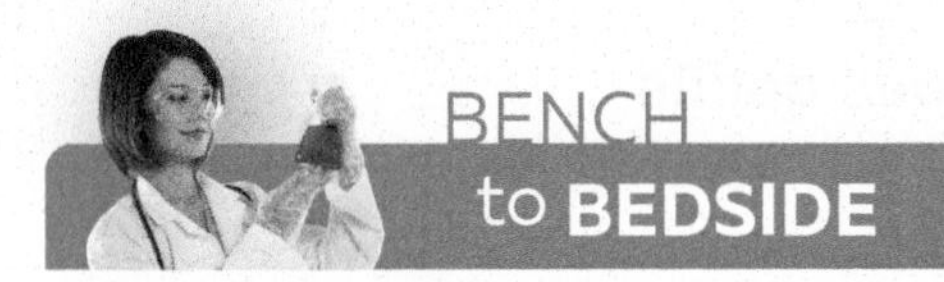

Using Excrement to Cure Chronic *Clostridioides difficile* Infections

Using feces as a curative agent for chronic diarrhea may sound outlandish, but as antibiotics are becoming less effective against *C. difficile,* fecal transplants are one of the alternative treatments being tested. In some cases, cure rates of 90 percent have been achieved, making these promising therapies.

The procedure involves a colonoscopy in which stool slurries from healthy donors are infused into the digestive tract of the person with the infection. If successful, the bacteria found in the donor feces can repopulate the intestines and help keep *C. difficile* in check. Donors are screened for various diseases to ensure no new pathogens get passed along with the transplant.

Research shows that a recipient's gut microbiota is indeed changed when donor fecal material is introduced. The procedure shows great promise, and researchers are working to develop easier administration methods, as well as specific microbial transplants that might one day replace the use of whole stool in the process. Fecal transplantation is also being tested on patients suffering from irritable bowel syndrome, inflammatory bowel disease, and metabolic syndrome.

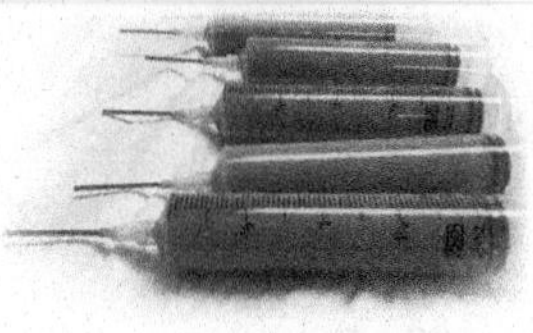
Stool samples for transplantation

Sources: Kelly, C. R., Yen, E. F., Grinspan, A. M., Kahn, S. A., Atreja, A., Lewis, J. D., ... & Laine, L. (2021). Fecal microbiota transplantation is highly effective in real-world practice: initial results from the FMT national registry. *Gastroenterology, 160*(1), 183–192.

Build Your Foundation (BYF) Quick Quiz: Visit the **Mastering Microbiology** Study Area to quiz yourself.

Build Your Foundation

11. What are the signs and symptoms of dental caries and periodontal disease and how do biofilms promote these conditions?
12. How does *Helicobacter pylori* cause ulcers?
13. Compare and contrast foodborne infections to food poisoning, including the common bacterial culprits of each.
14. What features distinguish disease caused by *Campylobacter* species, *Shigella* species, *E. coli*, *Salmonella* species, and *Listeria monocytogenes*?
15. What virulence factors do *Shigella* and *E. coli* O157:H7 have?
16. What are the features of cholera and the agent that causes it?
17. What disease is caused by *Clostridioides difficile* and why is it an increasing concern in healthcare settings?

19.5 PROTOZOAN AND HELMINTHIC DIGESTIVE SYSTEM INFECTIONS

Learning Outcomes

After reading this section, you should be able to:

19.18 Explain differences between protozoan and helminthic parasites and state how they are mainly transmitted.
19.19 Describe the causative agents and features of giardiasis, amoebiasis, and cryptosporidiosis.
19.20 Explain the typical life cycle of a parasitic protozoan that infects the GI tract.
19.21 Describe the major tapeworm and roundworm infections of the human GI tract.

Common protozoan infections include giardiasis, amoebiasis, and cryptosporidiosis.

Protozoans are single-celled eukaryotes. Although proper water and sewage management have made these parasites less of a burden in developed countries than in developing nations, they are by no means rare. Protozoan infections are underdiagnosed in developed countries due to lack of clinical suspicion of them, and the lack of sensitive detection tests. As climate change impacts the frequency and intensity of rainfall and as urban development shifts where animal and human populations settle, we'll inevitably see an increase in these diseases in all countries—wealthy and poor. The top three intestinal protozoans in developed countries are *Giardia lamblia*, *Entamoeba histolytica*, and *Cryptosporidium species*.

Giardiasis

"Montezuma's revenge," "Delhi belly," "Turkey trot," and "the Nepali quickstep" are all names given to traveler's diarrhea, which is occasionally caused by the protozoan parasite *Giardia lamblia*. However, this parasite is not restricted to just travelers. The CDC estimates that *Giardia lamblia* infects the intestines of about

DiseaseSnapshot • Giardiasis

Causative agent	*Giardia lamblia*
Epidemiology	Worldwide incidence; most common intestinal protozoan in the U.S.; anyone who drinks untreated water or has close contact with an infected person is at risk. Incubation is 1–2 weeks.
Transmission & prevention	Transmits via fecal–oral route, via contaminated fomites; or from direct contact with infected patient. Ingesting as few as 10 cysts is sufficient to establish infection. Prevention involves boiling water to deactivate cysts, or filtering water to remove them.
Signs & symptoms	Diarrhea, gas, abdominal cramps, nausea, vomiting; chronic infections can cause nutrient malabsorption
Pathogenesis & virulence factors	Cysts are resistant to chlorine, so treating water with that agent alone is ineffective. The cysts also survive outside the body for lengthy periods of time.
Diagnosis & treatment	Diagnosed via microscopic evaluation of feces for cysts or trophozoites. There are also molecular methods to detect *Giardia* antigens. Treatment involves supportive therapy, and drugs metronidazole or nitazoxanide for severe infections.

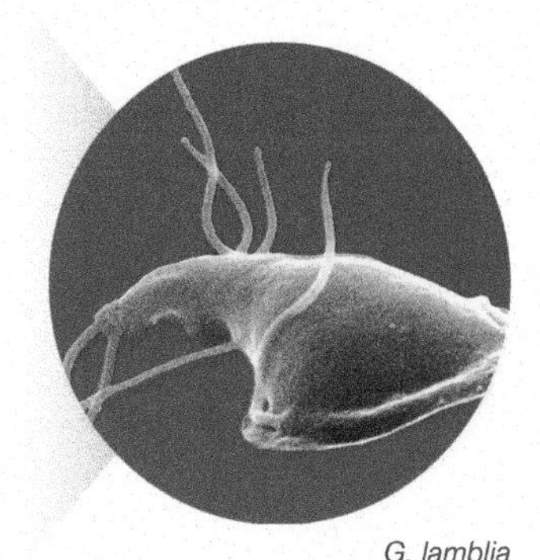

G. lamblia

2 percent of adults and 6 to 8 percent of children in developed countries around the world. It also lives in animals other than humans and is frequently seen in developing nations. In the United States, *Giardia* is the most common intestinal parasite; it is usually acquired by swallowing feces-contaminated food or water, but can also be transmitted from person to person or from animals to people. People at risk include backpackers and campers who drink untreated water from lakes, ponds, or streams; people in areas that have experienced hurricanes, flooding rains, or other natural disasters that compromise sewage and water management; and anyone who has close contact with an infected person.

Outside of its host, *Giardia* exists as a tough cyst form that resists chlorine disinfection used for municipal water treatment, UV light, and freezing. After being ingested, it passes unharmed through the stomach, and in the small intestines the cyst "hatches" to release two trophozoites. Trophozoites resemble flattened teardrop-shaped organisms with flagella. Using an adhesive disc, the protozoan attaches to epithelial cells that line the small intestines and leeches off the host (FIG. 19.15). Following this feeding stage, the trophozoites reproduce asexually. Some will then progress toward the colon and, while in transit, enter the cyst stage. Both cysts and trophozoites are expelled in the feces, but only the cysts survive and can go on to establish an infection in another host.

In addition to diarrhea, symptoms can include excessive flatulence ("gas"), which causes a feeling of bloating, and vomiting. Usually symptoms subside without treatment in 2–6 weeks. For severe symptoms, several drugs are available, such as metronidazole and nitazoxanide.

Amebiasis

Amebiasis, or amebic dysentery, affects people worldwide, but is most prevalent in tropical areas such as southeast Asia, southeast and west Africa, as well as Central and South America. Most cases in the United States are imported. The WHO estimates that 500 million people worldwide harbor the causative intestinal parasite, *Entamoeba histolytica*. Infection is via the fecal–oral route when contaminated food or water is ingested, but the amoeba can also be transmitted by direct contact with anyone that is infected and via feces-contaminated fomites.

This amoeba is similar to *Giardia lamblia* in that its life cycle consists of two stages: cyst and trophozoite. Ingested cysts become trophozoites in the intestine; there they primarily feed on bacteria. While 90 percent of infections are asymptomatic, when symptoms do develop they can range from mild diarrhea to severe dysentery. The dysentery form is characterized by profuse diarrhea containing mucus and blood, severe abdominal cramping, and a low-grade fever. Invasive disease develops when trophozoites release enzymes that kill epithelial cells and cause ulcers. When trophozoites enter the bloodstream, they can cause abscesses in the liver, lungs, and (very rarely) the brain.

Clinical CASE NCLEX HESI TEAS

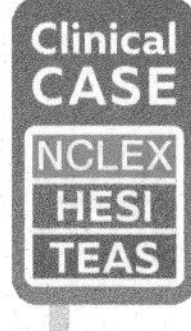

The Case of the Hurricane Health Hazard

Practice applying what you know clinically: visit the **Mastering Microbiology** Study Area to watch Part 3 and practice for future exams.

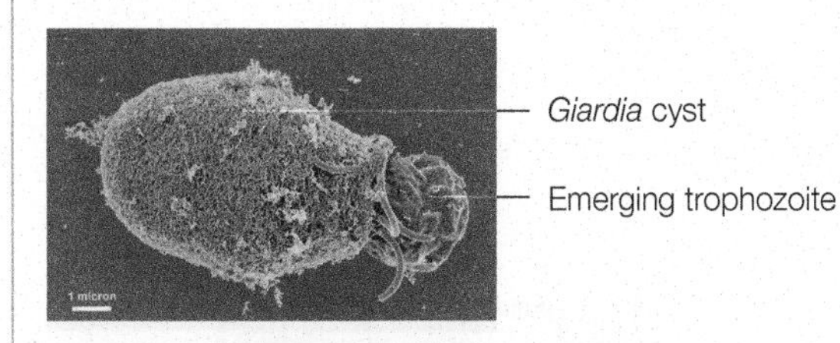

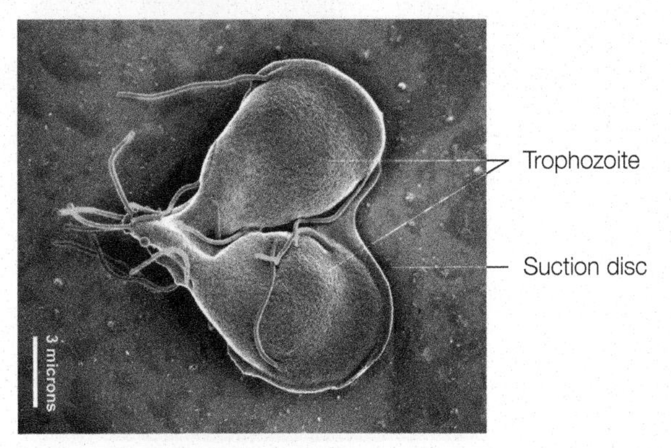

FIGURE 19.15 Two stages of *Giardia* development *Top*: Cyst stage of *Giardia* at the moment when the cyst is releasing a trophozoite. *Bottom*: Twin trophozoites that emerge from the cyst.

DiseaseSnapshot • Amebiasis

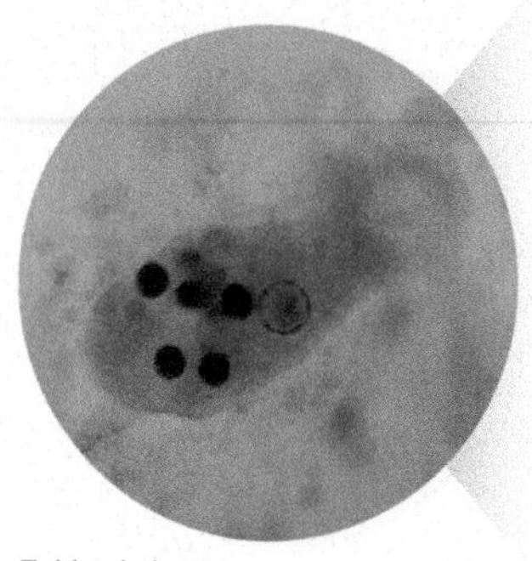

E. histolytica

Causative agent	*Entamoeba histolytica*
Epidemiology	Endemic in tropical areas with poor sanitation. Incubation period is 2–4 weeks.
Transmission & prevention	Transmits via the fecal–oral route, direct contact, or contaminated fomites. To prevent, avoid untreated water and ice. Boiling water deactivates cysts while filtering removes them.
Signs & symptoms	Dysentery form is characterized by profuse diarrhea containing mucus and blood, severe abdominal cramping, and low-grade fever. Invasive disease may develop when trophozoites release enzymes that kill epithelial cells and cause ulcers. If trophozoites enter the bloodstream they can cause abscesses in the liver, lungs, and (very rarely) the brain.
Pathogenesis & virulence factors	Trophozoites produce enzymes that lyse host epithelial cells and promote invasive disease. Cysts resist environmental stress to aid parasite transmission.
Diagnosis & treatment	Diagnosed via microscopic evaluation of feces for cysts or trophozoites. All infections should be treated because the protozoan can cause invasive disease. Metronidazole is prescribed first, followed by paromomycin.

Cryptosporidiosis

In April 1993, pharmacies in Milwaukee started selling out of antidiarrheal medications and emergency rooms were inundated with diarrhea cases. Eventually it was determined that about a quarter of Milwaukee's population was suffering from cryptosporidiosis. While the Milwaukee incident has gone down as the largest documented outbreak of waterborne infection in the United States, *Cryptosporidium* species (especially *C. parvum* and *C. hominis*) are an ever-present worldwide concern and tend to account for about 9 percent of infectious diarrhea in developed countries.[8] According to the U.S. Environmental Protection Agency, water-testing programs have revealed that about 65 to 97 percent of U.S. surface water (lakes, streams, etc.) contains *Cryptosporidium* oocysts. As such, it's not especially surprising that most people, even in developed countries, have been exposed to this parasite.

This water-associated apicomplexan parasite is transmitted by the fecal–oral route. Outbreaks are also linked to daycare centers and recreational water such as swimming pools, hot tubs, lakes, ponds, rivers, and streams. *Cryptosporidium* species naturally reside in the intestines of humans and animals, particularly cattle. Since the diarrhea of infected individuals and animals can contain millions of cells, spread in water or food by fecal contamination is relatively easy. Ingestion of as few as ten cells can result in infection. Most people who swallow the parasite show mild diarrheal symptoms for 1 to 2 weeks. However, for the immunocompromised, *cryptosporidiosis* can lead to more serious diarrhea and chronic infection that may lead to death.

Cryptosporidium enters its host as oocysts, spore-like structures that survive lengthy periods outside a mammalian host. Upon entering the intestines, the oocysts then release sickle-shaped sporozoites into the intestine. The sporozoites attach to epithelial cells to multiply, inciting an inflammatory response. The parasites then undergo asexual reproduction and later sexual reproduction to make macrogametes (female) and microgametes (male). The microgametes then fertilize the macrogametes to produce the oocysts, which are the infectious form of the parasite that gets excreted with feces (FIG. 19.16).

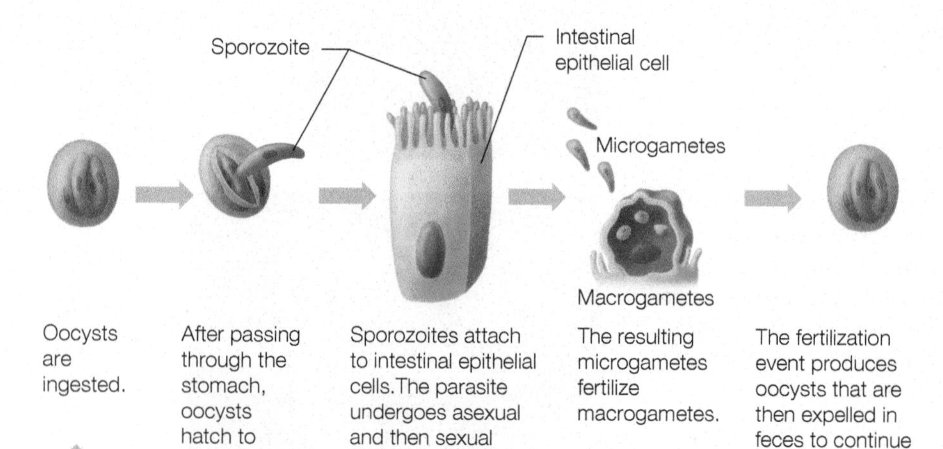

FIGURE 19.16 ***Cryptosporidium* life cycle**

[8] Fletcher, S. M., Stark, D., Harkness, J., & Ellis, J. (2012). Enteric protozoa in the developed world: A public health perspective. *Clinical Microbiology Reviews*, 25(3), 420–449.

DiseaseSnapshot • Cryptosporidiosis

Causative agent	*Cryptosporidium species* (especially *C. parvum* and *C. hominis*)
Epidemiology	Associated with drinking or swimming in untreated water; common in daycare settings; seasonal outbreaks mainly occur in summer. Incubation is 2–10 days (7 days average).
Transmission & prevention	Transmits via fecal–oral, direct contact, or via contaminated fomites. Prevent by avoiding untreated water and ice. Water filters can remove oocysts, while boiling deactivates them.
Signs & symptoms	Diarrhea, nausea, vomiting, fever, and stomach cramps
Pathogenesis & virulence factors	Oocysts have a protective layer that protects them as they pass through the stomach, and also allows them to resist environmental stress and to survive outside the host for long periods. Once in the intestine, sporozoites attach to host cells and build a protective membrane from modified host cell membrane materials. This covering allows them to blend in with host cells while still technically remaining outside the host cell.
Diagnosis & treatment	Diagnosis involves microscopic evaluation of stool for oocysts, or detecting antigens. Cases usually self-resolve, as long as the patient keeps hydrated. Nitazoxanide is prescribed for more severe infections or for immune-compromised patients.

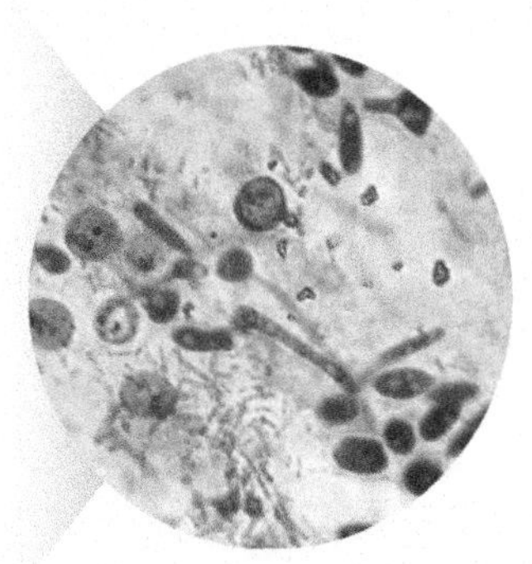

Cryptosporidium species

To prevent infection, water should be treated using filtration and UV radiation. Chlorination alone is not sufficient—hence outbreaks that have occurred in swimming pools and water systems.

Numerous helminths can infect the digestive tract and may migrate to other tissues.

Unlike protozoans, which are unciellular, helminths (or worms) are multicellular animals. In their adult form they can often be seen with the naked eye, but they are usually transmitted as microscopic eggs or larvae in either soil or water. Helminths are either round worms or flatworms, with flatworms including tapeworms and flukes. Many people living in developed nations tend to think that helminth infections only affect people living in poor and unhygienic settings. While helminths certainly present a greater burden of disease in developing countries, anyone can "get worms," and many people who live in developed countries have had some sort of worm infection in their lifetime.

Tapeworms

Tapeworms have flattened, segmented bodies with a **scolex** (head) at one end. The scolex has hook and/or sucker structures to help the worm attach to the intestinal lining in the definitive host. In their definitive host, they mature and release eggs as segments of the worm called **proglottids** break off from the distal end of the body—the end furthest from the scolex. The eggs (or even intact proglottids) are passed in the feces, at which point another host may ingest the eggs and become infected. Most tapeworm infections in humans are linked to *Hymenolepis nana*, *Taenia* species (*T. saginata*, *T. solium*, and *T. asiatica*), and *Diphyllobothrium latum*.

Hymenolepis nana These worms have a worldwide distribution and are the single most common cause of tapeworm infections in humans. In their adult form, *H. nana* worms are less than two inches long, which is why they are also called dwarf tapeworms. Most infections are transmitted person to person via the fecal–oral route when the worm's eggs are ingested in food or water contaminated with human feces. However, some cases are acquired through accidental ingestion of infected arthropods, such as fleas. Although anyone may be infected, most cases are in children. Infections with light worm loads are asymptomatic, but can last for years. When heavy worm loads develop, symptoms may include diarrhea, abdominal discomfort, nausea, and loss of appetite; children may also experience sleep disturbance, anal itching, and headache. Unlike other tapeworm infections that require an intermediate host, *H. nana*'s

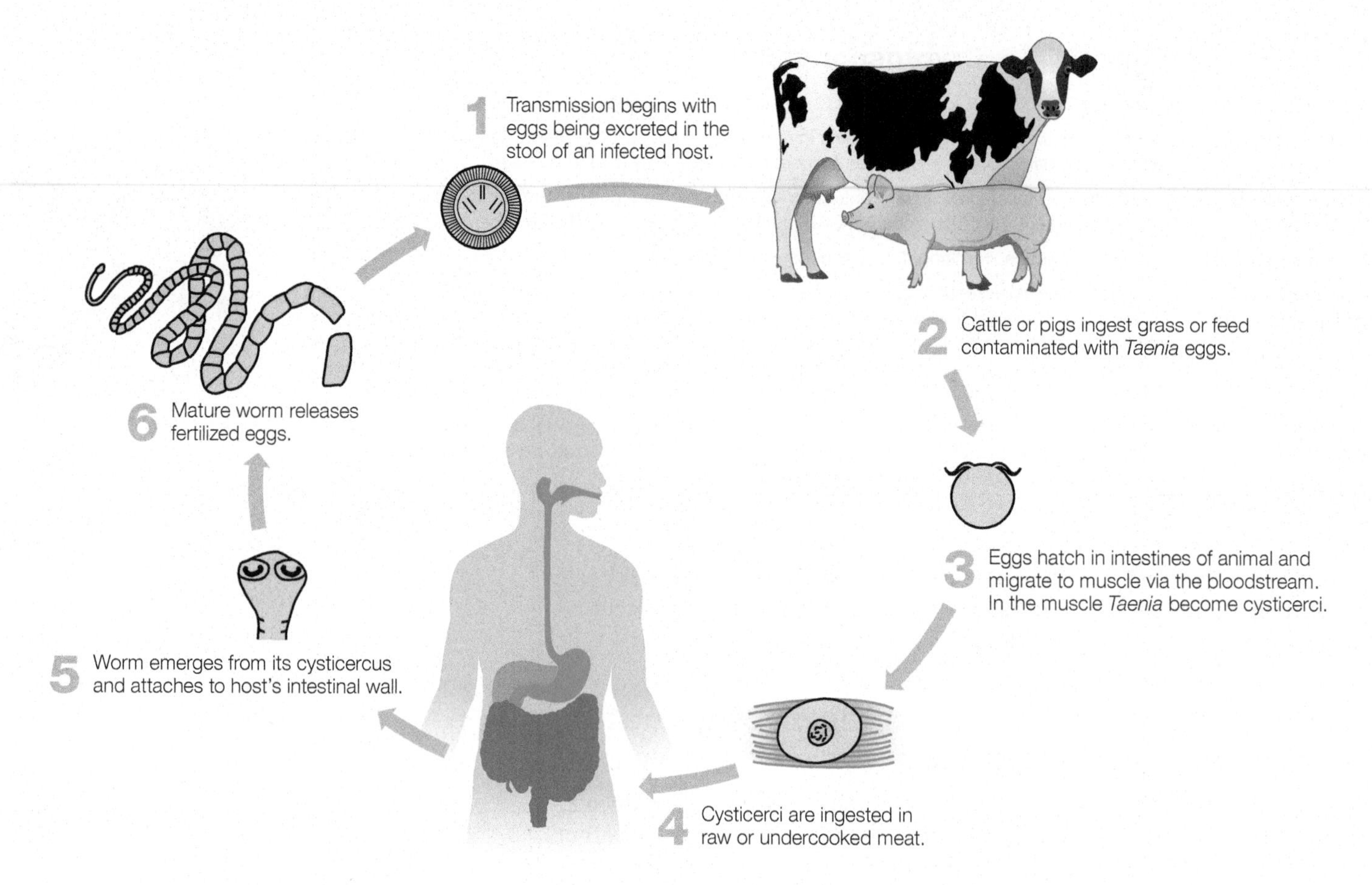

FIGURE 19.17 ***Taenia*** **tapeworm life cycle and transmission**

Critical Thinking *How would this figure be modified to depict the development of cysticercosis?*

eggs can hatch inside the same human host where the adult worm released them, which can easily lead to heavy worm loads.

***Taenia* species** Of the 32 recognized *Taenia* species, only three are known to cause human infections. *T. saginata* infects mainly cattle, while *T. solium* and *T. asiatica* infect pigs. These worms have a two-host life cycle, with humans being the definitive host where the adult worms thrive and release eggs, and cattle or pigs serving as intermediate hosts for the worm's larval stage. When sewage contaminates areas where animals graze or are housed, cattle or pigs are likely to ingest the worm eggs found in feces from the human host. The eggs then hatch in the pig's or cow's intestine. The hatched larvae migrate from the intestine to the animal's muscles. In the muscles, the larvae become *cysticerci* (singular = cysticercus)—larvae encased in a protective cyst-like structure that can survive for years in the animal's tissues. Humans are infected upon ingesting cysticerci in raw or undercooked meat. In the small intestine, the worm's scolex and neck emerge from the cysticercus so that the worm can attach to the human host's intestinal wall. After 5 to 12 weeks of thriving on the host for nutrients, the tapeworm reaches maturity and starts to release eggs (FIG. 19.17). Most infections are asymptomatic, but tapeworms can be very long, so they can potentially cause intestinal irritation and bowel obstruction.[9]

A complication of infection called **cysticercosis** occurs when humans ingest the *T. solium* eggs rather than the cysticerci in undercooked pork. Cysticercosis has not been linked to *T. saginata* (beef tapeworm) and it is still unclear if *T. asiatica* causes this form of infection. Exposure to *T. solium* eggs could occur if crops are irrigated with sewage-contaminated water, if drinking water is

[9] *T. saginata* is on average 5 meters or less in length (16 ft), but can grow to be a staggering 25 m (82 ft). *T. solium* and *T. asiatica* are usually 2 to 7 meters long (6.5 to 22 ft). All three species can live for years in the human intestine.

contaminated, or if food is prepared by someone who is a tapeworm carrier and very likely carries the eggs on their hands (thus vegetarians can also become infected). Humans are an accidental host of the larval stage; it doesn't benefit the parasite to enter humans because we are not a food source, so this form of infection is a dead end for the parasite.

Upon ingestion, the eggs hatch in the human host's intestines; the larvae then burrow through the intestinal epithelium and make their way via the bloodstream to human tissues, including the brain and eyes. When cysticerci form in the tissues they interfere with tissue function and cause severe and potentially deadly symptoms, especially when the brain is affected (neurocysticercosis). In developing nations it is a leading cause of adult-onset seizures. Symptomatic cases can be treated with antihelminthic drugs (such as praziquantel) and anti-inflammatory agents like corticosteroids. Surgery may be needed to remove cysticerci in certain parts of the brain.

The WHO estimates that at any given time there are 50–100 million people in the world with cysticercosis. However, the CDC estimates there are only about 1,000 cases per year in the United States, with most cases occurring in newly immigrated people from Latin America.

Diphyllobothrium latum Also known as fish tapeworm, this parasite requires three hosts: crustaceans, fish, and mammals. Eggs that are defecated by a mammalian host (humans, bears, large cats, etc.) make their way into freshwater and become ciliated embryos. Crustaceans, like copepods and water fleas, then eat the ciliated embryo. In the crustacean, the embryo matures into its first larval stage. Fish eat the infected crustaceans and the parasite migrates into the fish's flesh to form larval cysts. When humans (or other mammals) eat infected fish, the larval form becomes an immature worm that attaches to the intestinal wall. As it parasitizes its host for nutrients, it matures into an adult worm. In about 6 weeks the mature worm releases eggs in the feces to continue the infection cycle. *D. latum* is the largest tapeworm in humans; it averages 1 to 15 meters in length (about 3 to 49 ft). Other features of this infection are summarized in the Disease Snapshot table.

Pinworm and Hookworm

Pinworms (*Enterobius vermicularis*) and hookworms (mostly *Necator americanus*) are common roundworms that infect the human intestinal tract. The CDC estimates that at any given time at least half a billion people worldwide harbor hookworm. Like hookworm, pinworm also has a worldwide distribution, but pinworm is much more common in the United States and western Europe than hookworm. The CDC estimates that at any given time 40 million people in the United States are infected with pinworms.

Pinworm These white roundworms reside in the colon and are only about 4–12 millimeters long. When the infected person is asleep, the female worm migrates to the periphery of the anus to lay eggs. The sticky eggs cause intense anal itching. When an infected person scratches, the eggs readily stick to fingers. Because children are more likely to scratch their bottoms without concern for social norms or hand washing, they are a primary means of transmission. Infection spreads when someone swallows pinworm eggs acquired from contaminated surfaces or fingers. Surfaces such as bed linens, toys, clothing, and toilet seats can harbor eggs for up to three weeks. Disturbed sleep, teeth grinding, and an itchy anus are common symptoms. Prevention, diagnosis, and treatment are reviewed in the Disease Snapshot table.

Hookworm *Necator americanus* are soil-transmitted roundworms. When human feces harboring eggs is deposited in the soil by a person defecating outside, through poor sewage management, or by using human waste as fertilizer, the eggs hatch and mature into adult worms that burrow into bare feet and legs.

DiseaseSnapshot • Pinworm and Hookworm

Pinworm

Disease	Pinworm	Hookworm
Causative agent	*Enterobius vermicularis*	*Necator americanus*
Epidemiology	Worldwide; common among children in school settings; easily transmitted to family members in households with infected children. Eggs are infective within a few hours of being laid. It takes 1–2 months for an adult female to mature.	Worldwide, particularly in warm, moist climates; used to be endemic in the southeastern U.S. before most people wore shoes. Now mostly common in children. The process from skin penetration to egg release takes roughly 5 to 9 weeks; eggs hatch in soil within a few days after being deposited.
Transmission & prevention	Ingesting eggs that readily stick to hands or surfaces. Effective hand washing prevents infection.	Eggs are deposited into the soil and hatch into larvae; the larvae penetrate the skin, often through bare feet. Prevention includes sewage management and wearing shoes.
Signs & symptoms	Itchy anus, disturbed sleep, teeth grinding	Few symptoms unless a heavy worm load induces nutrient deficiencies
Pathogenesis & virulence factors	Adult worms exit partially from the anus to deposit eggs, causing an inflammatory response and itching that results in fingers picking up eggs and transferring them to individuals or surfaces. Eggs can remain infective on surfaces for several weeks.	Adult hookworms burrow into skin by secreting enzymes that break down skin proteins; the hookworm enters the bloodstream to eventually penetrate lungs. Coughing brings up the worm to be swallowed, and it enters the intestine, where it attaches and feeds on blood.
Diagnosis & treatment	Tape adhesion test detects eggs; the sticky side of clear tape is pressed against the perianal region and then examined for eggs. Treated with over-the-counter medications such as pyrantel pamoate. When one person is diagnosed, treat the whole household simultaneously.	Detected via microscopic identification of eggs in stool. Treated with albendazole and mebendazole.

The hookworm larvae then enter the bloodstream and make their way to the lungs, where they are coughed up and swallowed to enter the intestine. Here they attach to the intestinal wall to feed on the host's blood, and mature and mate to release fertilized eggs to continue the infection cycle. Symptoms are usually only detected in severe cases caused by a heavy worm load. In such cases, patients may exhibit iron and nutrient deficiencies; such deficiencies tend to have a greater impact on children by impairing mental and physical growth. Infections can be limited with proper sewage management to reduce human feces in the soil and by wearing shoes to prevent larvae from burrowing into the feet.

Ascariasis

According to the CDC, *Ascaris lumbricoides* is among the most prevalent helminthic infections in people. It is estimated to infect between 800 million and 1.2 billion people worldwide—that's roughly one in every four people—at any given time. This roundworm is most common in tropical and subtropical regions with poor sanitation measures or areas that use human fecal mater to fertilize crops. In the United States it is estimated that there are about 4 million people infected at any given time, with most cases occurring among impoverished populations in southeastern states, New Mexico, and Arizona.[10]

Eggs are found in soil contaminated by human feces. Eating unwashed fruit and vegetables, or eating with unwashed hands, can introduce eggs to the GI tract. *A. lumbricoides* eggs easily survive the stomach when ingested and make their way to the intestine. Once in the intestine, the eggs hatch and larvae

[10] Valentine, C. C., Hoffner, R. J., & Henderson, S. O. (2001). Three common presentations of ascariasis infection in an urban emergency department. *The Journal of Emergency Medicine*, 20(2), 135–139.

DiseaseSnapshot • Ascariasis

Causative agent	*Ascaris lumbricoides*
Epidemiology	Uncommon in the U.S.; common worldwide, with higher prevalence in subtropical and tropical areas that lack sanitation and good hygiene practices. Children are more likely to be affected than adults. It takes 2–3 months from egg ingestion to the production of infective eggs by a mature female worm.
Transmission & prevention	Fecal–oral; ingestion of eggs directly from soil or on contaminated hands or produce. Prevention includes sewage management, hand washing, and proper food handling.
Signs & symptoms	Usually asymptomatic; mild infection can cause abdominal pain, nausea, diarrhea, or bloody stools; with high worm load patient may develop aforementioned symptoms, plus fatigue, vomiting, weight loss, malnutrition, or bowel obstruction; worms can live 1 to 2 years in the host, at which point they die and may be excreted in a bowel movement or vomited up.
Pathogenesis & virulence factors	When eggs are swallowed, larvae hatch in the intestine to penetrate into the bloodstream; circulation brings the larvae to the lungs where they are then coughed up and swallowed; once larvae enter the intestines, they attach and develop into adult worms.
Diagnosis & treatment	Microscopic evaluation of feces to detect eggs. Treated with albendazole and mebendazole.

A. lumbricoides

migrate to the lungs via the bloodstream. For the 10–14 days they are in the lungs, they can cause coughing and wheezing. From the lungs, the larvae are coughed up and swallowed. Finally, they settle in the small intestine to mature, mate, and feed on intestinal contents. Unlike the other worms discussed to this point, *Ascaris* worms do not attach to the intestinal wall. Adult worms grow to about 30 cm (about 1 ft) in length and can be up to 6 mm wide (0.25 in). When these worms mate, the females can make up to 200,000 fertilized eggs per day for roughly a year. The eggs are infectious after spending a few days in the soil and they can remain infective for a decade or more.

Similar to the pinworm and hookworm, intestinal *A. lumbricoides* infections are usually asymptomatic. However, a high worm load, which is more common in children, tends to manifest with symptoms that can range from severe abdominal pain and malnutrition that stunts growth to an intestinal blockage as worms aggregate in the bowel. Most alarming is when the worms move out of the body through the anus, nose, or mouth—usually at night. Occasionally an aide worker, missionary, or soldier serving abroad will relate stories of seeing children with *Ascaris* worms extruding from their anus or mouth. The worms tend to survive as long as two years in the host, after which they die and may be passed in a bowel movement or vomited up.

Trichinellosis

Roundworms of the genus *Trichinella* (most notably, though not exclusively, *T. spiralis*) are responsible for **trichinellosis**, also known as trichinosis. Improved practices in the pork industry have made this infection rare in developed countries. But eating undercooked wild game, particularly from carnivores or omnivores like bears, increases the risk of infection.

Worldwide, this roundworm is acquired by eating raw or undercooked wild game meat or pork that contains the parasite. When an animal or human ingests raw meat contaminated with embedded larvae (termed cysts), the larvae mature into worms in the intestine. From the intestine, the larvae migrate into blood and embed in muscles, where they can live for years. Thus, this is a zoonotic infection that is normally transmitted from animals to humans, with the humans serving as an accidental host that is a dead end for the parasite, as we are not a food source. Symptoms of nausea, diarrhea, or vomiting develop within a few days of eating the contaminated meat. Two weeks later muscle pain, fever, headache, chills, and rash may occur. Symptom severity depends on

Disease**Snapshot** • **Trichinellosis**

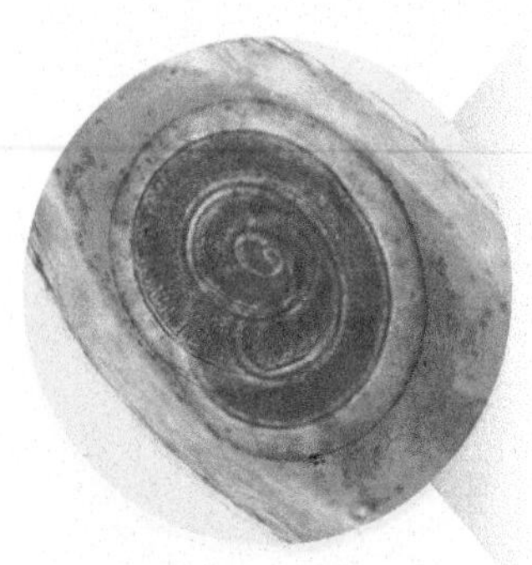

T. spiralis cyst

Causative agent	*Trichinella spiralis*
Epidemiology	Less common in the U.S. and other developed countries; found worldwide in pigs and in wild game such as cougars and bears. Abdominal symptoms develop 1–2 days after cyst ingestion. Further symptoms develop over the next 2–8 weeks as larvae encyst in tissues.
Transmission & prevention	Ingestion of meat that contains cysts, including wild game or pork. Freezing for long periods (over 20 days) can sometimes kill cysts, but is not completely reliable, especially in wild game. The best prevention method is to cook meat to well done.
Signs & symptoms	Nausea, diarrhea, and vomiting within the first couple of days after consumption. Two weeks after consumption, muscle pain, fever, swelling, fatigue, chills, and headache may appear.
Pathogenesis & virulence factors	Cysts are resistant to stomach acid and are only destroyed by sustained extremes in temperatures.
Diagnosis & treatment	Muscle biopsy can be done to detect cysts; analyzing patient's blood for antibodies to *T. spiralis*. Treatment is albendazole or mebendazole, and glucocorticosteroids.

how many cysts were ingested and where the larvae eventually form cysts in the human host. If the patient is exposed to a high load of worms, there may be cardiovascular and nervous system complications; death is also possible, but rare. Symptoms eventually disappear within a few months. Mild cases are rarely diagnosed and are often mistaken for a common illness like the flu.

Avoiding raw and undercooked meat is the way to prevent trichinellosis. Diagnosis is done through a *Trichinella* antibody test. A muscle biopsy may also be warranted. Like most other roundworm infections, mebendazole or albendazole may be prescribed and is most effective within the first several days of infection. Longer drug therapy may be warranted for infections that include encysted larvae hiding in the muscles.

Schistosomiasis

Schistosomiasis is caused by a number of related blood flukes, with the main ones being *Schistosoma haematobium*, *S. japonicum*, and *S. mansoni*. These flattened parasites are taxonomically considered a subgrouping of the flatworms. Although schistosomiasis is rare in the United States and all cases are imported, the CDC estimates that worldwide there are at least 200 million people infected with this blood fluke at any given time. Killing about 200,000 people per year, schistosomiasis is one of the most devastating parasitic infections in terms of global impact—second only to malaria.

FIGURE 19.18 ***Schistosoma* life cycle**

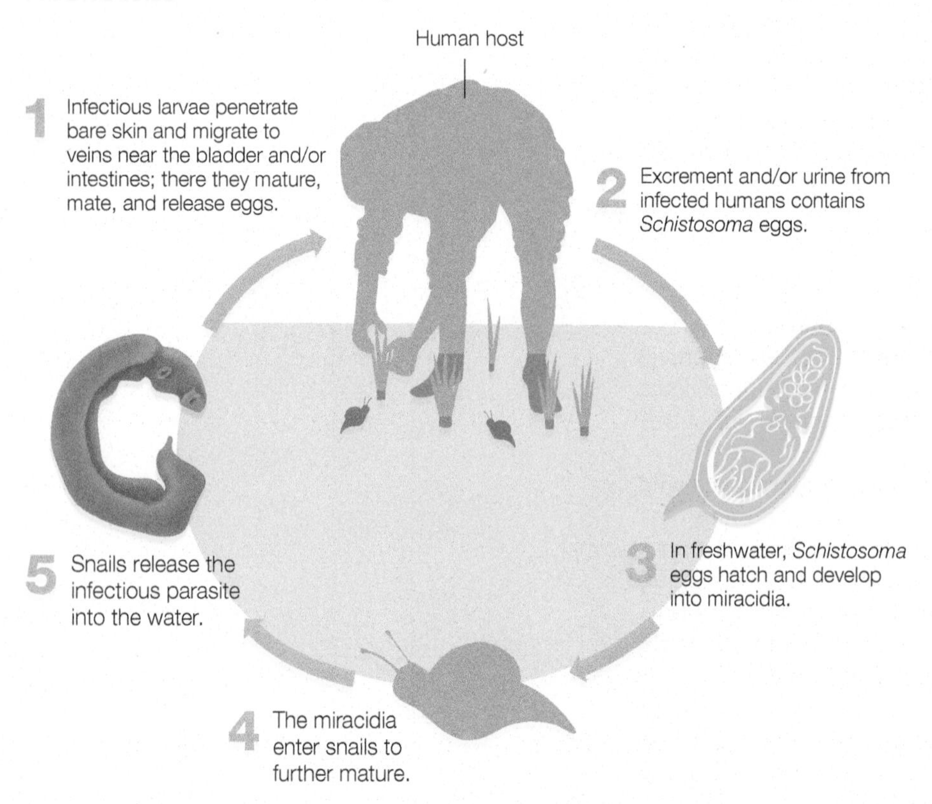

In order for transmission to occur, four participants are needed: *Schistosoma* parasites, humans, snails, and freshwater. When someone defecates or urinates in the water or if sewage pollutes a freshwater source, the deposited *Schistosoma* eggs mature into miracidia, which are a first larval stage. The ciliated, elliptical miracidia enter snails to further mature into a second-stage larval form that can infect humans. Snails release the infectious larval form of the parasite into the water, where it can then penetrate the skin of a human host (FIG. 19.18).

Once the larvae penetrate a person's skin, they lose their tails and migrate to various veins that serve the intestines or bladder—the final vein where they reside is based on which species

DiseaseSnapshot • Schistosomiasis

Causative agent	*Schistosoma masoni, S. japonicum,* and *S. haematobium*
Epidemiology	85% of the world's cases are in Africa; also endemic in the Middle East, China, Latin America, and southeast Asia. Incubation period is 14–84 days.
Transmission & prevention	Bare skin contact with contaminated freshwater via wading, bathing, swimming, or washing transmits the disease. Avoid wading, swimming, or washing in freshwater where schistosomiasis is endemic, to prevent disease. Proper sanitation measures; snail vector management/control and public education to limit defecation/urination in water can also prevent new cases.
Signs & symptoms	Itchy rash may develop when first infected, and within two months symptoms of fever, chills, cough, and muscle aches may arise.
Pathogenesis & virulence factors	Flukes coat themselves with proteins that mimic human proteins, thus avoiding antibody recognition.
Diagnosis & treatment	Detected with microscopy to identify eggs in feces or urine. There are also molecular methods to detect parasite antigens or DNA, and a patient blood test can detect antibodies to the parasite. Treatment is praziquantel.

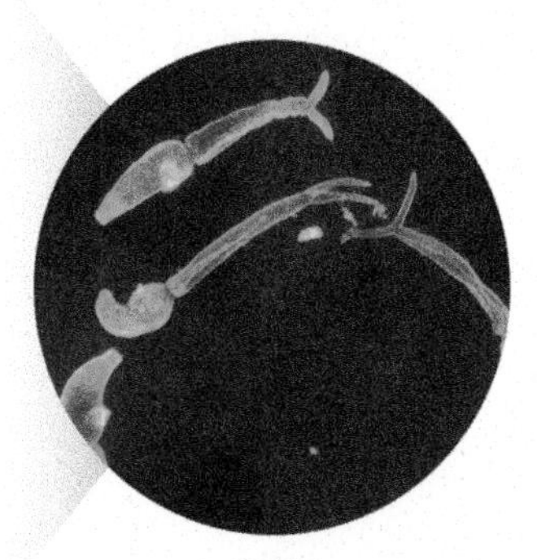

Schistosoma species

is being considered. While in the veins, the larvae mature into adult worms and mate. The resulting fertilized eggs are released into the blood and make their way into the intestinal tract and/or bladder. The infection cycle continues once eggs are deposited into freshwater via excreted feces or urine.

The course of infection is described as having three stages: migratory, acute, and chronic. The migratory phase occurs within a few days of initial infection and is characterized by a rash or itchy skin at the site of migration. Some people then progress to an acute phase. If this stage is going to develop, it tends to do so within 2 months of initial infection and coincides with the first wave of egg release. In the acute stage the patient develops fever, chills, cough, and muscle aches. The chronic stage is related to deposition of eggs in vessels and tissues. While eggs make their way toward the intestines or bladder, a number of them are swept up in the blood circulation and instead lodge in blood vessels and various tissues around the body to cause varied and potentially deadly symptoms. With chronic infection, additional symptoms of abdominal pain, enlarged liver, and swollen abdomen may be present.

Build Your Foundation

18. What are some key differences between protozoans and helminths and how are these parasites primarily transmitted?
19. What are the causative agents and features of giardiasis, amoebiasis, and cryptosporidiosis?
20. What is the typical life cycle of a parasitic protozoan that infects the human GI tract?
21. Describe the major tapeworm and roundworm infections of the human GI tract.

Build Your Foundation (BYF) Quick Quiz: Visit the **Mastering Microbiology** Study Area to quiz yourself.

VISUAL SUMMARY | *Digestive System Infections*

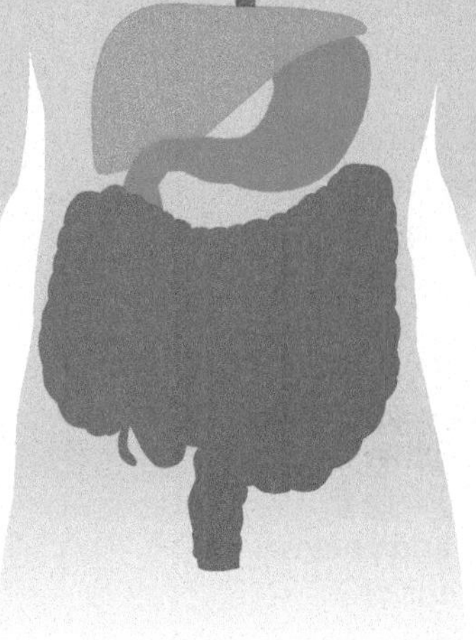

Vaccine available

Mouth and Salivary Glands

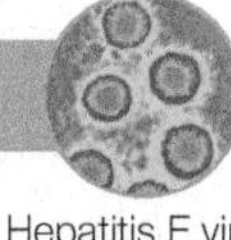

Dental caries caused by *S. mutans*

Viral

Mumps

- Salivary gland infection by mumps virus. (vaccine available)

Bacterial

Dental caries

Streptococcus mutans

- Bacteria in plaque release acids as they metabolize sugars; the acid erodes tooth enamel.

Periodontal disease

- Gums and underlying bone become inflamed as anaerobic bacteria infiltrate below the gum line; may lead to tooth loss.

Liver

Hepatitis E virus

Viral

Hepatitis

Hepatitis A virus (vaccine available)

- Acute viral infection that can cause jaundice and fatigue.
- Transmitted by fecal—oral route.

Hepatitis B virus (vaccine available)

- Infection can be acute or chronic; bloodborne agent.

Hepatitis C virus

- Chronic infection that leads to cirrhosis of the liver; bloodborne agent.

Hepatitis D virus

- Coinfecting virus with hepatitis B.

Hepatitis E virus

- Similar to hepatitis A symptoms; transmitted by fecal—oral route.

Stomach

H. pylori

Bacterial

Stomach ulcers

Helicobacter pylori

- Causes localized inflammation in stomach lining.

Intestines

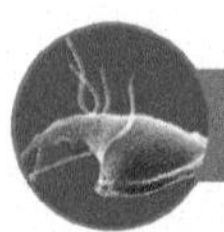

G. lamblia

Viral

Rotavirus (vaccine available)

- Diarrhea can cause severe dehydration in infants.

Norovirus

- Vomiting and diarrhea can last from 3 to 8 days.

Helminth

Tapeworms

Taenia species and *Diphyllobothrium latum*

- Transmission by eating under-cooked beef, pork, or fish containing larvae.

Hymenolepis nana

- Human-to-human transmission; most common tapeworm infection.

Roundworms

Necator americanus (hookworms)

- Mature adults in soil can burrow into bare feet and migrate to intestine.

Enterobius vermicularis (pinworms)

- Common in U.S.; worm's eggs are the infectious stage.

Trichinella spiralis (trichinellosis)

- Transmitted by eating under-cooked meat.

Ascaris lumbricoides

- Fecal—oral transmission by ingesting worm eggs on food or hands.

Flatworms

Schistosoma species

- Snails release infectious larvae into water; larvae penetrate human host's skin and migrate to GI tract or bladder.

Protozoan

Infectious diarrhea

Giardia lamblia (giardiasis)

- Fecal—oral transmission of parasite cysts.

Entamoeba histolytica (amebiasis)

- Fecal—oral transmission of parasite cysts.

Cryptosporidium species (cryptoporidosis)

- Fecal—oral transmission of parasite oocysts.

Bacterial

Food poisoning

Toxins from *Clostridium perfringens, Bacillus cereus,* and *Staphylococcus aureus*

- Enterotoxins induce diarrhea and/or vomiting and abdominal pain.

Foodborne infections

Campylobacter jejuni

- Neurological complications can follow infection.

Shigella species

- Human fecal—oral transmission through food, water, or fomites.

Escherichia coli O157:H7

- Fecal contamination of meat, fruits, or vegetables.

Salmonella enteric (serovars Enteritidis and Typhimurium)

- Often associated with poultry or seafood.

Infectious diarrhea

Vibrio cholerae (vaccine available)

- Fecal—oral transmission by improper sewage disposal.

Clostridioides difficile

- Resistant to several antibiotics and thrives in intestine if decrease in normal microbiota.

For all printed and additional Disease Snapshot tables, go to the Mastering Microbiology Study Area.

19 CHAPTER 19 OVERVIEW

19.1 Digestive System Anatomy and Defenses

- The digestive system's role is to acquire water, vitamins, and nutrients. The upper GI tract includes the mouth, pharynx, esophagus, and stomach. The lower GI tract consists of the small and large intestines, rectum, and anus. The digestive system's accessory organs include the salivary glands, liver, gallbladder, and pancreas.
- Mucosa-associated lymphoid tissue found along the digestive tract and normal microbiota provide protection against pathogens.

19.2 Gastrointestinal Infection Symptoms and Diagnostic Tools

- Digestive system infections are usually transmitted by the fecal–oral route.
- They may cause: diarrhea (loose stool), dysentery (diarrhea containing blood and/or pus), gastroenteritis (inflammation of the stomach and intestines), or dehydration.
- Differential and selective media, endoscopy, and colonoscopy are tools to help with diagnosing GI tract infections and causative agents.

19.3 Viral Digestive System Infections

- Mumps is an acute viral infection of the salivary glands that results in swelling of the jaw; it is prevented by the MMR vaccine.
- Rotavirus can cause severe dehydration in infants due to diarrhea. A vaccine is available.
- Norovirus is highly contagious among children and adults. Vomiting and diarrhea symptoms resolve within three to eight days.
- Hepatitis A virus is foodborne and waterborne; it causes an acute infection of the liver, but doesn't present a risk of chronic infection. It is vaccine preventable.
- Hepatitis B and C viruses are transmitted by contact with bodily fluids. They can cause chronic or acute infection. HBV is vaccine preventable, while HCV is usually curable with drugs.
- Hepatitis D virus coinfects with HBV; coinfected patients tend to have a more severe infection than patients with only HBV. Hepatitis E virus is transmitted by the fecal–oral route, causing symptoms similar to hepatitis A.

19.4 Bacterial Digestive System Infections

- Bacteria that release acid as they ferment sugar cause dental caries. *Streptococcus mutans* living in plaque (a biofilm on teeth) is a key agent of cavities. Periodontal disease is the result of biofilm buildup along the gums, which incites inflammatory responses.
- *Helicobacter pylori* causes stomach ulcers (damaged sections of the stomach lining).
- Most foodborne illnesses are due to feces-contaminated food or water. Foodborne infections are due to bacterial contamination, usually through fecal contamination. Foodborne infections are commonly caused by *Campylobacter jejuni*, *Shigella* species, *E. coli*, *Salmonella* species, and *Listeria monocytogenes*. Food poisoning is from toxins deposited by bacteria in food that lead to diarrhea and abdominal pain. *Clostridium perfringens, Bacillus cereus,* and *Staphylococcus aureus* are three leading agents of food poisoning.
- *Campylobacter jejuni*, which is commonly associated with ingestion of raw or undercooked poultry, is a leading cause of foodborne infection. Guillain-Barré syndrome is a neurological disease that can follow a *C. jejuni* infection.
- Children are most likely to be infected by *Shigella*, which is transmitted by feces-contaminated food, water, or surfaces. One species, *S. dysenteriae*, makes Shiga toxin and causes dysentery and may cause hemolytic uremic syndrome.
- There are six diarrheagenic *E. coli* pathotypes; all are transmitted via the fecal–oral route. *Escherichia coli* O157:H7 (also called Shiga toxin-producing *E. coli*) is the most common dysentery-associated serovar in the United States.
- The two most common diarrheagenic *Salmonella* serovars in the United States are Enteritidis and Typhimurium; most infections are linked to eating undercooked poultry products. Another serovar, Typhi, which causes typhoid fever, is no longer common in the United States, but is endemic in many developing nations.
- *Vibrio cholerae* causes cholera and is preventable with proper sewage disposal and treatment.
- *C. difficile* is now considered the most common healthcare-acquired infection. When antibiotics kill off the normal microbiota, *C. difficile,* which forms dormant endospores and can therefore naturally resist antibiotics, may thrive.

19.5 Protozoan and Helminthic Digestive System Infections

- Giardiasis is due to the ingestion of water or food contaminated with the protozoan *Giardia lamblia*.
- *Entamoeba histolytica* is a protozoan that causes amebiasis and is most prevalent in tropical areas of the world.
- *Cryptosporidium* species are protozoan parasites spread by feces-contaminated water usually found in lakes, ponds, and streams.
- Most tapeworm infections in humans are linked to *Hymenolepis nana*, which has a human-to-human transmission. *T. saginata* is contracted by eating undercooked beef. *T. solium* and *T. asiatica* are contracted by eating undercooked pork. Cysticercosis occurs when humans ingest the *T. solium* eggs rather than the cysticerci in undercooked pork. *Diphyllobothrium latum* is acquired from undercooked freshwater fish.
- The pinworm *Enterobius vermicularis* is an intestinal roundworm. Infected individuals get eggs on their hands upon scratching the perianal region and thereafter contaminate surfaces and food. The hookworm *Necator americanus* is a soil-transmitted roundworm that burrows into skin and eventually migrates to the intestines.
- *Ascaris lumbricoides* is one of the most common parasitic intestinal worms spread by fecal–oral transmission. High worm loads can cause bowel obstruction and malnutrition.
- *Trichinella spiralis* is a roundworm acquired by eating raw or undercooked wild game meat or pork.
- *Schistosoma* blood flukes in freshwater may burrow into skin and establish themselves in veins near the bowel or bladder. When eggs lodge in vessels and tissue, serious symptoms may develop.

THINK CLINICALLY | *Be S.M.A.R.T. About Cases*

COMPREHENSIVE CASE

The following case integrates basic principles from the chapter. Try to answer the case questions on your own. Don't forget to be S.M.A.R.T.* about your case study to help you interlink the scientific concepts you have just learned and apply your new understanding of digestive system infections to a case study.

*** The five-step method is shown in detail in the Chapter 1 Comprehensive Case on cholera. See pages 31–33. Refer back to this example to help you apply a SMART approach to other critical thinking questions and cases throughout the book.**

• • •

Jane is a pediatric nurse at the local hospital. She has twin daughters, 10 months old, who keep her extremely busy. One day, Jane noticed one of the twins acting unusually tired. She took the child's temperature and noted it was 101 degrees Fahrenheit. As Jane was changing the girls' diapers later that afternoon she noted the same twin showing loose and bloody stool. Alarmed, she immediately scheduled an appointment with her children's pediatrician for the next morning.

The pediatrician requested a stool sample from the next diaper change. She began talking to Jane about how to handle raw eggs and poultry but Jane interrupted her, explaining her professional background, and that she knew how intestinal pathogens could be transmitted through food. She was sure her food handling was not suspect. She mentioned that the previous owners of her house raised free-roaming chickens. The pediatrician did not prescribe any antibiotics for Jane's daughter, and sent them home waiting for culture results.

When Jane arrived home, she began to retrace all the places she and her twins had visited, as well as all the foods they had eaten for the past week. For the most part, they had stayed home and played either in the house or the yard. A babysitter comes to the home during the week to watch the twins while Jane goes to work. She usually feeds the twins from the same jars of baby food or cuts up soft fruit for them to feed themselves. Jane liked having the twins outside, but it was getting tricky to keep them on the blanket. They wanted to crawl everywhere. She remembered how she had caught the one twin a couple of days ago stuffing clover leaves into her mouth. She had tried to remove all the leaves; however, later that day she found another leftover leaf in the child's mouth.

• • •

CASE-BASED QUESTIONS

1. Why did the pediatrician assume Jane was mishandling raw eggs and poultry? Could there be other intestinal pathogen sources?
2. Is the pediatrician suspecting a bacterial or viral pathogen? What clues will help identify the causative pathogen?
3. Based on information provided, what is another piece of evidence that may support Jane's claim of safe food handling?
4. Was the pediatrician negligent in not prescribing antibiotics? Why or why not?
5. If you were an epidemiologist looking for the source of the pathogen, what samples would you collect at Jane's house?
6. How does the child chewing on clover leaves impact the list of suspected potential pathogens?
7. What other potential symptoms should Jane be looking for in her daughter over the next few days?

END OF CHAPTER QUESTIONS

Think Critically and Clinically

Questions highlighted in **orange** are opportunities to practice NCLEX, HESI, and TEAS-style questions.

1. Compare and contrast hepatitis A and B viruses.
2. Which of the following can occur even if the infectious agent is killed by the time it enters the host?
   a. Salmonellosis
   b. Cholera
   c. Giardiasis
   d. Food poisoning
   e. Mumps
3. Which one of the following measures is the most effective way to prevent viral infections of the digestive system?
   a. Antibiotics
   b. Thoroughly cooking food
   c. Decontaminating water
   d. Washing hands
   e. Disinfecting bathroom surfaces
4. How would an upper endoscopy help diagnose a *Helicobacter pylori* infection?

5. Undercooked poultry is commonly a source of infection caused by ____________ bacteria or by ____________ bacteria.

6. Match the preventative measure to the infectious agent it can help limit:

a. Thoroughly cooking pork	_____ Hepatitis C
b. Brushing teeth	_____ *Vibrio cholerae*
c. Wearing boots in freshwater	_____ *Trichinella spiralis*
d. Using a clean syringe for injection drugs	_____ *Schistosoma*
e. Ensuring fecal material is deposited away from drinking-water sources	_____ *Streptococcus mutans*

7. Name two ways *Shigella* can avoid host immune response.

8. Which of the following pathogens are not associated with foodborne infections?
   a. *Salmonella*
   b. *Shigella*
   c. *Campylobacter jejuni*
   d. *Bacillus cereus*
   e. *Escherichia coli*

9. Describe one way to prevent hookworm infection by *N. americanus*.

10. True or False: *Clostridioides difficile* cases are usually associated with a prior antibiotic therapy.

11. What is the main symptom of *Vibrio cholerae* infection and how is this infection best treated?

12. Why does dietary sugar increase the risk for developing dental caries?

13. What animals generally serve as the source of *Escherichia coli* O157:H7 for human infection?

14. A father suspects his son has intestinal worms due to his complaint of an itchy bottom. What habit is most likely responsible for his infection?
    a. Walking outside barefoot
    b. Eating without first washing hands
    c. Eating fruit that is not washed
    d. Eating undercooked meat
    e. Swimming in a nearby freshwater lake

15. Using the following list, construct a concept map or a chart connecting these terms according to the pathogen and correlated disease: fecal–oral, diarrhea, rotavirus, *Campylobacter jejuni*, bacterial, viral, foodborne transmission, antibiotics, vaccine.

16. List the symptoms associated with the various stages of Schistosomiasis.

17. True or False: *Giardia* infections in the United States are all imported from other countries.

18. What virulence factor is shared by *E. coli* O157:H7 and *Shigella*?

19. A lower endoscopy examination of a patient complaining of abdominal cramps and diarrhea reveals pus-laden patches along the intestinal wall. Which of the following infectious agents could be considered the pathogen responsible for this condition?
    a. *Clostridioides difficile*
    b. *Shigella*
    c. *Salmonella* serotype *Typhi*
    d. *Helicobacter pylori*
    e. *Campylobacter jejuni*

20. Larry goes to his local Red Cross center to donate blood for the first time. A few weeks later a Red Cross agent contacts him to ask him to come in for a confidential meeting. During the meeting Larry learns that his blood tested positive for hepatitis. He claims to be in excellent health and doesn't believe the diagnosis. What virus does Larry most likely have?
    a. Hepatitis A
    b. Hepatitis B
    c. Hepatitis C
    d. Hepatitis D
    e. Hepatitis E

21. Which of the following is most controversial for a nonseptic patient with *E. coli* O157:H7?
    a. Administering oral rehydration therapy
    b. Administering antibiotic therapy
    c. Administering fever-reducing medications
    d. Administering intravenous rehydration therapy
    e. Withholding agents like diphenoxylate-atropine that reduce GI tract motility

CRITICAL THINKING QUESTIONS

1. Why would a healthcare provider recommend yogurt to an individual who just finished a third round of antibiotics to treat a sinus infection? What GI tract infection are they most likely hoping to help this patient avoid by applying this dietary therapy?

2. With the increase in households raising their own chickens, what kind of disease would one expect to be on the rise? What can chicken owners do to protect themselves?

3. Two friends go out to dinner one night. A recent outbreak of beef-associated *E. coli* O157:H7 causes one friend to order a salad, while the other friend orders a steak, cooked medium. Which one is more likely to acquire the pathogen if it is transmitted by a careless chef preparing the food?

4. After going to the bathroom one night, Fred notices a long, thin, cream-colored, tube-shaped agent in his feces. Fred is a healthy 23-year-old who loves to travel; he recently returned from a two-month stay in Brazil. What is the most likely *specific* candidate for the excreted agent and how could he have acquired it? Should he seek medical help immediately?

5. *Helicobacter pylori* is defined as a microaerophilic neutralophile in spite of being found as a pathogen in the stomach. Explain how *H. pylori* can remain in the stomach in spite of the low pH. Also, describe two types of growth conditions you would need to provide if you were attempting to grow *H. pylori* from a patient sample.

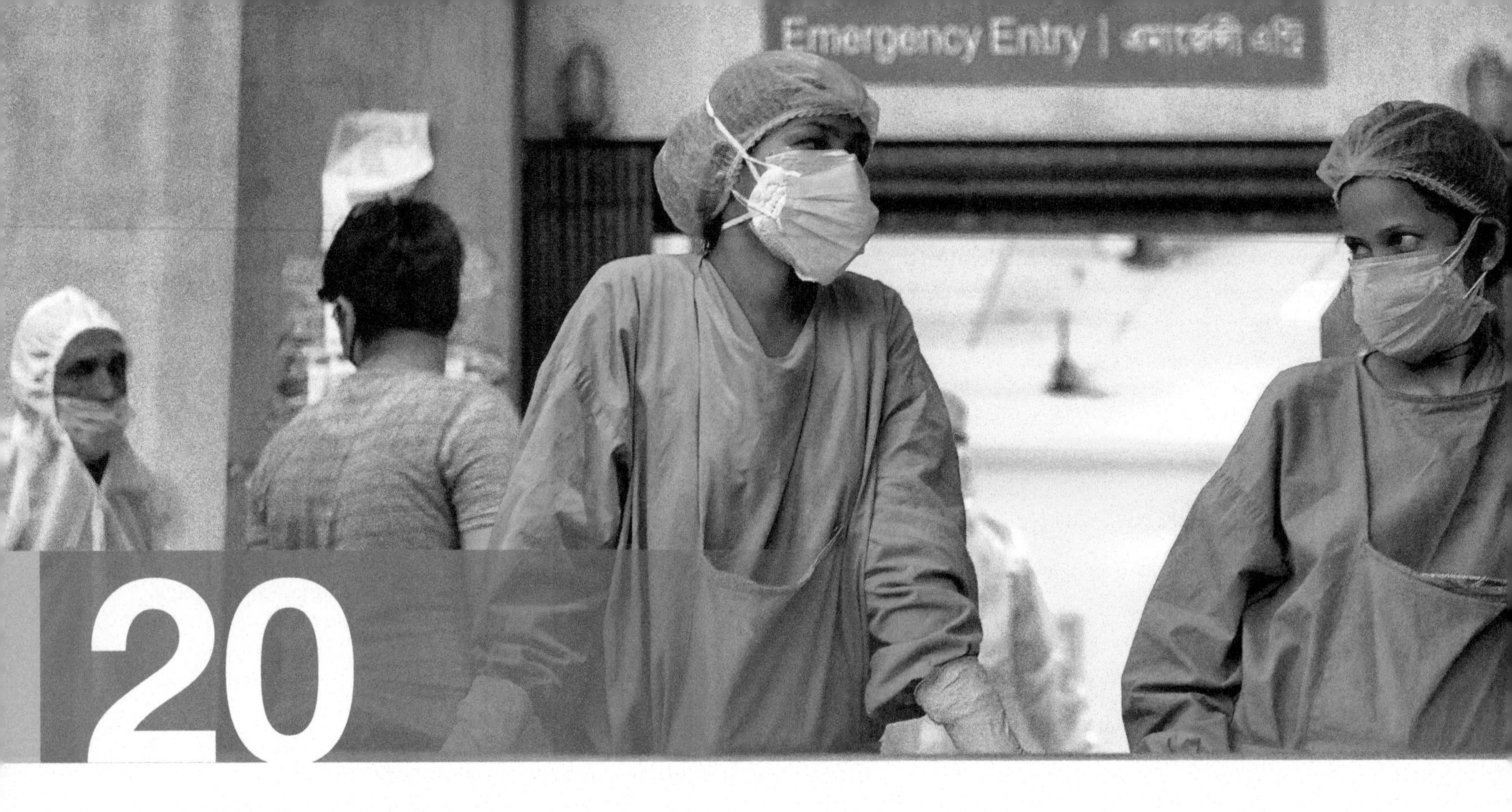

20

Urinary and Reproductive System Infections

What Will We Explore? Many pathogens infect the urinary and reproductive systems. According to the World Health Organization (WHO), every year about 357 million people of reproductive age contract chlamydia, gonorrhea, syphilis, or trichomoniasis—the four most common, curable, sexually transmitted infections. Millions of others contract treatable, but incurable, sexually transmitted infections such as human papilloma virus (HPV), herpes simplex virus 2 (also called human herpes virus 2), human immunodeficiency virus (HIV), and hepatitis B virus. This chapter focuses on the most common and noteworthy infections that target the urinary and/or reproductive systems.

Why Is It Important? Given their high prevalence, it's essential that healthcare providers understand the signs, symptoms, diagnostic tests, and treatments for these infections. Urinary tract infections (UTIs) are a common reason for a trip to the doctor. Depending on the infection type and patient health status, a UTI may be a self-limiting annoyance, or life threatening. The CDC estimates that the cost of treating just the eight most common sexually transmitted infections is almost $16 billion per year in the United States; this makes them a major healthcare priority. Infections in these body areas often vary in symptoms between sexes and from one individual to the next. Disease progression can range from asymptomatic and self-limiting, to causing infertility, pregnancy complications, renal failure, or even death. Unfortunately, the social stigma surrounding sexually transmitted infections persists; this presents challenges in prevention campaigns as well as hinders patient communication that is central to early diagnosis and treatment.

The WHO has called for a 90 percent global reduction in syphilis and gonorrhea by the year 2030. To reach this goal, it will take dedicated public health officials and clinicians who are well-versed in recognizing and treating these infections.

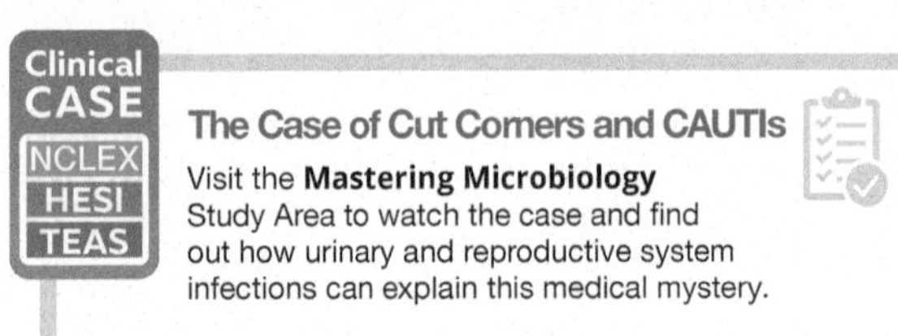

Healthcare workers discussing a patient's status
Kolkata, India

20.1 OVERVIEW OF THE URINARY AND REPRODUCTIVE SYSTEMS

Our urinary system includes kidneys, ureters, the bladder, and the urethra.

Learning Outcomes

After reading this section, you should be able to:

20.1 Name and describe the parts of the urinary system and discuss features that limit infection in this system.

20.2 Compare and contrast the anatomical differences between the male and female urinary system and the implications these differences may have.

20.3 Discuss the features of the urinary microbiome.

20.4 Discuss the structural and functional features of the male and female reproductive systems and describe features that help limit infection.

20.5 Outline what is meant by TORCH agents and discuss their implications in pregnancy.

20.6 Discuss the roles normal reproductive system microbiota have in both sexes.

The human urinary system consists of two kidneys, two ureters, the bladder, and the urethra (FIG. 20.1). Our **kidneys** produce a minimum of 500 ml of urine each day. They filter metabolic waste products such as urea and creatinine out of our blood and regulate the body's water and electrolyte balance. Our kidneys also help manage our blood pressure and blood pH.

Each kidney contains over a million microscopic units called **nephrons**, tiny functional units that secrete waste products into urine while retaining useful substances in the blood. As urine forms in each kidney, it's funneled to the associated **ureters.** The ureters convey urine from the kidneys to a smooth muscle sac called the **bladder.** The bladder stretches to hold up to 400 ml of urine. As the bladder fills, the stretching action triggers the nervous system to stimulate an urge to urinate.

Urine flows from the bladder to the exterior of the body through the **urethra.** In women the urethra is short and straight, while in men the urethra is longer, curved, and passes through the prostate gland. Also, in men the urethra serves as a part of the reproductive system when it transports semen out of the penis during an ejaculation. In women the urethra is associated with the external genitalia, just as it is in men, but it does not have a role in the reproductive system. The term **urogenital** describes infections and disorders that may jointly affect the urinary and reproductive systems. Some **sexually transmitted infections (STIs)** are described as urogenital infections because they can affect the urinary and reproductive systems at the same time. To clarify, STIs include diverse agents that can be transmitted by penetrative anal or vaginal sex, as well as nonpenetrative sexual contact such as oral sex or genital contact.

The anatomical differences between the male and female urethra help explain some differences in urinary system infection risks and progression in men versus women. For example, if the prostate gland, which has a role in the male reproductive system, becomes enlarged (**benign prostatic hypertrophy**) or inflamed (**prostatitis**), it can constrict the part of the urethra that passes through it—the prostatic urethra. This constriction makes it difficult for the bladder to empty during urination and can increase a man's risk for a **urinary**

CHEM • NOTE

Our liver makes **urea** as a normal part of amino acid breakdown. **Creatinine** is released as a normal byproduct of the constant tissue remodeling and maintenance that muscle tissues experience.

CHEM • NOTE

Electrolytes are ions that are dissolved in body fluids. The most common electrolytes in physiology are sodium, potassium, calcium, phosphate, magnesium, and chloride. Kidney failure is the most common cause of recurring electrolyte imbalances. (See Chapter 2 for more on electrolytes.)

CHEM • NOTE

Normal **blood pH** is 7.35–7.45. When blood pH drops below this range, the patient is said to have acidosis. When it's higher than this range, the patient is described as experiencing alkalosis. The kidneys regulate the levels of ions that affect blood pH.

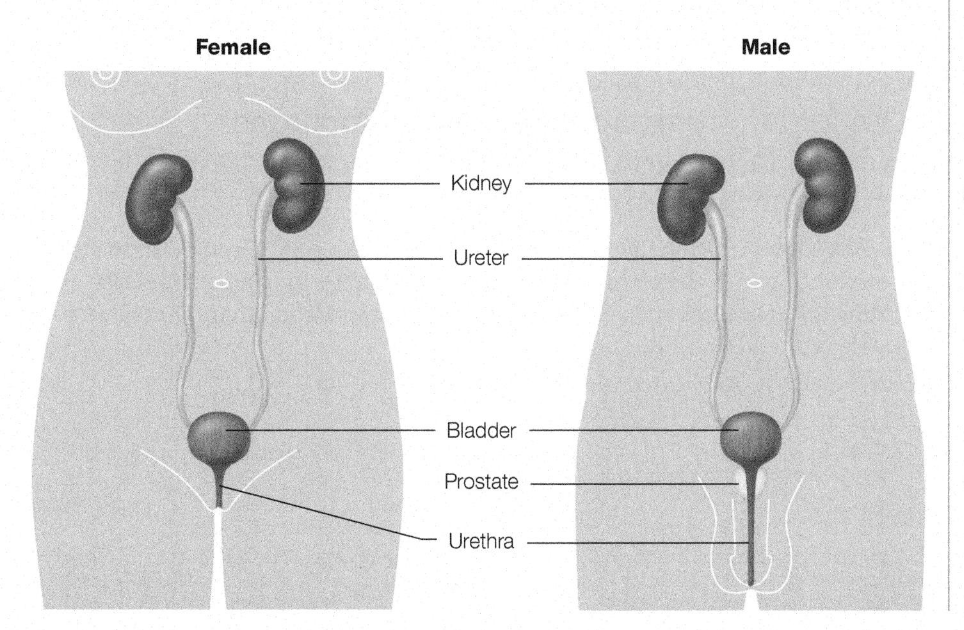

FIGURE 20.1 The urinary system The kidneys create urine, which passes through ureters to the bladder and then exits the body through the urethra.

Critical Thinking *Why would a patient with a urinary tract infection be advised to increase water intake?*

CLINICAL VOCABULARY

Benign prostatic hypertrophy a noncancerous enlargement of the prostate (also called benign prostatic hyperplasia); as the prostate enlarges it may constrict the urethra and make it difficult to fully void the bladder; BPH develops in about half of all men by age 50 and is almost universally present in men over age 75

Prostatitis an inflammation of the prostate gland; often caused by a bacterial infection, but may be due to other factors; it's the most common prostate issue in men under age 50

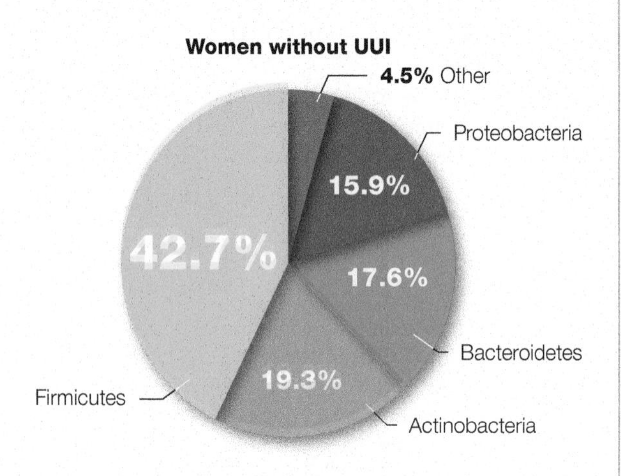

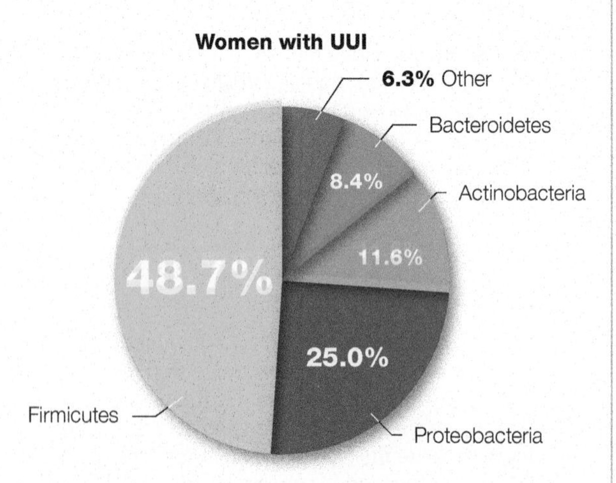

FIGURE 20.2 The urinary system microbiome in women with or without urinary urgency incontinence (UUI)

Source: Karstens, L., Asquith, M., Davin, S., Stauffer, P., Fair, D., Gregory, W. T., et al. (2016). Does the urinary microbiome play a role in urinary urgency incontinence and its severity? *Frontiers in Cellular and Infection Microbiology*, vol. 6, article 78.

tract infection (UTI), which is a general term that refers to an infection that affects the urinary system. However, UTIs are eight times more prevalent in women than in men.[1] This is mainly because the urethra in women is anatomically closer to the anus, making it easier for bacteria to get introduced into the urinary system. Also, in women the urethra is about 2 inches long, while in men it is about 8 inches long; this means pathogens have a much shorter journey to the bladder in women. Fortunately, the urinary system in men and women has factors that help limit infection. Usually urine has a pH near 6. This slightly acidic pH combined with urine's flushing action as it passes through the ureters and urethra help protect the urinary system from bacterial infections.

The Urinary System Microbiome

The urinary tract was traditionally considered a sterile environment. This misperception was largely based on the fact that urine from healthy patients rarely contains bacteria that grow using standard culture techniques. However, specialized culture techniques and molecular methods that detect bacterial genes have made it clear that there is a urinary microbiome. The urinary microbiome is diverse and varies significantly based on many factors, including age, sex, and general patient health. The types of microbes present also differ based on what part of the urinary system is being considered. For example, the bladder has a different microbiota profile than the urethra. This variability makes it difficult to pinpoint what a healthy urinary microbiome looks like. In general, *Lactobacillus* and *Streptococcus* are the most consistently detected genera in the healthy urinary tract. However, the urinary tract harbors a highly diverse microbiome that also includes other genera such as *Prevotella, Gardnerella, Peptoniphilus, Dialister, Finegoldia, Anaerococcus, Allisonella, Staphylococcus, Sneathia, Veillonella, Corynebacterium, Ureaplasma, Mycoplasma, Atopobium, Aerococcus, Gemella,* and *Enterococcus*.[2] It's likely that these normal microbial residents compete with potential pathogens to reduce infection incidence.

Many studies have compared the urinary microbiome of healthy men and women to the microbiome of patients with urinary tract pathologies such as incontinence (urine leakage), **urinary urgency incontinence** (UUI; a frequent urge to urinate accompanied by incontinence), urological cancers, prostatitis, and many others. These studies show that urinary microbiome shifts may contribute to disease, but the exact nature of those shifts and their significance is not yet fully understood. An example of such a study involved analyzing the urinary microbiome in healthy women to age-matched women with urinary urgency incontinence. The study revealed a shift in the microbiota phyla of UUI patients as compared to healthy controls (FIG. 20.2).[3]

The female reproductive system includes ovaries, fallopian tubes, uterus, cervix, vagina, and external genitalia.

FIG. 20.3 shows the female reproductive system. Let's walk through these structures starting with the **ovaries**, the organs that make eggs and also release a number of hormones such as estrogen, progesterone, and others that govern female reproductive physiology.

[1] Al-Badr, A., & Al-Shaikh, G. (2013). Recurrent urinary tract infections management in women: A review. *Sultan Qaboos University Medical Journal*, 13 (3), 359.

[2] Aragón, I. M., Herrera-Imbroda, B., Queipo-Ortuño, M. I., Castillo, E., Del Moral, J. S. G., Gómez-Millán, J., et al. (2016). The urinary tract microbiome in health and disease. *European Urology Focus*.

[3] Karstens, L., Asquith, M., Davin, S., Stauffer, P., Fair, D., Gregory, W. T., et al. (2016). Does the urinary microbiome play a role in urinary urgency incontinence and its severity? *Frontiers in Cellular and Infection Microbiology*, vol. 6, article 78.

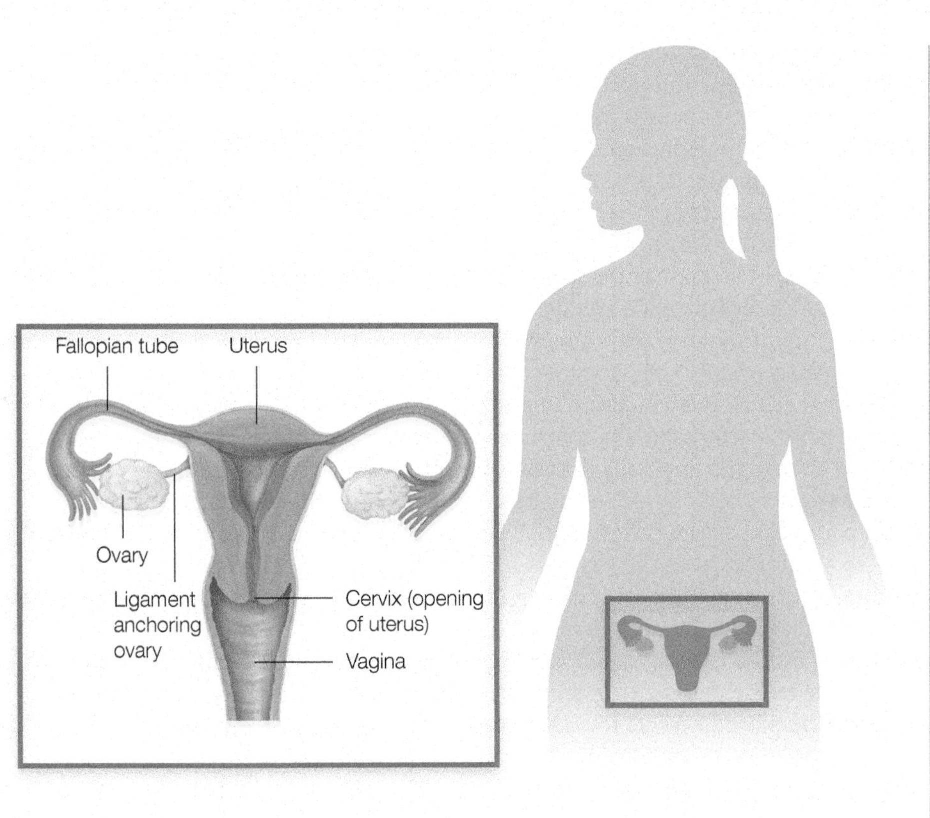

FIGURE 20.3 The female reproductive system
The uterus looks like an inverted pear. The cervix is the opening of the uterus; it is the "neck" of the pear-shaped uterus.

During ovulation, an egg is released from an ovary and is swept up into the adjacent **fallopian tube.** From there it's conveyed toward the **uterus** (womb)—a muscular, inverted-pear-shaped organ that houses the developing fetus during pregnancy and contracts to expel the baby during labor. As the egg journeys through the fallopian tube toward the uterus, sperm may fertilize it. If the egg is not fertilized, it eventually disintegrates. If it is fertilized, then a zygote forms and starts to undergo many rounds of cell division to form a ball of cells called a blastocyst. Eventually the blastocyst *implants*, or embeds itself, in the **endometrium,** a superficial layer of tissue that lines the uterus and is shed with each menstrual cycle (or period). During pregnancy the endometrial layer is not shed by menstruation. Instead, soon after implantation certain embryonic cells, along with select maternal uterine cells, collaborate to build the **placenta**—an organ that serves as an interface between maternal and fetal circulation and releases hormones that support pregnancy (FIG. 20.4). The umbilical cord links the baby to the placenta and allows the baby's blood supply to enter the placenta. On the maternal end, the mother's blood flows into the placenta from capillaries in the uterine wall.

If you interlace your fingers from your left and right hands you can envision how the capillaries from the maternally derived part of the placenta (the fingers of your right hand) interface with the placental capillaries that develop from the fetal side of the placenta (the fingers of your left hand). As fetal blood enters the placenta, it is brought into close proximity with maternal blood that has circulated into the placenta, but a healthy placenta does

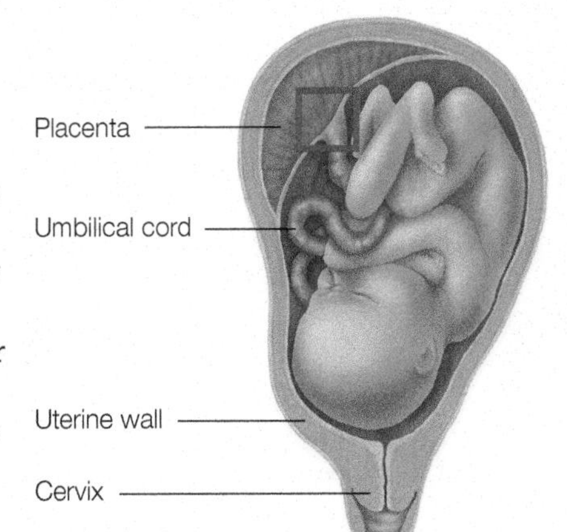

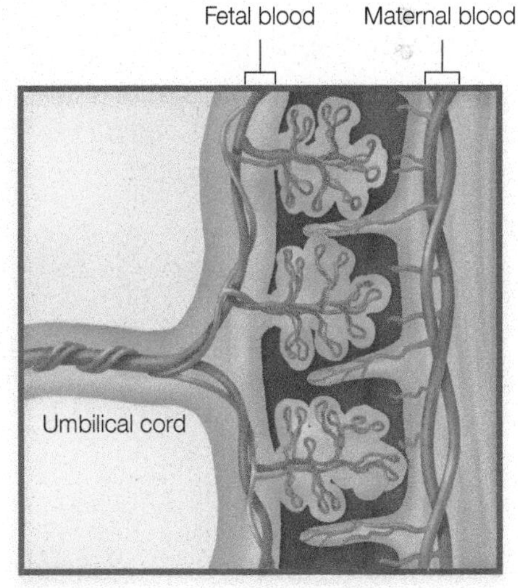

FIGURE 20.4 The placenta

CLINICAL VOCABULARY

Effacement this term describes how the opening to the uterus, the cervix, becomes thinner, softer, and shorter before the baby is delivered

Pap smear a routine test that collects cervical cells for microscopic analysis to detect precancerous and/or cancerous cells; Pap tests are the cheapest and easiest tool for early cervical cancer detection; human papilloma viruses may also be detected in cells collected by the Pap smear test

Parturition the action of giving birth; labor

Stillbirth in utero fetal death after 20 weeks gestation

Congenital defects birth defects or developmental issues present from birth

Spontaneous abortion the medical term for a miscarriage; noninduced embryonic or fetal death before 20 weeks' gestation

Neonatal herpes can present as skin, eye, and mouth lesions, neurological disorders, or a disseminated infection.

not allow maternal and fetal blood to mix. Instead, useful substances such as hormones, oxygen, and nutrients diffuse across the maternally formed capillaries of the placenta to enter fetal blood. At the same time, hormones, metabolic waste products, and carbon dioxide from the fetal bloodstream diffuse into the maternal circulation. The mother then exhales that collected carbon dioxide and eliminates waste for her baby through her own urinary system.

When it's time for the baby to be born, the **cervix**, which is the opening to the uterus that leads into the vagina, begins to open (or dilate) as well as soften and thin—a process called *effacement*. The outer margin of the cervix, which lies near the top of the vagina, is also the site that is swabbed during a **Pap smear** (*Papanikolaou smear*), a routine test that detects potentially cancerous changes in cervical cells (reviewed later when human papilloma viruses are discussed). During labor (*parturition*), the uterus contracts to push the baby through the cervix and toward the **vagina**, a tube-like structure that extends from the cervix to the exterior of the body. After the baby exits through the vagina, the placenta will be expelled as the "afterbirth." The vagina is also the site of standard intercourse and it serves as the exit site for menstrual fluids. The opening of the vagina is surrounded by the **vulva**, which collectively refers to a woman's external genitalia. The vulva consists of the **labia minora** (the pair of small lip-like folds that enfold the vagina and urethra), the **clitoris** (the erectile tissue that is analogous to the penis and is stimulated to induce orgasm), and the **labia majora** (the paired lip-like folds that enclose the female external genitalia).

Pathogens that Cross the Placenta (TORCH Agents)

Although the placenta is a selective barrier that protects the developing fetus, it's not a universal barrier. A number of pathogens can cross the placenta and affect the fetus. When this occurs, serious complications may develop. The baby could be stillborn, or may suffer a variety of *congenital defects* (physical and/or developmental abnormalities present from birth). Alternatively, the mother may suffer a *spontaneous abortion* (miscarriage). The acronym "**TORCH**" refers to pathogens that can be vertically transmitted to a developing fetus and cause congenital defects. The original TORCH acronym was developed in the 1970s and at first strictly referred to **T**oxoplasma gondii (reviewed in Chapter 18); **r**ubella virus (reviewed in Chapter 17); **c**ytomegalovirus (reviewed in Chapter 21); and **h**erpes simplex viruses (reviewed later in this chapter). At the time, these were the four most common agents known to be vertically transmitted and linked to congenital defects. Since the original acronym was developed, we have discovered many other pathogens that fit the TORCH criteria. In light of this, the "O" in TORCH has been reassigned to stand for "other infections." As a miscellaneous category, the "O" grouping now includes HIV, syphilis, parvovirus B19, listeriosis, varicella-zoster virus, coxsackievirus, enteroviruses, and most recently, Zika virus.[4]

Although each of these infections has its own epidemiological and clinical features, they all may cause devastating fetal effects. Early identification and rapid treatment (if available) can prevent many of the serious congenital abnormalities caused by TORCH infections. Therefore, TORCH infection screening is an important part of prenatal care. These tests usually involve blood work that detects maternal antibodies to TORCH pathogens. As discussed in earlier chapters, the antibody class detected gives clues about pathogen exposure time frames. IgG antibodies suggest a past exposure, while elevated IgM suggests an early infection stage.

[4] Schwartz, D. A. (2016). The origins and emergence of Zika virus, the newest TORCH infection: What's old is new again. *Archives of Pathology & Laboratory Medicine*, 141 (1), 18–25.

The female reproductive system has built-in innate protections, including a specialized microbiome.

The vagina has several features that help limit female reproductive tract infections. Vaginal epithelial cells constantly shed, with the top layer exfoliating about every four hours.[5] Vaginal and cervical mucus is enriched with antimicrobial peptides and lysozyme that combat bacteria; the sticky mucus also traps pathogens so they are shed along with the epithelial cells. Lastly, the vaginal epithelium releases glycogen, which lactic acid-producing bacteria of the normal microbiota metabolize. The next section reviews the important role these bacteria have in vaginal health.

Female Reproductive System Microbiome

The precise profiles of the vaginal microbiome vary from person to person and are influenced by a number of factors that include ethnicity, age, sexual activity, and pregnancy. Overall, more than 200 different taxa of bacteria have been identified in the vagina, but in general the vaginal microbiome in reproductive-age women is dominated by the *Lactobacillus* genus, with *L. iners*, *L. crispatus*, *L. gasseri*, and *L. jensenii* species being the most prevalent.[6] Most importantly, *Lactobacillus* species ferment host-secreted glycogen into lactic acid, which lowers the vaginal pH to about 3.5 to 4.5—an acidic pH that limits what microbes can thrive. These bacteria are also notable for making hydrogen peroxide, a substance that is especially toxic to anaerobic microbes, which lack catalase or other enzymes that eliminate toxic oxygen intermediates. Vaginal lactobacilli also make substances called **bacteriocins**, which are antimicrobial peptides that limit the growth of other competing bacteria in the same region. Antibiotic therapies are the most common triggers for a dysbiosis, or imbalance, in the normal vaginal microbiota. As lactic acid-producing bacteria die off, vaginal pH tends to increase; even a modest pH increase can foster the development of an opportunistic infection, especially by yeasts like *Candida albicans*.

Most studies have focused on reproductive-age women, so very little is known about the vaginal microbiome in prepubescent girls or postmenopausal women. It used to be thought that the hormonal shifts that led to *menarche* (the first onset of menstruation) promoted lactobacillus growth, but more recent data show that lactobacilli are already dominant members of the vaginal microbiota well before puberty onset.[7] That said, vaginal pH in healthy prepubescent girls remains higher (less acidic) than in healthy reproductive-age women. This may be due to estrogen in reproductively mature women stimulating glycogen production in the vaginal epithelium, which lactobacilli can then use to make lactic acid. Vaginal microbiome studies in postmenopausal women show decreases in lactobacillus populations compared with reproductive-age women, with corresponding increases in vaginal pH.

Although the vagina is certainly the most heavily colonized by normal microbiota, it is not the only place in the female reproductive system where microbiota have been identified (FIG. 20.5). Recent studies have revealed that every part of the female reproductive tract may harbor normal microbiota and

---

[5] Anderson, D. J., Marathe, J., & Pudney, J. (2014). The structure of the human vaginal stratum corneum and its role in immune defense. *American Journal of Reproductive Immunology, 71* (6), 618–623.

[6] Ravel, J., Gajer, P., Abdo, Z., Schneider, G. M., Koenig, S. S. K., et al. (2011). Vaginal microbiome of reproductive-age women. *Proceedings of the National Academy of Sciences of the United States of America, 108* (Suppl 1), 4680–4687.

[7] Hickey, R. J., Zhou, X., Settles, M. L., Erb, J., Malone, K., Hansmann, M. A., et al. (2015). Vaginal microbiota of adolescent girls prior to the onset of menarche resemble those of reproductive-age women. *MBio*, 6 (2), e00097–15.

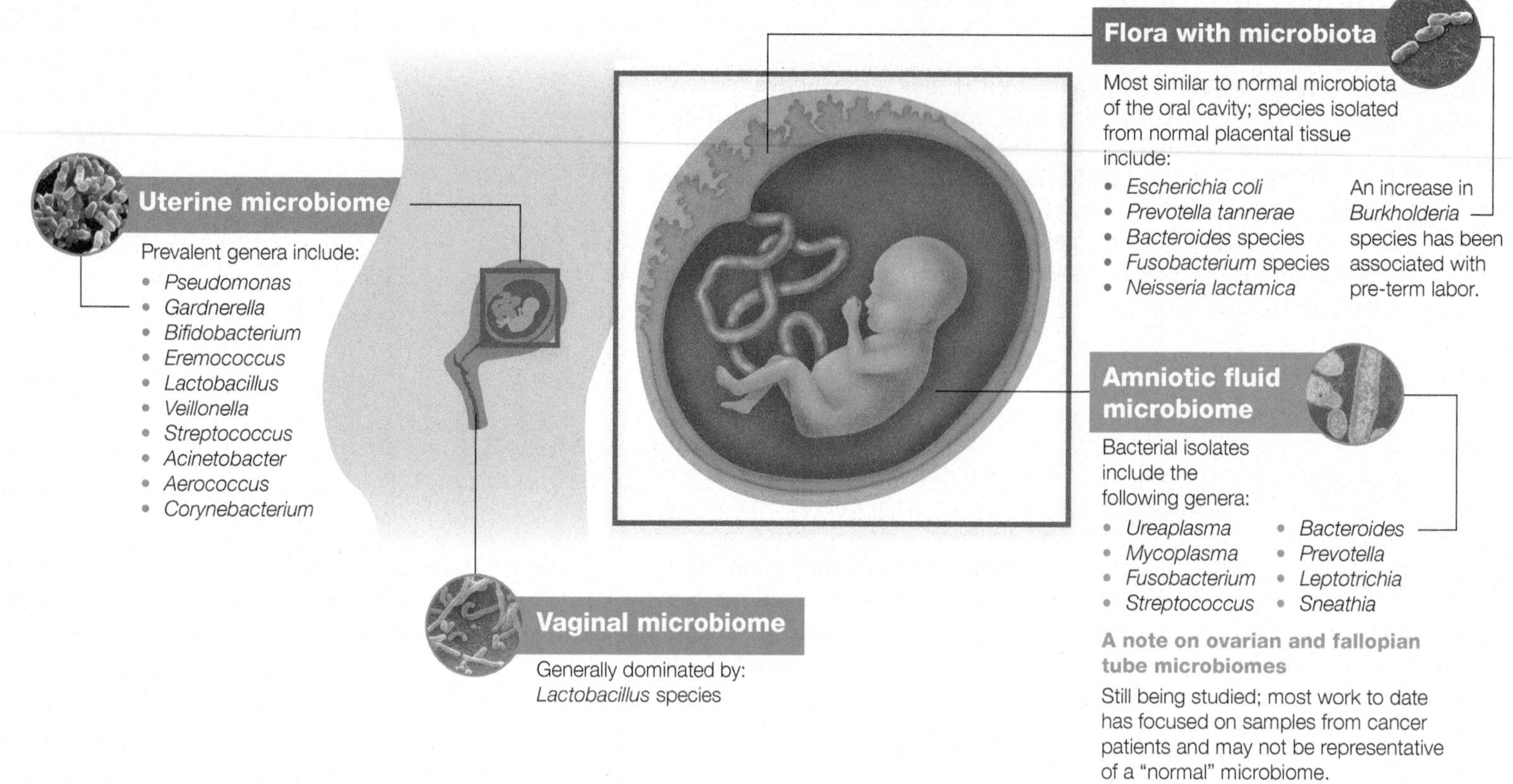

FIGURE 20.5 **Microbiome of the female reproductive system. Note, the placental microbiome findings are debated and studies are ongoing.**

even the previously embraced "sterile womb" notion is being questioned. Even in asymptomatic patients, bacteria have been isolated from ovaries, fallopian tubes, uterus, placenta, and amniotic fluid, suggesting that perhaps microbes are not strictly limited to the vagina.[8,9,10,11]

It is well established that intrauterine infections are a major risk factor for **pre-term labor**, which is labor onset before 37 weeks of gestation. It is estimated that at least 40 percent of pre-term deliveries are associated with an intrauterine infection.[12] But a more recent development is the idea that the normal microbiota profile of the female reproductive tract, not just infections, might impact the risk for pre-term labor. Initial studies suggest that decreased lactobacillus levels and increases in *Gardnerella* and/or *Ureaplasma* in the female reproductive tract may increase pre-term labor risk.[13] A separate study that

[8] Miles, S. M., Hardy, B. L., & Merrell, D. S. (2017). Investigation of the microbiota of the reproductive tract in women undergoing a total hysterectomy and bilateral salpingo-oopherectomy. *Fertility and Sterility, 107* (3), 813–820.

[9] Franasiak, J. M., Tao, X., Lonczak, A., Taylor, D., Treff, N., Hong, K. H., et al. (2014). Characterizing the uterine microbiome: Next generation sequencing of the V4 region of the 16S ribosomal gene. *Fertility and Sterility, 102* (3), e135–e136.

[10] Cao, B., Stout, M. J., Lee, I., & Mysorekar, I. U. (2014). Placental microbiome and its role in preterm birth. *NeoReviews*, 15 (12), e537–e545. http://doi.org/10.1542/neo.15-12-e537.

[11] DiGiulio, D. B. (2012, February). Diversity of microbes in amniotic fluid. In *Seminars in Fetal and Neonatal Medicine, 17* (1), 2–11. WB Saunders.

[12] Agrawal, V., & Hirsch, E. (2012, February). Intrauterine infection and preterm labor. In *Seminars in Fetal and Neonatal Medicine, 17* (1), 12–19. WB Saunders.

[13] DiGiulio, D. B., Callahan, B. J., McMurdie, P. J., Costello, E. K., Lyell, D. J., Robaczewska, A., et al. (2015). Temporal and spatial variation of the human microbiota during pregnancy. *Proceedings of the National Academy of Sciences, 112* (35), 11060–11065.

FIGURE 20.6 **Male reproductive system**

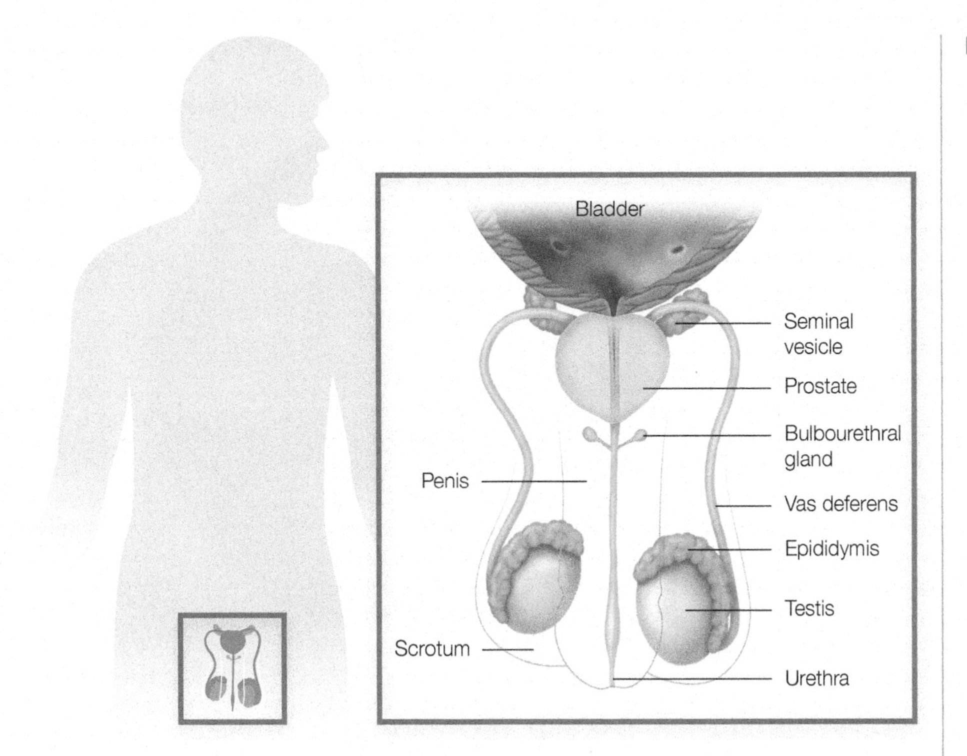

analyzed the placental microbiome suggested that an increase in *Burkholderia* species might increase pre-term labor risk.[14] Much work must still be done to validate findings and cement our understanding of what a "normal" reproductive system microbiome looks like, and what microbiome shifts may be linked to abnormal physiology.

The male reproductive system consists of the scrotum, testes, spermatic ducts, sex glands, and penis.

The male reproductive system consists of all the structures needed to make semen and convey it out of the penis during an ejaculation (FIG. 20.6). The **scrotum** is a pouch of skin that contains two **testicles** (testes), the organs that make sperm cells and a variety of hormones, including testosterone. The scrotum is located outside of the main body cavity because sperm develop best when slightly cooler than core body temperature. After sperm are made in the testes, they are conveyed to the adjacent **epididymis** for storage until they are ejaculated. The **vas deferens** transport sperm from the epididymis toward the prostate. Before sperm are released through ejaculation, fluids are added by the sex glands (the prostate gland, bulbourethral gland, and seminal vesicles) to make semen.

Semen is a liquid that in its final composition contains sperm and substances such as fructose and various proteins that nurture sperm and aid their journey through the female reproductive tract. The major volume of semen is not sperm—instead, the majority of semen volume comes from the fluids added by the sex glands. During an orgasm, semen is ejaculated through the urethra that passes down the length of the **penis**.

[14] Aagaard, K., Ma, J., Antony, K. M., Ganu, R., Petrosino, J., & Versalovic, J. (2014). The placenta harbors a unique microbiome. *Science Translational Medicine*, 6 (237), 237ra65.

FIGURE 20.7 **Semen microbiome** These data represent the main bacterial genera identified in semen from 46 different healthy men.

Source: Data are from Mändar, R., Punab, M., Korrovits, P., Türk, S., Ausmees, K., Lapp, E., et al. (2017). Seminal microbiome in men with and without prostatitis. *International Journal of Urology*, *24* (3), 211–216.

Critical Thinking *The data were collected for ejaculated samples as opposed to semen that was directly collected from the seminal vesicles. How might this complicate the data?*

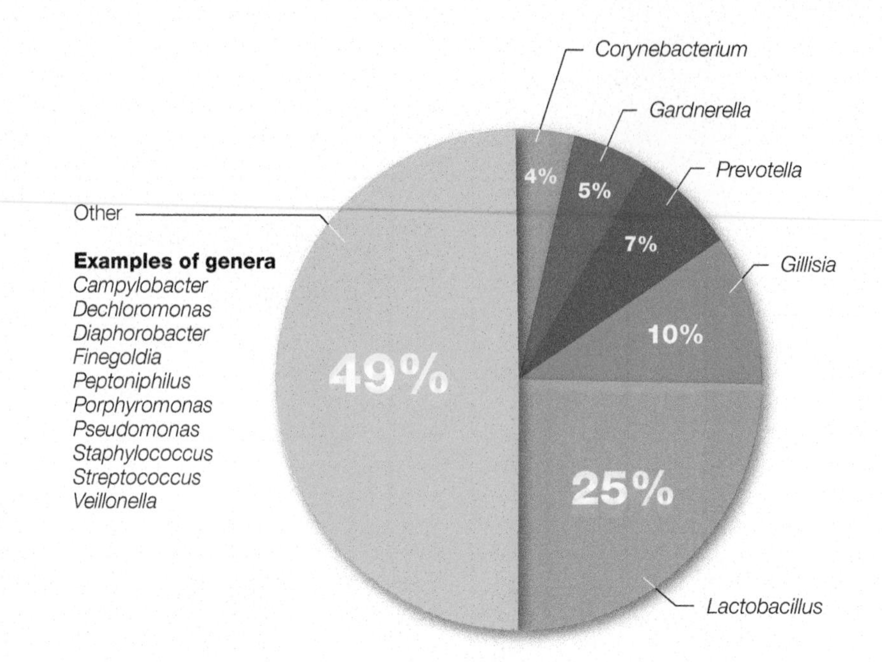

Male Reproductive System Microbiome

In men, the urethra serves a role in the urinary and reproductive system, so it's not surprising that there is some overlap between male urinary and reproductive microbiota. Interestingly, in mice the microbiota of the seminal vesicles seems to impact a number of health aspects. Although data obtained in animal models have not been thoroughly confirmed in humans, these studies serve as a proof of concept that can be further investigated. For example, studies in mice suggest that microbiome profiles in the seminal vesicles may impact the risk of developing prostate cancer.[15] The seminal fluid microbiome was characterized in mice and scientists mostly found members of the following phyla: Firmicutes (class Bacilli), Proteobacteria, Actinobacteria, Fusobacteria, Bacteroidetes (class Flavobacteria), and Acidobacteria.[16] In one study, over 500 species were unique to this habitat. It's likely that nutrients such as fructose that support sperm cells in semen also nurture a unique microbiome (FIG. 20.7). A lot of work remains to be done in order to determine how these preliminary findings in mice may apply to humans.

Build Your Foundation

1. What are the structures and functional features of each part of the urinary system and what features help limit infection?
2. Compare and contrast the male and female urinary systems.
3. What are the features of the urinary microbiome?
4. Name and describe each part of the male and female reproductive systems and describe the features that limit pathogens in these systems.
5. What does TORCH stand for, and what significance do TORCH agents have in pregnancy?
6. What are the roles of normal reproductive system microbiota in both sexes?

Build Your Foundation (BYF) Quick Quiz: Visit the **Mastering Microbiology** Study Area to quiz yourself.

[15] Shannon, B. A., Garrett, K. L. & Cohen, R. J. (2006). *Links between* Propionibacterium acnes *and prostate cancer*. *Future Oncology*, 2, 225–232, 10.2217/14796694.2.2.225.

[16] Javurek, Angela B., et al. (2016). Discovery of a novel seminal fluid microbiome and influence of estrogen receptor alpha genetic status. *Scientific Reports*, 6, 23–27.

Urinary tract infections are described by the part of the urinary tract affected.

Urinary tract infections (UTIs) can be described by the part of the urinary tract that is affected. **Urethritis** is inflammation of the urethra, while **cystitis** (sis TIE tis) is an inflammation of the bladder. Urethritis and cystitis are sometimes referred to as **lower UTIs**, based on the location of the affected organs within the urinary system. Although these conditions can develop from a number of factors, bacterial infections are the most common cause. Underlying conditions, or failure to treat lower urinary tract infections, can allow a lower UTI to progress to **ureteritis** (inflammation of the ureters) or **pyelonephritis**, an inflammation of one or both kidneys. Because of their location, infections that impact the ureters or kidneys are referred to as **upper UTIs**.

Generally speaking, the further "upstream" an infection goes in the urinary system, the more serious the situation becomes. While many cases of urethritis may be self-limiting, or only require a short course of antimicrobial drugs to cure, pyelonephritis can be fatal if not promptly treated. That's because bacteria that make their way up to the kidneys can use the vessels there as an entry point to the bloodstream, a scenario that may lead to *bacteremia* (bacteria in the blood), which can cause a dangerous systemic infection. Chronic pyelonephritis can also leave scar tissue in the kidneys that impairs function. When kidneys no longer effectively screen waste products from the blood, the patient is described as experiencing **renal failure**. Dialysis or kidney transplant may be the only options available to keep a patient alive once renal failure occurs.

Bacteria are the most common cause of urinary tract infections.

UTIs are a major healthcare issue; the National Kidney Foundation states that they account for about 10 million U.S. doctor's office visits each year. Clinically, UTIs can be categorized as uncomplicated or complicated. **Uncomplicated UTIs** occur in otherwise healthy people who have a normally structured urinary tract. These infections tend to readily resolve with primary drug therapies and are less likely to recur. **Complicated UTIs** tend to develop in people who are catheterized, have urinary tract malformations or obstructions, and/or have underlying health conditions that make it harder for them to fight off infections. These infections are much more likely to recur and may not respond to first-line drug therapies. Irrespective of whether the UTI is complicated or uncomplicated, bacteria are the leading cause of infection. The proximity of the urethra to the anus and genitalia means that millions of bacteria have fairly easy access to the urinary tract entry point and can work their way up toward the bladder and kidneys from there. Bacteria can also be introduced to the urinary tract by medical interventions, with urinary catheters being a particular concern.

Catheter-associated urinary tract infections (CAUTIs) account for at least one million cases of complicated UTIs each year and are among the most common type of healthcare-acquired infection.[17] CAUTIs are also the most common cause of secondary bloodstream infections that result when bacteria spread from the urinary tract into the bloodstream to cause a systemic infection. **Urinary catheters** allow urine to be directly drained from the bladder into a bag

[17] Flores-Mireles, A. L., Walker, J. N., Caparon, M., & Hultgren, S. J. (2015). Urinary tract infections: epidemiology, mechanisms of infection and treatment options. *Nature reviews. Microbiology*, 13(5), 269–284.

Learning Outcomes

After reading this section, you should be able to:

- **20.7** Differentiate complicated from uncomplicated urinary tract infections.
- **20.8** Define the terms urethritis, cystitis, ureteritis, and pyelonephritis.
- **20.9** Describe the signs, symptoms, risk factors, and potential long-term impacts of lower and upper UTIs and describe how UTIs are diagnosed.
- **20.10** Explain the infection-related challenges associated with urinary catheters and the significance of CAUTIs.
- **20.11** Summarize the features of causative agents of UTIs, with particular emphasis on uropathogenic *Escherichia coli* and *Staphylococcus saprophyticus*.
- **20.12** Describe the clinical and epidemiological features of leptospirosis.

CLINICAL VOCABULARY

Urethritis inflammation of the urethra

Cystitis inflammation of the bladder

Ureteritis inflammation of the ureter

Pyelonephritis inflammation of the kidneys

Renal failure a condition where kidneys cannot effectively filter waste from the blood or balance fluids

Uncomplicated UTI occurs in otherwise healthy individuals with normal urinary tract structure; tends to resolve quickly, without reccurrence, using first-line drugs

Complicated UTI usually occurs in people with catheters, urinary tract malformations/obstructions, or immune-compromised people; may not respond to first-line drug therapies, and tends to recur

Urinary catheter (Foley catheter) a tube inserted through the urethra into the bladder that allows urine to drain into a bag; frequently used on bed-bound or surgical patients

Clinical CASE
NCLEX
HESI
TEAS

The Case of Cut Corners and CAUTIs

Practice applying what you know clinically: visit the **Mastering Microbiology** Study Area to watch Part 2 and practice for future exams.

FIGURE 20.8 Urinary catheter

Critical Thinking *Aside from providing a surface for biofilm formation, how else might catheters increase the risk of a urinary tract infection?*

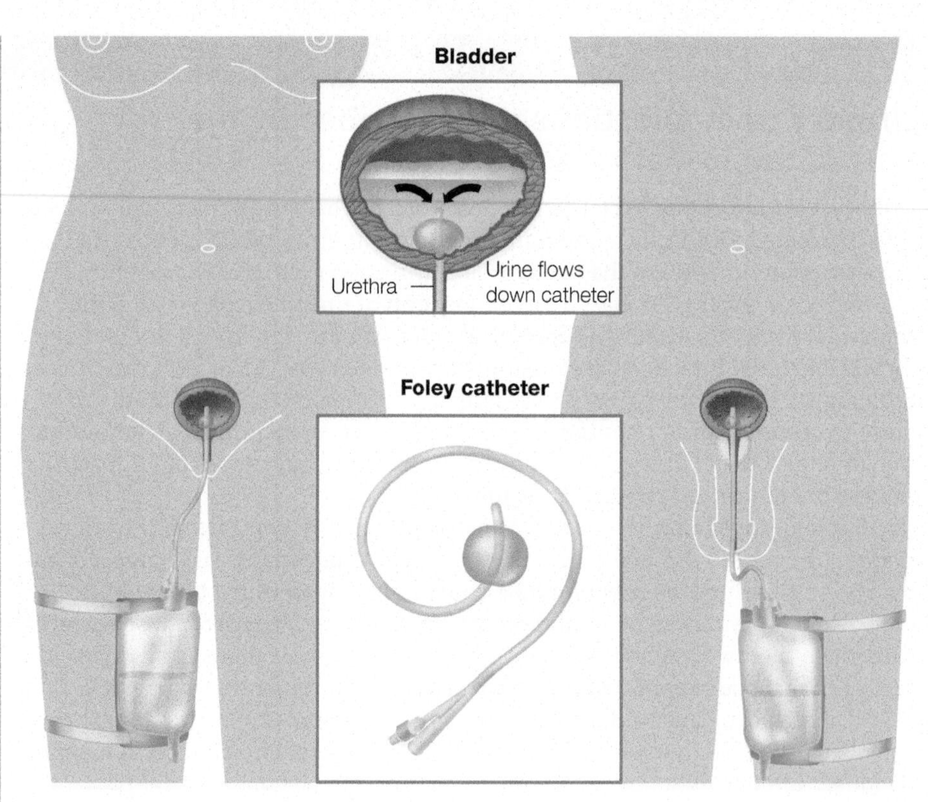

(FIG. 20.8). Although these indwelling devices are helpful in bed-bound patients, they provide a surface for biofilm formation and greatly increase the risk for UTIs. It's estimated that they cause 70 to 80 percent of complicated UTIs.[18]

Uropathogenic *Escherichia coli* (UPEC) and Other Enteric Bacteria That Cause UTIs

Enteric bacteria, which are bacteria that are normally found in the intestinal tract, cause the vast majority of UTIs—with the main culprit being **uropathogenic** *Escherichia coli* **(UPEC)**. These bacteria are small, flagellated, Gram-negative rods that may also take on a filamentous form when infecting a host. According to the Merck Manual, *Escherichia coli* cause 70–95 percent of all uncomplicated urinary tract infections and at least 50–65 percent of complicated UTIs. However, *Proteus mirabilis*, *Pseudomonas aeruginosa*, and *Klebsiella pneumoniae* are other enteric bacteria that are isolated in UTIs—though not nearly as frequently as UPEC.

Uropathogenic *E. coli* have diverse virulence mechanisms that help them successfully infect the urinary tract. For example, they have a number of adhesion factors such as pili and adhesins that allow them to attach to epithelial cells in the urinary tract and avoid being flushed out by a stream of urine. They also can invade bladder epithelial cells and take on an intracellular lifestyle. UPEC's ability to invade host cells means it can take shelter in a nutrient-rich environment while simultaneously hiding from host immune responses. Once inside host cells, the bacteria release siderophores to obtain iron, as well as proteases and toxins that can induce cell damage. If the bacteria make their way to the kidneys, which have numerous capillaries that bacteria can invade, then the infection may spread into the bloodstream.[19]

Renal Panel

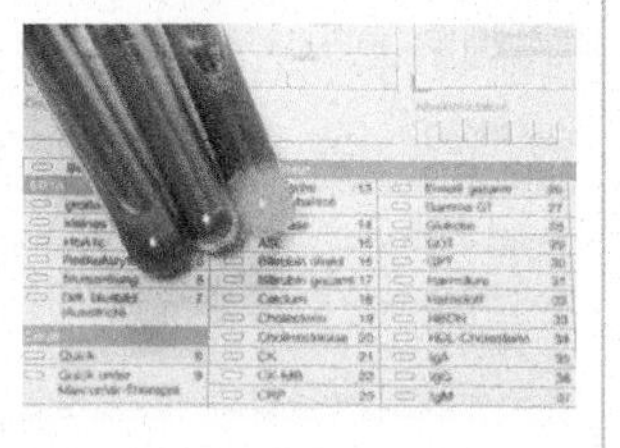

A renal panel, which is a collection of tests performed to assess kidney function, is one of the most commonly ordered diagnostics in health care. Kidney damage may result from any number of conditions, including infections, certain drugs, or other pre-existing health conditions such as inherited kidney disorders, diabetes, or autoimmune disorders like lupus.

Some tests can be performed immediately during a patient visit with urine dipstick strips. However, others, such as creatinine and blood urea nitrogen (BUN), require a blood sample and laboratory analysis.

Creatinine levels above 1.2 mg/dl in women and 1.4 mg/dl in men are typically considered abnormal, as are BUN levels above 20 mg/dl. Lower abnormal results may serve as an early indicator of kidney damage, while progressively higher numbers indicate that a patient is progressing toward—or already in—renal failure.

QUESTION 20.1

Would a blood creatinine level of 1.5 mg/dl in an elite male athlete be indicative of kidney disease? Explain your reasoning and state how you would follow up with the patient. (Hint: Consider the source of creatinine.)

[18] Gad, M. H., & AbdelAziz, H. H. (2021). Catheter-Associated Urinary Tract Infections in the Adult Patient Group: A Qualitative Systematic Review on the Adopted Preventative and Interventional Protocols From the Literature. *Cureus, 13*(7).

[19] Flores-Mireles, A. L., Walker, J. N., Caparon, M., & Hultgren, S. J. (2015). Urinary tract infections: Epidemiology, mechanisms of infection and treatment options. *Nature Reviews: Microbiology*, *13* (5), 269.

DiseaseSnapshot • Bacterial UTIs

	Enteric bacteria	Gram-positive bacteria
Causative agents	Uropathogenic *Escherichia coli* (UPEC) causes the majority of UTIs (up to 95 percent of uncomplicated UTIs and at least half of all complicated UTIs); occasionally other enteric bacteria such as *Proteus mirabilis, Klebsiella pneumoniae*, or *Pseudomonas aeruginosa* are responsible	*Staphylococcus saprophyticus* causes about 6 percent of all uncomplicated UTIs but is noted as much more common in UTIs that occur in young sexually active women; it's the main Gram-positive bacterium isolated in uncomplicated UTIs, followed by *Enterococcus faecalis*, group B Streptococcus (GBS), and *Staphylococcus aureus*
Epidemiology	UTIs are among the most common bacterial infections; risk factors include being female, being catheterized, having an underlying urological structural abnormality or disorder, or reduced immune function; incubation period is 3–8 days for *Escherichia coli* but is not well established for other enteric bacteria or for *S. saprophyticus*	
Transmission & prevention	Accidental transfer from anus or from genitalia of self or sexual partner; healthcare-acquired cases are almost always associated with catheters; prevention involves good hygiene and staying well hydrated so that you frequently urinate to flush out bacteria that may have entered the urethra	
Signs & symptoms	Common symptoms in lower UTIs include urgency to urinate; burning or painful sensation upon urination (dysuria), pus in the urine (pyuria), blood in the urine (hematuria), pain or pressure in lower abdomen (suprapubic tenderness/discomfort); upper UTIs, which include pyelonephritis, present with the same symptoms seen in lower UTIs, with the addition of fever, chills, flank pain (costovertebral angle tenderness), nausea, and vomiting	
Pathogenesis & virulence factors	Siderophores and abundant adhesion factors are key to helping bacteria invade the urinary tract; UPEC can also invade host urinary tract epithelial cells	*S. saprophyticus* has numerous adhesins and urease allows the bacterium to thrive in the bladder
Diagnosis & treatment	The fastest diagnostic is dipstick detection of nitrite or leukocyte esterase (Note: Nitrite test is noninformative for most Gram-positive infections); microscopy of urine may show presence of bacteria and white blood cells; bacteria can also be cultured from urine samples; in complicated UTIs, imaging tools like ultrasound, CT scan, or MRI are sometimes used to rule out kidney stones or urinary tract obstructions; treatment for uncomplicated cystitis may include nitrofurantoin, trimethoprim-sulfamethoxazole (TMP-SMX), or fosfomycin; fluoroquinolones such as ciprofloxacin and levofloxacin are second-line therapies; pyelonephritis treatment often includes fluoroquinolones or aminoglycosides like gentamicin or tobramycin	

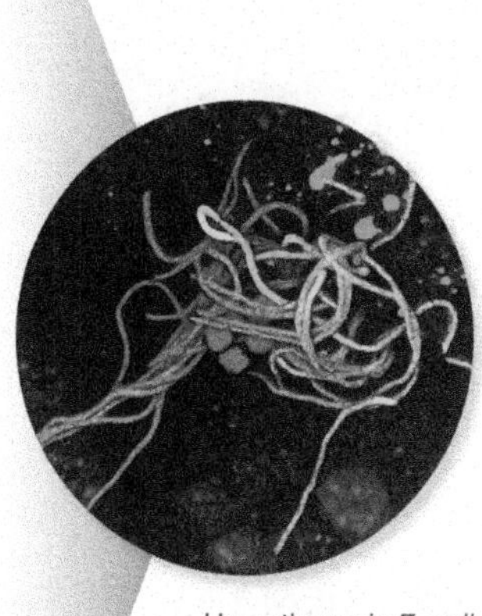

Uropathogenic *E. coli*

UTIs Caused by Gram-Positive Bacteria

Gram-positive bacteria can also cause UTIs. Usually these bacteria are introduced into the urethra via the skin or from the genitalia. The most common Gram-positive bacterium isolated in uncomplicated UTIs is *Staphylococcus saprophyticus*—followed in prevalence by *Enterococcus faecalis*, group B Streptococcus (GBS), and *Staphylococcus aureus*. About 6 percent of uncomplicated UTIs and 2 percent of complicated UTIs are caused by *S. saprophyticus*.[20] Although this bacterium seems relatively rare in terms of its likelihood of causing UTIs in the general population, it's a leading cause of UTIs in young sexually active women. It has been noted as causing up to 42 percent of uncomplicated UTIs in sexually active women who are 16 to 25 years old.[21] This bacterium is found in the male and female reproductive tract, including the external genitalia, of both sexes—making it easy to transmit to the urethra during sexual intercourse.

Emerging UTIs (Leptospirosis)

As climate change causes more coastal flooding and loss of natural habitat, it's feared that rats and other small mammals may increasingly migrate into residential areas. This means certain pathogens will have new opportunities to

[20] Flores-Mireles, A. L., Walker, J. N., Caparon, M., & Hultgren, S. J. (2015). Urinary tract infections: Epidemiology, mechanisms of infection and treatment options. *Nature Reviews: Microbiology*, 13 (5), 269.

[21] Raz, R., Colodner, R., & Kunin, C. M. (2005). Who are you—*Staphylococcus saprophyticus? Clinical Infectious Diseases*, 40 (6), 896–898.

DiseaseSnapshot • Leptospirosis

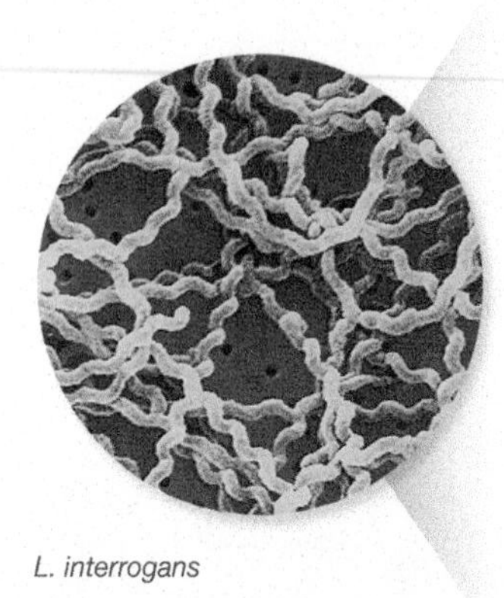

L. interrogans

Causative agent	*Leptospira interrogans* and other species (*L. borgpetersenii, L. santarosai, L. noguchii, L. weilli, L. kirschneri, L. alexanderi*); Gram-negative bacteria
Epidemiology	Worldwide distribution in developing countries with tropical and temperate climates; WHO estimates there are 873,000 caser per year with 48,600 deaths; incubation period ranges from 2 days to 4 weeks
Transmission & prevention	Zoonosis, mainly transmitted from rodents to humans, but is also found in livestock and dogs; most cases are from contact with water that is contaminated with urine from an infected animal; most outbreaks are related to recent flooding; no vaccine; avoid swimming in areas where leptopspirosis is endemic; be careful about the water you drink when in an endemic area; doxycycline is recommended for prophylaxis in epidemics
Signs & symptoms	Fever, headaches, vomiting, jaundice; can progress to renal failure, ocular disease, meningitis or pulmonary hemorrhage
Pathogenesis & virulence factors	Adhesion factors aid pathogen attachment to diverse host cells; pathogen releases a number of enzymes that damage host tissues; escape intracellular killing to live inside host cells, which allows them to avoid certain immune defenses and thrive off stolen host cell nutrients
Diagnosis & treatment	Detecting antibodies to the pathogen or PCR to detect bacterial DNA in blood, urine, or cerebrospinal fluid (CSF) samples; mild disease is treated with amoxicillin; penicillin G and ampicillin are indicated for severe disease; ceftriaxone is also used

CHEM • NOTE

Reduction reactions allow some bacteria to turn **nitrate** (NO_3^-) into **nitrite** (NO_2^-). Energy released during the reduction reactions is then channeled into making adenosine triphosphate (ATP), which is used to fuel biosynthesis reactions.

transmit to humans. This is the case for leptospirosis, an emerging disease caused by the *Leptospira interrogans* bacteria that can be transmitted to humans through animal excrement, especially that of rodents, livestock, and dogs. The long, slender, Gram-negative spirochetes usually enter the body through skin abrasions or across mucous membranes of the eyes, nose, or gastrointestinal tract. This means ingesting or bathing in contaminated water can lead to infection. Once in the body, the bacteria make their way through the bloodstream to the kidneys. Leptospirosis starts with general symptoms of fever, headache, and vomiting and may progress to renal failure, liver failure, meningitis, or respiratory distress. The World Health Organization estimates there are 873,000 cases worldwide each year, with 48,600 deaths. There are about 200 confirmed leptospirosis cases each year in the United States, with about half of all cases occurring in Hawaii. The globally increasing incidence of leptospirosis and concerns that people swimming in recreational water may be at increased risk led the CDC to list leptospirosis as a notifiable disease.

UTI Signs, Symptoms, and Diagnosis

Despite the diverse agents that can cause UTIs, they all tend to induce similar symptoms. Lower UTIs tend to cause frequent and urgent urination that may be painful (**dysuria**). **Pyuria** (pus in the urine) and/or **hematuria** (blood in the urine) may also develop. Sometimes the urine appears cloudy and may smell foul. The patient may also experience lower abdominal pain. When upper UTIs occur, additional signs and symptoms may include fever, nausea, vomiting, and severe abdominal and lower back pain near the waistline (*flank pain*).

The most common method for detecting a urinary infection is the **urine dipstick test** (FIG. 20.9). This test consists of a series of filter paper squares on a flexible plastic strip. The strip is dipped into a freshly collected **clean-catch urine sample** (midstream sample) and results are immediately read by comparing strip color changes to a key. A single strip tests for multiple factors, including infection markers such as white blood cells (WBCs) and nitrite. The test strip indirectly assesses WBC levels by checking for an enzyme called leukocyte esterase. High WBC levels would be reflected by higher esterase activity. An elevated urine nitrite level is another UTI indicator. Urine normally has high levels of nitrate, but certain bacteria reduce nitrate to nitrite. Therefore, an increase in nitrite levels suggests that there may be bacteria in the urine that are carrying out this chemical reaction. It should be noted that S. *saprophyticus*, which causes up to 6 percent of uncomplicated UTIs, cannot convert nitrate to nitrite.

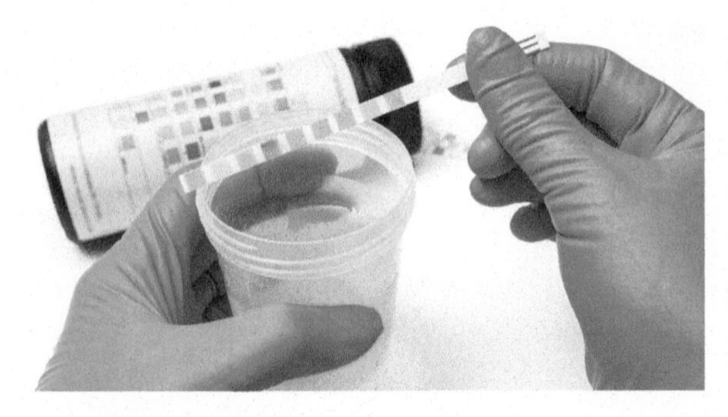

FIGURE 20.9 **Urine dipstick test** The dipstick test checks urine for signs of bacterial infection such as increased leukocytes and nitrites.

Critical Thinking *Assume a patient with a negative dipstick test has a UTI confirmed by bacteriological culturing. How could you explain the false negative dipstick results?*

If a dipstick test shows that the patient's urine contains infection markers such as nitrite and elevated white blood cell levels, then the urine sample may also be cultured in an effort to identify the causative agent. Urinary culturing is not usually necessary for uncomplicated UTIs, but is recommended in complicated infections.

CLINICAL VOCABULARY

Dysuria pain, discomfort, or a burning sensation during urination

Pyuria pus present in urine

Hematuria blood in the urine

Flank pain (or costovertebral angle tenderness) pain at the lower back near the natural waistline; often develops during a kidney infection

Clean-catch urine sample a way of collecting urine that limits contamination from the genitalia; the area around the opening of the urethra is wiped with an alcohol prep pad or sanitation wipe and then urine is collected midstream into a sterile sample cup

Viruses occasionally cause UTIs.

Symptomatic viral UTIs are extremely rare in people who have a normal immune system. Hantaviruses are one agent that may be considered in the differential diagnosis for an immune-competent patient who presents with an aseptic UTI (a case where bacterial or fungal agents cannot be cultured and the patient fails to improve with antimicrobial drugs). Hantaviruses were reviewed in Chapter 16 as the causative agents of a pulmonary syndrome and may also cause hemorrhagic fever with renal syndrome (HFRS).

HFRS symptoms tend to develop within two weeks of exposure. Early symptoms include a sudden and severe headache, flank pain, abdominal pain, fever, chills, nausea, blurred vision, and occasionally a rash. Later symptoms can include low blood pressure, leakage of blood into the abdomen, and acute kidney failure. HFRS is mainly treated with supportive therapies, renal dialysis, and the antiviral drug ribavirin. Even when treatment is applied, the CDC reports that hantavirus infections have a mortality rate of up to 45 percent.

Viral UTIs in Immune-Compromised Patients

Aside from hantavirus, when symptomatic viral UTIs occur they are mainly in immune-compromised patients. The incidence of viral UTIs varies in immune-compromised groups and seems highest in bone marrow transplant patients, people who are undergoing chemotherapy for blood cancers like leukemia or lymphoma, and organ transplant patients. Even in these patients, viral UTIs likely represent less than 25.5 percent of UTIs.[22]

Adenoviruses, cytomegalovirus, and BK virus cause most viral UTIs. Adenoviruses are mainly associated with cold-like symptoms (see Chapter 16), while cytomegalovirus is best known for causing mononucleosis (see Chapter 21). BK virus has not yet been discussed in the text; it is a *Polyomaviridae* family member and has a double-stranded DNA genome. It was named using the initials of the first patient in whom the virus was detected. Most people are infected with this virus in childhood and experience mild cold-like symptoms. The virus then goes latent in our body and tends to remain quiet the rest of our lives. In immune-compromised patients, the BK virus can re-emerge and cause cystitis. The virus is especially concerning in kidney transplant patients because it can attack transplanted kidney tissue.

The most common sign of a viral UTI is the sudden onset of **hemorrhagic cystitis** (bladder inflammation that is hallmarked by bleeding). Viral UTIs are diagnosed using molecular methods that can detect viral genes or viral antigens. Cidofovir is the most common treatment for viral UTIs.

CLINICAL VOCABULARY

Hemorrhagic cystitis a bladder inflammation accompanied by blood in the urine; a common hallmark sign of viral UTIs, but may also occur in fungal and bacterial UTIs

Fungi can cause UTIs.

The most common fungi that infect the urinary tract are *Candida* species (yeast)—especially *C. albicans*, which is also notorious for causing infections of the vagina and vulva (vulvovaginal candidiasis). *Candida albicans* causes about 7 percent of complicated UTIs and about 1 percent of uncomplicated UTIs.[23]

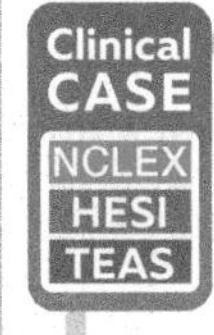

The Case of Cut Corners and CAUTIs

Practice applying what you know clinically: visit the **Mastering Microbiology** Study Area to watch Part 3 and practice for future exams.

[22] Paduch, D. A. (2007). Viral lower urinary tract infections. *Current Urology Reports*, 8 (4), 324–335.

[23] Flores-Mireles, A. L., Walker, J. N., Caparon, M., & Hultgren, S. J. (2015). Urinary tract infections: Epidemiology, mechanisms of infection and treatment options. *Nature Reviews: Microbiology*, 13 (5), 269.

CLINICAL VOCABULARY

Candiduria yeast in the urine

Fungal UTIs present with the same signs and symptoms seen in bacterial and viral UTIs. They are more common in catheterized patients and/or in patients who have underlying medical conditions like diabetes or reduced immune function. Patients who have unexplained **candiduria** (yeast in the urine) should be evaluated for underlying urological abnormalities. Fluconazole is the first-line drug for fungal UTIs. Yeast infections are covered in more depth at the end of this chapter, under reproductive system infections.

Build Your Foundation

Build Your Foundation (BYF) Quick Quiz: Visit the **Mastering Microbiology** Study Area to quiz yourself.

7. What differentiates complicated from uncomplicated urinary tract infections?
8. Define the terms urethritis, cystitis, ureteritis, and pyelonephritis.
9. What are the signs, symptoms, risk factors, and potential long-term impacts of lower and upper UTIs, and how are UTIs diagnosed?
10. What are the infection-related challenges associated with urinary catheters and the significance of CAUTIs?
11. What are the names and clinical features of the main causative agents of UTIs?
12. What are the main clinical and epidemiological features of leptospirosis?

20.3 REPRODUCTIVE SYSTEM VIRAL INFECTIONS

Learning Outcomes

After reading this section, you should be able to:

20.13 Describe the clinical and epidemiological features of genital herpes.

20.14 Compare and contrast HSV-1 and HSV-2.

20.15 Detail the consequences of neonatal herpes.

20.16 Explain the clinical and epidemiological features of human papilloma viruses (HPVs).

20.17 Describe the Pap smear method and its clinical applications in preventive care.

Many sexually transmitted pathogens do not target the reproductive system.

Sexually transmitted infections (STIs) include bacterial, viral, parasitic, and fungal pathogens. Before we delve into these infections it is important to clarify that the term STI simply refers to the *mode of transmission*. That means that while STIs may spread by sexual contact, it doesn't mean it's an infectious agent that will affect the reproductive system. For example, human immunodeficiency virus (HIV) is arguably one of the most infamous STIs, but sexual contact is only one way it spreads. Plus, HIV targets the immune system, not the reproductive system, which is why it is discussed in Chapter 21. Other agents that can be sexually transmitted but that don't target the reproductive system include hepatitis viruses, Zika virus, Ebola, and just about any other pathogen that spreads through body fluids. It's important to note that any STI that causes lesions/ulcers or pronounced inflammation that may lead to easily bleeding tissues is considered a risk factor for transmission and acquisition of HIV or other bloodborne, sexually transmissible infections like hepatitis. This is because lesions and bleeding tissues provide a readily accessible entry and exit point for bloodborne agents. The WHO estimates that the presence of ulcerative STIs increases the risk of HIV transmission by 10–50 percent in women and 50–300 percent in men.

Many infections that can be sexually transmitted, including HIV, chlamydia, gonorrhea, syphilis, chancroid, Zika, and hepatitis viruses, are nationally notifiable diseases. This means that healthcare providers are required to report confirmed cases to local and state health agencies. In this section we'll focus on pathogens that specifically target the reproductive system.

Genital herpes is an STI that affects the reproductive system and can cause serious neonatal complications.

Herpes simplex virus 2 (**HSV-2**, also known as human herpes virus 2) causes genital herpes. According to the CDC, in the United States about one in every six people between 14 and 49 years old have HSV-2. Another related virus,

HSV-1 (or human herpes virus 1), is the main cause of oral herpes lesions. (See Chapter 17 for more on oral herpes.) HSV-2 is usually contracted sexually, while HSV-1 is usually not. That said, unprotected oral sex *can* transmit HSV-1 to the genital area. On rare occasions, oral sex may transmit HSV-2 from the genitals to the mouth. However, oral HSV-2 infections shed very few virions.

The herpes virus enters epithelial cells after contact with an infected person's lesions. It usually causes an initial, severe outbreak of ulcers as quickly as three days after transmission. The early appearance of the lesions is sometimes called "dewdrop on a rose petal," describing the look of the clear pustule above a reddened base (FIG. 20.10). Herpes viruses are notorious for causing persistent latent infections. After the primary infection, the virus tends to go dormant in peripheral nerves near the area where lesions had developed. A number of events such as stress, illness, or hormone changes can reactivate latent viruses and lead to a new outbreak of sores on any area of the genitalia or perianal region. But lesions may form on other areas too. The sacral form of HSV-2 re-emerges as nongenital lesions located below the waist and can cause lesions on the buttocks, lower back, waist, or thighs; it is sometimes mistaken for shingles. In some people the virus rarely re-emerges, while others experience frequent outbreaks. (Chapter 6 further reviews the mechanisms of viral latency.)

FIGURE 20.10 Genital herpes lesion This photo shows the "dewdrop on a rose petal"—the initial rash caused by HSV-2.

Although the antiviral drug acyclovir can shorten outbreaks and suppress flare-ups to reduce the risk of transmission, there is not a cure for herpes. (A detailed discussion of antivirals is found in Chapter 6.) Condoms may not cover areas affected by a herpetic lesion and therefore may not provide full prevention. Also, even if people don't have active lesions they still may shed low levels of the virus and be infectious.

Genital herpes is rarely dangerous to immune-competent adults and most cases are so mild that people don't even know they're infected. The biggest threat herpes poses to adults is that lesions may increase a person's risk for contracting bloodborne STIs like HIV, because the herpetic lesions can provide a point of entry to the bloodstream. As already mentioned, *any* STI that generates

BENCH to BEDSIDE

Therapeutic Vaccine Development for Herpes Simplex Virus 2 (HSV-2)

Genital herpes has no cure, nor is there a vaccine for prevention. As of this book's publication, the only way patients can reduce genital herpes outbreaks is by taking daily oral antiviral drugs. However, advances have been made in the emerging field of immunotherapy that may mean that a shot three times a year could suppress outbreaks and reduce transmission. This sort of vaccine is called a therapeutic vaccine, because unlike other more commonly administered vaccines, it does not prevent disease. Instead, it treats infection.

The injectable formulation consists of parts of an HSV-2 glycoprotein, along with a specific protein that HSV-2-infected host cells make. To complete the injected cocktail, these components are mixed with an adjuvant—a substance that amplifies an immune response.

Initial trials were performed on HSV-2-infected guinea pigs. The injections were administered 14, 21, and 46 days after the guinea pigs were infected with HSV-2. After the animals had their first treatment, they experienced a 55 percent reduction in the number of lesions. After the third immunization, viral shedding fell below detectable levels for all animals in the study.

Now this suppressive therapy is in human clinical trials and the results look promising. After the three-injection series, study patients showed a 55 percent reduction in viral shedding and 60 percent reduction in lesions. Even a year after the last injection, viral shedding was reduced 52 percent compared to individuals in the control group. Although this immunotherapy is not a cure for genital herpes, is it a step in the right direction toward improving patient lives and reducing the incidence of outbreaks by providing a suppression therapy that would be much easier for patients to comply with than taking a daily oral medication. If FDA approved, this would be the first vaccine on the market designed for therapy, rather than prevention.

References: Skoberne, M., Cardin, R., Lee, A., Kazimirova, A., Zielinski, V., Garvie, D., et al. (2013). An adjuvanted herpes simplex virus 2 subunit vaccine elicits a T cell response in mice and is an effective therapeutic vaccine in guinea pigs. *Journal of Virology*, *87* (7), 3930–3942.

Kim, H. C., & Lee, H. K. (2020). Vaccines against Genital Herpes: Where Are We?. *Vaccines, 8*(3), 420.

DiseaseSnapshot • Genital Herpes

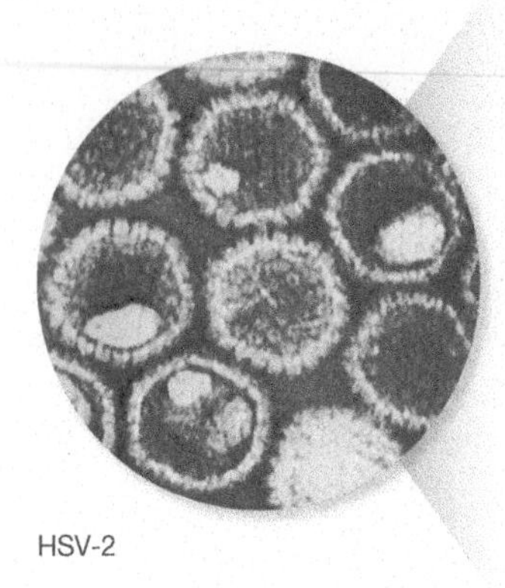

HSV-2

Causative agent	Herpes simplex virus 2 (HSV-2): enveloped, double-stranded DNA genome, *Herpesviridae* family
Epidemiology	One of the most common sexually transmitted viruses; incubation period is 3–7 days
Transmission & prevention	Mainly transmitted through sex with an infected person; occasionally transmitted through oral sex; virus can be shed even in the absence of active lesions; the only guaranteed way to prevent infection is abstinence from sex and oral sex with infected individuals; reduced contact during active outbreaks; and using condoms may reduce transmission, but lesions can also be on parts of the genitalia that are not covered by a condom
Signs & symptoms	Often asymptomatic or so mild that the person may not know they have been infected; occasionally HSV-2 may cause a mild flu-like illness upon the initial outbreak of lesions on the genitalia; recurrent outbreaks are also possible; recurrent lesions may be on the genitalia but they may develop elsewhere below the waist in a form called sacral herpes
Pathogenesis & virulence factors	Virus multiplication in tissues produces lesions and the host's immune responses are repressed; virus goes latent in host nerve cells near the initial site of infection and can reemerge
Diagnosis	Diagnosed by lesion appearance and medical history; detection of viral DNA by polymerase chain reaction or serological antibody detection methods may be used; there is no cure, but acyclovir can reduce frequency and severity of outbreaks

a lesion can increase the risk of infection with HIV or other bloodborne infections that can be sexually transmitted.

Neonatal Herpes

While herpes is not considered dangerous in adults, the story is different in newborns. According to the CDC, in the United States about one in every 3,500 newborns develops neonatal herpes. Roughly 90 percent of cases develop when HSV-2 is transmitted from a mother to her baby during delivery; only about 10 percent of neonatal herpes cases develop due to viral transmission across the placenta.[24] Mothers experiencing a first-ever herpes outbreak are more likely to transmit HSV-2 to their baby than are mothers who experience a re-emergence of infection. If the mother is experiencing a herpes outbreak (an initial or recurrent episode) and is at term with her pregnancy, then a cesarean section can reduce the risk of neonatal herpes.

Neonatal herpes usually presents with symptoms within the first month of life. It has three major clinical presentations: skin, eye, and mouth (SEM) manifestations, central nervous system (CNS) effects, and disseminated infections. Of these, SEM manifestations are the mildest and most common; SEM occurs in about 45 percent of neonatal herpes cases. Administering antiviral drugs early on can prevent the virus from spreading to the CNS or progressing to a full-blown **disseminated infection** (an infection that has spread throughout the body). About 30 percent of neonatal herpes cases progress to CNS involvement. Disseminated infections develop when the virus spreads from the mucous membranes and skin to the internal organs and CNS. Disseminated infections are the most dangerous form of neonatal herpes and occur in about 25 percent of cases. Even with antiviral drug treatment, around half of these infants will experience long-term neurological problems and about 1 in 4 affected infants die.[25]

Human papilloma viruses cause the most common STI in the world.

According to the CDC, **human papilloma viruses (HPVs)** cause the most common STI in the world. The CDC states that these viruses are so common that almost every sexually active person will be infected with some type of HPV in their lifetime. There are over 200 known subtypes of HPV; about 40 of them are

[24] Swetha, G. P., & David, W. (2014). Preventing HSV in the newborn. *Clinics in Perinatology*, (41), 4.
[25] Swetha, G. P., & David, W. (2014). Preventing HSV in the newborn. *Clinics in Perinatology*, (41), 4.

Disease**Snapshot** • Human Papilloma Viruses (HPVs)

Causative agent	Over 200 distinct types of human papilloma virus: nonenveloped, double-stranded DNA genome, *Papillomaviridae* family
Epidemiology	14 million people in the U.S. contract HPV every year, making it the most common STI in the U.S.; HPV types 16 and 18 account for approximately 70% of cervical cancers worldwide; HPV 6 and 11 are responsible for about 90% of genital warts; the incubation period varies based on HPV type; usually HPVs induce cellular changes in the host within 2 to 3 months after infection, but it may take years to see effects
Transmission & prevention	Sexual transmission mainly through oral, anal, vaginal sex; and nonpenetrative genital-to-genital contact; several HPV vaccines available, with the latest version providing protection against 9 types of HPV; risk reduced through consistent and correct use of condoms
Signs & symptoms	Certain subtypes cause genital warts and/or cancers (including penile, cervical, and oropharyngeal cancers)
Pathogenesis & virulence factors	HPV infects epithelial cells and stimulates them to divide, which is how warts and cancers may develop; the virus also has factors that prevent interferon and antigen presentation, which allows it to hide from the immune system and establish a persistent infection
Diagnosis & treatment	HPV is often detected through cervical cancer screening by a Pap smear; there are several nucleic acid-based tests to confirm viral presence in cervical cells; urogenital and anal warts can be surgically removed, physically ablated with heating or freezing, or treated with topical gels, creams, or ointments

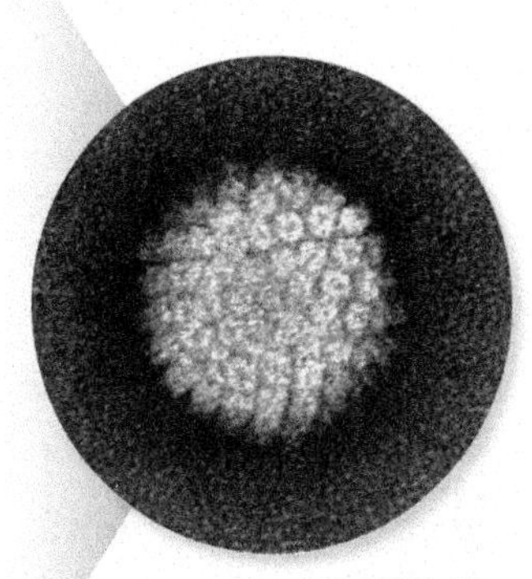

An HPV virion

sexually transmitted. Most people who get a sexually transmitted HPV infection will never develop any symptoms and the virus tends to self-clear in 9 out of 10 patients within two years. Those who cannot resolve the infection may experience outbreaks of genital warts and may have an increased risk of cancer. The sexually transmitted forms of HPV can cause anal and genital warts and some cause cancer—especially cervical, anal, vaginal, vulvar, penile, and oropharyngeal cancers (mouth/throat cancer). Types 6 and 11 are best known for causing warts, while types 16 and 18 are notorious for causing cancer. The CDC reports that overall, HPV causes about 90 percent of cervical cancers. HPVs invade the epithelial cells of the reproductive tract, with HPVs that cause cervical cancer showing a tropism (infection preference) for cervical cells. The CDC states that in the United States HPV causes 37,000 cancer cases per year in men and women. Cervical cancers are the most prevalent HPV-related cancer. The Pap smear is a cheap, easy, and noninvasive method for detecting HPV infections as well as precancerous and cancerous cervical cells (FIG. 20.11). According to the American Cancer Society, if cervical cancer is caught early it has an excellent prognosis, with a five-year survival rate of 93 percent. (For more on how HPV causes cancer, see Chapter 6.)

The FDA has approved three different vaccines for HPV since 2006. The most recent version is formulated to protect against nine different types of HPV, including the strains that are most commonly linked to cancers and genital warts. The CDC recommends boys and girls receive the HPV vaccine series at age 11 or 12.

The CDC reports that since vaccines against HPV were introduced in 2006, data has shown that teen girls in the United States have shown a 64 percent reduction in vaccine-type HPV infections.

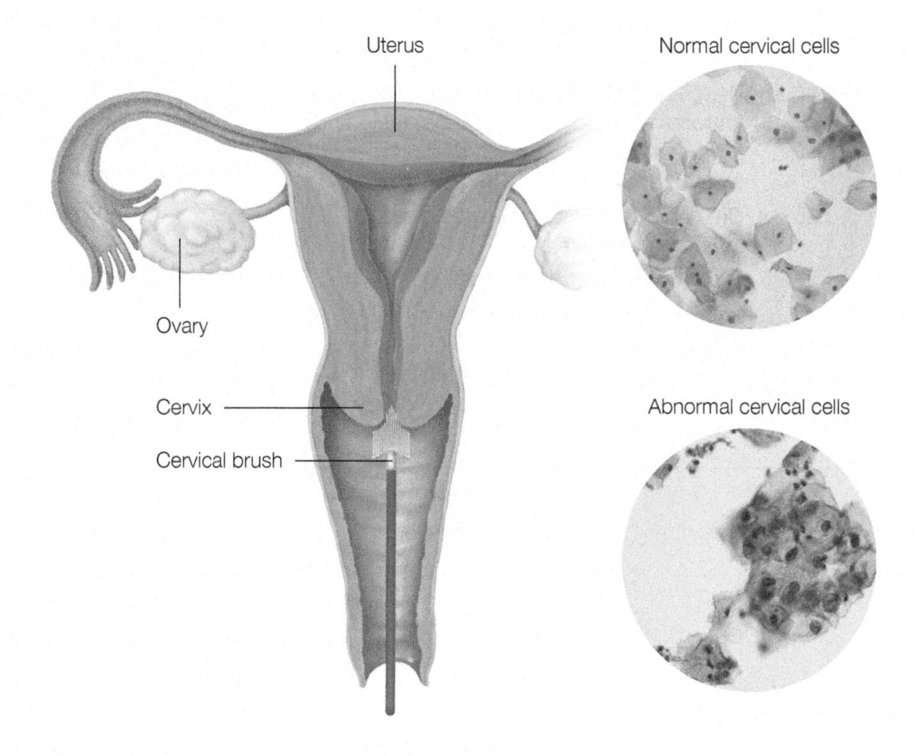

FIGURE 20.11 Pap smear to detect abnormal cervical cells The cervical cells collected by a Pap smear test are analyzed by microscopy to check for abnormal cells. Most cervical cancer cases are associated with HPV.

Critical Thinking *Why might douching (rinsing the vagina with a solution) or having sex within two days of having a Pap test diminish the test's sensitivity?*

Build Your Foundation

13. What are the clinical and epidemiological features of herpes simplex virus 2 and how can it be treated?
14. Compare and contrast herpes simplex virus 1 to herpes simplex virus 2.
15. What are the consequences of neonatal herpes?
16. What are the clinical and epidemiological features of human papilloma virus (HPV) infections?
17. What is a Pap smear and how is it useful in preventive care?

Build Your Foundation (BYF) Quick Quiz: Visit the **Mastering Microbiology** Study Area to quiz yourself.

20.4 REPRODUCTIVE SYSTEM BACTERIAL INFECTIONS

Learning Outcomes

After reading this section, you should be able to:

20.18 Explain the differences between vaginitis and vaginosis.

20.19 Describe the clinical features of vaginosis and explain how it can be diagnosed and treated.

20.20 Describe the clinical and epidemiological features of chlamydia, including potential complications in neonates and in untreated women.

20.21 Summarize the life cycle of *Chlamydia trachomatis*.

20.22 Describe the clinical and epidemiological features of lymphogranuloma venereum.

20.23 Describe the clinical and epidemiological features of gonorrhea infection, including potential complications in untreated women and neonates.

20.24 Describe the clinical features of pelvic inflammatory disease.

20.25 Name the agent that causes syphilis and describe the clinical features of each infection stage as well as the consequences of congenital syphilis.

20.26 Outline the clinical and epidemiological features of chancroid.

Vaginosis and vaginitis describe different vaginal conditions.

Not every affliction of the reproductive tract is sexually transmitted—indeed, some can be due to dysbiosis of normal microbiota. Such imbalances are common in the vagina. They can develop following an antibiotic treatment or even as hormone levels fluctuate with a woman's menstrual cycle or during pregnancy. An overgrowth of one type of microbe over another in the vagina is called **vaginosis**. When a patient experiences vaginosis, she may or may not simultaneously experience vaginal inflammation, or **vaginitis**.

It is entirely possible to have one scenario without the other. For example, an overgrowth of a benign bacterium from among the normal vaginal microbiota that doesn't trigger inflammation would lead to vaginosis. During vaginosis the patient may have symptoms like foul-smelling vaginal secretions or excessive secretions. While these manifestations are unpleasant and possibly embarrassing, they aren't painful like vaginitis. (Note that lack of pain doesn't mean lack of pathology; vaginosis has been linked to numerous issues like pre-term labor and infertility.) The flip side of this coin is a scenario where the patient could have vaginitis due to an adverse reaction to a spermicide or douching product, but there is no microbial involvement.

Vaginosis

Most women will experience **bacterial vaginosis** at least once in their lifetime. Bacterial vaginosis is *polymicrobial*, meaning there is not one causative agent. Sometimes polymicrobial diseases require a specific set of organisms to cause disease. In this case, the best indicator isn't the presence of one specific agent, or even several organisms, but the lack of *Lactobacillus*. In general, vaginosis features a decrease in the normal level of *Lactobacillus* species in the vagina, accompanied by an increase in mixed anaerobes, including *Gardnerella vaginalis*, *Bacteroides* (*Prevotella*) species, *Mobiluncus* species, *Ureaplasma urealyticum*, and *Mycoplasma hominis*.[26] While there are commercial test kits available for vaginosis, diagnosis is usually made based on a collection of signs called **Amsel's criteria**. These criteria are usually applied to assess reproductive-age women who have not yet reached menopause. Women who meet three of the four below criteria are 90 percent likely to have vaginosis.

1. Vaginal discharge has a greyish-white coloration and uniform texture (exact color and amount may vary).
2. An increase in vaginal pH above 4.5.

[26] Mastromarino, P., Vitali, B., & Mosca, L. (2013). Bacterial vaginosis: A review of clinical trials with probiotics. *New Microbiology*, 36 (3), 229–238.

Bacteria

Vaginal epithelial cell

Cell nucleus

Normal cell
Vaginal epithelial cell without adhered bacteria or yeast

Clue cell
Vaginal epithelial cell with bacteria or yeast adhered to it

FIGURE 20.12 Clue cells The presence of "clue cells" covered with bacteria is one vaginosis indicator.

3. Presence of "clue cells," which are epithelial cells with bacteria attached (FIG. 20.12).
4. A positive whiff test: after potassium hydroxide is added to a vaginal mucus smear, volatile amines develop and can be detected as a fishy odor (FIG. 12.13).

CHEM • NOTE

In the whiff test the patient's vaginal mucus is smeared onto a glass slide and drops of potassium hydroxide (KOH) are added. In vaginosis cases, the vaginal mucus is enriched with nitrogen-containing compounds that react with the added KOH to produce **volatile amines** that have a fishy odor. These amines are called "volatile" because they readily evaporate into the air at room temperature and can therefore be smelled.

Gram staining sampled vaginal secretions is still the gold standard for confirming vaginosis. Scoring standards called **Nugent's criteria** help evaluate Gram-stained preparations. As with Amsel's criteria, Nugent's scoring is most accurate when it is used to evaluate samples from reproductive-age women. In general, scoring is based on seeing abundant cocci and abundant Gram-negative (or Gram-variable) rods in the vaginal mucus, along with a decrease in long Gram-positive rods. Nugent's criteria also look for clue cells in the patient sample.

Although it is generally accepted that vaginosis is a microbiota dysbiosis as opposed to a sexually transmitted infection, there are some new studies that suggest that sexual intercourse could in part promote vaginosis. Some studies suggest that bacteriophages from a sexual partner's microbiome may be introduced to the vagina during intercourse. The transmitted bacteriophages then kill off the lactobacilli species in the vagina, which provides an opportunity for other organisms among the normal vaginal microbiota to overgrow.[27]

Chlamydia trachomatis is a common cause of bacterial STIs.

Chlamydia trachomatis is a Gram-negative bacterium that takes on a short rod or coccus shape, lives inside eukaryotic cells, and has a sophisticated taxonomy. This species is subdivided into two major groupings, or biovars: trachoma and lymphogranuloma venereum. When people talk about the disease **chlamydia**, they usually mean illnesses related to *Chlamydia trachomatis* of the trachoma biovar (sometimes denoted as *C. trachomatis* trachoma). Even within this biovar there are further divisions, or serovars. Some serovars target the eyes and cause an infection called trachoma that can lead to blindness (reviewed in Chapter 17). A separate set of *C. trachomatis* trachoma serovars (D through K) cause urogenital infections, which we collectively call chlamydia. A less

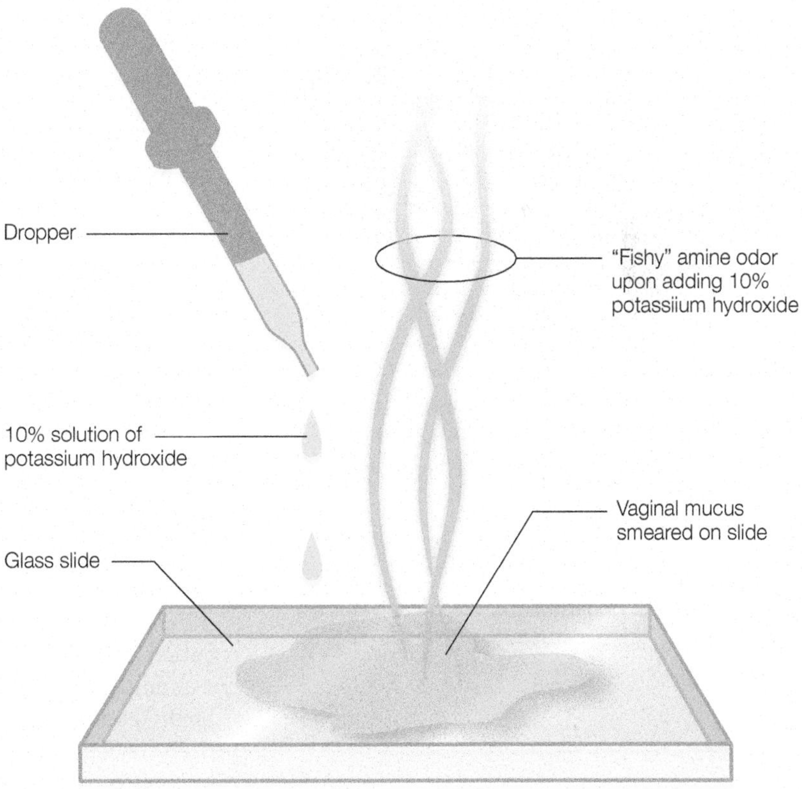

FIGURE 20.13 Whiff test One quick and easy way to detect vaginosis is the whiff test. In this test a swab is used to collect vaginal mucus from the patient. The sample is then smeared on a glass slide and a few drops of 10% potassium hydroxide solution are added to the smear. The generation of a fishy odor indicates possible vaginosis.

[27] Turovskiy, Y., Sutyak Noll, K., & Chikindas, M. L. (2011). The aetiology of bacterial vaginosis. *Journal of Applied Microbiology*, *110* (5), 1105–1128.

Disease**Snapshot** • Bacterial Vaginosis

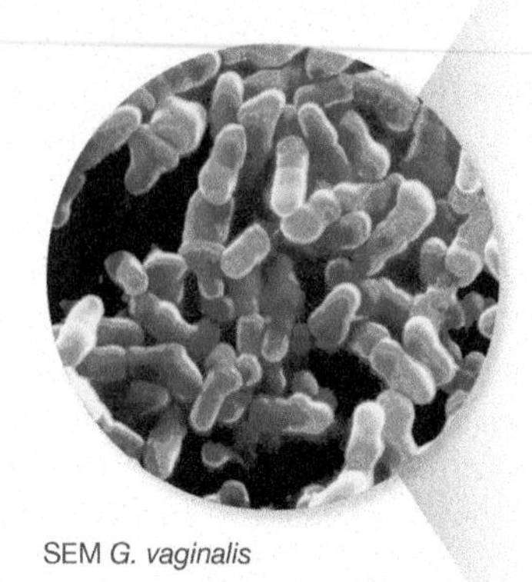
SEM *G. vaginalis*

Causative agents	Polymicrobial dysbiosis involving a decrease in vaginal lactobacilli and increases in mixed anaerobes such as *Gardnerella vaginalis, Bacteroides (Prevotella)* species, *Mobiluncus* species, *Ureaplasma urealyticum, and Mycoplasma hominis*
Epidemiology	Prevalence varies widely (5–50% depending on the population); the incubation period is variable and not well established
Transmission & prevention	Occurs after disturbance of vaginal flora, including but not limited to: douching, antibiotic use, hormonal shifts, etc.; avoid douching and other possible disturbances to the normal vaginal microbiome
Signs & symptoms	Whitish-grey homogenous vaginal discharge that may have a foul odor; while vaginosis does not usually lead to pain or inflammation in the patient, studies suggest it could also be a risk factor for pre-term labor and may contribute to infertility
Pathogenesis & virulence factors	Many bacteria enriched in vaginosis make nitrogen-rich compounds that are converted to foul- or fishy-smelling volatile amines; these organisms also have enzymes (mucinases, sialidases, and neuraminidases) that alter the normal vaginal discharge; some of the enriched organisms also have enzymes that cleave host IgG and IgM antibodies
Diagnosis & treatment	Amsel's criteria are accepted for diagnosis, but the gold standard remains confirmation using Nugent's criteria; nucleic acid diagnostic kits and detection of specialized enzymes are also available; treatments include oral or topical metronidazole or clindamycin

Ages 10–14
<1% of cases
Ages 40+
4% of cases
Ages 30–39
14%
of cases
Ages 20–24
37%
of cases
Ages 25–29
18%
of cases
Ages 15–19
27%
of cases

FIGURE 20.14 **Chlamydia cases in the U.S. by age group** These data are based on the incidence of new cases in 2019 in the United States and are rounded to the nearest percent.

Source: Data are from Centers for Disease Control and Prevention. Sexually Transmitted Disease Surveillance 2019. Atlanta: U.S. Department of Health and Human Services; 2021.

discussed though still clinically important biovar of *C. trachomatis* is the lymphogranuloma venereum grouping. This biovar causes the sexually transmitted infection **lymphogranuloma venereum.** First we'll review the more commonly recognized STI chlamydia and then we'll review the lesser-appreciated STI lymphogranuloma venereum.

Chlamydia

Among the most common STIs in the world, the CDC reports that in 2019 there were about 1.8 million cases of chlamydia in the United States (FIG. 20.14). However, many cases are asymptomatic, so they go undiagnosed and unreported and the true infection incidence is probably much higher. In men, the bacteria are a leading cause of **nongonococcal urethritis.** Recall that in men the urethra is part of the urinary and reproductive system, so this infection would affect both systems. In women the bacteria mainly target the cervix, but can spread to the uterus and fallopian tubes. As already mentioned, chlamydia infections are often asymptomatic; about 50 percent of men and 75 percent of women are asymptomatic. When symptoms do develop they can differ between the sexes. In men symptoms include dysuria, burning and itching around the opening of the penis, and swelling or pain in the testicles. There can also be a discharge from the penis. In women, symptoms may include bleeding/spotting between menstrual periods and/or after sex, dysuria, vaginal discharge with an odor, itching or burning around the vagina, pain during sexual intercourse, and pelvic pain that may be accompanied by fever.

Chlamydia hasn't yet developed many antibiotic-resistant strains, so it's relatively easy to cure. However, the fact that most cases are asymptomatic means they go undetected and can lead to complications. For women, complications include pelvic inflammatory disease (discussed more in later sections), a condition that increases risk of ectopic pregnancy, infertility, and pre-term birth. Women with an untreated infection can also transmit this bacterium to their newborn as the baby passes through the birth canal. This **perinatal exposure**, or exposure during delivery, may lead to bacterial conjunctivitis and/or neonatal pneumonia. Male complications include chronic **epididymitis** (swelling of the epididymis), which can also cause infertility.

***Chlamydia trachomatis* life cycle** The word *Chlamydia* comes from "cloak." The name ended up being appropriate, given that the bacteria have an intracellular lifestyle in which they hide from the immune system by entering our own cells to

CLINICAL VOCABULARY

Perinatal exposure describes a newborn's exposure to an agent during delivery

Epididymitis inflammation of the epididymis

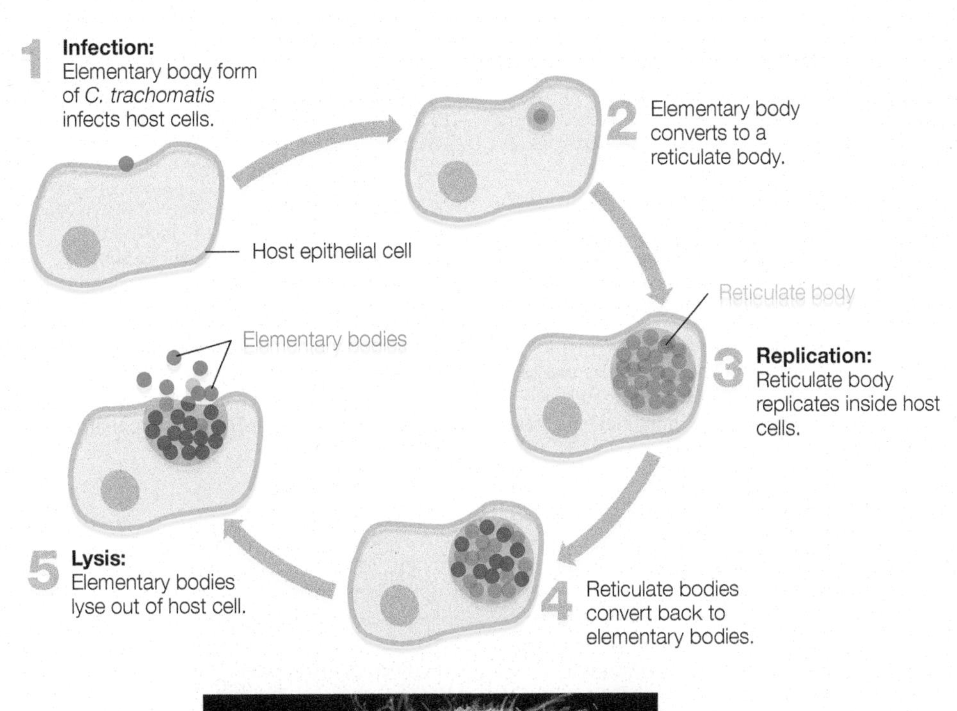

FIGURE 20.15 ***Chlamydia trachomatis* life cycle** This bacterium is an obligate intracellular pathogen that cycles between an elementary body form and a reticulate body form.

live and reproduce. Inside the "cloak" of our own cells, *C. trachomatis* undergoes a remarkable life cycle (FIG. 20.15). One form of the bacterium, the *elementary body*, is infectious but dormant. Another form is noninfectious but metabolically active, and called a *reticulate body*. Although *Chlamydia* cells don't have peptidoglycan, they have a membrane stabilized with lipopolysaccharide. This helps them bind host epithelial cells and then force the cells to take them up by phagocytosis. Once inside the host's cells, the bacteria steal the cell's ATP and become metabolically active reticulate bodies. The reticulate bodies then undergo binary fission until they are ready to burst out of the host cell to spread. Before they burst free of the host cell, they reconvert to their infectious elementary body form.

Lymphogranuloma Venereum

As previously described, lymphogranuloma venereum is the second main biovar (or grouping) of *Chlamydia trachomatis*. Although this pathogen is related to the bacteria that cause chlamydia, and is sexually transmitted, it is clinically distinct regarding its clinical profile and epidemiology. The infection progresses through three stages.

In the first stage a small painless lesion develops within the urethra, vagina, or on the cervix, or it may develop in the anus or on the external genitalia. If it forms in internal areas of the reproductive tract it is very easy to miss. In this early stage men may experience prostatitis and women may experience **cervicitis** (cervical inflammation). In the second stage the bacteria enter lymphatic vessels and travel to nearby lymph nodes in the groin region (*inguinal*

CLINICAL VOCABULARY

Cervicitis the opening of the uterus (the cervix) becomes inflamed

Inguinal lymph nodes lymph nodes located in the groin region

Buboes severe lymph node inflammation and swelling

Necrosis tissue death

Edema swelling

Fibrosis the formation of scar tissue

Disease**Snapshot** • Chlamydia and Lymphogranuloma venereum

C. *trachomatis*

Disease	Chlamydia	Lymphogranuloma venereum
Causative agent	*Chlamydia trachomatis* biovar trachoma (serovars D through K)	*Chlamydia trachomatis* biovar lymphogranuloma venereum
Epidemiology	An estimated 2.86 million infections occur annually in the U.S.; almost two-thirds of new infections occur in 15- to 24-year-olds; racial disparities and risk behaviors also influence incidence; incubation period remains unclear	Previously endemic in north and west Africa, now more common in Europe and North America among men who have sex with men; reliable data for incidence is lacking; incubation period is 3–30 days
Transmission & prevention	Transmitted by sexual contact; ejaculation is not required for transmission; newborns may be exposed to the bacterium in the birth canal in untreated mothers and may suffer neonatal conjunctivitis (see Chapter 17); can be prevented by condom use, abstinence, or monogamous sex after STI testing confirms noninfected state	Sexual transmission with higher risk associated with anal sex; prevention includes proper condom use, abstinence, or monogamous sex after STI testing confirms noninfected state
Signs & symptoms	10% of men and 5 to 30% of women are symptomatic; purulent discharge, pyuria, dysuria, and abdominal or pelvic pains are possible symptoms in women. Symptomatic men usually display mucoid or watery urethral discharge, dysuria, and sometimes testicular pain; rectal infections also present as mucoid discharge and local pain	Primary stage: painless lesions, swollen prostate, cervicitis; secondary stage: swollen lymph nodes, formation of buboes, sometimes fever; tertiary stage: often permanent edema and fibrosis of the lymph nodes, especially lymph nodes located near the genitals
Pathogenesis & virulence factors	Obligate intracellular pathogen, converting between two forms, elementary bodies and reticulate bodies; bacterium has tools to trigger host cells to take it up and tools to avoid destruction within the host cell; incites a host inflammatory response that is responsible for much of the tissue damage; repeated infections are possible	Binds epithelial cells, travels to lymph nodes where it invades macrophages; has an intracellular lifestyle that allows it to evade certain host immune defenses
Diagnosis & treatment	Diagnosis is usually confirmed using molecular methods that detect bacterial genes (NAAT: nucleic acid amplification test)—these tests are run on urine (for men) or endocervical swabs (for women); bacterial culture may also be performed; first-line therapies are azithromycin or doxycycline	Serologic testing for LGV is not widely available in the U.S., instead the chlamydial complement fixation test may be used to confirm diagnosis; first-line treatment is doxycycline; pregnant or lactating women are prescribed erythromycin or azithromycin

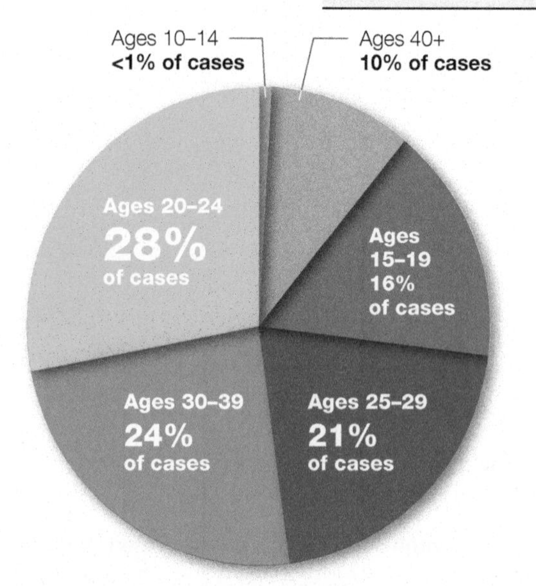

FIGURE 20.16 Gonorrhea incidence by age group These data are based on the incidence of new cases in 2019 in the United States and are rounded to the nearest percent.

Source: Centers for Disease Control and Prevention. Sexually Transmitted Disease Surveillance 2019. Atlanta: U.S. Department of Health and Human Services; 2021.

lymph nodes). Patients may develop a fever in this stage. Once in the lymph nodes, the bacteria infect macrophages. As the bacteria replicate, the affected lymph nodes become swollen and painful and develop into **buboes** that may have regions of **necrosis** and may rupture in up to a third of patients. The tertiary stage can lead to severe destruction of the genitalia as tissues experience progressive **edema** (swelling) and permanent **fibrosis** (scarring). As with biovar trachoma, lymphogranuloma venereum can be completely cured using antibiotics such as doxycycline.

Gonorrhea incidence is on the rise in the United States and increased antibiotic resistance is making it harder to treat.

According to the CDC, in 2019 about 600,000 people were diagnosed with gonorrhea (FIG. 20.16). The CDC suggests that the actual number of infections per year is probably closer to 820,000, but because the infection is often asymptomatic it often goes undiagnosed. Gonorrhea is an STI caused by the Gram-negative, intracellular diplococcus *Neisseria gonorrhoeae*. This bacterium can infect the genitals, rectum, and throat. Gonorrhea and chlamydia infections share a lot of features; both are caused by bacteria that have an intracellular lifestyle and both infections tend to be asymptomatic in women. However, in patients who do develop symptoms, chlamydia and gonorrhea are indistinguishable from each other based solely on signs and symptoms. Therefore, it's recommended that clinicians order tests for both when working up a patient diagnosis. Also, prior infection with *any* bacterial STI does not confer long-term immunity. Because people can repeatedly get chlamydia and/or gonorrhea (or *any* bacterial STI), all of a patient's sexual partners must be treated.

Men may develop dysuria and a white, yellow, or green discharge from the penis (FIG. 20.17). In men the bacteria can also occasionally cause epididymitis, prostatitis, and/or spread to the testicles; any of these complications could decrease fertility.

When women experience symptoms they tend to be mild and may be mistaken for a bladder infection or vaginal irritation. Possible symptoms include pain upon urination, increased vaginal discharge, and bleeding/spotting between periods and/or after sex.

Unfortunately, antimicrobial-resistant *Neisseria gonorrhoeae* strains are increasingly common. *N. gonorrhoeae* strains that resist drugs such as penicillin, tetracyclines, macrolides, quinolones, sulfonilamides with trimethoprim, and, more recently, extended-spectrum cephalosporins circulate in the general population. Due to increased antimicrobial resistance, the CDC now recommends that patients be prescribed a combination therapy of ceftriaxone and azithromycin. It is more difficult for organisms to evolve resistance mechanisms to two different classes of drugs administered together. Despite the relative ease of treatment, many gonorrhea cases are asymptomatic and therefore go untreated and may progress to pelvic inflammatory disease, which we discuss next.

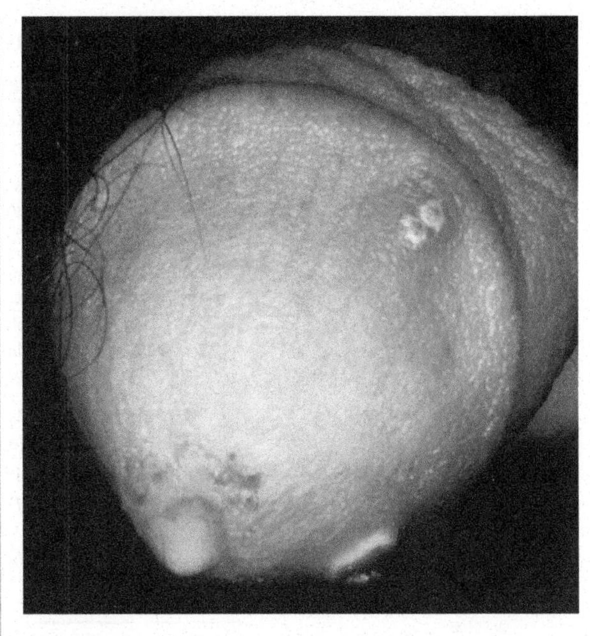

FIGURE 20.17 Gonorrhea discharge Although gonorrhea is often asymptomatic in women, infected men tend to develop dysuria and a whitish urethral discharge.

Critical Thinking *Given what you know about anatomy and the immune system, do you think some men with gonorrhea have pyuria? Support your response.*

Pelvic Inflammatory Disease

Gonorrhea and chlamydia frequently cause asymptomatic infections in women and often remain untreated. In such cases, **pelvic inflammatory disease (PID)** may develop. In PID the fallopian tubes can become inflamed (**salpingitis**) and abscesses in the fallopian tubes or ovaries may develop (FIG. 20.18). In some cases the abdominal cavity is also affected.

Diagnosis is sometimes challenging because symptoms can be mild and nonspecific. Preliminary diagnosis can be made with a pelvic exam, if uterine or cervical tenderness is present. Blood work and then biopsy, ultrasound, MRI, or laparoscopy can be used as follow-up procedures to confirm diagnosis. Once the causative pathogens have been identified, antibiotic therapy should be started. Although chlamydia and gonorrhea are the main culprits of PID, *Mycoplasma genitalium*, anaerobic bacteria associated with vaginosis, can also be involved.

CLINICAL VOCABULARY

Salpingitis inflammation of the fallopian tubes

Laparoscopy placement of fiber-optic instrument into the abdomen to view the organs and tissues there and/or to permit a surgical procedure with just a small incision

Ectopic pregnancy a pregnancy in which the fertilized egg implants outside the uterus; often in the fallopian tube

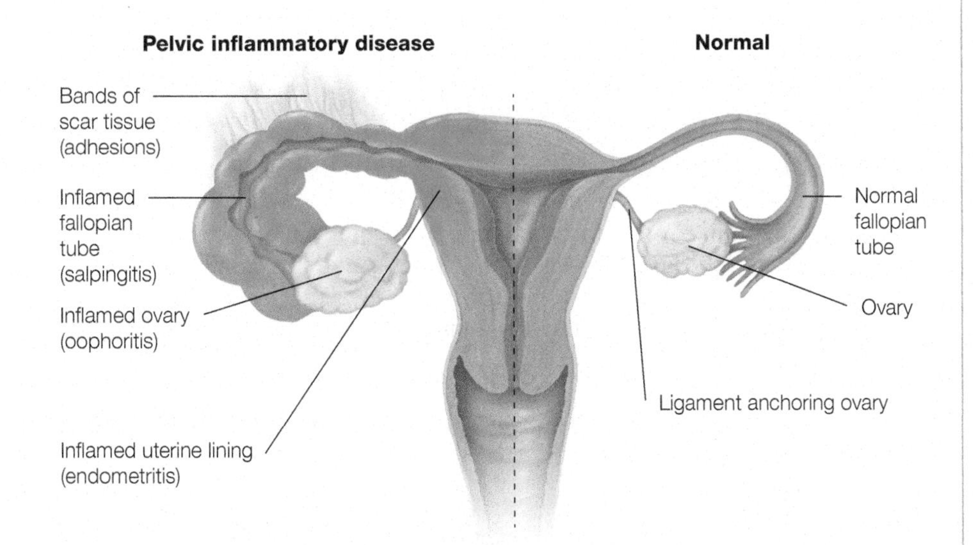

FIGURE 20.18 Pelvic inflammatory disease (PID) Untreated bacterial infections of the female reproductive system may advance to PID.

DiseaseSnapshot • Gonorrhea

N. gonorrhoeae

Causative agent	*Neisseria gonorrhoeae*; Gram-negative bacterium
Epidemiology	According to the WHO, globally there are around 106 million new cases every year; incubation period is 1 to 14 days
Transmission & prevention	Sexual contact; vertical transmission from mother to newborn may occur during vaginal delivery; prevention includes abstinence, monogamous sex after STI testing confirms noninfected state, or proper condom usage
Signs & symptoms	Men are often asymptomatic, but can experience dysuria (painful urination) and urethral discharge; women are often asymptomatic, but can experience dysuria, increased vaginal discharge, and bleeding/spotting between periods and/or after sex; women can also experience serious complications during pregnancy and pelvic inflammatory disease if gonorrhea is not treated; perinatal exposures can lead to eye infections that can result in blindness (ophthalmia neonatorum) and/or sepsis
Pathogenesis & virulence factors	Pili help with adhesion, then ciliated cells allow passage into deeper layers of the host epithelium; intracellular lifestyle and bacterial capsules help with immune evasion
Diagnosis & treatment	Urine (men) or endocervical swab samples are used for nucleic acid amplification testing (NAAT); *Neisseria gonorrhoeae* can also be cultured from swabs of the urethra or vagina; first-line treatment in nonpregnant patients is usually a combination therapy of ceftriaxone and azithromycin; pregnant patients may be given dual therapy consisting of ceftriaxone as a single injected dose and azithromycin as a single oral dose; newborns are given erythromycin ophthalmic ointment in each eye in a single application at birth

If the patient doesn't improve in 72 hours, alternative drugs should be considered, since no response to the medication is an indication that the microbes causing the infection could be resistant to that particular medication. Left untreated, the resulting swelling causes tissue damage that can lead to chronic pelvic pain or **ectopic pregnancy** (zygotes that implant in the fallopian tubes instead of the uterus), as well as miscarriages and infertility. The CDC estimates that around 1 million women experience PID each year. The disease is most common in young, sexually active women.

Neonatal Exposure

Perinatal exposure to a mother's infected cervix can cause neonatal gonnococcal infections. These infections include conjunctivitis (ophthalmia neonatorum; see Chapter 17) and sepsis. If untreated, neonatal infections cause permanent blindness, arthritis, meningitis, or even death. Sometimes ophthalmia neonatorum progresses so rapidly that the globe of the eye (eyeball) can break within 24 hours after birth. CDC guidelines call for screening all pregnant women for gonorrhea and administering prophylactic erythromycin eyedrops to all babies born in the United States, which has dramatically decreased the number of neonatal infections.

Syphilis killed millions for centuries, and remains a common but curable infection today.

Syphilis became prominent in Europe starting in the late 1400s. Originally known as "the French disease" because French troops spread it during an invasion of Italy, syphilis eventually took root in every portion of society, from the royal courts down through the peasants living in cities and small towns alike. For centuries this major killer had no effective treatment. Today, syphilis is curable with antibiotics like penicillin. The main groups at risk for the disease today are listed in FIG. 20.19.

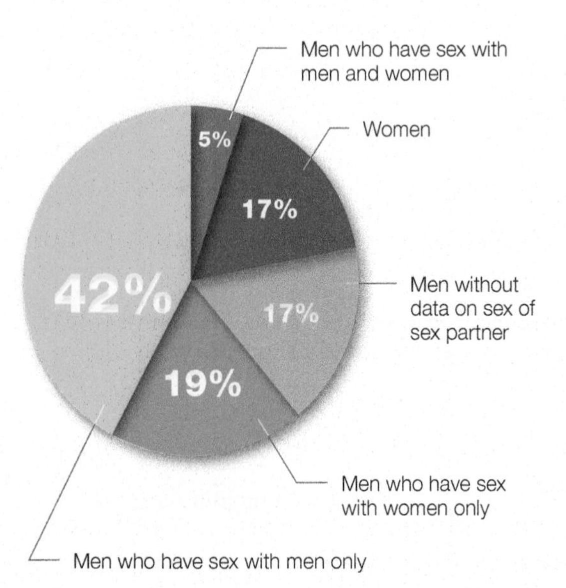

FIGURE 20.19 **Distribution of primary and secondary cases of syphilis by sex and sexual behavior** These data are based on new cases in the United States in 2019.

Source: Centers for Disease Control and Prevention. Sexually Transmitted Disease Surveillance 2019. Atlanta: U.S. Department of Health and Human Services; 2021.

Syphilis is caused by the bacterium *Treponema pallidum*. This Gram-negative spirochete has very few surface antigens that stimulate an immune response. And it can vary the types of antigens displayed on the cell surface as the disease progresses to help avoid immune detection. Because of this, it is sometimes called a "stealth pathogen." As with all bacterial STIs, one infection does not confer life-long protection; even after being cured with antibiotics, people can easily have a reinfection with syphilis. This means all of a patient's sexual partners must be treated to stop the spread.

Syphilis Stages

Untreated disease progresses in three phases. The infection starts with contact with an infected person's **chancres** (sores); generally, chancres are found on or around the penis, vagina, rectum, or anus, but they may also appear on the lips or mouth (FIG. 20.20). The **primary stage** of syphilis is characterized by a chancre appearing at the initial infection site, where the spirochete invades the new host's tissues and begins to multiply. In rabbit models, the bacteria are so invasive that they show up in the bloodstream of infected animals within hours of exposure. The bacteria also invade the central nervous system. The primary stage goes away in 4 to 6 weeks due to a strong cellular immune response.

Unfortunately, the immune system doesn't usually kill all the bacteria during the primary stage, since they are good at hiding from the immune system. During the **secondary stage** *T. pallidum* establishes a persistent infection. This occurs around three months after initial infection and usually involves a rash, which may appear on the skin or on mucous membranes in the mouth, vagina, or anus (Figure 20.20 *center*). A subset of patients experience **condylomata lata**, genital or anal warts. Swollen lymph nodes, sore throat, patchy hair loss, fever, muscle aches, or malaise may also occur. Sometimes people don't even notice these symptoms. These symptoms usually disappear within 3 months and the pathogen enters a latent period that lasts for months to years before it launches the final, tertiary stage.

The final stage, **tertiary syphilis**, can be fatal. It causes lesions on the skin and bones, heart damage, and neurological symptoms (Figure 20.20 *bottom*). Insomnia, vertigo, and seizures can be observed 5 to 10 years after infection. Partial paralysis, loss of sensation, memory impairment, and hallucinations can occur decades after infection.

Congenital Syphilis

Mother-to-fetus transmission may occur at any time during pregnancy, resulting in **congenital syphilis**. Antibiotic treatment of the mother can prevent fetal infection. If untreated, fetal death, birth defects, or infected infants may result. Babies born with syphilis show skin lesions and nasal discharge with blood in the primary stage, occurring 2 to 10 weeks after birth. Up to two years later, nasal, tooth, or palate deformations can be observed and many children experience blindness or deafness. In the United States routine screening has dramatically reduced cases of congenital syphilis. The usual cure for syphilis is two injections of Bicillin L-A, a type of penicillin.

Primary stage: Chancre appearing at the initial infection site

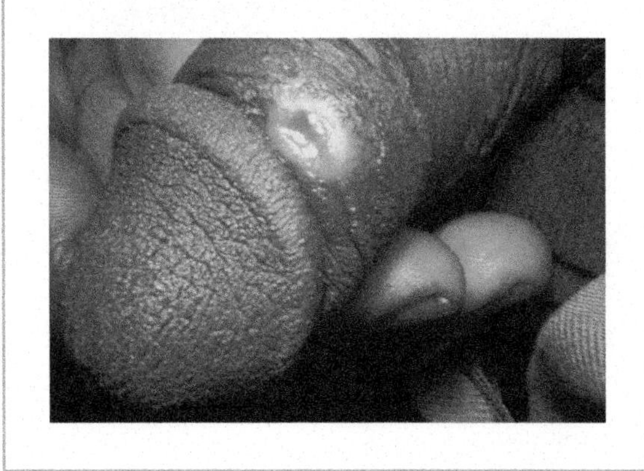

Second stage: Disseminated rash

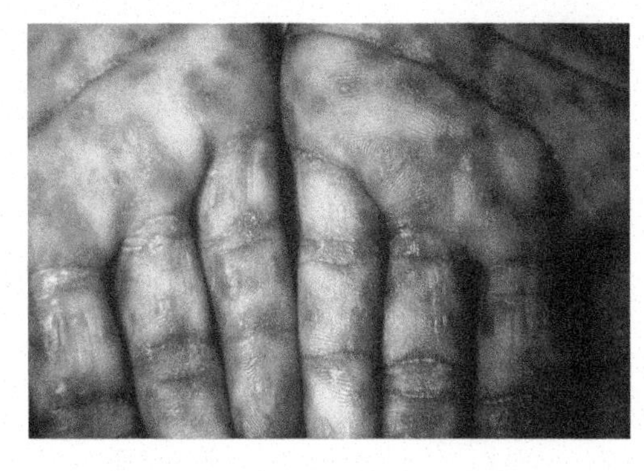

Tertiary stage: Gumma lesions on the skin and bones

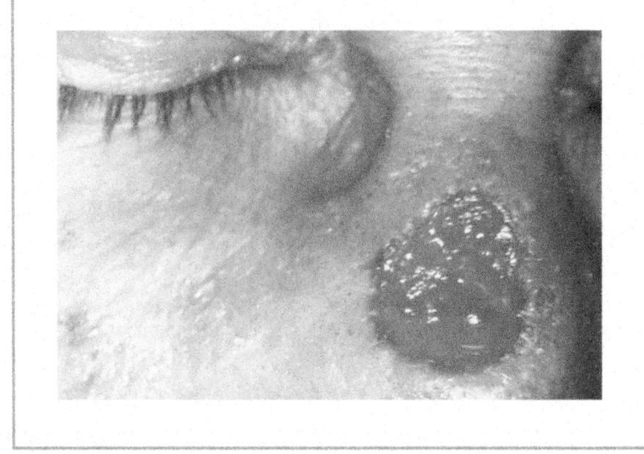

FIGURE 20.20 **Syphilis stages**

Chanchroid is an STI that is mainly found in developing nations.

Haemophilus ducreyi are Gram-negative nonmotile rods that cause a disease known as **chancroid**. It is rare in developed countries. Unlike many of the STIs discussed thus far, the disease is more common in men than in women. In developing nations, chancroid is a common cause of genital ulcers and genital lymph node buboes. Importantly, it is one of the most common cofactors for HIV transmission. This means that when someone is infected with chancroid it

CLINICAL VOCABULARY

Chancre an ulcer-like lesion that is often on the genitalia, but could develop elsewhere, mainly associated with syphilis

Condylomata lata wart-like genital lesions

Disease**Snapshot** • **Syphilis**

T. pallidum

Causative agent	*Treponema pallidum*; Gram-negative bacterium
Epidemiology	Most of the cases in the U.S. are among men who have sex with men; incubation period is 10–90 days, usually near 21 days after exposure
Transmission & prevention	Sexual transmission upon contact with a syphilitic chancre; prevention includes abstinence, monogamous sex after STI testing confirms noninfected state, and proper condom usage
Signs & symptoms	Primary stage features muscle aches, swollen lymph nodes, fever, and chancres; secondary stage is characterized by a rash, swollen lymph nodes, and sometimes genital warts; tertiary stage manifests as neurological symptoms, lesions on the skin and bones, and heart damage
Pathogenesis & virulence factors	Binds fibronectin on host cell surfaces for invasion; hyaluronidase aids in tissue invasion; hemolysins lyse red blood cells for iron; the immune system reaction to the pathogen plays an important role in the damage resulting from infection
Diagnosis & treatment	Direct observation of the spirochete by dark field microscopy, and nontreponemal and treponemal serologic antibody studies; two injections of Bicillin L-A, a type of penicillin, is the first-choice drug; azithromycin or doxycycline may also be prescribed

is easier for them to contract HIV if they are exposed; the characteristic chancroid lesion provides an avenue for the bloodborne pathogen HIV to be transmitted into a new host.

Unlike the often-asymptomatic STIs we've discussed so far, chancroid tends to have symptoms, and typically the ulcers are quite painful. This means that people usually seek treatment relatively quickly, and macrolides, cephalosporins, and fluoroquinolones can completely cure the disease.

Build Your Foundation

Build Your Foundation (BYF) Quick Quiz: Visit the **Mastering Microbiology** Study Area to quiz yourself.

18. What is the difference between vaginitis and vaginosis?
19. What are the clinical features of vaginosis and how can it be diagnosed and treated?
20. What are the clinical and epidemiological features of chlamydia and the potential complications in untreated women and neonates?
21. What stage of *Chlamydia trachomatis*'s life cycle is infectious and which stage is replicative?
22. What are the clinical end epidemiological features of lymphogranuloma venereum?
23. What are the clinical and epidemiological features of gonorrhea and the potential complications in untreated women and neonates?
24. What are the clinical features of pelvic inflammatory disease?
25. What causes syphilis, what are the clinical features of each infection stage, and what are the consequences of congenital syphilis?
26. What are the clinical and epidemiological features of chancroid?

20.5 REPRODUCTIVE SYSTEM EUKARYOTIC INFECTIONS

Learning Outcomes

After reading this section, you should be able to:

20.27 Identify the signs, symptoms, causes, treatments, and consequences of candidiasis.
20.28 Describe the clinical significance of structural morphogenesis in *Candida albicans*.
20.29 Explain the clinical and epidemiological features of trichomoniasis, including treatments.

Candidiasis is the most common fungal infection of the reproductive system.

The most common fungal infection of the reproductive system is *Candida albicans*. It can invade and colonize the urinary tract and reproductive organs and also become disseminated—spread to the bloodstream and cause a systemic infection. Systemic cases are usually healthcare-acquired infections related to surgery, catheters, or feeding tubes. *Candida* can also colonize the mouth, where it is known as "thrush." One of the most common infections caused by *C. albicans* is **vulvovaginal candidiasis**, commonly known as a vaginal yeast infection.

DiseaseSnapshot • Candidiasis

Causative agent	Usually *Candida albicans*
Epidemiology	According to the CDC at least 75% of women will develop vulvovaginal candidiasis at some point; infections may also occur in the male reproductive system, but are rare; incubation period is 2 to 5 days
Transmission & prevention	Causative yeast are usually part of the normal microbiota; sexual transmission to men occasionally occurs; wearing cotton underwear and taking probiotics while undergoing an antimicrobial therapy may help prevent infection
Signs & symptoms	Women experience vaginal itching, burning, and often "cottage cheese-like" discharge; men may experience an itchy rash on the penis and dysuria
Pathogenesis & virulence factors	In the hyphal form, yeast make virulence factors such as proteases that stimulate inflammation; shape shifting (or morphogenesis) from the budding yeast form to a hyphal form may also help the fungus evade host immune defenses
Diagnosis & treatment	Microscopic analysis of vaginal discharge reveals budding yeast or yeast in hyphal form; yeast can also be identified by culture methods; antifungal azole drugs are first-line treatments

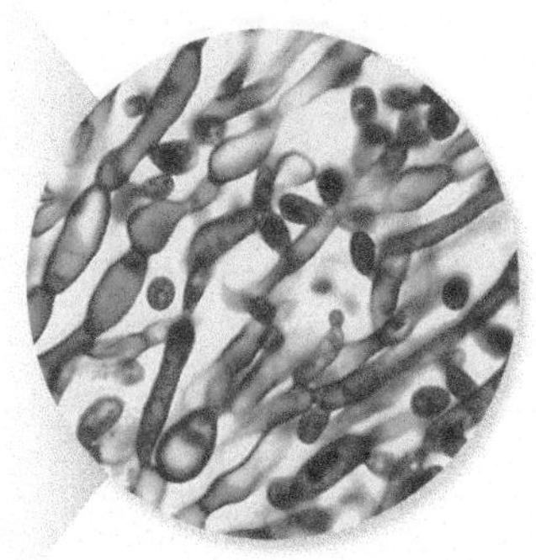

C. albicans

C. albicans is a normal resident of the vagina. A disruption in normal microbiota by antimicrobial treatments, hormonal changes, or other factors can lead to dysbiosis, where yeast can then overgrow and act as an opportunistic pathogen (FIG. 20.21). Vaginal itching, burning, pain, and a characteristic "cottage cheese" discharge are symptoms millions of women recognize. Occasionally other *Candida* species such as *C. glabrata* or *C. krusei* can cause candidiasis.

Yeast are dimorphic, meaning they grow as a single-celled yeast form, but can undergo **structural morphogenesis** (a switch in shape) to a filamentous or hyphal form in response to environmental changes. This switch from the yeast form to the hyphal form is required for a symptomatic infection to develop. It seems that altered microbiota profiles, increased estrogen levels, and/or increased local pH trigger this necessary shift. Scientists believe that hyphal yeast generate symptomatic infections because they secrete protein-destroying enzymes (proteases) and cause cellular damage that recruits the neutrophils responsible for the characteristic inflammation.

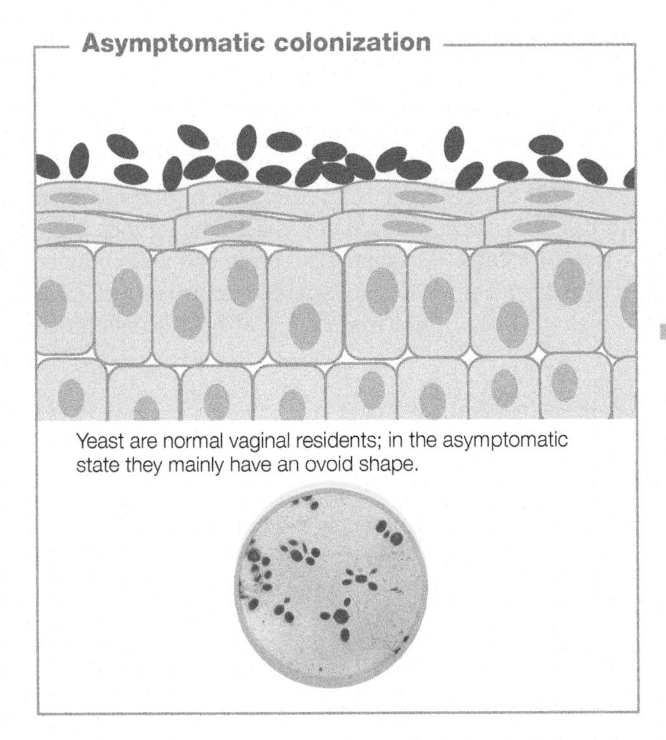

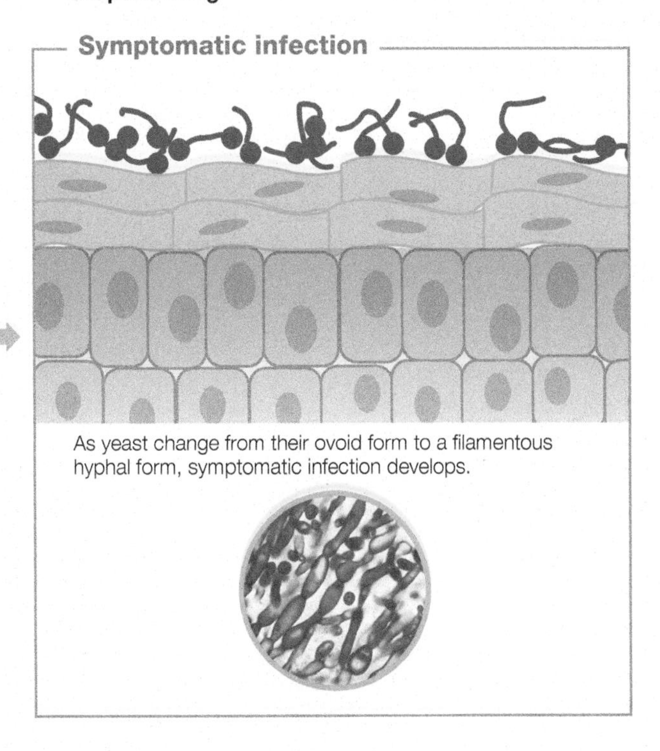

FIGURE 20.21 Vaginal yeast infections and yeast shape shifting

Disease**Snapshot** • Trichomoniasis

T. vaginalis

Causative agent	*Trichomonas vaginalis*
Epidemiology	No formal surveillance system exists, but the WHO estimates that globally 277 million people are infected; the CDC estimates that 3.7 million people in the U.S. are infected; data collected by the CDC also reveal that African-American women are especially burdened by this infection—they have infection rates almost ten times that of Hispanic and white women; incubation period is 4 to 28 days
Transmission & prevention	Sexually transmitted; can be prevented by abstinence, monogamous sex after STI testing confirms noninfection status, and proper condom use
Signs & symptoms	Usually asymptomatic; men may experience discharge from the penis, dysuria, and urethral discomfort after ejaculation; women may develop a green-tinged frothy vaginal discharge, dysuria, and itchiness, soreness, and burning in the vagina; infection during pregnancy can cause complications
Pathogenesis & virulence factors	The protozoa phagocytizes vaginal epithelia and bacteria; it can lyse red blood cells using pore-forming proteins; it has flagella for motility and makes a glycocalyx that helps it adhere to host epithelium
Diagnosis & treatment	Tests that detect *T. vaginalis* DNA or antigens are commonly used; wet-mount preparations can also confirm infection; mainstay treatments are a single dose of oral metronidazole or tinidazole

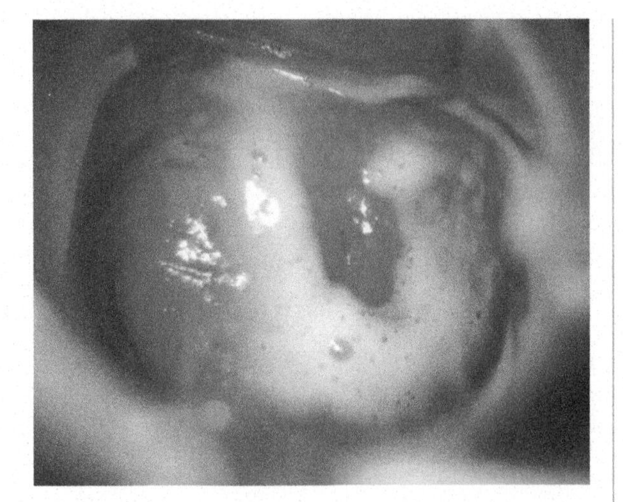

FIGURE 20.22 Trichomoniasis Discharge due to trichomoniasis is often frothy.

Vulvovaginal candidiasis is treatable with antifungal suppositories, creams, or pills. Unfortuantely, *C. albicans* can form biofilms that may help them resist antimicrobial treatments.

Caused by a parasite, trichomoniasis is often undiagnosed and underreported.

The CDC estimates that approximately 3.7 million people in the United States have trichomoniasis. But since only one-third of these people ever develop symptoms, most cases go undiagnosed. The infection is sexually transmitted and caused by a protist parasite known as *Trichomonas vaginalis*. *T. vaginalis* are flagellated cells with a long undulating membrane. They are highly motile in warm environments. Diagnosis can be made using a wet mount of vaginal discharge. To perform a wet mount, a swab of vaginal discharge is smeared onto a slide and a couple of drops of saline solution are added to adhere a cover slip. The preparation is then viewed under a microscope. If present, *T. vaginalis* are easily spotted swimming around in the wet mount preparation.

Men are usually asymptomatic and the disease self-resolves. When men do develop symptoms they can consist of a discharge from the penis, dysuria, and urethral discomfort after ejaculation. In women, the characteristic symptoms are a frothy, greenish discharge (FIG. 20.22), dysuria, and itchiness, soreness, and burning in the vagina. Trichomoniasis during pregnancy is associated with premature labor and low-birth-weight babies. *T. vaginalis* infection can also induce genital inflammation that makes it easier for HIV to establish an infection. All it takes to cure most people is a single dose of metronidazole, although resistance is on the rise.

Build Your Foundation

Build Your Foundation (BYF) Quick Quiz: Visit the **Mastering Microbiology** Study Area to quiz yourself.

27. What are the signs, symptoms, causes, treatments, and consequence of candidiasis?

28. What is the clinical significance of structural morphogenesis in *Candida albicans*?

29. What are the clinical and epidemiological features of trichomoniasis and how is it treated?

URINARY INFECTIONS

Bacterial

Urinary tract infections (UTIs) occur in men and women but are more common in women. They are often associated with urinary catheters.

- **Upper UTIs** include ureteritis (inflammation of the ureters) and pyelonephritis (kidney inflammation); characterized by flank pain and lower UTI symptoms; are usually more serious and can lead to renal failure
- **Lower UTIs** include urethritis and cystitis (bladder inflammation); are usually less serious than upper UTIs; cause urgency to urinate, painful urination (dysuria), pus in urine (pyuria), blood in urine (hematuria), pain in lower abdomen.

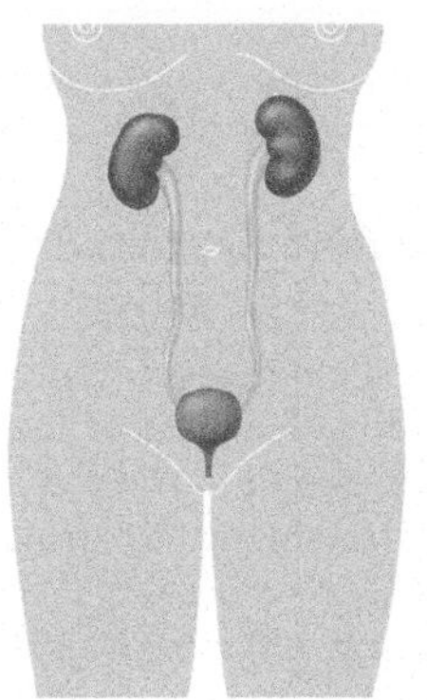

UPEC (*E. coli*)
- Most common cause of UTIs
- Invade host cells to live intracellulary, hiding from immune defenses
- Usually transmission is from anus, self-genitalia, or sexual partner

Staphylococcus saprophyticus
- Most common Gram-positive bacterium that is isolated in uncomplicated UTIs
- Causes up to 6% of all uncomplicated UTIs but up to 42% of uncomplicated UTIs in sexually active women age 16 to 25 years old

Leptospirosis
- Mainly transmitted by rats
- Spirochetes enter body through mucous membranes or cuts
- Symptoms include fever, headache, vomiting; can lead to renal failure

Leptospira interrogans

REPRODUCTIVE INFECTIONS

Viral

HSV-2

Genital Herpes (HSV-2)
- Contracted sexually
- Initial, severe outbreak of genital lesions
- Virus may reemerge causing new outbreaks
- No cure; antivirals suppress symptoms

Human Papilloma Viruses (HPV)

- Contracted sexually
- Cause genital warts
- Responsible for 90% of cervical cancers
- Vaccine available

Bacterial

C. trachomatis

Bacterial vaginosis
- Caused by a dysbiosis of the vaginal microbiota
- Characterized by a decrease in lactobacilli and an increase in mixed anaerobes of the vaginal microbiota

Chlamydia (*Chlamydia trachomatis*)
- Contracted sexually
- Among the most common STIs; easily cured with antibiotics
- Asymptomatic infections often go untreated and may lead to pelvic inflammatory disease (PID)

Gonorrhea (*Neisseria gonorrhoeae*)
- Contracted sexually; number of cases is increasing
- Antibiotic resistance is a growing concern
- Women are usually asymptomatic, causing undetected infections; complications include PID

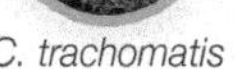

Syphilis (*Treponema pallidum*)
- Contracted sexually; curable with antibiotics
- Untreated disease has three stages: primary (chancre at infection site), secondary (persistent infection, rash), tertiary (lesions, heart damage, neurological symptoms, can be fatal)
- Congenital syphilis can cause serious fetal complications

Chanchroid (*Haemophilus ducreyi*)
- Contracted sexually
- More common in men than women
- Obvious symptoms include genital ulcers and buboes
- Common cofactor for HIV transmission

Fungal/Protozoan

T. vaginalis

Candidiasis (*Candida species*)
- Opportunistic fungal infection of the vagina
- Symptoms include vaginal itching, burning, and pain
- Can be treated with antifungals

Trichomoniasis (*Trichomonas vaginalis*)
- Contracted sexually
- Usually asymptomatic in men and women, but when symptomatic cases develop they tend to be in women
- Not transmitted to the fetus, but linked to premature labor and low-birth-weight babies

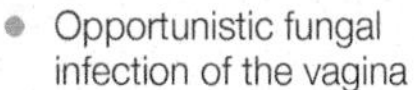

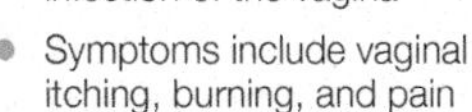

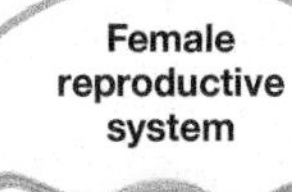

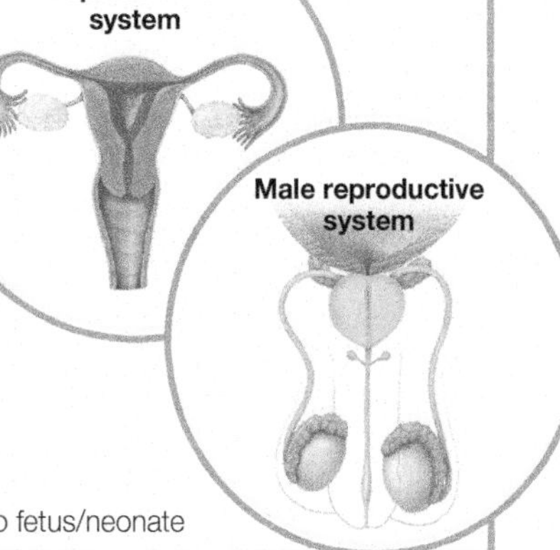

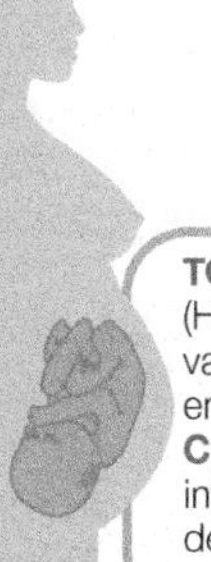

TORCH stands for **T**oxoplasmosis, **O**ther (HIV, syphilis, parvovirus B19, listeriosis, varicella-zoster virus, coxsackievirus, enteroviruses, and Zika virus), **R**ubella, **C**ytomegalovirus (CMV), and **H**erpes. These infections can be vertically transmitted to a developing fetus and cause devastating congenital defects.

Common manifestations: Discharge | Ulcers and lesions | Warts | Swollen lymph nodes | Passes to fetus/neonate

For all printed and additional Disease Snapshot tables, go to the Mastering Microbiology Study Area.

CHAPTER 20 OVERVIEW

20.1 Overview of the Urinary and Reproductive Systems

- The urinary system consists of two kidneys, two ureters, the bladder, and the urethra. The kidneys have nephrons that filter metabolic waste products such as urea and creatinine out of our blood and regulate the body's water and electrolyte balance. The ureters convey urine from our kidneys to the bladder. Urine is stored in the bladder until it exits the body through the urethra. Due to anatomical differences in the urinary tract, women have a greater risk than men for urinary tract infections (UTIs).
- The most consistently detected genera in the healthy urinary tract are *Lactobacillus* and *Streptococcus*.
- The major structures of the female reproductive system are paired ovaries and fallopian tubes, the uterus, vagina, and vulva. *Lactobacillus* bacteria dominate the normal vaginal microbiota and help limit vaginal infections. Every part of the female reproductive tract may harbor normal microbiota. Shifts in the female reproductive system microbiota could impact a patient's risk for ovarian cancer and are linked to pre-term labor.
- The placenta is an important interface between a mother and fetus. TORCH agents can be transmitted across the placenta or may be transmitted perinatally. These agents may cause congenital developmental and/or structural abnormalities or even death.
- Male reproductive structures include the testicles within the scrotum, the vas deferens, epididymis, sex glands (the prostate gland, bulbourethral gland, and seminal vesicles), and penis. The male reproductive system microbiome isn't well understood. Preliminary studies suggest that shifts in the male reproductive system microbiota could affect risks for pathologies like prostate cancer.

20.2 Urinary System Infections

- Upper UTIs include pyelonephritis and ureteritis. Lower UTIs include cystitis and urethritis. Urinary tract infections are divided into complicated or uncomplicated categories based on clinical criteria. Catheter-associated urinary tract infections (CAUTIs) are considered complicated UTIs and are among the most common healthcare-acquired infection.
- UPEC (uropathogenic *Escherichia coli*) causes most UTIs. *Staphylococcus saprophyticus* is the main Gram-positive agent that causes UTIs. *Leptospira interrogans* bacteria are zoonotic infections transmitted through animal urine to humans. These bacteria cross mucous membranes of the eyes, nose, or gastrointestinal tract, or can enter through skin abrasions.
- Clinical signs of a UTI include frequent and urgent urination that may be painful (dysuria), lower abdominal pain, pyuria (pus in the urine), and hematuria (blood in the urine). Sometimes the urine appears cloudy and may smell foul. In upper UTIs the patient may report all of the above, along with flank pain (costovertebral angle tenderness). Uncomplicated UTIs are typically diagnosed based on urine dipstick testing and the patient's medical history, signs, and symptoms. Uncomplicated UTIs are rarely confirmed by culture methods, but microbial culturing is recommended in complicated UTIs.
- Viral UTIs mainly occur in immune-compromised patients. Hantaviruses should be ruled out in patients who have a normal immune system but develop aseptic UTIs. The main agents of viral UTIs in immune-compromised patients are adenovirus, BK virus, and cytomegalovirus. Fungi like *Candida* species can also infect the urinary system.

20.3 Reproductive System Viral Infections

- Herpes simplex virus 2 causes genital herpes. Herpes simplex virus 1 usually infects the mouth but can infect the genitals. Herpes infections are characterized by occasional outbreaks/recurrence. Neonatal herpes can cause skin, eye, and mouth infections, as well as serious neurological damage.
- Human papilloma viruses (HPVs) are the most common sexually transmitted agents in the world. HPV can cause genital warts as well as cancers. There are vaccines to protect against the main cancer- and wart-associated HPVs. Pap smears are an important part of preventive care in women because they detect precancerous and cancerous cervical cells and may also help diagnose an HPV infection.

20.3 Reproductive System Bacterial Infections

- Bacterial vaginosis is a dysbiosis characterized by a decrease in lactobacilli and an increase in mixed anaerobes. Amsel's criteria and Nugent's criteria are used for diagnosis.
- *Chlamydial trachomatis* biovar trachoma causes chlamydia. Infections are often asymptomatic. When symptoms develop in men they can include a penile discharge, dysuria, burning and itching around the urethral opening in the penis, and testicular swelling or pain. In women, symptoms may include bleeding between menstrual periods and/or after sex, dysuria, vaginal discharge with an odor, itching or burning around the vagina, pain during sexual intercourse, and pelvic pain that may be accompanied by fever. Main complications include pelvic inflammatory disease (PID) in untreated women and perinatal transmission to a neonate. Male complications include chronic epididymitis, which can cause infertility.
- Lymphogranuloma venereum is caused by a different biovar of *C. trachomatis*. This biovar multiples inside macrophages and causes swelling and eventual necrosis in lymph nodes near the genitals.
- Gonorrhea is caused by *Neisseria gonorrhoeae* and is usually asymptomatic. If symptoms develop they are indistinguishable from a chlamydia infection. Complications are the same as seen in chlamydia. Antibiotic resistance is on the rise for gonorrhea.
- *Treponema pallidum* causes syphilis. Primary syphilis features chancres near the genitals. Secondary syphilis manifests after a persistent infection is established and features a rash and swollen lymph nodes. In tertiary syphilis, a patient may experience irreversible damage to the heart and nervous system. Congenital syphilis leads to neurological and physical complications.
- Chanchroid, caused by *Haemophilus ducreyi*, is characterized by painful genital ulcers and may also manifest as buboes.

20.4 Reproductive System Eukaryotic Infections

- *Candida albicans* is the main fungus associated with candidiasis. It's a common opportunistic infection that develops following an antimicrobial treatment in women. Symptoms including a thick, white, clumpy vaginal discharge, along with vaginal itching and pain. Yeasts change from ovoid to hyphal forms when they cause infections.
- *Trichomonas vaginalis* is a eukaryotic parasite that is a common, curable STI. Men are usually asymptomatic and the disease self-resolves. In women, the characteristic symptoms are a frothy, greenish discharge. Diagnosis can be made using a wet mount of vaginal discharge.

THINK CLINICALLY | *Be S.M.A.R.T. About Cases*

COMPREHENSIVE CASE

The following case integrates basic principles from the chapter. Try to answer the case questions on your own. Don't forget to be S.M.A.R.T.* about your case study to help you interlink the scientific concepts you have just learned and apply your new understanding of urinary and reproductive system infections to a case study.

*** The five-step method is shown in detail in the Chapter 1 Comprehensive Case on cholera. See pages 31–33. Refer back to this example to help you apply a SMART approach to other critical thinking questions and cases throughout the book.**

• • •

Kelly is a nurse practitioner at an obstetrics and gynecology practice. One day she sees a nonpregnant 23-year-old patient named Jessica who is complaining of dysuria, frequent urgency to urinate, spotting between periods, and increased vaginal discharge. In the process of taking a patient history, Kelly learns that Jessica has not received the HPV (human papilloma virus) vaccination series. She also learns that although Jessica and her boyfriend separated about two weeks ago, they had been together for six months and sexually intimate for the last 4 months. Jessica also reported she had been intimate once with another guy about two years ago. Kelly asked what form of birth control Jessica used, if any. Jessica said she took a birth control pill that she had been prescribed by her general practitioner, but that she and her partners did not consistently use condoms.

Kelly collects an endocervical swab to send for testing. She also performs a wet-mount analysis of the vaginal discharge. Lastly, she does a urine dipstick test, which reveals elevated white blood cells in the urine and a normal nitrite level. Ultimately, Kelly decides to treat Jessica with a single injection of ceftriaxone and send her home with a prescription for oral azithromycin. She tells Jessica that she suspects a chlamydia or gonorrhea infection. She advises Jessica to call the office if she doesn't feel better within a few days or if she develops new symptoms. She walks Jessica to the checkout area and asks the office staff to help her schedule a follow-up appointment and Pap smear for 14 to 20 days out to check for HPV. She also tells Jessica they will call her once they have the lab results back.

The lab results come back and are positive for chlamydia and negative for gonorrhea. Kelly calls Jessica and tells her that when she comes back in for her Pap smear they will retest her to ensure the infection has cleared. She also says it would be wise to check for a panel of other sexually transmitted infections (STIs) on her next visit. She advises Jessica to contact her ex-boyfriend and tell him that she was diagnosed with chlamydia and that he should also be tested. Jessica is shocked. She has so many questions, but she is overwhelmed and decides that instead of talking to the nurse practitioner she'll call you—her closest friend who just started nursing school.

• • •

CASE-BASED QUESTIONS

1. Jessica tells you that she's stunned that she was diagnosed with an STI; she was really convinced she had a urinary tract infection (UTI). What information aligns with her self-diagnosis?
2. What signs, symptoms, and medical history components would have led the nurse practitioner to suspect chlamydia and gonorrhea over a UTI?
3. Jessica asks you if this infection is proof that her ex-boyfriend cheated on her. How do you respond? Use information from the text to support your response.
4. What conditions would a wet-mount analysis of the vaginal discharge help eliminate from the list of possible urogenital conditions?
5. Jessica wonders if she should call both men that she's been intimate with in the past. What do you advise? Support your position with information from the chapter and the case.
6. Should Jessica go ahead and get the HPV vaccine if she is diagnosed with HPV, or is it too late for her to derive any benefits from the vaccine? Explain your response.

END OF CHAPTER QUESTIONS

Think Critically and Clinically

Questions highlighted in **orange** are opportunities to practice NCLEX, HESI, and TEAS-style questions.

1. Which of the following does not fall into our current definition of TORCH infections?
   a. Syphilis
   b. Trichomoniasis
   c. HIV
   d. Rubella
   e. Cytomegalovirus

2. Factors that contribute to vulvovaginal candidiasis include
   a. low vaginal pH.
   b. high nitrates in urine.
   c. increased blood bilirubin levels.
   d. increased estrogen levels.
   e. taking a fluconazole drug.

3. A patient you are assessing reports an increase in vaginal discharge that has a foul odor, but she does not report other symptoms. The patient has been in a monogamous relationship for 3 years. Which of the following would be the most useful for this patient? Select all that apply.
   a. Evaluate her urine for increased white blood cells.
   b. Perform a Gram stain on the vaginal discharge.
   c. Run a pregnancy test.
   d. Test for an STI.
   e. Perform a whiff test.
4. A 3-week-old infant with mouth, eye, and skin sores is brought to your clinic. What is the most likely diagnosis?
   a. Toxoplasmosis
   b. Herpes virus
   c. Syphilis
   d. Human papilloma virus
   e. Neonatal gonorrhea
5. How do lactobacilli limit infections in the vagina? Select all that apply.
   a. They competitively exclude potential pathogens.
   b. They make hydrogen peroxide.
   c. They ferment fructose to make lactic acid.
   d. They lower the vaginal pH, which limits pathogen growth.
   e. They increase the rate of vaginal mucus secretion.
6. A patient arrives in your clinic with buboes and a painful genital ulcer. Which drug do you recommend for treatment?
   a. Penicillin G
   b. Fluconazole
   c. Doxycycline
   d. Azithromycin
7. A male patient is complaining of frothy discharge and painful urination. After negative NAAT testing, you will *most likely* prescribe
   a. penicillin.
   b. tetracycline.
   c. a cephalosporin.
   d. metronidazole.
8. Uncomplicated UTIs are most common in
   a. pregnant women.
   b. catheterized patients.
   c. young women.
   d. homosexual men.
   e. people with urogenital disorders.
9. You see a corkscrew-like bacterial cell under dark field microscopy. If the sample came from __________, it is likely __________.
   a. genital lesions, leptospirosis
   b. urine, gonorrhea
   c. urine, syphilis
   d. genital lesions, syphilis
   e. vaginal discharge, chlamydia
10. Patients can contract chlamydia more than once because
    a. it is so commonly antibiotic resistant.
    b. long-term immunity is not established after infection.
    c. it hides in neurons and reactivates, lysing cells.
    d. it is sexually transmitted.
    e. all of the above apply.
11. Indicate if the statement is true or false and then correct each false statement so that it is true.
    a. Sexually transmitted infections always affect the reproductive tract.
    b. CAUTIs are usually classified as uncomplicated UTIs.
    c. *Candida* species are the leading cause of uncomplicated UTIs.
    d. Syphilis can be vertically transmitted from a mother to her fetus.
    e. Herpes simplex virus 2 can be transmitted to a partner without intercourse.
12. Match the following terms to their definition. Some terms may be matched to more than one definition and some definitions may not be used.

Cystitis	Inflammation of the urethra
Dysuria	Pus in the urine
Pyelonephritis	Painful urination
Hematuria	Lack of urination
Pyuria	Blood in the urine
	Inflammation of the kidneys
	Inflammation of the bladder
	Considered an upper UTI

CRITICAL THINKING QUESTIONS

1. What are some of the barriers to vaccination for human papilloma virus? Describe two different ways you would meet those challenges as a healthcare provider.
2. A 52-year-old man comes to your practice complaining of dysuria, urinary urgency, and a weak urine stream. Aside from this, he's in good health. What diagnostic measure(s) will be useful in this patient and what drug would likely help him? *Follow-up care:* After treatment he continues to experience a weak urine stream and urinary urgency, but his dysuria resolves. What condition should you consider?
3. Nugent's and Amsel's criteria are considered less reliable for diagnosing vaginosis in postmenopausal women. Why does this make sense?
4. Using the modern definition of TORCH and considering current vaccination recommendations, which TORCH agent can we expect to see less of in the next decade? Explain your reasoning.
5. Given what you learned in this chapter, why might it be helpful for pregnant women to secrete more glycogen than women who are not pregnant?
6. In some cases intentionally withholding STI information from a sexual partner and/or knowingly infecting a partner is punishable by law. A number of court cases have gone forward and prevailed on this matter. Herpes infections present a unique challenge in such cases. Why?

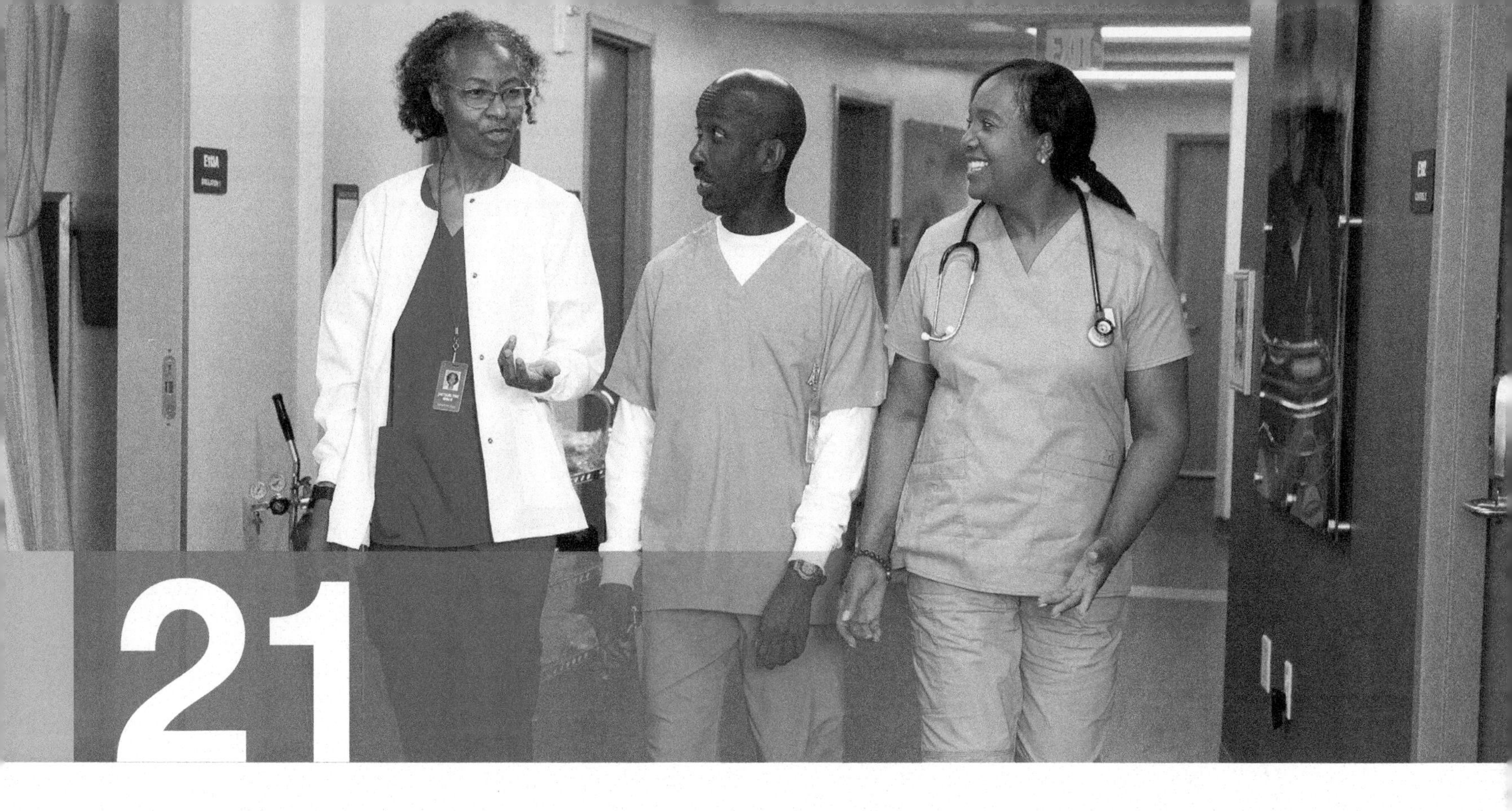

Cardiovascular and Lymphatic Infections

What Will We Explore? When it comes to battles between pathogens and us, localized infections such as strep throat are small skirmishes on the overall battlefield. However, with cardiovascular and lymphatic infections, the fight becomes a defense against a full-fledged invasion, as pathogens circulate through the entire body.

In this chapter, we explore the anatomy of the cardiovascular and lymphatic systems and the outcomes of infections that circulate through and attack these systems. As you might imagine, infection signs, symptoms, and consequences can vary significantly in their onset, progress, and prognosis. Some, like endocarditis, encroach slowly, whereas others, like Ebola, can be rapidly deadly.

Why Is It Important? Cardiovascular and lymphatic infections take an enormous toll on human health. From HIV infections to antimicrobial-resistant malaria, systemic infections incapacitate and kill millions worldwide. Historically, in the United States we've remained somewhat protected from many of the diseases discussed in this chapter. However, international travel and climate change have increased our risk for what we'd formerly considered "remote" disease threats. This makes it more important than ever before for healthcare workers to understand the sources and consequences of a wide array of cardiovascular and lymphatic infections.

Clinical CASE NCLEX HESI TEAS

The Case of the Bloated Belly

Visit the **Mastering Microbiology** Study Area to watch the case and find out how cardiovascular and lymphatic infections can explain this medical mystery.

(L-R) Jacquelyne Mack - CRDH, MS, MPA; Terry Thomas - BS, RMA; Carol Jenkins Neil - PhD, RN; Jacksonville, FL

21.1 OVERVIEW OF THE CARDIOVASCULAR AND LYMPHATIC SYSTEMS

Learning Outcomes

After reading this section, you should be able to:

21.1 Describe the basic structure of the heart, including the layers of the heart wall.

21.2 Trace how blood flows within the heart and describe the routes blood takes during pulmonary and systemic circulation.

21.3 Explain the basics of the lymphatic system—including what lymph is, how it enters tissues, and how it is collected for return to the venous blood supply.

21.4 Explain how the lymphatic system is interconnected with the cardiovascular system.

21.5 Describe sepsis and its progressive nature, name some sepsis risk factors, and discuss how sepsis is detected and managed.

Cardiovascular and lymphatic system infections are often referred to as systemic infections.

The cardiovascular and lymphatic systems are complex vascular networks that collaborate toward the distribution and collection of various biologically important substances throughout the body. Given the extent of the cardiovascular and lymphatic systems, infectious agents gaining access to these networks can readily spread throughout the body. This is why infections impacting the cardiovascular and lymphatic systems are often called **systemic infections**.

However, many initially localized infections can progress to systemic infections—that is, they can spread throughout the body—upon gaining access to cardiovascular or lymphatic vessels. Main risk factors for this potentially dangerous scenario include a patient not receiving appropriate or effective treatment for a localized infection. Antibiotic resistance and patient factors such as diabetes or compromised immunity also play roles. TABLE 21.1 lists select diseases that you've read about in earlier chapters that may progress to systemic forms. This chapter discusses some of the more common systemic infections facing healthcare providers today, as well as select emerging infectious agents.

TABLE 21.1 Infections Discussed in Other Chapters That May Become Systemic

Infection		Agent	Systemic Considerations	Main Book Coverage
Viral	Chickenpox	Varicella-zoster virus	Mainly diagnosed by skin lesions, though the disease has a systemic progression	Skin infections (Chapter 17)
	Measles	Measles virus	Mainly diagnosed by skin lesions, though the disease has a systemic progression	Skin infections (Chapter 17)
Bacterial	Anthrax	*Bacillus anthracis*	Cutaneous anthrax is main form, but pulmonary, gastrointestinal, and septicemic forms exist; septicemic form typically develops as a complication of pulmonary or other forms of anthrax	Skin infections (Chapter 17)
Fungal	Blastomycosis	*Blastomyces dermatitidis*	Primarily manifests as lung infection; occasionally spreads to bloodstream and can impact many organs; disseminated form is sometimes accompanied by skin lesions, which is why the agent's name includes "dermatitidis"	Respiratory infections (Chapter 16)
	Histoplasmosis	*Histoplasma capsulatum*	Mainly a lung infection; when spread to the bloodstream, multiple organs are impacted; often fatal if not treated	Respiratory infections (Chapter 16)
	Mucormycosis	*Rhizopus arrhizus* and other fungi	Six possible manifestations; lung and sinus forms are most common; spread from sinuses to brain is a possibly fatal complication; transmits by ingestion of fungi or inoculation into wounds	Respiratory infections (Chapter 16)
Parasitic	Toxoplasmosis	*Toxoplasma gondii*	Parasite enters the bloodstream and migrates to form cysts in various organs, with brain cysts being the most serious; protozoan may cross the placenta to cause severe birth defects or stillbirth; as the parasite disseminates through the body, mild flu-like symptoms may develop, but most immune-competent hosts will not experience symptoms	Urinary/reproductive system infections (Chapter 20)
	Amebiasis	*Entamoeba histolytica*	Most U.S. cases are imported; WHO estimates 500 million people wordwide harbor the causative intestinal parasite; trophozoite stage may invade the bloodstream to spread the parasite systemically	Digestive system infections (Chapter 19)

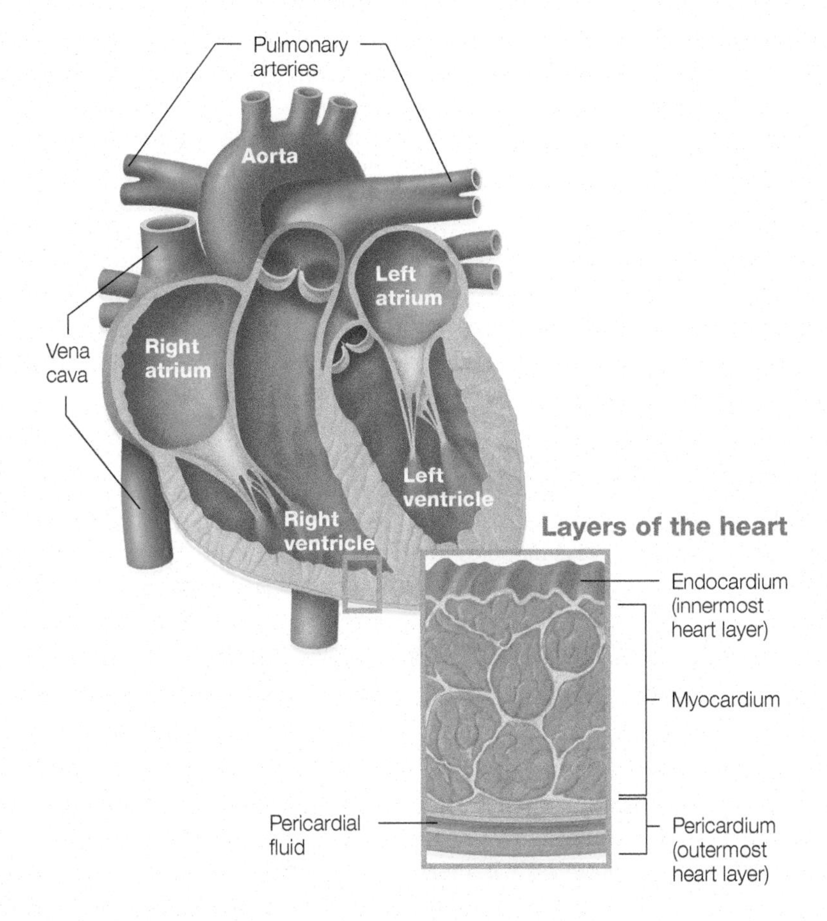

FIGURE 21.1 Heart anatomy

The cardiovascular system includes the heart and blood vessels.

The **heart** is a four-chambered muscle that constantly pumps blood through the body (FIG. 21.1). The two upper heart chambers are the **right atrium** and **left atrium** (plural: atria). The two lower chambers are the **right ventricle** and **left ventricle**. Carbon dioxide-laden blood enters the right atrium and is passed down to the right ventricle. From there, blood gets pumped through the pulmonary arteries to the lungs (a route often referred to as pulmonary circulation). Once in the lungs, blood is loaded with oxygen while carbon dioxide is removed. Next, blood returns to the left atrium and is then pumped into the left ventricle. The muscular wall of the left ventricle contracts to force blood out of the heart through the aorta with enough propulsion that it flows to the whole body (or what is often called systemic circulation). Heart valves open and close in coordination with heart chamber contractions; this ensures that blood flows in a single direction through the heart.

The heart consists of three layers: the outermost pericardium (of which the layer closest to the heart muscle is called the **epicardium**), the middle **myocardium**, which is the muscular tissue that makes up the bulk of the heart, and the innermost **endocardium**, which lines the heart's chambers. Inflammation of the myocardium is called **myocarditis**, whereas **endocarditis** is an inflammation of the endocardium. (We will discuss these conditions later in the chapter.) Enclosing the heart is the **pericardium**, a protective, double-layered sac that contains a thin layer of fluid called the **pericardial fluid**. Inflammation of the pericardium is known as **pericarditis**.

Blood delivers useful things like oxygen, nutrients, and immune system factors to our tissues while also collecting waste products like carbon dioxide. The liquid portion of blood is a fluid called plasma. Suspended in plasma we find red blood cells (erythrocytes), white blood cells (leukocytes), and a variety

CLINICAL VOCABULARY

Myocarditis inflammation of the heart muscle

Endocarditis inflammation of the inner lining of the heart

Pericarditis inflammation of the pericardium, which is a thin, double-membrane sac that encloses the heart

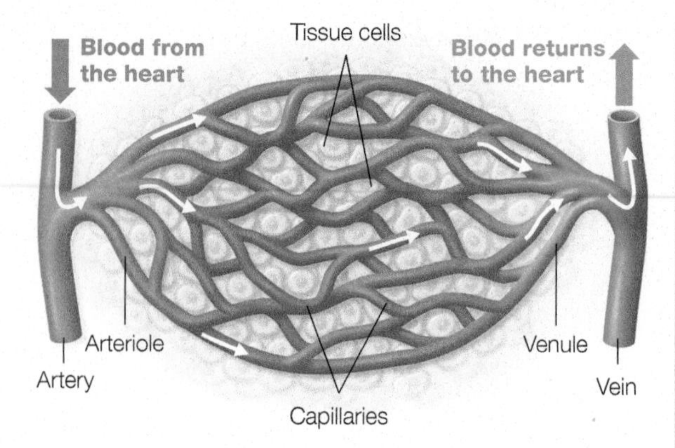

FIGURE 21.2 **Blood vessels of the cardiovascular system**

Blood fractions

Plasma

White blood cells and platelets

Red blood cells

Percent of blood microbiome in each fraction

0.03% in plasma

6.23% in red blood cells

93.74%

In white blood cells

Most common bacterial phyla in blood microbiome

Proteobacteria
Actinobacteria
Firmicutes
Bacteroidetes
Fusobacteria (in red blood cells)

FIGURE 21.3 **Blood microbiome**

CHEM • NOTE

Diffusion is the movement of a substance from an area of high concentration to an area of lower concentration, without an energy investment.

of hormones, clotting factors, nutrients, dissolved gases, and immune system proteins. (For more on blood, see Chapter 11.) We rely on our cardiovascular system's blood vessels to convey blood to our tissues.

Like the highways that crisscross a nation, blood vessels can be thought of as the roads that our blood takes through the body (FIG. 21.2). **Arteries** are muscular vessels; they conduct blood away from the heart (think "arteries away" to remember this point). The *aorta* is the largest artery of all. It conducts blood to other major arteries in our body that are responsible for systemic circulation—blood flow throughout the body. Major arteries branch off the aorta and then split off into smaller arterioles. These arterioles further branch into thin **capillaries** that directly supply blood to our organs and tissues and serve as the site of nutrient and gas exchange. At only one cell thick, capillary walls easily allow oxygen and nutrients to diffuse from the blood into surrounding tissues. Meanwhile, carbon dioxide and other metabolic waste products exit the same tissues, moving into capillaries. Oxygen-depleted, waste-rich blood eventually passes from capillaries to venules and finally to **veins**. Our systemic veins direct blood toward the *vena cava*, a large vein that conveys blood to the heart's right atrium. You can think of the phrase "veins visit," to recall that veins bring blood back to the heart.

Recent microbiome studies indicate that, contrary to common belief, normal blood isn't completely sterile.

Using techniques that detect bacterial DNA, researchers have shown that healthy blood donors harbor low levels of bacteria in their blood. It is thought that on a day-to-day basis, a small number of bacteria make their way into our bloodstream from some of the most heavily colonized areas of our body, such as our intestinal tract and mouth. Bacteria probably get into our blood by invading capillary-rich areas of our gums or crossing into the richly vascularized epithelium of our intestinal tract. That said, blood is cultured all the time in hospital microbiology labs, yet only patients with bloodborne infections generate positive cultures. With such frequent searches for bacteria in blood, how have they remained hidden?

It turns out that most of our blood microbiome is not freely floating in plasma—the part of blood samples tested by bacterial culture techniques. Instead, the microbiome is mainly dwelling *inside* our blood cells. When blood is separated into fractions, researchers find that a minuscule 0.03 percent of the microbiome is free floating in plasma. At such a low level, it's not surprising that blood cultures do not routinely detect these bacteria. It's also possible that the bacterial DNA detected in plasma is from dead bacteria (which of course would not be cultivable) and/or from bacteria that are not cultivable by existing methods. Looking inside blood cells tells a different story. Red blood cells harbor 6.23 percent of our blood microbiome, and 93.74 percent is tucked away inside our white blood cells (FIG. 21.3).[1] The precise role of the blood microbiome and its potential effects on health and disease are far from understood. However, the fact that blood is not sterile, even in healthy people, raises some intriguing possibilities for effects on our systemic health and disease. It also emphasizes the interconnectedness of our gut and oral microbiome with our systemic health.

Lymphatic vessels collect lymph from tissues and convey it back to the venous blood supply.

To briefly summarize, blood plasma diffuses out of capillaries into our tissues. Once plasma leaves the capillaries and enters the spaces around the cells in our tissues, it's called lymph. This protein-rich fluid constantly bathes all tissue cells.

[1] Castillo, D. J., Rifkin, R. F., Cowan, D. A., & Potgieter, M. (2019). The healthy human blood microbiome: fact or fiction? *Frontiers in Cellular and Infection Microbiology*, 9, 148.

FIGURE 21.4 Lymph capillaries in tissue The black arrows show the general flow of fluid located between cells in a tissue (also known as interstitial fluid) into lymph capillaries.

Critical Thinking *What happens when excess lymph accumulates in tissue and is not adequately collected by lymph capillaries?*

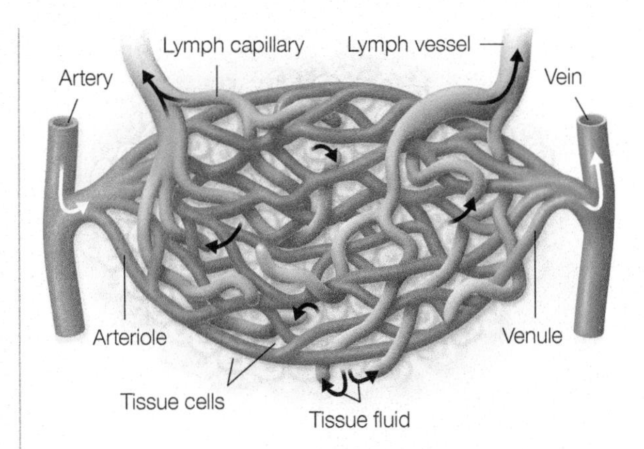

Our lymphatic vessels drain excess lymph from our tissues and shuttle it to the lymph nodes and other MALT tissue (mucosa-associated lymphoid tissue), where waste filtering and pathogen detection take place. If lymph is not collected, then it accumulates in the tissues and leads to swelling, or **edema**. To maintain our blood volume, lymphatic vessels will ultimately funnel lymph back toward our venous blood supply. FIG. 21.4 shows a schematic view of lymph capillaries, which collect fluid from the tissues (often called interstitial fluid) and return the collected fluid to the blood. (Chapter 11 contains the primary coverage of the lymphatic system's anatomy; in particular, see Figures 11.4 and 11.5 to review lymph flow and lymphatic tissues.)

Sepsis is a potentially deadly immune response syndrome.

Sepsis is *not* an infection; rather, this term describes a life-threatening organ dysfunction caused by a patient's dysregulated immune response to a toxin and/or a bacterial, fungal, viral, or parasitic infection. Although many sepsis cases originate from a systemic infection, it is entirely possible for a localized infection to trigger a septic response; technically, microbes do not need to enter the bloodstream or lymphatic vessels for sepsis to develop.[2] Sometimes, in fact, sepsis is referred to as *septicemia*, a term that emphasizes that sepsis is often triggered by a pathogen or a toxin spreading through the blood. Some clinicians have suggested that the term septicemia should be completely dropped from use because it incorrectly implies that sepsis is strictly caused by bloodborne agents.[3]

If not treated early, sepsis may progress to **septic shock**. In septic shock, a patient's blood pressure may drop to the point that blood is not adequately propelled into tissues and organs. As oxygen and nutrients are cut off, tissues can no longer perform cellular respiration, which leads to tissue death (necrosis) and organ failure. (See Chapter 8 to review cellular respiration and its role in making ATP that fuels life.) The formation of blood clots throughout the body—a scenario called **disseminated intravascular coagulation**—also blocks blood flow and promotes organ failure and tissue necrosis.

Bacteria or their toxins are frequently associated with sepsis. *Escherichia coli*, *Staphylococcus aureus*, and *Streptococcus* species are the most common pathogens isolated from septic patients. However, sepsis can develop from any type of infection—including those caused by fungal, parasitic, or even viral agents, including SARS-CoV-2, the virus that causes COVID-19.[4] In most sepsis cases, blood cultures are performed to isolate possible bacterial or fungal pathogens—but according to a study from the U.S. Centers for Disease Control and Prevention (CDC), in about one-third of sepsis patients, the causative agent is never identified.[5] Lung and kidney infections have the highest risk of spreading to the bloodstream and inducing sepsis (FIG. 21.5).

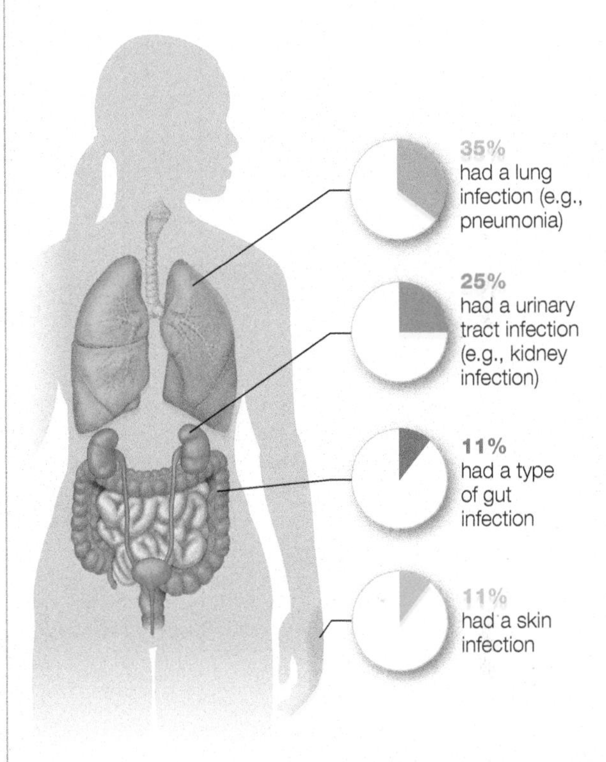

FIGURE 21.5 Infections associated with sepsis Note that 18% of sepsis cases fall into an "all other infections" category, which is not represented in the pie charts above.

Source: Centers for Disease Control and Prevention.

CLINICAL VOCABULARY

Sepsis overwhelming, body-wide inflammatory response to a pathogen or toxin; often due to a systemic infection, but localized infections may also trigger it; sepsis describes the immune response to an infection or toxin

Septic shock the most dangerous stage of sepsis that features tissue death and organ failure

Disseminated intravascular coagulation the formation of blood clots in the bloodstream; as the clots lodge in capillaries, they block blood flow to tissues

[2] Markwart, R., Saito, H., Harder, T., et al. (2020). Epidemiology and burden of sepsis acquired in hospitals and intensive care units: a systematic review and meta-analysis. *Intensive Care Medicine*, 1-16.

[3] Odeh, M. (1996). Sepsis, septicaemia, sepsis syndrome, and septic shock: The correct definition and use. *Postgraduate Medical Journal*, 72 (844), 66.

[4] Li, H., Liu, L., Zhang, D., et al. (2020). SARS-CoV-2 and viral sepsis: observations and hypotheses. *The Lancet*.

[5] CDC Morbidity and Mortality Weekly Report https://www.cdc.gov/mmwr/volumes/65/wr/mm6533e1.htm

DiseaseSnapshot • Sepsis

S. aureus

Causative agent	Sepsis is not an infection—it is a life-threatening organ dysfunction caused by a patient's dysregulated immune response to a toxin and/or a bacterial, fungal, viral, or parasitic infection.
Epidemiology	Over one million cases per year in the U.S.; mortality rate is about 38%; more common in patients who cannot fully fight off infections; recent surgery, invasive medical procedures, and severe tissue trauma increase sepsis risk.
Transmission & prevention	Not transmissible since it is generated by the host's immune response; best way to avoid sepsis is effective infection management.
Signs & symptoms	Diarrhea, vomiting, pale skin, sleepiness, decreased urine output, confusion/delirium, and lymphangitis may occur.
Pathogenesis & virulence factors	Pathogens with an enhanced ability to spread system-wide or make toxins are more likely to trigger sepsis; at more advanced stages, the patient exhibits severe hypotension, deregulated body temperature, and a dysregulated immune response that leads to tissue and organ damage.
Diagnosis & treatment	Sepsis diagnosis requires Sequential Organ Failure Assessment (SOFA) scores of 2 or more; septic shock diagnosis requires SOFA scores of 2 or more combined with severe hypotension and high blood lactate levels. Treatment goals include reducing system-wide inflammation, managing body temperature, stabilizing blood pressure, and increasing the patient's blood oxygen level; antimicrobial drugs may be administered if a nonviral infectious agent is suspected; severe sepsis may require life support.

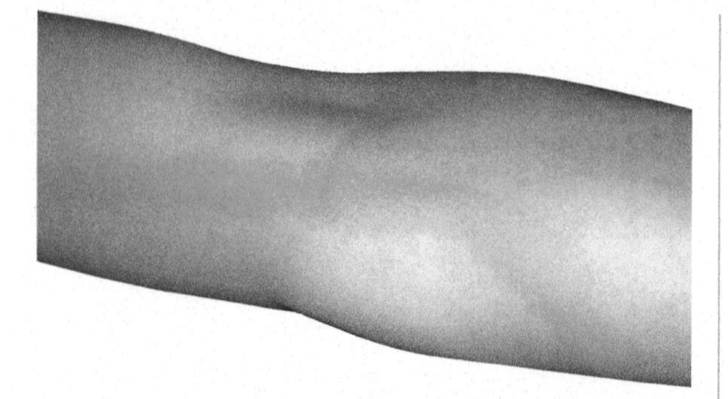

FIGURE 21.6 Lymphangitis

CLINICAL VOCABULARY

Lymphangitis inflammation of the lymphatic vessels that may be evident as red streaks in the skin that radiate from the infection source to the nearest lymph node

Intubation insertion of a flexible plastic tube into the airway

Analgesia pain management/relief

Sepsis Incidence and Risk Factors

Ultimately sepsis, no matter what triggers it, is a medical emergency because if it progresses to septic shock, then the patient has about a 38 percent chance of dying.[6] According to the CDC, sepsis occurs in over 1.5 million adults in the United States each year. When it comes to hospital charges, sepsis is among the most expensive health condition in the United States—annually it costs more than $62 billion to treat.[7]

Patients with chronic health issues such as cancer, diabetes, AIDS, and cardiovascular disease often struggle to fight off infections; this puts them at an increased risk for the agent spreading systemically and inducing sepsis. Recent surgery, invasive procedures, a wound, or blunt force trauma can also increase sepsis risk.

Diagnosis and Treatment

In 2016, the **Sequential Organ Failure Assessment (SOFA)** was established for diagnosing sepsis. The criteria for this assessment are tied to scoring levels; patients with a SOFA score of 2 or more are diagnosed as suffering sepsis—with higher scores being associated with a poorer prognosis (TABLE 21.2). Patients are diagnosed with septic shock if they have a SOFA score of two or more combined with severe hypotension (i.e., require drugs called vasopressors to maintain a mean arterial pressure of at least 65 mmHg) and a dangerously high blood lactate level.[8] Other signs and symptoms of sepsis and/or septic shock may include diarrhea, vomiting, pale skin, sleepiness, decreased urine output, confusion/delirium, and **lymphangitis**—a condition in which lymphatic vessels become inflamed and appear as red streaks that radiate from the initial infection site (often a wound on the extremities) to the nearest lymph node (FIG. 21.6).

[6] Vincent, J. L., Jones, G., David, S., Olariu, E., & Cadwell, K. K. (2019). Frequency and mortality of septic shock in Europe and North America: a systematic review and meta-analysis. *Critical Care*, 23(1), 1-11.

[7] Buchman, T. G., Simpson, S. Q., Sciarretta, K. L., et al. (2020). Sepsis Among Medicare Beneficiaries: 3. The Methods, Models, and Forecasts of Sepsis, 2012–2018. *Critical Care Medicine*, 48(3), 302.

[8] Recall from Chapter 8 that lactate is a fermentation byproduct; muscle cells only make this substance when they are not receiving enough oxygen to support ATP production. A blood lactate level greater than 2 mmol/L despite adequate fluid resuscitation is concerning. Fluid resuscitation is the practice of administering specialized intravenous fluids (e.g., intravenous crystalloid fluid) to increase intravascular volume, cardiac output, and oxygen delivery to tissues. A key indicator for fluid resuscitation is a blood lactate level of 4.0 mmol/L or more.

TABLE 21.2 A Summary of Sequential Organ Failure Assessment (SOFA) Criteria

	Scoring Criteria*	Notes
Respiratory Function	• $PaO_2/FiO_2 < 400$ mmHg (kPa) = 1 point • $PaO_2/FiO_2 < 300$ mmHg (kPa) = 2 points • $PaO_2/FiO_2 < 200$ mmHg (kPa) with respiratory ventilation = 3 points • $PaO_2/FiO_2 < 100$ mmHg (kPa) with respiratory ventilation = 4 points	As numbers worsen, respiratory system failure is suspected. (Note: PaO_2/FiO_2 is a ratio measurement that is used when a patient is on supplemental oxygen).
Kidney Function	• Creatinine 110 – 170 mmol/L = 1 point • Creatinine 171 – 299 mmol/L = 2 points • Creatinine 300 – 440 mmol/L or urine output < 500 mL/day = 3 points • Creatinine > 440 mmol/L or urine output < 200 mL/day = 4 points	Creatinine levels rise and urine output decreases as kidney function decreases.
Liver Function	• Bilirubin 20 – 32 mmol/L = 1 point • Bilirubin 33 – 101 mmol/L = 2 points • Bilirubin 102 – 204 mmol/L = 3 points • Bilirubin > 204 mmol/L = 4 points	Increased bilirubin levels reflect possible liver failure.
Cardiovascular Function	• MAP < 70 mmHg = 1 point • Dopamine < 5 or any dose of dobutamine = 2 points • Dopamine 5.1 – 15 mg/kg/minute or any dose of epinephrine or a dose of norepinephrine ≤ 0.1 μg/kg/minute = 3 points • Dopamine < 5 μg/kg/minute or any dose of epinephrine or a dose of norepinephrine > 0.1 μg/kg/minute = 3 points	Hypotension is a manifestation of cardiovascular pathology. Dopamine, dobutamine, epinephrine, and norepinephrine are all administered to raise blood pressure.
Low Platelet Level	• Platelets < 150×10^3/mL = 1 point • Platelets < 100×10^3/mL = 2 points • Platelets < 50×10^3/mL = 3 points • Platelets < 20×10^3/mL = 4 points	Platelet levels are monitored to detect dysregulated coagulation.
Glasgow Coma Scale Score	• 13 – 14 = 1 point • 10 – 12 = 2 points • 6 – 9 = 3 points • less than 6 = 4 points	The Glasgow Coma Scale is used to evaluate impaired consciousness and coma.

Abbreviations: PaO_2 = Partial pressure of oxygen dissolved in plasma; FiO_2 = Fraction of inspired oxygen; MAP = Mean arterial pressure

Managing sepsis mainly involves reducing system-wide inflammation, regulating body temperature, stabilizing blood pressure, and increasing blood oxygen levels. If sepsis is caused by a nonviral pathogen, then intravenous antimicrobial drugs may also be administered. Patients experiencing septic shock are often put on life support until they are fully stabilized and the condition starts to resolve. Life support care often includes intubation, mechanical ventilation, sedation, and analgesia (pain management/relief).

Build Your Foundation

1. Draw a diagram of a heart and label the chambers as well as the three layers of the heart wall. (LO 21.1)
2. List the route that blood takes as it flows through the heart and through pulmonary and systemic circuits. (LO 21.2)
3. What is lymph, how does it enter tissues, and how does it make its way back to the blood supply? (LO 21.3)
4. Why are the lymphatic and cardiovascular systems considered linked? (LO 21.4)
5. Discuss sepsis, its risk factors, detection, progression, and management. (LO 21.5)

Build Your Foundation (BYF) Quick Quiz: Visit the **Mastering Microbiology** Study Area to quiz yourself.

21.2 SYSTEMIC VIRAL INFECTIONS

Learning Outcomes

After reading this section, you should be able to:

21.6 Describe the features of the following: chikungunya, yellow fever, dengue, and Zika and discuss how they are prevented.

21.7 Describe the stages of dengue infection, and why the second stage is of concern.

21.8 Summarize the three cycles of yellow fever transmission.

21.9 Explain how chikungunya can be differentiated from dengue.

21.10 State the recommended wait periods that men and women should observe to avoid Zika-induced fetal complications.

21.11 Describe the typical symptoms of Ebola, Marburg, and Lassa hemorrhagic fevers, and discuss how these infections can be prevented.

21.12 Detail the features of Epstein-Barr virus (EBV) infection, including how it is transmitted and possible complications.

21.13 Explain the three stages of human immunodeficiency virus (HIV) infection and describe how antiretroviral drugs can be used to combat its progression.

21.14 Discuss the clinical and epidemiological features of human T lymphotropic virus (HTLV).

Vectorborne systemic viral infections include dengue fever, yellow fever, chikungunya, and Zika.

A number of arboviruses (arthropod-borne viruses) cause systemic infections in millions of patients worldwide. Key examples of these arboviruses include mosquito-borne viruses such as dengue, yellow fever, chikungunya, and Zika. At the time of this text's publication, only yellow fever and Ebola are vaccine preventable, which means controlling mosquito levels remains crucial for reducing the incidence of most arboviral infections. Given their link with mosquitoes, it's not surprising to learn that these infections are mainly perpetuated in tropical and subtropical regions.

The United States is relatively untouched by many of these illnesses, primarily thanks to mosquito abatement. Ready access to healthcare also helps limit complications in cases that happen to develop. Unfortunately, warming climate has begun to expand the year-round habitat available to mosquitoes, and certain species are expanding their territory. Of particular concern is the *Aedes aegypti* mosquito, which prefers human blood meals, thrives in urban areas, and bites throughout the day rather than just at dawn and dusk. *Aedes* mosquitoes transmit all four of the arboviruses discussed in the next sections. To date, *Aedes* mosquitoes have made their way into regions of the southernmost U.S. states, and will likely continue moving north over time.

Dengue Fever

Dengue fever is caused by four related though distinct dengue virus serotypes (DENV-1, DENV-2, DENV-3 and DENV-4). Recovery from infection with one serotype does not provide immunity to other serotypes; thus, a person may develop dengue fever more than once—with subsequent infections being more severe. Dengue fever was a geographically isolated disease with relatively low incidence until the 1950s, when epidemics emerged in the Philippines and Thailand. From these nations it slowly spread, and it is now endemic in about a hundred different countries that have tropical and subtropical climates, such as India and countries in South and Central America, the Caribbean, and sub-Saharan Africa. Cases typically increase with the onset of the rainy season, when mosquitoes proliferate. The World Health Organization (WHO) estimates that globally there are up to 400 million infections per year, with about 22,000 deaths. Most U.S. cases occur in travelers who get infected while traveling to dengue-prone areas. However, dengue fever is also endemic in parts of Mexico that border the United States. So far, mosquito control measures have prevented the disease from spreading much over the northern border. However, sporadic cases have occurred in Texas, Florida, and a few other states over the past decade. In 2020, the CDC recorded 332 dengue cases in the U.S. and 760 cases in U.S. territories.

CLINICAL VOCABULARY

Viremia a high number of virions in the blood

Anemia a condition characterized by a low red blood cell count or low hemoglobin level

Petechiae a small red or purple spot caused by bleeding under the skin

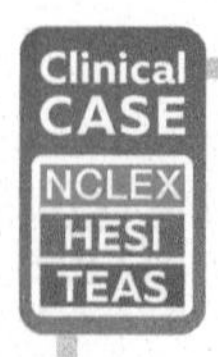

The Case of the Bloated Belly

Practice applying what you know clinically: visit the **Mastering Microbiology** Study Area to watch Part 2 and practice for future exams.

Signs, symptoms, and stages The first stage of dengue is called the febrile phase. It's characterized by fever and other signs and symptoms such as nausea, vomiting, rash, and body aches. **Viremia**, a high number of viruses in the blood, develops during this stage. Viremia promotes spread because if an infected person is bitten by mosquitoes during this stage, the insect takes up the virus in its blood meal and subsequently transmits it to new hosts.

Identifying dengue fever in its early stage allows timely supportive interventions to limit poor outcomes. Examples of common interventions include keeping the patient hydrated and monitoring for **anemia** development. When the fever stops, the second stage follows, and critical symptoms may develop within the next few days. Possible severe second-stage symptoms include extreme abdominal pain, persistent vomiting, and dehydration. Occasionally hemorrhagic features may also develop. The hemorrhagic form of dengue fever may feature

DiseaseSnapshot • Dengue Fever

Causative agent	Dengue virus types 1–4: enveloped, single-stranded RNA genome, *Flaviviridae* family
Epidemiology	Endemic in over 100 tropical and subtropical countries; potentially expanding into new areas where *Aedes aegypti* mosquitoes can thrive; incubation period of 4 to 7 days
Transmission & prevention	Vectorborne, usually by *Aedes* mosquitoes; no vaccine available to prevent primary infections, so prevention includes avoiding mosquito bites through insecticides and repellants
Signs & symptoms	Fever, body aches, rash, vomiting, or nausea; severe symptoms may also include extreme abdominal pain, bleeding from nose and mouth, or fluid accumulation that leads to shock
Pathogenesis & virulence factors	The virus enters through the mosquito bite to infect leukocytes; the presence of dengue virus nonstructural protein 1 (NS1) incites cytokine release and inflammatory response, which causes plasma leakage through capillaries that appears to aid the virus in its pathogenesis
Diagnosis & treatment	Preferred diagnostic test is RT-PCR to detect dengue virus genetic material (RNA) in the patient's blood; treatment goals include reducing fever and maintaining hydration; the vaccine Dengvaxia can prevent the severe disease that tends to develop during a subsequent infection; the vaccine is limited to young people who have recovered from a laboratory-confirmed dengue infection; in the U.S., Dengvaxia is only approved for administration where dengue is endemic (American Samoa, Guam, Puerto Rico and the U.S. Virgin Islands); vaccine is administered in 3 doses—each 6 months apart

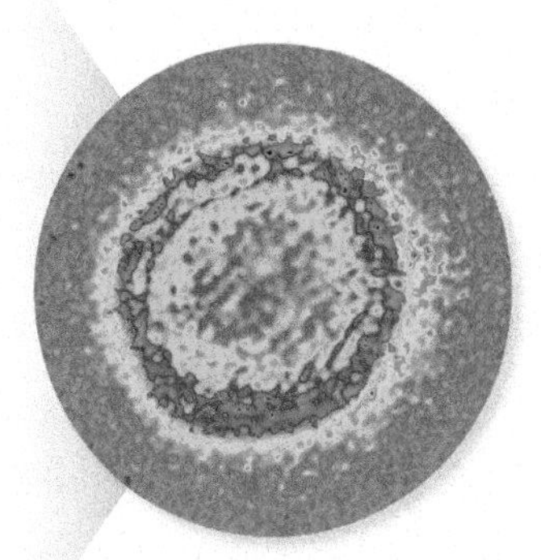

A dengue virus virion

bleeding from the nose or mouth, the appearance of small red patches on the skin (**petechiae**), and plasma leaking from capillaries into tissues. These hemorrhagic events can lead to a dangerous decrease in blood volume, which may lead to shock, organ failure, and eventually death if not mitigated. If the patient makes it through the second stage, then they proceed to a recovery phase.

Diagnosis A **tourniquet test** can indicate whether a patient may have dengue fever that has progressed to a hemorrhagic form. A blood pressure cuff is applied to the forearm of the patient for five minutes. If red patchy skin appears, it indicates that the patient's capillaries have become leaky in response to the viral infection (FIG. 21.7). For a definitive diagnosis, follow-up blood tests are necessary. Options include identification of dengue virus RNA, and an ELISA test that recognizes dengue viral proteins. The CDC requires that confirmed dengue fever be reported to the health department.

Treatment of dengue fever is simply administration of acetaminophen (Tylenol), hydration, and close observation during the critical stage. If a patient is showing symptoms of shock, then intravenous blood volume expanders may be administered to help raise blood pressure.

As mentioned earlier, there are four serotypes of dengue virus. Infection with one serotype confers lifelong immunity against re-infection with the same serotype, but patients get only a fleeting immune protection against other dengue serotypes. In fact, subsequent infection with a different serotype seems to be linked to a greater risk of developing a more severe case. Dengvaxia can

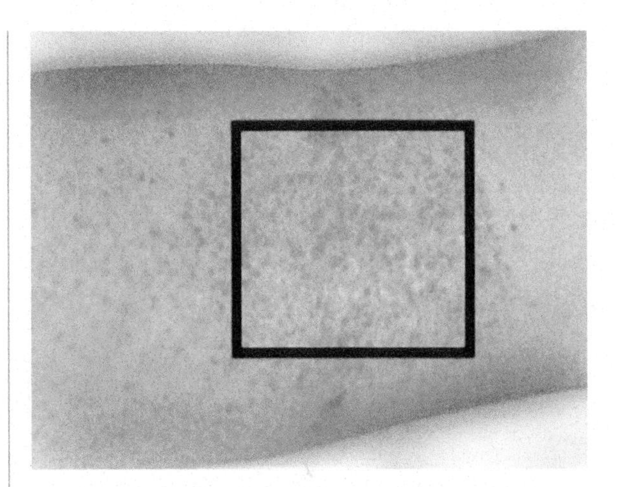

FIGURE 21.7 **Tourniquet test** Upon removing the blood pressure cuff, the skin is observed for petechiae. The test result is positive if more than 10 petechial spots develop in a square-inch region. The boxed region in this photograph reflects a positive result.

Critical Thinking *What part of the cardiovascular system would be involved in creating the petechial spots that result in a positive tourniquet test result?*

DiseaseSnapshot • Yellow Fever

Causative agent	Yellow fever virus: enveloped, single-stranded RNA genome, *Flaviviridae* family
Epidemiology	Endemic in sub-Saharan Africa, South America, and parts of Central America and Mexico; about 200,000 cases per year, with about 30,000 deaths; incubation period is 3–6 days
Transmission & prevention	Usually transmitted by *Aedes* or *Haemagogus* mosquitoes; prevented by avoiding mosquito bites; attenuated vaccine is also available for prevention
Signs & symptoms	Fever, chills, severe headache, and back pain; in severe cases jaundice, high fever, renal failure, shock, and organ failure may develop
Pathogenesis & virulence factors	The virus is deposited into tissue by mosquito bite; dendritic cells take up the virus and then move to the lymph nodes where the virus can kill dendritic cells, releasing more virus to infect macrophages; once the virus enters a host cell, infected macrophages migrate through blood and lymph, where they die and release more viruses to the patient's organs (particularly the liver)
Diagnosis & treatment	ELISA methods can detect patient antibodies to the virus or viral antigens in blood; supportive therapies for fever, pain, and maintaining hydration are the only treatments

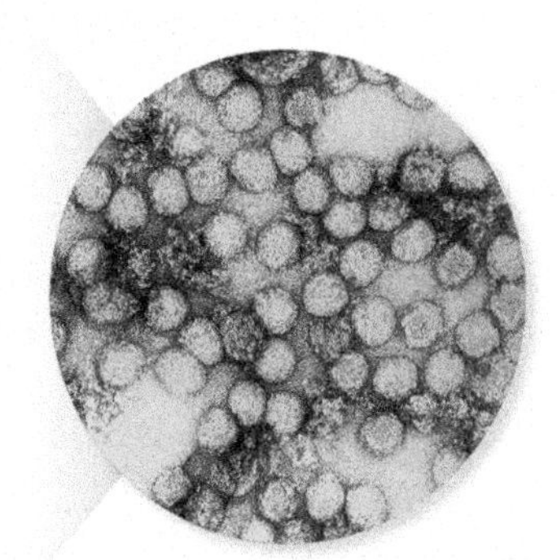

Yellow fever virus

prevent the severe disease that tends to develop during a subsequent dengue infection; this vaccine is limited to young people (ages 9-45 in many nations; ages 9-16 in the U.S.) who have recovered from a laboratory-confirmed dengue infection. Because it is not recommended to prevent primary dengue infections, Dengvaxia is not a preventative vaccine in the traditional sense. A vaccine against primary dengue infection is in clinical trials. However, studying the virus is difficult because there is no good animal model for infection.

Yellow Fever

It's believed that **yellow fever** sailed out of Africa and into Europe, the Americas, and the Caribbean via slave trade ships during the 17th and 18th centuries. The most notorious U.S. outbreak happened in Philadelphia during 1793, where about one in eight people died. Spread by mosquitoes of the *Aedes* and *Haemagogus* genera, yellow fever was largely eliminated from the United States and the Caribbean by intensive mosquito control and vaccination efforts. It remains endemic in sub-Saharan Africa, South America, and parts of Central America because of lapses in vaccination and poor mosquito control. According to the CDC, every year there are about 200,000 cases of yellow fever worldwide and approximately 30,000 deaths; about 90% of the cases occur in Africa.

Transmission The causative agent, a single-stranded RNA virus, infects a variety of primates, including humans and monkeys. Infected female mosquitoes can also pass the virus to their offspring. There are three cycles of transmission (FIG. 21.8):

- **Jungle:** Transmits primarily between monkeys via mosquitoes in forested, jungle areas.
- **Intermediate:** In areas with both villages and jungle, mosquitoes infect both humans and monkeys. This is the predominant transmission cycle in Africa.
- **Urban:** Human-to-human transmission via mosquitoes. Large epidemics are possible with urban transmission in areas of high population and low vaccination rates.

Signs, symptoms, pathogenesis, and prevention Most yellow fever infections are asymptomatic or mild. However, some patients develop fever, chills, severe headache, and back pain. Fatigue, nausea, and vomiting may also occur. About 15 percent of symptomatic patients develop more severe disease, including jaundice—the clinical feature that gives yellow fever its name. Extremely high fever and potentially deadly shock may also develop. Supportive therapy that includes rest, fluids, and pain relievers is the only treatment. Patients who survive infection generally have lifelong immunity against the virus.

Once deposited into tissue by a mosquito bite, the virus is taken up by local dendritic cells. Given the number of asymptomatic cases, it may be that dendritic cells and local innate immune responses can often prevent further infection progression. However, if viruses successfully infect dendritic cells then the infection can spread to other cells, including macrophages. As infected immune cells move through lymph and blood, viruses gain access to the liver and spleen where massive numbers of viruses can be made, resulting in viremia. As viruses spread through blood, cytokine release incites inflammatory responses, causing plasma to leak out of capillaries. This results in reduction in blood volume and may lead to shock and organ failure.

An attenuated yellow fever vaccine is available. Since the disease is rare in the United States, the vaccine is not routine. However, Thailand, Vietnam, the Philippines, parts of the Caribbean, and many African nations require that travelers who are over 1 year old present proof of yellow fever vaccination before they are granted entry. A single vaccine dose provides at least a decade of immunity. Other preventive measures include avoiding mosquito bites by using insect repellent and mosquito nets.

Jungle transmission cycle

Between monkeys and mosquitoes in forested areas

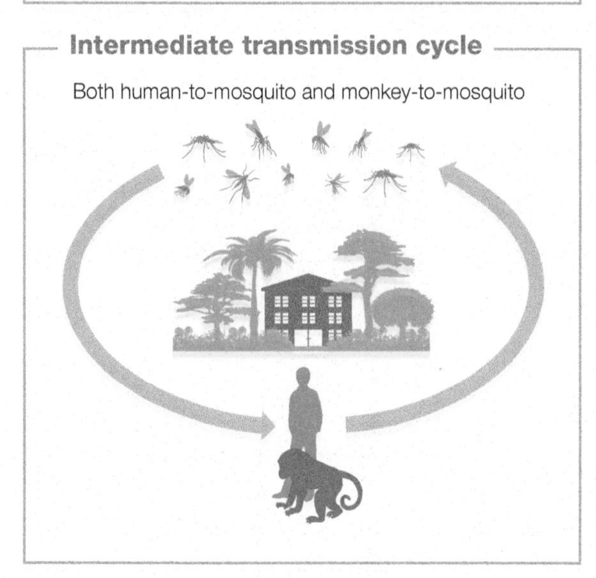

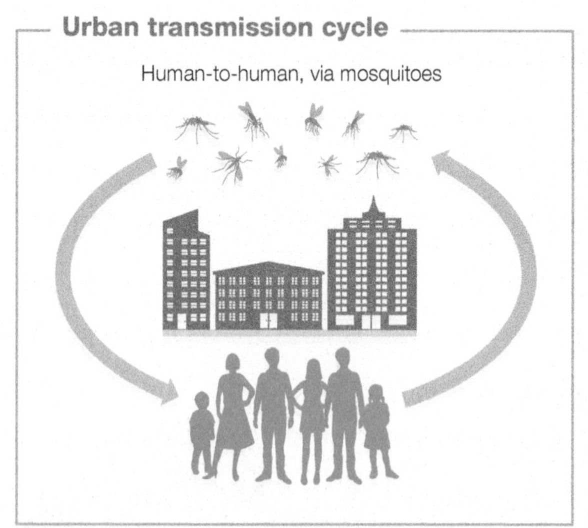

FIGURE 21.8 **Yellow fever transmission cycles**

CLINICAL VOCABULARY

Jaundice yellowing of the skin and eyes due to high bilirubin levels of the blood; a sign of blood, liver, or gallbladder problems when present in non-newborns

Disease**Snapshot** • **Chikungunya**

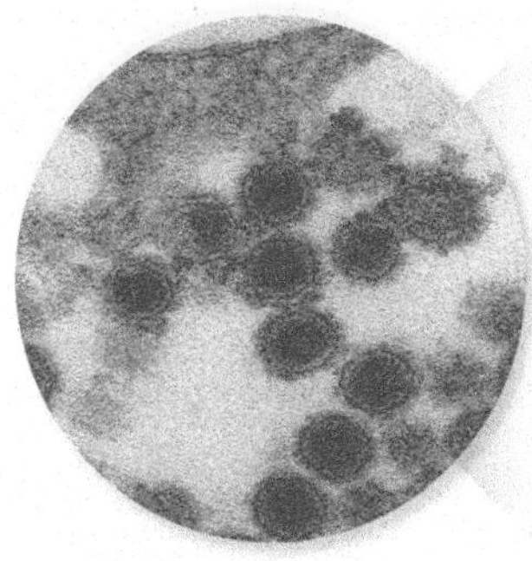
Chikungunya virus

Causative agent	Chikungunya virus: enveloped, single-stranded RNA genome, *Flaviviridae* family
Epidemiology	Outbreaks in Africa, Asia, Europe, Indian and Pacific Islands; slowly spreading through Americas; incubation period is 3 to 7 days
Transmission & prevention	Vectorborne, usually spread by *Aedes aegypti* mosquito; no vaccine; prevented by limiting mosquito populations through environmental interventions (spraying and removing standing water) and limiting mosquito bites by using insect repellent and bed netting
Signs & symptoms	Commonly features fever, muscle aches, headaches, joint pain, and rash
Pathogenesis & virulence factors	Virus spreads to multiple tissues and organs and induces direct cellular damage and inflammation
Diagnosis & treatment	Observation of symptoms, and blood tests that detect viral proteins, viral genetic material, or patient antibodies to the virus; supportive treatments to manage fever and maintain hydration

Chikungunya

Chikungunya (CHEE kun *goon* ya) is endemic in Central Africa. Genetic studies suggest that although the first human cases were described in Africa in the 1950s, the virus originated there nearly 500 years ago.[9] Ever since human cases started cropping up, the virus has slowly spread across the globe—mirroring the expanding territory of *Aedes* mosquito species. The virus has been detected in Asia, Europe, the Indian and Pacific islands, and—as of 2013—in the Americas. Within 5 years of being introduced to the Americas, there were at least 2 million nonimported chikungunya cases. In 2015, chikungunya virus disease became a nationally notifiable condition in the U.S. To date, mosquito abatement programs have prevented the virus from gaining much of a foothold—most cases in the U.S. are contracted while traveling abroad. According to the CDC, in 2019, 171 travel-related cases were recorded. In 2020, a year of greatly reduced travel due to the COVID-19 pandemic, there were only 18 cases reported. Annual global incidence varies greatly, from a few hundred thousand to over a million cases.

Most people develop symptoms within a week of infection. Generally, patients experience a sudden onset of a high fever (at least 102°F [39°C]) and **polyarthralgia,** which is extreme pain in multiple joints. Headaches, vomiting, and rash may also occur. The word chikungunya comes from the Kimakonde language; it translates to "that which bends up." This describes the bent and stiff posture of those suffering with the joint pain that is a hallmark of infection. Upon entering the host, the virus replicates in fibroblasts (cells that make collagen for our skin and connective tissues) and possibly macrophages. From there the virus spreads via the bloodstream and lymphatic system to infect multiple tissues, including skin, lymph nodes, spleen, joints, muscles, and tendons. In severe cases, the brain and liver may also become infected. Most infection symptoms resolve within about a week, but about 87 percent of patients experience ongoing joint and muscle pain that persists beyond three months.[10] Immune-compromised patients, the elderly, and infants are at a greater risk for complications such as hemorrhage, myocarditis, neurological disorders, hepatitis, and ocular disease (eye infections).

As of the publication of this text, there is no vaccine against chikungunya, and supportive care such as rest, fluids, and anti-inflammatory drugs is the main treatment. However, the search for a vaccine and treatments is actively under way. A specific antiviral therapy has not been approved for routine treatment.

CLINICAL VOCABULARY

Polyarthralgia aching/pain in multiple joints

[9] Deeba, F., Haider, M. S. H., Ahmed, A., Tazeen, A., et al. (2020). Global transmission and evolutionary dynamics of the Chikungunya virus. *Epidemiology & Infection*, 148.

[10] de Moraes, L., Cerqueira-Silva, T., Nobrega, V., et al. (2020). A clinical scoring system to predict long-term arthralgia in Chikungunya disease: A cohort study. *PLoS Neglected Tropical Diseases*, 14(7), e0008467.

DiseaseSnapshot • Zika

Zika virus

Causative agent	Zika virus: enveloped, single-stranded RNA genome, *Flaviviridae* family
Epidemiology	Identified in Africa, India, Southeast Asia, Pacific Islands, Central and South America, and more recently in Florida; incubation period is 3 to 10 days
Transmission & prevention	Transmitted by mosquitoes, bodily fluids, sexual contact, and vertically (pregnant mother to fetus); prevention involves mosquito abatement and condom use; to protect against vertical transmission, women should wait at least 8 weeks after showing symptoms to conceive, while men should wait 6 months after infection to ensure viral clearance
Signs & symptoms	Sudden fever, rash, joint pain, and eye infections (conjunctivitis); Guillain-Barré may be a complication
Pathogenesis & virulence factors	The virus may cross the placenta and infect fetal neural tissue
Diagnosis & treatment	Definitive diagnosis involves detecting patient IgM antibodies to the virus or viral RNA in a patient sample; the only treatments at this time are supportive therapies to alleviate discomfort, limit dehydration, and reduce fever

CLINICAL VOCABULARY

Microcephaly arrested fetal brain development that results in an abnormally small brain and head size

Coinfection with dengue virus The chikungunya virus is a member of the same viral family as dengue, so it's not surprising that they share certain features, including transmission and symptom features. Chikungunya and dengue coinfection can occur because these viruses are endemic in the same areas, and a single mosquito may harbor both pathogens. In the case of a suspected coinfection, clinicians attempt to first confirm the dengue infection because of the risk of shock and death during its secondary infection stage. Chikungunya diagnosis is mainly based on symptoms—if symptoms suggest an infection, then a definitive diagnosis can be made using blood tests that identify viral antigens, viral RNA, or patient antibodies to the virus.

Zika

In 2015, Brazil reported a dramatic increase in the number of babies born with **microcephaly**—an abnormally small head. Eventually, the phenomenon was linked to mothers contracting **Zika virus** during pregnancy. Zika virus was first identified in 1947; over the decades it spread to various nations in sub-Saharan Africa and South Asia before more recently arriving in Brazil and spreading through the Americas. The first U.S. Zika cases were in 2016.

Signs, symptoms, transmission, and clinical progress Most Zika virus infections are asymptomatic, and people who recover from Zika tend to have lifelong immunity to it. When signs and symptoms do occur, they last about a week and include fever, joint pain, red eyes, and a rash (FIG. 21.9). Because symptoms are similar to those caused by dengue and chikungunya viruses, definitive diagnosis requires molecular tests that identify the virus or antibodies made against it.

Mosquitoes transmit the virus to humans, but once it is in a person the virus can be spread through bodily fluids, via sexual contact, or transplacentally from a mother to her developing fetus. Because blood can harbor the virus, the CDC recommends that all donated blood be screened for Zika virus before it is used for a transfusion.

As of this text's publication, researchers are working to understand Zika's infection mechanisms and to develop a vaccine and antivirals to combat the virus. As of 2020, there are 18 different vaccine candidates in some stage of preclinical or clinical trial.[11] However, to date, there is no effective vaccine or treatment.

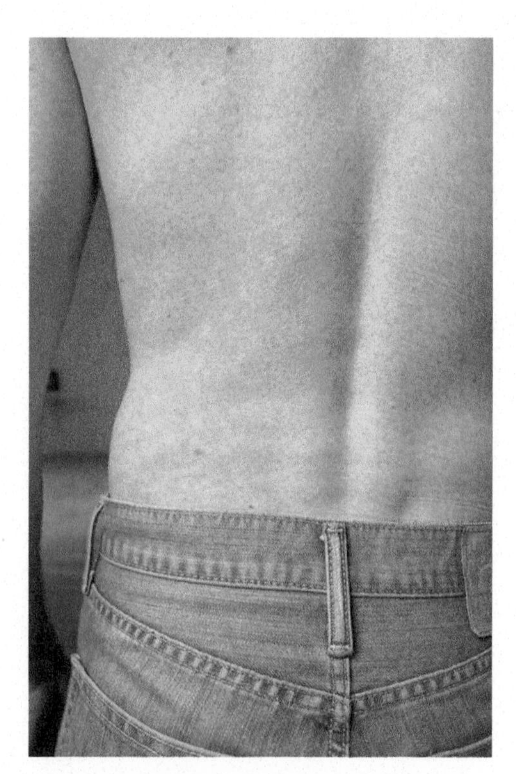

FIGURE 21.9 Zika rash The Zika rash resembles dengue and chikungunya rashes.

Critical Thinking *Why is it not surprising that Zika, dengue, and chikungunya have a large degree of symptom overlap?*

[11] Poland, G. A., Ovsyannikova, I. G., & Kennedy, R. B. (2019, December). Zika vaccine development: current status. *Mayo Clinic Proceedings* (Vol. 94, No. 12, pp. 2572-2586). Elsevier.

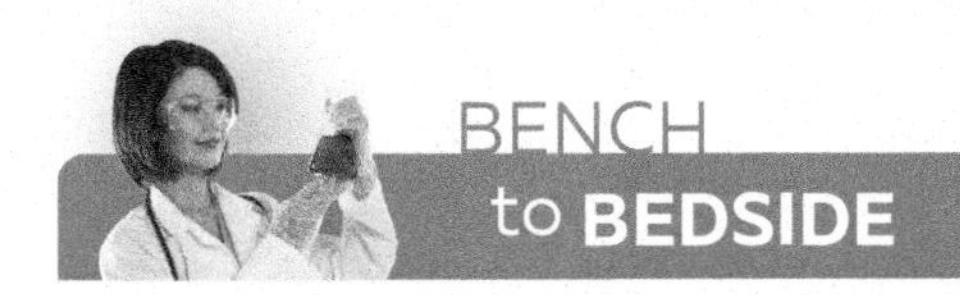

Eliminating *Aedes* Mosquitoes to Quell Vector-Transmitted Diseases

The applications of CRISPR, the gene editing technology discussed in Chapter 14, have been boundless. CRISPR has been used, for example, on mosquitoes—the disease-carrying vectors for dengue, malaria, Zika, yellow fever, and many other infectious diseases. The intent of one initiative was to create an "edited" *Aedes* mosquito population. Since only the female bites humans, CRISPR was used to genetically modify a population of mosquitoes to create females that would only produce males. Engineered mosquitoes could then be released into the environment to lower the *Aedes* mosquito population.

However, concerns were raised about this approach. First, some researchers worry that gene-editing a mosquito population could be the start of a slippery slope toward gene-editing more advanced organisms. Currently, there are no regulations in place to restrict scientists in the use of gene-modification tools in nonhuman organisms. Another concern is that once the release of gene-edited mosquitoes into a natural population occurs, there is no turning back. Removal of the population would be difficult, if not impossible, if damaging environmental effects were to result. Already, areas around the globe where vectorborne diseases are high host testing grounds for these genetically modified mosquitoes to be released. These test areas raise political and ethical issues. More discussion and research are being conducted to determine how best to balance human health with ethical and environmental concerns.

Source: Hall, A. B., et al. (2015). A male-determining factor in the mosquito *Aedes aegypti*. *Science, 348* (6240), 1268–1270.

Congenital defects and other complications Because most individuals have only mild or asymptomatic cases, Zika infections were largely overlooked and underreported in the past. However, with the observation that the virus can pass from a pregnant mother to her fetus (vertical transmission) and potentially result in microcephaly and other brain defects, Zika has been declared a growing public health threat. Another type of complication from Zika infection is Guillain-Barré syndrome, in which the infection damages the myelin sheath that surrounds nerves, resulting in muscle weakness and paralysis with varying degrees of permanence.

To avoid fetal complications, men and women who develop a Zika infection or may have been exposed to Zika and had an asymptomatic case are encouraged to delay conception. Women should wait at least eight weeks after infection before conceiving. The virus is detectable in semen for an unusually long period, so men should wait at least six months after infection to ensure viral clearance.

CLINICAL VOCABULARY

Hemorrhagic accompanied by bleeding or producing bleeding; a term often used to describe pathogens that interfere with host blood-clotting cascades

At least four viral families are known to cause hemorrhagic fevers.

Infections called **viral hemorrhagic fevers** induce high fever, chills, diarrhea, vomiting, and headache, as well as bleeding from the eyes, mouth, ears, skin pores, and internal organs. Mosquitoes, ticks, rodents, bats, and blood contact when slaughtering infected animals have all been known to transmit viral hemorrhagic fevers. Occasionally, infected humans transfer these viruses through blood or semen.

So far we have discussed dengue and yellow fever (viruses in the *Flaviviridae* family) as having some hemorrhagic manifestations in their most severe forms, but three other virus families also cause hemorrhagic fevers: the *Bunyaviridae* family, which includes hantavirus (see Chapter 16 for more on hantavirus); the *Arenaviridae* family that includes Lassa fever (discussed later); and the *Filoviridae* family that includes Ebola and Marburg viruses (discussed next).

Ebola

The **Ebola** virus is a single-stranded RNA virus. Scientists suspect that fruit bats, apes, or monkeys are the virus's natural reservoir. Contact with an infected animal's excrement can transmit the virus, as can handling or eating infected bushmeat (meat from nondomesticated animals such as monkeys and gorillas). Once in the human population, the virus spreads from person to person through contact with bodily fluids as well as sexual contact (FIG. 21.10).

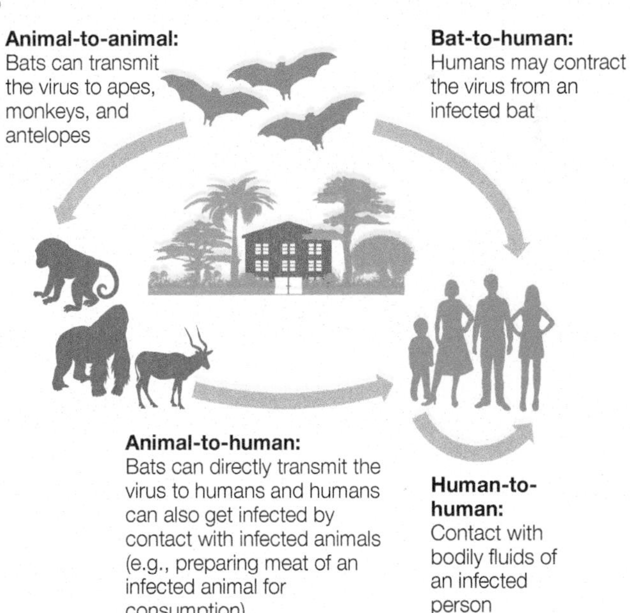

FIGURE 21.10 Ebola virus transmission Ebola cases can develop when Infected animals, such as fruit bats or nonhuman primates (e.g., apes, monkeys), and antelopes transmit Ebola viruses to humans. Human-to-human transmission is also observed.

Critical Thinking *What transmission route is most likely tied to an Ebola epidemic? Explain your reasoning.*

DiseaseSnapshot • Ebola Hemorrhagic Fever

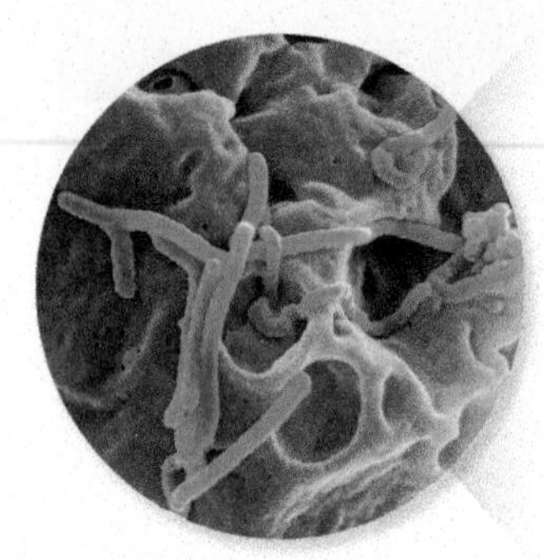

Ebola virus (red)

Causative agent	Ebola virus: enveloped, single-stranded RNA genome, *Filoviridae* family
Epidemiology	Mainly in Central and West Africa; incubation period of 2–21 days
Transmission & prevention	Avoid traveling to areas where outbreak is occurring; avoid contact with blood and bodily fluids of infected people; avoid areas inhabited by bats and where outbreaks have occurred previously; the vaccine Ervebo is FDA-approved to prevent Ebola in adults
Signs & symptoms	Flu-like symptoms progressing to rash and severe bleeding; death by shock
Pathogenesis & virulence factors	Ebola virus begins to infect macrophages and dendritic cells, then spreads to various other tissues, resulting in internal bleeding; the virus uses an envelope protein, GP, to dock on various cell surfaces
Diagnosis & treatment	Diagnosis involves detecting virus, viral components, or antiviral antibodies in a patient's blood; vaccine prevention is the best strategy; specific Ebola treatments have not been FDA-approved, although certain antibody preparations seemed to reduce morality rates in recent outbreaks

Pathophysiology and clinical course Ebola's exact infection mechanisms are still being investigated. We do know that, once the virus enters the host, it attacks macrophages and dendritic cells. An influx of cytokines then causes a massive inflammatory response. Ebola virus can infect almost every cell in the body; this means it impacts organs such as the spleen, kidneys, liver, and lungs. Because the virus targets antigen-presenting cells, the adaptive immune response is essentially stopped from fighting the infection. Antibodies are barely detected during Ebola infection.

Symptoms begin in a "dry phase" that is characterized by fever, intense weakness, sore throat, headache, and muscle pain. The "wet phase" is hallmarked by blood-tinged coughing, bloody diarrhea, and/or blood-laden vomit; a rash may also appear. Severe internal and external bleeding along with multiple organ failure and shock are the deadliest infection consequences. In 2019, Ervebo became the first vaccine approved by the U.S. Food and Drug Administration (FDA) for use in adults to prevent infections caused by the most prevalent circulating ebolavirus strain—Ebola Zaire. Approximately half of infected individuals who have not been vaccinated die, although recovery rates depend on whether supportive therapy is supplied in a timely manner, as well as the overall health of the individual prior to infection. Ebola survivors generally have lifelong immunity to the strain that infected them.

Recent outbreaks Ebolavirus has been recognized in equatorial Africa for more than 40 years; however, low population density, high mortality rates, and high awareness among health workers probably all helped limit initial outbreaks. In contrast, the epidemic from 2014-2016 in West Africa was the largest in history: 28,000 cases were recorded and 11,000 people died. This epidemic took place in highly populated areas where most healthcare workers were initially unfamiliar with Ebola. Cultural beliefs and traditional burial practices common to West Africa also played a part in the high transmission rates. Lack of adequate health care meant that many patients didn't receive the care necessary to improve their chance of survival.

The world's second largest Ebola outbreak was in 2018-2020 in the Democratic Republic of the Congo. During this outbreak approximately 3,481 people were infected, and 66 percent (2,299) died. In June of 2020, just as the 2018-2020 outbreak was declared over, a smaller wave of Ebola infections (130 infections with 55 deaths) emerged. Fortunately, it was extinguished by mid-November, 2020. Ervebo, the newly approved Ebola vaccine, was administered to over 40,000 people in the region and was essential to snuffing out the resurgence event. Until there is more widespread vaccination, sporadic Ebola cases are likely. As such, health agencies such as the WHO continue to monitor the region for Ebola, promote vaccination efforts, and treat sporadic Ebola infections as they arise.

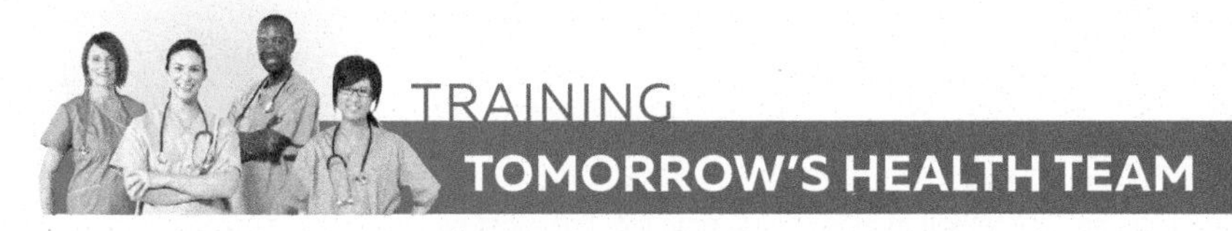

Handling Patients Infected with Hemorrhagic Fever

Ebola, Marburg, and Lassa viruses are classified as biosafety level 4 (BSL 4) agents. This means hemorrhagic fevers require the highest level of precautions among healthcare providers. Patients suspected or confirmed to have a hemorrhagic fever should be placed in isolation rooms. Since patients bleed as part of these diseases, and the blood harbors virus, the usual uniform of scrubs, gowns, masks, and thin gloves is inadequate protection. People who directly attend to patients should be trained in how to properly put on (don), wear, and remove (doff) personal protective equipment that includes a face mask, several layers of gloves, boots, and an airtight suit that pipes clean air from a compressed air canister or from outside the patient room. Special care must be taken when removing equipment, since it's easy to infect yourself, or contaminate areas outside the isolation room. Unfortunately, the specialized suits are bulky and hot to wear, which complicates patient care.

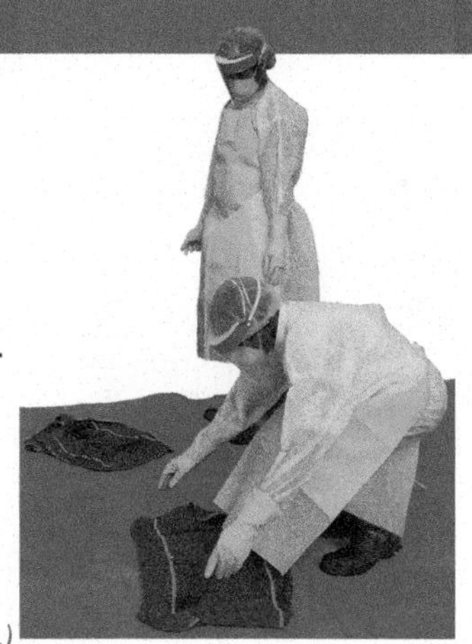

QUESTION 21.1

What features of Lassa, Ebola, and Marburg viruses make it appropriate for them to be classified as BSL level 4 agents rather than BSL 3 agents? (Hint: Refer to Chapter 10 to review BSL levels.)

Marburg

Marburg virus is similar to Ebola virus in its disease features and modes of transmission. However, there are fewer Marburg cases than Ebola—and so far, outbreaks seem to have a higher death rate (reaching up to 88 percent mortality in more recent outbreaks). Marburg cases occur sporadically, with the last one occurring in Uganda in 2014, with only one confirmed case. This infrequent occurrence means that many characteristics of this disease and virus are still unknown. So far, monkeys and fruit bats have been identified as hosts, but the natural reservoir remains unknown.

The virus enters the human population when people inhale bat droppings that have been stirred up in dust, or handle or eat infected bushmeat. Infected people can then transmit the virus to others through bodily fluids and sexual contact.

Symptoms begin as a sudden onset of headache, fever, and chills. Because these earlier stage symptoms are nondescript, the infection may initially be incorrectly diagnosed as malaria or typhoid fever. As the infection progresses, the patient develops nausea, vomiting, and abdominal pain, followed by severe bleeding from various body sites, jaundice, delirium, and shock within 5 to 7 days. Supportive therapy is the only treatment. There is not yet a vaccine for Marburg, but experimental vaccines are being studied.

Lassa

Lassa virus is a single-stranded RNA virus that causes **Lassa fever**. The virus is endemic to West Africa; most cases occur in Sierra Leone, Guinea, Liberia, and Nigeria. It was discovered in 1969 in Nigeria following the death of two missionary nurses in Lassa town, Borno State. Approximately 80 percent of those infected are asymptomatic or may have a slight fever, minor **malaise**, and weakness that is mistaken for a passing cold. This makes it hard to know the actual incidence of Lassa virus infections. The other 20 percent of cases show more severe symptoms, including bleeding of the gums, nose, or eyes, in addition to vomiting, facial swelling, and shock. Death can occur after two weeks of symptoms. Each year there are about 300,000 to 400,000 cases with approximately 5,000 deaths.[12]

Infected rats release the Lassa virus in their urine and feces. Thus, inhaling airborne dried rodent urine (often during sweeping), eating food contaminated with rodent droppings, or having rodent droppings enter a wound can all lead to

CLINICAL VOCABULARY

Malaise a general sense of illness or discomfort

[12] Okoro, O. A., Bamgboye, E., Dan-Nwafor, C., et al. (2020). Descriptive epidemiology of Lassa fever in Nigeria, 2012-2017. *The Pan African Medical Journal*, 37.

DiseaseSnapshot • Mononucleosis

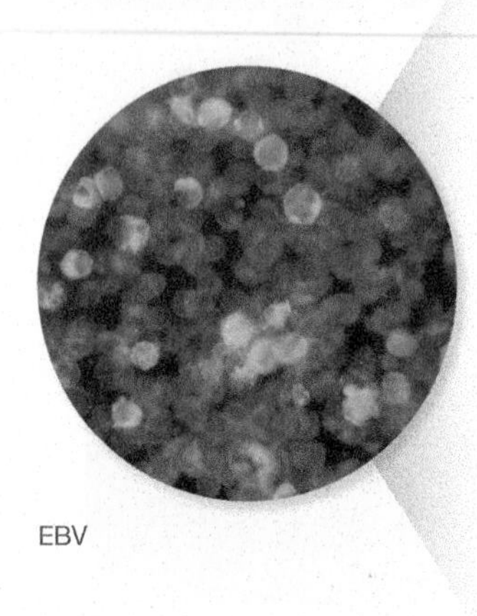

EBV

Causative agent	Mostly caused by Epstein-Barr virus (EBV): enveloped, double-stranded DNA genome, *Herpesviridae* family
Epidemiology	Almost everyone has an EBV infection at some point in life, but most cases are asymptomatic or mild; 25 percent of teens and young adults with EBV infection progress to mononucleosis; 4- to 6-week incubation period
Transmission & prevention	Mainly transmits via infected saliva, but contact with other body fluids can transmit EBV; no vaccine, so prevention revolves around not sharing utensils, cups, and toothbrushes and not kissing infected individuals who are shedding virus
Signs & symptoms	Fever, headache, extreme fatigue, severe sore throat, enlarged lymph nodes, splenomegaly, and possibly a rash
Pathogenesis & virulence factors	Virus enters tonsils at the back of the throat to infect nonactivated B cells; some infected B cells differentiate into memory cells and serve as a reserve for the latent virus; at any time the virus can become reactivated and new virions may be shed into saliva; only people who are immunocompromised tend to have symptoms during a reactivation
Complications	Endemic Burkitt's lymphoma is linked to EBV infection
Diagnosis & treatment	Diagnosis is based on symptoms and confirmed by an elevated lymphocyte count with abnormal monocyte-like cells present in the patient's blood and/or by detecting patient antibodies to the virus; treatment is rest, fluids, and over-the-counter pain relievers

infection. Initially dendritic cells are infected, which allows the virus to then spread to the lymphatic system. Hearing loss is a frequent complication—about one-third of symptomatic cases result in it. Also, women infected during pregnancy are 95 percent more likely to miscarry. When symptoms arise, administration of intravenous ribavirin within one week can decrease the risk of death, but it's not a cure.

Epstein-Barr virus is the most common cause of mononucleosis and can also lead to Burkitt's lymphoma.

Epstein-Barr virus (EBV) is a member of the *Herpesviridae* family and is one of the most widely transmitted infectious agents in the world. According to the CDC, at least 90 percent of the U.S. population experiences an EBV infection at some point, but most cases are mild or asymptomatic. EBV infection in childhood, for example, tends to cause only mild disease accompanied by a fever and sore throat.

CLINICAL VOCABULARY

Splenomegaly enlarged/swollen spleen

Mononucleosis In some cases, EBV infection leads to **mononucleosis** (or "mono"), a more severe form of infection. Mononucleosis induces extreme fatigue, swollen lymph nodes in the neck and armpit regions, severe sore throat, headache, splenomegaly (enlarged spleen), and occasionally a rash. Any age group may develop mononucleosis, but teens and young adults have the highest risk, with about 25 percent of infections progressing to mono.

Mononucleosis is transmitted through contact with infected bodily fluids—especially saliva, which is how it got the nickname, "the kissing disease." Sharing anything that contacts saliva, such as eating utensils or toothbrushes, or taking a sip of someone's drink, are common transmission modes. Once the virus enters the mouth, it invades the lymphatic system through the tonsils. The virus then infects nonactivated B cells. The virus tends to remain in B cells as they differentiate into memory B cells. This allows EBV to remain latent, and undergo occasional reactivation in a subset of B cells. When EBV reactivates, new viral particles are shed in the patient's saliva. However, unless the patient is immunocompromised, they are unlikely to have symptoms during a reactivation episode.

Mononucleosis is usually diagnosed based on symptoms and confirmed with blood work that measures leukocyte levels and detects antibodies to the virus. Infected patients tend to have a high lymphocyte count, and microscopic analysis reveals unusual lymphocytes that resemble monocytes—this is the "mono" referred to in the disease name (FIG. 21.11). The only treatment is supportive care.

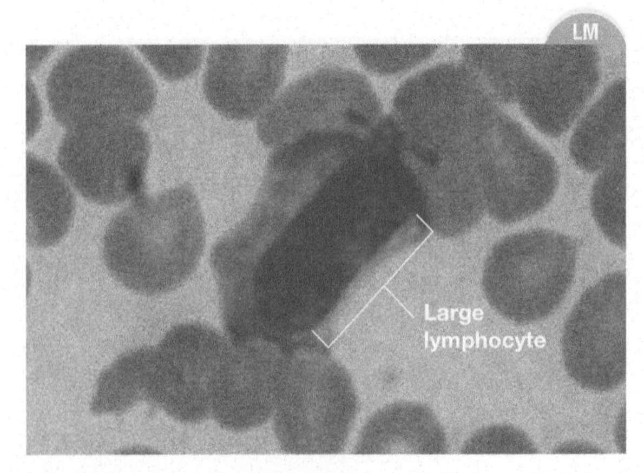

FIGURE 21.11 **Lymphocytes during mononucleosis** This blood sample shows a large unusual lymphocyte that is indicative of mononucleosis.

Critical Thinking *Why is a sore throat a typical symptom of mononucleosis?*

Burkitt's lymphoma The isolation of EBV in Burkitt's lymphoma—a type of lymphatic system cancer—was the first evidence that certain viruses can cause cancer in humans. Burkitt's lymphoma (BL) can be classified into three groups: sporadic, HIV associated, and endemic. Sporadic BL is rarely linked to EBV. In contrast, about 30 to 40 percent of HIV-associated BL is linked to EBV, whereas almost all endemic BL cases are linked to EBV.[13] Endemic BL mainly occurs in areas where malaria is endemic and coinfection with EBV is likely—such as countries of equatorial Africa. There is ongoing research to explore how malaria spurs an EBV infection to progress to Burkitt's lymphoma. Unlike sporadic BL disease, which is mainly in adults and usually features abdominal tumors, endemic BL is mainly seen in children and usually leads to tumors in the jaw or facial bones (FIG. 21.12); sometimes tumors may develop in the gastrointestinal tract, ovaries, and breast.

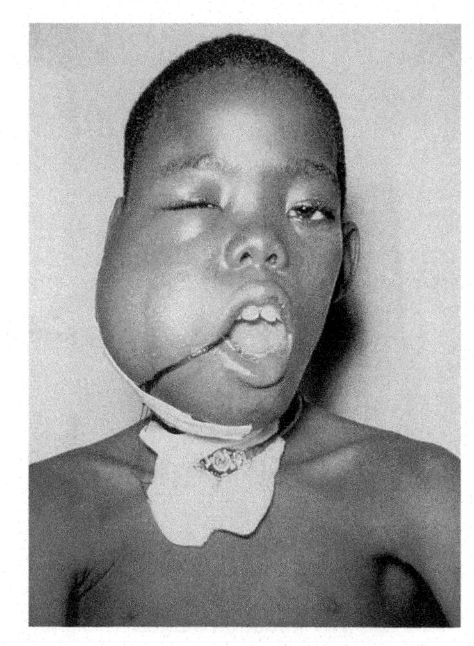

FIGURE 21.12 Endemic Burkitt's lymphoma

Chronic retroviral infections can lead to cancer and immunodeficiencies.

Retroviruses are unique from other viruses. As a part of their replication cycle, they use an enzyme called reverse transcriptase to convert their RNA genome into DNA. (See Figures 6.5 and 6.17 to review this concept.) Retroviruses permanently embed their genetic material in the host cell genome, forming a provirus (see Figure 6.17). Retroviruses are usually transmitted through sexual contact, with unprotected vaginal or anal sex accounting for the highest risk for transmission in men and women. However, any activity that leads to direct infected blood exposure (such as sharing needles for intravenous drug use) can cause infection. Retroviruses can also be vertically transmitted from mother to child through intrauterine infection or breast milk.

Retroviruses are notorious for infecting T helper cells, which are one type of $CD4^+$ immune system cell. Over months or years, retroviruses assault these immune system cells, eroding their numbers to the point that immunodeficiency develops, leaving the adaptive immune system ineffective at fighting pathogens or warding off cancer. Alternatively, retroviral infections can trigger lymphoma—a cancer of lymphocytes—where T cells grow normally. HIV and human lymphotropic viruses (HTLV) are retroviruses that infect humans.

Human Immunodeficiency Virus (HIV) Infection

Human immunodeficiency virus (HIV) originated in Africa and by the 1980s had emerged as a pandemic. The CDC reports that, in 2019 alone, 1.7 million new HIV cases were reported worldwide and about 38 million people were living with HIV. Approximately 25.4 million of those infected received medicines to treat HIV (up from 6.4 million receiving treatment in 2009).[14] According to the WHO, sub-Saharan Africa is the most affected region—about 25.7 million people there live with HIV infection.

CLINICAL VOCABULARY

Lymphoma cancer of cells associated with the lymphatic system; may result in consistently enlarged lymph nodes

HIV infection progression There are three broad stages in HIV infection—and patients can transmit the virus to others at any stage following initial infection. The first stage is **acute HIV infection/syndrome**. In this stage, the virus enters the body and circulates in the blood. It infects an array of white blood cells, but especially targets T helper cells. This infection stage is usually marked by flu-like symptoms that resolve within a few weeks. During this period there is enough virus in the blood that it can be detected using diagnostic tests. However, serological tests that look for patient antibodies to the virus may be negative. This is because the **seroconversion window** may not yet have passed—the period between infection and the point that the patient's immune system starts

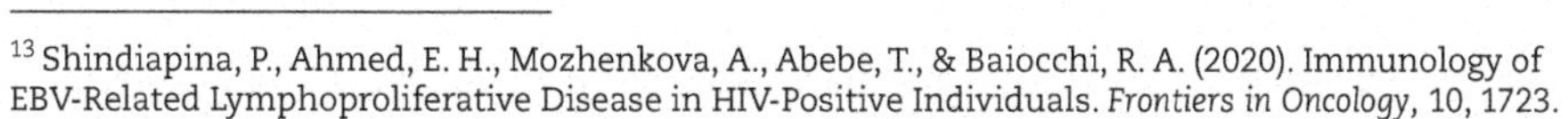

[13] Shindiapina, P., Ahmed, E. H., Mozhenkova, A., Abebe, T., & Baiocchi, R. A. (2020). Immunology of EBV-Related Lymphoproliferative Disease in HIV-Positive Individuals. *Frontiers in Oncology*, 10, 1723.

[14] Data from the UNAIDS Global HIV & AIDS statistics — 2020 fact sheet

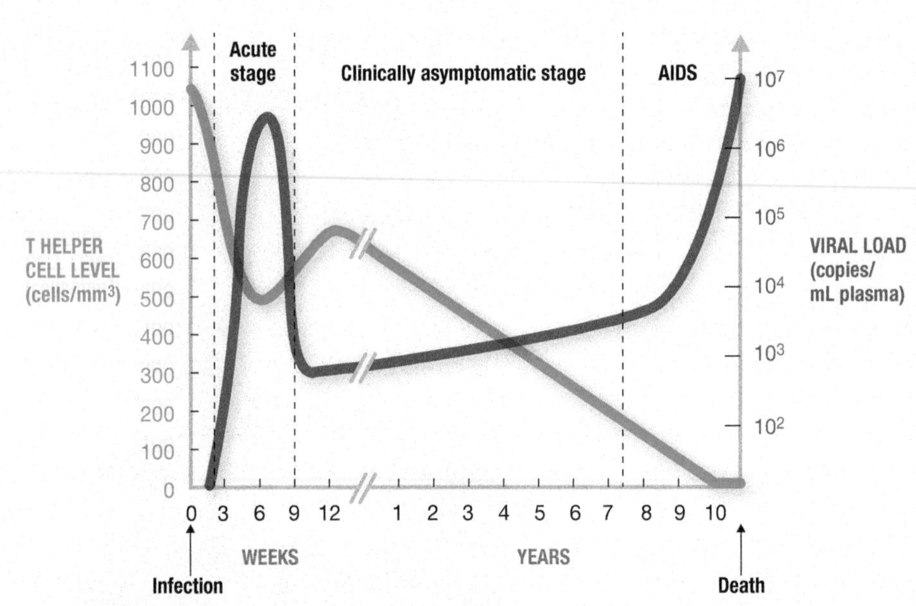

FIGURE 21.13 HIV infection stages This graph shows the relationship of T helper cell numbers (blue line) to HIV virus levels (teal line). In general, as HIV virus load increases, T helper cell levels decrease.

Critical Thinking *How might this graph look different if the patient was placed on an effective antiretroviral treatment?*

to make antibodies to the pathogen. However, most patients do not suspect HIV as the problem, so they don't seek testing.

Next, the **clinically asymptomatic stage** develops. As the name suggests, this stage is not marked by symptoms. It is sometimes called the dormancy stage, but research has made it clear that the virus is not dormant and instead is actively replicating to shed low levels of virions into the patient's blood. As the virus continues to slowly replicate, it infects new T helper cells. Each infected cell becomes a new factory that produces greater amounts of the virus. Depending on the individual, this clinical latency period can last from months to years. Eventually the third stage, **acquired immunodeficiency syndrome (AIDS)**, is reached (FIG. 21.13).

HIV-positive status versus AIDS Individuals in the first two stages of infection are generally referred to as "HIV positive." The normal range for T helper cells (a type of $CD4^+$ cell) is about 500 – 1,500 cells/mm^3 of blood. Patients with T helper cell levels below 200 cells/mm^3 typically progress to AIDS. In this stage, patients' immune systems are too ineffective to fight off infections and cancers. They tend to develop opportunistic infections and cancers such as Kaposi's sarcoma (FIG. 21.14) that are rare in immune competent patients. See TABLE 21.3 for a list of cancers and secondary infections commonly associated with AIDS.

CLINICAL VOCABULARY

Sarcoma cancer of bone cells or soft tissue cells (including blood vessels, muscles, tendons, joints, and fat)

Drug therapies When HIV infections first emerged, no treatments existed and the average time between diagnosis and death was only a year and a half. In the mid-1990s, the first **antiretroviral therapies (ART)** began to hit the market, improving patient outcomes and increasing life expectancy—often by many decades, as long as patients remain in treatment. Usually three or more drugs are prescribed in combination. This **antiretroviral regimen** usually includes two reverse transcriptase inhibitors. The third drug can be a protease inhibitor, integrase inhibitor, or another type of reverse transcriptase inhibitor. Because HIV

TABLE 21.3 Cancers and Secondary Infections Associated with AIDS

Disease	Pathogen	Disease Notes
Fungal pneumonia	*Candida* or *Histoplasmosis* species	AIDS patients with pneumonia have an increased risk of the infection becoming systemic and triggering sepsis
CMV retinitis	Cytomegalovirus (CMV)	Causes blindness
Genital/oral herpes	Herpes simplex virus 1 or 2	Immune system can no longer suppress these latent viruses so the patient has continuous outbreaks of mouth and/or genital ulcers
Tuberculosis (TB)	*Mycobacterium tuberculosis*	A healthy immune system tends to keep latent TB quiet; in AIDS the immune system cannot keep TB latent and it becomes an active and progressive infection; multidrug-resistant TB strains are emerging that are harder to combat
Kaposi's sarcoma	Human herpesvirus-8	Virus leads to cancer of cells that line lymph or blood vessels, resulting in blotchy skin lesions; called "AIDS defining cancer"
Recurrent *Salmonella* septicemia	*Salmonella* bacteria species	*Salmonella* bacteria are common in food, but usually at a level insufficient to cause infection in people with a healthy immune system; AIDS patients are easily infected by food and environmental *Salmonella* sources and are more likely to have a fatal infection
Mucormycosis	Fungal infection mainly caused by *Rhizopus* and *Mucor* species	Infection of the sinuses, lungs, intestine, and skin can move to the bloodstream

mutates easily, a patient's antiretroviral regimen must be adjusted regularly, with different drugs being rotated in or out of the therapy in response to the type of resistance the individual's HIV infection exhibits.

Antiretroviral drugs have saved millions of lives but are not without some drawbacks. As mentioned, people must take them for the rest of their lives. The high cost of antiretroviral therapy remains an issue for many patients. While access to antiretrovirals has improved in Africa, cost remains a constant challenge there. In addition, these drugs can interfere with other medications. Studies also indicate that people in long-term treatment for HIV are more likely to develop certain age-related conditions—cognitive impairment, osteoporosis, and kidney and liver disease, to name a few—decades before their healthy peers do. It's not fully understood if the HIV infection, the antiretrovirals, or some combination of both is what causes age-related issues with HIV infections.

Sometimes people who are HIV negative also benefit from antiretroviral drugs. For example, people who are in sexual relationships with HIV-positive partners can reduce their risk of contracting HIV by taking antiretroviral medications. This regimen is called **pre-exposure prophylaxis (PrEP)** and involves taking antiretroviral drugs daily. As of 2020, the fixed-dose combination of tenofovir disoproxil fumarate (TDF)/emtricitabine (FTC) is the only FDA-approved PrEP therapy. When properly taken, these drugs are 92 to 100 percent effective at preventing HIV infection.[15] PrEP should be used in combination with condoms, since prevention is not 100 percent in all cases and PrEP is not protective against other sexually transmitted infections. Another regimen, called **post-exposure prophylaxis (PEP)**, is used to prevent HIV in people who have an exposure through a sexual assault or an accidental HIV exposure—perhaps through a needle stick while caring for an HIV-positive patient. For optimal protection, PEP should ideally start within two hours of an exposure. This means if a healthcare worker has an accidental needle stick, they need to speak up right away. PEP normally consists of taking a 28-day course of three antiretroviral drugs from two different classes.

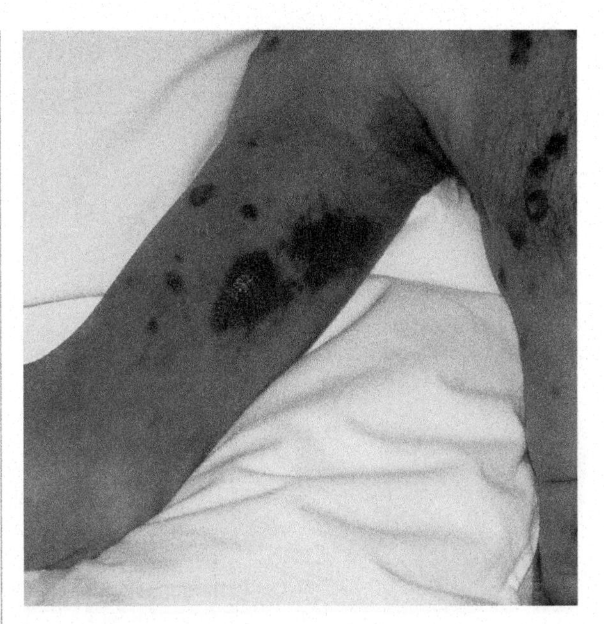

FIGURE 21.14 **Kaposi's sarcoma**

Vaccine development Despite decades of research and billions of dollars spent, an effective HIV vaccine remains elusive. One challenge is that the frequently mutating virus provides a difficult target. Another issue is that the virus infects some of the very immune cells that any vaccine is supposed to activate. As a result, traditional methods of vaccine formulation haven't worked as they would for infections that affect other cells. To date the most promising HIV vaccine trial is the RV144 study. This large clinical trial was conducted over three and a half years and enrolled about 16,000 adults. Researchers observed modest protective effects of the trial vaccine—there was a 31 percent decrease in HIV infection among vaccine recipients as compared to placebo groups.[16] The large-scale Mosaico trial results were originally expected by 2021, but because of the COVID-19 pandemic, the study has been delayed. Unfortunately the other large scale trial, Imbokodo, was terminated in August of 2021—data revealed that the trial vaccine, though well tolerated and safe, was only about 25 percent effective at preventing HIV infection.

Human T Lymphotropic Virus (HTLV) Infection

Human T lymphotropic viruses (HTLV) include HTLV-1, -2, -3, and -4. Here we will focus on HTLV-1 because it has the greatest clinical impact. HTLV-1 was the

---

[15] Tanner, M. R., Miele, P., Carter, W., Valentine, S. S., Dunville, R., Kapogiannis, B. G., & Smith, D. K. (2020). Preexposure Prophylaxis for Prevention of HIV Acquisition Among Adolescents: Clinical Considerations, 2020. *MMWR Recommendations and Reports*, 69(3), 1.

[16] The original clinical trial data are presented in this article: Rerks-Ngarm, S., Pitisuttithum, P., Nitayaphan, S., et al. (2009). Vaccination with ALVAC and AIDSVAX to prevent HIV-1 infection in Thailand. *New England Journal of Medicine*, 361(23), 2209-2220.

DiseaseSnapshot • HIV and HTLV

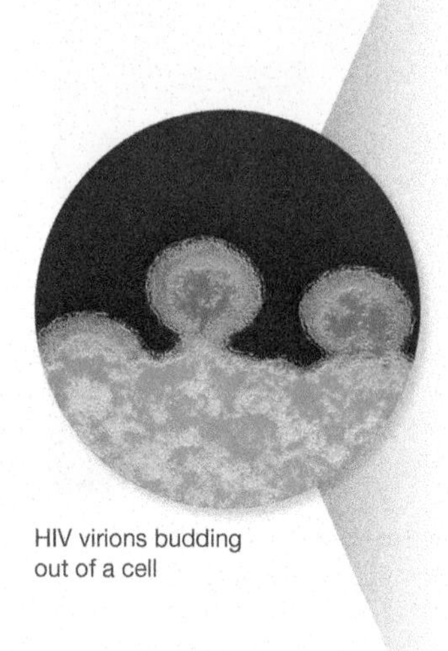

HIV virions budding out of a cell

Disease	Acquired Immunodeficiency Syndrome (AIDS)	Human T lymphotropic virus (HTLV)
Causative agent	Human immunodeficiency virus (HIV): enveloped, single-stranded RNA genome, *Retroviridae* family	HTLV-1, -2, -3, or -4: enveloped, single-stranded RNA genome, *Retroviridae* family
Epidemiology	Global, with highest incidence in eastern and southern Africa; 2- to 3-week incubation period	Endemic in Japan, sub-Saharan Africa, the Caribbean, and South America; 2- to 3-week incubation period
Transmission & prevention	Transmits through infected bodily fluids; mainly spreads through intravenous drug use, sexual contact, and vertical transmission; no vaccine, so prevention involves condom use and not sharing needles; post-exposure and pre-exposure drug prophylaxis therapies are also available	Transmits through infected bodily fluids; mainly spreads through intravenous drug use, sexual contact, and vertical transmission; no vaccine, so prevention involves condom use and not sharing needles
Signs & symptoms	Initial flu-like conditions followed with an asymptomatic period, ending with immunodeficiency, known as AIDS	Slight immunodeficiency is noted along with fever, fatigue, vomiting, and abdominal pain; may be asymptomatic
Pathogenesis & virulence factors	The retrovirus, after entering blood and/or lymph, infects lymphocytes, particularly T cells; upon integration into the host cell genome, the virus slowly produces more over a period of time (months to years); the reverse transcriptase gene can induce mutations to the virus that allow for avoidance of immune responses and antiviral medications	
Complications	As the virus kills T helper cells, immune function declines and patients develop diverse secondary/opportunistic infections and cancers	Can lead to T cell leukemia, lymphoma, and/or HTLV-1-associated myelopathy
Diagnosis & treatment	Molecular test can detect viral RNA or patient antibodies to the virus; combined antiretroviral medications must be taken for life	Molecular test can detect viral RNA or patient antibodies to the virus; chemotherapy treatment is used for HTLV-associated cancers

CLINICAL VOCABULARY

Leukemia cancer of the tissue that creates blood products (such as bone marrow)

first retrovirus ever discovered. Worldwide, about 15–20 million people are infected. HTLV-1 is endemic in Japan, sub-Saharan Africa, the Caribbean, and South America, but is fairly rare in the United States. However, its ability to pass from mother to child makes it more common in children born to immigrants from endemic areas. It is also considered endemic in certain Native American populations where up to 13 percent of people are infected. Infections may result in three outcomes:

1. Virus-infected T cells grow abnormally and result in adult T cell leukemia or lymphoma.
2. Infected T cells may produce abundant interferon, causing an inflammatory response to the central nervous system termed **HTLV-1-associated myelopathy (HAM)**.
3. Unchecked or opportunistic infections occur as a result of the decrease in functioning leukocytes.

HTLV is transmitted the same ways as HIV. After initial infection, viral load in blood remains low. Symptoms of fever, fatigue, vomiting, or abdominal pain are also minimal or nonexistent. In some cases, central nervous system inflammation triggers lower back pain or weakness/stiffness in the legs. Such nondescript early symptoms mean that infection often goes undiagnosed until the person develops cancer or donates blood (all blood is screened for HTLV, HIV, and other known bloodborne pathogens).

Blood tests that detect viral RNA or patient antibodies to the virus are used as definitive diagnostics. Currently, there is no treatment for HTLV. For patients afflicted with T cell leukemia, treatment is the same whether or not HTLV is detected.

Build Your Foundation

6. What are the clinical features of chikungunya, yellow fever, dengue, and Zika and how they are prevented? (LO 21.6)
7. What are the signs and symptoms of dengue fever, how is infection prevented, and why should the second stage be closely monitored? (LO 21.7)
8. Describe the three different transmission cycles of yellow fever. (LO 21.8)
9. What are the signs and symptoms of chikungunya virus infection and how can it be differentiated from a dengue virus infection? (LO 21.9)
10. What are the signs and symptoms of Zika virus and how can fetal complications be avoided? (LO 21.10)
11. What are the signs and symptoms associated with Ebola virus, Marburg virus, and Lassa virus infections and how can these infections be avoided? (LO 21.11)
12. What are the features of Epstein-Barr virus (EBV) infection, how is it transmitted, and what complications may it cause? (LO 21.12)
13. What are the features of each stage of HIV infection and what treatments exist to prevent and treat HIV infections? (LO 21.13)
14. What are the clinical and epidemiological features of human T lymphotropic virus (HTLV)? (LO 21.14)

Build Your Foundation (BYF) Quick Quiz: Visit the **Mastering Microbiology** Study Area to quiz yourself.

21.3 SYSTEMIC BACTERIAL INFECTIONS

Learning Outcomes

After reading this section, you should be able to:

21.15 Describe the causative agent and clinical features of plague.

21.16 Identify the risk factors and causes of endocarditis, and describe its possible complications.

21.17 Explain how tularemia spreads to humans, and why it's difficult to diagnose by symptoms.

21.18 Describe the clinical and epidemiological features of Lyme disease, Rocky Mountain spotted fever, ehrlichiosis, and anaplasmosis.

When bacteria directly invade the circulatory or lymphatic systems (or spread there from a previously localized infection), a systemic infection may develop. Sometimes catheters and intravenous lines allow bacteria this direct access to the circulatory and/or lymphatic systems. Flea or tick bites can also directly deposit bacteria into blood and lead to **bacteremia**—the presence of bacteria in circulating blood. No matter how they are initiated, systemic bacterial infections can result in significant inflammatory responses that may lead to sepsis and progress to septic shock. If bacteremia is suspected, a blood sample is collected in the hope of growing the causative bacterium so it can be identified and tested for antimicrobial sensitivity. (Chapter 7 reviewed how bacterial samples are collected and characterized; Chapter 15 covered antimicrobial sensitivity testing.)

Yersinia pestis continues to cause plague.

Say the word **plague**, and most people think of the Black Death, a Middle Ages pandemic notable for killing up to half of Europe's population. The Gram-negative bacterium *Yersinia pestis* causes bubonic, pneumonic, and septicemic plague. **Bubonic plague**, which affects the lymphatic system, is the most common form. It develops when a flea carrying *Y. pestis* bites a person. At the bite site, neutrophils and macrophages phagocytize the bacteria, but the pathogen has virulence factors that protect it from phagocyte-mediated death. Consequently, as the infected white blood cells make their normal journey through the lymphatic network, the bacteria have an opportunity to divide in lymph nodes. As the bacteria proliferate, buboes (swollen and painful lymph nodes) develop in the groin, neck, or armpit (axillary region). **Pneumonic plague** affects the lungs and has two development mechanisms; it can occur as bubonic plague spreads to the lung or it can be acquired via person-to person transmission. The person-to-person route is likely what led to widespread plague cases in the Middle Ages—as patients coughed up aerosols laden with the pathogen, others inhaled the pathogen and became infected. **Septicemic plague** develops when *Y. pestis* enters the bloodstream, which can occur as either bubonic plague or pneumonic plague advance.

CLINICAL VOCABULARY

Bacteremia the presence of bacteria in the blood

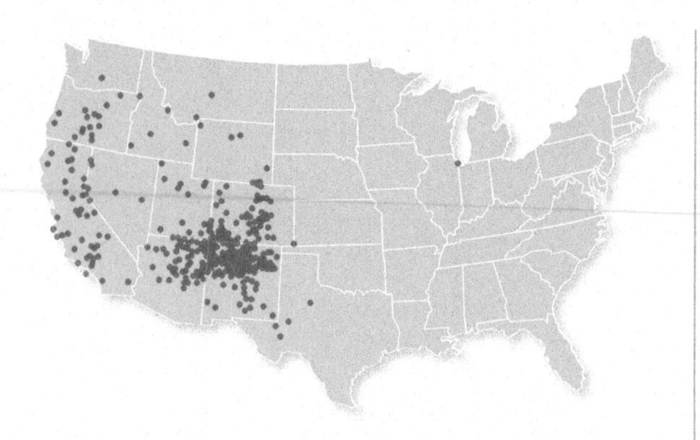

FIGURE 21.15 Plague incidence in the United States This map, based on data from the CDC, reflects plague cases in the United States from 1970 to 2018. The single case in Illinois was in a laboratory worker.

Critical Thinking *How could climate change theoretically alter the distribution of plague cases?*

CLINICAL VOCABULARY

Acral gangrene insufficient blood supply to the extremities that results in tissue death and blackening of skin

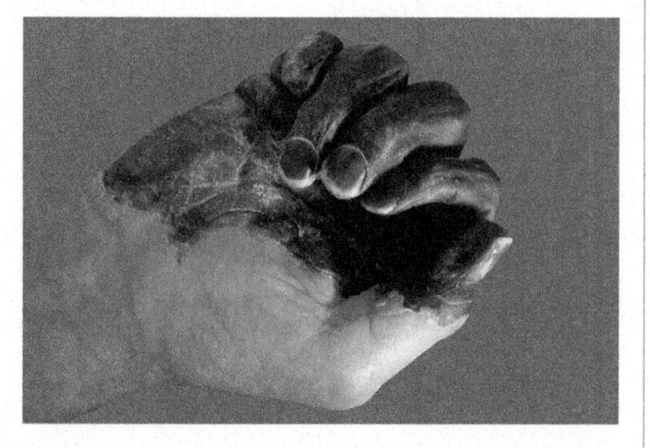

FIGURE 21.16 Acral gangrene in septicemic plague

Critical Thinking *What transmission precautions would healthcare providers need to observe when removing necrotic tissue such as that shown in the photograph? (Hint: Refer back to Chapter 10 to review transmission precautions.)*

CLINICAL VOCABULARY

Echocardiogram a sonogram of the heart; technique uses sound waves to create real-time images of the heart

Now curable with antibiotics, plague may seem like a disease that doesn't impact us at all in modern times. But according to the WHO, each year there are between 1,000 and 2,000 plague cases worldwide—including an average of about seven cases per year in the United States (FIG. 21.15). Because plague can develop anywhere that a flea can become infected by taking a blood meal from an infected animal (e.g., rats, mice, squirrels, prairie dogs, etc.), eradication is arguably impossible.

Signs, symptoms, and treatment Bubonic plague is characterized by flu-like signs and symptoms (e.g., fever, headache, chills, fatigue, weakness) and buboes. Although the name bubonic plague suggests that buboes are specific to this infectious disease, buboes develop in many infectious diseases and are not a specific diagnostic for any infection. Pneumonic plague has many of the same signs and symptoms of bubonic plague, with the addition of cough and possibly chest pain and shortness of breath. Without timely treatment, bubonic and pneumonic plague are likely to progress to septicemic plague. During septicemic plague, abdominal pain and septic shock may accompany the previously described signs and symptoms. As septic shock progresses, skin may turn black, particularly in fingers, toes, and the nose—which is why people living during the Middle Ages called it the "Black Death." This blackening is a result of tissue death (necrosis) and is called **acral gangrene** (FIG. 21.16). It develops because of disseminated intravascular coagulation. As clots lodge in capillaries, they cut off blood flow to tissues, which quickly die. Clot formation is triggered by a systemic inflammatory response that develops in response to bacteremia. Amputation is often necessary to remove dead, infected tissue. If left untreated, any form of plague can be fatal. Antibiotics are the most effective treatments for all three plague manifestations; gentamicin and ciprofloxacin are two that are typically prescribed.

Bacterial endocarditis can cause heart valve damage.

Bacterial endocarditis is an inflammation of the heart lining (the endocardium) and/or one or more of the heart's valves. People who have a normal, healthy heart rarely get endocarditis. Instead, it is seen mainly in people who have damaged, abnormal, or artificial heart valves or other heart defects. Typically, the left side of the heart, responsible for forcing oxygenated blood out of the heart and into the body for systemic circulation, is affected.

Any number of microorganisms, both bacterial and fungal, can be involved with endocarditis, but bacteria are the most common agents—particularly *Streptococcus* species and *Staphylococcus aureus*. The origin of the bacteremia leading to endocarditis may be infected gums, skin abscesses, urinary tract infections, or venous catheters. As they circulate, bacteria attach to the endocardium or valves, and begin to colonize the tissue. Fibrin and platelets will cover the colonized area and thus prevent white blood cells from accessing the infection (FIG. 21.17). Inflammation ensues as the colonization progresses.

Signs, symptoms, and treatment Signs and symptoms such as malaise, low-grade fever, chills, and weight loss can occur. Additional symptoms of heart murmur and abscesses in the heart tissue are likely. If the colonizing bacterial cells and/or fibrin/platelet matrix encasing the bacteria break off, several complications can occur. Petechiae, stroke, organ damage, and abscesses in other organs can follow as bacteria spread through the bloodstream. Finally, heart failure and death can result.

Since endocarditis can be deadly, there are preventive recommendations for certain patients. People who have artificial heart valves or certain heart defects and are undergoing oral or respiratory surgery should take antimicrobials before such procedures. This **prophylactic antibiotic therapy** will limit the risk of bacteria colonizing the heart if they enter the bloodstream. For those who have been diagnosed with endocarditis, ideally the infectious agent should be identified before prescribing antibiotics, but in practice this is rarely feasible.

DiseaseSnapshot • Bacterial Endocarditis

Causative agent	Many bacteria, with Gram-positive *Streptococci* and *Staphylococcus aureus* being the most prevalent; 1- to 2-week incubation period, but varies based on patient factors and causative agent
Epidemiology	People with heart defects or artificial heart valves are most susceptible, as are intravenous drug users who inject with dirty needles; subacute endocarditis can take months to develop, while acute endocarditis can take just days
Transmission & prevention	Not transmitted between humans; patients with certain heart defects or artificial valves should have prophylactic antimicrobial treatments before invasive procedures and even before certain noninvasive procedures like dental cleanings
Signs & symptoms	Subacute endocarditis presents with malaise, weight loss, low-grade fever, and chills, while severe symptoms include petechiae, stroke, and organ dysfunction
Pathogenesis & virulence factors	Bacterial cells enter blood through a variety of means and attach to heart endothelium or valves
Diagnosis & treatment	Blood culture to diagnose causative agent and echocardiogram to detect blockages; treatment depends on the infectious agent but is commonly vancomycin or teicoplanin with gentamicin

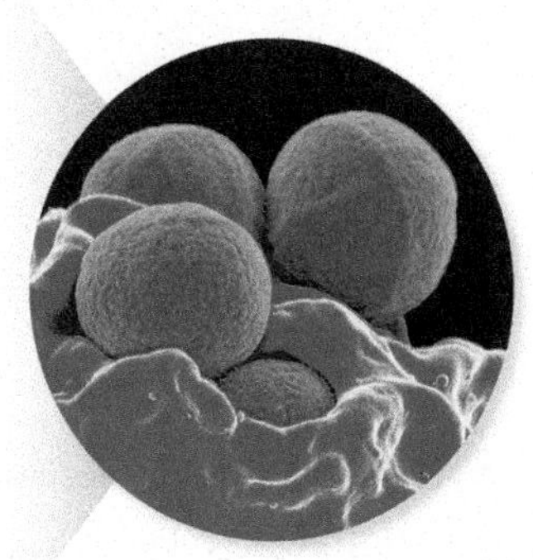

S. aureus (purple cells)

That said, every effort should be made to facilitate bacterial identification. To do this, blood should be drawn at three different time points over a minimum of one hour *before* antibiotics are given. To help the microbiology team differentiate bacteria present in the blood versus pathogens possibly on the skin, at each time point, blood should be taken from a different site on the patient's body. Furthermore, because microbes that cause endocarditis tend to thrive under lower oxygen levels, each collected sample should be placed in an anaerobic culture vial. Endocarditis is often treated with intravenous vancomycin or teicoplanin with gentamicin.

Highly infectious, tularemia is notable for being a potential bioterrorism agent.

Francisella tularensis is a Gram-negative coccobacillus bacterium that causes the disease **tularemia**. It is endemic in North America. According to the CDC, in recent decades incidence in the United States has tended to be less than 300 cases per year, making the disease extremely rare. If left untreated, tularemia has a mortality rate of 30 to 60 percent.

This zoonotic disease is typically spread by handling ill or dead animals infected with the bacterium; if the person has a cut or scratch, the bacteria can enter. Ticks and other insects can also become infected after feeding on ill animals, subsequently passing the bacteria to humans when they bite them. A variety of animals and birds can become infected; however, tularemia is usually associated with animals hunted by humans, such as wild rabbits. For this reason, the disease is sometimes called "rabbit fever."

Signs, symptoms, diagnosis, and treatment Tularemia infection can lead to headache, fever, chills, nausea and vomiting, and extreme weakness. Within one or two days, a small bump forms at the site of infection and expands into an ulcerated area (FIG. 21.18). Lymph nodes near the site of infection, such as those in the armpit or groin, also swell. If the bacterium is rubbed in the eye, then inflammation of the eye will occur with swelling of lymph glands near the ear. Ingesting contaminated food or water can result in infected tonsils, mouth ulcers, and a sore throat with swollen glands on the neck. Since early symptoms can vary based on how the bacteria are transmitted, the disease is often difficult to diagnose. Diagnosis is based on symptoms, lesions, and history of insect bites or contact with wild animals. The bacterium can be challenging and dangerous to grow in culture, so usually serological testing that detects patient antibodies to the bacterium or the bacterium itself confirms the diagnosis. Tularemia is treatable with antimicrobials such as streptomycin, gentamicin, doxycycline, and ciprofloxacin.

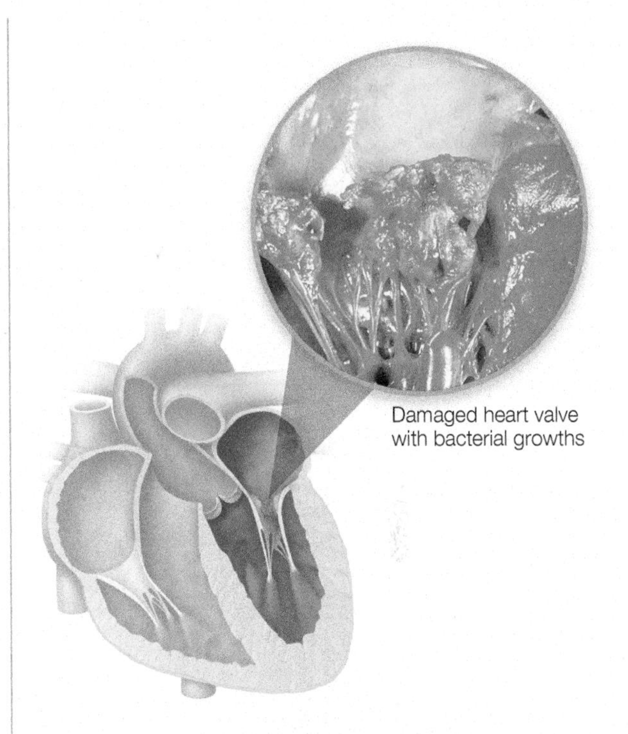

FIGURE 21.17 Bacterial endocarditis This photograph shows a heart valve that has been colonized by bacteria.

Critical Thinking *People with certain congenital heart defects or artificial valves are given antibiotics before they have a procedure that could introduce normal microbiota into the bloodstream—even before a noninvasive procedure like routine dental cleanings (scaling). It would be reasonable to assume that bacteria in the blood could be dangerous to anyone, so why aren't all patients put on prophylactic antibiotics for these procedures?*

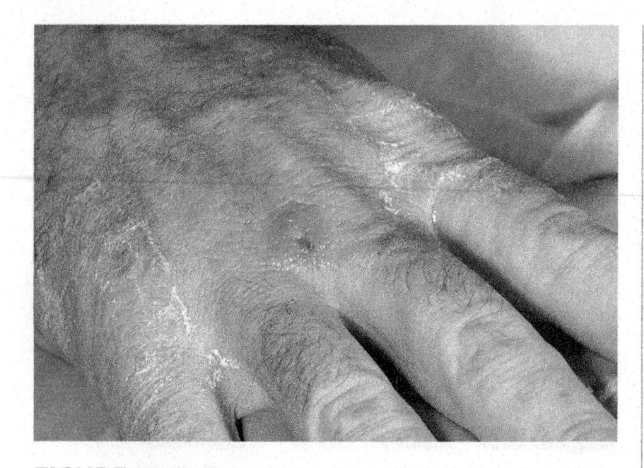

FIGURE 21.18 **Tularemia ulceration** A red ulcerated bump is one diagnostic sign for tularemia.

Critical Thinking *What features of this lesion help rule out a staphylococcal infection? (Hint: Refer to Chapter 17 to review skin lesions.)*

Bioterrorism threat *F. tularensis* is highly infectious—requiring only 10 cells to establish an infection—and it may cause pulmonary infections when inhaled. Although tularemia is not known to spread from person to person, the fact that it is so infectious, can be spread through the air, and is potentially deadly make it a biosafety level 3 agent. Unfortunately, these features also make it a candidate for weaponization. In fact, during the 20th century, *F. tularensis* was weaponized and stockpiled by the United States, Soviet Union, and Japan.[17] An international treaty ratified in the 1970s required nations to destroy all bioweapons and related research. However, the threat of tularemia being weaponized by terrorists remains.

Certain *Bartonella* species can cause systemic infections.

Bartonella species are known to cause infections such as trench fever (*Bartonella quintana* transmitted to humans by body lice), Carrión's disease (*Bartonella bacilliformis* transmitted to humans by infected sand flies), and **cat scratch disease (CSD)**. Of these infections, CSD is the most common and is therefore the focus of this section. The infection is named for the fact that cats can carry the Gram-negative bacterium *Bartonella henselae* and transmit it to people via a scratch. Most cats are infected either via bites from fleas carrying the bacterium or from the scratch of another infected cat. Notably, cats are usually asymptomatic carriers and, since about 30% of cats carry the bacterium, it's no wonder that CSD remains a common infection in humans. Kittens are especially notorious for spreading CSD; they love to play and, in their enthusiasm, they tend to inadvertently scratch their playmates (human and feline alike).

Signs, symptoms, diagnosis, and treatment Most cat scratches do not lead to CSD. People who do develop CSD will typically experience a fever; tender, enlarged lymph nodes; and a lesion at the scratch site within 1-3 weeks of sustaining a cat scratch (usually by a kitten). In the U.S., most cases are in children under age 15. There are no FDA-approved tests to detect this zoonotic infection; consequently, diagnosis is based entirely on clinical suspicion. This makes true disease incidence difficult to establish, although the CDC estimates that in the U.S. there are about 5 cases per 100,000 people under age 65. Fortunately, most CSD cases readily resolve without any treatment. In severe cases, in which the patient has excessively painful lymph nodes, or in the case of immune compromised patients, antibiotics such as azithromycin may be recommended.

Ticks are common vectors for transmitting diverse systemic bacterial infections to humans.

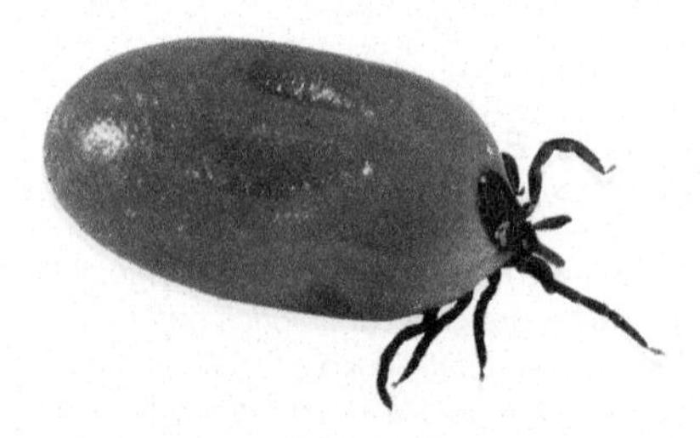

FIGURE 21.19 **Deer tick,** *Ixodes scapularis* Ticks take blood meals from infected animals and can serve as vectors that transmit bacteria to humans. Bacteria can remain in the tick for months.

Critical Thinking *What factors could lead to increased prevalence of Lyme disease?*

Diverse arthropod vectors transmit disease, with ticks being important vectors of bacterial infections (FIG. 21.19). In most cases a tick needs to stay on the host for 24 to 48 hours to regurgitate enough bacteria into the skin to cause infection, so checking your body for ticks after walking through fields and woods can limit disease transmission. Here we'll review Lyme disease, Rocky Mountain spotted fever, ehrlichiosis, and anaplasmosis.

Lyme Disease

Lyme disease is named after a Connecticut town where a large outbreak occurred in the 1970s. Caused by the Gram-negative spirochete *Borrelia burgdorferi*, the reservoir is the white-footed mouse, while the vector is deer

[17] Geissler, E. (2005). Alibek, tularaemia and the Battle of Stalingrad. *CBW Conventions Bulletin*, 69 (70), 10–15.

DiseaseSnapshot • Lyme Disease and Rocky Mountain Spotted Fever

Disease	Lyme Disease	Rocky Mountain Spotted Fever
Causative agent	*Borrelia burgdorferi*: Gram-negative spirochete bacterium	*Rickettsia rickettsii*: intracellular Gram-negative bacterium
Epidemiology	Endemic to the U.S. Northeast and Mid-Atlantic regions; also found in Europe, Russia, Asia, and Japan; 3- to 30-day incubation period	Endemic throughout the U.S., with 60 percent of cases developing in Arkansas, Missouri, North Carolina, Oklahoma, and Tennessee; 2- to 4-day incubation period
Transmission & prevention	Tick bite (usually deer ticks) transmits it; prevented by limiting tick exposure by covering arms and legs while outside, and wearing insect repellant	
Signs & symptoms	Characteristic bull's-eye rash appears in 75 percent of patients 3 to 30 days after tick bite, along with general malaise, headache, dizziness, fatigue, fever, chills, and joint and muscle pain	Most patients develop a non-itchy macule rash; a petechial rash develops in 35 percent of cases; other symptoms are fever, headache, nausea, vomiting, abdominal pain, muscle pain, reduced appetite, and conjunctivitis
Pathogenesis & virulence factors	Bacteria move from bite to lymphatic system; avoid immune detection by antigenic variation of outer membrane; don't need iron for metabolism (use manganese instead), so the body's iron-binding proteins don't impede growth; once in joints, can persist for years	Bacteria escape killing by host's phagocytes; damage blood vessels, which leads to tissue damage
Diagnosis & treatment	Diagnosed based on symptoms and serological testing for antibodies against the bacterium; several weeks of the antibiotic doxycycline is the usual early-stage treatment; other antibiotics may be used for late-stage treatments	Diagnosed based on symptoms and history of tick bites; confirmed by testing for antibodies to the bacterium; one week of doxycycline is a common treatment

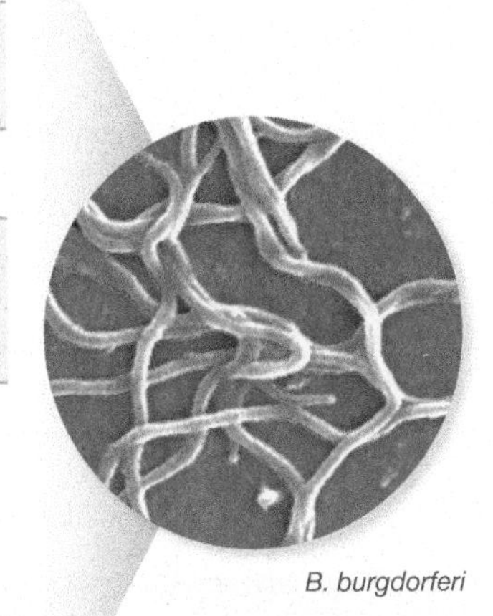

B. burgdorferi

ticks. Most U.S. states now report cases; however, case incidence is usually greatest in northeastern states. Though a definitive Lyme disease diagnosis is reported for only about 30,000 U.S. patients per year, the CDC estimates that about 470,000 Americans are infected each year.

The infection slowly spreads out from the bite via the lymphatic system. About 7 to 14 days after infection, a bull's-eye-shaped rash, termed **erythema migrans**, often develops around the tick bite area (FIG. 21.20). Fever, fatigue, and headache can accompany the rash. Diagnosing Lyme disease can be tricky because the characteristic rash is absent in about 25 percent of cases. The earlier treatment is started, the shorter the doxycycline drug course needed for recovery. If untreated, the bacteria can enter a latent stage where 60 percent of those infected exhibit *Lyme arthritis*—chronic arthritis due to a persistent inflammatory response. Other complications can include irregular heartbeat, myocarditis, and neurological problems such as facial palsy, balance issues, mood disturbances, and occasionally a schizophrenia-like illness.

Rocky Mountain Spotted Fever

Despite its name, tickborne **Rocky Mountain spotted fever (RMSF)** occurs in all parts of the United States. Like Lyme disease, it was named for the region where it was first detected and features a rash (FIG. 21.21). The extremely small Gram-negative bacteria *Rickettsia rickettsii* causes the disease. Following infection, these bacteria live intracellularly, particularly in endothelial cells that make up blood vessels. RMFS cases resemble other rickettsial spotted fevers (e.g., *Rickettsia parkeri* rickettsiosis, Pacific Coast tick fever, and rickettsialpox). Because commonly available serologic tests don't effectively differentiate among the various rickettsial diseases, RMSF is simply reported under the generalized category of Spotted Fever Rickettsiosis (SFR). According to the CDC, about 4,000-6,000 SFR cases are reported each year in the United States.

The initial RMSF symptom is fever. About 90 percent of those infected develop a rash 2 to 5 days after fever onset. The rash features numerous small, flat, pink/red, non-itchy spots that spread from the wrists, arms, and ankles inward to the torso as the disease progresses. Other symptoms include

CLINICAL VOCABULARY

Erythema migrans a slowly expanding rash surrounding a vector bite

Palsy paralysis, inability to move; often accompanied by uncontrolled shaking (tremors)

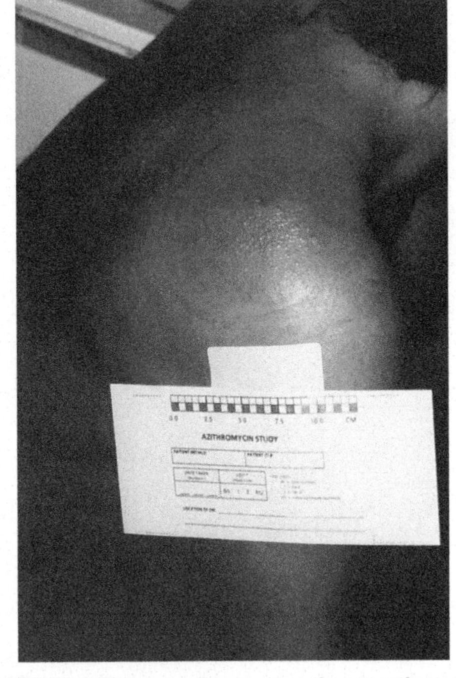

FIGURE 21.20 Lyme disease erythema migrans rash An outward migrating "bull's-eye" rash is characteristic of Lyme disease.

Critical Thinking *What does the rash pattern reflect about concentrations of the bacterium within the skin?*

TRAINING TOMORROW'S HEALTH TEAM

Lyme Disease and Rocky Mountain Spotted Fever on the Rise

Lyme disease and Rocky Mountain spotted fever cases continue to rise each year, according to CDC tracking statistics. RMSF is now spreading into Mexico. And a new species of *Borrelia* (*Borrelia mayonii*) has been identified. *B. mayonii* infections exhibit similar symptoms to classic Lyme disease. Compared to *Borrelia burgdorferi*, this new species can be found in unusually high numbers in the blood of an infected individual, may cause nausea/vomiting, and causes a more diffuse rash than traditional Lyme disease.

Thus far, *B. mayonii* has been identified only in the upper midwestern United States. Much is still to be discovered about whether or not this newly identified species will spread or remain an uncommon source.

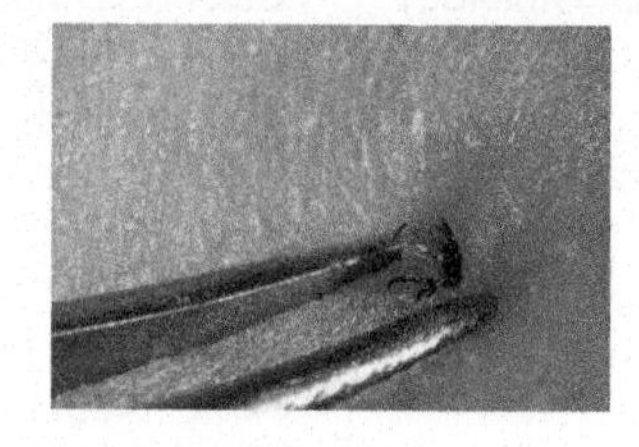

QUESTION 21.2

What are some pathogen factors that might lead B. mayonii *to emerge as a new, major source of Lyme disease?*

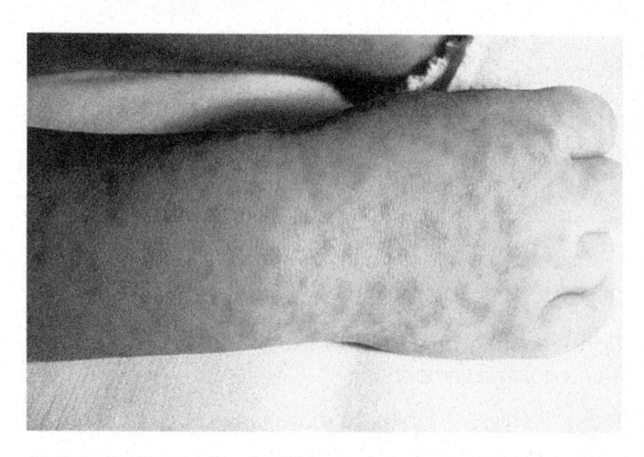

FIGURE 21.21 **Rocky Mountain spotted fever rash**

headache, vomiting, and muscle pain. Symptoms can vary from patient to patient, however, making diagnosis difficult.

Diagnosis and administration of the antibiotic doxycycline within the first five days of symptoms are critical because the infection can progress rapidly. If left untreated, RMSF may cause blood vessel damage that reduces circulation to the arms and legs—severe cases may require limb amputation. Vessel damage in 5 percent of cases can also cause fatal organ or brain bleeding. Untreated cases can be fatal in as few as eight days.

Ehrlichiosis and Anaplasmosis

Ehrlichiosis and **anaplasmosis** are systemic bacterial infections whose incidence is increasing. According to the CDC, both ehrlichiosis and anaplasmosis had fewer than 200 U.S. cases per year in the 1990s. However, the number of cases for each of these infections has steadily increased each year—by 2018, there were nearly 1,800 ehrlichiosis cases whereas anaplasmosis cases spiked to nearly 4,000 cases that year.

The lone star tick can carry *Ehrlichia chaffeensis* as well as other *Ehrlichia* species responsible for ehrlichiosis. Most cases are reported in the southeastern and south-central United States. The black-legged tick, which transmits *Anaplasma phagocytophilum*, is responsible for anaplasmosis. These ticks are more common in the northeastern and upper midwestern United States.

Ehrlichiosis and anaplasmosis are rarely fatal. Fever, chills, headache, and malaise develop soon after infection and a rash similar to that seen in RMSF may be present, especially in children. Unfortunately, the ticks that carry anaplasmosis may also harbor the bacteria that cause Lyme disease and RMSF. Since coinfection can occasionally occur, blood testing for more than one infectious agent is recommended if disease is suspected.

The Case of the Bloated Belly

Practice applying what you know clinically: visit the **Mastering Microbiology** Study Area to watch Part 3 and practice for future exams.

Build Your Foundation

15. What causes plague and what are the clinical features of plague? (LO 21.15)
16. What are the risk factors, causes, and possible complications of endocarditis? (LO 21.16)
17. What type of activities would expose you to tularemia, and why is it sometimes difficult to diagnose? (LO 21.17)
18. Compare and contrast the causative agents and major signs and symptoms of Lyme disease and Rocky Mountain spotted fever. (LO 21.18)
19. Which bacteria are responsible for ehrlichiosis and anaplasmosis, and how are the diseases similar? (LO 21.18)

BYF QUICK QUIZ

Build Your Foundation (BYF) Quick Quiz: Visit the **Mastering Microbiology** Study Area to quiz yourself.

21.4 SYSTEMIC FUNGAL INFECTIONS

Candidemia (invasive candidiasis) is the most common systemic fungal infection.

Learning Outcomes

After reading this section, you should be able to:

21.19 Identify the clinical features of candidiasis and discuss how it is contracted.

Fungi are typically not pathogenic; however, opportunistic fungi will take advantage of immunocompromised individuals who have underlying conditions such as cancer, diabetes mellitus, or AIDS, or who are taking medication that suppresses the immune system. That said, a few fungal species can cause primary infections in anyone, including immune-competent individuals. Here we focus on the main systemic fungal threat—*Candida*. (See Chapter 16 for coverage on opportunistic respiratory fungal infections that may become systemic, especially in AIDS patients.)

Candida species are yeasts. At least 20 different *Candida* species are pathogenic in humans, but the most common culprit is a member of our own normal microbiota—*Candida albicans*. It's abundant on skin and commonly found in the intestinal tract and vagina. The body has numerous mechanisms that help keep *Candida* numbers in check. However, antibiotic usage can disrupt the microbial balance, allowing *Candida* to proliferate. Usually this sort of microbiota disruption results in a local infection (see Chapter 20 for more on vaginal yeast infections). However, *C. albicans* infections can also become systemic.

Candidemia (or **invasive candidiasis**) develops when *Candida* invades the bloodstream. It is the most common systemic fungal infection in immunocompromised individuals and the fourth most common healthcare-acquired bloodstream infection in the United States. Surgery, central venous catheters, kidney dialysis, and intravenous lines are examples of interventions that increase the risk for candidemia. FIG. 21.22 shows an infected catheter site that indicates potential candidemia. Treatment with a broad-spectrum antibiotic that dramatically alters bacterial normal flora also increases susceptibility to invasive candidiasis.

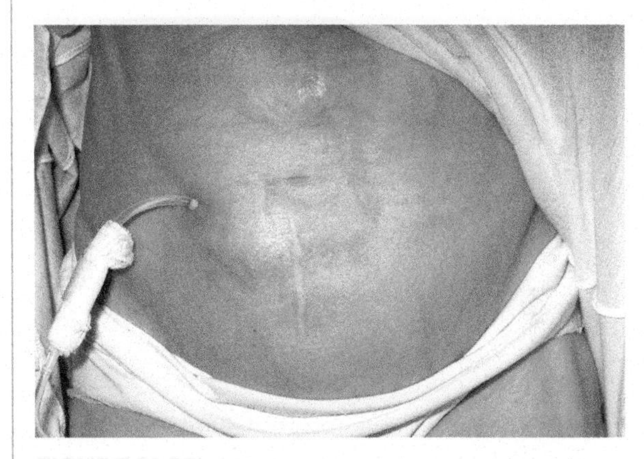

FIGURE 21.22 **Infected catheter exit site** The red area surrounding the catheter exit site indicates an infection.

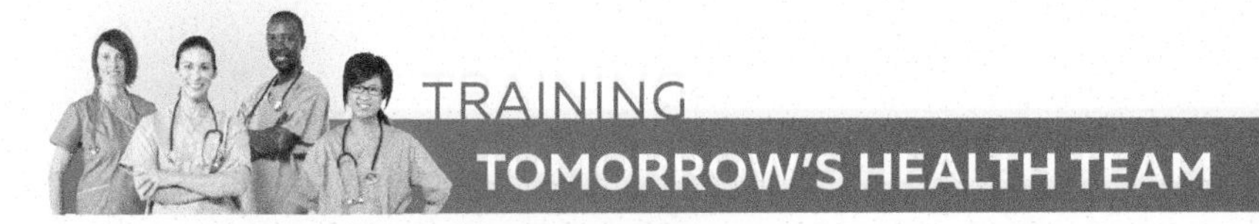

TRAINING TOMORROW'S HEALTH TEAM

Multidrug-Resistant Candidemia

A new *Candida* species has been identified as a source of hospital infections. *Candida auris* infection was first documented in Japan in 2009. Since then, numerous other countries have reported the invasive fungal infection. Between 2013 and 2016, seven cases in the United States were reported to the CDC; 2016 to 2017 saw 122 cases being reported in the United States, indicating a serious and rapidly emerging threat. The CDC reports that isolates within each region where *C. auris* is found are quite similar to one another, but differences between subtypes are found from region to region. This genetic profile suggests that *C. auris* is emerging spontaneously in multiple regions at roughly the same time, rather than spreading around the globe from a single point of origin.

C. auris isolates are often resistant to multiple drugs, including fluconazole. There have even been reports of the fungus resisting all three drugs that are used to treat candidemia—so far, such isolates have been rare. The agent is also easily misidentified, leading to incorrect treatments and prolonged infection. Another disturbing feature is that, unlike other yeast species, this pathogen can linger on surfaces such as beds, countertops, and infusion pumps and serve as a source of infection. The population at risk for *C. auris* is the same as that for *C. albicans*: the immunocompromised, as well as anyone hospitalized for long periods and whose treatment includes indwelling devices such as central venous catheters. The best way to accurately diagnose *C. auris* is through molecular methods rather than culturing.

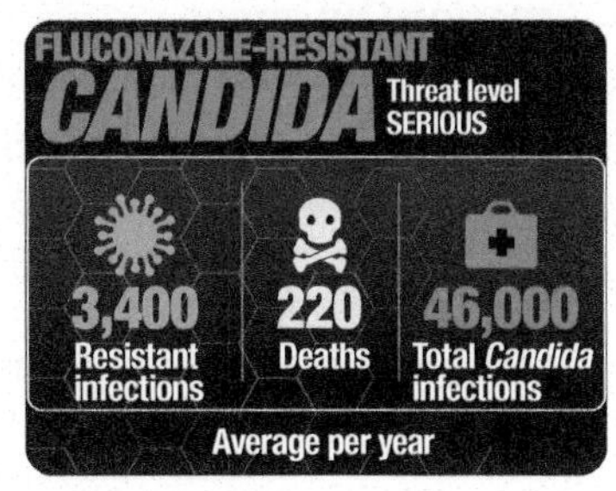

QUESTION 21.3

Since Candida auris *spontaneously arose in multiple areas of the globe, rather than spreading from a single geographic region, what disease factors are most likely at play with its emergence?*

DiseaseSnapshot • Invasive Candidiasis (or Candidemia)

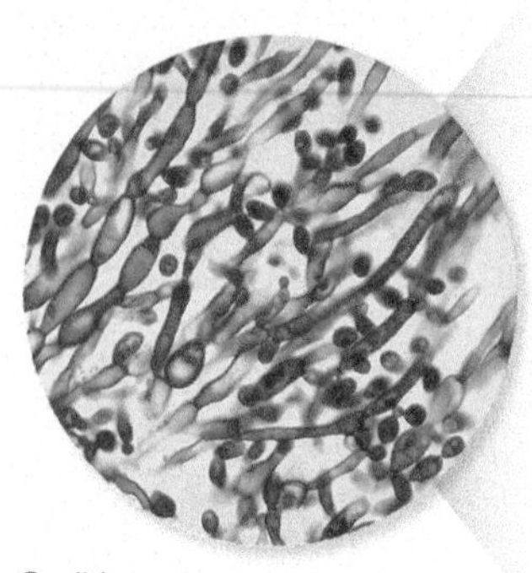

C. albicans

Causative Agent	*Candidia albicans*
Epidemiology	Mostly occurs in patients who are immunocompromised by another health condition or have extended hospital stays; incubation is days to weeks, depending on overall patient health status
Transmission & prevention	Exposure tends to occur in surgery, by direct contact with contaminated objects/hands, or via indwelling devices such as venous catheters; high-risk groups may be given prophylactic antifungal medication
Signs & symptoms	Fever and chills and possible progression to sepsis and death by septic shock
Pathogenesis & virulence factors	The yeast usually enters blood via a catheter or some other invasive procedure or device; it can exhibit antigen switching, whereby it changes the antigens it displays; it also evades immune responses by alternating between unicellular and filamentous forms
Diagnosis & treatment	Diagnosed via blood culture; treated with echinocandin drugs administered intravenously for at least 2 weeks after negative blood cultures; second-line drugs include fluconazole and liposomal amphotericin B; some strains are now resistant to first- and second-line drugs

Invasive candidiasis symptoms can range from a fever to sepsis, and even when treated, the mortality rate is approximately 25-40 percent.[18] Individuals more susceptible to fungal disease, such as AIDS patients, may be placed on antifungal medication as a preventive measure. For those in extended hospital stays, being aware of how long a catheter has been in place and alerting medical personnel if redness or pain occurs near the exit site, can help prevent the infection from becoming septic. Echinocandin is the first-line treatment; if the yeast is resistant to it, then fluconazole or liposomal amphotericin B may be used. To monitor how therapy is progressing, blood should be sampled and cultured at least every other day until growth is not detected. Provided there are no other complications, it is typically recommended that treatment continue for two weeks after the first negative blood culture.

Build Your Foundation (BYF) Quick Quiz: Visit the **Mastering Microbiology** Study Area to quiz yourself.

Build Your Foundation

20. What are candidemia's risk factors and clinical features? (LO 21.19)

21.5 SYSTEMIC PROTOZOAN INFECTIONS: MALARIA

Learning Outcomes

After reading this section, you should be able to:

21.20 Describe malaria's clinical and epidemiological features.

21.21 Differentiate between uncomplicated and complicated malaria.

21.22 Describe the main stages of *Plasmodium* development.

Malaria was described in ancient Chinese, Greek, and Sanskrit writings as long ago as 2700 B.C., but it wasn't until the late 1800s that the *Plasmodium* parasite was identified as the causative agent, and mosquitoes (*Anopheles* species) were identified as the vector. Although five *Plasmodium* species cause malaria, 95 percent of cases are caused by either *P. falciparum* or *P. vivax*, with *P. falciparum* causing the most deaths.[19]

By 1951, draining swamps, spraying to reduce mosquito populations, and educating the public to reduce standing water sources that serve as mosquito breeding sites helped eliminate malaria from the southern United States. Today, almost all malaria cases in the United States are imported. However, malaria endures as a major health problem in certain tropical and subtropical regions. Just over 90 percent of reported cases occur in Africa, with Southeast Asia and

[18] Tsay, S. V., Mu, Y., Williams, et al. (2020). Burden of Candidemia in the United States, 2017. *Clinical Infectious Diseases*.

[19] *P. falciparum*, *P. vivax*, *P. ovale*, *P. malariae*, and *P. knowlesi* all cause malaria. *P. falciparum* is mainly in Africa; *P. vivax* dominates in Asia and South America; *P. ovale* is primarily in West Africa; *P. malariae* is rare and is mainly isolated to Africa; *P. knowlesi* is very rare and is mainly in parts of Southeast Asia.

the eastern Mediterranean experiencing smaller yet significant outbreaks. According to the WHO, in 2019 alone there were approximately 229 million malaria cases worldwide, with about 409,000 deaths.

There are two forms of malaria.

Uncomplicated malaria is the milder version of the disease. It presents with three stages: a cold stage hallmarked by shivering and a sensation of cold; a hot stage with fever; and a sweating stage. Other symptoms that may develop in these stages include headaches, body aches, fatigue, nausea, and vomiting. Each stage is linked to a step in the protozoan's development (see more below). These malarial stages follow each other as a progression and altogether constitute an episode that lasts 6–10 hours. The 6- to 10-hour attack episode then repeats every two to three days.

The more severe form of disease, called **complicated malaria**, includes the aforementioned episodic attacks, but may be accompanied by anemia, low blood pressure (**hypotension**), low blood glucose (**hypoglycemia**), and/or excessive acidity of the blood (**acidosis**). Kidney failure, acute respiratory distress syndrome, and malarial infection of the brain are also possible in severe cases. Complicated malaria is more common in young or elderly patients, as well as those who are pregnant or immunocompromised. Improved treatment in endemic areas has reduced incidence, and related deaths have fallen by as much as 45 percent.

CLINICAL VOCABULARY

Hypotension abnormally low blood pressure that can lead to dizziness and fainting; typically a systolic pressure under 90 and a diastolic pressure under 60 (noted as 90/60)

Hypoglycemia low blood glucose level that may cause the patient to feel light headed and shaky

Acidosis a drop in arterial blood pH to 7.35 or lower; this condition can develop from respiratory or metabolic issues

The recurring attack episodes of malaria reflect the *Plasmodium* life cycle.

Plasmodium depends on both the *Anopheles* mosquito and humans to complete its complicated life cycle (FIG. 21.23). To summarize, the parasite passes as a *sporozoite* form from infected mosquitoes into the patient's bloodstream. Sporozoites then travel to the liver, where they reproduce. In the liver, the newly formed parasites mature into *merozoites*. They then lyse out of infected liver cells and reenter the bloodstream. Merozoites then infect red blood cells, where they further replicate and eventually burst out of the blood cell. The 6- to 10-hour attack episodes of malaria correspond to the merozoites bursting out of red blood cells. Some of the merozoites released from bursting red blood cells go on to infect other red blood cells, which leads to recurring attack episodes.

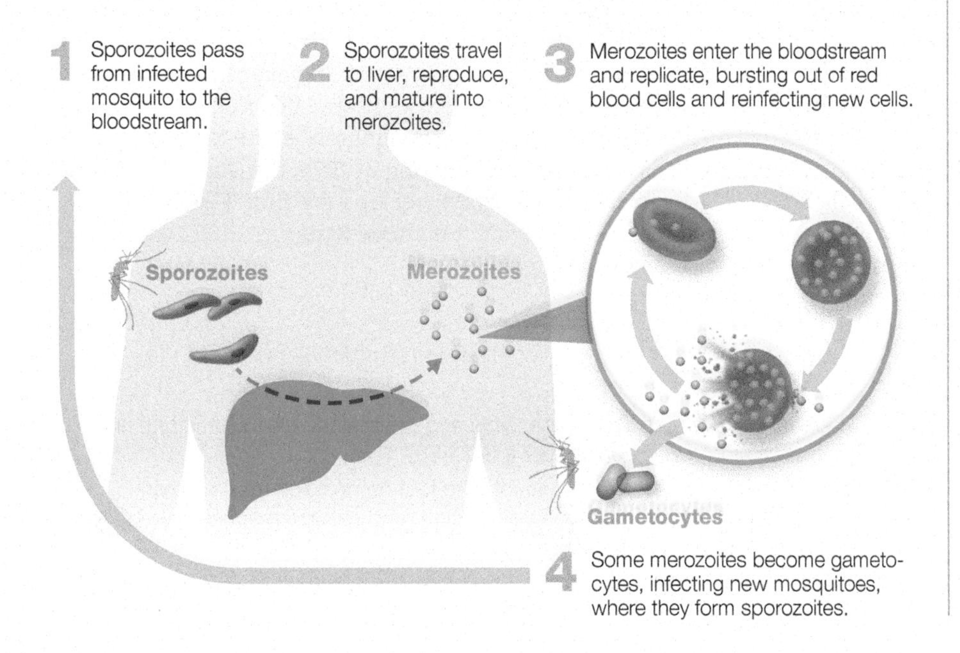

FIGURE 21.23 *Plasmodium* **life cycle in humans** *Plasmodium* species rely on humans and mosquitoes to complete their life cycle. The parasite's asexual replication phase is what leads to the 6- to 10-hour-long episodic attacks that characterize malaria.

Critical Thinking *Would you classify the mosquito as a biological or mechanical vector in this transmission cycle? Explain your reasoning.*

Disease**Snapshot** • Malaria

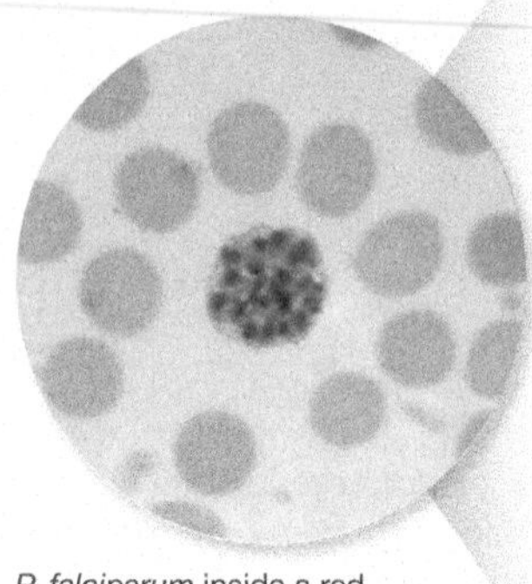

P. falciparum inside a red blood cell

Causative agent	Most cases are caused by *Plasmodium falciparum* or *P. vivax*
Epidemiology	Endemic in Africa, Southeast Asia, and the eastern Mediterranean area; typically 9- to 14-day incubation for *P. falciparum* and 12- to 18-day incubation for *P. vivax*
Transmission & prevention	Transmitted by bites from infected *Anopheles* mosquitoes; prevented through mosquito abatement (pesticides, repellents, and bed nets); oral prophylaxis such as doxycycline can be taken daily when traveling in malaria-endemic areas; Mosquirix™ vaccine is recommended to protect children against *P. falciparum*
Signs & symptoms	Episodic attacks of shivering, fever, and sweating last 6 to 10 hours and may be accompanied by headaches, body aches, fatigue, nausea, and vomiting; in complicated cases anemia, low blood pressure (hypotension), low blood glucose (hypoglycemia), and/or excessive acidity of the blood (acidosis), kidney failure, acute respiratory distress syndrome, and brain infections may develop
Pathogenesis & virulence factors	The parasite's asexual merozoite stage is responsible for malaria's characteristic recurrent episodic attacks; during the blood cell infection *Plasmodium* releases an enzyme that inactivates the trapping mechanisms of neutrophils
Diagnosis & treatment	Diagnosis is based on symptoms and microscopic observation of *Plasmodium* parasite in blood; treatments may include chloroquinine, quinine, and/or artemisinin-based drugs; usually combination therapies are recommended because of increasing drug resistance in *Plasmodium* species

Some released merozoites will go on to continue the asexual cycle of infecting red blood cells; others will develop into *gametocytes*. Mosquitoes that bite the infected patient take up these gametocytes. Once inside the mosquito, *Plasmodium* goes through another growth cycle that culminates in the formation of the sporozoites that can perpetuate the infection cycle.

Pathogen drug resistance and mosquito resistance to insecticides is increasing.

Current treatment options include some of the same drugs that have been used for centuries: artemisinin derivatives and chloroquine. Other antimalarial drugs such as doxycycline, mefloquine, and atovaquone are also used in combination. (Antimalarial drugs for active cases as well as prevention are reviewed in Chapter 15.) Unfortunately, in some regions, antimalarial drugs are often counterfeit and ineffective. Moreover, pathogen resistance to many malaria drugs is a problem, particularly with *P. falciparum* and *P. vivax* species. Travelers to areas where malaria is endemic are advised to carry a full drug regimen with them in case they get malaria. In addition, they should take preventive antimalarial drugs (malaria prophylaxis) daily while in endemic areas.

Malaria cases dropped somewhat in the 2000s when governments and aid groups in Africa widely disseminated insecticide-treated bed nets to cut down on mosquito bites during sleep (FIG. 21.24). Unfortunately, some mosquitoes are developing resistance to the insecticides. Starting in 2021, the GlaxoSmithKline Biologicals' RTS,S/AS01 vaccine (Mosquirix™) became the first WHO-recommended vaccine to protect children against malaria. Sanaria's PfSPZ vaccine is another promising vaccine candidate.

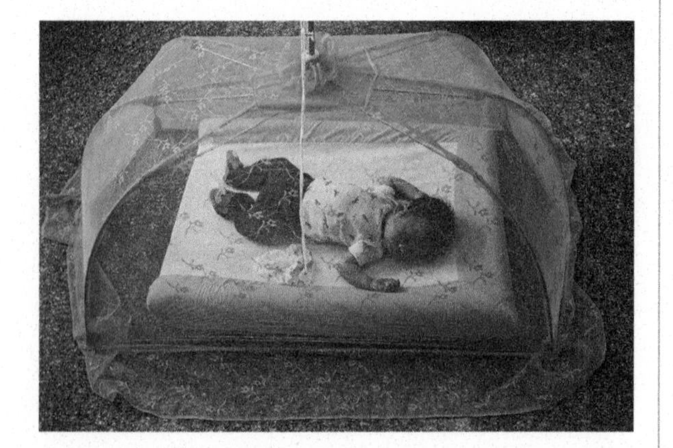

FIGURE 21.24 Bed nets Insecticide-treated bed nets cut down on mosquito bites while sleeping and help reduce mosquito-borne infections.

Build Your Foundation

21. Where in the world could you possibly acquire malaria and what signs and symptoms would you anticipate? (LO 21.20)
22. Assume you are assessing a patient who just came back from a safari in Africa and has anemia, severe hypotension, and recurrent episodes of fever and chills. What form of malaria would you suspect and why? (LO 21.21)
23. What leads to the episodic attacks featured in malaria? (LO 21.22)

BYF QUICK QUIZ

Build Your Foundation (BYF) Quick Quiz: Visit the **Mastering Microbiology** Study Area to quiz yourself.

VISUAL SUMMARY | *Cardiovascular and Lymphatic Infections*

Systemic Infections

May originate in the cardiovascular or lymphatic system, or may develop from a local infection that spreads to these systems.

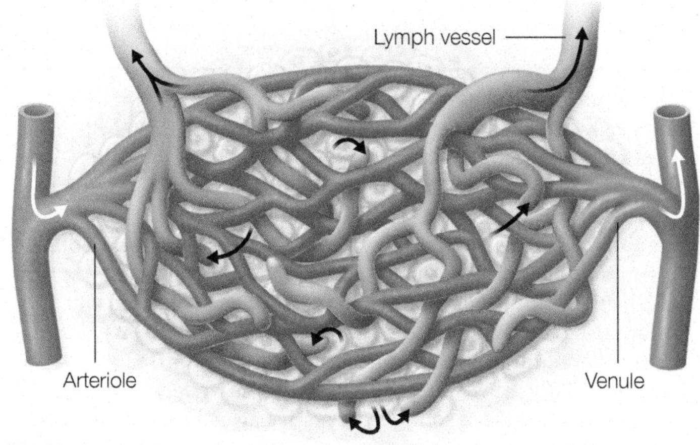

The black arrows show the collection of fluid from tissue spaces by lymph capillaries and vessels.

Vectors

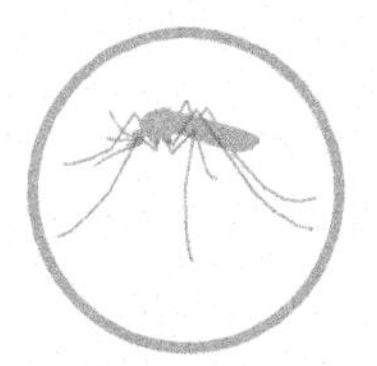

Mosquitoes
(Zika, malaria, dengue, and many more)

Ticks
(Lyme disease, tularemia, Rocky Mountain spotted fever)

Fleas
(Plague)

Systemic Bacterial Infections

Zoonotic bacteria

Plague, caused by *Yersinia pestis*, and tularemia, caused by *Francisella tularensis*, are relatively rare zoonotic systemic bacterial infections.

Endocarditis

Endocaditis impacts the heart lining and/or heart valves. It mainly occurs in people with heart defects or artificial heart valves.

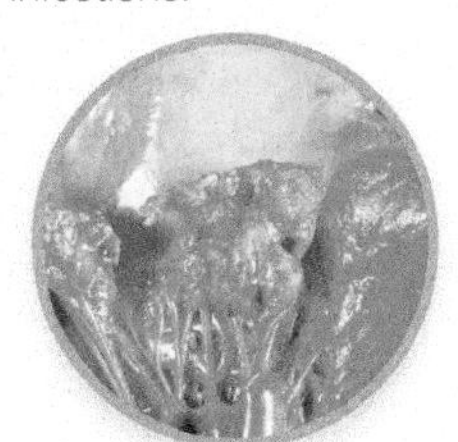

Bacterial endocarditis

Tickborne bacteria

Tickborne bacterial systemic infections often feature rashes. The most common examples include Lyme disease, caused by *Borrelia burgdorferi*, and Rocky Mountain spotted fever, caused by *Rickettsia rickettsii*.

Lyme disease rash

Sepsis

- Sepsis is a patient's overwhelming immune response to a toxin and/or a bacterial, fungal, viral, or parasitic infection.
- Sepsis can progress to septic shock, a potentially deadly scenario that features tissue death and organ failure.

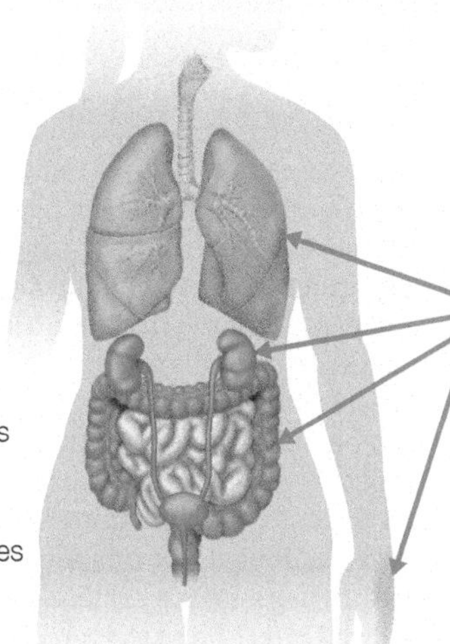

Systemic Viral Infections

Mosquito-borne viruses

Dengue, yellow fever, chikungunya, and Zika viruses belong to the same viral family, *Flaviviridae*. They share certain clinical features such as inducing high fever and aches. Dengue may lead to severe second-stage symptoms including hemorrhagic features. Zika can lead to birth defects such as microcephaly when vertically transmitted.

Microcephaly

Hemorrhagic fevers

Ebola, Marburg, and Lassa viruses cause hemorrhagic fevers that have high mortality rates. Most cases are in Africa.

Epstein-Barr virus

Epstein-Barr virus (EBV) is a member of the *Herpesviridae* family; it's the most common cause of mononucleosis, and can also lead to endemic Burkitt's lymphoma.

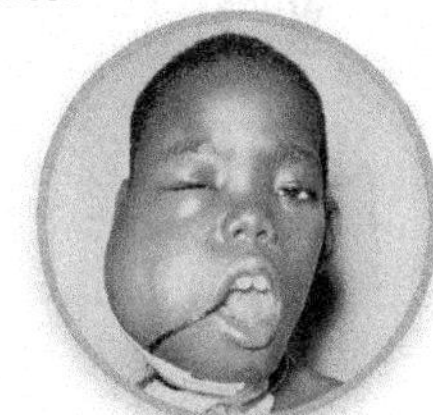

Burkitt's lymphoma

Retroviruses

- Human immunodeficiency virus (HIV) and human T lymphotropic virus (HTLV) are retroviruses that mainly target T cells. They are both transmitted by infected body fluids and can pass from mother to child.
- HIV can be treated and prevented with antiretroviral agents, but if not managed leads to suppressed immunity. People with suppressed immunity have a greater risk for infections (including systemic infections) and cancers like Kaposi sarcoma.
- HTLV can lead to T cell leukemia or lymphoma, HTLV-1-associated myelopathy, and immuno-suppression as T cell function is impacted.

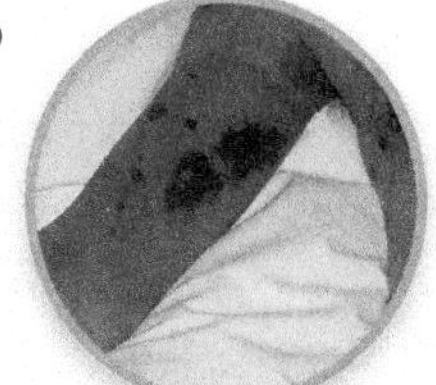

Kaposi sarcoma

For all printed and additional Disease Snapshot tables, go to the Mastering Microbiology Study Area.

21 CHAPTER 21 OVERVIEW

21.1 Overview of the Cardiovascular and Lymphatic Systems

- The cardiovascular and lymphatic systems are intertwined networks that circulate nutrients, oxygen, fluids, hormones, and immune system components to every part of the body. Pathogens that gain access to these systems cause systemic infections.
- The heart is a four-chambered organ. The heart wall consists of three main layers: the innermost endocardium, the muscular myocardium, and the outer pericardium. Veins convey blood to the heart and arteries conduct it away from the heart. Nutrient and gas exchange occurs in the tissues across capillaries.
- Lymphatic vessels collect lymph from tissues, ferry it to lymphatic tissues such as lymph nodes, and then return filtered lymph to the venous blood supply.
- Sepsis describes a patient's overwhelming immune response to a toxin and/or a bacterial, fungal, viral, or parasitic infection. Although many sepsis cases originate from a systemic infection, it is entirely possible for a localized infection to trigger a septic response. Sepsis progresses through stages and may lead to shock. In septic shock, severe low blood pressure and disseminated intravascular coagulation cause tissue death and organ failure.

21.2 Systemic Viral Infections

- Most systemic viral infections are caused by arboviruses that are transmitted by mosquitoes. Examples include dengue, yellow fever, chikungunya, and Zika viruses. Dengue virus causes dengue fever; the second stage of dengue can involve severe hemorrhagic symptoms. Yellow fever can be transmitted between primates and humans via a mosquito vector. Chikungunya virus causes polyarthralgia (extreme joint pain). Zika virus causes fever and rash in healthy individuals; vertical transmission can lead to microcephaly.
- Ebola and Marburg viruses are not caused by arboviruses; they are transmitted by contact with infected animals, bushmeat, or contaminated bodily fluids. Both infections are high-mortality hemorrhagic fevers that feature fever, chills, and severe internal bleeding. Lassa virus causes another example of hemorrhagic fever; it's transmitted by ingestion or inhalation of rodent feces or dried urine.
- Epstein-Barr virus is the main cause of mononucleosis and has been linked to endemic Burkitt's lymphoma.
- Human immunodeficiency virus (HIV) and human lymphotropic viruses (HTLV) are retroviruses that cause decreased immunity, resulting in secondary opportunistic infections and various cancers. There are three stages to HIV infection. Patients with T helper cell levels below 200 cells/mm^3 have progressed to AIDS. Antiretroviral therapies can help limit HIV progression. There are also drugs that can be used as pre-exposure and post-exposure HIV preventions.

21.3 Systemic Bacterial Infections

- Bacteria can cause systemic infections upon introduction to the circulatory or lymphatic systems—or, they can spread there from a previously localized infection. The presence of bacteria in the blood is called bacteremia.
- *Yersinia pestis* causes plague. Fleas transmit the bacterium from infected animals to people. There are three forms of plague: pneumonic plague, bubonic plague, and septicemic plague. Both pneumonic and bubonic plague can progress to the septicemic form, which can feature acral gangrene caused by disseminated intravascular coagulation.
- Bacterial endocarditis is an infection of heart valves due to bacteremia; it mainly affects patients with heart defects or artificial heart valves. The most common causative agents are *Streptococcus* species and *Staphylococcus aureus*.
- *Francisella tularensis* is a bacterium that causes tularemia—a zoonotic disease that is usually spread by bites from infected ticks and handling infected animals.
- Examples of tickborne systemic bacterial infections include Lyme disease, Rocky Mountain spotted fever, ehrlichiosis, and anaplasmosis. Lyme disease is often characterized by a bull's-eye rash. Rocky Mountain spotted fever is a rapidly progressing infection that features numerous small, flat, pink/red, non-itchy spots. Ehrlichiosis and anaplasmosis are emerging bacterial infections.

21.4 Systemic Fungal Infections

- Candidemia develops when yeast (*Candida* species) invade the bloodstream. *Candida albicans* is the most common cause. This fungal infection mainly affects immunocompromised individuals and it's a common healthcare-acquired infection. Surgery, central venous catheters, kidney dialysis, and intravenous lines are examples of interventions that increase the risk for candidemia.

21.5 Systemic Protozoan Infections: Malaria

- Malaria is a common systemic protozoan infection. Most cases are caused by *Plasmodium falciparum* or *P. vivax*. *Plasmodium*'s life cycle is dependent on mosquito and human hosts. Infection is characterized by episodic attacks that feature shivering, fever, and sweating. These attacks correlate with the parasite's asexual merozoite development.

SMART

THINK CLINICALLY | *Be S.M.A.R.T. About Cases*

COMPREHENSIVE CASE

The following case integrates basic principles from the chapter. Try to answer the case questions on your own. Don't forget to be S.M.A.R.T.* about your case study to help you interlink the scientific concepts you have just learned and apply your new understanding of cardiovascular and lymphatic infections to a case study.

***The five-step method is shown in detail in the Chapter 1 Comprehensive Case on cholera. See pages 31–33. Refer back to this example to help you apply a SMART approach to other critical thinking questions and cases throughout the book.**

• • •

Deja works at the CDC as a resource provider for the malaria hotline. Most calls she receives come from within the United States—either from healthcare providers seeking recommendations for managing an imported malaria case or from people planning travel to malaria-prone areas and requesting information on how to prevent malaria. But on Sunday night, Deja received a call from a college student named Kiara, who is two weeks into a summer-long internship in Sierra Leone with a nonprofit agency that digs wells in remote villages of West Africa.

Kiara complains of a high fever, severe body aches, and nausea. She's concerned that she might have malaria, and worried that if she needs to immediately see a doctor, there isn't one available in the village where she's staying.

Deja tries to calm and reassure Kiara, and then gathers a patient history. Kiara came to West Africa with enough antimalaria prophylaxis medication for her entire stay, but admits she has forgotten to take it "a few times" since her arrival. She doesn't sleep in a mosquito bed net, but applies DEET-based insect repellent each morning before going to work. Over the past day Kiara developed a faint pink rash on her torso. The fever and aches are persistent, rather than intermittent. She feels "awful all over," but doesn't report joint pain. She started feeling ill yesterday and got progressively worse throughout today. She doesn't have any bleeding from the nose or mouth.

Deja then asked if over the past week Kiara had contacted anyone who was bleeding or if any of her coworkers were ill. Kiara reported that none of her team members were sick, but that a man died in the village this week, after a long illness. Kiara paid her respects to the family members of the deceased, but did not have contact with the man himself, enter his house, or attend the funeral.

Deja does a computer search to help Kiara locate the nearest reliable medical clinic. The town Kiara mentioned as being three hours away does have a clinic. Deja asks Kiara to find a way to get to the clinic as soon as possible, and to call back if the clinic tries to prescribe any medication. She also requests that Kiara let her know what treatment is offered, if any. Kiara would need a blood test to make final confirmation of the disease Deja suspects.

• • •

CASE-BASED QUESTIONS

1. What diseases would Deja automatically suspect based solely on Kiara's fever, nausea, body aches, and West Africa location? How does adding a rash to the list of symptoms change your response?
2. What disease was Deja trying to ascertain risk of when she asked Kiara about contact with ill people showing symptoms of bleeding? What factor(s) in Kiara's situation makes asking about this particular disease important?
3. Is malaria the most likely disease that Kiara is experiencing? Support your response with evidence.
4. Based on her history and symptoms, what disease(s) does Kiara have the greatest probability of contracting? Explain your answer.
5. Why did Deja tell Kiara to get to a clinic as soon as possible? What complications does Deja fear?
6. Is there any danger of Kiara infecting her coworkers? Why or why not?

END OF CHAPTER QUESTIONS

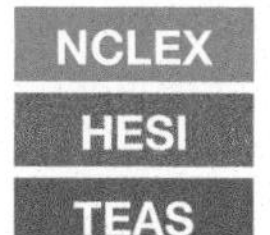

Think Critically and Clinically

Questions highlighted in **orange** are opportunities to practice NCLEX, HESI, and TEAS-style questions.

1. How are sepsis and septic shock related?
2. First label the following diseases as bacterial, viral, or protozoan, and then indicate which ones are vectorborne infections: dengue fever, AIDS, chikungunya, Lyme disease, malaria.
3. Which of the following microbes is most commonly associated with sepsis in immune-compromised individuals?
   a. *Borrelia burgdorferi*
   b. Human immunodeficiency virus
   c. Zika virus
   d. *Candida albicans*
   e. Epstein-Barr virus

4. If systemic capillaries become blocked due to clotting factors or invading microbes, what is the most likely immediate outcome?
   a. The heart cannot pump blood.
   b. Tissues die due to lack of nutrients and oxygen.
   c. Arteries will become blocked as well.
   d. Veins will begin to supply nutrients to tissue.
   e. The lungs will begin to malfunction without CO_2 being deposited.

5. What virus is responsible for mononucleosis and Burkitt's lymphoma?
   a. Dengue virus
   b. Yellow fever virus
   c. Human immunodeficiency virus
   d. *Plasmodium falciparum*
   e. Epstein-Barr virus

6. A patient in the United States presents with symptoms of fever, chills, nausea, and a rash. Which of the following questions would provide a clue as to whether dengue fever, chikungungya, or Zika should be included in the differential diagnosis?
   a. Have you recently traveled anywhere?
   b. Has your sexual partner exhibited these symptoms recently?
   c. Do you live near woods?
   d. Have you been hiking lately?
   e. Do you use intravenous drugs?

7. A tourniquet test can be used to identify which disease?
   a. Dengue fever
   b. Zika
   c. HTLV
   d. Lyme disease
   e. Rocky Mountain spotted fever

8. Name two similarities between Ebola and Marburg infections.

9. How is Lassa different from Ebola and Marburg, aside from being caused by a different virus?

10. Which of the following may transmit Zika virus? Select all that apply.
    a. Sexual contact
    b. Consuming undercooked meat
    c. Being bitten by a mosquito
    d. Fomites
    e. Coughs and sneezes

11. What effect would AIDS have on latent coinfections?
    a. None at all.
    b. Latent infections may become reactivated.
    c. Latent infections would turn into asymptomatic infections.
    d. Opportunistic pathogens could easily be reactivated.
    e. Latent infections would turn into opportunistic infections.

12. A reverse transcriptase antiviral medication is administered for human T cell leukemia virus (HTLV)–infected patients.
    a. True
    b. False

13. Why is endocarditis not transmissible from person to person?

14. Identify the zoonotic disease:
    a. HTLV lymphoma
    b. Systemic candidiasis
    c. Mucormycosis
    d. Plague
    e. Endocarditis

15. How does *Rickettsia rickettsii* damage blood vessels?
    a. It blocks blood vessels and causes vessel inflammation.
    b. It releases a toxin that destroys blood vessels.
    c. It causes sepsis and septic shock, which damages blood vessels.
    d. It avoids antibodies by binding to blood vessel surfaces.
    e. It infects the epithelial cells of the vessels.

16. How does draining standing water reduce the incidence of malaria, dengue fever, yellow fever, Zika, and chikungunya?

17. Which *Plasmodium* life stage infects the liver?
    a. Sporozoite
    b. Merozoite
    c. Gametocyte
    d. Sexual stage
    e. Vector stage

CRITICAL THINKING QUESTIONS

1. CDC estimates that 1 in 5 people infected with HIV are unaware of their infection. What HIV features are likely responsible for this lack of infection status?

2. How is coinfection with Rocky Mountain spotted fever and anaplasmosis possible in the same individual, and could you distinguish the two based on symptoms?

3. Which underlying systemic disease would you suspect if an apparently healthy individual was suffering from systemic candidiasis? Explain your reasoning.

4. If you entered a home where someone was suffering from Ebola, what types of items (clothing, cooking utensils, etc.) would you consider contaminated and handle only when wearing personal protective equipment to prevent transmission?

5. Explain how sepsis symptoms ultimately relate to cellular respiration.

APPENDIX A

Answers to End of Chapter Questions

CHAPTER 1

1. a. SEM, b. TEM, c. TEM, d. Both, e. Both, f. Both
2.

Pasteur	Disproved spontaneous generation using an S-necked flask
Koch	Developed postulates of disease
van Leeuwenhoek	The first to observe bacteria
Pasteur	The first to show that fermentation was caused by microbes
Lister	Implemented aseptic techniques in surgery
Nightingale	Practiced aseptic techniques in nursing
Semmelweis	Showed that hand washing decreased puerperal sepsis
Hooke	First to publish descriptions of cells

Francesco Redi (not used)
3. a. False. The Gram stain is a *differential* stain, b. True, c. True, d. False. Dark field microscopy *does not* require a stained sample, e. False. The acid-fast stain detects *mycolic acid* in the cell walls of certain bacteria, f. False. Gram-positive bacteria have a *thick* peptidoglycan layer in their cell wall; **4.** e; **5.** The 100 × oil objective would be in place; **6.** a. Observation, b. Observation, c. Conclusion, d. Conclusion, e. Conclusion, f. Conclusion, g. Observation, h. Observation, i. Conclusion, j. Conclusion, **7.** prokaryotic, Bacteria. Eukarya, eukaryotic; **8.** b; **9.** domains, kingdoms. species, second; **10.** Opportunistic pathogens require reduced resistance of the host, such as a compromised immune system, in order to establish an infection and cause disease. True pathogens can cause disease in a patient with a healthy immune system. Opportunistic pathogens like the yeast *Candida albicans* may also cause disease if there is a shift in normal microbiota that allows the less abundant opportunist to thrive in light of limited competition; **11.** strain; **12.** a. objective lens is closest to specimen, b. condenser sharpens the light into a precise cone, c. lamp provides a light source, d. fine focus knob allows for precision focusing; **13.** Bacterial endospores, endospore stain; **14.** a. Bright field, b. Differential interference contrast, c. Differential interference contrast, d. Bright field, e. Dark field, f. Phase contrast; **15.** b, c, d; **16.** a; **17.** Domain, Kingdom, Phylum, Class, Order, Family, Genus, Species; **18.** Knowing Gram property is important because certain drugs are more effective against Gram-positive versus Gram-negative bacteria and vice versa. **19.** b, c, a, d; **20.** mycolic acid; **21.** d; **22.** b.

CHAPTER 2

1. a. 2, b. 16 amu, c. K, d. 7, e. 125; **2.** H^+ and F^-; **3.** a. compound, b. compound, c. molecule, d. molecule, e. ion (cation), f. compound, g. ion (cation); **4.** a. 0.5 M (or 0.5 molar), b. 2%, c. 1,000 mg/dL, d. 0.001 M (or 0.001 molar); **5.** e; **6.** $C_2H_4O_2$; **7.** a. False. Isotopes are atoms *with the same number of protons and a different* number of neutrons., b. True, c. True, d. False. *Unequal* sharing of electrons leads to polar covalent bonds, e. True, f. True, g. True, h. False. Adding a base to a solution will *increase* the pH; **8.** H^+ ions (hydronium ions), decrease, OH^- ions (hydroxide ions), raise, basic
9.

$2\,Ca + O_2$	$\rightarrow 2\,CaO$	Synthesis reaction
$2\,HCl_2 + Ca(OH)_2$	$\rightarrow CaCl2 + 2H_2O$	Exchange reaction
$2\,AgCl(s)$	$\xrightarrow{Light} 2\,Ag\,(s) + Cl_2\,(g)$	Decomposition reaction

10. Complete the table:

Biomolecule	Basic Components	Type of Covalent Linkage between Monomer Units
Carbohydrates	Simple sugars	Glycosidic bonds
Lipid	Glycerol and fatty acids	Ester bonds
RNA	Ribonucleotides	Phosphodiester bonds
Proteins	Amino acids	Peptide bonds
DNA	Nucleotides (dNTPs)	Phosphodiester bonds

11. b.
12. Label the features of the periodic table box:

Magnesium ← a. Element name
12 ← b. Atomic number
Mg ← c. Chemical symbol (or simply symbol)
24.305 ← d. Atomic mass

13. 1,000; **14.** b; **15.** b.
16.

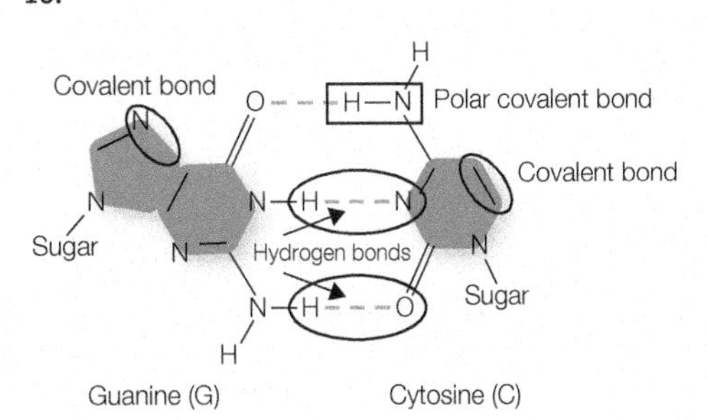

Answers will vary, but any of the dotted lines are hydrogen bonds and any of the solid lines are covalent bonds. Any hydrogen in the pictured molecule that is participating in a hydrogen bond has a polar covalent bond with a nitrogen or oxygen atom. **17.** a. dehydration synthesis reaction, b. neutralization reaction, c. hydrolysis reaction, d. neutralization reaction; **18.** a, b
19.

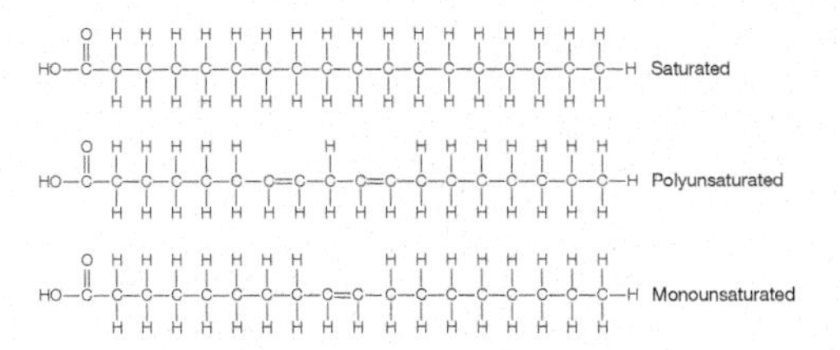

CHAPTER 3

1. b, d; **2.** a; **3.** b, e, f, g, j; **4.** a. FALSE. Prokaryotic cells have 70S ribosomes, b. FALSE. Prokaryotic cells asexually reproduce by binary fission, c. FALSE. Prokaryotic cell plasma membranes underlie the cell wall, d. TRUE, e. TRUE, f. FALSE. Fimbriae are used for adhesion, g. FALSE. Bacteria can be classified using the Gram stain; archaea cannot; **5.** b; **6.** a. amphitrichous, b. peritrichous, c. monotrichous; **7.** a. Shape: Coccus (or cocci), Arrangement: staph (or cluster); b. Shape: bacillus (or bacilli), Arrangement: Strep (or chain); c. Shape: spiral, Arrangement: single; **8.** a, e; **9.** Ribosome-Makes proteins; Cytoskeleton-Aids in cellular organization; Cell wall-Supports cell shape; Pili-Exchange genetic information; Flagella-Motility; Fimbriae-Adhesion; Capsule-Protection; Inclusion bodies-Nutrient storage
10.

Gram-positive	Gram-positive and Gram-negative	Gram-negative
• Thick peptidoglycan layer	• Single cell organisms	• Thin peptidoglycan layer
• Lack porins	• Prokaryotes	• Have porins
• Lack lapid A	• Bacteria	• Have lipid A
• No outer membrane	• Differentiated by Gram stain	• Have an outer membrane
• Teichoic acids	• Can include pathogens	• Lack teichoic acids
• Stain purple upon Gram-staining	• Both have peptidoglycan in their cell wall	• Stain pink upon Gram-staining
• Resist drying		• More susceptible to drying
• High penicillin susceptibility		• Low penicillin susceptibility
• More susceptible to anionic detergents		• Less susceptible to anionic detergents

11.

	Simple Diffu-sion	Facili-tated Diffusion	Active Trans-port
Requires energy?	NO	NO	YES
Requires a transporter of some kind?	NO	YES	YES
Moves substances from high to low concentration?	YES	YES	YES
Move substances from low to high concentration?	NO	NO	YES

12. b.
13.
- isotonic environment: Nothing
- hypertonic environment: Water would move out of the cell and cause the plasma membrane to draw away from the cell wall (plasmolysis)
- hypotonic environment: Water would move into the cell and because the cell wall is damaged (as described in the wording of the question) the cell would lyse or burst.

14. b; **15.** flagellin; **16.** a, c, h; **17.** a, c, e, f, h.

CHAPTER 4

1. a, b, c, e, f, h; **2.** a, h, i; **3.** a. False. Eukaryotic cells have 80S ribosomes OR Prokaryotic cells have 70S ribosomes, b. False. Eukaryotic cells asexually reproduce by mitosis OR Eukaryotic cells sexually reproduce by meiosis, c. True, d. False. Eukaryotic cells sometimes have a cell wall, e. True, f. False. Eukaryotic cells use flagella, cilia, or pseudopods for motility (they do not make fimbriae). (Note: A student may name *any* eukaryotic motility structure, or all of the motility structures, discussed in the chapter to correct the statement.) g. False. Yeast is a unicellular eukaryote; **4.** a; **5.** Flagellated Protozoans (Mastigophora); **6.** Exocytosis, endocytosis; **7.** Pinocytosis; **8.** a, d, e; **9.** Nucleus – Houses the cell's DNA; Endoplasmic reticulum – Modifies and packages proteins; can be rough or smooth; Golgi apparatus – Builds and packages lipids; Cilia – Motility; Mitochondria – Energy production; Glycocalyx – Adhesion and cell signaling; Ribosome – Makes proteins; Centrosome – Builds the cytoskeleton; Nucleolus – Initial site of ribosome subunit formation
10.

Prokaryotes
- Always unicellular
- No nucleus or membrane-bound organelles
- Almost always have a cell wall
- 70S ribosomes
- Usually a single circular chromosome
- Usually smaller than eukaryotes
- No sexual reproduction
- Divide asexually by binary fission
- Plasma membrane rarely contains sterols

Prokaryotes and Eukaryotes
- DNA genome
- Living
- Have plasma membrane
- Can divide
- Have ribosomes

Eukaryotes
- May be unicellular OR multicellular
- Have a nucleus and membrane-bound organelles
- Only plants, fungi, and certain protists have a cell wall
- 80S ribosomes
- Multiple linear chromosomes
- Usually larger than prokaryotes
- Divide asexually by mitosis
- Some divide sexually by meiosis
- Plasma membrane tends to contain sterols

11.

	Endocytosis	Exocytosis
Requires energy?	Yes	Yes
Starts at the plasma membrane?	Yes	No
Transports substances out of the cell?	No	Yes
Transports substances into the cell?	Yes	No
Includes phagocytosis?	Yes	No
Includes pinocytosis?	Yes	No

12. b, c; **13.** d; **14.** b, c, d; **15.** hyphae; **16.** a, c, d, e, g.

CHAPTER 5

1. Redundancy, triplet code, 64 different codons that encode 20 different amino acids, 2 stop signals, and one start signal; **2.** a. False. DNA is replicated in a 5′ to 3′ direction, b. False. DNA has an antiparallel arrangement, c. True, d. True, e. False. Eukaryotic mRNA requires processing before it is translated, f. True, g. True, h. False. DNA contains deoxyribonucleotides; RNA contains ribonucleotides; **3.** e; **4.** conjugation; F^+; **5.** Primase: Lays RNA primers to start DNA replication, Gyrase: Relieves the coiling tension in DNA as it is unwound, Helicase: Unwinds DNA helix, DNA polymerase I: Important for removing RNA primers on the lagging strand, DNA polymerase III: The primary enzyme that copies DNA, Single-strand DNA-binding protein: Keeps DNA strands separated during replication, RNA primer: Serves as a jump-start platform for DNA polymerase I; **6.** three; mRNA; translation; amino acids; stop codon (or nonsense codon); **7.** a. 5′-TGCATAG-GTCGTCGAGGTGGTT-3′, b. 5′-UGCAUAGGUC-GUCGAGGUGGUU-3′, c. This sequence does not have a start codon and therefore would not be translated; **8.** UV radiation: Physical mutagen, Transposons: Biological mutagen, Cigarette smoke: Chemical mutagen, Viruses: Biological mutagen, X-rays: Physical mutagen, Plasmids: Biological mutagen, Alcohol: Chemical mutagen; **9.** Ribonucleotides contain ribose instead of deoxyribose, which is found in deoxyribonucleotides. Ribonucleotides are used to build RNA, while deoxyribonucleotides are used to build DNA. Ribonucleotides will not incorporate thymine as a nitrogen base and instead use uracil; **10.** lactose; glucose; **11.** Gene expression requires transcription and translation, which is protein synthesis; **12.** If the promoter of a gene was deleted or severely mutated then most likely the gene would not be expressed; **13.** a, c, e; **14.** spliced; translated; intron; exons; spliceosome; **15.** a. RNA, b. DNA, c. RNA, d. DNA, e. RNA, f. DNA and RNA, g. DNA and RNA, h. RNA, i. DNA and RNA, j. DNA and RNA; **16.** a. Missense, b. Nonsense, c. Silent, d. Silent, e. Missense; **17.** d, f; **18.** a, c, d; **19.** b; **20.** competent; **21.** a, c; **22.** transcription and translation; **23.** a; **24.** a. 5′-AUGUAUUUUAUUACCGCAAGAUAA-3′, b. methionine-tyrosine-phenylalanine-isoleucine-threonine-alanine-arginine-(stop), c. AAA, d. New mRNA = 5′-AUGAUUUUAUUACCGCAAGAUAA-3′ New protein = methionine-isoleucine-leucine-leucine-proline-glutamine-aspartic acid
25.

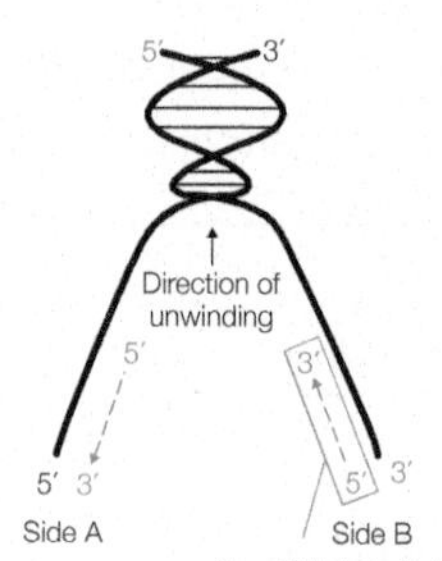

New DNA is built 5′ to 3′. As such, on Side B the new DNA is built in the same direction as unwinding (note arrow on new DNA and arrow for unwinding point in the same direction), while on Side A the new DNA is built in the opposite direction as unwinding.

DNA is built in a 5′ to 3′ direction. On the lagging strand DNA is built in the opposite direction of unwinding, while on the leading strand DNA is built in the same direction as unwinding. The parent strand dictates which direction the new DNA will be built since DNA must maintain its antiparallel orientation. The bottom of the parent strand on Side A says it is the 5′ end, so the partner strand (strand B) must be 3′ at the bottom end (for antiparallel orientation of the strands). Consequently the new DNA on Side B will be built in the same direction of the unwinding and serves as the leading strand. (See drawing for added clarification.)

CHAPTER 6

1. c; **2.** c; **3.** a, b, d; **4.** **Compare:** Both dsDNA viruses and $ssRNA^+$ viruses both direct protein production in a host cell.
Contrast: An $ssRNA^+$ genome can be directly translated by host cell ribosomes without a transcription step because the genome is essentially the same as mRNA. In contrast, dsDNA genomes must undergo transcription into mRNA before translation can occur. **5.** a; **6.** Phage conversion can provide new traits to a bacterium that may increase its chances for survival. **7.** persistent latent; **8.** Retrovirus – Human immunodeficiency virus (HIV); Lytic replication – T4 bacteriophage; Lysogenic replication – Lambda phage; Persistent infection – Human papilloma virus (HPV); Acute infection – Rhinoviruses; **9.** c, d; **10.** a, b, d, e, f; **11.** Uncoating is characterized by the release of the genome from the capsid after the virion enters the cell. Bacteriophages do not enter the host cell but inject the genome from outside the cell. **12.** b

CHAPTER 7

1. 2 hours; **2.** d; **3.** Single colonies theoretically arise from a single cell. As such, if the culture is pure, then all the colonies on the plate should exhibit the same features, such as shape, color, sheen, and texture; **4.** b, e; **5.** a.

6.

Death Phase	No population growth observed Rate of cell death exceeds cell division
Log Phase	A period of exponential population growth
Stationary Phase	No population growth observed Cells die and are generated at an equal rate Cells that can form spores prepare to do so
Lag Phase	Cells are acclimating to their environment No population growth observed

7. Some cells adapt to utilize nutrients from dead cells and withstand the accumulated waste;
8. c;
9.

Sterilization	A form of decontamination Autoclaving surgical equipment Treatment needed for critical equipment
Antisepsis	A form of decontamination Using a germicide on living tissue
Disinfection	A form of decontamination Wiping bedrails with a low-level germicide

10. a; **11.** a. False. Most pathogens would be considered mesophilic *neutralophiles*, b. True, c. True, d. True, e. False. Ionizing radiation is a form of *physical* microbial control, f. False. *Ultra*-pasteurization is a way to sterilize milk; **12.** e; **13.** d; **14.** Aerobic, anaerobic; **15.** e; **16.** c; **17.** decimal reduction time, thermal death point; **18.** a, c, d. **19.** a, d.
20. Huda could boil the water or if she brought along a filtration system, she could run the water through her filter.

CHAPTER 8

1.

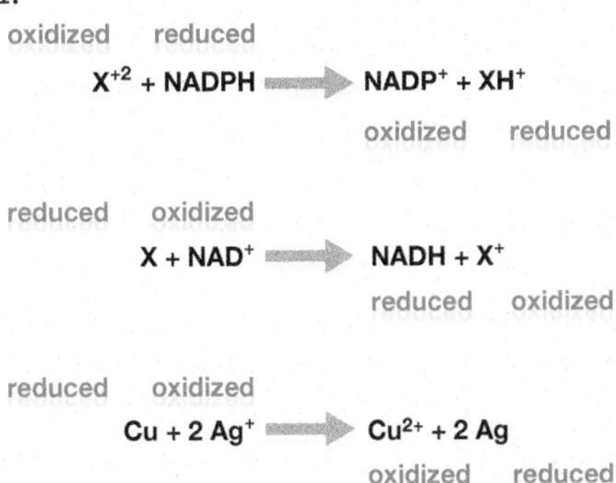

2. a. True., b. False. Substrate-level phosphorylation converts ADP to ATP, c. False. ATP is commonly used by cells but is not stored energy, d. True, e. False. Anabolic reactions are used to make ATP, f. False. In cellular respiration the most ATP is made by electron transport chains, g. True;

3.

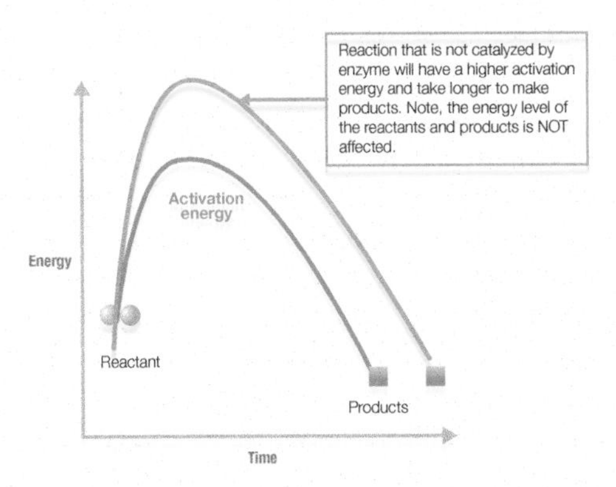

This is an exergonic reaction because the products are at a lower energy than the reactants, which means energy would have been released overall; **4.** Electron transport chain: Requires oxygen as a final electron acceptor AND Uses a proton gradient to make ATP, **Glycolysis:** Requires an investment of two ATP AND Produces two pyruvic acid molecules from glucose, **Krebs cycle:** Produces $FADH_2$ AND Decarboxylation and reduction reactions combine to catabolize acetyl-CoA, **Fermentation:** An organic molecule acts as the final electron acceptor, **Beta-oxidation:** Results in the production of acetyl-CoA from fatty acids, **Intermediate step:** Makes acetyl-CoA from pyruvic acid.

5.

Process	Cellular Location in Prokaryotic Cells	Cellular Location in Eukaryotic Cells	Uses Oxygen?
Glycolysis	Cytoplasm (also called the cytosol)	Cytoplasm (also called the cytosol)	No
Intermediate step	Cytoplasm	Mitochondria (in the mitochondrial matrix)	No
Krebs cycle	Cytoplasm	Mitochondria (in the mitochondrial matrix)	No
Aerobic electron transport chain	Plasma membrane	Mitochondria (in the inner mitochondrial membrane)	Yes

6. a. True, b. False. Sugars are not the only nutrients that can be fermented, c. True, d. False. There are hundreds of types of fermentation: homolactic, heterolactic, alcohol, mixed acid, and butanediol fermentation are just examples, e. False. Fermentation is not the same as anaerobic respiration.
7.

Anaerobic respiration
- Does not require oxygen
- Uses something other then oxygen as final electron acceptor (e.g. nitrate)
- Less efficient (tends to make less ATP per glucose than aerobic respiration)

Anaerobic respiration and Aerobic respiration
- Use glycolysis, the intermediate step, Krebs & electron transport chains
- Redox reactions
- Make ATP by substrate-level *and* oxidative phosphorylation
- Form a proton gradient to drive ATP production
- Use NAD^+ and FAD
- Make ATP
- Found in cells
- Catabolic
- Extract energy from nutrients

Aerobic respiration
- Requires oxygen
- Use oxygen as final electron acceptor
- More efficient (tends to make more ATP per glucose than anaerobic respiration)

8. c, d, f; **9.** Substrate-level phosphorylation; **10.** Photosynthesis; **11.** Cellular respiration; **12.** catalase test; negative; positive; observing bubbles after adding hydrogen peroxide; **13.** Because this is a glucose fermentation test, we can't make a conclusion about sucrose fermentation. However, we can conclude that glucose was not fermented.; **14.** a. Enzyme 4 is carrying out a redox reaction. Intermediate D is being oxidized and NAD^+ is being reduced, b. Enzyme 1, because it is earliest in the pathway, c. Substance E is the end product, d. Enzyme 2
15. Photoautotrophs: Harvest energy from light to make ATP; Make their own organic carbon-containing molecules from inorganic start materials. **Chemoheterotrophs:** Require an external source of organic carbon; Extract energy from nutrients to make ATP. **Mixotrophs:** Can switch between modes of metabolism.
16. A lipid made of glycerol and three 10-carbon fatty acid chains entering cellular respiration, 1 glucose molecule entering cellular respiration, 1 glucose molecule entering a fermentation pathway, 1 glucose molecule entering the Entner-Doudoroff pathway.

CHAPTER 9

1. Influenza: endemic; Tetanus: sporadic; Plague: sporadic; Common cold: endemic; Streptococcal pharyngitis: endemic; Botulism: sporadic; Pneumonia: endemic; **2.** a. False. Zoonotic diseases are transmitted from animals to humans, b. True, c. False. Communicable diseases are contagious, d. False. Koch's postulates of disease are used to study infectious diseases; **3.** a. Emerging: A new disease in a population, b. Epidemic: An outbreak of a disease, c. Chronic: A disease that has a slow progression and a slow onset, d. True pathogen: A pathogen that causes disease in a normally healthy host, e. Pandemic: Worldwide epidemic, f. Acute: A disease with rapid onset and rapid progression, g. Endemic: A disease with a common occurrence in a population, h. Opportunistic pathogen: A pathogen that usually only causes disease in a compromised host; **4.** Evaluate effectiveness of medical services; Link people to personal health services; Develop policies for community health; Identify and solve community health problems; Enforce laws that protect health and safety; Provide information on health issues;

5. Scenario 1: Maternal mortality rate; Scenario 2: Case fatality rate; Scenario 3: Infant mortality rate; Scenario 4: Cause-specific mortality rate. **6.**

Descriptive Epidemiology	Descriptive Epidemiology and Analytical Epidemiology	Analytical Epidemiology
• Who is infected • Where cases occur • When cases occur • *Does not* reveal the cause of the disease • Usually cheaper and less time consuming than other types of studies	Both are epidemiology studies	• What causes the disease • Why people get the disease • How to prevent or treat the disease • Tend to test hypotheses developed in descriptive epidemiology studies

7. c; **8.** a; **9.** c;
10.

- Transmission of HIV across the placenta: DIRECT
- Transmission of a pathogen through contaminated water: INDIRECT (vehicle transmission)
- Transmission of malaria by a mosquito to a human host: INDIRECT (vector transmission)
- Transmission of a pathogen through breast milk: DIRECT
- Transmission of rabies by a dog bite: DIRECT
- Transmission of a pathogen by touching a doorknob: INDIRECT (vehicle/fomite transmission)
- Transmission of a pathogen by a contaminated needle: INDIRECT (vehicle/fomite transmission)
- Transmission of a respiratory pathogen through airborne respiratory droplets: INDIRECT (droplet transmission);

11. a, c, e, f.

CHAPTER 10

1. c; **2.** d; **3.** a. Superantigen: type I, b. Hemolysins: type II, c. *Staphylococcus aureus* enterotoxins that cause food poisoning: type I, d. AB toxin: type III, e. Membrane-damaging toxins: type II, f. Phospholipases: type II; **4.** a; **5.** a. True, b. True, c. False. Gram-positive bacteria may produce exotoxins, d. False. Siderophores help pathogens obtain iron, e. True, f. False. The more toxic a substance is, the lower its LD_{50}, g. False. Pathogenicity is the ability of a microbe to cause disease, h. True; **6.** b; **7.** adhesins, stick to (or adhere to), invasins; **8.** A disease reservoir is any environment or organism where a pathogen tends to naturally thrive. *C. difficile* makes endospores, which are metabolically dormant structures. Because endospores can resist many disinfectants, and survive environmental stresses like drying out and nutrient limitations, they can remain outside a host for long periods. This makes it easy for them to be transmitted by fomites; **9.** toxins. the agent is not especially infectious, the toxin made is highly toxic; **10.** c.
11.

Description of Portal(s) of Entry	Microbe/pathology Example
Eye (optic) mucous membrane entry	Conjunctivitis
Respiratory system entry	Influenza
Transplacental entry	Congenital infections
Parenteral or transplacental entry	HIV

12. a; **13.**

Exotoxins	Exotoxins and Endotoxins	Endotoxins
• Made by Gram-negative and Gram-positive bacteria • Made by actively growing bacteria • Protein • Vaccines available for some • Can be neutralized in host	• Toxins • May cause fever	• Made by Gram-negative bacteria as they die or divide • Lipid • No vaccines • Not neutralized in host

14. Following adhesion, the pathogen has several options: It may stay on the surface of the host cell, pass through cells (or between cells) to invade deeper tissues, or enter cells to reside as an intracellular pathogen; **15.** b; **16.** b, d; **17.** mimicry, variation; **18.** a, b, f; **19.** It would directly attack host IgA antibodies and reduce immune function; **20.** The correct order from first to last is: Enter the host, Adhere to host tissues, Invade tissues and obtain nutrients, Evade immune defenses, Exit the host; **21.** This pathogen would be similar to the agent that causes tuberculosis, *Mycobacterium tuberculosis*, and would be a BSL-3 agent.

CHAPTER 11

1. Inflammation: combined action of second-line cellular and molecular; Neutrophils: second-line cellular; Skin: first line; Antimicrobial peptides: first line; Lysozyme: first line; Stomach acid: first line; Eosinophils: second-line cellular; Fever: combined action of second-line cellular and molecular; Complement proteins: second-line molecular; Mucus: first line; Iron-binding proteins: second-line molecular; Phagocytosis: second-line cellular; **2.** a. False: Localized heat, not fever, is a sign of inflammation, b. False: Monocytes and lymphocytes are agranulocytes. Granulocytes include neutrophils, eosinophils, basophils, and mast cells, c. True, d. False: The adaptive and innate immune system collaborate and function as a single immune system, e. True, f. True (once the cells mature into macrophages), g. False: Complement cascades share the same outcomes: opsonization, cytolysis, and inflammation (not fever); **3.** b, c; **4.** a, c, e; **5.** e, which is an eosinophil; **6.** classical; alternative; lectin; **7.** c; **8.** Basophil: granulocytes, innate cellular responders; Lymphocyte: agranulocytes, adaptive cellular responders; Mast cell: granulocytes, innate cellular responders; Monocyte: agranulocytes, innate cellular responders; Neutrophil: granulocytes, innate cellular responders; NK cell: agranulocytes, innate cellular responders; Macrophage: agranulocytes, innate cellular responders; Eosinophil: granulocytes, innate cellular responders; T cell: agranulocytes, adaptive cellular responders.
9.

Innate Immunity	Both Innate and Adaptive Immunity	Adaptive Immunity
• Older immunity found in all eukaryotes • Immediate response time • No immunological memory • Nonspecific immunity • Inborn protection • Consists of innate physical and chemical barriers as well as innate cellular and molecular defenses	• Distinguishes self from foreign • Effectively responds against diverse threats • Kills invaders	• Newer immunity only in vertebrates • Response time within 4–7 days • Has immunological memory, allowing for an anamnestic response • Tailors response to specific pathogen • Must be acquired (either passively or actively, artificially or naturally) • Consists of adaptive defenses (such as lymphocytes and antibodies)

10. e; **11.** b, c, e; **12.** Iron-binding proteins; ferritin; hemoglobin; lactoferrin; transferrin; **13.** c; **14.** a; **15.** Spleen: secondary lymphoid tissue; lymph node: secondary lymphoid tissue; adenoids: secondary lymphoid tissue; thymus: primary lymphoid tissue; tonsils: secondary lymphoid tissue; bone marrow: primary lymphoid tissue; **16.** Vascular changes including both vasodilation and increased vessel permeability contribute to the cardinal signs of inflammation. Vasodilation stimulates blood to enter the affected area. Since blood is red and warm, this leads to the early inflammation signs of redness and warmth. Swelling results as fluids exit leaky blood vessels to enter tissues. Furthermore, the pressure generated by swelling activates pain receptors, leading to pain; **17.** d; **18.** The innate immune system generalized responses (including cytokine production, phagocytosis, and inflammation) do not change based on the pathogen being fought. This is in contrast to the adaptive immune system with varying antibody specificities depending on the pathogen.; **19.** e; **20.** c; **21.** b, c; **22.** a, b, d; **23.** Relapsing: Characteristic of certain tick-transmitted bacterial infections; Quartan: A cyclical fever that manifests in a first and fourth day pattern; Sustained: A stably elevated body temperature with limited fluctuation; Pel-Ebstein: Fever lasting several days followed by nonfever state of similar length; Remittent: does not match definitions given.

CHAPTER 12

1. a. False: T cells are activated by antigen-presenting cells, b. True, c. True, d. False: IgM is the first antibody produced during the primary response, e. True; **2.** a, c, e; **3.** d; **4.** b, e;
5.

	MHC I	MHC II
B cell	Present	Present
Macrophage	Present	Present
Dendritic cell	Present	Present
Red blood cell	Absent	Absent
T_H cell	Present	Absent
T_C cell	Present	Absent

6. a, b; **7.** a, b, d; **8.** plasma cells; antibodies; **9.** T-dependent antigens are much more common than T-independent antigens. T-independent antigens are usually repetitive antigens such as polysaccharides. T-independent antigens bind to B cells along multiple B cell receptors and allow for direct B cell activation. In contrast, T-dependent antigens require a two-signal mechanism where T helper cells are needed for full activation. In the T-dependent antigen mechanism, B cells bind, internalize, and process the antigen. The antigen is then displayed along with MHC-II on the B cell surface where it interacts with T helper cells. Cytokines are then released by the T helper cell to allow for proliferation and differentiation of B cells; **10.** a; **11.** T cells undergo self-tolerance in the thymus; **12.** c; **13.** epitope; **14.** Plasma cells: Make antibodies, Lymphocytes; T helper cells: Have CD4, Activate T cells, Activated by MHC II bound to antigen, Lymphocytes, Make memory cells upon activation; T cytotoxic cells: Have CD8, Activated by MHC I bound to antigen, Lymphocytes, Make memory cells upon activation; APCs: Activate T cells, Present antigens to T cells; B cells: Make memory cells upon activation, Lymphocytes, Make antibodies (once activated), Present antigens to T cells; **15.** Opsonization is tagging antigens to enhance their chances of being phagocytized. IgG and IgA opsonize antigens; **16.** Requiring two activation signals prevents immune system misfires. Also, it allows for specialization of T cell populations, as production of T cell subpopulations depends on the type of signaling cascade triggered; **17.** d, f; **18.** Most abundant antibody in serum: IgG; Made as a dimer: IgA; Stimulates allergic responses: IgE; Does not cross the placenta: IgA, IgM, IgE, IgD; Considered a complement activator: IgG, IgA, IgM; Rare antibody that's poorly characterized: IgD; Main antibody in breast milk and mucus: IgA; Dominates the secondary immune response: IgG; Made early in the course of infection: IgM; Made in a primary immune response: IgM, IgG; **19.** a. Antigen-binding site; b. Light chain; c. Heavy chain; d. Covalent bonds (disulfide bridges); **20.** T_{reg}: Decreases the immune response upon threat removal, Have MHC I; T_H1: Activates T cytotoxic cells, Activates natural killer cells, Have MHC I; T_H2: Activates B cells, Have MHC I, Mainly supports the humoral response.

CHAPTER 13

1. d; **2.** b; **3.** a. False: Immunodefiencies are associated with increased cancer risk, b. False: Type I reactions are mediated by IgE antibodies, c. True, d. True, e. True, f. False: Immunotherapies reduce type I reactions; **4.** a, d; **5.** e; **6.**

Blood Type	Can Safely Donate to:	Can Safely Receive Blood From:	Antigens on Red Blood Cell Surface	Antibodies Present in Serum
A^+	A^+, AB^+	A^+, A^-, O^-, and O^+	A and Rh	B
AB^-	AB^+, AB^-	A^-, B^-, AB^-, and O^-	A and B	Rh
O^+ (worked out here for O^+, but could be any blood that is Rh^+)	A^+, B^+, AB^+, and O^+	O^- and O^+	Rh	A and B
O^-	All blood types (A^+, A^-, B^+, B^-, AB^+,AB^-, O^+, and O^-)	O^-	None of the A, B, or Rh antigens are present	A, B, and Rh

7. a; **8.** c; **9.** Mediated by IgG antibodies: types II, III; Mediated by T cells: IV; May be generated in response to pharmaceutical drugs: types I, II, III, IV; May be associated with autoimmunity: types II, III, IV; Require a sensitizing exposure: types I, IV; Mediated by IgE antibodies: type I; **10.** a, b, e; **11.** a; **12.** b, c, d; **13.** c, e; **14.** Managing autoimmunity usually involves suppressing the immune response and reducing inflammation that damages tissues.; **15.** a, b, d, and e.

CHAPTER 14

1. c; **2.** e; **3.** d; **4.** a; **5.** e; **6.** The purpose of this practice was to trigger an immune response to protect individuals from smallpox through artificial, active immunity; **7.** In 1998, a research article published in *The Lancet* (a medical journal) described a study of just 12 patients and claimed a correlation between the measles, mumps, and rubella (MMR) vaccine and the development of autism; **8.** No. In 2010, *The Lancet* fully retracted the MMR/autism study after it was revealed that the data was untruthful and there was a conflict of interest. Much larger subsequent studies found no link between vaccines and autism; **9.** Any form of active immunity (natural or artificial); memory; **10.** An inactivated toxin from a pathogen: d; The entire intact pathogen killed by heat or chemical methods: a; An antigen that elicits a strong immune response is joined to a pathogen part: e; Parts of a pathogen are made by another host cell via genetic engineering: c; A portion or piece of the pathogen such as a surface protein: b; **11.** a, c, d **12.** d, b, e, a, c; **13.** A direct ELISA typically allows for antigen identification in a sample. In this test, a solution containing antigens is added to microtiter plate wells, allowing antigens to stick to the bottom of the wells. Next, antibodies are specifically detected using a linked reporter enzyme. Any unbound antibodies are then washed out of the wells. Bound antibodies are detected by adding a substrate to the wells that interacts with the reporter system. A spectrophotometer can then detect the colorimetric or chemiluminescent signal. This test is not recommended for crude samples. Direct ELISAs can also be modified to determine if a particular antibody is present in a test sample. However, this is a challenging and impractical approach. The indirect ELISA is a more common tool for determining if a patient has antibodies to a particular pathogen. This form of ELISA requires two antibodies: one that will recognize the antigen bound to the bottom of the plate and one which is linked to an enzyme and serves as the detection antibody. To perform the test, the patient's serum is added to plates that have wells that were precoated with known antigens, and a rinse is performed. Next, detection antibody is added and unbound detection antibodies are rinsed away. Then, a substrate that interacts with the reporter enzymes is added and the resulting signal levels are then measured; **14.** c; 2^n (or 2^{15} in this case) = 32,768; **15.** Viral vectors such as adenovirus and retrovirus can be engineered to deliver the CRISPR-Cas9 system into human cells.

CHAPTER 15

1. c; **2.** b; **3.** inhibit DNA replication: b; inhibit transcription: c; inhibit translation: a, d; inhibit cell wall synthesis: e, f; disrupt plasma membrane integrity: g; block folic acid production: none; **4.** e; **5.** d; **6.** Many microbes have naturally developed resistance to antibiotics, which are naturally occurring compounds. Semisynthetic drugs are modified in ways that increase their stability and may overcome antimicrobial resistance mechanisms. Furthermore, it may be difficult to obtain mass quantities of an antibiotic but easier to make a semisynthetic derivative; **7.** d; **8.** d; **9.** a; **10.** Beta-lactamase targets the beta lactam ring, which is a chemical structure found in many drugs that target bacterial cell wall synthesis (penicillins, carbapenems, cephalosporins, and monobactams). Because glycopeptides lack the beta-lactam ring structure, they would not be inactivated by beta-lactamase and could remain a viable treatment option to fight beta-lactamase producing strains; **11.** b; **12.** d; **13.** b; **14.** a: True; b: False. The minimum bactericidal concentration is the minimum concentration of the drug that kills at least 99.9 percent of the bacteria present. c: False. The E-test *can not* reveal if a drug is bactericidal or bacteriostatic, d: True; e: True; **15.** Unlike bacteria, which are prokaryotic, fungi and protozoans are eukaryotic like our own cells. Consequently, fungi and protozoans share many physiological and structural similarities with human cells. This makes it difficult to develop selectively toxic dugs that target the microbe while sparing our own cells. The challenge in antiviral drug development lies in the fact that viruses rely on host cell machinery for replication. This makes it difficult to target viruses without adversely affecting our own cellular processes; **16.** a, e; **17.** Chloroquine: Antimalarial drug that treats nonresistant cases; Metronidazole: Treats nonmalarial protozoan infections; Quinine: Antimalarial drug that treats nonresistant cases; Praziquantel: Antihelminthic drug; Albendazole: Antihelminthic drug; ACT: Treats malaria that resists more common drugs; (not used) Treats MRSA infections; **18.** Broth dilution test.

CHAPTER 16

1. a. An inanimate object or material capable of carrying infectious microbes (such as a doorknob, cell phone, pacifier, and toy), b. A mucus-sweeping mechanism in the lower respiratory tract in which ciliated mucous membranes trap inhaled debris and sweep it toward the mouth to prevent lung entry, c. Macrophages residing in the aveoli (the air sacs) of the lungs, d. The likely outcome of disease, e. Genetically distinct variant within a species of bacterium or virus allowing for distinguishable antigenicity, f. Middle ear infection, g. Eardrum, h. A seasonal outbreak in which the number of cases of a particular disease has increased higher than would be expected in a defined community, geographical area, or season, i. Worldwide outbreaks, j. Minor changes in hemagglutinin and neuraminidase viral spikes as a result of random mutations, k. An indirect protection from an infectious disease received when a significant proportion of the population has been vaccinated against the pathogen, l. When a virus undergoes major genetic changes as a result of reassortment that cause major alterations in viral antigens, m. An antigen (such as M proteins) that can stimulate autoimmune reactions, n. Active but weakened viruses that do not cause disease in people with a normal immune system, o. A viral strain that has been chemically or physically deactivated and is incapable of causing disease, p. Fluid accumulation in the lungs, q. A case of pneumonia that has developed in a patient who was not recently hospitalized or admitted to some other healthcare or long-term-care facility such as a nursing home or rehabilitation center, r. A general term that describes pneumonia cases that develop at least 48 hours after admission to a healthcare facility, s. Illness that resembles pneumococcal pneumonia (pneumonia caused by

Streptococcus pneumoniae). Symptoms include high fever, sudden onset of severe chills, shortness of breath, productive cough, and chest pain when inhaling, t. Pneumonia-like illness that presented and progressed differently from pneumococcal pneumonia. Symptoms may include lower fever, little to no chills, shortness of breath, nonproductive cough, no chest pain when inhaling, runny nose, muscle aches, and joint aches, u. Fungal infections associated with distinct geographical locations, v. A fungus able to exist in two morphological forms (i.e., as a yeast and a mold), with the form determined by temperature; **2.** Paranasal sinuses, nasal cavity, pharynx (throat), and epiglottis; **3.** Larynx (voice box), trachea (windpipe), bronchus, and bronchiole; **4.** d; **5.** b; **6.** c; **7.** d; **8.** DTaP; Tdap; **9.** The three stages of pertussis are the catarrhal phase, the paroxysmal phase, and the convalescent stage. During the catarrhal phase, the patient experiences cold-like symptoms. The paroxysmal stage is characterized by intense coughing fits that leave the patient gasping for air and can also induce convulsions. Finally, the convalescent stage lasts up to three months and is characterized by less frequent coughing spells; **10.** First, there is wide variation in how well the vaccine protects adults. Secondly, tuberculosis is not endemic in the United States. The vaccine is only recommended in locations where tuberculosis is endemic. Finally, there is some evidence that the vaccine may interfere with the tuberculin skin test reactivity; **11.** a. True, b. False: Bacteria can be treated with antibiotics, c. False: Many studies have shown that there is not a statistical increase of colds in people exposed to cold/wet environments, d. True, e. False: The Centers for Disease Control and Prevention recommends the flu vaccine for children 6 months of age and older, f. False: Pneumonia is among the most common healthcare-acquired infections; **12.** Mycosis: Disease caused by fungi; Bacteremia: A clinical condition characterized by bacteria in the bloodstream; Dyspnea: Shortness of breath; Stridor: Wheezing or loud breathing associated with a blocked or narrowed airway; Antigenic drift: Often responsible for seasonal epidemics; Antigenic shift: Can lead to pandemics; Tachypnea: An above-normal number of breaths per minute; Consolidation: A clinical condition characterized by fluid in the alveoli; **13.** a. bacterium, b. fungus, c. bacterium, d. fungus, e. virus, f. viruses, g. fungus, h. bacterium, i. bacterium, j. bacterium, k. bacterium, l. bacterium, m. virus, n. bacterium;
14.

Latent Tuberculosis	Both Latent and Active Tuberculosis	Active Tuberculosis
• Not infectious • Symptoms of tuberculosis absent • Not detectable with either microscopy or culture analysis • Can progress to active tuberculosis • Granuloma remains intact	• Positive tuberculin skin test • Positive INF-gamma assay • Positive Xpert MTB/RIF assay • Needs treatment • Disease acquired when patient inhaled *Mycobacterium tuberculosis* and the bacteria entered alveolar macrophages	• Infectious • Symptoms of tuberculosis present • Detectable with microscopy and culture analysis • Granuloma breaks down

15. a, c, d, f; **16.** a, b, e; **17.** a, c, d.

CHAPTER 17

1. d; **2.** a; **3.** d; **4.** e; **5.** b; **6.** d; **7.** a; **8.** b, d, e; **9.** e; **10.** d; **11.** a, b; **12.** a, c; **13.** a; **14.** b; **15.** b; **16.** c; **17.** b; **18.** a; **19.** d;

20.

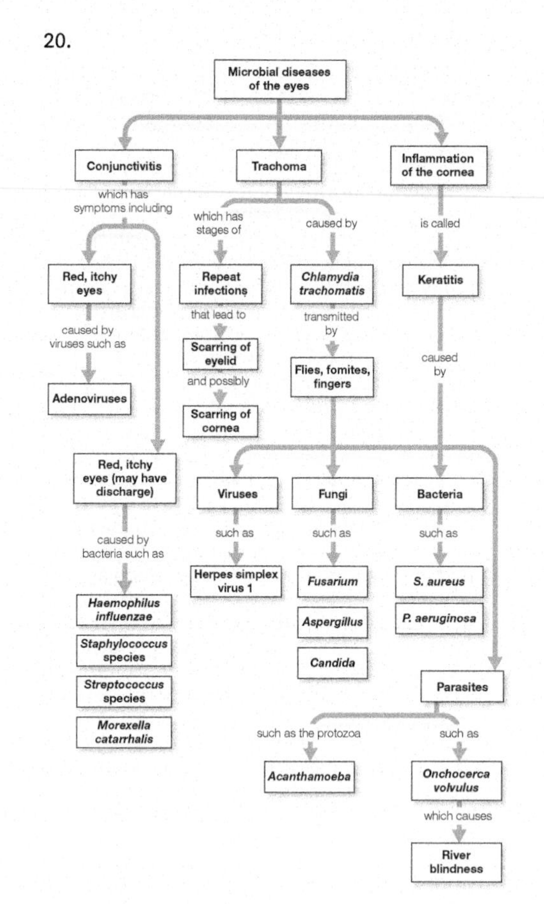

CHAPTER 18

1. d; **2.** c; **3.** c; **4.** d; **5.** a; **6.** c; **7.** b; **8.** c; **9.** a; **10.** c; **11.** a; **12.** a.

CHAPTER 19

1. Hepatitis A is an acute infection, while hepatitis B can be either acute or chronic. HAV is spread by the fecal–oral route, while HBV is spread through contact with bodily fluids from an infected person. HAV is a naked, single-stranded RNA virus, while HBV is an enveloped, double-stranded DNA virus. Both viruses cause similar acute-phase symptoms and both are preventable by vaccination; **2.** d; **3.** c; **4.** An upper endoscopy could identify an ulcer of the stomach, which is caused by *H. pylori*; **5.** *Campylobacter jejuni; Salmonella*; **6.** a. *Trichinella spiralis*, b. *Streptococcus mutans*, c. *Schistosoma*, d. Hepatitis C, e. *Vibrio cholera*; **7.** *Shigella* can move between neighboring epithelial cells by attaching actin filaments to one end of the bacterium and propelling itself into the next cell. This avoids being trapped in a cell targeted for killing by the immune response. *Shigella* can also remain in phagocytes without being digested and eventually be released; **8.** d; **9.** Wearing shoes outside to prevent contact with contaminated soil is the best way to limit infection; **10.** True; **11.** *V. cholerae* causes a severe watery diarrhea that is best treated with hydration therapy. If hydration therapy is not provided, then the patient can quickly become severely dehydrated, which may lead to shock and death; **12.** Sugars, especially sucrose, are fermented by bacteria such as *Streptococcus mutans* found in biofilms on teeth. This fermentation process releases acids that erode tooth enamel to cause cavities; **13.** Cattle; **14.** b;
15.

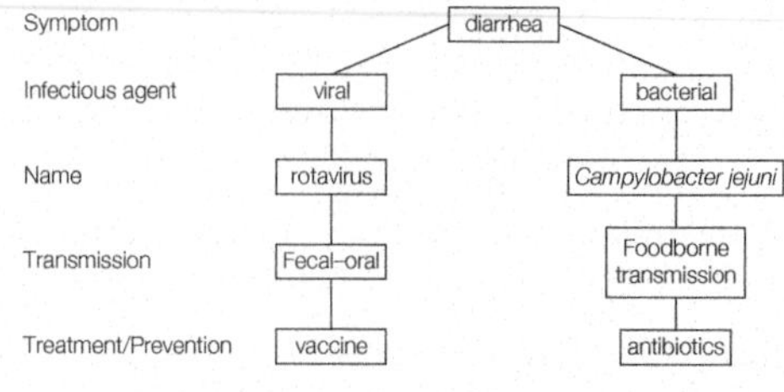

16. Schistosomiasis migratory phase symptoms include a rash or itchy skin at the site of migration. In the acute stage of infection, the patient develops fever, chills, cough, and muscle aches. The chronic stage is related to deposition of eggs in vessels and tissues and may be accompanied by abdominal pain, enlarged liver, and swollen abdomen; **17.** False; **18.** Both make Shiga toxin; **19.** a; **20.** c; **21.** b.

CHAPTER 20

1. b; **2.** d; **3.** b, e; **4.** b; **5.** a, b, d; **6.** d; **7.** d; **8.** c; **9.** d; **10.** b; **11.** a. False: Sexually transmitted infections may or may not affect the reproductive tract, b. False: CAUTIs are usually classified as complicated UTIs, c. False: UPEC are the leading cause of uncomplicated UTIs, d. True, e. True; **12.** Cystitis: Inflammation of the bladder; Dysuria: Painful urination; Pyelonephritis: Inflammation of the kidneys and considered an upper UTI; Hematuria: Blood in the urine; Pyuria: Pus in the urine. Not used: Lack of urination and inflammation of the urethra.

CHAPTER 21

1. Sepsis is the body's reaction to large amounts of circulating pathogenic microbes by inciting inflammation. Because inflammation causes plasma to leak into tissue body-wide, blood volume rapidly decreases and less blood flows to vital organs, which is termed septic shock; **2.** Dengue fever: viral and vectorborne; AIDS: viral; Chikungunya fever: viral and vectorborne; Lyme disease: bacterial and vectorborne; Malaria: protozoan and vectorborne; **3.** d; **4.** b; **5.** e; **6.** a; **7.** a; **8.** Both infections cause hemorrhagic fevers, both are of the filovirus family and thus both have negative RNA genomes, and both are transmitted by contact with primates and bats; **9.** Lassa is transmitted by ingestion or inhalation of the virus from rodent feces and urine; **10.** a, c; **11.** b; **12.** False; **13.** This type of infection is of the heart and stems from various kinds of bacteria that attach to heart valves or the lining of the heart. It is not an infection that can be necessarily shared by blood contact or other bodily fluids, since the infection does not remain systemic; **14.** d; **15.** e; **16.** Mosquitoes breed in standing water; thus, if you remove the breeding ground, the mosquito population will move away or diminish. Mosquitoes can carry Plasmodium species, which cause malaria; **17.** a.

APPENDIX B

Credits

Photo Credits

Clinical Case: SVStudio/Shutterstock
Bench to Bedside: Sjenner13/123RF
Training Tomorrow's Health Team: Ranplette/E+/Getty Images

CHAPTER 1 **Opener:** Brandi Bleak/Brandi Angel Photography. **1.1 (top to bottom):** www.rijksmuseum.nl, H. Lenthall, London, circa 1850s, PF-(bygone1)/Alamy Stock Photo, Mary Evans Picture Library/Alamy Stock Photo. **1.2 (top to bottom):** Ann Ronan Picture Library/Heritage-Images/The Print Collector/Alamy Stock Photo, Interfoto/Alamy Stock Photo. **page 5:** Mediscan/Alamy Stock Photo. **1.5:** *Les Remedes á tous maux: Le Tout Par Precaution* by Nicolas Guerard, the Younger, circa 1600s. Images from the History of Medicine (NLM). **Table 1.2 (left to right):** Eye of Science/Science Source, Frank DeLeo/NIAID, Nature's Geometry/Science Source, cparrphotos/Shutterstock, Dirk Ercken/Shutterstock, Greg Antipa/Science Source. **1.6:** James Cavallini/BSIP SA/Alamy Stock Photo. **Training Tomorrow's Health Team:** Steve Gschmeissner/Science Source. **page 14:** Scimat/Science Source. **Training Tomorrow's Health Team:** Pablo Paul/Alamy Stock Photo. **1.8:** Science Photo Library/Science Source/Science Source. **1.9:** Lourdes Norman-McKay. **1.10:** Image courtesy The Baker Company, Sanford, Maine, USA. **1.11:** Lourdes Norman-McKay. **1.12:** John Durham/Science Source. **1.13 (left to right):** William A. Clark/CDC, L. Brent Selinger/Pearson Education, Inc., Larry Stauffer, Oregon State Public Health Laboratory, CDC. **1.14:** toeytoey/Shutterstock, Blossom Tomorrow/Shutterstock. **Training Tomorrow's Health Team:** Andrey_Popov/Shutterstock. **1.15:** Rich Robison/Pearson Education. **1.16:** AkeSak/Shutterstock. **Table 1.3 (top to bottom):** Biophoto Associates/Science Source, Astrid & Hanns-Frieder Michler/Science Source, James W. Evarts/Science Source, Gerd Guenther/Science Source. **1.18 (left to right):** Peter Hawtin/University of Southampton/Science Source, Steve Gschmeilssner/Science Source. **Table 1.5 (top to bottom):** Larry Stauffer, Oregon State Public Health Laboratory/CDC, Stephanie Schuller/Science Source, Driscoll/Youngquist, & Baldeschwieler/Caltech/Science Source, Torunn Berge/Science Source. **Bench to Bedside:** Tabulae: *Anatomiam entozoorum illustrantes, congestae, nec non explicatione praeditae.* Delorieux for Johann Gottfried Bremser. **Visual Summary:** Michael Abbey/Science Source, Biophoto Associates/Science Source, John Durham/Science Source, Peter Hawtin/University of Southampton/Science Source, William A. Clark/CDC, L. Brent Selinger/Pearson Education, Inc., Larry Stauffer, Oregon State Public Health Laboratory/CDC, ASM/Science Source, Rich Robison/Pearson Education, Inc. **Critical Thinking Questions:** KuLouKu/Shutterstock, L. Brent Selinger/Pearson Education, Inc.

CHAPTER 2 **Opener:** Anthony Behar/Sipa USA/AP Images. **page 43:** Lourdes Norman-McKay. **2.11:** ThamKC/Shutterstock, Mallory Morrison Lann/Alamy Stock Photo. **Training Tomorrow's Health Team:** Neglected Tropical Diseases Study Group, University of the Philippines Manila. **2.17:** anat chant/ Shutterstock, Valentyn Volkov/Shutterstock. **page 55:** Kwangshin Kim/Science Source. **Training Tomorrow's Health Team (left to right):** From: *Transmissible Spongiform Encephalopathies Affecting Humans.* by G. B. Dudhatra et al. ISRN Infectious Diseases, vol. 2013, Article ID 387925, 11 pages, 2013. doi:10.5402/2013/387925 Fig. 2., Biophoto Associates/Science Source.

CHAPTER 3 **Opener:** Joseph Hwang. **3.3:** SPL/Science Source. **3.4:** Phasin Sudjai/123RF. **3.6:** Juergen Berger/Science Source. **Training Tomorrow's Health Team (left to right):** Steve Gschmeissner/Science Source, *Filamentation by Escherichia coli subverts innate defenses during urinary tract infection.* S. S. Justice et al. Proc Natl Acad Sci U S A. 2006 Dec 26;103(52):19884-9. Epub 2006 Dec 15; Fig. 3C. © 2006 National Academy of Sciences, U.S.A. **Table 3.1:** Chirawan Somsanuk/123RF. **3.13:** James Cavallini/Science Source. **Training Tomorrow's Health Team:** David M. Phillips/Science Source. **Training Tomorrow's Health Team:** Photoongraphy/Shutterstock. **3.19:** Alfred Pasieka/Science Source, A. Barry Dowsett/CAMR/Science Source, Eric V. Grave/Science Source, Linda M. Stannard/University of Cape Town/Science Source. **3.20:** From: *Periplasmic flagellar export apparatus protein, FliH, is involved in post-transcriptional regulation of FlaB, motility and virulence of the relapsing fever spirochete Borrelia hermsii.* C. Guyard et al. PLoS One. 2013 Aug 29;8(8):e72550. doi: 10.1371/journal.pone.0072550. eCollection 2013 Fig. 2E. **3.21:** Kwangshin Kim/Science Source. **3.22:** Xavier Nassif. **3.23:** Kari Lounatmaa/Science Source, Michael Abbey/Science Source. **Bench to Bedside:** Andrew Tingle. **3.24:** CNRI/Science Source. **3.25:** Protein Data Bank in Europe, EMD-2976. **3.26:** Duplication and segregation of the actin (MreB) cytoskeleton during the prokaryotic cell cycle. P. Vats P1 and L. Rothfield. *Proc Natl Acad Sci* U S A. 2007 Nov 6;104(45):17795-800. Epub 2007 Oct 31. Fig. 3A. © 2007 National Academy of Sciences, U.S.A. **3.27 (left to right):** From: Synthesis of High-Molecular-Weight Polyhydroxyalkanoates by Marine Photosynthetic Purple Bacteria. M. Higuchi-Takeuchi et al.*PLoS One.* 2016 Aug 11;11(8):e0160981. doi: 10.1371/journal.pone.0160981. eCollection 2016. Fig. 2, From: A comparison of methods to measure the magnetic moment of magnetotactic bacteria through analysis of their trajectories in external magnetic fields. R. Nadkarni et al. *PLoS One.* 2013 Dec 12;8(12):e82064. doi: 10.1371/journal.pone.0082064. eCollection 2013. Fig. 2B, Raul Gonzalez and Cheryl Kerfeld. **3.28 (left to right):** Larry Stauffer, Oregon State Public Health Laboratory, CDC, David M. Phillips/Science Source. **Training Tomorrow's Health Team:** Anna Rut Fridholm/Alamy Stock Photo (inset) Maurice Savage/Alamy Stock Photo. **page 94 (top to bottom):** A. Barry Dowsett/Science Source, William A. Clark/CDC, Richard Facklam/CDC, Kateryna Kon/Shutterstock, CDC.

CHAPTER 4 **Opener:** Mahesh Kumar A/AP Images. **4.5:** Callista Images/Cultura Creative (RF)/Alamy Stock Photo. **Training Tomorrow's Health Team:** Sherry Yates Young/123RF. **Table 4.3 (left to right):** David Scharf/Science Source, Andrew Syred/Science Source, SCHAEFER-NCI/Phanie/Alamy Stock Photo. **4.8:** L. Brent Selinger/Pearson Education, Inc. **4.9 (left to right, top to bottom):** Lucille K. Georg/CDC, Science Stock Photography/Science Source, Ed Reschke/Photolibrary/Getty Images, Biology Pics/Science Source, Olaf Speier/123RF. **4.10 (top to bottom):** Tatyana Okhitina/123RF, Steve Gschmeissner/Science Source. **4.11 (top to bottom):** Mark Conlin/Alamy Stock Photo, Jarrod Erbe Photography/Alamy Stock Photo. **4.12 (left to right):** Steve Gschmeissner/Science Source, Janice Haney Carr/CDC, Aaron J. Bell/Science Source, Sciencepics/Shutterstock. **Training Tomorrow's Health Team:** Allesalltag/Alamy Stock Photo. **4.14:** David M. Phillips/Science Source. **4.15:** Aaron J. Bell/Science Source. **4.19:** Professors Pietro M. Motta & Tomonori Naguro/Science Source. **Training Tomorrow's Health Team:** Scott Camazine/Alamy Stock Photo. **page 115:** SPL/Science Source. **Bench to Bedside:** Irzhanova Asel/Shutterstock. **Visual Summary:** Steve Gschmeissner/Science Source, Janice Haney Carr/CDC, Aaron J. Bell/Science Source, Sciencepics/Shutterstock, Science Stock Photography/Science Source, Ed Reschke/Photolibrary/Getty Images, Biology Pics/Science Source. **page 120:** Ed Reschke/Photolibrary/Getty Images

CHAPTER 5 **Opener:** Renata JM Henderson. **page 128:** Science Picture Co/Science Source. **Training Tomorrow's Health Team:** Dr. P. Marazzi/Science Source. **Training Tomorrow's Health Team:** Jack Sullivan/Alamy Stock Photo. **Bench to Bedside:** Daniel Schoenen/Image broker/Alamy Stock Photo. **page 134:** StudioMolekuul/Shutterstock. **5.13:** Elena Kiseleva/SPL/Science Source. **page 154:** Eye of Science/Science Source. **5.24:** Juergen Berger/Science Source. **5.26:** Leena Robinson/Shutterstock.

CHAPTER 6 **Opener:** Naveen Sharma/SOPA Images/Sipa/Newscom. **page 167:** Biozentrum/University of Basel/Science Source. **6.2 (top to bottom):** Linda M- Stannard/University of Cape Town/Science Source, Alfred Pasieka/Science Source, Frederick Murphy/CDC. **6.3.2:** Frederick Murphy/CDC. **Training Tomorrow's Health Team:** Timothy Nwachukwu/The New York Times/Getty Images. **page 178:** NIBSC/Science Source. **6.15:** NIBSC/Science Source. **Training Tomorrow's Health Team:** NIBSC/Science Source. **6.18:** Voisin/Phanie/Alamy Stock Photo. **6.19:** L. Brent Selinger/Pearson Education, Inc. **6.20:** Reuters/Alamy Stock Photo. **6.21:** Mauro Fermariello/Science Source. **Training Tomorrow's Health Team:** Luiscar74/Shutterstock. **Bench to Bedside:** Tommy E-Trenchard/Alamy Stock Photo. **6.24:** Sherif Zaki and Wun-Ju Shieh/CDC, Michael Abbey/Science Source. **Visual Summary:** Sherif Zaki and Wun-Ju Shieh/CDC.

CHAPTER 7 **Opener:** Kimberly S. Horn. **7.1:** Janice Haney Carr/CDC. **7.2:** Steve Gschmeissner/Science Source. **7.3:** From: T- Shomura, S- Amano & T- Niida, The Society for Actinomycetes Japan (http://atlas-actino-jp/). **7.6:** Maximilian Stock Ltd/Science Source. **7.7:** **(top to bottom)** NOAA PMEL EOI Program (http://www-pmel-noaa-gov/eoi/gallery/smoker-images-html), Wisanu boonrawd/123RF, Elnavegante/123RF. **7.8:** Adonis Villanueva/123RF. **7.9:** *Microbiology*- 1995 Dec;141 (Pt 12):3161-70- Phase and electron microscopic observations of osmotically induced wrinkling and the role of endocytotic vesicles in the plasmolysis of the Gram-negative cell wall- H Schwarz and A- L- Koch- Fig-7 (left) and 8 (right). **Training Tomorrow's Health Team:** Stem Jems/Science Source. **7.12:** Katarzyna BiaÅ‚asiewicz/123RF. **7.13 (left to right):** R. Weaver/ CDC, Hardy Diagnostics, www.HardyDiagnostics .com. **7.14:** L. Brent Selinger/Pearson Education, Inc. **7.15:** Lourdes Norman-McKay. **7.16 (left to right):** Hardy Diagnostics, www.HardyDiagnostics.com, AGE fotostock/Alamy Stock Photo. **7.17:** Lourdes Norman-McKay. **7.21:** ThermoFisher Scientific. **Training Tomorrow's Health Team:** Abir Roy Barman/ Shutterstock. **7.22:** Jasminko Ibrakovic/123RF. **7.23:** Hansen's Dairy, Hudson, Iowa. **7.24:** Newbie~commonswiki. **Bench to Bedside:** USDA. **7.26 (top to bottom):** Mr-Smith Chetanachan/123RF, Courtesy of Vestergaard Frandsen/Cooper-Hewitt Design Museum/MCT/Newscom. **Training Tomorrow's Health Team:** Marina Lohrbach/123RF. **7.27:** Soluscope Series 3. **page 233:** Chansom Pantip/123RF.

CHAPTER 8 **Opener:** Douglas Graham/Newscom. **8.9:** Evan Oto/Science Source. **Training Tomorrow's Health Team:** B.Boissonnet/BSIP SA/Alamy Stock Photo. **Training Tomorrow's Health Team:** Dr. M.A. Ansary/Science Source. **Training Tomorrow's Health Team:** Mile Atanasov/Alamy Stock Photo. **8.26:** Lourdes Norman-McKay. **Bench to Bedside:** John Sholtis/The Rockefeller University/AP Images. **8.32:** Lourdes Norman-McKay. **8.33:** Lourdes Norman-McKay. **8.34:** Lourdes Norman-McKay. **8.35 (top to bottom):** Tim Vernon, LTH NHS Trust/Science Source, CNRI/Science Source. **Visual Summary:** Lourdes Norman-McKay. **page 274:** Lourdes Norman-McKay.

CHAPTER 9 **Opener:** Rita Griffiths. **page 277:** Eye of Science/Science Source. **9.3:** Photo Researchers/ Science History Images/Alamy Stock Photo. **Training Tomorrow's Health Team:** Yuthakan Thaila/123RF, (inset) Dietmar Hoepfl/123RF. **Training Tomorrow's Health Team:** Alery Hache/AFP/Getty Images

CHAPTER 10 **Opener:** Brandi Bleak/Brandi Angel Photography. **10.1 (top to bottom):** Mediscan/Alamy Stock Photo, Janice Carr/CDC. **page 309:** Photo Researchers/Science History Images/Alamy Stock Photo. **page 311:** Mediscan/Alamy Stock Photo **Bench to Bedside:** Botox helping kids with cerebral palsy. Scott Hadow. *Northern Ontario Medical Journal* (www.nomj.ca/), June 1, 2011. **10.5 (left side, top to bottom):** Pathdoc/Shutterstock, James Cavallini/ BSIP/Alamy Stock Photo, WENN US/Alamy Stock Photo, Biomedical Imaging Unit, Southampton General Hospital/Science Source **(right side, top to bottom):** Mackoflower/123RF, NIAID, SrjT/ Shutterstock, Naveen kalwa/Alamy Stock Photo. **Training Tomorrow's Health Team:** Courtesy of Mary Ellen Rimsza, M.D. **10.6 (top to bottom):** WENN US/ Alamy Stock Photo, From: *Engineering out the risk for infection with urinary catheters*. D. G. Maki and P. A. Tambyah. Emerg Infect Dis. 2001 Mar-Apr;7(2):342-7. Fig. 2. **Training Tomorrow's Health Team:** Hxdbzxy/123RF. **10.10 (left side, top to bottom):** Dr P. Marazzi/Science Source, Rui Santos/123RF, Arve Bettum/123RF, Viacheslav Krisanov/123RF **(right side, top to bottom):** Teeramet thanomkiat/123RF, Fedorkondratenko/123RF, Dmitry Lobanov/123RF. **10.11:** M_Agency/Shutterstock. **10.12:** Yin Gang/ Xinhua/Alamy Stock Photo. **10.14:** The Venusian One/Shutterstock;

CHAPTER 11 **Opener:** Jordyn Henderson and Laurie Iliff. **11.6:** Chirawan Somsanuk/123RF, Scott Camazine/Alamy Stock Photo. **page 343:** Jarun Ontakrai/123RF. **Training Tomorrow's Health Team:** Chirawan Somsanuk/123RF. **11.7:** SPL/ Science Source. **11.8:** Eye of Science/Science Source. **Training Tomorrow's Health Team:** Marc Bruxelle/ Shutterstock. **Training Tomorrow's Health Team (top to bottom):** Neil Winokur/The LIFE Images Collection/Getty Images, PackStock/Alamy Stock Photo. **Bench to Bedside:** Alfio Scisetti/123RF. **11.16:** Thomas Perkins/Alamy Stock Photo. **page 359:** Chirawan Somsanuk/123RF.

CHAPTER 12 **Opener:** Jane Galligan. **12.4 (left to right):** James Cavallini/Science Source, ProImmune Limited 2017. All Rights Reserved. **Training Tomorrow's Health Team:** AlexLMX/ Shutterstock. **Training Tomorrow's Health Team:** Steve Gschmeissner/Science Source. **Training Tomorrow's Health Team:** Butaiump/Shutterstock. **Bench to Bedside:** Courtesy of Marty Ytreberg, Center for Modeling Complex Interactions (CMCI), Univ. of Idaho. (for more information on this research, see: Initiating a watch list for Ebola virus antibody escape mutations, C. R. Miller et al., *PeerJ*, 4:e1674-1-17 (2016) doi:10.7717/peerj.1674. and, New perspectives on Ebola virus evolution, C. J. Brown et al., *PLoS One*, 11:e0160410 (2016) doi:10.1371/journal. pone.0160410. **Visual Summary:** Thomas Perkins/ Alamy Stock Photo, Chirawan Somsanuk/123RF.

CHAPTER 13 **Opener:** Jennifer de La Cruz. **13.2:** Steve Gschmeissner/Science Source. **13.4 (top to bottom):** Mediscan/Alamy Stock Photo, Ana-Maria Tegzes/Alamy Stock Photo. **page 398:** CNRI/ Science Source. **13.7 (top to bottom):** Gorillaimages/ Shutterstock, Joseph Songco/Alamy Stock Photo. **Training Tomorrow's Health Team:** Daniel Heighton/Shutterstock. **Bench to Bedside:** Phanie/ Alamy Stock Photo. **Training Tomorrow's Health Team:** Robert Dant/Alamy Stock Photo. **Training Tomorrow's Health Team:** Scott Camazine/Alamy Stock Photo. **13.16:** Ezume Images/Shutterstock. **Visual Summary:** Joseph Songco/Alamy Stock Photo, Ezume Images/Shutterstock.

CHAPTER 14 **Opener:** MD Anderson. **14.1:** Gaston Melingue/Bridgeman Images. **14.2:** Science History Images/Alamy Stock Photo. **Training Tomorrow's Health Team:** The Photo Works/Alamy Stock Photo. **14.6:** Alessio Maddaluno/Shutterstock. **14.8:** Vincent Racaniello (microbe.tv | virology.ws | twitter.com/profvrr). **14.9:** Lab Photo/Shutterstock. **14.10:** magicoven/Shutterstock. **14.11:** Tierkel/CDC. **Training Tomorrow's Health Team:** Image courtesy of Quidel Corporation. **14.12:** SINITAR/Shuttertsock. **14.18:** Louis M. Staudt, National Cancer Institute. **Visual Summary:** Lab Photo/Shutterstock, Louis M. Staudt, National Cancer Institute.

CHAPTER 15 **Opener:** Nicholas Dyer. **15.1:** Wellcome Images/Science Source. **15.2:** Science History Images/Photo Researchers/Alamy Stock Photo. **15.5:** Electron microscope image by Peta Clode and Lyn Kirilak of the Centre for Microscopy, University of Western Australia, demonstrating RECCE® 327's antibiotic action. Courtesy of Recce Pharmaceuticals. **15.7:** Chang-Pooh24/Shutterstock. **Training Tomorrow's Health Team:** David Dorward/ NIAID. **15.8:** Science Source. **page 459:** Molekuul be/Shutterstock. **15.10:** Boucharlat/Science Source. **15.12:** Ted Croll/Science Source. **Bench to Bedside:** Kunal Mahto/Shutterstock. **page 464:** Maska82/123RF. **page 466:** Studiomode/Alamy Stock Photo. **15.14:** CDC. **15.15 (left to right):** From: NmcA carbapenem-hydrolyzing enzyme in Enterobacter cloacae in North America. S. Pottumarthy et al. *Emerg Infect Dis*. 2003 Aug;9(8):999-1002. Fig 1, Microrao, JJMMC, Davangere, Karnataka, India. **15.16:** Alessio Maddaluno/Shutterstock, **(inset):** Satirus/ Shutterstock. **Training Tomorrow's Health Team:** MBI/Shutterstock. **Visual Summary:** Wellcome Images/Science Source, Studiomode/Alamy Stock Photo, David Scharf/Science Source, CDC.

CHAPTER 16 **Opener:** Jackie Lewis. **Training Tomorrow's Health Team:** Arthit Buarapa/Alamy Stock Photo. **page 488:** BSIP SA/Alamy Stock Photo. **page 489:** Cultura Creative/Alamy Stock Photo. **page 491:** Frederick Murphy/CDC. **Training Tomorrow's Health Team:** Finnbarr Webster/Alamy Stock Photo. **page 494:** NIAID. **page 497:** Phanie/ Alamy Stock Photo. **16.7:** BSIP SA/Alamy Stock Photo. **16.8:** Scott Camazine/Alamy Stock Photo. **16.10 (top to bottom):** Biophoto Associates/Science Source, Chalie Chulapornsiri/Shutterstock. **16.11 (top to bottom):** CDC, Mediscan/Alamy Stock Photo. **page 501:** NIBSC/Science Source. **page 502:** Biophoto Associates/Science Source. **Training Tomorrow's Health Team:** BSIP SA/Alamy Stock Photo. **Bench to Bedside:** Corona Borealis Studio/ Shutterstock. **page 507:** James Cavallini/BSIP SA/ Alamy Stock Photo. **page 509:** Steve Gschmeissner/ Science Source. **page 510:** Janice Haney Carr/CDC. **page 511:** NIAID. **page 513:** James Scott. **page 514:** CDC. **page 515:** CDC. **page 515 (top to bottom):** VEM/ BSIP SA/Alamy Stock Photo, Dr Jeremy Burgess/ Science Source. **page 517:** CDC. **Visual Summary:** VEM/BSIP SA/Alamy Stock Photo, NIBSC/Science Source, Biophoto Associates/Science Source.

CHAPTER 17 **Opener:** Brandi Bleak/Brandi Angel Photography. **Table 17.1 (top to bottom):** CDC/J.D. Millar, MD, MPH, DTPH, Gustoimages/ Science Source, Jasmine Hartsook/Tetra Images/ Alamy Stock Photo, Dr P. Marazzi/Science Source, ntstudio/Shutterstock, Jean-Michel Girand/Science Source, CDC/Dr. Andre J. Lebrun. **Table 17.2 (top to bottom):** CDC/Dr. Thomas F. Sellers; Emory University, Milan Lipowski/Alamy Stock Photo, Mediscan/Alamy Stock Photo, Dr P. Marazzi/Science Source, Casa nayafana/Shutterstock. **17.3 (left to right):** Rehtse_c/Shutterstock, Dr P. Marazzi/ Science Source. **17.4:** Mediscan/Alamy Stock Photo. **page 529:** Kwangshin Kim/Mary Martin/Science Source. **17.5:** Sergii Chepulskyi/Shutterstock. **17.6 (top to bottom):** Dr P. Marazzi/Science Source, Mediscan/Alamy Stock Photo, CDC/Betty G. Partin. **page 531:** Alfred Pasieka/Science Source. **17.7:** Courtesy of Mary Ellen Rimsza, M.D.. **17.8 (top to bottom, left to right):** Lourdes Norman-McKay, Biophoto Associates/Science Source, Jes2u.photo/ Shutterstock, Anucha Cheechang/Shutterstock, Dr P. Marazzi/Science Source, CNRI/Science Source. **Training Tomorrow's Health Team:** Sherry Yates Young/123RF. **17.9 (left to right):** Zoonar GmbH/ Alamy Stock Photo, Zay Nyi Nyi/Shutterstock. **page 533:** Eye of Science/Science Source. **17.11 (left to right, top to bottom):** Zay Nyi Nyi/Shutterstock, DonyaHHI/Shutterstock, DermPics/Science Source, SunflowerMomma/Shutterstock, SPL/Science Source, Dr P. Marazzi/Science Source. **Bench to Bedside:** Eye of Science/Science Source. **17.12:** Keiji Hagiwara. **17.13:** © Smuszkiewicz et al; licensee BioMed Central Ltd. 2008. **page 539:** James Cavallini/ Science Source. **17.14:** Fujitani S, Moffett KS, Yu VL. Pseudomonas aeruginosa. In: www.antimicrobe. org. ESun Technologies LLC, 2015. **17.15:** Dr P. Marazzi/Science Source. **17.16:** Scott Camazine/ Science Source. **17.17:** CDC/F. Marc LaForce, M.D.. **page 541 (top to bottom):** Juergen Berger/Science Source, Eye of Science/Science Source. **17.18:** CDC/Dr. Martin. **page 542:** Stem Jems/Science Source. **17.19 (left to right, top to bottom):** CDC/ Dr. Lucille K. Georg, SPL/Science Source, Yopan90/ Shutterstock. **17.20 (left to right):** Andy Crump, TDR, World Health Organization/Science Source,

CDC/Dr. A.J. Sulzer. **17.22 (left to right):** ARZTSAMUI/Shutterstock, sruilk/Shutterstock. **page 547:** Biophoto Associates/Science Source. **17.23 (left to right):** Hugh Taylor, Murray McGavin. **17.24:** Andy Crump, TDR, World Health Organization/Science Source. **Visual Summary:** Kwangshin Kim/Science Source, Kwangshin Kim/Mary Martin/Science Source, Stem Jems/Science Source, Dr P. Marazzi/Science Source, Andy Crump/TDR, World Health Organization/Science Source, Hugh Taylor. **Exclusive to eText:** TisforThan/Shutterstock, Dr. M.A. Ansary/Science Source, CDC/Archil Navdarashvili, Georgia (Republic), CDC/O.T. Chambers, Kwarkot/Getty Images, Nenad Nedomacki/Shutterstock, 501room/Shutterstock, Shooter IZAZ/Shutterstock, Byron Ortiz/Shutterstock, Phadungsak photo/Shutterstock, NinaMalyna/Shutterstock, Gustoimages/Science Source, Mediscan/Alamy Stock Photo, Hercules Robinson/Alamy Stock Photo, CDC, sarocha wangdee/Shutterstock, Zay Nyi Nyi/Shutterstock, Dr. Harout Tanielian/Science Source/Shutterstock, Bachkova Natalia/Shutterstock, Rattana/Shutterstock, TisforThan/Shutterstock, Jodi Jacobson/Getty Images.

CHAPTER 18 **Opener:** Brandi Bleak/Brandi Angel Photography. **18.7:** Trappe Agency/Agencja Fotograficzna Caro/Alamy Stock Photo. **page 562, 563:** Chris Bjornberg/Science Source. **page 565:** Eye of Science/Science Source. **18.8:** Mediscan Agency/Alamy Stock Photo. **page 566:** James Cavallini/Science Source. **18.9:** CDC. **page 568:** SCIMAT/Science Source. **18.11 (top to bottom):** Biophoto Associates/Science Source, Medicshots/Alamy Stock Photo. **page 570:** Eye of Science/Science Source. **18.12:** Photo Researchers/Science History Images/Alamy Stock Photo. **Bench to Bedside:** Airelle-Joubert/Science Source. **page 573:** E- Gueho/CNRI/Science Source. **18.14:** CDC. **page 575:** Eye of Science/Science Source. **page 576:** Stanley B-Prusiner/UCSF. **Visual Summary:** Laguna Design/SPL/Getty Images, David Spears FRPS FRMS/Corbis Documentary/Getty Images, Chris Bjornberg/Science Source, Eye of Science/Science Source, Alfred Pasieka/Science Source.

CHAPTER 19 **Opener:** Vaani Bhatia. **19.2 (left to right, top to bottom):** James Cavallini/BSIP SA/Alamy Stock Photo, BSIP SA/Alamy Stock Photo, CDC/Dr. Holdeman, VEM/BSIP SA/Alamy Stock Photo, MedicalRF.com/Alamy Stock Photo. **19.3:** Lee Karen Stow/Alamy Stock Photo. **19.4:** CDC. **page 587:** VEM/BSIP SA/Alamy Stock Photo. **page 588 (top):** Erskine L. Palmer/CDC. **19.5:** Catherine Makison Booth, HSE's Health & Safety Laboratory, UK, © Crown Copyright, Health and Safety Executive, licensed under the Open Government Licence v3.0. **page 588 (bottom):** Catherine Makison Booth, HSE's Health & Safety Laboratory, UK, © Crown Copyright, Health and Safety Executive, licensed under the Open Government Licence v3.0. **19.6:** Thomas F. Sellers; Emory University/CDC. **19.7 (left to right):** Southern Illinois University/Science Source, CNRI/Science Source. **page 590:** Erskine L. Palmer/CDC. **page 591:** James Cavallini/BSIP SA/Alamy Stock Photo. **Training Tomorrow's Health Team:** Kerry Sheridan/AFP/Getty Images. **19.10:** Jean-Pierre Casteyde/CNRI/Science Source. **page 594:** David Scharf/Science Source. **19.11:** David M. Martin/Science Source. **page 595:** P. Hawtin/Science Source. **Training Tomorrow's Health Team:** Gastrolab/Science Source. **Training Tomorrow's Health Team:** Janice Haney Carr/CDC. **19.12:** Reins Bigacs/123RF, Janice Haney Carr/CDC, Michael Gray/123RF, CDC, Nataliia Pyzhova/123RF, Science History Images/Photo Researchers/Alamy Stock Photo. **page 598:** Alissa Eckert/CDC. **page 600:** Dan Higgins and Jennifer Oosthuizen/CDC. **page 601:** Janice Haney Carr/CDC. **page 602:** James Archer/CDC. **page 604:** CDC. **19.14 (top to bottom):** David M. Martin/Science Source, Gastrolab/Science Source. **page 605:** Janice Haney Carr/CDC. **Bench to Bedside:** Kelly, C. R., deLeon, L. and Jasutkar, N. (2012). Fecal microbiota transplantation for relapsing Clostridium difficile infection in 26 patients: methodology and results. *Journal of Clinical Gastroenterology*, 46(2), 145-149. **page 607:** Frederick Gandolfo, MD (retroflexions.com). **19.15:** Stan Erlandsen/CDC. **page 608:** Stan Erlandsen/CDC. **page 609:** Melvin and Greene/CDC. **page 612:** Kateryna Kon/Shutterstock. **page 613:** James Gathany/CDC. **page 614:** Science History Images/Photo Researchers/Alamy Stock Photo. **page 615:** CDC/Dr. Sulzer. **Visual Summary:** Dr Jean-Pierre Casteyde/CNRI/Science Source, P. Hawtin/Science Source, James Cavallini/BSIP SA/Alamy Stock Photo, Stan Erlandsen and Dennis Feely/CDC.

CHAPTER 20 **Opener:** Avishek Das/SOPA Images/SIPA/AP Images. **page 624:** J.D. Millar/CDC. **20.5 (top to bottom):** Janice Haney Carr/CDC, VEM/BSIP SA/Alamy Stock Photo, CNRI/Science Source, VEM/BSIP SA/Alamy Stock Photo. **Training Tomorrow's Health Team:** Henrik Dolle/Shutterstock. **page 631:** Filamentation by *Escherichia coli* subverts innate defenses during urinary tract infection. S. S. Justice et al. *Proc Natl Acad Sci USA*. 2006 Dec 26;103(52):19884-9. Epub 2006 Dec 15; Fig. 3C. **page 632:** Janice Haney Carr/CDC. **20.9:** Alexander Raths/Shutterstock. **20.10:** SneSivan/Shutterstock. **Bench to Bedside:** Kozak Dmytro/Shutterstock. **page 636:** Dr. Fred Murphy, Sylvia Whitfield/CDC. **page 637:** Laboratory of Tumor Virus Biology/NIH-Visuals Online. **20.11:** Komsan Loonprom/Shutterstock. **20.12:** Mike Miller/CDC. **page 640:** BSIP SA/Alamy Stock Photo. **20.15:** SPL/Science Source. **page 642:** Biomedical Imaging Unit, Southampton General Hospital/Science Source. **20.17:** CDC. **page 644:** Wiesner/CDC. **20.20 (top to bottom):** BSIP SA/Alamy Stock Photo, CDC, J. Pledger/CDC. **page 646:** Mediscan/Alamy Stock Photo. **page 647:** ASCP/CDC. **20.21 (left to right):** CDC, ASCP/CDC. **page 648:** Eye of Science/Science Source. **20.22:** Photo by E. J. Mayeaux, Jr. Used with permission from Richard P. Usatine. From Usatine RP, Sabella C, Smith M, Mayeaux EJ, Jr., Chumley HS, Appachi E. *The Color Atlas of Pediatrics*; 2015. © 2017 McGraw-Hill Education. All rights reserved. **Visual Summary:** Janice Haney Carr/CDC, Fred Murphy/CDC, Biomedical Imaging Unit, Southampton General Hospital/Science Source, Eye of Science/Science Source.

CHAPTER 21 **Opener:** Brandi Bleak/Brandi Angel Photography. **page 658:** Janice Haney Carr/CDC. **21.6:** SPL/Science Source. **page 661 (top):** James Cavallini/BSIP SA/Alamy Stock Photo. **21.7:** CDC. **page 661 (bottom):** BSIP SA/Alamy Stock Photo. **page 663:** Cynthia Goldsmith/CDC. **page 664:** Cynthia Goldsmith/CDC. **21.9:** CDC. **page 666:** National Institute of Allergy and Infectious Diseases (NIAID). **Training Tomorrow's Health Team:** Heidi Soeters/CDC. **page 668:** James Cavallini /BSIP SA/Alamy Stock Photo. **21.11:** Ed Uthman, MD. **21.12:** CDC/Robert S. Craig. **21.14:** SPL/Science Source. **page 672:** Paul M. Feorino/CDC. **21.16:** CDC/Christina Nelson, MD, MPH. **page 675:** NIBSC/Science Source. **21.17:** CDC/Dr. Edwin P. Ewing, Jr.. **21.18:** James Gathany/CDC. **21.19:** CDC/Dr. Gary Alpert - Urban Pests - Integrated Pest Management (IPM). **page 677:** Larry Stauffer, Oregon State Public Health Laboratory/CDC. **21.20:** CDC. **Training Tomorrow's Health Team:** Sahara Frost/Shutterstock. **21.21:** CDC. **21.22:** Mediscan Agency/Alamy Stock Photo. **Training Tomorrow's Health Team:** U. S. Department of Health and Human Services. **page 680:** NIAID. **page 682:** NIAID. **21.24:** Huaykwang/Shutterstock. **Visual Summary:** CDC/Dr. Edwin P. Ewing, Jr., James Gathany/CDC, Felipe Dana/AP Images, CDC/Robert S. Craig, SPL/Science Source.

TEXT AND ART CREDITS

CHAPTER 1 **Page 1:** 1665 Robert Hooke first formally described microbial life in his book *Micrographia*.

CHAPTER 4 **Figure 4.17:** Schematic image scraps from: http://classes.midlandstech.edu/carterp/courses/bio225/chap04/lecture5.htm © 2010 Pearson.

CHAPTER 7 **Figure 7.25:** http://air-purifier-reviewsite.com/featured/air-purifier-technologies-hepa-filter/.

CHAPTER 9 **Figure 9.8:** Data from Human papillomavirus genotype attribution invasive cervical cancer: a retrospective cross-sectional worldwide study. *The Lancet Oncology*, Volume 11, Issue 11, Pages 1048-1056, November 2010. **Page 292:** Florence Nightingale (1820-1910). **Figure 9.9:** Semmelweis's original data translated into English can be found in: Spirer, H. F., & Spirer, L. (1991). Death and numbers: Semmelweis the statistician. *Physicians for Social Responsibility Quarterly*, **1**, 43–52. **Figure 9.10:** Graph based on data from the Nation Healthcare Safety Network 2016 annual report (uses 2014 national data from acute care facilities).

CHAPTER 15 **Figure 15.13:** Data source: Structure—Activity Relationships of Polymyxin Antibiotics. http://www.ncbi.nlm.nih.gov/pmc/articles/PMC2907661/. **Figure 15.17:** From CDC: http://phil.cdc.gov/phil/details.asp image 16883.

CHAPTER 16 **Page 482:** Data from WHO. https://covid19.who.int/. **Figure 16.5:** Interactive flu map from CDC: http://gis.cdc.gov/grasp/fluview/main.html. **Figure 16.15:** Data source: CDC. **Figure 16.16:** From CDC: http://www.cdc.gov/fungal/diseases/coccidioidomycosis/causes.html. **Figure 16.17:** Data source: CDC.

CHAPTER 19 **Figure 19.8:** Data source: CDC.

CHAPTER 21 **Figure 21.15:** Case incidence map from CDC: https://www.cdc.gov/plague/maps/index.html. **Training Tomorrow's Health Team:** Biggest Threats, CDC. https://www.cdc.gov/drugresistance/biggest_threats.html.

APPENDIX C

ABRIDGED Microbiology in Nursing and Allied Health (MINAH) Undergraduate Curriculum Guidelines

These are abridged from the full MINAH guidelines, which can be found using the QR code located on the inside back cover of this textbook.

Impact of Microorganisms in Health and Disease

1. **The microbiome includes diverse cellular and acellular microbes that impact human health; a microbiome dysbiosis (imbalance) that changes the level, location, or diversity of our microbiota may lead to disease.**
   ASM Recommended Curriculum Guideline: 23
   NCLEX-RN Alignment: Basic Care & Comfort; Physiological Adaptation

2. **Microorganisms are everywhere and live in diverse and dynamic ecosystems, including the human body.**
   Microbes are classified into different taxonomic groups; understanding these groups supports infection management and informs patient care plans. Knowing what parts of the body microbes colonize is important in infection control and for diagnosing infectious diseases.
   ASM Recommended Curriculum Guideline: 20
   NCLEX-RN Alignment: Pharmacological & Parenteral Therapies; Safety & Infection Control

3. **Most bacteria in nature live in biofilm communities.**
   Biofilm production presents unique challenges to health care, such as providing a continuously available pathogen source for renewed infections and conferring resistance to antimicrobial agents. Implanted medical devices are a common location for biofilm formation.
   ASM Recommended Curriculum Guideline: 21
   NCLEX-RN Alignment: Pharmacological & Parenteral Therapies; Reduction of Risk Potential

4. **Microbes interact with human hosts in beneficial, neutral, or detrimental ways.**
   Nurses and allied health workers should understand the microbiological and epidemiological features of pathogens. Host factors impact infectious disease development.
   ASM Recommended Curriculum Guideline: 23
   NCLEX-RN Alignment: Pharmacological & Parenteral Therapies; Safety & Infection Control; Health Promotion & Maintenance; Basic Care & Comfort; Physiological Adaptation

5. **Humans use microorganisms and their products to make pharmaceuticals.**
   ASM Recommended Curriculum Guideline: 26
   NCLEX-RN Alignment: Health Promotion & Maintenance

Microbial Pathogenicity

6. **Pathogens have diverse virulence factors that influence their pathogenesis and impact treatment options and clinical management.**
   Understanding virulence factors and pathogenesis mechanisms allows healthcare workers to identify, properly treat, and reduce infectious disease transmission.
   ASM Recommended Curriculum Guidelines: 8, 9, 10, 23
   NCLEX-RN Alignment: Health Promotion & Maintenance; Safety & Infection Control

7. **Pathogens are continuously evolving and virulence is not a static property. Understanding mechanisms that impact pathogen evolution (e.g., vertical and horizontal genetic variation, mutations, recombination, etc.) is central to limiting pathogen evolution.**
   ASM Recommended Curriculum Guidelines: 2, 3, 15
   NCLEX-RN Alignment: Safety & Infection Control

Identifying and Managing Infectious Diseases

8. **Koch's postulates are used to identify the etiological agent of certain infectious diseases.**
   ASM Recommended Curriculum Guideline: 23
   NCLEX-RN Alignment: Management of Care

9. **A variety of methods are used to identify infectious agents.**
   Serology, diverse molecular methods, biochemical tests, and staining methods are used to definitively diagnose infections.
   ASM Recommended Curriculum Guideline: 34
   NCLEX-RN Alignment: Physiological Adaptation

10. **Vaccines are safe and effective tools for preventing disease.**
    Vaccines come in different formulations, and have different recommended administration schedules; they train the immune system and promote herd immunity against a particular pathogen. Nurses and other allied health workers must understand how vaccines work and be able to speak intelligibly about vaccines to all stakeholders.
    ASM Recommended Curriculum Guideline: 31
    NCLEX-RN Alignment: Reduction of Risk Potential; Communication & Documentation; Physiological Adaptation

Healthcare-Acquired Infections and Epidemiology

11. **Healthcare-acquired infections (HAIs, nosocomial infections) are costly and often have a poorer prognosis than community-acquired infections.**
    Standard/universal precautions, transmission precautions, surgical asepsis, and biosafety level precautions limit HAIs and allow healthcare workers to safely manage patients and safely collect/analyze patient samples.
    ASM Recommended Curriculum Guidelines: 23, 37
    NCLEX-RN Alignment: Safety & Infection Control; Health Promotion & Maintenance; Therapeutic Environment

12. **Tracking and reducing the incidence of healthcare-acquired infections is a collaborative effort that saves lives.**
    Epidemiologists use diverse surveillance techniques to monitor certain infectious diseases. Being familiar with emerging and re-emerging pathogens is important in managing potential outbreaks. An appreciation of nationally notifiable diseases is essential for compliance with reporting protocols.
    ASM Recommended Curriculum Guidelines: 23, 37
    NCLEX-RN Alignment: Safety & Infection Control; Health Promotion & Maintenance; Therapeutic Environment

13. **There are many strategies (such as quarantine, vector control, and patient education) to break the epidemiological triangle and prevent disease transmission.**
    ASM Recommended Curriculum Guideline: 23
    NCLEX-RN Alignment: Safety & Infection Control; Behavioral Intervention; Therapeutic Communication

Controlling Microbial Growth to Limit Disease

14. **A microbe's survival and growth in a given environment depends on its metabolic characteristics.**
    Understanding a pathogen's metabolic features helps in recognizing where they can thrive and their potential for introduction into humans.
    ASM Recommended Curriculum Guidelines: 11, 13
    NCLEX-RN Alignment: Physiological Adaptation; Basic Care & Comfort

15. **Microbial growth is controlled using physical, chemical, mechanical, and biological means.**
    Controlling microbial growth lowers healthcare-acquired infection (HAI) incidence. Specific and nonspecific immune defenses are biological controls against pathogens. An understanding of microbial control helps healthcare workers understand how critical, semicritical, and noncritical equipment should be managed, as well as how to properly prepare patient body sites for medical procedures like injections and surgery.
    ASM Recommended Curriculum Guidelines: 7, 8, 14
    NCLEX-RN Alignment: Pharmacological & Parenteral Therapies; Health Promotion & Maintenance; Physiological Adaptation; Basic Care & Comfort

16. **Antimicrobial compounds combat bacteria, fungi, helminths, protozoans, and viruses.**
    Understanding structural and functional microbial features allows us to develop new antimicrobial drugs and assess drug specificity mechanisms to limit adverse drug effects. The type of antimicrobial drug used to treat a particular pathogen depends on patient and microbe features.
    ASM Recommended Curriculum Guidelines: 14, 15
    NCLEX-RN Alignment: Safety & Infection Control

17. **Proper stewardship of antimicrobial drugs is essential to limit antimicrobial resistance.**
    Antimicrobial drug stewardship includes testing and tracking antimicrobial resistance; prescribing antimicrobial drugs only when truly needed; and promoting patient compliance with drug dosing regimens.
    ASM Recommended Curriculum Guidelines: 14, 15
    NCLEX-RN Alignment: Pharmacological & Parenteral Therapies; Safety & Infection Control; Psychosocial Integrity

Scientific Thinking and Critical Thinking Skills

18. **Applying the process of science is relevant to nursing.**
    Understanding the process of science is central to science literacy and fundamental to nursing practices. Analyzing and interpreting results from a variety of microbiological tests and applying analytical reasoning to solve problems are central to nursing practices.
    ASM Recommended Curriculum Guideline: 28
    NCLEX-RN Alignment: Nursing Process

19. **Using quantitative reasoning ties into nursing practice.**
Healthcare workers should: (1) be competent in drawing conclusions from charts and graphs related to patient medical history; (2) understand the metric system and scientific notation, as this terminology is used in patient medical history and is used in calculating medication dosages; (3) appreciate that microbe levels impact disease development and prognosis.
ASM Recommended Curriculum Guideline: 29
NCLEX-RN Alignment: Pharmaceutical & Parenteral Therapies; Reduction of Risk Potential

20. **The ability to communicate and collaborate with other disciplines is important for a cross-disciplinary healthcare team.**
Healthcare workers should be able to effectively communicate about microbiology-related topics in written and oral formats and effectively work as individuals and in groups.
ASM Recommended Curriculum Guideline: 30
NCLEX-RN Alignment: Communication & Documentation; Management of Care

21. **Understanding the relationship between science and society improves clinical practice and promotes the human aspect of medicine.**
Nurses should be able to identify and discuss ethical issues in microbiology, especially with regard to vaccines.
ASM Recommended Curriculum Guideline: 31
NCLEX-RN Alignment: Management of Care; Communication & Documentation; Psychosocial Integrity

Microbiology Laboratory Skills

22. **Aseptic technique is central to collecting clinical samples and to protecting healthcare providers and patients.**
ASM Recommended Curriculum Guidelines: 32, 34
NCLEX-RN Alignment: Management of Care; Safety & Infection Control; Health Promotion & Maintenance; Psychosocial Integrity; Basic Care & Comfort

23. **Microbiological and molecular lab techniques are key to identifying pathogens and implementing effective treatment options.**
ASM Recommended Curriculum Guidelines: 33, 34, 35, 36
NCLEX-RN Alignment: Management of Care

24. **All healthcare providers must understand protective procedures for handling infectious materials to prevent the spread of disease.**
Understanding biosafety levels and emergency procedures is central to safe nursing. Understanding proper biomedical waste management is important to reduce risk of pathogen exposure and to limit infections.
ASM Recommended Curriculum Guideline: 37
NCLEX-RN Alignment: Safety & Infection Control; Physiological Adaptation; Reduction of Risk Potential; Basic Care & Comfort

25. **The ability to document and report on experimental protocols, results, and conclusions is key to patient treatment.**
Nurses must accurately label specimens and keep records.
ASM Recommended Curriculum Guideline: 38
NCLEX-RN Alignment: Communication & Documentation

Glossary

A antigen The carbohydrate antigen present on the surface of red blood cells (erythrocytes) in people with type A or type AB blood.

AB toxins A class of exotoxin; B is the binding part of the toxin while the A is the active portion of the toxin that enters a target cell to exert its effect.

Acceptor or (A) site Site in a ribosome that accepts incoming charged t-RNAs.

Accessory organs Digestive system organs that do not directly contact food, but have essential roles in digestion; examples include salivary glands, liver, gallbladder, and pancreas.

Acetone-alcohol The decolorizing agent used in the Gram stain procedure; following the decolorizing step, Gram-positive bacteria remain purple while Gram-negative bacteria are left colorless.

Acetyl group A functional group with the chemical formula C_2H_3O; it's common in biology and found in acetylcholine (a neurotransmitter) and acetyl-CoA (a molecule that is fed into the Krebs cycle).

Acid-fast stain Differential staining procedure that distinguishes between cells with and without waxy mycolic acid cell walls; *Mycobacterium tuberculosis* and *Mycobacterium leprae* are examples of clinically important acid-fast bacteria.

Acidic dyes Negatively charged dyes that do not easily enter cells and are used to stain the background of a sample (e.g., negative stain procedure); examples of acidic dyes include nigrosin and India ink.

Acidophiles Organisms that grow at pH 1 (or less) to pH 5, and live in areas such as sulfur hot springs and volcanic vents; often these organisms are archaea that use the inorganic elements around them for energy and carbon sources.

Acidosis A condition of lower than normal blood pH; a drop in arterial blood pH to 7.35 or lower; this condition can develop from respiratory or metabolic issues.

Acids Contribute hydrogen ions (H^+) to an aqueous solution; acids lower the pH of a solution.

Acquired Immune Deficiency Syndrome (AIDS) A syndrome that develops when T helper cell populations are killed by the human immunodeficiency virus (HIV); HIV^+ patients are said to have progressed to AIDS once their T helper cell levels drop below 200 cells per cubic millimeter (mm^3).

Acquired resistance A form of antimicrobial drug resistance that develops due to genetic mutation or, in the case of bacteria, a form of antimicrobial drug resistance that develops when drug resistance genes are transferred or shared via horizontal gene transfer events such as conjugation and transformation.

Acral gangrene Tissue death (necrosis) characterized by a blackening of tissue; tends to develop because of disseminated intravascular coagulation that blocks blood flow to the extremities; most commonly develops in fingers, toes, and the nose.

Actin A protein that makes up microfilaments; associates with myosin to facilitate cell movement.

Activation energy The minimum amount of energy required to start a chemical reaction.

Active infection In this type of infection, the patient is symptomatic, meaning they have signs and symptoms of infection.

Active site The site of an enzyme that interacts with the substrate to generate a chemical reaction.

Active transport Transport of substances by the cell that requires energy to operate.

Acute A term used to describe rapid onset and progression of a disease or infection (e.g., acute disease = disease with a rapid onset and progression; acute infections = infections that run their course and are usually cleared by the host immune system).

Acute HIV infection/syndrome The first of three broad stages in HIV infection; usually marked by flu-like symptoms that resolve within a few weeks; in this stage of infection, the virus enters the body, circulates in the blood, and infects an array of white blood cells (especially T helper cells).

Acute phase Stage of infection occurring after the prodromal phase; in this phase the patient could experience the full-blown classical symptoms of the disease.

Acute respiratory distress syndrome (ARDS) This syndrome develops in critically ill COVID-19 patients; it is characterized by a rapidly progressing form of inflammatory lung injury that leads to a sudden and marked reduction in blood oxygenation.

Adaptive immune response/adaptive immunity Also called specific immunity or acquired immunity, these responses constitute our third and final line of our immune defenses; called into action when innate first- and second-line defenses fail to contain a threat; mature over time, are tailored to pathogens that are encountered, and may exhibit immunological memory.

Adenine (A) Purine nitrogenous base used to build nucleotides of DNA and RNA.

Adenosine diphosphate (ADP) Molecule made of adenine and two phosphate groups; formed when the last (terminal) phosphate group is removed from ATP by a dephosphorylation reaction.

Adenosine monophosphate (AMP) Molecule made of adenine and one phosphate group; formed when cells dephosphorylate ADP (remove the last/terminal phosphate group of ADP).

Adenosine triphosphate (ATP) A key molecule required by all cells to do cellular work; made by catabolic reactions and provides the energy for anabolic reactions.

Adenoviruses Non-enveloped DNA viruses; infect a variety of age groups; over 50 types of adenoviruses infect humans; most attack the respiratory tract and cause sore throat and cold-like symptoms; occasionally cause bronchitis, pneumonia, and croup.

Adhesins Virulence factors that pathogens use to stick/adhere to host cells in a specific or nonspecific manner.

Adjuvants Pharmacological additives that enhance the body's natural immune response to an antigen; often added to vaccines to encourage antigen uptake and processing by antigen-presenting cells.

Aerobes Microbes that have evolved to use oxygen in their metabolic pathways; usually have mechanisms to detoxify ROS (reactive oxygen species) so that they can safely use oxygen in their metabolism.

Aerobic respiratory chains Electron transport chains that use oxygen as a final electron acceptor.

Aerotaxis "Aero" means oxygen and "taxis" means movement; aerotaxis describes the movement of cells in response to oxygen gradient/level.

Aerotolerant anaerobes Organisms that tolerate atmospheric oxygen, even though they do not use it in their metabolic processes; like aerobes, these microbes have ways to deactivate ROS (reactive oxygen species).

African sleeping sickness Neurological disease that is fatal if not treated and is characterized by hallucinations and sleep disturbances; found in tropical regions of Africa; caused by the flagellated protozoan *Trypanosoma brucei*, which is transmitted by the blood-sucking tsetse fly.

Agglutinate To clump.

Agglutination A reaction in which antibodies bind antigens into a clump; these reactions are usually used for blood typing, to identify infections, and to diagnose noninfectious immune disorders such as autoimmune hemolytic anemia.

Agranulocytes White blood cells that lack cytoplasmic granules that are readily detectable by light microscopy; agranulocytes include monocytes/macrophages and lymphocytes (natural killer cells, B cells, and T cells).

Airborne transmission A type of indirect contact transmission in which the pathogen enters through the respiratory route as it is inhaled.

Albendazole Broad-spectrum antihelminthic drug that interferes with glucose uptake in worms by targeting microtubules; treats infections caused by roundworms such as *Ascaris*, hookworms, pinworms, *Trichinella*, and certain tapeworms.

Alcohol fermentation Biochemical process that converts pyruvic acid from glycolysis into ethanol and carbon dioxide; not carried out by human cells.

Alcohols Organic molecules that have a polar hydroxyl (R-OH) functional group that confers most of their chemical properties; examples include ethanol and isopropanol, that are most commonly used as antiseptics on skin and to disinfect small equipment like thermometers, scissors, and stethoscopes.

Aldehydes Substances, such as formaldehyde and glutaraldehyde, that act as high- or intermediate-level disinfectants based on their concentration; work by reacting with proteins and nucleic acid; used to sterilize surgical instruments, endoscopes, dialyzers, and anesthesia and respiratory equipment.

Alkaliphiles Microbes that grow in the basic pH range of 9–11.

Alkalosis A condition of higher than normal blood pH (above 7.45); can develop from respiratory or metabolic issues.

Allergen Any antigen that triggers IgE production and leads to allergy.

Allergen desensitization Occurs when the body shifts its immune responses to the allergen, by moving from a T helper 2–dominated immune response to a T helper 1–dominated response; the shift in T cells then alters the balance of antibodies made to the allergen from IgE to IgG.

Allergy A scenario in which IgE production has been triggered by an allergen and the immune system fights off a perceived threat that would otherwise be harmless.

Allografts Tissue or organ transplants that are similar to the host, but not genetically identical.

Allorecognition The process lymphocytes use to differentiate self from foreign MHCs (major histocompatibility complexes).

Allosteric activation Occurs when a regulatory molecule increases enzyme activity by binding to the enzyme's allosteric site.

Allosteric inhibition Occurs when a regulatory molecule binds to the enzyme's allosteric site and decreases the enzyme's activity.

Allosteric site A site on the enzyme other than the active site.

Allyamines Broad-spectrum antifungal agents (e.g., terbinafine and naftifine) commonly prescribed for athlete's foot, ringworm, and yeast infections; disrupt fungal plasma membranes by inhibiting enzymes that build ergosterol.

Alpha hemolytic (or α-hemolytic) Term applied to bacteria that do not lyse red blood cells and instead just oxidize hemoglobin, the oxygen-carrying component of blood; they turn blood agar a green color.

Alpha-helices (α-helices) One of two forms of secondary structure in proteins, these are spiral structures that have specified dimensions.

Alternative pathway Pathway of complement activation that is triggered when complement proteins directly interact with the invading agent.

Amebiasis Amebic dysentery caused by the intestinal parasite *Entamoeba histolytica*.

Ames test Test used to identify potential mutagens.

Amination Process in which cells make nonessential amino acids by adding an amine group (NH_2) to a biological intermediate.

Amines Organic molecules that have NH_2 functional groups.

Amino acid catabolism tests Tests that help identify certain enteric bacteria based on their amino acid catabolism profiles.

Amino acids Molecular building block of proteins; have a core structure consisting of an amine group (NH_2) and a carboxyl group (COOH).

Aminoglycosides Collection of mainly narrow spectrum antibacterial drugs that bind to the 30S ribosomal subunit to block protein synthesis; drug names tend to end in "mycin" or "micin."

Amphibolic pathways Metabolic pathways used for both breaking down and building substances; simultaneously function in anabolism and catabolism.

Amphipathic Term applied to molecules that have a region that interacts well with water (hydrophilic) and a region that interacts well with lipids or other water-repellent (hydrophobic) molecules; soaps and detergents are common examples.

Amphitrichous A term that describes cells with one or more flagella present at each end (pole) of the cell.

Amsel's criteria Collection of signs used in diagnosing vaginosis; these criteria are usually applied to assess reproductive-age women who have not yet reached menopause.

Anabolic pathways Metabolic pathways that combine energy and molecules to build new substances.

Anaerobes Organisms that do not use oxygen in their metabolic processes.

Anaerobic respiratory chains Electron transport chains that are used in anaerobic cellular respiration and require an inorganic substance other than oxygen as their final electron acceptor.

Analgesia Pain management/relief.

Analytical epidemiology An epidemiological approach that investigates what caused the disease (etiological agent), why people get the disease, and how the disease can be prevented or treated.

Anaphylaxis (or anaphylactic shock) A system-wide, potentially life-threatening response to an allergen.

Anaplasmosis Systemic bacterial infection caused by *Anaplasma phagocytophilum*; considered an emerging disease; ticks that transmit anaplasmosis may also harbor bacteria that cause Lyme disease and Rocky Mountain spotted fever.

Anemia A condition characterized by a low red blood cell count or low hemoglobin level.

Angiogenesis The process of building new blood vessels.

Anion Negative ion; has a negative charge as a result of gaining electrons.

Antibiogram A summary report of an organism's susceptibility to a panel of tested antimicrobial drugs.

Antibiotics Naturally occurring antimicrobial compounds.

Antibodies Also known as immunoglobulins (Ig), these are protein molecules secreted by plasma cells; antibodies can activate complement cascades, neutralize antigens, and promote phagocytosis of targeted antigens.

Anticodon loop Part of t-RNA that is complementary to a codon on mRNA; has a role in ensuring that the proper amino acid is being brought into the ribosome to be added to the growing protein chain.

Antigen Any molecule, that if presented in the right context, may stimulate an immune response.

Antigen masking Immune evasion tactic used by a pathogen in which it may conceal antigenic features so the immune system doesn't quickly mount an attack; accomplished in a number of ways, such as a pathogen coating itself with host molecules, so it may masquerade as part of the body.

Antigen-binding sites Tips of the "Y"-shaped antibody molecule that bind to antigens.

Antigenic drift Minor genetic changes in a virus; leads to influenza epidemics.

Antigenic mimicry Immune evasion tactic used by a pathogen in which it emulates (mimics) host molecules; for example, numerous bacteria make capsules that resemble host carbohydrates and thus do not as strongly stimulate an immune response.

Antigenic shift A major genetic reassortment that dramatically changes the virus; often leads to viral strains with new features; leads to influenza pandemics.

Antigenic variation Immune evasion tactic in which a pathogen undergoes periodic altering of the surface molecules that host immune cells used to recognize a previously met pathogen, so that a rapid immune response cannot result.

Antigen-presenting cells (APCs) Dendritic cells and certain other white blood cells that take up antigens, process them, and show them to T cells.

Antihistamines Substances that serve as anti-inflammatory drugs by blocking histamine's actions.

Antimalarial drugs Drugs for treating malaria; fall into six main classes: aminoquinolines, arylaminoalcohols, artemisinins, respiratory chain inhibitors, antifolates, and certain antibacterial drugs such as doxycycline and clindamycin.

Antimicrobial drugs Therapeutic compounds that kill microbes or inhibit their growth; categorized based on the type of pathogen they target: antibacterial drugs treat bacterial infections, antiviral drugs target viral infections, antifungal drugs work against fungal infections, and antiparasitic drugs treat protozoan and helminthic (worm) infections.

Antimicrobial peptides (AMPs) Proteins that act as chemical barriers by destroying a wide spectrum of viruses, parasites, bacteria, and fungi; thousands of different AMPs exist and they have diverse modes of action.

Antimicrobial resistance A situation in which a microbe, which could be a bacterium, virus, fungus, protozoan, or helminth, is not affected by a drug therapy that is intended to inhibit or eliminate the pathogen.

Antioxidants Compounds and enzymes that reduce the effects of reactive oxygen species; many aerobes rely on antioxidants to survive in aerobic conditions.

Antiparallel An arrangement of double-stranded DNA in which one strand runs 5' to 3' and the partner strand runs 3' to 5', allowing the complementary base pairs of DNA to properly associate.

Antiport A form of secondary active transport in which ions flow in the opposite direction of the target substance being transported.

Antiretroviral therapies (ART) Therapies for HIV infection that first hit the market in the mid-1990s; usually three or more drugs are prescribed in combination.

Antisense antivirals Drugs that block viral replication by preventing the host ribosome from translating viral mRNA.

Antiseptics A class of germicides that are applied to living tissue such as skin.

Antiserum A preparation of antibodies developed to neutralize specific toxins or venoms.

Antitoxin Preparations made using antibodies against toxins, such as botulinum, anthrax, and tetanus toxins.

Antivenom Antibody preparation that neutralizes the effects of venom from snakes, spiders, scorpions, or other venomous creatures; also called antivenins.

Antonie van Leeuwenhoek Refined earlier versions of the microscope and was the first to see bacteria.

Apoenzyme An enzyme that is inactive because it lacks a necessary cofactor.

Apoptosis Programmed cell death.

Aqueous solution A liquid mixture where water is the solvent (dissolving agent), and the dissolved substance is called the solute.

Archaea Prokaryotic unicellular microorganisms; nonpathogenic; often live in extreme environments.

Arginine (*arg*) operon An example of a repressible operon that regulates the production of the amino acid arginine; a pre-transcription regulation for gene expression that allows bacterial cells to stop making the amino acid arginine if it is already abundant.

Artemisinin-based combination therapies (ACT) Used to treat certain drug-resistant malaria cases; these therapies rely on combining an artemisinin-class drug (often artesunate) with one or more nonartemisinin drugs.

Arteries Muscular vessels that conduct blood away from the heart.

Artificially acquired active immunity Immunity that involves using vaccines or other man-made interventions to trigger an immune response in the patient; confers long-term protection against specific antigens.

Artificially acquired passive immunity Immunity in which the patient receives protective antibodies from a medical treatment (such as the administration of an antivenom); does not lead to long-term protection against specific antigens.

Aseptic culturing techniques Technique by which conditions are maintained to limit contaminants, so that only the microbes in a given sample are grown.

Aseptic technique Germ-free practices; term applied to techniques designed to prevent the introduction of contaminating microbes to a patient, a clinical sample, or others in the healthcare setting; methods that prevent healthcare-acquired infections by preventing the introduction of potentially dangerous microbes to a patient.

Asexual spores Fungal reproductive structures that arise from mitosis and do not result in genetic variation in the resulting offspring; two classes of asexual fungal spores are conidiospores and sporangiospores.

Asymmetric paralysis Paralysis that affects one side of the body more than another; may develop from a polio infection.

Asymptomatic carriers Individuals who harbor certain pathogens for extended periods without experiencing symptoms.

Asymptomatic case An infection that fails to generate symptoms; also called a subclinical case.

Atomic mass Determined by the mass of the protons and neutrons in the atom; the average mass of 6.022×10^{23} atoms, or one mole, of the given element.

Atomic number Number of protons in an atom.

Atoms The smallest units of elements.

Atopic asthma Allergy-based asthma.

Atopic dermatitis An inflamed and itchy skin condition also known as atopic eczema or allergy-based eczema.

ATP (adenosine triphosphate) Made by catabolic reaction; fuels anabolic reactions.

ATP synthase An enzyme in electron transport chains that captures the energy of the flowing protons and uses it to recharge ADP to ATP; found in the mitochondrial inner membrane of eukaryotes and in the plasma membrane of prokaryotes.

ATP–ADP cycle A cycle in which the terminal (or end) phosphate of ATP is removed to release energy and make ADP; the ADP is then recharged to ATP by adding a phosphate group.

Attenuated (or attenuated strains) Term applied to pathogens that lose virulence factors needed to cause disease; they are still infectious, but weakened to the point that they do not cause disease in an immune competent host; may be used to formulate certain vaccines.

Atypical bacterial pneumonia Refers to a respiratory infection that has symptoms different from typical bacterial pneumonia; *Mycoplasma pneumoniae* and *Chlamydophila pneumoniae* are the most common causative agents.

Autoantibodies Antibodies that bind to self-tissues.

Autoclave A machine that applies steam heat along with pressure to sterilize microbiological media and assorted medical or laboratory equipment.

Autografts Transplants from self; a self-skin graft from one part of the body to another location; lowest risk for a rejection reaction.

Autoimmune disorders Chronic conditions that develop when a patient's immune system attacks self-tissues.

Autoimmunity An inappropriate immune system attack against self-cells or self-tissues.

Autotrophs "Self-feeding" organisms, such as plants and other photosynthetic organisms, that use carbon fixation to convert inorganic carbon into organic carbon.

Axon The length of a neuron along which an electrical wave travels; axons are covered in myelin, a fatty layer that helps electrical impulses travel quickly along the axon's length; once an electrical wave reaches the end of the axon, neurotransmitters are released to convey the signal to a neighboring neuron.

Azoles Broad-spectrum antifungal drugs (examples: fluconazole and ketoconazole); disrupt fungal plasma membranes by inhibiting enzymes that build ergosterol.

B antigen A carbohydrate antigen found on red blood cells (erythrocytes) of people with type B or type AB blood.

B cell receptors (BCRs) Antigen recognition receptors on B cells.

B cells Adaptive immunity leukocytes that are produced and mature in the bone marrow, and are present in lymphoid tissues throughout the body; B cells are activated to plasma cells and coordinate the humoral response by making antibodies.

Bacilli (singular = bacillus) Cells that have a rod or cylindrical shape.

Bacitracin A polypeptide drug that targets bacterial cell walls and acts as a narrow-spectrum agent against Gram-positive bacteria; common antibacterial drug in topically applied ointments like Neosporin.

Bacteremia A situation in which bacteria spread to the bloodstream; usually confirmed by culturing bacteria from a blood sample.

Bacteria Prokaryotic unicellular microorganisms; pathogenic and nonpathogenic.

Bacterial endocarditis An inflammation of the heart lining (the endocardium) and/or one or more of the heart's valves.

Bacterial vaginosis A polymicrobial disease of the vagina involving overgrowth of bacteria; the best indicator of disease is the lack of *Lactobacillus* species.

Bactericidal Substance that kills bacteria; bactericidal drugs tend to target bacterial cell walls, cell membranes, or nucleic acids.

Bacteriocins Antimicrobial peptides that limit the growth of other competing bacteria in the same local area.

Bacteriophages (or phages) Viruses that infect bacteria.

Bacteriostatic Substances that prevent bacteria from growing; bacteriostatic drugs tend to target bacterial protein synthesis and metabolic pathways like folic acid production.

Barophiles Bacteria that can withstand high-barometric pressure environments, such as occurs in the deep sea.

Bases Release hydroxide ions (OH^-) in an aqueous solution; raise the pH of a solution.

Basic dyes Dyes used in staining that are mildly basic on the pH scale; frequently used basic dyes include methylene blue, crystal violet, safranin, and malachite green.

Basic reproduction number or **R_0 (R-naught)** A measure of a pathogen's transmissibility, or contagiousness—it represents the number of people that a single infected person is, on average, expected to infect in a population where all people are susceptible to infection and no prevention strategies have been applied.

Basophils Represent less than 1 percent of the leukocyte population; their overall circulating levels are rarely elevated except in certain blood cancers; these granulocytes release histamine (a molecule that stimulates inflammation); these cells combat parasitic infections and have roles in allergic responses.

Benign prostatic hypertrophy (BPH or benign prostatic hyperplasia) A noncancerous enlargement of the prostate; as the prostate enlarges it may constrict the urethra and make it difficult to fully void the bladder; BPH develops in about half of all men by age 50 and is almost universally present in men over age 75.

Beta hemolytic (or β-hemolytic) Term applied to pathogens (such as *Streptococcus pyogenes*) that make hemolysins and can therefore lyse red blood cells and generate a yellow zone around colonies growing on blood agar.

Beta-lactam antimicrobials (or β-lactam antimicrobials) A superfamily of antibacterial drugs that work by blocking bacterial cell wall construction; includes penicillins, cephalosporins, carbapenems, and monobactams.

Beta-lactam ring (or β-lactam ring) Four-sided ring that is part of the active chemical structure of penicillins, cephalosporins, carbapenems, and monobactams.

Beta-lactamase inhibitors Drugs such as clavulanate, sulbactam, or tazobactam that can be co-administered with certain beta-lactam antibiotics to combat beta-lactamases.

Beta-lactamases (or β-lactamases) Enzymes that are made by bacteria and that inactivate beta-lactam drugs.

Beta-oxidation (or β-oxidation) A process in which fatty acids are broken down, two carbons at a time, into acetyl-CoA molecules that then enter the Krebs cycle.

Beta-pleated sheets (or β-pleated sheets) One of two forms of secondary structure in proteins; accordion-like folds that have characteristic dimensions.

Binary fission An asexual form of reproduction used by most prokaryotic cells.

Binomial nomenclature system Two-name system that includes genus and species designations; developed by Carl Linnaeus.

Biochemical tests Tests that allow us to detect metabolic end products, intermediates, or particular enzymes and thus are extremely useful in identifying microbes.

Biofilms Sticky microbial communities made up of single or diverse species; they allow microbes to coordinate responses within an environment.

Biogenesis Idea that life emerges from existing life.
Biological safety cabinet An enclosed cabinet that maintains a specific flow of filtered air and also is readily decontaminated using UV light and surface cleaning with an antimicrobial solution.
Biological vectors Vector organisms that have a role in the pathogen's life cycle; common examples are ticks and mosquitos.
Bioremediation Approach that harnesses the power of microbes to help clean up toxic waste.
Biosafety level (BSL) Rankings to help hospitals and laboratories properly apply safety practices, such as appropriate safety gear and facility features, that limit exposure risks; the assignment of level, of which there are four, is based on numerous criteria, with the following being some key considerations: level of infectivity (not necessarily level of contagiousness), extent of disease caused and mortality rates, mode of transmission, and availability of preventions and treatments for the disease.
Biosynthesis The construction of biological molecules.
Biosynthetic reactions Also called anabolic reactions; reactions that use energy to build molecules.
Bladder A smooth muscle sac that receives urine from the kidneys via the ureters.
Blastomycosis A noncommunicable lung infection caused by the endemic fungus *Blastomyces dermatitidis*; also known as Chicago disease and Namekagon River fever.
Blood agar A common example of a differential medium; contains sheep red blood cells that serve as both a nutrient and a differentiating indicator for hemolytic bacteria.
Blood–brain barrier The blood vessels that deliver nutrients and oxygen to the central nervous system (CNS) and that are structurally specialized to protect the CNS from pathogens; because blood vessel cells are sealed much more tightly in the areas around our CNS than they are in other areas of the body, the blood–brain barrier allows only a few types of molecules to pass from capillaries into the CNS.
Blue light A short wavelength that has higher energy as compared with other visible light spectrum colors; closest to ultraviolet light in its energy; energy in the blue light range activates porphyrin, a light-absorbing molecule in *Propionibacterium acnes*, to make toxic oxygen species that kill the bacterium.
Bone marrow A spongy primary lymphoid tissue that is located inside bones and is a key site for red and white blood cell production.
Botulinum toxin A neurotoxin produced by *Clostridium botulinum* that causes flaccid (relaxed) paralysis.
Botulism A disease caused by *Clostridium botulinum*, a Gram-positive, rod-shaped anaerobe that makes a collection of powerful neurotoxins; the effects of these neurotoxins produce the disease.
Bright field microscopy The simplest and most common form of microscopy; in this technique a compound light microscope is used to see the specimen, which is illuminated from below with a solid cone of visible light; the sample appears as a darker contrasting image on a bright background.
Broad-spectrum drugs Drugs that are effective against both Gram-negative and Gram-positive bacterial cells.
Bronchitis Inflammation of the bronchi and/or bronchioles.
Broth dilution test A type of antimicrobial susceptibility test that can differentiate between bactericidal versus bacteriostatic actions; in this test each antibiotic is diluted in series and then added to a set amount of liquid growth medium containing a standardized amount of bacteria; after an incubation period, the level of growth for each diluted drug series is measured by assessing turbidity—the cloudiness of the sample; alternatively, there are colorimetric indicators of population growth.
Buboes Lymph nodes that become severely swollen and painful with regions of necrosis; in lymphogranuloma venereum these buboes may rupture.
Bubonic plague The most common form of plague; it infects the lymphatic system; characterized by the formation of buboes.
Budding A method of asexual reproduction in certain fungi and some bacteria.
Buffers Compounds that stabilize solution pH by absorbing or releasing H^+ ions.
Butanediol fermentation A type of fermentation in which pyruvic acid is converted to acidic products; detected by the Voges–Proskauer (VP) test.

Candidemia (or invasive candidiasis) A situation that develops when *Candida* yeast invades the bloodstream; the most common systemic fungal infection in immunocompromised individuals and the fourth most common healthcare-acquired bloodstream infection in the United States.
Candiduria The presence of yeast in the urine.
Capillaries Thin vessels that branch from arterioles and that directly supply blood to our organs and tissues and serve as the site of nutrient and gas exchange; at only one cell thick, capillary walls easily allow oxygen and nutrients to diffuse from the blood into surrounding tissues.
Capsid The protein shell that packages and protects the genome and also accounts for the bulk of a virion's mass.
Capsomeres Three-dimensional protein subunits that make up capsids.
Capsule A sticky carbohydrate-based structure made by some prokaryotes; a well-organized glycocalyx that is tightly associated with the cell wall; presence of a capsule often increases pathogenicity (ability to cause disease), since it promotes adhesion to host tissues, and provides some protection against host immune cells by interfering with phagocytosis.
Carbapenem-resistant Enterobacteriaceae (CRE) Certain genera of the Enterobacteriaceae, a large family of Gram-negative bacteria, that have developed resistance to carbapenems and have rendered even these last-resort drugs useless against some infections; *Klebsiella pneumonia* and *Escherichia coli* are among the more prevalent CRE.
Carbapenems Broad-spectrum beta-lactam drugs that target bacterial cell wall construction; usually well tolerated; drug names for group members tend to end with the suffix "penem."
Carbohydrate Also known as saccharides, carbohydrates consist of oxygen, hydrogen, and carbon; polar organic molecules consisting of one or more sugar monomers.
Carbohydrate catabolism The breakdown of carbohydrates to release energy; central to a cell's survival.
Carbolic acid (C_6H_6O) Also known as phenol; an organic molecule with antiseptic (degerming) and anesthetic (numbing) properties.
Carbon fixation Process by which some organisms are able to make their own organic carbon molecules from inorganic start materials; used by autotrophs to convert inorganic carbon into organic carbon.
Carcinogens Mutagens that cause a rate of mutation that promotes the development of cancers.
Cardinal signs of inflammation Redness, pain, localized heat (not fever), and swelling; occasionally a fifth feature, loss of function, also develops with severe inflammation that generates intense pain and/or swelling that limits use of the affected body part.
Carl Linnaeus Considered the father of taxonomy; developed the binomial nomenclature system, or two-name system, that includes genus and species designations.
Case reports Descriptive epidemiological studies that consist of individual or group records of a disease; provide an important connection between clinical medicine and epidemiology.
Catabolic pathways Metabolic pathways that break down substances and release energy.
Catalase test Uses hydrogen peroxide to detect whether an organism has the enzyme catalase, which protects organisms from reactive oxygen species (ROS).
Catalyst An organic or inorganic substance that is only needed in small amounts to make a reaction happen faster; it is not consumed or permanently changed by a reaction.
Catenation The ability of atoms of the same element to form long chains; carbon is a prime example of an element with catenation ability.
Catheter-associated urinary tract infections (CAUTIs) Infections that account for at least one million cases of complicated UTIs each year and are the most common type of healthcare-acquired infection; the most common cause of secondary bloodstream infections that result when bacteria spread from the urinary tract into the bloodstream to cause a systemic infection.
Cation Positive ion; has lost electrons and consequently has an overall positive charge; examples include magnesium (Mg^{2+}) and sodium (Na^+) ions.
Cationic Refers to a substance that has positively charged ions.
Cell count Enumerates the number of cells in a small portion of the sample; done using automated or manual procedures.
Cell signaling Collection of cellular communications that occur within a cell and between cells to manage cellular activities.
Cell wall A cell wall that lies just outside of the plasma membrane in most prokaryotic cells, fungi, plants and some other organisms; animal cells lack a cell wall.
Cellular respiration A collection of reactions that extract energy from foods using redox reactions and then transfer that energy into the bonds of ATP; occurs through the combined efforts of glycolysis, an intermediate step, the Krebs cycle, and the electron transport chain.
Cellular response Also called T cell–mediated immunity; one of two branches of the adaptive immune system that eliminates an identified antigen and remembers it so that next time adaptive responses are faster.
Cellulitis A deeper *Staphylococcus aureus* infection of the lower dermal and subcutaneous fat; mainly seen in adults and is characterized by red, swollen, and painful skin; may be accompanied

by fever, leukocytosis, and/or lymphangitis (inflamed lymphatic vessels).

Centers for Disease Control and Prevention (CDC) A federal health agency that falls under the U.S. Department of Health and Human Services; serves as a central source of epidemiological information.

Central dogma Refers to the general flow of genetic information from DNA to RNA to protein; the revised central dogma takes into account that RNA can be used as a template for DNA.

Central nervous system (CNS) Spinal cord and brain; the CNS integrates information received, and sends back an "action plan."

Centrifuge A machine that spins samples at high speed to separate out components in a liquid based on their density; when blood is centrifuged, the formed elements (platelets and all blood cells) are separated from plasma.

Centrosome Specialized organelle normally located near the nucleus; it is made up of two centrioles (barrel-shaped structures made of microtubules arranged in a cartwheel structure of nine triplets) and surrounding unstructured material.

Cephalosporins Broad-spectrum, bactericidal drugs that constitute the largest collection of beta-lactam drugs; can be recognized by their "cef" or "ceph" prefixes.

Cerebrospinal fluid (CSF) Colorless, watery fluid produced by structures within the brain and that flows in a space between the meninges called the subarachnoid space; cushions and nourishes the central nervous system; some infectious agents invade this fluid and attack the CNS.

Cervicitis Opening of the uterus (the cervix) becomes inflamed.

Cervix The opening to the uterus that leads into the vagina.

Chancre An ulcer-like lesion that is often on the genitalia, but could develop elsewhere; mainly associated with syphilis.

Chancroid Disease caused by the Gram-negative nonmotile rod *Haemophilus ducreyi*; rare in developed countries; more common in men than in women; one of the most common cofactors for HIV transmission.

Chaperones Smaller proteins that the cell makes to help fold larger proteins.

Chemical barriers Chemical factors may directly attack invaders or establish environments that limit pathogen survival in or on a particular tissue; examples include stomach acid and lysozyme in secretions.

Chemical bonds Forces that bind atoms in molecules and are typically classified as ionic or covalent; the types of bonds present in a given molecule depend on how electrons of the bonding participants interact; hydrogen bonds do not bind atoms into molecules and are an attractive force within or between molecules.

Chemical symbol An abbreviated letter notation that derives from the name of the element, which is often Greek or Latin.

Chemiosmosis The flow of a proton gradient across a membrane; this process occurs in respiratory chains to drive the phosphorylation reactions that ultimately recharge ADP to ATP.

Chemoattractants A diverse collection of cytokines and other signaling molecules that draw leukocytes to injured tissue; the recruitment phase of inflammation relies on chemoattractants.

Chemokines A class of cytokines that induce chemotaxis and act as signaling proteins that attract white blood cells to areas where they are needed; have important roles in wound healing, blood vessel formation and repair, lymphoid tissue development, and activation of innate and adaptive immune responses.

Chemostat A system of bacterial cell growth in which fresh growth medium is added at one end of the culturing device, while waste, nutrient-depleted medium, and excess cells are removed at another end of the system to maintain a constant culture volume; this maintains cellular growth at a constant rate instead of the culture experiencing all four growth phases.

Chemotaxis Cell movement in response to a chemical stimulus.

Chemotrophs Organisms that break down chemical compounds for energy; organisms that rely on energy found in the chemical bonds of their nutrients to make ATP.

Chikungunya Viral illness caused by the chikungunya virus, a single-stranded RNA virus; features a sudden onset of fever and polyarthralgia (extreme pain in multiple joints), headaches, vomiting, and rash may also occur.

Chlamydia In the context of disease, this term usually refers to urogenital infections or neonatal conjunctivitis caused by *Chlamydia trachomatis*; in the taxonomy context, the genus *Chlamydia* includes three pathogenic species (*C. psittac; C. trachomatis*; and *C. pneumoniae*) that are Gram-negative intracellular bacteria.

Chlamydophila pneumonia Also called chlamydia pneumonia; pneumonia caused by *Chlamydophila pneumoniae* (also called *Chlamydia pneumoniae*), a Gram-negative bacterium that lives inside host cells; a common cause of sinusitis, pneumonia, bronchitis, and pharyngitis.

Chloramphenicol A phenicol drug effective against Gram-positive and Gram-negative cocci and bacilli that include *Rickettsia*, *Mycoplasma*, and *Chlamydia* species; this antimicrobial has a narrow therapeutic index and is associated with bone marrow toxicity that results in aplastic anemia.

Chloroplasts An organelle that is structurally similar to mitochondria and that allow the eukaryotic cells that have them to harvest energy from sunlight using light-collecting pigments like chlorophylls and carotenoids.

Chloroquine An aminoquinoline drug that has been the most widely used antimalarial drug; malarial pathogen resistance to this drug is increasingly common.

Cholera toxin Potent exotoxin produced by *Vibrio cholerae*, a comma-shaped Gram-negative bacterium with a single flagellum; this exotoxin triggers intestinal epithelial cells to release electrolytes and water into the intestines, generating a watery, starchy-colored diarrhea or "rice water" stool.

Chromosomes Carefully packaged strands of DNA associated with organizational proteins; linear in eukaryotes and usually circular in prokaryotes.

Chronic carrier A patient in whom the pathogen can exist in a dormant (or latent) state and reactivate later; a chronic carrier patient may remain asymptomatic for long periods, only to have symptoms reemerge from time to time; even when chronic carriers are asymptomatic, they may infect others.

Chronic diseases Diseases with a slower onset and progression.

Chyme Mixture of food and gastric juices in the stomach; small amounts enter the small intestine.

Cilia Structurally similar to flagella except that they are much shorter and far more numerous on a cell; only found on eukaryotic cells; our airways are lined with ciliated cells that sweep debris-laden mucus toward the nose and mouth.

Cirrhosis Scarring of the liver; can be caused by hepatitis B virus infection.

Classical complement cascade The first complement pathway discovered; most commonly triggered by antibodies bound to an invading agent.

Clathrin A protein involved in clathrin-mediated endocytosis; this protein coats the inner surface of the plasma membrane where the receptor–ligand complex is located.

Clathrin-mediated endocytosis Most common form of receptor-mediated endocytosis in which a clathrin-coated vesicle containing the receptor–ligand complex forms and enters the cell, where it sheds its clathrin coat and fuses with an endosome, after which ligand and receptor separate from one another and are then delivered to their final destination.

Clean-catch urine sample Midstream sample of urine used for testing; a way of collecting urine that limits contamination from the genitalia; the area around the opening of the urethra is wiped with an alcohol prep pad or sanitation wipe and then urine is collected midstream into a sterile sample cup.

Clinically asymptomatic stage The second of three broad stages in HIV infection; as the name suggests, this stage is not marked by symptoms, however, the virus is not dormant and instead is actively replicating to shed low levels of virions into the patient's blood.

Clitoris Part of the vulva (female external genitalia), erectile tissue that is analogous to the penis and is stimulated to induce orgasm.

Clonal expansion A process in which activated B and T cells undergo multiple rounds of cell division to proliferate (increase their population).

Clones Genetically identical cells.

Cluster of differentiation (CD) proteins Specialized glycoproteins on the surfaces of T cells; these proteins enable us to tell the cytotoxic and helper T cells apart.

Coarse focus knob A part of the compound light microscope that allows the viewer to roughly focus the image by affecting the distance of the objective lens from the specimen.

Cocci (singular = coccus) Spherical prokaryotic cells.

Coccidioidomycosis (cock SID ee oy doh my *koh* sis) A noncommunicable lung disease caused by the dimorphic fungi *Coccidioides immitis* and *Coccidioides posadasii*.

Coccobacilli (singular = coccobacillus) Ovoid prokaryotic cells.

Codon Triplet of nucleotides found in messenger RNA (mRNA); read by ribosome during translation, codons can encode an amino acid, a start signal or a stop signal; there are 64 codons in nature.

Coenzyme A (CoA) A coenzyme that contains a derivative of pantothenic acid (vitamin B5); important in fat metabolism, and has a central role in the Krebs cycle.

Coenzymes Organic nonprotein cofactors that range from free molecules that can move about to factors anchored to the enzyme they assist; are often vitamins or are made from vitamins; some coenzymes collect electrons from one reaction and shuttle them to other reactions in the cell; common coenzymes that act as electron carriers in metabolism include NAD^+, $NADP^+$, FMN, and FAD.

Cofactors Additional components needed by enzymes to function.

Cohorting Rooming patients with the same active infection and no other infections; may be acceptable if approved by the infection control team.

Colistin (also called polymyxin E) A bactericidal narrow-spectrum polypeptide drug that targets the outer membrane of facultative anaerobes and aerobic Gram-negative cells.

Colitis Inflammation of the colon (the large intestine).

Collision theory A theory that proposes that the energy transferred during collisions of atoms and molecules can disturb the electron structures of atoms and molecules enough to make or break chemical bonds.

Colony Grouping of cells that developed from a single parent cell; cells in a colony are genetically identical to the parent cell; a clonal population of cells that form a small spot or mound that is visible on solid media.

Colony-forming units Measurement that represents the numerical data for plate counts; expressed as colony-forming units (CFU) per milliliter (or per gram).

Coma Prolonged state of unconsciousness that resists waking; can result from physical damage to CNS tissues.

Combination therapy Therapy that uses two or more drugs in combination; combination therapies are often used to decrease the likelihood that a pathogen will survive the therapy due to drug resistance.

Comedonal acne One of the most common skin afflictions; mainly caused by *Propionibacterium acnes* bacteria; mild form of acne.

Commensalism A type of symbiotic relationship that has no perceived benefit or cost to the host.

Communicable Pertaining to pathogens that transmit from human to human.

Community-acquired Obtained from a human, animal, or environmental source in a non-healthcare setting.

Community-acquired pneumonia (CAP) A case of pneumonia that develops within 48 hours of admission and does not meet healthcare-associated pneumonia (HCAP), hospital-acquired pneumonia (HAP) or ventilator-associated pneumonia (VAP) definitions.

Competitive inhibitors Substances that slow reactions by competing with a substrate for the target enzyme's active site.

Complement fixation assays Used to detect patient antibodies in a serum; faster and more sensitive techniques like ELISAs and some genetics methods have made these tests much less prevalent in modern diagnostic testing.

Complement system (or complement cascades) A system of over 30 different proteins that work together in a cascade fashion to protect us against infectious agents; complement proteins tend to be labeled with a "C" followed by a number (such as C1, C2, etc.).

Complex media Also called enriched media; contain a mixture of organic and inorganic nutrients that are not fully defined; instead, they contain more complex ingredients like blood, milk proteins, or yeast extract.

Complex viral structure Catchall structural category for viruses with less conventional capsids.

Complicated malaria A more severe form of the protozoan infection, malaria; includes the episodic attacks, but may be accompanied by anemia, low blood pressure (hypotension), low blood glucose (hypoglycemia), and/or excessive acidity of the blood (acidosis); kidney failure, acute respiratory distress syndrome, and malarial infection of the brain are also possible.

Complicated UTIs Urinary tract infections that tend to develop in people who are catheterized, have urinary tract malformations or obstructions, and/or have underlying health conditions that make it harder for them to fight off infections; these infections are much more likely to recur and may not respond to first-line drug therapies; bacteria are the leading cause of infection.

Compound Term used to describe molecules that are made of more than one type of element.

Compound light microscope The most common type of optical microscope and a basic tool found in microbiology labs.

Concentration Determined by the amount of solute dissolved in a specific volume of solvent.

Concentration gradient A gradation of a substance that forms when there is an unequal distribution of a dissolved substance in an environment; electrochemical gradients are generated when ions (charged atoms) are unequally distributed; electrochemical gradients drive many biological processes, including the spread or propagation of signals in our nervous system.

Conclusion Statement of whether the data supported or contradicted the hypothesis; an interpretation of what data mean; derived from observations.

Concussion Injury to the brain and/or spinal cord caused by jarring physical force; usually causes temporary CNS dysfunction that may or may not be accompanied by a brief period of unconsciousness.

Condenser A part of the compound light microscope consisting of lenses that sharpen light into a precise cone to illuminate the specimen; the condenser's iris diaphragm allows the viewer to modulate how much light is aimed at the specimen in order to improve contrast.

Condylomata lata Wart-like genital or anal lesions.

Congenital defects Birth defects or developmental issues present from birth.

Congenital rubella syndrome (CRS) Rubella infection in pregnant women, especially during the first trimester, that can lead to stillbirth or miscarriage or a variety of birth defects, including blindness, deafness, heart defects, and growth or mental retardation.

Congenital syphilis Syphilis resulting from mother-to-fetus transmission, which may occur at any time during pregnancy; if untreated, fetal death, birth defects, or infected infants may result; routine screening has dramatically reduced cases of congenital syphilis in the United States.

Conjugate vaccines Also called polysaccharide vaccines; these vaccines are created by linking polysaccharides to a protein antigen to stimulate an immune response in a vaccinated patient.

Conjugation A form of horizontal gene transfer that requires the formation of a hollow tube called a pilus to transfer genetic information from one bacterium to another.

Conjunctiva The epithelial membrane that covers the eyeball and lines the eyelids (it does not cover the cornea but surrounds it); pathogens that invade the conjunctiva cause inflammation called conjunctivitis.

Conjunctivitis Inflammation of the conjunctiva; the white sclera portion of the eye may become red or pink from inflamed blood vessels, which gives conjunctivitis its other name, "pink eye;" most forms of conjunctivitis are bacterial or viral and tend to be highly contagious.

Consolidation A term that refers to the apparent merging of the lung air sacs (alveoli) when fluid accumulates in the lungs; evident as white opaque areas by X-ray imaging.

Constant region The stem portion of the "Y"-shaped antibody molecule; this region does not directly participate in antigen binding and is the same (held constant) for a given antibody class.

Contact dermatitis Reaction that sometimes develops against a complete antigen and other times against haptens like pharmaceutical drugs, nickel, chromate, or the poison ivy toxin (pentadecacatechol) that have combined with proteins in skin cells; once T cells are sensitized, upon a secondary exposure to the same antigen, skin inflammation generates an extremely itchy (pruritic) red rash.

Contagious Pertaining to communicable diseases that are easily transmitted from one host to the next.

Continuous replication Characteristic of DNA replication along the leading strand template because the direction of replication on the leading strand is the same as unwinding.

Convalescent period Not fully better, but recovering; stage of infection that usually involves elimination of the pathogen from the body, but sometimes the host harbors a pathogen indefinitely; this is especially true for certain viruses.

Coronavirus disease 2019 (COVID-19) Respiratory infection caused by SARS-CoV-2.

Cornea A transparent tissue at the front of the eye that covers the iris; it has five to six layers of epithelial cells that can quickly be replaced if superficially damaged.

Correlation studies Also called ecological studies; descriptive epidemiological studies that search for possible associations between an exposure and the development of a disease in a population.

Co-stimulatory proteins Proteins on the surface of an antigen presenting cell (APC) that bind to proteins on the T cell's surface; cell membrane proteins on an APC and on a T cell that interact to provide the secondary activation signal.

Coulter counter A machine that counts the number of cells, both living and dead, as they pass through a thin tube; an electronic counter detects a cell as it passes and keeps a tally.

Covalent bond The electrostatic force of attraction between atoms that share one or more pairs of electrons.

Creatinine A molecule that is released as a normal byproduct of the constant tissue remodeling and maintenance that muscle tissues experience; elevated blood creatinine levels can indicate a kidney disorder.

CRISPR-Cas9 A gene-editing tool that was developed based on what is essentially a prokaryotic immune response against bacteriophages; this system locates a specific DNA sequence and cuts it out with surgical precision so that new DNA can be plugged into the cut site.

Critical equipment Equipment that comes into contact with sterile body sites or the vascular system and must therefore be sterilized; examples include surgical tools, implants, and urinary and cardiac catheters.

Cross-sectional studies Also called prevalence studies; descriptive epidemiological studies that evaluate exposure and the development of disease across a defined population at a single point in time; these studies often use surveys to determine prevalence of a disease.

Croup A combined inflammation of the larynx, trachea, bronchi, and bronchioles, most often caused by viruses; also called laryngotracheobronchitis; characterized by a barking cough and stridor.

Cryptococcosis A disease caused by infection with the yeast-like fungus *Cryptococcus neoformans*, which is inhaled from soil or bird droppings; once the fungus enters the lungs, it invades macrophages and grows inside them.

Crystal violet A purple dye used in the Gram stain technique.

Culture media Mixtures of nutrients that support organismal growth in an artificial setting.

Cutaneous anthrax Mildest form of the disease anthrax caused by *Bacillus anthracis*; develops when *B. anthracis* enters a wound through soil or by direct contact with infected animals or contaminated animal products; due to various exotoxins that these bacteria make, the infection is potentially deadly if not promptly treated with antibiotics.

Cutaneous leishmaniasis A protozoan infection caused by *Leishmania* in which skin ulcers, which are usually painless, form at bite sites and can persist for months or even years.

Cutaneous mycoses Fungal skin infections; common causative agents include *Candida* species and organisms that cause athlete's foot or ringworm; most cutaneous mycoses are superficial.

Cysticercosis A complication of *Taenia solium* infection that occurs when humans ingest the *T. solium* eggs rather than the cysticerci in undercooked pork; upon ingestion, the eggs hatch in the human host's intestines; the larvae then burrow through the intestinal epithelium and make their way via the bloodstream to human tissues, including the brain and eyes; when cysticerci form in the tissues they interfere with tissue function and cause severe and potentially deadly symptoms, especially when the brain is affected (neurocysticercosis).

Cystitis An inflammation of the bladder.

Cytolytic An agent that causes the lysis, or rupture, of a cell; usually applies to the ability of a virus to induce rapid host cell rupture after infection and viral replication.

Cytopathic effects Effects that occur as pathogens establish themselves in the host and damage host cells; these effects can kill the cell (cytocidal), or simply damage it (noncytocidal).

Cytoplasm Watery substance that bathes the internal structures of every cell; since prokaryotic cells lack membrane-bound organelles to compartmentalize the intracellular environment, most of their biochemical reactions occur in the cytoplasm.

Cytosine (C) Pyrimidine nitrogenous base that is an essential ingredient in nucleic acids.

Cytoskeleton A dynamic and responsive intracellular network of protein fibers that maintains cell shape, facilitates movement, protects against external forces that may otherwise deform the cell, and directs transport of vesicles, organelles, and other cellular cargo; it also coordinates cell division by moving chromosomes and organelles to developing daughter cells; eukaryotes and prokaryotes have a cytoskeleton.

Cytotoxic reactions (or cytolytic reactions) Develop in type II hypersensitivities; these reactions tend to trigger cell lysis; cytotoxic reactions can be a normal response to a foreign antigen or can be autoimmune based.

Deamination A reaction in which a molecule is stripped of its amine group (NH_2); deamination is performed as a part of amino acid catabolism.

Death phase Fourth and final growth phase of bacteria in which waste buildup and decreasing nutrients reach a critical point and the cells begin to die; during the death phase the rate of cell death is exponential and varies based on the starting factors and species being grown.

Decarboxylation reactions Reactions in which carbon dioxide is removed.

Decimal reduction time (DRT or *D value*) The time in minutes that it takes to kill 90 percent of a given microbial population at a set temperature.

Decontamination Interventions or treatments that remove or reduce microbial populations to render an object safe for handling.

Deep A format for solid or semisolid culture media; to make a deep, the test tube of culture media is placed in an upright position while the medium solidifies.

Defensins An important class of mammalian AMPs that rapidly kill invaders by inserting themselves into target cell membranes.

Defined media Also called synthetic media; media with a precisely known composition; each organic and inorganic component is completely known and quantified.

Degranulation An IgE–allergen interaction that causes the antibody-coated mast cells and basophils (also called IgE-primed mast cells or basophils) to release proinflammatory factors, such as histamine and leukotrienes, from their cytoplasmic granules.

Dehydration A clinical condition that develops due to excessive water and electrolyte loss; vomiting, diarrhea, and dysentery can lead to dehydration.

Dehydration synthesis Chemical reactions in which building a complex organic molecule requires bringing reactants together in such a way that water is released when a covalent bond is formed.

Delayed hypersensitivity reactions Also called type IV hypersensitivities; these reactions are T cell–mediated responses against self-antigens or otherwise harmless antigens; these reactions manifest slowly over 12–72 hours after the stimulating antigen is encountered.

Deletion mutations Mutations that occur when one or more nucleotides are removed from a DNA sequence.

Denatured Term applied to enzymes and other proteins that, through exposure to high temperatures or other conditions, lose their three-dimensional structure and become nonfunctional; denaturation can be reversible or irreversible.

Dendritic cell Named after the spindly dendrite extensions of the neurons (nerve cells) they resemble; highly phagocytic cells found in most body tissues, especially areas of the body that have contact with the environment, such as skin and body openings; an important type of antigen presenting cell.

Dengue fever Systemic infection caused by dengue viruses 1, 2, 3, or 4; the febrile phase of the infection is characterized by fever and other signs and symptoms such as nausea, vomiting, rash, and body aches, as well as viremia (virions circulate in the blood); second stage of infection is characterized by extreme abdominal pain, persistent vomiting, and dehydration, hemorrhagic features may also develop.

Dental caries Cavities; bacteria present in the mouth create acids that can eat away at the enamel to cause dental caries.

Dental plaque A complex biofilm created by the attachment of bacteria to the tooth enamel; *Streptococcus* and *Actinomyces* species are among the first to begin attaching to the tooth enamel to form plaque; the presence of plaque promotes tooth decay because it holds resident bacteria close to the tooth enamel.

Deoxyribonucleic acid (DNA) The genetic material in all cells; made of repeating subunits called nucleotides and exists as a double-stranded helical molecule.

Deoxyribonucleotides The nucleotides that make up DNA (adenine, guanine, cytosine, and thymine); differs in sugar type (deoxyribose) from ribonucleotides (ribose).

Dephosphorylation reactions Removal of a phosphoryl group from a molecule.

Dermatophytes A collection of fungal pathogens that infect the skin, hair, and nails and break down the protein keratin in these structures.

Dermis Thicker layer of skin beneath the epidermis; this skin layer contains blood vessels, nerves, hair follicles, sweat gland ducts, and sebaceous glands (oil glands).

Descriptive epidemiology An epidemiological approach that uncovers who is infected, where cases occur, and when cases occur; aims to describe the occurrence and distribution of disease so that hypotheses related to causes, prevention, or treatment can be developed and tested.

Detergents Substances such as detergents, that have polar and non-polar features (amphipathic molecules) that can remove water-soluble and water-insoluble substances; tend to have limited microbiocidal activity and instead work mostly as cleaning agents that reduce microbial counts by simply washing them away.

Diapedesis Also called transmigration or extravasation; a process whereby white blood cells dramatically change shape in order to squeeze out of a blood vessel during the recruitment phase of inflammation.

Diarrhea Frequent passing of loose or watery stool; risk factor for dehydration.

Differential media Specialized media that are formulated to allow us to visually distinguish one microbe from another based on how they metabolize media components.

Differential staining Staining technique that highlights differences in bacterial cell walls in order to discriminate between classes of cells; the two most common differential stains used in microbiology are the Gram stain and the acid-fast stain.

Differential white blood cell count (WBC differential) A rapid, inexpensive test that healthcare providers commonly use to help with diagnosis; determines if any leukocytes are over- or underrepresented in a patient's blood.

Differentiate To become specialized.

Diffusion The movement of molecules from an area of higher concentration to an area of lower concentration without an energy investment; the passive movement of substances from areas of high concentration to areas of low concentration (that is, down or along a concentration gradient) until they are uniformly distributed.

Diglycerides A lipid that has two fatty acid chains linked to glycerol.

Dimorphic Able to exist in two distinct forms; *Blastomyces dermatitidis* is an example of a dimorphic fungus; it lives as a spore-producing saprophyte in natural environments, and once inside the lungs, the saprophyte spores can germinate into a unicellular, yeast-like form that causes disease.

Diphtheria A bacterial illness caused by the Gram-positive rod *Corynebacterium diphtheria;* it can affect the upper respiratory tract or the skin; vaccine preventable infection.

Diplobacilli A cellular arrangement that results when bacilli (rod-shaped bacterial cells) divide to produce a pairing of cells.

Diplococci A cellular arrangement that results when cocci (or spherical-shaped bacterial cells) divide to produce a pairing of cells.

Dipole The asymmetric charge distribution between the participants in the covalent bond.

Direct contact transmission A form of transmission in which the host comes into physical contact with the source of the pathogen.

Direct ELISA A form of ELISA (enzyme-linked immunosorbent assay) that allows for the direct identification of antigens in a sample; it can also be modified to determine if a particular antibody is present in a test sample.

Disaccharides Two monosaccharides linked together by a glycosidic bond, which is the covalent bond formed between monosaccharides to build complex sugars.

Discontinuous replication Characteristic of DNA replication on the lagging strand, where chunks of the new DNA are constructed as Okazaki fragments.

Disinfectants Germicides used to treat inanimate objects.

Disinfection Decontamination measure that reduces microbial numbers.

Disseminated infection An infection that has spread throughout the body.

Disseminated intravascular coagulation In septic shock, the deregulated formation of blood clots throughout the body; this blood clot formation blocks blood flow and promotes organ failure and tissue necrosis (tissue death).

Disulfide bridges Class of covalent bond found in many proteins; these bonds form between sulfur-containing functional groups (thiol groups) of amino acids and add structural stability to proteins.

DNA methylation A type of epigenetic regulation that involves adding methyl groups (CH_3 groups) to DNA in order to prevent transcription.

DNA polymerase I An enzyme that replaces the RNA primer with DNA on the leading and lagging strand during DNA replication.

DNA polymerase III The main enzyme that copies/replicates DNA.

DNA replication The process by which a cell copies its genome before it divides.

DNA vaccines Relatively new vaccines that use a pathogen's DNA to stimulate an immune response; most DNA vaccines being developed focus on HIV or cancer.

DNAases These are DNA-degrading enzymes that break up DNA.

Domain The broadest taxonomic grouping of organisms; the three recognized domains include Bacteria, Archaea, and Eukarya.

Double immunodiffusion test Also called the Ouchterlony test; one of the two most common antigen–antibody precipitation methods used in clinical applications; a gel diffusion test that is useful for testing a patient sample for diverse antibodies or antigens all at one time.

Drug-induced liver injury (DILI) Injury to the liver from antimicrobial or other drugs; the most commonly cited reason for discontinuing drug development or removing a drug from the market.

Duration An epidemiological measure of how long an infection lasts.

Dysbiosis Microbiota disruption; an example is when a course of antibiotics kills off normal microbiota in the gut.

Dysentery A condition that may develop in severe cases of gastroenteritis; similar to diarrhea, but it's painful and is typically characterized by blood and mucus in the stool.

Dyspnea Shortness of breath.

Dysuria A symptom common in urinary tract infections; urgent urination that may be painful.

Ebola An infection caused by the Ebola virus, a single-stranded RNA virus that causes viral hemorrhagic fever; once in the human population, the virus spreads from person to person through contact with bodily fluids and through sexual contact.

Echinocandin drugs Drugs, such as caspofungin acetate, that inhibit fungal cell wall synthesis by targeting an enzyme that makes beta-glucan (β-glucan), a component of the fungal cell wall.

Echocardiogram (ECG or EKG) A sonogram of the heart; technique uses sound waves to create real-time images of the heart.

Ectopic pregnancy A pregnancy in which the fertilized egg implants outside the uterus; often in the fallopian tube.

Edema Generalized swelling, often in the extremities, in response to fluid accumulation in tissues.

Effacement A physiological process in which the opening to the uterus, the cervix, becomes thinner, softer, and shorter before the baby is delivered.

Effective reproduction number (R_e) As compared to R-naught (R_0) values, R_e values are more appropriate to consider in the midst of epidemics and pandemics. R_e values can change as host–pathogen interactions change; for example, as transmission precautions are applied or as pathogens evolve enhanced transmissibility.

Effector cells In adaptive immunity, cells that will engage in the response against the antigen.

Efflux pumps Plasma membrane–spanning proteins that actively pump drugs out of cells; these pumps tend to remove diverse classes of drugs, which makes them important contributors to multidrug resistance.

Ehrlichiosis Considered an emerging disease, this systemic bacterial infection is caused by *Ehrlichia chaffeensis* as well as other *Ehrlichia* species; it is mainly transmitted by the lone star tick; infection is characterized by fever, chills, headache, malaise and a rash.

Eicosanoids (or icosanoids) Vasoactive signaling molecules made from arachidonic acid, which is derived from cell membrane phospholipids; promote many physiological events to include inducing uterine contractions in childbirth, regulating stomach acid secretions, inducing vascular changes, stimulating pain receptors, and generating fever when released in the hypothalamus of the brain.

Electrolytes Name for ions that are freer to move around when ionic compounds dissolve in a solution; the most common electrolytes in physiology are sodium, potassium, calcium, phosphate, magnesium, and chloride.

Electron carriers Also called electron-carrier coenzymes; in cells, coenzymes often collect electrons from one reaction and shuttle them to other reactions in the cell; common coenzymes in metabolism include NAD^+, $NADP^+$, FMN, and FAD.

Electron shells Regions around the atomic nucleus where electrons are found; electron shells are organized into sub-shells and orbitals; each shell has a maximum number of electrons it can hold, with those closest to the nucleus tending to hold fewer electrons than the shells farther from the nucleus.

Electron transport chains Also called respiratory chains; these chains are collections of factors that pass electrons to one another in a series of redox reactions to generate a proton gradient that fuels the phosphorylation of ADP to ATP; electron transport chains can be aerobic or anaerobic.

Electronegativity The tendency of an atom to attract electrons; oxygen (O), nitrogen (N), and fluorine (F) are examples of highly electronegative elements that are commonly involved in polar covalent bonds.

Electrons Negatively charged particles around the nucleus of an atom; can be shared or transferred in reactions to form bonds.

Electrostatic forces The attraction forces that exist between positive and negative atoms or molecules; often seen as an attraction force between positive and negative atoms or molecules.

Elements Pure substances consisting of atoms that all have the same number of protons; elements make up ordinary matter.

Elongation Second step of the translation process, in which transfer RNAs (tRNAs) shuttle amino acids to ribosomes to translate mRNA to build proteins.

Emerging diseases Newer infectious diseases that exhibit an increasing incidence in populations.

Emerging pathogens Pathogens that previously caused only sporadic cases, but are increasingly common and/or exhibit an expanded geographical distribution; cause emerging diseases.

Emetic Term applied to toxins that trigger vomiting.

Empiric therapy A standard, accepted, or typical treatment based on clinical presentation in the absence of definitive or complete clinical data; when considering antimicrobial therapies, empiric treatments are often intentionally broad spectrum and later narrowed if the causative agent is identified and found to respond to a narrow-spectrum drug.

Encephalitis Inflammation of the brain; viruses are the most common cause; however, bacteria (and less commonly parasites and fungi) are also culprits.

Endemic Describes infections that are routinely detected in a population or region.

Endemic fungi Fungi that only grow in specific geographical locations because they require certain climate and soil conditions to thrive.

Endemic mycoses Fungal infections associated with distinct geographical locations.

Endergonic reactions Reactions that make products that have a higher final energy than the reactants; reactions that use more energy than they release; in biological systems, the energy released by exergonic reactions is used to fuel endergonic reactions.

Endocarditis Inflammation of the endocardium (the inner lining of the heart).

Endocardium The innermost layer of the three layers of the heart.

Endocytosis Importation mechanism primarily used by eukaryotic cells; it requires ATP and is used as a generalized (bulk or mass) transport mechanism, as well as specialized transport regulated by receptors on the cell's surface.

Endogenous source A scenario where the pathogen comes from the host's own body.

Endometrium A superficial layer of tissue that lines the uterus and is shed with each menstrual cycle (or period).

Endoplasmic reticulum (ER) An undulating series of interconnected membranous enclosures in eukaryotic cells, either rough or smooth in nature, that originate from the outer membrane of the nuclear envelope and that has essential roles in protein and lipid production.

Endosome A small vesicle with an acidic interior that fuses with the clathrin-coated vesicle during clathrin-mediated endocytosis; this fusion, through altering pH, allows the ligand and receptor to separate and be delivered to their final destinations.

Endospores Specialized dormant structures that certain bacteria such as *Bacillus* species and *Clostridium* species make in response to stressful conditions; metabolically inactive structures that allow certain bacterial cells to enter a dormant state; the structures are highly resistant to environmental stress such as starvation, heat, drying, freezing, radiation, or various chemicals.

Endosymbiotic theory Describes the evolution of eukaryotes as a series of sequential, cell-merging events between an ancient eukaryotic ancestor and certain prokaryotes.

Endotoxin The lipid A region of lipopolysaccharide (LPS); poisonous to us and other animals and is mainly released by Gram-negative bacteria when they die.

Enteric bacteria A collection of harmless bacteria as well as pathogenic species, which inhabit the intestinal tract of mammals.

Enteritis Inflammation of the intestines.

Enterotoxins Bacterial exotoxins that target the intestines; ingested enterotoxins may trigger intestinal inflammation that prevents water from absorbing into the intestine, causing diarrhea and abdominal pain; enterotoxins may also be emetic (trigger vomiting).

Envelope A lipid-based coating that surrounds the capsid of some animal viruses.

Enzyme-linked immunosorbent assay (ELISA) Among the most versatile and rapid diagnostic tests used today; ELISA techniques rely on antibody–antigen interactions along with a reporter enzyme (often horseradish peroxidase or alkaline phosphatase) that is attached to a detecting antibody—this is where the tests get the "enzyme-linked" part of their name.

Enzymes Protein catalysts that help chemical reactions occur under cellular conditions.

Enzyme–substrate complex Forms when an enzyme and its substrate come together.

Eosin methylene blue agar (EMB) A common medium with selective and differential capabilities; it contains the dyes eosin and methylene blue, which limit Gram-positive bacterial growth, while allowing Gram-negative bacteria to grow; also differentiates among Gram-negative species based on their ability to ferment the sugar lactose.

Eosinophils Account for less than 5 percent of the total white blood cell population; granulocytes that have granules packed with diverse enzymes and antimicrobial toxins; elevated eosinophil count (eosinophilia) suggests a parasitic infection, or possibly asthma or seasonal allergies.

Epicardium The layer of the pericardium that is closest to the heart muscle.

Epidemic A widespread disease outbreak in a particular region during a specific time frame.

Epidemiology Branch of medicine that aims to understand and prevent illness in communities.

Epidermis The outermost layer of the skin, made up of several layers of keratinocytes; consists of tightly packed dead cells.

Epididymis Structure adjacent to the testes; stores sperm until they are ejaculated.

Epididymitis Inflammation of the epididymis.

Epigenome The collection of all the chemical changes to the genome; the DNA methylation pattern of a genome.

Epiglottitis Inflammation and swelling of the epiglottis, which can block the airway; a medical emergency caused by certain infections—especially *Haemophilus influenzae* type b (Hib).

Episomally Describes the existence of viruses that do not integrate themselves into the host genome and whose genome remains outside the host's genome.

Epithelial cells Cells that are diverse in their structure and organization, but that all form essential barriers; epithelial cells are bound together into sheets that line each body cavity and body entrance, including the lining of the entire digestive tract, the outer layer of skin, and the airways.

Epitopes Also called antigenic determinants; parts or features of an antigen that B and T cells recognize.

Equilibrium Chemical term that describes the point at which the forward and reverse directions of a reaction occur at the same rate; no net change in product or reactant levels is realized, but this does not mean that product and reactant levels are equal.

Eradication A term applied when there are no longer any cases of a particular disease anywhere in the world.

Erysipelas A condition, more common in children and the elderly, in which impetigo spreads to surrounding skin and lymph nodes and involves extensive inflammation and pain; it is a much more serious condition than superficial impetigo and is accompanied by fever, chills, and an increase in white blood cells (leukocytosis) to fight off the infection; can be fatal if untreated.

Erythema migrans A slowly expanding rash surrounding a vector bite; in Lyme disease, a bull's-eye-shaped rash that often develops around the tick bite area about 7 to 14 days after infection.

Erythrocyte sedimentation rate (ESR) Hematological test that may be used in combination with other measures to monitor inflammation; ESR reflects how long it takes for red blood cells to separate out from plasma; it is reported as the volume of plasma in millimeters, present at the top of the tube after one hour (mm/hr).

Erythrocytes Red blood cells; formed elements in the blood that deliver oxygen to body tissues.

Essential amino acids Amino acids that cells cannot make and must get from their environment.

Essential nutrients Nutrients that are required to build new cells and can be found in the organic and inorganic compounds of a microbe's environment.

Ester bond A type of covalent bond that forms between acids and alcohols.

E-test An agar diffusion test for bacterial susceptibility (with modified protocols for testing fungi) that is conceptually similar to the Kirby-Bauer test, but instead of applying round disks infused with a set amount of an antimicrobial drug to the agar plate, strips infused with a variable gradient of drug are placed on the agar surface; it can reveal the minimal inhibitory concentration (MIC) of the drug.

Ethylene oxide A colorless gas that is a good sterilization method for temperature-sensitive materials and equipment susceptible to moisture; works by damaging proteins and nucleic acids.

Etiological agent Causative agent.

Exanthem subitum See roseola.

Excision repair A process in which specialized enzymes clip out damaged or mismatched nucleotides, and then DNA polymerase I lays down new nucleotides to repair the DNA.

Exergonic reactions Reactions that ultimately release more energy than they use; reactions that make products with a lower final energy than the reactants; in biological systems, the energy released by exergonic reactions is used to fuel endergonic reactions.

Exfoliative Describes the process of removing the outer skin layer; caused by certain toxins made by *Staphylococcus* strains that cause scalded skin syndrome.

Exit site (or E site) Region of an active ribosome that, together with the ribosome's peptidyl site (P) and acceptor site (A), coordinates the translation process.

Exocytosis A cellular exportation process that involves vesicles delivering their contents to the plasma membrane.

Exoenzymes Enzymes secreted by bacteria into their local environment that break down large macromolecules into smaller molecules.

Exogenous source A source of an infecting pathogen that is external to the host.

Exons In eukaryotic cells; certain segments of mRNA that are decoded to build a protein.

Exotoxin Toxic soluble proteins made by both Gram-positive and Gram-negative bacteria that affect a wide range of cells; these protein toxins are often named based on the organism that makes the toxin or the type of cells the toxin targets; usually classified into three main families based on their mode of action.

Experimental studies Analytical epidemiological studies that allow the researcher to change variables and determine the effect of the change on the outcome; used to determine the effectiveness of a treatment or preventive measure.

Exponential Describes a type of growth in which one cell begets two, the two yield four, those four become eight, and so on.

Extracellular Outside the cell.

Extreme thermophiles Organisms that prefer growth temperatures from 65–120°C; these organisms can live in boiling water and volcanic vents.

Extremophiles Organisms that live in extremes of pH, temperature, and/or salt and that are exposed to a combination of stresses.

Exudate A fluid discharge (can be pus-like or clear fluid) that is present in a given tissue; usually produced in response to inflammation; fluid that accumulates in the tissues as a result of vascular changes that accompany early inflammation.

Facultative anaerobes Group of microbes that spans both aerobic and anaerobic environments; can use oxygen for their metabolism, but if needed, they can perform anaerobic metabolism when oxygen is absent.

Fallopian tube Tube-like tissue adjacent to the ovary; during ovulation, an egg is released from an ovary and is swept up into the adjacent fallopian tube, from there it's conveyed toward the uterus (womb).

Fastidious Term applied to organisms that need multiple growth factors to thrive.

Fatty acid metabolism Fatty acids are broken down into a wide variety of organic molecules.

Fatty acids Organic molecules made up of long hydrocarbon chains with a carboxylic acid functional group; can be saturated or unsaturated.

Febrile seizures Convulsions, twitching, or shaking induced by fevers that could result in loss of consciousness; last from a few minutes to 15 minutes or longer.

Feedback inhibition A form of enzyme regulation that leads to the slowdown (or turnoff) of biochemical pathways in response to the accumulation of a pathway intermediate or end product.

Fermentation A biochemical process whose main goal is not to make ATP, but rather to sustain ATP production by glycolysis when respiratory chains are not available; works in combination with glycolysis but is not considered part of cellular respiration; a broad term that describes catabolic processes that certain cells use to break down nutrients (often carbohydrates) into simpler substances; acids and alcohols are common products of fermentation reactions.

Fermentation tests A collection of biochemical tests commonly used to identify an unknown bacterial specimen.

Fertility plasmid (or F plasmid) A bacterial plasmid, a copy of which can be transferred to a neighboring bacterial cell during conjugation.

Fever Also called pyrexia; an abnormally high systemic body temperature, different from the "local heat" that characterizes inflammation; infectious agents may trigger fevers, but certain drugs and a number of noninfectious pathologies also induce it.

Fibrosis The formation of scar tissue.

Fimbriae Short, bristle-like structures that extrude from the cell surface; only found on prokaryotic cells; made of protein, they tend to be numerous and cover the cell; their adhesive properties help prokaryotes stick to surfaces or to each other for establishing biofilms or for invading a host.

Fine focus knob The part of a compound light microscope that allows the viewer to carefully focus the image by affecting the distance of the objective lens from the specimen in small increments; allows for precision focusing of a light microscope.

First Law of Thermodynamics States that energy is not created or destroyed, it just changes forms.

First-generation drugs Early versions of a drug family; usually a term reserved for denoting early versions of cephalosporin drugs and sometimes penicillin drugs.

First-line defenses Defenses that attempt to prevent pathogen entry; traditionally subcategorized as mechanical, chemical, and physical barriers.

Fixed (heat fixed) In bacterial staining techniques, a treatment that adheres specimen cells to the slide by exposing it to heat or a chemical reagent.

Fixed macrophages Macrophages that reside in specific tissues; include Kupffer cells in the liver or alveolar dust cells in the lungs.

Flaccid (relaxed) paralysis A form of paralysis that develops when muscles cannot contract; paralysis produced by botulinum toxin.

Flagella Tail-like structures used by some cells for motility; while eukaryotic microbes tend to have only one flagellum located at a single pole of the cell, prokaryotes can have single or multiple flagella with diverse arrangements.

Flank pain (or costovertebral angle tenderness) Pain at the lower back near the natural waistline; often develops during a kidney infection.

Flavin adenine dinucleotide (FAD) A coenzyme that serves as an electron carrier and that is made from the B vitamin riboflavin.

Flavin mononucleotide (FMN) A coenzyme that serves as an electron carrier and that is made from the B vitamin riboflavin.

Florence Nightingale In the 1860s she established the use of aseptic techniques in nursing practices, which, along with other patient-care innovations, earned her the historical distinction of being the founder of modern nursing.

Flow cytometer A cell counter that uses a laser light to detect cells passing through a narrow channel; can differentiate one cell type from another by using different colored fluorescent labels for different cell types.

Flow cytometry A process by which cells can be sorted using a flow cytometer; overall process is useful for enumerating specific cells.

Flucytosine A drug that targets fungal DNA replication and also blocks transcription; usually administered in combination with amphotericin B for severe fungal infections.

Fluorescence A naturally occurring phenomenon where a substance absorbs energy, usually ultraviolet (UV) light that is invisible to the human eye, and then emits that energy as visible light.

Fluorescence-activated cell sorter (FACS) A special piece of equipment used in flow cytometry to count tagged cells and sort them from nontagged cells.

Fluorochromes Fluorescent dyes that can be used to stain samples so they will fluoresce when illuminated by a UV light microscope; can be natural or synthetic.

Folliculitis *Staphylococcus aureus* infection that is hallmarked by swollen, red, pus-filled hair follicles.

Fomite An inanimate object that can harbor pathogens; examples include doorknobs, used needles, toys in a busy childcare center, etc.

Fomite transmission Another name for vehicle transmission; transmission of an infectious disease by an inanimate object (fomite).

Foodborne bacterial infections Infection that results from ingesting food that harbors live bacterial pathogens that infect the GI tract; the most common bacterial causes of foodborne infections are *Campylobacter* species, *Shigella* species, *E. coli*, *Salmonella*, and *Listeria monocytogenes*.

Frameshift mutation A mutation in which the reading frame of an mRNA is altered due to nucleotides being inserted or deleted from the coding region of a genetic sequence.

Francesco Redi A 17th century scientist who performed experiments to test the hypothesis of spontaneous generation.

Functional groups In chemistry this term refers to specific groups of atoms that confer distinct chemical properties on a molecule; they often participate in chemical reactions.

Fungi Eukaryotic unicellular and multicellular microorganisms; can be pathogenic or nonpathogenic.

GALT Gut-associated lymphoid tissue; a secondary lymphoid tissue.

Gamma hemolytic Term applied to bacteria that do not lyse red blood cells.

Gamogony In apicomplexan protists, this is the sexual phase of reproduction in which meiosis produces male and female haploid gametes.

Gangrene Tissue necrosis due to lack of blood flow; usually due to an infection and generally occurs at an extremity.

Gastritis Inflammation of the stomach.

Gastroenteritis Inflammation of the stomach and intestines.

Gastrointestinal (GI) tract Also called the alimentary canal; the part of the digestive system that comes into direct contact with our food; the tube-like part of the digestive system that extends from the mouth to the anus and is divided into upper and lower segments.

Gel electrophoresis Method used to separate molecules (especially nucleic acids and proteins) based on their size.

Gell and Coombs classification system A system that describes four classes of hypersensitivity reactions—types I, II, III, and IV.

Gene A heritable unit of genetic material that determines a particular trait.

Gene expression Also called protein synthesis; a cellular process in which genetic information within a cell is read and used to create proteins.

Gene shuffling Mechanisms that generate an incredibly diverse repertoire of antigen receptors (TCRs and BCRs) during lymphocyte development.

Gene therapy The process of introducing genetic material that is not a human cell as a treatment.

Generalized transduction Carried out by bacteriophages, this is one example of a horizontal gene transfer mechanism that can convey new genes to a bacterial cell; in this process a bacteriophage randomly takes up a bacterial gene and carries it from the prior host to the new host bacterium.

Generation time The time it takes for a particular species of cell to divide.

Genetic code The set of rules that determines which codon represents a particular amino acid or start/stop signal; often displayed in a table format of 64 codons.

Genetic Information Nondiscrimination Act of 2008 (GINA) A piece of legislation that aims to protect patients from discrimination based on genetic information; as currently presented this act provides a number of loopholes for insurers and employers.

Genome The entire collection of genetic material in a cell or virus.

Genotype The genetic makeup of an organism.

Genus A taxonomic category comprised of related species; generally the first word in an organism's scientific name.

Germ theory of disease States that microbes cause infectious diseases.

Germicides Chemicals used to control microbial growth; two key classes of germicides are disinfectants, which are used to treat inanimate objects, and antiseptics, which are applied to living tissue such as skin.

Gingivitis Infection of the gums; marks start of periodontal disease.

Gluconeogenesis The biochemical process of building glucose from nonsugar starting materials.

Glycerol A short-chain alcohol; a component of triglycerides.

Glycocalyx A sticky extracellular layer made of carbohydrates; found in prokaryotes and eukaryotes; in prokaryotes this can be loosely or tightly associated with the cell wall.

Glycogen A branched polymer of glucose that is mainly stored in the liver.

Glycogenesis The production of glycogen, a branched polymer of glucose.

Glycolipids Fats that have added sugar groups.

Glycolysis Also called the Embden-Meyerhof-Parnas pathway; the first stage in aerobic and anaerobic carbohydrate catabolism in which energy is extracted from carbohydrates.

Glycopeptide drugs The term glycopeptide describes a peptide molecule (a short sequence of amino acids) that is associated with sugar components; are narrow-spectrum antibacterial compounds that interfere with bacterial cell wall construction; do not have a beta-lactam ring, so they are not susceptible to beta-lactamases; teicoplanin and vancomycin are examples; these drugs are effective against Gram-positive organisms and can be useful for fighting certain antibiotic-resistant bacteria, such as MRSA and *Clostridioides difficile*.

Glycoproteins Proteins that are covalently bonded to sugars; the sugars can be diverse and they often participate in cell signaling, binding, or other cellular actions.

Glycosidic bond The covalent bond formed between monosaccharides to build complex sugars.

Golden age of microbiology This period (approximately 1850–1920) was sparked by innovations in microscopes, the careful documentations made by earlier scientists, and new techniques to isolate and grow microbes; many of the techniques that spurred this turning point in biology are still used today.

Golgi apparatus An organelle in eukaryotic cells; resembles a series of disc-like, flattened sacs called cisternae that stack upon one another; coordinates with the endoplasmic reticulum (ER) to modify cellular proteins, build lipids, and further sort and distribute the finished products.

Graft-versus-host disease (GVHD) Mainly seen in bone marrow transplants; it is a scenario that can develop if donor and recipient's MHCs are not closely matched and the developing B and T cells in transplanted bone marrow mount an immune response against their new host.

Gram stain A staining procedure that allows us to classify bacteria as either Gram-positive or Gram-negative; following the Gram staining procedure, Gram-positive cells appear purple while Gram-negative bacteria appear pink; the final outcome of the Gram stain is based on the cell wall properties of the stained cells.

Granulocytes A name applied to white blood cells that contain cytoplasmic granules that are visible when stained and then viewed by light microscopy; examples include basophils, eosinophils, and neutrophils.

Granzymes Proteases (enzymes that break down proteins) that T cytotoxic cells release to induce apoptosis in a target cell.

Group A streptococci (GAS) Term often applied to *Streptococcus pyogenes* strains that can be invasive and deadly, thanks to their diverse virulence factors that break down tissues and facilitate spread.

Growth factors The necessary substances that a cell can't make on its own; an organism will not grow if these factors are missing from the environment.

Growth media Mixtures of nutrients that support organismal growth in an artificial setting.

Guanine (G) A purine nitrogenous base used to make nucleotides that are the building blocks of DNA and RNA.

Guillain-Barré syndrome (GBS) A rare but troubling complication associated with certain infectious agents such as *Campylobacter jejuni*; GBS starts as a tingling sensation in the legs and feet and eventually advances to paralysis; the symptoms are usually temporary, but individuals must be hospitalized.

Gyrase A specialized detangling enzyme that relieves the coiling tension that develops as the helix unwinds during DNA replication.

HA (hemagglutinin) spikes Glycoprotein structures on influenza's viral envelope that help the virus attach to host cells.

Haemophilus influenzae Gram-negative bacterium that is common among the normal microbiota of healthy people; *H. influenzae* type b (or Hib) causes bacterial meningitis and pneumonia; incidence of Hib infetcions has decreased as a result of the Hib vaccine.

***Haemophilus influenzae* (Hib) pneumonia** Pneumonia caused by *Haemophilus influenzae* type b.

Half-life The time it takes for half of a drug dose to be eliminated or deactivated by the body; can also refer to the time needed for radioactivity emitted by a radioisotope to decrease by half its original value.

Halogens Germicides, including chlorine and iodine compounds, that work by oxidizing cell components, especially cellular proteins and nucleic acids; chlorine bleach (sodium hypochlorite) is one of the most widely used halogen disinfectants.

Halophiles Organisms that thrive in high-salt environments.

Hand, foot, and mouth disease (HFMD) An infection mainly caused by coxsackievirus A16 and enterovirus 71 viruses that is common in infants and children but can also occur in adults; uncomfortable blister-like sores in the mouth (herpangina) and lesions on the hands and feet distinguish this disease.

Hansen's disease Also called leprosy; disease caused by the acid-fast bacterium *Mycobacterium leprae*, which infects the peripheral nervous system of the host.

Hanta pulmonary syndrome (HPS) An acute and potentially fatal respiratory illness that humans can acquire when they inhale airborne dust particles that contain hantavirus particles (also called sin nombre virus) shed in rodent urine or feces.

Haptens Incomplete antigens that are unable to stimulate an immune response (not immunogenic) unless they are linked to a more complex protein or polysaccharide.

Healthcare-acquired infection (HAI) Also known as nosocomial infections; an infection that a patient develops while receiving care in a healthcare setting.

Healthcare-acquired pneumonia A general term that describes pneumonia cases that develop at least 48 hours after admission to a healthcare facility; among the most common healthcare-acquired infections (HAIs).

Healthcare-associated infections/ healthcare-acquired infections/ HAIs/ nosocomial infections These terms describe any infection that develops as a result of medical interventions applied in a healthcare setting.

Healthcare-associated pneumonia (HCAP) Occurs in ambulatory patients who are not hospitalized but who have had extensive healthcare contact within the last 3 months.

Heart A four-chambered muscle that constantly pumps blood through the body.

Helical A term describing a hollow tube shape for viral capsids.

Helicase Enzyme in the primosome that unwinds DNA during DNA replication.

Helminths Eukaryotic multicellular parasitic roundworms and flatworms.

Hematopoiesis The process of stimulating the production of new blood cells and platelets.

Hematuria Blood in the urine.

Hemolysins Specialized proteins that can affect hemoglobin, red blood cells, white blood cells, and other cells; blood agar can be used to differentiate bacteria that make hemolysins (such as *Streptococcus pyogenes*) from bacteria that do not make hemolysins.

Hemolytic disease of the newborn (HDN) A disease in which an Rh^+ baby's red blood cells are lysed in response to maternal anti-Rh antibodies that cross the placenta; the lysis induces a severe and possibly fatal anemia in the baby.

Hemolytic transfusion reaction Immune response against incompatible transfused red blood cells in which red blood cells are lysed; this reaction could kill the patient.

Hemolytic uremic syndrome (HUS) Syndrome in which the kidneys are damaged by the Shiga toxin; HUS symptoms include blood in the urine (hematuria), bruising (purpura), and generalized swelling (edema) in the extremities; most HUS cases in the United States are due to Shiga toxin-producing *E. coli*.

Hemorrhagic Causing bleeding; a term often used to describe pathogens that interfere with host blood-clotting cascades.

Hemorrhagic cystitis A bladder inflammation accompanied by blood in the urine; common hallmark sign of viral urinary tract infections (UTIs), but may also occur in fungal and bacterial UTIs.

Hepatitis A general term meaning inflammation of the liver; there are infectious and noninfectious causes of hepatitis.

Hepatitis A virus (HAV) A nonenveloped, single-stranded RNA virus; vaccine against HAV was introduced in 1995; virus is typically spread by the fecal–oral route.

Hepatitis B virus (HBV) A double-stranded enveloped DNA virus; vaccine against HBV was introduced in 1991; transmission usually occurs through direct contact with bodily fluids or open sores from an infected individual—sexual contact with an infected partner, sharing HBV-contaminated needles, razors, or even sharing toothbrushes can lead to infection.

Hepatitis C virus (HCV) An enveloped virus, with a single-stranded RNA genome; most HCV cases are associated with shared needles, although some cases develop following direct contact with blood and bodily fluids from an HCV-infected patient; newer drug treatments cure up to 90 percent of those infected.

Hepatotoxic Liver-toxic; term applied to antimicrobial drugs that can induce liver damage and are leading agents of drug-induced liver injury.

Herd immunity Type of communal immunity that occurs when a pathogen won't find enough susceptible people in the community to persist, even if a small number of individuals there remain unvaccinated; herd immunity allows nonvaccinated individuals, such as premature babies and immune-compromised patients, to still reap the protective benefits of immunization provided a sufficient percentage of the rest of the population is vaccinated.

Heritable Passed from one generation to the next.

Herpangina Uncomfortable blisterlike sores in the mouth that occur in hand, foot, and mouth disease (HFMD).

Herpes gladitorum An HSV-1 infection common in wrestlers who come in contact with the virus through contaminated wrestling mats or direct contact with an infected wrestler.

Herpes simplex virus 2 (HSV-2) Also known as human herpes virus 2; causes genital herpes; after the primary infection, which results in a severe outbreak of ulcers as quickly as three days after transmission, the virus tends to go dormant in peripheral nerves near the area where lesions had developed; stress, illness, or hormone changes can reactivate latent viruses; the antiviral drug acyclovir can shorten outbreaks and suppress flare-ups to reduce the risk of transmission, but there is not a cure for herpes.

Herpes whitlow A rare condition that can occur when HSV-1 enters fingers via an abrasion; has been known to occur in healthcare providers that contact the virus.

Herpes zoster Also known as shingles; name derives from the Greek word for belt, indicating that the rash develops along a belt of nerves.

Heterolactic fermentation A type of fermentation that results in the production of equal quantities of lactic acid, ethanol, and carbon dioxide, with minor amounts of acidic end products such as acetic acid and formic acid; the pentose phosphate pathway is involved in heterolactic fermentation.

Heterotrophs Organisms, such as humans, that cannot fix carbon; they require an external source of organic carbon in order to live and grow.

High-frequency recombination (Hfr) strains Describes a scenario where a fertility plasmid merges or integrates into the bacterium's chromosome and ceases to be an independent plasmid in the cytoplasm.

Histamine A small amine molecule and potent inducer of vascular changes and inflammation; made by many cells, including mast cells, basophils, eosinophils, and platelets.

Histones Organizational proteins in eukaryotic chromosomes that help keep DNA from getting tangled.

Histoplasmosis Also called Darling's disease, cave disease, spelunker's lung, and Ohio Valley disease; an infectious disease caused by the dimorphic fungus *Histoplasma capsulatum*; exposure to the fungus is largely occupational, with poultry farmers and people who harvest guano (bat droppings) from caves for fertilizer at the greatest risk for infection.

Holoenzyme The functional form of the enzyme that includes the enzyme and any necessary cofactors.

Homolactic fermentation A type of fermentation in which the pyruvic acid made in glycolysis is reduced to lactic acid.

Horizontal gene transfer Occurs when genetic information is passed between cells by a process independent of cell division, and therefore separate from reproduction.

Hospital-acquired pneumonia (HAP) Describes new pneumonia cases that develop in hospitalized, nonintubated patients at least 48 hours after admission.

Host The organism targeted by a particular pathogen.

Host range The collection of species that a pathogen can infect.

Host–microbe interactions Dynamic give-and-take between microbe and host; not always damaging, in fact many host-microbe interactions are helpful.

HTLV-1-associated myelopathy (HAM) Outcome of HTLV infection in which infected T cells may produce abundant interferon, creating inflammatory responses to the central nervous system.

Human immunodeficiency virus (HIV) A virus that directly infects T helper cells; when enough T helper cells are killed, the patient is said to have developed AIDS (acquired immunodeficiency syndrome).

Human microbiome project A five-year project launched in 2008 that was a major initiative to catalog the microbes that comprise the normal human microbiota.

Human papilloma viruses (HPVs) Among the most well-known oncogenic viruses, with about 90 percent of all cervical cancer cases linked to HPV; over 200 known subtypes of HPV; about 40 of them are sexually transmitted; some HPVs just cause simple cutaneous (skin) warts and are not sexually transmitted.

Human parainfluenza virus (HPIV) infections Infections caused by HPIV (four types) that are characterized by cold-like symptoms but that, in infants or the elderly, can cause more severe illnesses like croup, bronchitis, and pneumonia.

Human pathogens Agents that cause disease in humans.

Human T-lymphotropic viruses (HTLV) A group of viruses that include HTLV-1, -2, -3, and -4; oncogenic retroviruses that can persist for more than a decade in host cells before emerging to cause leukemia or lymphoma.

Humoral response Also called B cell or antibody–mediated immunity; one of two branches of the adaptive immune system whose goal is to eliminate an identified antigen and remember it so that next time adaptive responses are faster.

Hydrogen bond A noncovalent electrostatic attraction between two or more molecules (*inter*molecular hydrogen bonds) or within a single large molecule (*intra*molecular hydrogen bonds); weaker than ionic or covalent bonds and do not bind atoms into molecules; hold complementary DNA strands together.

Hydrogenated A term used to describe the addition of hydrogen to reduce or saturate a carbon-carbon double bond; for example, hydrogenated fats have had the double bonds in their fatty acids reduced with hydrogen to saturate the double bonds.

Hydrolysis reaction A common process in biochemical pathways in which water is added to break the covalent bonds in complex molecules.

Hydrolytic enzymes Enzymes that carry out hydrolysis reactions; enzymes that use water to break chemical bonds; important for breaking down organic molecules in cells; these enzymes are commonly found inside lysosomes.

Hydrophilic Water loving; name applied to substances like sugar and other polar substances that are readily dissolved in water.

Hydrophobic Water fearing; name applied to substances that are not readily dissolved in water.

Hydrophobic interactions Occur between nonpolar molecules in a watery (aqueous) environment; oil droplets joining together in water is an example.

Hydroxychloroquine A drug used to treat malaria and certain autoimmune disorders; chemically related to the compound chloroquine (also used to treat malaria); investigated for efficacy against SARS-CoV-2, but in clinical trials it did not statistically reduce the severity or duration of COVID-19, nor did it prevent SARS-CoV-2 infection.

Hygiene hypothesis Hypothesis that aims to explain why autoimmunity and allergy are more common in residents of developed nations than in residents of developing countries; proposes that in developed countries, ultra-clean water and food combined with common antibiotic usage and decreased infection incidence has decreased the biodiversity of people's normal microbiota, thereby affecting how our immune system develops, increasing our risk for developing autoimmune disorders and allergies.

Hypersensitivities Inappropriate responses of the immune system to a threat; immune-based pathologies such as allergy and autoimmunity.

Hypertonic Solutions that contain more concentrated solute than a cell; cells will experience plasmolysis as water is drawn out of the cell into a hypertonic environment.

Hyphae (singular: hypha) A collection of tubular structures by which most fungi (aside from yeasts) grow.

Hypoglycemia Low blood glucose level that may cause the patient to feel light-headed and shaky.

Hypotension Abnormally low blood pressure that can lead to dizziness and fainting; typically a systolic pressure under 90 and a diastolic pressure under 60 (noted as 90/60).

Hypothesis A prediction based on prior experience or observation.

Hypotonic Solutions that contain a lower solute concentration than a cell; cells will take on water when exposed to a hypotonic solution and if the cell doesn't have a cell wall or some other tool for managing osmotic stress, it will lyse (burst).

Hypovolemic shock Low blood volume due to loss of blood or severe dehydration; can lead to organ failure due to lowered blood pressure.

Icosahedral Term describing a shape for capsids that is like three-dimensional polygons, but may appear fairly spherical—just as a soccer ball is spherical, yet made of multiple hexagon and pentagon shapes.

Ignaz Semmelweis Hungarian physician who, in the 1840s, first developed aseptic techniques; recommended hand washing to decrease mortality rates from childbed fever (puerperal sepsis), an infection that killed many women in childbirth before the antibiotics era.

Immersion oil Specialized oil formulated to have the same refractive index as glass, ensuring that the light that interacts with the specimen on the glass slide is smoothly funneled up toward the high-power objective lens instead of scattering when it reaches the air between the slide and lens; used in a variety of light microscopy techniques.

Immune Term applied when someone has specific protection conferred by adaptive immune responses against a particular pathogen.

Immune response A physiological process coordinated by the immune system to eliminate foreign substances (antigens).

Immune-privileged site Body sites in which transplants are the least likely to be rejected; immune-privileged sites include the eye, brain, uterus, and testicles.

Immunocompromised An individual who doesn't have a fully functioning immune system; can be associated with organ transplants, blood transfusion, and certain diseases including, but not limited to, cancer and HIV/AIDS.

Immunodeficiency The lack of a properly functioning immune system; can be primary (inborn) or secondary (acquired).

Immunofluorescence A technique in which a sample is exposed to fluorescent-tagged antibodies that recognize a specific target; tagged samples will glow when viewed under UV light; used for rapid identification of bacteria in blood cultures, virus identification in patient samples, and fast screening for pathogenic bacteria in food-processing plants.

Immunofluorescence assays (IFAs) Assays that detect antigens or antibodies in a patient sample; similar to ELISA (enzyme linked immunosorbent assays), but the detection antibody is linked to a fluorescent tag instead of an enzyme.

Immunofluorescence microscopy A protocol where fluorescent-tagged antibodies that recognize a specific antigen are incubated with tissue or cell samples, followed by observing the sample with a specialized UV light microscope.

Immunogenic Term applied to any antigen that can successfully trigger an immune response.

Immunoglobulin A (IgA) Represents up to 15 percent of our total antibodies; prevalent in mucus and in most bodily secretions such as tears, saliva, sweat, and milk.

Immunoglobulin D (IgD) A sparsely represented antibody mainly found on the surface of B cells; only small quantities can be found circulating in blood; its precise function remains unknown.

Immunoglobulin E (IgE) Present in very low concentrations, and mostly in lungs, skin, and mucous membranes; central to fighting parasites and mediating allergic responses.

Immunoglobulin G (IgG) The most abundant antibody in human blood (constitutes up to 85 percent of total antibodies); found in all bodily fluids, it is especially abundant in blood, lymph, cerebrospinal fluid, and peritoneal fluid (which lubricates organs and tissues of the abdominal cavity).

Immunoglobulin M (IgM) Present mainly in blood and accounts for up to 10 percent of total antibody population; usually made early in infection upon a primary antigen exposure, its presence indicates a recent exposure to a given antigen.

Immunoglobulins (Ig) Another name for antibodies; glycoprotein molecules made by plasma cells (activated B cells); essential participants in the humoral immune response.

Immunological memory Function of our memory cells that allows for a rapid reactivation of the cellular and humoral adaptive response if the same antigen is encountered again later; memory responses are so rapid and effective that we frequently will not even experience disease symptoms while our bodies eliminate the pathogen.

Immunotherapies Anticancer treatments that are geared at boosting immune defenses; these therapies can be prophylactic, meaning they try to prevent cancer, or therapeutic, meaning they treat existing cancer.

Impetigo A common and highly contagious skin infection; cases caused by *Staphylococcus aureus* are characterized by superficial pus-filled vesicles on reddened skin usually on the face, lips, or extremities (these itchy vesicles may rupture and ooze leading them to crust over into honey-colored lesions).

Inactivated vaccines Vaccines that consist of whole inactivated pathogens, such as killed bacteria or inactivated viruses, or parts of pathogens.

Inactivated viral strains Used in vaccines; viral strains that cannot cause disease in a human host because they have been physically or chemically deactivated.

Incidence rate An epidemiological measure of frequency that expresses the number of new cases in a defined population during a defined time frame.

Inclusion bodies Distinct collections of substances inside prokaryotic cells; consist of insoluble granules, or are sometimes bound in a membrane; bacteria typically build inclusions in times of excess, such as when grown on nutrient-rich media.

Incubation period Time between infection and the development of disease symptoms.

Indirect contact transmission Transmission in which a pathogen spreads to a host without direct physical contact with the source (source could be an animal, a human, or the environment).

Indirect ELISA An enzyme-linked immunosorbent assay for determining if a patient has antibodies to a particular pathogen; this form of ELISA requires two antibodies, the first antibody recognizes the antigen bound at the bottom of the plate wells, and the second antibody is linked to an enzyme and serves as the detection antibody.

Indirect methods Epidemiological study methods that rely on secondary reflections of overall population size.

Induced fit model A model describing how substrates and enzymes interact in the active site; both can change shape slightly upon interacting, which allows enzymes to slightly mold and position the substrate in a way that will encourage a reaction.

Induced mutations Genetic change prompted by a factor in an organism's environment; mutagens can lead to induced mutations.

Inducible operons Mainly used by bacterial cells as a means to regulate gene expression at the pre-transcriptional level; inducible operons are by default "off" unless certain conditions arise under which they are activated (induced) to allow transcription; a key example of an inducible operon is the lactose (or *lac*) operon.

Induration The hardening of a tissue often caused by inflammation; a hardened or fibrous mass in a normally elastic and soft tissue; a hardened area of the skin that develops following the injection site of a Mantoux test for tuberculosis; redness is also common around the induration, although only the size of the induration weighs into the test result.

Infectious disease An illness caused by a pathogen.

Infectious dose-50 (ID50) Term describing how many cells (bacterial, fungal, or parasitic) or virions are needed to establish an infection in 50 percent of exposed susceptible hosts; the more infectious the pathogen is, the lower its ID50.

Infectivity Term describing how good an infectious agent is at establishing an infection.

Inflammation An innate immune response that tends to develop when our tissues are damaged, either from physical factors like trauma or burns, or from infectious agents.

Inflammatory acne One of the most common skin afflictions, affecting 70–80 percent of teenagers and young adults at some point, mainly caused by *Propionibacterium acnes* bacteria; this is a moderate form of acne.

Influenza ("the flu") Viral infection in which most common causative virus strains infect the host by binding to ciliated cells in the upper respiratory tract; however, some strains (especially more virulent avian and swine strains) can directly affect the lungs; flu symptoms tend to mimic those of a severe cold and infection can be associated with dangerous complications like pneumonia.

Inguinal lymph nodes Lymph nodes located in the groin region.

Initiation The first step of the translation process in which a ribosome attaches to the mRNA and scans it until it encounters a start codon, at which point it adds the first amino acid.

Innate immunity Sometimes called non-specific immunity; an inborn, ancient protection existing in one form or another in all eukaryotic organisms; these generalized responses don't vary based on the pathogen being fought.

Inorganic molecules Molecules that lack carbon; even if carbon is present, inorganic molecules lacking the associated hydrogen; CO_2 is an example of an inorganic molecule that contains carbon, but lacks the associated hydrogen that would make it organic.

Insertion mutations Mutations that occur when a cell adds one or more nucleotides to a DNA sequence.

Integumentary system The largest body system; consists of skin, hair, nails, and associated glands; it blocks most microbes, but some have virulence factors that penetrate this essential barrier.

Interferon gamma release assays (IGRAs) Fast and relatively reliable way to detect tuberculosis (TB) in the early stages in vaccinated populations; IGRAs work by measuring how a patient's T cells respond to *Mycobacterium tuberculosis* antigens; T cells from a person who has TB (either a latent or an active infection) release more interferon gamma (INF-γ) T cells from someone who does not have TB.

Interferons Naturally occurring substances released by cells in response to viral infections; molecules that signal the immune system when pathogens or tumor cells are detected; especially well known for antiviral effects, they derive their name from their ability to "interfere" with viral replication; they are also important activators of innate and adaptive immune responses against bacteria and parasites.

Interleukins Diverse cytokines that have roles in activating adaptive and innate immune responses and stimulating the production of new blood cells and platelets.

Intermediate filaments Rope-like fibers of about 10 nm in diameter that mainly contribute tensile strength to the cytoskeleton in order to oppose external mechanical forces.

Interstitial fluid Also called extracellular fluid; plasma that exits the capillaries and seeps into the small spaces between tissue cells.

Intrinsic resistance Built-in qualities that help some bacteria naturally resist antimicrobial drugs; while intrinsic resistance rarely provides full protection from all antimicrobials, it does make certain pathogens harder to eliminate and it tends to limit what drugs can be used to combat those pathogens.

Introns Intervening sequences in mRNA that are clipped out before translation; genetic material that is not decoded to build the protein.

Intubation Insertion of a flexible plastic tube into the airway.

Invasins Local-acting factors, usually membrane-associated or secreted enzymes, that allow pathogens to invade host tissues.

Invasive aspergillosis Fungal disease having five forms, with the pulmonary (or lung) and rhinocerebral (or brain sinus) forms being the most common; although *Aspergillus* species are a common soil fungus throughout the world and the fungi make spores that are inhaled by most of us every day, invasive aspergillosis is almost exclusively seen in immune-compromised patients.

Invasive candidiasis See **candidemia**.

Iodine Serves as a mordant in the Gram stain technique; interacts with crystal violet to form an insoluble crystal violet iodine complex (CV-I complex) that is retained by Gram-positive cells following the decolorization step.

Ionic bond The electrostatic force of attraction that exists between oppositely charged ions (between cations and anions); type of bond formed when electrons are transferred from one atom to another.

Ionizing radiation Gamma rays released by radioactive substances and X-rays, which are electron beams, that generate reactive ions that kill microorganisms and inactivate viruses by damaging their nucleic acids.

Ions Charged atoms that have an unequal number of protons and electrons.

Iris diaphragm The part of a light microscope's condenser that allows the viewer to modulate how much light is aimed at the specimen in order to improve contrast.

Iron-binding proteins Proteins that bind free iron and thereby keep the amount of iron in circulation and within our tissues well below the amount microbes require for survival.

Isografts Transplanted tissue that is genetically identical to the host, these grafts are accepted as "self" and typically safe from immune rejection.

Isomerases Specialized cellular enzymes that perform isomerization reactions.

Isomerization reactions Occur in many cellular pathways; a reaction in which one isomer of a molecule is converted to another.

Isomers Molecules with the same molecular formula but different molecular structures.

Isoniazid A drug that interferes with mycolic acid construction in cell walls of acid-fast bacteria and is used in combination with other antibacterial drugs to treat tuberculosis.

Isotopes Atoms with the same number of protons, but different numbers of neutrons.

Isotype Class or subclass of immunoglobulin; isotypes include IgG, IgA, IgM IgE, and IgD.

Isotype switching A process that enables a given B cell to alter what class (subtype) of antibody it makes.

Jaundice Yellowing of the skin and eyes due to high bilirubin levels of the blood; a sign of blood, liver, or gallbladder problems when present in non-newborns.

Joseph Lister British surgeon whose work in the 1860s proved that sterilizing instruments, and sanitizing wounds with carbolic acid encouraged healing and prevented pus formation.

Julius Richard Petri Developed the petri dish, which when filled with the agar-solidified media, made it easier to isolate and observe bacteria.

Keratin A waterproofing protein that makes up the cells of most of the epidermis.

Keratitis Severe inflammation of the cornea; warrants immediate medical attention to prevent serious and irreversible damage to the eyes; in keratitis, the protective layers that cover the cornea are destroyed, leaving the cornea accessible to bacteria, viruses, parasites, or fungi.

Kidneys Paired organs of the human urinary system; our kidneys filter metabolic waste products such as urea and creatinine out of our blood and regulate the body's water and electrolyte balance; they also help manage our blood pressure and blood pH.

Kinases Specialized enzymes that use phosphorylation reactions to add phosphates to their targets.

Kingdoms Taxonomic groups that fall beneath the umbrella of domains; the number of designated kingdoms has fluctuated from five to eight; the older five-kingdom classification scheme includes Animalia, Plantae, Fungi, Protista, and Monera; the six-kingdom schematic, employed by this text, replaces Kingdom Monera with Kingdom Archaea and Kingdom Bacteria.

Kinins A class of pro-inflammatory polypeptides that induce vascular changes, stimulate pain receptors, and assist in blood-clotting cascades; they also amplify inflammation by triggering the production of numerous downstream signaling molecules, including eicosanoids.

Kirby-Bauer test Also called the disk diffusion test; this test is used to determine a bacterium's susceptibility to antimicrobial drugs; the diameter of the zone of inhibition that may develop around a drug infused disk is measured and compared to a standard table to ascertain if the bacterium is susceptible, resistant, or intermediate in its response to the tested drug.

Koch's postulates of disease Four principles that establish the criteria for determining the causative agent of an infectious disease: (1) The same organism must be present in every case of the disease; (2) The organism must be isolated from the diseased host and grown as a pure culture; (3) The isolated organism should cause the disease in question when it is introduced (inoculated) into a susceptible host (a host that can develop the disease); (4) The organism must then be re-isolated from the inoculated, diseased animal.

Krebs cycle Also called citric acid cycle or the tricarboxylic acid cycle; this cycle is a series of redox reactions and decarboxylation reactions that begins with the formation of citric acid from oxaloacetic acid and the acetyl-CoA made in the intermediate step; the Krebs cycle produces some ATP and a lot of the reduced cofactors NADH and $FADH_2$.

La Crosse virus A neuroinvasive arbovirus that causes around 80–100 cases of encephalitis in the United States each year; transmitted by mosquitos.

Labia majora The paired lip-like folds that enclose the female external genitalia; part of the vulva.

Labia minora The pair of small lip-like folds that enfold the vagina and urethra; part of the vulva.

Lacrimal system A system in the eye that produces and collects tears.

Lactose (*lac*) operon An example of an inducible operon that is induced, or actively transcribed, only when lactose is present and the cell's preferred food, glucose, is absent.

Lag phase First phase of a standard bacterial growth curve; occurs while bacterial cells adjust to their new environment; during this phase, cells alter their gene expression in response to their new setting; population growth is not usually seen in this phase.

Lagging strand The complementary strand of the parent DNA molecule that is copied/replicated in the opposite direction of helix unwinding; built in chunks called Okazaki fragments.

Laparoscopy Placement of a fiber-optic instrument into the abdomen to view the organs and tissues there and/or to permit a surgical procedure with just a small incision.

Laryngitis Inflammation of the larynx, which can cause temporary voice loss due to the swelling of the vocal cords.

Lassa fever Viral hemorrhagic fever caused by Lassa virus, a single-stranded RNA virus; contracted by inhaling airborne dried rodent urine (often during sweeping), eating food contaminated with rodent droppings, or having rodent droppings enter a wound.

Latency Ability of a pathogen to quietly exist inside a host; pathogens with this capability cause persistent or recurrent disease, staying under the immune system's "radar" and only reemerging in visible ways at opportune moments.

Latent infection (general latent infections) Infection in which the host does not have signs or symptoms (they are asymptomatic); if conditions are right, the agent can cause an active illness; infections that may be distinguished by flare-ups with intermittent periods of dormancy (latency).

Latent infection (specifically referring to latent tuberculosis) In latent tuberculosis the infected person has no symptoms of the disease and is not contagious.

Latex agglutination test In this test, a specific antigen is used to coat latex beads, the beads are then exposed to a sample that might contain antibodies that can recognize the antigen on the beads; if the beads clump (or agglutinate) then it indicates antibodies are present in the tested sample.

Law (scientific law) A precise statement, or mathematical formula, that predicts a specific occurrence under specific conditions.

Leading strand The side of the parent DNA molecule that is copied in the same direction as helix unwinding.

Lectin pathway A complement pathway that is independent of antibodies; becomes activated when mannose-binding lectin protein, a protein in our blood, associates with certain sugars on a microbe's surface; other than how it is initiated, the lectin pathway for complement activation is identical to the classical pathway.

Left atrium One of two upper heart chambers; heart chamber that receives oxygenated blood form the pulmonary veins.

Left ventricle One of two lower heart chambers; this chamber pumps blood to the whole body and is therefore responsible for systemic circulation.

Legionnaires' disease A pulmonary disease that came to attention in 1972, when more than 2,000 military veterans gathered at a hotel in Philadelphia for a convention; caused by the bacterium *Legionella pneumophila*.

Leishmaniasis Protozoan infection classified as a neglected tropical disease; caused by *Leishmania* species of protozoans that are transmitted by the bite of infected sand flies.

Leprosy See **Hansen's disease**.

Lesion Clinical term for any observable abnormality of the skin.

Lethal dose-50 (LD50) Term describing the amount of toxin needed to kill 50 percent of affected hosts that are not treated; the lethality of an infectious agent is usually expressed as mortality rate, rather than as LD50.

Leukemia Cancer of the tissue that creates blood components (such as bone marrow).

Leukocytes White blood cells; specialized cells of the immune system.

Leukocytosis Increase in leukocytes (white blood cells).

L-form A term usually reserved for bacteria that normally have a cell wall but have lost it in the course of their life cycle, or as a result of a mutation; upon losing their rigid cell wall, these bacteria assume various morphologies and are described as pleomorphic.

Ligase An enzyme used in DNA replication; this enzyme joins DNA segments at the junction between where DNA polymerase I replaced the RNA primer with DNA and the rest of the DNA strand; covalently links Okazaki fragments.

Lincosamides Broad-spectrum antimicrobial drugs that bind to the 50S subunit of prokaryotic ribosomes to block protein synthesis; work against a wide collection of aerobic and anaerobic Gram-positive as well as anaerobic Gram-negative bacteria and even some protozoans; clindamycin is an example.

Lipid Predominantly hydrophobic biomolecules that include fats, oils, waxes, and steroids; all lipids are organic molecules made up of carbon, hydrogen, and oxygen, but they lack the 2 to 1 ratio of hydrogen to oxygen seen in carbohydrates.

Lipid A The lipid portion of lipopolysaccharide (LPS).

Lipophilic molecules "Fat-loving" molecules that dissolve easily in fats or lipids; the hydrocarbon tails on fatty acids are a common example of lipophilic molecules; these are often synonymous with hydrophobic, or "water hating," molecules.

Lipopolysaccharide (LPS) A glycolipid that enriches the outer membrane of Gram-negative bacteria; consists of a lipid portion (lipid A or endotoxin) that is poisonous to animals.

Lipoproteins Lipids that are linked to proteins.

***Listeria* meningitis** Bacterial meningitis caused by the Gram-positive bacterium *Listeria monocytogenes;* accounts for 20 percent of all meningitis in neonates and in people over 60 years old; can develop from the preliminary gastrointestinal symptoms of listeriosis.

Lithotrophs Organisms that get reducing power from inorganic sources to fuel anabolic processes.

Live attenuated vaccines Vaccines in which pathogens have been altered so that they do not cause disease (are not pathogenic), but are still infectious; these vaccines are the closest to the actual agent encountered in nature and therefore tend to simulate potent immunological responses that are accompanied by long-lived memory.

Localized Term applied to an infection or inflammatory reaction that is restricted to a specific part of the body.

Localized anaphylaxis Allergic reaction that tends to feature isolated symptoms such as watery eyes, a runny nose, or a confined rash.

Lock-and-key model Now outdated, this model stated that an enzyme is like a lock that can only be opened by a specific key: its substrate.

Logarithimic (log) phase Second growth phase of bacteria, characterized by an upward-sloped line that results when the number of viable cells is plotted on a logarithmic scale as a function of time; this phase lasts as long as sufficient nutrients are available and metabolic wastes are not appreciably accumulating.

Lophotrichous A term applied to cells with a tuft or cluster of flagella at one pole (end) of the cell (*lopho* = tuft).

Louis Pasteur Scientist who, in the late 1800s, showed that biogenesis is responsible for the propagation of life; disproved spontaneous generation using his S-necked flask experiment.

Lower endoscopy A procedure that involves a colonoscope, which resembles an endoscope; the colonoscope is inserted into the anus and is used to view the rectum and the large intestine (colon).

Lower GI tract The portion of the digestive tract that consists of the small and large intestines, rectum, and anus.

Lower UTIs Urinary tract infections such as urethritis and cystitis; although these conditions can develop from a number of factors, bacterial infections are the most common cause.

Lumbar puncture Also called a spinal tap; a diagnostic procedure in which a needle is inserted between two vertebrae of the lower back to collect cerebrospinal fluid (CSF) for analysis; one of the most useful diagnostic tools for meningitis; results can reveal either purulent (pus-containing, usually associated with bacterial infection) or aseptic (without cultivable organisms, usually viral infections) meningitis.

Lyme disease Infection caused by the Gram-negative spirochete *Borrelia burgdorferi* and named after a Connecticut town where a large outbreak occurred in the 1970s; the reservoir is the white-footed mouse, while the vector is deer ticks; about 7 to 14 days after infection, a bull's-eye-shaped rash, termed erythema migrans, often develops around the tick bite area.

Lymph Interstitial fluid that enters lymphatic vessels; the lymphatic vessels transport lymph to lymph nodes, where it is screened for pathogens and filtered; from there, the fluid is eventually channeled into veins, where it rejoins the blood plasma.

Lymph nodes Bean-shaped secondary lymphoid organs; clustered in specific areas including in the neck, underarm, and groin; lymph nodes serve as filtering and screening centers for lymph before returning it to the bloodstream.

Lymphangitis A condition in which lymphatic vessels become inflamed and appear as red streaks that radiate from the initial infection site (often a wound on the extremities) to the nearest lymph node.

Lymphatic system A collection of tissues and organs that have roles in collecting, circulating, and filtering fluid in body tissues before such fluid is returned to the bloodstream.

Lymphocytes A subgroup of leukocytes that include natural killer (NK) cells, B cells, and T cells; account for about 25 percent of circulating leukocytes.

Lymphogranuloma venereum A sexually transmitted infection that progresses in three stages, the last of which may result in severe destruction of the genitalia; caused by the lymphogranuloma venereum grouping, a clinically important biovar of *Chlamydia trachomatis*.

Lymphoma Cancer of cells associated with the lymphatic system; may result in consistently enlarged lymph nodes.

Lysogenic pathway Replication pathway used by temperate phages.

Lysosomes Vesicle-like organelles that contain a wide variety of hydrolytic enzymes that break down substances engulfed by the cell during phagocytosis and receptor-mediated endocytosis; also act as garbage disposal tools for the cell.

Lysozyme An enzyme encoded by bacteriophages that breaks down host cell walls and causes bacterial cell lysis (bursting) once the newly assembled phages are mature; an enzyme found in secretions like mucus, saliva, sweat, and tears that acts as a chemical barrier by breaking down bacterial cell walls.

Lytic replication pathway The five-step pathway used by bacteriophages that kill the host cell as newly made bacteriophages are released.

Macrolide drugs A group of broad-spectrum drugs that include erythromycin, azithromycin ("Z-Pak" or "Z-Pack"), and clarithromycin; macrolides target the 50S subunit of prokaryotic ribosomes to block protein synthesis; these compounds are effective against various Gram-negative cocci as well as aerobic and anaerobic Gram-positive rods and cocci.

Macromolecules Large biomolecules built by a series of synthesis reactions and broken down by a series of decomposition reactions.

Macrophages Highly phagocytic cells that mature from monocytes and that destroy a wide range of pathogens; may be fixed or wandering. See **monocytes**.

Major histocompatibility complex (MHC) proteins Specialized "self-proteins" also known as human leukocyte antigens (HLAs); for T cells, ensuring self-tolerance involves screening the cells for their ability to recognize MHC proteins.

Malaise General sense of illness or discomfort; an overarching sense of feeling unwell or uncomfortable for reasons that may be unclear.

MALT (mucosa-associated lymphoid tissue) The majority of secondary lymphoid tissue in the body; this diffuse system of lymphoid tissue is found in all mucosal linings of our body and at all body entryways, playing a key role in finding and fighting harmful microbes; tonsils, the appendix, and Peyer's patches (lymphoid tissues found in the small intestine) are common examples.

Mannitol salt agar (MSA) A medium that is selective due to its high salt content; most bacteria can't grow on this medium; it also differentiates organisms based on their ability to ferment a sugar called mannitol.

Mannose-binding lectin In the lectin pathway of complement activation, a protein in the blood that associates with certain sugars on a microbe's surface.

Marburg virus A virus that causes viral hemorrhagic fever and that is similar to Ebola virus in its disease features and modes of transmission; so far, monkeys and fruit bats have been identified as hosts, but the natural reservoir remains unknown; inhaling bat droppings that have been stirred up in dust, or handling or eating infected bushmeat allows the virus to enter the human population; infected people can then transmit the virus to others through bodily fluids and sexual contact; symptoms begin as a sudden onset of headache, fever, and chills; as the infection progresses, the patient develops nausea, vomiting, and abdominal pain, followed by severe bleeding from various body sites, jaundice, delirium, and shock within 5 to 7 days.

Margination The first step by which cells exit blood vessels in the recruitment phase of inflammation; leukocytes slow down as they roll along vessel walls, and they eventually adhere to the vessel wall and stop rolling.

Mast cells A distinct population of leukocytes once thought to be basophils that resided in tissues instead of circulating in blood; similar to basophils in that they release histamine and have roles in allergies and fighting parasites; their ability to conduct phagocytosis, along with diverse enzymes and AMPs in their granules, also make them important in fighting bacteria; have key roles in activating the adaptive immune response.

Maximum and minimum temperatures The upper and lower temperatures that support a given microbe's growth.

Measles A viral infection that is highly contagious and spreads by the respiratory route; about ten days after exposure, unvaccinated individuals may develop a fever, sore throat, dry cough, spots in the mouth, and finally the characteristic maculopapular rash that begins on the face and spreads to the trunk and then to the extremities.

Measles, mumps, and rubella (MMR) vaccine A live attenuated vaccine that confers protection against measles, mumps, and rubella; the most common vaccine side effects are fever, rash, or swelling at the injection site; in very rare cases, there have been incidents of joint pain, febrile seizures (convulsions caused by fever), severe allergic reactions (less often than 1 in a million doses), or a decrease in platelet count.

Measures of association Epidemiological measures that tell us what factors may be linked with cases of the disease, and this in turn shows who might be at risk for developing an illness.

Measures of frequency Epidemiological measures that give information about the occurrence of a disease in a population during a certain period of time.

Mebendazole A common broad-spectrum antihelminthic drug that interferes with glucose uptake in worms by targeting microtubules. These drugs treat a wide collection of roundworms such as *Ascaris*, hookworms, pinworms, and *Trichinella* (an infection mainly associated with eating undercooked pork); also used against certain tapeworms.

Mechanical barriers First line defenses in which pathogens are rinsed, flushed, or trapped to limit their spread into the body.

Mechanical vector Vector organism that spreads disease without being integral to a pathogen's life cycle.

Meiosis A form of cell division that generates specialized cells, called gametes, that combine in sexual reproduction to make a genetically unique zygote.

Melanin The brown pigment in our skin that has antimicrobial properties and protects against ultraviolet radiation that depletes the skin of folic acid and causes skin damage.

Membrane attack complex (MAC) An attack complex made by complement proteins; it drills into cells, causing them to burst (undergo cytolysis).

Memory cells In adaptive immunity, cells that remain in lymphatic tissues to serve as a rapid recognition of the antigen if it's encountered again later.

Meninges Three layers of specialized tissue that encase the central nervous system; layers include the dura mater, arachnoid mater, and pia mater; these layers support and care for the CNS by supplying nutrients, removing waste, and protecting the CNS from physical shocks.

Meningitis Inflammation of any of the meninges; viruses are the most common cause of meningitis; however, bacteria (and less commonly parasites and fungi) are also culprits.

Meningococcal meningitis Bacterial meningitis caused by *Neisseria meningitides*, a Gram-negative aerobe that has a capsule that contributes to pathogenesis; there are several different serotypes based on the polysaccharides in the capsule; serotypes A, B, C, W, X, and Y are the groups implicated in meningococcal meningitis (this organism can also cause generalized bacteremia).

Meningoencephalitis Inflammation of the meninges (meningitis) occurring along with inflammation of the brain (encephalitis); the condition displays a variable combination of meningitis and encephalitis symptoms; viruses are the most common cause of both meningitis and encephalitis; however, bacteria (and less commonly parasites and fungi) are also culprits.

Merogony An asexual stage of reproduction in apicomplexans that entails daughter cells called merozoites being produced by repeated asexual cell division.

MERS-CoV An emerging coronavirus that causes Middle East Respiratory Syndrome (MERS); human-to-human transmission is possible (though not as likely as camel-to-human transmission), thus public health agencies around the world are carefully tracking this virus.

Mesophiles Organisms that prefer moderate temperatures and tend to grow best around 10°–50°C, a range that includes body temperature; most pathogens are part of the mesophilic temperature group and cover a broad range of the planet, from soil to streams to dwelling in eukaryotic organisms.

Messenger RNA (mRNA) Mostly linear RNA molecules that carry the genetic messages stored in DNA; made by RNA polymerase during transcription; mRNA contains the triplet code that is translated to build proteins.

Metabolism Collectively refers to the chemical reactions that organisms use to break down substances to release energy, as well as to reactions that use the released energy to build new substances.

Methyl red A pH indicator that turns red in acidic solutions.

Methyl red (MR) test A biochemical test that is often used in combination with the Voges-Proskauer (VP) test to identify enteric bacteria; detects the formation of acidic end products from glucose.

Methyl red/Voges-Proskauer (MRVP) Combination test designed to distinguish between mixed acid fermentation and butanediol fermentation.

Metronidazole One of the most commonly prescribed drugs (trade name: Flagyl) that treats nonmalarial protozoan infections; it is a member of the nitroimidazoles, it targets nucleic acids, and it is effective against *Toxoplasma gondi* (toxoplasmosis), *Trichomonas vaginalis* (trichomoniasis), and the intestinal protozoans *Giardia lamblia* (giardiasis) and *Entamoeba hystolytica* (amoebiasis).

MHC I–antigen complexes Combination of MHC I molecules and protein antigens bound to them; if the MHC I–antigen complex on the surface of cells is the same as that used to activate T cytotoxic cells, then the cell bearing the complex will be attacked by the cellular immune response.

MHC II-antigen complexes Association between MHC II proteins and antigens processed by antigen presenting cells (APCs); these MHC II-antigen complexes move to the APC cell's surface; displaying these complexes, the APC migrates from the inflammation site to lymphoid tissues and seeks out T helper cells to activate them.

Micelles Assemblies of amphipathic molecules where the hydrophobic portion of the molecule is positioned toward the center of the structure, while the hydrophilic region of the molecule faces the aqueous environment.

Microaerophiles Organisms that use only small amounts of atmospheric oxygen and live in low-oxygen settings where they can limit their exposure to reactive oxygen species (ROS) while still meeting their oxygen needs.

Microbes A term that encompasses living microorganisms such as bacteria, archaea, fungi, protists, and helminths, and nonliving/noncellular entities such as viruses and prions (infectious proteins).

Microbial growth Cell division that produces new (daughter) cells and increases the total cell population.

Microbiocidal Term applied to germicides that kill microbes.

Microbiostatic Term applied to germicides that inhibit microbial growth, but do not kill microbes.

Microcephaly Arrested fetal brain development that results in an abnormally small brain and head size; has occurred in babies whose mothers contracted Zika virus infection during pregnancy.

Microfilaments Fine fibers made of the protein **actin**; they tend to be 3–6 nm in diameter and associate with the motor protein **myosin** to facilitate movement.

Micrographs Pictures taken through a microscope.

Microorganisms Cellular, living microorganisms such as bacteria, archaea, fungi, protists, and helminths.

Microtubules Hollow tubes made of a protein called **tubulin** that tend to be about 25 nm in diameter, making microtubules the thickest of the three cytoskeleton fibers; microtubules serve as roadways in the cell.

Middle East Respiratory Syndrome (MERS) A respiratory disease primarily transmitted from dromedary camels to humans; it is caused by an emerging coronavirus, MERS-CoV.

Minimal inhibitory concentration (MIC) The lowest concentration of the antimicrobial drug that inhibits the microbe's growth.

Minimum bactericidal concentration (MBC) The minimum concentration of the drug that kills at least 99.9 percent of the bacteria present.

Missense mutation A mutation in which the meaning of the codon is changed in a way that the wrong amino acid is added to the growing protein.

Mitochondria (singular: mitochondrion) Organelles that make the most of a cell's adenosine triphosphate (ATP), which is the preferred molecule cells use to meet their energy needs; nicknamed the "powerhouse" of the cell.

Mitosis A form of asexual reproduction that is the most common way eukaryotic cells divide; generates two genetically identical offspring from one parent.

Mixed culture A culture with at least two characteristically different colonies.

Mixotrophs Microbes that use a variety of carbon sources and are also diverse in how they obtain energy and reducing power; they manage to survive dramatic changes in their surroundings by switching between metabolic modes.

Mole The average mass of 6.022×10^{23} atoms of a given substance.

Molecular formula Also called chemical formula; basically the atomic recipe of a molecule; molecular formulas reveal what elements are in a molecule as well as their ratios.

Molecule Formed when two or more atoms bond together.

Monobactams Drugs that have only one ring in their structure, unlike other beta-lactam drugs that usually have two rings; aztreonam is the lone example of a commercially available monobactam drug for human use.

Monocistronic An mRNA molecule that encodes only one protein; eukaryotic mRNAs are generally monocistronic.

Monoclonal antibody An antibody that recognizes just one epitope of a particular antigen; recommended for use as the detection antibody in ELISAs (enzyme-linked immunosorbent assays).

Monocytes When they mature, these cells are called macrophages; monocytes/macrophages are the largest agranular white blood cells and they account for up to 10 percent of circulating leukocytes; their levels can increase due to chronic infections and inflammation, autoimmune disorders, and certain cancers. See **Macrophages**.

Monoglyceride A name for a lipid that has only one fatty acid linked to glycerol.

Monomers In chemistry, this term refers to smaller foundational units that can be bonded together to form larger macromolecules; a single subunit or part of a greater whole.

Mononucleosis (or "mono") An infection caused by Epstein-Barr virus (EBV) that induces extreme fatigue, swollen lymph nodes in the neck and armpit regions, severe sore throat, headache, splenomegaly (enlarged spleen), and occasionally a rash; teens and young adults have the highest risk; EBV transmits through contact with infected bodily fluids—especially saliva, which is how it got the nickname, "the kissing disease;" once the virus enters the mouth, it invades the lymphatic system through the tonsils; the virus then infects nonactivated B cells.

Monosaccharide One sugar unit; the smallest unit of a carbohydrate; can be polymerized to build larger carbohydrates. See **polymerization**.

Monotrichous A term that describes cells with a single flagellum (mono = one).

Monounsaturated Term describing fats that have one double bond in their fatty acid chains.

Morbidity Presence of a disease in a population.

Mordants Chemicals that may be required in certain staining procedures to interact with a dye and fix, or trap, it on a treated specimen; mordants include iodine that is used in the Gram staining procedure. See **iodine**.

Morphology Physical traits such as shape, size, and arrangement that are used in classification of microbes.

Mortality rate An epidemiological measure of association that indicates the number of deaths during a specific time period.

MRSA Methicillin-resistant *Staphylococcus aureus* strains that have altered their cell wall construction enzymes in such a way that they do not associate with methicillin, penicillin, or a variety of other penicillin-based drugs; infections caused by these bacteria are now commonplace in healthcare settings and in the general community.

Mucociliary escalator A mechanism within airways whereby mucus's trapping action is complemented by the mucus being swept away from the lungs and toward the mouth by ciliated cells of the respiratory mucus membranes; this mechanism trap inhaled debris and sweeps it toward the mouth to prevent it from entering the lungs.

Mucocutaneous leishmaniasis A protozoan infection caused by *Leishmania* in which the lesions develop in mucous membranes of the nose or mouth; it can lead to severe and permanent disfigurement as the protozoan destroys these structures.

Mucormycosis A rare fungal disease having five forms, with the most common being the pulmonary and rhinocerebral forms; the fungi that cause mucormycosis are found in soil that is enriched with rotting wood material; most of us encounter the fungi that can cause mucormycosis in our everyday lives and never develop illness; although several fungi can cause the disease, most infections are associated with inhaling spores made by *Rhizopus arrhizus*.

Mucosa-associated lymphoid tissue (MALT) Lymphoid tissue found along the gastrointestinal tract, including tonsils at the back of the throat, the appendix, and Peyer's patches of the small intestine; cells in these lymphatic tissues "sample" the environment by phagocytosis, and present their findings to lymphocytes that may or may not need to initiate an immune response; if a response is warranted, then inflammation is an early defense, followed by activation of the adaptive immune response.

Muller Hinton agar Solidified, plated agar medium that is used in the Kirby-Bauer test because it supports the growth of most common bacterial pathogens while also allowing for fairly uniform diffusion of drugs through the agar from their point of application.

Multidrug-resistant (MDR) Term applied to bacterial strains that are resistant to multiple antimicrobial drugs.

Mumps An infection caused by the mumps virus, an RNA virus that spreads among humans through infected saliva; when the mumps virus enters the nose or mouth, it quickly replicates in the upper respiratory tract; thereafter, the virus enters the bloodstream to infect salivary glands, particularly the parotid glands; the incubation period of 2 to 3 weeks is often followed by dry mouth, fever, headache, and fatigue.

Mutagens Chemical, physical, or biological agents that increase the rate of mutation.

Mutations Errors in genetic material that can lead to changes in a protein's primary structure; changes in the genetic material of a cell or virus.

Mutualism A type of symbiotic relationship that helps the host.

Mycolic acid A substance in the waxy cell walls of acid-fast bacteria; the presence of mycolic acid can be detected by acid-fast staining; genera *Nocardia* and *Mycobacterium* are the best-known examples of acid-fast bacteria.

***Mycoplasma* pneumonia** The most common cause of atypical community-acquired pneumonia; studies suggest that up to 20 percent of infections are asymptomatic; it is estimated that about 2 million cases occur in the United States every year, with about 100,000 cases requiring hospitalization; the causative agent, *Mycoplasma pneumoniae*, is one of the smallest self-replicating organisms known to mankind; the virus-like characteristics of these bacteria (small size, lack of cell wall, parasitic nature) initially caused scientists to wrongly assume that they were viruses.

Mycoses Diseases caused by fungi.

Mycotoxins Potentially deadly toxin produced by fungi.

Myocarditis Inflammation of the myocardium (muscle layer of the heart).

Myocardium The middle layer of the three layers of the heart; the muscular tissue that makes up the bulk of the heart.

Myosin Motor protein that associates with **actin** to facilitate movement.

NA (neuraminidase) spikes Glycoprotein structures on influenza's viral envelope that help the virus attach to host cells.

Naked A term that describes virions that lack an envelope.

Narrow-spectrum drugs Drugs that target a limited range of bacteria; for example, bacitracin and vancomycin are mainly effective against Gram-positive bacterial species.

National Notifiable Diseases Surveillance System (NNDSS) A network of the CDC that depends upon local hospitals, laboratories, and private healthcare providers to monitor and report certain infectious and noninfectious diseases.

Natural killer cells (NK cells) Lymphocytes that have important roles in innate immune protection against viruses, bacteria, parasites, and even tumor cells.

Naturally acquired active immunity Immunity that involves contracting an infection that triggers the patient's immune system to make memory cells and antibodies that confer long-term protection thereafter.

Naturally acquired passive immunity Immunity that occurs when someone receives antibodies to an antigen through nonmedical means; the most common examples are maternal antibodies passing across the placenta to a baby in utero, and maternal antibodies moving into an infant through colostrum (an antibody- and mineral-rich fluid that precedes milk) and breast milk.

Necrosis Tissue death due to loss of blood flow; may be caused by infections, injury, or diseases that affect the vasculature (e.g., diabetes).

Necrotizing fasciitis Severe, yet rare, disease associated with *Streptococcus pyogenes* bacteria that break down flesh with their specialized tissue-degrading enzymes; referred to by the media as "flesh-eating disease."

Negative staining A technique using acidic dyes to stain the background of a specimen, giving to the sample a more true-to-life appearance, with fewer distortions of delicate cellular features. See **acidic dyes**.

Neonatal meningitis A form of bacterial meningitis that presents as acute meningitis in infants within the first 28 days of being born; it occurs when the bacteria are passed to a neonate from their mother during delivery; several different bacteria cause neonatal meningitis, including Group B *Streptococcus, Escherichia coli,* and *Listeria monocytogenes*.

Nephrons Microscopic functional units of the kidneys, over a million in each kidney, that secrete waste products into urine while retaining useful substances in the blood.

Nephrotoxic Kidney-toxic; a term applied to antimicrobial drugs that include a long list of agents, with aminoglycosides being prevalent on the list.

Neurons Specialized cells in the central nervous system and peripheral nervous system that transfer signals; neurons can be categorized as sensory (specializing in input, or afferent), motor (specializing in output, or efferent), or interneurons (allowing CNS and PNS communication).

Neutralization Occurs when OH^- and H^+ ions combine to form water.

Neutralophiles Organisms that grow best in a pH range of 5–8; they make up the majority of microorganisms, especially pathogens.

Neutrons Noncharged particles within the nucleus of an atom.

Neutropenia A lower than normal neutrophil count that may be caused by certain viral infections.

Neutrophils The most numerous white blood cells in circulation (40 to 70 percent); they are the first leukocytes recruited from the bloodstream to injured tissues; they release potent antimicrobial peptides (AMPs) and enzymes from their granules into the surrounding environment; these released factors destroy many microbes, especially bacteria, and stimulate inflammation; neutrophils also phagocytize foreign cells and viruses; an elevated neutrophil count may indicate an acute bacterial infection.

Nicotinamide adenine dinucleotide (NAD^+) A coenzyme that serves as an electron carrier and that is made from the B vitamin niacin (nicotinic acid).

Nicotinamide adenine dinucleotide phosphate ($NADP^+$) A coenzyme that serves as an electron carrier; similar in structure and function to NAD^+, but unlike NAD^+ it has a phosphate group; $NADP^+$ is reduced to NADPH, which is mainly used in anabolic reactions.

Nitrazoxanide An antiprotozoan drug mainly used to treat *Giardia* and *Cryptosporidium* as well as against certain parasitic worms; nitrazoxanide blocks anaerobic energy metabolism in protozoa.

Nitrogen base A component of nucleotides; includes adenine (A), guanine (G), cytosine (C), thymine (T), and uracil (U).

Nitrogen fixation Process by which some cells get their nitrogen directly from the atmosphere; an important step in converting atmospheric nitrogen from a gas form to a nongaseous form like ammonia that can be used by other cellular life.

Nodular cystic acne One of the most common skin afflictions, affecting 70–80 percent of teenagers and young adults at some point, mainly caused by *Propionibacterium acnes* bacteria; this is the most severe and scarring form of acne.

Noncommunicable Pertaining to zoonotic pathogens that do not spread from person to person.

Noncompetitive inhibitors Substances that do not compete with a substrate for the enzyme's active site but instead decrease enzyme activity by binding to the enzyme at a site other than the active site.

Noncritical equipment Equipment such as stethoscopes, blood pressure cuffs, and most surfaces in patient rooms that only contact patients' intact skin and therefore require less stringent disinfection than semicritical and critical equipment; low-level disinfectants are sufficient to remove microbes from noncritical surfaces.

Nonessential amino acids Amino acids that a cell can make and therefore does not have to obtain from its environment/food source.

Nongonococcal urethritis Infection of the urethra that is not caused by *Gonorrhea* bacteria; may be caused by diverse microbes such as *Ureaplasma urealyticum*, *Mycoplasma genitalium*, or *Chlamydia* bacteria.

Noninfectious diseases Illnesses not directly caused by pathogens.

Non-ionizing radiation A form of radiation that can change the bond structure of DNA, leading to severe mutations that ultimately result in cell death; an example is ultraviolet (UV) rays.

Nonsense mutation A mutation that causes a codon to go from encoding an amino acid to encoding a stop signal.

Nonsoluble Substances that do not dissolve in water.

Normal blood pH A pH range of 7.35–7.45; the kidneys regulate the levels of ions that affect blood pH.

Normal microbiota Normal human flora, including bacteria, archaea, and eukaryotic microbes.

Norovirus An RNA virus that is the leading cause of acute viral gastroenteritis in the United States; over half of all outbreaks occur in long-term-care facilities; the virus is extremely contagious, as only 20 particles are required for infection; feces and vomit serve to spread the virus via a fecal–oral route; symptoms—vomiting and diarrhea—develop within 12 to 48 hours of ingesting the virus; also known as Norwalk virus.

Nosocomial infections See **Healthcare-acquired infections (HAIs)**.

Notifiable diseases Diseases that the CDC recommends reporting to government health agencies for monitoring incidence; there are usually about 60 different infectious diseases (and a number of noninfectious diseases) on the CDC's notifiable diseases list.

Nucleic acids DNA and RNA; macromolecules that serve as the genetic material of cells and viruses; nucleic acids are polymers built from nucleotides.

Nucleoid A somewhat centralized region of a prokaryotic cell, though its boundaries are not distinct, since it isn't membrane enclosed; in addition to containing prokaryotic DNA, the region also includes some RNA and proteins.

Nucleoside analogs Antiviral drugs that block viral replication by blocking nucleic acid production.

Nucleoside reverse transcriptase inhibitors (NRTIs) Drugs that target reverse transcriptase enzymes and can be used to combat retroviruses like HIV.

Nucleotides The basic building blocks of the nucleic acids DNA and RNA; nucleotides have three basic parts: a phosphate, a sugar (deoxyribose in DNA or ribose in RNA), and one nitrogen base that can be adenine (A), guanine (G), cytosine (C), thymine (T), or uracil (U).

Nucleus The distinct, membrane-enclosed structure that is the command center of the cell and houses DNA, which ultimately orchestrates all cellular activities.

Nugent's criteria Scoring standards for diagnosing vaginoses based on Gram-stained sampled vaginal secretions; nugent's scoring criteria is most accurate when it is used to evaluate samples from reproductive-age women; in general, scoring is based on seeing abundant cocci and Gram-negative (or Gram-variable) rods in the vaginal mucus, along with a decrease in long Gram-positive rods; Nugent's criteria also look for clue cells in the patient sample.

O antigen The carbohydrate antigen on the surface of red blood cells (erythrocytes) of people with type O blood; shares a core structure with the type A and type B antigens which is why this antigen is not immunogenic in people that have type A or B blood.

Objective lens In a compound light microscope, the lens that is near the specimen.

Obligate aerobes Organisms, including humans, that have an absolute dependence on oxygen for cellular processes, and will die unless it is abundant.

Obligate anaerobes Organisms that do not use oxygen in their metabolism and that tend to die in aerobic environments because they can't eliminate reactive oxygen species (ROS).

Obligate intracellular pathogens Pathogens, which include viruses and certain bacteria and protozoans, that only replicate inside a host cell and therefore cannot be grown as independent pure cultures.

Observational studies Analytical epidemiological studies that do not involve administering an intervention, a treatment, or an exposure for a disease; they simply watch and collect data on cases that exist, used to exist, or develop over time in a tracked/monitored group; these studies are designed to directly measure risk for developing a certain disease when certain risk factors are present.

Observations Part of the scientific method; data collected and analyzed by the researcher.

Ocular lens In a compound light microscope, the lens that sits at the top of the microscope near the viewer's eyes.

Okazaki fragments Segments of DNA that are built on the lagging strand during DNA replication; these segments are glued together by ligase to form a continuous DNA strand.

Oncogenic viruses Viruses that can cause cancer.

Operator The part of the operon that the repressor binds to in order to block transcription.

Operons Collection of genes controlled by a shared regulatory element; mechanism used mainly by bacterial cells to regulate protein synthesis at the pre-transcriptional level.

Opportunistic pathogens Pathogens that only cause disease when their host, the targeted organism, is weakened in some way.

Opsonization Tagging of the invader with complement proteins so it stands out and is more readily cleared by phagocytic cells.

Optimal pH A pH above or below this value will alter enzyme structure, leading to a reduced reaction rate; also used to refer to the favored or best pH level for maximal growth of a particular microbe.

Optimal temperature The temperature where a given enzyme's activity is highest; also, this term is used to describe the temperature where cellular growth rate is highest.

Optimum pH See optimal pH.

Oral rehydration therapies Therapies that consist of a simple water-based solution enriched with electrolytes.

Organic carbon Carbon that is bonded to hydrogen; methane (CH_4), sugars, lipids, and proteins are all examples of organic carbon sources; inorganic carbon such as that found in carbon dioxide (CO_2) and carbon monoxide (CO) is not bonded to hydrogen.

Organic molecules Molecules that contain carbon and hydrogen.

Organotrophs Organisms that get reducing power from organic sources to fuel anabolic processes; the most common organic source of reducing power is glucose.

ORSA Oxacillin-resistant *Staphylococcus aureus* strain that is relatively common; rather than trying to go after the beta-lactam drugs, this resistant strain adopted a disconcertingly effective strategy—it altered the transpeptidase enzymes so that the majority of beta-lactam drugs can no longer target it; this means that the ORSA strain is resistant to almost all beta-lactam drugs, not just oxacillin, as its acronym implies.

Osmosis The net movement of water from an area of low solute (salt, sugar, or other dissolved substance) concentration to an area of higher solute concentration across a selectively permeable membrane; it does not require energy to occur; cells experience osmotic stress when they are placed in environments where their water balance is disrupted.

Osmotic stress A situation that may occur when a cell tries to maintain its water balance by counteracting the movement of water down its concentration gradient; such a scenario would occur when the solute concentration in the cell and its environment are unequal.

Otitis externa Swimmer's ear; *Pseudomonas aeruginosa* infection in which the outer ear canal is infected; in some cases the pinna (the flaplike tissue that most think of as the ear) can also become inflamed and exhibit pus-filled lesions.

Otitis media Middle ear infection, which may occur during a cold, when inflamed membranes can cause mucus to accumulate in the respiratory tract, providing an ideal environment for bacteria to grow; in children, otitis media is a frequent example of such an infection.

Ovaries The organs that make eggs and also release a number of hormones such as estrogen, progesterone, and others that govern female reproductive physiology.

Oxidase test A test that can detect if the electron carrier called cytochrome c oxidase is present.

Oxidation reactions Remove electrons from an atom.

Oxidative phosphorylation A chemical process involving a collection of redox reactions that strip electrons from a food source (such as carbohydrates, fats, or proteins) and eventually hand off those electrons to an electron transport chain to fuel the phosphorylation of ADP to ATP.

Oxygen radicals Oxygen atoms that have an unpaired electron in their outer shell that makes them reactive with nearby molecules; hydrogen peroxide and nitric oxide are examples of molecules that contain reactive oxygen; the immune system uses these molecules to combat invaders since they lethally damage targeted DNA and proteins.

Palisade An arrangement consisting of clusters of bacilli-shaped cells.

Palsy Paralysis, inability to move; often accompanied by uncontrolled shaking (tremors).

Pandemic Worldwide outbreak of disease.

Pap smear (*Papanikolaou smear*) A routine test that detects potentially cancerous changes in cervical cells.

Papillomaviruses Viruses that cause warts; they enter abrasions by direct or indirect contact and through contaminated fomites.

Para-aminobenzoic acid (PABA) The natural substrate of a bacterial enzyme in the pathway that builds folic acid; sulfa drugs chemically resemble PABA.

Paraesthesia A neurological symptom associated with a prickling or "pins-and-needles" sensation.

Parasite A term commonly used to describe organisms that benefit themselves at the expense of their host (in medical contexts this usually refers to protozoans and helminths).

Parasitism A type of symbiotic relationship that hurts the host.

Parenteral administration Injection or infusion of a drug; the drug may be injected into a vein (intravenously), into the muscle tissue (intramuscularly), or under the skin (subcutaneously).

Parenteral entry Type of entry in which a pathogen can bypass the skin and directly invade the underlying subcutaneous tissues, muscles, or bloodstream; bites, cuts, injections, and surgical incisions facilitate parenteral entry.

Paroxysm A sudden and violent attack (as in a paroxysm of coughing).

Parturition The action of giving birth; labor.

Passive transport Transport of substances without an energy investment by the cell.

Pasteurization Application of moderate heat, well below the liquid's boiling point, that eliminates pathogens and reduces (though does not fully eliminate) harmless microbes; commonly used to treat milk, juices, and wines to render these foods safe for consumption and slow spoilage.

Pathogenicity A term describing the general ability of an infectious agent to cause disease.

Pathogenicity islands Special gene groupings in certain regions in pathogen genomes that encode toxins, virulence factors, and resistance mechanisms; these gene groupings also reveal how some virulence factors are acquired through horizontal gene transfer such as bacteriophages, transposons, and conjugation events.

Pathogens Microbes that cause disease.

Pelvic inflammatory disease (PID) Inflammation of the female reproductive system; gonorrhea and chlamydia frequently cause asymptomatic infections in women that often remain untreated; in such cases, the fallopian tubes can become inflamed (salpingitis) and abscesses in the fallopian tubes or ovaries may develop; in some cases the abdominal cavity is also affected.

Penicillin drugs A family of natural and semisynthetic compounds that tend to end in the suffix "cillin;" they remain among the most widely prescribed and safest antimicrobials.

Penis Male reproductive organ; part of the external genitalia; during an orgasm, semen is ejaculated through the urethra that passes down the length of the penis.

Pentose phosphate pathway Pathway that uses a carbon-shuffling process to convert pentoses (five-carbon sugars) into trioses (three-carbon sugars) and hexoses (six-carbon sugars).

Peptide Short amino acid chain (usually less than 55 amino acids).

Peptide bonds The type of bond formed between the amino group of one amino acid and the carboxyl group of the neighboring amino acid; peptide bonds link amino acids together to form polypeptides, or proteins; this chemistry happens between amino acids in the P and A sites of the ribosome.

Peptidoglycan Also called murein; a core component of bacterial cell walls, this meshlike molecule includes protein and sugar units; built by alternating N-acetylglucoseamine and N-acetylmuramic acid residues; the long strands of these alternating residues are then cross-linked by short peptides to create a meshlike structure much like a chain-link fence.

Peptidyl (P) site The site on an active ribosome that, together with an exit site (E) and an acceptor site (A), coordinates the translation process.

Percutaneous Through the skin.

Perforins Substances released by T cytotoxic cells to form pores in a target cell.

Pericardial fluid A thin layer of fluid within the pericardium.

Pericarditis Inflammation of the pericardium, which is a thin, double-membrane sac that encloses the heart.

Pericardium The outermost layer of the heart consisting of a protective, double-layered sac that encloses the heart and that contains a thin layer of fluid called the pericardial fluid.

Perinatal exposure Describes a newborn's exposure to an agent during delivery.

Period of decline Stage of infection during which pathogen replication decreases and the patient begins to feel better; in this phase, patients may still be infectious even though they feel better.

Periodic table of elements Table in which all of the known elements are organized by atomic number, electron arrangements, and chemical properties.

Periodontitis Inflammation and eventual erosion of the bone that surrounds the tooth.

Peripheral nervous system (PNS) One of two major segments of the human nervous system; comprised of nerves, the PNS inputs and transmits information.

Periplasmic flagella Flagella located in the space between the plasma membrane and the cell wall; these unique flagella allow spirochetes to move with their distinct corkscrew motion.

Periplasmic space In bacterial cells, a gap that lies between the peptidoglycan layer and the outer membrane of the cell wall; this space is filled with a gel-like fluid that is enriched with various factors that have important roles in helping the bacterium obtain nutrients and neutralize potentially toxic substances.

Peritrichous A term that describes cells that have flagella distributed all over the cell surface.

Peroxisomes Vesicles that are found in most eukaryotic cells and that contain enzymes that break down certain fats and amino acids and protect the cell from hydrogen peroxide and other toxic oxygen intermediates.

Peroxygens A group of chemicals with strong oxidizing properties that easily convert into reactive oxygen species that attack proteins, nucleic acids, and other biomolecules; hydrogen peroxide and peracetic acid are two commonly used peroxygens.

Persistent infections Chronic or latent infections caused by persistent viruses that tend to remain in the host for long periods–from many weeks to a lifetime.

Personal protective equipment (PPE) Equipment worn to reduce a person's exposure to a hazard; in healthcare, examples of PPE include gloves, isolation gowns, and face shields.

Perspiration Sweat; a particularly effective weapon against microorganisms: The pH is too acidic for some, while the salt content is too high for others; sweat may also wash away microbes, or be wiped away, carrying microbes with it.

Pertussis Disease caused by the Gram-negative bacterium *Bordetella pertussis*; commonly referred to as whooping cough, due to the sound patients make as they try to catch a breath between long and intense coughing attacks.

Petechiae A small red or purple spot caused by bleeding under the skin.

pH A measure of a solution's acidity or alkalinity; it is based on the concentration of H^+ (hydrogen proton concentration); the lower the pH value, the higher the H^+ concentration; acids lower pH and bases raise pH; calculated based on the concentration of H^+ ions: $pH = -\log^{10}[H^+]$.

pH scale Scale used to describe the acidity and basicity of a solution; a pH of 7 is neutral, above 7 is basic, and below 7 is acidic.

Phage conversion The ability of prophages to confer new pathogenic properties to bacterial cells.

Phagocytes Means "cell eating;" specialized immune system cells, such as macrophages, that aim to destroy the targets they engulf.

Phagocytosis A specialized form of endocytosis where nondissolved, extracellular targets are engulfed by a cell and broken down.

Phagolysosome Fusion of a phagosome and a lysosome.

Pharyngitis Inflammation of the pharynx (throat); mainly caused by bacteria, viruses, or allergens.

Phenicols Structurally simple, broad-spectrum drugs that bind to the 50S ribosomal subunit; the main phenicol drug used in humans is chloramphenicol.

Phenols Have long been used as intermediate-level germicides; destroy bacterial cell walls and interact with proteins; are often used in medical settings for surface disinfection (bed rails, tables, and floors) and to disinfect semicritical and noncritical equipment.

Phenotype The physiological and physical traits of an organism.

Phosphatases Specialized enzymes that use dephosphorylation reactions to remove phosphate groups from their targets.

Phosphate groups (PO_4^{3-}) A type of functional group found in phospholipids, nucleic acids, ATP, and many other biologically important molecules; phosphate groups should not be confused with a phosphoryl group which has phosphorous bound to three oxygen atoms.

Phosphodiester bonds Bonds through which the sugar in one nucleotide bonds with a phosphate of another; creates the alternating pattern of sugar and phosphate seen in the backbone of DNA and RNA; covalent bonds that link nucleotides together in DNA and RNA.

Phospholipids Amphipathic molecules with a hydrophilic (water-loving) phosphate head group and hydrophobic (water-hating) fatty acid chains; molecules that make up the plasma membrane.

Phosphorylation reactions Reaction in which phosphoryl groups (PO_3^{2-}) are added to a molecule by a phosphorylation reaction.

Photophosphorylation Relies on the redox reactions of an electron transport chain; however, light energy is used to activate electrons, and only photosynthetic cells can perform photophosphorylation.

Phototaxis Movement of a cell in response to light.

Phototrophs Organisms that can harvest energy from light to make ATP.

Physical barriers Structures that act as a first line of defense by physically blocking pathogen entry; a key example is our skin and other epithelial barriers.

Pili Similar to fimbriae, except that they tend to be longer, more rigid, and less numerous; they are used to adhere to surfaces, move, and aid in gene transfer through conjugation.

Pilus A hollow tube that serves as a bridge for transferring a copy of the fertility plasmid from one cell to a cell lacking it during conjugation.

Pinocytosis Means "cell drinking;" describes endocytosis of dissolved substances in small vesicles.

Placenta An organ that serves as an interface between maternal and fetal circulation and releases hormones that support pregnancy; the umbilical cord links the baby to the placenta and allows the baby's blood supply to enter the placenta; on the maternal end, the mother's blood flows into the placenta from capillaries in the uterine wall.

Plague Term people use to refer to the Black Death, a Middle Ages pandemic notable for killing up to half the population in Europe; the causative agent is *Yersinia pestis*; there are three forms of plague: pneumonic plague, bubonic plague, and septicemic plague.

Planktonic Free-floating (bacteria).

Plaque assay Technique for determining the quantity of bacteriophages as well as how much of a given animal virus is present; the plaque assay can also be used to purify specific viruses.

Plaque reduction neutralization test (PRNT) A test that detects patient antibodies against a specific virus, such as Zika virus.

Plaque-forming units In a plaque assay, the units that express the quantity of bacteriophages in an initial volume of sample.

Plaques Areas that develop in a lawn of cultured cells as the cells are infected and lysed by a virus.

Plasma The liquid portion of blood.

Plasma cells In adaptive immunity, effector cells that make antibodies, which are a secreted form of the BCR that binds to the antigen that stimulated the activation event.

Plasma membrane An outer boundary surrounding all cells, whether prokaryotic or eukaryotic; also called the cytoplasmic membrane or cell membrane.

Plasmids Pieces of DNA that exist outside of the chromosomal DNA in bacteria and a number of eukaryotic cells.

Platelets Cell fragments that clump together and help stem blood loss when vessels are damaged; platelets also confine invading agents in mesh-like clots.

Pleomorphic Ability to take on different forms; a term used to describe the diverse shapes assumed by *Mycoplasma* (a genera of bacteria that lack a cell wall) and L-forms (bacteria that have been treated in a way that eliminates the cell wall).

Pneumococcal meningitis An infection caused by the Gram-positive diplococcus *Streptococcus pneumoniae*, this is the most common form of bacterial meningitis in the U.S.; incidence has decreased due to vaccination.

***Pneumocystis* pneumonia (PCP)** A serious fungal infection caused by *Pneumocystis jirovecii*; this infection is seen in immune-compromised patients; *P. jirovecii* is not native to the soil and it does not make spores; in fact, the natural reservoir for this fungus is not entirely known, but it is thought that asymptomatic human carriers may transmit the fungus; untreated cases are typically fatal, while up to 40 percent of treated cases are fatal.

Pneumonia Inflammation of the lung tissue, especially of the alveoli, which are specialized regions of the lung where gas exchange occurs; there are over 100 different microbes that can cause pneumonia when they invade the normally sterile lower tract, including certain bacteria, viruses, fungi, protists, or parasitic worms; in general, though, bacteria and viruses cause most pneumonia cases.

Pneumonic plague A form of plague that infects the lungs and is the only form of plague that is transmitted from person to person; may progress to the septicemic form.

Polar A term to describe molecules that have an unequal charge distribution in their covalent bonds (polar covalent bonds), and thus dipoles.

Polar covalent bonds A covalent bond in which electrons are not equally shared by the bonding atoms.

Polar molecules Molecules that tend to interact with water, while nonpolar molecules avoid water; phospholipids, the main components of plasma membranes, are built from glycerol, phosphate, and two fatty acids; the phosphate-containing region of a phospholipid is polar, while fatty acids are nonpolar. Also see **polar.**

Polio Common name for **poliomyelitis**.

Poliomyelitis A viral infection, more commonly known as **polio**, that affects the nervous system and is caused by the poliovirus, which belongs to the *Picornaviridae* family of very small RNA viruses that have only a protein capsid without an envelope; the poliovirus is cytolytic, causing rapid cell rupture after infection and viral replication; this lysis of infected neurons causes severe inflammation, giving the disease its name: inflammation of the myelin, or myelitis; if poliovirus reaches the CNS, it can attack different areas of the spine and brain, causing muscle weakness or paralysis; with development of vaccines against this pathogen, there has been a massive decrease in the incidence of polio, a result that is considered to be one of the triumphs of modern medicine.

Polyarthralgia Extreme pain in multiple joints.

Polycistronic An mRNA molecule that encodes more than one protein; prokaryotic mRNAs are commonly polycistronic.

Polyenes Agents (examples: amphotericin B and nystatin) that directly interact with ergosterols in fungal plasma membranes; these drugs cause targeted fungal plasma membranes to become leaky and lead to cell lysis; unfortunately, polyenes interact with cholesterol in animal cell membranes, which leads to a narrow therapeutic index and potential adverse side effects, including nephrotoxicity.

Polymer Large molecules (macromolecules) that consist of repeated smaller units (monomers).

Polymerase chain reaction (PCR) A DNA amplification technique that creates billions of copies of a target gene in just a few hours; PCR is sensitive enough to detect even a single viral particle in a sample; PCR's speed and sensitivity have made it an essential tool in clinical labs.

Polymerization A process in which building block subunits (or monomers) are linked to make a larger molecule, or polymer. See **monomer** and **polymer**.

Polymyxin B A bactericidal narrow-spectrum polypeptide drug that targets the outer membrane of facultative anaerobes and aerobic Gram-negative cells.

Polypeptides Long chains of amino acids (usually greater than 55 amino acids).

Polysaccharides This term directly translates as "many sugars;" it's another name for carbohydrates; polysaccharides can be linear or highly branched.

Polyunsaturated A term that describes fats that have more than one double bond in their fatty acid chains.

Pontiac fever One of two clinically distinct types of legionellosis caused by *L. pneumophila*; the other type is Legionnaires' disease.

Population Any defined group of people.

Porins Protein channels that form a pore in a cell membrane to facilitate the transport of substances such as amino acids, vitamins, and other nutrients to enter a cell, while excluding large molecules and a variety of substances that may be harmful to the cell (like certain drugs).

Portal of entry Any site that a pathogen uses to enter the host.

Portal of exit Any site that a pathogen uses to exit the host.

Post translational modifications Changes made to a protein after it has been translated; modifications such as trimming and adding various organic and/or inorganic factors are examples.

Postexposure prophylaxis (PEP) A prevention treatment applied after an exposure to limit infection.

Postherpetic neuralgia (PHN) A chronic pain condition that results from shingles.

Post-transcriptional Occurring after transcription; describes regulation that impacts how often mRNA is translated into protein.

Post-translational modifications Modifications made after the termination of translation that are often required for a protein to function and provide a way for cells to regulate gene product functionality.

Praziquantel An antihelminthic drug (trade name: Bilticide) for fluke and tapeworm infections; it essentially paralyzes these parasites, which release their attachment to the intestinal wall and are then expelled in the feces.

Pre-exposure prophylaxis (PrEP) Regimen of drugs for people who may be at risk for an exposure to a particular pathogen.

Pre-term labor Labor onset before 37 weeks of gestation; it is well established that intrauterine infections are a major risk factor for pre-term labor; it is estimated that at least 40 percent of pre-term deliveries are associated with an intrauterine infection, but a more recent development is the idea that the normal microbiota profile of the female reproductive tract, not just infections, might impact the risk for pre-term labor.

Pre-transcriptional regulation A process that helps cells control when and how often transcription occurs (regulates RNA production) to control protein synthesis/gene expression; a key example of this type of regulation for gene expression are operons.

Prevalence rate Sometimes just called prevalence; an epidemiological measure of frequency that describes morbidity in a given population during a specified time; to calculate the prevalence rate, take the total number of disease cases during a given time and divide it by the total number of people in the defined population during that same time.

Primary activation signal In the cellular immune response, the first of two activation signals needed to activate a T cell; this signal involves the T cell's TCR (T cell receptor) interacting with an MHC–antigen complex on an antigen presenting cell.

Primary amoebic meningoencephalitis (PAM) A rare infection caused by the amoeba *Naegleria fowleri* (often called "the brain-eating amoeba"); it's a thermophilic protist that lives in streams and lakes all around the world; once *N. fowleri* enters the nasal passages of a human host, the trophozoite can burrow into the nasal mucosa; from there, it travels up the olfactory nerves into the brain; initial symptoms include intense headaches, sore throat, vomiting, and fever; as the disease progresses, stiff neck, seizures, hallucinations, and coma can occur; patients usually die within a week or two after infection.

Primary immunodeficiency (congenital immunodeficiency) An inborn error that affects one or more immune system factors and leads to deficient immunity.

Primary lesions Lesions that are directly associated with a disease and are considered key features for diagnosing a variety of infections.

Primary lymphoid tissues Lymphoid tissues, including the thymus and bone marrow, that have roles in the production and maturation of leukocytes and other formed elements in blood.

Primary Stage (of syphilis) In syphilis, the stage that is characterized by a chancre appearing at the initial infection site, where the spirochete invades the new host's tissues and begins to multiply.

Primase An enzyme involved in DNA replication; found in the primosome; lays down RNA primers to jumpstart DNA replication.

Primosome A collection of factors, including primase and helicase, that is recruited to a specific point in the chromosome to start DNA replication.

Prions Infectious proteins that can cause a disease such as transmissible spongiform encephalopathies; prions are transmitted by transplant or ingestion; some prion diseases are inherited.

Prodromal phase Stage of infection in which the patient starts to feel run down and may have mild symptoms.

Products The substances generated as a result of a reaction.

Proglottids Segments of a tapeworm that break off from the distal end of the body—the end furthest from the scolex; the eggs released by the tapeworm (or even intact proglottids) are passed in the feces, at which point another host may ingest the eggs and become infected.

Prognosis The likely outcome of a disease or illness.

Prokaryotes Organisms that are all unicellular and lack a membrane-bound nucleus; they also lack other membrane-bound organelles, and have a much simpler genetic makeup than eukaryotic cells; bacteria and archaea are prokaryotes.

Prokaryotic species Cells that share physical characteristics, but also have at least 70 percent DNA similarity (based on the degree to which their DNA can stably pair with each other in solution); species are also at least 97 percent identical when genetic material in their ribosomes is compared (16s rRNA sequence similarity).

Promoter The part of the operon with which RNA polymerase associates to start transcription.

Prophage In the lysogenic pathway, the result of the phage genome being incorporated into the host cell genome; as the infected bacterial cell divides, it copies its own genome as well as the prophage's genome.

Prophylactic antibiotic therapy A preventive therapy in which antimicrobials are given to prevent a bacterial infection; a common example is administering antibiotics to people who have artificial heart valves or certain heart defects and who are undergoing oral or respiratory surgery; the goal in this case would be to limit the risk of bacteria colonizing the heart if they enter the bloodstream.

Proportion An epidemiological measure expressed as a percentage of a whole (80 out of 100 women with gonorrhea do not have symptoms, or 80 percent of women with gonorrhea do not have symptoms).

Prostatitis An inflammation of the prostate gland; often caused by a bacterial infection, but may be due to other factors; it's the most common prostate issue in men under age 50.

Proteases Specialized enzymes that can break peptide bonds by catalyzing hydrolysis reactions.

Proteasome Barrel-shaped cell component consisting of enzymes called proteases that break down proteins; sometimes referred to as "protein garbage disposals" because they break down proteins that have been tagged for destruction.

Protein folding The process whereby proteins take on higher order structure.

Protein self-assembly Most proteins aren't physiologically active until they are properly folded into a higher order structure; to achieve proper folding, most proteins enlist the help of specialized cellular tools called chaperones; proteins that self-assemble, like some capsid proteins, achieve their final form without the help of chaperones.

Protein synthesis A process in which genetic information within a cell is read and used to create gene products—proteins; also called gene expression.

Proteins Organic molecules that are made of amino acids; can have primary, secondary, tertiary, and quaternary structure.

Protists Eukaryotic unicellular and multicellular microorganisms; can be pathogenic while others are nonpathogenic.

Proton motive force A chemical gradient of protons (H^+ ions) that have accumulated on one side of the plasma membrane (or accumulated within the intermembrane space of the mitochondrion); sets the stage for chemiosmosis.

Protons Positively charged particles contained within the nucleus of an atom.

Protozoans Means "first animal;" a term of convenience to describe animal-like protists that are unicellular, lack a true cell wall, exhibit asexual and sexual reproduction, and typically live by heterotrophic means.

Provirus Formed when some viruses that cause persistent infections integrate their genome into the host cell; resembles the prophage made by temperate bacteriophages.

Pruritic Itchy.

Pseudomembranous colitis An inflammation of the colon that is accompanied by pus-filled nodules (pseudomembranous plaques); it is caused by *Clostridioides difficile* infection; symptoms usually develop toward the end of an antibiotic treatment—or soon after the antibiotic therapy is completed; mild cases feature diarrhea and moderate abdominal pain; serious cases are characterized by severe abdominal pain, fever, nausea, and abundant watery diarrhea that may progress to dysentery; in extreme but rare cases, toxic megacolon may develop.

Pseudopods Extensions of the protoplasm used for movement as "false feet" in amoeboid protozoans.

Psittacosis A zoonotic atypical pneumonia that transmits from infected birds to humans; *Chlamydophila psittaci*, the causative agent, are Gram-negative bacteria that live as a parasite of eukaryotic cells; the usual route of transmission is when people inhale dust that is stirred up from dried bird droppings found in birdcages.

Psychrophiles Organisms that can thrive between about −20°C and 10°C and that tend to live in environments that are consistently cold, like the Arctic.

Psychrotrophs Cold-tolerant organisms that grow at about 0–30°C, and are associated with foodborne illness because they grow at room temperature as well as in refrigerated and frozen foods.

Pulmonary edema Fluid accumulation in the lungs; can have a high mortality.

Pure culture Single-species of microbes in a sample; pure cultures do not tend to exist in natural settings.

Purified subunit vaccine Vaccine with pathogen components that were either harvested from a natural pathogen or purified from a genetically engineered expression system.

Purines A class of double-ring structured nitrogen bases; includes adenine (A) and guanine (G).

Purpura A rash of purple spots/bruising caused by red blood cell lysis and/or capillary destruction rather than by external physical trauma.

Pus In the resolution phase of inflammation, a substance that may form as dead tissue cells and leukocytes are collected into the fluid exudate made by earlier phases of inflammation.

Pyelonephritis An inflammation of one or both kidneys.

Pyrimidines A class of single-ring structured nitrogen bases; includes cytosine (C), thymine (T), and uracil (U).

Pyrogens Fever-inducing agents; many bacterial toxins act as pyrogens.

Pyuria Pus in the urine.

Q fever A respiratory illness caused by Gram-negative bacterium *Coxiella burnetii*.

Quarantine A period of confinement away from the general population for an animal or human that may have contracted an infectious disease; a measure intended to limit the spread of an infectious disease.

Quaternary ammonium compounds (QACs) Cationic detergents that have bactericidal activity and are sporostatic (inhibit spore germination); used as antiseptics on unbroken skin and certain mucous membranes and as disinfectants on noncritical equipment and surfaces.

Quinine An arylaminoalcohol compound that was one of the earliest treatments ever used for malaria, remains an important drug to fight uncomplicated and severe forms of malaria; malarial protozoans are increasingly able to resist quinine and quinine derivatives.

Quinolones Synthetic antimicrobials that target DNA replication enzymes; namely DNA gyrase and topoisomerases; the more modern and most commonly prescribed group of quinolones are called fluoroquinolones, so called because they contain a fluorine atom; examples include ciprofloxacin and levofloxacin.

Quorum sensing The collective sensing and responding to changes within a bacterial community.

R group Shorthand notation for the remainder or core of an organic molecule; this notation allows the reader to focus on the part of the molecule being discussed instead of including cumbersome chemical structures.

R plasmids These non-chromosomal genetic elements (resistance plasmids) include genes that confer resistance to antimicrobial drugs.

Rabies A classical zoonosis caused by the rabies virus, which is a member of the *Rhabdoviridae* (Greek for "rod-shaped") family of enveloped, single-stranded RNA viruses; usually transmitted to humans through an animal bite, although aerosolized bat droppings have also led to infection.

Radial immunodiffusion test One of the two most common antigen–antibody precipitation methods used in clinical applications; this test helps to quantify the amount of a particular antigen present in a sample. Also called the Mancini method.

Radiation High-energy waves that can serve as a disinfection or sterilization tool depending on the protocol applied.

Rash An eruption of lesions; may or may not be accompanied by other symptoms.

Rate An epidemiological measure that is used to measure the occurrence of an event over time.

Ratio An epidemiological measure that presents the occurrence of an event in one group as compared to another group (e.g., in the United States in people age 65 and older there are 0.77 male/female).

Reactants The ingredients of a chemical reaction.

Reaction rate Speed at which a reaction occurs.

Reactive oxygen species (ROS) Reactive intermediates that include superoxide ions (O_2^-) and hydrogen peroxide (H_2O_2), both of which can rapidly damage cellular proteins and DNA.

Real-time PCR A modified PCR method that uses fluorescence imaging to visualize DNA copies as they are made—making the data "real time," or immediate; sometimes called quantitative PCR (or qPCR) because it can quantify, or measure, how many copies of a particular gene were present in a sample from the start of the reaction.

Receptor-mediated endocytosis A highly specific importation tool in which target substances that are to be imported bind to specific cell-surface receptors.

Recombinant DNA (rDNA) Term referring to DNA that is generated or engineered by combining DNA from different organisms.

Recombinant subunit vaccines Vaccines that are purified from a genetically engineered expression system.

Recombinant vector vaccines A developing technology in which genetic material from the pathogen is packed inside a harmless virus or bacterium, which is then introduced into the body; if a virus is used, it will cause the host cells to make pathogen antigens to stimulate an immune response; in scenarios that employ engineered bacteria, harmless bacteria are genetically modified to make selected pathogen antigens that will stimulate an immune response.

Recombination An exchange of genetic material that leads to new genetic combinations; a process important in nature and also used as a molecular biology tool that allows researchers to alter the genetic landscape of a cell.

Red man syndrome A glycopeptide side effect that can be mistaken for a classical drug allergy because it is characterized by a red flush spreading over skin and itchiness.

Redox reactions A term used for oxidation and reduction reactions because they always occur as partners; you can remember "OIL RIG" to keep track of what happens to electrons in these reactions: Oxidation Is Loss of electrons; Reduction Is Gain of electrons.

Reduction reaction A reaction in which an atom or molecule gains electrons.

Reemerging diseases Diseases that were previously under control, but are now showing increased incidence; these include infections caused by well-known pathogens that have evolved increased virulence (capacity to cause disease).

Reemerging pathogen An infectious agent that was under control due to prevention or treatment strategies and is now resurfacing.

Refractive index The degree to which a substance bends light; a factor in the resolution of microscopes.

Regulators of complement activation A collection of proteins that turn off complement cascades after a threat passes.

Renal failure A condition where kidneys cannot effectively filter waste from the blood or balance fluids; dialysis or kidney transplant may be the only options available to keep a patient alive once renal failure occurs.

Replication fork During DNA replication, the immediate point where unwinding occurs and new DNA is built.

Reportable diseases The diseases on a state or local tracking list, usually including the diseases that the CDC has an interest in monitoring, as well as diseases that the local authorities want to monitor.

Repressible operons Operons that are on by default, meaning they are actively transcribed until they are switched off (repressed); an example is the arginine (*arg*) operon.

Repressor The part of the operon that blocks transcription.

Reservoir The animate or inanimate habitat where the pathogen is naturally found.

Resolution The ability to distinguish two distinct points as separate.

Respiratory droplets These small airborne particles laden with an infectious agent can be expelled by a sick person's coughs or sneezes and serve as a means for pathogens to enter the respiratory tract of another host.

Respiratory syncytial virus (RSV) An enveloped RNA virus that belongs to the *Paramyxoviridae* family—the same viral family as measles, rubella, and parainfluenza viruses; worldwide, RSV is the leading cause of acute lower respiratory tract infection in children under age five.

Restriction enzymes Enzymes that are used to generate compatible or sticky ends of complementary base pairs between the copied DNA and the plasmid to construct recombinant DNA; occur naturally in a variety of bacterial species and constitute a defense mechanism that allows a bacterium to cut up bacteriophage genomes when they enter a cell; generate specific ends after cutting the DNA sequence they identify.

Reverse transcriptase A virally encoded enzyme that enables retroviruses to use their single-stranded RNA genome to direct formation of DNA; special enzymes that carry out reverse transcription.

Reverse transcription PCR (RT-PCR) Another form of PCR that is useful for detecting RNA in a sample—such as the genome of an RNA virus; in RT-PCR, the enzyme reverse transcriptase is used in conjunction with primers to build DNA that is complementary to RNA molecules in a sample.

Reversible reaction Reactions in which the forward and reverse reactions are both possible; initially one reaction may occur at a higher rate than the other, but at equillibrium the forward and reverse reactions occur at the same rate.

Reversion mutation A base-substitution mutation that can be corrected by another base substitution (the error is corrected by another error).

Rh (rhesus) factor One of the most widely analyzed antigens on the red blood cell's surface; Rh factor is a protein.

Rh(D) immunoglobulin Immunoglobulin preparation (such as RhoGAM) administered to an Rh^- mother around the 28th week of pregnancy as a way to prevent hemolytic disease of the newborn by preventing Rh^- women from ever being sensitized to the Rh factor; this drug prevents the mother's body from making antibodies to the Rh factor that may be present on fetal red blood cells.

Ribonucleic acid (RNA) A nucleic acid made of repeating subunits called nucleotides, which are made up of a phosphate, a sugar (ribose), and one of four possible nitrogen bases (A, U, G, or C).

Ribonucleotides The nucleotides that make up RNA (adenine, guanine, cytosine, and uracil); differ in sugar type (ribose) from deoxyribonucleotides (deoxyribose).

Ribosomal RNA (rRNA) RNA molecules that fold up into elaborate three-dimensional structures and combine with proteins to form ribosomes.

Ribosomes Essential organelles for making proteins; eukaryotic ribosomes, like prokaryotic ribosomes, are made up of protein and ribosomal RNA (rRNA); ribosomes build proteins by linking together amino acids; prokaryotes have 70S ribosomes while eukaryotes have 80S ribosomes.

Riboswitches Built-in switches in mRNAs that act as protein synthesis controls.

Ribozymes Enzymes that are made of the nucleic acid RNA; they have a more limited range of substrates than do protein enzymes.

Rifampin Also called rifampicin; a key representative of the rifamycins; inhibits transcription by binding to RNA polymerase; this broad-spectrum antimicrobial is useful to combat mycobacterial species that can be especially tough to treat due to the challenges of getting drugs across their waxy mycolic acid–enriched cell wall.

Rifamycins A group of bactericidal drugs that were originally isolated from bacteria; now mainly produced as synthetic and semisynthetic compounds; rifampin is a key example.

Right atrium One of two upper heart chambers; receives oxygen-depleted blood from the superior and inferior vena cava.

Right ventricle One of two lower heart chambers; pumps blood to the lungs for re-oxygenation.

RNA polymerase The enzyme that performs transcription to make RNA using a DNA template.

RNA primers During DNA replication, these short segments of RNA are made by the enzyme primase and are required for DNA polymerase III to start replicating DNA.

RNA-dependent RNA polymerases (RdRPs) Virally encoded enzymes that transcribe the RNA genome into a readable mRNA format before translation.

Robert Hooke The first to observe eukaryotic cells, publishes *Micrographia*.

Robert Koch The first to prove a microbe causes disease (work on anthrax); in 1877, Robert Koch publishes his postulates of disease.

Rocky Mountain spotted fever (RMSF) A tickborne disease that is caused by the extremely small Gram-negative bacteria *Rickettsia rickettsii* and that occurs in all parts of the United States; rash features numerous small, flat, pink/red, non-itchy spots that spread from the wrists, arms, and ankles inward to the torso as the disease progresses; other symptoms include headache, vomiting, and muscle pain.

Roseola Mild or asymptomatic viral infection caused by human herpes viruses 6 and 7; manifestations, if they occur, include sudden high fever (up to 104°F) for a few days followed by a rash that covers most of the body; after a day or two the rash subsides.

Roseola infantum See **roseola**.

Rotavirus The leading cause of enteritis in children under five years old; this RNA virus also infects adults, but symptoms are usually mild and may go unnoticed; most cases are acquired through fecal–oral transmission, in which contaminated food or water is ingested.

Rubella Infection caused by the rubella virus, an enveloped, single-stranded RNA virus of the *Togaviridae* family; sometimes called German measles, but should not be confused with measles; the virus enters by the respiratory tract, spreads to the lymph nodes, and eventually spreads to the rest of the body by the bloodstream; infection is characterized by a red rash that lasts for three to seven days; rash starts at the face and then spreads downward; patients may also experience fever, swollen and tender lymph nodes, muscle and joint aches, and a runny or stuffy nose.

Safranin In the last step of the Gram stain, the smear is covered with safranin, a counterstain that stains decolorized Gram-negative cells pink.

Salmonellosis Foodborne infection caused by *Salmonella* bacteria, which are among the most common causative agents of diarrhea and dysentery; the CDC estimates that every year these Gram-negative rods are responsible for at least 1.2 million foodborne infections in the United States.

Salpingitis Inflammation of the fallopian tubes.

Salts Form when acids and bases react with each other; the acid contributes the anion of a salt, while the base contributes a cation.

Sandwich ELISA Enzyme-linked immunosorbent assay that allows us to detect a specific antigen in a sample, but that is more sensitive and overcomes the downsides of the direct ELISA; in its simplest form, the sandwich ELISA requires two antibodies: a capture antibody and a detection antibody; gets its name because the antigen is "sandwiched" between the two antibodies.

Saprobes Organisms, such as fungi, that feed on dead organic material to survive.

Sarcoma Cancer of bone cells or soft tissue cells (including blood vessels, muscles, tendons, joints, and fat).

Saturated Term describing lipids that do not have double bonds in their fatty acid chains and therefore have the maximum number of hydrogen-carbon bonds; because saturated fatty acids pack tightly together, lipids high in saturated fatty acids, like butter and lard, exist as solids or semisolids at room temperature.

Saturated fatty acids Fatty acids that lack double bonds in their hydrocarbon backbone; unsaturated fatty acids have double bonds in their hydrocarbon backbone, which limits how tightly they pack together, thereby keeping them fluid at colder temperatures.

Scarlet fever An illness caused by *Streptocccus pyogenes* strains that have gained the ability to produce an erythrogenic toxin as a result of infection by specific bacteriophages; these lysogenized bacterial strains generate easily recognizable disease features such as a red sandpaper-like rash and a reddened tongue that resembles the surface of a strawberry ("strawberry tongue"); fairly rare and occurs in less than 10 percent of streptococcal pharyngitis cases; most common in children under 10 years old.

Scientific method In its most basic form, the scientific method starts with a question that can be investigated.

Sclera White portion of the eye.

Scolex Head at one end of a tapeworm; the scolex has hook and/or sucker structures to help the worm attach to the intestinal lining in the definitive host.

Scrotum A pouch of skin that contains two testicles (testes), the organs that make sperm cells and a variety of hormones, including testosterone; located outside of the main body cavity because sperm develop best when slightly cooler than core body temperature.

Sebaceous glands Oil glands in skin.

Sebum An oily or waxy substance produced by the skin's sebaceous glands (oil glands) that is a combination of low-pH lipids and proteins that moisturize and further protect the skin.

Secondary activation signal In the cellular immune response, a signal that involves co-stimulatory proteins on the surface of the antigen presenting cell (APC) binding to co-stimulatory proteins on the T cell's surface.

Secondary immune response Response involving immunological memory that requires the coordinated activity of memory B and T cells.

Secondary immunodeficiencies (acquired immunodeficiencies) Immunodeficiencies that are much more common than primary immunodeficiencies; a person with this type of immunodeficiency doesn't have an inborn error in the immune system, but starts out with a normal immune system and then experiences a decline in immune system rigor.

Secondary lesions Lesions that are less strictly defined and that have diverse origins; may evolve from primary lesions, or from external forces like trauma or scratching; they are also seen during the progression of skin damage from certain infectious agents or allergic reactions.

Secondary lymphoid tissues Lymphoid tissues, including lymph nodes, spleen, and mucosa-associated lymphoid tissue (MALT), that filter lymph and sample surrounding body sites for antigens; leukocytes that reside in secondary lymphoid tissues are brought into contact with antigens to stimulate an immune response.

Secondary stage (specifically the secondary stage of syphilis) The stage at which *Treponema pallidum* establishes a persistent infection; this occurs around three months after initial infection and usually involves a rash, which may appear on the skin or on mucous membranes in the mouth, vagina, or anus; a subset of patients experience condylomata lata (genital or anal warts); swollen lymph nodes, sore throat, patchy hair loss, fever, muscle aches, or malaise may also occur; these symptoms usually disappear within 3 months and the pathogen enters a latent period that lasts for months to years before it launches the final, tertiary stage.

Second-generation drugs Drugs resulting from a round of chemical modification to first generation drugs.

Second-line defenses Defenses that primarily consist of assorted molecular factors and leukocytes (white blood cells)—specialized cells of the immune system; call on the body's lymphatic system.

Secretory vesicles A specific type of transport vesicle that shuttles materials to the cell surface for discharge from the cell; in some instances, the discharged materials are waste products; in other cases, the secreted substances are bioactive molecules.

Selective media Media that single out bacteria with specific properties, which is accomplished by including ingredients in the media that foster the growth of certain bacteria while suppressing the growth of others.

Selective toxicity Term applied to a drug, meaning that it inhibits or kills the targeted microbe without damaging host cells.

Self-tolerance A property of immune cells by which they will not attack normal self-cells; the ability to differentiate self from foreign and only attack foreign substances.

Semen A liquid that in its final composition contains sperm and substances such as fructose and various proteins that nurture sperm and aid their journey through the female reproductive tract; the major volume of semen is not sperm—instead, the majority of semen volume comes from the fluids added by the sex glands; during an orgasm, semen is ejaculated through the urethra that passes down the length of the penis.

Semiconservative Characteristic of the DNA replication process because it produces a hybrid molecule that is half new and half parent DNA.

Semicritical equipment Equipment that comes in contact with mucous membranes or non–intact skin and should be free of bacteria, fungi, and viruses; low numbers of endospores are not a threat; tubing for endoscopes or anesthesia or respiratory therapy is an example of semicritical equipment.

Semisynthetic antimicrobials Naturally occurring antibiotics that can be chemically modified to improve their pharmacological actions and/or stability.

Sensitivity A term that describes the ability of a diagnostic test to detect very low levels of the target to limit false negative results.

Sensitizing exposure An allergen exposure that triggers the immune system to produce IgE.

Sepsis Overwhelming, body-wide inflammatory response to a pathogen or toxin; often due to a systemic infection, but localized infections may also trigger it; technically, microbes do not need to enter the bloodstream or lymphatic vessels for sepsis to develop.

Septic shock Most dangerous stage of sepsis in which a patient's blood pressure may drop to the point that blood is not adequately propelled into tissues and organs; as oxygen and nutrients are cut off, tissues can no longer perform cellular respiration, which leads to tissue death (necrosis) and organ failure; the most dangerous stage of sepsis that features tissue death and organ failure.

Septicemic plague Form of plague that develops if *Yersinia pestis*, the bacterium that causes plague, enters the bloodstream; septicemic plague begins with fever, weakness, diarrhea, and chills; skin may turn black, particularly in fingers, toes, and the nose—which is why people living during the Middle Ages called it the "Black Death;" this blackening is a result of tissue death (necrosis) and is called acral gangrene; both pneumonic and bubonic plague can progress to the septicemic form.

Seroconversion period The period between infection and the point that the patient's immune system starts to make antibodies to the pathogen; also called the seroconversion window.

Serology The study of what is in a patient's serum; such study includes immunological tests; often the goal of serology is to determine if a patient has certain antigens and/or antibodies in their blood.

Serotype A group of related microbes that share common antigens (i.e., specific proteins on their surface) that are often used for detection and diagnosis.

Serovar Genetically distinct variants of the same species that are distinguished according to their different surface antigens.

Serum sickness Sickness that can be caused by antivenoms and antitoxins, as well as certain other medications, such as penicillin and sulfa drugs, in which the patient's immune system recognizes the administered substance as foreign; the patient then forms antibodies against the drug, antivenom, or antitoxin; as the antibodies bind to their targets, the immune complexes become lodged in vessels or joints, producing symptoms such as rash, fever, fatigue, achiness in joints and muscles, headache, labored breathing (dyspnea), and abdominal pain.

Severe acute respiratory syndrome (SARS) Caused by severe acute respiratory syndrome coronavirus (SARS-CoV), which was discovered in 2003; a highly infectious and dangerous respiratory illness.

Severe acute respiratory syndrome coronavirus-1 (SARS-CoV-1) Respiratory virus that originated in China and rapidly spread to at least 24 other countries causing the SARS outbreak of 2003.

Severe acute respiratory syndrome coronavirus-2 (SARS-CoV-2) Respiratory virus that originated in China and spread worldwide causing the COVID-19 pandemic.

Sexual spores Spores that arise from the union of complementary mating strains of fungi generated by meiosis and that do result in genetic variation; three types of sexual fungal spores exist: zygospores, ascospores, and basidiospores.

Sexually transmitted infections (STIs) Diverse agents that can be transmitted by penetrative anal or vaginal sex, as well as nonpenetrative sexual contact such as oral sex or genital contact; the term STI simply refers to the mode of transmission—that means that while STIs may spread by sexual contact, it doesn't mean it's an infectious agent that will affect the reproductive system.

Shiga toxin Toxin made by *S. dysenteriae*; it targets ribosomes to block protein synthesis and kill host cells; Shiga toxin's cytotoxic effects are largely why *S. dysenteriae* has such a high mortality rate compared with other *Shigella* species; in some patients, the Shiga toxin can damage the kidneys and induce hemolytic uremic syndrome (HUS).

Shingles Reactivation of the varicella-zoster virus within the nerves of someone who previously had chickenpox.

Siderophores Organic molecules, made by pathogenic bacteria, that pull iron from our iron-binding proteins.

Signs Objective indicators of disease that can be measured or verified; common signs include fever, rash, or blood in stool.

Silent mutations Mutations that do not change the amino acid sequence of a protein.

Simple staining Staining techniques that use just one dye; typically only size, shape, and cellular arrangement can be determined using simple stains.

Single-strand DNA-binding proteins Proteins that bind the template DNA strands during DNA replication to keep them separated until they are copied.

Sinusitis Inflammation of the sinus membranes caused by a wide variety of infectious agents and inhaled allergens like pollen that aggravate the sinuses and nasal passages.

Slants Solid media that are poured into test tubes and allowed to cool at an angle to create a slanted surface for inoculation.

Slime layer A type of glycocalyx that is fairly unorganized and loosely associated with the cell wall; a carbohydrate-rich layer around cells.

Small noncoding RNAs RNAs that impact protein synthesis and regulate gene expression at a post-transcriptional level.

Smear In bacterial staining techniques, a small amount of the sample that is placed on a glass slide.

Solubility Refers to a substance's ability to dissolve in a solvent, especially water.

Soluble A term used to describe substances that dissolve in water (the main liquid component of blood and tissues).

Solute The dissolved substance in an aqueous solution.

Solvent Dissolving agent in an aqueous solution.

Source The animate or inanimate habitat which disseminates the agent from the reservoir to new hosts.

Spastic paralysis A type of paralysis produced by tetanospasmin in which there is an inability to relax muscles; this in turn causes intense muscle spasms, drooling, sweating, and irritability.

Specialized transduction A type of horizontal gene transfer mechanism; a scenario in which a bacteriophage packs specific bacterial genes into a new phage particle and then conveys those genes to new bacterial host cells.

Specificity Term for when the diagnostic test reliably detects only the agent of interest without producing false positive results.

Spikes (or peplomers) Extensions in many viruses that may protrude from the viral capsid or, if present, from the viral envelope; spikes help viruses attach and gain entry to host cells.

Spirochetes Bacteria that resemble spiral-shaped bacteria, but move in a corkscrew-rotary motion due to a specialized periplasmic flagellum; spirochetes include the pathogens *Treponema pallidum*, the causative agent of syphilis, and *Borrelia burgdorferi*, the causative agent of Lyme disease.

Spleen A fist-sized secondary lymphoid organ located in the upper left part of the abdomen, just under the diaphragm; like lymph nodes, the spleen is a place where leukocytes look for invaders; unlike the lymph nodes, it filters blood rather than lymph; also has non–immune system functions, including disposal of damaged red blood cells.

Splenomegaly Enlarged/swollen spleen.

Spliceosome A complex in the nucleus that performs mRNA editing.

Splicing A form of RNA editing that involves clipping out specific sequences in RNA and joining the remaining parts of the molecule.

Spontaneous abortion The medical term for a miscarriage; noninduced embryonic or fetal death before 20 weeks of gestation.

Spontaneous generation Idea that life comes from nonliving items.

Spontaneous mutations Naturally occurring mutations, most of which are either neutral or harmful to a cell.

Sporadic Describes a disease that causes occasional infections in a population.

Spore formation A form of reproduction used by some fungi; in fungi, spores can be sexual or asexual.

Sporogony In apicomplexans, the zygote made by gamete fusion divides to make sporozoites.

Sporulation The process of forming an endospore during which a cell copies its genetic material and then packages it with the ribosomes and enzymes needed to return to a vegetative state into a compact structure called the spore core; it then surrounds the spore core with several heat- and chemical-resistant layers.

Stains Dyes used in microbiology that were first added simply to increase contrast so the sample was easier to see; eventually certain stains became an integral part of classifying microbes.

Standard precautions A more comprehensive expansion of universal precautions because they require that all patients are treated as potential sources of bloodborne or other infectious agents; standard precautions include handling precautions for all bodily fluids, including blood, urine, feces, sputum, and vomitus; for membranes, nonintact skin, and fresh tissues; and for all excretions or secretions except for sweat.

Staph An arrangement produced when cocci divide to form grapelike clusters of cells; members of the genus *Staphylococcus* exhibit such an arrangement.

Staphylococcal scalded skin syndrome A potentially serious condition caused by *Staphylococcus aureus*; looks as if the skin has been burned in boiling water; infant may present as fussy and tired, with malaise and/or fever. Red blistering skin is of significant concern; the fluid in the blisters does not contain bacteria, but rather the exfoliative toxins of certain *Staphylococcus* strains.

Start codon During the initiation phase of translation, this codon indicates where the ribosome is to add the first amino acid.

Stationary phase Third growth phase of bacteria in which the population growth rate slows and eventually levels off as the number of cells dying matches the number of cells dividing; as with the other phases, the length of this third phase also depends on factors such as the amount and type of nutrients in the culture and the species being grown.

Stella Star-shaped prokaryotic cells.

Sterilization Decontamination measure that eliminates all bacteria, viruses, and endospores.

Sterols (or steroid alcohols) Organic ring-structured compounds that have an alcohol functional group (OH group); the most common example of a sterol is cholesterol; sterol compounds are hydrophobic, which means they do not like interacting with water; they are plentiful in eukaryotic membranes where they have central roles in maintaining membrane stability and fluidity.

Sticky ends In building recombinant DNA, the compatible ends of complementary base pairs between the copied DNA and the plasmid, both of which are incubated with restriction enzymes to generate these compatible ends.

Stillbirth In utero fetal death after 20 weeks of gestation.

Stop codon A codon that signals that translation should end.

Strain Genetic variants of the same species; different strains of a species typically have a hallmark genetic trait that warrants a special designation; mutations and gene transfer often lead to new strains; strain names typically include numbers and/or letters after the species name; for example, a strain of *Escherichia coli* commonly found in laboratories is *E. coli* K-12.

Streak plate technique Technique that helps to isolate a specific species of microbe for study, accomplished by spreading the sample thinly enough on an agar plate, so that the various cells in the sample are sufficiently separated and can give rise to individual colonies.

Streptobacilli Bacilli that form longer chains (strep = chain).

Streptococcal pharyngitis (strep throat) Bacterial pharyngitis caused by the group A streptococcus (GAS) *S. pyogenes*.

Streptococcal toxic shock syndrome Severe, yet rare, disease associated with *Streptococcus pyogenes* bacteria in which bacterial toxins are released into the bloodstream and cause a drop in blood pressure and organ failure; cases may manifest with a red skin rash; different from staphylococcal toxic shock syndrome.

Streptococci Longer chains of spherical cells (strep = chain; cocci = sphere) that form as a result of cell division patterns; these bacteria have a beads-on-a-string appearance.

Stridor Wheezing or loud breathing associated with a blocked or narrowed airway.

Structural morphogenesis A switch in shape; for instance, yeasts are dimorphic, meaning they grow as a single-celled yeast form, but can undergo structural morphogenesis to a filamentous or hyphal form in response to environmental changes; this switch from the yeast form to the hyphal form is required for a symptomatic infection to develop.

Subcutaneous layer A layer beneath the dermis composed of fat cells, nerves, and blood vessels; this layer is closely associated with the skin, but is not technically considered a skin layer.

Subcutaneous mycoses Deeper dermal or muscle fungal infections that can be more serious than cutaneous mycoses; subcutaneous mycoses are often associated with wounds or abrasions.

Substitution mutations Mutations that occur when an incorrect nucleotide is added.

Substrate The molecule(s) that an enzyme chemically acts upon.

Subunit vaccines Vaccines that do not include whole pathogens, instead consisting of purified antigens or parts of the infectious agent; because subunit vaccines have fewer antigens than whole-agent or live attenuated vaccines, they require adjuvants to stimulate a strong immune response.

Sulfa drugs (or sulfonamides) Among the first synthetic antimicrobials used to combat bacterial infections; act as competitive inhibitors of folic acid production, are bacteriostatic, and have a broad spectrum of action.

Sulfamethoxazole A sulfa drug often administered in combination with another antifolate drug called trimethoprim.

Superantigens Include a variety of bacterial toxins such as staphylococcal enterotoxins, staphylococcal toxic shock toxin, and streptococcal exotoxins that generate the features of scarlet fever; superantigens are especially potent T helper cell activators; they overstimulate the immune system to cause massive inflammation that harms the host; also tend to cause fever (they are pyrogens).

Superbugs A name for resistant microbes; derives from the point that resistant microbes remain unaffected by the administered antimicrobial therapy and may readily increase their numbers in the patient and cause a superinfection that is difficult to treat.

Superinfection An infection caused by superbugs that is difficult to treat.

Supportive therapy Treatment to manage discomfort or symptoms, such as rest, fluids, or general pain medication.

Susceptibility testing Testing performed by clinical microbiology laboratories on bacteria (and to a lesser extent on fungi) to assess if the pathogen is likely to be treatable with a particular antimicrobial drug.

Susceptible Term applied when someone is not immune to a given pathogen and it may infect them; may also be used to describe a pathogen that will respond to a particular antimicrobial drug.

Susceptible host A host that can develop a disease in question.

Symbiotic relationship A relationship in which two or more organisms (microbes and humans) are closely connected; includes parasitism, mutualism, and commensalism.

Symmetric paralysis Paralysis that affects both sides of the body; can develop due to a poliovirus infection.

Symport Occurs when, in secondary active transport, ions flow in the same direction as the target substance.

Symptomatic case A case of infection in which a patient experiences the full-blown classical symptoms of the disease; also called a clinical infection.

Symptoms Indicators of disease that are sensed by the patient and are subjective rather than precisely measurable; pain, fatigue, and nausea are examples.

Syncytia Host cells in HSV-1 (herpes simplex virus-1) infection that have fused and become larger; these tissue changes are due to viral cytopathic effects on host cells.

Synergism The combined effectiveness of two drugs that is greater than the sum of their effectiveness when used alone.

Synthetic antimicrobials Antimicrobials that are wholly manufactured by chemical processes; they represent one avenue for making drugs that can overcome antibiotic-resistance mechanisms that naturally evolve as pathogens encounter our pharmacopeia of drugs.

Systemic Widespread throughout the body.

Systemic infection An infection that affects the whole body, usually carried in the blood or lymph; often describes infections that impact the cardiovascular and lymphatic systems.

T cell receptors (TCRs) Antigen recognition receptors on T cells.

T cells Adaptive immunity leukocytes that are produced in the bone marrow and mature in the thymus gland, and are present in lymphoid tissues throughout the body; T cells have roles in both the humoral and cellular branches of adaptive immunity.

T cytotoxic cells (T_C cells) T cells that directly destroy infected or cancerous cells.

T helper 1 (T_H1) cells T helper cells that mainly activate T cytotoxic cells, macrophages, and natural killer cells to destroy pathogens inside of host cells.

T helper 2 (T_H2) cells T helper cells that primarily stimulate B cells to make antibodies and are therefore key stimulators of humoral immune responses.

T helper cells (T_H cells) T cells that do not directly seek and destroy invaders—instead, they coordinate an adaptive immune response by stimulating other white blood cells.

T regulatory (T_{reg}) cells T cells that control functions of other white blood cells, including dendritic cells, mast cells, B cells, and other T cells to ensure that immune responses taper off once a threat subsides; this reduces risk for auto-inflammatory disorders and collateral host tissue damage from sustained inflammation.

Tachypnea An abnormally high number of breaths per minute.

***Taq* polymerase** An enzyme from the hot springs bacterium *Thermus aquaticus*; used as a heat-resistant DNA polymerase in polymerase chain reaction methods.

Tartar Calcified plaque that must be scraped off the teeth; also known as calculus.

Taxonomy The study of how organisms can be grouped by shared features; encompasses identifying, naming, and classifying organisms; modern classification schemes organize life forms by their shared characteristics, including physical and biochemical features, ecology, and gene sequences.

T-dependent antigens Antigens that require T helper cells (especially T_H2 cells) for full activation of B cells.

Temperate phages A category of bacteriophage that can carry out lysogenic replication and in so doing integrate their genetic material into a host bacterium's chromosome; can facilitate specialized transduction.

Termination Third step of the translation process in which the ribosome reaches a stop codon (or nonsense codon), signaling that translation should end.

Tertiary syphilis The final stage of syphilis, during which the disease can be fatal. It causes lesions on the skin and bones, heart damage, and neurological symptoms; insomnia, vertigo, and seizures can be observed 5 to 10 years after infection; partial paralysis, loss of sensation, memory impairment, and hallucinations can occur decades after infection.

Testicles Contained within the scrotum; the two organs that make sperm cells and a variety of hormones, including testosterone; also known as testes.

Tetanospasmin A toxin produced by *Clostridium tetani* that is taken up by peripheral motor neurons and transported to the spinal cord; once there, it becomes active and blocks the release of inhibitory neurotransmitters, preventing muscle relaxation; this leads to spastic paralysis (paralysis that develops when muscles cannot relax); the common name "lock jaw" comes from one of the effects of the toxin.

Tetanus Condition caused by *Clostridium tetani*.

Tetracyclines A family of chemically related broad spectrum bacteriostatic drugs that include tetracycline, demeclocycline, and doxycycline, with doxycycline being the most commonly used in a clinical setting; these drugs bind to the prokaryotic 30S ribosomal subunit.

Theory A hypothesis that has been proven through many studies with consistent, supporting conclusions.

Therapeutic drug monitoring (TDM) Monitoring used to ensure patient well-being and/or assess the therapeutic benefit of a drug; TDM may involve measuring drug concentrations in the bloodstream as well as monitoring other patient parameters such as renal and hepatic function; also used in late-phase clinical trials to ensure that the drug being developed is not adversely impacting the patient.

Therapeutic index (or therapeutic ratio) One measure of a drug's general safety; this is the ratio of the maximum tolerated or safe dose to the minimum effective or therapeutic dose.

Thermal death point The minimum temperature needed to kill all microbes in a sample within ten minutes.

Thermal death time The shortest period of time that a given temperature must be held to kill all microbes in a sample.

Thermocycler A machine used to perform polymerase chain reaction (PCR).

Thermophiles Organisms that prefer warm temperatures of roughly 40–75°C; they dwell in compost piles and hot springs.

Thioglycolate A reducing agent added to remove molecular oxygen from media by converting the molecular oxygen to water.

Thymine (T) Pyrimidine nitrogenous base that is an essential ingredient in nucleotides that build nucleic acids.

Thymus A butterfly-shaped organ that lies just behind the sternum (breastbone) near the heart and is classified as primary lymphoid tissue; maturation of T lymphocytes takes place here.

T-independent antigens Antigens that bind to B cells and instigate a direct activation cascade.

Tinea Ringworm infection; a fungal infection, not an actual worm.

Titer Amount of antibody present in the blood.

Topoisomerases Specialized detangling enzymes that relieve the coiling tension that develops as the helix unwinds during DNA replication.

TORCH An acronym that refers to pathogens that can be vertically transmitted to a developing fetus and cause congenital defects; the original TORCH acronym was developed in the 1970s and at first strictly referred to *Toxoplasma gondii*, rubella virus; cytomegalovirus, and herpes simplex viruses—at the time, these were the four most common agents known to be vertically transmitted and linked to congenital defects; since the original acronym was developed, we have discovered many other pathogens that fit the TORCH criteria; in light of this, the "O" in TORCH has been reassigned to stand for "other infections;" as a miscellaneous category, the "O" grouping now includes HIV, syphilis, parvovirus B19, listeriosis, varicella-zoster virus, coxsackievirus, enteroviruses, and most recently, Zika virus.

Tourniquet test Test that can indicate whether a patient may have dengue fever that has progressed to a hemorrhagic form; a blood pressure cuff is applied to the forearm of the patient for five minutes—if red patchy skin appears, it indicates that the patient's capillaries have become leaky due to the viral infection.

Toxemia Presence of bioactive toxins in the blood; often these toxins are made by bacterial pathogens; although many bacteria secrete deadly toxins, those of *Clostridium botulinum* and *Clostridium tetani* are among the most infamous causes of toxemia.

Toxic megacolon A condition that may develop with *Clostridioides difficile* infection, in which the large intestine can't expel gas and feces so it becomes distended and stressed; eventually a perforation, or a tear in the colon, may develop; as bacteria seep into the normally sterile abdominal cavity and cause inflammation, the patient may progress to septic shock and die.

Toxigenic Term applied to a microbe that makes toxins.

Toxins Molecules that, in small amounts, generate a range of adverse host effects such as tissue damage and suppressed immune response.

Toxoid vaccines A type of subunit vaccine; vaccines made with naturally occurring bacterial toxins that have been purified and inactivated.

Toxoplasmosis Disease caused by the protozoan *Toxoplasma gondii*; cats, both feral and household pets, often carry *T. gondii* and act as the definitive host, shedding the protozoans in their feces, making the disease toxoplasmosis a zoonosis; in addition to cat feces exposure, other important risk factors include eating raw or undercooked meat such as wild game, pork, and shellfish, because these organisms may have cysts in their tissues.

Tracheitis Inflammation and swelling of the trachea.

Transcription The first stage of protein synthesis, in which genes in DNA are copied into a new format, RNA.

Transduction Introduction of new genetic material into a bacterial cell by a virus.

Transfer RNA (tRNA) Cloverleaf-shaped RNA molecules that bring the correct amino acid to a ribosome to build proteins.

Transformation A type of horizontal gene transfer in which bacteria are genetically altered when they take up DNA from their environment.

Transient microbiota Temporary microbiota that do not persist as stable residents of our bodies; can be picked up through a handshake or contact with environmental surfaces; most acquired pathogens are transient microbiota; unlike normal microbiota, transient microbiota can be removed through hygiene—especially via proper hand-washing technique.

Transition state A brief yet critical point when reactants are chemically becoming products, but the reaction is not yet completed.

Translation The second stage of protein synthesis, a process that entails reading mRNA to build proteins.

Transmissible spongiform encephalopathy (TSE) A neurological disease caused by infectious proteins called prions that destroy brain tissue; there are many different TSEs; for instance, European sheep have a long history of a disease called scrapie, named for the way that affected sheep rub their skin raw on fences; there is also bovine spongiform encephalopathy (BSE), also known as mad cow disease.

Transpeptidation The step where reinforcing peptide linkages are made between sugar chains; drugs that block peptidoglycan production tend to do so by interfering with transpeptidation; this weakens the cell wall and leads to cell lysis, making these drugs bactericidal.

Transplacental entry Vertical transmission (passed from mother to child) by pathogens, meaning they cross the placenta to infect the fetus.

Transport vesicles Lipid bilayer sacs that move substances around the cell and make their way to diverse cellular destinations—including the plasma membrane.

Trichinellosis Zoonotic infection, also known as trichinosis, caused by roundworms of the genus *Trichinella* (most notably, though not exclusively, *T. spiralis*); worldwide, this roundworm is acquired by eating raw or undercooked wild game meat or pork that contains the parasite; symptoms of nausea, diarrhea, or vomiting develop within a few days of eating the contaminated meat; two weeks later muscle pain, fever, headache, chills, and rash may occur; symptom severity depends on how many cysts were ingested and where the larvae eventually form cysts in the human host.

Triglycerides Name for lipids that have three fatty acid chains bonded to glycerol.

Trimethoprim An antifolate drug that is often administered in combination with sulfamethoxazole and that targets a part of folic acid synthesis that is different from the step that sulfa drugs inhibit.

Trimethoprim-sulfamethoxazole (TMP/SMX) An antifolate drug combination (trade names: Septra, Bactrim) that works by blocking folate production in certain bacteria as well as in certain protozoans.

Tropism The preference of a pathogen for a specific host (and even a specific tissue within the host).

True pathogen A pathogen that does not require a weakened host to cause disease.

Tuberculosis (TB) An ancient infectious disease that historically was known as "consumption" because of the increasingly thin and wasted appearance of victims in the terminal stage of the disease; found throughout the world, but especially prevalent in Latin America, Africa, eastern Europe, and throughout Asia; worldwide, TB is the fourth leading cause of death from an infectious disease; usually caused by *Mycobacterium tuberculosis* (also called tubercle bacillus).

Tubulin Protein that constitutes microtubules.

Tularemia A rare but extremely infectious zoonotic disease caused by the Gram-negative bacterium *Francisella tularensis*; can exist as a free-living bacterium, but it usually prefers to live as a parasite of eukaryotic cells—especially macrophages; however, *F. tularensis* remains infectious for long periods of time outside of the host; many scientists consider it to be the most infectious pathogenic bacterium known to mankind, making it a potential biological weapon; when aerosolized, inhaled *F. tularensis* can cause pneumonic tularemia; if left untreated, has a mortality rate of 30 to 60 percent.

Tumor necrosis factor alpha (TNF-α) A factor made primarily by macrophages; it has received much clinical attention due to its ability to stimulate inflammation and kill tumor cells; TNF-α also stimulates fever.

Turbidity Cloudiness of a liquid culture, used to indirectly measure cell numbers.

Type II hypersensitivities Antibody-mediated reactions that involve IgG or IgM binding to nonsoluble antigens on the surface of a cell or within the extracellular environment, such as extracellular collagen in connective tissues.

Type III hypersensitivities Reactions that develop when IgG or IgM antibodies bind to soluble targets to make an excessive number of antibody–antigen complexes; the antibodies involved in type III reactions can be made as part of an autoimmune response or formed as a normal response to foreign antigens.

Type IV hypersensitivities Reactions that are T cell-mediated responses against self-antigens or otherwise harmless antigens; because type IV reactions manifest slowly over 12–72 hours after the stimulating antigen is encountered, they are often called delayed hypersensitivity reactions.

Typhoid fever Fever caused by *S. enteritidis* serotype Typhi, which is much more dangerous than the nontyphoid salmonellosis caused by the more common Enteritidis and Typhimurium serovars; the increased virulence of this serovar is primarily due to a toxin that its members make: the typhoid toxin; humans are the only known source and host for the bacterium, so typhoid fever is only spread from human to human; patients tend to exhibit fever, headache, a rose-colored spotted rash, constipation, and generalized abdominal pain that can last weeks; vomiting is usually absent; up to 5 percent of people who contract this infection may become asymptomatic carriers and continue to shed the bacteria in their feces and serve as a source of infection.

Typhoid toxin Toxin produced by *Salmonella enteritidis* serotype Typhi.

Ubiquitous fungi Fungi that grow in varied climates and under diverse conditions; some are even common members of the normal flora of the human body.

Ulcers Lesions that form in areas of the stomach where the lining is damaged by *Heliocbacter pylori* and the host inflammatory response; the epithelial layer of the stomach becomes exposed to stomach acid, and is further damaged, resulting in ulcers.

Uncoating Occurs when animal virus capsids enter the host cells, whereupon the capsid is then entirely or partially broken down, releasing the viral genome.

Uncomplicated malaria A typical version of the protozoan infection; a form of malaria that presents with three stages: a cold stage hallmarked by shivering and a sensation of cold; a hot stage with fever; and a sweating stage; other symptoms that may develop in these stages include headaches, body aches, fatigue, nausea, and vomiting; each stage is linked to a step in the protozoan's development; these malarial stages follow each other as a progression and altogether constitute an episode that lasts 6–10 hours; the 6- to 10-hour attack episode then repeats every two to three days.

Uncomplicated UTI Urinary tract infection (UTI) that occurs in otherwise healthy individuals with normal urinary tract structure; these UTIs tend to resolve quickly, respond to first-line drugs, and don't usually recur.

Unicellular Consisting of only one cell.

Universal precautions Older guidelines adopted by healthcare facilities in the wake of the first HIV cases in the 1980s to limit transmission of bloodborne pathogens such as HIV and hepatitis B and C; these guides applied when dealing with patients suspected or known to have a bloodborne infection. See **Standard precautions**.

Unsaturated Term describing lipids that have double bonds in one or more of the hydrocarbon chains of their fatty acids; the more double bonds present, the less saturated the fat; lipids rich in unsaturated fatty acids, like corn oil and olive oil, tend to have kinks and bends in their chains that prevent them from neatly stacking, which allows them to exist as liquids at room temperature.

Upper endoscopy Also called an upper GI endoscopy; a procedure in which an endoscope, a long flexible tube with a light and lens attached to the end, is inserted in the mouth and used to view the esophagus, stomach, and small intestine.

Upper GI tract The portion of the digestive tract that includes the mouth, pharynx, esophagus, and stomach.

Upper UTIs Urinary tract infections that impact the ureters or kidneys.

Uracil (U) Pyrimidine nitrogenous base that is an essential ingredient in nucleotides that make up nucleic acids.

Urea Our liver makes urea as a normal part of amino acid breakdown; the kidneys remove urea by excreting it into urine.

Ureters Convey urine from the kidneys to a smooth muscle sac called the bladder.

Urethra Conveys urine from the bladder to the exterior of the body; in women, the urethra is short and straight, while in men the urethra is longer, curved, and passes through the prostate gland; also, in men the urethra serves as a part of the reproductive system when it transports semen out of the penis during an ejaculation.

Urethritis Inflammation of the urethra.

Urinary catheter (Foley catheter) A tube inserted through the urethra into the bladder that allows urine to drain into a bag; frequently used on bed-bound or surgical patients; although these indwelling devices are helpful in bed-bound patients, they provide a surface for biofilm formation and greatly increase the risk for urinary tract infections.

Urinary tract infection (UTI) General term that refers to an infection that affects the urinary system; usually caused by bacteria.

Urinary urgency incontinence (UUI) A frequent urge to urinate accompanied by incontinence; may be linked to a shift in the microbiota phyla, as indicated by a study comparing women with UUI with healthy controls.

Urine dipstick test Most common method for detecting a urinary infection; this test consists of a series of filter paper squares on a flexible plastic strip; the strip is dipped into a freshly collected clean-catch urine sample (midstream sample) and results are immediately read by comparing strip color changes to a key; a single strip tests for multiple factors, including infection markers such as white blood cells (WBCs) and nitrite.

Urogenital Term that describes infections and disorders that may jointly affect the urinary and reproductive systems.

Uropathogenic *Escherichia coli* (UPEC) Enteric bacteria (bacteria that are normally found in the intestinal tract) that cause the vast majority of urinary tract infections (UTIs); small, flagellated, Gram-negative rods that may also take on a filamentous form when infecting a host; according to the Merck Manual, *Escherichia coli* cause 70–95 percent of all uncomplicated urinary tract infections and at least 50–65 percent of complicated UTIs.

Uterus In the female reproductive system, a muscular, inverted-pear-shaped organ that houses the developing fetus during pregnancy and contracts to expel the baby during labor; also called the womb.

Vacuoles A conglomerate of many vesicles that have merged to make a larger membranous sac; contain mainly water and various organic and inorganic substances such as nutrients, toxins, and even waste products.

Vagina A tube-like structure that extends from the cervix to the exterior of the body; also the site of standard intercourse; serves as the exit site for menstrual fluids.

Vaginitis Vaginal inflammation; may or may not simultaneously occur with vaginosis.

Vaginosis An overgrowth of one type of microbe over another in the vagina; marked by a lack of lactobacilli bacteria in the vagina.

Valence shell The outermost shell of electrons in an atom; valence electrons are the electrons found in the valence shell; for simplicity, they can be thought of as the electrons that typically participate in chemical reactions.

Valley fever Another name for coccidioidomycosis; one of the causative agents, *Coccidioides immitis*, is native to the semi-arid soil of California's San Joaquin Valley, which is why this name came about.

Van der Waals interactions Interactions that do not bind atoms into molecules and instead are considered electrostatic interactions between molecules.

Vancomycin-intermediate S. *aureus* (VISA) *Staphlococcus aureus* strain that is emerging, making *S. aureus* even more challenging to treat.

Vancomycin-resistant S. *aureus* (VRSA) *Staphlococcus aureus* strain that is emerging, making *S. aureus* even more challenging to treat.

Variable region The portion of an antibody that binds to the antigen; this region varies based on the epitope that activated the B cell producing the given antibody.

Variolation Procedure that originated in China in the 1400s to combat smallpox; in this procedure, the practitioner blew a powder made from the dried scabs of smallpox lesions into a healthy individual's nose; the resulting smallpox infections tended to be milder—with only a 1–2 percent mortality rate compared to the 30 percent mortality associated with naturally acquired infections.

Vas deferens Structure that transports sperm from the epididymis toward the prostate.

Vasoactive Term applied to a substance that is a potent inducer of vascular changes.

Vasodilatation Occurs in early inflammation, an increase in the diameter of undamaged blood vessels so that more blood flows to injured tissues.

Vector transmission A type of indirect contact transmission in which organisms such as arthropods and rodents spread infectious agents to other susceptible hosts; many zoonotic diseases require a vector for transmission to human hosts.

Vectors Organisms such as arthropods and rodents that spread infectious agents to other susceptible hosts.

Vegetative cells Actively growing cells that result when environmental conditions become favorable; bacterial spores germinate back into these cells when conditions are favorable.

Vehicle transmission A type of indirect contact transmission in which contaminated fomites (inanimate objects) convey the pathogen to a new host.

Veins Vessels that convey oxygen-depleted, waste-rich blood to the heart. Our systemic veins direct blood toward the vena cava, a large vein that conveys blood to the heart's right atrium.

Ventilator-associated pneumonia (VAP) A form of pneumonia that occurs more than 48 hours after a hospitalized patient is put on a mechanical ventilator.

Vertical gene transfer Occurs when cells pass their genetic information to the next generation (from parent cell to offspring) as a result of sexual or asexual cell division.

Vertical transmission A specialized form of direct contact transmission that occurs when the pathogen passes from mother to offspring during pregnancy (in utero), during delivery (transcervical), or through breast milk (postpartum).

Vessel permeability Increases in early inflammation, causing vessels to be slightly "leaky," and thus allowing blood plasma and the proteins it carries (including complement proteins) to enter tissues.

Viable plate count An important method for directly enumerating a population in which samples from a liquid broth culture are serially diluted and either spread on solid agar (spread plate method) or poured into a petri plate with melted agar media (pour plate method); after an incubation period, colonies are visible and can be counted.

Vibrio Comma-shaped prokaryotic cells.

Viral hemorrhagic fevers Infections that induce high fever, chills, diarrhea, vomiting, and headache, as well as bleeding from the eyes, mouth, ears, skin pores, and internal organs; mosquitoes, ticks, rodents, bats, and blood contact when slaughtering infected animals have all been known to transmit viral hemorrhagic fevers; occasionally, infected humans transfer these viruses through blood or semen; viruses that cause hemorrhagic fevers include the following: dengue and yellow fever (viruses in the *Flaviviridae* family); the *Bunyaviridae* family, which includes hantavirus; the *Arenaviridae* family that includes the virus causing Lassa fever; and the *Filoviridae* family that includes Ebola and Marburg viruses.

Viral strain A term used to refer to viral variants that have mutations with physiological consequences.

Viral titer Quantity of virus present in a given volume of sample.

Viral variant An isolate of a virus that has at least one mutation that makes it genetically distinct from the original form of the virus; the mutation (or mutations) may or may not change how the virus behaves; if the mutation *does* change how the virus behaves, then the term viral strain may be used. (See viral strain definition.)

Viremia Presence of virions in the blood.

Virion Single, infectious virus particle with an exterior protective protein capsid, packed with genetic material (DNA or RNA).

Virology The study of viruses.

Virulence Term describing severity of disease following infection; degree or extent of disease that a pathogen causes.

Virulence factors The ways pathogens overcome our defenses; they include features that help microbes adhere to host cells, invade host tissues, acquire nutrients, and evade immune defenses; also include toxins, substances with diverse ways of thwarting the immune system or damaging our cells.

Viruses Nonliving/noncellular entities that can infect animal, plant, or bacterial cells; can have a DNA or RNA genome; originally described as *filterable* infectious agents, reflecting tiny size compared with cells; today over 5,000 mammal-infecting viral species have been described; of these, at least 220 infect humans.

Visceral leishmaniasis A protozoan infection caused by *Leishmania* in which the protozoan spreads throughout the body; according to the World Health Organization, in developing countries where proper treatment is difficult, visceral leishmaniasis is usually fatal within two years.

Voges-Proskauer (VP) A test designed to detect acetoin, which is an intermediate of butanediol fermentation; often used in combination with the methyl red (MR) test.

Volatile amines Molecules that contain amine groups and easily evaporate into the air; responsible for the fishy odor generated by the whiff test.

Vulva External genitalia of females; includes the labia minora (the pair of small lip-like folds that enfold the vagina and urethra), the clitoris (the erectile tissue that is analogous to the penis and is stimulated to induce orgasm), and the labia majora (the paired lip-like folds that enclose the female external genitalia).

Vulvovaginal candidiasis Vaginal yeast infection; one of the most common infections caused by the fungus *Candida albicans*.

Walking pneumonia An infection caused by *Mycoplasma pneumoniae* in which the clinical findings point to a more severe illness than the symptoms may suggest; immunity to *M. pneumoniae* is not long lived, so a person can have walking pneumonia more than once.

Wandering macrophages Macrophages that roam through tissues.

Western blotting A protocol that detects specific proteins in a sample; like ELISA methods, Western blots reveal what proteins are present as well as their levels (although Western blots are not currently as sensitive as ELISA for quantifying proteins); unlike ELISA, Western blots provide added information about the size of the protein being detected.

Wheal and flare lesion Lesion that is produced in skin tests if the patient has an allergy to the tested allergen; the wheal is the raised, inflamed area of the lesion, and the flare is the flattened, reddened area.

Whiff test In this test, the patient's vaginal mucus is smeared onto a glass slide and drops of potassium hydroxide (KOH) are added; in vaginosis cases, the vaginal mucus is enriched with nitrogen-containing compounds that react with the added KOH to produce volatile amines that have a fishy odor.

Whole-agent vaccines Vaccines that contain the entire pathogen, which has been rendered inactive by heat, chemicals, or radiation; benefit of whole-agent vaccines is that the agent is essentially the same as what would be encountered in nature, but they can't cause disease in a weak host or mutate to cause disease.

Xenografts Interspecies transplants (for example, implanting pig heart valves into a human heart).

Yellow fever Systemic illness caused by the yellow fever virus, a single-stranded RNA virus that infects a variety of primates, including humans and monkeys; transmitted by mosquitoes; most yellow fever infections are asymptomatic or mild; about 15 percent of symptomatic patients develop more severe disease, including jaundice (the clinical feature that gives yellow fever its name) extremely high fever, and potentially deadly shock.

Zika virus Enveloped, single-stranded RNA genome virus that belongs to the *Flaviviridae* family of viruses; transmitted by *Aedes* mosquitoes and through sexual contact; can lead to microcephaly if transmitted to the fetus during pregnancy; an emerging virus that originated in Africa.

Zone of equivalence In the radial immunodiffusion test, the area where there is an equal or roughly equal amount of antigen and antibody molecules and where the precipitate is seen as a ring; the diameter of the ring is compared to a standard curve graph to get information about antigen concentration; the larger the ring, the more concentrated the antigen in the sample.

Zone of inhibition Develops in disk diffusion antimicrobial susceptibility testing; a clear zone that forms around a disk infused with a test drug.

Zoonosis Also called a zoonotic disease or zoonotic infection; an infection passed from animals to humans.

Index

NOTE: The letter "f" following a page number indicates a figure; "n" indicates a note; and "t" a table.

A

D

E

S

T